# RULES OF DIFFERENTIATION

## Basic Formulas

**1.** $\dfrac{d}{dx}(c) = 0$

**2.** $\dfrac{d}{dx}(cu) = cu\dfrac{du}{dx}$

**3.** $\dfrac{d}{dx}(u \pm v) = \dfrac{du}{dx} \pm \dfrac{dv}{dx}$

**4.** $\dfrac{d}{dx}(uv) = u\dfrac{dv}{dx} + v\dfrac{du}{dx}$

**5.** $\dfrac{d}{dx}\left(\dfrac{u}{v}\right) = \dfrac{v\dfrac{du}{dx} - u\dfrac{dv}{du}}{v^2}$

**6.** $\dfrac{d}{dx}f(g(x)) = f'(g(x))g'(x)$

**7.** $\dfrac{d}{dx}(u^n) = nu^{n-1}\dfrac{du}{dx}$

## Exponential and Logarithmic Functions

**8.** $\dfrac{d}{dx}(e^u) = e^u\dfrac{du}{dx}$

**9.** $\dfrac{d}{dx}(a^u) = (\ln a)a^u\dfrac{du}{dx}$

**10.** $\dfrac{d}{dx}\ln|u| = \dfrac{1}{u}\dfrac{du}{dx}$

**11.** $\dfrac{d}{dx}(\log_a u) = \dfrac{1}{u\ln a}\dfrac{du}{dx}$

## Trigonometric Functions

**12.** $\dfrac{d}{dx}(\sin u) = \cos u\dfrac{du}{dx}$

**13.** $\dfrac{d}{dx}(\cos u) = -\sin u\dfrac{du}{dx}$

**14.** $\dfrac{d}{dx}(\tan u) = \sec^2 u\dfrac{du}{dx}$

**15.** $\dfrac{d}{dx}(\csc u) = -\csc u\cot u\dfrac{du}{dx}$

**16.** $\dfrac{d}{dx}(\sec u) = \sec u\tan u\dfrac{du}{dx}$

**17.** $\dfrac{d}{dx}(\cot u) = -\csc^2 u\dfrac{du}{dx}$

## Inverse Trigonometric Functions

**18.** $\dfrac{d}{dx}(\sin^{-1} u) = \dfrac{1}{\sqrt{1-u^2}}\dfrac{du}{dx}$

**19.** $\dfrac{d}{dx}(\cos^{-1} u) = -\dfrac{1}{\sqrt{1-u^2}}\dfrac{du}{dx}$

**20.** $\dfrac{d}{dx}(\tan^{-1} u) = \dfrac{1}{1+u^2}\dfrac{du}{dx}$

**21.** $\dfrac{d}{dx}(\csc^{-1} u) = -\dfrac{1}{|u|\sqrt{u^2-1}}\dfrac{du}{dx}$

**22.** $\dfrac{d}{dx}(\sec^{-1} u) = \dfrac{1}{|u|\sqrt{u^2-1}}\dfrac{du}{dx}$

**23.** $\dfrac{d}{dx}(\cot^{-1} u) = -\dfrac{1}{1+u^2}\dfrac{du}{dx}$

## Hyperbolic Functions

**24.** $\dfrac{d}{dx}(\sinh u) = \cosh u\dfrac{du}{dx}$

**25.** $\dfrac{d}{dx}(\cosh u) = \sinh u\dfrac{du}{dx}$

**26.** $\dfrac{d}{dx}(\tanh u) = \operatorname{sech}^2 u\dfrac{du}{dx}$

**27.** $\dfrac{d}{dx}(\operatorname{csch} u) = -\operatorname{csch} u\coth u\dfrac{du}{dx}$

**28.** $\dfrac{d}{dx}(\operatorname{sech} u) = -\operatorname{sech} u\tanh u\dfrac{du}{dx}$

**29.** $\dfrac{d}{dx}(\coth u) = -\operatorname{csch}^2 u\dfrac{du}{dx}$

## Inverse Hyperbolic Functions

**30.** $\dfrac{d}{dx}(\sinh^{-1} u) = \dfrac{1}{\sqrt{1+u^2}}\dfrac{du}{dx}$

**31.** $\dfrac{d}{dx}(\cosh^{-1} u) = \dfrac{1}{\sqrt{u^2-1}}\dfrac{du}{dx}$

**32.** $\dfrac{d}{dx}(\tanh^{-1} u) = \dfrac{1}{1-u^2}\dfrac{du}{dx}$

**33.** $\dfrac{d}{dx}(\operatorname{csch}^{-1} u) = -\dfrac{1}{|u|\sqrt{u^2+1}}\dfrac{du}{dx}$

**34.** $\dfrac{d}{dx}(\operatorname{sech}^{-1} u) = -\dfrac{1}{u\sqrt{1-u^2}}\dfrac{du}{dx}$

**35.** $\dfrac{d}{dx}(\coth^{-1} u) = \dfrac{1}{1-u^2}\dfrac{du}{dx}$

## TABLE OF INTEGRALS

### Basic Forms

1. $\displaystyle\int u^n \, du = \frac{u^{n+1}}{n+1} + C, \quad n \neq -1$

2. $\displaystyle\int \frac{du}{u} = \ln|u| + C$

3. $\displaystyle\int \sin u \, du = -\cos u + C$

4. $\displaystyle\int \cos u \, du = \sin u + C$

5. $\displaystyle\int \tan u \, du = \ln|\sec u| + C$

6. $\displaystyle\int e^u \, du = e^u + C$

7. $\displaystyle\int a^u \, du = \frac{a^u}{\ln a} + C$

8. $\displaystyle\int \sec u \, du = \ln|\sec u + \tan u| + C$

9. $\displaystyle\int \csc u \, du = \ln|\csc u - \cot u| + C$

10. $\displaystyle\int \cot u \, du = \ln|\sin u| + C$

11. $\displaystyle\int \sec^2 u \, du = \tan u + C$

12. $\displaystyle\int \csc^2 u \, du = -\cot u + C$

13. $\displaystyle\int \sec u \tan u \, du = \sec u + C$

14. $\displaystyle\int \csc u \cot u \, du = -\csc u + C$

15. $\displaystyle\int \frac{du}{\sqrt{a^2 - u^2}} = \sin^{-1}\frac{u}{a} + C$

16. $\displaystyle\int \frac{du}{u\sqrt{u^2 - a^2}} = \frac{1}{a}\sec^{-1}\frac{u}{a} + C$

17. $\displaystyle\int \frac{du}{a^2 + u^2} = \frac{1}{a}\tan^{-1}\frac{u}{a} + C$

18. $\displaystyle\int \frac{du}{a^2 - u^2} = \frac{1}{2a}\ln\left|\frac{u+a}{u-a}\right| + C$

### Forms Involving $a + bu$

19. $\displaystyle\int \frac{u \, du}{a + bu} = \frac{1}{b^2}\left(a + bu - a\ln|a + bu|\right) + C$

20. $\displaystyle\int \frac{u^2 \, du}{a + bu}$

$\displaystyle\qquad = \frac{1}{2b^3}\left[(a + bu)^2 - 4a(a + bu) + 2a^2\ln|a + bu|\right] + C$

21. $\displaystyle\int \frac{u \, du}{(a + bu)^2} = \frac{a}{b^2(a + bu)} + \frac{1}{b^2}\ln|a + bu| + C$

22. $\displaystyle\int \frac{u^2 \, du}{(a + bu)^2} = \frac{1}{b^3}\left(a + bu - \frac{a^2}{a + bu} - 2a\ln|a + bu|\right) + C$

23. $\displaystyle\int \frac{du}{u(a + bu)} = \frac{1}{a}\ln\left|\frac{u}{a + bu}\right| + C$

24. $\displaystyle\int \frac{du}{u^2(a + bu)} = -\frac{1}{au} + \frac{b}{a^2}\ln\left|\frac{a + bu}{u}\right| + C$

25. $\displaystyle\int \frac{du}{u(a + bu)^2} = \frac{1}{a(a + bu)} - \frac{1}{a^2}\ln\left|\frac{a + bu}{u}\right| + C$

26. $\displaystyle\int \frac{du}{u^2(a + bu)^2} = -\frac{1}{a^2}\left[\frac{a + 2bu}{u(a + bu)} + \frac{2b}{a}\ln\left|\frac{u}{a + bu}\right|\right] + C$

### Forms Involving $\sqrt{a + bu}$

27. $\displaystyle\int u\sqrt{a + bu} \, du = \frac{2}{15b^2}(3bu - 2a)(a + bu)^{3/2} + C$

28. $\displaystyle\int \frac{u \, du}{\sqrt{a + bu}} = \frac{2}{3b^2}(bu - 2a)\sqrt{a + bu} + C$

29. $\displaystyle\int \frac{u^2 \, du}{\sqrt{a + bu}} = \frac{2}{15b^3}(8a^2 + 3b^2u^2 - 4abu)\sqrt{a + bu} + C$

30. $\displaystyle\int \frac{du}{u\sqrt{a + bu}} = \begin{cases} \dfrac{1}{\sqrt{a}}\ln\left|\dfrac{\sqrt{a + bu} - \sqrt{a}}{\sqrt{a + bu} + \sqrt{a}}\right| + C & \text{if } a > 0 \\[3mm] \dfrac{2}{\sqrt{-a}}\tan^{-1}\sqrt{\dfrac{a + bu}{-a}} + C & \text{if } a < 0 \end{cases}$

31. $\displaystyle\int \frac{\sqrt{a + bu}}{u} \, du = 2\sqrt{a + bu} + a\int \frac{du}{u\sqrt{a + bu}}$

32. $\displaystyle\int \frac{\sqrt{a + bu}}{u^2} \, du = -\frac{\sqrt{a + bu}}{u} + \frac{b}{2}\int \frac{du}{u\sqrt{a + bu}}$

33. $\displaystyle\int u^n\sqrt{a + bu} \, du$

$\displaystyle\qquad = \frac{2}{b(2n + 3)}\left[u^n(a + bu)^{3/2} - na\int u^{n-1}\sqrt{a + bu} \, du\right]$

34. $\displaystyle\int \frac{u^n \, du}{\sqrt{a + bu}} = \frac{2u^n\sqrt{a + bu}}{b(2n + 1)} - \frac{2na}{b(2n + 1)}\int \frac{u^{n-1} \, du}{\sqrt{a + bu}}$

**35.** $\displaystyle\int \frac{du}{u^n\sqrt{a + bu}} = -\frac{\sqrt{a + bu}}{a(n - 1)u^{n-1}} - \frac{b(2n - 3)}{2a(n - 1)}\int \frac{du}{u^{n-1}\sqrt{a + bu}}$

**36.** $\displaystyle\int \frac{\sqrt{a + bu}}{u^n}\, du = \frac{-1}{a(n - 1)}\left[\frac{(a + bu)^{3/2}}{u^{n-1}} + \frac{(2n - 5)b}{2}\int \frac{\sqrt{a + bu}}{u^{n-1}}\, du\right], \quad n \neq 1$

---

## Forms Involving $\sqrt{a^2 + u^2},\ a > 0$

**37.** $\displaystyle\int \sqrt{a^2 + u^2}\, du = \frac{u}{2}\sqrt{a^2 + u^2} + \frac{a^2}{2}\ln\left(u + \sqrt{a^2 + u^2}\right) + C$

**38.** $\displaystyle\int u^2\sqrt{a^2 + u^2}\, du = \frac{u}{8}(a^2 + 2u^2)\sqrt{a^2 + u^2}$
$$- \frac{a^4}{8}\ln\left(u + \sqrt{a^2 + u^2}\right) + C$$

**39.** $\displaystyle\int \frac{\sqrt{a^2 + u^2}}{u}\, du = \sqrt{a^2 + u^2} - a\ln\left|\frac{a + \sqrt{a^2 + u^2}}{u}\right| + C$

**40.** $\displaystyle\int \frac{\sqrt{a^2 + u^2}}{u^2}\, du = -\frac{\sqrt{a^2 + u^2}}{u} + \ln\left(u + \sqrt{a^2 + u^2}\right) + C$

**41.** $\displaystyle\int \frac{du}{\sqrt{a^2 + u^2}} = \ln\left(u + \sqrt{a^2 + u^2}\right) + C$

**42.** $\displaystyle\int \frac{u^2\, du}{\sqrt{a^2 + u^2}} = \frac{u}{2}\sqrt{a^2 + u^2} - \frac{a^2}{2}\ln\left(u + \sqrt{a^2 + u^2}\right) + C$

**43.** $\displaystyle\int \frac{du}{u\sqrt{a^2 + u^2}} = -\frac{1}{a}\ln\left|\frac{\sqrt{a^2 + u^2} + a}{u}\right| + C$

**44.** $\displaystyle\int \frac{du}{u^2\sqrt{a^2 + u^2}} = -\frac{\sqrt{a^2 + u^2}}{a^2 u} + C$

**45.** $\displaystyle\int \frac{du}{(a^2 + u^2)^{3/2}} = \frac{u}{a^2\sqrt{a^2 + u^2}} + C$

---

## Forms Involving $\sqrt{a^2 - u^2},\ a > 0$

**46.** $\displaystyle\int \sqrt{a^2 - u^2}\, du = \frac{u}{2}\sqrt{a^2 - u^2} + \frac{a^2}{2}\sin^{-1}\frac{u}{a} + C$

**47.** $\displaystyle\int u^2\sqrt{a^2 - u^2}\, du = \frac{u}{8}(2u^2 - a^2)\sqrt{a^2 - u^2} + \frac{a^4}{8}\sin^{-1}\frac{u}{a} + C$

**48.** $\displaystyle\int \frac{\sqrt{a^2 - u^2}}{u}\, du = \sqrt{a^2 - u^2} - a\ln\left|\frac{a + \sqrt{a^2 - u^2}}{u}\right| + C$

**49.** $\displaystyle\int \frac{\sqrt{a^2 - u^2}}{u^2}\, du = -\frac{1}{u}\sqrt{a^2 - u^2} - \sin^{-1}\frac{u}{a} + C$

**50.** $\displaystyle\int \frac{u^2\, du}{\sqrt{a^2 - u^2}} = -\frac{u}{2}\sqrt{a^2 - u^2} + \frac{a^2}{2}\sin^{-1}\frac{u}{a} + C$

**51.** $\displaystyle\int \frac{du}{u\sqrt{a^2 - u^2}} = -\frac{1}{a}\ln\left|\frac{a + \sqrt{a^2 - u^2}}{u}\right| + C$

**52.** $\displaystyle\int \frac{du}{u^2\sqrt{a^2 - u^2}} = -\frac{1}{a^2 u}\sqrt{a^2 - u^2} + C$

**53.** $\displaystyle\int (a^2 - u^2)^{3/2}\, du$
$$= -\frac{u}{8}(2u^2 - 5a^2)\sqrt{a^2 - u^2} + \frac{3a^4}{8}\sin^{-1}\frac{u}{a} + C$$

**54.** $\displaystyle\int \frac{du}{(a^2 - u^2)^{3/2}} = \frac{u}{a^2\sqrt{a^2 - u^2}} + C$

---

## Forms Involving $\sqrt{u^2 - a^2},\ a > 0$

**55.** $\displaystyle\int \sqrt{u^2 - a^2}\, du = \frac{u}{2}\sqrt{u^2 - a^2} - \frac{a^2}{2}\ln\left|u + \sqrt{u^2 - a^2}\right| + C$

**56.** $\displaystyle\int u^2\sqrt{u^2 - a^2}\, du$
$$= \frac{u}{8}(2u^2 - a^2)\sqrt{u^2 - a^2} - \frac{a^4}{8}\ln\left|u + \sqrt{u^2 - a^2}\right| + C$$

**57.** $\displaystyle\int \frac{\sqrt{u^2 - a^2}}{u}\, du = \sqrt{u^2 - a^2} - a\cos^{-1}\frac{a}{|u|} + C$

**58.** $\displaystyle\int \frac{\sqrt{u^2 - a^2}}{u^2}\, du = -\frac{\sqrt{u^2 - a^2}}{u} + \ln\left|u + \sqrt{u^2 - a^2}\right| + C$

**59.** $\displaystyle\int \frac{du}{\sqrt{u^2 - a^2}} = \ln\left|u + \sqrt{u^2 - a^2}\right| + C$

**60.** $\displaystyle\int \frac{u^2\, du}{\sqrt{u^2 - a^2}} = \frac{u}{2}\sqrt{u^2 - a^2} + \frac{a^2}{2}\ln\left|u + \sqrt{u^2 - a^2}\right| + C$

**61.** $\displaystyle\int \frac{du}{u^2\sqrt{u^2 - a^2}} = \frac{\sqrt{u^2 - a^2}}{a^2 u} + C$

**62.** $\displaystyle\int \frac{du}{(u^2 - a^2)^{3/2}} = -\frac{u}{a^2\sqrt{u^2 - a^2}} + C$

## Forms Involving sin *u*, cos *u*, tan *u*

**63.** $\displaystyle\int \sin^2 u\, du = \frac{1}{2}u - \frac{1}{4}\sin 2u + C$

**64.** $\displaystyle\int \cos^2 u\, du = \frac{1}{2}u + \frac{1}{4}\sin 2u + C$

**65.** $\displaystyle\int \tan^2 u\, du = \tan u - u + C$

**66.** $\displaystyle\int \sin^3 u\, du = -\frac{1}{3}(2 + \sin^2 u)\cos u + C$

**67.** $\displaystyle\int \cos^3 u\, du = \frac{1}{3}(2 + \cos^2 u)\sin u + C$

**68.** $\displaystyle\int \tan^3 u\, du = \frac{1}{2}\tan^2 u + \ln|\cos u| + C$

**69.** $\displaystyle\int \sin^n u\, du = -\frac{1}{n}\sin^{n-1} u \cos u + \frac{n-1}{n}\int \sin^{n-2} u\, du$

**70.** $\displaystyle\int \cos^n u\, du = \frac{1}{n}\cos^{n-1} u \sin u + \frac{n-1}{n}\int \cos^{n-2} u\, du$

**71.** $\displaystyle\int \tan^n u\, du = \frac{1}{n-1}\tan^{n-1} u - \int \tan^{n-2} u\, du$

**72.** $\displaystyle\int \sin au \sin bu\, du = \frac{\sin(a-b)u}{2(a-b)} - \frac{\sin(a+b)u}{2(a+b)} + C$

**73.** $\displaystyle\int \cos au \cos bu\, du = \frac{\sin(a-b)u}{2(a-b)} + \frac{\sin(a+b)u}{2(a+b)} + C$

**74.** $\displaystyle\int \sin au \cos bu\, du = -\frac{\cos(a-b)u}{2(a-b)} - \frac{\cos(a+b)u}{2(a+b)} + C$

**75.** $\displaystyle\int u \sin u\, du = \sin u - u \cos u + C$

**76.** $\displaystyle\int u \cos u\, du = \cos u + u \sin u + C$

**77.** $\displaystyle\int u^n \sin u\, du = -u^n \cos u + n\int u^{n-1}\cos u\, du$

**78.** $\displaystyle\int u^n \cos u\, du = u^n \sin u - n\int u^{n-1}\sin u\, du$

**79.** $\displaystyle\int \sin^n u \cos^m u\, du$

$$= -\frac{\sin^{n-1} u \cos^{m+1} u}{n+m} + \frac{n-1}{n+m}\int \sin^{n-2} u \cos^m u\, du$$

$$= \frac{\sin^{n+1} u \cos^{m-1} u}{n+m} + \frac{m-1}{n+m}\int \sin^n u \cos^{m-2} u\, du$$

## Forms Involving cot *u*, sec *u*, csc *u*

**80.** $\displaystyle\int \cot^2 u\, du = -\cot u - u + C$

**81.** $\displaystyle\int \cot^3 u\, du = -\frac{1}{2}\cot^2 u - \ln|\sin u| + C$

**82.** $\displaystyle\int \sec^3 u\, du = \frac{1}{2}\sec u \tan u + \frac{1}{2}\ln|\sec u + \tan u| + C$

**83.** $\displaystyle\int \csc^3 u\, du = -\frac{1}{2}\csc u \cot u + \frac{1}{2}\ln|\csc u - \cot u| + C$

**84.** $\displaystyle\int \cot^n u\, du = \frac{-1}{n-1}\cot^{n-1} u - \int \cot^{n-2} u\, du$

**85.** $\displaystyle\int \sec^n u\, du = \frac{1}{n-1}\tan u \sec^{n-2} u + \frac{n-2}{n-1}\int \sec^{n-2} u\, du$

**86.** $\displaystyle\int \csc^n u\, du = \frac{-1}{n-1}\cot u \csc^{n-2} u + \frac{n-2}{n-1}\int \csc^{n-2} u\, du$

## Forms Involving Inverse Trigonometric Functions

**87.** $\displaystyle\int \sin^{-1} u\, du = u \sin^{-1} u + \sqrt{1-u^2} + C$

**88.** $\displaystyle\int \cos^{-1} u\, du = u \cos^{-1} u - \sqrt{1-u^2} + C$

**89.** $\displaystyle\int \tan^{-1} u\, du = u \tan^{-1} u - \frac{1}{2}\ln(1+u^2) + C$

**90.** $\displaystyle\int u \sin^{-1} u\, du = \frac{2u^2-1}{4}\sin^{-1} u + \frac{u\sqrt{1-u^2}}{4} + C$

**91.** $\displaystyle\int u \cos^{-1} u\, du = \frac{2u^2-1}{4}\cos^{-1} u - \frac{u\sqrt{1-u^2}}{4} + C$

**92.** $\displaystyle\int u \tan^{-1} u\, du = \frac{u^2+1}{2}\tan^{-1} u - \frac{u}{2} + C$

**93.** $\displaystyle\int u^n \sin^{-1} u\, du = \frac{1}{n+1}\left[u^{n+1}\sin^{-1} u - \int \frac{u^{n+1}\, du}{\sqrt{1-u^2}}\right],$

$$n \neq -1$$

**94.** $\displaystyle\int u^n \cos^{-1} u\, du = \frac{1}{n+1}\left[u^{n+1}\cos^{-1} u + \int \frac{u^{n+1}\, du}{\sqrt{1-u^2}}\right],$

$$n \neq -1$$

**95.** $\displaystyle\int u^n \tan^{-1} u\, du = \frac{1}{n+1}\left[u^{n+1}\tan^{-1} u - \int \frac{u^{n+1}\, du}{\sqrt{1+u^2}}\right],$

$$n \neq -1$$

## Forms Involving Exponential and Logarithmic Functions

**96.** $\displaystyle\int ue^{au}\,du = \frac{1}{a^2}\,(au-1)e^{au} + C$

**97.** $\displaystyle\int u^n\,e^{au}\,du = \frac{1}{a}\,u^n\,e^{au} - \frac{n}{a}\int u^{n-1}e^{au}\,du$

**98.** $\displaystyle\int e^{au}\sin bu\,du = \frac{e^{au}}{a^2+b^2}\,(a\sin bu - b\cos bu) + C$

**99.** $\displaystyle\int e^{au}\cos bu\,du = \frac{e^{au}}{a^2+b^2}\,(a\cos bu + b\sin bu) + C$

**100.** $\displaystyle\int \frac{du}{1+be^{au}} = u - \frac{1}{a}\ln(1+be^{au}) + C$

**101.** $\displaystyle\int \ln u\,du = u\ln u - u + C$

**102.** $\displaystyle\int u^n\ln u\,du = \frac{u^{n+1}}{(n+1)^2}\,[(n+1)\ln u - 1] + C$

**103.** $\displaystyle\int \frac{1}{u\ln u}\,du = \ln|\ln u| + C$

## Forms Involving Hyperbolic Functions

**104.** $\displaystyle\int \sinh u\,du = \cosh u + C$

**105.** $\displaystyle\int \cosh u\,du = \sinh u + C$

**106.** $\displaystyle\int \tanh u\,du = \ln\cosh u + C$

**107.** $\displaystyle\int \coth u\,du = \ln|\sinh u| + C$

**108.** $\displaystyle\int \operatorname{sech} u\,du = \tan^{-1}|\sinh u| + C$

**109.** $\displaystyle\int \operatorname{csch} u\,du = \ln\left|\tanh\tfrac{1}{2}u\right| + C$

**110.** $\displaystyle\int \operatorname{sech}^2 u\,du = \tanh u + C$

**111.** $\displaystyle\int \operatorname{csch}^2 u\,du = -\coth u + C$

**112.** $\displaystyle\int \operatorname{sech} u \tanh u\,du = -\operatorname{sech} u + C$

**113.** $\displaystyle\int \operatorname{csch} u \coth u\,du = -\operatorname{csch} u + C$

## Forms Involving $\sqrt{2au^2 - u^2}$, $a>0$

**114.** $\displaystyle\int \sqrt{2au - u^2}\,du = \frac{u-a}{2}\sqrt{2au - u^2}$
$$+ \frac{a^2}{2}\cos^{-1}\left(\frac{a-u}{a}\right) + C$$

**115.** $\displaystyle\int u\sqrt{2au - u^2}\,du = \frac{2u^2 - au - 3a^2}{6}\sqrt{2au - u^2}$
$$+ \frac{a^3}{2}\cos^{-1}\left(\frac{a-u}{a}\right) + C$$

**116.** $\displaystyle\int \frac{\sqrt{2au - u^2}}{u}\,du = \sqrt{2au - u^2} + a\cos^{-1}\left(\frac{a-u}{a}\right) + C$

**117.** $\displaystyle\int \frac{\sqrt{2au - u^2}}{u^2}\,du = -\frac{2\sqrt{2au - u^2}}{u} - \cos^{-1}\left(\frac{a-u}{a}\right) + C$

**118.** $\displaystyle\int \frac{du}{\sqrt{2au - u^2}} = \cos^{-1}\left(\frac{a-u}{a}\right) + C$

**119.** $\displaystyle\int \frac{u\,du}{\sqrt{2au - u^2}} = -\sqrt{2au - u^2} + a\cos^{-1}\left(\frac{a-u}{a}\right) + C$

**120.** $\displaystyle\int \frac{u^2\,du}{\sqrt{2au - u^2}} = -\frac{(u+3a)}{2}\sqrt{2au - u^2}$
$$+ \frac{3a^2}{2}\cos^{-1}\left(\frac{a-u}{a}\right) + C$$

**121.** $\displaystyle\int \frac{du}{u\sqrt{2au - u^2}} = -\frac{\sqrt{2au - u^2}}{au} + C$

# CALCULUS

## EARLY TRANSCENDENTALS

## SOO T. TAN
STONEHILL COLLEGE

BROOKS/COLE
CENGAGE Learning™

Australia • Brazil • Japan • Korea • Mexico • Singapore • Spain • United Kingdom • United States

**Calculus: Early Transcendentals**
Soo T. Tan

Senior Acquisitions Editor: Liz Covello

Publisher: Richard Stratton

Senior Developmental Editor: Danielle Derbenti

Developmental Editor: Ed Dodd

Associate Editor: Jeannine Lawless

Editorial Assistant: Lauren Hamel

Media Editor: Maureen Ross

Marketing Manager: Jennifer Jones

Marketing Assistant: Erica O'Connell

Marketing Communications Manager: Mary Anne Payumo

Content Project Manager: Cheryll Linthicum

Creative Director: Rob Hugel

Art Director: Vernon Boes

Print Buyer: Becky Cross

Rights Acquisitions Account Manager, Text: Roberta Broyer

Rights Acquisitions Account Manager, Image: Amanda Groszko

Production Service: Martha Emry

Text Designer: Diane Beasley

Art Editors: Leslie Lahr, Lisa Torri, Martha Emry

Photo Researcher: Kathleen Olsen

Copy Editor: Barbara Willette

Illustrators: Precision Graphics, Matrix Art Services, Network Graphics

Cover Designer: Terri Wright

Cover Image: Nathan Fariss for *Popular Mechanics*

Compositor: Graphic World

For product information and technology assistance, contact us at **Cengage Learning Customer & Sales Support, 1-800-354-9706.** For permission to use material from this text or product, submit all requests online at **www.cengage.com/permissions.** Further permissions questions can be e-mailed to **permissionrequest@cengage.com.**

Library of Congress Control Number: 2009941227

Student Edition:
ISBN-13: 978-0-534-46554-4

ISBN-10: 0-534-46554-4

**Brooks/Cole**
20 Davis Drive
Belmont, CA 94002-3098
USA

Cengage Learning is a leading provider of customized learning solutions with office locations around the globe, including Singapore, the United Kingdom, Australia, Mexico, Brazil, and Japan. Locate your local office at **www.cengage.com/global.**

Cengage Learning products are represented in Canada by Nelson Education, Ltd.

To learn more about Brooks/Cole, visit **www.cengage.com/brookscole**
Purchase any of our products at your local college store or at our preferred online store **www.ichapters.com.**

Printed in Canada
1 2 3 4 5 6 7 13 12 11 10

To Olivia, Maxwell, Sasha, Isabella, and Ashley

## About the Author

**SOO T. TAN** received his S.B. degree from the Massachusetts Institute of Technology, his M.S. degree from the University of Wisconsin–Madison, and his Ph.D. from the University of California at Los Angeles. He has published numerous papers on optimal control theory, numerical analysis, and the mathematics of finance. He is also the author of a series of textbooks on applied calculus and applied finite mathematics.

*One of the most important lessons I have learned from my many years of teaching undergraduate mathematics courses is that most students, mathematics and non-mathematics majors alike, respond well when introduced to mathematical concepts and results using real-life illustrations.*

*This awareness led to the intuitive approach that I have adopted in all of my texts. As you will see, I try to introduce each abstract mathematical concept through an example drawn from a common, real-life experience. Once the idea has been conveyed, I then proceed to make it precise, thereby assuring that no mathematical rigor is lost in this intuitive treatment of the subject. Another lesson I learned from my students is that they have a much greater appreciation of the material if the applications are drawn from their fields of interest and from situations that occur in the real world. This is one reason you will see so many examples and exercises in my texts that are drawn from various and diverse fields such as physics, chemistry, engineering, biology, business, and economics. There are also many exercises of general and current interest that are modeled from data gathered from newspapers, magazines, journals, and other media. Whether it be global warming, brain growth and IQ, projected U.S. gasoline usage, or finding the surface area of the Jacqueline Kennedy Onassis Reservoir, I weave topics of current interest into my examples and exercises to keep the book relevant to all of my readers.*

# Contents

## Author's Commitment to Accuracy

As with all of my projects, accuracy is of paramount importance. For this reason, I solved every problem myself and wrote the solutions for the solutions manual. In this accuracy checking process, I worked very closely with several professors who contributed in different ways and at different stages throughout the development of the text and manual: Jason Aubrey (*University of Missouri*), Kevin Charlwood (*Washburn University*), Jerrold Grossman (*Oakland University*), Tao Guo (*Rock Valley College*), James Handley (*Montanta Tech of the University of Montana*), Selwyn Hollis (*Armstrong Atlantic State University*), Diane Koenig (*Rock Valley College*), Michael Montano (*Riverside Community College*), John Samons (*Florida Community College*), Doug Shaw (*University of Northern Iowa*), and Richard West (*Francis Marion University*).

## Accuracy Process

### First Round

- The first draft of the manuscript was reviewed by numerous calculus instructors, all of whom either submitted written reviews, participated in a focus group discussion, or class-tested the manuscript.

### Second Round

- The author provided revised manuscript to be reviewed by additional calculus instructors who went through the same steps as the first group and submitted their responses.
- Simultaneously, author Soo Tan was writing the solutions manual, which served as an additional check of his work on the text manuscript.

### Third Round

- Two calculus instructors checked the revised manuscript for accuracy while simultaneously checking the solutions manual, sending their corrections back to the author for inclusion.
- Additional groups of calculus instructors participated in focus groups and class testing of the revised manuscript.
- First drafts of the art were produced and checked for accuracy.
- The manuscript was edited by a professional copyeditor.
- Biographies were written by a calculus instructor and submitted for copyedit.

### Fourth Round

- Once the manuscript was declared final, a compositor created galley pages, whose accuracy was checked by several calculus instructors.
- Revisions were made to the art, and revised art proofs were checked for accuracy.
- Further class testing and live reviews were completed.
- Galley proofs were checked for consistency by the production team and carefully reviewed by the author.
- Biographies were checked and revised for accuracy by another calculus instructor.

### Fifth Round

- First round page proofs were distributed, proofread, and checked for accuracy again. As with galley proofs, these pages were carefully reviewed by the author with art seen in place with the exposition for the first time.
- The revised art was again checked for accuracy by the author and the production service.

### Sixth Round

- Revised page proofs were checked by a second proofreader and the author.

### Seventh Round

- Final page proofs were checked for consistency by the production team and the author performed his final review of the pages.

# Preface

**Throughout my teaching career** I have always enjoyed teaching calculus and helping students to see the elegance and beauty of calculus. So when I was approached by my editor to write this series, I welcomed the opportunity. Upon reflecting, I see that I started this project from a strong vantage point. I have written an *Applied Mathematics* series, and over the years I have gotten a lot of feedback from many professors and students using the books in the series. The wealth of suggestions that I gained from them coupled with my experience in the classroom served me well when I embarked upon this project.

In writing the *Calculus* series, I have constantly borne in mind two primary objectives: first, to provide the instructor with a book that is easy to teach from and yet has all the content and rigor of a traditional calculus text, and second, to provide students with a book that motivates their interest and at the same time is easy for them to read. In my experience, students coming to calculus for the first time respond best to an intuitive approach, and I try to use this approach by introducing abstract ideas with concrete, real-life examples that students can relate to, wherever appropriate. Often a simple real-life illustration can serve as motivation for a more complex mathematical concept or theorem. Also, I have tried to use a clear, precise, and concise writing style throughout the book and have taken special care to ensure that my intuitive approach does not compromise the mathematical rigor that is expected of an engineering calculus text.

In addition to the applications in mathematics, engineering, physics, and the other natural and social sciences, I have included many other examples and exercises drawn from diverse fields of current interest. The solutions to all the exercises in the book are provided in a separate manual. In keeping with the emphasis on conceptual understanding, I have included concept questions at the beginning of each exercise set. In each end-of-chapter review section I have also included fill-in-the-blank questions for a review of the concepts. I have found these questions to be an effective learning tool to help students master the definitions and theorems in each chapter. Furthermore, I have included many questions that ask for the interpretation of graphical, numerical, and algebraic results in both the examples and the exercise sets.

## Unique Approach to the Presentation of Limits

Finally, I have employed a unique approach to the introduction of the limit concept. Many calculus textbooks introduce this concept via the slope of a tangent line to a curve and then follow by relating the slope to the notion of the rate of change of one quantity with respect to another. In my text I do precisely the opposite: I introduce the limit concept by looking at the rate of change of the maglev (magnetic levitation train). This approach is more intuitive and captures the interest of the student from the very beginning—it shows immediately the relevance of calculus to the real world. I might add that this approach has worked very well for me not only in the classroom; it has also been received very well by the users of my applied calculus series. This intuitive approach (using the maglev as a vehicle) is carried into the introduction and explanation of some of the fundamental theorems in calculus, such as the Intermediate Value Theorem and the Mean Value Theorem. Consistently woven throughout the text, this idea permeates much of the text—from concepts in limits, to continuity, to integration, and even to inverse functions.

*Soo T. Tan*

## ■ Tan *Calculus* Series

The Tan *Calculus* series includes the following textbooks:

- *Calculus: Early Transcendentals* © 2011 (ISBN 0-534-46554-4)
- *Single Variable Calculus: Early Transcendentals* © 2011 (ISBN 0-534-46570-6)
- *Calculus* © 2010 (ISBN 0-534-46579-X)
- *Single Variable Calculus* © 2010 (ISBN 0-534-46566-8)
- *Multivariable Calculus* © 2010 (ISBN 0-534-46575-7)

## ■ Features

### An Intuitive Approach . . . Without Loss of Rigor

Beginning with each chapter opening vignette and carrying through each chapter, Soo Tan's intuitive approach links the abstract ideas of calculus with concrete, real-life examples. This intuitive approach is used to advantage to introduce and explain many important concepts and theorems in calculus, such as tangent lines, Rolles's Theorem, absolute extrema, increasing and decreasing functions, limits at infinity, and parametric equations. In this example from Chapter 5 the discussion of the area between two curves is motivated with a real-life illustration that is followed by the precise discussion of the mathematical concepts involved.

### ■ A Real-Life Interpretation

Two cars are traveling in adjacent lanes along a straight stretch of a highway. The velocity functions for Car A and Car B are $v = f(t)$ and $v = g(t)$, respectively. The graphs of these functions are shown in Figure 1.

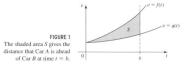 

**FIGURE 1**
The shaded area $S$ gives the distance that Car A is ahead of Car B at time $t = b$.

The area of the region under the graph of $f$ from $t = 0$ to $t = b$ gives the total distance covered by Car A in $b$ seconds over the time interval $[0, b]$. The distance covered by Car B over the same period of time is given by the area under the graph of $g$ on the interval $[0, b]$. Intuitively, we see that the area of the (shaded) region $S$ between the graphs of $f$ and $g$ on the interval $[0, b]$ gives the distance that Car A will be ahead of Car B at time $t = b$.

### ■ The Area Between Two Curves

Suppose $f$ and $g$ are continuous functions with $f(x) \geq g(x)$ for all $x$ in $[a, b]$, so that the graph of $f$ lies on or above that of $g$ on $[a, b]$. Let's consider the region $S$ bounded by the graphs of $f$ and $g$ between the vertical lines $x = a$ and $x = b$ as shown in Figure 2. To define the *area* of $S$, we take a regular partition of $[a, b]$,

$$a = x_0 < x_1 < x_2 < x_3 < \cdots < x_n = b$$

**FIGURE 2**
The region $S$ between the graphs of $f$ and $g$ on $[a, b]$

and form the Riemann sum of the function $f - g$ over $[a, b]$ with respect to this partition:

$$\sum_{k=1}^{n} [f(c_k) - g(c_k)] \Delta x$$

where $c_k$ is an evaluation point in the subinterval $[x_{k-1}, x_k]$ and $\Delta x = (b - a)/n$. The $k$th term of this sum gives the area of a rectangle with height $[f(c_k) - g(c_k)]$ and width $\Delta x$. As you can see in Figure 3, this area is an approximation of the area of the subregion of $S$ that lies between the graphs of $f$ and $g$ on $[x_{k-1}, x_k]$.

 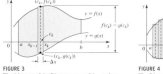 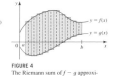

**FIGURE 3**
The $k$th term of the Riemann sum of $f - g$ gives the area of the $k$th rectangle of width $\Delta x$.

**FIGURE 4**
The Riemann sum of $f - g$ approximates the area of $S$.

## Unique Applications in the Examples and Exercises

Our relevant, unique applications are designed to illustrate mathematical concepts and at the same time capture students' interest.

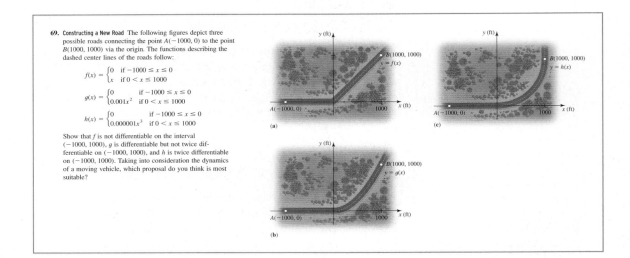

**69. Constructing a New Road** The following figures depict three possible roads connecting the point $A(-1000, 0)$ to the point $B(1000, 1000)$ via the origin. The functions describing the dashed center lines of the roads follow:

$$f(x) = \begin{cases} 0 & \text{if } -1000 \le x \le 0 \\ x & \text{if } 0 < x \le 1000 \end{cases}$$

$$g(x) = \begin{cases} 0 & \text{if } -1000 \le x \le 0 \\ 0.001x^2 & \text{if } 0 < x \le 1000 \end{cases}$$

$$h(x) = \begin{cases} 0 & \text{if } -1000 \le x \le 0 \\ 0.000001x^3 & \text{if } 0 < x \le 1000 \end{cases}$$

Show that $f$ is not differentiable on the interval $(-1000, 1000)$, $g$ is differentiable but not twice differentiable on $(-1000, 1000)$, and $h$ is twice differentiable on $(-1000, 1000)$. Taking into consideration the dynamics of a moving vehicle, which proposal do you think is most suitable?

## Connections

One particular example—the maglev (magnetic levitation) train—is used as a common thread throughout the development of calculus from limits through integration. The goal here is to show students the connection between the important theorems and concepts presented. Topics that are introduced through this example include the Intermediate Value Theorem, the Mean Value Theorem, the Mean Value Theorem for Definite Integrals, limits, continuity, derivatives, antiderivatives, initial value problems, inverse functions, and indeterminate forms.

### ■ A Real-Life Example

A prototype of a maglev (magnetic levitation train) moves along a straight monorail. To describe the motion of the maglev, we can think of the track as a coordinate line. From data obtained in a test run, engineers have determined that the maglev's displacement (directed distance) measured in feet from the origin at time $t$ (in seconds) is given by

$$s = f(t) = 4t^2 \qquad 0 \le t \le 30 \tag{1}$$

where $f$ is called the position function of the maglev. The position of the maglev at time $t = 0, 1, 2, 3, \ldots, 30$, measured in feet from its initial position, is

$$f(0) = 0, \qquad f(1) = 4, \qquad f(2) = 16, \qquad f(3) = 36, \qquad \ldots, \qquad f(30) = 3600$$

(See Figure 1.)

**FIGURE 1**
A maglev moving along an elevated monorail track

0    4         16                    36              3600    $s$ (ft)

## Precise Figures That Help Students Visualize the Concepts

Carefully constructed art helps the student to visualize the mathematical ideas under discussion.

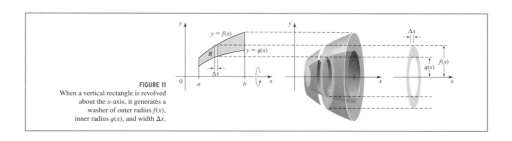

**FIGURE 11**
When a vertical rectangle is revolved about the *x*-axis, it generates a washer of outer radius $f(x)$, inner radius $g(x)$, and width $\Delta x$.

---

172    **Chapter 2**    The Derivative

### 2.1  CONCEPT QUESTIONS

**1. a.** Give a geometric and a physical interpretation of the expression

$$\frac{f(x + h) - f(x)}{h}$$

**b.** Give a geometric and a physical interpretation of the expression

$$\lim_{h \to 0} \frac{f(x + h) - f(x)}{h}$$

**2.** Under what conditions does a function fail to have a derivative at a number? Illustrate your answer with sketches.

### 2.1  EXERCISES

*In Exercises 1–14, use the definition of the derivative to find the derivative of the function. What is its domain?*

**1.** $f(x) = 5$
**2.** $f(x) = 2x + 1$
**3.** $f(x) = 3x - 4$
**4.** $f(x) = 2x^2 + x$
**5.** $f(x) = 3x^2 - x + 1$
**6.** $f(x) = x^3 - x$
**7.** $f(x) = 2x^3 + x - 1$
**8.** $f(x) = 2\sqrt{x}$
**9.** $f(x) = \sqrt{x + 1}$
**10.** $f(x) = \dfrac{1}{x}$
**11.** $f(x) = \dfrac{1}{x + 2}$
**12.** $f(x) = -\dfrac{2}{\sqrt{x}}$
**13.** $f(x) = \dfrac{3}{2x + 1}$
**14.** $f(x) = x + \sqrt{x}$

*In Exercises 15–20, find an equation of the tangent line to the graph of the function at the indicated point.*

| Function | Point |
|---|---|
| **15.** $f(x) = x^2 + 1$ | $(2, 5)$ |
| **16.** $f(x) = 3x^2 - 4x + 2$ | $(2, 6)$ |
| **17.** $f(x) = 2x^3$ | $(1, 2)$ |
| **18.** $f(x) = 3x^3 - x$ | $(-1, -2)$ |
| **19.** $f(x) = \sqrt{x - 1}$ | $(4, \sqrt{3})$ |
| **20.** $f(x) = \dfrac{2}{x}$ | $(2, 1)$ |

**21. a.** Find an equation of the tangent line to the graph of $f(x) = 2x - x^3$ at the point $(1, 1)$.
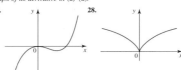 **b.** Plot the graph of $f$ and the tangent line in successively smaller viewing windows centered at $(1, 1)$ until the graph of $f$ and the tangent line appear to coincide.

**22. a.** In Example 6 we showed that $f(x) = |x|$ is not differentiable at $x = 0$. Plot the graph of $f$ using the viewing window $[-1, 1] \times [-1, 1]$. Then ZOOM IN using successively smaller viewing windows centered at $(0, 0)$. What can you say about the existence of a tangent line at $(0, 0)$?
**b.** Plot the graph of

$$f(x) = \begin{cases} x + 1 & \text{if } x \le 1 \\ \dfrac{2}{x} & \text{if } x > 1 \end{cases}$$

using the viewing window $[-2, 4] \times [-2, 3]$. Then ZOOM IN using successively smaller viewing windows centered at $(1, 2)$. Is $f$ differentiable at $x = 1$?

*In Exercises 23–26, find the rate of change of y with respect to x at the given value of x.*

**23.** $y = -2x^2 + x + 1$;   $x = 1$
**24.** $y = 2x^3 + 2$;   $x = 2$
**25.** $y = \sqrt{2x}$;   $x = 2$
**26.** $y = x^2 - \dfrac{1}{x}$;   $x = -1$

*In Exercises 27–30, match the graph of each function with the graph of its derivative in (a)–(d).*

**27.**

**28.**

### Concept Questions

Designed to test student understanding of the basic concepts discussed in the section, these questions encourage students to explain learned concepts in their own words.

### Exercises

Each exercise section contains an ample set of problems of a routine computational nature, followed by a set of application-oriented problems (many of them sourced) and true/false questions that ask students to explain their answer.

### Graphing Utility and CAS Exercises

Indicated by and **cas** icons next to the corresponding exercises, these exercises offer practice in using technology to solve problems that might be difficult to solve by hand. Sourced problems using real-life data are often included.

## Concept Review Questions

Beginning each end of chapter review, these questions give students a chance to check their knowledge of the basic definitions and concepts from the chapter.

## Review Exercises

Offering a solid review of the chapter material, these exercises contain routine computational exercises as well as applied problems.

## Problem-Solving Techniques

At the end of selected chapters the author discusses problem-solving techniques that provide students with the tools they need to make seemingly complex problems easier to solve.

---

Concept Review    1137

## CHAPTER 12 REVIEW

### CONCEPT REVIEW

*In Exercises 1–17, fill in the blanks.*

**1. a.** A function $f$ of two variables, $x$ and $y$, is a _____ that assigns to each ordered pair _____ in the domain of $f$, exactly one real number $f(x, y)$.
  **b.** The number $z = f(x, y)$ is called a _____ variable, and $x$ and $y$ are _____ variables. The totality of the numbers $z$ is called the _____ of the function $f$.
  **c.** The graph of $f$ is the set $S = $ _____.

**2. a.** The curves in the $xy$-plane with equation $f(x, y) = k$, where $k$ is a constant in the range of $f$, are called the _____ of $f$.
  **b.** A level surface of a function $f$ of three variables is the graph of the equation _____, where $k$ is a constant in the range of _____.

**3.** $\lim_{(x, y) \to (a, b)} f(x, y) = L$ means there exists a number _____ such that $f(x, y)$ can be made as close to _____ as we please by restricting $(x, y)$ to be sufficiently close to _____.

**4.** If $f(x, y)$ approaches $L_1$ as $(x, y)$ approaches $(a, b)$ along one path, and $f(x, y)$ approaches $L_2$ as $(x, y)$ approaches $(a, b)$ along another path with $L_1 \ne L_2$, then $\lim_{(x, y) \to (a, b)} f(x, y)$ _____ exist.

**5. a.** $f(x, y)$ is continuous at $(a, b)$ if $\lim_{(x, y) \to (a, b)} f(x, y) = $ _____.
  **b.** $f(x, y)$ is continuous on a region $R$ if $f$ is continuous at every point $(x, y)$ in _____.

**6. a.** A polynomial function is continuous _____; a rational function is continuous at all points in its _____.
  **b.** If $f$ is continuous at $(a, b)$ and $g$ is continuous at $f(a, b)$, then the composite function $h = g \circ f$ is continuous at _____.

**7. a.** The partial derivative of $f(x, y)$ with respect to $x$ is _____ if the limit exists. The partial derivative $(\partial f/\partial x)(a, b)$ gives the slope of the tangent line to the curve obtained by the intersection of the plane _____ and the graph of $z = f(x, y)$ at _____; it also measures the rate of change of $f(x, y)$ in the _____-direction with $y$ held _____ at _____.
  **b.** To compute $\partial f/\partial x$ where $f$ is a function of $x$ and $y$, treat _____ as a constant and differentiate with respect to _____ in the usual manner.

**8.** If $f(x, y)$ and its partial derivatives $f_x, f_y, f_{xy}$, and $f_{yx}$ are continuous on an open region $R$, then $f_{xy}(x, y) = $ _____ for all $(x, y)$ in $R$.

**9. a.** The total differential $dz$ of $z = f(x, y)$ is $dz = $ _____.
  **b.** If $\Delta z = f(x + \Delta x, y + \Delta y) - f(x, y)$, then $\Delta z \approx$ _____.
  **c.** $\Delta z = f_x(x, y) \, \Delta x + f_y(x, y) \, \Delta y + \varepsilon_1 \, \Delta x + \varepsilon_2 \, \Delta y$, where $\varepsilon_1$ and $\varepsilon_2$ are functions of _____ and _____ such that $\lim_{(\Delta x, \Delta y) \to (0, 0)} \varepsilon_1 = $ _____ and $\lim_{(\Delta x, \Delta y) \to (0, 0)} \varepsilon_2 = $ _____.
  **d.** The function $z = f(x, y)$ is differentiable at $(a, b)$ if $\Delta z$ can be expressed in the form $\Delta z = $ _____, where _____ and _____ as $(\Delta x, \Delta y) \to$ _____.

**10. a.** If $f$ is a function of $x$ and $y$, and $f_x$ and $f_y$ are continuous on an open region $R$, then $f$ is _____ in $R$.
  **b.** If $f$ is differentiable at $(a, b)$, then $f$ is _____ at $(a, b)$.

**11. a.** If $w = f(x, y)$, $x = g(t)$, and $y = h(t)$, then under suitable conditions the Chain Rule gives $dw/dt = $ _____.
  **b.** If $w = f(x, y)$, $x = g(u, v)$, and $y = h(u, v)$, then $\partial w/\partial u = $ _____.
  **c.** If $F(x, y) = 0$ _____, pr _____
  **d.** If $F(x, y, z) = $ _____ as _____ $z$ implicitly as _____ and _____.

**12. a.** If $f$ is a functi _____ vector, then th _____ of $\mathbf{u}$ is $D_{\mathbf{u}}f(x$ _____
  **b.** The directiona _____ change of $f$ at _____
  **c.** If $f$ is differen _____
  **d.** The gradient _____
  **e.** In terms of th _____

**13. a.** The maximum _____ occurs when t _____
  **b.** The minimum _____ occurs when t _____

**14. a.** $\nabla f$ is _____
  **b.** $\nabla F$ is _____
  **c.** The tangent p _____ point $P(a, b,$ _____ through $P(a,$ _____

**15. a.** If $f(x, y) \le f($ _____ ing $(a, b)$, the _____
  **b.** If $f(x, y) \ge f($ _____ has an _____

---

Review Exercises    1025

### REVIEW EXERCISES

*In Exercises 1–4, sketch the curve with the given vector equation, and indicate the orientation of the curve.*

**1.** $\mathbf{r}(t) = (2 + 3t)\mathbf{i} + (2t - 1)\mathbf{j}$
**2.** $\mathbf{r}(t) = t^3\mathbf{i} + t^2\mathbf{j}; \quad 0 \le t \le 2$
**3.** $\mathbf{r}(t) = (\cos t - 1)\mathbf{i} + (\sin t + 2)\mathbf{j} + 2\mathbf{k}$
**4.** $\mathbf{r}(t) = 2\cos t\mathbf{i} + 3\sin t\mathbf{j} + t^2\mathbf{k}; \quad 0 \le t \le 2\pi$
**5.** Find the domain of $\mathbf{r}(t) = \dfrac{1}{\sqrt{5 - t}}\mathbf{i} + \dfrac{\sin t}{t}\mathbf{j} + \ln(1 + t)\mathbf{k}$.
**6.** Find $\lim_{t \to 0^+} \mathbf{r}(t)$, where $\mathbf{r}(t) = \dfrac{\sqrt{t}}{1 + t^2}\mathbf{i} + \dfrac{t^2}{\sin t}\mathbf{j} + \dfrac{e^t - 1}{t}\mathbf{k}$.
**7.** Find the interval in which
$$\mathbf{r}(t) = \sqrt{t + 1}\mathbf{i} + \frac{e^t}{\sqrt{2 - t}}\mathbf{j} + \frac{t^2}{(t - 1)^2}\mathbf{k}$$
is continuous.
**8.** Find $\mathbf{r}'(t)$ if $\mathbf{r}(t) = \left[\int_0^t \cos^2 u \, du\right]\mathbf{i} + \left[\int_0^t \sin u \, du\right]\mathbf{j}$.

*In Exercises 9–12, find $\mathbf{r}'(t)$ and $\mathbf{r}''(t)$.*

**9.** $\mathbf{r}(t) = \sqrt{t}\mathbf{i} + t^2\mathbf{j} + \dfrac{1}{t + 1}\mathbf{k}$
**10.** $\mathbf{r}(t) = e^{-t}\mathbf{i} + t\cos t\mathbf{j} + t\sin t\mathbf{k}$
**11.** $\mathbf{r}(t) = (t^2 + 1)\mathbf{i} + 2t\mathbf{j} + \ln t\mathbf{k}$
**12.** $\mathbf{r}(t) = \langle t\sin t, t\cos t, e^{2t} \rangle$

*In Exercises 13 and 14, find parametric equations for the tangent line to the curve with the given parametric equations at the point with the given value of $t$.*

**13.** $x = t^2 + 1, \quad y = 2t - 3, \quad z = t^3 + 1; \quad t = 0$
**14.** $x = t\cos t - \sin t, \quad y = t\sin t + \cos t, \quad z = t^2; \quad t = \dfrac{\pi}{2}$

*In Exercises 15 and 16, evaluate the integral.*

**15.** $\displaystyle\int \left(\sqrt{t}\mathbf{i} + e^{-2t}\mathbf{j} + \frac{1}{t + 1}\mathbf{k}\right) dt$
**16.** $\displaystyle\int_1 (2t\mathbf{i} + t^2\mathbf{j} + t^{3/2}\mathbf{k}) \, dt$

*In Exercises 19 and 20, find the unit tangent and the unit normal vectors for the curve C defined by $\mathbf{r}(t)$ for the given value of $t$.*

**19.** $\mathbf{r}(t) = t\mathbf{i} + t^2\mathbf{j} + t^3\mathbf{k}; \quad t = 1$
**20.** $\mathbf{r}(t) = 2\cos t\mathbf{i} + 2\sin t\mathbf{j} + e^t\mathbf{k}; \quad t = 0$

*In Exercises 21 and 22, find the length of the curve.*

**21.** $\mathbf{r}(t) = 2\sin 2t\mathbf{i} + 2\cos 2t\mathbf{j} + 3t\mathbf{k}; \quad 0 \le t \le 2$
**22.** $\mathbf{r}(t) = \sqrt{2}t\mathbf{i} + \dfrac{1}{2}t^2\mathbf{j} + \ln t\mathbf{k}; \quad 1 \le t \le 2$

*In Exercises 23 and 24, find the curvature of the curve.*

**23.** $\mathbf{r}(t) = t\mathbf{i} + t^2\mathbf{j} + t^3\mathbf{k}$
**24.** $\mathbf{r}(t) = t\sin t\mathbf{i} + t\cos t\mathbf{j} + t\mathbf{k}$

*In Exercises 25 and 26, find the curvature of the plane curve, and determine the point on the curve at which the curvature is largest.*

**25.** $y = x - \dfrac{1}{4}x^2$        **26.** $y = e^{-x}$

*In Exercises 27 and 28, find the velocity, acceleration, and speed of the object with the given position vector.*

**27.** $\mathbf{r}(t) = 2t\mathbf{i} + e^{-2t}\mathbf{j} + \cos t\mathbf{k}$
**28.** $\mathbf{r}(t) = te^{-t}\mathbf{i} + \cos 2t\mathbf{j} + \sin 2t\mathbf{k}$

*In Exercises 29 and 30, find the velocity and position vectors of an object with the given acceleration and the given initial velocity and position.*

**29.** $\mathbf{a}(t) = t\mathbf{i} + \dfrac{1}{3}t^2\mathbf{j} + 3\mathbf{k}; \quad \mathbf{v}(0) = 2\mathbf{i} + 3\mathbf{j} + \mathbf{k}, \quad \mathbf{r}(0) = \mathbf{0}$
**30.** $\mathbf{a}(t) = e^t\mathbf{i} + e^{-t}\mathbf{j} + t\mathbf{k}; \quad \mathbf{v}(0) = 2\mathbf{i}, \quad \mathbf{r}(0) = \mathbf{i} + \mathbf{k}$

*In Exercises 31–34, find the scalar tangential and normal components of acceleration of a particle with the given position vector.*

**31.** $\mathbf{r}(t) = \mathbf{i} + t\mathbf{j} + t^2\mathbf{k}$
**32.** $\mathbf{r}(t) = 2\cos t\mathbf{i} + 3\sin t\mathbf{j} + t\mathbf{k}$
**33.** $\mathbf{r}(t) = \cos t\mathbf{i} + \sin 2t\mathbf{j}$
**34.** $\mathbf{r}(t) = \sqrt{2}t\mathbf{i} + e^t\mathbf{j} + e^{-t}\mathbf{k}$

**35. A Shot Put** In a track and field meet, a shot putter heaves a shot at an angle of 45° with the horizontal. As the shot leaves her hand, it is at a height of 7 ft and moving at a speed of 40 ft/sec. Set up a coordinate system so that the shot putter is at the origin.
  **a.** What is the position of the shot at time $t$?
  **b.** How far is her put?

*(continued)* ...ctor function $\mathbf{r}'(t)$ or ... $\mathbf{r}(0) = \mathbf{i} + 2\mathbf{j}$ ...**k**,

---

### PROBLEM-SOLVING TECHNIQUES

The following example shows that rewriting a function in an alternative form sometimes pays dividends.

**EXAMPLE** Find $f^{(n)}(x)$ if $f(x) = \dfrac{x}{x^2 - 1}$.

**Solution** Our first instinct is to use the Quotient Rule to compute $f'(x)$, $f''(x)$, and so on. The expectation here is either that the rule for $f^{(n)}$ will become apparent or that at least a pattern will emerge that will enable us to guess at the form for $f^{(n)}(x)$. But the futility of this approach will be evident when you compute the first two derivatives of $f$.

Let's see whether we can transform the expression for $f(x)$ before we differentiate. You can verify that $f(x)$ can be written as

$$f(x) = \frac{x}{x^2 - 1} = \frac{\frac{1}{2}(x - 1) + \frac{1}{2}(x + 1)}{(x + 1)(x - 1)} = \frac{1}{2}\left[\frac{1}{x + 1} + \frac{1}{x - 1}\right]$$

## CHALLENGE PROBLEMS

1. Find $\lim_{x \to 2} \dfrac{x^{10} - 2^{10}}{x^5 - 2^5}$.

2. Find the derivative of $y = \sqrt{x + \sqrt{x + \sqrt{x}}}$.

3. **a.** Verify that $\dfrac{2x + 1}{x^2 + x - 2} = \dfrac{1}{x + 2} + \dfrac{1}{x - 1}$.

   **b.** Find $f^{(n)}(x)$ if $f(x) = \dfrac{2x + 1}{x^2 + x - 2}$.

4. Find the values of $x$ for which $f$ is differentiable.
   **a.** $f(x) = \sin|x|$     **b.** $f(x) = |\sin x|$

5. Find $f^{(10)}(x)$ if $f(x) = \dfrac{1 + x}{\sqrt{1 - x}}$.

   **Hint:** Show that $f(x) = \dfrac{2}{\sqrt{1 - x}} - \sqrt{1 - x}$.

6. Find $f^{(n)}(x)$ if $f(x) = \dfrac{ax + b}{cx + d}$.

7. Suppose that $f$ is differentiable and $f(a + b) = f(a)f(b)$ for all real numbers $a$ and $b$. Show that $f'(x) = f'(0)f(x)$ for all $x$.

8. Suppose that $f^{(n)}(x) = 0$ for every $x$ in an interval $(a, b)$ and $f(c) = f'(c) = \cdots = f^{(n-1)}(c) = 0$ for some $c$ in $(a, b)$. Show that $f(x) = 0$ for all $x$ in $(a, b)$.

9. Let $F(x) = f(\sqrt{1 + x^2})$, where $f$ is a differentiable function. Find $F'(x)$.

10.  Determine the values of $b$ and $c$ such that the parabola $y = x^2 + bx + c$ is tangent to the graph of $y = \sin x$ at the point $\left(\frac{\pi}{6}, \frac{1}{2}\right)$. Plot the graphs of both functions on the same set of axes.

11. Suppose $f$ is defined on $(-\infty, \infty)$ and satisfies $|f(x) - f(y)| \leq (x - y)^2$ for all $x$ and $y$. Show that $f$ is a constant function.
    **Hint:** Look at $f'(x)$.

12. Use the definition of the derivative to find the derivative of $f(x) = \tan ax$.

13. Find $y''$ at the point $(1, -2)$ if
$$2x^2 + 2xy + xy^2 - 3x + 3y + 7 = 0$$

## Challenge Problems

Providing students with an opportunity to stretch themselves, the Challenge Problems develop their skills beyond the basics. These can be solved by using the techniques developed in the chapter but require more effort than the problems in the regular exercise sets do.

## Guidance When Students Need It

The caution icon advises students how to avoid common mistakes and misunderstandings. This feature addresses both student misconceptions and situations in which students often follow unproductive paths.

> ⚠ Theorem 1 states that a relative extremum of $f$ can occur only at a critical number of $f$. It is important to realize, however, that the converse of Theorem 1 is false. In other words, you may *not* conclude that if $c$ is a critical number of $f$, then $f$ must have a relative extremum at $c$. (See Example 3.)

## Biographies to Provide Historical Context

Historical biographies provide brief looks at the people who contributed to the development of calculus, focusing not only on their discoveries and achievements, but on their human side as well.

## Videos to Help Students Draw Complex Multivariable Calculus Artwork

Unique to this book, Tan's *Calculus* provides video lessons for the multivariable sections of the text that help students learn, step-by-step, how to draw the complex sketches required in multivariable calculus. Videos of these lessons will be available at the text's companion website.

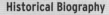

### Historical Biography

Sheila Terry/Photo Researchers, Inc.

**BLAISE PASCAL**
**(1623-1662)**

A great mathematician who was not acknowledged in his lifetime, Blaise Pascal came extremely close to discovering calculus before Leibniz (page 157) and Newton (page 179), the two people who are most commonly credited with the discovery. Pascal was something of a prodigy and published his first important mathematical discovery at the age of sixteen. The work consisted of only a single printed page, but it contained a vital step in the development of projective geometry and a proposition called *Pascal's mystic hexagram* that discussed a property of a hexagon inscribed in a conic section. Pascal's interests varied widely, and from 1642 to 1644 he worked on the first manufactured calculator, which he designed to help his father with his tax work. Pascal manufactured about 50 of the machines, but they proved too costly to continue production. The basic principle of Pascal's calculating machine was still used until the electronic age. Pascal and Pierre de Fermat (page 307) also worked on the mathematics in games of chance and laid the foundation for the modern theory of probability. Pascal's later work, *Treatise on the Arithmetical Triangle*, gave important results on the construction that would later bear his name, *Pascal's Triangle*.

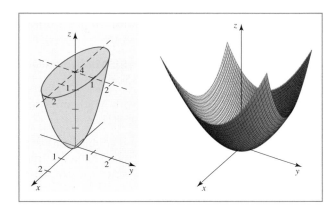

## ■ Instructor Resources

**Instructor's Solutions Manual for Single Variable Calculus:**
**Early Transcendentals** (ISBN 0-534-46572-2)

**Instructor's Solutions Manual for Multivariable Calculus** (ISBN 0-534-46578-1)
Prepared by Soo T. Tan

These manuals provide worked-out solutions to all problems in the text.

**PowerLecture CD** (ISBN 0-495-11482-0)
This comprehensive CD-ROM includes the *Instructor's Solutions Manual;* PowerPoint
slides with art, tables, and key definitions from the text; and ExamView computer-
ized testing, featuring algorithmically generated questions to create, deliver, and cus-
tomize tests. A static version of the test bank will also be available online.

**Solution Builder** (ISBN 0-534-41831-7)
The online Solution Builder lets instructors easily build and save personal solution
sets either for printing or for posting on password-protected class websites. Contact
your local sales representative for more information on obtaining an account for
this instructor-only resource.

**Enhanced WebAssign** (ISBN 0-495-39345-2)
Instant feedback and ease of use are just two reasons why WebAssign is the most
widely used homework system in higher education. WebAssign allows instructors to
assign, collect, grade, and record homework assignments via the Web. Now this
proven homework system has been enhanced to include links to textbook sections,
video examples, and problem-specific tutorials. Enhanced WebAssign is more than
a homework system—it is a complete learning system for math students.

## ■ Student Resources

**Student Solutions Manual for Single Variable Calculus:**
**Early Transcendentals** (ISBN 0-534-46573-0)

**Student Solutions Manual for Multivariable Calculus** (ISBN 0-534-46577-3)
Prepared by Soo T. Tan

Providing more in-depth explanations, this insightful resource includes fully worked-
out solutions for the answers to select exercises included at the back of the textbook,
as well as problem-solving strategies, additional algebra steps, and review for selected
problems.

**CalcLabs with Maple: Single Variable Calculus, 4e** by Phil Yasskin and Art
Belmonte (ISBN 0-495-56062-6)

**CalcLabs with Maple: Multivariable Calculus, 4e** by Phil Yasskin and Art
Belmonte (ISBN 0-495-56058-8)

**CalcLabs with Mathematica: Single Variable Calculus, 4e** by Selwyn Hollis
(ISBN 0-495-56063-4)

**CalcLabs with Mathematica: Multivariable Calculus, 4e** by Selwyn Hollis
(ISBN 0-495-82722-3)

Each of these comprehensive lab manuals helps students learn to effectively use the
technology tools that are available to them. Each lab contains clearly explained exercises and a variety of labs and projects to accompany the text.

# ■ Acknowledgments

I want to express my heartfelt thanks to the reviewers for their many helpful comments and suggestions at various stages during the development of this text. I also want to thank Kevin Charlwood, Jerrold Grossman, Tao Guo, James Handley, Selwyn Hollis, Diane Koenig, and John Samons, who checked the manuscript and text for accuracy; Richard West, Richard Montano, and again Kevin Charlwood for class testing the manuscript; and Andrew Bulman-Fleming for his help with the production of the solutions manuals. Additionally, I would like to thank Diane Koenig and Jason Aubrey for writing the biographies and also Doug Shaw and Richard West for their work on the projects. A special thanks to Tao Guo for his contribution to the content and accuracy of the solutions manuals.

I feel fortunate to have worked with a wonderful team during the development and production of this text. I wish to thank the editorial, production, and marketing staffs of Cengage Learning: Richard Stratton, Liz Covello, Cheryll Linthicum, Danielle Derbenti, Ed Dodd, Terri Mynatt, Leslie Lahr, Jeannine Lawless, Peter Galuardi, Lauren Hamel, Jennifer Jones, Angela Kim, and Mary Ann Payumo. My editor, Liz Covello, who joined the team this year, has done a great job working with me to finalize the product before publication. My development editor, Danielle Derbenti, as in the many other projects I have worked with her on, brought her enthusiasm and expertise to help me produce a better book. My production manager, Cheryll Linthicum, coordinated the entire project with equal enthusiasm and ensured that the production process ran smoothly from beginning to end. I also wish to thank Martha Emry, Barbara Willette, and Marian Selig for the excellent work they did in the production of this text. Martha spent countless hours working with me to ensure the accuracy and readability of the text and art. Without the help, encouragement, and support of all those mentioned above, I wouldn't have been able to complete this mammoth task.

I wish to express my personal appreciation to each of the following colleagues whose many suggestions have helped to make this a much improved book.

Arun Agarwal
*Grambling State University*

Mazenia Agustin
*Southern Illinois University–Edwardsville*

Mike Albanese
*Central Piedmont Community College*

Robert Andersen
*University of Wisconsin–Eau Claire*

Daniel Anderson
*George Mason University*

Joan Bell
*Northeastern State University*

David Bradley
*University of Maine–Orono*

Bob Bradshaw
*Ohlone College*

Paul Britt
*Louisiana State University*

Bob Buchanon
*Millersville University*

Christine Bush
*Palm Beach Community College*

Nick Bykov
*San Joaquin Delta College*

Janette Campbell
*Palm Beach Community College*

Kevin Charlwood
*Washburn University*

S.C. Cheng
*Creighton University*

Vladimir Cherkassky
*Diablo Valley College*

Charles Cooper
*University of Central Oklahoma*

Kyle Costello
*Salt Lake Community College*

Katrina Cunningham
*Southern University–Baton Rouge*

Eugene Curtin
*Texas State University*

Wendy Davidson
*Georgia Perimeter College–Newton*

Steven J. Davidson
*San Jacinto College*

John Davis
*Baylor University*

Ann S. DeBoever
*Catawba Valley Community College*

John Diamantopoulos
*Northwestern State University*

John Drost
*University of Wisconsin–Eau Claire*

Joe Fadyn
*Southern Polytechnic State University*

Tom Fitzkee
*Francis Marion University*

James Galloway
*Collin County Community College*

Jason Andrew Geary
*Harper College*

Don Goral
*Northern Virginia Community College*

Alan Graves
*Collin County Community College*

Elton Graves
*Rose-Hulman Institute*

Ralph Grimaldi
*Rose-Hulman Institute*

Ron Hammond
*Blinn College–Bryan*

James Handley
*Montana Tech of the University of Montana*

Patricia Henry
*Drexel University*

Irvin Hentzel
*Iowa State University*

Alfa Heryudono
*University of Massachusetts–Dartmouth*

Guy Hinman
*Brevard Community College–Melbourne*

Gloria Hitchcock
*Georgia Perimeter College–Newton*

Joshua Holden
*Rose-Hulman Institute*

Martin Isaacs
*University of Wisconsin–Madison*

Mic Jackson
*Earlham College*

Hengli Jiao
*Ferris State College*

Clarence Johnson
*Cuyahoga Community College–Metropolitan*

Cindy Johnson
*Heartland Community College*

Phil Johnson
*Appalachian State University*

Jack Keating
*Massasoit Community College*

John Khoury
*Brevard Community College–Melbourne*

Raja Khoury
*Collin County Community College*

Rethinasamy Kittappa
*Millersville University*

Carole Krueger
*University of Texas–Arlington*

Don Krug
*Northern Kentucky University*

Kouok Law
*Georgia Perimeter College–Lawrenceville*

Richard Leedy
*Polk Community College*

Suzanne Lindborg
*San Joaquin Delta College*

Tristan Londre
*Blue River Community College*

Ann M. Loving
*J. Sargeant Reynolds Community College*

Cyrus Malek
*Collin County Community College*

Robert Maynard
*Tidewater Community College*

Phillip McCartney
*Northern Kentucky University*

Robert McCullough
*Ferris State University*

Shelly McGee
*University of Findlay*

Rhonda McKee
*Central Missouri State University*

George McNulty
*University of South Carolina–Columbia*

Martin Melandro
*Sam Houston State University*

Mike Montano
*Riverside Community College*

Humberto Munoz
*Southern University–Baton Rouge*

Robert Nehs
*Texas Southern University*

## Board of Advisors

## ■ Note to the Student

The invention of calculus is one of the crowning intellectual achievements of mankind. Its roots can be traced back to the ancient Egyptians, Greeks, and Chinese. The invention of modern calculus is usually credited to both Gottfried Wilhelm Leibniz and Isaac Newton in the seventeenth century. It has widespread applications in many fields, including engineering, the physical and biological sciences, economics, business, and the social sciences. I am constantly amazed not only by the wonderful mathematical content in calculus but also by the enormous reach it has into every practical field of human endeavor. From studying the growth of a population of bacteria, to building a bridge, to exploring the vast expanses of the heavenly bodies, calculus has always played and continues to play an important role in these endeavors.

In writing this book, I have constantly kept you, the student, in mind. I have tried to make the book as interesting and readable as possible. Many mathematical concepts are introduced by using real-life illustrations. On the basis of my many years of teaching the subject, I am convinced that this approach makes it easier for you to understand the definitions and theorems in this book. I have also taken great pains to include as many steps in the examples as are needed for you to read through them smoothly. Finally, I have taken particular care with the graphical illustrations to ensure that they help you to both understand a concept and solve a problem.

The exercises in the book are carefully constructed to help you understand and appreciate the power of calculus. The problems at the beginning of each exercise set are relatively straightforward to solve and are designed to help you become familiar with the material. These problems are followed by others that require a little more effort on your part. Finally, at the end of each exercise set are problems that put the material you have just learned to good use. Here you will find applications of calculus that are drawn from many fields of study. I think you will also enjoy solving real-life problems of general interest that are drawn from many current sources, including magazines and newspapers. The answers often reveal interesting facts.

However interesting and exciting as it may be, reading a calculus book is not an easy task. You might have to go over the definitions and theorems more than once in order to fully understand them. Here you should pay careful attention to the conditions stated in the theorems. Also, it's a good idea to try to understand the definitions, theorems, and procedures as thoroughly as possible before attempting the exercises. Sometimes writing down a formula is a good way to help you remember it. Finally, if you study with a friend, a good test of your mastery of the material is to take turns explaining the topic you are studying to each other.

One more important suggestion: When you write out the solutions to the problems, make sure that you do so neatly, and try to write down each step to explain how you arrive at the solution. Being neat helps you to avoid mistakes that might occur through misreading your own handwriting (a common cause of errors in solving problems), and writing down each step helps you to work through the solution in a logical manner and to find where you went wrong if your answer turns out to be incorrect. Besides, good habits formed here will be of great help when you write reports or present papers in your career later on in life.

Finally, let me say that writing this book has been a labor of love, and I hope that I can convince you to share my love and enthusiasm for the subject.

*Soo T. Tan*

Clark County in Nevada—dominated by greater Las Vegas—was the fastest-growing metropolitan area in the United States from 1990 through the early 2000s. In this chapter, we will construct a mathematical model that can be used to describe how the population of Clark County grew over that period.

Jose Fuste/Raga/Corbis

# 0

# Preliminaries

**LINES PLAY AN** important role in calculus, albeit indirectly. So we begin our study of calculus by looking at the properties of lines in the plane. Next, we turn our attention to the discussion of functions. More specifically, we will see how functions can be combined to yield other functions; we will see how functions can be represented graphically; and finally, we will see how functions afford us a way to describe real-world phenomena in mathematical terms.

In this chapter we also look at some of the ways in which graphing calculators and computer algebra systems can help us in our study of calculus. Finally, we consider two families of very important functions: the exponential and logarithmic functions. They play an important role in both mathematics and its applications.

Ⓥ This symbol indicates that one of the following video types is available for enhanced student learning at **www.academic.cengage.com/login:**
- Chapter lecture videos
- Solutions to selected exercises

## 0.1   Lines

*The diagnostic tests that appear at the beginning of each section in Chapter 0 (other than Section 0.5) are designed to allow you to determine whether you should spend time reviewing the material in that section or should skip it and move on.*

### 0.1   SELF-CHECK DIAGNOSTIC TEST

1. Find an equation of the line passing through the points $(-1, 3)$ and $(2, -4)$.

2. Find an equation of the line that passes through the point $(3, -2)$ and is perpendicular to the line with equation $2x + 3y = 6$.

3. Determine whether the points $A(-1, 4)$, $B(1, 1)$, and $C(3, -4)$ lie on a straight line.

4. Find an equation of the line that has an $x$-intercept of 4 and a $y$-intercept of 6.

5. Find an equation of the line that is parallel to the line $3x + 4y = 6$ and passes through the point of intersection of the lines $4x - 5y = 1$ and $2x + 3y = -5$.

*Answers to Self-Check Diagnostic Test 0.1 can be found on page ANS 1.*

Figure 1a depicts a ladder leaning against a vertical wall, and Figure 1b depicts the trajectory of an aircraft flying along a straight line shortly after takeoff. How do we measure the steepness of the ladder (with respect to the ground) and the steepness of the flight path of the plane (with respect to the horizontal)? To answer these questions, we need to define the steepness or the slope of a straight line. (We will solve the problems posed here in Examples 2 and 3, respectively.)

**FIGURE 1**      (**a**) How steep is the ladder?                                    (**b**) How steep is the path of the plane?

### ■ Slopes of Lines

#### DEFINITION   Slope

Let $L$ be a nonvertical line in a coordinate plane. If $P_1(x_1, y_1)$ and $P_2(x_2, y_2)$ are any two distinct points on $L$, then the **slope** of $L$ is

$$m = \frac{\Delta y}{\Delta x} = \frac{y_2 - y_1}{x_2 - x_1} \tag{1}$$

(See Figure 2.) The slope of a vertical line is undefined.

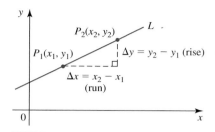

**FIGURE 2**
The slope of the line $L$ is
$$m = \frac{\Delta y}{\Delta x} = \frac{y_2 - y_1}{x_2 - x_1} = \frac{\text{rise}}{\text{run}}.$$

The quantity $\Delta y = y_2 - y_1$ ($\Delta y$ is read "delta $y$") measures the change in $y$ from $P_1$ to $P_2$ and is called the **rise**; the quantity $\Delta x = x_2 - x_1$ measures the change in $x$ from $P_1$ to $P_2$ and is called the **run.** Thus, the slope of a line is the ratio of its rise to its run.

Since the ratios of corresponding sides of similar triangles are equal, we see from Figure 3 that the slope of a line is independent of the two distinct points that are used to compute it; that is,

$$m = \frac{y_2 - y_1}{x_2 - x_1} = \frac{y_2' - y_1'}{x_2' - x_1'}$$

**FIGURE 3**
The slope of a nonvertical straight line is independent of the two distinct points used to compute it.

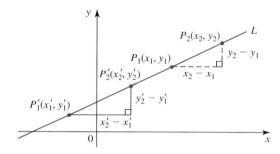

The slope of a straight line is a numerical measure of its steepness with respect to the positive $x$-axis. In fact, if we take $\Delta x = x_2 - x_1$ to be equal to 1 in Equation (1), then we see that

$$m = \frac{\Delta y}{\Delta x} = \frac{\Delta y}{1} = \Delta y = y_2 - y_1$$

gives the *change in y per unit change in x.*

Figure 4 shows four lines with different slopes. By taking a run of 1 unit to compute each slope, you can see that the larger the absolute value of the slope is, the larger the change in $y$ per unit change in $x$ is and, therefore, the steeper the line is. We also see that if $m > 0$, the line slants upward; if $m < 0$, the line slants downward; and finally, if $m = 0$, the line is horizontal.

**FIGURE 4**
The slope of a line is a numerical measure of its steepness.

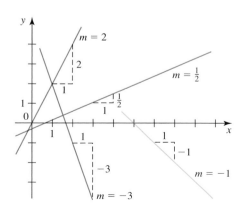

**EXAMPLE 1**   Find the slope of the line passing through (a) the points $P_1(1, 1)$ and $P_2(3, 5)$ and (b) the points $P_1(1, 3)$ and $P_2(3, 2)$.

**Solution**
**a.** Using Equation (1), we obtain the required slope as

$$m = \frac{5 - 1}{3 - 1} = 2$$

This tells us that $y$ increases by 2 units for each unit increase in $x$ (see Figure 5a).
**b.** Equation (1) gives the required slope as

$$m = \frac{2 - 3}{3 - 1} = -\frac{1}{2}$$

This tells us that $y$ decreases by $\frac{1}{2}$ unit for each unit increase in $x$ or, equivalently, $y$ decreases by 1 unit for each increase of 2 units in $x$ (see Figure 5b).

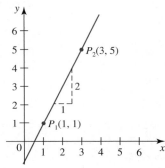    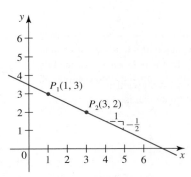

**FIGURE 5**       (a) The slope of the line is 2.        (b) The slope of the line is $-\frac{1}{2}$.

**Note**   In Example 1 we arbitrarily labeled the point $P_1(1, 1)$ and the point $P_2(3, 5)$. Suppose we had labeled the points $P_1(3, 5)$ and $P_2(1, 1)$ instead. Then Equation (1) would give

$$m = \frac{1 - 5}{1 - 3} = 2$$

as before. In general, relabeling the points $P_1$ and $P_2$ simply changes the sign of both the numerator and denominator of the ratio in Equation (1) and therefore does not change the value of $m$. Therefore, when we compute the slope of a line using Equation (1), it does not matter which point we label as $P_1$ and which point we label as $P_2$.

**EXAMPLE 2**   A 20-ft ladder leans against a wall with its top located 12 ft above the ground. What is the slope of the ladder?

**Solution**   The situation is depicted in Figure 6, where $x$ denotes the distance of the base of the ladder from the wall. By the Pythagorean Theorem we have

$$x^2 + 12^2 = 20^2$$

$$x^2 = 256$$

**FIGURE 6**
A ladder leaning against a wall

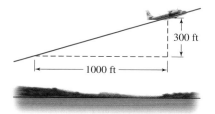

**FIGURE 7**
The flight path of the plane along a straight line

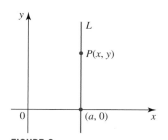

**FIGURE 8**
Every point on the vertical line L has an x-coordinate that is equal to a.

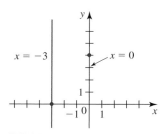

**FIGURE 9**
The graphs of the equations $x = -3$ and $x = 0$

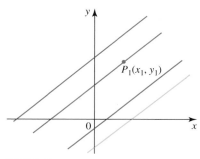

**FIGURE 10**
There are infinitely many lines with slope m but only one that passes through the point $P_1(x_1, y_1)$ with slope m.

or $x = 16$. The slope of the ladder is

$$\frac{12}{16} = \frac{3}{4} \qquad \frac{\text{rise}}{\text{run}}$$

**EXAMPLE 3** Shortly after takeoff a plane climbs along a straight path. The plane gains altitude at the rate of 300 ft for each 1000 ft it travels horizontally, that is, parallel to the ground. What is the slope of the trajectory of the plane? What is the altitude gained by the plane after traveling 5000 ft horizontally?

**Solution** The flight path is depicted in Figure 7. We see that the slope of the flight path of the plane is

$$\frac{300}{1000} = \frac{3}{10} \qquad \frac{\text{rise}}{\text{run}}$$

This tells us that the plane gains an altitude of $\frac{3}{10}$ ft for each foot traveled by the plane horizontally. Therefore, the altitude gained after traveling 5000 ft horizontally is

$$\frac{3}{10} \cdot 5000 = 1500$$

or 1500 ft.

## Equations of Vertical Lines

Let $L$ be a vertical line in the $xy$-plane. Then $L$ must intersect the axis at some point $(a, 0)$ as shown in Figure 8. If $P(x, y)$ is any point on $L$, then $x$ must be equal to $a$, whereas $y$ may take on any value, depending on the position of $P$. In other words, the only conditions on the coordinates of the point $(a, y)$ on $L$ are $x = a$ and $-\infty < y < \infty$. Conversely, we see that the set of all points $(x, y)$ where $x = a$ and $y$ is arbitrary is precisely the vertical line $L$. We have found an algebraic representation of a vertical line in a coordinate plane.

---

**DEFINITION    Equation of a Vertical Line**

An equation of the vertical line passing through the point $(a, b)$ is

$$x = a \qquad \qquad (2)$$

---

**EXAMPLE 4** The graph of $x = -3$ is the vertical line passing through $(-3, 0)$. An equation of the vertical line passing through $(0, 4)$ is $x = 0$. This is an equation of the $y$-axis (see Figure 9).

## Equations of Nonvertical Lines

If a line $L$ is nonvertical, then it has a well-defined slope $m$. But specifying the slope of a line alone is not enough to pin down a particular line, because there are infinitely many lines with a given slope (Figure 10). However, if we specify a point $P_1(x_1, y_1)$ through which a line $L$ passes in addition to its slope $m$, then $L$ is uniquely determined. To derive an equation of the line passing through a given point $P_1(x_1, y_1)$ and having slope $m$, let $P(x, y)$ be *any* point distinct from $P_1$ lying on $L$. Using Equation (1)

and the points $P_1(x_1, y_1)$ and $P(x, y)$, we can write the slope of $L$ as

$$\frac{y - y_1}{x - x_1}$$

But the slope of $L$ is $m$. So

$$\frac{y - y_1}{x - x_1} = m$$

or, upon multiplying both sides of the equation by $x - x_1$,

$$y - y_1 = m(x - x_1) \tag{3}$$

Observe that $x = x_1$ and $y = y_1$ also satisfy Equation (3), so all points on $L$ satisfy this equation. We leave it as an exercise to show that only the points that satisfy Equation (3) can lie on $L$.

Equation (3) is called the point-slope form of an equation of a line because it utilizes a point on the line and its slope.

---

**DEFINITION   Point-Slope Form of an Equation of a Line**

An equation of the line passing through the point $P_1(x_1, y_1)$ and having slope $m$ is

$$y - y_1 = m(x - x_1)$$

---

**EXAMPLE 5**  Find an equation of the line passing through the point $(2, 1)$ and having slope $m = -\frac{1}{2}$.

**Solution**   Using Equation (3) with $x_1 = 2$, $y_1 = 1$, and $m = -\frac{1}{2}$, we find

$$y - 1 = -\frac{1}{2}(x - 2)$$

or

$$y = -\frac{1}{2}x + 2$$

**EXAMPLE 6**  Find an equation of the line passing through the points $(-1, -2)$ and $(2, 3)$.

**Solution**   We first calculate the slope of the line, obtaining

$$m = \frac{3 - (-2)}{2 - (-1)} = \frac{5}{3}$$

Then using Equation (3) with $P_1(-1, -2)$ (the other point will also do, as you can verify) and $m = \frac{5}{3}$, we obtain

$$y - (-2) = \frac{5}{3}[x - (-1)]$$

$$y = \frac{5}{3}x + \frac{5}{3} - 2$$

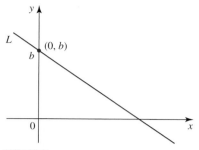

**FIGURE 11**
The line $L$ with $y$-intercept $b$ and slope $m$ has equation $y = mx + b$.

or

$$y = \frac{5}{3}x - \frac{1}{3}$$

A nonvertical line $L$ crosses the $y$-axis at some point $(0, b)$. The number $b$ is called the $y$-intercept of the line. (See Figure 11.) If we use the point $P_1(0, b)$ in Equation (3), we obtain

$$y - b = m(x - 0)$$

or

$$y = mx + b$$

which is called the **slope-intercept** form of an equation of a line.

---

**DEFINITION   Slope-Intercept Form of an Equation of a Line**

An equation of the line with slope $m$ and $y$-intercept $b$ is

$$y = mx + b \qquad\qquad (4)$$

---

**EXAMPLE 7**   Find an equation of the line with slope $\frac{3}{4}$ and $y$-intercept 4.

**Solution**   We use Equation (4) with $m = \frac{3}{4}$ and $b = 4$, obtaining the equation

$$y = \frac{3}{4}x + 4$$

## ■ The General Equation of a Line

An equation of the form

$$Ax + By + C = 0 \qquad\qquad (5)$$

where $A$, $B$, and $C$ are constants and $A$ and $B$ are not both zero, is called a **first-degree equation** in $x$ and $y$. You can verify the following result.

---

**THEOREM 1   General Equation of a Line**

Every first-degree equation in $x$ and $y$ has a straight line for its graph in the $xy$-plane; conversely, every straight line in the $xy$-plane is the graph of a first-degree equation in $x$ and $y$.

---

Because of this theorem, Equation (5) is often referred to as a **general equation of a line** or a **linear equation** in $x$ and $y$.

**EXAMPLE 8**   Find the slope of the line with equation $2x + 3y + 5 = 0$.

**Solution**   Rewriting the equation in the slope-intercept form by solving it for $y$ in terms of $x$, we obtain

$$3y = -2x - 5$$

or

$$y = -\frac{2}{3}x - \frac{5}{3}$$

Comparing this equation with Equation (4), we see immediately that the slope of the line is $m = -\frac{2}{3}$. ■

**Note**   Example 8 illustrates one advantage of writing an equation of a line in the slope-intercept form: The slope of the line is given by the coefficient of $x$. ■

## ■ Drawing the Graphs of Lines

We have already mentioned that the $y$-intercept of a straight line is the $y$-coordinate of the point $(0, b)$ at which the line crosses the $y$-axis. Similarly, the $x$-intercept of a straight line is the $x$-coordinate of the point $(a, 0)$ at which the line crosses the $x$-axis (see Figure 12). To find the $x$-intercept of a line $L$, we set $y = 0$ in the equation for $L$ because every point on the $x$-axis must have its $y$-coordinate equal to zero. Similarly, to find the $y$-intercept of $L$, we set $x = 0$. The easiest way to sketch a straight line is to find its $x$- and $y$-intercepts, when possible, as the following example shows.

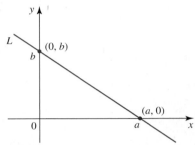

**FIGURE 12**
The $x$-intercept of $L$ is $a$, and the $y$-intercept of $L$ is $b$.

**EXAMPLE 9**   Sketch the graphs of

**a.** $2x + 3y - 6 = 0$     **b.** $x - 3y = 0$

**Solution**
**a.** Setting $y = 0$ gives the $x$-intercept as 3. Next, setting $x = 0$ gives the $y$-intercept as 2. Plotting the points $(3, 0)$ and $(0, 2)$ and drawing the line passing through them, we obtain the desired graph (see Figure 13a).
**b.** Setting $y = 0$ gives $x = 0$ as the $x$-intercept. Next, setting $x = 0$ gives $y = 0$ as the $y$-intercept. Thus, the line passes through the origin. In this situation we need to find another point through which the line passes. If we pick, say, $x = 3$ and substitute this value of $x$ into the equation $x - 3y = 0$ and solve the resulting equation for $y$, we obtain $y = 1$ as the $y$-coordinate. Plotting the points $(0, 0)$ and $(3, 1)$ and drawing the line through them, we obtain the desired graph (Figure 13b).

**FIGURE 13**

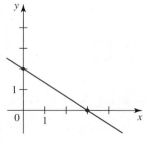

(a) The graph of $2x + 3y - 6 = 0$

(b) The graph of $x - 3y = 0$   ■

## ◼ Angles of Inclination

> **DEFINITION  Angle of Inclination**
>
> The **angle of inclination** of a line $L$ is the smaller angle $\phi$ (the Greek letter *phi*) measured in a counterclockwise direction from the direction of the positive $x$-axis to $L$ (see Figure 14).

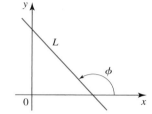

**FIGURE 14**
The angle of inclination is measured in a counterclockwise direction from the direction of the positive $x$-axis.

**Note**  The angle of inclination $\phi$ satisfies $0° \leq \phi < 180°$ or, in radian measure, $0 \leq \phi < \pi$. ◼

The relationship between the slope of a line and the angle of inclination of the line can be seen from examining Figure 15. Letting $m$ denote the slope of $L$ and $\phi$ its angle of inclination, we have

$$m = \tan \phi \tag{6}$$

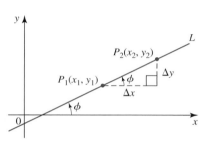

**FIGURE 15**

The slope of $L$ is $m = \dfrac{\Delta y}{\Delta x} = \tan \phi$.

**Notes**
1. Although Figure 15 illustrates Equation (6) for the case in which $0° \leq \phi < 90°$, it can be shown that the equation also holds when $90° < \phi < 180°$. We leave it as an exercise.
2. Observe that the angle of inclination of a vertical line is $90°$. Since $\tan 90°$ is undefined, we see that the slope of a vertical line is undefined, as was noted earlier. ◼

**EXAMPLE 10**  Refer to Example 3. Find the angle of the flight path of the plane. (Note: This angle is referred to as the *angle of climb.*)

**Solution**  From the result of Example 3 we see that $m = \frac{3}{10} = 0.3$. Therefore, the angle of climb, $\phi$, satisfies

$$\tan \phi = 0.3$$

from which we deduce that the angle of climb is

$$\phi = \tan^{-1} 0.3 \approx 0.29 \text{ rad}$$

or approximately $17°$. ◼

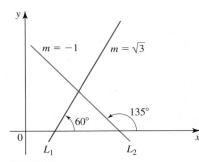

**FIGURE 16**
$L_1$ has slope $m = \sqrt{3}$, and $L_2$ has angle of inclination 135°.

**EXAMPLE 11**

**a.** Find the slope of a line whose angle of inclination is 60° ($\pi/3$ radians).
**b.** Find the angle of inclination of a line with slope $m = -1$.

**Solution**
**a.** Equation (6) immediately yields

$$m = \tan 60° = \sqrt{3}$$

as the slope of the line (see Figure 16).
**b.** Equation (6) gives

$$-1 = \tan \phi$$

and we see that $\phi = 3\pi/4$ radians, or 135° (see Figure 16).    ∎

## ■ Parallel Lines and Perpendicular Lines

Two lines are parallel if and only if they have the same angle of inclination (see Figure 17).

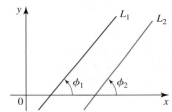

Slope of $L_1 = m_1 = \tan \phi_1$
Slope of $L_2 = m_2 = \tan \phi_2$

**FIGURE 17**
$L_1$ and $L_2$ are parallel if and only if their slopes are equal or both lines are vertical.

Therefore, using Equation (6), we have the following result.

---

**THEOREM 2**

Two nonvertical lines are parallel if and only if they have the same slope.

---

**Note**   If two lines are vertical, then they are parallel.    ∎

Suppose that $L_1$ and $L_2$ are two nonvertical perpendicular lines with slopes $m_1$ and $m_2$ and angles of inclination $\phi_1$ and $\phi_2$, respectively. The case in which $\phi_1$ is acute and $\phi_2$ is obtuse is shown in Figure 18.

Since $90° < \phi_2 < 180°$, $m_2$ is negative, so the length of the side $BC$ is $-m_2$. The two right triangles $\triangle ABC$ and $\triangle DAC$ are similar, and since the ratios of corresponding sides of similar triangles are equal, we have

$$\frac{m_1}{1} = \frac{1}{-m_2}$$

which may be rewritten as

$$m_1 = -\frac{1}{m_2} \qquad \text{or} \qquad m_1 m_2 = -1$$

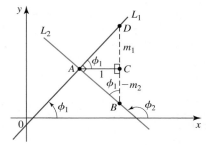

**FIGURE 18**
$\triangle ABC$ and $\triangle DAC$ are similar.

This argument can be reversed to prove the converse: The lines are perpendicular if $m_1 m_2 = -1$.

---

> **THEOREM 3  Slopes of Perpendicular Lines**
>
> Two nonvertical lines $L_1$ and $L_2$ with slopes $m_1$ and $m_2$, respectively, are perpendicular if and only if $m_1 m_2 = -1$ or, equivalently, if and only if
>
> $$m_1 = -\frac{1}{m_2} \qquad \text{or} \qquad m_2 = -\frac{1}{m_1} \qquad \text{(7)}$$
>
> Thus, the slope of each is the negative reciprocal of the slope of the other.

---

**Note**  If a line $L_1$ is vertical (and hence has no slope), then another line $L_2$ is perpendicular to it if and only if $L_2$ is horizontal (has zero slope), and vice versa.  ■

**EXAMPLE 12**  Find an equation of the line that passes through the point $(6, 7)$ and is perpendicular to the line with equation $2x + 3y = 12$.

**Solution**  First we find the slope of the given line by rewriting the equation in the slope-intercept form:

$$y = -\frac{2}{3}x + 4$$

From this we see that its slope is $-\frac{2}{3}$. Since the required line is perpendicular to the given line, its slope is

$$-\frac{1}{-\frac{2}{3}} = \frac{3}{2}$$

Therefore, using the point-slope form of an equation of a line with $m = \frac{3}{2}$ and $P_1(6, 7)$, we obtain the required equation as

$$y - 7 = \frac{3}{2}(x - 6)$$

or

$$y = \frac{3}{2}x - 2 \qquad ■$$

## ■ The Distance Formula

Another benefit that arises from using the Cartesian coordinate system is that the distance between any two points in the plane may be expressed solely in terms of their coordinates. Suppose, for example, that $(x_1, y_1)$ and $(x_2, y_2)$ are any two points in the plane (see Figure 19). Then the distance between these two points can be computed using the following formula.

---

> **Distance Formula**
>
> The distance $d$ between two points $P_1(x_1, y_1)$ and $P_2(x_2, y_2)$ in the plane is given by
>
> $$d = \sqrt{(x_2 - x_1)^2 + (y_2 - y_1)^2} \qquad \text{(8)}$$

---

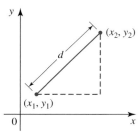

**FIGURE 19**

The distance $d$ between the points $(x_1, y_1)$ and $(x_2, y_2)$

In what follows, we give several applications of the distance formula.

**EXAMPLE 13**  Find the distance between the points $(-4, 3)$ and $(2, 6)$.

**Solution**  Let $P_1(-4, 3)$ and $P_2(2, 6)$ be points in the plane. Then, we have

$$x_1 = -4, \qquad y_1 = 3, \qquad x_2 = 2, \qquad y_2 = 6$$

Using Formula (8), we have

$$d = \sqrt{[2 - (-4)]^2 + (6 - 3)^2}$$
$$= \sqrt{6^2 + 3^2}$$
$$= \sqrt{45} = 3\sqrt{5} \qquad \blacksquare$$

**EXAMPLE 14**  Let $P(x, y)$ denote a point lying on the circle with radius $r$ and center $C(h, k)$. (See Figure 20.) Find a relationship between $x$ and $y$.

**Solution**  By the definition of a circle, the distance between $C(h, k)$ and $P(x, y)$ is $r$. Using Formula (8), we have

$$\sqrt{(x - h)^2 + (y - k)^2} = r$$

which, upon squaring both sides, gives the equation

$$(x - h)^2 + (y - k)^2 = r^2$$

that must be satisfied by the variables $x$ and $y$. $\qquad \blacksquare$

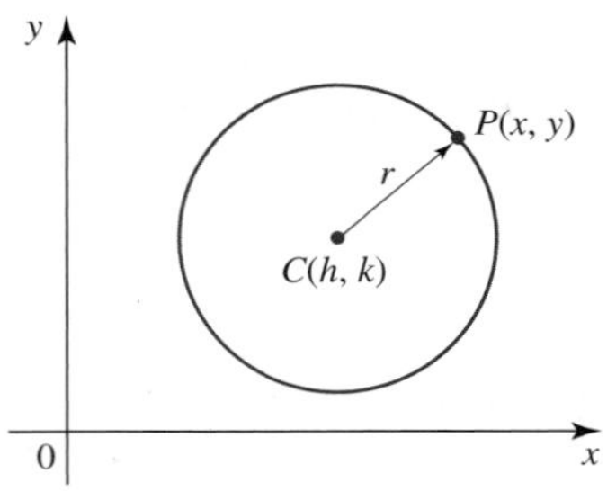

**FIGURE 20**
A circle with radius $r$ and center $C(h, k)$

A summary of the result obtained in Example 14 follows.

---

**Equation of a Circle**

An equation of the circle with center $C(h, k)$ and radius $r$ is given by

$$(x - h)^2 + (y - k)^2 = r^2 \qquad (9)$$

---

**EXAMPLE 15**  Find an equation of the circle with

**a.** Radius 2 and center $(-1, 3)$.
**b.** Radius 3 and center located at the origin.

**Solution**
**a.** We use Formula (9) with $r = 2$, $h = -1$, and $k = 3$, obtaining

$$[x - (-1)]^2 + (y - 3)^2 = 2^2 \qquad \text{or} \qquad (x + 1)^2 + (y - 3)^2 = 4$$

(See Figure 21a.)
**b.** Using Formula (9) with $r = 3$ and $h = k = 0$, we obtain

$$x^2 + y^2 = 3^2 \qquad \text{or} \qquad x^2 + y^2 = 9$$

(See Figure 21b.)

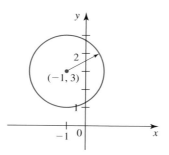

(a) The circle with radius 2 and center $(-1, 3)$

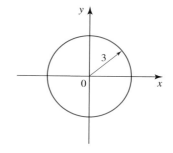

(b) The circle with radius 3 and center $(0, 0)$

**FIGURE 21**

## 0.1  EXERCISES

*In Exercises 1–4, find the slope of the line passing through the pair of points.*

**1.** $(1, -2)$ and $(2, 4)$

**2.** $(-4, -2)$ and $(-1, 3)$

**3.** $(1.2, 3.6)$ and $(3.2, 1.4)$

**4.** $(-3, -3)$ and $(-3, \sqrt{3})$

**5.** Refer to the figure below.

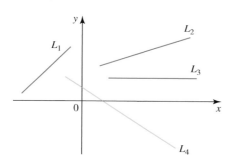

  **a.** Give the sign of the slope of each of the lines.
  **b.** List the lines in order of increasing slope.

**6.** Find the slope of each of the lines shown in the accompanying figure.

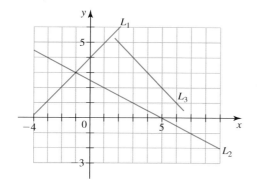

**7.** Find $a$ if the line passing through $(1, 3)$ and $(-4, a)$ has slope 5.

**8.** Find $a$ if the line passing through $(2, a)$ and $(-9, 3)$ has slope $-3$.

*In Exercises 9–14, find the slope of the line that has the angle of inclination.*

**9.** $45°$    **10.** $135°$    **11.** $30°$

**12.** $\dfrac{\pi}{4}$    **13.** $\dfrac{\pi}{3}$    **14.** $\dfrac{2\pi}{3}$

*In Exercises 15–20, find the angle of inclination of a line with the given slope. You may use a calculator.*

**15.** $-1$    **16.** $\dfrac{1}{2}$    **17.** $\sqrt{3}$

**18.** $10$    **19.** $-\dfrac{1}{\sqrt{3}}$    **20.** $20$

*In Exercises 21–24, sketch the line through the given point with the indicated slope.*

**21.** $(1, 2);\quad 3$    **22.** $(2, 3);\quad -2$

**23.** $(-1, -2);\quad -1$    **24.** $(-2, 3);\quad 4$

*In Exercises 25–28, determine whether the lines through the given pairs of points are parallel or perpendicular to each other.*

**25.** $(1, -2), (-3, -10)$ and $(1, 5), (-1, 1)$

**26.** $(4, 6), (4, -2)$ and $(-1, 5), (-1, 8)$

**27.** $(-2, 5), (4, 2)$ and $(-1, -2), (3, 6)$

**28.** $(-1, -2), (3, 4)$ and $(9, -6), (3, -2)$

**29.** If the line passing through the points $(-1, a)$ and $(3, -1)$ is parallel to the line passing through the points $(3, 6)$ and $(-5, a + 2)$, what must the value of $a$ be?

**30.** If the line passing through the points $(-2, 4)$ and $(1, a)$ is perpendicular to the line passing through the points $(a + 4, 8)$ and $(3, -4)$, what must the value of $a$ be?

**31.** The point $(-5, k)$ lies on the line passing through the point $(1, 3)$ and perpendicular to a line with slope 3. Find $k$.

**32.** Show that the triangle with vertices $A(-2, -8)$, $B(2, 2)$, and $C(-3, 4)$ is a right triangle.

**33.** A line passes through $(3, 4)$ and the midpoint of the line segment joining $(-1, 1)$ and $(3, 9)$. Show that this line is perpendicular to the line segment.

*In Exercises 34 and 35, determine whether the given points lie on a straight line.*

**34.** $A(-2, 1)$, $B(1, 7)$, and $C(4, 13)$

**35.** $A(-3, 6)$, $B(3, 3)$, and $C(6, 0)$

*In Exercises 36–41, write the equation in the slope-intercept form, and then find the slope and y-intercept of the corresponding lines.*

**36.** $2x - 3y - 12 = 0$

**37.** $-3x + 4y - 8 = 0$

**38.** $y + 4 = 0$

**39.** $Ax + By = C, \quad B \neq 0$

**40.** $\dfrac{x}{3} + \dfrac{y}{4} = 1$

**41.** $\sqrt{2}x - \sqrt{3}y = 4$

*In Exercises 42–45, find the angle of inclination of the line represented by the equation.*

**42.** $4x - 7y - 8 = 0$

**43.** $\sqrt{3}x - y + 4 = 0$

**44.** $x + \sqrt{3}y - 5 = 0$

**45.** $x + y - 8 = 0$

*In Exercises 46–59, find an equation of the line satisfying the conditions. Write your answer in the slope-intercept form.*

**46.** Is perpendicular to the $x$-axis and passes through the point $(\pi, \pi^2)$

**47.** Passes through $(4, -3)$ with slope 2

**48.** Passes through $(-3, 3)$ and has slope 0

**49.** Passes through $(2, 4)$ and $(3, 8)$

**50.** Passes through $(-1, -2)$ and $(3, -4)$

**51.** Passes through $(2, 5)$ and $(2, 28)$

**52.** Has slope $-2$ and $y$-intercept 3

**53.** Has slope 3 and $y$-intercept $-5$

**54.** Has $x$-intercept 3 and $y$-intercept $-5$

**55.** Passes through $(3, -5)$ and is parallel to the line with equation $2x + 3y = 12$

**56.** Is perpendicular to the line with equation $y = -3x - 5$ and has $y$-intercept 7

**57.** Passes through $(-2, -4)$ and is perpendicular to the line with equation $3x - y - 4 = 0$

**58.** Passes through $(3, -4)$ and is perpendicular to the line through $(-1, 2)$ and $(3, 6)$

**59.** Passes through $(2, 3)$ and has an angle of inclination of $\pi/6$ radians

*In Exercises 60–63, determine whether the pair of lines represented by the equations are parallel, perpendicular, or neither.*

**60.** $3x - 4y = 8 \quad$ and $\quad 6x - 8y = 10$

**61.** $x - 3 = 0 \quad$ and $\quad y - 5 = 0$

**62.** $2x - 3y - 12 = 0 \quad$ and $\quad 3x + 2y - 6 = 0$

**63.** $\dfrac{x}{a} + \dfrac{y}{b} = 1 \quad$ and $\quad \dfrac{x}{b} - \dfrac{y}{a} = 1$

*In Exercises 64–65, find the point of intersection of the lines with the given equations.*

**64.** $2x - y = 1 \quad$ and $\quad 3x + 2y = 12$

**65.** $x - 3y = -1 \quad$ and $\quad 4x + 3y = 11$

**66.** Find the distance between the points.
 **a.** $(1, 3)$ and $(4, 7)$
 **b.** $(1, 0)$ and $(4, 4)$
 **c.** $(-1, 3)$ and $(4, 9)$
 **d.** $(-2, 1)$ and $(10, 6)$

**67.** Find an equation of the circle that satisfies the conditions.
 **a.** Radius 5 and center $(2, -3)$
 **b.** Center at the origin and passes through $(2, 3)$
 **c.** Center $(2, -3)$ and passes through $(5, 2)$
 **d.** Center $(-a, a)$ and radius $2a$

**68.** Show that the two lines with equations $a_1x + b_1y + c_1 = 0$ and $a_2x + b_2y + c_2 = 0$, respectively, are parallel if and only if $a_1b_2 - b_1a_2 = 0$.

**69.** Show that an equation of the line $L$ that passes through the points $(a, 0)$ and $(0, b)$ with $a \neq 0$ and $b \neq 0$ can be written in the form

$$\frac{x}{a} + \frac{y}{b} = 1$$

This is called the **intercept form** of the equation of $L$.

**70.** Use the result of Exercise 69 to find an equation of the line with $x$-intercept 2 and $y$-intercept 5.

**71.** Use the result of Exercise 69 to find an equation of the line passing through the points $(-4, 0)$ and $(0, -1)$.

**72.** Find an equation of the line passing through $(5, 2)$ and the midpoint of the line segment joining $(-1, 1)$ and $(3, 9)$.

**73.** Find the distance from the point $(5, 3)$ to the line with equation $2x - y + 3 = 0$.
 **Hint:** Find the point of intersection of the given line and the line perpendicular to it that passes through $(5, 3)$.

**74.** The top of a ladder leaning against a wall is 9 ft above the ground. The slope of the ladder with respect to the ground is $3\sqrt{7}/7$. What is the length of the ladder?

**75.** A plane flying along a straight path loses altitude at the rate of 1000 ft for each 6000 ft covered horizontally. What is the angle of descent of the plane?

**76.** A plane flies along a straight line that has a slope of 0.22. If the plane gains altitude of 1000 ft over a certain period of time, what will be the horizontal distance covered by the plane over that period?

**77. Truss Bridges** Simple trusses are common in bridges. The following figure depicts such a truss superimposed on a coordinate system. Find an equation of the line containing the line segments (a) *OD*, (b) *AD*, (c) *AC*, and (d) *BC*.

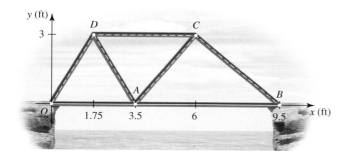

**78. Temperature Conversion** The relationship between the temperature in degrees Fahrenheit (°F) and the temperature in degrees Celsius (°C) is

$$F = \frac{9}{5}C + 32$$

  **a.** Sketch the line with the given equation.
  **b.** What is the slope of the line? What does it represent?
  **c.** What is the *F*-intercept of the line? What does it represent?

**79. Nuclear Plant Utilization** The United States is not building many nuclear plants, but the ones that it has are running full tilt. The output (as a percent of total capacity) of nuclear plants is described by the equation

$$y = 1.9467t + 70.082$$

where *t* is measured in years, with $t = 0$ corresponding to the beginning of 1990.
  **a.** Sketch the line with the given equation.
  **b.** What are the slope and the *y*-intercept of the line found in part (a)?
  **c.** Give an interpretation of the slope and the *y*-intercept of the line found in part (a).
  **d.** If the utilization of nuclear power continued to grow at the same rate and the total capacity of nuclear plants in the United States remained constant, by what year were the plants generating at maximum capacity?
  *Source: Nuclear Energy Institute.*

**80. Social Security Contributions** For wages less than the maximum taxable wage base, Social Security contributions by employees are 7.65% of the employee's wages.
  **a.** Find an equation that expresses the relationship between the wages earned (*x*) and the Social Security taxes paid (*y*) by an employee who earns less than the maximum taxable wage base.
  **b.** For each additional dollar that an employee earns, by how much is his or her Social Security contribution increased? (Assume that the employee's wages are less than the maximum taxable wage base.)
  **c.** What Social Security contributions will an employee who earns $75,000 (which is less than the maximum taxable wage base) be required to make?
  *Source: Social Security Administration.*

**81. Weight of Whales** The equation $W = 3.51L - 192$, $70 \le L \le 100$, which expresses the relationship between the length *L* (in feet) and the expected weight *W* (in British tons) of adult blue whales, was adopted in the late 1960s by the International Whaling Commission.
  **a.** What is the expected weight of an 80-ft whale?
  **b.** Sketch the straight line that represents the equation.

**82. The Narrowing Gender Gap** Since the founding of the Equal Employment Opportunity Commission and the passage of equal-pay laws, the gap between men's and women's earnings has continued to close gradually. At the beginning of 1990 ($t = 0$), women's wages were 68% of men's wages; and by the beginning of 2000 ($t = 10$), women's wages were projected to be 80% of men's wages. If this gap between women's and men's wages continued to narrow *linearly*, what percent of men's wages were women's wages at the beginning of 2004?
  *Source: Journal of Economic Perspectives.*

**83.** Show that only those points satisfying Equation (3) can lie on the line *L* passing through $P_1(x_1, y_1)$ with slope *m*.

**84.** Show that Equation (6) also holds when $90° < \phi < 180°$.

*In Exercises 85–88, determine whether the statement is true or false. If it is true, explain why. If it is false, explain why or give an example that shows it is false.*

**85.** Suppose the slope of a line *L* is $-\frac{1}{2}$ and *P* is a given point on *L*. If *Q* is the point on *L* lying 4 units to the left of *P*, then *Q* lies 2 units above *P*.

**86.** The line with equation $Ax + By + C = 0$, where $B \ne 0$, and the line with equation $ax + by + c = 0$, where $b \ne 0$, are parallel if $Ab - aB = 0$.

**87.** If the slope of the line $L_1$ is positive, then the slope of a line $L_2$ perpendicular to $L_1$ must be negative.

**88.** The lines with equations $ax + by + c_1 = 0$ and $bx - ay + c_2 = 0$, where $a \ne 0$ and $b \ne 0$, are perpendicular to each other.

## 0.2 | Functions and Their Graphs

---

**0.2  SELF-CHECK DIAGNOSTIC TEST**

**1.** If

$$f(x) = \begin{cases} \sqrt{-x} & \text{if } x < 0 \\ \sqrt{x} + 1 & \text{if } x \geq 0 \end{cases}$$

find $f(-4)$, $f(0)$, and $f(9)$.

**2.** If $f(x) = x^2 + 2x$, find and simplify $\dfrac{f(x + h) - f(x)}{h}$.

**3.** Find the domain of $f(x) = \dfrac{\sqrt{2x + 1}}{x^2 + x - 2}$.

**4.** Find the domain and range, and sketch the graph of

$$f(x) = \begin{cases} -2x + 1 & \text{if } x < 0 \\ 2x - 1 & \text{if } x \geq 0 \end{cases}$$

**5.** Determine whether $f(x) = \dfrac{2x^3 - x}{x^2 + 1}$ is odd, even, or neither.

*Answers to Self-Check Diagnostic Test 0.2 can be found on page ANS 1.*

---

### ■ Definition of a Function

In many situations, one quantity depends on another. For example:

- The area of a circle depends on its radius.
- The distance fallen by an object dropped from a building depends on the length of time it has fallen.
- The initial speed of a chemical reaction depends on the amount of substrate used.
- The size of the population of a certain culture of bacteria after the introduction of a bactericide depends on the time elapsed.
- The profit of a manufacturer depends on the company's level of production.

To describe these situations, we use the concept of a function.

---

**DEFINITION  Function**

A **function** $f$ from a set $A$ to a set $B$ is a rule that assigns to each element $x$ in $A$ one and only one element $y$ in $B$.

---

Let's consider an example that illustrates why there can be only one element $y$ in $B$ for each $x$ in $A$. Suppose that $A$ is the set of items on sale in a department store and $f$ is a "pricing" function that assigns to each item $x$ in $A$ its selling price $y$ in $B$. Then for each $x$ there should be exactly one $y$. Note that the definition does not preclude the possibility of more than one element in $A$ being associated with an element in $B$. In the context of our present example, this could mean that two or more items would have the same selling price.

The set $A$ is called the **domain** of the function. The element $y$ in $B$, called the value of $f$ at $x$, is written $f(x)$ and read "$f$ of $x$." The set of all values $y = f(x)$ as $x$ varies over the domain of $f$ is called the **range of $f$**. If $A$ and $B$ are subsets of the set of real numbers, then both $x$ and $f(x)$ are also real numbers. In this case we refer to the function $f$ as a *real-valued* function of a *real variable*.

We can think of a function $f$ as a machine or processor. In this analogy the domain of $f$ consists of the set of "inputs," the rule describes how the "inputs" are to be processed by the machine, and the range is made up of the set of "outputs" (see Figure 1).

**FIGURE 1**
A function machine

As an example, consider the function that associates with each nonnegative number $x$ its square root, $\sqrt{x}$. We can view this function as a square root extracting machine. Its domain is the set of all nonnegative numbers, and so is its range. Given the input 4, for example, the function extracts its square root $\sqrt{4}$ and yields the output 2.

Another way of viewing a function is to think of the function $f$ from a set $A$ to a set $B$ as a *mapping* or *transformation* that maps an element $x$ in $A$ onto its image $f(x)$ in $B$ (Figure 2). For example, the "square root" function is a function from the set of nonnegative real numbers to the set of real numbers. This function maps the number 4 onto the number 2, the number 7 onto the number $\sqrt{7}$, and so on.

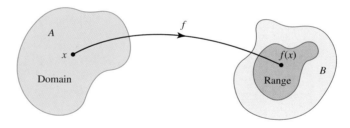

**FIGURE 2**
$f$ maps a point $x$ in its domain onto its image $f(x)$ in its range.

**Note** The range of $f$ is contained in the set $B$ but need not be equal to $B$. For example, consider the function $f$ that associates with each real number $x$ its square, $x^2$, from the set of real numbers $R$ to the set of real numbers $R$ (so $A = B = R$). Then the range of $f$ is the set of nonnegative numbers, a proper subset of $B$. ■

## ■ Describing Functions

Functions can be described in many ways. Earlier, we defined the square root function by giving a verbal description of the rule. Functions can also be described by giving a table of values describing the relationship between $x$ and $f(x)$. This method of describing a function is particularly effective when both the domain and the range of $f$ contain a small number of elements. For example, the function $f$ giving the Manhattan hotel occupancy rate in each of the years 1999 ($x = 0$) through 2006 can be defined by the data given in Table 1.

Here, the domain of $f$ is $A = \{0, 1, 2, 3, 4, 5, 6, 7\}$ and the range of $f$ is $B = \{74.5, 75.0, \ldots, 85.1\}$. Observe that we can also describe the rule for $f$ by writing $f(0) = 81.1$, $f(1) = 83.7$, $\ldots$, $f(7) = 85.1$.

**TABLE 1** The function $f$ giving the Manhattan hotel occupancy rate in year $x$

| $x$ (year) | $y = f(x)$ (percent) |
|:----------:|:--------------------:|
| 0 | 81.1 |
| 1 | 83.7 |
| 2 | 74.5 |
| 3 | 75.0 |
| 4 | 75.9 |
| 5 | 83.2 |
| 6 | 84.9 |
| 7 | 85.1 |

*Source:* PricewaterhouseCoopers LLP.

y (%)

3.5

3.0

2.5

2.0

1.5

1.0

0.5

0    1    2    3    *t* (months)

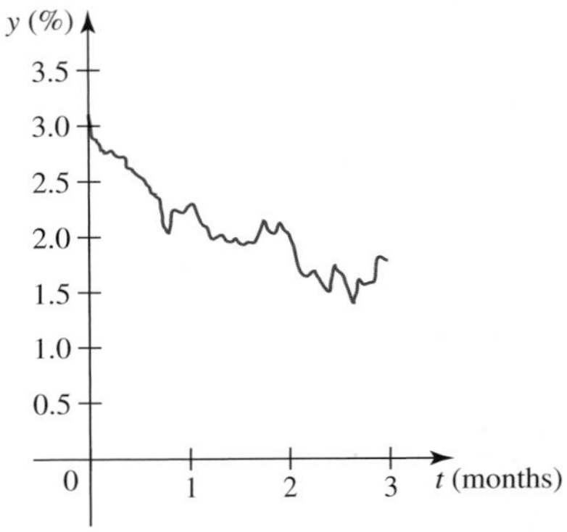

**FIGURE 3**
The function *f* gives the annual yield
for two-year Treasury notes in the first
three months of 2008.
*Source: Financial Times.*

A function can also be described graphically, as shown in Figure 3. Here, the function *f* gives the annual yield in percent for two-year Treasury notes, $f(t)$, for the first three months of 2008.

**EXAMPLE 1** The function *f* defined by the formula $y = \sqrt{x}$, or $f(x) = \sqrt{x}$, is just the square root function mentioned earlier. The domain of this function is the set of all values of *x* in the interval $[0, \infty)$. For example, if $x = 16$, then $f(16) = \sqrt{16} = 4$ is the square root of 16. The range of *f* consists of all the square roots of nonnegative numbers and is therefore the set of all numbers in $[0, \infty)$. (See Figure 4.)

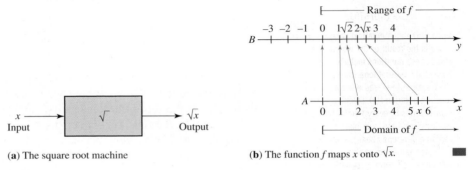

**FIGURE 4**   (a) The square root machine          (b) The function *f* maps *x* onto $\sqrt{x}$. ■

**Notes**
1. We often use letters other than *f* to denote a function. For example, we might speak of the area function *A*, the population function *P*, the function *F*, and so on.
2. Strictly speaking, it is improper to refer to a function *f* as $f(x)$ (recall that $f(x)$ is the value of *f* at *x*), but it is conventional to do so. ■

If a function *f* is described by an equation $y = f(x)$, we call *x* the **independent variable** and *y* the **dependent variable** because *y* (the value of *f* at *x*) is dependent upon the choice of *x*. Here, *x* represents a number in the *domain* of *f* and *y* the unique number in the *range* of *f* associated with *x*.

### ■ Evaluating Functions

Let's look again at the square root function *f* defined by the rule $f(x) = \sqrt{x}$. We could very well have defined this function by giving the rule as $f(t) = \sqrt{t}$ or $f(u) = \sqrt{u}$. In other words, it doesn't matter what letter we choose to represent the independent variable when describing the rule for a function. Indeed, we can describe the rule for *f* using the expression

$$f(\ ) = \sqrt{(\ )} = (\ )^{1/2}$$

To find the value of *f* at *x*, we simply insert *x* into the blank spaces inside the parentheses! As another example, consider the function *g* defined by the rule $g(x) = 2x^2 + x$. We can describe the rule for *g* by

$$g(\ ) = 2(\ )^2 + (\ )$$

obtained by replacing each *x* in the expression for $g(x)$ by a pair of parentheses. To find the value of *g* at $x = 2$, insert the number 2 in the blank spaces inside each pair of parentheses to obtain

$$g(2) = 2(2)^2 + 2 = 10$$

**EXAMPLE 2** Let $f(x) = x^2 + 2x - 1$. Find

**a.** $f(-1)$
**b.** $f(\pi)$
**c.** $f(t)$, where $t$ is a real number
**d.** $f(x + h)$, where $h$ is a real number
**e.** $f(2x)$

**Solution** We think of $f(x)$ as

$$f(\ ) = (\ )^2 + 2(\ ) - 1$$

Then

**a.** $f(-1) = (-1)^2 + 2(-1) - 1 = -2$
**b.** $f(\pi) = (\pi)^2 + 2(\pi) - 1 = \pi^2 + 2\pi - 1$
**c.** $f(t) = (t)^2 + 2(t) - 1 = t^2 + 2t - 1$
**d.** $f(x + h) = (x + h)^2 + 2(x + h) - 1 = x^2 + 2xh + h^2 + 2x + 2h - 1$
**e.** $f(2x) = (2x)^2 + 2(2x) - 1 = 4x^2 + 4x - 1$ ■

## ■ Finding the Domain of a Function

Sometimes the domain of a function is determined by the nature of a problem. For example, the domain of the function $A(r) = \pi r^2$ that gives the area of a circle in terms of its radius is the interval $(0, \infty)$, since $r$ must be positive.

**EXAMPLE 3** A man wants to enclose a vegetable garden in his backyard with a rectangular fence. If he has 100 ft of fencing with which to enclose his garden, find a function that gives the area of the garden in terms of its length $x$ (see Figure 5). (Assume that he uses all of the fencing.) What is the domain of this function?

**Solution** From Figure 5, we see that the perimeter of the rectangle, $(2x + 2y)$ ft, must be equal to 100 ft. Thus, we have the equation

$$2x + 2y = 100 \tag{1}$$

The area of the rectangle is given by

$$A = xy \tag{2}$$

Solving Equation (1) for $y$ in terms of $x$, we obtain $y = 50 - x$. Substituting this value of $y$ into Equation (2) yields

$$A = x(50 - x)$$
$$= -x^2 + 50x$$

Since the sides of the rectangle must be positive, we have $x > 0$ and $50 - x > 0$, which is equivalent to $0 < x < 50$. Therefore, the required function is

$$A(x) = -x^2 + 50x$$

with domain $(0, 50)$. ■

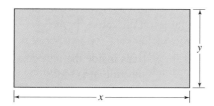

**FIGURE 5**
A rectangular garden with dimensions
$x$ ft by $y$ ft

Unless we specifically mention the domain of a function $f$, we will adopt the convention that the domain of $f$ is the set of all numbers for which $f(x)$ is a real number.

**EXAMPLE 4** Find the domain of each function:

**a.** $f(x) = \dfrac{2x + 1}{x^2 - x - 2}$     **b.** $f(x) = \dfrac{x + \sqrt{x + 1}}{2x - 1}$

**Solution**

**a.** Since division by zero is prohibited and the denominator of $f(x)$ is equal to zero if $x^2 - x - 2 = (x - 2)(x + 1) = 0$, or $x = -1$ or $x = 2$, we conclude that the domain of $f$ is the set of all numbers except $-1$ and 2. Equivalently, the domain of $f$ is the set $(-\infty, -1) \cup (-1, 2) \cup (2, \infty)$.

**b.** We begin by looking at the numerator of $f(x)$. Because the expression under the radical sign must be nonnegative, we see that $x + 1 \geq 0$, or $x \geq -1$. Next, since division by zero is not allowed, we see that $2x - 1 \neq 0$. But $2x - 1 = 0$ if $x = \frac{1}{2}$, so $x \neq \frac{1}{2}$. Therefore, the domain of $f$ is the set $\left[-1, \frac{1}{2}\right) \cup \left(\frac{1}{2}, \infty\right)$. ∎

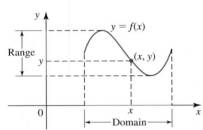

**FIGURE 6**
The graph of a function $f$

> **DEFINITION   Graph of a Function**
>
> The graph of a function $f$ is the set of all points $(x, y)$ such that $y = f(x)$, where $x$ lies in the domain of $f$.

The graph of $f$ provides us with a way of visualizing a function (see Figure 6).

**Note**   If the function $f$ is defined by the equation $y = f(x)$, then the domain of $f$ is the set of all $x$-values, and the range of $f$ is the set of all $y$-values. ∎

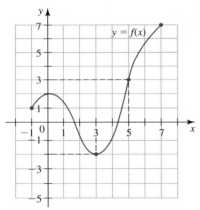

**FIGURE 7**
The graph of a function $f$

**EXAMPLE 5** The graph of a function $f$ is shown in Figure 7.

**a.** What is $f(3)$? $f(5)$?
**b.** What is the distance of the point $(3, f(3))$ from the $x$-axis? The point $(5, f(5))$ from the $x$-axis?
**c.** What is the domain of $f$? The range of $f$?

**Solution**

**a.** From the graph of $f$, we see that $y = -2$ when $x = 3$, and we conclude that $f(3) = -2$. Similarly, we see that $f(5) = 3$.
**b.** Since the point $(3, -2)$ lies below the $x$-axis, we see that the distance of the point $(3, f(3))$ from the $x$-axis is $-f(3) = -(-2) = 2$ units. The point $(5, f(5))$ lies above the $x$-axis, and its distance is $f(5)$, or 3 units.
**c.** Observe that $x$ may take on all values between $x = -1$ and $x = 7$, inclusive, so the domain of $f$ is $[-1, 7]$. Next, observe that as $x$ takes on all values in the domain of $f$, $y$ takes on all values between $-2$ and 7, inclusive. (You can see this by running your index finger along the $x$-axis from $x = -1$ to $x = 7$ and observing the corresponding values assumed by the $y$-coordinate of each point on the graph of $f$.) Therefore, the range of $f$ is $[-2, 7]$. ∎

**EXAMPLE 6** Sketch the graph of the function $f(x) = \dfrac{1}{x}$. What is the range of $f$?

**Solution**   The domain of $f$ is $(-\infty, 0) \cup (0, \infty)$. From the following table of values for $y = f(x)$ corresponding to some selected values of $x$, we obtain the graph of $f$ shown in Figure 8.

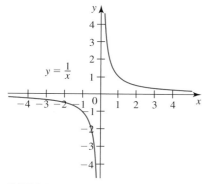

**FIGURE 8**
The graph of $f(x) = \dfrac{1}{x}$

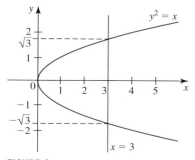

**FIGURE 9**
The number 3 has two images, $-\sqrt{3}$ and $\sqrt{3}$.

| $x$ | $\frac{1}{3}$ | $\frac{1}{2}$ | 1 | 2 | 3 | $-3$ | $-2$ | $-1$ | $-\frac{1}{2}$ | $-\frac{1}{3}$ |
|---|---|---|---|---|---|---|---|---|---|---|
| $y$ | 3 | 2 | 1 | $\frac{1}{2}$ | $\frac{1}{3}$ | $-\frac{1}{3}$ | $-\frac{1}{2}$ | $-1$ | $-2$ | $-3$ |

Setting $f(x) = y$ gives $1/x = y$, or $x = 1/y$, where $y \neq 0$. This shows that corresponding to any nonzero value of $y$ there is an $x$ in the domain of $f$ that is mapped onto $y$. So the range of $f$ is $(-\infty, 0) \cup (0, \infty)$. ∎

## The Vertical Line Test

Consider the equation $y^2 = x$. Solving for $y$ in terms of $x$, we obtain

$$y = \pm\sqrt{x} \tag{3}$$

Since each positive value of $x$ is associated with *two* values of $y$—for example, the number 3 is mapped onto the two images $-\sqrt{3}$ and $\sqrt{3}$—we see that the equation $y^2 = x$ does not define $y$ as a function of $x$. The graph of $y^2 = x$ is shown in Figure 9.

Note that the vertical line $x = 3$ intersects the graph of $y^2 = x$ at the *two* points $(3, -\sqrt{3})$ and $(3, \sqrt{3})$, verifying geometrically our earlier observation that the number $x = 3$ is associated with the two values $y = -\sqrt{3}$ and $y = \sqrt{3}$. These observations lead to the following criterion for determining when the graph of an equation is a function.

> **The Vertical Line Test**
>
> A curve in the $xy$-plane is the graph of a function $f$ defined by the equation $y = f(x)$ if and only if no vertical line intersects the curve at more than one point.

## Piecewise Defined Functions

In certain situations, a function is defined by several equations, each valid over a certain portion of the domain of the function.

**EXAMPLE 7** Sketch the graph of the absolute value function $f(x) = |x|$.

**Solution** We can plot a few points lying on the graph of $f$ and draw a suitable curve passing through them. Alternatively, we can proceed as follows. Recall that

$$|x| = \begin{cases} x & \text{if } x \geq 0 \\ -x & \text{if } x < 0 \end{cases}$$

This shows that the function $f(x) = |x|$ is defined piecewise over its domain $(-\infty, \infty)$. In the subdomain $[0, \infty)$ the rule for $f$ is $f(x) = x$. So the graph of $f$ coincides with that of $y = x$ for $x \geq 0$. But the latter is the right half of the line with equation $y = x$. In the subdomain $(-\infty, 0)$ the rule for $f$ is $f(x) = -x$, and we see that the graph of $f$ over this portion of its domain coincides with the left half of the line with equation $y = -x$. The graph of $f$ is sketched in Figure 10. ∎

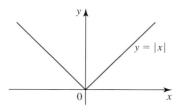

**FIGURE 10**
The graph of $f(x) = |x|$ consists of the left half of the line $y = -x$ and the right half of the line $y = x$.

**EXAMPLE 8** Sketch the graph of the function

$$f(x) = \begin{cases} x & \text{if } x < 1 \\ \frac{1}{4}x^2 - 1 & \text{if } x \geq 1 \end{cases}$$

**Solution**  The function $f$ is defined piecewise and has domain $(-\infty, \infty)$. In the subdomain $(-\infty, 1)$ the rule for $f$ is $f(x) = x$, so the graph of $f$ over this portion of its domain is the half-line with equation $y = x$. In the subdomain $[1, \infty)$ the rule for $f$ is $f(x) = \frac{1}{4}x^2 - 1$. To sketch the graph of $f$ over this subdomain, we use the following table.

| $x$ | 1 | 2 | 3 | 4 |
|---|---|---|---|---|
| $f(x) = \frac{1}{4}x^2 - 1$ | $-\frac{3}{4}$ | 0 | $\frac{5}{4}$ | 3 |

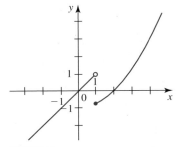

**FIGURE 11**

The graph of $f$ is shown in Figure 11.    ■

**Note**  Be sure that you use the correct equation when you evaluate a function that is defined piecewise. For instance, to find $f\left(\frac{1}{2}\right)$ in the preceding example, we note that $x = \frac{1}{2}$ lies in the subdomain $(-\infty, 1)$. So the correct rule here is $f(x) = x$ giving $f\left(\frac{1}{2}\right) = \frac{1}{2}$. To compute $f(5)$, we use the rule $f(x) = \frac{1}{4}x^2 - 1$, which gives $f(5) = \frac{21}{4}$.    ■

## ■ Even and Odd Functions

A function $f$ that satisfies $f(-x) = f(x)$ for every $x$ in its domain is called an **even function**. The graph of an even function is symmetric with respect to the $y$-axis (see Figure 12a). An example of an even function is $f(x) = x^2$, since $f(-x) = (-x)^2 = x^2 = f(x)$.

A function $f$ that satisfies $f(-x) = -f(x)$ for every $x$ in its domain is called an **odd function**. The graph of an odd function is symmetric with respect to the origin (see Figure 12b). An example of an odd function is $f(x) = x^3$, since $f(-x) = (-x)^3 = -x^3 = -f(x)$.

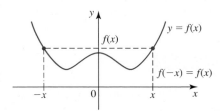    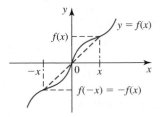

**FIGURE 12**        (a) $f$ is even.                                (b) $f$ is odd.

**EXAMPLE 9** Determine whether the function is even, odd, or neither even nor odd:

**a.** $f(x) = x^3 - x$        **b.** $g(x) = x^4 - x^2 + 1$        **c.** $h(x) = x - 2x^2$

**Solution**
**a.** $f(-x) = (-x)^3 - (-x) = -x^3 + x = -(x^3 - x) = -f(x)$. Therefore, $f$ is an odd function.

**b.** $g(-x) = (-x)^4 - (-x)^2 + 1 = x^4 - x^2 + 1 = g(x)$, and we see that $g$ is even.

**c.** $h(-x) = (-x) - 2(-x)^2 = -x - 2x^2$, which is neither equal to $h(x)$ nor $-h(x)$, and we conclude that $h$ is neither even nor odd.

The graphs of the functions $f$, $g$, and $h$ are shown in Figure 13.

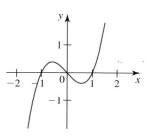
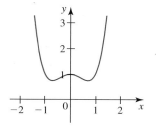
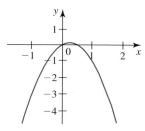

**FIGURE 13**  (**a**) $f(x) = x^3 - x$    (**b**) $g(x) = x^4 - x^2 + 1$    (**c**) $h(x) = x - 2x^2$

## 0.2  EXERCISES

**1.** If $f(x) = 3x + 4$, find $f(0)$, $f(-4)$, $f(a)$, $f(-a)$, $f(a + 1)$, $f(2a)$, $f(\sqrt{a})$, and $f(x + 1)$.

**2.** If $f(x) = 2x - 1$, find $f(-\sqrt{2})$, $f(t + 1)$, $f(2t - 1)$, $f\left(\dfrac{x}{2}\right)$, and $f(a + h)$.

**3.** If $g(x) = -x^2 + 2x$, find $g(-2)$, $g(\sqrt{3})$, $g(a^2)$, $g(a + h)$, and $\dfrac{1}{g(3)}$.

**4.** If $f(t) = \dfrac{2t^2}{\sqrt{t - 1}}$, find $f(2)$, $f(x + 1)$, and $f(2x - 1)$.

**5.** If $f(x) = 2x^3 - x$, find $f(-1)$, $f(0)$, $f(x^2)$, $f(\sqrt{x})$, and $f\left(\dfrac{1}{x}\right)$.

**6.** If $f(x) = \dfrac{\sqrt{x}}{x^2 + 1}$, find $f(4)$, $f(x + h)$, $f(x - h)$, and $f(x + 2h)$.

**7.** If
$$f(x) = \begin{cases} x^2 + 1 & \text{if } x \le 0 \\ \sqrt{x} & \text{if } x > 0 \end{cases}$$
find $f(-2)$, $f(0)$, and $f(1)$.

**8.** If
$$f(x) = \begin{cases} \dfrac{x}{x + 1} & \text{if } x < -1 \\ 1 + \sqrt{x + 1} & \text{if } x \ge -1 \end{cases}$$
find $f(-2)$, $f(-1)$, and $f(0)$.

**9.** If $f(x) = x^2$, find and simplify $\dfrac{f(x) - f(1)}{x - 1}$, where $x \ne 1$.

**10.** If $f(x) = 2x^2 + 1$, find and simplify $\dfrac{f(1 + h) - f(1)}{h}$, where $h \ne 0$.

**11.** If $f(x) = x - x^2$, find and simplify $\dfrac{f(x + h) - f(x)}{h}$, where $h \ne 0$.

**12.** If $f(x) = \sqrt{x}$, find and simplify $\dfrac{f(a + h) - f(a)}{h}$, where $h \ne 0$.

**13. a.** If $f(x) = x^2 - 2x + k$ and $f(1) = 3$, find $k$.
   **b.** If $g(t) = |t - 1| + k$ and $g(-1) = 0$, find $k$.

**14.** If $f(x) = ax^3 + b$, find $a$ and $b$ if it is known that $f(1) = 1$ and $f(2) = 15$.

*In Exercises 15–26, find the domain of the function.*

**15.** $f(x) = \dfrac{3x + 1}{x^2}$

**16.** $f(x) = \dfrac{2x + 1}{x - 1}$

**17.** $g(t) = \dfrac{t + 1}{2t^2 - t - 1}$

**18.** $h(x) = \sqrt{2x + 3}$

**19.** $f(x) = \sqrt{9 - x^2}$

**20.** $F(x) = \sqrt{x^2 - 2x - 3}$

**21.** $f(x) = \sqrt{x - 2} + \sqrt{4 - x}$

**22.** $f(x) = \dfrac{\sqrt{x - 1}}{x^2 - x - 6}$

**23.** $f(x) = \dfrac{\sqrt{x + 2} + \sqrt{2 - x}}{x^3 - x}$

**24.** $f(x) = \dfrac{\sqrt[3]{x^2 - x + 1}}{x^2 + 1}$

**25.** $f(x) = \dfrac{x^3 + 1}{x\sqrt{x^2 - 1}}$

**26.** $f(x) = \dfrac{1}{\sqrt{|x| - x}}$

**27.** Refer to the graph of the function $f$ in the following figure.

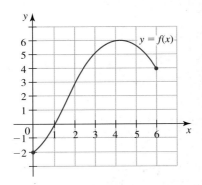

a. Find $f(0)$.
b. Find the value of $x$ for which (i) $f(x) = 3$ and (ii) $f(x) = 0$.
c. Find the domain of $f$.
d. Find the range of $f$.

**28.** Refer to the graph of the function $f$ in the following figure.

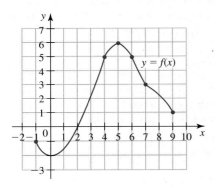

a. Find $f(7)$.
b. Find the values of $x$ corresponding to the point(s) on the graph of $f$ located at a height of 5 units above the $x$-axis.
c. Find the point on the $x$-axis at which the graph of $f$ crosses it. What is $f(x)$ at this point?
d. Find the domain and range of $f$.

*In Exercises 29–30, determine whether the point lies on the graph of the function.*

**29.** $P(3, 3); \quad f(x) = \dfrac{x + 1}{\sqrt{x^2 + 7}} + 2$

**30.** $P\left(-3, -\tfrac{1}{13}\right); \quad f(t) = \dfrac{|t + 1|}{t^3 + 2}$

*In Exercises 31–38, find the domain and sketch the graph of the function. What is its range?*

**31.** $f(x) = -2x + 1$

**32.** $f(x) = \dfrac{1}{2}x^2 + 1$

**33.** $g(x) = \sqrt{x - 1}$

**34.** $f(x) = |x| - 1$

**35.** $h(x) = \sqrt{x^2 - 1}$

**36.** $f(t) = \dfrac{|t - 1|}{t - 1}$

**37.** $f(x) = \begin{cases} -x + 1 & \text{if } x \le 1 \\ x^2 - 1 & \text{if } x > 1 \end{cases}$

**38.** $f(x) = \begin{cases} -x - 1 & \text{if } x < -1 \\ 0 & \text{if } -1 \le x \le 1 \\ x + 1 & \text{if } x > 1 \end{cases}$

*In Exercises 39–42, use the vertical line test to determine whether the curve is the graph of a function of x.*

**39.**
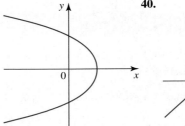

**40.**
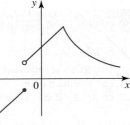

**41.**
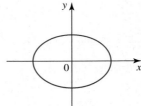

**42.**
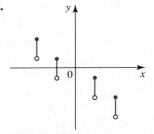

**43.** Refer to the curve for Exercise 39. Is it the graph of a function of $y$? Explain.

**44.** Refer to the curve for Exercise 40. Is it the graph of a function of $y$? Explain.

*In Exercises 45–48, determine whether the function whose graph is given, is even, odd, or neither.*

**45.**
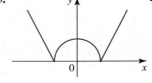

**46.**
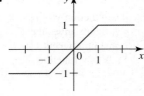

**47.** 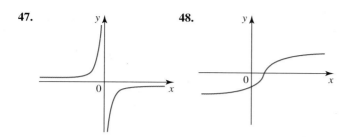   **48.**

*In Exercises 49–54, determine whether the function is even, odd, or neither.*

**49.** $f(x) = 1 - 2x^2$

**50.** $f(x) = \dfrac{x}{x^2 + 1}$

**51.** $f(x) = 2x^3 - 3x + 1$

**52.** $f(x) = 2x^{1/3} - 3x^2$

**53.** $f(x) = \dfrac{|x| + 1}{x^4 - 2x^2 + 3}$

**54.** $f(x) = \sqrt{x^2 + x + 1} - \sqrt{x^2 - x + 1}$

**55.** The following figure shows a portion of the graph of a function $f$ defined on the interval $[-2, 2]$. Sketch the complete graph of $f$ if it is known that (a) $f$ is even, (b) $f$ is odd.

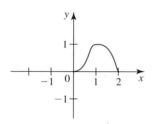

**56.** The following figure shows a portion of the graph of a function $f$ defined on the interval $[-2, 2]$.
  **a.** Can $f$ be odd? Explain. If so, complete the graph of $f$.
  **b.** Can $f$ be even? Explain. If so, complete the graph of $f$.

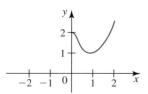

**57.** The function $y = f(t)$, whose graph is shown in the following figure, gives the distance the Jacksons were from their home on a recent trip they took from Boston to Niagara Falls as a function of time $t$ ($t = 0$ corresponds to 7 A.M.).

The 500-mi trip took a total of 8 hr. What does the graph tell us about the trip?

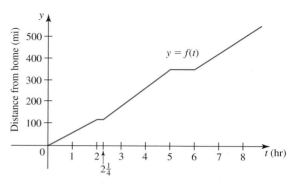

**58.** A plane departs from Logan Airport in Boston bound for Heathrow Airport in London, a 6-hr, 3267-mi flight. After takeoff, the plane climbs to a cruising altitude of 35,000 ft, which it maintains until its descent to the airport. While at its cruising altitude, the plane maintains a ground speed of 550 mph. Let $D = f(t)$ denote the distance (in miles) flown by the plane as a function of time (in hours), and let $A = g(t)$ denote the altitude (in feet) of the plane.
  **a.** Sketch a graph of $f$ that could describe the situation.
  **b.** Sketch a graph of $g$ that could describe the situation.

**59.** **Oxygen Content of a Pond** When organic waste is dumped into a pond, the oxidation process that takes place reduces the pond's oxygen content. However, given time, nature will restore the oxygen content to its natural level. Let $P = f(t)$ denote the oxygen content (as a percentage of its normal level) $t$ days after organic waste has been dumped into the pond. Sketch a graph of $f$ that could depict the process.

**60.** **The Gender Gap** The following graph shows the ratio of women's earnings to men's from 1960 through 2000.

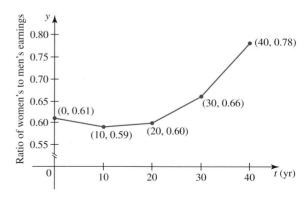

  **a.** Write the rule for the function $f$ giving the ratio of women's earnings to men's in year $t$, with $t = 0$ corresponding to 1960.
  **Hint:** The function $f$ is defined piecewise and is linear over each of four subintervals.

**b.** In what decade(s) was the gender gap expanding? Shrinking?

**c.** Refer to part (b). How fast was the gender gap (the ratio per year) expanding or shrinking in each of these decades?

*Source:* U.S. Bureau of Labor Statistics.

**61. Prevalence of Alzheimer's Patients** On the basis of a study conducted in 1997, the percentage of the U.S. population by age afflicted with Alzheimer's disease is given by the function

$$P(x) = 0.0726x^2 + 0.7902x + 4.9623 \qquad 0 \le x \le 25$$

where $x$ is measured in years, with $x = 0$ corresponding to age 65. What percentage of the U.S. population at age 65 is expected to have Alzheimer's disease? At age 90?

*Source:* Alzheimer's Association.

**62. U.S. Health Care Information Technology Spending** As health care costs increase, payers are turning to technology and outsourced services to keep a lid on expenses. The amount of health care information technology spending by payer is approximated by

$$S(t) = -0.03t^3 + 0.2t^2 + 0.23t + 5.6 \qquad 0 \le t \le 4$$

where $S(t)$ is measured in billions of dollars and $t$ is measured in years with $t = 0$ corresponding to 2004. What was the amount spent by payers on health care IT in 2004? What amount was spent by payers in 2008?

*Source:* U.S. Department of Commerce.

**63. Hotel Rates** The average daily rate of U.S. hotels from 2001 through 2006 is approximated by the function

$$f(t) = \begin{cases} 82.95 & \text{if } 1 \le t \le 3 \\ 0.95t^2 - 3.95t + 86.25 & \text{if } 3 < t \le 6 \end{cases}$$

where $f(t)$ is measured in dollars and $t = 1$ corresponds to 2001.

**a.** What was the average daily rate of U.S. hotels from 2001 through 2003?

**b.** What was the average daily rate of U.S. hotels in 2004? In 2005? In 2006?

**c.** Sketch the graph of $f$.

*Source:* Smith Travel Research.

**64. Postal Regulations** In 2007 the postage for packages sent by first-class mail was raised to $1.13 for the first ounce or fraction thereof and 17¢ for each additional ounce or fraction thereof. Any parcel not exceeding 13 oz may be sent by first-class mail. Letting $x$ denote the weight of a parcel in ounces and letting $f(x)$ denote the postage in dollars, complete the following description of the "postage function" $f$:

$$f(x) = \begin{cases} 1.13 & \text{if } 0 < x \le 1 \\ 1.30 & \text{if } 1 < x \le 2 \\ \vdots & \\ ? & \text{if } 12 < x \le 13 \end{cases}$$

**a.** What is the domain of $f$?

**b.** Sketch the graph of $f$.

**65. Harbor Cleanup** The amount of solids discharged from the Massachusetts Water Resources Authority sewage treatment plant on Deer Island (near Boston Harbor) is given by the function

$$f(t) = \begin{cases} 130 & \text{if } 0 \le t \le 1 \\ -30t + 160 & \text{if } 1 < t \le 2 \\ 100 & \text{if } 2 < t \le 4 \\ -5t^2 + 25t + 80 & \text{if } 4 < t \le 6 \\ 1.25t^2 - 26.25t + 162.5 & \text{if } 6 < t \le 10 \end{cases}$$

where $f(t)$ is measured in tons/day and $t$ is measured in years, with $t = 0$ corresponding to 1989.

**a.** What amount of solids were discharged per day in 1989? In 1992? In 1996?

**b.** Sketch the graph of $f$.

*Source:* Metropolitan District Commission.

**66. Rising Median Age** Increased longevity and the aging of the baby boom generation—those born between 1946 and 1965—are the primary reasons for a rising median age. The median age (in years) of the U.S. population from 1900 through 2000 is approximated by the function

$$f(t) = \begin{cases} 1.3t + 22.9 & \text{if } 0 \le t \le 3 \\ -0.7t^2 + 7.2t + 11.5 & \text{if } 3 < t \le 7 \\ 2.6t + 9.4 & \text{if } 7 < t \le 10 \end{cases}$$

where $t$ is measured in decades, with $t = 0$ corresponding to the beginning of 1900.

**a.** What was the median age of the U.S. population at the beginning of 1900? At the beginning of 1950? At the beginning of 1990?

**b.** Sketch the graph of $f$.

*Source:* U.S. Census Bureau.

**67.** Suppose a function has the property that whenever $x$ is in the domain of $f$, then so is $-x$. Show that $f$ can be written as the sum of an even function and an odd function.

**68.** Prove that a nonzero *polynomial function*

$$f(x) = a_n x^n + a_{n-1} x^{n-1} + \cdots + a_2 x^2 + a_1 x + a_0$$

where $n$ is a nonnegative integer and $a_0, a_1, \ldots, a_n$ are real numbers with $a_n \ne 0$, can be expressed as the sum of an even function and an odd function.

*In Exercises 69–72, determine whether the statement is true or false. If it is true, explain why. If it is false, explain why or give an example that shows that it is false.*

**69.** If $a = b$, then $f(a) = f(b)$.

**70.** If $f(a) = f(b)$, then $a = b$.

**71.** If $f$ is a function, then $f(a + b) = f(a) + f(b)$.

**72.** A curve in the $xy$-plane can be simultaneously the graph of a function of $x$ and the graph of a function of $y$.

## 0.3 The Trigonometric Functions

### 0.3 SELF-CHECK DIAGNOSTIC TEST

1. Given that $\sec \theta = \frac{5}{3}$ and $0 \le \theta < \frac{\pi}{2}$, find $\tan \theta$.

2. Determine whether the function $f(x) = \dfrac{\sin 2x}{\sqrt{1 + \cos^2 x} + 1}$ is even, odd, or neither.

3. Verify the identity $\dfrac{\cot x - 1}{1 - \tan x} = \cot x$.

4. Using the substitution $x = a \sin \theta$, where $a > 0$ and $-\frac{\pi}{2} \le \theta \le \frac{\pi}{2}$, express $a^2 - x^2$ in terms of $\theta$.

5. Solve the equation $\cos \theta - 2 \sin^2 \theta + 1 = 0$, where $0 \le \theta < 2\pi$.

*Answers to Self-Check Diagnostic Test 0.3 can be found on page ANS 2.*

In this section we review the basic properties of the trigonometric functions and their graphs. The emphasis is placed on those topics that we will use later in calculus.

### ■ Angles

An angle in the plane is generated by rotating a ray about its endpoint. The starting position of the ray is called the initial side of the angle, the final position of the ray is called the terminal side, and the point of intersection of the two sides is called the **vertex** of the angle (see Figure 1a).

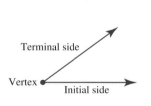

(a) An angle

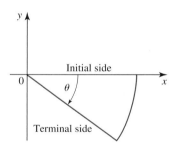

(b) A positive angle in standard position

(c) A negative angle in standard position

**FIGURE 1**

In a rectangular coordinate system an angle $\theta$ (the Greek *theta*) is in **standard position** if its vertex is centered at the origin and its initial side coincides with the positive $x$-axis. An angle is **positive** if it is generated by a counterclockwise rotation and **negative** if it is generated by a clockwise rotation (Figure 1b–c).

### ■ Radian Measure of Angles

We can express the magnitude of an angle in either degrees or radians. In calculus, however, we prefer to use the radian measure of an angle because it simplifies our work.

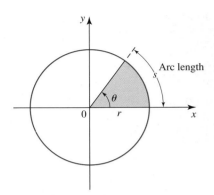

**FIGURE 2**

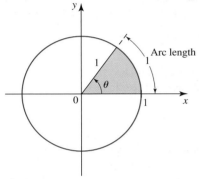

**FIGURE 3**
The unit circle $x^2 + y^2 = 1$

**DEFINITION    Radian Measure of an Angle**

If $s$ is the length of the arc subtended by a central angle $\theta$ in a circle of radius $r$, then

$$\theta = \frac{s}{r} \qquad (1)$$

is the **radian measure** of $\theta$ (see Figure 2).

For convenience we often work with the **unit circle,** that is, the circle of radius 1 centered at the origin. On the unit circle, an angle of 1 radian is subtended by an arc of length 1 (see Figure 3). To specify the units of measure for the angle $\theta$ in Figure 3, we write $\theta = 1$ radian or $\theta = 1$. By convention, if the unit of measure is not specifically stated, we assume that it is radians.

Since the circumference of the unit circle is $2\pi$ and the central angle subtended by one complete revolution is $360°$, we see that

$$2\pi \text{ radians (rad)} = 360°$$

or

$$1 \text{ rad} = \left(\frac{180}{\pi}\right)^{\circ} \qquad (2)$$

and

$$1° = \frac{\pi}{180} \text{ rad} \qquad (3)$$

These relationships suggest the following useful conversion rules.

**Converting Degrees and Radians**

To convert degrees to radians, multiply by $\dfrac{\pi}{180}$.

To convert radians to degrees, multiply by $\dfrac{180}{\pi}$.

**EXAMPLE 1**   Convert each of the following to radian measure:

**a.** $60°$      **b.** $300°$      **c.** $-225°$

**Solution**

**a.** $60 \cdot \dfrac{\pi}{180} = \dfrac{\pi}{3}$, or $\dfrac{\pi}{3}$ rad

**b.** $300 \cdot \dfrac{\pi}{180} = \dfrac{5\pi}{3}$, or $\dfrac{5\pi}{3}$ rad

**c.** $-225 \cdot \dfrac{\pi}{180} = -\dfrac{5\pi}{4}$, or $-\dfrac{5\pi}{4}$ rad

**EXAMPLE 2** Convert each of the following to degree measure:

**a.** $\dfrac{\pi}{3}$ rad    **b.** $\dfrac{3\pi}{4}$ rad    **c.** $-\dfrac{7\pi}{4}$ rad

**Solution**

**a.** $\dfrac{\pi}{3} \cdot \dfrac{180}{\pi} = 60$, or $60°$

**b.** $\dfrac{3\pi}{4} \cdot \dfrac{180}{\pi} = 135$, or $135°$

**c.** $-\dfrac{7\pi}{4} \cdot \dfrac{180}{\pi} = -315$, or $-315°$ ∎

More than one angle may have the same initial and terminal sides. We call such angles **coterminal.** For example the angle $4\pi/3$ has the same initial and terminal sides as the angle $\theta = -2\pi/3$ (see Figure 4).

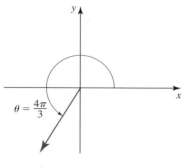    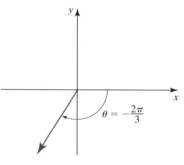

**FIGURE 4**
Coterminal angles    **(a)** $\theta = \dfrac{4\pi}{3}$    **(b)** $\theta = -\dfrac{2\pi}{3}$

An angle may be greater than $2\pi$ rad. For example, an angle of $3\pi$ rad is generated by rotating a ray in a counterclockwise direction through one and a half revolutions (Figure 5a). Similarly, an angle of $-5\pi/2$ radians is generated by rotating a ray in a clockwise direction through one and a quarter revolutions (Figure 5b).

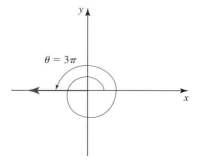    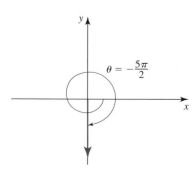

**FIGURE 5**
Angles generated by more
than one revolution    **(a)** $\theta = 3\pi$    **(b)** $\theta = -\dfrac{5\pi}{2}$

The radian and degree measures of several common angles are given in Table 1. Be sure that you familiarize yourself with these values.

**TABLE 1**

| Degrees | 0° | 30° | 45° | 60° | 90° | 120° | 135° | 150° | 180° | 270° | 360° |
|---------|-----|-----|-----|-----|-----|------|------|------|------|------|------|
| Radians | 0 | $\dfrac{\pi}{6}$ | $\dfrac{\pi}{4}$ | $\dfrac{\pi}{3}$ | $\dfrac{\pi}{2}$ | $\dfrac{2\pi}{3}$ | $\dfrac{3\pi}{4}$ | $\dfrac{5\pi}{6}$ | $\pi$ | $\dfrac{3\pi}{2}$ | $2\pi$ |

By rewriting Equation (1), $\theta = s/r$, we obtain the following formula, which gives the length of a circular arc.

---

**Length of a Circular Arc**

$$s = r\theta \tag{4}$$

---

Another related formula that we will use later in calculus gives the area of a circular sector.

---

**Area of a Circular Sector**

$$A = \frac{1}{2}r^2\theta \tag{5}$$

---

**Note**   In Equations (4) and (5) $\theta$ must be expressed in radians.   ■

**EXAMPLE 3**   What is the length of the arc subtended by $\theta = 7\pi/6$ radians in a circle of radius 3? What is the area of the circular sector determined by $\theta$?

**Solution**   To find the length of the arc, we use Equation (4) to obtain

$$s = 3\left(\frac{7\pi}{6}\right) = \frac{7\pi}{2}$$

The area of the sector is obtained by using Equation (5). Thus,

$$A = \frac{1}{2}r^2\theta = \frac{1}{2}(3)^2\left(\frac{7\pi}{6}\right)$$

$$= \frac{21\pi}{4}$$
   ■

## ■ The Trigonometric Functions

Two approaches are generally used to define the six trigonometric functions. We summarize each approach here.

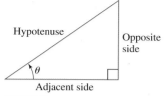
**FIGURE 6**

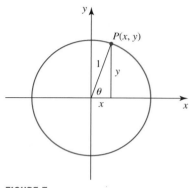
**FIGURE 7**
The unit circle

## THE TRIGONOMETRIC FUNCTIONS

### The Right Triangle Definition

For an acute angle $\theta$ (see Figure 6),

$$\sin \theta = \frac{\text{opp}}{\text{hyp}} \qquad \cos \theta = \frac{\text{adj}}{\text{hyp}} \qquad \csc \theta = \frac{\text{hyp}}{\text{opp}}$$

$$\sec \theta = \frac{\text{hyp}}{\text{adj}} \qquad \tan \theta = \frac{\text{opp}}{\text{adj}} \qquad \cot \theta = \frac{\text{adj}}{\text{opp}}$$

### The Unit Circle Definition

Let $\theta$ denote an angle in standard position, and let $P(x, y)$ denote the point where the terminal side of $\theta$ meets the unit circle. (See Figure 7.) Then

$$\sin \theta = y \qquad\qquad \cos \theta = x$$

$$\csc \theta = \frac{1}{y}, \quad y \neq 0 \qquad\qquad \sec \theta = \frac{1}{x}, \quad x \neq 0$$

$$\tan \theta = \frac{y}{x}, \quad x \neq 0 \qquad\qquad \cot \theta = \frac{x}{y}, \quad y \neq 0$$

Referring to the point $P(x, y)$ on the unit circle (Figure 7), we see that the coordinates of $P$ can also be written in the form

$$x = \cos \theta \qquad \text{and} \qquad y = \sin \theta \qquad\qquad (6)$$

**Note**   $\tan \theta$ and $\sec \theta$ are not defined when $x = 0$. Also, $\csc \theta$ and $\cot \theta$ are not defined when $y = 0$. ▪

Table 2 lists the values of the trigonometric functions of certain angles. Since these values occur very frequently in problems involving trigonometry, you will find it helpful to memorize them. The right triangles shown in Figure 8 can be used to help jog your memory.

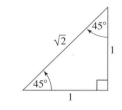

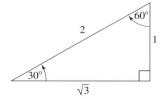
**FIGURE 8**

**TABLE 2**

| $\theta$ (radians) | $\theta$ (degrees) | $\sin \theta$ | $\cos \theta$ | $\tan \theta$ |
|:---:|:---:|:---:|:---:|:---:|
| $\dfrac{\pi}{6}$ | $30°$ | $\dfrac{1}{2}$ | $\dfrac{\sqrt{3}}{2}$ | $\dfrac{\sqrt{3}}{3}$ |
| $\dfrac{\pi}{4}$ | $45°$ | $\dfrac{\sqrt{2}}{2}$ | $\dfrac{\sqrt{2}}{2}$ | $1$ |
| $\dfrac{\pi}{3}$ | $60°$ | $\dfrac{\sqrt{3}}{2}$ | $\dfrac{1}{2}$ | $\sqrt{3}$ |

The sign of a trigonometric function of an angle $\theta$ is determined by the quadrant in which the terminal side of $\theta$ lies. Figure 9 shows a helpful way of remembering the functions that are positive in each quadrant. The signs of the other functions are easy to remember, since they are all negative.

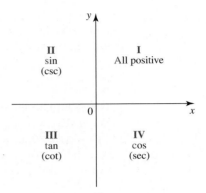

**FIGURE 9**
The trigonometric functions that are positive in each quadrant can be remembered with the mnemonic device ASTC: **A**ll **S**tudents **T**ake **C**alculus. The functions that are not listed in each quadrant are negative.

To evaluate the trigonometric functions in quadrants other than the first quadrant, we use a reference angle. A **reference angle** for an angle $\theta$ is the acute angle formed by the $x$-axis and the terminal side of $\theta$. Reference angles for each quadrant are depicted in Figure 10.

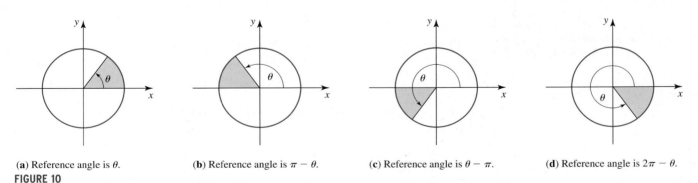

**(a)** Reference angle is $\theta$.  **(b)** Reference angle is $\pi - \theta$.  **(c)** Reference angle is $\theta - \pi$.  **(d)** Reference angle is $2\pi - \theta$.
**FIGURE 10**

The next example illustrates how we find the trigonometric functions of an angle.

**EXAMPLE 4** Find the sine, cosine, and tangent of $5\pi/4$.

**Solution** We first determine the reference angle for the given angle. As is indicated in Figure 11, the reference angle is $(5\pi/4) - \pi = \pi/4$, or $45°$. Since $\sin 45° = \sqrt{2}/2$ and the sine is negative in Quadrant III, we conclude that $\sin(5\pi/4) = -\sqrt{2}/2$. Similarly, since $\cos 45° = \sqrt{2}/2$ and the cosine is negative in Quadrant III, we conclude that $\cos(5\pi/4) = -\sqrt{2}/2$. Finally, since $\tan 45° = 1$ and the tangent is positive in Quadrant III, we conclude that $\tan(5\pi/4) = 1$. ∎

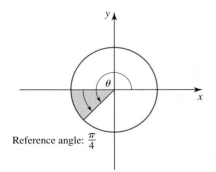

Reference angle: $\frac{\pi}{4}$

**FIGURE 11**
The reference angle for $\theta = 5\pi/4$ is $\pi/4$, or $45°$.

The values of the trigonometric functions that we found in Example 4 are *exact*. The *approximate* value of any trigonometric function can be found by using a calculator. If you use a calculator, be sure to set the mode correctly. For example, to find

$\sin(5\pi/4)$, first set the calculator in radian mode and then enter $\sin(5\pi/4)$. The result will be

$$\sin\frac{5\pi}{4} \approx -0.7071068$$

The number of digits in your answer will depend on the calculator that you use. As we saw in Example 4, the *exact* value of $\sin(5\pi/4)$ is $-\sqrt{2}/2$. Notice that we do not need to use reference angles when we use a calculator.

## ◼ Graphs of the Trigonometric Functions

Referring once again to the unit circle, which is reproduced in Figure 12, we see that an angle of $2\pi$ rad corresponds to one complete revolution on the unit circle. Since $P(x, y) = (\cos\theta, \sin\theta)$ is the point where the terminal side of $\theta$ intersects the unit circle, we see that the values of $\sin\theta$ and $\cos\theta$ repeat themselves in subsequent revolutions.

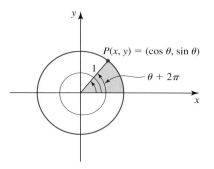

**FIGURE 12**
The $x$ and $y$ coordinates of the point $P$ are the same for $\theta$ and $\theta + 2\pi$.

Therefore,

$$\sin(\theta + 2\pi) = \sin\theta \qquad \text{and} \qquad \cos(\theta + 2\pi) = \cos\theta \tag{7a}$$

and

$$\sin(\theta + 2n\pi) = \sin\theta \qquad \text{and} \qquad \cos(\theta + 2n\pi) = \cos\theta \tag{7b}$$

for every real number $\theta$ and every integer $n$, and we say that the sine and cosine functions are periodic with period $2\pi$.

More generally, we have the following definition of a periodic function.

> **DEFINITION   Periodic Function**
> A function $f$ is **periodic** if there is a number $p > 0$ such that
> $$f(x + p) = f(x)$$
> for all $x$ in the domain of $f$. The smallest such number $p$ is called the **period** of $f$.

The graphs of the six trigonometric functions are shown in Figure 13. Note that we have denoted the independent variable by $x$ instead of $\theta$. Here, the real number $x$ denotes the radian measure of an angle. As their graphs indicate, the six trigonometric functions are all periodic. The sine and cosine functions, as well as their reciprocals, the

Domain: $(-\infty, \infty)$
Range: $[-1, 1]$
Period: $2\pi$

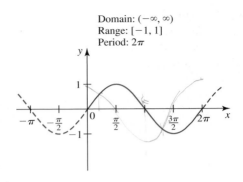

**(a)** $y = \sin x$

Domain: $(-\infty, \infty)$
Range: $[-1, 1]$
Period: $2\pi$

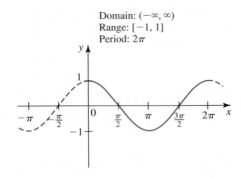

**(b)** $y = \cos x$

Domain: $x \neq \frac{\pi}{2} + n\pi$
Range: $(-\infty, \infty)$
Period: $\pi$

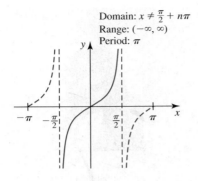

**(c)** $y = \tan x$

Domain: $x \neq n\pi$
Range: $(-\infty, -1] \cup [1, \infty)$
Period: $2\pi$

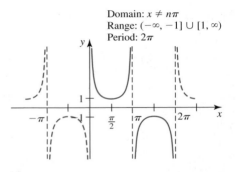

**(d)** $y = \csc x$

Domain: $x \neq \frac{\pi}{2} + n\pi$
Range: $(-\infty, -1] \cup [1, \infty)$
Period: $2\pi$

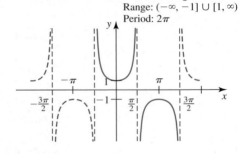

**(e)** $y = \sec x$

Domain: $x \neq n\pi$
Range: $(-\infty, \infty)$
Period: $\pi$

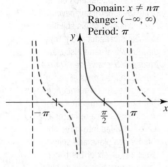

**(f)** $y = \cot x$

**FIGURE 13**
Graphs of the six trigonometric functions

cosecant and secant functions, have period $2\pi$. The period of the tangent and cotangent functions, however, is $\pi$.

Let's look more closely at the graphs shown in Figure 13a–b. Notice that the graphs of $y = \sin x$ and $y = \cos x$ oscillate between $y = -1$ and $y = 1$. In general, the graphs of the functions $y = A \sin x$ and $y = A \cos x$ oscillate between $y = -A$ and $y = A$, and we say that their amplitude is $|A|$. The graphs of $y = 4 \sin x$ and $y = \frac{1}{4} \sin x$ are shown in Figure 14a–b. Observe that the factor 4 in $y = 4 \sin x$ has the effect of "stretching" the graph of $y = \sin x$ between the values of $-4$ and 4, whereas the factor $\frac{1}{4}$ in $y = \frac{1}{4} \sin x$ has the effect of "compressing" the graph between $-\frac{1}{4}$ and $\frac{1}{4}$.

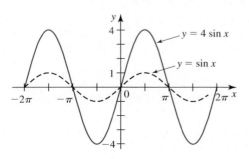

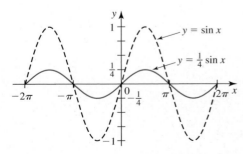

**FIGURE 14**

**(a)** The graph of $y = 4 \sin x$ superimposed upon the graph of $y = \sin x$

**(b)** The graph of $y = \frac{1}{4} \sin x$ superimposed upon the graph of $y = \sin x$

Next, let's compare the graphs of $y = \cos 2x$ and $y = \cos(x/2)$ with the graph of $y = \cos x$ (see Figure 15a–b). Notice here that the factor of 2 has the effect of "speeding up" the graph of the cosine: The period is decreased from $2\pi$ to $\pi$. In contrast, the factor of $\frac{1}{2}$ has the effect of "slowing down" the graph of the cosine: The period is increased from $2\pi$ to $4\pi$. In general, the period of both $y = \sin Bx$ and $y = \cos Bx$ is $2\pi/|B|$ if $B \neq 0$.

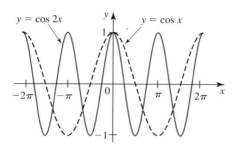

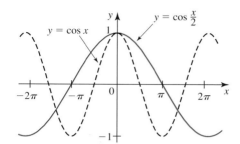

**(a)** The graph of $y = \cos 2x$ superimposed upon the graph of $y = \cos x$

**(b)** The graph of $y = \cos \frac{x}{2}$ superimposed upon the graph of $y = \cos x$

**FIGURE 15**

We now summarize these definitions.

---

**DEFINITION**  **Period and Amplitude of $A \sin Bx$ and $A \cos Bx$**

The graphs of

$$f(x) = A \sin Bx \qquad \text{and} \qquad f(x) = A \cos Bx$$

where $A \neq 0$ and $B \neq 0$, have period $2\pi/|B|$ and **amplitude** $|A|$.

---

**EXAMPLE 5**  Sketch the graph of $y = 3 \sin \frac{1}{2} x$.

**Solution**  The function $y = 3 \sin \frac{1}{2} x$ has the form $y = A \sin Bx$, where $A = 3$ and $B = \frac{1}{2}$. This tells us that the amplitude of the graph is 3 and the period is $2\pi/|\frac{1}{2}| = 4\pi$. Using the graph of the sine curve, we sketch the graph of $y = 3 \sin \frac{1}{2} x$ over one period $[0, 4\pi]$. (See Figure 16.) Next, the periodic properties of the sine function allow us to extend the graph in either direction by completing another cycle as shown.

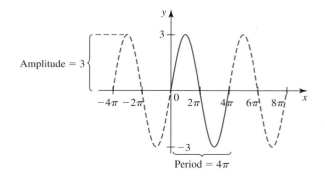

**FIGURE 16**
The graph of $y = 3 \sin \frac{1}{2} x$ has amplitude 3 and period $4\pi$.

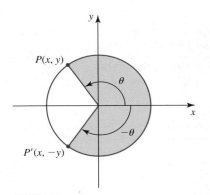

**FIGURE 17**
The angles $\theta$ and $-\theta$ have the same magnitude but opposite signs.

## The Trigonometric Identities

By comparing the angles $\theta$ and $-\theta$ in Figure 17, we see that the points $P$ and $P'$ have the same $x$-coordinates and that their $y$-coordinates differ only in sign. Thus,

$$\cos \theta = x = \cos(-\theta) \tag{8}$$

and

$$\sin \theta = y = -\sin(-\theta) \tag{9}$$

We conclude that the cosine function is even and the sine function is odd. Similarly, we can show that the cosecant, tangent, and cotangent functions are odd, while the secant function is even. These results are also confirmed by the symmetry of the graph of each function (see Figure 13).

Equations such as Equations (8) and (9) that express a relationship between trigonometric functions are called **trigonometric identities.** Each identity holds true for every value of $\theta$ in the domain of the specified trigonometric functions.

Referring once again to the point $P(x, y)$ on the unit circle (see Figure 7), we see that the equation $x^2 + y^2 = 1$ can also be written in the form

$$\cos^2 \theta + \sin^2 \theta = 1 \tag{10}$$

**Note**   Recall that $\sin^2 \theta = (\sin \theta)^2$. In general, $(\sin \theta)^n$ is usually written $\sin^n \theta$. The same convention applies to the other trigonometric functions.   ■

The **addition** and **subtraction formulas** for the sine and cosine are

$$\sin(A \pm B) = \sin A \cos B \pm \cos A \sin B \tag{11}$$

and

$$\cos(A \pm B) = \cos A \cos B \mp \sin A \sin B \tag{12}$$

If we let $A = B$ in Formulas (11) and (12), we obtain the **double-angle** formulas

$$\sin 2A = 2 \sin A \cos A \tag{13}$$

and

$$\cos 2A = \cos^2 A - \sin^2 A \tag{14a}$$
$$= 2 \cos^2 A - 1 \tag{14b}$$
$$= 1 - 2 \sin^2 A \tag{14c}$$

Solving (14b) and (14c) for $\cos^2 A$ and $\sin^2 A$, respectively, we obtain the **half-angle formulas**

$$\cos^2 A = \frac{1}{2} (1 + \cos 2A) \tag{15}$$

and

$$\sin^2 A = \frac{1}{2} (1 - \cos 2A) \tag{16}$$

These and several other trigonometric identities are summarized in Table 3.

**TABLE 3**   Trigonometric Identities

| Pythagorean identities | Half-angle formulas | Addition and subtraction formulas |
|---|---|---|
| $\cos^2 \theta + \sin^2 \theta = 1$ | $\cos^2 A = \frac{1}{2}(1 + \cos 2A)$ | $\sin(A \pm B) = \sin A \cos B \pm \cos A \sin B$ |
| $\tan^2 \theta + 1 = \sec^2 \theta$ | $\sin^2 A = \frac{1}{2}(1 - \cos 2A)$ | $\cos(A \pm B) = \cos A \cos B \mp \sin A \sin B$ |
| $\cot^2 \theta + 1 = \csc^2 \theta$ | | |

| Double-angle formulas | Cofunctions of complementary angles |
|---|---|
| $\sin 2A = 2 \sin A \cos A$ | $\sin \theta = \cos\left(\frac{\pi}{2} - \theta\right)$ |
| $\cos 2A = \cos^2 A - \sin^2 A = 2 \cos^2 A - 1$ | $\cos \theta = \sin\left(\frac{\pi}{2} - \theta\right)$ |
| $\qquad\qquad\quad = 1 - 2 \sin^2 A$ | |

A more complete list of trigometric identities can be found in the reference pages at the back of the book.

**EXAMPLE 6**   Find the solutions of the equation $\cos 2x - \cos x = 0$ that lie in the interval $[0, 2\pi]$.

**Solution**   Using the identity (14b), we make the substitution $\cos 2x = 2 \cos^2 x - 1$, obtaining

$$\cos 2x - \cos x = 0$$
$$(2 \cos^2 x - 1) - \cos x = 0$$
$$2 \cos^2 x - \cos x - 1 = 0$$
$$(2 \cos x + 1)(\cos x - 1) = 0$$
$$2 \cos x + 1 = 0 \qquad \text{or} \qquad \cos x - 1 = 0$$

Thus,

$$\cos x = -\frac{1}{2} \qquad \text{or} \qquad \cos x = 1$$

and $x = 2\pi/3$, $4\pi/3$, $0$, and $2\pi$ are the solutions in the interval $[0, 2\pi]$.  ■

## 0.3   EXERCISES

*In Exercises 1–8, convert each angle to radian measure.*

**1.** $150°$   **2.** $210°$   **3.** $330°$   **4.** $405°$

**5.** $-120°$   **6.** $-225°$   **7.** $-75°$   **8.** $-495°$

*In Exercises 9–16, convert each angle to degree measure.*

**9.** $\dfrac{\pi}{3}$   **10.** $\dfrac{3\pi}{4}$   **11.** $\dfrac{5\pi}{6}$   **12.** $\dfrac{9\pi}{4}$

**13.** $-\dfrac{\pi}{2}$   **14.** $-\dfrac{11\pi}{6}$   **15.** $-\dfrac{13\pi}{4}$   **16.** $-\dfrac{11\pi}{3}$

*In Exercises 17–24, find the exact value of the trigonometric functions at the indicated angle.*

**17.** $\sin \theta$, $\cos \theta$, and $\tan \theta$ for $\theta = \pi/3$

**18.** $\sin \theta$, $\cos \theta$, and $\csc \theta$ for $\theta = -\pi/4$

**19.** $\cos x$, $\tan x$, and $\sec x$ for $x = 2\pi/3$

**20.** $\sin x$, $\cot x$, and $\csc x$ for $x = 5\pi/6$

**21.** $\sin \alpha$, $\tan \alpha$, and $\csc \alpha$ for $\alpha = \pi$

**22.** $\cos \alpha$, $\cot \alpha$, and $\csc \alpha$ for $\alpha = -3\pi/2$

**23.** $\csc t$, $\sec t$, and $\cot t$ for $t = 17\pi/6$

**24.** $\sin t$, $\tan t$, and $\cot t$ for $t = -11\pi/3$

**25.** Given that $\sin \theta = \frac{3}{5}$ and $\frac{\pi}{2} \le \theta \le \pi$, find the five other trigonometric functions of $\theta$.

**26.** Given that $\cot \theta = -\frac{5}{3}$ and $\frac{\pi}{2} \le \theta \le \pi$, find the five other trigonometric functions of $\theta$.

**27.** If $f(x) = \sin x$, find $f(0)$, $f\left(\frac{\pi}{4}\right)$, $f\left(-\frac{\pi}{3}\right)$, $f(3\pi)$, and $f\left(a + \frac{\pi}{2}\right)$.

**28.** If

$$f(x) = \begin{cases} 2 + \sqrt{1 - x} & \text{if } x \le 1 \\ 2 \cos 2\pi x & \text{if } x > 1 \end{cases}$$

find $f(0)$, $f(1)$, and $f(2)$.

*In Exercises 29 and 30, find the domain of the function.*

**29.** $f(t) = \sqrt{\sin t - 1}$      **30.** $f(x) = \dfrac{x}{2 + \sin x}$

*In Exercises 31–32, determine whether the functions are even, odd, or neither.*

**31. a.** $y = 2 \sin x$      **32. a.** $y = \cot x$

     **b.** $y = -\dfrac{\cos^2 x}{x}$      **b.** $y = 2 \sin \dfrac{x}{2}$

     **c.** $y = -\csc x$      **c.** $y = 2 \sec x$

*In Exercises 33–42, verify the identity.*

**33.** $\sec t - \cos t = \tan t \sin t$

**34.** $2 \csc 2u = \sec u \csc u$

**35.** $\dfrac{\sin y}{\csc y} + \dfrac{\cos y}{\sec y} = 1$

**36.** $(\sin x)(\csc x - \sin x) = \cos^2 x$

**37.** $\tan A + \tan B = \dfrac{\sin(A + B)}{\cos A \cos B}$

**38.** $\dfrac{\cos \theta \tan \theta + \sin \theta}{\tan \theta} = 2 \cos \theta$

**39.** $\csc t - \sin t = \cos t \cot t$

**40.** $\sin 3t = 3 \sin t - 4 \sin^3 t$

**41.** $\sin 2\theta = 2 \sin^3 \theta \cos \theta + 2 \sin \theta \cos^3 \theta$

**42.** $\tan \dfrac{t}{2} = \dfrac{1 - \cos t}{\sin t}$

*In Exercises 43 and 44, find the domain and sketch the graph of the function. What is its range?*

**43.** $h(\theta) = 2 \sin \pi\theta$      **44.** $f(t) = |\cos t|$

*In Exercises 45–58, determine the amplitude and the period for the function. Sketch the graph of the function over one period.*

**45.** $y = \sin(x - \pi)$      **46.** $y = \cos(x + \pi)$

**47.** $y = \sin\left(x + \frac{\pi}{2}\right)$      **48.** $y = \cos\left(x - \frac{\pi}{4}\right)$

**49.** $y = \cos x + 2$      **50.** $y = 2 - \sin x$

**51.** $y = 2 \sin\left(2x + \frac{\pi}{2}\right)$      **52.** $y = \cos\left(2x + \frac{\pi}{4}\right)$

**53.** $y = 2 \sin x \cos x$      **54.** $y = \cos^2 x - \sin^2 x$

**55.** $y = -2 \cos 3x$      **56.** $y = -3 \sin(-4x)$

**57.** $y = 3 \cos 2x$      **58.** $y = -3 \sin(\pi x + \pi)$

*In Exercises 59–66, find the solutions of the equation in $[0, 2\pi)$.*

**59.** $\sin 2x = 1$      **60.** $\tan 2\theta = -1$

**61.** $\cos t + 2 \sec t = -3$      **62.** $\tan^2 x - \sec x - 1 = 0$

**63.** $\cos^2 x - \sin x \cos x = 0$      **64.** $\csc^2 x - \cot x - 1 = 0$

**65.** $2 \cos^2 x - 3 \cos x + 1 = 0$

**66.** $(\sin 2x)(\sin x) = 0$

**67.** After takeoff, an airplane climbs at an angle of $20°$ at a speed of 200 ft/sec. How long does it take for the airplane to reach an altitude of 10,000 ft?

**68.** A man located at a point $A$ on one bank of a river that is 1000 ft wide observed a woman jogging on the opposite bank. When the jogger was first spotted, the angle between the river bank and the man's line of sight was $30°$. One minute later, the angle was $40°$. How fast was the woman running if she maintained a constant speed?

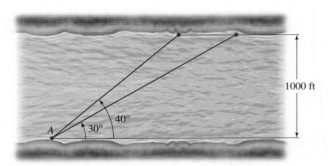

*In Exercises 69–74, determine whether the statement is true or false. If it is true, explain why. If it is false, explain why or give an example that shows it is false.*

**69.** The graph of $y = \cos x$ is the same as the graph of $y = \cos(-x)$.

**70.** The product $y = (\sin x)(\cos x)$ is an odd function of $x$.

**71.** The graph of $y = \cos(x + \pi)$ is the same as the graph of $y = -\cos x$.

**72.** The graph of $y = \cos\left(x + \frac{\pi}{4}\right)$ is the same as the graph of $y = -\sin\left(x - \frac{\pi}{4}\right)$.

**73.** The graph of $y = \csc x$ is symmetric with respect to the $y$-axis.

**74.** The function $y = \sin^2 x$ is an odd function.

# 0.4 Combining Functions

## ■ Arithmetic Operations on Functions

Many functions are built up from other, and generally simpler, functions. Consider, for example, the function $h$ defined by $h(x) = x + (1/x)$. Note that the value of $h$ at $x$ is the sum of two terms. The first term, $x$, may be viewed as the value of the function $f$ defined by $f(x) = x$ at $x$, and the second term, $1/x$, may be viewed as the value of the function $g$ defined by $g(x) = 1/x$ at $x$. These observations suggest that $h$ can be viewed as the *sum* of the functions $f$ and $g$, $f + g$, defined by

$$(f + g)(x) = f(x) + g(x) = x + \frac{1}{x}$$

The domain of $f + g$ is $(-\infty, 0) \cup (0, \infty)$, the intersection of the domains of $f$ and $g$. Note that the plus sign on the left side of this equation denotes an operation (addition in this case) on two *functions*.

Since the value of $h = f + g$ at $x$ is the sum of the values of $f$ and $g$ at $x$, we see that the graph of $h$ can be obtained from the graphs of $f$ and $g$ by adding the $y$-coordinates of $f$ and $g$ at $x$ to obtain the corresponding $y$-coordinate of $h$ at $x$. This technique is used to sketch the graph of $h$, the sum of $f(x) = x$ and $g(x) = 1/x$, discussed above (see Figure 1). We show the graph of $h$ only in the first quadrant.

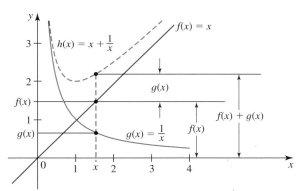

**FIGURE 1**
The graphs of $f$, $g$, and $h$

The difference, product, and quotient of two functions are defined in a similar manner.

---

**DEFINITION** Operations on Functions

Let $f$ and $g$ be functions with domains $A$ and $B$, respectively. Then their sum $f + g$, difference $f - g$, product $fg$, and quotient $f/g$ are defined as follows:

$$(f + g)(x) = f(x) + g(x) \quad \text{with domain } A \cap B \tag{1a}$$

$$(f - g)(x) = f(x) - g(x) \quad \text{with domain } A \cap B \tag{1b}$$

$$(fg)(x) = f(x)g(x) \quad \text{with domain } A \cap B \tag{1c}$$

$$\left(\frac{f}{g}\right)(x) = \frac{f(x)}{g(x)} \quad \text{with domain } \{x \mid x \in A \cap B \text{ and } g(x) \neq 0\} \tag{1d}$$

---

**EXAMPLE 1** Let $f$ and $g$ be functions defined by $f(x) = \sqrt{x}$ and $g(x) = \sqrt{3 - x}$. Find the domain and the rule for each of the functions $f + g$, $f - g$, $fg$, and $f/g$.

**Solution** The domain of $f$ is $[0, \infty)$, and the domain of $g$ is $(-\infty, 3]$. Therefore, the domain of $f + g$, $f - g$, and $fg$ is

$$[0, \infty) \cap (-\infty, 3] = [0, 3]$$

The rules for these functions are

$$(f + g)(x) = f(x) + g(x) = \sqrt{x} + \sqrt{3 - x} \qquad \text{By Equation (1a)}$$

$$(f - g)(x) = f(x) - g(x) = \sqrt{x} - \sqrt{3 - x} \qquad \text{By Equation (1b)}$$

and

$$(fg)(x) = f(x)g(x) = \sqrt{x}\sqrt{3 - x} = \sqrt{3x - x^2} \qquad \text{By Equation (1c)}$$

For the domain of $f/g$ we must exclude the value of $x$ for which $g(x) = \sqrt{3 - x} = 0$ or $x = 3$. Therefore, $f/g$ is defined by

$$\left(\frac{f}{g}\right)(x) = \frac{f(x)}{g(x)} = \frac{\sqrt{x}}{\sqrt{3 - x}} = \sqrt{\frac{x}{3 - x}} \qquad \text{By Equation (1d)}$$

with domain $[0, 3)$. ∎

**Notes**

1. To determine the domain of the product or quotient of two functions, begin by examining the domains of the functions to be combined. One common mistake is to try to deduce the domain of the combined function by studying its rule. For example, suppose $f(x) = \sqrt{x}$ and $g(x) = 2\sqrt{x}$. Then, if $h = fg$, we have $h(x) = f(x)g(x) = (\sqrt{x})(2\sqrt{x}) = 2x$. On the basis of the rule for $h$ alone, we might be tempted to conclude that its domain is $(-\infty, \infty)$. But bearing in mind that $h$ is a product of the functions $f$ with domain $[0, \infty)$ and $g$ with domain $[0, \infty)$, we see that the domain of $h$ is $[0, \infty)$.

2. Equations (1a–d) can be extended to the case involving more than two functions. For example, $fg - h$ is just the function with rule

$$(fg - h)(x) = f(x)g(x) - h(x) \qquad ∎$$

## ■ Composition of Functions

There is another way in which certain functions are built up from simpler functions. For example, consider the function $h(x) = \sqrt{2x + 1}$. Let $f$ be the function defined by $f(x) = 2x + 1$, and let $g$ be the function defined by $g(x) = \sqrt{x}$. Then

$$h(x) = \sqrt{2x + 1} = \sqrt{f(x)} = g(f(x))$$

In other words, the value of $h$ at $x$ can be obtained by *evaluating* the function $g$ at $f(x)$. This method of combining two functions is called **composition.** More specifically, we say that the function $h$ is the **composition** of $g$ and $f$, and we denote it by $g \circ f$ (read "$g$ circle $f$").

---

**DEFINITION**   **Composition of Two Functions**

Given two functions $g$ and $f$, the composition of $g$ and $f$, denoted by $g \circ f$, is the function defined by

$$(g \circ f)(x) = g(f(x)) \tag{2}$$

The domain of $g \circ f$ is the set of all $x$ in the domain of $f$ for which $f(x)$ is in the domain of $g$.

---

Figure 2 shows an interpretation of the composition $g \circ f$, in which the functions $f$ and $g$ are viewed as machines. Notice that the output of $f$, $f(x)$, must lie in the domain of $g$ for $f(x)$ to be an input for $g$.

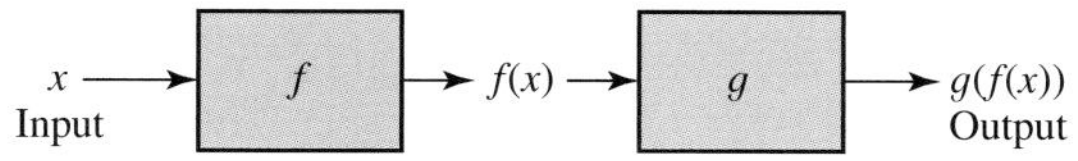

**FIGURE 2**
The output of $f$ is the input for $g$ (in this order).

$$x \longrightarrow \boxed{f} \longrightarrow f(x) \longrightarrow \boxed{g} \longrightarrow g(f(x))$$
Input                                                                Output

Figure 3 shows how the composition $g \circ f$ can be viewed in terms of transformations or mappings. The point $x$ in the domain of $g \circ f$ is mapped onto the image $f(x)$ that lies in the domain of $g$. The function $g$ then maps $f(x)$ onto its image $g(f(x))$. Thus, we may view the function $g \circ f$ as a transformation that maps a point $x$ in its domain onto its image $g(f(x))$ in two steps: from $x$ to $f(x)$ via the function $f$, then from $f(x)$ to $g(f(x))$ via the function $g$.

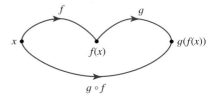

**FIGURE 3**
$g \circ f$ maps $x$ onto $g(f(x))$ in two steps: via $f$, then via $g$.

**EXAMPLE 2**   Let $f$ and $g$ be functions defined by $f(x) = x + 1$ and $g(x) = \sqrt{x}$. Find the functions $g \circ f$ and $f \circ g$. What is the domain of $g \circ f$?

**Solution**   The rule for $g \circ f$ is found by evaluating $g$ at $f(x)$. Thus,

$$(g \circ f)(x) = g(f(x)) = \sqrt{f(x)} = \sqrt{x + 1}$$

To find the domain of $g \circ f$, recall that $f(x)$ must lie in the domain of $g$. Since the domain of $g$ consists of all nonnegative numbers and the range of $f$ is the set of all numbers $f(x) = x + 1$, we require that $x + 1 \geq 0$ or $x \geq -1$. Therefore, the domain of $g \circ f$ is $[-1, \infty)$. Note that all $x$ are in the domain of $f$.

The rule for $f \circ g$ is found by evaluating $f$ at $g(x)$. Thus,

$$(f \circ g)(x) = f(g(x)) = g(x) + 1 = \sqrt{x} + 1$$

We leave it to you to show that the domain of $f \circ g$ is $[0, \infty)$.   ■

**Note**   In general, $g \circ f \neq f \circ g$, as was demonstrated in Example 2. Thus, the order in which functions are composed is important. For example, in the composition $g \circ f$, remember that $f$ is applied first, followed by $g$.   ■

**EXAMPLE 3**   Let $f(x) = \sin x$ and $g(x) = 1 - 2x$. Find the functions $g \circ f$ and $f \circ g$. What are their domains?

**Solution**   $(g \circ f)(x) = g(f(x)) = 1 - 2f(x) = 1 - 2 \sin x$. Since the range of $f$ is $[-1, 1]$ and this interval lies in $(-\infty, \infty)$, the domain of $g$, we see that the domain of $g \circ f$ is given by the domain of $f$, namely, $(-\infty, \infty)$. Next,

$$(f \circ g)(x) = f(g(x)) = f(1 - 2x) = \sin(1 - 2x)$$

The range of $g$ is $(-\infty, \infty)$, and this is also the domain of $f$. So the domain of $f \circ g$ is given by the domain of $g$, namely, $(-\infty, \infty)$.   ■

**EXAMPLE 4**   Find two functions $f$ and $g$ such that $F = g \circ f$ if $F(x) = (x + 2)^4$.

**Solution**   The expression $(x + 2)^4$ can be evaluated in two steps. First, given any value of $x$, add 2 to it. Second, raise this result to the fourth power. This suggests that we take

$$f(x) = x + 2 \qquad \text{Remember that } f \text{ is applied first in } g \circ f.$$

and

$$g(x) = x^4$$

Then

$$(g \circ f)(x) = g(f(x)) = [f(x)]^4 = (x + 2)^4 = F(x)$$

so $F = g \circ f$, as required.   ■

**Note**   There is always more than one way to write a function as a composition of functions. In Example 4 we could have taken $f(x) = (x + 2)^2$ and $g(x) = x^2$. However, there is usually a "natural" way of decomposing a complicated function.   ■

Composite functions play an important role in describing practical situations in which one variable quantity depends on another, which in turn depends on a third, as the following example shows.

**EXAMPLE 5**   **Oil Spills**   In calm waters, the oil spilling from the ruptured hull of a grounded tanker spreads in all directions. Assuming that the area polluted is a circle and that its radius is increasing at the rate of 2 ft/sec, find the area as a function of time.

**Solution**   The circular polluted area is described by the function $g(r) = \pi r^2$, where $r$ is the radius of the circle, measured in feet. Next, the radius of the circle is described by the function $f(t) = 2t$, where $t$ is the time elapsed, measured in seconds. Therefore, the required function $A$ describing the polluted area as a function of time is $A = g \circ f$ defined by

$$A(t) = (g \circ f)(t) = g(f(t)) = \pi[f(t)]^2 = \pi(2t)^2 = 4\pi t^2$$   ■

The composition of functions can be extended to include the composition of three or more functions. For example, the composite function $h \circ g \circ f$ is found by applying $f$, $g$, and $h$ in that order. Thus,

$$(h \circ g \circ f)(x) = h(g(f(x)))$$

**EXAMPLE 6** Let $f(x) = x - (\pi/2)$, $g(x) = 1 + \cos^2 x$, and $h(x) = \sqrt{x}$. Find $h \circ g \circ f$.

**Solution** $(h \circ g \circ f)(x) = h(g(f(x))) = \sqrt{g(f(x))}$. But

$$g(f(x)) = 1 + \cos^2[f(x)] = 1 + \cos^2\left(x - \tfrac{\pi}{2}\right)$$

So

$$(h \circ g \circ f)(x) = \sqrt{1 + \cos^2\left(x - \tfrac{\pi}{2}\right)} \qquad \blacksquare$$

**EXAMPLE 7** Suppose $F(x) = \dfrac{1}{\sqrt{2x + 3} + 1}$. Find functions $f$, $g$, and $h$ such that $F = h \circ g \circ f$.

**Solution** The rule for $F$ says that as a first step, we multiply $x$ by 2 and add 3 to it. This suggests that we take $f(x) = 2x + 3$. Next, we take the square root of this result and add 1 to it. This suggests that we take $g(x) = \sqrt{x} + 1$. Finally, we take the reciprocal of the last result, so let $h(x) = 1/x$. Then

$$F(x) = (h \circ g \circ f)(x) = h(g(f(x)))$$

$$= h(g(2x + 3)) = h(\sqrt{2x + 3} + 1) = \frac{1}{\sqrt{2x + 3} + 1} \qquad \blacksquare$$

## ■ Graphs of Transformed Functions

Sometimes it is possible to obtain the graph of a relatively complicated function by transforming the graph of a simpler but related function. We will describe some of these transformations here.

## 1. Vertical Translations

The graph of the function $g$ defined by $g(x) = f(x) + c$, where $c$ is a positive constant, is obtained from the graph of $f$ by shifting the latter vertically upward by $c$ units (see Figure 4). This follows by observing that for each $x$ in the domain of $g$ (which is the same as the domain of $f$) the point $(x, f(x) + c)$ on the graph of $g$ lies precisely $c$ units above the point $(x, f(x))$ on the graph of $f$. Similarly, the graph of the function $g$ defined by $g(x) = f(x) - c$, where $c$ is a positive constant, is obtained from the graph of $f$ by shifting the latter vertically downward by $c$ units (see Figure 4). These results are also evident if you think of $g$ as the sum of the function $f$ and the constant function $h(x) = c$ and use the graphical interpretation of the sum of two functions described earlier.

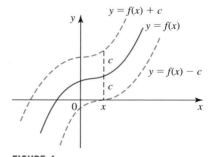

**FIGURE 4**
The graphs of $y = f(x) + c$ and $y = f(x) - c$, where $c > 0$, are obtained by translating the graph of $y = f(x)$ vertically upward and downward, respectively.

## 2. Horizontal Translations

The graph of the function $g$ defined by $g(x) = f(x + c)$, where $c$ is a positive constant, is obtained from the graph of $f$ by shifting the latter horizontally to the left by $c$ units (see Figure 5a). To see this, observe that the number $x + c$ lies $c$ units to the right of $x$. Therefore, for each $x$ in the domain of $g$, $(x, f(x + c))$ on the graph of $g$ has precisely the same $y$-coordinate as the point on the graph of $f$ located $c$ units to the *right* of $x$ (measured horizontally). Similarly, the graph of the function $g(x) = f(x - c)$, where $c$ is a positive constant, is obtained from the graph of $y = f(x)$ by shifting the latter horizontally to the right by $c$ units (see Figure 5b). We summarize these results in Table 1.

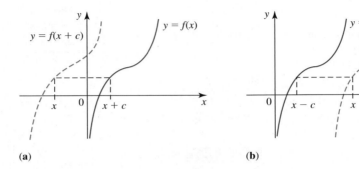

**FIGURE 5**
The graphs of $y = f(x + c)$ and $y = f(x - c)$, where $c > 0$, are obtained by shifting the graph of $y = f(x)$ horizontally to the left and right, respectively.

**(a)**                    **(b)**

**TABLE 1**    Vertical and Horizontal Translations

If $c > 0$, then we have the following:

| Function $g$ | The graph of $g$ is obtained by shifting the graph of $f$ |
|---|---|
| $g(x) = f(x) + c$ | Upward by a distance of $c$ units |
| $g(x) = f(x) - c$ | Downward by a distance of $c$ units |
| $g(x) = f(x + c)$ | To the left by a distance of $c$ units |
| $g(x) = f(x - c)$ | To the right by a distance of $c$ units |

## 3. Vertical Stretching and Compressing

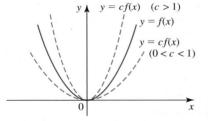

**FIGURE 6**
The graph of $y = cf(x)$ is obtained from the graph of $y = f(x)$ by stretching it (if $c > 1$) or compressing it (if $0 < c < 1$).

The graph of the function $g$ defined by $g(x) = cf(x)$, where $c$ is a constant with $c > 1$, is obtained from the graph of $f$ by stretching the latter vertically by a factor of $c$. This can be seen by observing that for each $x$ in the domain of $g$ (and therefore in the domain of $f$), the point $(x, cf(x))$ on the graph of $g$ has a $y$-coordinate that is $c$ times as large as the $y$-coordinate of the point $(x, f(x))$ on the graph of $f$ (see Figure 6). Similarly, if $0 < c < 1$ then the graph of $g$ is obtained from that of $f$ by compressing the latter vertically by a factor of $1/c$ (see Figure 6).

## 4. Horizontal Stretching and Compressing

The graph of the function $g$ defined by $g = f(cx)$, where $c$ is a constant with $0 < c < 1$, is obtained from the graph of $f$ by stretching the graph of the latter horizontally by a factor of $1/c$ (see Figure 7). To see this, observe that if $x > 0$, the number $cx$ lies to the left of $x$. Therefore, for each $x$ in the domain of $g$, the point $(x, g(x)) = (x, f(cx))$ on the graph of $g$ has precisely the same $y$-coordinate as the point on the graph of $f$ located at the point with $x$-coordinate $cx$. (We leave it to you to analyze the case in which $x < 0$.) Similarly if $c > 1$, then the graph of $g$ is obtained from that of $f$ by compressing the latter horizontally by a factor of $c$. We summarize these results in Table 2.

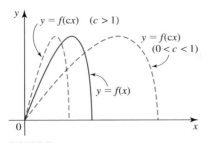

**FIGURE 7**
The graph of $y = f(cx)$ is obtained from the graph of $y = f(x)$ by compressing it if $c > 1$ and stretching it if $0 < c < 1$.

**TABLE 2**   Vertical and Horizontal Stretching and Compressing

**a.** If $c > 1$ then we have the following:

| Function $g$ | The graph of $g$ is obtained by |
|---|---|
| $g(x) = cf(x)$ | Stretching the graph of $f$ vertically by a factor of $c$ |
| $g(x) = f(cx)$ | Compressing the graph of $f$ horizontally by a factor of $c$ |

**b.** If $0 < c < 1$, then we have the following:

| Function $g$ | The graph of $g$ is obtained by |
|---|---|
| $g(x) = cf(x)$ | Compressing the graph of $f$ vertically by a factor of $1/c$ |
| $g(x) = f(cx)$ | Stretching the graph of $f$ horizontally by a factor of $1/c$ |

## 5. Reflecting

The graph of the function defined by $g(x) = -f(x)$ is obtained from the graph of $f$ by reflecting the latter with respect to the $x$-axis (see Figure 8a). This follows from the observation that for each $x$ in the domain of $g$, the point $(x, -f(x))$ on the graph of $g$ is the mirror reflection of the point $(x, f(x))$ with respect to the $x$-axis. Similarly, the graph of $g(x) = f(-x)$ is obtained from the graph of $f$ by reflecting the latter with respect to the $y$-axis (see Figure 8b). These results are summarized in Table 3.

**FIGURE 8**
The graphs of $y = -f(x)$ and $y = f(-x)$ are obtained from the graph of $y = f(x)$ by reflecting it with respect to the $x$-axis and with respect to the $y$-axis, respectively.

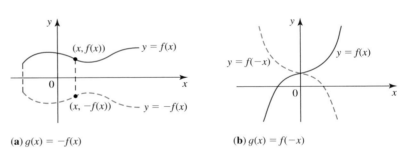

**(a)** $g(x) = -f(x)$

**(b)** $g(x) = f(-x)$

**TABLE 3**   Reflecting

| Function $g$ | The graph of $g$ is obtained by reflecting the graph of $f$ |
|---|---|
| $g(x) = -f(x)$ | With respect to the $x$-axis |
| $g(x) = f(-x)$ | With respect to the $y$-axis |

**EXAMPLE 8**   By translating the graph of $y = x^2$, sketch the graphs of $y = x^2 + 2$, $y = x^2 - 2$, $y = (x + 2)^2$, and $y = (x - 2)^2$.

**Solution**   The graph of $y = x^2$ is shown in Figure 9a. The graph of $y = x^2 + 2$ is obtained from the graph of $y = x^2$ by translating the latter vertically upward by 2 units (see Figure 9b). The graph of $y = x^2 - 2$ is obtained by translating the graph of $y = x^2$ vertically downward by 2 units (see Figure 9c). The graph of $y = (x + 2)^2$ is obtained by translating the graph of $y = x^2$ horizontally to the left by 2 units (see Figure 9d). Finally, the graph of $y = (x - 2)^2$ is obtained by translating the graph of $y = x^2$ to the right by 2 units (see Figure 9e).

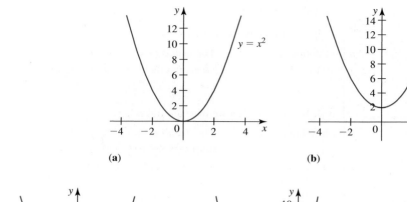

(a)

(b)

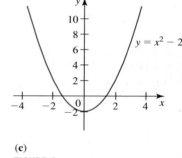

(c)

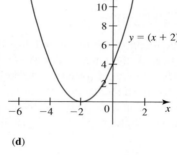

(d)

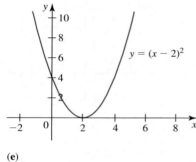

(e)

**FIGURE 9**

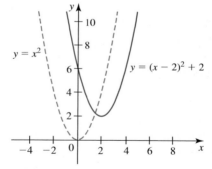

**FIGURE 10**
The graph of $y = (x - 2)^2 + 2$ can be obtained by shifting the graph of $y = x^2$.

**EXAMPLE 9** Sketch the graph of the function $f$ defined by $f(x) = x^2 - 4x + 6$.

**Solution** By completing the square, we can rewrite the given equation in the form

$$y = [x^2 - 4x + (-2)^2] + 6 - (-2)^2$$
$$= (x - 2)^2 + 2$$

We see that the required graph can be obtained from the graph of $y = x^2$ by shifting it 2 units to the right and 2 units upward (see Figure 10). Compare this with Example 8.

**EXAMPLE 10** By stretching or compressing the graph of $y = \sin x$, sketch the graphs of $y = 2 \sin x$, $y = \frac{1}{2} \sin x$, $y = \sin 2x$, and $y = \sin(x/2)$.

**Solution** The graph of $y = \sin x$ is shown in Figure 11a. The graph of $y = 2 \sin x$ is obtained from the graph of $y = \sin x$ by stretching the latter vertically by a factor of 2 (see Figure 11b). The graph of $y = \frac{1}{2} \sin x$ is obtained by compressing the graph of $y = \sin x$ vertically by a factor of 2 (see Figure 11c). The graph of $y = \sin 2x$ is obtained from the graph of $y = \sin x$ by compressing the graph of the latter horizontally by a factor of 2. In fact, the period of $\sin x$ is $2\pi$, whereas the period of $\sin 2x$ is $\pi$ (see Figure 11d). Finally, the graph of $y = \sin(x/2)$ is obtained from the graph of $y = \sin x$ by stretching the latter horizontally by a factor of 2 (see Figure 11e).

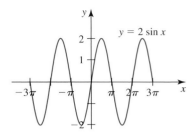

**(a)**

**(b)** Vertical stretching

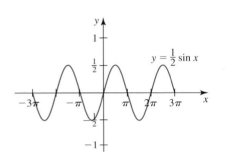

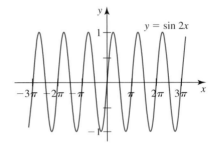

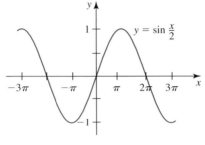

**(c)** Vertical compression

**(d)** Horizontal compression

**(e)** Horizontal stretching

**FIGURE 11**

**EXAMPLE 11**  By reflecting the graph of $y = \sqrt{x}$, sketch the graphs of $y = -\sqrt{x}$ and $y = \sqrt{-x}$.

**Solution**  The graph of $y = \sqrt{x}$ is shown in Figure 12a. To obtain the graph of $y = -\sqrt{x}$, we reflect the graph of $y = \sqrt{x}$ with respect to the $x$-axis (see Figure 12b). To obtain the graph of $y = \sqrt{-x}$, we reflect the graph of $y = \sqrt{x}$ with respect to the $y$-axis (see Figure 12c).

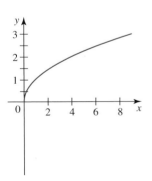

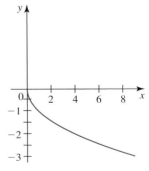

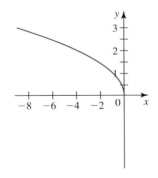

**(a)** The graph of $y = \sqrt{x}$

**(b)** The graph of $y = -\sqrt{x}$

**(c)** The graph of $y = \sqrt{-x}$

**FIGURE 12**

The next example involves the use of another transformation of interest.

**EXAMPLE 12**

**a.** Explain how you can obtain the graph of $y = |f(x)|$ given the graph of $y = f(x)$.
**b.** Use the method you devised in part (a) to sketch the graph of $y = ||x| - 1|$.

**Solution**

**a.** By the definition of the absolute value, we have

$$|f(x)| = \begin{cases} f(x) & \text{if } f(x) \geq 0 \\ -f(x) & \text{if } f(x) < 0 \end{cases}$$

So to obtain the graph of $y = |f(x)|$ from that of $y = f(x)$ (Figure 13a), we retain the portion of the graph of $y = f(x)$ that lies above the axis and reflect the portion of the graph of $y = f(x)$ that lies below the $x$-axis with respect to the $x$-axis (see Figure 13b).

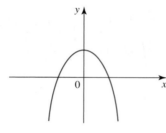

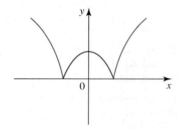

**FIGURE 13**

(**a**) $y = f(x)$

(**b**) $y = |f(x)|$

**b.** We begin by sketching the graph of $y = |x|$ as shown in Figure 14a. Next, we sketch the graph of $y = |x| - 1$ by translating the graph of $y = |x|$ vertically downward by 1 unit (see Figure 14b). Finally, using the method of part (a), we obtain the desired graph (see Figure 14c).

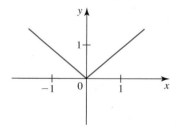

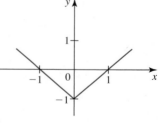

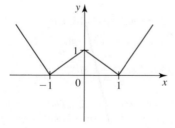

(**a**) $y = |x|$

(**b**) $y = |x| - 1$

(**c**) $y = ||x| - 1|$

**FIGURE 14**

---

## 0.4 EXERCISES

*In Exercises 1–4, find* (a) $f + g$, (b) $f - g$, (c) $fg$, *and* (d) $f/g$. *What is the domain of the function?*

**1.** $f(x) = 3x$, $g(x) = x^2 - 1$

**2.** $f(x) = x^2 + 1$, $g(x) = 1 + \sqrt{x}$

**3.** $f(x) = \sqrt{x + 1}$, $g(x) = \sqrt{x - 1}$

**4.** $f(x) = \dfrac{1}{x + 1}$, $g(x) = \dfrac{x}{x - 1}$

*In Exercises 5–8, find f ∘ g and g ∘ f, and give their domains.*

**5.** $f(x) = x^2$, $g(x) = 2x + 3$

**6.** $f(x) = \sqrt{x}$, $g(x) = 1 - x^2$

**7.** $f(x) = \dfrac{1}{x}$, $g(x) = \dfrac{x + 1}{x - 1}$

**8.** $f(x) = \sqrt{x + 1}$, $g(x) = \dfrac{1}{x - 1}$

*In Exercises 9–10, evaluate h(2), where h = g ∘ f.*

**9.** $f(x) = \sqrt[3]{x^2 - 1}$, $g(x) = 3x^3 + 1$

**10.** $f(x) = \dfrac{\pi x}{4}$, $g(x) = 2 \sin x + 3 \cos x$

**11.** Let

$$f(x) = \begin{cases} x + 1 & \text{if } x < 0 \\ x - 1 & \text{if } x \geq 0 \end{cases}$$

and let $g(x) = x^2$. Find

**a.** $g \circ f$, and sketch its graph.

**b.** $f \circ g$, and sketch its graph.

**12.** Suppose the function $f$ is defined on the interval $[0, 1]$. Find the domain of $h$ if (a) $h(x) = f(2x + 3)$ and (b) $h(x) = f(2x^2)$.

**13.** Let $f(x) = x + 2$ and $g(x) = 2x^2 + \sqrt{x}$. Find

**a.** $(g \circ f)(0)$     **b.** $(g \circ f)(2)$

**c.** $(f \circ g)(4)$     **d.** $(g \circ g)(1)$

**14.** Let $f(x) = \dfrac{\pi}{2} - x$ and $g(x) = \dfrac{2 \sin x}{1 + \cos x}$. Find

**a.** $g(f(0))$     **b.** $(g \circ f)\left(\dfrac{\pi}{2}\right)$

**c.** $f\left(g\left(\dfrac{\pi}{2}\right)\right)$     **d.** $(f \circ f)\left(\dfrac{\pi}{2}\right)$

*In Exercises 15–16, find f ∘ g ∘ h.*

**15.** $f(x) = \sqrt{x}$, $g(x) = 2x + 1$, $h(x) = x^2 - 1$

**16.** $f(x) = \dfrac{1}{x}$, $g(x) = a - bx$, $h(x) = \cos x$

*In Exercises 17–22, find functions f and g such that h = g ∘ f. (Note: The answer is not unique.)*

**17.** $h(x) = (3x^2 + 4)^{3/2}$

**18.** $h(x) = |x^2 - 2x + 3|$

**19.** $h(x) = \dfrac{1}{\sqrt{x^2 - 4}}$

**20.** $h(x) = \sqrt{2x + 1} + \dfrac{1}{\sqrt{2x + 1}}$

**21.** $h(t) = \sin(t^2)$

**22.** $h(t) = \dfrac{\tan t}{1 + \cot t}$

*In Exercises 23–24, find functions f, g, and h such that F = f ∘ g ∘ h. (Note: The answer is not unique.)*

**23. a.** $F(x) = \sqrt{1 - \sqrt{x}}$     **b.** $F(x) = \sin^3(2x + 3)$

**24. a.** $F(x) = \dfrac{1}{(2x^2 + x + 3)^3}$

**b.** $F(x) = \dfrac{\sqrt{x + 1} - 1}{\sqrt{x + 1} + 1}$

**25.** Use the following table to evaluate each composite function.

**a.** $(f \circ g)(1)$     **b.** $(g \circ f)(2)$

**c.** $f(g(2))$     **d.** $g(f(0))$

**e.** $f(f(2))$     **f.** $g(g(1))$

| $x$ | 0 | 1 | 2 | 3 | 4 | 5 |
|------|---|------------|---|---|---|---|
| $f(x)$ | 1 | $\sqrt{2}$ | 2 | 4 | 3 | 1 |
| $g(x)$ | 2 | 3 | 5 | 6 | 7 | 9 |

**26.** Use the graphs of $f$ and $g$ to estimate the values of $(g \circ f)(x)$ for $x = -2, -1, 0, 1, 2$, and 3. Then use these values to make a rough sketch of the graph of $g \circ f$.

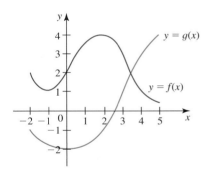

*In Exercises 27–30 the graph of f is given. Match the other graphs with the given function(s).*

**27.** $y = f(x) + 1$, $y = f(x) - 1$

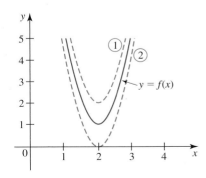

**28.** $y = f(x + 2)$,   $y = f(x - 2)$

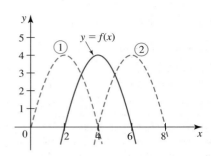

**29.** $y = f(2x)$,   $y = f\left(\dfrac{x}{2}\right)$

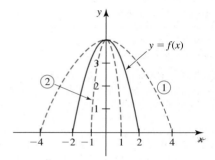

**30.** $y = f(-x)$,   $y = -f(x)$,   $y = 2f(x)$,   $y = \dfrac{1}{2}f(-x)$

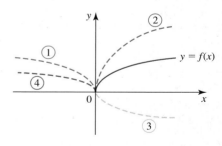

*In Exercises 31–40, the graph of the function f is to be transformed as described. Find the function for the transformed graph.*

**31.** $f(x) = x^3 + x - 1$; shifted vertically upward by 3 units

**32.** $f(x) = x + \sqrt{x + 1}$; shifted vertically downward by 2 units

**33.** $f(x) = x + \dfrac{1}{\sqrt{x}}$; shifted horizontally to the left by 3 units

**34.** $f(x) = \dfrac{\sin x}{1 + \cos x}$; shifted horizontally to the right by 4 units

**35.** $f(x) = \dfrac{\sqrt{x}}{x^2 + 1}$; stretched vertically by a factor of 3

**36.** $f(x) = \sqrt{x^2 + 4}$; compressed vertically by a factor of 2

**37.** $f(x) = x \sin x$; stretched horizontally by a factor of 2

**38.** $f(x) = 5 \sin 4x$; compressed horizontally by a factor of 3

**39.** $f(x) = \sqrt{4 - x^2}$; shifted horizontally to the right by 2 units, compressed horizontally by a factor of 2, and shifted vertically upward by 1 unit

**40.** $f(x) = \sqrt{x} + 1$; shifted horizontally to the left by 1 unit, compressed horizontally by a factor of 3, stretched vertically by a factor of 3, and shifted vertically downward by 2 units

**41.** The graph of the function $f$ follows.

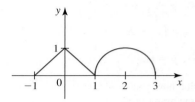

Use it to sketch the following graphs.
**a.** $y = f(x) + 1$      **b.** $y = f(x + 2)$
**c.** $y = 2f(x)$      **d.** $y = f(2x)$
**e.** $y = -f(x)$      **f.** $y = f(-x)$
**g.** $y = 2f(x - 1) + 2$      **h.** $y = -2f(x + 1) + 3$

**42.** The graph of the function $f$ follows.

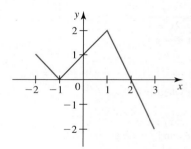

Use it to sketch the following graphs.

**a.** $y = f(x - 1)$      **b.** $y = f\left(\dfrac{x}{2}\right)$

**c.** $y = |f(x)|$      **d.** $y = \dfrac{|f(x)|}{f(x)}$

**e.** $y = f(-x)$      **f.** $y = \dfrac{(|f(x)| + f(x))}{2}$

**g.** $y = -2f(-x) + 1$

*In Exercises 43–54, sketch the graph of the first function by plotting points if necessary. Then use transformation(s) to obtain the graph of the second function.*

**43.** $y = x^2$,   $y = x^2 - 2$

**44.** $y = x^2$,   $y = (x - 2)^2$

**45.** $y = \dfrac{1}{x}, \quad y = \dfrac{1}{x-1}$

**46.** $y = \sqrt{x}, \quad y = 2\sqrt{x-1} + 1$

**47.** $y = |x|, \quad y = 2|x+1| - 1$

**48.** $y = |x|, \quad y = |2x-1| + 1$

**49.** $y = x^2, \quad y = 2x^2 - 4x + 1$

**50.** $y = x^2, \quad y = |x^2 - 1|$

**51.** $y = \sin x, \quad y = 2\sin\dfrac{x}{2}$

**52.** $y = \cos x, \quad y = \dfrac{1}{2}\cos\left(x - \dfrac{\pi}{4}\right)$

**53.** $y = x^2, \quad y = |x^2 - 2x - 1|$

**54.** $y = \tan x, \quad y = \tan\left(x + \dfrac{\pi}{3}\right)$

**55. a.** Describe how you would construct the graph of $f(|x|)$ from the graph of $y = f(x)$.
    **b.** Use the result of part (a) to sketch the graph of $y = \sin|x|$.

**56.** Find $f(x)$ if $f(x+1) = 2x^2 + 7x + 4$.

**57. a.** If $f(x) = x - 1$ and $h(x) = 2x + 3$, find a function $g$ such that $h = g \circ f$.
    **b.** If $g(x) = 3x + 4$ and $h(x) = 4x - 8$, find a function $f$ such that $h = g \circ f$.

**58.** Let $g(x) = \dfrac{x+1}{2x-1}$, and let $h(x) = \dfrac{2x+2}{4x+1}$. Find a function $f$ such that $h = g \circ f$.

**59.** Let $f(x) = 2x^2 + x$, and let $h(x) = 6x^2 + 3x - 1$. Find a function $g$ such that $h = g \circ f$.

**60.** Determine whether $h = g \circ f$ is even, odd, or neither, given that
    **a.** both $g$ and $f$ are even.
    **b.** $g$ is even and $f$ is odd.
    **c.** $g$ is odd and $f$ is even.
    **d.** both $g$ and $f$ are odd.

**61.** Let $f$ be a function defined by $f(x) = \sqrt{x} + \sin x$ on the interval $[0, 2\pi]$.
    **a.** Find an even function $g$ defined on the interval $[-2\pi, 2\pi]$ such that $g(x) = f(x)$ for all $x$ in $[0, 2\pi]$.
    **b.** Find an odd function $h$ defined on the interval $[-2\pi, 2\pi]$ such that $h(x) = f(x)$ for all $x$ in $[0, 2\pi]$.

**62. a.** Show that if a function $f$ is defined at $-x$ whenever it is defined at $x$, then the function $g$ defined by $g(x) = f(x) + f(-x)$ is an even function and the function $h$ defined by $h(x) = f(x) - f(-x)$ is an odd function.
    **b.** Use the result of part (a) to show that any function $f$ defined on an interval $(-a, a)$ can be written as a sum of an even function and an odd function.

**c.** Rewrite the function

$$f(x) = \dfrac{x+1}{x-1} \qquad -1 < x < 1$$

as a sum of an even function and an odd function.

**63. Spam Messages** The total number of email messages per day (in billions) between 2003 and 2007 is approximated by

$$f(t) = 1.54t^2 + 7.1t + 31.4 \qquad 0 \le t \le 4$$

where $t$ is measured in years, with $t = 0$ corresponding to 2003. Over the same period the total number of spam messages per day (in billions) is approximated by

$$g(t) = 1.21t^2 + 6t + 14.5 \qquad 0 \le t \le 4$$

    **a.** Find the rule for the function $D = f - g$. Compute $D(4)$, and explain what it measures.
    **b.** Find the rule for the function $P = g/f$. Compute $P(4)$, and explain what it means.
*Source: Technology Review.*

**64. Global Supply of Plutonium** The global stockpile of plutonium for military applications between 1990 ($t = 0$) and 2003 ($t = 13$) stood at a constant 267 tons. On the other hand, the global stockpile of plutonium for civilian use was $2t^2 + 46t + 733$ tons in year $t$ over the same period.
    **a.** Find the function $f$ giving the global stockpile of plutonium for military use from 1990 through 2003 and the function $g$ giving the global stockpile of plutonium for civilian use over the same period.
    **b.** Find the function $h$ giving the total global stockpile of plutonium between 1990 and 2003.
    **c.** What was the total global stockpile of plutonium in 2003?
*Source: Institute for Science and International Security.*

**65. Motorcycle Deaths** Suppose that the fatality rate (deaths per 100 million miles traveled) of motorcyclists is given by $g(x)$, where $x$ is the percentage of motorcyclists who wear helmets. Next, suppose that the percentage of motorcyclists who wear helmets at time $t$ ($t$ measured in years) is $f(t)$, where $t = 0$ corresponds to the year 2000.
    **a.** If $f(0) = 0.64$ and $g(0.64) = 26$, find $(g \circ f)(0)$, and interpret your result.
    **b.** If $f(6) = 0.51$ and $g(0.51) = 42$, find $(g \circ f)(6)$, and interpret your result.
    **c.** Comment on the results of parts (a) and (b).
*Source: NHTSA.*

**66. Fighting Crime** Suppose that the reported serious crimes (crimes that include homicide, rape, robbery, aggravated assault, burglary, and car theft) that end in arrests or in the identification of suspects is $g(x)$ percent, where $x$ denotes the total number of detectives. Next, suppose that the total number of detectives in year $t$ is $f(t)$, where $t = 0$ corresponds to 2001.

**a.** If $f(1) = 406$ and $g(406) = 23$, find $(g \circ f)(1)$, and interpret your result.

**b.** If $f(6) = 326$ and $g(326) = 18$, find $(g \circ f)(6)$, and interpret your result.

**c.** Comment on the results of parts (a) and (b).

*Source:* Boston Police Department.

**67. Overcrowding of Prisons** The 1980s saw a trend toward old-fashioned punitive deterrence of crime in contrast to the more liberal penal policies and community-based corrections that were popular in the 1960s and early 1970s. As a result, prisons became more crowded, and the gap between the number of people in prison and the prison capacity widened. The number of prisoners (in thousands) in federal and state prisons is approximated by the function

$$N(t) = 3.5t^2 + 26.7t + 436.2 \qquad 0 \le t \le 10$$

where $t$ is measured in years, with $t = 0$ corresponding to 1983. The number of inmates for which prisons were designed is given by

$$C(t) = 24.3t + 365 \qquad 0 \le t \le 10$$

where $C(t)$ is measured in thousands and $t$ has the same meaning as before.

**a.** Find an expression that shows the gap between the number of prisoners and the number of inmates for which the prisons were designed at any time $t$.

**b.** Find the gap at the beginning of 1983 and at the beginning of 1986.

*Source:* U.S. Department of Justice.

**68. Hotel Occupancy Rate** The occupancy rate of the all-suite Wonderland Hotel, located near an amusement park, is given by the function

$$r(t) = \frac{10}{81}t^3 - \frac{10}{3}t^2 + \frac{200}{9}t + 55 \qquad 0 \le t \le 11$$

where $t$ is measured in months and $t = 0$ corresponds to the beginning of January. Management has estimated that the monthly revenue (in thousands of dollars) is approximated by the function

$$R(r) = -\frac{3}{5000}r^3 + \frac{9}{50}r^2 \qquad 0 \le r \le 100$$

where $r$ (percent) is the occupancy rate.

**a.** What is the hotel's occupancy rate at the beginning of January? At the beginning of July?

**b.** What is the hotel's monthly revenue at the beginning of January? At the beginning of July?

*In Exercises 69–74, determine whether the statement is true or false. If it is true, explain why. If it is false, explain why or give an example that shows it is false.*

**69.** If $f$ and $g$ are both linear functions of $x$, then so are $f \circ g$ and $g \circ f$.

**70.** If $f$ is a polynomial function of $x$ and $g$ is a rational function, then $g \circ f$ and $f \circ g$ are rational functions.

**71.** If $f$ and $g$ are both even (odd), then $f + g$ is even (odd).

**72.** If $f$ is even and $g$ is odd, then $f + g$ is neither even nor odd.

**73.** If $f$ and $g$ are both even, then $fg$ is even.

**74.** If $f$ and $g$ are both odd, then $fg$ is odd.

## 0.5 Graphing Calculators and Computers

The graphing calculator and the computer are indispensable tools in helping us to solve complex mathematical problems. In this book we will use them to help us explore ideas and concepts in calculus both graphically and numerically. But the amount and accuracy of the information obtained by using a graphing utility depend on the experience and sophistication of the user. As you progress through this text, you will see that the more knowledge of calculus you gain, the more effective the graphing utility will prove to be as a tool for problem solving. But there are pitfalls in using the graphing utility, and we will point them out when the opportunity arises.

In this section we will look at some basic capabilities of the graphing calculator and the computer that we will use later.

### ■ Finding a Suitable Viewing Window

The first step in plotting the graph of a function with a graphing utility is to select a suitable viewing window $[a, b] \times [c, d]$ that displays the portion of the graph of the function in the rectangular set $\{(x, y) \mid a \le x \le b, c \le y \le d\}$. For example, you might first plot the graph using the *standard viewing window* $[-10, 10] \times [-10, 10]$. If necessary, you then might adjust the viewing window by enlarging it, reducing it, or even changing it altogether to obtain a sufficiently complete view of the graph or at least the portion of the graph that is of interest.

**EXAMPLE 1** Plot the graph of $f(x) = 2x^2 - 4x - 5$ in the standard viewing window $[-10, 10] \times [-10, 10]$.

**Solution** The graph of $f$, shown in Figure 1a, is a parabola. Figure 1b shows a typical window screen, and Figure 1c shows a typical equation screen.

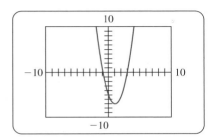

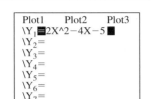

(**a**) The graph of $f(x) = 2x^2 - 4x - 5$ in $[-10, 10] \times [-10, 10]$

(**b**) A window screen on a graphing calculator

(**c**) An equation screen on a graphing calculator

**FIGURE 1**

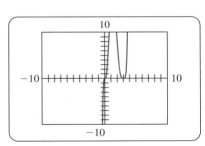

**FIGURE 2**
An incomplete sketch of
$f(x) = x^3(x - 3)^4$ on
$[-10, 10] \times [-10, 10]$

**EXAMPLE 2** Let $f(x) = x^3(x - 3)^4$.

**a.** Plot the graph of $f$ in the standard viewing window.
**b.** Plot the graph of $f$ in the window $[-1, 5] \times [-40, 40]$.

**Solution**
**a.** The graph of $f$ in the standard viewing window is shown in Figure 2. Since the graph does not appear to be complete, we need to adjust the viewing window.
**b.** The graph of $f$ in the window $[-1, 5] \times [-40, 40]$, shown in Figure 3, is an improvement over the previous graph. (Later, we will be able to show that the figure does in fact give a rather complete view of the graph of $f$.)

## Evaluating a Function

A graphing utility can be used to find the value of a function with minimal effort, as the following example shows.

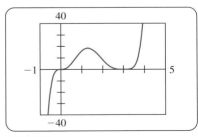

**FIGURE 3**
A more complete sketch of
$f(x) = x^3(x - 3)^4$ is shown by using
the window $[-1, 5] \times [-40, 40]$.

**EXAMPLE 3** Let $f(x) = x^3 - 4x^2 + 4x + 2$.

**a.** Plot the graph of $f$ in the standard viewing window.
**b.** Find $f(3)$ using a calculator, and verify your result by direct computation.
**c.** Find $f(4.215)$.

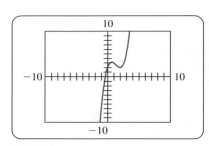

**FIGURE 4**
The graph of
$f(x) = x^3 - 4x^2 + 4x + 2$ in
the standard viewing window

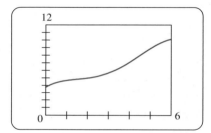

**FIGURE 5**
The graph of $f$ in the viewing window
$[0, 6] \times [0, 12]$

**Solution**
a. The graph of $f$ is shown in Figure 4.
b. Using the evaluation function of the graphing utility and the value 3 for $x$, we find $y = 5$. This result is verified by computing

$$f(3) = 3^3 - 4(3)^2 + 4(3) + 2 = 27 - 36 + 12 + 2 = 5$$

c. Using the evaluation function of the graphing utility and the value 4.215 for $x$, we find $y = 22.679738375$. Thus, $f(4.215) = 22.679738375$. The efficacy of the graphing utility is clearly demonstrated here! ■

**EXAMPLE 4  Number of Alzheimer's Patients**  The number of patients with Alzheimer's disease in the United States is approximated by

$$f(t) = -0.0277t^4 + 0.3346t^3 - 1.1261t^2 + 1.7575t + 3.7745 \qquad 0 \le t \le 6$$

where $f(t)$ is measured in millions and $t$ is measured in decades, with $t = 0$ corresponding to the beginning of 1990.

a. Use a graphing utility to plot the graph of $f$ in the viewing window $[0, 6] \times [0, 12]$.
b. What is the anticipated number of Alzheimer's patients in the United States at the beginning of 2010 ($t = 2$)? At the beginning of 2030 ($t = 4$)?

**Solution**
a. The graph of $f$ is shown in Figure 5.
b. Using the evaluation function of the graphing utility and the value 2 for $x$, we see that the anticipated number of Alzheimer's patients at the beginning of 2010 is given by $f(2) = 5.0187$, or approximately 5 million. The anticipated number of Alzheimer's patients at the beginning of 2030 is given by $f(4) = 7.1101$, or approximately 7.1 million. ■

### ■ Finding the Zeros of a Function

There will be many occasions when we need to find the zeros of a function. This task is greatly simplified if we use a graphing calculator or a computer algebra system (CAS).

**EXAMPLE 5**  Let $f(x) = x^3 + x + 1$. Find the zero of $f$ using (a) a graphing calculator and (b) a CAS.

**Solution**
a. The graph of $f$ in the window $[-2, 2] \times [-5, 5]$ is shown in Figure 6. Using TRACE and ZOOM or the function for finding the zero of a function, we find the zero to be approximately $-0.6823278$.
b. In Maple we use the command

$$\text{solve(x\textasciicircum3+x+1=0,x);}$$

and in Mathematica we use the command

$$\text{Solve[x\textasciicircum3+x+1==0,x]}$$

to obtain the solution $x \approx -0.682328$. ■

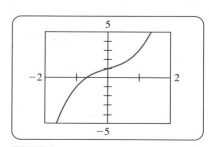

**FIGURE 6**
The graph of $f$ intersects the $x$-axis at $x \approx -0.6823278$.

### ■ Finding the Point(s) of Intersection of Two Graphs

A graphing calculator or a CAS can be used to find the point(s) of intersection of the graphs of two functions. Although the points of intersection of the graphs of the functions $f$ and $g$ can be found by finding the zeros of the function $f - g$, it is often more illuminating to proceed as in Example 6.

**EXAMPLE 6** Find the points of intersection of the graphs of $f(x) = 0.3x^2 - 1.4x - 3$ and $g(x) = -0.4x^2 + 0.8x + 6.4$.

**Solution** The graphs of both $f$ and $g$ in the standard viewing window are shown in Figure 7a. Using **TRACE** and **ZOOM** or the function for finding the points of intersection of two graphs on your graphing utility, we find the point(s) of intersection, accurate to four decimal places, to be $(-2.4158, 2.1329)$ (Figure 7b) and $(5.5587, -1.5125)$ (Figure 7c).

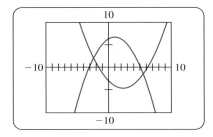

(**a**) The graphs of $f$ and $g$ in the standard viewing window

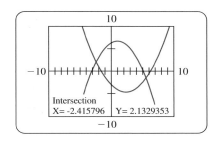

(**b**) An intersection screen

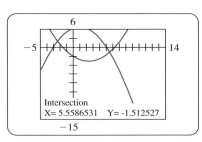

(**c**) An intersection screen

**FIGURE 7**

### ■ Constructing Functions from a Set of Data

A graphing calculator or a CAS can often be used to find the function that fits a given set of data points "best" in some sense. For example, if the points corresponding to the given data are scattered about a straight line, then we use linear regression to obtain a function that approximates the data at hand. If the points seem to be scattered about a parabola (the graph of a quadratic function), then we use second-degree polynomial regression, and so on.

We will exploit these capabilities of graphing calculators and computer algebra systems in Section 0.6, where we will see how "mathematical models" are constructed from raw data. The solution to the following example is obtained by using linear regression. (Consult the manual that accompanies your calculator for instructions for using linear regression. If you are using a CAS, consult your HELP menu for instructions.)

**EXAMPLE 7**

**a.** Use a graphing calculator or computer algebra system to find a linear function whose graph fits the following data "best" in the sense of least squares:

| $x$ | 1 | 2 | 3 | 4 | 5 |
|---|---|---|---|---|---|
| $y$ | 3 | 5 | 5 | 7 | 8 |

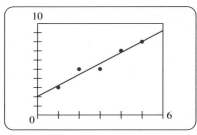

**FIGURE 8**
The scatter diagram and least-squares
line for the data set.

**b.** Plot the data points $(x, y)$ for the values of $x$ and $y$ given in the table (the graph is
called a *scatter diagram*) and the graph of the least-squares line (called the
*regression line*) on the same set of axes.

**Solution**

**a.** We first enter the data and then use the linear regression function on the calculator or computer to obtain the graph shown in Figure 8. We also find that the
equation of the least-squares regression line is $y = 1.2x + 2$.

**b.** See Figure 8.

## 0.5 EXERCISES

 *In Exercises 1–4, plot the graph of the function f in* (a) *the
standard viewing window and* (b) *the indicated window.*

**1.** $f(x) = x^3 - 20x^2 + 8x - 10$; $[-20, 20] \times [-1200, 100]$

**2.** $f(x) = x^4 - 2x^2 + 8$; $[-2, 2] \times [6, 10]$

**3.** $f(x) = x\sqrt{4 - x^2}$; $[-3, 3] \times [-2, 2]$

**4.** $f(x) = \dfrac{4}{x^2 - 8}$; $[-5, 5] \times [-5, 5]$

*In Exercises 5–16, plot the graph of the function f in an appro-
priate viewing window. (Note: The answer is not unique.)*

**5.** $f(x) = 2x^4 - 3x^3 + 5x^2 - 20x + 40$

**6.** $f(x) = -2x^4 + 5x^2 - 4$    **7.** $f(x) = \dfrac{x^3}{x^3 + 1}$

**8.** $f(x) = \dfrac{2x^4 - 3x}{x^2 - 1}$    **9.** $f(x) = \sqrt[3]{x} - \sqrt[3]{x + 1}$

**10.** $f(x) = \dfrac{5x}{x - 1} + 5x$    **11.** $f(x) = x^2 \sin \dfrac{1}{x}$

**Hint:** Stay close to the origin.

**12.** $f(x) = \dfrac{1}{2 + \cos x}$    **13.** $f(x) = \dfrac{\sin \sqrt{x}}{\sqrt{x}}$

**14.** $f(x) = \dfrac{1}{2} \sin 2x + \cos x$    **15.** $f(x) = x + 0.01 \sin 50x$

**16.** $f(x) = x^2 - 0.1x$

*In Exercises 17–22, find the zero(s) of the function f to five
decimal places.*

**17.** $f(x) = 2x^3 - 3x + 2$    **18.** $f(x) = x^3 - 9x + 4$

**19.** $f(x) = x^4 - 2x^3 + 3x - 1$    **20.** $f(x) = 2x^4 - 4x^2 + 1$

**21.** $f(x) = \sin 2x - x^2 + 1$    **22.** $f(x) = x^2 - 2x - 2 \sin x + 1$

*In Exercises 23–28, find the point(s) of intersection of the graphs of
the functions. Express your answers accurate to five decimal places.*

**23.** $f(x) = 0.3x^2 - 1.7x - 3.2$; $g(x) = -0.4x^2 + 0.9x + 6.7$

**24.** $f(x) = -0.3x^2 + 0.6x + 3.2$; $g(x) = 0.2x^2 - 1.2x - 4.8$

**25.** $f(x) = 0.3x^3 - 1.8x^2 + 2.1x - 2$; $g(x) = 2.1x - 4.2$

**26.** $f(x) = -0.2x^3 + 1.2x^2 - 1.2x + 2$;
$g(x) = -0.2x^2 + 0.8x + 2.1$

**27.** $f(x) = 2 \sin x$; $g(x) = 2 - \dfrac{1}{2}x^2$

**28.** $f(x) = \sin^2 x$; $g(x) = \sqrt{x^2 - x^4}$

**29.** Let $f(x) = x + \dfrac{1}{100} \sin 100x$.

   **a.** Plot the graph of $f$ using the viewing window
   $[-10, 10] \times [-10, 10]$.

   **b.** Plot the graph of $f$ using the viewing window
   $[-0.1, 0.1] \times [-0.1, 0.1]$.

   **c.** Explain why the two displays obtained in parts (a) and (b)
   taken together give a complete description of the graph of $f$.

**30. a.** Plot the graph of $f(x) = \cos(\sin x)$. Is $f$ odd or even?

   **b.** Verify your answer to part (a) analytically.

**31. a.** Plot the graph of $f(x) = x/x$ and $g(x) = 1$.

   **b.** Are the functions $f$ and $g$ identical? Why or why not?

**32. a.** Plot the graph of $f(x) = \sqrt{x}\sqrt{x - 1}$ using the viewing
   window $[-5, 5] \times [-5, 5]$.

   **b.** Plot the graph of $g(x) = \sqrt{x(x - 1)}$ using the viewing
   window $[-5, 5] \times [-5, 5]$.

   **c.** In what interval are the functions $f$ and $g$ identical?

   **d.** Verify your observation in part (c) analytically.

**33.** Let $f(x) = 2x^3 - 5x^2 + x - 2$ and $g(x) = 2x^3$.

   **a.** Plot the graph of $f$ and $g$ using the same viewing
   window: $[-5, 5] \times [-5, 5]$.

   **b.** Plot the graph of $f$ and $g$ using the same viewing window:
   $[-50, 50] \times [-100{,}000, 100{,}000]$.

   **c.** Explain why the graphs of $f$ and $g$ that you obtained in
   part (b) seem to coalesce as $x$ increases or decreases
   without bound.

   **Hint:** Write $f(x) = 2x^3\left(1 - \dfrac{5}{2x} + \dfrac{1}{2x^2} - \dfrac{1}{x^3}\right)$ and study its

   behavior for large values of $x$.

**V** Videos for selected exercises are available online at **www.academic.cengage.com/login**.

**34.** Let $f(x) = \left(1 + \dfrac{1}{x}\right)^x$, where $x > 0$.

    **a.** Plot the graph of $f$ using the window $[0, 10] \times [0, 3]$, and then using the window $[0, 100] \times [0, 3]$. Does $f(x)$ appear to approach a unique number as $x$ gets larger and larger?

    **b.** Use the evaluation function of your graphing utility to fill in the accompanying table. Use the table of values to estimate, accurate to five decimal places, the number that $f(x)$ seems to approach as $x$ increases without bound.

    **Note:** We will see in Section 2.8 that this number, written $e$, is given by 2.71828 . . .

| $x$ | $f(x)$ |
|---|---|
| 10 | |
| 100 | |
| 1000 | |
| 10,000 | |
| 100,000 | |
| 1,000,000 | |
| 10,000,000 | |
| 100,000,000 | |
| 1,000,000,000 | |

# 0.6 Mathematical Models

## 0.6 SELF-CHECK DIAGNOSTIC TEST

**1.** Give an example of each of the following.

    **a.** a linear function

    **b.** a polynomial function of degree 4

    **c.** a rational function

    **d.** a power function

    **e.** an algebraic function

    **f.** a trigonometric function

**2.** The book value of an asset at time $t$ (measured in years) being depreciated linearly over a period of $n$ years is given by

$$V(t) = C - \frac{C - S}{n} t$$

where $C$ and $S$ (in dollars) give the initial and scrap value of the asset, respectively.

    **a.** What is the $V$-intercept? Interpret your result.

    **b.** By how much is the asset being depreciated annually?

**3.** By cutting away identical squares from each corner of a square piece of cardboard with sides 12 in. long and then folding up the resulting flaps, an open box can be made. If the square cutaways have dimensions $x$ in. by $x$ in., find a function giving the volume of the resulting box.

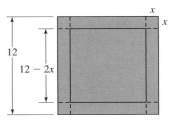

*Answers to Self-Check Diagnostic Test 0.6 can be found on page ANS 6.*

Mathematical modeling is a process that enables us to use mathematics as a tool to analyze and understand real-world phenomena. The four steps in this process are illustrated in Figure 1.

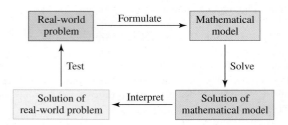

FIGURE 1

1. **Formulate.** Given a real-world problem, our first task is to formulate the problem using the language of mathematics. This mathematical description of the real-world phenomenon is called a **mathematical model.** The many techniques that are used in constructing mathematical models range from theoretical consideration of the problem on the one extreme to an interpretation of data associated with the problem on the other. For example, the mathematical model that gives the accumulated amount at any time after a certain sum of money has been deposited in the bank can be derived theoretically (see Section 0.8, pp. 87–89). On the other hand, the mathematical models in Examples 2 and 3 of this section are constructed by requiring that they fit the data associated with the problem "best" according to some specified criterion. In calculus we are primarily concerned with how one (dependent) variable depends on one or more (independent) variables. Consequently, most of our mathematical models will involve functions of one or more variables or equations defining these functions (implicitly).

2. **Solve.** Once a mathematical model has been constructed, we can use the appropriate mathematical techniques, which we will develop throughout this text, to solve the problem.

3. **Interpret.** Bearing in mind that the solution obtained in Step 2 is just the solution of the mathematical model, we need to interpret these results in the context of the original real-world problem.

4. **Test.** Some mathematical models of real-world applications describe the situations with complete accuracy. For example, the model describing a deposit in a bank account gives the exact accumulated amount in the account at any time. But other mathematical models give, at best, an approximate description of the real-world problem. In such cases we need to test the accuracy of the model by observing how well it describes the original real-world problem and how well it predicts past and/or future behavior. If the results are unsatisfactory, then we might have to reconsider the assumptions that were made in the construction of the model or, in the worst case, return to Step 1.

## ■ Modeling with Functions

Many real-world phenomena, such as the speed at which a screwdriver falls after being accidentally dropped from a building under construction, the speed of a chemical reaction, the population of a certain strain of bacteria, the life expectancy of a female infant at birth in a certain country, and the demand for a product, can be modeled by an appropriate function.

In what follows, we will recall some familiar functions and give examples of real-world phenomena that are modeled by using these functions.

## ■ Polynomial Functions

A **polynomial function of degree $n$** is a function of the form

$$f(x) = a_n x^n + a_{n-1} x^{n-1} + \cdots + a_2 x^2 + a_1 x + a_0 \qquad a_n \neq 0$$

where $n$ is a nonnegative integer and the numbers $a_0, a_1, \ldots, a_n$ are constants called the **coefficients** of the polynomial function. For example, the functions

$$f(x) = 2x^5 - 3x^4 + \frac{1}{2}x^3 + \sqrt{2}x^2 - 6$$

$$g(x) = 0.001x^3 - 0.2x^2 + 10x + 200$$

are polynomial functions of degree 5 and 3, respectively. Observe that a polynomial function is defined for every value of $x$, so its domain is $(-\infty, \infty)$.

A polynomial function of degree 1 ($n = 1$) has the form

$$y = f(x) = a_1 x + a_0 \qquad a_1 \neq 0$$

and is an equation of a straight line in the slope-intercept form with slope $m = a_1$ and $y$-intercept $b = a_0$ (see Section 0.1). For this reason a polynomial function of degree 1 is called a **linear function.**

Linear functions are used extensively in mathematical modeling for two important reasons. First, some models are linear by nature. For example, the formula for converting temperature from Celsius (°C) to Fahrenheit (°F) is $F = \frac{9}{5}C + 32$, and $F$ is a linear function of $C$ for $C$ in any feasible prescribed domain (see Figure 2a). Second, some natural phenomena exhibit linear characteristics over a small range of values and can therefore be modeled by a linear function that is restricted to a small interval. For example, according to Hooke's Law, the magnitude of a force $F$ required to stretch a spring by an elongation $x$ beyond its unstretched length is given by $F = kx$, provided that the elongation $x$ is not too great. If stretched beyond a certain point, called the elastic limit, the spring will become permanently deformed and will not return to its natural length when the force is removed. The constant $k$ is called the *spring constant* or the *stiffness* of the spring. In this instance we have to restrict our interest to the portion of the graph that is linear (see Figure 2b).

**FIGURE 2**
The graph of a linear function and the graph of a function that is linear over a small interval

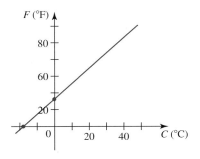

**(a)** $F$ is linear in $C$.

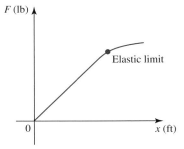

**(b)** $F$ is linear for small values of $x$.

In the following example we assume that Hooke's Law applies.

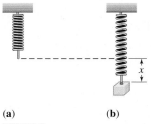

**(a)**                    **(b)**

**FIGURE 3**
The spring in part (a) is stretched by an elongation of $x$ feet beyond its natural length by a weight in part (b).

**EXAMPLE 1** **Force Required to Stretch a Spring**    A force of 3.18 lb is required to stretch a spring by 2.4 in. beyond its unstretched length (see Figure 3).

**a.** Use Hooke's Law to find a mathematical model that describes the force $F$ required to stretch the spring by $x$ feet beyond its unstretched length.
**b.** What is the spring constant?
**c.** Find the force required to stretch the spring by 1.8 in. beyond its unstretched length.

**Solution**
**a.** By Hooke's Law, $F = kx$, where $k$ is the spring constant. Next, using the given data, we find

$$3.18 = 0.2k \qquad \text{2.4 in. is equal to 0.2 ft}$$

from which we deduce that $k = 15.90$. Therefore, the required mathematical model is $F = 15.9x$.
**b.** From the result of part (a) we see that the spring constant is 15.9 lb/ft.
**c.** We first note that 1.8 in. is equal to 0.15 ft. Then, using the model obtained in part (a), we see that the required force is

$$F = (15.9)(0.15) = 2.385$$

or approximately 2.39 lb.    ■

   In Example 1 the model was constructed by using the data obtained from one measurement. In practice, one normally takes a set of measurements and then uses these data to construct a mathematical model. This practice generally results in a more accurate model.

**EXAMPLE 2** **Force Required to Stretch a Spring**    Table 1 gives the force $F$ required to stretch the spring (Example 1) by an elongation $x$ ft beyond its unstretched length. As Hooke's Law predicts, the data points in the *scatter plot* associated with these data appear to lie close to a straight line passing through the origin (see Figure 4).

**TABLE 1**

| $x$ (ft) | 0 | 0.1 | 0.2 | 0.3 | 0.4 | 0.5 |
|---|---|---|---|---|---|---|
| $F$ (lb) | 0 | 1.68 | 3.18 | 4.84 | 6.36 | 8.02 |

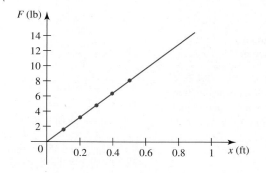

**FIGURE 4**
The data points are scattered about a line through the origin.

To find a mathematical model based on these data, we use the *method of least squares* to find a function of the form $f(x) = kx$ (as suggested by Hooke's Law) that fits the data "best" in the sense of least squares. (See Exercises 3.7, Problems 71 and 72.) We obtain the function

$$f(x) = 16.02x$$

as the required model. Incidentally, this model also tells us that the spring constant is approximately 16.02 lb/ft. ◼

**Notes**

1. If you use the linear least-squares regression program that is built into most graphing calculators and computers to find a mathematical model using the data in Example 2, you will obtain a different model, namely, $g(x) = 15.94x + 0.028$. This occurs because the program finds the "best" fit for the data (in the sense of least squares) using the most general linear function, that is, one having the form $f(x) = ax + b$.

2. Since $F$ must be equal to zero if $x$ is equal to zero, we see that the class of functions chosen to fit the data should have the form $f(x) = ax$, that is, with $b = 0$. Therefore, the model $F = 16.02x$ that we found in Example 2 should be regarded as being a more accurate mathematical model than the model suggested by $g(x) = 15.94x + 0.028$, in which $g(0) = 0.028 \neq 0$. As a consequence, we should accept the spring constant to be 16.02 lb/ft found in Example 2 rather than the figure of 15.94 that is found by using the function $g$ as the model. ◼

A polynomial function of degree 2 has the form

$$y = f(x) = a_2 x^2 + a_1 x + a_0 \qquad a_2 \neq 0$$

or, more simply, $y = ax^2 + bx + c$ and is called a **quadratic function.** The graph of a quadratic function is a parabola (see Figure 5). The parabola opens upward if $a > 0$ and downward if $a < 0$. To see this, we rewrite

$$f(x) = ax^2 + bx + c = x^2 \left( a + \frac{b}{x} + \frac{c}{x^2} \right) \qquad x \neq 0$$

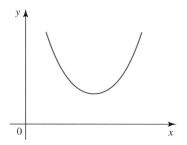

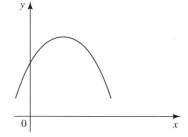

**FIGURE 5**
The graph of a quadratic function is a parabola.

(a) If $a > 0$, the parabola opens upward.

(b) If $a < 0$, the parabola opens downward.

Observe that if $x$ is large in absolute value, then the expression inside the parentheses is close to $a$, so $f(x)$ behaves like $ax^2$ for large values of $x$. Therefore, for large values of $x$, $y = f(x)$ is large and positive if $a > 0$ (the parabola opens upward) and is large in magnitude and negative if $a < 0$ (the parabola opens downward). The highest point on a parabola that opens downward or the lowest point on a parabola that opens upward is called the *vertex* of the parabola. The vertex of the parabola with equation $y = ax^2 + bx + c$, where $a \neq 0$, is $(-b/(2a), f(-b/(2a)))$ since $y = f(x)$. You can verify this fact by using the method of completing the square (see Exercise 30).

Quadratic functions serve as mathematical models for many phenomena. For example, Newton's Second Law of Motion can be used to show that the distance covered by a falling object dropped near the surface of the earth is given by $D = \frac{1}{2}gt^2$ where $g$, the gravitational constant at sea level at the equator, is approximately 32.088 ft/sec². In fact, a model for this motion can be found, experimentally, as the following example shows.

**EXAMPLE 3** A steel ball is dropped from a height of 10 ft. The distance covered by the ball at intervals of one tenth of a second is measured and recorded in Table 2. A scatter plot of the data is shown in Figure 6. You can see from the figure that the points associated with the data do lie close to a parabola with equation $y = at^2$ for some constant $a$, as was suggested earlier.

**TABLE 2**

| Time (sec)    | 0.0 | 0.1    | 0.2    | 0.3    | 0.4    | 0.5    | 0.6    | 0.7    |
|---------------|-----|--------|--------|--------|--------|--------|--------|--------|
| Distance (ft) | 0   | 0.1608 | 0.6416 | 1.4444 | 2.5672 | 4.0108 | 5.7760 | 7.8614 |

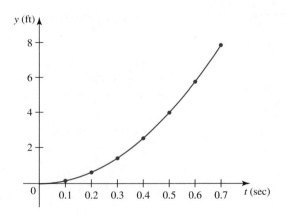

**FIGURE 6**

To find a mathematical model to describe this motion, we use the method of least squares to find a function of the form $y = at^2$ that fits the data "best." We obtain the function

$$y = 16.044t^2$$

(See Exercises 3.7, Problems 73 and 74.) On the basis of this model, the ball will hit the ground when $y = 10$. Solving the equation $16.044t^2 = 10$ gives $t \approx \pm 0.7895$. Rejecting the negative root, we conclude that the ball will hit the ground approximately 0.79 sec after it is dropped. Thus, a complete description of the mathematical model for this motion is

$$D = 16.044t^2 \qquad 0 \leq t \leq 0.79$$

where $D$ is the distance covered by the ball after $t$ sec.

**Notes**

1. Observe that even though the function $f(t) = 16.044t^2$ is defined on $(-\infty, \infty)$, we need to restrict its domain to the interval $[0, 0.79]$ to obtain a mathematical model for the motion of the ball. Once the ball reaches the ground, the function $f$ no longer describes its motion.

2. If you use the quadratic regression program that is found in most graphing calculators and in computers, you will find the quadratic model

$$D = 16.0425t^2 + 0.00075t + 0.000075$$

which is not very satisfactory, since we know that $D = 0$ when $t = 0$. Besides, as you will be able to confirm later, this model implies that the ball started out with an initial velocity of 0.00075 ft/sec. But we know that the steel ball had an initial velocity of 0 ft/sec. ∎

A polynomial of degree three is called a **cubic polynomial,** one of degree four is called a **quartic polynomial,** and one of degree five is called a **quintic** polynomial. In general, the higher the degree of the polynomial function, the more its graph wiggles. Figure 7a–c shows the graph of a cubic, a quartic, and a quintic, respectively.

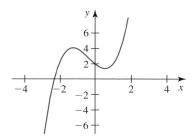

**(a)** $y = x^3 + x^2 - 2x + 2$
(a cubic)

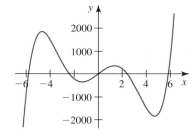

**(b)** $y = x^4 - 6x^2 + x + 2$
(a quartic)

**(c)** $y = 2x^5 - 80x^3 + 400x$
(a quintic)

**FIGURE 7**

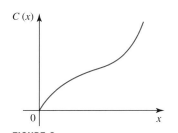

**FIGURE 8**
A total cost function is often modeled by using a cubic function.

Cubic polynomials lend themselves to modeling some phenomena in business and economics. For example, let $C(x)$ denote the total cost incurred when $x$ units of a certain commodity are produced. A typical graph of the function $C$ is shown in Figure 8. As the level of production $x$ increases, the cost per unit drops, so $C$ increases but at a slower pace. However, a level of production is soon reached at which the cost per unit begins to increase dramatically (because of overtime, a shortage of raw materials, and breakdown of machinery due to excessive stress and strain), so $C$ continues to increase at a faster pace. The graph of a cubic polynomial can exhibit precisely the characteristics just described.

The following example shows how we can use a quartic function to describe the assets of the Social Security system.

**EXAMPLE 4** **Social Security Trust Fund Assets** The projected assets of the Social Security trust fund (in trillions of dollars) from 2008 through 2040 are given in Table 3. The scatter plot associated with these data is shown in Figure 9a, where $t = 0$ corresponds to 2008. A mathematical model giving the approximate value of the assets in the trust fund $A(t)$ (in trillions of dollars) in year $t$ is

$$A(t) = -0.00000268t^4 - 0.000356t^3 + 0.00393t^2 + 0.2514t + 2.4094$$

The graph of $A$ is shown in Figure 9b.

**TABLE 3**

| Year | 2008 | 2011 | 2014 | 2017 | 2020 | 2023 | 2026 | 2029 | 2032 | 2035 | 2038 | 2040 |
|------|------|------|------|------|------|------|------|------|------|------|------|------|
| Assets | $2.4 | $3.2 | $4.0 | $4.7 | $5.3 | $5.7 | $5.9 | $5.6 | $4.9 | $3.6 | $1.7 | 0 |

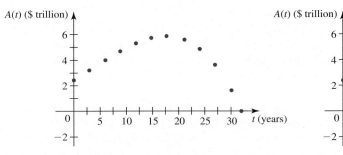

(**a**) Scatter plot           (**b**) Graph of $A$

**FIGURE 9**

*Source:* Social Security Administration.

**a.** The first baby boomers will turn 65 in 2011. What will the assets of the Social Security system trust fund be at that time? The last of the baby boomers will turn 65 in 2029. What will the assets of the trust fund be at that time?

**b.** Unless payroll taxes are increased significantly and/or benefits are scaled back dramatically, it is only a matter of time before the assets of the current system are depleted. Use the graph of the function $A(t)$ to estimate the year in which the current Social Security system is projected to go broke.

**Solution**

**a.** The assets of the Social Security trust fund in 2011 ($t = 3$) will be

$$A(3) = -0.00000268(3)^4 - 0.000356(3)^3$$
$$+ 0.00393(3)^2 + 0.2514(3) + 2.4094 \approx 3.19$$

or approximately $3.19 trillion. The assets of the trust fund in 2029 ($t = 21$) will be

$$A(21) = -0.00000268(21)^4 - 0.000356(21)^3$$
$$+ 0.00393(21)^2 + 0.2514(21) + 2.4094 \approx 5.60$$

or approximately $5.60 trillion.

**b.** From Figure 9b we see that the graph of $A$ crosses the $t$-axis at approximately $t = 32$. So unless the current system is changed, it is projected to go broke in 2040. (At this time the first of the baby boomers will be 94, and the last of the baby boomers will be 76.) ■

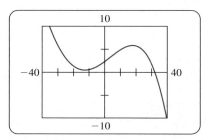

**FIGURE 10**
The graph of $f$ in the viewing window $[-40, 40] \times [-10, 10]$

**Note** Observe that the model in Example 4 utilizes only a small portion of the graph of $f$, as is often the case in practice. A more complete picture of the graph of $f$ is shown in Figure 10. ■

## ■ Power Functions

A **power function** is a function of the form $f(x) = x^a$, where $a$ is a real number. If $a$ is a nonnegative integer, then $f$ is just a polynomial function of degree $a$ with one term (a monomial). Examples of other power functions are

$$f(x) = x^{-2} = \frac{1}{x^2}, \quad f(x) = x^{-1} = \frac{1}{x}, \quad f(x) = x^{1/2} = \sqrt{x}, \quad \text{and} \quad f(x) = x^{1/3} = \sqrt[3]{x}$$

whose graphs are shown in Figure 11.

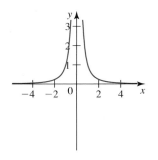

(**a**) $f(x) = x^{-2}$

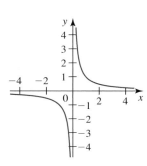

(**b**) $f(x) = x^{-1}$

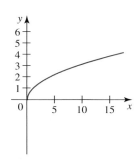

(**c**) $f(x) = x^{1/2}$

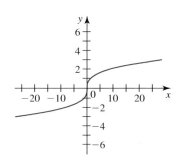

(**d**) $f(x) = x^{1/3}$

**FIGURE 11**
The graphs of some power functions

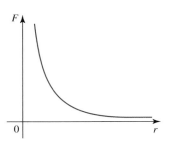

**FIGURE 12**
The magnitude of a gravitational force $F$

Power functions serve as mathematical models in many fields of study. For example, according to Newton's Law of Gravitation, the force exerted by a particle of mass $m_1$ on another particle of mass $m_2$ a distance $r$ away is directed toward $m_1$ and has magnitude

$$F = \frac{Gm_1 m_2}{r^2}$$

where $G$ is the universal gravitational constant. The graph of $F$ is similar to that of $f(x) = x^{-2}$ for $x > 0$ (see Figure 12).

## ■ Rational Functions

A **rational function** is a quotient of two polynomials. Examples of rational functions are

$$f(x) = \frac{3x^3 + x^2 - x + 1}{x - 2} \quad \text{and} \quad g(x) = \frac{x^2 + 1}{x^2 - 1}$$

In general, a rational function has the form

$$f(x) = \frac{P(x)}{Q(x)}$$

where $P$ and $Q$ are polynomial functions. The domain of a rational function is the set of all real numbers except the zeros of $Q$, that is, the roots of the equation $Q(x) = 0$. Thus, the domain of $f$ is $\{x \mid x \neq 2\}$, and the domain of $g$ is $\{x \mid x \neq \pm 1\}$. A mathematical model involving a rational function is suggested by the experiments conducted by A.J. Clark on the response $R(x)$ of a frog's heart muscle to the injection of $x$ units of

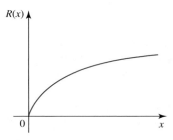

**FIGURE 13**

The graph of $R(x) = \dfrac{100x}{b + x}$

acetylcholine (as a percentage of the maximum possible effect of the drug). His results show that $R$ has the form

$$R(x) = \frac{100x}{b + x} \qquad x \geq 0$$

where $b$ is a positive constant that depends on the particular frog (see Figure 13).

## ■ Algebraic Functions

**Algebraic functions** are functions that can be expressed as sums, differences, products, quotients, or roots of polynomial functions. By definition, rational functions are algebraic functions. The function

$$f(x) = 2x^3 - 3\sqrt{x} + \frac{x\sqrt[3]{x^2 + 1}}{x(x + \sqrt{x})}$$

is another example of an algebraic function. The following example from the special theory of relativity involves an algebraic function.

**EXAMPLE 5**   **Special Theory of Relativity**   According to the special theory of relativity, the relativistic mass of a particle moving with a speed $v$ is

$$m = f(v) = \frac{m_0}{\sqrt{1 - \dfrac{v^2}{c^2}}}$$

where $m_0$ is the rest mass (the mass at zero speed) and $c = 2.9979 \times 10^8$ m/sec is the speed of light in a vacuum. What is the speed of a particle whose relativistic mass is twice that of its rest mass?

**Solution**   We solve the equation

$$2m_0 = \frac{m_0}{\sqrt{1 - \dfrac{v^2}{c^2}}}$$

for $v$, obtaining

$$2 = \frac{1}{\sqrt{1 - \dfrac{v^2}{c^2}}}$$

$$\sqrt{1 - \frac{v^2}{c^2}} = \frac{1}{2}$$

$$1 - \frac{v^2}{c^2} = \frac{1}{4}$$

$$\frac{v^2}{c^2} = \frac{3}{4}$$

$$v = \frac{\sqrt{3}}{2} c$$

or approximately 0.866 times the speed of light (approximately $2.596 \times 10^8$ m/sec). ■

## ■ Trigonometric Functions

Trigonometric functions were reviewed in Section 0.3. The characteristics of the trigonometric functions make them suitable for modeling phenomena that exhibit cyclical, or almost cyclical, behavior such as the motion of sound waves, the vibration of strings, and the motion of a simple pendulum.

**EXAMPLE 6** **Average Temperature** Table 4 gives the average monthly temperature in degrees Fahrenheit recorded in Boston.

**TABLE 4**

| Month | Jan. | Feb. | March | April | May | June | July | Aug. | Sept. | Oct. | Nov. | Dec. |
|---|---|---|---|---|---|---|---|---|---|---|---|---|
| **Temp (°F)** | 28.6 | 30.3 | 38.6 | 48.1 | 58.2 | 67.7 | 73.5 | 71.9 | 64.8 | 54.8 | 45.3 | 33.6 |

*Source: The Boston Globe.*

To find a model describing the average temperature $T$ in month $t$, we assume that $T$ is a sine function with period 12 and amplitude given by $\frac{1}{2}(73.5 - 28.6) = 22.45$. A possible model is

$$T = 51.05 + 22.45 \sin\left[\tfrac{\pi}{6}(t - 4.3)\right]$$

where $t = 1$ corresponds to January. The graph of $T$ is shown in Figure 14.

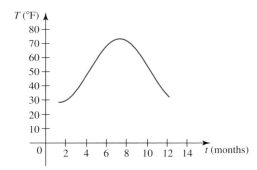

**FIGURE 14**
A model of the average temperature in Boston is
$T = 51.05 + 22.45 \sin\left[\tfrac{\pi}{6}(t - 4.3)\right].$

Other functions, such as *exponential* and *logarithmic* functions, also play an important role in modeling and will be studied in later chapters.

## ■ Constructing Mathematical Models

We close this section by showing how some mathematical models can be constructed by using elementary geometric and algebraic arguments.

The following guidelines can be used to construct mathematical models.

> **Guidelines for Constructing Mathematical Models**
>
> **1.** Assign a letter to each variable mentioned in the problem. If appropriate, draw and label a figure.
> **2.** Find an expression for the quantity that is being sought.
> **3.** Use the conditions given in the problem to write the quantity being sought as a function $f$ of one variable. Note any restrictions to be placed on the domain of $f$ from physical considerations of the problem.

**EXAMPLE 7**　**Enclosing an Area**　The owner of Rancho Los Feliz has 3000 yd of fencing with which to enclose a rectangular piece of grazing land along the straight portion of a river. Fencing is not required along the river. Letting $x$ denote the width of the rectangle, find a function $f$ in the variable $x$ giving the area of the grazing land if she uses all of the fencing (see Figure 15).

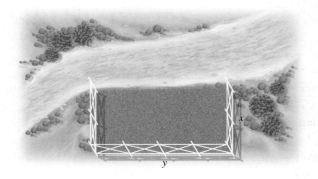

**FIGURE 15**
The rectangular grazing land
has width $x$ and length $y$.

**Solution**

1. This information is given in the statement of the problem.
2. The area of the rectangular grazing land is $A = xy$. Next, observe that the amount of fencing is $2x + y$ and that this must be equal to 3000, since all the fencing is to be used; that is,

$$2x + y = 3000$$

3. From the equation we see that $y = 3000 - 2x$. Substituting this value of $y$ into the expression for $A$ gives

$$A = xy = x(3000 - 2x) = 3000x - 2x^2$$

   Finally, observe that both $x$ and $y$ must be positive, since they represent the width and length of a rectangle, respectively. Thus, $x > 0$ and $y > 0$, but the latter is equivalent to $3000 - 2x > 0$, or $x < 1500$. So the required function is $f(x) = 3000x - 2x^2$ with domain $0 < x < 1500$. ∎

**EXAMPLE 8**　**Charter Flight Revenue**　If exactly 200 people sign up for a charter flight, Leisure World Travel Agency charges $300 per person. However, if more than 200 people sign up for the flight (assume that this is the case), then each fare is reduced by $1 for each additional person. Letting $x$ denote the number of passengers above 200, find a function giving the revenue realized by the company.

**Solution**

1. This information is given.
2. If there are $x$ passengers above 200, then the number of passengers signing up for the flight is $200 + x$. Furthermore, the fare will be $(300 - x)$ dollars per passenger.
3. The revenue will be

$$R = (200 + x)(300 - x) \qquad \text{number of passengers} \cdot \text{fare per passenger}$$

$$= -x^2 + 100x + 60{,}000$$

   Clearly, $x$ must be positive, and $300 - x > 0$, or $x < 300$. So the required function is $f(x) = -x^2 + 100x + 60{,}000$ with domain $(0, 300)$. ∎

## 0.6 EXERCISES

In Exercises 1 and 2, *classify each function as a polynomial function (state its degree), a power function, a rational function, an algebraic function, a trigonometric function, or other.*

**1. a.** $f(x) = 2x^3 - 3x^2 + x - 4$

  **b.** $f(x) = \sqrt[3]{x^2}$

  **c.** $g(x) = \dfrac{x}{x^2 - 4}$

  **d.** $f(t) = 3t^{-2} - 2t^{-1} + 4$

  **e.** $h(x) = \dfrac{\sqrt{x} + 1}{\sqrt{x} - 1}$

  **f.** $f(x) = \sin x + \cos x$

**2. a.** $f(t) = 2t^4 - 3t^2 - 2\sqrt{t}$

  **b.** $g(x) = 2\sqrt{1 - x^2}$

  **c.** $f(x) = \dfrac{\sqrt{x}}{x^2 + 1}$

  **d.** $h(x) = \dfrac{\sin x}{1 + \tan x}$

  **e.** $f(x) = \tan 2x$

  **f.** $h(x) = (3x - 1)^2 + 4$

**3. Instant Messaging Accounts** The number of enterprise instant messaging (IM) accounts is approximated by the function

$$N(t) = 2.96t^2 + 11.37t + 59.7 \qquad 0 \le t \le 5$$

where $N(t)$ is measured in millions and $t$ is measured in years with $t = 0$ corresponding to 2006.
  **a.** How many enterprise IM accounts were there in 2006?
  **b.** How many enterprise IM accounts are there projected to be in 2010?
*Source: The Radical Group.*

**4. Average Single-Family Property Tax** On the basis of data from 298 of Massachusetts' 351 cities and towns, the average single-family tax bill from 1997 through 2007 in that state is approximated by the function

$$T(t) = 7.26t^2 + 91.7t + 2360 \qquad 0 \le t \le 10$$

where $T(t)$ is measured in dollars and $t$ is measured in years with $t = 0$ corresponding to 1997.
  **a.** What was the average property tax on a single-family home in Massachusetts in 1997?
  **b.** If the trend continued, what would the average property tax be in 2010?
*Source: Massachusetts Department of Revenue.*

**5. Testosterone Use** Fueled by the promotion of testosterone as an antiaging elixir, use of the hormone by middle-aged and older men grew dramatically. The total number of prescriptions for testosterone from 1999 through 2002 is given by

$$N(t) = -35.8t^3 + 202t^2 + 87.8t + 648 \qquad 0 \le t \le 3$$

where $N(t)$ is measured in thousands and $t$ is measured in years with $t = 0$ corresponding to 1999. Find the total number of prescriptions for testosterone in 1999, 2000, 2001, and 2002.
*Source: IMS Health.*

**6. Aging Drivers** The number of driver fatalities due to car crashes, based on the number of miles driven, begins to climb after the driver is past age 65 years. Aside from declining ability as one ages, the older driver is more fragile. The number of driver fatalities per 100 million vehicle miles driven is approximately

$$N(x) = 0.0336x^3 - 0.118x^2 + 0.215x + 0.7 \qquad 0 \le x \le 7$$

where $x$ denotes the age group of drivers, with $x = 0$ corresponding to those aged 50–54 years, $x = 1$ corresponding to those aged 55–59, $x = 2$ corresponding to those aged 60–64, . . . , and $x = 7$ corresponding to those aged 85–89. What is the driver fatality rate per 100 million vehicle miles driven for an average driver in the 50–54 age group? In the 85–89 age group?
*Source: U.S. Department of Transportation.*

**7. Obese Children in the United States** The percentage of obese children aged 12–19 years in the United States is approximately

$$P(t) = \begin{cases} 0.04t + 4.6 & \text{if } 0 \le t < 10 \\ -0.01005t^2 + 0.945t - 3.4 & \text{if } 10 \le t \le 30 \end{cases}$$

where $t$ is measured in years, with $t = 0$ corresponding to the beginning of 1970. What was the percentage of obese children aged 12–19 years at the beginning of 1970? At the beginning of 1985? At the beginning of 2000?
*Source: Centers for Disease Control and Prevention.*

**8. Rwandan Genocide** The population of Rwanda in millions from 1990 through 2002 is approximated by the function

$$P(t) = \begin{cases} 0.17t + 6.99 & \text{if } 0 \le t < 3 \\ -0.9t + 10.2 & \text{if } 3 \le t < 5 \\ 0.7t + 2.2 & \text{if } 5 \le t < 7 \\ 0.12t + 6.26 & \text{if } 7 \le t \le 12 \end{cases}$$

where $t$ is measured in years, with $t = 0$ corresponding to 1990. The genocide that the majority Hutus committed against the Tutsis and moderate Hutus resulted in almost a million deaths and mass migration of the population out of the country. Eventually, most of the refugees returned to the country.
  **a.** Sketch the graph of the population function $P$.
  **b.** What was the population in 1993? In 1995? In 2002?
  **c.** In what year was the population of Rwanda at the lowest level?
  **d.** Did the population eventually recover to at least its previous level?
*Source: CIA World Factbook.*

---

Ⓥ Videos for selected exercises are available online at **www.academic.cengage.com/login**.

9. **Linear Depreciation** In computing income tax, businesses are allowed by law to depreciate certain assets, such as buildings, machines, furniture, and automobiles, over a period of time. The linear depreciation method, or straight-line method, is often used for this purpose. Suppose an asset has an initial value of $\$C$ and is to be depreciated linearly over $n$ years with a scrap value of $\$S$. Show that the book value of the asset at any time $t$, where $0 \leq t \leq n$, is given by the linear function

$$V(t) = C - \frac{C - S}{n}t$$

**Hint:** Find an equation of the straight line that passes through the points $(0, C)$ and $(n, S)$. Then rewrite the equation in the slope-intercept form.

10. **Cricket Chirping and Temperature** Entomologists have discovered that a linear relationship exists between the number of chirps of crickets of a certain species and the air temperature. When the temperature is 70°F, the crickets chirp at the rate of 120 times/min; when the temperature is 80°F, they chirp at the rate of 160 times/min.
   a. Find an equation giving the relationship between the air temperature $t$ and the number of chirps per minute, $N$, of the crickets.
   b. Find $N$ as a function of $t$, and use this formula to determine the rate at which the crickets chirp when the temperature is 102°F.

11. **Reaction of a Frog to a Drug** Experiments conducted by A.J. Clark suggest that the response $R(x)$ of a frog's heart muscle to the injection of $x$ units of acetylcholine (as a percent of the maximum possible effect of the drug) can be approximated by the rational function

$$R(x) = \frac{100x}{b + x} \qquad x \geq 0$$

where $b$ is a positive constant that depends on the particular frog.
   a. If a concentration of 40 units of acetylcholine produces a response of 50% for a certain frog, find the response function for this frog.
   b. Using the model found in part (a), find the response of the frog's heart muscle when 60 units of acetylcholine are administered.

12. **Outsourcing of Jobs** According to a study conducted in 2003, the total number of U.S. jobs (in millions) that were projected to leave the country by year $t$, where $t = 0$ corresponds to 2000, is

$$N(t) = 0.0018425(t + 5)^{2.5} \qquad 0 \leq t \leq 15$$

How many jobs were projected to be outsourced in 2005? In 2010?
*Source:* Forrester Research.

13. **Online Video Viewers** As broadband Internet grows more popular, video services such as YouTube will continue to expand. The number of online video viewers (in millions) is projected to grow according to the rule

$$N(t) = 52t^{0.531} \qquad 1 \leq t \leq 10$$

where $t = 1$ corresponds to 2003.
   a. Sketch the graph of $N$.
   b. How many online video viewers will there be in 2010?
*Source:* eMarketer.com.

14. **Cost, Revenue, and Profit Functions** A manufacturer of indoor-outdoor thermometers has fixed costs (executive salaries, rent, etc.) of $\$F$/month, where $F$ is a positive constant. The cost for manufacturing its product is $\$c$/unit, and the product sells for $\$s$/unit.
   a. Write a function $C(x)$ that gives the total cost incurred by the manufacturer in producing $x$ thermometers/month.
   b. Write a function $R(x)$ that gives the total revenue realized by the manufacturer in selling $x$ thermometers.
   c. Write a function $P(x)$ that gives the total monthly profit realized by the manufacturer in selling $x$ thermometers/month.
   d. Refer to your answer in part (c). Find $P(0)$, and interpret your result.
   e. How many thermometers should the manufacturer produce per month to have a break-even operation?
   **Hint:** Solve $P(x) = 0$.

 15. **Global Warming** The increase in carbon dioxide in the atmosphere is a major cause of global warming. The Keeling Curve, named after Dr. Charles David Keeling, a professor at Scripps Institution of Oceanography, gives the average amount of carbon dioxide ($CO_2$) measured in parts per million volume (ppmv), in the atmosphere from 1958 ($t = 1$) through 2007 ($t = 50$). (Even though data were available for every year in this time interval, we will construct the curve only on the basis of the following randomly selected data points.)

| Year | 1958 | 1970 | 1974 | 1978 | 1985 | 1991 | 1998 | 2003 | 2007 |
|---|---|---|---|---|---|---|---|---|---|
| Amount | 315 | 325 | 330 | 335 | 345 | 355 | 365 | 375 | 380 |

   a. Use a graphing utility to find a second-degree polynomial regression model for the data.
   b. Plot the graph of the function $f$ that you found in part (a), using the viewing window $[1, 50] \times [310, 400]$.
   c. Use the model to estimate the average amount of atmospheric carbon dioxide in 1980 ($t = 23$).
   d. Assume that the trend continues, and use the model to predict the average amount of atmospheric carbon dioxide in 2010.
*Source:* Scripps Institution of Oceanography.

**16.** **Population Growth in Clark County**  Clark County in Nevada, dominated by greater Las Vegas, is one of the fastest-growing metropolitan areas in the United States. The population of the county from 1970 through 2000 is given in the following table.

| Year | 1970 | 1980 | 1990 | 2000 |
|------|------|------|------|------|
| Population | 273,288 | 463,087 | 741,459 | 1,375,765 |

**a.** Use a graphing utility to find a third-degree polynomial regression model for the data. Let $t$ be measured in years, with $t = 0$ corresponding to the beginning of 1970.
**b.** Plot the graph of the function $f$ that you found in part (a), using the viewing window $[0, 30] \times [0, 1,500,000]$.
**c.** Compare the values of $f$ at $t = 0, 10, 20,$ and 30 with the given data.
*Source:* U.S. Census Bureau.

**17.** **Hiring Lobbyists**  Many public entities such as cities, counties, states, utilities, and Indian tribes are hiring firms to lobby Congress. One goal of such lobbying is to place earmarks—money directed at a specific project—into appropriation bills. The amount (in millions of dollars) spent by public entities on lobbying from 1998 through 2004 is shown in the following table.

| Year | 1998 | 1999 | 2000 | 2001 | 2002 | 2003 | 2004 |
|------|------|------|------|------|------|------|------|
| Amount | 43.4 | 51.7 | 62.5 | 76.3 | 92.3 | 101.5 | 107.7 |

**a.** Use a graphing utility to find a third-degree polynomial regression model for the data, letting $t = 0$ correspond to 1998.
**b.** Plot the scatter diagram and the graph of the function $f$ that you found in part (a).
**c.** Compare the values of $f$ at $t = 0, 3,$ and 6 with the given data.
*Source:* Center for Public Integrity.

**18.** **Measles Deaths**  Measles is still a leading cause of vaccine-preventable death among children, but because of improvements in immunizations, measles deaths have dropped globally. The following table gives the number of measles deaths (in thousands) in sub-Saharan Africa from 1999 through 2005.

| Year | 1999 | 2001 | 2003 | 2005 |
|------|------|------|------|------|
| Number | 506 | 338 | 250 | 126 |

**a.** Use a graphing utility to find a third-degree polynomial regression model for the data, letting $t = 0$ correspond to 1999.
**b.** Plot the scatter diagram and the graph of the function $f$ that you found in part (a).
**c.** Compute the values of $f$ for $t = 0, 2,$ and 6.
**d.** How many measles deaths were there in 2004?
*Source:* Centers for Disease Control and Prevention, World Health Organization.

**19.** **Nicotine Content of Cigarettes**  Even as measures to discourage smoking have been growing more stringent in recent years, the nicotine content of cigarettes has been rising, making it more difficult for smokers to quit. The following table gives the average amount of nicotine in cigarette smoke from 1999 through 2004.

| Year | 1999 | 2000 | 2001 | 2002 | 2003 | 2004 |
|------|------|------|------|------|------|------|
| Yield per cigarette (mg) | 1.71 | 1.81 | 1.85 | 1.84 | 1.83 | 1.89 |

**a.** Use a graphing utility to find a fourth-degree polynomial regression model for the data. Let $t = 0$ correspond to 1999.
**b.** Plot the graph of the function $f$ that you found in part (a), using the viewing window $[0, 5] \times [1, 3]$.
**c.** Compute the values of $f(t)$ for $t = 0, 1, 2, 3, 4,$ and 5.
*Source:* Massachusetts Tobacco Control Program.

**20.** **Periods of Planets**  The following table gives the mean distance $D$ between a planet and the sun measured in astronomical units (an AU is the mean distance between the earth and the sun), and its period $T$, measured in years, of some planets of the solar system.

| Planet | $D$ | $T$ |
|--------|-----|-----|
| Mercury | 0.39 | 0.24 |
| Venus | 0.72 | 0.62 |
| Earth | 1.00 | 1.00 |
| Mars | 1.52 | 1.88 |
| Jupiter | 5.20 | 11.9 |
| Saturn | 9.54 | 29.5 |

**a.** Use a graphing utility to find a power regression model, $T(D)$, for the data.
**b.** Does the model that you obtained in part (a) confirm Kepler's Third Law of Planetary Motion? (The squares of the periods of the planets are proportional to the cubes of their mean distances from the sun.)

**21. Enclosing an Area**  Patricia wishes to have a rectangular-shaped garden in her backyard. She has 80 ft of fencing with which to enclose her garden. Letting $x$ denote the width of the garden, find a function $f$ in the variable $x$ that gives the area of the garden. What is its domain?

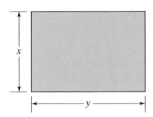

**22. Enclosing an Area**  Ramon wishes to have a rectangular-shaped garden in his backyard. But Ramon wants his garden to have an area of 250 ft$^2$. Letting $x$ denote the width of the garden, find a function $f$ in the variable $x$ that gives the length of the fencing required to construct the garden. What is the domain of the function?

**23. Packaging**  By cutting away identical squares from each corner of a rectangular piece of cardboard and folding up the resulting flaps, an open box can be made. If the cardboard is 15 in. long and 8 in. wide and the square cutaways have dimensions of $x$ in. by $x$ in., find a function that gives the volume of the resulting box.

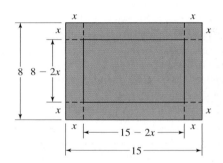

**24. Construction Costs**  A rectangular box is to have a square base and a volume of 20 ft$^3$. The material for the base costs 30¢/ft$^2$, the material for the sides costs 10¢/ft$^2$, and the material for the top costs 20¢/ft$^2$. Letting $x$ denote the length of one side of the base, find a function in the variable $x$ that gives the cost of materials for constructing the box.

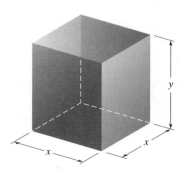

**25. Area of a Norman Window**  A Norman window has the shape of a rectangle surmounted by a semicircle. Suppose a Norman window is to have a perimeter of 28 ft. Find a function in the variable $x$ that gives the area of the window.

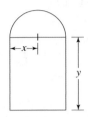

**26. Yield of an Apple Orchard**  An apple orchard has an average yield of 36 bushels of apples per tree if tree density is 22 trees per acre. For each unit increase in tree density, the yield decreases by 2 bushels per tree. Letting $x$ denote the number of trees beyond 22 per acre, find a function of $x$ that gives the yield of apples.

**27. Book Design**  A book designer decided that the pages of a book should have 1-in. margins at the top and bottom and $\frac{1}{2}$-in. margins on the sides. She further stipulated that each page should have an area of 50 in.$^2$. Find a function in the variable $x$, giving the area of the printed page (see the figure). What is the domain of the function?

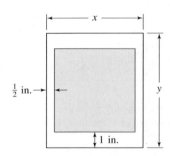

**28. Profit of a Vineyard**  Phillip, the proprietor of a vineyard, estimates that if 10,000 bottles of wine are produced this season, then the profit will be $5 per bottle. But if more than 10,000 bottles are produced, then the profit per bottle for the entire lot will drop by $0.0002 for each bottle sold. Assume that at least 10,000 bottles of wine are produced and sold, and let $x$ denote the number of bottles produced and sold above 10,000.
  **a.** Find a function $P$ giving the profit in terms of $x$.
  **b.** What is the profit that Phillip can expect from the sale of 16,000 bottles of wine from his vineyard?

**29. Charter Revenue**  The owner of a luxury motor yacht that sails among the 4000 Greek islands charges $600 per person per day if exactly 20 people sign up for the cruise. However, if more than 20 people (up to the maximum capacity of 90) sign up for the cruise, then each fare is reduced by $4 per day for each additional passenger. Assume at least 20 people sign up for the cruise, and let $x$ denote the number of passengers above 20.

**a.** Find a function $R$ giving the revenue per day realized from the charter.

**b.** What is the revenue per day if 60 people sign up for the cruise?

**c.** What is the revenue per day if 80 people sign up for the cruise?

**30.** Show that the vertex of the parabola $f(x) = ax^2 + bx + c$, where $a \neq 0$, is $(-b/(2a), f(-b/(2a)))$.

Hint: Complete the square.

---

## 0.7    Inverse Functions

### 0.7   SELF-CHECK DIAGNOSTIC TEST

**1.** Determine whether $f(x) = |x + 1| - |x|$ is one-to-one.

**2.** Find $f^{-1}(1)$ if $f(x) = \dfrac{1}{2} + \sqrt{x - \dfrac{3}{4}}, \quad x \geq \dfrac{3}{4}$.

**3.** Find the inverse of $f(x) = 3x - 2$. Then sketch the graph of $f$ and $f^{-1}$ on the same set of axes.

**4.** Find the exact value of each of the following:

   **a.** $\tan^{-1}(-1)$

   **b.** $\cot^{-1}(-\sqrt{3})$

**5.** Write $\tan(\sin^{-1} x)$ in algebraic form.

*Answers to Self-Check Diagnostic Test 0.7 can be found on page ANS 6.*

### ▇ The Inverse of a Function

A prototype of a maglev (magnetic levitation train) moves along a straight monorail. To describe the motion of the maglev, we can think of the track as a coordinate line. From data obtained in a test run, engineers have determined that the displacement (directed distance) of the maglev measured in feet from the origin at time $t$ (in seconds) is given by

$$s = f(t) = 4t^2 \qquad 0 \leq t \leq 30 \qquad \textbf{(1)}$$

where $f$ is called the position function of the maglev (see Figure 1).

**FIGURE 1**
A maglev moving along an elevated monorail track

The domain of this position function is [0, 30], and the graph of $f$ is shown in Figure 2. Formula (1) enables us to compute algebraically the position of the maglev at any given time $t$. Geometrically, we can find the position of the maglev at any given time $t$ by following the path indicated in Figure 2, which associates the given time $t$ with the desired position $f(t)$.

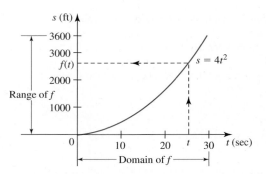

**FIGURE 2**
Each $t$ in the domain of $f$ is
associated with the (unique)
position $s = f(t)$ of the maglev.

Now consider the reverse problem: Knowing the position function of the maglev, can we find some way of obtaining the time it takes for the maglev to reach a given position? Geometrically, this problem is easily solved: Locate the point on the $s$-axis corresponding to the given position. Follow the path considered earlier but traced in the *opposite* direction. This path associates the given position $s$ with the desired time $t$.

Algebraically, we can obtain a formula for the time $t$ it takes for the maglev to get to the position $s$ by solving Equation (1) for $t$ in terms of $s$. Thus,

$$t = \frac{1}{2}\sqrt{s}$$

(we reject the negative root, since $t$ lies in $[0, 30]$). Observe that the function $g$ defined by

$$t = g(s) = \frac{1}{2}\sqrt{s}$$

has domain $[0, 3600]$ (the range of $f$) and range $[0, 30]$ (the domain of $f$). The graph of $g$ is shown in Figure 3.

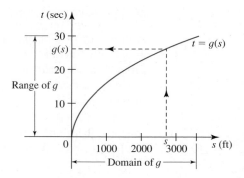

**FIGURE 3**
Each $s$ in the domain of $g$ is associated
with the (unique) time $t = g(s)$.

The functions $f$ and $g$ have the following properties:

**1.** The domain of $g$ is the range of $f$ and vice versa.
**2.** They satisfy the relationships

$$(g \circ f)(t) = g[f(t)] = \frac{1}{2}\sqrt{f(t)} = \frac{1}{2}\sqrt{4t^2} = t$$

and

$$(f \circ g)(t) = f[g(t)] = 4[g(t)]^2 = 4\left(\frac{1}{2}\sqrt{t}\right)^2 = t$$

In other words, one undoes what the other does. This is to be expected because $f$ maps $t$ onto $s = f(t)$ and $g$ maps $s = f(t)$ back onto $t$.

The functions $f$ and $g$ are said to be *inverses* of each other. More generally, we have the following definition.

---

**DEFINITION    Inverse Functions**

A function $g$ is the inverse of the function $f$ if

$$f[g(x)] = x \text{ for every } x \text{ in the domain of } g$$

and

$$g[f(x)] = x \text{ for every } x \text{ in the domain of } f$$

Equivalently, $g$ is the inverse of $f$ if the following condition is satisfied:

$$y = f(x) \qquad \text{if and only if} \qquad x = g(y)$$

for every $x$ in the domain of $f$ and for every $y$ in its range.

---

**Note**   The inverse of $f$ is normally denoted by $f^{-1}$ (read "$f$ inverse"), and we will use this notation throughout the text.   ■

 Do not confuse $f^{-1}(x)$ with $[f(x)]^{-1} = \dfrac{1}{f(x)}$.

**EXAMPLE 1**   Show that the functions $f(x) = x^{1/3}$ and $g(x) = x^3$ are inverses of each other.

**Solution**   First, observe that the domain and range of both $f$ and $g$ are $(-\infty, \infty)$. Therefore, both the composite functions $f \circ g$ and $g \circ f$ are defined. Next, we compute

$$(f \circ g)(x) = f[g(x)] = [g(x)]^{1/3} = (x^3)^{1/3} = x$$

and

$$(g \circ f)(x) = g[f(x)] = [f(x)]^3 = (x^{1/3})^3 = x$$

Since $f[g(x)] = x$ for all $x$ in $(-\infty, \infty)$, and $g[f(x)] = x$ for all $x$ in $(-\infty, \infty)$, we conclude that $f$ and $g$ are inverses of each other. In short, $f^{-1}(x) = x^3$.   ■

## Interpreting Our Results

We can view $f$ as a cube root extracting machine and $g$ as a "cubing" machine. In this light, it is easy to see that one function does undo what the other does. So $f$ and $g$ are indeed inverses of each other.

## ■ The Graphs of Inverse Functions

The graphs of $f(x) = x^{1/3}$ and $f^{-1}(x) = x^3$ are shown in Figure 4. They seem to suggest that the graphs of inverse functions are mirror images of each other with respect to the line $y = x$. This is true in general, as we will now show.

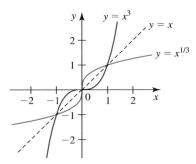

**FIGURE 4**
The functions $y = x^{1/3}$ and $y = x^3$ are inverses of each other.

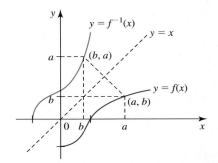

**FIGURE 5**
The graph of $f^{-1}$

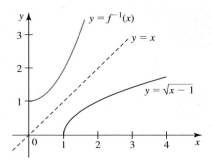

**FIGURE 6**
The graph of $f^{-1}$ is obtained by reflecting the graph of $f$ with respect to the line $y = x$.

Suppose that $(a, b)$ is any point on the graph of a function $f$. (See Figure 5.) Then $b = f(a)$, and we have

$$f^{-1}(b) = f^{-1}[f(a)] = a$$

This shows that $(b, a)$ is on the graph of $f^{-1}$ (Figure 5). Similarly, we can show that if $(b, a)$ lies on the graph of $f^{-1}$, then $(a, b)$ must be on the graph of $f$. But the point $(b, a)$, as you can see in Figure 5, is the reflection of the point $(a, b)$ with respect to the line $y = x$. We have proved the following.

---

**The Graphs of Inverse Functions**

The graph of $f^{-1}$ is the reflection of the graph of $f$ with respect to the line $y = x$ and vice versa.

---

**EXAMPLE 2**　Sketch the graph of $f(x) = \sqrt{x - 1}$. Then reflect the graph of $f$ with respect to the line $y = x$ to obtain the graph of $f^{-1}$.

**Solution**　The graphs of both $f$ and $f^{-1}$ are sketched in Figure 6.　■

## ■ Which Functions Have Inverses

Does every function have an inverse? Consider, for example, the function $f$ defined by $y = x^2$ with domain $(-\infty, \infty)$ and range $[0, \infty)$. From the graph of $f$ shown in Figure 7, you can see that each value of $y$ in the range of $[0, \infty)$ of $f$ is associated with exactly *two* numbers $x = \pm\sqrt{y}$ in the domain $(-\infty, \infty)$ of $f$ (except for $y = 0$). This implies that $f$ does not have an inverse, since the uniqueness requirement of a function cannot be satisfied in this case. Observe that any horizontal line $y = c$, where $c > 0$, intersects the graph of $f$ at more than one point.

Next, consider the function $g$ defined by the same rule as that of $f$, namely, $y = x^2$, but with domain restricted to $[0, \infty)$. From the graph of $g$ shown in Figure 8, you can see that each value of $y$ in the range $[0, \infty)$ of $g$ is mapped onto exactly *one* number $x = \sqrt{y}$ in the domain $[0, \infty)$ of $g$. Thus, in this case we can define the inverse function of $g$, from the range $[0, \infty)$ of $g$, onto the domain $[0, \infty)$ of $g$. To find the rule for $g^{-1}$, we solve the equation $y = x^2$ for $x$ in terms of $y$. Thus, $x = \sqrt{y}$, since $x \geq 0$, so $g^{-1}(y) = \sqrt{y}$, or, since $y$ is a dummy variable, we can write $g^{-1}(x) = \sqrt{x}$. Also, observe that every horizontal line intersects the graph of $g$ at no more than one point.

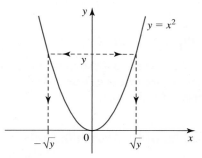

**FIGURE 7**
Each value of $y$ is associated with two values of $x$.

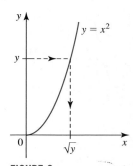

**FIGURE 8**
Each value of $y$ is associated with exactly one value of $x$.

Our analysis of the functions $f$ and $g$ reveals the following important difference between the two functions that enables $g$ to have an inverse but not $f$. Observe that $f$ takes on the same value twice; that is, there are two values of $x$ that are mapped onto each value of $y$ (except $y = 0$). On the other hand, $g$ never takes on the same value more than once; that is, any two values of $x$ have different images. The function $g$ is said to be *one-to-one*.

---

**DEFINITION   One-to-One Function**

A function $f$ with domain $D$ is **one-to-one** if no two numbers in $D$ have the same image; that is,

$$f(x_1) \neq f(x_2) \qquad \text{whenever} \qquad x_1 \neq x_2$$

---

Geometrically, a function is one-to-one if every horizontal line intersects its graph at no more than one point. This is called the **horizontal line test.**

We have the following important theorem concerning the existence of an inverse function.

---

**THEOREM 1   The Existence of an Inverse Function**

A function has an inverse if and only if it is one-to-one.

---

You will be asked to prove this theorem in Exercise 60.

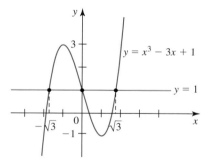

**FIGURE 9**
$f$ is not one-to-one because it fails the horizontal line test.

**EXAMPLE 3** Determine whether the function has an inverse.

**a.** $f(x) = x^{1/3}$      **b.** $f(x) = x^3 - 3x + 1$

**Solution**

**a.** Refer to Figure 4, page 75. Using the horizontal line test, we see that $f$ is one-to-one on $(-\infty, \infty)$. Therefore, $f$ has an inverse on $(-\infty, \infty)$.

**b.** The graph of $f$ is shown in Figure 9. Observe that the horizontal line $y = 1$ intersects the graph of $f$ at three points, so $f$ does not pass the horizontal line test. Therefore, $f$ is not one-to-one. In fact, the three points $x = -\sqrt{3}$, $0$, and $\sqrt{3}$ are mapped onto the point 1. Therefore, by Theorem 1, $f$ does not have an inverse.

## Finding the Inverse of a Function

Before looking at the next example, let's summarize the steps for finding the inverse of a function, assuming that it exists.

---

**Guidelines for Finding the Inverse of a Function**

**1.** Write $y = f(x)$.
**2.** Solve this equation for $x$ in terms of $y$ (if possible).
**3.** Interchange $x$ and $y$ to obtain $y = f^{-1}(x)$.

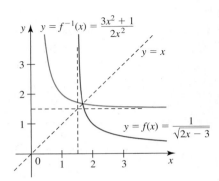

**FIGURE 10**
The graphs of $f$ and $f^{-1}$. Notice that they are reflections of each other with respect to the line $y = x$.

**EXAMPLE 4** Find the inverse of the function defined by $f(x) = \dfrac{1}{\sqrt{2x - 3}}$.

**Solution**   The graph of $f$ shown in Figure 10 shows that $f$ is one-to-one and so $f^{-1}$ exists. To find the rule for this inverse, write

$$y = \frac{1}{\sqrt{2x - 3}}$$

and then solve the equation for $x$:

$$y^2 = \frac{1}{2x - 3} \qquad \text{Square both sides.}$$

$$2x - 3 = \frac{1}{y^2} \qquad \text{Take reciprocals.}$$

$$2x = \frac{1}{y^2} + 3 = \frac{3y^2 + 1}{y^2}$$

and

$$x = \frac{3y^2 + 1}{2y^2}$$

Finally, interchanging $x$ and $y$, we obtain

$$y = \frac{3x^2 + 1}{2x^2}$$

giving the rule for $f^{-1}$ as

$$f^{-1}(x) = \frac{3x^2 + 1}{2x^2}$$

The graphs of both $f$ and $f^{-1}$ are shown in Figure 10.    ■

## Inverse Trigonometric Functions

Generally speaking, the trigonometric functions, being periodic, are not one-to-one and, therefore, do not have inverse functions. For example, you can see by examining the graph of $y = \sin x$ shown in Figure 11 that this function is not one-to-one, since it fails the horizontal line test. But observe that by restricting the domain of the function $f(x) = \sin x$ to the interval $\left[-\frac{\pi}{2}, \frac{\pi}{2}\right]$, it is one-to-one and its range is $[-1, 1]$ (Figure 12a). So, by Theorem 1, $f$ has an inverse function with domain $[-1, 1]$ and range $\left[-\frac{\pi}{2}, \frac{\pi}{2}\right]$. This function is called the **inverse sine function** or **arcsine function**

**FIGURE 11**
The horizontal line cuts the graph of $y = \sin x$ at infinitely many points, so the sine function is not one-to-one.

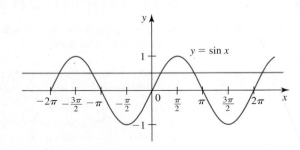

and is denoted by arcsin or $\sin^{-1}$. Thus,

$$y = \sin^{-1} x \qquad \text{if and only if} \qquad \sin y = x$$

where $-1 \leq x \leq 1$ and $-\frac{\pi}{2} \leq y \leq \frac{\pi}{2}$. (The graph of $y = \sin^{-1} x$ is shown in Figure 13a.)

Similarly, by suitably restricting the domains of the other five trigonometric functions, each function can also be made one-to-one, and therefore, each function also has an inverse. Figure 12 shows the graphs of the six trigonometric functions and their restricted domains.

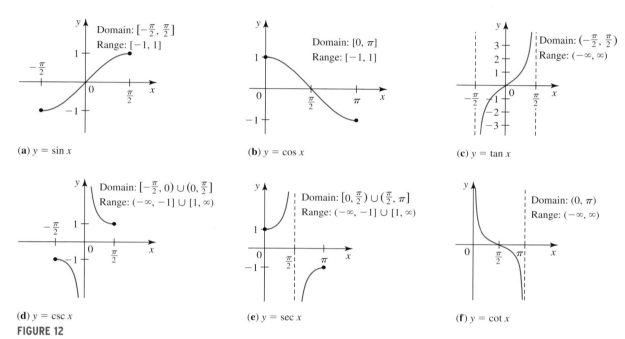

(a) $y = \sin x$

(b) $y = \cos x$

(c) $y = \tan x$

(d) $y = \csc x$

(e) $y = \sec x$

(f) $y = \cot x$

**FIGURE 12**

When restricted to the indicated domains, each of the six trigonometric functions is one-to-one.

With these restrictions the corresponding trigonometric inverse functions are defined as follows.

**DEFINITION  Inverse Trigonometric Functions**

|  |  | Domain |  |
|---|---|---|---|
| $y = \sin^{-1} x$ if and only if $x = \sin y$ | | $[-1, 1]$ | (2a) |
| $y = \cos^{-1} x$ if and only if $x = \cos y$ | | $[-1, 1]$ | (2b) |
| $y = \tan^{-1} x$ if and only if $x = \tan y$ | | $(-\infty, \infty)$ | (2c) |
| $y = \csc^{-1} x$ if and only if $x = \csc y$ | | $(-\infty, -1] \cup [1, \infty)$ | (2d) |
| $y = \sec^{-1} x$ if and only if $x = \sec y$ | | $(-\infty, -1] \cup [1, \infty)$ | (2e) |
| $y = \cot^{-1} x$ if and only if $x = \cot y$ | | $(-\infty, \infty)$ | (2f) |

The graphs of the six inverse trigonometric functions are shown in Figures 13a–13f.

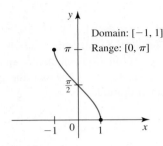

**(a)** $y = \sin^{-1} x$

Domain: $[-1, 1]$
Range: $\left[-\frac{\pi}{2}, \frac{\pi}{2}\right]$

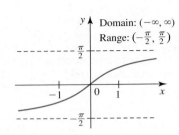

**(b)** $y = \cos^{-1} x$

Domain: $[-1, 1]$
Range: $[0, \pi]$

**(c)** $y = \tan^{-1} x$

Domain: $(-\infty, \infty)$
Range: $\left(-\frac{\pi}{2}, \frac{\pi}{2}\right)$

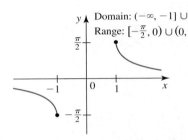

**(d)** $y = \csc^{-1} x$

Domain: $(-\infty, -1] \cup [1, \infty)$
Range: $\left[-\frac{\pi}{2}, 0\right) \cup \left(0, \frac{\pi}{2}\right]$

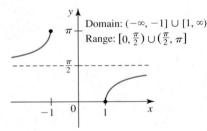

**(e)** $y = \sec^{-1} x$

Domain: $(-\infty, -1] \cup [1, \infty)$
Range: $\left[0, \frac{\pi}{2}\right) \cup \left(\frac{\pi}{2}, \pi\right]$

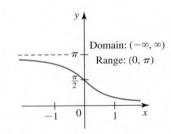

**(f)** $y = \cot^{-1} x$

Domain: $(-\infty, \infty)$
Range: $(0, \pi)$

**FIGURE 13**

**EXAMPLE 5** Evaluate

**a.** $\sin^{-1} \dfrac{1}{2}$    **b.** $\cos^{-1}\left(-\dfrac{\sqrt{3}}{2}\right)$    **c.** $\tan^{-1} \sqrt{3}$    **d.** $\cos^{-1} 0.6$

**Solution**
**a.** Let $y = \sin^{-1} \frac{1}{2}$. Then by Formula (2a), $\sin y = \frac{1}{2}$. Since $y$ must lie in the interval $\left[-\frac{\pi}{2}, \frac{\pi}{2}\right]$, we see that $y = \pi/6$. Therefore,

$$\sin^{-1} \frac{1}{2} = \frac{\pi}{6}$$

**b.** Let $y = \cos^{-1}(-\sqrt{3}/2)$ so that, by Formula (2b), $\cos y = -\sqrt{3}/2$. Since $y$ must be in the interval $[0, \pi]$, we see that $y = 5\pi/6$. Therefore,

$$\cos^{-1}\left(-\frac{\sqrt{3}}{2}\right) = \frac{5\pi}{6}$$

**c.** Let $y = \tan^{-1} \sqrt{3}$ so that $\tan y = \sqrt{3}$. Since $y$ must lie in the interval $\left(-\frac{\pi}{2}, \frac{\pi}{2}\right)$, we see that $y = \pi/3$. Therefore,

$$\tan^{-1} \sqrt{3} = \frac{\pi}{3}$$

**d.** Here, we use a calculator to find

$$\cos^{-1} 0.6 \approx 0.9273$$

Remember to set the calculator in the *radian mode*.

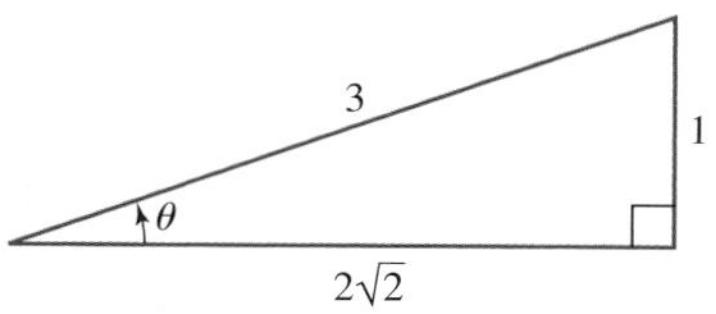

**FIGURE 14**
The right triangle associated with the equation $\theta = \sin^{-1}\frac{1}{3}$

**EXAMPLE 6** Evaluate $\cot\left(\sin^{-1}\frac{1}{3}\right)$.

**Solution** Let $\theta = \sin^{-1}\frac{1}{3}$. Then $\theta$ is the angle in the right triangle with opposite side of length 1 and hypotenuse of length 3. (See Figure 14.) Therefore, by the Pythagorean Theorem the length of the adjacent side of the right triangle is

$$\sqrt{9-1} = 2\sqrt{2}$$

and

$$\cot\left(\sin^{-1}\frac{1}{3}\right) = \cot\theta = \frac{2\sqrt{2}}{1} = 2\sqrt{2} \qquad \blacksquare$$

Recall that if $f$ and $f^{-1}$ are inverses of each other, then

$$f(f^{-1}(x)) = x \qquad \text{and} \qquad f^{-1}(f(x)) = x$$

For the trigonometric functions sine, cosine, and tangent (and similarly for the other three trigonometric functions) these relationships translate into the following properties.

---

**Inverse Properties of Trigonometric Functions**

| | | | |
|---|---|---|---|
| $\sin(\sin^{-1}x) = x$ | for | $-1 \le x \le 1$ | **(3a)** |
| $\sin^{-1}(\sin x) = x$ | for | $-\frac{\pi}{2} \le x \le \frac{\pi}{2}$ | **(3b)** |
| $\cos(\cos^{-1}x) = x$ | for | $-1 \le x \le 1$ | **(3c)** |
| $\cos^{-1}(\cos x) = x$ | for | $0 \le x \le \pi$ | **(3d)** |
| $\tan(\tan^{-1}x) = x$ | for | $-\infty < x < \infty$ | **(3e)** |
| $\tan^{-1}(\tan x) = x$ | for | $-\frac{\pi}{2} < x < \frac{\pi}{2}$ | **(3f)** |

---

⚠ Remember that these properties hold only for the specified values of $x$. For example, $\sin^{-1}(\sin \pi) = \sin^{-1}(0) = 0$, but a careless application of the property $\sin^{-1}(\sin x) = x$ with $x = \pi$—which does not lie in the interval $\left[-\frac{\pi}{2}, \frac{\pi}{2}\right]$—leads to the erroneous result $\sin^{-1}(\sin \pi) = \pi$.

**EXAMPLE 7** Evaluate

**a.** $\sin(\sin^{-1}0.7)$ **b.** $\cos^{-1}(\cos(3\pi/2))$

**Solution**
**a.** Since 0.7 lies in the interval $[-1, 1]$, we conclude, by Formula (3a), that

$$\sin(\sin^{-1}0.7) = 0.7$$

**b.** Notice that $3\pi/2$ does not lie in the interval $[0, \pi]$, so we may not use Formula (3d). But observe that $\cos(3\pi/2) = 0$, and since 0 lies in the interval $[-1, 1]$, we have

$$\cos^{-1}\left(\cos\frac{3\pi}{2}\right) = \cos^{-1}0 = \frac{\pi}{2} \qquad \blacksquare$$

## 0.7   CONCEPT QUESTIONS

**1. a.** What is a one-to-one function? Give an example.
   **b.** Explain how the horizontal line test is used to determine whether a curve in the plane is the graph of a one-to-one function. Illustrate with a figure.

**2.** Suppose that $f$ is a one-to-one function with domain $[a, b]$ and range $[c, d]$.
   **a.** How is $f^{-1}$ defined?
   **b.** What are the domain and range of $f^{-1}$? Illustrate with a figure.

**3.** Suppose that $f$ is a one-to-one function defined by $y = f(x)$.
   **a.** Describe how to find the rule for $f^{-1}$. Give an example.

**b.** Describe the relationship between the graph of $f$ and that of $f^{-1}$.

**4.** For each of the following inverse trigonometric functions, (a) give its definition, (b) give its domain and range, and (c) sketch its graph:
   **(i)** $f(x) = \sin^{-1} x$    **(ii)** $f(x) = \cos^{-1} x$    **(iii)** $f(x) = \tan^{-1} x$

**5.** For each of the following inverse trigonometric functions, (a) give its definition, (b) give its domain and range, and (c) sketch its graph:
   **(i)** $f(x) = \csc^{-1} x$    **(ii)** $f(x) = \sec^{-1} x$    **(iii)** $f(x) = \cot^{-1} x$

## 0.7   EXERCISES

*In Exercises 1–6, show that f and g are inverses of each other by verifying that $f[g(x)] = x$ and $g[f(x)] = x$.*

**1.** $f(x) = \dfrac{1}{3}x^3; \quad g(x) = \sqrt[3]{3x}$

**2.** $f(x) = \dfrac{1}{x}; \quad g(x) = \dfrac{1}{x}$

**3.** $f(x) = 2x + 3; \quad g(x) = \dfrac{x - 3}{2}$

**4.** $f(x) = x^2 + 1 \ (x \le 0); \quad g(x) = -\sqrt{x - 1}$

**5.** $f(x) = 4(x + 1)^{2/3}$, where $x \ge -1$;

   $g(x) = \dfrac{1}{8}(x^{3/2} - 8)$, where $x \ge 0$

**6.** $f(x) = \dfrac{1 + x}{1 - x}; \quad g(x) = \dfrac{x - 1}{x + 1}$

*In Exercises 7–10, you are given the graph of a function f. Determine whether f is one-to-one.*

**7.**

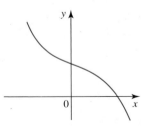

**8.**

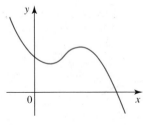

**9.**

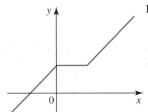

**10.**

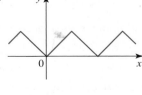

*In Exercises 11–14, determine whether the function is one-to-one.*

**11.** $f(x) = 4x - 3$

**12.** $f(x) = -x^2 + 2x - 3$

**13.** $f(x) = \sqrt{1 - x}$

**14.** $f(x) = -x^4 + 16$

**15.** Suppose that $f$ is a one-to-one function such that $f(2) = 5$. Find $f^{-1}(5)$.

**16.** Suppose that $f$ is a one-to-one function such that $f(3) = 7$. Find $f[f^{-1}(7)]$.

*In Exercises 17–20, find $f^{-1}(a)$ for the function f and the real number a.*

**17.** $f(x) = x^3 + x - 1; \quad a = -1$

**18.** $f(x) = 2x^5 + 3x^3 + 2; \quad a = 2$

**19.** $f(x) = \dfrac{3}{\pi}x + \sin x; \quad -\dfrac{\pi}{2} < x < \dfrac{\pi}{2}; \quad a = 1$

**20.** $f(x) = 2 + \tan\left(\dfrac{\pi x}{2}\right), \quad -1 < x < 1; \quad a = 2$

**21.** The graph of $f$ is given. Sketch the graph of $f^{-1}$ on the same set of axes.

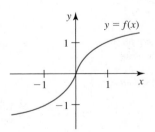

**22.** The graph of the inverse of a function $f$, $f^{-1}$, is given. Sketch the graph of $f$ on the same set of axes.

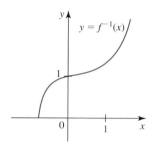

*In Exercises 23–28, find the inverse of f. Then sketch the graphs of f and $f^{-1}$ on the same set of axes.*

**23.** $f(x) = x^3 + 1$

**24.** $f(x) = 2\sqrt{x} + 3$

**25.** $f(x) = \sqrt{9 - x^2}$, $x \geq 0$

**26.** $f(x) = x^{3/5} + 1$

**27.** $f(x) = \sin(2x - 1)$, $\frac{1}{2}\left(1 - \frac{\pi}{2}\right) \leq x \leq \frac{1}{2}\left(1 + \frac{\pi}{2}\right)$

**28.** $f(x) = \cot^{-1}\left(\frac{x}{3}\right)$, $0 < x < 3\pi$

 *In Exercises 29–32, find the inverse of f. Then use a graphing utility to plot the graphs of f and $f^{-1}$ using the same viewing window.*

**29.** $f(x) = \sqrt[3]{x - 1}$

**30.** $f(x) = 1 - \dfrac{1}{x}$

**31.** $f(x) = \dfrac{x}{x^2 + 1}$, $-\frac{1}{2} \leq x \leq \frac{1}{2}$

**32.** $f(x) = \dfrac{x}{\sqrt{x^2 + 1}}$, $-1 \leq x \leq 1$

**33.** Let

$$f(x) = \begin{cases} 2x - 1 & \text{if } x < 1 \\ \sqrt{x} & \text{if } 1 \leq x < 4 \\ \dfrac{1}{2}x^2 - 6 & \text{if } x \geq 4 \end{cases}$$

Find $f^{-1}(x)$, and state its domain.

**34. a.** Show that $f(x) = -x^2 + x + 1$ on $\left[\frac{1}{2}, \infty\right)$ and $g(x) = \frac{1}{2} + \sqrt{\frac{5}{4} - x}$ on $\left(-\infty, \frac{5}{4}\right)$ are inverses of each other.

**b.** Solve the equation $-x^2 + x + 1 = \frac{1}{2} + \sqrt{\frac{5}{4} - x}$.
Hint: Use the result of part (a).

*In Exercises 35–48, find the exact value of the given expression.*

**35.** $\sin^{-1} 0$

**36.** $\cos^{-1} 0$

**37.** $\sin^{-1} \dfrac{1}{2}$

**38.** $\cos^{-1} \dfrac{1}{2}$

**39.** $\tan^{-1} \sqrt{3}$

**40.** $\cot^{-1}(-1)$

**41.** $\sin^{-1}\left(\dfrac{\sqrt{3}}{2}\right)$

**42.** $\cos^{-1}\left(-\dfrac{1}{\sqrt{2}}\right)$

**43.** $\sec^{-1} 2$

**44.** $\csc^{-1} \sqrt{2}$

**45.** $\sin^{-1}\left(-\dfrac{1}{2}\right)$

**46.** $\tan^{-1}\left(-\dfrac{1}{\sqrt{3}}\right)$

**47.** $\sin\left(\sin^{-1} \dfrac{1}{\sqrt{2}}\right)$

**48.** $\cos\left(\sin^{-1} \dfrac{1}{2}\right)$

*In Exercises 49–54, write the expression in algebraic form.*

**49.** $\cos(\sin^{-1} x)$

**50.** $\sin(\cos^{-1} x)$

**51.** $\tan(\tan^{-1} x)$

**52.** $\sec(\sin^{-1} x)$

**53.** $\cot(\sec^{-1} x)$

**54.** $\csc(\cot^{-1} x)$

**55. Temperature Conversion** The formula $F = f(C) = \frac{9}{5}C + 32$, where $C \geq -273.15$, gives the temperature $F$ (in degrees) on the Fahrenheit scale as a function of the temperature $C$ (in degrees) on the Celsius scale.
**a.** Find a formula for $f^{-1}$, and interpret your result.
**b.** What is the domain of $f^{-1}$?

**56. Motion of a Hot Air Balloon** A hot air balloon rises vertically from the ground so that its height after $t$ sec is $h = \frac{1}{2}t^2 + \frac{1}{2}t$, where $h$ is measured in feet and $0 \leq t \leq 60$.
**a.** Find the inverse of the function $f(t) = \frac{1}{2}t^2 + \frac{1}{2}t$ and explain what it represents.
**b.** Use the result of part (a) to find the time when the balloon is between an altitude of 120 ft and 210 ft.

**57. Aging Population** The population of Americans age 55 and over as a percent of the total population is approximated by the function

$$f(t) = 10.72(0.9t + 10)^{0.3} \qquad 0 \leq t \leq 25$$

where $t$ is measured in years and $t = 0$ corresponds to the year 2000.
**a.** Find the rule for $f^{-1}$.
**b.** Evaluate $f^{-1}(25)$, and interpret your result.
*Source: U.S. Census Bureau.*

**58. Special Theory of Relativity** According to the special theory of relativity, the relativistic mass of a particle moving with speed $v$ is

$$m = f(v) = \dfrac{m_0}{\sqrt{1 - \dfrac{v^2}{c^2}}}$$

where $m_0$ is the rest mass (the mass at zero speed) and $c = 2.9979 \times 10^8$ m/sec is the speed of light in a vacuum.
**a.** Find $f^{-1}$, and interpret your result.
**b.** What is the speed of a particle when its relativistic mass is four times its rest mass?

**59.** Prove that if $f$ has an inverse, then $(f^{-1})^{-1} = f$.

**60.** Prove that a function has an inverse if and only if it is one-to-one.

*In Exercises 61–66, determine whether the statement is true or false. If it is true, explain why it is true. If it is false, explain why or give an example to show why it is false.*

**61.** If $f$ is one-to-one on $(-\infty, \infty)$, then $f^{-1}(f(a)) = a$ if $a$ is a real number.

**62.** The function $f(x) = 1/x^2$ has an inverse on any interval $(a, b)$, where $a < b$, not containing the origin.

**63.** If $f(x) = a_{2n+1}x^{2n+1} + a_{2n-1}x^{2n-1} + \cdots + a_1 x$, where $a_1$, $a_3, \ldots, a_{2n+1}$ are nonnegative numbers $(a_{2n+1} \neq 0)$, then $f^{-1}$ exists.

**64.** $\sin^{-1} x = \dfrac{1}{\sin x}$

**65.** $\cot^{-1} x = \dfrac{\cos^{-1} x}{\sin^{-1} x}$

**66.** $(\sin^{-1} x)^2 + (\cos^{-1} x)^2 = 1$

---

## 0.8 Exponential and Logarithmic Functions

---

**0.8 SELF-CHECK DIAGNOSTIC TEST**

**1.** Simplify the expression $e^{\ln x} + \ln e^{2x}$.

**2.** Solve the equation $\dfrac{200}{1 + 3e^{-0.3t}} = 100$.

**3.** Find the domain of $f(x) = \dfrac{e^{1/x}}{1 + \ln x}$.

**4.** Find the inverse of $f(x) = 2\ln(x + 1)$. What is its domain?

**5.** Express $\ln x + \dfrac{1}{2}\ln(x + 1) - \ln \cos x$ as a single logarithm.

*Answers to Self-Check Diagnostic Test 0.8 can be found on page ANS 7.*

---

### ■ Exponential Functions and Their Graphs

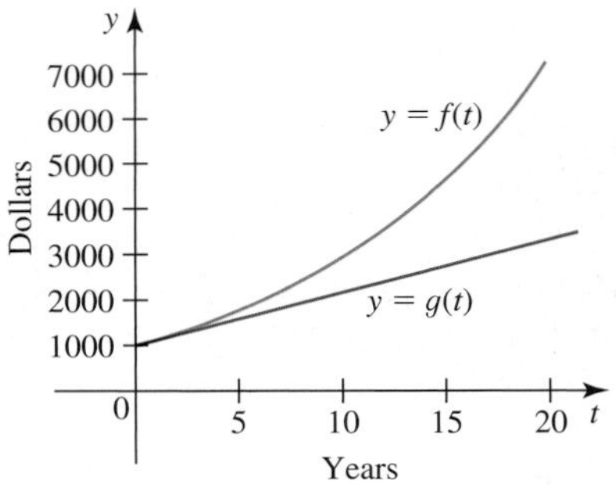

Suppose you deposit a sum of $1000 in an account earning interest at the rate of 10% per year *compounded continuously* (the way most financial institutions compute interest). Then, the accumulated amount at the end of $t$ years ($0 \le t \le 20$) is described by the function $f$, whose graph appears in Figure 1.* This function is called an *exponential function*. Observe that the graph of $f$ rises rather slowly at first but very rapidly as time goes by. For purposes of comparison, we have also shown the graph of the function $y = g(t) = 1000(1 + 0.10t)$, giving the accumulated amount for the same principal ($1000) but earning *simple* interest at the rate of 10% per year. The moral of the story: It is never too early to save.

Recall that if $a$ is a real number and $n$ is a positive integer, then

$$a^n = \underbrace{a \cdot a \cdots \cdot a}_{n \text{ factors}}$$

**FIGURE 1**
Under continuous compounding, a sum of money grows exponentially.

In the expression $a^n$, $a$ is called the base and $n$ is the exponent or power to which the base is raised. Also, by definition, $a^0 = 1$, and if $n$ is a positive integer, then

$$a^{-n} = \frac{1}{a^n}$$

---

*We will discuss simple and compound interest later in this section. Continuous compound interest will be discussed in Section 3.5.

If $p/q$ is a rational number, where $p$ and $q$ are integers with $q > 0$, then we define the expression $a^{p/q}$ with rational exponent by

$$a^{p/q} = \sqrt[q]{a^p} = (\sqrt[q]{a})^p$$

To define expressions with irrational exponents such as $2^{\sqrt{2}}$, we proceed as follows. Observe that $\sqrt{2} = 1.414213\ldots$. So $\sqrt{2}$ can be approximated successively and with increasing accuracy by the rational numbers

$$1.4, \quad 1.41, \quad 1.414, \quad 1.4142, \quad 1.41421, \quad 1.414213, \quad \ldots$$

Thus, we can expect that $2^{\sqrt{2}}$ may be approximated by the numbers

$$2^{1.4}, \quad 2^{1.41}, \quad 2^{1.414}, \quad 2^{1.4142}, \quad 2^{1.41421}, \quad 2^{1.414213}, \quad \ldots$$

In fact, from Table 1 we see that as $x$ approaches $\sqrt{2}$, the corresponding values of $2^x$ approach the number $2.665143\ldots$. It can be shown that this number is unique, and we define it to be $2^{\sqrt{2}}$. Furthermore, Table 1 suggests that correct to five decimal places,

$$2^{\sqrt{2}} \approx 2.66514$$

**TABLE 1**

| $x$ | 1.4 | 1.41 | 1.414 | 1.4142 | 1.41421 | 1.414213 |
|-----|-----|------|-------|--------|---------|----------|
| $2^x$ | 2.639015 ... | 2.657371 ... | 2.664749 ... | 2.665119 ... | 2.665137 ... | 2.665143 ... |

Similarly, we can define $2^x$, where $x$ is an irrational number. In fact, this procedure can be used to define $a^x$, where $a$ is any positive number and $x$ is an irrational number. In this manner, we see that the number $a^x$ can be defined for *all* real numbers $x$.

Computations involving exponentials are facilitated by the following laws of exponents.

---

### LAWS OF EXPONENTS

If $a$ and $b$ are positive numbers and $x$ and $y$ are real numbers, then

**a.** $a^x a^y = a^{x+y}$      **b.** $\dfrac{a^x}{a^y} = a^{x-y}$      **c.** $(a^x)^y = a^{xy}$

**d.** $(ab)^x = a^x b^x$      **e.** $\left(\dfrac{a}{b}\right)^x = \dfrac{a^x}{b^x}$

---

**EXAMPLE 1**

**a.** $(2^{1/3})(2^{3/5}) = 2^{(1/3)+(2/5)} = 2^{11/15}$      **b.** $\dfrac{3^{1/2}}{3^{1/3}} = 3^{(1/2)-(1/3)} = 3^{1/6}$

**c.** $(2^x)^3 = 2^{3x}$      **d.** $(4x^3)^{-1/2} = (4^{-1/2})(x^{-3/2}) = \dfrac{1}{2x^{3/2}}$      ■

Since the number $a^x$ ($a > 0$) is defined for all real numbers $x$, we can define a function $f$ with the rule given by

$$f(x) = a^x$$

where $a$ is a positive constant and $a \neq 1$. The domain of $f$ is $(-\infty, \infty)$. This function is called an **exponential function with base $a$.** Examples of exponential functions are

$$f(x) = 2^x, \quad f(x) = \left(\frac{1}{2}\right)^x, \quad \text{and} \quad f(x) = \pi^x$$

An alternative and more rigorous definition of exponential functions is given in Appendix C.

 Do not confuse an exponential function with a power function such as $f(x) = x^2$, encountered in Section 0.6. In the case of the power function the base is a variable, and its exponent is a constant.

**EXAMPLE 2**    Sketch the graphs of $f(x) = 2^x$, $g(x) = 3^x$, $h(x) = \left(\frac{1}{2}\right)^x$, and $F(x) = \left(\frac{1}{3}\right)^x$ on the same set of axes.

**Solution**    We first construct a table of values for each of the functions (see Table 2). With the help of Table 2 we obtain the graphs of $f$, $g$, $h$, and $F$ shown in Figure 2.

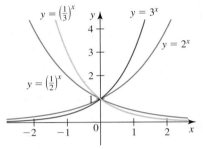

**FIGURE 2**
The graphs of $f(x) = 2^x$, $g(x) = 3^x$, $h(x) = \left(\frac{1}{2}\right)^x$, and $F(x) = \left(\frac{1}{3}\right)^x$

**TABLE 2**

| $x$ | $-4$ | $-3$ | $-2$ | $-1$ | 0 | 1 | 2 | 3 | 4 |
|---|---|---|---|---|---|---|---|---|---|
| $2^x$ | $\frac{1}{16}$ | $\frac{1}{8}$ | $\frac{1}{4}$ | $\frac{1}{2}$ | 1 | 2 | 4 | 8 | 16 |
| $3^x$ | $\frac{1}{81}$ | $\frac{1}{27}$ | $\frac{1}{9}$ | $\frac{1}{3}$ | 0 | 3 | 9 | 27 | 81 |
| $\left(\frac{1}{2}\right)^x$ | 16 | 8 | 4 | 2 | 1 | $\frac{1}{2}$ | $\frac{1}{4}$ | $\frac{1}{8}$ | $\frac{1}{16}$ |
| $\left(\frac{1}{3}\right)^x$ | 81 | 27 | 9 | 3 | 0 | $\frac{1}{3}$ | $\frac{1}{9}$ | $\frac{1}{27}$ | $\frac{1}{81}$ |

The graphs of $f(x) = 2^x$, $g(x) = 3^x$, $h(x) = \left(\frac{1}{2}\right)^x$, and $F(x) = \left(\frac{1}{3}\right)^x$ obtained in Example 2 are special cases of the graphs of $f(x) = a^x$, obtained by setting $a = 2, 3, \frac{1}{2}$, and $\frac{1}{3}$, respectively. In general, the exponential function $y = a^x$ with $a > 1$ has a graph similar to that of $y = 2^x$ or $y = 3^x$, whereas the graph of $y = a^x$ for $0 < a < 1$ is similar to that of $y = \left(\frac{1}{2}\right)^x$. If $a = 1$, then the function $y = a^x$ reduces to the constant function $y = 1$. The graphs of $y = a^x$ for each of these three cases are shown in Figure 3. Observe that all the graphs pass through the point $(0, 1)$ because $a^0 = 1$. Also, as suggested by Figure 2, the larger $a$ is ($a > 1$), the faster the graph of $f(x) = a^x$ rises for $x > 0$.

The properties of exponential functions are summarized below.

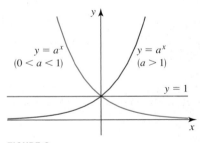

**FIGURE 3**
The graph of $y = a^x$ rises from left to right if $a > 1$, is constant if $a = 1$, and falls from left to right if $0 < a < 1$.

> **Properties of Exponential Functions**
>
> The exponential function $f(x) = a^x$ ($a > 0$, $a \neq 1$) has the following properties.
>
> 1. Its domain is $(-\infty, \infty)$.
> 2. Its range is $(0, \infty)$.
> 3. Its graph passes through the point $(0, 1)$.
> 4. Its graph rises from left to right on $(-\infty, \infty)$ if $a > 1$ and falls from left to right on $(-\infty, \infty)$ if $a < 1$.

**EXAMPLE 3**    Sketch the graph of the function $f(x) = 1 - 2^x$, and find its domain and range.

**Solution**    The required graph is obtained by first reflecting the graph of $y = 2^x$ (see Figure 4a) to obtain the graph of $y = -2^x$ (see Figure 4b) and then translating this graph upward by 1 unit. The resulting graph is shown in Figure 4c. The domain of $f$ is $(-\infty, \infty)$ and its range is $(-\infty, 1)$.

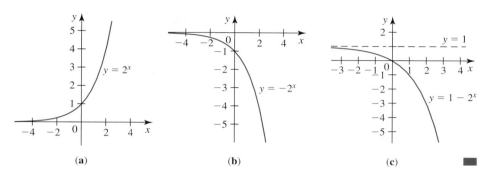

**FIGURE 4**    (a)    (b)    (c)

## ■ The Natural Exponential Function

Of all the possible choices for the base of an exponential function, there is one that plays an important role in calculus. This base, denoted by the letter $e$, is the irrational number whose value, correct to five decimal places, is given by

$$e \approx 2.71828$$

As you will see later on, the use of $e$ for the base of an exponential function enables us to express some of the formulas of calculus in the simplest form possible. The rationale for this choice of the base will be given in Section 2.2.

The function $f(x) = e^x$ is called the **natural exponential function.** Since the number $e$ lies between 2 and 3, we expect the graph of $y = e^x$ to lie between the graphs of $y = 2^x$ and $y = 3^x$, as we will see in Example 4. In the definition of the exponential function $f(x) = a^x$, the base $a$ can be any positive constant. But as mentioned earlier, the choice of $e$ as the base of the exponential function will lead to much simpler calculations in our work ahead. We will give a precise definition of $e$ in Section 2.8.

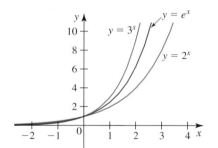

**FIGURE 5**
The graph of $y = e^x$ lies between the graphs of $y = 2^x$ and $y = 3^x$.

**EXAMPLE 4**  Sketch the graphs of $f(x) = 2^x$, $g(x) = e^x$, and $h(x) = 3^x$ on the same set of axes.

**Solution**  The values of $f(x)$ and $h(x)$ for selected values of $x$ were found in Example 2. With the aid of a calculator we obtain the following table. The graphs of $f$, $g$, and $h$ are shown in Figure 5.

| $x$ | $-3$ | $-2$ | $-1$ | 0 | 1 | 2 | 3 |
|---|---|---|---|---|---|---|---|
| $e$ | 0.05 | 0.14 | 0.37 | 1 | 2.72 | 7.39 | 20.09 |

## ■ Compound Interest

An important application of exponential functions is found in computations involving interest—charges on borrowed money.

*Simple interest* is interest that is computed on the original principal only. Thus, if $I$ denotes the interest on a principal $P$ (in dollars) at an interest rate of $r$ per year for $t$ years, then

$$I = Prt$$

The **accumulated amount** $A$, the sum of the principal and interest after $t$ years, is given by

$$A = P + I = P + Prt$$

$$= P(1 + rt) \tag{1}$$

and is a linear function of $t$.

In contrast to simple interest, earned interest that is periodically added to the principal and thereafter itself earns interest at the same rate is called **compound interest.** To find a formula for the accumulated amount, suppose that $P$ dollars (the principal) is deposited in a bank for a term of $t$ years, earning interest at the rate of $r$ per year (called the **nominal** or **stated rate**) compounded annually. Then, using Equation (1), we see that the accumulated amount at the end of the first year is

$$A_1 = P(1 + rt)$$

To find the accumulated amount $A_2$ at the end of the second year, we use Equation (1) again, this time with $P = A_1$, since the principal *and* interest now earn interest over the second year. We obtain

$$A_2 = A_1(1 + rt) = P(1 + rt)(1 + rt) = P(1 + rt)^2$$

Continuing, we see that the accumulated amount $A$ after $t$ years is

$$A = P(1 + r)^t \tag{2}$$

Equation (2) was derived under the assumption that interest was compounded *annually*. In practice, however, interest is usually compounded more than once a year. The interval of time between successive interest calculations is called the **conversion period.**

If interest at a nominal rate of $r$ per year is compounded $m$ times a year on a principal of $P$ dollars, then the simple interest rate per conversion period is

$$i = \frac{r}{m} \qquad \frac{\text{annual interest rate}}{\text{number of periods per year}}$$

For example, if the nominal rate is 8% per year ($r = 0.08$) and interest is compounded quarterly ($m = 4$), then

$$i = \frac{r}{m} = \frac{0.08}{4} = 0.02$$

or 2% per period.

To find a general formula for the accumulated amount when a principal of $P$ dollars is deposited in a bank for a term of $t$ years and earns interest at the (nominal) rate of $r$ per year compounded $m$ times per year, we proceed as before, using Equation (2) repeatedly with the interest rate $i = r/m$. We see that the accumulated amount at the end of each period is as follows:

First period: $\quad A_1 = P(1 + i)$

Second period: $\quad A_2 = A_1(1 + i) = [P(1 + i)](1 + i) = P(1 + i)^2$

$\qquad \vdots \qquad\qquad\qquad\qquad \vdots$

$n$th period: $\quad A_n = A_{n-1}(1 + i) = [P(1 + i)^{n-1}](1 + i) = P(1 + i)^n$

But there are $n = mt$ periods in $t$ years (number of conversion periods times the term). Therefore, the accumulated amount at the end of $t$ years is given by

$$A = P\left(1 + \frac{r}{m}\right)^{mt} \tag{3}$$

TABLE 3   The accumulated amount $A$ after 3 years when interest is converted $m$ times/year

| $m$ | $A$ (dollars) |
|-----|---------------|
| 1 | 1259.71 |
| 2 | 1265.32 |
| 4 | 1268.24 |
| 12 | 1270.24 |
| 365 | 1271.22 |

**EXAMPLE 5**   Find the accumulated amount after 3 years if $1000 is invested at 8% per year compounded annually, semiannually, quarterly, monthly, and daily. (Assume that there are 365 days in a year.)

**Solution**   We use Equation (3) with $P = 1000$, $r = 0.08$, and $m = 1, 2, 4, 12$, and 365 in succession to obtain the results summarized in Table 3.   ▪

## ■ Logarithmic Functions

If you examine the graph of the exponential function $f(x) = a^x$ where $a > 0$ and $a \neq 1$ (see Figure 3), you will see that it passes the horizontal line test, and so the function $f$ is one-to-one and therefore possesses an inverse function $f^{-1}$. This function is called the **logarithmic function with base $a$.** The graph of $f^{-1}(x) = \log_a x$ is obtained by reflecting the graph of $f(x) = a^x$ about the line $y = x$. The graph of $y = \log_a x$ for the case $a > 1$ is given in Figure 6.

The function $f^{-1}$ is called the **logarithmic function with base $a$** and is denoted by $\log_a$. Using the definition of an inverse function given in Section 0.7,

$$f^{-1}(x) = y \qquad \text{if and only if} \qquad f(y) = x$$

we are led to the following:

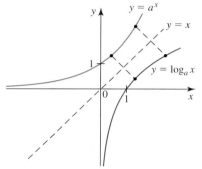

**FIGURE 6**
The graphs of $f^{-1}(x) = \log_a x$ and $f(x) = a^x$ are mirror reflections about the line $y = x$.

> $$\log_a x = y \qquad \text{if and only if} \qquad a^y = x$$

Thus, if $x > 0$, then $\log_a x$ is the exponent to which $a$ must be raised to obtain $x$. Also because $f(x) = a^x$ and $g(x) = \log_a x$ are inverses of each other, we have

> $$a^{\log_a x} = x \qquad \text{for all } x \text{ in } (0, \infty)$$
>
> and
>
> $$\log_a(a^x) = x \qquad \text{for all } x \text{ in } (-\infty, \infty)$$

A summary of the properties of logarithmic functions follows.

**Properties of Logarithmic Functions**

The logarithmic function $f(x) = \log_a x$ ($a > 0$, $a \neq 1$) has the following properties.

1. Its domain is $(0, \infty)$.
2. Its range is $(-\infty, \infty)$.
3. Its graph passes through the point $(1, 0)$.
4. Its graph rises from left to right on $(0, \infty)$ if $a > 1$ and falls from left to right if $a < 1$.

The graphs of $y = \log_a x$ for different bases $a$ are shown in Figure 7.

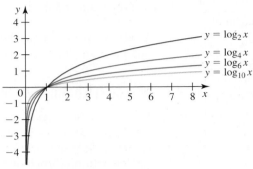

**FIGURE 7**
The graphs of $y = \log_a x$ for $a = 2, 4, 6,$ and 10

As in the case of exponentials, computations involving logarithms are facilitated by the following laws of logarithms. (These laws are proved in Appendix C.)

---

**LAWS OF LOGARITHMS**

If $x$ and $y$ are positive numbers, then

**a.** $\log_a xy = \log_a x + \log_a y$

**b.** $\log_a \dfrac{x}{y} = \log_a x - \log_a y$

**c.** $\log_a x^r = r \log_a x$    where $r$ is any real number

**d.** $\log_a 1 = 0$

**e.** $\log_a a = 1$

---

**EXAMPLE 6**    Use the laws of logarithms to evaluate $\log_2 40 - \log_2 5$.

**Solution**    We have

$$\log_2 40 - \log_2 5 = \log_2 \frac{40}{5} \qquad \text{Use Law b.}$$

$$= \log_2 8 = \log_2 2^3$$

$$= 3 \log_2 2 \qquad \text{Use Law c.}$$

$$= 3(1) = 3 \qquad \text{Use Law e.} \qquad \blacksquare$$

Before turning to another example, we mention that the two widely used systems of logarithms are the system of **common logarithms,** which uses the number 10 as the base, and the system of **natural logarithms,** which uses the number $e$ as the base. It is standard practice to write **log** for $\log_{10}$ and **ln** for $\log_e$. As in the case of exponentials, the use of natural logarithms rather than logarithms with other bases leads to simpler expressions.

**EXAMPLE 7**   Expand and simplify the following expressions:

**a.** $\log_2 \dfrac{x^2 + 1}{2^x}$        **b.** $\ln \dfrac{x^2\sqrt{x^2 - 1}}{e^x}$

**Solution**

**a.** $\log_2 \dfrac{x^2 + 1}{2^x} = \log_2 (x^2 + 1) - \log_2 2^x$        Use Law b.

$\qquad\qquad\quad = \log_2 (x^2 + 1) - x \log_2 2$        Use Law c.

$\qquad\qquad\quad = \log_2 (x^2 + 1) - x$        Use Law e.

**b.** $\ln \dfrac{x^2\sqrt{x^2 - 1}}{e^x} = \ln \dfrac{x^2(x^2 - 1)^{1/2}}{e^x}$        Rewrite.

$\qquad\qquad\quad = \ln x^2 + \dfrac{1}{2}\ln(x^2 - 1) - x \ln e$        Use Laws a, b, and c.

$\qquad\qquad\quad = 2 \ln x + \dfrac{1}{2}\ln(x^2 - 1) - x \ln e$        Use Law c.

$\qquad\qquad\quad = 2 \ln x + \dfrac{1}{2}\ln(x^2 - 1) - x$        Use Law e.   ■

## Properties Relating the Natural Exponential and the Natural Logarithmic Functions

The following properties follow as an immediate consequence of the definition of the natural logarithm of a number.

---

**Properties Relating $e^x$ and $\ln x$**

$$e^{\ln x} = x \qquad x > 0 \tag{4}$$

$$\ln e^x = x \qquad \text{for any real number} \tag{5}$$

---

The relationships expressed in Equations (4) and (5) are useful in solving equations that involve exponentials and logarithms.

**EXAMPLE 8**   Solve the equation $2e^{x+2} = 5$.

**Solution**   We first divide both sides of the equation by 2 to obtain

$$e^{x+2} = \frac{5}{2} = 2.5$$

Next, taking the natural logarithm of each side of the equation and using Equation (5), we have

$$\ln e^{x+2} = \ln 2.5$$

$$x + 2 = \ln 2.5$$

$$x = -2 + \ln 2.5 \approx -1.08 \qquad\qquad ■$$

**EXAMPLE 9**    Solve the equation $2 \ln(3x - 5) = 15$.

**Solution**    We have

$$2 \ln(3x - 5) = 15$$

$$\ln(3x - 5) = 7.5$$

$$3x - 5 = e^{7.5}$$

$$x = \frac{1}{3}(e^{7.5} + 5) \approx 604.347$$    ■

## ■ Change of Base Formula

As we mentioned earlier, it is sometimes preferable to use one base rather than another when solving a problem. More specifically, we mentioned that we often use natural logarithms to simplify formulas in calculus. The following formula enables us to write the logarithms with any base in terms of natural logarithms.

---

**Change of Base Formula**

If $a$ is a positive number and $a \neq 1$, then

$$\log_a x = \frac{\ln x}{\ln a}$$

---

**PROOF**    Let $y = \log_a x$. Then $x = a^y$. Taking the natural logarithm of both sides of this equation gives $\ln x = \ln a^y = y \ln a$, and so, solving for $y$, we obtain

$$y = \frac{\ln x}{\ln a}$$

and this proves the result.    ■

**EXAMPLE 10**    Evaluate $\log_9 7$ correct to five decimal places.

**Solution**    We have

$$\log_9 7 = \frac{\ln 7}{\ln 9} \approx 0.88562$$    ■

## 0.8    CONCEPT QUESTIONS

1. Define the number $e$. What is its approximate value?
2. Define the natural exponential function $f(x) = e^x$. What are its domain and range?
3. State the laws of exponents.
4. What is the relationship between the graph of $f(x) = e^x$ and that of $g(x) = \ln x$? Sketch the graphs on the same set of axes.

5. Define the natural logarithmic function $f(x) = \ln x$. What are its domain and range?
6. State the laws of logarithms.

# 0.8  EXERCISES

*In Exercises 1–4, given that* $\ln 2 \approx 0.6931$, $\ln 3 \approx 1.0986$, *and* $\ln 5 \approx 1.6094$, *use the laws of logarithms to approximate each expression.*

**1. a.** $\ln 6$  **b.** $\ln \dfrac{3}{2}$

**2. a.** $\ln \dfrac{20}{\sqrt{3}}$  **b.** $\ln\left(\dfrac{15}{2}\right)^{1/3}$

**3. a.** $\ln 30$  **b.** $\ln 7.5$

**4. a.** $\ln \dfrac{1}{125}$  **b.** $\ln \dfrac{5}{9}$

*In Exercises 5–10, use the laws of logarithms to expand the expression.*

**5.** $\ln \dfrac{2\sqrt{3}}{5}$  **6.** $\ln \dfrac{xy}{z}$

**7.** $\ln \dfrac{x^{1/3}y^{2/3}}{z^{1/2}}$  **8.** $\ln\left(x^2\sqrt{x^2 + 1}\right)$

**9.** $\ln\left(\dfrac{x + 1}{x - 1}\right)^{1/3}$  **10.** $\ln\left[\sqrt{x}\,|\cos x|\,(x + 1)^{-1/3}\right]$

*In Exercises 11–14, use the laws of logarithms to write the expression as the logarithm of a single quantity.*

**11.** $\ln 4 + \ln 6 - \ln 12$  **12.** $\ln(x^2 - 1) - 2\ln(x + 1)$

**13.** $3\ln 2 - \dfrac{1}{2}\ln(x + 1)$

**14.** $\dfrac{1}{2}\left[2\ln(x + 1) + \ln x - \ln(x - 1)\right]$

*In Exercises 15–18, simplify the expression.*

**15. a.** $\ln e^3$  **b.** $\ln e^{x^2}$

**16. a.** $\ln \sqrt{e}$  **b.** $\ln e^{\sqrt{e}}$

**17. a.** $e^{2\ln 3}$  **b.** $e^{\ln \sqrt{x}}$

**18. a.** $\ln e^{x^2 + 1}$  **b.** $e^{2\ln x + \cos x}$

*In Exercises 19–26, find the domain of the function.*

**19.** $f(x) = xe^{-x}$  **20.** $g(x) = \dfrac{x}{1 - e^x}$

**21.** $h(t) = \sqrt{2^t - 1}$  **22.** $f(x) = \sin^{-1}(|x| - 3)$

**23.** $f(x) = \ln(2x + 1)$  **24.** $g(x) = \ln(-x)$

**25.** $g(x) = \ln(\cos x)$  **26.** $h(x) = \ln\left(\dfrac{x + 1}{x - 1}\right)$

*In Exercises 27–32, solve the equation.*

**27. a.** $e^{\ln x} = 2$  **b.** $\ln e^{-2x} = 3$

**28. a.** $\ln(2x + 1) = 3$  **b.** $\ln x^2 = 5$

**29. a.** $2e^{x+2} = 5$  **b.** $\ln \sqrt{x + 1} = 1$

**30. a.** $\ln x + \ln(x - 1) = \ln 2$  **b.** $2e^{-0.2x} - 2 = 8$

**31. a.** $\dfrac{50}{1 + 4e^{0.2x}} = 20$  **b.** $e^{2x} - 5e^x + 6 = 0$

**32. a.** $\ln(x + \sqrt{x^2 + 1}) = 2$  **b.** $x^{1/\ln x} - x^2 + 1 = 0$

*In Exercises 33–36, determine whether f is even, odd, or neither even nor odd.*

**33.** $f(x) = \dfrac{1 + e^{kx}}{1 - e^{kx}}$  **34.** $f(x) = \dfrac{2^x}{(1 + 2^x)^2}$

**35.** $f(x) = \ln \dfrac{1 - x}{1 + x}$  **36.** $f(x) = \ln(x + \sqrt{1 + x^2})$

*In Exercises 37–42, use the graph of* $y = \ln x$ *as an aid to sketch the graph of the function.*

**37.** $f(x) = 2\ln x$  **38.** $g(x) = -\ln x$

**39.** $y = 1 + \ln x$  **40.** $f(x) = \ln 2x$

**41.** $g(x) = \ln(x + 1)$  **42.** $h(x) = \ln|x|$

 **43. a.** Plot the graphs of $f(x) = \ln x + \ln(x - 1)$ and $g(x) = \ln x(x - 1)$ using the same viewing window.
  **b.** For what values of $x$ is $f = g$? Prove your assertion.

 **44. a.** For what values of $x$ is $f = g$ if $f(x) = \ln\sqrt{x/(x - 1)}$ and $g(x) = \frac{1}{2}[\ln x - \ln(x - 1)]$?
  **b.** Verify the result of part (a) graphically by plotting the graphs of $f$ and $g$.

*In Exercises 45–48, show that the functions are inverses of each other. Sketch the graphs of each pair of functions on the same set of axes.*

**45.** $f(x) = e^{2x}$  and  $g(x) = \ln \sqrt{x}$

**46.** $f(x) = e^{-x}$  and  $g(x) = -\ln x$

**47.** $f(x) = e^{x/2}$  and  $g(x) = 2\ln x$

**48.** $f(x) = e^{x-1}$  and  $g(x) = 1 + \ln x$

 *In Exercises 49–52, find the inverse of f. Then use a graphing utility to plot the graphs of f and* $f^{-1}$ *on the same set of axes.*

**49.** $f(x) = e^x + 1$  **50.** $f(x) = \ln(2x + 3)$

**51.** $f(x) = \dfrac{e^x + 1}{e^x - 1}$  **52.** $f(x) = 2^{\ln x}$

**53. a.** Plot the graph of $f(x) = \tan^{-1}(\tan x)$ using the viewing window $[-10, 10] \times [-2, 2]$.
  **b.** Is $f$ periodic? Prove your assertion.

**54.** Sketch the graph of $f(x) = x^{1/\log x}$.

**55.** Are the functions $f(x) = x$ and $g(x) = e^{\ln x}$ identical? Explain.

56. **Over-100 Population** On the basis of data obtained from the Census Bureau, the number of Americans over age 100 years is expected to be

$$P(t) = 0.07e^{0.54t} \qquad 0 \le t \le 4$$

where $P(t)$ is measured in millions and $t$ is measured in decades, with $t = 0$ corresponding to the beginning of 2000. What was the population of Americans over age 100 years at the beginning of 2000? What will it be at the beginning of 2030?

*Source*: U.S. Census Bureau.

57. **World Population Growth** After its fastest rate of growth ever during the 1980s and 1990s, the rate of growth of world population is expected to slow dramatically, in the twenty-first century. The function

$$G(t) = 1.58e^{-0.213t}$$

gives the projected average percent population growth/decade in the $t$th decade, with $t = 1$ corresponding to the beginning of 2000. What will the projected average percent population growth rate be at the beginning of 2020 ($t = 3$)?

*Source*: U.S. Census Bureau.

58. **Epidemic Growth** During a flu epidemic the number of children in the Woodhaven Community School System who contracted influenza by the $t$th day is given by

$$N(t) = \frac{5000}{1 + 99e^{-0.8t}}$$

($t = 0$ corresponds to the date when data were first collected.) How many students were stricken by the flu on the first day?

59. **Blood Alcohol Level** The percentage of alcohol in a person's bloodstream $t$ hr after drinking 8 fluid oz of whiskey is given by

$$A(t) = 0.23te^{-0.4t} \qquad 0 \le t \le 12$$

What is the percentage of alcohol in a person's bloodstream after $\frac{1}{2}$ hr? After 8 hr?

*Source*: Encyclopedia Britannica.

60. **Von Bertalanffy Functions** The mass $W(t)$ (in kilograms) of the average female African elephant at age $t$ (in years) can be approximated by a *von Bertalanffy function*

$$W(t) = 2600(1 - 0.51e^{-0.075t})^3$$

   a. What is the mass of a newborn female elephant?
   b. If a female elephant has a mass of 1600 kg, what is her approximate age?

61. **Death Due to Strokes** Before 1950, little was known about strokes. By 1960, however, risk factors such as hypertension had been identified. In recent years, CAT scans used as a diagnostic tool have helped to prevent strokes. As a result, deaths due to strokes have fallen dramatically. The function

$$N(t) = 130.7e^{-0.1155t^2} + 50 \qquad 0 \le t \le 6$$

gives the number of deaths per 100,000 people from the beginning of 1950 to the beginning of 2010, where $t$ is measured in decades, with $t = 0$ corresponding to the beginning of 1950.
   a. How many deaths due to strokes per 100,000 people were there at the beginning of 1950?
   b. If the trend continues, how many deaths due to strokes per 100,000 people will there be at the beginning of 2010?

*Source*: American Heart Association, Centers for Disease Control and Prevention, and National Institutes of Health.

62. **Length of Fish** The length (in centimeters) of a typical Pacific halibut $t$ years old is approximately

$$f(t) = 200(1 - 0.956e^{-0.18t})$$

   a. Plot the graph of $f$ using the viewing window $[0, 20] \times [0, 200]$. What is the maximum length that a typical Pacific halibut can attain?
   b. What is the approximate length of a typical 10-year-old Pacific halibut?

63. **Annuities** At the time of retirement, Christine expects to have a sum of $500,000 in her retirement account. Her accountant pointed out to her that if she made withdrawals in monthly installments amounting to $x$ dollars per year ($x > 25,000$), assuming that the account earns interest at the rate of 5% per year compounded continuously, then the time required to deplete her savings would be $T$ years, where

$$T = f(x) = 20 \ln\left(\frac{x}{x - 25,000}\right) \qquad x > 25,000$$

   a. Plot the graph of $f$, using the viewing window $[25,000, 50,000] \times [0, 100]$.
   b. How much should Christine plan to withdraw from her retirement account each year if she wants it to last for 25 years?

64. **Growth of a Tumor** The rate at which a tumor grows with respect to time is given by

$$R = Ax \ln \frac{B}{x}$$

for $0 < x < B$, where $A$ and $B$ are positive constants and $x$ is the radius of the tumor. Plot the graph of $R$ for the case $A = B = 10$.

65. **Atmospheric Pressure** In the troposphere (lower part of the atmosphere), the atmospheric pressure $p$ is related to the height $y$ from the earth's surface by the equation

$$\ln\left(\frac{p}{p_0}\right) = \frac{Mg}{R\alpha} \ln\left(\frac{T_0 - \alpha y}{T_0}\right)$$

where $p_0$ is the pressure at the earth's surface, $T_0$ is the temperature at the earth's surface, $M$ is the molecular mass for air, $g$ is the constant of acceleration due to gravity, $R$ is the ideal gas constant, and $\alpha$ is called the lapse rate of temperature. Find $p$ for $y = 6194$ m (the altitude at the summit of Mount McKinley), taking $M = 28.8 \times 10^{-3}$ kg/mol, $T_0 = 300$ K, $g = 9.8$ m/sec$^2$, $R = 8.314$ J/mol $\cdot$ K, and $\alpha = 0.006$ K/m. Explain why mountaineers experience difficulty in breathing at very high altitudes.

66. **A Sliding Chain** A chain of length 6 m is held on a table with 1 m of the chain hanging down from the table. Upon release, the chain slides off the table. Assuming that there is no friction, the end of the chain that initially was 1 m from the edge of the table is given by the function

$$s(t) = \frac{1}{2}\left(e^{\sqrt{g/6}\,t} + e^{-\sqrt{g/6}\,t}\right)$$

where $g = 9.8$ m/sec$^2$ and $t$ is measured in seconds. Find the time it takes for the end of the chain to move 1 m.

67. **Increase in Juvenile Offenders** The number of youths aged 15 to 19 years increased by 21% between 1994 and 2005, pushing up the crime rate. According to the National Council on Crime and Delinquency, the number of violent crime arrests of juveniles under age 18 in year $t$ is given by

$$f(t) = -0.438t^2 + 9.002t + 107 \qquad 0 \le t \le 13$$

where $f(t)$ is measured in thousands and $t$ in years, with $t = 0$ corresponding to the beginning of 1989. According to the same source, if trends such as inner-city drug use and wider availability of guns continues, then the number of violent crime arrests of juveniles under age 18 in year $t$ will be given by

$$g(t) = \begin{cases} -0.438t^2 + 9.002t + 107 & \text{if } 0 \le t < 4 \\ 99.456e^{0.07824t} & \text{if } 4 \le t \le 13 \end{cases}$$

where $g(t)$ is measured in thousands and $t = 0$ corresponds to the beginning of 1989. Compute $f(11)$ and $g(11)$, and interpret your results.
*Source:* National Council on Crime and Delinquency.

68. **Percent of Females in the Labor Force** Based on data from the U.S. Census Bureau, the following model giving the percent of the total female population in the civilian labor force, $P(t)$, at the beginning of the $t$th decade ($t = 0$ corresponds to the year 1900) was constructed.

$$P(t) = \frac{74}{1 + 2.6e^{-0.166t + 0.04536t^2 - 0.0066t^3}} \qquad 0 \le t \le 12$$

What was the percent of the total female population in the civilian labor force at the beginning of 2010?
*Source:* U.S. Census Bureau.

69. **An Extinction Situation** The number of saltwater crocodiles in a certain area of northern Australia $t$ years from now is given by

$$P(t) = \frac{300e^{-0.024t}}{5e^{-0.024t} + 1}$$

a. How many crocodiles were in the population initially?
b. Plot the graph of $P$ in the viewing window $[0, 200] \times [0, 70]$.
**Note:** This phenomenon is referred to as an extinction situation.

70. **Income of American Families** On the basis of data from the Census Bureau, it is estimated that the number of American families $y$ (in millions) who earned $x$ thousand dollars in 1990 is given by the equation

$$y = 0.1584xe^{-0.0000016x^3 + 0.00011x^2 - 0.04491x} \qquad x > 0$$

Plot the graph of the equation in the viewing window $[0, 150] \times [0, 2]$.
*Source:* House Budget Committee, House Ways and Means Committee, and U.S. Census Bureau.

71. Find the accumulated amount after 5 years on an investment of $5000 earning interest at the rate of 10% per year compounded (a) annually, (b) semiannually, (c) quarterly, (d) monthly, and (e) daily.

72. Find the accumulated amount after 10 years on an investment of $10,000 earning interest at the rate of 12% per year compounded (a) annually, (b) semiannually, (c) quarterly, (d) monthly, and (e) daily.

73. **Pension Funds** The managers of a pension fund have invested $1.5 million in U.S. government certificates of deposit that pay interest at the rate of 5.5% per year compounded semiannually over a period of 10 years. At the end of this period, how much will the investment be worth?

74. **Retirement Funds** Five and a half years ago, Chris invested $10,000 in a retirement fund that grew at the rate of 10.82% per year compounded quarterly. What is his account worth today?

*In Exercises 75–80, determine whether the statement is true or false. If it is true, explain why it is true. If it is false, give an example to show why it is false.*

75. The inverse of $f(x) = e^{x/2}$ is $f^{-1}(x) = 2 \ln x$.

76. $f(x) = \dfrac{\cos x}{e^x}$ is not defined at $x = 0$.

77. $e^{3 \ln x} = x^3$ on $(0, \infty)$

78. $\ln a - \ln b = \ln(a - b)$ for all positive numbers $a > b > 0$.

79. $(\ln x)^3 = 3 \ln x$ for all $x$ in $(0, \infty)$.

80. The domain of $f(x) = \ln|x|$ is $(-\infty, 0) \cup (0, \infty)$.

# CHAPTER 0 REVIEW

## REVIEW EXERCISES

*In Exercises 1–4, find the slope of the line satisfying the given condition.*

**1.** Passes through the points $(-1, 3)$ and $(2, -4)$

**2.** Has the same slope as the line $2x + 3y = 8$

**3.** Has the same slope as the line perpendicular to the line $-2x + 4y = -6$

**4.** Has an angle of inclination of $120°$

*In Exercises 5–10, find an equation of the line satisfying the conditions.*

**5.** Passes through $(-2, -4)$ and is parallel to the $x$-axis

**6.** Passes through $(1, 3)$ and has slope $-4$

**7.** Passes through $(-2, 3)$ and $(4, -5)$

**8.** Passes through $(2, 3)$ and is parallel to the line $3x + 4y - 8 = 0$

**9.** Passes through $(-1, 3)$ and is parallel to the line passing through the points $(-3, 4)$ and $(2, 1)$

**10.** Passes through $(-2, -4)$ and is perpendicular to the line $2x - 3y - 24 = 0$

**11.** Find an equation of the line passing through the point $(2, -1)$ and the point of intersection of the lines $x + 2y = 3$ and $2x - 3y = 13$.

**12. Dial-up Internet Households** The number of U.S. dial-up Internet households stood at 42.5 million at the beginning of 2004 and was projected to decline at the rate of 3.9 million households per year for the next 6 years.
   **a.** Find a linear function $f$ giving the projected U.S. dial-up Internet households (in millions) in year $t$, where $t = 0$ corresponds to the beginning of 2004.
   **b.** What is the projected number of U.S. dial-up Internet households at the beginning of 2010?
   *Source:* Strategy Analytics, Inc.

**13. Satellite TV Subscribers** The following table gives the number of satellite TV subscribers in the United States (in millions) from 1998 through 2005 ($x = 0$ corresponds to 1998).

| Year, $x$ | 0 | 1 | 2 | 3 | 4 | 5 | 6 | 7 |
|-----------|-----|------|------|------|------|------|------|------|
| Number, $y$ | 8.5 | 11.1 | 15.0 | 17.0 | 18.9 | 21.5 | 24.8 | 27.4 |

   **a.** Plot the number of satellite TV subscribers in the United States ($y$) versus the year ($x$).

   **b.** Draw the line $L$ through the points $(0, 8.5)$ and $(7, 27.4)$.
   **c.** Find an equation of the line $L$.
   **d.** Assuming that this trend continues, estimate the number of satellite TV subscribers in the United States in 2006.
   *Sources:* National Cable & Telecommunications Association, Federal Communications Commission.

**14.** If $f(x) = x^2 + x + 1$, find and simplify $\dfrac{f(x + h) - f(x)}{h}$.

**15.** If $f(x) = \tan x$, find $f(0)$, $f\left(\frac{\pi}{6}\right)$, $f\left(\frac{\pi}{4}\right)$, $f\left(\frac{\pi}{3}\right)$, and $f(\pi)$.

**16.** Let $f(x) = \begin{cases} \sqrt{-x} & \text{if } x \leq 0 \\ x^2 + x & \text{if } x > 0 \end{cases}$
   Find
   **a.** $f(-4)$
   **b.** $f(1)$
   **c.** $\dfrac{f(-1 - h) - f(-1)}{h}$, $\quad h > 0$
   **d.** $\dfrac{f(2 + h) - f(2)}{h}$, $\quad h > 0$

*In Exercises 17–23, find the domain of the function.*

**17.** $f(x) = \dfrac{x}{x^2 - 4}$

**18.** $g(x) = \sqrt{x^2 - 4}$

**19.** $h(x) = \dfrac{\sqrt{x - 1}}{x(x - 2)}$

**20.** $f(x) = \sec \pi x$

**21.** $f(x) = \dfrac{\sin x}{2 - \cos x}$

**22.** $f(x) = \tan^{-1}(e^{-x})$

**23.** $h(t) = \ln(e^t - 1)$

*In Exercises 24–25, find the domain and sketch the graph of the function. What is its range?*

**24.** $f(x) = \sqrt{1 - x}$

**25.** $g(t) = |\sin t| + 1$

*In Exercises 26–29, determine whether the function is even, odd, or neither.*

**26.** $f(x) = -3x^7 + 4x^3 - 2x$

**27.** $g(x) = \dfrac{\sin x}{x}$, $\quad x \neq 0$

**28.** $f(x) = x\dfrac{e^x + 1}{e^x - 1}$

**29.** $f(x) = x^3 \ln\left(\dfrac{1 + x}{1 - x}\right)$, $\quad -1 < x < 1$

**30.** Convert the angle to radian measure.
    **a.** 120°      **b.** 450°      **c.** −225°

**31.** Convert the angle to degree measure.
    **a.** $\dfrac{11\pi}{6}$ radians  **b.** $-\dfrac{5\pi}{2}$ radians  **c.** $-\dfrac{7\pi}{4}$ radians

**32.** If $f(x) = \cos x$, find $f(0), f\left(\frac{\pi}{4}\right), f\left(-\frac{\pi}{4}\right), f(3\pi)$ and $f\left(a + \frac{\pi}{2}\right)$.

**33.** Find all values of $\theta$ that satisfy the equation over the interval $[0, 2\pi)$.
    **a.** $\cos \theta = \dfrac{1}{2}$      **b.** $\cot \theta = -\sqrt{3}$

**34.** Verify the identity.
    **a.** $(\sec \theta + \tan \theta)(1 - \sin \theta) = \cos \theta$
    **b.** $\dfrac{\sec \theta - \cos \theta}{\tan \theta} = \sin \theta$

**35.** Find the solutions of the equation in $[0, 2\pi)$.
    **a.** $\cot^2 x - \cot x = 0$    **b.** $\sin x + \sin 2x = 0$

**36.** If $f(x) = 2x + 3$ and $g(x) = \dfrac{x}{2x^2 - 1}$, find the functions $f + g, g - f, fg, f/g,$ and $g/f$.

**37.** Find $g \circ f$ if $f(x) = x^2 - 1$ and $g(x) = \sqrt{x + 1}$. What is its domain?

**38.** Find functions $f$ and $g$ such that $h = g \circ f$, where $h(x) = \cos^2(\pi x)$.

**39.** Find functions $f$, $g$, and $h$ such that $F = f \circ g \circ h$ if $F(x) = \cos^2(1 + \sqrt{x + 2})$.

**40.** If $f(x) = 2x$ and $h(x) = 4x^2 - 1$, find $g$ such that $h = g \circ f$.

*In Exercises 41–54, solve the equation for x.*

**41.** $\ln x = \dfrac{2}{5}$             **42.** $e^x = 3$

**43.** $\log_3 x = 2$         **44.** $\log_8(x - 3) = \dfrac{2}{3}$

**45.** $e^{\sqrt{x}} = 4$           **46.** $e^{x^2} = 15$

**47.** $2 + 3e^{-x} = 6$      **48.** $\ln x = -1 + \ln(x + 2)$

**49.** $\ln x + \ln(x - 2) = 0$  **50.** $\dfrac{50}{1 + 4e^{0.2x}} = 20$

**51.** $3^{2x} - 12 \cdot 3^x + 27 = 0$  **52.** $\ln x^e = 2$

**53.** $\tan^{-1} x = 1$       **54.** $\cos^{-1}(\sin x) = 0$

*In Exercises 55 and 56, solve the equation for x in terms of y.*

**55.** $y = e^{2x} + 2$      **56.** $y = \dfrac{e^x - e^{-x}}{2}$

*In Exercises 57 and 58, expand the expression. Assume all variables are positive.*

**57.** $\ln x^3 \sqrt{y/z^2}$      **58.** $\ln \dfrac{\sqrt{x}}{y\sqrt[3]{x^2 + y^2}}$

*In Exercises 59 and 60, write the expression as a single logarithm.*

**59.** $2 \ln x + \ln \dfrac{x^3}{y^2} - 4 \ln \sqrt{x + y}$

**60.** $3 \ln x - \dfrac{1}{3} \ln(yz) + 6 \ln \sqrt{xy}$

*In Exercises 61–66, use a transformation to sketch the graph of the function.*

**61.** $y = x^3 - 2$           **62.** $y = 3(x + 2)^2$

**63.** $y = 2 - \sqrt{x}$         **64.** $y = \dfrac{1}{x + 1}$

**65.** $y = 3 \cos \dfrac{x}{2}$        **66.** $y = |\sin x|$

**67.** Plot the graph of $f(x) = x^5 - 3x^2 + x - 1$.

**68.** Plot the graph of $f(x) = x^3 - 0.01x^2$.

**69.** Use a calculator or computer to find the zeros of $f(x) = x^5 - 4x^3 + x^2 - x + 1$ accurate to five decimal places.

**70.** Find the point(s) of intersection of the graphs of $f(x) = \cos^2 x$ and $g(x) = 0.1x^2$ accurate to five decimal places.

**71.** Find the zero(s) of $f(x) = 2x^5 - 3x^3 + x^2 - 2$ accurate to five decimal places.

**72.** Find the point(s) of intersection of the graphs of $f(x) = \sin 2x$ and $g(x) = 3x^2 - 2$ accurate to four decimal places.

**73. Clark's Rule** Clark's Rule is a method for calculating pediatric drug dosages on the basis of a child's weight. If $a$ denotes the adult dosage (in milligrams) and $w$ is the weight of the child (in pounds), then the child's dosage is given by

$$D(w) = \frac{aw}{150}$$

If the adult dose of a substance is 500 mg, how much should a child who weighs 35 lb receive?

**74. Population Growth** A study prepared for a Sunbelt town's chamber of commerce projected that the population of the town in the next 3 years will grow according to the rule

$$P(t) = 50{,}000 + 30t^{3/2} + 20t$$

where $P(t)$ denotes the population $t$ months from now. By how much will the population increase during the next 9 months? The next 16 months?

**75. Thurstone Learning Curve** Psychologist L.L. Thurstone discovered the following model for the relationship between the learning time $T$ and the length of a list $n$:

$$T = f(n) = An\sqrt{n - b}$$

where $A$ and $b$ are constants that depend on the person and the task. Suppose that for a certain person and a certain task, $A = 4$ and $b = 4$. Compute $f(4), f(5), \ldots, f(12)$, and use this information to sketch the graph of the function $f$. Interpret your results.

**76. Forecasting Sales** The annual sales of Crimson Drug Store are expected to be given by

$$S_1(t) = 2.3 + 0.4t$$

million dollars $t$ years from now, whereas the annual sales of Cambridge Drug Store are expected to be given by

$$S_2(t) = 1.2 + 0.6t$$

million dollars $t$ years from now. When will the annual sales of Cambridge first surpass the annual sales of Crimson?

**77. Oil Spills** The oil spilling from the ruptured hull of a grounded tanker spreads in all directions in calm waters. Suppose that the area polluted after $t$ sec is a circle of radius $r$ and the radius is increasing at the rate of 2 ft/sec.
a. Find a function $f$ giving the area polluted in terms of $r$.
b. Find a function $g$ giving the radius of the polluted area in terms of $t$.
c. Find a function $h$ giving the area polluted in terms of $t$.
d. What is the size of the polluted area 30 sec after the hull was ruptured?

**78. Film Conversion Prices** PhotoMart transfers movie films to DVDs. The fees charged for this service are shown in the following table. Find a function $C$ relating the cost $C(x)$ to the number of feet $x$ of film transferred. Sketch the graph of the function $C$.

| Length of film, $x$ (ft) | Cost for conversion ($) |
|:---:|:---:|
| $1 < x \leq 100$ | 5.00 |
| $100 < x \leq 200$ | 9.00 |
| $200 < x \leq 300$ | 12.50 |
| $300 < x \leq 400$ | 15.00 |
| $x > 400$ | $7.00 + 0.02x$ |

**79. Packaging** An open box is made from a square piece of cardboard by cutting away identical squares from each corner of

the cardboard and folding up the resulting flaps. The length of one side of the cardboard is 10 in. Let the square cutaways have dimensions $x$ in. by $x$ in.
a. Draw and label an appropriate figure.
b. Find a function of $x$ giving the volume of the resulting box.
c. What is the volume of the box if the cutaway is 1 in. by 1 in.?

**80.** A closed cylindrical can has a volume of 54 in.$^3$. Find a function $S$ giving the total area of the cylindrical can in terms of $r$, the radius of the base. What is the total surface area of a closed cylindrical can of radius 4 in.?

**81.** A man wishes to construct a cylindrical barrel with a capacity of $32\pi$ ft$^3$. The cost of the material for the side of the barrel is \$4/ft$^2$, and the cost of the material for the top and bottom is \$8/ft$^2$.
a. Draw and label an appropriate figure.
b. Find a function in terms of the radius of the barrel giving the total cost for constructing the barrel.
c. What is the total cost for constructing a barrel of radius 2 ft?

**82. Linear Depreciation** A farmer purchases a new machine for \$10,000. The machine is to have a salvage value of \$2000 after 5 years. Assuming linear depreciation, find a function giving the book value $V$ of the machine after $t$ years, where $0 \leq t \leq 5$.

**83. Cost of Housing** The Brennans are planning to buy a house 4 years from now. Housing experts in their area have estimated that the cost of a home will increase at a rate of 3% per year during that 4-year period. If their predictions are correct, how much can the Brennans expect to pay for a house that currently costs \$300,000?

**84. Yahoo! in Europe** Yahoo! is putting more emphasis on Western Europe, where the number of online households is expected to grow steadily. In a study conducted in 2004, the number of online households (in millions) in Western Europe was projected to be

$$N(t) = 34.68 + 23.88 \ln(1.05t + 5.3) \qquad 0 \leq t \leq 2$$

where $t = 0$ corresponds to the beginning of 2004. What was the projected number of online households in Western Europe at the beginning of 2005?
*Source:* Jupiter Research.

A maglev is a train that uses electromagnetic force to levitate, guide, and propel it. Compared to the more conventional steel-wheel and track trains, the maglev has the potential to reach very high speeds, perhaps 600 mph. In Section 1.1 we use the maglev as a vehicle to help us introduce the concept of the *limit* of a function. Specifically, we will see how the limit concept enables us to find the velocity of the maglev knowing only its position as a function of time. Then, generalizing, we use the limit to define the *derivative* of a function, the fundamental tool in differential calculus, which we will use to solve many practical problems in the ensuing chapters.

# 1 Limits

**THE NOTION OF** a limit permeates much of our work in calculus. We begin with an intuitive introduction to limits. We then develop techniques that will allow us to find limits much more easily than would be the case if we had to use the definition. The limit of a function allows us to define a very important property of functions: that of continuity. Finally, the limit plays a central role in the study of the rate of change of one quantity with respect to another—the central theme of calculus.

Ⓥ This symbol indicates that one of the following video types is available for enhanced student learning at **www.academic.cengage.com/login:**
  • Chapter lecture videos         • Solutions to selected exercises

## 1.1    An Intuitive Introduction to Limits

### ■ A Real-Life Example

A prototype of a maglev (magnetic levitation train) moves along a straight monorail. To describe the motion of the maglev, we can think of the track as a coordinate line. From data obtained in a test run, engineers have determined that the maglev's displacement (directed distance) measured in feet from the origin at time $t$ (in seconds) is given by

$$s = f(t) = 4t^2 \qquad 0 \le t \le 30 \tag{1}$$

where $f$ is called the position function of the maglev. The position of the maglev at time $t = 0, 1, 2, 3, \ldots, 30$, measured in feet from its initial position, is

$$f(0) = 0, \qquad f(1) = 4, \qquad f(2) = 16, \qquad f(3) = 36, \qquad \ldots, \qquad f(30) = 3600$$

(See Figure 1.)

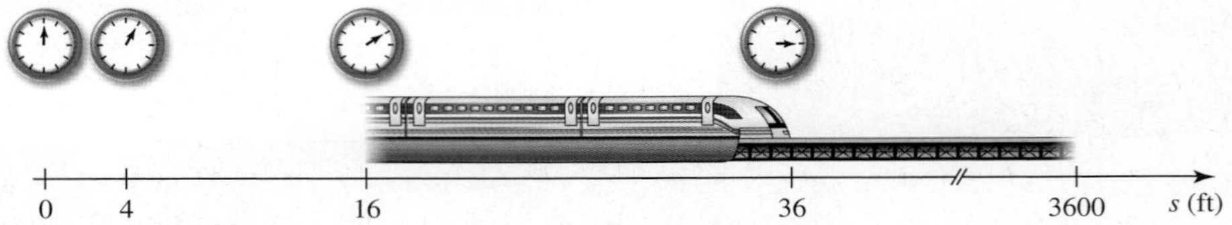

**FIGURE 1**
A maglev moving along an elevated monorail track

It appears that the maglev is accelerating over the time interval [0, 30] and, therefore, that its velocity varies over time. This raises the following question: Can we find the velocity of the maglev at *any* time in the interval (0, 30) using only Equation (1)? To be more specific, can we find the velocity of the maglev when, say, $t = 2$?

For a start, let's see what quantities we can compute. We can certainly compute the position of the maglev for some selected values of $t$ by using Equation (1), as we did earlier. Using these values of $f$, we can then compute the *average velocity* of the maglev over any interval of time. For example, to compute the average velocity of the train over the time interval [2, 4], we first compute the **displacement** of the train over that interval, $f(4) - f(2)$, and then divide this quantity by the time elapsed. Thus,

$$\frac{\text{displacement}}{\text{time elapsed}} = \frac{f(4) - f(2)}{4 - 2} = \frac{4(4)^2 - 4(2)^2}{2} = \frac{64 - 16}{2} = 24$$

or 24 ft/sec. Although this is not quite the velocity of the maglev at $t = 2$, it does provide us with an approximation of its velocity at that time.

Can we do better? Intuitively, the smaller the time interval we pick (with $t = 2$ as the left endpoint), the more closely the average velocity over that time interval will approximate the actual velocity of the maglev at $t = 2$.*

Now let's describe this process in general terms. Let $t > 2$. Then the average velocity of the maglev over the time interval [2, $t$] is given by

$$v_{\text{av}} = \frac{f(t) - f(2)}{t - 2} = \frac{4t^2 - 4(2)^2}{t - 2} = \frac{4(t^2 - 4)}{t - 2} \tag{2}$$

---

*Actually, any interval containing $t = 2$ will do.

By choosing the values of $t$ closer and closer to 2, we obtain a sequence of numbers that gives the average velocities of the maglev over smaller and smaller time intervals. As we observed earlier, this sequence of numbers should approach the *instantaneous velocity* of the train at $t = 2$.

Let's try some sample calculations. Using Equation (2) and taking the sequence $t = 2.5, 2.1, 2.01, 2.001,$ and $2.0001$, which approaches 2, we find

$$\text{The average velocity over [2, 2.5] is } \frac{4(2.5^2 - 4)}{2.5 - 2} = 18 \text{ ft/sec}$$

$$\text{The average velocity over [2, 2.1] is } \frac{4(2.1^2 - 4)}{2.1 - 2} = 16.4 \text{ ft/sec}$$

and so forth. These results are summarized in Table 1. From the table we see that the average velocity of the maglev seems to approach the number 16 as it is computed over smaller and smaller time intervals. These computations suggest that the instantaneous velocity of the train at $t = 2$ is 16 ft/sec.

**TABLE 1**   The average velocity of the maglev

| $t$ | 2.5 | 2.1 | 2.01 | 2.001 | 2.0001 |
|---|---|---|---|---|---|
| $v_{av}$ **over [2, $t$]** | 18 | 16.4 | 16.04 | 16.004 | 16.0004 |

**Note**   We cannot obtain the instantaneous velocity for the maglev at $t = 2$ by substituting $t = 2$ into Equation (2) because this value of $t$ is not in the domain of the average velocity function. ▪

## ▪ Intuitive Definition of a Limit

Consider the function $g$ defined by

$$g(t) = \frac{4(t^2 - 4)}{t - 2}$$

which gives the average velocity of the maglev (see Equation (2)). Suppose that we are required to determine the value that $g(t)$ approaches as $t$ approaches the (fixed) number 2. If we take a sequence of values of $t$ approaching 2 from the right-hand side, as we did earlier, we see that $g(t)$ approaches the number 16. Similarly, if we take a sequence of values of $t$ approaching 2 from the left, such as $t = 1.5, 1.9, 1.99, 1.999,$ and 1.9999, we obtain the results in Table 2.

**TABLE 2**   The values of $g$ as $t$ approaches 2 from the left

| $t$ | 1.5 | 1.9 | 1.99 | 1.999 | 1.9999 |
|---|---|---|---|---|---|
| $g(t)$ | 14 | 15.6 | 15.96 | 15.996 | 15.9996 |

Observe that $g(t)$ approaches the number 16 as $t$ approaches 2—this time from the left-hand side. In other words, as $t$ approaches 2 from *either* side of 2, $g(t)$ approaches 16. In this situation we say that the *limit* of $g(t)$ as $t$ approaches 2 is 16, written

$$\lim_{t \to 2} g(t) = \lim_{t \to 2} \frac{4(t^2 - 4)}{t - 2} = 16$$

The graph of the function $g$, shown in Figure 2, confirms this observation.

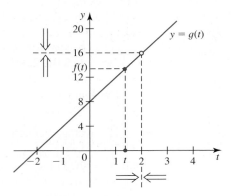

**FIGURE 2**
As $t$ approaches 2, $g(t)$ approaches 16.

**Note** Observe that the number 2 does not lie in the domain of $g$. (For this reason the point $(2, 16)$ is not on the graph of $g$, and we indicate this by an open circle on the graph.) Notice, too, that the existence or nonexistence of $g(t)$ at $t = 2$ plays no role in our computation of the limit. ∎

---

**DEFINITION** **Limit of a Function at a Number**

Let $f$ be a function defined on an open interval containing $a$, with the possible exception of $a$ itself. Then the limit of $f(x)$ as $x$ approaches $a$ is the number $L$, written

$$\lim_{x \to a} f(x) = L \qquad (3)$$

if $f(x)$ can be made as close to $L$ as we please by taking $x$ to be sufficiently close to $a$.

---

**EXAMPLE 1** Use the graph of the function $f$ shown in Figure 3 to find the given limit, if it exists.

**a.** $\lim_{x \to 1} f(x)$    **b.** $\lim_{x \to 3} f(x)$    **c.** $\lim_{x \to 5} f(x)$    **d.** $\lim_{x \to 7} f(x)$    **e.** $\lim_{x \to 10} f(x)$

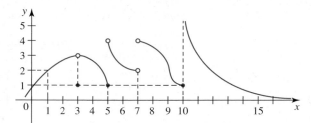

**FIGURE 3**
The graph of the function $f$

**Solution**
**a.** The values of $f$ can be made as close to 2 as we please by taking $x$ to be sufficiently close to 1. So $\lim_{x \to 1} f(x) = 2$.
**b.** The values of $f$ can be made as close to 3 as we please by taking $x$ to be sufficiently close to 3. So $\lim_{x \to 3} f(x) = 3$. Observe that $f(3) = 1$, but this has no bearing on the answer.

**c.** No matter how close $x$ is to 5, there are values of $f$, corresponding to values of $x$ smaller than 5, that are close to 1; and there are values of $f$, corresponding to values of $x$ greater than 5, that are close to 4. In other words, there is no *unique* number that $f(x)$ approaches as $x$ approaches 5. Therefore, $\lim_{x \to 5} f(x)$ does not exist. Observe that $f(5) = 1$, but, again, this has no bearing on the existence or nonexistence of the limit.

**d.** No matter how close $x$ is to 7, there are values of $f$ that are close to 2 (corresponding to values of $x$ less than 7) and values of $f$ that are close to 4 (corresponding to values of $x$ greater than 7). So $\lim_{x \to 7} f(x)$ does not exist. Observe that $x = 7$ is not in the domain of $f$, but this does not affect our answer.

**e.** As $x$ approaches 10 from the right, $f(x)$ increases without bound. Therefore, $f(x)$ cannot approach a unique number as $x$ approaches 10, and $\lim_{x \to 10} f(x)$ does not exist. Here, $f(10) = 1$, but this fact plays no role in our determination of the limit.   ■

**Note**   Example 1 shows that when we evaluate the limit of a function $f$ as $x$ approaches $a$, it is immaterial whether $f$ is defined at $a$. Furthermore, even if $f$ is defined at $a$, the value of $f$ at $a$, $f(a)$, has no bearing on the existence or the value of the limit in question.   ■

**EXAMPLE 2**   Find $\lim_{x \to 2} f(x)$ if it exists, where $f$ is the piecewise-defined function

$$f(x) = \begin{cases} 4x + 8 & \text{if } x \neq 2 \\ 4 & \text{if } x = 2 \end{cases}$$

**Solution**   From the graph of $f$ shown in Figure 4, we see that $\lim_{x \to 2} f(x) = 16$. If you compare the function $f$ with the function $g$ discussed earlier (page 102), you will see that the values of $f$ are identical to the values of $g$ except at $x = 2$ (Figures 2 and 4). Thus, the limits of $f(x)$ and $g(x)$ as $x$ approaches 2 are equal, as expected. We can see why the graphs of the two functions coincide everywhere except at $x = 2$ by writing

$$g(x) = \frac{4(x^2 - 4)}{x - 2} \qquad \text{Use } x \text{ instead of } t.$$

$$= \frac{4(x + 2)(x - 2)}{x - 2}$$

$$= 4(x + 2) \qquad \text{Assume that } x \neq 2.$$

which is equivalent to the rule defining $f$ when $x \neq 2$.   ■

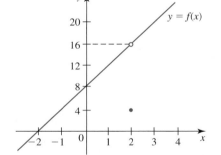

**FIGURE 4**
The graph of $f$ coincides with the graph of the function $g$ shown in Figure 2, except at $x = 2$.

**EXAMPLE 3**   **The Heaviside Function**   The Heaviside function $H$ (the unit step function) is defined by

$$H(t) = \begin{cases} 0 & \text{if } t < 0 \\ 1 & \text{if } t \geq 0 \end{cases}$$

This function, named after Oliver Heaviside (1850–1925), can be used to describe the flow of current in a DC electrical circuit that is switched on at time $t = 0$. Show that $\lim_{t \to 0} H(t)$ does not exist.

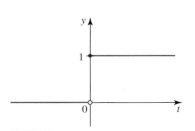

**FIGURE 5**
$\lim_{t \to 0} H(t)$ does not exist.

**Solution**   The graph of $H$ is shown in Figure 5. You can see from the graph that no matter how close $t$ is to 0, $H(t)$ takes on the value 1 or 0, depending on whether $t$ is to the right or to the left of 0. Therefore, $H(t)$ cannot approach a unique number $L$ as $t$ approaches 0, and we conclude that $\lim_{t \to 0} H(t)$ does not exist.   ■

## ■ One-Sided Limits

Let's reexamine the Heaviside function. We have shown that $\lim_{t \to 0} H(t)$ does not exist, but what can we say about the behavior of $H(t)$ at values of $t$ that are close to but greater than 0? If you look at Figure 5 again, it is evident that as $t$ approaches 0 through positive values (from the right of 0), $H(t)$ approaches 1. In this situation we say that the right-hand limit of $H$ as $t$ approaches 0 is 1, written

$$\lim_{t \to 0^+} H(t) = 1$$

More generally, we have the following:

---

**DEFINITION    Right-Hand Limit of a Function**

Let $f$ be a function defined for all values of $x$ close to but greater than $a$. Then the right-hand limit of $f(x)$ as $x$ approaches $a$ is equal to $L$, written

$$\lim_{x \to a^+} f(x) = L \qquad (4)$$

if $f(x)$ can be made as close to $L$ as we please by taking $x$ to be sufficiently close to but greater than $a$.

---

**Note**   Equation (4) is just Equation (3) with the further restriction $x > a$.    ■

The left-hand limit of a function is defined in a similar manner.

---

**DEFINITION    Left-Hand Limit of a Function**

Let $f$ be a function defined for all values of $x$ close to but less than $a$. Then the left-hand limit of $f(x)$ as $x$ approaches $a$ is equal to $L$, written

$$\lim_{x \to a^-} f(x) = L \qquad (5)$$

if $f(x)$ can be made as close to $L$ as we please by taking $x$ to be sufficiently close to but less than $a$.

---

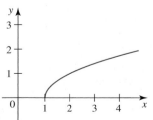

**FIGURE 6**
The right-hand limit of $f(x) = \sqrt{x - 1}$ as $x$ approaches 1 is 0.

For the function $H$ of Example 3 we have $\lim_{t \to 0^-} H(t) = 0$.

The right-hand and left-hand limits of a function, $\lim_{x \to a^+} f(x)$ and $\lim_{x \to a^-} f(x)$, are often referred to as **one-sided limits,** whereas $\lim_{x \to a} f(x)$ is called a **two-sided limit.**

For some functions it makes sense to look only at one-sided limits. Consider, for example, the function $f$ defined by $f(x) = \sqrt{x - 1}$, whose domain is $[1, \infty)$. Here it makes sense to talk only about the right-hand limit of $f(x)$ as $x$ approaches 1. Also, from Figure 6, we see that $\lim_{x \to 1^+} f(x) = 0$.

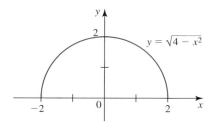

**FIGURE 7**
We can approach −2 only from the
right and 2 only from the left.

**EXAMPLE 4**   Let $f(x) = \sqrt{4 - x^2}$. Find $\lim_{x \to -2^+} f(x)$ and $\lim_{x \to 2^-} f(x)$.

**Solution**   The graph of $f$ is the upper semicircle shown in Figure 7. From this graph we see that $\lim_{x \to -2^+} f(x) = 0$ and $\lim_{x \to 2^-} f(x) = 0$.   ■

Theorem 1 gives the connection between one-sided limits and two-sided limits.

---

**THEOREM 1    Relationship Between One-Sided and Two-Sided Limits**

Let $f$ be a function defined on an open interval containing $a$, with the possible exception of $a$ itself. Then

$$\lim_{x \to a} f(x) = L \quad \text{if and only if} \quad \lim_{x \to a^-} f(x) = \lim_{x \to a^+} f(x) = L \qquad (6)$$

---

Thus, the (two-sided) limit exists if and only if the one-sided limits exist and are equal.

**EXAMPLE 5**   Sketch the graph of the function $f$ defined by

$$f(x) = \begin{cases} 3 - x & \text{if } x < 1 \\ 1 & \text{if } x = 1 \\ 2 + \sqrt{x - 1} & \text{if } x > 1 \end{cases}$$

Use your graph to find $\lim_{x \to 1^-} f(x)$, $\lim_{x \to 1^+} f(x)$,  and $\lim_{x \to 1} f(x)$.

**Solution**   From the graph of $f$, shown in Figure 8, we see that

$$\lim_{x \to 1^-} f(x) = 2 \quad \text{and} \quad \lim_{x \to 1^+} f(x) = 2$$

Since the one-sided limits are equal, we conclude that $\lim_{x \to 1} f(x) = 2$. Notice that $f(1) = 1$, but this has no effect on the value of the limit.   ■

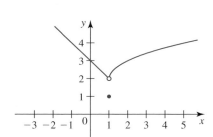

**FIGURE 8**
$\lim_{x \to 1^-} f(x) = \lim_{x \to 1^+} f(x) = \lim_{x \to 1} f(x) = 2$

**EXAMPLE 6**   Let $f(x) = \dfrac{\sin x}{x}$. Use your calculator to complete the following table.

| $x$ | $\pm 1$ | $\pm 0.5$ | $\pm 0.1$ | $\pm 0.05$ | $\pm 0.01$ | $\pm 0.005$ | $\pm 0.001$ |
|---|---|---|---|---|---|---|---|
| $\dfrac{\sin x}{x}$ | | | | | | | |

Then sketch the graph of $f$, and use your graph to guess at the value of $\lim_{x \to 0^-} f(x)$, $\lim_{x \to 0^+} f(x)$, and $\lim_{x \to 0} f(x)$.

**Solution**   Using a calculator, we obtain Table 3. (Remember to use radian mode!) The graph of $f$ is shown in Figure 9. We find

$$\lim_{x \to 0^-} f(x) = 1, \qquad \lim_{x \to 0^+} f(x) = 1, \qquad \text{and so} \qquad \lim_{x \to 0} f(x) = 1$$

We will prove in Section 1.2 that our guesses here are correct.

**TABLE 3**

| $x$ | $\dfrac{\sin x}{x}$ |
|---|---|
| $\pm 1$ | 0.841470985 |
| $\pm 0.5$ | 0.958851077 |
| $\pm 0.1$ | 0.998334166 |
| $\pm 0.05$ | 0.999583385 |
| $\pm 0.01$ | 0.999983333 |
| $\pm 0.005$ | 0.999995833 |
| $\pm 0.001$ | 0.999999833 |

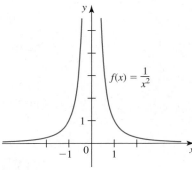

**FIGURE 9**
The graph of $f(x) = \dfrac{\sin x}{x}$

**EXAMPLE 7**   Let $f(x) = \dfrac{1}{x^2}$. Evaluate the limit, if it exists.

**a.** $\displaystyle\lim_{x\to 0^-} f(x)$     **b.** $\displaystyle\lim_{x\to 0^+} f(x)$     **c.** $\displaystyle\lim_{x\to 0} f(x)$

**Solution**   Some values of the function are listed in Table 4, and the graph of $f$ is shown in Figure 10.

### Historical Biography

Bettmann/Corbis

**JOHN WALLIS**
**(1616–1703)**

The first mathematician to use the symbol ∞ to indicate infinity, John Wallis contributed to the earliest forms, notations, and terms of calculus and other areas of mathematics. Born November 23, 1616, in the borough of Ashford, in Kent, England, Wallis attended boarding school as a child, and his exceptional mathematical ability was evident at an early age. He mastered arithmetic in two weeks and was able to solve a problem such as the square root of a 53-digit number to 17 places without notation. Considered to be the most influential British mathematician before Isaac Newton (page 202), Wallis published his first major work, *Arithmetica Infinitorum*, in 1656. It became a standard reference and is still recognized as a monumental text in British mathematics.

**TABLE 4**

| $x$ | $\dfrac{1}{x^2}$ |
|---|---|
| $\pm 1$ | 1 |
| $\pm 0.5$ | 4 |
| $\pm 0.1$ | 100 |
| $\pm 0.05$ | 400 |
| $\pm 0.01$ | 10,000 |
| $\pm 0.001$ | 1,000,000 |

$f(x) = \dfrac{1}{x^2}$

**FIGURE 10**
As $x \to 0$ from the left (or from the right), $f(x)$ increases without bound.

**a.** As $x$ approaches 0 from the left, $f(x)$ increases without bound and does not approach a unique number. Therefore, $\lim_{x\to 0^-} f(x)$ does not exist.
**b.** As $x$ approaches 0 from the right, $f(x)$ increases without bound and does not approach a unique number. Therefore, $\lim_{x\to 0^+} f(x)$ does not exist.
**c.** From the results of parts (a) and (b) we conclude that $\lim_{x\to 0} f(x)$ does not exist.

**Note**   Even though the limit $\lim_{x\to 0} f(x)$ does not exist, we write $\lim_{x\to 0} (1/x^2) = \infty$ to indicate that $f(x)$ increases without bound as $x$ approaches 0. We will study "infinite limits" in Section 3.5.

## ■ Using Graphing Utilities to Evaluate Limits

In Example 6 we employed both a numerical and a graphical approach to help us conjecture that

$$\lim_{x \to 0} \frac{\sin x}{x} = 1$$

Either or both of these approaches can often be used to estimate the limit of a function as $x$ approaches a specified value. But there are pitfalls in using graphing utilities, as the following examples show.

**EXAMPLE 8**  Use a graphing utility to find

$$\lim_{x \to 0} \frac{\sqrt{x + 4} - 2}{x}$$

**Solution**  We first investigate the problem numerically by constructing a table of values of $f(x) = (\sqrt{x + 4} - 2)/x$ corresponding to values of $x$ that approach 0 from either side of 0. Table 5a shows the values of $f$ for $x$ close to but to the left of 0, and Table 5b shows the values of $f$ for $x$ close to but to the right of 0.

If you look at $f$ evaluated at the first nine values of $x$ shown in each column, we are tempted to conclude that the required limit is $\frac{1}{4}$. But how do we reconcile this result with the last two values of $f$ in each column? Upon reflection we see that this discrepancy can be attributed to a phenomenon known as *loss of significance*.

**TABLE 5**  Values of $f$ for $x$ close to 0

| $x$ | $\dfrac{\sqrt{x + 4} - 2}{x}$ | $x$ | $\dfrac{\sqrt{x + 4} - 2}{x}$ |
|---|---|---|---|
| $-0.001$ | 0.250015627 | 0.001 | 0.249984377 |
| $-0.0001$ | 0.250001562 | 0.0001 | 0.249998438 |
| $-10^{-5}$ | 0.25000016 | $10^{-5}$ | 0.24999984 |
| $-10^{-6}$ | 0.25 | $10^{-6}$ | 0.25 |
| $-10^{-7}$ | 0.25 | $10^{-7}$ | 0.25 |
| $-10^{-8}$ | 0.25 | $10^{-8}$ | 0.25 |
| $-10^{-9}$ | 0.25 | $10^{-9}$ | 0.25 |
| $-10^{-10}$ | 0.25 | $10^{-10}$ | 0.25 |
| $-10^{-11}$ | 0.25 | $10^{-11}$ | 0.25 |
| $-10^{-12}$ | 0.3 | $10^{-12}$ | 0.2 |
| $-10^{-13}$ | 0 | $10^{-13}$ | 0 |

**(a)** $x$ approaches 0 from the left.     **(b)** $x$ approaches 0 from the right.

When $x$ is very small, the computed values of $\sqrt{x + 4}$ are very close to 2. For $x = -10^{-13}$ or $x = 10^{-13}$ (and values that are smaller in absolute value) the calculator rounds off the value of $\sqrt{x + 4}$ to 2 and gives the value of $f(x)$ as 0. Figures 11a–b show the graphs of $f$ using the viewing windows $[-2, 2] \times [0.2, 0.3]$ and $[-10^{-3}, 10^{-3}] \times [0.2, 0.3]$, respectively. Both these graphs reinforce the earlier observation that the required limit is $\frac{1}{4}$. The graph of $f$ using the viewing window $[-10^{-11}, 10^{-11}] \times [0.24995, 0.25005]$, shown in Figure 11c, proves to be of no help because of the problem with loss of significance stated earlier.

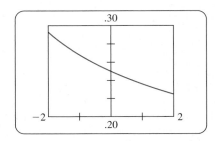

(a) $[-2, 2] \times [0.2, 0.3]$

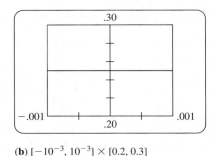

(b) $[-10^{-3}, 10^{-3}] \times [0.2, 0.3]$

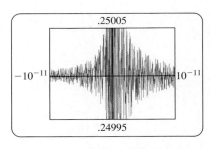

(c) $[-10^{-11}, 10^{-11}] \times [0.24995, 0.25005]$

**FIGURE 11**
The graphs of $f(x) = \dfrac{\sqrt{x+4}-2}{x}$ in different viewing windows

Having recognized the source of the difficulty, how can we remedy the situation? Let's find another expression for $f(x)$ that does not involve subtracting numbers that are so close to each other that it results in a loss of significance. Rationalizing the numerator, we obtain

$$f(x) = \frac{\sqrt{x+4}-2}{x} = \frac{\sqrt{x+4}-2}{x} \cdot \frac{\sqrt{x+4}+2}{\sqrt{x+4}+2}$$

$$= \frac{(x+4)-4}{x(\sqrt{x+4}+2)} \qquad (a+b)(a-b) = a^2 - b^2$$

$$= \frac{1}{\sqrt{x+4}+2} \qquad x \neq 0$$

Observe that the use of the last expression avoids the pitfalls that we encountered with the original expression. We leave it as an exercise to show that both a numerical analysis and a graphical analysis of

$$\lim_{x \to 0} \frac{1}{\sqrt{x+4}+2}$$

suggest that a good guess for

$$\lim_{x \to 0} \frac{\sqrt{x+4}-2}{x} = \lim_{x \to 0} \frac{1}{\sqrt{x+4}+2}$$

is $\frac{1}{4}$, a result that can be proved analytically using the techniques to be developed in the next section.

**EXAMPLE 9**    Find $\displaystyle \lim_{x \to 0} \sin \frac{1}{x}$.

**Solution**    Let $f(x) = \sin(1/x)$. The graph of $f$ using the viewing window $[-1, 1] \times [-1.2, 1.2]$ does not seem to be of any help to us in finding the required limit (see Figure 12). To obtain a more accurate graph of $f(x) = \sin(1/x)$, note that the sine function is bounded by $-1$ and $1$. Thus, the graph of $f$ lies between the horizontal lines $y = -1$ and $y = 1$. Next, observe that the sine function has period $2\pi$. Since $1/x$ increases without bound (decreases without bound) as $x$ approaches 0 from the right (from the left), we see that $\sin(1/x)$ undergoes more and more cycles as $x$ approaches 0. Thus, the graph of $f(x) = \sin(1/x)$ oscillates between $-1$ and 1, as shown in Figure 13. Therefore, it seems reasonable to conjecture that the limit does not exist. Indeed, we can demonstrate this conclusion by constructing Table 6.

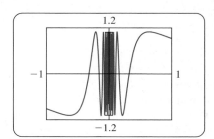

**FIGURE 12**
The graph of $f(x) = \sin(1/x)$ in the viewing window $[-1, 1] \times [-1.2, 1.2]$

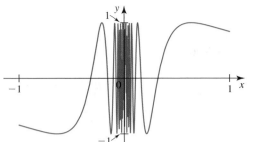

**FIGURE 13**
The graph of $f(x) = \sin(1/x)$

**TABLE 6**

| $x$ | $\dfrac{2}{\pi}$ | $\dfrac{2}{3\pi}$ | $\dfrac{2}{5\pi}$ | $\dfrac{2}{7\pi}$ | $\dfrac{2}{9\pi}$ | $\dfrac{2}{11\pi}$ | $\cdots$ |
|---|---|---|---|---|---|---|---|
| $\sin\dfrac{1}{x}$ | 1 | $-1$ | 1 | $-1$ | 1 | $-1$ | $\cdots$ |

Note that the values of $x$ approach 0 from the right. From the table we see that no matter how close $x$ is to 0 (from the right), there are values of $f$ corresponding to these values of $x$ that are equal to 1 or $-1$. Therefore, $f(x)$ cannot approach any fixed number as $x$ approaches 0. A similar result is true if the values of $x$ approach 0 from the left. This shows that

$$\lim_{x \to 0} \sin \frac{1}{x}$$

does not exist. ■

## 1.1  CONCEPT QUESTIONS

1. Explain what is meant by the statement $\lim_{x \to 2} f(x) = 3$.
2. **a.** If $\lim_{x \to 3} f(x) = 5$, what can you say about $f(3)$? Explain.
   **b.** If $f(2) = 6$, what can you say about $\lim_{x \to 2} f(x)$? Explain.

3. Explain what is meant by the statement $\lim_{x \to 3^-} f(x) = 2$.
4. Suppose $\lim_{x \to 1^-} f(x) = 3$ and $\lim_{x \to 1^+} f(x) = 4$.
   **a.** What can you say about $\lim_{x \to 1} f(x)$? Explain.
   **b.** What can you say about $f(1)$? Explain.

## 1.1  EXERCISES

*In Exercises 1–6, use the graph of the function f to find each limit.*

**1.**

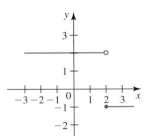

**a.** $\lim_{x \to 2^-} f(x)$
**b.** $\lim_{x \to 2^+} f(x)$
**c.** $\lim_{x \to 2} f(x)$

**2.**

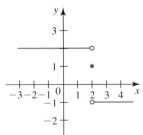

**a.** $\lim_{x \to 2^-} f(x)$
**b.** $\lim_{x \to 2^+} f(x)$
**c.** $\lim_{x \to 2} f(x)$

**3.**

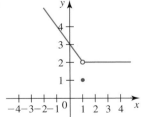

**a.** $\lim_{x \to 1^-} f(x)$
**b.** $\lim_{x \to 1^+} f(x)$
**c.** $\lim_{x \to 1} f(x)$

**4.**

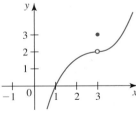

**a.** $\lim_{x \to 3^-} f(x)$
**b.** $\lim_{x \to 3^+} f(x)$
**c.** $\lim_{x \to 3} f(x)$

Ⓥ Videos for selected exercises are available online at **www.academic.cengage.com/login**.

**5.**

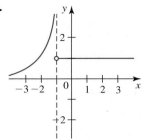

**a.** $\lim\limits_{x \to -1^-} f(x)$   **b.** $\lim\limits_{x \to -1^+} f(x)$   **c.** $\lim\limits_{x \to -1} f(x)$

**6.**

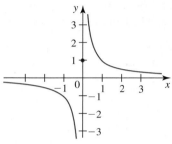

**a.** $\lim\limits_{x \to 0^-} f(x)$   **b.** $\lim\limits_{x \to 0^+} f(x)$   **c.** $\lim\limits_{x \to 0} f(x)$

**7.** Use the graph of the function $f$ to determine whether each statement is true or false. Explain.

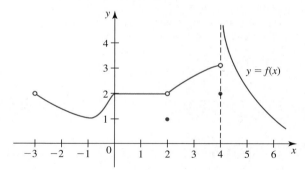

**a.** $\lim\limits_{x \to -3^+} f(x) = 2$   **b.** $\lim\limits_{x \to 0} f(x) = 2$

**c.** $\lim\limits_{x \to 2} f(x) = 1$   **d.** $\lim\limits_{x \to 4^-} f(x) = 3$

**e.** $\lim\limits_{x \to 4^+} f(x)$ does not exist   **f.** $\lim\limits_{x \to 4} f(x) = 2$

**8.** Use the graph of the function $f$ to determine whether each statement is true or false. Explain.

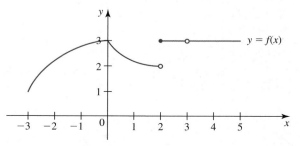

**a.** $\lim\limits_{x \to -3^+} f(x) = 1$   **b.** $\lim\limits_{x \to 0} f(x) = f(0)$

**c.** $\lim\limits_{x \to 2^-} f(x) = 2$   **d.** $\lim\limits_{x \to 2^+} f(x) = 3$

**e.** $\lim\limits_{x \to 3} f(x)$ does not exist   **f.** $\lim\limits_{x \to 5^-} f(x) = 3$

*In Exercises 9–16, complete the table by computing $f(x)$ at the given values of x, accurate to five decimal places. Use the results to guess at the indicated limit, if it exists.*

**9.** $\lim\limits_{x \to 1} \dfrac{x - 1}{x^2 - 3x + 2}$

| $x$ | 0.9 | 0.99 | 0.999 | 1.001 | 1.01 | 1.1 |
|------|-----|------|-------|-------|------|-----|
| $f(x)$ | | | | | | |

**10.** $\lim\limits_{x \to 1} \dfrac{x - 1}{x^2 + x - 2}$

| $x$ | 0.9 | 0.99 | 0.999 | 1.001 | 1.01 | 1.1 |
|------|-----|------|-------|-------|------|-----|
| $f(x)$ | | | | | | |

**11.** $\lim\limits_{x \to 2} \dfrac{\sqrt{x + 2} - 2}{x - 2}$

| $x$ | 1.9 | 1.99 | 1.999 | 2.001 | 2.01 | 2.1 |
|------|-----|------|-------|-------|------|-----|
| $f(x)$ | | | | | | |

**12.** $\lim\limits_{x \to 0} \dfrac{\sqrt{3 + x} - \sqrt{3 - x}}{x}$

| $x$ | $-0.1$ | $-0.01$ | $-0.001$ | 0.001 | 0.01 | 0.1 |
|------|--------|---------|----------|-------|------|-----|
| $f(x)$ | | | | | | |

**13.** $\lim\limits_{x \to 2} \dfrac{\dfrac{1}{\sqrt{2 + x}} - \dfrac{1}{2}}{x - 2}$

| $x$ | 1.9 | 1.99 | 1.999 | 2.001 | 2.01 | 2.1 |
|------|-----|------|-------|-------|------|-----|
| $f(x)$ | | | | | | |

**14.** $\lim\limits_{x \to 3} \dfrac{3\sqrt{x + 1} - 2x}{x(x - 3)}$

| $x$ | 2.9 | 2.99 | 2.999 | 3.001 | 3.01 | 3.1 |
|------|-----|------|-------|-------|------|-----|
| $f(x)$ | | | | | | |

**15.** $\lim\limits_{x\to 0}\dfrac{e^x - 1}{x}$

| $x$ | $-0.1$ | $-0.01$ | $-0.001$ | $0.001$ | $0.01$ | $0.1$ |
|---|---|---|---|---|---|---|
| $f(x)$ | | | | | | |

**16.** $\lim\limits_{x\to 0}\dfrac{\tan^{-1} 2x}{\ln(1 + 2x)}$

| $x$ | $-0.1$ | $-0.01$ | $-0.001$ | $0.001$ | $0.01$ | $0.1$ |
|---|---|---|---|---|---|---|
| $f(x)$ | | | | | | |

*In Exercises 17–22, sketch the graph of the function f and evaluate* (a) $\lim\limits_{x\to a^-} f(x)$, (b) $\lim\limits_{x\to a^+} f(x)$, *and* (c) $\lim\limits_{x\to a} f(x)$ *for the given value of a.*

**17.** $f(x) = \begin{cases} x - 1 & \text{if } x \le 3 \\ -2x + 8 & \text{if } x > 3 \end{cases}$;   $a = 3$

**18.** $f(x) = \begin{cases} 2x - 4 & \text{if } x < 4 \\ x - 2 & \text{if } x \ge 4 \end{cases}$;   $a = 4$

**19.** $f(x) = \begin{cases} -e^{-x} & \text{if } x \ne 0 \\ 1 & \text{if } x = 0 \end{cases}$;   $a = 0$

**20.** $f(x) = \begin{cases} x^2 - 1 & \text{if } x \ne 0 \\ 1 & \text{if } x = 0 \end{cases}$;   $a = 0$

**21.** $f(x) = \begin{cases} x & \text{if } x < 1 \\ 2 & \text{if } x = 1 \\ -x + 2 & \text{if } x > 1 \end{cases}$;   $a = 1$

**22.** $f(x) = \begin{cases} x^2 - 1 & \text{if } x < 1 \\ 2 & \text{if } x = 1 \\ \ln x & \text{if } x > 1 \end{cases}$;   $a = 1$

*The symbol $\llbracket\ \rrbracket$ denotes the greatest integer function defined by $\llbracket x \rrbracket$ = the greatest integer n such that $n \le x$. For example, $\llbracket 2.8 \rrbracket = 2$, and $\llbracket -2.7 \rrbracket = -3$. In Exercises 23–28, use the graph of the function to find the indicated limit, if it exists.*

**23.** $\lim\limits_{x\to 3^-} \llbracket x \rrbracket$

**24.** $\lim\limits_{x\to 3^+} \llbracket x \rrbracket$

**25.** $\lim\limits_{x\to -1^+} \llbracket x \rrbracket$

**26.** $\lim\limits_{x\to -1} \llbracket x \rrbracket$

**27.** $\lim\limits_{x\to 3.1} \llbracket x \rrbracket$

**28.** $\lim\limits_{x\to 2.4} \llbracket 2x \rrbracket$

**29.** Let

$$f(x) = \begin{cases} 0 & \text{if } x \le 0 \\ \sin\dfrac{1}{x} & \text{if } x > 0 \end{cases}$$

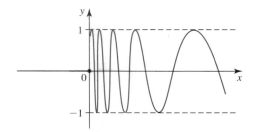

(As $x$ approaches 0 from the right, $y$ oscillates more and more.) Use the figure and construct a table of values to guess at $\lim\limits_{x\to 0^+} f(x)$, $\lim\limits_{x\to 0^-} f(x)$,  and $\lim\limits_{x\to 0} f(x)$. Justify your answer.

**30.** Let

$$f(x) = \begin{cases} 0 & \text{if } x = 0 \\ x\sin\dfrac{1}{x} & \text{if } x \ne 0 \end{cases}$$

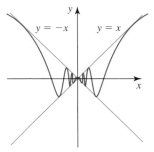

Use the figure, and construct a table of values to guess at $\lim\limits_{x\to 0^+} f(x)$, $\lim\limits_{x\to 0^-} f(x)$, and $\lim\limits_{x\to 0} f(x)$. Justify your answer.

**31.** Let

$$f(x) = \begin{cases} \dfrac{1}{x} & \text{if } x < 0 \\ \sin x & \text{if } 0 \le x < \pi \\ 0 & \text{if } x \ge \pi \end{cases}$$

**a.** Sketch the graph of $f$.
**b.** Find all values of $x$ in the domain of $f$ at which the limit of $f$ exists.
**c.** Find all values of $x$ in the domain of $f$ at which the left-hand limit of $f$ exists.
**d.** Find all values of $x$ in the domain of $f$ at which the right-hand limit of $f$ exists.

**32.** Let

$$f(x) = \begin{cases} -x^2 & \text{if } x < 0 \\ \tan x & \text{if } 0 \le x < \frac{\pi}{2} \\ \ln\left(x - \frac{\pi}{2} + 1\right) & \text{if } x \ge \frac{\pi}{2} \end{cases}$$

**a.** Sketch the graph of $f$.

**b.** Find all values of $x$ in the domain of $f$ at which the limit of $f$ exists.

**c.** Find all values of $x$ in the domain of $f$ at which the left-hand limit of $f$ exists.

**d.** Find all values of $x$ in the domain of $f$ at which the right-hand limit of $f$ exists.

**33. The Heaviside Function** A generalization of the unit step function or Heaviside function $H$ of Example 3 is the function $H_c$ defined by

$$H_c(t - t_0) = \begin{cases} 0 & \text{if } t < t_0 \\ c & \text{if } t \ge t_0 \end{cases}$$

where $c$ is a constant and $t_0 \ge 0$. Show that if $c \ne 0$, then $\lim_{t \to t_0} H_c(t - t_0)$ does not exist.

**34. The Square-Wave Function** The square-wave function $f$ can be expressed in terms of the Heaviside function (Exercise 33) as follows:

$$f(t) = H_k(t) - H_k(t - k) + H_k(t - 2k)$$
$$- H_k(t - 3k) + H_k(t - 4k) - \cdots$$

Referring to the following figure, show that $\lim_{t \to nk} f(t)$ does not exist for $n = 1, 2, 3, \dots$.

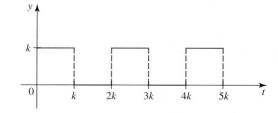

**35.**  Let $f(h) = (1 + h)^{1/h}$, and assume that $\lim_{h \to 0} (1 + h)^{1/h}$ exists. (We will establish this in Section 2.8.) Find its value to four decimal places of accuracy by computing $f(h)$ for $h = 0.1, 0.01, 0.001, 0.0001, 0.00001, 0.000001$, and $0.0000001$.

**36.**  Let $f(\theta) = (\tan \theta - \theta)/\theta^3$. By computing $f(\theta)$ for $\theta = \pm 0.1$, $\pm 0.01$, and $\pm 0.001$, accurate to five decimal places, guess at $\lim_{\theta \to 0} (\tan \theta - \theta)/\theta^3$.

 *In Exercises 37–42, plot the graph of f. Then zoom-in to guess at the specified limit (if it exists).*

**37.** $f(x) = \dfrac{2x^2 - x - 6}{x - 2}$;   $\lim_{x \to 2} f(x)$

**38.** $f(x) = \dfrac{x - 3}{\sqrt{x + 1} - 2}$;   $\lim_{x \to 3} f(x)$

**39.** $f(x) = \dfrac{x^3 + x^2 - 3x + 1}{|x - 1|}$;   $\lim_{x \to 1} f(x)$

**40.** $f(x) = \dfrac{\tan x}{x}$;   $\lim_{x \to 0} f(x)$

**41.** $f(x) = \dfrac{\sin^{-1}\sqrt{x}}{1 - \cos \sqrt[4]{x}}$;   $\lim_{x \to 0^+} f(x)$

**42.** $f(x) = \dfrac{e^{\tan 3x} - 1}{\ln(1 + \sin 2x)}$;   $\lim_{x \to 0} f(x)$

*In Exercises 43–46, determine whether the statement is true or false. If it is true, explain why. If it is false, explain why or give an example that shows it is false.*

**43.** If $\lim_{x \to a} f(x) = c$, then $f(a) = c$.

**44.** If $f$ is defined at $a$, then $\lim_{x \to a} f(x)$ exists.

**45.** If $\lim_{x \to a} f(x) = \lim_{x \to a} g(x)$, then $f(a) = g(a)$.

**46.** If both $\lim_{x \to a^+} f(x)$ and $\lim_{x \to a^-} f(x)$ exist, then $\lim_{x \to a} f(x)$ exists.

---

## 1.2   Techniques for Finding Limits

### ■ Computing Limits Using the Laws of Limits

In Section 1.1 we used tables of functional values and graphs of functions to help us guess at the limit of a function, if it exists. This approach, however, is useful only in suggesting whether the limit exists and what its value might be for simple functions. In practice, the limit of a function is evaluated by using the laws of limits that we now introduce.

> **LAW 1   Limit of a Constant Function $f(x) = c$**
>
> If $c$ is a real number, then
>
> $$\lim_{x \to a} c = c$$

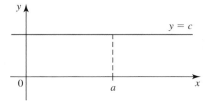

**FIGURE 1**
For the constant function $f(x) = c$,
$\lim_{x \to a} f(x) = c$.

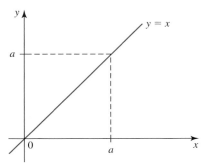

**FIGURE 2**
If $f$ is the identity function $f(x) = x$,
then $\lim_{x \to a} f(x) = a$.

You can see this intuitively by studying the graph of the constant function $f(x) = c$ shown in Figure 1. You will be asked to prove this law in Exercise 15, Section 1.3.

**EXAMPLE 1**   $\lim_{x \to 2} 5 = 5$, $\lim_{x \to -1} 3 = 3$, and $\lim_{x \to 0} 2\pi = 2\pi$.   ■

---

**LAW 2   Limit of the Identity Function $f(x) = x$**

$$\lim_{x \to a} x = a$$

---

Again, you can see this intuitively by examining the graph of the identity function $f(x) = x$. (See Figure 2.) You will also be asked to prove this law in Exercise 16, Section 1.3.

**EXAMPLE 2**   $\lim_{x \to 4} x = 4$, $\lim_{x \to 0} x = 0$, and $\lim_{x \to -\pi} x = -\pi$.   ■

The following limit laws allow us to find the limits of functions algebraically.

---

**LIMIT LAWS**

If $\lim_{x \to a} f(x) = L$ and $\lim_{x \to a} g(x) = M$, then

**LAW 3   Sum Law**

$$\lim_{x \to a}[f(x) \pm g(x)] = L \pm M$$

**LAW 4   Product Law**

$$\lim_{x \to a}[f(x)g(x)] = LM$$

**LAW 5   Constant Multiple Law**

$$\lim_{x \to a}[cf(x)] = cL, \quad \text{for every } c$$

**LAW 6   Quotient Law**

$$\lim_{x \to a}\frac{f(x)}{g(x)} = \frac{L}{M}, \quad \text{provided that } M \neq 0$$

**LAW 7   Root Law**

$$\lim_{x \to a}\sqrt[n]{f(x)} = \sqrt[n]{L}, \quad \text{provided that } n \text{ is a positive integer,} \\ \text{and } L > 0 \text{ if } n \text{ is even}$$

---

In words, these laws say the following:

**3.** The limit of the sum (difference) of two functions is the sum (difference) of their limits.
**4.** The limit of the product of two functions is the product of their limits.
**5.** The limit of a constant times a function is the constant times the limit of the function.

**6.** The limit of a quotient of two functions is the quotient of their limits, provided that the limit of the denominator is not zero.

**7.** The limit of the *n*th root of a function is the *n*th root of the limit of the function, provided that *n* is a positive integer and $L > 0$ if *n* is even.

(We will prove the Sum Law in Section 1.3. The other laws are proved in Appendix B.)

Although the Sum Law and the Product Law are stated for two functions, they are also valid for any finite number of functions. For example, if

$$\lim_{x \to a} f_1(x) = L_1, \qquad \lim_{x \to a} f_2(x) = L_2, \qquad \ldots, \qquad \lim_{x \to a} f_n(x) = L_n$$

then

$$\lim_{x \to a}[f_1(x) + f_2(x) + \cdots + f_n(x)] = L_1 + L_2 + \cdots + L_n$$

and

$$\lim_{x \to a}[f_1(x)f_2(x) \cdots f_n(x)] = L_1 L_2 \cdots L_n \tag{1}$$

If we take $f_1(x) = f_2(x) = \cdots = f_n(x) = f(x)$, then Equation (1) gives the following result for powers of *f*.

---

**LAW 8**    If *n* is a positive integer and $\lim_{x \to a} f(x) = L$, then $\lim_{x \to a}[f(x)]^n = L^n$.

---

Next, if we take $f(x) = x$, then Equation (1) and Law 8 give the following result.

---

**LAW 9**    $\lim_{x \to a} x^n = a^n$, where *n* is a positive integer.

---

**EXAMPLE 3**    Find $\lim_{x \to 2}(2x^3 - 4x^2 + 3)$.

**Solution**

$$
\begin{aligned}
\lim_{x \to 2}(2x^3 - 4x^2 + 3) &= \lim_{x \to 2} 2x^3 - \lim_{x \to 2} 4x^2 + \lim_{x \to 2} 3 & \text{Law 3}\\
&= 2 \lim_{x \to 2} x^3 - 4 \lim_{x \to 2} x^2 + \lim_{x \to 2} 3 & \text{Law 5}\\
&= 2(2)^3 - 4(2)^2 + 3 & \text{Law 9}\\
&= 3 &\blacksquare
\end{aligned}
$$

## ■ Limits of Polynomial and Rational Functions

The method of solution that we used in Example 3 can be used to prove the following.

---

**LAW 10**    **Limits of Polynomial Functions**

If $p(x) = a_n x^n + a_{n-1}x^{n-1} + \cdots + a_0$ is a polynomial function, then

$$\lim_{x \to a} p(x) = p(a)$$

Thus, the limit of a polynomial function as $x$ approaches $a$ is equal to the value of the function at $a$.

**PROOF**   Applying the (generalized) sum law and the constant multiple law repeatedly, we find

$$\lim_{x \to a} p(x) = \lim_{x \to a}(a_n x^n + a_{n-1}x^{n-1} + \cdots + a_0)$$

$$= a_n(\lim_{x \to a} x^n) + a_{n-1}(\lim_{x \to a} x^{n-1}) + \cdots + \lim_{x \to a} a_0$$

Next, using Laws 1, 2, and 9, we obtain

$$\lim_{x \to a} p(x) = a_n a^n + a_{n-1}a^{n-1} + \cdots + a_0 = p(a)$$

In light of this, we could have solved the problem posed in Example 3 as follows:

$$\lim_{x \to 2}(2x^3 - 4x^2 + 3) = 2(2)^3 - 4(2)^2 + 3 = 3 \qquad \blacksquare$$

**EXAMPLE 4**   Find $\lim_{x \to -1}(3x^2 + 2x + 1)^5$.

**Solution**

$$\lim_{x \to -1} (3x^2 + 2x + 1)^5 = [\lim_{x \to -1} (3x^2 + 2x + 1)]^5 \qquad \text{Law 8}$$

$$= [3(-1)^2 + 2(-1) + 1]^5 \qquad \text{Law 10}$$

$$= 2^5 = 32 \qquad \blacksquare$$

The following result follows from the Quotient Law for limits and Law 10.

---

**LAW 11   Limits of Rational Functions**

If $f$ is a rational function defined by $f(x) = P(x)/Q(x)$, where $P(x)$ and $Q(x)$ are polynomial functions and $Q(a) \neq 0$, then

$$\lim_{x \to a} f(x) = f(a) = \frac{P(a)}{Q(a)}$$

---

Thus, the limit of a rational function as $x$ approaches $a$ is equal to the value of the function at $a$ provided the denominator is not zero at $a$.

**PROOF**   Since $P$ and $Q$ are polynomial functions, we know from Law 10 that

$$\lim_{x \to a} P(x) = P(a) \qquad \text{and} \qquad \lim_{x \to a} Q(x) = Q(a)$$

Since $Q(a) \neq 0$, we can apply the Quotient Law to conclude that

$$\lim_{x \to a} f(x) = \lim_{x \to a} \frac{P(x)}{Q(x)} = \frac{\lim_{x \to a} P(x)}{\lim_{x \to a} Q(x)} = \frac{P(a)}{Q(a)} = f(a) \qquad \blacksquare$$

**EXAMPLE 5**   Find $\displaystyle \lim_{x \to 3} \frac{4x^2 - 3x + 1}{2x - 4}$.

**Solution**   Using Law 11, we obtain

$$\lim_{x \to 3} \frac{4x^2 - 3x + 1}{2x - 4} = \frac{4(3)^2 - 3(3) + 1}{2(3) - 4} = \frac{28}{2} = 14 \qquad \blacksquare$$

**EXAMPLE 6**   Find $\displaystyle \lim_{x \to 1} \sqrt[3]{\frac{2x + 14}{x^2 + 1}}$.

**Solution**

$$\lim_{x \to 1} \sqrt[3]{\frac{2x + 14}{x^2 + 1}} = \sqrt[3]{\lim_{x \to 1} \frac{2x + 14}{x^2 + 1}} \qquad \text{Law 7}$$

$$= \sqrt[3]{\frac{2(1) + 14}{1^2 + 1}} \qquad \text{Law 11}$$

$$= \sqrt[3]{8} = 2 \qquad \blacksquare$$

Lest you think that we can *always* find the limit of a function by substitution, consider the following example.

**EXAMPLE 7**   Find $\displaystyle \lim_{x \to 2} \frac{x^2 - 4}{x - 2}$.

**Solution**   Because the denominator of the rational expression is 0 at $x = 2$, we cannot find the limit by direct substitution. However, by factoring the numerator, we obtain

$$\frac{x^2 - 4}{x - 2} = \frac{(x + 2)(x - 2)}{x - 2}$$

so if $x \neq 2$, we can cancel the common factors. Thus,

$$\frac{x^2 - 4}{x - 2} = x + 2 \qquad x \neq 2$$

In other words, the values of the function $f$ defined by $f(x) = (x^2 - 4)/(x - 2)$ coincide with the values of the function $g$ defined by $g(x) = x + 2$ for all values of $x$ except $x = 2$. Since the limit of $f(x)$ as $x$ approaches 2 depends only on the values of $x$ other than 2, we can find the required limit by evaluating the limit of $g(x)$ as $x$ approaches 2 instead. Thus,

$$\lim_{x \to 2} \frac{x^2 - 4}{x - 2} = \lim_{x \to 2} (x + 2) = 2 + 2 = 4 \qquad \blacksquare$$

In certain instances the technique that we used in Example 7 can be applied to find the limit of a quotient in which both the numerator and denominator of the quotient approach 0 as $x$ approaches $a$. The trick here is to use the appropriate algebraic manipulations that will enable us to replace the original function by one that is identical to that function except perhaps at $a$. The limit is then found by evaluating this function at $a$.

**Notes**

**1.** If the numerator does not approach 0 but the denominator does, then the limit of the quotient does not exist. (See Example 7 in Section 1.1.)

**2.** A function whose limit at $a$ can be found by evaluating it at $a$ is said to be continuous at $a$. (We will study continuous functions in Section 1.4.)    ■

**EXAMPLE 8**    Find $\displaystyle\lim_{x \to -3} \frac{x^2 + 2x - 3}{x^2 + 4x + 3}$.

**Solution**    Notice that both the numerator and the denominator of the quotient approach 0 as $x$ approaches $-3$, so Law 6 is not applicable. Instead, we proceed as follows:

$$\lim_{x \to -3} \frac{x^2 + 2x - 3}{x^2 + 4x + 3} = \lim_{x \to -3} \frac{(x + 3)(x - 1)}{(x + 3)(x + 1)}$$

$$= \lim_{x \to -3} \frac{x - 1}{x + 1} \qquad x \neq -3$$

$$= \frac{-3 - 1}{-3 + 1} = 2 \qquad\qquad ■$$

**EXAMPLE 9**    Find $\displaystyle\lim_{x \to 0} \frac{\sqrt{1 + x} - 1}{x}$.

**Solution**    Both the numerator and the denominator of the quotient approach 0 as $x$ approaches 0, so we cannot evaluate the limit using Law 6. Let's rationalize the numerator of the quotient by multiplying both the numerator and the denominator by $\sqrt{1 + x} + 1$. Thus,

$$\lim_{x \to 0} \frac{\sqrt{1 + x} - 1}{x} = \lim_{x \to 0} \frac{\sqrt{1 + x} - 1}{x} \cdot \frac{\sqrt{1 + x} + 1}{\sqrt{1 + x} + 1}$$

$$= \lim_{x \to 0} \frac{(\sqrt{1 + x} - 1)(\sqrt{1 + x} + 1)}{x(\sqrt{1 + x} + 1)}$$

$$= \lim_{x \to 0} \frac{1 + x - 1}{x(\sqrt{1 + x} + 1)} \qquad \text{Difference of two squares}$$

$$= \lim_{x \to 0} \frac{1}{\sqrt{1 + x} + 1} = \frac{1}{2} \qquad x \neq 0 \qquad ■$$

All of the limit laws stated for two-sided limits in this section also hold true for one-sided limits.

**EXAMPLE 10**    Let

$$f(x) = \begin{cases} -x + 3 & \text{if } x < 2 \\ \sqrt{x - 2} + 1 & \text{if } x \geq 2 \end{cases}$$

Find $\lim_{x \to 2} f(x)$ if it exists.

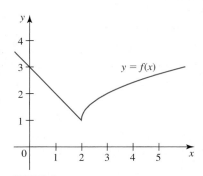

**FIGURE 3**
$\lim_{x \to 2^-} f(x) = \lim_{x \to 2^+} f(x) = 1$, so
$\lim_{x \to 2} f(x) = 1$.

**Solution**   The function $f$ is defined piecewise. For $x \geq 2$ the rule for $f$ is $f(x) = \sqrt{x - 2} + 1$. Letting $x$ approach 2 from the right, we obtain

$$\lim_{x \to 2^+}(\sqrt{x - 2} + 1) = \lim_{x \to 2^+} \sqrt{x - 2} + \lim_{x \to 2^+} 1 \qquad \text{Sum Law}$$

$$= 0 + 1 = 1$$

For $x < 2$, $f(x) = -x + 3$, and

$$\lim_{x \to 2^-}(-x + 3) = \lim_{x \to 2^-}(-x) + \lim_{x \to 2^-} 3 \qquad \text{Sum Law}$$

$$= -2 + 3 = 1$$

The right-hand and left-hand limits are equal. Therefore, the limit exists and

$$\lim_{x \to 2} f(x) = 1$$

The graph of $f$ is shown in Figure 3.   ■

The next example involves the **greatest integer** function defined by $f(x) = [\![x]\!]$, where $[\![x]\!]$ is the greatest integer $n$ such that $n \leq x$. For example, $[\![3]\!] = 3$, $[\![2.4]\!] = 2$, $[\![\pi]\!] = 3$, $[\![-4.6]\!] = -5$, $[\![-\sqrt{2}]\!] = -2$, and so on. As an aid to finding the value of the greatest integer function, think of "rounding down."

**EXAMPLE 11**   Show that $\lim_{x \to 2} [\![x]\!]$ does not exist.

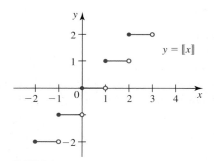

**FIGURE 4**
The graph of $y = [\![x]\!]$

**Solution**   The graph of the greatest integer function is shown in Figure 4. Observe that if $2 \leq x < 3$, then $[\![x]\!] = 2$, and therefore,

$$\lim_{x \to 2^+} [\![x]\!] = \lim_{x \to 2^+} 2 = 2$$

Next, observe that if $1 \leq x < 2$, then $[\![x]\!] = 1$, so

$$\lim_{x \to 2^-} [\![x]\!] = \lim_{x \to 2^-} 1 = 1$$

Since these one-sided limits are not equal, we conclude by Theorem 1, Section 1.1, that $\lim_{x \to 2} [\![x]\!]$ does not exist.   ■

## Limits of Trigonometric Functions

So far, we have dealt with limits involving algebraic functions. The following theorem tells us that if $a$ is a number in the domain of a trigonometric function, then the limit of that function as $x$ approaches $a$ can be found by substitution.

> **THEOREM 1   Limits of Trigonometric Functions**
>
> Let $a$ be a number in the domain of the given trigonometric function. Then
>
> **a.** $\lim_{x \to a} \sin x = \sin a$     **b.** $\lim_{x \to a} \cos x = \cos a$
>
> **c.** $\lim_{x \to a} \tan x = \tan a$     **d.** $\lim_{x \to a} \cot x = \cot a$
>
> **e.** $\lim_{x \to a} \sec x = \sec a$     **f.** $\lim_{x \to a} \csc x = \csc a$

The proofs of Theorem 1a and Theorem 1b are sketched in Exercises 97 and 98. The proofs of the other parts follow from Theorems 1a and 1b and the limit laws.

**EXAMPLE 12**  Find

**a.** $\displaystyle\lim_{x\to\pi/2} x \sin x$     **b.** $\displaystyle\lim_{x\to\pi/4} (2x^2 + \cot x)$

**Solution**

**a.** $\displaystyle\lim_{x\to\pi/2} x \sin x = \left(\lim_{x\to\pi/2} x\right)\left(\lim_{x\to\pi/2} \sin x\right) = \frac{\pi}{2} \sin \frac{\pi}{2} = \frac{\pi}{2}$

**b.** $\displaystyle\lim_{x\to\pi/4} (2x^2 + \cot x) = \lim_{x\to\pi/4} 2x^2 + \lim_{x\to\pi/4} \cot x$

$$= 2\left(\frac{\pi}{4}\right)^2 + \cot \frac{\pi}{4}$$

$$= \frac{\pi^2}{8} + 1 = \frac{\pi^2 + 8}{8}$$  ▪

## ■ The Squeeze Theorem

The techniques that we have developed so far do not work in all situations. For example, they cannot be used to find

$$\lim_{x\to 0} x^2 \sin \frac{1}{x}$$

For limits such as this we use the Squeeze Theorem.

---

**THEOREM 2   The Squeeze Theorem**

Suppose that $f(x) \le g(x) \le h(x)$ for all $x$ in an open interval containing $a$, except possibly at $a$, and

$$\lim_{x\to a} f(x) = L = \lim_{x\to a} h(x)$$

Then

$$\lim_{x\to a} g(x) = L$$

---

The Squeeze Theorem says that if $g(x)$ is squeezed between $f(x)$ and $h(x)$ near $a$ and both $f(x)$ and $h(x)$ approach $L$ as $x$ approaches $a$, then $g(x)$ must approach $L$ as well (see Figure 5). A proof of this theorem is given in Appendix B.

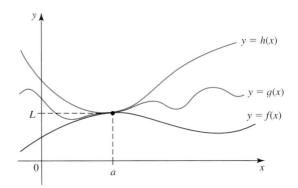

**FIGURE 5**
An illustration of the Squeeze Theorem

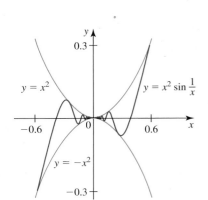

**FIGURE 6**

$$\lim_{x \to 0} g(x) = \lim_{x \to 0} x^2 \sin \frac{1}{x} = 0$$

**EXAMPLE 13**  Find $\lim\limits_{x \to 0} x^2 \sin \dfrac{1}{x}$.

**Solution**  Since $-1 \le \sin t \le 1$ for every real number $t$, we have

$$-1 \le \sin \frac{1}{x} \le 1$$

for every $x \ne 0$. Therefore,

$$-x^2 \le x^2 \sin \frac{1}{x} \le x^2 \qquad x \ne 0$$

Let $f(x) = -x^2$, $g(x) = x^2 \sin(1/x)$, and $h(x) = x^2$. Then $f(x) \le g(x) \le h(x)$. Since

$$\lim_{x \to 0} f(x) = \lim_{x \to 0}(-x^2) = 0 \qquad \text{and} \qquad \lim_{x \to 0} h(x) = \lim_{x \to 0} x^2 = 0$$

the Squeeze Theorem implies that

$$\lim_{x \to 0} g(x) = \lim_{x \to 0} x^2 \sin \frac{1}{x} = 0$$

(See Figure 6.)

The property of limits given in Theorem 3 will be used later. (Its proof is given in Appendix B.)

---

**THEOREM 3**

Suppose that $f(x) \le g(x)$ for all $x$ in an open interval containing $a$, except possibly at $a$, and

$$\lim_{x \to a} f(x) = L \qquad \text{and} \qquad \lim_{x \to a} g(x) = M$$

Then

$$L \le M$$

---

The Squeeze Theorem can be used to prove the following important result, which will be needed in our work later on.

---

**THEOREM 4**

$$\lim_{\theta \to 0} \frac{\sin \theta}{\theta} = 1$$

---

**PROOF**  First, suppose that $0 < \theta < \frac{\pi}{2}$. Figure 7 shows a sector of a circle of radius 1. From the figure we see that

$$\text{Area of } \triangle OAB = \frac{1}{2}(1)(\sin \theta) = \frac{1}{2} \sin \theta \qquad \tfrac{1}{2} \text{base} \cdot \text{height}$$

$$\text{Area of sector } OAB = \frac{1}{2}(1)^2 \theta = \frac{1}{2} \theta \qquad \tfrac{1}{2} r^2 \theta$$

$$\text{Area of } \triangle OAC = \frac{1}{2}(1)(\tan \theta) = \frac{1}{2} \tan \theta \qquad \tfrac{1}{2} \text{base} \cdot \text{height}$$

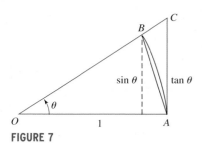

**FIGURE 7**

Since $0 <$ area of $\triangle OAB <$ area of sector $OAB <$ area of $\triangle OAC$, we have

$$0 < \frac{1}{2}\sin\theta < \frac{1}{2}\theta < \frac{1}{2}\tan\theta$$

Multiplying through by $2/(\sin\theta)$ and keeping in mind that $\sin\theta > 0$ and $\cos\theta > 0$ for $0 < \theta < \frac{\pi}{2}$, we obtain

$$1 < \frac{\theta}{\sin\theta} < \frac{1}{\cos\theta}$$

or, upon taking reciprocals,

$$\cos\theta < \frac{\sin\theta}{\theta} < 1 \tag{2}$$

If $-\frac{\pi}{2} < \theta < 0$, then $0 < -\theta < \frac{\pi}{2}$, and Inequality (2) gives

$$\cos(-\theta) < \frac{\sin(-\theta)}{-\theta} < 1$$

or, since $\cos(-\theta) = \cos\theta$ and $\sin(-\theta) = -\sin\theta$, we have

$$\cos\theta < \frac{\sin\theta}{\theta} < 1$$

which is just Inequality (2). Therefore, Inequality (2) holds whenever $\theta$ lies in the intervals $\left(-\frac{\pi}{2}, 0\right)$ and $\left(0, \frac{\pi}{2}\right)$.

   Finally, let $f(\theta) = \cos\theta$, $g(\theta) = (\sin\theta)/\theta$, and $h(\theta) = 1$, and observe that

$$\lim_{\theta\to0} f(\theta) = \lim_{\theta\to0} \cos\theta = 1$$

and

$$\lim_{\theta\to0} h(\theta) = \lim_{\theta\to0} 1 = 1$$

Then the Squeeze Theorem implies that

$$\lim_{\theta\to0} g(\theta) = \lim_{\theta\to0} \frac{\sin\theta}{\theta} = 1 \qquad\blacksquare$$

**EXAMPLE 14**   Find $\displaystyle\lim_{x\to0} \frac{\sin 2x}{3x}$.

**Solution**   We first rewrite

$$\frac{\sin 2x}{3x} \qquad \text{as} \qquad \left(\frac{2}{3}\right)\frac{\sin 2x}{2x}$$

Then, making the substitution $\theta = 2x$ and observing that $\theta \to 0$ as $x \to 0$, we find

$$\lim_{x\to0} \frac{\sin 2x}{3x} = \lim_{\theta\to0}\left(\frac{2}{3}\right)\frac{\sin\theta}{\theta}$$

$$= \frac{2}{3}\lim_{\theta\to0}\frac{\sin\theta}{\theta}$$

$$= \frac{2}{3} \qquad \text{Use Theorem 4.} \qquad\blacksquare$$

**EXAMPLE 15** Find $\lim\limits_{x \to 0} \dfrac{\tan x}{x}$.

**Solution**

$$\lim_{x \to 0} \frac{\tan x}{x} = \lim_{x \to 0} \left( \frac{\sin x}{x} \cdot \frac{1}{\cos x} \right)$$

$$= \left( \lim_{x \to 0} \frac{\sin x}{x} \right) \left( \lim_{x \to 0} \frac{1}{\cos x} \right)$$

$$= (1)(1)$$

$$= 1$$

Theorem 5 is a consequence of Theorem 4.

---

**THEOREM 5**

$$\lim_{\theta \to 0} \frac{\cos \theta - 1}{\theta} = 0$$

---

**PROOF** We use the identity $\sin^2 x = \frac{1}{2}(1 - \cos 2x)$ to write

$$1 - \cos \theta = 2 \sin^2 \left( \frac{\theta}{2} \right) \qquad \text{Let } x = \frac{\theta}{2}.$$

Then

$$\lim_{\theta \to 0} \frac{\cos \theta - 1}{\theta} = \lim_{\theta \to 0} \left( \frac{-2 \sin^2 \left( \frac{\theta}{2} \right)}{\theta} \right)$$

$$= \lim_{\theta \to 0} \left( -\sin \frac{\theta}{2} \right) \left( \frac{\sin \frac{\theta}{2}}{\frac{\theta}{2}} \right)$$

$$= -\left( \lim_{\theta \to 0} \sin \frac{\theta}{2} \right) \left( \lim_{\theta \to 0} \frac{\sin \frac{\theta}{2}}{\frac{\theta}{2}} \right) \qquad \text{Note: } \frac{\theta}{2} \to 0 \text{ as } \theta \to 0.$$

$$= 0 \cdot 1 = 0$$

---

## 1.2 CONCEPT QUESTIONS

1. State the Sum, Product, Constant Multiple, Quotient, and Root Laws for limits at a number.

2. Find the limit and state the limit law that you use at each step.

   **a.** $\lim\limits_{x \to 2} (3x^2 - 2x + 1)$   **b.** $\lim\limits_{x \to 3} \dfrac{x^2 + 4}{2x + 3}$

3. Find the limit and state the limit law that you use at each step.

   **a.** $\lim\limits_{x \to 4} \sqrt{x}(2x^2 + 1)$   **b.** $\lim\limits_{x \to 1} \left( \dfrac{2x^2 + x + 5}{x^4 + 1} \right)^{3/2}$

4. State the Squeeze Theorem in your own words, and give a graphical interpretation.

## 1.2  EXERCISES

*In Exercises 1–22, find the indicated limit.*

**1.** $\lim\limits_{t\to 2}(3t + 4)$

**2.** $\lim\limits_{x\to 2}(3x^2 + 2x - 8)$

**3.** $\lim\limits_{h\to -1}(h^4 - 2h^3 + 2h - 1)$

**4.** $\lim\limits_{x\to 2}(x^2 + 1)(2x^2 - 4)$

**5.** $\lim\limits_{x\to 1}(3x^2 - 4x + 2)^4$

**6.** $\lim\limits_{t\to 3}(2t - 1)^2(t^2 - 2t)^3$

**7.** $\lim\limits_{x\to 1}\dfrac{x - 2}{x^2 + x + 1}$

**8.** $\lim\limits_{t\to -1}\dfrac{t^3 - 1}{t^3 - 2t + 4}$

**9.** $\lim\limits_{x\to 2}\left(\sqrt{2x^3} - \sqrt{2}x\right)$

**10.** $\lim\limits_{x\to 3}\sqrt{2x^3 - 3x + 7}$

**11.** $\lim\limits_{x\to -1^+}(x^3 - 2x^2 - 5)^{2/3}$

**12.** $\lim\limits_{x\to -2}(x + 3)^2\sqrt{4x^2 - 8}$

**13.** $\lim\limits_{x\to 0^+}\dfrac{1 + \sqrt{x}}{\sqrt{x + 4}}$

**14.** $\lim\limits_{t\to 4}t^{-1/2}(t^2 - 3t + 4)^{3/2}$

**15.** $\lim\limits_{u\to -2}\sqrt[3]{\dfrac{3u^2 + 2u}{3u^3 - 3}}$

**16.** $\lim\limits_{w\to 0}\dfrac{\sqrt{w + 1} - \sqrt{w^2 + 4}}{(w + 2)^2 - (w + 1)^2}$

**17.** $\lim\limits_{x\to 1}\sin\dfrac{\pi x}{2}$

**18.** $\lim\limits_{x\to \pi/4}(x\tan x)$

**19.** $\lim\limits_{x\to \pi/4}\dfrac{\sin x}{x}$

**20.** $\lim\limits_{x\to 0}\dfrac{\sec 2x}{\sqrt{x + 4}}$

**21.** $\lim\limits_{x\to \pi}\sqrt{2 + \cos x}$

**22.** $\lim\limits_{x\to \pi/4}\dfrac{\tan^2 x}{1 + \cos x}$

*In Exercises 23–28, you are given that $\lim_{x\to a}f(x) = 2$, $\lim_{x\to a}g(x) = 4$, and $\lim_{x\to a}h(x) = -1$. Find the indicated limit.*

**23.** $\lim\limits_{x\to a}[2f(x) + 3g(x)]$

**24.** $\lim\limits_{x\to a}\dfrac{f(x) + g(x)}{2h(x)}$

**25.** $\lim\limits_{x\to a}\dfrac{f(x)}{\sqrt{g(x)}}$

**26.** $\lim\limits_{x\to a}\dfrac{f(x)g(x)}{\sqrt{g(x)} + 5}$

**27.** $\lim\limits_{x\to a}\{[h(x)]^2 - f(x)g(x)\}$

**28.** $\lim\limits_{x\to a}\dfrac{\sqrt[3]{f(x)g(x)}}{\sqrt{f(x)g(x) + 1}}$

*In Exercises 29 and 30, suppose that $\lim_{x\to -2}f(x) = 2$ and $\lim_{x\to -2}g(x) = 3$. Find the indicated limit.*

**29.** $\lim\limits_{x\to -2}[xf(x) + (x^2 + 1)g(x)]$

**30.** $\lim\limits_{x\to -2}\dfrac{xf(x)}{1 + x^2}$

*In Exercises 31–36, use the graphs of f and g that follow to find the indicated limit, if it exists. If the limit does not exist, explain why.*

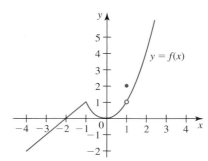

The graph of $f$

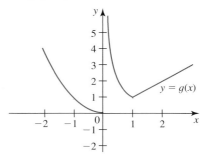

The graph of $g$

**31.** $\lim\limits_{x\to -1}[f(x) + g(x)]$

**32.** $\lim\limits_{x\to 0}[f(x) - g(x)]$

**33.** $\lim\limits_{x\to 1}[f(x)g(x)]$

**34.** $\lim\limits_{x\to 2}\dfrac{f(x)}{g(x)}$

**35.** $\lim\limits_{x\to 0^-}[2f(x) + 3g(x)]$

**36.** $\lim\limits_{x\to 0^+}\dfrac{f(x)}{g(x)}$

**37.** Is the following argument correct?

$$f(x) = \frac{x^2 - 9}{x + 3} = \frac{(x + 3)(x - 3)}{x + 3} = x - 3$$

Therefore, $\lim_{x\to -3}f(x) = f(-3) = -6$. Explain your answer.

**38.** Is the following argument correct?

$$\lim_{x\to -3}\frac{x^2 - 9}{x + 3} = \lim_{x\to -3}\frac{(x + 3)(x - 3)}{x + 3} = \lim_{x\to -3}(x - 3) = -6$$

Explain your answer. Compare it with Exercise 37.

**39.** Give an example to illustrate the following: If $\lim_{x \to a} f(x) = L \neq 0$ and $\lim_{x \to a} g(x) = 0$, then $\lim_{x \to a} [f(x)/g(x)]$ does not exist.

**40.** Give examples to illustrate the following: If $\lim_{x \to a} f(x) = 0$ and $\lim_{x \to a} g(x) = 0$, then $\lim_{x \to a} [f(x)/g(x)]$ might or might not exist.

*In Exercises 41–76, find the limit, if it exists.*

**41.** $\lim\limits_{x \to 2} \dfrac{x^2 - 4}{x - 2}$

**42.** $\lim\limits_{x \to 5} \dfrac{5 - x}{x^2 - 25}$

**43.** $\lim\limits_{t \to 1} \dfrac{t + 1}{(t - 1)^2}$

**44.** $\lim\limits_{x \to 2^+} \dfrac{x + 1}{x - 2}$

**45.** $\lim\limits_{x \to 1} \dfrac{x^2 + 2x - 3}{x^2 - 1}$

**46.** $\lim\limits_{x \to 2} \dfrac{x^2 - x - 2}{x - 2}$

**47.** $\lim\limits_{x \to -1} \sqrt{\dfrac{2 + x - x^2}{x^2 + 4x + 3}}$

**48.** $\lim\limits_{x \to -5} \sqrt{\dfrac{x^2 - 25}{2x^2 + 6x - 20}}$

**49.** $\lim\limits_{t \to 0} \dfrac{2t^3 + 3t^2}{3t^4 - 2t^2}$

**50.** $\lim\limits_{t \to 1} \dfrac{3t^3 + 4t + 1}{(t - 1)(2t^2 + 1)}$

**51.** $\lim\limits_{x \to 1} \dfrac{x^3 - 1}{x - 1}$

**52.** $\lim\limits_{v \to 2} \dfrac{v^4 - 16}{v^2 - 4}$

**53.** $\lim\limits_{t \to 1} \dfrac{\sqrt{t} - 1}{t - 1}$

**54.** $\lim\limits_{x \to 4} \dfrac{x - 4}{\sqrt{x} - 2}$

**55.** $\lim\limits_{t \to 1} \dfrac{\sqrt{t} + 1}{t - 1}$

**56.** $\lim\limits_{t \to 0} \dfrac{t}{\sqrt{2t + 1} - 1}$

**57.** $\lim\limits_{x \to 0} \dfrac{\sqrt{x + 3} - \sqrt{3}}{x}$

**58.** $\lim\limits_{h \to 0} \dfrac{\sqrt{a + h} - \sqrt{a}}{h}$

**59.** $\lim\limits_{x \to 1} \dfrac{\sqrt{5 - x} - 2}{\sqrt{2 - x} - 1}$

**60.** $\lim\limits_{h \to 0} \dfrac{(2 + h)^{-1} - 2^{-1}}{h}$

**61.** $\lim\limits_{x \to 7^-} [\![x]\!]$

**62.** $\lim\limits_{x \to -5^+} [\![x]\!]$

**63.** $\lim\limits_{x \to 2^-} (x - [\![x]\!])$

**64.** $\lim\limits_{x \to 3^+} [\![x + 1]\!]$

**65.** $\lim\limits_{x \to 0} \dfrac{\sin x}{3x}$

**66.** $\lim\limits_{x \to 0} \dfrac{\sin 2x}{x}$

**67.** $\lim\limits_{h \to 0^+} \dfrac{\cos^{-1}(1 - h)}{\sqrt{h}}$

**68.** $\lim\limits_{x \to 0} \dfrac{\tan 2x}{3x}$

Hint: Let $x = \cos^{-1}(1 - h)$.

**69.** $\lim\limits_{x \to 0} \dfrac{\tan^2 x}{x}$

**70.** $\lim\limits_{x \to 0} \dfrac{\cos x - 1}{\sin x}$

**71.** $\lim\limits_{\theta \to 0} \dfrac{\cos \theta - 1}{\theta^2}$

**72.** $\lim\limits_{x \to 0} \dfrac{x}{1 - \cos^2 x}$

**73.** $\lim\limits_{x \to \pi/4} \dfrac{\sin x - \cos x}{1 - \tan x}$

**74.** $\lim\limits_{\theta \to 0} \dfrac{\theta}{\cos\left(\theta - \frac{\pi}{2}\right)}$

**75.** $\lim\limits_{x \to 0} \dfrac{\tan^{-1} x}{x}$

**76.** $\lim\limits_{x \to 0^+} \sqrt{\dfrac{\tan x - \sin x}{x^2}}$

Hint: Let $u = \tan^{-1} x$.

**77.** Find $\lim\limits_{x \to \pi/2} \dfrac{\cos x}{x - \frac{\pi}{2}}$.

Hint: Let $t = x - (\pi/2)$.

**78.** Find $\lim\limits_{x \to \pi/2} \dfrac{\sin\left(x - \frac{\pi}{2}\right)}{2x - \pi}$.

Hint: Let $t = 2x - \pi$.

 **79.** Let $f(x) = \dfrac{x - 1}{\sqrt[3]{x + 7} - 2}$.

    **a.** Plot the graph of $f$, and use it to estimate the value of $\lim_{x \to 1} f(x)$.

    **b.** Construct a table of values of $f(x)$ accurate to three decimal places, and use it to estimate $\lim_{x \to 1} f(x)$.

    **c.** Find the exact value of $\lim_{x \to 1} f(x)$ analytically.

    Hint: Make the substitution $x + 7 = t^3$, and observe that $t \to 2$ as $x \to 1$.

 **80.** Let $f(x) = \dfrac{x + 2}{\sqrt[4]{x + 18} - 2}$.

    **a.** Plot the graph of $f$, and use it to estimate the value of $\lim_{x \to -2} f(x)$.

    **b.** Construct a table of values of $f(x)$ accurate to three decimal places, and use it to estimate $\lim_{x \to -2} f(x)$.

    **c.** Find the exact value of $\lim_{x \to -2} f(x)$ analytically.

    Hint: Make the substitution $x + 18 = t^4$, and observe that $t \to 2$ as $x \to -2$.

**81. Special Theory of Relativity** According to the special theory of relativity, when force and velocity are both along a straight line, resulting in straight-line motion, the magnitude of the acceleration of a particle acted upon by the force is

$$a = f(v) = \frac{F}{m}\left(1 - \frac{v^2}{c^2}\right)^{3/2}$$

where $v$ is its speed, $F$ is the magnitude of the force, $m$ is the mass of the particle at rest, and $c$ is the speed of light.

    **a.** Find the domain of $f$, and use this result to explain why we may consider only $\lim_{v \to c^-} f(v)$.

    **b.** Find $\lim_{v \to c^-} f(v)$, and interpret your result.

**82. Special Theory of Relativity** According to the special theory of relativity, the speed of a particle is

$$v = c\sqrt{1 - \left(\frac{E_0}{E}\right)^2}$$

where $E_0 = m_0 c^2$ is the rest energy and $E$ is the total energy.

    **a.** Find the domain of $v$, use this result to explain why we may consider only $\lim_{E \to E_0^+} v$, and interpret your result.

    **b.** Find $\lim_{E \to E_0^+} v$, and interpret your result.

 **83.** Use the Squeeze Theorem to find $\lim_{x \to 0} x \sin(1/x)$. Verify your result visually by plotting the graphs of $f(x) = -x$, $g(x) = x \sin(1/x)$, and $h(x) = x$ in the same window.

**84.**  Use the Squeeze Theorem to find $\lim_{x\to 0^+} \sqrt{x}\cos(1/x^2)$. Verify your result visually.
Hint: See Exercise 83.

**85.** Let
$$f(x) = \begin{cases} x + 2 & \text{if } x < -1 \\ x^2 + 2x + 3 & \text{if } x > -1 \end{cases}$$
  **a.** Find $\lim_{x\to -1^-} f(x)$ and $\lim_{x\to -1^+} f(x)$.
  **b.** Does $\lim_{x\to -1} f(x)$ exist? Why?

**86.** Let
$$f(x) = \begin{cases} \dfrac{x^3 - 16}{x} & \text{if } x < -2 \\ -x^2 - 4x + 8 & \text{if } x > -2 \end{cases}$$
Does $\lim_{x\to -2} f(x)$ exist? If so, what is its value?

**87.** Let
$$f(x) = \begin{cases} -x^5 + x^3 + x + 1 & \text{if } x < 0 \\ 2 & \text{if } x = 0 \\ x^2 + \sqrt{x + 1} & \text{if } x > 0 \end{cases}$$
Find $\lim_{x\to 0^-} f(x)$ and $\lim_{x\to 0^+} f(x)$. Does $\lim_{x\to 0} f(x)$ exist? Justify your answer.

**88.** Let
$$f(x) = \begin{cases} \sqrt{1 - x} + 2 & \text{if } x < 1 \\ 1 & \text{if } x = 1 \\ 1 + x^{3/2} & \text{if } x > 1 \end{cases}$$
Find $\lim_{x\to 1^-} f(x)$ and $\lim_{x\to 1^+} f(x)$. Does $\lim_{x\to 1} f(x)$ exist? Justify your answer.

**89.** Let
$$f(x) = \begin{cases} [\![x]\!] & \text{if } x < 2 \\ \sqrt{x - 2} + 1 & \text{if } x \geq 2 \end{cases}$$
Does $\lim_{x\to 2} f(x)$ exist? If so, what is its value?

**90.** Let
$$f(x) = \begin{cases} |x| & \text{if } x \leq 1 \\ [\![x]\!] & \text{if } x > 1 \end{cases}$$
Does $\lim_{x\to 1} f(x)$ exist? If so, what is its value?

**91.** Let
$$f(x) = \begin{cases} x^2 & \text{if } x \text{ is rational} \\ -x^2 & \text{if } x \text{ is irrational} \end{cases}$$
Show that $\lim_{x\to 0} f(x) = 0$.

**92. The Dirichlet Function** The function
$$f(x) = \begin{cases} 1 & \text{if } x \text{ is rational} \\ 0 & \text{if } x \text{ is irrational} \end{cases}$$
is called the *Dirichlet function*. For example, $f\left(\frac{1}{2}\right) = 1$, $f\left(\frac{20}{21}\right) = 1, f(\sqrt{2}) = 0$, and $f(-\pi) = 0$. Show that for every $a$, $\lim_{x\to a} f(x)$ does not exist.

**93.** Show by means of an example that $\lim_{x\to a}[f(x) + g(x)]$ may exist even though neither $\lim_{x\to a} f(x)$ nor $\lim_{x\to a} g(x)$ exists. Does this example contradict the Sum Law of limits?

**94.** Show by means of an example that $\lim_{x\to a}[f(x)g(x)]$ may exist even though neither $\lim_{x\to a} f(x)$ nor $\lim_{x\to a} g(x)$ exists. Does this example contradict the Product Law of limits?

**95.** Suppose that $f(x) < g(x)$ for all $x$ in an open interval containing $a$, except possibly at $a$, and that both $\lim_{x\to a} f(x)$ and $\lim_{x\to a} g(x)$ exist. Does it follow that $\lim_{x\to a} f(x) < \lim_{x\to a} g(x)$? Explain.

**96.** The following figure shows a sector of radius 1 and angle $\theta$ satisfying $0 < \theta < \frac{\pi}{2}$.

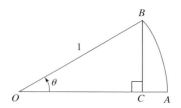

  **a.** From the inequality $|BC| < \text{arc } AB$, deduce that $0 < \sin \theta < \theta$.
  **b.** Use the Squeeze Theorem to prove that $\lim_{\theta\to 0^+} \sin \theta = 0$.
  **c.** Use the result of part (a) to show that if $-\frac{\pi}{2} < \theta < 0$, then $\lim_{\theta\to 0^-} \sin \theta = 0$. Conclude that $\lim_{\theta\to 0} \sin \theta = 0$.
  **d.** Use the result of part (c) and the trigonometric identity $\sin^2 \theta + \cos^2 \theta = 1$ to show that $\lim_{\theta\to 0} \cos \theta = 1$.

**97.** Use the result of Exercise 96 to prove that $\lim_{x\to a} \sin x = \sin a$.
Hint: It suffices to show that $\lim_{h\to 0} \sin(a + h) = \sin a$. Use the addition formula for the sine function.

**98.** Show that $\lim_{x\to a} \cos x = \cos a$. (See the hint for Exercise 97.)

*In Exercises 99–102, determine whether the statement is true or false. If it is true, explain why. If it is false, explain why or give an example that shows it is false.*

**99.** $\lim_{x\to 2}\left(\dfrac{3x}{x - 2} - \dfrac{2}{x - 2}\right) = \lim_{x\to 2}\dfrac{3x}{x - 2} - \lim_{x\to 2}\dfrac{2}{x - 2}$.

**100.** $\lim_{x\to 1}\dfrac{x^2 + 3x - 4}{x^2 - 2x - 3} = \dfrac{\lim_{x\to 1} x^2 + 3x - 4}{\lim_{x\to 1} x^2 - 2x - 3}$

**101.** If $\lim_{x\to a}[f(x) - g(x)]$ exists, then $\lim_{x\to a} f(x)$ and $\lim_{x\to a} g(x)$ also exist.

**102.** If $f(x) \leq g(x) \leq h(x)$ for all $x$ in an open interval containing $a$, except possibly at $a$, and both $\lim_{x\to a} f(x)$ and $\lim_{x\to a} h(x)$ exist, then $\lim_{x\to a} g(x)$ exists.

## 1.3   A Precise Definition of a Limit

### ■ Precise Definition of a Limit

The definition of the limit of a function given in Section 1.1 is intuitive. In this section we give precise meaning to phrases such as "$f(x)$ can be made as close to $L$ as we please" and "by taking $x$ to be sufficiently close to $a$." We will focus our attention on the (two-sided) limit

$$\lim_{x \to a} f(x) = L \tag{1}$$

where $a$ and $L$ are real numbers. (The precise definition of one-sided limits is given in Exercise 28.)

Let's begin by investigating how we might establish the result

$$\lim_{x \to 2}(2x - 1) = 3 \tag{2}$$

with some degree of mathematical rigor. Here, $f(x) = 2x - 1$, $a = 2$, and $L = 3$. We need to show that "$f(x)$ can be made as close to 3 as we please by taking $x$ to be sufficiently close to 2."

Our first step is to establish what we mean by "$f(x)$ is close to 3." For a start, suppose that we invite a challenger to specify some sort of "tolerance." For example, our challenger might declare that $f(x)$ is close to 3 provided that $f(x)$ differs from 3 by no more than 0.1 unit. Recalling that $|f(x) - 3|$ measures the distance from $f(x)$ to 3, we can rephrase this statement by saying that $f(x)$ is close to 3 provided that

$$|f(x) - 3| < 0.1 \qquad \text{Equivalently, } 2.9 < f(x) < 3.1. \tag{3}$$

(See Figure 1.)

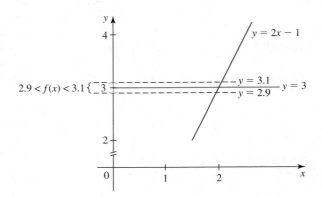

**FIGURE 1**
All the values of $f$ satisfying
$2.9 < f(x) < 3.1$ are "close" to 3.

Now let's show that Inequality (3) is satisfied by all $x$ that are "sufficiently close to 2." Because $|x - 2|$ measures the distance from $x$ to 2, what we need to do is to show that there exists some positive number, call it $\delta$ (delta), such that

$$0 < |x - 2| < \delta \qquad \text{implies that} \qquad |f(x) - 3| < 0.1$$

(The first half of the first inequality precludes the possibility of $x$ taking on the value 2. Remember that when we evaluate the limit of a function at a number $a$, we are not concerned with whether $f$ is defined at $a$ or its value there if it is defined.) To find $\delta$, consider

$$|f(x) - 3| = |(2x - 1) - 3| = |2x - 4| = |2(x - 2)|$$
$$= 2|x - 2|$$

Now, $2|x - 2| < 0.1$ holds whenever

$$|x - 2| < \frac{0.1}{2} = 0.05 \qquad \qquad (4)$$

Therefore, if we pick $\delta = 0.05$, then $0 < |x - 2| < \delta$ implies that Inequality (4) holds. This in turn implies that

$$|f(x) - 3| = 2|x - 2| < 2(0.05) = 0.1$$

as we set out to show. (See Figure 2.)

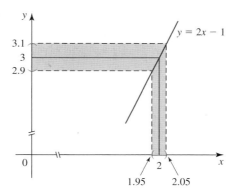

**FIGURE 2**
Whenever $x$ satisfies $|x - 2| < 0.05$,
$f(x)$ satisfies $|f(x) - 3| < 0.1$.

Have we established Equation (2)? The answer is a resounding no! What we have demonstrated is that by restricting $x$ to be sufficiently close to 2, $f(x)$ can be made "close to 3" as measured by the norm, or tolerance, specified by one particular challenger. Another challenger might specify that "$f(x)$ is close to 3" if the tolerance is $10^{-20}$! If you retrace these last steps, you can show that corresponding to a tolerance of $10^{-20}$, we can make $|f(x) - 3| < 10^{-20}$ by requiring that $0 < |x - 2| < 5 \times 10^{-21}$. (Choose $\delta = 5 \times 10^{-21}$.)

To handle *all* such possible notions of closeness that could arise, suppose that a tolerance is given by specifying a number $\varepsilon$ (epsilon) that may be *any* positive number whatsoever. Can we show that $f(x)$ is close to 3 (with tolerance $\varepsilon$) by restricting $x$ to be sufficiently close to 2? In other words, given any number $\varepsilon > 0$, can we find a number $\delta > 0$ such that

$$|f(x) - 3| < \varepsilon \qquad \text{whenever} \qquad 0 < |x - 2| < \delta$$

All we have to do to answer these questions is to repeat the earlier computations with $\varepsilon$ in place of 0.1. Consider

$$|f(x) - 3| = |(2x - 1) - 3| = |2x - 4| = 2|x - 2|$$

Now,

$$2|x - 2| < \varepsilon \qquad \text{provided that} \qquad |x - 2| < \frac{\varepsilon}{2}$$

Therefore, if we pick $\delta = \varepsilon/2$, then $0 < |x - 2| < \delta$ implies that $|x - 2| < \varepsilon/2$, which implies that

$$|f(x) - 3| = 2|x - 2| < 2\left(\frac{\varepsilon}{2}\right) = \varepsilon$$

Now, because $\varepsilon$ is arbitrary, we have indeed shown that "$f(x)$ can be made as close to 3 as we please" by restricting $x$ to be sufficiently close to 2.

This analysis suggests the following precise definition of a limit.

> **DEFINITION (Precise)    Limit of a Function at a Number**
>
> Let $f$ be a function defined on an open interval containing $a$ with the possible exception of $a$ itself. Then the limit of $f(x)$ as $x$ approaches $a$ is the number $L$, written
>
> $$\lim_{x \to a} f(x) = L$$
>
> if for every number $\varepsilon > 0$, we can find a number $\delta > 0$ such that
>
> $$0 < |x - a| < \delta \qquad \text{implies that} \qquad |f(x) - L| < \varepsilon$$

## ■ A Geometric Interpretation

Here is a geometric interpretation of the definition. Let $\varepsilon > 0$ be given. Draw the lines $y = L + \varepsilon$ and $y = L - \varepsilon$. Since $|f(x) - L| < \varepsilon$ is equivalent to $L - \varepsilon < f(x) < L + \varepsilon$, $\lim_{x \to a} f(x) = L$ exists provided that we can find a number $\delta$ such that if we restrict $x$ to lie in the interval $(a - \delta, a + \delta)$ with $x \neq a$, then the graph of $y = f(x)$ lies inside the band of width $2\varepsilon$ determined by the lines $y = L + \varepsilon$ and $y = L - \varepsilon$. (See Figure 3.) You can see from Figure 3 that once a number $\delta > 0$ has been found, then any number smaller than $\delta$ will also satisfy the requirement.

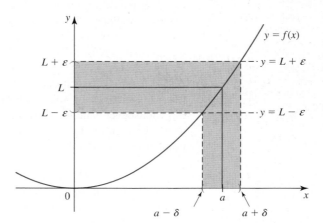

**FIGURE 3**
If $x \in (a - \delta, a)$ or $(a, a + \delta)$, then $f(x)$ lies in the band defined by $y = L + \varepsilon$ and $y = L - \varepsilon$.

## ■ Some Illustrative Examples

**EXAMPLE 1**    Prove that $\displaystyle\lim_{x \to 2} \frac{4(x^2 - 4)}{x - 2} = 16$. (Recall that this limit gives the instantaneous velocity of the maglev at $x = 2$ as described in Section 1.1.)

## Historical Biography

### SOPHIE GERMAIN
(1776-1831)

Overcoming great adversity, Sophie Germain won acknowledgment for her mathematical works from some of the most prominent mathematicians of her day. Born in 1776 in Paris to a prosperous bourgeois family, Germain was able to devote herself to research without financial concerns but also without the education accorded to women of the aristocracy. Germain became interested in geometry, an interest that her family deemed inappropriate for a woman. In an effort to prevent her studying at night, her family confiscated her candles and left her bedroom fire unlit in order to keep her in her bed. Determined, Germain would wait until the family was asleep, wrap herself in quilts, and study through the night by the light of contraband candles. Despite having to study alone and to teach herself Latin in order to read the mathematics of Newton (page 202) and Euler (page 19), Germain eventually made important breakthroughs in the fields of number theory and the theory of elasticity. She anonymously entered a paper into a contest sponsored by the French Academy of Sciences. She won the prize and became the first woman not related to a member by marriage to attend Academie des Sciences meetings and the first woman invited to attend sessions at the Institut de France.

**FIGURE 4**

If we pick $\delta = \varepsilon/4$, then

$$0 < |x - 2| < \delta \Rightarrow$$

$$\left| \frac{4(x^2 - 4)}{x - 2} - 16 \right| < \varepsilon.$$

**Solution**   Let $\varepsilon > 0$ be given. We must show that there exists a $\delta > 0$ such that

$$\left| \frac{4(x^2 - 4)}{x - 2} - 16 \right| < \varepsilon$$

whenever $0 < |x - 2| < \delta$. To find $\delta$, consider

$$\left| \frac{4(x^2 - 4)}{x - 2} - 16 \right| = \left| \frac{4(x - 2)(x + 2)}{x - 2} - 16 \right|$$

$$= |4(x + 2) - 16| = |4x - 8| \qquad x \neq 2$$

$$= 4|x - 2|$$

Therefore,

$$\left| \frac{4(x^2 - 4)}{x - 2} - 16 \right| = 4|x - 2| < \varepsilon$$

whenever

$$|x - 2| < \frac{1}{4}\varepsilon$$

So we may take $\delta = \varepsilon/4$. (See Figure 4.)

By reversing the steps, we see that if $0 < |x - 2| < \delta$, then

$$\left| \frac{4(x^2 - 4)}{x - 2} - 16 \right| = 4|x - 2| < 4\left(\frac{1}{4}\varepsilon\right) = \varepsilon$$

Thus,

$$\lim_{x \to 2} \frac{4(x^2 - 4)}{x - 2} = 16$$

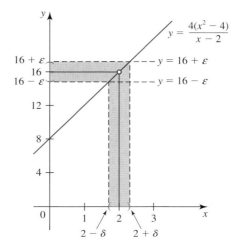

**EXAMPLE 2**   Prove that $\lim_{x \to 2} x^2 = 4$.

**Solution**   Let $\varepsilon > 0$ be given. We must show that there exists a $\delta > 0$ such that

$$|x^2 - 4| < \varepsilon$$

whenever $|x - 2| < \delta$. To find $\delta$, consider

$$|x^2 - 4| = |(x + 2)(x - 2)|$$
$$= |x + 2||x - 2| \qquad (5)$$

At this stage, one might be tempted to set

$$|x + 2||x - 2| < \varepsilon$$

and then divide both sides of this inequality by $|x + 2|$ to obtain

$$|x - 2| < \frac{\varepsilon}{|x + 2|}$$

and conclude that we may take

$$\delta = \frac{\varepsilon}{|x + 2|}$$

But this approach will not work because $\delta$ *cannot depend on* $x$. Let us begin afresh with Equation (5). On the basis of the experience just gained, we should obtain an upper bound for the quantity $|x + 2|$; that is, we want to find a positive number $k$ such that $|x + 2| < k$ for all $x$ "close to 2." As we observed earlier, once a $\delta$ has been found that satisfies our requirement, then any number smaller than $\delta$ will also do. This allows us to agree beforehand to take $\delta \le 1$ (or any other positive constant); that is, we will consider only those values of $x$ that satisfy $|x - 2| < 1$; that is $-1 < x - 2 < 1$, or $1 < x < 3$. Adding 2 to each side of this last inequality, we have $1 + 2 < x + 2 < 3 + 2$; $3 < x + 2 < 5$; thus, $|x + 2| < 5$. So $k = 5$, and Equation (5) gives

$$|x^2 - 4| = |x + 2||x - 2| < 5|x - 2|$$

Now

$$5|x - 2| < \varepsilon$$

whenever $|x - 2| < \varepsilon/5$. Therefore, if we take $\delta$ to be the smaller of the numbers 1 and $\varepsilon/5$, we are guaranteed that $|x - 2| < \delta$ implies that

$$|x^2 - 4| < 5|x - 2| < 5\left(\frac{\varepsilon}{5}\right) = \varepsilon$$

This proves the assertion (see Figure 5).

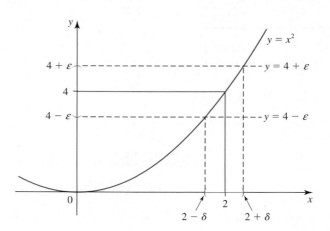

**FIGURE 5**
If we pick $\delta$ to be the smaller of 1 and $\varepsilon/5$, then $|x - 2| < \delta \Rightarrow |x^2 - 4| < \varepsilon$.

**EXAMPLE 3** Let

$$f(x) = \begin{cases} 1 & \text{if } x \geq 0 \\ -1 & \text{if } x < 0 \end{cases}$$

Prove that $\lim_{x \to 0} f(x)$ does not exist.

**Solution**  Suppose that the limit exists. We will show that this assumption leads to a contradiction. It will follow, therefore, that the opposite is true, namely, the limit does not exist.

So suppose that there exists a number $L$ such that

$$\lim_{x \to 0} f(x) = L$$

Then, for every $\varepsilon > 0$ there exists a $\delta > 0$ such that

$$|f(x) - L| < \varepsilon \qquad \text{whenever} \qquad 0 < |x - 0| < \delta$$

In particular, if we take $\varepsilon = 1$, there exists a $\delta > 0$ such that

$$|f(x) - L| < 1 \qquad \text{whenever} \qquad 0 < |x - 0| < \delta$$

If we take $x = -\delta/2$, which lies in the interval defined by $0 < |x - 0| < \delta$, we have

$$\left| f\left(-\frac{\delta}{2}\right) - L \right| = |-1 - L| < 1$$

This inequality is equivalent to

$$-1 < -1 - L < 1$$
$$0 < -L < 2$$

or

$$-2 < L < 0$$

Next, if we take $x = \delta/2$, which also lies in the interval defined by $0 < |x - 0| < \delta$, we have

$$\left| f\left(\frac{\delta}{2}\right) - L \right| = |1 - L| < 1$$

This inequality is equivalent to

$$-1 < 1 - L < 1$$
$$-2 < -L < 0$$

or

$$0 < L < 2$$

But the number $L$ cannot satisfy both the inequalities

$$-2 < L < 0 \qquad \text{and} \qquad 0 < L < 2$$

simultaneously. This contradiction proves that $\lim_{x \to 0} f(x)$ does not exist.  ▪

We end this section by proving the Sum Law for limits.

**EXAMPLE 4**   Prove the Sum Law for limits: If $\lim_{x \to a} f(x) = L$ and $\lim_{x \to a} g(x) = M$, then $\lim_{x \to a}[f(x) + g(x)] = L + M$.

**Solution**   Let $\varepsilon > 0$ be given. We must show that there exists a $\delta > 0$ such that

$$|[f(x) + g(x)] - (L + M)| < \varepsilon$$

whenever $0 < |x - a| < \delta$. But by the Triangle Inequality,*

$$|[f(x) + g(x)] - (L + M)| = |(f(x) - L) + (g(x) - M)|$$

$$\leq |f(x) - L| + |g(x) - M| \tag{6}$$

and this suggests that we consider the bounds for $|f(x) - L|$ and $|g(x) - M|$ separately.

Since $\lim_{x \to a} f(x) = L$, we can take $\varepsilon/2$, which is a positive number, and be guaranteed that there exists a $\delta_1 > 0$ such that

$$|f(x) - L| < \frac{\varepsilon}{2} \qquad \text{whenever} \qquad 0 < |x - a| < \delta_1 \tag{7}$$

Similarly, since $\lim_{x \to a} g(x) = M$, we can find a $\delta_2 > 0$ such that

$$|g(x) - M| < \frac{\varepsilon}{2} \qquad \text{whenever} \qquad 0 < |x - a| < \delta_2 \tag{8}$$

If we take $\delta$ to be the smaller of the two numbers $\delta_1$ and $\delta_2$ so that $\delta$ is itself positive, then both Inequalities (7) and (8) hold simultaneously if $0 < |x - a| < \delta$. Therefore, by Inequality (6)

$$|[f(x) + g(x)] - (L + M)| \leq |f(x) - L| + |g(x) - M|$$

$$< \frac{\varepsilon}{2} + \frac{\varepsilon}{2} = \varepsilon$$

whenever $0 < |x - a| < \delta$, and this proves the Sum Law.    ■

---

*The Triangle Inequality $|a + b| \leq |a| + |b|$ is proved in Appendix A.

## 1.3   CONCEPT QUESTIONS

1. State the precise definition of $\lim_{x \to 2}(x^3 + 5) = 13$.
2. Write the precise definition of $\lim_{x \to a} f(x) = L$ without using absolute values.
3. Use the figure to find a number $\delta$ such that $|x^2 - 1| < \frac{1}{2}$ whenever $|x - 1| < \delta$.

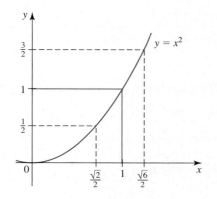

4. Use the figure to find a number $\delta$ such that $|\frac{1}{x} - 1| < \frac{1}{4}$ whenever $|x - 1| < \delta$.

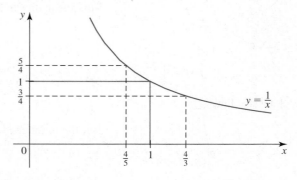

## 1.3   EXERCISES

*In Exercises 1–10 you are given* $\lim_{x \to a} f(x) = L$ *and a toler-ance* $\varepsilon$. *Find a number* $\delta$ *such that* $|f(x) - L| < \varepsilon$ *whenever* $0 < |x - a| < \delta$.

**1.** $\lim\limits_{x \to 2} 3x = 6$;   $\varepsilon = 0.01$

**2.** $\lim\limits_{x \to -1} 2x = -2$;   $\varepsilon = 0.001$

**3.** $\lim\limits_{x \to 1}(2x + 3) = 5$;   $\varepsilon = 0.01$

**4.** $\lim\limits_{x \to -2}(3x - 2) = -8$;   $\varepsilon = 0.05$

**5.** $\lim\limits_{x \to 3} \dfrac{x^2 - 9}{x - 3} = 6$;   $\varepsilon = 0.02$

**6.** $\lim\limits_{x \to -2} \dfrac{x^2 - 4}{x + 2} = -4$;   $\varepsilon = 0.005$

**7.** $\lim\limits_{x \to 3} 2x^2 = 18$;   $\varepsilon = 0.01$

**8.** $\lim\limits_{x \to 4} \sqrt{x} = 2$;   $\varepsilon = 0.01$

**9.** $\lim\limits_{x \to 2} \dfrac{x^2 + 4}{x + 2} = 2$;   $\varepsilon = 0.01$

**10.** $\lim\limits_{x \to 2} \dfrac{1}{x} = \dfrac{1}{2}$;   $\varepsilon = 0.05$

*In Exercises 11–22, use the precise definition of a limit to prove that the statement is true.*

**11.** $\lim\limits_{x \to 2} 3 = 3$

**12.** $\lim\limits_{x \to -2} \pi = \pi$

**13.** $\lim\limits_{x \to 3} 2x = 6$

**14.** $\lim\limits_{x \to -2}(2x - 3) = -7$

**15.** $\lim\limits_{x \to a} c = c$

**16.** $\lim\limits_{x \to a} x = a$

**17.** $\lim\limits_{x \to 1} 3x^2 = 3$

**18.** $\lim\limits_{x \to 2}(x^2 - 2) = 2$

**19.** $\lim\limits_{x \to 2} \dfrac{x^2 - 4}{x - 2} = 4$

**20.** $\lim\limits_{x \to 0} \dfrac{x^2 + 2x}{x} = 2$

**21.** $\lim\limits_{x \to 9} \sqrt{x} = 3$

**22.** $\lim\limits_{x \to 0}(x^3 + 1) = 1$

**23.** Let
$$f(x) = \begin{cases} -1 & \text{if } x < 0 \\ 1 & \text{if } x \geq 0 \end{cases}$$
Prove that $\lim_{x \to 0} f(x)$ does not exist.

**24.** Let
$$g(x) = \begin{cases} -1 + x & \text{if } x < 0 \\ 1 + x & \text{if } x \geq 0 \end{cases}$$
Prove that $\lim_{x \to 0} g(x)$ does not exist.

**25.** Prove that $\lim_{x \to 0} H(x)$ does not exist, where $H$ is the Heaviside function
$$H(x) = \begin{cases} 0 & \text{if } x < 0 \\ 1 & \text{if } x \geq 0 \end{cases}$$

**26.** Let
$$f(x) = \begin{cases} 0 & \text{if } x \text{ is rational} \\ 1 & \text{if } x \text{ is irrational} \end{cases}$$
Prove that $\lim_{x \to 0} f(x)$ does not exist.

**27.** Prove the Constant Multiple Law for limits: If $\lim_{x \to a} f(x) = L$ and $c$ is a constant, then $\lim_{x \to a}[cf(x)] = cL$.

**28.** The precise definition of the left-hand limit, $\lim_{x \to a^-} f(x) = L$, may be stated as follows: For every number $\varepsilon > 0$ there exists a number $\delta > 0$ such that $|f(x) - L| < \varepsilon$ whenever $a - \delta < x < a$. Similarly, for the right-hand limit, $\lim_{x \to a^+} f(x) = L$ if for every number $\varepsilon > 0$ there exists a number $\delta > 0$ such that $|f(x) - L| < \varepsilon$ whenever $a < x < a + \delta$. Explain, with the aid of figures, why these definitions are appropriate.

**29.** Use the definition in Exercise 28 to prove that $\lim_{x \to 2^-} \sqrt[4]{4 - x^2} = 0$.

**30.** Use the definition in Exercise 28 to prove that $\lim_{x \to 2^+} \sqrt{x - 2} = 0$.

*In Exercises 31–34, determine whether the statement is true or false. If it is true, explain why. If it is false, explain why or give an example that shows it is false.*

**31.** The limit of $f(x)$ as $x$ approaches $a$ is $L$ if there exists a number $\varepsilon > 0$ such that for all $\delta > 0$, $|f(x) - L| < \varepsilon$ whenever $0 < |x - a| < \delta$.

**32.** If $\lim_{x \to a} f(x) = L$, then given the number 0.01, there exists a $\delta > 0$ such that $0 < |x - a| < \delta$ implies that $|f(x) - L| < 0.01$.

**33.** The limit of $f(x)$ as $x$ approaches $a$ is $L$ if for all $\varepsilon > 0$, there exists a $\delta > 0$ such that $|f(x) - L| < \varepsilon$ whenever $0 < |x - a| < \delta$.

**34.** The limit of $f(x)$ as $x$ approaches $a$ is $L$ if for all $\delta > 0$, there exists an $\varepsilon > 0$, such that $|f(x) - L| < \varepsilon$ whenever $0 < |x - a| < \delta$.

## 1.4    Continuous Functions

### ■ Continuous Functions

The graph of the function

$$s = f(t) = 4t^2 \qquad 0 \le t \le 30$$

giving the position of the maglev at any time $t$ (discussed in Section 1.1) is shown in Figure 1. Observe that the curve has no holes or jumps. This tells us that the displacement of the maglev must vary continuously with respect to time—it cannot vanish at any instant of time, and it cannot skip a stretch of the track to reappear and resume its motion somewhere else. The function $s$ is an example of a *continuous function*. Observe that you can draw the graph of this function without lifting your pencil from the paper.

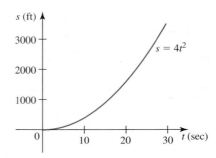

**FIGURE 1**
$s = f(t) = 4t^2$ gives the position of the maglev at any time $t$.

Functions that are *discontinuous* also occur in practical applications. Consider, for example, the Heaviside function $H$ defined by

$$H(t) = \begin{cases} 0 & \text{if } t < 0 \\ 1 & \text{if } t \ge 0 \end{cases}$$

and first introduced in Example 3 in Section 1.1. You can see from the graph of $H$ that it has a jump at $t = 0$ (Figure 2). If we think of $H$ as describing the flow of current in an electrical circuit, then $t = 0$ corresponds to the time at which the switch is turned on. The function $H$ is discontinuous at 0.

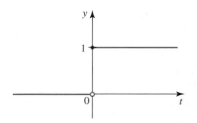

**FIGURE 2**
The Heaviside function is discontinuous at $t = 0$.

### ■ Continuity at a Number

We now give a formal definition of continuity.

> **DEFINITION    Continuity at a Number**
>
> Let $f$ be a function defined on an open interval containing all values of $x$ close to $a$. Then $f$ **is continuous at $a$** if
>
> $$\lim_{x \to a} f(x) = f(a) \qquad (1)$$

If we write $x = a + h$ and note that $x$ approaches $a$ as $h$ approaches 0, we see that the condition for $f$ to be continuous at $a$ is equivalent to

$$\lim_{h \to 0} f(a + h) = f(a) \qquad (2)$$

Briefly, $f$ is continuous at $a$ if $f(x)$ gets closer and closer to $f(a)$ as $x$ approaches $a$. Equivalently, $f$ is continuous at $a$ if proximity of $x$ to $a$ implies proximity of $f(x)$ to $f(a)$. (See Figure 3.)

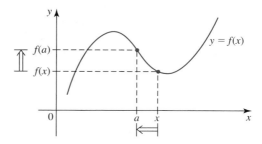

**FIGURE 3**
As $x$ approaches $a, f(x)$ approaches $f(a)$.

If $f$ is defined for all values of $x$ close to $a$ but Equation (1) is not satisfied, then $f$ is **discontinuous at $a$** or $f$ has a **discontinuity at $a$.**

**Note**  It is implicit in Equation (1) that $f(a)$ is defined and the $\lim_{x \to a} f(x)$ exists. However, for emphasis we sometimes define continuity at $a$ by requiring that the following three conditions hold: (1) $f(a)$ is defined, (2) $\lim_{x \to a} f(x)$ exists, and (3) $\lim_{x \to a} f(x) = f(a)$.   ■

**EXAMPLE 1**  Use the graph of the function shown in Figure 4 to determine whether $f$ is continuous at 0, 1, 2, 3, 4, and 5.

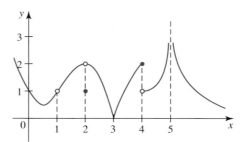

**FIGURE 4**
The graph of $f$

**Solution**  The function $f$ is continuous at 0 because

$$\lim_{x \to 0} f(x) = 1 = f(0)$$

It is discontinuous at 1 because $f(1)$ is not defined. It is discontinuous at 2 because

$$\lim_{x \to 2} f(x) = 2 \neq 1 = f(2)$$

Since

$$\lim_{x \to 3} f(x) = 0 = f(3)$$

we see that $f$ is continuous at 3. Next, we see that $\lim_{x \to 4} f(x)$ does not exist, so $f$ is not continuous at 4. Finally, because $\lim_{x \to 5} f(x)$ does not exist, we see that $f$ is discontinuous at 5.   ■

Refer to the function $f$ in Example 1. The discontinuity at 1 and at 2, where the limit exists, is called a **removable discontinuity** because $f$ can be made continuous at each of these numbers by defining or redefining it there. For example, if we define $f(1) = 1$, then $f$ is made continuous at 1; if we redefine $f(2)$ by specifying that $f(2) = 2$, then $f$ is also made continuous at 2.

The discontinuity at 4 is called a **jump discontinuity,** whereas the discontinuity at 5 is called an **infinite discontinuity.** Because the limit does not exist at a jump or at an infinite discontinuity, the discontinuity cannot be removed by defining or redefining the function at the number in question.

**EXAMPLE 2** Let

$$f(x) = \begin{cases} \dfrac{x^2 - x - 2}{x - 2} & \text{if } x \neq 2 \\ 1 & \text{if } x = 2 \end{cases}$$

Show that $f$ has a removable discontinuity at 2. Redefine $f$ at 2 so that it is continuous everywhere.

**Solution**   First, let's find the limit of $f(x)$ as $x$ approaches 2:

$$\lim_{x \to 2} \frac{x^2 - x - 2}{x - 2} = \lim_{x \to 2} \frac{(x - 2)(x + 1)}{x - 2}$$

$$= \lim_{x \to 2}(x + 1) = 3$$

Because $\lim_{x \to 2} f(x) = 3 \neq 1 = f(2)$, we see that $f$ is discontinuous at 2. We can remove this discontinuity and thus render $f$ continuous everywhere by redefining the value of $f$ at 2 to be equal to 3. (See Figure 5.)

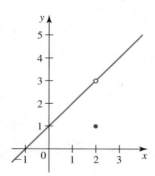

(a) $f$ has a removable discontinuity at 2.

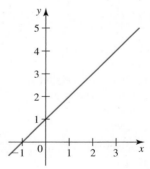

(b) $f$ is continuous at 2.

**FIGURE 5**
The discontinuity at 2 is removed by redefining $f$ at $x = 2$.

## ■ Continuity at an Endpoint

When we defined continuity, we assumed that $f(x)$ was defined for all values of $x$ close to $a$. Sometimes $f(x)$ is defined only for those values of $x$ that are greater than or equal to $a$ or for values of $x$ that are less than or equal to $a$. For example, $f(x) = \sqrt{x}$ is defined for $x \geq 0$, and $g(x) = \sqrt{3 - x}$ is defined for $x \leq 3$. The following definition covers these situations.

> **DEFINITION** Continuity from the Right and from the Left
>
> A function $f$ is **continuous from the right at $a$** if
>
> $$\lim_{x \to a^+} f(x) = f(a) \tag{3a}$$
>
> A function $f$ is **continuous from the left at $a$** if
>
> $$\lim_{x \to a^-} f(x) = f(a) \tag{3b}$$
>
> (See Figure 6.)

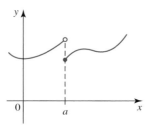 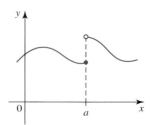

**FIGURE 6**

(a) $f$ is continuous from the right at $a$.

(b) $f$ is continuous from the left at $a$.

**EXAMPLE 3** **The Heaviside Function** Consider the Heaviside function $H$ defined by

$$H(t) = \begin{cases} 0 & \text{if } t < 0 \\ 1 & \text{if } t \geq 0 \end{cases}$$

Determine whether $H$ is continuous from the right at 0 and/or from the left at 0.

**Solution** Because

$$\lim_{t \to 0^+} H(t) = \lim_{t \to 0^+} 1 = 1$$

and this is equal to $H(0) = 1$, $H$ is continuous from the right at 0. Next, because

$$\lim_{t \to 0^-} H(t) = \lim_{t \to 0^-} (0) = 0$$

and this is not equal to $H(0) = 1$, $H$ is not continuous from the left at 0. (See Figure 7.)

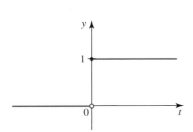

**FIGURE 7**
The Heaviside function $H$ is continuous from the right at the number 0.

**Note** It follows from the definition of continuity that a function $f$ is continuous at $a$ if and only if $f$ is simultaneously continuous from the right and from the left at $a$.

## ■ Continuity on an Interval

You might have noticed that continuity is a "local" concept; that is, we say that $f$ is continuous at a number. The following definition tells us what it means to say that a function is continuous on an interval.

### DEFINITION    Continuity on Open and Closed Intervals

A function *f* **is continuous on an open interval** $(a, b)$ if it is continuous at every number in the interval. A function *f* **is continuous on a closed interval** $[a, b]$ if it is continuous on $(a, b)$ and is also continuous from the right at *a* and from the left at *b*. A function *f* **is continuous on a half-open interval** $[a, b)$ or $(a, b]$ if *f* is continuous on $(a, b)$ and *f* is continuous from the right at *a* or *f* is continuous from the left at *b*, respectively.

**EXAMPLE 4**  Show that the function *f* defined by $f(x) = \sqrt{4 - x^2}$ is continuous on the closed interval $[-2, 2]$.

**Solution**  We first show that *f* is continuous on $(-2, 2)$. Let *a* be any number in $(-2, 2)$. Then, using the laws of limits, we have

$$\lim_{x \to a} f(x) = \lim_{x \to a} \sqrt{4 - x^2} = \sqrt{\lim_{x \to a}(4 - x^2)} = \sqrt{4 - a^2} = f(a)$$

and this proves the assertion.

Next, let us show that *f* is continuous from the right at $-2$ and from the left at 2. Again, by invoking the limit properties, we see that

$$\lim_{x \to -2^+} f(x) = \lim_{x \to -2^+} \sqrt{4 - x^2} = \sqrt{\lim_{x \to -2^+}(4 - x^2)} = 0 = f(-2)$$

and

$$\lim_{x \to 2^-} f(x) = \lim_{x \to 2^-} \sqrt{4 - x^2} = \sqrt{\lim_{x \to 2^-}(4 - x^2)} = 0 = f(2)$$

and this proves the assertion. Therefore, *f* is continuous on $[-2, 2]$. The graph of *f* is shown in Figure 8.

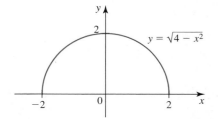

**FIGURE 8**
The function $f(x) = \sqrt{4 - x^2}$ is continuous on $[-2, 2]$.

### THEOREM 1    Continuity of a Sum, Product, and Quotient

If the functions *f* and *g* are continuous at *a*, then the following functions are also continuous at *a*.

**a.** $f \pm g$
**b.** $fg$
**c.** $cf$,   where *c* is any constant
**d.** $\dfrac{f}{g}$,   if $g(a) \neq 0$

We will prove Theorem 1b and leave some of the other parts as exercises. (See Exercises 94–95.)

### PROOF OF THEOREM 1b

Since *f* and *g* are continuous at *a*, we have

$$\lim_{x \to a} f(x) = f(a) \qquad \text{and} \qquad \lim_{x \to a} g(x) = g(a)$$

By the Product Law for limits,

$$\lim_{x \to a}[f(x)g(x)] = \lim_{x \to a} f(x) \cdot \lim_{x \to a} g(x) = f(a)g(a)$$

so $fg$ is continuous at $a$.  ■

**Note**  As in the case of the Sum Law and the Product Law, Theorems 1a and 1b can be extended to the case involving finitely many functions.  ■

The following theorem is an immediate consequence of Laws 10 and 11 for limits from Section 1.2.

---

**THEOREM 2  Continuity of Polynomial and Rational Functions**

**a.** A polynomial function is continuous on $(-\infty, \infty)$.
**b.** A rational function is continuous on its domain.

---

**EXAMPLE 5**  Find the values of $x$ for which the function

$$f(x) = x^8 - 3x^4 + x + 4 + \frac{x + 1}{(x + 1)(x - 2)}$$

is continuous.

**Solution**  We can think of the function $f$ as the sum of the polynomial function $g(x) = x^8 - 3x^4 + x + 4$ and the rational function $h(x) = (x + 1)/[(x + 1)(x - 2)]$. By Theorem 2 we see that $g$ is continuous on $(-\infty, \infty)$, whereas $h$ is continuous everywhere except at $-1$ and $2$. Therefore, $f$ is continuous on $(-\infty, -1)$, $(-1, 2)$, and $(2, \infty)$.  ■

If you examine the graphs of the sine and cosine functions, you can see that they are continuous on $(-\infty, \infty)$. You will be asked to provide a rigorous demonstration of this in Exercises 92 and 93. Since the other trigonometric functions are defined in terms of these two functions, the continuity of the other trigonometric functions can be determined from them.

---

**THEOREM 3  Continuity of Trigonometric Functions**

The functions $\sin x$, $\cos x$, $\tan x$, $\sec x$, $\csc x$, and $\cot x$ are continuous at every number in their respective domain.

---

For example, since $\tan x = (\sin x)/(\cos x)$, we see that $\tan x$ is continuous everywhere except at the values of $x$ where $\cos x = 0$; that is, except at $\pi/2 + n\pi$, where $n$ is an integer. In other words, $f(x) = \tan x$ is continuous on

$$\ldots, \left(-\frac{3\pi}{2}, -\frac{\pi}{2}\right), \quad \left(-\frac{\pi}{2}, \frac{\pi}{2}\right), \quad \left(\frac{\pi}{2}, \frac{3\pi}{2}\right), \quad \ldots$$

**EXAMPLE 6** Find the values at which the following functions are continuous.

**a.** $f(x) = x \cos x$      **b.** $g(x) = \dfrac{\sqrt{x}}{\sin x}$

**Solution**

**a.** Since the functions $x$ and $\cos x$ are continuous everywhere, we conclude that $f$ is continuous on $(-\infty, \infty)$.

**b.** The function $\sqrt{x}$ is continuous on $[0, \infty)$. The function $\sin x$ is continuous everywhere and has zeros at $n\pi$, where $n$ is an integer. It follows from Theorem 1d, that $g$ is continuous at all positive values of $x$ that are not integral multiples of $\pi$; that is, $g$ is continuous on $(0, \pi)$, $(\pi, 2\pi)$, $(2\pi, 3\pi), \ldots$ ∎

Because of the *reflective property* of inverse functions, we might expect that $f$ and $f^{-1}$ have similar properties. Thus, if $f$ is continuous on its domain, then we might expect $f^{-1}$ to be continuous on its domain. We give a proof of this in Appendix B. As a consequence of this result, Theorem 3, and the continuity of the exponential function (by the very way it is defined), we have the following.

---

**THEOREM 4**    **Continuity of Inverse Functions, Inverse Trigonometric Functions, Exponential Functions, and Logarithmic Functions**

If $f$ is continuous on its domain, then $f^{-1}$ is continuous on its domain. Also, the functions

$$\sin^{-1} x, \quad \cos^{-1} x, \quad \tan^{-1} x, \quad \sec^{-1} x, \quad \csc^{-1} x, \quad \cot^{-1} x, \quad a^x, \quad \text{and} \quad \log_a x$$

are continuous on their respective domains.

---

**EXAMPLE 7** Find the values at which the function

$$f(x) = \frac{\tan^{-1} x + e^x}{(\log x)\sqrt{2 - x}}$$

is continuous.

**Solution**    Since both the functions $\tan^{-1} x$ and $e^x$ are continuous on $(-\infty, \infty)$, we see that $\tan^{-1} x + e^x$ is continuous on $(-\infty, \infty)$. Next, $\log x$ is continuous on $(0, \infty)$, and since we require that $2 - x > 0$, or $x < 2$, we see that $f$ is continuous on $(0, 2)$. ∎

## ◼ Continuity of Composite Functions

The following theorem shows us how to compute the limit of a composite function $f \circ g$ where $f$ is continuous.

---

**THEOREM 5**    **Limit of a Composite Function**

If the function $f$ is continuous at $L$ and $\lim_{x \to a} g(x) = L$, then

$$\lim_{x \to a} f(g(x)) = f(L)$$

Intuitively, Theorem 5 is plausible because as $x$ approaches $a$, $g(x)$ approaches $L$. Since $f$ is continuous at $L$, proximity of $g(x)$ to $L$ implies proximity of $f(g(x))$ to $f(L)$, which is what the theorem asserts. Theorem 5 is proved in Appendix B.

**Note** Theorem 5 states that the limit symbol can be moved through a continuous function. Thus,

$$\lim_{x \to a} f(g(x)) = f(\lim_{x \to a} g(x)) = f(L) \qquad \blacksquare$$

It follows from Theorem 5 that compositions of continuous functions are also continuous.

---

**THEOREM 6    Continuity of Composite Functions**

If the function $g$ is continuous at $a$ and the function $f$ is continuous at $g(a)$, then the composition $f \circ g$ is continuous at $a$.

---

**PROOF**    We compute

$$\lim_{x \to a}(f \circ g)(x) = \lim_{x \to a} f(g(x))$$

$$= f(\lim_{x \to a} g(x)) \qquad \text{Theorem 5}$$

$$= f(g(a)) \qquad \text{Since } g \text{ is continuous at } a$$

$$= (f \circ g)(a)$$

which is precisely the condition for $f \circ g$ to be continuous at $a$.    $\blacksquare$

### EXAMPLE 8

**a.** Show that $h(x) = |x|$ is continuous everywhere.
**b.** Use the result of part (a) to evaluate

$$\lim_{x \to 1} \left| \frac{-x^2 - x + 2}{x - 1} \right|$$

**Solution**

**a.** Since $|x| = \sqrt{x^2}$ for all $x$, we can view $h$ as $h = f \circ g$, where $g(x) = x^2$ and $f(x) = \sqrt{x}$. Now $g$ is continuous on $(-\infty, \infty)$, and $g(x) \geq 0$ for all $x$ in $(-\infty, \infty)$. Also, $f$ is continuous on $[0, \infty)$. Therefore, Theorem 6 says that $h = f \circ g$ is continuous on $(-\infty, \infty)$.

**b.** By the continuity of the absolute value function established in part (a) and Theorem 5, we find

$$\lim_{x \to 1} \left| \frac{-x^2 - x + 2}{x - 1} \right| = \left| \lim_{x \to 1} \frac{-x^2 - x + 2}{x - 1} \right|$$

$$= \left| \lim_{x \to 1} \frac{-(x - 1)(x + 2)}{x - 1} \right|$$

$$= \left| \lim_{x \to 1} (-1)(x + 2) \right| = |-3| = 3. \qquad \blacksquare$$

**Historical Biography**

**MARIN MERSENNE**
(1588-1648)

Father Marin Mersenne was a close friend of Descartes (page 6), Fermat (page 348), and many other mathematicians, scientists, and philosophers of the early 1600s. Referred to as the "correspondent extraordinaire," he is best remembered for his extensive exchanges of letters with the brightest European scholars of the time. Through Mersenne the French mathematicians learned of one another's thoughts on newly developed mathematical concepts. Mersenne's name is also preserved in connection with prime numbers of the form $2^p - 1$, where $p$ is prime. The search for such primes continues today through the Great Internet Mersenne Prime Search (GIMPS). In 2008 a German electrical engineer discovered the largest known Mersenne prime: $2^{37,156,667} - 1$. This number is 11,185,272 digits long!

**EXAMPLE 9**    Find the intervals where the following functions are continuous.

**a.** $f(x) = \cos(\sqrt{3}x + 4)$          **b.** $g(x) = x^2 \sin \dfrac{1}{x}$

**Solution**

**a.** We can view $f$ as a composition, $g \circ h$, of the functions $g(x) = \cos x$ and $h(x) = \sqrt{3}x + 4$. Since each of these functions is continuous everywhere, we conclude that $f$ is continuous on $(-\infty, \infty)$.

**b.** The function $f(x) = \sin(1/x)$ is the composition of the functions $h(x) = \sin x$ and $k(x) = 1/x$. Since $h$ is continuous everywhere and $k$ is continuous everywhere except at 0, Theorem 6 says that the function $f = h \circ k$ is continuous on $(-\infty, 0)$ and $(0, \infty)$. Also, the function $F(x) = x^2$ is continuous everywhere. Therefore, we conclude by Theorem 1b that $g$, which is the product of $F$ and $f$, is continuous on $(-\infty, 0)$ and $(0, \infty)$. The graph of $g$ is shown in Figure 9.

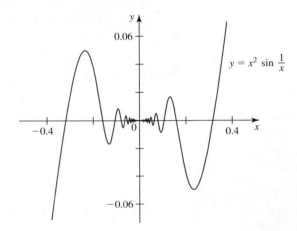

**FIGURE 9**
$g$ is continuous everywhere except at 0.

## ▮ Intermediate Value Theorem

Let's look again at our model of the motion of the maglev on a straight stretch of track. We know that the train cannot vanish at any instant of time, and it cannot skip portions of the track and reappear someplace else. To put it another way, the train cannot occupy the positions $s_1$ and $s_2$ without at least, at some time, occupying every intermediate position (Figure 10). To state this fact mathematically, recall that the position of the maglev as a function of time is described by

$$s = f(t) = 4t^2 \qquad 0 \le t \le 30$$

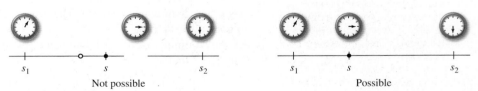

**FIGURE 10**
Position of the maglev

Suppose that the position of the maglev is $s_1$ at some time $t_1$ and that its position is $s_2$ at some time $t_2$. (See Figure 11.) Then if $\bar{s}$ is any number between $s_1$ and $s_2$ giving an intermediate position of the maglev, there must be at least one $\bar{t}$ between $t_1$ and $t_2$ giving the time at which the train is at $\bar{s}$; that is, $f(\bar{t}) = \bar{s}$.

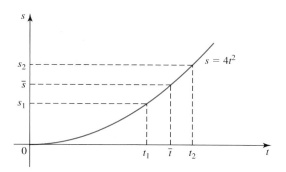

**FIGURE 11**
If $s_1 \leq \bar{s} \leq s_2$, then there must be at least one $\bar{t}$, where $t_1 \leq \bar{t} \leq t_2$, such that $f(\bar{t}) = \bar{s}$.

This discussion carries the gist of the Intermediate Value Theorem.

---

**THEOREM 7    The Intermediate Value Theorem**

If $f$ is a continuous function on a closed interval $[a, b]$ and $M$ is any number between $f(a)$ and $f(b)$, inclusive, then there is at least one number $c$ in $[a, b]$ such that $f(c) = M$. (See Figure 12.)

---

**FIGURE 12**
If $f$ is continuous on $[a, b]$ and $f(a) \leq M \leq f(b)$, then there is at least one $c$, where $a \leq c \leq b$ such that $f(c) = M$.

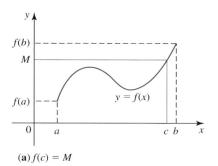

**(a)** $f(c) = M$

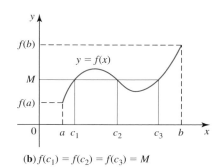

**(b)** $f(c_1) = f(c_2) = f(c_3) = M$

To illustrate the Intermediate Value Theorem, let's look at the example involving the motion of the maglev again (see Figure 1 in Section 1.1). Notice that the initial position of the train is $f(0) = 0$ and that the position at the end of its test run is $f(30) = 3600$. Furthermore, the function $f$ is continuous on $[0, 30]$. So the Intermediate Value Theorem guarantees that if we arbitrarily pick a number between 0 and 3600, say, 400, giving the position of the maglev, there must be a $\bar{t}$ between 0 and 30 at which time the train is at the position $\bar{s} = 400$. To find the value of $\bar{t}$, we solve the equation $f(\bar{t}) = \bar{s}$, or

$$4\bar{t}^2 = 400$$

giving $\bar{t} = 10$. (Note that $\bar{t}$ must lie between 0 and 30.)

 Remember that when you use Theorem 7, the function $f$ must be continuous. The conclusion of the Intermediate Value Theorem might not hold if $f$ is not continuous (see Exercise 70).

The next theorem is an immediate consequence of the Intermediate Value Theorem. It not only tells us when a *zero of a function* $f$ (root of the equation $f(x) = 0$) exists but also provides the basis for a method of approximating it.

**THEOREM 8    Existence of Zeros of a Continuous Function**

If $f$ is a continuous function on a closed interval $[a, b]$ and $f(a)$ and $f(b)$ have opposite signs, then the equation $f(x) = 0$ has at least one solution in the interval $(a, b)$ or, equivalently, the function $f$ has at least one zero in the interval $(a, b)$. (See Figure 13.)

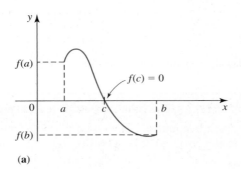

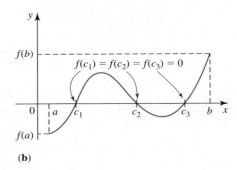

**FIGURE 13**
If $f(a)$ and $f(b)$ have opposite signs, there must be at least one number $c$, where $a < c < b$, such that $f(c) = 0$.

(a)

(b)

**EXAMPLE 10** Let $f(x) = x^3 + x - 1$. Since $f$ is a polynomial, it is continuous everywhere. Observe that $f(0) = -1$ and $f(1) = 1$, so Theorem 8 guarantees the existence of at least one root of the equation $f(x) = 0$ in $(0, 1)$.* We can locate the root more precisely by using Theorem 8 once again as follows: Evaluate $f(x)$ at the midpoint of $[0, 1]$. Thus,

$$f(0.5) = -0.375$$

Because $f(0.5) < 0$ and $f(1) > 0$, Theorem 8 now tells us that a root must lie in $(0.5, 1)$. Repeat the process: Evaluate $f(x)$ at the midpoint of $[0.5, 1]$, which is

$$\frac{0.5 + 1}{2} = 0.75$$

Thus,

$$f(0.75) = 0.171875$$

Because $f(0.5) < 0$ and $f(0.75) > 0$, Theorem 8 tells us that a root is in $(0.5, 0.75)$. This process can be continued. Table 1 summarizes the results of our computations through nine steps. From Table 1 we see that the root is approximately 0.68, accurate to two decimal places. By continuing the process through a sufficient number of steps, we can obtain as accurate an approximation to the root as we please. ∎

**Note**    The process of finding the root of $f(x) = 0$ used in Example 10 is called the method of bisection. It is crude but effective. Later, we will look at a more efficient method, called the Newton-Raphson method, for finding the roots of $f(x) = 0$. ∎

**TABLE 1**

| Step | Root of $f(x) = 0$ lies in |
|------|------------------------------|
| 1 | $(0, 1)$ |
| 2 | $(0.5, 1)$ |
| 3 | $(0.5, 0.75)$ |
| 4 | $(0.625, 0.75)$ |
| 5 | $(0.625, 0.6875)$ |
| 6 | $(0.65625, 0.6875)$ |
| 7 | $(0.671875, 0.6875)$ |
| 8 | $(0.6796875, 0.6875)$ |
| 9 | $(0.6796875, 0.68359375)$ |

*It can be shown that $f$ has exactly one zero in $(0, 1)$ (see Exercise 90).

## 1.4 CONCEPT QUESTIONS

1. Explain what it means for a function $f$ to be continuous (a) at a number $a$, (b) from the right at $a$, and (c) from the left at $a$. Give examples.
2. Explain what it means for a function $f$ to be continuous (a) on an open interval $(a, b)$ and (b) on a closed interval $[a, b]$. Give examples.
3. Determine whether each function $f$ is continuous or discontinuous. Explain your answer.
   a. $f(t)$ gives the altitude of an airplane at time $t$.
   b. $f(t)$ measures the total amount of rainfall at time $t$ over the past 24 hr at the municipal airport.

   c. $f(t)$ is the price of admission for an adult at a movie theater as a function of time on a weekday.
   d. $f(t)$ is the speed of a pebble at time $t$ when it is dropped from a height of 6 ft into a swimming pool.
4. a. Suppose that $\lim_{x \to a^-} f(x)$ and $\lim_{x \to a^+} f(x)$ both exist and $f$ is discontinuous at $a$. Under what conditions does $f$ have a removable discontinuity at $a$?
   b. Suppose that $f$ is continuous from the left at $a$ and continuous from the right at $a$. What can you say about the continuity of $f$ at $a$? Explain.

## 1.4 EXERCISES

In Exercises 1–6, use the graph to determine where the function is discontinuous.

1.
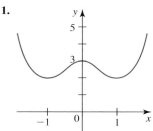

2.
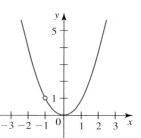

3.
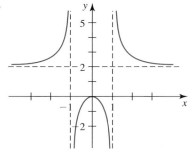

4.
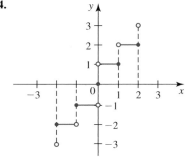

5.
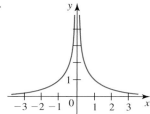

6.
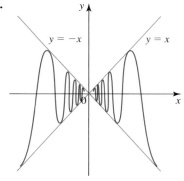

In Exercises 7–26, find the numbers, if any, where the function is discontinuous.

7. $f(x) = 2x^3 - 3x^2 + 4$

8. $f(x) = \dfrac{3}{x^2 + 1}$

9. $f(x) = \dfrac{e^x}{x - 2}$

10. $f(x) = \dfrac{\cos x}{x^2 - 1}$

11. $f(x) = \dfrac{x - 2}{x^2 - 4}$

12. $f(x) = \dfrac{x + 1}{x^2 - 2x - 3}$

13. $f(x) = \dfrac{x^2 - 3x + 2}{x^2 - 2x}$

14. $f(x) = |x^3 - 2x + 1|$

**15.** $f(x) = \left| \dfrac{x + 2}{x^2 + 2x} \right|$

**16.** $f(x) = \tan^{-1} x + \dfrac{x + 1}{|x + 1|}$

**17.** $f(x) = x - [\![x]\!]$

**18.** $f(x) = [\![x - 2]\!]$

**19.** $f(x) = \begin{cases} \tan^{-1} \left| \dfrac{1}{x - 5} \right| & \text{if } x \neq 5 \\ \dfrac{\pi}{2} & \text{if } x = 5 \end{cases}$

**20.** $f(x) = \begin{cases} x + 2 & \text{if } x < 3 \\ \ln(x - 2) + 5 & \text{if } x \geq 3 \end{cases}$

**21.** $f(x) = \begin{cases} \dfrac{x^2 - 1}{x + 1} & \text{if } x \neq -1 \\ 1 & \text{if } x = -1 \end{cases}$

**22.** $f(x) = \begin{cases} \dfrac{x^2 + x - 6}{x - 2} & \text{if } x \neq 2 \\ 5 & \text{if } x = 2 \end{cases}$

**23.** $f(x) = \begin{cases} e^{1/x} & \text{if } x \neq 0 \\ 1 & \text{if } x = 0 \end{cases}$

**24.** $f(x) = \begin{cases} -|x| + 1 & \text{if } x \neq 0 \\ 0 & \text{if } x = 0 \end{cases}$

**25.** $f(x) = \sec 2x$

**26.** $f(x) = \cot \pi x$

**27.** Let

$$f(x) = \begin{cases} x + 2 & \text{if } x \leq 1 \\ kx^2 & \text{if } x > 1 \end{cases}$$

Find the value of $k$ that will make $f$ continuous on $(-\infty, \infty)$.

**28.** Let

$$f(x) = \begin{cases} \dfrac{x^2 - 4}{x + 2} & \text{if } x \neq -2 \\ k & \text{if } x = -2 \end{cases}$$

Find the value of $k$ that will make $f$ continuous on $(-\infty, \infty)$.

**29.** Let

$$f(x) = \begin{cases} ax + b & \text{if } x < 1 \\ 4 & \text{if } x = 1 \\ 2ax - b & \text{if } x > 1 \end{cases}$$

Find the values of $a$ and $b$ that will make $f$ continuous on $(-\infty, \infty)$.

**30.** Let

$$f(x) = \begin{cases} kx + \ln(x - 3)^2 & \text{if } x \leq 2 \\ 4ke^{x-2} - 3 & \text{if } x > 2 \end{cases}$$

Find the value of $k$ that will make $f$ continuous on $(-\infty, \infty)$.

**31.** Let

$$f(x) = \begin{cases} \dfrac{\sin 2x}{x} & \text{if } x \neq 0 \\ c & \text{if } x = 0 \end{cases}$$

Find the value of $c$ that will make $f$ continuous on $(-\infty, \infty)$.

**32.** Let

$$f(x) = \begin{cases} x \cot kx & \text{if } x < 0 \\ x^2 + c & \text{if } x \geq 0 \end{cases}$$

Find the value of $c$ that will make $f$ continuous at $x = 0$.

*In Exercises 33–36, determine whether the function is continuous on the closed interval.*

**33.** $f(x) = \sqrt{16 - x^2}, \quad [-4, 4]$

**34.** $g(x) = \ln(x + 3) + \sqrt{4 - x^2}, \quad [-2, 1]$

**35.** $f(x) = \begin{cases} x + 1 & \text{if } x < 0 \\ 2 - x & \text{if } x \geq 0 \end{cases}, \quad [-2, 4]$

**36.** $h(t) = \dfrac{1}{t^2 - 9}, \quad [-2, 2]$

*In Exercises 37–48, find the interval(s) where f is continuous.*

**37.** $f(x) = (3x^3 + 2x^2 + 1)^4$

**38.** $f(x) = \sqrt{x}(x - 5)^4$

**39.** $f(x) = \sqrt{x^2 + x + 1}$

**40.** $h(x) = \sqrt{x} + \dfrac{1}{\sqrt{x}}$

**41.** $f(x) = e^{\sqrt{9 - x^2}}$

**42.** $f(x) = \ln|x^2 - 4|$

**43.** $f(x) = \dfrac{1}{x\sqrt{9 - x^2}}$

**44.** $f(x) = \dfrac{1}{x} + \dfrac{3\sqrt{x}}{(x - 2)^2}$

**45.** $f(x) = \sin e^{\sqrt{x}}$

**46.** $f(x) = \dfrac{\ln(x + 1)}{e^x - 2}$

**47.** $f(x) = \tan^{-1} \dfrac{1}{x - 2}$

**48.** $f(x) = \dfrac{2 \cos x}{5 + 2 \sin x}$

**49.** Find $\displaystyle\lim_{x \to 2} \left| \dfrac{x^2 + x - 6}{x - 2} \right|$

**50.** Find $\displaystyle\lim_{x \to 1} \sin^{-1} \left[ \dfrac{x^2 + x - 2}{6(x - 1)} \right]$

*In Exercises 51–56, define the function at a so as to make it continuous at a.*

**51.** $f(x) = \dfrac{3x^3 - 2x}{5x}, \quad a = 0$

**52.** $f(x) = \dfrac{2x^3 + x - 3}{x - 1}, \quad a = 1$

**53.** $f(x) = \dfrac{\sqrt{x + 1} - 1}{x}, \quad a = 0$

**54.** $f(x) = \dfrac{4 - x}{2 - \sqrt{x}}, \quad a = 4$

**55.** $f(x) = \dfrac{\tan x}{x}, \quad a = 0$

**56.** $f(x) = \dfrac{e^{-x} \sin^2 x}{1 - \cos x}, \quad a = 0$

*In Exercises 57 and 58, let $f(x) = x(1 - x^2)$, and let g be the signum (or sign) function defined by*

$$g(x) = \begin{cases} -1 & \text{if } x < 0 \\ 0 & \text{if } x = 0 \\ 1 & \text{if } x > 0 \end{cases}$$

**57.** Show that $f \circ g$ is continuous on $(-\infty, \infty)$. Does this contradict Theorem 6?

**58.** Sketch the graph of the function $g \circ f$, and determine where $g \circ f$ is continuous.

*In Exercises 59–62, use the Intermediate Value Theorem to find the value of $c$ such that $f(c) = M$.*

**59.** $f(x) = x^2 - x + 1$ on $[-1, 4]$;   $M = 7$

**60.** $f(x) = x^2 - 4x + 6$ on $[0, 3]$;   $M = 3$

**61.** $f(x) = x^3 - 2x^2 + x - 2$ on $[0, 4]$;   $M = 10$

**62.** $f(x) = \dfrac{x - 1}{x + 1}$ on $[-4, -2]$;   $M = 2$

*In Exercises 63–66, use Theorem 8 to show that there is at least one root of the equation in the given interval.*

**63.** $x^3 - 2x - 1 = 0$;   $(0, 2)$

**64.** $x^4 - 2x^3 - 3x^2 + 7 = 0$;   $(1, 2)$

**65.** $e^{-x} - \ln x = 0$;   $(1, 2)$

**66.** $x^4 - 2x^3 = \sqrt{x - 1}$;   $(2, 3)$

**67.** Let $f(x) = x^2$. Use the Intermediate Value Theorem to prove that there is a number $c$ in the interval $[0, 2]$ such that $f(c) = 2$. (This proves the existence of the number $\sqrt{2}$.)

**68.** Let $f(x) = x^5 - 3x^2 + 2x + 5$.
   **a.** Show that there is at least one number $c$ in the interval $[0, 2]$ such that $f(c) = 12$.
    **b.** Use a graphing utility to find all values of $c$ accurate to five decimal places.
   **Hint:** Find the point(s) of intersection of the graphs of $f$ and $g(x) = 12$.

**69.** Let $f(x) = \frac{1}{2}x^2 - \cos \pi x + 1$.
   **a.** Show that there is at least one number $c$ in the interval $[0, 1]$ such that $f(c) = \sqrt{2}$.
    **b.** Use a graphing utility to find all values of $c$ accurate to five decimal places.
   **Hint:** Find the point(s) of intersection of the graphs of $f$ and $g(x) = \sqrt{2}$.

**70.** Let
$$f(x) = \begin{cases} -x^2 & \text{if } x \le 0 \\ x + 1 & \text{if } x > 0 \end{cases}$$

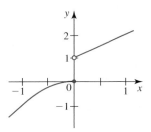

   **a.** Show that $f$ is not continuous on $[-1, 1]$.
   **b.** Show that $f$ does not take on all values between $f(-1)$ and $f(1)$.

**71.** Let
$$f(x) = \begin{cases} -x + 2 & \text{if } -2 \le x < 0 \\ -(x^2 + 2) & \text{if } 0 \le x \le 2 \end{cases}$$

Does $f$ have a zero in the interval $[-2, 2]$? Explain your answer.

 **72.** Use the method of bisection to approximate the root of the equation $x^3 - x + 1 = 0$ accurate to two decimal places. (Refer to Example 10.)

 **73.** Use the method of bisection to approximate the root of the equation $x^5 + 2x - 7 = 0$ accurate to two decimal places.

**74. Acquisition of Failing S&L's** The Tri-State Savings and Loan Company acquired two ailing financial institutions in 2009. One of them was acquired at time $t = T_1$, and the other was acquired at time $t = T_2$. ($t = 0$ corresponds to the beginning of 2009.) The following graph shows the total amount of money on deposit with Tri-State. Explain the significance of the discontinuities of the function at $T_1$ and $T_2$.

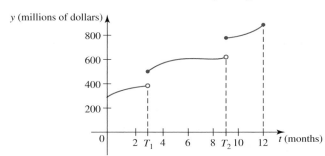

**75. Colliding Billiard Balls** While moving at a constant speed of $v$ m/sec, billiard ball $A$ collides with another stationary ball $B$ at time $t_1$, hitting it "dead center." Suppose that at the moment of impact, ball $A$ comes to rest. Draw graphs depicting the speeds of ball $A$ and ball $B$ (neglect friction).

**76. Action of an Impulse on an Object** An object of mass $m$ is at rest at the origin on the $x$-axis. At $t = t_0$ it is acted upon by an impulse $P_0$ for a very short duration of time. The position of the object is given by
$$x = f(t) = \begin{cases} 0 & \text{if } 0 \le t < t_0 \\ \dfrac{P_0(t - t_0)}{m} & \text{if } t \ge t_0 \end{cases}$$

Sketch the graph of $f$, and interpret your results.

**77.** Joan is looking straight out of a window of an apartment building at a height of 32 ft from the ground. A boy throws a tennis ball straight up by the side of the building where the window is located. Suppose the height of the ball (measured in feet) from the ground at time $t$ (in sec) is $h(t) = 4 + 64t - 16t^2$.
   **a.** Show that $h(0) = 4$ and $h(2) = 68$.
   **b.** Use the Intermediate Value Theorem to conclude that the ball must cross Joan's line of sight at least once.
   **c.** At what time(s) does the ball cross Joan's line of sight? Interpret your results.

**78. A Mixture Problem** A tank initially contains 10 gal of brine with 2 lb of salt. Brine with 1.5 lb of salt per gallon enters the tank at the rate of 3 gal/min, and the well-stirred mixture leaves the tank at the rate of 4 gal/min. It can be shown that the amount of salt in the tank after $t$ min is $x$ lb, where

$$x = f(t) = 1.5(10 - t) - 0.0013(10 - t)^4 \qquad 0 \le t \le 3$$

Show that there is at least one instant of time between $t = 0$ and $t = 3$ when the amount of salt in the tank is 5 lb.

**Note:** We will find the times(s) when the amount of salt in the tank is 5 lb in Example 4 of Section 3.9.

**79. Elastic Curve of a Beam** The following figure shows the *elastic curve* (the dashed curve in the figure) of a beam of length $L$ ft carrying a concentrated load of $W_0$ lb at its center. An equation of the curve is

$$y = f(x)$$

$$= \begin{cases} \dfrac{W_0}{48EI}(3L^2x - 4x^3) & \text{if } 0 \le x < \frac{L}{2} \\[2mm] \dfrac{W_0}{48EI}(4x^3 - 12Lx^2 + 9L^2x - L^3) & \text{if } \frac{L}{2} \le x \le L \end{cases}$$

where the product $EI$ is a constant called the *flexural rigidity* of the beam. Show that the function $y = f(x)$ describing the elastic curve is continuous on $[0, L]$.

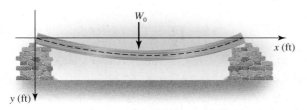

**80. Newton's Law of Attraction** The magnitude of the force exerted on a particle of mass $m$ by a thin homogeneous spherical shell of radius $R$ is

$$F(r) = \begin{cases} 0 & \text{if } r < R \\[2mm] \dfrac{GMm}{r^2} & \text{if } r \ge R \end{cases}$$

where $M$ is the mass of the shell, $r$ is the distance from the center of the shell to the particle, and $G$ is the gravitational constant.

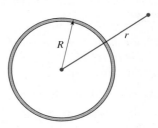

**a.** What is the force exerted on a particle just inside the shell? Just outside the shell?

**b.** Sketch the graph of $F$. Is $F$ a continuous function of $r$?

**81.** A couple leaves their house at 6 P.M. on Friday for a weekend escape to their mountain cabin, where they arrive at 8 P.M. On the return trip, the couple leaves the cabin at 6 P.M. on Sunday and reverses the route they took on Friday, arriving home at 8 P.M. Use the Intermediate Value Theorem to show that there is a location on the route that the couple will pass at the same time of day on both days.

**82. a.** Suppose that $f$ is continuous at $a$ and $g$ is discontinuous at $a$. Prove that the sum $f + g$ is discontinuous at $a$.

**b.** Suppose that $f$ and $g$ are both discontinuous at $a$. Is the sum $f + g$ necessarily discontinuous at $a$? Explain.

**83. a.** Suppose that $f$ is continuous at $a$ and $g$ is discontinuous at $a$. Is the product $fg$ necessarily discontinuous at $a$? Explain.

**b.** Suppose that $f$ and $g$ are both discontinuous at $a$. Is the product $fg$ necessarily discontinuous at $a$? Explain.

**84. The Dirichlet Function** The Dirichlet function is defined by

$$f(x) = \begin{cases} 0 & \text{if } x \text{ is rational} \\ 1 & \text{if } x \text{ is irrational} \end{cases}$$

Show that $f$ is discontinuous at every real number.

**85.** Show that every polynomial equation of the form

$$a_{2n+1}x^{2n+1} + a_{2n}x^{2n} + \cdots + a_2x^2 + a_1x + a_0 = 0$$

with real coefficients and $a_{2n+1} \ne 0$ has at least one real root.

**86.** Suppose that $f$ is continuous on $[a, b]$ and has a finite number of zeros $x_1, x_2, \ldots, x_n$ in $(a, b)$, satisfying $a < x_1 < x_2 < \cdots < x_n < b$. Show that $f(x)$ has the same sign within each of the intervals $(a, x_1), (x_1, x_2), \ldots, (x_n, b)$.

**87.** Let $g$ be a continuous function on an interval $[a, b]$ and suppose $a \le g(x) \le b$ whenever $a \le x \le b$. Show that the equation $x = g(x)$ has at least one solution $c$ in the interval $[a, b]$. Give a geometric interpretation.

**Hint:** Apply the Intermediate Value Theorem to the function $f(x) = x - g(x)$.

In Exercises 88 and 89, *plot the graph of f. Then use the graph to determine where the function is continuous. Verify your answer analytically.*

**88.** $f(x) = \begin{cases} \dfrac{x+1}{x\sqrt{1-x}} & \text{if } x < 1 \\ 2 & \text{if } x = 1 \\ \dfrac{x^4+1}{x^2} & \text{if } x > 1 \end{cases}$

**89.** $f(x) = \dfrac{|\sin x|}{\sin x}$

**90.** Show that $f(x) = x^3 + x - 1$ has exactly one zero in $(0, 1)$.

**91.** Show that there is at least one root of the equation $\sin x - x + 2 = 0$ in the interval $\left(0, \frac{3\pi}{2}\right)$.

**92.** Prove that $f(x) = \sin x$ is continuous everywhere.
**Hint:** Use the result of Exercise 97 in Section 1.2.

**93.** Prove that $f(x) = \cos x$ is continuous everywhere.

**94.** Prove that if $f$ and $g$ are continuous at $a$, then $f - g$ is continuous at $a$.

**95.** Prove that if $f$ and $g$ are continuous at $a$ with $g(a) \neq 0$, then $f/g$ is continuous at $a$.

In Exercises 96–100, *determine whether the statement is true or false. If it is true, explain why. If it is false, explain why or give an example that shows it is false.*

**96.** If $|f|$ is continuous at $a$, then $f$ is continuous at $a$.

**97.** If $f$ is discontinuous at $a$, then $f^2$ is continuous at $a$.

**98.** If $f$ is defined on the interval $[a, b]$ with $f(a)$ and $f(b)$ having opposite signs, then $f$ must have at least one zero in $(a, b)$.

**99.** If $f$ is continuous and $f + g$ is continuous, then $g$ is continuous.

**100.** If $f$ is continuous on the interval $(1, 5)$, then $f$ is continuous on the interval $(2, 4)$.

## 1.5   Tangent Lines and Rates of Change

### ◼ An Intuitive Look

One of the two problems that played a fundamental role in the development of calculus is the tangent line problem: How do we find the tangent line at a given point on a curve? (See Figure 1a.) To gain an intuitive feeling for the notion of the tangent line to a curve, think of the curve as representing a stretch of roller coaster track, and imagine that you are sitting in a car at the point $P$ and looking straight ahead. Then the tangent line $T$ to the curve at $P$ is just the line parallel to your line of sight (Figure 1b).

Observe that the slope of the tangent line $T$ at the point $P$ appears to reflect the "steepness" of the curve at $P$. In other words, the slope of the tangent line at the point $P(x, f(x))$ on the graph of $y = f(x)$ provides us with a natural yardstick for measuring the rate of change of one quantity ($y$) with respect to another quantity ($x$).

Let's see how this intuitive observation bears out in a specific example. The function $s = f(t) = 4t^2$ gives the position of a maglev moving along a straight track at time $t$. We have drawn the tangent line $T$ to the graph of $s$ at the point $(2, 16)$ in Figure 2. Observe that the slope of $T$ is $32/2 = 16$. This suggests that the quantity $s$ is changing at the rate of 16 units per unit change in $t$; that is, the velocity of the maglev at $t = 2$ is 16 ft/sec. You might recall that this was the figure we arrived at in our calculations in Section 1.1!

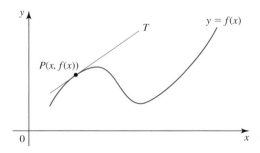

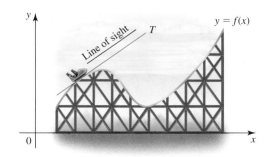

**FIGURE 1**   (a) $T$ is the tangent line to the curve at $P$.        (b) The line of sight is parallel to $T$.

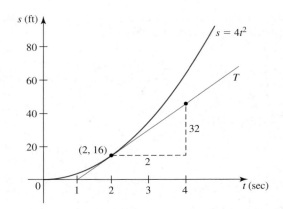

**FIGURE 2**
The position of the maglev at time $t$

## ■ Estimating the Rate of Change of a Function from Its Graph

**EXAMPLE 1   Automobile Fuel Economy**   According to a study by the U.S. Department of Energy and the Shell Development Company, a typical car's fuel economy as a function of its speed is described by the graph of the function $f$ shown in Figure 3. Assuming that the rate of change of the function $f$ at any value of $x$ is given by the slope of the tangent line at the point $P(x, f(x))$, use the graph of $f$ to estimate the rate of change of a typical car's fuel economy, measured in miles per gallon (mpg), when a car is driven at 20 mph and when it is driven at 60 mph.

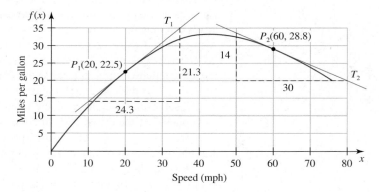

**FIGURE 3**
The fuel economy of a typical car
*Source:* U.S. Department of Energy
and Shell Development Company.

**Solution**   The slope of the tangent line $T_1$ to the graph of $f$ at $P_1(20, 22.5)$ is approximately

$$\frac{21.3}{24.3} \approx 0.88 \qquad \frac{\text{rise}}{\text{run}}$$

This tells us that the quantity $f(x)$ is increasing at the rate of approximately 0.9 unit per unit change in $x$ when $x = 20$. In other words, when a car is driven at a speed of 20 mph, its fuel economy typically increases at the rate of approximately 0.9 mpg per 1 mph increase in the speed of the car. The slope of the tangent line $T_2$ to the graph of $f$ at $P_2(60, 28.8)$ is

$$-\frac{14}{30} \approx -0.47$$

This says that the quantity $y$ is decreasing at the rate of approximately 0.5 unit per unit change in $x$ when $x = 60$. In other words, when a car is driven at a speed of 60 mph, its fuel economy typically decreases at the rate of approximately 0.5 mpg per 1 mph increase in the speed of the car.   ■

## ■ More Examples Involving Rates of Change

The discovery of the relationship between the problem of finding the slope of the tangent line and the problem of finding the rate of change of one quantity with respect to another spurred the development in the seventeenth century of the branch of calculus called **differential calculus** and made it an indispensable tool for solving practical problems. A small sample of the types of problems that we can solve using differential calculus follows:

- Finding the velocity (rate of change of position with respect to time) of a sports car moving along a straight road
- Finding the rate of change of the harmonic distortion of a stereo amplifier with respect to its power output
- Finding the rate of growth of a bacteria population with respect to time
- Finding the rate of change of the Consumer Price Index with respect to time
- Finding the rate of change of a company's profit (loss) with respect to its level of sales

## ■ Defining a Tangent Line

The main purpose of Example 1 was to illustrate the relationship between tangent lines and rates of change. Ideally, the solution to a problem should be analytic and not rely, as in Example 1, on how accurately we can draw a curve and estimate the position of its tangent lines. So our first task will be to give a more precise definition of a tangent line to a curve. After that, we will devise an analytical method for finding an equation of such a line.

Let $P$ and $Q$ be two distinct points on a curve, and consider the secant line passing through $P$ and $Q$. (See Figure 4.) If we let $Q$ move along the curve toward $P$, then the secant line rotates about $P$ and approaches the fixed line $T$. We define $T$ to be the tangent line at $P$ on the curve.

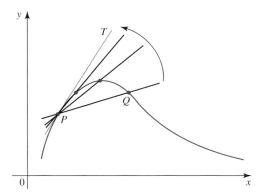

**FIGURE 4**
As $Q$ approaches $P$ along the curve, the secant lines approach the tangent line $T$.

Let's make this notion more precise: Suppose that the curve is the graph of a function $f$ defined by $y = f(x)$. (See Figure 5.) Let $P(a, f(a))$ be a point on the graph of $f$, and let $Q$ be a point on the graph of $f$ distinct from $P$. Then the $x$-coordinate of $Q$ has the form $x = a + h$, where $h$ is some appropriate nonzero number. If $h > 0$, then $Q$ lies to the right of $P$; and if $h < 0$, then $Q$ lies to the left of $P$. The corresponding $y$-coordinate of $Q$ is $y = f(a + h)$. In other words, we can specify $Q$ in the usual manner by writing $Q(a + h, f(a + h))$. Observe that we can make $Q$ approach $P$ along the graph of $f$ by letting $h$ approach 0. This situation is illustrated in Figure 5b. (You are encouraged to sketch your own figures for the case $h < 0$.)

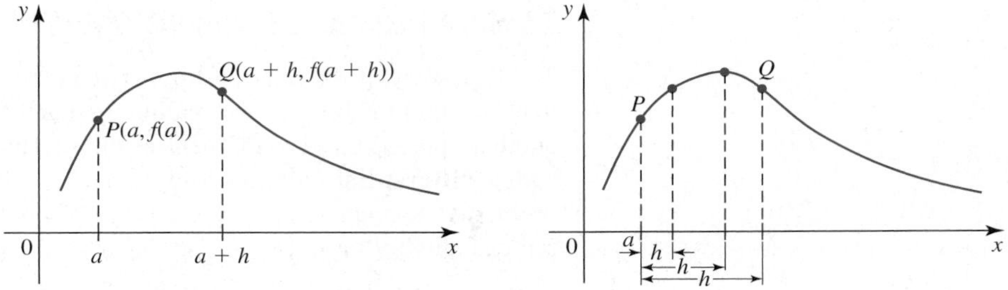

**FIGURE 5**    (**a**) The points $P(a, f(a))$ and $Q(a + h, f(a + h))$    (**b**) As $h$ approaches 0, $Q$ approaches $P$.

Next, using the formula for the slope of a line, we can write the slope of the secant line passing through $P(a, f(a))$ and $Q(a + h, f(a + h))$ as

$$m_{\text{sec}} = \frac{f(a + h) - f(a)}{(a + h) - a} = \frac{f(a + h) - f(a)}{h} \tag{1}$$

The expression on the right-hand side of Equation (1) is called a **difference quotient.**

As we observed earlier, if we let $h$ approach 0, then $Q$ approaches $P$ and the secant line passing through $P$ and $Q$ approaches the tangent line $T$. This suggests that if the tangent line does exist at $P$, then its slope $m_{\text{tan}}$ should be the limit of $m_{\text{sec}}$ obtained by letting $h$ approach zero. This leads to the following definition.

---

**DEFINITION    Tangent Line**

Let $P(a, f(a))$ be a point on the graph of a function $f$. Then the **tangent line** at $P$ (if it exists) on the graph of $f$ is the line passing through $P$ and having slope

$$m_{\text{tan}} = \lim_{h \to 0} \frac{f(a + h) - f(a)}{h} \tag{2}$$

---

**Notes**
1. If the limit in Equation (2) does not exist, then $m_{\text{tan}}$ is undefined.
2. If the limit in Equation (2) exists, then we can find an equation of the tangent line at $P$ by using the point-slope form of an equation of a line. Thus, $y - f(a) = m_{\text{tan}}(x - a)$.  ■

**EXAMPLE 2**    Find the slope and an equation of the tangent line to the graph of $f(x) = x^2$ at the point $P(1, 1)$.

**Solution**    To find the slope of the tangent line at the point $P(1, 1)$, we use Equation (2) with $a = 1$, obtaining

$$m_{\text{tan}} = \lim_{h \to 0} \frac{f(1 + h) - f(1)}{h} = \lim_{h \to 0} \frac{(1 + h)^2 - 1^2}{h}$$

$$= \lim_{h \to 0} \frac{(1 + 2h + h^2) - 1}{h} = \lim_{h \to 0} \frac{2h + h^2}{h}$$

$$= \lim_{h \to 0} (2 + h) = 2$$

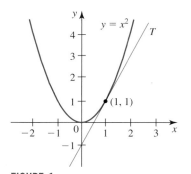

**FIGURE 6**
$T$ is the tangent line at the point $P(1, 1)$ on the graph of $y = x^2$.

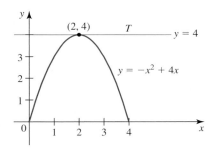

**FIGURE 7**
The tangent line at the point $(2, 4)$ is horizontal.

To find an equation of the tangent line, we use the point-slope form of an equation of a line to obtain

$$y - 1 = 2(x - 1)$$

or

$$y = 2x - 1$$

The graphs of $f$ and the tangent line at $(1, 1)$ are sketched in Figure 6.   ■

**EXAMPLE 3**   Find the slope and an equation of the tangent line to the graph of the equation $y = -x^2 + 4x$ at the point $P(2, 4)$.

**Solution**   The slope of the tangent line at the point $P(2, 4)$ is found by using Equation (2) with $a = 2$ and $f(x) = -x^2 + 4x$. We have

$$m_{\tan} = \lim_{h \to 0} \frac{f(2 + h) - f(2)}{h} = \lim_{h \to 0} \frac{[-(2 + h)^2 + 4(2 + h)] - [-(2)^2 + 4(2)]}{h}$$

$$= \lim_{h \to 0} \frac{-4 - 4h - h^2 + 8 + 4h + 4 - 8}{h} = \lim_{h \to 0} -\frac{h^2}{h}$$

$$= \lim_{h \to 0} (-h) = 0$$

An equation of the tangent line at $P(2, 4)$ is

$$y - 4 = 0(x - 2) \qquad \text{or} \qquad y = 4$$

The graphs of $f$ and the tangent line at $(2, 4)$ are sketched in Figure 7.   ■

The solution in Example 3 is fully expected if we recall that the graph of the equation $y = -x^2 + 4x$ is a parabola with vertex at $(2, 4)$. At the vertex the tangent line is horizontal, and therefore its slope is zero.

## ■ Tangent Lines, Secant Lines, and Rates of Change

As we observed earlier, there seems to be a connection between the slope of the tangent line at a given point $P(a, f(a))$ on the graph of a function $f$ and the rate of change of $f$ when $x = a$. Let's show that this is true.

Consider the function $f$ whose graph is shown in Figure 8a. You can see from Figure 8a that as $x$ changes from $a$ to $a + h$, $f(x)$ changes from $f(a)$ to $f(a + h)$. (We call $h$ the *increment* in $x$.) The ratio of the change in $f(x)$ to the change in $x$ measures the average rate of change of $f$ over the interval $[a, a + h]$.

---

**DEFINITION**   **Average Rate of Change of a Function**
The **average rate of change of a function** $f$ over the interval $[a, a + h]$ is

$$\frac{f(a + h) - f(a)}{h} \tag{3}$$

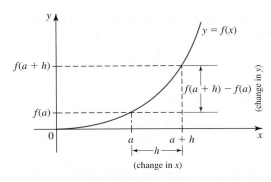

**FIGURE 8** **(a)** The average rate of change of $f$ over $[a, a + h]$ is given by

$$\frac{f(a + h) - f(a)}{h}$$

**(b)** $m_{\text{sec}} = \dfrac{f(a + h) - f(a)}{h}$

Figure 8b depicts the graph of the same function $f$. The slope of the secant line passing through the points $P(a, f(a))$ and $Q(a + h, f(a + h))$ is

$$m_{\text{sec}} = \frac{f(a + h) - f(a)}{(a + h) - a} = \frac{f(a + h) - f(a)}{h}$$

But this is just Equation (1). Comparing the expression in (3) and that on the right-hand side of Equation (1), we conclude that the *average rate of change of f with respect to x over the interval* $[a, a + h]$ *has the same value as the slope of the secant line passing through the points* $(a, f(a))$ *and* $(a + h, f(a + h))$.

Next, by letting $h$ approach zero in the expression in (3), we obtain the (instantaneous) rate of change of $f$ at $a$.

---

**DEFINITION   Instantaneous Rate of Change of a Function**

The **(instantaneous) rate of change of a function** $f$ **with respect to** $x$ **at** $a$ is

$$\lim_{h \to 0} \frac{f(a + h) - f(a)}{h} \qquad (4)$$

if the limit exists.

---

But this expression also gives the slope of the tangent line to the graph of $f$ at $P(a, f(a))$. Thus, we conclude that *the instantaneous rate of change of f with respect to x at a has the same value as the slope of the tangent line at the point* $(a, f(a))$.

Our earlier calculations suggested that the instantaneous velocity of the maglev at $t = 2$ is 16 ft/sec. We now verify this assertion.

**EXAMPLE 4** The position function of the maglev at time $t$ is $s = f(t) = 4t^2$, where $0 \le t \le 30$. Then the average velocity of the maglev over the time interval $[2, 2 + h]$ is given by the average rate of change of the position function $s$ over $[2, 2 + h]$, where $h > 0$ and $2 + h$ lies in the interval $(2, 30)$. Using the expression in (3) with $a = 2$, we see that the average velocity is given by

$$\frac{f(2 + h) - f(2)}{h} = \frac{4(2 + h)^2 - 4(2)^2}{h} = \frac{16 + 16h + 4h^2 - 16}{h} = 16 + 4h$$

Next, using the expression in (4), we see that the instantaneous velocity of the maglev at $t = 2$ is given by

$$v = \lim_{h \to 0} \frac{f(2 + h) - f(2)}{h} = \lim_{h \to 0}(16 + 4h) = 16$$

or 16 ft/sec, as observed earlier.

## 1.5  CONCEPT QUESTIONS

*For Questions 1 and 2, refer to the following figure.*

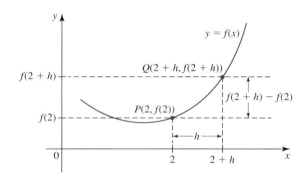

1. Let $P(2, f(2))$ and $Q(2 + h, f(2 + h))$ be points on the graph of a function $f$.
   a. Find an expression for the slope of the secant line passing through $P$ and $Q$.
   b. Find an expression for the slope of the tangent line passing through $P$.
2. Refer to Question 1.
   a. Find an expression for the average rate of change of $f$ over the interval $[2, 2 + h]$.
   b. Find an expression for the instantaneous rate of change of $f$ at 2.
   c. Compare your answers for parts (a) and (b) with those of Question 1.

## 1.5  EXERCISES

1. **Traffic Flow** Opened in the late 1950s, the Central Artery in downtown Boston was designed to move 75,000 vehicles per day. The following graph shows the average speed of traffic flow in miles per hour versus the number of vehicles moved per day. Estimate the rate of change of the average speed of traffic flow when the number of vehicles moved per day is 100,000 and when it is 200,000. (According to our model, there will be permanent gridlock when we reach 300,000 cars per day!)

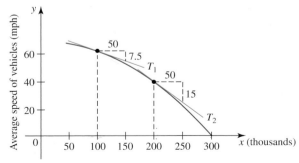

*Source: The Boston Globe.*

**Note:** Since 2003 the city of Boston has ameliorated the situation with the "Big Dig."

2. **Forestry** The following graph shows the volume of wood produced in a single-species forest. Here, $f(t)$ is measured in cubic meters per hectare, and $t$ is measured in years. By computing the slopes of the respective tangent lines, estimate the rate at which the wood grown is changing at the beginning of year 10 and at the beginning of year 30.

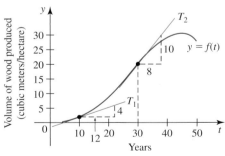

*Source: The Random House Encyclopedia.*

3. **TV-Viewing Patterns** The graph on the following page shows the percentage of U.S. households watching television during a 24-hr period on a weekday ($t = 0$ corresponds to 6 A.M.). By computing the slopes of the respective tangent lines,

estimate the rate of change of the percentage of households watching television at 4 P.M. and 11 P.M.

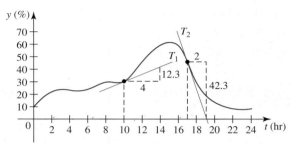

*Source:* A.C. Nielsen Company.

**4. Crop Yield**  Productivity and yield of cultivated crops are often reduced by insect pests. The following graph shows the relationship between the yield of a certain crop, $f(x)$, as a function of the density of aphids $x$. (Aphids are small insects that suck plant juices.) Here, $f(x)$ is measured in kilograms per 4000 square meters, and $x$ is measured in hundreds of aphids per bean stem. By computing the slopes of the respective tangent lines, estimate the rate of change of the crop yield with respect to the density of aphids if the density is 200 aphids per bean stem and if it is 800 aphids per bean stem.

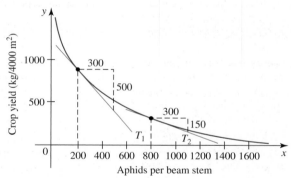

*Source:* The *Random House Encyclopedia.*

**5.** The velocities of car $A$ and car $B$, starting out side by side and traveling along a straight road, are given by $v_A = f(t)$ and $v_B = g(t)$, respectively, where $v$ is measured in feet per second and $t$ is measured in seconds.

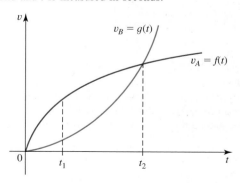

**a.** What can you say about the velocity and acceleration of the two cars at $t_1$? (Acceleration is the rate of change of velocity.)
**b.** What can you say about the velocity and acceleration of the two cars at $t_2$?

**6. Effect of a Bactericide on Bacteria**  In the figure below, $f(t)$ gives the population $P_1$ of a certain bacteria culture at time $t$ after a portion of bactericide $A$ was introduced into the population at $t = 0$. The graph of $g(t)$ gives the population $P_2$ of a similar bacteria culture at time $t$ after a portion of bactericide $B$ was introduced into the population at $t = 0$.

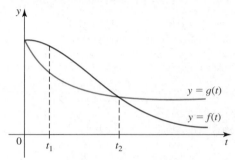

**a.** Which population is decreasing faster at $t_1$?
**b.** Which population is decreasing faster at $t_2$?
**c.** Which bactericide is more effective in reducing the population of bacteria in the short run? In the long run?

*In Exercises 7–14, (a) use Equation (1) to find the slope of the secant line passing through the points $(a, f(a))$ and $(a + h, f(a + h))$; (b) use the results of part (a) and Equation (2) to find the slope of the tangent line at the point $(a, f(a))$; and (c) find an equation of the tangent line to the graph of f at the point $(a, f(a))$.*

| Function | $(a, f(a))$ | Function | $(a, f(a))$ |
|---|---|---|---|
| **7.** $f(x) = 5$ | $(1, 5)$ | **8.** $f(x) = 2x + 3$ | $(1, 5)$ |
| **9.** $f(x) = 2x^2 - 1$ | $(2, 7)$ | **10.** $f(x) = x^2 - x$ | $(2, 2)$ |
| **11.** $f(x) = x^3$ | $(2, 8)$ | **12.** $f(x) = x^3 + x$ | $(2, 10)$ |
| **13.** $f(x) = \dfrac{1}{x}$ | $(1, 1)$ | **14.** $f(x) = \dfrac{1}{x + 1}$ | $\left(1, \frac{1}{2}\right)$ |

*In Exercises 15–20, find the instantaneous rate of change of the given function when $x = a$.*

**15.** $f(x) = 2x^2 + 1$;  $a = 1$

**16.** $g(x) = x^2 - x + 2$;  $a = -1$

**17.** $H(x) = x^3 + x$;  $a = 2$

**18.** $f(x) = \sqrt{x}$;  $a = 4$

**19.** $f(x) = \dfrac{2}{x} + x$;  $a = 1$

**20.** $f(x) = \dfrac{1}{x - 2}$;  $a = 1$

*In Exercises 21–24, the position function of an object moving along a straight line is given by $s = f(t)$. The **average velocity** of the object over the time interval $[a, b]$ is the average rate of change of $f$ over $[a, b]$; its (instantaneous) **velocity at $t = a$** is the rate of change of $f$ at $a$.*

**21.** The position of a car at any time $t$ is given by $s = f(t) = \frac{1}{4}t^2$, $0 \le t \le 10$, where $s$ is given in feet and $t$ in seconds.
   **a.** Find the average velocity of the car over the time intervals $[2, 3]$, $[2, 2.5]$, $[2, 2.1]$, $[2, 2.01]$, and $[2, 2.001]$.
   **b.** Find the velocity of the car at $t = 2$.

**22. Velocity of a Car**  Suppose the distance $s$ (in feet) covered by a car moving along a straight road after $t$ sec is given by the function $s = f(t) = 2t^2 + 48t$.
   **a.** Calculate the average velocity of the car over the time intervals $[20, 21]$, $[20, 20.1]$, and $[20, 20.01]$.
   **b.** Calculate the (instantaneous) velocity of the car when $t = 20$.
   **c.** Compare the results of part (a) with those of part (b).

**23. Velocity of a Ball Thrown into the Air**  A ball is thrown straight up with an initial velocity of 128 ft/sec, so its height (in feet) after $t$ sec is given by $s = f(t) = 128t - 16t^2$.
   **a.** What is the average velocity of the ball over the time intervals $[2, 3]$, $[2, 2.5]$, and $[2, 2.1]$?
   **b.** What is the instantaneous velocity at time $t = 2$?
   **c.** What is the instantaneous velocity at time $t = 5$? Is the ball rising or falling at this time?
   **d.** When will the ball hit the ground?

**24.** During the construction of a high-rise building, a worker accidentally dropped his portable electric screwdriver from a height of 400 ft. After $t$ sec the screwdriver had fallen a distance of $s = f(t) = 16t^2$ ft.
   **a.** How long did it take the screwdriver to reach the ground?
   **b.** What was the average velocity of the screwdriver during the time it was falling?
   **c.** What was the velocity of the screwdriver at the time it hit the ground?

**25.** A hot air balloon rises vertically from the ground so that its height after $t$ seconds is $h(t) = \frac{1}{2}t^2 + \frac{1}{2}t$ feet, where $0 \le t \le 60$.
   **a.** What is the height of the balloon after 40 sec?
   **b.** What is the average velocity of the balloon during the first 40 sec of its flight?
   **c.** What is the velocity of the balloon after 40 sec?

**26. Average Velocity of a Helicopter**  A helicopter lifts vertically from its pad and reaches a height of $h(t) = 0.2t^3$ feet after $t$ sec, where $0 \le t \le 12$.
   **a.** How long does it take for the helicopter to reach an altitude of 200 ft?
   **b.** What is the average velocity of the helicopter during the time it takes to attain this height?
   **c.** What is the velocity of the helicopter when it reaches this height?

**27. a.** Find the average rate of change of the area of a circle with respect to its radius $r$ as $r$ increases from $r = 1$ to $r = 2$.
   **b.** Find the rate of change of the area of a circle with respect to $r$ when $r = 2$.

**28. a.** Find the average rate of change of the volume of a sphere with respect to its radius $r$ as $r$ increases from $r = 1$ to $r = 2$.
   **b.** Find the rate of change of the volume of a sphere with respect to $r$ when $r = 2$.

**29. Demand for Tents**  The quantity demanded of the Sportsman $5 \times 7$ tents, $x$, is related to the unit price, $p$, by the function

$$p = f(x) = -0.1x^2 - x + 40$$

where $p$ is measured in dollars and $x$ is measured in units of a thousand.
   **a.** Find the average rate of change in the unit price of a tent if the quantity demanded is between 5000 and 5050 tents; between 5000 and 5010 tents.
   **b.** What is the rate of change of the unit price if the quantity demanded is 5000?

**30.** At a temperature of 20°C, the volume $V$ (in liters) of 1.33 g of $O_2$ is related to its pressure $p$ (in atmospheres) by the formula $V(p) = 1/p$.
   **a.** What is the average rate of change of $V$ with respect to $p$ as $p$ increases from $p = 2$ to $p = 3$?
   **b.** What is the rate of change of $V$ with respect to $p$ when $p = 2$?

**31. Average Velocity of a Motorcycle**  The distance $s$ (in feet) covered by a motorcycle traveling in a straight line at any time $t$ (in seconds) is given by the function

$$s(t) = -0.1t^3 + 2t^2 + 24t$$

Calculate the motorcycle's average velocity over the time interval $[2, 2 + h]$ for $h = 1, 0.1, 0.01, 0.001, 0.0001$, and 0.00001, and use your results to guess at the motorcycle's instantaneous velocity at $t = 2$.

**32. Rate of Change of a Cost Function**  The daily total cost $C(x)$ incurred by Trappee and Sons for producing $x$ cases of TexaPep hot sauce is given by

$$C(x) = 0.000002x^3 + 5x + 400$$

Calculate

$$\frac{C(100 + h) - C(100)}{h}$$

for $h = 1, 0.1, 0.01, 0.001$, and 0.0001, and use your results to estimate the rate of change of the total cost function when the level of production is 100 cases per day.

**33. a.** Plot the graph of

$$g(h) = \frac{(2 + h)^3 - 8}{h}$$

using the viewing window $[-1, 1] \times [0, 20]$.

**b.** Zoom-in to find $\lim_{h \to 0} g(h)$.

**c.** Verify analytically that the limit found in part (b) is $\lim_{h \to 0} \dfrac{f(2 + h) - f(2)}{h}$ where $f(x) = x^3$.

**34.** Use the technique of Exercise 33a–b to find

$$\lim_{h \to 0} \frac{f(8 + h) - f(8)}{h} \text{ if } f(x) = \sqrt[3]{x}, \text{ using the viewing}$$

window $[-1, 1] \times [0, 0.1]$.

*In Exercises 35–40 the expression gives the (instantaneous) rate of change of a function f at some number a. Identify f and a.*

**35.** $\lim\limits_{h \to 0} \dfrac{(1 + h)^5 - 1}{h}$

**36.** $\lim\limits_{h \to 0} \dfrac{2\sqrt[4]{16 + h} - 4}{h}$

**37.** $\lim\limits_{h \to 0} \left[ \dfrac{(4 + h)^2 - 16}{h} + \dfrac{\sqrt{4 + h} - 2}{h} \right]$

**38.** $\lim\limits_{h \to 0} \dfrac{2^{3+h} - 8}{h}$

**39.** $\lim\limits_{x \to 1} \dfrac{x^4 - 1}{x - 1}$

**40.** $\lim\limits_{x \to \pi/2} \dfrac{\sin x - 1}{x - \frac{\pi}{2}}$

*In Exercises 41–44, determine whether the statement is true or false. If it is true, explain why. If it is false, explain why or give an example that shows it is false.*

**41.** The slope of the secant line passing through the points $(a, f(a))$ and $(b, f(b))$ measures the average rate of change of $f$ over the interval $[a, b]$.

**42.** A tangent line to the graph of a function may intersect the graph at infinitely many points.

**43.** There may be more than one tangent line at a given point on the graph of a function.

**44.** The slope of the tangent line to the graph of $f$ at the point $(a, f(a))$ is given by

$$\lim_{x \to a} \frac{f(x) - f(a)}{x - a}$$

# CHAPTER 1 REVIEW

## CONCEPT REVIEW

*In Exercises 1–8, fill in the blanks.*

**1. a.** The statement $\lim_{x \to a} f(x) = L$ means that there exists a number _____ such that the values of _____ can be made as close to _____ as we please by taking $x$ to be sufficiently close to _____.

**b.** The statement $\lim_{x \to a^+} f(x) = L$ is similar to $\lim_{x \to a} f(x) = L$, but here we require that $x$ lie to the _____ of $a$.

**c.** $\lim_{x \to a} f(x) = L$ if and only if $\lim_{x \to a^-} f(x)$ and $\lim_{x \to a^+} f(x)$ both _____ and are equal to _____.

**d.** The precise meaning of $\lim_{x \to a} f(x) = L$ is that given any number _____, there exists a number _____ such that $0 < |x - a| < \delta$ implies $|f(x) - L| < \varepsilon$.

**2. a.** If $\lim_{x \to a} f(x) = L$ and $\lim_{x \to a} g(x) = M$, then the Sum Law states _____, the Product Law states _____, the Constant Multiple Law states _____, the Quotient Law states _____ $M \neq 0$, and the Root Law states _____ provided $L > 0$, if $n$ is even.

**b.** If $p(x)$ is a polynomial function, then $\lim_{x \to a} p(x) = $ _____ for every real number $a$.

**c.** If $r(x)$ is a rational function, then $\lim_{x \to a} r(x) = r(a)$, provided that $a$ is in the domain of _____.

**3.** Suppose that $f(x) \leq g(x) \leq h(x)$ for all $x$ in an interval containing $a$, except possibly at $a$, and that $\lim_{x \to a} f(x) = \lim_{x \to a} h(x) = L$. Then the Squeeze Theorem says that _____.

**4. a.** If $\lim_{x \to a} f(x) = f(a)$, then $f$ is said to be _____ at $a$.

**b.** If $f$ is discontinuous at $a$ but it can be made continuous at $a$ by defining or redefining $f$ at $a$, then $f$ has a _____ discontinuity at $a$.

**c.** If $\lim_{x \to a^-} f(x) = L$ and $\lim_{x \to a^+} f(x) = M$ and $L \neq M$, then $f$ has a _____ discontinuity at $a$.

**d.** If $\lim_{x \to a^-} f(x) = f(a)$, then $f$ is continuous from the _____ at $a$.

**5. a.** A polynomial function is continuous on _____.

**b.** A rational function is continuous on _____ _____.

**c.** The composition of two continuous functions is a _____ function.

**6. a.** Suppose that $f$ is continuous on $[a, b]$ and $f(a) \leq M \leq f(b)$. Then the Intermediate Value

Theorem guarantees the existence of at least one number $c$ in _____ such that _____.

**b.** If $f$ is continuous on $[a, b]$ and $f(a)f(b) < 0$, then there must be at least one solution of the equation _____ in the interval _____.

**7. a.** The tangent line at $P(a, f(a))$ to the graph of $f$ is the line passing through $P$ and having slope _____.

**b.** If the slope of the tangent line at $P(a, f(a))$ is $m_{tan}$, then an equation of the tangent line at $P$ is _____.

**8. a.** The slope of the secant line passing through $P(a, f(a))$ and $Q(a + h, f(a + h))$ and the average rate of change of $f$ over the interval $[a, a + h]$ are both given by _____.

**b.** The slope of the tangent line at $P(a, f(a))$ and the instantaneous rate of change of $f$ at $a$ are both given by _____.

# REVIEW EXERCISES

_In Exercises 1 and 2, use the graph of the function f to find_ (a) $\lim_{x \to a^-} f(x)$, (b) $\lim_{x \to a^+} f(x)$, _and_ (c) $\lim_{x \to a} f(x)$ _for the given value of a._

**1.** $a = 4$

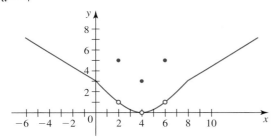

**2.** $a = 0$

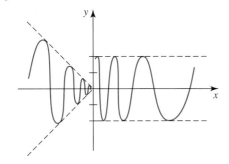

_In Exercises 3–6, sketch the graph of f, and evaluate_ (a) $\lim_{x \to a^-} f(x)$, (b) $\lim_{x \to a^+} f(x)$, _and_ (c) $\lim_{x \to a} f(x)$ _for the given value of a._

**3.** $f(x) = \begin{cases} -x + 5 & \text{if } x \le 3 \\ 2x - 4 & \text{if } x > 3 \end{cases}$;  $a = 3$

**4.** $f(x) = \begin{cases} \dfrac{|x - 2|}{x - 2} & \text{if } x \ne 2 \\ 2 & \text{if } x = 2 \end{cases}$;  $a = 2$

**5.** $f(x) = \begin{cases} -x + 2 & \text{if } x < 2 \\ \sqrt{x - 2} & \text{if } x \ge 2 \end{cases}$;  $a = 2$

**6.** $f(x) = \begin{cases} -\dfrac{1}{2}x + 1 & \text{if } x < 0 \\ 0 & \text{if } x = 0; \\ -\dfrac{1}{x^2} & \text{if } x > 0 \end{cases}$  $a = 0$

_In Exercises 7–28, find the indicated limit if it exists._

**7.** $\lim_{h \to 3}(4h^2 - 2h + 4)$

**8.** $\lim_{x \to 2}(x^3 + 1)(x^2 - 1)$

**9.** $\lim_{x \to 3}\sqrt{x^2 + 2x - 3}$

**10.** $\lim_{t \to 1}\dfrac{t^2 - 1}{1 - t}$

**11.** $\lim_{x \to 5}(x^2 + 2)^{2/3}$

**12.** $\lim_{x \to 3}\dfrac{27 - x^3}{x - 3}$

**13.** $\lim_{y \to 0}\dfrac{2y^2 + 1}{y^3 - 2y^2 + y + 2}$

**14.** $\lim_{x \to 3^+}\dfrac{x + 1}{x - 3}$

**15.** $\lim_{x \to 3}\dfrac{2x^2 - 5x - 3}{3x^2 - 10x + 3}$

**16.** $\lim_{x \to 4}\dfrac{x - 4}{\sqrt{x} - 2}$

**17.** $\lim_{h \to 0}\dfrac{(4 + h)^{-1} - 4^{-1}}{h}$

**18.** $\lim_{x \to 2^+}\dfrac{x - 2}{|x - 2|}$

**19.** $\lim_{x \to 3^-}\sqrt{9 - x^2}$

**20.** $\lim_{x \to 3^+}\dfrac{1 + \sqrt{2x - 6}}{x - 2}$

**21.** $\lim_{x \to 0}\dfrac{2 \sin 3x}{x}$

**22.** $\lim_{x \to 0} x \cot 2x$

**23.** $\lim_{x \to 0^+}\dfrac{\cos x}{\sqrt{x}}$

**24.** $\lim_{x \to 0^+}\sqrt{x} \sin \dfrac{1}{x}$

**25.** $\lim_{x \to 0}\ln(x^2 + 1)$

**26.** $\lim_{x \to \pi/2} e^{\sin x}$

**27.** $\lim_{x \to 0} e^{\sqrt{x}/(x+1)}$

**28.** $\lim_{x \to 0}\ln(\sec x + x)$

**29.** Prove that $\lim_{x \to 0^+} x^2 \cos(1/\sqrt{x}) = 0$.

**30.** Suppose that $1 - x^2 \le f(x) \le 1 + x^2$ for all $x$. Find $\lim_{x \to 0} f(x)$.

*In Exercises 31 and 32, use the graph of the function f to determine where the function is discontinuous.*

**31.**

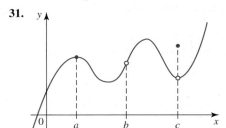

**32.**

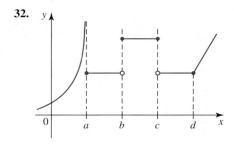

*In Exercises 33 and 34, use the graph of the function f to determine whether f is continuous on the given interval(s). Justify your answer.*

**33.**

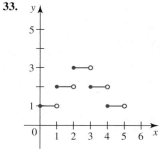

**a.** [1, 2)
**b.** (0, 1)
**c.** (3, 5)

**34.**

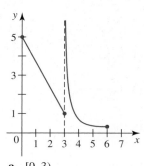

**a.** [0, 3)
**b.** [0, 3]
**c.** [2, 6]
**d.** (3, 6]

*In Exercises 35–42, find the numbers, if any, where the function is discontinuous.*

**35.** $f(x) = x^2 + 3x + \sqrt{-x}$

**36.** $g(x) = \dfrac{3|x - 1|}{x^2 + x - 6}$

**37.** $f(t) = \dfrac{(t + 2)^{1/2}}{(t + 1)^{1/2}}$

**38.** $h(x) = \dfrac{1}{\cos x}$

**39.** $f(x) = \dfrac{1}{\sin x}$

**40.** $f(x) = \begin{cases} x^2 + 1 & \text{if } x < 0 \\ -x + 1 & \text{if } x \geq 0 \end{cases}$

**41.** $f(x) = \dfrac{\ln x}{1 - \sqrt{x}}$

**42.** $f(x) = \dfrac{e^x}{\sqrt{1 - x}}$

**43.** Let

$$f(x) = \begin{cases} \dfrac{x^2 - 2x}{x^2 - 4} & \text{if } x \neq 2 \\ c & \text{if } x = 2 \end{cases}$$

Find the value of $c$ such that $f$ will be continuous at 2.

**44.** True or false? The square of a discontinuous function is also a discontinuous function. Justify your answer.

*In Exercises 45–48, show that the equation has at least one zero in the given interval.*

**45.** $x^4 + x - 5 = 0$; (1, 2)

**46.** $\sin x - x + 1 = 0$; $\left(0, \frac{3\pi}{2}\right)$

**47.** $x \ln x = 1$; (1, e)

**48.** $e^{-x} - x = 0$; (0, 1)

**49.** Let

$$f(x) = \begin{cases} -(x^2 + 1) & \text{if } -2 \leq x < 0 \\ x^2 + 1 & \text{if } 0 \leq x \leq 2 \end{cases}$$

Is there a number $c$ in $[-2, 2]$ such that $f(c) = 0$? Why?

**50.** Find where the function

$$f(x) = \begin{cases} x \sin \dfrac{1}{x} & \text{if } x \neq 0 \\ 0 & \text{if } x = 0 \end{cases}$$

is continuous.

*In Exercises 51 and 52, use the precise definition of the limit to prove the statement.*

**51.** $\lim_{x \to -1}(2x + 3) = 1$

**52.** $\lim_{x \to 0} \sqrt[3]{x} = 0$

**53.** According to the special theory of relativity, the Lorentz contraction formula $L = L_0\sqrt{1 - (v^2/c^2)}$ gives the relationship between the length $L$ of an object moving with a speed $v$ relative to an observer and its length $L_0$ at rest. Here, $c$ is the speed of light.
**a.** Find the domain of $L$, and use the result to explain why one may consider only $\lim_{v \to c^-} L$.
**b.** Evaluate $\lim_{v \to c^-} L$, and interpret your result.

**54.** **Temperature Changes** The following graph shows the air temperature over a 24-hr period on a certain day in November in Chicago, with $t = 0$ corresponding to 12 midnight. Using the given data, compute the slopes of the respective tangent lines, and estimate the rate of change of the temperature at 8 A.M. and at 6 P.M.

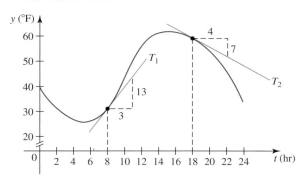

**55.** The position of an object moving along a straight line is $s(t) = 2t^2 + t + 1$, where $s(t)$ is measured in feet and $t$ is measured in seconds.
   **a.** Find the average velocity of the object over the time intervals $[1, 2]$, $[1, 1.5]$, $[1, 1.1]$, and $[1, 1.01]$.
   **b.** Find the instantaneous velocity of the object when $t = 1$.

**56. Gravitational Force** The magnitude of the gravitational force exerted by the earth on a particle of mass $m$ at a distance $r$ from the center of the earth is

$$F(r) = \begin{cases} \dfrac{GMmr}{R^3} & \text{if } r < R \\[2ex] \dfrac{GMm}{r^2} & \text{if } r \geq R \end{cases}$$

where $M$ is the mass of the earth, $R$ is its radius, and $G$ is the gravitational constant.
   **a.** Is $F$ a continuous function of $r$?
   **b.** Sketch the graph of $F$.

## PROBLEM-SOLVING TECHNIQUES

In this very first example in Problem-Solving Techniques, we illustrate the efficacy of the *method of substitution*. When the right substitution is used, a problem which at first glance seems impossible to solve, or as in this case, difficult to solve, is often reduced to one that is familiar or is much easier to solve. In the Problem-Solving Techniques sections throughout this book, we will showcase other problem-solving techniques.

**EXAMPLE** Evaluate $\displaystyle\lim_{x \to 1} \frac{3x - 3}{\sqrt[3]{x + 7} - 2}$.

**Solution** The obvious approach is to use the Quotient Law for limits. But since the numerator and the denominator approach zero as $x$ approaches 1, the law is not applicable.

Drawing from experience in solving such problems, we might attempt to rationalize the denominator. Although this can be done directly, it is better to transform the expression into a simpler one. A reasonable *substitution* is to put

$$t = \sqrt[3]{x + 7}$$

so $t^3 = x + 7$ or $x = t^3 - 7$. Observe that as $x$ approaches 1, $t$ approaches 2. Therefore,

$$\lim_{x \to 1} \frac{3x - 3}{\sqrt[3]{x + 7} - 2} = \lim_{t \to 2} \frac{3(t^3 - 7) - 3}{t - 2} = \lim_{t \to 2} \frac{3t^3 - 24}{t - 2} = \lim_{t \to 2} \frac{3(t^3 - 2^3)}{t - 2}$$

$$= \lim_{t \to 2} \frac{3(t - 2)(t^2 + 2t + 4)}{t - 2}$$

$$= \lim_{t \to 2} 3(t^2 + 2t + 4) = 36$$

## CHALLENGE PROBLEMS

**1.** Find $\lim\limits_{x \to 0} \dfrac{\sqrt[3]{x+1} - 1}{x}$.

**2. a.** Find $\lim\limits_{x \to 1^-} \dfrac{x^2 - 1}{|x - 1|}$ and $\lim\limits_{x \to 1^+} \dfrac{x^2 - 1}{|x - 1|}$.

   **b.** Find $\lim\limits_{x \to 0^+} \dfrac{|\sin x|}{\sin x}$ and $\lim\limits_{x \to 0^-} \dfrac{|\sin x|}{\sin x}$.

**3.** Find $\lim\limits_{x \to \pi/2} \dfrac{\cos x}{1 - \dfrac{4x^2}{\pi^2}}$.

**4.** Let $P\left(c, \sqrt{a^2 - c^2}\right)$ be a point on the upper half of the circle $x^2 + y^2 = a^2$ and located in the first quadrant, and let $Q\left(c + h, \sqrt{a^2 - (c + h)^2}\right)$ be another point on the circle in the same quadrant.

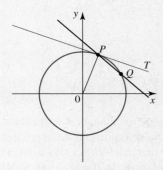

   **a.** Find an expression for the slope $m_{\sec}$ of the secant line passing through $P$ and $Q$.

   **b.** Evaluate $\lim_{h \to 0} m_{\sec}$, and show that this limit is the slope of the tangent line $T$ to the circle at $P$.

   **c.** How would you establish a similar result for the case in which $P$ and $Q$ both lie in the third quadrant?

**5.** An $n$-sided regular polygon is inscribed in a circle of radius $R$, and another is circumscribed in the same circle. The figure below illustrates the case in which $n = 6$.

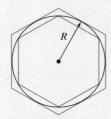

   **a.** Show that the perimeter of the circumscribing polygon is $2Rn \tan(\pi/n)$ and the perimeter of the inscribing polygon is $2Rn \sin(\pi/n)$

   **b.** Use the Squeeze Theorem and the results of part (a) to show that the circumference of a circle of radius $R$ is $2\pi R$.

**6.** Find the values of $x$ at which the function is discontinuous.

   **a.** $f(x) = [\![\sqrt{x}]\!]$

   **b.** $g(x) = [\![x]\!] + [\![-x]\!]$

**7.** A function $f$ is defined by

$$f(x) = \begin{cases} \dfrac{\tan^2 x}{1 - \cos x} & \text{if } x \neq 0 \\ c & \text{if } x = 0 \end{cases}$$

Determine the value of $c$ such that $f$ is continuous at 0.

**8.** Show that

$$f(x) = \begin{cases} \tan x \cos \dfrac{1}{x} & \text{if } x \neq 0 \\ 0 & \text{if } x = 0 \end{cases}$$

is continuous at 0.

**9.** Let $f$ be a continuous function with domain $[1, 3]$ and range $[0, 4]$ satisfying $f(1) = 0$ and $f(3) = 4$. Show that there is at least one point $c$ in $(1, 3)$ such that $f(c) = c$. The point $c$ is called a *fixed point* of $f$.

**10.** Let $f(x) = \dfrac{1}{1 - x}$. Determine where the composite function $g = f \circ f \circ f$ defined by $g(x) = f\{f[f(x)]\}$ is discontinuous.

**11.** Determine where the composite function $h = f \circ g$ defined by $f(x) = \dfrac{1}{x^2 - x - 2}$ and $g(x) = \dfrac{1}{x - 1}$ is discontinuous.

**12.** Let $f$ be a polynomial function of even degree, and suppose that there is a number $c$ such that $f(c)$ and the leading coefficient of $f$ have opposite signs. Show that $f$ must have at least two real zeros.

**13.** Suppose that $a$, $b$, and $c$ are positive and that $A < B < C$. Show that the equation

$$\frac{a}{x - A} + \frac{b}{x - B} + \frac{c}{x - C} = 0$$

has a root between $A$ and $B$, and a root between $B$ and $C$.

**14.** Suppose that $f$ is continuous on an interval $(a, b)$ and that $x_1, x_2, \ldots, x_n$ are any $n$ numbers in $(a, b)$. Show that there exists a number $c$ in $(a, b)$ such that

$$f(c) = \frac{1}{n} [f(x_1) + f(x_2) + \cdots + f(x_n)]$$

The photograph shows a space shuttle being launched from Cape Kennedy. Suppose a spectator watches the launch from an observation deck located at a known distance from the launch pad. If the speed of the shuttle at a certain instant of time is known, can we find the speed at which the distance between the shuttle and the spectator is changing? The derivative allows us to answer questions such as this.

Matt Stroshane/Getty Images

# 2

# The Derivative

**IN THIS CHAPTER** we introduce the notion of the derivative of a function. The derivative is the principal tool that we use to solve problems in differential calculus. We also develop rules of differentiation that will enable us to calculate, with relative ease, the derivatives of complicated functions. The rest of the chapter will be devoted to applications of the derivative.

**V** This symbol indicates that one of the following video types is available for enhanced student learning at **www.academic.cengage.com/login:**
- Chapter lecture videos
- Solutions to selected exercises

## 2.1 | The Derivative

### ■ The Derivative

In Section 1.5 we saw that the slope of the tangent line to the graph of a function $y = f(x)$ at the point $(a, f(a))$ has the same value as the rate of change of the quantity $y$ with respect to $x$ at the number $a$. Both values are given by

$$\lim_{h \to 0} \frac{f(a + h) - f(a)}{h}$$

provided that the limit exists. Recall that in deriving this expression, the number $a$ was fixed but otherwise arbitrary. Therefore, if we simply replace the constant $a$ by the variable $x$, we obtain a formula that gives us the slope of the tangent line at *any* point $(x, f(x))$ on the graph of $f$ as well as the rate of change of the quantity $y$ with respect to $x$ for any value of $x$. The resulting function is called the *derivative of $f$*, since it is derived from the function $f$.

---

**DEFINITION** **The Derivative**

The derivative of a function $f$ with respect to $x$ is the function $f'$ defined by the rule

$$f'(x) = \lim_{h \to 0} \frac{f(x + h) - f(x)}{h} \tag{1}$$

The domain of $f'$ consists of all values of $x$ for which the limit exists.

---

Two interpretations of the derivative follow.

1. **Geometric Interpretation of the Derivative:** The derivative $f'$ of a function $f$ is a measure of the slope of the tangent line to the graph of $f$ at any point $(x, f(x))$, provided that the derivative exists.
2. **Physical Interpretation of the Derivative:** The derivative $f'$ of a function $f$ measures the instantaneous rate of change of $f$ at $x$.

(See Figure 1.)

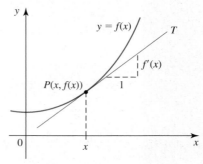

**FIGURE 1**
$f'(x)$ is the slope of $T$ at $P$; $f(x)$ is changing at the rate of $f'(x)$ units per unit change in $x$ at $x$.

## Using the Derivative to Describe the Motion of the Maglev

Let's look at these two interpretations of the derivative via an example involving the motion of the maglev. Once again, recall that the position $s$ of the maglev at any time $t$ is

$$s = f(t) = 4t^2 \qquad 0 \le t \le 30$$

The derivative of the function $f$ is

$$
\begin{aligned}
f'(t) &= \lim_{h \to 0} \frac{f(t + h) - f(t)}{h} \\
&= \lim_{h \to 0} \frac{4(t + h)^2 - 4t^2}{h} \\
&= \lim_{h \to 0} \frac{4t^2 + 8th + 4h^2 - 4t^2}{h} \\
&= \lim_{h \to 0} \frac{h(8t + 4h)}{h} = \lim_{h \to 0} (8t + 4h) \\
&= 8t
\end{aligned}
$$

Thus, the rate of change of the position of the maglev with respect to time, at time $t$, as well as the slope of the tangent line at the point $(t, f(t))$ on the graph of $f$, is given by

$$f'(t) = 8t \qquad 0 < t < 30$$

So in this setting, $f'$ is just the velocity function giving the velocity of the maglev at any time $t$. In particular, the velocity of the maglev when $t = 2$ is

$$f'(2) = 8(2) = 16$$

or 16 ft/sec. Equivalently, the slope of the tangent line to the graph of $f$ at the point $P(2, 16)$ is 16. The graph of $f'$ is sketched in Figure 2.

From the velocity curve we see that the velocity of the maglev is steadily increasing with respect to time. We can even say more. Because the equation $v = 8t$ is a linear equation in the slope-intercept form with slope 8, we see that $v$ is increasing at the rate of 8 units per unit change in $t$. Put another way, the maglev is accelerating at the constant rate of 8 ft/sec/sec, usually abbreviated 8 ft/sec$^2$. (Acceleration is the rate of change of velocity.)

Starting from just a formula giving the position of the maglev, we have now been able to give a complete description of the motion of the maglev, albeit just for this particular situation.

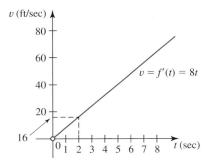

**FIGURE 2**
The graph of $v = f'(t) = 8t$ gives the velocity of the maglev at any time $t$ and is called a velocity curve.

## Differentiation

The process of finding the derivative of a function is called **differentiation.** We can view this process as an operation on a function $f$ to produce another function $f'$. For example, if we let $D_x$ denote the **differential operator,** then the process of differentiation can be written

$$D_x f = f' \qquad \text{or} \qquad D_x f(x) = f'(x)$$

Differentiation is always performed with respect to the independent variable. (Remember that we are concerned with the rate of change of the dependent variable with respect to the independent variable.) Therefore, if the independent variable is $t$, we write $D_t$ instead of $D_x$. Another notation, and one that we will adopt, is

$$\frac{d}{dx}$$

which is read "dee dee x of." For example

$$\frac{d}{dx}f = D_x f = f' \quad \text{or} \quad \frac{d}{dx}f(x) = D_x f(x) = f'(x) \qquad \text{\small $f'(x)$ is read "$f$ prime of $x$."}$$

If we denote the dependent variable by $y$ so that $y = f(x)$, then the derivative is written

$$\frac{dy}{dx}$$

(read "dee y, dee x") or, in an even more abbreviated form, as $y'$ (read "y prime").

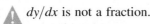 $dy/dx$ is not a fraction.

The value of the derivative of $f$ at $a$ is denoted by $f'(a)$. If the dependent variable is denoted by a letter such as $y$, then the value of the derivative at $a$ is denoted by

$$\frac{dy}{dx}\bigg|_{x=a}$$

(read "dy/dx evaluated at x = a"). For example, since the position of the maglev is denoted by the letter $s$, where $s = f(t) = 4t^2$, the velocity of the maglev when $t = 2$ may be written as $f'(2) = 16$ or

$$\frac{ds}{dt}\bigg|_{t=2} = 8t\bigg|_{t=2} = 16$$

## ■ Finding the Derivative of a Function

**EXAMPLE 1** Let $y = \sqrt{x}$.

**a.** Find $dy/dx$, and determine its domain.
**b.** How fast is $y$ changing at $x = 4$?
**c.** Find the slope and an equation of the tangent line to the graph of the equation $y = \sqrt{x}$ at the point where $x = 4$.

**Solution**  Here, $f(x) = \sqrt{x}$.

**a.** $\dfrac{dy}{dx} = \lim\limits_{h \to 0} \dfrac{f(x + h) - f(x)}{h} = \lim\limits_{h \to 0} \dfrac{\sqrt{x + h} - \sqrt{x}}{h}$

$\qquad = \lim\limits_{h \to 0} \dfrac{(\sqrt{x + h} - \sqrt{x})(\sqrt{x + h} + \sqrt{x})}{h(\sqrt{x + h} + \sqrt{x})}$  \qquad Rationalize the numerator.

$\qquad = \lim\limits_{h \to 0} \dfrac{(x + h) - x}{h(\sqrt{x + h} + \sqrt{x})} = \lim\limits_{h \to 0} \dfrac{h}{h(\sqrt{x + h} + \sqrt{x})}$

$\qquad = \lim\limits_{h \to 0} \dfrac{1}{\sqrt{x + h} + \sqrt{x}} = \dfrac{1}{2\sqrt{x}}$

The domain of $dy/dx$ is $(0, \infty)$.

**b.** The rate of change of $y$ with respect to $x$ at $x = 4$ is

$$\left.\frac{dy}{dx}\right|_{x=4} = \left.\frac{1}{2\sqrt{x}}\right|_{x=4} = \frac{1}{2\sqrt{4}} = \frac{1}{4}$$

or $\frac{1}{4}$ unit per unit change in $x$.

**c.** The slope $m$ of the tangent line to the graph of $y = \sqrt{x}$ at the point where $x = 4$ has the same value as the rate of change of $y$ with respect to $x$ at $x = 4$. From the result of part (b), we find $m = \frac{1}{4}$. Next, when $x = 4$, $y = \sqrt{4} = 2$, giving $(4, 2)$ as the point of tangency. Finally, using the point-slope form of an equation of a line, we find

$$y - 2 = \frac{1}{4}(x - 4)$$

or $y = \frac{1}{4}x + 1$ as an equation of the tangent line.

The graph of $y = \sqrt{x}$ and the tangent line at $(4, 2)$ are sketched in Figure 3.

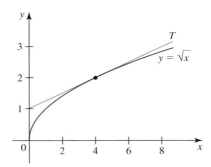

**FIGURE 3**
$T$ is the tangent line to the graph of $y = \sqrt{x}$ at $(4, 2)$.

**EXAMPLE 2** Let $f(x) = 2x^3 + x$.

**a.** Find $f'(x)$.
**b.** What is the slope of the tangent line to the graph of $f$ at $(2, 18)$?
**c.** How fast is $f$ changing when $x = 2$?

**Solution**

**a.** $f'(x) = \lim\limits_{h \to 0} \dfrac{f(x + h) - f(x)}{h} = \lim\limits_{h \to 0} \dfrac{[2(x + h)^3 + (x + h)] - (2x^3 + x)}{h}$

$= \lim\limits_{h \to 0} \dfrac{(2x^3 + 6x^2h + 6xh^2 + 2h^3 + x + h) - (2x^3 + x)}{h}$

$= \lim\limits_{h \to 0} \dfrac{h(6x^2 + 6xh + 2h^2 + 1)}{h} = \lim\limits_{h \to 0} (6x^2 + 6xh + 2h^2 + 1)$

$= 6x^2 + 1$

**b.** The required slope is given by

$$f'(2) = 6(2)^2 + 1 = 25$$

**c.** From the result of part (b), we see that $f$ is changing at the rate of 25 units per unit change in $x$ when $x = 2$.

**EXAMPLE 3**  Find $\dfrac{dy}{dx}$ if $y = \dfrac{1}{x + 1}$.

**Solution**    If we write $y = f(x)$, then

$$\frac{dy}{dx} = f'(x) = \lim_{h \to 0} \frac{f(x + h) - f(x)}{h}$$

$$= \lim_{h \to 0} \frac{\dfrac{1}{(x + h) + 1} - \dfrac{1}{x + 1}}{h}$$

$$= \lim_{h \to 0} \frac{\dfrac{x + 1 - (x + h + 1)}{(x + h + 1)(x + 1)}}{h} \qquad \text{Simplify the numerator.}$$

$$= \lim_{h \to 0} -\frac{1}{(x + h + 1)(x + 1)} = -\frac{1}{(x + 1)^2}$$

## ■ Using the Graph of $f$ to Sketch the Graph of $f'$

It was a simple matter to sketch the graph of the derivative function $f'$ in the example describing the motion of a maglev, because we were able to obtain the formula $f'(t) = 8t$ from the position function $f$ for the maglev. The next example shows how we can make a rough sketch of the graph of $f'$ using only the graph of $f$. The method that is used is based on the geometric interpretation of $f'$.

**EXAMPLE 4**  **The Trajectory of a Projectile**    The graph of the function $f$ shown in Figure 4 gives the ballistic trajectory of a projectile that starts from the origin and is confined to move in the $xy$-plane. Use this graph to draw the graph of $f'$. Then use it to estimate the rate at which the altitude of the projectile ($y$) is changing with respect to $x$ (the distance traveled horizontally by the projectile) when $x = 5000$ and when $x = 16,000$.

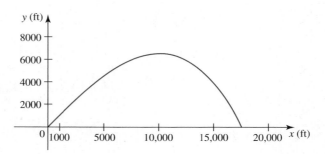

**FIGURE 4**
The trajectory of a projectile

**Solution**    First we estimate the slopes $f'(x)$ of the tangent lines (drawn by sight) to some points on the graph of $f$ using the techniques of Example 1 in Section 1.5. The results are shown in Figure 5a. Next, we plot the points $(x, f'(x))$ on the $xy'$-coordinate system placed directly below the $xy$-coordinate system. Finally, we draw a smooth curve through these points, obtaining the graph of $f'$ shown in Figure 5b. From the graph of $f'$ we see that the altitude of the projectile is increasing at the rate of approximately

0.7 ft/ft when $x = 5000$, and it is decreasing at the rate of approximately 1.3 ft/ft when $x = 16,000$.

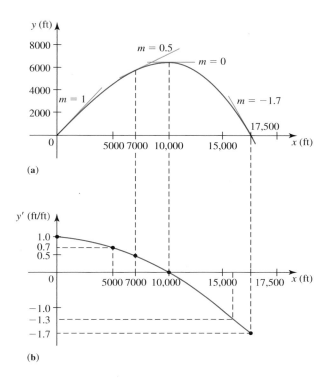

(a)

(b)

**FIGURE 5**
The graphs of $f$ and $f'$

## Differentiability

A function is said to be **differentiable** at a number if it has a derivative at that number. As we will soon see, a function may fail to be differentiable at one or more numbers in its domain. This should not surprise us because the derivative is the limit of a function, and we have already seen that the limit of a function does not always exist as we approach a number.

Loosely speaking, a function $f$ does not have a derivative at $a$ if the graph of $f$ does not have a tangent line at $a$, or if the tangent line does exist, then it is vertical.

In this text we will deal only with functions whose derivatives fail to exist at a finite number of values of $x$. Typically, these values correspond to points where the graph of $f$ has a discontinuity, a corner, or a vertical tangent. These situations are illustrated in the following examples.

**EXAMPLE 5** Show that the Heaviside function

$$H(t) = \begin{cases} 0 & \text{if } t < 0 \\ 1 & \text{if } t \geq 0 \end{cases}$$

which is discontinuous at 0, is not differentiable at 0 (Figure 6).

**Solution** Let's show that the (left-hand) limit

$$\lim_{h \to 0^-} \frac{H(0 + h) - H(0)}{h} \qquad h < 0$$

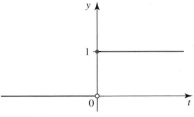

**FIGURE 6**
The Heaviside function is not differentiable at 0.

does not exist. This, in turn, will imply that

$$H'(0) = \lim_{h \to 0} \frac{H(0 + h) - H(0)}{h}$$

does not exist; that is, $H$ does not have a derivative at 0. Now

$$\lim_{h \to 0^-} \frac{H(h) - H(0)}{h} = \lim_{h \to 0^-} \frac{0 - 1}{h} = \infty \qquad \text{Since } h < 0$$

so $H'(0)$ does not exist, as asserted. ■

The next example shows that if $f$ has a sharp corner at $a$, then $f$ is not differentiable at $a$.

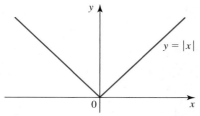

**FIGURE 7**
The function $f(x) = |x|$ is continuous everywhere and has a corner at 0.

**EXAMPLE 6** Show that the function $f(x) = |x|$ is differentiable everywhere except at 0.

**Solution** The graph of $f$ is shown in Figure 7. To prove that $f$ is not differentiable at 0, we will show that $f'(0)$ does not exist by demonstrating that the one-sided limits of the quotient

$$\frac{f(0 + h) - f(0)}{h} = \frac{f(h) - f(0)}{h} = \frac{|h| - 0}{h} = \frac{|h|}{h}$$

as $h$ approaches 0 are not equal. First, suppose $h > 0$. Then $|h| = h$, so

$$\lim_{h \to 0^+} \frac{|h|}{h} = \lim_{h \to 0^+} \frac{h}{h} = \lim_{h \to 0^+} 1 = 1$$

Next, if $h < 0$, then $|h| = -h$, and therefore,

$$\lim_{h \to 0^-} \frac{|h|}{h} = \lim_{h \to 0^-} \frac{-h}{h} = \lim_{h \to 0^-} (-1) = -1$$

Therefore,

$$f'(0) = \lim_{h \to 0} \frac{f(0 + h) - f(0)}{h} = \lim_{h \to 0} \frac{|h|}{h}$$

does not exist, and $f$ is not differentiable at 0.

To show that $f$ is differentiable at all other numbers, we rewrite $f(x)$ in the form

$$f(x) = |x| = \begin{cases} -x & \text{if } x < 0 \\ x & \text{if } x \geq 0 \end{cases}$$

and then differentiate $f(x)$ to obtain

$$f'(x) = \begin{cases} -1 & \text{if } x < 0 \\ 1 & \text{if } x > 0 \end{cases}$$

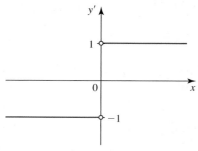

**FIGURE 8**
$f'(0)$ is not defined; therefore, $f$ is not differentiable at 0.

Geometrically, this result is evident if you consider the graph of $f$, which consists of two rays (Figure 7). The slope of the half-line to the left of the origin is $-1$, and the slope of the half-line to the right of the origin is 1. The graph of $f'$ is shown in Figure 8. ■

The graph of a function $f$ has a **vertical tangent line** $x = a$ at $a$, if $f$ is continuous at $a$ and

$$\lim_{x \to a} f'(x) = -\infty \qquad \text{or} \qquad \lim_{x \to a} f'(x) = \infty$$

The next example shows that the function $f$ is not differentiable at $a$ because the graph of $f$ has a vertical tangent line at $a$.

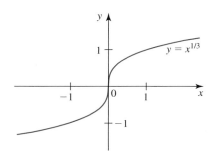

**FIGURE 9**
The graph of $f$ has a vertical tangent line at $(0, 0)$.

**EXAMPLE 7** Show that the function $f(x) = x^{1/3}$ is not differentiable at 0.

**Solution** We compute

$$\lim_{h \to 0} \frac{f(0 + h) - f(0)}{h} = \lim_{h \to 0} \frac{f(h) - f(0)}{h}$$

$$= \lim_{h \to 0} \frac{h^{1/3} - 0}{h} = \lim_{h \to 0} \frac{1}{h^{2/3}} = \infty$$

This shows that $f$ is not differentiable at 0. (See Figure 9.)

## ◼ Differentiability and Continuity

Examples 6 and 7 show that a function can be continuous at a number yet not be differentiable there. The next theorem shows that the requirement that a function is differentiable at a number is stronger than the requirement that it be continuous there.

### THEOREM 1

If $f$ is differentiable at $a$, then $f$ is continuous at $a$.

**PROOF** If $x$ is in the domain of $f$ and $x \neq a$, then we can write

$$f(x) - f(a) = \frac{f(x) - f(a)}{x - a} (x - a)$$

We have

$$\lim_{x \to a} [f(x) - f(a)] = \lim_{x \to a} \frac{f(x) - f(a)}{x - a} \cdot (x - a)$$

$$= \lim_{x \to a} \frac{f(x) - f(a)}{x - a} \cdot \lim_{x \to a} (x - a)$$

$$= f'(a) \cdot 0 = 0$$

So

$$\lim_{x \to a} f(x) = \lim_{x \to a} [f(a) + (f(x) - f(a))]$$

$$= \lim_{x \to a} f(a) + \lim_{x \to a} [f(x) - f(a)] = f(a) + 0 = f(a)$$

and this shows that $f$ is continuous at $a$, as asserted.

## 2.1   CONCEPT QUESTIONS

**1. a.** Give a geometric and a physical interpretation of the expression

$$\frac{f(x + h) - f(x)}{h}$$

**b.** Give a geometric and a physical interpretation of the expression

$$\lim_{h \to 0} \frac{f(x + h) - f(x)}{h}$$

**2.** Under what conditions does a function fail to have a derivative at a number? Illustrate your answer with sketches.

## 2.1   EXERCISES

*In Exercises 1–14, use the definition of the derivative to find the derivative of the function. What is its domain?*

**1.** $f(x) = 5$

**2.** $f(x) = 2x + 1$

**3.** $f(x) = 3x - 4$

**4.** $f(x) = 2x^2 + x$

**5.** $f(x) = 3x^2 - x + 1$

**6.** $f(x) = x^3 - x$

**7.** $f(x) = 2x^3 + x - 1$

**8.** $f(x) = 2\sqrt{x}$

**9.** $f(x) = \sqrt{x + 1}$

**10.** $f(x) = \dfrac{1}{x}$

**11.** $f(x) = \dfrac{1}{x + 2}$

**12.** $f(x) = -\dfrac{2}{\sqrt{x}}$

**13.** $f(x) = \dfrac{3}{2x + 1}$

**14.** $f(x) = x + \sqrt{x}$

*In Exercises 15–20, find an equation of the tangent line to the graph of the function at the indicated point.*

| Function | Point |
|---|---|
| **15.** $f(x) = x^2 + 1$ | $(2, 5)$ |
| **16.** $f(x) = 3x^2 - 4x + 2$ | $(2, 6)$ |
| **17.** $f(x) = 2x^3$ | $(1, 2)$ |
| **18.** $f(x) = 3x^3 - x$ | $(-1, -2)$ |
| **19.** $f(x) = \sqrt{x - 1}$ | $(4, \sqrt{3})$ |
| **20.** $f(x) = \dfrac{2}{x}$ | $(2, 1)$ |

**21. a.** Find an equation of the tangent line to the graph of $f(x) = 2x - x^3$ at the point $(1, 1)$.

 **b.** Plot the graph of $f$ and the tangent line in successively smaller viewing windows centered at $(1, 1)$ until the graph of $f$ and the tangent line appear to coincide.

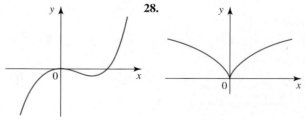

**22. a.** In Example 6 we showed that $f(x) = |x|$ is not differentiable at $x = 0$. Plot the graph of $f$ using the viewing window $[-1, 1] \times [-1, 1]$. Then **ZOOM IN** using successively smaller viewing windows centered at $(0, 0)$. What can you say about the existence of a tangent line at $(0, 0)$?

**b.** Plot the graph of

$$f(x) = \begin{cases} x + 1 & \text{if } x \le 1 \\ \dfrac{2}{x} & \text{if } x > 1 \end{cases}$$

using the viewing window $[-2, 4] \times [-2, 3]$. Then **ZOOM IN** using successively smaller viewing windows centered at $(1, 2)$. Is $f$ differentiable at $x = 1$?

*In Exercises 23–26, find the rate of change of y with respect to x at the given value of x.*

**23.** $y = -2x^2 + x + 1; \quad x = 1$

**24.** $y = 2x^3 + 2; \quad x = 2$

**25.** $y = \sqrt{2x}; \quad x = 2$

**26.** $y = x^2 - \dfrac{1}{x}; \quad x = -1$

*In Exercises 27–30, match the graph of each function with the graph of its derivative in (a)–(d).*

**27.**

**28.**

**29.**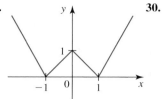

**30.**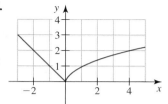

**(a)**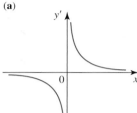

**(b)**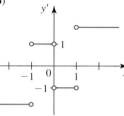

**(c)**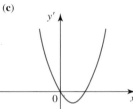

**(d)**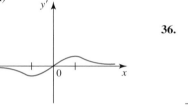

*In Exercises 31–36, sketch the graph of the derivative f' of the function f whose graph is given.*

**31.**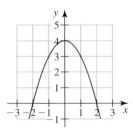

**32.**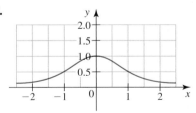

**33.**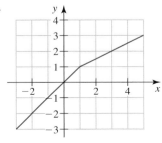

**34.**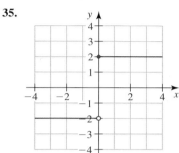

**35.**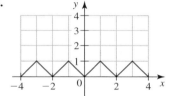

**36.**

**37. Air Temperature and Altitude** The air temperature at a height of $h$ feet from the surface of the earth is $T = f(h)$ degrees Fahrenheit.
   **a.** Give a physical interpretation of $f'(h)$. Give units.
   **b.** Generally speaking, what do you expect the sign of $f'(h)$ to be?
   **c.** If you know that $f'(1000) = -0.05$, estimate the change in the air temperature if the altitude changes from 1000 ft to 1001 ft.

**38. Advertising and Revenue** Suppose that the total revenue realized by the Odyssey Travel Agency is $R = f(x)$ thousand dollars if $x$ thousand dollars are spent on advertising.
   **a.** What does

$$\frac{f(b) - f(a)}{b - a} \qquad 0 < a < b$$

   measure? What are the units?
   **b.** What does $f'(x)$ measure? Give units.
   **c.** Given that $f'(20) = 3$, what is the approximate change in the revenue if Odyssey increases its advertising budget from $20,000 to $21,000?

**39. Production Costs** Suppose that the total cost in manufacturing $x$ units of a certain product is $C(x)$ dollars.
   **a.** What does $C'(x)$ measure? Give units.
   **b.** What can you say about the sign of $C'$?
   **c.** Given that $C'(1000) = 20$, estimate the additional cost to be incurred by the company in producing the 1001st unit of the product.

**40. Range of a Projectile** A projectile is fired from a cannon that makes an angle of $\theta$ degrees with the horizontal. If the muzzle velocity is constant, then the range in feet of the projectile is a function of $\theta$, that is, $R = f(\theta)$.

   **a.** What is the physical meaning of $f'(\theta)$? Give units.

   **b.** What can you say about the sign of $f'(\theta)$, where $0° < \theta < 90°$?

   **c.** Given that $f(40) = 10{,}000$ and $f'(40) = 20$, estimate the range of a projectile if it is fired at an angle of elevation of 41°.

**41.** Let $f(x) = x^2 - 2x + 1$.

   **a.** Find the derivative $f'$ of $f$.

   **b.** Find the point on the graph of $f$ where the tangent line to the curve is horizontal.

   **c.** Sketch the graph of $f$ and the tangent line to the curve at the point found in part (b).

   **d.** What is the rate of change of $f$ at this point?

**42.** Let $f(x) = \dfrac{1}{x - 1}$.

   **a.** Find the derivative $f'$ of $f$.

   **b.** Find an equation of the tangent line to the curve at the point $\left(-1, -\frac{1}{2}\right)$.

   **c.** Sketch the graph of $f$ and the tangent line to the curve at the point $\left(-1, -\frac{1}{2}\right)$.

*In Exercises 43–48, use the graph of the function $f$ to find the value(s) of $x$ at which $f$ is not differentiable.*

**43.**

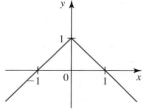

**44.**

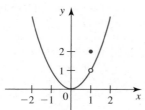

**45.**

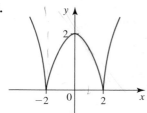

**46.**

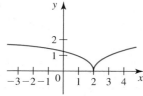

**47.**

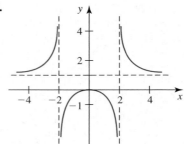

**48.**

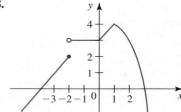

*In Exercises 49–52, show that the function is continuous but not differentiable at the given value of $x$.*

**49.** $f(x) = \begin{cases} x + 2 & \text{if } x \le 0 \\ 2 - 3x & \text{if } x > 0 \end{cases}; \quad x = 0$

**50.** $f(x) = \begin{cases} x + 1 & \text{if } x \le 0 \\ x^2 + 1 & \text{if } x > 0 \end{cases}; \quad x = 0$

**51.** $f(x) = |2x - 1|; \quad x = \dfrac{1}{2}$

**52.** $f(x) = \begin{cases} x \sin \dfrac{1}{x} & \text{if } x \ne 0 \\ 0 & \text{if } x = 0 \end{cases}; \quad x = 0$

**53. R & D Expenditure** The graph of the function $f$ shown in the figure gives the Department of Energy budget for research and development for solar, wind, and other renewable energy sources over a 12-year period. Use the slopes of $f$ at the indicated values of $t$ and the technique of Example 4 to sketch the graph of $f'$. Then use the graph of $f'$ to estimate the rate of change of the budget when $t = 1$ and when $t = 5$.

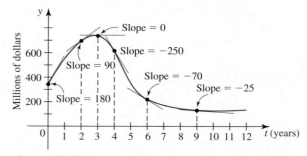

*Source:* U.S. Department of Energy.

**54. Velocity of a Model Car** The graph of the function $f$ shown in the figure gives the position $s = f(t)$ of a model car moving along a straight line as a function of time. Use the technique of Example 4 to sketch the velocity curve for the car. (Recall that the velocity of an object is given by the rate of change (derivative) of its position.) Then use the graph of $f'$ to estimate the velocity of the car at $t = 5$ and $t = 12$.

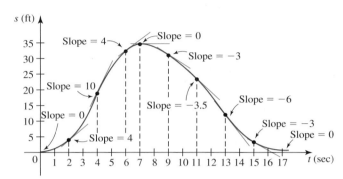

**55.** Let $f(x) = x^3$. For each real number $h \neq 0$, define

$$g(x) = \frac{(x + h)^3 - x^3}{h}$$

**a.** For each fixed value of $h$, what does $g(x)$ measure?

**b.** What function do you expect $g(x)$ to approach as $h$ approaches zero?

 **c.** Verify your answer to part (b) visually by plotting the graph of the function you guessed at in part (b) and the graph of the function $g(x)$ for $h = 1, 0.5$, and $0.1$ in a common viewing window.

**56.** Let $f(x) = x^3 - x$.

**a.** Find $f'(x)$.

 **b.** Plot the graphs of $f'$ and $g$, where

$$g(x) = \frac{[(x + 0.01)^3 - (x + 0.01)] - (x^3 - x)}{0.01}$$

using a common viewing window. Is the result expected? Explain.

**57.** Let $f(x) = |x^3|$.

**a.** Sketch the graph of $f$.

**b.** For what values of $x$ is $f$ differentiable?

**c.** Find a formula for $f'(x)$.

**58.** Let $f(x) = x|x|$.

**a.** Sketch the graph of $f$.

**b.** For what values of $x$ is $f$ differentiable?

**c.** Find a formula for $f'(x)$.

**59.** Suppose that $g(x) = |x - a| f(x)$, where $f$ is a continuous function and $f(a) \neq 0$. Show that $g$ is continuous at $a$ but not differentiable at $a$.

**60.** Let

$$f(x) = \begin{cases} x^{1/3} \sin \dfrac{1}{x} & \text{if } x \neq 0 \\ 0 & \text{if } x = 0 \end{cases}$$

**a.** Show that $f$ is continuous at 0, but not differentiable at 0.

 **b.** Plot the graph of $f$ using the viewing window $[-0.5, 0.5] \times [-0.1, 0.1]$.

**61.** Let

$$f(x) = \begin{cases} x^2 \sin \dfrac{1}{x} & \text{if } x \neq 0 \\ 0 & \text{if } x = 0 \end{cases}$$

**a.** Show that $f$ is differentiable at 0. What is $f'(0)$?

**b.** Plot the graph of $f$ using the viewing window $[-0.5, 0.5] \times [-0.1, 0.1]$.

**62.** A function $f$ is called *periodic* if there exists a number $T > 0$ such that $f(x + T) = f(x)$ for all $x$ in the domain of $f$. Prove that the derivative of a differentiable periodic function with period $T$ is also a periodic function with period $T$.

**63.** Show that if $f'(x)$ exists, then

$$\lim_{h \to 0} \frac{f(x + nh) - f[x + (n - 1)h]}{h} = f'(x) \qquad n \neq 0, 1$$

**64.** Use the result of Exercise 63 to find the derivative of (a) $f(x) = \sqrt{x}$ by taking $n = 2$ and (b) $f(x) = \dfrac{1}{x + 1}$ by taking $n = 3$. (Compare with Examples 1 and 3.)

*In Exercises 65–70, determine whether the statement is true or false. If it is true, explain why it is true. If it is false, explain why or give an example to show why it is false.*

**65.** If $f$ is differentiable at $x = 3$, then the slope of the tangent line to the graph of $f$ at the point $(3, f(3))$ is

$$\lim_{h \to 0} \frac{f(3 + h) - f(3)}{h}$$

**66.** If $f$ is differentiable at $a$, and $g$ is not differentiable at $a$, then the product $fg$ is not differentiable at $a$.

**67.** If both $f$ and $g$ are not differentiable at $a$, then the product $fg$ is not differentiable at $a$.
**Hint:** Consider $f(x) = |x|$ and $g(x) = |x|$.

**68.** If both $f$ and $g$ are not differentiable at $a$, then the sum $f + g$ is not differentiable at $a$.

**69.** The domain of $f'$ is the same as that of $f$.

**70.** If $n$ is a positive integer, then there exists a function $f$ such that $f$ is differentiable everywhere except at $n$ numbers.

## 2.2 Basic Rules of Differentiation

### ■ Some Basic Rules

Up to now we have computed the derivative of a function using its definition. But as you have seen, this process is tedious even for relatively simple functions. In this section we will develop some rules of differentiation that will simplify the process of finding the derivative of a function.

> **THEOREM 1    Derivative of a Constant Function**
>
> If $c$ is a constant, then
>
> $$\frac{d}{dx}(c) = 0$$

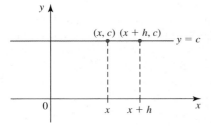

**FIGURE 1**
The slope of the graph of $f(x) = c$ is zero at every point. Hence, $f'(x) = 0$.

**PROOF**    Let $f(x) = c$. Then

$$f'(x) = \lim_{h \to 0} \frac{f(x + h) - f(x)}{h} = \lim_{h \to 0} \frac{c - c}{h} = \lim_{h \to 0} 0 = 0$$  ■

This result is also evident geometrically (see Figure 1). The tangent line to a straight line at any point on the line must coincide with the line itself. Since the constant function $f$ defined by $f(x) = c$ is a horizontal line with slope 0, any tangent line to $f$ must also have slope 0. Hence, $f'(x) = 0$ for every $x$.

### EXAMPLE 1

**a.** If $f(x) = 19$, then $f'(x) = \dfrac{d}{dx}(19) = 0$.

**b.** If $f(x) = -\pi^2$, then $f'(x) = \dfrac{d}{dx}(-\pi^2) = 0$.  ■

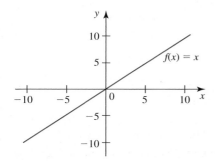

**FIGURE 2**
The graph of $f(x) = x$ is the line with slope 1. Hence, $f'(x) = 1$.

Next, we turn our attention to the rule for differentiating power functions $f(x) = x^n$ with positive integral exponents $n$. For the special case in which $n = 1$, we have $f(x) = x$. Its derivative is

$$f'(x) = \lim_{h \to 0} \frac{f(x + h) - f(x)}{h} = \lim_{h \to 0} \frac{(x + h) - x}{h} = \lim_{h \to 0} 1 = 1$$

This result is also evident geometrically because the graph of $y = x$ is the line with slope 1 (see Figure 2) and hence $f'(x) = 1$ for every $x$. That is,

$$\frac{d}{dx}(x) = 1 \tag{1}$$

We now state the general rule for finding the derivative of $f(x) = x^n$, where $n$ is a positive integer.

---

**THEOREM 2    The Power Rule**

If $n$ is a positive integer and $f(x) = x^n$, then

$$f'(x) = \frac{d}{dx}(x^n) = nx^{n-1}$$

---

**PROOF**   Let $f(x) = x^n$. Then

$$f'(x) = \lim_{h \to 0} \frac{f(x+h) - f(x)}{h} = \lim_{h \to 0} \frac{(x+h)^n - x^n}{h}$$

Now observe that

$$a^n - b^n = (a - b)(a^{n-1} + a^{n-2}b + \cdots + ab^{n-2} + b^{n-1})$$

which can be verified by simply expanding the expression on the right-hand side. If we use this equation with $a$ replaced by $x + h$ and $b$ replaced by $x$, then we can write

$$f'(x) = \lim_{h \to 0} \frac{[(x+h) - x][(x+h)^{n-1} + (x+h)^{n-2}x + \cdots + (x+h)x^{n-2} + x^{n-1}]}{h}$$

$$= \lim_{h \to 0} [(x+h)^{n-1} + (x+h)^{n-2}x + \cdots + (x+h)x^{n-2} + x^{n-1}]$$

$$= \underbrace{x^{n-1} + x^{n-1} + \cdots + x^{n-1} + x^{n-1}}_{n \text{ terms}}$$

$$= nx^{n-1}$$

Theorem 2 can also be proved by using the Binomial Theorem (see Exercise 73).

**EXAMPLE 2**

**a.** If $f(x) = x^{10}$, then $f'(x) = \dfrac{d}{dx}(x^{10}) = 10x^{10-1} = 10x^9$.

**b.** If $g(u) = u^3$, then $g'(u) = \dfrac{d}{du}(u^3) = 3u^{3-1} = 3u^2$.

Although Theorem 2 was stated for the case in which the power $n$ is a positive integer, the Power Rule is true for all real numbers $n$. For example, if we apply the more general rule *formally* to finding the derivative of $f(x) = \sqrt{x} = x^{1/2}$, we find

$$f'(x) = \frac{d}{dx}(x^{1/2}) = \frac{1}{2}x^{-1/2} = \frac{1}{2\sqrt{x}}$$

a result that we obtained in Example 1, Section 2.1, using the definition of the derivative.

We will demonstrate the validity of the Power Rule for negative integers $n$ in Section 2.3. The rule will be extended to include rational powers $n$ in Section 2.7. Finally, we will prove the general version of the Power Rule, where $n$ may be any real number, in Section 2.8. But for now, we will assume that the Power Rule is *valid for all real numbers* and use it in our work.

**THEOREM 3    The Power Rule (General Version)**

If $n$ is any real number, then

$$\frac{d}{dx}(x^n) = nx^{n-1}$$

**EXAMPLE 3**

**a.** If $f(x) = \dfrac{1}{x^3}$, then $f'(x) = \dfrac{d}{dx}\left(\dfrac{1}{x^3}\right) = \dfrac{d}{dx}(x^{-3}) = -3x^{-3-1} = -3x^{-4} = -\dfrac{3}{x^4}$.

**b.** If $y = x^{3/2}$, then $\dfrac{dy}{dx} = \dfrac{d}{dx}(x^{3/2}) = \dfrac{3}{2}x^{(3/2)-1} = \dfrac{3}{2}x^{1/2} = \dfrac{3\sqrt{x}}{2}$.

**c.** If $g(x) = x^{0.12}$, then $g'(x) = \dfrac{d}{dx}(x^{0.12}) = 0.12x^{0.12-1} = 0.12x^{-0.88} = \dfrac{0.12}{x^{0.88}}$.    ■

The next theorem tells us that *the derivative of a constant times a function is equal to the constant times the derivative of the function.*

**THEOREM 4    The Constant Multiple Rule**

If $f$ is a differentiable function and $c$ is a constant, then

$$\frac{d}{dx}[cf(x)] = cf'(x)$$

**PROOF**    Let $F(x) = cf(x)$. Then

$$F'(x) = \lim_{h \to 0} \frac{F(x+h) - F(x)}{h} = \lim_{h \to 0} \frac{cf(x+h) - cf(x)}{h}$$

$$= \lim_{h \to 0} c\left[\frac{f(x+h) - f(x)}{h}\right]$$

$$= c\lim_{h \to 0} \frac{f(x+h) - f(x)}{h} \qquad \text{Constant Multiple Law for limits}$$

$$= cf'(x) \qquad\qquad\qquad ■$$

**EXAMPLE 4**

**a.** If $f(x) = 3x^5$, then $f'(x) = \dfrac{d}{dx}(3x^5) = 3\dfrac{d}{dx}(x^5) = 3(5x^4) = 15x^4$.

**b.** If $y = -2u^3$, then $\dfrac{dy}{du} = \dfrac{d}{du}(-2u^3) = -2\dfrac{d}{du}(u^3) = -2(3u^2) = -6u^2$.    ■

The next theorem says that *the derivative of the sum of two functions is the sum of their derivatives.*

**THEOREM 5    The Sum Rule**

If $f$ and $g$ are differentiable functions, then

$$\frac{d}{dx}[f(x) + g(x)] = f'(x) + g'(x)$$

**PROOF**    Let $F(x) = f(x) + g(x)$. Then

$$F'(x) = \lim_{h \to 0} \frac{F(x + h) - F(x)}{h}$$

$$= \lim_{h \to 0} \frac{[f(x + h) + g(x + h)] - [f(x) + g(x)]}{h}$$

$$= \lim_{h \to 0} \left[ \frac{f(x + h) - f(x)}{h} + \frac{g(x + h) - g(x)}{h} \right]$$

$$= \lim_{h \to 0} \frac{f(x + h) - f(x)}{h} + \lim_{h \to 0} \frac{g(x + h) - g(x)}{h} \quad \text{Sum Law for limits}$$

$$= f'(x) + g'(x) \qquad \blacksquare$$

**Notes**

1. Since $f(x) - g(x)$ can be written as $f(x) + [-g(x)]$, Theorem 5 implies that

$$\frac{d}{dx}[f(x) - g(x)] = \frac{d}{dx}[f(x)] + \frac{d}{dx}[-g(x)]$$

$$= \frac{d}{dx}[f(x)] - \frac{d}{dx}[g(x)] \qquad \text{By Theorem 4 with } c = -1$$

$$= f'(x) - g'(x)$$

and we see that Theorem 5 also applies to the difference of two functions.

2. The Sum (Difference) Rule is valid for any finite number of functions. For example, if $f$, $g$, and $h$ are differentiable at $x$, then so is $f + g - h$, and

$$\frac{d}{dx}[f(x) + g(x) - h(x)] = f'(x) + g'(x) - h'(x) \qquad \blacksquare$$

**EXAMPLE 5**    Find the derivative of $f(x) = 4x^5 + 2x^4 - 3x^2 + 6x + 1$.

**Solution**    Using the generalized Sum Rule, we find that

$$f'(x) = \frac{d}{dx}(4x^5 + 2x^4 - 3x^2 + 6x + 1)$$

$$= \frac{d}{dx}(4x^5) + \frac{d}{dx}(2x^4) - \frac{d}{dx}(3x^2) + \frac{d}{dx}(6x) + \frac{d}{dx}(1)$$

$$= 4\frac{d}{dx}(x^5) + 2\frac{d}{dx}(x^4) - 3\frac{d}{dx}(x^2) + 6\frac{d}{dx}(x) + \frac{d}{dx}(1)$$

$$= 4(5x^4) + 2(4x^3) - 3(2x) + 6(1) + 0$$

$$= 20x^4 + 8x^3 - 6x + 6 \qquad \blacksquare$$

**EXAMPLE 6** Find the derivative of $y = \dfrac{x^3 - 2x^2 + x - 4}{2\sqrt{x}}$.

**Solution** Using the generalized Sum Rule, we find

$$\frac{dy}{dx} = \frac{d}{dx}\left(\frac{x^3 - 2x^2 + x - 4}{2x^{1/2}}\right)$$

$$= \frac{d}{dx}\left(\frac{1}{2}x^{5/2} - x^{3/2} + \frac{1}{2}x^{1/2} - 2x^{-1/2}\right)$$

$$= \frac{1}{2}\left(\frac{5}{2}x^{3/2}\right) - \frac{3}{2}x^{1/2} + \frac{1}{2}\left(\frac{1}{2}x^{-1/2}\right) - 2\left(-\frac{1}{2}x^{-3/2}\right)$$

$$= \frac{5}{4}x^{3/2} - \frac{3}{2}x^{1/2} + \frac{1}{4}x^{-1/2} + x^{-3/2}$$ ■

**EXAMPLE 7** Find the points on the graph of $f(x) = x^4 - 2x^2 + 2$ where the tangent line is horizontal.

**Solution** At a point on the graph of $f$ where its tangent line is horizontal, the derivative of $f$ is zero. So we begin by finding

$$f'(x) = \frac{d}{dx}(x^4 - 2x^2 + 2) = 4x^3 - 4x = 4x(x^2 - 1)$$

Setting $f'(x) = 0$ leads to $4x(x^2 - 1) = 0$, giving $x = -1$, $0$, or $1$. Substituting each of the numbers into $f(x)$ gives the points $(-1, 1)$, $(0, 2)$, and $(1, 1)$ as the required points. (See Figure 3.) ■

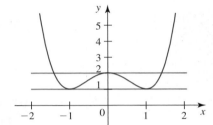

**FIGURE 3**
The graph of $f(x) = x^4 - 2x^2 + 2$ has horizontal tangent lines at $(-1, 1)$, $(0, 2)$, and $(1, 1)$.

**EXAMPLE 8** **Carbon Monoxide in the Atmosphere** The projected average global atmospheric concentration of carbon monoxide is approximated by

$$f(t) = 0.88t^4 - 1.46t^3 + 0.7t^2 + 2.88t + 293 \qquad 0 \le t \le 4$$

where $t$ is measured in 40-year intervals with $t = 0$ corresponding to the beginning of 1860 and $f(t)$ is measured in parts per million by volume. How fast was the projected average global atmospheric concentration of carbon monoxide changing at the beginning of the year 1900 ($t = 1$) and at the beginning of 2000 ($t = 3.5$)?
*Source:* Meadows et al., "Beyond the Limits."

**Solution** The rate at which the concentration of carbon monoxide is changing at time $t$ is given by

$$f'(t) = \frac{d}{dt}(0.88t^4 - 1.46t^3 + 0.7t^2 + 2.88t + 293)$$

$$= 3.52t^3 - 4.38t^2 + 1.4t + 2.88$$

parts/million/(40 years). Therefore, the rate at which the concentration of carbon monoxide was changing at the beginning of 1900 was

$$f'(1) = 3.52(1) - 4.38(1) + 1.4(1) + 2.88 = 3.42$$

or approximately 3.4 parts/million/(40 years). At the beginning of the year 2000, it was

$$f'(3.5) = 3.52(3.5)^3 - 4.38(3.5)^2 + 1.4(3.5) + 2.88 = 105.045$$

or approximately 105 parts/million/(40 years). ■

## ■ The Derivative of the Natural Exponential Function

We now turn our attention to exponential functions. To find the derivative of $f(x) = a^x$, where $a > 0$ and $a \neq 1$, we use the definition of the derivative to write

$$f'(x) = \lim_{h \to 0} \frac{f(x + h) - f(x)}{h} = \lim_{h \to 0} \frac{a^{x+h} - a^x}{h}$$

$$= \lim_{h \to 0} \frac{a^x a^h - a^x}{h} = \lim_{h \to 0} \frac{a^x(a^h - 1)}{h}$$

Since $a^x$ does not depend on $h$, it can be treated as a constant with respect to the limiting process, and we have

$$f'(x) = a^x \lim_{h \to 0} \frac{a^h - 1}{h} \tag{2}$$

Thus, the derivative of $f(x) = a^x$ exists for all values of $x$ provided that

$$\lim_{h \to 0} \frac{a^h - 1}{h}$$

exists. If we put $x = 0$ in Equation (2), we obtain

$$f'(0) = \lim_{h \to 0} \frac{a^h - 1}{h} \tag{3}$$

so

$$f'(x) = f'(0)a^x \tag{4}$$

Equation (4) tells us that the derivative of $f(x) = a^x$ is a constant multiple of itself, the constant being the slope of the tangent line to the graph of $f$ at 0. Now, it can be shown that the limit in Equation (3) exists for all $a > 0$. Before proceeding further, let us consider the cases in which $a = 2$ and $a = 3$.

**TABLE 1**  The values of $\dfrac{a^h - 1}{h}$ for $a = 2$ and $a = 3$ correct to four decimal places

| $h$ | $-0.1$ | $-0.01$ | $-0.001$ | $-0.0001$ | 0.0001 | 0.001 | 0.01 | 0.1 |
|---|---|---|---|---|---|---|---|---|
| $\dfrac{2^h - 1}{h}$ | 0.6697 | 0.6908 | 0.6929 | 0.6931 | 0.6932 | 0.6934 | 0.6956 | 0.7177 |
| $\dfrac{3^h - 1}{h}$ | 1.0404 | 1.0926 | 1.0980 | 1.0986 | 1.0987 | 1.0992 | 1.1047 | 1.1612 |

From Table 1 we see that if $a = 2$, then

$$f'(x) = \frac{d}{dx}(2^x) = f'(0)2^x = \lim_{h \to 0} \frac{2^h - 1}{h} \cdot 2^x \approx (0.69)2^x$$

and if $a = 3$, then

$$f'(x) = \frac{d}{dx}(3^x) = f'(0)3^x = \lim_{h \to 0} \frac{3^h - 1}{h} \cdot 3^x \approx (1.10)3^x$$

These calculations suggest that it might be possible to pick a number between 2 and 3 for which $f'(0) = 1$. The choice of this number for the base of the exponential function will lead to the simplest formula for the derivative of the function. This number is denoted by the letter $e$ and is the same number mentioned in Section 0.8 when we introduced the natural exponential function.

> **DEFINITION    The Number $e$**
>
> The number $e$ is the number such that
>
> $$\lim_{h \to 0} \frac{e^h - 1}{h} = 1$$

Putting $a = e$ in Equation (2) leads to the following formula:

$$f'(x) = \frac{d}{dx}(e^x) = e^x \lim_{h \to 0} \frac{e^h - 1}{h} = e^x \cdot (1) = e^x$$

In other words, the derivative of the natural exponential function is equal to itself.

> **THEOREM 6    Derivative of $e^x$**
>
> $$\frac{d}{dx} e^x = e^x$$

**EXAMPLE 9**   Let $f(x) = e^x + x$.

**a.** Find the derivative of $f$.
**b.** Find an equation of the tangent line to the graph of $f$ at the point where $x = 0$.

**Solution**

**a.** $f'(x) = \dfrac{d}{dx}(e^x + x) = \dfrac{d}{dx}(e^x) + \dfrac{d}{dx}(x) = e^x + 1$

**b.** The slope of the tangent line at the point where $x = 0$ is

$$f'(0) = e^x + 1 \Big|_{x=0} = 2$$

The $y$-coordinate of the point of tangency is $f(0) = e^0 + 0 = 1$. So a required equation is

$$y - 1 = 2(x - 0) \qquad \text{or} \qquad y = 2x + 1$$

The graph of $f$ and the tangent line are shown in Figure 4.    ∎

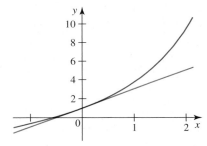

**FIGURE 4**
The graph of $f$ and the tangent line at $(0, 1)$

## 2.2    CONCEPT QUESTIONS

1. State the rule of differentiation and explain it in your own words.
   **a.** The Power Rule
   **b.** The Constant Multiple Rule
   **c.** The Sum Rule
2. State the derivative of the function.
   **a.** $f(x) = c$,   $c$ a constant
   **b.** $f(x) = x^n$,   $n$ a real number
   **c.** $f(x) = e^x$

3. **a.** Give a definition of the number $e$.
   **b.** Explain why it is more desirable to use the natural exponential function $f(x) = e^x$ than the more general exponential function $f(x) = a^x$ in calculus.
4. If $f'(2) = 3$ and $g'(2) = -2$, find
   **a.** $h'(2)$ if $h(x) = 2f(x)$
   **b.** $F'(2)$ if $F(x) = 2f(x) - 4g(x)$

# 2.2   EXERCISES

*In Exercises 1–32, find the derivative of the function.*

**1.** $f(x) = 2.718$

**2.** $f(x) = 3x + 4$

**3.** $f(x) = 3x^2$

**4.** $f(x) = -2x^3 - 3e^2$

**5.** $f(x) = x^{2.1}$

**6.** $f(x) = 9x^{1/3}$

**7.** $f(x) = 3\sqrt{x} + 2e^x$

**8.** $f(u) = \dfrac{2}{\sqrt{u}}$

**9.** $f(x) = 7x^{-12}$

**10.** $f(x) = 0.3x^{-1.2}$

**11.** $f(x) = x^2 - 2x + 8$

**12.** $g(x) = -\dfrac{1}{3}x^2 + \sqrt{2}x$

**13.** $f(r) = \pi r^2 + 2\pi r$

**14.** $y = -\dfrac{1}{3}(x^3 + 2x^2 + x - 1)$

**15.** $f(x) = 0.03x^2 - 0.4x + 10$

**16.** $f(x) = 0.002x^3 - 0.05x^2 + 0.1x + 0.1e^x - 20$

**17.** $g(x) = x^2(2x^3 - 3x^2 + x + 4)$

**18.** $H(u) = (2u)^3 - 3u + 7$

**19.** $f(x) = \dfrac{x^3 - 4x^2 + 3}{x}$

**20.** $h(t) = \dfrac{t^5 - 3t^3 + 2t^2 e^t}{2t^2}$

**21.** $f(x) = 4x^4 - 3x^{5/2} + 2$

**22.** $f(x) = 5x^{4/3} - \dfrac{2}{3}x^{3/2} + x^2 - 3x + 1$

**23.** $f(x) = 3x^{-1} + 4x^{-2}$

**24.** $f(x) = -\dfrac{1}{3}(x^{-3} - x^6)$

**25.** $f(t) = \dfrac{4}{t^4} - \dfrac{3}{t^3} + \dfrac{2}{t}$

**26.** $f(x) = \dfrac{5}{x^3} - \dfrac{2}{x^2} - \dfrac{1}{x} + 200$

**27.** $A = 0.001x^2 - 0.4x + 5 + \dfrac{200}{x}$

**28.** $y = 0.0002t^3 - 0.4t^2 + 4 + \dfrac{100}{t^2}$

**29.** $f(x) = 2x - 5\sqrt{x} + e^{x+1}$

**30.** $f(t) = 2t^2 + \sqrt{t^3}$

**31.** $y = \sqrt[3]{x} + \dfrac{1}{\sqrt{x}}$

**32.** $f(u) = \dfrac{1}{\sqrt{u}} - \dfrac{3}{\sqrt[3]{u}}$

**33.** Let $f(x) = 2x^3 - 4x$. Find

   **a.** $f'(-2)$       **b.** $f'(0)$       **c.** $f'(2)$

**34.** Let $f(x) = 4x^{5/4} + 2x^{3/2} + x$. Find

   **a.** $f'(0)$               **b.** $f'(16)$

 *In Exercises 35–38, (a) find an equation of the tangent line to the graph of the function at the indicated point, and (b) use a graphing utility to plot the graph of the function and the tangent line on the same screen.*

**35.** $f(x) = 2x^2 - 3x + 4;\quad (2, 6)$

**36.** $f(x) = -\dfrac{5}{3}x^2 + 2x + 2;\quad \left(-1, -\dfrac{5}{3}\right)$

**37.** $f(x) = x^4 - 3x^3 + 2x^2 - x + e^{x-1};\quad (1, 0)$

**38.** $f(x) = \sqrt{x} + \dfrac{1}{\sqrt{x}};\quad \left(4, \dfrac{5}{2}\right)$

*In Exercises 39–42, find the point(s) on the graph of the function at which the tangent line has the indicated slope.*

**39.** $f(x) = 2x^3 + 3x^2 - 12x - 10;\quad m_{\tan} = 0$

**40.** $g(x) = \dfrac{1}{3}x^3 - \dfrac{1}{2}x^2 - x + 1;\quad m_{\tan} = -1$

**41.** $h(t) = 2t + \dfrac{1}{t};\quad m_{\tan} = -2$

**42.** $F(s) = \dfrac{2s + 1}{s};\quad m_{\tan} = -\dfrac{1}{9}$

*A straight line perpendicular to and passing through a point of tangency of the tangent line is called a **normal line** to the curve. In Exercises 43 and 44, (a) find the equations of the tangent line and the normal line to the curve at the given point, and (b) use a graphing utility to plot the graph of the function, the tangent line, and the normal line on the same screen.*

**43.** The curve $y = x^3 - 3x + 1$ at the point $(2, 3)$.

**44.** The curve $y = 2x + (1/\sqrt{x})$ at the point $(1, 3)$.

**45.** Find the value(s) of $x$ at which $y = 2x - (9/x)$ is increasing at the rate of 3 units per unit change in $x$.

**46.** Let $f(x) = \frac{2}{3}x^3 + x^2 - 12x + 6$. Find the values of $x$ for which

   **a.** $f'(x) = -12$    **b.** $f'(x) = 0$    **c.** $f'(x) = 12$

**47.** Let $f(x) = \frac{1}{4}x^4 - \frac{1}{3}x^3 - x^2$. Find the point(s) on the graph of $f$ where the slope of the tangent line is equal to

   **a.** $-2x$       **b.** $0$       **c.** $10x$

**48.** Find the points on the graph of $y = \frac{1}{3}x^3 - 2x + 5$ at which the tangent line is parallel to the line $y = 2x + 3$.

**49.** Find the points on the graph of $y = \frac{1}{3}x^3 - 2x + 5$ at which the tangent line is perpendicular to the line $y = x + 2$.

**50.** Given that the line $y = 2x$ is tangent to the graph of $y = x^2 + c$, find $c$.

**51.** Find equations of the lines passing through the point $(3, 2)$ that are tangent to the parabola $y = x^2 - 2x$.

   **Hint:** Find two expressions for the slope of a tangent line.

**52.** Find an equation of the normal line to the parabola $y = x^2 - 6x + 11$ that is perpendicular to the line passing through the point $(1, 0)$ and the vertex of the parabola. (Refer to the directions given for Exercise 43.)

*In Exercises 53–56, find the limit by evaluating the derivative of a suitable function at an appropriate value of x. (Hint: Use the definition of the derivative.)*

**53.** $\lim\limits_{h \to 0} \dfrac{(1 + h)^3 - 1}{h}$

**54.** $\lim\limits_{x \to 1} \dfrac{x^5 - 1}{x - 1}$

Hint: Let $h = x - 1$.

**55.** $\lim\limits_{h \to 0} \dfrac{3(2 + h)^2 - (2 + h) - 10}{h}$

**56.** $\lim\limits_{t \to 0} \dfrac{(8 + t)^{1/3} - 2}{t}$

*In Exercises 57 and 58, write the expression as a derivative of a function of x.*

**57.** $\lim\limits_{h \to 0} \dfrac{2(x + h)^7 - (x + h)^2 - 2x^7 + x^2}{h}$

**58.** $\lim\limits_{h \to 0} \dfrac{\dfrac{1}{x + h} + \sqrt{x + h} - \dfrac{1}{x} - \sqrt{x}}{h}$

**59. Temperature Changes** The temperature (in degrees Fahrenheit) on a certain day in December in Minneapolis is given by

$$T = -0.05t^3 + 0.4t^2 + 3.8t + 19.6 \qquad 0 \le t \le 12$$

where $t$ is measured in hours and $t = 0$ corresponds to 6 A.M. Determine the time of day when the temperature is increasing at the rate of 2.05°F/hr.

**60. Traffic Flow** Opened in the late 1950s, the Central Artery in downtown Boston was designed to move 75,000 vehicles per day. Suppose that the average speed of traffic flow $S$ in miles per hour is related to the number of vehicles $x$ (in thousands) moved per day by the equation

$$S = -0.00075x^2 + 67.5 \qquad 50 < x < 300$$

Find the rate of change of the average speed of traffic flow when the number of vehicles moved per day is 100,000; 200,000. (Compare with Exercise 1 in Section 1.5.)
*Source: The Boston Globe.*

**61. Spending on Medicare** On the basis of the current eligibility requirement, a study conducted in 2004 showed that federal spending on entitlement programs, particularly Medicare, would grow enormously in the future. The study predicted that spending on Medicare, as a percentage of the gross domestic product (GDP), will be

$$P(t) = 0.27t^2 + 1.4t + 2.2 \qquad 0 \le t \le 5$$

percent in year $t$, where $t$ is measured in decades with $t = 0$ corresponding to the year 2000.

**a.** How fast will the spending on Medicare, as a percentage of the GDP, be growing in 2010? In 2020?

**b.** What will the predicted spending on Medicare, as a percentage of the GDP, be in 2010? In 2020?
*Source: Congressional Budget Office.*

**62. Effect of Stopping on Average Speed** According to data from a study by General Motors, the average speed of a trip, $A$ (in miles per hour), is related to the number of stops per mile made on that trip, $x$, by the equation

$$A = \frac{26.5}{x^{0.45}}$$

Compute $dA/dx$ for $x = 0.25$ and $x = 2$, and interpret your results.
*Source: General Motors.*

**63. Health-Care Spending** Health-care spending per person by the private sector comprising payments by individuals, corporations, and their insurance companies is approximated by the function

$$f(t) = 2.48t^2 + 18.47t + 509 \qquad 0 \le t \le 6$$

where $f(t)$ is measured in dollars and $t$ is measured in years with $t = 0$ corresponding to the beginning of 1994. The corresponding government spending—including expenditures for Medicaid, Medicare, and other federal, state, and local government public health care—is

$$g(t) = -1.12t^2 + 29.09t + 429 \qquad 0 \le t \le 6$$

where $g(t)$ is measured in dollars and $t$ in years.

**a.** Find a function that gives the difference between private and government health-care spending per person at any time $t$.

**b.** How fast was the difference between private and government expenditures per person changing at the beginning of 1995? At the beginning of 2000?
*Source: Health Care Financing Administration.*

 **64. Fuel Economy of Cars** According to data obtained from the U.S. Department of Energy and the Shell Development Company, a typical car's fuel economy depends on the speed it is driven and is approximated by the function

$$f(x) = 0.00000310315x^4 - 0.000455174x^3$$
$$+ 0.00287869x^2 + 1.25986x \qquad 0 \le x \le 75$$

where $x$ is measured in miles per hour and $f(x)$ is measured in miles per gallon (mpg).

**a.** Use a graphing utility to graph the function $f$ on the interval $[0, 75]$.

**b.** Use a calculator or computer to find the rate of change of $f$ when $x = 20$ and when $x = 50$.

**c.** Interpret your results.
*Source: U.S. Department of Energy and the Shell Development Company.*

**65. Prevalence of Alzheimer's Patients** The projected number of Alzheimer's patients in the United States is given by

$$f(t) = -0.02765t^4 + 0.3346t^3 - 1.1261t^2$$
$$+ 1.7575t + 3.7745 \qquad 0 \le t \le 6$$

where $f(t)$ is measured in millions and $t$ in decades, with $t = 0$ corresponding to the beginning of 1990.
a. Use a graphing utility to graph the function $f$ on the interval [0, 6].
b. Use a calculator or computer to find the rate at which the number of Alzheimer's patients in the United States is anticipated to be changing at the beginning of 2010? At the beginning of 2020? At the beginning of 2030?
c. Interpret your results.
*Source:* Alzheimer's Association.

**66. Hedge Fund Assets** A hedge fund is a lightly regulated pool of professionally managed money. The assets (in billions of dollars) of hedge funds from the beginning of 1999 ($t = 0$) through the beginning of 2004 are given in the following table.

| Year | 1999 | 2000 | 2001 | 2002 | 2003 | 2004 |
|------|------|------|------|------|------|------|
| Assets (billions of dollars) | 472 | 517 | 594 | 650 | 817 | 950 |

a. Use the regression capability of a calculator or computer to find a third-degree polynomial function for the data, letting $t = 0$ correspond to the beginning of 1999.
b. Plot the graph of the function found in part (a).
c. Use a calculator or computer to find the rate at which the assets of hedge funds were increasing at the beginning of 2000. At the beginning of 2003.
*Sources:* Hennessee Group, Institutional Investor.

**67. Population Decline** Political and social upheaval stemming from Russia's difficult transition from communism to capitalism is expected to contribute to the decline of the country's population well into the next century. The following table shows the total population at the beginning of each year.

| Year | 1985 | 1990 | 1995 | 2000 | 2005 |
|------|------|------|------|------|------|
| Population (millions) | 143.3 | 147.9 | 147.8 | 145.5 | 143.8 |

| Year | 2010 | 2015 | 2020 | 2025 | 2030 |
|------|------|------|------|------|------|
| Population (millions) | 141.2 | 137.5 | 133.2 | 128.7 | 123.3 |

a. Use the regression capability of a calculator or computer to find a fourth-degree polynomial function for the data, letting $t = 0$ correspond to the beginning of 1985, where $t$ is measured in 5-year intervals.

b. Plot the graph of the function found in part (a).
c. Use a calculator or computer to find the rate at which the population was changing at the beginning of 1985? At the beginning of 1995? At the beginning of 2030?
*Sources:* Population Reference Bureau, United Nations.

**68. Newton's Law of Gravitation** According to Newton's Law of Gravitation, the magnitude $F$ (in newtons) of the force of attraction between two bodies of masses $M$ and $m$ kilograms is

$$F = \frac{GmM}{r^2}$$

where $G$ is a constant and $r$ is the distance between the two bodies in meters. What is the rate of change of $F$ with respect to $r$?

**69. Period of a Satellite** The period of a satellite in a circular orbit of radius $r$ is given by

$$T = \frac{2\pi r}{R}\sqrt{\frac{r}{g}}$$

where $R$ is the earth's radius and $g$ is the constant of acceleration. Find the rate of change of the period with respect to the radius of the orbit.

**70. Coast Guard Launch** In the figure the $x$-axis represents a straight shoreline. A spectator located at the point $P(2.5, 0)$ observes a Coast Guard launch equipped with a search light execute a turn. The path of the launch is described by the parabola $y = -2.5x^2 + 10$ ($x$ and $y$ are measured in hundreds of feet). Find the distance between the launch and the spectator at the instant of time the bow of the launch is pointed directly at the spectator.

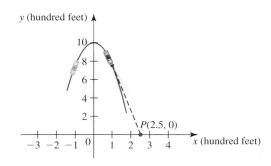

**71.** Determine the constants $A$, $B$, and $C$ such that the parabola $y = Ax^2 + Bx + C$ passes through the point $(-1, 0)$ and is tangent to the line $y = x$ at the point where $x = 1$.

**72.** Let

$$f(x) = \begin{cases} x^2 & \text{if } x \le a \\ Ax + B & \text{if } x > a \end{cases}$$

Find the values of $A$ and $B$ such that $f$ is continuous and differentiable at $a$.

**73.** Prove the Power Rule $f'(x) = nx^{n-1}$ ($n$, a positive integer) using the Binomial Theorem

$$(a + b)^n = a^n + na^{n-1}b$$
$$+ \frac{n(n-1)}{2}a^{n-2}b^2 + \cdots + nab^{n-1} + b^n$$

Compute

$$f'(x) = \lim_{h \to 0} \frac{f(x+h) - f(x)}{h} = \lim_{h \to 0} \frac{(x+h)^n - x^n}{h}$$

using the substitution $a = x$ and $b = h$.

**74.** Show that

$$f(x) = \begin{cases} \dfrac{e^x - 1}{x} & \text{if } x \neq 0 \\ 1 & \text{if } x = 0 \end{cases}$$

is continuous on $(-\infty, \infty)$.

**Hint:** Use the definition of the derivative.

*In Exercises 75–78, determine whether the statement is true or false. If it is true, explain why it is true. If it is false, explain why or give an example to show why it is false.*

**75.** If $f(x) = x^{2n}$, where $n$ is an integer, then $f'(x) = 2nx^{2(n-1)}$.

**76.** If $f(x) = 2^x$, then $f'(x) = x \cdot 2^{x-1}$ by the Power Rule.

**77.** If $f$ and $g$ are differentiable, then

$$\frac{d}{dx}[2f(x) - 5g(x)] = 2f'(x) - 5g'(x)$$

**78.** If $g(x) = f(x^2)$, where $f$ is differentiable, then $g'(x) = f'(x^2)$.

## 2.3  The Product and Quotient Rules

In this section we study two more rules of differentiation: the **Product Rule** and the **Quotient Rule.** We also consider higher-order derivatives.

### The Product and Quotient Rules

In general, the derivative of the product of two functions is *not* equal to the product of their derivatives. The following rule tells us how to differentiate a product of two functions.

---

**THEOREM 1    The Product Rule**

If $f$ and $g$ are differentiable functions, then

$$\frac{d}{dx}[f(x)g(x)] = f(x)g'(x) + g(x)f'(x)$$

---

**PROOF**    Let $F(x) = f(x)g(x)$. Then

$$F'(x) = \lim_{h \to 0} \frac{F(x+h) - F(x)}{h} = \lim_{h \to 0} \frac{f(x+h)g(x+h) - f(x)g(x)}{h}$$

If we add the quantity $[-f(x+h)g(x) + f(x+h)g(x)]$, which is equal to zero, to the numerator, we obtain

$$F'(x) = \lim_{h \to 0} \frac{f(x+h)g(x+h) - \mathbf{f(x+h)g(x)} + \mathbf{f(x+h)g(x)} - f(x)g(x)}{h}$$

$$= \lim_{h \to 0}\left\{ f(x+h)\left[\frac{g(x+h) - g(x)}{h}\right] + g(x)\left[\frac{f(x+h) - f(x)}{h}\right] \right\}$$

$$= \lim_{h \to 0} f(x+h) \cdot \lim_{h \to 0} \frac{g(x+h) - g(x)}{h} + \lim_{h \to 0} g(x) \cdot \lim_{h \to 0} \frac{f(x+h) - f(x)}{h} \qquad \textbf{(1)}$$

Since $f$ is assumed to be differentiable at $x$, Theorem 1 of Section 2.1 tells us that it is continuous there, so

$$\lim_{h \to 0} f(x + h) = f(x)$$

Also, because $g(x)$ does not involve $h$, it is constant with respect to the limiting process and

$$\lim_{h \to 0} g(x) = g(x)$$

Therefore, Equation (1) reduces to

$$F'(x) = f(x)g'(x) + g(x)f'(x)$$

In words, the Product Rule states that *the derivative of the product of two functions is the first function times the derivative of the second, plus the second function times the derivative of the first.*

### EXAMPLE 1

**a.** Find the derivative of $f(x) = xe^x$.
**b.** How fast is $f$ changing when $x = 1$?

**Solution**
**a.** Using the Product Rule, we find

$$f'(x) = \frac{d}{dx}(xe^x) = x\frac{d}{dx}(e^x) + e^x\frac{d}{dx}(x)$$

$$= xe^x + e^x \cdot 1 = (x + 1)e^x$$

**b.** When $x = 1$, $f$ is changing at the rate of $f'(1) = 2e$ units per unit change in $x$.

**EXAMPLE 2** Suppose that $g(x) = (x^2 + 1)f(x)$ and it is known that $f(2) = 3$ and $f'(2) = -1$. Evaluate $g'(2)$.

**Solution** Using the Product Rule, we find

$$g'(x) = \frac{d}{dx}[(x^2 + 1)f(x)] = (x^2 + 1)\frac{d}{dx}[f(x)] + f(x)\frac{d}{dx}(x^2 + 1)$$

$$= (x^2 + 1)f'(x) + 2xf(x)$$

Therefore,

$$g'(2) = (2^2 + 1)f'(2) + 2(2)f(2)$$

$$= (5)(-1) + 4(3) = 7$$

Just as the derivative of a product of two functions is not the product of their derivatives, the derivative of a quotient of two functions is not the quotient of their derivatives! Rather, we have the following rule.

**THEOREM 2    The Quotient Rule**

If $f$ and $g$ are differentiable functions and $g(x) \neq 0$, then

$$\frac{d}{dx}\left[\frac{f(x)}{g(x)}\right] = \frac{g(x)f'(x) - f(x)g'(x)}{[g(x)]^2}$$

**PROOF**    Let $F(x) = \dfrac{f(x)}{g(x)}$. Then

$$F'(x) = \lim_{h \to 0} \frac{F(x + h) - F(x)}{h}$$

$$= \lim_{h \to 0} \frac{\dfrac{f(x + h)}{g(x + h)} - \dfrac{f(x)}{g(x)}}{h}$$

$$= \lim_{h \to 0} \frac{f(x + h)g(x) - f(x)g(x + h)}{hg(x + h)g(x)}$$

Subtracting and adding $f(x)g(x)$ in the numerator yield

$$F'(x) = \lim_{h \to 0} \frac{f(x + h)g(x) - \mathbf{f(x)g(x)} + \mathbf{f(x)g(x)} - f(x)g(x + h)}{hg(x + h)g(x)}$$

$$= \lim_{h \to 0} \frac{g(x)\left[\dfrac{f(x + h) - f(x)}{h}\right] - f(x)\left[\dfrac{g(x + h) - g(x)}{h}\right]}{g(x + h)g(x)}$$

$$= \frac{\displaystyle\lim_{h \to 0} g(x) \cdot \lim_{h \to 0} \frac{f(x + h) - f(x)}{h} - \lim_{h \to 0} f(x) \cdot \lim_{h \to 0} \frac{g(x + h) - g(x)}{h}}{\displaystyle\lim_{h \to 0} g(x + h) \cdot \lim_{h \to 0} g(x)} \quad (2)$$

As in the proof of the Product Rule, we see that

$$\lim_{h \to 0} g(x) = g(x) \qquad \text{and} \qquad \lim_{h \to 0} f(x) = f(x)$$

and, because $g$ is continuous at $x$,

$$\lim_{h \to 0} g(x + h) = g(x)$$

Therefore, Equation (2) is

$$F'(x) = \frac{g(x)f'(x) - f(x)g'(x)}{[g(x)]^2} \qquad \blacksquare$$

As an aid to remembering the Quotient Rule, observe that it has the following form:

$$\frac{d}{dx}\left[\frac{f(x)}{g(x)}\right] = \frac{\text{(denominator)(derivative of numerator)} - \text{(numerator)(derivative of denominator)}}{\text{(square of denominator)}}$$

⚠ Because of the presence of the minus sign in the numerator, the order of the terms is important!

**EXAMPLE 3** Find the derivative of $f(x) = \dfrac{2x^2 + x}{x^3 - 1}$.

**Solution** Using the Quotient Rule, we have

$$f'(x) = \frac{(x^3 - 1)\dfrac{d}{dx}(2x^2 + x) - (2x^2 + x)\dfrac{d}{dx}(x^3 - 1)}{(x^3 - 1)^2}$$

$$= \frac{(x^3 - 1)(4x + 1) - (2x^2 + x)(3x^2)}{(x^3 - 1)^2}$$

$$= \frac{(4x^4 + x^3 - 4x - 1) - (6x^4 + 3x^3)}{(x^3 - 1)^2}$$

$$= -\frac{2x^4 + 2x^3 + 4x + 1}{(x^3 - 1)^2} \qquad \blacksquare$$

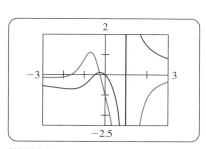

**FIGURE 1**
The graph of $f$ is shown in blue, and the graph of $f'$ is shown in red.

**Note** Figure 1 shows the graph of $f$ and $f'$ in the same viewing window. Observe that the graph of $f$ has horizontal tangent lines at the points where $x \approx -1.63$ and $x \approx -0.24$, the approximate roots of $f'(x) = 0$. $\qquad \blacksquare$

**EXAMPLE 4** Find an equation of the tangent line to the graph of

$$f(x) = \frac{e^x}{x + 1}$$

at the point where $x = 1$.

**Solution** The slope of the tangent line at any point on the graph of $f$ is given by

$$f'(x) = \frac{(x + 1)\dfrac{d}{dx}(e^x) - e^x \dfrac{d}{dx}(x + 1)}{(x + 1)^2}$$

$$= \frac{(x + 1)e^x - e^x(1)}{(x + 1)^2} = \frac{xe^x}{(x + 1)^2}$$

In particular, the slope of the required tangent line is

$$f'(1) = \frac{(1)e^1}{(1 + 1)^2} = \frac{e}{4}$$

Also, when $x = 1$,

$$y = f(1) = \frac{e^1}{1 + 1} = \frac{e}{2}$$

Therefore, the point of tangency is $\left(1, \frac{1}{2}e\right)$, and an equation of the tangent line is

$$y - \frac{1}{2}e = \frac{1}{4}e(x - 1) \qquad \text{or} \qquad y = \frac{1}{4}e(x + 1)$$

The graph of $f$ and the tangent line to the graph at $\left(1, \frac{1}{2}e\right)$ are shown in Figure 2. $\qquad \blacksquare$

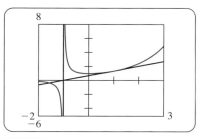

**FIGURE 2**
The graph of $f$ and the tangent line to the graph of $f$ at $\left(1, \frac{1}{2}e\right)$

**EXAMPLE 5** **Rate of Change of DVD Sales**    The sales (in millions of dollars) of a DVD recording of a hit movie $t$ years from the date of release are given by

$$S(t) = \frac{5t}{t^2 + 1} \qquad t \ge 0$$

a. Find the rate at which the sales are changing at time $t$.
b. How fast are the sales changing at the time the DVDs are released ($t = 0$)? Two years from the date of release?

**Solution**

a. The rate at which the sales are changing at time $t$ is given by $S'(t)$. Using the Quotient Rule, we obtain

$$S'(t) = \frac{d}{dt}\left[\frac{5t}{t^2 + 1}\right] = 5\frac{d}{dt}\left[\frac{t}{t^2 + 1}\right]$$

$$= 5\left[\frac{(t^2 + 1)(1) - t(2t)}{(t^2 + 1)^2}\right]$$

$$= 5\left[\frac{t^2 + 1 - 2t^2}{(t^2 + 1)^2}\right] = \frac{5(1 - t^2)}{(t^2 + 1)^2}$$

b. The rate at which the sales are changing at the time the DVDs are released is given by

$$S'(0) = \frac{5(1 - 0)}{(0 + 1)^2} = 5$$

That is, they are increasing at the rate of $5 million per year.

Two years from the date of release, the sales are changing at the rate of

$$S'(2) = \frac{5(1 - 4)}{(4 + 1)^2} = -\frac{3}{5} = -0.6$$

That is, they are decreasing at the rate of $600,000 per year. The graph of the function $S$ is shown in Figure 3.

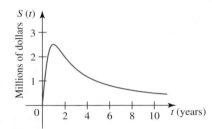

**FIGURE 3**
After a spectacular rise, the sales begin to taper off.

You may have observed that the domain of the function $S$ in Example 5 is restricted, for practical reasons, to the interval $[0, \infty)$. Since the definition of the derivative of a function $f$ at a number $a$ requires that $f$ be defined in an open interval containing $a$, the derivative of $S$ is not, strictly speaking, defined at 0. But notice that the function $S$ can, in fact, be defined for all values of $t$, and hence it makes sense to calculate $S'(0)$. You will encounter situations such as this throughout the book, especially in exercises pertaining to real-world applications. The nature of the functions appearing in these applications obviates the necessity to consider "one-sided" derivatives.

### ■ Extending the Power Rule

The Quotient Rule can be used to extend the Power Rule to include the case in which $n$ is a negative integer.

**THEOREM 3    The Power Rule for Integral Powers**

If $f(x) = x^n$, where $n$ is any integer, then

$$\frac{d}{dx}(x^n) = nx^{n-1}$$

**PROOF**   If $n$ is a positive integer, then the formula holds by Theorem 2 of Section 2.2. If $n = 0$, the formula gives

$$\frac{d}{dx}(x^0) = \frac{d}{dx}(1) = 0$$

which is true by Theorem 1 of Section 2.2. Next, suppose $n < 0$. Then $-n > 0$, and therefore, there is a positive integer $m$ such that $n = -m$. Write

$$f(x) = x^n = x^{-m} = \frac{1}{x^m}$$

Since $m > 0$, $x^m$ can be differentiated using Theorem 2 of Section 2.2. Applying the Quotient Rule, we have

$$f(x) = \frac{d}{dx}(x^n) = \frac{d}{dx}\left(\frac{1}{x^m}\right)$$

$$= \frac{x^m \dfrac{d}{dx}(1) - 1 \cdot \dfrac{d}{dx}(x^m)}{x^{2m}} \qquad \text{Use the Quotient Rule.}$$

$$= \frac{0 - mx^{m-1}}{x^{2m}}$$

$$= -mx^{-m-1}$$

$$= nx^{n-1} \qquad \text{Substitute } n = -m.$$

### Higher-Order Derivatives

The derivative $f'$ of a function $f$ is itself a function. As such, we may consider differentiating the function $f'$. The derivative of $f'$, if it exists, is denoted by $f''$ and is called the **second derivative** of $f$. Continuing in this fashion, we are led to the third, fourth, fifth, and higher-order derivatives of $f$, whenever they exist. Notations for the first, second, third, and in general, the $n$th derivative of $f$ are

$$f', \quad f'', \quad f''', \quad \ldots, \quad f^{(n)}$$

or

$$\frac{d}{dx}[f(x)], \quad \frac{d^2}{dx^2}[f(x)], \quad \frac{d^3}{dx^3}[f(x)], \quad \ldots, \quad \frac{d^n}{dx^n}[f(x)]$$

or

$$D_x f(x), \quad D_x^2 f(x), \quad D_x^3 f(x), \quad \ldots, \quad D_x^n f(x)$$

respectively.

If we denote the dependent variable by $y$, so that $y = f(x)$, then its first $n$ derivatives are also written

$$y', \quad y'', \quad y''', \quad \ldots, \quad y^{(n)}$$

or

$$\frac{dy}{dx}, \quad \frac{d^2y}{dx^2}, \quad \frac{d^3y}{dx^3}, \quad \ldots, \quad \frac{d^ny}{dx^n}$$

or

$$D_x y, \quad D_x^2 y, \quad D_x^3 y, \quad \ldots, \quad D_x^n y$$

respectively.

**EXAMPLE 6** Find the derivatives of all orders of $f(x) = x^4 - 3x^3 + x^2 - 2x + 8$.

**Solution** We have

$$f'(x) = 4x^3 - 9x^2 + 2x - 2$$

$$f''(x) = \frac{d}{dx}f'(x) = 12x^2 - 18x + 2$$

$$f'''(x) = \frac{d}{dx}f''(x) = 24x - 18$$

$$f^{(4)}(x) = \frac{d}{dx}f'''(x) = 24$$

$$f^{(5)}(x) = \frac{d}{dx}f^{(4)}(x) = 0$$

and

$$f^{(6)}(x) = f^{(7)}(x) = \cdots = 0 \qquad \blacksquare$$

**EXAMPLE 7** Find the third derivative of $y = \dfrac{1}{x}$.

**Solution** Rewriting the given equation in the form $y = x^{-1}$, we find

$$y' = \frac{d}{dx}(x^{-1}) = -x^{-2}$$

$$y'' = \frac{d}{dx}(-x^{-2}) = (-1)(-2x^{-3}) = 2x^{-3}$$

and hence

$$y''' = \frac{d}{dx}(2x^{-3}) = 2(-3x^{-4}) = -6x^{-4} = -\frac{6}{x^4} \qquad \blacksquare$$

Just as the first derivative $f'(x)$ of a function $f$ at any point $x$ gives the rate of change of $f(x)$ at that point, the second derivative $f''(x)$ of $f$, which is the derivative of $f'$ at $x$, gives the rate of change of $f'(x)$ at $x$. The third derivative $f'''(x)$ of $f$ gives the rate of change of $f''(x)$ at $x$, and so on. For example, if $P = f(t)$ gives the population of a certain city at time $t$, then $P'$ gives the rate of change of the population of the city at time $t$ and $P''$ gives the rate of change of the rate of change of the population at time $t$.

A geometric interpretation of the second derivative of a function will be given in Chapter 3, and applications of higher-order derivatives will be given in Chapter 8.

## 2.3 CONCEPT QUESTIONS

1. State the rule of differentiation and explain it in your own words.
   a. The Product Rule
   b. The Quotient Rule

2. If $f(1) = 3$, $g(1) = 2$, $f'(1) = -1$, and $g'(1) = 4$, find
   a. $h'(1)$ if $h(x) = f(x)g(x)$
   b. $F'(1)$ if $F(x) = \dfrac{f(x)}{g(x)}$

## 2.3 EXERCISES

*In Exercises 1–6, use the Product Rule to find the derivative of each function.*

1. $f(x) = x^2 e^x$

2. $f(x) = \sqrt{x} e^x$

3. $f(t) = \sqrt{t}(t + 2)e^t$

4. $f(x) = \dfrac{e^x}{x}$

5. $f(x) = e^{2x} + 2e^x + 4$

6. $f(w) = \dfrac{\sqrt{w} e^w + w^2}{2w}$

*In Exercises 7–12, use the Quotient Rule to find the derivative of each function.*

7. $f(x) = \dfrac{x}{x - 1}$

8. $g(x) = \dfrac{2x}{x^2 + 1}$

9. $h(x) = \dfrac{2x + 1}{3x - 2}$

10. $P(t) = \dfrac{e^t}{1 + t}$

11. $F(x) = \dfrac{x}{1 + xe^x}$

12. $f(s) = \dfrac{s^2 - 4}{s + 1}$

*In Exercises 13 and 14, find the derivative of each function in two ways.*

13. $F(x) = (x + 2)(x^2 - x + 1)$

14. $h(t) = \dfrac{t^5 - 3t^3 + 2t^2}{2t^2}$

*In Exercises 15–30, find the derivative of each function.*

15. $f(x) = (x + 2e^x)(x - e^x)$

16. $f(t) = (1 + \sqrt{t})(e^t - 3)$

17. $f(x) = \dfrac{2\sqrt{x}}{x^2 + 1}$

18. $f(x) = \dfrac{e^x + 1}{e^x - 1}$

19. $y = \dfrac{2x^2}{x^2 + x + 1}$

20. $y = \dfrac{2t - 1}{t^2 - 3t + 2}$

21. $f(x) = \dfrac{e^x(x + 1)}{x - 2}$

22. $f(r) = \dfrac{1 - re^r}{1 + e^r}$

23. $f(x) = \dfrac{1 + \dfrac{1}{x}}{x + 2}$

24. $y = \dfrac{1 + \dfrac{1}{x}}{1 - \dfrac{1}{x}}$

25. $f(x) = \dfrac{x + \sqrt{3x}}{3x - 1}$

26. $f(x) = \dfrac{x}{x^2 - 4} - \dfrac{x - 1}{x^2 + 4}$

27. $F(x) = \dfrac{ax + b}{cx + d}$,   $a, b, c, d$ constants

28. $g(t) = \dfrac{at^2}{t^2 + b}$,   $a, b$ constants

29. $f(x) = \dfrac{x + e^x}{1 - xe^x}$

30. $g(t) = (2t + 1)\left(t - 1 + \dfrac{2}{t - 1}\right)$

*In Exercises 31–34, find the derivative of the function and evaluate $f'(x)$ at the given value of $x$.*

31. $f(x) = (2x - 1)(x^2 + 3)$;   $x = 1$

32. $f(x) = \dfrac{2x + 1}{2x - 1}$;   $x = 2$

33. $f(x) = (\sqrt{x} + 2x)(x^{3/2} - x)$;   $x = 4$

34. $f(x) = \dfrac{x}{x^4 - 2x^2 - 1}$;   $x = -1$

*In Exercises 35 and 36, find the point(s) on the graph of f where the tangent line is horizontal.*

**35.** $f(x) = xe^{-x}$

**36.** $f(x) = \dfrac{x}{x^2 + 1}$

*In Exercises 37 and 38, find the point(s) on the graph of the function at which the tangent line has the indicated slope.*

**37.** $f(x) = (e^x + 1)(2 - x);\quad m_{\tan} = -1$

**38.** $F(s) = \dfrac{2s + 1}{s - 2};\quad m_{\tan} = -\dfrac{1}{5}$

 *In Exercises 39–42, (a) find an equation of the tangent line to the graph of the function at the indicated point, and (b) use a graphing utility to plot the graph of the function and the tangent line on the same screen.*

**39.** $y = \dfrac{e^x}{1 + x};\quad \left(1, \frac{1}{2}e\right)$

**40.** $y = \dfrac{2x}{x^2 + 1};\quad (-1, -1)$

**41.** $y = x^2 + 1 + \dfrac{3}{x - 1};\quad (-2, 4)$

**42.** $f(x) = \dfrac{\sqrt{x} - 1}{\sqrt{x} + 1};\quad \left(4, \frac{1}{3}\right)$

 *The straight line perpendicular to and passing through the point of tangency of the tangent line is called the normal line to the curve. In Exercises 43 and 44, (a) find the equations of the tangent line and the normal line to the curve at the given point, and (b) use a graphing utility to plot the graph of the function, the tangent line, and the normal line on the same screen.*

**43.** The curve $f(x) = (x^3 + 1)(3x^2 - 4x + 2)$ at the point $(1, 2)$.

**44.** The curve $y = 1/(1 + x^2)$ at the point $\left(1, \frac{1}{2}\right)$.

*In Exercises 45–48, suppose that f and g are functions that are differentiable at $x = 1$ and that $f(1) = 2$, $f'(1) = -1$, $g(1) = -2$, and $g'(1) = 3$. Find $h'(1)$.*

**45.** $h(x) = f(x)g(x)$

**46.** $h(x) = (x^2 + 1)g(x)$

**47.** $h(x) = \dfrac{xf(x)}{x + g(x)}$

**48.** $h(x) = \dfrac{f(x)g(x)}{f(x) - g(x)}$

*In Exercises 49 and 50, find the limit by evaluating the derivative of a suitable function at an appropriate value of x. (Hint: Use the definition of the derivative.)*

**49.** $\displaystyle\lim_{t \to 0} \dfrac{1 - (1 + t)^2}{t(1 + t)^2}$

**50.** $\displaystyle\lim_{x \to 1} \dfrac{(x + 1)^2 - 4}{x - 1}$

*In Exercises 51–54, find $f''(x)$.*

**51.** $f(x) = x^8 - x^4 + 2x^2 + 1$

**52.** $f(x) = x^3 e^x$

**53.** $f(x) = \dfrac{e^x}{x}$

**54.** $f(x) = \dfrac{x + 1}{x - 1}$

*In Exercises 55–58, find $y''$.*

**55.** $y = x^3 - 2x^2 + 1$

**56.** $y = e^x\left(x + \dfrac{1}{x}\right)$

**57.** $y = x^{5/2} e^{-x}$

**58.** $y = \dfrac{x}{x^2 + 1}$

**59.** Find

a. $f''(2)$ if $f(x) = 4x^3 - 2x^2 + 3$

b. $y''\Big|_{x=1}$ if $y = 2x^3 - \dfrac{1}{x}$

**60.** Find

a. $f'''(0)$ if $f(x) = 8x^7 - 6x^5 + 4x^3 - x$

b. $y'''\Big|_{x=1}$ if $y = xe^x$

**61.** Find the derivatives of all order of $f(x) = 2x^4 - 4x^2 + 1$.

**62. Newton's Second Law of Motion** Consider a particle moving along a straight line. Newton's Second Law of Motion states that the external force $F$ acting on the particle is equal to the rate of change of its momentum. Thus,

$$F = \dfrac{d}{dt}(mv)$$

where $m$, the mass of the particle, and $v$, the velocity of the particle, are both functions of time.

a. Use the Product Rule to show that

$$F = m\dfrac{dv}{dt} + v\dfrac{dm}{dt}$$

b. Use the results of part (a) to show that if the mass of a particle is constant, then $F = ma$, where $a$ is the acceleration of the particle.

**63. Formaldehyde Levels** A study on formaldehyde levels in 900 homes indicates that emissions of various chemicals can decrease over time. The formaldehyde level (parts per million) in an average home in the study is given by

$$f(t) = \dfrac{0.055t + 0.26}{t + 2}\qquad 0 \le t \le 12$$

where $t$ is the age of the house in years. How fast is the formaldehyde level of the average house dropping when the house is new? At the beginning of its fourth year?

*Source: Bonneville Power Administration.*

**64. Oxygen Content of a Pond** When organic waste is dumped into a pond, the oxidization process that takes place reduces the pond's oxygen content. However, given time, nature will restore the oxygen content to its natural level. Suppose that the oxygen content $t$ days after organic waste has been dumped into a pond is given by

$$f(t) = 100\left(\dfrac{t^2 + 10t + 100}{t^2 + 20t + 100}\right)$$

where $f(t)$ is the percentage of the oxygen content of the pond prior to dumping.

**a.** Derive a general expression that gives the rate of change of the pond's oxygen level at any time $t$.

**b.** How fast is the oxygen content of the pond changing one day after organic waste has been dumped into the pond? Ten days after? Twenty days after?

**c.** Interpret your results.

 **65. Importance of Time in Treating Heart Attacks** According to the American Heart Association, the treatment benefit for heart attacks depends on the time until treatment and is described by the function

$$f(t) = \frac{-16.94t + 203.28}{t + 2.0328} \qquad 0 \le t \le 12$$

where $t$ is measured in hours and $f(t)$ is expressed as a percent.

**a.** Use a graphing utility to graph the function $f$ using the viewing window $[0, 13] \times [0, 120]$.

**b.** Use the numerical derivative capability of a graphing utility to find the derivative of $f$ when $t = 0$ and $t = 2$.

**c.** Interpret the results obtained in part (b).

*Source:* American Heart Association.

**66. Cylinder Pressure** The pressure $P$, volume $V$, and temperature $T$ of a gas in a cylinder are related by the van der Waals equation

$$P = \frac{kT}{V - b} + \frac{ab}{V^2(V - b)} - \frac{a}{V(V - b)}$$

where $a$, $b$, and $k$ are constants. If the temperature of the gas is kept constant, find $dP/dV$.

**67. Constructing a New Road** The following figures depict three possible roads connecting the point $A(-1000, 0)$ to the point $B(1000, 1000)$ via the origin. The functions describing the dashed center lines of the roads follow:

$$f(x) = \begin{cases} 0 & \text{if } -1000 \le x \le 0 \\ x & \text{if } 0 < x \le 1000 \end{cases}$$

$$g(x) = \begin{cases} 0 & \text{if } -1000 \le x \le 0 \\ 0.001x^2 & \text{if } 0 < x \le 1000 \end{cases}$$

$$h(x) = \begin{cases} 0 & \text{if } -1000 \le x \le 0 \\ 0.000001x^3 & \text{if } 0 < x \le 1000 \end{cases}$$

Show that $f$ is not differentiable on the interval $(-1000, 1000)$, $g$ is differentiable but not twice differentiable on $(-1000, 1000)$, and $h$ is twice differentiable on $(-1000, 1000)$. Taking into consideration the dynamics of a moving vehicle, which proposal do you think is most suitable?

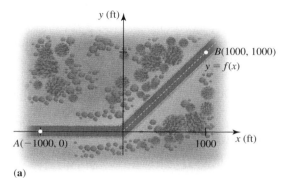

**(a)**

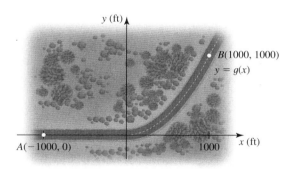

**(b)**

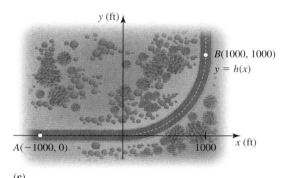

**(c)**

**68. Obesity in America** The body mass index (BMI) measures body weight in relation to height. A BMI of 25 to 29.9 is considered overweight, a BMI of 30 or more is considered obese, and a BMI of 40 or more is morbidly obese. The percent of the U.S. population that is obese is approximated by the function

$$P(t) = 0.0004t^3 + 0.0036t^2 + 0.8t + 12 \qquad 0 \le t \le 13$$

where $t$ is measured in years, with $t = 0$ corresponding to the beginning of 1991. Show that the rate of change of the rate of change of the percent of the U.S. population that is deemed obese was positive from 1991 to 2004. What does this mean?

*Source:* Centers for Disease Control and Prevention.

**69.** Find $f''(x)$ if $f(x) = |x^3|$. Does $f''(0)$ exist?

**70. a.** Use the Product Rule twice to prove that if $h = uvw$, where $u$, $v$, and $w$ are differentiable functions, then

$$h' = u'vw + uv'w + uvw'$$

**b.** Use the result of part (a) to find the derivative of

$$h(x) = (2x + 5)(x + 3)(x^2 + 4)$$

*In Exercises 71–76, determine whether the statement is true or false. If it is true, explain why it is true. If it is false, explain why or give an example to show why it is false.*

**71.** If $f$ and $g$ are differentiable, then

$$\frac{d}{dx}[f(x)g(x)] = f'(x)g'(x)$$

**72.** If $f$ is differentiable, then $\dfrac{d}{dx}[xf(x)] = f(x) + xf'(x)$.

**73.** If $f$ and $g$ have second derivatives, then

$$\frac{d}{dx}[f(x)g'(x) - f'(x)g(x)] = f(x)g''(x) - f''(x)g(x)$$

**74.** If $P$ is a polynomial function of degree $n$, then $P^{(n+1)}(x) = 0$.

**75.** If $g(x) = [f(x)]^2$, where $f$ is differentiable, then $g'(x) = 2f(x)f'(x)$.

**76.** If $g(x) = [f(x)]^{-2}$, where $f$ is differentiable, then

$$g'(x) = -\frac{2f'(x)}{[f(x)]^3}$$

## 2.4 The Role of the Derivative in the Real World

In this section we will see how the derivative can be used to solve real-world problems. Our first example calls for interpreting the derivative as a measure of the slope of a tangent line to the graph of a function. Before we look at the example, however, we make the following observation: If $\alpha$ denotes the angle that the tangent line to the graph of $f$ at $P(x, f(x))$ makes with the positive $x$-axis, then $\tan \alpha = dy/dx$ or, equivalently, $\alpha = \tan^{-1}(dy/dx)$. (See Figure 1.)

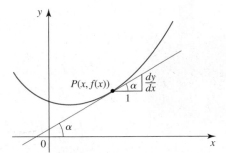

**FIGURE 1**

$$\tan \alpha = \frac{dy}{dx}$$

**EXAMPLE 1** **Flight Path of a Plane** After taking off from a runway, an airplane continues climbing for 10 sec before turning to the right. Its flight path during that time period can be described by the curve in the $xy$-plane with equation

$$y = -1.06x^3 + 1.61x^2 \qquad 0 \le x \le 0.6$$

where $x$ is the distance along the ground in miles, $y$ is the height above the ground in miles, and the point at which the plane leaves the runway is located at the origin. Find the angle of climb of the airplane when it is at the point on the flight path where $x = 0.5$. (See Figure 2.)

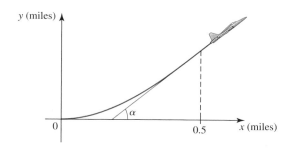

**FIGURE 2**
The flight path of the airplane

**Solution**  The required angle of climb, $\alpha$, is given by

$$\tan \alpha = \left. \frac{dy}{dx} \right|_{x=0.5}$$

But

$$\frac{dy}{dx} = \frac{d}{dx}(-1.06x^3 + 1.61x^2) = -3.18x^2 + 3.22x$$

so

$$\left. \frac{dy}{dx} \right|_{x=0.5} = \left. (-3.18x^2 + 3.22x) \right|_{x=0.5} = 0.815$$

Therefore, $\tan \alpha = 0.815$ and $\alpha = \tan^{-1} 0.815 \approx 39.18°$, giving the required angle of climb of the airplane as approximately $39°$.  ■

We now turn our attention to real-world problems that require the interpretation of the derivative as a measure of the rate of change of one quantity with respect to another.

## ■ Motion Along a Line

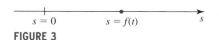

**FIGURE 3**
The position of a moving body at any time $t$ is at the point $s = f(t)$ on the coordinate line.

An example of motion along a straight line was encountered in Section 1.1, where we studied the motion of a maglev. In considering such motion, we assume that it takes place along a coordinate line. Then the position of a moving body may be specified by giving its coordinate $s$. Since $s$ varies with time $t$, we write $s = f(t)$, where the function $f(t)$ is called the **position function** of the body (see Figure 3). As we saw in Section 1.1, the (instantaneous) velocity of a body at any time $t$ is the rate of change of the position function $f$ with respect to $t$.

> **DEFINITION  Velocity**
> If $s = f(t)$, where $f$ is the position function of a body moving on a coordinate line, then the velocity of the body at time $t$ is given by
> $$v(t) = \frac{ds}{dt} = f'(t)$$

The function $v(t)$ is called the **velocity function** of the body.

Observe that if $v(t) > 0$ at a given time $t$, then $s$ is increasing, and the body is moving in the positive direction along the coordinate line at that instant of time (Figure 4a). Similarly, if $v(t) < 0$, then the body is moving in the negative direction at that instant of time (Figure 4b).

**FIGURE 4**

(a) If $v(t) > 0$, then the body is moving in the positive direction.

(b) If $v(t) < 0$, then the body is moving in the negative direction.

Sometimes we merely need to know how fast a body is moving and are not concerned with its direction of motion. In this instance we are asking for the magnitude of the velocity, or the *speed,* of the body.

---

**DEFINITION    Speed**

If $v(t)$ is the velocity of a body at any time $t$, then the speed of the body at time $t$ is given by

$$|v(t)| = |f'(t)| = \left|\frac{ds}{dt}\right|$$

---

**EXAMPLE 2**    The position of a particle moving along a straight line is given by

$$s = f(t) = 2t^3 - 15t^2 + 24t \qquad t \geq 0$$

where $t$ is measured in seconds and $s$ in feet.

**a.** Find an expression giving the velocity of the particle at any time $t$. What are the velocity and speed of the particle when $t = 2$?
**b.** Determine the position of the particle when it is stationary.
**c.** When is the particle moving in the positive direction? In the negative direction?

**Solution**
**a.** The required velocity of the particle is given by

$$v(t) = f'(t) = \frac{d}{dt}(2t^3 - 15t^2 + 24t)$$

$$= 6t^2 - 30t + 24 = 6(t^2 - 5t + 4)$$

$$= 6(t - 1)(t - 4)$$

The velocity of the particle when $t = 2$ is

$$v(2) = 6(2 - 1)(2 - 4) = -12$$

or $-12$ ft/sec. The speed of the particle when $t = 2$ is $|v(2)| = 12$ ft/sec. In short, the particle is moving in the negative direction at a speed of 12 ft/sec.

**b.** The particle is stationary when its velocity is equal to zero. Setting $v(t) = 0$ gives

$$v(t) = 6(t - 1)(t - 4) = 0$$

and we see that the particle is stationary at $t = 1$ and $t = 4$. Its position at $t = 1$ is given by

$$f(1) = 2(1)^3 - 15(1)^2 + 24(1) = 11 \quad \text{or} \quad 11 \text{ ft}$$

Its position at $t = 4$ is given by

$$f(4) = 2(4)^3 - 15(4)^2 + 24(4) = -16 \quad \text{or} \quad -16 \text{ ft}$$

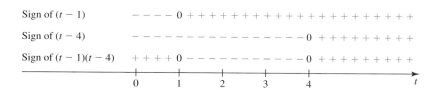

**FIGURE 5**
The sign diagram for
determining the sign of $v(t)$

**c.** The particle is moving in the positive direction when $v(t) > 0$ and is moving in the negative direction when $v(t) < 0$. From the sign diagram shown in Figure 5, we see that $v(t) = 6(t - 1)(t - 4)$ is positive in the intervals $(0, 1)$ and $(4, \infty)$ and negative in $(1, 4)$. We conclude that the particle is moving to the right in the time intervals $(0, 1)$ and $(4, \infty)$ and to the left in the time interval $(1, 4)$.

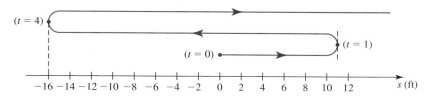

**FIGURE 6**
A schematic showing the
position of the particle

A schematic of the motion of the particle is shown in Figure 6. The graph of the position function $s = f(t) = 2t^3 - 15t^2 + 24t$ is shown in Figure 7. Try to explain the motion of the particle in terms of this graph.    ■

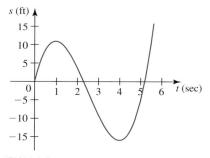

**FIGURE 7**
The graph of $s = 2t^3 - 15t^2 + 24t$
gives the position of the particle versus
time $t$. (Do not confuse this with the
path of the particle.)

If a body moves along a coordinate line, the acceleration of the body is the rate of change of its velocity, and the jerk of the body is the rate of change of its acceleration.

**DEFINITIONS    Acceleration, Jerk**

If $f(t)$ and $v(t)$ are the position and velocity functions, respectively, of a body moving on a coordinate line, then the **acceleration** of the body at time $t$ is

$$a(t) = v'(t) = f''(t)$$

and the **jerk** of the body at time $t$ is

$$j(t) = a'(t) = v''(t) = f'''(t)$$

The jerk function $j(t)$ is of particular interest to safety engineers of automobile companies who are constantly performing jerk tests on various components of motor vehicles. Large jerk conditions in automobiles not only lead to discomfort but may also cause harm to the occupants, including whiplash.

**EXAMPLE 3**    Consider the motion of the particle of Example 2 with position function

$$s = f(t) = 2t^3 - 15t^2 + 24t \qquad t \geq 0$$

where $t$ is measured in seconds and $s$ in feet.

**a.** Find the acceleration function of the particle. What is the acceleration of the particle when $t = 2$?
**b.** When is the acceleration zero? Positive? Negative?
**c.** Find the jerk function of the particle.

**Solution**

**a.** From the solution to Example 2 we have $v(t) = 6t^2 - 30t + 24$. Therefore,

$$a(t) = v'(t) = \frac{d}{dt}(6t^2 - 30t + 24)$$

$$= 12t - 30 = 6(2t - 5)$$

In particular, the acceleration of the particle when $t = 2$ is

$$a(2) = 6[2(2) - 5] \qquad \text{or} \qquad -6 \text{ ft/sec}^2$$

In other words, the particle is decelerating at 6 ft/sec$^2$ when $t = 2$.

**b.** The acceleration of the particle is zero when $a(t) = 0$, or

$$6(2t - 5) = 0$$

giving $t = \frac{5}{2}$. Since $2t - 5 < 0$ when $t < \frac{5}{2}$ and $2t - 5 > 0$ when $t > \frac{5}{2}$, we also conclude that the acceleration is negative for $0 < t < \frac{5}{2}$ and positive for $t > \frac{5}{2}$.

**c.** Using the result of part (b), we find

$$j(t) = \frac{d}{dt}[a(t)] = \frac{d}{dt}(12t - 30) = 12$$

or 12 ft/sec$^3$. ■

---

**EXAMPLE 4** **The Velocity of Exploding Fireworks** In a fireworks display, a shell is launched vertically upward from the ground, reaching a height (in feet) of

$$s = -16t^2 + 256t$$

after $t$ sec. The shell is designed to burst when it reaches its maximum altitude, simultaneously igniting a cluster of explosives.

**a.** At what time after the launch will the shell burst?
**b.** What will the altitude of the shell be at the instant it explodes?

**Solution**

**a.** At its maximum altitude the velocity of the shell is zero. But the velocity of the shell at any time $t$ is

$$v(t) = \frac{ds}{dt} = \frac{d}{dt}(-16t^2 + 256t)$$

$$= -32t + 256 = -32(t - 8)$$

which is equal to zero when $t = 8$. Therefore, the shell will burst 8 sec after it has been launched.

**b.** The altitude of the shell at the instant it explodes will be

$$s = -16(8)^2 + 256(8) = 1024$$

or 1024 ft. A schematic of the motion of the shell and the graph of the function $s = -16t^2 + 256t$ are shown in Figure 8.

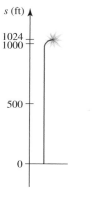
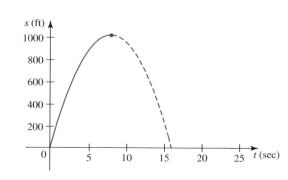

**FIGURE 8**

**(a)** Schematic of the position of the shell

**(b)** Graph of the function $s = -16t^2 + 256t$
(The portion of interest is drawn with a solid line.)

## ■ Marginal Functions in Economics

The derivative is an indispensable tool in the study of the rate of change of one economic quantity with respect to another. Economists refer to this field of study as **marginal analysis**. The following example will help to explain the use of the adjective **marginal**.

**EXAMPLE 5**   **Cost Functions**   Suppose that the total cost in dollars incurred per week by the Polaraire Corporation in manufacturing $x$ refrigerators is given by the total cost function

$$C(x) = -0.2x^2 + 200x + 9000 \qquad 0 \le x \le 400$$

**a.** What is the cost incurred in manufacturing the 201st refrigerator?
**b.** Find the rate of change of $C$ with respect to $x$ when $x = 200$.

### Solution
**a.** The cost incurred in manufacturing the 201st refrigerator is the difference between the total cost incurred in manufacturing the first 201 units and the total cost incurred in manufacturing the first 200 units. Thus, the cost is

$$C(201) - C(200) = [-0.2(201)^2 + 200(201) + 9000]$$

$$- [-0.2(200)^2 + 200(200) + 9000]$$

$$= 41119.8 - 41000 = 119.8$$

or $119.80.
**b.** The rate of change of $C$ with respect to $x$ is

$$C'(x) = \frac{d}{dx}(-0.2x^2 + 200x + 9000)$$

$$= -0.4x + 200$$

In particular, when $x = 200$, we find

$$C'(200) = -0.4(200) + 200 = 120$$

In other words, when the level of production is 200 units, the total cost function is increasing at the rate of $120 per refrigerator. ■

## Historical Biography

### ISAAC NEWTON
### (1643-1727)

Born three months after the death of his father, Isaac Newton was small and unhealthy at birth. His mother nursed him back to health, and when he was three, she remarried and sent him to be raised by his maternal grandmother. At the age of twelve, Newton began grammar school, where he excelled, learning Latin along with his other studies. When he was sixteen, his mother called him home to take care of the family farm, but Newton was inattentive to the animals and a poor farmer. Newton's uncle and the schoolmaster at the grammar school convinced Newton's mother to let him return to his studies, and in 1661 he was admitted to Cambridge University. There he read the works of the great mathematicians Euclid, Descartes (page 6), Galileo, and Kepler (page 885) and he attained his bachelor degree in 1665. Starting that same year, the plague shut Cambridge down for two years, and Newton spent the time working on the foundation for calculus, which he called "fluxional method." Newton's geometric approach to calculus was not published until 1689, several years after publication of Leibniz's (page 179) paper which presented the same topic with a more algebraic approach. However, Newton had made his work known to a small group of mathematicians in 1668, and a debate broke out as to who had developed calculus first. This caused great animosity between Newton and Leibniz, which lasted until Leibniz's death in 1716.

If you compare the results of parts (a) and (b) of Example 5, you will notice that $C'(200)$ is a pretty good approximation to $C(201) - C(200)$, the cost incurred in manufacturing an additional refrigerator when the level of production is already 200 units. To see why, let's recall the definition of the derivative of a function and write

$$C'(200) = \lim_{h \to 0} \frac{C(200 + h) - C(200)}{h}$$

Next, the definition of the limit tells us that if $h$ is small, then

$$C'(200) \approx \frac{C(200 + h) - C(200)}{h}$$

In particular, by taking $h = 1$, we see that

$$C'(200) \approx \frac{C(200 + 1) - C(200)}{1} = C(201) - C(200)$$

as we wished to show.

Economists call the cost incurred in producing an additional unit of a commodity, given that the plant is already operating at a certain level $x = a$, the marginal cost. But as we have just seen, this quantity may be suitably approximated by $C'(a)$, where $C$ is the total cost function associated with the process. Furthermore, as you can see from the computations in Example 5, it is often much easier to calculate $C'(a)$ than to calculate $C(a + 1) - C(a)$. For this reason, economists prefer to work with $C'$ rather than $C$ in marginal analysis. The derivative $C'$ of the total cost function is called the **marginal cost function.**

The other marginal functions in economics are defined in a similar manner and have similar meanings. For example, the marginal revenue function $R'$ is the derivative of the total revenue function $R$, and $R'$ gives an approximation of the change in revenue that results when sales are increased by one unit from $x = a$ to $x = a + 1$.

A summary of these definitions follows.

| Function | Marginal Function |
|---|---|
| $C$, cost function | $C'$, marginal cost function |
| $R$, revenue function | $R'$, marginal revenue function |
| $P$, profit function | $P'$, marginal profit function |
| $\overline{C}$, average cost function | $\overline{C}'$, marginal average cost function |

**Note**   $\overline{C}(x) = C(x)/x$, the total cost incurred in producing $x$ units of a commodity divided by the number of units produced.    ■

**EXAMPLE 6** **Marginal Revenue**   Suppose the weekly revenue realized through the sale of $x$ Pulsar cell phones is

$$R(x) = -0.000078x^3 - 0.0016x^2 + 80x \qquad 0 \le x \le 800$$

dollars.

**a.** Find the marginal revenue function.
**b.** If the company currently sells 200 phones per week, by how much will the revenue increase if sales increase by one phone per week?

**Solution**

**a.** The marginal revenue function is

$$R'(x) = \frac{d}{dx}(-0.000078x^3 - 0.0016x^2 + 80x)$$

$$= -0.000234x^2 - 0.0032x + 80$$

**b.** The company's revenue will increase by approximately

$$R'(200) = -0.000234(200)^2 - 0.0032(200) + 80$$

$$= 70$$

or $70.

### ■ Other Applications

We close this section by looking at a few more examples involving applications of the derivative in fields as diverse as engineering and the social sciences.

Engineering
If the shape of an electric power line strung between two transmission towers is described by the graph of $y = f(x)$, then the (acute) angle $\alpha$ that the cable makes with the horizontal at any point $P(x, f(x))$ on the cable is given by

$$\alpha = \left| \tan^{-1}\left(\frac{dy}{dx}\right) \right|$$

(See Figure 9.)

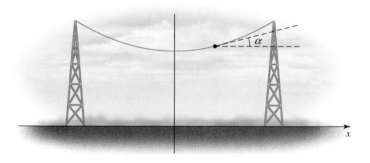

**FIGURE 9**
The shape of the cable is described by $y = f(x)$.

Meteorology
If $P(h)$ is the atmospheric pressure at an altitude $h$, then $P'(h)$ gives the rate of change of the atmospheric pressure with respect to altitude at an altitude $h$.

Chemistry
Certain proteins, known as enzymes, serve as catalysts for chemical reactions in living things. If $V(x)$ gives the initial speed (in moles per liter per second) at which a chemical reaction begins as a function of $x$, the amount of substrate (the substance being acted upon, measured in moles per liter), then $dV/dx$ measures the rate of change of the initial speed at which the reaction begins with respect to the amount of substrate, when the amount of substrate is $x$ moles/liter.

Biology
If $R(I)$ denotes the rate of production in photosynthesis, where $I$ is the light intensity, then $dR/dI$ measures the rate of change of the rate of production with respect to light intensity, when the light intensity is $I$.

| | |
|---|---|
| **Epidemiology** | If $p(t)$ stands for the percentage of infected students in a university in week $t$, then $dp/dt$ gives the rate of change of the percentage of infected students with respect to time at time $t$. |
| **Life Sciences** | If $A(t)$ gives the amount of radioactive substance remaining after $t$ years, then $A'(t)$ gives the rate of decay of that substance with respect to time at time $t$. |
| **Medicine** | If $C(t)$ gives the concentration of a drug in a patient's bloodstream $t$ hours after injection, then $C'(t)$ measures the rate at which the concentration of the drug is changing with respect to time at time $t$. |
| **Business** | If $S(x)$ is the total sales of a company when the amount spent on advertising its products and services is $x$, then $S'(x)$ measures the rate of change of the sales level with respect to the amount spent on advertising when the expenditure is $x$. |
| **Demographics** | If $P(t)$ gives the population of the United States in year $t$, then $P'(t)$ gives the rate of change of the population with respect to time at time $t$. |

## 2.4 CONCEPT QUESTIONS

1. Let $f(t)$ denote the position of an object moving along a coordinate line, where $f(t)$ is measured in feet and $t$ in seconds. Explain each of the following in terms of $f$:
   a. average velocity
   b. velocity
   c. speed
   d. acceleration
   e. jerk

2. Suppose that $P$ is a profit function giving the total profit $P(x)$ in dollars resulting from the sale of $x$ units of a certain commodity. What does $P'(a)$ measure if $a$ is a given level of sales?

3. The following figure shows the cross section of a narrow tube of radius $a$ immersed in water. Because of a surface-tension phenomenon called *capillarity*, the water rises until it reaches an equilibrium height. The curved liquid surface is called a *meniscus*, and the angle $\theta$ at which it meets the

inner wall of the tube is called the contact angle. If the meniscus is described by the function $y = f(x)$, what is the contact angle $\theta$?

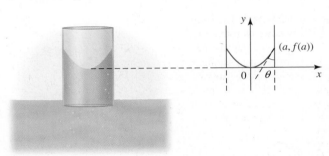

**(a)** Cross section of the tube and the meniscus

**(b)** The meniscus is described by $y = f(x)$.

## 2.4 EXERCISES

*In Exercises 1–8, $s(t)$ is the position function of a body moving along a coordinate line; $s(t)$ is measured in feet and $t$ in seconds, where $t \geq 0$. Find the position, velocity, and speed of the body at the indicated time.*

1. $s(t) = 1.86t^2$; $t = 2$ (free fall on Mars)

2. $s(t) = 2t^3 - 3t^2 + 4t + 1$; $t = 1$

3. $s(t) = 2t^4 - 8t^2 + 4$; $t = 1$

4. $s(t) = \dfrac{t}{t+1}$; $t = 0$      5. $s(t) = \dfrac{2t}{t^2+1}$; $t = 2$

6. $s(t) = te^{-t}$; $t = 2$      7. $s(t) = (t^2 - 1)^2$; $t = 1$

8. $s(t) = \dfrac{t^3}{t^3 + 1}$; $t = 1$

*In Exercises 9–16, $s(t)$ is the position function of a body moving along a coordinate line, where $t \geq 0$, and $s(t)$ is measured in feet and $t$ in seconds. (a) Determine the times(s) and the position(s) when the body is stationary. (b) When is the body moving in the positive direction? In the negative direction? (c) Sketch a schematic showing the position of the body at any time t.*

9. $s(t) = 2t + 3$      10. $s(t) = 4 - t^2$

11. $s(t) = 8 + 2t - t^2$      12. $s(t) = \dfrac{1}{3}t^3 - \dfrac{3}{2}t^2 + 1$

13. $s(t) = 2t^4 - 8t^3 + 8t^2 + 1$

14. $s(t) = (t^2 - 1)^2$

15. $s(t) = \dfrac{2t}{t^2 + 1}$

16. $s(t) = \dfrac{t^3}{t^3 + 1}$

*In Exercises 17–20, s(t) is the position function of a body moving along a coordinate line, where t ≥ 0, and s(t) is measured in feet and t in seconds.* (a) *Find the acceleration of the body.* (b) *When is the acceleration zero? Positive? Negative?*

17. $s(t) = 2t^3 - 9t^2 + 12t - 2$

18. $s(t) = t^4 - 2t^2 + 2$

19. $s(t) = \dfrac{2t}{t^2 + 1}$

20. $s(t) = \dfrac{t^3}{t^3 + 1}$

*In Exercises 21 and 22, s(t) is the position function of a body moving along a coordinate line, where t ≥ 0. If the mass of the body is 20 kg and s(t) and t are measured in meters and seconds, respectively, find* (a) *the momentum (mv) of the body and* (b) *the kinetic energy ($\frac{1}{2}mv^2$) of the body at the indicated times.*

21. $s(t) = 2t^2 - 3t + 1; \quad t = 2$

22. $s(t) = t^3 - 3t^2 + 1; \quad t = 1$

23. **Tiltrotor Plane** The tiltrotor plane takes off and lands vertically, but its rotors tilt forward for conventional cruising. The figure depicts the graph of the position function of a tiltrotor plane during a test flight in the vertical takeoff and landing mode. Answer the following questions pertaining to the motion of the plane at each of the times $t_0$, $t_1$, and $t_2$:
    a. Is the plane ascending, stationary, or descending?
    b. Is the acceleration positive, zero, or negative?

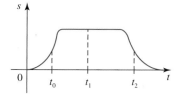

(a) The graph of the position function of a tiltrotor plane          (b) A tiltrotor plane

24. **Explosion of a Gas Main** An explosion caused by the ignition of a leaking underground gas main blew a manhole cover vertically into the air. The height of the manhole cover $t$ seconds after the explosion was $s = 24t - 16t^2$ ft.
    a. How high did the manhole cover go?
    b. What was the velocity of the manhole cover when it struck the ground?

25. **Diving** The position of a diver executing a high dive from a 10-m platform is described by the position function
$$s(t) = -4.9t^2 + 2t + 10 \qquad t \geq 0$$
where $t$ is measured in seconds and $s$ in meters.

a. When will the diver hit the water?
b. How fast will the diver be traveling at that time? (Ignore the height of the diver and his outstretched arms.)

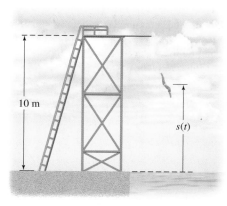

26. **Stopping Distance of a Sports Car** A test of the stopping distance (in feet) of a sports car was conducted by the editors of an auto magazine. For a particular test, the position function of the car was
$$s(t) = 88t - 12t^2 - \frac{1}{6}t^3$$
where $t$ is measured in seconds and $t = 0$ corresponds to the time when the brakes were first applied.
    a. What was the car's velocity when the brakes were first applied?
    b. What was the car's stopping distance for that particular test?
    c. What was the jerk at time $t$? At the time when the brakes were first applied?

27. **Flight of a VTOL Aircraft** In a test flight of McCord Aviation's experimental VTOL (vertical takeoff and landing) aircraft, the altitude of the aircraft operating in the vertical takeoff mode was given by the position function
$$h(t) = \frac{1}{64}t^4 - \frac{1}{2}t^3 + 4t^2 \qquad 0 \leq t \leq 16$$
where $h(t)$ is measured in feet and $t$ is measured in seconds.
    a. Find the velocity function.
    b. What was the velocity of the VTOL at $t = 0$, $t = 8$, and $t = 16$? Interpret your results.
    c. What was the maximum altitude attained by the VTOL during the test flight?

28. **Rotating Fluid** If a right circular cylinder of radius $a$ is filled with water and rotated about its vertical axis with a constant angular velocity $\omega$, then the water surface assumes a shape whose cross section in a plane containing the vertical axis is a parabola. If we choose the $xy$-system so that the $y$-axis is the axis of rotation and the vertex of the parabola passes

through the origin of the coordinate system, then the equation of the parabola is

$$y = \frac{\omega^2 x^2}{2g}$$

where $g$ is the acceleration due to gravity. Find the angle $\alpha$ that the tangent line to the water level makes with the $x$-axis at any point on the water level. What happens to $\alpha$ as $\omega$ increases? Interpret your result.

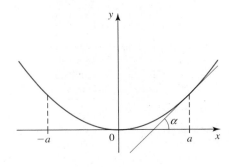

29. **Motion of a Projectile** A projectile is fired from a cannon located on a horizontal plane. If we think of the cannon as being located at the origin $O$ of an $xy$-coordinate system, then the path of the projectile is

$$y = \sqrt{3}x - \frac{x^2}{400}$$

where $x$ and $y$ are measured in feet.

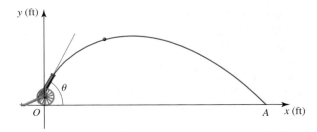

a. Find the value of $\theta$ (the angle of elevation of the gun).
b. At what point on the trajectory is the projectile traveling parallel to the ground?
c. What is the maximum height attained by the projectile?
d. What is the range of the projectile (the distance $OA$ along the $x$-axis)?
e. At what angle with respect to the $x$-axis does the projectile hit the ground?

30. **Deflection of a Beam** A horizontal uniform beam of length $L$ is supported at both ends and bends under its own weight $w$ per unit length. Because of its elasticity, the beam is distorted in shape, and the resulting distorted axis of symmetry (shown dashed in the figure) is called the elastic curve. It

can be shown that an equation for the elastic curve is

$$y = \frac{w}{24EI}(x^4 - 2Lx^3 + L^3x)$$

where the product $EI$ is a constant called the *flexural rigidity*.

**(a)** The distorted beam

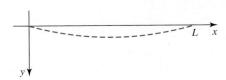

**(b)** The elastic curve in the $xy$-plane (The positive direction of the $y$-axis is directed downward.)

a. Find the angle that the elastic curve makes with the positive $x$-axis at each end of the beam in terms of $w$, $E$, and $I$.
b. Show that the angle that the elastic curve makes with the horizontal at $x = L/2$ is zero.
c. Find the deflection of the beam at $x = L/2$. (We will show that the deflection is maximal in Section 3.1, Exercise 74.)

31. **Flight Path of an Airplane** The path of an airplane on its final approach to landing is described by the equation $y = f(x)$ with

$$f(x) = 4.3404 \times 10^{-10}x^3 - 1.5625 \times 10^{-5}x^2 + 3000$$
$$0 \le x \le 24{,}000$$

where $x$ and $y$ are both measured in feet.
a. Plot the graph of $f$ using the viewing window [0, 24000] × [0, 3000].
b. Find the maximum angle of descent during the landing approach.
**Hint:** When is $dy/dx$ smallest?

32. **Middle-Distance Race** As they round the corner into the final (straight) stretch of the bell lap of a middle-distance race, the positions of the two leaders of the pack, $A$ and $B$, are given by

$$s_A(t) = 0.063t^2 + 23t + 15 \qquad t \ge 0$$

and

$$s_B(t) = 0.298t^2 + 24t \qquad t \ge 0$$

respectively, where the reference point (origin) is taken to be the point located 300 feet from the finish line and $s$ is measured in feet and $t$ in seconds. It is known that one of the two runners, $A$ and $B$, was the winner of the race and the other was the runner-up.

a. Show that $B$ won the race.
b. At what point from the finish line did $B$ overtake $A$?
c. By what distance did $B$ beat $A$?
d. What was the speed of each runner as he crossed the finish line?

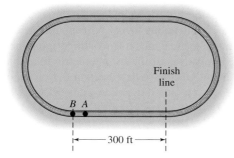

**33. Acceleration of a Car**  A car starting from rest and traveling in a straight line attains a velocity of

$$v(t) = \frac{110t}{2t + 5}$$

feet per second after $t$ sec. Find the initial acceleration of the car and its acceleration 10 sec after starting from rest.

**34. Marginal Cost of Producing Compact Discs**  The weekly total cost in dollars incurred by the BMC Recording Company in manufacturing $x$ compact discs is

$$C(x) = 4000 + 3x - 0.0001x^2 \qquad 0 \le x \le 10,000$$

a. What is the actual cost incurred by the company in producing the 2001st disc? The 3001st disc?
b. What is the marginal cost when $x = 2000$? When $x = 3000$?

**35. Marginal Cost of Producing Microwave Ovens**  A division of Ditton Industries manufactures the "Spacemaker" model microwave oven. Suppose that the daily total cost (in dollars) of manufacturing $x$ microwave ovens is

$$C(x) = 0.0002x^3 - 0.06x^2 + 120x + 6000$$

What is the marginal cost when $x = 200$? Compare the result with the actual cost incurred by the company in manufacturing the 201st oven.

**36. Marginal Average Cost of Producing Television Sets**  The Advance Visual Systems Corporation manufactures a 19-inch LCD HDTV. The weekly total cost incurred by the company in manufacturing $x$ sets is

$$C(x) = 0.000002x^3 - 0.02x^2 + 120x + 70,000$$

dollars.
a. Find the average cost function $\overline{C}(x)$ and the marginal average cost function $\overline{C}'(x)$.
b. Compute $\overline{C}'(5000)$ and $\overline{C}'(10,000)$, and interpret your results.

**37. Marginal Revenue of an Airline**  The Commuter Air Service realizes a revenue of

$$R(x) = 10,000x - 100x^2$$

dollars per month when the price charged per passenger is $x$ dollars.
a. Find the marginal revenue function $R'$.
b. Compute $R'(49)$, $R'(50)$, and $R'(51)$. What do your results seem to imply?

**38. Marginal Profit in Producing Television Sets**  The Advance Visual Systems Corporation realizes a total profit of

$$P(x) = -0.000002x^3 + 0.016x^2 + 80x - 70,000$$

dollars per week from the manufacture and sale of $x$ units of their 26-in. LCD HDTVs.
a. Find the marginal profit function $P'$.
b. Compute $P'(2000)$ and interpret your result.

**39. Optics**  The equation

$$\frac{1}{f} = \frac{1}{p} + \frac{1}{q}$$

sometimes called a **lens-maker's equation,** gives the relationship between the focal length $f$ of a thin lens, the distance $p$ of the object from the lens, and the distance $q$ of its image from the lens. We can think of the eye as an optical system in which the ciliary muscle constantly adjusts the curvature of the cornea-lens system to focus the image on the retina. Assume that the distance from the cornea to the retina is 2.5 cm.

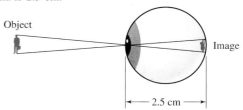

a. Find the focal length of the cornea-lens system if an object located 50 cm away is to be focused on the retina.
b. What is the rate of change of the focal length with respect to the distance of the object when the object is 50 cm away?

**40. Gravitational Force**  The magnitude of the gravitational force exerted by the earth on a particle of mass $m$ at a distance $r$ from the center of the earth is

$$F(r) = \begin{cases} \dfrac{GMmr}{R^2} & \text{if } r < R \\[2mm] \dfrac{GMm}{r^2} & \text{if } r \ge R \end{cases}$$

where $M$ is the mass of the earth, $R$ is its radius, and $G$ is the gravitational constant.
a. Compute $F'(r)$ for $r < R$, and interpret your result.
b. Compute $F'(r)$ for $r > R$, and interpret your result.

## 2.5    Derivatives of Trigonometric Functions

Many real-world problems are modeled using trigonometric functions. The motion of a pendulum, for example, is periodic (or almost periodic) and can be described by using a combination of sine and cosine functions. The motion of a shock absorber in a car can also be described by using a combination of trigonometric functions and exponential functions. You will see many other applications involving trigonometric functions throughout the book. To analyze the mathematical models involving trigonometric functions, we need to be able to find the derivatives of the trigonometric functions.

Before starting this section, you might wish to review Section 0.3 on trigonometric functions. Keep in mind that all angles are measured in radians, unless otherwise stated.

### ■ Derivatives of Sines and Cosines

Our first result tells us how to find the derivative of $\sin x$.

---

**THEOREM 1    Derivative of $\sin x$**

$$\frac{d}{dx}(\sin x) = \cos x$$

---

**PROOF**    Let $f(x) = \sin x$. Then

$$f'(x) = \lim_{h \to 0} \frac{f(x + h) - f(x)}{h} \qquad \text{Definition of the derivative}$$

$$= \lim_{h \to 0} \frac{\sin(x + h) - \sin x}{h}$$

$$= \lim_{h \to 0} \frac{\sin x \cos h + \cos x \sin h - \sin x}{h} \qquad \begin{array}{l} \text{Expand } \sin(x + h) \text{ using the} \\ \text{Angle Addition Formula.} \end{array}$$

$$= \lim_{h \to 0} \left[ \frac{\sin x \cos h - \sin x}{h} + \frac{\cos x \sin h}{h} \right]$$

$$= \lim_{h \to 0} \left[ (\sin x)\left( \frac{\cos h - 1}{h} \right) + (\cos x)\left( \frac{\sin h}{h} \right) \right]$$

$$= \left( \lim_{h \to 0} \sin x \right)\left( \lim_{h \to 0} \frac{\cos h - 1}{h} \right) + \left( \lim_{h \to 0} \cos x \right)\left( \lim_{h \to 0} \frac{\sin h}{h} \right)$$

Use the Sum and Product Laws for limits.

But $\lim_{h \to 0} \sin x = \sin x$ and $\lim_{h \to 0} \cos x = \cos x$ because these expressions do not involve $h$ and thus remain constant with respect to the limiting process. From Section 1.2 we have

$$\lim_{h \to 0} \frac{\cos h - 1}{h} = 0$$

and

$$\lim_{h \to 0} \frac{\sin h}{h} = 1$$

Using these results, we see that

$$f'(x) = (\sin x)(0) + (\cos x)(1) = \cos x$$

The relationship between the function $f(x) = \sin x$ and its derivative $f'(x) = \cos x$ can be seen by sketching the graphs of both functions (see Figure 1). Here, we interpret $f'(x)$ as the slope of the tangent line to the graph of $f$ at the point $(x, f(x))$.

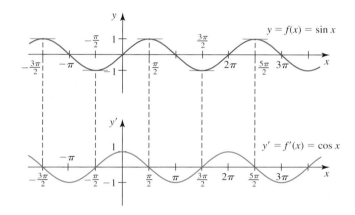

**FIGURE 1**
The graphs of $f(x) = \sin x$ and its derivative $f'(x) = \cos x$

**EXAMPLE 1** Find $f'(x)$ if $f(x) = x^2 \sin x$.

**Solution** Using the Product Rule and Theorem 1, we obtain

$$f'(x) = \frac{d}{dx}(x^2 \sin x) = x^2 \frac{d}{dx}(\sin x) + (\sin x)\frac{d}{dx}(x^2)$$

$$= x^2 \cos x + 2x \sin x$$

---

**THEOREM 2   Derivative of cos x**

$$\frac{d}{dx}(\cos x) = -\sin x$$

---

The proof of this rule is similar to the proof of Theorem 1 and is left as an exercise (Exercise 47).

## Derivatives of Other Trigonometric Functions

The remaining trigonometric functions are defined in terms of the sine and cosine functions. Thus,

$$\tan x = \frac{\sin x}{\cos x}, \qquad \csc x = \frac{1}{\sin x}, \qquad \sec x = \frac{1}{\cos x}, \qquad \text{and} \qquad \cot x = \frac{\cos x}{\sin x}$$

Therefore, their derivatives can be found by using Theorems 1 and 2 and the Quotient Rule. For example,

$$\frac{d}{dx}(\tan x) = \frac{d}{dx}\left(\frac{\sin x}{\cos x}\right)$$

$$= \frac{(\cos x)\dfrac{d}{dx}(\sin x) - (\sin x)\dfrac{d}{dx}(\cos x)}{\cos^2 x} \qquad \text{Quotient Rule}$$

$$= \frac{(\cos x)(\cos x) - (\sin x)(-\sin x)}{\cos^2 x}$$

$$= \frac{\cos^2 x + \sin^2 x}{\cos^2 x}$$

$$= \frac{1}{\cos^2 x} = \sec^2 x$$

that is,

$$\frac{d}{dx}(\tan x) = \sec^2 x$$

A complete list of the rules for differentiating trigonometric functions follows. The proofs of the remaining three rules are left as exercises. (See Exercises 48–50.)

---

**THEOREM 3    Rules for Differentiating Trigonometric Functions**

$$\frac{d}{dx}(\sin x) = \cos x \qquad \frac{d}{dx}(\cos x) = -\sin x \qquad \frac{d}{dx}(\tan x) = \sec^2 x$$

$$\frac{d}{dx}(\csc x) = -\csc x \cot x \qquad \frac{d}{dx}(\sec x) = \sec x \tan x \qquad \frac{d}{dx}(\cot x) = -\csc^2 x$$

---

**Note**  As an aid to remembering the signs of the derivatives of the trigonometric functions, observe that those functions beginning with a "c" (cos $x$, csc $x$, and cot $x$) have a minus sign attached to their derivatives.  ■

**EXAMPLE 2**  Differentiate $y = (\sec x)(x + \tan x)$.

**Solution**  Using the Product Rule and Theorem 3, we have

$$\frac{dy}{dx} = \frac{d}{dx}[(\sec x)(x + \tan x)]$$

$$= (\sec x)\frac{d}{dx}(x + \tan x) + (x + \tan x)\frac{d}{dx}(\sec x)$$

$$= (\sec x)(1 + \sec^2 x) + (x + \tan x)(\sec x \tan x)$$

$$= (\sec x)(1 + \sec^2 x + x \tan x + \tan^2 x)$$

$$= (\sec x)(2 + x \tan x + 2\tan^2 x) \qquad \sec^2 x = 1 + \tan^2 x \qquad ■$$

**EXAMPLE 3** Find the derivative of $y = \dfrac{\sin x}{1 - \cos x}$.

**Solution**    Using the Quotient Rule and Theorems 1 and 2, we obtain

$$\frac{dy}{dx} = \frac{d}{dx}\left(\frac{\sin x}{1 - \cos x}\right)$$

$$= \frac{(1 - \cos x)\dfrac{d}{dx}(\sin x) - (\sin x)\dfrac{d}{dx}(1 - \cos x)}{(1 - \cos x)^2}$$

$$= \frac{(1 - \cos x)(\cos x) - (\sin x)(\sin x)}{(1 - \cos x)^2}$$

$$= \frac{\cos x - \cos^2 x - \sin^2 x}{(1 - \cos x)^2} = \frac{\cos x - 1}{(1 - \cos x)^2}$$

$$= \frac{1}{\cos x - 1} \qquad\blacksquare$$

**EXAMPLE 4** Find an equation of the tangent line to the graph of $y = x \sin x$ at the point where $x = \pi/2$.

**Solution**    The slope of the tangent line at any point $(x, y)$ on the graph of $y = x \sin x$ is given by

$$\frac{dy}{dx} = \frac{d}{dx}(x \sin x) = x\frac{d}{dx}(\sin x) + (\sin x)\frac{d}{dx}(x)$$

$$= x \cos x + \sin x$$

In particular, the slope of the tangent line at the point where $x = \pi/2$ is

$$\frac{dy}{dx}\bigg|_{x=\pi/2} = (x \cos x + \sin x)\bigg|_{x=\pi/2}$$

$$= \frac{\pi}{2}\cos\frac{\pi}{2} + \sin\frac{\pi}{2}$$

$$= \frac{\pi}{2}(0) + 1 = 1$$

The $y$-coordinate of the point of tangency is

$$y\bigg|_{x=\pi/2} = x \sin x\bigg|_{x=\pi/2}$$

$$= \frac{\pi}{2}\sin\frac{\pi}{2} = \frac{\pi}{2}$$

Using the point-slope form of an equation of a line, we find that

$$y - \frac{\pi}{2} = x - \frac{\pi}{2} \qquad \text{or} \qquad y = x$$

The graph of $y = x \sin x$ and the tangent line are shown in Figure 2.

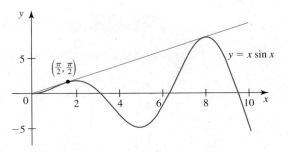

**FIGURE 2**
An equation of the tangent line to the graph of $y = x \sin x$ at $\left(\frac{\pi}{2}, \frac{\pi}{2}\right)$ is $y = x$.

**EXAMPLE 5** **Simple Harmonic Motion** Suppose that a flexible spring is attached vertically to a rigid support (Figure 3a). If a weight is attached to the free end of the spring, it will settle in a certain equilibrium position (Figure 3b). Suppose that the weight is pulled downward (a positive direction) and released from rest from a position that is 3 units below the equilibrium position at time $t = 0$ (Figure 3c). Then, in the absence of opposing forces such as air resistance, the weight will oscillate back and forth about the equilibrium position. This motion is referred to as *simple harmonic motion*.

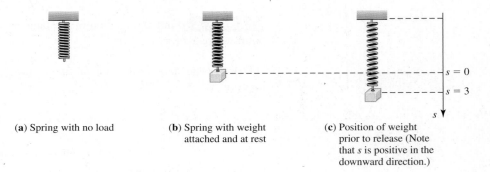

(**a**) Spring with no load   (**b**) Spring with weight attached and at rest   (**c**) Position of weight prior to release (Note that $s$ is positive in the downward direction.)

**FIGURE 3**

Suppose that for a particular spring and weight, the motion is described by the equation

$$s = 3 \cos t \qquad t \geq 0$$

(See Figure 4.)

**a.** Find the velocity and acceleration functions describing the motion.
**b.** Find the values of $t$ when the weight passes the equilibrium position.
**c.** What are the velocity and acceleration of the weight at these values of $t$?

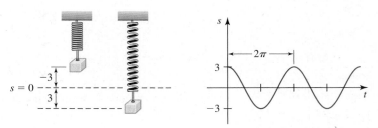

(**a**) Extreme positions of the weight   (**b**) The graph of the function $s = 3 \cos t$ describing the simple harmonic motion of the weight

**FIGURE 4**

**Solution**

**a.** The velocity of the weight at any time $t > 0$ is

$$v(t) = \frac{ds}{dt} = \frac{d}{dt}(3\cos t) = -3\sin t$$

and its acceleration at any time $t > 0$ is

$$a(t) = \frac{dv}{dt} = \frac{d}{dt}(-3\sin t) = -3\cos t$$

**b.** When $s = 0$, the weight is at the equilibrium position. Solving the equation

$$s = 3\cos t = 0$$

we see that the required values of $t$ are $t = \pi/2 + n\pi$, where $n = 0, 1, 2, \ldots$.

**c.** Using the results of parts (a) and (b), we then calculate the velocity and acceleration of the weight as it passes the equilibrium position:

| $t$ | $\dfrac{\pi}{2}$ | $\dfrac{3\pi}{2}$ | $\dfrac{5\pi}{2}$ | $\dfrac{7\pi}{2}$ | $\cdots$ |
|---|---|---|---|---|---|
| $v(t)$ | $-3$ | $3$ | $-3$ | $3$ | $\cdots$ |
| $a(t)$ | $0$ | $0$ | $0$ | $0$ | $\cdots$ |

## 2.5  CONCEPT QUESTIONS

1. State the rules for differentiating $\sin x$, $\cos x$, $\tan x$, $\csc x$, $\sec x$, and $\cot x$.

2. Find

   **a.** $\displaystyle\lim_{h\to 0}\frac{\cos(a+h)-\cos a}{h}$    **b.** $\displaystyle\lim_{h\to 0}\frac{\sec\left(\frac{\pi}{4}+h\right)-\sqrt{2}}{h}$

## 2.5  EXERCISES

*In Exercises 1–22, find the derivative of the function.*

**1.** $f(x) = 4\cos x - 2x + 1$     **2.** $g(x) = x + \tan x$

**3.** $h(t) = 3\tan t - 4\sec t$     **4.** $y = \sqrt{x}\sin x$

**5.** $f(u) = e^u \cot u$     **6.** $g(v) = e^v \sin v - 2v\csc v$

**7.** $s = \sin x \cos x$     **8.** $f(t) = \sec t \tan t$

**9.** $f(\theta) = \cos\theta(1 + \sec\theta)$     **10.** $g(x) = \dfrac{\cos x}{1 + x}$

**11.** $g(x) = e^{-x}\sin x$     **12.** $y = \dfrac{\cos\theta}{1 - \sin\theta}$

**13.** $y = \dfrac{x}{1 + \sec x}$     **14.** $f(x) = \dfrac{\cot x}{1 + \csc x}$

**15.** $f(x) = e^x \sec x + e^{-x}\cot x$     **16.** $y = \cos 2x$

**17.** $f(x) = \dfrac{1 + \sin x}{1 - \cos x}$     **18.** $y = \dfrac{\sin x \cos x}{1 + \csc x}$

**19.** $h(\theta) = \dfrac{\sin\theta + \cos\theta}{\sin\theta - \cos\theta}$     **20.** $s = \dfrac{1 - \tan t}{1 + \cot t}$

**21.** $f(x) = e^x \sin^2 x$     **22.** $y = \dfrac{a\sin t}{1 + b\cos t}$

*In Exercises 23–28, find the second derivative of the function.*

**23.** $f(x) = e^x \sin x$     **24.** $g(x) = \sec x$

**25.** $y = 3\cos x - x\sin x$

**26.** $h(t) = (t^2 + 1)\sin t$

**27.** $y = \sqrt{x}\cos x$     **28.** $w = \dfrac{\cos\theta}{\theta}$

 *In Exercises 29–32, (a) find an equation of the tangent line to the graph of the function at the indicated point, and (b) use a graphing utility to plot the graph of the function and the tangent line on the same screen.*

**29.** $f(x) = \sin x$; $\left(\frac{\pi}{6}, \frac{1}{2}\right)$     **30.** $f(x) = \tan x$; $\left(\frac{\pi}{4}, 1\right)$

**31.** $f(x) = \sec x$; $\left(\frac{\pi}{3}, 2\right)$     **32.** $f(x) = \dfrac{\sin x}{x}$; $\left(\frac{\pi}{2}, \frac{2}{\pi}\right)$

Ⓥ Videos for selected exercises are available online at **www.academic.cengage.com/login**.

*In Exercises 33–36, find the rate of change of y with respect to x at the indicated value of x.*

**33.** $y = x^2 \sec x; \quad x = \dfrac{\pi}{4}$

**34.** $y = \csc x - 2 \cos x; \quad x = \dfrac{\pi}{6}$

**35.** $y = \dfrac{\sin x}{1 - \cos x}; \quad x = \dfrac{\pi}{2}$  **36.** $y = \dfrac{e^x \tan x}{\sec x}; \quad x = 0$

*In Exercises 37–40, find the x-coordinate(s) of the point(s) on the graph of the function at which the tangent line has the indicated slope.*

**37.** $f(x) = \sin x; \quad m_{\tan} = 1$  **38.** $g(x) = x + \sin x; \quad m_{\tan} = 1$

**39.** $h(x) = \csc x; \quad m_{\tan} = 0$  **40.** $f(x) = \cot x; \quad m_{\tan} = -2$

**41.** Let $f(x) = \sin x$. Compute $f^{(n)}(x)$ for $n = 1, 2, 3, \ldots$. Then use your results to show that $|f^{(n)}(x)| \leq 1$ for all $x$. In other words, the values of the sine function as well as all of its derivatives lie between $-1$ and $1$.

**42.** Repeat Exercise 41 with the function $f(x) = \cos x$.

**43. Simple Harmonic Motion** The position function of a body moving along a coordinate line is

$$s(t) = 2 \sin t + 3 \cos t \qquad t \geq 0$$

where $t$ is measured in seconds and $s(t)$ in feet. Find the position, velocity, speed, and acceleration of the body when $t = \pi/2$.

**44. Pure Resonance** Refer to Example 5. Suppose that the system shown in the figure is initially at rest in the equilibrium position. Further, suppose that starting at $t = 0$, the system is subjected to an external driving force that has the same frequency as the natural frequency of the system. Then the resulting motion of the body is described by the position function

$$s(t) = \sin t - t \cos t \qquad t \geq 0$$

(The frequency is just the reciprocal of the period of the position function, in this case, $1/(2\pi)$.)

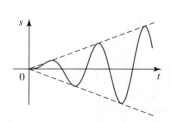

**(a)** The support is subject to an up-and-down motion whose frequency is the same as the natural frequency of the system.

**(b)** The resulting motion is one in which the amplitude of the wave gets larger and larger.

**a.** By computing $s(t)$ for $t = n\pi$, where $n = 1, 2, 3, \ldots$, show that $|s(t)|$ gets larger and larger as $t$ increases. This implies that the "amplitude" of the motion becomes unbounded.

**b.** What is the velocity of the body when $t = \pi/2 + n\pi$, $n = 0, 1, 2, 3, \ldots$?

**c.** What is the acceleration of the body when $t = n\pi$, $n = 1, 2, 3, \ldots$?

This phenomenon is called **pure resonance**. A mechanical system subjected to resonance will necessarily fail. For example, a singer hitting the "right note" can induce acoustic vibrations that will lead to the shattering of a wine glass.

**45.** Evaluate $\displaystyle\lim_{h \to 0} \dfrac{\dfrac{1}{\sin(x + h)} - \dfrac{1}{\sin x}}{h}$.

**46.** Evaluate $\displaystyle\lim_{h \to 0} \dfrac{\tan\left(\dfrac{\pi}{4} + h\right) - 1}{h}$.

**47.** Prove $\dfrac{d}{dx} (\cos x) = -\sin x$.

**48.** Prove $\dfrac{d}{dx} (\csc x) = -\csc x \cot x$.

**49.** Prove $\dfrac{d}{dx} (\cot x) = -\csc^2 x$.

**50.** Prove $\dfrac{d}{dx} (\sec x) = \sec x \tan x$.

*In Exercises 51–52, determine whether the statement is true or false. If it is true, explain why it is true. If it is false, give an example to show why it is false.*

**51.** If $f(x) = \dfrac{1 - \sin^2 2x}{\cos^2 2x}$, $\left(x \neq \dfrac{\pi}{4} + \dfrac{n\pi}{2}, n \text{ is an integer}\right)$ then $f'(x) = 0$.

**52.** If $f(x) = \cos(x + h)$, where $h$ is a constant, then $f'(x) = -\sin(x + h)$.

# 2.6 | The Chain Rule

## ■ Composite Functions

Suppose that we wish to differentiate the function $F$ defined by

$$F(x) = (x^2 + 1)^{120}$$

If we use only the rules of differentiation developed so far, then a possible approach might be to expand $F(x)$ using the binomial theorem and differentiate the resulting expression term by term. But the amount of work involved would be prodigious! How about the function $G$ defined by $G(x) = \sqrt{2x^2 - 1}$?

You can convince yourself that the same differentiation rules cannot be applied directly to compute $G'(x)$. Observe that both $F$ and $G$ are composite functions. For example, $F$ is the composition of $g(x) = x^{120}$ and $f(x) = x^2 + 1$. Thus,

$$F(x) = (g \circ f)(x) = g[f(x)]$$
$$= [f(x)]^{120} = (x^2 + 1)^{120}$$

and $G$ is the composition of $g(x) = \sqrt{x}$ and $f(x) = 2x^2 - 1$. Thus,

$$G(x) = (g \circ f)(x) = g[f(x)]$$
$$= \sqrt{f(x)} = \sqrt{2x^2 - 1}$$

Notice that each of the component functions $f$ and $g$ is easily differentiated by using the rules of differentiation already available to us. The question, then, is whether we can take advantage of this fact to compute the derivatives of the more complicated composite functions $F$ and $G$. We will return to these examples later. But for now, let's turn our attention to the general problem of finding the derivative $h'$ of a composite function $h$.

## ■ The Chain Rule

For each $x$ in the domain of $h = g \circ f$, let $u = f(x)$ and $y = g(u) = g[f(x)]$. Then, as illustrated in Figure 1, we see that the composite function $h$ maps the number $x$ onto the number $y$ in one step. Alternatively, we see that $x$ is also mapped onto $y$ in two steps—via $f$ ($x$ onto $u$) then via $g$ ($u$ onto $y$). Since it might be too difficult to compute $h' = (g \circ f)'$ directly, the following question arises: Can we find $h'$ by somehow combining $g'$ and $f'$?

**FIGURE 1**
The function $h$ is composed of the functions $g$ and $f$: $h(x) = g[f(x)]$.

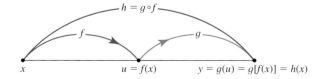

Since $u$ is a function of $x$, we can compute the derivative of $u$ with respect to $x$, $du/dx = f'(x)$. Next, $y$ is a function of $u$, and we can compute the derivative of $y$ with respect to $u$, $dy/du = g'(u)$. Because $h$ is composed of $g$ and $f$, it seems reasonable to expect that $h'$, or $dy/dx$, must be a combination of $f'$ and $g'$ ($du/dx$ and $dy/du$). But how should we combine them?

Consider the following argument: Interpreting the derivative of a function as the rate of change of that function, suppose that $u = f(x)$ changes twice as fast as $x$ [$f'(x) = du/dx = 2$], and $y = g(u)$ changes three times as fast as $u$ [$g'(u) = dy/du = 3$]. Then we would expect $y = h(x)$ to change six times as fast as $x$; that is,

$$h'(x) = g'(u)f'(x) = (3)(2) = 6$$

or, equivalently,

$$\frac{dy}{dx} = \frac{dy}{du} \cdot \frac{du}{dx} = (3)(2) = 6$$

Although it is far from being a proof, this argument does suggest how $f'(x)$ and $g'(u) = g'[f(x)]$ should be combined to obtain $h'(x)$ (that is, how $du/dx$ and $dy/du$ should be combined to obtain $dy/dx$). We simply multiply them together.

The rule for calculating the derivative of a composite function follows.

---

**THEOREM 1   The Chain Rule**

If $f$ is differentiable at $x$ and $g$ is differentiable at $f(x)$, then the composition $h = g \circ f$ defined by $h(x) = g[f(x)]$ is differentiable at $x$, and

$$h'(x) = g'[f(x)]f'(x) \qquad \text{(a)}$$

Also, if we write $u = f(x)$ and $y = g(u) = g[f(x)]$, then

$$\frac{dy}{dx} = \frac{dy}{du} \cdot \frac{du}{dx} \qquad \text{(b)}$$

---

The proof of the Chain Rule is given in Appendix B.

**Notes**

1. The "Inside-Outside" Rule: If we label the composite function $h(x) = g[f(x)]$ in the following way

"inside function"

↓

$$h(x) = g[f(x)]$$

↑

"outside function"

then $h'(x)$ is just the derivative of the "outside function" evaluated at the "inside function" times the derivative of the "inside function."

2. When written in the form of Theorem 1b, the Chain Rule can be remembered by observing that if we "cancel" the $du$'s on the right of the equation, we do obtain $dy/dx$. ∎

## ■ Applying the Chain Rule

**EXAMPLE 1**   Find $F'(x)$ if $F(x) = (x^2 + 1)^{120}$.

**Solution**   As we observed earlier, $F$ can be viewed as the composite function defined by $F(x) = g[f(x)]$, where $f(x) = x^2 + 1$ and $g(x) = x^{120}$, or $g(u) = u^{120}$ (remember that $x$ and $u$ are dummy variables). The derivative of the "outside function" is

$$g'(u) = \frac{d}{du}[u^{120}] = 120u^{119}$$

which, when evaluated at $f(x) = x^2 + 1$, yields

$$g'[f(x)] = g'(x^2 + 1) = 120(x^2 + 1)^{119}$$

The derivative of the "inside function" is

$$f'(x) = \frac{d}{dx}(x^2 + 1) = 2x$$

Using Theorem 1a, we obtain

$$F'(x) = g'[f(x)]f'(x) = 120(x^2 + 1)^{119} \cdot (2x)$$
$$= 240x(x^2 + 1)^{119}$$

**Alternative Solution**   Let $u = f(x) = x^2 + 1$ and $y = g(u) = u^{120}$. Then, using Theorem 1b, we find

$$F'(x) = \frac{dy}{dx} = \frac{dy}{du} \cdot \frac{du}{dx}$$

$$= 120u^{119} \cdot (2x) = 240xu^{119}$$

$$= 240x(x^2 + 1)^{119} \qquad ■$$

**EXAMPLE 2**   Find $G'(x)$ if $G(x) = \sqrt{2x^2 - 1}$.

**Solution**   We view $G(x)$ as $G(x) = g[f(x)]$, where $f(x) = 2x^2 - 1$ and $g(x) = \sqrt{x}$ (so $g(u) = \sqrt{u}$). Now

$$g'(u) = \frac{d}{du}\left[\sqrt{u}\right] = \frac{1}{2\sqrt{u}}$$

$$g'[f(x)] = \frac{1}{2\sqrt{f(x)}} = \frac{1}{2\sqrt{2x^2 - 1}}$$

and

$$f'(x) = \frac{d}{dx}(2x^2 - 1) = 4x$$

Therefore, if we use Theorem 1a, we obtain

$$G'(x) = g'[f(x)]f'(x) = \frac{1}{2\sqrt{2x^2 - 1}} \cdot (4x)$$

$$= \frac{2x}{\sqrt{2x^2 - 1}}$$

**Alternative Solution**   Let $u = f(x) = 2x^2 - 1$ and $y = g(u) = \sqrt{u}$. Then, using Theorem 1b, we find

$$G'(x) = \frac{dy}{dx} = \frac{dy}{du} \cdot \frac{du}{dx}$$

$$= \frac{1}{2\sqrt{u}} \cdot (4x) = \frac{1}{2\sqrt{2x^2 - 1}} \cdot (4x)$$

$$= \frac{2x}{\sqrt{2x^2 - 1}}$$   ■

**EXAMPLE 3**   Find $\dfrac{dy}{dx}$ if $y = u^3 - u^2 + u + 1$ and $u = x^3 + 1$.

**Solution**   In this situation it is more convenient to use Theorem 1b. Thus,

$$\frac{dy}{dx} = \frac{dy}{du} \cdot \frac{du}{dx} = \frac{d}{du}(u^3 - u^2 + u + 1) \cdot \frac{d}{dx}(x^3 + 1)$$

$$= (3u^2 - 2u + 1)(3x^2)$$

If we wish, we could write $dy/dx$ in terms of $x$ as follows:

$$\frac{dy}{dx} = [3(x^3 + 1)^2 - 2(x^3 + 1) + 1](3x^2)$$

$$= 3x^2(3x^6 + 4x^3 + 2)$$   ■

**Note**   Of course, we could have worked Example 3 using Theorem 1a. In this event, simply observe that $y = g(u) = u^3 - u^2 + u + 1$ and $u = f(x) = x^3 + 1$.   ■

## ■ The General Power Rule

Although we have used the Chain Rule in its most general form to help us find the derivatives of the functions in the previous examples, in many situations we need only use a special version of the rule. For example, some functions, such as those in Examples 1 and 2, have the form $y = [f(x)]^n$. These functions are called **generalized power functions.**

To find a formula for computing the derivative of the generalized power function $y = [f(x)]^n$, where $n$ is an integer, let $u = f(x)$ so that $y = u^n$. Using the Chain Rule, we find

$$\frac{dy}{dx} = \frac{dy}{du} \cdot \frac{du}{dx}$$

$$= nu^{n-1} \cdot f'(x)$$

$$= n[f(x)]^{n-1}f'(x)$$

---

**THEOREM 2    General Power Rule**

Let $y = u^n$, where $u = f(x)$ is a differentiable function and $n$ is a real number. Then

$$\frac{dy}{dx} = nu^{n-1}\frac{du}{dx}$$

Equivalently,

$$\frac{dy}{dx} = n[f(x)]^{n-1} \cdot f'(x)$$

---

Before looking at another example, let's rework Example 1 using Theorem 2. We have $F(x) = (x^2 + 1)^{120}$, which is a generalized power function with $f(x) = x^2 + 1$. Therefore, using the General Power Rule, we obtain

$$F'(x) = \frac{d}{dx}(x^2 + 1)^{120}$$

$$= \underbrace{120(x^2 + 1)^{119}}_{nu^{n-1}} \cdot \underbrace{\frac{d}{dx}(x^2 + 1)}_{\frac{du}{dx}}$$

$$= 120(x^2 + 1)^{119}(2x) = 240x(x^2 + 1)^{119}$$

as before.

**EXAMPLE 4** Find $\dfrac{dy}{dx}$ if $y = \dfrac{1}{(2x^4 - x^2 + 1)^3}$.

**Solution** If we rewrite the given equation as $y = (2x^4 - x^2 + 1)^{-3}$, then an application of the General Power Rule gives

$$\frac{dy}{dx} = -3(2x^4 - x^2 + 1)^{-4}\frac{d}{dx}(2x^4 - x^2 + 1)$$

$$= -3(2x^4 - x^2 + 1)^{-4}(8x^3 - 2x)$$

$$= \frac{6x(1 - 4x^2)}{(2x^4 - x^2 + 1)^4}$$

Observe that the graph of $y$ has horizontal tangents at the points where $x = -\frac{1}{2}$, 0, and $\frac{1}{2}$ and that these are the numbers on the $x$-axis where the graph of $y'$ crosses the axis. (See Figure 2.) ■

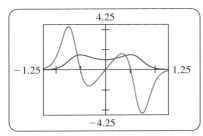

**FIGURE 2**
The graph of $y$ is shown in blue, and the graph of $y'$ is shown in red.

**EXAMPLE 5** How fast is $y = \left(\dfrac{2t - 1}{t^2 + 1}\right)^5$ changing when $t = 1$?

**Solution** The rate of change of $y$ at any value of $t$ is given by $dy/dt$. To find $dy/dt$, we use the General Power Rule, obtaining

$$\frac{dy}{dt} = \frac{d}{dt}\left(\frac{2t - 1}{t^2 + 1}\right)^5$$

$$= \underbrace{5\left(\frac{2t - 1}{t^2 + 1}\right)^4}_{nu^{n-1}} \cdot \underbrace{\frac{d}{dt}\left(\frac{2t - 1}{t^2 + 1}\right)}_{\frac{du}{dt}}$$

$$= 5\left(\frac{2t - 1}{t^2 + 1}\right)^4\left[\frac{(t^2 + 1)\frac{d}{dt}(2t - 1) - (2t - 1)\frac{d}{dt}(t^2 + 1)}{(t^2 + 1)^2}\right]$$

Use the Quotient Rule.

$$= 5\left(\frac{2t - 1}{t^2 + 1}\right)^4\left[\frac{(t^2 + 1)(2) - (2t - 1)(2t)}{(t^2 + 1)^2}\right]$$

$$= -\frac{10(t^2 - t - 1)(2t - 1)^4}{(t^2 + 1)^6}$$

In particular, when $t = 1$

$$\frac{dy}{dt}\bigg|_{t=1} = -\frac{10(-1)(1)}{2^6} = \frac{5}{32}$$

Therefore, $y$ is increasing at the rate of $\frac{5}{32}$ units per unit change in $t$ when $t = 1$. ■

## ■ The Chain Rule and Trigonometric Functions

In Section 2.5 we learned how to find the derivative of trigonometric functions such as $f(x) = \sin x$. How do we differentiate the function $F(x) = \sin(x^2 - \pi)$? Observe that $F$ is the composition $g \circ f$ of the functions $g$ and $f$ defined by $g(x) = \sin x$ and $f(x) = x^2 - \pi$. Therefore, an application of the Chain Rule yields

$$F'(x) = \frac{d}{dx}[\sin(x^2 - \pi)]$$

$$= \underbrace{\cos(x^2 - \pi)}_{g'[f(x)]} \cdot \underbrace{\frac{d}{dx}(x^2 - \pi)}_{f'(x)} \qquad f(x) = x^2 - \pi, \quad g(x) = \sin x$$

$$= \cos(x^2 - \pi) \cdot (2x) = 2x\cos(x^2 - \pi)$$

Another approach to differentiating generalized trigonometric functions is to derive the appropriate formulas using the Chain Rule. For example, we can find the formula for differentiating the generalized sine function $y = \sin[f(x)]$ by letting $u = f(x)$ so that $y = \sin u$ and then applying the Chain Rule to obtain

$$\frac{dy}{dx} = \frac{dy}{du} \cdot \frac{du}{dx}$$

$$= \frac{d}{du}(\sin u) \cdot \frac{du}{dx}$$

$$= (\cos u)f'(x)$$

$$= \cos[f(x)] \cdot f'(x)$$

In a similar manner we obtain the following rules.

---

**THEOREM 3    Derivatives of Generalized Trigonometric Functions**

$$\frac{d}{dx}(\sin u) = \cos u \cdot \frac{du}{dx} \qquad\qquad \frac{d}{dx}(\cos u) = -\sin u \cdot \frac{du}{dx}$$

$$\frac{d}{dx}(\tan u) = \sec^2 u \cdot \frac{du}{dx} \qquad\qquad \frac{d}{dx}(\csc u) = -\csc u \cot u \cdot \frac{du}{dx}$$

$$\frac{d}{dx}(\cot u) = -\csc^2 u \cdot \frac{du}{dx} \qquad\qquad \frac{d}{dx}(\sec u) = \sec u \tan u \cdot \frac{du}{dx}$$

**EXAMPLE 6** Find the slope of the tangent line to the graph of $y = 3 \cos x^2$ at the point where $x = \sqrt{\pi/2}$.

**Solution** The slope of the tangent line at any point on the graph is given by $dy/dx$. To find $dy/dx$, we use Theorem 3, obtaining

$$\frac{dy}{dx} = \frac{d}{dx}(3 \cos x^2)$$

$$= 3\frac{d}{dx}(\cos x^2) \qquad \text{Constant Multiple Rule}$$

$$= 3\underbrace{(-\sin x^2)}_{-\sin f(x)}\underbrace{(2x)}_{f'(x)}$$

$$= -6x \sin x^2$$

In particular, the slope of the tangent line to the graph of the given equation at the point where $x = \sqrt{\pi/2}$ is

$$\frac{dy}{dx}\bigg|_{x=\sqrt{\pi/2}} = -6\left(\sqrt{\frac{\pi}{2}}\right)\sin\frac{\pi}{2} = -6\sqrt{\frac{\pi}{2}} \qquad \blacksquare$$

⚠ Do not confuse $\cos x^2$ with $(\cos x)^2$, usually written as $\cos^2 x$.

**EXAMPLE 7** Find an equation of the tangent line at the point on the graph of $y = x^2 \sin 3x$, where $x = \pi/2$.

**Solution** The slope of the tangent line at any point $(x, y)$ on the graph of $y = x^2 \sin 3x$ is given by $dy/dx$. Using the Product Rule and Theorem 3, we obtain

$$\frac{dy}{dx} = \frac{d}{dx}(x^2 \sin 3x)$$

$$= x^2\frac{d}{dx}(\sin 3x) + (\sin 3x)\frac{d}{dx}(x^2)$$

$$= x^2(\cos 3x) \cdot \frac{d}{dx}(3x) + 2x \sin 3x$$

$$= 3x^2 \cos 3x + 2x \sin 3x$$

In particular, the slope of the tangent line at the point where $x = \pi/2$ is

$$\frac{dy}{dx}\bigg|_{x=\pi/2} = 3\left(\frac{\pi}{2}\right)^2 \cos\frac{3\pi}{2} + 2\left(\frac{\pi}{2}\right)\sin\frac{3\pi}{2} = 0 + \pi(-1) = -\pi$$

The point of tangency has $y$-coordinate given by

$$y\bigg|_{x=\pi/2} = x^2 \sin 3x\bigg|_{x=\pi/2} = \left(\frac{\pi}{2}\right)^2 \sin\frac{3\pi}{2} = -\frac{\pi^2}{4}$$

Therefore, an equation of the required tangent line is

$$y - \left(-\frac{\pi^2}{4}\right) = -\pi\left(x - \frac{\pi}{2}\right) \qquad \text{or} \qquad y = -\pi x + \frac{\pi^2}{4}$$

The graph of $f$ and its tangent line at $\left(\frac{\pi}{2}, -\frac{\pi^2}{4}\right)$ are shown in Figure 3. $\blacksquare$

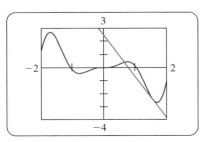

**FIGURE 3**

Next we consider an example in which the Chain Rule is applied more than once to differentiate a function.

**EXAMPLE 8** Find $\dfrac{dy}{dx}$ if $y = \tan^3(3x^2 + 1)$.

**Solution**

$$\frac{dy}{dx} = \frac{d}{dx}[\tan^3(3x^2 + 1)] = \frac{d}{dx}[\tan(3x^2 + 1)]^3$$

$$= 3[\tan(3x^2 + 1)]^2 \cdot \frac{d}{dx}[\tan(3x^2 + 1)] \qquad \text{Use the General Power Rule.}$$

$$= 3\tan^2(3x^2 + 1) \cdot \sec^2(3x^2 + 1) \cdot \frac{d}{dx}(3x^2 + 1) \qquad \text{Use Theorem 3.}$$

$$= 3\tan^2(3x^2 + 1) \cdot \sec^2(3x^2 + 1) \cdot 6x$$

$$= 18x\tan^2(3x^2 + 1)\sec^2(3x^2 + 1) \qquad\blacksquare$$

**Notes**

1. The function in Example 8 can be viewed as a composition of three functions. For example, letting $f(x) = 3x^2 + 1$, $g(u) = \tan u$, and $h(w) = w^3$, you can see that $g \circ f$ is defined by

$$(g \circ f)(x) = g[f(x)] = \tan[f(x)] = \tan(3x^2 + 1)$$

Now, if we compose $h$ with $g \circ f$, we obtain

$$[h \circ (g \circ f)](x) = h[(g \circ f)(x)] = h[\tan(3x^2 + 1)] = \tan^3(3x^2 + 1)$$

and this is just the expression for $y$.

2. Suppose that $y = [h \circ (g \circ f)](x) = h\{g[f(x)]\}$. In this case the Chain Rule is

$$\frac{dy}{dx} = h'\{g[f(x)]\}g'[f(x)]f'(x)$$

Equivalently, if we let $u = f(x)$ and $v = g(u)$, then

$$\frac{dy}{dx} = \frac{dy}{dv} \cdot \frac{dv}{du} \cdot \frac{du}{dx}$$

Incidentally, the reason for calling Theorem 1 the Chain Rule becomes clear if we look at the larger "chain" that results when we have the composition of many functions. $\blacksquare$

## ■ The Derivative of $e^u$

If we apply the Chain Rule to the function $y = e^{f(x)}$ by letting $u = f(x)$ (the "inside function"), we have

$$\frac{dy}{dx} = \frac{dy}{du}\frac{du}{dx}$$

$$= \frac{d}{du}(e^u)\frac{du}{dx}$$

$$= e^u\frac{du}{dx}$$

So we have the following.

---

**THEOREM 4**  **Derivative of the Generalized Natural Exponential Function**

If $u$ is a differentiable function of $x$, then

$$\frac{d}{dx}(e^u) = e^u \frac{du}{dx}$$

---

**EXAMPLE 9**  Differentiate $y = e^{-x \cos x}$.

**Solution**  Using Theorem 4, we have

$$\frac{dy}{dx} = \frac{d}{dx}(e^{-x \cos x}) = e^{-x \cos x} \frac{d}{dx}(-x \cos x)$$

$$= e^{-x \cos x}(-\cos x + x \sin x) = (x \sin x - \cos x)e^{-x \cos x} \quad \blacksquare$$

## ■ The Derivative of $a^u$

Thanks to the Chain Rule, we are also able to find the formula for differentiating more general exponential functions of the form $a^u$, where $a > 0$. We begin by recalling from Section 0.8, that if $a > 0$, then $a = e^{\ln a}$. So

$$f(x) = a^x = (e^{\ln a})^x = e^{(\ln a)x}$$

and using the Chain Rule, we obtain

$$f'(x) = \frac{d}{dx}(a^x) = \frac{d}{dx}\left(e^{(\ln a)x}\right) = e^{(\ln a)x} \cdot \frac{d}{dx}(\ln a)x$$

$$= e^{(\ln a)x} \cdot \ln a = a^x \ln a$$

To find a formula for computing the derivative of the generalized exponential function $y = a^{f(x)}$, where $f$ is a differentiable function, let $u = f(x)$ so that $y = a^u$. Then, using the Chain Rule, we find

$$\frac{dy}{dx} = \frac{dy}{du}\frac{du}{dx} = a^u \ln a \frac{du}{dx}$$

---

**THEOREM 5**  **The Derivatives of $a^x$ and $a^u$**

Let $u$ be a differentiable function of $x$ and $a > 0$, $a \neq 1$. then

**a.** $\dfrac{d}{dx}a^x = a^x \ln a$      **b.** $\dfrac{d}{dx}a^u = a^u \ln a \dfrac{du}{dx}$

---

**EXAMPLE 10**  Find the derivative of

**a.** $f(x) = 2^x$      **b.** $g(x) = 3^{\sqrt{x}}$      **c.** $y = 10^{\cos 2x}$

**Solution**

**a.** $f'(x) = \dfrac{d}{dx}2^x = (\ln 2)2^x$

**b.** $g'(x) = \dfrac{d}{dx} 3^{\sqrt{x}} = (\ln 3)3^{\sqrt{x}} \dfrac{d}{dx} x^{1/2} = (\ln 3)3^{\sqrt{x}} \left(\dfrac{1}{2} x^{-1/2}\right) = \dfrac{(\ln 3)3^{\sqrt{x}}}{2\sqrt{x}}$

**c.** $\dfrac{dy}{dx} = \dfrac{d}{dx} 10^{\cos 2x} = (\ln 10)10^{\cos 2x} \dfrac{d}{dx} \cos 2x = (\ln 10)10^{\cos 2x}(-\sin 2x)\dfrac{d}{dx}(2x)$

$\qquad = -2(\ln 10)(\sin 2x)10^{\cos 2x}$ ■

In Section 2.2 we saw that if $f(x) = a^x$ ($a > 0$), then $f'(x) = f'(0)a^x$, where $f'(0)$, the slope of the tangent line to the graph of $f$ at $(0, 1)$, is $\lim_{h\to 0}(a^h - 1)/h$. In view of Theorem 5 we now know that

$$\lim_{h\to 0} \frac{a^h - 1}{h} = \ln a$$

For example, if $a = 2$, then $\lim_{h\to 0}(2^h - 1)/h = \ln 2 \approx 0.69$, and this agrees with our estimate,

$$\frac{d}{dx}(2^x) \approx (0.69)2^x$$

obtained in Section 2.2. Finally, observe that if we put $a = e$, then

$$\lim_{h\to 0} \frac{e^h - 1}{h} = \ln e = 1$$

as expected.

**EXAMPLE 11**  **A Spring System**   The equation of motion of a weight attached to a spring and a dashpot damping device is

$$x(t) = e^{-t}\left(-2\cos 3t - \frac{2}{3}\sin 3t\right)$$

where $x(t)$, measured in feet, is the displacement from the equilibrium position of the spring system and $t$ is measured in seconds. (See Figure 4.) Find the initial position and the initial velocity of the weight.

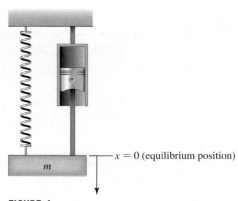

$x = 0$ (equilibrium position)

$m$

**FIGURE 4**
The system in equilibrium (The positive direction is downward.)

**Solution**  The initial position of the spring system is given by

$$x(0) = e^0 \left( -2 \cos 0 - \frac{2}{3} \sin 0 \right) = -2$$

This tells us that the spring system is 2 ft above the equilibrium position.

The velocity of the spring system at any time $t$ is given by

$$v(t) = \frac{d}{dt} \left[ e^{-t} \left( -2 \cos 3t - \frac{2}{3} \sin 3t \right) \right]$$

$$= -e^{-t} \left( -2 \cos 3t - \frac{2}{3} \sin 3t \right) + e^{-t} (6 \sin 3t - 2 \cos 3t)$$

Use the Product Rule.

$$= \frac{20}{3} e^{-t} \sin 3t$$

In particular, its initial velocity is

$$v(0) = \frac{20}{3} e^0 \sin 0 = 0$$

that is, it is released from rest.

**EXAMPLE 12**  **Path of a Boat**  A boat leaves the point $O$ (the origin) located on one bank of a river traveling with a constant speed of 20 mph and always heading toward a dock at the point $A(1000, 0)$, which is directly due east of the origin (see Figure 5). The river flows north at a constant speed of 5 mph. It can be shown that the path of the boat is

$$y = 500 \left[ \left( \frac{1000 - x}{1000} \right)^{3/4} - \left( \frac{1000 - x}{1000} \right)^{5/4} \right] \qquad 0 \le x \le 1000$$

Find $dy/dx$ when $x = 100$ and when $x = 900$. Interpret your results.

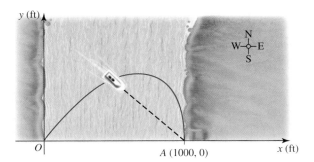

**FIGURE 5**
The path of the boat

**Solution**  We find

$$\frac{dy}{dx} = 500 \left[ \frac{3}{4} \left( \frac{1000 - x}{1000} \right)^{-1/4} \left( -\frac{1}{1000} \right) - \frac{5}{4} \left( \frac{1000 - x}{1000} \right)^{1/4} \left( -\frac{1}{1000} \right) \right]$$

$$= \frac{1}{2} \left[ \frac{5}{4} \left( \frac{1000 - x}{1000} \right)^{1/4} - \frac{3}{4} \left( \frac{1000 - x}{1000} \right)^{-1/4} \right]$$

So

$$\left.\frac{dy}{dx}\right|_{x=100} = \frac{1}{2}\left[\frac{5}{4}\left(\frac{9}{10}\right)^{1/4} - \frac{3}{4}\left(\frac{10}{9}\right)^{1/4}\right] \approx 0.22$$

This tells us that at the point on the path where $x = 100$, the boat is drifting north at the rate of 0.22 ft/ft in the $x$-direction. Next,

$$\left.\frac{dy}{dx}\right|_{x=900} = \frac{1}{2}\left[\frac{5}{4}\left(\frac{1}{10}\right)^{1/4} - \frac{3}{4}(10)^{1/4}\right] \approx -0.32$$

This tells us that at the point on the path where $x = 900$, the boat is drifting south at the rate of 0.32 ft/ft in the $x$-direction. ∎

## 2.6 CONCEPT QUESTIONS

1. State the Chain Rule for differentiating the composite function $h = g \circ f$. Explain it in your own words.
2. **a.** State the rule for differentiating the generalized power function $g(x) = [f(x)]^n$, where $n$ is any real number.
   **b.** State the rule for differentiating the generalized trigonometric function
   $$h(x) = \sec[f(x)]$$

3. Suppose the population $P$ of a certain bacteria culture is given by $P = f(T)$, where $T$ is the temperature of the medium. Further, suppose that the temperature $T$ is a function of time $t$ in seconds—that is, $T = g(t)$. Give an interpretation of each of the following quantities:
   **a.** $\dfrac{dP}{dT}$ **b.** $\dfrac{dT}{dt}$ **c.** $\dfrac{dP}{dt}$ **d.** $(f \circ g)(t)$ **e.** $f'(g(t))g'(t)$

## 2.6 EXERCISES

*In Exercises 1–6, identify the "inside function" $u = f(x)$ and the "outside function" $y = g(u)$. Then find $dy/dx$ using the Chain Rule.*

**1.** $y = (2x + 4)^3$

**2.** $y = \sqrt{x^2 - 4}$

**3.** $y = \dfrac{1}{\sqrt[3]{x^2 + 1}}$

**4.** $y = 2 \sin \pi x$

**5.** $y = \sqrt{e^x + \cos x}$

**6.** $y = \sec \sqrt{x}$

*In Exercises 7–64, find the derivative of the function.*

**7.** $f(x) = (2x + 1)^5$

**8.** $g(x) = (3x^2 + x - 1)^{4/3}$

**9.** $y = e^{x^2 - x}$

**10.** $f(t) = e^{\sqrt{t}}$

**11.** $y = \left(t + \dfrac{2}{t}\right)^6$

**12.** $f(x) = \left(\dfrac{x^2 + 3}{x}\right)^{-2}$

**13.** $h(u) = u^3(2u^2 - 1)^4$

**14.** $h(x) = (2x - 1)^2(x^2 + 1)^3$

**15.** $f(x) = x^2 e^{-2x}$

**16.** $g(t) = \sqrt{t - 2} + \sqrt{4 - t}$

**17.** $g(u) = u\sqrt{1 - u^2}$

**18.** $f(x) = \dfrac{x}{\sqrt{x^2 - x - 2}}$

**19.** $f(t) = \dfrac{2t + 3}{(t + 2t^2)^3}$

**20.** $f(x) = (x^2 + \sqrt{x})^6$

**21.** $y(s) = \left(1 + \sqrt{1 + s^2}\right)^5$

**22.** $f(x) = \left(\dfrac{x + 2}{x - 3}\right)^{3/2}$

**23.** $g(x) = \dfrac{e^{2x}}{1 + e^{-x}}$

**24.** $h(t) = \dfrac{e^t - e^{-t}}{e^t + e^{-t}}$

**25.** $f(x) = \sin 3x$

**26.** $g(x) = e^{-2x} \cos 3x$

**27.** $g(t) = \tan(\pi t - 1)$

**28.** $y = \cot(2x + 1)$

**29.** $f(x) = \sin^3 x$

**30.** $y = \cos(x^3)$

**31.** $f(x) = \sin 2x + \tan \sqrt{x}$

**32.** $y = \cos(x^2 - 3x + 1) + \tan\left(\dfrac{2}{x}\right)$

**33.** $f(x) = \sin^3 x + \cos^3 x$

**34.** $f(x) = \tan^2 x + \cot x^2$

**35.** $f(x) = (1 + \sin^2 3x)^{2/3}$

**36.** $z = (1 + \csc^2 x)^4$

**37.** $h(x) = (x^2 - \sec \pi x)^{-3}$

**38.** $g(x) = \tan^2(x^2 + x)$

**39.** $y = \sqrt{1 + 2 \cos x}$

**40.** $f(x) = \sqrt{2 + 3 \tan 2x}$

**41.** $f(x) = \dfrac{1 + \cos 3x}{1 - \cos 3x}$

**42.** $y = \dfrac{x + \sin 2x}{2 + \cos 3x}$

**43.** $y = e^{\cos x}$

**44.** $f(x) = x \sin \dfrac{1}{x}$

**45.** $f(x) = \sqrt{\sin 2x - \cos 2x}$

**46.** $g(t) = \sqrt{t + \tan 3t}$

**47.** $y = \sin^2\left(\dfrac{1 + x}{1 - x}\right)$

**48.** $y = \sec^3\left(\dfrac{\sqrt{x}}{1 + x}\right)$

**49.** $f(x) = \dfrac{\cos 2x}{\sqrt{1 + x^2}}$

**50.** $f(t) = \dfrac{\cot 2t}{1 + t^2}$

**V** Videos for selected exercises are available online at **www.academic.cengage.com/login.**

**51.** $y = \sec^2 x \tan 3x$

**52.** $y = x \tan^2(2x + 3)$

**53.** $f(x) = \sin(\sin x)$

**54.** $g(t) = \tan(\cos 2t)$

**55.** $f(x) = \cos^3(\sin \pi x)$

**56.** $y = e^{\cos x^2} \tan(e^{2x} + x)$

**57.** $y = x(5^{3x})$

**58.** $f(u) = 2^{u^2}$

**59.** $h(x) = (2^x + 3^{-x})^6$

**60.** $f(x) = x^e + e^x$

**61.** $g(x) = x^e e^x$

**62.** $f(x) = \dfrac{2^{3x}}{x}$

**63.** $y = 2^{\cot x}$

**64.** $g(x) = \dfrac{2^x}{\sqrt{3^x + 1}}$

*In Exercises 65–68, find the second derivative of the function.*

**65.** $f(x) = x(2x^2 - 1)^4$

**66.** $g(x) = \dfrac{1}{(2x + 1)^2}$

**67.** $f(t) = \sin^2 t - \sin t^2$

**68.** $f(x) = e^{-x} \tan e^x$

 *In Exercises 69–72, (a) find an equation of the tangent line to the graph of the function at the indicated point, and (b) use a graphing utility to plot both the graph of the function and the tangent line on the same screen.*

**69.** $f(x) = x\sqrt{x^2 + 1}$; $(1, \sqrt{2})$

**70.** $g(x) = \left(\dfrac{x - 1}{x + 1}\right)^3$; $(2, \frac{1}{27})$

**71.** $f(x) = xe^{-x}$; $(1, e^{-1})$

**72.** $h(t) = 2 \cos^2 \pi t$; $(\frac{1}{4}, 1)$

**73.** Suppose that $F = g \circ f$ and $f(2) = 5$, $f'(2) = 4$, and $g'(5) = 75$. Find $F'(2)$.

**74.** Suppose that $F(x) = g[f(x)]$ and $f(3) = 16$, $f'(3) = 6$, and $g'(16) = \frac{1}{8}$. Find $F'(3)$.

**75.** Let $F(x) = f[f(x)]$. Does it follow that $F'(x) = [f'(x)]^2$?

**76.** Suppose that $h = g \circ f$. Does it follow that $h' = g' \circ f'$?

**77.** Find an equation of the line tangent to the graph of $y = 2^x + 1$ at the point $(0, 2)$.

**78.** Find an equation of the line tangent to the graph of $y = \dfrac{e^{-2x}}{x^2 + 1}$ at the point where $x = 0$.

*In Exercises 79–80, refer to the following graph.*

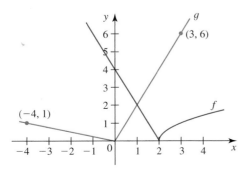

**79.** The graphs of $f$ and $g$ are shown in the figure. Let $F(x) = g[f(x)]$ and $G(x) = f[g(x)]$. Find $F'(1)$, $G'(-1)$, and $F'(2)$. If a derivative does not exist, explain why.

**80.** The graphs of $f$ and $g$ are shown in the figure. Find
   **a.** $h'(1)$ if $h = g \circ g$
   **b.** $H'(1)$ if $H = f \circ f$
   **c.** $G'(1)$ if $G(x) = f(x^2 - 1)$
   **d.** $F'(\frac{\pi}{6})$ if $F(x) = f(2 \sin x)$

*In Exercises 81–84, find $F'(x)$. Assume that all functions are differentiable.*

**81.** $F(x) = a[f(\sin x)] + b[g(\cos x)]$, where $a$ and $b$ are real numbers

**82.** $F(x) = a \sin[f(x)] + b \cos[g(x)]$

**83.** $F(x) = f(x^2 + 1) + g(x^2 - 1)$

**84.** $F(x) = f(x^a) + [f(x)]^a$, where $a$ is a real number

**85.** Find $F''(2)$ if $F(x) = x^2 f(2x)$ and it is known that $f(4) = -2$, $f'(4) = 1$, and $f''(4) = -1$.

**86.** Suppose that $f$ has second-order derivatives and $g(x) = xf(x^2 + 1)$. Find $g''(x)$ in terms of $f(x), f'(x)$, and $f''(x)$.

**87.** The graph of the function

$$f(x) = \dfrac{|x|}{\sqrt{2 - x^2}}$$

is called a bullet-nose curve.
   **a.** What is the derivative of $f$ for $x \neq 0$? Find the equations of the tangent lines to the graph of $f$ at $(-1, 1)$ and $(1, 1)$.
    **b.** Plot the graph of $f$ and the tangent lines found in part (a) using the same viewing window.

**88.** Refer to Exercise 87. Explain why $f$ is not differentiable at 0.

**89. Aging Population** The population of Americans age 55 years and over as a percent of the total population is approximated by the function

$$f(t) = 10.72(0.9t + 10)^{0.3} \qquad 0 \leq t \leq 20$$

where $t$ is measured in years, with $t = 0$ corresponding to the year 2000. At what rate was the percent of Americans age 55 years and over changing at the beginning of 2000? At what rate will the percent of Americans age 55 years and over be changing at the beginning of 2010? What will be the percent of the population of Americans age 55 years and over at the beginning of 2010?
*Source:* U.S. Census Bureau.

**90. Accumulation Years** People from their mid-40s to their mid-50s are in the prime investing years. Demographic studies of this type are of particular importance to financial institutions. The function

$$N(t) = 34.4(1 + 0.32125t)^{0.15} \qquad 0 \leq t \leq 12$$

gives the projected number of people in this age group in the United States (in millions) in year $t$, where $t = 0$ corresponds to the beginning of 1996.

**a** How large was this segment of the population projected to be at the beginning of 2005?

**b.** How fast was this segment of the population growing at the beginning of 2005?

*Source:* U.S. Census Bureau.

**91. World Population Growth** After its fastest rate of growth ever during the 1980s and 1990s, the rate of growth of world population is expected to slow dramatically, in the twenty-first century. The function

$$G(t) = 1.58e^{-0.213t}$$

gives the projected average percent population growth/decade in the $t$th decade, with $t = 1$ corresponding to the beginning of 2000.

**a.** What will the projected average population growth rate be at the beginning of 2020 ($t = 3$)?

**b.** How fast will the projected average population growth rate be changing at the beginning of 2020?

*Source:* U.S. Census Bureau.

**92. Blood Alcohol Level** The percentage of alcohol in a person's bloodstream $t$ hr after drinking 8 fluid oz of whiskey is given by

$$A(t) = 0.23te^{-0.4t} \qquad 0 \le t \le 12$$

How fast is the percentage of alcohol in a person's bloodstream changing after $\frac{1}{2}$ hr? After 8 hr?

*Source:* Encyclopedia Britannica.

**93. Radioactivity** The radioactive element polonium decays according to the law

$$Q(t) = Q_0 \cdot 2^{-(t/140)}$$

where $Q_0$ is the initial amount and $t$ is measured in days.

**a.** If the amount of polonium left after 280 days is 20 mg, what was the initial amount present?

**b.** How fast is the amount of polonium changing at any time $t$?

**94. Forensic Science** Forensic scientists use the following formula to determine the time of death of accident or murder victims. If $T$ denotes the temperature of a body $t$ hr after death, then

$$T = T_0 + (T_1 - T_0)(0.97)^t$$

where $T_0$ is the air temperature and $T_1$ is the body temperature (in degrees Fahrenheit) at the time of death. John Doe was found murdered at midnight in his house, when the room temperature was 70°F. Assume that his body temperature at the time of death was 98.6°F.

 **a.** Plot the graph of $T$ using the viewing window [0, 40] × [70, 100].

**b.** How fast was the temperature of John Doe's body dropping 2 hr after his death?

**c.** If the temperature of John Doe's body was 80°F when it was found, when was he killed? Solve the problem analytically, and then verify it using a graphing calculator.

**95. Simple Harmonic Motion** The position function of a body moving along a coordinate line is

$$s(t) = \frac{1}{2}\cos 2t + \frac{3}{4}\sin 2t \qquad t \ge 0$$

where $s(t)$ is measured in feet and $t$ in sec. Find the position, velocity, speed, and acceleration of the body when $t = \pi/4$.

**96. Predator-Prey Population Model** The wolf population in a certain northern region is estimated to be

$$P_W(t) = 9000 + 1000 \sin \frac{\pi t}{24}$$

in month $t$, and the caribou population in the same region is given by

$$P_C(t) = 36{,}000 + 12{,}000 \cos \frac{\pi t}{24}$$

Find the rate of change of each population when $t = 12$.

**97. Stock Prices** The closing price (in dollars) per share of the stock of Tempco Electronics on the $t$th day it was traded is approximated by

$$P(t) = 20 + 12 \sin\left(\frac{\pi t}{30}\right) - 6 \sin\left(\frac{\pi t}{15}\right)$$

$$+ 4 \sin\left(\frac{\pi t}{10}\right) - 3 \sin\left(\frac{2\pi t}{15}\right) \qquad 0 \le t \le 20$$

where $t = 0$ corresponds to the time the stock was first listed on a major stock exchange. What was the rate of change of the stock's price at the close of the fifteenth day of trading? What was the closing price on that day?

**98. Shortage of Nurses** The projected number of nurses (in millions) from the year 2000 through 2015 is given by

$$N(t) = \begin{cases} 1.9 & \text{if } 0 \le t < 5 \\ \sqrt{0.123t + 2.995} & \text{if } 5 \le t \le 15 \end{cases}$$

where $t = 0$ corresponds to the year 2000, while the projected number of nursing jobs (in millions) over the same period is

$$J(t) = \begin{cases} \sqrt{0.129t + 4} & \text{if } 0 \le t < 10 \\ \sqrt{0.4t + 1.29} & \text{if } 10 \le t \le 15 \end{cases}$$

**a.** Let $G = J - N$ be the function giving the gap between the demand and the supply of nurses from the year 2000 through 2015. Find $G'$.

**b.** How fast was the gap between the demand and the supply of nurses changing in 2008? In 2012?

*Source:* Department of Health and Human Services.

**99. Potential Energy** A commonly used potential-energy function for the interaction of two molecules is the Lennard-Jones 6-12 potential, given by

$$u(r) = u_0\left[\left(\frac{\sigma}{r}\right)^{12} - \left(\frac{\sigma}{r}\right)^{6}\right]$$

where $u_0$ and $\sigma$ are constants. The force corresponding to this potential is $F(r) = -u'(r)$. Find $F(r)$.

**100. Mass of a Body Moving Near the Speed of Light** According to the special theory of relativity, the mass $m$ of a body moving at a speed $v$ is given by

$$m = \frac{m_0}{\sqrt{1 - \dfrac{v^2}{c^2}}}$$

where $m_0$ is the mass of the body at rest and $c \approx 3 \times 10^8$ m/sec is the speed of light. How fast is the mass of an electron changing with respect to its speed when its speed is $0.999c$? The rest mass of an electron is $9.11 \times 10^{-31}$ kg.

**101. Motion Along a Line** A body moves along a coordinate line in such a way that its position function at any time $t$ is given by

$$s(t) = t\sqrt{1 - t^2} \qquad 0 \le t \le 1$$

where $s(t)$ is measured in feet and $t$ in seconds. Find the velocity and acceleration of the body when $t = \frac{1}{2}$.

**102. Surface Area of a Cone** The lateral surface area of a right circular cone is

$$S = \pi r\sqrt{r^2 + h^2}$$

where $r$ is the radius of the base and $h$ is the height.
**a.** What is the rate of change of the lateral surface area with respect to the height if the radius is constant?
**b.** What is the rate of change of the lateral surface area with respect to the radius if the height is constant?

**103. A Sliding Chain** A chain of length 6 m is held on a table with 1 m of the chain hanging down from the table. Upon release, the chain slides off the table. Assuming that there is no friction, the end of the chain that initially was 1 m from the edge of the table is given by the function

$$s(t) = \frac{1}{2}\left(e^{\sqrt{g/6}\,t} + e^{-\sqrt{g/6}\,t}\right)$$

where $g = 9.8$ m/sec$^2$ and $t$ is measured in seconds.
**a.** Find the time it takes for the chain to slide off the table.
**b.** What is the speed of the chain at the instant of time when it slides off the table?

**104. Percent of Females in the Labor Force** Based on data from the U.S. Census Bureau, the following model giving the percent of the total female population in the civilian labor force, $P(t)$, at the beginning of the $t$th decade ($t = 0$ corresponds to the year 1900) was constructed.

$$P(t) = \frac{74}{1 + 2.6e^{-0.166t + 0.04536t^2 - 0.0066t^3}} \qquad 0 \le t \le 11$$

**a.** What was the percent of the total female population in the civilian labor force at the beginning of 2000?
**b.** What was the growth rate of the percentage of the total female population in the civilian labor force at the beginning of 2000?

*Source:* U.S. Census Bureau.

**105. Electric Current in a Circuit** The following figure shows an R-C series circuit comprising a variable resistor, a capacitor, and an electromotive force. If the resistance at any time $t$ is given by $R = k_1 + k_2t$ ohms, where $k_1 > 0$ and $k_2 > 0$, the capacitance is $C$ farads, and the electromotive force is a constant $E$ volts, then the charge at any time $t$ is given by

$$q(t) = EC + (q_0 - EC)\left(\frac{k_1}{k_1 + k_2t}\right)^{1/(Ck_2)}$$

coulombs where the constant $q_0$ is the charge at $t = 0$. What is the current $i(t)$ at any time $t$?
**Hint:** $i(t) = dq/dt$.

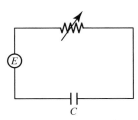

**106. Simple Harmonic Motion** The equation of motion of a body executing simple harmonic motion is given by

$$x(t) = A\sin(\omega t + \phi)$$

where $x$ (in feet) is the displacement of the body, $A$ is the amplitude, $\omega = \sqrt{k/m}$, $k$ is a constant, and $m$ (in slugs) is the mass of the body. Find expressions for the velocity and acceleration of the body at time $t$.

**107. Potential of a Charged Disk** The potential on the axis of a uniformly charged disk is

$$V(r) = \frac{\sigma}{2\varepsilon_0}\left(\sqrt{r^2 + R^2} - r\right)$$

where $\varepsilon_0$ and $\sigma$ are constants. The force corresponding to this potential is $F(r) = -V'(r)$. Find $F(r)$.

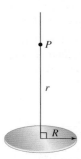

**108. Electric Potential** Suppose that a ring-shaped conductor of radius $a$ carries a total charge $Q$. Then the electrical potential at the point $P$, a distance $x$ from the center and along the line perpendicular to the plane of the ring through its center, is given by

$$V(x) = \frac{1}{4\pi\varepsilon_0} \frac{Q}{\sqrt{x^2 + a^2}}$$

where $\varepsilon_0$ is a constant called the permittivity of free space. The magnitude of the electric field induced by the charge at the point $P$ is $E = -dV/dx$, and the direction of the field is along the $x$-axis. Find $E$.

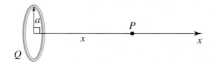

**109. Motion of a Conical Pendulum** A metal ball is attached to a string of length $L$ ft and is whirled in a horizontal circle as shown in the figure. The speed of the ball is $v = \sqrt{Lg \sec \theta \sin^2 \theta}$ ft/sec, where $\theta$ is the angle the string makes with the vertical.

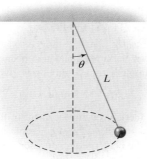

**a.** Show that

$$\frac{dv}{d\theta} = \frac{\sqrt{Lg}(\tan^2 \theta + 2)}{2\sqrt{\sec \theta}}$$

and interpret your result.
**b.** Find $v$ and $dv/d\theta$ if $L = 4$ and $\theta = \pi/6$ rad. (Take $g = 32$ ft/sec$^2$.)
**c.** Evaluate $\lim_{\theta \to \pi/2^-} v$, and interpret your result.
**d.** Plot the graph of $v$ for $0 \le \theta < \frac{\pi}{2}$ to verify the result of part (c) visually.

**110. Orbit of a Satellite** An artificial satellite moves around the earth in an elliptic orbit. Its distance $r$ from the center of the earth is approximated by

$$r = a\left[1 - e \cos M - \frac{e^2}{2}(\cos 2M - 1)\right]$$

where $M = (2\pi/P)(t - t_n)$. Here, $t$ is time and $a$, $e$, $P$, and $t_n$ are constants measuring the semimajor axis of the orbit, the eccentricity of the orbit, the period of orbiting, and the time taken by the satellite to pass the perigee, respectively. Find $dr/dt$, the radial velocity of the satellite.

**111.** Find $f''(x)$ if

$$f(x) = \begin{cases} x^2 \sin \dfrac{1}{x} & \text{if } x \ne 0 \\ 0 & \text{if } x = 0 \end{cases}$$

Does $f''(0)$ exist?

**112.** Suppose that $u$ is a differentiable function of $x$ and $f(x) = |u|$. Show that

$$f'(x) = \frac{u'u}{|u|} \qquad u \ne 0$$

Hint: $|u| = \sqrt{u^2}$.

*In Exercises 113–116, use the result of Exercise 112 to find the derivative of the function.*

**113.** $f(x) = |x + 1|$

**114.** $g(x) = x|x^2 + x|$

**115.** $h(x) = |\sin x|$

**116.** $f(x) = \dfrac{|x|}{x^2}$

**117.** Let $f(x) = \sqrt{|(x - 1)(x - 2)|}$.
**a.** Find $f'(x)$.
Hint: See Exercise 112.
**b.** Sketch the graph of $f$ and $f'$.

**118.** A function is called *even* if $f(-x) = f(x)$ for all $x$ in the domain of $f$; it is called *odd* if $f(-x) = -f(x)$ for all $x$ in the domain of $f$. Prove that the derivative of a differentiable even function is an odd function and that the derivative of a differentiable odd function is an even function.

*In Exercises 119–122, determine whether the statement is true or false. If it is true, explain why it is true. If it is false, give an example to show why it is false.*

**119.** If $f$ has a second-order derivative at $x$, $g$ has a second-order derivative at $f(x)$, and $h(x) = g[f(x)]$, then $h''(x) = g''[f(x)]f''(x)$.

**120.** If $f$ is differentiable and $h(t) = f(a + bt) + f(a - bt)$, then $h'(t) = bf'(a + bt) - bf'(a - bt)$.

**121.** If $f$ is differentiable, then
$$\frac{d}{dx}f\left(\frac{1}{x}\right) = -\frac{f'\left(\frac{1}{x}\right)}{x^2} \qquad x \neq 0$$

**122.** If $f$ is differentiable and $h = (f \circ f)^2$, then $h' = 2(f \circ f)(f' \circ f)f'$.

## 2.7 Implicit Differentiation

### ■ Implicit Functions

Up to now, the functions we have dealt with are represented by equations of the form $y = f(x)$, in which the dependent variable $y$ has been expressed explicitly in terms of the independent variable $x$. Sometimes, however, a function $f$ is defined implicitly by an equation $F(x, y) = 0$. For example, the equation

$$x^2y + y - \cos x + 1 = 0 \tag{1}$$

defines $y$ as a function of $x$. (Here, $F(x, y) = x^2y + y - \cos x + 1$.) In fact, if we solve the equation for $y$ in terms of $x$, we obtain the explicit representation

$$y = f(x) = \frac{\cos x - 1}{x^2 + 1} \tag{2}$$

You can verify that Equation (2) satisfies Equation (1); that is,

$$x^2 f(x) + f(x) - \cos x + 1 = 0$$

Suppose we are given Equation (1) and we wish to find $dy/dx$. An obvious approach would be to first find an explicit representation for the function $f$, such as Equation (2), and then differentiate this expression in the usual manner to obtain $dy/dx = f'(x)$.

How about the equation

$$4x^4 + 8x^2y^2 - 25x^2y + 4y^4 = 0 \tag{3}$$

whose graph is shown in Figure 1? The Vertical Line Test shows that Equation (3) does not define $y$ as a function of $x$. But with suitable restrictions on $x$ and $y$, Equation (3) does define $y$ as a function of $x$ implicitly. Figure 2 shows the graphs (the solid curves) of two such functions, $f$ and $g$. In this instance we would be hard pressed to find explicit representations for the functions $f$ and $g$. So how do we go about computing $dy/dx$ in this case?

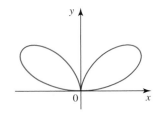

**FIGURE 1**
The graph of
$$4x^4 + 8x^2y^2 - 25x^2y + 4y^4 = 0$$
is a bifolium.

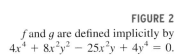

**FIGURE 2**
$f$ and $g$ are defined implicitly by
$4x^4 + 8x^2y^2 - 25x^2y + 4y^4 = 0.$

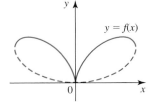

**(a)** The graph of $f$

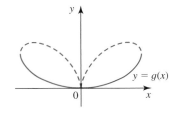

**(b)** The graph of $g$

Thanks to the Chain Rule, there exists a method for finding the derivative of a function directly from the equation defining it implicitly. This method is called **implicit differentiation** and will be demonstrated in the next several examples.

## ■ Implicit Differentiation

**EXAMPLE 1**

**a.** Find $dy/dx$ if $x^2 + y^2 = 4$.
**b.** Find an equation of the tangent line to the graph of $x^2 + y^2 = 4$ at the point $(1, \sqrt{3})$.
**c.** Solve part (b) again, this time using an explicit representation of a function.

**Solution**
**a.** Differentiating both sides of the equation with respect to $x$, we obtain

$$\frac{d}{dx}(x^2 + y^2) = \frac{d}{dx}(4)$$

$$\frac{d}{dx}(x^2) + \frac{d}{dx}(y^2) = 0 \qquad \text{Use the Sum Rule for derivatives.}$$

To carry out the differentiation of the term $y^2$, we note that $y$ is a function of $x$. Writing $y = f(x)$ to remind us of this, we see that

$$\frac{d}{dx}(y^2) = \frac{d}{dx}[f(x)]^2 \qquad \text{Write } y = f(x).$$

$$= 2f(x)f'(x) \qquad \text{Use the Chain Rule.}$$

$$= 2y\frac{dy}{dx} \qquad \text{Return to using } y \text{ instead of } f(x).$$

Therefore, the equation

$$\frac{d}{dx}(x^2 + y^2) = \frac{d}{dx}(4)$$

is equivalent to

$$2x + 2y\frac{dy}{dx} = 0$$

Solving for $dy/dx$ yields

$$\frac{dy}{dx} = -\frac{x}{y}$$

**b.** Using the result of part (a), we see that the slope of the required tangent line is

$$\left.\frac{dy}{dx}\right|_{(1,\sqrt{3})} = -\left.\frac{x}{y}\right|_{(1,\sqrt{3})} = -\frac{1}{\sqrt{3}} \qquad \left.\frac{dy}{dx}\right|_{(a,b)} \text{means } \frac{dy}{dx} \text{ evaluated at } x = a \text{ and } y = b.$$

Using the slope-intercept form of an equation of a line, we see that an equation of the tangent line is

$$y - \sqrt{3} = -\frac{1}{\sqrt{3}}(x - 1)$$

$$\sqrt{3}y - 3 = -(x - 1) \qquad \text{or} \qquad x + \sqrt{3}y - 4 = 0$$

**c.** Solving the equation $x^2 + y^2 = 4$ for $y$ in terms of $x$ gives the functions

$$y = f(x) = \sqrt{4 - x^2} \quad \text{and} \quad y = g(x) = -\sqrt{4 - x^2}$$

among others. The graph of $f$ is the upper semicircle centered at the origin with radius 2 (here, $y \geq 0$), whereas the graph of $g$ is the lower semicircle (here, $y \leq 0$). (See Figure 3.) Since the point $(1, \sqrt{3})$ lies on the upper semicircle, we will work with the function

$$f(x) = \sqrt{4 - x^2} = (4 - x^2)^{1/2}$$

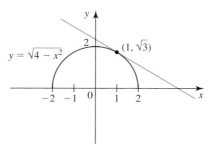

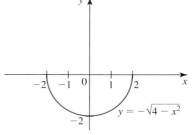

The graph of $f$

The graph of $g$

**FIGURE 3**

The graphs of $f(x) = \sqrt{4 - x^2}$
and $g(x) = -\sqrt{4 - x^2}$

Differentiating $f(x)$ with the help of the Chain Rule gives

$$f'(x) = \frac{1}{2}(4 - x^2)^{-1/2}\frac{d}{dx}(4 - x^2)$$

$$= \frac{1}{2}(4 - x^2)^{-1/2}(-2x)$$

$$= -\frac{x}{\sqrt{4 - x^2}}$$

$$= -\frac{x}{y}$$

and this gives the slope of the tangent line at any point $(x, y)$ on the graph of $f$. In particular, the slope of the tangent line at $(1, \sqrt{3})$ is

$$f'(1) = -\frac{1}{\sqrt{3}}$$

as before. Continuing, we find that an equation of the tangent line is $x + \sqrt{3}y - 4 = 0$, as obtained earlier. ∎

**Notes**

**1.** You can verify that

$$g'(x) = \frac{x}{\sqrt{4 - x^2}} = -\frac{x}{y} = \frac{dy}{dx}$$

so $-x/y$ is the derivative of both $f$ and $g$.

**2.** Even when it is possible to find an explicit representation for $f$, it still can be easier to find $f'(x)$ by implicit differentiation. (See Example 1.)

**3.** In general, if $dy/dx$ is found by implicit differentiation, the expression for $dy/dx$ will usually involve both $x$ and $y$. ∎

Guidelines for differentiating a function implicitly follow.

> **Finding $dy/dx$ by Implicit Differentiation**
>
> Suppose that a function $y = f(x)$ is defined implicitly via an equation in $x$ and $y$. To compute $dy/dx$:
>
> 1. Differentiate both sides of the equation with respect to $x$. Make sure that the derivative of any term involving $y$ includes the factor $dy/dx$.
> 2. Solve the resulting equation for $dy/dx$ in terms of $x$ and $y$.

**EXAMPLE 2** Find $\dfrac{dy}{dx}$ if $y^4 - 2y^3 + x^3y^2 - \cos x = 8$.

**Solution** Differentiating both sides of the given equation with respect to $x$, we obtain

$$\frac{d}{dx}(y^4 - 2y^3 + x^3y^2 - \cos x) = \frac{d}{dx}(8)$$

$$\frac{d}{dx}(y^4) - \frac{d}{dx}(2y^3) + \frac{d}{dx}(x^3y^2) - \frac{d}{dx}(\cos x) = 0$$

or

$$\frac{d}{dx}(y^4) - 2\frac{d}{dx}(y^3) + x^3\frac{d}{dx}(y^2) + y^2\frac{d}{dx}(x^3) - \frac{d}{dx}(\cos x) = 0,$$

where we have used the Product Rule to differentiate the term $x^3y^2$. Next, recalling that $y$ is a function of $x$, we apply the Chain Rule to the first three terms on the left, obtaining

$$4y^3\frac{dy}{dx} - 6y^2\frac{dy}{dx} + 2x^3y\frac{dy}{dx} + 3x^2y^2 + \sin x = 0$$

$$(4y^3 - 6y^2 + 2x^3y)\frac{dy}{dx} = -3x^2y^2 - \sin x$$

$$\frac{dy}{dx} = -\frac{3x^2y^2 + \sin x}{2y(2y^2 - 3y + x^3)}$$

**EXAMPLE 3** Find $y'$ if $e^x \cos y - e^{-y} \sin x = \pi$.

**Solution** Differentiating both sides of the given equation with respect to $x$, we obtain

$$\frac{d}{dx}(e^x \cos y - e^{-y} \sin x) = \frac{d}{dx}(\pi)$$

$$\frac{d}{dx}(e^x \cos y) - \frac{d}{dx}(e^{-y} \sin x) = 0$$

$$e^x\frac{d}{dx}(\cos y) + (\cos y)\frac{d}{dx}(e^x) - e^{-y}\frac{d}{dx}(\sin x) - (\sin x)\frac{d}{dx}(e^{-y}) = 0$$

$$e^x(-\sin y)y' + (\cos y)e^x - e^{-y}(\cos x) - (\sin x)e^{-y}(-y') = 0$$

$$(e^{-y} \sin x - e^x \sin y)y' = e^{-y} \cos x - e^x \cos y$$

and so

$$y' = \frac{e^{-y}\cos x - e^x \cos y}{e^{-y}\sin x - e^x \sin y}$$ ■

If we wish to find $dy/dx$ at a specific point $(a, b)$ on the graph of a function defined implicitly by an equation, we need not find a general expression for $dy/dx$, as illustrated in Example 4.

**EXAMPLE 4** Find $\dfrac{dy}{dx}$ at the point $\left(\frac{\pi}{2}, \pi\right)$ if $x \sin y - y \cos 2x = 2x$.

**Solution** Differentiating both sides of the equation with respect to $x$, we obtain

$$\frac{d}{dx}(x \sin y - y \cos 2x) = \frac{d}{dx}(2x)$$

$$\frac{d}{dx}(x \sin y) - \frac{d}{dx}(y \cos 2x) = 2$$

Using the Product Rule on each term on the left, we have

$$x\frac{d}{dx}(\sin y) + (\sin y)\frac{d}{dx}(x) - y\frac{d}{dx}(\cos 2x) - (\cos 2x)\frac{d}{dx}(y) = 2$$

Next, using the Chain Rule on the first, third, and fourth terms on the left, we obtain

$$(x \cos y)\frac{dy}{dx} + \sin y - y(-\sin 2x)\frac{d}{dx}(2x) - (\cos 2x)\frac{dy}{dx} = 2$$

or

$$(x \cos y)\frac{dy}{dx} + \sin y + 2y \sin 2x - (\cos 2x)\frac{dy}{dx} = 2$$

Replacing $x$ by $\pi/2$ and $y$ by $\pi$ in the last equation gives

$$\left(\frac{\pi}{2}\cos \pi\right)\frac{dy}{dx} + \sin \pi + 2\pi \sin \pi - (\cos \pi)\frac{dy}{dx} = 2$$

$$-\frac{\pi}{2} \cdot \frac{dy}{dx} + \frac{dy}{dx} = 2$$

or

$$\frac{dy}{dx} = \frac{2}{1 - \dfrac{\pi}{2}} = \frac{4}{2 - \pi}$$ ■

**EXAMPLE 5** Find an equation of the tangent line to the bifolium

$$4x^4 + 8x^2y^2 - 25x^2y + 4y^4 = 0$$

at the point $(2, 1)$.

**Solution**    The slope of the tangent line to the bifolium at any point $(x, y)$ is given by $dy/dx$. To compute $dy/dx$, we differentiate both sides of the equation with respect to $x$ to obtain

$$\frac{d}{dx}(4x^4 + 8x^2y^2 - 25x^2y + 4y^4) = \frac{d}{dx}(0)$$

$$\frac{d}{dx}(4x^4) + \frac{d}{dx}(8x^2y^2) - \frac{d}{dx}(25x^2y) + \frac{d}{dx}(4y^4) = 0$$

Using the Product Rule on the second and third terms on the left, we find

$$16x^3 + 8x^2\frac{d}{dx}(y^2) + y^2\frac{d}{dx}(8x^2) - 25x^2\frac{d}{dx}(y) - y\frac{d}{dx}(25x^2) + \frac{d}{dx}(4y^4) = 0$$

With the aid of the Chain Rule, we obtain

$$16x^3 + 16x^2y\frac{dy}{dx} + 16xy^2 - 25x^2\frac{dy}{dx} - 50xy + 16y^3\frac{dy}{dx} = 0$$

By substituting $x = 2$ and $y = 1$ into the last equation, we obtain

$$16(8) + 16(4)\frac{dy}{dx} + 32 - 25(4)\frac{dy}{dx} - 100 + 16\frac{dy}{dx} = 0$$

or

$$\frac{dy}{dx} = 3$$

Using the slope-intercept form for an equation of a line, we see that an equation of the tangent line is

$$y - 1 = 3(x - 2) \qquad \text{or} \qquad y = 3x - 5$$

(See Figure 4.)    ■

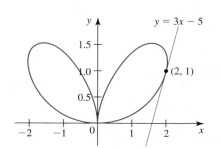

**FIGURE 4**

The graph of

$4x^4 + 8x^2y^2 - 25x^2y + 4y^4 = 0$

The slope of the curve at the point $(2, 1)$ is

$$\left.\frac{dy}{dx}\right|_{(2,\,1)} = 3$$

## ■ Derivatives of Inverse Functions

Because of the *reflective property* of inverse functions, we might expect that $f$ and $f^{-1}$ have similar properties. More specifically, we might expect that if $f$ is differentiable, then so is $f^{-1}$. The next theorem shows us how to compute the derivative of an inverse function, assuming that it exists.

---

**THEOREM 1    The Derivative of an Inverse Function**

Let $f$ be differentiable on its domain and have an inverse function $g = f^{-1}$. Then the derivative of $g$ is given by

$$g'(x) = \frac{1}{f'[g(x)]} \tag{4}$$

provided that $f'[g(x)] \neq 0$.

---

The proof of Theorem 1 is given in Appendix B.

**Note** If we write $y = f^{-1}(x) = g(x)$, then $x = f(y)$, and we can write Equation (4) in the form

$$\frac{dy}{dx} = \frac{1}{\frac{dx}{dy}} \tag{5}$$

**EXAMPLE 6** Let $f(x) = x^2$ for $x$ in $[0, \infty)$.

**a.** Show that the point $(2, 4)$ lies on the graph of $f$.
**b.** Find $g'(4)$, where $g$ is the inverse of $f$.

**Solution**
**a.** Since $f(2) = 4$, we conclude that the point $(2, 4)$ does lie on the graph of $f$.
**b.** Since $f'(x) = 2x$, Equation (4) gives

$$g'(4) = \frac{1}{f'[g(4)]} = \frac{1}{f'(2)} = \frac{1}{2x}\Big|_{x=2} = \frac{1}{2(2)} = \frac{1}{4}$$

## ■ Derivatives of Inverse Trigonometric Functions

The rules for differentiating the inverse trigonometric functions follow. Here, $u = g(x)$ is a differentiable function of $x$.

---

**Derivatives of Inverse Trigonometric Functions**

$$\frac{d}{dx}(\sin^{-1} u) = \frac{1}{\sqrt{1 - u^2}} \frac{du}{dx} \qquad \frac{d}{dx}(\csc^{-1} u) = -\frac{1}{|u|\sqrt{u^2 - 1}} \frac{du}{dx}$$

$$\frac{d}{dx}(\cos^{-1} u) = -\frac{1}{\sqrt{1 - u^2}} \frac{du}{dx} \qquad \frac{d}{dx}(\sec^{-1} u) = \frac{1}{|u|\sqrt{u^2 - 1}} \frac{du}{dx}$$

$$\frac{d}{dx}(\tan^{-1} u) = \frac{1}{1 + u^2} \frac{du}{dx} \qquad \frac{d}{dx}(\cot^{-1} u) = -\frac{1}{1 + u^2} \frac{du}{dx}$$

---

**PROOF** We will prove the first of these formulas and leave the proofs of the others as an exercise. Let $y = \sin^{-1} x$ so that $\sin y = x$ for $-\frac{\pi}{2} \le y \le \frac{\pi}{2}$. Differentiating the latter equation implicitly with respect to $x$, we obtain

$$(\cos y) \frac{dy}{dx} = 1$$

or

$$\frac{dy}{dx} = \frac{1}{\cos y}$$

Now $\cos y \ge 0$, since $-\frac{\pi}{2} \le y \le \frac{\pi}{2}$, so we can write

$$\cos y = \sqrt{1 - \sin^2 y} = \sqrt{1 - x^2} \qquad \text{Recall that } x = \sin y.$$

Therefore,

$$\frac{dy}{dx} = \frac{1}{\cos y} = \frac{1}{\sqrt{1-x^2}} \qquad -1 < x < 1$$

Finally, if $u$ is a differentiable function of $x$, then the Chain Rule gives

$$\frac{d}{dx} \sin^{-1} u = \frac{1}{\sqrt{1-u^2}} \frac{du}{dx}$$

∎

**EXAMPLE 7** Find the derivative of

**a.** $f(x) = \cos^{-1} 3x$

**b.** $g(x) = \tan^{-1} \sqrt{2x+3}$

**c.** $y = \sec^{-1} e^{-2x}$

**Solution**

**a.** $f'(x) = \dfrac{d}{dx} \cos^{-1} 3x$ $\qquad u = 3x$

$$= -\frac{1}{\sqrt{1-(3x)^2}} \cdot \frac{d}{dx}(3x) = -\frac{3}{\sqrt{1-9x^2}}$$

**b.** $g'(x) = \dfrac{d}{dx} \tan^{-1}(2x+3)^{1/2}$ $\qquad u = (2x+3)^{1/2}$

$$= \frac{1}{1 + [(2x+3)^{1/2}]^2} \cdot \frac{d}{dx}(2x+3)^{1/2}$$

$$= \frac{1}{1 + 2x + 3} \cdot \frac{1}{2}(2x+3)^{-1/2} \frac{d}{dx}(2x)$$

$$= \frac{1}{2(x+2)\sqrt{2x+3}}$$

**c.** $\dfrac{dy}{dx} = \dfrac{d}{dx} \sec^{-1} e^{-2x}$ $\qquad u = e^{-2x}$

$$= \frac{1}{e^{-2x}\sqrt{(e^{-2x})^2 - 1}} \frac{d}{dx} e^{-2x}$$

$$= \frac{-2e^{-2x}}{e^{-2x}\sqrt{e^{-4x} - 1}} = -\frac{2}{\sqrt{e^{-4x} - 1}}$$

∎

**EXAMPLE 8** **Videographing a Moving Boat** A boat is cruising at a constant speed of 20 ft/sec along a course that is parallel to a straight shoreline and 100 ft from it. A spectator standing on the shore begins to videograph the boat as soon as it passes him. Let $\theta(t)$ denote the angle of the spectator's camera at time $t$, where $t$ is measured in seconds and $t = 0$ corresponds to the time that the boat passes him (see Figure 5).

**a.** Find an expression for $\theta$ as a function of $t$.

**b.** Find the rate at which the videographer must rotate his camera in order to keep the boat in frame.

**Solution**

**a.** Referring to Figure 5, we find

$$\theta(t) = \tan^{-1}\left(\frac{20t}{100}\right)$$

$$= \tan^{-1}\left(\frac{t}{5}\right)$$

**b.** In order to keep the boat in frame, the spectator must rotate his camera at the rate given by

$$\frac{d\theta}{dt} = \frac{d}{dt}\tan^{-1}\left(\frac{t}{5}\right)$$

$$= \frac{\frac{d}{dt}\left(\frac{t}{5}\right)}{1 + \left(\frac{t}{5}\right)^2} = \frac{\frac{1}{5}}{1 + \frac{t^2}{25}}$$

$$= \frac{5}{25 + t^2}$$

radians/sec.

## Derivatives of Rational Powers of $x$

In Section 2.3 we proved that

$$\frac{d}{dx}(x^n) = nx^{n-1}$$

for integral values of $n$. Using implicit differentiation, we can now prove that this formula holds for rational powers of $x$. Thus, if $r$ is a rational number, then

$$\frac{d}{dx}(x^r) = rx^{r-1}$$

**PROOF**   Let $y = x^r$. Since $r$ is a rational number, it can be written in the form $r = m/n$, where $m$ and $n$ are integers with $n \neq 0$. Thus,

$$y = x^{m/n}$$

or

$$y^n = x^m$$

Using the Chain Rule to differentiate both sides of this equation with respect to $x$, we obtain

$$\frac{d}{dx}(y^n) = \frac{d}{dx}(x^m)$$

$$ny^{n-1}\frac{dy}{dx} = mx^{m-1}$$

$$\frac{dy}{dx} = \frac{m}{n}x^{m-1}y^{-n+1}$$

$$= \frac{m}{n}x^{m-1}(x^{m/n})^{-n+1} \qquad \text{Replace } y \text{ by } x^{m/n}.$$

$$= \frac{m}{n}x^{m-1}x^{-m+(m/n)}$$

$$= \frac{m}{n}x^{m-1-m+(m/n)}$$

$$= \frac{m}{n}x^{(m/n)-1} = rx^{r-1} \qquad \text{Replace } \frac{m}{n} \text{ by } r.$$ ∎

## 2.7    CONCEPT QUESTIONS

1. **a.** Suppose that the equation $F(x, y) = 0$ defines $y$ as a function of $x$. Explain how implicit differentiation can be used to find $dy/dx$.
   **b.** What is the role of the Chain Rule in implicit differentiation?

2. Suppose that the equation $x\, g(y) + y\, f(x) = 0$, where $f$ and $g$ are differentiable functions, defines $y$ as a function of $x$. Find an expression for $dy/dx$.

3. Write the derivative with respect to $x$ of (a) $\sin^{-1} u$, (b) $\cos^{-1} u$, and (c) $\tan^{-1} u$.

## 2.7    EXERCISES

*In Exercises 1–22, find dy/dx by implicit differentiation.*

**1.** $2x^2 + y^2 = 4$

**2.** $y^2 - 3y = 2x$

**3.** $xy^2 + yx^2 - 2 = 0$

**4.** $x^2y + 2xy^2 - x + 3 = 0$

**5.** $x^3 - 2y^3 - y = x + 2$

**6.** $x^3y^2 - 2x^2y + 2x = 3$

**7.** $\dfrac{1}{x} + \dfrac{1}{y} = 1$

**8.** $\dfrac{x^3}{y} + \dfrac{y^2}{x^2} = 3$

**9.** $\dfrac{xy}{x^2 + y^2} = x + 1$

**10.** $\dfrac{x + y}{x - y} = y^2 + 1$

**11.** $(x + 1)^2 + (y - 2)^2 = 9$

**12.** $(2x^2 + 3y^2)^{5/2} = x$

**13.** $\sqrt{x} + \sqrt{y} = 1$

**14.** $\sqrt{xy} = x^2 + 2y^2$

**15.** $y^2 = \sin(x + y)$

**16.** $x + y^2 = \cos xy$

**17.** $\tan^2(x^3 + y^3) = xy$

**18.** $x = \sec 2y$

**19.** $\sqrt{1 + \cos^2 y} = xy$

**20.** $x + y^2 = \cot xy$

**21.** $xe^{2y} - x^3 + 2y = 5$

**22.** $e^{xy} - x^2 + y^2 = 5$

*In Exercises 23–26, use implicit differentiation to find an equation of the tangent line to the curve at the indicated point.*

**23.** $x^2 + 4y^2 = 4;\quad \left(1, \frac{-\sqrt{3}}{2}\right)$    **24.** $x^2y + y^3 = 2;\quad (-1, 1)$

**25.** $x^{2/3} + y^{2/3} = 2;\quad (1, -1)$    **26.** $y = \sin xy;\quad \left(\frac{\pi}{2}, 1\right)$

*In Exercises 27–30, find the rate of change of y with respect to x at the given values of x and y.*

**27.** $xy^2 - x^2y - 2 = 0;\quad x = 1, y = -1$

**28.** $x^{2/3} + y^{2/3} = 5;\quad x = 1, y = 8$

**29.** $x \csc y = 2;\quad x = 1, y = \dfrac{\pi}{6}$

**30.** $\tan(x + 2y) - \sin x = 1;\quad x = 0, y = \dfrac{\pi}{8}$

**31.** Find an equation of the tangent line to the curve $e^y + xy = e$ at $(0, 1)$.

**32.** Find an equation of the tangent line to the curve $xe^y + 2x + y = 3$ at $(1, 0)$.

*In Exercises 33–36, find $d^2y/dx^2$ in terms of x and y.*

**33.** $xy + x^3 = 4$

**34.** $x^3 - y^3 = 8$

**35.** $\sin x + \cos y = 1$

**36.** $\tan y - xy = 0$

**37.** Suppose that $f(x) = x^2$ for x in $[0, \infty)$, and let g be the inverse of f.
   **a.** Compute $g'(x)$ using Equation (4).
   **b.** Find $g'(x)$ by first computing $g(x)$.

**38.** Let $f(x) = x^{1/3}$, and let g be the inverse of f.
   **a.** Find $g'(x)$ using Equation (4).
   **b.** Find $g'(x)$ by first computing $g(x)$.

*In Exercises 39–48, let g denote the inverse of the function f.*
*(a) Show that the point (a, b) lies on the graph of f. (b) Find $g'(b)$.*

**39.** $f(x) = 2x + 1; \quad (2, 5)$

**40.** $f(x) = x^3 + x + 2; \quad (1, 4)$

**41.** $f(x) = x^5 + 2x^3 + x - 1; \quad (0, -1)$

**42.** $f(x) = \dfrac{x + 1}{2x - 1}; \quad (1, 2)$

**43.** $f(x) = (x^3 + 1)^3; \quad (1, 8)$

**44.** $f(x) = 2 - \sqrt[3]{x + 1}; \quad (7, 0)$

**45.** $f(x) = \dfrac{1}{1 + x^2}$, where $x \geq 0; \quad \left(2, \frac{1}{5}\right)$

**46.** $f(x) = \dfrac{1}{\sqrt{x^2 + 1}}$, where $x \geq 0; \quad \left(1, \frac{\sqrt{2}}{2}\right)$

**47.** Suppose that g is the inverse of a function f. If $f(2) = 4$ and $f'(2) = 3$, find $g'(4)$.

**48.** Suppose that g is the inverse of a differentiable function f and $H = g \circ g$. If $f(4) = 3$, $g(4) = 5$, $f'(4) = \frac{1}{2}$, and $f'(5) = 2$, find $H'(3)$.

*In Exercises 49–72, find the derivative of the function.*

**49.** $f(x) = \sin^{-1} 3x$

**50.** $g(x) = \cos^{-1}(2x - 1)$

**51.** $f(x) = \tan^{-1} x^2$

**52.** $f(t) = \sin^{-1} \sqrt{2t + 1}$

**53.** $g(t) = t \tan^{-1} 3t$

**54.** $y = \sin^{-1}\left(\dfrac{1}{x}\right)$

**55.** $f(u) = \sec^{-1} 2u$

**56.** $g(\theta) = \dfrac{\sec^{-1} \theta}{\theta}$

**57.** $h(x) = \sin^{-1} x + 2 \cos^{-1} x$

**58.** $f(x) = \sin^{-1} 2x + \cos^{-1} 3x$

**59.** $g(x) = \tan^{-1} x + x \cot^{-1} x$

**60.** $y = \sec^{-1} x + \csc^{-1} x$

**61.** $y = (x^2 + 1) \tan^{-1} x$

**62.** $f(x) = \tan^{-1} \sqrt{3x + 1}$

**63.** $g(t) = \tan^{-1}\left(\dfrac{t - 1}{t + 1}\right)$

**64.** $f(x) = \cos^{-1}(\sin 2x)$

**65.** $y = \tan^{-1}(\sin 2x)$

**66.** $h(\theta) = \tan^{-1}\left(\dfrac{\cos \theta}{2}\right)$

**67.** $f(x) = \sin^{-1}(e^{2x})$

**68.** $y = e^{\tan^{-1} 2t}$

**69.** $h(x) = \cot(\cos^{-1} x^2)$

**70.** $y = \sin^{-1}\left(\dfrac{\sin x}{1 + \cos x}\right)$

**71.** $f(\theta) = (\sec^{-1} \theta)^{-1}$

**72.** $f(x) = x \tan x \sec^{-1} x$

*In Exercises 73–78, find an equation of the tangent line to the given curve at the indicated point.*

**73.** $\dfrac{x^2}{4} + \dfrac{y^2}{9} = 1; \quad \left(-1, \frac{3\sqrt{3}}{2}\right)$

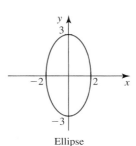

Ellipse

**74.** $\dfrac{x^2}{9} - \dfrac{y^2}{4} = 1; \quad \left(5, \frac{8}{3}\right)$

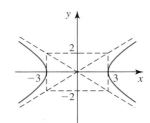

Hyperbola

**75.** $y^2 - xy^2 - x^3 = 0; \quad \left(\frac{1}{2}, \frac{1}{2}\right)$

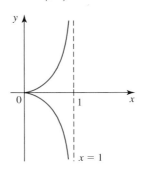

Cissoid of Diocles

**76.** $2y^2 - x^3 - x^2 = 0$;   (1, 1)

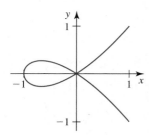

Tschirnhausen's cubic

**77.** $2(x^2 + y^2)^2 = 25(x^2 - y^2)$;   (3, 1)

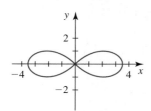

Lemniscate

**78.** $x^2 y^2 = (y + 1)^2(4 - y^2)$;   $(-2\sqrt{3}, 1)$

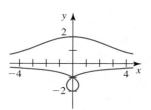

The Conchoid of Nicomedes

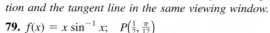 *In Exercises 79 and 80, find an equation of the tangent line to the graph of the function at the indicated point. Graph the function and the tangent line in the same viewing window.*

**79.** $f(x) = x \sin^{-1} x$;   $P\left(\frac{1}{2}, \frac{\pi}{12}\right)$

**80.** $f(x) = \sec^{-1} 2x$;   $P\left(\frac{\sqrt{2}}{2}, \frac{\pi}{4}\right)$

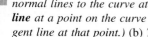 *In Exercises 81–84, (a) find the equations of the tangent and the normal lines to the curve at the indicated point. (The **normal line** at a point on the curve is the line perpendicular to the tangent line at that point.) (b) Then use a graphing utility to plot the curve and the tangent and normal lines on the same screen.*

**81.** $4xy - 9 = 0$;   $\left(3, \frac{3}{4}\right)$

**82.** $x^2 + y^2 = 9$;   $(-1, 2\sqrt{2})$

**83.** $4x^3 - 3xy^2 - 5xy - 8y^2 + 9x = -38$;   $(-2, 3)$

**84.** $x^5 - 2xy + y^5 = 0$;   (1, 1)

**85.** The graph of the equation $x^3 + y^3 = 3xy$ is called the **folium of Descartes.**

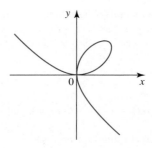

**a.** Find $y'$.

**b.** Find an equation of the tangent line to the folium at the point in the first quadrant where it intersects the line $y = x$.

**c.** Find the points on the folium where the tangent line is horizontal.

**86.** The curve with equation $x^{2/3} + y^{2/3} = 4$ is called an astroid. Find an equation of the tangent line to the curve at the point $(3\sqrt{3}, 1)$.

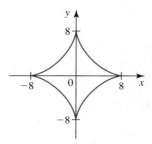

**87.** Water flows from a tank of constant cross-sectional area 50 ft$^2$ through an orifice of constant cross-sectional area $\frac{1}{4}$ ft$^2$ located at the bottom of the tank. Initially, the height of the water in the tank was 20 ft, and $t$ sec later it was given by the equation

$$2\sqrt{h} + \frac{1}{25}t - 2\sqrt{20} = 0 \qquad 0 \le t \le 50\sqrt{20}$$

How fast was the height of the water decreasing when its height was 9 ft?

**88. Watching a Rocket Launch** At a distance of 2000 ft from the launch site, a spectator is observing a rocket 120-ft long being launched vertically. Let $\theta$ be her viewing angle of the rocket, and let $y$ denote the altitude (measured in feet) of the rocket. (Neglect the height of the spectator.)

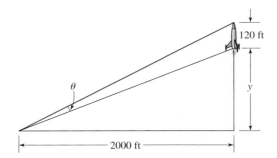

120 ft

$\theta$

$y$

2000 ft

**a.** Show that

$$\tan \theta = \frac{240{,}000}{y^2 + 120y + 4{,}000{,}000}$$

**b.** What is the viewing angle when the rocket is on the launching pad? When it is at an altitude of 10,000 feet?

**c.** Find the rate of change of the viewing angle when the rocket is at an altitude of 10,000 feet.

**d.** What happens to the viewing angle when the rocket is at a very great altitude?

*Two curves are said to be **orthogonal** if their tangent lines are perpendicular at each point of intersection of the curves. In Exercises 89–92, show that the curves with the given equations are orthogonal.*

**89.** $x^2 + 2y^2 = 6$, $\quad x^2 = 4y$

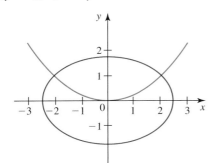

**90.** $x^2 - y^2 = 3$, $\quad xy = 2$

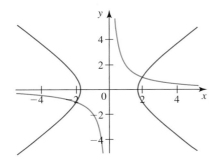

**91.** $x^2 + 3y^2 = 4$, $\quad y = x^3$

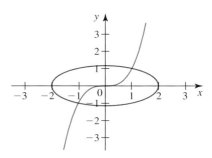

**92.** $y - x = \dfrac{\pi}{2}$, $\quad x = \cos y$

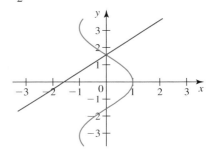

*Two families of curves are **orthogonal trajectories** of each other if every curve of one family is orthogonal to every curve in the other family. In Exercises 93–96, (a) show that the given families of curves are orthogonal to each other, and (b) sketch a few members of each family on the same set of axes.*

**93.** $x^2 + y^2 = c^2$, $\quad y = kx$, $\quad c, k$ constants

**94.** $x^2 + y^2 = cx$, $\quad x^2 + y^2 = ky$, $\quad c, k$ constants

**95.** $2x^2 + y^2 = c$, $\quad y^2 = kx$, $\quad c, k$ constants

**96.** $9x^2 + 4y^2 = c^2$, $\quad y^9 = kx^4$, $\quad c, k$ constants

**97. The Path of Steepest Descent** The contour lines of a **topographic** or **contour map** are curves that connect the contiguous points of the same altitude. The figure gives the contour map of a hill. Suppose that you start at the point $A$ and you want to get to the point $B$ by taking the shortest path.

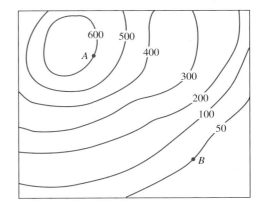

600  500

$A$

400

300

200

100

50

$B$

**a.** Explain why the direction that you start out with at $A$ should be perpendicular to the tangent line to the contour line passing through $A$.

**b.** Using the observation made in part (a), explain why the desired path should be the curve that is orthogonal to the contour lines. Sketch this path from $A$ to $B$. This path is called the *path of steepest descent*.

**98. Isobars** are curves on a weather map that connect points having the same air pressure. The figure shows a family of isobars.

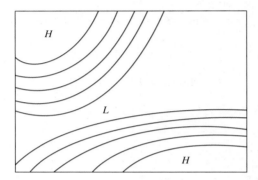

**a.** Sketch several members of the family of orthogonal trajectories of the family of isobars.

**b.** Use the fact that air flows from regions of high air pressure to those of lower air pressure to give an interpretation of the role of the orthogonal family.

**99.** A 20-ft ladder leaning against a wall begins to slide. How fast is the angle between the ladder and the wall changing at the instant of time when the bottom of the ladder is 12 ft from the wall and sliding away from the wall at the rate of 5 ft/sec?

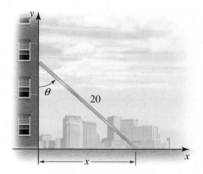

**100.** A trough of length $L$ feet has a cross section in the shape of a semicircle with radius $r$ feet. When the trough is filled with water to a level that is $h$ feet as measured from the top of the trough, the volume of the water is

$$V = L\left[\frac{1}{2}\pi r^2 - r^2\sin^{-1}\left(\frac{h}{r}\right) - h\sqrt{r^2 - h^2}\right]$$

Suppose that a trough with $L = 10$ and $r = 1$ springs a leak at the bottom and that at a certain instant of time, $h = 0.4$ ft and $dV/dt = -0.2$ ft$^3$/sec. Find the rate at which $h$ is changing at that instant of time.

**101.** Verify each differentiation formula.

**a.** $\dfrac{d}{dx}\cos^{-1}u = -\dfrac{1}{\sqrt{1 - u^2}}\dfrac{du}{dx}$

**b.** $\dfrac{d}{dx}\tan^{-1}u = \dfrac{1}{1 + u^2}\dfrac{du}{dx}$

**c.** $\dfrac{d}{dx}\csc^{-1}u = -\dfrac{1}{|u|\sqrt{u^2 - 1}}\dfrac{du}{dx}$

**d.** $\dfrac{d}{dx}\sec^{-1}u = \dfrac{1}{|u|\sqrt{u^2 - 1}}\dfrac{du}{dx}$

**e.** $\dfrac{d}{dx}\cot^{-1}u = -\dfrac{1}{1 + u^2}\dfrac{du}{dx}$

*In Exercises 102–104, determine whether the statement is true or false. If it is true, explain why it is true. If it is false, explain why or give an example to show why it is false.*

**102.** If $f$ and $g$ are differentiable and $f(x)g(y) = 0$, then

$$\frac{dy}{dx} = -\frac{f'(x)g(y)}{f(x)g'(y)} \qquad f(x) \neq 0 \quad \text{and} \quad g'(y) \neq 0$$

**103.** If $f$ and $g$ are differentiable and $f(x) + g(y) = 0$, then

$$\frac{dy}{dx} = -\frac{f'(x)}{g'(y)}$$

**104.** $\dfrac{d}{dx}[\cos^{-1}(\cos x)] = 1$ for all $x$ in $(0, \pi)$.

## 2.8 Derivatives of Logarithmic Functions

### ■ The Derivatives of Logarithmic Functions

The method of differentiation developed in Section 2.7 can be used to help us find the derivatives of logarithmic functions. The fact that logarithmic functions are differentiable can be demonstrated with mathematical rigor, but we will not do so here. However, if we recall that the graph of $y = \log_a x$ is the reflection of the graph of $y = a^x$ with respect to the line $y = x$, then it seems plausible that the differentiability of $a^x$ would imply the differentiability of $\log_a x$.

To find the derivative of $y = \log_a x$, where $a > 0$, $a \neq 1$, we first recall that the equation is equivalent to

$$a^y = x$$

Differentiating the last equation implicitly with respect to $x$ yields

$$a^y (\ln a) \frac{dy}{dx} = 1$$

so

$$\frac{dy}{dx} = \frac{1}{a^y \ln a} = \frac{1}{x \ln a}$$

provided that the derivative exists. If we put $a = e$, we obtain

$$\frac{dy}{dx} = \frac{1}{e^y \ln e} = \frac{1}{x}$$

Also, using the Chain Rule, we find that if $u$ is a differentiable function of $x$, then

$$\frac{d}{dx} \ln u = \frac{1}{u} \frac{du}{dx} \qquad \text{and} \qquad \frac{d}{dx} \log_a u = \frac{1}{u \ln a} \frac{du}{dx}$$

Let's summarize these results.

---

**THEOREM 1 The Derivatives of Logarithmic Functions**

Let $u$ be a differentiable function of $x$, and let $a > 0$, where $a \neq 1$. Then

**a.** $\dfrac{d}{dx} \ln x = \dfrac{1}{x}$ 

**b.** $\dfrac{d}{dx} \ln u = \dfrac{1}{u} \dfrac{du}{dx}$

**c.** $\dfrac{d}{dx} \log_a x = \dfrac{1}{x \ln a}$ 

**d.** $\dfrac{d}{dx} \log_a u = \dfrac{1}{u \ln a} \dfrac{du}{dx}$

---

**EXAMPLE 1** Find the derivative of

**a.** $f(x) = \ln(2x^2 + 1)$     **b.** $g(x) = x^2 \log 2x$     **c.** $y = \ln \cos x$

**Solution**

**a.** $f'(x) = \dfrac{d}{dx} \ln(2x^2 + 1) = \dfrac{1}{2x^2 + 1} \dfrac{d}{dx}(2x^2 + 1) = \dfrac{4x}{2x^2 + 1}$

**b.** $g'(x) = \dfrac{d}{dx}(x^2 \log 2x) = x^2 \dfrac{d}{dx}(\log 2x) + (\log 2x)\dfrac{d}{dx}(x^2)$   Use the Product Rule.

$$= x^2\left[\frac{1}{(2x)\ln 10}\right]\frac{d}{dx}(2x) + (\log 2x)(2x) = x\left(\frac{1}{\ln 10} + 2\log 2x\right)$$

**c.** $\dfrac{dy}{dx} = \dfrac{d}{dx}\ln\cos x = \dfrac{1}{\cos x}\dfrac{d}{dx}(\cos x) = -\dfrac{\sin x}{\cos x} = -\tan x$   ∎

**EXAMPLE 2**  Find the derivative of $y = \ln(e^{2x} + e^{-2x})$

**Solution**  Using the rule for differentiating a logarithmic function gives

$$\frac{dy}{dx} = \frac{d}{dx}\ln(e^{2x} + e^{-2x})$$

$$= \frac{1}{e^{2x} + e^{-2x}}\frac{d}{dx}(e^{2x} + e^{-2x})$$

$$= \frac{1}{e^{2x} + e^{-2x}}(2e^{2x} - 2e^{-2x})$$

$$= \frac{2(e^{2x} - e^{-2x})}{e^{2x} + e^{-2x}}$$   ∎

If an expression contains a logarithm, it may be helpful to use the laws of logarithms to simplify the expression *before* differentiating, as illustrated in Examples 3 and 4.

**EXAMPLE 3**  Find the derivative of $f(x) = \ln\sqrt{x^2 + 1}$.

**Solution**  We first rewrite the given expression as

$$f(x) = \ln(x^2 + 1)^{1/2} = \frac{1}{2}\ln(x^2 + 1)$$

Differentiating this function, we obtain

$$f'(x) = \frac{d}{dx}\left[\frac{1}{2}\ln(x^2 + 1)\right] = \frac{1}{2}\frac{d}{dx}[\ln(x^2 + 1)]$$

$$= \frac{1}{2}\cdot\frac{1}{x^2 + 1}\frac{d}{dx}(x^2 + 1) = \frac{1}{2}\cdot\frac{1}{x^2 + 1}(2x) = \frac{x}{x^2 + 1}$$   ∎

**EXAMPLE 4**  Find the rate of change of

$$f(x) = \ln\left[\frac{x^2(2x^2 + 1)^3}{\sqrt{5 - x^2}}\right]$$

when $x = 1$.

**Solution**  The rate of change of $f(x)$ for any value of $x$ is given by $f'(x)$. To find $f'(x)$, we first rewrite

$$f(x) = \ln\left[\frac{x^2(2x^2 + 1)^3}{(5 - x^2)^{1/2}}\right] = 2\ln x + 3\ln(2x^2 + 1) - \frac{1}{2}\ln(5 - x^2)$$

Then we have

$$f'(x) = \frac{2}{x} + \frac{3}{2x^2 + 1}\frac{d}{dx}(2x^2 + 1) - \frac{1}{2(5 - x^2)}\frac{d}{dx}(5 - x^2)$$

$$= \frac{2}{x} + \frac{12x}{2x^2 + 1} + \frac{x}{5 - x^2}$$

from which we see that the rate of change of $f(x)$ at $x = 1$ is

$$f'(1) = 2 + \frac{12}{3} + \frac{1}{4}$$

or $\frac{25}{4}$ units per unit change in $x$. ∎

## ■ Logarithmic Differentiation

Having seen how the laws of logarithms can help simplify the work involved in differentiating logarithmic expressions, we now look at a procedure that takes advantage of these same laws to help us differentiate functions that at first blush do not necessarily involve logarithms. This method, called **logarithmic differentiation,** is especially useful for differentiating functions involving products, quotients, and/or powers that can be simplified by using logarithms.

**EXAMPLE 5**  Find the derivative of $y = \dfrac{(2x - 1)^3}{\sqrt{3x + 1}}$.

**Solution**  We begin by taking the logarithm on both sides of the equation, getting

$$\ln y = \ln \frac{(2x - 1)^3}{(3x + 1)^{1/2}}$$

or

$$\ln y = 3 \ln(2x - 1) - \frac{1}{2} \ln(3x + 1) \qquad \text{Use the laws of logarithms.}$$

Next, we differentiate implicitly with respect to $x$, obtaining

$$\frac{1}{y}(y') = \frac{3}{2x - 1}(2) - \frac{1}{2(3x + 1)}(3)$$

$$= \frac{6}{2x - 1} - \frac{3}{2(3x + 1)} = \frac{6(2)(3x + 1) - 3(2x - 1)}{2(2x - 1)(3x + 1)}$$

$$= \frac{15(2x + 1)}{2(2x - 1)(3x + 1)}$$

Multiplying both sides of this equation by $y$ gives

$$y' = \frac{15(2x + 1)}{2(2x - 1)(3x + 1)} \cdot y$$

$$= \frac{15(2x + 1)}{2(2x - 1)(3x + 1)} \cdot \frac{(2x - 1)^3}{(3x + 1)^{1/2}} \qquad \text{Substitute for } y.$$

$$= \frac{15(2x + 1)(2x - 1)^2}{2(3x + 1)^{3/2}}$$  ∎

Here is a summary of this procedure.

---

**Finding $dy/dx$ by Logarithmic Differentiation**

Suppose we are given the equation $y = f(x)$. To compute $dy/dx$:

1. Take the logarithm of both sides of the equation, and use the laws of logarithms to simplify the resulting equation.
2. Differentiate implicitly with respect to $x$.
3. Solve the equation found in Step 2 for $dy/dx$.
4. Substitute for $y$.

---

## ■ The General Version of the Power Rule

As was promised in Section 2.2, we will now prove that the Power Rule holds for all exponents (Theorem 3). But before we prove this, we need the following result.

If $f(x) = \ln|x|$, where $x \neq 0$, then

$$f'(x) = \frac{d}{dx} \ln|x| = \frac{1}{x} \tag{1}$$

**PROOF**  We have

$$f(x) = \ln|x| = \begin{cases} \ln x & \text{if } x > 0 \\ \ln(-x) & \text{if } x < 0 \end{cases}$$

So

$$f'(x) = \begin{cases} \dfrac{1}{x} & \text{if } x > 0 \\ \dfrac{-1}{-x} = \dfrac{1}{x} & \text{if } x < 0 \end{cases}$$

■

We now prove the Power Rule for all real exponents. If $n$ is any real number, then

$$\frac{d}{dx}(x^n) = nx^{n-1}$$

**PROOF**  Let $y = x^n$. Then

$$\ln|y| = \ln|x|^n = n\ln|x|$$

Using the Chain Rule and Equation (1), we obtain

$$\frac{y'}{y} = \frac{n}{x}$$

or

$$y' = \frac{ny}{x} = \frac{nx^n}{x} = nx^{n-1}$$

■

### ■ The Number $e$ as a Limit

In Section 0.8 we mentioned that $e \approx 2.71828$, correct to five decimal places. We are now in the position to give the exact value of $e$, albeit in the form of a limit. If we use the definition of the derivative as a limit to compute $f'(1)$, where $f(x) = \ln x$, we obtain

$$f'(1) = \lim_{h \to 0} \frac{f(1 + h) - f(1)}{h}$$

$$= \lim_{h \to 0} \frac{\ln(1 + h) - \ln 1}{h} = \lim_{h \to 0} \frac{\ln(1 + h)}{h} \qquad \ln 1 = 0$$

$$= \lim_{h \to 0} \ln(1 + h)^{1/h}$$

$$= \ln\left[\lim_{h \to 0} (1 + h)^{1/h}\right] \qquad \text{Use the continuity of ln.}$$

But

$$f'(1) = \left[\frac{d}{dx} \ln x\right]_{x=1} = \left[\frac{1}{x}\right]_{x=1} = 1$$

so

$$\ln\left[\lim_{h \to 0} (1 + h)^{1/h}\right] = 1$$

or

$$\lim_{h \to 0} (1 + h)^{1/h} = e \qquad\qquad\qquad (2)$$

Table 1 shows that $e \approx 2.71828$, correct to five decimal places, as was mentioned earlier.

**TABLE 1**   Table of values of $(1 + x)^{1/x}$

| $x$ | $(1 + x)^{1/x}$ | $x$ | $(1 + x)^{1/x}$ |
|---|---|---|---|
| $-0.1$ | 2.867972 | 0.1 | 2.593742 |
| $-0.01$ | 2.732000 | 0.01 | 2.704814 |
| $-0.001$ | 2.719642 | 0.001 | 2.716924 |
| $-0.0001$ | 2.718418 | 0.0001 | 2.718146 |
| $-0.00001$ | 2.718295 | 0.00001 | 2.718268 |
| $-0.000001$ | 2.718283 | 0.000001 | 2.718280 |

## 2.8  CONCEPT QUESTIONS

1. State the rule for differentiating (a) $f(x) = \ln x$ and (b) $f(x) = \log_a x$.
2. Let $u$ be a differentiable function of $x$. State the rule for differentiating (a) $f(x) = \ln u$ and (b) $f(x) = \log_a u$.
3. **a.** If $f(x) = \ln|x|$, what is $f'(x)$?
   **b.** If $f(x) = \log_a|x|$, where $a > 0$, $a \neq 1$, what is $f'(x)$?
4. Give the steps used in logarithmic differentiation.
5. Give a definition of the number $e$ as a limit.

## 2.8   EXERCISES

*In Exercises 1–26, differentiate the function.*

**1.** $f(x) = \ln(2x + 3)$

**2.** $g(x) = \ln(x^2 + 4)^2$

**3.** $h(x) = \ln\sqrt{x}$

**4.** $y = \sqrt{\ln x}$

**5.** $g(u) = \ln\dfrac{u}{u + 1}$

**6.** $g(t) = t \ln 2t$

**7.** $y = x(\ln x)^2$

**8.** $f(x) = \ln\left(x + \sqrt{x^2 - 1}\right)$

**9.** $g(x) = \dfrac{\ln x}{x + 1}$

**10.** $y = \ln\left(\dfrac{x - 1}{x + 1}\right)^{2/3}$

**11.** $f(x) = \ln(\ln x)$

**12.** $h(t) = \dfrac{\ln t}{\ln 2t}$

**13.** $f(x) = \ln(x \ln x)$

**14.** $f(x) = \ln[x \ln(x + 2)]$

**15.** $g(x) = \sin(\ln x)$

**16.** $h(t) = t \sin(\ln 2t)$

**17.** $f(x) = x^2 \ln \cos x$

**18.** $g(\theta) = \ln|\tan 3\theta|$

**19.** $h(u) = \ln|\sec u|$

**20.** $f(x) = \sec[\ln(2x + 3)]$

**21.** $g(t) = \ln\left|\dfrac{\sin t + 1}{\cos t + 2}\right|$

**22.** $g(x) = \ln\sqrt{\dfrac{x \cos x}{(2x + 1)^3}}$

**23.** $f(x) = \log_2(x^2 + x + 1)$

**24.** $h(x) = \log_3|2x - 1|$

**25.** $f(t) = \log\sqrt{t^2 + 1}$

**26.** $y = x^2 \log_2\sqrt{x^2 - 1}$

**27.** Find an equation of the tangent line to the graph of $y = x \ln x$ at $(1, 0)$.

**28.** Find an equation of the tangent line to the curve $y - \ln(x^2 + y^2) = 0$ at $(1, 0)$.

*In Exercises 29–40, use logarithmic differentiation to find the derivative of the function.*

**29.** $y = (2x + 1)^2(3x^2 - 4)^3$

**30.** $y = \dfrac{x^2\sqrt{2x - 4}}{(x + 1)^2}$

**31.** $y = \sqrt[3]{\dfrac{x - 1}{x^2 + 1}}$

**32.** $y = \dfrac{\sin^2 x}{x^2\sqrt{1 + \tan x}}$

**33.** Find $y''$ if $y = x^x$.

**34.** Find $y'$ if $y = x^{x^x}$.

**35.** $y = 3^x$

**36.** $y = x^{x^2}$

**37.** $y = (x + 2)^{1/x}$

**38.** $y = (x^2 + x)^{\sqrt{x}}$

**39.** $y = (\sqrt{\cos x})^x$

**40.** $y = \sin x^{\tan x}$

*In Exercises 41–44, use implicit differentiation to find $dy/dx$.*

**41.** $\ln y - x \ln x = -1$

**42.** $\ln xy - y^2 = 5$

**43.** $\tan^{-1}\left(\dfrac{y}{x}\right) - \ln\sqrt{x^2 + y^2} = 0$

**44.** $\ln(x + y) - \cos y - x^2 = 0$

**45.** **Flight of a Rocket**   A rocket having mass $M$ kg and carrying fuel of mass $m$ kg takes off vertically from the earth's surface. The fuel is burned at the constant rate of $a$ kg/sec, and the gas is expelled at a constant velocity of $b$ m/sec relative to the rocket, where $a > 0$ and $b > 0$. If the external force acting on the rocket is a constant gravitational field, then the height of the rocket $t$ seconds after liftoff is

$$x = bt + \frac{b}{a}(M + m - at)\ln\left(\frac{M + m - at}{M + m}\right) - \frac{1}{2}gt^2$$

$$0 \le t \le \tfrac{m}{a}$$

a. Find expressions for the velocity and acceleration of the rocket at any time $t$ after liftoff.

b. What are the velocity and acceleration of the rocket at burnout (that is, when $t = m/a$).

**46.** **Distance Traveled by a Motorboat**   The distance $x$ (in feet) traveled by a motorboat moving in a straight line $t$ sec after the engine of the moving boat has been cut off is given by

$$x = \frac{1}{k}\ln(v_0kt + 1)$$

where $k$ is a constant and $v_0$ is the speed of the boat at $t = 0$.

a. Find expressions for the velocity and acceleration of the boat at any time $t$ after the engine has been cut off.

b. Show that the acceleration of the boat is in the direction opposite to that of its velocity and is directly proportional to the square of its velocity.

c. Use the results of part (a) to show that the velocity of the boat after traveling a distance of $x$ ft is given by

$$v = v_0 e^{-kx}$$

**47.** **Strain on Vertebrae**   The strain (percentage of compression) on the lumbar vertebral disks in an adult human as a function of the load $x$ (in kilograms) is given by

$$f(x) = 7.2956 \ln(0.0645012x^{0.95} + 1)$$

What is the rate of change of the strain with respect to the load when the load is 100 kg? When the load is 500 kg?
*Source:* Benedek and Villars, *Physics with Illustrative Examples from Medicine and Biology.*

**48.** **Predator-Prey Model**   The relationship between the number of rabbits $y(t)$ and the number of foxes $x(t)$ at any time $t$ is given by

$$-C \ln y + Dy = A \ln x - Bx + E$$

where $A$, $B$, $C$, $D$, and $E$ are constants. This relationship is based on a model by Lotka (1880–1949) and Volterra (1860–1940) for analyzing the ecological balance between two species of animals, one of which is a prey species and the other of which is a predator species. Use implicit differentiation to find the relationship between the rate of change of the rabbit population in terms of the rate of change of the fox population.

**49. Force Exerted by an Electric Charge** An electric charge $Q$ is distributed uniformly along a line of length $2a$, lying along the $y$-axis, as shown in the figure. A point charge $q$ lies on the $x$-axis, at a distance $x$ from the origin. It can be shown that the magnitude of the total force $F$ that $Q$ exerts on $q$ (in the direction of the $x$-axis) is $F = -q\, dV/dx$, where

$$V(x) = \frac{1}{4\pi\varepsilon_0}\frac{Q}{2a}\ln\frac{\sqrt{a^2 + x^2} + a}{\sqrt{a^2 + x^2} - a} \qquad \varepsilon_0,\text{ a constant}$$

Show that

$$F = \frac{qQ}{4\pi\varepsilon_0}\frac{1}{x\sqrt{x^2 + a^2}}$$

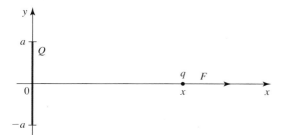

A line of charge with length $2a$ and total charge $Q$ exerts an electrostatic force on the point charge $q$.

**50. Rate of a Catalytic Chemical Reaction** A catalyst is a substance that either accelerates a chemical reaction or is necessary for the reaction to occur. Suppose that an enzyme $E$ (a catalyst) combines with a substrate $S$ (a reacting chemical) to form an intermediate product $X$, which then produces a product $P$ and releases the enzyme. If initially there are $x_0$ moles per liter of $S$ and there is no $P$, then on the basis of the theory of Michaelis and Menten, the concentration of $P$, $p(t)$, after $t$ hours is given by the equation

$$Vt = p - k\ln\left(1 - \frac{p}{x_0}\right)$$

where the constant $V$ is the maximum possible speed of the reaction and the constant $k$ is called the **Michaelis constant** for the reaction. Find the rate of change of the formation of the product $P$ in this reaction.

**51. Atmospheric Pressure** In the troposphere (lower part of the atmosphere), the atmospheric pressure $p$ is related to the height $y$ from the earth's surface by the equation

$$\ln\left(\frac{p}{p_0}\right) = \frac{Mg}{R\alpha}\ln\left(\frac{T_0 - \alpha y}{T_0}\right)$$

where $p_0$ is the pressure at the earth's surface, $T_0$ is the temperature at the earth's surface, $M$ is the molecular mass for air, $g$ is the constant of acceleration due to gravity, $R$ is the ideal gas constant, and $\alpha$ is called the lapse rate of temperature.

**a.** Find $p$ for $y = 8882$ m (the altitude at the summit of Mount Everest), taking $M = 28.8 \times 10^{-3}$ kg/mol, $T_0 = 300$ K, $g = 9.8$ m/sec$^2$, $R = 8.314$ J/mol $\cdot$ K, and $\alpha = 0.006$ K/m. Explain why mountaineers experience difficulty in breathing at very high altitudes.

**b.** Find the rate of change of the atmospheric pressure with respect to altitude when $y = 8882$ m.
**Hint:** Use logarithmic differentiation.

*In Exercises 52–56, determine whether the statement is true or false. If it is true, explain why it is true. If it is false, give an example to show why it is false.*

**52.** The function $f(x) = 1/(\ln x)$ is continuous on $(1, \infty)$.

**53.** If $f(x) = \ln 5$, then $f'(x) = \frac{1}{5}$.

**54.** $\displaystyle\lim_{x\to 0}\frac{a^x - 1}{x} = \ln a$, where $a > 0$

**55.** $\displaystyle\frac{d}{dx}\log_a\sqrt{x} = \frac{1}{(\ln a)\sqrt{x}}$

**56.** $\displaystyle\lim_{x\to 0}\frac{\log(3 + x) - \log 3}{x} = \frac{1}{3\ln 10}$

## 2.9   Related Rates

### Related Rates Problems

The following is a typical related rates problem: Suppose that $x$ and $y$ are two quantities that depend on a third quantity $t$ and that we know the relationship between $x$ and $y$ in the form of an equation. Can we find a relationship between $dx/dt$ and $dy/dt$? In particular, if we know one of the rates of change at a specific value of $t$, say, $dx/dt$, can we find the other rate, $dy/dt$, at that value of $t$?

As an example, consider this problem from the field of aviation: Suppose that $x(t)$ and $y(t)$ describe the $x$- and $y$-coordinates at time $t$ of a plane pulling out of a shallow dive (Figure 1). The flight path of the plane is described by the equation

$$y^2 - x^2 = 160{,}000 \qquad\qquad (1)$$

where $x$ and $y$ are both measured in feet.

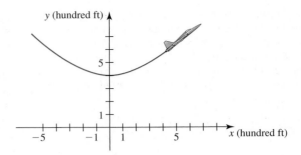

**FIGURE 1**
The flight path of a plane
pulling out of a shallow dive

Suppose that $x$ and $y$ are both differentiable functions of $t$, where $t$ is measured in seconds. Then differentiating both sides of Equation (1) implicitly with respect to $t$, we obtain

$$2y \frac{dy}{dt} - 2x \frac{dx}{dt} = 0$$

giving a relationship between the variables $x$ and $y$ and their rates of change $dx/dt$ and $dy/dt$. Now suppose that $dx/dt = 500$ at the point where $x = 300$ and $y = 500$. At that instant of time,

$$2(500) \frac{dy}{dt} - 2(300)(500) = 0$$

or $dy/dt = 300$. This says that the plane's altitude is increasing at the rate of 300 ft/sec.

## ■ Solving Related Rates Problems

In the last example we were given the relationship between $x$ and $y$ in the form of an equation. In certain related rates problems we must first identify the variables and then find a relationship between them before solving the problem. The following guidelines can be used to solve these problems.

---

**Guidelines for Solving a Related Rates Problem**

1. Draw a diagram, and label the variable quantities.
2. Write down the *given* values of the variables and their rates of change with respect to $t$.
3. Find an equation that relates the variables.
4. Differentiate both sides of this equation implicitly with respect to $t$.
5. Replace the variables and derivative in the resulting equation by the values found in Step 2, and solve this equation for the required rate of change.

---

**EXAMPLE 1    The Speed of a Rocket During Liftoff**    At a distance of 12,000 feet from the launch site, a spectator is observing a rocket being launched vertically. What is the speed of the rocket at the instant when the distance of the rocket from the spectator is 13,000 ft and is increasing at the rate of 480 ft/sec?

**Solution**

**Step 1**    Let $y =$ the altitude of the rocket and $z =$ the distance of the rocket from the spectator at any time $t$. (See Figure 2.)

**Step 2**   We are given that at a certain instant of time

$$z = 13,000 \quad \text{and} \quad \frac{dz}{dt} = 480$$

and are asked to find $dy/dt$ at that time.

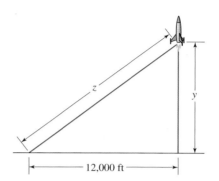

**FIGURE 2**
We want to find the speed of
the rocket when $z = 13,000$ ft
and $dz/dt = 480$ ft/sec.

**Step 3**   Applying the Pythagorean Theorem to the right triangle in Figure 2, we find
that

$$z^2 = y^2 + 12,000^2 \tag{2}$$

**Step 4**   Differentiating Equation (2) implicitly with respect to $t$, we obtain

$$2z\frac{dz}{dt} = 2y\frac{dy}{dt} \tag{3}$$

**Step 5**   Using Equation (2) we see that if $z = 13,000$, then

$$y = \sqrt{13,000^2 - 12,000^2} = 5000$$

Finally, substituting $z = 13,000$, $y = 5000$, and $dz/dt = 480$ in Equation (3),
we find

$$2(13,000)(480) = 2(5000)\frac{dy}{dt} \quad \text{and} \quad \frac{dy}{dt} = 1248$$

Therefore, the rocket is rising at the rate of 1248 ft/sec.   ■

Don't replace the variables in Equation (2) found in Step 3 by their values before
differentiating this equation. Look at Steps 3–5 in Example 1 once again, and make
sure you understand that this substitution takes place *after* the differentiation.

**EXAMPLE 2**   **Televising a Rocket Launch**   A major network is televising the launching of the rocket described in Example 1. A camera tracking the liftoff of the rocket is
located at point $A$, as shown in Figure 3, where $\phi$ denotes the angle of elevation of the
camera at $A$. When the rocket is 13,000 ft from the camera and this distance is increasing at the rate of 480 ft/sec, how fast is $\phi$ changing?

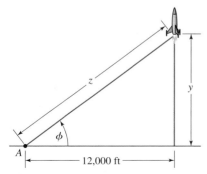

**FIGURE 3**
A television camera tracking a rocket
launch

**Solution**   We are given that at a certain instant of time,

$$z = 13,000 \quad \text{and} \quad \frac{dz}{dt} = 480$$

and are asked to find $d\phi/dt$ at that time. From Figure 3 we see that

$$\cos \phi = \frac{12{,}000}{z}$$

Differentiating this equation implicitly with respect to $t$, we obtain

$$(-\sin \phi)\frac{d\phi}{dt} = -\frac{12{,}000}{z^2} \cdot \frac{dz}{dt} \qquad \qquad \textbf{(4)}$$

Now when $z = 13{,}000$, we find that $y = 5000$ (the same value that was obtained in Example 1). Therefore, at this instant of time,

$$\sin \phi = \frac{5{,}000}{13{,}000} = \frac{5}{13}$$

Finally, substituting $z = 13{,}000$, $\sin \phi = 5/13$, and $dz/dt = 480$ into Equation (4), we obtain

$$-\frac{5}{13}\frac{d\phi}{dt} = -\frac{12{,}000}{13{,}000^2}(480)$$

from which we deduce that

$$\frac{d\phi}{dt} \approx 0.0886$$

Therefore, the angle of elevation of the camera is increasing at the rate of approximately 0.09 rad/sec, or about 5°/sec.  ■

**EXAMPLE 3**   Water is poured into a conical funnel at the constant rate of 1 in.³/sec and flows out at the rate of $\frac{1}{2}$ in.³/sec (Figure 4a). The funnel is a right circular cone with a height of 4 in. and a radius of 2 in. at the base (Figure 4b). How fast is the water level changing when the water is 2 in. high?

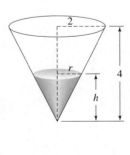

(a) Water is poured into a
    conical funnel.

(b) We want to find the rate
    at which the water level
    is rising when $h = 2$.

**FIGURE 4**

**Solution**

**Step 1**    Let

$$V = \text{the volume of the water in the funnel}$$

$$h = \text{the height of the water in the funnel}$$

and

$$r = \text{the radius of the surface of the water in the funnel}$$

at any time $t$ (in seconds).

**Step 2**    We are given that

$$\frac{dV}{dt} = 1 - \frac{1}{2} = \frac{1}{2} \qquad \text{Rate of flow in minus rate of flow out}$$

and are asked to find $dh/dt$ when $h = 2$.

**Step 3**    The volume of water in the funnel is equal to the volume of the shaded cone in Figure 4b. Thus,

$$V = \frac{1}{3} \pi r^2 h$$

but we need to express $V$ in terms of $h$ alone. To do this, we use similar triangles and deduce that

$$\frac{r}{h} = \frac{2}{4} \qquad \text{or} \qquad r = \frac{h}{2} \qquad \text{Ratio of corresponding sides}$$

Substituting this value of $r$ into the expression for $V$, we obtain

$$V = \frac{1}{3} \pi \left(\frac{h}{2}\right)^2 h = \frac{1}{12} \pi h^3$$

**Step 4**    Differentiating this last equation implicitly with respect to $t$, we obtain

$$\frac{dV}{dt} = \frac{1}{4} \pi h^2 \frac{dh}{dt}$$

**Step 5**    Finally, substituting $dV/dt = \frac{1}{2}$ and $h = 2$ into this equation gives

$$\frac{1}{2} = \frac{1}{4} \pi (2^2) \frac{dh}{dt}$$

or

$$\frac{dh}{dt} = \frac{1}{2\pi} \approx 0.159$$

and we see that the water level is rising at the rate of 0.159 in./sec.    ■

**EXAMPLE 4**    A passenger ship and an oil tanker left port sometime in the morning; the former headed north, and the latter headed east. At noon the passenger ship was 40 mi from port and moving at 30 mph, while the oil tanker was 30 mi from port and moving at 20 mph. How fast was the distance between the two ships changing at that time?

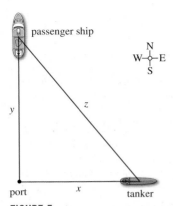

**FIGURE 5**
We want to find $dz/dt$, the rate at which the distance between the two ships is changing at a certain instant of time.

**Solution**

**Step 1**    Let

$$x = \text{the distance of the oil tanker from port}$$

$$y = \text{the distance of the passenger ship from port}$$

and

$$z = \text{the distance between the two ships}$$

(See Figure 5.)

**Step 2**    We are given that at noon,

$$x = 30, \quad y = 40, \quad \frac{dx}{dt} = 20, \quad \text{and} \quad \frac{dy}{dt} = 30$$

and we are required to find $dz/dt$ at that time.

**Step 3**    Applying the Pythagorean Theorem to the right triangle in Figure 5, we find that

$$z^2 = x^2 + y^2 \tag{5}$$

**Step 4**    Differentiating Equation (5) implicitly with respect to $t$, we obtain

$$2z\frac{dz}{dt} = 2x\frac{dx}{dt} + 2y\frac{dy}{dt}$$

or

$$z\frac{dz}{dt} = x\frac{dx}{dt} + y\frac{dy}{dt}$$

**Step 5**    Using Equation (5) with $x = 30$ and $y = 40$, we have

$$z^2 = 30^2 + 40^2 = 2500$$

or $z = 50$. Finally substituting $x = 30$, $y = 40$, $z = 50$, $dx/dt = 20$, and $dy/dt = 30$ into the last equation of Step 4, we find

$$50\frac{dz}{dt} = (30)(20) + (40)(30)$$

and

$$\frac{dz}{dt} = 36$$

Therefore, at noon on the day in question, the ships are moving apart at the rate of 36 mph.    ■

## 2.9    CONCEPT QUESTIONS

**1.** What is a related rates problem?

**2.** Give the steps involved in solving a related rates problem.

## 2.9 EXERCISES

*In Exercises 1–6, an equation relating the variables x and y, the values of x and y, and the value of either dx/dt or dy/dt at a particular instant of time are given. Find the value of the rate of change that is not specified.*

1. $x^2 + y^2 = 25$;  $x = 3, y = -4, \dfrac{dx}{dt} = 2$;  $\dfrac{dy}{dt} = ?$

2. $y^3 - 2x^3 = -10$;  $x = 1, y = -2, \dfrac{dy}{dt} = -1$;  $\dfrac{dx}{dt} = ?$

3. $x^2 y = 8$;  $x = 2, y = 2, \dfrac{dx}{dt} = 3$;  $\dfrac{dy}{dt} = ?$

4. $y^2 + xy + x^2 - 1 = 0$;  $x = 1, y = -1, \dfrac{dy}{dt} = -2$;  $\dfrac{dx}{dt} = ?$

5. $\sin^2 x + \cos y = 1$;  $x = \dfrac{\pi}{4}, y = \dfrac{\pi}{3}, \dfrac{dx}{dt} = \dfrac{\sqrt{3}}{2}$;  $\dfrac{dy}{dt} = ?$

6. $4x \cos y - \pi \tan x = 0$;  $x = \dfrac{\pi}{6}, y = \dfrac{\pi}{6}, \dfrac{dx}{dt} = 1$;  $\dfrac{dy}{dt} = ?$

7. The volume $V$ of a cube with sides of length $x$ inches is changing with respect to time $t$ (in seconds).
   a. Find a relationship between $dV/dt$ and $dx/dt$.
   b. When the sides of the cube are 10 in. long and increasing at the rate of 0.5 in./sec, how fast is the volume of the cube increasing?

8. The volume of a right circular cylinder of radius $r$ and height $h$ is $V = \pi r^2 h$. Suppose that the radius and height of the cylinder are changing with respect to time $t$.
   a. Find a relationship between $dV/dt$, $dr/dt$, and $dh/dt$.
   b. At a certain instant of time, the radius and height of the cylinder are 2 in. and 6 in. and are increasing at the rate of 0.1 in./sec and 0.3 in./sec, respectively. How fast is the volume of the cylinder increasing?

9. A point moves along the curve $2x^2 - y^2 = 2$. When the point is at $(3, -4)$, its $x$-coordinate is increasing at the rate of 2 units per second. How fast is its $y$-coordinate changing at that instant of time?

10. A point moves along the curve $3y + 4y^2 + 3x = 4$. When the point is at $(1, -1)$, its $x$-coordinate is increasing at the rate of 3 units per second. How fast is its $y$-coordinate changing at that instant of time?

11. **Motion of a Particle** A particle moves along the curve defined by $y = \frac{1}{6}x^3 - x$. Determine the values of $x$ at which the rate of change of its $y$-coordinate is (a) less than, (b) equal to, and (c) greater than that of its $x$-coordinate.

12. **Rectilinear Motion** The velocity of a particle moving along the $x$-axis is proportional to the square root of the distance, $x$, covered by the particle. Show that the force acting on the particle is constant.
   **Hint:** Use Newton's Second Law of Motion, which states that the force is proportional to the rate of change of momentum.

13. **Oil Spill** In calm waters, the oil spilling from the ruptured hull of a grounded tanker spreads in all directions. Assuming that the polluted area is circular, determine how fast the area is increasing when the radius of the circle is 60 ft and is increasing at the rate of $\frac{1}{2}$ ft/sec?

14. **Blowing a Soap Bubble** Carlos is blowing air into a spherical soap bubble at the rate of 8 cm³/sec. How fast is the radius of the bubble changing when the radius is 10 cm? How fast is the surface area of the bubble changing at that time?

15. If a spherical snowball melts at a rate that is proportional to its surface area, show that its radius decreases at a constant rate.

16. **Speed of a Race Car** A race car is moving along a track described by the equation

$$x^4 - 4x^2 + 2x^2 y^2 + 4y^2 + y^4 = 0$$

where both $x$ and $y$ are measured in miles. How fast is the car moving in the $y$-direction ($dy/dt$), when $dx/dt = -20$ (mph) and the car is at the point in the first quadrant with coordinate $x = 1$?

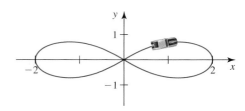

17. The base of a 13-ft ladder that is leaning against a wall begins to slide away from the wall. When the base is 12 ft from the wall and moving at the rate of 8 ft/sec, how fast is the top of the ladder sliding down the wall?

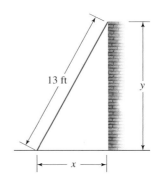

18. A 20-ft ladder leaning against a wall begins to slide. How fast is the top of the ladder sliding down the wall at the instant of time when the bottom of the ladder is 12 ft from the wall and sliding away from the wall at the rate of 5 ft/sec?

**19. Demand for Compact Discs** The demand equation for the Olympus recordable compact disc is

$$100x^2 + 9p^2 = 3600$$

where $x$ represents the number (in thousands) of 50-packs demanded per week when the unit price is $p$ dollars. How fast is the quantity demanded increasing when the unit price per 50-pack is $14 and the selling price is dropping at the rate of 10¢ per 50-pack per week?

**20.** Let $V$ denote the volume of a rectangular box of length $x$ inches, width $y$ inches, and height $z$ inches. Suppose that the sides of the box are changing with respect to time $t$.
   **a.** Find a relationship between $dV/dt$, $dx/dt$, $dy/dt$, and $dz/dt$.
   **Hint:** Write $V = x(yz)$, and use the Product Rule.
   **b.** At a certain instant of time, the length, width, and height of the box are 3, 5, and 10 in., respectively. If the length, width, and height of the box are increasing at the rate of 0.2, 0.3, and 0.1 in./sec, respectively, how fast is the volume of the box increasing?

**21. Baseball Diamond** The sides of a square baseball diamond are 90 ft long. When a player who is between the second and third base is 60 ft from second base and heading toward third base at a speed of 22 ft/sec, how fast is the distance between the player and home plate changing?

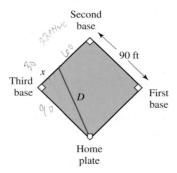

Second base

90 ft

Third base

$x$

First base

$D$

Home plate

**22. Docking a Boat** A boat is pulled into a dock by means of a rope attached to the bow of the boat and passing through a pulley on the dock. The pulley is located at a point on the dock that is 2 m higher than the bow of the boat. If the rope is being pulled in at the rate of 1 m/sec, how fast is the boat approaching the dock when it is 12 m from the dock?

**23. Tracking the Path of a Submarine** The position $P(x, y)$ of a submarine moving in an $xy$-plane is described by the equation

$$y = 10^{-10}x^3(x - 2000) \qquad 0 \le x \le 1500$$

where both $x$ and $y$ are measured in feet (see the figure). How fast is the depth of the submarine changing when it is at the position $(1000, -100)$ and its speed in the $x$-direction is 50 ft/sec?

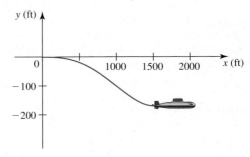

**24. Length of a Shadow** A man who is 6 ft tall walks away from a streetlight that is 15 ft from the ground at a speed of 4 ft/sec. How fast is the tip of his shadow moving along the ground when he is 30 ft from the base of the light pole?

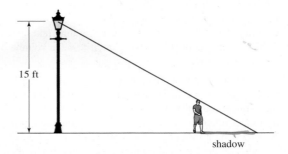

15 ft

shadow

**25.** A coffee pot that has the shape of a circular cylinder of radius 4 in. is being filled with water flowing at a constant rate. At what rate is the water flowing into the coffee pot when the water level is rising at the rate of 0.4 in./sec?

$h$

26. A car leaves an intersection traveling west. Its position 4 sec later is 20 ft from the intersection. At the same time, another car leaves the same intersection heading north so that its position 4 sec later is 28 ft from the intersection. If the speeds of the cars at that instant of time are 9 ft/sec and 11 ft/sec, respectively, find the rate at which the distance between the two cars is changing.

27. A car leaves an intersection traveling east. Its position $t$ sec later is given by $x = t^2 + t$ ft. At the same time, another car leaves the same intersection heading north, traveling $y = t^2 + 3t$ ft in $t$ sec. Find the rate at which the distance between the two cars will be changing 5 sec later.

28. A police cruiser hunting for a suspect pulls over and stops at a point 20 ft from a straight wall. The flasher on top of the cruiser revolves at a constant rate of 90 deg/sec, and the light beam casts a spot of light as it strikes the wall. How fast is the spot of light moving along the wall at a point 30 ft from the point on the wall closest to the cruiser?

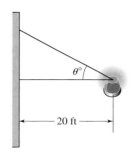

29. At 8:00 A.M. ship $A$ is 120 km due east of ship $B$. Ship $A$ is moving north at 20 km/hr, and ship $B$ is moving east at 25 km/hr. How fast is the distance between the two ships changing at 8:30 A.M.?

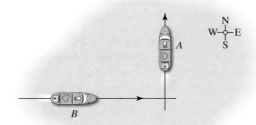

30. Two ships leave the same port at noon. Ship $A$ moves north at 18 km/hr, and ship $B$ moves northeast at 20 km/hr. How fast is the distance between them changing at 1 P.M.?

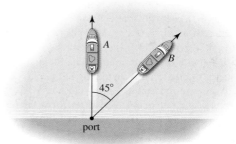

31. **Adiabatic Process** In an adiabatic process (one in which no heat transfer takes place), the pressure $P$ and volume $V$ of an ideal gas such as oxygen satisfy the equation $P^5 V^7 = C$, where $C$ is a constant. Suppose that at a certain instant of time, the volume of the gas is 4L, the pressure is 100 kPa, and the pressure is decreasing at the rate of 5 kPa/sec. Find the rate at which the volume is changing.

32. **Electric Circuit** The voltage $V$ in volts (V) in an electric circuit is related to the current $I$ in amperes (A) and the resistance $R$ in ohms ($\Omega$) by the equation $V = IR$. When $V = 12$, $I = 2$, $V$ is increasing at the rate of 2 V/sec, and $I$ is increasing at the rate of $\frac{1}{2}$ A/sec, how fast is the resistance changing?

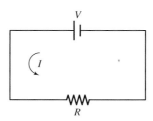

33. **Mass of a Moving Particle** The mass $m$ of a particle moving at a velocity $v$ is related to its rest mass $m_0$ by the equation

$$m = \frac{m_0}{\sqrt{1 - \dfrac{v^2}{c^2}}}$$

where $c$ (2.98 × 10⁸ m/sec) is the speed of light. Suppose that an electron of mass 9.11 × 10⁻³¹ kg is being accelerated in a particle accelerator. When its velocity is 2.92 × 10⁸ m/sec and its acceleration is 2.42 × 10⁵ m/sec², how fast is the mass of the electron changing?

**34. Variable Resistors** Two rheostats (variable resistors) are connected in parallel as shown in the figure. If the resistances of the rheostats are $R_1$ and $R_2$ ohms ($\Omega$), then the single resistor that could replace this combination has resistance $R$, called the equivalent resistance, and is given by

$$\frac{1}{R} = \frac{1}{R_1} + \frac{1}{R_2}$$

Suppose that at a certain instant of time the first rheostat has a resistance of 60 $\Omega$ that is increasing at the rate of 2 $\Omega$/sec, while the second rheostat has a resistance of 90 $\Omega$ that is decreasing at the rate of 3 $\Omega$/sec. How fast is the resistance of the equivalent resistor changing at that time?

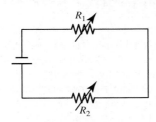

**35. Coast Guard Patrol Search Mission** The pilot of a Coast Guard patrol aircraft on a search mission had just spotted a disabled fishing trawler and decided to go in for a closer look. Flying in a straight line at a constant altitude of 1000 ft and at a constant speed of 264 ft/sec, the aircraft passed directly over the trawler. How fast was the aircraft receding from the trawler when the aircraft was 1500 ft from the trawler?

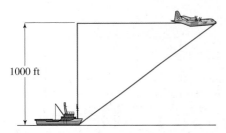

1000 ft

**36. Tracking a Plane with Radar** Shortly after taking off, a plane is climbing at an angle of 30° and traveling at a constant speed of 600 ft/sec as it passes over a ground radar tracking station. At that instant of time, the altitude of the plane is 1000 ft. How fast is the distance between the plane and the radar station increasing at that instant of time?

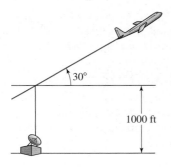

30°

1000 ft

**37.** A piston is attached to a crankshaft of radius 3 in. by means of a 7-in. connecting rod (see Figure a).
   **a.** Let $x$ denote the position of the piston (Figure b). Use the law of cosines to find an equation relating $x$ to $\theta$.
   **b.** If the crankshaft rotates counterclockwise at a constant rate of 60 rev/sec, what is the velocity of the piston when $\theta = \pi/3$?

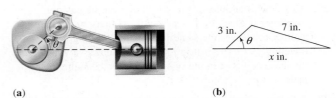

3 in.    7 in.

$\theta$

$x$ in.

(a)                         (b)

**38.** An aircraft carrier is sailing due east at a constant speed of 30 ft/sec. When the aircraft carrier is at the origin ($t = 0$), a plane is launched from its deck with a flight path that is described by the graph of $y = 0.001x^2$ where $y$ is the altitude of the plane (in feet). Ten seconds later, when the plane is at the point (1000, 1000) and $dx/dt = 500$ ft/sec, how fast is the distance between the plane and the aircraft carrier changing?

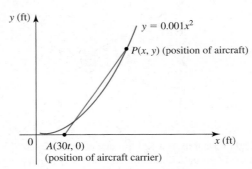

$y$ (ft)

$y = 0.001x^2$

$P(x, y)$ (position of aircraft)

0

$A(30t, 0)$
(position of aircraft carrier)

$x$ (ft)

**39.** As a tender leaves an offshore oil rig, traveling in a straight line and at a constant velocity of 20 mph, a helicopter approaches the oil rig in a direction perpendicular to the direction of motion of the tender. The helicopter, flying at a constant altitude of 100 ft, approaches the rig at a constant velocity of 60 mph. When the helicopter is 1000 ft (measured horizontally) from the rig and the tender is 200 ft from the rig, how fast is the distance between the helicopter and the tender changing? (Recall that 60 mi/hr $=$ 88 ft/sec.)

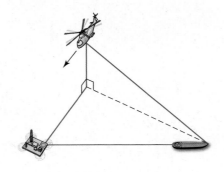

**40.** The following figure shows the cross section of a swimming pool that is 30 ft wide. When the pool is being filled with water at the rate of 600 gal/min and the depth at the deep end is 4 ft, how fast is the water level rising? (1 gal = 0.1337 ft³.)

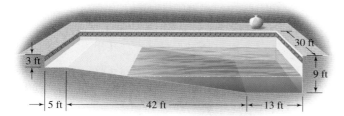

**41.** A hole is to be drilled into a block of Plexiglas. The 1-in. drill bit is shown in Figure (a), and the cross section of the Plexiglas block is shown in Figure (b). The drill press operator drives the drill bit into the Plexiglas at a constant speed of 0.05 in./sec. At what rate is the Plexiglas being removed 10 sec after the drill bit first makes contact with the block of Plexiglas?

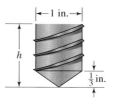
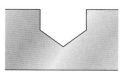

(a) Cross section of drill bit   (b) Cross section of Plexiglas block

**Hint:** First show that the amount of material removed when the drill bit is $h$ in. from the top surface of the Plexiglas block is $V = [\pi(9h - 2)]/36$.

**42. Home Mortgage Payments** The Garcias are planning to buy their first home within the next several months and estimate that they will need a home mortgage loan of $250,000 to be amortized over 30 years. At an interest rate of $r$ per year, compounded monthly, the Garcias' monthly repayment $P$ (in dollars) can be computed by using the formula

$$P = \frac{250{,}000r}{12\left[1 - \left(1 + \dfrac{r}{12}\right)^{-360}\right]}$$

**a.** If the interest rate is currently 7% per year and they secure the rate right now, what will the Garcias' monthly repayment on the mortgage be?

**b.** If the interest rate is currently increasing at the rate of $\frac{1}{4}\%$ per month, how fast is the monthly repayment on a mortgage loan of $250,000 increasing? Interpret your result.

# Differentials and Linear Approximations

The Jacksons are planning to buy a house in the near future and estimate that they will need a 30-year fixed-rate mortgage of $240,000. If the interest rate increases from the present rate of 7% per year compounded monthly to 7.3% per year compounded monthly between now and the time the Jacksons decide to secure the loan, approximately how much more per month will their mortgage be? (You will be asked to answer this question in Exercise 42.)

Questions like this, in which we wish to *estimate* the change in the dependent variable (monthly mortgage payment) corresponding to a small change in the independent variable (interest rate per year), occur in many real-life applications. Here are a few more examples:

- An engineer would like to know the changes in the gaps between the rails in a railroad track due to expansions caused by small fluctuations in temperature.
- A chemist would like to know how a small increase in the amount of a catalyst will affect the initial speed at which a chemical reaction begins.
- An economist would like to know how a small increase in a country's capital expenditure will affect the country's gross domestic product.
- A bacteriologist would like to know how a small increase in the amount of a bactericide will affect a population of bacteria.

- A businesswoman would like to know how raising the unit price of a product by a small amount will affect her profits.
- A sociologist would like to know how a small increase in the amount of capital investment in a housing project will affect the crime rate.

To calculate these changes and their approximate effect, we need the concept of the *differential* of a function.

## ■ Increments

Let $x$ denote a variable quantity and suppose that $x$ changes from $x_1$ to $x_2$. Then the change in $x$, called the **increment in $x$,** is denoted by the symbol $\Delta x$ (delta $x$). Thus,

$$\Delta x = x_2 - x_1 \qquad \text{Final value minus initial value} \qquad (1)$$

For example, if $x$ changes from 2 to 2.1, then $\Delta x = 2.1 - 2 = 0.1$; and if $x$ changes from 2 to 1.9, then $\Delta x = 1.9 - 2 = -0.1$.

Sometimes it is more convenient to express the change in $x$ in a slightly different manner. For example, if we solve Equation (1) for $x_2$, we find $x_2 = x_1 + \Delta x$, where $\Delta x$ is an increment in $x$. Observe that $\Delta x$ plays precisely the role that $h$ played in our earlier discussions.

Now, suppose that two quantities, $x$ and $y$, are related by an equation $y = f(x)$, where $f$ is some function. If $x$ changes from $x$ to $x + \Delta x$, then the corresponding change in $y$, or the **increment in $y$,** is denoted by $\Delta y$. It is the value of $f(x)$ at $x + \Delta x$ minus the value of $f(x)$ at $x$; that is,

$$\Delta y = f(x + \Delta x) - f(x) \qquad (2)$$

(See Figure 1.)

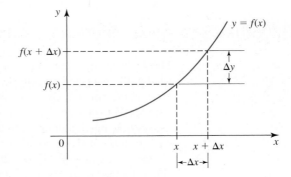

**FIGURE 1**
An increment of $\Delta x$ in $x$
induces an increment of
$\Delta y = f(x + \Delta x) - f(x)$ in $y$.

**EXAMPLE 1**  Suppose that $y = 2x^3 - x + 1$. Find $\Delta x$ and $\Delta y$ when (a) $x$ changes from 3 to 3.01 and (b) $x$ changes from 3 to 2.98.

**Solution**

**a.** Here, $\Delta x = 3.01 - 3 = 0.01$. Next, letting $f(x) = 2x^3 - x + 1$, we see that

$$\Delta y = f(x + \Delta x) - f(x) = f(3.01) - f(3)$$

$$= [2(3.01)^3 - 3.01 + 1] - [2(3)^3 - 3 + 1]$$

$$= 0.531802$$

**b.** Here, $\Delta x = 2.98 - 3 = -0.02$. Also,

$$\Delta y = f(x + \Delta x) - f(x) = f(2.98) - f(3)$$
$$= [2(2.98)^3 - 2.98 + 1] - [2(3)^3 - 3 + 1]$$
$$= -1.052816$$

## ◼ Differentials

To find a quick and simple way of estimating the change in $y$, $\Delta y$, due to a small change in $x$, $\Delta x$, let's look at the graph in Figure 2.

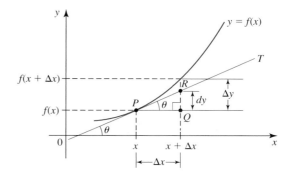

**FIGURE 2**
If $\Delta x$ is small, $dy$ is a good approximation of $\Delta y$.

We can see that the tangent line $T$ lies close to the graph of $f$ near the point of tangency at $P$. Therefore, if $\Delta x$ is small, the $y$-coordinate of the point $R$ on $T$ is a good approximation of $f(x + \Delta x)$. Equivalently, the quantity $dy$ is a good approximation of $\Delta y$.

Now consider the right triangle $\triangle PQR$. We have

$$\frac{dy}{\Delta x} = \tan \theta$$

or $dy = (\tan \theta)\Delta x$. But the derivative of $f$ gives the slope of the tangent line $T$, so we have $\tan \theta = f'(x)$. Therefore,

$$dy = f'(x)\Delta x$$

The quantity $dy$ is called the *differential* of $y$.

---

**DEFINITION** **Differential**

Let $y = f(x)$ where $f$ is a differentiable function. Then

1. The **differential $dx$** of the independent variable $x$ is $dx = \Delta x$, where $\Delta x$ is an increment in $x$.
2. The **differential $dy$** of the dependent variable $y$ is

$$dy = f'(x)\Delta x = f'(x)\,dx \tag{3}$$

---

**Notes**

1. For the independent variable $x$, there is no difference between the differential $dx$ and the increment $\Delta x$; both measure the change in $x$ from $x$ to $x + \Delta x$.

2. For the dependent variable $y$, the differential $dy$ is an *approximation* of the change in $y$, $\Delta y$, corresponding to a small change in $x$ from $x$ to $x + \Delta x$.
3. The differential $dy$ depends on both $x$ and $dx$. However, if $x$ is fixed, then $dy$ is a *linear* function of $dx$.    ■

Later, we will show that the approximation of $\Delta y$ by $dy$ is very good when $dx$, or $\Delta x$, is small. First, let's look at some examples.

**EXAMPLE 2**    Consider the equation $y = 2x^3 - x + 1$ of Example 1. Use the differential $dy$ to approximate $\Delta y$ when (a) $x$ changes from 3 to 3.01 and (b) $x$ changes from 3 to 2.98. Compare your results with those of Example 1.

**Solution**    Let $f(x) = 2x^3 - x + 1$. Then

$$dy = f'(x)\, dx = (6x^2 - 1)\, dx$$

**a.** Here, $x = 3$ and $dx = 3.01 - 3 = 0.01$. Therefore,

$$dy = [6(3^2) - 1](0.01) = 0.53$$

and we obtain the approximation

$$\Delta y \approx 0.53 \qquad \text{From Example 1 we know that the actual value of } \Delta y \text{ is } 0.531802.$$

**b.** Here, $x = 3$ and $dx = 2.98 - 3 = -0.02$. Therefore,

$$dy = [6(3)^2 - 1](-0.02) = -1.06$$

and we obtain the approximation

$$\Delta y \approx -1.06 \qquad \text{From Example 1 we know that the actual value of } \Delta y \text{ is } -1.052816. \quad ■$$

**EXAMPLE 3**    **Estimating Fuel Costs of Operating an Oil Tanker**    The total cost incurred in operating an oil tanker on an 800-mi run, traveling at an average speed of $v$ mph, is estimated to be

$$C(v) = \frac{1{,}000{,}000}{v} + 200v^2$$

dollars. Find the approximate change in the total operating cost if the average speed is increased from 10 mph to 10.5 mph.

**Solution**    Letting $v = 10$ and $dv = 0.5$, we find

$$\Delta C \approx dC = C'(10)\, dv$$

$$= -\frac{1{,}000{,}000}{v^2} + 400v \Big|_{v=10} \cdot (0.5)$$

$$= (-10{,}000 + 4000)(0.5) \approx -3000$$

So the total operating costs decrease by approximately \$3000.    ■

**EXAMPLE 4** The Rings of Neptune

**a.** A planetary ring has an inner radius of $r$ units and an outer radius of $R$ units, where $(R - r)$ is small in comparison to $r$ (see Figure 3a). Use differentials to estimate the area of the ring.

**b.** Observations including those of Voyager I and II showed that Neptune's ring system is considerably more complex than had been believed. For one thing, it is made up of a large number of distinguishable rings rather than one continuous great ring, as had previously been thought (see Figure 3b). The outermost ring, 1989N1R, has an inner radius of approximately 62,900 km (measured from the center of the planet) and a radial width of approximately 50 km. Using these data, estimate the area of the ring.

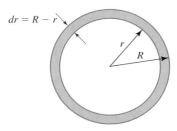

$dr = R - r$

$r$

$R$

(a) The area of the ring can be approximated by the circumference of the inner circle times the thickness.

**FIGURE 3**

NASA

(b) Neptune and its rings

**Solution**

**a.** Since the area of a circle of radius $x$ is $A = f(x) = \pi x^2$, we have

$$\pi R^2 - \pi r^2 = f(R) - f(r)$$

$$= \Delta A$$

Remember that $\Delta A =$ change in $f$ when $x$ changes from $x = r$ to $x = R$.

$$\approx dA$$

$$= f'(r)\, dr$$

where $dr = R - r$. So we see that the area of the ring is approximately $f'(r)\, dr = 2\pi r(R - r)$ square units. In words, the area of the ring is approximately equal to

circumference of the inner circle $\times$ thickness of the ring

**b.** Applying the results of part (a) with $r = 62,900$ and $dr = 50$, we find that the area of the ring is approximately $2\pi(62,900)(50)$, or 19,760,618 sq km, which is approximately 4% of the earth's surface.  ■

## ■ Error Estimates

An important application of differentials lies in the calculation of error propagation. For example, suppose that the quantities $x$ and $y$ are related by the equation $y = f(x)$, where $f$ is some function; then a small error $\Delta x$ or $dx$ incurred in measuring the quantity $x$ results in an error $\Delta y$ in the calculated value of $y$.

**EXAMPLE 5** **Estimating the Surface Area of the Moon**   Assume that the moon is a perfect sphere, and suppose that we have measured its radius and found it to be 1080 mi with a possible error of 0.05 mi. Estimate the maximum error in the computed surface area of the moon.

**Solution**   The surface area of a sphere of radius $r$ is

$$S = 4\pi r^2$$

We are given that the error in $r$ is $\Delta r = 0.05$ mi and are required to find the error $\Delta S$ in $S$. But if $\Delta r$ (equivalently, $dr$) is small, then

$$\Delta S \approx dS = f'(r)\Delta r = 8\pi r \, dr \qquad \text{Let } f(r) = 4\pi r^2. \qquad \textbf{(4)}$$

Substituting $r = 1080$ and $dr = \Delta r = 0.05$ in Equation (4), we obtain

$$\Delta S \approx 8\pi(1080)(0.05) \approx 1357.17$$

Therefore, the maximum error in the calculated area is approximately 1357 mi$^2$. ∎

In Example 5 we calculated the **error $\Delta q$ of a quantity $q$.** There are two other common error measurements. They are

$$\frac{\Delta q}{q}, \quad \text{the \textbf{relative error} in the measurement}$$

and

$$\frac{\Delta q}{q}(100), \quad \text{the \textbf{percentage error} in the measurement}$$

The error, relative error, and percentage error are often approximated by

$$dq, \quad \frac{dq}{q}, \quad \text{and} \quad \frac{dq}{q}(100)$$

respectively.

The relative errors made when the surface area of the moon was calculated in Example 5 are given by

$$\text{relative error in } r \approx \frac{dr}{r} = \frac{0.05}{1080} \approx 0.0000463$$

and

$$\text{relative error in } S \approx \frac{dS}{S} = \frac{8\pi r}{4\pi r^2} \, dr = \frac{2}{r} \, dr \approx 0.0000926$$

A summary of these results and the approximate percentage errors follows.

| Variable | Error | Approximate relative error | Approximate percentage error |
|:---:|:---:|:---:|:---:|
| $r$ | 0.05 | 0.0000463 | 0.00463% |
| $S$ | 1357.17 | 0.0000926 | 0.00926% |

**Note**  Example 5 illustrates why the relative error is so important. The (absolute) error in $S$ is 1357.17 mi². By itself, the error appears to be rather large (a little larger than the area of the state of Rhode Island). But when the error is compared to the area of the moon (approximately 14,657,415 mi²), it is a relatively small number.  ■

**EXAMPLE 6**  The edge of a cube was measured and found to be 3 in. with a maximum possible error of 0.02 in. Find the approximate maximum percentage error that would be incurred in computing the volume of the cube using this measurement.

**Solution**  Let $x$ denote the length of an edge of the cube. Then the volume of the cube is $V = x^3$. The error in the measurement of its volume is approximated by the differential

$$dV = 3x^2 dx \qquad \text{Let } f(x) = x^3, \text{ so } f'(x)\, dx = 3x^2 dx.$$

But we are given that

$$|dx| \leq 0.02 \qquad \text{and} \qquad x = 3$$

so

$$|dV| = 3x^2 |dx| \leq 3(3)^2(0.02) = 0.54$$

Therefore, the approximate maximum percentage error that would be incurred in computing the volume of the cube is

$$\frac{|dV|}{V}(100) = \frac{0.54}{3^3}(100) = \frac{54}{27} = 2$$

or 2%.  ■

## ■ Linear Approximations

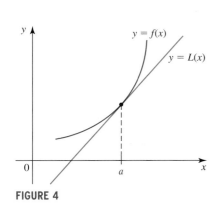

**FIGURE 4**

As you can see in Figure 4, the graph of $f$ lies very close to its tangent line near the point of tangency. This suggests that the values of $f(x)$ for $x$ near $a$ can be approximated by the corresponding values of $L(x)$, where $L$ is the linear function describing the tangent line.

The function $L$ can be found by using the point-slope form of the equation of a line. Indeed, the slope of the tangent line at $(a, f(a))$ is $f'(a)$, and an equation of the tangent line is

$$y - f(a) = f'(a)(x - a)$$

or

$$y = L(x) = f(a) + f'(a)(x - a)$$

Next, if we replace $x$ by $a$ in Equation (2) and let $\Delta x = x - a$, then

$$\Delta y = f(x) - f(a)$$

so

$$f(x) - f(a) \approx dy = f'(a)\Delta x = f'(a)(x - a) \qquad \text{By Equation (3)}$$

or

$$f(x) \approx f(a) + f'(a)(x - a) \tag{5}$$

provided that $\Delta x$ is small or, equivalently, $x$ is close to $a$. But the expression on the right of Equation (5) is $L(x)$. So $f(x) \approx L(x)$ for $x$ near $a$. The approximation in Equation (5) is called the **linear approximation** of $f$ at $a$. The linear function $L$ defined by

$$L(x) = f(a) + f'(a)(x - a) \tag{6}$$

whose graph is the tangent line to the graph of $f$ at $(a, f(a))$, is called the **linearization** of $f$ at $a$. Observe that the linearization of $f$ gives an approximation of $f$ over a *small interval containing a.*

### EXAMPLE 7

**a.** Find the linearization of $f(x) = \sqrt{x}$ at $a = 4$.
**b.** Use the result of part (a) to approximate the numbers $\sqrt{3.9}$, $\sqrt{3.98}$, $\sqrt{4}$, $\sqrt{4.04}$, $\sqrt{4.8}$, $\sqrt{6}$, and $\sqrt{8}$. Compare the results with the actual values obtained with a calculator.

**Solution**
**a.** Here, $a = 4$. Since

$$f'(x) = \frac{1}{2} x^{-1/2} = \frac{1}{2\sqrt{x}}$$

we find $f'(4) = \frac{1}{4}$. Also, $f(4) = 2$. Using Equation (6), we see that the required linearization of $f$ is

$$L(x) = f(4) + f'(4)(x - 4)$$

or

$$L(x) = 2 + \frac{1}{4}(x - 4) = \frac{1}{4}x + 1$$

(See Figure 5.)

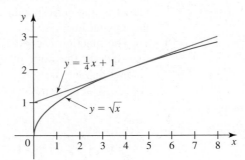

**FIGURE 5**
The linear approximation of
$f(x) = \sqrt{x}$ by $L(x) = \frac{1}{4}x + 1$

**b.** Using the result of part (a), we see that

$$\sqrt{3.9} = f(3.9) \approx L(3.9) = \frac{1}{4}(3.9) + 1 = 1.975$$

We obtain the other approximations in a similar manner. The results are summarized in the following table. You can see from the table that the approximations of $f(x)$ by $L(x)$ are good if $x$ is close to 4 but are less accurate if $x$ is farther away from 4.

| Number | $x$ | $L(x)$ | $f(x)$ (actual value) |
|--------|-----|--------|------------------------|
| $\sqrt{3.9}$ | 3.9 | 1.975 | 1.97484177... |
| $\sqrt{3.98}$ | 3.98 | 1.995 | 1.99499373... |
| $\sqrt{4}$ | 4 | 2 | 2.00000000... |
| $\sqrt{4.04}$ | 4.04 | 2.01 | 2.00997512... |
| $\sqrt{4.8}$ | 4.8 | 2.2 | 2.19089023... |
| $\sqrt{6}$ | 6 | 2.5 | 2.44948974... |
| $\sqrt{8}$ | 8 | 3 | 2.82842712... |

## ■ Error in Approximating $\Delta y$ by $dy$

Through several numerical examples we have seen how closely the (true) increment $\Delta y = f(x + \Delta x) - f(x)$, where $y = f(x)$, is approximated by the differential $dy$. Let's demonstrate that this is no accident. We start by computing the error in the approximation

$$\Delta y - dy = [f(x + \Delta x) - f(x)] - f'(x)\Delta x$$

$$= \left[\frac{f(x + \Delta x) - f(x)}{\Delta x}\right]\Delta x - f'(x)\Delta x$$

$$= \left[\frac{f(x + \Delta x) - f(x)}{\Delta x} - f'(x)\right]\Delta x$$

For fixed $x$, the quantity in brackets depends only on $\Delta x$. Furthermore, because

$$\frac{f(x + \Delta x) - f(x)}{\Delta x}$$

approaches $f'(x)$ as $\Delta x$ approaches 0, the bracketed quantity approaches 0 as $\Delta x$ approaches 0. Let's denote this quantity, which is a function of $\Delta x$, by $\varepsilon(\Delta x)$.* Then we have

$$\Delta y - dy = \varepsilon(\Delta x)\Delta x$$

Therefore, if $\Delta x$ is small, then

$$\Delta y - dy = (\text{small number})(\text{small number})$$

and is a *very* small number, which accounts for the closeness of the approximation.

---

*We could have called this function of $\Delta x$, $g$ or $h$, say, but in mathematical literature the Greek letter $\varepsilon$ is often used to denote a small quantity. Since the functional value $\varepsilon(\Delta x)$ is small when $\Delta x$ is small, for emphasis we chose the letter $\varepsilon$ to denote that function.

## ■ 2.10 CONCEPT REVIEW

1. If $y = f(x)$, what is the differential of $x$? Write an expression for the differential $dy$.

2. Let $y = f(x)$. What is the relationship between the actual change in $y$, $\Delta y$, when $x$ changes from $x$ to $x + \Delta x$ and the differential $dy$ of $f$ at $x$? Illustrate this relationship graphically.

## 2.10 EXERCISES

**1.** Let $y = x^2 + 1$.
   **a.** Find $\Delta x$ and $\Delta y$ if $x$ changes from 2 to 2.02.
   **b.** Find the differential $dy$, and use it to approximate $\Delta y$ if $x$ changes from 2 to 2.02.
   **c.** Compute $\Delta y - dy$, the error in approximating $\Delta y$ by $dy$.

**2.** Let $y = 2x^3 - x$.
   **a.** Find $\Delta x$ and $\Delta y$ if $x$ changes from 2 to 1.97.
   **b.** Find the differential $dy$, and use it to approximate $\Delta y$ if $x$ changes from 2 to 1.97.
   **c.** Compute $\Delta y - dy$, the error in approximating $\Delta y$ by $dy$.

**3.** Let $w = \sqrt{2u + 3}$.
   **a.** Find $\Delta u$ and $\Delta w$ if $u$ changes from 3 to 3.1.
   **b.** Find the differential $dw$, and use it to approximate $\Delta w$ if $u$ changes from 3 to 3.1.
   **c.** Compute $\Delta w - dw$, the error in approximating $\Delta w$ by $dw$.

**4.** Let $y = 1/x$.
   **a.** Find $\Delta x$ and $\Delta y$ if $x$ changes from 1 to 1.02.
   **b.** Find the differential $dy$, and use it to approximate $\Delta y$ if $x$ changes from 1 to 1.02.
   **c.** Compute $\Delta y - dy$, the error in approximating $\Delta y$ by $dy$.

*In Exercises 5–18, find the differential of the function at the indicated number.*

**5.** $f(x) = 2x^2 - 3x + 1; \quad x = 1$

**6.** $f(x) = x^4 - 2x^3 + 3; \quad x = 0$

**7.** $f(x) = 2x^{1/4} + 3x^{-1/2}; \quad x = 1$

**8.** $f(x) = \sqrt{2x^2 + 1}; \quad x = 2$

**9.** $f(x) = x^2(3x - 1)^{1/3}; \quad x = 3$

**10.** $f(x) = \dfrac{x^2}{x^3 - 1}; \quad x = -1$

**11.** $f(x) = 2 \sin x + 3 \cos x; \quad x = \dfrac{\pi}{4}$

**12.** $f(x) = x \tan x; \quad x = \dfrac{\pi}{4}$

**13.** $f(x) = (1 + 2 \cos x)^{1/2}; \quad x = \dfrac{\pi}{2}$

**14.** $f(x) = \sin^2 x; \quad x = \dfrac{\pi}{6}$

**15.** $f(x) = e^x + \ln(1 + x); \quad x = 0$

**16.** $f(x) = \ln(2 \cos x + x); \quad x = 0$

**17.** $f(x) = x^3 e^{1-x}; \quad x = 1$

**18.** $f(x) = 2^{-x^2}; \quad x = 1$

*In Exercises 19–22 find the linearization $L(x)$ of the function at a.*

**19.** $f(x) = x^3 + 2x^2; \quad a = 1$

**20.** $f(x) = \sqrt{2x + 3}; \quad a = 3$

**21.** $f(x) = \ln x; \quad a = 1$

**22.** $f(x) = \sin x; \quad a = \dfrac{\pi}{4}$

**23.** Find the linearization of $f(x) = \sqrt{x + 3}$ at $a = 1$, and use it to approximate the numbers $\sqrt{3.9}$ and $\sqrt{4.1}$. Plot the graphs of $f$ and $L$ on the same set of axes.

**24.** Find the linearization $L(x)$ of $f(x) = \sqrt[3]{1 - x}$ at $a = 0$, and use it to approximate the numbers $\sqrt[3]{0.95}$ and $\sqrt[3]{1.05}$. Plot the graphs of $f$ and $L$ on the same set of axes.

*In Exercises 25–28, find the linearization of a suitable function, and then use it to approximate the number.*

**25.** $1.002^3$

**26.** $\sqrt{63.8}$

**27.** $\sqrt[5]{31.08}$

**28.** $\sin 0.1$

**29.** The side of a cube is measured with a maximum possible error of 2%. Use differentials to estimate the maximum percentage error in its computed volume.

**30.** **Estimating the Area of a Ring of Neptune** The ring 1989N2R of the planet Neptune has an inner radius of approximately 53,200 km (measured from the center of the planet) and a radial width of 15 km. Use differentials to estimate the area of the ring.

**31.** **Effect of Advertising on Profits** The relationship between the quarterly profits of the Lyons Realty Company, $P(x)$, and the amount of money $x$ spent on advertising per quarter is described by the function

$$P(x) = -\frac{1}{8}x^2 + 7x + 32 \qquad 0 \le x \le 50$$

where both $P(x)$ and $x$ are measured in thousands of dollars. Use differentials to estimate the increase in profits when the amount spent on advertising each quarter is increased from $24,000 to $26,000.

**32. Construction of a Storage Tank** A storage tank for propane gas has the shape of a right circular cylinder with hemispherical ends. The length of the cylinder is 6 ft, and the radius of each hemisphere is $r$ ft.

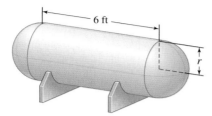

6 ft

$r$

**a.** Show that the volume of the tank is $\frac{2}{3}\pi r^2(2r + 9)$ ft$^3$.
**b.** If the tank were constructed with a radius of 4.1 ft instead of a specified radius of 4 ft, what would be the approximate percentage error in its volume?

**33. Unclogging Arteries** Research done in the 1930s by the French physiologist Jean Poiseuille showed that the resistance $R$ of a blood vessel of length $l$ and radius $r$ is $R = kl/r^4$, where $k$ is a constant. Suppose that a dose of the drug TPA increases $r$ by 10%. How will this affect the resistance $R$? (Assume that $l$ is constant.)

**34. Period of a Pendulum** The period of a simple pendulum is given by

$$T = 2\pi\sqrt{\frac{L}{g}}$$

where $L$ is the length of the pendulum in feet, $g$ is the constant of acceleration due to gravity, and $T$ is measured in seconds. Suppose that the length of a pendulum was measured with a maximum error of $\frac{1}{2}\%$. What will be the maximum percentage error in measuring its period?

**35. Period of a Satellite** The period of a satellite in a circular orbit of radius $r$ is given by

$$T = \frac{2\pi r}{R}\sqrt{\frac{r}{g}}$$

where $R$ is the earth's mean radius and $g$ is the constant of acceleration. Estimate the percentage change in the period if the radius of the orbit increases by 2%.

**36. Surface Area of a Horse** Animal physiologists use the formula

$$S = kW^{2/3}$$

to calculate the surface area of an animal (in square meters) from its mass $W$ (in kilograms), where $k$ is a constant that depends on the animal under consideration. Suppose that a physiologist calculates the surface area of a horse ($k = 0.1$). If the estimated mass of the horse is 280 kg with a maximum error in measurement of 0.5 kg, determine the maximum percentage error in the calculation of the horse's surface area.

**37. Child-Langmuir Law** In a vacuum diode a steady current $I$ flows between the cathode with potential 0 and anode which is held at a positive potential $V_0$. The Child-Langmuir Law states that $I = kV_0^{3/2}$, where $k$ is a constant. Use differentials to estimate the percentage change in the current corresponding to a 10% increase in the positive potential.

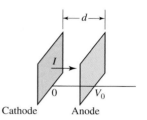

Cathode    Anode

**38. Effect of Price Increase on Quantity Demanded** The quantity $x$ demanded per week of the Alpha Sports Watch (in thousands) is related to its unit price of $p$ dollars by the equation

$$x = f(p) = 10\sqrt{\frac{50 - p}{p}} \qquad 0 < p \le 50$$

Use differentials to find the decrease in the quantity of watches demanded per week if the unit price is increased from \$40 to \$42.

**39. Range of an Artillery Shell** The range of an artillery shell fired at an angle of $\theta°$ with the horizontal is

$$R = \frac{1}{32}v_0^2 \sin 2\theta$$

in feet, where $v_0$ is the muzzle speed of the shell. Suppose that the muzzle speed of a shell is 80 ft/sec and the shell is fired at an angle of 29.5° instead of the intended 30°. Estimate how far short of the target the shell will land.

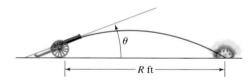

$\theta$

$R$ ft

**40. Range of an Artillery Shell** The range of an artillery shell fired at an angle of $\theta°$ with the horizontal is

$$R = \frac{v_0^2}{g} \sin 2\theta$$

in feet, where $v_0$ is the muzzle speed of the shell and $g = 32$ ft/sec$^2$ is the constant of acceleration due to gravity. Suppose the angle of elevation of the cannon is set at 45°. Because of variations in the amount of charge in a shell, the muzzle speed of a shell is subject to a maximum error of 0.1%. Calculate the effect this will have on the range of the shell.

**41. Forecasting Commodity Crops** Government economists in a certain country have determined that the demand equation for soybeans is given by

$$p = f(x) = \frac{55}{2x^2 + 1}$$

where the unit price $p$ is expressed in dollars per bushel and $x$, the quantity demanded per year, is measured in billions of bushels. The economists are forecasting a harvest of 2.2 billion bushels for the year, with a possible error of 10% in their forecast. Determine the corresponding error in the predicted price per bushel of soybeans.

**42. Financing a Home** The Jacksons are considering the purchase of a house in the near future and estimate that they will need a loan of $240,000. Their monthly repayment for a 30-year conventional mortgage with an interest rate of $r$ per year compounded monthly will be

$$P = \frac{20{,}000r}{1 - \left(1 + \dfrac{r}{12}\right)^{-360}}$$

dollars.

a. Find the differential of $P$.
b. If the interest rate increases from the present rate of 7% per year to 7.2% per year between now and the time the Jacksons decide to secure the loan, approximately how much more per month will their mortgage payment be? How much more will it be if the interest rate increases to 7.3% per year?

**43. Period of a Communications Satellite** According to Kepler's Third Law, the period $T$ (in days) of a satellite moving in a circular orbit $x$ mi above the surface of the earth is given by

$$T = 0.0588\left(1 + \frac{x}{3959}\right)^{3/2}$$

Suppose that a communications satellite is moving in a circular orbit 22,000 mi above the earth's surface. Because of friction, the satellite drops down to a new orbit 21,500 mi above the earth's surface. Estimate the decrease in the period of the satellite to the nearest one-hundredth hour.

**44. Effect of an Earthquake on a Structure** To study the effect an earthquake has on a structure, engineers look at the way a beam bends when subjected to an earth tremor. The equation

$$D = a - a\cos\left(\frac{\pi h}{2L}\right) \qquad 0 \le h \le L$$

where $L$ is the length of a beam and $a$ is the maximum deflection from the vertical, has been used by engineers to calculate the deflection $D$ at a point on the beam $h$ ft from the ground. Suppose that a 10-ft vertical beam has a maximum deflection of $\frac{1}{2}$ ft when subjected to an external force. Using

differentials, estimate the difference in the deflection between the point midway on the beam and the point $\frac{1}{10}$ ft above it.

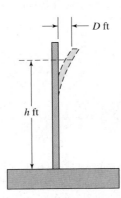

**45. Relative Error in Measuring Electric Current** When measuring an electric current with a tangent galvanometer, we use the formula

$$I = k \tan \phi$$

where $I$ is the current, $k$ is a constant that depends on the instrument, and $\phi$ is the angle of deflection of the pointer. Find the relative error in measuring the current $I$ due to an error in reading the angle $\phi$. At what position of the pointer can one obtain the most reliable results?

**46. Percentage Error in Measuring Height** From a point on level ground 150 ft from the base of a derrick, Jose measures the angle of elevation to the top of the derrick as 60°. If Jose's measurements are subject to a maximum error of 1%, find the percentage error in the measured height of the derrick.

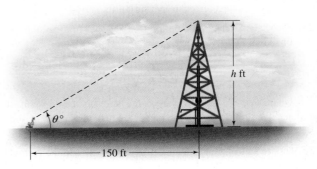

**47. Heights of Children** For children between the ages of 5 and 13 years, the Ehrenberg equation

$$\ln W = \ln 2.4 + 1.84h$$

gives the relationship between the weight $W$ (in kilograms) and the height $h$ (in meters) of a child. Use differentials to estimate the change in the weight of a child who grows from 1 m to 1.1 m.

In Exercises 48–51, *determine whether the statement is true or false. If it is true, explain why it is true. If it is false, explain why or give an example to show why it is false.*

**48.** If $y = ax + b$, where $a$ and $b$ are constants, then $\Delta y = dy$.

**49.** If $f$ is differentiable at $a$ and $x$ is close to $a$, then $f(x) \approx f(a) + f'(a)(x - a)$.

**50.** If $h = g \circ f$, where $g$ and $f$ are differentiable everywhere, then $h(x + \Delta x) \approx g(f(x)) + g'(f(x))f'(x)\Delta x$.

**51.** If $y = f(x)$ and $f'(x) > 0$, then $\Delta y \geq dy$.

# CHAPTER 2 REVIEW

## CONCEPT REVIEW

In Exercises 1–14, *fill in the blanks.*

**1. a.** The derivative of a function with respect to $x$ is the function $f'$ defined by the rule _____.
  **b.** The domain of $f'$ consists of all values of $x$ for which the _____ exists.
  **c.** The number $f'(a)$ gives the slope of the _____ _____ to the graph of $f$ at _____.
  **d.** The number $f'(a)$ also measures the rate of change of _____ with respect to _____ at _____.
  **e.** If $f$ is differentiable at $a$, then an equation of the tangent line to the graph of $f$ at $(a, f(a))$ is _____.

**2. a.** A function might not be differentiable at a _____. For example, the function _____ fails to be differentiable at _____.
  **b.** If a function $f$ is differentiable at $a$, then $f$ is _____ at $a$. The converse is false. For example, the function _____ is continuous at _____ but is not differentiable at _____.

**3. a.** If $c$ is a constant, then $\dfrac{d}{dx}(c) = $ _____.
  **b.** If $n$ is any real number, then $\dfrac{d}{dx}(x^n) = $ _____.

**4.** If $f$ and $g$ are differentiable functions and $c$ is a constant, then the Constant Multiple Rule states that _____, the Sum Rule states that _____, the Product Rule states that _____, and the Quotient Rule states that _____.

**5.** If $y = f(x)$, where $f$ is a differentiable function, and $\alpha$ denotes the angle that the tangent line to the graph of $f$ at $(x, f(x))$ makes with the positive $x$-axis, then $\tan \alpha = $ _____.

**6.** Suppose that $f(t)$ gives the position of an object moving on a coordinate line.
  **a.** The velocity of the object is given by _____, its acceleration is given by _____, and its jerk is given by _____. The speed of the object is given by _____.
  **b.** The object is moving in the positive direction if $v(t)$ _____ and in the negative direction if $v(t)$ _____. It is stationary if $v(t) = $ _____.

**7.** If $C$, $R$, $P$, and $\bar{C}$ denote the total cost function, the total revenue function, the profit function, and the average cost function, respectively, then the marginal total cost function is given by _____, the marginal total revenue function by _____, the marginal profit function by _____, and the marginal average cost function by _____.

**8.** If $f$ is differentiable at $x$ and $g$ is differentiable at $f(x)$, then the function $h = g \circ f$ is differentiable at _____, and $h'(x) = $ _____.

**9. a.** The General Power Rule states that $\dfrac{d}{dx}[f(x)]^n = $ _____.
  **b.** If $f$ is differentiable, then $\dfrac{d}{dx}[\sin f(x)] = $ _____, $\dfrac{d}{dx}[\cos f(x)] = $ _____, $\dfrac{d}{dx}[\tan f(x)] = $ _____, $\dfrac{d}{dx}[\sec f(x)] = $ _____, $\dfrac{d}{dx}[\csc f(x)] = $ _____, and $\dfrac{d}{dx}[\cot f(x)] = $ _____.

**10. a.** If $u$ is a differentiable function of $x$, then $\dfrac{d}{dx}e^u = $ _____.
  **b.** If $a > 0$ and $a \neq 1$, then $a^x = $ _____; if $u$ is a differentiable function of $x$, then $\dfrac{d}{dx}a^u = $ _____.

**11.** If $u$ is a differentiable function of $x$, then $\dfrac{d}{dx}\log_a|u| = $ _____.

**12.** Suppose that a function $y = f(x)$ is defined implicitly by an equation in $x$ and $y$. To find $dy/dx$, we differentiate _____ _____ of the equation with respect to $x$ and then solve the resulting equation for $dy/dx$. The derivative of a term involving $y$ includes _____ as a factor.

**13.** In a related rates problem we are given a relationship between a variable $x$ and a variable _____ that depend on a third variable $t$. Knowing the values of $x$, $y$, and $dx/dt$ at $a$, we want to find _____ at _____.

**14.** Let $y = f(t)$ and $x = g(t)$. If $x^2 + y^2 = 4$, then $dx/dt = $ _____. If $xy = 1$, then $dy/dt = $ _____.

**15. a.** If a variable quantity $x$ changes from $x_1$ to $x_2$, then the increment in $x$ is $\Delta x = $ _____.
**b.** If $y = f(x)$ and $x$ changes from $x$ to $x + \Delta x$, then the increment in $y$ is $\Delta y = $ _____.

**16.** If $y = f(x)$, where $f$ is a differentiable function, then the differential $dx$ of $x$ is $dx = $ _____, where _____ is an increment in _____, and the differential $dy$ of $y$ is $dy = $ _____.

## REVIEW EXERCISES

*In Exercises 1 and 2, use the definition of the derivative to find the derivative of the function.*

**1.** $f(x) = x^2 - 2x - 4$

**2.** $f(x) = 2x^3 - 3x + 2$

*In Exercises 3 and 4, sketch the graph of $f'$ for the function $f$ whose graph is given.*

**3.**

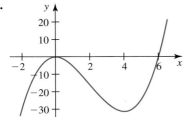

**4.**

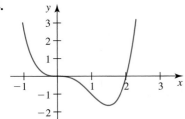

**5.** The amount of money on fixed deposit at the end of 5 years in a bank paying interest at the rate of $r$ per year is given by $A = f(r)$ (dollars).
**a.** What does $f'(r)$ measure? Give units.
**b.** What is the sign of $f'(r)$? Explain.
**c.** If you know that $f'(6) = 60{,}775.31$, estimate the change in the amount after 5 years if the interest rate changes from 6% to 7% per year.

**6.** Use the graph of the function $f$ to find the value(s) of $x$ at which $f$ is not differentiable.

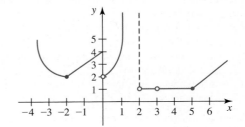

**7.** Find a function $f$ and a number $a$ such that

$$\lim_{h \to 0} \frac{3(4 + h)^{3/2} - 24}{h} = f'(a)$$

**8.** Evaluate $\displaystyle\lim_{h \to 0} \frac{2(1 + h)^3 + (1 + h)^2 - 3}{h}$.

*In Exercises 9–64, find the derivative of the function.*

**9.** $f(x) = \dfrac{1}{3}x^6 - 2x^4 + x^2 - 5$

**10.** $g(x) = 2x^4 + 3x^{1/2} - x^{-1/3} + x^{-4}$

**11.** $s = 2t^2 - \dfrac{4}{t} + \dfrac{2}{\sqrt{t}}$

**12.** $f(x) = \dfrac{x + 1}{x - 1}$

**13.** $g(t) = \dfrac{t - 1}{2t + 1}$

**14.** $h(x) = \dfrac{x}{2x^2 + 3}$

**15.** $h(u) = \dfrac{\sqrt{u}}{u^2 + 1}$

**16.** $u = \dfrac{t^2}{1 - \sqrt{t}}$

**17.** $g(\theta) = \cos \theta - 2 \sin \theta$

**18.** $f(x) = x \tan x + \sec x$

**19.** $f(x) = x \sin x + x^2 \cos x$

**20.** $y = \dfrac{1 - \sin x}{1 + \sin x}$

**21.** $h(t) = \dfrac{t \cos t}{1 + \tan t}$

**22.** $f(x) = (1 + 2x)^7$

**23.** $y = (t^3 + 2t + 1)^{-3/2}$

**24.** $g(t) = \left(t^2 + \dfrac{1}{t^2}\right)^3$

**25.** $f(s) = s(s^3 + s + 1)^{3/2}$

**26.** $y = \left(\dfrac{1 + t^2}{1 - t^2}\right)^{3/2}$

**27.** $y = \dfrac{2t}{\sqrt{t + 1}}$

**28.** $h(x) = \dfrac{1 + x}{(2x^2 + 1)^2}$

**29.** $f(x) = \cos(2x + 1)$

**30.** $g(t) = t^2 \sin(\pi t + 1)$

**31.** $y = x^2 + \dfrac{\sin 2x}{x}$

**32.** $h(x) = \sec\left(\dfrac{x + 1}{x - 1}\right)$

**33.** $u = \tan \dfrac{2}{x}$

**34.** $v = \sec 2x + \tan 3x$

**35.** $w = \cot^3 x$

**36.** $f(x) = \tan(x^2 + 1)^{-1/2}$

**37.** $f(\theta) = \dfrac{\cos \theta}{\theta^2}$

**38.** $y = \dfrac{\sin(2x + 1)}{2x + 1}$

**39.** $y = \ln \sqrt{x + 1}$

**40.** $y = \ln \dfrac{x(x - 1)}{x + 2}$

**41.** $y = \sqrt{x} \ln x$

**42.** $y = x^2 e^{\sqrt{x}}$

**43.** $y = e^{-x}(\cos 2x + 3 \sin 2x)$

**44.** $y = \dfrac{\sqrt{x}(x + 2)^3}{\sqrt{x + 3}}$

**45.** $x \ln y + y \ln x = 3$

**46.** $\ln(x - y) + \sin y - x^2 = 0$

**47.** $y = \ln(x^2 e^{-2x})$

**48.** $y = \ln(\tan x)$

**49.** $y = \dfrac{e^x}{\sqrt{1 + e^{-x}}}$

**50.** $y = \ln|\sec 2x + \tan 2x|$

**51.** $y = e^{\csc x}$

**52.** $y = x \cdot 3^{x^2 + 1}$

**53.** $y = e^{e^x}$

**54.** $y = (2e)^{x/2}$

**55.** $ye^{-x} + xe^{y^2} = 8$

**56.** $y = 2e^{\sec^{-1} x}$

**57.** $y = 3^{x \cot x}$

**58.** $y = x^2 \ln(x + \sin^{-1} x)$

**59.** $y = x \sec^{-1} x$

**60.** $y = \tan(\cos^{-1} 2x)$

**61.** $y = \tan^{-1} \sqrt{x^2 + 1}$

**62.** $y = \sin^{-1}\left(\dfrac{x + 1}{x + 2}\right)$

**63.** $y = \tan^{-1}(\cos^{-1} \sqrt{x})$

**64.** $y = (\sin x)^{\cos x}$

*In Exercises 65 and 66, find $f'(a)$.*

**65.** $f(x) = \dfrac{\sqrt{x}}{x^2 + 1}$;  $a = 4$

**66.** $f(x) = \sin(\cos x)$;  $a = \dfrac{\pi}{4}$

*In Exercises 67–76, find the second derivative of the function.*

**67.** $y = x^3 + x^2 - \dfrac{1}{x}$

**68.** $g(x) = \dfrac{1}{3x + 1}$

**69.** $y = x\sqrt{2x - 1}$

**70.** $f(x) = \dfrac{x - 1}{x^2 + 1}$

**71.** $f(x) = \cos^2 x$

**72.** $f(x) = \sin \dfrac{1}{x}$

**73.** $y = \cot \dfrac{\theta}{2}$

**74.** $u = \cos(\pi - 2t) + \sin(\pi + 2t)$

**75.** $f(t) = t \cot t$

**76.** $h(x) = x^2 \cos \dfrac{1}{x}$

*In Exercises 77 and 78, find $f''(a)$.*

**77.** $f(x) = \sqrt{2x + 1}$;  $a = 4$

**78.** $f(x) = x \tan x$;  $a = \dfrac{\pi}{4}$

*In Exercises 79 and 80, suppose that $f$ and $g$ are functions that are differentiable at $x = 2$ and that $f(2) = 3$, $f'(2) = -1$, $g(2) = 2$, and $g'(2) = 4$. Find $h'(2)$.*

**79.** $h(x) = f(x)g(x)$

**80.** $h(x) = \dfrac{f(x)}{g(x)}$

*In Exercises 81 and 82, find $h'(x)$ in terms of $f$, $g$, $f'$, and $g'$.*

**81.** $h(x) = \sqrt[3]{\dfrac{f(x)}{g(x)}}$

**82.** $h(x) = g[\sin f(x)]$

*In Exercises 83–92, find $dy/dx$ by implicit differentiation.*

**83.** $3x^2 - 2y^2 = 6$

**84.** $x^3 + 3xy^2 + y^3 = 1$

**85.** $\dfrac{1}{x^2} + \dfrac{1}{y^2} = 1$

**86.** $x\sqrt{y} + y\sqrt{x} - 1 = 0$

**87.** $(x + y)^3 + x^3 + y^3 = 0$

**88.** $x \sin x + y \cos y = 3$

**89.** $\cos(x + y) + x \sin y = 1$

**90.** $\csc x + x \cot y = 1$

**91.** $\sec xy = 8$

**92.** $\cos^2 x + \sin^2 y = 1$

*In Exercises 93 and 94, write the expression as a function of $x$.*

**93.** $\displaystyle\lim_{h \to 0} \dfrac{2(x + h)^5 - (x + h)^3 - 2x^5 + x^3}{h}$

**94.** $\displaystyle\lim_{h \to 0} \dfrac{\sqrt{x + h} + \dfrac{1}{x + h} - \sqrt{x} - \dfrac{1}{x}}{h}$

*In Exercises 95–100, find the differential of the function at the indicated number.*

**95.** $f(x) = \sqrt{x} + \dfrac{1}{\sqrt{x}};\quad x = 4$

**96.** $f(x) = \sqrt{2x^2 + x + 1};\quad x = 1$

**97.** $f(x) = x(2x^2 - 1)^{1/3};\quad x = 1$

**98.** $f(x) = \sec^2 x;\quad x = \dfrac{\pi}{4}$  **99.** $f(x) = x \sin \dfrac{1}{x};\quad x = \dfrac{6}{\pi}$

**100.** $f(x) = \dfrac{\tan x}{1 + \cot x};\quad x = \dfrac{\pi}{4}$

**101.** Find an equation of the tangent line to the graph of $y = (2x)/(\ln x)$ at the point $(e, 2e)$.

**102.** Find an equation of the tangent line to the graph of $y = xe^{-x}$ that is parallel to the line $x - y + 3 = 0$.

*In Exercises 103 and 104, find equations of the tangent line and normal line to the curve at the indicated point.*

**103.** $x^2 + 5xy + y^2 - 7 = 0;\quad (1, 1)$

**104.** $x + \sqrt{xy} + y = 6;\quad (2, 2)$

**105.** Find $d^2y/dx^2$ by implicit differentiation, given $x^3 + y^3 = 1$.

**106.** Find $d^2y/dx^2$ by implicit differentiation, given $\sin 2x + \cos 2y = 1$.

**107.** Find the linearization $L(x)$ of $f(x) = \cos^2 x$ at $\pi/6$.

**108.** Find the linearization of a suitable function, and then use it to approximate $\sqrt[3]{0.00096}$.

**109.** Let $f(x) = x^2 + 1$.
 a. Find the point on the graph of $f$ at which the slope of the tangent line is equal to 2.
 b. Find an equation of the tangent line of part (a).

**110.** Let $f(x) = 2x^3 - 3x^2 - 16x + 3$.
 a. Find the points on the graph of $f$ at which the slope of the tangent line is equal to $-4$.
 b. Find the equation(s) of the tangent line(s) of part (a).

**111.** Let $y = \dfrac{\sec x}{1 + \tan x}$. How fast is $y$ changing when $x = \pi/4$?

**112.** The position of a particle moving along a coordinate line is
$$s(t) = t^3 - 12t + 1 \qquad t \geq 0$$
where $s(t)$ is measured in feet and $t$ in seconds.
 a. Find the velocity and acceleration functions of the particle.
 b. Determine the times(s) when the particle is stationary.
 c. When is the particle moving in the positive direction and when is it moving in the negative direction?
 d. Construct a schematic showing the position of the body at any time $t$.
 e. What is the total distance traveled by the particle in the time interval $[0, 3]$?

**113.** The position function of a body moving along a coordinate line is
$$s(t) = 10t^{2/3} - t^{5/3} \qquad t \geq 0$$
where $s(t)$ is measured in feet and $t$ in seconds. Find the velocity and acceleration functions for the body.

**114.** The position function of a particle moving along a coordinate line is
$$s(t) = 5\cos\left(t + \frac{\pi}{4}\right) \qquad t \geq 0$$
where $s(t)$ is measured in feet and $t$ in seconds.
 a. Find the velocity and acceleration functions for the particle.
 b. At what time does the particle first reach the origin?
 c. What are the velocity and acceleration of the particle when it first reaches the origin?

**115. Velocity of Blood** The velocity (in centimeters per second) of blood $r$ cm from the central axis of an artery is given by
$$v(r) = k(R^2 - r^2)$$
where $k$ is a constant and $R$ is the radius of the artery. Suppose that $k = 1000$ and $R = 0.2$. Find $v(0.1)$, and $v'(0.1)$ and interpret your results.

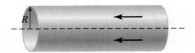

**116. Traffic Flow** The average speed of traffic flow on a stretch of Route 106 between 6 A.M. and 10 A.M. on a typical weekday is approximated by the function
$$f(t) = 20t - 45t^{0.45} + 50 \qquad 0 \leq t \leq 4$$
where $f(t)$ is measured in miles per hour and $t$ is measured in hours with $t = 0$ corresponding to 6 A.M. How fast is the average speed of traffic flow changing at 7 A.M.? At 8 A.M.?

**117. Surface Area of a Human Body** An empirical formula by E.F. Dubois relates the surface area $S$ of a human body (in square meters) to its mass $W$ in kilograms and its height $H$ in centimeters. The formula given by
$$S = 0.007184W^{0.425}H^{0.725}$$
is used by physiologists in metabolism studies. Suppose that a man is 1.83 m tall. How fast does his surface area change with respect to his mass when his mass is 80 kg?

**118.** Refer to Exercise 117. If the measurement of the mass of the man is subject to a maximum error of 0.5 kg, what is the percentage error in the calculation of the man's surface area?

**119. Number of Hours of Daylight** The number of hours of daylight on a particular day of the year in Boston is approximated by the function

$$f(t) = 3 \sin\left[\frac{2\pi}{365}(t - 79)\right] + 12$$

where $t = 0$ corresponds to January 1. Compute $f'(79)$, and interpret your result.

**120. Projected Profit** The management of the company that makes Long Horn Barbeque Sauce estimates that the daily profit from the production and sale of $x$ cases of sauce is

$$P(x) = -0.000002x^3 + 6x - 350$$

dollars. Management forecasts that they will sell, on average, 900 cases of the sauce per day in the next several months. If the forecast is subject to a maximum error of 10%, find the corresponding error in the company's projected average daily profit.

**121.** The volume of a circular cone is $V = \pi r^2 h/3$, where $r$ is the radius of the base and $h$ is the height.

**a.** What is the rate of change of the volume with respect to the height if the radius is constant?
**b.** What is the rate of change of the volume with respect to the radius if the height is constant?

**122. Depreciation of Equipment** For assets such as machines, whose market values drop rapidly in the early years of usage, businesses often use the double declining balance method. In practice, a business firm normally employs the double declining balance method for depreciating such assets for a certain number of years and then switches over to the linear method. The double declining balance formula is

$$V(n) = C\left(1 - \frac{2}{N}\right)^n$$

where $V(n)$ denotes the book value of the assets at the end of $n$ years and $N$ is the number of years over which the asset is depreciated.

**a.** Find $V'(n)$.
**b.** What is the relative rate of change of $V(n)$?

**123.** Show that if the equation of motion of an object is $x(t) = ae^t + be^{-t}$, where $a$ and $b$ are constants, then its acceleration is numerically equal to the distance covered by the object.

**124.** The equation of motion of a mass attached to a spring and a dashpot damping device is

$$x(t) = e^{-2t}(2 \cos 4t - 3 \sin 4t)$$

where $x(t)$, measured in feet, is the displacement from the equilibrium position of the spring system and $t$ is measured in seconds. Find expressions for the velocity and acceleration of the mass.

**125.** Given the equation $x^2 - y^2 = 9$, where $x$ and $y$ are both functions of $t$, find $dy/dt$ if $x = 5$, $y = 4$, and $dx/dt = 3$.

**126.** Given the equation $\sin 2x + \cos 2y = 1$, where $x$ and $y$ are both functions of $t$, find $dx/dt$ if $x = \pi/2$, $y = 0$, and $dy/dt = -1$.

**127. Watching a Boat Race** A spectator is watching a rowing race from the edge of a riverbank. The lead boat is moving in a straight line that is 120 ft from the river bank. If the boat is moving at a constant speed of 20 ft/sec, how fast will the boat be moving away from the spectator when it is 50 ft past her?

**128. Watching a Space Shuttle Launch** At a distance of 6000 ft from the launch site, a spectator is observing a space shuttle being launched. If the space shuttle lifts off vertically, at what rate is the distance between the spectator and the space shuttle changing with respect to the angle of elevation $\theta$ at the instant when the angle is 30° and the shuttle is traveling at 600 mph (880 ft/sec)?

## PROBLEM-SOLVING TECHNIQUES

The following example shows that rewriting a function in an alternative form sometimes pays dividends.

**EXAMPLE**    Find $f^{(n)}(x)$ if $f(x) = \dfrac{x}{x^2 - 1}$.

**Solution**    Our first instinct is to use the Quotient Rule to compute $f'(x), f''(x)$, and so on. The expectation here is either that the rule for $f^{(n)}$ will become apparent or that at least a pattern will emerge that will enable us to guess at the form for $f^{(n)}(x)$. But the futility of this approach will be evident when you compute the first two derivatives of $f$.

Let's see whether we can transform the expression for $f(x)$ before we differentiate. You can verify that $f(x)$ can be written as

$$f(x) = \frac{x}{x^2 - 1} = \frac{\frac{1}{2}(x - 1) + \frac{1}{2}(x + 1)}{(x + 1)(x - 1)} = \frac{1}{2}\left[\frac{1}{x + 1} + \frac{1}{x - 1}\right]$$

There is actually a systematic method for obtaining the last expression for $f(x)$. It is called *partial fraction decomposition* and will be taken up in Section 6.4. Differentiating, we obtain

$$f'(x) = \frac{1}{2}\frac{d}{dx}\left[\frac{1}{x + 1} + \frac{1}{x - 1}\right]$$

$$= \frac{1}{2}\left[\frac{d}{dx}(x + 1)^{-1} + \frac{d}{dx}(x - 1)^{-1}\right]$$

$$= \frac{1}{2}[(-1)(x + 1)^{-2} + (-1)(x - 1)^{-2}]$$

$$f''(x) = \frac{1}{2}[(-1)(-2)(x + 1)^{-3} + (-1)(-2)(x - 1)^{-3}]$$

$$f'''(x) = \frac{1}{2}[(-1)(-2)(-3)(x + 1)^{-4} + (-1)(-2)(-3)(x - 1)^{-4}]$$

$$= \frac{1}{2}[(-1)^3 3!(x + 1)^{-4} + (-1)^3 3!(x - 1)^{-4}]$$

$$\vdots$$

$$f^{(n)}(x) = \frac{(-1)^n n!}{2}\left[\frac{1}{(x + 1)^{n+1}} + \frac{1}{(x - 1)^{n+1}}\right]$$

where $n! = n(n - 1)(n - 2) \cdots (1)$ and $0! = 1$.    ■

## CHALLENGE PROBLEMS

**1.** Find $\lim\limits_{x \to 2} \dfrac{x^{10} - 2^{10}}{x^5 - 2^5}$.

**2.** Find the derivative of $y = \sqrt{x + \sqrt{x + \sqrt{x}}}$.

**3. a.** Verify that $\dfrac{2x + 1}{x^2 + x - 2} = \dfrac{1}{x + 2} + \dfrac{1}{x - 1}$.

   **b.** Find $f^{(n)}(x)$ if $f(x) = \dfrac{2x + 1}{x^2 + x - 2}$.

**4.** Find the values of $x$ for which $f$ is differentiable.
   **a.** $f(x) = \sin |x|$          **b.** $f(x) = |\sin x|$

**5.** Find $f^{(10)}(x)$ if $f(x) = \dfrac{1 + x}{\sqrt{1 - x}}$.

   **Hint:** Show that $f(x) = \dfrac{2}{\sqrt{1 - x}} - \sqrt{1 - x}$.

**6.** Find $f^{(n)}(x)$ if $f(x) = \dfrac{ax + b}{cx + d}$.

**7.** Suppose that $f$ is differentiable and $f(a + b) = f(a)f(b)$ for all real numbers $a$ and $b$. Show that $f'(x) = f'(0)f(x)$ for all $x$.

**8.** Suppose that $f^{(n)}(x) = 0$ for every $x$ in an interval $(a, b)$ and $f(c) = f'(c) = \cdots = f^{(n-1)}(c) = 0$ for some $c$ in $(a, b)$. Show that $f(x) = 0$ for all $x$ in $(a, b)$.

**9.** Let $F(x) = f(\sqrt{1 + x^2})$, where $f$ is a differentiable function. Find $F'(x)$.

 **10.** Determine the values of $b$ and $c$ such that the parabola $y = x^2 + bx + c$ is tangent to the graph of $y = \sin x$ at the point $(\frac{\pi}{6}, \frac{1}{2})$. Plot the graphs of both functions on the same set of axes.

**11.** Suppose $f$ is defined on $(-\infty, \infty)$ and satisfies $|f(x) - f(y)| \le (x - y)^2$ for all $x$ and $y$. Show that $f$ is a constant function.
   **Hint:** Look at $f'(x)$.

**12.** Use the definition of the derivative to find the derivative of $f(x) = \tan ax$.

**13.** Find $y''$ at the point $(1, -2)$ if
$$2x^2 + 2xy + xy^2 - 3x + 3y + 7 = 0$$

**14.** Prove that the function $f(x) = |\ln x|$ is not differentiable at $x = 1$.

**15.** Let $g = f^{-1}$ be the inverse function of $f$. Show that if $f$ has derivatives of order 3, then
$$g''' = \dfrac{3(f'')^2 - f'f'''}{(f')^5} \qquad f' \ne 0$$

**16.** Let $f$ be positive and differentiable. Prove that the graphs of $y = f(x)$ and $y = f(x) \sin ax$ are tangent to each other at their points of intersection.

**17.** Let $f$ be defined by
$$f(x) = \begin{cases} \dfrac{2x}{3 + e^{1/x}} & \text{if } x \ne 0 \\ 0 & \text{if } x = 0 \end{cases}$$

Is $f$ differentiable at $x = 0$? Explain.
   **Hint:** Use the definition of the derivative.

**18.** Find $y'$ if $y = \log_{f(x)} g(x)$, where $f$ and $g$ are differentiable functions with $f(x) > 0$ and $g(x) > 0$ for all values of $x$.

Antarctic glaciers are calving into the ocean with greater frequency as a result of global warming. A major cause of global warming is the increase of carbon dioxide in the atmosphere. We can use the derivative to help us study the rate of change of the average amount of atmospheric $CO_2$.

Marco Simoni/Getty Images

# 3

# Applications of the Derivative

**IN THIS CHAPTER** we continue to explore the power of the derivative of a function as a tool for solving problems. We will see how the first and second derivatives of a function can be used to help us sketch the graph of the function. We will also see how the derivative of a function can help us find the maximum and minimum values of the function. Determining these values is important because many practical problems call for finding one or both of these extreme values. For example, an engineer might be interested in finding the maximum horsepower a prototype engine can deliver, and a businesswoman might be interested in the level of production of a certain commodity that will minimize the unit cost of producing that commodity.

**V** This symbol indicates that one of the following video types is available for enhanced student learning at **www.academic.cengage.com/login:**
- Chapter lecture videos
- Solutions to selected exercises

## 3.1  Extrema of Functions

### ■ Absolute Extrema of Functions

The graph of the function $f$ in Figure 1 gives the altitude of a hot-air balloon over the time interval $I = [a, d]$. The point $(c, f(c))$, the lowest point on the graph of $f$, tells us that the hot-air balloon attains its minimum altitude, $f(c)$, at time $t = c$. The smallest value attained by $f$ for all values of $t$ in the domain $I$ of $f$, $f(c)$, is called the *absolute minimum value* of $f$ on $I$. Similarly, the point $(d, f(d))$, the highest point on the graph of $f$, tells us that the balloon attains its maximum altitude, $f(d)$, at time $t = d$. The largest value attained by $f$ for all values of $t$ in $I$ is called the *absolute maximum value* of $f$ on $I$.

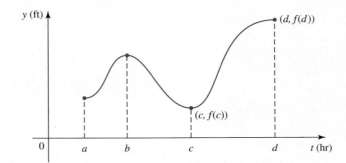

**FIGURE 1**
The altitude $f(t)$ of a hot-air balloon for $a \leq t \leq d$

More generally, we have the following definitions.

> **DEFINITIONS    Extrema of a Function $f$**
>
> A function $f$ has an **absolute maximum** at $c$ if $f(x) \leq f(c)$ for all $x$ in the domain $D$ of $f$. The number $f(c)$ is called the **maximum value** of $f$ on $D$. Similarly, $f$ has an **absolute minimum** at $c$ if $f(x) \geq f(c)$ for all $x$ in $D$. The number $f(c)$ is called the **minimum value** of $f$ on $D$. The absolute maximum and absolute minimum values of $f$ on $D$ are called the **extreme values,** or **extrema,** of $f$ on $D$.

**EXAMPLE 1**  Find the extrema of the function, if any, by examining its graph.

**a.** $f(x) = x^2$    **b.** $g(x) = -x^2$    **c.** $h(x) = \dfrac{1}{x}$    **d.** $k(x) = \dfrac{x}{\sqrt{x^2 + 7}}$

**Solution**    The graphs of the functions $f$, $g$, $h$, and $k$ are shown in Figure 2.

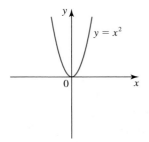

**(a)** $f$ has a minimum at 0.

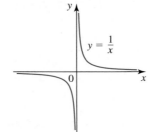

**(c)** $h$ has no extrema.

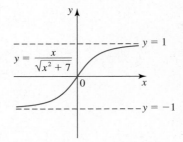

**(b)** $g$ has a maximum at 0.

**(d)** $k$ has no extrema.

**FIGURE 2**

**a.** $f$ has a minimum value of 0 at 0. Next, since the values of $f$ are not bounded above, $f$ has no maximum value.

**b.** $g$ has a maximum value of 0 at 0. Also, because the values of $g$ are not bounded below, $g$ has no minimum value.

**c.** The values of $h$ are neither bounded above nor bounded below, so $h$ has no absolute extrema.

**d.** As $x$ gets larger and larger, $k(x)$ gets closer and closer to 1. But this value is never attained; that is, a real number $c$ does not exist such that $k(c) = 1$. Therefore, $k$ has no maximum value. Similarly, you can show that $k$ has no minimum value.

**EXAMPLE 2** Find the extrema of the function:

**a.** $f(x) = x^2$   $-1 < x < 2$

**b.** $g(x) = x^2$   $-1 \le x \le 2$

**Solution**

**a.** The graph of $f$ is shown in Figure 3a. We see that $f$ has a minimum value of 0 at 0. Next, observe that as $x$ approaches 2 through values less than 2, $f(x)$ increases and approaches 4. But $f$ never attains the value 4. Therefore, $f$ does not have a maximum.

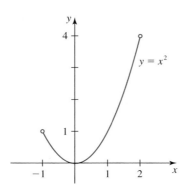

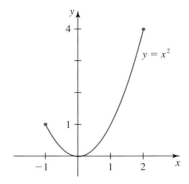

**FIGURE 3**       (a) $f$ has a minimum at 0.                    (b) $g$ has a minimum at 0 and a maximum at 2.

**b.** The graph of $g$ is shown in Figure 3b. As before, we see that $g$ has a minimum value of 0 at 0. Next, because 2 lies in the domain of $g$, we see that $g$ does attain a largest value, namely, $g(2) = 4$.

## ■ Relative Extrema of Functions

If you refer once again to the graph of the function $f$ giving the altitude of a hot-air balloon over the interval $[a, d]$ shown in Figure 4, you will see that the point $(b, f(b))$ is the highest point on the graph of $f$ when compared to *neighboring points*. (For example, it is the highest point when compared to the points $(t, f(t))$, where $a < t < c$.) This tells us that $f(b)$ is the highest altitude attained by the balloon when considered over a small time interval containing $t = b$. The value $f(b)$ is called a *relative* (or *local*) *maximum value of $f$*.

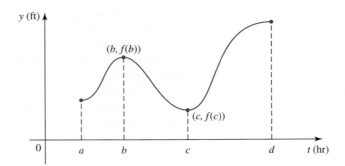

**FIGURE 4**
The altitude of a hot-air
balloon for $a \le t \le d$

Similarly, the point $(c, f(c))$ is the lowest point on the graph of $f$ when compared to points nearby. (For example, it is the lowest point when compared to the points $(t, f(t))$, where $b < t < d$.) This tells us that the balloon attains the lowest altitude at $t = c$ when considered over a small time interval containing $t = c$. The value $f(c)$ is called a *relative* (or *local*) *minimum value* of $f$. Recall that $f(c)$ also happens to be the (absolute) minimum value of $f$, as we observed earlier.

More generally, we have the following definition.

---

**DEFINITIONS    Relative Extrema of a Function**

A function $f$ has a **relative** (or **local**) **maximum** at $c$ if $f(c) \ge f(x)$ for all values of $x$ in some open interval containing $c$. Similarly, $f$ has a **relative** (or **local**) **minimum** at $c$ if $f(c) \le f(x)$ for all values of $x$ in some open interval containing $c$.

---

The function $f$ whose graph is shown in Figure 5 has a relative maximum at $a$ and at $c$ and a relative minimum at $b$ and at $d$. The graph of $f$ suggests that at a point corresponding to a relative extremum of $f$, either the tangent line is horizontal or it does not exist. Put another way, the values of $x$ that correspond to these points are precisely the numbers in the domain of $f$ at which $f'$ is zero or $f'$ does not exist.

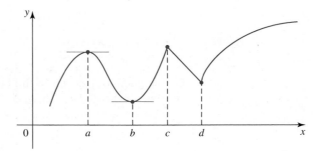

**FIGURE 5**
The function $f$ has relative extrema
at $a$, $b$, $c$, and $d$. The tangent lines
at $a$ and $b$ are horizontal. There
are no tangent lines at $c$ and $d$.

These observations suggest the following theorem, which tells us where the relative extrema of a function may occur.

---

**THEOREM 1    Fermat's Theorem**

If $f$ has a relative extremum at $c$, then either $f'(c) = 0$ or $f'(c)$ does not exist.

**PROOF** First, suppose that $f$ has a relative maximum at $c$. If $f$ is not differentiable at $c$, then there is nothing to prove. So let's suppose that $f'(c)$ exists. Since $f$ has a relative maximum at $c$, there exists an open interval, $I$, such that $f(x) \leq f(c)$ for all $x$ in $I$. This implies that if we pick $h$ to be positive and sufficiently small (so that $c + h$ lies in $I$), then

$$f(c + h) \leq f(c) \quad \text{or} \quad f(c + h) - f(c) \leq 0$$

Multiplying both sides of the latter inequality by $1/h$, where $h > 0$, we obtain

$$\frac{f(c + h) - f(c)}{h} \leq 0$$

Taking the right-hand limit of both sides of this inequality gives

$$\lim_{h \to 0^+} \frac{f(c + h) - f(c)}{h} \leq \lim_{h \to 0^+} 0 = 0 \qquad \text{By Theorem 3 of Section 1.2}$$

Since $f'(c)$ exists, we have

$$f'(c) = \lim_{h \to 0} \frac{f(c + h) - f(c)}{h} = \lim_{h \to 0^+} \frac{f(c + h) - f(c)}{h}$$

and we have shown that $f'(c) \leq 0$.

Next, we pick $h$ to be negative and sufficiently small (so that $c + h$ lies in $I$). Then

$$f(c + h) \leq f(c) \quad \text{or} \quad f(c + h) - f(c) \leq 0$$

Upon multiplying this last inequality by $1/h$ and reversing the direction of the inequality (because $1/h < 0$), we have

$$f'(c) = \lim_{h \to 0} \frac{f(c + h) - f(c)}{h} = \lim_{h \to 0^-} \frac{f(c + h) - f(c)}{h} \geq 0$$

Thus, we have shown that $f'(c) \leq 0$ and $f'(c) \geq 0$, simultaneously. Therefore, $f'(c) = 0$. This proves the theorem for the case in which $f$ has a relative maximum at $c$. The case in which $f$ has a relative minimum at $c$ can be proved in a similar manner (see Exercise 90). ∎

The values of $x$ at which $f'$ is zero or $f'$ does not exist are given a special name.

---

**DEFINITION** Critical Number of $f$

A **critical number** of a function $f$ is any number $c$ in the domain of $f$ at which $f'(c) = 0$ or $f'(c)$ does not exist.

---

Theorem 1 states that a relative extremum of $f$ can occur only at a critical number of $f$. It is important to realize, however, that the converse of Theorem 1 is false. In other words, you may *not* conclude that if $c$ is a critical number of $f$, then $f$ must have a relative extremum at $c$. (See Example 3.)

**EXAMPLE 3** Show that zero is a critical number of each of the functions $f(x) = x^3$ and $g(x) = x^{1/3}$ but that neither function has a relative extremum at 0.

**Solution**    The graphs of $f$ and $g$ are shown in Figure 6. Since $f'(x) = 3x^2 = 0$ if $x = 0$, we see that 0 is a critical number of $f$. But observe that $f(x) < 0$ if $x < 0$ and $f(x) > 0$ if $x > 0$, and this tells us that $f$ cannot have a relative extremum at 0.

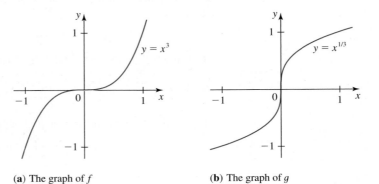

**FIGURE 6**
Both $f$ and $g$ have 0 as a critical number, but neither function has a relative extremum at 0.

(a) The graph of $f$          (b) The graph of $g$

Next, we compute

$$g'(x) = \frac{1}{3}x^{-2/3} = \frac{1}{3x^{2/3}}$$

Note that $g'$ is not defined at 0, but $g$ is; so 0 is a critical number of $g$. Observe that $g(x) < 0$ if $x < 0$ and $g(x) > 0$ if $x > 0$, so $g$ cannot have a relative extremum at 0. ∎

**EXAMPLE 4**    Find the critical numbers of $f(x) = x - 3x^{1/3}$.

**Solution**    The derivative of $f$ is

$$f'(x) = 1 - x^{-2/3} = \frac{x^{2/3} - 1}{x^{2/3}}$$

Observe that $f'$ is not defined at 0 and also $f'(x) = 0$ if $x = \pm 1$. Therefore, the critical numbers of $f$ are $-1$, 0, and 1. ∎

We will develop a systematic method for finding the relative extrema of a function in Section 3.3. For the rest of this section we will develop techniques for finding the extrema of continuous functions defined on closed intervals.

## Finding the Extreme Values of a Continuous Function on a Closed Interval

As you saw in the preceding examples, an arbitrary function might or might not have a maximum value or a minimum value. But there is an important case in which the extrema always exist for a function. The conditions are spelled out in Theorem 2.

**THEOREM 2    The Extreme Value Theorem**

If $f$ is continuous on a closed interval $[a, b]$, then $f$ attains an absolute maximum value $f(c)$ for some number $c$ in $[a, b]$ and an absolute minimum value $f(d)$ for some number $d$ in $[a, b]$.

In certain applications, not only is a function continuous on a closed interval [a, b], but it is also differentiable, with the possible exception of a finite set of numbers, on the open interval (a, b). In such cases, the following procedure can be used to find the extrema of the function.

> **Guidelines for Finding the Extrema of a Continuous Function f on [a, b]**
> 1. Find the critical numbers of f that lie in (a, b).
> 2. Compute the value of f at each of these critical numbers, and also compute f(a) and f(b).
> 3. The absolute maximum value of f and the absolute minimum value of f are precisely the largest and the smallest numbers found in Step 2.

This procedure can be justified as follows: If an extremum of f occurs at a number in the open interval (a, b), then it must also be a relative extremum of f; hence it must occur at a critical number of f. Otherwise, the extremum of f must occur at one or both of the endpoints of the interval [a, b]. (See Figure 7.)

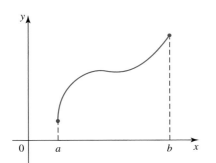

(**a**) The extreme values of f occur at the endpoints.

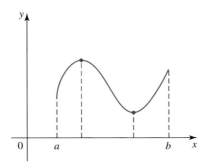

(**b**) The extreme values of f occur at critical numbers.

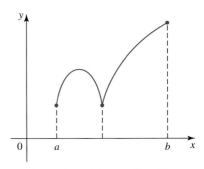

(**c**) The absolute minimum value of f occurs at both an endpoint and a critical number of f, whereas the absolute maximum value of f occurs at an endpoint.

**FIGURE 7**
f is continuous on [a, b].

**EXAMPLE 5** Find the extreme values of the function $f(x) = 3x^4 - 4x^3 - 8$ on $[-1, 2]$.

**Solution** Since f is a polynomial function, it is continuous everywhere; in particular, it is continuous on the closed interval $[-1, 2]$. Therefore, we can use the Extreme Value Theorem.

First, we find the critical numbers of f in $(-1, 2)$:

$$f'(x) = 12x^3 - 12x^2$$
$$= 12x^2(x - 1)$$

Observe that $f'$ is continuous on $(-1, 2)$. Next, setting $f'(x) = 0$ gives $x = 0$ or $x = 1$. Therefore, 0 and 1 are the only critical numbers of f in $(-1, 2)$.

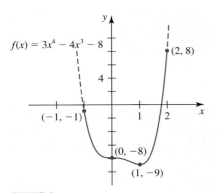

$f(x) = 3x^4 - 4x^3 - 8$

(2, 8)

(−1, −1)

(0, −8)

(1, −9)

**FIGURE 8**
The maximum value of $f$ is 8, and the minimum value is −9.

Next, we compute $f(x)$ at these critical numbers as well as at the endpoints −1 and 2. These values are shown in the following table.

| $x$ | −1 | 0 | 1 | 2 |
|-----|----|----|----|----|
| $f(x)$ | −1 | −8 | −9 | 8 |

From the table we see that $f$ attains the absolute maximum value of 8 at 2 and the absolute minimum value of −9 at 1. The graph of $f$ shown in Figure 8 confirms our results. (You don't need to draw the graph to solve the problem.) ∎

**EXAMPLE 6**  Find the extreme values of the function $f(x) = 2\cos x - x$ on $[0, 2\pi]$.

**Solution**   The function $f$ is continuous everywhere; in particular, it is continuous on the closed interval $[0, 2\pi]$. Therefore, the Extreme Value Theorem is applicable.
   First, we find the critical numbers of $f$ in $(0, 2\pi)$. We have

$$f'(x) = -2\sin x - 1$$

Observe that $f'$ is continuous on $(0, 2\pi)$. Setting $f'(x) = 0$ gives

$$-2\sin x - 1 = 0$$

$$\sin x = -\frac{1}{2}$$

Thus, $x = 7\pi/6$ or $11\pi/6$. (Remember $x$ lies in $(0, 2\pi)$.) So $7\pi/6$ and $11\pi/6$ are the only critical numbers of $f$ in $(0, 2\pi)$.
   Next, we compute the values of $f$ at these critical numbers as well as at the endpoints 0 and $2\pi$. These values are shown in the following table.

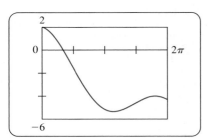

**FIGURE 9**
The graph of $f(x) = 2\cos x - x$ on $[0, 2\pi]$

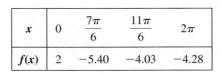

| $x$ | 0 | $\dfrac{7\pi}{6}$ | $\dfrac{11\pi}{6}$ | $2\pi$ |
|-----|----|----|----|----|
| $f(x)$ | 2 | −5.40 | −4.03 | −4.28 |

From the table we see that $f$ attains the absolute maximum value of 2 at 0 and the absolute minimum value of approximately −5.4 at $7\pi/6$. The graph of $f$ shown in Figure 9 confirms our results. ∎

## ■ An Optimization Problem

The solution to many practical problems involves finding the absolute maximum or the absolute minimum of a function. If we know that the function to be optimized is continuous on a closed interval, then the techniques of this section can be used to solve the problem, as illustrated in the following example.

**EXAMPLE 7** **Maximum Deflection of a Beam** Figure 10 depicts a beam of length $L$ and uniform weight $w$ per unit length that is rigidly fixed at one end and simply supported at the other. An equation of the elastic curve (the dashed curve in the figure) is

$$y = \frac{w}{48EI}(2x^4 - 5Lx^3 + 3L^2x^2)$$

where the product $EI$ is a constant called the *flexural rigidity* of the beam. Show that the maximum deflection (the displacement of the elastic curve from the $x$-axis) occurs at $x = (15 - \sqrt{33})L/16 \approx 0.578L$ and has a magnitude of approximately $0.0054wL^4/(EI)$.

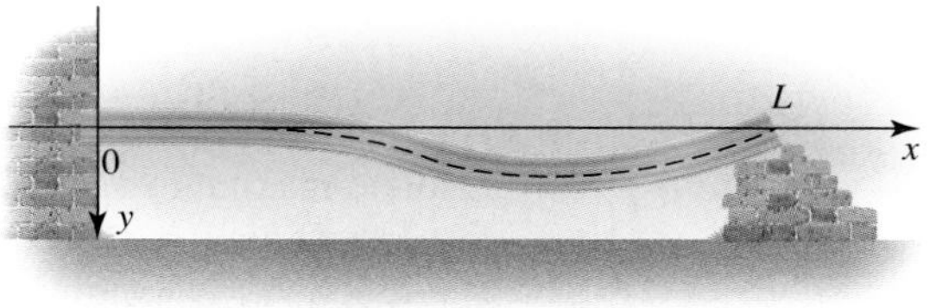

**FIGURE 10**
The beam is rigidly fixed at $x = 0$ and simply supported at $x = L$. Note the orientation of the $y$-axis.

**Solution** We wish to find the value of $x$ on the closed interval $[0, L]$ at which the function $f$ defined by

$$f(x) = \frac{w}{48EI}(2x^4 - 5Lx^3 + 3L^2x^2)$$

attains its absolute maximum value. Since $f$ is continuous on $[0, L]$, this value must be attained at a critical number of $f$ in $(0, L)$ or at an endpoint of the interval. To find the critical numbers of $f$, we compute

$$f'(x) = \frac{w}{48EI}(8x^3 - 15Lx^2 + 6L^2x)$$

$$= \frac{w}{48EI}x(8x^2 - 15Lx + 6L^2)$$

Setting $f'(x) = 0$ gives $x = 0$ or

$$x = \frac{15L \pm \sqrt{225L^2 - 192L^2}}{16}$$

$$= \frac{15L \pm \sqrt{33}L}{16}$$

Because $(15 + \sqrt{33})L/16 > L$, we see that the sole critical number of $f$ in $(0, L)$ is $x = (15 - \sqrt{33})L/16 \approx 0.578L$. Evaluating $f$ at $0$, $0.578L$, and $L$, we obtain the following table of values.

| $f(0)$ | $f(0.578L)$ | $f(L)$ |
|--------|-------------|--------|
| $0$ | $\dfrac{0.0054wL^4}{EI}$ | $0$ |

We conclude that the maximum deflection occurs at $x = (15 - \sqrt{33})L/16 \approx 0.578L$ and has a magnitude of approximately $0.0054wL^4/(EI)$. ■

Our final example shows how a graphing utility can be used to approximate the maximum and minimum values of a continuous function defined on a closed interval. But to obtain the *exact* values, we must solve the problem analytically.

**EXAMPLE 8**  Let $f(x) = 2 \sin x + \sin 2x$.

**a.** Use a graphing utility to plot the graph of $f$ using the viewing window $\left[0, \frac{3\pi}{2}\right] \times [-3, 3]$. Find the approximate absolute maximum and absolute minimum values of $f$ on the interval $\left[0, \frac{3\pi}{2}\right]$.

**b.** Obtain the exact absolute maximum and absolute minimum values of $f$ analytically.

**Solution**

**a.** The required graph is shown in Figure 11. From the graph we see that the absolute maximum value of $f$ is approximately 2.6 obtained when $x \approx 1$. The absolute minimum value of $f$ is $-2$ obtained when $x = 3\pi/2$.

**b.** The function $f$ is continuous everywhere and, in particular, on the interval $\left[0, \frac{3\pi}{2}\right]$. We find

$$f'(x) = 2 \cos x + 2 \cos 2x$$
$$= 2 \cos x + 2(\cos^2 x - \sin^2 x) \qquad \cos 2x = \cos^2 x - \sin^2 x$$
$$= 2 \cos x + 2(\cos^2 x - 1 + \cos^2 x) \qquad \sin^2 x = 1 - \cos^2 x$$
$$= 2(2 \cos^2 x + \cos x - 1)$$

Since

$$2 \cos^2 x + \cos x - 1 = (2 \cos x - 1)(\cos x + 1) = 0$$

if $\cos x = -1$ or $\frac{1}{2}$, we see that $x = \pi/3$ or $\pi$. From the following table we see that the absolute maximum value of $f$ is $3\sqrt{3}/2$ and the absolute minimum value of $f$ is $-2$.

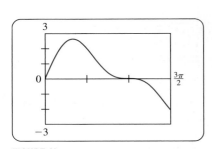

**FIGURE 11**

| $x$ | 0 | $\dfrac{\pi}{3}$ | $\pi$ | $\dfrac{3\pi}{2}$ |
|---|---|---|---|---|
| $f(x)$ | 0 | $\dfrac{3\sqrt{3}}{2}$ | 0 | $-2$ |

# 3.1  CONCEPT QUESTIONS

**1.** Explain each of the following terms: (a) absolute maximum value of a function $f$; (b) relative maximum value of a function $f$. Illustrate each with an example.

**2. a.** What is a critical number of a function $f$?
   **b.** Explain the role of a critical number in determining the relative extrema of a function.

**3. a.** Explain the Extreme Value Theorem in your own words.
   **b.** Describe a procedure for finding the extrema of a continuous function $f$ on a closed interval $[a, b]$.

## 3.1   EXERCISES

*In Exercises 1–6, you are given the graph of a function f defined on the indicated domain. Find the absolute maximum and absolute minimum values of f (if they exist) and where they are attained.*

**1.** *f* defined on (0, 2]

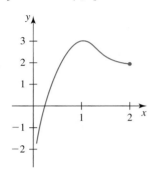

**2.** *f* defined on (−∞, ∞)

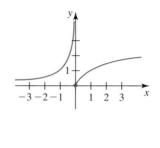

**3.** *f* defined on (−∞, ∞)

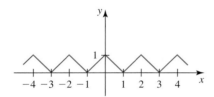

**4.** *f* defined on (−2, ∞)

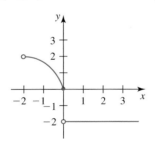

**5.** *f* defined on [0, 5]

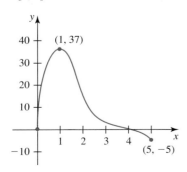

**6.** *f* defined on (−1, ∞)

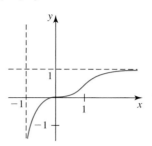

*In Exercises 7–24, sketch the graph of the function and find its absolute maximum and absolute minimum values, if any.*

**7.** $f(x) = 2x + 3$ on $[-1, \infty)$   **8.** $g(x) = -3x + 2$ on $(-1, 2]$

**9.** $h(t) = t^2 - 1$ on $(-1, 0)$   **10.** $f(t) = t^2 - 1$ on $[-1, 0)$

**11.** $g(x) = x^2 + 1$ on $(0, \infty)$   **12.** $h(x) = x^2 + 1$ on $(-2, 1]$

**13.** $f(x) = x^2 - 4x + 3$ on $(-\infty, \infty)$

**14.** $g(x) = 2x^2 - 3x + 1$ on $[0, 1)$

**15.** $f(x) = \dfrac{1}{x}$ on $(0, 1]$   **16.** $g(x) = \dfrac{1}{x}$ on $(-1, 1)$

**17.** $f(x) = |x|$ on $[-2, 1)$   **18.** $g(x) = |2x - 1|$ on $(0, 2]$

**19.** $f(t) = 2\sin t$ on $\left(0, \frac{3\pi}{2}\right)$   **20.** $h(t) = \cos \pi t$ on $\left[\frac{1}{4}, 1\right)$

**21.** $f(x) = e^x$ on $(-\infty, 1]$   **22.** $g(x) = \ln x$ on $(0, e)$

**23.** $f(x) = \begin{cases} x & \text{if } -1 \le x \le 0 \\ 2 - x & \text{if } 0 < x \le 2 \end{cases}$

**24.** $f(x) = \begin{cases} \sqrt{4 - x^2} & \text{if } -2 \le x < 0 \\ -\sqrt{4 - x^2} & \text{if } 0 \le x \le 2 \end{cases}$

*In Exercises 25–40, find the critical number(s), if any, of the function.*

**25.** $f(x) = 2x + 3$   **26.** $g(x) = 4 - 3x$

**27.** $f(x) = 2x^2 + 4x$   **28.** $h(t) = 6t^2 - t - 2$

**29.** $f(x) = x^3 - 6x + 2$   **30.** $g(t) = 2t^3 + 3t^2 - 12t + 4$

**31.** $h(x) = x^4 - 4x^3 + 12$

**32.** $g(t) = 3t^4 + 4t^3 - 12t^2 + 8$

**33.** $f(x) = x^{2/3}$   **34.** $g(t) = 4t^{1/3} + 3t^{4/3}$

**35.** $h(u) = \dfrac{u}{u^2 + 1}$   **36.** $g(x) = \dfrac{x^2}{x^2 + 3}$

**37.** $f(t) = \cos^2(2t)$   **38.** $g(\theta) = 2\sin\theta - \cos 2\theta$

**39.** $f(x) = e^{x^2} - \pi$   **40.** $g(t) = t^2 \ln t$

*In Exercises 41–60, find the absolute maximum and absolute minimum values, if any, of the function.*

**41.** $f(x) = x^2 - x - 2$ on $[0, 2]$

**42.** $f(x) = -x^2 + 4x + 3$ on $[-1, 3]$

**43.** $h(x) = x^3 + 3x^2 + 1$ on $[-3, 2]$

**44.** $f(t) = -2t^3 + 3t^2 + 12t + 3$ on $[-2, 3]$

**45.** $g(x) = 3x^4 + 4x^3 + 1$ on $[-2, 1]$

**46.** $f(x) = 2x^4 - \dfrac{8}{3}x^3 - 8x^2 + 12$ on $[-2, 3]$

**47.** $f(x) = \dfrac{x}{x^2 + 1}$ on $[-1, 2]$ **48.** $g(u) = \dfrac{\sqrt{u}}{u^2 + 1}$ on $[0, 2]$

**49.** $g(v) = \dfrac{v}{v - 1}$ on $[2, 4]$ **50.** $f(x) = 2x + \dfrac{1}{x}$ on $[-1, 3]$

**51.** $f(x) = x - 2\sqrt{x}$ on $[0, 9]$ **52.** $f(t) = \dfrac{1}{8}t^2 - 4\sqrt{t}$ on $[0, 9]$

**53.** $f(x) = x^{2/3}(x^2 - 4)$ on $[-1, 2]$

**54.** $g(x) = x\sqrt{4 - x^2}$ on $[0, 2]$

**55.** $f(x) = 2 + 3 \sin 2x$ on $\left[0, \frac{\pi}{2}\right]$

**56.** $g(x) = \cos x - \sin x$ on $[0, 2\pi]$

**57.** $f(x) = xe^{-x}$; $[-1, 2]$ **58.** $f(x) = e^{2x} - e^x$; $[-2, 0]$

**59.** $f(x) = x \ln x - x$; $\left[\frac{1}{2}, 2\right]$ **60.** $f(x) = \dfrac{\ln x + 1}{x}$; $\left[\frac{1}{2}, 3\right]$

**61. Maximizing Profit** The total daily profit in dollars realized by the TKK Corporation in the manufacture and sale of $x$ dozen recordable DVDs is given by the total profit function

$$P(x) = -0.000001x^3 + 0.001x^2 + 5x - 500$$
$$0 \le x \le 2000$$

Find the level of production that will yield a maximum daily profit.

**62. Reaction to a Drug** The strength of a human body's reaction to a dosage $D$ of a certain drug is given by

$$R = D^2\left(\frac{k}{2} - \frac{D}{3}\right)$$

where $k$ is a positive constant. Show that the maximum reaction is achieved if the dosage is $k$ units.

**63. Traffic Flow** The average speed of traffic flow on a stretch of Route 124 between 6 A.M. and 10 A.M. on a typical weekday is approximated by the function

$$f(t) = 20t - 40\sqrt{t} + 50 \qquad 0 \le t \le 4$$

where $f(t)$ is measured in miles per hour and $t$ is measured in hours, with $t = 0$ corresponding to 6 A.M. At what time in the morning is the average speed of traffic flow highest? At what time in the morning is it lowest?

**64. Foreign-Born Medical Residents** The percentage of foreign-born medical residents in the United States from the beginning of 1910 to the beginning of 2000 is approximated by the function

$$P(t) = 0.04363t^3 - 0.267t^2 - 1.59t + 14.7 \qquad 0 \le t \le 9$$

where $t$ is measured in decades with $t = 0$ corresponding to the beginning of 1910. Show that the percentage of foreign-born medical residents was lowest in early 1970.
*Source: Journal of the American Medical Association.*

**65. Brain Growth and IQs** In a study conducted at the National Institute of Mental Health, researchers followed the development of the cortex, the thinking part of the brain, in 307 children. Using repeated magnetic resonance imaging scans from childhood to the late teens, they measured the thickness (in millimeters) of the cortex of children of age $t$ years with the highest IQs: 121 to 149. These data lead to the model

$$S(t) = 0.000989t^3 - 0.0486t^2 + 0.7116t + 1.46$$
$$5 \le t \le 19$$

Show that the cortex of children with superior intelligence reaches maximum thickness around age 11.
*Source: Nature.*

**66. Brain Growth and IQs** Refer to Exercise 65. The researchers at the institute also measured the thickness (also in millimeters) of the cortex of children of age $t$ years who were of average intelligence. These data lead to the model

$$A(t) = -0.00005t^3 - 0.000826t^2 + 0.0153t + 4.55$$
$$5 \le t \le 19$$

Show that the cortex of children with average intelligence reaches maximum thickness at age 6.
*Source: Nature.*

**67. Maximizing Revenue** The quantity demanded per month of the Peget wristwatch is related to the unit price by the demand equation

$$p = \frac{50}{0.01x^2 + 1} \qquad 0 \le x \le 20$$

where $p$ is measured in dollars and $x$ is measured in units of a thousand. How many watches must be sold by the manufacturer to maximize its revenue?
**Hint:** Recall that the revenue $R = px$.

**68. Poiseuille's Law** According to Poiseuille's Law, the velocity (in centimeters per second) of blood $r$ cm from the central axis of an artery is given by

$$v(r) = k(R^2 - r^2) \qquad 0 \le r \le R$$

where $k$ is a constant and $R$ is the radius of the artery. Show that the flow of blood is fastest along the central axis. Where is the flow of blood slowest?

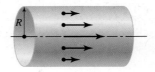

**69. Chemical Reaction** In an autocatalytic chemical reaction the product formed acts as a catalyst for the reaction. If $Q$ is the amount of the original substrate that is present initially and $x$ is the amount of catalyst formed, then the rate of change of the chemical reaction with respect to the amount of catalyst present in the reaction is

$$R(x) = kx(Q - x) \qquad 0 \le x \le Q$$

where $k$ is a constant. Show that the rate of the chemical reaction is greatest at the point at which exactly half of the original substrate has been transformed.

**70. Velocity of Airflow During a Cough** When a person coughs, the trachea (windpipe) contracts, allowing air to be expelled at a maximum velocity. It can be shown that the velocity $v$ of airflow during a cough is given by

$$v = f(r) = kr^2(R - r) \qquad 0 \le r \le R$$

where $r$ is the radius of the trachea in centimeters during a cough, $R$ is the normal radius of the trachea in centimeters, and $k$ is a constant that depends on the length of the trachea. Find the radius for which the velocity of airflow is greatest.

**71. A Mixture Problem** A tank initially contains 10 gal of brine with 2 lb of salt. Brine with 1.5 lb of salt per gallon enters the tank at the rate of 3 gal/min, and the well-stirred mixture leaves the tank at the rate of 4 gal/min. It can be shown that the amount of salt in the tank after $t$ min is $x$ lb, where

$$x = f(t) = 1.5(10 - t) - 0.0013(10 - t)^4 \qquad 0 \le t \le 10$$

What is the maximum amount of salt present in the tank at any time?

**72. Air Pollution** According to the South Coast Air Quality Management district, the level of nitrogen dioxide, a brown gas that impairs breathing, that is present in the atmosphere between 7 A.M. and 2 P.M. on a certain May day in downtown Los Angeles is approximated by

$$I(t) = 0.03t^3(t - 7)^4 + 60.2 \qquad 0 \le t \le 7$$

where $I(t)$ is measured in pollutant standard index (PSI) and $t$ is measured in hours, with $t = 0$ corresponding to 7 A.M. Determine the time of day when the PSI is the lowest and when it is the highest.
*Source: The Los Angeles Times.*

**73. Office Rents** After the economy softened, the sky-high office space rents of the late 1990s started to come down to earth. The function $R$ gives the approximate price per square foot in dollars, $R(t)$, of prime space in Boston's Back Bay and Financial District from the beginning of 1997 ($t = 0$) to the beginning of 2002 ($t = 5$), where

$$R(t) = -0.711t^3 + 3.76t^2 + 0.2t + 36.5 \qquad 0 \le t \le 5$$

Show that the office space rents peaked at about the middle of the year 2000. What was the highest office space rent during the period in question?
*Source: Meredith & Grew Inc./Oncor.*

**74. Maximum Deflection of a Beam** A uniform beam of length $L$ ft and negligible weight rests on supports at both ends. When subjected to a uniform load of $w_0$ lb/ft, it bends and has the *elastic curve* (the dashed curve in the figure below) described by the equation

$$y = \frac{w_0}{24EI}(x^4 - 2Lx^3 + L^3x) \qquad 0 \le x \le L$$

where the product $EI$ is a constant called the *flexural rigidity* of the beam. Show that the maximum deflection of the beam occurs at the midpoint of the beam and that its value is $5w_0L^4/(384EI)$.

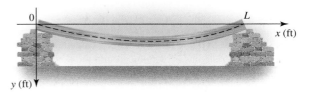

**75. Use of Diesel Engines** Diesel engines are popular in cars in Europe, where fuel prices are high. The percentage of new vehicles in Western Europe equipped with diesel engines is approximated by the function

$$f(t) = 0.3t^4 - 2.58t^3 + 8.11t^2 - 7.71t + 23.75 \qquad 0 \le t \le 4$$

where $t$ is measured in years, with $t = 0$ corresponding to the beginning of 1996.

 **a.** Plot the graph of $f$ using the viewing window $[0, 4] \times [0, 40]$.

**b.** What was the lowest percentage of new vehicles equipped with diesel engines for the period in question?
*Source: German Automobile Industry Association.*

**76. Federal Debt** According to data obtained from the Congressional Budget Office, the national debt (in trillions of dollars) is given by the function

$$f(t) = 0.0022t^3 - 0.0465t^2 + 0.506t + 3.27 \qquad 0 \le t \le 20$$

where $t$ is measured in years, with $t = 0$ corresponding to the beginning of 1990.

**a.** Plot the graph of $f$ using the viewing window $[0, 20] \times [0, 14]$.

**b.** When was the federal debt at the highest level over the period under consideration? What was that level?

*Source:* Congressional Budget Office.

**77.** A cylindrical tank of height $h$ is filled with water. Suppose a jet of water flows through an orifice on the tank. According to Torricelli's law, the velocity of flow of the jet of water is given by $V = \sqrt{2gx}$ where $g$ is the gravitational constant. It can be shown that the range $R$ (in feet) of the jet of water is given by $R = 2\sqrt{x(h - x)}$. Where should the orifice be located so that the jet of water will have the maximum range?

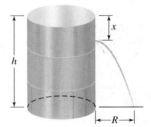

**78. Water Pollution** When organic waste is dumped into a pond, the oxidation process that takes place reduces the pond's oxygen content. However, given time, nature will restore the oxygen content to its natural level. In the accompanying graph, $P(t)$ gives the oxygen content (as a percentage of its normal level) $t$ days after organic waste has been dumped into the pond. Suppose that the oxygen content $t$ days after the organic waste has been dumped into the pond is given by

$$P(t) = 100\left(\frac{t^2 + 10t + 100}{t^2 + 20t + 100}\right)$$

percent of its normal level. Find the coordinates of the point $P$, and explain its significance.

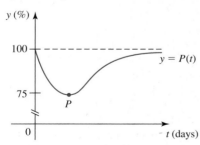

**79. Path of a Boat** A boat leaves the point $O$ (the origin) located on one bank of a river, traveling with a constant speed of 20 mph and always heading toward a dock located at the point $A(1000, 0)$, which is due east of the origin (see the figure). The river flows north at a constant speed of 5 mph. It can be shown that the path of the boat is

$$y = 500\left[\left(\frac{1000 - x}{1000}\right)^{3/4} - \left(\frac{1000 - x}{1000}\right)^{5/4}\right]$$

$$0 \le x \le 1000$$

Find the maximum distance the boat has drifted north during its trip.

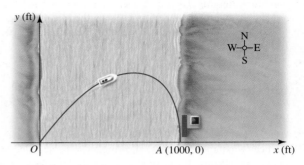

**80. A Motorcyclist's Turn** A motorcyclist weighing 180 lb traveling at a constant speed of 30 mph executes a turn on a road described by the graph of $y = 100e^{0.01x}$, where $-200 \le x \le 50$. It can be shown that the magnitude of the normal force acting on the motorcyclist is approximately

$$F = \frac{10,890e^{0.1x}}{(1 + 100e^{0.2x})^{3/2}}$$

pounds. Find the maximum force acting on the motorcyclist as he makes the turn.

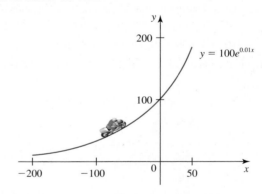

**81. Construction of an AC Transformer** In constructing an AC transformer, a cross-shaped iron core is inserted into a coil (see the figure). If the radius of the coil is $a$, find the values of $x$ and $y$ such that the iron core has the largest surface area.

**Hint:** Let $x = a \cos \theta$ and $y = a \sin \theta$. Then maximize the function

$$S = 4xy + 4y(x - y) = 8xy - 4y^2$$
$$= 4a^2(\sin 2\theta - \sin^2 \theta)$$

on the interval $0 \le \theta \le \frac{\pi}{4}$.

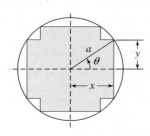

**82.** A body of mass $m$ moves in an elliptical path with a constant angular speed $\omega$ (see the figure). It can be shown that the force acting on the body is always directed toward the origin and has magnitude given by

$$F = m\omega^2 \sqrt{a^2 \cos^2 \omega t + b^2 \sin^2 \omega t} \qquad t \geq 0$$

where $a$ and $b$ are constants with $a > b$. Find the points on the path where the force is greatest and where it is smallest. Does your result agree with your intuition?

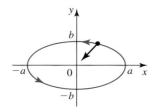

**83.** The object shown in the figure is a crate full of office equipment that weighs $W$ lb. Suppose you try to move the crate by tying a rope around it and pulling on the rope at an angle $\theta$ to the horizontal. Then the magnitude $F$ of the force that is required to set the crate in motion is

$$F = \frac{\mu W}{\mu \sin \theta + \cos \theta} \qquad 0 \leq \theta \leq \frac{\pi}{2}$$

where $\mu$ is a constant called the *coefficient of static friction*.
**a.** Find the angle $\theta$ at which $F$ is minimized.
**b.** What is the magnitude of the force found in part (a)?
 **c.** Suppose $W = 60$ and $\mu = 0.4$. Plot the graph of $F$ as a function of $\theta$ on the interval $\left[0, \frac{\pi}{2}\right]$. Then verify the result obtained in parts (a) and (b) for this special case.

**84.** A uniform beam of length 3 ft and negligible weight is supported at both ends. When subjected to a concentrated load $W$ at a distance 1 ft from one end, it bends and has the elastic curve (the dashed curve in the figure) described by the equation

$$y = \begin{cases} \dfrac{W}{9EI}(5x - x^3) & \text{if } 0 \leq x \leq 1 \\ \dfrac{W}{18EI}(x^3 - 9x^2 + 19x - 3) & \text{if } 1 < x \leq 3 \end{cases}$$

where the product $EI$ is a constant called the *flexural rigidity* of the beam. Find the maximum deflection of the beam.
**Hint:** Maximize $y = f(x)$ over each interval [0, 1] and [1, 3] separately. Then combine your results.

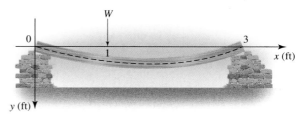

**85.** Let

$$f(x) = \begin{cases} -x & \text{if } -1 \leq x < 0 \\ x - 1 & \text{if } 0 \leq x \leq 1 \end{cases}$$

Show that $f$ is discontinuous at $x = 0$ but attains an absolute maximum value and an absolute minimum value on $[-1, 1]$. Does this contradict the Extreme Value Theorem?

**86.** Let

$$f(x) = \begin{cases} x^2 + 1 & \text{if } -1 < x \leq 2 \\ 2 & \text{if } 2 < x < 4 \end{cases}$$

Show that $f$ attains an absolute maximum value and an absolute minimum value on the *open* interval $(-1, 4)$. Does this contradict the Extreme Value Theorem?

**87.** Show that the function $f(x) = x^3 + x + 1$ has no relative extrema on $(-\infty, \infty)$.

**88.** Find the critical numbers of the greatest integer function $f(x) = \lfloor x \rfloor$.

**89.** Find the absolute maximum value and the absolute minimum value (if any) of the function $g(x) = x - \lfloor x \rfloor$, where $f(x) = \lfloor x \rfloor$ is the greatest integer function.

**90. a.** Suppose $f$ has a relative minimum at $c$. Show that the function $g$ defined by $g(x) = -f(x)$ has a relative maximum at $c$.
**b.** Use the result of (a) to prove Theorem 1 for the case in which $f$ has a relative minimum at $c$.

In Exercises 91–94, *plot the graph of $f$ and use the graph to estimate the absolute maximum and absolute minimum values of $f$ in the given interval.*

**91.** $f(x) = -0.02x^5 - 0.3x^4 + 2x^3 - 6x + 4$ on $[-2, 2]$
**92.** $f(x) = 0.3x^6 - 2x^4 + 3x^2 - 3$ on $[0, 2]$
**93.** $f(x) = \dfrac{0.2x^2}{3x^4 + 2x^2 + 1}$ on $[0, 4]$
**94.** $f(x) = \dfrac{x + \cos x}{1 + 0.5 \sin x}$ on $[0, 2]$

*In Exercises 95–98,* (a) *plot the graph of f in the given viewing window and find the approximate absolute maximum and absolute minimum values accurate to three decimal places, and* (b) *obtain the exact absolute maximum and absolute minimum values of f analytically.*

**95.** $f(x) = \dfrac{1}{2}x^4 - \dfrac{3}{2}x + 2$ on $[-1, 2] \times [0, 8]$

**96.** $f(x) = x - \sqrt{1 - x^2}$ on $[-1, 1] \times [-2, 2]$

**97.** $f(x) = \dfrac{x + 1}{\sqrt{x} + 1}$ on $[0, 1] \times [0.8, 1]$

**98.** $f(x) = 2 \sin x - x$ on $\left[0, \frac{\pi}{2}\right] \times [0, 1]$

*In Exercises 99–102, determine whether the statement is true or false. If it is true, explain why it is true. If it is false, explain why or give an example to show why it is false.*

**99.** If $f'(c) = 0$, then $f$ has a relative maximum or a relative minimum at $c$.

**100.** If $f$ has a relative minimum at $c$, then $f'(c) = 0$.

**101.** If $f$ is defined on the closed interval $[a, b]$, then $f$ has an absolute minimum value in $[a, b]$.

**102.** If $f$ is continuous on the interval $(a, b)$, then $f$ attains an absolute minimum value at some number $c$ in $(a, b)$.

---

## 3.2 The Mean Value Theorem

### ■ Rolle's Theorem

The graph of the function $f$ shown in Figure 1 gives the *depth* of a radical new twin-piloted submarine during a test dive. The submarine is on the surface at $t = a$ $[f(a) = 0]$ when it commences its dive. It resurfaces at $t = b$ $[f(b) = 0]$, the end of the test run. As you can see from the graph of $f$, there is at least one point on the graph of $f$ at which the tangent line to the curve is horizontal.

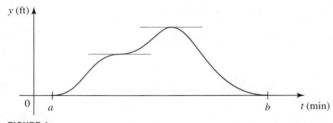

**FIGURE 1**
$f(t)$ gives the depth of the submarine at time $t$.

We can convince ourselves that there must exist at least one such point on the graph of $f$ through the following intuitive argument: Since we know that the submarine returned to the surface, there must be at least one point on the graph of $f$ that corresponds to the time when the submarine stops diving and begins to resurface. The tangent line to the graph of $f$ at this point must be horizontal.

A mathematical description of this phenomenon is contained in Rolle's Theorem, named in honor of the French mathematician Michel Rolle (1652–1719).

**THEOREM 1  Rolle's Theorem**

Let $f$ be continuous on $[a, b]$ and differentiable on $(a, b)$. If $f(a) = f(b)$, then there exists at least one number $c$ in $(a, b)$ such that $f'(c) = 0$.

**PROOF**   Let $f(a) = f(b) = d$. There are two cases to consider (see Figure 2).

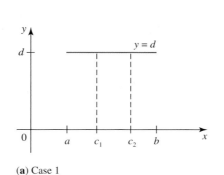

(a) Case 1

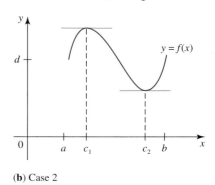

(b) Case 2

**FIGURE 2**
Geometric interpretations
of Rolle's Theorem

**Case 1**   $f(x) = d$ for all $x$ in $[a, b]$ (see Figure 2a).
In this case, $f'(x) = 0$ for all $x$ in $(a, b)$, so $f'(c) = 0$ for any number $c$ in $(a, b)$.

**Case 2**   $f(x) \neq d$ for at least one $x$ in $[a, b]$ (see Figure 2b).
In this case there must be a number $x$ in $(a, b)$ where $f(x) > d$ or $f(x) < d$. First, suppose that $f(x) > d$. Since $f$ is continuous on $[a, b]$, the Extreme Value Theorem implies
that $f$ attains an absolute maximum value at some number $c$ in $[a, b]$. The number $c$
cannot be an endpoint because $f(a) = f(b) = d$, and we have assumed that $f(x) > d$
for some number $x$ in $(a, b)$. Therefore, $c$ must be in $(a, b)$. Since $f$ is differentiable on
$(a, b)$, $f'(c)$ exists, and by Fermat's Theorem $f'(c) = 0$.
   The proof for the case in which $f(x) < d$ is similar and is left as an exercise (Exercise 40).   ∎

**EXAMPLE 1**   Let $f(x) = x^3 - x$ for $x$ in $[-1, 1]$.

**a.** Show that $f$ satisfies the hypotheses of Rolle's Theorem on $[-1, 1]$.
**b.** Find the number(s) $c$ in $(-1, 1)$ such that $f'(c) = 0$ as guaranteed by Rolle's
Theorem.

**Solution**
**a.** The polynomial function $f$ is continuous and differentiable on $(-\infty, \infty)$. In particular, it is continuous on $[-1, 1]$ and differentiable on $(-1, 1)$. Furthermore,

$$f(-1) = (-1)^3 - (-1) = 0 \quad \text{and} \quad f(1) = 1^3 - 1 = 0$$

and the hypotheses of Rolle's theorem are satisfied.
**b.** Rolle's Theorem guarantees that there exists at least one number $c$ in $(-1, 1)$
such that $f'(c) = 0$. But $f'(x) = 3x^2 - 1$, so to find $c$, we solve

$$3c^2 - 1 = 0$$

obtaining $c = \pm\sqrt{3}/3$. In other words, there are two numbers, $c_1 = -\sqrt{3}/3$ and
$c_2 = \sqrt{3}/3$, in $(-1, 1)$ for which $f'(c) = 0$ (Figure 3).   ∎

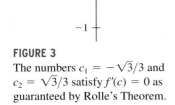

**FIGURE 3**
The numbers $c_1 = -\sqrt{3}/3$ and
$c_2 = \sqrt{3}/3$ satisfy $f'(c) = 0$ as
guaranteed by Rolle's Theorem.

**EXAMPLE 2**   **A Real-Life Illustration of Rolle's Theorem**   During a test dive of a prototype of a twin-piloted submarine, the depth in feet of the submarine at time $t$ in minutes is given by $h(t) = t^3(t - 7)^4$, where $0 \leq t \leq 7$.

**a.** Use Rolle's Theorem to show that there is some instant of time $t = c$ between 0
and 7 when $h'(c) = 0$.
**b.** Find the number $c$ and interpret your results.

**Solution**

**a.** The polynomial function $h$ is continuous on $[0, 7]$ and differentiable on $(0, 7)$. Furthermore, $h(0) = 0$ and $h(7) = 0$, so the hypotheses of Rolle's Theorem are satisfied. Therefore, there exists at least one number $c$ in $(0, 7)$ such that $h'(c) = 0$.

**b.** To find the value of $c$, we first compute

$$h'(t) = 3t^2(t - 7)^4 + t^3(4)(t - 7)^3$$
$$= t^2(t - 7)^3[3(t - 7) + 4t]$$
$$= 7t^2(t - 7)^3(t - 3)$$

Setting $h'(t) = 0$ gives $t = 0$, 3, or 7. Since 3 is the only number in the interval $(0, 7)$ such that $h'(3) = 0$, we see that $c = 3$. Interpreting our results, we see that the submarine is on the surface initially (since $h(0) = 0$) and returns to the surface again after 7 minutes (since $h(7) = 0$). The vertical component of the velocity of the submarine is zero at $t = 3$, at which time the submarine attains the greatest depth of $h(3) = 3^3(3 - 7)^4 = 6912$ ft. The graph of $h$ is shown in Figure 4. ∎

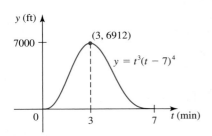

**FIGURE 4**
The submarine is at a depth of $h(t)$ feet at time $t$ minutes.

Rolle's Theorem is a special case of a more general result known as the Mean Value Theorem.

---

**THEOREM 2    The Mean Value Theorem**

Let $f$ be continuous on $[a, b]$ and differentiable on $(a, b)$. Then there exists at least one number $c$ in $(a, b)$ such that

$$f'(c) = \frac{f(b) - f(a)}{b - a} \tag{1}$$

---

To interpret this theorem geometrically, notice that the quotient in Equation (1) is just the slope of the secant line passing through the points $P(a, f(a))$ and $Q(b, f(b))$ lying on the graph of $f$ (Figure 5). The quantity $f'(c)$ on the left, however, gives the slope of the tangent line to the graph of $f$ at $x = c$. The Mean Value Theorem tells us that under suitable conditions on $f$, there is always at least one point $(c, f(c))$ on the graph of $f$ for $a < c < b$ such that the tangent line to the graph of $f$ at this point is parallel to the secant line passing through $P$ and $Q$. Observe that if $f(a) = f(b)$, then Theorem 2 reduces to Rolle's Theorem.

**FIGURE 5**
The tangent line $T$ at $(c, f(c))$ is parallel to the secant line $S$ through $P$ and $Q$.

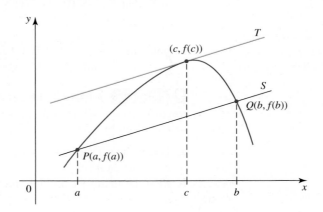

**PROOF**   If you examine Figure 5, you will see that the vertical distance between the graph of $f$ and the secant line $S$ passing through $P$ and $Q$ is maximal at $x = c$. This observation gives a clue to the proof of the Mean Value Theorem: Find a function whose absolute value gives the vertical distances between the graph of $f$ and the secant line. Then optimize this function.

Now an equation of the secant line can be found by using the point-slope form of the equation of a line with slope $[f(b) - f(a)]/(b - a)$ and the point $(b, f(b))$. Thus,

$$y - f(b) = \frac{f(b) - f(a)}{b - a} \cdot (x - b)$$

or

$$y = f(b) + \frac{f(b) - f(a)}{b - a} \cdot (x - b)$$

Define the function $D$ by

$$D(x) = f(x) - \left[ f(b) + \frac{f(b) - f(a)}{b - a} \cdot (x - b) \right] \tag{2}$$

Notice that $|D(x)|$ gives the vertical distance between the graph of $f$ and the secant line through $P$ and $Q$ (Figure 6). The function $D$ is continuous on $[a, b]$ and differentiable on $(a, b)$, so we can use Rolle's Theorem on $D$. First, we note that $D(a) = D(b) = 0$. Therefore, there exists at least one number $c$ in $(a, b)$ such that $D'(c) = 0$. But

$$D'(x) = f'(x) - \frac{f(b) - f(a)}{b - a}$$

so $D'(c) = 0$ implies that

$$0 = f'(c) - \frac{f(b) - f(a)}{b - a}$$

or

$$f'(c) = \frac{f(b) - f(a)}{b - a}$$

as was to be shown.

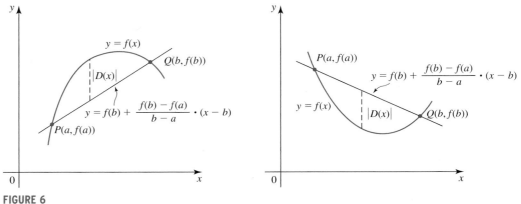

**FIGURE 6**
$|D(x)|$ gives the vertical distance between the graph of $f$ and the secant line passing through $P$ and $Q$.

**EXAMPLE 3** Let $f(x) = x^3$.

a. Show that $f$ satisfies the hypotheses of the Mean Value Theorem on $[-1, 1]$.
b. Find the number(s) $c$ in $(-1, 1)$ that satisfy Equation (1) as guaranteed by the Mean Value Theorem.

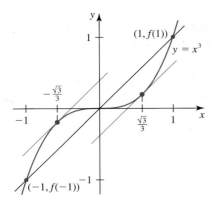

**FIGURE 7**
The numbers $c_1 = -\sqrt{3}/3$ and $c_2 = \sqrt{3}/3$ satisfy Equation (1), as guaranteed by the Mean Value Theorem.

**Solution**

a. $f$ is a polynomial function, so it is continuous and differentiable on $(-\infty, \infty)$. In particular, $f$ is continuous on $[-1, 1]$ and differentiable on $(-1, 1)$. So the hypotheses of the Mean Value Theorem are satisfied.
b. $f'(x) = 3x^2$, so $f'(c) = 3c^2$. With $a = -1$ and $b = 1$, Equation (1) gives

$$\frac{f(1) - f(-1)}{1 - (-1)} = f'(c)$$

or

$$\frac{1 - (-1)}{1 - (-1)} = 3c^2$$

$$1 = 3c^2$$

and $c = \pm\sqrt{3}/3$. So there are two numbers, $c_1 = -\sqrt{3}/3$ and $c_2 = \sqrt{3}/3$, in $(-1, 1)$ that satisfy Equation (1). (See Figure 7.)  ∎

The next example gives an interpretation of the Mean Value Theorem in a real-life setting.

**EXAMPLE 4** **The Mean Value Theorem and the Maglev**    The position of a maglev moving along a straight, elevated monorail track is given by $s = f(t) = 4t^2$, $0 \le t \le 30$, where $s$ is measured in feet and $t$ is measured in seconds. Then the average velocity of the maglev during the first 4 sec of the run is

$$\frac{f(4) - f(0)}{4 - 0} = \frac{64 - 0}{4} = 16 \qquad (3)$$

or 16 ft/sec. Next, since $f$ is continuous on $[0, 4]$ and differentiable on $(0, 4)$, the Mean Value Theorem guarantees that there is a number $c$ in $(0, 4)$ such that

$$\frac{f(4) - f(0)}{4 - 0} = f'(c) \qquad (4)$$

But $f'(t) = 8t$, so using Equation (3), we see that Equation (4) is equivalent to

$$16 = 8c$$

or $c = 2$. Since $f'(t)$ measures the instantaneous velocity of the maglev at any time $t$, the Mean Value Theorem tells us that at some time $t$ between $t = 0$ and $t = 4$ (in this case, $t = 2$) the maglev must attain an instantaneous velocity equal to the average velocity of the maglev over the time interval $[0, 4]$.  ∎

## Some Consequences of the Mean Value Theorem

An important application of the Mean Value Theorem is to establish other mathematical results. For example, we know that the derivative of a constant function is zero. Now we can show that the converse is also true.

---

### THEOREM 3

If $f'(x) = 0$ for all $x$ in an interval $(a, b)$, then $f$ is constant on $(a, b)$.

---

**PROOF** Suppose that $f'(x) = 0$ for all $x$ in $(a, b)$. To prove that $f$ is constant on $(a, b)$, it suffices to show that $f$ has the same value at every pair of numbers in $(a, b)$. So let $x_1$ and $x_2$ be arbitrary numbers in $(a, b)$ with $x_1 < x_2$. Since $f$ is differentiable on $(a, b)$, it is also differentiable on $(x_1, x_2)$ and continuous on $[x_1, x_2]$. Therefore, the hypotheses of the Mean Value Theorem are satisfied on the interval $[x_1, x_2]$. Applying the theorem, we see that there exists a number $c$ in $(x_1, x_2)$ such that

$$f'(c) = \frac{f(x_2) - f(x_1)}{x_2 - x_1} \qquad (5)$$

But by hypothesis, $f'(x) = 0$ for all $x$ in $(a, b)$, so $f'(c) = 0$. Therefore, Equation (5) implies that $f(x_2) - f(x_1) = 0$, or $f(x_1) = f(x_2)$; that is, $f$ has the same value at any two numbers in $(a, b)$. This completes the proof. ■

---

### COROLLARY TO THEOREM 3

If $f'(x) = g'(x)$ for all $x$ in an interval $(a, b)$, then $f$ and $g$ differ by a constant on $(a, b)$; that is, there exists a constant $c$ such that $f(x) = g(x) + c$ for all $x$ in $(a, b)$.

---

**PROOF** Let $h(x) = f(x) - g(x)$. Then

$$h'(x) = f'(x) - g'(x) = 0$$

for every $x$ in $(a, b)$. By Theorem 3, $h$ is constant; that is, $f - g$ is constant on $(a, b)$. Thus, $f(x) - g(x) = c$ for some constant $c$ and $f(x) = g(x) + c$ for all $x$ in $(a, b)$. ■

**EXAMPLE 5** Prove the identity $\sin^{-1} x + \cos^{-1} x = \pi/2$.

**Solution** Let $f(x) = \sin^{-1} x + \cos^{-1} x$ for all $x$ in $[-1, 1]$. Then $f(-1) = f(1) = \pi/2$ by direct computation. For $-1 < x < 1$ we have

$$f'(x) = \frac{1}{\sqrt{1 - x^2}} - \frac{1}{\sqrt{1 - x^2}} = 0$$

Therefore, by Theorem 3, $f(x)$ is constant on $(-1, 1)$; that is, there exists a constant $C$ such that

$$\sin^{-1} x + \cos^{-1} x = C$$

To determine the value of $C$, we put $x = 0$, giving

$$\sin^{-1} 0 + \cos^{-1} 0 = C \quad \text{or} \quad C = 0 + \frac{\pi}{2}$$

Thus, $\sin^{-1} x + \cos^{-1} x = \pi/2$. ■

## ■ Determining the Number of Zeros of a Function

Our final example brings together two important theorems—the Intermediate Value Theorem and Rolle's Theorem—to help us determine the number of zeros of a function $f$ in a given interval $[a, b]$.

**EXAMPLE 6** Show that the function $f(x) = x^3 + x + 1$ has exactly one zero in the interval $[-2, 0]$.

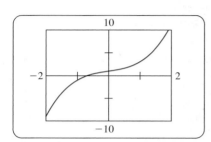

**FIGURE 8**
The graph shows the zero of $f$.

**Solution** First, observe that $f$ is continuous on $[-2, 0]$ and that $f(-2) = -9$ and $f(0) = 1$. Therefore, by the Intermediate Value Theorem, there must exist at least one number $c$ that satisfies $-2 < c < 0$ such that $f(c) = 0$. In other words, $f$ has at least one zero in $(-2, 0)$.

To show that $f$ has exactly one zero, suppose, on the contrary, that $f$ has at least two distinct zeros, $x_1$ and $x_2$. Without loss of generality, suppose that $x_1 < x_2$. Then $f(x_1) = f(x_2) = 0$. Because $f$ is differentiable on $(x_1, x_2)$, an application of Rolle's Theorem tells us that there exists a number $c$ between $x_1$ and $x_2$ such that $f'(c) = 0$. But $f'(x) = 3x^2 + 1 \geq 1$ can never be zero in $(x_1, x_2)$. This contradiction establishes the result.

The graph of $f$ is shown in Figure 8. ■

## 3.2 CONCEPT QUESTIONS

1. State Rolle's Theorem and give a geometric interpretation of it.
2. State the Mean Value Theorem, and give a geometric interpretation of it.
3. Refer to the graph of $f$.
   a. Sketch the secant line through the points $(0, 3)$ and $(9, 8)$. Then draw all lines parallel to this secant line that are tangent to the graph of $f$.
   b. Use the result of part (a) to estimate the values of $c$ that satisfy the Mean Value Theorem on the interval $[0, 9]$.

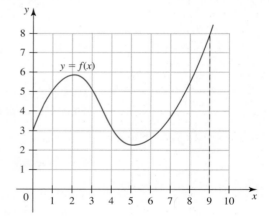

## 3.2 EXERCISES

*In Exercises 1–8, verify that the function satisfies the hypotheses of Rolle's Theorem on the given interval, and find all values of $c$ that satisfy the conclusion of the theorem.*

1. $f(x) = x^2 - 4x + 3$; $\quad [1, 3]$
2. $g(x) = x^3 - 9x$; $\quad [-3, 3]$
3. $f(x) = x^3 + x^2 - 2x$; $\quad [-2, 0]$
4. $h(x) = x^3(x - 7)^4$; $\quad [0, 7]$
5. $f(x) = x\sqrt{1 - x^2}$; $\quad [-1, 1]$
6. $f(t) = t^{2/3}(6 - t)^{1/3}$; $\quad [0, 6]$
7. $h(t) = \sin^2 t$; $\quad [0, \pi]$
8. $f(x) = \cos 2x - 1$; $\quad [0, \pi]$

*In Exercises 9–16, verify that the function satisfies the hypotheses of the Mean Value Theorem on the given interval, and find all values of c that satisfy the conclusion of the theorem.*

**9.** $f(x) = x^2 + 1$;  [0, 2]

**10.** $f(x) = x^3 - 2x^2$;  [-1, 2]

**11.** $h(x) = \dfrac{1}{x}$;  [1, 3]

**12.** $g(t) = \dfrac{t}{t-1}$;  [-2, 0]

**13.** $h(x) = x\sqrt{2x+1}$;  [0, 4]

**14.** $f(x) = \sin x$;  $\left[0, \frac{\pi}{2}\right]$

**15.** $f(x) = e^{-x/2}$;  [0, 4]

**16.** $g(x) = \ln x$;  [1, 3]

**17. Flight of an Aircraft** A commuter plane takes off from the Los Angeles International Airport and touches down 30 min later at the Ontario International Airport. Let $A(t)$ (in feet) be the altitude of the plane at time $t$ (in minutes), where $0 \le t \le 30$. Use Rolle's Theorem to explain why there must be at least one number $c$ with $0 < c < 30$ such that $A'(c) = 0$. Interpret your result.

**18. Breaking the Speed Limit** A trucker drove from Bismarck to Fargo, a distance of 193 mi, in 2 hr and 55 min. Use the Mean Value Theorem to show that the trucker must have exceeded the posted speed limit of 65 mph at least once during the trip.

**19. Test Flights** In a test flight of the McCord Terrier, an experimental VTOL (vertical takeoff and landing) aircraft, it was determined that $t$ sec after takeoff, when the aircraft was operated in the vertical takeoff mode, its altitude was

$$h(t) = \frac{1}{16}t^4 - t^3 + 4t^2 \qquad 0 \le t \le 8$$

Use Rolle's Theorem to show that there exists a number $c$ satisfying $0 < c < 8$ such that $h'(c) = 0$. Find the value of $c$, and explain its significance.

**20. Hotel Occupancy** The occupancy rate of the all-suite Wonderland Hotel, located near a theme park, is given by the function

$$r(t) = \frac{10}{81}t^3 - \frac{10}{3}t^2 + \frac{200}{9}t + 56 \qquad 0 \le t \le 12$$

where $t$ is measured in months with $t = 0$ corresponding to the beginning of January. Show that there exists a number $c$ that satisfies $0 < c < 12$ such that $r'(c) = 0$. Find the value of $c$, and explain its significance.

**21.** Let $f(x) = |x| - 1$. Show that there is no number $c$ in $(-1, 1)$ such that $f'(c) = 0$ even though $f(-1) = f(1) = 0$. Why doesn't this contradict Rolle's Theorem?

**22.** Let $f(x) = 1 - x^{2/3}$, $a = -1$, and $b = 8$. Show that there is no number $c$ in $(a, b)$ such that

$$f'(c) = \frac{f(b) - f(a)}{b - a}$$

Doesn't this contradict the Mean Value Theorem? Explain.

**23.** Let

$$f(x) = \begin{cases} x^2 & \text{if } x < 1 \\ 2 - x & \text{if } x \ge 1 \end{cases}$$

Does $f$ satisfy the hypotheses of the Mean Value Theorem on [0, 2]? Explain.

**24.** Prove that $f(x) = 4x^3 - 4x + 1$ has at least one zero in the interval (0, 1).

**Hint:** Apply Rolle's Theorem to the function $g(x) = x^4 - 2x^2 + x$ on [0, 1].

**25.** Prove that $f(x) = x^5 + 6x + 4$ has exactly one zero in $(-\infty, \infty)$.

**26.** Prove that the equation $x^7 + 6x^5 + 2x - 6 = 0$ has exactly one real root.

**27.** Prove that the function $f(x) = x^5 - 12x + c$, where $c$ is any real number, has at most one zero in [0, 1].

**28.** Use the Mean Value Theorem to prove that $|\sin a - \sin b| \le |a - b|$ for all real numbers $a$ and $b$.

**29.** Suppose that the equation

$$a_n x^n + a_{n-1} x^{n-1} + \cdots + a_1 x = 0$$

has a positive root $r$. Show that the equation

$$n a_n x^{n-1} + (n-1) a_{n-1} x^{n-2} + \cdots + a_1 = 0$$

has a positive root smaller than $r$.

**Hint:** Use Rolle's Theorem.

**30.** Suppose $f'(x) = c$, where $c$ is a constant, for all values of $x$. Show that $f$ must be a linear function of the form $f(x) = cx + d$ for some constant $d$.

**Hint:** Use the corollary to Theorem 3.

**31.** Let $f(x) = x^4 - 4x - 1$.

**a.** Use Rolle's Theorem to show that $f$ has exactly two distinct zeros.

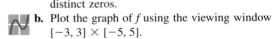  **b.** Plot the graph of $f$ using the viewing window $[-3, 3] \times [-5, 5]$.

**32.** Let

$$f(x) = \begin{cases} x \sin \dfrac{\pi}{x} & \text{if } x > 0 \\ 0 & \text{if } x = 0 \end{cases}$$

Use Rolle's Theorem to prove that $f$ has infinitely many critical numbers in the interval (0, 1). Plot the graph of $f$ using the viewing window $[0, 1] \times [-1, 1]$.

**33.** Prove the formula

$$\cos^2 x = \frac{1 + \cos 2x}{2}$$

**34.** Prove the formula

$$\cos^{-1} \frac{1 - x^2}{1 + x^2} = 2 \tan^{-1} x$$

for $0 \le x < \infty$.

**35.** Suppose that $f$ and $g$ are continuous on an interval $[a, b]$ and differentiable on the interval $(a, b)$. Furthermore, suppose that $f(a) = g(a)$ and $f'(x) < g'(x)$ for $a < x < b$. Prove that $f(x) < g(x)$ for $a < x < b$.
Hint: Apply the Mean Value Theorem to the function $h = f - g$.

**36.** Let $f(x) = Ax^2 + Bx + C$, and let $[a, b]$ be an arbitrary interval. Show that the number $c$ in the Mean Value Theorem applied to the function $f$ lies at the midpoint of the interval $[a, b]$.

**37.** Let $f(x) = 2(x - 1)(x - 2)(x - 3)(x - 4)$. Prove that $f'$ has exactly three real zeros.

**38.** A real number $c$ such that $f(c) = c$ is called a *fixed point* of the function $f$. Geometrically, a fixed point of $f$ is a point that is mapped by $f$ onto itself. Prove that if $f$ is differentiable and $f'(x) \neq 1$ for all $x$ in an interval $I$, then $f$ has at most one fixed point in $I$.

**39.** Use the result of Exercise 38 to show that $f(x) = \sqrt{x + 6}$ has exactly one fixed point in the interval $(0, \infty)$. What is the fixed point?

**40.** Complete the proof of Rolle's Theorem by considering the case in which $f(x) < d$ for some number $x$ in $(a, b)$.

**41.** Let $f$ be continuous on $[a, b]$ and differentiable on $(a, b)$. Put $h = b - a$.
  **a.** Use the Mean Value Theorem to show that there exists at least one number $\theta$ that satisfies $0 < \theta < 1$ such that
$$\frac{f(a + h) - f(a)}{h} = f'(a + \theta h)$$
  **b.** Find $\theta$ in the formula in part (a) for the function $f(x) = x^2$.

**42.** Let $f(x) = x^4 - 2x^3 + x - 2$.
  **a.** Show that $f$ satisfies the hypotheses of Rolle's Theorem on the interval $[-1, 2]$.
  **b.** Use a calculator or a computer to estimate all values of $c$ accurate to five decimal places that satisfy the conclusion of Rolle's Theorem.
  **c.** Plot the graph of $f$ and the (horizontal) tangent lines to the graph of $f$ at the point(s) $(c, f(c))$ for the values of $c$ found in part (b).

**43.** Let $f(x) = x^4 - 2x^2 + 2$.
  **a.** Use a calculator or a computer to estimate all values of $c$ accurate to three decimal places that satisfy the conclusion of the Mean Value Theorem for $f$ on the interval $[0, 2]$.
  **b.** Plot the graph of $f$, the secant line passing through the points $(0, 2)$ and $(2, 10)$, and the tangent line to the graph of $f$ at the point(s) $(c, f(c))$ for the value(s) of $c$ found in part (a).

**44.** Let $f(x) = x^2 \sin x$.
  **a.** Show that $f$ satisfies the hypotheses of Rolle's Theorem on the interval $[0, \pi]$.
  **b.** Use a calculator or a computer to estimate all value(s) of $c$ accurate to five decimal places that satisfy the conclusion of Rolle's Theorem.
  **c.** Plot the graph of $f$ and the (horizontal) tangent lines to the graph of $f$ at the point(s) $(c, f(c))$ for the value(s) of $c$ found in part (b).

**45.** Let $f(x) = \sin \sqrt{x}$.
  **a.** Use a calculator or a computer to estimate all values of $c$ accurate to three decimal places that satisfy the conclusion of the Mean Value Theorem for $f$ on the interval $\left[0, \frac{\pi^2}{4}\right]$.
  **b.** Plot the graph of $f$, the secant line passing through the points $(0, 0)$ and $\left(\frac{\pi^2}{4}, 1\right)$, and the tangent line to the graph of $f$ at the point(s) $(c, f(c))$ for the value(s) of $c$ found in part (b).

**46.** Prove the inequality
$$\frac{x}{x + 1} < \ln(1 + x) < x$$
for $x > 0$.
Hint: Use the Mean Value Theorem.

*In Exercises 47–52, determine whether the statement is true or false. If it is true, explain why it is true. If it is false, explain why or give an example to show why it is false.*

**47.** Suppose that $f$ is continuous on $[a, b]$ and differentiable on $(a, b)$. If $f'(c) = 0$ for at least one $c$ in $(a, b)$, then $f(a) = f(b)$.

**48.** Suppose that $f$ is continuous on $[a, b]$ but is not differentiable on $(a, b)$. Then there does not exist a number $c$ in $(a, b)$ such that
$$f'(c) = \frac{f(b) - f(a)}{b - a}$$

**49.** If $f'(x) = 0$ for all $x$, then $f$ is a constant function.

**50.** If $|f'(x)| \leq 1$ for all $x$, then
$$|f(x_1) - f(x_2)| \leq |x_1 - x_2|$$
for all numbers $x_1$ and $x_2$.

**51.** There does not exist a continuous function defined on the interval $[2, 5]$ and differentiable on $(2, 5)$ satisfying $|f(5) - f(2)| \leq 6$ on $[2, 5]$ and $|f'(x)| > 2$ for all $x$ in $(2, 5)$.

**52.** If $f$ is continuous on $[1, 3]$, differentiable on $(1, 3)$, and satisfies $f(1) = 2$, $f(3) = 5$, then there exists a number $c$ satisfying $1 < c < 3$, such that $f'(c) = \frac{3}{2}$.

## 3.3 Increasing and Decreasing Functions and the First Derivative Test

### ▪ Increasing and Decreasing Functions

Among the important factors in determining the structural integrity of an aircraft is its age. Advancing age makes the parts of a plane more likely to crack. The graph of the function $f$ in Figure 1 is referred to as a "bathtub curve" in the airline industry. It gives the fleet damage rate (damage due to corrosion, accident, and metal fatigue) of a typical fleet of commercial aircraft as a function of the number of years of service.

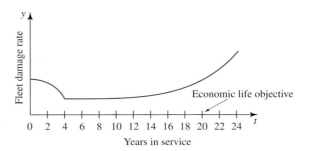

**FIGURE 1**

The "bathtub curve" gives the number of planes in a fleet that are damaged as a function of the age of the fleet.

The function is *decreasing* on the interval $(0, 4)$, showing that the fleet damage rate is dropping as problems are found and corrected during the initial shakedown period. The function is *constant* on the interval $(4, 10)$, reflecting that planes have few structural problems after the initial shakedown period. Beyond this, the function is *increasing,* reflecting an increase in structural defects due mainly to metal fatigue.

These intuitive notions involving increasing and decreasing functions can be described mathematically as follows.

---

**DEFINITIONS   Increasing and Decreasing Functions**

A function $f$ is **increasing** on an interval $I$, if for every pair of numbers $x_1$ and $x_2$ in $I$,

$$x_1 < x_2 \quad \text{implies that} \quad f(x_1) < f(x_2) \qquad \text{See Figure 2a.}$$

$f$ is **decreasing** on $I$ if, for every pair of numbers $x_1$ and $x_2$ in $I$,

$$x_1 < x_2 \quad \text{implies that} \quad f(x_1) > f(x_2) \qquad \text{See Figure 2b.}$$

$f$ is **monotonic** on $I$ if it is either increasing or decreasing on $I$.

---

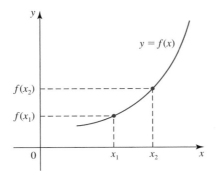

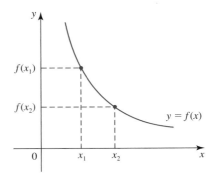

**FIGURE 2**    **(a)** $f$ is increasing on $I$.    **(b)** $f$ is decreasing on $I$.

Since the derivative of a function measures the rate of change of that function, it lends itself naturally as a tool for determining the intervals where a differentiable function is increasing or decreasing. As you can see in Figure 3, if the graph of $f$ has tangent lines with positive slopes over an interval, then the function is increasing on that interval. Similarly, if the graph of $f$ has tangent lines with negative slopes over an interval, then the function is decreasing on that interval. Also, we know that the slope of the tangent line at $(x, f(x))$ and the rate of change of $f$ at $x$ are given by $f'(x)$. Therefore, $f$ is increasing on an interval where $f'(x) > 0$ and decreasing on an interval where $f'(x) < 0$.

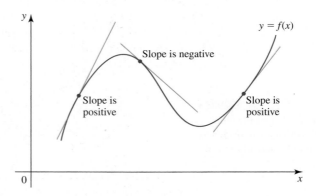

**FIGURE 3**
$f$ is increasing on an interval where $f'(x) > 0$ and decreasing on an interval where $f'(x) < 0$.

These intuitive observations lead to the following theorem.

---

**THEOREM 1**   Suppose $f$ is differentiable on an open interval $(a, b)$.
**a.** If $f'(x) > 0$ for all $x$ in $(a, b)$, then $f$ is increasing on $(a, b)$.
**b.** If $f'(x) < 0$ for all $x$ in $(a, b)$, then $f$ is decreasing on $(a, b)$.
**c.** If $f'(x) = 0$ for all $x$ in $(a, b)$, then $f$ is constant on $(a, b)$.

---

**PROOF**
**a.** Let $x_1$ and $x_2$ be any two numbers in $(a, b)$ with $x_1 < x_2$. Since $f$ is differentiable on $(a, b)$, it is continuous on $[x_1, x_2]$ and differentiable on $(x_1, x_2)$. By the Mean Value Theorem, there exists a number $c$ in $(x_1, x_2)$ such that

$$f'(c) = \frac{f(x_2) - f(x_1)}{x_2 - x_1}$$

or, equivalently,

$$f(x_2) - f(x_1) = f'(c)(x_2 - x_1) \tag{1}$$

Now, $f'(c) > 0$ by assumption, and $x_2 - x_1 > 0$ because $x_1 < x_2$. Therefore, $f(x_2) - f(x_1) > 0$, or $f(x_1) < f(x_2)$. This shows that $f$ is increasing on $(a, b)$.
**b.** The proof of (b) is similar and is left as an exercise (see Exercise 66).
**c.** This was proved in Theorem 3 in Section 3.2.    ■

Theorem 1 enables us to develop a procedure for finding the intervals where a function is increasing, decreasing, or constant. In this connection, recall that a function can only change sign as we move across a zero or a number at which the function is discontinuous.

**Determining the Intervals Where a Function Is Increasing or Decreasing**

1. Find all the values of $x$ for which $f'(x) = 0$ or $f'(x)$ does not exist. Use these values of $x$ to partition the domain of $f$ into open intervals.
2. Select a test number $c$ in each interval found in Step 1, and determine the sign of $f'(c)$ in that interval.
   **a.** If $f'(c) > 0$, then $f$ is increasing on that interval.
   **b.** If $f'(c) < 0$, then $f$ is decreasing on that interval.
   **c.** If $f'(c) = 0$, then $f$ is constant on that interval.

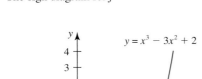

**FIGURE 4**
The sign diagram for $f'$

**EXAMPLE 1** Determine the intervals where the function $f(x) = x^3 - 3x^2 + 2$ is increasing and where it is decreasing.

**Solution** We first compute

$$f'(x) = 3x^2 - 6x = 3x(x - 2)$$

from which we see that $f'$ is continuous everywhere and has zeros at 0 and 2. These zeros of $f'$ partition the domain of $f$ into the intervals $(-\infty, 0)$, $(0, 2)$, and $(2, \infty)$. To determine the sign of $f'(x)$ on each of these intervals, we evaluate $f'(x)$ at a convenient test number in each interval. These results are summarized in the following table.

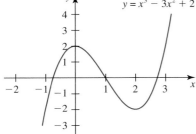

**FIGURE 5**
$f$ is increasing on $(-\infty, 0)$, decreasing on $(0, 2)$, and increasing on $(2, \infty)$.

| Interval | Test number $c$ | $f'(c)$ | Sign of $f'(c)$ |
|----------|-----------------|---------|-----------------|
| $(-\infty, 0)$ | $-1$ | 9 | $+$ |
| $(0, 2)$ | 1 | $-3$ | $-$ |
| $(2, \infty)$ | 3 | 9 | $+$ |

Using these results, we obtain the sign diagram for $f'(x)$ shown in Figure 4. We conclude that $f$ is increasing on $(-\infty, 0)$ and $(2, \infty)$ and decreasing on $(0, 2)$. The graph of $f$ is shown in Figure 5. ∎

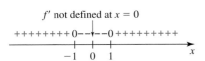

**FIGURE 6**
The sign diagram for $f'$

**EXAMPLE 2** Determine the intervals where the function $f(x) = x + 1/x$ is increasing and where it is decreasing.

**Solution** The derivative of $f$ is

$$f'(x) = 1 - \frac{1}{x^2} = \frac{x^2 - 1}{x^2} = \frac{(x + 1)(x - 1)}{x^2}$$

from which we see that $f'(x)$ is continuous everywhere except at $x = 0$ and has zeros at $x = -1$ and $x = 1$. These values of $x$ partition the domain of $f$ into the intervals $(-\infty, -1)$, $(-1, 0)$, $(0, 1)$, and $(1, \infty)$. By evaluating $f'(x)$ at each of the test numbers $x = -2, -\frac{1}{2}, \frac{1}{2}$, and 2, we find

$$f'(-2) = \frac{3}{4}, \qquad f'\left(-\frac{1}{2}\right) = -3, \qquad f'\left(\frac{1}{2}\right) = -3, \qquad \text{and} \qquad f'(2) = \frac{3}{4}$$

giving us the sign diagram of $f'(x)$ shown in Figure 6. We conclude that $f$ is increasing on $(-\infty, -1)$ and $(1, \infty)$ and decreasing on $(-1, 0)$ and $(0, 1)$. The graph of $f$ is shown in Figure 7.* Note that $f'(x)$ does not change sign as we move across the point of discontinuity. ∎

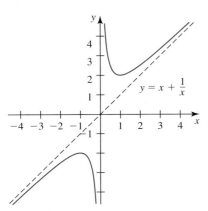

**FIGURE 7**
The graph of $f$

---

*The graph of $f$ approaches the dashed line as $x \to \pm\infty$. The dashed line is called a *slant asymptote* and will be discussed in Section 3.6.

### ■ Finding the Relative Extrema of a Function

We will now see how the derivative of a function $f$ can be used to help us find the relative extrema of $f$. If you examine Figure 8, you can see that the graph of $f$ is *rising* to the left of the relative maximum that occurs at $b$ and *falling* to the right of it. Likewise, at the relative minima of $f$ at $a$ and $d$, you can see that the graph of $f$ is *falling* to the left of these critical numbers and *rising* to the right of them. Finally, look at the behavior of the graph of $f$ at the critical numbers $c$ and $e$. These numbers do not give rise to relative extrema. Notice that $f$ is either increasing or decreasing on *both* sides of these critical numbers.

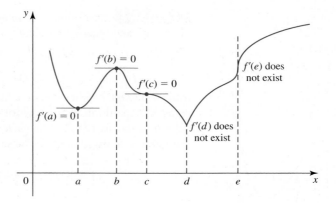

**FIGURE 8**
$a$, $b$, $c$, $d$, and $e$ are critical numbers of $f$, but only the critical numbers $a$, $b$, and $d$ give rise to relative extrema.

This discussion leads to the following theorem.

---

**THEOREM 2**   **The First Derivative Test**

Let $c$ be a critical number of a continuous function $f$ in the interval $(a, b)$ and suppose that $f$ is differentiable at every number in $(a, b)$ with the possible exception of $c$ itself.

**a.** If $f'(x) > 0$ on $(a, c)$ and $f'(x) < 0$ on $(c, b)$, then $f$ has a *relative maximum* at $c$ (Figure 9a).

**b.** If $f'(x) < 0$ on $(a, c)$ and $f'(x) > 0$ on $(c, b)$, then $f$ has a *relative minimum* at $c$ (Figure 9b).

**c.** If $f'(x)$ has the same sign on $(a, c)$ and $(c, b)$, then $f$ does not have a relative extremum at $c$ (Figure 9c).

---

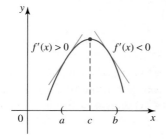

**(a)** Relative maximum at $c$

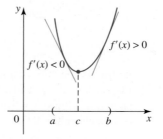

**(b)** Relative minimum at $c$

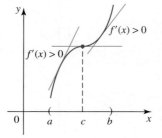

**(c)** No relative extrema at $c$

**FIGURE 9**

**PROOF** We will prove part (a) and leave the other two parts for you to prove (see Exercise 67). Suppose $f'$ changes sign from positive to negative as we pass through $c$. Then there are numbers $a$ and $b$ such that $f'(x) > 0$ for all $x$ in $(a, c)$ and $f'(x) < 0$ for all $x$ in $(c, b)$. By Theorem 1 we see that $f$ is increasing on $(a, c)$ and decreasing on $(c, b)$. Therefore, $f(x) \le f(c)$ for all $x$ in $(a, b)$. We conclude that $f$ has a relative maximum at $c$. ∎

The following procedure for finding the relative extrema of a continuous function is based on Theorem 2.

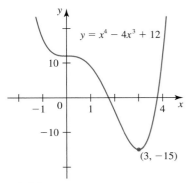

**FIGURE 10**
The sign diagram of $f'$

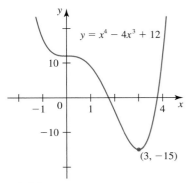

**FIGURE 11**
The graph of $f$

> **Finding the Relative Extrema of a Function**
>
> 1. Find the critical numbers of $f$.
> 2. Determine the sign of $f'(x)$ to the left and to the right of each critical number.
>    a. If $f'(x)$ changes sign from *positive* to *negative* as we move across a critical number $c$, then $f(c)$ is a relative maximum value.
>    b. If $f'(x)$ changes sign from *negative* to *positive* as we move across a critical number $c$, then $f(c)$ is a relative minimum value.
>    c. If $f'(x)$ does not change sign as we move across a critical number $c$, then $f(c)$ is not a relative extremum.

**EXAMPLE 3** Find the relative extrema of $f(x) = x^4 - 4x^3 + 12$.

**Solution** The derivative of $f$,

$$f'(x) = 4x^3 - 12x^2 = 4x^2(x - 3)$$

is continuous everywhere. Therefore, the zeros of $f'$, which are 0 and 3, are the only critical numbers of $f$. The sign diagram of $f'$ is shown in Figure 10. Since $f'$ has the same sign on $(-\infty, 0)$ and $(0, 3)$, the First Derivative Test tells us that $f$ does not have a relative extremum at 0. Next, we note that $f'$ changes sign from negative to positive as we move across 3, so 3 does give rise to a relative minimum of $f$. The relative minimum value of $f$ is $f(3) = -15$. The graph of $f$ is shown in Figure 11 and confirms these results. ∎

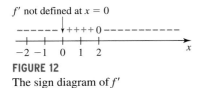

**FIGURE 12**
The sign diagram of $f'$

**EXAMPLE 4** Find the relative extrema of $f(x) = 15x^{2/3} - 3x^{5/3}$.

**Solution** The derivative of $f$ is

$$f'(x) = 10x^{-1/3} - 5x^{2/3} = 5x^{-1/3}(2 - x) = \frac{5(2 - x)}{x^{1/3}}$$

Note that $f'$ is discontinuous at 0 and has a zero at 2, so 0 and 2 are critical numbers of $f$. Referring to the sign diagram of $f'$ (Figure 12) and using the First Derivative Test, we conclude that $f$ has a relative minimum at 0 and a relative maximum at 2. The relative minimum value is $f(0) = 0$, and the relative maximum value is

$$f(2) = 15(2)^{2/3} - 3(2)^{5/3} \approx 14.29$$

The graph of $f$ is shown in Figure 13. ∎

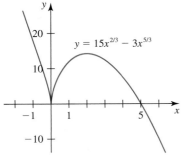

**FIGURE 13**
The graph of $f$

**EXAMPLE 5** **Motion of a Projectile** A projectile starts from the origin of the $xy$-coordinate system, and its motion is confined to the $xy$-plane. Suppose the trajectory of the projectile is

$$y = f(x) = 1.732x - 0.000008x^2 - 0.000000002x^3 \qquad 0 \le x \le 27,496$$

where $y$ measures the height in feet and $x$ measures the horizontal distance in feet covered by the projectile.

**a.** Find the interval where $y$ is increasing and the interval where $y$ is decreasing.
**b.** Find the relative extrema of $f$.
**c.** Interpret the results obtained in part (a) and part (b).

**Solution**
**a.** Observe that

$$\frac{dy}{dx} = 1.732 - 0.000016x - 0.000000006x^2$$

is continuous everywhere. Setting $dy/dx = 0$ gives

$$0.000000006x^2 + 0.000016x - 1.732 = 0$$

Using the quadratic formula to solve this equation, we obtain

$$x = \frac{-0.000016 \pm \sqrt{(0.000016)^2 - 4(0.000000006)(-1.732)}}{2(0.000000006)}$$

$$\approx -18,376 \text{ or } 15,709$$

We reject the negative root, since $x$ must be nonnegative. So the critical number of $y$ is approximately 15,709. From the sign diagram for $f'$ shown in Figure 14, we see that $y$ is increasing on $(0, 15,709)$ and decreasing on $(15,709, 27,496)$.

**b.** From part (a) we see that $y$ has a relative maximum at $x \approx 15,709$ with value

$$y \approx 1.732x - 0.000008x^2 - 0.000000002x^3 |_{x=15,709} \approx 17,481$$

**c.** After leaving the origin, the projectile gains altitude as it travels downrange. It reaches a maximum altitude of approximately 17,481 ft after it has traveled approximately 15,709 ft downrange. From this point on, the missile descends until it strikes the ground (after traveling approximately 27,496 ft horizontally). The trajectory of the projectile is shown in Figure 15. ∎

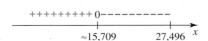

**FIGURE 14**
The sign diagram of $f'$

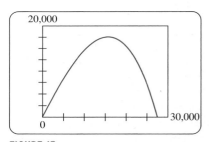

**FIGURE 15**
The trajectory of the projectile

## 3.3 CONCEPT QUESTIONS

**1.** Explain each of the following statements: (a) $f$ is increasing on an interval $I$, (b) $f$ is decreasing on an interval $I$, and (c) $f$ is monotonic on an interval $I$.

**2.** Describe a procedure for determining where a function is increasing and where it is decreasing.

**3.** Describe a procedure for finding the relative extrema of a function.

## 3.3 EXERCISES

In Exercises 1–6 you are given the graph of a function f.
(a) Determine the intervals on which f is increasing, constant,
or decreasing. (b) Find the relative maxima and relative
minima, if any, of f.

**1.**

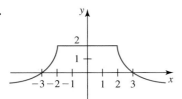

**2.**

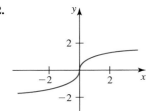

**3.**

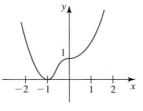

**4.**

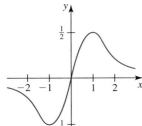

**5.**

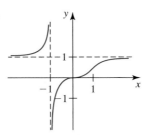

**6.**

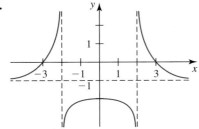

In Exercises 7 and 8 you are given the graph of the derivative f'
of a function f. (a) Determine the intervals on which f is increasing, constant, or decreasing. (b) Find the x-coordinates of the
relative maxima and relative minima of f.

**7.**

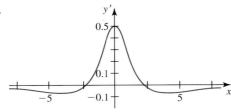

**8.**

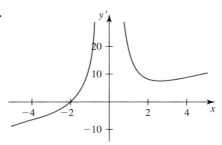

In Exercises 9–38, (a) find the intervals on which f is increasing
or decreasing, and (b) find the relative maxima and relative minima of f.

**9.** $f(x) = x^2 - 2x$

**10.** $f(x) = -x^2 + 4x + 2$

**11.** $f(x) = x^3 - 6x + 1$

**12.** $f(x) = -x^3 + 3x^2 + 1$

**13.** $f(x) = 2x^3 + 3x^2 - 12x + 5$

**14.** $f(x) = x^3 - 3x^2 - 9x + 6$

**15.** $f(x) = x^4 - 4x^3 + 6$

**16.** $f(x) = -x^4 + 2x^2 + 1$

**17.** $f(x) = x^{1/3} - 1$

**18.** $f(x) = x^{1/3} - x^{2/3}$

**19.** $f(x) = x^2(x - 2)^3$

**20.** $f(x) = x^3(x - 6)^4$

**21.** $f(x) = x + \dfrac{1}{x}$

**22.** $f(x) = \dfrac{x}{x - 1}$

**23.** $f(x) = \dfrac{x^2}{x - 1}$

**24.** $f(x) = \dfrac{x}{x^2 + 1}$

**25.** $f(x) = \dfrac{2x - 3}{x^2 - 4}$

**26.** $f(x) = \dfrac{x^2 - 3x + 2}{x^2 + 2x + 1}$

**27.** $f(x) = x^{2/3}(x - 3)$

**28.** $f(x) = x\sqrt{4 - x}$

**29.** $f(x) = x\sqrt{x - x^2}$

**30.** $f(x) = \dfrac{x}{\sqrt{x^2 - 1}}$

**31.** $f(x) = x - 2 \sin x, \quad 0 < x < 2\pi$

**32.** $f(x) = x - \cos x, \quad 0 < x < 2\pi$

**33.** $f(x) = \cos^2 x, \quad 0 < x < 2\pi$

**34.** $f(x) = \sin^2 2x, \quad 0 < x < \pi$

**35.** $f(x) = x^2 e^{-x}$

**36.** $f(x) = x^2 - \ln x$

**37.** $f(x) = \dfrac{2x}{\ln x}$

**38.** $f(x) = \ln(e^x + e^{-x} - 2)$

In Exercises 39 and 40, find the relative extrema of the function.

**39.** $f(x) = \sin^{-1} x - 2x$

**40.** $f(x) = 3 \tan^{-1} x - 2x$

**41. The Boston Marathon** The graph of the function $f$ shown in the accompanying figure gives the elevation of that part of the Boston Marathon course that includes the notorious Heartbreak Hill. Determine the intervals (stretches of the course) where the function $f$ is increasing (the runner is laboring), where it is constant (the runner is taking a breather), and where it is decreasing (the runner is coasting).

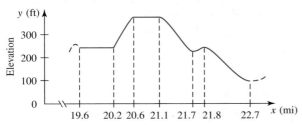

*Source: The Boston Globe.*

**42. The Flight of a Model Rocket** The altitude (in feet) attained by a model rocket $t$ sec into flight is given by the function

$$h(t) = 0.1t^2(t - 7)^4 \qquad 0 \le t \le 7$$

When is the rocket ascending, and when is it descending? What is the maximum altitude attained by the rocket?

**43. Morning Traffic Rush** The speed of traffic flow on a certain stretch of Route 123 between 6 A.M. and 10 A.M. on a typical weekday is approximated by the function

$$f(t) = 20t - 40\sqrt{t} + 52 \qquad 0 \le t \le 4$$

where $f(t)$ is measured in miles per hour and $t$ is measured in hours, with $t = 0$ corresponding to 6 A.M. Find the interval where $f$ is increasing, the interval where $f$ is decreasing, and the relative extrema of $f$. Interpret your results.

**44. Air Pollution** The amount of nitrogen dioxide, a brown gas that impairs breathing, that is present in the atmosphere on a certain day in May in the city of Long Beach is approximated by

$$A(t) = \frac{136}{1 + 0.25(t - 4.5)^2} + 28 \qquad 0 \le t \le 11$$

where $A(t)$ is measured in pollutant standard index (PSI) and $t$ is measured in hours with $t = 0$ corresponding to 7 A.M. When is the PSI increasing, and when is it decreasing? At what time is the PSI highest, and what is its value at that time?

*Source: The Los Angeles Times.*

**45. Finding the Lowest Average Cost** A subsidiary of the Electra Electronics Company manufactures an MP3 player. Management has determined that the daily total cost of producing these players (in dollars) is given by

$$C(x) = 0.0001x^3 - 0.08x^2 + 40x + 5000$$

When is the average cost function $\overline{C}$, defined by $\overline{C}(x) = C(x)/x$, decreasing, and when is it increasing?

At what level of production is the average cost lowest? What is the average cost corresponding to this level of production?

**Hint:** $x = 500$ is a root of the equation $\overline{C}'(x) = 0$.

**46. Cantilever Beam** The figure below depicts a cantilever beam clamped at the left end ($x = 0$) and free at its right end ($x = L$). If a constant load $w$ is uniformly distributed along its length, then the deflection $y$ is given by

$$y = \frac{w}{24EI}(x^4 - 4Lx^3 + 6L^2x^2)$$

where the product $EI$ is a constant called the *flexural rigidity* of the beam. Show that $y$ is increasing on the interval $(0, L)$ and, therefore, that the maximum deflection of the beam occurs at $x = L$. What is the maximum deflection?

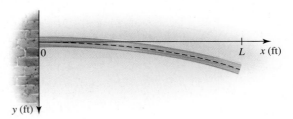

The beam is fixed at $x = 0$ and free at $x = L$.
(Note that the positive direction of $y$ is downward.)

**47. Water Level in a Harbor** The water level in feet in Boston Harbor during a certain 24-hr period is approximated by the formula

$$H = 4.8 \sin\left(\frac{\pi}{6}(t - 10)\right) + 7.6 \qquad 0 \le t \le 24$$

where $t = 0$ corresponds to 12 A.M. When is the water level rising and when is it falling? Find the relative extrema of $H$ and interpret your results.

*Source: SMG Marketing Group.*

**48. Spending on Fiber-Optic Links** U.S. telephone company spending on fiber-optic links to homes and businesses from the beginning of 2001 to the beginning of 2006 is approximated by

$$S(t) = -2.315t^3 + 34.325t^2 + 1.32t + 23 \qquad 0 \le t \le 5$$

billion dollars in year $t$, where $t$ is measured in years with $t = 0$ corresponding to the beginning of 2001.

**a.** Plot the graph of $S$ in the viewing window $[0, 5] \times [0, 600]$.

**b.** Plot the graph of $S'$ in the viewing window $[0, 5] \times [0, 175]$. What conclusion can you draw from your result?

**c.** Verify your result analytically.

*Source: RHK, Inc.*

**49. Surgeries in Physicians' Offices** Driven by technological advances and financial pressures, the number of surgeries performed in physicians' offices nationwide has been increasing over the years. The function

$$f(t) = -0.00447t^3 + 0.09864t^2 + 0.05192t + 0.8$$
$$0 \le t \le 15$$

gives the number of surgeries (in millions) performed in physicians' offices in year $t$, with $t = 0$ corresponding to the beginning of 1986.

 **a.** Plot the graph of $f$ in the viewing window $[0, 15] \times [0, 10]$.

**b.** Prove that $f$ is increasing on the interval $[0, 15]$.

*Source:* SMG Marketing Group.

**50. Age of Drivers in Crash Fatalities** The number of crash fatalities per 100,000 vehicle miles of travel (based on 1994 data) is approximated by the model

$$f(x) = \frac{15}{0.08333x^2 + 1.91667x + 1} \qquad 0 \le x \le 11$$

where $x$ is the age of the driver in years, with $x = 0$ corresponding to age 16. Show that $f$ is decreasing on $(0, 11)$ and interpret your result.

*Source:* National Highway Traffic Safety Administration.

**51. Sales of Functional Food Products** The sales of functional food products—those that promise benefits beyond basic nutrition—have risen sharply in recent years. The sales (in billions of dollars) of foods and beverages with herbal and other additives is approximated by the function

$$S(t) = 0.46t^3 - 2.22t^2 + 6.21t + 17.25 \qquad 0 \le t \le 4$$

where $t$ is measured in years, with $t = 0$ corresponding to the beginning of 1997.

 **a.** Plot the graph of $S$ in the viewing window $[0, 4] \times [15, 40]$.

**b.** Show that sales were increasing over the 4-year period beginning in 1997.

*Source:* Frost & Sullivan.

**52. Halley's Law** Halley's Law states that the barometric pressure (in inches of mercury) at an altitude of $x$ miles above sea level is approximated by

$$p(x) = 29.92e^{-0.2x} \qquad x \ge 0$$

**a.** If a hot-air balloonist measures the barometric pressure as 20 in. of mercury, what is the balloonist's altitude?

**b.** If the barometric pressure is decreasing at the rate of 1 in./hr at that altitude, how fast is the balloon rising?

**53. Polio Immunization** Polio, a once-feared killer, declined markedly in the United States in the 1950s after Jonas Salk developed the inactivated polio vaccine and mass immunization of children took place. The number of polio cases in the United States from the beginning of 1959 to the beginning of 1963 is approximated by the function

$$N(t) = 5.3e^{0.095t^2 - 0.85t} \qquad 0 \le t \le 4$$

where $N(t)$ gives the number of polio cases (in thousands) and $t$ is measured in years with $t = 0$ corresponding to the beginning of 1959.

**a.** Show that the function $N$ is decreasing over the time interval under consideration.

**b.** How fast was the number of polio cases decreasing at the beginning of 1959? At the beginning of 1962?

**Note:** Since the introduction of the oral vaccine developed by Dr. Albert B. Sabin in 1963, polio in the United States has, for all practical purposes, been eliminated.

**54.** Find the intervals where $f(x) = e^{-x^2/2}$ is increasing and where it is decreasing.

**55.** Find the intervals where $f(x) = (\log x)/x$ is increasing and where it is decreasing.

**56.** Prove that the function $f(x) = 2x^5 + x^3 + 2x$ is increasing everywhere.

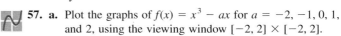 **57. a.** Plot the graphs of $f(x) = x^3 - ax$ for $a = -2, -1, 0, 1$, and 2, using the viewing window $[-2, 2] \times [-2, 2]$.

**b.** Use the results of part (a) to guess at the values of $a$ such that $f$ is increasing on $(-\infty, \infty)$.

**c.** Prove your conjecture analytically.

**58.** Find the values of $a$ such that $f(x) = \cos x - ax + b$ is decreasing everywhere.

**59.** Show that the equation $x + \sin x = b$ has no positive root if $b < 0$ and has one positive root if $b > 0$.

**Hint:** Show that $f(x) = x + \sin x - b$ is increasing and that $f(0) > 0$ if $b < 0$ and $f(0) < 0$ if $b > 0$.

**60.** Prove that $x < \tan x$ if $0 < x < \frac{\pi}{2}$.

**Hint:** Let $f(x) = \tan x - x$ and show that $f$ is increasing on $\left(0, \frac{\pi}{2}\right)$.

**61.** Prove that $2x/\pi < \sin x < x$ if $0 < x < \frac{\pi}{2}$.

**Hint:** Show that $f(x) = (\sin x)/x$ is decreasing on $\left(0, \frac{\pi}{2}\right)$.

**62.** Let $f(x) = -2x^2 + ax + b$. Determine the constants $a$ and $b$ such that $f$ has a relative maximum at $x = 2$ and the relative maximum value is 4.

**63.** Let $f(x) = ax^3 + 6x^2 + bx + 4$. Determine the constants $a$ and $b$ such that $f$ has a relative minimum at $x = -1$ and a relative maximum at $x = 2$.

**64.** Let $f(x) = (ax + b)/(cx + d)$, where $a$, $b$, $c$, and $d$ are constants. Show that $f$ has no relative extrema if $ad - bc \ne 0$.

**65.** Let

$$f(x) = \begin{cases} \dfrac{1}{x^2} & \text{if } x < 0 \\ x^2 & \text{if } x \ge 0 \end{cases}$$

Show that $f$ has a relative minimum at 0, although its first derivative does not change sign as we move across $x = 0$. Does this contradict the First Derivative Test?

**66.** Prove part (b) of Theorem 1.

**67.** Prove parts (b) and (c) of Theorem 2.

**68.** Prove that $x - x^3/6 < \sin x < x$ if $x > 0$.

**Hint:** To prove the left inequality, let $f(x) = \sin x - x + x^3/6$, and show that $f$ is increasing on the interval $(0, \infty)$.

**69.** Let $f(x) = 3x^5 - 8x^3 + x$.

**a.** Plot the graph of $f$ using the viewing window $[-2, 2] \times [-6, 6]$. Can you determine from the graph of $f$ the intervals where $f$ is increasing or decreasing?

**b.** Plot the graph of $f$ using the viewing window $[-0.5, 0.5] \times [-0.5, 0.5]$. Using this graph and the result of part (a), determine the intervals where $f$ is increasing and where $f$ is decreasing.

**70.** Let

$$f(x) = \begin{cases} \frac{1}{2}x + x^2 \sin \frac{1}{x} & \text{if } x \neq 0 \\ 0 & \text{if } x = 0 \end{cases}$$

**a.** Plot the graph of $f$. Use ZOOM to obtain successive magnifications of the graph in the neighborhood of the origin. Can you see that $f$ is not monotonic on any interval containing the origin?

**b.** Prove the observation made in part (a).

**71.** Let

$$f(x) = \begin{cases} \left(2 - \sin \frac{1}{x}\right)|x| & \text{if } x \neq 0 \\ 0 & \text{if } x = 0 \end{cases}$$

**a.** Plot the graph of $f$. Use ZOOM to obtain successive magnifications of the graph in the neighborhood of

the origin. Can you see that $f$ has a relative minimum at 0 but is not monotonic to the left or to the right of $x = 0$?

**b.** Prove the observation made in part (a).

**Hint:** For $x > 0$, show that $f'(x) > 0$ if $x = 1/(2n\pi)$ and $f'(x) < 0$ if $x = 1/((2n + 1)\pi)$.

**72. a.** Show that $e^x \geq 1 + x$ if $x \geq 0$.

**b.** Show that $e^x \geq 1 + x + x^2/2$ if $x \geq 0$.

**Hint:** Show that $f(x) = e^x - 1 - x - x^2/2$ is increasing for $x \geq 0$.

*In Exercises 73–78, determine whether the statement is true or false. If it is true, explain why it is true. If it is false, explain why or give an example to show why it is false.*

**73.** If $f$ and $g$ are increasing on an interval $I$, then $f + g$ is also increasing on $I$.

**74.** If $f$ is increasing on an interval $I$ and $g$ is decreasing on the same interval $I$, then $f - g$ is increasing on $I$.

**75.** If $f$ and $g$ are increasing functions on an interval $I$, then their product $fg$ is also increasing on $I$.

**76.** If $f$ and $g$ are positive on an interval $I$, $f$ is increasing on $I$, and $g$ is decreasing on $I$, then the quotient $f/g$ is increasing on $I$.

**77.** If $f$ is increasing on an interval $(a, b)$, then $f'(x) \geq 0$ for every $x$ in $(a, b)$.

**78.** $f(x) = \cos^{-1} x$ is a decreasing function.

---

## 3.4 Concavity and Inflection Points

### ■ Concavity

The graphs of the position functions $s_1$ and $s_2$ of two cars $A$ and $B$ traveling along a straight road are shown in Figure 1. Both graphs are rising, reflecting the fact that both cars are moving forward, that is, moving with positive velocities.

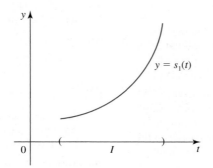

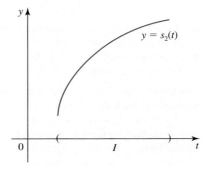

**FIGURE 1**  (a) $s_1$ is increasing on $I$.  (b) $s_2$ is increasing on $I$.

Observe, however, that the graph shown in Figure 1a opens upward, whereas the graph shown in Figure 1b opens downward. How do we interpret the way the curves bend in terms of the motion of the cars? To answer this question, let's look at the slopes of the tangent lines at various points on each graph (Figure 2).

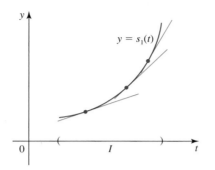

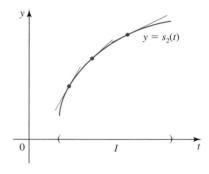

**FIGURE 2**
The slopes of the tangent lines to the graph of $s_1$ are increasing, whereas those to the graph of $s_2$ are decreasing.

(**a**) The graph of $s_1$ is concave upward.

(**b**) The graph of $s_2$ is concave downward.

In Figure 2a you can see that the slopes of the tangent lines to the graph increase as $t$ increases. Since the slope of the tangent line at the point $(t, s_1(t))$ measures the velocity of car $A$ at time $t$, we see not only that the car is moving forward, but also that its velocity is increasing on the time interval $I$. In other words, car $A$ is accelerating over the interval $I$. A similar analysis of the graph in Figure 2b shows that car $B$ is moving forward as well but decelerating over the time interval $I$.

We can describe the way a curve bends using the notion of concavity.

> **DEFINITIONS   Concavity of the Graph of a Function**
> Suppose $f$ is differentiable on an open interval $I$. Then
>
> **a.** the graph of $f$ is **concave upward** on $I$ if $f'$ is increasing on $I$.
> **b.** the graph of $f$ is **concave downward** on $I$ if $f'$ is decreasing on $I$.

**Note**   It can be shown that if the graph of $f$ is concave upward on an open interval $I$, then it lies above all of its tangent lines (Figure 2a), and if the graph is concave downward on $I$, then it lies below all of its tangent lines (Figure 2b). A proof of this is given in Appendix B.   ◼

Figure 3 shows the graph of a function that is concave upward on the intervals $(a, b)$, $(c, d)$, and $(d, e)$ and concave downward on $(b, c)$ and $(e, g)$.

**FIGURE 3**
The interval $[a, g]$ is divided into subintervals showing where the graph of $f$ is concave upward and where it is concave downward.

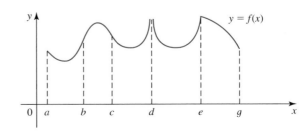

If a function $f$ has a second derivative $f''$, we can use it to determine the intervals of concavity of the graph of $f$. Indeed, since the second derivative of $f$ measures the rate of change of the first derivative of $f$, we see that $f'$ is increasing on an open interval $(a, b)$ if $f''(x) > 0$ for all $x$ in $(a, b)$ and that $f'$ is decreasing on $(a, b)$ if $f''(x) < 0$ for all $x$ in $(a, b)$. Thus, we have the following result.

---

**THEOREM 1**

Suppose $f$ has a second derivative on an open interval $I$.

**a.** If $f''(x) > 0$ for all $x$ in $I$, then the graph of $f$ is concave upward on $I$.
**b.** If $f''(x) < 0$ for all $x$ in $I$, then the graph of $f$ is concave downward on $I$.

---

The following procedure, based on the conclusions of Theorem 1, can be used to determine the intervals of concavity of a function.

---

**Determining the Intervals of Concavity of a Function**

**1.** Find all values of $x$ for which $f''(x) = 0$ or $f''(x)$ does not exist. Use these values of $x$ to partition the domain of $f$ into open intervals.
**2.** Select a test number $c$ in each interval found in Step 1 and determine the sign of $f''(c)$ in that interval.
   **a.** If $f''(c) > 0$, the graph of $f$ is concave upward on that interval.
   **b.** If $f''(c) < 0$, the graph of $f$ is concave downward on that interval.

---

**Note**   In developing this procedure, we have once again used the fact that a function (in this case, the function $f''$) can change sign only as we move across a zero or a number at which the function is discontinuous.   ■

**FIGURE 4**
The sign diagram of $f''$

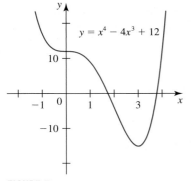

**FIGURE 5**
The graph of $f$ is concave upward on $(-\infty, 0)$ and on $(2, \infty)$ and concave downward on $(0, 2)$.

**EXAMPLE 1**   Determine the intervals where the graph of $f(x) = x^4 - 4x^3 + 12$ is concave upward and the intervals where it is concave downward.

**Solution**   We first calculate the second derivative of $f$:

$$f'(x) = 4x^3 - 12x^2$$
$$f''(x) = 12x^2 - 24x = 12x(x - 2)$$

Next, we observe that $f''$ is continuous everywhere and has zeros at 0 and 2. Using this information, we draw the sign diagram of $f''$ (Figure 4). We conclude that the graph of $f$ is concave upward on $(-\infty, 0)$ and on $(2, \infty)$ and concave downward on $(0, 2)$. The graph of $f$ is shown in Figure 5. Observe that the concavity of the graph of $f$ changes from upward to downward at the point $(0, 12)$ and from downward to upward at the point $(2, -4)$.   ■

**EXAMPLE 2**   Determine the intervals where the graph of $f(x) = x^{2/3}$ is concave upward and where it is concave downward.

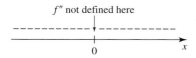

**FIGURE 6**
The sign diagram of $f''$

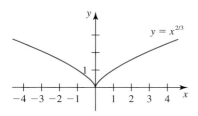

**FIGURE 7**
The graph of $f$ is concave downward on $(-\infty, 0)$ and on $(0, \infty)$.

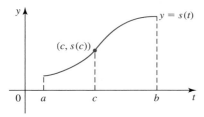

**FIGURE 8**
The point $(c, s(c))$ at which the concavity of the graph of $s$ changes is called an inflection point of $s$.

**Solution** We find

$$f'(x) = \frac{2}{3} x^{-1/3}$$

and

$$f''(x) = -\frac{2}{9} x^{-4/3} = -\frac{2}{9x^{4/3}}$$

Observe that $f''$ is continuous everywhere except at 0. From the sign diagram of $f''$ shown in Figure 6, we conclude that the graph of $f$ is concave downward on $(-\infty, 0)$ and on $(0, \infty)$ (Figure 7). ■

## ◼ Inflection Points

The graph of the position function $s$ of a car traveling along a straight road is shown in Figure 8. Observe that the graph of $s$ is concave upward on $(a, c)$ and concave downward on $(c, b)$. Interpreting the graph, we see that the car is accelerating for $a < t < c$ ($s''(t) > 0$ for $t$ in $(a, c)$) and decelerating for $c < t < b$ ($s''(t) < 0$ for $t$ in $(c, b)$). Its acceleration is zero when $t = c$, at which time the car also attains the maximum velocity in the time interval $(a, b)$. The point $(c, s(c))$ on the graph of $s$ at which the concavity changes is called an *inflection point* or *point of inflection* of $s$.

More generally, we have the following definition.

> **DEFINITION Inflection Point**
>
> Let the function $f$ be continuous on an open interval containing the point $c$, and suppose the graph of $f$ has a tangent line at $P(c, f(c))$. If the graph of $f$ changes from concave upward to concave downward (or vice versa) at $P$, then the point $P$ is called an **inflection point** of the graph of $f$.

Observe that the graph of a function crosses its tangent line at a point of inflection (Figure 9).

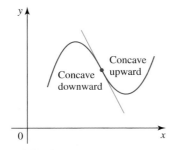
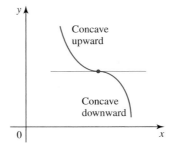
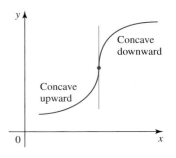

**FIGURE 9**
At a point of inflection the graph of a function crosses its tangent line.

The following procedure can be used to find the inflection points of a function that has a second derivative, except perhaps at isolated numbers.

> **Finding Inflection Points**
> 1. Find all numbers $c$ in the domain of $f$ for which $f''(c) = 0$ or $f''(c)$ does not exist. These numbers give rise to candidates for inflection points.
> 2. Determine the sign of $f''(x)$ to the left and to the right of each number $c$ found in Step 1. If the sign of $f''(x)$ changes, then the point $P(c, f(c))$ is an inflection point of $f$, provided that the graph of $f$ has a tangent line at $P$.

**FIGURE 10**
The sign diagram of $f''$

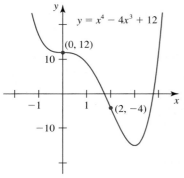

**FIGURE 11**
$(0, 12)$ and $(2, -4)$ are inflection points.

**EXAMPLE 3** Find the points of inflection of $f(x) = x^4 - 4x^3 + 12$.

**Solution** We compute

$$f'(x) = 4x^3 - 12x^2 \qquad \text{and} \qquad f''(x) = 12x^2 - 24x = 12x(x - 2)$$

We see that $f''$ is continuous everywhere and has zeros at 0 and 2. These numbers give rise to candidates for the inflection points of $f$. From the sign diagram of $f''$ shown in Figure 10, we see that $f''(x)$ changes sign from positive to negative as we move across 0. Therefore, the point $(0, 12)$ is an inflection point of $f$. Also, $f''(x)$ changes sign from negative to positive as we move across 2, so $(2, -4)$ is also an inflection point of $f$. These inflection points are shown in Figure 11, where the graph of $f$ is sketched. ■

**EXAMPLE 4** Find the points of inflection of $f(x) = (x - 1)^{1/3}$.

**Solution** We find

$$f'(x) = \frac{1}{3}(x - 1)^{-2/3}$$

and

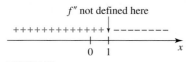

**FIGURE 12**
The sign diagram of $f''$

$$f''(x) = -\frac{2}{9}(x - 1)^{-5/3} = -\frac{2}{9(x - 1)^{5/3}}$$

We see that $f''$ is continuous everywhere except at 1, where it is not defined. Furthermore, $f''$ has no zeros, so 1 gives rise to the only candidate for an inflection point of $f$. From the sign diagram of $f''$ shown in Figure 12, we see that $f''(x)$ does change sign from positive to negative as we move across 1. Therefore, $(1, 0)$ is indeed an inflection point of $f$. Observe that the graph of $f$ has a vertical tangent line at that point. (See Figure 13.) ■

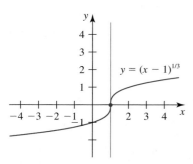

**FIGURE 13**
$f$ has an inflection point at $(1, 0)$.

⚠ Remember that the numbers where $f''(x) = 0$ or where $f''$ is discontinuous give rise only to *candidates* for inflection points of $f$. For example, you can show that if $f(x) = x^4$, then $f''(0) = 0$, but the point $(0, 0)$ is not an inflection point of $f$ (Figure 14). Also, if $g(x) = x^{2/3}$, then $g''$ is discontinuous at 0, as we saw in Example 2, but the point $(0, 0)$ is not an inflection point of $g$ (Figure 15).

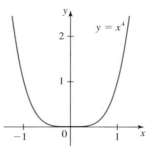

**FIGURE 14**
$f''(0) = 0$, but $(0, 0)$ is not an inflection point of $f$.

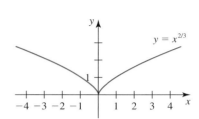

**FIGURE 15**
$g''$ is discontinuous at 0, but $(0, 0)$ is not an inflection point of $g$.

Examples 5 and 6 provide us with two practical interpretations of the inflection point of a function.

**EXAMPLE 5** **Test Dive of a Submarine** Refer to Example 2 in Section 3.2. Recall that the depth (in feet) at time $t$ (measured in minutes) of the prototype of a twin-piloted submarine is given by

$$h(t) = t^3(t - 7)^4 \qquad 0 \le t \le 7$$

Find the inflection points of $h$, and explain their significance.

**Solution** We have

$$h'(t) = 3t^2(t - 7)^4 + t^3(4)(t - 7)^3 = t^2(t - 7)^3(3t - 21 + 4t)$$

$$= 7t^2(t - 3)(t - 7)^3$$

$$h''(t) = \frac{d}{dt}[7(t^3 - 3t^2)(t - 7)^3]$$

$$= 7[(3t^2 - 6t)(t - 7)^3 + (t^3 - 3t^2)(3)(t - 7)^2]$$

$$= 7[3t(t - 2)(t - 7)^3 + 3t^2(t - 3)(t - 7)^2]$$

$$= 21t(t - 7)^2[(t - 2)(t - 7) + t(t - 3)]$$

$$= 42t(t - 7)^2(t^2 - 6t + 7)$$

Observe that $h''$ is continuous everywhere and, therefore, on $[0, 7]$. Setting $h''(t) = 0$ gives $t = 0$, $t = 7$ or $t^2 - 6t + 7 = 0$. Using the quadratic formula to solve the last equation, we obtain

$$t = \frac{6 \pm \sqrt{36 - 28}}{2} = 3 \pm \sqrt{2}$$

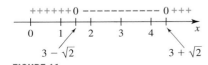

**FIGURE 16**
The sign diagram for $h''$

Since both of these roots lie inside the interval $(0, 7)$, they give rise to candidates for the inflection points of $h$. From the sign diagram of $h''$ we see that $t = 3 - \sqrt{2} \approx 1.59$ and $t = 3 + \sqrt{2} \approx 4.41$ do indeed give rise to inflection points of $h$ (Figure 16). The graph of $h$ is reproduced in Figure 17.

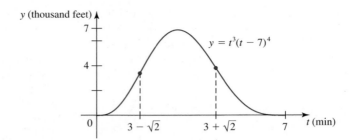

**FIGURE 17**
The graph of $h$ has inflection points at $(3 - \sqrt{2}, h(3 - \sqrt{2}))$ and $(3 + \sqrt{2}, h(3 + \sqrt{2}))$.

To interpret our results, observe that the graph of $h$ is concave upward on $(0, 3 - \sqrt{2})$. This says that the submarine is accelerating downward to a depth of $h(3 - \sqrt{2}) \approx 3427$ ft over the time interval $(0, 1.6)$. (Verify!) The graph of $f$ is concave downward on $(3 - \sqrt{2}, 3 + \sqrt{2})$, and this says that the submarine is decelerating downward from $t \approx 1.6$ to its lowest point. Then it is accelerating upward until $t \approx 4.4$. From $t \approx 4.4$ until $t = 7$, the submarine decelerates upward until it reaches the surface, 7 min after the start of the test dive. The rate of descent of the submarine is greatest at $t = 3 - \sqrt{2} \approx 1.6$ and is approximately $h'(3 - \sqrt{2})$, or 3951 ft/min. Also the rate of ascent of the submarine is greatest at $t = 3 + \sqrt{2} \approx 4.4$ and is approximately $-h'(3 + \sqrt{2})$, or 3335 ft/min.    ■

**EXAMPLE 6**    **Effect of Advertising on Revenue**    The total annual revenue $R$ of the Odyssey Travel Agency, in thousands of dollars, is related to the amount of money $x$ that the agency spends on advertising its services by the formula

$$R = -0.01x^3 + 1.5x^2 + 200 \qquad 0 \le x \le 100$$

where $x$ is measured in thousands of dollars. Find the inflection point of $R$ and interpret your results.

**Solution**

$$R' = -0.03x^2 + 3x$$

and

$$R'' = -0.06x + 3$$

which is continuous everywhere. Setting $R'' = 0$ gives $x = 50$, and this number gives rise to a candidate for an inflection point of $R$. Moreover, because $R'' > 0$ for $0 < x < 50$ and $R'' < 0$ for $50 < x < 100$, we see that the point $(50, 2700)$ is an inflection point of the function $R$. The graph of $R$ appears in Figure 18.

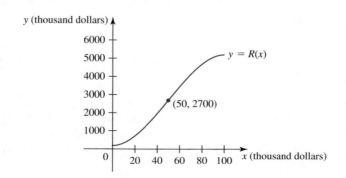

**FIGURE 18**
The graph of $R$ has an inflection point at $x = 50$.

To interpret these results, observe that the revenue of the agency increases rather slowly at first. As the amount spent on advertising increases, the revenue increases rapidly, reflecting the effectiveness of the company's ads. But a point is soon reached beyond which any additional advertising expenditure results in increased revenue but at a slower rate of increase. This level of expenditure is commonly referred to as the *point of diminishing returns* and corresponds to the $x$-coordinate of the inflection point of $R$. ∎

## The Second Derivative Test

The second derivative of a function can often be used to help us determine whether a critical number gives rise to a relative extremum. Suppose that $c$ is a critical number of $f$ and suppose that $f''(c) < 0$. Then the graph of $f$ is concave downward on some interval $(a, b)$ containing $c$. Intuitively, we see that $f(c)$ must be the largest value of $f(x)$ for all $x$ in $(a, b)$. In other words, $f$ has a relative maximum at $c$ (Figure 19a). Similarly, if $f''(c) > 0$ at a critical number $c$, then $f$ has a relative minimum at $c$ (Figure 19b).

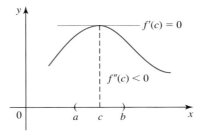
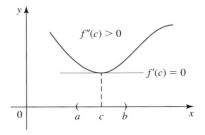

**FIGURE 19**       **(a)** $f$ has a relative maximum at $c$.       **(b)** $f$ has a relative minimum at $c$.

These observations suggest the following theorem.

---

**THEOREM 2    The Second Derivative Test**

Suppose that $f$ has a continuous second derivative on an interval $(a, b)$ containing a critical number $c$ of $f$.

**a.** If $f''(c) < 0$, then $f$ has a relative maximum at $c$.
**b.** If $f''(c) > 0$, then $f$ has a relative minimum at $c$.
**c.** If $f''(c) = 0$, then the test is inconclusive.

---

**PROOF**   We will give an outline of the proof for (a). The proof for (b) is similar and will be omitted. So suppose that $f''(c) < 0$. Then the continuity of $f''$ implies that $f''(x) < 0$ on some open interval $I$ containing $c$. This means that the graph of $f$ is concave downward on $I$. Therefore, the graph of $f$ lies below its tangent line at the point $(c, f(c))$. (See the note on page 315.) But this tangent line is horizontal because $f'(c) = 0$, and this shows that $f(x) \leq f(c)$ for all $x$ in $I$ (Figure 19a). So $f$ has a relative maximum at $c$ as asserted. ∎

**EXAMPLE 7** Find the relative extrema of $f(x) = x^3 - 3x^2 - 24x + 32$ using the Second Derivative Test.

**Solution**

$$f'(x) = 3x^2 - 6x - 24 = 3(x - 4)(x + 2)$$

Setting $f'(x) = 0$, we see that $-2$ and $4$ are critical numbers of $f$. Next, we compute

$$f''(x) = 6x - 6 = 6(x - 1)$$

Evaluating $f''(x)$ at the critical number $-2$, we find

$$f''(-2) = 6(-2 - 1) = -18 < 0$$

and the Second Derivative Test implies that $-2$ gives rise to a relative maximum of $f$. Also

$$f''(4) = 6(4 - 1) = 18 > 0$$

so $4$ gives rise to a relative minimum of $f$. The graph of $f$ is shown in Figure 20. ■

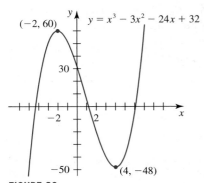

**FIGURE 20**
$f$ has a relative maximum at $(-2, 60)$ and a relative minimum at $(4, -48)$.

**EXAMPLE 8** **Watching a Helicopter Take Off** A spectator standing 200 ft from a helicopter pad watches a helicopter take off. The helicopter rises vertically with a constant acceleration of 8 ft/sec$^2$ and reaches a height (in feet) of $h(t) = 4t^2$ after $t$ sec, where $0 \leq t \leq 10$. (See Figure 21.) As the helicopter rises, $d\theta/dt$ increases, slowly at first, then faster, and finally it slows down again. The spectator perceives the helicopter to be rising at the greatest speed when $d\theta/dt$ is maximal. Determine the height of the helicopter at the moment the spectator perceives it to be rising at the greatest speed.

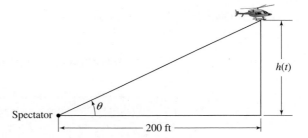

**FIGURE 21**
The helicopter attains a height of $h(t) = 4t^2$ after $t$ sec.

Spectator

$\theta$

200 ft

$h(t)$

**Solution** The angle of elevation of the spectator's line of sight at time $t$ is

$$\theta(t) = \tan^{-1}\left(\frac{h(t)}{200}\right) = \tan^{-1}\left(\frac{4t^2}{200}\right) = \tan^{-1}\left(\frac{t^2}{50}\right)$$

Therefore,

$$\frac{d\theta}{dt} = \frac{1}{1 + \left(\dfrac{t^2}{50}\right)^2} \cdot \frac{d}{dt}\left(\frac{t^2}{50}\right) = \frac{2500}{2500 + t^4} \cdot \frac{2t}{50}$$

$$= \frac{100t}{2500 + t^4}$$

To find when $d\theta/dt$ is maximal, we first compute

$$\frac{d^2\theta}{dt^2} = \frac{(2500 + t^4)100 - 100t(4t^3)}{(2500 + t^4)^2} = \frac{100(2500 - 3t^4)}{(2500 + t^4)^2}$$

Then, setting $d^2\theta/dt^2 = 0$ gives $t = (2500/3)^{1/4} \approx 5.37$ as the sole critical number of $d\theta/dt$. Using either the First or Second Derivative Test, we can show that this critical number gives rise to a maximum for $d\theta/dt$. The height of the helicopter at this instant of time is

$$h(\sqrt[4]{2500/3}) = 4(\sqrt[4]{2500/3})^2 = 4\sqrt{2500/3} \approx 115.47$$

or approximately 115 ft. ▪

**Note**   The point $(5.37, 115.47)$ in Example 8 is an inflection point of the graph of $h$.

▬

The Second Derivative Test is not useful if $f''(c) = 0$ at a critical number $c$. For example, each of the functions $f(x) = -x^4$, $g(x) = x^4$, and $h(x) = x^3$ has a critical number 0. Notice that $f''(0) = g''(0) = h''(0) = 0$; but as you can see from the graphs of these functions (Figure 22), $f$ has a relative maximum at 0, $g$ has a relative minimum at 0, and $h$ has no extremum at 0.

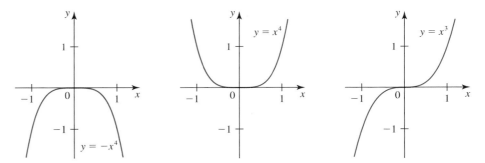

**FIGURE 22**
The Second Derivative Test is not useful when the second derivative is zero at a critical number $c$.

What are the pros and cons of using the First Derivative Test (FDT) and the Second Derivative Test (SDT) to determine the relative extrema of a function? First, because the SDT can be used only when $f''$ exists, it is less versatile than the FDT. For example, the SDT cannot be used to show that $f(x) = x^{2/3}$ has a relative minimum at 0. Furthermore, the SDT is inconclusive if $f''$ is equal to zero at a critical number of $f$, whereas the FDT always yields positive conclusions. The SDT is also inconvenient to use when $f''$ is difficult to compute. However, on the plus side, the SDT is easy to apply if $f''$ is easy to compute. (See Example 7.) Also, the conclusions of the SDT are often used in theoretical work.

## ▮ The Roles of $f'$ and $f''$ in Determining the Shape of a Graph

Let's summarize our discussion of the properties of the graph of a function $f$ that are determined by its first and second derivatives: The first derivative $f'$ tells us where $f$ is increasing and where $f$ is decreasing, whereas the second derivative $f''$ tells us where the graph of $f$ is concave upward and where it is concave downward. Each of these properties is determined by the signs of $f'$ and $f''$ in the interval of interest and is reflected in the shape of the graph of $f$. Table 1 gives the characteristics of the graph of $f$ for the various possible combinations of the signs of $f'$ and $f''$.

**TABLE 1**

| Signs of $f'$ and $f''$ | Properties of the graph of $f$ | General shape of the graph of $f$ |
|---|---|---|
| $f'(x) > 0$ <br> $f''(x) > 0$ | $f$ increasing <br> $f$ concave upward | |
| $f'(x) > 0$ <br> $f''(x) < 0$ | $f$ increasing <br> $f$ concave downward | |
| $f'(x) < 0$ <br> $f''(x) > 0$ | $f$ decreasing <br> $f$ concave upward | |
| $f'(x) < 0$ <br> $f''(x) < 0$ | $f$ decreasing <br> $f$ concave downward | |

## 3.4 CONCEPT QUESTIONS

1. Explain what it means for the graph of a function $f$ to be (a) concave upward and (b) concave downward on an open interval $I$. Given that $f$ has a second derivative on $I$ (except at isolated numbers), how do you determine where the graph of $f$ is concave upward and where it is concave downward?

2. What is an inflection point of the graph of a function $f$? How do you find the inflection points of the graph of a function $f$ whose rule is given?

3. State the Second Derivative Test. What are the pros and cons of using the First Derivative Test and the Second Derivative Test?

## 3.4 EXERCISES

*In Exercises 1–6 you are given the graph of a function f. Determine the intervals where the graph of f is concave upward and where it is concave downward. Find all inflection points of f.*

**1.**

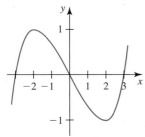

**2.**

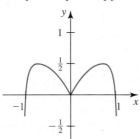

**4.**

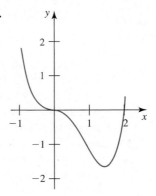

**3.**

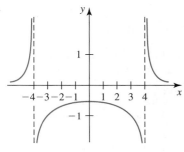

**5.**

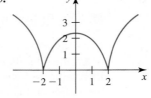

**6.**

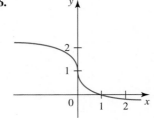

*In Exercises 7 and 8 you are given the graph of the second derivative $f''$ of a function $f$. (a) Determine the intervals where the graph of $f$ is concave upward and the intervals where it is concave downward. (b) Find the x-coordinates of the inflection points of $f$.*

**7.**

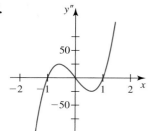

**8.**

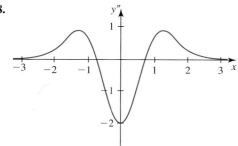

*In Exercises 9–10, determine which graph—(a), (b), or (c)—is the graph of the function $f$ with the specified properties. Explain.*

**9.** $f'(0)$ is undefined, $f$ is decreasing on $(-\infty, 0)$, the graph of $f$ is concave downward on $(0, 3)$, and $f$ has an inflection point at $x = 3$.

**(a)**

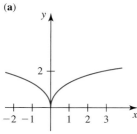

**(b)**

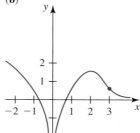

**(c)**

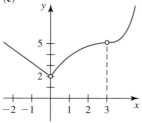

**10.** $f$ is decreasing on $(-\infty, 2)$ and increasing on $(2, \infty)$, the graph of $f$ is concave upward on $(1, \infty)$, and $f$ has inflection points at $x = 0$ and $x = 1$.

**(a)**

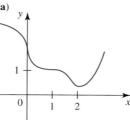

**(b)**

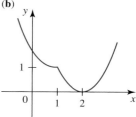

**(c)**

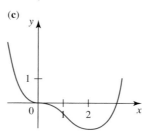

*In Exercises 11–32, determine where the graph of the function is concave upward and where it is concave downward. Also, find all inflection points of the function.*

**11.** $f(x) = x^3 - 6x$

**12.** $g(x) = x^3 - 6x^2 + 2x + 3$

**13.** $f(t) = t^4 - 2t^3$

**14.** $h(x) = 3x^4 + 4x^3 + 1$

**15.** $f(x) = 1 + 3x^{1/3}$

**16.** $g(x) = 2x - x^{1/3}$

**17.** $h(t) = \dfrac{1}{3}t^2 + \dfrac{3}{5}t^{5/3}$

**18.** $f(x) = x - \sqrt{1 - x^2}$

**19.** $h(x) = \sqrt{x^2 - x^4}$

**20.** $g(x) = x + \dfrac{1}{x}$

**21.** $h(x) = x^2 + \dfrac{1}{x^2}$

**22.** $f(x) = \dfrac{x}{x + 1}$

**23.** $f(u) = \dfrac{u}{u^2 - 1}$

**24.** $f(x) = \dfrac{x^2 - 9}{1 - x^2}$

**25.** $f(x) = \sin 2x, \quad 0 \le x \le \pi$

**26.** $g(x) = \cos^2 x, \quad 0 \le x \le 2\pi$

**27.** $h(t) = \sin t + \cos t, \quad 0 \le t \le 2\pi$

**28.** $f(x) = x - \sin x, \quad 0 \le x \le 4\pi$

**29.** $f(x) = \tan 2x, \quad -\pi \le x \le \pi$

**30.** $g(x) = x^2 e^{-x}$

**31.** $h(x) = \ln|x|$

**32.** $f(x) = e^{\sin x}, \quad -\dfrac{\pi}{2} \le x \le \dfrac{\pi}{2}$

In Exercises 33–36, plot the graph of f, and find (a) the approximate intervals where the graph of f is concave upward and where it is concave downward and (b) the approximate coordinates of the point(s) of inflection accurate to 1 decimal place.

**33.** $f(x) = x^5 - 2x^4 + 3x^2 - 5x + 4$

**34.** $f(x) = \dfrac{x^3 + x^2 - x + 1}{x^3 + 1}$    **35.** $f(x) = \dfrac{x}{\sqrt{x^2 + 1}}$

**36.** $f(x) = \cos(\sin x)$  $-2 < x < 2$

In Exercises 37–48, find the relative extrema, if any, of the function. Use the Second Derivative Test, if applicable.

**37.** $h(t) = \dfrac{1}{3}t^3 - 2t^2 - 5t - 10$

**38.** $h(x) = 2x^3 + 3x^2 - 12x - 2$

**39.** $f(x) = x^4 - 4x^3$    **40.** $f(x) = 2x^4 - 8x + 4$

**41.** $f(t) = 2t + \dfrac{1}{t}$    **42.** $h(t) = e^t - t - 1$

**43.** $g(t) = t - 2\ln t$    **44.** $f(x) = x(\ln x)^2$

**45.** $f(x) = \sin x + \cos x$,  $0 < x < \frac{\pi}{2}$

**46.** $f(x) = \sin^2 x$,  $0 < x < \frac{3\pi}{2}$

**47.** $f(x) = 2\sin x + \sin 2x$,  $0 < x < \pi$

**48.** $f(x) = x^2 \ln x$

In Exercises 49 and 50, find the point(s) of inflection of the graph of the function.

**49.** $f(x) = \sin^{-1} x$    **50.** $f(x) = (\tan^{-1} x)^2$

In Exercises 51–54, sketch the graph of a function having the given properties.

**51.** $f(0) = 0, f'(0) = 0$
$f'(x) < 0$ on $(-\infty, 0)$
$f'(x) > 0$ on $(0, \infty)$
$f''(x) > 0$ on $(-1, 1)$
$f''(x) < 0$ on $(-\infty, -1) \cup (1, \infty)$

**52.** $f(0) = -1, f(-1) = f(1) = 0$
$f'(0)$ does not exist
$f'(x) < 0$ on $(-\infty, 0)$
$f'(x) > 0$ on $(0, \infty)$
$f''(x) < 0$ on $(-\infty, 0) \cup (0, \infty)$

**53.** $f(-1) = 0, f'(-1) = 0$
$f(0) = 1, f'(0) = 0$
$f'(x) < 0$ on $(-\infty, -1)$
$f'(x) > 0$ on $(-1, \infty)$
$f''(x) > 0$ on $\left(-\infty, -\frac{2}{3}\right) \cup (0, \infty)$
$f''(x) < 0$ on $\left(-\frac{2}{3}, 0\right)$

**54.** $f(-1) = f(1) = 2, f'(-1) = f'(1) = 0$
$f'(x) < 0$ on $(-\infty, -1) \cup (0, 1)$
$f'(x) > 0$ on $(-1, 0) \cup (1, \infty)$
$\lim_{x \to 0} f(x) = \infty$
$f''(x) > 0$ on $(-\infty, 0) \cup (0, \infty)$

**55. Effect of Advertising on Bank Deposits** The CEO of the Madison Savings Bank used the graphs on the following page to illustrate what effect a projected promotional campaign would have on its deposits over the next year. The functions $D_1$ and $D_2$ give the projected amount of money on deposit with the bank over the next 12 months with and without the proposed promotional campaign, respectively.
  **a.** Determine the signs of $D_1'(t)$, $D_2'(t)$, $D_1''(t)$, and $D_2''(t)$ on the interval $(0, 12)$.
  **b.** What can you conclude about the rate of change of the growth rate of the money on deposit with the bank with and without the proposed promotional campaign?

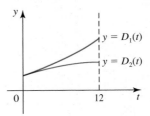

**56. Assembly Time of a Worker** In the following graph, $N(t)$ gives the number of satellite radios assembled by the average worker by the $t$th hour, where $t = 0$ corresponds to 8 A.M. and $0 \le t \le 4$. The point $P$ is an inflection point of $N$.
  **a.** What can you say about the rate of change of the rate of the number of satellite radios assembled by the average worker between 8 A.M. and 10 A.M.? Between 10 A.M. and 12 P.M.?
  **b.** At what time is the rate at which the satellite radios are being assembled by the average worker greatest?

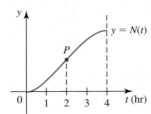

**57. Water Pollution** When organic waste is dumped into a pond, the oxidation process that takes place reduces the pond's oxygen content. However, given time, nature will restore the oxygen content to its natural level. In the following graph, $P(t)$ gives the oxygen content (as a percentage of its normal level) $t$ days after organic waste has been dumped into the pond. Explain the significance of the inflection point $Q$.

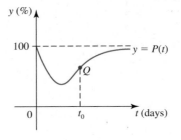

**58. Effect of Budget Cuts on Drug-Related Crimes** A police commissioner used the following graphs to illustrate what effect a budget cut would have on crime in the city. The number $N_1(t)$ gives the projected number of drug-related crimes in the next 12 months. The number $N_2(t)$ gives the projected number of drug-related crimes in the same time frame if next year's budget is cut.

   **a.** Explain why $N'_1(t)$ and $N'_2(t)$ are both positive on the interval $(0, 12)$.

   **b.** What are the signs of $N''_1(t)$ and $N''_2(t)$ on the interval $(0, 12)$?

   **c.** Interpret the results of part (b).

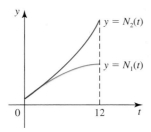

**59.** In the figure below, water is poured into the vase at a constant rate (in appropriate units), and the water level rises to a height of $f(t)$ units at time $t$ as measured from the base of the vase. Sketch the graph of $f$, and explain its shape, indicating where it is concave upward and concave downward. Indicate the inflection point on the graph, and explain its significance.

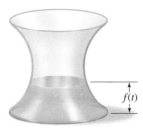

**60.** In the figure below, water is poured into an urn at a constant rate (in appropriate units), and the water level rises to a height of $f(t)$ units at time $t$ as measured from the base of the urn. Sketch the graph of $f$, and explain its shape, indicating where it is concave upward and concave downward. Indicate the inflection point on the graph, and explain its significance.

**61. Effect of Smoking Bans** The sales (in billions of dollars) in restaurants and bars in California from the beginning of 1993 ($t = 0$) to the beginning of 2000 ($t = 7$) are approximated by the function

$$S(t) = 0.195t^2 + 0.32t + 23.7 \qquad 0 \le t \le 7$$

   **a.** Show that the sales in restaurants and bars continued to rise after smoking bans were implemented in restaurants in 1995 and in bars in 1998.

   **Hint:** Show that $S$ is increasing on the interval $(2, 7)$.

   **b.** What can you say about the rate at which the sales were rising after smoking bans were implemented?

*Source:* California Board of Equalization.

**62. Global Warming** The increase in carbon dioxide in the atmosphere is a major cause of global warming. Using data obtained by Charles David Keeling, professor at Scripps Institution of Oceanography, the average amount of $CO_2$ in the atmosphere from 1958 through 2007 is approximated by

$$A(t) = 0.010716t^2 + 0.8212t + 313.4$$

where $t = 1$ corresponds to the beginning of 1958 and $1 \le t \le 50$.

   **a.** What can you say about the rate of change of the average amount of atmospheric $CO_2$ from 1958 through 2007?

   **b.** What can you say about the rate of the rate of change of the average amount of atmospheric $CO_2$ from 1958 through 2007?

*Source:* Scripps Institution of Oceanography.

**63. Population Growth in Clark County** Clark County in Nevada, which is dominated by greater Las Vegas, is one of the fastest-growing metropolitan areas in the United States. The population of the county from 1970 through 2000 is approximated by the function

$$P(t) = 44,560t^3 - 89,394t^2 + 234,633t + 273,288$$
$$0 \le t \le 3$$

where $t$ is measured in decades, with $t = 0$ corresponding to the beginning of 1970.

   **a.** Show that the population of Clark County was always increasing over the time period in question.

   **b.** Show that the population of Clark County was increasing at the slowest pace some time around the middle of August 1976.

*Source:* U.S. Census Bureau.

**64. Air Pollution** The level of ozone, an invisible gas that irritates and impairs breathing, that was present in the atmosphere on a certain day in May in the city of Riverside is approximated by

$$A(t) = 1.0974t^3 - 0.0915t^4 \qquad 0 \le t \le 11$$

where $A(t)$ is measured in pollutant standard index (PSI) and $t$ is measured in hours, with $t = 0$ corresponding to 7 A.M. Use the Second Derivative Test to show that the function $A$ has a relative maximum at approximately $t = 9$. Interpret your results.

*Source: The Los Angeles Times.*

**65. Women's Soccer** Starting with the youth movement that took hold in the 1970s and buoyed by the success of the U.S. national women's team in international competition in recent years, girls and women have taken to soccer in ever-growing numbers. The function

$$N(t) = -0.9307t^3 + 74.04t^2 + 46.8667t + 3967$$
$$0 \leq t \leq 16$$

gives the number of participants in women's soccer in year $t$ with $t = 0$ corresponding to the beginning of 1985.
**a.** Verify that the number of participants in women's soccer has been increasing from 1985 through 2000.
**b.** Show that the number of participants in women's soccer has been growing at an increasing rate from 1985 through 2000.
*Source:* NCCA News.

**66. Surveillance Cameras** Research reports indicate that surveillance cameras at major intersections dramatically reduce the number of drivers who barrel through red lights. The cameras automatically photograph vehicles that drive into intersections after the light turns red. Vehicle owners are then mailed citations instructing them to pay a fine or sign an affidavit that they were not driving at the time. The function

$$N(t) = 6.08t^3 - 26.79t^2 + 53.06t + 69.5$$
$$0 \leq t \leq 4$$

gives the number, $N(t)$, of U.S. communities using surveillance cameras at intersections in year $t$ with $t = 0$ corresponding to the beginning of 2003.
**a.** Show that $N$ is increasing on $(0, 4)$.
**b.** When was the number of communities using surveillance cameras at intersections increasing least rapidly? What was the rate of increase?
*Source:* Insurance Institute for Highway Safety.

**67. Measles Deaths** Measles is still a leading cause of vaccine-preventable death among children, but because of improvements in immunizations, measles deaths have dropped globally. The function

$$N(t) = -2.42t^3 + 24.5t^2 - 123.3t + 506$$
$$0 \leq t \leq 6$$

gives the number of measles deaths (in thousands) in sub-Saharan Africa in year $t$ with $t = 0$ corresponding to the beginning of 1999.
**a.** What was the number of measles deaths in 1999? In 2005?
**b.** Show that $N'(t) < 0$ on $(0, 6)$. What does this say about the number of measles deaths from 1999 through 2005?
**c.** When was the number of measles deaths decreasing most rapidly? What was the rate of measles death at that instant of time?
*Source:* Centers for Disease Control and World Health Organization.

**68. Epidemic Growth** During a flu epidemic the number of children in the Woodhaven Community School System who contracted influenza by the $t$th day is given by

$$N(t) = \frac{5000}{1 + 99e^{-0.8t}}$$

($t = 0$ corresponds to the date when data were first collected.)
**a.** How fast was the flu spreading on the third day ($t = 2$)?
**b.** When was the flu being spread at the fastest rate?

**69. Von Bertalanffy Functions** The mass $W(t)$ (in kilograms) of the average female African elephant at age $t$ (in years) can be approximated by a *von Bertalanffy function*

$$W(t) = 2600(1 - 0.51e^{-0.075t})^3$$

**a.** How fast does a newborn female elephant gain weight? A 1600 kg female elephant?
**b.** At what age does a female elephant gain weight at the fastest rate?

**70. Death Due to Strokes** Before 1950, little was known about strokes. By 1960, however, risk factors such as hypertension had been identified. In recent years, CAT scans used as a diagnostic tool have helped to prevent strokes. As a result, deaths due to strokes have fallen dramatically. The function

$$N(t) = 130.7e^{-0.1155t^2} + 50 \qquad 0 \leq t \leq 6$$

gives the number of deaths per 100,000 people from the beginning of 1950 to the beginning of 2010, where $t$ is measured in decades, with $t = 0$ corresponding to the beginning of 1950.
**a.** How fast was the number of deaths due to strokes per 100,000 people changing at the beginning of 1950? At the beginning of 1960? At the beginning of 1970? At the beginning of 1980?
**b.** When was the decline in the number of deaths due to strokes per 100,000 people greatest?
*Source:* American Heart Association, Centers for Disease Control and Prevention, and National Institutes of Health.

**71. Oxygen Content of a Pond** Refer to Exercise 57. When organic waste is dumped into a pond, the oxidation process that takes place reduces the pond's oxygen content. However, given time, nature will restore the oxygen content to its natural level. Suppose that the oxygen content $t$ days after the organic waste has been dumped into the pond is given by

$$f(t) = 100\left(\frac{t^2 + 10t + 100}{t^2 + 20t + 100}\right)$$

percent of its normal level. Show that an inflection point of $f$ occurs at $t = 20$.

**72.** Find the intervals where $f(x) = \log_3 |x|$ is concave upward or where it is concave downward.

**73. a.** Determine where the graph of $f(x) = 2 - |x^3 - 1|$ is concave upward and where it is concave downward.
  **b.** Does the graph of $f$ have an inflection point at $x = 1$? Explain.
  **c.** Sketch the graph of $f$.

**74.** Show that the graph of the function $f(x) = x|x|$ has an inflection point at $(0, 0)$ but $f''(0)$ does not exist.

**75.** Find the values of $c$ such that the graph of

$$f(x) = x^4 + 2x^3 + cx^2 + 2x + 2$$

is concave upward everywhere.

**76.** Find conditions on the coefficients $a$, $b$, and $c$ such that the graph of $f(x) = ax^4 + bx^3 + cx^2 + dx + e$ has inflection points.

 **77.** If the graph of a function $f$ is concave upward on an open interval $I$, must the graph of the function $f^2$ also be concave upward on $I$?
  **Hint:** Study the function $f(x) = x^2 - 1$ on $(-1, 1)$. Plot the graphs of $f$ and $f^2$ on the same set of axes.

**78.** Suppose $f$ is twice differentiable on an open interval $I$. If $f$ is positive and the graph of $f$ is concave upward on $I$, show that the graph of the function $f^2$ is also concave upward. (Compare with Exercise 77.)

**79.** Show that a polynomial function of odd degree greater than or equal to three has at least one inflection point.

**80.** Show that the graph of a polynomial function of the form

$$f(x) = a_{2n}x^{2n} + a_{2n-2}x^{2n-2} + \cdots + a_2x^2 + a_0$$

where $n$ is a positive integer and the coefficients $a_0, a_2, \ldots, a_{2n}$ are positive, is concave upward everywhere and that $f$ has an absolute minimum.

**81.** Suppose that the point $(a, f(a))$ is a point of inflection of the graph of $y = f(x)$. Prove that the number $a$ gives rise to a relative extremum of the function $f'$.

**82. a.** Suppose that $f''$ is continuous and $f'(a) = f''(a) = 0$, but $f'''(a) \neq 0$. Show that the graph of $f$ has an inflection point at $a$.
  **b.** Find the relative maximum and minimum values of

$$f(x) = \cos x - 1 + \frac{x^2}{2} - \frac{x^3}{6}$$

 **c.** Verify the result of part (b) by plotting the graph of $f$ using the viewing window $[-2, 2] \times [-1.5, 1.5]$.

*In Exercises 83–86, determine whether the statement is true or false. If it is true, explain why it is true. If it is false, explain why or give an example to show why it is false.*

**83.** If $f$ has an inflection point at $a$, then $f'(a) = 0$.

**84.** If $f''(x)$ exists everywhere except at $x = a$ and $f''(x)$ changes sign as we move across $a$, then $f$ has an inflection point at $a$.

**85.** A polynomial function of degree 3 has exactly one inflection point.

**86.** If the graph of a function $f$ that has a second derivative is concave upward on an open interval $I$, then the graph of the function $-f$ is concave downward.

## 3.5 Limits Involving Infinity; Asymptotes

### Infinite Limits

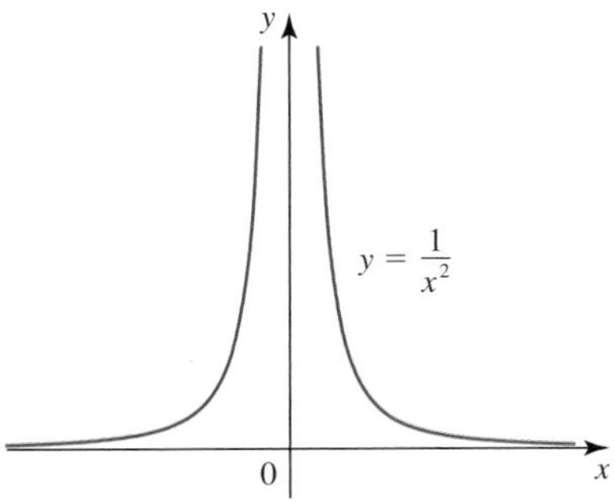

**FIGURE 1**
$f(x)$ gets larger and larger without bound as $x$ gets closer and closer to 0.

In Section 1.1 we were concerned primarily with whether or not the functional values of $f$ approach a number $L$ as $x$ approaches a number $a$. Even if $f(x)$ does not approach a (finite) limit, there are situations in which it is useful to describe the behavior of $f(x)$ as $x$ approaches $a$. Recall that the function $f(x) = 1/x^2$ does not have a limit as $x$ approaches 0 because $f(x)$ becomes arbitrarily large as $x$ gets arbitrarily close to 0. (See Example 7 in Section 1.1.) The graph of $f$ is reproduced in Figure 1. We described this behavior by writing

$$\lim_{x \to 0} \frac{1}{x^2} = \infty$$

with the understanding that this is not a limit in the usual sense.

More generally, we have the following definitions concerning the behavior of functions whose values become unbounded as $x$ approaches $a$.

> ### DEFINITIONS Infinite Limits
>
> Let $f$ be a function defined on an open interval containing $a$ with the possible exception of $a$ itself. Then
>
> $$\lim_{x \to a} f(x) = \infty$$
>
> if all the values of $f$ can be made arbitrarily large (as large as we please) by taking $x$ sufficiently close to but not equal to $a$. Similarly,
>
> $$\lim_{x \to a} f(x) = -\infty$$
>
> if all the values of $f$ can be made as large in absolute value and negative as we please by taking $x$ sufficiently close to but not equal to $a$.

These definitions are illustrated graphically in Figure 2.

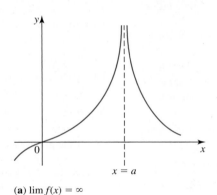

**FIGURE 2**

$f$ has an infinite limit as $x$ approaches $a$.

**(a)** $\lim_{x \to a} f(x) = \infty$  |  **(b)** $\lim_{x \to a} f(x) = -\infty$

Similar definitions can be given for the one-sided limits

$$\lim_{x \to a^-} f(x) = \infty \qquad \lim_{x \to a^+} f(x) = \infty$$

$$\lim_{x \to a^-} f(x) = -\infty \qquad \lim_{x \to a^+} f(x) = -\infty \tag{1}$$

(see Figure 3). The expression $\lim_{x \to a} f(x) = \infty$ is read "the limit of $f(x)$ as $x$ approaches $a$ is infinity." The expression $\lim_{x \to a} f(x) = -\infty$ is read "the limit of $f(x)$ as $x$ approaches $a$ is negative infinity."

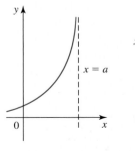
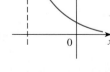
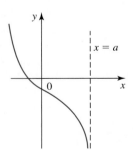
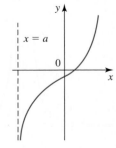

**(a)** $\lim_{x \to a^-} f(x) = \infty$  |  **(b)** $\lim_{x \to a^+} f(x) = \infty$  |  **(c)** $\lim_{x \to a^-} f(x) = -\infty$  |  **(d)** $\lim_{x \to a^+} f(x) = -\infty$

**FIGURE 3**

$f$ has one-sided infinite limits as $x$ approaches $a$.

⚠ The "infinite limits" that are defined here are not limits in the sense defined in Section 1.1. They are simply expressions used to indicate the direction (positive or negative) taken by the unbounded values of $f(x)$ as $x$ approaches $a$.

## ▮ Vertical Asymptotes

Each vertical line $x = a$ shown in Figures 2a–b and 3a–d is called a *vertical asymptote* of the graph of $f$. Note that an asymptote does not constitute part of the graph of $f$, but it is a useful aid for sketching the graph of $f$.

> **DEFINITION    Vertical Asymptote**
>
> The line $x = a$ is a **vertical asymptote** of the graph of a function $f$ if at least one of the following statements is true:
>
> $$\lim_{x \to a^-} f(x) = \infty \quad (\text{or} -\infty); \quad \lim_{x \to a^+} f(x) = \infty \quad (\text{or} -\infty); \quad \lim_{x \to a} f(x) = \infty \quad (\text{or} -\infty)$$

It follows from the above definition that $x = 0$ (the $y$-axis) is a vertical asymptote of the graph of $f(x) = 1/x^2$. (See Figure 1.) Another example of a function whose graph has a vertical asymptote is the natural logarithmic function $y = \ln x$. From the graph of $y = \ln x$ shown in Figure 4, we see that

$$\lim_{x \to 0^+} \ln x = -\infty$$

So $x = 0$ is a vertical asymptote of $f(x) = \ln x$. In fact, it is true that $x = 0$ is a vertical asymptote of $f(x) = \log_a x$ if $a > 1$. (See Figures 6 and 7 in Section 0.8.)

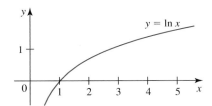

**FIGURE 4**
The graph of the natural logarithmic function $y = \ln x$

**EXAMPLE 1**  Find $\displaystyle\lim_{x \to 1^-} \frac{1}{x-1}$ and $\displaystyle\lim_{x \to 1^+} \frac{1}{x-1}$, and the vertical asymptote of the graph of $f(x) = \dfrac{1}{x-1}$.

**Solution**  From the graph of $f(x) = 1/(x - 1)$ shown in Figure 5, we see that

$$\lim_{x \to 1^-} \frac{1}{x-1} = -\infty \quad \text{and} \quad \lim_{x \to 1^+} \frac{1}{x-1} = \infty$$

The line $x = 1$ is a vertical asymptote of the graph of $f$.

| $x$ | $f(x)$ |
|-----|--------|
| 0.9 | $-10$ |
| 0.99 | $-100$ |
| 0.999 | $-1000$ |

| $x$ | $f(x)$ |
|-----|--------|
| 1.1 | 10 |
| 1.01 | 100 |
| 1.001 | 1000 |

$$\lim_{x \to 1^-} \frac{1}{x-1} = -\infty \quad \text{and} \quad \lim_{x \to 1^+} \frac{1}{x-1} = \infty$$

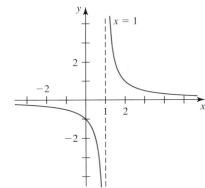

**FIGURE 5**

**Alternative Solution**    Observe that if $x$ is close to but less than 1, then $(x - 1)$ is a small negative number. The numerator, however, remains constant with value 1. Therefore, $1/(x - 1)$ is a number that is large in absolute value and negative. Consequently, as $x$ approaches 1 from the left, $1/(x - 1)$ becomes larger and larger in absolute value and negative; that is,

$$\lim_{x \to 1^-} \frac{1}{x - 1} = -\infty$$

Similarly, if $x$ is close to but greater than 1, then $(x - 1)$ is a small positive number, and we see that $1/(x - 1)$ is a large positive number. Thus,

$$\lim_{x \to 1^+} \frac{1}{x - 1} = \infty$$    ∎

**EXAMPLE 2**    **Special Theory of Relativity**    According to Einstein's special theory of relativity, the mass $m$ of a particle moving with speed $v$ is

$$m = f(v) = \frac{m_0}{\sqrt{1 - \dfrac{v^2}{c^2}}} \tag{2}$$

where $c$ is the speed of light (approximately $3 \times 10^8$ m/sec) and $m_0$ is the rest mass.

**a.** Evaluate $\lim_{v \to c^-} f(v)$.

**b.** Sketch the graph of $f$, and interpret your result.

**Solution**

**a.** Observe that as $v$ approaches $c$ from the left, $v^2/c^2$ approaches 1 through values less than 1 and $1 - (v^2/c^2)$ approaches zero. Thus, the denominator of Equation (2) approaches zero through positive values, and the numerator remains constant, so $f(v)$ increases without bound. Thus, we have

$$\lim_{v \to c^-} f(v) = \lim_{v \to c^-} \frac{m_0}{\sqrt{1 - \dfrac{v^2}{c^2}}} = \infty$$

**b.** From the result of part (a) we see that $v = c$ is a vertical asymptote of the graph of $f$. The graph of $f$ is shown in Figure 6. This mathematical model tells us that the mass of a particle grows without bound as its speed approaches the speed of light. This is why the speed of light is called the "ultimate speed."    ∎

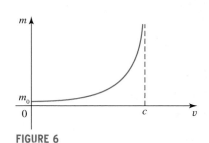

**FIGURE 6**

If a function $f$ is the quotient of two functions, $g$ and $h$, that is,

$$f(x) = \frac{g(x)}{h(x)}$$

then the zeros of the denominator $h(x)$ provide us with candidates for the vertical asymptotes of the graph of $f$, as the following example shows.

**EXAMPLE 3**    Find the vertical asymptotes of the graph of

$$f(x) = \frac{x}{x^2 - x - 2}$$

**Solution**   By factoring the denominator, we can rewrite $f(x)$ in the form

$$f(x) = \frac{x}{(x+1)(x-2)}$$

Notice that the denominator of $f(x)$ is equal to zero when $x = -1$ or $x = 2$. The lines $x = -1$ and $x = 2$ are candidates for vertical asymptotes of the graph of $f$. To see whether $x = -1$ is, in fact, a vertical asymptote of the graph of $f$, let's evaluate

$$\lim_{x \to -1^-} f(x)$$

If $x$ is close to but less than $-1$, then $(x + 1)$ is a small negative number. Furthermore, $(x - 2)$ is close to $-3$, so $[(x + 1)(x - 2)]$ is a small positive number. Also, the numerator of $f(x)$ is close to $-1$ when $x$ is close to $-1$. Therefore, $x/[(x + 1)(x - 2)]$ is a number that is large in absolute value and negative. Thus,

$$\lim_{x \to -1^-} \frac{x}{(x+1)(x-2)} = -\infty$$

We conclude that $x = -1$ is a vertical asymptote of the graph of $f$. We leave it to you to show that

$$\lim_{x \to -1^+} \frac{x}{(x+1)(x-2)} = \infty$$

which also confirms that $x = -1$ is a vertical asymptote of the graph of $f$.

Next, notice that if $x$ is close to but less than 2, then $(x - 2)$ is a small negative number. Furthermore, $(x + 1)$ is close to 3, so $[(x + 1)(x - 2)]$ is a small negative number. Also, the numerator of $f(x)$ is close to 2 when $x$ is close to 2. Therefore,

$$\lim_{x \to 2^-} \frac{x}{(x+1)(x-2)} = -\infty$$

We conclude that $x = 2$ is also a vertical asymptote of the graph of $f$. We leave it to you to show that

$$\lim_{x \to 2^+} \frac{x}{(x+1)(x-2)} = \infty$$

The graph of $f$ is shown in Figure 7. Don't worry about sketching it at this time. We will study curve sketching in Section 3.6.

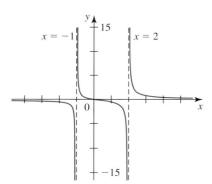

**FIGURE 7**
The graph of

$$y = \frac{x}{x^2 - x - 2}$$

has a vertical asymptote at $x = -1$ and another at $x = 2$.

**EXAMPLE 4**   Find the vertical asymptotes of the graph of $f(x) = \tan x$.

**Solution**   We write

$$f(x) = \tan x = \frac{\sin x}{\cos x}$$

Since $\cos x = 0$ if $x = (2n + 1)\pi/2$, where $n$ is an integer, we see that the vertical lines $x = (2n + 1)\pi/2$ are candidates for vertical asymptotes of the graph of $f$. Consider the line $x = \pi/2$, where $n = 0$. If $x$ is close to but less than $\pi/2$, then $\sin x$ is close to 1, but $\cos x$ is positive and close to 0. Therefore, $(\sin x)/(\cos x)$ is positive and large. Thus,

$$\lim_{x \to (\pi/2)^-} \tan x = \infty$$

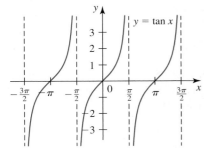

**FIGURE 8**
The lines $x = (2n + 1)\pi/2$ ($n$, an integer) are vertical asymptotes of the graph of $f$.

Next, if $x$ is close to but greater than $\pi/2$, then $\sin x$ is close to 1, and $\cos x$ is negative and close to 0. Therefore, $(\sin x)/(\cos x)$ is negative and large in absolute value. Thus,

$$\lim_{x \to (\pi/2)^+} \tan x = -\infty$$

This shows that the line $x = \pi/2$ is a vertical asymptote of the graph of $f$. Similarly, you can show that the lines $x = (2n + 1)\pi/2$, where $n$ is an integer, are vertical asymptotes of the graph of $f$ (see Figure 8).    ∎

## Limits at Infinity

Up to now we have studied the limit of a function as $x$ approaches a finite number $a$. Sometimes we wish to know whether $f(x)$ approaches a unique number as $x$ increases without bound. Consider, for example, the function $P$ giving the number of fruit flies (*Drosophila melanogaster*) in a container under controlled laboratory conditions as a function of time $t$. The graph of $P$ is shown in Figure 9. You can see from the graph of $P$ that as $t$ increases without bound (tends to infinity), $P(t)$ approaches the number 400. This number, called the *carrying capacity* of the environment, is determined by the amount of living space and food available, as well as other environmental factors.

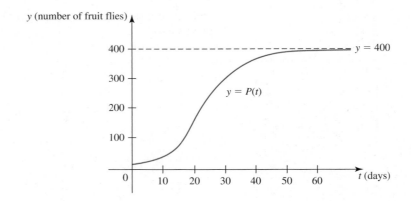

**FIGURE 9**
The graph of $P(t)$ gives the population of fruit flies in a laboratory experiment.

More generally, we have the following intuitive definition of the limit of a function at infinity.

---

**DEFINITION    Limit of a Function at Infinity**

Let $f$ be a function that is defined on an interval $(a, \infty)$. Then the limit of $f(x)$ as $x$ approaches infinity (increases without bound) is the number $L$, written

$$\lim_{x \to \infty} f(x) = L$$

if all the values of $f$ can be made arbitrarily close to $L$ by taking $x$ to be sufficiently large.

---

This definition is illustrated graphically in Figure 10.

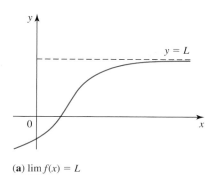 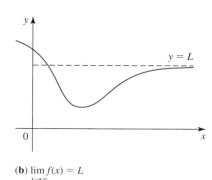

**FIGURE 10**  (a) $\lim\limits_{x \to \infty} f(x) = L$    (b) $\lim\limits_{x \to \infty} f(x) = L$

We define the limit at negative infinity in a similar manner.

---

**DEFINITION**  **Limit of a Function at Negative Infinity**

Let $f$ be a function that is defined on an interval $(-\infty, a)$. Then the limit of $f(x)$ as $x$ approaches negative infinity (decreases without bound) is the number $L$, written

$$\lim_{x \to -\infty} f(x) = L$$

if all the values of $f$ can be made arbitrarily close to $L$ by taking $x$ to be sufficiently large in absolute value and negative. (See Figure 11.)

---

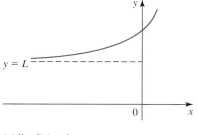 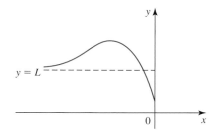

**FIGURE 11**  (a) $\lim\limits_{x \to -\infty} f(x) = L$    (b) $\lim\limits_{x \to -\infty} f(x) = L$

## ■ Horizontal Asymptotes

Each horizontal line $y = L$ shown in Figures 10a–b and 11a–b is called a *horizontal asymptote* of the graph of $f$.

---

**DEFINITION**  **Horizontal Asymptote**

The line $y = L$ is a **horizontal asymptote** of the graph of a function $f$ if

$$\lim_{x \to \infty} f(x) = L \quad \text{or} \quad \lim_{x \to -\infty} f(x) = L$$

(or both).

---

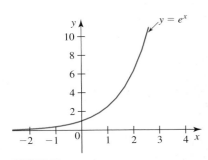

**FIGURE 12**
The graph of the natural exponential function $y = e^x$

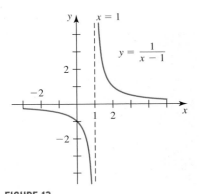

**FIGURE 13**
$\lim\limits_{x \to \infty} \dfrac{1}{x-1} = 0$, $\lim\limits_{x \to -\infty} \dfrac{1}{x-1} = 0$,
and, therefore, $y = 0$ is a horizontal asymptote of the graph of $f$.

An example of a function whose graph has a horizontal asymptote is the natural exponential function $y = e^x$. From the graph of $y = e^x$ shown in Figure 12, we see that

$$\lim_{x \to -\infty} e^x = 0$$

and so $y = 0$ (the $x$-axis) is a horizontal asymptote of $f(x) = e^x$. (Also see Figure 5, Section 0.8.) In fact, it is true that $y = 0$ is a horizontal asymptote of $f(x) = a^x$ provided $a > 1$.

**EXAMPLE 5**   Find $\lim\limits_{x \to \infty} \dfrac{1}{x-1}$, $\lim\limits_{x \to -\infty} \dfrac{1}{x-1}$, and the horizontal asymptote of the graph

of $f(x) = \dfrac{1}{x-1}$.

**Solution**   We have

$$\lim_{x \to \infty} \frac{1}{x-1} = 0 \quad \text{and} \quad \lim_{x \to -\infty} \frac{1}{x-1} = 0$$

We conclude that $y = 0$ is a horizontal asymptote of $f$ (Figure 13).   ■

The following theorem is useful for evaluating limits at infinity. We also point out that the laws of limits in Section 1.2 are valid if we replace $x \to a$ by $x \to -\infty$ or $x \to \infty$.

**THEOREM 1**

Let $r > 0$ be a rational number. Then

$$\lim_{x \to \infty} \frac{1}{x^r} = 0$$

Also, if $x^r$ is defined for all $x$, then

$$\lim_{x \to -\infty} \frac{1}{x^r} = 0$$

**EXAMPLE 6**   Let $f(x) = \dfrac{2x^2 - x + 1}{3x^2 + 2x - 1}$. Find $\lim_{x \to \infty} f(x)$ and $\lim_{x \to -\infty} f(x)$, and find all horizontal asymptotes of the graph of $f$.

**Solution**   If we divide both the numerator and denominator by $x^2$, the highest power of $x$ in the denominator, we obtain

$$\lim_{x \to \infty} \frac{2x^2 - x + 1}{3x^2 + 2x - 1} = \lim_{x \to \infty} \frac{2 - \dfrac{1}{x} + \dfrac{1}{x^2}}{3 + \dfrac{2}{x} - \dfrac{1}{x^2}} = \frac{\lim\limits_{x \to \infty}\left(2 - \dfrac{1}{x} + \dfrac{1}{x^2}\right)}{\lim\limits_{x \to \infty}\left(3 + \dfrac{2}{x} - \dfrac{1}{x^2}\right)}$$

$$= \frac{\lim\limits_{x \to \infty} 2 - \lim\limits_{x \to \infty} \dfrac{1}{x} + \lim\limits_{x \to \infty} \dfrac{1}{x^2}}{\lim\limits_{x \to \infty} 3 + \lim\limits_{x \to \infty} \dfrac{2}{x} - \lim\limits_{x \to \infty} \dfrac{1}{x^2}} = \frac{2 - 0 + 0}{3 + 0 - 0} = \frac{2}{3}$$

In a similar manner, we can show that

$$\lim_{x \to -\infty} \frac{2x^2 - x + 1}{3x^2 + 2x - 1} = \frac{2}{3}$$

We conclude that $y = \frac{2}{3}$ is a horizontal asymptote of the graph of $f$.  ∎

**EXAMPLE 7**  Find the horizontal asymptotes of the graph of the function

$$f(x) = \frac{3x}{\sqrt{x^2 + 1}}$$

**Solution**  First, let's investigate $\lim_{x \to \infty} f(x)$. We may assume that $x > 0$. In this case $\sqrt{x^2} = x$. Dividing the numerator and the denominator by $x$, the highest power of $x$ in the denominator, we find

$$f(x) = \frac{\frac{1}{x}(3x)}{\frac{1}{x}\sqrt{x^2 + 1}} = \frac{3}{\frac{1}{\sqrt{x^2}}\sqrt{x^2 + 1}}$$

$$= \frac{3}{\sqrt{\frac{1}{x^2}(x^2 + 1)}} = \frac{3}{\sqrt{1 + \frac{1}{x^2}}}$$

Therefore,

$$\lim_{x \to \infty} f(x) = \lim_{x \to \infty} \frac{3}{\sqrt{1 + \frac{1}{x^2}}} = \frac{\lim_{x \to \infty} 3}{\lim_{x \to \infty} \sqrt{1 + \frac{1}{x^2}}}$$

$$= \frac{3}{\sqrt{\lim_{x \to \infty} 1 + \lim_{x \to \infty} \frac{1}{x^2}}} = \frac{3}{\sqrt{1 + 0}} = 3$$

We conclude that $y = 3$ is a horizontal asymptote of the graph of $f$. Next, we investigate $\lim_{x \to -\infty} f(x)$. In this case we may assume that $x < 0$. Then $\sqrt{x^2} = |x| = -x$. Dividing both the numerator and the denominator of $f(x)$ by $-x$, we obtain

$$f(x) = \frac{-\frac{1}{x}(3x)}{-\frac{1}{x}\sqrt{x^2 + 1}} = \frac{-3}{\frac{1}{\sqrt{x^2}}\sqrt{x^2 + 1}}$$

$$= \frac{-3}{\sqrt{\frac{1}{x^2}(x^2 + 1)}} = \frac{-3}{\sqrt{1 + \frac{1}{x^2}}}$$

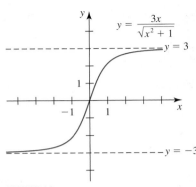

**FIGURE 14**
$y = 3$ and $y = -3$ are horizontal asymptotes of the graph of $f$.

Therefore,

$$\lim_{x \to -\infty} f(x) = \lim_{x \to -\infty} \frac{-3}{\sqrt{1 + \dfrac{1}{x^2}}} = -3$$

and we see that $y = -3$ is also a horizontal asymptote of the graph of $f$. The graph of $f$ is sketched in Figure 14. ■

### The Number $e$ as a Limit at $\infty$

In Section 2.8 we saw that

$$e = \lim_{h \to 0} (1 + h)^{1/h} \tag{3}$$

We can obtain an alternative expression for $e$ by putting $n = 1/h$ so that $h = 1/n$. Observe that $n \to \infty$ as $h \to 0$. Making this substitution in Equation (3), we have

$$e = \lim_{n \to \infty} \left(1 + \frac{1}{n}\right)^n \tag{4}$$

Equation (4) gives an expression for $e$ as a limit at infinity. (See Exercise 69.)

### Interest Compounded Continuously

In Section 0.8 we saw that when a principal of $P$ dollars earns compound interest at the rate of $r$ per year converted $m$ times a year, then the accumulated amount at the end of $t$ years is given by

$$A = P\left(1 + \frac{r}{m}\right)^{mt} \tag{5}$$

Also, the results of Example 5 in Section 0.8 suggest that as interest is converted more and more frequently, the accumulated amount over a fixed term seems to increase—but ever so slowly. This raises the question: Does the accumulated amount grow without bound, or does it approach a limit as interest is computed more and more frequently?

To answer this question, we let $m$ approach infinity in Equation (5), obtaining

$$A = \lim_{m \to \infty} P\left(1 + \frac{r}{m}\right)^{mt}$$

$$= \lim_{m \to \infty} P\left[\left(1 + \frac{r}{m}\right)^m\right]^t$$

If we make the substitution $u = m/r$ and observe that $u \to \infty$ as $m \to \infty$, then

$$A = \lim_{u \to \infty} P\left[\left(1 + \frac{1}{u}\right)^{ur}\right]^t = \lim_{u \to \infty} P\left[\left(1 + \frac{1}{u}\right)^u\right]^{rt}$$

$$= P\left[\lim_{u \to \infty} \left(1 + \frac{1}{u}\right)^u\right]^{rt}$$

But the limit in this expression is equal to the number $e$ (see Equation (4)). Therefore,

$$A = Pe^{rt} \tag{6}$$

Equation (6) gives the accumulated amount of $P$ dollars over a term of $t$ years and earning interest at the rate of $r$ per year **compounded continuously.**

**EXAMPLE 8** Find the accumulated amount after 3 years if $1000 is invested at 8% per year compounded continuously.

**Solution** We use Equation (6) with $P = 1000$, $r = 0.08$, and $t = 3$, obtaining

$$A = 1000e^{(0.08)(3)} \approx 1271.25$$

or $1271.25. ∎

## ◼ Infinite Limits at Infinity

The notation

$$\lim_{x \to \infty} f(x) = \infty$$

is used to indicate that $f(x)$ becomes arbitrarily large as $x$ increases without bound (approaches infinity). For example,

$$\lim_{x \to \infty} x^3 = \infty$$

(See Figure 15.) Similarly, we can define

$$\lim_{x \to \infty} f(x) = -\infty, \qquad \lim_{x \to -\infty} f(x) = \infty, \qquad \lim_{x \to -\infty} f(x) = -\infty$$

For example, an examination of Figure 15 once again will confirm that

$$\lim_{x \to -\infty} x^3 = -\infty$$

| $x$ | $f(x) = x^3$ |
|------|------|
| $-1$ | $-1$ |
| $-10$ | $-1000$ |
| $-100$ | $-1000000$ |
| $-1000$ | $-1000000000$ |

| $x$ | $f(x) = x^3$ |
|------|------|
| $1$ | $1$ |
| $10$ | $1000$ |
| $100$ | $1000000$ |
| $1000$ | $1000000000$ |

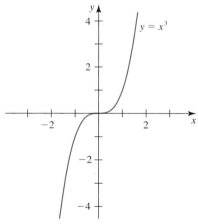

**FIGURE 15**
$\lim\limits_{x \to \infty} x^3 = \infty$ and $\lim\limits_{x \to -\infty} x^3 = -\infty$

**EXAMPLE 9** Find $\lim_{x \to \infty}(2x^3 - x^2 + 1)$ and $\lim_{x \to -\infty}(2x^3 - x^2 + 1)$.

**Solution** We rewrite

$$2x^3 - x^2 + 1 = x^3\left(2 - \frac{1}{x} + \frac{1}{x^3}\right)$$

and note that if $x$ is very large, then $\left(2 - \dfrac{1}{x} + \dfrac{1}{x^3}\right)$ is close to 2 and $x^3$ is very large. This shows that

$$\lim_{x \to \infty} (2x^3 - x^2 + 1) = \infty$$

Next, note that if $x$ is large in absolute value and negative, so is $x^3$. Furthermore, $\left(2 - \dfrac{1}{x} + \dfrac{1}{x^3}\right)$ is close to 2. Therefore, $x^3\left(2 - \dfrac{1}{x} + \dfrac{1}{x^3}\right)$ is numerically very large and negative. So

$$\lim_{x \to -\infty} (2x^3 - x^2 + 1) = -\infty \qquad \blacksquare$$

**EXAMPLE 10** Find $\displaystyle\lim_{x \to -\infty} \dfrac{x^2 + 1}{x - 2}$.

**Solution**　Dividing both the numerator and the denominator by $x$ (the largest power of $x$ in the denominator), we obtain

$$\lim_{x \to -\infty} \frac{x^2 + 1}{x - 2} = \lim_{x \to -\infty} \frac{x + \dfrac{1}{x}}{1 - \dfrac{2}{x}}$$

If $x$ is very large in absolute value and negative, then the denominator of this last expression is close to 1, whereas the numerator is large in absolute value and negative. Thus, the quotient is large in absolute value and negative. We conclude that

$$\lim_{x \to -\infty} \frac{x^2 + 1}{x - 2} = -\infty \qquad \blacksquare$$

## ■ Precise Definitions

We begin by giving a precise definition of an infinite limit as $x$ approaches a number $a$.

---

**DEFINITION**　**Infinite Limit**

Let $f$ be a function defined on an open interval containing $a$, with the possible exception of $a$ itself. We write

$$\lim_{x \to a} f(x) = \infty$$

if for every number $M > 0$ we can find a number $\delta > 0$ such that for all $x$ satisfying

$$0 < |x - a| < \delta$$

then $f(x) > M$.

---

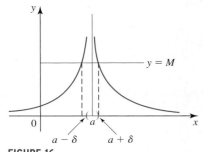

**FIGURE 16**
If $x \in (a - \delta, a) \cup (a, a + \delta)$, then $f(x) > M$.

For a geometric interpretation, let $M > 0$ be given. Draw the line $y = M$ shown in Figure 16. You can see that there exists a $\delta > 0$ such that whenever $x$ lies in the interval $(a - \delta, a + \delta)$, the graph of $y = f(x)$ lies above the line $y = M$. You can also see from the figure that once you have found a number $\delta > 0$ called for in the definition, then any positive number smaller than $\delta$ will also satisfy the requirement in the definition.

**EXAMPLE 11** Prove that $\lim\limits_{x \to 0} \dfrac{1}{x^2} = \infty$.

**Solution** Let $M > 0$ be given. We want to show that there exists a $\delta > 0$ such that

$$\frac{1}{x^2} > M$$

whenever $0 < |x - 0| < \delta$. To find $\delta$, consider

$$\frac{1}{x^2} > M$$

$$x^2 < \frac{1}{M}$$

or

$$|x| < \frac{1}{\sqrt{M}}$$

This suggests that we may take $\delta$ to be $1/\sqrt{M}$ or any positive number less than or equal to $1/\sqrt{M}$. Reversing the steps, we see that if $0 < |x| < \delta$, then

$$x^2 < \delta^2$$

so

$$\frac{1}{x^2} > \frac{1}{\delta^2} \geq M$$

Therefore,

$$\lim\limits_{x \to 0} \frac{1}{x^2} = \infty$$　　　　■

The precise definition of $\lim_{x \to a} f(x) = -\infty$ is similar to that of $\lim_{x \to a} f(x) = \infty$.

---

**DEFINITION** **Infinite Limit**

Let $f$ be a function defined on an open interval containing $a$, with the possible exception of $a$ itself. We write

$$\lim\limits_{x \to a} f(x) = -\infty$$

if for every number $N < 0$, we can find a number $\delta > 0$ such that for all $x$ satisfying

$$0 < |x - a| < \delta$$

then $f(x) < N$.

---

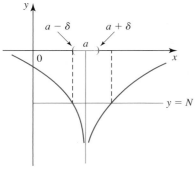

**FIGURE 17**
If $x \in (a - \delta, a) \cup (a, a + \delta)$, then $f(x) < N$.

(See Figure 17 for a geometric interpretation.)

The precise definitions for one-sided infinite limits are similar to the previous definitions. For example, in defining

$$\lim_{x \to a^-} f(x) = \infty$$

we must restrict $x$ so that $x < a$. Otherwise, the definition is similar to that for

$$\lim_{x \to a} f(x) = \infty$$

We now turn our attention to the precise definition of the limit of a function at infinity.

> **DEFINITION    Limit at Infinity**
>
> Let $f$ be a function defined on an interval $(a, \infty)$. We write
>
> $$\lim_{x \to \infty} f(x) = L$$
>
> if for every number $\varepsilon > 0$ there exists a number $N$ such that for all $x$ satisfying $x > N$ then $|f(x) - L| < \varepsilon$.

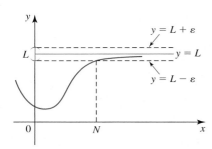

**FIGURE 18**
If $x > N$, then $f(x)$ lies in the band defined by $y = L - \varepsilon$ and $y = L + \varepsilon$.

As Figure 18 illustrates, the definition states that given any number $\varepsilon > 0$, we can find a number $N$ such that $x > N$ implies that all the values of $f$ lie inside the band of width $2\varepsilon$ determined by the lines $y = L - \varepsilon$ and $y = L + \varepsilon$.

Finally, infinite limits at infinity can also be defined precisely. For example, the precise definition of $\lim_{x \to \infty} f(x) = \infty$ follows.

> **DEFINITION    Infinite Limit at Infinity**
>
> Let $f$ be a function defined on an interval $(a, \infty)$. We write
>
> $$\lim_{x \to \infty} f(x) = \infty$$
>
> if for every number $M > 0$ there exists a number $N$ such that for all $x$ satisfying $x > N$, then $f(x) > M$.

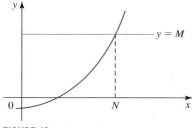

**FIGURE 19**
If $x > N$, then $f(x) > M$.

Figure 19 gives a geometric illustration of this definition. The precise definitions for $\lim_{x \to \infty} f(x) = -\infty$, $\lim_{x \to -\infty} f(x) = \infty$, and $\lim_{x \to -\infty} f(x) = -\infty$ are similar.

## 3.5    CONCEPT QUESTIONS

1. Explain what is meant by the statements
   (a) $\lim_{x \to 3} f(x) = \infty$ and (b) $\lim_{x \to 2^-} f(x) = -\infty$.
2. Explain what is meant by the statements
   (a) $\lim_{x \to -\infty} f(x) = 2$ and (b) $\lim_{x \to \infty} f(x) = -5$.
3. Explain the following terms in your own words:
   **a.** Vertical asymptote
   **b.** Horizontal asymptote

4. **a.** How many vertical asymptotes can the graph of a function $f$ have? Explain using graphs.
   **b.** How many horizontal asymptotes can the graph of a function $f$ have? Explain, using graphs.
5. State the precise definition of
   (a) $\lim_{x \to 2} \dfrac{3}{(x - 2)^2} = \infty$ and (b) $\lim_{x \to \infty} \dfrac{2x^2 + x + 1}{3x^2 + 4} = \dfrac{2}{3}$.

# 3.5   EXERCISES

*In Exercises 1–6, use the graph of the function f to find the given limits.*

**1. a.** $\displaystyle\lim_{x\to 0^-} f(x)$   **b.** $\displaystyle\lim_{x\to 0^+} f(x)$

**c.** $\displaystyle\lim_{x\to\infty} f(x)$   **d.** $\displaystyle\lim_{x\to -\infty} f(x)$

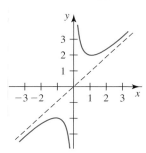

**2. a.** $\displaystyle\lim_{x\to 0^-} f(x)$   **b.** $\displaystyle\lim_{x\to 0^+} f(x)$

**c.** $\displaystyle\lim_{x\to\infty} f(x)$   **d.** $\displaystyle\lim_{x\to -\infty} f(x)$

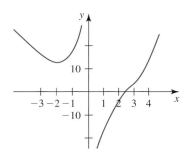

**3. a.** $\displaystyle\lim_{x\to 0} f(x)$   **b.** $\displaystyle\lim_{x\to -\infty} f(x)$   **c.** $\displaystyle\lim_{x\to\infty} f(x)$

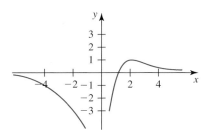

**4. a.** $\displaystyle\lim_{x\to -\infty} f(x)$   **b.** $\displaystyle\lim_{x\to\infty} f(x)$

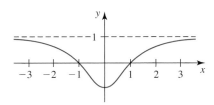

**5.** $\displaystyle\lim_{x\to 2n\pi} f(x)$ for $n = 0, 1, 2, \dots$

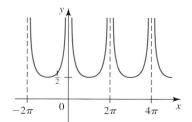

**6. a.** $\displaystyle\lim_{x\to -\infty} f(x)$   **b.** $\displaystyle\lim_{x\to\infty} f(x)$

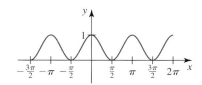

*In Exercises 7–36, find the limit.*

**7.** $\displaystyle\lim_{x\to -1^-} \frac{1}{x+1}$

**8.** $\displaystyle\lim_{t\to -3^+} \frac{t}{t+3}$

**9.** $\displaystyle\lim_{x\to 1^-} \frac{1+x}{1-x}$

**10.** $\displaystyle\lim_{x\to 1^+} \frac{x+1}{1-x}$

**11.** $\displaystyle\lim_{u\to 4^+} \frac{u^2+1}{u-4}$

**12.** $\displaystyle\lim_{t\to 1} \frac{t^3}{(t^2-1)^2}$

**13.** $\displaystyle\lim_{x\to 0^+} \frac{x+1}{\sqrt{x}(x-1)^2}$

**14.** $\displaystyle\lim_{x\to -1^+}\left(\frac{1}{x}-\frac{1}{x+1}\right)$

**15.** $\displaystyle\lim_{x\to 0^+} \frac{1}{\sin x}$

**16.** $\displaystyle\lim_{x\to 0^+} \cot 2x$

**17.** $\displaystyle\lim_{x\to 0^+} \frac{1}{1-e^{(1/\ln x)}}$

**18.** $\displaystyle\lim_{x\to(\pi/2)^-} \frac{2e^{\tan x}}{2x-\pi}$

**19.** $\displaystyle\lim_{t\to -(3/2)^-} \sec \pi t$

**20.** $\displaystyle\lim_{x\to\infty} \frac{x+1}{x-5}$

**21.** $\displaystyle\lim_{x\to -\infty} \frac{3x+4}{2x-3}$

**22.** $\displaystyle\lim_{x\to\infty} \frac{2x^2-1}{4x^2+1}$

**23.** $\displaystyle\lim_{x\to -\infty} \frac{1-2x^2}{x^3+1}$

**24.** $\displaystyle\lim_{x\to -\infty} \frac{2x^3+x^2+3}{x+1}$

**25.** $\displaystyle\lim_{x\to\infty}\left(\frac{x^3}{3x^2-2}-\frac{x^2}{3x+1}\right)$

**26.** $\displaystyle\lim_{x\to -\infty} \frac{x^4+1}{x^3+1}$

**27.** $\displaystyle\lim_{x\to\infty} \frac{-2x^4}{3x^4-3x^2+x+1}$

**28.** $\displaystyle\lim_{x\to\infty}\left(1+\frac{1}{x}\right)\left(\frac{x^2+1}{x^2-1}\right)$

**29.** $\displaystyle\lim_{t\to\infty}\left(\frac{t+1}{2t-1}+\frac{2t^2-1}{1-3t^2}\right)$

**30.** $\displaystyle\lim_{s\to -\infty}\left(\frac{s}{s+1}-\frac{s^2}{2s^2+1}\right)$

**31.** $\lim\limits_{x\to\infty} \dfrac{2x}{\sqrt{3x^2 + 1}}$

**32.** $\lim\limits_{t\to-\infty} \dfrac{2t^2}{\sqrt{t^4 + t^2}}$

**33.** $\lim\limits_{x\to\infty} \dfrac{2e^x + 1}{3e^x + 2}$

**34.** $\lim\limits_{t\to-\infty} \dfrac{e^{-t} - 2e^{2t}}{e^{-2t} + 3e^{2t}}$

**35.** $\lim\limits_{t\to\infty} \left(\dfrac{3t^2 + 1}{2t^2 - 1}\right)e^{-0.1t}$

**36.** $\lim\limits_{x\to\infty} \tan^{-1}(\ln x)$

**37.** Let

$$f(x) = \begin{cases} \dfrac{1}{x} & \text{if } x < 0 \\ 1 & \text{if } x > 0 \end{cases}$$

Find $\lim_{x\to 0^-} f(x)$, $\lim_{x\to 0^+} f(x)$, $\lim_{x\to-\infty} f(x)$, and $\lim_{x\to\infty} f(x)$.

**38.** Let

$$f(x) = \begin{cases} \dfrac{1}{\pi}x^2 + x & \text{if } x < 0 \\ \sin 2x & \text{if } x \ge 0 \end{cases}$$

(See the graph of $f$.) Find $\lim_{x\to-\infty} f(x)$ and $\lim_{x\to\infty} f(x)$.

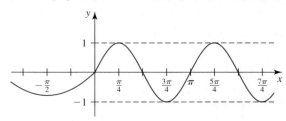

**39.** Let $f(x) = \dfrac{\sin x}{x}$.

   **a.** Show that $-\dfrac{1}{x} \le \dfrac{\sin x}{x} \le \dfrac{1}{x}$, for $x > 0$.

   **b.** Use the results of (a) and the Squeeze Theorem (which also holds for limits at infinity) to find $\lim\limits_{x\to\infty} \dfrac{\sin x}{x}$.

   **c.** Plot the graphs of $f(x) = -\dfrac{1}{x}$, $g(x) = \dfrac{\sin x}{x}$, and $h(x) = \dfrac{1}{x}$ using the viewing window $[0, 20] \times \left[-\frac{1}{2}, \frac{1}{2}\right]$.

**40.** Let $g(x) = \dfrac{\cos x^2}{\sqrt{x}}$. Find $\lim_{x\to\infty} g(x)$.

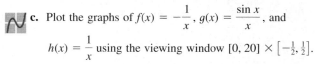 In Exercises 41–44, (a) *find an approximate value of the limit by plotting the graph of an appropriate function f,* (b) *find an approximate value of the limit by constructing a table of values of f,* and (c) *find the exact value of the limit.*

**41.** $\lim\limits_{x\to\infty} x\left(\sqrt{x^2 + 1} - x\right)$

**42.** $\lim\limits_{x\to-\infty} \left(x + \sqrt{x^2 + 5x}\right)$

**43.** $\lim\limits_{x\to\infty} \left(\sqrt{2x^2 + 3x + 4} - \sqrt{2x^2 + x + 1}\right)$

**44.** $\lim\limits_{x\to\infty} \dfrac{\sqrt{3x + 2} - \sqrt{3x}}{\sqrt{2x + 1} - \sqrt{2x}}$

*In Exercises 45–48 you are given the graph of a function f. Find the horizontal and vertical asymptotes of the graph of f.*

**45.**

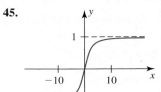

**46.**

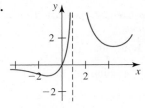

**47.**

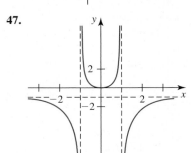

**48.**

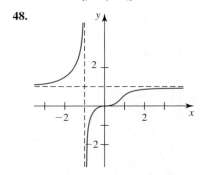

*In Exercises 49–56, find the horizontal and vertical asymptotes of the graph of the function. Do not sketch the graph.*

**49.** $f(x) = \dfrac{1}{x + 2}$

**50.** $g(x) = \dfrac{x}{x + 1}$

**51.** $h(x) = \dfrac{x - 1}{x + 1}$

**52.** $f(t) = \dfrac{t^2}{t^2 - 4}$

**53.** $f(x) = \dfrac{2x}{x^2 - x - 6}$

**54.** $h(x) = \dfrac{e^x}{e^x - 2}$

**55.** $f(t) = \dfrac{t^2 - 2}{t^2 - 4}$

**56.** $f(x) = \dfrac{2x^3}{\sqrt{3x^6 + 2}}$

*In Exercises 57–60, sketch the graph of a function having the given properties.*

**57.** $f(0) = 0, f'(0) = 1, f''(x) > 0$ on $(-\infty, 0), f''(x) < 0$ on $(0, \infty), \lim_{x\to-\infty} f(x) = -1, \lim_{x\to\infty} f(x) = 1$

**58.** $f(0) = \pi/2, f'(0)$ does not exist, $f(-1) = f(1) = 0,$ $f''(x) > 0$ on $(-\infty, 0) \cup (0, \infty),$ $\lim_{x\to-\infty} f(x) = \lim_{x\to\infty} f(x) = -\pi/2$

**59.** Domain of $f$ is $(-\infty, -1) \cup (1, \infty)$, $f(-2) = -1$,
$f'(-2) = 0$, $f''(x) < 0$ on $(-\infty, -1) \cup (1, \infty)$,
$\lim_{x \to -1^-} f(x) = -\infty$, $\lim_{x \to 1^+} f(x) = -\infty$,
$\lim_{x \to \infty} f(x) = -\infty$

**60.** $f(2) = 3$, $f'(2) = 0$, $f'(x) < 0$ on $(-\infty, 0) \cup (2, \infty)$,
$f'(x) > 0$ on $(0, 2)$, $\lim_{x \to 0^-} f(x) = -\infty$, $\lim_{x \to 0^+} f(x) = -\infty$,
$\lim_{x \to -\infty} f(x) = \lim_{x \to \infty} f(x) = 1$, $f''(x) < 0$ on
$(-\infty, 0) \cup (0, 3)$, $f''(x) > 0$ on $(3, \infty)$

**61. Chemical Pollution** As a result of an abandoned chemical dump leaching chemicals into the water, the main well of a town has been contaminated with trichloroethylene, a cancer-causing chemical. A proposal submitted by the town's board of health indicates that the cost, measured in millions of dollars, of removing $x$ percent of the toxic pollutant is given by

$$C(x) = \frac{0.5x}{100 - x}$$

    **a.** Evaluate $\lim_{x \to 100^-} C(x)$, and interpret your results.
   **b.** Plot the graph of $C$ using the viewing window $[0, 100] \times [0, 10]$.

**62. Driving Costs** A study of driving costs of a 2008 medium-sized sedan found that the average cost (car payments, gas, insurance, upkeep, and depreciation) is given by the function

$$C(x) = \frac{1735.2}{x^{1.72}} + 38.6$$

where $C(x)$ is measured in cents per mile and $x$ denotes the number of miles (in thousands) the car is driven in a year. Compute $\lim_{x \to \infty} C(x)$, and interpret your results.
*Source:* American Automobile Association.

**63. City Planning** A major developer is building a 5000-acre complex of homes, offices, stores, schools, and churches in the rural community of Marlboro. As a result of this development, the planners have estimated that Marlboro's population (in thousands) $t$ years from now will be given by

$$P(t) = \frac{25t^2 + 125t + 200}{t^2 + 5t + 40}$$

    **a.** What will the population of Marlboro be in the long run?
      Hint: Find $\lim_{t \to \infty} P(t)$.
   **b.** Plot the graph of $P$ using the viewing window $[0, 20] \times [0, 30]$.

**64. Oxygen Content of a Pond** When organic waste is dumped into a pond, the oxidation process that takes place reduces the pond's oxygen content. However, given time, nature will restore the oxygen content to its natural level. Suppose that the oxygen content $t$ days after the organic waste has been dumped into the pond is given by

$$f(t) = 100 \left( \frac{t^2 + 10t + 100}{t^2 + 20t + 100} \right)$$

percent of its normal level.

    **a.** Evaluate $\lim_{t \to \infty} f(t)$ and interpret your result.
   **b.** Plot the graph of $f$ using the viewing window $[0, 200] \times [70, 100]$.

**65. Terminal Velocity** A skydiver leaps from the gondola of a hot-air balloon. As she free-falls, air resistance, which is proportional to her velocity, builds up to a point at which it balances the force due to gravity. The resulting motion may be described in terms of her velocity as follows: Starting at rest (zero velocity), her velocity increases and approaches a constant velocity, called the *terminal velocity*. Sketch a graph of her velocity $v$ versus time $t$.

**66. Terminal Velocity** A skydiver leaps from a helicopter hovering high above the ground. Her velocity $t$ sec later and before deploying her parachute is given by

$$v(t) = 52[1 - (0.82)^t]$$

where $v(t)$ is measured in meters per second.

    **a.** Complete the following table, giving her velocity at the indicated times.

| $t$ (sec) | 0 | 10 | 20 | 30 | 40 | 50 | 60 |
|---|---|---|---|---|---|---|---|
| $v(t)$ (m/sec) | | | | | | | |

   **b.** Plot the graph of $v$ using the viewing window $[0, 60] \times [0, 60]$.
    **c.** What is her terminal velocity?
      Hint: Evaluate $\lim_{t \to \infty} v(t)$.

**67. Mass of a Moving Particle** The mass $m$ of a particle moving at a speed $v$ is related to its rest mass $m_0$ by the equation

$$m = \frac{m_0}{\sqrt{1 - \dfrac{v^2}{c^2}}}$$

where $c$, a constant, is the speed of light. Show that

$$\lim_{v \to c^-} \frac{m_0}{\sqrt{1 - \dfrac{v^2}{c^2}}} = \infty$$

thus proving that the line $v = c$ is a vertical asymptote of the graph of $m$ versus $v$. Make a sketch of the graph of $m$ as a function of $v$.

**68. Special Theory of Relativity** According to the special theory of relativity

$$v = c\sqrt{1 - \left( \frac{E_0}{E} \right)^2}$$

where $E_0 = m_0 c^2$ is the rest energy and $E$ is the total energy.

    **a.** Find $\lim_{E \to \infty} v$.
    **b.** Sketch the graph of $v$.
    **c.** What do your results say about the speed of light?

**69.** Complete the following table to show that Equation (4),

$$e = \lim_{n \to \infty} \left(1 + \frac{1}{n}\right)^n$$

appears to be valid.

| $n$ | 1 | 10 | $10^2$ | $10^3$ | $10^4$ | $10^5$ | $10^6$ |
|---|---|---|---|---|---|---|---|
| $\left(1 + \dfrac{1}{n}\right)^n$ | | | | | | | |

**70.** Find the accumulated amount after 5 years on an investment of $5000 earning interest at the rate of 10% per year compounded continuously.

**71.** Find the accumulated amount after 10 years on an investment of $10,000 earning interest at the rate of 12% per year compounded continuously.

**72. Annual Return of an Investment** A group of private investors purchased a condominium complex for $2.1 million and sold it 6 years later for $4.4 million. Find the annual rate of return (compounded continuously) on their investment.

**73. Establishing a Trust Fund** The parents of a child wish to establish a trust fund for the child's college education. If they need an estimated $96,000 8 years from now and they are able to invest the money at 8.5% compounded continuously in the interim, how much should they set aside in trust now?

**74. Effect of Inflation on Salaries** Mr. Gilbert's current annual salary is $75,000. Ten years from now, how much will he need to earn to retain his present purchasing power if the rate of inflation over that period is 5% per year? Assume that inflation is compounded continuously.

 **75.** Let $f(x) = \sqrt{3x + \sqrt{x}} - \sqrt{3x - \sqrt{x}}$.
a. Plot the graph of $f$, and use it to estimate $\lim_{x \to \infty} f(x)$ to one decimal place.
b. Use a table of values to estimate $\lim_{x \to \infty} f(x)$.
c. Find the exact value of $\lim_{x \to \infty} f(x)$ analytically.

 **76.** Let

$$f(x) = \sqrt[3]{x^3 + 2x^2 + 3x - 1} - \sqrt[3]{x^3 - 3x^2 + x - 4}$$

a. Plot the graph of $f$, and use it to estimate $\lim_{x \to \infty} f(x)$ to one decimal place.
b. Use a table of values to estimate $\lim_{x \to \infty} f(x)$.
c. Find the exact value of $\lim_{x \to \infty} f(x)$ analytically.

**77. Escape Velocity** An object is projected vertically upward from the earth's surface with an initial velocity $v_0$ of magnitude less than the *escape velocity* (the velocity that a projectile should have in order to break free of the earth forever). If

only the earth's influence is taken into consideration, then the maximum height reached by the rocket is

$$H = \frac{v_0^2 R}{2gR - v_0^2}$$

where $R$ is the radius of the earth and $g$ is the acceleration due to gravity.
a. Show that the graph of $H$ has a vertical asymptote at $v_0 = \sqrt{2gR}$, and interpret your result.
b. Use the result of part (a) to find the escape velocity. Take the radius of the earth to be 4000 mi ($g = 32$ ft/sec$^2$).
c. Sketch the graph of $H$ as a function of $v_0$.

**78.** Determine the constants $a$ and $b$ such that

$$\lim_{x \to \infty} \left(\frac{2x^2 + 3}{x + 1} - ax - b\right) = 0$$

**79.** Let

$$P(x) = \frac{a_n x^n + a_{n-1} x^{n-1} + \cdots + a_0}{b_m x^m + b_{m-1} x^{m-1} + \cdots + b_0}$$

where $a_n \neq 0$, $b_m \neq 0$, and $m$, $n$, are positive integers. Show that

$$\lim_{x \to \infty} P(x) = \begin{cases} \pm\infty & \text{if } n > m \\ \dfrac{a_n}{b_m} & \text{if } n = m \\ 0 & \text{if } n < m \end{cases}$$

**80.** Prove that $\lim_{x \to \infty} f(x) = \lim_{t \to 0^+} f(1/t)$.

**81.** Use the result of Exercise 80 to find $\lim_{x \to \infty} x \sin(1/x)$.

**82.** a. Show that $\lim_{x \to \infty}(x^a/e^x) = 0$ for any fixed number $a$. Thus, $e^x$ eventually grows faster than any power of $x$.
   **Hint:** Use the result of part (b) of Exercise 72, Section 3.3, to show that if $a = 1$, then $\lim_{x \to \infty}(x/e^x) = 0$. For the general case, introduce the variable $y$ defined by $x = ay$ if $a > 0$.

 b. Plot the graph of $f(x) = (x^{10}/e^x)$ using the viewing window $[0, 40] \times [0, 460{,}000]$, thus verifying the result of part (a) for the special case in which $a = 10$.
c. Find the value of $x$ at which the graph of $f(x) = e^x$ eventually overtakes that of $g(x) = x^{10}$.

*In Exercises 83–88, use the appropriate precise definition to prove the statement.*

**83.** $\lim_{x \to 0} \dfrac{2}{x^4} = \infty$

**84.** $\lim_{x \to 0^+} \dfrac{1}{\sqrt{x}} = \infty$

**85.** $\lim_{x \to 0^-} \dfrac{1}{x} = -\infty$

**86.** $\lim_{x \to \infty} \dfrac{x}{x^2 + 1} = 0$

**87.** $\lim_{x \to -\infty} \dfrac{x}{x + 1} = 1$

**88.** $\lim_{x \to \infty} 3x = \infty$

In Exercises 89–94, *determine whether the given statement is true or false. If it is true, explain why it is true. If it is false, explain why or give an example to show why it is false.*

**89.** $\lim\limits_{x \to 2} \dfrac{1}{x-2} = \infty$

**90.** $\lim\limits_{x \to \infty} c = c$ for any real number $c$.

**91.** If $y = L$ is a horizontal asymptote of the graph of the function $f$, then the graph of $f$ cannot intersect $y = L$.

**92.** If the denominator of a rational function $f$ is equal to zero at $a$, then $x = a$ is a vertical asymptote of the graph of $f$.

**93.** The graph of a function can have two distinct horizontal asymptotes.

**94.** If $f$ is defined on $(0, \infty)$ and $\lim\limits_{x \to 0^+} f(x) = L$, then $\lim\limits_{x \to \infty} f(1/x) = 1/L$.

## 3.6 Curve Sketching

### ■ The Graph of a Function

We have seen on many occasions how the graph of a function can help us to visualize the properties of the function. From a practical point of view, the graph of a function also gives, at one glance, a complete summary of all the information captured by the function.

Consider, for example, the graph of the function giving the Dow-Jones Industrial Average (DJIA) on Black Monday: October 19, 1987 (Figure 1). Here, $t = 0$ corresponds to 9:30 A.M., when the market was open for business, and $t = 6.5$ corresponds to 4 P.M., the closing time. The following information can be gleaned from studying the graph.

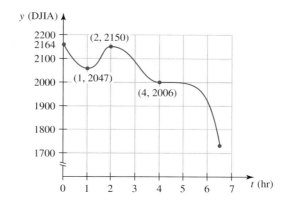

**FIGURE 1**
The Dow-Jones Industrial
Average on Black Monday
*Source: The Wall Street Journal.*

The graph is *decreasing* rapidly from $t = 0$ to $t = 1$, reflecting the sharp drop in the index in the first hour of trading. The point $(1, 2047)$ is a *relative minimum* point of the function, and this turning point coincides with the start of an aborted recovery. The short-lived rally, represented by the portion of the graph that is *increasing* on the interval $(1, 2)$, quickly fizzled out at $t = 2$ (11:30 A.M.). The *relative maximum* point $(2, 2150)$ marks the highest point of the recovery. The function is decreasing on the rest of the interval. The point $(4, 2006)$ is an *inflection point* of the function; it shows that there was a temporary respite at $t = 4$ (1:30 P.M.). However, selling pressure continued unabated, and the DJIA continued to fall until the closing bell. Finally, the graph also shows that the index opened at the high of the day ($f(0) = 2164$ is the *absolute maximum* of the function) and closed at the low of the day ($f\left(\frac{13}{2}\right) = 1739$ is the *absolute minimum* of the function), a drop of 508 points, or approximately 23%, from the previous close.

## Guide to Curve Sketching

A systematic approach to sketching the graph of a function $f$ begins with an attempt to gather as much information as possible about $f$. The following guidelines provide us with a step-by-step procedure for doing this.

> **Guidelines for Curve Sketching**
> 1. Find the domain of $f$.
> 2. Find the $x$- and $y$-intercepts of $f$.
> 3. Determine whether the graph of $f$ is symmetric with respect to the $y$-axis or the origin.
> 4. Determine the behavior of $f$ for large absolute values of $x$.
> 5. Find the asymptotes of the graph of $f$.
> 6. Find the intervals where $f$ is increasing and where $f$ is decreasing.
> 7. Find the relative extrema of $f$.
> 8. Determine the concavity of the graph of $f$.
> 9. Find the inflection points of $f$.
> 10. Sketch the graph of $f$.

**EXAMPLE 1** Sketch the graph of the function $f(x) = 2x^3 - 3x^2 - 12x + 12$.

**Solution**   First, we obtain the following information about $f$.
1. Since $f$ is a polynomial function of degree 3, the domain of $f$ is $(-\infty, \infty)$.
2. By setting $x = 0$, we see that the $y$-intercept is 12. Since the cubic equation $2x^3 - 3x^2 - 12x + 12 = 0$ is not readily solved, we will not attempt to find the $x$-intercept.*
3. Since $f(-x) = -2x^3 - 3x^2 + 12x + 12$ is not equal to $f(x)$ or $-f(x)$, the graph of $f$ is not symmetric with respect to the $y$-axis or the origin.
4. Since
$$\lim_{x \to -\infty} f(x) = -\infty \qquad \text{and} \qquad \lim_{x \to \infty} f(x) = \infty$$
we see that $f$ decreases without bound as $x$ decreases without bound and $f$ increases without bound as $x$ increases without bound.
5. Because $f$ is a polynomial function (a rational function whose denominator is 1 and is therefore never zero), we see that the graph of $f$ has no vertical asymptotes. From part (4), we see that the graph of $f$ has no horizontal asymptotes.
6. $f'(x) = 6x^2 - 6x - 12 = 6(x^2 - x - 2) = 6(x + 1)(x - 2)$ and is continuous everywhere. Setting $f'(x) = 0$ gives $-1$ and 2 as critical numbers. The sign diagram for $f'$ shows that $f$ is increasing on $(-\infty, -1)$ and on $(2, \infty)$ and decreasing on $(-1, 2)$. (See Figure 2.)
7. From the results of part (6), we see that $-1$ and 2 are critical numbers of $f$. Furthermore, from the sign diagram of $f'$, we see that $f$ has a relative maximum at $-1$ with value
$$f(-1) = 2(-1)^3 - 3(-1)^2 - 12(-1) + 12 = 19$$
and a relative minimum at 2 with value
$$f(2) = 2(2)^3 - 3(2)^2 - 12(2) + 12 = -8$$

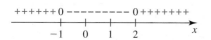

**FIGURE 2**
The sign diagram for $f'$

---
*If the equation $f(x) = 0$ is difficult to solve, disregard finding the $x$-intercepts.

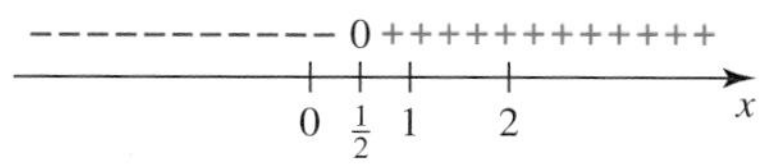

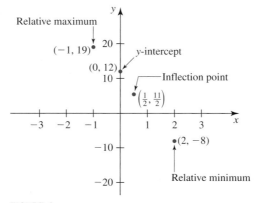

**FIGURE 3**
The sign diagram for $f''$

8. $f''(x) = 12x - 6 = 6(2x - 1)$

Setting $f''(x) = 0$ gives $x = \frac{1}{2}$. The sign diagram for $f''$ shows that the graph of $f$ is concave downward on $\left(-\infty, \frac{1}{2}\right)$ and concave upward on $\left(\frac{1}{2}, \infty\right)$. (See Figure 3.)

9. From the results of part (8) we see that $f$ has an inflection point when $x = \frac{1}{2}$. Next,

$$f\left(\tfrac{1}{2}\right) = 2\left(\tfrac{1}{2}\right)^3 - 3\left(\tfrac{1}{2}\right)^2 - 12\left(\tfrac{1}{2}\right) + 12 = \tfrac{11}{2}$$

so $\left(\frac{1}{2}, \frac{11}{2}\right)$ is the inflection point of $f$.

10. The following table summarizes this information.

| | |
|---|---|
| **Domain** | $(-\infty, \infty)$ |
| **Intercepts** | $y$-intercept: 12 |
| **Symmetry** | None |
| **End behavior** | $\displaystyle\lim_{x \to -\infty} f(x) = -\infty$ and $\displaystyle\lim_{x \to \infty} f(x) = \infty$ |
| **Asymptotes** | None |
| **Intervals where $f$ is $\nearrow$ or $\searrow$** | $\nearrow$ on $(-\infty, -1)$ and on $(2, \infty)$; $\searrow$ on $(-1, 2)$ |
| **Relative extrema** | Rel. max. at $(-1, 19)$; rel. min. at $(2, -8)$ |
| **Concavity** | Downward on $\left(-\infty, \frac{1}{2}\right)$; upward on $\left(\frac{1}{2}, \infty\right)$ |
| **Point of inflection** | $\left(\frac{1}{2}, \frac{11}{2}\right)$ |

We begin by plotting the intercepts, the inflection point, and the relative extrema of $f$ as shown in Figure 4. Then, using the rest of the information, we complete the graph of $f$ as shown in Figure 5.

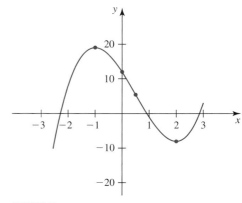

**FIGURE 4**
First plot the $y$-intercept, the relative extrema, and the inflection point.

**FIGURE 5**
The graph of $y = 2x^3 - 3x^2 - 12x + 12$ ■

---

**EXAMPLE 2** Sketch the graph of the function $f(x) = \dfrac{x^2}{x^2 - 1}$.

**Solution**

1. The denominator of the rational function $f$ is equal to zero if $x^2 - 1 = (x + 1)(x - 1) = 0$, that is, if $x = -1$ or $x = 1$. Therefore, the domain of $f$ is $(-\infty, -1) \cup (-1, 1) \cup (1, \infty)$.

2. Setting $x = 0$ gives 0 as the $y$-intercept. Next, setting $f(x) = 0$ gives $x^2 = 0$, or $x = 0$. So the $x$-intercept is 0.

**3.** $f(-x) = \dfrac{(-x)^2}{(-x)^2 - 1} = \dfrac{x^2}{x^2 - 1} = f(x)$

and this shows that the graph of $f$ is symmetric with respect to the $y$-axis.

**4.** $\displaystyle\lim_{x \to -\infty} \dfrac{x^2}{x^2 - 1} = \lim_{x \to \infty} \dfrac{x^2}{x^2 - 1} = 1$

**5.** Because the denominator of $f(x)$ is equal to zero at $-1$ and 1, the lines $x = -1$ and $x = 1$ are candidates for the vertical asymptotes of the graph of $f$. Since

$$\lim_{x \to -1^-} \dfrac{x^2}{x^2 - 1} = \infty \qquad \text{and} \qquad \lim_{x \to 1^-} \dfrac{x^2}{x^2 - 1} = -\infty$$

we see that $x = -1$ and $x = 1$ are indeed vertical asymptotes. From part (4) we see that $y = 1$ is a horizontal asymptote of the graph of $f$.

**6.** $f'(x) = \dfrac{(x^2 - 1)\dfrac{d}{dx}(x^2) - x^2 \dfrac{d}{dx}(x^2 - 1)}{(x^2 - 1)^2}$

$= \dfrac{(x^2 - 1)(2x) - x^2(2x)}{(x^2 - 1)^2} = -\dfrac{2x}{(x^2 - 1)^2}$

Notice that $f'$ is continuous everywhere except at $\pm 1$ and that it has a zero when $x = 0$. The sign diagram of $f'$ is shown in Figure 6.

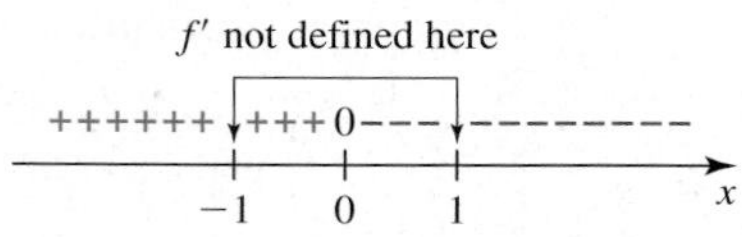

*f'* not defined here

++++++ +++0--- ---------
  −1   0   1   *x*

**FIGURE 6**
The sign diagram for *f'*

From the diagram we see that $f$ is increasing on $(-\infty, -1)$ and on $(-1, 0)$ and decreasing on $(0, 1)$ and on $(1, \infty)$.

**7.** From the results of part (6) we see that 0 is a critical number of $f$. The numbers $-1$ and 1 are not in the domain of $f$ and, therefore, are not critical numbers of $f$. Also, from Figure 6 we see that $f$ has a relative maximum at $x = 0$. Its value is $f(0) = 0$.

**8.** $f''(x) = \dfrac{d}{dx}\left[\dfrac{-2x}{(x^2 - 1)^2}\right]$

$= \dfrac{(x^2 - 1)^2(-2) - (-2x)(2)(x^2 - 1)(2x)}{(x^2 - 1)^4}$

$= \dfrac{2(x^2 - 1)[-(x^2 - 1) + 4x^2]}{(x^2 - 1)^4} = \dfrac{2(3x^2 + 1)}{(x^2 - 1)^3}$

Notice that $f''$ is continuous everywhere except at $\pm 1$ and that $f''$ has no zeros. From the sign diagram of $f''$ shown in Figure 7, we see that the graph of $f$ is concave upward on $(-\infty, -1)$ and on $(1, \infty)$ and concave downward on $(-1, 1)$.

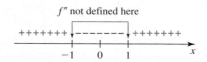

*f''* not defined here

**FIGURE 7**
The sign diagram for *f''*

**9.** $f$ has no inflection points. Remember that $-1$ and $1$ are not in the domain of $f$.

**10.** The following table summarizes this information.

| Domain | $(-\infty, -1) \cup (-1, 1) \cup (1, \infty)$ |
|---|---|
| Intercepts | $x$- and $y$-intercepts: 0 |
| Symmetry | With respect to the $y$-axis |
| Asymptotes | Vertical: $x = -1$ and $x = 1$<br>Horizontal: $y = 1$ |
| End behavior | $\displaystyle \lim_{x \to -\infty} \frac{x^2}{x^2 - 1} = \lim_{x \to \infty} \frac{x^2}{x^2 - 1} = 1$ |
| Intervals where $f$ is ↗ or ↘ | ↗ on $(-\infty, -1)$ and on $(-1, 0)$; ↘ on $(0, 1)$ and on $(1, \infty)$ |
| Relative extrema | Rel. max. at $(0, 0)$ |
| Concavity | Downward on $(-1, 1)$; upward on $(-\infty, -1)$ and on $(1, \infty)$ |
| Point of inflection | None |

We begin by plotting the relative maximum of $f$ and drawing the asymptotes of the graph of $f$ as shown in Figure 8. In this case, plotting a few additional points will ensure a more accurate graph. For example, from the table

| $x$ | $\frac{1}{2}$ | $\frac{3}{2}$ | 2 |
|---|---|---|---|
| $f(x)$ | $-\frac{1}{3}$ | $\frac{9}{5}$ | $\frac{4}{3}$ |

we see that the points $\left(\frac{1}{2}, -\frac{1}{3}\right)$, $\left(\frac{3}{2}, \frac{9}{5}\right)$, and $\left(2, \frac{4}{3}\right)$ and, by symmetry, $\left(-\frac{1}{2}, -\frac{1}{3}\right)$, $\left(-\frac{3}{2}, \frac{9}{5}\right)$, and $\left(-2, \frac{4}{3}\right)$ lie on the graph of $f$. Finally, using the rest of the information about $f$, we sketch its graph as shown in Figure 9.

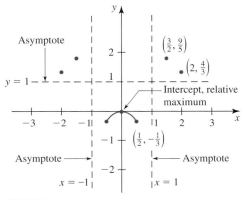

**FIGURE 8**
First plot the $y$-intercept, relative maximum, and asymptotes. Then plot a few additional points.

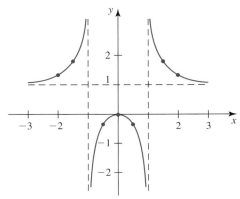

**FIGURE 9**
The graph of $f(x) = \dfrac{x^2}{x^2 - 1}$

**EXAMPLE 3** Sketch the graph of the function $f(x) = \dfrac{1}{1 + \sin x}$.

**Solution**

1. The denominator of $f(x)$ is equal to zero if $1 + \sin x = 0$; that is, if $\sin x = -1$ or $x = (3\pi/2) + 2n\pi$ $(n = 0, \pm 1, \pm 2, \ldots )$. Therefore, the domain of $f$ is $\cdots \left(-\frac{\pi}{2}, \frac{3\pi}{2}\right) \cup \left(\frac{3\pi}{2}, \frac{7\pi}{2}\right) \cup \cdots$.

2. Setting $x = 0$ gives 1 as the $y$-intercept. Since $y \neq 0$, there are no $x$-intercepts.

3. $f(-x) = \dfrac{1}{1 + \sin(-x)} = \dfrac{1}{1 - \sin x}$     $\sin(-x) = -\sin x$

   and is equal to neither $f(x)$ nor $-f(x)$. Therefore, $f$ is not symmetric with respect to the $y$-axis or the origin.

4. $\displaystyle\lim_{x \to -\infty} \left[\dfrac{1}{1 + \sin x}\right]$ and $\displaystyle\lim_{x \to \infty} \left[\dfrac{1}{1 + \sin x}\right]$ do not exist.

5. The denominator of $f(x)$ is equal to zero when $1 + \sin x = 0$, that is, when $x = (3\pi/2) + 2n\pi$ $(n = 0, \pm 1, \pm 2, \ldots )$ (see part (1)). Since

$$\lim_{x \to (3\pi/2) + 2n\pi} \left[\dfrac{1}{1 + \sin x}\right] = \infty$$

we see that the lines $x = (3\pi/2) + 2n\pi$ $(n = 0, \pm 1, \pm 2, \ldots )$ are vertical asymptotes of the graph of $f$. From part (4) we see that there are no horizontal asymptotes.

6. $f'(x) = \dfrac{d}{dx} (1 + \sin x)^{-1}$

   $= -(1 + \sin x)^{-2}(\cos x)$     Use the Chain Rule.

   $= -\dfrac{\cos x}{(1 + \sin x)^2}$

   Notice that $f'$ is continuous everywhere except at $x = (3\pi/2) + 2n\pi$ $(n = 0, \pm 1, \pm 2, \ldots )$ and has zeros at $x = (\pi/2) + 2n\pi$ $(n = 0, \pm 1, \pm 2, \ldots )$. The sign diagram of $f'$ is shown in Figure 10. We see that $f$ is increasing on $\cdots \left(-\frac{3\pi}{2}, -\frac{\pi}{2}\right), \left(\frac{\pi}{2}, \frac{3\pi}{2}\right),$ and on $\left(\frac{5\pi}{2}, \frac{7\pi}{2}\right) \cdots$ and decreasing on $\cdots \left(-\frac{5\pi}{2}, -\frac{3\pi}{2}\right), \left(-\frac{\pi}{2}, \frac{\pi}{2}\right),$ and on $\left(\frac{3\pi}{2}, \frac{5\pi}{2}\right) \cdots$.

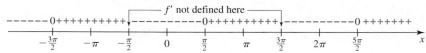

**FIGURE 10**
The sign diagram for $f'$

7. From the results of part (6) we see that $(\pi/2) + 2n\pi$ $(n = 0, \pm 1, \pm 2, \ldots )$ are critical numbers of $f$. From Figure 10 we see that these numbers give rise to the relative minima of $f$, each with value $\frac{1}{2}$, since

$$f\left(\tfrac{\pi}{2} + 2n\pi\right) = \dfrac{1}{1 + \sin\left(\frac{\pi}{2} + 2n\pi\right)} = \dfrac{1}{1 + \sin\frac{\pi}{2}} = \dfrac{1}{2}$$

**8.** $f''(x) = \dfrac{d}{dx}[-(\cos x)(1 + \sin x)^{-2}]$

$= (\sin x)(1 + \sin x)^{-2} - (\cos x)(-2)(1 + \sin x)^{-3}(\cos x)$

$= (1 + \sin x)^{-3}[(\sin x)(1 + \sin x) + 2\cos^2 x]$

$= \dfrac{\sin x + \sin^2 x + 2\cos^2 x}{(1 + \sin x)^3}$

$= \dfrac{\sin x + \sin^2 x + 2(1 - \sin^2 x)}{(1 + \sin x)^3} = -\dfrac{\sin^2 x - \sin x - 2}{(1 + \sin x)^3}$

$= -\dfrac{(\sin x - 2)(\sin x + 1)}{(1 + \sin x)^3} = \dfrac{2 - \sin x}{(1 + \sin x)^2}$

Because $|\sin x| \le 1$ for all values of $x$, we see that $f''(x) > 0$ whenever it is defined. From the sign diagram of $f''$ shown in Figure 11, we conclude that the graph of $f$ is concave upward on $\cdots \left(-\frac{5\pi}{2}, -\frac{\pi}{2}\right), \left(-\frac{\pi}{2}, \frac{3\pi}{2}\right)$, and on $\left(\frac{3\pi}{2}, \frac{7\pi}{2}\right) \cdots$.

**FIGURE 11**
The sign diagram for $f''$

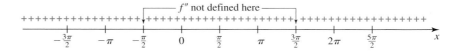

**9.** $f$ has no inflection points.

**10.** The following table summarizes this information.

| | |
|---|---|
| **Domain** | $\cdots \left(-\frac{\pi}{2}, \frac{3\pi}{2}\right) \cup \left(\frac{3\pi}{2}, \frac{7\pi}{2}\right) \cup \cdots$ |
| **Intercept** | $y$-intercept: 1 |
| **Symmetry** | None (with respect to the $y$-axis or the origin) |
| **End behavior** | $\displaystyle\lim_{x \to -\infty}\left[\dfrac{1}{1 + \sin x}\right]$ and $\displaystyle\lim_{x \to \infty}\left[\dfrac{1}{1 + \sin x}\right]$ do not exist. |
| **Asymptotes** | Vertical: $x = \frac{3\pi}{2} + 2n\pi \ (n = 0, \pm 1, \pm 2, \dots )$ |
| **Intervals where $f$ is $\nearrow$ or $\searrow$** | $\nearrow$ on $\cdots \left(-\frac{3\pi}{2}, -\frac{\pi}{2}\right)$ and on $\left(\frac{\pi}{2}, \frac{3\pi}{2}\right) \cdots$ <br> $\searrow$ on $\cdots \left(-\frac{\pi}{2}, \frac{\pi}{2}\right)$ and on $\left(\frac{3\pi}{2}, \frac{5\pi}{2}\right) \cdots$ |
| **Relative extrema** | Rel. min: $\cdots \left(-\frac{3\pi}{2}, \frac{1}{2}\right), \left(\frac{\pi}{2}, \frac{1}{2}\right), \left(\frac{5\pi}{2}, \frac{1}{2}\right) \cdots$ |
| **Concavity** | Upward on $\cdots \left(-\frac{5\pi}{2}, -\frac{\pi}{2}\right)$ and on $\left(-\frac{\pi}{2}, \frac{3\pi}{2}\right) \cdots$ |
| **Point of inflection** | None |

The graph of $f$ is shown in Figure 12.

**FIGURE 12**
The graph of $f(x) = \dfrac{1}{1 + \sin x}$

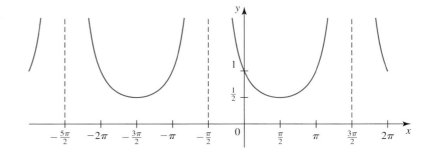

**EXAMPLE 4** Sketch the graph of the function $f(x) = e^{-x^2}$.

**Solution** First, we obtain the following information on the function $f$.

1. The domain of $f$ is $(-\infty, \infty)$.
2. Setting $x = 0$ gives 1 as the $y$-intercept. Next, since $e^{-x^2} = 1/e^{x^2}$ is never zero, there are no $x$-intercepts.
3. Since

$$f(-x) = e^{-(-x)^2} = e^{-x^2} = f(x)$$

   we see that the graph of $f$ is symmetric with respect to the $y$-axis.
4. and 5. Since

$$\lim_{x \to -\infty} e^{-x^2} = \lim_{x \to -\infty} \frac{1}{e^{x^2}} = 0 = \lim_{x \to \infty} e^{-x^2}$$

   we see that $y = 0$ is a horizontal asymptote of the graph of $f$.

6. $f'(x) = \dfrac{d}{dx} e^{-x^2} = e^{-x^2} \dfrac{d}{dx} (-x^2) = -2xe^{-x^2}$

   Setting $f'(x) = 0$ gives $x = 0$. The sign diagram of $f'$ shows that $f$ is increasing on $(-\infty, 0)$ and decreasing on $(0, \infty)$. (See Figure 13.)
7. From the results of Step 6 we see that 0 is the sole critical number of $f$. Furthermore, from the sign diagram of $f'$, we see that $f$ has a relative maximum at $x = 0$ with value $f(0) = e^0 = 1$.

8. $f''(x) = \dfrac{d}{dx}\left[-2xe^{-x^2}\right]$

   $= -2e^{-x^2} - 2xe^{-x^2}(-2x)$     Use the Product Rule and the Chain Rule.

   $= 2(2x^2 - 1)e^{-x^2}$

   Setting $f''(x) = 0$ gives $2x^2 - 1 = 0$ or $x = \pm\sqrt{2}/2$. The sign diagram of $f''$ shows that $f$ is concave upward on $\left(-\infty, -\frac{\sqrt{2}}{2}\right)$ and on $\left(\frac{\sqrt{2}}{2}, \infty\right)$ and concave downward on $\left(-\frac{\sqrt{2}}{2}, \frac{\sqrt{2}}{2}\right)$. (See Figure 14.)

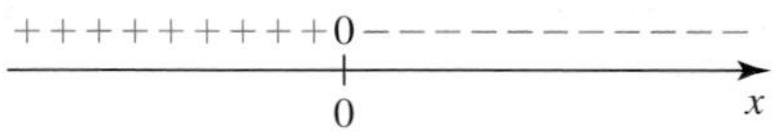

**FIGURE 13**
The sign diagram for $f'$

**FIGURE 14**
The sign diagram for $f''$

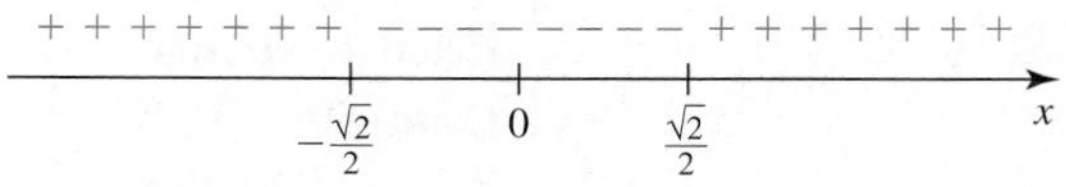

9. From the results of Step 8 we see that $f$ has inflection points at $x = \pm\sqrt{2}/2$. Since $f\left(-\frac{\sqrt{2}}{2}\right) = e^{-1/2} = f\left(\frac{\sqrt{2}}{2}\right)$, we see that $\left(-\frac{\sqrt{2}}{2}, e^{-1/2}\right)$ and $\left(\frac{\sqrt{2}}{2}, e^{-1/2}\right)$ are inflection points of $f$.
10. The graph of $f(x) = e^{-x^2}$ is sketched in Figure 15. ■

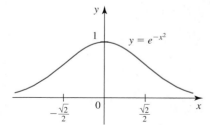

**FIGURE 15**
The graph of $y = e^{-x^2}$

## Slant Asymptotes

The graph of a function $f$ may have an asymptote that is neither vertical nor horizontal but slanted. We call the line with equation $y = mx + b$ a **slant** or **oblique (right)** **asymptote** of the graph of $f$ if

$$\lim_{x \to \infty} \frac{f(x)}{x} = m \quad \text{and} \quad \lim_{x \to \infty} [f(x) - mx] = b \qquad (1)$$

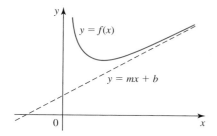

**FIGURE 16**
The graph of $f$ has a slant asymptote.

Observe that the second equation in (1) is equivalent to the statement $\lim_{x\to\infty}[f(x) - mx - b] = 0$. Since $|f(x) - mx - b|$ measures the vertical distance between the graph of $f(x)$ and the line $y = mx + b$, the second equation in (1) simply states that the graph of $f$ approaches the line with equation $y = mx + b$ as $x$ approaches infinity. (See Figure 16.)

Similarly, if

$$\lim_{x\to -\infty}\frac{f(x)}{x} = m \qquad \text{and} \qquad \lim_{x\to -\infty}[f(x) - mx] = b \tag{2}$$

then the line $y = mx + b$ is called a **slant (left) asymptote** of the graph of $f$. Note that a horizontal asymptote of the graph of $f$ may be considered a special case of a slant asymptote where $m = 0$.

Before looking at the next example, we point out that the graph of a rational function has a slant asymptote if the degree of its numerator exceeds the degree of its denominator by 1 or more. In fact, if the degree of the numerator exceeds the degree of the denominator by 1, the slant asymptote is a straight line, as the next example shows; if it exceeds the denominator by 2, then the slant asymptote is parabolic, and so forth.

**EXAMPLE 5** Find the slant asymptotes of the graph of $f(x) = \dfrac{2x^2 - 3}{x - 2}$.

**Solution** We compute

$$\lim_{x\to\infty}\frac{f(x)}{x} = \lim_{x\to\infty}\frac{\dfrac{2x^2 - 3}{x - 2}}{x} = \lim_{x\to\infty}\frac{2x - \dfrac{3}{x}}{x - 2}$$

$$= \lim_{x\to\infty}\frac{2 - \dfrac{3}{x^2}}{1 - \dfrac{2}{x}} \qquad \text{Divide the numerator and the denominator by } x.$$

$$= 2$$

Next, taking $m = 2$, we compute

$$\lim_{x\to\infty}[f(x) - mx] = \lim_{x\to\infty}\left(\frac{2x^2 - 3}{x - 2} - 2x\right)$$

$$= \lim_{x\to\infty}\frac{2x^2 - 3 - 2x^2 + 4x}{x - 2}$$

$$= \lim_{x\to\infty}\frac{4x - 3}{x - 2} = \lim_{x\to\infty}\frac{4 - \dfrac{3}{x}}{1 - \dfrac{2}{x}} = 4$$

So, taking $b = 4$, we see that the line with equation $y = 2x + 4$ is a slant asymptote of the graph of $f$. You can show that the computations using the equations in (2) lead to the same conclusion (see Exercise 43), so $y = 2x + 4$ is the only slant asymptote of the graph of $f$. The graph of $f$ is sketched in Figure 17.

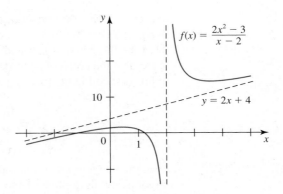

**FIGURE 17**

$y = 2x + 4$ is a slant asymptote of the graph of $f$.

## ◼ Finding Relative Extrema Using a Graphing Utility

Although we found the relative extrema of the functions in the previous examples analytically, these relative extrema can also be found with the aid of a graphing utility. For instance, the relative extrema of the graphs of the functions in Examples 3 and 5 are easily identified (see Figures 18 and 19).

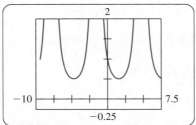

**FIGURE 18**
The graph of the function $f(x) = \dfrac{1}{1 + \sin x}$

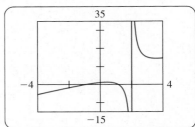

**FIGURE 19**
The graph of the function $f(x) = \dfrac{2x^2 - 3}{x - 2}$

For more complicated functions, however, it could prove to be rather difficult to find their relative extrema by using only a graphing utility. Consider, for example, the function

$$f(x) = \frac{(x - 2)(x - 3)}{(x + 2)(x + 3)}$$

The graph of $f$ in the viewing window $[-10, 10] \times [-10, 10]$ is shown in Figure 20.

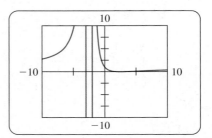

**FIGURE 20**

A cursory examination of the graph seems to indicate that $f$ has no relative extrema, at least for $x$ in the interval $(-10, 10)$.

Let's look at the problem analytically. We compute

$$f'(x) = \frac{d}{dx}\left[\frac{(x-2)(x-3)}{(x+2)(x+3)}\right] = \frac{d}{dx}\left(\frac{x^2 - 5x + 6}{x^2 + 5x + 6}\right)$$

$$= \frac{(x^2 + 5x + 6)(2x - 5) - (x^2 - 5x + 6)(2x + 5)}{(x+2)^2(x+3)^2} = \frac{10(x^2 - 6)}{(x+2)^2(x+3)^2}$$

$$= \frac{10(x - \sqrt{6})(x + \sqrt{6})}{(x+2)^2(x+3)^2}$$

We see that $f$ has two critical numbers, $-\sqrt{6}$ and $\sqrt{6}$. Note that $f'$ is discontinuous at $-2$ and $-3$, but because these numbers are not in the domain of $f$, they do not qualify as critical numbers. From the sign diagram of $f'$ (Figure 21) we see that $f$ has a relative maximum at $-\sqrt{6}$ with value $f(-\sqrt{6}) \approx -97.99$ and a relative minimum at $\sqrt{6}$ with value $f(\sqrt{6}) \approx -0.01$. These calculations tell us that we need to adjust the viewing window to see the relative maximum of $f$. Figure 22a shows the graph of $f$ using the viewing window $[-5, 5] \times [-300, 100]$. A close-up of the relative maximum is shown in Figure 22b, where the viewing window $[-3, -2] \times [-300, 100]$ is used. The point $(-\sqrt{6}, f(-\sqrt{6}))$ can be estimated by using the function for finding the maximum on your graphing utility.

$f'$ not defined here

**FIGURE 21**
The sign diagram for $f'$

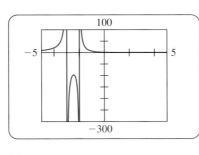

$-300$

(a)

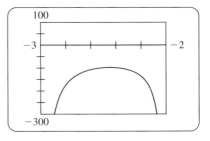

$-300$

(b)

**FIGURE 22**
The relative maximum at $-\sqrt{6}$ an be seen in part (a). The same relative maximum is shown in close-up view in part (b).

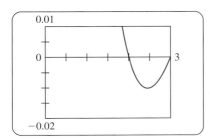

**FIGURE 23**
The relative minimum at $\sqrt{6}$ can be seen using the viewing window $[0, 3] \times [-0.02, 0.01]$.

Our calculations also indicate that there is a relative minimum at $\sqrt{6}$ with value $f(\sqrt{6}) \approx -0.01$. This relative minimum point $(\sqrt{6}, f(\sqrt{6}))$ shows up when we use the viewing window $[0, 3] \times [-0.02, 0.01]$. (See Figure 23.) You can use the graphing utility to approximate the point $(\sqrt{6}, f(\sqrt{6}))$.

This example shows that a combination of analytical and graphical techniques sometimes forms a powerful team when it comes to solving calculus problems.

## 3.6  CONCEPT QUESTIONS

1. Give the guidelines for sketching a curve.
2. Let $f(x) = x^2 + 1/x^2$.
   a. Show that if $|x|$ is very large, then $f(x)$ behaves like $g(x) = x^2$.

b. Show that $\lim_{x \to 0} f(x) = \infty$.

c. Use the guidelines for curve sketching and the results of parts (a) and (b) to sketch the graph of $f$.

## 3.6  EXERCISES

*In Exercises 1–4, use the information summarized in the table to sketch the graph of f.*

**1.** $f(x) = x^3 - 3x^2 + 1$

| Domain | $(-\infty, \infty)$ |
|---|---|
| Intercepts | $y$-intercept: 1 |
| Symmetry | None |
| Asymptotes | None |
| Intervals where $f$ is ↗ or ↘ | ↗ on $(-\infty, 0)$ and on $(2, \infty)$; ↘ on $(0, 2)$ |
| Relative extrema | Rel. max. at $(0, 1)$; rel. min. at $(2, -3)$ |
| Concavity | Downward on $(-\infty, 1)$; upward on $(1, \infty)$ |
| Point of inflection | $(1, -1)$ |

**2.** $f(x) = \dfrac{1}{9}(x^4 - 4x^3)$

| Domain | $(-\infty, \infty)$ |
|---|---|
| Intercepts | $x$-intercepts: 0, 4 <br> $y$-intercept: 0 |
| Symmetry | None |
| Asymptotes | None |
| Intervals where $f$ is ↗ or ↘ | ↗ on $(3, \infty)$; ↘ on $(-\infty, 3)$ |
| Relative extrema | Rel. min. at $(3, -3)$ |
| Concavity | Downward on $(0, 2)$; upward on $(-\infty, 0)$ and on $(2, \infty)$ |
| Points of inflection | $(0, 0)$ and $\left(2, -\frac{16}{9}\right)$ |

**3.** $f(x) = \dfrac{4x - 4}{x^2}$

| Domain | $(-\infty, 0) \cup (0, \infty)$ |
|---|---|
| Intercepts | $x$-intercept: 1 |
| Symmetry | None |
| Asymptotes | $x$-axis; $y$-axis |
| Intervals where $f$ is ↗ or ↘ | ↗ on $(0, 2)$; ↘ on $(-\infty, 0)$ and on $(2, \infty)$ |
| Relative extrema | Rel. max. at $(2, 1)$ |
| Concavity | Downward on $(-\infty, 0)$ and on $(0, 3)$; upward on $(3, \infty)$ |
| Point of inflection | $\left(3, \frac{8}{9}\right)$ |

**4.** $f(x) = x - 3x^{1/3}$

| Domain | $(-\infty, \infty)$ |
|---|---|
| Intercepts | $x$-intercepts: $\pm 3\sqrt{3}$, 0; <br> $y$-intercept: 0 |
| Symmetry | With respect to the origin |
| Asymptotes | None |
| Intervals where $f$ is ↗ or ↘ | ↗ on $(-\infty, -1)$ and on $(1, \infty)$; ↘ on $(-1, 1)$ |
| Relative extrema | Rel. max. at $(-1, 2)$; rel. min. at $(1, -2)$ |
| Concavity | Downward on $(-\infty, 0)$ upward on $(0, \infty)$ |
| Point of inflection | $(0, 0)$ |

*In Exercises 5–38, sketch the graph of the function using the curve-sketching guidelines on page 348.*

**5.** $f(x) = 4 - 3x - 2x^3$

**6.** $f(x) = x^3 - 3x^2 + 2$

**7.** $f(x) = x^3 - 6x^2 + 9x + 2$

**8.** $y = 2t^3 - 15t^2 + 36t - 20$

**9.** $f(x) = 2x^3 - 9x^2 + 12x - 3$

**10.** $f(t) = 3t^4 + 4t^3$

**11.** $g(x) = x^4 + 2x^3 - 2$

**12.** $f(x) = (x - 2)^4 + 1$

**13.** $f(x) = 4x^5 - 5x^4$

**14.** $g(x) = \dfrac{1}{2}x - \sqrt{x}$

**15.** $y = (x + 2)^{3/2} + 1$

**16.** $f(t) = \sqrt{t^2 - 4}$

**17.** $f(x) = \dfrac{x}{x + 1}$

**18.** $g(x) = \dfrac{x + 1}{x - 1}$

**19.** $h(x) = \dfrac{x}{x^2 - 9}$

**20.** $f(t) = \dfrac{t}{t^2 + 1}$

**21.** $f(x) = \dfrac{x^2}{x^2 + 1}$

**22.** $g(x) = \dfrac{x^2 - 9}{x^2 - 4}$

**23.** $h(x) = \dfrac{1}{x^2 - x - 2}$

**24.** $f(x) = x\sqrt{9 - x^2}$

**25.** $f(x) = x - \sin x, \quad 0 \le x \le 2\pi$

**26.** $g(x) = 2 \sin x + \sin 2x, \quad 0 \le x \le 2\pi$

**27.** $f(x) = \dfrac{1}{1 - \cos x}, \quad -2\pi < x < 2\pi$

**28.** $y = \cos^2 x, \quad -\pi \le x \le \pi$

**29.** $g(x) = \dfrac{\sin x}{1 + \sin x}$

**30.** $f(x) = 2x - \tan x, \quad -\frac{\pi}{2} < x < \frac{\pi}{2}$

**31.** $f(x) = xe^{-x}$

**32.** $f(x) = xe^x$

**33.** $f(x) = \dfrac{e^x - e^{-x}}{2}$

**34.** $f(x) = e^x - x$

**35.** $f(x) = x + \ln x$

**36.** $f(x) = x \ln x - x$

**37.** $f(x) = \ln(x^2 + 1)$

**38.** $f(x) = \ln(\cos x), \quad -\frac{\pi}{2} < x < \frac{\pi}{2}$

*In Exercises 39–42, find the slant asymptotes of the graphs of the function. Then sketch the graph of the function.*

**39.** $g(u) = \dfrac{u^3 + 1}{u^2 - 1}$

**40.** $h(x) = \dfrac{x^3 + 1}{x(x + 1)}$

**41.** $f(x) = \dfrac{x^2 - 2x - 3}{2x - 2}$

**42.** $f(x) = e^{-x} + x$

**43.** Refer to Example 5. Show that

$$\lim_{x \to -\infty} \frac{f(x)}{x} = 2 \quad \text{and} \quad \lim_{x \to -\infty} [f(x) - 2x] = 4$$

so $y = 2x + 4$ is a (left) slant asymptote of the graph of $f(x) = \dfrac{2x^2 - 3}{x - 2}$.

 **44.** Find the (right) slant asymptote and the (left) slant asymptote of the graph of the function $f(x) = \sqrt{1 + x^2} + 2x$. Plot the graph of $f$ together with the slant asymptotes.

**45. Worker Efficiency** An efficiency study showed that the total number of cell phones assembled by the average worker at Alpha Communications $t$ hours after starting work at 8 A.M. is given by

$$N(t) = -\frac{1}{2}t^3 + 3t^2 + 10t \qquad 0 \le t \le 4$$

Sketch the graph of the function $N$, and interpret your result.

**46. Crime Rate** The number of major crimes per 100,000 people committed in a city from the beginning of 2002 to the beginning of 2009 is approximated by the function

$$N(t) = -0.1t^3 + 1.5t^2 + 80 \qquad 0 \le t \le 7$$

where $N(t)$ denotes the number of crimes per 100,000 people committed in year $t$ and $t = 0$ corresponds to the beginning of 2002. Enraged by the dramatic increase in the crime rate, the citizens, with the help of the local police, organized Neighborhood Crime Watch groups in early 2007 to combat this menace. Sketch the graph of the function $N$, and interpret your results. Is the Neighborhood Crime Watch program working?

 **47. Air Pollution** The level of ozone, an invisible gas that irritates and impairs breathing, that is present in the atmosphere on a certain day in June in the city of Riverside is approximated by

$$S(t) = 1.0974t^3 - 0.0915t^4 \qquad 0 \le t \le 11$$

where $S(t)$ is measured in Pollutant Standard Index (PSI) and $t$ is measured in hours with $t = 0$ corresponding to 7 A.M. Plot the graph of $S$, and interpret your results.
*Source: The Los Angeles Times.*

**48. Production Costs** The total daily cost in dollars incurred by the TKK Corporation in manufacturing $x$ multipacks of DVDs is given by the function

$$f(x) = 0.000001x^3 - 0.003x^2 + 5x + 500$$

$$0 \le x \le 3000$$

Plot the graph of $f$, and interpret your results.

**49. A Mixture Problem** A tank initially contains 10 gal of brine with 2 lb of salt. Brine with 1.5 lb of salt per gallon enters the tank at the rate of 3 gal/min, and the well-stirred mixture leaves the tank at the rate of 4 gal/min. It can be shown that the amount of salt in the tank after $t$ min is $x$ lb, where

$$x = f(t) = 1.5(10 - t) - 0.0013(10 - t)^4 \qquad 0 \leq t \leq 10$$

Plot the graph of $f$, and interpret your result.

**50. Traffic Flow Analysis** The speed of traffic flow in miles per hour on a stretch of Route 123 between 6 A.M. and 10 A.M. on a typical workday is approximated by the function

$$f(t) = 20t - 40\sqrt{t} + 52 \qquad 0 \leq t \leq 4$$

where $t$ is measured in hours and $t = 0$ corresponds to 6 A.M. Sketch the graph of $f$ and interpret your results.

**51. Einstein's Theory of Special Relativity** The mass of a particle moving at a velocity $v$ is related to its rest mass $m_0$ by the equation

$$m = f(v) = \frac{m_0}{\sqrt{1 - \dfrac{v^2}{c^2}}}$$

where $c$ is the speed of light. Sketch the graph of the function $f$, and interpret your results.

**52. Absorption of Drugs** A liquid carries a drug into an organ of volume $V$ cm$^3$ at the rate of $a$ cm$^3$/sec and leaves at the same rate. The concentration of the drug in the entering liquid is $c$ g/cm$^3$. Letting $x(t)$ denote the concentration of the drug in the organ at any time $t$, we have $x(t) = c(1 - e^{-at/V})$, where $a$ is a positive constant that depends on the organ.
**a.** Show that $x$ is an increasing function on $(0, \infty)$.
**b.** Sketch the graph of $x$.

**53. Harbor Water Level** The water level (in feet) at Boston Harbor during a certain 24-hour period is approximated by the function

$$H = f(t) = 4.8 \sin\left(\frac{\pi}{6}(t - 10)\right) + 7.6 \qquad 0 \leq t \leq 24$$

where $t = 0$ corresponds to 12 A.M. Plot the graph of $f$, and interpret your results.

**54. Chemical Mixtures** Two chemicals react to form another chemical. Suppose that the amount of the chemical formed in time $t$ (in hours) is given by

$$x(t) = \frac{15\left[1 - \left(\frac{2}{3}\right)^{3t}\right]}{1 - \frac{1}{4}\left(\frac{2}{3}\right)^{3t}}$$

where $x(t)$ is measured in pounds.

**a.** Plot the graph of $x$ using the viewing window $[0, 10] \times [0, 16]$.
**b.** Find the rate at which the chemical is formed when $t = 1$.
**c.** How many pounds of the chemical are formed eventually?

*In Exercises 55–58, plot the graph of the function.*

**55.** $f(t) = \dfrac{\sqrt{t^2 + 1}}{t - 1}$

**56.** $f(x) = \dfrac{x^2 + x}{3x^2 + x - 1}$

**57.** $g(t) = t^2 + 3 \sin 2t, \quad -2\pi < t < 2\pi$

**58.** $h(x) = 2 \sin x + 3 \cos 2x + \sin 3x, \quad -2\pi < x < 2\pi$

**59. Snowfall Accumulation** The snowfall accumulation at Logan Airport $t$ hr after a 33-hr snowstorm in Boston in 1995 is given in the following table.

| Hour | 0 | 3 | 6 | 9 | 12 | 15 |
|---|---|---|---|---|---|---|
| Inches | 0.1 | 0.4 | 3.6 | 6.5 | 9.1 | 14.4 |

| Hour | 18 | 21 | 24 | 27 | 30 | 33 |
|---|---|---|---|---|---|---|
| Inches | 19.5 | 22 | 23.6 | 24.8 | 26.6 | 27 |

By using the logistic curve-fitting capability of a graphing calculator, it can be verified that a regression model for this data is given by

$$f(t) = \frac{26.71}{1 + 31.74e^{-0.24t}}$$

where $t$ is measured in hours, $t = 0$ corresponds to noon of February 6, and $f(t)$ is measured in inches.
**a.** Plot the scatter diagram and the graph of the function $f$ using the viewing window $[0, 36] \times [0, 30]$.
**b.** How fast was the snowfall accumulating at midnight on February 6? At noon on February 7?
**c.** At what time during the storm was the snowfall accumulating at the greatest rate? What was the rate of accumulation?

*Source: The Boston Globe.*

**60. Worldwide PC Shipments** The number of worldwide PC shipments (in millions of units) from 2005 through 2009, according to data from the International Data Corporation, are given in the following table.

| Year | 2005 | 2006 | 2007 | 2008 | 2009 |
|---|---|---|---|---|---|
| PCs | 207.1 | 226.2 | 252.9 | 283.3 | 302.4 |

By using the logistic curve-fitting capability of a graphing calculator, it can be verified that a regression model for this data is given by

$$f(t) = \frac{544.65}{1 + 1.65e^{-0.1846t}}$$

where $t$ is measured in years and $t = 0$ corresponds to 2005.
**a.** Plot the scatter diagram and the graph of the function $f$ using the viewing window $[0, 4] \times [200, 300]$.
**b.** How fast were the worldwide PC shipments increasing in 2006? In 2008?
*Source:* International Data Corporation.

 **61. Flight Path of a Plane** The function

$$f(x) = \begin{cases} 0 & \text{if } 0 \le x < 1 \\ -0.0411523x^3 + 0.679012x^2 \\ \quad - 1.23457x + 0.596708 & \text{if } 1 \le x < 10 \\ 15 & \text{if } 10 \le x \le 11 \end{cases}$$

where both $x$ and $f(x)$ are measured in units of 1000 ft, describes the flight path of a plane taking off from the origin and climbing to an altitude of 15,000 ft. Plot the graph of $f$ to visualize the trajectory of the plane.

**62.** Let

$$f(x) = \frac{x^{2n} - 1}{x^{2n} + 1}$$

 **a.** Plot the graphs of $f$ for $n = 1, 5, 10, 100,$ and 1000. Do these graphs approach a "limiting" graph as $n$ approaches infinity?
**b.** Can you prove this result analytically?

## 3.7 Optimization Problems

We first encountered optimization problems in Section 3.1. There, we solved certain problems by finding the absolute maximum value or the absolute minimum value of a continuous function on a *closed, bounded interval*. Thanks to the Extreme Value Theorem, we saw that these problems always have a solution.

In practice, however, there are optimization problems that are solved by finding the absolute extremum value of a continuous function on an *arbitrary interval*. If the interval is not closed, there is no guarantee that the function to be optimized has an absolute maximum value or an absolute minimum value on that interval (see Example 1 in Section 3.1). Thus, for these problems, a solution might not exist. But if the function to be maximized (minimized) has exactly one relative maximum (relative minimum) inside that interval, then there is a solution to the problem. In fact, as Figure 1 suggests, the relative extremum value at a critical number turns out to be the absolute extremum value of the function on the interval. Thus, the solutions to such problems are found by finding the relative extreme values of the function in that interval.

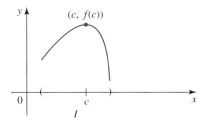

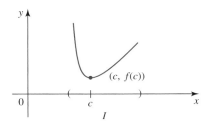

**FIGURE 1**
$f$ has only one critical number on an interval $I$.

**(a)** The relative maximum value $f(c)$ is the absolute maximum value.

**(b)** The relative minimum value $f(c)$ is the absolute minimum value.

Before proceeding further, let us summarize this important observation.

**Guidelines for Finding the Absolute Extrema of a Continuous Function $f$ on an Arbitrary Interval**

Suppose that a continuous function $f$ has only one critical number $c$ in an interval $I$.

1. Use the First Derivative Test or the Second Derivative Test to ascertain whether $f$ has a relative maximum (minimum) value at $c$.
2. **a.** If $f$ has a relative maximum value at $c$, then the number $f(c)$ is also the absolute maximum value of $f$ on $I$.
   **b.** If $f$ has a relative minimum value at $c$, then the number $f(c)$ is also the absolute minimum value of $f$ on $I$.

Armed with these guidelines and the guidelines for finding the absolute extrema of functions on closed intervals, we are ready to tackle a large class of optimization problems.

## ▪ Formulating Optimization Problems

If you reexamine the optimization problems in Section 3.1, you will see that the functions to be optimized were given to you. More often than not, we first need to find an appropriate function and then optimize it. The following guidelines can be used to formulate these optimization problems.

**Guidelines for Solving Optimization Problems**

1. Assign a letter to each variable. Draw and label a figure (if appropriate).
2. Find an expression for the quantity to be maximized or minimized.
3. Use the conditions given in the problem to express the quantity to be optimized as a function $f$ of one variable. Note any restrictions to be placed on the domain of $f$.
4. Optimize the function $f$ over its domain using the guidelines of Section 3.1 and the guidelines on this page.

**FIGURE 2**
The area of the rectangle is $A = xy$.

**EXAMPLE 1**  **A Fencing Problem**  A man has 100 ft of fencing to enclose a rectangular garden in his backyard. Find the dimensions of the garden of largest area he can have if he uses all of the fencing.

**Solution**

Step 1  Let $x$ and $y$ denote the length and width of the garden (in feet) and let $A$ denote its area (see Figure 2).

Step 2  The area of the rectangle is

$$A = xy \tag{1}$$

and is the quantity to be maximized.

Step 3  The perimeter of the rectangle is $(2x + 2y)$ ft, and this must be equal to 100 ft. Therefore, we have the equation

$$2x + 2y = 100 \tag{2}$$

relating the variables $x$ and $y$. Solving Equation (2) for $y$ in terms of $x$, we have

$$y = 50 - x \qquad (3)$$

which, when substituted into Equation (1), yields

$$A = x(50 - x)$$
$$= -x^2 + 50x$$

(Remember, the function to be optimized must involve just one variable.) Because the sides of the rectangle must be positive, $x > 0$ and $y = 50 - x > 0$, giving us the inequality $0 < x < 50$. Thus, the problem is reduced to that of finding the value of $x$ in $(0, 50)$ at which $f(x) = -x^2 + 50x$ attains the largest value.

**Step 4** To find the critical number(s) of $f$, we compute

$$f'(x) = -2x + 50 = -2(x - 25)$$

Setting $f'(x) = 0$, yields 25 as the only critical number of $f$. Since $f''(x) = -2 < 0$, we see, by the Second Derivative Test, that $f$ has a relative maximum at $x = 25$. But 25 is the only critical number in $(0, 50)$, so we conclude that $f$ attains its largest value of $f(25) = 625$ at $x = 25$. From Equation (3) the corresponding value of $y$ is 25. Thus, the man would have a garden of maximum area ($625 \text{ ft}^2$) if it were in the form of a square with sides of length 25 ft. ■

**EXAMPLE 2** **Finding the Maximum Area** Find the dimensions of the rectangle of greatest area that has its base on the $x$-axis and is inscribed in the parabola $y = 9 - x^2$.

**Solution**

**Step 1** Consider the rectangle of width $2x$ and height $y$ as shown in Figure 3. Let $A$ denote its area.

**Step 2** The area of the rectangle is $A = 2xy$ and is the quantity to be maximized.

**Step 3** Because the point $(x, y)$ lies on the parabola, it must satisfy the equation of the parabola; that is, $y = 9 - x^2$. Therefore,

$$A = 2xy$$
$$= 2x(9 - x^2)$$
$$= -2x^3 + 18x$$

Furthermore, $y > 0$ implies that $9 - x^2 > 0$ or, equivalently, $-3 < x < 3$. Also, $x > 0$, since the side of a rectangle must be positive. Therefore, the problem is equivalent to the problem of finding the value of $x$ in $(0, 3)$ for which $f(x) = -2x^3 + 18x$ attains the largest value.

**Step 4** To find the critical numbers of $f$, we compute

$$f'(x) = -6x^2 + 18 = -6(x^2 - 3)$$

Setting $f'(x) = 0$ yields $x = \pm\sqrt{3}$. We consider only the critical number $\sqrt{3}$, since $-\sqrt{3}$ lies outside the interval $(0, 3)$. Since $f''(x) = -12x$ and $f''(\sqrt{3}) = -12\sqrt{3} < 0$, we see, by the Second Derivative Test, that $f$ has a relative maximum at $x = \sqrt{3}$. Since $f$ has only one critical number in $(0, 3)$, we see that $f$ attains its largest value at $x = \sqrt{3}$. Substituting this value of $x$ into $y = 9 - x^2$ gives $y = 6$. Thus, the dimensions of the desired rectangle are $2\sqrt{3}$ by 6 and its area is $12\sqrt{3}$. ■

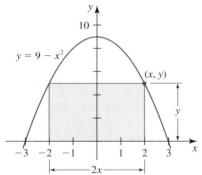

**FIGURE 3**
The area of the rectangle is
$2xy = 2x(9 - x^2)$.

**EXAMPLE 3** **Minimizing the Cost of Laying Cable** In Figure 4, the point $S$ gives the location of a power relay station on a straight coast, and the point $E$ gives the location of a marine biology experimental station on an island. The point $Q$ is located 7 mi west of the point $S$, and the point $Q$ is 3 mi south of the point $E$. A cable is to be laid connecting the relay station with the experimental station. If the cost of running the cable along the shoreline is \$10,000/mi and the cost of running the cable under water is \$30,000/mi, where should the point $P$ be located to minimize the cost of laying the cable?

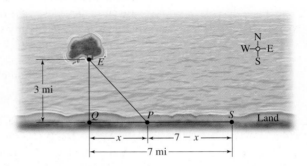

**FIGURE 4**
The cable connects the marine biology station at $E$ to the power relay station at $S$. The cable from $E$ to $P$ will be laid under water, and the cable from $P$ to $S$ will be laid over land.

**Solution**

**Step 1**   It is clear that the point $P$ should lie between $Q$ and $S$, inclusive. Let $x$ denote the distance between $P$ and $Q$ (in miles), and let $C$ denote the cost of laying the cable (in thousands of dollars).

**Step 2**   The length of the cable to be laid under water is given by the distance between $E$ and $P$. Using the Pythagorean Theorem, we find that this length is $\sqrt{x^2 + 9}$ mi. So the cost of laying the cable under water is $30\sqrt{x^2 + 9}$ thousand dollars. Next, we see that the length of cable to be laid over land is $(7 - x)$ mi. So the cost of laying this stretch of the cable is $10(7 - x)$ thousand dollars. Therefore, the total cost incurred in laying the cable is

$$C = 30\sqrt{x^2 + 9} + 10(7 - x)$$

thousand dollars, and this is the quantity to be minimized.

**Step 3**   Because the distance between $Q$ and $S$ is 7 mi, we see that $x$ must satisfy the constraint $0 \le x \le 7$. So the problem is that of finding the value of $x$ in $[0, 7]$ at which $f(x) = 30\sqrt{x^2 + 9} + 10(7 - x)$ attains the smallest value.

**Step 4**   Observe that $f$ is continuous on the closed interval $[0, 7]$. So the absolute minimum value of $f$ must be attained at an endpoint of $[0, 7]$ or at a critical number of $f$ in the interval. To find the critical numbers of $f$, we compute

$$f'(x) = \frac{d}{dx}[30(x^2 + 9)^{1/2} + 10(7 - x)]$$

$$= (30)\left(\frac{1}{2}\right)(x^2 + 9)^{-1/2}(2x) - 10$$

$$= 10\left[\frac{3x}{\sqrt{x^2 + 9}} - 1\right]$$

Setting $f'(x) = 0$ gives

$$\frac{3x}{\sqrt{x^2 + 9}} - 1 = 0$$

$$3x = \sqrt{x^2 + 9}$$

$$9x^2 = x^2 + 9$$

$$8x^2 = 9$$

or

$$x = \pm\frac{3}{2\sqrt{2}} = \pm\frac{3\sqrt{2}}{4} \approx \pm 1.06$$

We reject the root $-3\sqrt{2}/4$ because it lies outside the interval $[0, 7]$. We are left with $x = 3\sqrt{2}/4$ as the only critical number of $f$. Finally, from the following table we see that $f(x)$ attains its smallest value of 154.85 at $x = 3\sqrt{2}/4 \approx 1.06$. We conclude that the cost of laying the cable will be minimized (approximately \$155,000) if the point $P$ is located at a distance of approximately 1.06 miles from $Q$.

| $f(0)$ | $f(3\sqrt{2}/4)$ | $f(7)$ |
|--------|------------------|--------|
| 160 | 154.85 | 228.47 |

**EXAMPLE 4** **Packaging** The Betty Moore Company requires that its beef stew containers have a capacity of 64 in.$^3$, have the shape of right circular cylinders, and be made of aluminum. Determine the radius and height of the container that requires the least amount of metal.

**FIGURE 5**
We want to minimize the amount of material used to construct the container.

**Solution**

**Step 1**  Let $r$ and $h$ denote the radius and height, respectively, of a container (Figure 5). The amount of aluminum required to construct a container is given by the total surface area of the cylinder, which we denote by $S$.

**Step 2**  The area of the base or top of the cylinder is $\pi r^2$ in.$^2$, and the area of its lateral surface is $2\pi rh$ in.$^2$. Therefore,

$$S = 2\pi r^2 + 2\pi rh \tag{4}$$

and this is the quantity to be minimized.

**Step 3**  The requirement that the volume of the container be 64 in.$^3$ translates into the equation

$$\pi r^2 h = 64 \tag{5}$$

Solving Equation (5) for $h$ in terms of $r$, we obtain

$$h = \frac{64}{\pi r^2} \tag{6}$$

which, when substituted into Equation (4), yields

$$S = 2\pi r^2 + 2\pi r\left(\frac{64}{\pi r^2}\right)$$

$$= 2\pi r^2 + \frac{128}{r}$$

The domain of $S$ is $(0, \infty)$. The problem has been reduced to one of finding the value of $r$ in $(0, \infty)$ at which $f(r) = 2\pi r^2 + (128/r)$ attains the smallest value.

**Step 4**   Observe that $f$ is continuous on $(0, \infty)$. Following the guidelines given at the beginning of this section, we first find the critical number of $f$,

$$f'(r) = 4\pi r - \frac{128}{r^2}$$

Setting $f'(r) = 0$ gives

$$4\pi r - \frac{128}{r^2} = 0$$

$$4\pi r^3 - 128 = 0$$

$$r^3 = \frac{32}{\pi}$$

or

$$r = \left(\frac{32}{\pi}\right)^{1/3} \approx 2.17$$

as the only critical number of $f$.

To see whether this critical number gives rise to a relative extremum of $f$, we use the Second Derivative Test. Now

$$f''(r) = 4\pi + \frac{256}{r^3}$$

so

$$f''\left(\left(\frac{32}{\pi}\right)^{1/3}\right) = 4\pi + \frac{256}{\dfrac{32}{\pi}} = 12\pi > 0$$

Therefore, $f$ has a relative minimum value at $r = (32/\pi)^{1/3}$. Finally, because $f$ has only one critical number in $(0, \infty)$, we conclude that $f$ attains the absolute minimum value at this number. Using Equation (6), we find that the corresponding value of $h$ is

$$h = \frac{64}{\pi\left(\dfrac{32}{\pi}\right)^{2/3}} = \frac{64}{\pi\left(\dfrac{32}{\pi}\right)^{2/3}} \cdot \frac{\left(\dfrac{32}{\pi}\right)^{1/3}}{\left(\dfrac{32}{\pi}\right)^{1/3}}$$

$$= \frac{64}{\pi\left(\dfrac{32}{\pi}\right)} \cdot \left(\frac{32}{\pi}\right)^{1/3}$$

$$= 2r$$

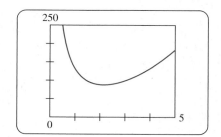

**FIGURE 6**

The graph of $S = 2\pi r^2 + \dfrac{128}{r}$

Thus, the required container has a radius of approximately 2.17 in. and a height twice the size of its radius, or approximately 4.34 in. The graph of $S$ is shown in Figure 6. ■

**EXAMPLE 5** **Finding the Minimum Distance** Figure 7 shows an aerial view of a race-track composed of two sides of a rectangle and two semicircles. It also shows the position $P$ of a spectator watching a race from the roof of his car. Find the point $Q$ on the track that is closest to the spectator. What is the distance between these two points?

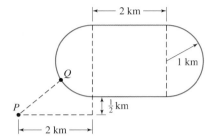

**FIGURE 7**
The diagram shows the position of a spectator, $P$, in relation to a racetrack.

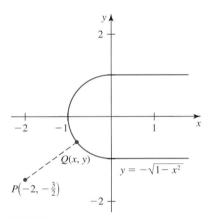

**FIGURE 8**
We want to minimize the distance between $P$ and $Q$.

**Solution**

**Step 1** Clearly, the required point must lie on the lower left semicircular stretch of the racetrack. Let us set up a rectangular coordinate system as shown in Figure 8. To find an equation describing this curve, begin with the equation $x^2 + y^2 = 1$ of the circle with center at the origin and radius 1. Solving for $y$ in terms of $x$ and observing that both $x$ and $y$ must be nonpositive, we are led to the following representation of the curve:

$$y = -\sqrt{1 - x^2} \qquad -1 \le x \le 0 \qquad (7)$$

Next, let $D$ denote the distance between $P\left(-2, -\frac{3}{2}\right)$ and a point $Q(x, y)$ lying on the curve described by Equation (7).

**Step 2** Using the distance formula, we see that the distance $D$ between $P$ and $Q$ is given by

$$D = \sqrt{(x + 2)^2 + \left(y + \frac{3}{2}\right)^2}$$

Thus,

$$D^2 = (x + 2)^2 + \left(y + \frac{3}{2}\right)^2$$

$$= x^2 + 4x + 4 + y^2 + 3y + \frac{9}{4} \qquad (8)$$

Since $D$ is minimal if and only if $D^2$ is minimal, we will minimize $D^2$ instead of $D$.

**Step 3** Substituting Equation (7) into Equation (8), we obtain

$$D^2 = x^2 + 4x + 4 + (1 - x^2) - 3\sqrt{1 - x^2} + \frac{9}{4}$$

$$= 4x - 3\sqrt{1 - x^2} + \frac{29}{4}$$

So the problem is reduced to that of finding the value of $x$ in $[-1, 0]$ at which $f(x) = 4x - 3\sqrt{1 - x^2} + (29/4)$ attains the smallest value.

**Step 4** Observe that $f$ is continuous on $[-1, 0]$. So the absolute minimum value of $f$ must be attained at an endpoint of $[-1, 0]$ or at a critical number of $f$ in that interval. To find the critical numbers of $f$, we compute

$$f'(x) = \frac{d}{dx}\left[4x - 3(1 - x^2)^{1/2} + \frac{29}{4}\right]$$

$$= 4 - 3\left(\frac{1}{2}\right)(1 - x^2)^{-1/2}(-2x) = 4 + \frac{3x}{\sqrt{1 - x^2}}$$

Setting $f'(x) = 0$ and solving for $x$, we obtain

$$4 + \frac{3x}{\sqrt{1 - x^2}} = 0$$

$$3x = -4\sqrt{1 - x^2}$$

$$9x^2 = 16(1 - x^2)$$

$$25x^2 = 16$$

or $x = \pm\frac{4}{5}$. Only $-\frac{4}{5}$ is a solution of $f'(x) = 0$; so it is the only critical number of interest. Finally, from the following table

| $f(-1)$ | $f\left(-\frac{4}{5}\right)$ | $f(0)$ |
|---|---|---|
| $\frac{13}{4} = 3.25$ | $\frac{9}{4} = 2.25$ | $\frac{17}{4} = 4.25$ |

we see that $f$ attains its smallest value of 2.25 at $x = -\frac{4}{5}$. Using Equation (7), we find that the corresponding value of $y$ is

$$y = -\sqrt{1 - \left(-\frac{4}{5}\right)^2} = -\frac{3}{5}$$

We conclude that the point $\left(-\frac{4}{5}, -\frac{3}{5}\right)$ is the point on the track closest to the spectator. The distance between the spectator and the point is

$$\sqrt{f\left(-\frac{4}{5}\right)} = \sqrt{4\left(-\frac{4}{5}\right) - 3\sqrt{1 - \left(-\frac{4}{5}\right)^2} + \frac{29}{4}} = \sqrt{\frac{9}{4}} = \frac{3}{2}$$

or 1.5 km. ■

**EXAMPLE 6** **Minimizing Length** Figure 9a depicts a cross section of a high-rise building. A ladder from a fire engine to the front wall of the building must clear the canopy, which extends 12 ft from the building. Find the length of the shortest ladder that will enable the firefighters to accomplish this task.

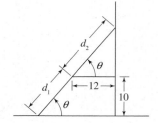

**FIGURE 9**        (a) The ladder touches the edge of the canopy.            (b) The length of the ladder is $L = d_1 + d_2$.

### Solution

**Step 1**    Let $L$ denote the length of the ladder, and let $\theta$ be the angle the ladder makes with the horizontal.

**Step 2** From Figure 9b we see that

$$L = d_1 + d_2$$
$$= 10 \csc \theta + 12 \sec \theta \qquad \csc \theta = \frac{d_1}{10} \quad \text{and} \quad \sec \theta = \frac{d_2}{12}$$

and this is the quantity to be minimized.

**Step 3** The domain of $L$ is $\left(0, \frac{\pi}{2}\right)$. So the problem is to find the value of $\theta$ in $\left(0, \frac{\pi}{2}\right)$ for which $f(\theta) = 10 \csc \theta + 12 \sec \theta$ has the smallest value.

**Step 4** Observe that $f$ is continuous on $\left(0, \frac{\pi}{2}\right)$. Following the guidelines given at the beginning of this section, we first find the critical numbers of $f$. Thus,

$$f'(\theta) = -10 \csc \theta \cot \theta + 12 \sec \theta \tan \theta$$

Setting $f'(\theta) = 0$ gives

$$12 \sec \theta \tan \theta = 10 \csc \theta \cot \theta$$

$$12\left(\frac{1}{\cos \theta}\right)\left(\frac{\sin \theta}{\cos \theta}\right) = 10\left(\frac{1}{\sin \theta}\right)\left(\frac{\cos \theta}{\sin \theta}\right)$$

$$\frac{\sin^3 \theta}{\cos^3 \theta} = \frac{10}{12}$$

$$\tan^3 \theta = \frac{5}{6}$$

or $\theta = \tan^{-1} \sqrt[3]{5/6} \approx 0.76$. The sign diagram for $f'$ shown in Figure 10 tells us that $f$ has a relative minimum value at $\tan^{-1} \sqrt[3]{5/6}$. Since $f$ has only one critical number in $\left(0, \frac{\pi}{2}\right)$, this value is also the absolute minimum value of $f$. Finally, $f(0.76) \approx 31.07$, so we conclude that the ladder must be at least 31.1 ft long. ∎

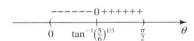

**FIGURE 10**
The sign diagram for $f'$

## 3.7 CONCEPT QUESTIONS

1. Give the procedure for finding the absolute extrema of a continuous function $f$ on (a) a closed interval and on (b) an arbitrary interval in which $f$ possesses only one critical number at which an extremum occurs.

2. Give the guidelines for solving optimization problems.

## 3.7 EXERCISES

1. Find two positive numbers whose sum is 100 and whose product is a maximum.

2. Find two numbers whose difference is 50 and whose product is a minimum.

3. The product of two positive numbers is 54. Find the numbers if the sum of the first number plus the square of the second number is as small as possible.

4. The sum of a positive number and its reciprocal is to be as small as possible. What is the number?

5. Find the dimensions of a rectangle with a perimeter of 100 m that has the largest possible area.

6. Find the dimensions of a rectangle of area 144 ft² that has the smallest possible perimeter.

7. **A Fencing Problem** A rancher has 400 ft of fencing with which to enclose two adjacent rectangular parts of a corral. What are the dimensions of the parts if the area enclosed is to be as large as possible and she uses all of the fencing available?

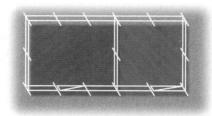

8. **A Fencing Problem** The owner of the Rancho Grande has 3000 yd of fencing with which to enclose a rectangular piece of grazing land situated along the straight portion of a river. If fencing is not required along the river, what are the dimensions of the largest area he can enclose? What is the area?

9. **Packaging** An open box is made from a rectangular piece of cardboard of dimensions 16 × 10 in. by cutting out identical squares from each corner and bending up the resulting flaps. Find the dimensions of the box with the largest volume that can be made.

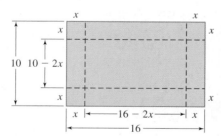

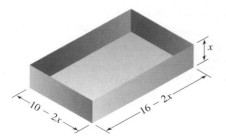

10. **Packaging** If an open box is made from a metal sheet 10 in. square by cutting out identical squares from each corner and bending up the resulting flaps, determine the dimensions of the box with the largest volume that can be made.

11. **Packaging** An open box constructed from a tin sheet has a square base and a volume of 216 in.$^3$. Find the dimensions of the box, assuming that the minimum amount of material was used in its construction.

12. **Satisfying Postal Regulations** Postal regulations specify that a parcel sent by priority mail may have a combined length and girth of no more than 108 in. Find the dimensions of a rectangular package that has a square cross section and largest volume that may be sent by priority mail. What is the volume of such a package?

**Hint:** The length plus the girth is $4x + l$.

13. **Satisfying Postal Regulations** Postal regulations specify that a package sent by priority mail may have a combined length and girth of no more than 108 in. Find the dimensions of a cylindrical package with the greatest volume that may be sent by priority mail. What is the volume of such a package?

**Hint:** The length plus the girth is $2\pi r + l$.

14. **Packaging** A container for a soft drink is in the form of a right circular cylinder. If the container is to have a capacity of 12 fluid ounces (fl oz), find the dimensions of the container that can be constructed with a minimum of material. **Hint:** 1 fl oz ≈ 1.805 in.$^3$.

15. **Designing a Loudspeaker** The rectangular enclosure for a loudspeaker system is to have an internal volume of 2.4 ft$^3$. For aesthetic reasons the height of the enclosure is to be 1.5 times its width. If the top, bottom, and sides of the enclosure are to be constructed of veneer costing 80 cents per square foot and the front and rear are to be constructed of particle board cost-

ing 40 cents per square foot, find the dimensions of the enclosure that can be constructed at a minimum cost.

16. **Book Publishing** A production editor at Weston Publishers decided that the pages of a book should have a 1-in. margin at the top and the bottom, and a $\frac{1}{2}$-in. margin on each side of the page. She further stipulated that each page of the book should have an area of 50 in.$^2$. Determine the dimensions of the page that will result in the maximum printed area on the page.

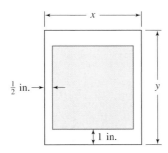

*In Exercises 17–20, find the dimensions of the shaded region so that its area is maximized.*

17.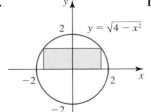

18.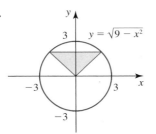

19.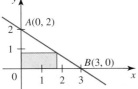

20.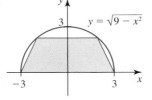

21. Find the point on the line $y = 2x + 5$ that is closest to the origin.

22. Find the points on the hyperbola $x^2/4 - y^2/9 = 1$ that are closest to the point $(0, 3)$.

 23. Find the approximate location of the points on the hyperbola $xy = 1$ that are closest to the point $(-1, 1)$.

24. Let $P$ be a point lying on the axis of the parabola $y^2 = 2px$ at a distance $a$ from its vertex. Find the $x$-coordinate(s) of the point(s) on the parabola that are closest to $P$.

25. Find the dimensions of the rectangle of maximum possible area that can be inscribed in a semicircle of radius 4.

26. Find the point on the graph of $x + y = 1$ that is closest to $(3, 2)$.

 27. Find the point on the graph of the parabola $y = 4 - x^2$ that is closest to the point $(-3, -4)$.

28. **Optimal Driving Speed** A truck gets $600/x$ miles per gallon (mpg) when driven at a constant speed of $x$ mph, where $40 \le x \le 80$. If the price of fuel is $2.80/gal and the driver is paid $12/hr, at what speed is it most economical for the trucker to drive?

29. **Maximizing Yield** An apple orchard has an average yield of 36 bushels of apples per tree if tree density is 22 trees per acre. For each unit increase in tree density, the yield decreases by 2 bushels per tree. How many trees per acre should be planted to maximize the yield?

30. **Packaging** A rectangular box is to have a square base and a volume of 20 ft$^3$. If the material for the base costs $0.30 per square foot, the material for the sides costs $0.10 per square foot, and the material for the top costs $0.20 per square foot, determine the dimensions of the box that can be constructed at minimum cost.

31. **Packaging** A rectangular box having a top and square base is to be constructed at a cost of $2. If the material for the bottom costs $0.30 per square foot, the material for the top costs $0.20 per square foot, and the material for the sides costs $0.15 per square foot, find the dimensions and volume of the box of maximum volume that can be constructed.

32. **Maximizing Revenue** If exactly 200 people sign up for a charter flight, the operators of a charter airline charge $300 for a round-trip ticket. However, if more than 200 people sign up for the flight, then each fare is reduced by $1 for each additional person. Assuming that more than 200 people sign up, determine how many passengers will result in a maximum revenue for the travel agency. What is the maximum revenue? What would the fare per person be in this case?

33. **Optimal Subway Fare** A city's Metropolitan Transit Authority (MTA) operates a subway line for commuters from a certain suburb to downtown. Currently, an average of 6000 passengers a day take the trains, paying a fare of $3.00 per ride. The board of the MTA, contemplating raising the fare to $3.50 per ride to generate a larger revenue, engages the services of a consulting firm. The firm's study reveals that for each $0.50 increase in fare, the ridership will be reduced by an average of 1000 passengers a day. Therefore, the consulting firm recommends that the MTA stick to the current fare of $3.00 per ride, which already yields a maximum revenue. Show that the consultants are correct.

34. **Strength of a Beam** A wooden beam has a rectangular cross section of height $h$ and width $w$. The strength $S$ of the beam is directly proportional to its width and the square of its

height. Find the dimensions of the cross section of such a beam of maximum strength that can be cut from a round log of diameter 24 in.

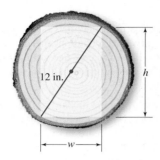

**Hint:** $S = kh^2w$, where $k$ is the constant of proportionality.

**35. Stiffness of a Beam** The stiffness $S$ of a wooden beam with a rectangular cross section is proportional to its width $w$ and the cube of its height $h$. Find the dimensions of the cross section of the beam of maximum stiffness that can be cut from a round log of diameter 23 in.
**Hint:** $S = kwh^3$, where $k$ is the constant of proportionality.

**36. Maximizing Drainage Capacity** The cross section of a drain is a trapezoid as shown in the figure. The sides and the bottom of the trapezoid each have length 5 ft. Determine the angle $\theta$ such that the drain will have a maximal cross-sectional area.

**37. Designing a Conical Figure** A cone is constructed by cutting out a sector of central angle $\theta$ from a circular sheet of radius 12 in. and then gluing the edges of the remaining piece together. Find the value of $\theta$ that will result in a cone of maximal volume. What is the maximal volume?

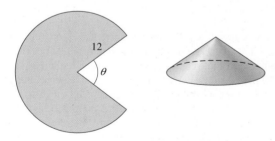

**38. A Norman Window** A Norman window has the shape of a rectangle surmounted by a semicircle. Find the dimensions of a Norman window of perimeter 28 ft that will admit the greatest possible amount of light.

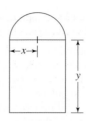

**39. Designing a Grain Silo** A grain silo has the shape of a right circular cylinder surmounted by a hemisphere. If the silo is to have a volume of $504\pi$ ft$^3$, determine the radius and height of the silo that requires the least amount of material to build.

**Hint:** The volume of the silo is $\pi r^2h + \frac{2}{3}\pi r^3$, and the surface area of the silo (including the floor) is $\pi(3r^2 + 2rh)$.

**40. Racetrack Design** The figure below depicts a racetrack with ends that are semicircular. The length of the track is 1760 ft $\left(\frac{1}{3}\text{ mi}\right)$. Find $l$ and $r$ so that the area of the rectangular portion of the region enclosed by the racetrack is as large as possible. What is the area enclosed by the track in this case?

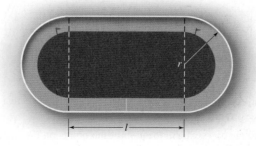

**41. Packaging** A container of capacity 64 in.$^3$ is to be made in the form of a right circular cylinder. The top and the bottom of the can are to be cut from squares, whereas the side is to be made by bending a rectangular sheet so that the ends match. Find the radius and height of the can that can be constructed with the least amount of material.

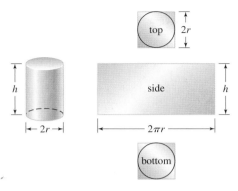

**42. A Fencing Problem** Joan has 50 ft of interlocking stone available for fencing off a flower bed in the form of a circular sector. Find the radius of the circle that will yield a flower bed with the largest area if Joan uses all of the stone.

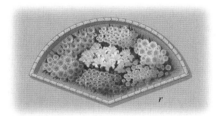

**43. Constructing a Marina** The figure below shows the position of two islands located off a straight stretch of coastal highway. A marina is to be constructed at the point $M$ on the highway to serve both island communities. Determine the location of $M$ if the total distance from both the islands to $M$ is as small as possible.

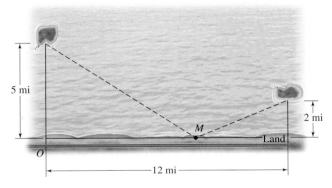

**44. Flights of Birds** During daylight hours some birds fly more slowly over water than over land because some of their

energy is expended in overcoming the downdrafts of air over open bodies of water. Suppose a bird that flies at a constant speed of 4 mph over water and 6 mph over land starts its journey at the point $E$ on an island and ends at its nest $N$ on the shore of the mainland, as shown in the figure. Find the location of the point $P$ that allows the bird to complete its journey in the minimum time (solve for $x$).

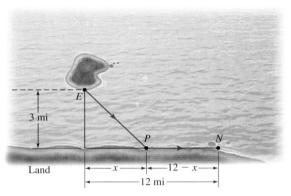

**45. Avoiding a Collision** Upon spotting a disabled and stationary boat, the driver of a speedboat took evasive action. Suppose that the disabled boat is located at the point $(0, 2)$ in an $xy$-coordinate system (both scales measured in miles) and the path of the speedboat is described by the graph of $f(x) = (x + 1)/\sqrt{x}$.
   **a.** Find an expression $D(x)$ that gives the distance between the speedboat and the disabled boat.
   **b.** Plot the graph of $D$, and use it to determine how close the speedboat came to the disabled boat before it changed its path.

**46. Minimizing Costs** Suppose that the cost incurred in operating a cruise ship for 1 hr is $a + bv^3$ dollars, where $a$ and $b$ are positive constants and $v$ is the ship's speed in miles per hour. At what speed should the ship be operated between two ports to minimize the cost?

**47. Maximum Power Output** Suppose that the source of current in an electric circuit is a battery. Then the power output $P$ (in watts) obtained if the circuit has a resistance of $R$ ohms is given by

$$P = \frac{E^2 R}{(R + r)^2}$$

where $E$ is the electromotive force in volts and $r$ is the internal resistance of the battery in ohms. If $E$ and $r$ are constant, find the value of $R$ that will result in the greatest power output. What is the maximum power output?

**48. Optimal Inventory Control** The equation

$$A(q) = \frac{km}{q} + cm + \frac{hq}{2}$$

gives the annual cost of ordering and storing (as yet unsold) merchandise. Here, $q$ is the size of each order, $k$ is the cost of placing each order, $c$ is the unit cost of the product, $m$ is the number of units of the product sold per year, and $h$ is the annual cost for storing each unit. Determine the size of each order such that the annual cost $A(q)$ is as small as possible.

**49. Velocity of a Wave** In deep water a wave of length $L$ travels with a velocity

$$v = k\sqrt{\frac{L}{C} + \frac{C}{L}}$$

where $k$ and $C$ are positive constants. Find the length of the wave that has a minimum velocity.

**50.** Show that the isosceles triangle of maximum area that can be inscribed in a circle of fixed radius $a$ is equilateral.

**51.** Show that the rectangle of maximum area that can be inscribed in a circle of fixed radius $a$ is a square.

**52.** Find the dimensions of the cylinder of largest volume that will fit inside a right circular cone of radius 3 in. and height 5 in. Assume that the axis of the cylinder coincides with the axis of the cone.

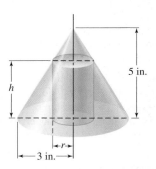

**53.** A right circular cylinder is inscribed in a cone of height $H$ and base radius $R$ so that the axis of the cylinder coincides with the axis of the cone. Determine the dimensions of the cylinder with the largest lateral surface area.

**54.** Find the radius and height of a right circular cylinder with the largest possible lateral surface area that can be inscribed in a sphere of radius $a$.

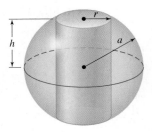

**55.** Find an equation of the line passing through the point $(1, 2)$ such that the area of the triangle formed by this line and the positive coordinate axes is as small as possible.

**56. Range of a Projectile** The range of an artillery shell fired at an angle of $\theta°$ with the horizontal is

$$R = \frac{v_0^2}{g} \sin 2\theta$$

feet, where $v_0$ is the muzzle velocity of the shell in feet per second, and $g$ is the constant of acceleration due to gravity ($32$ ft/sec$^2$). Find the angle of elevation of the gun that will give it a maximum range.

**57. Optimal Illumination** A hobbyist has set up a railroad track on a circular table to display a recently acquired model railroad locomotive. The radius of the track is 5 ft, and the display is to be illuminated by a light source suspended from an 8-ft ceiling located directly above the center of the table (see the figure). How high above the table should the light source be placed in order to achieve maximum illumination on the railroad track?

Hint: The intensity of light at $P$ is proportional to the cosine of the angle that the incident light makes with the vertical and inversely proportional to the square of the distance $r$ between $P$ and the light source.

**58. Cells of a Honeycomb** The accompanying figure depicts a single prism-shaped cell in a honeycomb. The front end of the prism is a regular hexagon, and the back is formed by the sides of the cell coming together at a point. It can be shown that the surface area of a cell is given by

$$S(\theta) = 6ab + \frac{3}{2}b^2\left(\frac{\sqrt{3} - \cos\theta}{\sin\theta}\right) \qquad 0 < \theta < \frac{\pi}{2}$$

where $\theta$ is the angle between one of the (three) upper surfaces and the altitude. The lengths of the sides of the hexagon, $b$, and the altitude, $a$, are both constants.

**a.** Show that the surface area is minimized if $\cos\theta = 1/\sqrt{3}$, or $\theta \approx 54.7°$. (Measurements of actual honeycombs have confirmed that this is, in fact, the angle found in beehives.)

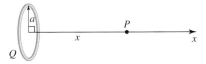 **b.** Using a graphing utility, verify the result of part (a) by finding the absolute minimum of

$$f(\theta) = \frac{\sqrt{3} - \cos\theta}{\sin\theta} \qquad 0 < \theta < \frac{\pi}{2}$$

**59. Maximizing Length** A metal pipe of length 16 ft is to be carried horizontally around a corner from a hallway 8 ft wide into a hallway 4 ft wide. Can this be done?

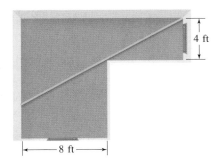

**Hint:** Find the length of the largest pipe that can be carried horizontally around the corner.

**60. Distance Between Two Aircraft** Two aircraft approach each other, each flying at a speed of 500 mph and at an altitude of 35,000 ft. Their paths are straight lines that intersect at an angle of 120°. At a certain instant of time, one aircraft is 200 mi from the point of intersection of their paths, while the other is 300 mi from it. At what time will the aircraft be closest to each other, and what will that distance be?

**61. Storing Radioactive Waste** A cylindrical container for storing radioactive waste is to be constructed from lead and have a thickness of 6 in. (see the figure). If the volume of the outside cylinder is to be $16\pi$ ft³, find the radius and the height of the inside cylinder that will result in a container of maximum storage capacity.

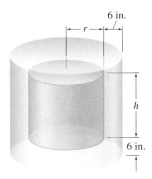

**Hint:** Show that the storage capacity (inside volume in ft³) is given by

$$V(r) = \pi r^2\left[\frac{16}{(r + \frac{1}{2})^2} - 1\right] \qquad 0 \le r \le \frac{7}{2}$$

**62. Electrical Force of a Conductor** A ring-shaped conductor of radius $a$ carrying a total charge $Q$ induces an electrical force of magnitude

$$F = \frac{Q}{4\pi\varepsilon_0} \cdot \frac{x}{(x^2 + a^2)^{3/2}}$$

where $\varepsilon_0$ is a constant called the permittivity of free space, at a point $P$, a distance $x$ from the center, along the line perpendicular to the plane of the ring through its center. Find the value of $x$ for which $F$ is greatest.

**63. Energy Expended by a Fish** It has been conjectured that the total energy expended by a fish swimming a distance of $L$ ft at a speed of $v$ ft/sec relative to the water and against a current flowing at the rate of $u$ ft/sec ($u < v$) is given by

$$E(v) = \frac{aLv^3}{v - u}$$

where $E$ is measured in foot-pounds (ft-lb) and $a$ is a constant.

**a.** Find the speed at which the fish must swim to minimize the total energy expended.

**b.** Sketch the graph of $E$.

**Note:** This result has been verified by biologists.

**64.** A poster of height 36 in. is mounted on a wall so that its lower edge is 12 in. above the eye level of an observer. How far from the wall should the observer stand so that the viewing angle $\theta$ subtended at his eye by the poster is as large as possible?

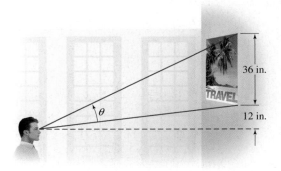

**65.** A restaurateur has a choice of a site for a restaurant to be constructed between two jetties. The two jetties lie along a straight stretch of a coastal highway and are 1000 ft apart. How far from the longer jetty should the restaurant be located in order to have the largest unobstructed view of the ocean?

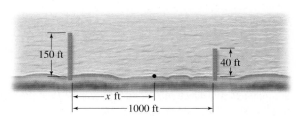

**66.** An observer stands on a straight path that is parallel to a straight test track. At $t = 0$ a Formula 1 car is directly opposite her and 200 ft away. As she watches, the car moves with a constant acceleration of 20 ft/sec², so it is at a distance of $10t^2$ ft from the starting position after $t$ sec, where $0 \le t \le 15$. As the car moves, $d\theta/dt$ increases, slowly at first, then faster, and finally it slows down again. The observer perceives the car to be moving at the fastest speed when $d\theta/dt$ is maximal. Determine the position of the car at the moment she perceives it to be moving at the fastest speed.

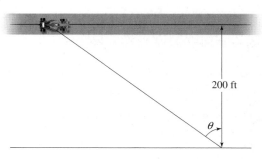

**67.** A woman is on a lake in a rowboat located one mile from the closest point $P$ of a straight shoreline (see the figure). She wishes to get to a point $Q$, 10 miles along the shore from $P$, by rowing to a point $R$ between $P$ and $Q$ and then walking the rest of the distance. If she can row at a speed of 3 mph and walk at a speed of 4 mph, how should she pick the point $R$ to get to $Q$ as quickly as possible? How much time does she require?

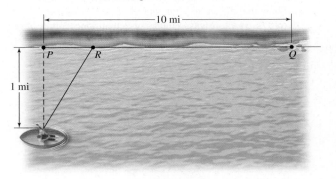

**68. Resonance** A spring system comprising a weight attached to a spring and a dashpot damping device (see the accompanying figure) is acted on by an oscillating external force. Its motion for large values of $t$ is described by the equation

$$x(t) = \frac{F}{(\omega^2 - \gamma^2)^2 + 4\lambda^2\gamma^2} \sin(\gamma t + \theta)$$

where $F$, $\omega$, $\lambda$, and $\theta$ are constants. ($F$ is the amplitude of the external force, $\theta$ is a phase angle, $\gamma$ is associated with the frequency of the external force, and $\omega$ and $\lambda$ are associated with the stiffness of the spring and the degree of resistance of the dashpot damping device, respectively.) Show that the amplitude of the motion of the system

$$g(\gamma) = \frac{F}{\sqrt{(\omega^2 - \gamma^2)^2 + 4\lambda^2\gamma^2}}$$

has a maximum value at $\gamma_1 = \sqrt{\omega^2 - 2\lambda^2}$. When the frequency of the external force is $\sqrt{\omega^2 - 2\lambda^2}/2\pi$, the system is said to be in **resonance.** The figure below shows a typical **resonance curve.**

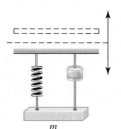
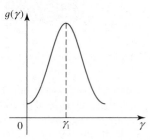

The external force imparts an oscillatory vertical motion on the support.

**69. Snell's Law of Refraction** The following figure shows the path of a ray of light traveling in air from the source $A$ to the point $C$ and then from $C$ to the point $B$ in water. Let $v_1$ denote the velocity of light in air, and let $v_2$ denote the velocity of light in water. Use Fermat's Principle, which states that a ray of light will travel from one point to another in the least time, to prove that

$$\frac{\sin\theta_1}{v_1} = \frac{\sin\theta_2}{v_2}$$

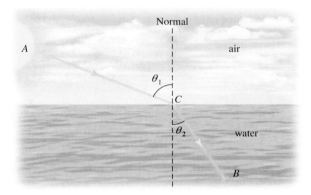

**70. Flow of Blood** Suppose that some of the fluid flowing along a pipe of radius $R$ is diverted to a pipe of smaller radius $r$ attached to the former at an angle $\theta$ (see the figure). Such is the case when blood flowing along an artery is pumped into an arteriole. What should the angle $\theta$ be so that the energy loss due to friction in moving the fluid is minimal? Solve the problem via the steps below.

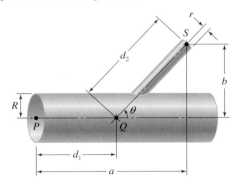

**a.** Use Poiseuille's Law, which states that the loss of energy due to friction in nonturbulent flow is proportional to the length of the path and inversely proportional to the fourth power of the radius, to show that the energy loss in moving the fluid from $P$ to $S$ via $Q$ is

$$E = \frac{kd_1}{R^4} + \frac{kd_2}{r^4}$$

where $k$ is a constant.

**b.** Suppose $a$ and $b$ are fixed. Find $d_1$ and $d_2$ in terms of $a$ and $b$. Then use this result together with the result from part (a) to show that

$$E = k\left[\frac{a - b\cot\theta}{R^4} + \frac{b\csc\theta}{r^4}\right]$$

**c.** Using the technique of this section, show that $E$ is minimized when

$$\theta = \cos^{-1}\frac{r^4}{R^4}$$

**71. Least Squares Approximation** Suppose we are given $n$ data points

$$P_1(x_1, y_1), P_2(x_2, y_2), \ldots, P_n(x_n, y_n)$$

that are scattered about the graph of a straight line with equation $y = ax$ (see the figure). The error in approximating $y_i$ by the value of the function $f(x) = ax$ at $x_i$ is

$$[y_i - f(x_i)] \qquad 1 \le i \le n$$

**a.** Show that the sum of the squares of the errors in approximating $y_i$ by $f(x_i)$ for $1 \le i \le n$ is

$$g(a) = (y_1 - ax_1)^2 + (y_2 - ax_2)^2 + \cdots + (y_n - ax_n)^2$$

**b.** Show that $g$ is minimized if

$$a = \frac{x_1y_1 + x_2y_2 + \cdots + x_ny_n}{x_1^2 + x_2^2 + \cdots + x_n^2}$$

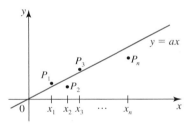

**Note:** The straight line with equation $y = ax$, where $a$ is the number found in part (b), is called the least-squares, or regression, line associated with the given data and is a line that fits the data "best" in the sense of least squares.

 **72. Calculating a Spring Constant** The following table gives the force required to stretch a spring by an elongation $x$ beyond its unstretched length.

| $x$ (ft) | 0 | 0.1 | 0.2 | 0.3 | 0.4 | 0.5 |
|---|---|---|---|---|---|---|
| Force, $y$ (lb) | 0 | 1.68 | 3.18 | 4.84 | 6.36 | 8.02 |

**a.** Use the result of Exercise 71 to find the straight line $y = ax$ that fits the data "best" in the sense of least squares.

**b.** Plot the data points and the least squares line found in part (a) on the same set of axes.

**c.** Using Hooke's Law, which states that $F = kx$, where $k$ is the spring constant, what does the result of part (b) give as the spring constant?

**73. Least Squares Approximation** Suppose we are given $n$ data points

$$P_1(x_1, y_1), P_2(x_2, y_2), \ldots, P_n(x_n, y_n)$$

that are scattered about the graph of a parabola with equation $y = ax^2$ (see the figure). The error in approximating $y_i$ by the value of the function $f(x) = ax^2$ at $x_i$ is

$$[y_i - f(x_i)] \qquad 1 \le i \le n$$

**a.** Show that the sum of the squares of the errors in approximating $y_i$ by $f(x_i)$ for $1 \le i \le n$ is

$$g(a) = \left(y_1 - ax_1^2\right)^2 + \left(y_2 - ax_2^2\right)^2 + \cdots + \left(y_n - ax_n^2\right)^2$$

**b.** Show that $g$ is minimized if

$$a = \frac{x_1^2 y_1 + x_2^2 y_2 + \cdots + x_n^2 y_n}{x_1^4 + x_2^4 + \cdots + x_n^4}$$

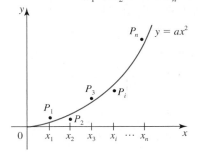

**Note:** The curve with equation $y = ax^2$, where $a$ is the number found in part (b), is called the least-squares curve associated with the data given and is a curve that fits the data "best" in the sense of least squares.

**74. Calculating the Constant of Acceleration** A steel ball is dropped from a height of 10 ft. The distance covered by the ball at intervals of one tenth of a second is measured and recorded in the following table.

| Time $t$ (sec)    | 0.0 | 0.1    | 0.2    | 0.3    |
|-------------------|-----|--------|--------|--------|
| Distance $y$ (ft) | 0   |        | 0.1608 | 0.6416 | 1.4444 |

| Time $t$ (sec)    | 0.4    | 0.5    | 0.6    | 0.7    |
|-------------------|--------|--------|--------|--------|
| Distance $y$ (ft) | 2.5672 | 4.0108 | 5.7760 | 7.8614 |

**a.** Use the result of Exercise 73 to find the parabola $y = at^2$ that fits the data "best" in the sense of least squares.
**b.** Plot the data points and the least-squares curve found in part (a) on the same set of axes.
**c.** Using the fact that a free-falling object acted upon only by the force of gravity covers a distance of $s = \frac{1}{2} gt^2$ ft after $t$ sec, what does the result of part (a) give as the constant of acceleration $g$?

---

## 3.8 Indeterminate Forms and l'Hôpital's Rule

In Section 1.1 we encountered the limit

$$\lim_{t \to 2} \frac{4(t^2 - 4)}{t - 2} \tag{1}$$

when we attempted to find the velocity of the maglev at time $t = 2$, and in Section 1.2 we studied the limit

$$\lim_{x \to 0} \frac{\sin x}{x} \tag{2}$$

Observe that both the numerator and the denominator of expression (1) approach zero as $t$ approaches 2. Similarly, both the numerator and the denominator of expression (2) also approach zero as $x$ approaches zero.

More generally, if $\lim_{x \to a} f(x) = 0$ and $\lim_{x \to a} g(x) = 0$, then the limit

$$\lim_{x \to a} \frac{f(x)}{g(x)}$$

is called an **indeterminate form of the type $0/0$.** As the name implies, the undefined expression $0/0$ does not provide us with a definitive answer concerning the existence of the limit or its value, if the limit exists.

Recall that we evaluated the limit in (1) through algebraic sleight of hand. Thus,

$$\lim_{t \to 2} \frac{4(t^2 - 4)}{t - 2} = \lim_{t \to 2} \frac{4(t + 2)(t - 2)}{t - 2} = \lim_{t \to 2} 4(t + 2) = 16$$

This method, however, will not work in evaluating the limit in (2). In Section 1.2 we used a geometric argument to show that

$$\lim_{x \to 0} \frac{\sin x}{x} = 1$$

These examples raise the following question: Given an indeterminate form of the type $0/0$, is there a more general and efficient method for resolving whether the limit

$$\lim_{x \to a} \frac{f(x)}{g(x)}$$

exists, and if so, what is the limit?

## ■ The Indeterminate Forms $0/0$ and $\infty/\infty$

To gain insight into the nature of an indeterminate form of the type $0/0$, let's consider the following limits:

**a.** $\lim_{x \to 0^+} \frac{x^2}{x}$     **b.** $\lim_{x \to 0^+} \frac{2x}{3x}$     **c.** $\lim_{x \to 0^+} \frac{x}{x^2}$

Each of these limits is an indeterminate form of the type $0/0$. We can evaluate each limit as follows:

**a.** $\lim_{x \to 0^+} \frac{x^2}{x} = \lim_{x \to 0^+} x = 0$

**b.** $\lim_{x \to 0^+} \frac{2x}{3x} = \lim_{x \to 0^+} \frac{2}{3} = \frac{2}{3}$

**c.** $\lim_{x \to 0^+} \frac{x}{x^2} = \lim_{x \to 0^+} \frac{1}{x} = \infty$

Let's examine each limit in greater detail. In (a) the numerator $f_1(x) = x^2$ goes to zero faster than the denominator $g_1(x) = x$, when $x$ is close to zero. So it is plausible that the ratio $f_1(x)/g_1(x)$ should approach 0 as $x$ approaches 0. In (b) the numerator $f_2(x) = 2x$ goes to zero at $(2x)/(3x) = \frac{2}{3}$ the rate at which $g_2(x) = 3x$ goes to zero, so the answer seems reasonable. Finally, in (c) the denominator $g_3(x) = x^2$ goes to zero faster than the numerator $f_3(x) = x$, and consequently, we expect the ratio to "blow up."

These three examples show that the existence or nonexistence of the limit as well as the value of the limit depend on how fast the numerator $f(x)$ and the denominator $g(x)$ go to zero. This observation suggests the following technique for evaluating these indeterminate forms: Because both $f(x)$ and $g(x)$ go to 0 as $x$ approaches 0, we cannot determine the limit of the quotient by using the Quotient Rule for limits. So we might consider the limit of the ratio of their *derivatives, f'(x)* and $g'(x)$, since the derivatives measure how fast $f(x)$ and $g(x)$ change. In other words, it might be plausible that if both $f(x) \to 0$ and $g(x) \to 0$ as $x \to 0$, then

$$\lim_{x \to 0} \frac{f(x)}{g(x)} = \lim_{x \to 0} \frac{f'(x)}{g'(x)}$$

Let's try this on the limits in (1) and (2). For the limit in (1) we have

$$\lim_{t \to 2} \frac{4(t^2 - 4)}{t - 2} = \lim_{t \to 2} \frac{\frac{d}{dt}[4(t^2 - 4)]}{\frac{d}{dt}(t - 2)} = \lim_{t \to 2} \frac{8t}{1} = 16$$

which is the value we obtained before! For the limit in (2) we find

$$\lim_{x \to 0} \frac{\sin x}{x} = \lim_{x \to 0} \frac{\dfrac{d}{dx}(\sin x)}{\dfrac{d}{dx}(x)} = \lim_{x \to 0} \frac{\cos x}{1} = 1$$

which we demonstrated in Section 1.2.

   This method, which we have arrived at intuitively, is given validity by the theorem known as l'Hôpital's Rule. The theorem is named after the French mathematician Guillaume Francois Antoine de l'Hôpital (1661–1704), who published the first calculus text in 1696. But before stating l'Hôpital's Rule, we need to define another type of indeterminate form.

   If $\lim_{x \to a} f(x) = \pm\infty$ and $\lim_{x \to a} g(x) = \pm\infty$, then the limit

$$\lim_{x \to a} \frac{f(x)}{g(x)}$$

is said to be an indeterminate form of the type $\infty/\infty$, $-\infty/\infty$, $\infty/-\infty$, or $-\infty/-\infty$. To see why this limit is an indeterminate form, we simply write

$$\lim_{x \to a} \frac{f(x)}{g(x)} = \lim_{x \to a} \frac{\dfrac{1}{g(x)}}{\dfrac{1}{f(x)}}$$

which has the form $0/0$ and, therefore, is indeterminate. We refer to each of these limits as an **indeterminate form of the type $\infty/\infty$**, since the sign provides no useful information.

---

### THEOREM 1    l'Hôpital's Rule

Suppose that $f$ and $g$ are differentiable on an open interval $I$ that contains $a$, with the possible exception of $a$ itself, and $g'(x) \neq 0$ for all $x$ in $I$. If $\lim_{x \to a} \dfrac{f(x)}{g(x)}$ is an indeterminate form of the type $0/0$ or $\infty/\infty$, then

$$\lim_{x \to a} \frac{f(x)}{g(x)} = \lim_{x \to a} \frac{f'(x)}{g'(x)}$$

provided that the limit on the right-hand side exists or is infinite.

---

We will prove this theorem in Appendix B.

⚠ The expression $f'(x)/g'(x)$ is the *ratio* of the derivatives of $f(x)$ and $g(x)$—it is *not obtained* from $f/g$ by using the Quotient Rule.

**Notes**
1. l'Hôpital's Rule is also valid for one-sided limits as well as limits at infinity or negative infinity; that is, we can replace "$x \to a$" by any of the symbols $x \to a^+$, $x \to a^-$, $x \to \infty$, or $x \to -\infty$.

**2.** Before applying l'Hôpital's Rule, check to see that the limit does have one of the indeterminate forms. For example, $\cos x \to 1$ as $x \to 0^+$, so

$$\lim_{x \to 0^+} \frac{\cos x}{x} = \infty$$

If we had applied l'Hôpital's Rule to evaluate the limit without first ascertaining that it had an indeterminate form, we would have obtained the erroneous result

$$\lim_{x \to 0^+} \frac{\cos x}{x} = \lim_{x \to 0^+} \frac{-\sin x}{1} = 0 \qquad \blacksquare$$

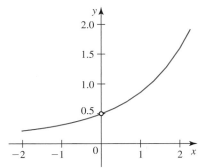

**FIGURE 1**
The graph of $y = \dfrac{e^x - 1}{2x}$ gives a visual confirmation of the result of Example 1.

**EXAMPLE 1** Evaluate $\lim\limits_{x \to 0} \dfrac{e^x - 1}{2x}$.

**Solution** We have an indeterminate form of the type $0/0$. Applying l'Hôpital's Rule, we obtain

$$\lim_{x \to 0} \frac{e^x - 1}{2x} = \lim_{x \to 0} \frac{\dfrac{d}{dx}(e^x - 1)}{\dfrac{d}{dx}(2x)} = \lim_{x \to 0} \frac{e^x}{2} = \frac{1}{2}$$

(See Figure 1.) $\blacksquare$

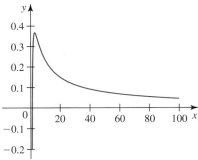

**FIGURE 2**
The graph of $y = \dfrac{\ln x}{x}$ shows that $y \to 0$ as $x \to \infty$.

**EXAMPLE 2** Evaluate $\lim\limits_{x \to \infty} \dfrac{\ln x}{x}$.

**Solution** We have an indeterminate form of the type $\infty/\infty$. Applying l'Hôpital's Rule, we obtain

$$\lim_{x \to \infty} \frac{\ln x}{x} = \lim_{x \to \infty} \frac{\dfrac{d}{dx}(\ln x)}{\dfrac{d}{dx}(x)} = \lim_{x \to \infty} \frac{1}{x} = 0$$

(See Figure 2.) $\blacksquare$

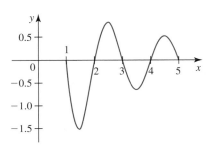

**FIGURE 3**
The graph of $y = \dfrac{\sin \pi x}{(x - 1)^{1/2}}$ shows that
$$\lim_{x \to 1^+} \frac{\sin \pi x}{(x - 1)^{1/2}} = 0.$$

**EXAMPLE 3** Evaluate $\lim\limits_{x \to 1^+} \dfrac{\sin \pi x}{\sqrt{x - 1}}$.

**Solution** We have an indeterminate form of the type $0/0$. Applying l'Hôpital's Rule, we obtain

$$\lim_{x \to 1^+} \frac{\sin \pi x}{(x - 1)^{1/2}} = \lim_{x \to 1^+} \frac{\pi \cos \pi x}{\frac{1}{2}(x - 1)^{-1/2}}$$

$$= \lim_{x \to 1^+} 2\pi(\cos \pi x)\sqrt{x - 1}$$

$$= 0$$

(See Figure 3.) $\blacksquare$

Sometimes we need to apply l'Hôpital's Rule more than once to resolve a limit involving an indeterminate form. This is illustrated in the next two examples.

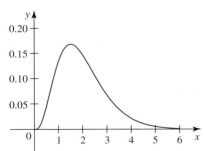

**FIGURE 4**
The graph of $y = \dfrac{x^3}{e^{2x}}$ shows that $y \to 0$ as $x \to \infty$.

**EXAMPLE 4** Evaluate $\displaystyle\lim_{x\to\infty} \frac{x^3}{e^{2x}}$.

**Solution** Applying l'Hôpital's Rule (three times), we obtain

$$\lim_{x\to\infty} \frac{x^3}{e^{2x}} = \lim_{x\to\infty} \frac{3x^2}{2e^{2x}} \qquad \text{Type: } \infty/\infty$$

$$= \lim_{x\to\infty} \frac{6x}{4e^{2x}} \qquad \text{Type: } \infty/\infty$$

$$= \lim_{x\to\infty} \frac{6}{8e^{2x}} = 0$$

(See Figure 4.) ■

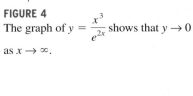
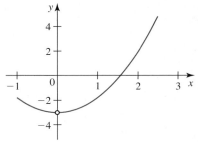

**FIGURE 5**
The graph of $y = \dfrac{x^3}{x - \tan x}$ shows that $y \to -3$ as $x \to 0$. Note that $y$ is not defined at $x = 0$.

**EXAMPLE 5** Evaluate $\displaystyle\lim_{x\to 0} \frac{x^3}{x - \tan x}$.

**Solution** We have an indeterminate form of the type $0/0$. Using l'Hôpital's Rule, repeatedly, we obtain

$$\lim_{x\to 0} \frac{x^3}{x - \tan x} = \lim_{x\to 0} \frac{3x^2}{1 - \sec^2 x} \qquad \text{Type: } 0/0$$

$$= \lim_{x\to 0} \frac{6x}{-2 \sec^2 x \tan x} \qquad \text{Type: } 0/0$$

$$= \lim_{x\to 0} \frac{6}{-4 \sec^2 x \tan^2 x - 2 \sec^4 x} = \frac{6}{-2} = -3$$

(See Figure 5.) ■

## ■ The Indeterminate Forms $\infty - \infty$ and $0 \cdot \infty$

If $\lim_{x\to a} f(x) = \infty$ and $\lim_{x\to a} g(x) = \infty$, then the limit

$$\lim_{x\to a}[f(x) - g(x)]$$

is said to be an **indeterminate form of the type** $\infty - \infty$. An indeterminate form of this type can often be expressed as one of the type $0/0$ or $\infty/\infty$ by algebraic manipulation. This is illustrated in the following example.

**EXAMPLE 6** Evaluate $\displaystyle\lim_{x\to 0^+} \left( \frac{1}{x} - \frac{1}{e^x - 1} \right)$.

**Solution** We have an indeterminate form of the type $\infty - \infty$. By writing the expression in parentheses as a single fraction, we obtain an indeterminate form of

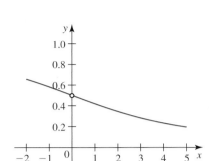

**FIGURE 6**
The graph of $y = \dfrac{1}{x} - \dfrac{1}{e^x - 1}$ shows

that $\displaystyle\lim_{x \to 0^+}\left(\dfrac{1}{x} - \dfrac{1}{e^x - 1}\right) = \dfrac{1}{2}$.

the type 0/0. This enables us to evaluate the resulting expression using l'Hôpital's Rule:

$$\lim_{x \to 0^+}\left(\frac{1}{x} - \frac{1}{e^x - 1}\right) = \lim_{x \to 0^+}\frac{e^x - x - 1}{x(e^x - 1)} \qquad \text{Type: } 0/0$$

$$= \lim_{x \to 0^+}\frac{e^x - 1}{e^x - 1 + xe^x} \qquad \text{Apply l'Hôpital's Rule.}$$

$$= \lim_{x \to 0^+}\frac{e^x}{(x + 2)e^x} = \frac{1}{2} \qquad \text{Apply l'Hôpital's Rule again.}$$

(See Figure 6.) ∎

If $\lim_{x \to a} f(x) = 0$ and $\lim_{x \to a} g(x) = \pm\infty$, then $\lim_{x \to a} f(x)g(x)$ is said to be an **indeterminate form of the type 0 · ∞.** An indeterminate form of this type also can be expressed as one of the type 0/0 or ∞/∞ by algebraic manipulation, as illustrated in the following example.

**EXAMPLE 7** Evaluate $\lim_{x \to 0^+} x \ln x$.

**Solution** We have an indeterminate form of the type $0 \cdot \infty$. By writing

$$x \ln x = \frac{\ln x}{\dfrac{1}{x}}$$

the given limit can be cast in an indeterminate form of the type ∞/∞. Then, applying l'Hôpital's Rule, we obtain

$$\lim_{x \to 0^+} x \ln x = \lim_{x \to 0^+}\frac{\ln x}{\dfrac{1}{x}} \qquad \text{Type: } \infty/\infty$$

$$= \lim_{x \to 0^+}\frac{\dfrac{1}{x}}{-\dfrac{1}{x^2}} = \lim_{x \to 0^+}(-x) = 0$$

(See Figure 7.) ∎

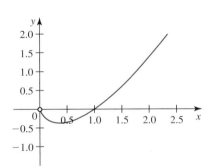

**FIGURE 7**
The graph of $y = x \ln x$ gives a visual verification of the result in Example 7.

## ■ The Indeterminate Forms $0^0$, $\infty^0$, and $1^\infty$

The limit

$$\lim_{x \to a}[f(x)]^{g(x)}$$

is said to be an **indeterminate form of the type**

$$\mathbf{0^0} \text{ if } \lim_{x \to a} f(x) = 0 \text{ and } \lim_{x \to a} g(x) = 0$$

$$\boldsymbol{\infty^0} \text{ if } \lim_{x \to a} f(x) = \infty \text{ and } \lim_{x \to a} g(x) = 0$$

$$\mathbf{1^\infty} \text{ if } \lim_{x \to a} f(x) = 1 \text{ and } \lim_{x \to a} g(x) = \pm\infty$$

These indeterminate forms can usually be converted to indeterminate forms of the type $0 \cdot \infty$ by taking logarithms or by using the identity

$$[f(x)]^{g(x)} = e^{g(x)\ln f(x)}$$

**EXAMPLE 8**    Evaluate $\lim_{x \to 0^+} x^x$.

**Solution**    We have an indeterminate form of the type $0^0$. Let

$$y = x^x$$

Then

$$\ln y = \ln x^x = x \ln x$$

and using the result of Example 7, we obtain

$$\lim_{x \to 0^+} \ln y = \lim_{x \to 0^+} x \ln x = 0$$

Finally, using the identity $y = e^{\ln y}$ and the continuity of the exponential function, we have

$$\lim_{x \to 0^+} x^x = \lim_{x \to 0^+} y = \lim_{x \to 0^+} e^{\ln y} = e^{\lim_{x \to 0^+} \ln y} = e^0 = 1$$

(See Figure 8.)

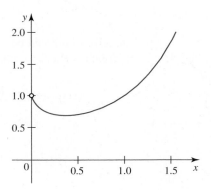

**FIGURE 8**
The graph of $y = x^x$
shows that $\lim_{x \to 0^+} x^x = 1$.

**EXAMPLE 9**    Evaluate $\lim_{x \to 0^+} \left(\dfrac{1}{x}\right)^{\sin x}$.

**Solution**    We have an indeterminate form of the type $\infty^0$. Let

$$y = \left(\frac{1}{x}\right)^{\sin x}$$

Then

$$\ln y = \ln\left(\frac{1}{x}\right)^{\sin x} = (\sin x)\ln \frac{1}{x}$$

and

$$\lim_{x \to 0^+} \ln y = \lim_{x \to 0^+} (\sin x)\ln \frac{1}{x}$$

This last limit is an indeterminate form of the type $0 \cdot \infty$. By writing

$$(\sin x)\ln\left(\frac{1}{x}\right) = \frac{\ln \dfrac{1}{x}}{\dfrac{1}{\sin x}}$$

we can transform it into an indeterminate form of the type $\infty/\infty$ and hence use l'Hôpital's Rule. We have

$$\lim_{x \to 0^+} \ln y = \lim_{x \to 0^+} \frac{\ln \dfrac{1}{x}}{\dfrac{1}{\sin x}} \qquad \text{Type: } \infty/\infty$$

$$= \lim_{x \to 0^+} -\frac{\ln x}{\dfrac{1}{\sin x}} \qquad \text{Rewrite } \ln\left(\dfrac{1}{x}\right).$$

$$= \lim_{x \to 0^+} \frac{-\dfrac{1}{x}}{-\dfrac{\cos x}{\sin^2 x}} = \lim_{x \to 0^+} \frac{\sin^2 x}{x \cos x} \qquad \text{Apply l'Hôpital's Rule.}$$

$$= \lim_{x \to 0^+} \left(\frac{\sin x}{x}\right)(\tan x) = 0 \qquad \lim_{x \to 0^+} \frac{\sin x}{x} = 1$$

Therefore,

$$\lim_{x \to 0^+} \left(\frac{1}{x}\right)^{\sin x} = \lim_{x \to 0^+} y = \lim_{x \to 0^+} e^{\ln y} = e^{\lim_{x \to 0^+} \ln y} = e^0 = 1$$

since the exponential function is continuous. (See Figure 9.)

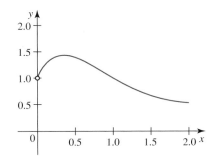

**FIGURE 9**

The graph of $y = \left(\dfrac{1}{x}\right)^{\sin x}$

shows that $\lim\limits_{x \to 0^+} \ln y = 1$.

**EXAMPLE 10** Evaluate $\lim\limits_{x \to \infty} \left(1 + \dfrac{1}{x}\right)^x$.

**Solution** We have an indeterminate form of the type $1^\infty$. Let

$$y = \left(1 + \frac{1}{x}\right)^x$$

Then

$$\ln y = \ln\left(1 + \frac{1}{x}\right)^x = x \ln\left(1 + \frac{1}{x}\right)$$

so

$$\lim_{x \to \infty} \ln y = \lim_{x \to \infty} x \ln\left(1 + \frac{1}{x}\right)$$

has an indeterminate form of the type $0 \cdot \infty$. Rewriting and using l'Hôpital's Rule, we obtain

$$\lim_{x \to \infty} \ln y = \lim_{x \to \infty} x \ln\left(1 + \frac{1}{x}\right) \qquad \text{Type: } 0 \cdot \infty$$

$$= \lim_{x \to \infty} \frac{\ln\left(1 + \frac{1}{x}\right)}{\frac{1}{x}} \qquad \text{Type: } 0/0$$

$$= \lim_{x \to \infty} \left[\frac{\left(\frac{1}{1 + \frac{1}{x}}\right)\left(-\frac{1}{x^2}\right)}{-\frac{1}{x^2}}\right] \qquad \text{Apply l'Hôpital's Rule.}$$

$$= \lim_{x \to \infty} \frac{1}{1 + \frac{1}{x}} = 1$$

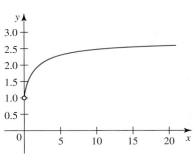

**FIGURE 10**
The graph of $y = \left(1 + \dfrac{1}{x}\right)^x$ shows that
$y \to e \approx 2.718$ as $x \to \infty$.

Therefore,

$$\lim_{x \to \infty}\left(1 + \frac{1}{x}\right)^x = \lim_{x \to \infty} y = \lim_{x \to \infty} e^{\ln y} = e^{\lim_{x \to \infty} \ln y} = e^1 = e$$

since the exponential function is continuous. (See Figure 10.) ■

## 3.8 CONCEPT QUESTIONS

*In Exercises 1–8, evaluate the limit or classify the type of indeterminate form to which it gives rise.*

**1.** $\lim_{x \to a} \dfrac{f(x)}{g(x)}$ if $\lim_{x \to a} f(x) = 1$ and $\lim_{x \to a} g(x) = \infty$

**2.** $\lim_{x \to a} \dfrac{f(x)}{g(x)}$ if $\lim_{x \to a} f(x) = 4$ and $\lim_{x \to a} g(x) = 0$, where $g(x) > 0$

**3.** $\lim_{x \to a^+} \dfrac{f(x)}{x - a}$ if $\lim_{x \to a} f(x) = \infty$

**4.** $\lim_{x \to a}[f(x) - g(x)]$ if $\lim_{x \to a} f(x) = \infty$ and $\lim_{x \to a} g(x) = -\infty$

**5.** $\lim_{x \to 3}\left[\dfrac{f(x)}{(x - 3)^2} + \dfrac{2}{|x - 3|}\right]$ if $\lim_{x \to 3} f(x) = 8$

**6.** $\lim_{x \to \infty}[f(x)]^{g(x)}$ if $\lim_{x \to \infty} f(x) = 0$ and $\lim_{x \to \infty} g(x) = \infty$ where $f(x) \geq 0$

**7.** $\lim_{x \to \infty}\left[\dfrac{x}{\sqrt{x^2 + 5}}\right]^{f(x)}$ if $\lim_{x \to \infty} f(x) = \infty$

**8.** $\lim_{x \to a}\left[\dfrac{2}{f(x)}\right]^{g(x)}$ if $\lim_{x \to a} f(x) = 0$ and $\lim_{x \to a} g(x) = 0$ where $f(x) > 0$

**9. a.** State l'Hôpital's Rule.
   **b.** Explain how l'Hôpital's Rule can be used to evaluate
   **(i)** $\lim_{x \to a} f(x)g(x)$ if $\lim_{x \to a} f(x) = \infty$ and $\lim_{x \to a} g(x) = 0$
   **(ii)** $\lim_{x \to a}[f(x) - g(x)]$ if $\lim_{x \to a} f(x) = -\infty$ and $\lim_{x \to a} g(x) = -\infty$
   **(iii)** $\lim_{x \to a}[f(x)]^{g(x)}$ if $\lim_{x \to a} f(x) = 0$ and $\lim_{x \to a} g(x) = 0$, where $f(x) > 0$

## 3.8  EXERCISES

*In Exercises 1–56, evaluate the limit using l'Hôpital's Rule if appropriate.*

1. $\lim\limits_{x\to1}\dfrac{x-1}{x^2-1}$

2. $\lim\limits_{x\to-1}\dfrac{x^2-2x-3}{x+1}$

3. $\lim\limits_{x\to2}\dfrac{x^3-8}{x-2}$

4. $\lim\limits_{x\to1}\dfrac{x^7-1}{x^4-1}$

5. $\lim\limits_{x\to0}\dfrac{e^x-1}{x^2+x}$

6. $\lim\limits_{x\to1}\dfrac{\ln x}{x-1}$

7. $\lim\limits_{t\to\pi}\dfrac{\sin t}{\pi-t}$

8. $\lim\limits_{x\to0}\dfrac{e^x-1}{x+\sin x}$

9. $\lim\limits_{\theta\to0}\dfrac{\tan 2\theta}{\theta}$

10. $\lim\limits_{x\to0}\dfrac{\sin 2x}{x}$

11. $\lim\limits_{x\to\infty}\dfrac{x+\cos x}{2x+1}$

12. $\lim\limits_{\theta\to0}\dfrac{\theta+\sin\theta}{\tan\theta}$

13. $\lim\limits_{x\to0}\dfrac{\sin x-x\cos x}{\tan^3 x}$

14. $\lim\limits_{u\to\pi}\dfrac{2\sin^2 u}{1+\cos u}$

15. $\lim\limits_{x\to\infty}\dfrac{\sqrt{x}}{\ln x}$

16. $\lim\limits_{x\to\infty}\dfrac{e^x}{x^4}$

17. $\lim\limits_{x\to\infty}\dfrac{(\ln x)^3}{x^2}$

18. $\lim\limits_{x\to1}\dfrac{x^{1/2}-x^{1/3}}{x-1}$

19. $\lim\limits_{x\to\infty}\dfrac{\ln(1+e^x)}{x^2}$

20. $\lim\limits_{x\to1}\dfrac{a^{\ln x}-x}{\ln x}$

21. $\lim\limits_{x\to-1}\dfrac{\sqrt{x+2}+x}{\sqrt[3]{2x+1}+1}$

22. $\lim\limits_{x\to0}\dfrac{\ln(x^2+1)}{\cos x-1}$

23. $\lim\limits_{x\to0^+}\dfrac{e^{x^2}+x-1}{1-\sqrt{1-x^2}}$

24. $\lim\limits_{x\to0}\dfrac{\ln(1+x)-\tan x}{x^2}$

25. $\lim\limits_{x\to0}\dfrac{\sin x-x}{e^x-e^{-x}-2x}$

26. $\lim\limits_{x\to0}\dfrac{e^{x^2}-1}{1-\cos x}$

27. $\lim\limits_{x\to0}\dfrac{\sin^{-1}(2x)}{x}$

28. $\lim\limits_{x\to0}\dfrac{2x}{\tan^{-1}(3x)}$

29. $\lim\limits_{x\to0}\left(\cot x-\dfrac{1}{x}\right)$

30. $\lim\limits_{x\to0}\dfrac{\sin^{-1}x-x}{\tan^{-1}x-x}$

31. $\lim\limits_{x\to0}\dfrac{(\sin x)^2}{1-\sec x}$

32. $\lim\limits_{t\to\pi/2}(\tan t-\sec t)$

33. $\lim\limits_{x\to0^+}\left(\dfrac{1}{x}-\dfrac{1}{1-\cos x}\right)$

34. $\lim\limits_{x\to0}\left(\dfrac{e^x-\cos x+\tan x}{x+\tan x+\sin x}\right)$

35. $\lim\limits_{x\to1}\left(\dfrac{1}{\ln x}-\dfrac{1}{x-1}\right)$

36. $\lim\limits_{x\to\pi/2}[(\pi-2x)\sec x]$

37. $\lim\limits_{x\to0^+}[\csc x\cdot\ln(1-\sin x)]$

38. $\lim\limits_{x\to0^+}x^{\sin x}$

39. $\lim\limits_{x\to\infty}\left(\dfrac{1}{x}\right)e^{-x}$

40. $\lim\limits_{x\to0^+}(x-\sin x)^{\sqrt{x}}$

41. $\lim\limits_{x\to0^+}(1-\cos x)^{\tan x}$

42. $\lim\limits_{x\to\infty}(\ln x)^{1/x}$

43. $\lim\limits_{x\to\infty}(e^{2x}+1)^{1/x}$

44. $\lim\limits_{x\to\infty}(x^2+e^x)^{1/x}$

45. $\lim\limits_{x\to\frac{\pi}{2}^-}(\tan x)^{\cos x}$

46. $\lim\limits_{x\to\infty}x^{\tan(1/x)}$

47. $\lim\limits_{x\to\infty}\left(1+\dfrac{1}{x}\right)^{x^3}$

48. $\lim\limits_{x\to\infty}\left(1-\dfrac{1}{x}\right)^{x}$

49. $\lim\limits_{x\to\infty}\left(\dfrac{2x+1}{2x-1}\right)^{\sqrt{x}}$

50. $\lim\limits_{x\to\frac{\pi}{2}^-}(\sin x)^{\tan x}$

51. $\lim\limits_{x\to\infty}(x-\sqrt{x^2+1})$

52. $\lim\limits_{x\to\infty}(2\tan^{-1}x-\pi)\ln x$

53. $\lim\limits_{x\to\infty}\dfrac{2\tan^{-1}x-\pi}{e^{1/x^2}-1}$

54. $\lim\limits_{x\to0}\dfrac{a^x-b^x}{x}$, $a,b>0$

55. $\lim\limits_{x\to0^+}\dfrac{\ln x}{2+3\ln(\sin x)}$

56. $\lim\limits_{x\to\infty}(a^{1/x}-1)x$

*In Exercises 57 and 58, l'Hôpital's Rule is used incorrectly. Find where the error is made, and give the correct solution.*

57. $\lim\limits_{x\to1}\dfrac{x^5-1}{x^2-1}=\lim\limits_{x\to1}\dfrac{5x^4}{2x}=\lim\limits_{x\to1}\dfrac{20x^3}{2}=10$

58. $\lim\limits_{x\to0}\dfrac{e^{3x}+x-1}{e^x-1}=\lim\limits_{x\to0}\dfrac{3e^{3x}+1}{e^x}=\lim\limits_{x\to0}\dfrac{9e^{3x}}{e^x}=9$

59. **Continuous Compound Interest Formula** See Section 3.5. Use l'Hôpital's Rule to derive the continuous compound interest formula

$$A=Pe^{rt}$$

where $A$ is the accumulated amount, $P$ is the principal, $t$ is the time in years, and $r$ is the nominal interest rate per year compounded continuously, from the compound interest formula

$$A=P\left(1+\dfrac{r}{m}\right)^{mt}$$

where $r$ is the nominal interest rate per year compounded $m$ times per year.

60. **Velocity of a Ballast** A ballast of mass $m$ slugs is dropped from a hot-air balloon with an initial velocity of $v_0$ ft/sec. If the ballast is subjected to air resistance that is directly proportional to its instantaneous velocity, then its velocity at time $t$ is

$$v(t)=\dfrac{mg}{k}+\left(v_0-\dfrac{mg}{k}\right)e^{-kt/m}$$

feet per second, where $k>0$ is the constant of proportionality and $g$ is the constant of acceleration. Find an expression for the velocity of the ballast at any time $t$, assuming that there is no air resistance. **Hint:** Find $\lim\limits_{k\to0}v(t)$.

Videos for selected exercises are available online at **www.academic.cengage.com/login**.

**61. Current in a Circuit** A series $RL$ circuit including a resistor $R$ and inductance $L$ is shown in the schematic. Suppose that the electromotive force $E(t)$ is $V$ volts, the resistance is $R$ ohms, and the inductance is $L$ henries, where $V$, $R$, and $L$ are positive constants. Then the current at time $t$ is given by

$$I(t) = \frac{V}{R}\left(1 - e^{-Rt/L}\right)$$

amperes. Using l'Hôpital's Rule, evaluate $\lim_{R \to 0^+} I$ to find an expression for the current in a circuit in which the resistance is 0 ohms.

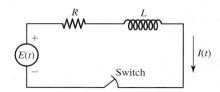

**62. Resonance** Refer to Section 2.5. A weight of mass $m$ is attached to a spring suspended from a support. The weight is then set in motion by an oscillatory force $f(t) = F_0 \sin \omega t$ acting on the support. Here, $F_0$ and $\omega$ are positive constants, and $t$ is time. In the absence of frictional and damping forces, the position of the weight from its equilibrium position at time $t$ is given by

$$x(t) = \frac{F_0(-\omega \sin \omega_0 t + \omega_0 \sin \omega t)}{\omega_0(\omega_0^2 - \omega^2)}$$

with $\omega_0 = \sqrt{k/m}$, where $k$ is the spring constant. Show that if $\omega$ approaches $\omega_0$, the resulting oscillations of the mass increase without bound. This phenomenon is known as **pure resonance.**

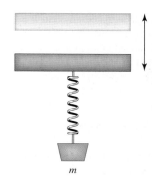

$m$

**63. Bimolecular Reaction** In a bimolecular reaction $A + B \to M$, $a$ moles per liter of $A$ and $b$ moles per liter of $B$ are combined. The number of moles per liter that have reacted after time $t$ is given by

$$x = \frac{ab[1 - e^{(b-a)kt}]}{a - be^{(b-a)kt}} \qquad a \neq b$$

where the positive number $k$ is called the *velocity constant*. Find an expression for $x$ if $a = b$, and find $\lim_{t \to \infty} x$. Interpret your results.

**64. a.** Prove that $\lim\limits_{x \to \infty} \dfrac{x^k}{e^x} = 0$ for every positive constant $k$. This shows that the natural exponential function approaches infinity faster than any power function.

**b.** Prove that $\lim\limits_{x \to \infty} \dfrac{\ln x}{x^k} = 0$ for every positive constant $k$. This shows that the natural logarithmic function approaches infinity slower than any power function.

**65.** Prove that $\lim\limits_{x \to \pi/2} \dfrac{\tan x}{\sec x} = 1$. Can l'Hôpital's Rule be used to compute this limit?

**66.** Show that

$$\lim_{x \to 0} \frac{x^2 \sin\left(\dfrac{1}{x}\right)}{\sin x} = 0$$

Can l'Hôpital's Rule be used to compute this limit?

**67.** Use l'Hôpital's Rule to show that if $f'$ is continuous, then

$$\lim_{h \to 0} \frac{f(x + h) - f(x - h)}{2h} = f'(x)$$

and that if $f''$ is continuous, then

$$\lim_{h \to 0} \frac{f(x + h) - 2f(x) + f(x - h)}{h^2} = f''(x)$$

**68.** Find the slant asymptotes of the graph of $f(x) = x \tan^{-1} x$. Then sketch the graph of the function.
**Hint:** See Section 3.6.

 *In Exercises 69–72, plot the graph the function and use it to guess at the limit. Verify your result using l'Hôpital's Rule.*

**69.** $\lim\limits_{x \to 0} \dfrac{e^x - e^{-2x}}{\ln(1 + x)}$

**70.** $\lim\limits_{x \to 0^+} \left(\dfrac{1}{x}\right)^{\tan x}$

**71.** $\lim\limits_{x \to 1}\left(\dfrac{1}{\ln x} - \dfrac{1}{x - 1}\right)$

**72.** $\lim\limits_{x \to 0} \dfrac{1}{x}[(1 + x)^{1/x} - e]$

*In Exercises 73 and 74, determine whether the statement is true or false. If it is true, explain why it is true. If it is false, give an example to show why it is false.*

**73.** If $\lim\limits_{x \to a} f(x) = 0$ and $\lim\limits_{x \to a} g(x) = 0$, then

$$\lim_{x \to a} \frac{f(x)}{g(x)} = \lim_{x \to a} \frac{d}{dx}\left[\frac{f(x)}{g(x)}\right].$$

**74.** $\lim\limits_{x \to \pi^+} \dfrac{\sin x}{1 - \cos x} = \lim\limits_{x \to \pi^+} \dfrac{\cos x}{\sin x} = \infty$

## 3.9 Newton's Method

There are many occasions when we need to find one or more zeros of a function $f$. For example, the $x$-intercepts of a function $f$ are precisely the values of $x$ that satisfy $f(x) = 0$; the critical numbers of $f$ include the roots of the equation $f'(x) = 0$; and the $x$-coordinates of the candidates for the inflection points of $f$ include the roots of the equation $f''(x) = 0$.

If $f$ is a linear or quadratic function, the zeros are easily found. But, if $f$ is a polynomial function of degree three or higher, the task of finding its zeros is difficult unless the polynomial is easily factored.* We encounter similar difficulties when we try to solve transcendental equations such as

$$x - \tan x = 0 \qquad x > 0$$

In such situations, we have to settle for *approximations* of the roots of the equation.

Actually, in many practical applications we often have a rough idea as to where the zero(s) of interest are located. We can also determine the approximate location of a zero by making a rough sketch of the graph of $f$ and noting where the graph crosses the $x$-axis. Now, once a crude approximation of the desired zero has been found, a procedure is needed that will yield an approximation of the zero to the desired accuracy. Newton's method (also called the Newton-Raphson method) provides us with one such procedure.

### ■ Newton's Method

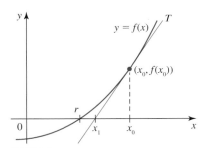

**FIGURE 1**
Starting with an initial estimate $x_0$, Newton's method gives a better approximation $x_1$ to the root $r$.

Suppose $f$ has a zero at $r$ that we want to approximate (see Figure 1). Let $x_0$ be an approximation of $r$ obtained, for example, from a rough sketch of the graph of $f$ or by using the Intermediate Value Theorem, and let $T$ denote the tangent line to the graph of $f$ at $(x_0, f(x_0))$.

Because of the proximity of the points on $T$ to the graph of $f$ near $(x_0, f(x_0))$, one may expect that if $x_0$ is close to $r$, then the $x$-intercept of $T$ (call it $x_1$) will be close to $r$ as well. As it turns out, $x_1$ often provides us with an even better approximation to $r$ than $x_0$ does (such is the situation depicted in Figure 1).

To find a formula for $x_1$ in terms of $x_0$, we first find an equation for $T$. Since the slope of $T$ is $f'(x_0)$ and its point of tangency is $(x_0, f(x_0))$, we see that such an equation is

$$y - f(x_0) = f'(x_0)(x - x_0) \tag{1}$$

Now, if $f'(x_0) \neq 0$, then setting $y = 0$ in Equation (1) and solving for $x$ gives the $x$-intercept of $T$, $x_1$. Thus,

$$x_1 = x_0 - \frac{f(x_0)}{f'(x_0)} \tag{2}$$

---

*Although formulas exist for solving third- and fourth-degree polynomial equations, they are seldom used because of their complexity. No formula exists for finding the roots of a general polynomial equation of degree five or higher. This fact was demonstrated by the Norwegian mathematician Niels Henrik Abel (1802–1829).

If we repeat the process, this time letting $x_1$ play the role of $x_0$, we obtain yet another approximation of $r$:

$$x_2 = x_1 - \frac{f(x_1)}{f'(x_1)}$$

Continuing in this manner, we generate a sequence of approximations $x_1, x_2, \ldots x_n$, $x_{n+1}, \ldots$, with

$$x_{n+1} = x_n - \frac{f(x_n)}{f'(x_n)} \tag{3}$$

provided that $f'(x_n) \neq 0$.

The sequence obtained through the repetitive use, or *iteration*, of Equation (3) frequently converges to the root $r$ of $f(x) = 0$. Our discussion leads us to the following algorithm for finding an approximation to the root $r$ of $f(x) = 0$.

**The Newton Algorithm**

1. Pick an initial estimate $x_0$ of the root $r$.
2. Generate a sequence of estimates using the iterative formula

$$x_{n+1} = x_n - \frac{f(x_n)}{f'(x_n)} \qquad n = 0, 1, 2, \ldots \tag{4}$$

3. Compute $|x_n - x_{n+1}|$. If this number is less than a prescribed number, stop. The required approximation to the root $r$ is $x_{n+1}$.

**Note**    The initial estimate $x_0$ is normally taken to be a guess of the root $r$. For example, a rough sketch of the graph of $f$ could reveal what a good choice of $x_0$ might be.

## ■ Applying Newton's Method

**EXAMPLE 1**    In Example 9 in Section 1.4, we saw that a zero of $f(x) = x^3 + x - 1$ lies in the interval $(0, 1)$. Use Newton's method with $x_0 = 0.5$ to obtain an approximation of this root. Stop the iteration when $|x_n - x_{n+1}| < 0.000001$.

**Solution**    We have

$$f(x) = x^3 + x - 1$$

and

$$f'(x) = 3x^2 + 1$$

so the iterative formula (4) becomes

$$x_{n+1} = x_n - \frac{f(x_n)}{f'(x_n)} = x_n - \frac{x_n^3 + x_n - 1}{3x_n^2 + 1}$$

$$= \frac{2x_n^3 + 1}{3x_n^2 + 1} \tag{5}$$

Letting $n = 0$ in Equation (5) and using $x_0 = 0.5$, we obtain

$$x_1 = \frac{2(0.5)^3 + 1}{3(0.5)^2 + 1} \approx 0.714285714$$

Next, with $n = 1$ and the value of $x_1$ just obtained, we find

$$x_2 \approx \frac{2(0.714285714)^3 + 1}{3(0.714285714)^2 + 1} \approx 0.683179724$$

Continuing with the iteration, we obtain

$$x_3 \approx 0.682328423 \qquad \text{and} \qquad x_4 \approx 0.682327804$$

Since

$$|x_3 - x_4| \approx 0.000000619 < 0.000001$$

the process is terminated, and we find the required root to be approximately 0.682328. Note that $f(0.682328) \approx 4.7 \times 10^{-7}$ is very close to zero.  ∎

**EXAMPLE 2**  Use Newton's method to find an approximation to the root of the equation $\cos x - x = 0$ accurate to eight decimal places.

**Solution**  By writing the given equation in the form

$$\cos x = x$$

we see that the root of the equation is just the $x$-coordinate of the point of intersection of the graphs of $y = \cos x$ and $y = x$. This observation enables us to obtain an initial estimate of the root of $\cos x - x = 0$ graphically (Figure 2).

From Figure 2 we see that $x_0 = 0.5$ is a reasonable approximation. Writing

$$f(x) = \cos x - x$$

we have

$$f'(x) = -\sin x - 1$$

and the required iterative formula is

$$x_{n+1} = x_n - \frac{f(x_n)}{f'(x_n)} = x_n - \frac{\cos x_n - x_n}{-\sin x_n - 1}$$

$$= \frac{x_n \sin x_n + \cos x_n}{\sin x_n + 1}$$

With $x_0 = 0.5$ we obtain the sequence

$$x_1 \approx 0.755222417$$

$$x_2 \approx 0.739141666$$

$$x_3 \approx 0.739085134$$

$$x_4 \approx 0.739085133$$

$$x_5 \approx 0.739085133$$

Therefore, the root of $\cos x - x = 0$ is approximately 0.73908513.  ∎

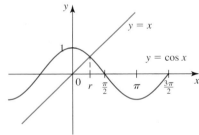

**FIGURE 2**
Our initial estimate of the root of the equation $\cos x = x$ is $x_0 = 0.5$.

**EXAMPLE 3** **Approximating the Square Root of a Positive Number**   Observe that the positive root of the positive number $A$ can be found by solving the equation $x^2 - A = 0$. Thus, an approximation of $\sqrt{A}$ can be obtained by using Newton's method to solve the equation $f(x) = x^2 - A = 0$.

**a.** Find the iterative formula for solving the equation $x^2 - A = 0$.
**b.** Compute $\sqrt{2}$ using this formula with $x_0 = 1$, terminating the process when two successive approximations differ by less than 0.00001.

**Solution**
**a.** We have $f(x) = x^2 - A$ and $f'(x) = 2x$, so by Equation (4)

$$x_{n+1} = x_n - \frac{f(x_n)}{f'(x_n)} = x_n - \frac{x_n^2 - A}{2x_n} = \frac{x_n^2 + A}{2x_n}$$

**b.** With $A = 2$ and $x_0 = 1$, we find

$$x_1 = \frac{1^2 + 2}{2(1)} = 1.5$$

$$x_2 = \frac{(1.5)^2 + 2}{2(1.5)} \approx 1.416667$$

$$x_3 = \frac{(1.416667)^2 + 2}{2(1.416667)} \approx 1.414216$$

$$x_4 = \frac{(1.414216)^2 + 2}{2(1.414216)} \approx 1.414214$$

Since $|x_3 - x_4| = 0.000002 < 0.00001$, we terminate the process. The sequence that was generated converges to $\sqrt{2}$, which is one of the two roots of the equation $x^2 - 2 = 0$. Note that the value of $\sqrt{2}$ to six places is 1.414214.    ■

**EXAMPLE 4** **A Mixture Problem**   A tank initially contains 10 gal of brine with 2 lb of salt. Brine with 1.5 lb of salt per gallon enters the tank at the rate of 3 gal/min, and the well-stirred mixture leaves the tank at the rate of 4 gal/min (see Figure 3). It can be shown that the amount of salt in the tank after $t$ min is $x$ lb, where

$$x = f(t) = 1.5(10 - t) - 0.0013(10 - t)^4 \qquad 0 \le t \le 10$$

Find the time(s) $t$ when the amount of salt in the tank is 5 lb.

**Solution**   We solve the equation

$$1.5(10 - t) - 0.0013(10 - t)^4 = 5$$

or

$$(10 - t)^4 - 1153.846154(10 - t) + 3846.153846 = 0$$

Let's make the substitution $u = 10 - t$. Then the above equation becomes

$$u^4 - 1153.846154u + 3846.153846 = 0$$

Put

$$g(u) = u^4 - 1153.846154u + 3846.153846$$

**FIGURE 3**
Brine enters the tank at the rate of 3 gal/min, and the mixture exits at the rate of 4 gal/min.

Then

$$g'(u) = 4u^3 - 1153.846154$$

and the Newton iteration formula becomes

$$u_{n+1} = u_n - \frac{g(u_n)}{g'(u_n)} = u_n - \frac{u_n^4 - 1153.846154u_n + 3846.153846}{4u_n^3 - 1153.846154}$$

$$= \frac{3u_n^4 - 3846.153846}{4u_n^3 - 1153.846154}$$

A rough sketch of the graph of $g$ on the interval $[0, 10]$ shows that $g$ has zeros near $u = 3$ and $u = 9$. Taking $u_0 = 2$ as an initial guess, we obtain the following sequence:

$$u_1 \approx 3.38563, \qquad u_2 \approx 3.45677, \qquad u_3 \approx 3.45713, \qquad u_4 \approx 3.45713$$

Therefore, $t \approx 10 - 3.45713 = 6.54287 \approx 6.54$.

Next, taking $u_0 = 8$ as an initial guess, we obtain

$$u_1 = 9.44116, \quad u_2 = 9.03541, \quad u_3 = 8.98780, \quad u_4 = 8.98716, \quad u_5 = 8.98716$$

Therefore, $t = 10 - 8.98716 = 1.01284 \approx 1.01$.

So 5 lb of salt are in the tank approximately 1 min and approximately 6.5 min after the brine enters the tank.

## ■ When Newton's Method Does Not Work

Now that we have seen how effective Newton's method can be for finding the zeros of a function, we wish to point out that there are situations in which the method fails and that care must be exercised in applying it. Figure 4a illustrates a situation in which $f'(x_n) = 0$ for some $n$ (in this case, $n = 2$). Since the iterative formula (4) involves division by $f'(x_n)$, it should be clear why the method fails to work in this case. However, we are sometimes able to salvage the situation by choosing a different initial estimate $x_0$ (see Figure 4b).

The situation shown in Figure 5 is more serious, and Newton's method will not work for any choice of the initial estimate $x_0$ other than the actual zero of the function $f(x)$. As you can see from the figure, the sequence $x_1, x_2, x_3, \ldots$ actually moves farther and farther away with each iteration, and thus the method fails.

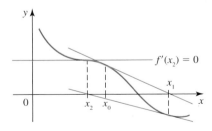

**(a)** Newton's method fails to work because $f'(x_2) = 0$.

**FIGURE 4**

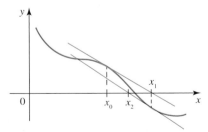

**(b)** The situation in part (a) is remedied by selecting a different initial estimate $x_0$.

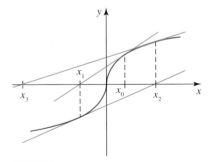

**FIGURE 5**

Newton's method fails here because the sequence of estimates diverges.

## 3.9 CONCEPT QUESTIONS

1. Give a geometric description of Newton's method for finding the zeros of a function $f$. Illustrate graphically.
2. Describe Newton's algorithm for finding the zero of a function $f$.

3. Does Newton's method always work for any choice of the initial estimate $x_0$ of the root of $f(x) = 0$? Explain graphically.

## 3.9 EXERCISES

*In Exercises 1–4, use Newton's method to find the zero(s) of f to four decimal places by solving the equation $f(x) = 0$. Use the initial estimate(s) $x_0$.*

1. $f(x) = -x^3 - 2x + 2$, $x_0 = 1$

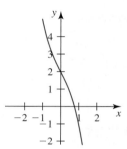

2. $f(x) = 2x^3 - 15x^2 + 36x - 20$, $x_0 = 1$

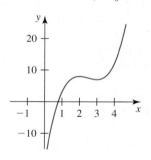

3. $f(x) = \dfrac{3}{2}x^4 - 2x^3 - 6x^2 + 8$, $x_0 = 1$ and $x_0 = 3$

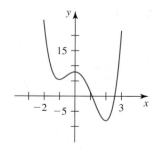

4. $f(x) = x - \sqrt{1 - x^2}$, $x_0 = 0.5$

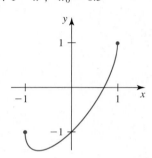

*In Exercises 5–8, use Newton's method to find the point of intersection of the graphs to four decimal places of accuracy by solving the equation $f(x) - g(x) = 0$. Use the initial estimate $x_0$ for the x-coordinate.*

5. $f(x) = x^2$, $g(x) = \sin x$, $x_0 = 1$

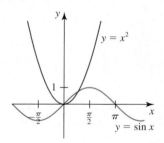

6. $f(x) = \tan x$, $g(x) = 1 - x$, $x_0 = 1$

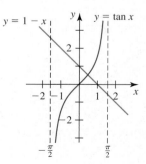

**7.** $f(x) = \dfrac{1}{2}\cos x,\ g(x) = x,\quad x_0 = 0.5$

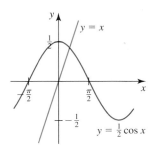

**8.** $f(x) = \sin x,\ g(x) = \dfrac{1}{5}x,\quad x_0 = 2$

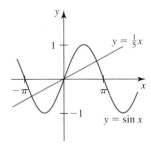

Hint: Use symmetry.

*In Exercises 9–14, use Newton's method to approximate the indicated zero of the function. Continue with the iteration until two successive approximations differ by less than 0.0001.*

**9.** The zero of $f(x) = x^3 + x - 4$ between $x = 0$ and $x = 2$. Take $x_0 = 1$.

**10.** The zero of $f(x) = x^3 + 2x^2 + x - 6$ between $x = 1$ and $x = 2$. Take $x_0 = 1.5$.

**11.** The zero of $f(x) = x^5 + x - 1$ between $x = 0$ and $x = 1$. Take $x_0 = 0.5$.

**12.** The zero of $f(x) = x^5 + 2x^4 + 2x - 4$ between $x = 0$ and $x = 1$. Take $x_0 = 0.5$.

**13.** The zero of $f(x) = 5x + \cos x - 5$ between $x = 0$ and $x = 1$. Take $x_0 = 0.5$.

**14.** The zero of $f(x) = x - \sin x - 0.5$ between $x = 0$ and $x = 2$. Take $x_0 = 1$.

 *In Exercises 15–18, approximate the zero of the function in the indicated interval to six decimal places.*

**15.** $f(x) = x^3 + 3x^2 - 3$ in $[-2, 0]$

**16.** $f(x) = x^3 - x - 1$ in $[1, 2]$

**17.** $f(x) = \cos x - x$ in $\left[0, \dfrac{\pi}{2}\right]$

**18.** $f(x) = 2x - \sin x - 2$ in $[0, \pi]$

*In Exercises 19 and 20, use Newton's method to find the roots of the equation correct to five decimal places.*

**19.** $x \ln x - 1 = 0$

**20.** $\ln x + x - 3 = 0$

**21.** What can you say about the sequence of approximations obtained using Newton's method if your initial estimate, through a stroke of luck, happens to be the root you are seeking?

**22.** Let $f(x) = 2x^3 - 9x^2 + 12x - 2$. Use the Intermediate Value Theorem to prove that $f$ has a zero between $x = 0$ and $x = 1$, and then use Newton's method to find it.

**23.** Let $f(x) = x^3 - 3x - 1$. Use the Intermediate Value Theorem to prove that $f$ has a zero between $x = 1$ and $x = 2$, and then use Newton's method to find it.

*In Exercises 24–27, estimate the value of the radical accurate to four decimal places by using three iterations of Newton's method to solve the equation $f(x) = 0$ with initial estimate $x_0$.*

**24.** $\sqrt{3};\quad f(x) = x^2 - 3;\ x_0 = 1.5$

**25.** $\sqrt{6};\quad f(x) = x^2 - 6;\ x_0 = 2.5$

**26.** $\sqrt[3]{7};\quad f(x) = x^3 - 7;\ x_0 = 2$

**27.** $\sqrt[4]{20};\quad f(x) = x^4 - 20;\ x_0 = 2.1$

**28.** Consider the equation $xe^x = 2$.
  **a.** Show that this equation has one positive root in the interval $(0, 1)$.
  **b.** Use Newton's method to compute the root accurate to five decimal places.

**29.** Use Newton's method to solve the equation
$$320(t + 10e^{-0.1t}) - 13{,}200 = 0$$
accurate to five decimal places.

**30.** Use Newton's method to obtain an approximation of the root of $\cos^{-1} x - x = 0$ accurate to three decimal places.

**31.** Use Newton's method to find the point of intersection of the graphs of $y = \tan^{-1} x$ and $y = \cos^{-1} x$ accurate to three decimal places.

**32.** **Approximating the kth Root of a Positive Number**
  **a.** Apply Newton's method to the solution of the equation $f(x) = x^k - A = 0$ to show that an approximation of $\sqrt[k]{A}$ can be found by using the iteration
$$x_{n+1} = \frac{1}{k}\left[(k-1)x_n + \frac{A}{x_n^{k-1}}\right]$$
  **b.** Use this iteration to find $\sqrt[10]{50}$ accurate to four decimal places.

**33** The graph of $f(x) = x - \sqrt{1 - x^2}$ accompanies Exercise 4. Explain why $x_0 = 1$ cannot be used as an initial estimate for solving the equation $f(x) = 0$ using Newton's method. Can you explain this analytically? How about the initial estimate $x_0 = 0$?

**34.** The temperature at 6 A.M. on a certain December day in Chicago was 15.6°F. As a cold front moved in gradually, the temperature in the next $t$ hours was given by

$$T(t) = -0.05t^3 + 0.4t^2 + 3.8t + 15.6$$

degrees Fahrenheit, where $0 \le t \le 15$. At what time was the temperature 0°F?

**35. Tracking a Submarine** A submarine traveling along a path described by the equation $y = x^2 + 1$ (both $x$ and $y$ are measured in miles) is being tracked by a sonobuoy (sound detector) located at the point $(3, 0)$ in the figure.

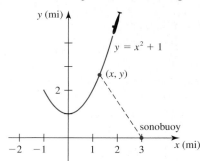

**a.** Show that the submarine is closest to the sonobuoy if $x$ satisfies the equation $2x^3 + 3x - 3 = 0$.
   **Hint:** Minimize the square of the distance between the points $(3, 0)$ and $(x, y)$.
**b.** Use Newton's method to solve the equation $2x^3 + 3x - 3 = 0$.
**c.** What is the distance between the submarine and the sonobuoy at the closest point of approach?
 **d.** Plot the graph of the function $g(x)$ giving the distance between the submarine and the sonobuoy using the viewing window $[0, 1] \times [2, 4]$. Then use it to verify the result that you obtained in parts (b) and (c).

**36. Finding the Position of a Planet** As shown in the accompanying figure, the position of a planet that revolves about the sun with an elliptical orbit can be located by calculating the central angle $\theta$. Suppose the central angle sustained by a planet on a certain day satisfies the equation $\theta - 0.5 \sin \theta = 1$. Using Newton's method, find an approximation of $\theta$ to five decimal places.
   **Hint:** A rough sketch of the graphs of $y = x - 1$ and $y = \frac{1}{2} \sin x$ will show that an initial estimate of $\theta$ may be taken to be $\theta_0 = 1.5$ (radians).

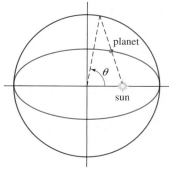

**37. Loan Amortization** The size of the monthly repayment $k$ that amortizes a loan of $A$ dollars in $N$ years at an interest rate of $r$ per year, compounded monthly, on the unpaid balance is given by

$$k = \frac{Ar}{12\left[1 - \left(1 + \dfrac{r}{12}\right)^{-12N}\right]}$$

Show that $r$ can be found by performing the iteration

$$r_{n+1} = r_n - \frac{Ar_n + 12k\left[\left(1 + \dfrac{r_n}{12}\right)^{-12N} - 1\right]}{A - 12Nk\left(1 + \dfrac{r_n}{12}\right)^{-12N-1}}$$

**Hint:** Apply Newton's method to solve the equation

$$Ar + 12k\left[\left(1 + \frac{r}{12}\right)^{-12N} - 1\right] = 0$$

**38. Financing a Home** Refer to Exercise 37. The McCoys secured a loan of $360,000 from a bank to finance the purchase of a house. They have agreed to repay the loan in equal monthly installments of $2106 over 25 years. The bank charges interest at the rate of $r$ per year, compounded monthly. Find $r$.

**39. Financing a Home** Refer to Exercise 37. The Wheatons borrowed a sum of $200,000 from a bank to help finance the purchase of a house. The bank charges interest at the rate of $r$ per year on the unpaid balance, compounded monthly. The Wheatons have agreed to repay the loan in equal monthly installments of $1287.40 over 30 years. What is the rate of interest $r$ charged by the bank?

**40. a.** Show that Newton's method fails when it is used to find the zero of $f(x) = x^{1/3}$ with any initial guess $x_0 \ne 0$.
 **b.** Illustrate the result graphically in the special case in which $x_0 = 1$ by plotting the graph of $f$ at the points $(x_0, f(x_0))$ and $(x_1, f(x_1))$. Use the viewing window $[-5, 5] \times [-2, 2]$.

**41.** For a concrete interpretation of a situation similar to that depicted in Figure 4a, consider the function

$$f(x) = x^3 - 1.5x^2 - 6x + 2$$

**a.** Show that Newton's method fails to work if we choose $x_0 = -1$ or $x_0 = 2$ for an initial estimate.
**b.** Using the initial estimates $x_0 = -2.5$, $x_0 = 1$, and $x_0 = 2.5$, show that the three roots of $f(x) = 0$ are $-2$, $0.313859$, and $3.186141$, respectively.
 **c.** Plot the graph of $f$ using the viewing window $[-3, 4] \times [-10, 7]$.

## CHAPTER 3   REVIEW

## CONCEPT REVIEW

*In Exercises 1–9, fill in the blanks.*

**1. a.** A function $f$ has an absolute maximum at $c$ if _____ for all $x$ in the domain $D$ of $f$. The number $f(c)$ is called the _____ _____ _____ of $f$ on $D$.

   **b.** A function $f$ has a relative minimum at $c$ if _____ for all values of $x$ in some _____ _____ containing $c$.

**2. a.** A critical number of a function $f$ is any number in the _____ of $f$ at which $f'(c)$ _____ or $f'(c)$ _____ _____ _____.

   **b.** If $f$ has a relative extremum at $c$, then $c$ must be a _____ _____ of $f$.

   **c.** If $c$ is a critical number of $f$, then $f$ may or may not have a _____ _____ at $c$.

**3.** The Extreme Value Theorem states that if $f$ is _____ on a closed interval $[a, b]$, then $f$ attains an _____ _____ _____ $f(c)$ for some number $c$ in $[a, b]$ and an _____ _____ _____ $f(d)$ for some number $d$ in $[a, b]$.

**4.** Suppose that $f$ is continuous on $[a, b]$ and differentiable on $(a, b)$.

   **a.** If $f(a) = f(b)$, then Rolle's Theorem states that there exists at least one number $c$ in _____ such that _____.

   **b.** The Mean Value Theorem states that there exists at least one number $c$ in $(a, b)$ such that $f'(c) =$ _____.

**5. a.** A function $f$ is increasing on an interval $I$ if for every pair of numbers $x_1$ and $x_2$ in $I$, $x_1 < x_2$ implies that _____.

   **b.** A function $f$ is monotonic if it is either _____ or _____ on $I$.

   **c.** If $f$ is differentiable on an open interval $(a, b)$ and if $f'(x)$ _____ for all $x$ in $(a, b)$, then $f$ is decreasing on $(a, b)$.

**6. a.** The graph of a differentiable function $f$ is concave upward on an interval $I$ if _____ is increasing on $I$.

   **b.** If $f$ has a second derivative on an open interval $I$ and $f''(x)$ _____ on $I$, then the graph of $f$ is concave upward on $I$.

   **c.** Suppose that the graph of $f$ has a tangent line at $P(c, f(c))$ and the graph of $f$ changes concavity from _____ to _____ or vice versa at $P$; then $P$ is called an inflection point of the graph of $f$.

   **d.** Suppose that $f$ has a continuous second derivative on an interval $(a, b)$, containing a critical number $c$ of $f$. If $f''(c) < 0$, then $f$ has a _____ _____ at $c$. If $f''(c) = 0$, then $f$ may or may not have a _____ _____ at $c$.

**7. a.** The statement $\lim_{x \to a} f(x) = \infty$ means that all the _____ of $f$ can be made _____ _____ by taking $x$ sufficiently close but not equal to _____.

   **b.** The statement $\lim_{x \to \infty} f(x) = L$ means that all the _____ of $f$ can be made arbitrarily close _____ _____ by taking $x$ _____ _____ _____.

   **c.** The statement $\lim_{x \to -\infty} f(x) = \infty$ means that all the values of $f$ can be made _____ _____ as $x$ _____ without bound.

**8. a.** The line $x = a$ is a vertical asymptote of the graph of $f$ if at least one of the following is true: _____, _____, or _____.

   **b.** The line $y = L$ is a horizontal asymptote of the graph of $f$ if _____ or _____.

**9. a.** The precise definition of $\lim_{x \to a} f(x) = \infty$ is: Given any number _____, we can find a number _____ such that $0 < |x - a| < \delta$ implies that _____.

   **b.** The precise definition of $\lim_{x \to -\infty} f(x) = \infty$ means for every number _____, there exists a number _____ such that _____ implies that _____.

## REVIEW EXERCISES

*In Exercises 1–14, find the absolute maximum value and the absolute minimum value, if any, of the function.*

**1.** $f(x) = -x^2 + 4x - 3$ on $[-1, 3]$

**2.** $g(x) = \dfrac{1}{3}x^3 - x^2 + 1$ on $[0, 2]$

**3.** $h(x) = x^3 - 6x^2$ on $[2, 5]$

**4.** $f(t) = \dfrac{t}{t^2 + 1}$ on $[0, 5]$

**5.** $f(x) = 4x - \dfrac{1}{x^2}$ on $[1, 3]$

**6.** $g(x) = x\sqrt{1 - x^2}$ on $[-1, 1]$

**7.** $f(x) = -2x^3 + 9x^2 - 12x + 6$ on $(0, 3]$

8. $g(x) = x^3 - 2x^2 - 4x + 4$ on $(-1, 3)$

9. $f(x) = \cos x - \sin x$ on $[0, 2\pi]$

10. $f(x) = \sin 2x - 2 \sin x$ on $[-\pi, \pi]$

11. $f(x) = \dfrac{x}{2} - \sin x$ on $\left(0, \dfrac{\pi}{2}\right)$

12. $f(x) = x \tan x$ on $\left(-\dfrac{\pi}{2}, \dfrac{\pi}{2}\right)$

13. $f(x) = \dfrac{\ln x}{x}$ on $[1, 5]$

14. $f(x) = \tan^{-1} x - \frac{1}{2} \ln x$ on $\left[\frac{1}{\sqrt{3}}, \sqrt{3}\right]$

*In Exercises 15–20, verify that the function satisfies the hypotheses of the Mean Value Theorem on the given interval, and find all values of c that satisfy the conclusion of the theorem.*

15. $f(x) = x^3$;  $[-2, 1]$

16. $g(x) = \sqrt{x}$;  $[0, 4]$

17. $h(x) = x + \dfrac{1}{x}$;  $[1, 3]$

18. $f(x) = \dfrac{1}{\sqrt{x + 1}}$;  $[0, 3]$

19. $f(x) = x - \sin x$;  $\left[0, \frac{\pi}{2}\right]$

20. $g(x) = \cos 2x$;  $\left[0, \frac{\pi}{2}\right]$

21. Let $f(x) = \dfrac{x}{x + 1}$. Show that there is no value of $c$ in $[-2, 0]$ such that

$$f'(c) = \frac{f(0) - f(-2)}{0 - (-2)}$$

Why doesn't this contradict the Mean Value Theorem?

22. Let $f(x) = |x - 1|$. Show that there is no value $c$ in $[0, 2]$ such that

$$f'(c) = \frac{f(2) - f(0)}{2 - 0}$$

Why doesn't this contradict the Mean Value Theorem?

*In Exercises 23–32, (a) find the intervals where the function f is increasing and where it is decreasing, (b) find the relative extrema of f, (c) find the intervals where the graph of f is concave upward and where it is concave downward, and (d) find the inflection points, if any, of f.*

23. $f(x) = \dfrac{1}{3}x^3 - x^2 + x - 6$

24. $f(x) = (x - 2)^3$

25. $f(x) = x^4 - 2x^2$

26. $f(x) = x + \dfrac{4}{x}$

27. $f(x) = \dfrac{x^2}{x - 1}$

28. $f(x) = \sqrt{x - 1}$

29. $f(x) = (1 - x)^{1/3}$

30. $f(x) = x\sqrt{x - 1}$

31. $f(x) = \dfrac{2x}{x + 1}$

32. $f(x) = -\dfrac{1}{1 + x^2}$

33. Prove that the equation $x^5 + 5x - 2 = 0$ has only one real root.

34. Prove that the equation $x^5 + 3x^3 + x - 2 = 0$ has exactly one real root.

35. Find the intervals where $f(x) = 2x^2 - \ln x$ is increasing and where it is decreasing.

36. Find the intervals where $f(x) = x^2 e^{-x}$ is increasing and where it is decreasing.

*In Exercises 37–46, find the limit.*

37. $\displaystyle\lim_{x \to 2^-} \frac{x^2}{x - 2}$

38. $\displaystyle\lim_{x \to -2^+} \frac{1 - 2x}{x^2 - 4}$

39. $\displaystyle\lim_{x \to 3} \frac{2 - 3x}{(x - 3)^2}$

40. $\displaystyle\lim_{x \to 2^-} \frac{x^2 - x + 1}{x^2 - x - 2}$

41. $\displaystyle\lim_{x \to 0^+} \frac{\sqrt{x}}{\sin x}$

42. $\displaystyle\lim_{x \to (\pi/2)^-} \frac{2x + \sin x}{(\tan x)^2}$

43. $\displaystyle\lim_{x \to \infty} \frac{1 - 2x + 3x^2}{x^2 - 1}$

44. $\displaystyle\lim_{x \to \infty} \frac{x + 4}{\sqrt{x^2 - 1}}$

45. $\displaystyle\lim_{x \to -\infty} \frac{\sqrt{2 - x}}{x + 2}$

46. $\displaystyle\lim_{x \to -\infty} \frac{x}{\sin x - 2x}$

*In Exercises 47 and 48, find the horizontal and vertical asymptotes of the graph of each function. Do not sketch the graph.*

47. $f(x) = \dfrac{1}{2x + 3}$

48. $f(x) = \dfrac{x^2 + x}{x(x - 1)}$

*In Exercises 49–64, use the guidelines on page 348 to sketch the graph of the function.*

49. $f(x) = x^2 - 4x + 3$

50. $f(x) = 2x^3 - 6x^2 + 6x + 3$

51. $g(x) = 3x^4 - 4x^3$

52. $h(x) = 2x + \dfrac{3}{x}$

53. $f(x) = x^2 + \dfrac{1}{x}$

54. $f(x) = \dfrac{2}{1 + x^2}$

55. $f(x) = \dfrac{x^2}{x^2 - 1}$

56. $f(x) = \dfrac{x^2}{x^4 + 1}$

57. $f(x) = (1 - x)^{1/3}$

58. $g(x) = x + \cos x$,  $0 \le x \le 2\pi$

59. $h(x) = 2 \sin x - \sin 2x$,  $0 \le x \le 2\pi$

60. $f(x) = \dfrac{1}{1 - \sin x}$,  $-\dfrac{3\pi}{2} < x < \dfrac{5\pi}{2}$

61. $f(x) = x \ln x$

62. $f(x) = 2 - e^{-x}$

63. $f(x) = \dfrac{3}{1 + e^{-x}}$

64. $f(x) = \sin^{-1}\left(\dfrac{1}{x}\right)$

*In Exercises 65–74, evaluate the limit.*

**65.** $\lim\limits_{x\to 1} \dfrac{x^3 - 2x^2 + x}{x^5 - 1}$

**66.** $\lim\limits_{x\to 2} \dfrac{\sqrt{x} - \sqrt{2}}{x - 2}$

**67.** $\lim\limits_{x\to 0} \dfrac{\sin 2x}{\sin 3x}$

**68.** $\lim\limits_{x\to \infty} \dfrac{e^{2x}}{e^x + x^2}$

**69.** $\lim\limits_{x\to \infty} e^{-x} \cos x$

**70.** $\lim\limits_{x\to (\pi/2)^-} (\cos x)^{\tan x}$

**71.** $\lim\limits_{x\to 0} \left( \csc x - \dfrac{1}{x} \right)$

**72.** $\lim\limits_{x\to (\pi/2)^-} (\sin x)^{\tan x}$

**73.** $\lim\limits_{x\to 0^+} \left( \dfrac{1}{x} - \dfrac{1}{e^x - 1} \right)$

**74.** $\lim\limits_{x\to 0^+} x^n \ln x$

**75.** Find the accumulated amount after 4 years if $5000 is invested at 8% per year compounded continuously.

**76. Effect of Inflation on Salaries** Omar's current annual salary is $70,000. How much will he need to earn 10 years from now in order to retain his present purchasing power if the rate of inflation over that period is 6% per year? Assume that inflation is continuously compounded.

**77. Spread of an Epidemic** The incidence (number of new cases per day) of a contagious disease that is spreading in a population of $M$ people is given by

$$R(x) = kx(M - x)$$

where $k$ is a positive constant and $x$ denotes the number of people already infected. Show that the incidence $R$ is greatest when half the population is infected.

**78. Maximizing Profit** The management of the company that makes Long Horn Barbecue Sauce estimates that the daily profit from the production and sale of $x$ cases (each case contains 24 bottles of the sauce) is

$$P(x) = -0.000002x^3 + 6x - 350$$

dollars. Determine the largest possible daily profit the company can realize.

**79. Senior Workforce** The percent of women age 65 years and older in the workforce from the beginning of 1970 to the beginning of 2000 is approximated by the function

$$P(t) = -0.0002t^3 + 0.018t^2 - 0.36t + 10 \qquad 0 \le t \le 30$$

where $t$ is measured in years, with $t = 0$ corresponding to the beginning of 1970.
  **a.** Find the interval where $P$ is decreasing and the interval where $P$ is increasing.
  **b.** Find the absolute minimum of $P$.
  **c.** Interpret the results of parts (a) and (b).
  *Source:* U.S. Census Bureau.

**80.** Find the point on the graph of $y = x^2 + 1$ that is closest to the point (3, 1).

**81.** Find the dimensions of the rectangle of maximum area, with sides parallel to the coordinate axes, that can be inscribed in the ellipse with equation

$$\frac{x^2}{100} + \frac{y^2}{16} = 1$$

**82.** The sum of two numbers is 8. Find the numbers if the sum of their cubes is as small as possible.

**83.** Find the point on the parabola $y = (x - 3)^2$ that is closest to the origin.

**84.** Find the point on the parabola $y = x^2$ that is closest to the point (5, −1).

**85. Demand for Electricity** The demand for electricity from 1 A.M. through 7 P.M. on August 1, 2006, in Boston is described by the function

$$D(t) = -11.3975t^3 + 285.991t^2 - 1467.73t + 23{,}755$$
$$0 \le t \le 18$$

where $D(t)$ is measured in megawatts (MW), with $t = 0$ corresponding to 1 A.M. Driven overwhelmingly by air-conditioning and refrigeration systems, the demand for electricity reached a new record high that day.
   **a.** Plot the graph of $D$ in the viewing window $[0, 18] \times [20{,}000, 30{,}000]$.
  **b.** Show that the demand for electricity did not exceed the system capacity of 31,000 MW, thus negating the necessity for imposing rolling blackouts if electricity demand were to exceed supply.
  *Source:* ISO New England.

**86. Sickouts** In a sickout by pilots of American Airlines in February 1999, the number of canceled flights from February 6 ($t = 0$) through February 14 ($t = 8$) is approximated by the function

$$N(t) = 1.2576t^4 - 26.357t^3 + 127.98t^2 + 82.3t + 43$$
$$0 \le t \le 8$$

where $t$ is measured in days. The sickout ended after the union was threatened with millions of dollars in fines.
   **a.** Plot the graph of $N$ in the viewing window $[0, 8] \times [0, 1200]$.
  **b.** Show that the number of canceled flights was increasing at the fastest rate on February 8.
  **c.** Estimate the maximum number of canceled flights in a day during the sickout.
  *Source:* Associated Press.

**87. Air Inhaled During Respiration** Suppose that the volume of air inhaled by a person during respiration is given by

$$V(t) = \frac{6}{5\pi} \left( 1 - \cos \frac{\pi t}{2} \right)$$

liters at time $t$ (in seconds). At what time is the rate of flow of air at a maximum? At a minimum?

**88. Path of an Acrobatic Plane** In a fly-by, the path of an acrobatic plane may be described by the equation

$$y = 200(e^{0.01x} + e^{-0.02x})$$

where $x$ and $y$ are both measured in feet.

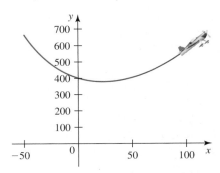

How close to the ground does the plane get?

**89. Absorption of Drugs** Jane took 100 mg of a drug in the morning and another 100 mg of the same drug at the same time the following morning. The amount of the drug in her body $t$ days after the first dosage was taken is given by

$$A(t) = \begin{cases} 100e^{-1.4t} & \text{if } 0 \le t < 1 \\ 100(1 + e^{1.4})e^{-1.4t} & \text{if } t \ge 1 \end{cases}$$

a. How fast was the amount of drug in Jane's body changing after 12 hr $\left(t = \frac{1}{2}\right)$? After 2 days?

b. When was the amount of drug in Jane's body a maximum?

c. What was the maximum amount of drug in Jane's body?

**90.** A box with an open top is to be constructed from a square piece of cardboard, 8 in. wide, by cutting out a square from each of the four corners and bending up the sides. What is the largest volume of such a box?

**91. Packaging** A closed rectangular box with a volume of 4 ft³ is to be constructed. The length of the base of the box will be twice as long as its width. The material for the top and bottom of the box costs 30 cents per square foot, and the material for the sides of the box costs 20 cents per square foot. Find the dimensions of the least expensive box that can be constructed.

**92. Minimizing Construction Costs** A man wishes to construct a cylindrical barrel with a capacity of $32\pi$ ft³. The cost per square foot of the material for the side of the barrel is half that of the cost per square foot for the top and bottom. Help him find the dimensions of the barrel that can be constructed at a minimum cost in terms of material used.

**93. Maximizing Light Intensity** A light is suspended over the center of a 2 ft by 2 ft table. The intensity of the light striking a point on the table is directly proportional to the sine of the angle the path of the light makes with the table and inversely proportional to the square of the distance between

the point and the light. How high above the table should the light be positioned to maximize the light intensity at the corners of the table?

**94. Minimizing Length** Two towers, one 120 ft high, the other 300 ft high, and standing 500 ft apart, are to be stayed by two wires (among others) running from the top of the towers to the ground between them. Where should the stake on the ground be placed if we want to minimize the length of wire used?

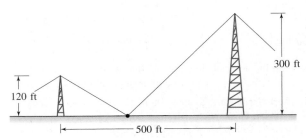

*In Exercises 95–97, use three iterations of Newton's method to approximate the indicated zero of the given function accurate to 4 decimal places.*

**95.** The zero of $f(x) = x^3 + 2x + 2$ between $x = -1$ and $x = 0$. Take $x_0 = -0.5$.

**96.** The zero of $f(x) = x^4 + x - 4$ between $x = 0$ and $x = 2$. Take $x_0 = 1$.

**97.** The zero of $f(x) = x^2 - \sin x$ between $x = \pi/6$ and $x = \pi/2$. Take $x_0 = \pi/4$.

**98.** On what interval is the quadratic function

$$f(x) = ax^2 + bx + c \qquad a \ne 0$$

increasing? On what interval is $f$ decreasing?

**99.** Let $f(x) = x^2 + ax + b$. Determine the constants $a$ and $b$ such that $f$ has a relative minimum at $x = 2$ and the relative minimum value is 7.

**100.** Find the values of $c$ such that the graph of

$$f(x) = x^4 + 2x^3 + cx^2 + 2x + 2$$

is concave upward everywhere.

**101.** Let $f(x) = ax^6 + bx^4 + cx^2 + d$, where $a$, $b$, $c$, and $d$ are positive constants. Can the graph of $f$ have any inflection points? Explain.

**102.** Let

$$f(x) = \begin{cases} x^3 + 1 & \text{if } x \ne 0 \\ 2 & \text{if } x = 0 \end{cases}$$

a. Compute $f'(x)$, and show that it does not change sign as we move across $x = 0$.

b. Show that $f$ has a relative maximum at $x = 0$. Does this contradict the first derivative test? Explain your answer.

## PROBLEM-SOLVING TECHNIQUES

As a first step in analyzing functions involving absolute values, we often rewrite the function as an equivalent piecewise defined function. This is illustrated in the following example.

**EXAMPLE** Find the absolute minimum value of the function $f(x) = |x|^3 + |x - 2|^3$.

**Solution**  We write

$$f(x) = |x|^3 + |x - 2|^3 = \begin{cases} -x^3 - (x - 2)^3 & \text{if } x < 0 \\ x^3 - (x - 2)^3 & \text{if } 0 \leq x < 2 \\ x^3 + (x - 2)^3 & \text{if } x \geq 2 \end{cases}$$

Differentiating, we obtain

$$f'(x) = \begin{cases} -3x^2 - 3(x - 2)^2 & \text{if } x < 0 \\ 3x^2 - 3(x - 2)^2 & \text{if } 0 \leq x < 2 \\ 3x^2 + 3(x - 2)^2 & \text{if } x \geq 2 \end{cases}$$

Since $\lim_{x \to 0^-} f'(x) = -12 = \lim_{x \to 0^+} f'(x)$, we see that $f'$ exists at 0. Similarly, you can show that $f'$ exists at 2. We see that $f'(x) < 0$ for $x < 0$ and $f'(x) > 0$ for $x > 2$. Next, observe that for $0 < x < 2, f'(x) = 0$ if

$$3x^2 - 3(x - 2)^2 = 3x^2 - 3x^2 + 12x - 12 = 12(x - 1) = 0$$

so $x = 1$ is a critical number of $f$. The sign diagram of $f'$ follows.

The sign diagram for $f'$

From the sign diagram for $f'$ we see that $f$ has a relative minimum at $x = 1$, where $f$ takes on the value $f(1) = 2$. Therefore, the absolute minimum value of $f$ is 2. ∎

## CHALLENGE PROBLEMS

1. Find the absolute minimum value of $f(x) = |x|^3 + |x - 1|^3$.

2. Find the relative extrema of

$$f(x) = \frac{(x + a)(x + b)}{(x - a)(x - b)} \qquad a + b \neq 0$$

3. Find the highest and the lowest points on the graph of the equation $x^2 + xy + y^2 = 3$.

4. Use the Mean Value Theorem to prove the inequalities

$$nx^{n-1}(y - x) \leq y^n - x^n \leq ny^{n-1}(y - x)$$

for $0 \leq x \leq y$, where $n$ is a natural number.

5. **a.** Show that if $0 \leq p \leq 1$ and $x \geq 0$, then
$(1 + x)^p \leq 1 + x^p$.

**b.** Use the result of part (a) to show that $(a + b)^p < a^p + b^p$ for all positive numbers $a$ and $b$ and $0 \leq p \leq 1$.

**c.** Use the result of part (b) to show that $\sqrt[n]{a + b} \leq \sqrt[n]{a} + \sqrt[n]{b}$ if $n \geq 1$.

6. Find the maximum of $x + y$ defined on the part of the *eight curve* $x^4 = x^2 - y^2$ that lies in the first quadrant.

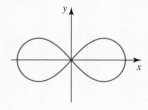

7. Show that the equation $x^5 - 5x + c = 0$ does not have two roots between 0 and 1 for any $c$.

8. The equation

$$v^3 - \frac{Tv^2}{2\alpha} + 1 = 0$$

is called *Rivlin's equation* and arises in the study of incompressible material. Specifically, $v$ gives the factor by which a rubber cube is stretched in two directions and the factor by which it is contracted in the other when the material is pulled on all faces with a force $T$. The positive constant $\alpha$ is analogous to the spring constant for a spring. Show that Rivlin's equation has no positive root if $T < 3\sqrt[3]{2}\alpha$ and two positive roots if $T > 3\sqrt[3]{2}\alpha$.

9. Prove that between any two zeros of the polynomial function

$$f(x) = a_n x^n + a_{n-1} x^{n-1} + \cdots + a_1 x + a_0$$

there exists a zero of the polynomial function

$$g(x) = na_n x^{n-1} + \cdots + 2a_2 x + a_1$$

10. A generalization of the Mean Value Theorem is Cauchy's Theorem: If $f$ and $g$ are continuous on $[a, b]$ and differentiable on $(a, b)$, $g'(x) \neq 0$ for all $x$ in $(a, b)$, and $g(a) \neq g(b)$, then there exists a number $c$ in $(a, b)$ such that

$$\frac{f(b) - f(a)}{g(b) - g(a)} = \frac{f'(c)}{g'(c)}$$

   a. Explain what is wrong with the following proof of Cauchy's Theorem: Since all the conditions of the Mean Value Theorem are satisfied by $f$ and $g$, there exists a number $c$ in $(a, b)$ such that

   $$\frac{f(b) - f(a)}{b - a} = f'(c) \quad \text{and} \quad \frac{g(b) - g(a)}{b - a} = g'(c)$$

   Therefore, dividing the first expression by the second gives the desired result.

   b. Prove Cauchy's Theorem by applying Rolle's Theorem to the function

   $$h(x) = f(x) - f(a) - \frac{f(b) - f(a)}{g(b) - g(a)} [g(x) - g(a)]$$

11. Prove that if $a_0, a_1, a_2, \ldots, a_n$ are real numbers such that

$$\frac{a_0}{1} + \frac{a_1}{2} + \frac{a_2}{3} + \cdots + \frac{a_n}{n+1} = 0$$

then the polynomial function $f(x) = a_0 + a_1 x + a_2 x^2 + \cdots + a_n x^n$ has at least one zero in the interval $(0, 1)$.

12. Suppose that $F$ is continuous on $[a, b]$ and that it has a derivative $f$ for all $x$ in $(a, b)$, and suppose that $x_i$, where $0 \leq i \leq n$, are real numbers satisfying

$$x_0 = a < x_1 < x_2 < \cdots < x_n = b$$

Show that there exist numbers $c_1$ in $[x_0, x_1]$, $c_2$ in $[x_1, x_2], \ldots, c_n$ in $[x_{n-1}, x_n]$ such that

$$F(b) - F(a) = f(c_1)(x_1 - x_0) + f(c_2)(x_2 - x_1)$$
$$+ \cdots + f(c_n)(x_n - x_{n-1})$$

13. Let $f$ be defined and have a continuous derivative of order $(n - 1)$, $f^{(n-1)}(x)$, on an interval $[a, b]$ and a derivative of order $n$, $f^{(n)}(x)$, on the interval $(a, b)$. Furthermore, let

$$f(x_0) = f(x_1) = \cdots = f(x_n)$$
$$x_0 = a < x_1 < x_2 < \cdots < x_n = b$$

Show that there exists at least one number $c$ in $(a, b)$ such that $f^{(n)}(c) = 0$.

14. Prove the inequality $x/(x + 1) \leq \ln(x + 1) \leq x$ for $x \geq 0$.

15. Prove the inequality $\dfrac{x}{1 + x^2} < \tan^{-1} x < x$ for $x > 0$.

16. Let $f(x) = x^x (1 - x)^{1-x}$ for $0 < x < 1$.
    a. Evaluate $\lim_{x \to 0^+} f(x)$ and $\lim_{x \to 1^-} f(x)$.
    b. Find the absolute extrema of $f$ on $(0, 1)$.

17. Evaluate $\lim_{x \to \infty} \left( \sqrt[n]{x^n + a_{n-1} x^{n-1} + \cdots + a_0} - \sqrt[n]{x^n + b_{n-1} x^{n-1} + \cdots + b_0} \right)$.

In Chapter 2, we saw how the derivative of a function enabled us to calculate the velocity of the maglev knowing only its position function. In this chapter, we will see how the knowledge of the velocity of the maglev at time $t$ will enable us to calculate its position at any time $t$. The tool used here is the antiderivative of a function. As it turns out, the derivative of a function and the antiderivative of the function are intimately related—one of the fundamental results of this chapter.

Liu Jin/Getty Images

# 4  Integration

**IN THIS CHAPTER** we begin the study of the other major branch of calculus, known as *integral calculus*. Historically, the development of integral calculus, like the development of differential calculus, was motivated by a geometric problem. In this case the problem is that of finding the area of a region in the plane.

The principal tool in the study of integral calculus is the *definite integral*, which, as in the case of the derivative, is defined by using the notion of a limit. As we shall see in the ensuing chapters, the concept of the integral allows us to solve not only the area problem, but also other geometric problems, such as finding the lengths of curves and the volumes and surface areas of solids. The integral also proves to be an all-important tool in solving problems in physics, chemistry, biology, engineering, economics, and other fields.

Although the two branches of calculus seem at first sight to be unconnected, they are, in fact, intimately related. This relationship is established via the Fundamental Theorem of Calculus, which is the main result of this chapter. This theorem also simplifies the calculations involved in solving many problems.

**V** This symbol indicates that one of the following video types is available for enhanced student learning at **www.academic.cengage.com/login:**
- Chapter lecture videos
- Solutions to selected exercises

## 4.1  Indefinite Integrals

### ■ Antiderivatives

Let's return to the example involving the motion of the maglev (see Figure 1). In Chapter 2 we discussed the following problem: *If we know the position of the maglev at all times t, can we find its velocity at any time t?* As it turns out, if the position of the maglev is described by the position function $f$, then its velocity at any time $t$ is given by $f'(t)$. Here $f'$, the velocity function of the maglev, is just the derivative of $f$.

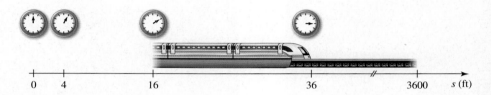

**FIGURE 1**       0    4                16                              36              3600       $s$ (ft)

Now, in Chapters 4 and 5 we will consider precisely the opposite problem: *If we know the velocity of the maglev at all times t, can we find its position at any time t?* Stated another way, if we know the velocity function $f'$ of the maglev, can we find its position function $f$? To solve this problem, we need the concept of an *antiderivative* of a function.

> **DEFINITION   Antiderivative**
>
> A function $F$ is an **antiderivative** of a function $f$ on an interval $I$ if $F'(x) = f(x)$ for all $x$ in $I$.

Thus, an antiderivative of a function $f$ is a function $F$ whose derivative is $f$.

**EXAMPLE 1**  Show that $F_1(x) = x^3$, $F_2(x) = x^3 + 1$, and $F_3(x) = x^3 - \pi$ are antiderivatives of $f(x) = 3x^2$. How about the function $G(x) = x^3 + C$, where $C$ is any constant?

**Solution**   You can easily verify that $F_1'(x) = F_2'(x) = F_3'(x) = 3x^2 = f(x)$ for all $x$ in $(-\infty, \infty)$. Therefore, by the definition of an antiderivative, $F_1$, $F_2$, and $F_3$ are all antiderivatives of $f$, as was asserted.

Next, we find

$$G'(x) = \frac{d}{dx}(x^3 + C)$$

$$= 3x^2 + 0 = 3x^2 = f(x)$$

so $G$ is also an antiderivative of $f$.                                        ■

Example 1 suggests the following more general result: If $F$ is an antiderivative of $f$ on $I$, then so is every function of the form $G(x) = F(x) + C$, where $C$ is an arbitrary constant. To prove this, we find

$$G'(x) = \frac{d}{dx}[F(x) + C] = \frac{d}{dx}[F(x)] + \frac{d}{dx}(C) = F'(x) + 0 = F'(x) = f(x)$$

Are there any antiderivatives of $f$ other than those that are obtained in this manner? To answer this question, suppose that $H$ is any other antiderivative of $f$ on $I$. Then

$$F'(x) = H'(x) = f(x)$$

Since two functions having the same derivative on an interval differ only by a constant, (by the corollary to the Mean Value Theorem, page 298), we have $H(x) - F(x) = C$, where $C$ is a constant. Equivalently, $H(x) = F(x) + C$.

---

**THEOREM 1**

If $F$ is an antiderivative of $f$ on an interval $I$, then *every* antiderivative of $f$ on $I$ has the form

$$G(x) = F(x) + C$$

where $C$ is a constant.

---

**EXAMPLE 2** Let $f(x) = 1$.

**a.** Show that $F(x) = x$ is an antiderivative of $f$ on $(-\infty, \infty)$.
**b.** Find all antiderivatives of $f$ on $(-\infty, \infty)$.

**Solution**
**a.** $F'(x) = 1 = f(x)$, and this proves that $F$ is an antiderivative of $f$.
**b.** By Theorem 1 the antiderivatives of $f$ have the form $G(x) = x + C$, where $C$ is an arbitrary constant.

Figure 2 shows the graphs of some antiderivatives of $f(x) = 1$. These graphs constitute part of a family of infinitely many parallel lines, each having slope 1. This result is expected, because an antiderivative $G$ of $f$ satisfies $G'(x) = f(x) = 1$, and there are infinitely many straight lines that have slope 1. The antiderivatives $G(x) = x + C$, where $C$ is a constant, are precisely the functions representing this family of straight lines.

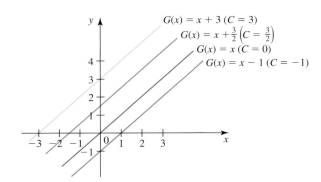

**FIGURE 2**
The graphs of the antiderivatives $G$ of $f(x) = 1$ constitute a family of straight lines, each with slope 1.

### ■ The Indefinite Integral

The process of finding all antiderivatives of a function is called **antidifferentiation** or **integration.** We can view this process as an operation on a function $f$ to produce the entire family of antiderivatives of $f$. The integral operator is denoted by the integral sign $\int$, and the process of integration is indicated by the expression

$$\int f(x)\, dx = F(x) + C$$

which is read "the **indefinite integral** of $f(x)$ with respect to $x$ equals $F(x)$ plus $C$." The function $f$ to be integrated is called the **integrand.** The differential $dx$ reminds us that the integration is performed with respect to the variable $x$. The function $F$ is an antiderivative of $f$, and the constant $C$ is called a **constant of integration.** Using this notation, the result of Example 2 is written

$$\int 1\, dx = x + C$$

### ■ Basic Rules of Integration

Because integration and differentiation are, in a sense, reverse operations, we can discover many of the rules of integration by guessing at an antiderivative $F$ of an integrand $f$ and then verifying that $F$ is an antiderivative of $f$ by demonstrating that $F'(x) = f(x)$. For example, to find the indefinite integral of $f(x) = x^n$, we first recall the Power Rule for differentiating $f(x) = x^n$. Thus,

$$f'(x) = \frac{d}{dx}(x^n) = nx^{n-1}$$

In writing the derivative $nx^{n-1}$, we followed these steps:

**Step 1**  Diminish the power of $x^n$ by 1 to obtain $x^{n-1}$.
**Step 2**  Multiply $x^{n-1}$ by the "old" power $n$ to obtain $nx^{n-1}$.

Now, if we reverse the operation in each step, we have

**Step 1**  Increase the power of $x^n$ by 1 to obtain $x^{n+1}$.

**Step 2**  Divide $x^{n+1}$ by the "new" power $n + 1$ to obtain $\dfrac{x^{n+1}}{n+1}$.

This argument suggests that

$$\int x^n\, dx = \frac{x^{n+1}}{n+1} + C \qquad n \neq -1$$

To verify that this formula is correct, we compute

$$\frac{d}{dx}\left[\frac{x^{n+1}}{n+1} + C\right] = \frac{n+1}{n+1}x^{(n+1)-1} = x^n$$

In a similar manner we obtain the following integration formulas by studying the corresponding differentiation formulas.

## Basic Integration Formulas

| Differentiation Formula | Integration Formula | |
|---|---|---|
| 1. $\dfrac{d}{dx}(C) = 0$ | $\displaystyle\int 0 \, dx = C$ | |
| 2. $\dfrac{d}{dx}(x^n) = nx^{n-1}$ | $\displaystyle\int x^n \, dx = \dfrac{x^{n+1}}{n+1} + C$ | $n \neq -1$ |
| 3. $\dfrac{d}{dx}(\sin x) = \cos x$ | $\displaystyle\int \cos x \, dx = \sin x + C$ | |
| 4. $\dfrac{d}{dx}(\cos x) = -\sin x$ | $\displaystyle\int \sin x \, dx = -\cos x + C$ | |
| 5. $\dfrac{d}{dx}(\tan x) = \sec^2 x$ | $\displaystyle\int \sec^2 x \, dx = \tan x + C$ | |
| 6. $\dfrac{d}{dx}(\sec x) = \sec x \tan x$ | $\displaystyle\int \sec x \tan x \, dx = \sec x + C$ | |
| 7. $\dfrac{d}{dx}(\csc x) = -\csc x \cot x$ | $\displaystyle\int \csc x \cot x \, dx = -\csc x + C$ | |
| 8. $\dfrac{d}{dx}(\cot x) = -\csc^2 x$ | $\displaystyle\int \csc^2 x \, dx = -\cot x + C$ | |
| 9. $\dfrac{d}{dx}(e^x) = e^x$ | $\displaystyle\int e^x \, dx = e^x + C$ | |
| 10. $\dfrac{d}{dx}(\ln|x|) = \dfrac{1}{x}$ | $\displaystyle\int \dfrac{dx}{x} = \ln|x| + C$ | |
| 11. $\dfrac{d}{dx}(a^x) = a^x \ln a$ | $\displaystyle\int a^x \, dx = \dfrac{a^x}{\ln a} + C$ | $a > 0, a \neq 1$ |

**Note**   The formulas for integrals such as $\int \tan x \, dx$ and $\int \sec x \, dx$ are not as easily found. We will learn how to find formulas for such integrals later on.   ■

**EXAMPLE 3**   Using Formula 2 for integration, we see that

**a.** $\displaystyle\int 1 \, dx = \int x^0 \, dx = \dfrac{x^{0+1}}{0+1} + C = x + C$     Here, $n = 0$.

**b.** $\displaystyle\int x^2 \, dx = \dfrac{x^{2+1}}{2+1} + C = \dfrac{1}{3}x^3 + C$

**c.** $\displaystyle\int \dfrac{1}{x^3} \, dx = \int x^{-3} \, dx = \dfrac{x^{-3+1}}{-3+1} + C = -\dfrac{1}{2x^2} + C$     Here, $n = -3$.

**d.** $\displaystyle\int x^{1/4} \, dx = \dfrac{x^{1/4+1}}{\frac{1}{4}+1} + C = \dfrac{4}{5}x^{5/4} + C$     Here, $n = \frac{1}{4}$.   ■

**Note**   We can check our answers by differentiating each indefinite integral and showing that the result is equal to the integrand. Thus, to verify the result of Example 3d, we compute

$$\dfrac{d}{dx}\left(\dfrac{4}{5}x^{5/4} + C\right) = \dfrac{4}{5} \cdot \dfrac{5}{4}x^{5/4-1} = x^{1/4}$$   ■

---

**Rules of Integration**

1. $\displaystyle\int c\,f(x)\,dx = c\int f(x)\,dx,$   where $c$ is a constant

2. $\displaystyle\int [f(x) \pm g(x)]\,dx = \int f(x)\,dx \pm \int g(x)\,dx$

---

Thus, the indefinite integral of a constant multiple of a function is equal to the constant multiple of the indefinite integral, and the indefinite integral of the sum (difference) of two functions is equal to the sum (difference) of their indefinite integrals.

Also, Rule 2 is valid for any finite number of functions; that is,

$$\int [f_1(x) \pm \cdots \pm f_n(x)]\,dx = \int f_1(x)\,dx \pm \int f_2(x)\,dx \pm \cdots \pm \int f_n(x)\,dx$$

**EXAMPLE 4**   Find

**a.** $\displaystyle\int 2x^3\,dx$     **b.** $\displaystyle\int (2x + 3\sin x)\,dx$     **c.** $\displaystyle\int\left(3e^x - \frac{2}{x}\right)dx$

**Solution**

**a.** $\displaystyle\int 2x^3\,dx = 2\int x^3\,dx$     Rule 1

$$= 2\left(\frac{1}{4}x^4 + C_1\right)$$     Formula 2

$$= \frac{1}{2}x^4 + 2C_1$$

where $C_1$ is a constant of integration. Since $C_1$ is arbitrary, so is $2C_1$, and we can write $2C_1 = C$, where $C$ is an arbitrary number. Therefore,

$$\int 2x^3\,dx = \frac{1}{2}x^4 + C$$

**b.** $\displaystyle\int (2x + 3\sin x)\,dx = \int 2x\,dx + \int 3\sin x\,dx$     Rule 2

$$= 2\int x\,dx + 3\int \sin x\,dx$$     Rule 1

$$= 2\left(\frac{1}{2}x^2 + C_1\right) + 3(-\cos x + C_2)$$     Formulas 2 and 4

$$= x^2 + 2C_1 - 3\cos x + 3C_2$$

$$= x^2 - 3\cos x + C$$

where $C = 2C_1 + 3C_2$ is an arbitrary constant.

**c.** $\displaystyle\int\left(3e^x - \frac{2}{x}\right)dx = \int 3e^x\,dx - \int\frac{2}{x}\,dx$     Rule 2

$$= 3\int e^x\,dx - 2\int\frac{1}{x}\,dx$$     Rule 1

$$= 3(e^x + C_1) - 2(\ln|x| + C_2)$$

$$= 3e^x + 3C_1 - 2\ln|x| - 2C_2$$

$$= 3e^x - \ln x^2 + C$$

where $C = 3C_1 - 2C_2$.

From now on, we will use a single letter $C$ to represent any combination of constants of integration.

**EXAMPLE 5**  Find $\int (3x^5 - 2x^3 + 2 - 3x^{-1/3}) \, dx$.

**Solution**  Using the generalized sum rule, we have

$$\int (3x^5 - 2x^3 + 2 - 3x^{-1/3}) \, dx = 3\int x^5 \, dx - 2\int x^3 \, dx + 2\int 1 \, dx - 3\int x^{-1/3} \, dx$$

$$= 3\left(\frac{x^6}{6}\right) - 2\left(\frac{x^4}{4}\right) + 2x - 3\left(\frac{x^{2/3}}{\frac{2}{3}}\right) + C$$

$$= \frac{1}{2}x^6 - \frac{1}{2}x^4 + 2x - \frac{9}{2}x^{2/3} + C$$

Sometimes we need to rewrite the integrand in a different form before integrating, as is illustrated in the next example.

**EXAMPLE 6**  Find

**a.** $\int (x + 1)(x^2 - 2) \, dx$      **b.** $\int \frac{2x^2 - 1}{x^2} \, dx$      **c.** $\int \frac{\sin t}{\cos^2 t} \, dt$

**Solution**

**a.** $\int (x + 1)(x^2 - 2) \, dx = \int (x^3 + x^2 - 2x - 2) \, dx = \frac{1}{4}x^4 + \frac{1}{3}x^3 - x^2 - 2x + C$

**b.** $\int \frac{2x^2 - 1}{x^2} \, dx = \int \left(2 - \frac{1}{x^2}\right) dx = \int (2 - x^{-2}) \, dx = 2x + x^{-1} + C = 2x + \frac{1}{x} + C$

**c.** $\int \frac{\sin t}{\cos^2 t} \, dt = \int \frac{1}{\cos t} \cdot \frac{\sin t}{\cos t} \, dt = \int \sec t \tan t \, dt = \sec t + C$

## Differential Equations

Let's return to the problem posed at the beginning of the section: Given the derivative of a function, $f'$, can we find the function $f$? As an example, suppose that we are given the function

$$f'(x) = 2x - 1 \tag{1}$$

and we wish to find $f(x)$. From what we now know, we can find $f$ by integrating Equation (1). Thus,

$$f(x) = \int f'(x) \, dx = \int (2x - 1) \, dx = x^2 - x + C \tag{2}$$

where $C$ is an arbitrary constant. So there are infinitely many functions having the derivative $f'$; these functions differ from each other by a constant.

Equation (1) is called a *differential equation.* In general, a **differential equation** is an equation that involves the derivative or differential of an unknown function. (In Equation (1) the unknown function is $f$.) A **solution** of a differential equation on an interval $I$ is any function that satisfies the differential equation on $I$. Thus, Equation (2) gives all solutions of the differential equation (1) on $(-\infty, \infty)$ and is, accordingly, called the **general solution** of the differential equation $f'(x) = 2x - 1$.

The graphs of $f(x) = x^2 - x + C$ for selected values of $C$ are shown in Figure 3. These graphs have one property in common: For any fixed value of $x$, the tangent lines to these graphs have the same slope. This follows because any member of the family $f(x) = x^2 - x + C$ must have the same slope at $x$, namely, $2x - 1$. (We will study differential equations in greater depth in Chapter 8.)

Although there are infinitely many solutions to the differential equation $f'(x) = 2x - 1$, we can obtain a **particular solution** by specifying the value that the function must assume at a certain value of $x$. For example, suppose we stipulate that the solution $f(x) = x^2 - x + C$ must satisfy the condition $f(1) = 3$. Then, we find that

$$f(1) = 1 - 1 + C = 3$$

and $C = 3$. Thus, the particular solution is $f(x) = x^2 - x + 3$ (see Figure 3). The condition $f(1) = 3$ is an example of an *initial condition.* More generally, an **initial condition** is a condition that is imposed on the value of $f$ at a number $x = a$. Geometrically, this means that the graph of the particular solution passes through the point $(a, f(a))$.

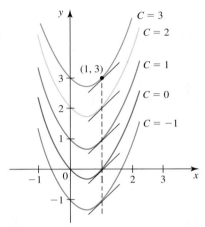

**FIGURE 3**
The graphs of some functions having the derivative $f'(x) = 2x - 1$

### ■ Initial Value Problems

In an **initial value problem** we are required to find a function satisfying (1) a differential equation and (2) one or more initial conditions. The following are examples.

---

**EXAMPLE 7**   **Finding the Position of a Maglev**   In a test run of a maglev along a straight elevated monorail track, data obtained from reading its speedometer indicated that the velocity of the maglev at time $t$ can be described by the velocity function

$$v(t) = 8t \qquad 0 \le t \le 30$$

Find the position function of the maglev. Assume that the maglev is initially located at the origin of a coordinate line.

**Solution**   Let $s(t)$ denote the position of the maglev at time $t$, where $0 \le t \le 30$. Then $s'(t) = v(t)$. So we have the initial value problem

$$\begin{cases} s'(t) = 8t \\ s(0) = 0 \end{cases}$$

Integrating both sides of the differential equation $s'(t) = 8t$, we obtain

$$s(t) = \int s'(t)\, dt = \int 8t\, dt = 4t^2 + C$$

where $C$ is an arbitrary constant. To evaluate $C$, we use the initial condition $s(0) = 0$ to write

$$s(0) = 4(0) + C = 0 \qquad \text{or} \qquad C = 0$$

Therefore, the required position function is $s(t) = 4t^2$, where $0 \le t \le 30$.   ■

**EXAMPLE 8**   **Describing the Path of a Pop-Up**   In a baseball game, one of the batters hit a pop-up. Suppose that the initial velocity of the ball was 96 ft/sec and the initial height of the ball was 4 ft from the ground.

**a.** Find the position function giving the height of the ball at any time $t$.
**b.** How high did the ball go?
**c.** How long did the ball stay in the air after being struck?

**Solution**

**a.** Let $s(t)$ denote the position of the ball at time $t$, and let $t = 0$ represent the (initial) time when the ball was struck. The only force acting on the ball during the motion is the force of gravity; taking the acceleration due to this force as $-32$ ft/sec², we see that $s$ must satisfy

$$s''(t) = -32$$

When $t = 0$,

$$s(0) = 4 \qquad \text{Initial height was 4 ft.}$$

and

$$s'(0) = 96 \qquad \text{Initial velocity was 96 ft/sec.}$$

To solve this initial value problem, we integrate the differential equation $s''(t) = -32$ with respect to $t$, obtaining

$$s'(t) = \int s''(t)\, dt = \int -32\, dt = -32t + C_1$$

where $C_1$ is an arbitrary constant. To determine the value of $C_1$, we use the initial condition $s'(0) = 96$. We find

$$s'(0) = -32(0) + C_1 = 96$$

which gives $C_1 = 96$. Therefore,

$$s'(t) = -32t + 96$$

Integrating again, we have

$$s(t) = \int s'(t)\, dt = \int (-32t + 96)\, dt = -16t^2 + 96t + C_2$$

where $C_2$ is an arbitrary constant. To evaluate $C_2$, we use the initial condition $s(0) = 4$ to obtain

$$s(0) = -16(0) + 96(0) + C_2 = 4 \qquad \text{or} \qquad C_2 = 4$$

Therefore, the required position function is

$$s(t) = -16t^2 + 96t + 4$$

**b.** At the highest point, the velocity of the ball is zero. But from part (a) the velocity of the ball at any time $t$ is $v(t) = s'(t) = -32t + 96$. So setting $v(t) = 0$, we obtain $-32t + 96 = 0$, or $t = 3$. Substituting this value of $t$ into the position function gives

$$s(3) = -16(3^2) + 96(3) + 4$$

or 148 ft as the maximum height attained by the ball. (See Figure 4.)

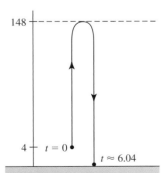

**FIGURE 4**
The ball attains a maximum height of 148 ft and stays in the air approximately 6 sec.

**c.** The ball hits the ground when $s(t) = 0$. Solving this equation, we have

$$-16t^2 + 96t + 4 = 0$$

or

$$4t^2 - 24t - 1 = 0$$

Next, using the quadratic formula, we obtain

$$t = \frac{24 \pm \sqrt{576 + 16}}{8} = \frac{24 \pm 4\sqrt{37}}{8} \approx -0.04 \quad \text{or} \quad 6.04$$

Since $t$ must be positive, we see that the ball hit the ground when $t \approx 6.04$. Therefore, after the ball was struck, it remained in the air for approximately 6 sec.    ■

## 4.1    CONCEPT QUESTIONS

1. What is an antiderivative of a function $f$? Give an example.
2. If $f'(x) = g'(x)$ for all $x$ in an interval $I$, what is the relationship between $f$ and $g$?
3. What is the difference between an antiderivative of $f$ and the indefinite integral of $f$?

4. Define each of the following:
   **a.** A differential equation
   **b.** A solution of a differential equation
   **c.** A general solution of a differential equation
   **d.** A particular solution of a differential equation
   **e.** An initial value problem

## 4.1    EXERCISES

In Exercises 1–30, find the indefinite integral, and check your answer by differentiation.

**1.** $\displaystyle\int (x + 2)\,dx$

**2.** $\displaystyle\int (6x^2 - 2x + 1)\,dx$

**3.** $\displaystyle\int (3 - 2x + x^2)\,dx$

**4.** $\displaystyle\int (x^3 - 2x^2 + x + 1)\,dx$

**5.** $\displaystyle\int (2x^9 + 3e^x + 4)\,dx$

**6.** $\displaystyle\int (2x^{2/3} - 4x^{1/3} + 4)\,dx$

**7.** $\displaystyle\int \left(\sqrt{x} + \frac{3}{\sqrt{x}}\right)dx$

**8.** $\displaystyle\int x^{2/3}(x - 1)\,dx$

**9.** $\displaystyle\int (e^t + t^e)\,dt$

**10.** $\displaystyle\int \left(\frac{2}{1 + x^2} + \frac{1}{x}\right)dx$

**11.** $\displaystyle\int \frac{3}{\sqrt{u}}\,du$

**12.** $\displaystyle\int \frac{x^2 + 1}{x^2}\,dx$

**13.** $\displaystyle\int \frac{3x^4 - 2x^2 + 1}{x^4}\,dx$

**14.** $\displaystyle\int \frac{t^2 - 2\sqrt{t} + 1}{t^2}\,dt$

**15.** $\displaystyle\int \frac{x^2 - 2x + 3}{\sqrt{x}}\,dx$

**16.** $\displaystyle\int 10^x\,dx$

**17.** $\displaystyle\int \frac{x^2 2^x + x}{x^2}\,dx$

**18.** $\displaystyle\int (2t + 3\cos t)\,dt$

**19.** $\displaystyle\int (3\sin x - 4\cos x)\,dx$

**20.** $\displaystyle\int (\csc\theta\cot\theta - 3\sec^2\theta)\,d\theta$

**21.** $\displaystyle\int (\csc^2 x + \sqrt{x})\,dx$

**22.** $\displaystyle\int \sec u\,(\tan u + \sec u)\,du$

**23.** $\displaystyle\int \frac{\cos x}{1 - \cos^2 x}\,dx$

**24.** $\displaystyle\int \frac{\sin 2x}{\cos x}\,dx$

**25.** $\displaystyle\int \frac{1 - 2\cot^2 x}{\cos^2 x}\,dx$

**26.** $\displaystyle\int \frac{\cos 2x}{\cos x - \sin x}\,dx$

**27.** $\displaystyle\int \frac{1}{\sin^2 x \cos^2 x}\,dx$

Hint: Use the identity $\sin^2 x + \cos^2 x = 1$.

**28.** $\displaystyle\int \tan^2 x\,dx$

Hint: Rewrite the integrand.

**29.** $\displaystyle\int \frac{dx}{1 - \sin x}$

**30.** $\displaystyle\int \cot^2 x\,dx$

In Exercises 31–34, (a) *find the indefinite integral, and* (b) *plot the graphs of the antiderivatives corresponding to* $C = -2, -1, 0, 1,$ *and* 2 *(C is the constant of integration).*

**31.** $\displaystyle\int (x - 3)\, dx$

**32.** $\displaystyle\int (3x^2 + x - 1)\, dx$

**33.** $\displaystyle\int (2x + \sin x)\, dx$

**34.** $\displaystyle\int (1 + \sec^2 x)\, dx$

In Exercises 35–46, *find f by solving the initial value problem.*

**35.** $f'(x) = 2x + 1, \quad f(1) = 3$

**36.** $f'(x) = 3x^2 - 6x, \quad f(2) = 4$

**37.** $f'(x) = \dfrac{1}{\sqrt{x}}, \quad f(4) = 2$

**38.** $f'(x) = 1 + \dfrac{1}{x^2}, \quad f(1) = 2$

**39.** $f'(x) = e^x + \sin x, \quad f(0) = 0$

**40.** $f'(t) = \sec^2 t + 2\cos t, \quad f\left(\dfrac{\pi}{4}\right) = \sqrt{2}$

**41.** $f''(x) = 6; \quad f(1) = 4, \quad f'(1) = 2$

**42.** $f''(x) = 2x + 1; \quad f(0) = 5, \quad f'(0) = 1$

**43.** $f''(x) = 6x^2 + 6x + 2; \quad f(-1) = \dfrac{1}{2}, \quad f'(-1) = 2$

**44.** $f''(x) = \dfrac{1}{x^3}; \quad f(1) = 1, \quad f'(1) = \dfrac{1}{2}$

**45.** $f''(t) = t^{-3/2}; \quad f(4) = 1, \quad f'(4) = 3$

**46.** $f''(t) = 2\sin t + 3\cos t; \quad f\left(\dfrac{\pi}{2}\right) = 1, \quad f'\left(\dfrac{\pi}{2}\right) = 2$

**47.** Find the function $f$ given that the slope of the tangent line to the graph of $f$ at any point $(x, f(x))$ is $x^2 - 2x + 3$ and the graph of $f$ passes through the point $(1, 2)$.

**48.** Find the function $f$ given that it satisfies $f''(x) = 36x^2 + 24x$ and its graph has a horizontal tangent line at the point $(0, 1)$.

In Exercises 49–50, *identify which of the two graphs 1 and 2 is the graph of the function f and the graph of its antiderivative. Give a reason for your choice.*

**49.**

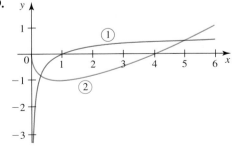

**50.**

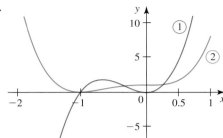

In Exercises 51–56, *find the position function of a particle moving along a coordinate line that satisfies the given condition(s).*

**51.** $v(t) = 6t^2 - 4t + 1, \quad s(1) = -1$

**52.** $v(t) = 2\sin t - 3\cos t, \quad s(0) = -2$

**53.** $a(t) = 6t - 4, \quad s(0) = -2, \quad v(0) = 4$

**54.** $a(t) = -6t^2 + 4t + 8, \quad s(1) = \dfrac{7}{6}, \quad v(1) = 4$

**55.** $a(t) = \sin t - 2\cos t, \quad s(0) = 3, \quad v(0) = 0$

**56.** $a(t) = 6\sin t, \quad s(\pi) = \pi, \quad v(0) = -4$

**57. Velocity of a Maglev** The velocity of a maglev is $v(t) = 0.2t + 3$ (ft/sec), where $0 \leq t \leq 120$. At $t = 0$ the maglev is at the station. Find the function that gives the position of the maglev at time $t$ assuming that the motion takes place along a straight stretch of the track.

**58.** A ball is thrown straight up from a height of 3 ft with an initial velocity of 40 ft/sec. How high will the ball go? (Take $g = 32$ ft/sec$^2$.)

**59. Ballast Dropped from a Balloon** A ballast is dropped from a stationary hot-air balloon that is at an altitude of 400 ft. Find (a) an expression for the altitude of the ballast after $t$ seconds, (b) the time when it strikes the ground, and (c) its velocity when it strikes the ground. (Disregard air resistance and take $g = 32$ ft/sec$^2$.)

ballast

**60. Ballast Dropped from a Balloon** Refer to Exercise 59. Suppose that the hot-air balloon is rising vertically with a velocity of 16 ft/sec at an altitude of 128 ft when the ballast is dropped. How long will it take for the ballast to strike the ground? What will its impact velocity be?

**61.** A particle located at the point $x = x_0$ on a coordinate line is given an initial velocity of $v_0$ ft/sec and a constant accelera-tion of $a$ ft/sec$^2$. Show that its position at any time $t$ is

$$x = x_0 + v_0 t + \frac{1}{2} a t^2$$

**62.** Refer to Exercise 61. Show that the velocity $v$ of the particle at any time $t$ satisfies

$$v^2 = v_0^2 + 2a(x - x_0)$$

 **63. Flight of a Model Rocket** A model rocket is fired vertically upward from a height of $s_0$ ft above the ground with a velocity of $v_0$ ft/sec. If air resistance is negligible, show that its height (in feet) after $t$ seconds is given by

$$s(t) = -16t^2 + v_0 t + s_0$$

(Take $g = 32$ ft/sec$^2$.)

**64.** Kaitlyn drops a stone into a well. Approximately 4.22 sec later, she hears the splash made by the impact of the stone in the water. How deep is the well? (The speed of sound is approximately 1128 ft/sec.)

**65. Jumping While on Mars** The acceleration due to gravity on Mars is approximately 3.72 m/sec$^2$. If an astronaut jumps straight up on the surface of the planet with an initial veloc-ity of 4 m/sec, what height will she attain? Find the compa-rable height that she would jump on the earth. (The constant of acceleration due to gravity on the earth is 9.8 m/sec$^2$.)

**66. Acceleration of a Car** A car traveling along a straight road at 66 ft/sec accelerated to a speed of 88 ft/sec over a distance of 440 ft. What was the acceleration of the car, assuming that the acceleration was constant?

**67. Stopping Distance of a Car** To what constant deceleration would a car moving along a straight road be subjected if the car were brought to rest from a speed of 88 ft/sec in 9 sec? What would the stopping distance be?

**68. Acceleration of a Car** A car traveling along a straight road at a constant speed was subjected to a constant acceleration of 12 ft/sec$^2$. It reached a speed of 60 mph after traveling 242 ft. What was the speed of the car just prior to the acceleration?

**69. Crossing the Finish Line** After rounding the final turn in the bell lap, two runners emerged ahead of the pack. When run-ner $A$ is 200 ft from the finish line, his speed is 22 ft/sec, a speed that he maintains until he crosses the line. At that instant of time, runner $B$, who is 20 ft behind runner $A$ and running at a speed of 20 ft/sec, begins to spurt. Assuming that runner $B$ sprints with a constant acceleration, what min-imum acceleration will enable him to cross the finish line ahead of runner $A$?

**70. Velocity of a Car** Two cars, side by side, start from rest and travel along a straight road. The velocity of car $A$ is given by $v = f(t)$, and the velocity of car $B$ is given by $v = g(t)$. The graphs of $f$ and $g$ are shown in the following figure. Are the cars still side by side after $T$ sec? If not, which car is ahead of the other? Justify your answer.

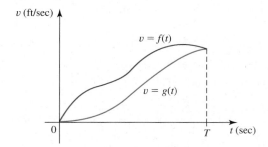

**71. Bank Deposits** Madison Finance opened two branches on September 1 ($t = 0$). Branch $A$ is located in an established industrial park, and branch $B$ is located in a fast-growing new development. The net rates at which money was deposited into branch $A$ and branch $B$ in the first 180 busi-ness days are given by the graphs of $f$ and $g$, respectively. Which branch has a larger amount on deposit at the end of 180 business days? Justify your answer.

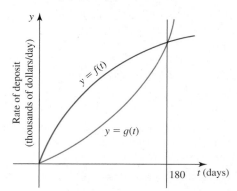

**72. Collision of Two Particles** Two points $A$ and $B$ are located 100 ft apart on a straight line. A particle moves from $A$ toward $B$ with an initial velocity of 10 ft/sec and an acceleration of $\frac{1}{2}$ ft/sec$^2$. Simultaneously, a particle moves from $B$ toward $A$ with an initial velocity of 5 ft/sec and an acceleration of $\frac{3}{4}$ ft/sec$^2$. When will the two particles collide? At what distance from $A$ will the collision take place?

**73. Revenue** The monthly marginal revenue of Commuter Air Service is $R'(x) = 10,000 - 200x$ dollars per passenger, where $x$ stands for the fare per passenger. Find the monthly total revenue function $R$ if $R(0) = 0$.

**74. Total Cost Function** The weekly marginal cost of the Electra Electronics Company in producing its Zephyr laser jet printers is given by

$$C'(x) = 0.000006x^2 - 0.04x + 1000$$

dollars per printer, where $x$ stands for the number of printers manufactured. Find the weekly total cost function $C$ if the fixed cost of the company is $120,000 per week.

75. **Risk of Down Syndrome** The rate at which the risk of Down syndrome is changing is approximated by the function

$$r(x) = 0.004641x^2 - 0.3012x + 4.9 \qquad 20 \le x \le 45$$

where $r(x)$ is measured in percent of all births per year and $x$ is the maternal age at delivery.

   **a.** Find a function $f$ giving the risk as a percentage of all births when the maternal age at delivery is $x$ years, given that the risk of Down syndrome at age 30 is 0.14% of all births.

   **b.** What is the risk of Down syndrome when the maternal age at delivery is 40 years? 45 years?

*Source: New England Journal of Medicine.*

76. **Online Ad Sales** In a study conducted in 2004, it was found that the share of online advertisement worldwide, as a percentage of the total ad market, was expected to grow at the rate of

$$R(t) = -0.033t^2 + 0.3428t + 0.07 \qquad 0 \le t \le 6$$

percent per year at time $t$ (in years), with $t = 0$ corresponding to the beginning of 2000. The online ad market at the beginning of 2000 was 2.9% of the total ad market.

   **a.** What is the projected online ad market share at any time $t$?

   **b.** What was the projected online ad market share at the beginning of 2006?

*Source: Jupiter Media Metrix, Inc.*

77. **Ozone Pollution** The rate of change of the level of ozone, an invisible gas that is an irritant and impairs breathing, present in the atmosphere on a certain May day in the city of Riverside is given by

$$R(t) = 3.2922t^2 - 0.366t^3 \qquad 0 < t < 11$$

(measured in pollutant standard index per hour). Here, $t$ is measured in hours, with $t = 0$ corresponding to 7 A.M. Find the ozone level $A(t)$ at any time $t$, assuming that at 7 A.M. it is 34.

*Source: The Los Angeles Times.*

78. **U.S. Sales of Organic Milk** The sales of organic milk from 1999 through 2004 grew at the rate of approximately

$$R(t) = 3t^3 - 17.9445t^2 + 28.7222t + 26.632$$
$$0 \le t \le 5$$

million dollars per year, where $t$ is measured in years with $t = 0$ corresponding to 1999. Sales of organic milk in 1999 totaled $108 million.

   **a.** Find an expression giving the total sales of organic milk by year $t$, where $0 \le t \le 5$.

   **b.** According to this model, what were the total sales of organic milk in 2004?

*Source: Resource, Inc.*

79. **Water Level of a Tank** A tank has a constant cross-sectional area of 50 ft² and an orifice of constant cross-sectional area of $\frac{1}{2}$ ft² located at the bottom of the tank. If the tank is filled with water to a height of $h$ ft and allowed to drain, then the height of the water decreases at a rate that is described by the equation

$$\frac{dh}{dt} = -\frac{1}{25}\left(\sqrt{20} - \frac{t}{50}\right) \qquad 0 \le t \le 50\sqrt{20}$$

Find an expression for the height of the water at any time $t$ if its height initially is 20 ft.

80. **The Elastic Curve of a Beam** A horizontal, uniform beam of length $L$, supported at its ends, bends under its own weight, $w$ per unit length. The *elastic curve* of the beam (the shape that it assumes) has equation $y = f(x)$ satisfying

$$EIy'' = \frac{wx^2}{2} - \frac{wLx}{2}$$

where $E$ and $I$ are positive constants that depend on the material and the cross section of the beam.

**(a)** The distorted beam

**(b)** The elastic curve in the $xy$-plane (The positive direction of the $y$-axis is downward.)

   **a.** Find an equation of the elastic curve.
      **Hint:** $y = 0$ at $x = 0$ and at $x = L$.

   **b.** Show that the maximum deflection of the beam occurs at $x = L/2$.

*In Exercises 81–86, determine whether the statement is true or false. If it is true, explain why it is true. If it is false, explain why or give an example to show why it is false.*

**81.** $\int f'(x)\,dx = f(x) + C$

**82.** $\int f(x)g(x)\,dx = F(x)G(x) + C,$ where $F' = f$ and $G' = g$

**83.** $\int x f(x)\,dx = x\int f(x)\,dx = x F(x) + C,$ where $F' = f$

**84.** If $F$ and $G$ are antiderivatives of $f$ and $g$, respectively, then

$$\int [2f(x) - 3g(x)]\,dx = 2F(x) - 3G(x) + C$$

**85.** If $R(x) = P(x)/Q(x)$ is a rational function, then

$$\int R(x)\,dx = \frac{\displaystyle\int P(x)\,dx}{\displaystyle\int Q(x)\,dx}$$

**86.** $\int \left[\int f(x)\,dx\right]dx = G(x) + C_1 x + C_2,$

where $G' = F$ and $F' = f$

**87.** $\int e^{\ln x}\,dx = \dfrac{1}{2}\,x e^{\ln x} + C$

---

## 4.2 Integration by Substitution

In this section we introduce a technique of integration that will enable us to integrate a large class of functions. This method of integration, like the integration formulas of Section 4.1, is obtained by reversing a differentiation rule—in this case the Chain Rule.

### How the Method of Substitution Works

Consider the indefinite integral

$$\int 2x\sqrt{x^2 + 3}\,dx \tag{1}$$

You can convince yourself that this integral cannot be evaluated as it stands by using any of the integration formulas that we are now familiar with.

Let's try to simplify the indefinite integral (1) by making a change of variable from $x$ to a new variable $u$ as follows. Write

$$u = x^2 + 3$$

with differential

$$du = 2x\,dx$$

If we *formally* substitute these quantities into the indefinite integral (1), we obtain

$$\int 2x\sqrt{x^2 + 3}\,dx = \int \sqrt{x^2 + 3}\,(2x\,dx) = \int \sqrt{u}\,du$$

$$\underset{\text{Rewriting}}{\uparrow} \qquad\qquad \underset{\begin{cases} u = x^2 + 3 \\ du = 2x\,dx \end{cases}}{\uparrow}$$

Now the integral is in a form that is easily integrated by using Formula (2) of Section 4.1. Thus,

$$\int \sqrt{u}\,du = \int u^{1/2}\,du = \frac{2}{3}u^{3/2} + C$$

Finally, replacing $u$ by $x^2 + 3$, we see that

$$\int 2x\sqrt{x^2 + 3}\, dx = \frac{2}{3}(x^2 + 3)^{3/2} + C$$

To verify that this solution is correct, we compute

$$\frac{d}{dx}\left[\frac{2}{3}(x^2 + 3)^{3/2} + C\right] = \frac{2}{3} \cdot \frac{3}{2}(x^2 + 3)^{1/2}(2x) \qquad \text{Use the Chain Rule.}$$

$$= 2x\sqrt{x^2 + 3}$$

which is the integrand. This proves the assertion.

## ■ The Technique of Integration by Substitution

As was shown in the preceding example, it is sometimes possible to transform one indefinite integral into another that is easier to integrate by using a suitable change of variable. Now, by letting $f(x) = \sqrt{x}$ and $g(x) = x^2 + 3$, so that $g'(x) = 2x$, we can see that the indefinite integral (1) is a special case of an indefinite integral of the form

$$\int f(g(x))g'(x)\, dx \qquad (2)$$

Let's show that the integral (2) can always be rewritten in the form

$$\int f(u)\, du \qquad (3)$$

where $u$ is differentiable on an interval $(a, b)$ and $f$ is continuous on the range of $u$. Suppose that $F$ is an antiderivative of $f$. By the Chain Rule we have

$$\frac{d}{dx}[F(g(x))] = F'(g(x))g'(x)$$

or, equivalently,

$$\int F'(g(x))g'(x)\, dx = F(g(x)) + C$$

Writing $F' = f$ and making the substitution $u = g(x)$, we have

$$\int f(g(x))g'(x)\, dx = F(u) + C = \int F'(u)\, du = \int f(u)\, du$$

as was to be shown.

Before looking at an example, let's summarize the steps used in this method.

---

**Integration by Substitution: Evaluating $\int f(g(x))g'(x)\, dx$**

**Step 1**  Let $u = g(x)$, where $g(x)$ is part of the integrand, usually, the "inside function" of the composite function $f(g(x))$.

**Step 2**  Compute $du = g'(x)\, dx$.

**Step 3**  Use the substitution $u = g(x)$ and $du = g'(x)\, dx$ to transform the integral into one that involves *only* $u$: $\int f(u)\, du$.

**Step 4**  Find the resulting integral.

**Step 5**  Replace $u$ by $g(x)$ so that the final solution is in terms of $x$.

**EXAMPLE 1** Find $\int x^2(x^3 + 2)^4 \, dx$.

**Solution**

**Step 1**  If you examine the integrand, you will see that it involves the composite function $(x^3 + 2)^4$, with "inside function" $g(x) = x^3 + 2$. So let us choose $u = x^3 + 2$.

**Step 2**  We compute $du = 3x^2 \, dx$.

**Step 3**  Making the substitution $u = x^3 + 2$ and $du = 3x^2 \, dx$ or $x^2 dx = \frac{1}{3} \, du$, we obtain

$$\int x^2(x^3 + 2)^4 \, dx = \int (x^3 + 2)^4 \, x^2 \, dx = \int u^4 \left( \frac{1}{3} \, du \right) = \frac{1}{3} \int u^4 \, du$$

$$\uparrow$$
$$\text{Rewriting}$$

an integral involving only the variable $u$.

**Step 4**  We find

$$\frac{1}{3} \int u^4 \, du = \frac{1}{3} \left( \frac{u^5}{5} \right) + C = \frac{1}{15} u^5 + C$$

**Step 5**  Replacing $u$ by $x^3 + 2$, we find

$$\int x^2(x^3 + 2)^4 \, dx = \frac{1}{15} (x^3 + 2)^5 + C$$    ∎

**EXAMPLE 2** Find $\int \dfrac{dx}{(2x - 4)^3}$.

**Solution**    First rewrite the integral in the form

$$\int (2x - 4)^{-3} \, dx$$

**Step 1**  In the composite function $(2x - 4)^{-3}$, the "inside function" is $g(x) = 2x - 4$. So let $u = 2x - 4$.

**Step 2**  We find $du = 2 \, dx$.

**Step 3**  Substituting $u = 2x - 4$ and $du = 2 \, dx$ or $dx = \frac{1}{2} \, du$ into the integral yields

$$\int (2x - 4)^{-3} \, dx = \int u^{-3} \left( \frac{1}{2} \, du \right) = \frac{1}{2} \int u^{-3} \, du$$

an integral involving only the variable $u$.

**Step 4**  We integrate

$$\frac{1}{2} \int u^{-3} \, du = \frac{1}{2} \left( \frac{u^{-2}}{-2} \right) + C = -\frac{1}{4u^2} + C$$

**Step 5**  Replacing $u$ by $2x - 4$ gives

$$\int \frac{dx}{(2x - 4)^3} = -\frac{1}{4(2x - 4)^2} + C$$    ∎

**EXAMPLE 3**   Find $\int (x + 1)\sqrt{2x - 1}\, dx$.

**Solution**   First, rewrite the integral in the form

$$\int (x + 1)(2x - 1)^{1/2}\, dx$$

**Step 1**   Examining the integrand, we spot the composite function $(2x - 1)^{1/2}$, which has the "inside function" $g(x) = 2x - 1$. So let $u = 2x - 1$.

**Step 2**   We find $du = 2\, dx$.

**Step 3**   We use the substitution $u = 2x - 1$ and $du = 2\, dx$ or $dx = \frac{1}{2}\, du$. Because of the factor $x + 1$ in the integrand, we need to solve $u = 2x - 1$ for $x$, obtaining $x = \frac{1}{2}u + \frac{1}{2}$. Therefore,

$$\int (x + 1)(2x - 1)^{1/2}\, dx = \int \left(\frac{1}{2}u + \frac{1}{2} + 1\right)u^{1/2}\left(\frac{1}{2}\, du\right) = \frac{1}{2}\int \left(\frac{1}{2}u^{3/2} + \frac{3}{2}u^{1/2}\right) du$$

$$= \frac{1}{4}\int (u^{3/2} + 3u^{1/2})\, du$$

an integral involving only the variable $u$.

**Step 4**   Integrating, we find

$$\frac{1}{4}\int (u^{3/2} + 3u^{1/2})\, du = \frac{1}{4}\left(\frac{2}{5}u^{5/2} + 2u^{3/2}\right) + C$$

$$= \frac{1}{10}u^{5/2} + \frac{1}{2}u^{3/2} + C$$

**Step 5**   Replacing $u$ by $2x - 1$, we have

$$\int (x + 1)\sqrt{2x - 1}\, dx = \frac{1}{10}(2x - 1)^{5/2} + \frac{1}{2}(2x - 1)^{3/2} + C \qquad \blacksquare$$

Now that we are familiar with this procedure, we will drop the practice of labeling the steps as we work through the next several examples.

**EXAMPLE 4**   Find

**a.** $\int \sin 5x\, dx$   **b.** $\displaystyle \int \frac{\cos \sqrt{x}}{\sqrt{x}}\, dx$

**Solution**

**a.** Let $u = 5x$, so that $du = 5\, dx$ or $dx = \frac{1}{5}\, du$. Substituting these quantities into the integral gives

$$\int (\sin u)\frac{1}{5}\, du = \frac{1}{5}\int \sin u\, du = -\frac{1}{5}\cos u + C = -\frac{1}{5}\cos 5x + C$$

**b.** Let $u = \sqrt{x} = x^{1/2}$. (Here, $\sqrt{x}$ is the "inside function" of the composite function $y = \cos \sqrt{x}$.) Then

$$du = \frac{1}{2}x^{-1/2}\, dx = \frac{dx}{2\sqrt{x}} \qquad \text{or} \qquad \frac{dx}{\sqrt{x}} = 2\, du$$

Substituting these quantities into the integral yields

$$\int (\cos u)\, 2\, du = 2\int \cos u\, du = 2\sin u + C = 2\sin \sqrt{x} + C \qquad \blacksquare$$

**EXAMPLE 5**  Find $\int \sin^3 x \cos x \, dx$.

**Solution**   The integrand contains the composite function $y = \sin^3 x = (\sin x)^3$. So let's put $u = \sin x$. Then $du = \cos x \, dx$, so

$$\int \sin^3 x \cos x \, dx = \int u^3 du = \frac{1}{4} u^4 + C = \frac{1}{4} \sin^4 x + C \qquad \blacksquare$$

**EXAMPLE 6**  Find

**a.** $\int \dfrac{e^{2/x}}{x^2} \, dx$     **b.** $\int \dfrac{1}{2x + 1} \, dx$

**Solution**
**a.** Let $u = 2/x$, so that

$$du = -\frac{2}{x^2} \, dx \qquad \text{or} \qquad \frac{dx}{x^2} = -\frac{1}{2} \, du$$

Making these substitutions, we obtain

$$\int \frac{e^{2/x}}{x^2} \, dx = -\frac{1}{2} \int e^u \, du = -\frac{1}{2} e^u + C = -\frac{1}{2} e^{2/x} + C$$

**b.** Let $u = 2x + 1$, so that $du = 2 \, dx$ or $dx = \frac{1}{2} du$. Making these substitutions, we have

$$\int \frac{1}{2x + 1} \, dx = \frac{1}{2} \int \frac{1}{u} \, du = \frac{1}{2} \ln|u| + C$$

$$= \frac{1}{2} \ln|2x + 1| + C \qquad \blacksquare$$

**EXAMPLE 7**  Find $\int \tan x \, dx$.

**Solution**   We first rewrite

$$\int \tan x \, dx = \int \frac{\sin x}{\cos x} \, dx$$

Then we use the substitution $u = \cos x$, so that $du = -\sin x \, dx$ or $\sin x \, dx = -du$, giving

$$\int \frac{\sin x}{\cos x} \, dx = -\int \frac{1}{u} \, du = -\ln|u| + C$$

Therefore,

$$\int \tan x \, dx = -\ln|\cos x| + C \quad \text{or} \quad \ln|\sec x| + C \qquad \blacksquare$$

We can obtain the formula for $\int \cot x \, dx$ in a similar manner by observing that $\cot x = (\cos x)/(\sin x)$.

**EXAMPLE 8** Find $\int \sec x \, dx$.

**Solution** Multiplying both the numerator and denominator of the integrand by $\sec x + \tan x$ gives

$$\int \sec x \, dx = \int \sec x \, \frac{\sec x + \tan x}{\sec x + \tan x} \, dx = \int \frac{\sec^2 x + \sec x \tan x}{\sec x + \tan x} \, dx$$

Next, use the substitution

$$u = \sec x + \tan x \qquad \text{so that} \qquad du = (\sec x \tan x + \sec^2 x) \, dx$$

This gives

$$\int \sec x \, dx = \int \frac{1}{u} \, du = \ln|u| + C = \ln|\sec x + \tan x| + C \qquad \blacksquare$$

We can find the formula for $\int \csc x \, dx$ by using the same technique. The results of Examples 7 and 8 are summarized below.

---

**Integrals of Trigonometric Functions**

$$\int \tan u \, du = \ln|\sec u| + C \tag{4a}$$

$$\int \cot u \, du = \ln|\sin u| + C \tag{4b}$$

$$\int \sec u \, du = \ln|\sec u + \tan u| + C \tag{4c}$$

$$\int \csc u \, du = \ln|\csc u - \cot u| + C \tag{4d}$$

---

**EXAMPLE 9** Find $\int x \sec x^2 \, dx$.

**Solution** Let $u = x^2$, so that $du = 2x \, dx$ or $x \, dx = \frac{1}{2} du$. Making these substitutions, we obtain

$$\int x \sec x^2 \, dx = \frac{1}{2} \int \sec u \, du = \frac{1}{2} \ln|\sec u + \tan u| + C$$

$$= \frac{1}{2} \ln|\sec x^2 + \tan x^2| + C \qquad \blacksquare$$

By reversing the rules of differentiation for inverse trigonometric functions, we obtain the following formulas.

**Integrals Involving Inverse Trigonometric Functions**

$$\int \frac{1}{\sqrt{1 - u^2}}\, du = \sin^{-1} u + C \qquad\qquad \text{(5a)}$$

$$\int \frac{1}{1 + u^2}\, du = \tan^{-1} u + C \qquad\qquad \text{(5b)}$$

$$\int \frac{1}{|u|\sqrt{u^2 - 1}}\, du = \sec^{-1} |u| + C \qquad\qquad \text{(5c)}$$

**EXAMPLE 10**  Find

**a.** $\displaystyle\int \frac{1}{\sqrt{1 - 9x^2}}\, dx$      **b.** $\displaystyle\int \frac{1}{\sqrt{4 - x^2}}\, dx$

**Solution**

**a.** Comparing the integral with Formula (5a) suggests the substitution $u = 3x$. Then $du = 3\, dx$ or $dx = \frac{1}{3}\, du$. Therefore,

$$\int \frac{1}{\sqrt{1 - 9x^2}}\, dx = \int \frac{1}{\sqrt{1 - (3x)^2}}\, dx$$

$$= \frac{1}{3} \int \frac{1}{\sqrt{1 - u^2}}\, du$$

$$= \frac{1}{3} \sin^{-1} u + C = \frac{1}{3} \sin^{-1}(3x) + C$$

**b.** Once again, comparing the integral with Formula (5a) suggests that we write

$$\int \frac{1}{\sqrt{4 - x^2}}\, dx = \int \frac{1}{2\sqrt{1 - \left(\dfrac{x}{2}\right)^2}}\, dx$$

Next, we let $u = x/2$ so that $du = \frac{1}{2}\, dx$ or $dx = 2\, du$. Then

$$\int \frac{1}{\sqrt{4 - x^2}}\, dx = \frac{1}{2}\,(2)\int \frac{1}{\sqrt{1 - u^2}}\, du$$

$$= \sin^{-1} u + C = \sin^{-1}\left(\frac{x}{2}\right) + C \qquad\blacksquare$$

**EXAMPLE 11**  Find

**a.** $\displaystyle\int \frac{e^x}{e^{2x} + 1}\, dx$      **b.** $\displaystyle\int \frac{1}{x\sqrt{x^4 - 16}}\, dx$

**Solution**

**a.** Let $u = e^x$ so that $du = e^x\, dx$. Then

$$\int \frac{e^x}{e^{2x} + 1}\, dx = \int \frac{1}{u^2 + 1}\, du$$

$$= \tan^{-1} u + C = \tan^{-1} e^x + C$$

**b.** Let $u = x^2$. Then $du = 2x\,dx$, or $dx = \dfrac{du}{2x}$. Making these substitutions, we find

$$\int \frac{1}{x\sqrt{x^4 - 16}}\,dx = \frac{1}{2}\int \frac{1}{x^2\sqrt{u^2 - 16}}\,du$$

$$= \frac{1}{2}\int \frac{1}{u\sqrt{u^2 - 16}}\,du \qquad \text{Replace } x^2 \text{ by } u.$$

$$= \frac{1}{2}\cdot\frac{1}{4}\int \frac{1}{u\sqrt{\left(\dfrac{u}{4}\right)^2 - 1}}\,du$$

Next, we let $v = u/4$ so that $dv = \frac{1}{4}\,du$ or $du = 4\,dv$. Then

$$\int \frac{1}{x\sqrt{x^4 - 16}}\,dx = \left(\frac{1}{8}\right)4\int \frac{1}{4v\sqrt{v^2 - 1}}\,dv$$

$$= \frac{1}{8}\int \frac{1}{v\sqrt{v^2 - 1}}\,dv$$

$$= \frac{1}{8}\sec^{-1}v + C = \frac{1}{8}\sec^{-1}\left(\frac{x^2}{4}\right) + C \qquad \blacksquare$$

**EXAMPLE 12 A Falling Ballast** A ballast is dropped from a balloon at a height of 10,000 ft. The velocity at any time $t$ (until it reaches the ground) is given by

$$v(t) = 320(e^{-0.1t} - 1)$$

where the velocity is measured in feet per second and $t$ is measured in seconds.

**a.** Find an expression for the height $s(t)$ of the ballast at any time $t$. ($s(t)$ is measured from the ground.)
**b.** What is the terminal velocity of the ballast?
**c.** Estimate the time it takes for the ballast to hit the ground.

**Solution**
**a.** The velocity of the ballast is

$$s'(t) = v(t) = 320(e^{-0.1t} - 1)$$

Therefore, its height at any time $t$ is

$$s(t) = \int v(t)\,dt = \int 320(e^{-0.1t} - 1)\,dt$$

$$= 320\left(-\frac{1}{0.1}e^{-0.1t} - t\right) + C \qquad \text{Use integration by substitution with } u = -0.1t.$$

$$= -320(t + 10e^{-0.1t}) + C$$

To determine $C$, we use the initial condition $s(0) = 10{,}000$, obtaining

$$s(0) = -3200 + C = 10{,}000 \qquad \text{or} \qquad C = 13{,}200$$

Therefore,

$$s(t) = -320(t + 10e^{-0.1t}) + 13{,}200$$

**b.** The terminal velocity is given by

$$\lim_{t \to \infty} v(t) = \lim_{t \to \infty} 320(e^{-0.1t} - 1)$$

$$= 320 \lim_{t \to \infty}(e^{-0.1t} - 1) = -320$$

or $-320$ ft/sec.

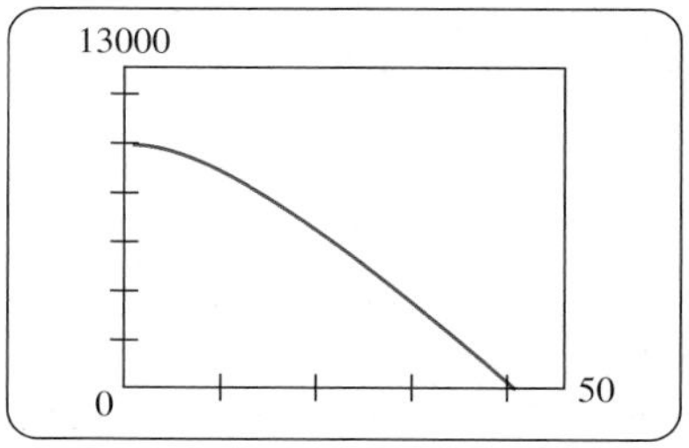

**c.** The ballast hits the ground when $s(t) = 0$. Using the result of part (a), we have

$$-320(t + 10e^{-0.1t}) + 13{,}200 = 0 \qquad (6)$$

This equation is not easily solved for $t$, but we can obtain an approximation of $t$ by observing that if $t$ is large, then the term $10e^{-0.1t}$ is relatively small in comparison to $t$. So, dropping this term, we solve the equation

$$-320t + 13{,}200 = 0$$

getting $t \approx 41$. Therefore, the ballast hits the ground approximately 41 sec after it has been jettisoned. The graph of $s(t) = -320(t + 10e^{-0.1t}) + 13{,}200$ is shown in Figure 1.

**FIGURE 1**
The graph of
$s(t) = -320(t + 10e^{-0.1t}) + 13{,}200$

### Notes

**1.** Let's show that the approximation obtained in part (c) of Example 12 is reasonably accurate by computing the position of the ballast 41 sec after it was jettisoned. Thus,

$$s(41) = -320(41 + 10e^{-4.1}) + 13{,}200 \approx 27$$

or 27 ft above the ground. If greater accuracy is required, one can use Newton's method to solve the equation $s(t) = 0$. (See Exercise 29, Section 3.9.)

**2.** We can also estimate the time of impact of the ballast by using a graphing utility to find the zero of $f$. We find $t \approx 41.086$, accurate to three decimal places.

**EXAMPLE 13   Flight of a Projectile**   Suppose that a projectile is launched vertically upward from the earth's surface with an initial velocity equal to the *escape velocity.* (The escape velocity is the velocity the projectile must attain to escape from the earth's gravitational pull.) Then, if we neglect the gravitational influence of the sun and other planets, the rotation of the earth, and air resistance, it can be shown that the differential equation governing the motion of the projectile is

$$\frac{dt}{dx} = \frac{\sqrt{x + R}}{R\sqrt{2g}}$$

where $t$ is the time in seconds, $x$ is the distance of the projectile from the surface of the earth, $R$ is the radius of the earth (approximately 4000 mi), and $g$ is the gravitational constant of acceleration (approximately 32 ft/sec$^2$). (See Figure 2.) Find the time it takes for the projectile to travel a distance of $x$ miles. How long would it take the projectile to cover 100,000 mi?

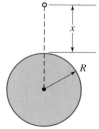

**FIGURE 2**
The projectile is launched vertically upward from the earth's surface.

**Solution**   Integrating the given equation with respect to $x$, we obtain

$$t = \int \frac{dt}{dx}\, dx = \int \frac{\sqrt{x + R}}{R\sqrt{2g}}\, dx = \frac{1}{R\sqrt{2g}} \int (x + R)^{1/2}\, dx$$

Let $u = x + R$, so that $du = dx$. Then

$$t = \frac{1}{R\sqrt{2g}} \int u^{1/2}\, du = \frac{2u^{3/2}}{3R\sqrt{2g}} + C = \frac{2(x + R)^{3/2}}{3R\sqrt{2g}} + C$$

where $C$ is the constant of integration. To determine the value of $C$, we use the condition that $t = 0$ when $x = 0$. This gives

$$0 = \frac{2R^{3/2}}{3R\sqrt{2g}} + C \qquad \text{or} \qquad C = -\frac{2R^{3/2}}{3R\sqrt{2g}}$$

Therefore,

$$t = \frac{2(x + R)^{3/2}}{3R\sqrt{2g}} - \frac{2R^{3/2}}{3R\sqrt{2g}} = \frac{2}{3R\sqrt{2g}}[(x + R)^{3/2} - R^{3/2}]$$

and this is the time it takes for the projectile to cover a distance $x$. Finally, to find the time it takes for the projectile to travel 100,000 mi, we substitute

$$R = 4000(5280) = 21,120,000 \quad \text{(ft)}$$

and

$$x = 100,000(5280) = 528,000,000 \quad \text{(ft)}$$

into the expression for $t$, obtaining

$$t = \frac{2}{3(21,120,000)\sqrt{64}}(549,120,000^{3/2} - 21,120,000^{3/2}) \approx 50,389.2$$

So it takes approximately 50,000 sec, or 14 hr, to cover 100,000 mi.  ■

## 4.2  CONCEPT QUESTIONS

1. Explain how the method of integration by substitution works by showing the steps that are used to find $\int f(g(x))g'(x)\, dx$.

2. Explain why the method of substitution works for the integral $\int x^2 \cos(x^3 + 1)\, dx$. Does it work for $\int x \cos(x^3 + 1)\, dx$?

## 4.2  EXERCISES

*In Exercises 1–6, find the integral using the indicated substitution.*

1. $\displaystyle \int (2x + 3)^5\, dx, \quad u = 2x + 3$

2. $\displaystyle \int x^2 \sqrt{x^3 + 2}\, dx, \quad u = x^3 + 2$

3. $\displaystyle \int \frac{x}{\sqrt{x^2 + 1}}\, dx, \quad u = x^2 + 1$

4. $\displaystyle \int e^{-3x}\, dx, \quad u = -3x$

5. $\displaystyle \int \tan^3 x \sec^2 x\, dx, \quad u = \tan x$

6. $\displaystyle \int \frac{\sin x}{\cos^2 x}\, dx, \quad u = \cos x$

*In Exercises 7–72, find the indefinite integral.*

7. $\displaystyle \int 2x(x^2 + 1)^4\, dx$

8. $\displaystyle \int x^2(2x^3 - 1)^4\, dx$

9. $\displaystyle \int (2x - 4)^{3/5}\, dx$

10. $\displaystyle \int (1 - 3x)^{1.4}\, dx$

11. $\displaystyle \int 3x(2x^2 + 3)^5\, dx$

12. $\displaystyle \int x^2(2x^3 - 1)^{-4}\, dx$

13. $\displaystyle \int \sqrt{1 - 2x}\, dx$

14. $\displaystyle \int 2x\sqrt[3]{1 - 4x^2}\, dx$

15. $\displaystyle \int s^3(s^4 - 1)^{3/2}\, ds$

16. $\displaystyle \int x^{-1/3}\sqrt{x^{2/3} - 1}\, dx$

17. $\displaystyle \int xe^{-x^2}\, dx$

18. $\displaystyle \int (x^2 - \tfrac{1}{3})e^{x^3 - x}\, dx$

**19.** $\displaystyle\int (e^x + e^{-x})^2 \, dx$

**20.** $\displaystyle\int \sqrt{1 + 2e^x} \, e^x \, dx$

**21.** $\displaystyle\int 2^{-x} \, dx$

**22.** $\displaystyle\int (x + 1)3^{x^2 + 2x} \, dx$

**23.** $\displaystyle\int \frac{e^{-1/x}}{x^2} \, dx$

**24.** $\displaystyle\int \frac{\sqrt{1 + u^{-1}}}{u^2} \, du$

**25.** $\displaystyle\int 2 \cos \frac{x}{2} \, dx$

**26.** $\displaystyle\int x \sin x^2 \, dx$

**27.** $\displaystyle\int x \cos \pi x^2 \, dx$

**28.** $\displaystyle\int x^2 \sec^2 x^3 \, dx$

**29.** $\displaystyle\int \sin \pi x \cos \pi x \, dx$

**30.** $\displaystyle\int \cot^3 x \csc^2 x \, dx$

**31.** $\displaystyle\int \tan 3x \sec 3x \, dx$

**32.** $\displaystyle\int \sqrt{\sin \theta} \cos \theta \, d\theta$

**33.** $\displaystyle\int \frac{\sin u^{-1}}{u^2} \, du$

**34.** $\displaystyle\int \frac{\sin x}{(1 + \cos x)^3} \, dx$

**35.** $\displaystyle\int \frac{\csc^2 3x}{\cot^3 3x} \, dx$

**36.** $\displaystyle\int \frac{\sin \sqrt{x}}{\sqrt{x}} \, dx$

**37.** $\displaystyle\int \frac{\cos 2t}{\sqrt{2 + \sin 2t}} \, dt$

**38.** $\displaystyle\int (\csc^2 x)(\cot x - 1)^3 \, dx$

**39.** $\displaystyle\int \frac{\sec x \tan x}{(\sec x - 1)^2} \, dx$

**40.** $\displaystyle\int \sec^2(x + 1)\sqrt{1 + \tan(x + 1)} \, dx$

**41.** $\displaystyle\int \sin^2 \pi x \, dx$

Hint: $\sin^2 \theta = \dfrac{1 - \cos 2\theta}{2}$

**42.** $\displaystyle\int \frac{1 + \sin x}{\cos^2 x} \, dx$

**43.** $\displaystyle\int \frac{e^{-x}}{1 + e^{-x}} \, dx$

**44.** $\displaystyle\int \frac{3^x}{1 + 3^x} \, dx$

**45.** $\displaystyle\int \frac{\sqrt{\log x}}{x} \, dx$

**46.** $\displaystyle\int e^x \sin e^x \, dx$

**47.** $\displaystyle\int \frac{e^x \ln(e^x + 1)}{e^x + 1} \, dx$

**48.** $\displaystyle\int e^{\sin x} \cos x \, dx$

**49.** $\displaystyle\int \frac{1}{2x + 3} \, dx$

**50.** $\displaystyle\int \frac{1}{x^{2/3}(x^{1/3} + 1)} \, dx$

**51.** $\displaystyle\int \frac{1}{x \ln x} \, dx$

**52.** $\displaystyle\int \frac{\sqrt{1 + \ln x}}{x} \, dx$

**53.** $\displaystyle\int \frac{\cos x}{1 + \sin x} \, dx$

**54.** $\displaystyle\int \frac{\sec^2 3x}{4 - \tan 3x} \, dx$

**55.** $\displaystyle\int (\sec \theta + \cos \theta) \, d\theta$

**56.** $\displaystyle\int \frac{\sin 2x}{1 + \sin^2 x} \, dx$

**57.** $\displaystyle\int \frac{1 + \ln x}{2 + x \ln x} \, dx$

**58.** $\displaystyle\int \frac{(\ln x)\sqrt{1 + \ln x}}{x} \, dx$

**59.** $\displaystyle\int \frac{1}{\sqrt{16 - x^2}} \, dx$

**60.** $\displaystyle\int \frac{1}{x\sqrt{9x^2 - 1}} \, dx$

**61.** $\displaystyle\int \frac{1}{x\sqrt{x^4 - 81}} \, dx$

**62.** $\displaystyle\int \frac{1}{t\sqrt{t^6 - 16}} \, dt$

**63.** $\displaystyle\int \frac{x^2 - 1}{x^2 + 1} \, dx$

**64.** $\displaystyle\int \frac{\cos 3x}{1 + \sin^2 3x} \, dx$

**65.** $\displaystyle\int \frac{\sin x}{\sqrt{4 - \cos^2 x}} \, dx$

**66.** $\displaystyle\int \frac{dx}{|x|(\sec^{-1} x)^3 \sqrt{x^2 - 1}}$

**67.** $\displaystyle\int \frac{e^{\tan^{-1} x}}{1 + x^2} \, dx$

**68.** $\displaystyle\int \frac{1}{(x + 1)\sqrt{(x + 1)^2 - 9}} \, dx$

**69.** $\displaystyle\int \frac{1}{\sqrt{x}\,(4 + x)} \, dx$

**70.** $\displaystyle\int \frac{e^x}{\sqrt{1 - e^{2x}}} \, dx$

**71.** $\displaystyle\int \frac{1}{4 + (x - 2)^2} \, dx$

**72.** $\displaystyle\int \frac{1}{x[9 + (\ln x)^2]} \, dx$

*In Exercises 73–78, find the indefinite integral.*

**73.** $\displaystyle\int x\sqrt{x - 4} \, dx$

**74.** $\displaystyle\int x^2(1 - x)^7 \, dx$

**75.** $\displaystyle\int x^3(x^2 + 1)^{5/2} \, dx$

**76.** $\displaystyle\int \frac{x + 1}{(\sqrt{x} - 1)^{3/2}} \, dx$

**77.** $\displaystyle\int \frac{dx}{\sqrt{x} + \sqrt{x + 1}}$

Hint: First rationalize the denominator of the integrand.

**78.** $\displaystyle\int \frac{\sqrt{a^2 - x^2}}{x^4} \, dx$

Hint: Let $x = 1/t$.

**79.** Find the function $f$ given that its derivative is $f'(x) = x\sqrt{1 + x^2}$ and that its graph passes through the point $(0, 1)$.

**80.** The slope of the tangent line at any point on the graph of $f$ is $\dfrac{x}{(2x^2 + 1)^{3/2}}$, and the graph of $f$ passes through the point $\left(2, -\frac{1}{6}\right)$. Find $f$.

**81.** **Rectilinear Motion** A body moves along a coordinate line in such a way that its velocity at any time $t$, where $0 \leq t \leq 4$, is given by

$$v(t) = t\sqrt{16 - t^2}$$

Find its position function if the body is initially located at the origin.

**82.** **Population Growth** The population of a certain city is projected to grow at the rate of

$$r(t) = 400\left(1 + \frac{2t}{\sqrt{24 + t^2}}\right) \qquad 0 \leq t \leq 5$$

people per year $t$ years from now. The current population is 60,000. What will be the population 5 years from now?

83. **Life Expectancy of a Female**  Suppose that in a certain country the life expectancy at birth of a female is changing at the rate of

$$g'(t) = \frac{5.45218}{(1 + 1.09t)^{0.9}}$$

years per year. Here, $t$ is measured in years, with $t = 0$ corresponding to the beginning of 1900. Find an expression $g(t)$ giving the life expectancy at birth (in years) of a female in that country if the life expectancy at the beginning of 1900 is 50.02 years. What is the life expectancy at birth of a female born at the beginning of 2000 in that country?

84. **Revenue**  The weekly marginal revenue of a company selling $x$ units (in lots of 100) of a portable hair dryer is given by

$$R'(x) = \frac{225 - 10x}{\sqrt{225 - 5x}}$$

dollars per lot. Find the weekly total revenue function.
**Hint:** $R(0) = 0$.

85. **Respiratory Cycle**  Suppose that the rate at which air is inhaled by a person during respiration is

$$r(t) = \frac{3}{5} \sin \frac{\pi t}{2}$$

liters per second, at time $t$. Find $V(t)$, the volume of inhaled air in the lungs at any time $t$. Assume that $V(0) = 0$.

86. **Revenue**  The total revenue of McMenamy's Fish Shanty at a popular summer resort is changing at the rate of approximately

$$R'(t) = 2\left(5 - 4 \cos \frac{\pi t}{6}\right) \qquad 0 \le t \le 12$$

thousand dollars per week, where $t$ is measured in weeks, with $t = 0$ corresponding to the beginning of June. Find the total revenue $R$ of the Shanty at the end of $t$ weeks after its opening on June 1.

87. **Simple Harmonic Motion**  The acceleration function of a body moving along a coordinate line is

$$a(t) = -4 \cos 2t - 3 \sin 2t \qquad t \ge 0$$

Find its velocity and position functions at any time $t$ if the body is located at the origin and has an initial velocity of $\frac{3}{2}$ m/sec.

88. **Special Theory of Relativity**  According to Einstein's special theory of relativity, the mass of a particle is given by

$$m = \frac{m_0}{\sqrt{1 - \dfrac{v^2}{c^2}}}$$

where $m_0$ is the rest mass of the particle, $v$ is its velocity, and $c$ is the speed of light. Suppose that a particle starts from rest at $t = 0$ and moves along a straight line under the action of a constant force $F$. Then, according to Newton's second law of motion, the equation of motion is

$$F = m_0 \frac{d}{dt} \left( \frac{v}{\sqrt{1 - \dfrac{v^2}{c^2}}} \right)$$

Find the velocity and position functions of the particle. What happens to the velocity of the particle as time goes by?

*In Exercises 89 and 90, determine whether the statement is true or false. If it is true, explain why it is true. If it is false, explain why or give an example to show why it is false.*

89. If $f$ is continuous, then $\int x f(x^2) \, dx = \frac{1}{2} \int f(u) \, du$, where $u = x^2$.

90. If $f$ is continuous, then $\int f(ax + b) \, dx = \int f(x) \, dx$.

---

## 4.3   Area

### ■ An Intuitive Look

Consider a car moving on a straight road with a velocity function given by

$$v(t) = 44 \qquad 0 \le t \le 10$$

where $t$ is measured in seconds and $v(t)$ in feet per second. Since $v(t) > 0$, it also gives the speed of the car over this time interval. The distance traveled by the car between $t = 1$ and $t = 5$ is

$$(44)(5 - 1) \qquad \text{constant speed} \cdot \text{time elapsed}$$

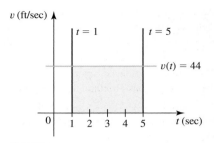

**FIGURE 1**
The distance traveled by the car can be represented by the area of the rectangular region.

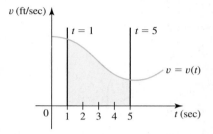

**FIGURE 2**
The distance covered by the car is given by the "area" of the shaded region.

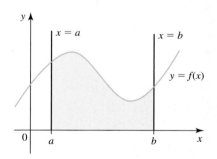

**FIGURE 3**
The shaded region is the area under the graph of $f$ on $[a, b]$.

or 176 ft. If you examine the graph of $v$ shown in Figure 1, you will see that this distance is just the area of the rectangular region bounded above by the graph of $v$, below by the $t$-axis, and to the left and right by the vertical lines $t = 1$ and $t = 5$, respectively.

Suppose that the same car moves along a straight road but this time with a velocity function $v$ that is positive but not necessarily constant over an interval of time. What is the distance traveled by the car between $t = 1$ and $t = 5$? We might be tempted to conjecture that it is given by the "area" of the region bounded above by the graph of $v$, below by the $t$-axis, and to the left and right by the vertical lines $t = 1$ and $t = 5$, respectively (see Figure 2). Later, we will show that this is indeed the case.

This example raises two questions:

1.  What do we mean by the "area" of a region such as the one shown in Figure 2?
2.  How do we find the area of such a region?

## ▇ The Area Problem

Here, we have touched upon the second fundamental problem in calculus: How do we find the area of the region bounded above by the graph of a *nonnegative* function $f$, below by the $x$-axis, and to the left and right by the vertical lines $x = a$ and $x = b$, as shown in Figure 3? We refer to the area of this region as the **area under the graph of $f$** on the interval $[a, b]$.

## ▇ Defining the Area of the Region Under the Graph of a Function

When we defined the slope of the tangent line to the graph of a function at a point on the graph, we first approximated it with the slopes of secant lines (quantities that we could compute). We then took the limit of these approximations to give us the slope of the tangent line. We will now adopt a parallel approach to define the area of the region under the graph of a function.

The idea here is to approximate the area of a region by using the sums of the areas of rectangles (quantities that we can compute).* We can then find the desired area by taking the limit of these sums. Let's begin by looking at a specific example.

**EXAMPLE 1**    Consider the region $S$ bounded above by the parabola $f(x) = x^2$, below by the $x$-axis, and to the left and right by the vertical lines $x = 0$ and $x = 1$, respectively (see Figure 4). As you can see, the area $A$ of the region $S$ can be approximated by the area $A_1$ of the rectangle $R_1$ with base lying on the interval $[0, 1]$ and height given by the value of $f(x) = x^2$ evaluated at the midpoint of $[0, 1]$. Thus,

$$A \approx A_1 = 1 \cdot f\left(\frac{1}{2}\right) = (1)\left(\frac{1}{2}\right)^2 = \frac{1}{4}$$

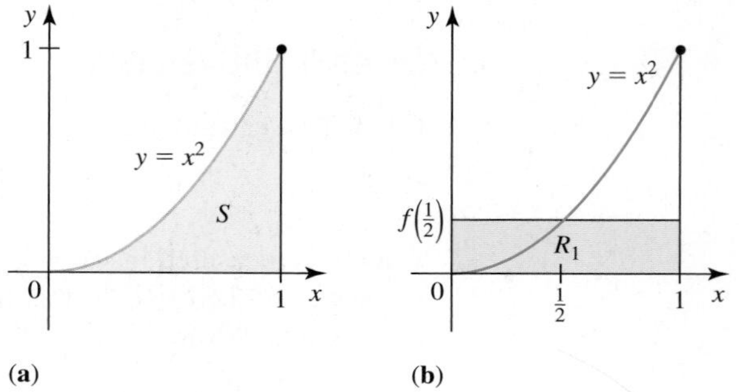

**FIGURE 4**
The area of the region $S$ in part (a) is approximated by the area of the rectangle $R_1$ in part (b).

**(a)**

**(b)**

---

*Until a formal definition of area is given, the term *area* will refer to our intuitive notion of area.

We have used the midpoint of the interval $[0, 1]$ to compute the height of the approximating rectangle because it seems to be a logical choice. But you should convince yourself that any other point in the interval, including the endpoints, would also serve our purpose. Of course, the approximation obtained will be different depending on your choice.

Can we do better? Let's divide the interval $[0, 1]$ into two subintervals $\left[0, \frac{1}{2}\right]$ and $\left[\frac{1}{2}, 1\right]$, each of (equal) length $\frac{1}{2}$. Figure 5a shows the region $S$ expressed as the union of two nonoverlapping subregions $S_1$ and $S_2$ with bases lying on the subintervals $\left[0, \frac{1}{2}\right]$ and $\left[\frac{1}{2}, 1\right]$, respectively. Figure 5b shows the rectangle $R_1$ with base lying on $\left[0, \frac{1}{2}\right]$ and height $f\left(\frac{1}{4}\right)$, the value of $f$ evaluated at the midpoint of $\left[0, \frac{1}{2}\right]$, and the rectangle $R_2$ with base lying on $\left[\frac{1}{2}, 1\right]$ and height $f\left(\frac{3}{4}\right)$, where $x = \frac{3}{4}$ is the midpoint of $\left[\frac{1}{2}, 1\right]$. If we approximate the area of $S_1$ by the area of $R_1$ and the area of $S_2$ by the area of $R_2$ and denote the sum of the areas of the two rectangles by $A_2$, we obtain

$$A \approx A_2 = \frac{1}{2}f\left(\frac{1}{4}\right) + \frac{1}{2}f\left(\frac{3}{4}\right)$$

$$= \frac{1}{2}\left(\frac{1}{4}\right)^2 + \frac{1}{2}\left(\frac{3}{4}\right)^2$$

$$= \frac{1}{2}\left(\frac{1}{16} + \frac{9}{16}\right) = \frac{5}{16}$$

or $0.3125$. Continuing with this process, we divide the interval $[0, 1]$ into four subintervals of equal length $\frac{1}{4}$ using the five points

$$x_0 = 0, \qquad x_1 = \frac{1}{4}, \qquad x_2 = \frac{1}{2}, \qquad x_3 = \frac{3}{4}, \qquad \text{and} \qquad x_4 = 1$$

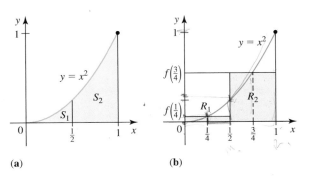

**FIGURE 5**

The subregions $S_1$ and $S_2$ in part (a) are approximated by the rectangles $R_1$ and $R_2$ in part (b).

**(a)**　　　　**(b)**

The resulting subintervals are

$$\left[0, \tfrac{1}{4}\right], \qquad \left[\tfrac{1}{4}, \tfrac{1}{2}\right], \qquad \left[\tfrac{1}{2}, \tfrac{3}{4}\right], \qquad \text{and} \qquad \left[\tfrac{3}{4}, 1\right]$$

Figure 6a shows the region $S$ expressed as the union of four nonoverlapping subregions $S_1$, $S_2$, $S_3$, and $S_4$ with bases lying on these subintervals. The midpoints of the subintervals are

$$c_1 = \frac{1}{8}, \qquad c_2 = \frac{3}{8}, \qquad c_3 = \frac{5}{8}, \qquad \text{and} \qquad c_4 = \frac{7}{8}$$

respectively. The rectangles $R_1$, $R_2$, $R_3$, and $R_4$ with bases lying on these subintervals and having heights evaluated at their respective midpoints are shown in Figure 6b.

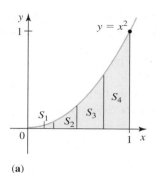

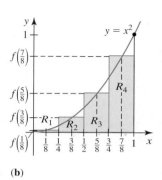

**(a)**                                                        **(b)**

**FIGURE 6**

The area of subregion $S_i$ in part (a) is approximated by the area
of rectangle $R_i$ for $1 \le i \le 4$.

Approximating the area of the subregion $S_i$ by the area of the rectangle $R_i$, where $1 \le i \le 4$, and letting $A_4$ denote the sum of the areas of the four rectangles, we obtain yet another approximation of the area $A$ of $S$:

$$
\begin{aligned}
A \approx A_4 &= \frac{1}{4}f\left(\frac{1}{8}\right) + \frac{1}{4}f\left(\frac{3}{8}\right) + \frac{1}{4}f\left(\frac{5}{8}\right) + \frac{1}{4}f\left(\frac{7}{8}\right) \\
&= \frac{1}{4}\left(\frac{1}{8}\right)^2 + \frac{1}{4}\left(\frac{3}{8}\right)^2 + \frac{1}{4}\left(\frac{5}{8}\right)^2 + \frac{1}{4}\left(\frac{7}{8}\right)^2 \\
&= \frac{1}{4}\left(\frac{1}{64} + \frac{9}{64} + \frac{25}{64} + \frac{49}{64}\right) = \frac{21}{64}
\end{aligned}
$$

or 0.328125.

We can keep going. Figure 7a shows what happens if we use eight rectangles to approximate the area of the region $S$, and Figure 7b shows the situation if sixteen rectangles are used.

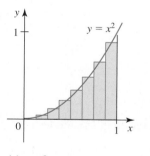

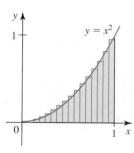

**(a)** $n = 8$                            **(b)** $n = 16$

**FIGURE 7**

As the number of rectangles used increases, the approximation
of the area of the region $S$ seems to improve.

With the aid of a computer we can find the approximations of the area $A$ of the region $S$ using $n$ approximating rectangles. In the following table, $A_n$ denotes the approximation of $A$ when $n$ rectangles are used. The results include approximations obtained earlier ($n = 1, 2, 4$) and are rounded off to seven decimal places.

| $n$ | $A_n$ |
|---|---|
| 1 | 0.25 |
| 2 | 0.3125 |
| 4 | 0.328125 |
| 8 | 0.3320313 |
| 16 | 0.3330078 |
| 50 | 0.3333000 |
| 100 | 0.3333250 |
| 500 | 0.3333330 |
| 1000 | 0.3333333 |

These results seem to suggest that $A_n$ approaches $\frac{1}{3}$ as $n$ gets larger and larger; that is, $\lim_{n \to \infty} A_n = \frac{1}{3}$. This in turn suggests that we could take the area of the region $S$ to be $\frac{1}{3}$. ∎

## ■ Sigma Notation

Before confirming this result, we will digress a little to introduce a notation that will provide us with a shorthand method for writing sums involving a large number of terms. The notation uses the uppercase Greek letter sigma $\Sigma$ and is accordingly called *sigma notation*.

---

**DEFINITION   Sigma Notation**

The sum of the $n$ terms $a_1, a_2, a_3, \ldots, a_n$ is abbreviated $\displaystyle\sum_{k=1}^{n} a_k$. Thus,

$$\sum_{k=1}^{n} a_k = a_1 + a_2 + a_3 + \cdots + a_k + \cdots + a_n$$

The variable $k$ is called the **index of summation,** the term $a_k$ is called the **$k$th term** of the sum, and the numbers $n$ and 1 are called the **upper** and **lower limits of summation,** respectively.

---

The sum $\sum_{k=1}^{n} a_k$ is read "the sum of $a_k$ where $k$ runs from 1 to $n$."

**EXAMPLE 2**   Write each of the following sums in expanded form:

**a.** $\displaystyle\sum_{k=1}^{5} k$   **b.** $\displaystyle\sum_{k=1}^{10} k^2$   **c.** $\displaystyle\sum_{k=1}^{20} \frac{1}{(k+1)^2}$   **d.** $\displaystyle\sum_{k=1}^{15} (-1)^k k^3$   **e.** $\displaystyle\sum_{k=1}^{10} \sin\left(\frac{k\pi}{4}\right)$

**Solution**

**a.** $\displaystyle\sum_{k=1}^{5} k = 1 + 2 + 3 + 4 + 5$   Here, $a_k = k$, so $a_1 = 1$, $a_2 = 2$, $a_3 = 3$, $a_4 = 4$, and $a_5 = 5$.

**b.** $\displaystyle\sum_{k=1}^{10} k^2 = 1^2 + 2^2 + 3^2 + \cdots + 10^2$    Here, $a_k = k^2$, so $a_1 = 1^2$, $a_2 = 2^2$, ....

**c.** $\displaystyle\sum_{k=1}^{20} \frac{1}{(k+1)^2} = \frac{1}{2^2} + \frac{1}{3^2} + \frac{1}{4^2} + \cdots + \frac{1}{21^2}$    Here, $a_k = \dfrac{1}{(k+1)^2}$.

**d.** $\displaystyle\sum_{k=1}^{15} (-1)^k k^3 = (-1)^1 1^3 + (-1)^2 2^3 + (-1)^3 3^3 + \cdots + (-1)^{15} 15^3$

$$= -1 + 2^3 - 3^3 + \cdots - 15^3$$

**e.** $\displaystyle\sum_{k=1}^{10} \sin\left(\frac{k\pi}{4}\right) = \sin\frac{\pi}{4} + \sin\frac{2\pi}{4} + \sin\frac{3\pi}{4} + \cdots + \sin\frac{10\pi}{4}$    ■

So far, we have used $k$ as the index of summation, but any letter will do. For example, each of the following

$$\sum_{k=1}^{5} a_k, \qquad \sum_{i=1}^{5} a_i, \qquad \text{and} \qquad \sum_{j=1}^{5} a_j$$

represents the sum $a_1 + a_2 + a_3 + a_4 + a_5$. Sometimes it is more convenient to use a lower limit of summation other than 1. For example, we can write

$$\sum_{k=2}^{6} (2k+1) = 5 + 7 + 9 + 11 + 13$$

which is equivalent to

$$\sum_{k=1}^{5} (2k+3) = 5 + 7 + 9 + 11 + 13$$

Also, if the upper and lower indices of summation are the same, then the sum consists of just one term. For example,

$$\sum_{k=1}^{1} \frac{1}{k} = \frac{1}{1} = 1 \qquad k \text{ runs from 1 to 1.}$$

In the next example, keep in mind that the upper limit of summation $n$ is constant *with respect to the summation.*

**EXAMPLE 3**    Write each of the following sums in expanded form:

**a.** $\displaystyle\sum_{k=1}^{n} \frac{1}{n}(2k-1)$    **b.** $\displaystyle\sum_{k=1}^{n}\left(1 + \frac{k}{n}\right)^3\left(\frac{1}{n}\right)$    **c.** $\displaystyle\sum_{k=1}^{n-1} \sin\left(\frac{k\pi}{n}\right)$

**Solution**

**a.** $\displaystyle\sum_{k=1}^{n} \frac{1}{n}(2k-1) = \frac{1}{n}(2-1) + \frac{1}{n}(4-1) + \frac{1}{n}(6-1) + \cdots + \frac{1}{n}(2n-1)$

$$= \frac{1}{n} + \frac{3}{n} + \frac{5}{n} + \cdots + \frac{2n-1}{n}$$

**b.** $\displaystyle\sum_{k=1}^{n}\left(1+\frac{k}{n}\right)^{3}\left(\frac{1}{n}\right) = \left(1+\frac{1}{n}\right)^{3}\left(\frac{1}{n}\right) + \left(1+\frac{2}{n}\right)^{3}\left(\frac{1}{n}\right)$

$$+ \left(1+\frac{3}{n}\right)^{3}\left(\frac{1}{n}\right) + \cdots + \left(1+\frac{n}{n}\right)^{3}\left(\frac{1}{n}\right)$$

$$= \left(\frac{1}{n}\right)\left[\left(1+\frac{1}{n}\right)^{3} + \left(1+\frac{2}{n}\right)^{3} + \left(1+\frac{3}{n}\right)^{3} + \cdots + 2^{3}\right]$$

**c.** $\displaystyle\sum_{k=1}^{n-1}\sin\left(\frac{k\pi}{n}\right) = \sin\frac{\pi}{n} + \sin\frac{2\pi}{n} + \sin\frac{3\pi}{n} + \cdots + \sin\frac{(n-1)\pi}{n}$

## Summation Formulas

The following rules are useful in manipulating sums written using sigma notation.

---

**Rules of Summation**

**1.** $\displaystyle\sum_{k=1}^{n} ca_k = c\sum_{k=1}^{n} a_k,$   where $c$ is a constant

**2.** $\displaystyle\sum_{k=1}^{n} (a_k + b_k) = \sum_{k=1}^{n} a_k + \sum_{k=1}^{n} b_k$

**3.** $\displaystyle\sum_{k=1}^{n} (a_k - b_k) = \sum_{k=1}^{n} a_k - \sum_{k=1}^{n} b_k$

---

**PROOF**   All three rules can be proved by writing the respective sums in expanded form. For example, to prove Rule 1, we write

$$\sum_{k=1}^{n} ca_k = ca_1 + ca_2 + \cdots + ca_n = c(a_1 + a_2 + \cdots + a_n)$$

Use the distributive property.

$$= c\sum_{k=1}^{n} a_k$$

The proof of Rules 2 and 3 are left as exercises. ▬

**EXAMPLE 4**   Use the rules of summation to expand each sum:

**a.** $\displaystyle\sum_{k=1}^{10} 3k^{2}$   **b.** $\displaystyle\sum_{k=2}^{8} (k + 3k^{3})$

**Solution**

**a.** $\displaystyle\sum_{k=1}^{10} 3k^{2} = 3\sum_{k=1}^{10} k^{2} = 3(1^{2} + 2^{2} + 3^{2} + \cdots + 10^{2})$

**b.** $\displaystyle\sum_{k=2}^{8} (k + 3k^{3}) = \sum_{k=2}^{8} k + \sum_{k=2}^{8} 3k^{3} = \sum_{k=2}^{8} k + 3\sum_{k=2}^{8} k^{3}$

$$= (2 + 3 + \cdots + 8) + 3(2^{3} + 3^{3} + \cdots + 8^{3})$$ ▬

The following summation formulas will be used later.

**THEOREM 1    Summation Formulas**

**a.** $\displaystyle\sum_{k=1}^{n} c = nc,$   $c$ a constant

**b.** $\displaystyle\sum_{k=1}^{n} k = \frac{n(n + 1)}{2}$

**c.** $\displaystyle\sum_{k=1}^{n} k^2 = \frac{n(n + 1)(2n + 1)}{6}$

**d.** $\displaystyle\sum_{k=1}^{n} k^3 = \left[\frac{n(n + 1)}{2}\right]^2$

We will omit the proofs.

**EXAMPLE 5**   Use Theorem 1 to evaluate each sum:

**a.** $\displaystyle\sum_{k=1}^{10} 3$    **b.** $\displaystyle\sum_{k=1}^{20} k$    **c.** $\displaystyle\sum_{k=1}^{50} k^2$

**Solution**

**a.** $\displaystyle\sum_{k=1}^{10} 3 = \underbrace{3 + 3 + 3 + \cdots + 3}_{10 \text{ terms}} = 10(3) = 30$    Use Theorem 1a.

**b.** $\displaystyle\sum_{k=1}^{20} k = 1 + 2 + 3 + \cdots + 20 = \frac{20(20 + 1)}{2} = 210$    Use Theorem 1b.

**c.** $\displaystyle\sum_{k=1}^{50} k^2 = 1^2 + 2^2 + 3^2 + \cdots + 50^2$

$\qquad = \dfrac{50(50 + 1)(2 \cdot 50 + 1)}{6} = 42{,}925$    Use Theorem 1c.

**EXAMPLE 6**   Evaluate $\displaystyle\sum_{k=1}^{10} 3k^2(2k + 1).$

**Solution**

$$\sum_{k=1}^{10} 3k^2(2k + 1) = \sum_{k=1}^{10} (6k^3 + 3k^2)$$

$$= 6\sum_{k=1}^{10} k^3 + 3\sum_{k=1}^{10} k^2$$

$$= 6\left[\frac{10(10 + 1)}{2}\right]^2 + 3\left[\frac{10(10 + 1)(20 + 1)}{6}\right]$$

$$= 18{,}150 + 1155 = 19{,}305$$

**EXAMPLE 7** Evaluate

$$\lim_{n\to\infty} \sum_{k=1}^{n} \left[ \left( \frac{k}{n} \right)^2 + 2 \right]\left( \frac{4}{n} \right)$$

**Solution**

$$\lim_{n\to\infty} \sum_{k=1}^{n} \left[ \left( \frac{k}{n} \right)^2 + 2 \right]\left( \frac{4}{n} \right)$$

$$= \lim_{n\to\infty} \sum_{k=1}^{n} \left( \frac{4k^2}{n^3} + \frac{8}{n} \right)$$

$$= \lim_{n\to\infty} \left[ \frac{4}{n^3} \sum_{k=1}^{n} k^2 + \frac{8}{n} \sum_{k=1}^{n} 1 \right] \qquad \text{Remember that } n \text{ is constant with respect to the summations.}$$

$$= \lim_{n\to\infty} \left[ \frac{4}{n^3} \cdot \frac{n(n+1)(2n+1)}{6} + \frac{8}{n} \cdot n \right] \qquad \text{Use Theorems 1a and 1c.}$$

$$= \lim_{n\to\infty} \left[ \frac{2}{3}\left( \frac{n}{n} \right)\left( \frac{n+1}{n} \right)\left( \frac{2n+1}{n} \right) + 8 \right]$$

$$= \lim_{n\to\infty} \left[ \frac{2}{3}\left( 1 + \frac{1}{n} \right)\left( 2 + \frac{1}{n} \right) + 8 \right]$$

$$= \frac{2}{3}(1)(2) + 8 = \frac{28}{3} \qquad\qquad\blacksquare$$

## ◼ An Intuitive Look at Area (Continued)

We are now ready to resume the discussion of the area concept.

**EXAMPLE 8** Following the procedure of Example 1, we can obtain an expression, $A_n$, for approximating the area of the region under the graph of $f(x) = x^2$ on the interval $[0, 1]$ using $n$ rectangles. Then, by letting $n$ take on increasingly larger values, we will show that

$$\lim_{n\to\infty} A_n = \frac{1}{3}$$

To find such an expression, let's divide the interval $[0, 1]$ into $n$ subintervals of equal length $1/n$ using the $(n + 1)$ points

$$x_0 = 0, \quad x_1 = \frac{1}{n}, \quad x_2 = \frac{2}{n}, \quad x_3 = \frac{3}{n}, \quad \dots, \quad x_k = \frac{k}{n}, \quad \dots, \quad x_n = 1$$

The subintervals are

$$\underbrace{\left[ 0, \frac{1}{n} \right],}_{\text{1st subinterval}} \quad \underbrace{\left[ \frac{1}{n}, \frac{2}{n} \right],}_{\text{2nd subinterval}} \quad \underbrace{\left[ \frac{2}{n}, \frac{3}{n} \right],}_{\text{3rd subinterval}} \quad \dots, \quad \underbrace{\left[ \frac{k-1}{n}, \frac{k}{n} \right],}_{k\text{th subinterval}} \quad \dots, \quad \underbrace{\left[ \frac{n-1}{n}, 1 \right]}_{n\text{th subinterval}}$$

Next, we note that the midpoints of these subintervals are

$$c_1 = \frac{1}{2n}, \quad c_2 = \frac{3}{2n}, \quad c_3 = \frac{5}{2n}, \quad \dots, \quad c_k = \frac{2k-1}{2n}, \quad \dots, \quad c_n = \frac{2n-1}{2n}$$

so the heights of the $n$ corresponding rectangles are

$$f\left(\frac{1}{2n}\right), \quad f\left(\frac{3}{2n}\right), \quad f\left(\frac{5}{2n}\right), \quad \cdots, \quad f\left(\frac{2k-1}{2n}\right), \quad \cdots, \quad f\left(\frac{2n-1}{2n}\right)$$

(See Figure 8.) Letting $A_n$ denote the sum of the areas of the $n$ rectangles, we have

$$A_n = \frac{1}{n}f\left(\frac{1}{2n}\right) + \frac{1}{n}f\left(\frac{3}{2n}\right) + \frac{1}{n}f\left(\frac{5}{2n}\right) + \cdots + \frac{1}{n}f\left(\frac{2k-1}{2n}\right) + \cdots + \frac{1}{n}f\left(\frac{2n-1}{2n}\right)$$

$$= \frac{1}{n}\left[f\left(\frac{1}{2n}\right) + f\left(\frac{3}{2n}\right) + f\left(\frac{5}{2n}\right) + \cdots + f\left(\frac{2k-1}{2n}\right) + \cdots + f\left(\frac{2n-1}{2n}\right)\right] \qquad \text{Factor out } \frac{1}{n}.$$

$$= \frac{1}{n}\sum_{k=1}^{n} f\left(\frac{2k-1}{2n}\right) \qquad \text{Use sigma notation.}$$

$$= \frac{1}{n}\sum_{k=1}^{n} \left(\frac{2k-1}{2n}\right)^2 \qquad f(x) = x^2$$

$$= \frac{1}{n}\sum_{k=1}^{n} \left(\frac{4k^2 - 4k + 1}{4n^2}\right) \qquad \text{Expand the expression following the summation sign.}$$

$$= \frac{1}{4n^3}\sum_{k=1}^{n} (4k^2 - 4k + 1) \qquad n \text{ is constant with respect to summation.}$$

$$= \frac{1}{4n^3}\left[4\sum_{k=1}^{n} k^2 - 4\sum_{k=1}^{n} k + \sum_{k=1}^{n} 1\right] \qquad \text{Use the rules of summation.}$$

$$= \frac{1}{4n^3}\left[\frac{4n(n+1)(2n+1)}{6} - \frac{4n(n+1)}{2} + n\right] \qquad \text{Use Theorems 1a, 1b, and 1c.}$$

$$= \frac{1}{4n^3} \cdot \frac{n(4n^2 - 1)}{3}$$

$$= \frac{4n^2 - 1}{12n^2}$$

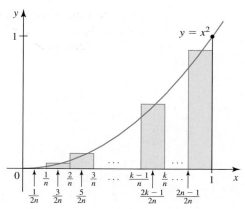

**FIGURE 8**
The area of the first rectangle
is $\frac{1}{n} \cdot f\left(\frac{1}{2n}\right)$, the area of
the second rectangle is
$\frac{1}{n} \cdot f\left(\frac{3}{2n}\right), \ldots$, and the
area of the $n$th rectangle
is $\frac{1}{n} \cdot f\left(\frac{2n-1}{2n}\right)$.

By letting $n$ take on the values 4, 10, and 100, for example, we see that

$$A_4 = \frac{4(4)^2 - 1}{12(4)^2} = 0.328125 \qquad \text{Compare this with Example 1.}$$

$$A_{10} = \frac{4(10)^2 - 1}{12(10)^2} = 0.3325$$

and

$$A_{100} = \frac{4(100)^2 - 1}{12(100)^2} = 0.333325$$

Our computations seem to show that $A_n$ approaches $\frac{1}{3}$ as $n$ gets larger and larger. This result is confirmed by the following calculation:

$$\lim_{n \to \infty} A_n = \lim_{n \to \infty} \frac{4n^2 - 1}{12n^2}$$

$$= \lim_{n \to \infty} \left( \frac{1}{3} - \frac{1}{12n^2} \right)$$

$$= \frac{1}{3}$$

The results of Example 8 suggest that we *define* the area of the region $S$ under the graph of $f(x) = x^2$ on the interval $[0, 1]$ to be $\frac{1}{3}$.

## Defining the Area of the Region Under the Graph of a Function

Example 8 paves the way to defining the area of the region under the graph of a continuous nonnegative function $f$ on an interval $[a, b]$. (See Figure 9.) We begin by partitioning the interval $[a, b]$ using $n + 1$ equally spaced points

$$a = x_0 < x_1 < x_2 < x_3 < \cdots < x_{n-1} < x_n = b$$

This is called a **regular partition** of $[a, b]$. The resulting subintervals are

$$[x_0, x_1], \qquad [x_1, x_2], \qquad [x_2, x_3], \qquad \ldots, \qquad [x_{n-1}, x_n]$$

with $x_0 = a$ and $x_n = b$. The width of each subinterval is

$$\Delta x = \frac{b - a}{n}$$

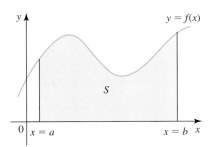

**FIGURE 9**
The region $S$ under the graph of $f$ on $[a, b]$.

This partitioning leads to the subdivision of the region $S$ into $n$ nonoverlapping subregions $S_1, S_2, S_3, \ldots, S_n$, where $S_1$ is the subregion under the graph of $f$ on $[x_0, x_1]$, $S_2$ is the subregion under the graph of $f$ on $[x_1, x_2]$, and so on. (See Figure 10.)

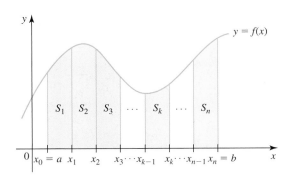

**FIGURE 10**
The region $S$ is the union of $n$ nonoverlapping subregions.

Next, we approximate the area of the subregion $S_1$ by the area of the rectangle $R_1$ with base $[x_0, x_1]$ and height $f(c_1)$, where $c_1$ is an arbitrarily chosen point in the subinterval $[x_0, x_1]$. (See Figure 11.) Thus,

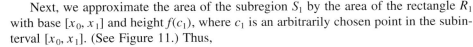

$$\text{area of } S_1 \approx \text{area of } R_1 = f(c_1)\Delta x$$

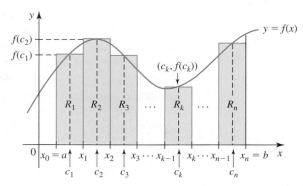

**FIGURE 11**

The area of the subregion $S_k$ in Figure 10 is approximated by the area of the rectangle $R_k$.

Similarly, we approximate the area of the subregion $S_2$ by the area of the rectangle $R_2$ with base $[x_1, x_2]$ and height $f(c_2)$, where $c_2$ is an arbitrary point in $[x_1, x_2]$. Thus,

$$\text{area of } S_2 \approx \text{area of } R_2 = f(c_2)\Delta x$$

In general, we approximate the area of the subregion $S_k$ by the area of the rectangle $R_k$ with base $[x_{k-1}, x_k]$ and height $f(c_k)$, where $c_k$ is an arbitrary point in $[x_{k-1}, x_k]$. Thus,

$$\text{area of } S_k \approx \text{area of } R_k = f(c_k)\Delta x$$

If we denote the area of the region $S$ by $A$ and the sum of the areas of the $n$ rectangles by $A_n$, then, intuitively, we see that

$$A \approx A_n = f(c_1)\Delta x + f(c_2)\Delta x + \cdots + f(c_n)\Delta x = \sum_{k=1}^{n} f(c_k)\Delta x$$

If we let the number of partition points, $n$, increase, then the number of subregions increases, and, as shown in Figure 12, the approximations seem to improve.

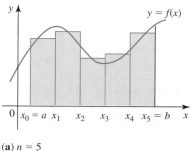

**FIGURE 12**

As $n$ increases the approximation seems to improve.

**(a)** $n = 5$

**(b)** $n = 10$

This observation suggests that we define the area $A$ of the region $S$ as follows.

---

**DEFINITION** **Area of the Region $S$ Under the Graph of a Function**

Let $f$ be a continuous, nonnegative function defined on an interval $[a, b]$. Suppose that $[a, b]$ is divided into $n$ subintervals of equal length $\Delta x = (b - a)/n$ by means of $(n + 1)$ equally spaced points

$$a = x_0 < x_1 < x_2 < \cdots < x_n = b$$

Then the area of the region $S$ that lies under the graph of $f$ on $[a, b]$ is

$$A = \lim_{n \to \infty} A_n = \lim_{n \to \infty} \sum_{k=1}^{n} f(c_k)\Delta x \qquad (1)$$

where $c_k$ lies in the $k$th subinterval $[x_{k-1}, x_k]$.

Because of the supposition that $f$ is continuous, it can be shown that the limit (1) in the definition always exists, regardless of how the points $c_k$ in $[x_{k-1}, x_k]$, where $1 \le k \le n$, are chosen. In Exercises 47 and 48 you will be asked to compute the area of the region under the graph of $f(x) = x^2$ on $[0, 1]$ choosing $c_k$ to be (1) the left endpoint of the subinterval $[x_{k-1}, x_k]$, that is, $c_k = x_{k-1}$, for $1 \le k \le n$, and (2) the right endpoint of the subinterval $[x_{k-1}, x_k]$, that is, $c_k = x_k$, for $1 \le k \le n$. You will see that the results are indeed the same as those obtained in Example 8, where $c_k$ was chosen to be the midpoint of the subinterval $[x_{k-1}, x_k]$.

**EXAMPLE 9** Find the area of the region under the graph of $f(x) = 4 - x^2$ on the interval $[-2, 1]$.

**Solution** Observe that $f$ is continuous and nonnegative on $[-2, 1]$. The region under consideration is shown in Figure 13. If we partition the interval $[-2, 1]$ into $n$ subintervals of equal length by means of $(n + 1)$ points, then the width of each subinterval is

$$\Delta x = \frac{b - a}{n} = \frac{1 - (-2)}{n} = \frac{3}{n}.$$

and the partition points are

$$x_0 = -2, \qquad x_1 = -2 + \frac{3}{n}, \qquad x_2 = -2 + 2\left(\frac{3}{n}\right), \qquad \ldots,$$

$$x_k = -2 + k\left(\frac{3}{n}\right), \qquad \ldots, \qquad x_n = 1$$

Since $f$ is continuous, we have a free hand at picking $c_k$ in $[x_{k-1}, x_k]$. So let's pick $c_k$ to be the right endpoint of the subinterval; that is,

$$c_k = x_k = -2 + \frac{3k}{n}$$

Using the definition of the area of the region under a graph of a function, we find the required area to be

$$A = \lim_{n \to \infty} \sum_{k=1}^{n} f(c_k)\Delta x$$

$$= \lim_{n \to \infty} \sum_{k=1}^{n} f\left(-2 + \frac{3k}{n}\right)\left(\frac{3}{n}\right)$$

$$= \lim_{n \to \infty} \sum_{k=1}^{n} \left[4 - \left(-2 + \frac{3k}{n}\right)^2\right]\left(\frac{3}{n}\right) \qquad f(x) = 4 - x^2$$

$$= \lim_{n \to \infty} \frac{3}{n} \sum_{k=1}^{n} \left(4 - 4 + \frac{12k}{n} - \frac{9k^2}{n^2}\right)$$

$$= \lim_{n \to \infty} \frac{3}{n} \left[\frac{12}{n} \sum_{k=1}^{n} k - \frac{9}{n^2} \sum_{k=1}^{n} k^2\right]$$

$$= \lim_{n \to \infty} \frac{3}{n} \left[\frac{12}{n} \cdot \frac{n(n + 1)}{2} - \frac{9}{n^2} \cdot \frac{n(n + 1)(2n + 1)}{6}\right] \qquad \text{Use Theorems 1b and 1c.}$$

$$= \lim_{n \to \infty} \left[18\left(1 + \frac{1}{n}\right) - \frac{9}{2}\left(1 + \frac{1}{n}\right)\left(2 + \frac{1}{n}\right)\right]$$

$$= 18 - \frac{9}{2}(2) = 9 \qquad \blacksquare$$

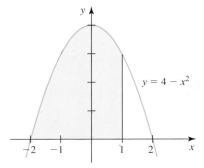

**FIGURE 13**
The region under the graph of $f(x) = 4 - x^2$ on $[-2, 1]$

**Note** In Exercise 49 you will be asked to solve Example 9 again, this time choosing the midpoint for $c_k$. Of course, you should obtain the same answer. ▪

## ■ Area and Distance

We now show that if $v$ is a (continuous) velocity function of a car traveling in a straight line and $v(t) \geq 0$ on $[a, b]$, then the distance covered by the car between $t = a$ and $t = b$ is numerically equal to the area of the region under the graph of the velocity function on $[a, b]$. (See Figure 14.) Let's divide the time interval $[a, b]$ into $n$ subintervals each of equal length $\Delta t = (b - a)/n$ by means of $(n + 1)$ equally spaced points

$$t_0 = a, \quad t_1 = a + \Delta t, \quad t_2 = a + 2(\Delta t), \quad \ldots, \quad t_k = a + k(\Delta t), \quad \ldots, \quad t_n = b$$

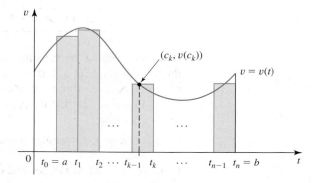

**FIGURE 14**
$v$ is the velocity function on $[a, b]$.

Observe that if $n$ is large, then the time intervals $[t_0, t_1], [t_1, t_2], \ldots, [t_{n-1}, t_n]$ are uniformly small.

Let's focus our attention on the first subinterval $[t_0, t_1]$. Because $v$ is continuous, we see that the speed of the car does not vary appreciably in that interval and can be approximated by the *constant* speed $v(c_1)$, where $c_1$ is an arbitrary point in $[t_0, t_1]$.* Therefore, the distance covered by the car from $t = t_0$ to $t = t_1$ may be approximated by

$$v(c_1)\Delta t \qquad \text{distance = constant speed · time elapsed}$$

In a similar manner we see that the distance covered by the car from $t_1$ to $t_2$ is approximately

$$v(c_2)\Delta t$$

where $c_2$ is an arbitrary point in $[t_1, t_2]$. Continuing, we see that the distance covered by the car from $t_{k-1}$ to $t_k$ is approximately

$$v(c_k)\Delta t$$

where $c_k$ is an arbitrary point in $[t_{k-1}, t_k]$. Therefore, the distance traveled by the car from $t = a$ to $t = b$ is approximately

$$v(c_1)\Delta t + v(c_2)\Delta t + \cdots + v(c_n)\Delta t = \sum_{k=1}^{n} v(c_k)\Delta t \qquad (2)$$

---

*Recall that if a function $f$ is continuous at $t$, then a small change in $t$ implies a small change in $f(t)$.

As $n$ gets larger and larger, the length of the time subintervals gets smaller and smaller. Intuitively, we expect that the approximations will improve. It seems reasonable, therefore, to define the distance covered by the car to be

$$\lim_{n \to \infty} \sum_{k=1}^{n} v(c_k) \Delta t$$

But as you can see from Figure 14, this quantity also gives the area of the region under the graph of $v$ on $[a, b]$.

**EXAMPLE 10** **Distance Covered by a Cyclist** The speed of a cyclist is measured at 4-sec intervals over a 32-sec time span and recorded in the following table.

| Time (sec)      | 0 | 4 | 8 | 12 | 16 | 20 | 24 | 28 | 32 |
|-----------------|---|---|---|----|----|----|----|----|----|
| Speed (ft/sec)  | 2 | 4 | 6 | 10 | 12 | 14 | 10 | 8  | 6  |

If we let $v$ denote the velocity function associated with the motion of the cyclist over the time interval $[0, 32]$, then the values of $v$ are available to us only at a discrete set of numbers, even though $v$ is clearly a continuous function defined on the interval. Using Equation (2), find the approximate distance $D$ covered by the cyclist from $t = 0$ to $t = 32$ using

**a.** Eight ($n = 8$) rectangles and choosing $c_k$ to be the left endpoint of the $k$th subinterval

**b.** Eight ($n = 8$) rectangles and choosing $c_k$ to be the right endpoint of the $k$th subinterval

**c.** Four ($n = 4$) rectangles and choosing $c_k$ to be the midpoint of the $k$th subinterval.

**Solution** An approximation to the graph of $v$ is shown in Figure 15. (Remember that we know the values of $v$ at only $t = 0, 4, 8, \ldots, 32$.)

**a.** Using eight rectangles with $t_0 = 0$, $t_1 = 4$, $t_2 = 8, \ldots, t_8 = 32$ and $c_1 = 0$, $c_2 = 4$, $c_3 = 8, \ldots, c_8 = 28$, we see that the required approximate distance is

$$D = \sum_{k=1}^{8} v(c_k) \Delta t = \sum_{k=1}^{8} v(t_{k-1}) \Delta t = v(t_0) \cdot 4 + v(t_1) \cdot 4 + \cdots + v(t_7) \cdot 4$$

$$= v(0) \cdot 4 + v(4) \cdot 4 + v(8) \cdot 4 + \cdots + v(28) \cdot 4$$

$$= 2 \cdot 4 + 4 \cdot 4 + 6 \cdot 4 + 10 \cdot 4 + 12 \cdot 4 + 14 \cdot 4 + 10 \cdot 4 + 8 \cdot 4$$

$$= 264$$

or 264 ft.

**b.** Using the same partition as in part (a) and

$$c_1 = 4, \qquad c_2 = 8, \qquad c_3 = 12, \qquad \ldots, \qquad c_8 = 32$$

we find

$$D = \sum_{k=1}^{8} v(c_k) \Delta t = \sum_{k=1}^{8} v(t_k) \Delta t = v(t_1) \cdot 4 + v(t_2) \cdot 4 + \cdots + v(t_8) \cdot 4$$

$$= v(4) \cdot 4 + v(8) \cdot 4 + v(12) \cdot 4 + \cdots + v(32) \cdot 4$$

$$= 4 \cdot 4 + 6 \cdot 4 + 10 \cdot 4 + 12 \cdot 4 + 14 \cdot 4 + 10 \cdot 4 + 8 \cdot 4 + 6 \cdot 4$$

$$= 280$$

or 280 ft.

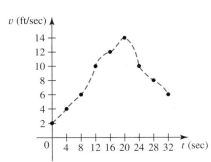

**FIGURE 15**
An approximation of the graph of $v$ on $[0, 32]$

**c.** We use four rectangles with $t_0 = 0$, $t_1 = 8$, $t_2 = 16$, $t_3 = 24$, and $t_4 = 32$ and

$$c_1 = 4, \qquad c_2 = 12, \qquad c_3 = 20, \qquad \text{and} \qquad c_4 = 28$$

obtaining

$$
\begin{aligned}
D &= \sum_{k=1}^{4} v(c_k)\Delta t = v(c_1) \cdot 8 + v(c_2) \cdot 8 + v(c_3) \cdot 8 + v(c_4) \cdot 8 \\
&= v(4) \cdot 8 + v(12) \cdot 8 + v(20) \cdot 8 + v(28) \cdot 8 \\
&= 4 \cdot 8 + 10 \cdot 8 + 14 \cdot 8 + 8 \cdot 8 \\
&= 288
\end{aligned}
$$

or 288 ft. ∎

## 4.3   CONCEPT QUESTIONS

**1.** Let $f(x) = x + 1$ on the interval $[1, 5]$. Divide the interval $[1, 5]$ into four subintervals of length 1 using the points

$$x_0 = 1, \quad x_1 = 2, \quad x_2 = 3, \quad x_3 = 4, \quad \text{and} \quad x_4 = 5$$

Write the sum $\Sigma_{k=1}^{4} f(c_k)\Delta x$ to approximate the area of the region $S$ under the graph of $f$ on $[1, 5]$, choosing $c_k$ in the subinterval $[x_{k-1}, x_k]$, where $1 \le k \le 4$, to be (a) the left endpoint of the subinterval, $c_k = x_{k-1}$; (b) the right endpoint of the subinterval; $c_k = x_k$; and (c) the midpoint of the

subinterval, $c_k = \frac{1}{2}(x_{k-1} + x_k) = x_{k-1} + \frac{1}{2}\Delta x$. Sketch the graph of $f$ and the approximating rectangles for parts (a)–(c).

**2.** Refer to Exercise 1. Find the area of the region $S$ under the graph of $f$ in $[1, 5]$ by calculating $\lim_{n\to\infty} \Sigma_{k=1}^{n} f(c_k)\Delta x$, where $c_k$ is chosen to be (a) the left endpoint, (b) the right endpoint, and (c) the midpoint of the subinterval $[x_{k-1}, x_k]$, where $1 \le k \le n$. Verify your result by using elementary geometry to find the area of the region $S$.

## 4.3   EXERCISES

*In Exercises 1–12 you are given a function f, an interval $[a, b]$, the number n of subintervals into which $[a, b]$ is divided (each of length $\Delta x = (b - a)/n$), and the point $c_k$ in $[x_{k-1}, x_k]$, where $1 \le k \le n$. (a) Sketch the graph of f and the rectangles with base on $[x_{k-1}, x_k]$ and height $f(c_k)$, and (b) find the approximation $\Sigma_{k=1}^{n} f(c_k)\Delta x$ of the area of the region S under the graph of f on $[a, b]$.*

**1.** $f(x) = x$,   $[0, 1]$,   $n = 5$,   $c_k$ is the left endpoint

**2.** $f(x) = x$,   $[1, 4]$,   $n = 6$,   $c_k$ is the midpoint

**3.** $f(x) = 2x + 3$,   $[0, 4]$,   $n = 5$,   $c_k$ is the right endpoint

**4.** $f(x) = 3 - 2x$,   $[0, 1]$,   $n = 5$,   $c_k$ is the left endpoint

**5.** $f(x) = 8 - 2x$,   $[1, 3]$,   $n = 4$,   $c_k$ is the midpoint

**6.** $f(x) = x^2$,   $[0, 1]$,   $n = 5$,   $c_k$ is the right endpoint

**7.** $f(x) = x^2$,   $[1, 3]$,   $n = 4$,   $c_k$ is the midpoint

**8.** $f(x) = 4 - x^2$,   $[0, 2]$,   $n = 8$,   $c_k$ is the left endpoint

**9.** $f(x) = 16 - x^2$,   $[1, 3]$,   $n = 5$,   $c_k$ is the right endpoint

**10.** $f(x) = \sqrt{x}$,   $[0, 4]$,   $n = 8$,   $c_k$ is the left endpoint

**11.** $f(x) = \dfrac{1}{x}$,   $[1, 2]$,   $n = 10$,   $c_k$ is the left endpoint

**12.** $f(x) = \cos x$,   $\left[0, \frac{\pi}{2}\right]$,   $n = 4$,   $c_k$ is the midpoint

*In Exercises 13–20, expand and then evaluate the sum.*

**13.** $\displaystyle\sum_{k=1}^{10} 1$

**14.** $\displaystyle\sum_{k=1}^{5} 2k$

**15.** $\displaystyle\sum_{k=1}^{5} (2k - 1)$

**16.** $\displaystyle\sum_{k=1}^{5} k(k + 1)$

**17.** $\displaystyle\sum_{k=1}^{5} k^2$

**18.** $\displaystyle\sum_{k=1}^{5} \frac{1}{k}$

**19.** $\displaystyle\sum_{k=1}^{4} \sqrt{k}$

**20.** $\displaystyle\sum_{k=1}^{4} k \sin \frac{k\pi}{2}$

*In Exercises 21–30, rewrite the sum using sigma notation. Do not evaluate.*

**21.** $2 + 4 + 6 + 8 + \cdots + 60$

**22.** $2 \cdot 1 + 2 \cdot 2 + 2 \cdot 3 + \cdots + 2 \cdot 10$

**23.** $3 + 5 + 7 + 9 + \cdots + 23$

**24.** $\dfrac{1}{5} + \dfrac{2}{5} + \dfrac{3}{5} + \dfrac{4}{5} + \cdots + \dfrac{8}{5}$

**25.** $\left[2\left(\dfrac{1}{5}\right)+1\right]+\left[2\left(\dfrac{2}{5}\right)+1\right]+\left[2\left(\dfrac{3}{5}\right)+1\right]$

$\qquad +\left[2\left(\dfrac{4}{5}\right)+1\right]+\left[2\left(\dfrac{5}{5}\right)+1\right]$

**26.** $\left[\left(\dfrac{1}{4}\right)^2-1\right]\left(\dfrac{1}{4}\right)+\left[\left(\dfrac{2}{4}\right)^2-1\right]\left(\dfrac{1}{4}\right)$

$\qquad +\left[\left(\dfrac{3}{4}\right)^2-1\right]\left(\dfrac{1}{4}\right)+\left[\left(\dfrac{4}{4}\right)^2-1\right]\left(\dfrac{1}{4}\right)$

**27.** $\left[2\left(\dfrac{1}{n}\right)^3-1\right]\left(\dfrac{1}{n}\right)+\left[2\left(\dfrac{2}{n}\right)^3-1\right]\left(\dfrac{1}{n}\right)$

$\qquad +\left[2\left(\dfrac{3}{n}\right)^3-1\right]\left(\dfrac{1}{n}\right)+\cdots+\left[2\left(\dfrac{n}{n}\right)^3-1\right]\left(\dfrac{1}{n}\right)$

**28.** $\left[\sqrt{\dfrac{0}{n}}+1\right]\left(\dfrac{1}{n}\right)+\left[\sqrt{\dfrac{1}{n}}+1\right]\left(\dfrac{1}{n}\right)$

$\qquad +\left[\sqrt{\dfrac{2}{n}}+1\right]\left(\dfrac{1}{n}\right)+\cdots+\left[\sqrt{\dfrac{n-1}{n}}+1\right]\left(\dfrac{1}{n}\right)$

**29.** $\dfrac{1}{n}\sin\left(1+\dfrac{1}{n}\right)+\dfrac{1}{n}\sin\left(1+\dfrac{2}{n}\right)$

$\qquad +\dfrac{1}{n}\sin\left(1+\dfrac{3}{n}\right)+\cdots+\dfrac{1}{n}\sin\left(1+\dfrac{n}{n}\right)$

**30.** $\dfrac{1}{n}\sec^2\left(1+\dfrac{1}{n}\right)+\dfrac{1}{n}\sec^2\left(1+\dfrac{2}{n}\right)$

$\qquad +\dfrac{1}{n}\sec^2\left(1+\dfrac{3}{n}\right)+\cdots+\dfrac{1}{n}\sec^2\left(1+\dfrac{n}{n}\right)$

*In Exercises 31–38, use the rules of summation and the summation formulas to evaluate the sum.*

**31.** $\displaystyle\sum_{k=1}^{10}(2k+1)$

**32.** $\displaystyle\sum_{k=1}^{8}(3-k^2)$

**33.** $\displaystyle\sum_{k=1}^{10}k(k-2)$

**34.** $\displaystyle\sum_{k=1}^{40}k(k^2-k)$

**35.** $\displaystyle\sum_{k=1}^{10}k(2k+1)^2$

**36.** $\displaystyle\sum_{k=1}^{n}\dfrac{1}{n^2}(2k+1)$

**37.** $\displaystyle\sum_{k=1}^{n}(2k+1)^2$

**38.** $\displaystyle\sum_{k=1}^{n}\dfrac{1}{n}\left(1+\dfrac{k}{n}\right)^2$

*In Exercises 39–44, evaluate the limit after first finding the sum (as a function of n) using the summation formulas.*

**39.** $\displaystyle\lim_{n\to\infty}\sum_{k=1}^{n}\dfrac{2k}{n^2}$

**40.** $\displaystyle\lim_{n\to\infty}\sum_{k=1}^{n}\dfrac{1}{n^3}(2k+1)^2$

**41.** $\displaystyle\lim_{n\to\infty}\sum_{k=1}^{n}\left(\dfrac{k}{n}+2\right)\left(\dfrac{3}{n}\right)$

**42.** $\displaystyle\lim_{n\to\infty}\sum_{k=1}^{n}\left[1+2\left(\dfrac{k}{n}\right)^2\right]\left(\dfrac{2}{n}\right)$

**43.** $\displaystyle\lim_{n\to\infty}\sum_{k=1}^{n}\left(1+\dfrac{2k}{n}\right)^2\left(\dfrac{1}{n}\right)$

**44.** $\displaystyle\lim_{n\to\infty}\sum_{k=1}^{n}\left(1+\dfrac{2k-1}{2n}\right)\left(\dfrac{1}{n}\right)$

*In Exercises 45–52, use the definition of area (page 438) to find the area of the region under the graph of f on [a, b] using the indicated choice of $c_k$.*

**45.** $f(x)=2x+1$, $[0,2]$, $\quad c_k$ is the left endpoint

**46.** $f(x)=3x-1$, $[1,3]$, $\quad c_k$ is the midpoint

**47.** $f(x)=x^2$, $[0,1]$, $\quad c_k$ is the left endpoint

**48.** $f(x)=x^2$, $[0,1]$, $\quad c_k$ is the right endpoint

**49.** $f(x)=4-x^2$, $[-2,1]$, $\quad c_k$ is the midpoint

**50.** $f(x)=x-x^2$, $[-2,1]$, $\quad c_k$ is the right endpoint

**51.** $f(x)=x^2+2x+2$, $[-1,1]$, $\quad c_k$ is the right endpoint

**52.** $f(x)=2x-x^3$, $[0,1]$, $\quad c_k$ is the right endpoint

**cas** *In Exercises 53–56, (a) express the area of the region under the graph of the function f over the interval as the limit of a sum (use the right endpoints), (b) use a computer algebra system (CAS) to find the sum obtained in part (a) in compact form, and (c) evaluate the limit of the sum found in part (b) to obtain the exact area of the region.*

**53.** $f(x)=x^4$; $\quad[0,2]$

**54.** $f(x)=x^5$; $\quad[0,2]$

**55.** $f(x)=x^4+2x^2+x$; $\quad[2,5]$

**56.** $f(x)=\sin x$; $\quad\left[0,\dfrac{\pi}{2}\right]$

**57.** A regular *n*-sided polygon is inscribed in a circle of radius *r* as shown in the figure with $n=6$.

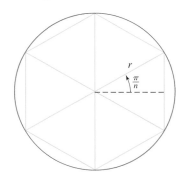

   **a.** Show that the area of the polygon is
$A_n=\frac{1}{2}nr^2\sin(2\pi/n)$.

   **b.** Evaluate $\lim_{n\to\infty}A_n$ to obtain the area of the circle $A=\pi r^2$.

   **Hint:** Use the result $\displaystyle\lim_{x\to0}\dfrac{\sin x}{x}=1$.

**58.** Refer to Exercise 57.

   **a.** Show that the perimeter of the polygon is
$C_n=2nr\sin(\pi/n)$.

   **b.** Evaluate $\lim_{n\to\infty}C_n$ to obtain the circumference of the circle $C=2\pi r$.

**59. Real Estate** Figure (a) shows a vacant lot with a 100-ft frontage in a development. To estimate its area, we introduce a coordinate system so that the $x$-axis coincides with the edge of the straight road forming the lower boundary of the property, as shown in Figure (b). Then, thinking of the upper boundary of the property as the graph of a continuous function $f$ over the interval $[0, 100]$, we see that the problem is mathematically equivalent to that of finding the area of the region under the graph of $f$ on $[0, 100]$. To estimate the area of the lot using the sum of the areas of rectangles, we divide the interval $[0, 100]$ into five equal subintervals of length 20 ft. Then, using surveyor's equipment, we measure the distance from the midpoint of each of these subintervals to the upper boundary of the property. These measurements give the values of $f(x)$ at $x = 10, 30, 50, 70,$ and $90$. What is the approximate area of the lot?

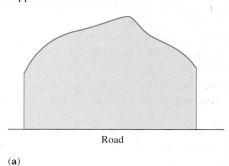

Road

**(a)**

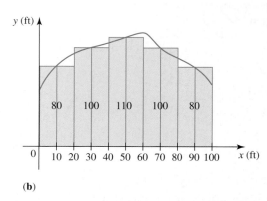

| | | | | | |
|---|---|---|---|---|---|
| 80 | 100 | 110 | 100 | 80 | |

**(b)**

**60. Hot-Air Balloon** The rate of ascent or descent of a hot-air balloon is measured at certain instants of time from $t = 0$ to $t = 45$ as summarized in the following table. The dashed curve in the figure is an estimate of the graph of the velocity function on the time interval $[0, 45]$.

| Time (sec) | 0 | 3 | 6 | 10 | 16 | 24 | 32 | 38 | 43 | 45 |
|---|---|---|---|---|---|---|---|---|---|---|
| Velocity (ft/sec) | 5 | 10 | 16 | 18 | 20 | 18 | 14 | 17 | 14 | 15 |

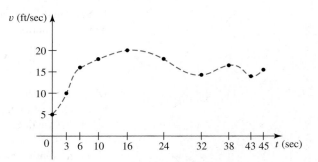

Using nine rectangles determined by the 10 points

$$t_0 = 0, \qquad t_1 = 3, \qquad t_2 = 6, \qquad \ldots, \qquad t_9 = 45$$

and choosing $c_k$ to be the left endpoint of the $k$th subinterval, estimate the total height gained by the balloon over the time period from $t = 0$ to $t = 45$.

**Note:** Here, the partition points are not spaced equally apart, so the subintervals are not of equal length.

**61.** Prove that $\sum_{k=1}^{n}(a_k + b_k) = \sum_{k=1}^{n} a_k + \sum_{k=1}^{n} b_k$.

**62.** Prove that $\sum_{k=1}^{n}(a_k - b_k) = \sum_{k=1}^{n} a_k - \sum_{k=1}^{n} b_k$.

*In Exercises 63–66, determine whether the statement is true or false. If it is true, explain why it is true. If it is false, explain why or give an example to show why it is false.*

**63.** If $f$ is a nonnegative function such that $f(x)$ is strictly positive for some value of $x$ in $[a, b]$, $[a, b]$ is partitioned into $n$ subintervals of equal length, and $c_k$ lies in the $k$th subinterval $[x_{k-1}, x_k]$, then $\sum_{k=1}^{n} f(c_k)\Delta x$ must be strictly positive.
**Hint:** Study the Dirichlet function of Exercise 92 in Section 1.2.

**64.** $\displaystyle\sum_{k=1}^{n} (ca_k - db_k) = c\sum_{k=1}^{n} a_k - d\sum_{k=1}^{n} b_k,$
where $c$ and $d$ are constants

**65.** $\displaystyle\left(\sum_{k=1}^{n} a_k\right)\left(\sum_{k=1}^{n} b_k\right) = \sum_{k=1}^{n} a_k b_k$

**66.** $\displaystyle\sum_{k=1}^{n} (a_k - b_k)^2 = \sum_{k=1}^{n} a_k^2 - \sum_{k=1}^{n} b_k^2$

## 4.4 The Definite Integral

### ■ Definition of the Definite Integral

In Section 4.3 we saw that the area of the region under the graph of a continuous, non-negative function $f$ on an interval $[a, b]$ is defined by a limit of the form

$$\lim_{n \to \infty} \sum_{k=1}^{n} f(c_k)\Delta x = \lim_{n \to \infty}[f(c_1)\Delta x + f(c_2)\Delta x + \cdots + f(c_n)\Delta x] \tag{1}$$

where $\Delta x = (b - a)/n$ and $c_k$ is in $[x_{k-1}, x_k]$. We also saw that the distance covered by an object moving along a straight line with a positive velocity is found by evaluating a similar limit.

In this section we will look at limits defined by Equation (1) in which $f$ may take on both positive and negative values. We will give a geometric interpretation for this general case later on. We will also interpret such limits in terms of the position of an object that moves with both positive and negative velocities. Looking ahead, we will see that limits of this type arise when we try to find the length and mass of a curved wire, the center of mass of a body, the volume of a solid, the area of a surface, the pressure exerted by a fluid against the wall of a container, the amount of oil consumed over a certain period of time, the net sales of a department store over a certain period, and the total number of AIDS cases diagnosed over a certain period of time, just to name a few applications.

In the following definition we will assume, as before, that $f$ is continuous. This allows for a relatively simple development of the material ahead of us.

> **DEFINITION** **Definite Integral**
>
> Let $f$ be a continuous function defined on an interval $[a, b]$. Suppose that $[a, b]$ is divided into $n$ subintervals of equal length $\Delta x = (b - a)/n$ by means of $(n + 1)$ equally spaced points
>
> $$a = x_0 < x_1 < x_2 < \cdots < x_n = b$$
>
> Let $c_1, c_2, \ldots, c_n$ be arbitrary points in the respective subintervals with $c_k$ lying in the $k$th subinterval $[x_{k-1}, x_k]$. Then the **definite integral of $f$ on $[a, b]$**, denoted by $\int_a^b f(x)\, dx$, is
>
> $$\int_a^b f(x)\, dx = \lim_{n \to \infty} \sum_{k=1}^{n} f(c_k)\Delta x \tag{2}$$

We also say that $f$ is **integrable on $[a, b]$** if the limit (2) exists. The process of evaluating a definite integral is called **integration.** The number $a$ in the definition is called the **lower limit of integration,** and the number $b$ is called the **upper limit of integration.** Together, the numbers $a$ and $b$ are referred to as the **limits of integration.** As in the case of the indefinite integral, the function $f$ to be integrated is called the **integrand.**

The sum $\sum_{k=1}^{n} f(c_k)\Delta x$ in the definition is called a **Riemann sum** in honor of the German mathematician Bernhard Riemann (1826–1866). Actually, this sum is a special case of a more general form of a Riemann sum in which no assumption is made requiring that $f$ be continuous on $[a, b]$ or that the interval be partitioned in such a way that the resulting subintervals have equal length. For completeness we will discuss this general case at the end of this section.

**Notes**

1. The assumption that $f$ is continuous on $[a, b]$ guarantees that the definite integral always exists. In other words, the limit in Equation (2) exists and is unique for all choices of the *evaluation* points $c_k$. Furthermore, if $f$ is nonnegative, then the definite integral gives the area of the region under the graph of $f$ on $[a, b]$ since the limit in Equation (2) reduces to the limit in Equation (1), page 438, in Section 4.3.

2. The symbol $\int$ in the definition of the definite integral is the same as that used to denote the indefinite integral of a function. (Remember that the definite integral is a number, in contrast to the indefinite integral, which is a family of functions (the antiderivatives of $f$).) ∎

**EXAMPLE 1** Compute the Riemann sum for $f(x) = 4 - x^2$ on $[-1, 3]$ using five subintervals ($n = 5$) and choosing the evaluation points to be the midpoints of the subintervals.

**Solution** Here, $a = -1$, $b = 3$, and $n = 5$. So the length of each subinterval is

$$\Delta x = \frac{b - a}{n} = \frac{3 - (-1)}{5} = \frac{4}{5}$$

The partition points are

$$x_0 = -1, \qquad x_1 = -1 + \frac{4}{5} = -\frac{1}{5}, \qquad x_2 = -1 + 2\left(\frac{4}{5}\right) = \frac{3}{5},$$

$$x_3 = \frac{7}{5}, \qquad x_4 = \frac{11}{5}, \qquad \text{and} \qquad x_5 = 3$$

The midpoints of the subintervals are given by $c_k = \frac{1}{2}(x_k + x_{k-1})$, or

$$c_1 = -\frac{3}{5}, \qquad c_2 = \frac{1}{5}, \qquad c_3 = 1, \qquad c_4 = \frac{9}{5}, \qquad \text{and} \qquad c_5 = \frac{13}{5}$$

(See Figure 1.)

Therefore, the required Riemann sum is

$$\sum_{k=1}^{5} f(c_k)\Delta x = f(c_1)\Delta x + f(c_2)\Delta x + f(c_3)\Delta x + f(c_4)\Delta x + f(c_5)\Delta x$$

$$= \left[ f\left(-\frac{3}{5}\right) + f\left(\frac{1}{5}\right) + f(1) + f\left(\frac{9}{5}\right) + f\left(\frac{13}{5}\right) \right]\Delta x$$

$$= \left(\frac{4}{5}\right)\left\{ \left[ 4 - \left(-\frac{3}{5}\right)^2 \right] + \left[ 4 - \left(\frac{1}{5}\right)^2 \right] + [4 - (1)^2] \right.$$

$$\left. + \left[ 4 - \left(\frac{9}{5}\right)^2 \right] + \left[ 4 - \left(\frac{13}{5}\right)^2 \right] \right\}$$

$$= \left(\frac{4}{5}\right)(3.64 + 3.96 + 3 + 0.76 - 2.76)$$

$$= 6.88 \qquad \blacksquare$$

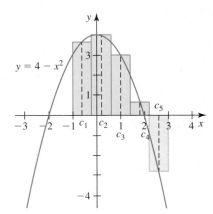

**FIGURE 1**
The positive terms of the Riemann sum are associated with the rectangles that lie above the $x$-axis; the negative term is associated with the rectangle that lies below the $x$-axis.

$y = 4 - x^2$

The Riemann sum computed in Example 1 is the sum of five terms. As you can see in Figure 1, these terms are associated with the areas of the five rectangles shown. The positive terms give the areas of the rectangles that lie above the $x$-axis, while the negative term is the negative of the area of the rectangle that lies below the $x$-axis.

**EXAMPLE 2** Evaluate $\displaystyle\int_{-1}^{3} (4 - x^2)\, dx$.

**Solution**   Here, $a = -1$ and $b = 3$. Furthermore, $f(x) = 4 - x^2$ is continuous on $[-1, 3]$, so $f$ is integrable on $[-1, 3]$. To evaluate the given definite integral, let's subdivide the interval $[-1, 3]$ into $n$ equal subintervals of length

$$\Delta x = \frac{b - a}{n} = \frac{3 - (-1)}{n} = \frac{4}{n}$$

The partition points are

$$x_0 = -1, \qquad x_1 = -1 + \frac{4}{n}, \qquad x_2 = -1 + 2\left(\frac{4}{n}\right), \qquad \dots,$$

$$x_{k-1} = -1 + (k-1)\left(\frac{4}{n}\right), \qquad x_k = -1 + k\left(\frac{4}{n}\right), \qquad \dots, \qquad x_n = 3$$

Next, we pick $c_k$ to be the right endpoint of the subinterval $[x_{k-1}, x_k]$ so that

$$c_k = x_k = -1 + k\left(\frac{4}{n}\right) = -1 + \frac{4k}{n}$$

Then

$$\int_{-1}^{3} (4 - x^2)\, dx = \int_{-1}^{3} f(x)\, dx = \lim_{n \to \infty} \sum_{k=1}^{n} f(c_k)\Delta x$$

$$= \lim_{n \to \infty} \sum_{k=1}^{n} f\left(-1 + \frac{4k}{n}\right)\left(\frac{4}{n}\right)$$

$$= \lim_{n \to \infty} \sum_{k=1}^{n} \left[4 - \left(-1 + \frac{4k}{n}\right)^2\right]\left(\frac{4}{n}\right) \qquad f(x) = 4 - x^2$$

$$= \lim_{n \to \infty} \left(\frac{4}{n}\right)\sum_{k=1}^{n} \left(3 + \frac{8k}{n} - \frac{16k^2}{n^2}\right)$$

$$= \lim_{n \to \infty} \left[\frac{4}{n}\sum_{k=1}^{n} 3 + \frac{32}{n^2}\sum_{k=1}^{n} k - \frac{64}{n^3}\sum_{k=1}^{n} k^2\right]$$

$$= \lim_{n \to \infty} \left[\frac{4}{n}(3n) + \frac{32}{n^2}\cdot\frac{n(n+1)}{2} - \frac{64}{n^3}\cdot\frac{n(n+1)(2n+1)}{6}\right]$$

$$= \lim_{n \to \infty} \left[12 + 16\left(1 + \frac{1}{n}\right) - \frac{32}{3}\left(1 + \frac{1}{n}\right)\left(2 + \frac{1}{n}\right)\right]$$

$$= 12 + 16 - \frac{64}{3} = \frac{20}{3} = 6\frac{2}{3}$$

(Compare this with the approximate value of $\int_{-1}^{3} (4 - x^2)\, dx$ obtained in Example 1.)   ∎

**EXAMPLE 3** Show that $\displaystyle\int_{a}^{b} x\, dx = \frac{1}{2}(b^2 - a^2)$.

**Solution**   Let's subdivide the interval $[a, b]$ into $n$ subintervals of length

$$\Delta x = \frac{b - a}{n}$$

The partition points are

$$x_0 = a, \qquad x_1 = a + \frac{b-a}{n}, \qquad x_2 = a + 2\left(\frac{b-a}{n}\right), \qquad \ldots,$$

$$x_k = a + k\left(\frac{b-a}{n}\right), \qquad \ldots, \qquad x_n = b$$

Next we choose the evaluation point $c_k$ to be the right endpoint of the subinterval $[x_{k-1}, x_k]$, where $1 \le k \le n$; that is, we pick $c_k = x_k$ for each $1 \le k \le n$. Then

$$\int_a^b x \, dx = \lim_{n \to \infty} \sum_{k=1}^{n} f(c_k)\Delta x$$

$$= \lim_{n \to \infty} \sum_{k=1}^{n} \left[ a + \left(\frac{b-a}{n}\right)k \right]\left(\frac{b-a}{n}\right)$$

$$= (b-a)\lim_{n \to \infty} \frac{1}{n} \sum_{k=1}^{n} \left[ a + \left(\frac{b-a}{n}\right)k \right]$$

$$= (b-a)\lim_{n \to \infty} \frac{1}{n} \left[ \sum_{k=1}^{n} a + \left(\frac{b-a}{n}\right)\sum_{k=1}^{n} k \right]$$

$$= (b-a)\lim_{n \to \infty} \frac{1}{n} \left[ na + \left(\frac{b-a}{n}\right) \cdot \frac{n(n+1)}{2} \right]$$

$$= (b-a)\lim_{n \to \infty} \left[ a + \left(\frac{b-a}{2}\right) \cdot \frac{n(n+1)}{n^2} \right]$$

$$= (b-a)\left[ a + \left(\frac{b-a}{2}\right)\lim_{n \to \infty} \frac{n+1}{n} \right]$$

$$= (b-a)\left( a + \frac{b-a}{2} \right) = (b-a)\left( \frac{2a+b-a}{2} \right)$$

$$= \frac{1}{2}(b-a)(b+a) = \frac{1}{2}(b^2 - a^2) \qquad \blacksquare$$

**EXAMPLE 4**   Divide the interval $[2, 5]$ into $n$ subintervals of equal length, and let $c_k$ be any point in $[x_{k-1}, x_k]$. Write

$$\lim_{n \to \infty} \sum_{k=1}^{n} \sqrt{1 + (c_k)^2} \, \Delta x$$

as an integral.

**Solution**   Comparing the given expression with Equation (2), we see that it is the limit of a Riemann sum of the function $f(x) = \sqrt{1 + x^2}$ on the interval $[2, 5]$. Next, since $f$ is continuous on $[2, 5]$, the limit exists, so by Equation (2),

$$\lim_{n \to \infty} \sum_{k=1}^{n} \sqrt{1 + (c_k)^2} \, \Delta x = \int_2^5 \sqrt{1 + x^2} \, dx \qquad \blacksquare$$

## ■ Geometric Interpretation of the Definite Integral

As was pointed out earlier, if $f$ is a continuous, nonnegative function on $[a, b]$, then the definite integral $\int_a^b f(x) \, dx$ gives the area of the region under the graph of $f$ on $[a, b]$. (See Figure 2.)

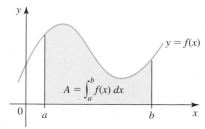

**FIGURE 2**
If $f(x) \ge 0$ on $[a, b]$, then $\int_a^b f(x) \, dx$ gives the area of the region under the graph of $f$ on $[a, b]$.

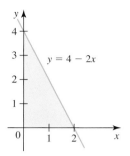

**FIGURE 3**
$\int_0^2 (4 - 2x)\, dx$ = area of the triangle.

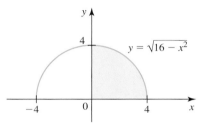

**FIGURE 4**
$f(x) = \sqrt{16 - x^2}$ represents the upper semicircle.

**EXAMPLE 5**  Evaluate the definite integral by interpreting it geometrically:

**a.** $\displaystyle\int_0^2 (4 - 2x)\, dx$     **b.** $\displaystyle\int_0^4 \sqrt{16 - x^2}\, dx$

**Solution**

**a.** The graph of the integrand $f(x) = 4 - 2x$ on $[0, 2]$ is the straight line segment shown in Figure 3. Since $f(x) \geq 0$ on $[0, 2]$, we can interpret the integral as the area of the triangle shown. Thus,

$$\int_0^2 (4 - 2x)\, dx = \frac{1}{2}(2)(4) = 4 \qquad \text{area} = \frac{1}{2}\,\text{base} \cdot \text{height}$$

**b.** The integrand $f(x) = \sqrt{16 - x^2}$ is the positive root obtained by solving the equation $x^2 + y^2 = 16$ for $y$, which represents the circle of radius 4 centered at the origin; therefore, it represents the upper semicircle shown in Figure 4. Since $f(x) \geq 0$ on $[0, 4]$, we can interpret the integral as the area of that part of the circle lying in the first quadrant. Since this area is $\frac{1}{4}\pi(4^2) = 4\pi$, we see that

$$\int_0^4 \sqrt{16 - x^2}\, dx = 4\pi \qquad\blacksquare$$

Next we look at a geometric interpretation of the definite integral for the case in which $f$ assumes both positive and negative values on $[a, b]$. Consider a typical Riemann sum of the function $f$,

$$\sum_{k=1}^{n} f(c_k)\Delta x$$

corresponding to a partition $P$ with points of subdivision

$$a = x_0 < x_1 < x_2 < \cdots < x_{k-1} < x_k < \cdots < x_{n-1} < x_n = b$$

and evaluation points $c_k$ in $[x_{k-1}, x_k]$. The sum consists of $n$ terms in which a positive term corresponds to the area of a rectangle of height $f(c_k)$ lying above the $x$-axis, and a negative term corresponds to the area of a rectangle of height $-f(c_k)$ lying below the $x$-axis. (See Figure 5, where $n = 6$.)

**FIGURE 5**
The positive (negative) terms in the Riemann sum are associated with the areas of the rectangles that lie above (below) the $x$-axis.

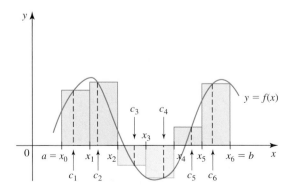

As $n$ gets larger and larger, the sums of the areas of the rectangles lying above the $x$-axis seem to give a better and better approximation of the area of the region lying above the $x$-axis. Similarly, the sums of the area of the rectangles lying below the *x*-axis seem to give a better and better approximation of the area of the region lying below the *x*-axis. (See Figure 6, where $n = 12$.)

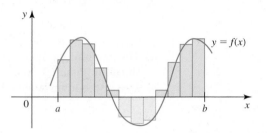

**FIGURE 6**
Approximating $\int_a^b f(x)\, dx$
with 12 rectangles

This observation suggests that we interpret the definite integral

$$\int_a^b f(x)\, dx = \lim_{n \to \infty} \sum_{k=1}^{n} f(c_k)\Delta x$$

as a difference of areas. Specifically,

$$\int_a^b f(x)\, dx = \text{area of } S_1 - \text{area of } S_2 + \text{area of } S_3$$

where $S_2$ is the region lying *above* the graph of *f* and below the *x*-axis. (See Figure 7.) More generally,

$$\int_a^b f(x)\, dx = \text{areas of the regions above } [a, b] - \text{areas of the regions below } [a, b]$$

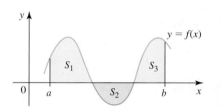

**FIGURE 7**
$\int_a^b f(x)\, dx = \text{area of } S_1 -$
area of $S_2 + $ area of $S_3$

### ■ The Definite Integral and Displacement

In Section 4.3 we showed that if $v(t)$ is a nonnegative velocity function of a car traveling in a straight line, then the distance covered by the car between $t = a$ and $t = b$ is given by the area of the region under the graph of the velocity function on the time interval $[a, b]$. Since the area of the region under the graph of a nonnegative function $v(t)$ on $[a, b]$ is just the definite integral of *v* on $[a, b]$, we can write

$$\int_a^b v(t)\, dt = \text{displacement of the car between } t = a \text{ and } t = b$$

If we denote the position of the car at any time *t* by $s(t)$, then its position at $t = a$ is $s(a)$. So we can then write its final position at $t = b$ as

$$s(b) = s(a) + \int_a^b v(t)\, dt$$

(See Figure 8.)

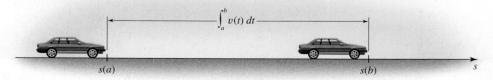

**FIGURE 8**
The position of the car at $t = b$
is $s(b) = s(a) + \int_a^b v(t)\, dt$.

Now suppose that $v(t)$ assumes both positive and negative values on $[a, b]$. (See Figure 9.)

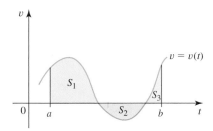

**FIGURE 9**
The area of $S_1$ and the area of $S_3$ give the distance the car moves in the positive direction, whereas the area of $S_2$ gives the distance it moves in the negative direction.

Then

$$\int_a^b v(t)\, dt = \text{area of the regions above } [a, b] - \text{area of the region below } [a, b]$$

$$= \text{distance covered by the car in the positive direction} - \text{distance covered by the car in the negative direction}$$

$$= \text{displacement of the car between } t = a \text{ and } t = b$$

In other words, the final position of the car at $t = b$ is

$$s(b) = s(a) + \int_a^b v(t)\, dt$$

as before.

**EXAMPLE 6** The velocity function of a car moving along a straight road is given by $v(t) = t - 20$ for $0 \le t \le 40$, where $v(t)$ is measured in feet per second and $t$ in seconds. Show that at $t = 40$ the car will be in the same position as it was initially.

**Solution** The graph of $v$ is shown in Figure 10. We have

$$\int_0^{40} v(t)\, dt = \text{area of } S_2 - \text{area of } S_1$$

$$= \frac{1}{2}(20)(20) - \frac{1}{2}(20)(20)$$

$$= 200 - 200 = 0$$

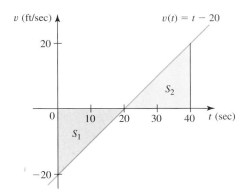

**FIGURE 10**
The area of $S_1$ is equal to the area of $S_2$.

Therefore,

$$s(40) = s(0) + \int_0^{40} v(t)\, dt = s(0)$$

so the net change in the position of the car is zero, as was to be shown.

We interpret this result as follows: The car moves a total of 200 ft in the negative direction in the first 20 sec and then moves a total of 200 ft in the positive direction in the next 20 sec, resulting in no net change in its position.

**Alternative Solution** Let $s(t)$ denote the position of the car at any time $t$. Then

$$\frac{ds}{dt} = v(t)$$

But $v(t) = t - 20$, so

$$\frac{ds}{dt} = t - 20$$

Integrating with respect to $t$, we have

$$s(t) = \int (t - 20)\, dt$$

$$= \frac{1}{2}t^2 - 20t + C \qquad \text{$C$ an arbitrary constant}$$

The position of the car at $t = 0$ is $s(0)$, and this condition gives

$$s(0) = \frac{1}{2}(0) - 20(0) + C \qquad \text{or} \qquad C = s(0)$$

Therefore, the position of the car at any time $t$ is

$$s(t) = \frac{1}{2}t^2 - 20t + s(0)$$

In particular, the position of the car at $t = 40$ is

$$s(40) = \frac{1}{2}(40^2) - 20(40) + s(0) = s(0)$$

its position at $t = 0$, as was to be shown.    ■

**Note**   The method used in the alternative solution of the problem in Example 6 hints at the relationship between the definite integral of a function and the indefinite integral of the function. We will exploit this relationship in the next section.    ■

## ■ Properties of the Definite Integral

When we defined the definite integral $\int_a^b f(x)\, dx$, we assumed that $a < b$. We now extend the definition to cover the cases $a = b$ and $a > b$.

---

**DEFINITIONS   Two Special Definite Integrals**

**1.** $\displaystyle \int_a^a f(x)\, dx = 0$

**2.** $\displaystyle \int_a^b f(x)\, dx = -\int_b^a f(x)\, dx, \quad$ if $a > b$

---

The first definition is compatible with the definition of the definite integral if we observe that here,

$$\Delta x = \frac{b - a}{n} = \frac{a - a}{n} = 0$$

The second definition is also compatible with the definition by observing that if we interchange $a$ and $b$, then the sign of the resulting Riemann sum changes because

$$\Delta x = \frac{b - a}{n} = -\frac{a - b}{n}$$

**EXAMPLE 7** Evaluate the definite integral:

**a.** $\displaystyle\int_{2}^{2} (x^2 - 2x + 4)\, dx$     **b.** $\displaystyle\int_{3}^{-1} (4 - x^2)\, dx$

**Solution**

**a.** $\displaystyle\int_{2}^{2} (x^2 - 2x + 4)\, dx = 0$

**b.** $\displaystyle\int_{3}^{-1} (4 - x^2)\, dx = -\int_{-1}^{3} (4 - x^2)\, dx = -6\frac{2}{3}$ using the result of Example 2.  ■

In the expression $\int_{a}^{b} f(x)\, dx$ the variable of integration, $x$, is a *dummy variable* in the sense that it may be replaced by any other letter without changing the value of the integral. As an illustration, the results of Example 2 may be written

$$\int_{-1}^{3} (4 - x^2)\, dx = \int_{-1}^{3} (4 - u^2)\, du = \int_{-1}^{3} (4 - s^2)\, ds = 6\frac{2}{3}$$

Suppose that $c > 0$. Interpreting $\int_{a}^{b} c\, dx$ as the area of the region under the graph of $f(x) = c$ on $[a, b]$ gives

$$\int_{a}^{b} c\, dx = c(b - a)$$

(See Figure 11.)

We will now look at some properties of the definite integral that will prove helpful later on when we evaluate integrals. Here we assume, as we did earlier, that all of the functions under consideration are continuous.

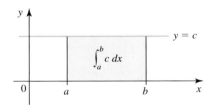

**FIGURE 11**
If $c > 0$, then interpreting $\int_{a}^{b} c\, dx$ as the area of the region under the graph of $f(x) = c$ on $[a, b]$ gives $\int_{a}^{b} f(x)\, dx = c(b - a)$.

---

**The Definite Integral of a Constant Function**

If $c$ is a real number, then

$$\int_{a}^{b} c\, dx = c(b - a) \tag{3}$$

---

The special case where $c > 0$ was discussed earlier.

**EXAMPLE 8** Evaluate $\displaystyle\int_{2}^{7} 3\, dx$.

**Solution** We use Equation (3) with $c = 3$, $a = 2$, and $b = 7$, obtaining

$$\int_{2}^{7} 3\, dx = 3(7 - 2) = 15$$  ■

The next two properties of the definite integral are analogous to the rules of integration for indefinite integrals (see Section 4.1).

**Properties of the Definite Integral**

**1. Sum (Difference)**

$$\int_a^b [f(x) \pm g(x)]\, dx = \int_a^b f(x)\, dx \pm \int_a^b g(x)\, dx$$

**2. Constant Multiple**

$$\int_a^b cf(x)\, dx = c\int_a^b f(x)\, dx, \quad \text{where } c \text{ is any constant}$$

Property 1 states that the integral of the sum (difference) is the sum (difference) of the integrals. Property 2 states that the integral of a constant times a function is equal to the constant times the integral of the function. Thus, a constant (and only a constant!) can be moved in front of the integral sign. These properties are derived by using the corresponding limit laws. For example, to prove Property 2, we use the definition of the definite integral to write

$$\int_a^b cf(x)\, dx = \lim_{n\to\infty} \sum_{k=1}^n cf(c_k)\Delta x$$

$$= c \lim_{n\to\infty} \sum_{k=1}^n f(c_k)\Delta x \qquad \text{Constant Multiple Law for limits}$$

$$= c \int_a^b f(x)\, dx$$

**EXAMPLE 9**  Use the result $\int_0^1 x^2\, dx = \dfrac{1}{3}$ of Example 8 in Section 4.3 to evaluate

**a.** $\displaystyle\int_0^1 (x^2 - 4)\, dx$      **b.** $\displaystyle\int_0^1 5x^2\, dx$

**Solution**

**a.** $\displaystyle\int_0^1 (x^2 - 4)\, dx = \int_0^1 x^2\, dx - \int_0^1 4\, dx$     Property 1

$$= \frac{1}{3} - 4(1)$$

$$= -\frac{11}{3}$$

**b.** $\displaystyle\int_0^1 5x^2\, dx = 5\int_0^1 x^2\, dx$     Property 2

$$= 5\left(\frac{1}{3}\right) = \frac{5}{3}$$

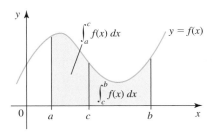

**FIGURE 12**

$$\int_a^b f(x)\, dx = \int_a^c f(x)\, dx + \int_c^b f(x)\, dx$$

Suppose that $f$ is continuous and nonnegative on $[a, b]$. Then $\int_a^b f(x)\, dx$ gives the area of the region under the graph of $f$ on $[a, b]$. Next, if $a < c < b$, then $\int_a^c f(x)\, dx$ and $\int_c^b f(x)\, dx$ give the area of the region under the graph of $f$ on $[a, c]$ and $[c, b]$, respectively. Therefore, as you can see in Figure 12,

$$\int_a^b f(x)\, dx = \int_a^c f(x)\, dx + \int_c^b f(x)\, dx$$

This observation suggests the following property of definite integrals.

---

**Property of the Definite Integral**

**3.** If $c$ is any number in $[a, b]$, then

$$\int_a^b f(x)\, dx = \int_a^c f(x)\, dx + \int_c^b f(x)\, dx$$

---

**Note**   The conclusion of Property 3 holds for *any* three numbers $a$, $b$, and $c$.

**EXAMPLE 10**   Suppose that $\int_1^6 f(x)\, dx = 8$ and $\int_4^6 f(x)\, dx = 5$. What is $\int_1^4 f(x)\, dx$?

**Solution**   Using Property 3, we have

$$\int_1^6 f(x)\, dx = \int_1^4 f(x)\, dx + \int_4^6 f(x)\, dx$$

from which we see that

$$\int_1^4 f(x)\, dx = \int_1^6 f(x)\, dx - \int_4^6 f(x)\, dx = 8 - 5 = 3$$

The next three properties of the definite integral involve inequalities.

---

**Properties of the Definite Integral**

**4.** If $f(x) \geq 0$ on $[a, b]$, then

$$\int_a^b f(x)\, dx \geq 0$$

**5.** If $f(x) \geq g(x)$ on $[a, b]$, then

$$\int_a^b f(x)\, dx \geq \int_a^b g(x)\, dx$$

**6.** If $m \leq f(x) \leq M$ on $[a, b]$, then

$$m(b - a) \leq \int_a^b f(x)\, dx \leq M(b - a)$$

---

The plausibility of Property 4 stems from the observation that the area of the region under the graph of a nonnegative function is nonnegative. Also, if we assume that $g$ and therefore $f$ are both nonnegative on $[a, b]$, then Property 5 is a statement that the area of the region under the graph of $f$ is larger than the area of the region under the graph of $g$. (See Figure 13.) The plausibility of Property 6 is suggested by Figure 14, where $m$ and $M$ are the absolute minimum and absolute maximum values of $f$, respectively, on $[a, b]$: The area of the region under the graph of $f$ on $[a, b]$, $\int_a^b f(x)\, dx$, is greater than the area of the rectangle with height $m$, $m(b - a)$, and smaller than the area of the rectangle with height $M$, $M(b - a)$.

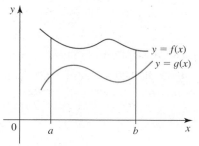

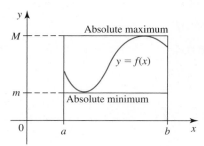

**FIGURE 13**
If $f(x) \geq g(x)$ on $[a, b]$, then the area of the region under the graph of $f$ is greater than the area of the region under the graph of $g$.

**FIGURE 14**
The area of the region under the graph of $f$ is greater than or equal to $m(b - a)$ and less than or equal to $M(b - a)$.

It should be mentioned that all of the properties of the definite integral can be proved with mathematical rigor and without any assumption regarding the sign of $f(x)$ (see Exercise 62).

**EXAMPLE 11** Use Property 6 to estimate $\displaystyle\int_0^1 e^{-\sqrt{x}}\, dx$.

**Solution** The integrand $f(x) = e^{-\sqrt{x}}$ is decreasing on $[0, 1]$. Therefore, its absolute maximum value occurs at $x = 0$ (the left endpoint of the interval), and its absolute minimum value occurs at $x = 1$ (the right endpoint of the interval). If we take $m = f(1) = e^{-1}$, $M = f(0) = 1$, $a = 0$, and $b = 1$, then Property 6 gives

$$e^{-1}(1 - 0) \leq \int_0^1 e^{-\sqrt{x}}\, dx \leq 1(1 - 0)$$

$$e^{-1} \leq \int_0^1 e^{-\sqrt{x}}\, dx \leq 1$$

Since $e^{-1} \approx 0.3679$, we have the estimate

$$0.367 \leq \int_0^1 e^{-\sqrt{x}}\, dx \leq 1$$ ■

## More General Definition of the Definite Integral

As was pointed out earlier, the points that make up a partition of an interval $[a, b]$ need not be chosen to be equally spaced. In general, a **partition of $[a, b]$** is any set $P = \{x_0, x_1, \ldots, x_n\}$ satisfying

$$a = x_0 < x_1 < x_2 < \cdots < x_{n-1} < x_n = b$$

The subintervals corresponding to this partition of $[a, b]$ are

$$[x_0, x_1], \qquad [x_1, x_2], \qquad \ldots, \qquad [x_{k-1}, x_k], \qquad \ldots, \qquad [x_{n-1}, x_n]$$

The length of the $k$th subinterval is

$$\Delta x_k = x_k - x_{k-1}$$

Figure 15 shows one possible partition of $[a, b]$.

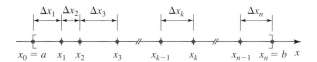

**FIGURE 15**
A possible partition of $[a, b]$

The length of the largest subinterval, denoted by $\|P\|$, is called the **norm** of $P$. For example, in the partition shown in Figure 16,

$$\Delta x_1 = \frac{1}{4}, \qquad \Delta x_2 = \frac{1}{4}, \qquad \Delta x_3 = \frac{1}{2}, \qquad \Delta x_4 = \frac{1}{4},$$

$$\Delta x_5 = \frac{1}{8}, \qquad \Delta x_6 = \frac{1}{8}, \qquad \Delta x_7 = \frac{1}{4}, \qquad \text{and} \qquad \Delta x_8 = \frac{1}{4}$$

so its norm is $\frac{1}{2}$.

**FIGURE 16**
A possible partition of $[0, 2]$

If the $(n + 1)$ points of a partition of $[a, b]$ are chosen to be equally spaced so that the resulting $n$ subintervals have equal length, then the partition is **regular.** In a regular partition, the norm satisfies

$$\|P\| = \Delta x = \frac{b - a}{n}$$

For a general partition $P$,

$$\|P\| \geq \frac{b - a}{n} \qquad \text{or} \qquad n \geq \frac{b - a}{\|P\|}$$

**FIGURE 17**
As the number of subintervals approach infinity, $\|P\|$ does not approach 0.

From this inequality we see that as the norm of a partition approaches 0, the number of subintervals approach infinity. The converse, however, is false. For example, the partition $P$ of the interval $[0, 1]$ in Figure 17 is given by

$$0 < \frac{1}{2} < \frac{3}{4} < \frac{7}{8} < \cdots < 1 - \frac{1}{2^{n-1}} < 1 - \frac{1}{2^n} < 1$$

has norm $\frac{1}{2}$ for any positive integer $n$. Therefore, $n \to \infty$ does not imply that $\|P\| \to 0$. But for a regular partition,

$$\|P\| \to 0 \quad \text{if and only if} \quad n \to \infty$$

a fact that we will use shortly.

We are now in a position to give a more general definition of the definite integral, but first we observe that a function $f$ is **bounded** on an interval $[a, b]$ if there exists some positive real number $M$ such that $|f(x)| \leq M$ for all $x$ in $[a, b]$.

---

**DEFINITION**    **Definite Integral (General Definition)**

Let $f$ be a bounded function defined on an interval $[a, b]$. Then the **definite integral of $f$ on $[a, b]$**, denoted by $\displaystyle\int_a^b f(x)\, dx$, is

$$\int_a^b f(x)\, dx = \lim_{\|P\| \to 0} \sum_{k=1}^n f(c_k) \Delta x \qquad (4)$$

if the limit exists for *all* partitions $P$ of $[a, b]$ and *all* choices of $c_k$ in $[x_{k-1}, x_k]$.

---

It can be shown that if $f$ is continuous on $[a, b]$, then the definite integral of $f$ on $[a, b]$ always exists. Therefore, the limit (4) exists for all choices of $P$ and $c_k$. In particular, the limit exists if we choose a regular partition, as was done in our earlier presentation. In fact, for regular partitions, $\|P\| \to 0$ if and only if $n \to \infty$. So the limit (4) is equivalent to

$$\int_a^b f(x)\, dx = \lim_{n \to \infty} \sum_{k=1}^n f(c_k) \Delta x$$

which is the definition of the definite integral given earlier.

Finally, we note the following precise definition of the definite integral.

---

**DEFINITION**    **Precise Definition of the Definite Integral**

The definite integral of $f$ on $[a, b]$ is

$$\int_a^b f(x)\, dx$$

if for every number $\varepsilon > 0$ there exists a number $\delta > 0$ such that for every partition $P$ of $[a, b]$ with $\|P\| < \delta$ and every choice of points $c_k$ in $[x_{k-1}, x_k]$, the inequality

$$\left| \sum_{k=1}^n f(c_k) \Delta x_k - \int_a^b f(x)\, dx \right| < \varepsilon$$

holds.

---

## 4.4  CONCEPT QUESTIONS

1. What is a Riemann sum of a continuous function $f$ on an interval $[a, b]$? Illustrate graphically the case in which $f$ assumes both positive and negative values on $[a, b]$.

2. Define the definite integral of a continuous function on the interval $[a, b]$. Give a geometric interpretation of $\int_a^b f(x)\, dx$ for the case in which (a) $f$ is nonnegative on $[a, b]$ and (b) $f$ assumes both positive and negative values on $[a, b]$. Illustrate your answers graphically.

3. The following figure depicts the graph of the velocity function $v$ of an object traveling along a coordinate line over the time interval $[a, b]$. The numbers $c$, $d$, and $e$ satisfy $a < c < d < e < b$. The areas of the regions $S_1$, $S_2$, $S_3$, and $S_4$ are $A_1$, $A_2$, $A_3$, and $A_4$ respectively. Assume that the object is located at the origin at $t = a$.

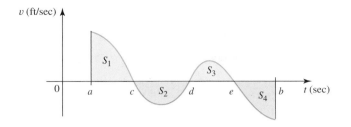

**a.** Write the displacement of the object at $t = c$, $t = d$, $t = e$, and $t = b$ (i) in terms of $A_1$, $A_2$, $A_3$, and $A_4$ and (ii) in terms of definite integrals.

**b.** Write the distances covered by the object over the time intervals $[a, d]$ and $[a, b]$. Express your answer using $A_1$, $A_2$, $A_3$, and $A_4$ and also using definite integrals.

## 4.4   EXERCISES

**1.** The graph of a function $f$ on the interval $[0, 8]$ is shown in the figure. Compute the Riemann sum for $f$ on $[0, 8]$ using four subintervals of equal length and choosing the evaluation points to be (a) the left endpoints, (b) the right endpoints, and (c) the midpoints of the subintervals.

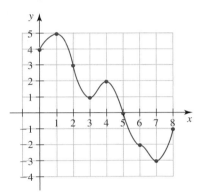

**2.** The graph of a function $g$ on the interval $[-2, 4]$ is shown in the figure. Compute the Riemann sum for $g$ on $[-2, 4]$ using six subintervals of equal length and choosing the evaluation points to be (a) the left endpoints, (b) the right endpoints, and (c) the midpoints of the subintervals.

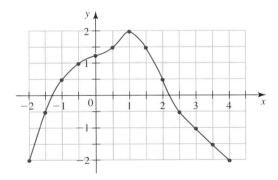

*In Exercises 3–6 you are given a function $f$ defined on an interval $[a, b]$, the number $n$ of subintervals of equal length $\Delta x = (b - a)/n$, and the evaluation points $c_k$ in $[x_{k-1}, x_k]$. (a) Sketch the graph of $f$ and the rectangles associated with the Riemann sum for $f$ on $[a, b]$, and (b) find the Riemann sum.*

**3.** $f(x) = 2x - 3$,   $[0, 2]$,   $n = 4$,   $c_k$ is the midpoint

**4.** $f(x) = -2x + 1$,   $[-1, 2]$,   $n = 6$,   $c_k$ is the left endpoint

**5.** $f(x) = \sqrt{x} - 1$,   $[0, 3]$,   $n = 6$,   $c_k$ is the right endpoint

**6.** $f(x) = 2 \sin x$,   $\left[0, \frac{5\pi}{4}\right]$,   $n = 5$,   $c_k$ is the right endpoint

*In Exercises 7–12, use Equation (2) to evaluate the integral.*

**7.** $\displaystyle\int_0^2 x \, dx$

**8.** $\displaystyle\int_{-1}^2 x^2 \, dx$

**9.** $\displaystyle\int_{-1}^3 (x - 2) \, dx$

**10.** $\displaystyle\int_{-1}^1 (2x + 1) \, dx$

**11.** $\displaystyle\int_1^2 (3 - 2x^2) \, dx$

**12.** $\displaystyle\int_{-2}^1 (x^3 + 2x) \, dx$

*In Exercises 13–16, the given expression is the limit of a Riemann sum of a function $f$ on $[a, b]$. Write this expression as a definite integral on $[a, b]$.*

**13.** $\displaystyle\lim_{n \to \infty} \sum_{k=1}^n (4c_k - 3)\Delta x$,   $[-3, -1]$

**14.** $\displaystyle\lim_{n \to \infty} \sum_{k=1}^n 2c_k(1 - c_k)^2 \Delta x$,   $[0, 3]$

**15.** $\displaystyle\lim_{n \to \infty} \sum_{k=1}^n \frac{2c_k}{c_k^2 + 1} \Delta x$,   $[1, 2]$

**16.** $\displaystyle\lim_{n \to \infty} \sum_{k=1}^n c_k(\cos c_k)\Delta x$,   $\left[0, \frac{\pi}{2}\right]$

*In Exercises 17 and 18, express the integral as a limit of a Riemann sum using a regular partition. Do not evaluate the limit.*

**17.** $\displaystyle\int_{-2}^1 (1 + x^3)^{1/3} \, dx$

**18.** $\displaystyle\int_1^4 \left(5 \ln x - \frac{1}{2}x^2\right) dx$

**19.** Use the graph of $f$ shown in the figure to evaluate the integral by interpreting it geometrically.

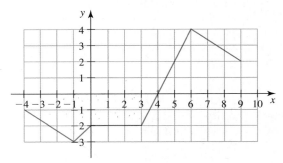

**a.** $\displaystyle\int_{-4}^{-1} f(x)\,dx$      **b.** $\displaystyle\int_{-1}^{4} f(x)\,dx$

**c.** $\displaystyle\int_{4}^{9} f(x)\,dx$      **d.** $\displaystyle\int_{-4}^{9} f(x)\,dx$

**20.** The graph of $f$ shown in the figure consists of straight line segments and a semicircle. Evaluate each integral by interpreting it geometrically.

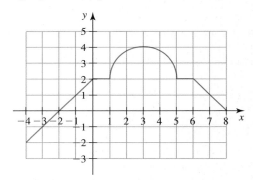

**a.** $\displaystyle\int_{-4}^{1} f(x)\,dx$      **b.** $\displaystyle\int_{1}^{5} f(x)\,dx$

**c.** $\displaystyle\int_{5}^{8} f(x)\,dx$      **d.** $\displaystyle\int_{-4}^{8} f(x)\,dx$

*In Exercises 21–28 you are given a definite integral $\int_a^b f(x)\,dx$. Make a sketch of $f$ on $[a, b]$. Then use the geometric interpretation of the integral to evaluate it.*

**21.** $\displaystyle\int_{-2}^{4} 3\,dx$      **22.** $\displaystyle\int_{-2}^{3} (2x + 1)\,dx$

**23.** $\displaystyle\int_{0}^{3} (-3x + 6)\,dx$      **24.** $\displaystyle\int_{-1}^{2} |x|\,dx$

**25.** $\displaystyle\int_{-1}^{2} |x - 1|\,dx$      **26.** $\displaystyle\int_{-2}^{2} \sqrt{4 - x^2}\,dx$

**27.** $\displaystyle\int_{0}^{3} -\sqrt{9 - x^2}\,dx$      **28.** $\displaystyle\int_{0}^{2} \sqrt{-x^2 + 2x}\,dx$

**29.** Given that $\int_0^2 f(x)\,dx = 3$ and $\int_2^5 f(x)\,dx = -1$, evaluate the following integrals.

**a.** $\displaystyle\int_{0}^{5} f(x)\,dx$      **b.** $\displaystyle\int_{5}^{2} f(x)\,dx$

**c.** $\displaystyle\int_{0}^{2} 2f(x)\,dx$      **d.** $\displaystyle\int_{2}^{5} [f(x) - 4]\,dx$

**30.** Given that $\int_1^3 f(x)\,dx = 4$ and $\int_3^6 f(x)\,dx = 2$, evaluate the following integrals.

**a.** $\displaystyle\int_{3}^{1} 2f(x)\,dx$      **b.** $\displaystyle\int_{6}^{1} f(x)\,dx$

**c.** $\displaystyle\int_{3}^{1} -2f(x)\,dx$      **d.** $\displaystyle\int_{2}^{2} 3f(x)\,dx$

**31.** Given that $\int_{-1}^{3} f(x)\,dx = 5$ and $\int_{-1}^{3} g(x)\,dx = -2$, evaluate the following integrals.

**a.** $\displaystyle\int_{-1}^{3} [f(x) + g(x)]\,dx$

**b.** $\displaystyle\int_{-1}^{3} [g(x) - f(x)]\,dx$

**c.** $\displaystyle\int_{-1}^{3} [3f(x) - 2g(x)]\,dx$

**32.** Given that $\int_{-2}^{2} f(x)\,dx = 3$ and $\int_0^2 f(x)\,dx = 2$, evaluate the following integrals.

**a.** $\displaystyle\int_{2}^{0} f(x)\,dx$      **b.** $\displaystyle\int_{-2}^{0} [f(x) + 3]\,dx$

**c.** $\displaystyle\int_{2}^{0} 3f(x)\,dx - \int_{0}^{-2} 2f(x)\,dx$

**33.** Evaluate $\int_2^2 \sqrt[3]{x^2 + x + 1}\,dx$.

**34.** Evaluate $\int_2^5 f(x)\,dx$ if it is known that $\int_5^2 f(x)\,dx = -10$.

**35.** Show that $\int_0^\pi \cos 2x\,dx = 0$ by interpreting the definite integral geometrically.

**36.** Show that
$$\int_0^x \sqrt{a^2 - t^2}\,dt = \frac{1}{2}x\sqrt{a^2 - x^2} + \frac{a^2}{2}\sin^{-1}\left(\frac{x}{a}\right)$$
$$0 < x \le a$$
by interpreting the definite integral geometrically.

*In Exercises 37–40, use the properties of the integral to prove the inequality without evaluating the integral.*

**37.** $\displaystyle\int_{0}^{1} \frac{\sqrt{x^3 + x}}{x^2 + 1}\,dx \ge 0$      **38.** $\displaystyle\int_{0}^{1} x^2\,dx \le \int_{0}^{1} \sqrt{x}\,dx$

**39.** $\displaystyle\int_{0}^{\pi/4} \sin^2 x \cos x\,dx \le \int_{0}^{\pi/4} \sin^2 x\,dx$

**40.** $\displaystyle\int_0^{\pi/2} \cos x \, dx \le \int_0^{\pi/2} (x^2 + 1) \, dx$

*In Exercises 41–46, use Property 6 of the definite integral to estimate the definite integral.*

**41.** $\displaystyle\int_1^2 \sqrt{1 + 2x^3} \, dx$

**42.** $\displaystyle\int_1^3 \frac{1}{x} \, dx$

**43.** $\displaystyle\int_{-1}^2 (x^2 - 2x + 2) \, dx$

**44.** $\displaystyle\int_0^2 \frac{x^2 + 5}{x^2 + 2} \, dx$

**45.** $\displaystyle\int_0^1 e^{-x^2} \, dx$

**46.** $\displaystyle\int_{\pi/4}^{\pi/2} x \sin x \, dx$

 **47. a.** Plot the graph of $f(x) = x\sqrt{x^4 + 1}$ on the interval $[-1, 1]$.
   **b.** Prove that the area of the region above the $x$-axis is equal to the area of the region below the $x$-axis.
   **c.** Use the result of part (b) to show that
   $$\int_{-1}^1 x\sqrt{x^4 + 1} \, dx = 0.$$

 **48. a.** Plot the graph of $f(x) = \sin^3 x$ on the interval $[0, 2\pi]$.
   **b.** Prove that the area of the region above the $x$-axis is equal to the area of the region below the $x$-axis.
   **Hint:** Look at $f(\pi + t)$ for $0 \le t \le \pi$.
   **c.** Use the result of part (b) to show that $\int_0^{2\pi} \sin^3 x \, dx = 0$.

**49.** Suppose that $f$ is continuous on $[a, b]$ and $f(x) \le 0$ on $[a, b]$. Prove that $\int_a^b f(x) \, dx \le 0$.

**50.** Suppose that $f$ is continuous on $[a, b]$. Prove that
$$\left| \int_a^b f(x) \, dx \right| \le \int_a^b |f(x)| \, dx$$
**Hint:** $-|f(x)| \le f(x) \le |f(x)|$.

**51.** Use the result of Exercise 50 to show that
$\left| \int_a^b x \sin 2x \, dx \right| \le \frac{1}{2}(b^2 - a^2)$, where $0 \le a < b$.
**Hint:** Use the result of Example 3.

**52.** Suppose that $f$ is continuous and increasing and its graph is concave upward on the interval $[a, b]$. Give a geometric argument to show that
$$(b - a)f(a) \le \int_a^b f(x) \, dx \le \frac{1}{2}(b - a)[f(a) + f(b)]$$

 **53. a.** Plot the graphs of $f(x) = \dfrac{x}{\sqrt{1 + x^5}}$ and $g(x) = x$ using the viewing window $[0, 1] \times [0, 1]$.
   **b.** Prove that $0 \le f(x) \le g(x)$.
   **c.** Use the result of part (b) and Property 5 to show that
   $$0 \le \int_0^1 \frac{x}{\sqrt{1 + x^5}} \, dx \le \frac{1}{2}$$
   **Hint:** Use the result of Example 3.

 **54. a.** Plot the graphs of $f(x) = \sin x$ and $g(x) = x$ using the viewing window $\left[0, \frac{\pi}{2}\right] \times [0, 2]$.
   **b.** Prove that $0 \le f(x) \le g(x)$.

**c.** Use the result of part (b) and Property 5 to show that
$$0 \le \int_{\pi/6}^{\pi/4} \sin x \, dx \le \frac{5\pi^2}{288}$$
**Hint:** Use the result of Example 3.

*In Exercises 55 and 56, use Property 5 to prove the inequality.*

**55.** $\displaystyle\int_2^4 \sqrt{x^4 + x} \, dx \ge \frac{56}{3}$

**56.** $\displaystyle\int_0^{\pi/4} x \sin x \, dx \le \frac{\pi^3}{192}$

**57.** Estimate the integral $\int_0^1 \sqrt{1 + x^2} \, dx$ using (a) Property 6 of the definite integral and (b) the result of Exercise 52. Which estimate is better? Explain.

**58.** Show that $\int_a^b x^2 \, dx = \frac{1}{3}(b^3 - a^3)$.

**59.** Find the constant $b$ such that $\int_0^b (2\sqrt{x} - x) \, dx$ is as large as possible. Explain your answer.

**60.** Define the function $F$ by $F(x) = \int_{-1}^x (t^4 - 2t^3) \, dt$ for $x$ in $[-1, 2]$.
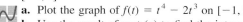 **a.** Plot the graph of $f(t) = t^4 - 2t^3$ on $[-1, 2]$.
   **b.** Use the result of part (a) to find the interval where $F$ is increasing and where $F$ is decreasing on $(-1, 2)$.

**61.** Determine whether the Dirichlet function
$$f(x) = \begin{cases} 1 & \text{if } x \text{ is rational} \\ 0 & \text{if } x \text{ is irrational} \end{cases}$$
is integrable on the interval $[0, 1]$. Explain.

**62.** Prove Properties 4, 5, and 6 of the definite integral.

*In Exercises 63–70, determine whether the statement is true or false. If it is true, explain why it is true. If it is false, explain why or give an example to show why it is false.*

**63.** If $f$ and $g$ are continuous on $[a, b]$ and $c$ is constant, then
$$\int_a^b [f(x) + cg(x)] \, dx = \int_a^b f(x) \, dx + c \int_a^b g(x) \, dx$$

**64.** If $f$ and $g$ are continuous on $[a, b]$, then
$$\int_a^b f(x) g(x) \, dx = \left[ \int_a^b f(x) \, dx \right]\left[ \int_a^b g(x) \, dx \right]$$

**65.** If $f$ is continuous on $[a, b]$, then $\int_a^b xf(x) \, dx = x \int_a^b f(x) \, dx$.

**66.** If $f$ is continuous on $[a, b]$ and $\int_a^b f(x) \, dx > 0$, then $f$ must be positive on $[a, b]$.

**67.** If $f$ is continuous and decreasing on $[a, b]$, then
$(b - a)f(b) \le \int_a^b f(x) \, dx \le (b - a)f(a)$.

**68.** If $f$ is nonnegative and continuous on $[a, b]$ and $a < c < d < b$, then $\int_c^d f(x) \, dx \le \int_a^b f(x) \, dx$.

**69.** $\displaystyle\int_1^3 \frac{dx}{x - 2} = -\int_3^1 \frac{dx}{x - 2}$

**70.** $\displaystyle\int_{-2}^2 \frac{dx}{x} = \ln|x| \Big|_{-2}^{2} = \ln|2| - \ln|-2| = \ln 2 - \ln 2 = 0$

## 4.5    The Fundamental Theorem of Calculus

### ■ How Are Differentiation and Integration Related?

In Section 4.4 we defined the definite integral of a function by taking the limit of its Riemann sums. But as we saw, the actual process of finding the definite integral of a function based on this definition turned out to be rather tedious even for simple functions. This is reminiscent of the process of finding the derivative of a function by finding the limit of the difference quotient of the function. Fortunately, there are better and easier ways of evaluating definite integrals.

In this section we will look at what is undoubtedly the most important theorem in calculus. Because it establishes the relationship between differentiation and integration, it is called the **Fundamental Theorem of Calculus.** It was discovered independently by Sir Isaac Newton (1643–1727) in England and by Gottfried Wilhelm Leibniz (1646–1716) in Germany. Before looking at this theorem, we need the results of the following theorem.

### ■ The Mean Value Theorem for Definite Integrals

Suppose that the velocity of a maglev traveling along a straight track is $v(t)$ ft/sec for $t$ between $t = a$ and $t = b$, where $t$ is measured in seconds. What is the average velocity of the maglev over the time interval $[a, b]$?

To answer this question, let's assume that $v$ is continuous on $[a, b]$. We begin by partitioning the interval $[a, b]$ into $n$ equal subintervals of length

$$\Delta t = \frac{b - a}{n}$$

by means of equally spaced points

$$a = t_0 < t_1 < t_2 < \cdots < t_n = b$$

Next, we choose the evaluation points $c_1, c_2, \ldots, c_n$ lying in the subintervals $[t_0, t_1]$, $[t_1, t_2], \ldots, [t_{n-1}, t_n]$, respectively, and compute the velocities of the maglev at these points:

$$v(c_1), \quad v(c_2), \quad \ldots, \quad v(c_n)$$

The average of these $n$ numbers

$$\frac{v(c_1) + v(c_2) + \cdots + v(c_n)}{n} = \frac{1}{n} \sum_{k=1}^{n} v(c_k)$$

gives an approximation of the average velocity of the maglev over $[a, b]$. Since

$$n = \frac{b - a}{\Delta t}$$

we can rewrite the expression in the form

$$\frac{1}{n} \sum_{k=1}^{n} v(c_k) = \frac{1}{\dfrac{b - a}{\Delta t}} \sum_{k=1}^{n} v(c_k) = \frac{1}{b - a} \sum_{k=1}^{n} v(c_k) \Delta t$$

By letting $n$ get larger and larger, we are approximating the average velocity of the maglev using measurements of its velocity at more and more points over smaller and smaller time intervals. Intuitively, the approximations should improve with increasing $n$. This suggests that we define the average velocity of the maglev over the time interval $[a, b]$ to be

$$\lim_{n \to \infty} \frac{1}{b - a} \sum_{k=1}^{n} v(c_k) \Delta t$$

But by the definition of the definite integral, we have

$$\lim_{n \to \infty} \frac{1}{b - a} \sum_{k=1}^{n} v(c_k) \Delta t = \frac{1}{b - a} \lim_{n \to \infty} \sum_{k=1}^{n} v(c_k) \Delta t$$

$$= \frac{1}{b - a} \int_{a}^{b} v(t) \, dt$$

Thus, we are led to define the **average velocity** of the maglev over the time interval $[a, b]$ to be

$$\frac{1}{b - a} \int_{a}^{b} v(t) \, dt$$

More generally, we have the following definition of the average value of a function $f$ over an interval $[a, b]$.

---

**DEFINITION**    **Average Value of a Function**

If $f$ is integrable on $[a, b]$, then the **average value of $f$** over $[a, b]$ is the number

$$f_{av} = \frac{1}{b - a} \int_{a}^{b} f(x) \, dx \qquad \qquad (1)$$

---

If we assume that $f$ is nonnegative, then we have the following geometric interpretation for the average value of a function over $[a, b]$. Referring to Figure 1, we see that $f_{av}$ is the height of the rectangle with base lying on the interval $[a, b]$ and having the same area as the area of the region under the graph of $f$ on $[a, b]$.

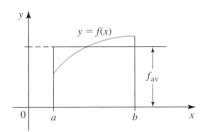

**FIGURE 1**
The area of the rectangle is
$(b - a)f_{av} = \int_{a}^{b} f(x) \, dx$ = area
of the region under the graph of $f$.

Returning to the example involving the motion of the maglev, we see that if we assume that $v(t) \geq 0$ on $[a, b]$, then the distance covered by the maglev over the time period $[a, b]$ is $\int_{a}^{b} v(t) \, dt$, the area of the region under the graph of $v$ on $[a, b]$. But this area is equal to $(b - a)v_{av}$, where $v_{av}$ is the average value of the velocity function $v$. Thus, we can cover the distance traveled by the maglev at a speed of $v(t)$ ft/sec from $t = a$ to $t = b$ by traveling at a *constant* speed, namely, at the average speed $v_{av}$ ft/sec over the same time interval.

**EXAMPLE 1**   Find the average value of $f(x) = 4 - x^2$ over the interval $[-1, 3]$.

**Solution**   Using Equation (1) with $a = -1$, $b = 3$, and $f(x) = 4 - x^2$, we find

$$f_{av} = \frac{1}{b - a} \int_a^b f(x) \, dx$$

$$= \frac{1}{3 - (-1)} \int_{-1}^3 (4 - x^2) \, dx$$

$$= \frac{1}{4}\left(\frac{20}{3}\right) \qquad \text{Use the result of Example 2 in Section 4.4.}$$

$$= \frac{5}{3} \qquad \blacksquare$$

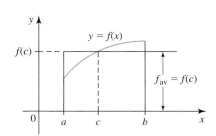

**FIGURE 2**

$$f_{av} = \frac{1}{b - a} \int_a^b f(x) \, dx$$

If you look at Figure 1 again, you will see that there is a number $c$ on $[a, b]$ such that $f(c) = f_{av}$. (See Figure 2.)

The following theorem guarantees that $f_{av}$ is always attained at (at least) one number in an interval $[a, b]$ if $f$ is continuous.

---

**THEOREM 1   The Mean Value Theorem for Integrals**

If $f$ is continuous on $[a, b]$, then there exists a number $c$ in $[a, b]$ such that

$$f(c) = \frac{1}{b - a} \int_a^b f(x) \, dx$$

---

**PROOF**   Since $f$ is continuous on the interval $[a, b]$, the Extreme Value Theorem tells us that $f$ attains an absolute minimum value $m$ at some number in $[a, b]$ and an absolute maximum value $M$ at some number in $[a, b]$. So $m \leq f(x) \leq M$ for all $x$ in $[a, b]$.

By Property 6 of integrals we have

$$m(b - a) \leq \int_a^b f(x) \, dx \leq M(b - a)$$

If $b > a$, then, upon dividing by $(b - a)$, we obtain

$$m \leq \frac{1}{b - a} \int_a^b f(x) \, dx \leq M$$

Because the number

$$\frac{1}{b - a} \int_a^b f(x) \, dx$$

lies between $m$ and $M$, the Intermediate Value Theorem guarantees the existence of at least one number $c$ in $[a, b]$ such that

$$f(c) = \frac{1}{b - a} \int_a^b f(x) \, dx$$

as was to be shown.   $\blacksquare$

**EXAMPLE 2**  Find the value of $c$ guaranteed by the Mean Value Theorem for Integrals for $f(x) = 4 - 2x$ on the interval $[0, 2]$.

**Solution**    The function $f(x) = 4 - 2x$ is continuous on the interval $[0, 2]$. Therefore, the Mean Value Theorem for Integrals states that there is a number $c$ in $[0, 2]$ such that

$$\frac{1}{b - a} \int_a^b f(x) \, dx = f(c)$$

where $a = 0$ and $b = 2$. Thus,

$$\frac{1}{2 - 0} \int_0^2 (4 - 2x) \, dx = 4 - 2c$$

but

$$\int_0^2 (4 - 2x) \, dx = 4 \qquad \text{See Example 5a in Section 4.4.}$$

So we have

$$\frac{1}{2}(4) = 4 - 2c$$

or $c = 1$. (See Figure 3.)    ∎

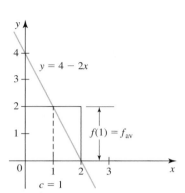

**FIGURE 3**
The number $c = 1$ in $[0, 2]$ gives $f(c) = f_{av}$ as guaranteed by the Mean Value Theorem for Integrals.

## The Fundamental Theorem of Calculus, Part I

Suppose that $f$ is a continuous, nonnegative function defined on the interval $[a, b]$. If $x$ is any number in $[a, b]$, let us put

$$A(x) = \int_a^x f(t) \, dt$$

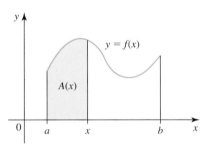

**FIGURE 4**
$A(x) = \int_a^x f(t) \, dt$ gives the area of the region under the graph of $f$ on $[a, x]$.

(We use the dummy variable $t$ because we are using $x$ to denote the upper limit of integration.) Since $f$ is nonnegative, we can interpret $A(x)$ to be the area of the region under the graph of $f$ on $[a, x]$, as shown in Figure 4. Since the number $A(x)$ is unique for each $x$ in $[a, b]$, we see that $A$ is a function of $x$ with domain $[a, b]$.

Let's look at a specific example. Suppose that $f(x) = x$ on the interval $[0, 1]$. If we use the result of Example 3 in Section 4.4, with $a = 0$ and $b = x$, we obtain

$$A(x) = \int_0^x t \, dt = \frac{1}{2}x^2 \qquad 0 \le x \le 1$$

This result is also evident if you refer to Figure 5 and interpret the integral $\int_0^x t \, dt$ as the area of the shaded triangle. Observe that

$$A'(x) = \frac{d}{dx} \int_0^x t \, dt = \frac{d}{dx}\left(\frac{1}{2}x^2\right) = x = f(x)$$

so $A(x)$ is an antiderivative of $f(x) = x$. Now if this result,

$$\frac{d}{dx} \int_a^x f(t) \, dt = f(x)$$

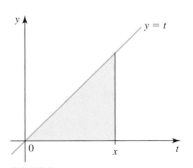

**FIGURE 5**
The area of the triangle is $\frac{1}{2}(x)(x) = \frac{1}{2}x^2$.

is true for all continuous functions $f$, then it is quite astounding because it provides a link between the processes of differentiation and integration. Roughly speaking, this equation says that differentiation undoes what integration does: The two operations are inverses of one another. Thus, the two seemingly unrelated problems of differential calculus (that of finding the slope of a tangent line to a curve) and integral calculus (that of finding the area of the region bounded by a curve) are indeed intimately related.

As it turns out, the result is true. Because of its importance, it is called the Fundamental Theorem of Calculus.

---

**THEOREM 2    The Fundamental Theorem of Calculus, Part 1**

If $f$ is continuous on $[a, b]$, then the function $F$ defined by

$$F(x) = \int_a^x f(t)\, dt \qquad a \leq x \leq b$$

is differentiable on $(a, b)$, and

$$F'(x) = \frac{d}{dx} \int_a^x f(t)\, dt = f(x) \tag{2}$$

---

**PROOF**    Fix $x$ in $(a, b)$, and suppose that $x + h$ is in $(a, b)$, where $h \neq 0$. Then

$$F(x + h) - F(x) = \int_a^{x+h} f(t)\, dt - \int_a^x f(t)\, dt$$

$$= \int_a^x f(t)\, dt + \int_x^{x+h} f(t)\, dt - \int_a^x f(t)\, dt \qquad \text{By Property 3}$$

$$= \int_x^{x+h} f(t)\, dt$$

By the Mean Value Theorem for Integrals there exists a number $c$ between $x$ and $x + h$ such that

$$\int_x^{x+h} f(t)\, dt = f(c) \cdot h$$

Therefore,

$$\frac{F(x + h) - F(x)}{h} = \frac{1}{h} \int_x^{x+h} f(t)\, dt = \frac{f(c) \cdot h}{h} = f(c)$$

Next, observe that as $h$ approaches 0, the number $c$, which is squeezed between $x$ and $x + h$, approaches $x$, and by continuity, $f(c)$ approaches $f(x)$. Therefore,

$$F'(x) = \lim_{h \to 0} \frac{F(x + h) - F(x)}{h} = \lim_{h \to 0} \frac{1}{h} \int_x^{x+h} f(t)\, dt = \lim_{h \to 0} f(c) = f(x)$$

which is the desired result.    ■

**EXAMPLE 3**    Find the derivative of the function:

**a.** $F(x) = \displaystyle\int_{-1}^x \frac{1}{1 + t^2}\, dt$    **b.** $G(x) = \displaystyle\int_x^3 \sqrt{1 + t^2}\, dt$

**Solution**

**a.** The integrand

$$f(t) = \frac{1}{1 + t^2}$$

is continuous everywhere. Using the Fundamental Theorem of Calculus, Part 1, we find

$$F'(x) = \frac{d}{dx} \int_{-1}^{x} \frac{1}{1 + t^2} \, dt = f(x) = \frac{1}{1 + x^2}$$

**b.** The integrand $\sqrt{1 + t^2}$ is continuous everywhere. Therefore,

$$G'(x) = \frac{d}{dx} \int_{x}^{3} \sqrt{1 + t^2} \, dt = \frac{d}{dx} \left[ -\int_{3}^{x} \sqrt{1 + t^2} \, dt \right] \qquad \int_{a}^{b} f(x) \, dx = -\int_{b}^{a} f(x) \, dx$$

$$= -\frac{d}{dx} \int_{3}^{x} \sqrt{1 + t^2} \, dt$$

$$= -\sqrt{1 + x^2}$$ ■

**EXAMPLE 4**  If $y = \displaystyle\int_{0}^{x^3} \cos t^2 \, dt$, what is $\dfrac{dy}{dx}$?

**Solution**   Notice that the upper limit of integration is not $x$, so the Fundamental Theorem of Calculus, Part 1, is not applicable as the problem now stands. Let's put

$$u = x^3 \qquad \text{so} \qquad \frac{du}{dx} = 3x^2$$

Using the Chain Rule and the Fundamental Theorem of Calculus, Part 1, we have

$$\frac{dy}{dx} = \frac{dy}{du} \cdot \frac{du}{dx} = \left[ \frac{d}{du} \int_{0}^{u} \cos t^2 \, dt \right] \cdot \frac{du}{dx}$$

$$= (\cos u^2)(3x^2) = 3x^2 \cos x^6$$ ■

## ■ Fundamental Theorem of Calculus, Part 2

The following theorem, which is a consequence of Part 1 of the Fundamental Theorem of Calculus, shows how to evaluate a definite integral by finding an antiderivative of the integrand, rather than relying on evaluating the limit of a Riemann sum, thus simplifying the task greatly.

**THEOREM 3   The Fundamental Theorem of Calculus, Part 2**

If $f$ is continuous on $[a, b]$, then

$$\int_{a}^{b} f(x) \, dx = F(b) - F(a) \tag{3}$$

where $F$ is any antiderivative of $f$, that is, $F' = f$.

**PROOF** Let $G(x) = \int_a^x f(t)\, dt$. By Theorem 2 we know that $G$ is an antiderivative of $f$. If $F$ is any other antiderivative of $f$, then Theorem 1 in Section 4.1 tells us that $F$ and $G$ differ by a constant. In other words, $F(x) = G(x) + C$. To determine $C$, we put $x = a$ to obtain

$$F(a) = G(a) + C = \int_a^a f(t)\, dt + C = C \qquad \int_a^a f(x)\, dx = 0$$

Therefore, evaluating $F$ at $b$, we have

$$F(b) = G(b) + C = \int_a^b f(t)\, dt + F(a)$$

from which we conclude that

$$F(b) - F(a) = \int_a^b f(x)\, dx \qquad\qquad \blacksquare$$

When applying the Fundamental Theorem of Calculus, it is convenient to use the notation

$$\big[F(x)\big]_a^b = F(b) - F(a) \qquad \text{“}F(x)\text{ evaluated at } b \text{ minus } F(x) \text{ evaluated at } a.\text{”}$$

For example, by using this notation, Equation (3) is written

$$\int_a^b f(x)\, dx = \big[F(x)\big]_a^b = F(b) - F(a)$$

Also, by the Fundamental Theorem of Calculus, if $F(x) + C$ is any antiderivative of $f$, then

$$\int_a^b f(x)\, dx = \big[F(x) + C\big]_a^b$$
$$= [F(b) + C] - [F(a) + C]$$
$$= F(b) - F(a) = \big[F(x)\big]_a^b$$

This result shows that we can drop the constant of integration when we use the Fundamental Theorem of Calculus.

From now on, thanks to the Fundamental Theorem of Calculus, Part 2, we can use our knowledge for finding antiderivatives to help us evaluate definite integrals.

**EXAMPLE 5** Evaluate

**a.** $\displaystyle\int_1^2 (x^3 - 2x^2 + 1)\, dx$     **b.** $\displaystyle\int_0^4 2\sqrt{x}\, dx$     **c.** $\displaystyle\int_0^{\pi/2} \cos x\, dx$

**Solution**

**a.** $\displaystyle\int_1^2 (x^3 - 2x^2 + 1)\, dx = \left[\frac{1}{4}x^4 - \frac{2}{3}x^3 + x\right]_1^2$

$$= \left(4 - \frac{16}{3} + 2\right) - \left(\frac{1}{4} - \frac{2}{3} + 1\right) = \frac{1}{12}$$

**b.** $\displaystyle\int_0^4 2\sqrt{x}\, dx = \int_0^4 2x^{1/2}\, dx = \left[\frac{4}{3}x^{3/2}\right]_0^4 = \frac{4}{3}(4)^{3/2} - \frac{4}{3}(0) = \frac{32}{3}$

**c.** $\displaystyle\int_0^{\pi/2} \cos x \, dx = \left[ \sin x \right]_0^{\pi/2} = 1 - 0 = 1$ ∎

The next example shows how to evaluate the definite integral of a function that is defined piecewise.

**EXAMPLE 6** Evaluate $\displaystyle\int_{-2}^2 f(x) \, dx$, where

$$f(x) = \begin{cases} -x^2 + 1 & \text{if } x < 0 \\ x^3 + 1 & \text{if } x \geq 0 \end{cases}$$

**Solution** The graph of $f$ is shown in Figure 6. Observe that $f$ is continuous on $[-2, 2]$. Since $f$ is defined by different rules for $x$ in the two subintervals $[-2, 0)$ and $[0, 2]$, we use Property 3 of definite integrals to write

$$\int_{-2}^2 f(x) \, dx = \int_{-2}^0 f(x) \, dx + \int_0^2 f(x) \, dx$$

$$= \int_{-2}^0 (-x^2 + 1) \, dx + \int_0^2 (x^3 + 1) \, dx$$

$$= \left[ -\frac{1}{3} x^3 + x \right]_{-2}^0 + \left[ \frac{1}{4} x^4 + x \right]_0^2$$

$$= 0 - \left( \frac{8}{3} - 2 \right) + (4 + 2) - 0 = \frac{16}{3}$$ ∎

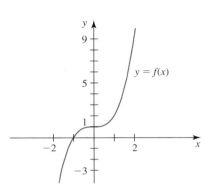

**FIGURE 6**

$\displaystyle\int_{-2}^2 f(x) \, dx = \int_{-2}^0 f(x) \, dx + \int_0^2 f(x) \, dx$

## ▇ Evaluating Definite Integrals Using Substitution

The next two examples show how the method of substitution can be used to help us evaluate definite integrals.

**EXAMPLE 7** Evaluate $\displaystyle\int_0^2 x\sqrt{x^2 + 4} \, dx$.

**Solution** *Method I:* Consider the corresponding indefinite integral

$$I = \int x\sqrt{x^2 + 4} \, dx = \int x(x^2 + 4)^{1/2} \, dx$$

Let $u = x^2 + 4$, so that $du = 2x \, dx$ or $x \, dx = \frac{1}{2} du$. Substituting these quantities into the integral gives

$$I = \int \frac{1}{2} u^{1/2} \, du = \frac{1}{3} u^{3/2} + C = \frac{1}{3} (x^2 + 4)^{3/2} + C$$

Armed with the knowledge of the antiderivative of the function $f(x) = x\sqrt{x^2 + 4}$, we can evaluate the given integral as follows:

$$\int_0^2 x\sqrt{x^2 + 4} \, dx = \left[ \frac{1}{3} (x^2 + 4)^{3/2} \right]_0^2 = \frac{1}{3} (8)^{3/2} - \frac{1}{3} (4)^{3/2} = \frac{8}{3} (2\sqrt{2} - 1)$$

**Solution** *Method II: Changing the Limits of Integration* As before, we make the substitution $u = x^2 + 4$, so that $du = 2x \, dx$ or $x \, dx = \frac{1}{2} du$. Next, we make the following intuitive observation: The given integral has lower and upper limits of integration 0 and 2, respectively, and hence a *range of integration* given by the interval $[0, 2]$. In making the substitution $u = x^2 + 4$, the original integral is transformed into another integral in which the integration is carried out with respect to the new variable $u$.

To obtain the new limits of integration, we note that if $x = 0$, then $u = 0 + 4 = 4$. This gives the lower limit of integration when integrating with respect to $u$. Similarly, if $x = 2$, then $u = 4 + 4 = 8$, and this gives the upper limit of integration. Thus, the range of integration when the integration is performed with respect to $u$ is $[4, 8]$. In view of this, we can write

$$\int_0^2 x(x^2 + 4)^{1/2}\, dx = \int_4^8 \frac{1}{2} u^{1/2}\, du = \left[\frac{1}{3} u^{3/2}\right]_4^8$$

$$= \frac{1}{3}(8)^{3/2} - \frac{1}{3}(4)^{3/2} = \frac{8}{3}(2\sqrt{2} - 1)$$

as was obtained earlier.    ■

**EXAMPLE 8**  Evaluate $\displaystyle\int_0^{\pi/4} \cos^3 2x \sin 2x\, dx$.

**Solution**   Let $u = \cos 2x$, so that $du = -2 \sin 2x\, dx$ or $\sin 2x\, dx = -\frac{1}{2} du$. Also, if $x = 0$, then $u = 1$, and if $x = \pi/4$, then $u = 0$, giving 1 and 0 as the lower and upper limits of integration with respect to $u$. Making these substitutions, we obtain

$$\int_0^{\pi/4} \cos^3 2x \sin 2x\, dx = \int_1^0 u^3\left(-\frac{1}{2}\, du\right)$$

$$= -\frac{1}{8} u^4 \Big|_1^0$$

$$= 0 - \left(-\frac{1}{8}\right) = \frac{1}{8}$$    ■

**Note**   Do not let the fact that the limits of integration with respect to $u$ run from 1 to 0 alarm you. This is not uncommon when we integrate using the method of substitution. Of course,

$$\int_1^0 u^3\left(-\frac{1}{2}\, du\right) = -\int_0^1 u^3\left(-\frac{1}{2}\, du\right) \qquad \int_a^b f(x)\, dx = -\int_b^a f(x)\, dx$$

as you can verify.    ■

**EXAMPLE 9**  **Drug Concentration in the Bloodstream**   The concentration of a certain drug (in mg/cc) in a patient's bloodstream $t$ hr after injection is

$$C(t) = \frac{0.2t}{t^2 + 1}$$

Determine the average concentration of the drug in the patient's bloodstream over the first 4 hr after the drug is injected.

**Solution**   The average concentration of the drug over the time interval $[0, 4]$ is given by

$$A = \frac{1}{4 - 0} \int_0^4 C(t)\, dt = \frac{1}{4} \int_0^4 \frac{0.2t}{t^2 + 1}\, dt$$

To evaluate this definite integral, we make the substitution

$$u = t^2 + 1 \qquad \text{so that} \qquad du = 2t \, dt \qquad \text{or} \qquad t \, dt = \frac{du}{2}$$

Observe that when $t = 0$, $u = 0^2 + 1 = 1$, and when $t = 4$, $u = 4^2 + 1 = 17$, giving $u = 1$ and $u = 17$ as the lower and upper limits of integration with respect to $u$, respectively. We have

$$A = \frac{1}{20} \int_0^4 \frac{t}{t^2 + 1} \, dt$$

$$= \frac{1}{20} \left( \frac{1}{2} \right) \int_1^{17} \frac{1}{u} \, du = \left[ \frac{1}{40} \ln u \right]_1^{17} = \frac{1}{40} (\ln 17 - \ln 1)$$

or approximately 0.071 mg/cc. ■

## Definite Integrals of Odd and Even Functions

The following theorem makes use of the symmetry properties of the integrand to help us evaluate a definite integral.

---

**THEOREM 4  Integrals of Odd and Even Functions**

Suppose that $f$ is continuous on $[-a, a]$.

**a.** If $f$ is even, then $\displaystyle\int_{-a}^a f(x) \, dx = 2 \int_0^a f(x) \, dx$.

**b.** If $f$ is odd, then $\displaystyle\int_{-a}^a f(x) \, dx = 0$.

---

**PROOF**  We write

$$\int_{-a}^a f(x) \, dx = \int_{-a}^0 f(x) \, dx + \int_0^a f(x) \, dx = -\int_0^{-a} f(x) \, dx + \int_0^a f(x) \, dx \qquad \textbf{(4)}$$

For the integral

$$\int_0^{-a} f(x) \, dx$$

let's make the substitution $u = -x$, so that $du = -dx$. Also, if $x = 0$, then $u = 0$, and if $x = -a$, then $u = a$. So

$$\int_0^{-a} f(x) \, dx = \int_0^a f(-u)(-du) = -\int_0^a f(-x) \, dx$$

Therefore, Equation (4) can be written as

$$\int_{-a}^a f(x) \, dx = \int_0^a f(-x) \, dx + \int_0^a f(x) \, dx = \int_0^a [f(-x) + f(x)] \, dx \qquad \textbf{(5)}$$

If $f$ is even, then $f(-x) = f(x)$, so, using Equation (5), we have

$$\int_{-a}^a f(x) \, dx = \int_0^a [f(x) + f(x)] \, dx = 2 \int_0^a f(x) \, dx$$

If $f$ is odd, then $f(-x) = -f(x)$, so Equation (5) gives

$$\int_{-a}^{a} f(x)\,dx = \int_{0}^{a} [-f(x) + f(x)]\,dx = 0$$ ∎

Figure 7 gives a geometric interpretation of Theorem 4. In Figure 7a the area of the region under the graph of the nonnegative function $f$ from $-a$ to $0$ is the same as that under the graph of $f$ from $0$ to $a$, so the area of the region under the graph of $f$ from $-a$ to $a$ is equal to twice that from $0$ to $a$. But each of these areas is given by an appropriate integral, leading to the first result in the theorem. In Figure 7b the area of the region above the graph of $f$ and under the $x$-axis from $-a$ to $0$ is equal to the area of the region under the graph of $f$ from $0$ to $a$; the former is given by the *negative* of the integral from $0$ to $a$.

**FIGURE 7**
The integral of (a) an even function
and (b) an odd function

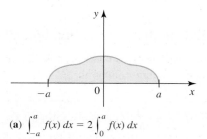

(a) $\displaystyle\int_{-a}^{a} f(x)\,dx = 2\int_{0}^{a} f(x)\,dx$      (b) $\displaystyle\int_{-a}^{a} f(x)\,dx = 0$

**EXAMPLE 10**  Evaluate

**a.** $\displaystyle\int_{-1}^{1} (x^2 + 2)\,dx$      **b.** $\displaystyle\int_{-2}^{2} \frac{\sin x}{\sqrt{1 + x^2}}\,dx$

**Solution**

**a.** Here, $f(-x) = (-x)^2 + 2 = x^2 + 2 = f(x)$, so $f$ is even. Therefore, by Theorem 4,

$$\int_{-1}^{1} (x^2 + 2)\,dx = 2\int_{0}^{1} (x^2 + 2)\,dx = 2\left(\frac{1}{3}x^3 + 2x\right)\Big|_{0}^{1} = 2\left(\frac{1}{3} + 2\right) = \frac{14}{3}$$

**b.** Here,

$$f(-x) = \frac{\sin(-x)}{\sqrt{1 + (-x)^2}} = -\frac{\sin x}{\sqrt{1 + x^2}} = -f(x)$$

so $f$ is odd. Therefore, by Theorem 4,

$$\int_{-2}^{2} \frac{\sin x}{\sqrt{1 + x^2}}\,dx = 0$$ ∎

## ◼ The Definite Integral as a Measure of Net Change

In real-world applications we are often interested in the net change of a quantity over a period of time. For example, suppose that $P$ is a function giving the population, $P(t)$, of a city at time $t$. Then the *net change* in the population over the period from $t = a$ to $t = b$ is given by

$$P(b) - P(a) \qquad \text{Population at } t = b \text{ minus population at } t = a$$

If $P$ has a continuous derivative $P'$ on $[a, b]$, then we can invoke the Fundamental Theorem of Calculus, Part 2, to write

$$P(b) - P(a) = \int_a^b P'(t) \, dt \qquad P \text{ is an antiderivative of } P'.$$

Thus, if we know the *rate of change* of the population at any time $t$, then we can calculate the net change in the population from $t = a$ to $t = b$ by evaluating an appropriate definite integral.

**EXAMPLE 11** **Population Growth in Clark County** Clark County in Nevada, dominated by Las Vegas, was one of the fastest-growing metropolitan areas in the United States. From 1970 through 2000 the population was growing at the rate of

$$R(t) = 133{,}680t^2 - 178{,}788t + 234{,}633 \qquad 0 \le t \le 4$$

people per decade, where $t = 0$ corresponds to the beginning of 1970. What was the net change in the population over the decade from the beginning of 1980 to the beginning of 1990?

*Source:* U.S. Census Bureau.

**Solution** The net change in the population over the decade from the beginning of 1980 ($t = 1$) to the beginning of 1990 ($t = 2$) is given by $P(2) - P(1)$, where $P$ denotes the population in the county at time $t$. But $P' = R$, so

$$\begin{aligned}
P(2) - P(1) &= \int_1^2 P'(t) \, dt = \int_1^2 R(t) \, dt \\
&= \int_1^2 (133{,}680t^2 - 178{,}788t + 234{,}633) \, dt \\
&= \left[ 44{,}560t^3 - 89{,}394t^2 + 234{,}633t \right]_1^2 \\
&= [44{,}560(2^3) - 89{,}394(2^2) + 234{,}633(2)] \\
&\quad - [44{,}560 - 89{,}394 + 234{,}633] \\
&= 278{,}371
\end{aligned}$$

so the net change is 278,371 people. ■

More generally, we have the following result. We assume that $f$ has a continuous derivative, even though the integrability of $f'$ is sufficient.

---

**Net Change Formula**

The net change in a function $f$ over an interval $[a, b]$ is given by

$$f(b) - f(a) = \int_a^b f'(x) \, dx \tag{6}$$

provided that $f'$ is continuous on $[a, b]$.

---

As another example of the net change of a function, let's consider the motion of an object along a straight line. Suppose that the position function and the velocity function of the object are $s$ and $v$, respectively. Since $s'(t) = v(t)$, Equation (6) gives

$$s(b) - s(a) = \int_a^b s'(t)\, dt = \int_a^b v(t)\, dt$$

the net change in the position of the object over the time interval $[a, b]$. This net change of position is the *displacement* of the object between $t = a$ and $t = b$. (Recall that this result was also discussed in Section 4.4.)

To calculate the distance covered by the object between $t = a$ and $t = b$, we observe that if $v(t) \geq 0$ on an interval $[c, d]$, then the distance covered by the object between $t = c$ and $t = d$ is given by its displacement $\int_c^d v(t)\, dt$. On the other hand, if $v(t) \leq 0$ on an interval $[c, d]$, then the distance covered by the object between $t = c$ and $t = d$ is given by the negative of its displacement, that is, by $-\int_c^d v(t)\, dt$. But $-\int_c^d v(t)\, dt = \int_c^d -v(t)\, dt$. Since

$$|v(t)| = \begin{cases} v(t) & \text{if } v(t) \geq 0 \\ -v(t) & \text{if } v(t) < 0 \end{cases}$$

we see that in either case the distance covered by the object is obtained by integrating the *speed* $|v(t)|$ of the object. Therefore, the distance covered by an object between $t = a$ and $t = b$ is

$$\int_a^b |v(t)|\, dt \tag{7}$$

Figure 8 gives a geometric interpretation of the displacement of an object and the distance covered by an object.

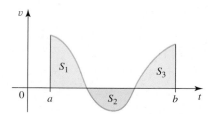

**FIGURE 8**
Displacement is $\int_a^b v(t)\, dt =$ area of $S_1$ − area of $S_2$ + area of $S_3$, and distance covered is $\int_a^b |v(t)|\, dt =$ area of $S_1$ + area of $S_2$ + area of $S_3$.

**EXAMPLE 12** A car moves along a straight road with velocity function

$$v(t) = t^2 + t - 6 \qquad 0 \leq t \leq 10$$

where $v(t)$ is measured in feet per second.

**a.** Find the displacement of the car between $t = 1$ and $t = 4$.
**b.** Find the distance covered by the car during this period of time.

**Solution**
**a.** Using Equation (6), we see that the displacement is

$$s(4) - s(1) = \int_1^4 v(t)\, dt = \int_1^4 (t^2 + t - 6)\, dt$$

$$= \left[ \frac{1}{3}t^3 + \frac{1}{2}t^2 - 6t \right]_1^4 = 10\frac{1}{2}$$

That is, at $t = 4$ the car is $10\frac{1}{2}$ ft to the right of its position at $t = 1$.

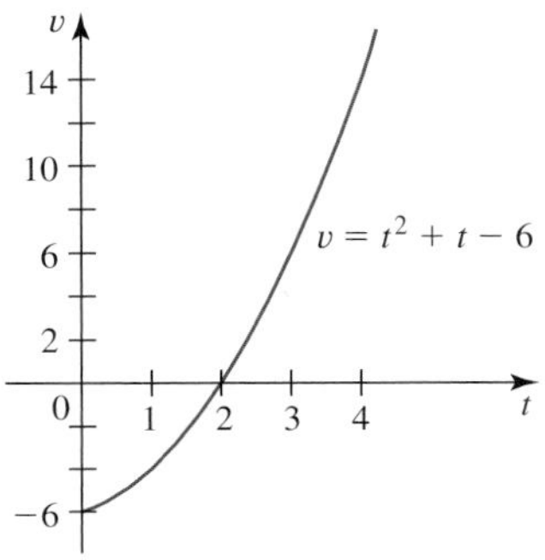


**FIGURE 9**
$v(t) \le 0$ if $t \in [1, 2]$, and
$v(t) \ge 0$ if $t \in [2, 4]$.

**b.** Writing $v(t) = t^2 + t - 6 = (t-2)(t+3)$, we see that $v(t) \le 0$ on [1, 2] and $v(t) \ge 0$ on [2, 4]. (See Figure 9.) Using the integral in (7), we see that the distance covered by the car between $t = 1$ and $t = 4$ is given by

$$\begin{aligned}\int_1^4 |v(t)|\,dt &= \int_1^2 (-v(t))\,dt + \int_2^4 v(t)\,dt\\ &= \int_1^2 (-t^2 - t + 6)\,dt + \int_2^4 (t^2 + t - 6)\,dt\\ &= \left[-\frac{1}{3}t^3 - \frac{1}{2}t^2 + 6t\right]_1^2 + \left[\frac{1}{3}t^3 + \frac{1}{2}t^2 - 6t\right]_2^4\\ &= 14\frac{5}{6}\end{aligned}$$

or $14\frac{5}{6}$ ft.

## 4.5 CONCEPT QUESTIONS

1. Define the average value of a function $f$ over an interval $[a, b]$. Give a geometric interpretation.
2. State the Mean Value Theorem for Integrals. Give a geometric interpretation.
3. State both parts of the Fundamental Theorem of Calculus.
4. State the Net Change Formula, and use it to answer the following:
   **a.** If water is flowing through a pipe at the rate of $R$ ft³/min, what does $\int_{t_1}^{t_2} R(t)\,dt$ measure, where $t_1$ and $t_2$ are measured in minutes with $t_1 < t_2$?
   **b.** If an object is moving along a straight line with an acceleration of $a(t)$ ft/sec², what does $\int_{t_1}^{t_2} a(t)\,dt$ measure if $t_1 < t_2$?
5. Suppose that a particle moves along a coordinate line with a velocity of $v(t)$ ft/sec. Explain the difference between $\int_a^b v(t)\,dt$ and $\int_a^b |v(t)|\,dt$.

## 4.5 EXERCISES

1. Let $F(x) = \int_2^x t^2\,dt$.
   **a.** Use Part 1 of the Fundamental Theorem of Calculus to find $F'(x)$.
   **b.** Use Part 2 of the Fundamental Theorem of Calculus to integrate $\int_2^x t^2\,dt$ to obtain an alternative expression for $F(x)$.
   **c.** Differentiate the expression for $F(x)$ found in part (b), and compare the result with that obtained in part (a). Comment on your result.
2. Repeat Exercise 1 with $G(x) = \int_0^x \sqrt{3t+1}\,dt$.

*In Exercises 3–14, find the derivative of the function.*

3. $F(x) = \int_0^x \sqrt{3t+5}\,dt$
4. $G(x) = \int_{-1}^x t\sqrt{t^2+1}\,dt$
5. $g(x) = \int_2^x \frac{1}{t^2+1}\,dt$
6. $h(x) = \int_x^3 \frac{t}{\sqrt{t+1}}\,dt$
7. $F(x) = \int_x^{\pi} \sin 2t\,dt$
8. $G(x) = \int_0^{x^2} t\sin t\,dt$
9. $g(x) = \int_2^{\sqrt{x}} \frac{\sin t}{t}\,dt$
10. $h(x) = \int_0^{x^2} \sin t^2\,dt$
11. $F(x) = \int_1^{\cos x} \frac{t^2}{t+1}\,dt$
12. $G(x) = \int_{\sqrt{x}}^5 \frac{\sin t^2}{t}\,dt$
13. $f(x) = \int_{x^2}^{x^3} \ln t\,dt;\quad x > 0$
14. $g(x) = \int_{1/x}^{e^x} \tan^{-1} t\,dt$

*In Exercises 15–36, evaluate the integral.*

15. $\int_{-3}^2 4\,dx$
16. $\int_{-2}^0 (2x-3)\,dx$
17. $\int_{-1}^1 (t^2-4)\,dt$
18. $\int_0^2 (2-4u+u^2)\,du$
19. $\int_{-2}^1 (3t+2)^2\,dt$
20. $\int_1^2 \frac{3}{x^3}\,dx$
21. $\int_1^3 \frac{x^2-x+3}{x}\,dx$
22. $\int_1^3 \frac{\ln x}{x}\,dx$

V Videos for selected exercises are available online at www.academic.cengage.com/login.

**23.** $\displaystyle\int_1^4 \frac{1}{\sqrt{x}}\, dx$

**24.** $\displaystyle\int_1^2 \frac{3x^4 - 2x^2 + 1}{2x^2}\, dx$

**25.** $\displaystyle\int_4^9 \frac{x-1}{\sqrt{x}}\, dx$

**26.** $\displaystyle\int_1^0 (t^{1/2} - t^{5/2})\, dt$

**27.** $\displaystyle\int_2^0 \sqrt{x}(x+1)(x-2)\, dx$

**28.** $\displaystyle\int_0^{\pi/2} (\sin x + 1)\, dx$

**29.** $\displaystyle\int_{\pi/6}^{\pi/4} \sec^2 t\, dt$

**30.** $\displaystyle\int_{\pi/6}^{\pi/4} \csc\theta \cot\theta\, d\theta$

**31.** $\displaystyle\int_0^{\pi} \sin 2x \cos x\, dx$

**32.** $\displaystyle\int_0^{\pi} |\cos x|\, dx$

**33.** $\displaystyle\int_{\pi/4}^{\pi/3} \frac{dx}{\sin^2 x \cos^2 x}$

**34.** $\displaystyle\int_0^{\pi} \sqrt{\sin x - \sin^3 x}\, dx$

**35.** $\displaystyle\int_{-1}^1 f(x)\, dx$ where $f(x) = \begin{cases} -x+1 & \text{if } x \le 0 \\ 2x^2+1 & \text{if } x > 0 \end{cases}$

**36.** $\displaystyle\int_{-\pi}^{\pi/2} f(x)\, dx$ where $f(x) = \begin{cases} x^2+1 & \text{if } x \le 0 \\ \cos x & \text{if } x > 0 \end{cases}$

*In Exercises 37–62, evaluate the integral.*

**37.** $\displaystyle\int_0^1 (3 - 2x)^4\, dx$

**38.** $\displaystyle\int_0^2 (t+1)^{0.2}\, dt$

**39.** $\displaystyle\int_1^2 8t(t^2 - 1)^7\, dt$

**40.** $\displaystyle\int_1^5 \sqrt{2x - 1}\, dx$

**41.** $\displaystyle\int_1^4 \sqrt[3]{5 - u}\, du$

**42.** $\displaystyle\int_1^2 \frac{x}{2x^2 - 1}\, dx$

**43.** $\displaystyle\int_1^4 \frac{1}{\sqrt{x}(\sqrt{x} + 1)^2}\, dx$

**44.** $\displaystyle\int_{\pi/4}^{\pi/2} \sin 2x\, dx$

**45.** $\displaystyle\int_{-1}^0 \frac{1}{1 + e^{-2x}}\, dx$

**46.** $\displaystyle\int_0^{\pi/4} \frac{e^{\tan x}}{\cos^2 x}\, dx$

**47.** $\displaystyle\int_{\pi/2}^{\pi} \cos\left(\frac{1}{2}x\right) dx$

**48.** $\displaystyle\int_0^{\pi/2} \sqrt{\cos\theta}\, \sin\theta\, d\theta$

**49.** $\displaystyle\int_{-1/4}^{1/4} \sec \pi t \tan \pi t\, dt$

**50.** $\displaystyle\int_{\pi/6}^{\pi/2} \csc^2\theta \cot\theta\, d\theta$

**51.** $\displaystyle\int_{1/\pi}^{2/\pi} \frac{\sin\frac{1}{x}}{x^2}\, dx$

**52.** $\displaystyle\int_{-\pi}^{\pi} \frac{x^2 \sin x}{\sqrt{1 + x^2}}\, dx$

**53.** $\displaystyle\int_0^1 xe^{x^2}e^{e^{x^2}}\, dx$

**54.** $\displaystyle\int_1^e \frac{\ln x}{x}e^{(\ln x)^2}\, dx$

**55.** $\displaystyle\int_0^1 (3^t + t^3)\, dt$

**56.** $\displaystyle\int_1^4 \frac{3^{\sqrt{x}}}{\sqrt{x}}\, dx$

**57.** $\displaystyle\int_0^{1/4} \frac{1}{\sqrt{1 - 4x^2}}\, dx$

**58.** $\displaystyle\int_0^{1/2} \frac{1}{1 + 4x^2}\, dx$

**59.** $\displaystyle\int_0^{4\sqrt{3}} \frac{1}{x^2 + 16}\, dx$

**60.** $\displaystyle\int_0^1 \frac{e^{2x}}{1 + e^{4x}}\, dx$

**61.** $\displaystyle\int_0^1 \frac{x^3}{1 + x^8}\, dx$

**62.** $\displaystyle\int_0^{\sqrt{3}/2} \frac{\sin^{-1} x}{\sqrt{1 - x^2}}\, dx$

**63. a.** Prove that $0 \le \displaystyle\int_0^1 \frac{x^5}{\sqrt[3]{1 + x^4}}\, dx \le \frac{1}{6}$.

**cas** **b.** Use a calculator or a computer to find the value of the integral accurate to five decimal places.

**64. a.** Prove that $0 \le \displaystyle\int_0^1 \frac{dx}{\sqrt{4 - 3x + x^2}} \le \frac{2}{3}$.

**cas** **b.** Use a calculator or a computer to find the value of the integral accurate to five decimal places.

*In Exercises 65–70, find the area of the region under the graph of $f$ on $[a, b]$.*

**65.** $f(x) = x^2 - 2x + 2$; $[-1, 2]$

**66.** $f(x) = \dfrac{1}{x^2}$; $[1, 2]$

**67.** $f(x) = 2 + \sqrt{x + 1}$; $[0, 3]$

**68.** $f(x) = \sec^2 x$; $\left[0, \frac{\pi}{4}\right]$

**69.** $f(x) = e^{-x/2}$; $[-1, 2]$

**70.** $f(x) = \dfrac{1}{4 + x^2}$; $[0, 1]$

**71.** Let $f(x) = -2x^4 + x^2 + 2x$.

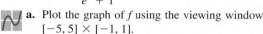

**a.** Plot the graph of $f$.
**b.** Find the $x$-intercepts of $f$ accurate to three decimal places.
**c.** Use the results of parts (a) and (b) to find the area of the region under the graph of $f$ and above the $x$-axis.

**72.** Let $f(x) = \dfrac{e^x - 1}{e^x + 1}$.

**a.** Plot the graph of $f$ using the viewing window $[-5, 5] \times [-1, 1]$.
**b.** Find the area of the region under the graph of $f$ over the interval $[0, \ln 3]$.
**c.** Verify your answer to part (b) using a calculator or a computer.

*In Exercises 73–76, evaluate the limit by interpreting it as the limit of a Riemann sum of a function on the interval $[a, b]$.*

**73.** $\displaystyle\lim_{n\to\infty} \frac{1}{n^5} \sum_{k=1}^n k^4$; $[0, 1]$

**74.** $\displaystyle\lim_{n\to\infty} \frac{1}{n} \sum_{k=1}^n \left(\frac{k}{n}\right)^{1/3}$; $[0, 1]$

**75.** $\displaystyle\lim_{n\to\infty} \frac{2}{n} \sum_{k=1}^n \left(2 + \frac{2k}{n}\right)^2$; $[2, 4]$

**76.** $\displaystyle\lim_{n\to\infty} \frac{\pi}{2n} \sum_{k=1}^n \cos\left(\frac{k\pi}{2n}\right)$; $\left[0, \frac{\pi}{2}\right]$

In Exercises 77–80, find the average value $f_{av}$ of the function over the indicated interval.

**77.** $f(x) = 2x^2 - 3x;$ $[-1, 2]$

**78.** $f(x) = 1 + \sqrt{x};$ $[0, 4]$

**79.** $f(x) = \dfrac{x}{\sqrt{x^2 + 1}};$ $[0, 3]$

**80.** $f(x) = \sin x;$ $[0, \pi]$

In Exercises 81–84, (a) find the number c whose existence is guaranteed by the Mean Value Theorem for Integrals for the function f on [a, b], and (b) sketch the graph of f on [a, b] and the rectangle with base on [a, b] that has the same area as that of the region under the graph of f.

**81.** $f(x) = x^2 + 2x;$ $[0, 1]$

**82.** $f(x) = x^3;$ $[0, 2]$

**83.** $f(x) = \sqrt{x + 3};$ $[1, 6]$

**84.** $f(x) = \cos x;$ $\left[-\frac{\pi}{3}, \frac{\pi}{3}\right]$

**85. Distance Covered by a Car** A car moves along a straight road with velocity function

$$v(t) = 2t^2 + t - 6 \qquad 0 \le t \le 8$$

where $v(t)$ is measured in feet per second.
**a.** Find the displacement of the car between $t = 0$ and $t = 3$.
**b.** Find the distance covered by the car during this period of time.

**86. Projected U.S. Gasoline Use** The White House wants to cut gasoline use from 140 billion gallons per year in 2007 to 128 billion gallons per year in 2017. But estimates by the Department of Energy's Energy Information Agency suggest that this will not happen. In fact, the agency's projection of gasoline use from the beginning of 2007 to the beginning of 2017 is given by

$$A(t) = 0.014t^2 + 1.93t + 140 \qquad 0 \le t \le 10$$

where $A(t)$ is measured in billions of gallons per year and $t$ is in years with $t = 0$ corresponding to the beginning of 2007.
**a.** According to the agency's projection, what will be the gasoline consumption at the beginning of 2017?
**b.** What will be the average consumption per year from the beginning of 2007 to the beginning of 2017?
Source: U.S. Department of Energy, Energy Information Agency.

**87. Air Purification** To test air purifiers, engineers ran a purifier in a smoke-filled 10-ft × 20-ft room. While conducting a test for a certain brand of air purifier, it was determined that the amount of smoke in the room was decreasing at the rate of $R(t)$ percent of the (original) amount of smoke per minute, $t$ min after the start of the test, where $R$ is given by

$$R(t) = 0.00032t^4 - 0.01872t^3 + 0.3948t^2 - 3.83t + 17.63$$
$$0 \le t \le 20$$

How much smoke was left in the room 5 min after the start of the test? How much smoke was left in the room 10 min after the start of the test?
Source: Consumer Reports

**88. Voltage in AC Circuits** The voltage in an AC circuit is given by

$$V = V_0 \sin \omega t$$

**a.** Show that the average (mean) voltage from $t = 0$ to $t = \pi/\omega$ (a half-cycle) is $V_{av} = (2/\pi)V_0$, which is $2/\pi$ (about $\frac{2}{3}$) times the maximum voltage $V_0$.
**b.** Show that the average voltage over a complete cycle is 0. Explain.

**89.** If $a$ feet of fencing are used to enclose a rectangular garden, show that the average area of such a garden is $a^2/24$ ft$^2$.

**90. Average Acceleration of a Car** A car moves along a straight road with velocity function $v(t)$ and acceleration function $a(t)$. The average acceleration of the car over the time interval $[t_1, t_2]$ is

$$\bar{a} = \frac{v(t_2) - v(t_1)}{t_2 - t_1}$$

Show that $\bar{a}$ is equal to the average value of $a(t)$ on $[t_1, t_2]$.

**91. Velocity of a Falling Hammer** During the construction of a high-rise apartment building, a construction worker accidentally drops a hammer that falls vertically a distance of $h$ ft. The velocity of the hammer after falling a distance of $x$ ft is $v = \sqrt{2gx}$ ft/sec, where $0 \le x \le h$. Show that the average velocity of the hammer over this path is $\bar{v} = \frac{2}{3}\sqrt{2gh}$.

**92. Flow of Water in a Canal** Water at a depth of $x$ ft in a wide rectangular canal flows at a velocity of

$$v = v_0 - 20\sqrt{hs}\left(\frac{x}{h}\right)^2$$

feet per second, where $v_0$ is the velocity of the water on the surface, $h$ is the depth of the canal, and $s$ is its gradient. Find the average velocity of flow in a cross section of the canal.

**93. Flow of Blood in an Artery** The velocity (in centimeters per second) of blood $r$ cm from the central axis of an artery is given by $v(r) = k(R^2 - r^2)$, where $k$ is a constant and $R$ is the radius of the artery. Suppose that $k = 1000$ and $R = 0.2$. Find the average velocity of the blood across a cross section of the artery.

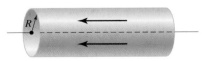

**94. Newton's Law of Cooling** A bottle of white wine at room temperature (68°F) is placed in a refrigerator at 4 P.M. Its temperature after $t$ hr is changing at the rate of $-18e^{-0.6t}$ °F/hr. By how many degrees will the temperature of the wine have dropped by 7 P.M.? What will the temperature of the wine be at 7 P.M.?

**95. Air Pollution** According to the South Coast Air Quality Management District, the level of nitrogen dioxide, a brown gas that impairs breathing, present in the atmosphere on a certain June day in downtown Los Angeles is approximated by

$$A(t) = 0.03t^3(t - 7)^4 + 62.7 \qquad 0 \le t \le 7$$

where $A(t)$ is measured in pollutant standard index and $t$ is measured in hours with $t = 0$ corresponding to 7 A.M. What is the average level of nitrogen dioxide present in the atmosphere from 7 A.M. to 2 P.M. on that day?
*Source: The Los Angeles Times.*

**96. Water Level in Boston Harbor** The water level (in feet) in Boston Harbor during a certain 24-hr period is approximated by the formula

$$H = 4.8 \sin\left[\frac{\pi}{6}(t - 10)\right] + 7.6 \qquad 0 \le t \le 24$$

where $t = 0$ corresponds to 12 A.M. What is the average water level in Boston Harbor over the 24-hr period on that day?

**97 Predator-Prey Populations** The wolf and caribou populations in a certain northern region are given by

$$P_1(t) = 8000 + 1000 \sin\frac{\pi t}{24}$$

and

$$P_2(t) = 40{,}000 + 12{,}000 \cos\frac{\pi t}{24}$$

respectively, at time $t$, where $t$ is measured in months. What are the average wolf and caribou populations over the time interval $[0, 6]$?

**98. Daylight Hours in Chicago** The number of hours of daylight at any time $t$ in Chicago is approximated by

$$L(t) = 2.8 \sin\left[\frac{2\pi}{365}(t - 79)\right] + 12$$

where $t$ is measured in days and $t = 0$ corresponds to January 1. What is the daily average number of hours of daylight in Chicago over the year? Over the summer months from June 21 ($t = 171$) through September 20 ($t = 262$)?

**99. Global Warming** The increase in carbon dioxide in the atmosphere is a major cause of global warming. Using data obtained by Dr. Charles David Keeling, professor at Scripps Institution of Oceanography, the average amount of carbon dioxide in the atmosphere from 1958 through 2007 is approximated by

$$A(t) = 0.010716t^2 + 0.8212t + 313.4 \qquad 1 \le t \le 50$$

where $A(t)$ is measured in parts per million volume (ppmv) and $t$ in years with $t = 1$ corresponding to the beginning of 1958. Find the average amount of carbon dioxide in the atmosphere from 1958 through 2007.
*Source: Scripps Institution of Oceanography.*

**100. Depreciation: Double Declining Balance Method** Suppose that a tractor purchased at a price of $60,000 is to be depreciated by the *double declining balance method* over a 10-year period. It can be shown that the rate at which the book value will be decreasing is given by

$$R(t) = 13{,}388.61e^{-0.22314t} \qquad 0 \le t \le 10$$

dollars per year at year $t$. Find the amount by which the book value of the tractor will depreciate over the first 5 years of its life.

**101. Canadian Oil-Sands Production** The production of oil (in millions of barrels per day) extracted from oil sands in Canada is projected to be

$$P(t) = \frac{4.76}{1 + 4.11e^{-0.22t}} \qquad 0 \le t \le 15$$

where $t$ is measured in years, with $t = 0$ corresponding to the beginning of 2005. What will the total oil production of oil from oil sands be over the years from the beginning of 2005 until the beginning of 2020 ($t = 15$)?
*Source: Canadian Association of Petroleum Producers.*

**102. Average Temperature** A homogenous hollow metallic ball of inner radius $r_1$ and outer radius $r_2$ is in thermal equilibrium. The temperature $T$ at a distance $r$ from the center of the ball is given by

$$T = T_1 + \frac{r_1 r_2 (T_2 - T_1)}{(r_1 - r_2)}\left(\frac{1}{r} - \frac{1}{r_1}\right) \qquad r_1 \le r \le r_2$$

where $T_1$ is the temperature on the inner surface and $T_2$ is the temperature on the outer surface. Find the average temperature of the ball in a radial direction between $r = r_1$ and $r = r_2$.

**103. Motion of a Submersible** A submersible moving in a straight line through water is subjected to a resistance $R$ that is proportional to its velocity. Suppose that the submersible travels with its engine shut off. Then the time it takes for the submersible to slow down from a velocity of $v_1$ to a velocity of $v_2$ is

$$T = -\int_{v_1}^{v_2} \frac{m}{kv}\, dv$$

where $m$ is the mass of the submersible and $k$ is a constant. Find the time it takes the submersible to slow down from a velocity of 16 ft/sec to 8 ft/sec if its mass is 1250 slugs and $k = 20$ (slug/sec).

 **104. Lengths of Infants** Medical records of infants delivered at Kaiser Memorial Hospital show that the percentage of infants whose length at birth is between 19 and 21 in. is given by

$$P = 100 \int_{19}^{21} \frac{1}{2.6\sqrt{2\pi}} e^{-(1/2)[(x - 20)/2.6]^2}\, dx$$

Use a calculator or computer to estimate $P$.

**105. Serum Cholesterol Levels** The percentage of a current Mediterranean population with serum cholesterol levels between 160 and 180 mg/dL is estimated to be

$$P = \sqrt{\frac{2}{\pi}} \int_{160}^{180} e^{-(1/2)[(x-160)/50]^2} \, dx$$

Estimate $P$.

 **106. Absorption of Drugs** The concentration of a drug in an organ at any time $t$, in seconds) is given by

$$C(t) = \begin{cases} 0.3t - 18(1 - e^{-t/60}) & \text{if } 0 \le t \le 20 \\ 18e^{-t/60} - 12e^{-(t-20)/60} & \text{if } t > 20 \end{cases}$$

where $C(t)$ is measured in grams per cubic centimeter (g/cm$^3$). Find the average concentration of the drug in the organ over the first 30 sec after it is administered.

**107.** Prove that

$$\int_{-1/2}^{1/2} 2^{\cos x} \, dx = 2 \int_0^{1/2} 2^{\cos x} \, dx$$

**108.** Find $dx/dy$ if

$$\int_0^x \sqrt{3 + 2\cos t} \, dt + \int_0^y \sin t \, dt = 0$$

**109.** Find the $x$-coordinates of the relative extrema of the function

$$F(x) = \int_0^x \frac{\sin t}{t} \, dt \qquad x > 0$$

**110.** Find all functions $f$ on $[0, 1]$ such that $f$ is continuous on $[0, 1]$ and

$$\int_0^x f(t) \, dt = \int_x^1 f(t) \, dt \quad \text{for every } x \in (0, 1)$$

**111.** If $f(x) = \int_2^x \frac{dt}{\sqrt{1 + t^3}}$, where $x > -1$, what is $(f^{-1})'(0)$?

**112.** Let

$$f(x) = \begin{cases} 1 - x & \text{if } 0 \le x \le 1 \\ x - 1 & \text{if } 1 < x \le 3 \end{cases}$$

   **a.** Find $F(x) = \int_0^x f(t) \, dt$.
   **b.** Plot the graph of $F$, and show that it is continuous on $[0, 3]$.
   **c.** Where is $f$ differentiable? Where is $F$ differentiable?

**113.** Evaluate $\displaystyle\lim_{h \to 0} \frac{1}{h} \int_2^{2+h} \sqrt{5 + t^2} \, dt$.

**114.** Evaluate $\displaystyle\int_{-1}^1 \frac{2x^5 + x^4 - 3x^3 + 2x^2 + 8x + 1}{x^2 + 1} \, dx$.

**115.** Evaluate $\displaystyle\int_{-\pi/4}^{\pi/4} (\cos x + 1) \tan^3 x \, dx$.

**116.** Show that

$$\int_{-1}^1 \sqrt{x^2 + 1} \sec x \, dx = 2 \int_0^1 \sqrt{x^2 + 1} \sec x \, dx$$

**117. a.** Show that $\int_0^\pi x f(\sin x) \, dx = (\pi/2) \int_0^\pi f(\sin x) \, dx$.
    **Hint:** Use the substitution $x = \pi - u$.
   **b.** Use the result of part (a) to evaluate $\int_0^\pi x \sin x \, dx$.

**118. a.** If $f$ is even, what can you say about $\int_{-\pi}^\pi f(x) \cos nx \, dx$ and $\int_{-\pi}^\pi f(x) \sin nx \, dx$ if $n$ is an integer? Explain.
   **b.** If $f$ is odd, what can you say about $\int_{-\pi}^\pi f(x) \cos nx \, dx$ and $\int_{-\pi}^\pi f(x) \sin nx \, dx$? Explain.

**119.** Use the identity

$$\frac{\sin\left(n + \frac{1}{2}\right) x}{2 \sin \frac{x}{2}} = \frac{1}{2} + \cos x + \cos 2x + \cdots + \cos nx$$

to show that

$$\int_0^\pi \frac{\sin\left(n + \frac{1}{2}\right) x}{\sin \frac{x}{2}} \, dx = \pi$$

**120. a.** Show that if $f$ is a continuous function, then

$$\int_0^a f(x) \, dx = \int_0^a f(a - x) \, dx$$

   and give a geometric interpretation of this result.
   **b.** Use the result of part (a) to prove that

$$\int_0^\pi \frac{\sin 2kx}{\sin x} \, dx = 0$$

   where $k$ is an integer.
    **c.** Plot the graph of

$$f(x) = \frac{\sin 2kx}{\sin x}$$

   for $k = 1, 2, 3$, and 4. Do these graphs support the result of part (b)?
   **d.** Prove that the graph of

$$f(x) = \frac{\sin 2kx}{\sin x}$$

   on $[0, \pi]$ is antisymmetric with respect to the line $x = \pi/2$ by showing that $f\left(x + \frac{\pi}{2}\right) = -f\left(x - \frac{\pi}{2}\right)$ for $0 \le x \le \frac{\pi}{2}$, and use this result to explain part (b).

**121.** A car travels along a straight road in such a way that the average velocity over *any* time interval $[a, b]$ is equal to the average of its velocities at $a$ and at $b$.
   **a.** Show that its velocity $v(t)$ satisfies

$$\int_a^b v(t) \, dt = \frac{1}{2} [v(a) + v(b)](b - a) \qquad (1)$$

   **b.** Show that $v(t) = ct + d$ for some constants $c$ and $d$.
    **Hint:** Differentiate Equation (1) with respect to $a$ and with respect to $b$.

**122.** Let $f$ be continuous on $(-\infty, \infty)$. Show that

$$\int_a^b f(x + h)\, dx = \int_{a+h}^{b+h} f(x)\, dx$$

**123.** Let $f$ be continuous on $(-\infty, \infty)$, and let $c$ be a constant. Show that

$$\int_{ca}^{cb} f(x)\, dx = c \int_a^b f(cx)\, dx$$

*In Exercises 124–128, determine whether the statement is true or false. If it is true, explain why it is true. If it is false, explain why or give an example to show why it is false.*

**124.** Assuming that the integral exists, then
$\int_{-a}^a f(x^2)\, dx = 2 \int_0^a f(x^2)\, dx$.

**125.** Assuming that the integral exists and that $f$ is even, then
$\int_{-a}^a f(x^3)\, dx = 2 \int_0^a f(x^3)\, dx$.

**126.** Assuming that the integral exists and that $f$ is even and $g$ is odd, then

$$\int_{-a}^a f(x)[g(x)]^2\, dx = 2 \int_0^a f(x)[g(x)]^2\, dx$$

**127.** If $F$ is defined by $F(x) = \int_0^x \sqrt[3]{1 + t^2}\, dt$, then $F'$ has an inverse on $(0, \infty)$.

**128.** $\displaystyle\int_1^2 \frac{e^{-x}}{x}\, dx < 0$

## 4.6 Numerical Integration

### ■ Approximating Definite Integrals

Table 1 gives the daily consumption of oil in the United States in millions of barrels, in two-year intervals from 1987 through the year 2007. Suppose that we want to determine the average daily consumption of oil over the period in question. From our earlier work, we know that the solution is obtained by computing

$$\frac{1}{20} \int_0^{20} f(t)\, dt$$

where $f(t)$ is the oil consumption in year $t$ and $t = 0$ corresponds to 1987. But the problem here is that we do not know the algebraic rule defining the integrand $f$ for all values of $t$ in $[0, 20]$. We are given its values only at a discrete set of points in that interval! Here, the Fundamental Theorem of Calculus cannot be used to help us evaluate the integral, since we cannot find an antiderivative of $f$. Other situations also arise (for example, $f(t) = \sin t^2$) in which, although the integrand of a definite integral is defined algebraically, we are not able to find its antiderivative in terms of elementary functions. In each of these situations the best we can do is to obtain an approximation to the definite integral. (We will return to the problem of finding the average daily consumption of petroleum in Example 5.)

**TABLE 1**

| Year | 1987 | 1989 | 1991 | 1993 | 1995 | 1997 | 1999 | 2001 | 2003 | 2005 | 2007 |
|------|------|------|------|------|------|------|------|------|------|------|------|
| **Consumption** | 16.7 | 17.3 | 16.7 | 17.2 | 17.7 | 18.6 | 19.5 | 19.6 | 20.0 | 20.8 | 20.7 |

*Source:* U.S. Energy Information Administration.

A Riemann sum gives us a good approximation of a definite integral of an integrable function if the norm of the partition is sufficiently small. But there are better methods for finding approximate values of definite integrals. In this section we will look at two such methods.

## The Trapezoidal Rule

The Trapezoidal Rule uses the sum of the areas of trapezoids to approximate the definite integral $\int_a^b f(x)\, dx$. To derive this rule, let's assume that $f$ is continuous and nonnegative on $[a, b]$.* We begin by subdividing the interval $[a, b]$ into $n$ subintervals, each of equal length $\Delta x = (b - a)/n$. Because $f$ is nonnegative, the definite integral $\int_a^b f(x)\, dx$ gives the area of the region $R$ under the graph of $f$ on $[a, b]$. (See Figure 1.) This area is given by the sum of the areas of the nonoverlapping subregions $R_1, R_2, \ldots, R_n$, where $R_1$ represents the region under the graph of $f$ on $[x_0, x_1]$, $R_2$ represents the region under the graph of $f$ on $[x_1, x_2], \ldots$, and $R_n$ represents the region under the graph of $f$ on $[x_{n-1}, x_n]$.

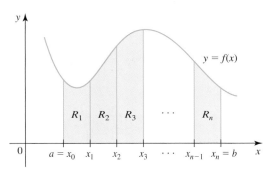

**FIGURE 1**

$\int_a^b f(x)\, dx =$ the sum of the areas of the subregions $R_1, R_2, \ldots, R_n$

The basis for the Trapezoidal Rule lies in the approximation of each of the subregions $R_1, R_2, \ldots, R_n$ by a suitable trapezoid. For example, consider the subregion $R_1$ reproduced in Figure 2. You can see that the area of $R_1$ may be approximated by the area of the trapezoid having width $\Delta x = x_1 - x_0$ and parallel sides of lengths $f(x_0)$ and $f(x_1)$. The area of this trapezoid is

$$\left[\frac{f(x_0) + f(x_1)}{2}\right]\Delta x \qquad \text{average of the lengths of the parallel sides · the width}$$

Similarly, the area of the subregion $R_2$ may be approximated by the area of the trapezoid having width $\Delta x$ and sides of length $f(x_1)$ and $f(x_2)$:

$$\left[\frac{f(x_1) + f(x_2)}{2}\right]\Delta x$$

Finally, the area of the subregion $R_n$ is approximately

$$\left[\frac{f(x_{n-1}) + f(x_n)}{2}\right]\Delta x$$

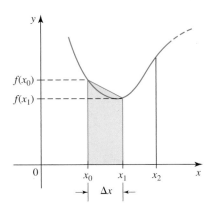

**FIGURE 2**
The area of $R_1$ is approximated by the area of a trapezoid.

---

*Actually, the nonnegativity condition is not necessary, but this assumption will simplify the derivation of the Trapezoidal Rule.

Therefore, the area of the region $R$ is approximated by the sum of the areas of the $n$ trapezoids:

$$\int_a^b f(x)\,dx \approx \Delta x \left[ \frac{f(x_0) + f(x_1)}{2} + \frac{f(x_1) + f(x_2)}{2} + \cdots + \frac{f(x_{n-1}) + f(x_n)}{2} \right]$$

$$= \frac{\Delta x}{2} [f(x_0) + 2f(x_1) + 2f(x_2) + \cdots + 2f(x_{n-1}) + f(x_n)]$$

where $\Delta x = (b - a)/n$.

**The Trapezoidal Rule**

$$\int_a^b f(x)\,dx \approx \frac{\Delta x}{2} [f(x_0) + 2f(x_1) + 2f(x_2) + \cdots + 2f(x_{n-1}) + f(x_n)] \quad \textbf{(1)}$$

where $\Delta x = (b - a)/n$ and $x_i = a + i\Delta x$, for $0 \le i \le n$.

**EXAMPLE 1**   Use the Trapezoidal Rule with $n = 10$ to approximate $\displaystyle\int_1^2 \frac{dx}{x}$.

**Solution**   Here, $a = 1$, $b = 2$, and $n = 10$, so

$$\Delta x = \frac{b - a}{n} = \frac{2 - 1}{10} = 0.1$$

Also,

$$x_0 = 1, \quad x_1 = 1.1, \quad x_2 = 1.2, \quad x_3 = 1.3, \quad \ldots, \quad x_9 = 1.9, \quad x_{10} = 2$$

The Trapezoidal Rule yields

$$\int_1^2 \frac{dx}{x} \approx \frac{0.1}{2} \left[ 1 + 2\left(\frac{1}{1.1}\right) + 2\left(\frac{1}{1.2}\right) + 2\left(\frac{1}{1.3}\right) + \cdots + 2\left(\frac{1}{1.9}\right) + \frac{1}{2} \right]$$

$$\approx 0.693771$$

If we use Formula (10) of Section 4.1, we see that

$$\int_1^2 \frac{dx}{x} = \ln 2 \qquad \text{Natural logarithm of 2}$$

$$\approx 0.693147$$

Thus, the Trapezoidal Rule with $n = 10$ yields an approximation with an error of approximately 0.000624.    ■

### ■ The Error in the Trapezoidal Rule

The **error** in the approximation of $I = \int_a^b f(x)\,dx$ by the Trapezoidal Rule is defined to be $E_n = I - T_n$, where

$$T_n = \frac{\Delta x}{2} [f(x_0) + 2f(x_1) + 2f(x_2) + \cdots + 2f(x_{n-1}) + f(x_n)]$$

An upper bound for this error follows.

**Error Bound for the Trapezoidal Rule**

If $f''$ is continuous on $[a, b]$, then the error $E_n$ in approximating $\int_a^b f(x)\,dx$ by the Trapezoidal Rule satisfies

$$|E_n| \le \frac{M(b - a)^3}{12n^2} \tag{2}$$

where $M$ is a positive number such that $|f''(x)| \le M$ for all $x$ in $[a, b]$.

**Note**   Observe that as $n \to \infty$, $E_n \to 0$, as our intuition tells us.   ∎

**EXAMPLE 2**   Find an upper bound for the error in the approximation of $\int_1^2 \dfrac{dx}{x}$ using the Trapezoidal Rule with $n = 10$ (see Example 1).

**Solution**   Here $f(x) = 1/x$. So

$$f'(x) = -\frac{1}{x^2} \quad \text{and} \quad f''(x) = \frac{2}{x^3}$$

Since $f''$ is positive and decreasing on $(1, 2)$, it attains its maximum value at the left endpoint of the interval. So if we take $M = f''(1) = 2$, then $|f''(x)| \le 2$ for all $x$ in $[1, 2]$. Finally, using Inequality (2) with this value of $M$ and $a = 1$, $b = 2$, and $n = 10$, we obtain

$$|E_{10}| \le \frac{2(2 - 1)^3}{12(10)^2} = \frac{1}{600} \approx 0.0016667$$

The actual error is approximately 0.000624 (as we computed in Example 1), and this is less than the upper bound that we just found.   ∎

**EXAMPLE 3**   Use the Trapezoidal Rule to approximate $\int_0^1 \sin x^2\,dx$ with an error that is less than 0.01.

**Solution**   First, we determine the number of subintervals, $n$, required in the Trapezoidal Rule to ensure that the error will be less than 0.01. To find the value of $M$ called for in Inequality (2), we compute the second derivative of $f(x) = \sin x^2$. Thus,

$$f'(x) = 2x \cos x^2$$

and

$$f''(x) = 2(\cos x^2 - 2x^2 \sin x^2)$$

Using the triangle inequality, we have

$$|f''(x)| \le 2|\cos x^2| + 4x^2|\sin x^2| \le 2 + 4 = 6$$

because $0 \le x \le 1$ and both $|\cos x^2|$ and $|\sin x^2|$ cannot exceed 1. Therefore, we can take $M = 6$.

Using Inequality (2) with $a = 0$, $b = 1$, and $M = 6$ and observing the requirement that the error in the approximation be less than 0.01, we have

$$\frac{6(1 - 0)^3}{12n^2} < 0.01, \qquad n^2 > \frac{100}{2}, \qquad \text{or} \qquad n > \sqrt{50} \approx 7.07$$

So by taking $n = 8$, the smallest integer exceeding 7.07, we are guaranteed that the error in the approximation will be smaller than that prescribed.

To obtain the approximation, we compute

$$\Delta x = \frac{b - a}{n} = \frac{1 - 0}{8} = \frac{1}{8}$$

Then with

$$x_0 = 0, \qquad x_1 = \frac{1}{8}, \qquad x_2 = \frac{1}{4}, \qquad x_3 = \frac{3}{8},$$

$$x_4 = \frac{1}{2}, \qquad x_5 = \frac{5}{8}, \qquad x_6 = \frac{3}{4}, \qquad x_7 = \frac{7}{8}, \qquad \text{and} \qquad x_8 = 1$$

the Trapezoidal Rule yields

$$\int_0^1 \sin x^2 \, dx \approx \frac{\frac{1}{8}}{2}\left[ f(0) + 2f\left(\frac{1}{8}\right) + 2f\left(\frac{1}{4}\right) + 2f\left(\frac{3}{8}\right) + 2f\left(\frac{1}{2}\right)\right.$$

$$\left. + 2f\left(\frac{5}{8}\right) + 2f\left(\frac{3}{4}\right) + 2f\left(\frac{7}{8}\right) + f(1) \right]$$

$$= \frac{1}{16}\left( \sin 0 + 2 \sin \frac{1}{64} + 2 \sin \frac{1}{16} + 2 \sin \frac{9}{64} + 2 \sin \frac{1}{4} \right.$$

$$\left. + 2 \sin \frac{25}{64} + 2 \sin \frac{9}{16} + 2 \sin \frac{49}{64} + \sin 1 \right)$$

$$\approx 0.3117 \qquad \blacksquare$$

## ◼ Simpson's Rule

Before deriving Simpson's Rule, let's look at the two methods that we currently have for approximating the definite integral $\int_a^b f(x) \, dx$ from a fresh point of view. Let $f$ be a continuous, nonnegative function on $[a, b]$.* Suppose that the interval $[a, b]$ is partitioned into $n$ subintervals of equal length by means of $n + 1$ equally spaced points $x_0 = a, x_1, x_2, \ldots, x_n = b$, where $n$ is a positive integer, so that the length of each subinterval is $\Delta x = (b - a)/n$. (See Figure 3.)

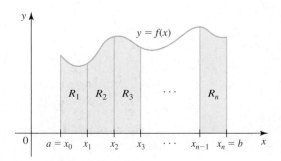

**FIGURE 3**
The area of the region under the graph of $f$ on $[a, b]$ is given by the sum of the areas of the subregions $R_1, R_2, \ldots, R_n$.

---

*Again, the nonnegativity condition is not necessary for our results but will be assumed here to simplify the ensuing discussion.

Figure 4 shows the approximation of $\int_a^b f(x)\,dx$ using a Riemann sum consisting of $n$ terms that are just the areas of the $n$ rectangles shown shaded. Here is another view of that method: Approximate $f(x)$ on $[x_0, x_1]$ by the *constant* function $y = f(p_1)$, where $p_1$ is any point in $[x_0, x_1]$; approximate $f(x)$ on $[x_1, x_2]$ by the constant function $y = f(p_2)$, where $p_2$ is any point in $[x_1, x_2]$; and so on. Then the area of the region under the graph of $f$ on $[a, b]$, $\int_a^b f(x)\,dx$, is approximated by the area of the region under the graph of the approximating "step function" $S(x)$ on $[a, b]$.

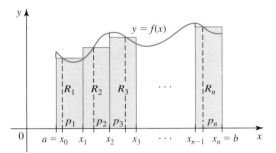

$$S(x) = \begin{cases} f(p_1) & \text{for } x_0 \le x < x_1 \\ f(p_2) & \text{for } x_1 \le x < x_2 \\ \vdots \\ f(p_n) & \text{for } x_{n-1} \le x \le x_n \end{cases}$$

**FIGURE 4**
$f$ is approximated by the step function $S$.

Next, Figure 5 shows the approximation of $\int_a^b f(x)\,dx$ using the Trapezoidal Rule. Another view of this method: Approximate $f(x)$ on $[x_0, x_1]$ by the *linear* function whose graph is the line passing through the two points $(x_0, f(x_0))$ and $(x_1, f(x_1))$; approximate $f(x)$ on $[x_1, x_2]$ by the linear function whose graph is the line passing through the points $(x_1, f(x_1))$ and $(x_2, f(x_2))$; and so on. Then the area of the region under the graph of $f$ on $[a, b]$, $\int_a^b f(x)\,dx$, is approximated by the area of the region under the graph of the approximating "polygonal function" $P(x)$ on $[a, b]$ whose graph is a *polygonal* curve.

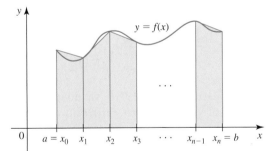

$$P(x) = \begin{cases} f(x_0) + \dfrac{f(x_1) - f(x_0)}{x_1 - x_0}(x - x_0) & \text{for } x_0 \le x < x_1 \\[2mm] f(x_1) + \dfrac{f(x_2) - f(x_1)}{x_2 - x_1}(x - x_1) & \text{for } x_1 \le x < x_2 \\ \vdots \\ f(x_{n-1}) + \dfrac{f(x_n) - f(x_{n-1})}{x_n - x_{n-1}}(x - x_{n-1}) & \text{for } x_{n-1} \le x \le x_n \end{cases}$$

**FIGURE 5**
$f$ is approximated by the "polygonal function" $P$.

A natural extension of the method used to approximate $\int_a^b f(x)\,dx$ is to approximate sections of the graph of $f$ by sections of the graphs of second-degree polynomials (parts of parabolas). We begin by showing that the area of the region under the parabola $y = ax^2 + bx + c$ on $[-h, h]$ is

$$A = \frac{h}{3}(y_0 + 4y_1 + y_2) \tag{3}$$

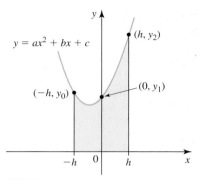

**FIGURE 6**
The area of the region under the parabola $y = ax^2 + bx + c$ on $[-h, h]$ is $A = \frac{h}{3}(y_0 + 4y_1 + y_2)$.

where $y_0$, $y_1$, and $y_2$ are the $y$-coordinates of the points lying on the parabola with $x$-coordinates $-h$, $0$, and $h$, respectively. (See Figure 6.)

The area under the parabola $y = ax^2 + bx + c$ on $[-h, h]$ is

$$A = \int_{-h}^{h} (ax^2 + bx + c)\, dx = \left[\frac{a}{3}x^3 + \frac{b}{2}x^2 + cx\right]_{-h}^{h}$$

$$= \frac{2ah^3}{3} + 2ch = \frac{h}{3}(2ah^2 + 6c)$$

Since the parabola passes through the three points $(-h, y_0)$, $(0, y_1)$, and $(h, y_2)$, the equation $y = ax^2 + bx + c$ must be satisfied at each of these points. This yields the following system of three equations:

$$ah^2 - bh + c = y_0$$

$$c = y_1$$

$$ah^2 + bh + c = y_2$$

from which we obtain

$$c = y_1$$

$$ah^2 - bh = y_0 - y_1$$

$$ah^2 + bh = y_2 - y_1$$

Adding the last two equations gives

$$2ah^2 = y_0 - 2y_1 + y_2$$

These expressions for $2ah^2$ and $c$ enable us to express $A$ in terms of $y_0$, $y_1$, and $y_2$ as follows:

$$A = \frac{h}{3}(2ah^2 + 6c) = \frac{h}{3}(y_0 - 2y_1 + y_2 + 6y_1) = \frac{h}{3}(y_0 + 4y_1 + y_2)$$

as was to be shown.

To derive Simpson's Rule for approximating $\int_a^b f(x)\, dx$, we divide the interval $[a, b]$ into an *even* number of subintervals of width $\Delta x = (b - a)/n$. If we approximate the area under the region of the graph of $f$ on $[x_0, x_2]$ by the area of the region under the parabola passing through the three points $(x_0, f(x_0))$, $(x_1, f(x_1))$, and $(x_2, f(x_2))$ on $[x_0, x_2]$ (Figure 7) and use Equation (3) with $h = \Delta x$, $y_0 = f(x_0)$, $y_1 = f(x_1)$, and $y_2 = f(x_2)$, we obtain

$$\int_{x_0}^{x_2} f(x)\, dx \approx \frac{\Delta x}{3}[f(x_0) + 4f(x_1) + f(x_2)]$$

In a similar manner we approximate the area of the region under the graph of $f$ on $[x_2, x_4]$ by the area of the region under the parabola passing through the three points $(x_2, f(x_2))$, $(x_3, f(x_3))$, and $(x_4, f(x_4))$ to obtain

$$\int_{x_2}^{x_4} f(x)\, dx \approx \frac{\Delta x}{3}[f(x_2) + 4f(x_3) + f(x_4)]$$

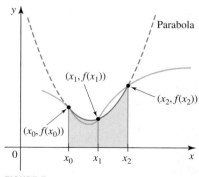

**FIGURE 7**
Simpson's Rule approximates portions of the area of the region under the curve by the area of the regions under parabolas.

Continuing, we approximate the area of the region under the graph of $f$ on $[a, b]$ using the sum of the areas of the regions under $n/2$ parabolas. Thus,

$$\int_a^b f(x)\, dx = \int_{x_0}^{x_2} f(x)\, dx + \int_{x_2}^{x_4} f(x)\, dx + \cdots + \int_{x_{n-2}}^{x_n} f(x)\, dx$$

$$\approx \frac{\Delta x}{3}[f(x_0) + 4f(x_1) + f(x_2)] + \frac{\Delta x}{3}[f(x_2) + 4f(x_3) + f(x_4)] + \cdots$$

$$+ \frac{\Delta x}{3}[f(x_{n-2}) + 4f(x_{n-1}) + f(x_n)]$$

$$= \frac{\Delta x}{3}[f(x_0) + 4f(x_1) + 2f(x_2) + 4f(x_3) + \cdots$$

$$+ 2f(x_{n-2}) + 4f(x_{n-1}) + f(x_n)]$$

Let's summarize this result.

---

**Simpson's Rule**

$$\int_a^b f(x)\, dx \approx \frac{\Delta x}{3}[f(x_0) + 4f(x_1) + 2f(x_2) + 4f(x_3) + \cdots$$

$$+ 2f(x_{n-2}) + 4f(x_{n-1}) + f(x_n)] \tag{4}$$

where $\Delta x = (b - a)/n$, and $n$ is even.

---

Simpson's rule is named after the English mathematician Thomas Simpson (1710–1761).

### ■ The Error in Simpson's Rule

If we denote the expression on the right of the approximation in (4) by $S_n$, then the error in approximating $I = \int_a^b f(x)\, dx$ by Simpson's Rule is defined to be $I - S_n$. An upper bound for this error follows. We omit the proof, which can be found in more advanced textbooks.

---

**Error Bound for Simpson's Rule**

If $f^{(4)}$ is continuous on $[a, b]$, then the error $E_n$ in approximating $\int_a^b f(x)\, dx$ by Simpson's Rule satisfies

$$|E_n| \le \frac{M(b - a)^5}{180n^4} \tag{5}$$

where $M$ is a positive number such that $|f^4(x)| \le M$ for all $x$ in $[a, b]$.

---

**EXAMPLE 4**  Use Simpson's Rule to approximate

$$\int_0^2 \frac{1}{\sqrt{x + 1}}\, dx$$

with an error that is less than 0.001.

**Solution**    We first determine the number of subintervals, $n$, required in Simpson's Rule to guarantee that the error will be less than 0.001. To find the value of $M$ that is called for in Inequality (5), we compute the fourth derivative of $f(x) = (x + 1)^{-1/2}$. Thus,

$$f'(x) = -\frac{1}{2}(x + 1)^{-3/2}, \qquad f''(x) = \frac{3}{4}(x + 1)^{-5/2}, \qquad f'''(x) = -\frac{15}{8}(x + 1)^{-7/2}$$

and

$$f^{(4)}(x) = \frac{105}{16}(x + 1)^{-9/2} = \frac{105}{16(x + 1)^{9/2}}$$

Because $f^{(4)}$ is decreasing on $(0, 2)$, we see that $|f^{(4)}(x)| \le |f^{(4)}(0)| = \frac{105}{16}$. So we may take $M = \frac{105}{16}$. Using Inequality (5) with $a = 0$, $b = 2$, and $M = \frac{105}{16}$ and observing the requirement that the error in the approximation be less than 0.001, we have

$$\frac{105(2 - 0)^5}{16(180n^4)} < 0.001, \qquad n^4 > \frac{3500}{3}, \qquad \text{or} \qquad n > \left(\frac{3500}{3}\right)^{1/4} \approx 5.84$$

So by taking $n = 6$ (which is even), we are guaranteed that the error in the approximation will be less than 0.001.

To obtain the approximation, we compute

$$\Delta x = \frac{2 - 0}{6} = \frac{1}{3}$$

Then, with

$$x_0 = 0, \qquad x_1 = \frac{1}{3}, \qquad x_2 = \frac{2}{3}, \qquad x_3 = 1, \qquad x_4 = \frac{4}{3}, \qquad x_5 = \frac{5}{3}, \qquad \text{and} \qquad x_6 = 2$$

Simpson's Rule yields

$$\int_0^2 \frac{dx}{\sqrt{x + 1}} \approx \frac{\frac{1}{3}}{3}\left[f(0) + 4f\left(\frac{1}{3}\right) + 2f\left(\frac{2}{3}\right) + 4f(1) + 2f\left(\frac{4}{3}\right) + 4f\left(\frac{5}{3}\right) + f(2)\right]$$

$$= \frac{1}{9}\left(1 + \frac{4}{\sqrt{\frac{4}{3}}} + \frac{2}{\sqrt{\frac{5}{3}}} + \frac{4}{\sqrt{2}} + \frac{2}{\sqrt{\frac{7}{3}}} + \frac{4}{\sqrt{\frac{8}{3}}} + \frac{1}{\sqrt{3}}\right)$$

$$\approx 1.46421$$

The actual value of the definite integral to six decimal places is 1.464102 and can be found by using the method of substitution.    ■

In the next example we solve the oil consumption problem posed at the beginning of this section.

**EXAMPLE 5**    **U.S. DAILY OIL CONSUMPTION**    Table 2 gives the daily consumption of oil in the United States, in millions of barrels, measured in two-year intervals, from the beginning of 1987 to the beginning of 2007. Use Simpson's Rule to estimate the average daily consumption of oil over the period in question.

**TABLE 2**

| Year | 1987 | 1989 | 1991 | 1993 | 1995 | 1997 | 1999 | 2001 | 2003 | 2005 | 2007 |
|------|------|------|------|------|------|------|------|------|------|------|------|
| **Consumption** | 16.7 | 17.3 | 16.7 | 17.2 | 17.7 | 18.6 | 19.5 | 19.6 | 20.0 | 20.8 | 20.7 |

*Source:* U.S. Energy Information Administration

**Solution** The average daily consumption from the beginning of 1987 to the beginning of 2007 is given by

$$\frac{1}{20} \int_0^{20} f(t)\, dt$$

where $t$ is measured in years, with $t = 0$ corresponding to 1987. Using Simpson's Rule with $a = 0$, $b = 20$, and $n = 10$, so that $\Delta t = (20 - 0)/10 = 2$, we have $t_0 = 0$, $t_1 = 2$, $t_3 = 4, \ldots, t_{10} = 20$. Therefore,

$$\frac{1}{20} \int_0^{20} f(t)\, dt \approx \left(\frac{1}{20}\right)\left(\frac{2}{3}\right)[f(0) + 4f(2) + 2f(4) + 4f(6) + \cdots + 4f(18) + f(20)]$$

$$= \frac{1}{30}[16.7 + 4(17.3) + 2(16.7) + 4(17.2) + 2(17.7) + 4(18.6)$$

$$+ 2(19.5) + 4(19.6) + 2(20.0) + 4(20.8) + 20.7]$$

$$= 18.64$$

and we conclude that the average daily oil consumption in the United States from 1987 through 2007 is approximately 18.6 million barrels per day. ∎

## 4.6 CONCEPT QUESTIONS

1. Describe (a) the Trapezoidal Rule and (b) Simpson's Rule. What are the main differences in approximating $\int_a^b f(x)\, dx$ using a Riemann sum, using the Trapezoidal Rule, and using Simpson's Rule?

2. Explain why $n$ can be odd or even in the Trapezoidal Rule, but it must be even in Simpson's Rule.

3. Explain, without alluding to the error formulas, why the Trapezoidal Rule gives the exact value of $\int_a^b f(x)\, dx$ if $f$ is a linear function and why Simpson's Rule gives the exact value of the integral if $f$ is a quadratic function.

## 4.6 EXERCISES

In Exercises 1–8, use (a) the Trapezoidal Rule and (b) Simpson's Rule to approximate the integral. Compare your results with the exact value of the integral.

1. $\int_0^2 x^2\, dx$; $\quad n = 4$

2. $\int_1^3 (x^2 - 1)\, dx$; $\quad n = 6$

3. $\int_1^2 x^3\, dx$; $\quad n = 6$

4. $\int_1^2 \frac{1}{x^2}\, dx$; $\quad n = 4$

5. $\int_0^2 x\sqrt{2x^2 + 1}\, dx$; $\quad n = 6$

6. $\int_0^1 e^{-x}\, dx$; $\quad n = 6$

7. $\int_0^1 xe^{-x^2}\, dx$; $\quad n = 6$

8. $\int_0^{\pi/2} \cos 2x\, dx$; $\quad n = 6$

In Exercises 9–14, use the Trapezoidal Rule to approximate the integral with answers rounded to four decimal places.

9. $\int_0^1 \frac{dx}{2x + 1}$; $\quad n = 7$

10. $\int_1^3 \sqrt{x^2 + 1}\, dx$; $\quad n = 5$

11. $\int_0^2 e^{-x^2}\, dx$; $\quad n = 4$

12. $\int_0^1 \cos x^2\, dx$; $\quad n = 6$

13. $\int_1^2 \sqrt{x} \sin x\, dx$; $\quad n = 5$

14. $\int_2^4 \frac{dx}{\ln x}$; $\quad n = 6$

In Exercises 15–20, use Simpson's Rule to approximate the integral with answers rounded to four decimal places.

15. $\int_0^2 \sqrt{x^3 + 1}\, dx$; $\quad n = 6$

16. $\int_0^1 \frac{dx}{x^2 + 1}$; $\quad n = 4$

17. $\int_{-1}^1 \sqrt{x^2 + 1}\, dx$; $\quad n = 6$

18. $\int_1^2 x^{-1/2}e^x\, dx$; $\quad n = 4$

**19.** $\int_0^{\pi/2} \sqrt{1 + \sin^2 x}\, dx; \quad n = 6$

**20.** $\int_2^4 \dfrac{dx}{\ln x}; \quad n = 6$

*In Exercises 21–28, find a bound on the error in approximating the integral using* (a) *the Trapezoidal Rule and* (b) *Simpson's Rule with n subintervals.*

**21.** $\int_{-1}^2 x^3\, dx; \quad n = 6$

**22.** $\int_0^1 \dfrac{dx}{x + 1}; \quad n = 8$

**23.** $\int_1^4 \sqrt{x}\, dx; \quad n = 8$

**24.** $\int_0^2 \dfrac{dx}{\sqrt{x + 1}}; \quad n = 6$

**25.** $\int_0^{\pi/2} x \sin x\, dx; \quad n = 6$

**26.** $\int_0^1 \cos x^2\, dx; \quad n = 6$

**27.** $\int_1^2 \ln x^2\, dx; \quad n = 8$

**28.** $\int_0^1 \cot^{-1} x\, dx; \quad n = 10$

**cas** *In Exercises 29–34, use a calculator or a computer and the error formula for the Trapezoidal Rule to find n such that the error in the approximation of the integral using the Trapezoidal Rule is less than 0.001.*

**29.** $\int_1^2 \dfrac{dx}{x}$

**30.** $\int_0^2 \dfrac{dx}{x^2 + 1}$

**31.** $\int_1^2 e^{1/x}\, dx$

**32.** $\int_0^2 \sqrt{x^2 + 1}\, dx$

**33.** $\int_0^{\pi/2} x \cos x\, dx$

**34.** $\int_0^1 \cos x^2\, dx$

**cas** *In Exercises 35–40, use a calculator or a computer and the error formula for Simpson's Rule to find n such that the error in the approximation of the integral using Simpson's Rule is less than 0.001.*

**35.** $\int_1^4 2x^{3/2}\, dx$

**36.** $\int_1^3 \dfrac{dx}{x}$

**37.** $\int_0^2 \dfrac{dx}{x^2 + 4}$

**38.** $\int_1^2 e^{x^2}\, dx$

**39.** $\int_0^{\pi/2} x \sin x\, dx$

**40.** $\int_0^1 \sin x^2\, dx$

**41. Velocity of a Sports Car** The velocity function for a sports car traveling on a straight road is given by

$$v(t) = \dfrac{80t^3}{t^3 + 100} \quad 0 \le t \le 16$$

where $t$ is measured in seconds and $v(t)$ in feet per second. Use Simpson's Rule with $n = 8$ to estimate the average velocity of the car over the time interval $[0, 16]$.

**42. Air Pollution** The amount of nitrogen dioxide, a brown gas that impairs breathing, present in the atmosphere on a certain May day in the city of Long Beach is approximated by

$$A(t) = \dfrac{136}{1 + 0.25(t - 4.5)^2} + 28 \quad 0 \le t \le 11$$

where $A(t)$ is measured in pollutant standard index and $t$ is measured in hours, with $t = 0$ corresponding to 7 A.M. Use the Trapezoidal Rule with $n = 11$ to find the approximate average level of nitrogen dioxide present in the atmosphere from 7 A.M. to 6 P.M. on that day.

*Source: The Los Angeles Times.*

**43. U.S. Strategic Petroleum Reserves** According to data from the American Petroleum Institute, the U.S. Strategic Petroleum Reserves from the beginning of 1981 to the beginning of 1990 can be approximated by the function

$$S(t) = \dfrac{613.7t^2 + 1449.1}{t^2 + 6.3} \quad 0 \le t \le 9$$

where $S(t)$ is measured in millions of barrels and $t$ in years, with $t = 0$ corresponding to the beginning of 1981. Using the Trapezoidal Rule with $n = 9$, estimate the average petroleum reserves from the beginning of 1981 to the beginning of 1990.

*Source: American Petroleum Institute.*

**44. Velocity of an Attack Submarine** The following data give the velocity of an attack submarine taken at 10-min intervals during a submerged trial run.

| Time $t$ (hr) | 0 | $\frac{1}{6}$ | $\frac{1}{3}$ | $\frac{1}{2}$ | $\frac{2}{3}$ | $\frac{5}{6}$ | 1 |
|---|---|---|---|---|---|---|---|
| Velocity $v$ (mph) | 14.2 | 24.3 | 40.2 | 45.0 | 38.5 | 27.6 | 12.8 |

Use Simpson's Rule to estimate the distance traveled by the submarine during the 1-hr submerged trial run.

**45. Flow of Water in a River** At a certain point, a river is 78 ft wide and its depth, measured at 6-ft intervals across the river, is recorded in the following table.

| $x$ | 0 | 6 | 12 | 18 | 24 | 30 | 36 |
|---|---|---|---|---|---|---|---|
| $y$ | 0.8 | 2.6 | 5.8 | 6.2 | 8.2 | 10.1 | 10.8 |

| $x$ | 42 | 48 | 54 | 60 | 66 | 72 | 78 |
|---|---|---|---|---|---|---|---|
| $y$ | 9.8 | 7.6 | 6.4 | 5.2 | 3.9 | 2.4 | 1.4 |

Here, $x$ denotes the distance (in feet) from one bank of the river, and $y$ (in feet) is the corresponding depth. If the average rate of flow through this section of the river is 4 ft/sec, estimate the rate of the volume of flow of water in the river. Use the Trapezoidal Rule with $n = 13$.

**46. Measuring Cardiac Output** Eight milligrams of a dye are injected into a vein leading to an individual's heart. The concentration of the dye in the aorta (in milligrams per liter) measured at 2-sec intervals is shown in the accompanying table. Use Simpson's Rule with $n = 12$ and the formula

$$R = \frac{60D}{\displaystyle\int_0^{24} C(t)\, dt}$$

to estimate the person's cardiac output, where $D$ is the quantity of dye injected in milligrams, $C(t)$ is the concentration of the dye in the aorta, and $R$ is measured in liters per minute.

| $t$ | 0 | 2 | 4 | 6 | 8 | 10 | 12 |
|------|---|---|-----|-----|-----|-----|-----|
| $C(t)$ | 0 | 0 | 2.8 | 6.1 | 9.7 | 7.6 | 4.8 |

| $t$ | 14 | 16 | 18 | 20 | 22 | 24 |
|------|-----|-----|-----|-----|-----|----|
| $C(t)$ | 3.7 | 1.9 | 0.8 | 0.3 | 0.1 | 0 |

*In Exercises 47–48, determine whether the statement is true or false. If it is true, explain why it is true. If it is false, explain why or give an example to show why it is false.*

**47.** If $f$ is a polynomial of degree greater than one, then the error $E_n$ in approximating $\int_a^b f(x)\, dx$ by the Trapezoidal Rule must be nonzero.

**48.** If $f$ is nonnegative and concave upward on $[a, b]$ and $A$ is an approximation of $\int_a^b f(x)\, dx$ using the Trapezoidal Rule, then $A > \int_a^b f(x)\, dx$.

# CHAPTER 4 REVIEW

## CONCEPT REVIEW

*In Exercises 1–10, fill in the blanks.*

**1. a.** A function $F$ is an antiderivative of $f$ on an interval if _____ for all $x$ in $I$.
   **b.** If $F$ is an antiderivative of $f$ on an interval $I$, then every antiderivative of $f$ on $I$ has the form _____.

**2. a.** $\int c f(x)\, dx = $ _____.
   **b.** $\int [f(x) \pm g(x)]\, dx = $ _____.

**3. a.** A differential equation is an equation that involves the derivative or differential of an _____ function.
   **b.** A solution of a differential equation on an interval $I$ is any _____ that satisfies the differential equation.

**4.** If we let $u = g(x)$, then $du = $ _____, and the substitution transforms the integral $\int f(g(x))g'(x)\, dx$ into the integral _____ involving only $u$.

**5. a.** If $f$ is continuous and nonnegative on an interval $[a, b]$, then the area of the region under the graph of $f$ on $[a, b]$ is given by _____.
   **b.** If $f$ is continuous on an interval $[a, b]$, then $\int_a^b f(x)\, dx$ is equal to the area(s) of the regions lying above the $x$-axis and bounded by the graph of $f$ on $[a, b]$ _____ the

area(s) of the regions lying below the $x$-axis and bounded by the graph of $f$ on $[a, b]$.

**6. a.** If $f$ is continuous on $[a, b]$, then the average value of $f$ over $[a, b]$ is the number $f_{av} = $ _____.
   **b.** If $f$ is a continuous and nonnegative function on $[a, b]$, then the number $f_{av}$ may be thought of as the _____ of the rectangle with base lying on the interval $[a, b]$ and having the same _____ as the area of the region under the graph of $f$ on $[a, b]$.

**7. a.** The Fundamental Theorem of Calculus, Part 1, states that if $f$ is continuous on $(a, b)$, then $F(x) = \int_a^x f(t)\, dt$ is differentiable on $(a, b)$, and $F'(x) = $ _____.
   **b.** The Fundamental Theorem of Calculus, Part 2, states that if $f$ is continuous on $[a, b]$, then $\int_a^b f(x)\, dx = $ _____, where $F$ is an _____ of $f$.
   **c.** The net change in a function $f$ over an interval $[a, b]$ is given by $f(b) - f(a) = $ _____, provided that $f'$ is continuous on $[a, b]$.

**8.** Let $f$ be continuous on $[-a, a]$. If $f$ is even, then $\int_{-a}^a f(x)\, dx = $ _____, and if $f$ is odd, then $\int_{-a}^a f(x)\, dx = $ _____.

**9.** The Mean Value Theorem for Integrals states that if $f$ is continuous on $[a, b]$, then there exists at least one point $c$ in $[a, b]$ such that _____.

**10. a.** The Trapezoidal Rule states that $\int_a^b f(x)\,dx \approx$ _____, where $\Delta x = (b - a)/n$. The error $E_n$ in approximating $\int_a^b f(x)\,dx$ by the Trapezoidal Rule satisfies $|E_n| \le$

_____, where $M$ is a positive number such that $|f''(x)| \le M$ for all $x$ in $[a, b]$.

**b.** Simpson's Rule states that $\int_a^b f(x)\,dx \approx$ _____, where $\Delta x = (b - a)/n$ and $n$ is _____. The error $E_n$ in approximating $\int_a^b f(x)\,dx$ by Simpson's Rule satisfies $|E_n| \le$ _____, where $M$ is a positive number such that $|f^{(4)}(x)| \le M$ for all $x$ in $[a, b]$.

## REVIEW EXERCISES

*In Exercises 1–32, find the indefinite integral.*

**1.** $\displaystyle\int (2x^3 - 4x^2 + 3x + 4)\,dx$

**2.** $\displaystyle\int \frac{x^5 - 3x + 2}{x^3}\,dx$

**3.** $\displaystyle\int (x^{5/3} - 2x^{2/5})\,dx$

**4.** $\displaystyle\int x^{1/3}(2x^2 - 3x + 1)\,dx$

**5.** $\displaystyle\int \left( x^{2/3} - \frac{2}{x^4} + 3 \right)\,dx$

**6.** $\displaystyle\int \frac{x^2 - x + \sqrt{x}}{\sqrt[3]{x}}\,dx$

**7.** $\displaystyle\int (1 + 2t)^3\,dt$

**8.** $\displaystyle\int \frac{(1 + x)^2}{\sqrt{x}}\,dx$

**9.** $\displaystyle\int (3t - 4)^8\,dt$

**10.** $\displaystyle\int \sqrt[3]{2u + 1}\,du$

**11.** $\displaystyle\int (x + x^{-1})^2\,dx$

**12.** $\displaystyle\int 2x^3\sqrt{x^4 + 1}\,dx$

**13.** $\displaystyle\int \frac{3x + 1}{(3x^2 + 2x)^3}\,dx$

**14.** $\displaystyle\int (\pi^2 + \sqrt{x} + 1)\,dx$

**15.** $\displaystyle\int \cos^4 t \sin t\,dt$

**16.** $\displaystyle\int \frac{\sec^2 x}{\sqrt{\tan x}}\,dx$

**17.** $\displaystyle\int \frac{\cos\theta}{\sqrt{1 - \sin\theta}}\,d\theta$

**18.** $\displaystyle\int \frac{\cos^3\theta + 1}{\cos^2\theta}\,d\theta$

**19.** $\displaystyle\int x\csc x^2 \cot x^2\,dx$

**20.** $\displaystyle\int \sec 3x \tan 3x\,dx$

**21.** $\displaystyle\int \frac{1}{5x - 3}\,dx$

**22.** $\displaystyle\int \frac{\cos x}{2 + 3\sin x}\,dx$

**23.** $\displaystyle\int \frac{2x + 1}{3x + 2}\,dx$

**24.** $\displaystyle\int \frac{(\ln x)^3}{x}\,dx$

**25.** $\displaystyle\int t \cdot 2^{t^2}\,dt$

**26.** $\displaystyle\int \frac{e^x}{e^x - 1}\,dx$

**27.** $\displaystyle\int \frac{\sin(\ln x)}{x}\,dx$

**28.** $\displaystyle\int \frac{e^{1/x}}{x^2}\,dx$

**29.** $\displaystyle\int \frac{\tan\sqrt{x}}{\sqrt{x}}\,dx$

**30.** $\displaystyle\int \frac{\sin^{-1}x}{\sqrt{1 - x^2}}\,dx$

**31.** $\displaystyle\int \frac{\tan^{-1}2x}{1 + 4x^2}\,dx$

**32.** $\displaystyle\int \frac{\sec t \tan t}{1 + \sec t}\,dt$

*In Exercises 33–46, evaluate the definite integral.*

**33.** $\displaystyle\int_0^2 (3x + 5)\,dx$

**34.** $\displaystyle\int_{-1}^1 \sqrt[3]{8x}\,dx$

**35.** $\displaystyle\int_1^2 \left( \frac{1}{x^2} - \frac{1}{x^3} \right)\,dx$

**36.** $\displaystyle\int_0^2 t^2\sqrt{t^3 + 1}\,dt$

**37.** $\displaystyle\int_0^4 \frac{1}{\sqrt{1 + 2x}}\,dx$

**38.** $\displaystyle\int_0^{\sqrt{2}} \frac{x}{\sqrt{x^2 + 2}}\,dx$

**39.** $\displaystyle\int_0^1 (x + 1)(2x + 3)^2\,dx$

**40.** $\displaystyle\int_1^4 \frac{(\sqrt{x} + 1)^5}{\sqrt{x}}\,dx$

**41.** $\displaystyle\int_0^{\pi/8} \frac{\sin 2x}{\cos^2 2x}\,dx$

**42.** $\displaystyle\int_0^{\pi/2} \sin\theta\sqrt{2 + 7\cos\theta}\,d\theta$

**43.** $\displaystyle\int_{\pi/6}^{\pi/4} (\csc\theta + \cot\theta)(1 - \cos\theta)\,d\theta$

**44.** $\displaystyle\int_0^1 \frac{x}{2x^2 + 1}\,dx$

**45.** $\displaystyle\int_1^2 \frac{x^3 - 2x + 1}{x^2}\,dx$

**46.** $\displaystyle\int_0^1 \frac{e^{3x}}{1 + e^{3x}}\,dx$

*In Exercises 47 and 48, use the properties of integrals to prove the inequality.*

**47.** $\displaystyle\int_0^1 \sqrt{1 + e^{4x}}\,dx \ge \frac{1}{2}(e^2 - 1)$

**48.** $1 < \displaystyle\int_0^1 e^{x^2}\,dx < e$

*In Exercises 49–52, find the average value of the function over the given interval.*

**49.** $f(x) = x^3$; $[0, 2]$

**50.** $f(x) = \dfrac{1}{\sqrt{x} + 1}$; $[1, 3]$

**51.** $f(x) = \dfrac{\ln x^2}{x}$; $[1, 2]$

**52.** $f(x) = \sin^2 x \cos x$; $\left[0, \frac{\pi}{2}\right]$

*In Exercises 53 and 54, find f'(x).*

**53.** $f(x) = \int_0^{x^2} \dfrac{e^t}{t^2 + 1}\, dt$

**54.** $f(x) = \int_{\ln x}^{\sqrt{x}} \sin t^2\, dt, \quad$ where $x > 0$

**55.** Find the derivative of $F(x) = \int_{x^2}^{x^3} \sin t\, dt$.

**56.** Find the area of the region under the graph of $y = xe^{-x^2}$ on $[0, 4]$.

**57.** Evaluate $\int_0^1 \sqrt{1 + x^2}\, dx$ by using the Trapezoidal Rule with $n = 5$.

**58.** Evaluate $\int_1^2 \dfrac{\sqrt{x}}{1 + x^2}\, dx$ by using Simpson's Rule with $n = 4$.

*In Exercises 59 and 60, find a bound on the error in approximating each definite integral using* (a) *the Trapezoidal Rule and* (b) *Simpson's Rule with n subintervals.*

**59.** $\int_0^2 \dfrac{1}{\sqrt{1 + x}}\, dx; \quad n = 8$

**60.** $\int_1^3 \ln x\, dx; \quad n = 10$

**61.** Find the function $f$ given that its derivative is $f'(x) = \sqrt{x} + \sin x$ and that its graph passes through the point $(0, 2)$.

**62.** A ball is thrown straight up from the ground with an initial velocity of 64 ft/sec. How long will it take for the ball to reach its highest point, and what will its height be at this point?

**63.** An electric drill rolls off the edge of a 128-ft-tall structure under construction.
 **a.** Find the position of the electric drill after $t$ sec.
 **b.** Determine when it strikes the ground.
 **c.** What is its velocity at impact?

**64.** A stone is dropped from a height of $h$ ft above the ground. Show that the speed at which the stone strikes the ground is $\sqrt{2gh}$ ft/sec, where $g$ is the gravitational constant.

**65.** A car traveling along a straight road undergoes constant deceleration that reduces its speed from 44 ft/sec to 22 ft/sec in 8 sec. How far will the car travel if it is brought to rest from 44 ft/sec at that rate of deceleration?

**66. Traffic Flow** The traffic department of a certain city estimates that $t$ years from now the number of vehicles (in thousands) in the city will be

$$0.2t^4 + 4t + 84$$

Find the estimated average number of motor vehicles in the city over the next 5 years.

**67.** An electromotive force (emf), $E$, is given by

$$E = E_0 \sin \frac{2\pi t}{T}$$

where $T$ is the period in seconds and $E_0$ is the amplitude of the emf. Find the average value of the emf over one period (the time interval $[0, T]$).

**68. Total Cost** The weekly marginal cost of the Advanced Visuals Systems Corporation in manufacturing its 42-in. plasma television sets is given by

$$C'(x) = 0.000006x^2 - 0.04x + 120$$

dollars per set, where $x$ stands for the number of sets manufactured. Find the weekly total cost function $C$ if the fixed cost of the company is \$70,000 per week.

**69. Total Profit** The weekly marginal profit of the Advanced Visuals Systems Corporation is given by

$$P'(x) = -0.000006x^2 - 0.04x + 200$$

dollars per set where $x$ stands for the number of 42-in. plasma television sets sold. Find the weekly total profit function $P$ if $P(0) = -80,000$.

**70. Hotel Occupancy Rate** The occupancy rate at a large hotel in Maui in month $t$ is described by the function

$$R(t) = 60 + 37 \sin^2\left(\frac{\pi t}{12}\right) \qquad 0 \le t \le 12$$

where $t$ is measured in months, $v(t)$ is measured in percent, and $t = 0$ corresponds to the beginning of June. Find the average occupancy rate of the hotel over a 1-year period.

Hint: $\sin^2 x = \dfrac{1 - \cos 2x}{2}$

**71. TV on Mobile Phones** The number of people watching TV on mobile phones is expected to grow at the rate of

$$N'(t) = \frac{5.4145}{\sqrt{1 + 0.91t}}\, dt \qquad 0 \le t \le 4$$

million per year. The number of people watching TV on mobile phones at the beginning of 2007 ($t = 0$) was 11.9 million.
 **a.** Find an expression giving the number of people watching TV on mobile phones in year $t$.
 **b.** According to this projection, how many people will be watching TV on mobile phones at the beginning of 2011?
 *Source:* International Data Corporation, U.S. forecast.

**72. Net Investment Flow** The net investment flow of a giant conglomerate, which is the rate of capital formation, is projected to be $t\sqrt{2t^2 + 1}$ million dollars per year $t$ years from now. Find the accruement on the company's capital stock after 2 years; that is, compute

$$\int_0^2 t\sqrt{2t^2 + 1}\, dt$$

**73. Respiratory Cycles** The volume of air inhaled by a person during respiration is given by

$$V(t) = \frac{6}{5\pi}\left(1 - \cos\frac{\pi t}{2}\right)$$

liters at time $t$ (in seconds). What is the average volume of air inhaled by a person over one cycle from $t = 0$ to $t = 4$?

**74. Average Temperature** The average daily temperature (in degrees Fahrenheit) on the $t$th day at a tourist resort in Cameron Highlands is approximated by

$$T = 62 - 18\cos\frac{2\pi(t - 23)}{365}$$

($t = 0$ corresponds to the beginning of the year). What is the average daily temperature in Cameron Highlands over the year?

**75. Average Temperature** The following graph shows the daily mean temperatures recorded during a certain month at Frazer's Hill. Using (a) the Trapezoidal Rule and

(b) Simpson's Rule with $n = 10$, estimate the average temperature during that month.

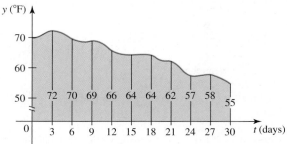

**76.** Show that $\dfrac{d}{dx}\displaystyle\int_x^c f(t)\,dt = -f(x)$.

**77. a.** Prove that

$$0.5 < \int_0^{1/2} \frac{dx}{\sqrt{1 - x^{2n}}} < 0.524 \qquad n > 1$$

 **b.** Use a computer or a calculator to find the value of the integral in part (a) with $n = 6$ accurate to six decimal places.

## PROBLEM-SOLVING TECHNIQUES

The following example introduces a technique for transforming a definite integral into another having the same value as the original integral. This technique is then used to evaluate a definite integral without the need to find an antiderivative associated with the integral.

### EXAMPLE

**a.** Show that if $f$ is a continuous function, then

$$\int_0^a f(x)\,dx = \int_0^a f(a - x)\,dx$$

**b.** Use the result of part (a) to show that $\int_0^{\pi/2} \sin^m x\,dx = \int_0^{\pi/2} \cos^m x\,dx$.

**c.** Use the result of part (b) to evaluate $\int_0^{\pi/2} \sin^2 x\,dx$ and $\int_0^{\pi/2} \cos^2 x\,dx$.

#### Solution

**a.** Let us evaluate the integral on the right-hand side by using the substitution $u = a - x$, so that $du = -dx$. To obtain the limits of integration, observe that if $x = 0$, then $u = a$, and if $x = a$, then $u = 0$. Substituting, we obtain

$$\int_0^a f(a - x)\,dx = -\int_a^0 f(u)\,du = \int_0^a f(u)\,du = \int_0^a f(x)\,dx$$

This proves the assertion.

There is a simple geometric explanation for this result. It stems from the fact that the graph of $f$ on the interval $[0, a]$ is the mirror image of the graph of $f(a - x)$ on the same interval with respect to the vertical line $x = a/2$. In fact, if the point $A$ lies on the $x$-axis and has $x$-coordinate $x$, then the point $A'$ that is

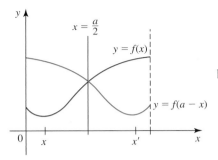

**FIGURE 1**
The graphs of $f(x)$ and $f(a - x)$ are mirror images with respect to $x = a/2$.

its mirror image with respect to the line $x = a/2$ has $x$-coordinate $x' = a - x$. Therefore, $f(a - x') = f(a - (a - x)) = f(x)$. (See Figure 1.) Since congruent figures have equal areas, the result follows from interpreting definite integrals as areas.

**b.** Using the result of part (a), we see that

$$\int_0^{\pi/2} \sin^m x \, dx = \int_0^{\pi/2} \sin^m \left( \frac{\pi}{2} - x \right) dx$$

$$= \int_0^{\pi/2} \left( \sin \frac{\pi}{2} \cos x - \cos \frac{\pi}{2} \sin x \right)^m dx$$

$$= \int_0^{\pi/2} \cos^m x \, dx$$

**c.** Using the result of part (b) with $m = 2$, we have

$$I = \int_0^{\pi/2} \sin^2 x \, dx = \int_0^{\pi/2} \cos^2 x \, dx$$

Therefore,

$$2I = \int_0^{\pi/2} \sin^2 x \, dx + \int_0^{\pi/2} \cos^2 x \, dx$$

$$= \int_0^{\pi/2} (\sin^2 x + \cos^2 x) \, dx = \int_0^{\pi/2} dx = \frac{\pi}{2}$$

and hence $I = \pi/4$. ∎

## CHALLENGE PROBLEMS

**1.** Evaluate $\displaystyle\int_a^b \frac{|x|}{x} \, dx$, where $a < b$.

**2.** Show that $\displaystyle\int_0^x [\![t]\!] \, dt = \frac{[\![x]\!]([\![x]\!] - 1)}{2} + [\![x]\!](x - [\![x]\!])$, where $[\![x]\!]$ is the greatest integer function.

**3.** Evaluate $\displaystyle\int_0^{10\pi} \sqrt{1 - \cos 2x} \, dx$.

**4.** By interpreting the integral geometrically, evaluate

$$\int_{\sqrt{2}/2}^{\sqrt{3}/2} \sqrt{1 - x^2} \, dx$$

**5.** Evaluate $\displaystyle\int_{-1}^1 \frac{3x^6 - 2x^5 + 4x^3 - 3x^2 + 5x}{x^2 + 1} \, dx$.

**6.** Find $\displaystyle\int \frac{1}{\sin^2 x \cos^4 x} \, dx$.

**7.** Find $\displaystyle\int \frac{dx}{1 + 2x \cos a + x^2}$, where $0 < |a| < \pi$.

**Hint:** Use the substitution $u = \dfrac{x + \cos a}{\sin a}$.

**8.** Evaluate

$$\lim_{x \to 0} \frac{\displaystyle\int_0^{x^3} \tan t^{1/3} \, dt}{\displaystyle\int_0^{2x^2} t \, dt}$$

**9.** Evaluate $\displaystyle\lim_{b \to a} \frac{1}{b - a} \int_a^b f(x) \, dx$, where $f$ is a continuous function.

**10.** Evaluate $\displaystyle\lim_{n \to \infty} \sum_{k=1}^n \frac{n}{n^2 + k^2}$.

**Hint:** Relate the limit to the limit of a Riemann sum of an appropriate function.

**11. a.** Show that $\int_a^b f(x) \, dx = \int_a^b f(a + b - x) \, dx$, and give a geometric interpretation of the result.

**b.** Use the result of part (a) to show that $\int_0^\pi f(\sin x) \cos x \, dx = 0$.

**12.** Show that $\int_0^t f(x)g(t - x) \, dx = \int_0^t g(x)f(t - x) \, dx$.

**13. a.** Suppose that $f$ is continuous and $g$ and $h$ are differentiable. Show that

$$\frac{d}{dx} \int_{g(x)}^{h(x)} f(t)\,dt = f[h(x)]h'(x) - f[g(x)]g'(x)$$

**b.** Use the result of part (a) to find $g'(x)$ if

$$g(x) = \int_{1/x}^{\sqrt{x}} \sin t^2\,dt \qquad x > 0$$

**14.** Prove that if $f$ and $g$ are continuous functions on $[a, b]$, then

$$\left| \int_a^b f(x)g(x)\,dx \right| \le \sqrt{\int_a^b [f(x)]^2\,dx \int_a^b [g(x)]^2\,dx}$$

This is known as Schwarz's inequality.
**Hint:** Consider the function $F(x) = [f(x) - tg(x)]^2$, where $t$ is a real number.

**15. a.** Use Schwarz's inequality (see Exercise 14) to prove that

$$\int_0^1 \sqrt{1 + x^3}\,dx < \frac{\sqrt{5}}{2}$$

**b.** Is this estimate better than the one obtained by using the Mean Value Theorem for Integrals?

**16.** Find the values of $x$ at which

$$F(x) = \int_0^{x^2} \frac{t^2 - 5t + 4}{t^2 + 1}\,dt$$

has relative extrema.

**17.** Suppose that $f$ is continuous on an interval $[a, b]$. Show that

$$\lim_{n \to \infty} \frac{1}{n} \sum_{k=1}^n f\left[a + \frac{k(b-a)}{n}\right] = \frac{1}{b-a} \int_a^b f(x)\,dx$$

**18. a.** Prove that

$$\int_a^b f(x)\,dx = (b-a) \int_0^1 f[(b-a)t + a]\,dt$$

Thus, an integral with interval of integration $[a, b]$ can be transformed into one with interval of integration $[0, 1]$ by means of the substitution $x = (b-a)t + a$.

**b.** Use the result of part (a) to evaluate

$$\int_{-3}^{-4} \cos(x + 4)^2\,dx + 3\int_{1/3}^{2/3} \cos\left[9\left(x - \frac{2}{3}\right)^2\right]dx$$

**19.** Suppose that $f$ is a continuous periodic function with period $p$.
**a.** Prove that if $a$ is any real number, then

$$\int_0^a f(x)\,dx = \int_p^{a+p} f(x)\,dx$$

**b.** Use the result of part (a) to show that if $a$ is any real number, then

$$\int_0^p f(x)\,dx = \int_a^{a+p} f(x)\,dx$$

**20.** Let $f$ be continuous on an interval $[-a, a]$.
**a.** Show that $\int_{-a}^a f(x^2)\,dx = 2\int_0^a f(x^2)\,dx$.
**b.** What can you say about $\int_{-a}^a f(x^2)\sin x\,dx$?

**21.** Let $f$ be continuous on an interval $[a, b]$ and satisfy $\int_a^x f(t)\,dt = \int_x^b f(t)\,dt$ for all $x$ in $[a, b]$. Show that $f(x) = 0$ on $[a, b]$.

**22.** The Fresnel function $S$ is defined by the integral

$$S(x) = \int_0^x \sin\left(\frac{\pi t^2}{2}\right)dt$$

**a.** Sketch the graphs of $f(x) = \sin(\pi x^2/2)$ and $S(x)$ on the same set of axes for $0 \le x \le 3$. Interpret your results.
**b.** Sketch the graph of $S$ on the interval $[-10, 10]$.

**23.** Find all continuous, nonnegative functions $f$ defined on $[0, b]$, where $b > 0$, satisfying the equation $[f(x)]^2 = 2\int_0^x f(t)\,dt$.

**24. a.** Prove that $e^{-R\sin x} < e^{(-2R/\pi)x}$ on the interval $\left(0, \frac{\pi}{2}\right)$, where $R > 0$.
**Hint:** Show that $f(x) = \sin x/x$ is decreasing on $\left(0, \frac{\pi}{2}\right)$.
**b.** Use the result of part (a) to prove the inequality

$$\int_0^{\pi/2} e^{-R\sin x}\,dx < \frac{\pi}{2R}(1 - e^{-R}) \qquad R > 0$$

Ambient Images/Alamy

The photograph shows the Jacqueline Kennedy Onassis Reservoir (formerly the Central Park Reservoir). Built between 1858 and 1862, it is located between 86th Street and 96th Street in the borough of Manhattan in New York City. In this chapter we will use calculus to help us estimate the surface area of the reservoir.

# 5

# Applications of the Definite Integral

**IN THIS CHAPTER** we continue to exploit the integral as a tool for solving a variety of problems. More specifically, we will use the techniques of integration to find the areas of regions between curves, the volumes of solids, the arc lengths of plane curves, and the areas of surfaces. We will also show how the integral is used to compute the work done by a force acting on an object and the force exerted on an object by a hydrostatic force. Finally, we will use integration to find the moments and the centers of mass of thin plates.

**V** This symbol indicates that one of the following video types is available for enhanced student learning at **www.academic.cengage.com/login:**
- Chapter lecture videos
- Solutions to selected exercises

## 5.1 Areas Between Curves

### A Real-Life Interpretation

Two cars are traveling in adjacent lanes along a straight stretch of a highway. The velocity functions for Car $A$ and Car $B$ are $v = f(t)$ and $v = g(t)$, respectively. The graphs of these functions are shown in Figure 1.

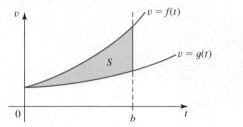
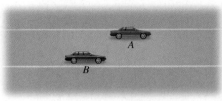

**FIGURE 1**
The shaded area $S$ gives the distance that Car $A$ is ahead of Car $B$ at time $t = b$.

The area of the region under the graph of $f$ from $t = 0$ to $t = b$ gives the total distance covered by Car $A$ in $b$ seconds over the time interval $[0, b]$. The distance covered by Car $B$ over the same period of time is given by the area under the graph of $g$ on the interval $[0, b]$. Intuitively, we see that the area of the (shaded) region $S$ between the graphs of $f$ and $g$ on the interval $[0, b]$ gives the distance that Car $A$ will be ahead of Car $B$ at time $t = b$.

Since the area of the region under the graph of $f$ on $[0, b]$ is

$$\int_0^b f(t)\, dt$$

and the area of the region under the graph of $g$ on $[0, b]$ is

$$\int_0^b g(t)\, dt$$

we see that the area of the region $S$ is given by

$$\int_0^b f(t)\, dt - \int_0^b g(t)\, dt = \int_0^b [f(t) - g(t)]\, dt$$

Therefore, the distance that Car $A$ will be ahead of Car $B$ at $t = b$ is

$$\int_0^b [f(t) - g(t)]\, dt$$

This example suggests that some applied problems can be solved by finding the area of a region between two curves, which in turn can be found by evaluating an appropriate definite integral. Let's make this notion more precise.

### The Area Between Two Curves

Suppose $f$ and $g$ are continuous functions with $f(x) \geq g(x)$ for all $x$ in $[a, b]$, so that the graph of $f$ lies on or above that of $g$ on $[a, b]$. Let's consider the region $S$ bounded by the graphs of $f$ and $g$ between the vertical lines $x = a$ and $x = b$ as shown in Figure 2. To define the *area* of $S$, we take a regular partition of $[a, b]$,

$$a = x_0 < x_1 < x_2 < x_3 < \cdots < x_n = b$$

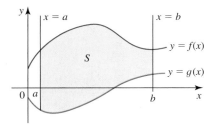

**FIGURE 2**
The region $S$ between the graphs of $f$ and $g$ on $[a, b]$

and form the Riemann sum of the function $f - g$ over $[a, b]$ with respect to this partition:

$$\sum_{k=1}^{n} [f(c_k) - g(c_k)]\Delta x$$

where $c_k$ is an evaluation point in the subinterval $[x_{k-1}, x_k]$ and $\Delta x = (b - a)/n$. The $k$th term of this sum gives the area of a rectangle with height $[f(c_k) - g(c_k)]$ and width $\Delta x$. As you can see in Figure 3, this area is an approximation of the area of the subregion of $S$ that lies between the graphs of $f$ and $g$ on $[x_{k-1}, x_k]$.

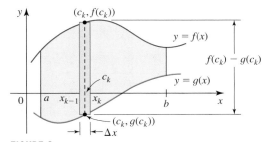

**FIGURE 3**
The $k$th term of the Riemann sum of $f - g$ gives the area of the $k$th rectangle of width $\Delta x$.

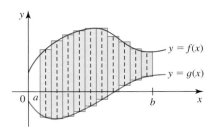

**FIGURE 4**
The Riemann sum of $f - g$ approximates the area of $S$.

Therefore, the Riemann sum provides us with an approximation of what we might intuitively think of as the area of $S$ (see Figure 4). As $n$ gets larger and larger, we might expect the approximation to get better and better. This suggests that we *define* the area $A$ of $S$ by

$$A = \lim_{n \to \infty} \sum_{k=1}^{n} [f(c_k) - g(c_k)]\Delta x \qquad (1)$$

Since $f - g$ is continuous on $[a, b]$, the limit in Equation (1) exists and is equal to the definite integral of $f - g$ from $a$ to $b$. This leads us to the following definition of the area $A$ of $S$.

---

**DEFINITION  Area of a Region Between Two Curves**

Let $f$ and $g$ be continuous on $[a, b]$, and suppose that $f(x) \geq g(x)$ for all $x$ in $[a, b]$. Then the area of the region between the graphs of $f$ and $g$ and the vertical lines $x = a$ and $x = b$ is

$$A = \int_a^b [f(x) - g(x)]\, dx \qquad (2)$$

---

**Notes**

**1.** If $g(x) = 0$ for all $x$ in $[a, b]$, then the region $S$ is just the region under the graph of $f$ on $[a, b]$, and its area is

$$\int_a^b [f(x) - 0]\, dx = \int_a^b f(x)\, dx$$

as expected (see Figure 5a).

**2.** If $f(x) = 0$ for all $x$ in $[a, b]$, then the region $S$ lies on or below the $x$-axis, and its area is

$$\int_a^b [0 - g(x)]\, dx = -\int_a^b g(x)\, dx$$

This shows that we can interpret the definite integral of a negative function as the *negative* of the area of the region *above* the graph of $g$ on $[a, b]$ (see Figure 5b).

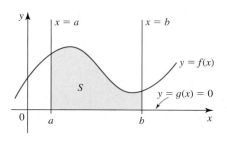

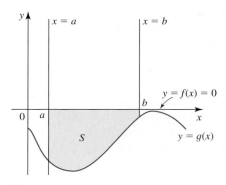

**FIGURE 5**

**(a)** If $g(x) = 0$ on $[a, b]$, then $\int_a^b f(x)\, dx$ gives the area of $S$.

**(b)** If $f(x) = 0$ on $[a, b]$, then $-\int_a^b g(x)\, dx$ gives the area of $S$.

The following guidelines are useful in setting up the integral in Equation (2).

**Finding the Area Between Two Curves**

**1.** Sketch the region between the graphs of $f$ and $g$ on $[a, b]$.

**2.** Draw a representative rectangle with height $[f(x) - g(x)]$ and width $\Delta x$ and note that its area is

$$\Delta A = [f(x) - g(x)]\Delta x$$

**3.** Observe that the height of the rectangle, $[f(x) - g(x)]$, is the integrand in Equation (2). The width $\Delta x$ reminds us to integrate with respect to $x$. Thus,

$$A = \int_a^b [f(x) - g(x)]\, dx$$

(See Figure 6.)

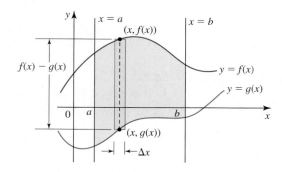

**FIGURE 6**
The area of the vertical rectangle
is $\Delta A = [f(x) - g(x)]\Delta x$.

**EXAMPLE 1** Find the area of the region between the graphs of $y = e^x$ and $y = x$ and the vertical lines $x = 0$ and $x = 1$.

**Solution** First, we make a sketch of the region and draw a representative rectangle. (See Figure 7.) Observe that the graph of $y = e^x$ lies above that of $y = x$. Therefore, if we let $f(x) = e^x$ and $g(x) = x$, then $f(x) \geq g(x)$ on $[0, 1]$. Also, from the figure we see that the area of the vertical rectangle is

$$\Delta A = [f(x) - g(x)]\Delta x = (e^x - x)\Delta x \qquad \text{(upper function − lower function)}\Delta x$$

So the area of the required region is

$$A = \int_a^b [f(x) - g(x)]\, dx = \int_0^1 (e^x - x)\, dx$$

$$= \left[ e^x - \frac{1}{2}x^2 \right]_0^1 = \left( e - \frac{1}{2} \right) - (1 - 0)$$

$$= e - \frac{3}{2} \approx 1.22$$

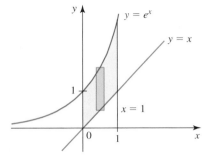

**FIGURE 7**
The graph of $y = e^x$ lies above that of $y = x$ on $[0, 1]$.

**EXAMPLE 2** Find the area of the region bounded by the graphs of $y = 2 - x^2$ and $y = -x$.

**Solution** We first make a sketch of the desired region and draw a representative rectangle. (See Figure 8.) The points of intersection of the two graphs are found by solving the equations $y = 2 - x^2$ and $y = -x$ simultaneously. Substituting the second equation into the first yields

$$-x = 2 - x^2$$

$$x^2 - x - 2 = 0$$

$$(x + 1)(x - 2) = 0$$

giving $x = -1$ and $x = 2$ as the $x$-coordinates of the points of intersection. We can think of the region in question as being bounded by the vertical lines $x = -1$ and $x = 2$. This gives the limits of integration as $a = -1$ and $b = 2$ in Equation (2). Next, if we let $f(x) = 2 - x^2$ and $g(x) = -x$, then $f(x) \geq g(x)$ on $[-1, 2]$, and the representative rectangle has area

$$\Delta A = [f(x) - g(x)]\Delta x = [(2 - x^2) - (-x)]\Delta x = (-x^2 + x + 2)\Delta x$$

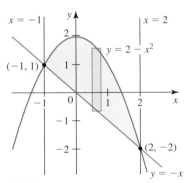

**FIGURE 8**
The graph of $f(x) = 2 - x^2$ lies above that of $g(x) = -x$ on $[-1, 2]$.

Therefore, the area of the required region is

$$A = \int_a^b [f(x) - g(x)]\, dx = \int_{-1}^2 (-x^2 + x + 2)\, dx$$

$$= \left[ -\frac{1}{3}x^3 + \frac{1}{2}x^2 + 2x \right]_{-1}^2$$

$$= \left( -\frac{8}{3} + 2 + 4 \right) - \left( \frac{1}{3} + \frac{1}{2} - 2 \right) = \frac{27}{6} \quad \text{or} \quad 4\frac{1}{2}$$

**EXAMPLE 3**   Refer to Figure 9. Find the area of the region enclosed by the graphs of

$$y = \frac{1}{4}x^2 \qquad \text{and} \qquad y = \frac{8}{x^2 + 4}$$

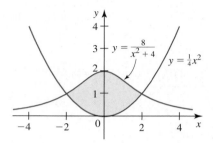

**FIGURE 9**
The region bounded by the graphs
of $y = \dfrac{8}{x^2 + 4}$ and $y = \dfrac{x^2}{4}$

**Solution**   We first find the $x$-coordinates of the points of intersection of the two graphs by solving the system

$$\begin{cases} y = \frac{1}{4}x^2 \\ y = \dfrac{8}{x^2 + 4} \end{cases}$$

simultaneously. We have

$$\frac{1}{4}x^2 = \frac{8}{x^2 + 4}$$

$$x^4 + 4x^2 - 32 = 0$$

$$(x^2 + 8)(x^2 - 4) = 0$$

giving $x = \pm 2$. Next, observing that the graph of $f(x) = 8/(x^2 + 4)$ lies above that of $g(x) = x^2/4$ on $[-2, 2]$, we find the required area to be

$$A = \int_{-2}^{2} \left( \frac{8}{x^2 + 4} - \frac{x^2}{4} \right) dx$$

$$= 2\int_{0}^{2} \left( \frac{8}{x^2 + 4} - \frac{x^2}{4} \right) dx \qquad \text{The integrand is even.}$$

$$= 2\left[ 4 \tan^{-1} \frac{x}{2} - \frac{1}{12}x^3 \right]_{0}^{2}$$

$$= 2\left( 4 \tan^{-1} 1 - \frac{8}{12} \right) = 2\pi - \frac{4}{3} \qquad \blacksquare$$

## ■ Integrating with Respect to $y$

Sometimes it is easier to find the area of a region by integrating with respect to $y$ rather than with respect to $x$. Consider, for example, the region $S$ bounded by the graphs of $x = f(y)$ and $x = g(y)$, where $f(y) \geq g(y)$, and the horizontal lines $y = c$ and $y = d$, where $c \leq d$, as shown in Figure 10.

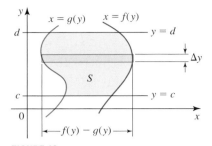

**FIGURE 10**
The region $S$ is bounded on the left by the graph of $x = g(y)$ and on the right by that of $x = f(y)$ on $[c, d]$.

Observe that the condition $f(y) \geq g(y)$ implies that the graph of $f$ lies to the right of the graph of $g$. Considering the horizontal rectangle of length $[f(y) - g(y)]$ and width $\Delta y$, we see that its area is

$$\Delta A = [f(y) - g(y)]\Delta y$$

This suggests that the area of $S$ is

$$A = \int_c^d [f(y) - g(y)]\, dy \qquad (3)$$

Since a rigorous derivation of Equation (3) proceeds along lines that are virtually identical to that of Equation (2), it will be omitted here.

**EXAMPLE 4** Find the area of the region of Example 3 by integrating with respect to $y$.

**Solution** We view the region $S$ as being bounded by the graphs of the functions $f(y) = y + 2$ (solve $y = x - 2$ for $x$), $g(y) = y^2$, and the horizontal lines $y = -1$ and $y = 2$. See Figure 11. Observe that $f(y) \geq g(y)$ for $y$ in $[-1, 2]$. The area of the representative horizontal rectangle is

$$\Delta A = [f(y) - g(y)]\Delta y = [(y + 2) - y^2]\Delta y = (y + 2 - y^2)\Delta y$$
$$\text{(right function } - \text{ left function)}\Delta y$$

This implies that

$$A = \int_{-1}^{2} (y + 2 - y^2)\, dy = \left[\frac{1}{2}y^2 + 2y - \frac{1}{3}y^3\right]_{-1}^{2}$$

$$= \left(2 + 4 - \frac{8}{3}\right) - \left(\frac{1}{2} - 2 + \frac{1}{3}\right) = \frac{9}{2} \quad \text{or} \quad 4\frac{1}{2} \quad \blacksquare$$

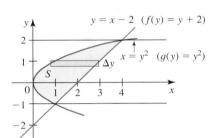

**FIGURE 11**
The horizontal rectangle has area $[f(y) - g(y)]\Delta y$.

**Note** Sometimes we prefer to use Equation (3) instead of Equation (2) or vice versa. In general, the choice of the formula depends on the shape of the region. Often one would integrate with respect to the variable that results in the minimal splitting of the region. But sometimes the use of one formula leads to an integral(s) that is difficult to evaluate, in which case the other formula should be used. $\blacksquare$

## What Happens When the Curves Intertwine?

Sometimes we are required to find the area of a region $S$ between two curves in which the graph of one function $f$ lies above that of another function $g$ for some values of $x$ ($f(x) \geq g(x)$) and lies below it for other values of $x$ ($f(x) \leq g(x)$). You will be asked to give a physical interpretation of a problem involving precisely such a situation in Exercise 48.

To find the area of the region $S$, we divide it into subregions $S_1, S_2, \ldots, S_n$, each of which is described by the sole condition $f(x) \geq g(x)$ or $f(x) \leq g(x)$. Figure 12 illustrates the case in which $n = 3$. We then use the guidelines developed earlier to calculate the area of each subregion. Adding up these areas gives the area of $S$. Thus, the area of the region $S$ shown in Figure 12 between the graphs of $f$ and $g$ and between the vertical lines $x = a$ and $x = b$ is

$$A = \int_a^c [f(x) - g(x)]\, dx + \int_c^d [g(x) - f(x)]\, dx + \int_d^b [f(x) - g(x)]\, dx$$

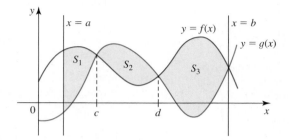

**FIGURE 12**
The region $S$ is the union of $S_1$, where $f(x) \geq g(x)$; $S_2$, where $f(x) \leq g(x)$, and $S_3$, where $f(x) \geq g(x)$.

Since

$$|f(x) - g(x)| = \begin{cases} f(x) - g(x) & \text{if } f(x) \geq g(x) \\ g(x) - f(x) & \text{if } f(x) \leq g(x) \end{cases}$$

we can also write $A$ in the abbreviated form

$$A = \int_a^b |f(x) - g(x)|\, dx \qquad (4)$$

When using Equation (4), however, we still need to determine the subintervals of $[a, b]$ where $f(x) \geq g(x)$ and/or where $g(x) \geq f(x)$ and write $A$ as the sum of integrals giving the areas of the subregions on these subintervals.

**EXAMPLE 5** Find the area of the region $S$ bounded by the graphs of $y = \cos x$ and $y = (2/\pi)x - 1$ and the vertical lines $x = 0$ and $x = \pi$.

**Solution** The region $S$ is shown in Figure 13. To find the points of intersection of the graphs of $y = \cos x$ and $y = (2/\pi)x - 1$, we solve the two equations simultaneously. Substituting the first equation into the second, we obtain

$$\cos x = \frac{2}{\pi}x - 1$$

By inspecting the graphs, we see that $x = \pi/2$ is the only solution of the equation. Therefore, the point of intersection is $\left(\frac{\pi}{2}, 0\right)$. Let $f(x) = \cos x$ and $g(x) = (2/\pi)x - 1$. Referring to Figure 13, we see that the areas $A_1$ and $A_2$ of the subregions $S_1$ and $S_2$ are

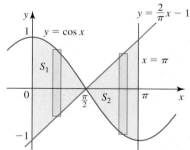

**FIGURE 13**
The area of $S$ is the sum of the areas of $S_1$ and $S_2$.

$$A_1 = \int_0^{\pi/2} [f(x) - g(x)]\, dx \qquad f(x) \geq g(x)$$

$$= \int_0^{\pi/2} \left[ \cos x - \left( \frac{2}{\pi}x - 1 \right) \right] dx = \int_0^{\pi/2} \left( \cos x - \frac{2}{\pi}x + 1 \right) dx$$

$$= \left[ \sin x - \frac{1}{\pi}x^2 + x \right]_0^{\pi/2} = 1 - \frac{1}{\pi}\left( \frac{\pi}{2} \right)^2 + \frac{\pi}{2} = \frac{4\pi - \pi^2 + 2\pi^2}{4\pi} = \frac{4 + \pi}{4}$$

and

$$A_2 = \int_{\pi/2}^{\pi} [g(x) - f(x)] \, dx \qquad g(x) \ge f(x)$$

$$= \int_{\pi/2}^{\pi} \left( \frac{2}{\pi} x - 1 - \cos x \right) dx$$

$$= \left[ \frac{1}{\pi} x^2 - x - \sin x \right]_{\pi/2}^{\pi} = \left[ \frac{1}{\pi} (\pi^2) - \pi - 0 \right] - \left[ \frac{1}{\pi} \left( \frac{\pi}{2} \right)^2 - \frac{\pi}{2} - 1 \right]$$

$$= \pi - \pi - \frac{\pi}{4} + \frac{\pi}{2} + 1 = \frac{4 + \pi}{4}$$

Therefore, the required area is

$$A = A_1 + A_2 = \frac{4 + \pi}{4} + \frac{4 + \pi}{4} = \frac{4 + \pi}{2} = 2 + \frac{\pi}{2} \qquad ■$$

The following example, drawn from the field of study known as the *theory of elasticity,* gives yet another physical interpretation of the area between two curves.

**EXAMPLE 6   Elastic Hysteresis**   Figure 14 shows a stress–strain curve for a sample of vulcanized rubber that has been stretched to seven times its original length. The function $f$ whose graph is the upper curve gives the relationship between the stress and the strain as the load (the stress) is applied to the material. Because the material is elastic, the rubber returns to its original length when the load is removed. However, when the load is decreased, the graph of $f$ is not retraced. Instead, the stress–strain curve given by the graph of the function $g$ is obtained.

**FIGURE 14**
A stress–strain curve for a sample of vulcanized rubber: The upper curve shows what happens when the load is applied, and the lower curve shows what happens when the load is decreased.

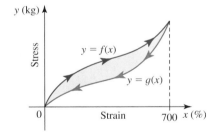

The lack of coincidence of the curves for increasing and decreasing stress is known as *elastic hysteresis.* The graphs of $f$ and $g$ on the interval $[0, 700]$ form the *hysteresis loop* for the material. It can be shown that the area of the region enclosed by the hysteresis loop is proportional to the energy dissipated within the rubber. Thus, the elastic hysteresis of the rubber is given by

$$\int_0^{700} [f(x) - g(x)] \, dx \qquad \text{Since } f(x) \ge g(x) \text{ on } [0, 700]$$

Certain types of rubber have large hysteresis, and these materials are often used as vibration absorbers. Most of the internal energy is dissipated in the form of heat, thereby minimizing the transmission of the energy of vibration to the mediums to which the machinery is mounted.   ■

## 5.1     CONCEPT QUESTIONS

1. Write an expression that gives the area of the region completely enclosed by the graphs of $f$ and $g$ in Figures 1 and 2 in terms of (a) a single integral and (b) two integrals.

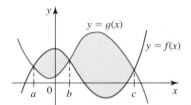

**FIGURE 1**

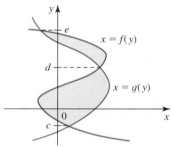

**FIGURE 2**

2. Two cars start out side by side moving down a straight road. The velocity functions for Car $A$ and Car $B$ are $f$ and $g$, respectively. Their graphs are shown in Figure 3. Suppose that $f(t)$ and $g(t)$ are measured in feet per second and $t$ in seconds, where $t$ lies in the interval [0, 10]. Answer the following questions using definite integral(s) if appropriate.

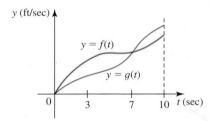

**FIGURE 3**

a. By what distance is Car $A$ ahead of Car $B$ after 3 sec? After 7 sec? After 10 sec?
b. Is one car always ahead of the other after the start of motion?
c. What is the greatest distance between the two cars over the 10-sec interval?

## 5.1     EXERCISES

*In Exercises 1–6, find the area of the shaded region.*

1.

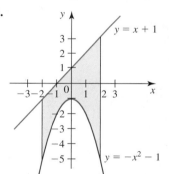

2.

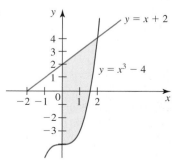

3.

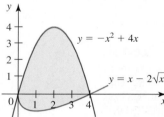

4.

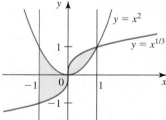

**5.**

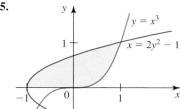

**6.**

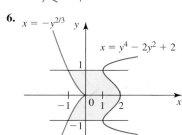

**7. Oil Production Shortfall** Energy experts disagree about when global oil production will begin to decline. In the following figure, the function $f$ gives the annual world oil production in billions of barrels from 1980 to 2050 according to the U.S. Department of Energy projection. The function $g$ gives the world oil production in billions of barrels per year over the same period according to longtime petroleum geologist Colin Campbell. Find an expression in terms of definite integrals involving $f$ and $g$ giving the shortfall in the total oil production over the period in question heeding Campbell's dire warnings.

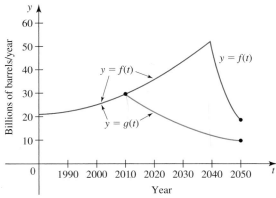

*Source:* U.S. Department of Energy and Colin Campbell.

**8. Rate of Change of Revenue** The rate of change of the revenue of Company $A$ over the (time) interval $[0, T]$ is $f(t)$ dollars per week, whereas the rate of change of the revenue of Company $B$ over the same period is $g(t)$ dollars per week. Suppose the graphs of $f$ and $g$ are as depicted in the following figure. Find an expression in terms of definite integrals involving $f$ and $g$ giving the additional revenue that Company $B$ will have over Company $A$ in the period $[0, T]$.

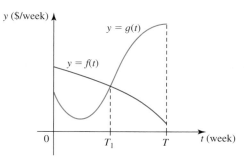

*In Exercises 9–40, sketch the region bounded by the graphs of the given equations and find the area of that region.*

**9.** $y = x^2 + 3$, $\quad y = x + 1$, $\quad x = -1$, $\quad x = 1$

**10.** $y = x^3 + 1$, $\quad y = x - 1$, $\quad x = -1$, $\quad x = 1$

**11.** $y = x^2 - 2x - 1$, $\quad y = -e^x - 1$, $\quad x = -1$, $\quad x = 1$

**12.** $y = \dfrac{1}{\sqrt{1 - x^2}}$, $\quad y = e^{-x}$, $\quad x = 0$, $\quad x = \dfrac{1}{2}$

**13.** $y = -x^2 + 4$, $\quad y = 3x + 4$

**14.** $y = x^2 - 4x$, $\quad y = -x + 4$

**15.** $y = x^2 - 4x + 3$, $\quad y = -x^2 + 2x + 3$

**16.** $y = (x - 2)^2$, $\quad y = 4 - x^2$

**17.** $y = x$, $\quad y = x^3$

**18.** $y = x^2$, $\quad y = x^4$

**19.** $y = \dfrac{x}{x^2 + 1}$, $\quad y = -\dfrac{1}{2}x^2$, $\quad x = 1$

**20.** $y = x^3 - 6x^2 + 9x$, $\quad y = x^2 - 3x$

**21.** $y = \sqrt{x}$, $\quad y = -\dfrac{1}{2}x + 1$, $\quad x = 1$, $\quad x = 4$

**22.** $y = 2\sqrt{x} - x$, $\quad y = -\sqrt{x}$

**23.** $y = \dfrac{1}{x^2}$, $\quad y = x^2$, $\quad x = 3$

**24.** $y = 2x$, $\quad y = x\sqrt{x + 1}$

**25.** $y = -x^2 + 6x + 5$, $\quad y = x^2 + 5$

**26.** $y = x\sqrt{4 - x^2}$, $\quad y = 0$

**27.** $y = \dfrac{x}{\sqrt{16 - x^2}}$, $\quad y = 0$, $\quad x = 3$

**28.** $x = y^2 + 1$, $\quad x = 0$, $\quad y = -1$, $\quad y = 2$

**29.** $x = y^2$, $\quad x = y - 3$, $\quad y = -1$, $\quad y = 2$

**30.** $x = y^2$, $\quad x = 2y + 3$

**31.** $y = -x^3 + x$, $\quad y = x^4 - 1$

**32.** $\sqrt{x} + \sqrt{y} = 1$, $\quad x + y = 1$

**33.** $y = |x|, \quad y = x^2 - 2$

**34.** $y = 2^x, y = 2^{-x}, x = -2,$ and $x = 2.$

**35.** $y = \sin 2x, \quad y = \cos x, \quad x = \dfrac{\pi}{6}, \quad x = \dfrac{\pi}{2}$

**36.** $y = \cos 2x, \quad y = \sin x, \quad x = 0, \quad x = \dfrac{3\pi}{2}$

**37.** $y = \sec^2 x, \quad y = 2, \quad x = -\dfrac{\pi}{4}, \quad x = \dfrac{\pi}{4}$

**38.** $y = \sec^2 x, \quad y = \cos x, \quad x = -\dfrac{\pi}{3}, \quad x = \dfrac{\pi}{3}$

**39.** $y = 2 \sin x + \sin 2x, \quad y = 0, \quad x = 0, \quad x = \pi$

**40.** $x = \sin y + \cos 2y, \quad x = 0, \quad y = 0, \quad y = \dfrac{\pi}{2}$

**41.** Find the area of the region in the first quadrant bounded by the parabolas $y = x^2$ and $y = \frac{1}{4}x^2$ and the line $y = 2$.

**42.** Find the area of the region enclosed by the curve $y^2 = x^2(1 - x^2)$.

*In Exercises 43 and 44, use integration to find the area of the triangle with the given vertices.*

**43.** $(0, 0), (1, 6), (4, 2)$

**44.** $(-2, 4), (0, -2), (6, 2)$

*In Exercises 45 and 46, find the area of the region bounded by the given curves (a) using integration with respect to x and (b) using integration with respect to y.*

**45.** $y = x^3, \quad y = 2x + 4, \quad x = 0$

**46.** $y = \sqrt{x}, \quad y = \dfrac{1}{2}x, \quad y = 1, \quad y = 2$

**47. Effect of Advertising on Revenue** In the accompanying figure, the function $f$ gives the rate of change of Odyssey Travel's revenue with respect to the amount $x$ it spends on advertising with its current advertising agency. By engaging the services of a different advertising agency, Odyssey expects its revenue to grow at the rate given by the function $g$. Give an interpretation of the area $A$ of the region $S$, and find an expression for $A$ in terms of a definite integral involving $f$ and $g$.

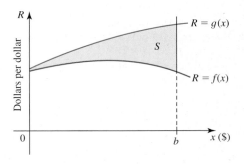

**48.** Two cars start out side by side and travel along a straight road. The velocity of Car A is $f(t)$ ft/sec, and the velocity of Car B is $g(t)$ ft/sec over the interval $[0, T]$, where $0 < T_1 < T$. Furthermore, suppose that the graphs of $f$ and $g$ are as depicted in the figure. Let $A_1$ and $A_2$ denote the areas of the regions shown shaded.

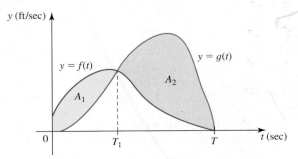

**a.** Write the number

$$\int_{T_1}^{T} [g(t) - f(t)] \, dt - \int_0^{T_1} [f(t) - g(t)] \, dt$$

in terms of $A_1$ and $A_2$.

**b.** What does the number obtained in part (a) represent?

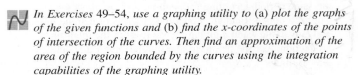

*In Exercises 49–54, use a graphing utility to (a) plot the graphs of the given functions and (b) find the x-coordinates of the points of intersection of the curves. Then find an approximation of the area of the region bounded by the curves using the integration capabilities of the graphing utility.*

**49.** $y = x^2, \quad y = 4 - x^4$

**50.** $y = x^3 - 3x^2 + 1, \quad y = x^2 - 4$

**51.** $y = x^3 - 4x^2, \quad y = x^3 - 9x$

**52.** $y = x^4 - 2x^2 + 2, \quad y = 4 - x^2$

**53.** $y = x^2, \quad y = \sin x$

**54.** $y = \cos x, \quad y = |x|$

**55. Turbocharged Engine Versus Standard Engine** In tests conducted by *Auto Test Magazine* on two identical models of the Phoenix Elite, one equipped with a standard engine and the other with a turbocharger, it was found that the acceleration of the former (in ft/sec$^2$) is given by

$$a = f(t) = 4 + 0.8t \qquad 0 \le t \le 12$$

$t$ sec after starting from rest at full throttle, whereas the acceleration of the latter (in ft/sec$^2$) is given by

$$a = g(t) = 4 + 1.2t + 0.03t^2 \qquad 0 \le t \le 12$$

How much faster is the turbocharged model moving than the model with the standard engine at the end of a 10-sec test run at full throttle?

**56. Velocity of Dragsters** Two dragsters start out side by side. The velocity of Dragster A, $V_A$, and the velocity of Dragster B,

$V_B$, for the first 8 sec of the race are shown in the following table, where $V_A$ and $V_B$ are measured in feet per second. Use Simpson's Rule with $n = 8$ to estimate how far Dragster $A$ is ahead of Dragster $B$ 8 sec after the start of the race.

| $t$ (sec) | 0 | 1 | 2 | 3 | 4 | 5 | 6 | 7 | 8 |
|---|---|---|---|---|---|---|---|---|---|
| $V_A$ (ft/sec) | 0 | 22 | 46 | 70 | 94 | 118 | 142 | 166 | 190 |
| $V_B$ (ft/sec) | 0 | 20 | 44 | 66 | 88 | 112 | 138 | 160 | 182 |

57. **Surface Area of the Jacqueline Kennedy Onassis Reservoir** The reservoir located in Central Park in New York City has the shape depicted in the figure below. The measurements shown were taken at 206-ft intervals. Use Simpson's Rule with $n = 10$ to estimate the surface area of the reservoir.

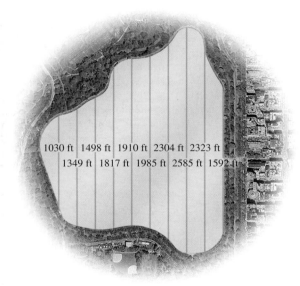

1030 ft  1498 ft  1910 ft  2304 ft  2323 ft
1349 ft  1817 ft  1985 ft  2585 ft  1592 ft

*Source: The Boston Globe*

58. **Estimating the Rate of Flow of a River** A stream is 120 ft wide. The following table gives the depths of the river measured across a section of the river in intervals of 6 ft. Here, $x$ denotes the distance from one bank of the river, and $y$ denotes the corresponding depth (in feet). The average rate of flow of the river across this section of the river is 4.2 ft/sec. Use Simpson's Rule to estimate the rate of flow of the river.

| $x$ (ft) | 0 | 6 | 12 | 18 | 24 | 30 | 36 | 42 | 48 | 54 | 60 |
|---|---|---|---|---|---|---|---|---|---|---|---|
| $y$ (ft) | 0.8 | 1.2 | 3.0 | 4.1 | 5.8 | 6.6 | 6.8 | 7.0 | 7.2 | 7.4 | 7.8 |

| $x$ (ft) | 66 | 72 | 78 | 84 | 90 | 96 | 102 | 108 | 114 | 120 |
|---|---|---|---|---|---|---|---|---|---|---|
| $y$ (ft) | 7.6 | 7.4 | 7.0 | 6.6 | 6.0 | 5.1 | 4.3 | 3.2 | 2.2 | 1.1 |

59. **Profit Functions** The weekly total marginal cost incurred by the Advance Visuals Systems Corporation in manufacturing $x$ 19-inch LCD HDTVs is

$$C'(x) = 0.000006x^2 - 0.04x + 120$$

dollars per set. The weekly marginal revenue realized by the company from the sale of $x$ sets is

$$R'(x) = -0.008x + 200$$

dollars per set.

 **a.** Plot the graphs of $C'$ and $R'$ using the viewing window $[0, 10{,}000] \times [0, 300]$.

**b.** Find the area of the region bounded by the graphs of $C'$ and $R'$ and the vertical lines $x = 2000$ and $x = 5000$. Interpret your result.

60. Find the area of the region bounded by the curve $y^2 = x^3 - x^2$ and the line $x = 2$.

61. Find the area of the region bounded by the graph of $f(x) = \sqrt{x}$, the $y$-axis, and the tangent line to the graph of $f$ at $(1, 1)$.

62. Find the number $a$ such that the area of the region bounded by the graph of $x = (y - 1)^2$ and the line $x = a$ is $\frac{9}{2}$.

63. Find the area of the region bounded by the $x$-axis and the graph of $f(x) = x^4 - 2x^3$ and to the right of the vertical line that passes through the point at which $f$ attains its absolute minimum.

64. The area of the region in the right half plane bounded by the $y$-axis, the parabola $y = -x^2 - 2x + 3$, and a line tangent to the parabola is $\frac{8}{3}$. Find the coordinates of the point of tangency.

 65. The region $S$ is bounded by the graphs of $y = \sqrt{x}$, the $x$-axis, and the line $x = 4$.
   **a.** Find $a$ such that the line $x = a$ divides $S$ into two subregions of equal area.
   **b.** Find $b$ such that the line $y = b$ divides $S$ into two subregions of equal area.

66. Find the value of $c$ such that the parabola $y = cx^2$ divides the region bounded by the parabola $y = \frac{1}{9}x^2$, and the lines $y = 2$, and $x = 0$ into two subregions of equal area.

67. Let $A(x)$ denote the area of the region in the first quadrant completely enclosed by the graphs of $f(x) = x^m$ and $g(x) = x^{1/m}$, where $m$ is a positive integer.
   **a.** Find an expression for $A(m)$.
   **b.** Evaluate $\lim_{m \to 1} A(m)$ and $\lim_{m \to \infty} A(m)$. Give a geometric interpretation.
    **c.** Verify your observations in part (b) by plotting the graphs of $f$ and $g$.

**68.** Let $f(x) = \dfrac{1}{x^2 + 1}$ and $g(x) = |x|$.

  **a.** Plot the graphs of $f$ and $g$ using the viewing window $[-1, 1] \times [0, 1.5]$. Find the points of intersection of the graphs of $f$ and $g$ accurate to three decimal places.

  **b.** Use a calculator or computer and the result of part (a) to find the area of the region bounded by the graphs of $f$ and $g$.

**69.** The curve with equation $y^2 - 4x^3 + 4x^4 = 0$ is called a **piriform.**

  **a.** Plot the curve using the viewing window $[-1, 1] \times [-1, 1]$.

  **b.** Find the area of the region enclosed by the curve accurate to five decimal places.

**70.** The curve with equation $4y^2 - 4xy^2 - x^2 - x^3 = 0$ is called a **right strophoid.**

  **a.** Plot the curve using the viewing window $[-1.5, 1.5] \times [-0.5, 0.5]$.

  **b.** Find the area of the region enclosed by the loop of the curve.

*In Exercises 71–74, determine whether the statement is true or false. If it is true, explain why it is true. If it is false, explain why or give an example to show why it is false.*

**71.** If $A$ denotes the area bounded by the graphs of $f$ and $g$ on $[a, b]$, then

$$A^2 = \int_a^b [f(x) - g(x)]^2 \, dx$$

**72.** If $f$ and $g$ are continuous on $[a, b]$ and $\int_a^b [f(t) - g(t)] \, dt > 0$, then $f(t) \geq g(t)$ for all $t$ in $[a, b]$.

**73.** Two cars start out traveling side by side along a straight road at $t = 0$. Twenty seconds later, Car $A$ is 30 ft behind Car $B$. If $v_1$ and $v_2$ are continuous velocity functions for Car $A$ and Car $B$, respectively, where $v_1(t)$ and $v_2(t)$ are measured in feet per second, then

$$\int_0^{20} v_2(t) \, dt = \int_0^{20} v_1(t) \, dt + 30$$

**74.** Suppose that the acceleration of Car $A$ and Car $B$ along a straight road are $a_1(t)$ ft/sec$^2$ and $a_2(t)$ ft/sec$^2$, respectively, over the time interval $[t_1, t_2]$, where $a_1$ and $a_2$ are continuous functions with $a_1(t) \geq a_2(t)$ on $[t_1, t_2]$. Then at time $t = t_2$, Car $A$ will be traveling $\int_{t_1}^{t_2} [a_1(t) - a_2(t)] \, dt$ ft/sec faster than Car $B$. (Assume that $t$ is measured in seconds.)

## 5.2   Volumes: Disks, Washers, and Cross Sections

In Section 5.1 we saw the role played by the definite integral in finding the area of plane regions. In the next two sections we will see how the definite integral can be used to help us find the volumes of solids such as those shown in Figure 1.

**FIGURE 1**    **(a)** Wine barrel      **(b)** Pyramid          **(c)** Pontoon

Figure 1c depicts a pontoon for a seaplane. In designing a pontoon, the engineer needs to know the volume of water displaced by the part of the pontoon that lies below the waterline in order to determine the buoyancy of the pontoon (Archimedes' Principle).

## ■ Solids of Revolution

A **solid of revolution** is a solid obtained by revolving a region in the plane about a line in the plane. The line is called the **axis of revolution.** For example, if the region $R$ under the graph of $f$ on the interval $[a, b]$ shown in Figure 2a is revolved about the $x$-axis, we obtain the solid of revolution $S$ shown in Figure 2b. Here, the axis of revolution of the solid is the $x$-axis.

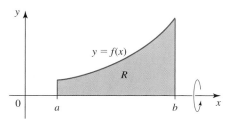 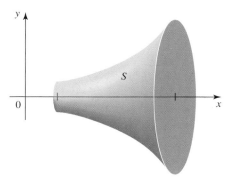

**FIGURE 2**      (**a**) Region $R$ under the graph of $f$              (**b**) Solid obtained by revolving $R$ about the $x$-axis

## ■ The Disk Method

To define the volume of a solid of revolution and to devise a method for computing it, let's consider the solid $S$ generated by the region $R$ shown in Figure 3a. Let $P = \{x_0, x_1, \ldots, x_n\}$ be a regular partition of $[a, b]$. This partition divides the region $R$ into $n$ nonoverlapping subregions $R_1, R_2, \ldots, R_n$. When these regions are revolved about the $x$-axis, they give rise to the $n$ nonoverlapping solids $S_1, S_2, \ldots, S_n$, whose union is $S$. (See Figure 3b.)

**FIGURE 3**
A partition of $[a, b]$ produces $n$ subregions $R_1, R_2, \ldots, R_n$ that are revolved about the $x$-axis to obtain the $n$ solids $S_1, S_2, \ldots, S_n$ that together form $S$. (Here $n = 8$.)

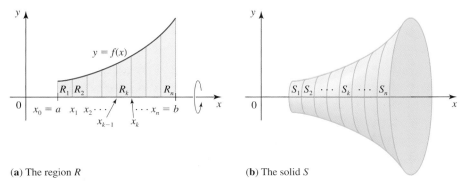

(**a**) The region $R$                              (**b**) The solid $S$

Let's concentrate on the part of the solid of revolution that is generated by the region $R_k$ under the graph of $f$ on the interval $[x_{k-1}, x_k]$. This region is shown enlarged for the sake of clarity in Figure 4. If $c_k$ is an evaluation point in $[x_{k-1}, x_k]$, then the area of $R_k$ is approximated by the rectangle of height $f(c_k)$ and width $\Delta x = (b - a)/n$. If this rectangle is revolved about the $x$-axis, it generates the disk $D_k$ having radius $f(c_k)$ and width $\Delta x$; therefore, its volume is

$$\Delta V_k = \pi[f(c_k)]^2 \Delta x \qquad \pi(\text{radius})^2 \cdot \text{width}$$

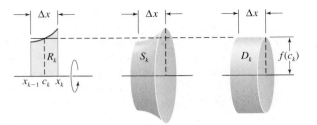

**FIGURE 4**
The region $R_k$, shown shaded, is approximated by the rectangle. The volume of $S_k$ is approximated by the volume of the disk $D_k$.

The $k$th region and the approximating rectangle

The $k$th solid of revolution

The $k$th disk

The volume of $D_k$ provides us with an approximation of the volume of $S_k$. Therefore, by approximating the volume of each solid $S_1, S_2, \dots, S_n$ with the volume of a corresponding disk $D_1, D_2, \dots, D_n$, we see that the volume $V$ of $S$ is approximated by the sum of the volumes of these disks. (See Figure 5.) Thus,

$$V \approx \sum_{k=1}^{n} \Delta V_k = \sum_{k=1}^{n} \pi [f(c_k)]^2 \, \Delta x$$

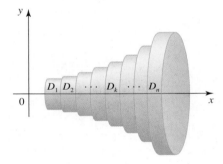

**FIGURE 5**
The volume $V$ of the solid of revolution $S$ is approximated by the sum of the volume of the $n$ disks $D_1, \dots, D_n$.

Recognizing this sum to be the Riemann sum of the function $\pi f^2$ on the interval $[a, b]$, we see that

$$\lim_{n \to \infty} \sum_{k=1}^{n} \pi [f(c_k)]^2 \, \Delta x = \int_{a}^{b} \pi [f(x)]^2 \, dx$$

---

**DEFINITION**   **Volume of a Solid of Revolution**
(Region revolved about the $x$-axis)

Let $f$ be a continuous nonnegative function on $[a, b]$, and let $R$ be the region under the graph of $f$ on the interval $[a, b]$. The volume of the solid of revolution generated by revolving $R$ about the $x$-axis is

$$V = \lim_{n \to \infty} \sum_{k=1}^{n} \pi [f(c_k)]^2 \, \Delta x = \int_{a}^{b} \pi [f(x)]^2 \, dx \qquad \textbf{(1)}$$

---

Just as we were able to recall the formulas for finding the area under a curve by looking at the area of a representative rectangle, so can we recall Formula (1) by looking at the volume of the disk obtained by revolving a representative rectangle about the $x$-axis.

We proceed as follows: Having made a sketch of the region $R$ under the graph of $y = f(x)$ on $[a, b]$, draw a representative vertical rectangle of height $f(x)$, or $y$, corresponding to a value of $x$ in $[a, b]$, and width $\Delta x$. (See Figure 6.) We can regard this disk with volume

$$\Delta V = \pi[f(x)]^2 \Delta x = \pi y^2 \Delta x \qquad \pi(\text{radius})^2 \cdot \text{width}$$

as representing an element of volume of a solid. Now observe that the expression next to $\Delta x$, $\pi y^2$, is *the integrand* in Formula (1).

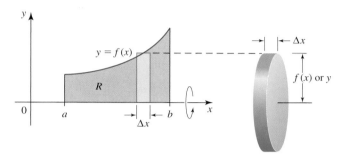

**FIGURE 6**
If a representative vertical rectangle is revolved about the $x$-axis, it generates a disk of radius $f(x)$, or $y$, and width $\Delta x$.

---

**Volume by Disk Method** (Region revolved about the $x$-axis)

$$V = \pi \int_a^b [f(x)]^2 \, dx = \pi \int_a^b y^2 \, dx \qquad f \geq 0$$

---

From now on, when we introduce a notion and/or derive a formula through the use of Riemann sums, we will often use the heuristic approach of looking at a representative element associated with the general term of the Riemann sum (without the subscripts) to help us recall the appropriate formula.

**EXAMPLE 1**   Find the volume of the solid obtained by revolving the region under the graph of $y = \sqrt{x}$ on $[0, 2]$ about the $x$-axis.

**Solution**   From the graph of $y = \sqrt{x}$ sketched in Figure 7a, we see that the radius of the representative disk corresponding to a particular value of $x$ in $[0, 2]$ (the height of the representative rectangle) is $y$, or $\sqrt{x}$. Therefore, the volume of the disk is

$$\Delta V = \pi y^2 \Delta x \qquad \text{Here } y = f(x) = \sqrt{x}.$$
$$= \pi(\sqrt{x})^2 \Delta x = \pi x(\Delta x)$$

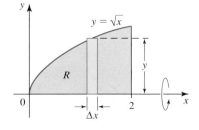

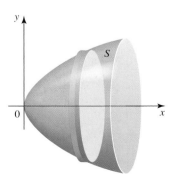

**FIGURE 7**
If $R$ is revolved about the $x$-axis, we obtain the solid of revolution $S$.

(**a**) The region $R$

(**b**) The solid $S$

Summing the volumes of the disks and taking the limit, we find that the volume of the solid is

$$V = \int_0^2 \pi x \, dx = \pi \int_0^2 x \, dx$$

$$= \frac{1}{2} \pi x^2 \bigg|_0^2 = \frac{1}{2} \pi (4 - 0) \quad \text{or} \quad 2\pi \qquad \blacksquare$$

**EXAMPLE 2**  By revolving the region under the graph of $y = \sqrt{r^2 - x^2}$ on $[-r, r]$, show that the volume of a sphere of radius $r$ is $V = \frac{4}{3} \pi r^3$.

**Solution**  The graph of $y = \sqrt{r^2 - x^2}$ is a semicircle, as shown in Figure 8a. We can see that the radius of a representative disk is $y$, the height of the vertical rectangle. Therefore, the volume of the disk is

$$\Delta V = \pi y^2 \, \Delta x$$

$$= \pi (r^2 - x^2) \Delta x \qquad \text{Since } y = \sqrt{r^2 - x^2}$$

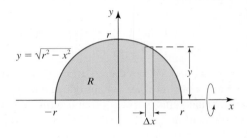

**FIGURE 8**
By revolving the region $R$ about the $x$-axis, we obtain the sphere of radius $r$.

(a) The region $R$          (b) The solid $S$

Summing the volumes of the disks and taking the limit, we obtain the required volume as

$$V = \int_{-r}^{r} \pi (r^2 - x^2) \, dx$$

$$= \pi \int_{-r}^{r} (r^2 - x^2) \, dx$$

$$= 2\pi \int_0^r (r^2 - x^2) \, dx \qquad \text{Use the symmetry of the region.}$$

$$= 2\pi \left[ r^2 x - \frac{1}{3} x^3 \right]_0^r$$

$$= 2\pi \left( r^3 - \frac{1}{3} r^3 \right)$$

$$= \frac{4}{3} \pi r^3 \qquad \blacksquare$$

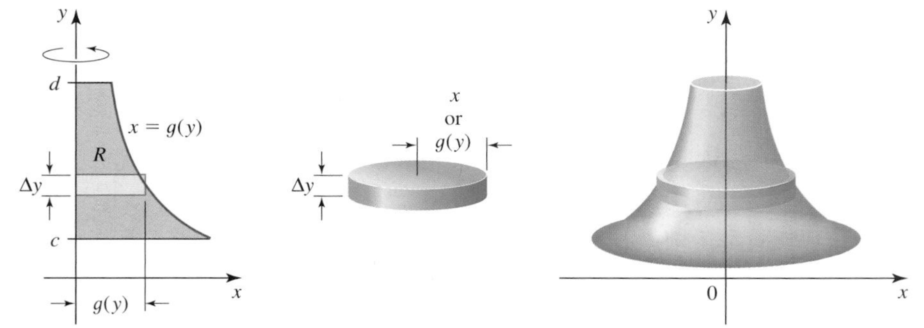

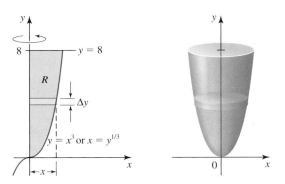

**FIGURE 9**
If a representative horizontal
rectangle is revolved about the
$y$-axis, it generates a disk of
radius $g(y)$, or $x$, and width $\Delta y$
and hence volume $\Delta V = \pi x^2 \Delta y$.

Formula (1) is used to find the volume of a solid of revolution when the axis of revolution is the $x$-axis. To derive a formula for the volume $V$ of a solid of revolution obtained by revolving a region about the $y$-axis, consider the region $R$ bounded by the graphs of $x = g(y)$, $x = 0$, $y = c$, and $y = d$ as shown in Figure 9.

If $R$ is revolved about the $y$-axis, then a representative horizontal rectangle (perpendicular to the axis of revolution) with length $x$, or $g(y)$, and width $\Delta y$ generates a disk with volume

$$\Delta V = \pi[g(y)]^2 \Delta y = \pi x^2 \Delta y$$

Summing the volumes of the disks and taking the limit, we obtain the following formula.

---

**Volume by Disk Method** (Region revolved about the $y$-axis)

$$V = \pi \int_c^d [g(y)]^2 \, dy = \pi \int_c^d x^2 \, dy \qquad g \geq 0$$

---

**EXAMPLE 3** Find the volume of the solid obtained by revolving the region bounded by the graphs of $y = x^3$, $y = 8$, and $x = 0$ about the $y$-axis.

**Solution** The region $R$ in question together with the solid generated by revolving that region about the $y$-axis is shown in Figure 10. A representative horizontal rectangle sweeps out a disk of radius $x$ and width $\Delta y$. Therefore, its volume is

$$\Delta V = \pi x^2 \Delta y$$
$$= \pi(y^{1/3})^2 \Delta y \qquad \text{Solve } y = x^3 \text{ for } x.$$
$$= \pi y^{2/3} \Delta y$$

**FIGURE 10**
If a horizontal rectangle is revolved
about the $y$-axis, it generates a disk of
radius $g(y) = y^{1/3}$, or $x$, and width $\Delta y$.

If we sum the volume of these disks and take the limit, we find that the required volume is

$$V = \pi \int_0^8 y^{2/3}\, dy$$

$$= \frac{3}{5}\pi y^{5/3}\Big|_0^8 = \frac{3}{5}\pi(8^{5/3}) \quad \text{or} \quad \frac{96\pi}{5} \qquad \blacksquare$$

## The Washer Method

Let $R$ be the region between the graphs of the functions $f$ and $g$ and between the vertical lines $x = a$ and $x = b$, where $f(x) \geq g(x) \geq 0$ on $[a, b]$. If $R$ is revolved about the $x$-axis, we obtain a solid of revolution with a hole in it. (See Figure 11.) Observe that when a representative vertical rectangle between the curves is revolved about the $x$-axis, the resultant element of volume of the solid has the shape of a washer with outer radius $f(x)$ and inner radius $g(x)$. Therefore, the volume of this element is

$$\Delta V = \pi[f(x)]^2\, \Delta x - \pi[g(x)]^2\, \Delta x$$

$$\underbrace{\qquad}_{\pi(\text{outer radius})^2 \,\cdot\, \text{width}} - \underbrace{\qquad}_{\pi(\text{inner radius})^2 \,\cdot\, \text{width}}$$

$$= \pi\{[f(x)]^2 - [g(x)]^2\}\Delta x$$

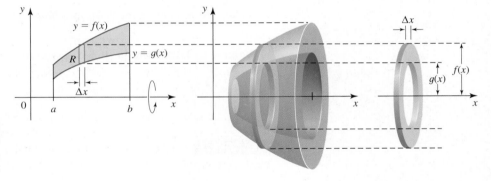

**FIGURE 11**
When a vertical rectangle is revolved about the $x$-axis, it generates a washer of outer radius $f(x)$, inner radius $g(x)$, and width $\Delta x$.

Summing the volumes of the washers and taking the limit, we see that the volume $V$ of the solid $S$ is given by the following.

> **Volume by Washer Method** (Region revolved about the $x$-axis)
>
> $$V = \pi \int_a^b \{[f(x)]^2 - [g(x)]^2\}\, dx \qquad f \geq g \geq 0$$

**EXAMPLE 4**    Find the volume of the solid obtained by revolving the region bounded by $y = \sqrt{x}$ and $y = x$ about the $x$-axis.

**Solution**    The region bounded by $y = \sqrt{x}$ and $y = x$ is shown in Figure 12. The curves $y = \sqrt{x}$ and $y = x$ intersect at $(0, 0)$ and $(1, 1)$, as may be verified by solving the equations simultaneously. The outer and inner radius of the washer generated by the representative vertical rectangle shown are $f(x) = \sqrt{x}$ and $g(x) = x$, respectively. Therefore, its volume is

$$\Delta V = \pi\{[f(x)]^2 - [g(x)]^2\}\Delta x$$

$$= \pi(x - x^2)\Delta x$$

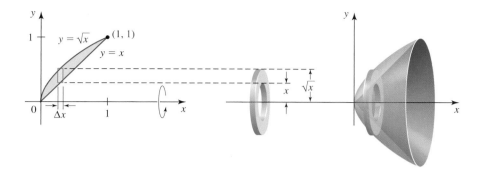

Summing the volumes of the washers and taking the limit, we find that the required volume is

$$V = \int_0^1 \pi(x - x^2)\, dx$$

$$= \pi \int_0^1 (x - x^2)\, dx$$

$$= \pi\left[\frac{1}{2}x^2 - \frac{1}{3}x^3\right]_0^1 = \pi\left(\frac{1}{2} - \frac{1}{3}\right) \quad \text{or} \quad \frac{\pi}{6}$$

**EXAMPLE 5** Find the volume of the solid generated by revolving the region of Example 4 about the line $y = 2$.

**Solution** The region and the resulting solid of revolution are shown in Figure 13. If a representative vertical rectangle is revolved about the line $y = 2$, the resultant solid is a washer with outer radius $2 - x$, inner radius $2 - \sqrt{x}$, and width $\Delta x$. Therefore, its volume is

$$\Delta V = \pi[(2 - x)^2 - (2 - \sqrt{x})^2]\Delta x$$
$$\underset{\pi[(\text{outer radius})^2 - (\text{inner radius})^2]\Delta x}{}$$

$$= \pi(x^2 - 5x + 4\sqrt{x})\Delta x$$

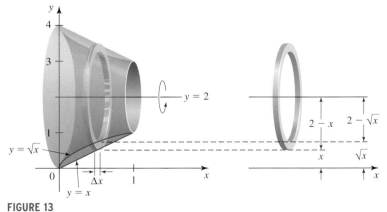

**FIGURE 13**
If a vertical rectangle is revolved about the line $y = 2$, it generates a washer with outer radius $2 - x$, inner radius $2 - \sqrt{x}$, and width $\Delta x$.

Summing the volumes of the washers and taking the limit, we find that the required volume is

$$V = \int_0^1 \pi(x^2 - 5x + 4\sqrt{x})\,dx$$

$$= \pi\left[\frac{1}{3}x^3 - \frac{5}{2}x^2 + \frac{8}{3}x^{3/2}\right]_0^1$$

$$= \pi\left(\frac{1}{3} - \frac{5}{2} + \frac{8}{3}\right) \quad \text{or} \quad \frac{\pi}{2}$$

$\blacksquare$

**EXAMPLE 6**   Find the volume of the solid generated by revolving the region of Example 4 about the $y$-axis.

**Solution**   The region together with the solid of revolution is shown in Figure 14. When a horizontal rectangle is revolved about the $y$-axis, the resultant solid is a washer with outer radius $y$, inner radius $y^2$, and width $\Delta y$. Therefore, the volume of the solid is

$$\Delta V = \pi(y^2 - y^4)\Delta y$$

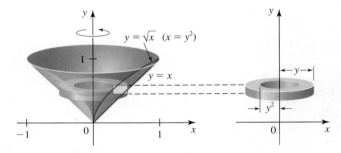

**FIGURE 14**

If a horizontal rectangle is revolved about the $y$-axis, it generates a washer with outer radius $y$, inner radius $y^2$, and width $\Delta y$.

Summing the volumes of the washers and taking the limit, we find that the volume of the solid is

$$V = \int_0^1 \pi(y^2 - y^4)\,dy = \pi\int_0^1 (y^2 - y^4)\,dy$$

$$= \pi\left[\frac{1}{3}y^3 - \frac{1}{5}y^5\right]_0^1 = \pi\left(\frac{1}{3} - \frac{1}{5}\right) \quad \text{or} \quad \frac{2\pi}{15}$$

$\blacksquare$

## ■ The Method of Cross Sections

We now turn to the more general problem of defining the volume of an irregularly shaped object. Consider, for example, the solid that is the part of a pontoon that lies below the waterline. The side view of one such pontoon is shown in Figure 15. A cross section of the pontoon (by a plane perpendicular to the $x$-axis) at the point $x$ is shown on the right.

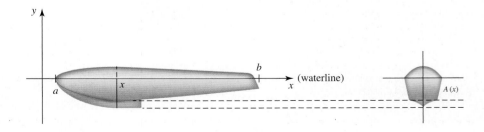

**FIGURE 15**

$A(x)$ is the area of a cross section of a pontoon at $x$.

To find the volume of the pontoon, let's take a regular partition $P = \{x_0, x_1, \ldots, x_n\}$ of the interval $[a, b]$. The planes that are perpendicular to the $x$-axis at the partition points will slice the pontoon into "slabs" much like the way one slices a loaf of bread. The volume $\Delta V$ of the $k$th slab between $x = x_{k-1}$ and $x = x_k$ is approximated by the volume of the cylinder with *constant* cross-sectional area $A(c_k)$ and height $\Delta x$, where $c_k$ lies in $[x_{k-1}, x_k]$. (See Figure 16.) Thus,

$$\Delta V \approx A(c_k)\Delta x$$

**FIGURE 16**
The volume of the $k$th "slab"
is approximately $A(c_k)\Delta x$.

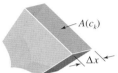

If we add up these $n$ terms, we obtain an approximation to the volume $V$ of the pontoon. We can expect the approximations to get better and better as $n \to \infty$. Recognizing this sum to be the Riemann sum of the function $A(x)$ on the interval $[a, b]$, we are led to the following definition.

---

**DEFINITION    Volume of a Solid with Known Cross Section**

Let $S$ be a solid bounded by planes that are perpendicular to the $x$-axis at $x = a$ and $x = b$. If the cross-sectional area of $S$ at any point $x$ in $[a, b]$ is $A(x)$, where $A$ is continuous on $[a, b]$, then the **volume** of $S$ is

$$V = \lim_{n \to \infty} \sum_{k=1}^{n} A(c_k)\Delta x = \int_a^b A(x)\, dx \qquad (2)$$

---

**EXAMPLE 7**    A solid has a circular base of radius 2. Parallel cross sections of the solid perpendicular to its base are equilateral triangles. What is the volume of the solid?

**Solution**    Suppose that the base of the solid is bounded by the circle with equation $x^2 + y^2 = 4$. The solid is shown in Figure 17a, where we have highlighted a typical cross section. To find the area of the cross section, observe that the base of the triangular cross section is $2y$, as shown in Figure 17b. Using the Pythagorean Theorem, we see that the height of the cross section is $\sqrt{3}y$ (see Figure 17c). Therefore, the area $A(x)$ of a typical cross section is

$$A(x) = \frac{1}{2}(2y)(\sqrt{3}y) = \sqrt{3}y^2 = \sqrt{3}(4 - x^2) \qquad y^2 = 4 - x^2$$

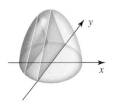
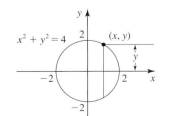
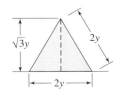

**FIGURE 17**        **(a)** The solid                    **(b)** The base of a cross section          **(c)** A cross section

Using Formula (2), we see that the volume of the solid is

$$V = \int_{-2}^{2} A(x)\, dx = \int_{-2}^{2} \sqrt{3}(4 - x^2)\, dx = 2\int_{0}^{2} \sqrt{3}(4 - x^2)\, dx$$

The integrand is even.

$$= 2\sqrt{3}\left[ 4x - \frac{1}{3}x^3 \right]_{0}^{2} = \frac{32\sqrt{3}}{3}$$

■

**EXAMPLE 8** Find the volume of a right pyramid with a square base of side $b$ and height $h$.

**Solution** Let's place the center of the base of the pyramid at the origin as shown in Figure 18a. A typical cross section of the pyramid perpendicular to the $y$-axis is a square of dimension $2x$ by $2x$. From Figure 18b we see by similar triangles that

$$\frac{x}{\dfrac{b}{2}} = \frac{h - y}{h}$$

or

$$x = \frac{b}{2h}(h - y)$$

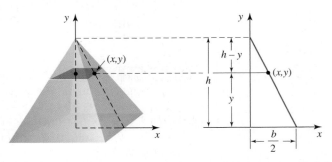

**FIGURE 18**    **(a)** A right pyramid    **(b)** A side view of the pyramid

Therefore, the area of the cross section is

$$A(y) = (2x)(2x) = 4x^2 = \frac{b^2}{h^2}(h - y)^2$$

The pyramid lies between $y = 0$ and $y = h$. Therefore, its volume is

$$V = \int_{0}^{h} A(y)\, dy = \int_{0}^{h} \frac{b^2}{h^2}(h - y)^2\, dy$$

$$= \left[ -\frac{b^2}{3h^2}(h - y)^3 \right]_{0}^{h} = 0 - \left( -\frac{b^2}{3h^2} \right)(h^3) = \frac{1}{3}b^2 h$$

■

**EXAMPLE 9** The external fuel tank for a space shuttle has a shape that may be obtained by revolving the region under the curve

$$f(x) = \begin{cases} 4\sqrt{10} & \text{if } -120 \le x \le 10 \\ \dfrac{1}{5}\sqrt{x}(30-x) & \text{if } 10 < x \le 30 \end{cases}$$

from $x = -120$ to $x = 30$ about the $x$-axis (Figure 19) where all measurements are given in feet. The tank carries liquid hydrogen for fueling the shuttle's three main engines. Estimate the capacity of the tank (231 cubic inches = 1 gallon).

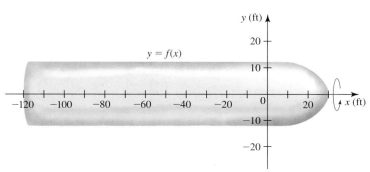

**FIGURE 19**
The solid of revolution obtained by revolving the region under the curve $y = f(x)$ about the $x$-axis

**Solution**  The volume of the tank is given by

$$V = \pi \int_{-120}^{30} [f(x)]^2 \, dx$$

$$= \pi \int_{-120}^{10} (4\sqrt{10})^2 \, dx + \pi \int_{10}^{30} \left[ \frac{1}{5}\sqrt{x}(30-x) \right]^2 \, dx$$

$$= 160\pi \int_{-120}^{10} dx + \frac{\pi}{25} \int_{10}^{30} x(30-x)^2 \, dx$$

$$= 160\pi x \Big|_{-120}^{10} + \frac{\pi}{25} \int_{10}^{30} (900x - 60x^2 + x^3) \, dx$$

$$= 160\pi(130) + \frac{\pi}{25} \left( 450x^2 - 20x^3 + \frac{1}{4}x^4 \right) \Big|_{10}^{30}$$

$$= 20{,}800\pi + \frac{\pi}{25} \left\{ \left[ 450(30)^2 - 20(30)^3 + \frac{1}{4}(30)^4 \right] \right.$$

$$\left. - \left[ 450(10)^2 - 20(10)^3 + \frac{1}{4}(10)^4 \right] \right\}$$

$$= 20{,}800\pi + 1600\pi = 22{,}400\pi$$

or approximately 70,372 cubic feet. Therefore, its capacity is approximately $(70{,}372)(12^3)/231$, or 526,419, gallons. ∎

## 5.2 CONCEPT QUESTIONS

1. Write the integral that gives the volume of a solid of revolution using (a) the disk method and (b) the washer method. Illustrate each case graphically by drawing the region $R$, indicating the axis of revolution, and drawing a representative rectangle that helps you to derive the formula.

2. Write the integral that gives the volume of a solid using the method of cross sections.

## 5.2 EXERCISES

*In Exercises 1–12, find the volume of the solid that is obtained by revolving the region about the indicated axis or line.*

1.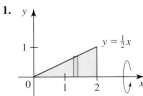

2.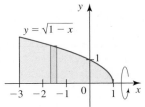

3.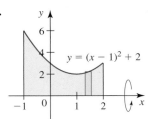

4.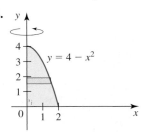

5.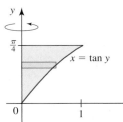

6.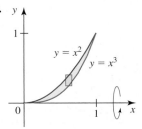

7.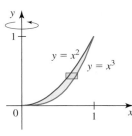

8.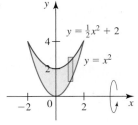

9.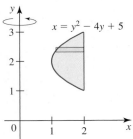

10.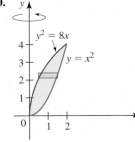

11.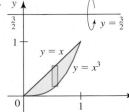

12.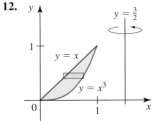

*In Exercises 13–34, find the volume of the solid generated by revolving the region bounded by the graphs of the equations and/or inequalities about the indicated axis.*

13. $y = x^2$, $y = 0$, $x = 2$; the $x$-axis

14. $y = x^3$, $y = 0$, $x = 1$; the $x$-axis

15. $y = -x^2 + 2x$, $y = 0$; the $x$-axis

16. $y = \sqrt{x - 1}$, $y = 0$, $x = 2$, $x = 5$; the $x$-axis

17. $y = e^x$, $y = 0$, $x = 0$, $x = 1$; the $x$-axis

18. $y = \dfrac{e^{x/2}}{(1 + e^x)^{3/2}}$, $y = 0$, $x = 0$, $x = 2$; the $x$-axis

19. $x = \dfrac{1}{y}$, $x = 0$, $y = 1$, $y = 2$; the $y$-axis

20. $x = y^{3/2}$, $x = 0$, $y = 1$; the $y$-axis

21. $x = \sqrt{4 - y^2}$, $x = 0$, $y = 0$; the $y$-axis

22. $x = -y^2 + 2y$, $x = 0$; the $y$-axis

**23.** $x^2 - y^2 = 4$,   $x \geq 0$,   $y = -2$,   $y = 2$;   the $y$-axis

**24.** $y = \dfrac{1}{\sqrt{x}(x^2 - 4)^{1/4}}$,   $y = 0$,   $x = 3$,   $x = 4$;   the $y$-axis

**25.** $x = y\sqrt{4 - y^2}$,   $x = 0$;   the $y$-axis

**26.** $y = \sqrt{\sin x}$,   $y = 0$,   $x \leq \dfrac{\pi}{2}$;   the $x$-axis

**27.** $y = \cos x$,   $x = 0$,   $y = 0$,   $x = \dfrac{\pi}{2}$;   the $x$-axis

**28.** $y = x^2$,   $y = x$;   the $x$-axis

**29.** $y = x^2$,   $y = \sqrt{x}$;   the $x$-axis

**30.** $y = x^2$,   $y = 2 - x^2$;   the $x$-axis

**31.** $y = \ln x$,   $y = 0$,   $y = 1$,   $x = 0$;   the $y$-axis

**32.** $y = \sin^{-1} x$,   $y = 0$,   $y = \dfrac{\pi}{2}$,   $x = 0$;   the $y$-axis

**33.** $x^2 + y^2 = 1$,   $y^2 = \dfrac{3}{2}x$,   $y \geq 0$;   the $x$-axis;

   (the smaller region)

**34.** $x^2 + y^2 = 1$,   $y^2 = \dfrac{3}{2}x$;   the $y$-axis;   (the smaller region)

 *In Exercises 35 and 36 use a graphing utility to (a) plot the graphs of the given functions and (b) find the approximate x-coordinates of the points of intersection of the graphs. Then find an approximation of the volume of the solid obtained by revolving the region bounded by the graphs of the functions about the x-axis.*

**35.** $y = \dfrac{1}{2}x^5$,   $y = 2x^2 - x^3$   **36.** $y = x^5$,   $y = \sin(x^2)$

*In Exercises 37–42, find the volume of the solid generated by revolving the region bounded by the graphs of the equations about the indicated line.*

**37.** $y = -x^2 + 2x$,   $y = 0$;   the line $y = 2$

**38.** $y = x$,   $y = x^2$;   the line $y = 2$

**39.** $y = 4 - x^2$,   $y = 0$;   the line $y = 5$

**40.** $y = x^2$,   $y = \dfrac{1}{2}x^2 + 2$;   the line $y = 5$

**41.** $x = y^2 - 4y + 5$,   $x = 2$;   the line $x = -1$

**42.** $y = x^2$,   $y^2 = 8x$;   the line $x = 2$

*In Exercises 43–46, sketch a plane region, and indicate the axis about which it is revolved so that the resulting solid of revolution has the volume given by the integral. (The answer is not unique.)*

**43.** $\pi \displaystyle\int_0^{\pi/2} \sin^2 x \, dx$     **44.** $\pi \displaystyle\int_0^1 y^{2/3} \, dy$

**45.** $\pi \displaystyle\int_0^1 (x^2 - x^4) \, dx$     **46.** $\pi \displaystyle\int_0^1 [(-1)^2 - (x^2 - 1)^2] \, dx$

**47.** Find the volume of the solid generated by revolving the region enclosed by the graph of $x^{1/2} + y^{1/2} = a^{1/2}$ and the coordinate axes about the $x$-axis.

**48.** **a.** Find the volume of the solid (a prolate spheroid) generated by revolving the upper half of the ellipse $9x^2 + 25y^2 = 225$ about the $x$-axis.
   **b.** Find the volume of the solid (an oblate spheroid) generated by revolving the right half of the ellipse $9x^2 + 25y^2 = 225$ about the $y$-axis.

**49.** Find the volume of the solid obtained by revolving the region enclosed by the curve $y^2 = \frac{1}{4}(2x^3 - x^4)$, where $y \geq 0$, about the $x$-axis.

**50.** Find the volume of the solid generated by revolving the region enclosed by the astroid $x^{2/3} + y^{2/3} = a^{2/3}$ about the $x$-axis.

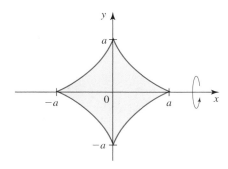

**51.** The function $f$ is defined by

$$f = \begin{cases} \sqrt{x} & \text{if } 0 \leq x \leq 1 \\ x^2 - 2x + 2 & \text{if } 1 < x \leq 2 \end{cases}$$

Find the volume of the solid generated by revolving the region under the graph of $f$ on $[0, 2]$ about the $x$-axis.

**52.** Verify the formula for the volume of a right circular cone by finding the volume of the solid obtained by revolving the triangular region with vertices $(0, 0)$, $(0, r)$, and $(h, 0)$ about the $x$-axis.

**53.** Find the volume of a frustum of a right circular cone with height $h$, lower base radius $R$, and upper radius $r$.

**54.** Verify the formula for the volume of a sphere of radius $r$ by finding the volume of the solid obtained by revolving the region bounded by the graph of $x^2 + y^2 = r^2$, $x \geq 0$, and the $y$-axis about the $y$-axis.

**55.** Find the volume of a cap of height $h$ formed from a sphere of radius $r$.

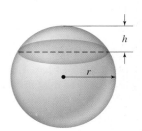

**56. Newton's Wine Barrel** Find the capacity of a wine barrel with the shape of a solid that is obtained by revolving the region bounded by the graphs of $x = R - ky^2$, $x = 0$, $y = -h/2$, and $y = h/2$ about the $y$-axis.

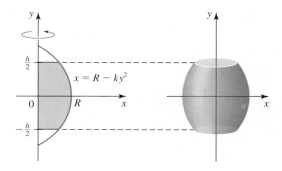

*In Exercises 57–60, find the volume of the solid with the given base R and the indicated shape of every cross section taken perpendicular to the x-axis.*

**57.** Cross section: a square

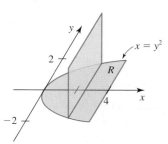

**58.** Cross section: a semicircle

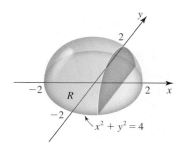

**59.** Cross section: an equilateral triangle

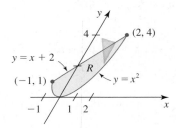

**60.** Cross section: a quarter circle

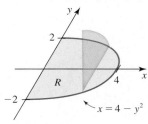

**61.** The curve defined by $y^4 = 1 - |x/2|^4$ is called a **hyperellipse.**
 **a.** Plot the curve using the viewing window $[-3, 3] \times [-2, 2]$.
   **b.** Estimate the volume $V$ of the solid obtained by revolving the region enclosed by the hyperellipse for $y \geq 0$ about the $x$-axis.
   **c.** Use a calculator or computer to find $V$ accurate to four decimal places.
   **Hint:** The hyperellipse is almost rectangular in shape.

**62.** A solid has a circular base of radius 2, and its parallel cross sections perpendicular to its base are rectangles of height 2. Find the volume of the solid.

**63.** The curve defined by $2y^2 - x^3 - x^2 = 0$ is called a **Tschirnhausen's cubic.**
 **a.** Plot the curve using the viewing window $[-1.5, 1.5] \times [-1.5, 1.5]$.
   **b.** Find the volume of the solid obtained by revolving the region enclosed by the loop of the curve about the $x$-axis.

**64.** Find the volume of the solid obtained by revolving the region bounded by the bullet-nosed curve $y = \dfrac{|x|}{\sqrt{2 - x^2}}$ about the $y$-axis for $0 \leq y \leq 12$.

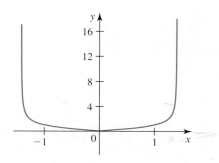

**65.** A solid has a circular base of radius 2, and its parallel cross sections perpendicular to its base are isosceles right triangles oriented so that the endpoints of the hypotenuse of a triangle lie on the circle. Find the volume of the solid.

**66.** The base of a solid is the region bounded by the graphs of $y = 4 - x^2$ and $y = 0$. The cross sections perpendicular to the $y$-axis are equilateral triangles. Find the volume of the solid.

**67.** The base of a wooden wedge is in the form of a semicircle with radius $a$, and its top is a plane that passes through the diameter of the base and makes a $45°$ angle with the plane of the base. Find the volume of the wedge.

**68.** The axes of two right cylinders, each of radius $r$, intersect at right angles. Find the volume of the resulting solid that is common to both cylinders. (The figure shows one eighth of the solid.)

**69. Cavalieri's Theorem** Cavalieri's Theorem states that if two solids have equal altitudes and all cross sections parallel to their bases and at equal distance from their bases have the same area, then the solids have the same volume.

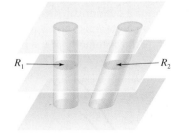

**(a)** area of $R_1$ = area of $R_2$        **(b)** An oblique circular cylinder

    **a.** Prove Cavalieri's Theorem.

    **b.** Use Cavalieri's Theorem to find the volume of the oblique circular cylinder shown in part (b) of the figure.

**70. Capacity of a Fuel Tank** The external fuel tank for a fighter aircraft is 8 m long. The areas of the cross sections in square meters measured from the front to the back of the tank at 1-m intervals are summarized in the following table.

| $x$ (distance from front) | 0 | 1 | 2 | 3 | 4 |
|---|---|---|---|---|---|
| $A(x)$ | 0 | 0.3041 | 0.6206 | 0.8937 | 0.8937 |

| $x$ (distance from front) | 5 | 6 | 7 | 8 |
|---|---|---|---|---|
| $A(x)$ | 0.8937 | 0.6206 | 0.3041 | 0 |

Use Simpson's Rule to estimate the capacity (in liters) of the fuel tank.

**71. The Volume of a Pontoon** A pontoon is 12 ft long. The areas of the cross sections in square feet measured from the blueprint at intervals of 2 ft from the front to the back of the part of the pontoon that is under the waterline are summarized in the following table.

| $x$ | 0 | 2 | 4 | 6 | 8 | 10 | 12 |
|---|---|---|---|---|---|---|---|
| $A(x)$ | 0 | 3.82 | 4.78 | 3.24 | 2.64 | 1.80 | 0 |

Use Simpson's Rule to estimate the volume of the pontoon.

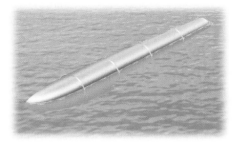

**72. a.** Let $S$ be a solid bounded by planes that are perpendicular to the $x$-axis at $x = 0$ and $x = h$. If the cross-sectional area of $S$ at any point $x$ in $[0, h]$ is $A(x)$, where $A$ is a polynomial of degree less than or equal to three, show that the volume of the solid is

$$V = \frac{h}{6}\left[A(0) + 4A\left(\frac{h}{2}\right) + A(h)\right]$$

    **b.** Use the result of part (a) to verify the result of Exercise 54.

## 5.3 Volumes Using Cylindrical Shells

In Section 5.2 we saw how the volume of a solid of revolution can be found by using the method of disks or the method of washers. Sometimes these methods are difficult or inconvenient to use. For example, suppose that we want to find the volume of the solid generated by revolving the region $R$ bounded by the graphs of the equations $y = -x^3 + 3x^2$, $y = 0$, $x = 0$, and $x = 3$ about the $y$-axis. (See Figure 1.)

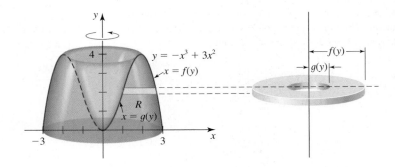

**FIGURE 1**
The washer generated by revolving the representative horizontal rectangle about the $y$-axis has outer radius $f(y)$ and inner radius $g(y)$.

As you can see from the figure, $f(y)$ is the outer radius and $g(y)$ is the inner radius of the washer generated by revolving a representative horizontal rectangle about the $y$-axis. Therefore, the volume of the solid is given by

$$\pi \int_0^b \{[f(y)]^2 - [g(y)]^2\}\, dy$$

where $b$ is the maximum value of $F(x) = -x^3 + 3x^2$ on $[0, 3]$. Using the techniques of Section 3.1, we can show that $b = 4$. So finding the interval of integration does not, at least in this case, pose much difficulty. But finding the functions $f$ and $g$ is an entirely different matter. Here, we need to solve the cubic equation $x^3 - 3x^2 + y = 0$ for $x$, a far more complicated task. Fortunately, there is another method that will allow us to find the volumes of such solids with relative ease. We will complete the solution of this problem in Example 1 after introducing the *method of cylindrical shells*.

### ■ The Method of Cylindrical Shells

As the name suggests, the method of cylindrical shells makes use of the volumes of cylindrical shells (or tubes) to approximate the volume of a solid of revolution. We begin with the derivation of an expression for the volume of a cylindrical shell.

Suppose a shell has outer radius $r_2$, inner radius $r_1$, and height $h$ as shown in Figure 2. The volume $V$ of the shell can be found by subtracting the volume $V_1$ of the inner cylinder from the volume $V_2$ of the outer cylinder. Thus,

$$V = V_2 - V_1$$
$$= \pi r_2^2 h - \pi r_1^2 h = \pi (r_2^2 - r_1^2) h$$
$$= \pi (r_2 + r_1)(r_2 - r_1) h$$
$$= 2\pi \left( \frac{r_2 + r_1}{2} \right) (r_2 - r_1) h$$

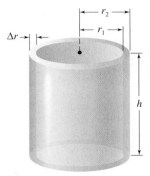

**FIGURE 2**
A cylindrical shell of outer radius $r_2$, inner radius $r_1$, and height $h$

This last equation can be written in the form

$$V = 2\pi r h \, \Delta r \tag{1}$$

where $r = (r_1 + r_2)/2$ is the average radius of the shell and $\Delta r = r_2 - r_1$ is the thickness of the shell. Formula (1) may also be written in the following form.

---

**Volume of a Cylindrical Shell**

$$V = 2\pi(\text{average radius})(\text{height})(\text{thickness})$$

---

Now let $R$ be the region under the graph of $f$ on the interval $[a, b]$, where $a \geq 0$, shown in Figure 3a. If this region $R$ is revolved about the $y$-axis, we obtain the solid shown in Figure 3b.

**FIGURE 3**

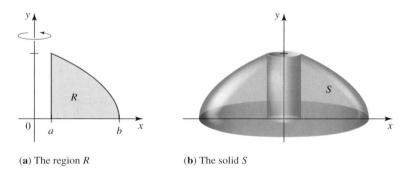

(a) The region $R$          (b) The solid $S$

Let $P = \{x_0, x_1, x_2, \ldots, x_n\}$ be a regular partition of the interval $[a, b]$, and let $c_k$ be the midpoint of the subinterval $[x_{k-1}, x_k]$; that is,

$$c_k = \frac{1}{2}(x_k + x_{k-1})$$

If the vertical rectangle with base $[x_{k-1}, x_k]$ and height $f(c_k)$ is revolved about the $y$-axis, we obtain a cylindrical shell with average radius $c_k$, height $f(c_k)$, and thickness $\Delta x = (b - a)/n$. (See Figure 4.) Therefore, by Formula (1) the volume of the shell is

$$\Delta V_k = 2\pi c_k f(c_k)\Delta x$$

**FIGURE 4**
If the vertical rectangle in (a)
is revolved about the $y$-axis, we
obtain the cylindrical shell (b). The
volume of $S$ is approximated by the
volumes of the nested shells (c).

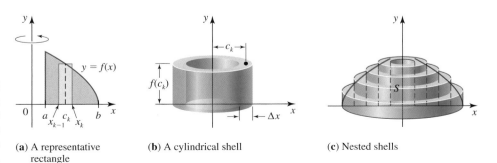

(a) A representative      (b) A cylindrical shell      (c) Nested shells
    rectangle

The volume $V$ of $S$ is approximated by the sum of the volumes of these shells (Figure 4c). Thus,

$$V \approx \sum_{k=1}^{n} \Delta V_k = \sum_{k=1}^{n} 2\pi c_k f(c_k) \Delta x$$

Recognizing this sum to be the Riemann sum of the function $2\pi x f(x)$ on the interval $[a, b]$, we see that

$$\lim_{n \to \infty} \sum_{k=1}^{n} 2\pi c_k f(c_k) \Delta x = \int_{a}^{b} 2\pi x f(x) \, dx = 2\pi \int_{a}^{b} x f(x) \, dx$$

This discussion leads to the following definition.

---

**Method of Cylindrical Shells** (Region revolved about the $y$-axis)

Let $f$ be a continuous nonnegative function on $[a, b]$, where $0 \le a \le b$, and let $R$ be the region under the graph of $f$ on the interval $[a, b]$. The volume $V$ of the solid of revolution generated by revolving $R$ about the $y$-axis is

$$V = \lim_{n \to \infty} \sum_{k=1}^{n} 2\pi c_k f(c_k) \Delta x = \int_{a}^{b} 2\pi x f(x) \, dx \tag{2}$$

---

As before, there is a convenient aid to help us recall this method. Draw a representative vertical rectangle of height $f(x)$, or $y$, and width $\Delta x$. Here we pick $x$ to be the midpoint of the base of the rectangle. Observe that this rectangle is parallel to the axis of revolution. When this rectangle is revolved about the $y$-axis, it generates a cylindrical shell of radius $x$, height $f(x)$, and thickness $\Delta x$. (See Figure 5.) Therefore, its volume is

$$\Delta V = 2\pi x f(x) \Delta x$$

Summing the volume of the shells and taking the limit, we obtain

$$V = \int_{a}^{b} 2\pi x f(x) \, dx$$

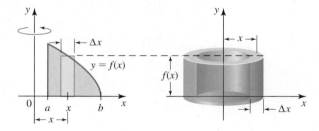

**FIGURE 5**
If a representative vertical rectangle is revolved about the $y$-axis, it generates a cylindrical shell of radius $x$, height $f(x)$, and thickness $\Delta x$ and hence volume $\Delta V = 2\pi x f(x) \Delta x$.

## ▪ Applying the Method of Cylindrical Shells

**EXAMPLE 1** The region under the graph of $y = -x^3 + 3x^2$ on $[0, 3]$ is revolved about the $y$-axis. Find the volume of the resulting solid.

**Solution** The region and the resulting solid of revolution are shown in Figure 6. If a representative vertical rectangle is revolved about the $y$-axis, the resulting cylindrical shell has an average radius $x$, height $-x^3 + 3x^2$, and thickness $\Delta x$. Therefore, its volume is

$$\Delta V = 2\pi x(-x^3 + 3x^2)\Delta x$$

$$= 2\pi(-x^4 + 3x^3)\Delta x$$

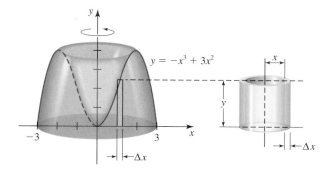

**FIGURE 6**
If a representative rectangle is revolved about the $y$-axis, it generates a cylindrical shell of volume $\Delta V = 2\pi xy\,\Delta x$.

Summing the volumes of the cylindrical shells and taking the limit, we find that the volume of the solid is

$$V = \int_0^3 2\pi(-x^4 + 3x^3)\,dx = 2\pi\int_0^3 (-x^4 + 3x^3)\,dx$$

$$= 2\pi\left[-\frac{1}{5}x^5 + \frac{3}{4}x^4\right]_0^3$$

$$= 2\pi\left(-\frac{243}{5} + \frac{243}{4}\right) \quad \text{or} \quad \frac{243\pi}{10}$$

Sometimes one method is preferable to another. In the next example the method of cylindrical shells is more convenient to use than the method of washers.

**EXAMPLE 2**   Let $R$ be the region bounded by the graphs of $y = x^2 + 1$, $y = -x + 1$, and $x = 1$. Find the volume of the solid that is obtained by revolving $R$ about the $y$-axis using (a) the method of washers and (b) the method of cylindrical shells.

**Solution**   The region $R$ is shown in Figure 7a.

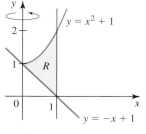

**FIGURE 7**
If each of the horizontal rectangles in part (b) is revolved about the $y$-axis, the resulting solid is a washer. If the vertical rectangle in part (c) is revolved about the $y$-axis, the resulting solid is a cylindrical shell.

(a) The region $R$          (b) The method of washers          (c) The method of shells

**a.** To use the method of washers, we regard the region $R$ as being made up of two subregions $R_1$ and $R_2$. (See Figure 7b.) Observe that if a representative horizontal rectangle lying in $R_1$ is revolved about the $y$-axis, we obtain a washer with outer radius $x = 1$ and inner radius $x = 1 - y$ (obtained by solving the equation $y = -x + 1$ for $x$). Therefore, its volume is

$$\Delta V_1 = \pi\{[f(y)]^2 - [g(y)]^2\}\Delta y$$

$$= \pi[1 - (1 - y)^2]\Delta y \qquad \text{Here, } f(y) = 1 \text{ and } g(y) = 1 - y.$$

$$= \pi(2y - y^2)\Delta y$$

Summing the volumes of the washers and taking the limit, we see that the volume of the solid obtained by revolving the subregion $R_1$ about the $y$-axis is

$$V_1 = \int_0^1 \pi(2y - y^2)\, dy$$

Similarly, we see that if a representative horizontal rectangle lying in $R_2$ is revolved about the $y$-axis, we obtain a washer with outer radius $x = 1$ and inner radius $x = \sqrt{y - 1}$ (obtained by solving the equation $y = x^2 + 1$ for $x$), where $x \geq 0$. Therefore, its volume is

$$\Delta V_2 = \pi\{[f(y)]^2 - [g(y)]^2\}\Delta y$$

$$= \pi[1 - (\sqrt{y - 1})^2]\Delta y \qquad \text{Here, } f(y) = 1 \text{ and } g(y) = \sqrt{y - 1}.$$

$$= \pi(2 - y)\Delta y$$

Summing the volumes of the washers and taking the limit, we see that the volume of the solid obtained by revolving the subregion $R_2$ about the $y$-axis is

$$V_2 = \int_1^2 \pi(2 - y)\, dy$$

Therefore, the required volume is

$$V_1 + V_2 = \int_0^1 \pi(2y - y^2)\, dy + \int_1^2 \pi(2 - y)\, dy$$

$$= \pi\left[y^2 - \frac{1}{3}y^3\right]_0^1 + \pi\left[2y - \frac{1}{2}y^2\right]_1^2$$

$$= \pi\left(1 - \frac{1}{3}\right) + \pi\left\{\left[2(2) - \frac{1}{2}(4)\right] - \left[2 - \frac{1}{2}\right]\right\} = \frac{7\pi}{6}$$

**b.** If a representative vertical rectangle is revolved about the $y$-axis, the resulting cylindrical shell has an average radius of $x$, height $[(x^2 + 1) - (-x + 1)]$, or $x^2 + x$, and thickness $\Delta x$ (Figure 7c). Therefore, its volume is

$$\Delta V = 2\pi x(x^2 + x)\Delta x$$

$$= 2\pi(x^3 + x^2)\Delta x$$

Summing the volumes of the cylindrical shells and taking the limit, we find that the volume of the solid is

$$V = \int_0^1 2\pi(x^3 + x^2)\, dx = 2\pi\int_0^1 (x^3 + x^2)\, dx$$

$$= 2\pi\left[\frac{1}{4}x^4 + \frac{1}{3}x^3\right]_0^1 = 2\pi\left(\frac{1}{4} + \frac{1}{3}\right) \quad \text{or} \quad \frac{7\pi}{6} \qquad \blacksquare$$

**Note**   Figure 7 reveals, once again, an intrinsic difference between the method of washers and the method of cylindrical shells. In the method of washers a representative rectangle is always *perpendicular* to the axis of revolution of the solid. In the method of cylindrical shells a representative rectangle is always *parallel* to the axis of revolution.

## ■ Shells Generated by Revolving a Region About the *x*-axis

The method of cylindrical shells can also be used to find the volume of a solid obtained by revolving a region about the *x*-axis. For example, suppose that the region $R$ bounded by the graphs of $x = f(y)$, $x = 0$, $y = c$, and $y = d$, where $f \geq 0$ and $c \leq d$, is revolved about the *x*-axis. (See Figure 8.) Then the volume of the resulting solid is given by the following formula.

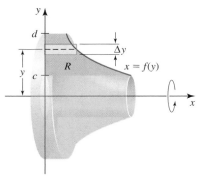

**FIGURE 8**
If a horizontal rectangle is revolved about the *x*-axis, it generates a cylindrical shell of volume $\Delta V = 2\pi y f(y) \Delta y$.

---

**Volume by Cylindrical Shells** (Region revolved about the *x*-axis)

$$V = \int_c^d 2\pi y f(y)\, dy \tag{3}$$

---

Equation (3) follows from Equation (2) if we interchange the roles of $x$ and $y$. It also follows from this observation: The solid generated by revolving the representative rectangle shown in Figure 8 is a cylindrical shell of average radius $y$, height $f(y)$, thickness $\Delta y$, and therefore volume $\Delta V = 2\pi y f(y) \Delta y$. Summing the volumes of the shells and taking the limit, we obtain

$$V = \int_c^d 2\pi y f(y)\, dy$$

**EXAMPLE 3**   Let $R$ be the region bounded by the graphs of $x = -y^2 + 6y$ and $x = 0$. Find the volume of the solid obtained by revolving $R$ about the *x*-axis.

**Solution**   The region $R$ is shown in Figure 9. If a representative horizontal rectangle is revolved about the *x*-axis, the resulting cylindrical shell has an average radius of $y$, a height of $x$, or $-y^2 + 6y$, and a thickness $\Delta y$. Therefore, its volume is

$$\Delta V = 2\pi y(-y^2 + 6y)\Delta y$$
$$= 2\pi(-y^3 + 6y^2)\Delta y$$

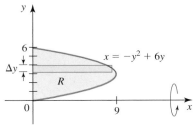

**FIGURE 9**
If a horizontal rectangle is revolved about the *x*-axis, it generates a cylindrical shell of volume $\Delta V = 2\pi y x \, \Delta y$.

Summing the volumes of the cylindrical shells and taking the limit, we find the volume of the solid to be

$$V = \int_0^6 2\pi(-y^3 + 6y^2)\, dy = 2\pi \int_0^6 (-y^3 + 6y^2)\, dy$$
$$= 2\pi\left[ -\frac{1}{4}y^4 + 2y^3 \right]_0^6 = 2\pi(-324 + 432) = 216\pi$$

**EXAMPLE 4** Let $R$ be the region bounded by the graphs of the equations $y = 4 - x^2$ and $y = -x + 2$. Find the volume of the solid obtained by revolving $R$ about the line $x = 4$.

**Solution** The region $R$ is shown in Figure 10. If a representative vertical rectangle is revolved about the line $x = 4$, it generates a cylindrical shell of average radius $4 - x$, height $(4 - x^2) - (-x + 2)$ or $-x^2 + x + 2$, and thickness $\Delta x$. Therefore, its volume is $\Delta V = 2\pi(4 - x)(-x^2 + x + 2)\Delta x$. Summing the volumes of the cylindrical shells and taking the limit, we find the volume of the solid to be

$$V = \int_{-1}^{2} 2\pi(4 - x)(-x^2 + x + 2)\, dx = 2\pi \int_{-1}^{2} (x^3 - 5x^2 + 2x + 8)\, dx$$

$$= 2\pi \left[ \frac{1}{4}x^4 - \frac{5}{3}x^3 + x^2 + 8x \right]_{-1}^{2}$$

$$= 2\pi \left[ \left( 4 - \frac{40}{3} + 4 + 16 \right) - \left( \frac{1}{4} + \frac{5}{3} + 1 - 8 \right) \right] = \frac{63\pi}{2}$$

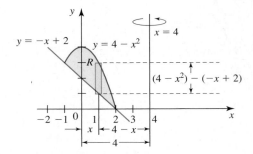

**FIGURE 10**
If a vertical rectangle is revolved about the line $x = 4$, it generates a cylindrical shell with average radius $4 - x$, height $(4 - x^2) - (-x + 2)$, and thickness $\Delta x$.

## 5.3 CONCEPT QUESTIONS

1. Let $S$ be the solid obtained by revolving the region shown in the figure about the $y$-axis.
   a. Sketch representative horizontal rectangles, and use them to help you set up the integrals giving the volume of $S$ using the disk and/or washer method.
   b. Sketch a representative vertical rectangle, and use it to help you set up an integral giving the volume of $S$ using the shell method.
   c. Find the volume of $S$. Which method is easier?

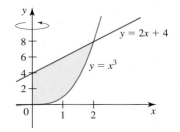

2. Let $S$ be the solid that is obtained by revolving the region shown in the figure about the $y$-axis.
   a. Is it desirable to use the disk method to find the volume of $S$? Explain.
   b. Use the shell method to find the volume of $S$.

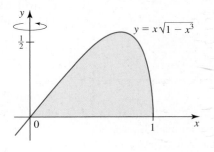

## 5.3   EXERCISES

*In Exercises 1–6, use the method of cylindrical shells to find the volume of the solid generated by revolving the region about the indicated axis or line.*

**1.**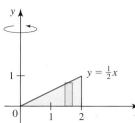

**2.**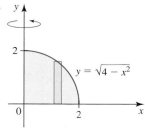

**3.**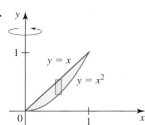

**4.**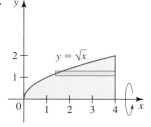

**5.**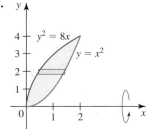

**6.**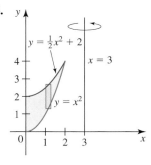

*In Exercises 7–22, use the method of cylindrical shells to find the volume of the solid generated by revolving the region bounded by the graphs of the equations and/or inequalities about the indicated axis. Sketch the region and a representative rectangle.*

**7.** $y = x^2$,   $y = 0$,   $x = 2$;   the $y$-axis

**8.** $y = x^3$,   $y = 0$,   $x = 1$;   the $y$-axis

**9.** $y = -x^2 + 2x$,   $y = 0$;   the $y$-axis

**10.** $y = \sqrt{x - 1}$,   $y = 0$,   $x = 5$;   the $y$-axis

**11.** $y = \dfrac{1}{x}$,   $y = 0$,   $x = 1$,   $x = 2$;   the $y$-axis

**12.** $y = e^{-x^2}$,   $y = 0$,   $x = 0$,   $x = 1$;   the $y$-axis

**13.** $y = \dfrac{1}{x^2 + 1}$,   $x = 0$,   $x = 2$;   the $y$-axis

**14.** $y = \dfrac{1}{x(x^2 + 4)}$,   $x = 1$,   $x = 2$;   the $y$-axis

**15.** $x = \sqrt{9 - y^2}$,   $x = 0$,   $y = 0$;   the $x$-axis

**16.** $y = 3^{x^2}$,   $y = 0$,   $x = 0$,   $x = 1$;   the $y$-axis

**17.** $y = x^2 + 1$,   $x \geq 0$,   $y = 5$;   the $y$-axis

**18.** $y = x$,   $y = \dfrac{1}{2}x^2$;   the $y$-axis

**19.** $y = \sqrt{1 - x^2}$,   $y = -x + 1$;   the $y$-axis

**20.** $y = \sin x^2$,   $y = 0$,   $x = 0$,   $x = \sqrt{\pi}$;   the $y$-axis

**21.** $y^2 = \dfrac{3}{2}x$;   $x^2 + y^2 = 1$,   $y \geq 0$;   the $x$-axis; (the smaller region)

**22.** $y = \sqrt{x - 1}$,   $y = x - 1$;   the $y$-axis

*In Exercises 23 and 24, use a graphing utility to (a) plot the graphs of the given functions, (b) find the approximate x-coordinates of the points of intersection of the graphs, and (c) find an approximation of the volume of the solid obtained by revolving the region bounded by the graphs of the functions about the y-axis.*

**23.** $y = x$,   $y = x^5 - x^2$;   $x \geq 0$

**24.** $y = \sin x$,   $y = x^2$

*In Exercises 25–30, use the method of disks or washers, or the method of cylindrical shells to find the volume of the solid generated by revolving the region bounded by the graphs of the equations about the indicated axis. Sketch the region and a representative rectangle.*

**25.** $y = \sqrt{x}$,   $y = x - 2$,   $y = 0$;   the $x$-axis

**26.** $y = (x - 1)^2$,   $y = x + 1$;   the $x$-axis

**27.** $y = x^2$,   $y = 2x - 1$,   $y = 4$;   the $y$-axis

**28.** $y = \sqrt{1 - x^2}$,   $y = -x + 1$;   the $x$-axis

**29.** $y = 2x^2$,   $y = x + 1$,   $y = 0$;   the $x$-axis

**30.** $y = \sqrt{9 - x^2}$,   $y = \dfrac{2}{3}\sqrt{9 - x^2}$,   $x \geq 0$;   the $y$-axis

*In Exercises 31–36, find the volume of the solid generated by revolving the region bounded by the graphs of the equations about the indicated line. Sketch the region and a representative rectangle.*

**31.** $y = x$, $y = 0$, $x = 2$; the line $x = 4$

**32.** $y = x^2 + 1$, $y = 0$, $x = 0$, $x = 2$; the line $x = 3$

**33.** $y = 4 - x^2$, $y = 0$; the line $x = -2$

**34.** $y = \sqrt{x}$, $y = 0$, $x = 4$; the line $y = 2$

**35.** $y = \sqrt{x - 1}$, $y = x - 1$; the line $x = 3$

**36.** $y = x$, $y = x^2$; the line $y = 2$

*In Exercises 37–40, sketch a plane region and indicate the axis about which it is revolved so that the resulting solid of revolution (found using the shell method) is given by the integral. (Answers may not be unique.)*

**37.** $2\pi \displaystyle\int_0^\pi x \sin x \, dx$

**38.** $2\pi \displaystyle\int_0^1 y^{4/3} \, dy$

**39.** $2\pi \displaystyle\int_0^1 y(y^{1/3} - y) \, dy$

**40.** $2\pi \displaystyle\int_0^1 (x + 1)x^2 \, dx$

**41.** Verify the formula for the volume of a right circular cone by applying the method of cylindrical shells to find the volume of the solid obtained by revolving the triangular region with vertices $(0, 0)$, $(h, 0)$, and $(0, r)$ about the $x$-axis.

**42.** Verify the formula for the volume of a sphere of radius $r$ by applying the method of cylindrical shells to find the volume of the solid obtained by revolving the semicircular region $x^2 + y^2 = r^2$, where $x \geq 0$, about the $y$-axis.

**43.** Use the method of cylindrical shells to find the volume of the ellipsoid obtained by revolving the elliptical region enclosed by the graph of

$$\frac{x^2}{a^2} + \frac{y^2}{b^2} = 1 \qquad x \geq 0$$

about the $y$-axis.

**44.** Find the volume of the solid that remains after a circular hole of radius $a$ is bored through the center of a solid sphere of radius $r > a$.

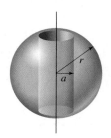

**45.** A torus (a doughnut-shaped object) is formed by revolving the circle $x^2 + y^2 = a^2$ about the vertical line $x = b$, where $0 < a < b$. Find its volume.

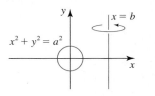

**46.** Find the volume of the solid obtained by revolving the region bounded by the graphs of $y = \sin x^2$ and $y = \cos x^2$ on $\left[0, \frac{\pi}{2}\right]$ about the $y$-axis.

**47. Volume of Liquid in a Rotating Container** A cylindrical container of radius 2 ft and height 4 ft is partially filled with a liquid. When the container is rotated about its axis of symmetry at a constant angular speed $\omega$, the surface of the liquid assumes a parabolic cross section. Suppose that the parabola is given by

$$y = 2 + \frac{\omega^2 x^2}{2g}$$

Find the volume of the liquid in the rotating container if the liquid reaches a height of 3 ft on the side of the container.

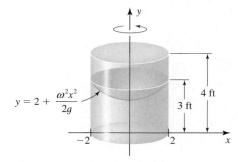

**48. The Clepsydra or Water Clock** A container having the shape of a solid of revolution obtained by revolving the graph of $y = kx^4$, $k > 0$, about the $y$-axis is made with a transparent material (see the following figure). A small hole is drilled in the bottom of the container to allow water to flow out.
  **a.** Find the volume $V(h)$ of water in the container as a function of $h$, the height of the water at time $t$.
  **b.** Use the Chain Rule and Torricelli's Law—which states that the rate of flow of water is $dV/dt = kA\sqrt{h}$, where $k$ is a negative constant, $A$ is the area of the hole at the bottom of the container, and $h$ is the height of the water—to show that the water level in the container drops at a *constant* rate.
  **c.** Explain why this property allows us to construct a water clock.

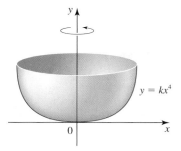

**49. Capacity of an Artificial Lake** A circular artificial lake has a diameter of 4000 ft. The following figure gives the depth of the water measured in 200-ft intervals starting from the center of the lake. Assume that readings taken along other radial directions produce similar data, so that the capacity of the lake can be approximated by the volume of the solid obtained by revolving the region bounded above by the $x$-axis and below by the graph of $y = f(x)$ about the $y$-axis. Use Simpson's Rule to approximate this capacity.

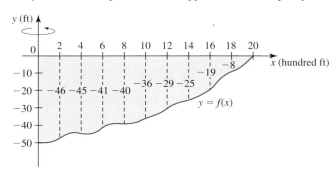

**50. Land Reclamation** A hill will be leveled, and the earth will be used in a land reclamation project that includes the construction of additional landing strips for an existing airport. The hill resembles the solid of revolution obtained by revolving the region under the graph of the function $f$ on [0, 240] about the $y$-axis. Use Simpson's Rule to approximate the amount of earth that can be recovered.

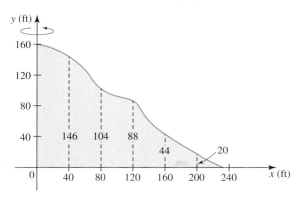

## 5.4  Arc Length and Areas of Surfaces of Revolution

Upon leaving port, an oil tanker sails along a course given by the curve $C$ shown in Figure 1, where the port is taken to be located at the origin of a coordinate system. What is the distance traveled by the tanker when it reaches a point on the course that is located 4 mi to the east and 2 mi to the north of the port?

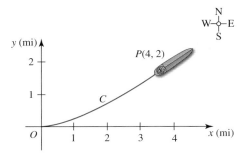

**FIGURE 1**
The curve $C$ gives the course taken by an oil tanker.

Intuitively, we see that this distance is given by the length of the curve $C$ between the points $O$ and $P$. So to answer this question, we must (a) define what we mean by the length of a curve and (b) devise a way of computing it. (We will solve this problem in Example 1.)

## Definition of Arc Length

Suppose that $C$ is the graph of a continuous function $f$ on a closed interval $[a, b]$. Let $P = \{x_0, x_1, \ldots, x_n\}$ be a regular partition of $[a, b]$. If $y_k = f(x_k)$, then the points $P_k(x_k, y_k)$ divide $C$ into $n$ arcs that we denote by $\overparen{P_0P_1}, \overparen{P_1P_2}, \ldots, \overparen{P_{n-1}P_n}$. (See Figure 2.)

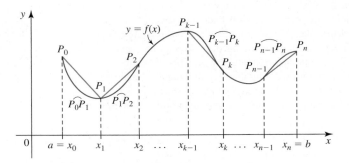

**FIGURE 2**
The graph of $f$ on $[a, b]$ is the union of the arcs $\overparen{P_0P_1}, \overparen{P_1P_2}, \ldots, \overparen{P_{n-1}P_n}$.

Since these arcs are disjoint (except for their endpoints), we see that the length $L$ of $C$ from $P_0$ to $P_n$ is just the sum of the lengths of these arcs. Now the length of the arc $\overparen{P_{k-1}P_k}$ can be approximated by the length $d(P_{k-1}P_k)$ of the line segment joining $P_{k-1}$ and $P_k$ (shown in red in Figure 2). Therefore, approximating the length of each arc with the length of the corresponding line segment, we see that

$$L \approx \sum_{k=1}^{n} d(P_{k-1}P_k)$$

This approximation improves as $n$ gets larger and larger. This observation suggests that we define the length of $C$ as follows.

---

**DEFINITION    Arc Length of a Curve**

Let $f$ be a continuous function defined on $[a, b]$, and let $P = \{x_0, x_1, \ldots, x_n\}$ be a regular partition of $[a, b]$. The **arc length of the graph** of $f$ from $P(a, f(a))$ to $Q(b, f(b))$ is

$$L = \lim_{n \to \infty} \sum_{k=1}^{n} d(P_{k-1}P_k) \tag{1}$$

if the limit exists.

---

**Note**    $L$ is also called the arc length of the graph of $f$ on the interval $[a, b]$.    ∎

## Length of a Smooth Curve

A function $f$ is **smooth** on an interval if its derivative $f'$ is continuous on that interval. The continuity of $f'$ implies that a small change in $x$ produces a small change in the slope $f'(x)$ of the tangent line to the graph of $f$ at any point $(x, f(x))$. Consequently, the graph of $f$ cannot have an abrupt change in direction. In other words, the graph of $f$ has no cusps or corners and is a **smooth curve**. (See Figure 3.)

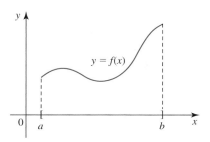
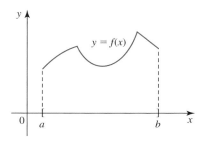

**FIGURE 3**    (a) The function $f$ is smooth.                    (b) The function $f$ is not smooth.

The length of the graph of a smooth function can be found by integration. To derive a formula for finding the length $L$ of such a graph, suppose $f$ is a smooth function defined on the closed interval $[a, b]$ and $P = \{x_0, x_1, \ldots, x_n\}$ is a regular partition of $[a, b]$. Then by Equation (1),

$$L = \lim_{n \to \infty} \sum_{k=1}^{n} d(P_{k-1} P_k)$$

Using the distance formula, we have

$$d(P_{k-1} P_k) = \sqrt{(x_k - x_{k-1})^2 + (y_k - y_{k-1})^2}$$
$$= \sqrt{(x_k - x_{k-1})^2 + [f(x_k) - f(x_{k-1})]^2} \qquad y_k = f(x_k)$$

Applying the Mean Value Theorem to $f$ on the interval $[x_{k-1}, x_k]$, we see that

$$f(x_k) - f(x_{k-1}) = f'(c_k)(x_k - x_{k-1})$$

where $c_k$ is a number in the interval $(x_{k-1}, x_k)$. Therefore,

$$d(P_{k-1} P_k) = \sqrt{(x_k - x_{k-1})^2 + [f'(c_k)(x_k - x_{k-1})]^2}$$
$$= \sqrt{\{1 + [f'(c_k)]^2\}(x_k - x_{k-1})^2}$$
$$= \sqrt{1 + [f'(c_k)]^2}\, \Delta x \qquad \Delta x = (b - a)/n$$

So

$$L = \lim_{n \to \infty} \sum_{k=1}^{n} d(P_{k-1} P_k)$$

$$= \lim_{n \to \infty} \sum_{k=1}^{n} \sqrt{1 + [f'(c_k)]^2}\, \Delta x$$

Recognizing this expression as the Riemann sum of the continuous function $g(x) = \sqrt{1 + [f'(x)]^2}$ leads to the following result.

---

**The Arc Length Formula**

Let $f$ be smooth on $[a, b]$. Then the arc length of the graph of $f$ from $P(a, f(a))$ to $Q(b, f(b))$ is

$$L = \int_a^b \sqrt{1 + [f'(x)]^2}\, dx \qquad (2)$$

---

**Note** If the equation defining the function $f$ is expressed in the form $y = f(x)$, then Equation (2) is sometimes written

$$L = \int_a^b \sqrt{1 + \left(\frac{dy}{dx}\right)^2}\, dx \qquad (3)$$

■

**EXAMPLE 1** **Distance Traveled by a Tanker** The graph $C$ of the equation $y = \frac{1}{4}x^{3/2}$ gives the course taken by an oil tanker after leaving port, which is taken to be located at the origin of a coordinate system. (See Figure 4.) Find the distance traveled by the tanker when it reaches a point on the course that is located 4 mi to the east and 2 mi to the north of the port.

**Solution** The required distance is given by the length $L$ of the curve $C$ from $x = 0$ to $x = 4$. To use Equation (3), we first find

$$\frac{dy}{dx} = \frac{d}{dx}\left(\frac{1}{4}x^{3/2}\right) = \frac{3}{8}x^{1/2}$$

and

$$1 + \left(\frac{dy}{dx}\right)^2 = 1 + \left(\frac{3}{8}x^{1/2}\right)^2 = 1 + \frac{9}{64}x$$

Then

$$L = \int_0^4 \sqrt{1 + \left(\frac{dy}{dx}\right)^2}\, dx = \int_0^4 \sqrt{1 + \frac{9}{64}x}\, dx$$

$$= \left(\frac{64}{9}\right)\left(\frac{2}{3}\right)\left(1 + \frac{9}{64}x\right)^{3/2}\Big|_0^4$$

$$= \frac{128}{27}\left[\left(1 + \frac{9}{16}\right)^{3/2} - 1\right] = \frac{128}{27}\left(\frac{125}{64} - 1\right) = \frac{122}{27} \approx 4.52$$

So the oil tanker will have traveled approximately 4.52 mi when it reaches the point in question. ■

**EXAMPLE 2** Find the length of the graph $f(x) = \frac{1}{3}x^3 + \frac{1}{4x}$ on the interval $[1, 3]$.

**Solution** The graph of $f$ is sketched in Figure 5.
We first find

$$f'(x) = \frac{d}{dx}\left[\frac{1}{3}x^3 + \frac{1}{4}x^{-1}\right] = x^2 - \frac{1}{4x^2}$$

Using Equation (2) with

$$1 + [f'(x)]^2 = 1 + \left(x^2 - \frac{1}{4x^2}\right)^2 = 1 + \left(\frac{4x^4 - 1}{4x^2}\right)^2$$

$$= 1 + \frac{16x^8 - 8x^4 + 1}{16x^4} = \frac{16x^8 + 8x^4 + 1}{16x^4}$$

$$= \frac{(4x^4 + 1)^2}{16x^4}$$

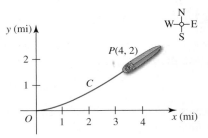

y (mi)

P(4, 2)

C

O    1    2    3    4    x (mi)

N
W—E
S

**FIGURE 4**
The course taken by the oil tanker

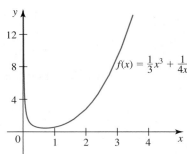

y

12

8

4

0    1    2    3    4    x

$f(x) = \frac{1}{3}x^3 + \frac{1}{4x}$

**FIGURE 5**
The graph of $f(x) = \frac{1}{3}x^3 + \frac{1}{4x}$

we see that the required length is

$$L = \int_1^3 \sqrt{1 + [f'(x)]^2}\, dx = \int_1^3 \sqrt{\frac{(4x^4 + 1)^2}{16x^4}}\, dx = \int_1^3 \frac{4x^4 + 1}{4x^2}\, dx$$

$$= \int_1^3 \left(x^2 + \frac{1}{4}x^{-2}\right) dx = \left[\frac{1}{3}x^3 - \frac{1}{4x}\right]_1^3 = \left(9 - \frac{1}{12}\right) - \left(\frac{1}{3} - \frac{1}{4}\right) = \frac{53}{6} \quad ∎$$

**EXAMPLE 3** Find the arc length of the graph of $f(x) = \ln(2\cos x)$ between the adjacent points of the intersection of the graph with the $x$-axis.

**Solution** To find the $x$-coordinates of the points of intersection of the graph of $f$ with the $x$-axis, we set $f(x) = 0$ or

$$\ln(2\cos x) = 0$$

Solving this equation we find $2\cos x = 1$, $\cos x = \frac{1}{2}$, or $x = \pm\frac{\pi}{3} \pm 2n\pi$ ($n = 0, 1, 2, \ldots$). The graph of $f$ is shown in Figure 6.

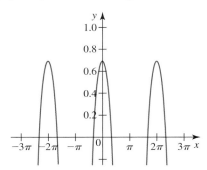

**FIGURE 6**
The graph of $f(x) = \ln(2\cos x)$

Making use of symmetry, we see that the required arc length is

$$L = 2\int_0^{\pi/3} \sqrt{1 + \left(\frac{dy}{dx}\right)^2}\, dx$$

But

$$\frac{dy}{dx} = \frac{d}{dx}[\ln(2\cos x)] = \frac{\frac{d}{dx}(2\cos x)}{2\cos x} = \frac{-2\sin x}{2\cos x} = -\tan x$$

and so

$$L = 2\int_0^{\pi/3} \sqrt{1 + (-\tan x)^2}\, dx = 2\int_0^{\pi/3} \sqrt{\sec^2 x}\, dx$$

$$= 2\int_0^{\pi/3} \sec x\, dx \qquad \text{Since } \sec x > 0 \text{ on } \left[0, \frac{\pi}{4}\right]$$

Using Formula (4c), Section 4.2, we have

$$L = 2\ln|\sec x + \tan x|\Big|_0^{\pi/3} = 2\ln(2 + \sqrt{3}) - 2\ln 1$$

$$= 2\ln(2 + \sqrt{3}) \quad ∎$$

By interchanging the roles of $x$ and $y$ in Equation (2), we obtain the following formula for finding the arc length of the graph of a smooth function defined by $x = g(y)$ on the interval $[c, d]$. (See Figure 7.)

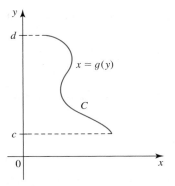

**FIGURE 7**
The curve $C$ is the graph of $x = g(y)$ for $c \le y \le d$.

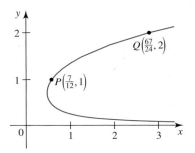

**FIGURE 8**
The graph of the function
$$x = g(y) = \frac{1}{3}y^3 + \frac{1}{4y}$$

---

**Arc Length: Integrating with Respect to $y$**

$$L = \int_c^d \sqrt{1 + [g'(y)]^2}\, dy = \int_c^d \sqrt{1 + \left(\frac{dx}{dy}\right)^2}\, dy \qquad (4)$$

**EXAMPLE 4** Find the length of the graph of $x = \dfrac{1}{3}y^3 + \dfrac{1}{4y}$ from $P\left(\frac{7}{12}, 1\right)$ to $Q\left(\frac{67}{24}, 2\right)$.

**Solution** The graph of $x = g(y)$ is shown in Figure 8. Here $x$ is a function of $y$, so we use Equation (4). First, we compute

$$1 + \left(\frac{dx}{dy}\right)^2 = 1 + \left(y^2 - \frac{1}{4y^2}\right)^2$$

$$= 1 + y^4 - \frac{1}{2} + \frac{1}{16y^4} = y^4 + \frac{1}{2} + \frac{1}{16y^4}$$

$$= \left(y^2 + \frac{1}{4y^2}\right)^2$$

Then observing that $y$ runs from $y = 1$ to $y = 2$ and using Equation (4), we find that the required length is

$$L = \int_1^2 \sqrt{\left(y^2 + \frac{1}{4y^2}\right)^2}\, dy = \int_1^2 \left(y^2 + \frac{1}{4y^2}\right) dy$$

$$= \left[\frac{1}{3}y^3 - \frac{1}{4y}\right]_1^2 = \left[\left(\frac{8}{3} - \frac{1}{8}\right) - \left(\frac{1}{3} - \frac{1}{4}\right)\right] = \frac{59}{24} \qquad \blacksquare$$

### ■ The Arc Length Function

Suppose that $C$ is the graph of a smooth function $f$ defined by $y = f(x)$ on the closed interval $[a, b]$. If $x$ is a point in $[a, b]$, we can use Equation (2) to express the length of the arc of the graph of $f$ from $P(a, f(a))$ to $Q(x, f(x))$. (See Figure 9.) Denoting this length by $s(x)$ (since it depends on $x$), we have

$$s(x) = \int_a^x \sqrt{1 + [f'(t)]^2}\, dt$$

This equation enables us to define the following function.

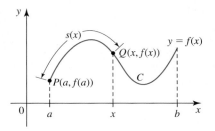

**FIGURE 9**
$s(x)$ is the length of the arc of the graph of $f$ from $P(a, f(a))$ to $Q(x, f(x))$.

---

**DEFINITION** **Arc Length Function**

Let $f$ be smooth on $[a, b]$. The **arc length function** $s$ for the graph of $f$ is defined by

$$s(x) = \int_a^x \sqrt{1 + [f'(t)]^2}\, dt \qquad (5)$$

with domain $[a, b]$.

If we use the Fundamental Theorem of Calculus, Part 1, to differentiate Equation (5), we obtain

$$s'(x) = \frac{d}{dx} \int_a^x \sqrt{1 + [f'(t)]^2}\, dt = \sqrt{1 + [f'(x)]^2} = \sqrt{1 + \left(\frac{dy}{dx}\right)^2} \tag{6}$$

The quantity $ds = s'(x)\, dx$ is the *differential of arc length*. In view of Equation (6) we can express $ds$ in the following forms.

---

**Arc Length Differentials**

$$ds = \sqrt{1 + \left(\frac{dy}{dx}\right)^2}\, dx \tag{7}$$

or, equivalently,

$$(ds)^2 = (dy)^2 + (dx)^2 \tag{8}$$

---

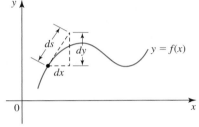

**FIGURE 10**
The relationship between $ds$, $dy$, and $dx$ follows from the Pythagorean Theorem.

Figure 10 gives a geometric interpretation of the differential of arc length in terms of the differentials $dx$ and $dy$. Observe that if $dx = \Delta x$ is small, then $ds$ affords a good approximation of the arc length of the graph of $f$ corresponding to the change, $\Delta x$, in $x$.

**EXAMPLE 5**  Use differentials to obtain an approximation of the arc length of the graph of $y = 2x^2 + x$ from $P(1, 3)$ to $Q(1.1, 3.52)$.

**Solution**  Using Equation (7), we find

$$ds = \sqrt{1 + \left(\frac{dy}{dx}\right)^2}\, dx = \sqrt{1 + (4x + 1)^2}\, dx$$

Letting $x = 1$ and $dx = 0.1$, we obtain the approximation

$$ds \approx \sqrt{1 + 5^2}(0.1) = 0.1\sqrt{26} \approx 0.51 \qquad \blacksquare$$

**Note**  The expression in Equation (8) provides us with an easy way of recalling the formula for the arc length $L$ of the graph of a function $f$ on $[a, b]$. From

$$(ds)^2 = (dy)^2 + (dx)^2$$

we see that

$$ds = \sqrt{1 + \left(\frac{dy}{dx}\right)^2}\, dx$$

and

$$s(x) = \int_a^x ds$$

Therefore,

$$L = s(b) = \int_a^b ds = \int_a^b \sqrt{1 + \left(\frac{dy}{dx}\right)^2}\, dx = \int_a^b \sqrt{1 + [f'(x)]^2}\, dx \qquad \blacksquare$$

## ■ Surfaces of Revolution

A **surface of revolution** is a surface that is obtained by revolving the graph of a continuous function about a line. For example, if the graph $C$ of the function $f$ on the interval $[a, b]$ shown in Figure 11a is revolved about the $x$-axis, we obtain the surface of revolution $S$ shown in Figure 11b.

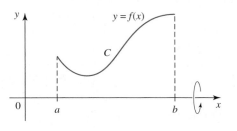

**FIGURE 11**
If we revolve the graph of $f$
about the $x$-axis in (a), we
obtain the surface $S$ in (b).　**(a)**　**(b)**

**FIGURE 12**
The frustum of a cone obtained
by cutting off its top using a
plane parallel to its base

Our immediate objective is to devise a formula for finding the surface area of $S$. To do this, we need the formula for the lateral surface area of a frustum of a right circular cone (Figure 12).

If the upper and lower radii of a frustum are $r_1$ and $r_2$, respectively, and its slant height is $l$, then the surface area $S$ of the frustum is

$$S = 2\pi r l \tag{9}$$

where $r = \frac{1}{2}(r_1 + r_2)$ is the average radius of the frustum. (You will be asked to establish this formula in Exercise 59.)

Next, consider the surface $S$ generated by revolving the graph $C$ of a smooth nonnegative function $f$ about the $x$-axis from $x = a$ to $x = b$. Let $P = \{x_0, x_1, \ldots, x_n\}$ be a regular partition of $[a, b]$. If $y_k = f(x_k)$, then the points $P_k(x_k, y_k)$ divide $C$ into $n$ disjoint (except at their endpoints) arcs $\overparen{P_0P_1}, \overparen{P_1P_2}, \ldots, \overparen{P_{n-1}P_n}$ whose union is $C$ (Figure 13a). The surface $S$ is the union of the surfaces $S_1, S_2, \ldots, S_n$ obtained by revolving these arcs about the $x$-axis (Figure 13b).

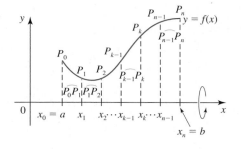

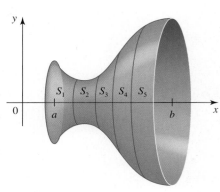

**FIGURE 13**
(a) A partition of $[a, b]$ produces $n$
arcs $\overparen{P_0P_1}, \overparen{P_1P_2}, \ldots, \overparen{P_{n-1}P_n}$,
which, when revolved about the
$x$-axis, give $n$ surfaces $S_1, S_2, \ldots, S_n$,
which together form $S$. (b) Here, $n = 5$.　**(a)**　**(b)**

Let's concentrate on the part of the surface generated by the arc of the graph of $f$ on the interval $[x_{k-1}, x_k]$. This arc is shown in Figure 14. If $\Delta x$ is small, then the arc $\overset{\frown}{P_{k-1}P_k}$ may be approximated by the line segment joining $P_{k-1}$ and $P_k$. This suggests that the surface area of the frustum that is generated by revolving this line segment about the $x$-axis will provide us with a good approximation of the surface area of $S_k$. (See Figure 14.)

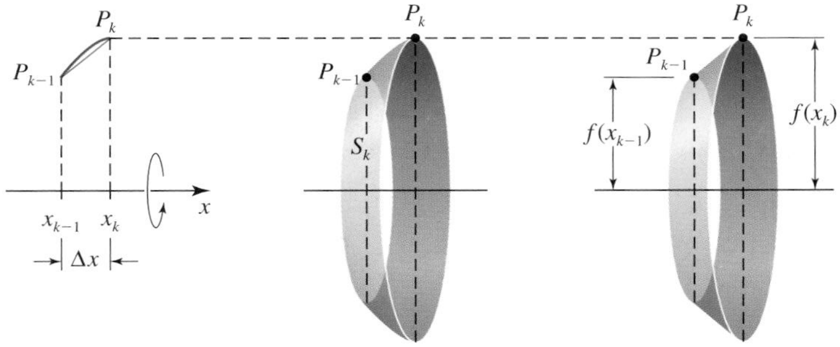

**FIGURE 14**
In part (a) the arc $\overset{\frown}{P_{k-1}P_k}$ is approximated by the line segment joining $P_{k-1}$ to $P_k$. So the area of the surface generated by $\overset{\frown}{P_{k-1}P_k}$ in (b) is approximated by the lateral surface area of the frustum generated by the line segment in (c).

(a)                    (b)                    (c)

Since the frustum has an average radius of $r = \frac{1}{2}[f(x_{k-1}) + f(x_k)]$ and a slant height of $l = d(P_{k-1}P_k)$, Formula (9) tells us that its surface area is

$$\Delta S = 2\pi \left[ \frac{f(x_{k-1}) + f(x_k)}{2} \right] d(P_{k-1}P_k)$$

But as in the computations leading to Equation (2), we have

$$d(P_{k-1}P_k) = \sqrt{1 + [f'(c_k)]^2}\, \Delta x$$

where $c_k$ is a number in the interval $(x_{k-1}, x_k)$. Also, if $\Delta x$ is small, the continuity of $f$ implies that $f(x_{k-1}) \approx f(c_k)$ and $f(x_k) \approx f(c_k)$. Therefore,

$$\Delta S = 2\pi \left[ \frac{f(c_k) + f(c_k)}{2} \right] \sqrt{1 + [f'(c_k)]^2}\, \Delta x$$

Approximating the area of each surface $S_k$ by the area of the corresponding frustum, we see that*

$$S \approx \sum_{k=1}^{n} 2\pi f(c_k) \sqrt{1 + [f'(c_k)]^2}\, \Delta x$$

This approximation can be expected to improve as $n$ gets larger and larger. Finally, recognizing this sum to be the Riemann sum of the function $g(x) = 2\pi f(x)\sqrt{1 + [f'(x)]^2}$ on the interval $[a, b]$, we see that

$$\lim_{n \to \infty} \sum_{k=1}^{n} 2\pi f(c_k) \sqrt{1 + [f'(c_k)]^2}\, \Delta x = \int_{a}^{b} 2\pi f(x) \sqrt{1 + [f'(x)]^2}\, dx$$

This discussion leads to the following definition.

---

*It is conventional to denote the area of a surface by $S$, and we will do so here even though we have used this very letter to denote the surface of revolution itself.

---

**DEFINITION** **Surface Area of a Surface of Revolution**

Let $f$ be a nonnegative smooth function on $[a, b]$. The **surface area** of the surface obtained by revolving the graph of $f$ about the $x$-axis is

$$S = 2\pi \int_a^b f(x)\sqrt{1 + [f'(x)]^2}\, dx \tag{10}$$

---

**Note** If we use Equation (7), then we can write Equation (10) in the form

$$S = 2\pi \int_a^b y\, ds \tag{11}$$

which is the *arc length differential form* of Equation (10). ■

This formula can be remembered as follows: If $\Delta x$ is small, the differential of arc length, $ds$, gives an approximation of the slant height of the frustum of a cone of average radius approximated by $y$ (or $f(x)$). So $2\pi y\, ds$ represents an element of area of the surface (Figure 15). By summing and taking the limit, we then obtain Equation (11).

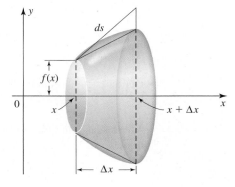

**FIGURE 15**
If $\Delta x$ is small, $ds$ approximates the slant height of the frustum, and $f(x)$ approximates the average height of the frustum.

**EXAMPLE 6** Find the area of the surface obtained by revolving the graph of $f(x) = \sqrt{x}$ on the interval $[0, 2]$ about the $x$-axis.

**Solution** The graph of $f$ and the resulting surface of revolution are shown in Figure 16. We have

$$f'(x) = \frac{1}{2\sqrt{x}}$$

Using Equation (10), we find that the required area is given by

$$S = 2\pi \int_0^2 f(x)\sqrt{1 + [f'(x)]^2}\, dx$$

$$= 2\pi \int_0^2 \sqrt{x}\sqrt{1 + \left(\frac{1}{2\sqrt{x}}\right)^2}\, dx = 2\pi \int_0^2 \sqrt{x}\sqrt{1 + \frac{1}{4x}}\, dx$$

$$= \pi \int_0^2 \sqrt{4x + 1}\, dx$$

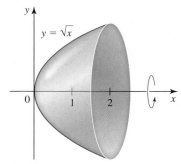

**FIGURE 16**
The graph of $y = \sqrt{x}$ on $[0, 2]$ and the resulting surface of revolution obtained by revolving the graph about the $x$-axis

We evaluate this integral using the method of substitution with $u = 4x + 1$, so that $du = 4\,dx$ or $dx = \frac{1}{4}\,du$. The lower and upper limits of integration with respect to $u$ are 1 and 9, respectively. We obtain

$$S = \frac{\pi}{4}\int_1^9 \sqrt{u}\,du = \frac{\pi}{4}\left[\frac{2}{3}u^{3/2}\right]_1^9 = \frac{\pi}{4}\left(18 - \frac{2}{3}\right) = \frac{13\pi}{3}$$    ∎

By interchanging the roles of $x$ and $y$ in Equation (10), we obtain the following formula for finding the area of the surface obtained by revolving the graph of a smooth function defined by $x = g(y)$ on the interval $[c, d]$ about the $y$-axis.

---

**Surface Area: Integrating with Respect to $y$**

$$S = 2\pi\int_c^d g(y)\sqrt{1 + [g'(y)]^2}\,dy = 2\pi\int_c^d x\,ds \qquad (12)$$

---

**EXAMPLE 7**   Find the area of the surface obtained by revolving the graph of $x = y^3$ on the interval $[0, 1]$ about the $y$-axis.

**Solution**   Here, $x = g(y) = y^3$ and so $g'(y) = 3y^2$. Therefore, Equation (12) gives the required surface area as

$$S = 2\pi\int_0^1 g(y)\sqrt{1 + [g'(y)]^2}\,dy$$

$$= 2\pi\int_0^1 y^3\sqrt{1 + (3y^2)^2}\,dy = 2\pi\int_0^1 y^3\sqrt{1 + 9y^4}\,dy$$

To evaluate the integral, we use the method of substitution with $u = 1 + 9y^4$ so that $du = 36y^3\,dy$. The lower and upper limits of integration are 1 and 10, respectively. We obtain

$$S = \frac{2\pi}{36}\int_1^{10}\sqrt{u}\,du = \frac{\pi}{18}\left[\frac{2}{3}u^{3/2}\right]_1^{10} = \frac{\pi}{18}\left(\frac{2}{3}10^{3/2} - \frac{2}{3}\right) = \frac{\pi}{27}(10\sqrt{10} - 1)$$

The surface is shown in Figure 17.    ∎

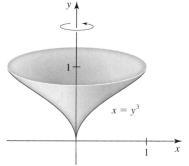

**FIGURE 17**
The graph of $x = y^3$ on $[0, 1]$ and the surface obtained by revolving it about the $y$-axis

---

## 5.4   CONCEPT QUESTIONS

**1. a.** Write an integral that gives the arc length of (1) a smooth function $y = f(x)$ on the interval $[a, b]$ and (2) a smooth function $x = g(y)$ on the interval $[c, d]$.

   **b.** Write two different integrals that give the arc length $L$ of the curve defined by the equation $y = x^{2/3} - 1$ from the point $P(0, -1)$ to the point $Q(5\sqrt{5}, 4)$. Which integral would you choose to compute $L$? Explain. Then use your choice of integral to compute $L$.

**2.** Write the formulas for finding the surface area of a surface of revolution obtained by (a) revolving the graph of a nonnegative smooth function $y = f(x)$ on the interval $[a, b]$ about the $x$-axis and (b) revolving the graph of a smooth function $x = g(y)$ on the interval $[c, d]$ about the $y$-axis.

## 5.4 EXERCISES

*In Exercises 1–4, find the arc length of the graph from A to B.*

**1.**
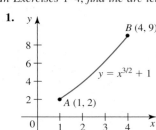

**2.**
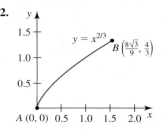

**3.**
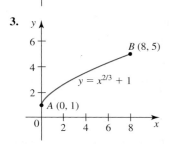

**4.**
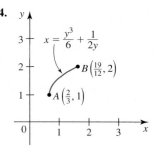

*In Exercises 5 and 6, find the length of the line segment joining the two given points by finding the equation of the line and using Equation (2). Then check your answer by using the distance formula.*

**5.** $(0, 0)$ and $(3, 8)$

**6.** $(-1, -2)$ and $(3, 6)$

*In Exercises 7–18, find the arc length of the graph of the given equation from P to Q or on the specified interval.*

**7.** $y = -2x + 3$;  $P(-1, 5), Q(2, -1)$

**8.** $y = \dfrac{2}{3}x^{3/2} - 1$;  $P\left(4, \frac{13}{3}\right), Q(9, 17)$

**9.** $y = 2(x - 1)^{3/2}$;  $P(1, 0), Q(5, 16)$

**10.** $x = \dfrac{1}{4}y^4 + \dfrac{1}{8y^2}$;  $P\left(\frac{3}{8}, 1\right), Q\left(\frac{129}{32}, 2\right)$

**11.** $y = \dfrac{2}{3}(x^2 + 1)^{3/2}$;  $[1, 4]$

**12.** $y = (2 - x^{2/3})^{3/2}$;  $[1, 2]$

**13.** $(y + 3)^2 = 4(x + 2)^3$;  $P(-2, -3), Q(2, 13)$

**14.** $y = \dfrac{x^3}{3} + \dfrac{1}{4x}$;  $[1, 3]$

**15.** $y = \frac{1}{2}(e^x + e^{-x})$;  $[0, \ln 2]$

**16.** $y = \dfrac{1}{2}\left[x\sqrt{x^2 - 1} - \ln\left(x + \sqrt{x^2 - 1}\right)\right]$;  $[1, 3]$

**17.** $y = \ln \cos x$;  $\left[0, \frac{\pi}{4}\right]$

**18.** $y = \sqrt{4 - x^2}$;  $[0, 2]$

*In Exercises 19–24, write an integral giving the arc length of the graph of the equation from P to Q or over the indicated interval. (Do not evaluate the integral.)*

**19.** $y = x^2$;  $P(-1, 1), Q(2, 4)$

**20.** $y = x^3 - 1$;  $[0, 1]$

**21.** $y = \dfrac{1}{x^2 + 1}$;  $P\left(-1, \frac{1}{2}\right), Q\left(2, \frac{1}{5}\right)$

**22.** $y = \cos x$;  $[0, \pi]$

**23.** $y = \tan x$;  $P(0, 0), Q\left(\frac{\pi}{4}, 1\right)$

**24.** $x = \sec y$;  $P\left(\sqrt{2}, -\frac{\pi}{4}\right), Q(1, 0)$

*In Exercises 25–28, (a) plot the graph of the function f, (b) write an integral giving the arc length of the graph of the function over the indicated interval, and (c) find the arc length of the curve accurate to four decimal places.*

**25.** $f(x) = 2x^3 - x^4$;  $[0, 2]$   **26.** $f(x) = \sqrt{x^2 - x^4}$;  $[0, 1]$

**27.** $f(x) = x - 2\sqrt{x}$;  $[0, 4]$   **28.** $f(x) = \dfrac{x^2}{1 + x^4}$;  $[0, 1]$

**29.** The graph of the equation $x^{2/3} + y^{2/3} = a^{2/3}$, where $a > 0$, shown in the following figure, is called an astroid. Find the arc length of the astroid.
**Hint:** By symmetry the arc length is equal to 8 times the length of the curve joining $P$ to $Q(a, 0)$. To find the coordinates of $P$, find the point of intersection of the astroid with the line $y = x$.

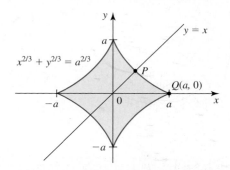

**30.** Use the fact that the circumference of a circle of radius 1 is $2\pi$ to evaluate the integral

$$\int_0^{\sqrt{2}/2} \frac{dx}{\sqrt{1 - x^2}}$$

**Hint:** Interpret $\int_0^{\sqrt{2}/2}\sqrt{1 + (y')^2}\, dx$, where $y = \sqrt{1 - x^2}$.

*In Exercises 31 and 32, use differentials to approximate the arc length of the graph of the equation from P to Q.*

**31.** $y = x^3 + 1$;  $P(1, 2), Q(1.2, 2.728)$

**32.** $y = \sqrt{x} + 1$;  $P(4, 3), Q(4.3, 3.074)$

V Videos for selected exercises are available online at **www.academic.cengage.com/login**.

*In Exercises 33–43, find the area of the surface obtained by revolving the given curve about the indicated axis.*

**33.** $y = \dfrac{1}{2}x + 2$ for $0 \le x \le 2$;   *x*-axis

**34.** $y = \sqrt{x}$ on $[4, 9]$;   *x*-axis

**35.** $y = x^3$ on $[0, 1]$;   *x*-axis

**36.** $y = x^{1/3}$ on $[1, 8]$;   *y*-axis

**37.** $y = 4 - x^2$ on $[0, 2]$;   *y*-axis

**38.** $x = \dfrac{1}{6}y^3 + \dfrac{1}{2y}$ for $1 \le y \le 2$;   *y*-axis

**39.** $2x + 3y = 6$ for $-2 \le y \le 1$;   *y*-axis

**40.** $y = \dfrac{1}{4}x^4 + \dfrac{1}{8x^2}$ on $[1, 2]$;   *x*-axis

**41.** $y = \dfrac{1}{2\sqrt{2}}\sqrt{x^2 - x^4}$ on $[0, 1]$;   *x*-axis

**42.** $x = \dfrac{1}{3}\sqrt{y(3 - y)^2}$ on $0 \le y \le 3$;   *y*-axis

**43.** $y = \dfrac{1}{2}(e^x + e^{-x})$ on $[0, \ln 2]$;   *x*-axis.

*In Exercises 44 and 45, write an integral giving the area of the surface obtained by revolving the curve about the x-axis. (Do not evaluate the integral.)*

**44.** $y = \dfrac{1}{x}$ on $[1, 2]$

**45.** $y = \sin x$ on $\left[0, \dfrac{\pi}{2}\right]$

 **46. a.** Plot the graph of $f(x) = \tan^{-1} x$ and the graph of the secant line passing through $(0, 0)$ and $\left(1, \dfrac{\pi}{4}\right)$.
    **b.** Use the Pythagorean Theorem to estimate the arc length of the graph of $f$ on the interval $[0, 1]$.
    **c.** Use a calculator or a computer to find the arc length of the graph of $f(x) = \tan^{-1} x$.

 **47.** Refer to Exercise 25. Use a calculator or computer to find the area of the surface formed by revolving the graph of $f(x) = 2x^3 - x^4$, where $0 \le x \le 2$, about the *x*-axis, accurate to four decimal places.

 **48.** Refer to Exercise 26. Use a calculator or computer to find the area of the surface formed by revolving the graph of $f(x) = \sqrt{x^2 - x^4}$, where $0 \le x \le 1$, about the *x*-axis, accurate to four decimal places.

**49.** Verify that the lateral surface area of a right circular cone of height $h$ and base radius $r$ is $S = \pi r\sqrt{r^2 + h^2}$ by evaluating a definite integral.
**Hint:** The cone is generated by revolving the region bounded by $y = (h/r)x$, $y = h$, and $x = 0$ about the *y*-axis.

**50.** Verify that the surface area of a sphere of radius $r$ is $S = 4\pi r^2$ by evaluating a definite integral.
**Hint:** Generate this sphere by revolving the semicircle $x^2 + y^2 = r^2$, where $y \ge 0$, about the *x*-axis.

**51.** Find the area of the surface obtained by revolving the graph of $y = \sqrt{4 - x^2}$ on $[0, 1]$ about the *x*-axis. This surface is called a **spherical zone.**

**52.** Find the area of the spherical zone formed by revolving the graph of $y = \sqrt{r^2 - x^2}$ on $[a, b]$, where $0 < a < b < r$, about the *x*-axis.

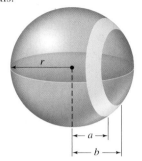

**53. A Pursuit Curve** The graph $C$ of the function

$$y = \dfrac{2}{3}\left(1 - \dfrac{x}{2}\right)^{3/2} - 2\left(1 - \dfrac{x}{2}\right)^{1/2} + \dfrac{4}{3}$$

gives the path taken by Boat $A$ as it pursues and eventually intercepts Boat $B$ ($x = 2$). Initially, Boat $A$ was at the origin, and Boat $B$ was at the point $(2, 0)$, heading due north. Find the distance traveled by Boat $A$ during the pursuit.

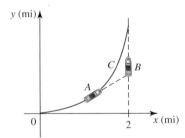

 **54. Motion of a Projectile** Refer to Exercise 29 in Section 2.4. A projectile is fired from a cannon located on a horizontal plane. If we think of the cannon as being located at the origin $O$ of an $xy$-coordinate system, then the path of the projectile is

$$y = \sqrt{3}x - \dfrac{x^2}{400}$$

where $x$ and $y$ are measured in feet. Estimate the distance traveled by the projectile in the air.

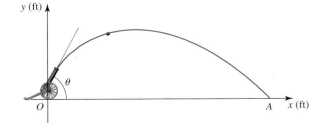

**55. Flight Path of an Airplane** The path of an airplane on its final approach to landing is described by the equation $y = f(x)$ with

$$f(x) = 4.3403 \times 10^{-10}x^3 - 1.5625 \times 10^{-5}x^2 + 3000$$
$$0 \le x \le 24{,}000$$

where $x$ and $y$ are both measured in feet. Estimate the distance traveled by the airplane during the landing approach.

**56. Area of a Roof** A hangar is 100 ft long and has a uniform cross section that is described by the equation $y = 10 - 0.0001x^4$, where both $x$ and $y$ are measured in feet. Estimate the area of the roof of the hangar.

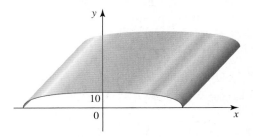

**57. Manufacturing Corrugated Sheets** A manufacturer of aluminum roofing products makes corrugated sheets as shown in the figure. The cross section of the corrugated sheets can be described by the equation

$$y = \sin\left(\frac{\pi x}{10}\right) \qquad 0 \le x \le 30$$

where $x$ and $y$ are measured in inches. If the corrugated sheets are made from flat sheets of aluminum using a stamping machine that does not stretch the metal, find the width $w$ of a flat aluminum sheet that is needed to make a 30-in. panel.

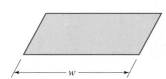
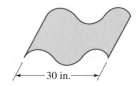

**58.** Let $f$ be a smooth nonnegative function on $[a, b]$. Show that the area of the surface obtained by revolving the graph of $f$ about the line $y = L$ is given by

$$S = 2\pi \int_a^b |f(x) - L|\sqrt{1 + [f'(x)]^2}\, dx$$

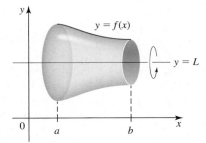

**59.** Show that the lateral surface area of a frustrum of a right circular cone of upper and lower radii $r_1$ and $r_2$, respectively, and slant height $l$ is $S = 2\pi r l$, where $r = \frac{1}{2}(r_1 + r_2)$.

**60.** Let $L$ denote the length of the graph of $y = f(x)$ connecting the points $(0, 0)$ and $(l, 0)$, and let $D = L - l$ (see the figure). Show that

$$\frac{1}{2}\int_0^l \frac{(y')^2}{\sqrt{1 + (y')^2}}\, dx \le D \le \frac{1}{2}\int_0^l (y')^2\, dx$$

assuming that $y'$ is continuous on $(0, l)$.

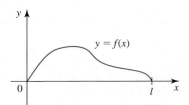

---

## 5.5    Work

The term *work,* as used in physics and engineering, is the transference of energy that results when the application of a force causes a body to move. Scientists and engineers need to know precisely how much energy is required to perform certain tasks. For example, a rocket scientist needs to know the amount of energy required to put an artificial satellite into an orbit around the earth, and a power engineer needs to know the amount of energy derived from water flowing through a dam.

### ■ Work Done by a Constant Force

We begin by defining the work done by a *constant* force in moving an object along a straight line.

> **DEFINITION** **Work Done by a Constant Force**
>
> The work $W$ done by a constant force $F$ in moving a body a distance $d$ in the direction of the force is
>
> $$W = Fd \qquad \text{work} = \text{force} \cdot \text{distance}$$

The unit of work in any system is the unit of force times the unit of distance. In the English system the unit of force is the pound (lb), the unit of distance is the foot (ft), and so the unit of work is the foot-pound (ft-lb). In the International System of Units, abbreviated SI (for Système international d'unités), the unit of force is the newton (N), the unit of distance is the meter (m), and so the unit of work is the newton-meter (N-m). A newton-meter is also called a *joule* (J).

**EXAMPLE 1**

**a.** Find the work done in lifting a 25-lb object 4 ft off the ground.
**b.** Find the work done in lifting a 2.4-kg package 0.8 m off the ground. (Take $g = 9.8$ m/sec$^2$.)

**Solution**
**a.** The force $F$ required to do the job is 25 lb (the weight of the object). Therefore, the work done by the force is

$$W = Fd = 25(4) = 100$$

or 100 ft-lb.
**b.** The magnitude of the force required is $F = mg = (2.4)(9.8) = 23.52$, or 23.52 N. So the work done is

$$W = Fd = (23.52)(0.8) \approx 18.8$$

or 18.8 J.

## Work Done by a Variable Force

Suppose that a body moves along the $x$-axis in the positive direction from $x = a$ to $x = b$ under the action of a force $F(x)$ that depends on $x$. Suppose also that the function $F$ is continuous on the interval $[a, b]$ with the graph depicted in Figure 1. Next, let $P = \{x_0, x_1, \ldots, x_n\}$ be a regular partition of $[a, b]$.

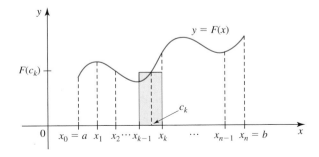

**FIGURE 1**
The graph of a variable force defined by the function $F$

Let's concentrate on the subinterval $[x_{k-1}, x_k]$. If $\Delta x = (b - a)/n$ is small, then the continuity of $F$ guarantees that the values of $F(x)$ at any two points in $[x_{k-1}, x_k]$

do not differ by much. Therefore, if $c_k$ is any point in $[x_{k-1}, x_k]$, we can approximate $F(x)$ by $F(c_k)$ for all $x$ in $[x_{k-1}, x_k]$. Physically, we are saying that the force $F(x)$ is approximately constant when measured over a small distance. So if we assume that $F(x) = F(c_k)$ in $[x_{k-1}, x_k]$, then the work done by $F$ in moving the body along the $x$-axis from $x = x_{k-1}$ to $x = x_k$ is

$$\Delta W_k \approx F(c_k)\Delta x \qquad \text{constant force · distance}$$

(This is the area of a rectangle of height $F(c_k)$ and width $\Delta x$.) It follows that the work $W$ done by $F$ in moving the body from $x = a$ to $x = b$ is

$$W \approx \sum_{k=1}^{n} F(c_k)\Delta x$$

Intuitively, we see that the approximation improves as $n$ gets larger and larger. This suggests that we define the work done by $F$ by taking the limit of the sum

$$\sum_{k=1}^{n} F(c_k)\Delta x$$

as $n \to \infty$. But this sum is just the Riemann sum of $F$ on the interval $[a, b]$. Therefore, our discussion leads to the following definition.

---

**DEFINITION** **Work Done by a Variable Force**

Suppose that a force $F$, where $F$ is continuous on $[a, b]$, acts on a body moving it along the $x$-axis. Then the **work** done by the force in moving the body from $x = a$ to $x = b$ is

$$W = \lim_{n\to\infty} \sum_{k=1}^{n} F(c_k)\Delta x = \int_a^b F(x)\, dx \qquad (1)$$

---

**Note** When we derived Equation (1), we assumed that $b > a$. This condition is not necessary and may be dropped. ■

**EXAMPLE 2** Find the work done by the force $F(x) = 3x^2 + x$ (measured in pounds) in moving a particle along the $x$-axis from $x = 2$ to $x = 4$ (measured in feet).

**Solution** Here, $F(x) = 3x^2 + x$, so the work done by $F$ in moving the body from $x = 2$ to $x = 4$ is

$$W = \int_2^4 F(x)\, dx = \int_2^4 (3x^2 + x)\, dx = \left[ x^3 + \frac{1}{2}x^2 \right]_2^4 = 72 - 10 = 62$$

or 62 ft-lb. ■

## ▪ Hooke's Law

As another application of Equation (1), let's find the work done in stretching or compressing a spring. Recall that **Hooke's Law** states that the force $F$ required to stretch or compress a spring $x$ units past its natural length is proportional to $x$. That is,

$$F(x) = kx$$

where $k$, the constant of proportionality, is called the **spring constant,** or the **stiffness.** Hooke's Law is valid provided that $|x|$ is not too large.

**EXAMPLE 3**  A force of 30 N is required to stretch a spring 4 cm beyond its natural length of 18 cm. Find the work required to stretch the spring from a length of 20 cm to a length of 24 cm.

**Solution**  Suppose that the spring is placed on the $x$-axis with the free end at the origin as shown in Figure 2. According to Hooke's Law, the force $F(x)$ required to stretch the spring $x$ meters beyond its natural length is $F(x) = kx$. Since a 30-N force is required to stretch the spring 4 cm, or 0.04 m, beyond its natural length, we see that

$$30 = k(0.04) \qquad \text{or} \qquad k = 750$$

that is, 750 N/m. Therefore, $F(x) = 750x$ for this spring. Using Equation (1), we find that the work required to stretch the spring from 20 cm to 24 cm is

$$W = \int_{0.02}^{0.06} 750x \, dx = 750 \left[ \frac{1}{2} x^2 \right]_{0.02}^{0.06}$$

$$= 375[(0.06)^2 - (0.02)^2] = 1.2$$

or 1.2 J.  ■

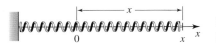

(a) The unstretched spring

(b) The spring stretched $x$ units beyond its natural length

**FIGURE 2**

## ■ Moving Nonrigid Matter

The next two examples involve the computation of the work involved in moving nonrigid matter, such as the evacuation of fluid from a container and the hoisting of an object.

**EXAMPLE 4**  A tank has the shape of an inverted right circular cone with a base of radius 5 ft and a height of 12 ft. If the tank is filled with water to a height of 8 ft, find the work required to empty the tank by pumping the water over the top of the tank. (Water weighs 62.4 lb/ft$^3$.)

**Solution**  We think of the tank as being placed on a coordinate system with its vertex at the origin and its axis along the $y$-axis as shown in Figure 3a. Think of the water as being subdivided into slabs by planes perpendicular to the $y$-axis from $y = 0$ to $y = 8$. The volume $\Delta V$ of a representative slab is approximated by a disk of radius $x$ and width $\Delta y$, that is,

$$\Delta V \approx \pi x^2 \, \Delta y$$

You can see how to express $x$ in terms of $y$ by referring to Figure 3b. By similar triangles,

$$\frac{x}{5} = \frac{y}{12} \qquad \text{or} \qquad x = \frac{5}{12} y$$

**FIGURE 3**
We wish to find the amount of work required to pump all of the water out of the top of the conical tank.

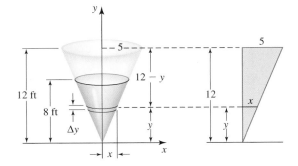

(a)

(b)

so

$$\Delta V \approx \pi \left( \frac{5}{12} y \right)^2 \Delta y = \frac{25\pi}{144} y^2 \, \Delta y$$

Since water weighs 62.4 lb/ft$^3$, or $62\frac{2}{5}$ lb/ft$^3$, the weight of a representative slab (the force required to lift this slab) is

$$\Delta F \approx 62 \frac{2}{5} \cdot \frac{25\pi}{144} y^2 \, \Delta y = \frac{65\pi}{6} y^2 \, \Delta y$$

Since this slab is transported a distance of approximately $(12 - y)$ ft, the work done by the force is

$$\Delta W \approx \Delta F(12 - y)$$

$$= \frac{65\pi}{6} y^2 (12 - y) \, \Delta y$$

Finally, summing the work done in lifting each slab to the top of the tank and taking the limit, we see that the work required to empty the tank is

$$W = \int_0^8 \frac{65\pi}{6} y^2 (12 - y) \, dy$$

$$= \frac{65\pi}{6} \int_0^8 (12y^2 - y^3) \, dy$$

$$= \frac{65\pi}{6} \left[ 4y^3 - \frac{1}{4} y^4 \right]_0^8 = \frac{65\pi}{6} (1024)$$

or approximately 34,851 ft-lb. ∎

**EXAMPLE 5** A ship's anchor, weighing 800 lb, is attached to a chain that weighs 10 lb per running foot. Find the work done by the winch if the anchor is pulled in from a height of 20 ft. (See Figure 4.)

**Solution** The work done by the winch is $W = W_A + W_C$, where $W_A$ is the work required to hoist the anchor to the top of the ship and $W_C$ is the work required to pull the cable to the top of the ship. To find $W_A$, observe that the force required to lift the anchor is 800 lb and that it will be applied over a distance of 20 ft. Therefore,

$$W_A = (800)(20) = 16,000$$

or 16,000 ft-lb.

**FIGURE 4**
The anchor is hoisted to the top of the ship by means of a cable chain.

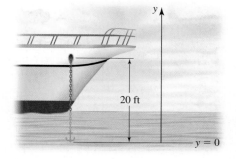

(a)

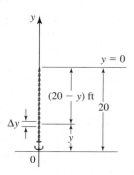

(b)

To find $W_C$, think of the chain as being subdivided into pieces. The length of a representative piece is $\Delta y$ ft, and its weight is $10\,\Delta y$ lb (weight per running foot times length). This element is to be lifted a distance of approximately $(20 - y)$ ft, so the work required is

$$\Delta W_C \approx 10\,\Delta y(20 - y)$$

Summing the work done in lifting each piece of the chain to the top and taking the limit, we see that the work required is

$$W_C = \int_0^{20} 10(20 - y)\, dy$$

$$= 10\left[20y - \frac{1}{2}y^2\right]_0^{20} = 2000$$

or 2000 ft-lb. So the work required to pull in the anchor from a height of 20 ft is

$$W = W_A + W_C = 16{,}000 + 2{,}000 = 18{,}000$$

or 18,000 ft-lb.                                                                                          ■

**Note**   A 2-horsepower winch with a capacity of 1100 ft-lb/sec can pull in this anchor in approximately 16 sec.                                                                               ■

## ■ Work Done by an Expanding Gas

**EXAMPLE 6**   Figure 5 shows the cross section of a cylindrical casing of internal radius $r$. When the confined gas expands, the resulting increase in pressure exerts a force against the piston, moving it and thus causing work to be done. If the confined gas has a pressure of $p$ lb/in.$^2$ and the gas expands from a volume of $V_0$ in.$^3$ to $V_1$ in.$^3$, show that the work done by the expanding gas is

$$W = \int_{V_0}^{V_1} p\, dV$$

**FIGURE 5**
Work is done by the expanding gas against the piston.

**Solution**   Draw the $x$-axis parallel to the side of the casing as shown in Figure 6, and suppose that the piston has initial and final positions $x = a$ and $x = b$, respectively.

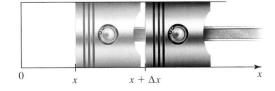

**FIGURE 6**
The work done by the expanding gas in moving the piston from $x$ to $x + \Delta x$ is $\Delta W$, which is approximately $p(x)(\pi r^2)\Delta x$.

The force exerted by the expanding gas against the piston head at $x$ $(a < x < b)$ is

$$F(x) = p(x)(\pi r^2) \qquad \text{pressure} \cdot \text{area}$$

so the work done by the force in moving the piston a distance of $\Delta x$ from $x$ to $x + \Delta x$ is

$$\Delta W \approx p(x)(\pi r^2)\Delta x \qquad \text{constant force} \cdot \text{distance}$$

Summing the work done by the force in moving the piston over each of the subintervals in the interval $[a, b]$ and taking the limit, we see that the work done is

$$W = \int_a^b p(x)\pi r^2 \, dx$$

To express this integral in terms of the volume of the gas, observe that the volume $V$ of the gas is related to $x$ by $V = \pi r^2 x$, so $dV = \pi r^2 \, dx$. Furthermore, observe that when $x = a$, $V = V_0$, and when $x = b$, $V = V_1$. Therefore,

$$W = \int_{V_0}^{V_1} p \, dV \qquad \blacksquare$$

## 5.5 CONCEPT QUESTIONS

1. **a.** A force of 3 lb moves an object along a coordinate line from $x = 0$ to $x = 10$ ($x$ is measured in feet). What is the work done by the force on the object?
   **b.** A force of magnitude 3 lb acts on an object in the negative direction with respect to a coordinate line as the object moves from $x = 0$ to $x = 10$ ($x$ is measured in feet). What is the work done by the force on the object?
   **c.** As an object moves in the coordinate plane from the point $A(0, 0)$ to the point $B(10, 0)$ along the $x$-axis, a

force of magnitude 5 lb acts on the body in the positive $y$-direction. What is the work done by the force on the object? Explain.

2. **a.** Can the work done on a body by a force be negative? Explain with an example.
   **b.** A force acts on an object situated on a coordinate line. If the work done by the force on the object is 0 ft-lb, does this mean that the force has magnitude 0 and/or the distance moved by the object is 0 ft? Explain.

## 5.5 EXERCISES

1. Find the work done in lifting a 50-lb sack of potatoes to a height of 4 ft above the ground.

2. How much work is done in lifting a 4-kg bag of rice to a height of 1.5 m above the ground?

3. A particle moves a distance of 100 ft along a straight line. As it moves, it is acted upon by a constant force of magnitude 5 lb in a direction opposite to that of the motion. What is the work done by the force?

4. An engine crane is used to raise a 400-lb engine a vertical distance of 2 ft so that it can be placed in an engine dolly. Find the work done by the crane.

5. Find the work done by the force $F(x) = 2x - 1$ (measured in pounds) in moving an object along the $x$-axis from $x = -2$ to $x = 4$ ($x$ is measured in feet).

6. Find the work done by the force $f(x) = 4/x^2$ (measured in pounds) in moving a particle along the $x$-axis from $x = 1$ to $x = 6$ ($x$ is measured in feet).

7. When a particle is at the point $x$ on the $x$-axis, it is acted upon by a force of $x^2 + 2x$ newtons. Find the work done by the force in moving the particle from the origin to the point $x = 3$ ($x$ is measured in meters).

8. A particle moves along the $x$-axis from $x = 1$ to $x = 3$. As it moves, it is acted upon by a force $F(x) = -3x^2 + x$. If $x$ is measured in meters and $F(x)$ is measured in newtons, find the work done by the force.

9. When a particle is at the point $x$ on the $x$-axis, it is acted upon by a force of $\sin \pi x$ newtons. Find the work done by the force in moving the particle from $x = 1$ to $x = 2$ ($x$ is measured in meters).

**10.** A force of 8 lb is required to stretch a spring 2 in. beyond its natural length. Find the work required to stretch the spring 3 in. beyond its natural length.

**11.** A force of 20 N is required to stretch a spring 3 cm beyond its natural length of 24 cm. Find the work required to stretch the spring from 30 to 35 cm.

**12.** Suppose that it takes 3 J of work to stretch a spring 5 cm beyond its natural length. How much work is required to stretch the spring from 2 cm beyond its natural length to 4 cm beyond its natural length?

**13.** A spring has a natural length of 8 in. If it takes a force of 14 lb to compress the spring to a length of 6 in., how much work is required to compress the spring from its natural length to 7 in.?

**14.** A chain with length 5 m and mass 30 kg is lying on the ground. Find the work done in pulling one end of the chain vertically upward to a height of 2 m.

**15.** A chain weighing 5 lb/ft hangs vertically from a winch located 12 ft above the ground, and the free end of the chain is just touching the ground. Find the work done by the winch in pulling in the whole chain.

**16.** A chain weighing 5 lb/ft hangs vertically from a winch located 16 ft above the ground, and the free end of the chain is 3 ft from the ground. Find the work done by the winch in pulling in 4 ft of the chain.

**17.** A steel girder weighing 200 lb is hoisted from ground level to the roof of a 60-ft building using a chain that weighs 2 lb/running foot. Find the work done.

**18.** An aquarium has the shape of a rectangular tank of length 4 ft, width 2 ft, and height 3 ft. If the tank is filled with water weighing 62.4 lb/ft$^3$, find the work required to empty the tank by pumping the water over the top of the tank.

**19.** A tank having the shape of a right-circular cylinder with a radius of 4 ft and a height of 6 ft is filled with water weighing 62.4 lb/ft$^3$. Find the work required to empty the tank by pumping the water over the top of the tank.

**20. Leaking Bucket** A bucket weighing 4 lb when empty and attached to a rope of negligible weight is used to draw water from a well that is 30 ft deep. Initially, the bucket contains 40 lb of water, but as it is pulled up at a constant rate of 2 ft/sec, the water leaks out of the bucket at the rate of 0.2 lb/sec. Find the work done in pulling the bucket to the top of the well.

**21. Leaking Bucket** A bucket weighing 4 lb when empty and attached to a rope of negligible weight is used to draw water from a well that is 40 ft deep. Initially, the bucket contains 40 lb of water and is pulled up at a constant rate of 2 ft/sec.

Halfway up, the bucket springs a leak and begins to lose water at the rate of 0.2 lb/sec. Find the work done in pulling the bucket to the top of the well.

**22.** A tank having the shape of a right circular cylinder with a radius of 5 ft and a height of 6 ft is filled with water weighing 62.4 lb/ft$^3$. Find the work required to empty the tank by pumping the water out of the tank through a pipe that extends to a height of 2 ft beyond the top of the tank.

**23.** A tank has the shape of an inverted right circular cone with a base radius of 2 m and a height of 5 m. If the tank is filled with water to a height of 3 m, find the work required to empty the tank by pumping the water over the top of the tank. (The mass of water is 1000 kg/m$^3$.)

**24.** Consider the tank described in Exercise 23. If water is pumped in through the bottom of the tank, find the work required to fill the empty tank to a depth of 2 m.

**25. Emptying a Storage Tank** A gasoline storage tank in the shape of a right cylinder of radius 3 ft and length 12 ft is buried in the ground in a horizontal position. If the top of the tank is 4 ft below the surface, find the work required to empty a full tank of gasoline weighing 42 lb/ft$^3$ by pumping it through a pipe that extends to a height of 2 ft above the ground.

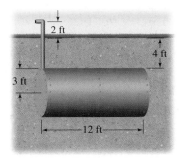

**26. Emptying a Trough** An 8-ft-long trough has ends that are equilateral triangles with sides that are 2 ft long. If the trough is full of water weighing 62.4 lb/ft$^3$, find the work required to empty it by pumping the water through a pipe that extends 1 ft above the top of the trough.

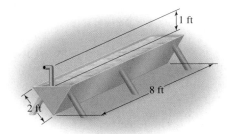

**27. Emptying a Trough** An 8-ft-long trough has ends that are semi-circles of radius 2 ft. If the trough is full of water weighing 62.4 lb/ft$^3$, find the work required to empty it by pumping the water through a pipe that extends 1 ft above the top of the trough.

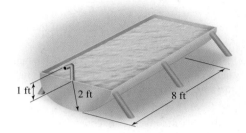

**28.** A boiler has the shape of a (lower) hemisphere of radius 5 ft. If it is filled with water weighing 62.4 lb/ft$^3$, find the work required to empty the boiler by pumping the water over the top of the boiler.

**29.** Refer to Example 6. Suppose that the pressure $P$ and volume $V$ of the steam in a steam engine are related by the law $PV^{1.4} = 100{,}000$, where $P$ is measured in pounds per square inch and $V$ is measured in cubic inches. Find the work done by the steam as it expands from a volume of 100 in.$^3$ to a volume of 400 in.$^3$.

**30.** Refer to Example 6. The pressure $P$ and volume $V$ of the steam in a steam engine are related by the equation $PV^{1.2} = k$, where $k$ is a constant. If the initial pressure of the steam is $P_0$ lb/in.$^2$ and its initial volume is $V_0$ in.$^3$, find an expression for the work done by the steam as it expands to a volume of four times its initial volume.

**31. Launching a Rocket** Newton's Law of Gravitation states that two bodies having masses $m_1$ and $m_2$ attract each other with a force

$$F = G\frac{m_1 m_2}{r^2}$$

where $G$ is the gravitational constant and $r$ is the distance between the two bodies. Assume that the mass of the earth is $5.97 \times 10^{24}$ kg and is concentrated at the center of the earth, the radius of the earth is $6.37 \times 10^6$ m, and $G = 6.67 \times 10^{-11}$ N-m$^2$/kg$^2$. Find the work required to launch a rocket of mass 500,000 kg vertically upwards to a height of 10,000 km.

**32. Launching a Rocket** Show that the work $W$ required to launch a rocket of mass $m$ from the ground vertically upward to a height $h$ is given by the formula

$$W = \frac{mgRh}{R + h}$$

where $R$ is the radius of the earth.

**Hint:** Use Newton's Law of Gravitation given in Exercise 31 and follow these steps: (i) Let $m$ and $M$ denote the mass of the rocket and the earth, respectively, so that $F = GmM/r^2$, where $R \leq r \leq R + h$. At $r = R$ the force will be the weight of the rocket, that is, $mg = GmM/R^2$. Therefore, $G = gR^2/M$, so $F = mgR^2/r^2$. (ii) $W = \int_R^{R+h} F \, dr$.

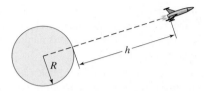

**33. Launching a Lunar Landing Module** A lunar landing module with a weight of 20,000 lb, as measured on the earth, is to be launched vertically upward from the surface of the moon to a height of 20 mi. Taking the radius of the moon to be 1100 mi and its gravitational force to be one sixth that of the earth, find the work required to accomplish the task. **Hint:** See Exercise 32.

**34. Work Done by a Repulsive Charge** Coulomb's Law states that the force exerted on two point charges $q_1$ and $q_2$ separated by a distance $r$ is given by

$$F = \frac{1}{4\pi\varepsilon_0}\frac{q_1 q_2}{r^2}$$

where $\varepsilon_0$ is a constant known as the permittivity of free space. Suppose that an electrical charge $q_1$ is concentrated at the origin of the coordinate line and that it repulses a like charge $q_2$ from the point $x = a$ to the point $x = b$. Show that the work $W$ done by the repulsive force is given by

$$W = \frac{q_1 q_2}{4\pi\varepsilon_0}\left(\frac{1}{a} - \frac{1}{b}\right)$$

**35. Work Done by a Repulsive Charge** An electric charge $Q$ distributed uniformly along a ring-shaped conductor of radius $a$ repulses a like charge $q$ along the line perpendicular to the plane of the ring, through its center. The magnitude of the force acting on the charge $q$ when it is at the point $x$ is given by

$$F = \frac{1}{4\pi\varepsilon_0} \cdot \frac{qQx}{(x^2 + R^2)^{3/2}}$$

and the force acts in the direction of the positive $x$-axis. Find the work done by the force of repulsion in moving the charge $q$ from $x = a$ to $x = b$.

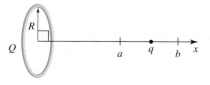

**36. Work Done by an Expanding Gas** In Example 6 we showed that the work done by an expanding gas against a piston as its volume expands from $V_0$ to $V_1$ is given by

$$W = \int_{V_0}^{V_1} p \, dV$$

where $p$ is the pressure of the gas. If the pressure and volume of a gas are related by the equation $pV = k$, where $k$ is a positive constant, show that $W = k \ln(V_1/V_0)$.

As the gas expands, work is done by the expanding gas against the piston.

**37. Work Done by an Expanding Gas** Refer to Exercise 36. At high pressure, the relationship between the volume $V$ and pressure $P$ of gases is approximated by the van der Waals equation:

$$\left( P + \frac{an^2}{V^2} \right)(V - nb) = nRT$$

where $R$ is the gas constant, $n$ is the number of moles, and $a$ and $b$ are constants having different values for different gases. (In the special case in which $a = b = 0$, we have the ideal gas equation.) Calculate the work done by a van der Waals gas when it undergoes isothermal expansion ($T =$ constant) from a volume of $V_0$ to a volume of $V_1$. Reconcile your result with that of Exercise 36 when $a = b = 0$ (that is, when expansion occurs under normal pressure).

**38.** The following table shows the force $F(x)$ (in pounds) exerted on an object as it is moved along a coordinate axis from $x = 0$ to $x = 10$ ($x$ is measured in feet). Use Simpson's Rule to estimate the work done by the force.

| $x$ (ft) | 0 | 1 | 2 | 3 | 4 | 5 |
|---|---|---|---|---|---|---|
| $F(x)$ (lb) | 0 | 0.69 | 1.61 | 2.28 | 2.88 | 3.20 |

| $x$ (ft) | 6 | 7 | 8 | 9 | 10 |
|---|---|---|---|---|---|
| $F(x)$ (lb) | 3.58 | 3.95 | 4.20 | 4.38 | 4.64 |

**39. Work and Kinetic Energy** A force $F(x)$ acts on a body of mass $m$ moving it along a coordinate axis. Show that the work done by the force in moving the body from $x = x_1$ to $x = x_2$ is

$$W = \int_{x_1}^{x_2} F(x) \, dx = \frac{1}{2} m v_2^2 - \frac{1}{2} m v_1^2$$

where $v_1$ and $v_2$ are the velocities of the body when it is at $x = x_1$ and $x = x_2$, respectively.
**Hint:** Use Newton's Second Law of Motion $\left( F = m \dfrac{dv}{dt} \right)$ and the Chain Rule to write

$$\frac{dv}{dt} = \frac{dv}{dx} \cdot \frac{dx}{dt} = v \frac{dv}{dx}$$

The quantity $\frac{1}{2} m v^2$ is the *kinetic energy* of a body of mass $m$ moving with a velocity $v$. Thus, the work done by the force is equal to the net change in the kinetic energy of the body.

**40.** Refer to Exercise 39. A 4-kg block is attached to a horizontal spring with a spring constant of 400 N/m. The spring is compressed 5 cm from equilibrium and released from rest. Find the speed of the block when the spring is at its equilibrium position.

# 5.6 Fluid Pressure and Force

Whether designing a hydroelectric dam, an aquarium, or a submarine, an engineer must consider the *pressure* exerted by the water on the walls or surfaces of the object. (See Figure 1.)

**FIGURE 1**

## ■ Fluid Pressure

Consider a thin horizontal plate of area $A$ ft$^2$ submerged to a depth of $h$ ft in a liquid of weight density $\delta$ lb/ft$^3$ (Figure 2a). The force acting on the surface of the plate is just the weight of the column of liquid above it (Figure 2b). Since the volume of this column of liquid is $Ah$ ft$^3$ and its weight density is $\delta$ lb/ft$^3$, we see that the force exerted on the plate by the liquid is given by

$$F = \delta A h \qquad \text{weight density} \cdot \text{volume}$$

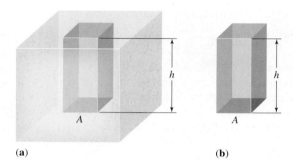

**FIGURE 2**
The force exerted by the fluid on the horizontal plate (a) is the weight of the column of liquid above it.

**(a)**          **(b)**

The pressure exerted by the liquid on the horizontal plate is

$$P = \delta h \qquad \text{force divided by area}$$

in lb/ft$^2$.

**EXAMPLE 1** A rectangular fish aquarium has a base measuring 2 ft by 4 ft. (See Figure 3.) Find the pressure and the force exerted on the base of the tank when the tank is filled with water to a height of $1\frac{1}{2}$ ft. (The weight density of water is 62.4 lb/ft$^3$.)

**FIGURE 3**
A rectangular fish aquarium with base 2 ft × 4 ft

**Solution** The pressure exerted by the water on the base of the tank is

$$P = \delta h = (62.4)(1.5) = 93.6$$

or 93.6 lb/ft$^2$.

Since the area of the base of the tank is $(4)(2)$ or 8 ft$^2$, we see that the force exerted on the base of the tank is

$$F = PA = (93.6)(8) \qquad \text{pressure} \cdot \text{area}$$

$$= 748.8$$

or 748.8 lb.

In the study of hydrostatics we are guided by the following important physical law: *The pressure at any point in a liquid is the same in all directions.* Thus, the water pressure at a point on the wall of a swimming pool $h$ ft from the surface of the water is the same as that at a point located away from the sides of the pool and $h$ ft from the surface of the water. (See Figure 4.) The pressure is $\delta h$ lb/ft$^2$ and is the same in every direction. This physical law, known as Pascal's Principle, is named after the French mathematician Blaise Pascal (1623–1662).

**FIGURE 4**

A cross section of a swimming pool

Referring once again to Figure 4, you can see that as we move vertically downward along the wall of the swimming pool, the depth of the water increases, and, therefore, the water pressure on the wall increases as well. Thus, unlike the case of a thin *horizontal* plate, in which the pressure is constant at every point on the plate, we have here a situation in which the pressure varies as we proceed down the vertical wall. How do we find the force exerted by the water against the wall?

To answer this question, let's consider the more general situation in which a thin vertical plate is submerged in a liquid with weight density $\delta$ lb/ft$^3$ as shown in Figure 5.

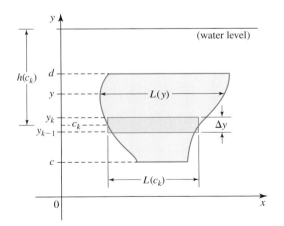

**FIGURE 5**

$L(y)$ gives the length of the vertical plate at $y$.

Let $P = \{y_0, y_1, \ldots, y_n\}$ be a regular partition of $[c, d]$, and let $c_k$ be any point in $[y_{k-1}, y_k]$. If $\Delta y = (d - c)/n$ is small, then the depth of the $k$th (representative) rectangular strip, shown shaded in Figure 5, is approximately $h(c_k)$. Its length is approximately $L(c_k)$, where $L(y)$ is the horizontal length of the plate at $y$. Therefore, the force exerted by the liquid on this representative rectangular strip is

$$\Delta F_k \approx \delta h(c_k) L(c_k) \Delta y \qquad \text{pressure} \cdot \text{area}$$

and so the sum

$$\sum_{k=1}^{n} \Delta F_k \approx \sum_{k=1}^{n} \delta h(c_k)L(c_k)\Delta y$$

provides us with an approximation of the force $F$ exerted by the liquid on the vertical plate. Recognizing this sum to be a Riemann sum of the function $g(y) = \delta h(y)L(y)$ on $[c, d]$, we have the following definition.

---

**DEFINITION    Force Exerted by a Fluid**

The **force $F$ exerted by a fluid** of constant weight density $\delta$ on one side of a submerged vertical plate from $y = c$ to $y = d$, where $c \le d$, is given by

$$F = \lim_{n\to\infty} \sum_{k=1}^{n} \delta h(c_k)L(c_k)\Delta y = \delta \int_{c}^{d} h(y)L(y)\, dy \qquad (1)$$

where $h(y)$ is the depth of the fluid at $y$ and $L(y)$ is the horizontal length of the plate at $y$.

---

**EXAMPLE 2    Fluid Pressure**    The vertical wall on the deep end of a rectangular swimming pool is 20 ft wide and 8 ft high. If water in the swimming pool is filled to a height of 7 ft as measured from the bottom of the wall, find the force exerted on the wall by the water. (The weight density of water is 62.4 lb/ft$^3$.)

**Solution**    Imagine that the wall is placed on a coordinate system with the bottom of the wall lying along the $x$-axis, as shown in Figure 6. Here, the width of the wall is constant, so the length of the thin horizontal strip at $y$ is $L(y) = 20$. The depth of the fluid at $y$ is $h(y) = 7 - y$. Therefore, the force exerted by the water on the wall is

$$F = \delta \int_{c}^{d} h(y)L(y)\, dy$$

$$= 62.4 \int_{0}^{7} (7 - y)(20)\, dy = 1248 \int_{0}^{7} (7 - y)\, dy$$

$$= 1248 \left[ 7y - \frac{1}{2}y^2 \right]_{0}^{7} = 30{,}576$$

or 30,576 lb.

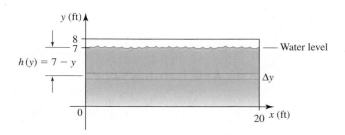

**FIGURE 6**
We want to find the force exerted on the wall of a swimming pool by the water. Here, the width of the wall is constant.

**EXAMPLE 3** **Fluid Pressure** The vertical gate of a dam has the shape of a trapezoid as shown in Figure 7. What is the force on the gate when the surface of the water is 2 ft above the top of the gate?

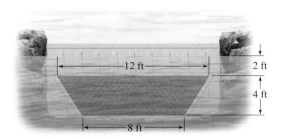

**FIGURE 7**
We want to find the force exerted on the gate of the dam by the water. Here, the width of the gate varies.

**Solution** Let's introduce a coordinate system so that the *x*-axis coincides with the water level, as shown in Figure 8. The length of the horizontal strip is $L(y) = 8 + 2t$. To find $t$, refer to Figure 8b. By similar triangles we have

$$\frac{t}{2} = \frac{6 + y}{4} \quad \text{or} \quad t = \frac{1}{2}(6 + y)$$

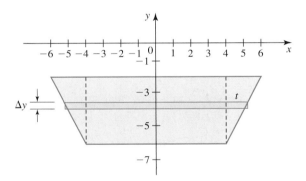

**FIGURE 8**    (a) The length of the strip is $8 + 2t$.    (b) We use similar triangles to find $t$.

So

$$L(y) = 8 + 2t = 8 + 6 + y = 14 + y$$

The depth of the fluid at $y$ is $h(y) = -y$. Therefore, the force exerted by the water on the gate is

$$F = \delta \int_c^d h(y)L(y)\, dy = 62.4 \int_{-6}^{-2} (-y)(14 + y)\, dy$$

$$= 62.4 \left[ -7y^2 - \frac{1}{3}y^3 \right]_{-6}^{-2}$$

$$= 62.4 \left[ \left( -28 + \frac{8}{3} \right) - (-252 + 72) \right] = 9651.2$$

or 9651.2 lb.

**EXAMPLE 4** The viewing port of a modern submersible used in oceanographic research has a radius of 1 ft. If the vertical viewing port is 100 ft under water as measured from its center, find the force exerted on it by the water.

**Solution** Let's choose a coordinate system so that its origin coincides with the center of the viewing port. Then the viewing port is described by the equation $x^2 + y^2 = 1$. (See Figure 9.)

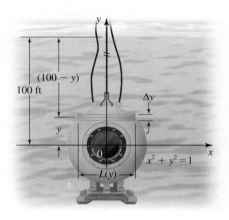

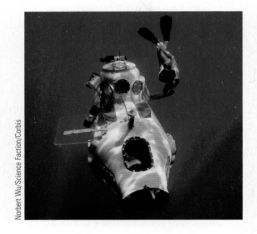

Norbert Wu/Science Faction/Corbis

**FIGURE 9**
The viewing port of a modern submersible

The length of a thin horizontal strip at $y$ is $L(y) = 2x = 2\sqrt{1 - y^2}$, and the depth of the fluid at $y$ is $h(y) = 100 - y$. Therefore, the force exerted by the water on the viewing port is

$$F = \delta \int_c^d h(y)L(y)\, dy = 62.4 \int_{-1}^1 (100 - y)(2)\sqrt{1 - y^2}\, dy$$

$$= 12{,}480 \int_{-1}^1 \sqrt{1 - y^2}\, dy - 124.8 \int_{-1}^1 y\sqrt{1 - y^2}\, dy$$

The second integral on the right is zero because the integrand is an odd function (see Theorem 4 in Section 4.5). To evaluate the first integral, observe that it represents the area of a semicircular disk with radius 1. Therefore,

$$F = 12{,}480 \int_{-1}^1 \sqrt{1 - y^2}\, dy$$

$$= 12{,}480 \left(\frac{1}{2}\pi\right)(1)^2 = 6240\pi \approx 19{,}604$$

or 19,604 lb.

## 5.6 CONCEPT QUESTIONS

1. Explain Pascal's Principle.
2. **a.** A thin vertical plate is submerged in a fluid of constant weight density $\delta$. Its length at a depth of $y$ is $L(y)$ for $c \le y \le d$. Write an integral giving the force exerted on one side of the plate.

   **b.** If the plate described in part (a) is submerged in the fluid so that the plate is parallel to the surface of the liquid at a depth of $h$ feet, what is the force exerted on the plate?

## 5.6 EXERCISES

1. An aquarium is 3 ft long, 1 ft wide, and 1 ft deep. If the aquarium is filled with water, find the force exerted by the water (a) on the bottom of the aquarium, (b) on the longer side of the aquarium, and (c) on the shorter side of the aquarium.

2. A rectangular swimming pool is 40 ft long, 15 ft wide, and 9 ft deep. If the pool is filled with water to a depth of 8 ft, find the force exerted by the water (a) on the bottom of the pool and (b) on one end of the pool.

*In Exercises 3–10, you are given the shape of the vertical ends of a trough that is completely filled with water. Find the force exerted by the water on one end of the trough.*

**3.**

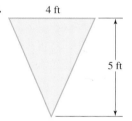

4 ft, 5 ft

**4.**

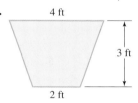

4 ft, 3 ft, 2 ft

**5.**

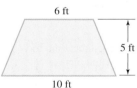

6 ft, 5 ft, 10 ft

**6.**

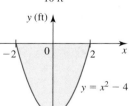

3 ft

**7.**

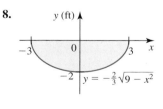

$y = x^2 - 4$

**8.**

$y = -\frac{2}{3}\sqrt{9 - x^2}$

**9.**

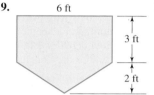

6 ft, 3 ft, 2 ft

**10.**

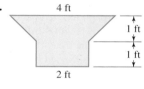

4 ft, 1 ft, 1 ft, 2 ft

*In Exercises 11–14, a vertical plate is submerged in water (the surface of the water coincides with the x-axis). Find the force exerted by the water on the plate.*

**11.**

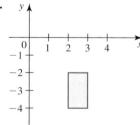

**12.**

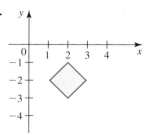

**13.**

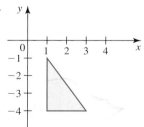

**14.**

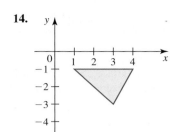

**15.** A trough has vertical ends that are equilateral triangles with sides of length 2 ft. If the trough is filled with water to a depth of 1 ft, find the force exerted by the water on one end of the trough.

**16.** A trough has vertical ends that are trapezoids with parallel sides of length 4 ft (top) and 2 ft (bottom) and a height of 3 ft. If the trough is filled with water to a depth of 2 ft, find the force exerted by the water on one end of the trough.

**17.** A cylindrical drum of diameter 4 ft and length 8 ft is lying on its side, submerged in water 12 ft deep. Find the force exerted by the water on one end of the drum.

**18.** A cylindrical oil storage tank of diameter 4 ft and length 8 ft is lying on its side. If the tank is half full of oil that weighs 50 lb/ft$^3$, find the force exerted by the oil on one end of the tank.

**19.** The first figure shows a vertical dam with a parabolic gate, and the second figure shows an enlargement of the parabolic gate.

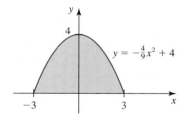

**a.** Find the force exerted by the water on the gate when the water is 10 ft deep.
**b.** The gate is designed to withstand twice the force that the water will exert on it under flood conditions (when the water level is 20 ft deep). What is this force?

**20.** Redo Exercise 19 for a semicircular gate as shown in the figure.

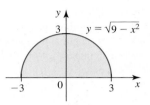

**21.** A rectangular tank has width 2 ft, height 3 ft, and length 6 ft. It is filled with equal volumes of water and oil. The oil has a weight density of 50 lb/ft$^3$ and floats on the water. Find the force exerted by the mixture on one end of the tank.

**22.** A rectangular swimming pool is 25 ft wide, 60 ft long, and 4 ft deep at the shallow end and 9 ft deep at the deep end. Its bottom is an inclined plane. If the pool is completely filled with water, find the force exerted by the water on each side of the pool.

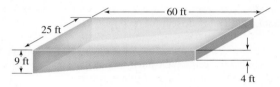

**23.** Refer to Exercise 22. Find the force exerted by the water on the bottom of the pool.

**24.** A vertical plate is submerged in water as shown in the figure below. The widths of the plate taken at $\frac{1}{2}$-ft intervals are recorded in the following table.

| Depth (ft) | 2.0 | 2.5 | 3.0 | 3.5 | 4.0 | 4.5 | 5.0 |
|---|---|---|---|---|---|---|---|
| Width of plate (ft) | 0 | 2.9 | 3.4 | 3.0 | 2.2 | 1.7 | 0 |

Use Simpson's Rule with $n = 6$ to estimate the force exerted by the water on one side of the plate.

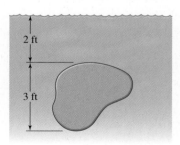

**FIGURE 1**
The center of mass of a circular plate is located at the center of the plate.

As every juggler knows, many objects will remain in equilibrium if supported at a certain point. As an example, for a homogeneous circular plate, this point is located at the center of the plate and is called the center of mass of the plate. (See Figure 1.)

The knowledge of the location of the center of mass of a body or a system of bodies is important in physics and engineering. As a matter of practical interest, every motorist knows that a car wheel must be balanced when a new tire is installed. Because of defects in the tire-manufacturing process, a wheel is seldom balanced when a new tire is installed; that is, the center of mass of the wheel is not located at "dead center." An unbalanced wheel causes the car to shimmy.

Before we learn how to find the center of mass of plane regions, we need to recall some basic notions from physics.

### ■ Measures of Mass

The **mass** of a body is the quantity of matter in the body. In the English system the unit of mass is the **slug;** in the international system (SI) the unit of mass is the **kilogram;** and in the centimeter-gram-second system (cgs) the unit of mass is the **gram.** On the surface of the earth, where the constant of acceleration due to gravity, $g$, is approximately 32 ft/sec$^2$ in the English system, 9.8 m/sec$^2$ in the international system, and 980 cm/sec$^2$ in the cgs system, Newton's Second Law of Motion ($F = ma$) tells us that

a body of mass $m$ slugs has a weight of $32m$ pounds,

a body of mass $m$ kilograms has a weight of $9.8m$ newtons, and

a body of mass $m$ grams has a weight of $980m$ dynes.

### ■ Center of Mass of a System on a Line

Consider a simple system consisting of two particles of mass $m_1$ and $m_2$ connected by a rod of negligible mass. If we place this system on a fulcrum as shown in Figure 2, then equilibrium is achieved if

$$m_1 d_1 = m_2 d_2 \tag{1}$$

where $d_1$ and $d_2$ are the distances (called *moment arms*) between the particles and the fulcrum. The quantity $m_1 d_1$, called the *moment* of $m_1$ about the fulcrum, is a measure of the tendency of $m_1$ to rotate the system about the fulcrum (in this case in the counterclockwise direction). On the other hand, the moment $m_2 d_2$ is a measure of the tendency of $m_2$ to rotate the system about the fulcrum (in a clockwise direction). Balance is achieved when these moments are equal, that is, when Equation (1) holds.

**FIGURE 2**
The condition for equilibrium of the system is $m_1 d_1 = m_2 d_2$.

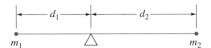

We can use Equation (1) to derive a formula for calculating the center of mass of the system. Place the system on a coordinate line, and suppose that the coordinates of $m_1$, $m_2$, and the fulcrum are $x_1$, $x_2$, and $\bar{x}$, respectively, as shown in Figure 3. You can see immediately that the distance between $m_1$ and the fulcrum is $d_1 = \bar{x} - x_1$ and that the distance between $m_2$ and the fulcrum is $d_2 = x_2 - \bar{x}$. Therefore, Equation (1) gives

$$m_1(\bar{x} - x_1) = m_2(x_2 - \bar{x})$$

$$m_1\bar{x} + m_2\bar{x} = m_1x_1 + m_2x_2$$

and

$$\bar{x} = \frac{m_1x_1 + m_2x_2}{m_1 + m_2} \tag{2}$$

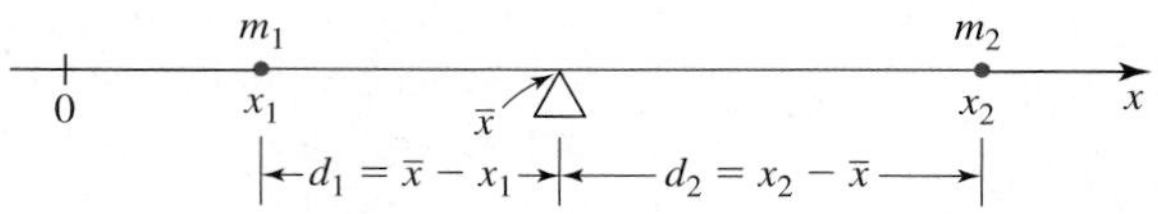

**FIGURE 3**
The system placed on a coordinate line

The numbers $m_1x_1$ and $m_2x_2$ in Equation (2) are called the *moments* of the masses $m_1$ and $m_2$ about the origin.

In general, if $m$ is a mass located at the point $x$ on a coordinate line, then $mx$ is called the **moment of the mass $m$ about the origin.** If you think of the mass $m$ as being connected to the origin by a rod of negligible mass, then $mx$ measures the tendency of $m$ to rotate the rod about the origin. Observe that Equation (2) says that to find the coordinate of the center of mass of a system comprising two masses, $m_1$ and $m_2$, add the moments of the masses about the origin and divide the sum by the total mass $m_1 + m_2$. A similar analysis of a system comprising $n$ particles located on a coordinate line, as shown in Figure 4, leads to the definition of the center of mass of that system.

**FIGURE 4**
A system of $n$ masses connected by a rod of negligible mass on a coordinate line

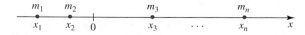

---

**DEFINITION**    **The Center of Mass of a System of $n$ Masses on a Line**

Let $S$ denote a system of $n$ masses $m_1, m_2, \ldots, m_n$ located at $x_1, x_2, \ldots, x_n$, lying on a line, respectively, and let $m = \sum_{k=1}^{n} m_k$ denote the total mass of the system.

1. The **moment of $S$ about the origin** is

$$M = \sum_{k=1}^{n} m_k x_k \tag{3a}$$

2. The **center of mass of $S$** is located at

$$\bar{x} = \frac{M}{m} = \frac{1}{m} \sum_{k=1}^{n} m_k x_k \tag{3b}$$

---

**Note**    If we write Equation (3b) in the form $m\bar{x} = M$, we obtain the following interpretation of the center of mass: Think of the total mass of the system as being concentrated at the center of mass $\bar{x}$. Then the moment of this mass about the origin will be the same as the moment of the system about the origin. ∎

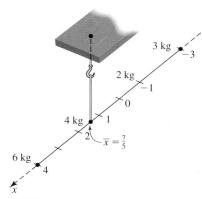

**FIGURE 5**
If the system is suspended at $\bar{x} = \frac{7}{5}$, it will hang in equilibrium horizontally.

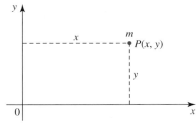

**FIGURE 6**
A particle of mass $m$ located at the point $P(x, y)$

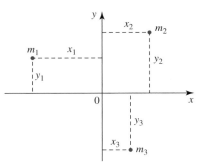

**FIGURE 7**
A system with three masses ($n = 3$)

**EXAMPLE 1**   Find the center of mass of a system of four objects located at the points $-3$, $-1$, 2, and 4, on the $x$-axis ($x$ in meters), with masses 3, 2, 4, and 6 kilograms, respectively.

**Solution**   Using Equation (3b) with $m_1 = 3$, $m_2 = 2$, $m_3 = 4$, and $m_4 = 6$ and $x_1 = -3$, $x_2 = -1$, $x_3 = 2$, and $x_4 = 4$ gives the coordinate of the center of mass of the system as

$$\bar{x} = \frac{3(-3) + 2(-1) + 4(2) + 6(4)}{3 + 2 + 4 + 6} = \frac{21}{15} = \frac{7}{5} \quad \text{or} \quad 1.4 \text{ m} \quad \blacksquare$$

## Interpreting Our Results

Think of the four masses as being connected by a rod of length 7 m and of mass very small in comparison to the mass of the four given objects. If the system is suspended by a string at $\bar{x}$, then the system will hang in equilibrium horizontally. (See Figure 5.)

## ■ Center of Mass of a System in the Plane

Consider a particle of mass $m$ located at the point $P(x, y)$ in a coordinate plane. (See Figure 6.) If you think of this mass as being connected to the $x$-axis by a rod perpendicular to the axis and of negligible mass, then the quantity $my$ measures the tendency of the mass $m$ to rotate the system about the $x$-axis. This quantity is called the **moment of the mass $m$ about the $x$-axis** and is denoted by $M_x$. Similarly, we define the **moment of the mass $m$ about the $y$-axis** to be $M_y = mx$. To find the moments about the $x$- and $y$-axes of a system comprising $n$ particles in the plane, we simply add the respective moments of each mass. (See Figure 7.)

This leads to the following definition.

---

**DEFINITION   The Center of Mass of a System of $n$ Particles in a Plane**

Let $S$ denote a system of $n$ particles with masses $m_1, m_2, \ldots, m_n$ located at the points $(x_1, y_1), (x_2, y_2), \ldots, (x_n, y_n)$, respectively, and let $m = \sum_{k=1}^{n} m_k$ denote the total mass of the system.

1. The **moment of $S$ about the $x$-axis** is

$$M_x = \sum_{k=1}^{n} m_k y_k \tag{4a}$$

2. The **moment of $S$ about the $y$-axis** is

$$M_y = \sum_{k=1}^{n} m_k x_k \tag{4b}$$

3. The **center of mass of $S$** is located at the point $(\bar{x}, \bar{y})$ where

$$\bar{x} = \frac{M_y}{m} = \frac{1}{m} \sum_{k=1}^{n} m_k x_k \quad \text{and} \quad \bar{y} = \frac{M_x}{m} = \frac{1}{m} \sum_{k=1}^{n} m_k y_k \tag{4c}$$

---

**Note**   The center of mass of a system of $n$ particles in the plane is the point at which, if the total mass $m$ of the system were concentrated, it would generate the same moments as the system. This can be seen by writing Equation (4c) in the form $m\bar{x} = M_y$ and $m\bar{y} = M_x$.   ■

**EXAMPLE 2** Find the center of mass of a system comprising three particles with masses 2, 3, and 5 slugs, located at the points $(-2, 2)$, $(4, 6)$, and $(2, -3)$, respectively. (Assume that all distances are measured in feet.)

**Solution** We first compute the moments

$$M_y = 2(-2) + 3(4) + 5(2) = 18$$

or 18 slug-ft, and

$$M_x = 2(2) + 3(6) + 5(-3) = 7$$

or 7 slug-ft. Since $m = 2 + 3 + 5 = 10$, an application of Equation (4c) yields

$$\bar{x} = \frac{M_y}{m} = \frac{18}{10} = \frac{9}{5} \quad \text{and} \quad \bar{y} = \frac{M_x}{m} = \frac{7}{10}$$

feet. Therefore, the center of mass of the system is located at $\left(\frac{9}{5}, \frac{7}{10}\right)$. ■

## Interpreting Our Results

Think of the three particles as being connected by rods of negligible mass to the center of mass, $P$. If the system is suspended by a string at $P$, then it will rest in a horizontal position, much like a mobile. (See Figure 8.)

## ■ Center of Mass of Laminas

We now turn our attention to the problem of finding the center of mass of a lamina (a thin, flat plate). We will assume that the laminas we consider are homogeneous, that is, that they have uniform mass density $\rho$ (the Greek letter rho), where $\rho$ is a positive constant.

Let's begin by assuming that the lamina $L$ has the shape of the region $R$ under the graph of a continuous nonnegative function $f$ on the interval $[a, b]$, as shown in Figure 9.

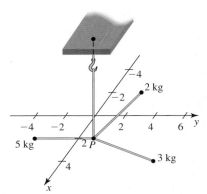

**FIGURE 8**
When the system is suspended by a string at $P$, it will rest in a horizontal position.

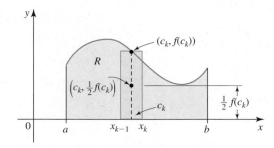

**FIGURE 9**
The lamina has the shape of the region $R$. The $k$th lamina is highlighted.

Let $P = \{x_0, x_1, \ldots, x_n\}$ be a regular partition of $[a, b]$. This partition divides $R$ into $n$ nonoverlapping subregions $R_1, \ldots, R_n$, each of which is again a lamina. The $k$th subregion is approximated by the $k$th rectangle of width $\Delta x = (b - a)/n$ and height $f(c_k)$, where $c_k$ is the midpoint of the $k$th subinterval $[x_{k-1}, x_k]$, that is, $c_k = (x_k + x_{k-1})/2$. The area of the $k$th rectangle is $f(c_k)\Delta x$, so the mass of the $k$th lamina is approximately

$$\rho f(c_k)\Delta x \qquad \text{density} \cdot \text{area}$$

Next, since the center of mass of a rectangular lamina is located at its center, we conclude that the center of mass of the $k$th lamina is located at the point $\left(c_k, \frac{1}{2}f(c_k)\right)$.

This tells us that the moment arm of the $k$th lamina with respect to the $y$-axis is $c_k$ and, therefore, that the moment of the $k$th lamina about the $y$-axis is

$$[\rho f(c_k)\Delta x]c_k = \rho c_k f(c_k)\Delta x \qquad \text{mass} \cdot \text{moment arm}$$

Adding the moments of the $n$ laminas and taking the limit of the associated Riemann sum as $n \to \infty$ lead to the following definition of the moment of $L$ about the $y$-axis:

$$M_y = \lim_{n \to \infty} \sum_{k=1}^{n} \rho c_k f(c_k)\Delta x = \rho \int_a^b xf(x)\,dx$$

Similarly, by observing that the moment arm of the $k$th rectangle about the $x$-axis is $\frac{1}{2}f(c_k)$, we see that the moment of $L$ about the $x$-axis may be defined as

$$M_x = \lim_{n \to \infty} \sum_{k=1}^{n} \rho \cdot \frac{1}{2}[f(c_k)]^2 \Delta x = \rho \int_a^b \frac{1}{2}[f(x)]^2\,dx$$

(See Figure 9.) Finally, the mass of $L$ may be defined as

$$m = \lim_{n \to \infty} \sum_{k=1}^{n} \rho f(c_k)\Delta x = \rho \int_a^b f(x)\,dx$$

---

**DEFINITION**  **Moments and Center of Mass of a Lamina**

Let $L$ denote a lamina of constant mass density $\rho$, and suppose that $L$ has the shape of the region $R$ under the graph of a nonnegative continuous function $f$ on $[a, b]$.

1. The **mass** of $L$ is

$$m = \rho \int_a^b f(x)\,dx = \rho A \tag{5a}$$

   where $A = \int_a^b f(x)\,dx$ is the area of $R$.
2. The **moments of $L$ about the $x$- and the $y$-axis** are

$$M_x = \rho \int_a^b \frac{1}{2}[f(x)]^2\,dx \tag{5b}$$

   and

$$M_y = \rho \int_a^b xf(x)\,dx \tag{5c}$$

3. The **center of mass** of $L$ is located at $(\bar{x}, \bar{y})$, where

$$\bar{x} = \frac{M_y}{m} = \frac{1}{A}\int_a^b xf(x)\,dx \qquad \text{and} \qquad \bar{y} = \frac{M_x}{m} = \frac{1}{A}\int_a^b \frac{1}{2}[f(x)]^2\,dx \tag{5d}$$

---

**EXAMPLE 3**  A lamina $L$ of uniform area density $\rho$ has the shape of the region $R$ under the graph of $f(x) = x^2$ on $[0, 2]$. (See Figure 10.) Find the mass of $L$, the moments of $L$ about each of the coordinate axes, and the center of mass of $L$.

**Solution**  Using Equation (5a), we find that the mass of the lamina is

$$m = \rho \int_0^2 x^2\,dx = \rho\left[\frac{1}{3}x^3\right]_0^2 = \frac{8\rho}{3}$$

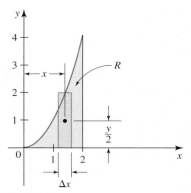

**FIGURE 10**
The center of mass of a representative
rectangle is $\left(x, \frac{y}{2}\right)$.

To find the moment $M_x$ of $L$ about the $x$-axis, we can use Equation (5b), or we can proceed as follows: Draw a representative rectangle of width $\Delta x$ and height $y$ (Figure 10). The moment arm of this rectangle with respect to the $x$-axis is

$$\frac{y}{2} = \frac{f(x)}{2} = \frac{x^2}{2}$$

and the mass of the representative rectangle is

$$\rho y\, \Delta x = \rho f(x)\Delta x = \rho x^2\, \Delta x \qquad \text{density} \cdot \text{area}$$

So the moment of this element about the $y$-axis is

$$\left(\frac{x^2}{2}\right)\rho x^2\, \Delta x \qquad \text{moment arm} \cdot \text{mass}$$

Summing and taking the limit of the Riemann sum, we have

$$M_x = \int_0^2 \left(\frac{x^2}{2}\right)\rho x^2\, dx = \frac{\rho}{2}\int_0^2 x^4\, dx = \left[\frac{\rho}{10}x^5\right]_0^2 = \frac{16\rho}{5}$$

To find $M_y$, we can use Equation (5c) or proceed as before, observing that the moment arm of the representative rectangle is $x$. Thus,

$$M_y = \int_0^2 x\rho x^2\, dx = \rho\int_0^2 x^3\, dx = \left[\frac{\rho}{4}x^4\right]_0^2 = 4\rho$$

Finally, using Equation (5d), we see that the coordinates of the center of mass of $L$ are

$$\bar{x} = \frac{M_y}{m} = \frac{4\rho}{\dfrac{8\rho}{3}} = \frac{3}{2} \quad \text{and} \quad \bar{y} = \frac{M_x}{m} = \frac{\dfrac{16\rho}{5}}{\dfrac{8\rho}{3}} = \frac{6}{5} \qquad \blacksquare$$

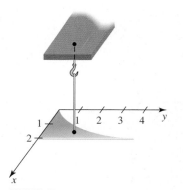

**FIGURE 11**
If the lamina is suspended at $\left(\frac{3}{2}, \frac{6}{5}\right)$,
it will hang horizontally.

## Interpreting Our Results

If the lamina is suspended by a string at $\left(\frac{3}{2}, \frac{6}{5}\right)$, it will hang in equilibrium horizontally. (See Figure 11.)

Observe that Equation (5d) does not involve $\rho$, the density of the lamina. This is always true when the lamina has *uniform* density. In other words, the center of mass of such a lamina depends only on the shape of the region $R$ that it occupies in the plane and not on its density. The point $(\bar{x}, \bar{y})$ at which the center of mass of such a lamina is located is called the **centroid** of the region $R$.

**EXAMPLE 4** Find the centroid of the region $R$ under the graph of $y = \sqrt{x}$ on the interval $[0, 4]$. (See Figure 12.)

**Solution** The area of the region $R$ is

$$A = \int_0^4 \sqrt{x}\, dx = \left[\frac{2}{3}x^{3/2}\right]_0^4 = \frac{16}{3}$$

Using Equation (5d), we have

$$\bar{x} = \frac{1}{A}\int_0^4 x f(x)\, dx = \frac{3}{16}\int_0^4 x^{3/2}\, dx = \frac{3}{16}\left[\frac{2}{5}x^{5/2}\right]_0^4 = \left(\frac{3}{16}\right)\left(\frac{64}{5}\right) = \frac{12}{5}$$

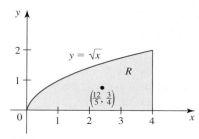

**FIGURE 12**
The centroid of the region $R$ is $\left(\frac{12}{5}, \frac{3}{4}\right)$.

and

$$\bar{y} = \frac{1}{A} \int_0^4 \frac{1}{2} [f(x)]^2 \, dx = \frac{3}{16} \int_0^4 \frac{1}{2} x \, dx = \frac{3}{16} \left[ \frac{1}{4} x^2 \right]_0^4 = \left( \frac{3}{16} \right)(4) = \frac{3}{4}$$

Therefore, the centroid of $R$ is $\left( \frac{12}{5}, \frac{3}{4} \right)$. ◼

We can use the following heuristic argument to derive the formulas for the centroid of a region $R$ between the graphs of two functions. Suppose that $R$ is bounded by the graphs of two continuous functions $f$ and $g$, where $f(x) \geq g(x)$ on an interval $[a, b]$, and to the left and right by the lines $x = a$ and $x = b$. (See Figure 13.) Then let $L$ be a lamina of uniform density that has the shape of $R$. Draw a representative rectangle of width $\Delta x$ and height $[f(x) - g(x)]$ and, therefore, area $\Delta A = [f(x) - g(x)]\Delta x$. Its mass is

$$\rho \, \Delta A = \rho [f(x) - g(x)] \Delta x \qquad \text{density} \cdot \text{area}$$

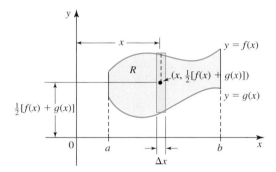

**FIGURE 13**
The region $R$ lies between the graphs of $f$ and $g$ on $[a, b]$.

Therefore, the mass of $R$ is

$$m = \rho \int_a^b [f(x) - g(x)] \, dx = \rho A$$

where $A$ is the area of $R$.

Next, the moment of the rectangle about the $x$-axis is

$$\frac{1}{2} [f(x) + g(x)]\rho [f(x) - g(x)]\Delta x \qquad \text{moment arm} \cdot \text{mass}$$

and this gives

$$M_x = \rho \int_a^b \left[ \frac{f(x) + g(x)}{2} \right] [f(x) - g(x)] \, dx = \frac{\rho}{2} \int_a^b \{ [f(x)]^2 - [g(x)]^2 \} \, dx$$

The moment of the rectangle about the $y$-axis is

$$x\rho [f(x) - g(x)]\Delta x$$

and this gives

$$M_y = \rho \int_a^b x[f(x) - g(x)] \, dx$$

Since the center of mass of $L$ (also called the **centroid of $R$**) is given by $(\bar{x}, \bar{y})$, where $\bar{x} = M_y / m$ and $\bar{y} = M_x / m$, we have the following result.

---

**DEFINITION**   **The Centroid of a Region Between Two Curves**

Let $R$ be a region bounded by the graphs of two continuous functions $f$ and $g$ on $[a, b]$, where $f(x) \geq g(x)$ for all $x$ in $[a, b]$. Then the centroid $(\bar{x}, \bar{y})$ of $R$ is given by

$$\bar{x} = \frac{1}{A} \int_a^b x[f(x) - g(x)]\, dx \tag{6a}$$

and

$$\bar{y} = \frac{1}{A} \int_a^b \frac{1}{2} \{[f(x)]^2 - [g(x)]^2\}\, dx \tag{6b}$$

where

$$A = \int_a^b [f(x) - g(x)]\, dx$$

---

**EXAMPLE 5**   Find the centroid of the region bounded by the graphs of $y = x^2 - 3$ and $y = -x^2 + 2x + 1$.

**Solution**   The region $R$ in question is shown in Figure 14. The points of intersection of the two graphs are $(-1, -2)$ and $(2, 1)$. If we let $f(x) = -x^2 + 2x + 1$ and $g(x) = x^2 - 3$, then $f(x) \geq g(x)$ on $[-1, 2]$, so the area $A$ of $R$ is

$$A = \int_{-1}^2 [f(x) - g(x)]\, dx$$

$$= \int_{-1}^2 (-2x^2 + 2x + 4)\, dx = \left[ -\frac{2}{3}x^3 + x^2 + 4x \right]_{-1}^2 = 9$$

Next, using Equations (6a) and (6b), we have

$$\bar{x} = \frac{1}{A} \int_{-1}^2 x[f(x) - g(x)]\, dx = \frac{1}{9} \int_{-1}^2 (-2x^3 + 2x^2 + 4x)\, dx$$

$$= \frac{1}{9} \left[ -\frac{1}{2}x^4 + \frac{2}{3}x^3 + 2x^2 \right]_{-1}^2 = \frac{1}{2}$$

$$\bar{y} = \frac{1}{A} \int_{-1}^2 \frac{1}{2} [(-x^2 + 2x + 1)^2 - (x^2 - 3)^2]\, dx = \frac{1}{9} \int_{-1}^2 (-2x^3 + 4x^2 + 2x - 4)\, dx$$

$$= \frac{1}{9} \left[ -\frac{1}{2}x^4 + \frac{4}{3}x^3 + x^2 - 4x \right]_{-1}^2 = -\frac{1}{2} \qquad \blacksquare$$

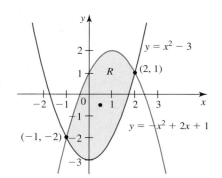

**FIGURE 14**
The region $R$ bounded by the graphs of $y = x^2 - 3$ and $y = -x^2 + 2x + 1$ has centroid $\left( \frac{1}{2}, -\frac{1}{2} \right)$.

## The Theorem of Pappus

Suppose that a solid of revolution is obtained by revolving a plane region $R$ about a line. The Theorem of Pappus enables us to find the volume of the solid in terms of the centroid of the region. (See Figure 15.) The theorem is named after the Greek mathematician Pappus of Alexandria, who lived in the fourth century A.D.

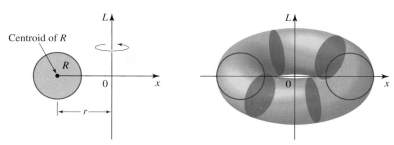

**FIGURE 15**
The volume of the solid generated by revolving $R$ about $L$ is $(2\pi r)$(area of $R$).

**The Theorem of Pappus**

Let $R$ be a plane region that lies entirely on one side of a line $L$ in the same plane. If $r$ is the distance between the centroid of $R$ and the line $L$, then the volume $V$ of the solid of revolution obtained by revolving $R$ about $L$ is given by

$$V = 2\pi r A$$

where $A$ is the area of $R$.

Note that $2\pi r$ is the distance traveled by the centroid as the region $R$ is revolved about the line $L$.

**EXAMPLE 6** **Volume of a Torus** A torus (a doughnut-shaped solid) is formed by revolving a circular region of radius $a$ about a line lying in the same plane as the circle and at a distance $b$ $(b > a)$ from the center of the circle. (See Figure 16.) Find the volume of the torus.

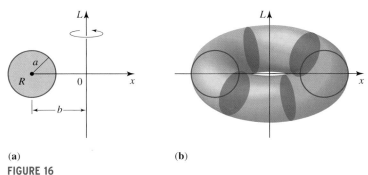

(a)                              (b)

**FIGURE 16**
If the circular region $R$ in part (a) is revolved about $L$, the resulting solid of revolution is a torus (b).

**Solution**  The centroid of the circular region is the center of the circle. So the distance traveled by the centroid during one revolution of the circular region is $2\pi b$. Since the area of the region is $\pi a^2$, the Theorem of Pappus says that the volume of the torus is

$$V = 2\pi b A = (2\pi b)(\pi a^2) = 2\pi^2 a^2 b$$

## 5.7 CONCEPT QUESTIONS

**1. a.** Let $S$ be a system of $n$ masses $m_1, m_2, \ldots, m_n$ located at $x_1, x_2, \ldots, x_n$ on a coordinate line, respectively. What is the center of mass of $S$?

**b.** Let $S$ be a system of $n$ masses $m_1, m_2, \ldots, m_n$ located at $(x_1, y_1), (x_2, y_2), \ldots, (x_n, y_n)$ in the plane. What is the moment of $S$ about the $x$-axis? About the $y$-axis? What is the center of mass of $S$?

**2.** Let $L$ denote a lamina having the shape of a region $R$ under the graph of a nonnegative continuous function $f$ on $[a, b]$ and having uniform mass density $\rho$. What is the center of mass of $L$?

**3.** Let $R$ be a region bounded by the graphs of two continuous functions $f$ and $g$ on $[a, b]$, where $f(x) \geq g(x)$. What is the centroid of $R$?

**4.** State the Theorem of Pappus.

## 5.7 EXERCISES

*In Exercises 1–4, find the center of mass of the system comprising masses $m_k$ located at the points $x_k$ on a coordinate line. Assume that mass is measured in kilograms and distance is measured in meters.*

**1.** $m_1 = 2$, $m_2 = 4$, $m_3 = 6$; $x_1 = -3$, $x_2 = -1$, $x_3 = 4$

**2.** $m_1 = 3$, $m_2 = 1$, $m_3 = 5$, $m_4 = 6$; $x_1 = -4$, $x_2 = -1$, $x_3 = 1$, $x_4 = 3$

**3.** $m_1 = 4$, $m_2 = 3$, $m_3 = 2$, $m_4 = 4$, $m_5 = 8$; $x_1 = -5$, $x_2 = -3$, $x_3 = -2$, $x_4 = 2$, $x_5 = 4$

**4.** $m_1 = 6$, $m_2 = 4$, $m_3 = 5$, $m_4 = 8$, $m_5 = 4$; $x_1 = -4$, $x_2 = -2$, $x_3 = 0$, $x_4 = 3$, $x_5 = 6$

*In Exercises 5–8, find the center of mass of the system comprising masses $m_k$ located at the points $P_k$ in a coordinate plane. Assume that mass is measured in grams and distance is measured in centimeters.*

**5.** $m_1 = 4$, $m_2 = 3$, $m_3 = 5$; $P_1(-3, -2)$, $P_2(-1, 2)$, $P_3(2, 4)$

**6.** $m_1 = 2$, $m_2 = 4$, $m_3 = 1$; $P_1(-2, 2)$, $P_2(2, 1)$, $P_3(3, -1)$

**7.** $m_1 = 3$, $m_2 = 4$, $m_3 = 6$, $m_4 = 5$; $P_1(-3, -2)$, $P_2(-2, 3)$, $P_3(2, 3)$, $P_4(4, -2)$

**8.** $m_1 = 4$, $m_2 = 1$, $m_3 = 2$, $m_4 = 5$; $P_1(-2, 3)$, $P_2(-1, 4)$, $P_3(1, 4)$, $P_4(4, -3)$

*In Exercises 9–24, find the centroid of the region bounded by the graphs of the given equations.*

**9.** $y = -\dfrac{2}{3}x + 2$, $y = 0$, $x = 0$

**10.** $y = x^2$, $y = 0$, $x = 1$, $x = 2$

**11.** $y = 4 - x^2$, $y = 0$

**12.** $y = \sqrt{x}$, $y = 0$, $x = 1$, $x = 4$

**13.** $y = |x|\sqrt{1 - x^2}$, $y = 0$, $x = -1$, $x = 1$

**14.** $y = x^3$, $y = 0$, $x = 3$

**15.** $y = 2x - x^2$, $y = 0$

**16.** $y = \sqrt{1 - x^2}$, $y = 0$

**17.** $y = x^{2/3}$, $y = 0$, $x = 8$

**18.** $y = x^{2/3}$, $y = 4$, $x = 0$

**19.** $y = x^2$, $y = \sqrt{x}$

**20.** $y = x^3$, $y = x$, $x = 0$, $x = 1$

**21.** $y = x^3$, $y = \sqrt[3]{x}$, $x = 0$, $x = 1$

**22.** $y = \dfrac{1}{x^3}$, $y = 0$, $x = 1$, $x = 2$

**23.** $y = 6 - x^2$, $y = 3 - 2x$

**24.** $y = -x^2 + 3$, $y = x^2 - 2x - 1$

*In Exercises 25–28, find the centroid of the region shown in the figure.*

**25.**

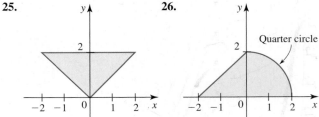

**26.**

Quarter circle

**27.**

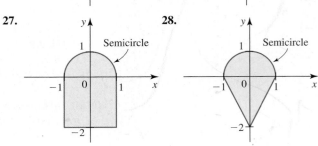

Semicircle

**28.**

Semicircle

*In Exercises 29–32, find the centroid of the region shown in the figure. (You can solve the problem without using integration.)*

**29.**

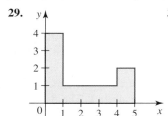

**30.**

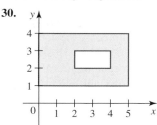

**31.**          **32.**

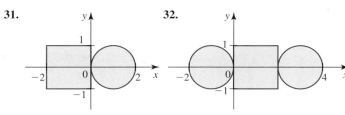

**33.** Find the center of mass of the lamina of Exercise 31 if the density of the circular lamina is twice that of the square lamina.

**34.** Find the center of mass of the lamina of Exercise 32 if the density of the circular laminae is 3 times that of the square lamina.

**35.** Find the centroid of the region bounded by the graphs of $x/a + y/b = 1$, $x = 0$, and $y = 0$.

**36.** Find the centroid of the region bounded by the graph of the equation $x^{1/2} + y^{1/2} = a^{1/2}$ and the coordinate axes.

**37.** Prove that the centroid of a triangular region is located at the point of intersection of the medians of the triangle.
Hint: Suppose that the vertices of the triangle are located at $(0, 0)$, $(a, 0)$, and $(b, h)$.

**38.** Find the centroid of the region bounded by the graphs of $y = \sqrt{1 - x^2}$ and $y = 1 - x$.

**39.** Find the centroid of the region bounded by the graphs of $y = 1/x$, $y = 0$, $x = 1$, and $x = 2$.

 **40.** Use a calculator or computer to find the centroid of the region $R$ under the graph of $y = \tan^{-1} x$ on $[0, 1]$ accurate to three decimal places.

*In Exercises 41–44, use the Theorem of Pappus to find the volume of the given solid.*

**41.** The torus formed by revolving the region bounded by the circle $(x - 4)^2 + y^2 = 9$ about the y-axis

**42.** A cone of radius $r$ and height $h$

**43.** The solid obtained by revolving the region bounded by the graphs of $y = 4 - x^2$, $y = 4$, and $x = 2$ about the y-axis

**44.** The solid obtained by revolving the region bounded by the graphs of $y = \sqrt{x - 2}$, $y = 0$, and $x = 6$ about the y-axis

**45.** Use the Theorem of Pappus to find the centroid of the region bounded by the upper semicircle $y = \sqrt{R^2 - x^2}$ and the x-axis.

**46.** Use the Theorem of Pappus to show that the y-coordinate of the centroid of a triangular region is located at the point that is one third of the distance along the altitude from the base of the triangle.
Hint: Suppose the vertices of the triangle are located at $(0, 0)$, $(a, 0)$, and $(b, h)$.

*In Exercises 47 and 48, C is a curve that is the graph of a continuous function $y = f(x)$ on the interval $[a, b]$, and the moments $M_x$ and $M_y$ of C about the x- and y-axis are defined by $M_x = \int_a^b y \, ds$ and $M_y = \int_a^b x \, ds$, respectively, where $ds = \sqrt{1 + (y')^2} \, dx$ is the element of arc length. The coordinates of the centroid of C are $\bar{x} = M_y/L$ and $\bar{y} = M_x/L$, where L is the arc length of C. Find the centroid of C.*

**47.** $C: y = \sqrt{a^2 - x^2}$, $\quad -a \le x \le a$ (upper semicircle)

**48.** $C: x^{2/3} + y^{2/3} = a^{2/3}$, $\quad 0 \le x \le a, y \ge 0$ (astroid in the first quadrant)

 **49.** Find the centroid of the region under the graph of $y = \sin \pi x$ on the interval $[0, 1]$. Find the exact values of $\bar{x}$ and $\bar{y}$.

 **50.** Find the centroid of the region under the graph of $y = 1/(1 + x^2)$ on the interval $[-1, 1]$.

## 5.8 Hyperbolic Functions

Figure 1 depicts a uniform flexible cable, such as a telephone or power line, suspended between two poles. The shape assumed by the cable is called a *catenary,* from the Latin word *catena,* which means "chain." Figure 1b shows the path taken by a heat-seeking missile as it locks onto and intercepts an aircraft. We assume here that the aircraft is flying along a straight line at a constant height and at a constant speed and that the missile, also flying at a constant speed is always pointed at the aircraft. The trajectory of the missile is called a *pursuit curve.*

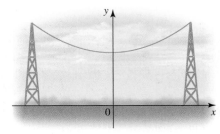

(a) The hanging cable takes the shape of a catenary.

(b) The trajectory of the missile is called a pursuit curve.

**FIGURE 1**

The analysis of problems such as these involves combinations of exponential functions of the form $e^{-cx}$ and $e^{cx}$, where $c$ is a constant. Because combinations of these functions arise so frequently in mathematics and its applications, they have been given special names. These combinations—the **hyperbolic sine,** the **hyperbolic cosine,** the **hyperbolic tangent,** and so on—are referred to as **hyperbolic functions** and are so called because they have many properties in common with the trigonometric functions.

**DEFINITIONS** **The Hyperbolic Functions**

$$\sinh x = \frac{e^x - e^{-x}}{2} \qquad \cosh x = \frac{e^x + e^{-x}}{2} \qquad \tanh x = \frac{\sinh x}{\cosh x}$$

$$\operatorname{csch} x = \frac{1}{\sinh x}, \quad x \neq 0 \qquad \operatorname{sech} x = \frac{1}{\cosh x} \qquad \coth x = \frac{\cosh x}{\sinh x}, \quad x \neq 0$$

**Note** The expression sinh $x$ is pronounced "cinch $x$," and cosh $x$ is pronounced "kosh $x$," which rhymes with "gosh $x$." ■

### ■ The Graphs of the Hyperbolic Functions

The graph of $y = \sinh x$ can be drawn by first sketching the graphs of $y = \frac{1}{2} e^x$ and $y = -\frac{1}{2} e^{-x}$ and then adding the $y$-coordinates of the points on these graphs corresponding to each $x$ to obtain the $y$-coordinates of the points on $y = \sinh x$ (Figure 2a). Similarly, the graph of $y = \cosh x$ can be drawn by first sketching the graphs of $y = \frac{1}{2} e^x$ and $y = \frac{1}{2} e^{-x}$ and then adding the $y$-coordinates of the points on these graphs corresponding to each $x$ to obtain the $y$-coordinates of the points on $y = \cosh x$ (Figure 2b).

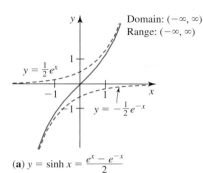
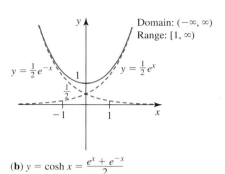

**FIGURE 2**
The graphs of the hyperbolic sine and cosine functions

(**a**) $y = \sinh x = \dfrac{e^x - e^{-x}}{2}$

(**b**) $y = \cosh x = \dfrac{e^x + e^{-x}}{2}$

The graphs of the other four hyperbolic functions are shown in Figure 3.

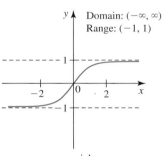
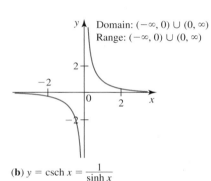

(**a**) $y = \tanh x = \dfrac{\sinh x}{\cosh x}$

(**b**) $y = \operatorname{csch} x = \dfrac{1}{\sinh x}$

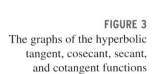
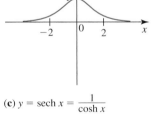

**FIGURE 3**
The graphs of the hyperbolic tangent, cosecant, secant, and cotangent functions

(**c**) $y = \operatorname{sech} x = \dfrac{1}{\cosh x}$

(**d**) $y = \coth x = \dfrac{1}{\tanh x}$

## ■ Hyperbolic Identities

The hyperbolic functions satisfy certain identities that look very much like those satisfied by trigonometric functions. For example, the analog of $\sin(-x) = -\sin x$ is $\sinh(-x) = -\sinh x$. To prove this identity, we simply compute

$$\sinh(-x) = \frac{e^{(-x)} - e^{-(-x)}}{2} = \frac{e^{-x} - e^{x}}{2} = -\frac{e^{x} - e^{-x}}{2} = -\sinh x$$

A list of frequently used hyperbolic identities is given in Table 1.

**TABLE 1** Hyperbolic Identities

| | |
|---|---|
| $\sinh(-x) = -\sinh x$ | $\cosh(-x) = \cosh x$ |
| $\cosh^2 x - \sinh^2 x = 1$ | $\text{sech}^2 x = 1 - \tanh^2 x$ |
| $\sinh(x + y) = \sinh x \cosh y + \cosh x \sinh y$ | $\cosh(x + y) = \cosh x \cosh y + \sinh x \sinh y$ |
| $\sinh 2x = 2 \sinh x \cosh x$ | $\cosh 2x = \cosh^2 x + \sinh^2 x$ |
| $\cosh^2 x = \frac{1}{2}(1 + \cosh 2x)$ | $\sinh^2 x = \frac{1}{2}(-1 + \cosh 2x)$ |

We will prove the identity $\cosh^2 x - \sinh^2 x = 1$ in Example 1. The proofs of the others will be left as exercises.

**EXAMPLE 1** Prove the identity $\cosh^2 x - \sinh^2 x = 1$.

**Solution** We compute

$$\cosh^2 x - \sinh^2 x = \left( \frac{e^x + e^{-x}}{2} \right)^2 - \left( \frac{e^x - e^{-x}}{2} \right)^2$$

$$= \frac{e^{2x} + 2 + e^{-2x}}{4} - \frac{e^{2x} - 2 + e^{-2x}}{4}$$

$$= \frac{4}{4} = 1$$

and this establishes the identity. ∎

## Derivatives and Integrals of Hyperbolic Functions

Since the hyperbolic functions are defined in terms of $e^x$ and $e^{-x}$, their derivatives are easily computed. For example,

$$\frac{d}{dx}(\sinh x) = \frac{d}{dx}\left( \frac{e^x - e^{-x}}{2} \right) = \frac{e^x + e^{-x}}{2} = \cosh x$$

Similarly, we can show that

$$\frac{d}{dx}(\cosh x) = \sinh x$$

Then, using these results, we can compute

$$\frac{d}{dx}(\tanh x) = \frac{d}{dx}\frac{\sinh x}{\cosh x} = \frac{\cosh x \dfrac{d}{dx}(\sinh x) - \sinh x \dfrac{d}{dx}(\cosh x)}{\cosh^2 x}$$

$$= \frac{\cosh^2 x - \sinh^2 x}{\cosh^2 x} = \frac{1}{\cosh^2 x} = \text{sech}^2 x$$

Following are the differentiation formulas together with the corresponding integration formulas for the six hyperbolic functions. We have assumed that $u = g(x)$, where $g$ is a differentiable function, and we have used the Chain Rule. The proofs of these formulas are left as exercises.

**Derivatives and Integrals of Hyperbolic Functions**

$$\frac{d}{dx}(\sinh u) = (\cosh u)\frac{du}{dx} \qquad \int \cosh u \, du = \sinh u + C$$

$$\frac{d}{dx}(\cosh u) = (\sinh u)\frac{du}{dx} \qquad \int \sinh u \, du = \cosh u + C$$

$$\frac{d}{dx}(\tanh u) = (\operatorname{sech}^2 u)\frac{du}{dx} \qquad \int \operatorname{sech}^2 u \, du = \tanh u + C$$

$$\frac{d}{dx}(\operatorname{csch} u) = -(\operatorname{csch} u \coth u)\frac{du}{dx} \qquad \int \operatorname{csch} u \coth u \, du = -\operatorname{csch} u + C$$

$$\frac{d}{dx}(\operatorname{sech} u) = -(\operatorname{sech} u \tanh u)\frac{du}{dx} \qquad \int \operatorname{sech} u \tanh u \, du = -\operatorname{sech} u + C$$

$$\frac{d}{dx}(\coth u) = -(\operatorname{csch}^2 u)\frac{du}{dx} \qquad \int \operatorname{csch}^2 u \, du = -\coth u + C$$

**EXAMPLE 2**

**a.** $\dfrac{d}{dx}\sinh(x^2 + 1) = \cosh(x^2 + 1)\dfrac{d}{dx}(x^2 + 1) = 2x\cosh(x^2 + 1)$

**b.** $\dfrac{d}{dx}\cosh^2(\ln 2x) = 2\cosh(\ln 2x)\dfrac{d}{dx}\cosh(\ln 2x)$

$$= 2\cosh(\ln 2x)\sinh(\ln 2x)\frac{d}{dx}\ln 2x$$

$$= \frac{2}{x}\cosh(\ln 2x)\sinh(\ln 2x) \qquad \blacksquare$$

**EXAMPLE 3**  Find $\int \cosh^2 3x \sinh 3x \, dx$.

**Solution**  Let $u = 3x$ so that $du = 3\,dx$ or $dx = \frac{1}{3}\,du$. Then

$$\int \cosh^2 3x \sinh 3x \, dx = \frac{1}{3}\int \cosh^2 u \sinh u \, du$$

Next, let $v = \cosh u$ so that $dv = \sinh u \, du$. Then

$$\frac{1}{3}\int \cosh^2 u \sinh u \, du = \frac{1}{3}\int v^2 \, dv = \frac{1}{9}v^3 + C$$

So

$$\int \cosh^2 3x \sinh 3x \, dx = \frac{1}{9}\cosh^3 3x + C \qquad \blacksquare$$

■ **Inverse Hyperbolic Functions**

If you examine Figures 2a and 3a, you will notice that both $\sinh x$ and $\tanh x$ are *one-to-one* on $(-\infty, \infty)$ and hence have inverse functions that we denote by $\sinh^{-1} x$ and $\tanh^{-1} x$, respectively. Also, an examination of Figure 2b shows that $\cosh x$ is one-to-one on $[0, \infty)$, so, if restricted to this domain, it has an inverse, $\cosh^{-1} x$. By examining the graphs of the other hyperbolic functions and making the necessary restrictions on their domains, we are able to define the other inverse hyperbolic functions.

---

**DEFINITIONS** **Inverse Hyperbolic Functions**

|  | | | Domain |
|---|---|---|---|
| $y = \sinh^{-1} x$ | if and only if | $x = \sinh y$ | $(-\infty, \infty)$ |
| $y = \cosh^{-1} x$ | if and only if | $x = \cosh y$ | $[1, \infty)$ |
| $y = \tanh^{-1} x$ | if and only if | $x = \tanh y$ | $(-1, 1)$ |
| $y = \operatorname{csch}^{-1} x$ | if and only if | $x = \operatorname{csch} y$ | $(-\infty, 0) \cup (0, \infty)$ |
| $y = \operatorname{sech}^{-1} x$ | if and only if | $x = \operatorname{sech} y$ | $(0, 1]$ |
| $y = \coth^{-1} x$ | if and only if | $x = \coth y$ | $(-\infty, -1) \cup (1, \infty)$ |

---

The graphs of $y = \sinh^{-1} x$, $y = \cosh^{-1} x$, and $y = \tanh^{-1} x$ are shown in Figure 4.

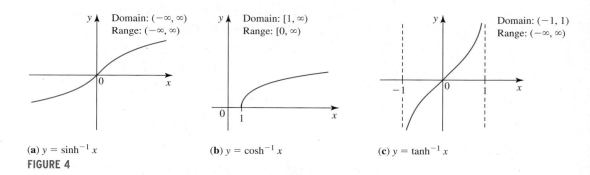

Domain: $(-\infty, \infty)$
Range: $(-\infty, \infty)$

Domain: $[1, \infty)$
Range: $[0, \infty)$

Domain: $(-1, 1)$
Range: $(-\infty, \infty)$

**(a)** $y = \sinh^{-1} x$   **(b)** $y = \cosh^{-1} x$   **(c)** $y = \tanh^{-1} x$

**FIGURE 4**

Since the hyperbolic functions are defined in terms of exponential functions, it seems natural that the inverse hyperbolic functions should be expressible in terms of logarithmic functions.

**EXAMPLE 4** Show that $\sinh^{-1} x = \ln\left(x + \sqrt{x^2 + 1}\right)$.

**Solution** Let $y = \sinh^{-1} x$. Then

$$x = \sinh y = \frac{e^y - e^{-y}}{2}$$

or

$$e^y - 2x - e^{-y} = 0$$

On multiplying both sides of this equation by $e^y$, we obtain

$$e^{2y} - 2xe^y - 1 = 0$$

which is a quadratic in $e^y$. Using the quadratic formula, we have

$$e^y = \frac{2x \pm \sqrt{4x^2 + 4}}{2} = x \pm \sqrt{x^2 + 1}$$

Only the root $x + \sqrt{x^2 + 1}$ is admissible. To see why, observe that $e^y > 0$, but $x - \sqrt{x^2 + 1} < 0$, since $x < \sqrt{x^2 + 1}$. Therefore, we have

$$e^y = x + \sqrt{x^2 + 1}$$

so

$$y = \ln\left(x + \sqrt{x^2 + 1}\right)$$

that is,

$$\sinh^{-1} x = \ln\left(x + \sqrt{x^2 + 1}\right) \qquad \blacksquare$$

Proceeding in a similar manner, we can obtain the representations of the other five inverse hyperbolic functions in terms of logarithmic functions. Three such representations follow.

---

**Representations of Inverse Hyperbolic Functions
in Terms of Logarithmic Functions**

|  | Domain |
|---|---|
| $\sinh^{-1} x = \ln\left(x + \sqrt{x^2 + 1}\right)$ | $(-\infty, \infty)$ |
| $\cosh^{-1} x = \ln\left(x + \sqrt{x^2 - 1}\right)$ | $[1, \infty)$ |
| $\tanh^{-1} x = \dfrac{1}{2} \ln\left(\dfrac{1 + x}{1 - x}\right)$ | $(-1, 1)$ |

---

## ■ Derivatives of Inverse Hyperbolic Functions

The derivatives of the inverse hyperbolic functions can be found by differentiating the function in question directly. For example,

$$\frac{d}{dx} \sinh^{-1} x = \frac{d}{dx} \ln\left(x + \sqrt{x^2 + 1}\right)$$

$$= \frac{1}{x + \sqrt{x^2 + 1}}\left[1 + \frac{1}{2}(x^2 + 1)^{-1/2}(2x)\right]$$

$$= \frac{1}{x + \sqrt{x^2 + 1}} \cdot \frac{\sqrt{x^2 + 1} + x}{\sqrt{x^2 + 1}}$$

$$= \frac{1}{\sqrt{x^2 + 1}}$$

Alternatively, we may proceed as follows:

$$y = \sinh^{-1} x \qquad \text{if and only if} \qquad x = \sinh y$$

Differentiating this last equation implicitly with respect to $x$, we obtain

$$\frac{d}{dx}(x) = \frac{d}{dx}(\sinh y)$$

$$1 = (\cosh y)\frac{dy}{dx}$$

or

$$\frac{dy}{dx} = \frac{1}{\cosh y} = \frac{1}{\sqrt{\sinh^2 y + 1}} = \frac{1}{\sqrt{x^2 + 1}}$$

as before.

Using techniques such as these, we obtain the following formulas for differentiating the inverse hyperbolic functions (once again, $u = g(x)$, where $g$ is a differentiable function).

---

**Derivatives of Inverse Hyperbolic Functions**

$$\frac{d}{dx}\sinh^{-1} u = \frac{1}{\sqrt{u^2 + 1}}\frac{du}{dx} \qquad \frac{d}{dx}\cosh^{-1} u = \frac{1}{\sqrt{u^2 - 1}}\frac{du}{dx}$$

$$\frac{d}{dx}\tanh^{-1} u = \frac{1}{1 - u^2}\frac{du}{dx} \qquad \frac{d}{dx}\operatorname{csch}^{-1} u = -\frac{1}{|u|\sqrt{u^2 + 1}}\frac{du}{dx}$$

$$\frac{d}{dx}\operatorname{sech}^{-1} u = -\frac{1}{u\sqrt{1 - u^2}}\frac{du}{dx} \qquad \frac{d}{dx}\coth^{-1} u = \frac{1}{1 - u^2}\frac{du}{dx}$$

---

**EXAMPLE 5** Find the derivative of $y = x^2 \operatorname{sech}^{-1} 3x$.

**Solution** We have

$$\frac{dy}{dx} = \operatorname{sech}^{-1} 3x \cdot \frac{d}{dx}(x^2) + x^2\frac{d}{dx}\operatorname{sech}^{-1} 3x \qquad \text{Use the Product Rule.}$$

$$= 2x\operatorname{sech}^{-1} 3x - x^2\left[\frac{1}{3x\sqrt{1 - 9x^2}}\right]\frac{d}{dx}(3x)$$

$$= 2x\operatorname{sech}^{-1} 3x - \frac{x}{\sqrt{1 - 9x^2}}$$

■

## ■ An Application

**EXAMPLE 6** **Length of a Power Line** A power line is suspended between two towers as depicted in Figure 5. The shape of the cable is a *catenary* with equation

$$y = 80\cosh\frac{x}{80} \qquad -100 \le x \le 100$$

where $x$ is measured in feet. Find the length of the cable.

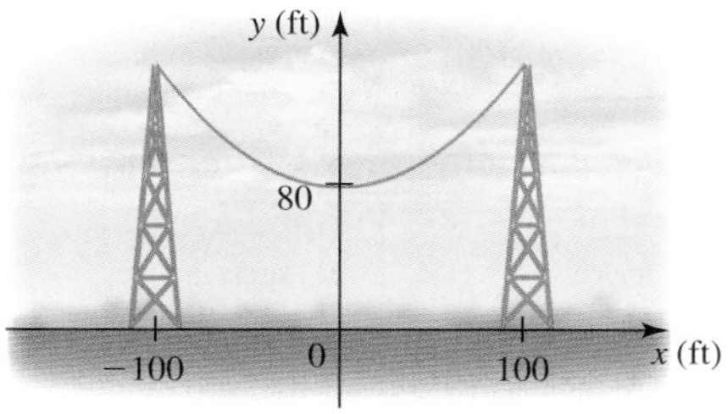

**FIGURE 5**
The shape of the hanging
cable is a catenary.

**Solution** Taking advantage of the symmetry of the situation, we see that the required length is given by

$$L = 2\int_0^{100} \sqrt{1 + \left(\frac{dy}{dx}\right)^2}\, dx$$

But

$$\frac{dy}{dx} = \frac{d}{dx}\left[80 \cosh \frac{x}{80}\right] = 80 \sinh \frac{x}{80} \cdot \frac{d}{dx}\left(\frac{x}{80}\right) = \sinh \frac{x}{80}$$

So

$$\sqrt{1 + \left(\frac{dy}{dx}\right)^2} = \sqrt{1 + \sinh^2\left(\frac{x}{80}\right)} = \sqrt{1 + \cosh^2\left(\frac{x}{80}\right) - 1}$$

$$= \sqrt{\cosh^2\left(\frac{x}{80}\right)} = \cosh \frac{x}{80}$$

Therefore,

$$L = 2\int_0^{100} \cosh \frac{x}{80}\, dx$$

$$= 2\left[80 \sinh \frac{x}{80}\right]_0^{100} \qquad \text{Use the substitution } u = \frac{x}{80}.$$

$$= 160 \sinh \frac{100}{80} = 160 \sinh \frac{5}{4}$$

or approximately 256 ft. ∎

## 5.8 CONCEPT QUESTIONS

1. Define (a) sinh $x$, (b) cosh $x$, and (c) tanh $x$.
2. State the derivative of (a) sinh $x$, (b) cosh $x$, and (c) tanh $x$.
3. Write an antiderivative of (a) sech$^2 u$, (b) sech $u$ tanh $u$, and (c) csch$^2 u$.

4. Define (a) csch$^{-1} x$, (b) sech$^{-1} x$, and (c) coth$^{-1} x$.
5. Write (a) sinh$^{-1} x$, (b) cosh$^{-1} x$, and (c) tanh$^{-1} x$ in terms of logarithmic functions.
6. State the derivative of (a) sinh$^{-1} u$, (b) cosh$^{-1} u$, and (c) tanh$^{-1} u$ with respect to $x$.

# 5.8 EXERCISES

*In Exercises 1–6, find the value of the expression accurate to four decimal places.*

1. **a.** $\sinh 2$    **b.** $\cosh 4$    **c.** $\operatorname{sech} 3$

2. **a.** $\operatorname{csch} 3$    **b.** $\tanh(-2)$    **c.** $\coth 5$

3. **a.** $\cosh 0$    **b.** $\operatorname{sech}(-1)$    **c.** $\operatorname{csch}(\ln 2)$

4. **a.** $\sinh^{-1} 1$    **b.** $\cosh^{-1} 2$    **c.** $\operatorname{sech}^{-1} \dfrac{1}{3}$

5. **a.** $\operatorname{csch}^{-1} 2$    **b.** $\operatorname{csch}^{-1}(-2)$    **c.** $\coth^{-1} \dfrac{3}{2}$

6. **a.** $\tanh^{-1}\left(-\dfrac{1}{2}\right)$    **b.** $\cosh^{-1}(\ln 5)$    **c.** $\tanh^{-1}(\sinh 0)$

*In Exercises 7–16, prove the identity.*

7. $\cosh(-x) = \cosh x$

8. $\tanh(-x) = -\tanh x$

9. $\operatorname{sech}^2 x + \tanh^2 x = 1$

10. $\sinh^2 x = \dfrac{\cosh 2x - 1}{2}$

11. $\cosh^2 x = \dfrac{1 + \cosh 2x}{2}$

12. $\sinh 2x = 2 \sinh x \cosh x$

13. $\cosh 2x = \cosh^2 x + \sinh^2 x$

14. $\sinh(x + y) = \sinh x \cosh y + \cosh x \sinh y$

15. $\cosh(x + y) = \cosh x \cosh y + \sinh x \sinh y$

16. $\tanh(x + y) = \dfrac{\tanh x + \tanh y}{1 + \tanh x \tanh y}$

17. If $\sinh x = \frac{4}{3}$, find the values of the other hyperbolic functions at $x$.

18. If $\cosh x = \frac{5}{4}$, find the values of the other hyperbolic functions at $x$.

*In Exercises 19–54, find the derivative of the function.*

19. $f(x) = \sinh 3x$

20. $f(x) = \cosh(2x + 1)$

21. $g(x) = \tanh(1 - 3x)$

22. $h(x) = \operatorname{sech}(x^2)$

23. $f(t) = e^t \sinh t$

24. $y = \coth \dfrac{1}{x}$

25. $F(x) = \ln(\cosh x)$

26. $y = \ln(\sinh 3x)$

27. $g(u) = \tanh(\cosh u^2)$

28. $h(s) = \coth(\cosh 2s)$

29. $f(t) = \cosh^2(3t^2 + 1)$

30. $f(x) = \sinh 2x \cosh 4x$

31. $g(v) = v \sinh v^2$

32. $F(t) = \cosh \sqrt{2t^2 + 1}$

33. $f(x) = \tanh(e^{2x} + 1)$

34. $y = \cosh \sqrt[3]{x^2 + 1}$

35. $f(x) = (\cosh x - \sinh x)^{2/3}$

36. $y = e^{\sinh 2t}$

37. $g(x) = \tanh^{-1}(\cosh x)$

38. $f(x) = \sqrt{2 + \coth 3x}$

39. $f(x) = \dfrac{\sinh x}{1 + \cosh x}$

40. $g(x) = \dfrac{\sinh x}{x}$

41. $y = \dfrac{\cosh^{-1} t}{1 + \tanh 2t}$

42. $f(x) = e^{-x} \operatorname{sech} 2x$

43. $f(x) = \sinh^{-1} 3x$

44. $g(x) = \tanh^{-1} \dfrac{x}{2}$

45. $y = \sqrt{\cosh^{-1} 2x}$

46. $f(x) = \operatorname{sech}^{-1} x^3$

47. $f(x) = \operatorname{sech}^{-1} \sqrt{2x + 1}$

48. $y = e^x \operatorname{sech}^{-1} x$

49. $y = x \cosh^{-1} x^2$

50. $g(x) = \ln(\tanh^{-1} x)$

51. $f(x) = \operatorname{sech}^{-1} \sqrt{x}$

52. $h(x) = \cosh^{-1}(\sinh x)$

53. $y = \sqrt{9x^2 - 1} - 3 \cosh^{-1} 3x$

54. $y = 2x \coth^{-1} 2x - \ln \sqrt{1 - 4x^2}$

*In Exercises 55–62, find the given integral.*

55. $\displaystyle\int \cosh(2x + 3)\,dx$

56. $\displaystyle\int \dfrac{\sinh \sqrt{x}}{\sqrt{x}}\,dx$

57. $\displaystyle\int \sqrt{\sinh x}\, \cosh x\,dx$

58. $\displaystyle\int \tanh x\,dx$

59. $\displaystyle\int \coth 3x\,dx$

60. $\displaystyle\int \operatorname{sech}^2(3x - 1)\,dx$

61. $\displaystyle\int \dfrac{\sinh x}{1 + \cosh x}\,dx$

62. $\displaystyle\int \dfrac{\operatorname{sech}\left(\dfrac{1}{x}\right)\tanh\left(\dfrac{1}{x}\right)}{x^2}\,dx$

63. Find the volume of the solid obtained by revolving the region under the graph of the catenary $y = a \cosh(x/a)$ on the interval $[-b, b]$ ($b > 0$) about the $x$-axis.

64. The arc of the catenary $y = a \cosh(x/a)$ for $x$ between $x = 0$ and $x = b$ is revolved about the $x$-axis. Show that the surface area $S$ and the volume $V$ of the resulting solid of revolution are related by the formula $S = 2V/a$.

65. Refer to Figure 5. Suppose that the cable has a constant weight density of $W$ lb/ft. Then the tension on the cable is

$$T = T_0 \cosh \dfrac{Wx}{T_0} \qquad -b \le x \le b$$

where $T_0$ is the tension at the lowest point. Find the average tension on the cable.

**66.** The velocity of a body of mass $m$ falling from rest through a viscous medium is given by

$$v(t) = \sqrt{\frac{mg}{k}} \tanh\left(\sqrt{\frac{gk}{m}}\, t\right)$$

where $g$ is the acceleration of gravity and $k$ is a positive constant that depends on the viscosity of the medium.
**a.** Find $\lim_{t \to \infty} v(t)$.
**b.** Plot the graph of $v$ taking $m = 2$, $g = 32$, and $k = 8$.
**Note:** This limiting velocity of the body is called the *terminal velocity*.

 **67. Damped Harmonic Motion** The equation of motion of a weight attached to a spring and a dashpot damping device is

$$x(t) = -\frac{1}{\sqrt{2}}\, e^{-4t} \sinh 2\sqrt{2}t$$

where $x(t)$, measured in feet, is the displacement from the equilibrium position of the spring system and $t$ is measured in seconds.
**a.** Find the initial position and the initial velocity of the weight.
**b.** Plot the graph of $x(t)$.

The system in equilibrium (The positive direction is downward.)

 **68. Heat-Seeking Missiles** In a test conducted on a heat-seeking Missile $A$, the target missile $B$, which is initially at a distance of $b$ miles from Missile $A$, is launched vertically upward. Assume that Missile $A$ travels at a constant speed $v_A$, that Missile $B$ travels at a constant speed $v_B$ ($v_A > v_B$), and that Missile $A$, which is launched from the origin, is always pointed at Missile $B$. Then the trajectory of Missile $A$ is

$$y = \frac{b}{2}\left[\frac{\left(1 - \dfrac{x}{b}\right)^{1+c}}{1+c} - \frac{\left(1 - \dfrac{x}{b}\right)^{1-c}}{1-c}\right] + \frac{bc}{1-c^2}$$

where $c = v_B/v_A$.

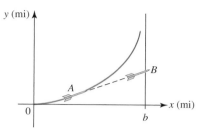

The trajectory of Missile $A$ is a pursuit curve.

**a.** Find the point at which Missile $A$ intercepts Missile $B$.
**b.** Show that

$$\frac{dy}{dx} = -\sinh\left[c \ln\left(1 - \frac{x}{b}\right)\right]$$

**c.** Suppose that $b = 1$ and $c = \frac{1}{2}$. Show that the distance $D$ traveled by Missile $A$ for the intercept is $1\frac{1}{3}$ mi.
**Hint:** $D = \displaystyle\int_0^1 \sqrt{1 + \left(\frac{dy}{dx}\right)^2}\, dx$
**d.** Plot the graph of the trajectory of the heat-seeking missile taking $b = 1$ and $c = \frac{1}{2}$.

**69.** The minimum-surface-of-revolution problem may be stated as follows: Of all curves joining two fixed points, find the one that, when revolved about the $x$-axis, will generate a surface of minimum area. It can be shown that the solution to the problem is a catenary. The resulting surface of revolution is called a ***catenoid***. Suppose a catenary described by the equation

$$y = \cosh x \qquad a \le x \le b$$

is revolved about the $x$-axis. Find the surface area of the resulting catenoid.

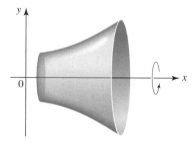

**Hint:** Use the identity $\cosh^2 x = \dfrac{1 + \cosh 2x}{2}$.
**Note:** A soap bubble formed by two parallel circular rings that are close to each other is an example of a catenoid.

**70.** Find the volume of the solid of revolution that is obtained by revolving the region bounded by the graph of $y = (x^2 - 1)^{3/4}$, the x-axis, and the lines $x = 1$ and $x = 2$, about the x-axis. **Hint:** Use the substitution $x = \cosh u$.

**71.** Find the centroid of the region under the graph of $f(x) = \cosh x$ on $[-a, a]$.

**72.** A power line is suspended between two towers that are 200 ft apart, as shown in the figure. The shape of the cable is a *catenary* with equation

$$y = 80 \cosh \frac{x}{80} \qquad -100 \le x \le 100$$

where x is measured in feet. What is the angle $\theta$ that the line makes with the pole?

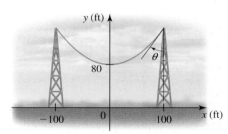

**73.** Prove that $\dfrac{d}{dx} \cosh u = (\sinh u) \dfrac{du}{dx}$.

**74.** Prove that $\dfrac{d}{dx} \operatorname{csch} u = -(\operatorname{csch} u \coth u) \dfrac{du}{dx}$.

**75.** Prove that $\dfrac{d}{dx} \operatorname{sech} u = -(\operatorname{sech} u \tanh u) \dfrac{du}{dx}$.

**76.** Prove that $\dfrac{d}{dx} \coth u = -(\operatorname{csch}^2 u) \dfrac{du}{dx}$.

*In Exercises 77–80, determine whether the statement is true or false. If it is true, explain why it is true. If it is false, explain why or give an example to show why it is false.*

**77.** $(\sinh x + \cosh x)^3 > 0$ for all x in $(-\infty, \infty)$.

**78.** $\dfrac{d}{dx} (\coth^2 x - \operatorname{csch}^2 x)^5 = 0$

**79.** $\displaystyle\int_{-\pi}^{\pi} (\cos x)\sinh x \, dx = 0$

**80.** $\displaystyle\int_{-3}^{3} x^2 \operatorname{sech} x \, dx = 2 \int_{0}^{3} x^2 \operatorname{sech} x \, dx$

# CHAPTER 5 REVIEW

## CONCEPT REVIEW

*In Exercises 1–12, fill in the blanks.*

**1. a.** If f and g are continuous on $[a, b]$ and $f(x) \ge g(x)$ for all x in $[a, b]$, then the area of the region between the graphs of f and g and the vertical lines $x = a$ and $x = b$ is $A = $ _____.

   **b.** If f and g are continuous on $[a, b]$, then the area of the region bounded by the graphs of f and g and the vertical lines $x = a$ and $x = b$ is $A = $ _____.

**2. a.** If f is a continuous nonnegative function on $[a, b]$, then the volume of the solid obtained by revolving the region R under the graph of f on $[a, b]$ about the x-axis is $V = $ _____.

   **b.** If g is a continuous nonnegative function on $[c, d]$, then the volume of the solid obtained by revolving the region R between the graph of $x = g(y)$ and the y-axis on $[c, d]$ about the y-axis is $V = $ _____.

**3.** If f and g are continuous on $[a, b]$ and $f(x) \ge g(x) \ge 0$ for all x in $[a, b]$, then the volume of the solid obtained by revolving the region between the graphs of f and g on $[a, b]$ about the x-axis is $V = $ _____.

**4.** If S is a solid bounded by planes that are perpendicular to the x-axis at $x = a$ and $x = b$ and the cross sectional area of S at any point x in $[a, b]$ is $A(x)$, where A is continuous on $[a, b]$, then the volume of S is $V = $ _____.

**5. a.** If f is a continuous nonnegative function on $[a, b]$ and R is the region under the graph of f on $[a, b]$, then the volume of the solid obtained by revolving R about the y-axis is $V = $ _____.

   **b.** If the region R bounded by the graphs of $x = f(y)$, where f is nonnegative, $x = 0$, $y = c$, and $y = d$, is revolved about the x-axis, where $0 \le c \le d$, then the volume of the resulting solid is $V = $ _____.

**6. a.** If f is smooth on $[a, b]$, then the arc length of the graph of f from $P(a, f(a))$ to $Q(b, f(b))$ is $L = $ _____.

   **b.** If g is smooth on $[c, d]$, then the arc length of the graph of $x = g(y)$ from $P(g(c), c)$ to $Q(g(d), d)$ is $L = $ _____.

**7.** If f is smooth on $[a, b]$, then the arc length function s for the graph of f is defined by $s(x) = $ _____ with domain _____. The arc length differential is $ds = $ _____ or $(ds)^2 = $ _____.

8. **a.** If $f$ is a nonnegative smooth function on $[a, b]$, then the surface area of the surface obtained by revolving the graph of $f$ about the $x$-axis is $S =$ _____.

   **b.** If $g$ is a smooth function on $[c, d]$ with $g(y) \geq 0$, then the surface area of the surface obtained by revolving the graph of $g$ about the $y$-axis is $S =$ _____.

9. If $F$ is continuous on $[a, b]$, then the work done by the force $F(x)$ in moving a body from $x = a$ to $x = b$ is $W =$ _____.

10. The force $F$ exerted by a fluid of constant weight density $\delta$ on one side of a submerged vertical plate from $y = c$ to $y = d$, where $c \leq d$, is given by $F =$ _____, where $L(y)$ is the horizontal length of the plate at $y$ and $h(y)$ is the depth of the fluid at $y$.

11. If $L$ denotes a lamina of constant mass density $\rho$ and $L$ has the shape of the region $R$ under the graph of a nonnegative continuous function $f$ on $[a, b]$, then

   **a.** The mass of $L$ is _____.

   **b.** The moments of $L$ about the $x$-axis and $y$-axis are $M_x =$ _____ and $M_y =$ _____.

   **c.** The center of mass of $L$ is located at $(\bar{x}, \bar{y})$, where $\bar{x} =$ _____ and $\bar{y} =$ _____.

12. If $R$ is the region bounded by the graphs of two continuous functions $f$ and $g$ on $[a, b]$, where $f(x) \geq g(x)$ for all $x$ in $[a, b]$, then the centroid $(\bar{x}, \bar{y})$ of $R$ is given by $\bar{x} =$ _____, and $\bar{y} =$ _____, where $A =$ _____.

## REVIEW EXERCISES

*In Exercises 1–16, sketch the region bounded by the graphs of the equations and find the area of the region.*

1. $y = \dfrac{1}{x^3}$,   $y = 0$,   $x = 1$,   $x = 2$

2. $y = \dfrac{1}{\sqrt{x}}$,   $y = 0$,   $x = 1$,   $x = 2$

3. $y = x^2 + 2$,   $y = x + 1$,   $x = 0$,   $x = 1$

4. $y = 2x^3 - 1$,   $y = x - 1$,   $x = 1$,   $x = 2$

5. $y = 2x^2 + 2x - 3$,   $y = 3x^2 + 2x - 4$

6. $y = x^3$,   $x = y^3$

7. $y = \sqrt{x - 1}$,   $y = 2$,   $y = 0$,   $x = 0$

8. $y = (x + 1)^3$,   $y = x + 1$

9. $x = (y - 1)^2$,   $x = 1$

10. $x = \sqrt{y - 1}$,   $x = y$,   $x = 0$,   $y = 2$

11. $x = y^2 - 1$,   $x = 1 - y$

12. $y = \cos x$,   $y = 1 - \dfrac{2}{\pi} x$,   $x = -\dfrac{\pi}{2}$,   $x = \dfrac{\pi}{2}$

13. $y = \cos x$,   $y = -\sin x$,   $x = 0$,   $x = \dfrac{\pi}{2}$

14. $y = \sec^2 x$,   $y = \sin x$,   $x = -\dfrac{\pi}{4}$,   $x = \dfrac{\pi}{4}$

15. $\sqrt{x} + \sqrt{y} = 1$,   $x + y = 1$

16. $(y - x)^2 = x^3$,   $x = 1$

17. $y = e^{2x}$,   $y = -\dfrac{1}{x}$,   $x = 1$;   $x = 2$

*In Exercises 18–26, find the volume of the solid generated by revolving the region bounded by the graphs of the equations about the indicated line.*

18. $y = x - x^2$,   $y = 0$;   the $x$-axis

19. $y = \sqrt{x + 1}$,   $y = 0$,   $x = 3$;   the $x$-axis

20. $y = x^2$,   $y = x^3$;   the line $y = 2$

21. $y = x^{1/3}$,   $y = x$;   the $y$-axis

22. $y = \dfrac{1}{\sqrt{1 + x^2}}$,   $y = 0$,   $x = 0$,   $x = 1$;   the $y$-axis

23. $y = x^2$,   $y = x^3$;   the line $x = 2$

24. $y = 1 - x^2$,   $y = -x + 1$;   the line $x = -2$

25. $y = \cos x^2$,   $x \geq 0$,   $y = 0$;   the $y$-axis

26. $y = \dfrac{\ln x}{x^2}$,   $y = 0$,   $x = e$;   the $y$-axis

*In Exercises 27–32, find $dy/dx$.*

27. $y = \ln(\sinh 2x)$

28. $y = e^{\tanh 3x}$

29. $y = \sinh^{-1}(\tanh x)$

30. $y = \sin^{-1}\left(\dfrac{x + 1}{x + 2}\right)$

31. $y = \tan^{-1}(\cos^{-1}\sqrt{x})$

32. $x \cosh y + e^{\sinh y} = 10$

33. Find $\displaystyle\int \sinh^2 x \, dx$.

34. Find $\displaystyle\int \cosh^3 x \, dx$.

35. Find $\displaystyle\int \sinh 2t \, dt$.

36. If $f(x) = \displaystyle\int_{\ln x}^{\sqrt{x}} \sinh t \, dt$, where $x > 0$, find $f'(x)$.

**37.** Find the area of the region completely enclosed by the parabola $y = x^2 - 6x + 11$ and the line passing through the point $(1, 0)$ and the vertex of the parabola.

**38.** The base of a solid is a circular disk of radius 2, and the cross sections perpendicular to the base are isosceles right triangles with the hypotenuse lying on the base. Find the volume of the solid.

**39.** A monument stands 50 m high. A horizontal cross section $x$ m from the top is an equilateral triangle with sides $x/5$ m. What is the volume of the monument?

**40.** Write an integral giving the arc length of the graph of the function $f(x) = x^2 + x^3$ on $[0, 1]$. Do not evaluate the integral.

**41.** Find the length of the graph of $y = (9 - x^{2/3})^{3/2}$ on the interval $[1, 27]$.

**42.** Find the length of the graph of

$$y = \int_1^x \sqrt{t^{3/2} - 1}\, dt \qquad 1 \le x \le 16$$

**43.** Find the arc length of the graph of $y = \frac{1}{4}x^2 - \frac{1}{2}\ln x$ on $[1, 2]$.

**44.** Find the arc length of the graph of $y = \ln \dfrac{e^x + 1}{e^x - 1}$ on $[1, 2]$.

**45.** Find the area of the surface obtained by revolving the portion of the graph of $y = \frac{1}{2}x^2$ that lies below $y = \frac{3}{2}$ about the $y$-axis.

 **46.** Find the area of the surface formed by revolving the graph of $y = 2\sqrt{x} - x$, $0 \le x \le 2$, about the $x$-axis, accurate to four decimal places.

**47.** Write an integral giving the area of the surface obtained by revolving the graph of $f(x) = x^2 + \dfrac{1}{x}$ on $[1, 2]$ about the $x$-axis. Do not evaluate the integral.

**48.** The region under the graph of $y = x^2/(1 + x^4)$ on the interval $[0, 1]$ is revolved about the $y$-axis. Find the volume of the resulting solid.

**49.** Consider the portion of the unit circle lying in the first quadrant.

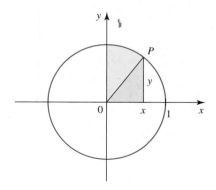

**a.** By considering the area of the shaded region, show that

$$\int_0^x \sqrt{1 - t^2}\, dt = \frac{1}{2}\sin^{-1} x + \frac{x}{2}\sqrt{1 - x^2}$$

and hence

$$\sin^{-1} x = 2\int_0^x \sqrt{1 - t^2}\, dt - x\sqrt{1 - x^2}$$

**b.** By differentiating the last equation in part (a) with respect to $x$, show that

$$\frac{d}{dx}(\sin^{-1} x) = \frac{1}{\sqrt{1 - x^2}}$$

**50.** When a particle is at the point $x$ on the $x$-axis, it is acted upon by a force of $x + \cos \pi x$ dynes. Find the work done by the force in moving the particle from $x = 1$ to $x = 2$, where $x$ is measured in centimeters.

**51.** A force of 6 lb is required to stretch a spring $1\frac{1}{2}$ in. beyond its natural length. Find the work required to stretch the spring 2 in. beyond its natural length.

**52.** A 1200-lb elevator at a construction site is suspended by a cable that weighs 10 lb/ft. How much work is done in raising the elevator from the ground to a height of 20 ft?

**53.** A tank having the shape of an inverted right circular cone with a base radius of 10 ft and a height of 15 ft is filled with water weighing 62.4 lb/ft$^3$. Find the work required to empty the tank by pumping the water over the rim of the tank.

**54.** A semicircular plate of radius $r$ is submerged vertically in a liquid weighing 50 lb/ft$^3$ in such a way that the diameter of the plate is flush with the surface of the liquid. Find the force exerted by the liquid on one side of the plate.

**55.** A rectangular swimming pool is 50 ft long, 20 ft wide, and 3 ft deep at the shallow end and 7 ft deep at the deep end. Its bottom is an inclined plane. If the pool is completely filled with water, find the force exerted by the water on each vertical wall of the swimming pool.

*In Exercises 56–59, find the centroid of the region bounded by the graphs of the equations.*

**56.** $y = \sqrt{x}$, $\quad y = \dfrac{1}{2}x$

**57.** $y = x^2 - 2x$, $\quad y = 0$

**58.** $y = \sqrt{9 - x^2}$, $\quad y = 0$

**59.** $y = 2x^2 - 4x$, $\quad y = 2x - x^2$

## PROBLEM-SOLVING TECHNIQUES

In the following example we invoke a rule of differentiation to solve a problem involving integration.

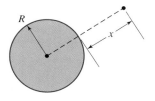

**EXAMPLE** **Height Reached by a Projectile**  A projectile is launched vertically upward from the earth's surface with an initial velocity $v_0$ of magnitude less than the escape velocity. If only the earth's influence is taken into consideration, then the differential equation governing its motion is

$$\frac{d^2x}{dt^2} = -\frac{gR^2}{(x+R)^2}$$

where $t$ is in seconds, $x$ is the distance of the projectile from the surface of the earth in miles, $R$ is the radius of the earth, and $g$ is the constant of acceleration due to gravity. Show that the maximum height reached by the projectile is $v_0^2 R/(2gR - v_0^2)$.

**Solution**  At first glance, we are tempted to integrate the given equation with respect to $t$. But the right-hand side of the equation is a function of the unknown variable $x$! The trick here is to use the Chain Rule to rewrite the left-hand side of the differential equation. Thus,

$$\frac{d^2x}{dt^2} = \frac{d}{dt}\left(\frac{dx}{dt}\right) = \frac{dv}{dt} = \frac{dv}{dx} \cdot \frac{dx}{dt} = \frac{dv}{dx} v$$

The given equation now reads

$$v\frac{dv}{dx} = -\frac{gR^2}{(x+R)^2}$$

Integrating both sides with respect to $x$, we obtain

$$\int v\frac{dv}{dx}\,dx = -gR^2\int (x+R)^{-2}dx$$

$$\frac{1}{2}v^2 = \frac{gR^2}{x+R} + C$$

or

$$v^2 = \frac{2gR^2}{x+R} + C$$

We have used the substitution $u = x + R$ to find the integral on the right-hand side. To find $C$, we use the initial condition $v = v_0$ if $x = 0$, obtaining

$$v_0^2 = \frac{2gR^2}{R} + C \quad \text{or} \quad C = v_0^2 - 2gR$$

Therefore,

$$v^2 = \frac{2gR^2}{x+R} + v_0^2 - 2gR$$

At the maximum height, $v = 0$, so we have

$$0 = \frac{2gR^2}{x + R} + v_0^2 - 2gR$$

$$\frac{2gR^2}{x + R} = 2gR - v_0^2$$

$$x + R = \frac{2gR^2}{2gR - v_0^2}$$

$$x = \frac{2gR^2}{2gR - v_0^2} - R = \frac{2gR^2 - 2gR^2 + v_0^2 R}{2gR - v_0^2}$$

$$= \frac{v_0^2 R}{2gR - v_0^2}$$

as was to be shown.    ■

## CHALLENGE PROBLEMS

1.  The figure shows the region bounded by the parabola $y = x^2 - 2x + 2$, the line tangent to it at the point $P(a, b)$, and the $y$-axis. If the area of the region is 9 square units, what are the values of $a$ and $b$?

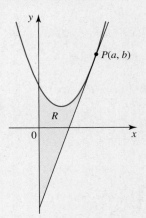

2.  Find the area of the region completely enclosed by the parabolas $x = y^2$ and $x = \frac{3}{4}y^2 + 1$.

3.  Let $R$ be the region bounded by the graph of the function $f(x) = x\sqrt{1 - x^2}$ and the positive $x$-axis. Find the parabola $y = cx^2$ that divides $R$ into two subregions of equal area.

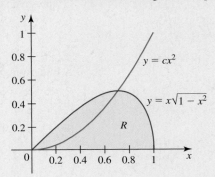

4.  A trough has a cross section in the form of a parabola with base $a$ ft and depth $h$ ft. What is the average depth of the trough?

5.  Let $f$ be a nonnegative, continuous function on the interval $[0, \infty)$. Suppose that the arc length of the graph of $f$ from $x = 0$ to $x = b$ is $b + \frac{2}{3}b^3$ and that $f(0) = \frac{2}{3}$. Find the function $f$.

6. A solid has a circular base of radius $R$, and its parallel cross sections perpendicular to its base are parabolas of height $h$. Find the volume of the solid.

7. Find the area of the region bounded by the graph of the function $f(x) = x^3 - 6x^2$, the $x$-axis, and the vertical lines passing through the inflection point and the relative minimum of $f$.

8. A semicircle of radius $a$ is revolved about an axis parallel to the straight edge of the semicircle and located at a distance $b > a$ from the edge. Find the volume of the resulting solid of revolution.

9. The region bounded by the hyperbola $x^2/a^2 - y^2/b^2 = 1$, the line $bx - 2ay = 0$, and the $x$-axis is revolved about the $x$-axis. What is the volume of the resulting solid of revolution?

10. A solid ball of mass $M$ and radius $R$ rotates with an angular velocity $\omega$ about an axis through its center. Calculate the work required to stop the ball.
    **Hint:** Calculate the kinetic energy $\left(\frac{1}{2} mv^2\right)$ of the ball.

11. The rate at which water evaporates from a pond is proportional to its surface area. Show that the depth of the water decreases at a constant rate and is not dependent on the shape of the pond.
    **Hint:** If $V(t)$ denotes the volume of water in the pond at time $t$ and $A(x)$ the surface area of the pond when the water has depth $x$, then $dV/dt = -kA(x)$, where $k$ is the constant of proportionality.

12. **Buffon's Needle Problem** A needle of length $l$ is dropped onto a board that is covered with parallel lines spaced at a distance $w$ units apart where $w > l$. What is the probability that the needle will intersect one of the lines? To solve the problem, refer to the figure and observe that the needle intersects one of the lines if and only if $|(l/2) \cos \theta| > y$, where $y$ is the distance from the center of the needle to the nearest line.

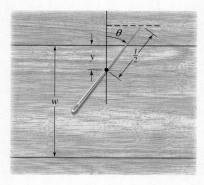

We can think of the set of all positions assumed by the needle as being associated with the rectangular region $S = \{(y, \theta) \mid 0 \le y \le \frac{w}{2} \text{ and } 0 \le \theta \le 2\pi\}$. The set of all positions assumed by the needle when it intersects a line can then be associated with the region $R = \{(y, \theta) \mid 0 \le y < |(l/2) \cos \theta| \text{ and } 0 \le \theta \le 2\pi\}$. We define the probability $p$ that the needle intersects a line to be the ratio of all "favorable" outcomes to "all" outcomes. Thus,

$$p = \frac{\text{area of } R}{\text{area of } S}$$

Show that $p = 2l/(\pi w)$.

Brand X/Alamy

There is a minimum speed that a rocket must attain in order to escape from the gravitational field of a planet. This speed is called the *escape velocity* for the planet. In this chapter, we will learn how to calculate the escape velocity for the earth.

# 6 Techniques of Integration

**UP TO NOW** we have relied on the basic integration formulas and the method of substitution to help us evaluate integrals. In this chapter we will look at some techniques of integration that will enable us to evaluate the integrals of more complicated functions.

We begin by introducing the method of integration by parts, which, like the method of substitution, is a general technique of integration. We then look at special methods for integrating trigonometric functions and rational functions. We also see how a table of integrals and a computer algebra system can help us to evaluate very general integrals.

Finally, we look at *improper integrals*, integrals in which the interval of integration is infinite or the integrand is unbounded (or both).

**V** This symbol indicates that one of the following video types is available for enhanced student learning at **www.academic.cengage.com/login:**
- Chapter lecture videos
- Solutions to selected exercises

## 6.1    Integration by Parts

As we have seen, a rule of integration can often be found by reversing a corresponding rule of differentiation. In this section we will look at a method of integration that is obtained by reversing the Product Rule for differentiation.

### ■ The Method of Integration by Parts

Recall that the Product Rule states that if $f$ and $g$ are differentiable functions, then

$$\frac{d}{dx}[f(x)g(x)] = f(x)g'(x) + g(x)f'(x)$$

If we integrate both sides of this equation with respect to $x$, we obtain

$$\int \frac{d}{dx}[f(x)g(x)]\,dx = \int [f(x)g'(x) + g(x)f'(x)]\,dx$$

or

$$f(x)g(x) = \int f(x)g'(x)\,dx + \int g(x)f'(x)\,dx$$

which may be written in the form

$$\int f(x)g'(x)\,dx = f(x)g(x) - \int g(x)f'(x)\,dx \qquad (1)$$

Formula (1) is called the **formula for integration by parts.** We use this formula to express one integral in terms of another that is easier to integrate.

Formula (1) can be simplified by using differentials. Let $u = f(x)$ and $v = g(x)$ so that $du = f'(x)\,dx$ and $dv = g'(x)\,dx$. Substituting these quantities into Formula (1) leads to the following version of the formula for integration by parts.

---

**Integration by Parts Formula**

$$\int u\,dv = uv - \int v\,du \qquad (2)$$

---

**EXAMPLE 1**    Find $\int xe^x\,dx$.

**Solution**    Let's use Formula (2) by choosing

$$u = x \qquad \text{and} \qquad dv = e^x\,dx$$

so that

$$du = dx \qquad \text{and} \qquad v = \int e^x\,dx = e^x$$

*Any* antiderivative will do—see the *Note* following the example.

This gives

$$\int xe^x \, dx = uv - \int v \, du$$

$$= xe^x - \int e^x \, dx$$

$$= xe^x - e^x + C = (x - 1)e^x + C$$ ■

**Notes**

1. In finding $v$ from the expression for $dv$, we don't need to include the constant of integration (that is, we may take $C = 0$). To see why, suppose that we replace $v$ in Formula (2) by $v + C$. Then we obtain

$$\int u \, dv = u(v + C) - \int (v + C) \, du = uv + Cu - \int v \, du - \int C \, du$$

$$= uv + Cu - \int v \, du - Cu = uv - \int v \, du$$

In other words, the constant $C$ "drops out."

2. The success of the method of integration by parts depends on a judicious choice of $u$ and $dv$. For instance, had we chosen $u = e^x$ and $dv = x \, dx$ in Example 1, then $du = e^x \, dx$ and $v = x^2/2$ and Formula (2) would have yielded

$$\int xe^x \, dx = uv - \int v \, du$$

$$= \frac{1}{2} x^2 e^x - \int \frac{1}{2} x^2 e^x \, dx$$

Since the integral on the right-hand side of this equation is more complicated than the original integral, we have not made a good choice of $u$ and $dv$. ■

Our original choice of $u$ and $dv$ in Example 1 suggests the following general guidelines.

**Guidelines for Choosing $u$ and $dv$**

Choose $u$ and $dv$ so that

1. $du$ is simpler than $u$ (if possible).
2. $dv$ is easily integrated.

**EXAMPLE 2**  Find $\int x \ln x \, dx$.

**Solution**  Let

$$u = \ln x \qquad \text{and} \qquad dv = x \, dx$$

so that

$$du = \frac{1}{x} \, dx \qquad \text{and} \qquad v = \int x \, dx = \frac{1}{2} x^2$$

Then Formula (2) gives

$$\int x \ln x \, dx = \frac{1}{2} x^2 \ln x - \int \frac{1}{2} x^2 \left( \frac{1}{x} \right) dx$$

$$= \frac{1}{2} x^2 \ln x - \frac{1}{4} x^2 + C = \frac{1}{4} x^2 (2 \ln x - 1) + C \qquad ■$$

Sometimes we need to apply the integration by parts formula more than once to find an integral, as illustrated in the next two examples.

**EXAMPLE 3**  Find $\int x^2 \sin x \, dx$.

**Solution**  Let

$$u = x^2 \qquad \text{and} \qquad dv = \sin x \, dx$$

so that

$$du = 2x \, dx \qquad \text{and} \qquad v = \int \sin x \, dx = -\cos x$$

Then Formula (2) yields

$$\int x^2 \sin x \, dx = -x^2 \cos x + \int 2x \cos x \, dx \qquad \text{(3)}$$

Observe that the integral on the right-hand side, although not readily integrable, is simpler than the original integral. In fact, the power in $x$ in the integrand is 1 instead of 2. This suggests that integrating by parts again might be a move in the right direction. So let's apply the formula once again to evaluate $\int 2x \cos x \, dx$. Let

$$u = 2x \qquad \text{and} \qquad dv = \cos x \, dx$$

so that

$$du = 2 \, dx \qquad \text{and} \qquad v = \int \cos x \, dx = \sin x$$

It follows from Formula (2) that

$$\int 2x \cos x \, dx = 2x \sin x - \int 2 \sin x \, dx = 2x \sin x + 2 \cos x + C \qquad \text{(4)}$$

Finally, substituting Equation (4) into Equation (3) gives

$$\int x^2 \sin x \, dx = -x^2 \cos x + 2x \sin x + 2 \cos x + C \qquad ■$$

**EXAMPLE 4**  Find $\int e^x \sin 2x \, dx$.

**Solution**  Let

$$u = e^x \qquad \text{and} \qquad dv = \sin 2x \, dx$$

so that

$$du = e^x \, dx \qquad \text{and} \qquad v = \int \sin 2x \, dx = -\frac{1}{2} \cos 2x$$

(In this case the choice $u = \sin 2x$ and $dv = e^x\, dx$ will work equally well.) Substituting this value into Formula (2) yields

$$\int e^x \sin 2x\, dx = -\frac{1}{2} e^x \cos 2x + \frac{1}{2} \int e^x \cos 2x\, dx \qquad (5)$$

The integral on the right-hand side is not readily integrable. But notice that it is certainly no more complicated than the original integral. So let's integrate by parts again and see where this leads us. Let

$$u = e^x \qquad \text{and} \qquad dv = \cos 2x\, dx$$

so that

$$du = e^x\, dx \qquad \text{and} \qquad v = \int \cos 2x\, dx = \frac{1}{2} \sin 2x$$

On using Formula (2), we find that

$$\int e^x \cos 2x\, dx = \frac{1}{2} e^x \sin 2x - \frac{1}{2} \int e^x \sin 2x\, dx \qquad (6)$$

Substituting Equation (6) into Equation (5) yields

$$\int e^x \sin 2x\, dx = -\frac{1}{2} e^x \cos 2x + \frac{1}{4} e^x \sin 2x - \frac{1}{4} \int e^x \sin 2x\, dx$$

Since the integral on the right-hand side is, except for the constant of integration, a constant multiple of the (original) integral on the left side, we can combine them to yield

$$\frac{5}{4} \int e^x \sin 2x\, dx = -\frac{1}{2} e^x \cos 2x + \frac{1}{4} e^x \sin 2x + C_1$$

so the required result is

$$\int e^x \sin 2x\, dx = -\frac{2}{5} e^x \cos 2x + \frac{1}{5} e^x \sin 2x + C \qquad C = \frac{4}{5} C_1$$

$$= \frac{1}{5} e^x (\sin 2x - 2 \cos 2x) + C \qquad \blacksquare$$

**Note**  We leave it for you to show that if we had chosen $u = \cos 2x$ and $dv = e^x\, dx$ in finding the integral on the right-hand side of Equation (5), then our final result would have been $\int e^x \sin 2x\, dx = \int e^x \sin 2x\, dx$. This is certainly a true statement, but it is of no help to us in evaluating the given integral. $\blacksquare$

**EXAMPLE 5**  Evaluate $\displaystyle\int_0^{\pi/4} \sec^3 x\, dx$.

**Solution**  We first find the indefinite integral

$$\int \sec^3 x\, dx = \int \sec x \cdot \sec^2 x\, dx$$

Let

$$u = \sec x \qquad \text{and} \qquad dv = \sec^2 x\, dx$$

so that

$$du = \sec x \tan x\, dx \qquad \text{and} \qquad v = \int \sec^2 x\, dx = \tan x$$

Using Formula (2), we obtain

$$\int \sec^3 x \, dx = \sec x \tan x - \int \tan^2 x \sec x \, dx$$

$$= \sec x \tan x - \int (\sec^2 x - 1)\sec x \, dx \qquad \sec^2 x = 1 + \tan^2 x$$

$$= \sec x \tan x - \int \sec^3 x \, dx + \int \sec x \, dx$$

$$= \sec x \tan x + \ln|\sec x + \tan x| - \int \sec^3 x \, dx$$

Combining the integrals, we obtain

$$2\int \sec^3 x \, dx = \sec x \tan x + \ln|\sec x + \tan x| + C_1$$

or

$$\int \sec^3 x \, dx = \frac{1}{2} \sec x \tan x + \frac{1}{2} \ln|\sec x + \tan x| + C \qquad C = \frac{1}{2} C_1$$

Finally, using this result, we find

$$\int_0^{\pi/4} \sec^3 x \, dx = \left[ \frac{1}{2} \sec x \tan x + \frac{1}{2} \ln|\sec x + \tan x| \right]_0^{\pi/4}$$

$$= \left[ \frac{1}{2} (\sqrt{2})(1) + \frac{1}{2} \ln|\sqrt{2} + 1| \right] - \left[ \frac{1}{2} (1)(0) + \frac{1}{2} \ln 1 \right]$$

$$= \frac{1}{2} [\sqrt{2} + \ln(\sqrt{2} + 1)] \approx 1.148$$

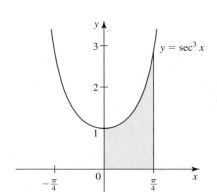

**FIGURE 1**
The area of the shaded region is given by $\int_0^{\pi/4} \sec^3 x \, dx$.

Since $f(x) = \sec^3 x$ is positive on $\left[0, \frac{\pi}{4}\right]$, we can interpret the integral in this example as the area of the region under the graph of $f$ on $\left[0, \frac{\pi}{4}\right]$. (See Figure 1.) ■

An alternative method for evaluating a definite integral using integration by parts is based on the following formula. Here we assume that both $f'$ and $g'$ are continuous. Then the Fundamental Theorem of Calculus, Part 2, gives

$$\int_a^b f(x)g'(x) \, dx = f(x)g(x) \Big|_a^b - \int_a^b g(x)f'(x) \, dx$$

Letting $u = f(x)$ and $v = g(x)$ and keeping in mind that the limits of integration are stated for $x$, we have the following.

---

**Integration by Parts Formula for a Definite Integral**

$$\int_a^b u \, dv = \left[ uv \right]_a^b - \int_a^b v \, du \qquad \qquad (7)$$

We illustrate the use of this formula in the next example.

**EXAMPLE 6** Find the centroid of the region under the graph of $f(x) = \ln x$ on $[1, e]$.

**Solution** The region $R$ under consideration is shown in Figure 2. The area of $R$ is given by

$$A = \int_1^e \ln x \, dx$$

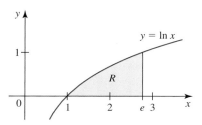

**FIGURE 2**
The region $R$

We integrate by parts, letting

$$u = \ln x \quad \text{and} \quad dv = dx$$

so that

$$du = \frac{1}{x} \, dx \quad \text{and} \quad v = x$$

Using Formula (7), we obtain

$$A = \Big[ x \ln x \Big]_1^e - \int_1^e dx$$

$$= (e \ln e - \ln 1) - x \Big|_1^e$$

$$= e - (e - 1) \qquad \text{\small ln } e = 1 \quad \text{and} \quad \text{ln } 1 = 0$$

$$= 1$$

Then, using Equation (6) of Section 5.7, we find

$$\bar{x} = \frac{1}{A} \int_a^b x f(x) \, dx = \frac{1}{1} \int_1^e x \ln x \, dx$$

$$= \Big[ \frac{1}{4} x^2 (2 \ln x - 1) \Big]_1^e \qquad \text{\small Use the result of Example 2.}$$

$$= \frac{1}{4} e^2 (2 \ln e - 1) - \frac{1}{4} (2 \ln 1 - 1) = \frac{1}{4} (e^2 + 1)$$

and

$$\bar{y} = \frac{1}{A} \int_a^b \frac{1}{2} [f(x)]^2 \, dx = \frac{1}{1} \int_1^e \frac{1}{2} (\ln x)^2 \, dx$$

$$= \frac{1}{2} \int_1^e (\ln x)^2 \, dx$$

We integrate by parts, letting

$$u = (\ln x)^2 \quad \text{and} \quad dv = dx$$

so that

$$du = \frac{2 \ln x}{x} \, dx \quad \text{and} \quad v = x$$

We obtain

$$\bar{y} = \frac{1}{2}\left\{\left[x(\ln x)^2\right]_1^e - \int_1^e 2 \ln x \, dx\right\}$$

$$= \frac{1}{2}\left\{\left[e(\ln e)^2 - 1(\ln 1)^2\right] - 2\int_1^e \ln x \, dx\right\}$$

$$= \frac{1}{2}e - 1$$

because $\int_1^e \ln x \, dx = 1$, as we saw earlier. Therefore, the centroid of $R$ is located at the point $\left(\frac{1}{4}(e^2 + 1), \frac{1}{2}e - 1\right)$. ∎

## Reduction Formulas

We can use the integration by parts formula to derive **reduction formulas** for evaluating certain integrals. These formulas enable us to express such integrals in terms of integrals whose integrands involve lower powers. In the next two examples we derive these formulas and then show how they are used.

**EXAMPLE 7** Find a reduction formula for $\int \sin^n x \, dx$, where $n \geq 2$ is an integer.

**Solution** We first rewrite

$$\int \sin^n x \, dx = \int \sin^{n-1} x \sin x \, dx$$

and then integrate by parts, letting

$$u = \sin^{n-1} x \qquad \text{and} \qquad dv = \sin x \, dx$$

so that

$$du = (n-1)\sin^{n-2} x \cos x \, dx \qquad \text{and} \qquad v = -\cos x$$

This gives

$$\int \sin^n x \, dx = uv - \int v \, du$$

$$= -\sin^{n-1} x \cos x + (n-1)\int \sin^{n-2} x \cos^2 x \, dx$$

Since $\cos^2 x = 1 - \sin^2 x$, we can write

$$\int \sin^n x \, dx = -\sin^{n-1} x \cos x + (n-1)\int \sin^{n-2} x \, dx - (n-1)\int \sin^n x \, dx$$

Transposing the last term on the right to the left-hand side gives

$$n\int \sin^n x \, dx = -\sin^{n-1} x \cos x + (n-1)\int \sin^{n-2} x \, dx$$

or

$$\int \sin^n dx = -\frac{1}{n}\sin^{n-1} x \cos x + \frac{n-1}{n}\int \sin^{n-2} x \, dx$$

∎

**EXAMPLE 8** Use the reduction formula obtained in Example 7 to find $\int \sin^4 x \, dx$.

**Solution**  We use the reduction formula with $n = 4$ to obtain

$$\int \sin^4 x \, dx = -\frac{1}{4} \sin^3 x \cos x + \frac{3}{4} \int \sin^2 x \, dx$$

Applying the reduction formula once more to the integral on the right-hand side with $n = 2$, we have

$$\int \sin^2 x \, dx = -\frac{1}{2} \sin x \cos x + \frac{1}{2} \int dx$$

$$= -\frac{1}{2} \sin x \cos x + \frac{1}{2} x + C_1$$

Therefore,

$$\int \sin^4 x \, dx = -\frac{1}{4} \sin^3 x \cos x + \frac{3}{4} \left( -\frac{1}{2} \sin x \cos x + \frac{1}{2} x + C_1 \right)$$

$$= -\frac{1}{4} \sin^3 x \cos x - \frac{3}{8} \sin x \cos x + \frac{3}{8} x + C$$

where $C = \frac{3}{4} C_1$.

## 6.1  CONCEPT QUESTIONS

1. Write the formula for integration by parts for (a) indefinite integrals and (b) definite integrals.

2. Explain how you would choose $u$ and $dv$ when using the integration by parts formula. Illustrate your answer with the integral $\int xe^{-x} \, dx$.

## 6.1  EXERCISES

*In Exercises 1–44, find or evaluate the integral.*

1. $\int xe^{2x} \, dx$

2. $\int xe^{-x} \, dx$

3. $\int x \sin x \, dx$

4. $\int x \cos 2x \, dx$

5. $\int x \ln 2x \, dx$

6. $\int x^3 \ln x \, dx$

7. $\int x^2 e^{-x} \, dx$

8. $\int t^3 e^t \, dt$

9. $\int x^2 \cos x \, dx$

10. $\int x^2 \sin 2x \, dx$

11. $\int \tan^{-1} x \, dx$

12. $\int \sin^{-1} x \, dx$

13. $\int \sqrt{t} \ln t \, dt$

14. $\int \frac{\ln t}{\sqrt{t}} \, dt$

15. $\int x \sec^2 x \, dx$

16. $\int e^{-x} \sin x \, dx$

17. $\int e^{2x} \cos 3x \, dx$

18. $\int_0^1 x \tan^{-1} x \, dx$

19. $\int u \sin(2u + 1) \, du$

20. $\int \theta \csc^2 \theta \, d\theta$

21. $\int x \tan^2 x \, dx$

22. $\int \cos(\ln x) \, dx$

23. $\int \sqrt{x} \cos \sqrt{x} \, dx$

24. $\int \csc^3 \theta \, d\theta$

25. $\int \sec^5 \theta \, d\theta$

26. $\int x \sinh x \, dx$

**27.** $\displaystyle\int x^3 \sinh x \, dx$

**28.** $\displaystyle\int x \cosh 2x \, dx$

**29.** $\displaystyle\int e^{-x} \ln(e^x + 1) \, dx$

**30.** $\displaystyle\int (x^2 - 1)\cos x \, dx$

**31.** $\displaystyle\int \frac{\ln x}{\sqrt{1 - x}} \, dx$

**32.** $\displaystyle\int_0^1 (t - 1)e^{-2t} \, dt$

**33.** $\displaystyle\int_1^e x^2 \ln x \, dx$

**34.** $\displaystyle\int_0^2 \ln(x + 1) \, dx$

**35.** $\displaystyle\int_0^{1/2} \cos^{-1} x \, dx$

**36.** $\displaystyle\int_0^\pi x \sin 2x \, dx$

**37.** $\displaystyle\int_{\sqrt{e}}^e x^{-2} \ln x \, dx$

**38.** $\displaystyle\int_0^{\pi/2} e^{2x} \cos x \, dx$

**39.** $\displaystyle\int_1^2 x \sec^{-1} x \, dx$

**40.** $\displaystyle\int_0^{\pi/2} (x + x \cos x) \, dx$

**41.** $\displaystyle\int_0^1 \ln(1 + t^2) \, dt$

**42.** $\displaystyle\int_0^{\pi^2/4} \sin \sqrt{x} \, dx$

**43.** $\displaystyle\int_{\pi/4}^{\pi/3} \frac{\theta}{\sin^2 \theta} \, d\theta$

**44.** $\displaystyle\int_0^1 \tan^{-1} \sqrt{x} \, dx$

**45.** Find the area of the region under the graph of $y = (\ln x)^2$ on the interval $[1, e]$.

**46.** Find the area of the region under the graph of $y = \dfrac{xe^x}{(1 + x)^2}$ on the interval $[0, 1]$.

**47.** Find the area of the region bounded by the graphs of $y = \tan^{-1} x$ and $y = (\pi/4)x$.
Hint: The graphs intersect at $(0, 0)$ and $\left(1, \frac{\pi}{4}\right)$.

**48.** Find the area of the region bounded by the graph of $y = e^{-x} \cos x$ and the x-axis for $x$ in the interval $\left[0, \frac{3\pi}{2}\right]$.

**49.** Let $f(x) = x\sqrt{x + 1}$ and $g(x) = 1 - x^2$.
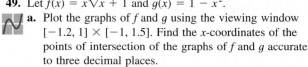
**a.** Plot the graphs of $f$ and $g$ using the viewing window $[-1.2, 1] \times [-1, 1.5]$. Find the x-coordinates of the points of intersection of the graphs of $f$ and $g$ accurate to three decimal places.
**b.** Use the result of part (a) and integration by parts to find the approximate area of the region bounded by the graphs of $f$ and $g$.

**50.** Let $f(x) = e^{-x} \sin x$ and $g(x) = -\sqrt{x} \cos \sqrt{x}$.
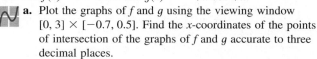
**a.** Plot the graphs of $f$ and $g$ using the viewing window $[0, 3] \times [-0.7, 0.5]$. Find the x-coordinates of the points of intersection of the graphs of $f$ and $g$ accurate to three decimal places.
**b.** Use the result of part (a) and integration by parts to find the approximate area of the region bounded by the graphs of $f$ and $g$.

**51.** The region under the graph of $y = \sqrt{\cos^{-1} x}$ on the interval $[0, 1]$ is revolved about the x-axis. Find the volume of the solid generated.

**52.** The region bounded by the graphs of $y = \ln x$, $y = 0$, $x = 1$, and $x = e$ is revolved about the x-axis. Find the volume of the solid generated.

**53.** The region bounded by the graphs of $y = e^{x/2} \cos x$, $x = 0$, $y = 0$, and $x = \pi/2$ is revolved about the x-axis. Find the volume of the solid generated.

**54.** Find the volume of the solid generated by revolving the region enclosed by the graphs of $y = \sin x$, $x = 0$, and $y = 1$ about the y-axis.

**55.** The region bounded by the graphs of $y = \sin x$, $y = 0$, $x = 0$, and $x = \pi$ is revolved about the line $x = -1$. Find the volume of the solid generated.

**56.** Find the centroid of the region $R$ bounded by the graphs of $y = \sin x$ and $y = (2/\pi)x$ on $\left[0, \frac{\pi}{2}\right]$.

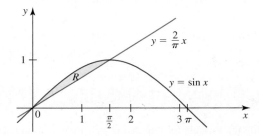

**57.** **Energy Production** To satisfy increased worldwide demand, the Metro Mining Company plans to increase its production of steam coal, the boiler-firing fuel used for generating electricity. Currently 20 million metric tons are produced per year; however, the company plans to increase production by $2te^{-0.05t}$ million metric tons per year, where $t$ is measured in years, for the next 10 years. Find a function that describes the company's total production of steam coal at the end of $t$ years. How much coal will the company have produced over the next 10 years if its production goals are met?

**58.** **Alcohol-Related Traffic Accidents** The number of alcohol-related accidents in a certain state, $t$ months after the passage of a series of strict anti-drunk-driving laws, has been decreasing at the rate of $R(t) = 20 + te^{0.1t}$ accidents per month. There were 882 alcohol-related accidents for the year before enactment of the laws. Determine how many alcohol-related accidents were expected to occur during the first year after passage of the laws.

**59.** **Damped Harmonic Motion** Consider the system shown in the accompanying figure. Here, the weight is attached to a spring and a dashpot damping device. Suppose that at $t = 0$, the weight is set in motion from its equilibrium position so that its velocity at any time $t$ is

$$v(t) = 3e^{-4t}(1 - 4t)$$

Find the position function $x(t)$ of the body.

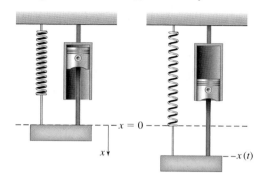

**(a)** System in equilibrium position     **(b)** System in motion

**60. A Mixture Problem** Two tanks are connected in tandem as shown in the figure. Each tank contains 60 gal of water. Starting at time $t = 0$, brine containing 3 lb/gal of salt flows into Tank 1 at the rate of 2 gal/min. The mixture then enters and leaves Tank 2 at the same rate. The mixtures in both tanks are stirred uniformly. It can be shown that the amount of salt in Tank 2 after $t$ min is given by

$$A(t) = 180(1 - e^{-t/30}) - 6te^{-t/30}$$

where $A(t)$ is measured in pounds.
**a.** What is the initial amount of salt in Tank 2?
**b.** What is the amount of salt in Tank 2 after 3 hr?
**c.** What is the average amount of salt in Tank 2 over the first 3 hr?

Tank 2

Tank 1

**61. The Charge in an Electric Current** The following figure shows an $LRC$ series electrical circuit comprising an inductor, a resistor, and a capacitor with inductance $L$ in henries, resistance $R$ in ohms, and capacitance $C$ in farads, respectively.

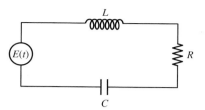

Here $E(t)$ is the electromotive force in volts. Suppose that the current in the system at time $t$ is

$$i(t) = -\frac{200}{3}e^{-20t}\sin 60t$$

amperes. Find the charge $q(t)$ on the capacitor at any time $t$ if the initial charge in the capacitor is $q_0$ coulombs.

Hint: $i = \dfrac{dq}{dt}$

**62. Damped Harmonic Motion** Refer to Exercise 59. Suppose that $t = 0$. The weight is set in motion from a point $\frac{1}{2}$ ft below the equilibrium position so that its velocity at any time $t$ is

$$v(t) = e^{-2t}(\cos 4t - 3\sin 4t)$$

Find the position function of the body.

**63. A Rocket Launch** A rocket with a mass $m$ (including fuel) is launched vertically upward from the surface of the earth ($t = 0$). If the fuel is consumed at a constant rate $r$ during the interval $0 \le t \le T$ and the speed of the exhaust gas relative to the rocket is a constant $s$, then the velocity of the rocket at time $t$ is given by

$$v(t) = v_0 - gt - s\ln\left(1 - \frac{r}{m}t\right)$$

where $v_0$ is its initial velocity and $g$ is the gravitational constant. Find the height $h(t)$ of the rocket at any time $t$ before burnout ($t = T$).

**64. Growth of HMOs** The membership of the Cambridge Community Health Plan, a health maintenance organization, is projected to grow at the rate of $9\sqrt{t+1}\ln\sqrt{t+1}$ thousand people per year $t$ years from now. If the HMO's current membership is 50,000, what will the membership be 5 years from now?

**65. Diffusion** A cylindrical membrane with inner radius $r_1$ cm and outer radius $r_2$ cm containing a chemical solution is introduced into a salt bath with constant concentration $c_2$ moles/L. If the concentration of the chemical inside the membrane is kept constant at a different concentration of $c_1$ moles/L, then the concentration of chemical across the membrane will be given by

$$c(r) = \left(\frac{c_1 - c_2}{\ln r_1 - \ln r_2}\right)(\ln r - \ln r_2) + c_2 \qquad r_1 < r < r_2$$

moles/L. Find the average concentration of the chemical across the membrane from $r = r_1$ to $r = r_2$.

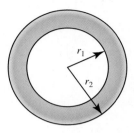

**66. Mechanical Resonance** Refer to Example 5 of Section 2.5. Suppose that an external force is applied to the spring so that the velocity of the weight at time $t$ is $v = 8t \sin 2t$.
  **a.** Find the position function of the weight if it is at the equilibrium position at $t = 0$.
  **b.** Plot the graph of the position function found in part (a).
  **c.** Show that $\lim_{t\to\infty} |s(t)| = \infty$ and hence the motion is one of resonance (see Exercise 44 of Section 2.5).

**67.** Suppose that $f''$ is continuous on $[1, 3]$ and $f(1) = 2$, $f(3) = -1, f'(1) = 2$, and $f'(3) = 5$. Evaluate $\int_1^3 x f''(x) \, dx$.

**68.** Consider the following "proof" that $0 = 1$. Integrate $\int (dx)/x$ by parts by letting $u = 1/x$ and $dv = dx$ so that $du = (-1/x^2) \, dx$ and $v = x$. This gives

$$\int \frac{dx}{x} = uv - \int v \, du$$

$$= \left(\frac{1}{x}\right)x - \int x\left(-\frac{1}{x^2}\right) dx = 1 + \int \frac{dx}{x}$$

Therefore, $0 = 1$. What is wrong with this argument?

*In Exercises 69 and 70, determine whether the statement is true or false. If it is true, explain why it is true. If it is false, explain why or give an example to show why it is false.*

**69.** $\displaystyle\int e^x f'(x) \, dx = e^x f(x) - \int e^x f(x) \, dx$

**70.** $\displaystyle\int uv \, dw = uvw - \int uw \, dv - \int vw \, du$

---

## 6.2 | Trigonometric Integrals

In this section we develop techniques for evaluating integrals involving combinations of trigonometric functions. Examples of such integrals are

$$\int \sin^5 x \cos^2 x \, dx, \qquad \int \csc 4x \cot^4 x \, dx, \qquad \text{and} \qquad \int \sin 5x \cos 4x \, dx$$

As you will see, these techniques rely on the use of the appropriate trigonometric identities.

### ■ Integrals of the Form $\int \sin^m x \cos^n x \, dx$

We begin by looking at integrals of the form

$$\int \sin^m x \cos^n x \, dx \qquad\qquad (1)$$

where

    **1.** $m$ and/or $n$ is an odd positive integer.
    **2.** $m$ and $n$ are both even nonnegative integers.

Examples 1 and 2 illustrate how an integral belonging to category 1 is evaluated, and Example 3 shows how to evaluate an integral belonging to category 2.

**EXAMPLE 1** Find $\int \sin^5 x \cos^2 x \, dx$.

**Solution** Here $m$ (the power of $\sin x$) is an odd positive integer. Let's write

$$\sin^5 x = (\sin^4 x)(\sin x) \qquad \text{Retain a factor of } \sin x.$$

$$= (\sin^2 x)^2 \sin x$$

$$= (1 - \cos^2 x)^2 \sin x \qquad \begin{array}{l}\text{Use the identity } \sin^2 x + \cos^2 x = 1 \text{ to convert} \\ \text{the other factor to a function of } \cos x.\end{array}$$

$$= (1 - 2\cos^2 x + \cos^4 x)\sin x$$

Then

$$\int \sin^5 x \cos^2 x \, dx = \int \cos^2 x (1 - 2\cos^2 x + \cos^4 x)\sin x \, dx$$

$$= \int (\cos^2 x - 2\cos^4 x + \cos^6 x)\sin x \, dx$$

If we make the substitution $u = \cos x$, then $du = -\sin x \, dx$, so

$$\int \sin^5 x \cos^2 x \, dx = \int (u^2 - 2u^4 + u^6)(-du)$$

$$= -\int (u^2 - 2u^4 + u^6) \, du$$

$$= -\left( \frac{1}{3}u^3 - \frac{2}{5}u^5 + \frac{1}{7}u^7 \right) + C$$

$$= -\frac{1}{3}\cos^3 x + \frac{2}{5}\cos^5 x - \frac{1}{7}\cos^7 x + C \qquad ■$$

**EXAMPLE 2** Find $\int \sin^4 x \cos^3 x \, dx$.

**Solution** Here $n$ (the power of $\cos x$) is an odd positive integer. Let's write

$$\cos^3 x = (\cos^2 x)(\cos x) \qquad \text{Retain a factor of } \cos x.$$

$$= (1 - \sin^2 x)\cos x \qquad \begin{array}{l}\text{Use the identity } \sin^2 x + \cos^2 x = 1 \text{ to con-} \\ \text{vert the other factor to a function of } \sin x.\end{array}$$

Then

$$\int \sin^4 x \cos^3 x \, dx = \int \sin^4 x (1 - \sin^2 x)\cos x \, dx$$

$$= \int (\sin^4 x - \sin^6 x)\cos x \, dx$$

Let $u = \sin x$ so that $du = \cos x \, dx$. Then

$$\int \sin^4 x \cos^3 x \, dx = \int (u^4 - u^6) \, du$$

$$= \frac{1}{5}u^5 - \frac{1}{7}u^7 + C$$

$$= \frac{1}{5}\sin^5 x - \frac{1}{7}\sin^7 x + C \qquad ■$$

**EXAMPLE 3**  Find $\int \sin^4 x \, dx$.

**Solution**  Here, $m = 4$ and $n = 0$. So both $m$ and $n$ are even nonnegative integers. In this case we use the half-angle formula for $\sin^2 x$:

$$\sin^2 x = \frac{1}{2}(1 - \cos 2x)$$

to write

$$\sin^4 x = (\sin^2 x)^2$$

$$= \left[\frac{1}{2}(1 - \cos 2x)\right]^2$$

$$= \frac{1}{4}(1 - 2\cos 2x + \cos^2 2x)$$

Applying the half-angle formula,

$$\cos^2 x = \frac{1}{2}(1 + \cos 2x)$$

to $\cos^2 2x$ in the last equation leads to

$$\sin^4 x = \frac{1}{4}\left(1 - 2\cos 2x + \frac{1}{2} + \frac{1}{2}\cos 4x\right)$$

$$= \frac{1}{4}\left(\frac{3}{2} - 2\cos 2x + \frac{1}{2}\cos 4x\right)$$

Therefore,

$$\int \sin^4 x \, dx = \frac{1}{4}\int \left(\frac{3}{2} - 2\cos 2x + \frac{1}{2}\cos 4x\right) dx$$

$$= \frac{3}{8}x - \frac{1}{4}\sin 2x + \frac{1}{32}\sin 4x + C \qquad ■$$

In general, we have the following guidelines for evaluating integrals of the form $\int \sin^m x \cos^n x \, dx$.

---

**Guidelines for Evaluating $\int \sin^m x \cos^n x \, dx$**

1. *If the power of* $\sin x$ *is odd and positive* ($m = 2k + 1$), retain a factor of $\sin x$, and use the identity $\sin^2 x = 1 - \cos^2 x$ to write

$$\int \sin^{2k+1} x \cos^n x \, dx = \int (\sin^2 x)^k \cos^n x \sin x \, dx$$

$$= \int (1 - \cos^2 x)^k \cos^n x \sin x \, dx$$

Then integrate using the substitution $u = \cos x$.

2. *If the power of* cos *x is odd and positive* ($n = 2k + 1$), *retain a factor of* cos *x, and use the identity* $\cos^2 x = 1 - \sin^2 x$ *to write*

$$\int \sin^m x \cos^{2k+1} x \, dx = \int \sin^m x \, (\cos^2 x)^k \cos x \, dx$$

$$= \int \sin^m x \, (1 - \sin^2 x)^k \cos x \, dx$$

*Then integrate using the substitution* $u = \sin x$.

3. *If the powers of* sin *x and* cos *x are both even and nonnegative,* use the half-angle formulas (repeatedly, if necessary) to write

$$\sin^2 x = \frac{1}{2}(1 - \cos 2x) \qquad \text{and} \qquad \cos^2 x = \frac{1}{2}(1 + \cos 2x)$$

**EXAMPLE 4** Evaluate $\displaystyle\int_0^{\pi/2} \sin^3 x \cos^{1/2} x \, dx$.

**Solution**  The power of sin $x$ is odd and positive. So we retain a factor of sin $x$. Thus,

$$\int_0^{\pi/2} \sin^3 x \cos^{1/2} x \, dx = \int_0^{\pi/2} \sin^2 x \cos^{1/2} x \sin x \, dx$$

$$= \int_0^{\pi/2} (1 - \cos^2 x)\cos^{1/2} x \sin x \, dx$$

$$= \int_0^{\pi/2} (\cos^{1/2} x - \cos^{5/2} x)\sin x \, dx$$

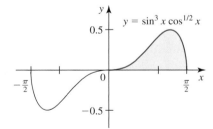

**FIGURE 1**
The area of the shaded region is given by $\int_0^{\pi/2} \sin^3 x \cos^{1/2} x \, dx$.

Let $u = \cos x$ so that $du = -\sin x \, dx$. Note that when $x = 0$, $u = \cos 0 = 1$, the lower limit of integration with respect to $u$; when $x = \pi/2$, $u = \cos(\pi/2) = 0$, the upper limit of integration with respect to $u$. Making these substitutions, we obtain

$$\int_0^{\pi/2} \sin^3 x \cos^{1/2} x \, dx = \int_1^0 (u^{1/2} - u^{5/2})(-du) = -\int_1^0 (u^{1/2} - u^{5/2}) \, du$$

$$= \left[ -\frac{2}{3} u^{3/2} + \frac{2}{7} u^{7/2} \right]_1^0 = \left[ 0 - \left( -\frac{2}{3} + \frac{2}{7} \right) \right] = \frac{8}{21}$$

Since $f(x) = \sin^3 x \cos^{1/2} x$ is nonnegative on $\left[ 0, \frac{\pi}{2} \right]$, we can interpret the integral in this example as the area of the region under the graph of $f$ on $\left[ 0, \frac{\pi}{2} \right]$. (See Figure 1.) ∎

## ■ Integrals of the Form $\int \tan^m x \sec^n x \, dx$ and $\int \cot^m x \csc^n x \, dx$

The techniques for evaluating integrals of the form

$$\int \tan^m x \sec^n x \, dx$$

are developed in a similar manner. We have the following guidelines for evaluating such integrals.

**Guidelines for Evaluating $\int \tan^m x \sec^n x \, dx$**

1. *If the power of* $\tan x$ *is odd and positive* $(m = 2k + 1)$, retain a factor of $\sec x \tan x$ and use the identity $\tan^2 x = \sec^2 x - 1$ to write

$$\int \tan^{2k+1} x \sec^n x \, dx = \int (\tan^2 x)^k \sec^{n-1} x \sec x \tan x \, dx$$

$$= \int (\sec^2 x - 1)^k \sec^{n-1} x \sec x \tan x \, dx$$

   Then integrate using the substitution $u = \sec x$.

2. *If the power of* $\sec x$ *is even and positive* $(n = 2k, k \geq 2)$, retain a factor of $\sec^2 x$ and use the identity $\sec^2 x = 1 + \tan^2 x$ to write

$$\int \tan^m x \sec^{2k} x \, dx = \int \tan^m x (\sec^2 x)^{k-1} \sec^2 x \, dx$$

$$= \int \tan^m x (1 + \tan^2 x)^{k-1} \sec^2 x \, dx$$

   Then integrate using the substitution $u = \tan x$.

The guidelines for evaluating $\int \cot^m x \csc^n x \, dx$ are similar to those for evaluating $\int \tan^m x \sec^n x \, dx$.

**EXAMPLE 5** Find $\int \tan^3 x \sec^7 x \, dx$.

**Solution** Here, $m$ (the power of $\tan x$) is an odd positive integer. Let's retain the factor $\sec x \tan x$ from the integrand and write

$$\tan^3 x \sec^7 x = \tan^2 x \sec^6 x (\sec x \tan x)$$

$$= (\sec^2 x - 1)\sec^6 x (\sec x \tan x) \qquad \text{Use the identity } \tan^2 x = \sec^2 x - 1 \text{ to convert the other factors to a function of } \sec x.$$

$$= (\sec^8 x - \sec^6 x)\sec x \tan x$$

Then

$$\int \tan^3 x \sec^7 x \, dx = \int (\sec^8 x - \sec^6 x)\sec x \tan x \, dx$$

Let $u = \sec x$ so that $du = \sec x \tan x \, dx$. Then

$$\int \tan^3 x \sec^7 x \, dx = \int (u^8 - u^6) \, du$$

$$= \frac{1}{9} u^9 - \frac{1}{7} u^7 + C$$

$$= \frac{1}{9} \sec^9 x - \frac{1}{7} \sec^7 x + C \qquad \blacksquare$$

**EXAMPLE 6**  Evaluate $\displaystyle\int_0^{\pi/4} \sqrt{\tan x}\, \sec^6 x\, dx$.

**Solution**  Here, $n$ (the power of sec $x$) is an even positive integer. So let's retain a factor of $\sec^2 x$. Thus,

$$\int_0^{\pi/4} \sqrt{\tan x}\, \sec^6 x\, dx = \int_0^{\pi/4} \tan^{1/2} x\, \sec^4 x\, \sec^2 x\, dx$$

$$= \int_0^{\pi/4} (\tan^{1/2} x)(1 + \tan^2 x)^2 \sec^2 x\, dx \qquad \sec^2 x = 1 + \tan^2 x$$

$$= \int_0^{\pi/4} (\tan^{1/2} x + 2\tan^{5/2} x + \tan^{9/2} x)\sec^2 x\, dx$$

Let $u = \tan x$ so that $du = \sec^2 x\, dx$. The lower and upper limits of integration with respect to $u$ are $u = 0$ (set $x = 0$) and $u = 1$ (set $x = \pi/4$), respectively. Making these substitutions, we obtain

$$\int_0^{\pi/4} \sqrt{\tan x}\, \sec^6 x\, dx = \int_0^1 (u^{1/2} + 2u^{5/2} + u^{9/2})\, du$$

$$= \left[\frac{2}{3} u^{3/2} + \frac{4}{7} u^{7/2} + \frac{2}{11} u^{11/2}\right]_0^1$$

$$= \frac{2}{3} + \frac{4}{7} + \frac{2}{11} = \frac{328}{231} \qquad\blacksquare$$

Integrals of the form $\int \cot^m x \csc^n x\, dx$ may be evaluated in a similar manner, as the following example illustrates.

**EXAMPLE 7**  Evaluate $\int \cot^5 x \csc^5 x\, dx$.

**Solution**  Here, the power of cot $x$ is an odd positive integer. So we retain the factor $\csc x \cot x$ from the integrand. Thus,

$$\int \cot^5 x \csc^5 x\, dx = \int \cot^4 x\, (\csc^4 x)(\csc x \cot x)\, dx$$

$$= \int (\csc^2 x - 1)^2 \csc^4 x \csc x \cot x\, dx \qquad \cot^2 x = \csc^2 x - 1$$

$$= \int (\csc^8 x - 2\csc^6 x + \csc^4 x)\csc x \cot x\, dx$$

Let $u = \csc x$ so that $du = -\csc x \cot x\, dx$. Then

$$\int \cot^5 x \csc^5 x\, dx = -\int (u^8 - 2u^6 + u^4)\, du$$

$$= -\frac{1}{9} u^9 + \frac{2}{7} u^7 - \frac{1}{5} u^5 + C$$

$$= -\frac{1}{9} \csc^9 x + \frac{2}{7} \csc^7 x - \frac{1}{5} \csc^5 x + C \qquad\blacksquare$$

## ◼ Converting to Sines and Cosines

For integrals involving powers of trigonometric functions that are not covered by the formulas just considered, we are sometimes able to evaluate the integral by converting the integrand to an expression involving sines and cosines, as the following example illustrates.

**EXAMPLE 8** Find $\displaystyle\int \frac{\tan x}{\sec^2 x}\, dx$.

**Solution** We have

$$\int \frac{\tan x}{\sec^2 x}\, dx = \int \left(\frac{\sin x}{\cos x}\right)\cos^2 x\, dx = \int \sin x \cos x\, dx$$

$$= \frac{1}{2}\sin^2 x + C \qquad \text{Let } u = \sin x.$$

## ◼ Integrals of the Form $\int \sin mx \sin nx\, dx$, $\int \sin mx \cos nx\, dx$, and $\int \cos mx \cos nx\, dx$

Integrals in which the integrand is a product of sines and cosines of two *different* angles can be evaluated with the help of the following identities.

---

**Trigonometric Identities**

$$\sin mx \sin nx = \frac{1}{2}[\cos(m - n)x - \cos(m + n)x] \tag{2a}$$

$$\sin mx \cos nx = \frac{1}{2}[\sin(m - n)x + \sin(m + n)x] \tag{2b}$$

$$\cos mx \cos nx = \frac{1}{2}[\cos(m - n)x + \cos(m + n)x] \tag{2c}$$

---

**EXAMPLE 9** Find $\int \sin 4x \cos 5x\, dx$.

**Solution** Using Formula (2b), we have

$$\int \sin 4x \cos 5x\, dx = \int \frac{1}{2}[\sin(-x) + \sin 9x]\, dx$$

$$= \frac{1}{2}\int (-\sin x + \sin 9x)\, dx$$

$$= \frac{1}{2}\left(\cos x - \frac{1}{9}\cos 9x\right) + C$$

## 6.2 CONCEPT QUESTIONS

1. Explain how you would find $\int \sin^m x \cos^n x\, dx$ if (a) $m$ is odd and positive, (b) $n$ is odd and positive, and (c) $m$ and $n$ are both nonnegative and even.

2. Explain how you would find $\int \tan^m x \sec^n x\, dx$ if (a) $m$ is odd and positive and (b) $n$ is even and positive.

3. Explain how you would find $\int \cot^m x \csc^n x\, dx$ if (a) $m$ is odd and positive and (b) $n$ is even and positive.

4. Explain how you would find (a) $\int \sin mx \cos nx\, dx$, (b) $\int \sin mx \sin nx\, dx$, and (c) $\int \cos mx \cos nx\, dx$.

## 6.2 EXERCISES

*In Exercises 1–48, find or evaluate the integral.*

**1.** $\int \sin^3 x \cos x \, dx$

**2.** $\int \sin^3 x \cos^2 x \, dx$

**3.** $\int \cos^3 2x \sin^5 2x \, dx$

**4.** $\int_0^{\pi/2} \sqrt{\cos x} \sin^3 x \, dx$

**5.** $\int \sin^3 x \, dx$

**6.** $\int \cos^3 2x \, dx$

**7.** $\int_0^{\pi} \cos^2 \frac{x}{2} \, dx$

**8.** $\int \cos^4 x \, dx$

**9.** $\int_0^1 \sin^4 \pi x \, dx$

**10.** $\int \sin^2 2x \cos^4 2x \, dx$

**11.** $\int \sin^2\left(\frac{x}{2}\right)\cos^2\left(\frac{x}{2}\right) dx$

**12.** $\int \cos^4 x \sin^4 x \, dx$

**13.** $\int_0^{\pi} \sin^2 x \cos^4 x \, dx$

**14.** $\int \sin^6 u \, du$

**15.** $\int x \cos^4(x^2) \, dx$

**16.** $\int \theta \sin^2(\theta^2)\cos^2(\theta^2) \, d\theta$

**17.** $\int x \sin^2 x \, dx$

**18.** $\int x \cos^2 x \, dx$

Hint: Integrate by parts.

Hint: Integrate by parts.

**19.** $\int_0^{\pi/4} \tan^2 x \, dx$

**20.** $\int \tan^3(\pi - x) \, dx$

**21.** $\int \tan^5 \frac{x}{2} \, dx$

**22.** $\int \tan^5 x \sec^3 x \, dx$

**23.** $\int \sec^2(\pi x)\tan^3(\pi x) \, dx$

**24.** $\int_0^{\pi/4} \sec^2 x \tan^2 x \, dx$

**25.** $\int \sec^4 3x \tan^2 3x \, dx$

**26.** $\int \sec^4(\pi - x)\tan(\pi - x) \, dx$

**27.** $\int \sec^4 \theta \sqrt{\tan \theta} \, d\theta$

**28.** $\int \sec^4\left(\frac{x}{2}\right)\tan^4\left(\frac{x}{2}\right)dx$

**29.** $\int \cot^2 2x \, dx$

**30.** $\int_{\pi/4}^{\pi/2} \cot^3 x \, dx$

**31.** $\int \csc^3 x \, dx$

**32.** $\int \csc^5 x \, dx$

Hint: Integrate by parts.

Hint: Integrate by parts.

**33.** $\int \csc^4 t \, dt$

**34.** $\int \csc^4 \theta \cot^4 \theta \, d\theta$

**35.** $\int \cot^6 t \, dt$

**36.** $\int \cot^3 x \csc^4 x \, dx$

**37.** $\int_0^{\pi/4} \frac{1}{\cos^4 x} \, dx$

**38.** $\int (1 + \cot x)^2 \csc x \, dx$

**39.** $\int_0^{\pi/2} \sin x \cos 2x \, dx$

**40.** $\int \sin 3\theta \sin 4\theta \, d\theta$

**41.** $\int \cos 2\theta \cos 4\theta \, d\theta$

**42.** $\int \frac{\sin^3 x}{\sec^2 x} \, dx$

**43.** $\int \cos^2 2\theta \cot 2\theta \, d\theta$

**44.** $\int_0^{\pi/2} \frac{\sin t}{1 + \cos t} \, dt$

**45.** $\int \frac{\tan^3 \sqrt{t} \sec^2 \sqrt{t}}{\sqrt{t}} \, dt$

**46.** $\int \frac{1}{\csc x \cot^2 x} \, dx$

**47.** $\int \frac{1 - \tan^2 x}{\sec^2 x} \, dx$

**48.** $\int \frac{\cos 2\theta}{\cos \theta + \sin \theta} \, d\theta$

**49.** Find the average value of $f(x) = \cos^2 x$ over the interval $[0, 2\pi]$.

**50.** Find the average value of $f(x) = \cos^2 x \sin^3 x$ over the interval $[0, \pi]$.

**51.** Find the area of the region under the graph of $y = \sin^2 \pi x$ on the interval $[0, 1]$.

**52.** Find the area of the region bounded by the graphs of $y = \sin^4 x$, $y = \cos^4 x$, $x = 0$, and $x = \pi/4$.

**53.** Let $f(x) = \sin^4 x$ and $g(x) = 1 - x^2$.
   **a.** Plot the graphs of $f$ and $g$ using the viewing window $[-1, 1] \times [0, 1]$. Find the $x$-coordinates of the points of intersection of the graphs of $f$ and $g$ accurate to three decimal places.
   **b.** Use the result of part (a) and the method of this section to find the approximate area of the region bounded by the graphs of $f$ and $g$.

**54.** Let $f(x) = \cos 2x \cos 4x$ and $g(x) = \sqrt{x}$.
   **a.** Plot the graphs of $f$ and $g$ using the viewing window $[0, \frac{1}{2}] \times [0, 1]$. Find the $x$-coordinate of the point of intersection of the graphs of $f$ and $g$ accurate to three decimal places.
   **b.** Use the result of part (a) and the method of this section to find the approximate area of the region bounded by the graphs of $f$ and $g$ and the $y$-axis.

**55.** The region under the graph of $y = \tan^2 x$ on the interval $\left[0, \frac{\pi}{4}\right]$ is revolved about the $x$-axis. Find the volume of the solid generated.

**56.** The region under the graph of $y = \sin^3 x$ on the interval $[0, \pi]$ is revolved about the $y$-axis. Find the volume of the solid generated.

**57.** Find the centroid of the region $R$ under the graph of $y = \cos x$ on the interval $\left[0, \frac{\pi}{2}\right]$.

**58.** Find the centroid of the region $R$ bounded by the graphs of $y = \cos x$ and $y = 1 - (2/\pi)x$ on $\left[0, \frac{\pi}{2}\right]$.

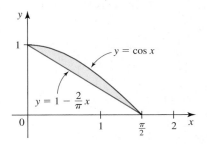

**59.** The velocity of a particle $t$ sec after leaving the origin, moving along a coordinate line, is $v(t) = \sin^3 \pi t$ ft/sec. What is the distance traveled by the particle during the first 6 sec? What is its position at $t = 6$?

**60.** Find the volume of the solid generated by revolving the region bounded by the graphs of $y = \sin x$ and $y = (2/\pi)x$ about the $x$-axis.

**61.** Find the volume of the solid generated by revolving the region under the graph of $f(x) = (\sin x)/(\cos^3 x)$ on $\left[0, \frac{\pi}{4}\right]$ about the $x$-axis.

**62. Interval Training** As part of her speed training, a long-distance runner runs in spurts for a minute. Suppose that she runs along a straight line so that her velocity $t$ sec after passing a marker is given by the velocity function

$$v(t) = 5 + 100 \sin^2\left(\frac{\pi}{16}t\right)\cos^2\left(\frac{\pi}{16}t\right) \qquad 0 \le t \le 60$$

where $v(t)$ is measured in feet per second. Find the position function as measured from the marker.

**63. Alternating Current Intensity** Find the average value of the alternating current intensity described by

$$I = I_0 \cos(\omega t + \alpha)$$

where $I_0$, $\omega$, and $\alpha$ are constants, over the time interval $\left[0, \frac{\pi}{\omega}\right]$.

**64. Electromotive Force** An electromotive force (emf), $E$, is given by

$$E = E_0 \sin \frac{2\pi t}{T}$$

where $T$ is the period in seconds and $E_0$ is the amplitude of the emf. Find the average value of the square of the emf, $E^2$, over the time interval $[0, T]$.

**65. Heat Generated by an Alternating Current** According to the Joule-Lenz Law, the amount of heat generated by an alternating current

$$I = I_0 \sin\left(\frac{2\pi t}{T} - \phi\right)$$

flowing in a conductor with resistance $R$ ohms from $t = T_1$ to $t = T_2$ is given by

$$Q = 0.24R\int_{T_1}^{T_2} I^2 \, dt$$

joules. Find the amount of heat generated during a cycle (from $t = 0$ to $t = T$).

**66. Fabricating Corrugated Metal Sheets**  A certain brand of corrugated metal sheets comes in 20 in. $\times$ 48 in. sizes. The cross section of the sheets can be described by the graph of

$$y = \sin \frac{\pi}{2} t \qquad 0 \le t \le 20$$

Use a calculator or computer to find the approximate length of the flat metal sheet before fabrication.

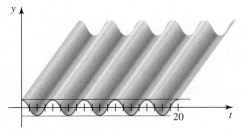

**67.**  Plot the graph of $f(x) = e^{-x} \sin(\pi x/2)$ for $0 \le x \le 2$. Then use a calculator or computer to approximate the volume of the solid generated by revolving the region under the graph of $f$ on $[0, 2]$ about (a) the $x$-axis and (b) the $y$-axis.

**68.**  Refer to Exercise 62. Use a calculator or computer to approximate the distance traveled by the athlete in her 60-sec speed exercise.

**69.** Prove that if $m$ and $n$ are positive integers, then

$$\int_{-\pi}^{\pi} \sin mx \cos nx \, dx = 0$$

**70.** Prove that if $m$ and $n$ are positive integers, then

$$\int_{-\pi}^{\pi} \sin mx \sin nx \, dx = \begin{cases} 0 & \text{if } m \ne n \\ \pi & \text{if } m = n \end{cases}$$

**71.** Prove that if $m$ and $n$ are positive integers, then

$$\int_{-\pi}^{\pi} \cos mx \cos nx \, dx = \begin{cases} 0 & \text{if } m \ne n \\ \pi & \text{if } m = n \end{cases}$$

**72. Finite Fourier Series** Prove that if

$$f(x) = \frac{a_0}{2} + \sum_{k=1}^{n} (a_k \cos kx + b_k \sin kx)$$

$$= \frac{a_0}{2} + a_1 \cos x + b_1 \sin x + a_2 \cos 2x$$

$$+ b_2 \sin 2x + \cdots + a_n \cos nx + b_n \sin nx$$

then

$$a_0 = \frac{1}{\pi} \int_{-\pi}^{\pi} f(x)\, dx, \qquad a_k = \frac{1}{\pi} \int_{-\pi}^{\pi} f(x) \cos kx\, dx$$

and

$$b_k = \frac{1}{\pi} \int_{-\pi}^{\pi} f(x) \sin kx\, dx \qquad k = 1, 2, \ldots, n$$

**Hint:** Use the results of Exercises 69–71.

# 6.3 Trigonometric Substitutions

Figure 1a depicts an aerial view of a scenic drive along the coast. We can approximate the part of the road between Point $A$ and Point $B$ by the graph of $f(x) = \frac{1}{2} x^2$ on the interval $[0, 1]$, where both $x$ and $f(x)$ are measured in miles (see Figure 1b). How far is the drive between $A$ and $B$?

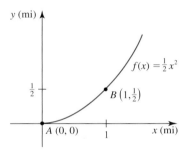

**FIGURE 1**      (a) A stretch of coastal road      (b) The graph of the function describing the road

To answer this question, we need to find the arc length $L$ of the graph of $f$ from $A(0, 0)$ to $B\left(1, \frac{1}{2}\right)$. Now, from Section 5.4 we know that the required arc length is given by

$$L = \int_0^1 \sqrt{1 + [f'(x)]^2}\, dx = \int_0^1 \sqrt{1 + x^2}\, dx \tag{1}$$

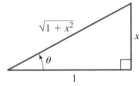

**FIGURE 2**
The right triangle associated with the integrand of Equation (1)

To evaluate this integral, observe that the integrand can be written in the form $\sqrt{1^2 + x^2}$ —the square root of the sum of two squares. This brings to mind the Pythagorean Theorem. In fact, we can think of this quantity as being associated with the length of the hypotenuse of the right triangle shown in Figure 2. This suggests that we try the substitution

$$\tan \theta = \frac{x}{1} \qquad \text{or} \qquad x = \tan \theta$$

We then have

$$\sqrt{1 + x^2} = \sqrt{1 + \tan^2 \theta} = \sqrt{\sec^2 \theta}$$
$$= \sec \theta$$

provided that $\frac{-\pi}{2} < \theta < \frac{\pi}{2}$. Proceeding with the substitution $x = \tan \theta$ so that $dx = \sec^2 \theta \, d\theta$, we see that the indefinite integral

$$\int \sqrt{1 + x^2} \, dx = \int \sec \theta \, (\sec^2 \theta) \, d\theta$$

$$= \int \sec^3 \theta \, d\theta$$

$$= \frac{1}{2} \left( \sec \theta \tan \theta + \ln|\sec \theta + \tan \theta| \right) + C$$

See Example 5 in Section 6.1.

Finally, referring to Figure 2 again, we see that

$$\sec \theta = \sqrt{1 + x^2} \qquad \text{hypotenuse/adjacent}$$

so

$$\int \sqrt{1 + x^2} \, dx = \frac{1}{2} \left( \sqrt{1 + x^2} \cdot x + \ln|\sqrt{1 + x^2} + x| \right) + C$$

Therefore, the distance of the drive between $A$ and $B$ is given by

$$L = \int_0^1 \sqrt{1 + x^2} \, dx = \frac{1}{2} \left[ x\sqrt{1 + x^2} + \ln\left(\sqrt{1 + x^2} + x\right) \right]_0^1$$

$$= \frac{1}{2} \left[ \sqrt{2} + \ln(\sqrt{2} + 1) \right] \approx 1.148$$

or approximately 1.15 mi.

## ■ Trigonometric Substitution

The techniques that we used in solving our introductory example involve **trigonometric substitution.** In general, this method can be used to evaluate integrals involving the radicals

$$\sqrt{a^2 - x^2}, \qquad \sqrt{a^2 + x^2}, \qquad \text{and} \qquad \sqrt{x^2 - a^2}$$

where $a > 0$.

The key to this technique lies in making an appropriate trigonometric substitution using one of the trigonometric identities

$$\cos^2 \theta = 1 - \sin^2 \theta$$

and

$$\sec^2 \theta = 1 + \tan^2 \theta$$

to transform the given integral into one that is radical free. The resulting *trigonometric integral* can then be evaluated by using the techniques developed earlier. Finally, the answer is written in terms of the original variable by converting from $\theta$'s to $x$'s.

The trigonometric substitutions for evaluating integrals involving the indicated radicals are listed in Table 1.

Note that in each case the restriction on $\theta$ ensures that the function $g$ in the substitution $x = g(\theta)$ is *one-to-one* and, therefore, has an inverse. This enables us to solve for $\theta$ in terms of $x$ and, hence, to express the answer in terms of the original variable $x$.

**TABLE 1**   Trigonometric Substitutions

| For integrals involving | Use the substitution | Use the identity | Right triangle associated with the substitution |
|---|---|---|---|
| $\sqrt{a^2 - x^2}, a > 0$ | $x = a \sin \theta, -\frac{\pi}{2} \le \theta \le \frac{\pi}{2}$ | $1 - \sin^2 \theta = \cos^2 \theta$ | 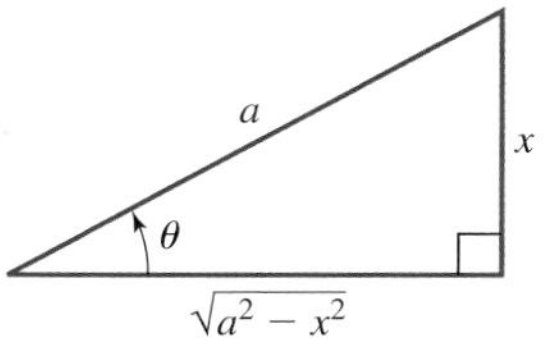 |
| $\sqrt{a^2 + x^2}, a > 0$ | $x = a \tan \theta, -\frac{\pi}{2} < \theta < \frac{\pi}{2}$ | $1 + \tan^2 \theta = \sec^2 \theta$ | 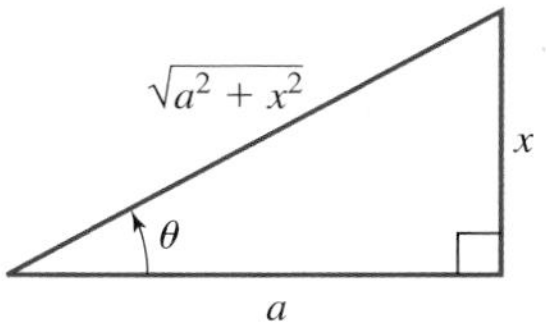 |
| $\sqrt{x^2 - a^2}, a > 0$ | $x = a \sec \theta, 0 \le \theta < \frac{\pi}{2}$   or   $\frac{\pi}{2} < \theta \le \pi$ | $\sec^2 \theta - 1 = \tan^2 \theta$ | 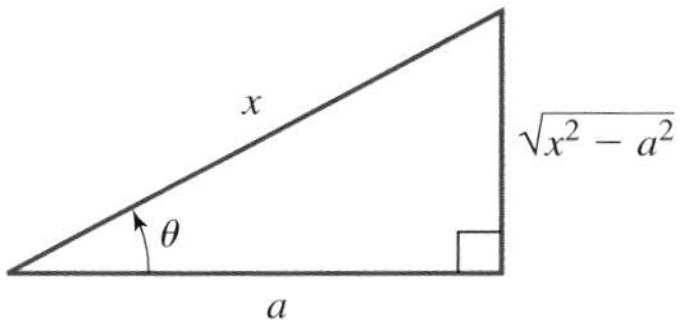 |

**EXAMPLE 1**   Find $\displaystyle\int \frac{x^2}{\sqrt{9 - x^2}}\, dx$.

**Solution**   Note that the integrand involves a radical of the form $\sqrt{a^2 - x^2}$, where $a = 3$. This suggests that we use the trigonometric substitution

$$x = 3 \sin \theta \qquad \text{so that} \qquad dx = 3 \cos \theta\, d\theta$$

where $-\frac{\pi}{2} < \theta < \frac{\pi}{2}$. In this example we have the further restriction $\theta \ne \pm \pi/2$ to ensure that $x \ne \pm 3$ (the integrand is not defined at these points). Making these substitutions, we have

$$\int \frac{x^2}{\sqrt{9 - x^2}}\, dx = \int \frac{9 \sin^2 \theta}{\sqrt{9 - 9 \sin^2 \theta}}\, (3 \cos \theta\, d\theta)$$

$$= 9 \int \sin^2 \theta\, d\theta$$

$$= \frac{9}{2} \int (1 - \cos 2\theta)\, d\theta \qquad \text{Use a half-angle formula.}$$

$$= \frac{9}{2} \left( \theta - \frac{1}{2} \sin 2\theta \right) + C$$

To express this result in terms of the original variable $x$, observe that $\sin \theta = x/3$ implies that $\theta = \sin^{-1}(x/3)$. Next, observe that $\sin 2\theta = 2 \sin \theta \cos \theta$. With the help of Figure 3, we find

$$\sin 2\theta = 2(\sin \theta)(\cos \theta) = 2 \left( \frac{x}{3} \right) \left( \frac{\sqrt{9 - x^2}}{3} \right)$$

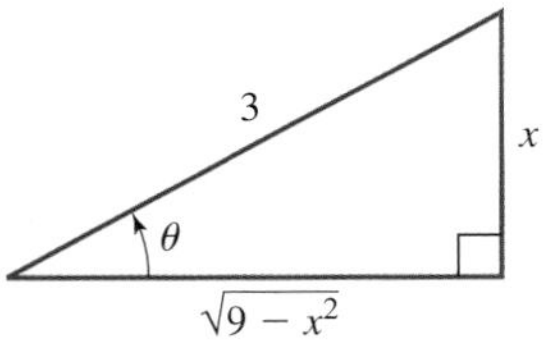

**FIGURE 3**
The right triangle associated with the substitution $x = 3 \sin \theta$

Therefore,

$$\int \frac{x^2}{\sqrt{9-x^2}}\,dx = \frac{9}{2}\left[\sin^{-1}\left(\frac{x}{3}\right) - \frac{1}{9}x\sqrt{9-x^2}\right] + C$$

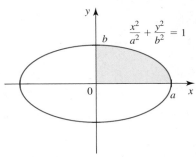

$$\frac{x^2}{a^2} + \frac{y^2}{b^2} = 1$$

**FIGURE 4**
The area enclosed by the ellipse is four times its area in the first quadrant.

**EXAMPLE 2** Find the area enclosed by the ellipse $\dfrac{x^2}{a^2} + \dfrac{y^2}{b^2} = 1$.

**Solution** The ellipse is shown in Figure 4. By symmetry we see that the area $A$ enclosed by the ellipse is just four times its area in the first quadrant. Next, to find the function describing the ellipse in this quadrant, we solve the given equation for $y$. Thus,

$$\frac{y^2}{b^2} = 1 - \frac{x^2}{a^2} = \frac{a^2 - x^2}{a^2} \qquad \text{or} \qquad y = \pm\frac{b}{a}\sqrt{a^2 - x^2}$$

Since $y > 0$ in this quadrant, the required function is $f(x) = (b/a)\sqrt{a^2 - x^2}$ for $x$ in the interval $[0, a]$. Therefore, the desired area $A$ is given by

$$A = 4\int_0^a \frac{b}{a}\sqrt{a^2 - x^2}\,dx = \frac{4b}{a}\int_0^a \sqrt{a^2 - x^2}\,dx$$

To evaluate this integral, we let

$$x = a\sin\theta \qquad \text{so that} \qquad dx = a\cos\theta\,d\theta$$

Note that when $x = 0$, $\sin\theta = 0$, so $\theta = 0$ is the lower limit of integration with respect to $\theta$; when $x = a$, $\sin\theta = 1$, giving $\theta = \pi/2$ as the upper limit of integration. Also,

$$\sqrt{a^2 - x^2} = \sqrt{a^2 - a^2\sin^2\theta} = a\sqrt{1 - \sin^2\theta} = a\sqrt{\cos^2\theta} = a|\cos\theta| = a\cos\theta$$

since $0 \le \theta \le \frac{\pi}{2}$. Therefore,

$$A = \frac{4b}{a}\int_0^a \sqrt{a^2 - x^2}\,dx = \frac{4b}{a}\int_0^{\pi/2} a\cos\theta \cdot a\cos\theta\,d\theta$$

$$= 4ab\int_0^{\pi/2} \cos^2\theta\,d\theta$$

$$= 4ab\int_0^{\pi/2} \frac{1}{2}(1 + \cos 2\theta)\,d\theta \qquad \text{Use a half-angle formula.}$$

$$= 2ab\left[\theta + \frac{1}{2}\sin 2\theta\right]_0^{\pi/2} = 2ab\left[\left(\frac{\pi}{2} + 0\right) - 0\right]$$

or $\pi ab$.

**Note** For a circle, $a = b = r$, where $r$ is the radius of the circle, and the result of Example 2 gives $\pi r^2$ as the area of the circle, as expected.

**EXAMPLE 3** Find $\displaystyle\int \frac{1}{(4 + x^2)^{3/2}}\,dx$.

**Solution** Observe that the denominator of the integrand can be written as $\left(\sqrt{4 + x^2}\right)^3$ and thus involves a radical of the form $\sqrt{a^2 + x^2}$ with $a = 2$. Hence, we make the substitution

$$x = 2\tan\theta \qquad \text{so that} \qquad dx = 2\sec^2\theta\,d\theta$$

Then

$$\sqrt{4 + x^2} = \sqrt{4 + 4\tan^2\theta} = 2\sqrt{1 + \tan^2\theta} = 2\sqrt{\sec^2\theta} = 2\sec\theta$$

Therefore,

$$\int \frac{1}{(4 + x^2)^{3/2}}\,dx = \int \frac{1}{(2\sec\theta)^3} \cdot 2\sec^2\theta\,d\theta$$

$$= \frac{1}{4}\int \cos\theta\,d\theta$$

$$= \frac{1}{4}\sin\theta + C$$

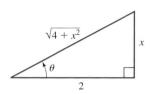

**FIGURE 5**
The right triangle associated
with the substitution $x = 2\tan\theta$

Finally, from the right triangle associated with the substitution $x = 2\tan\theta$, we see that $\sin\theta = x/\sqrt{4 + x^2}$. (See Figure 5.) Therefore,

$$\int \frac{1}{(4 + x^2)^{3/2}}\,dx = \frac{x}{4\sqrt{4 + x^2}} + C$$

■

**EXAMPLE 4**  Find $\displaystyle\int \frac{\sqrt{x^2 - 16}}{x}\,dx$.

**Solution**  Here, the integrand involves a radical of the form $\sqrt{x^2 - a^2}$, where $a = 4$. So we make the substitution

$$x = 4\sec\theta \qquad \text{so that} \qquad dx = 4\sec\theta\tan\theta\,d\theta$$

Then

$$\sqrt{x^2 - 16} = \sqrt{16\sec^2\theta - 16} = 4\sqrt{\sec^2\theta - 1} = 4\sqrt{\tan^2\theta} = 4\tan\theta$$

Therefore,

$$\int \frac{\sqrt{x^2 - 16}}{x}\,dx = \int \frac{4\tan\theta}{4\sec\theta} \cdot 4\sec\theta\tan\theta\,d\theta$$

$$= 4\int \tan^2\theta\,d\theta$$

$$= 4\int (\sec^2\theta - 1)\,d\theta = 4\int \sec^2\theta\,d\theta - 4\int d\theta$$

$$= 4\tan\theta - 4\theta + C$$

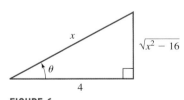

**FIGURE 6**
The right triangle associated with
the substitution $x = 4\sec\theta$

Since $x = 4\sec\theta$ or $\sec\theta = x/4$, we see that $\theta = \sec^{-1}(x/4)$. Furthermore, by inspecting the right triangle associated with the substitution, we see that

$$\tan\theta = \frac{\sqrt{x^2 - 16}}{4}$$

(See Figure 6.) Therefore,

$$\int \frac{\sqrt{x^2 - 16}}{x}\,dx = \sqrt{x^2 - 16} - 4\sec^{-1}\left(\frac{x}{4}\right) + C$$

■

Sometimes we can use the technique of completing the square to rewrite an integrand that involves a quadratic expression in the appropriate form before making a trigonometric substitution. This is illustrated in the next example.

**EXAMPLE 5**  Find $\displaystyle\int \frac{dx}{\sqrt{x^2 + 4x + 7}}$.

**Solution**  By completing the square for the expression under the radical sign, we obtain

$$x^2 + 4x + 7 = [x^2 + 4x + (2)^2] + 7 - 4 = (x + 2)^2 + 3$$

If we let $u = x + 2$, then

$$x^2 + 4x + 7 = u^2 + 3 \qquad \text{and} \qquad du = dx$$

so we can write

$$\int \frac{dx}{\sqrt{x^2 + 4x + 7}} = \int \frac{du}{\sqrt{u^2 + 3}}$$

Observe that the integrand $\sqrt{u^2 + 3}$ has the form $\sqrt{u^2 + a^2}$, where $a = \sqrt{3}$. This suggests that we make the substitution

$$u = \sqrt{3}\tan\theta \qquad \text{so that} \qquad du = \sqrt{3}\sec^2\theta\, d\theta$$

Then

$$\sqrt{u^2 + 3} = \sqrt{3\tan^2\theta + 3} = \sqrt{3}\sqrt{\tan^2\theta + 1} = \sqrt{3}\sqrt{\sec^2\theta} = \sqrt{3}\sec\theta$$

Therefore,

$$\int \frac{du}{\sqrt{u^2 + 3}} = \int \frac{\sqrt{3}\sec^2\theta}{\sqrt{3}\sec\theta}\, d\theta = \int \sec\theta\, d\theta$$

$$= \ln|\sec\theta + \tan\theta| + C$$

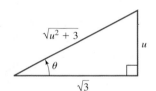

**FIGURE 7**
The right triangle associated with the substitution $u = \sqrt{3}\tan\theta$

From $u = \sqrt{3}\tan\theta$ we see that $\tan\theta = u/\sqrt{3}$. Also, from the right triangle associated with the substitution $u = \sqrt{3}\tan\theta$, we see that $\sec\theta = \sqrt{u^2 + 3}/\sqrt{3}$. (See Figure 7.) Therefore,

$$\int \frac{du}{\sqrt{u^2 + 3}} = \ln\left|\frac{\sqrt{u^2 + 3}}{\sqrt{3}} + \frac{u}{\sqrt{3}}\right| + \ln C_1$$

so

$$\int \frac{dx}{\sqrt{x^2 + 4x + 7}} = \ln\left|\sqrt{x^2 + 4x + 7} + x + 2\right| + C \qquad C = \ln\left(\frac{C_1}{\sqrt{3}}\right) \quad \blacksquare$$

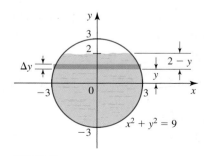

**FIGURE 8**
One end of a cylindrical oil storage tank

**EXAMPLE 6**  **Hydrostatic Force on a Window**  A cylindrical oil storage tank of radius 3 ft and length 10 ft is lying on its side. If the tank is filled to a height of 5 ft with oil having a weight density of 50 lb/ft³, find the force exerted by the oil on one end of the tank.

**Solution**  Let's introduce a coordinate system in such a way that the end of the tank (a disk of radius 3 ft) has its center at the origin. Then the circle in question is described by the equation $x^2 + y^2 = 9$. (See Figure 8.) The length of a representative horizontal rectangle is $L(y) = 2x = 2\sqrt{9 - y^2}$. Therefore, the area of a horizontal rectangle is

$$\Delta A = L(y)\Delta y = 2\sqrt{9 - y^2}\,\Delta y$$

The pressure exerted by the oil on the rectangle is

$$\delta(2 - y) = 50(2 - y) \qquad \text{density} \cdot \text{depth}$$

Therefore, the force exerted by the oil on the rectangle is

$$\delta(2 - y)\Delta A = 50(2 - y)(2)\sqrt{9 - y^2}\,\Delta y \qquad \text{pressure} \cdot \text{area}$$

$$= 100(2 - y)\sqrt{9 - y^2}\,\Delta y$$

Summing the forces on these rectangles and taking the limit, we find that the force exerted by the oil on the end of the storage tank is

$$F = \int_{-3}^{2} 100(2 - y)\sqrt{9 - y^2}\,dy$$

$$= 200\int_{-3}^{2}\sqrt{9 - y^2}\,dy - 100\int_{-3}^{2} y\sqrt{9 - y^2}\,dy$$

The second integral can be evaluated by making the substitution $u = 9 - y^2$, whereas the first integral can be evaluated by using the trigonometric substitution $y = 3\sin\theta$. We will leave the evaluation of these integrals as an exercise (Exercise 58). You will find that the force is approximately 2890 lb. ∎

## 6.3 CONCEPT QUESTIONS

1. What substitution would you use to find an integral whose integrand involves the expression (a) $\sqrt{a^2 - x^2}$, $a > 0$; (b) $\sqrt{a^2 + x^2}$, $a > 0$; and (c) $\sqrt{x^2 - a^2}$, $a > 0$?

2. How would you find an integral whose integrand involves the expression $\sqrt{ax^2 + bx + c}$?

## 6.3 EXERCISES

In Exercises 1–32, find or evaluate the integral using an appropriate trigonometric substitution.

1. $\displaystyle\int \frac{x}{\sqrt{9 - x^2}}\,dx$

2. $\displaystyle\int \frac{\sqrt{4 - x^2}}{x^2}\,dx$

3. $\displaystyle\int x\sqrt{4 - x^2}\,dx$

4. $\displaystyle\int \frac{1}{x^2\sqrt{1 - x^2}}\,dx$

5. $\displaystyle\int \frac{1}{x\sqrt{4 + x^2}}\,dx$

6. $\displaystyle\int x^3\sqrt{1 + x^2}\,dx$

7. $\displaystyle\int \frac{1}{x^2\sqrt{x^2 + 4}}\,dx$

8. $\displaystyle\int \frac{1}{x^3\sqrt{x^2 - 4}}\,dx$

9. $\displaystyle\int x^3\sqrt{1 - x^2}\,dx$

10. $\displaystyle\int x^3\sqrt{4 - x^2}\,dx$

11. $\displaystyle\int \frac{x^3}{\sqrt{x^2 + 9}}\,dx$

12. $\displaystyle\int_{0}^{3/4} \frac{x^2}{\sqrt{9 - 4x^2}}\,dx$

13. $\displaystyle\int \frac{1}{(x^2 - 9)^{3/2}}\,dx$

14. $\displaystyle\int (4 - x^2)^{3/2}\,dx$

15. $\displaystyle\int \frac{\sqrt{16x^2 - 9}}{x}\,dx$

16. $\displaystyle\int \frac{x^2}{\sqrt{3 - x^2}}\,dx$

17. $\displaystyle\int \frac{\sqrt{1 - x^2}}{x^4}\,dx$

18. $\displaystyle\int \frac{\sqrt{9x^2 + 4}}{x^4}\,dx$

19. $\displaystyle\int \frac{1}{x\sqrt{9x^2 + 4}}\,dx$

20. $\displaystyle\int \frac{1}{(9 + x^2)^2}\,dx$

21. $\displaystyle\int_{-\sqrt{3}}^{\sqrt{3}}\sqrt{4 - x^2}\,dx$

22. $\displaystyle\int_{2}^{4} \frac{\sqrt{x^2 - 4}}{x^4}\,dx$

23. $\displaystyle\int_{1}^{\sqrt{3}} \frac{1}{(1 + x^2)^{3/2}}\,dx$

24. $\displaystyle\int \frac{2x + 3}{\sqrt{1 - x^2}}\,dx$

25. $\displaystyle\int e^x\sqrt{4 - e^{2x}}\,dx$

26. $\displaystyle\int e^t\sqrt{1 + e^{2t}}\,dt$

27. $\displaystyle\int_{1/2}^{\sqrt{3}/2} \frac{dx}{x\sqrt{1 - x^2}}$

28. $\displaystyle\int \sqrt{4x - x^2}\,dx$

**29.** $\displaystyle\int \frac{1}{\sqrt{2t - t^2}}\, dt$

**30.** $\displaystyle\int \frac{t^2}{\sqrt{4t - t^2}}\, dt$

**31.** $\displaystyle\int \frac{1}{(x^2 + 4x + 8)^2}\, dx$

**32.** $\displaystyle\int \frac{1}{(3 - 2x - x^2)^{5/2}}\, dx$

**33.** Find the area of the region under the graph of
$$y = \frac{1}{x\sqrt{4 - x^2}}$$ on the interval $[1, \sqrt{2}]$.

**34.** Find the area of the region enclosed by the hyperbola $16x^2 - 9y^2 = 144$ and the line $x = 5$.

**35.** Find the average value of the positive $y$-coordinates of the ellipse $\dfrac{x^2}{a^2} + \dfrac{y^2}{b^2} = 1$.

**36.** The region under the graph of $y = \dfrac{x}{(9 + x^2)^{1/4}}$ on the interval $[0, 4]$ is revolved about the $x$-axis. Find the volume of the resulting solid.

**37.** The region under the graph of $y = \dfrac{x}{\sqrt{16 - x^2}}$ on the interval $[0, 2]$ is revolved about the $y$-axis. Find the volume of the resulting solid.

**38.** The graph of $y = e^x$ between $x = 0$ and $x = 1$ is revolved about the $x$-axis. Find the surface area of the resulting solid.

**39.** Find the arc length of the graph of $y = \ln 2x$ on the interval $[1, \sqrt{3}]$.

**40.** Find the arc length of the graph of $y = -\frac{1}{2}x^2 + 2x$ from $P(0, 0)$ to $Q(2, 2)$.

**41. Force Exerted on a Viewing Port**  The circular viewing port of a modern submersible used in oceanographic research has a radius of 1 ft. If the viewing port is partially submerged so that three fourths of it is under water, find the force exerted on it by the water. Note that the density of sea water is 64 lb/ft³.

**42.** Find the area of the region enclosed by the parabola $y = \frac{1}{4}x^2$ and the witch of Agnesi: $y = \dfrac{8}{x^2 + 4}$.

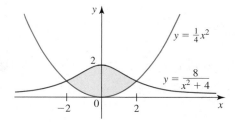

**43.** Let $f(x) = \dfrac{1}{(4 + x^2)^{3/2}}$ and $g(x) = 1 - x^2$.

    **a.** Plot the graphs of $f$ and $g$ using the viewing window $[-1.5, 1.5] \times [0, 1.2]$. Find the $x$-coordinates of the points of intersection of the graphs of $f$ and $g$ accurate to three decimal places.

    **b.** Use the result of part (a) and trigonometric substitution to find the approximate area of the region bounded by the graphs of $f$ and $g$.

**44.** Let $f(x) = x^3\sqrt{1 - x^2}$ and $g(x) = \dfrac{0.2}{(1 + x^2)^{3/2}}$.

    **a.** Plot the graphs of $f$ and $g$ using the viewing window $[0, 1.2] \times [0, 0.4]$. Find the $x$-coordinates of the points of intersection of the graphs of $f$ and $g$ accurate to three decimal places.

    **b.** Use the result of part (a) and trigonometric substitution to find the approximate area of the region bounded by the graphs of $f$ and $g$.

**45.** Find the surface area of the ellipsoid formed by revolving the ellipse $\dfrac{x^2}{a^2} + \dfrac{y^2}{b^2} = 1$, $a > b$, about the $x$-axis.

**46.** Find the force exerted by a liquid of constant weight density $\delta$ on a vertical ellipse $\dfrac{x^2}{a^2} + \dfrac{y^2}{b^2} = 1$ whose center is submerged in the liquid to a depth $h$, where $h \geq b$.

**47. Air Pollution**  The amount of nitrogen dioxide, a brown gas that impairs breathing, present in the atmosphere on a certain day in May in the city of Long Beach is approximated by
$$A(t) = \frac{544}{4 + (t - 4.5)^2} + 28 \qquad 0 \leq t \leq 11$$
where $A(t)$ is measured in pollutant standard index (PSI) and $t$ is measured in hours with $t = 0$ corresponding to 7 A.M. What is the average amount of the pollutant present in the atmosphere between 7 A.M. and noon on that day in the city?
*Source: The Los Angeles Times.*

**48. Work Done by a Repulsive Charge**  An electric charge $Q$ distributed uniformly along a line of length $2c$ lying along the $y$-axis repulses a like charge $q$ from the point $x = a$ ($a > 0$) to the point $x = b$, where $b > a$. The magnitude of the force acting on the charge $q$ when it is at the point $x$ is given by
$$F(x) = \frac{1}{4\pi\varepsilon_0} \frac{qQ}{x\sqrt{x^2 + c^2}}$$
and the force acts in the direction of the positive $x$-axis. Find the work done by the force of repulsion.

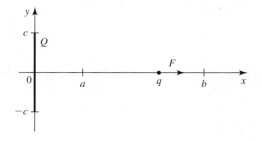

**49. Work Done by a Magnetic Field** The force of a circular electric current acting on a small magnet with its axis perpendicular to the plane of the circle and passing through its center has magnitude given by

$$F = \frac{k}{(x^2 + a^2)^{3/2}} \qquad 0 < x < \infty$$

where $a$ is the radius of the circle, $k$ is a constant, and $x$ is the distance from the center of the circle to the magnet in the direction along the axis. Find the work done by the magnetic field in moving the magnet along the axis from $x = 0$ to $x = 2a$.

The electric current $I$ flowing in the loop establishes a magnetic field that acts on the magnet.

**50. Average Illumination** Two lamp posts, each $h$ ft tall, are located $d$ ft apart. (See the figure.) If the intensity of each light is $I$ lumens, find the average intensity of light along the straight line connecting the bases of the lamp posts.

**Hint:** The intensity of light at a point $P$ from the base of the lamp post is proportional to the cosine of the angle that the incident light makes with the vertical and inversely proportional to the square of the distance between $P$ and the light source.

*In Exercises 51–54, use a trigonometric substitution to derive the formula.*

**51.** $\displaystyle\int \sqrt{a^2 - u^2}\, du = \frac{u}{2} \sqrt{a^2 - u^2} + \frac{a^2}{2} \sin^{-1} \frac{u}{a} + C$

**52.** $\displaystyle\int \frac{\sqrt{a^2 + u^2}}{u}\, du = \sqrt{a^2 + u^2} - a \ln\left| \frac{a + \sqrt{a^2 + u^2}}{u} \right| + C$

**53.** $\displaystyle\int \frac{du}{u\sqrt{a^2 + u^2}} = -\frac{1}{a} \ln\left| \frac{\sqrt{a^2 + u^2} + a}{u} \right| + C$

**54.** $\displaystyle\int \frac{\sqrt{u^2 - a^2}}{u^2}\, du = -\frac{\sqrt{u^2 - a^2}}{u} + \ln\left| u + \sqrt{u^2 - a^2} \right| + C$

**55. a.** Use trigonometric substitution to show that

$$\int \frac{dx}{\sqrt{x^2 + a^2}} = \ln\left( x + \sqrt{x^2 + a^2} \right) + C$$

**b.** Use integration by parts and the result of part (a) to find

$$\int \frac{x \tan^{-1} x}{\sqrt{1 + x^2}}\, dx$$

**56.** Evaluate

$$\int_0^{\pi/4} \frac{dx}{a^2 \cos^2 x + b^2 \sin^2 x} \qquad a > 0, \quad b > 0$$

**Hint:** Use the substitution $u = \tan x$.

**57.** Prove that

$$\int_0^x \sqrt{a^2 - u^2}\, du = \frac{1}{2} x\sqrt{a^2 - x^2} + \frac{a^2}{2} \sin^{-1} \frac{x}{a} + C$$

where $0 \leq x \leq a$, by interpreting the integral geometrically.

**58.** Refer to Example 6. Show that

$$F = 200 \int_{-3}^{2} \sqrt{9 - y^2}\, dy - 100 \int_{-3}^{2} y\sqrt{9 - y^2}\, dy \approx 2890$$

---

## 6.4 The Method of Partial Fractions

### ■ Partial Fractions

In algebra we learned how to combine two or more rational expressions (fractions) into a single expression by putting them together over a common denominator. For example,

$$\frac{2}{x - 3} - \frac{1}{x + 1} = \frac{2(x + 1) - (x - 3)}{(x - 3)(x + 1)} = \frac{x + 5}{(x - 3)(x + 1)} \qquad \textbf{(1)}$$

Sometimes, however, it is advantageous to reverse the process, that is, to express a complicated expression as a sum or difference of simpler ones. As an example, suppose that we wish to evaluate the integral

$$\int \frac{x+5}{x^2 - 2x - 3}\, dx \tag{2}$$

Thanks to Equation (1), we can write the integrand in the form

$$\frac{x+5}{x^2 - 2x - 3} = \frac{x+5}{(x-3)(x+1)} = \frac{2}{x-3} - \frac{1}{x+1} \tag{3}$$

so that upon integrating both sides with respect to $x$, we obtain

$$\int \frac{x+5}{x^2 - 2x - 3}\, dx = \int \left( \frac{2}{x-3} - \frac{1}{x+1} \right) dx$$

$$= 2 \ln|x-3| - \ln|x+1| + C$$

The expression on the right-hand side of Equation (3) is called the *partial fraction decomposition* of $(x+5)/(x^2 - 2x - 3)$, and each of the terms is called a *partial fraction*. The technique of integration that we have used to evaluate the integrand in (2) is called the **method of partial fractions** and can be used to integrate any rational function.

Suppose that $f$ is a rational function defined by

$$f(x) = \frac{P(x)}{Q(x)}$$

where $P$ and $Q$ are polynomials. If the degree of $P$ is greater than or equal to the degree of $Q$, we can use long division to express $f(x)$ in the form

$$f(x) = S(x) + \frac{R(x)}{Q(x)} \tag{4}$$

where $S$ is a *polynomial* and the degree of $R$ is *less than* that of $Q$. For example, if

$$f(x) = \frac{x^3 - 4x^2 + 3x - 5}{x^2 - 1}$$

then using long division, we can write

$$f(x) = x - 4 + \frac{4x - 9}{x^2 - 1}$$

Now suppose that we want to integrate $f$. Using Equation (4), we have

$$\int f(x)\, dx = \int S(x)\, dx + \int \frac{R(x)}{Q(x)}\, dx$$

The first integral on the right is easily evaluated since its integrand is a polynomial. To evaluate the second integral, we decompose $R(x)/Q(x)$ into a sum of partial fractions and integrate the resulting expression term by term. That $R(x)/Q(x)$ can be so decomposed is guaranteed by the following results from algebra, which we state without proof:

1. Every polynomial $Q$ can be factored into a product of linear factors (of the form $ax + b$) and irreducible quadratic factors (of the form $ax^2 + bx + c$ where $b^2 - 4ac < 0$).

**2.** Every rational function $R(x)/Q(x)$ where the degree of $R$ is *less than* the degree of $Q$ can be decomposed into a sum of partial fractions of the form

$$\frac{A}{(ax + b)^k} \quad \text{or} \quad \frac{Ax + B}{(ax^2 + bx + c)^k}$$

The form the partial fraction decomposition of the rational function $R(x)/Q(x)$ takes depends on the form of $Q(x)$ and can be illustrated through examining four cases.

---

**Case 1: Distinct Linear Factors**

If

$$\frac{R(x)}{Q(x)} = \frac{R(x)}{(a_1x + b_1)(a_2x + b_2) \cdots (a_nx + b_n)}$$

where all the factors $a_kx + b_k$, $k = 1, 2, \ldots, n$, are distinct, then there exist constants $A_1, A_2, \ldots, A_n$ such that

$$\frac{R(x)}{Q(x)} = \frac{A_1}{a_1x + b_1} + \frac{A_2}{a_2x + b_2} + \cdots + \frac{A_n}{a_nx + b_n}$$

---

**EXAMPLE 1**  Find $\int \dfrac{4x^2 - 4x + 6}{x^3 - x^2 - 6x}\, dx$.

**Solution**  The degree of the numerator of the integrand is less than that of the denominator, and no long division is required in this case. The denominator can be written in the form $x(x - 3)(x + 2)$, a product of three distinct linear factors. Therefore, a partial fraction decomposition of the form

$$\frac{4x^2 - 4x + 6}{x(x - 3)(x + 2)} = \frac{A}{x} + \frac{B}{x - 3} + \frac{C}{x + 2}$$

exists. To determine $A$, $B$, and $C$, we multiply both sides of the equation by $x(x - 3)(x + 2)$, obtaining

$$4x^2 - 4x + 6 = A(x - 3)(x + 2) + Bx(x + 2) + Cx(x - 3)$$

If we expand the terms on the right and collect like powers of $x$, the equation can be written in the form

$$4x^2 - 4x + 6 = (A + B + C)x^2 + (-A + 2B - 3C)x - 6A$$

Because the two polynomials are equal, the coefficients of like powers of $x$ must be equal. Equating, in turn, the coefficients of $x^2$, $x^1$, and $x^0$ leads to the following system of linear equations in $A$, $B$, and $C$:

$$A + B + C = 4$$
$$-A + 2B - 3C = -4$$
$$-6A = 6$$

Solving this system, we find $A = -1$, $B = 2$, and $C = 3$. Therefore, the partial fraction decomposition of the integrand is

$$\frac{4x^2 - 4x + 6}{x^3 - x^2 - 6x} = -\frac{1}{x} + \frac{2}{x - 3} + \frac{3}{x + 2}$$

## Historical Biography

**JOHANN BERNOULLI**
(1667-1748)

Despite his father's wish for him to become a merchant, Johann Bernoulli studied mathematics privately with his brother Jacob, who was a professor of mathematics at the University of Basel. Johann became so engrossed in mathematics that by the age of 25 he had composed two textbooks on calculus. These texts were not published until much later, and in the intervening years he became a tutor to Guillaume François Antoine, Marquis de l'Hôpital (page 380). l'Hôpital commissioned Bernoulli to sell his mathematical findings in exchange for a regular salary. In fact, one of Johann Bernoulli's greatest contributions to calculus was presented in l'Hôpital's book *Analyse des infiniment petits pour l'intelligence des lignes courbes* (1696), and this contribution has since been known as l'Hôpital's Rule. Bernoulli carried on an extensive correspondence with Liebniz (page 179) and was a firm supporter of Leibniz's methods over those of Isaac Newton (page 202). Among Bernoulli's many contributions to calculus is the method of partial fractions, which aids in the integration of some rational functions. Johann Bernoulli attained great fame during his lifetime and was known as the "Archimedes of his age." Mathematical talent ran deep in the Bernoulli family; in addition to his older brother Jacob, three of Johann Bernoulli's sons became mathematicians.

Finally, integrating both sides of this equation gives

$$\int \frac{4x^2 - 4x + 6}{x^3 - x^2 - 6x} \, dx = \int \left( -\frac{1}{x} + \frac{2}{x - 3} + \frac{3}{x + 2} \right) dx$$

$$= -\ln|x| + 2\ln|x - 3| + 3\ln|x + 2| + k$$

where $k$ is the constant of integration. ∎

**Note** There is another way of finding the coefficients $A$, $B$, and $C$ in Example 1. Our starting point is the equation

$$4x^2 - 4x + 6 = A(x - 3)(x + 2) + Bx(x + 2) + Cx(x - 3)$$

which holds for *all* values of $x$. If we let $x = 0$, then the second and third terms on the right are equal to zero, giving $6 = -6A$ or $A = -1$. Next, letting $x = 3$, so that the first and third terms are equal to zero, we find that $30 = 15B$, giving $B = 2$. Finally, letting $x = -2$ gives $30 = 10C$ or $C = 3$. ∎

**EXAMPLE 2** Find $\displaystyle \int \frac{4x^3 + x}{2x^2 + x - 3} \, dx$.

**Solution** Since the degree of the numerator of the integrand is greater than that of the denominator, we use long division to write

$$\frac{4x^3 + x}{2x^2 + x - 3} = 2x - 1 + \frac{8x - 3}{2x^2 + x - 3} \tag{5}$$

Next, we decompose $(8x - 3)/(2x^2 + x - 3)$ into a sum of partial fractions. Factoring, we see that $(2x^2 + x - 3) = (2x + 3)(x - 1)$ is a product of two distinct linear factors. Therefore,

$$\frac{8x - 3}{2x^2 + x - 3} = \frac{8x - 3}{(2x + 3)(x - 1)} = \frac{A}{2x + 3} + \frac{B}{x - 1}$$

Multiplying through by $(2x + 3)(x - 1)$ gives

$$8x - 3 = A(x - 1) + B(2x + 3)$$

If we let $x = 1$, then $5 = 5B$ or $B = 1$. Next, letting $x = -\frac{3}{2}$ yields $-15 = -\frac{5}{2}A$, or $A = 6$. Therefore,

$$\frac{8x - 3}{2x^2 + x - 3} = \frac{6}{2x + 3} + \frac{1}{x - 1}$$

Substituting the right-hand side of this equation into Equation (5) and integrating both sides of the resulting expression with respect to $x$, we get the desired result:

$$\int \frac{4x^3 + x}{2x^2 + x - 3} \, dx = \int \left( 2x - 1 + \frac{6}{2x + 3} + \frac{1}{x - 1} \right) dx$$

$$= x^2 - x + 3\ln|2x + 3| + \ln|x - 1| + k$$

Observe that we have used the substitution $u = 2x + 3$ to evaluate the integral of the third term on the right and the substitution $u = x - 1$ to evaluate the last term on the right. ∎

---

**Case 2: Repeated Linear Factors**

If $Q(x)$ contains a factor $(ax + b)^r$ with $r > 1$, then the partial fraction decomposition of $R(x)/Q(x)$ contains a sum of $r$ partial fractions of the form

$$\frac{A_1}{ax + b} + \frac{A_2}{(ax + b)^2} + \cdots + \frac{A_r}{(ax + b)^r}$$

where each $A_k$ is a real number.

---

For example,

$$\frac{2x^4 - 3x^2 + x - 4}{x(x - 1)(2x + 3)^3} = \frac{A}{x} + \frac{B}{x - 1} + \frac{C}{2x + 3} + \frac{D}{(2x + 3)^2} + \frac{E}{(2x + 3)^3}$$

**EXAMPLE 3** Find $\displaystyle\int \frac{2x^2 + 3x + 7}{x^3 + x^2 - x - 1}\, dx$.

**Solution** The degree of the numerator of the integrand is less than that of the denominator, and no long division is necessary. Note that

$$Q(x) = x^3 + x^2 - x - 1 = x^2(x + 1) - (x + 1) = (x + 1)(x^2 - 1)$$

$$= (x - 1)(x + 1)^2$$

Since $-1$ is a zero of multiplicity 2 (here, $r = 2$), the partial fraction decomposition of the integrand has the form

$$\frac{2x^2 + 3x + 7}{(x + 1)^2(x - 1)} = \frac{A}{x + 1} + \frac{B}{(x + 1)^2} + \frac{C}{x - 1}$$

Multiplying both sides of this equation by $(x + 1)^2(x - 1)$, we obtain

$$2x^2 + 3x + 7 = A(x + 1)(x - 1) + B(x - 1) + C(x + 1)^2$$

If we let $x = 1$, then we obtain $12 = 4C$, which yields $C = 3$. Next, letting $x = -1$, we have $6 = -2B$, so $B = -3$. Finally, to determine $A$, we let $x = 0$ (which is the most convenient choice), obtaining $7 = -A - B + C$. Using the values of $B$ and $C$ that we obtained earlier, we see that $A = -B + C - 7 = -1$. Therefore,

$$\int \frac{2x^2 + 3x + 7}{x^3 + x^2 - x - 1}\, dx = \int \left( -\frac{1}{x + 1} - \frac{3}{(x + 1)^2} + \frac{3}{x - 1} \right) dx$$

$$= -\ln|x + 1| + \frac{3}{x + 1} + 3\ln|x - 1| + k$$

$$= \frac{3}{x + 1} + \ln\left| \frac{(x - 1)^3}{x + 1} \right| + k \qquad \blacksquare$$

Recall that a quadratic expression $ax^2 + bx + c$ is **irreducible** if it cannot be written as a product of linear factors with real roots. For example, $3x^2 + x + 1$ is irreducible.

**Case 3: Distinct Irreducible Quadratic Factors**

If

$$\frac{R(x)}{Q(x)} = \frac{R(x)}{(a_1x^2 + b_1x + c_1)(a_2x^2 + b_2x + c_2) \cdots (a_nx^2 + b_nx + c_n)}$$

where all the factors $a_kx^2 + b_kx + c_k$, $k = 1, 2, \ldots, n$, are distinct and irreducible, then there exist constants $A_1, A_2, \ldots, A_n, B_1, B_2, \ldots, B_n$ such that

$$\frac{R(x)}{Q(x)} = \frac{A_1x + B_1}{a_1x^2 + b_1x + c_1} + \frac{A_2x + B_2}{a_2x^2 + b_2x + c_2} + \cdots + \frac{A_nx + B_n}{a_nx^2 + b_nx + c_n}$$

For example,

$$\frac{3x^3 + 8x^2 + 7x + 5}{(x^2 + 1)(x^2 + 2x + 2)} = \frac{Ax + B}{x^2 + 1} + \frac{Cx + D}{x^2 + 2x + 2}$$

**EXAMPLE 4** Find $\displaystyle\int \frac{x^4 + 3x^3 + 14x^2 + 14x + 41}{(x^2 + 4)(x^2 + 2x + 5)} \, dx$.

**Solution** Since the degree of the numerator is not less than the degree of the denominator, we use long division to write

$$\frac{x^4 + 3x^3 + 14x^2 + 14x + 41}{(x^2 + 4)(x^2 + 2x + 5)} = \frac{x^4 + 3x^3 + 14x^2 + 14x + 41}{x^4 + 2x^3 + 9x^2 + 8x + 20}$$

$$= 1 + \frac{x^3 + 5x^2 + 6x + 21}{(x^2 + 4)(x^2 + 2x + 5)}$$

Notice that the quadratic $x^2 + 2x + 5$ is irreducible because its discriminant

$$b^2 - 4ac = 2^2 - 4(1)(5) = -16 < 0$$

Since the quadratic factors are distinct, we can write

$$\frac{x^3 + 5x^2 + 6x + 21}{(x^2 + 4)(x^2 + 2x + 5)} = \frac{Ax + B}{x^2 + 4} + \frac{Cx + D}{x^2 + 2x + 5}$$

Multiplying both sides of the equation by $(x^2 + 4)(x^2 + 2x + 5)$ gives

$$x^3 + 5x^2 + 6x + 21 = (Ax + B)(x^2 + 2x + 5) + (Cx + D)(x^2 + 4)$$

$$= (A + C)x^3 + (2A + B + D)x^2$$

$$+ (5A + 2B + 4C)x + (5B + 4D)$$

Equating the coefficients of like powers of $x$ yields the system

$$A + \qquad C \quad = 1$$
$$2A + B + \qquad D = 5$$
$$5A + 2B + 4C \qquad = 6$$
$$5B + \qquad 4D = 21$$

The solution of the system is $A = 0$, $B = 1$, $C = 1$, and $D = 4$. Therefore,

$$\int \frac{x^4 + 3x^3 + 14x^2 + 14x + 41}{(x^2 + 4)(x^2 + 2x + 5)} \, dx = \int \left( 1 + \frac{1}{x^2 + 4} + \frac{x + 4}{x^2 + 2x + 5} \right) dx$$

$$= x + \frac{1}{2} \tan^{-1} \left( \frac{x}{2} \right) + \int \frac{x + 4}{x^2 + 2x + 5} \, dx$$

To evaluate the integral on the right, we complete the square in the denominator of the integrand. Thus, $x^2 + 2x + 5 = (x + 1)^2 + 4$. Next, using the substitution $u = x + 1$ so that $du = dx$ and $x = u - 1$, we obtain

$$\int \frac{x + 4}{x^2 + 2x + 5} \, dx = \int \frac{x + 4}{(x + 1)^2 + 4} \, dx = \int \frac{(u - 1) + 4}{u^2 + 4} \, du = \int \frac{u + 3}{u^2 + 4} \, du$$

$$= \int \frac{u}{u^2 + 4} \, du + \int \frac{3}{u^2 + 4} \, du$$

$$= \frac{1}{2} \ln(u^2 + 4) + \frac{3}{2} \tan^{-1} \left( \frac{u}{2} \right) + C_1$$

$$= \frac{1}{2} \ln(x^2 + 2x + 5) + \frac{3}{2} \tan^{-1} \left( \frac{x + 1}{2} \right) + C_1$$

So

$$\int \frac{x^4 + 3x^3 + 14x^2 + 14x + 41}{(x^2 + 4)(x^2 + 2x + 5)} \, dx$$

$$= x + \frac{1}{2} \tan^{-1} \left( \frac{x}{2} \right) + \frac{1}{2} \ln(x^2 + 2x + 5) + \frac{3}{2} \tan^{-1} \left( \frac{x + 1}{2} \right) + C \quad \blacksquare$$

---

**Case 4: Repeated Irreducible Quadratic Factors**

If $Q(x)$ contains a factor $(ax^2 + bx + c)^r$ with $r > 1$, where $ax^2 + bx + c$ is irreducible, then the partial fraction decomposition of $R(x)/Q(x)$ contains a sum of $r$ partial fractions of the form

$$\frac{A_1 x + B_1}{ax^2 + bx + c} + \frac{A_2 x + B_2}{(ax^2 + bx + c)^2} + \cdots + \frac{A_r x + B_r}{(ax^2 + bx + c)^r}$$

where $A_k$ and $B_k$ are real numbers.

---

For example,

$$\frac{x^4 - 3x^3 + x + 1}{x(x - 1)^2(x^2 + 1)(x^2 + x + 1)^2}$$

$$= \frac{A}{x} + \frac{B}{x - 1} + \frac{C}{(x - 1)^2} + \frac{Dx + E}{x^2 + 1} + \frac{Fx + G}{(x^2 + x + 1)} + \frac{Hx + I}{(x^2 + x + 1)^2}$$

**EXAMPLE 5** Find $\displaystyle\int \frac{x^3 - 2x^2 + 3x + 2}{x(x^2 + 1)^2}\, dx.$

**Solution**   The partial fraction decomposition of the integrand has the form

$$\frac{x^3 - 2x^2 + 3x + 2}{x(x^2 + 1)^2} = \frac{A}{x} + \frac{Bx + C}{x^2 + 1} + \frac{Dx + E}{(x^2 + 1)^2}$$

Multiplying both sides of the equation by $x(x^2 + 1)^2$ gives

$$
\begin{aligned}
x^3 - 2x^2 + 3x + 2 &= A(x^2 + 1)^2 + (Bx + C)x(x^2 + 1) + (Dx + E)x \\
&= A(x^4 + 2x^2 + 1) + B(x^4 + x^2) + C(x^3 + x) + Dx^2 + Ex \\
&= (A + B)x^4 + Cx^3 + (2A + B + D)x^2 + (C + E)x + A
\end{aligned}
$$

Equating the coefficients of like powers of $x$ yields the system

$$
\begin{aligned}
A + B &&&&&= 0 \\
&& C &&&= 1 \\
2A + B + && && D &= -2 \\
&& C && + E &= 3 \\
A &&&&&= 2
\end{aligned}
$$

The solution of this system is $A = 2$, $B = -2$, $C = 1$, $D = -4$, and $E = 2$. Therefore,

$$
\begin{aligned}
\int \frac{x^3 - 2x^2 + 3x + 2}{x(x^2 + 1)^2}\, dx \\
= \int \left( \frac{2}{x} + \frac{-2x + 1}{x^2 + 1} + \frac{-4x + 2}{(x^2 + 1)^2} \right) dx \\
= 2\int \frac{dx}{x} - 2\int \frac{x}{x^2 + 1}\, dx + \int \frac{dx}{x^2 + 1} - 4\int \frac{x}{(x^2 + 1)^2}\, dx + 2\int \frac{1}{(x^2 + 1)^2}\, dx \\
= 2\ln|x| - \ln(x^2 + 1) + \tan^{-1} x + \frac{2}{x^2 + 1} + 2\int \frac{1}{(x^2 + 1)^2}\, dx
\end{aligned}
$$

To find the integral on the right-hand side, we make the substitution

$$x = \tan\theta \qquad \text{so that} \qquad dx = \sec^2\theta\, d\theta$$

Also, $x^2 + 1 = \tan^2\theta + 1 = \sec^2\theta$. So

$$
\begin{aligned}
\int \frac{1}{(x^2 + 1)^2}\, dx &= \int \frac{1}{\sec^4\theta} \cdot \sec^2\theta\, d\theta = \int \cos^2\theta\, d\theta \\
&= \frac{1}{2}\int (1 + \cos 2\theta)\, d\theta = \frac{1}{2}\left( \theta + \frac{1}{2}\sin 2\theta \right) + k \\
&= \frac{1}{2}(\theta + \sin\theta\cos\theta) + k \qquad \sin 2\theta = 2\sin\theta\cos\theta \\
&= \frac{1}{2}\left( \tan^{-1} x + \frac{x}{\sqrt{x^2 + 1}} \cdot \frac{1}{\sqrt{x^2 + 1}} \right) + k \qquad \text{See Figure 1.} \\
&= \frac{1}{2}\left( \tan^{-1} x + \frac{x}{x^2 + 1} \right) + k
\end{aligned}
$$

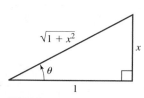

**FIGURE 1**
The right triangle associated
with the substitution $x = \tan\theta$

Therefore, the desired result is

$$\int \frac{x^3 - 2x^2 + 3x + 2}{x(x^2 + 1)^2} dx = \ln \frac{x^2}{x^2 + 1} + \tan^{-1} x + \frac{2}{x^2 + 1} + \tan^{-1} x + \frac{x}{x^2 + 1} + K$$

$$= \ln \frac{x^2}{x^2 + 1} + 2 \tan^{-1} x + \frac{x + 2}{x^2 + 1} + K \qquad \blacksquare$$

**Note**   Certain integrals involving rational functions can be more easily evaluated by using the method of substitution. For example, the integral

$$\int \frac{6x^2 + 4x + 2}{x(x^2 + x + 1)} dx$$

can be evaluated by letting

$$u = x(x^2 + x + 1) = x^3 + x^2 + x$$

Then $du = (3x^2 + 2x + 1)\, dx$, so

$$\int \frac{6x^2 + 4x + 2}{x(x^2 + x + 1)} dx = \int \frac{2}{u} du$$

$$= 2 \ln|u| + C$$

$$= \ln(x^3 + x^2 + x)^2 + C$$

However, such integrals rarely occur in practice. $\qquad \blacksquare$

**EXAMPLE 6**   **Waste Disposal**   When organic waste is dumped into a pond, the oxidization process that takes place reduces the pond's oxygen content. However, in time, nature will restore the oxygen content to its natural level. Suppose that the oxygen content $t$ days after organic waste has been dumped into a pond is given by

$$f(t) = 100 \left( \frac{t^2 + 10t + 100}{t^2 + 20t + 100} \right)$$

percent of its normal level. Find the average content of oxygen in the pond over the first 10 days after organic waste has been dumped into it.

**Solution**   The average content is given by

$$C = \frac{1}{10} \int_0^{10} f(t)\, dt = 10 \int_0^{10} \frac{t^2 + 10t + 100}{t^2 + 20t + 100} dt$$

Note that the degree of the numerator in the integrand is the same as that of the denominator, so long division is required here. Carrying through with the division, we find

$$\frac{t^2 + 10t + 100}{t^2 + 20t + 100} = 1 - \frac{10t}{t^2 + 20t + 100}$$

Next, observe that $t^2 + 20t + 100 = (t + 10)^2$. Therefore, we can write

$$\frac{10t}{t^2 + 20t + 100} = \frac{10t}{(t + 10)^2} = \frac{A}{t + 10} + \frac{B}{(t + 10)^2}$$

where $A$ and $B$ are real numbers to be determined. Multiplying both sides of this equation by $(t + 10)^2$, we obtain

$$10t = A(t + 10) + B$$

Equating the coefficients of like powers of $x$ yields $A = 10$ and $10A + B = 0$. Substituting $A = 10$ into the second equation then gives $B = -100$. Therefore,

$$C = 10 \int_0^{10} \left( 1 - \frac{10}{t + 10} + \frac{100}{(t + 10)^2} \right) dt$$

$$= 10 \left[ t - 10 \ln(t + 10) - \frac{100}{t + 10} \right]_0^{10}$$

$$= 10[(10 - 10 \ln 20 - 5) - (0 - 10 \ln 10 - 10)] = 80.69$$

or approximately 81%.

## 6.4  CONCEPT QUESTIONS

Let $f(x) = P(x)/Q(x)$ be a rational function in which the degree of $P$ is less than the degree of $Q$. What is the form of the partial fraction decomposition of $f$:

1. If $Q$ has only distinct linear factors?

2. If $Q$ contains a factor $(ax + b)^r$ with $r > 1$ that is repeated?
3. If $Q$ contains a factor $(ax^2 + bx + c)^r$ with $r = 1$ that is not repeated?

## 6.4  EXERCISES

In Exercises 1–6, write the form of the partial fraction decomposition of the rational expression. Do not find the numerical values of the constants.

1. a. $\dfrac{3}{x(x - 5)}$  b. $\dfrac{2x}{(x + 1)(3x - 2)}$

2. a. $\dfrac{2x + 1}{x^2 - x - 2}$  b. $\dfrac{x - 4}{x^2 + 4x + 3}$

3. a. $\dfrac{2x^2 - 1}{x^3 + x^2}$  b. $\dfrac{7}{x^2 + 3x - 4}$

4. a. $\dfrac{2x + 1}{x^3 + x}$  b. $\dfrac{8x}{x^3 - 5x^2}$

5. a. $\dfrac{x^3 - 2x + 1}{x^4 - 16}$  b. $\dfrac{x^2 - x - 27}{2x^3 - x^2 + 8x - 4}$

6. a. $\dfrac{2x^3 - 3x - 5}{x^2(x + 1)^3}$  b. $\dfrac{2x^4 - 3x^2 + 8x + 1}{(x - 1)^2(x^2 + x + 1)^3}$

In Exercises 7–51, find or evaluate the integral.

7. $\displaystyle\int \dfrac{dx}{x(x - 4)}$

8. $\displaystyle\int \dfrac{3x + 2}{x(x - 2)}\, dx$

9. $\displaystyle\int \dfrac{t + 3}{t(t + 1)}\, dt$

10. $\displaystyle\int \dfrac{2x - 1}{2x^2 - x}\, dx$

11. $\displaystyle\int_3^4 \dfrac{1}{x^2 - 4}\, dx$

12. $\displaystyle\int \dfrac{1}{4x^2 - 9}\, dx$

13. $\displaystyle\int \dfrac{x - 1}{x^2 - x - 2}\, dx$

14. $\displaystyle\int_0^1 \dfrac{2u + 3}{u^2 + 4u + 3}\, du$

15. $\displaystyle\int \dfrac{2x^2 + 3x + 6}{(x + 3)(x^2 - 4)}\, dx$

16. $\displaystyle\int \dfrac{x^2 + 2x + 8}{x^3 - 4x}\, dx$

17. $\displaystyle\int \dfrac{2x^2 + x - 1}{x^2 - x}\, dx$

18. $\displaystyle\int_2^3 \dfrac{x^3 - 2x + 7}{x^2 + x - 2}\, dx$

19. $\displaystyle\int \dfrac{2x^2 - 3x + 3}{x^3 - 2x^2 + x}\, dx$

20. $\displaystyle\int \dfrac{x^4 - 3x^2 - 3x - 2}{x^3 - x^2 - 2x}\, dx$

21. $\displaystyle\int \dfrac{4x^2 + 3x + 2}{x^3 + x^2}\, dx$

22. $\displaystyle\int_2^4 \dfrac{3x - 5}{(x - 1)^2}\, dx$

23. $\displaystyle\int \dfrac{v^3 + 1}{v(v - 1)^3}\, dv$

24. $\displaystyle\int_1^2 \dfrac{x^2 + 10x - 36}{x(x - 3)^2}\, dx$

25. $\displaystyle\int \dfrac{x^3 - x + 2}{x^3 + 2x^2 + x}\, dx$

26. $\displaystyle\int \dfrac{4x^2}{(x^2 - 4)^2}\, dx$

27. $\displaystyle\int \dfrac{dx}{x(x^2 - 1)^2}$

28. $\displaystyle\int \dfrac{x^2}{(x^2 + 4x + 3)^2}\, dx$

29. $\displaystyle\int \dfrac{6x^2 + 28x + 28}{x^3 + 4x^2 + x - 6}\, dx$

30. $\displaystyle\int \dfrac{x^2 + 16x + 7}{x^3 - x^2 + x + 3}\, dx$

**31.** $\displaystyle\int \frac{x^3 + 3}{(x + 1)(x^2 + 1)} \, dx$

**32.** $\displaystyle\int \frac{2r^2 - 3r + 4}{(r^2 + 2)^2} \, dr$

**33.** $\displaystyle\int \frac{5x^3 - 3x^2 + 7x - 3}{(x^2 + 1)^2} \, dx$

**34.** $\displaystyle\int \frac{13x + 4}{(x - 2)(x^2 + 2x + 2)} \, dx$

**35.** $\displaystyle\int \frac{8 - 3x}{(x + 1)(x^2 - 4x + 6)} \, dx$

**36.** $\displaystyle\int \frac{x^2 + 1}{x^3 - 1} \, dx$

**37.** $\displaystyle\int \frac{x}{x^3 + 1} \, dx$

**38.** $\displaystyle\int \frac{x^2 - x - 21}{2x^3 - x^2 + 8x - 4} \, dx$

**39.** $\displaystyle\int_0^1 \frac{3x^3 + 5x^2 + 5x + 1}{(x + 1)^2(x^2 + 1)} \, dx$

**40.** $\displaystyle\int \frac{3x - x^2}{(x^2 + 1)(x^2 + 2)} \, dx$

**41.** $\displaystyle\int \frac{3x^2 + x + 2}{(x^2 + x + 1)^2} \, dx$

**42.** $\displaystyle\int \frac{t^4}{t^4 - 1} \, dt$

**43.** $\displaystyle\int \frac{3x^2 - x + 4}{(2x^3 - x^2 + 8x + 4)^2} \, dx$

**44.** $\displaystyle\int \frac{\cos x}{\sin^2 x - \sin x - 6} \, dx$

**45.** $\displaystyle\int \frac{\sin x}{\cos^3 x + \cos^2 x} \, dx$

**46.** $\displaystyle\int \frac{\sec^2 \theta}{\tan \theta \, (\tan \theta - 1)} \, d\theta$

**47.** $\displaystyle\int \frac{e^t}{(e^t - 1)(e^t + 2)} \, dt$

**48.** $\displaystyle\int \frac{e^x}{e^{2x} + 2e^x - 8} \, dx$

**49.** $\displaystyle\int \frac{e^{4t}}{(e^t + 2)(e^{2t} - 1)} \, dt$

**50.** $\displaystyle\int \frac{dx}{e^x(1 + e^{2x})}$

**51.** $\displaystyle\int \frac{x^{1/3}}{1 + x} \, dx$

**Hint:** Let $u = x^{1/3}$.

**52.** An integral of the form $\int R(\sin x, \cos x) \, dx$, where $R$ is a rational function of $\sin x$ and $\cos x$, can be converted into an integral involving an ordinary rational function of $u$ by means of the substitution $u = \tan (x/2)$. Prove this by showing that if $u = \tan (x/2)$, where $-\pi < x < \pi$, then

$$\cos x = \frac{1 - u^2}{1 + u^2}, \qquad \sin x = \frac{2u}{1 + u^2}, \qquad dx = \frac{2}{1 + u^2} \, du$$

**Hint:** Sketch a right triangle.

*In Exercises 53–60, use the result of Exercise 52 to find the integral.*

**53.** $\displaystyle\int \frac{1}{1 + \cos x} \, dx$

**54.** $\displaystyle\int \frac{1}{3 \sin x - 4 \cos x} \, dx$

**55.** $\displaystyle\int \frac{1}{5 + \sin x - 3 \cos x} \, dx$

**56.** $\displaystyle\int \frac{1}{\sin x \, (2 + \cos x - 2 \sin x)} \, dx$

**57.** $\displaystyle\int_0^{\pi/2} \frac{1}{1 + \cos x + \sin x} \, dx$

**58.** $\displaystyle\int_0^{\pi/4} \frac{\tan x}{1 + \cos x} \, dx$

**59.** $\displaystyle\int \frac{1}{1 + \tan x} \, dx$

**60.** $\displaystyle\int_0^{\pi/6} \frac{\sin x}{\cos x \, (1 + \sin x)} \, dx$

**61.** Find the area of the region under the graph of $y = \dfrac{1}{x(x + 1)}$ on the interval $[1, 2]$.

**62.** Find the area of the region under the graph of $y = \dfrac{x^3}{x^3 + 1}$ on the interval $[0, 2]$.

**63.** Let $f(x) = \dfrac{x + 3}{x(x + 1)}$ and $g(x) = \ln x$.

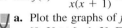

   **a.** Plot the graphs of $f$ and $g$ using the viewing window $[0, 3] \times [-1, 3]$. Find the $x$-coordinate of the point of intersection of the graphs of $f$ and $g$ accurate to three decimal places.

   **b.** Use the result of part (a) to find the approximate area of the region bounded by the graphs of $f$ and $g$ and the vertical line $x = 1$.

**64.** Let $f(x) = \dfrac{x}{x^3 + 1}$ and $g(x) = \dfrac{1}{3} x^3$.

   **a.** Plot the graphs of $f$ and $g$ using the viewing window $[0, 2] \times \left[-\frac{1}{2}, 1\right]$. Find the $x$-coordinate of the point of intersection of the graphs of $f$ and $g$ accurate to three decimal places.

   **b.** Use the result of part (a) to find the approximate area of the region enclosed by the graphs of $f$ and $g$.

**65.** The region under the graph of $y = \dfrac{1}{x(x + 1)}$ on the interval $[1, 2]$ is revolved about the $x$-axis. Find the volume of the resulting solid.

**66.** The region under the graph of $y = \dfrac{2x}{x^2 + 1}$ on the interval $[0, 2]$ is revolved about the $x$-axis. Find the volume of the resulting solid.

**67.** Find the length of the graph of $y = 2 \ln\!\left(\dfrac{4}{4 - x^2}\right)$ from $A(0, 0)$ to $B\!\left(1, 2 \ln \frac{4}{3}\right)$.

**68.** Find the centroid of the region under the graph of
$$y = \frac{2x}{x^2 + 1} \quad \text{on } [0, 2].$$

**69.** Let $I = \displaystyle\int \frac{x^2 + 1}{x^4 + 6x^3 + 12x^2 + 11x + 6}\, dx.$

a. Find $I$.

Hint: One root of $x^4 + 6x^3 + 12x^2 + 11x + 6 = 0$ is $-2$.

**cas** b. Use a CAS to find the partial fraction decomposition of

$$f(x) = \frac{x^2 + 1}{x^4 + 6x^3 + 12x^2 + 11x + 6}$$

c. Use a CAS to find $I$.

**cas** **70.** Let

$$I = \int \frac{8x^5 - 3x^4 + 2x^2 - 1}{36x^6 - 108x^5 + 105x^4 - 72x^3 + 58x^2 - 12x + 9}\, dx$$

a. Use a CAS to find the partial fraction decomposition of

$$f(x) = \frac{8x^5 - 3x^4 + 2x^2 - 1}{36x^6 - 108x^5 + 105x^4 - 72x^3 + 58x^2 - 12x + 9}$$

b. Use a CAS to find $I$.

**71. City Planning** A major corporation is building a 4325-acre complex of homes, offices, stores, schools, and churches in the rural community of Glen Cove. As a result of this development the planners have estimated that Glen Cove's population (in thousands) $t$ years from now will be given by

$$P(t) = \frac{3t^2 + 130t + 270}{t^2 + 6t + 45}$$

What will the average population of Glen Cove be over the next 10 years?

**72. Work Done in Moving a Charged Particle** Suppose that a particle of charge $+1$ is placed on a coordinate line between two

particles, each of charge $-1$, as shown in the figure. Then, according to Coulomb's Law, there is an electrical force acting on the particle of charge $+1$ given by

$$F(x) = k\left[\frac{2x - 3}{x^2(x - 3)^2}\right]$$

where $k$ is a positive constant. Find the work required to move the particle of charge $+1$ along the coordinate line from $x = 1$ to $x = 2$.

In Exercises 73–76, determine whether the statement is true or false. If it is true, explain why it is true. If it is false, explain why or give an example to show why it is false.

**73.** $\dfrac{x^3 + 2x}{(x + 1)(x - 2)}$ can be written in the form $\dfrac{A}{x + 1} + \dfrac{B}{x - 2}.$

**74.** $\dfrac{4x^2 - 15x - 1}{x(x^2 - 4x - 5)}$ can be written in the form

$$\frac{A}{x} + \frac{B}{x - 5} + \frac{C}{x + 1}.$$

**75.** $\dfrac{1}{x(x - 1)^2}$ can be written in the form $\dfrac{A}{x} + \dfrac{B}{(x - 1)^2}.$

**76.** $\dfrac{4x^3 - x^2 + 4x + 2}{(x^2 + 1)^2}$ can be written in the form

$$\frac{A}{x^2 + 1} + \frac{B}{(x^2 + 1)^2}.$$

---

## 6.5 Integration Using Tables of Integrals and a CAS; a Summary of Techniques

The techniques of integration that we have developed so far enable us to integrate a wide variety of functions. But in practice, there are many functions for which these techniques will not work or, if they do, work inefficiently. Other techniques have been developed that enable us to integrate many complicated functions. By using these techniques, extensive lists of integration formulas have been compiled. A small sample of such formulas can be found in the Table of Integrals on the reference pages at the back of this book. These formulas are grouped according to the following basic forms of the integrand: $a + bu$, $\sqrt{a + bu}$, $\sqrt{a^2 \pm u^2}$, $\sqrt{u^2 - a^2}$, $\sqrt{2au - u^2}$, trigonometric, inverse trigonometric, exponential, logarithmic, and hyperbolic functions.

### ■ Using a Table of Integrals

The Table of Integrals provides us with a quick and convenient way of integrating complicated functions. The idea is to match the integrand of the integral to be found with the integrand of an appropriate integral appearing in the table (whose antiderivative is known). Sometimes we need to recast the given integral by making an appropriate substitution or by using the integration by parts formula before we can use the Table of Integrals.

**EXAMPLE 1** Use the Table of Integrals to find $\displaystyle\int \frac{3x}{\sqrt{2+x}}\,dx$.

**Solution** We first write

$$\int \frac{3x}{\sqrt{2+x}}\,dx = 3\int \frac{x}{\sqrt{2+x}}\,dx$$

Scanning the Table of Integrals for integrands involving $\sqrt{a+bu}$, we see that Formula 28,

$$\int \frac{u}{\sqrt{a+bu}}\,du = \frac{2}{3b^2}(bu - 2a)\sqrt{a+bu} + C$$

is the proper choice. With $a = 2$, $b = 1$, and $u = x$ we obtain

$$\int \frac{3x}{\sqrt{2+x}}\,dx = 3\left[\frac{2}{3}(x-4)\sqrt{2+x}\right] + C$$

$$= 2(x-4)\sqrt{2+x} + C \qquad \blacksquare$$

**EXAMPLE 2** Use the Table of Integrals to find $\displaystyle\int \frac{\sqrt{3-4x^2}}{x^2}\,dx$.

**Solution** Looking at the Table of Integrals for integrands involving $\sqrt{a^2 - u^2}$, we find that Formula 49,

$$\int \frac{\sqrt{a^2 - u^2}}{u^2}\,du = -\frac{1}{u}\sqrt{a^2 - u^2} - \sin^{-1}\frac{u}{a} + C$$

is closest to the form of the given integral. Comparison of the two integrands suggests that we make the substitution $u = 2x$ and $du = 2\,dx$, obtaining

$$\int \frac{\sqrt{3-4x^2}}{x^2}\,dx = \int \frac{\sqrt{3-u^2}}{(u/2)^2}\left(\frac{du}{2}\right) = 2\int \frac{\sqrt{3-u^2}}{u^2}\,du$$

Then, using Formula 49 with $a = \sqrt{3}$, we obtain

$$\int \frac{\sqrt{3-4x^2}}{x^2}\,dx = 2\int \frac{\sqrt{3-u^2}}{u^2}\,du = 2\left[-\frac{1}{u}\sqrt{3-u^2} - \sin^{-1}\frac{u}{\sqrt{3}}\right] + C$$

$$= -\frac{\sqrt{3-4x^2}}{x} - 2\sin^{-1}\left(\frac{2x}{\sqrt{3}}\right) + C \qquad \blacksquare$$

**EXAMPLE 3** Use the Table of Integrals to find $\int x^3 \cos x\,dx$.

**Solution** Looking in the section of the Table of Integrals for integrands involving trigonometric functions, we find Formula 78, a reduction formula.

$$\int u^n \cos u\,du = u^n \sin u - n\int u^{n-1} \sin u\,du$$

Using the formula with $n = 3$, we obtain

$$\int x^3 \cos x\,dx = x^3 \sin x - 3\int x^2 \sin x\,dx$$

Next, using Formulas 77 and 76, we obtain

$$\int x^3 \cos x \, dx = x^3 \sin x - 3\left[ -x^2 \cos x + 2 \int x \cos x \, dx \right]$$

$$= x^3 \sin x + 3x^2 \cos x - 6(\cos x + x \sin x) + C$$

$$= x^3 \sin x + 3x^2 \cos x - 6x \sin x - 6 \cos x + C \qquad \blacksquare$$

## ▣ Using Tables of Integrals to Evaluate Definite Integrals

**EXAMPLE 4**   Use the Table of Integrals to evaluate $\displaystyle\int_0^{\pi/2} \frac{\sin 2x}{\sqrt{3 - 2\cos x}} \, dx$.

**Solution**   Let's begin by evaluating the corresponding indefinite integral, which can also be rewritten as

$$\int \frac{\sin 2x}{\sqrt{3 - 2\cos x}} \, dx = \int \frac{2\sin x \cos x}{\sqrt{3 - 2\cos x}} \, dx$$

No formula in the Table of Integrals has either of these forms, so let's consider making a substitution. Letting $u = \cos x$, so that $du = -\sin x \, dx$, we find

$$\int \frac{\sin 2x}{\sqrt{3 - 2\cos x}} \, dx = -2 \int \frac{\cos x \,(-\sin x)}{\sqrt{3 - 2\cos x}} \, dx = -2 \int \frac{u}{\sqrt{3 - 2u}} \, du$$

Looking in the Table of Integrals for integrands involving $\sqrt{a + bu}$ leads to Formula 28,

$$\int \frac{u}{\sqrt{a + bu}} \, du = \frac{2}{3b^2} (bu - 2a)\sqrt{a + bu} + C$$

with $a = 3$ and $b = -2$. We obtain

$$-2 \int \frac{u}{\sqrt{3 - 2u}} \, du = -2 \left( \frac{2}{12} \right)(-2u - 6)\sqrt{3 - 2u} + C = \frac{2}{3}(u + 3)\sqrt{3 - 2u} + C$$

Therefore,

$$\int \frac{\sin 2x}{\sqrt{3 - 2\cos x}} \, dx = \frac{2}{3}(u + 3)\sqrt{3 - 2u} + C$$

$$= \frac{2}{3}(\cos x + 3)\sqrt{3 - 2\cos x} + C \quad \text{\small Since } u = \cos x$$

So

$$\int_0^{\pi/2} \frac{\sin 2x}{\sqrt{3 - 2\cos x}} \, dx = \left[ \frac{2}{3}(\cos x + 3)\sqrt{3 - 2\cos x} \right]_0^{\pi/2}$$

$$= \frac{2}{3}(3)\sqrt{3 - 0} - \frac{2}{3}(1 + 3)\sqrt{3 - 2} = 2\sqrt{3} - \frac{8}{3}$$

$$= \frac{6\sqrt{3} - 8}{3} \qquad \blacksquare$$

**EXAMPLE 5**   The region $R$ under the graph of $y = \cos^{-1} x$ on the interval $[0, 1]$ is revolved about the $y$-axis. Find the volume of the resulting solid.

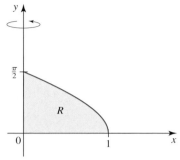

**FIGURE 1**
The region $R$ under $y = \cos^{-1} x$ on $[0, 1]$

**Solution** The region $R$ is shown in Figure 1. Using the method of cylindrical shells, we see that the required volume is

$$V = 2\pi \int_0^1 x \cos^{-1} x \, dx$$

We use Formula 91 from the Table of Integrals to evaluate this integral, obtaining

$$V = 2\pi \int_0^1 x \cos^{-1} x \, dx = 2\pi \left[ \frac{2x^2 - 1}{4} \cos^{-1} x - \frac{x\sqrt{1 - x^2}}{4} \right]_0^1$$

$$= 2\pi \left[ \frac{1}{4} \cos^{-1} 1 - \left( -\frac{1}{4} \cos^{-1} 0 \right) \right] = 2\pi \left[ \frac{1}{4}(0) + \frac{1}{4} \left( \frac{\pi}{2} \right) \right]$$

$$= \frac{\pi^2}{4}$$

## Graphing Calculators and CAS

Most of the graphing calculators that are available today will perform numerical integration; that is, the calculator will give a numerical approximation to the value of a definite integral. The more sophisticated graphing calculators, such as the TI-89 or TI-92, will even do symbolic integration; that is, the calculator will find an antiderivative of a given function.

A computer equipped with the appropriate software, such as *Mathematica, Maple,* or *Derive,* can be used to perform both tasks. If you use these programs, bear in mind that the commands are different for different programs and, more important, the answers may appear in different forms even though they are equivalent.

**EXAMPLE 6** Find $\int x(x^2 + 3)^6 \, dx$ using (a) the TI-89 and (b) CAS with *Maple* and *Mathematica*.

**Solution**
**a.** Using the TI-89:

$$\int x(x^2 + 3)^6 \, dx = \frac{(x^2 + 3)^7}{14}$$

**b.** Using *Mathematica*:

$$\int x(x^2 + 3)^6 \, dx = \frac{729x^2}{2} + \frac{729x^4}{2} + \frac{405x^6}{2} + \frac{135x^8}{2} + \frac{27x^{10}}{2} + \frac{3x^{12}}{2} + \frac{x^{14}}{14}$$

Using *Maple*:

$$\int x(x^2 + 3)^6 \, dx = \frac{1}{14}x^{14} + \frac{3}{2}x^{12} + \frac{27}{2}x^{10} + \frac{135}{2}x^8 + \frac{405}{2}x^6 + \frac{729}{2}x^4 + \frac{729}{2}x^2$$

Note that none of these programs include the constant of integration in their answers. Of course, we can find the integral under consideration by using the method of substitution. You can easily verify that your result is the same as that obtained by using the TI-89. The output obtained by using both *Maple* and *Mathematica* is in a more cumbersome form, but it is equivalent to the more compact answer obtained by using the TI-89. You can see this by expanding the latter using the Binomial Theorem.

**EXAMPLE 7** Find $\displaystyle\int \frac{\cos^4 x}{\sin^3 x}\, dx$ by using (a) the TI-89 and (b) CAS with *Maple* and *Mathematica*.

**Solution**

**a.** Using the TI-89:

$$\int \frac{\cos^4 x}{\sin^3 x}\, dx$$

$$= \frac{3(\sin^2 x) \cdot \ln(|\cos x + 1|) - 3(\sin^2 x \cdot \ln(|\sin x|)) - (2(\sin^2 x) + 1) \cdot \cos x}{2 \sin^2 x}$$

**b.** Using *Mathematica*:

$$\int \frac{\cos^4 x}{\sin^3 x}\, dx = -\cos x - \frac{1}{8}\csc\left(\frac{x}{2}\right)^2 + \frac{3}{2}\ln\left(\cos\frac{x}{2}\right) - \frac{3}{2}\ln\left(\sin\frac{x}{2}\right) + \frac{1}{8}\sec\left(\frac{x}{2}\right)^2$$

Using *Maple*:

$$\int \frac{\cos^4 x}{\sin^3 x}\, dx = -\frac{\cos(x)^5}{2\sin(x)^2} - \frac{1}{2}\cos(x)^3 - \frac{3}{2}\cos(x) - \frac{3}{2}\ln(\csc(x) - \cot(x)) \quad \blacksquare$$

## ■ Summary of Integration Techniques

Our first table gives a summary of the basic integration formulas that were covered in this and the previous chapters.

**BASIC INTEGRATION FORMULAS**

| | | |
|---|---|---|
| **1.** $\displaystyle\int u^n\, du = \frac{u^{n+1}}{n+1} + C, \quad n \neq -1$ | **9.** $\displaystyle\int \sec u \tan u\, du = \sec u + C$ | **17.** $\displaystyle\int \frac{du}{1+u^2} = \tan^{-1} u + C$ |
| **2.** $\displaystyle\int \frac{1}{u}\, du = \ln|u| + C$ | **10.** $\displaystyle\int \csc u \cot u\, du = -\csc u + C$ | **18.** $\displaystyle\int \sinh u\, du = \cosh u + C$ |
| **3.** $\displaystyle\int e^u\, du = e^u + C$ | **11.** $\displaystyle\int \sec u\, du = \ln|\sec u + \tan u| + C$ | **19.** $\displaystyle\int \cosh u\, du = \sinh u + C$ |
| **4.** $\displaystyle\int a^u\, du = \frac{a^u}{\ln a} + C$ | **12.** $\displaystyle\int \csc u\, du = -\ln|\csc u + \cot u| + C$ | **20.** $\displaystyle\int \text{sech}^2 u\, du = \tanh u + C$ |
| **5.** $\displaystyle\int \sin u\, du = -\cos u + C$ | **13.** $\displaystyle\int \tan u\, du = \ln|\sec u| + C$ | **21.** $\displaystyle\int \text{csch } u \coth u\, du = -\text{csch } u + C$ |
| **6.** $\displaystyle\int \cos u\, du = \sin u + C$ | **14.** $\displaystyle\int \cot u\, du = \ln|\sin u| + C$ | **22.** $\displaystyle\int \text{sech } u \tanh u\, du = -\text{sech } u + C$ |
| **7.** $\displaystyle\int \sec^2 u\, du = \tan u + C$ | **15.** $\displaystyle\int \frac{du}{\sqrt{1-u^2}} = \sin^{-1} u + C$ | **23.** $\displaystyle\int \text{csch}^2 u\, du = -\coth u + C$ |
| **8.** $\displaystyle\int \csc^2 u\, du = -\cot u + C$ | **16.** $\displaystyle\int \frac{du}{u\sqrt{u^2-1}} = \sec^{-1}|u| + C$ | |

The next table lists the methods of integration developed in Chapter 4 (Integration by Substitution) and this chapter.

## METHODS OF INTEGRATION

| Integration | Method of integration | Section |
|---|---|---|
| **1.** $\displaystyle\int f(g(x))g'(x)\,dx$ | Use the substitution $u = g(x)$. | Section 4.2 |
| **2.** $\displaystyle\int f(x)g'(x)\,dx$ | Use the integration by parts formula: $$\int f(x)g'(x)\,dx = f(x)g(x) - \int g(x)f'(x)\,dx \quad\text{or}\quad \int u\,dv = uv - \int v\,du$$ | Section 6.1 |

**Note:** Apply the method to integrals of the form $\int P(x)e^{ax}\,dx$, $\int P(x)\sin ax\,dx$, $\int P(x)\cos ax\,dx$, where $P(x)$ is a polynomial, $\int \ln x\,dx$, $\int \sin^{-1} x\,dx$, $\int \tan^{-1} x\,dx$, $\int \sec^m x\,dx$ ($m > 0$ and $m$ odd), $\int e^{ax}\cos bx\,dx$, $\int e^{ax}\sin bx$, and so on.

| | | |
|---|---|---|
| **3. a.** $\displaystyle\int \sin^m x \cos^n x\,dx$, where $m$ or $n$ is a positive integer | **a.** If $m$ is odd and positive, use the substitution $u = \cos x$. <br> **b.** If $n$ is odd and positive, use the substitution $u = \sin x$. <br> **c.** If $m$ and $n$ are even and nonnegative use the formulas $$\sin^2 x = \frac{1 - \cos 2x}{2}, \quad \cos^2 x = \frac{1 + \cos 2x}{2}$$ | |
| **b.** $\displaystyle\int \tan^m x \sec^n x\,dx$, where $m$ or $n$ is a positive integer. | **a.** If $m$ is odd and positive, use the substitution $u = \sec x$. <br> **b.** If $n$ is even and positive, use the substitution $u = \tan x$. | |

**Note:** Also try converting the integrand to one involving sines and cosines.

| | Use the identities: | |
|---|---|---|
| **c.** $\displaystyle\int \sin mx \sin nx\,dx$ | $\sin mx \sin nx = \frac{1}{2}[\cos(m - n)x - \cos(m + n)x]$ | |
| $\displaystyle\int \sin mx \cos nx\,dx$ | $\sin mx \cos nx = \frac{1}{2}[\sin(m - n)x + \sin(m + n)x]$ | |
| $\displaystyle\int \cos mx \cos nx\,dx$ | $\cos mx \cos nx = \frac{1}{2}[\cos(m - n)x + \cos(m + n)x]$ | |

| | | |
|---|---|---|
| **4.** $\displaystyle\int f(x)\,dx$, where $f$ involves | | Section 6.3 |
| $\sqrt{a^2 - x^2}$ <br> $\sqrt{a^2 + x^2}$ | Use the substitution $x = a \sin\theta$, where $-\frac{\pi}{2} \le \theta \le \frac{\pi}{2}$. <br> Use the substitution $x = a \tan\theta$, where $-\frac{\pi}{2} < \theta < \frac{\pi}{2}$. | |
| $\sqrt{x^2 - a^2}$ | Use the substitution $x = a \sec\theta$, where $0 \le \theta < \frac{\pi}{2}$ or $\frac{\pi}{2} < \theta \le \pi$. | |

| | | |
|---|---|---|
| **5.** $\displaystyle\int \frac{P(x)}{Q(x)}\,dx$, where $\deg P < \deg Q$ and $Q(x) = (p_1 x + q_1)^k (p_2 x + q_2)^l \cdots (ax^2 + bx + c)^m \cdots$ | Write the integrand as a sum of partial fractions: $$\frac{P(x)}{Q(x)} = \frac{A_1}{p_1 x + q_1} + \frac{A_2}{(p_1 x + q_1)^2} + \cdots + \frac{A_k}{(p_1 x + q_1)^k}$$ $$+ \frac{B_1}{p_2 x + q_2} + \frac{B_2}{(p_2 x + q_2)^2} + \cdots + \frac{B_l}{(p_2 x + q_2)^l}$$ $$+ \cdots + \frac{M_1 x + N_1}{ax^2 + bx + c} + \frac{M_2 x + N_2}{(ax^2 + bx + c)^2} + \cdots$$ $$+ \frac{M_m x + N_m}{(ax^2 + bx + c)^m} + \cdots$$ | Section 6.4 |

**EXAMPLE 8** Indicate the method of integration that you would use to find the integral. Explain how you arrive at your choice.

**a.** $\displaystyle\int x^2(1 - x)^{30}\, dx$  **b.** $\displaystyle\int \frac{x \sin^{-1} x}{\sqrt{1 - x^2}}\, dx$  **c.** $\displaystyle\int \sin x \sin 2x \cos 3x\, dx$

**d.** $\displaystyle\int \frac{\cos^4 x}{\sin^3 x}\, dx$  **e.** $\displaystyle\int \frac{x + 4}{(x - 1)(x^2 + 1)^2}\, dx$

**Solution**

**a.** We use the substitution $u = 1 - x$ so that $du = -dx$. Then

$$\int x^2(1 - x)^{30}\, dx = -\int (1 - u)^2 u^{30}\, du = -\int (1 - 2u + u^2)u^{30}\, du$$

$$= -\int (u^{30} - 2u^{31} + u^{32})\, du$$

which is easily integrated.

**b.** The integrand involves $\sin^{-1} x$, so we try the method of integration by parts with

$$u = \sin^{-1} x \quad \text{and} \quad dv = \frac{x}{\sqrt{1 - x^2}}\, dx$$

so that

$$du = \frac{1}{\sqrt{1 - x^2}}\, dx \quad \text{and} \quad v = \int \frac{x}{\sqrt{1 - x^2}}\, dx = -\sqrt{1 - x^2}$$

We obtain

$$\int \frac{x \sin^{-1} x}{\sqrt{1 - x^2}}\, dx = -(\sin^{-1} x)\sqrt{1 - x^2} + \int \frac{\sqrt{1 - x^2}}{\sqrt{1 - x^2}}\, dx$$

$$= -(\sin^{-1} x)\sqrt{1 - x^2} + x + C$$

**c.** We use the trigonometric identities of Section 6.2. Thus,

$$\sin x \sin 2x \cos 3x = [(\sin x)(\sin 2x)]\cos 3x$$

$$= \frac{1}{2}(\cos x - \cos 3x)\cos 3x$$

$$= \frac{1}{2}[(\cos x)(\cos 3x) - (\cos 3x)(\cos 3x)]$$

$$= \frac{1}{4}(\cos 2x + \cos 4x - 1 - \cos 6x)$$

So

$$\int \sin x \sin 2x \cos 3x\, dx = \frac{1}{4}\int (\cos 2x + \cos 4x - \cos 6x - 1)\, dx$$

which is readily integrated.

**d.** We rewrite

$$I = \int \frac{\cos^4 x}{\sin^3 x} \, dx = \int \frac{\cos^4 x \sin x}{\sin^4 x} \, dx$$

Letting $u = \cos x$, we have $du = -\sin x \, dx$, and this gives

$$I = \int \frac{\cos^4 x \sin x}{(\sin^2 x)^2} \, dx = \int \frac{\cos^4 x \sin x}{(1 - \cos^2 x)^2} \, dx$$

$$= -\int \frac{u^4}{(1 - u^2)^2} \, du$$

To complete the solution, we first perform long division and then use the method of partial fractions.

**e.** The integrand is a rational function whose numerator has degree less than that of the denominator. So we use the method of partial fractions. The form of the decomposition is

$$\frac{A}{x - 1} + \frac{Bx + C}{x^2 + 1} + \frac{Dx + E}{(x^2 + 1)^2}$$

## 6.5 EXERCISES

*In Exercises 1–36, use the Table of Integrals to evaluate the integral.*

**1.** $\displaystyle\int x\sqrt{1 + 2x} \, dx$

**2.** $\displaystyle\int \frac{x}{\sqrt{2 + 3x}} \, dx$

**3.** $\displaystyle\int \frac{x^2}{(1 + 2x)^2} \, dx$

**4.** $\displaystyle\int \frac{1}{x\sqrt{4 + x}} \, dx$

**5.** $\displaystyle\int \frac{\sqrt{3 + 2x}}{x^2} \, dx$

**6.** $\displaystyle\int \frac{x^2}{\sqrt{9 + 4x^2}} \, dx$

**7.** $\displaystyle\int \frac{1}{x\sqrt{3 + 2x^2}} \, dx$

**8.** $\displaystyle\int x^2\sqrt{4 - 3x^2} \, dx$

**9.** $\displaystyle\int \frac{\sqrt{2 - x^2}}{x} \, dx$

**10.** $\displaystyle\int \frac{\sqrt{9 - 2x^2}}{x^2} \, dx$

**11.** $\displaystyle\int \frac{\sqrt{x^2 - 3}}{x} \, dx$

**12.** $\displaystyle\int \frac{1}{x^2\sqrt{x^2 - 5}} \, dx$

**13.** $\displaystyle\int \frac{e^x}{(1 - e^{2x})^{3/2}} \, dx$

**14.** $\displaystyle\int x^4 \sin x \, dx$

**15.** $\displaystyle\int x \cos^{-1} 2x \, dx$

**16.** $\displaystyle\int \csc^5 \theta \, d\theta$

**17.** $\displaystyle\int x^3 \sin(x^2 + 1) \, dx$

**18.** $\displaystyle\int x^2 \tan^{-1} 3x \, dx$

**19.** $\displaystyle\int_0^1 \sin^{-1}\sqrt{x} \, dx$

Hint: Let $u = \sqrt{x}$.

**20.** $\displaystyle\int \frac{\cos^{-1}\sqrt{x}}{\sqrt{x}} \, dx$

Hint: Let $u = \sqrt{x}$.

**21.** $\displaystyle\int e^{-2x} \sin 3x \, dx$

**22.** $\displaystyle\int e^{2x} \sin^{-1} e^x \, dx$

**23.** $\displaystyle\int x^3 e^{-2x} \, dx$

**24.** $\displaystyle\int \frac{1}{\sqrt{1 + e^{2x}}} \, dx$

**25.** $\displaystyle\int \frac{\sin x}{1 + \cos^2 x} \, dx$

**26.** $\displaystyle\int \frac{\sec^3\sqrt{x}}{\sqrt{x}} \, dx$

**27.** $\displaystyle\int x^3 \ln 5x \, dx$

**28.** $\displaystyle\int \frac{1}{x \ln\sqrt{x}} \, dx$

**29.** $\displaystyle\int \sqrt{6x - x^2} \, dx$

**30.** $\displaystyle\int \frac{\sqrt{4x - 2x^2}}{x} \, dx$

**31.** $\displaystyle\int \frac{x^2}{\sqrt{8x - 3x^2}} \, dx$

**32.** $\displaystyle\int e^{2x} \ln(1 + e^{2x}) \, dx$

**33.** $\displaystyle\int_1^{e^2} \frac{\ln t}{t\sqrt{1 + \ln t}} \, dt$

**34.** $\displaystyle\int_0^{\pi/4} \frac{1}{a^2 \cos^2 x + b^2 \sin^2 x} \, dx \quad a > 0, b > 0$

Hint: Let $u = \tan x$.

**35.** $\displaystyle\int e^{\cos x} \sin 2x \, dx$

**36.** $\displaystyle\int e^{2x} \ln(1 + e^x) \, dx$

**37.** Find the area of the region under the graph of $y = x^2 \ln x$ on the interval $[1, e]$.

**38.** Find the length of the graph of $f(x) = \ln x$ from $A(1, 0)$ to $B(e, 1)$.

**39.** The region under the graph of $y = \cos^2 x$ on $\left[0, \frac{\pi}{2}\right]$ is revolved about the $x$-axis. Find the volume of the resulting solid of revolution.

**40.** The region under the graph of $y = \sin^{-1} x$ on $[0, 1]$ is revolved about the $y$-axis. Find the volume of the resulting solid of revolution.

**41.** Find the centroid of the region under the graph of $y = \cos^2 x$ on $\left[0, \frac{\pi}{2}\right]$.

**42.** Find the work done by the force $F(x) = x^2/(1 + 2x)^2$ (measured in pounds) in moving a particle along the $x$-axis from $x = 0$ to $x = 4$ (measured in feet).

**43. Theme Park Attendance** The management of Astro World ("The Amusement Park of the Future") estimates that visitors enter the park $t$ hours after opening time at 8 A.M. at the rate of

$$R(t) = \frac{60}{(2 + t^2)^{3/2}}$$

thousand people per hour. Determine the number of visitors admitted by noon.

**44. Voter Registration** The number of voters in a certain district of a city is expected to grow at the rate of

$$R(t) = \frac{3000}{\sqrt{4 + t^2}}$$

people per year $t$ years from now. If the number of voters at present is 20,000, how many voters will be in the district 5 years from now?

**45. Growth of Fruit Flies** On the basis of data collected during an experiment, a biologist found that the number of fruit flies (*Drosophila melanogaster*) with a limited food supply could be approximated by the exponential model

$$N(t) = \frac{1000}{1 + 24e^{-0.02t}}$$

where $t$ denotes the number of days since the beginning of the experiment. Find the average number of fruit flies in the colony in the first 10 days of the experiment and in the first 20 days.

**46. Average Mass of an Electron** According to the special theory of relativity, the mass $m$ of a particle moving at a velocity $v$ is given by

$$m = \frac{m_0}{\sqrt{1 - \dfrac{v^2}{c^2}}}$$

where $m_0$ is the mass of the body at rest and $c = 3 \times 10^8$ m/sec is the speed of light. If an electron is accelerated from a speed of $v_1$ m/sec to a speed of $v_2$ m/sec, find an expression for the average mass of the electron between $v = v_1$ and $v = v_2$.

**47.** Find the area of the surface generated by revolving the graph of $y = x^2$ for $0 \le x \le 1$ about the $x$-axis.

**48.** Find the centroid of the region enclosed by the graph of $y^2 = x^3 - x^4$.

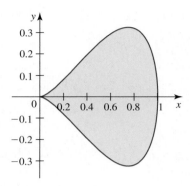

*In Exercises 49–52, verify the integration formula.*

**49.** $\displaystyle \int \frac{\sqrt{a^2 - u^2}}{u^2}\, du = -\frac{1}{u}\sqrt{a^2 - u^2} - \sin^{-1}\frac{u}{a} + C$

**50.** $\displaystyle \int \frac{1}{u^2(a + bu)}\, du = -\frac{1}{au} + \frac{b}{a^2}\ln\left|\frac{a + bu}{u}\right| + C$

**51.** $\displaystyle \int u^n \tan^{-1} u\, du = \frac{1}{n + 1}\left[u^{n+1}\tan^{-1}u - \int \frac{u^{n+1}}{1 + u^2}\, du\right],$
$$n \ne -1$$

**52.** $\displaystyle \int u^n \ln u\, du = \frac{u^{n+1}}{(n + 1)^2}[(n + 1)\ln u - 1] + C, \quad n \ne -1$

**cas** *In Exercises 53–62, use a CAS to find the integral.*

**53.** $\displaystyle \int x\sqrt{x + 2}\, dx$

**54.** $\displaystyle \int \frac{x}{\sqrt{1 + 2x}}\, dx$

**55.** $\displaystyle \int \frac{x + 1}{x\sqrt{x + 2}}\, dx$

**56.** $\displaystyle \int \frac{x^2 + x + 1}{x^3 + 1}\, dx$

**57.** $\displaystyle \int \cos^4 x\, dx$

**58.** $\displaystyle \int \tan^5 x\, dx$

**59.** $\displaystyle \int x^5 e^x\, dx$

**60.** $\displaystyle \int x^3 e^{-2x}\, dx$

**61.** $\displaystyle \int \frac{e^{2x}}{\sqrt{e^x + 1}}\, dx$

**62.** $\displaystyle \int x\sin^{-1} x\, dx$

*In Exercises 63–100, find or evaluate the integral.*

**63.** $\displaystyle \int \frac{x}{\sqrt[3]{2 - x}}\, dx$

**64.** $\displaystyle \int \frac{t^3}{\sqrt{1 - t^2}}\, dt$

**65.** $\displaystyle \int \frac{\cos\dfrac{1}{x}}{x^2}\, dx$

**66.** $\displaystyle \int \sqrt{1 + 2\cos^2 x}\,\sin 2x\, dx$

**67.** $\displaystyle\int_0^{1/2} \frac{x+1}{\sqrt{1-x^2}}\,dx$

**68.** $\displaystyle\int \frac{x+3}{\sqrt{5-4x-x^2}}\,dx$

**69.** $\displaystyle\int_0^1 \frac{x}{x^4+3}\,dx$

**70.** $\displaystyle\int_0^{1/\sqrt{2}} \frac{(\sin^{-1}x)^2}{\sqrt{1-x^2}}\,dx$

**71.** $\displaystyle\int \frac{dx}{x\sqrt{1+(\ln x)^2}}$

**72.** $\displaystyle\int e^{\sqrt{x}}\,dx$

**73.** $\displaystyle\int_1^e \frac{\sqrt{\ln x+3}}{x}\,dx$

**74.** $\displaystyle\int \frac{e^x}{\sqrt{1-e^x}}\,dx$

**75.** $\displaystyle\int x^2(3^{x^3+1})\,dx$

**76.** $\displaystyle\int \tan^{-1}x\,dx$

**77.** $\displaystyle\int x\sin^{-1}x\,dx$

**78.** $\displaystyle\int_1^e \sin(\ln x)\,dx$

**79.** $\displaystyle\int_2^{\sqrt{5}} \sqrt{x^2-4}\,dx$

**80.** $\displaystyle\int x^2 e^{3x}\,dx$

**81.** $\displaystyle\int_1^2 \frac{\ln x}{x^2}\,dx$

**82.** $\displaystyle\int \sin^2 x\cos^5 x\,dx$

**83.** $\displaystyle\int x\tan^2 x\,dx$

**84.** $\displaystyle\int e^x \sin^2 x\,dx$

**85.** $\displaystyle\int_0^{\pi/3} \sqrt{1-\cos x}\,dx$

**86.** $\displaystyle\int \sin 3x\cos 4x\,dx$

**87.** $\displaystyle\int \frac{dx}{x+1+\sqrt{x+1}}$

**88.** $\displaystyle\int \frac{dx}{1+\tan x}$

**89.** $\displaystyle\int \frac{dx}{\sqrt{x^2-6x}}$

**90.** $\displaystyle\int_0^{\pi/2} \frac{\sin x}{1+\sqrt{\cos x}}\,dx$

**91.** $\displaystyle\int \cot^4(2x)\,dx$

**92.** $\displaystyle\int \frac{\sqrt{9-4x^2}}{x}\,dx$

**93.** $\displaystyle\int \frac{\sqrt{x^2+9}}{x}\,dx$

**94.** $\displaystyle\int_0^{\pi/4} \tan^{3/2}x\sec^4 x\,dx$

**95.** $\displaystyle\int \frac{dx}{x^3-1}$

**96.** $\displaystyle\int_0^1 \frac{x^3}{(x+1)^2(x^2+x+1)}\,dx$

**97.** $\displaystyle\int \frac{dx}{x^4+x^2+1}$

**98.** $\displaystyle\int \frac{x^4}{(1-x)^3}\,dx$

**99.** $\displaystyle\int xe^{x^2+e^{x^2}}\,dx$

**100.** $\displaystyle\int_0^{\pi/4} \frac{\sin x}{1-4\cos^2 x}\,dx$

# 6.6  Improper Integrals

In defining the definite integral $\int_a^b f(x)\,dx$, we required that the interval of integration $[a, b]$ be finite and that $f$ be bounded. In many applications, one or both of these conditions do not hold. In this section we will extend the concept of the definite integral to include these cases:

**1.** The interval of integration is infinite (Figure 1a).
**2.** $f$ is unbounded (Figure 1b).

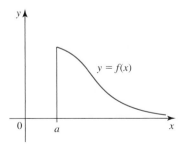

**(a)** The interval of integration $[a, \infty)$ is infinite.

**(b)** $f$ is unbounded on $[a, b]$ because it has an infinite discontinuity at $c$: $f(x) \longrightarrow \infty$ as $x \rightarrow c^{-}$.

**FIGURE 1**

Integrals that have infinite intervals of integration or unbounded integrands are called **improper integrals.**

## ■ Infinite Intervals of Integration

Suppose that we want to find the area $A$ of the unbounded region under the graph of $f(x) = 1/x^2$ on the interval $[1, \infty)$ as shown in Figure 2a. Because the interval $[1, \infty)$ is infinite, the definition of the integral that we have used thus far is not applicable, and a new approach to solving the problem is required. But observe that if $b > 1$, then $A$ can be approximated by the area $A(b)$ of the region under the graph of $f$ on $[1, b]$ (Figure 2b).

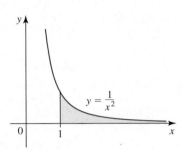

**FIGURE 2**

The shaded area in part (a) is approximated by the shaded area in part (b).

**(a)** The area of $A$ of the region under the graph of $y = 1/x^2$ on $[1, \infty)$.

**(b)** The area $A(b)$ of the region under the graph of $y = 1/x^2$ on $[1, b]$.

The approximation seems to get better and better as $b$ gets larger and larger (see Figure 3). Since $[1, b]$ is finite, we see that

$$A(b) = \int_1^b f(x)\, dx = \int_1^b \frac{1}{x^2}\, dx = -\frac{1}{x}\Big|_1^b = -\frac{1}{b} + 1$$

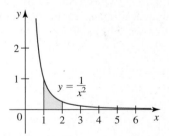
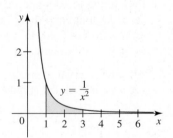

**(a)** Area of region under the graph of $f$ on $[1, 2]$

**(b)** Area of region under the graph of $f$ on $[1, 3]$

**(c)** Area of region under the graph of $f$ on $[1, 4]$

**FIGURE 3**

As $b$ increases, the approximation of $A$ by the definite integral improves.

Letting $b \to \infty$, we obtain

$$\lim_{b \to \infty} A(b) = \lim_{b \to \infty}\left(-\frac{1}{b} + 1\right) = 1$$

This suggests that we *define* the area $A$ to be 1 and write

$$A = \int_1^\infty \frac{1}{x^2}\, dx = \lim_{b \to \infty} \int_1^b \frac{1}{x^2}\, dx = 1$$

This example shows how we can define an integral over an infinite interval as the limit of integrals over finite intervals. More precisely, we have the following definitions. (Note that $f$ need not be positive in the interval under consideration.)

---

**DEFINITIONS**  **Improper Integrals with Infinite Limits of Integration**

**1.** If $f$ is continuous on $[a, \infty)$, then

$$\int_a^\infty f(x)\,dx = \lim_{b\to\infty} \int_a^b f(x)\,dx \tag{1}$$

provided that the limit exists.

**2.** If $f$ is continuous on $(-\infty, b]$, then

$$\int_{-\infty}^b f(x)\,dx = \lim_{a\to-\infty} \int_a^b f(x)\,dx \tag{2}$$

provided that the limit exists.

**3.** If $f$ is continuous on $(-\infty, \infty)$, then

$$\int_{-\infty}^\infty f(x)\,dx = \int_{-\infty}^c f(x)\,dx + \int_c^\infty f(x)\,dx \tag{3}$$

where $c$ is any real number, provided that both improper integrals on the right-hand side exist.

**Convergence and Divergence**

Each improper integral in Equation (1) and Equation (2) is **convergent** if the limit exists and **divergent** if the limit does not exist. The improper integral on the left-hand side in Equation (3) is **convergent** if both improper integrals on the right are convergent and **divergent** if one or both of the improper integrals on the right is divergent.

---

**EXAMPLE 1**  Evaluate $\displaystyle\int_1^\infty \frac{1}{x}\,dx$.

**Solution**  By Equation (1) we have

$$\int_1^\infty \frac{1}{x}\,dx = \lim_{b\to\infty} \int_1^b \frac{1}{x}\,dx = \lim_{b\to\infty} \Big[\ln x\Big]_1^b$$

$$= \lim_{b\to\infty} (\ln b - \ln 1) = \infty$$

Therefore, the given improper integral is divergent.

Let's compare the integral $\int_1^\infty (1/x)\,dx$ of Example 1 with the integral $\int_1^\infty (1/x^2)\,dx$ that we considered earlier. If we interpret each integral as the area of the region under the graph of a function on the infinite interval $[1, \infty)$, then the result $\int_1^\infty (1/x^2)\,dx = 1$ tells us that the area under the graph of $y = 1/x^2$ is equal to 1 and hence finite, whereas the result $\int_1^\infty (1/x)\,dx = \infty$ tells us that the area under the graph of $y = 1/x$ is infinite. Observe that the graphs of $y = 1/x^2$ and $y = 1/x$ are similar. (See Figure 4.) Both $1/x^2$ and $1/x$ approach zero as $x$ approaches infinity, but $1/x^2$ approaches zero faster than $1/x$ does.

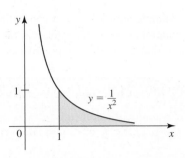 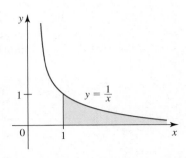

**FIGURE 4**     **(a)** The unbounded region has *finite* area.     **(b)** The unbounded region has *infinite* area.

These examples reveal the fine line between convergence and divergence of an improper integral. But a word of caution: It is not even necessary for $f(x)$ to approach zero as $x$ approaches infinity for an integral $\int_a^\infty f(x)\,dx$ to converge (see Challenge Problem 16 at the end of this chapter).

**EXAMPLE 2** Find the values of $p$ for which $\displaystyle\int_1^\infty \frac{1}{x^p}\,dx$ is convergent.

**Solution** From the result of Example 1 we see that the integral is divergent if $p = 1$. So let's assume that $p \neq 1$. We have

$$\int_1^\infty \frac{1}{x^p}\,dx = \lim_{b\to\infty} \int_1^b x^{-p}\,dx$$

$$= \lim_{b\to\infty} \left[\frac{x^{-p+1}}{-p+1}\right]_1^b$$

$$= \frac{1}{1-p} \lim_{b\to\infty} \left[\frac{1}{b^{p-1}} - 1\right]$$

If $p < 1$, then $1 - p > 0$, so

$$\lim_{b\to\infty} \frac{1}{b^{p-1}} = \lim_{b\to\infty} b^{1-p} = \infty$$

Therefore, the integral diverges. If $p > 1$, then $p - 1 > 0$, so

$$\lim_{b\to\infty} \frac{1}{b^{p-1}} = 0$$

Therefore, the integral converges to $1/(p - 1)$. To summarize

$$\int_1^\infty \frac{1}{x^p}\,dx = \begin{cases} \dfrac{1}{p-1} & \text{if } p > 1 \\[2mm] \text{diverges} & \text{if } p \leq 1 \end{cases}$$

**EXAMPLE 3** Evaluate

**a.** $\displaystyle\int_{-1}^\infty e^{-x}\,dx$      **b.** $\displaystyle\int_0^\infty \cos x\,dx$

## Solution

**a.** $\int_{-1}^{\infty} e^{-x}\,dx = \lim_{b\to\infty}\int_{-1}^{b} e^{-x}\,dx = \lim_{b\to\infty}\left[-e^{-x}\right]_{-1}^{b} = \lim_{b\to\infty}(-e^{-b} + e^{1}) = e$

**b.** $\int_{0}^{\infty} \cos x\,dx = \lim_{b\to\infty}\int_{0}^{b} \cos x\,dx = \lim_{b\to\infty}\left[\sin x\right]_{0}^{b} = \lim_{b\to\infty}(\sin b - 0)$

Since $\lim_{b\to\infty}\sin b$ does not exist, we conclude that the given integral is divergent. (To see why, just examine the graph of $y = \sin x$.)

**EXAMPLE 4** Evaluate $\int_{-\infty}^{0} xe^{x}\,dx$.

**Solution** By Equation (2) we have

$$\int_{-\infty}^{0} xe^{x}\,dx = \lim_{a\to-\infty}\int_{a}^{0} xe^{x}\,dx$$

From the result of Example 1 in Section 6.1 we have

$$\int xe^{x}\,dx = xe^{x} - \int e^{x}\,dx = (x - 1)e^{x} + C$$

Therefore,

$$\int_{-\infty}^{0} xe^{x}\,dx = \lim_{a\to-\infty}\int_{a}^{0} xe^{x}\,dx = \lim_{a\to-\infty}\left[(x - 1)e^{x}\right]_{a}^{0}$$

$$= \lim_{a\to-\infty}\left[-1 - (a - 1)e^{a}\right]$$

To evaluate the limit on the right-hand side, note that

$$\lim_{a\to-\infty} e^{a} = 0$$

and, by l'Hôpital's Rule,

$$\lim_{a\to-\infty} ae^{a} = \lim_{a\to-\infty}\frac{a}{e^{-a}} \qquad \text{Indeterminate form: } -\infty/\infty$$

$$= \lim_{a\to-\infty}\frac{1}{-e^{-a}} = 0$$

Therefore,

$$\int_{-\infty}^{0} xe^{x}\,dx = \lim_{a\to-\infty}(-1 - ae^{a} + e^{a})$$

$$= \lim_{a\to-\infty}(-1) - \lim_{a\to-\infty} ae^{a} + \lim_{a\to-\infty} e^{a}$$

$$= -1 - 0 + 0 = -1$$

**EXAMPLE 5** Evaluate $\int_{-\infty}^{\infty}\frac{1}{1 + x^{2}}\,dx$, and interpret your result geometrically.

**Solution**   By Equation (3) we have

$$\int_{-\infty}^{\infty} \frac{1}{1 + x^2}\, dx = \int_{-\infty}^{0} \frac{1}{1 + x^2}\, dx + \int_{0}^{\infty} \frac{1}{1 + x^2}\, dx \qquad \text{For convenience we have chosen } c = 0.$$

$$= \lim_{a \to -\infty} \int_{a}^{0} \frac{1}{1 + x^2}\, dx + \lim_{b \to \infty} \int_{0}^{b} \frac{1}{1 + x^2}\, dx$$

$$= \lim_{a \to -\infty} \left[ \tan^{-1} x \right]_{a}^{0} + \lim_{b \to \infty} \left[ \tan^{-1} x \right]_{0}^{b}$$

$$= \lim_{a \to -\infty} \left( \tan^{-1} 0 - \tan^{-1} a \right) + \lim_{b \to \infty} \left( \tan^{-1} b - \tan^{-1} 0 \right)$$

$$= \left[ 0 - \left( -\frac{\pi}{2} \right) \right] + \left( \frac{\pi}{2} - 0 \right) = \pi$$

Because the integrand $f(x) = 1/(1 + x^2)$ is nonnegative on $(-\infty, \infty)$, we can interpret the value of the improper integral as the area $(\pi)$ of the region under the graph of $f$ on $(-\infty, \infty)$. (See Figure 5.)  ∎

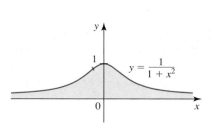

**FIGURE 5**
The area of the region under the graph
of $y = \dfrac{1}{1 + x^2}$ on $(-\infty, \infty)$ is $\pi$.

**EXAMPLE 6   A Rocket Launch**   Find the work done in launching a rocket weighing $P$ pounds, vertically upward from the surface of the earth so that the rocket completely escapes the earth's gravitational field.

**Solution**   According to Newton's Law of Gravitation, the rocket is attracted to the earth by a force $F(x)$ given by

$$F(x) = \frac{GmM}{x^2}$$

where $m$ is the mass of the rocket, $M$ is the mass of the earth, $x$ is the distance between the rocket and the center of the earth, and $G$ is the universal gravitational constant. Writing $k = GmM$, we have

$$F(x) = \frac{k}{x^2} \qquad R \le x < \infty$$

where $R$ is the radius of the earth. Since the rocket weighs $P$ pounds on the surface of the earth, we have

$$F(R) = \frac{k}{R^2} = P$$

This gives $k = PR^2$, and therefore

$$F(x) = \frac{PR^2}{x^2}$$

(See Figure 6.) Therefore, the work required to propel the rocket to an infinite height (to escape the earth's gravitational field) is

$$W = \int_{R}^{\infty} F(x)\, dx = \int_{R}^{\infty} \frac{PR^2}{x^2}\, dx$$

$$= \lim_{b \to \infty} \int_{R}^{b} \frac{PR^2}{x^2}\, dx = \lim_{b \to \infty} \left[ -\frac{PR^2}{x} \right]_{R}^{b}$$

$$= \lim_{b \to \infty} \left( -\frac{PR^2}{b} + \frac{PR^2}{R} \right) = PR$$

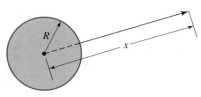

**FIGURE 6**
The force attracting the rocket to the
earth when it is at a distance $x$ is
$F = PR^2/x^2$, where $R \le x < \infty$.

For example, if the rocket weighs 20 tons (40,000 lb) on the ground and the radius of the earth is approximately 4000 mi (21,120,000 ft), then the work required is $W \approx 40{,}000 \times 21{,}120{,}000$ or $8.448 \times 10^{11}$ ft-lb.   ■

## ■ Improper Integrals with Infinite Discontinuities

As we mentioned earlier, there is another kind of improper integral: those having integrands that are unbounded on the interval of integration (Figure 1b). To see how we define this type of integral, consider the problem of finding the area $A$ of the unbounded region under the graph of $f(x) = 1/\sqrt{x}$ on the interval $(0, 4]$ shown in Figure 7a.

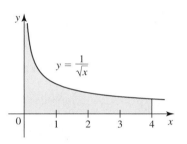
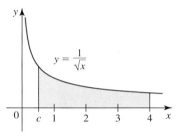

(**a**) The area $A$ of the region under
   the graph of $y = 1/\sqrt{x}$ on $(0, 4]$

(**b**) The area $A(c)$ of the region under
   the graph of $y = 1/\sqrt{x}$ on $[c, 4]$

**FIGURE 7**
The area of the shaded region in part (a) is approximated by the area of the shaded region in part (b).

Because the integrand is unbounded on the interval $(0, 4]$ (that is, $1/\sqrt{x} \to \infty$ as $x \to 0^+$), the definition of the integral given in Chapter 4 cannot be used to find $A$. But observe that if $c$ is any number such that $0 < c < 4$, then $A$ can be approximated by the area $A(c)$ of the region under the graph of $f$ on $[c, 4]$ (Figure 7b). Observe that the approximation appears to get better and better as $c$ approaches 0 from the right. Since $f(x) = 1/\sqrt{x}$ is bounded on the finite interval $[c, 4]$, we see that

$$A(c) = \int_c^4 f(x)\, dx = \int_c^4 \frac{1}{\sqrt{x}}\, dx = 2\sqrt{x}\,\Big|_c^4 = 4 - 2\sqrt{c}$$

Letting $c \to 0^+$, we obtain

$$\lim_{c \to 0^+} A(c) = \lim_{c \to 0^+} (4 - 2\sqrt{c}) = 4$$

This suggests that we *define* the area $A$ to be 4 and write

$$A = \int_0^4 \frac{1}{\sqrt{x}}\, dx = \lim_{c \to 0^+} \int_c^4 \frac{1}{\sqrt{x}}\, dx = 4$$

This example shows how we can define an integral whose integrand has an infinite discontinuity at a point as the limit of integrals whose integrands are bounded. More precisely, we have the following definitions. (Again, note that $f$ need not be positive in the interval under consideration.)

> **DEFINITIONS**   **Improper Integrals Whose Integrands Have Infinite Discontinuities**
>
> **1.** If $f$ is continuous on $[a, b)$ and $f$ has an infinite discontinuity at $b$, then
> $$\int_a^b f(x)\, dx = \lim_{c \to b^-} \int_a^c f(x)\, dx \tag{4}$$
> provided that the limit exists (Figure 8a).
> **2.** If $f$ is continuous on $(a, b]$ and $f$ has an infinite discontinuity at $a$, then
> $$\int_a^b f(x)\, dx = \lim_{c \to a^+} \int_c^b f(x)\, dx \tag{5}$$
> provided that the limit exists (Figure 8b).
> **3.** If $f$ has an infinite discontinuity at $c$, where $a < c < b$, but $f$ is continuous elsewhere on $[a, b]$, then
> $$\int_a^b f(x)\, dx = \int_a^c f(x)\, dx + \int_c^b f(x)\, dx \tag{6}$$
> provided that both improper integrals on the right exist (Figure 8c).
>
> **Convergence and Divergence**
>
> Each improper integral in Equations (4) and (5) is **convergent** if the limit exists and **divergent** if the limit does not exist. The improper integral on the left in Equation (6) is **convergent** if both improper integrals on the right are convergent and **divergent** if one or both improper integrals on the right is divergent.

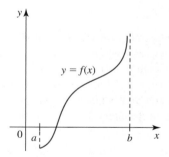

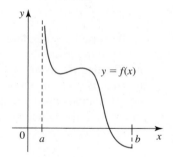

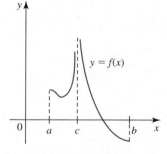

(a) $f$ has an infinte discontinuity at $b$.

(b) $f$ has an infinite discontinuity at $a$.

(c) $f$ has an infinite discontinuity at $c$.

**FIGURE 8**

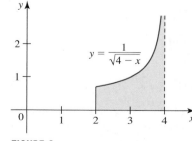

**FIGURE 9**
The area of the region under the graph of $y = 1/\sqrt{4 - x}$ on $[2, 4)$ is $2\sqrt{2}$.

**EXAMPLE 7**  Evaluate $\displaystyle\int_2^4 \frac{1}{\sqrt{4 - x}}\, dx$, and interpret your result geometrically.

**Solution**   The integrand $f(x) = 1/\sqrt{4 - x}$ has an infinite discontinuity at $x = 4$, as shown in Figure 9. Using Equation (4), we have

$$\int_2^4 \frac{1}{\sqrt{4 - x}}\, dx = \lim_{c \to 4^-} \int_2^c \frac{1}{\sqrt{4 - x}}\, dx$$

$$= \lim_{c \to 4^-} \left[ -2\sqrt{4 - x} \right]_2^c \qquad \text{Integrate using the substitution } u = 4 - x.$$

$$= \lim_{c \to 4^-} \left( -2\sqrt{4 - c} + 2\sqrt{2} \right) = 2\sqrt{2}$$

Since the integrand is positive on $[2, 4]$, we can interpret the value of the improper integral as the area of the region under the graph of $f$ on $[2, 4]$. ∎

**EXAMPLE 8** Evaluate $\displaystyle\int_0^1 \frac{dx}{x^2}$.

**Solution** The integrand $1/x^2$ has an infinite discontinuity at $x = 0$. Using Equation (5), we have

$$\int_0^1 \frac{dx}{x^2} = \lim_{a \to 0^+} \int_a^1 \frac{dx}{x^2} = \lim_{a \to 0^+}\left[-\frac{1}{x}\right]_a^1 = \lim_{a \to 0^+}\left(-1 + \frac{1}{a}\right) = \infty$$

and we conclude that the given improper integral is divergent. ∎

**EXAMPLE 9** Evaluate $\displaystyle\int_0^1 \ln x \, dx$.

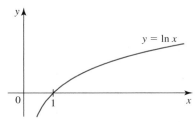

FIGURE 10
The integrand $f(x) = \ln x$ approaches $-\infty$ as $x$ approaches 0 from the right.

**Solution** The integrand has an infinite discontinuity at $x = 0$. (See Figure 10.) Therefore, we write

$$\int_0^1 \ln x \, dx = \lim_{a \to 0^+} \int_a^1 \ln x \, dx$$

$$= \lim_{a \to 0^+}\left[x \ln x - x\right]_a^1 \qquad \begin{array}{l}\text{Integrate by parts with } u = \ln x\\ \text{and } dv = dx.\end{array}$$

$$= \lim_{a \to 0^+} (0 - 1 - a \ln a + a)$$

To evaluate the limit on the right, we apply l'Hôpital's Rule, obtaining

$$\lim_{a \to 0^+} a \ln a = \lim_{a \to 0^+} \frac{\ln a}{\dfrac{1}{a}} = \lim_{a \to 0^+} \frac{\dfrac{1}{a}}{-\dfrac{1}{a^2}} = \lim_{a \to 0^+} (-a) = 0$$

Therefore,

$$\int_0^1 \ln x \, dx = \lim_{a \to 0^+} (-1 - a \ln a + a) = -1 - 0 + 0 = -1 \qquad ∎$$

**EXAMPLE 10** Evaluate $\displaystyle\int_{-1}^1 \frac{dx}{x^2}$.

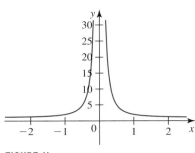

FIGURE 11
The integrand $f(x) = 1/x^2$ approaches $\infty$ as $x$ approaches 0.

**Solution** The integrand $f(x) = 1/x^2$ has an infinite discontinuity at $x = 0$. (See Figure 11.) Using Equation (6), we have

$$\int_{-1}^1 \frac{dx}{x^2} = \int_{-1}^0 \frac{dx}{x^2} + \int_0^1 \frac{dx}{x^2}$$

Now, using the result of Example 8, we see that the second integral on the right is divergent; that is,

$$\int_0^1 \frac{dx}{x^2} = \infty$$

Therefore, the given improper integral is divergent. Note that it is not necessary to evaluate the first integral on the right. ∎

**Note** If we had not realized that $f(x) = 1/x^2$ has an infinite discontinuity at $x = 0$, then we might have proceeded as follows:

$$\int_{-1}^{1} \frac{dx}{x^2} = -\frac{1}{x}\Big|_{-1}^{1} = -1 + (-1) = -2$$

giving a *wrong* answer. After all, a positive integrand could not possibly yield an integral whose value is negative! ∎

**EXAMPLE 11** **Length of a Pursuit Curve** The graph $C$ of the equation

$$y = \frac{1}{3}\sqrt{x}(x - 3) + \frac{2}{3}$$

gives the path taken by a coast guard patrol boat (Boat $A$) as it pursued and eventually intercepted boat $B$ that was suspected of carrying contraband. (See Figure 12.) Initially, the patrol boat was at point $P$, and Boat $B$ was at the origin, heading north. At the time of interception both boats were at point $Q$. Find the distance traveled by the patrol boat during the pursuit.

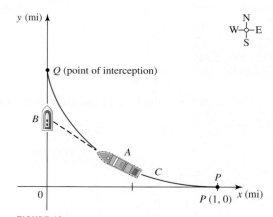

**FIGURE 12**
The pursuit curve $C$ gives the path taken by patrol boat $A$.

**Solution** The distance traveled by the patrol boat is given by the length $L$ of the curve $C$ from $x = 0$ to $x = 1$. To use Equation (5), we first compute

$$\frac{dy}{dx} = \frac{d}{dx}\left[\frac{1}{3}x^{3/2} - x^{1/2} + \frac{2}{3}\right]$$

$$= \frac{1}{2}x^{1/2} - \frac{1}{2}x^{-1/2} = \frac{1}{2}(x^{1/2} - x^{-1/2})$$

and

$$1 + \left(\frac{dy}{dx}\right)^2 = 1 + \frac{1}{4}(x^{1/2} - x^{-1/2})^2 = 1 + \frac{1}{4}(x - 2 + x^{-1})$$

$$= \frac{4x + x^2 - 2x + 1}{4x} = \frac{x^2 + 2x + 1}{4x} = \frac{(x + 1)^2}{4x}$$

Then

$$L = \int_0^1 \sqrt{1 + \left(\frac{dy}{dx}\right)^2}\, dx = \int_0^1 \sqrt{\frac{(x+1)^2}{4x}}\, dx = \frac{1}{2}\int_0^1 \frac{x+1}{\sqrt{x}}\, dx$$

$$= \frac{1}{2}\int_0^1 x^{1/2}\, dx + \frac{1}{2}\int_0^1 x^{-1/2}\, dx$$

The second integral on the right has an infinite discontinuity at $x = 0$. So we write

$$L = \frac{1}{2}\int_0^1 x^{1/2}\, dx + \frac{1}{2}\lim_{t\to0^+}\int_t^1 x^{-1/2}\, dx$$

$$= \left(\frac{1}{2}\right)\left(\frac{2}{3}x^{3/2}\right)\Big|_0^1 + \frac{1}{2}\lim_{t\to0^+}\left[2x^{1/2}\right]_t^1$$

$$= \frac{1}{3} + \frac{1}{2}\lim_{t\to0^+}(2 - 2t^{1/2}) = \frac{1}{3} + 1 = \frac{4}{3}$$

Therefore, the patrol boat traveled $\frac{4}{3}$ miles from the time Boat $B$ was spotted until the time it was intercepted. ■

The next example involves both an infinite limit of integration and an infinite discontinuity.

**EXAMPLE 12** Evaluate $\displaystyle\int_0^\infty \frac{e^{-\sqrt{x}}}{\sqrt{x}}\, dx$.

**Solution** We write

$$\int_0^\infty \frac{e^{-\sqrt{x}}}{\sqrt{x}}\, dx = \int_0^1 \frac{e^{-\sqrt{x}}}{\sqrt{x}}\, dx + \int_1^\infty \frac{e^{-\sqrt{x}}}{\sqrt{x}}\, dx$$

$$= \lim_{t\to0^+}\int_t^1 \frac{e^{-\sqrt{x}}}{\sqrt{x}}\, dx + \lim_{b\to\infty}\int_1^b \frac{e^{-\sqrt{x}}}{\sqrt{x}}\, dx$$

$$= \lim_{t\to0^+}\left[-2e^{-\sqrt{x}}\right]_t^1 + \lim_{b\to\infty}\left[-2e^{-\sqrt{x}}\right]_1^b$$

$$= \lim_{t\to0^+}\left(-2e^{-1} + 2e^{-\sqrt{t}}\right) + \lim_{b\to\infty}\left(-2e^{-\sqrt{b}} + 2e^{-1}\right)$$

$$= -2e^{-1} + 2 + 2e^{-1} = 2$$ ■

## ■ A Comparison Test for Improper Integrals

Sometimes it is impossible to find the exact value of an improper integral. In such instances we need to determine whether the integral is convergent or divergent. If we can ascertain that the improper integral is convergent, then we can proceed to obtain a sufficiently accurate approximation of its value, which, in practice, is all that is required. The following theorem is stated without proof, but its plausibility should be evident by examining Figure 13.

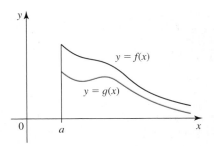

**FIGURE 13**
The function $f$ dominates the function $g$ on $[a, \infty)$.

> **THEOREM 1  A Comparison Test for Improper Integrals**
>
> Let $f$ and $g$ be continuous, and suppose that $f(x) \geq g(x) \geq 0$ for all $x \geq a$; that is, $f$ dominates $g$ on $[a, \infty)$.
>
> **a.** If $\displaystyle\int_a^\infty f(x)\, dx$ is convergent, then so is $\displaystyle\int_a^\infty g(x)\, dx$.
>
> **b.** If $\displaystyle\int_a^\infty g(x)\, dx$ is divergent, then so is $\displaystyle\int_a^\infty f(x)\, dx$.

Before looking at the next example, let's note that the functions that we have dealt with up until now have been functions such as polynomial, rational, power, exponential, logarithmic, trigonometric, and inverse trigonometric functions or functions obtained from this list by combining them using the operations of addition, subtraction, multiplication, division, and composition. Such functions are called *elementary functions*.

**EXAMPLE 13** Show that $\displaystyle\int_0^\infty e^{-x^2}\, dx$ is convergent.

**Solution** We cannot evaluate the integral directly because it turns out that the antiderivative of $e^{-x^2}$ is not an elementary function. To show that this integral is convergent, let's write

$$\int_0^\infty e^{-x^2}\, dx = \int_0^1 e^{-x^2}\, dx + \int_1^\infty e^{-x^2}\, dx$$

Observe that the first integral on the right is a proper integral, and therefore, it has a finite value, even though we don't know what that value is. For the second integral we note that $x^2 \geq x$ for $x \geq 1$, so $e^{-x^2} \leq e^{-x}$ on $[1, \infty)$. (See Figure 14.) Now

$$\int_1^\infty e^{-x}\, dx = \lim_{b \to \infty} \int_1^b e^{-x}\, dx = \lim_{b \to \infty} \left[-e^{-x}\right]_1^b = \lim_{b \to \infty}(-e^{-b} + e^{-1}) = \frac{1}{e}$$

So if we take $f(x) = e^{-x}$ and $g(x) = e^{-x^2}$, the Comparison Test tells us that $\int_1^\infty e^{-x^2}\, dx$ is convergent. Therefore, $\int_0^\infty e^{-x^2}\, dx$ is convergent. ∎

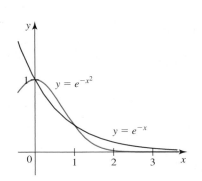

**FIGURE 14**
We use the Comparison Test to show that
$$\int_0^\infty e^{-x^2}\, dx = \int_0^1 e^{-x^2}\, dx + \int_1^\infty e^{-x^2}\, dx$$
is convergent.

## 6.6  CONCEPT QUESTIONS

1. Define the following improper integrals:

   **a.** $\displaystyle\int_{-\infty}^b f(x)\, dx$   **b.** $\displaystyle\int_a^\infty f(x)\, dx$   **c.** $\displaystyle\int_{-\infty}^\infty f(x)\, dx$

2. Define the improper integral $\displaystyle\int_a^b f(x)\, dx$ if
   **a.** $f$ has an infinite discontinuity at $a$.
   **b.** $f$ has an infinite discontinuity at $b$.
   **c.** $f$ has an infinite discontinuity at $c$, where $a < c < b$.

3. State the Comparison Test for improper integrals.

## 6.6 EXERCISES

*In Exercises 1–6, find the area of the shaded region, if it exists.*

**1.**

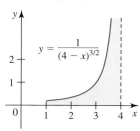

**2.**

**3.**

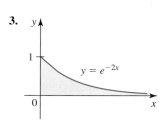

**4.**

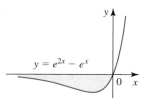

**5.**

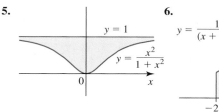

**6.**

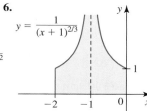

*In Exercises 7–42, determine whether the improper integral converges or diverges, and if it converges, find its value.*

**7.** $\displaystyle\int_1^\infty \frac{1}{x^3}\,dx$

**8.** $\displaystyle\int_1^\infty \frac{1}{x^{0.99}}\,dx$

**9.** $\displaystyle\int_1^\infty \frac{1}{x^{1.01}}\,dx$

**10.** $\displaystyle\int_0^\infty \frac{1}{(x+1)^2}\,dx$

**11.** $\displaystyle\int_1^\infty \frac{1}{(x+2)^{3/2}}\,dx$

**12.** $\displaystyle\int_2^\infty \frac{1}{\sqrt[3]{x-1}}\,dx$

**13.** $\displaystyle\int_1^\infty e^{-2x}\,dx$

**14.** $\displaystyle\int_e^\infty \frac{1}{x\ln^2 x}\,dx$

**15.** $\displaystyle\int_0^\infty \sin x\,dx$

**16.** $\displaystyle\int_0^\infty e^{-x}\sin x\,dx$

**17.** $\displaystyle\int_0^\infty \frac{x}{1+x^2}\,dx$

**18.** $\displaystyle\int_{-\infty}^0 \frac{1}{x^2+2x+5}\,dx$

**19.** $\displaystyle\int_{-\infty}^\infty \frac{1}{x^2+4}\,dx$

**20.** $\displaystyle\int_{-\infty}^\infty xe^{-x^2}\,dx$

**21.** $\displaystyle\int_{-\infty}^\infty \frac{e^x}{1+e^{2x}}\,dx$

**22.** $\displaystyle\int_{-\infty}^\infty \cos^2 x\,dx$

**23.** $\displaystyle\int_{-\infty}^\infty \frac{x}{(x^2+1)^{3/2}}\,dx$

**24.** $\displaystyle\int_{-\infty}^\infty e^{-|x|}\,dx$

**25.** $\displaystyle\int_0^1 \frac{1}{x^{2/3}}\,dx$

**26.** $\displaystyle\int_{-2}^1 \frac{1}{x^2}\,dx$

**27.** $\displaystyle\int_{-8}^1 \frac{1}{\sqrt[3]{x}}\,dx$

**28.** $\displaystyle\int_0^2 \frac{1}{2x-3}\,dx$

**29.** $\displaystyle\int_1^4 \frac{1}{(4-x)^{2/3}}\,dx$

**30.** $\displaystyle\int_0^2 \frac{1}{x^2-2x}\,dx$

**31.** $\displaystyle\int_0^4 \frac{1}{\sqrt{x}-1}\,dx$

**32.** $\displaystyle\int_0^e \ln x\,dx$

**33.** $\displaystyle\int_0^1 x\ln x\,dx$

**34.** $\displaystyle\int_0^\pi \sec^2 x\,dx$

**35.** $\displaystyle\int_{\pi/6}^{\pi/2} \frac{\cos x}{\sqrt{1-\sin x}}\,dx$

**36.** $\displaystyle\int_0^{\pi/2} \tan^2 x\,dx$

**37.** $\displaystyle\int_1^\infty \frac{\ln x}{x^{3/2}}\,dx$

**38.** $\displaystyle\int_0^\infty \frac{\sqrt{\tan^{-1}x}}{1+x^2}\,dx$

**39.** $\displaystyle\int_{-\infty}^\infty \frac{1}{x^{4/3}}\,dx$

**40.** $\displaystyle\int_0^\infty \frac{1}{e^x-1}\,dx$

**41.** $\displaystyle\int_0^1 \frac{\ln x}{\sqrt{x}}\,dx$

**42.** $\displaystyle\int_1^\infty \frac{dx}{x\sqrt{x^2-1}}$

*In Exercises 43–48, use the Comparison Test to determine whether the integral is convergent or divergent by comparing it with the second integral.*

**43.** $\displaystyle\int_1^\infty \frac{1}{1+x^2}\,dx; \quad \int_1^\infty \frac{1}{x^2}\,dx$

**44.** $\displaystyle\int_1^\infty \frac{1}{\sqrt{x^3+1}}\,dx; \quad \int_1^\infty \frac{1}{x^{3/2}}\,dx$

**45.** $\displaystyle\int_1^\infty \frac{\cos^2 x}{x^2}\,dx; \quad \int_1^\infty \frac{1}{x^2}\,dx$

**46.** $\displaystyle\int_1^\infty \frac{dx}{x+\sin^2 x}; \quad \int_1^\infty \frac{1}{1+x}\,dx$

**47.** $\displaystyle\int_1^\infty \frac{2+\cos x}{\sqrt{x}}\,dx; \quad \int_1^\infty \frac{1}{\sqrt{x}}\,dx$

**48.** $\displaystyle\int_1^\infty \frac{1}{\sqrt{1+x^2+x^4}}\,dx; \quad \int_1^\infty \frac{1}{x^2}\,dx$

**49.** Evaluate $\int_0^\infty x^5 e^{-x^2}\,dx$.

**50.** Find the area of the region bounded by the graph of $y = 1/\left(\frac{1}{2}x^2 - x + 1\right)$ and the $x$-axis.

**51.** Find the volume of the solid obtained by revolving the region under the graph of $y = 2\left(\dfrac{1}{x^2} - \dfrac{1}{x^4}\right)$ on $[1, \infty)$ about the $x$-axis.

**52.** Find the area of the surface obtained by revolving the graph of $y = e^{-x}$ on $[0, \infty)$ about the $x$-axis.

**53.** Find the volume of the solid obtained by revolving the region under the graph of $y = e^{-x}$ on $[0, \infty)$ about the $x$-axis.

**54.** Find the volume of the solid obtained by revolving the region under the graph of $y = e^{-x}$ on $[0, \infty)$ about the $y$-axis.

**55.** Find the area of the region bounded by the graphs of $y = 1/\sqrt{1 - x^2}$, $y = 0$, $x = 0$, and $x = 1$.

**56. Gabriel's Horn** The solid obtained by revolving the unbounded region under the graph of $f(x) = 1/x$ on the interval $[1, \infty)$ about the $x$-axis is called *Gabriel's Horn*. Show that this solid has a finite volume but an infinite surface area. Thus, Gabriel's Horn describes a can that does not hold enough paint to cover its outside surface!

**Hint:** The surface area is

$$S = 2\pi \int_1^\infty \frac{\sqrt{1 + x^4}}{x^3}\, dx$$

Use the substitution $u = x^2$, and integrate using Formula 40 from the Table of Integrals.

**57. Cissoid of Diocles** Find the area of the region bounded by the *cissoid of Diocles*

$$y = \pm \frac{x^{3/2}}{\sqrt{1 - x}}\, dx$$

and its asymptote $x = 1$.

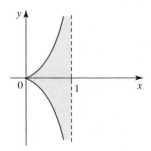

**Hint:** The area is

$$2 \int_0^1 \frac{x^{3/2}}{\sqrt{1 - x}}\, dx$$

Use the substitution $u = \sqrt{x}$ followed by the substitution $u = \sin \theta$.

**58.** Find the length of the astroid $x^{2/3} + y^{2/3} = a^{2/3}$, where $a > 0$.

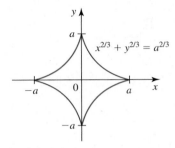

**59. Work Done by a Repulsive Force** An electric charge $Q$ located at the origin of a coordinate line repulses a like charge $q$ from the point $x = a$, where $a > 0$, an infinite distance to the right. Find the work done by the force of repulsion.
**Hint:** The magnitude of force acting on the charge $q$ when it is at the point $x$ is given by

$$F(x) = \frac{1}{4\pi\varepsilon_0} \frac{qQ}{x^2}$$

**60. Elastic Deformation of a Long Beam** The graph $C$ of the function

$$y = \frac{P\alpha}{2k} e^{-\alpha|x|}(\cos \alpha x + \sin \alpha|x|)$$

where $\alpha$ and $k$ are constants, gives the shape of a beam of infinite length lying on an elastic foundation and acted upon by a concentrated load $P$ applied to the beam at the origin. Before application of the force, the beam lies on the $x$-axis. Find the potential energy of elastic deformation $W$ using the formula

$$W = Ee \int_0^\infty (y'')^2\, dx$$

where $E$ and $e$ are constants.

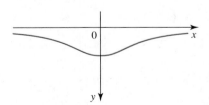

**Note:** This model provides a good approximation in working with long beams.

**61. Work Done by a Repulsive Charge** An electric charge $Q$ distributed uniformly along a line of length $2c$ lying along the $y$-axis repulses a like charge $q$ from the point $x = a$, where $a > 0$, an infinite distance to the right. The magnitude of the force acting on the charge $q$ when it is at the point $x$ is given by

$$F(x) = \frac{1}{4\pi\varepsilon_0} \frac{qQ}{x\sqrt{x^2 + c^2}}$$

and the force acts in the direction of the positive $x$-axis. Find the work done by the force of repulsion.

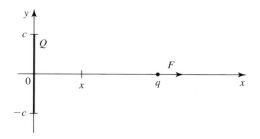

**62. Escape Velocity of a Rocket** The *escape velocity* $v_0$ is the minimum speed a rocket must attain in order to escape from the gravitational field of a planet. Use Newton's Law of Gravitation to find the escape velocity for the earth (see Exercise 32 in Section 5.5).

**Hint:** The work required to launch a rocket from the surface of the earth upward to escape from the earth's gravitational field is

$$W = \int_R^\infty \frac{mgR^2}{r^2} \, dr$$

Equate $W$ with the initial kinetic energy $\frac{1}{2}mv_0^2$ of the rocket.

**63. Capital Value of Property** The *capital value* (present sale value) $CV$ of a property that can be rented on a perpetual basis of $R$ dollars annually is given by

$$CV \approx \int_0^\infty Re^{-it} \, dt$$

where $i$ is the prevailing interest rate per year compounded continuously.
**a.** Show that $CV \approx R/i$.
**b.** Find the capital value of a property that can be rented out for \$10,000 annually when the prevailing interest rate is 10% per year.

**64. Average Power in AC Circuits** If $f$ is defined on $[0, \infty)$, then the average value of $f$ over $[0, \infty)$ is defined to be

$$f_{av} = \lim_{b \to \infty} \frac{1}{b} \int_0^b f(x) \, dx$$

Suppose that the voltage and current in an AC circuit are

$$V = V_0 \cos \omega t \quad \text{and} \quad I = I_0 \cos(\omega t + \phi)$$

so that the voltage and current differ by an angle $\phi$. Then the power output is $P = VI$. Show that the average power output is $P_{av} = \frac{1}{2} I_0 V_0 \cos \phi$.
**Note:** The factor $\cos \phi$ is called the *power factor*. When $V$ and $I$ are in phase ($\phi = 0°$), the average power output is $\frac{1}{2} IV$, but when $V$ and $I$ are out of phase ($\phi = 90°$), then the average power output is zero.

**cas 65. Serum Cholesterol Population Study** The percentage of a current Mediterranean population with serum cholesterol levels at or above 200 mg/dL is estimated to be

$$P = \frac{1}{20\sqrt{2\pi}} \int_{200}^\infty e^{(-1/2)[(x - 160)/20]^2} \, dx$$

Use a CAS to find $P$.

**66.** Find the arc length of the loop defined by $3y^2 = x(x - 1)^2$ from $x = 0$ to $x = 1$.

**67.** Find the value of the constant $C$ for which

$$\int_1^\infty \left( \frac{1}{\sqrt{x}} - \frac{C}{\sqrt{x + 1}} \right) dx$$

converges. Then evaluate the integral for this value of $C$.

**68.** Let $I = \displaystyle\int_0^\infty \frac{x^2}{x^4 + 1} \, dx$.
**a.** Use the substitution $u = 1/x$ to show that

$$I = \frac{1}{2} \int_0^\infty \frac{x^2 + 1}{x^4 + 1} \, dx = \frac{1}{2} \int_0^\infty \frac{1 + \dfrac{1}{x^2}}{x^2 + \dfrac{1}{x^2}} \, dx$$

**b.** Use the substitution $v = x - \dfrac{1}{x}$ to show that

$$I = \frac{1}{2} \int_{-\infty}^\infty \frac{dv}{v^2 + 2}.$$

**c.** Use the result of part (b) to show that

$$\int_0^\infty \frac{x^2}{x^4 + 1} \, dx = \frac{\sqrt{2}\,\pi}{4}$$

**69.** Find the values of $p$ for which the integral $\int_0^1 1/x^p \, dx$ converges and the values of $p$ for which it diverges.

**70.** Consider the integral $I = \displaystyle\int_3^\infty \frac{dx}{\sqrt{x(x - 1)(x - 2)}}$.
**a.** Plot the graphs of $f(x) = \dfrac{1}{\sqrt{x(x - 1)(x - 2)}}$ and $g(x) = \dfrac{3\sqrt{2}}{2x^{3/2}}$ using the viewing window $[0, 6] \times [0, 1.8]$ to see that $f(x) \le g(x)$ for all $x$ in $(3, \infty)$.
**b.** Prove the assertion in part (a).
**c.** Prove that $I$ converges.

**71.** Prove that $\displaystyle\int_0^1 \frac{\sin \dfrac{1}{\sqrt{x}}}{\sqrt{x}} \, dx$ converges.

**72.** Observe that $\int_0^\infty e^{-x^2} \, dx = \int_0^4 e^{-x^2} \, dx + \int_4^\infty e^{-x^2} \, dx$.
**a.** Show that $\int_4^\infty e^{-x^2} \, dx \le 10^{-7}$ so that $\int_0^\infty e^{-x^2} \, dx \approx \int_0^4 e^{-x^2} \, dx$.
**b.** Use a calculator or computer to obtain an estimate for $\int_0^\infty e^{-x^2} \, dx$.

In Exercises 73 and 74, (a) *find a "test integral" to be used in determining the convergence or divergence of the improper integral,* (b) *verify the result of part* (a) *by plotting the graphs of both integrands in the same viewing window, and* (c) *determine the convergence or divergence of the integral.*

**73.** $\displaystyle\int_1^\infty \frac{\sqrt{t^3 - t^2 + 1}}{t^5 + t + 2}\, dt$   **74.** $\displaystyle\int_1^\infty \frac{1 - 4\sin 2x}{x^3 + x^{1/3}}\, dx$

*Let* $f(t)$ *be continuous for* $t > 0$. *The* **Laplace transform** *of* $f$ *is the function* $F$ *defined by*

$$F(s) = \int_0^\infty f(t)e^{-st}\, dt$$

*provided that the integral exists. In Exercises 75–79, use this definition.*

**75.** Find the Laplace transform of $f(t) = 1$.

**76.** Find the Laplace transform of $f(t) = e^{at}$, where $a$ is a constant.

**77.** Find the Laplace transform of $f(t) = t$.

**78.** Show that the Laplace transform of $f(t) = \cos \omega t$ is

$$F(s) = \frac{s}{s^2 + \omega^2}.$$

**79.** Suppose that $f'$ is continuous for $t > 0$ and $f$ satisfies the condition $\lim_{t \to \infty} e^{-st} f(t) = 0$. Show that the Laplace transform of $f'(t)$ for $t > 0$, denoted by $G$, satisfies $G(s) = sF(s) - f(0)$, where $s > 0$ and $F$ is the Laplace transform of $f$.

In Exercises 80–87, *determine whether the statement is true or false. If it is true, explain why it is true. If it is false, explain why or give an example to show why it is false.*

**80.** If $f$ is continuous on $(-\infty, \infty)$, then $\int_{-\infty}^{\infty} f(x)\, dx = \lim_{t \to \infty} \int_{-t}^{t} f(x)\, dx$.

**81.** If $f$ is continuous on $[0, \infty)$ and $\lim_{x \to \infty} f(x) = 0$, then $\int_0^\infty f(x)\, dx$ is convergent.

**82.** If $\int_a^\infty [f(x) + g(x)]\, dx$ is convergent, then $\int_a^\infty f(x)\, dx$ and $\int_a^\infty g(x)\, dx$ must both be convergent.

**83.** If $\int_a^\infty f(x)\, dx$ and $\int_a^\infty g(x)\, dx$ are both convergent, then $\int_a^\infty [f(x) + g(x)]\, dx$ is convergent.

**84.** If both $\int_a^\infty f(x)\, dx$ and $\int_a^\infty g(x)\, dx$ are divergent, then $\int_a^\infty [f(x) + g(x)]\, dx$ must also be divergent.

**85.** If $f(x) \le g(x)$ for all $x$ in $[a, \infty)$ and $\int_a^\infty g(x)\, dx$ diverges, then $\int_a^\infty f(x)\, dx$ may converge.

**86.** If $f(x) \le g(x)$ for all $x$ in $[a, \infty)$ and $\int_a^\infty f(x)\, dx$ converges, then $\int_a^\infty g(x)\, dx$ also converges.

**87.** Suppose that $f$ is continuous on $[a, b)$ and $f$ has an infinite discontinuity at $b$. Furthermore, suppose that $\int_c^b f(x)\, dx$ is convergent, where $c$ is a number between $a$ and $b$. Then $\int_a^b f(x)\, dx$ is convergent.

# CHAPTER 6 REVIEW

## CONCEPT REVIEW

In Exercises 1–8, *fill in the blanks.*

**1.** The integration by parts formula is obtained by reversing the _____ Rule. The formula for indefinite integrals is $\int u\, dv =$ _____. In choosing $u$ and $dv$, we want $du$ to be simpler than _____ and $dv$ to be _____ _____. The formula for definite integrals is $\int_a^b f(x)g'(x)\, dx =$ _____.

**2.** To integrate $\sin^m x \cos^n x$, where $m$ and $n$ are positive integers, we use the substitution (a) $u =$ _____ if $m$ is _____ and (b) $u =$ _____ if $n$ is _____. If $m$ and $n$ are both even and nonnegative, we use the half-angle formulas $\sin^2 x =$ _____ and $\cos^2 x =$ _____.

**3.** To integrate $\tan^m x \sec^n x$, where $m$ and $n$ are positive integers, we use the substitution (a) $u =$ _____ if $m$ is _____ and (b) $u =$ _____ if $n$ is _____.

**4.** To integrate $\sin mx \sin nx$, we use the identity $\sin mx \sin nx =$ _____; to integrate $\sin mx \cos nx$, we use the identity $\sin mx \cos nx =$ _____; to integrate $\cos mx \cos nx$, we use the identity $\cos mx \cos nx =$ _____.

**5. a.** If an integral involves $\sqrt{a^2 - x^2}$, we use the substitution $x =$ _____.
   **b.** If an integral involves $\sqrt{a^2 + x^2}$, we use the substitution $x =$ _____.
   **c.** If an integral involves $\sqrt{x^2 - a^2}$, we use the substitution $x =$ _____.

**6.** The method of partial fractions is used to integrate _____ functions. As a first step, the integrand $f(x) = P(x)/Q(x)$ should be written as $f(x) = S(x) + R(x)/Q(x)$ where the degree of $R$ is _____ than the degree of $Q$. $R(x)/Q(x)$ is decomposed into a sum of partial fractions involving _____ and irreducible _____ factors. As an example, the form of the decomposition for $\dfrac{2x^4 + 3x^2 + 8x + 5}{(x-1)^3(x^2+x+1)^2}$ is _____. The integral of $f$ is then found by _____ this last expression.

**7.** The improper integrals are defined by $\int_{-\infty}^b f(x)\,dx =$ _____; $\int_a^\infty f(x)\,dx =$ _____; $\int_{-\infty}^\infty f(x)\,dx =$ _____. If $\lim_{x\to b^-} f(x) = \pm\infty$, where $a < b$, then the improper integral $\int_a^b f(x)\,dx =$ _____. If $f$ is continuous on $[a, b]$ except that $f$ has an infinite discontinuity at $c$, where $a < c < b$, then $\int_a^b f(x)\,dx =$ _____.

**8.** If $f$ and $g$ are continuous and $f(x) \geq g(x) \geq 0$ for all $x \geq a$, then if $\int_a^\infty f(x)\,dx$ converges, _____ converges and if $\int_a^\infty g(x)\,dx$ diverges, _____ diverges.

## REVIEW EXERCISES

*In Exercises 1–42, evaluate or find the integral.*

**1.** $\displaystyle\int \frac{2x}{x+1}\,dx$

**2.** $\displaystyle\int x^2 \cos 3x\,dx$

**3.** $\displaystyle\int \frac{x^3}{\sqrt{9-x^2}}\,dx$

**4.** $\displaystyle\int \frac{\cos^3 x}{\sin x}\,dx$

**5.** $\displaystyle\int x^2 \ln x\,dx$

**6.** $\displaystyle\int \frac{2x-1}{x(x^2-4)}\,dx$

**7.** $\displaystyle\int \frac{1}{1-\cos\theta}\,d\theta$

**8.** $\displaystyle\int \frac{1}{1-\sin x}\,dx$

**9.** $\displaystyle\int \frac{x+1}{x^4+6x^3+9x^2}\,dx$

**10.** $\displaystyle\int \sec^3 x \tan^5 x\,dx$

**11.** $\displaystyle\int \sqrt{x^2-4}\,dx$

**12.** $\displaystyle\int_0^1 \cos^4 \pi x\,dx$

**13.** $\displaystyle\int \theta \sin^{-1}\theta\,d\theta$

**14.** $\displaystyle\int \sqrt{4+x^2}\,dx$

**15.** $\displaystyle\int_1^{e^\pi} \cos(\ln x)\,dx$

**16.** $\displaystyle\int \frac{x^2+4x}{x^3-x^2+x-1}\,dx$

**17.** $\displaystyle\int \frac{x+2}{(x^2+x)(x^2+1)}\,dx$

**18.** $\displaystyle\int_1^\infty \sin(\ln x)\,dx$

**19.** $\displaystyle\int \sec^4 2x \tan^6 2x\,dx$

**20.** $\displaystyle\int_0^2 \sqrt{4x-x^2}\,dx$

**21.** $\displaystyle\int \frac{\cos x}{1+\cos x}\,dx$

**22.** $\displaystyle\int e^x \cos 2x\,dx$

**23.** $\displaystyle\int \frac{(\ln x)^3}{x}\,dx$

**24.** $\displaystyle\int \frac{1}{3\sin\theta+4\cos\theta}\,d\theta$

**25.** $\displaystyle\int \frac{1}{x\sqrt{4x-1}}\,dx$

**26.** $\displaystyle\int \tan^3 x \sec^3 x\,dx$

**27.** $\displaystyle\int \sec^2 x \ln(\tan x)\,dx$

**28.** $\displaystyle\int \frac{1}{x^2+4x+20}\,dx$

**29.** $\displaystyle\int \sin x \cos 3x\,dx$

**30.** $\displaystyle\int \cosh^{-1} x\,dx$

**31.** $\displaystyle\int \frac{1}{\sqrt{1-(2x+3)^2}}\,dx$

**32.** $\displaystyle\int \frac{\sqrt{x^2+4}}{x^2}\,dx$

**33.** $\displaystyle\int \sin^2 t \cos^4 t\,dt$

**34.** $\displaystyle\int x \cos^2 x\,dx$

**35.** $\displaystyle\int \frac{\sqrt{x}}{\sqrt{x}-1}\,dx$

**36.** $\displaystyle\int (x+1)e^{2x}\,dx$

**37.** $\displaystyle\int x \cos^{-1} 2x\,dx$

**38.** $\displaystyle\int_1^4 \frac{e^{1/x}}{x^2}\,dx$

**39.** $\displaystyle\int e^{-x} \cosh x\,dx$

**40.** $\displaystyle\int \csc^4 2x\,dx$

**41.** $\displaystyle\int \frac{1}{\sqrt{4x^2+4x+10}}\,dx$

**42.** $\displaystyle\int_0^1 \frac{(\sin^{-1} x)^{3/2}}{\sqrt{1-x^2}}\,dx$

*In Exercises 43–50, evaluate the integral or show that it is divergent.*

**43.** $\displaystyle\int_{-\infty}^0 e^x\,dx$

**44.** $\displaystyle\int_0^\infty \frac{1}{(x+1)^{3/2}}\,dx$

**45.** $\displaystyle\int_{-\infty}^\infty \frac{x}{1+x^2}\,dx$

**46.** $\displaystyle\int_0^3 \frac{1}{\sqrt{3-x}}\,dx$

**47.** $\displaystyle\int_{-8}^1 \frac{1}{\sqrt[3]{x}}\,dx$

**48.** $\displaystyle\int_e^\infty \frac{1}{x \ln^4 x}\,dx$

**49.** $\displaystyle\int_1^e \frac{1}{x(\ln x)^{1/3}}\,dx$

**50.** $\displaystyle\int_0^2 \frac{x}{\sqrt{4-x^2}}\,dx$

*In Exercises 51–56, use the Table of Integrals to find the integral.*

**51.** $\displaystyle\int x^2\sqrt{3+x^2}\,dx$

**52.** $\displaystyle\int e^{2x}\sqrt{5+2e^x}\,dx$

**53.** $\displaystyle\int \frac{dx}{(x+1)\ln(1+x)}$

**54.** $\displaystyle\int (\ln x)^3\,dx$

**55.** $\displaystyle\int \sec^4 x\,dx$

**56.** $\displaystyle\int \frac{\tan x}{\sqrt{1+2\cos x}}\,dx$

**57.** Find $\int e^x f(x)\, dx + \int f'(x)\, e^x\, dx$, where $f'$ is continuous.

**58.** Find the area of the region under the graph of

$$y = \frac{\sqrt{4 + x^2}}{x^2}$$

on the interval $[1, 2]$.

**59.** Find the area of the region bounded by the graphs of $y = 1/x^{2/3}$, $y = 0$, $x = -1$, and $x = 1$.

**60.** Find the area of the region bounded by the graphs of $y = \sin^2 x$, $y = \sin^3 x$, $x = 0$, and $x = \pi$.

**61.** Find the area of the region enclosed by the ellipse $9x^2 + 4y^2 = 36$.

**62.** Let $I = \displaystyle\int_{-4}^{4} \frac{x^2}{\sqrt{16 - x^2}}\, dx$.

 **a.** Plot the graph of $f(x) = \dfrac{x^2}{\sqrt{16 - x^2}}$ using the viewing window $[-5, 5] \times [0, 20]$.

    **b.** Evaluate $I$ using the Table of Integrals.

**63.** Consider the integral $I = \displaystyle\int_{1}^{\infty}\left(1 - \cos\frac{2}{x}\right) dx$.

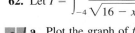 **a.** Plot the graphs of $f(x) = 1 - \cos(2/x)$ and $g(x) = (2/x^2)$ using the viewing window $[0, 5] \times [-1, 3]$ to see that $f(x) \le g(x)$ for all $x$ in $(0, \infty)$.

    **b.** Prove the assertion in part (a).

    **c.** Prove that $I$ converges.

**64.** The region under the graph of $y = \tan x$ on the interval $\left[0, \frac{\pi}{4}\right]$ is revolved about the x-axis. Find the volume of the resulting solid.

**65.** The region under the graph of $y = x \ln x$ on the interval $[1, e]$ is revolved about the y-axis. Find the volume of the resulting solid.

**66.** The region under the graph of $y = \tan^{-1} x$ on the interval $[0, 1]$ is revolved about the y-axis. Find the volume of the resulting solid.

**67.** Find the length of the graph of $y = \frac{1}{2}x^2$ from $(0, 0)$ to $\left(\sqrt{3}, \frac{3}{2}\right)$.

**68.** Use the Comparison Test to determine whether the integral

$$\int_{1}^{\infty} \frac{1 + 2\cos x}{x^3 + \sqrt{x}}\, dx$$

is convergent.

**69.** **Velocity of a Dragster** The velocity of a dragster $t$ sec after leaving the starting line is $v(t) = 80te^{-0.2t}$ ft/sec. What is the distance traveled by the dragster during the first 10 sec?

**70.** **Drug Concentration in the Bloodstream** The concentration of a certain drug (in mg/mL) in the bloodstream of a patient $t$ hours after it has been administered is given by

$$C(t) = 2te^{-t/3}$$

What is the average concentration of the drug in the patient's bloodstream over the first 12 hr after administration of the drug?

# PROBLEM-SOLVING TECHNIQUES

The following example shows that by making a suitable substitution, we can sometimes evaluate a definite integral whose indefinite integral cannot be expressed in terms of elementary functions, that is, as a sum, difference, product, quotient, or composition of the functions we have studied thus far.

**EXAMPLE 1** Evaluate $I = \displaystyle\int_{0}^{\pi} \frac{x \sin x}{1 + \cos^2 x}\, dx$.

**Solution** Let $u = \pi - x$ or $x = \pi - u$. Then $du = -dx$. Furthermore, if $x = 0$, then $u = \pi$, and if $x = \pi$, then $u = 0$. Making these substitutions, we have

$$I = \int_{0}^{\pi} \frac{x \sin x}{1 + \cos^2 x}\, dx = -\int_{\pi}^{0} \frac{(\pi - u)\sin(\pi - u)}{1 + \cos^2(\pi - u)}\, du$$

$$= \int_{0}^{\pi} \frac{(\pi - u)\sin(\pi - u)}{1 + \cos^2(\pi - u)}\, du$$

Next, we observe that

$$\sin(\pi - u) = \sin \pi \cos u - \cos \pi \sin u = \sin u$$

$$\cos(\pi - u) = \cos \pi \cos u + \sin \pi \sin u = -\cos u$$

and this leads to

$$I = \int_0^\pi \frac{(\pi - u)\sin u}{1 + \cos^2 u}\, du = \pi \int_0^\pi \frac{\sin u}{1 + \cos^2 u}\, du - \int_0^\pi \frac{u \sin u}{1 + \cos^2 u}\, du$$

But the second integral on the right-hand side is the same as $I$. Therefore, we have

$$2I = \pi \int_0^\pi \frac{\sin u}{1 + \cos^2 u}\, du$$

or

$$I = \frac{\pi}{2} \int_0^\pi \frac{\sin u}{1 + \cos^2 u}\, du$$

In this form, $I$ is easily evaluated. In fact, letting $t = \cos u$ so that $dt = -\sin u\, du$ and observing that if $u = 0$, then $t = 1$, and if $u = \pi$, then $t = -1$, we have

$$I = -\frac{\pi}{2} \int_1^{-1} \frac{dt}{1 + t^2} = -\frac{\pi}{2} \tan^{-1} t \Big|_1^{-1} = -\frac{\pi}{2} [\tan^{-1}(-1) - \tan^{-1}(1)]$$

$$= -\frac{\pi}{2}\left(-\frac{\pi}{4} - \frac{\pi}{4}\right) = \frac{\pi^2}{4} \qquad\qquad \blacksquare$$

If you look at the result of finding the integral of the form $\int P(x)e^{ax}\, dx$, where $P$ is a polynomial function and $a$ is a constant, you will see that $\int P(x)e^{ax}\, dx = Q(x)e^{ax} + C$, where $Q$ is a polynomial having the same degree as that of $P$. A similar observation reveals that using the integration by parts formula gives

$$\int P(x)\sin ax\, dx = P_1(x)\sin ax + Q_1(x)\cos ax + C$$

and

$$\int P(x)\cos ax\, dx = P_2(x)\cos ax + Q_2(x)\sin ax + C$$

where $P_1$, $P_2$, $Q_1$, and $Q_2$ are polynomial functions having the same degree as that of $P$.

These observations reduce the problem of finding integrals of the aforementioned form to one of solving an algebraic problem: that of solving a system of equations for the "undetermined" coefficients of a polynomial or polynomials.

**EXAMPLE 2**  Find $I = \int (2x^3 - 3x^2 + 8)e^{2x}\, dx$.

**Solution**  $\int (2x^3 - 3x^2 + 8)e^{2x}\, dx = (Ax^3 + Bx^2 + Dx + E)e^{2x} + C$. Differentiating both sides of the equation with respect to $x$ yields

$$(2x^3 - 3x^2 + 8)e^{2x} = (3Ax^2 + 2Bx + D)e^{2x} + 2(Ax^3 + Bx^2 + Dx + E)e^{2x}$$

$$2x^3 - 3x^2 + 8 = 2Ax^3 + (3A + 2B)x^2 + (2B + 2D)x + (D + 2E)$$

Since this equation holds for all values of $x$, the coefficients of like terms must be equal. This observation leads to the system

$$2A \qquad\qquad = 2$$
$$3A + 2B \qquad\quad = -3$$
$$2B + 2D \quad = 0$$
$$D + 2E = 8$$

Solving this system, we find $A = 1$, $B = -3$, $D = 3$, and $E = \frac{5}{2}$. So

$$\int (2x^3 - 3x^2 + 8)\, e^{2x}\, dx = \left(x^3 - 3x^2 + 3x + \frac{5}{2}\right)e^{2x} + C \qquad\blacksquare$$

## CHALLENGE PROBLEMS

1. Evaluate $\displaystyle\lim_{n\to\infty} \frac{\sqrt[n]{n!}}{n}$.

   Hint: Take logarithms and interpret the sum as an integral.

2. Find $\int \ln(\sqrt{1-x} + \sqrt{1+x})\, dx$.

3. **a.** Show that

$$f(x) = \begin{cases} x \ln\left(1 + \dfrac{1}{x}\right) & \text{if } 0 < x \le 1 \\ 0 & \text{if } x = 0 \end{cases}$$

   is continuous on $[0, 1]$.
   **b.** Evaluate $\int_0^1 f(x)\, dx$.

4. Let $I_n = \int (a^2 - x^2)^n\, dx$, where $n > 0$.

   **a.** Show that $I_n = \dfrac{x(a^2 - x^2)^n}{2n + 1} + \dfrac{2na^2}{2n + 1} I_{n-1}$.

   **b.** Use the result of part (a) to show that

$$\int \sqrt{a^2 - x^2}\, dx = \frac{x}{2}\sqrt{a^2 - x^2} + \frac{a^2}{2}\arcsin\frac{x}{a} + C$$

5. Find

$$\lim_{n\to\infty}\left(\frac{1}{\sqrt{4n^2 - 1}} + \frac{1}{\sqrt{4n^2 - 2^2}} + \cdots + \frac{1}{\sqrt{4n^2 - n^2}}\right)$$

   Hint: Interpret the sum as an integral.

6. Find

$$\int \frac{dx}{(1 + \sqrt{x})\sqrt{x - x^2}}$$

   Hint: Use the substitution $u = \sin^{-1}\sqrt{x}$.

7. Prove that $\left|\displaystyle\int_0^a \frac{\cos bx}{1 + x^2}\, dx\right| < \frac{\pi}{2}$.

8. Find the area of the region lying between the cissoid

$$y^2 = \frac{x^3}{2a - x} \quad \text{and its asymptote.}$$

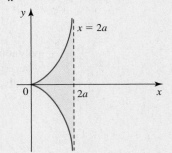

9. Prove that

$$\int_0^a (a^2 - x^2)^n\, dx = \frac{2^{2n}a^{2n+1}(n!)^2}{(2n + 1)!}$$

   where $n$ is a positive integer.
   Hint: Denote the integral by $I_n$, and show that

$$I_n = a^2\,\frac{2n}{2n + 1}\, I_{n-1}$$

10. Find the area enclosed by the ellipse with equation $2x^2 + \sqrt{3}xy + y^2 = 20$ as shown in the figure.

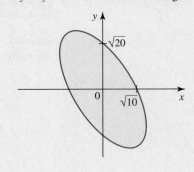

**11.** Show that

$$\int_0^1 x^m (\ln x)^n \, dx = \frac{(-1)^n n!}{(m+1)^{n+1}}$$

where $n$ is a positive integer and $m > -1$.

**12.** Show that $\displaystyle \int_0^\infty \frac{\ln x}{1 + x^2} \, dx = 0$.

**Hint:** Write

$$\int_0^\infty \frac{\ln x}{1 + x^2} \, dx = \int_0^1 \frac{\ln x}{1 + x^2} \, dx + \int_1^\infty \frac{\ln x}{1 + x^2} \, dx$$

and use the substitution $u = 1/x$ on the second integral on the right.

**13.** Suppose that $f$ is continuous on $(-\infty, \infty)$ and $\int_{-\infty}^\infty |f(x)| \, dx$ exists.
   **a.** Show that $\int_{-\infty}^\infty f(x) \, dx$ exists.
   **b.** If $g$ is continuous and bounded on $(-\infty, \infty)$, that is, there exists a positive number $M$ such that $|g(x)| \le M$ for all $x$ in $(-\infty, \infty)$, show that $\int_{-\infty}^\infty f(x) \, g(x) \, dx$ exists.
   **c.** Use the result of part (b) to show that $\displaystyle \int_{-\infty}^\infty \frac{\sin x}{x^2 + 1} \, dx = 0$.

**14.** Find the area of the surface obtained by revolving the ellipse $4x^2 + y^2 = 4$ about the $y$-axis.

**15.** Consider the Dirichlet integral $\displaystyle \int_0^\infty \frac{\sin x}{x} \, dx$, which may be written as the sum of the integrals

$$I_1 = \int_0^{\pi/2} \frac{\sin x}{x} \, dx \quad \text{and} \quad I_2 = \int_{\pi/2}^\infty \frac{\sin x}{x} \, dx$$

   **a.** Show that $I_1$ exists and is finite.
   **b.** Show that $I_2$ converges.
   **c.** Conclude that the Dirichlet integral converges.

**16.** The integrals $\int_0^\infty \sin x^2 \, dx$ and $\int_0^\infty \cos x^2 \, dx$ are called Fresnel integrals and are used in the study of light diffraction. Show that a Fresnel integral is convergent.
   **Hint:** Use the substitution $u = x^2$ and write the resulting integral as the sum of two integrals as in Exercise 15.
   **Note:** A Fresnel integral shows that an improper integral can converge even though the integrand does not approach zero as $x$ approaches infinity.

**17. a.** Use the result of Exercise 16 to show that the integral $\int_0^\infty 2x \cos x^4 \, dx$ converges.
   **Hint:** Use the substitution $x^2 = u$.
   **b.** Show that the integrand $f(x) = 2x \cos x^4$ is unbounded.
   **Note:** This integral shows that an improper integral can converge even though the integrand is unbounded.

**18. The Path of a Water Skier** A water skier is pulled along by means of a 40-ft tow rope attached to a boat. Initially, the boat is located at the origin and the skier is located at the point $(40, 0)$. As the boat moves along the $y$-axis, the tow rope is kept taut at all times. The path followed by the skier is a curve called a *tractrix* and has the property that the rope is tangent to the curve.

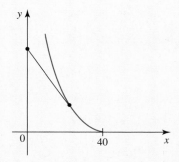

   **a.** Show that the path followed by the skier is the graph of $y = f(x)$, where $y$ satisfies the equation

$$\frac{dy}{dx} = -\frac{\sqrt{1600 - x^2}}{x}$$

   **b.** Solve the equation in part (a) to show that the path followed by the skier is

$$y = -\sqrt{1600 - x^2} + 40 \ln\left[\frac{40 + \sqrt{1600 - x^2}}{x}\right]$$

**19.** Refer to Exercise 18. Find the distance covered by the water skier after the boat has traveled 100 ft from its starting point, which is located at the origin.

Once a skydiver jumps out of a plane, the force of gravity acts on the skydiver, accelerating her fall to earth. But air resistance builds up quickly as she falls and soon matches the force due to gravity. The result is that her rate of fall approaches a constant (maximum) rate, called the *terminal velocity*. We will see how the motion of the skydiver is described by the solution of a differential equation.

Steve Fitchett/Getty Images

# 7  Differential Equations

**A DIFFERENTIAL EQUATION** is one that involves the derivative, or differential, of one or more unknown functions. In this chapter we give a brief introduction to the all-important field of differential equations by looking at *first-order* differential equations and their applications. The applications of differential equations are many and varied and appear in virtually every field of study. Examples are the study of motion, population growth, radioactive decay, calculations involving compound interest, electrostatic and electromagnetic fields, carbon dating, chemical reactions, concentration of a drug in the bloodstream, and the spread of a disease. We also take a brief look at an important application involving a system of differential equations: the predator-prey problem, which deals with how one population (predator) affects another (prey).

V This symbol indicates that one of the following video types is available for enhanced student learning at **www.academic.cengage.com/login:**
• Chapter lecture videos         • Solutions to selected exercises

## 7.1 Differential Equations: Separable Equations

We first encountered differential equations in Section 4.1, and you might want to review the material there before proceeding.

### ■ First-order Differential Equations and Solutions

Recall that a **differential equation** is an equation that involves the derivative or differential of an unknown function. The **order** of a differential equation is the order of the highest derivative that occurs in the equation. Thus, a **first-order differential equation** is one that involves only derivatives of order one. If we solve the equation for the derivative, it can be written in the form

$$\frac{dy}{dx} = f(x, y) \tag{1}$$

In Section 4.1 we considered first-order differential equations of the form

$$\frac{dy}{dx} = f(x)$$

where $f$ is a function of $x$ alone and, therefore, can be solved by integration. In fact, as we have seen, the general solution of this equation is

$$y = \int f(x)\, dx + C$$

where $C$ is an arbitrary constant.

In the general case in which $f$ involves both $x$ and $y$, the solution is not so easily obtained, and more sophisticated methods of solution are needed. Before going further, let's recall that a **solution** of the differential equation (1) is a differentiable function $y = y(x)$ defined on an open interval $I$ such that $y$ satisfies the equation on $I$. Thus, for Equation (1) a **solution** is a function $y(x)$ that satisfies

$$\frac{d}{dx}[y(x)] = f(x, y(x))$$

for all $x$ in the interval.

**EXAMPLE 1** Show that the function $y = x + 1 + Ce^x$, where $C$ is an arbitrary constant, is a solution of the differential equation $y' = y - x$.

**Solution** The function $y = x + 1 + Ce^x$ is defined and differentiable on the interval $(-\infty, \infty)$. To show that the differential equation is satisfied, we compute

$$y' = \frac{d}{dx}(x + 1 + Ce^x) = 1 + Ce^x$$

Next, substituting the expression for $y$ into the right-hand side of the differential equation gives

$$y - x = x + 1 + Ce^x - x = 1 + Ce^x$$

which is the same as the left-hand side of the differential equation for all values of $x$ in $(-\infty, \infty)$. Therefore, $y = x + 1 + Ce^x$ is a solution of the given differential equation on $(-\infty, \infty)$. ■

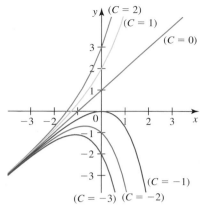

**FIGURE 1**
Some solution curves of $y' = y - x$

In general, a first-order differential equation will have a solution involving one arbitrary constant. Such a solution is called the *general solution* of the differential equation. For example, the solution $y = x + 1 + Ce^x$ is called the general solution of the equation $y' = y - x$. Graphically, the general solution of the differential equation represents a family of curves called the **solution** or **integral curves** of the differential equation. Figure 1 shows six solution curves of the differential equation for selected values of the parameter $C$.

We can obtain a *particular solution* of a differential equation by choosing a particular value of the arbitrary constant. This is usually done by requiring that the differential equation satisfy a side condition $y(x_0) = y_0$. Geometrically, the solution of the **initial-value problem**

$$\begin{cases} \dfrac{dy}{dx} = f(x, y) \\ y(x_0) = y_0 \end{cases}$$

is the solution curve of the differential equation that passes through the point $(x_0, y_0)$.

**EXAMPLE 2** Solve the initial-value problem

$$\begin{cases} y' = y - x \\ y(0) = 0 \end{cases}$$

**Solution** In Example 1 we saw that $y = x + 1 + Ce^x$ is a one-parameter solution of the equation $y' = y - x$. To determine $C$, we use the initial condition $y(0) = 0$, or $y = 0$, when $x = 0$. We obtain

$$0 = 0 + 1 + Ce^0 \qquad \text{or} \qquad C = -1$$

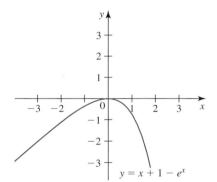

**FIGURE 2**
The particular solution of $y' = y - x$
satisfying $y(0) = 0$

Therefore, the required solution is $y = x + 1 - e^x$. The graph of this function is shown in Figure 2. If you compare this with the family of solution curves of this differential equation shown in Figure 1, you will see that the particular solution obtained here is the solution curve that passes through the origin $(0, 0)$. ∎

## ■ The Laws of Natural Growth and Decay

When free of constraints, certain quantities in nature grow or decay at a rate that is proportional to their current size. Examples of such phenomena are

- the growth of a population of bacteria under ideal conditions,
- the decay of a radioactive substance, and
- the discharge of an electrical condenser.

Even quantities that occur outside the realm of nature sometimes exhibit this type of growth or decay. For example, the accumulated amount of money on deposit with a bank, earning interest at a fixed rate compounded continuously, grows in this manner (see Example 9).

We can describe these phenomena mathematically. Suppose that the size or magnitude of a quantity $y$ at any time $t$ is given by $y(t)$.* Since the rate of change of $y$ with respect to $t$, $dy/dt$, is proportional to its size $y$ at any time $t$, we have

$$\frac{dy}{dt} = ky \tag{2}$$

---

*We use the letter $y$ to denote the function in question.

where $k$ is a constant. If $k > 0$, this equation is called the **law of natural growth;** and if $k < 0$, it is called the **law of natural decay.** Also, since the equation involves the derivative of an unknown function, it is a differential equation. If the amount present initially is $y_0$, then we have the initial-value problem

$$\begin{cases} \dfrac{dy}{dt} = ky \\ y(0) = y_0 \end{cases} \tag{3}$$

The differential equation $dy/dt = ky$ is an example of a **first-order separable differential equation.** It is separable because $dy/dt$ can be written as a function of $t$ times a function of $y$. In general, a first-order separable differential equation in $x$ and $y$ is one that can be written in the form

$$\frac{dy}{dx} = g(x)h(y) \tag{4}$$

where $g$ is a function of $x$ alone and $h$ is a function of $y$ alone. Equivalently, a separable equation is one that can be written in differential form as

$$G(x)\,dx + H(y)\,dy = 0 \tag{5}$$

where $G$ is a function of $x$ alone and $H$ is a function of $y$ alone. For example, the differential equation in system (3) is easily seen to be separable if we put $g(t) = k$ and $h(y) = y$.

As another example, the equation

$$(x^2 + 1)\,dx + \frac{1}{y}\,dy = 0$$

has the form of the differential equation in (5) with $G(x) = x^2 + 1$ and $H(y) = 1/y$, so it is separable. On the other hand, the differential equation

$$\frac{dy}{dx} = xy^2 + 2$$

is *not* separable, nor is the equation

$$(x + y)\,dx + xy\,dy = 1$$

We will return to the solution of the initial-value problem (3) later on.

## ▪ The Method of Separation of Variables

First-order separable differentiable equations can be solved by using the *method of separation of variables.* If $h(y) \neq 0$, we write differential equation (4) in the form

$$\frac{1}{h(y)}\frac{dy}{dx} = g(x)$$

When it is written in this form, the variables $x$ and $y$ are said to be separated. Integrating both sides of the equation with respect to $x$ then gives

$$\int \frac{1}{h(y)}\frac{dy}{dx}\,dx = \int g(x)\,dx$$

or

$$\int \frac{1}{h(y)}\,dy = \int g(x)\,dx \tag{6}$$

Carrying out the integration on each side of Equation (6) with respect to the appropriate variable gives the solution to the differential equation expressed implicitly by an equation in $x$ and $y$. In some cases we may be able to solve for $y$ explicitly in terms of $x$.

To justify the method of separation of variables, let's consider the separable Equation (4) in the general form:

$$\frac{dy}{dx} = g(x)h(y)$$

If $h(y) \neq 0$, we may rewrite the equation in the form

$$\frac{1}{h(y)} \frac{dy}{dx} - g(x) = 0$$

Now, suppose that $H$ is an antiderivative of $1/h$ and $G$ is an antiderivative of $g$. Using the chain rule, we see that

$$\frac{d}{dx}[H(y) - G(x)] = H'(y)\frac{dy}{dx} - G'(x) = \frac{1}{h(y)}\frac{dy}{dx} - g(x)$$

Therefore,

$$\frac{d}{dx}[H(y) - G(x)] = 0$$

and so

$$H(y) - G(x) = C \qquad C, \text{ a constant}$$

But the last equation is equivalent to

$$H(y) = G(x) + C \qquad \text{or} \qquad \int \frac{dy}{h(y)} = \int g(x)\,dx$$

which is precisely Equation (6).

**EXAMPLE 3** Solve the differential equation $\dfrac{dy}{dx} = \dfrac{y}{x}$.

**Solution** First, observe that $y = 0$ is a solution of the separable equation. To find other solutions, assume that $y \neq 0$, separate variables, and integrate each side of the resulting equation with respect to the appropriate variable to obtain

$$\int \frac{dy}{y} = \int \frac{dx}{x}$$

or

$$\ln|y| = \ln|x| + \ln|C_1|$$

where $C_1$ is a nonzero but otherwise arbitrary constant. We write $\ln|C_1|$ rather than an arbitrary constant $C_2$ because we can then more readily apply the laws of logarithms. Proceeding, we have

$$\ln|y| = \ln(|C_1| \cdot |x|) = \ln|C_1 x|$$

or, upon exponentiating using the base $e$,

$$|y| = |C_1 x| \qquad C_1 \neq 0$$

$$y = \pm C_1 x$$

Since $y = 0$ is also a solution, we see that the general solution can be written in the form $y = Cx$, where $C$ is an arbitrary constant. ∎

**EXAMPLE 4**  Solve the differential equation $y' = \dfrac{xy}{x^2 + 1}$.

**Solution**  First, observe that $y = 0$ is a solution of the differential equation. Next, write the given equation in the form

$$\frac{dy}{dx} = \left( \frac{x}{x^2 + 1} \right) y = g(x) h(y)$$

which is separable. Separating variables and integrating, we have

$$\int \frac{dy}{y} = \int \frac{x}{x^2 + 1} \, dx$$

$$\ln|y| = \frac{1}{2} \ln(x^2 + 1) + \ln|C_1| \qquad C_1 \neq 0$$

where $\ln|C_1|$ represents the combined constants of integration. We have

$$\ln|y| = \ln(x^2 + 1)^{1/2} + \ln|C_1|$$

$$= \ln|C_1 \sqrt{x^2 + 1}|$$

or

$$y = \pm C_1 \sqrt{x^2 + 1}$$

Since $y = 0$ is also a solution of the differential equation, we conclude that the general solution is

$$y = C \sqrt{x^2 + 1}$$

where $C$ is an arbitrary constant. Figure 3 shows six solution curves of the differential equation for selected values of $C$. ∎

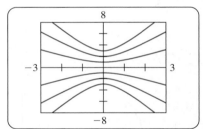

**FIGURE 3**
Solution curves for $y' = \dfrac{xy}{x^2 + 1}$

for $C = \pm 1, \pm 2, \pm 3$

**EXAMPLE 5**  Find the particular solution of the differential equation

$$ye^x \, dx + (y^2 - 1) \, dy = 0$$

that satisfies the condition $y(0) = 1$.

**Solution**  The equation is separable. By inspection we see that $y = 0$ is a solution of the differential equation. But this solution does not satisfy the initial condition $y(0) = 1$ and is rejected. Next, suppose that $y \neq 0$. Separating variables and integrating, we have

$$\int \frac{y^2 - 1}{y} \, dy = -\int e^x \, dx$$

$$\int \left( y - \frac{1}{y} \right) dy = -\int e^x \, dx$$

$$\frac{1}{2} y^2 - \ln|y| = -e^x + C_1$$

$$y^2 - \ln y^2 = -2e^x + C_2 \qquad C_2 = 2C_1$$

Using the condition $y(0) = 1$, we have

$$1 - \ln 1 = -2 + C_2 \qquad \text{or} \qquad C_2 = 3$$

Therefore, the required solution is

$$y^2 - \ln y^2 = -2e^x + 3 \qquad \blacksquare$$

Note that the solution of the differential equation in Example 3 is obtained in the form of an implicit equation in $x$ and $y$.

We now turn to the solution of the initial-value problem (3) posed earlier:

$$\begin{cases} \dfrac{dy}{dt} = ky \\ y(0) = y_0 \end{cases}$$

The differential equation $dy/dt = ky$ is separable. Separating variables and integrating, we obtain

$$\int \frac{dy}{y} = \int k\,dt$$

$$\ln|y| = kt + C_1$$

$$|y| = e^{kt+C_1} = C_2 e^{kt} \qquad C_2 = e^{C_1}$$

$$y = \pm C_2 e^{kt}$$

$$= Ce^{kt} \qquad C = \pm C_2$$

Using the initial condition $y(0) = y_0$ gives $y_0 = C$. Therefore, the solution is $y = y_0 e^{kt}$.

---

**THEOREM 1   Natural Law of Exponential Growth (Decay)**

The initial-value problem

$$\begin{cases} \dfrac{dy}{dt} = ky \qquad k, \text{ a constant} \\ y(0) = y_0 \end{cases}$$

has the unique solution $y = y_0 e^{kt}$.

---

The solution curves for the initial-value problem are shown in Figure 4.

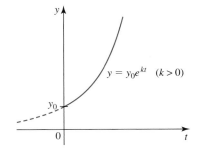

(a) Exponential growth $(k > 0)$

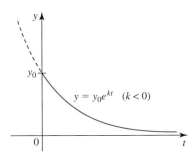

(b) Exponential decay $(k < 0)$

**FIGURE 4**
Graphs of $y = y_0 e^{kt}$

**EXAMPLE 6**  **Growth of Bacteria**   The population of bacteria in a culture grows at a rate that is proportional to the number present. Suppose that there are 1000 bacteria present initially in a culture and 3000 present 2 hr later. How many bacteria will there be in the culture after 4 hr?

**Solution**   Let $y = y(t)$ denote the number of bacteria present in the culture after $t$ hr. Then $y(0) = 1000$ and $y(2) = 3000$. Since the rate of growth of bacteria in the culture is proportional to the number present, the quantity $y$ satisfies the initial-value problem

$$\begin{cases} \dfrac{dy}{dt} = ky \\ y(0) = 1000 \end{cases}$$

By Theorem 1,

$$y(t) = y_0 e^{kt} = 1000 e^{kt}$$

Next, using the condition $y(2) = 3000$, we have

$$y(2) = 1000 e^{2k} = 3000$$

$$e^{2k} = \frac{3000}{1000} = 3$$

or

$$e^{k} = (e^{2k})^{1/2} = 3^{1/2}$$

Therefore, the number of bacteria present after $t$ hr is

$$y(t) = 1000 e^{kt} = 1000(e^{k})^{t} = 1000(3^{t/2})$$

In particular, the number of bacteria present in the culture after 4 hr is

$$y(4) = 1000(3^{4/2}) = 9000$$

Notice that it is not necessary to determine the value of $k$, the *growth constant*, which depends on the strain of the bacteria. However, if desired, its value can be found by solving the equation

$$e^{2k} = 3$$

Thus,

$$\ln e^{2k} = \ln 3$$

$$2k \ln e = \ln 3$$

$$2k = \ln 3$$

$$k = \frac{1}{2} \ln 3 \approx 0.5493$$

The graph of $y = 1000(3^{t/2})$, or $y = 1000 e^{0.5493t}$, is shown in Figure 5.    ■

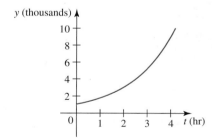

**FIGURE 5**
The graph of $y = 1000 e^{0.5493t}$ shows the population of bacteria at time $t$.

**EXAMPLE 7**  **Radioactive Decay**   Radioactive substances decay at a rate that is proportional to the amount present. The *half-life* of a substance is the time required for a given amount to be reduced by one-half. It is known that the half-life of radium-226 is approximately 1602 years. Suppose that initially there are 100 mg of radium in a sample.

**a.** Find a formula that gives the amount of radium-226 present after $t$ years.
**b.** Find the amount of radium-226 present after 1000 years.
**c.** How long will it take for the radium-226 to be reduced to 40 mg?

**Solution**

**a.** Let $y(t)$ denote the amount of radium-226 present in the sample after $t$ years. Then $y(0) = 100$ and $y(1602) = 50$. Since the rate of decay of the radium is proportional to the amount present, the quantity $y$ satisfies the initial-value problem

$$\begin{cases} \dfrac{dy}{dt} = ky \\ y(0) = 100 \end{cases}$$

By Theorem 1,

$$y(t) = y_0 e^{kt} = 100 e^{kt}$$

Next, we use the condition $y(1602) = 50$ to write

$$y(1602) = 100 e^{1602k} = 50$$

$$e^{1602k} = \frac{1}{2}$$

or

$$e^k = \left(\frac{1}{2}\right)^{1/1602}$$

Therefore, the amount of radium-226 present after $t$ years is

$$y(t) = 100 e^{kt} = 100(e^k)^t = 100\left(\frac{1}{2}\right)^{t/1602}$$

**b.** The amount of radium-226 present after 1000 years is given by

$$y(1000) = 100\left(\frac{1}{2}\right)^{1000/1602} \approx 64.88$$

or approximately 64.9 mg.

**c.** We want to find the value of $t$ such that $y(t) = 40$; that is,

$$100\left(\frac{1}{2}\right)^{t/1602} = 40$$

$$\left(\frac{1}{2}\right)^{t/1602} = \frac{40}{100} = \frac{2}{5}$$

Taking the natural logarithm on both sides, we obtain

$$\ln\left(\frac{1}{2}\right)^{t/1602} = \ln\frac{2}{5}$$

$$\frac{t}{1602}\ln\left(\frac{1}{2}\right) = \ln\frac{2}{5}$$

$$t = 1602\left(\frac{\ln\frac{2}{5}}{\ln\frac{1}{2}}\right) \approx 2118$$

So it will take approximately 2118 years for the radium-226 to decay to 40 mg. The graph of $y = 100\left(\frac{1}{2}\right)^{t/1602}$ is shown in Figure 6.

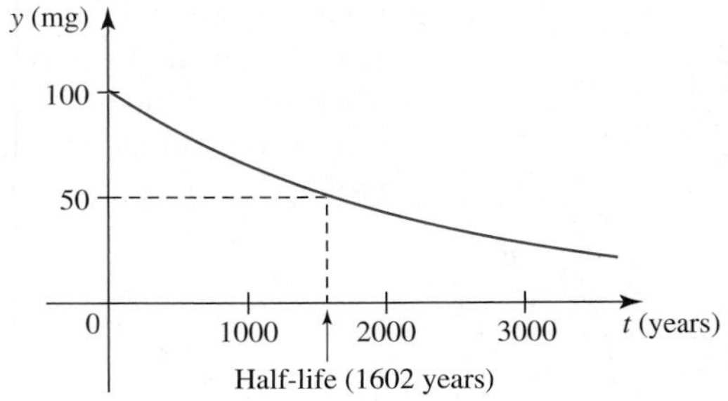

**FIGURE 6**
The graph of $y = 100\left(\frac{1}{2}\right)^{t/1602}$ shows
how the radium-226 decays.

**EXAMPLE 8    Newton's Law of Cooling**    Newton's Law of Cooling states that the temperature of a body drops at a rate that is proportional to the difference between the temperature of the body and the temperature of the surrounding medium. An apple pie is taken out of an oven at a temperature of 200°F and placed on the counter in a room where the temperature is 70°F. The temperature of the pie is 160°F after 15 min.

**a.** What is the temperature of the pie after 30 min?
**b.** How long will it take for the pie to cool to 120°F?

**Solution**

**a.** Let $y(t)$ denote the temperature of the apple pie $t$ min after it was placed on the counter. Then Newton's Law of Cooling gives

$$\frac{dy}{dt} = k(y - 70)$$

where $k$ is the constant of proportionality. The initial temperature of the pie is 200°F, and this translates into the condition $y(0) = 200$. So we have the initial-value problem

$$\begin{cases} \dfrac{dy}{dt} = k(y - 70) \\ y(0) = 200 \end{cases}$$

Observe that the differential equation here does not have the same form as that in Theorem 1. Let's define the function $u$ by $u(t) = y(t) - 70$. Then

$$\frac{du}{dt} = \frac{dy}{dt}$$

Making these substitutions, we obtain

$$\frac{du}{dt} = ku$$

Using Theorem 1, we obtain the solution

$$u(t) = u(0)e^{kt}$$

$$y(t) - 70 = [y(0) - 70]e^{kt} \qquad \text{Recall } u(t) = y(t) - 70.$$

or

$$y(t) = 70 + (200 - 70)e^{kt} = 70 + 130e^{kt}$$

Next, we use the condition $y(15) = 160$ to write

$$y(15) = 70 + 130e^{15k} = 160$$

$$130e^{15k} = 90$$

or

$$e^k = \left(\frac{9}{13}\right)^{1/15}$$

Therefore, the temperature of the pie after $t$ min is

$$y(t) = 70 + 130\left(\frac{9}{13}\right)^{t/15}$$

In particular, the temperature of the pie after 30 min is

$$y(30) = 70 + 130\left(\frac{9}{13}\right)^{30/15} = 70 + 130\left(\frac{9}{13}\right)^2 \approx 132.3$$

or approximately 132°F.

**b.** We need to find the value of $t$ for which $y(t) = 120$, that is,

$$70 + 130e^{kt} = 120$$

$$130e^{kt} = 50$$

$$(e^k)^t = \frac{5}{13}$$

$$\left(\frac{9}{13}\right)^{t/15} = \frac{5}{13} \qquad \text{Use the value of } e^k \text{ from part (a).}$$

Taking the natural logarithm on both sides, we obtain

$$\frac{t}{15}\ln\frac{9}{13} = \ln\frac{5}{13}$$

or

$$t = \frac{15\ln\frac{5}{13}}{\ln\frac{9}{13}} \approx 39$$

So it will take approximately 39 min for the pie to cool to 120°F. The graph of $y(t) = 70 + 130\left(\frac{9}{13}\right)^{t/15}$ is shown in Figure 7.

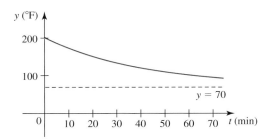

**FIGURE 7**
The graph of $y(t) = 70 + 130\left(\frac{9}{13}\right)^{t/15}$ gives the temperature of the pie as a function of time.

**Note** The differential equation $dy/dt = k(y - 70)$ in Example 8 is separable and can be solved directly without using Theorem 1.

**EXAMPLE 9** **Continuously Compounded Interest** Suppose that money deposited into a bank grows at a rate that is proportional to the amount accumulated. If the amount on deposit initially is $P$ dollars, find an expression for the accumulated amount $A$ after $t$ years. Reconcile your result with the continuous compound interest formula, $A = Pe^{rt}$.

**Solution** Since the rate of growth of the money is proportional to the amount present, we have the initial-value problem

$$\begin{cases} \dfrac{dA}{dt} = kA \\ A(0) = P \end{cases}$$

By Theorem 1,

$$A(t) = A(0)e^{kt} = Pe^{kt}$$

Therefore, the accumulated amount after $t$ years is given by

$$A(t) = Pe^{kt}$$

dollars.

If we compare this result with the formula $A = Pe^{rt}$, we see that the formulas are identical when the growth constant $k$ is taken to be equal to $r$, the nominal interest rate. This shows that money deposited into a bank with interest compounded continuously grows according to the law of natural growth. ■

Suppose all the curves of one family intersect all the curves of another family at right angles. Then the curves of the first family are said to be **orthogonal trajectories** of the other family, and vice versa. For example, the straight lines passing through the origin are the orthogonal trajectories of the concentric circles with center at the origin and vice versa. (See Figure 8.) Orthogonal trajectories occur in physics and engineering. For example, in electrostatics and electromagnetics the field lines are orthogonal to the equipotential curves.

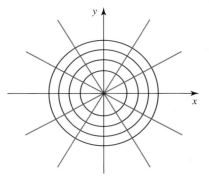

**FIGURE 8**
The concentric circles are the orthogonal trajectories of the lines, and vice versa.

**EXAMPLE 10** Find the orthogonal trajectories of the family of curves given by $y = Cx^2$, where $C$ is an arbitrary constant.

**Solution** First, recall that the slope of the tangent line to a curve in the given family at a point $(x, y)$ is given by $dy/dx$. To find $dy/dx$, we differentiate the given equation to obtain

$$\frac{dy}{dx} = 2Cx$$

Next, we eliminate $C$ from this equation. Solving for $C$ in the given equation gives $C = y/x^2$. Substituting this value of $C$ into the equation for $dy/dx$, we obtain

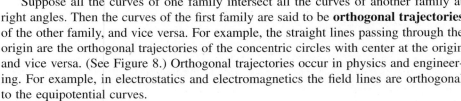

$$\frac{dy}{dx} = 2\left(\frac{y}{x^2}\right)x = \frac{2y}{x}$$

Since the required family is orthogonal to the given family, the slope of the tangent line to each member of the required family is given by the negative reciprocal of $2y/x$. Therefore, the orthogonal trajectories satisfy the differential equation

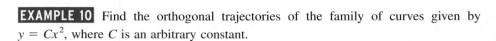

$$\frac{dy}{dx} = -\frac{x}{2y}$$

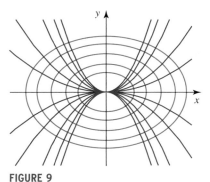

**FIGURE 9**
Each parabola is orthogonal to each
ellipse, and vice versa.

Separating variables and integrating with respect to the appropriate variable, we obtain

$$\int 2y \, dy + \int x \, dx = 0$$

$$y^2 + \frac{1}{2}x^2 = k$$

where $k$ is an arbitrary constant. We recognize this family to be a family of ellipses. (See Figure 9.) ∎

## 7.1 CONCEPT QUESTIONS

**1. a.** What is a first-order separable differential equation?
   **b.** Is the equation $f(x)g(y)dx + F(x)G(y)dy = 0$ separable? Explain.

**2. a.** Is $x\dfrac{dy}{dx} + y = e^y$ a separable differential equation? Explain.
   **b.** Is $(x^2 + y^2)dx + xy^3 dy = 4$ a separable differential equation? Explain.

**3.** Describe a method for solving a separable differential equation.

**4. a.** Write a differential equation to describe the natural law of exponential growth (decay).
   **b.** Write a differential equation to describe Newton's Law of Cooling.

## 7.1 EXERCISES

**1.** Show that $y = \dfrac{1}{2} + \dfrac{3}{x^2}$ is a solution of the differential equation $xy' + 2y = 1$ on any interval that does not contain $x = 0$.

**2.** Show that $y = Ce^{-2x} + e^x$ is a solution of the differential equation $y' + 2y = 3e^x$ on $(-\infty, \infty)$.

**3.** Show that $y = x^4 + 3x^3$ is a solution of the initial-value problem $xy' - 3y = x^4$, $y(1) = 4$ on $(-\infty, \infty)$.

**4.** Show that $y = \sin x - \cos x$ is a solution of the initial-value problem $\cos x \dfrac{dy}{dx} + y \sin x = 1$, $y(0) = -1$ on the interval $(-\infty, \infty)$.

**5.** Assume that the differential equation $y' = 3y$ has a solution of the form $y = Ce^{mx}$.
   **a.** Find the value of $m$.
   **b.** Plot the solution curves of $y' = 3y$ on the same set of axes for $C = -3, -2, -1, 0, 1, 2,$ and 3.
   **c.** Find the solution of $y' = 3y$ that satisfies the initial condition $y(0) = 2$. Is the solution curve for this solution among those in part (b)?

**6.** Suppose that a solution of the second-order differential equation $y'' - y' - 2y = 0$ has the form $y = e^{mx}$.
   **a.** Find an equation that $m$ must satisfy.
   **b.** Solve the equation found in part (a).
   **c.** Write two solutions of the differential equation.
   **d.** Verify the results of part (c) directly.

*In Exercises 7 and 8, the general solution of a differential equation is given. (a) Find the particular solution that satisfies the given initial condition. (b) Plot the solution curves corresponding to the given values of C. Indicate the solution curve that corresponds to the solution found in part (a).*

**7.** $x\dfrac{dy}{dx} + y = x^3$,  $y = \dfrac{C}{x} + \dfrac{x^3}{4}$;  $y(1) = \dfrac{5}{4}$;
$C = -2, -1, 0, 1, 2$

**8.** $y\dfrac{dy}{dx} - e^{2x} = 0$,  $y^2 = e^{2x} + C$;  $y(0) = 1$;
$C = -2, -1, 0, 1, 2$

 Videos for selected exercises are available online at **www.academic.cengage.com/login**.

*In Exercises 9–18, solve the differential equation.*

**9.** $\dfrac{dy}{dx} = \dfrac{2y}{x}$

**10.** $\dfrac{dy}{dx} = \dfrac{x+1}{y^2}$

**11.** $\dfrac{dy}{dx} = x^2 y$

**12.** $\dfrac{dy}{dx} = -\dfrac{xy}{x+1}$

**13.** $y' = \dfrac{2y+3}{x^2}$

**14.** $y' = e^{x-y}$

**15.** $\cos y \dfrac{dy}{dx} = \sec^2 x$

**16.** $(1-\cos\theta)\dfrac{dr}{d\theta} = r\sin\theta$

**17.** $xy' = y^2 \ln x$

**18.** $\dfrac{dy}{dt} = 1 + t + y + ty$

*In Exercises 19–26, solve the initial-value problem.*

**19.** $\dfrac{dy}{dx} = 3xy - 2x, \quad y(0) = 1$

**20.** $\dfrac{dy}{dx} = xe^{-y}, \quad y(0) = 1$

**21.** $y' = x^2 y^{-1/2}, \quad y(1) = 1$

**22.** $\dfrac{dy}{dx} = \dfrac{y^2}{x-2}, \quad y(3) = 1$

**23.** $y' = 3x^2 e^{-y}, \quad y(0) = 1$

**24.** $\sin^2 y\,dx + \cos^2 x\,dy = 0, \quad y\left(\dfrac{\pi}{4}\right) = \dfrac{\pi}{4}$

**25.** $\dfrac{dI}{dt} + 2I = 4, \quad I(0) = 0$

**26.** $\cos\theta \dfrac{du}{d\theta} = u\tan\theta, \quad u\left(\dfrac{\pi}{3}\right) = 2$

**27.** Find an equation defining a function $f$ given that (a) the slope of the tangent line to the graph of $f$ at any point $P(x, y)$ on the graph is given by
$$\frac{dy}{dx} = \frac{3x^2}{2y}$$
and (b) the graph of $f$ passes through the point $(1, 3)$.

**28.** Find an equation of a curve given that it passes through the point $\left(2, \dfrac{2\sqrt{5}}{3}\right)$ and that the slope of the tangent line to the curve at any point $P(x, y)$ is given by
$$\frac{dy}{dx} = -\frac{4x}{9y}$$

 *In Exercises 29–32, find the orthogonal trajectories of the family of curves. Use a graphing utility to draw several members of each family on the same set of axes.*

**29.** $xy = c$

**30.** $y^2 = cx^3$

**31.** $y = ce^x$

**32.** $y = \ln(cx)$

**33.** Find the constant $a$ such that the curves $x^2 + ay^2 = C_1$ and $y^3 = C_2 x$ are orthogonal trajectories of each other.

**34. Growth of Bacteria** The population of bacteria in a culture grows at a rate that is proportional to the number present. Initially, there are 600 bacteria, and after 3 hr there are 10,000 bacteria.
   **a.** What is the number of bacteria after $t$ hr?
   **b.** What is the number of bacteria after 5 hr?
   **c.** When will the number of bacteria reach 24,000?

**35. Growth of Bacteria** The population of bacteria in a culture grows at a rate that is proportional to the number present. After 2 hr there are 800 bacteria present. After 4 hr there are 3200 bacteria present. How many bacteria were there initially?

**36. Growth of Bacteria** The population of bacteria in a certain culture grows at a rate that is proportional to the number present. If the original population increases by 50% in $\frac{1}{2}$ hr, how long will it take for the population to triple in size?

**37. Lambert's Law of Absorption** According to Lambert's Law of Absorption, the percentage of incident light $L$, absorbed in passing through a thin layer of material $x$, is proportional to the thickness of the material. For a certain material, if $\frac{1}{2}$ in. of the material reduces the light to half of its intensity, how much additional material is needed to reduce the intensity to one fourth of its initial value?

**38. Savings Accounts** An amount of money deposited in a savings account grows at a rate proportional to the amount present. (Thus it earns interest compounded continuously (see Example 9).) Suppose that $10,000 is deposited in a fixed account earning interest at the rate of 10% compounded continuously.
   **a.** What is the accumulated amount after 5 years?
   **b.** How long does it take for the original deposit to double in value?

**39. Chemical Reactions** In a certain chemical reaction a substance is converted into another substance at a rate proportional to the square of the amount of the first substance present at any time $t$. Initially ($t = 0$), 50 g of the first substance was present; 1 hr later, only 10 g of it remained. Find an expression that gives the amount of the first substance present at any time $t$. What is the amount present after 2 hr?

**40. Radioactive Decay** Phosphorus-32 has a half-life of 14.3 days. If 100 g of this substance is present initially, find the amount present after $t$ days. What amount will be left after 7.1 days? How fast is the phosphorus-32 decaying when $t = 7.1$?

**41. Nuclear Fallout** Strontium-90, a radioactive isotope of strontium, is present in the fallout resulting from nuclear explosions. It is especially hazardous to animal life, including humans, because when contaminated food is ingested, the strontium-90 is absorbed into the bone structure. Its half-life is 28.9 years. If the amount of strontium-90 in a certain area is found to be four times the "safe" level, find how much time must elapse before an "acceptable" level is reached.

**42. Carbon-14 Dating** Wood deposits recovered from an archeological site contain 20% of the carbon-14 they originally contained. How long ago did the tree from which the wood was obtained die? (The half-life of carbon C-14 is 5730 years.)

**43. Carbon-14 Dating** Skeletal remains of the so-called Pittsburgh Man unearthed in Pennsylvania had lost 82% of the carbon-14 they originally contained. Determine the approximate age of the bones. (The half-life of carbon C-14 is 5730 years.)

**44. Newton's Law of Cooling** Newton's Law of Cooling states that the rate at which the temperature of an object changes is directly proportional to the difference in temperature between the object and that of the surrounding medium. A horseshoe that has been heated to a temperature of 600°C is immersed in a large tank of water at a (constant) temperature of 30°C at time $t = 0$. Two minutes later, the temperature of the horseshoe is reduced to 70°C. Derive an expression that gives the temperature of the horseshoe at any time $t$. What is the temperature of the horseshoe 3 min after it has been immersed in the water?

**45. Newton's Law of Cooling** Newton's Law of Cooling states that the rate at which the temperature of an object changes is directly proportional to the difference in temperature of the object and that of the surrounding medium. A cup of coffee is prepared with boiling water (212°F) and left to cool on a counter in a room where the temperature is 72°F. If the temperature of the coffee is 140°F after 2 min, when will the coffee will be cool enough to drink (say, 110°F)?

**46. Newton's Law of Heating** A thermometer is taken from a room where the temperature is 70°F to a patio. After 1 min the thermometer reads 50°F, and after 2 min it reads 40°F. What is the outdoor temperature?

**47. Motion of a Motorboat** A motorboat is traveling at a speed of 12 mph in calm water when its motor is cut off. Twenty seconds later, the boat's speed drops to 8 mph. Assuming that the water resistance on the boat is directly proportional to the speed of the boat, what will its speed be 2 min after the motor was cut off?

**48. Learning Curves** The American Court Reporter Institute finds that the average student taking elementary machine shorthand will progress at a rate given by

$$\frac{dQ}{dt} = k(80 - Q)$$

in a 20-week course, where $k$ is a positive constant and $Q(t)$ measures the number of words of dictation a student can take per minute after $t$ weeks in the course. If the average student can take 50 words of dictation per minute after 10 weeks in the course, how many words per minute can the average student take after completing the course?

**49. Effect of Immigration on Population Growth** Suppose that a country's population at any time $t$ grows in accordance with the rule

$$\frac{dP}{dt} = kP + I$$

where $P$ denotes the population at any time $t$, $k$ is a positive constant reflecting the natural growth rate of the population, and $I$ is a constant giving the (constant) rate of immigration into the country.

**a.** If the total population of the country at time $t = 0$ is $P_0$, find an expression for the population at any time $t$.

**b.** The population of the United States in the year 1980 ($t = 0$) was 226.5 million. Suppose that the natural growth rate is 0.8% annually ($k = 0.008$) and that net immigration is allowed at the rate of 0.5 million people per year ($I = 0.5$). What will the U.S. population be in 2010?

**50. Chemical Reaction Rates** Two chemical solutions, one containing $N$ molecules of chemical $A$ and another containing $M$ molecules of chemical $B$, are mixed together at time $t = 0$. The molecules from the two chemicals combine to form another chemical solution containing $y$ $AB$ molecules. The rate at which the $AB$ molecules are formed, $dy/dt$, is called the *reaction rate* and is jointly proportional to $(N - y)$ and $(M - y)$. Thus,

$$\frac{dy}{dt} = k(N - y)(M - y)$$

where $k$ is a constant. (We assume that the temperature of the chemical mixture remains constant during the interaction.) Solve this differential equation with the side condition $y(0) = 0$, assuming that $N - y > 0$ and $M - y > 0$.
**Hint:** Use the identity

$$\frac{1}{(N - y)(M - y)} = \frac{1}{M - N}\left(\frac{1}{N - y} - \frac{1}{M - y}\right)$$

**51. A Falling Raindrop** As a raindrop falls, it picks up more moisture, and as a result, its mass increases. Suppose that the rate of change of its mass is directly proportional to its current mass.

**a.** Using Newton's Law of Motion, $\dfrac{d}{dt}(mv) = F = mg$, where $m(t)$ is the mass of the raindrop at time $t$, $v$ is its velocity (positive direction is downward), and $g$ is the acceleration due to gravity, derive the (differential) equation of motion of the raindrop.

**b.** Solve the differential equation of part (a) to find the velocity of the raindrop at time $t$. Assume that $v(0) = 0$.

**c.** Find the *terminal velocity* of the raindrop, that is, find $\lim_{t \to \infty} v(t)$.

**52. Discharging Water from a Tank** A container that has a constant cross section $A$ is filled with water to height $H$. The water is discharged through an opening of cross section $B$ at the base of the container. By using Torricelli's Law, it can be shown that the height $h$ of the water at time $t$ satisfies the initial-value problem

$$\frac{dh}{dt} = -\frac{B}{A}\sqrt{2gh} \qquad h(0) = H$$

a. Find an expression for $h$.
b. Find the time $T$ it takes for the tank to empty.
c. Find $T$ if $A = 4$ (ft²), $B = 1$ (in.²), $H = 16$ (ft), and $g = 32$ (ft/sec²).

**53. Doomsday Equation** Suppose that the population $P$ satisfies the differential equation $dP/dt = kP^{1.01}$, where $k$ is a positive constant and $P(0) = 1$.
a. Solve the initial-value problem.
**cas** b. Suppose that $k = 0.1$. Plot the graph of $P(t)$.
c. Why is $dP/dt = kP^{1.01}$ called the "doomsday equation"?

**54. Stefan's Law** Stefan's Law states that the rate of change of the temperature $T$ of a body is directly proportional to the difference of the fourth power of $T$ and the fourth power of the temperature of the surrounding medium $T_m$. Thus,

$$\frac{dT}{dt} = k(T^4 - T_m^4)$$

where $k$ is the constant of proportionality. Stefan's Law holds over a greater temperature range than does Newton's Law of Cooling. Show that $T$ is given by the implicit equation

$$\ln\left(\frac{T + T_m}{T - T_m}\right) + 2\tan^{-1}\left(\frac{T}{T_m}\right) = -4T_m^3 kt + C \qquad T > T_m$$

**55. Concentration of a Drug in the Bloodstream** Suppose that the rate at which the concentration of a drug in the bloodstream decreases is proportional to the concentration at time $t$. Initially, there is no drug in the bloodstream. At time $t = 0$ a drug having a concentration of $C_0$ g/mL is introduced into the bloodstream.
a. What is the concentration of drug in the bloodstream at the end of $T$ hr?
b. If at time $T$ another dosage having the concentration of $C_0$ g/mL is infused into the bloodstream, what is the concentration of the drug at the end of $2T$ hr?
c. If the process were continued, what would the concentration of the drug be at the end of $NT$ hr?

d. Find the concentration of the drug in the bloodstream in the long run.
Hint: Evaluate $\lim_{N \to \infty} y(NT)$, where $y(NT)$ denotes the concentration of the drug at the end of $NT$ hr.

**56. Spread of Disease** A simple mathematical model in epidemiology for the spread of a disease assumes that the rate at which the disease spreads is jointly proportional to the number of infected people and the number of uninfected people. Suppose that there are a total of $N$ people in the population, of whom $N_0$ are infected initially. Show that the number of infected people after $t$ weeks, $x(t)$, is given by

$$x(t) = \frac{N}{1 + \left(\dfrac{N - N_0}{N_0}\right)e^{-kNt}}$$

where $k$ is a positive constant.

**57. Spread of Disease** Refer to Exercise 56. Suppose that there are 8000 students in a college and 400 students had contracted the flu at the beginning of the week.
a. If 1200 had contracted the flu at the end of the week, how many will have contracted the flu at the end of 2, 3, and 4 weeks?
b. How long does it take for 80% of the student population to become infected?
 c. Plot the graph of the function $x(t)$.

**58. Von Bertalanffy Growth Model** The von Bertalanffy growth model is used to predict the length of commercial fish. The model is described by the differential equation

$$\frac{dx}{dt} = k(L - x)$$

where $x(t)$ is the length of the fish at time $t$, $k$ is a positive constant called the von Bertalanffy growth rate, and $L$ is the maximum length of the fish.
a. Find $x(t)$ given that the length of the fish at $t = 0$ is $x_0$.
b. At the time the larvae hatch, the North Sea haddock are about 0.4 cm long, and the average haddock grows to a length of 10 cm after 1 year. Find an expression for the length of the North Sea haddock at time $t$.
 c. Plot the graph of $x$. Take $L = 100$ (cm).
d. On average, the haddock that are caught today are between 40 cm and 60 cm long. What are the ages of the haddock that are caught?

*In Exercises 59–62, determine whether the statement is true or false. If it is true, explain why. If it is false, explain why or give an example that shows it is false.*

**59.** If $f$ is a solution of a first-order differential equation, then so is $cf$, where $c$ is a constant.

**60.** The differential equation $y' = x^2 - y^2$ is separable.

**61.** The differential equation $y' = xy + 2x - y - 2$ is separable.

**62.** The curves $cx^2 + y^2 = 1$ and $x^2 + y^2 - \ln y^2 = k$ are orthogonal to each other.

## 7.2 Direction Fields and Euler's Method

In Section 7.1 we considered differential equations with solutions that could be found analytically. Armed with these solutions, we were able to draw the solution curves of these first-order differential equations.

In this section we will describe a way to visualize the general solution of a differential equation without actually solving the equation. This is especially useful when we are unable to find an exact solution of the equation. We will also describe a method for constructing an approximation for the solution curve of an initial-value problem.

### ■ Direction Fields

Suppose that we are given a first-order differential equation of the form

$$y' = f(x, y) \tag{1}$$

If $(a, b)$ is any point on a solution curve of Equation (1), then the slope of the tangent line to this curve at $(a, b)$ is given by

$$y'|_{(a, b)} = f(a, b)$$

(See Figure 1a.) If we retain a small portion of the tangent line at $(a, b)$, then we have a small line segment called a **lineal element** that indicates the direction of the solution curve at that point (Figure 1b).

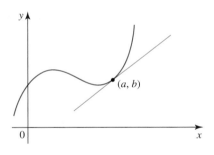

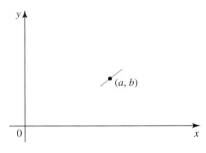

**FIGURE 1**
A short tangent line at $(a, b)$ gives the direction of the solution curve at $(a, b)$.

(**a**) The tangent line at $(a, b)$

(**b**) A lineal element at $(a, b)$

If $f$ is defined in a region in the $xy$-plane, then we can draw a lineal element at each point $(x, y)$ in the region. A set of lineal elements drawn at various points is called a **slope field** or **direction field** of the differential equation. For example, the direction field for the differential equation $y' = y - x$ (see Example 1 in Section 7.1) is shown in Figure 2.

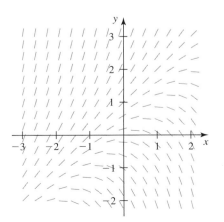

**FIGURE 2**
A direction field for $y' = y - x$

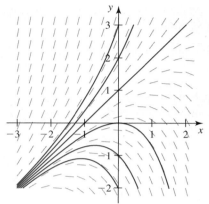

**FIGURE 3**
A few solution curves of $y' = y - x$ superimposed over its direction field

Since the small line segments represent tangent lines to the solution curves of the differential equation at these points, we see that a direction field indicates the general shape of the solution curves. Figure 3 shows a few solution curves of the differential equation $y' = y - x$ superimposed over its direction field.

**EXAMPLE 1** Consider the differential equation $y' = x + 2y$.

a. Sketch the lineal elements at $(-1, 0)$, $(-1, 1)$, $(0, 0)$, $(1, -1)$, $(1, 1)$, $(1, 2)$, $(2, -1)$, and $(2, 1)$.
b. Use a calculator or computer to draw a direction field with more lineal elements.
c. Use the slope field to sketch the solution that passes through the point $(0, 1)$.

**Solution**

a. Here, $f(x, y) = x + 2y$. The slope of the lineal element at $(-1, 0)$ is

$$f(-1, 0) = -1 + 2(0) = -1$$

We summarize the results of the other calculations in the table.

| $(x, y)$ | $(-1, 0)$ | $(-1, 1)$ | $(0, 0)$ | $(1, -1)$ | $(1, 1)$ | $(1, 2)$ | $(2, -1)$ | $(2, 1)$ |
|---|---|---|---|---|---|---|---|---|
| $y' = f(x, y)$ | $-1$ | $1$ | $0$ | $-1$ | $3$ | $5$ | $0$ | $4$ |

The lineal elements for the given ordered pairs are shown in Figure 4.

b. The required direction field is shown in Figure 5. Note that the lineal elements that we obtained in part (a) are contained in the direction field, as expected.
c. To sketch the solution curve passing through the point $(0, 1)$, we draw a curve starting out at $(0, 1)$ and extend it first to the right, then to the left, always requiring that it be parallel to nearby lineal elements as we proceed. The resulting approximating solution curve is shown in Figure 6 superimposed over the direction field of the differential equation.

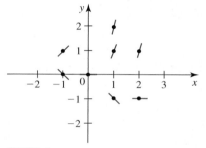

**FIGURE 4**
The lineal elements at selected points

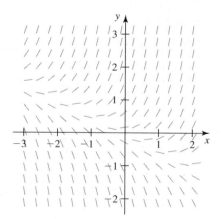

**FIGURE 5**
The direction field for $y' = x + 2y$

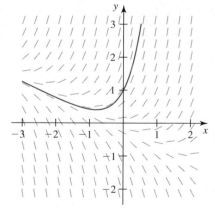

**FIGURE 6**
The solution curve of $y' = x + 2y$ passing through $(0, 1)$

In physical applications involving differential equations, the direction fields of the equations can shed much light on the nature of their solution. Suppose that an object of constant mass $m$ falls vertically downward under the influence of gravity. Assum-

**FIGURE 7**
Here, $x(t)$ is the position of
the object, $v = dx/dt$, and the
positive direction is downward.

ing that air resistance is proportional to the velocity of the object at any instant during the fall, then according to Newton's Second Law of Motion,

$$F = ma = m\frac{dv}{dt}$$

where $F$ is the net force acting on the object in the positive (downward) direction. (See Figure 7.) But $F$ is given by the weight of the object $mg$ (acting downward) minus the air resistance $kv$ (acting upward), where $k > 0$ is the constant of proportionality; that is, $F = mg - kv$. Therefore, the equation of motion is

$$m\frac{dv}{dt} = mg - kv \tag{2}$$

**EXAMPLE 2**   **A Parachute Jump**   A paratrooper and his equipment have a combined weight of 192 lb. At the instant that the parachute is deployed, he is traveling vertically downward at a speed of 30 ft/sec. Assume that air resistance is proportional to the instantaneous velocity with constant of proportionality $k = 4$ and that $g = 32$ (ft/sec$^2$).

**a.** Write an equation of motion, and draw a direction field associated with it.
**b.** Sketch the solution curve superimposed over the direction field.
**c.** What velocity does the paratrooper approach as $t$ increases without bound?

(The velocity found in part (c) is called the terminal velocity.)

**Solution**
**a.** Here, $m = \frac{192}{32}$, or 6 slugs. Therefore, using Equation (2), we have the equation of motion

$$6\frac{dv}{dt} = 192 - 4v \qquad \text{or} \qquad \frac{dv}{dt} = 32 - \frac{2}{3}v$$

At $t = 0$, $v = 30$, so we have the initial-value problem

$$\begin{cases} \dfrac{dv}{dt} = 32 - \dfrac{2}{3}v \\ v(0) = 30 \end{cases}$$

A direction field for the differential equation $v' = 32 - \frac{2}{3}v$ is shown in Figure 8.

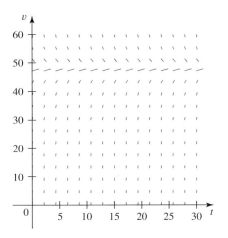

**FIGURE 8**
A direction field for $v' = 32 - \frac{2}{3}v$

**b.** The solution curve of the initial-value problem superimposed over the direction field is shown in Figure 9.

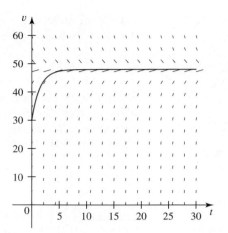

**c.** From the solution curve of part (b), we see that as $t \to \infty$, $v(t)$ seems to approach 48, and we conclude that the terminal velocity of the paratrooper is 48 ft/sec. ■

## ■ Euler's Method

The technique that we used to sketch the solution curves of a differential equation helps to reveal the nature of the solution of the equation, but it does not provide us with an accurate solution of the problem. We now turn our attention to a method that provides us with a more accurate approximation to the solution of the initial-value problem

$$\frac{dy}{dx} = F(x, y) \qquad y(x_0) = y_0 \tag{3}$$

**Euler's method,** named after the Swiss mathematician Leonard Euler (1707–1783), calls for approximating the actual solution $y = f(x)$ at certain selected values of $x$. The values of $f$ between two adjacent values of $x$ are then found by linear interpolation. This situation is depicted geometrically in Figure 10. As you can see, the actual solution of the differential equation is approximated by a suitable polygonal curve.

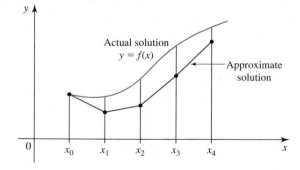

**FIGURE 10**
In using Euler's method, the actual solution curve of the differential equation is approximated by a polygonal curve.

To describe the method, let $h$ be a small positive number, and let

$$x_n = x_0 + nh \qquad n = 1, 2, 3, \dots$$

That is,

$$x_1 = x_0 + h, \qquad x_2 = x_0 + 2h, \qquad x_3 = x_0 + 3h, \qquad \dots$$

Observe that the points $x_0$, $x_1$, $x_2$, $x_3$, ... are spaced evenly apart, and the distance between any two adjacent points is $h$ units.

We begin by finding an approximation $y_1$ to the value of the actual solution, $f(x_1)$, at $x = x_1$. Observe that the *initial* condition $y(x_0) = y_0$ tells us that the point $(x_0, y_0)$ lies on the solution curve. Euler's method calls for approximating the part of the graph of $f$ on the interval $[x_0, x_1]$ by the straight-line segment that is tangent to the graph of $f$ at $(x_0, y_0)$. (See Figure 11.) To find an equation of this straight-line segment, observe that the slope of this line segment is equal to $F(x_0, y_0)$. So using the point-slope form of an equation of a line, we see that the required equation is

$$y - y_0 = F(x_0, y_0)(x - x_0)$$

or

$$y = y_0 + F(x_0, y_0)(x - x_0)$$

Therefore, the approximation $y_1$ to $f(x_1)$ is obtained by replacing $x$ by $x_1$. Thus,

$$y_1 = y_0 + F(x_0, y_0)(x_1 - x_0)$$
$$= y_0 + F(x_0, y_0)h \qquad \text{Since } x_1 - x_0 = h$$

This situation is depicted in Figure 11.

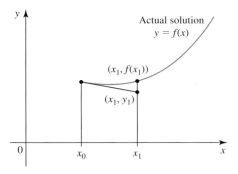

**FIGURE 11**
$y_1 = y_0 + hF(x_0, y_0)$ is an approximation of $f(x_1)$.

Next, to find an approximation $y_2$ to the value of the actual solution, $f(x_2)$, at $x = x_2$, we repeat the above procedure, this time taking the slope of the straight-line segment on $[x_1, x_2]$ to be $F(x_1, y_1)$. We obtain

$$y_2 = y_1 + hF(x_1, y_1)$$

(See Figure 12.)

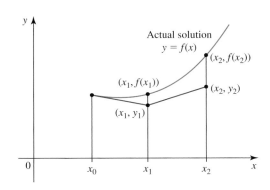

**FIGURE 12**
$y_2 = y_1 + hF(x_1, y_1)$ is the number used to approximate $f(x_2)$.

Continuing in this manner, we see that $y_1, y_2, \ldots, y_n$ can be found by the general formula

$$y_n = y_{n-1} + hF(x_{n-1}, y_{n-1}) \qquad n = 1, 2, \ldots$$

We now summarize this procedure.

> **Euler's Method**
>
> Suppose that we are given the differential equation
>
> $$\frac{dy}{dx} = F(x, y)$$
>
> subject to the initial condition $y(x_0) = y_0$, and we wish to find an approximation of $y(b)$, where $b$ is a number greater than $x_0$ and $n$ is a positive integer. Compute
>
> $$h = \frac{b - x_0}{n}$$
>
> $$\begin{aligned}
> x_1 &= x_0 + h & &\text{and} & y_0 &= y(x_0) \\
> x_2 &= x_0 + 2h & & & y_1 &= y_0 + hF(x_0, y_0) \\
> x_3 &= x_0 + 3h & & & y_2 &= y_1 + hF(x_1, y_1) \\
> &\;\;\vdots & & & &\;\;\vdots \\
> x_n &= x_0 + nh = b & & & y_n &= y_{n-1} + hF(x_{n-1}, y_{n-1})
> \end{aligned}$$
>
> Then $y_n$ gives an approximation of the true value $y(b)$ of the solution to the initial-value problem at $x = b$.

**EXAMPLE 3**  Use Euler's method with (a) $n = 5$ and (b) $n = 10$ to approximate the solution of the initial-value problem

$$y' = -2xy^2 \qquad y(0) = 1$$

on the interval $[0, 0.5]$. Find the actual solution of the initial-value problem. Finally, sketch the graphs of the approximate solutions and the actual solution for $0 \le x \le 0.5$ on the same set of axes.

**Solution**

**a.** Here, $x_0 = 0$ and $b = 0.5$. Taking $n = 5$, we find

$$h = \frac{0.5 - 0}{5} = 0.1$$

and $x_0 = 0$, $x_1 = 0.1$, $x_2 = 0.2$, $x_3 = 0.3$, $x_4 = 0.4$, and $x_5 = b = 0.5$. Also,

$$F(x, y) = -2xy^2 \qquad \text{and} \qquad y_0 = y(0) = 1$$

Therefore

$$y_0 = y(0) = 1$$

$$y_1 = y_0 + hF(x_0, y_0) = 1 + 0.1(-2)(0)(1)^2 = 1$$

$$y_2 = y_1 + hF(x_1, y_1) = 1 + 0.1(-2)(0.1)(1)^2 = 0.98$$

$$y_3 = y_2 + hF(x_2, y_2) = 0.98 + 0.1(-2)(0.2)(0.98)^2 \approx 0.9416$$

$$y_4 = y_3 + hF(x_3, y_3) = 0.9416 + 0.1(-2)(0.3)(0.9416)^2 \approx 0.8884$$

$$y_5 = y_4 + hF(x_4, y_4) = 0.8884 + 0.1(-2)(0.4)(0.8884)^2 \approx 0.8253$$

**b.** Here, $x_0 = 0$ and $b = 0.5$. Taking $n = 10$, we find

$$h = \frac{0.5 - 0}{10} = 0.05$$

and $x_0 = 0$, $x_1 = 0.05$, $x_2 = 0.10, \ldots, x_9 = 0.45$, and $x_{10} = 0.5 = b$. Proceeding as in part (a), we obtain the approximate solutions listed in the following table.

| $x_n$ | 0.00 | 0.05 | 0.10 | 0.15 | 0.20 | 0.25 | 0.30 | 0.35 | 0.40 | 0.45 | 0.50 |
|-------|------|------|------|------|------|------|------|------|------|------|------|
| $y_n$ | 1.0000 | 1.0000 | 0.9950 | 0.9851 | 0.9705 | 0.9517 | 0.9291 | 0.9032 | 0.8746 | 0.8440 | 0.8119 |

To obtain the actual solution of the differential equation, we separate variables, obtaining

$$\frac{dy}{y^2} = -2x\, dx$$

Integrating each side of the last equation with respect to the appropriate variable, we have

$$\int \frac{dy}{y^2} = \int -2x\, dx$$

or

$$-\frac{1}{y} = -x^2 + C_1$$

$$\frac{1}{y} = x^2 + C \qquad C = -C_1$$

$$y = \frac{1}{x^2 + C}$$

Using the condition $y(0) = 1$, we have

$$\frac{1}{0 + C} = 1 \qquad \text{or} \qquad C = 1$$

Therefore, the required solution is given by

$$y = \frac{1}{x^2 + 1}$$

The graphs of the approximate solutions and the actual solution are sketched in Figure 13.

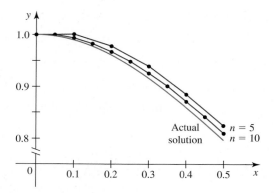

**FIGURE 13**
The approximate solutions
and the actual solution to
the initial-value problem

**EXAMPLE 4** **Parachute Jump** Refer to Example 2. There, we showed that the differential equation describing the motion of a paratrooper was

$$\frac{dv}{dt} = 32 - \frac{2}{3}v \qquad v(0) = 30$$

where $v$ is the velocity of the paratrooper at time $t$. Use Euler's method with $h = 0.2$ to estimate the velocity of the paratrooper 2 sec after his parachute was deployed.

**Solution** Here, $F(t, v) = 32 - \frac{2}{3}v$. With a step size of 0.2 ($n = 10$), we find

$$t_0 = 0, \qquad t_1 = 0.2, \qquad t_2 = 0.4, \qquad t_3 = 0.6, \qquad t_4 = 0.8, \qquad t_5 = 1.0,$$
$$t_6 = 1.2, \qquad t_7 = 1.4, \qquad t_8 = 1.6, \qquad t_9 = 1.8, \qquad t_{10} = 2.0$$

Therefore,

$$v_0 = 30$$

$$v_1 = v_0 + hF(t_0, v_0) = 30 + 0.2\left[32 - \frac{2}{3}(30)\right] = 32.4$$

$$v_2 = v_1 + hF(t_1, v_1) = 32.4 + 0.2\left[32 - \frac{2}{3}(32.4)\right] = 34.48$$

$$v_3 = v_2 + hF(t_2, v_2) = 34.48 + 0.2\left[32 - \frac{2}{3}(34.48)\right] \approx 36.28267$$

$$v_4 = v_3 + hF(t_3, v_3) = 36.28267 + 0.2\left[32 - \frac{2}{3}(36.28267)\right] \approx 37.84498$$

$$v_5 = v_4 + hF(t_4, v_4) = 37.84498 + 0.2\left[32 - \frac{2}{3}(37.84498)\right] \approx 39.19898$$

$$v_6 = v_5 + hF(t_5, v_5) = 39.19898 + 0.2\left[32 - \frac{2}{3}(39.19898)\right] \approx 40.37245$$

Carrying on, we find

$$v_7 \approx 41.38946, \qquad v_8 \approx 42.27087, \qquad v_9 \approx 43.03475, \qquad \text{and} \qquad v_{10} \approx 43.69678$$

So his velocity 2 sec after deployment of the parachute is $v_{10} \approx 43.70$ ft/sec.

## 7.2   CONCEPT QUESTIONS

**1. a.** What is the direction field of the differential equation
$y' = f(x, y)$?

**b.** Explain how you would use the direction field of part (a)
to sketch a solution curve passing through the point
$(x_0, y_0)$.

**2.** Explain how Euler's method is used to approximate the
solution of an initial-value problem.

## 7.2   EXERCISES

*In Exercises 1–4, match the differential equation with the direction field labeled* (a)–(d). *Give a reason for your choice.*

**1.** $y' = 1 - \dfrac{y}{2}$

**2.** $y' = 1 + xy$

**3.** $y' = 2x + y$

**4.** $y' = \sin x \cos y$

**(a)**

**(b)**

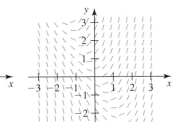

**(c)**

**(d)**

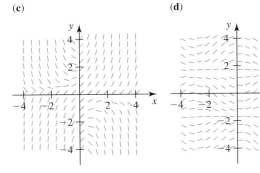

*In Exercises 5–8 a direction field for the differential equation is given. Sketch the solution curves that satisfy the initial condition.*

**5.** $y' = x^2 - y$
  **a.** $y(-2) = 0$   **b.** $y(0) = 0$   **c.** $y(0) = 1$   **d.** $y(1) = 0$

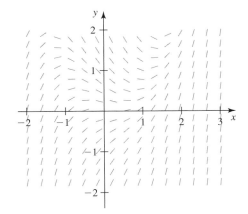

**6.** $y' = 1 - \dfrac{1}{4}y$
  **a.** $y(0) = 1$   **b.** $y(0) = 4$   **c.** $y(0) = 6$

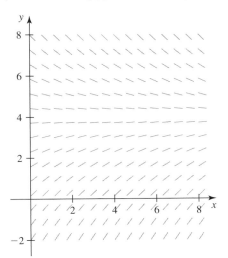

**7.** $y' = x^2 + y^2$
  **a.** $y(0) = 0$    **b.** $y(0) = 1$    **c.** $y(0) = 2$

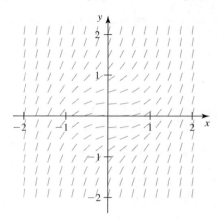

**8.** $y' = \sin x \sin y$
  **a.** $y(0) = -1$    **b.** $y(0) = 0$    **c.** $y(0) = 1$

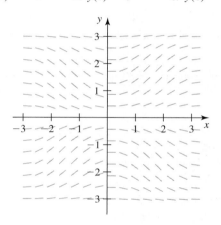

**cas** *In Exercises 9–16, use a computer algebra system (CAS) to draw a direction field for the differential equation. Then sketch approximate solution curves passing through the given points by hand superimposed over the direction field. Compare your sketch with the solution curve obtained by using a CAS.*

**9.** $y' = y$
  **a.** $(0, -1)$    **b.** $(0, 0)$    **c.** $(0, 1)$

**10.** $y' = y - 2$
  **a.** $(0, 1)$    **b.** $(0, 2)$    **c.** $(0, 4)$

**11.** $y' = x + y + 1$
  **a.** $(0, -2)$    **b.** $(0, 0)$    **c.** $(0, 1)$

**12.** $y' = \dfrac{1}{4}x^2 + y$
  **a.** $(0, -2)$    **b.** $(0, 1)$    **c.** $(1, 3)$

**13.** $y' = -\dfrac{x}{y}$
  **a.** $(-1, 1)$    **b.** $(2, 0)$    **c.** $(0, 4)$

**14.** $y' = x(2 - y)$
  **a.** $(0, -1)$    **b.** $(0, 2)$    **c.** $(0, 4)$

**15.** $y' = \cos x - y \tan x$
  **a.** $(0, -1)$    **b.** $(0, 0)$    **c.** $(0, 1)$

**16.** $y' = e^{x-y}$
  **a.** $(0, 0)$    **b.** $(0, 1)$    **c.** $(2, 1)$

*In Exercises 17–26, use Euler's method with (a) $n = 4$ and (b) $n = 6$ to estimate $y(b)$, where $y$ is the solution of the initial-value problem (accurate to two decimal places).*

**17.** $y' = x + y$,  $y(0) = 1$;  $b = 1$

**18.** $y' = x - 2y$,  $y(0) = 1$;  $b = 2$

**19.** $y' = 2x - y + 1$,  $y(0) = 2$;  $b = 2$

**20.** $y' = 2xy$,  $y(0) = 1$;  $b = 0.5$

**21.** $y' = -2xy^2$,  $y(0) = 1$;  $b = 0.5$

**22.** $y' = 1 + xy^2$,  $y(0) = 1$;  $b = 0.8$

**23.** $y' = \sqrt{x + y}$,  $y(0) = 1$;  $b = 1.5$

**24.** $y' = (x^2 + y^2)^{-1}$,  $y(0) = 1$;  $b = 1$

**25.** $y' = \dfrac{x}{y}$,  $y(0) = 1$;  $b = 1$

**26.** $y' = xy^{1/3}$,  $y(0) = 1$;  $b = 1$

*In Exercises 27 and 28, (a) sketch a few solution curves of the differential equation on the direction field, (b) solve the initial-value problem, and (c) sketch the solution curve found in part (b) on the direction field.*

**27.** $\dfrac{dy}{dx} = -\dfrac{x}{y}$,  $y(2) = 2\sqrt{3}$

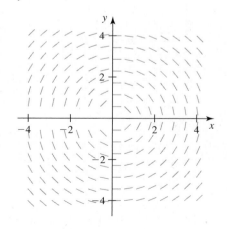

**28.** $\dfrac{dy}{dx} = y + xy, \quad y(0) = 1$

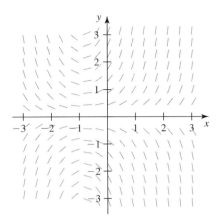

**cas 29. Gompertz Growth Curves** The differential equation
$P' = P(a - b \ln P)$, where $a$ and $b$ are constants,
is called a **Gompertz differential equation.** This
differential equation occurs in the study of population
growth and the growth of tumors.
**a.** Take $a = b = 1$ in the Gompertz differential equation,
and use a CAS to draw a direction field for the differential equation.
**b.** Use the direction field of part (a) to sketch the approximate curves for solutions satisfying the initial conditions
$P(0) = 1$ and $P(0) = 4$.
**c.** What can you say about $P(t)$ as $t$ tends to infinity? If the
limit exists, what is its approximate value?

**cas 30. Restricted Population Growth** The differential equation
$P' = P(a - bP)$, where $a$ and $b$ are constants, is called a
**logistic equation.** This differential equation is used in the
study of restricted population growth.
**a.** Take $a = 2$ and $b = 1$, and use a CAS to draw a direction field for the differential equation.
**b.** Use the direction field of part (a) to sketch the approximate solution curves passing through the points $(0, -1)$,
$(0, 1)$, and $(0, 3)$.
**c.** Suppose that $P(0) = c$. For what values of $c$ does
$\lim_{t \to \infty} P(t)$ exist? What is the value of the limit?

**31. Parachute Jump** A skydiver, together with her parachute and
equipment, have a combined weight of 160 lb. At the instant
of deployment of the parachute, she is falling vertically
downward at a speed of 30 ft/sec. Suppose that the air
resistance varies directly as the instantaneous velocity and
that the air resistance is 30 lb when her velocity is 30 ft/sec.
**a.** Use Euler's method with $n = 10$ to estimate her velocity
2 sec after deployment of her parachute.
**b.** Find the exact solution of the separable differential equation, and compute $v(2)$. Compare the answers obtained in
parts (a) and (b).

**32. R-C Series Circuit** The figure shows an R-C series circuit containing a resistor with a resistance of $R$ ohms, and a capacitor with a capacitance of $C$ farads. The voltage drop across
the capacitor is $Q(t)/C$, where $Q$ is the charge (in coulombs)
in the capacitor. Using Kirchhoff's Second Law, we have

$$RI + \dfrac{Q}{C} = E(t)$$

where $E(t)$ is the electromotive force (emf) in volts. But
$I = dQ/dt$, and this gives

$$R\dfrac{dQ}{dt} + \dfrac{1}{C}Q = E(t)$$

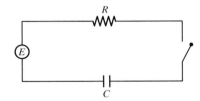

Suppose that an emf of 100 volts is applied to an R-C series
circuit in which the resistance is 50 ohms and the capacitance is 0.001 farad.
**a.** Draw a direction field for this differential equation.
**b.** Sketch the solution curve passing through the point
$(0, 0)$.
**c.** Using Euler's method with $n = 10$, estimate the charge
0.1 sec after the switch is closed.
**d.** Find the charge $Q(t)$ at time $t$ by solving the separable
differential equation if the initial charge is 0 coulomb.
Sketch the graph of $Q$, and compare this result with that
obtained in part (b).

*In Exercises 33–36, determine whether the statement is true or
false. If it is true, explain why. If it is false, explain why or give
an example that shows it is false.*

**33.** At each point $(x, y)$ on a solution curve of the differential
equation $y' = f(x, y)$, a small line segment that contains the
point $(x, y)$ and has slope $f(x, y)$ is drawn. The result is a
direction field of the differential equation.

**34.** The lineal elements in the direction field of a differential
equation constitute parts of the solution curve of the differential equation.

**35.** The lineal elements in the direction field of a differential
equation of the form $y' = f(y)$ at the point $(x, y_0)$ are parallel to each other for all values of $x$ and each fixed $y_0$.

**36.** The lineal elements in the direction field of a differential
equation of the form $y' = f(x)$ at the point $(x_0, y)$ are parallel to each other for all values of $y$ and each fixed $x_0$.

In Section 7.1 we considered a model for population growth in which the rate of change of the population at any time is proportional to the current population. Thus,

$$\frac{dP}{dt} = kP \tag{1}$$

where $P(t)$ is the population at time $t$, and $k$, the positive constant of proportionality, is the *growth constant*. Unfortunately, this model describing *unrestricted growth* is not very realistic. In the real world, one might expect that the population would grow rapidly at first and then eventually slow down because of overcrowding, scarcity of food, and other environmental factors. Indeed, one might expect that the population would eventually stabilize at a level that is compatible with the life-support capacity of the environment. In this section we will study a population model that exhibits precisely these characteristics.

### The Logistic Model

We can rewrite Equation (1) in the form

$$\frac{\dfrac{dP}{dt}}{P} = k$$

This tells us that the *relative growth* rate of the population in the unrestricted growth model is a (positive) constant $k$. Suppose that the population cannot exceed a number $L$, called the **carrying capacity** of the environment. Then it is reasonable to assume that the relative growth rate of the population starts out at $k$ when $P$ is small and approaches zero when $P$ is close to $L$. In other words, we want a model of the form

$$\frac{\dfrac{dP}{dt}}{P} = f(P)$$

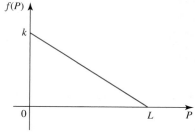

$f(P)$

$k$

$0$        $L$        $P$

**FIGURE 1**
The graph of the linear function $f$
satisfying $f(0) = k$ and $f(L) = 0$

where $f$ satisfies $f(0) = k$ and $f(L) = 0$. The simplest function $f$ satisfying these conditions is the linear function whose graph is the straight line passing through the points $(0, k)$ and $(L, 0)$. (See Figure 1.) You can verify that the desired function is

$$f(P) = k\left(1 - \frac{P}{L}\right)$$

This discussion leads to the following model for restricted population growth, known as the **logistic differential equation:**

$$\frac{dP}{dt} = kP\left(1 - \frac{P}{L}\right) \tag{2}$$

Observe that if $P$ is small relative to $L$, then $P/L$ is small and $dP/dt \approx kP$; that is, the logistic model behaves like the unrestricted growth model. But as $P$ approaches $L$, then $P/L$ approaches 1, and the rate of growth of $P$, $dP/dt$, approaches 0. Thus, the logistic differential equation exhibits both the property of rapid growth initially and that of saturation eventually. Also, note that if the (initial) population $P$ exceeds the carrying capacity $L$, then $1 - (P/L)$ is negative and $dP/dt < 0$, so the population decreases.

The following example of the logistic differential equation verifies these properties graphically.

**EXAMPLE 1**    **Logistic Growth Function**    Sketch the direction field for the logistic differential equation with $k = 0.05$ and $L = 1000$. Then draw the approximate solution curves of the equation satisfying the initial conditions $P(0) = 100$, $P(0) = 1400$, and $P(0) = 1000$ superimposed upon the direction field.

**Solution**    The logistic differential equation under consideration is

$$\frac{dP}{dt} = 0.05P\left(1 - \frac{P}{1000}\right)$$

Using a graphing utility, we obtain the direction field for this equation shown in Figure 2a. Note that the slopes are the same along any horizontal line. This occurs because the logistic differential equation is *autonomous;* that is, $P'$ depends on $P$ alone. The solution curves satisfying the initial conditions $P(0) = 100$, $P(0) = 1400$, and $P(0) = 1000$ are shown in Figure 2b–d.

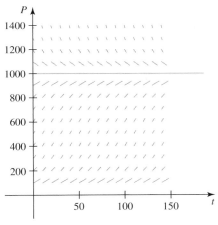

**(a)** Direction field for $P' = 0.05P\left(1 - \frac{P}{1000}\right)$

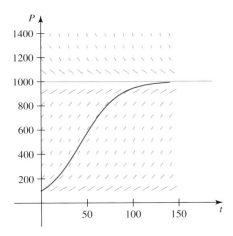

**(b)** Solution curve with $P(0) = 100$

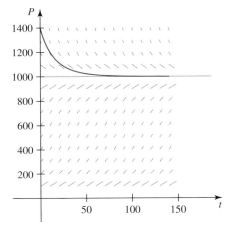

**(c)** Solution curve with $P(0) = 1400$

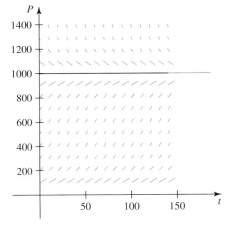

**(d)** Solution curve with $P(0) = 1000$

**FIGURE 2**

Note that in the two cases in which the initial populations do not begin at 1000, the carrying capacity of the environment, both populations tend to 1000 as $t$ increases without bound. But in the case in which the initial population is 1000, the population remains steady at that level for all values of $t$.   ■

## ■ Analytic Solution of the Logistic Differential Equation

The logistic differential equation (2) is separable and can be solved by using the method of Section 7.1.

**EXAMPLE 2**   Solve the logistic differential equation

$$\frac{dP}{dt} = kP\left(1 - \frac{P}{L}\right) \qquad P(0) = P_0$$

**Solution**   First, $P = 0$ and $P = L$ are solutions, as you can verify by substituting these values into the differential equation. Next, suppose that $P \neq 0$ and $P \neq L$. Observe that the equation is separable. Separating variables leads to

$$\frac{dP}{P\left(1 - \dfrac{P}{L}\right)} = k \, dt$$

Integrating each side of this equation with respect to the appropriate variable, we obtain

$$\int \frac{dP}{P\left(1 - \dfrac{P}{L}\right)} = \int k \, dt \qquad \text{or} \qquad \int \frac{L}{P(L - P)} \, dP = k \int dt \tag{3}$$

To find the integral on the left-hand side, we use partial fraction decomposition (see Section 6.4) to write

$$\frac{L}{P(L - P)} = \frac{1}{P} + \frac{1}{L - P}$$

This leads to

$$\int \left(\frac{1}{P} + \frac{1}{L - P}\right) dP = \int k \, dt$$

$$\ln|P| - \ln|L - P| = kt + C_1$$

$$\ln|L - P| - \ln|P| = -kt - C_1 \qquad \text{Multiply each side by } -1.$$

$$\ln\left|\frac{L - P}{P}\right| = -kt - C_1$$

$$\left|\frac{L - P}{P}\right| = e^{-kt - C_1} = e^{-C_1}e^{-kt} = C_2 e^{-kt} \qquad C_2 = e^{-C_1}$$

or

$$\frac{L - P}{P} = Ce^{-kt} \tag{4}$$

## Historical Biography

### JEAN LE ROND D'ALEMBERT
(1717-1783)

Jean le Rond d'Alembert was a man of varied interests, and he is remembered as a mathematician, physicist, and philosopher. In mathematics, d'Alembert is best known for his method of solving the so-called wave equation, an important differential equation that is used to describe the behavior of a large class of waves. Abandoned as an infant on the steps of St. Jean Baptiste le Rond, near Notre-Dame de Paris, d'Alembert was baptized and put in foster care to be raised by Madame Rousseau, the wife of a glazier. d'Alembert's biological parents were later found to be an artillery general and the socially prominent sister of a Cardinal. However, d'Alembert always regarded his foster mother as his real mother, and he continued to live with her until he reached the age of 47. d'Alembert's schooling was paid for by his biological father, and he excelled, becoming a celebrated mathematician. He wrote more than 1500 works, including the famous *Discours préliminaire* (1751) for Denis Diderot's *Encyclopedie*, an introduction that explained the structure and philosophy of the articles in the *Encyclopedie* and gave a thorough review of the intellectual history behind those articles and the philosophy of the French Enlightenment.

where $C = \pm C_2$. We can solve for $P$ in Equation (4) as follows:

$$\frac{L}{P} - 1 = Ce^{-kt}$$

$$\frac{L}{P} = 1 + Ce^{-kt}$$

and

$$P = \frac{L}{1 + Ce^{-kt}}$$

To determine $C$, we use the initial condition $P(0) = P_0$, where $P_0$ is the initial population. Putting $t = 0$ and $P = P_0$ in Equation (4) yields

$$\frac{L - P_0}{P_0} = Ce^0 = C$$

Thus, the solution of the initial-value problem is

$$P(t) = \frac{L}{1 + \left(\frac{L}{P_0} - 1\right)e^{-kt}} \tag{5}$$

Note that

$$\lim_{t\to\infty} P(t) = \lim_{t\to\infty} \frac{L}{1 + \left(\frac{L}{P_0} - 1\right)e^{-kt}} = L \tag{6}$$

as expected. The graph of Equation (5) is called the **logistic curve.**

## Logistic Curve

Example 1 suggests the shape of the logistic curve. We are now in the position to confirm this observation analytically.

We begin by determining the intervals where $P$ is increasing and where it is decreasing. To do this, we could compute $P'(t)$ from Equation (5), but this would be tedious and unnecessary. Instead, we will work with Equation (2), which expresses $P'$ in terms of $P$. Observe that

$$P' = kP\left(1 - \frac{P}{L}\right) \tag{7}$$

is a continuous function of $P$ on $(-\infty, \infty)$ and has zeros at $P = 0$ and $P = L$. The sign diagram for $P'$ is shown in Figure 3.

**FIGURE 3**
The sign diagram for $P'$

$$0++++++0-----------$$

Observe that this sign diagram is not the same as the sign diagrams that we encountered in Chapter 3. Here, $P$ is the *dependent variable,* and it lies on the *vertical* axis.

From the sign diagram for $P'$ we conclude that $P$ is increasing for $0 < P < L$ and $P$ is decreasing for $P > L$. Next, we compute

$$P'' = \frac{d}{dt}\left[kP\left(1 - \frac{P}{L}\right)\right] = k\frac{d}{dt}\left(P - \frac{P^2}{L}\right) \qquad \text{Use Equation (7).}$$

$$= k\left(1 - \frac{2P}{L}\right)P' = k^2 P\left(1 - \frac{2P}{L}\right)\left(1 - \frac{P}{L}\right)$$

**FIGURE 4**

The sign diagram for $P''$

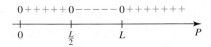

Observe that $P''$, as a function of $P$, is continuous on $(-\infty, \infty)$ and has zeros at $P = 0$, $L/2$, and $L$. The sign diagram of $P''$ is shown in Figure 4. From the sign diagram for $P''$ we conclude that the graph of $P$ is concave upward for $0 < P < \frac{L}{2}$ and $P > L$ and is concave downward for $\frac{L}{2} < P < L$. Also, $P$ has an inflection point at $P = L/2$. We have two cases. Referring to the sign diagrams for $P'$ and $P''$, we see the following:

1. If $0 < P < L$, then $P$ is increasing, concave upward for $0 < P < \frac{L}{2}$, and concave downward for $\frac{L}{2} < P < L$. Also, the graph of $P$ has an inflection point at $P = L/2$. (See Figure 5.)
2. If $P > L$, then $P$ is decreasing and concave upward.

It can be shown, although we will not do so here, that none of these curves can cross the horizontal lines $P = 0$ and $P = L$.

Let's summarize our results.

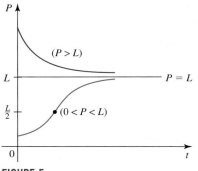

**FIGURE 5**

Two possible logistic curves

Suppose that a population at any time $t$ satisfies the logistic differential equation (2) and that the initial population at $t = 0$ is $P_0$.

1. If $P_0 = 0$, the population stays at zero at all times
2. If $0 < P_0 < L$, then the population increases and approaches the limiting value $L$, called the carrying capacity of the environment, asymptotically. The population increases most rapidly at the instant of time when it reaches $\frac{1}{2}L$.
3. If $P_0 = L$, the population at any later time remains at $L$.
4. If $P_0 > L$, then the population decreases and approaches the carrying capacity $L$ asymptotically.

In Exercise 9 you will be asked to show that the time referred to in part (2) when the population increases most rapidly is given by

$$T = \frac{\ln\left(\dfrac{L}{P_0} - 1\right)}{k} \qquad (8)$$

**Note** The constant solutions $P = 0$ and $P = L$ are called **equilibrium solutions.** ■

**EXAMPLE 3** **Logistic Growth Function**   Refer to Example 1. Suppose that the population $P(t)$ satisfies the logistic differential equation

$$\frac{dP}{dt} = 0.05P\left(1 - \frac{P}{1000}\right)$$

**a.** What is the population at any time $t$ if the initial population is 1000?
**b.** What is the population at any time $t$ if the initial population is 1400?
**c.** What is the population at any time $t$ if the initial population is 100?

**Solution**   Using Equation (5) with $k = 0.05$ and $L = 1000$, we see that the population at time $t$ is

$$P(t) = \frac{L}{1 + \left(\dfrac{L}{P_0} - 1\right)e^{-kt}} = \frac{1000}{1 + \left(\dfrac{1000}{P_0} - 1\right)e^{-0.05t}} \tag{9}$$

**a.** Here, $P_0 = 1000$, so Equation (9) gives

$$P(t) = \frac{1000}{1 + \left(\frac{1000}{1000} - 1\right)e^{-0.05t}} = 1000$$

That is, the population stays at the equilibrium level for all $t$. (See Figure 6a.)
**b.** Here, $P_0 = 1400$, so Equation (9) gives

$$P(t) = \frac{1000}{1 + \left(\frac{1000}{1400} - 1\right)e^{-0.05t}} = \frac{1000}{1 - \frac{2}{7}e^{-0.05t}}$$

(See Figure 6b.) The population decreases to 1000 asymptotically.
**c.** Here, $P_0 = 100$, so Equation (9) gives

$$P(t) = \frac{1000}{1 + \left(\frac{1000}{100} - 1\right)e^{-0.05t}} = \frac{1000}{1 + 9e^{-0.05t}}$$

(See Figure 6c.) The population increases to 1000 asymptotically.

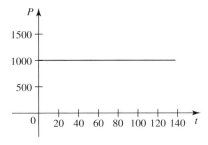

(**a**) $P(0) = 1000$

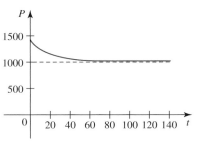

(**b**) $P(0) = 1400$

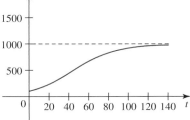

(**c**) $P(0) = 100$

**FIGURE 6**
Logistic curves for $\dfrac{dP}{dt} = 0.05P\left(1 - \dfrac{P}{1000}\right)$

**EXAMPLE 4** **Logistic Growth Function**   Refer to Example 3. Suppose that the population $P(t)$ satisfies the logistic differential equation

$$\frac{dP}{dt} = 0.05P\left(1 - \frac{P}{1000}\right)$$

where $t$ is measured in days and the initial population is 100.

**a.** What is the population at $t = 30$? At $t = 50$?
**b.** At what time is the population increasing most rapidly?
**c.** At what time is the population equal to 800?

**Solution**

**a.** The population at any time $t$ was obtained in part (c) of Example 3. We have

$$P(t) = \frac{1000}{1 + 9e^{-0.05t}}$$

The population at $t = 30$ is

$$P(30) = \frac{1000}{1 + 9e^{-0.05(30)}} \approx 332$$

The population at $t = 50$ is

$$P(50) = \frac{1000}{1 + 9e^{-0.05(50)}} \approx 575$$

**b.** The population increases most rapidly at the instant it reaches half the carrying capacity of the environment. Therefore, we can find the required time by solving the equation

$$500 = \frac{1000}{1 + 9e^{-0.05t}}$$

for $t$. Alternatively, we can use Equation (8) to find

$$T = \frac{\ln\left(\dfrac{L}{P_0} - 1\right)}{k} = \frac{\ln\left(\dfrac{1000}{100} - 1\right)}{0.05} \approx 43.944$$

So the population increases most rapidly on approximately the 44th day.

**c.** The required time is found by solving

$$800 = \frac{1000}{1 + 9e^{-0.05t}}$$

for $t$. We have

$$1 + 9e^{-0.05t} = \frac{1000}{800} = \frac{5}{4}$$

$$9e^{-0.05t} = \frac{1}{4}$$

$$e^{-0.05t} = \frac{1}{36}$$

$$-0.05t = \ln\frac{1}{36} = -\ln 36 \qquad \text{Take the natural logarithms.}$$

$$t = \frac{\ln 36}{0.05} \approx 71.67$$

So the population reaches 800 when $t$ is approximately 72, or approximately after 72 days.

The graph of $P$ is shown in Figure 7.

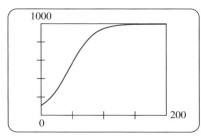

**FIGURE 7**
The graph of $P(t) = \dfrac{1000}{1 + 9e^{-0.05t}}$

**EXAMPLE 5** **Rate of Growth of a Fish Population**   A fish farm is stocked with 100 fish. Suppose that the fish population satisfies the logistic equation and that the carrying capacity of the pond is 2000.

a.  Find an expression for the fish population after $t$ years if the number of fish increased to 250 in the first year.
b.  How long will it take for the fish population to reach 1000?

**Solution**

a.  Using Equation (5), we see that the population after $t$ years is

$$P(t) = \frac{L}{1 + \left(\dfrac{L}{P_0} - 1\right)e^{-kt}}$$

Here, $L = 2000$ and $P_0 = P(0) = 100$, so

$$P(t) = \frac{2000}{1 + \left(\dfrac{2000}{100} - 1\right)e^{-kt}} = \frac{2000}{1 + 19e^{-kt}}$$

To determine $k$, we use the condition $P(1) = 250$. This leads to

$$P(1) = \frac{2000}{1 + 19e^{-k}} = 250$$

$$1 + 19e^{-k} = \frac{2000}{250} = 8$$

$$19e^{-k} = 7$$

$$e^{-k} = \frac{7}{19}$$

$$k = -\ln\frac{7}{19} = 0.9985$$

so

$$P(t) = \frac{2000}{1 + 19e^{-0.9985t}}$$

b.  We solve the equation $P(t) = 1000$ for $t$; that is,

$$1000 = \frac{2000}{1 + 19e^{-0.9985t}}$$

$$1 + 19e^{-0.9985t} = 2$$

$$e^{-0.9985t} = \frac{1}{19}$$

$$-0.9985t = \ln\left(\frac{1}{19}\right)$$

$$t \approx 2.949$$

So it takes approximately 2.9 years for the fish population to reach 1000.

The graph of $P$ is shown in Figure 8.

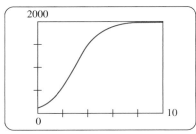

**FIGURE 8**
The graph of $P(t) = \dfrac{2000}{1 + 19e^{-0.9985t}}$

## 7.3 CONCEPT QUESTIONS

1. Consider the logistic differential equation $\dfrac{dP}{dt} = kP\left(1 - \dfrac{P}{L}\right)$.

   a. What does $k$ represent? What does $L$ represent?
   b. Write two constant solutions of the equation, and explain their meaning.
   c. What can you say about the rate of change of the population if the initial population is greater than $L$? If it is greater than zero but less than $L$? Interpret your answers.

2. a. Verify by direct computation that

   $$P(t) = \frac{L}{1 + \left(\dfrac{L}{P_0} - 1\right)e^{-kt}}$$

   is a solution of the initial-value problem

   $$\frac{dP}{dt} = kP\left(1 - \frac{P}{L}\right) \qquad P(0) = P_0$$

   b. Describe the solution of the logistic differential equation of part (a) with the aid of a graph. Assume that $P_0 > 0$.

## 7.3 EXERCISES

*In Exercises 1–4, a logistic differential equation describing population growth is given. Use the equation to find (a) the growth constant and (b) the carrying capacity of the environment.*

1. $\dfrac{dP}{dt} = 0.02P\left(1 - \dfrac{P}{1000}\right)$

2. $\dfrac{dP}{dt} = 0.03P - 0.000006P^2$

3. $\dfrac{dP}{dt} = P\left(0.5 - \dfrac{P}{1000}\right)$

4. $150{,}000\,\dfrac{dP}{dt} = 3P(2000 - P)$

5. A direction field of a logistic differential equation describing population growth is shown in the figure.

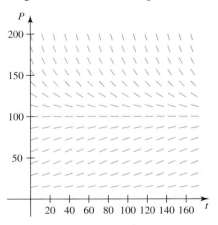

   a. What is the carrying capacity of the environment?
   b. What are the constant solutions?
   c. Sketch the solution curve with an initial population of 200.
   d. Sketch the solution curve with an initial population of 100.
   e. Sketch the solution curve with an initial population of 10.

6. A direction field of a modified logistic differential equation describing population growth is shown in the figure.

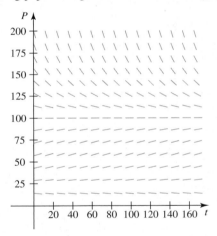

   a. What is the carrying capacity of the environment?
   b. What are the constant solutions?
   c. Sketch the solution curve with an initial population of 120.
   d. Sketch the solution curve with an initial population of 60.
   e. Sketch the solution curve with an initial population of 10.

*In Exercises 7 and 8, use the given logistic equation to find (a) the growth constant, (b) the carrying capacity of the environment, and (c) the initial population.*

7. $P(t) = \dfrac{8000}{2 + 798e^{-0.02t}}$

8. $P(t) = \dfrac{100e^{0.2t}}{e^{0.2t} + 19}$

**9.** Consider the logistic differential equation $\dfrac{dP}{dt} = kP\left(1 - \dfrac{P}{L}\right)$.

   **a.** Show that $P(t)$ grows most rapidly when $P = L/2$.

   **b.** Show that $P(t)$ grows most rapidly at time

$$T = \frac{\ln\left(\dfrac{L}{P_0} - 1\right)}{k}$$

   where $P_0$ is the initial population.

**10. Spread of an Epidemic** During a flu epidemic the number of children in the Woodbridge Community School System who contracted influenza after $t$ days was given by

$$Q(t) = \frac{1000}{1 + 199e^{-0.8t}}$$

   **a.** How many children were stricken by the flu after the first day?

   **b.** How many children had the flu after 10 days?

   **c.** How many children eventually contracted the disease?

**11. Lay Teachers at Roman Catholic Schools** The change from religious to lay teachers at Roman Catholic schools has been partly attributed to the decline in the number of women and men entering religious orders. The percentage of teachers who are lay teachers is given by

$$f(t) = \frac{98}{1 + 2.77e^{-t}} \qquad 0 \le t \le 4$$

   where $t$ is measured in decades, with $t = 0$ corresponding to the beginning of 1960.

   **a.** What percentage of teachers were lay teachers at the beginning of 1990?

   **b.** Find the year when the percentage of lay teachers was increasing most rapidly.

   *Source:* National Catholic Education Association and the Department of Education.

**12. People Living with HIV** On the basis of data compiled by the World Health Organization, it is estimated that the number of people living with HIV worldwide from 1985 through 2006 is

$$N(t) = \frac{39.88}{1 + 18.94e^{-0.2957t}} \qquad 0 \le t \le 21$$

   where $N(t)$ is measured in millions and $t$ in years with $t = 0$ corresponding to the beginning of 1985.

   **a.** How many people were living with HIV worldwide at the beginning of 1985? At the beginning of 2005?

   **b.** Assuming that the trend continued, how many people were living with HIV worldwide at the beginning of 2008?

   *Source:* World Health Organization.

**13. Growth of a Fruit Fly Population** Initially, there were 10 fruit flies (*Drosophila melanogaster*) in an experiment. Because

of a limit to be placed on the amount of food available, the maximum population of fruit flies was estimated to be 100.

   **a.** Suppose that the pattern of growth followed the logistic curve and that the population was 34 after 30 days. Find an expression for the fruit fly population $t$ days after the start of the experiment.

   **b.** How long did it take the population to reach 80?

**14. Rate of Growth for a Plant** The rate of growth of a certain type of plant is described by a logistic differential equation. Botanists have estimated the maximum theoretical height of such plants to be 30 in. At the beginning of an experiment, the height of a plant was 5 in., and the plant grew to 12 in. after 20 days.

   **a.** Find an expression for the height of the plant after $t$ days.

   **b.** What was the height of the plant after 30 days?

   **c.** How long did it take for the plant to reach 80% of its maximum theoretical height?

**15. Logistic Growth Function** Consider the logistic growth function

$$P(t) = \frac{L}{1 + \left(\dfrac{L}{P_0} - 1\right)e^{-kt}}$$

   Suppose that the population is $P_1$ when $t = t_1$ and $P_2$ when $t = t_2$. Show that the value of $k$ is

$$k = \frac{1}{t_2 - t_1}\ln\left[\frac{P_2(L - P_1)}{P_1(L - P_2)}\right]$$

**16. Logistic Growth Function** The carrying capacity of a colony of fruit flies (*Drosophila melanogaster*) is 600. The population of fruit flies after 14 days is 76, and the population after 21 days is 167. What is the value of the growth constant $k$? **Hint:** Use the result of Exercise 15.

**17. Rate of Growth of a Fish Population** Let $P(t)$ denote the population of a certain species of fish in a lake, where $t$ is measured in weeks. Then $P$ can be described by the modified logistic differential equation

$$\frac{dP}{dt} = kP\left(1 - \frac{P}{L}\right) - c$$

   where $k$ is the growth rate, $L$ is the carrying capacity of the environment, and $c$ is the constant rate at which fish are being removed because of fishing. Suppose that $k = 0.2$, $L = 800$, and $c = 30$.

   **a.** Draw a direction field for the resulting differential equation.

   **b.** Use the direction field that was obtained in part (a) to find the equilibrium solutions. Verify your results algebraically.

   **c.** Sketch the solution curves for the solutions with initial populations of 100, 300, and 700, and describe what happens to the fish population in each case.

**18. Gompertz Growth Curves** The **Gompertz differential equation,** a model for restricted population growth, is obtained by modifying the logistic differential equation and is given by

$$\frac{dP}{dt} = cP \ln\left(\frac{L}{P}\right)$$

where $c$ is a constant and $L$ is the carrying capacity of the environment.
   **a.** Find the equilibrium solution of the differential equation.
   **b.** Illustrate graphically the solutions of the equation with initial conditions $P(0) = P_0$, where (i) $P_0 > L$, (ii) $P_0 = L$, and (iii) $0 < P_0 < L$.

**19. Gompertz Growth Curves** Refer to Exercise 18. Consider the Gompertz differential equation with $L = 1000$ and $c = 0.02$.
   **a.** Draw a direction field for the differential equation.
   **b.** Identify the equilibrium solution.
   **c.** Plot the solution curve with initial conditions $P(0) = 1200$ and $P(0) = 100$.

**20. Gompertz Growth Curves** Refer to Exercises 18 and 19. Consider the Gompertz differential equation

$$\frac{dP}{dt} = cP \ln\left(\frac{L}{P}\right)$$

where $c$ is a positive constant and $L$ is the carrying capacity of the environment.
   **a.** Solve the differential equation.
   **b.** Find $\lim_{t\to\infty} P(t)$.
   **c.** Show that $P(t)$ is increasing most rapidly when $P = L/e$.
   **d.** Show that $P(t)$ is increasing most rapidly when

$$t = \frac{\ln \ln\left(\dfrac{L}{P_0}\right)}{c}$$

**21. A Goldfish Population** Refer to Exercise 20. A population of 20 goldfish was introduced into a pond that has an estimated carrying capacity of 200 fish. After 1 month, the population

of goldfish had grown to 80. If the pattern of growth of the population followed the Gompertz curve, how many goldfish were in the pond after 3 months?

**22. Cyclical Models** Some populations are subject to seasonal fluctuations. The population in a vacation resort serves as one example. A model for describing such situations is the differential equation

$$\frac{dP}{dt} = (k \cos t)P$$

where $k$ is a constant and $t$ is measured in months.
   **a.** Find the solution of the differential equation subject to $P(0) = P_0$.

   **b.** Let $k = 0.2$, and plot the graphs of $P$ for $P_0 = 400$, $500$, and $600$.
   **c.** What happens to $P(t)$ for large values of $t$?

*In Exercises 23–26, determine whether the statement is true or false. If it is true, explain why. If it is false, explain why or give an example that shows it is false.*

**23.** If $P$ is the solution of the initial-value problem

$$P' = 0.2P\left(1 - \frac{P}{100}\right), \quad P(0) = 150, \text{ then } P(t) \text{ is}$$

decreasing on the interval $(0, \infty)$.

**24.** If $P$ is the solution of the initial-value problem

$$P' = 0.3P\left(1 - \frac{P}{20}\right), \quad P(0) = 0, \text{ then } P(t) \text{ is}$$

increasing on the interval $(0, \infty)$.

**25.** If $P$ is the solution of the initial-value problem

$$P' = 0.5P\left(1 - \frac{P}{50}\right), \quad P(0) = 10, \text{ then the graph}$$

of $P$ has an inflection point.

**26.** If $P$ is the solution of the initial-value problem

$$P' = 0.02P\left(1 - \frac{P}{1000}\right), \quad P(0) = 1000, \text{ then}$$

$\lim_{t\to\infty} P(t) = 1000$.

---

## 7.4 First-Order Linear Differential Equations

We now consider another class of first-order differential equations. A **first-order linear differential equation** is one that can be written in the form

$$\frac{dy}{dx} + P(x)y = Q(x) \tag{1}$$

where $P$ and $Q$ are continuous functions of $x$ on a given interval. The equation is so named because it is *linear* in the unknown function and its derivative. A linear equation written in the form of Equation (1) is said to be in **standard form.** For example, the differential equation

$$x\frac{dy}{dx} - y - x^3 = 0$$

is a linear equation, since it is linear in both $y$ and $dy/dx$. By dividing through by $x$ and rearranging terms, we obtain the equation in standard form, namely,

$$\frac{dy}{dx} - \frac{1}{x}y = x^2$$

Here, $P(x) = -\dfrac{1}{x}$ and $Q(x) = x^2$. On the other hand, the equations

$$y\frac{dy}{dx} + 2y = e^x \qquad \text{and} \qquad \frac{dy}{dx} + 2\cos y = x^3$$

are not linear because of the nonlinear term $y(dy/dx)$ in the first equation and the nonlinear term $\cos y$ in the second equation.

## ■ Method of Solution

First-order linear differential equations can be solved by multiplying both sides of the equation

$$\frac{dy}{dx} + P(x)y = Q(x)$$

by a suitable function $u(x)$ that transforms the equation into one that can be solved by integration. To find $u(x)$, let's consider the equation that is obtained by putting $Q(x) = 0$. The resulting equation

$$\frac{dy}{dx} + P(x)y = 0 \tag{2}$$

is called a *homogeneous* linear equation. Observe that Equation (2) is a separable equation and, therefore, can be solved using the method of separation of variables. We find

$$\int \frac{dy}{y} = -\int P(x)\,dx$$

$$\ln|y| = -\int P(x)\,dx + C_1$$

$$|y| = e^{C_1}e^{-\int P(x)\,dx}$$

$$y = Ce^{-\int P(x)\,dx} \qquad C = \pm e^{C_1} \tag{3}$$

The solution of the homogeneous equation associated with Equation (1) points the way to solving the nonhomogeneous equation (1) itself. We rewrite Equation (3) in the form

$$ye^{\int P(x)\,dx} = C$$

Let's differentiate this last equation using the Fundamental Theorem of Calculus, Part 1. Thus,

$$\frac{d}{dx}\left[ye^{\int^x P(t)\,dt}\right] = \frac{d}{dx}(C) \qquad \text{Rewrite the integral using the dummy variable } t.^*$$

$$\frac{dy}{dx}e^{\int^x P(t)\,dt} + ye^{\int^x P(t)\,dt} \cdot \frac{d}{dx}\int^x P(t)\,dt = 0 \qquad \text{Use the Product Rule and the Chain Rule.}$$

$$\frac{dy}{dx}e^{\int^x P(t)\,dt} + ye^{\int^x P(t)\,dt} \cdot P(x) = 0 \qquad \text{Use the Fundamental Theorem of Calculus, Part 1.}$$

---

*Henceforth, we will usually write $e^{\int^x P(t)\,dt}$ in the form $e^{\int P(x)\,dx}$ to conform with the more standard practice.

or

$$e^{\int^x P(t)\,dt}\left[\frac{dy}{dx} + P(x)y\right] = 0$$

Observe that the expression within the square brackets is just the expression on the left-hand side of Equation (1). This suggests that by multiplying both sides of Equation (1) by $e^{\int P(x)\,dx}$, we have

$$e^{\int P(x)\,dx}\left[\frac{dy}{dx} + P(x)y\right] = e^{\int P(x)\,dx}Q(x)$$

which in turn can be written in the form

$$\frac{d}{dx}\left[ye^{\int P(x)\,dx}\right] = e^{\int P(x)\,dx}Q(x) \tag{4}$$

The expression on the left of Equation (4) is easily integrated because it is the derivative of a function. Since the function on the right does not involve the unknown function, it can also be integrated. Thus, the function $u$ that we are seeking is

$$u(x) = e^{\int P(x)\,dx}$$

This function is called an **integrating factor** because multiplying Equation (1) by $u$, as we have just seen, enables us to solve the problem by integration. Before looking at an example, let's summarize the steps in solving a linear differential equation.

---

**Solving a First-Order Linear Differential Equation**

1. Rewrite the equation in standard form $y' + P(x)y = Q(x)$ if necessary.
2. Find an integrating factor $u(x) = e^{\int P(x)\,dx}$.
3. Multiply both sides of the equation $y' + P(x)y = Q(x)$ by $u(x)$. The resulting equation can be written in the form

$$\frac{d}{dx}(yu) = uQ$$

Unknown ——————— Integrating
function                factor

which can then be integrated.

---

**EXAMPLE 1**  Solve the equation $x\dfrac{dy}{dx} - y - x^3 = 0$, where $x > 0$.

**Solution**

**Step 1**    We rewrite the equation in standard form

$$\frac{dy}{dx} - \frac{1}{x}y = x^2 \tag{5}$$

and identify $P(x) = -1/x$ and $Q(x) = x^2$

**Step 2**    We find

$$u(x) = e^{\int P(x)\,dx} = e^{-\int (1/x)\,dx} = e^{-\ln x} = e^{\ln x^{-1}} = \frac{1}{x}$$

**Step 3**    Multiplying both sides of Equation (5) by $u(x) = 1/x$, we obtain

$$\frac{1}{x}\frac{dy}{dx} - \frac{1}{x^2}y = x$$

which can be written in the form

$$\frac{d}{dx}\left[\frac{1}{x}y\right] = x$$

Integrating
factor ———↑↑——— Unknown
function

Integrating both sides with respect to $x$, we obtain

$$\frac{1}{x}y = \int x\,dx = \frac{1}{2}x^2 + C$$

so $y = \frac{1}{2}x^3 + Cx$. ∎

**Note**    If we integrate Equation (4), we can obtain a formula for the solution to the problem. Thus,

$$ye^{\int^{x} P(t)\,dt} = \int^{x} e^{\int^{u} P(t)\,dt}Q(u)\,du + C$$

or

$$y = e^{-\int^{x} P(t)\,dt}\int^{x} e^{\int^{u} P(t)\,dt}Q(u)\,du + Ce^{-\int^{x} P(t)\,dt}$$

But it is preferable to use the method of solution just described. ∎

**Historical Biography**

**JACOB BERNOULLI**
**(1654-1705)**

Going against his father's wish that he enter the ministry, Jacob Bernoulli followed his personal interests and studied mathematics and astronomy. He eventually founded a school for science and mathematics, where he lectured on mathematics and mechanics and carried out experiments in physics. He was the first in his mathematically talented family to pursue a career in mathematics. He was followed by his younger brother Johann and three of Johann's sons. Jacob's own two children did not pursue careers in mathematics or the sciences. In 1687, Jacob Bernoulli was named professor of mathematics at the University of Basel, a seat that he held until his death in 1705. He was among the first mathematicians of his time to fully understand the newly developing calculus. One of his many contributions to mathematics was a method to solve differential equations of the form $y' + P(x)y = Q(x)y^n$, a type of differential equation that is now known as Bernoulli's differential equation (see Exercise 23 in this section). Bernoulli also made important contributions to the theory of probability and to the study of mechanics.

**EXAMPLE 2**    Solve the initial-value problem

$$\begin{cases} x^2 y' + 3xy = e^{-x^2} & x > 0 \\ y(1) = 0 \end{cases}$$

**Solution**    First, we rewrite the differential equation in standard form by dividing both sides by $x^2$, obtaining

$$y' + \frac{3}{x}y = \frac{e^{-x^2}}{x^2} \tag{6}$$

An integrating factor is

$$u(x) = e^{\int (3/x)\,dx} = e^{3\ln x} = e^{\ln x^3} = x^3$$

Multiplying both sides of Equation (6) by $x^3$ gives

$$x^3 y' + 3x^2 y = xe^{-x^2}$$

which can be rewritten in the form

$$\frac{d}{dx}(x^3 y) = xe^{-x^2}$$

Integrating both sides with respect to $x$ yields

$$x^3 y = \int x e^{-x^2} \, dx = -\frac{1}{2} e^{-x^2} + C$$

$$y = -\frac{e^{-x^2}}{2x^3} + \frac{C}{x^3}$$

Since $y(1) = 0$, we have

$$-\frac{e^{-1}}{2} + C = 0 \qquad \text{or} \qquad C = \frac{1}{2e}$$

Therefore, the required solution is

$$y = -\frac{e^{-x^2}}{2x^3} + \frac{1}{2ex^3} \qquad\blacksquare$$

Our first application of first-order linear differential equations is a *mixture problem*.

**FIGURE 1**

**EXAMPLE 3** **Mixture Problem**    A tank initially contains 400 gal of water in which 50 lb of salt has been dissolved. Brine containing 2 lb of salt per gallon enters the tank at the rate of 3 gal/min. The well-stirred solution drains from the tank at the rate of 5 gal/min. How much salt is in the tank after 30 min? (See Figure 1.)

**Solution**    Let $y(t)$ denote the amount of salt in the tank (in pounds) at time $t$ (in minutes). Then the rate of change of the amount of salt at time $t$ is

$$\frac{dy}{dt} = (\text{rate of salt entering}) - (\text{rate of salt exiting})$$

$$\text{rate of salt entering} = (\text{concentration of brine entering}) \cdot (\text{rate of flow in})$$

$$= (2)(3) = 6 \text{ lb/min} \qquad \frac{\text{lb}}{\text{gal}} \cdot \frac{\text{gal}}{\text{min}} = \frac{\text{lb}}{\text{min}}$$

$$\text{rate of salt exiting} = (\text{concentration of brine in tank}) \cdot (\text{rate of flow out})$$

$$= \left( \frac{\text{amount of salt in the tank at time } t}{\text{volume of brine in the tank at time } t} \right) \cdot (\text{rate of flow out})$$

But the volume of brine at time $t$ is given by

$$(\text{initial volume}) + (\text{net change in volume})$$

$$= (\text{initial volume}) + (\text{rate of flow in minus rate of flow out})t$$

$$= 400 + (3 - 5)t = 400 - 2t$$

So the rate of salt exiting is

$$\frac{y}{400 - 2t} \cdot 5 = \frac{5y}{400 - 2t} \qquad 400 - 2t > 0$$

Therefore,

$$\frac{dy}{dt} = 6 - \frac{5y}{400 - 2t}$$

The condition that there are 50 lb of salt in the tank initially translates into the initial condition $y(0) = 50$. The mathematical formulation has led to the initial-value problem

$$\begin{cases} \dfrac{dy}{dt} = 6 - \dfrac{5y}{400 - 2t} & 0 \le t < 200 \\ y(0) = 50 \end{cases}$$

To solve the first-order linear differential equation, we first write it in standard form

$$\frac{dy}{dt} + \frac{5}{400 - 2t} y = 6 \tag{7}$$

and identify $P(t) = 5/(400 - 2t)$. An integrating factor is

$$u(t) = e^{\int [5/(400-2t)] \, dt} = e^{-(5/2)\ln(400-2t)} = (400 - 2t)^{-5/2}$$

Multiplying both sides of Equation (7) by $u(t)$, we obtain

$$(400 - 2t)^{-5/2} \frac{dy}{dt} + (400 - 2t)^{-5/2}\left(\frac{5}{400 - 2t}\right) y = 6(400 - 2t)^{-5/2}$$

which we can write as

$$\frac{d}{dt}[(400 - 2t)^{-5/2} y] = 6(400 - 2t)^{-5/2}$$

Integrating both sides with respect to $t$ yields

$$(400 - 2t)^{-5/2} y = 6 \int (400 - 2t)^{-5/2} \, dt$$

$$= 6\left(-\frac{1}{2}\right)\left(-\frac{2}{3}\right)(400 - 2t)^{-3/2} + C$$

or

$$y = 4(200 - t) + C(400 - 2t)^{5/2}$$

To determine the value of $C$, we use the initial condition $y = 50$ when $t = 0$, giving

$$4(200) + C(400)^{5/2} = 50$$

or

$$C = -\frac{750}{400^{5/2}}$$

Therefore,

$$y = 4(200 - t) - \frac{750}{400^{5/2}} (400 - 2t)^{5/2}$$

The amount of salt in the tank after 30 min is given by

$$y(30) = 4(200 - 30) - \frac{750}{400^{5/2}} (400 - 60)^{5/2} \approx 180.42$$

or approximately 180 lb. ■

First-order linear differential equations also arise in the analysis of electrical circuits. For example, suppose that we are given an electric circuit consisting of a battery

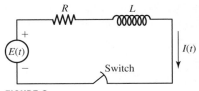

**FIGURE 2**
A single-loop series circuit

or generator having an electromotive force (emf) of $E(t)$ volts in series with a resistor having a resistance of $R$ ohms and an inductor having an inductance of $L$ henries. (See Figure 2.) According to Kirchhoff's Second Law, the emf that is supplied ($E$) is equal to the voltage drop across the inductor ($L\,dI/dt$) plus the voltage drop across the resistor ($RI$), where $I(t)$ is the current in amperes at time $t$. Thus, the differential equation for the circuit is a first-order linear equation

$$L\frac{dI}{dt} + RI = E \tag{8}$$

**EXAMPLE 4**   **Electric Circuits**   A 12-volt battery is connected in series with a 10-ohm resistor and an inductor of 2 henries. If the switch is closed at time $t = 0$, determine (a) the current at time $t$, (b) the current $\frac{1}{10}$ second after the switch is closed, and (c) the current after a long time.

**Solution**
**a.** We put $L = 2$, $R = 10$, and $E = 12$ in Equation (8) and let $I(0) = 0$ to obtain the initial-value problem

$$\begin{cases} 2\dfrac{dI}{dt} + 10I = 12 \\ I(0) = 0 \end{cases}$$

Rewriting the first-order linear equation in standard form, we obtain

$$\frac{dI}{dt} + 5I = 6$$

An integrating factor for this equation is

$$u(t) = e^{\int 5\,dt} = e^{5t}$$

Multiplying the differential equation by $u(t)$ gives

$$e^{5t}\frac{dI}{dt} + 5e^{5t}I = 6e^{5t}$$

$$\frac{d}{dt}\,(e^{5t}I) = 6e^{5t}$$

$$e^{5t}I = \int 6e^{5t}dt = \frac{6}{5}e^{5t} + C$$

$$I(t) = \frac{6}{5} + Ce^{-5t}$$

Since $I(0) = 0$, we have $\frac{6}{5} + C = 0$, or $C = -\frac{6}{5}$. Therefore,

$$I(t) = \frac{6}{5}(1 - e^{-5t})$$

**b.** The current $\frac{1}{10}$ second after the switch is closed is given by

$$I\left(\frac{1}{10}\right) = \frac{6}{5}(1 - e^{-5/10}) \approx 0.47$$

or approximately 0.5 amp.

**c.** The current after a long time (called the **steady-state current**) is given by

$$\lim_{t \to \infty} I(t) = \lim_{t \to \infty} \frac{6}{5}(1 - e^{-5t})$$

$$= \frac{6}{5} - \frac{6}{5} \lim_{t \to \infty} e^{-5t}$$

$$= \frac{6}{5} - 0 = \frac{6}{5}$$

or 1.2 amp.

The graph of $I$ is shown in Figure 3.

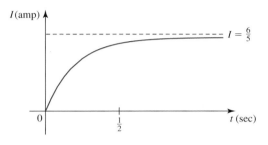

**FIGURE 3**
The current $I(t)$ approaches the
steady-state current as $t \to \infty$.

**EXAMPLE 5**  **Electric Circuits**   Suppose that the battery in the circuit of Example 4 is replaced by a generator having an emf of $E(t) = 20e^{-2t}$ volts. Find $I(t)$. What is the maximum current in the circuit?

**Solution**   The only difference between this problem and that of Example 4 is that $E = 12$ is replaced by $E = 20e^{-2t}$. We have

$$2\frac{dI}{dt} + 10I = 20e^{-2t} \qquad \text{or} \qquad \frac{dI}{dt} + 5I = 10e^{-2t}$$

We use the same integrating factor as before, obtaining

$$e^{5t}\frac{dI}{dt} + 5e^{5t}I = e^{5t} \cdot 10e^{-2t} = 10e^{3t}$$

$$\frac{d}{dt}(e^{5t}I) = 10e^{3t}$$

$$e^{5t}I = \int 10e^{3t}\,dt = \frac{10}{3}e^{3t} + C$$

$$I(t) = \frac{10}{3}e^{-2t} + Ce^{-5t}$$

Since $I(0) = 0$, we have $\frac{10}{3} + C = 0$ or $C = -\frac{10}{3}$. Therefore,

$$I(t) = \frac{10}{3}e^{-2t} - \frac{10}{3}e^{-5t}$$

To find the maximum current in the circuit, we set

$$I'(t) = -\frac{20}{3}e^{-2t} + \frac{50}{3}e^{-5t} = 0$$

obtaining

$$\frac{20}{3} e^{-2t} = \frac{50}{3} e^{-5t}$$

$$e^{3t} = \frac{50}{20}$$

$$\ln e^{3t} = \ln \frac{5}{2}$$

$$3t = \ln \frac{5}{2}$$

$$t \approx \frac{1}{3} \ln \frac{5}{2} \approx 0.3054$$

This is the only critical value of $I$ on the interval $[0, \infty)$. Since $I(0) = 0$ and

$$\lim_{t \to \infty} I(t) = \lim_{t \to \infty} \left( \frac{10}{3} e^{-2t} - \frac{10}{3} e^{-5t} \right) = 0$$

we see that the maximum current occurs at $t \approx 0.3054$ and has a value of approximately

$$I(0.3054) \approx \frac{10}{3} e^{-2(0.3054)} - \frac{10}{3} e^{-5(0.3054)} \approx 1.09$$

that is, approximately 1.1 amp. The graph of $I$ is shown in Figure 4.

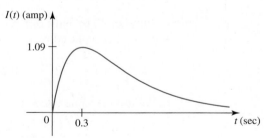

**FIGURE 4**

The graph of $I(t) = \dfrac{10}{3} e^{-2t} - \dfrac{10}{3} e^{-5t}$

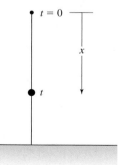

**FIGURE 5**
The positive direction is downward.

Our final example looks at an application of first-order linear differential equations to the motion of an object. Suppose that an object of constant mass $m$ falls vertically downward under the influence of gravity. If we assume that air resistance is proportional to the speed of the object at any instant during the fall, then according to Newton's second law of motion, $F = ma = m(dv/dt)$, where $F$ is the net force acting on the object in the positive (downward) direction. (See Figure 5.) But $F$ is given by the weight of the object $mg$ (acting downward) minus the air resistance $kv$ (acting upward), where $k > 0$ is the constant of proportionality; that is, $F = mg - kv$. Therefore, the equation of motion is

$$m \frac{dv}{dt} = mg - kv \tag{9}$$

**EXAMPLE 6** **Parachute Jump**  A paratrooper and his equipment have a combined weight of 192 lb. At the instant that the parachute is deployed, he is traveling vertically downward at a speed of 30 ft/sec. Assume that air resistance is proportional to the instantaneous velocity with constant of proportionality $k = 4$.

**a.** Determine the velocity and position of the paratrooper at any time.
**b.** Find his limiting velocity by evaluating $\lim_{t \to \infty} v(t)$.

**Solution**
**a.** Here, $m = \frac{192}{32}$, or 6 slugs. Therefore, using Equation (9) we have the equation of motion

$$6 \frac{dv}{dt} = 192 - 4v \qquad \text{or} \qquad \frac{dv}{dt} = 32 - \frac{2}{3}v$$

At $t = 0$, $v = 30$, so we have the initial-value problem

$$\begin{cases} \dfrac{dv}{dt} + \dfrac{2}{3}v = 32 \\ v(0) = 30 \end{cases}$$

An integrating factor for this equation is

$$u(t) = e^{\int (2/3)\, dt} = e^{(2/3)t}$$

Multiplying the differential equation by $u(t)$ gives

$$e^{(2/3)t} \frac{dv}{dt} + \frac{2}{3} e^{(2/3)t} v = 32 e^{(2/3)t}$$

$$\frac{d}{dt}\left(e^{(2/3)t} v\right) = 32 e^{(2/3)t}$$

$$e^{(2/3)t} v = 32 \int e^{(2/3)t}\, dt = 48 e^{(2/3)t} + C$$

$$v(t) = 48 + Ce^{-(2/3)t}$$

Since $v(0) = 30$, we have $v(0) = 48 + C = 30$, or $C = -18$. Therefore,

$$v(t) = 48 - 18e^{-(2/3)t}$$

The position of the paratrooper at time $t$ is

$$x(t) = \int v(t)\, dt = \int (48 - 18e^{-(2/3)t})\, dt = 48t + 27e^{-(2/3)t} + C_1$$

Since $x(0) = 0$, we have $27 + C_1 = 0$, or $C_1 = -27$. Therefore,

$$x(t) = 48t + 27e^{-(2/3)t} - 27$$

(See Figure 6 on the following page.)
**b.** The paratrooper's limiting (terminal) velocity is

$$\lim_{t \to \infty} v(t) = \lim_{t \to \infty}(48 - 18e^{-(2/3)t}) = 48 - \lim_{t \to \infty} 18e^{-(2/3)t} = 48$$

or 48 ft/sec. (See Figure 7 on the following page.)

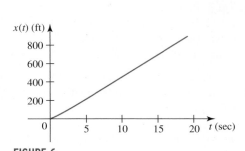

**FIGURE 6**
The position of the paratrooper at time $t$

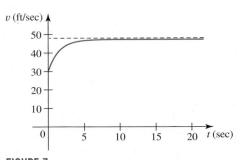

**FIGURE 7**
The velocity of the paratrooper at time $t$    ■

## 7.4    CONCEPT QUESTIONS

**1. a.** Write a first-order linear differential equation in standard form.

**b.** Is the differential equation $a_0(x)y' + a_1(x)y = g(x)$, where $a_0(x) \neq 0$, a first-order linear differential equation?

**2. a.** What is an integrating factor for a first-order linear differential equation?

**b.** Describe a method for solving a first-order linear differential equation.

## 7.4    EXERCISES

*In Exercises 1–4, determine whether the differential equation is linear.*

**1.** $\dfrac{dy}{dx} + xy^2 = \cos x$

**2.** $x^2y' + e^x y = 4$

**3.** $y \cos y + \dfrac{1}{x}\dfrac{dy}{dx} - \ln x = 0$

**4.** $y^2 \dfrac{dx}{dy} + 3x = \tan y$

*In Exercises 5–16, solve the differential equation.*

**5.** $\dfrac{dy}{dx} + 2y = e^{2x}$

**6.** $x\dfrac{dy}{dx} + 3y = 2$

**7.** $xy' + y = x^3$

**8.** $y \sin x + y' \cos x = 1$

**9.** $\dfrac{dy}{dx} - \dfrac{2y}{x} = x^2 \cos 3x$

**10.** $\dfrac{dy}{dx} + y \cot x = \cos x$

**11.** $xy' - y = 2x(\ln x)^2$

**12.** $(\cos y - xe^y)\, dy = e^y\, dx$
    **Hint:** Consider $x = f(y)$.

**13.** $(t + 1)\dfrac{dy}{dt} + y = t, \quad t > -1$

**14.** $xy' + (1 + x)y = e^{-x}(1 + \cos 2x)$

**15.** $xy' + (2x + 1)y = xe^{-2x}$

**16.** $\dfrac{dy}{dx} = \dfrac{y}{x + y^3}$
    **Hint:** Consider $x = f(y)$.

*In Exercises 17–22, solve the initial-value problem.*

**17.** $\dfrac{dy}{dx} + y = 1, \quad y(0) = -1$

**18.** $xy' - 3y = x^4, \quad y(1) = 5$

**19.** $\dfrac{dy}{dx} + 2xy = x, \quad y(0) = 1$

**20.** $\dfrac{dr}{d\theta} = \theta - \dfrac{r}{3\theta}, \quad r(1) = 1$

**21.** $x(x + 1)y' + xy = \ln x, \quad y(1) = \dfrac{1}{2}$

**22.** $(1 + e^x)\dfrac{dy}{dx} + e^x y = \sin x, \quad y(0) = \dfrac{1}{2}$

**23.** The equation

$$\dfrac{dy}{dx} + P(x)y = Q(x)y^n$$

where $n$ is a constant, is called *Bernoulli's differential equation*.

**a.** Show that the Bernoulli equation reduces to a linear equation if $n = 0$ or 1.

**b.** Show that if $n \neq 0$ or 1, then changing the dependent variable from $y$ to $v$ using the transformation $v = y^{1-n}$ reduces the Bernoulli equation to the linear equation

$$\dfrac{dv}{dx} + (1 - n)P(x)v = (1 - n)Q(x)$$

**24.** Use the method of Exercise 23 to solve $y' - y = xy^2$.

**25.** Use the method of Exercise 23 to solve the initial-value problem

$$x^2 y' - 2xy = 4y^3 \qquad y(1) = \sqrt{3}$$

**26. a.** Show that the differential equation

$$\frac{dy}{dx} + P(x)y = Q(x)y \ln y$$

can be solved by using the transformation $y = e^v$.
   **b.** Use the result of part (a) to solve $xy' - 2x^2 y = y \ln y$.

**27.** The slope of the tangent line to the graph of a function $y = f(x)$ at the point $(x, y)$ is $1 + y/x$. If the graph passes through the point $(1, 1)$, find $f$.

**28. Mixture Problem** A tank initially holds 16 gal of water in which 4 lb of salt has been dissolved. Brine that contains 6 lb of salt per gallon enters the tank at the rate of 2 gal/min, and the well-stirred mixture leaves at the same rate.
   **a.** Find a function that gives the amount of salt in the tank at time $t$.
   **b.** Find the amount of salt in the tank after 5 min.
   **c.** How much salt is in the tank after a long time?

**29. Mixture Problem** A tank initially holds 30 gal of pure water. Brine that contains 3 lb of salt per gallon enters the tank at the rate of 2 gal/min, and the well-stirred mixture leaves at the same rate.
   **a.** How much salt is in the tank at any time $t$?
   **b.** When will the tank hold 80 lb of salt?

**30. Mixture Problem** A tank initially holds 10 gal of water in which 2 lb of salt has been dissolved. Brine containing 1.5 lb of salt per gallon enters the tank at the rate of 2 gal/min, and the well-stirred mixture leaves at the rate of 3 gal/min.
   **a.** Find the amount of salt $y(t)$ in the tank at time $t$.
   **b.** Find the amount of salt in the tank after 10 min.
   **c.** Plot the graph of $y$.
   **d.** At what time is the amount of salt in the tank greatest? How much salt is in the tank at that time?

**31. Mixture Problem** A tank initially holds 40 gal of pure water. Brine that contains 2 lb of salt per gallon enters the tank at the rate of 1.5 gal/min, and the well-stirred mixture leaves at the rate of 2 gal/min.
   **a.** Find the amount of salt in the tank at time $t$.
   **b.** Find the amount of salt in the tank after 20 min.
   **c.** Find the amount of salt when the tank holds 20 gal of brine.
   **d.** Find the maximum amount of salt present.

**32. Electric Circuit** A 24-volt battery is connected in series with a 20-ohm resistor and an inductor of 4 henries. If the switch is closed at time $t = 0$, determine (a) the current at time $t$, (b) the current after 0.2 sec, and (c) the current after a long time.

**33. Electric Circuit** The figure shows an electric circuit consisting of a battery or generator of $E$ volts in series with a resistor of $R$ ohms and a capacitor of $C$ farads. The voltage drop across the capacitor is $Q/C$ where $Q$ is the charge (in coulombs), so by Kirchhoff's Second Law,

$$RI + \frac{1}{C}Q = E$$

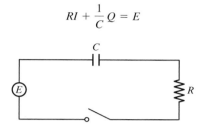

But $I = dQ/dt$, so we have the differential equation

$$R\frac{dQ}{dt} + \frac{1}{C}Q = E$$

Suppose that a circuit consists of a battery having a constant emf of 12 volts in series with a resistor of 10 ohms and a capacitor of 0.02 farad. The charge on the capacitor at $t = 0$ is 0.05 coulomb. Find the charge and the current at time $t$ after the switch is closed.

**34. Electric Circuit** Suppose that the battery in the electric circuit of Exercise 33 is replaced by a generator having an emf of $E(t) = 30e^{-2t} + 10e^{-6t}$ volts. If the charge in the capacitor is 0 coulomb at $t = 0$, find the maximum charge on the capacitor.

**35. Falling Weight** An 8-lb weight is dropped from rest from a cliff. Assume that air resistance is equal to the weight's instantaneous velocity.
   **a.** Find the velocity of the weight at time $t$.
   **b.** What is the velocity of the weight after 1 sec?
   **c.** How long does it take for the weight to reach a speed of 4 ft/sec?

**36. Parachute Jump** A skydiver and his equipment have a combined weight of 192 lb. At the instant that his parachute is deployed, he is traveling vertically downward at a speed of 112 ft/sec. Assume that air resistance is proportional to the instantaneous velocity with a constant of proportionality of $k = 12$. Determine the position and velocity of the skydiver $t$ sec after his parachute is deployed. What is his limiting velocity?

**37. Sinking Boat** As a boat weighing 1000 lb sinks in water from rest, it is acted upon by a buoyant force of 200 lb and a force of water resistance in pounds that is numerically equal to $100v$, where $v$ is in feet per second. Find the distance traveled by the boat after 4 sec. What is its limiting velocity?

**38.** An object of mass $m$ is thrown vertically upward with an initial velocity of $v_0$. Air resistance is proportional to its instantaneous velocity with constant of proportionality $k$. Show that the maximum height attained by the object is

$$\frac{mv_0}{k} - \frac{m^2 g}{k^2} \ln\left(1 + \frac{kv_0}{mg}\right)$$

**39. Electric Circuit** An electromotive force of $E_0 \cos \omega t$ volts, where $E_0$ and $\omega$ are constants, is applied to a series circuit consisting of a resistor of constant resistance $R$ ohms and an inductor of constant inductance $L$ henries. If we use Ohm's Law, the current $I(t)$, where $t$ is time in seconds, satisfies the first-order linear differential equation

$$L\frac{dI}{dt} + RI = E_0 \cos \omega t$$

If the current $I$ is 0 ampere initially, find an expression for the current at any time $t$.

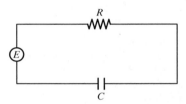

**40. Market Equilibrium** The quantity demanded of a certain commodity, $d(t)$, is related to its unit price $p(t)$, in dollars, by the *demand equation* $d(t) = 40 - p(t) + 2p'(t)$, where $t$ denotes time. The quantity of the commodity made available by the supplier, $s(t)$, is related to the unit price $p(t)$, in dollars, by the *supply equation* $s(t) = 22 + 2p(t) + 3p'(t)$. Both $d(t)$ and $s(t)$ are measured in units of a thousand. *Market equilibrium* prevails when the demand is equal to the supply.
  **a.** If market equilibrium prevails, find the *equilibrium price* at time $t$ if the price of the commodity is 10 dollars at $t = 0$.
  **b.** What happens to the price as $t \to \infty$?
  **Note:** If $\lim_{t \to \infty} p(t)$ exists, we say that there is price stability.

**41.** Suppose that $y_1$ is a solution of $y' + P(x)y = f(x)$ and $y_2$ is a solution of $y' + P(x)y = g(x)$. Show that $c_1 y_1 + c_2 y_2$ is a solution of $y' + P(x)y = c_1 f(x) + c_2 g(x)$ for all constants $c_1$ and $c_2$.

**42. a.** Find the general solution of

$$y' + \frac{2}{x}y = \frac{e^x}{x} \quad \text{and} \quad y' + \frac{2}{x}y = \frac{e^{-x}}{x}$$

  **b.** Use the result of Exercise 41 to write down the general solution of

$$y' + \frac{2}{x}y = \frac{2e^x}{x} - \frac{3e^{-x}}{x}$$

*In Exercises 43–47, determine whether the statement is true or false. If it is true, explain why it is true. If it is false, explain why or give an example to show why it is false.*

**43.** $y^2 \dfrac{dx}{dy} + e^y x = y \cos y$ is a first-order linear differential equation.

**44.** A first-order differential equation can be both separable and linear.

**45.** An integrating factor for the equation $a_0(x)y' + a_1(x)y = f(x)$ is $e^{\int [a_1(x)/a_0(x)]\, dx}$.

**46.** If $y_1$ is a solution of the homogeneous equation $y' + Py = 0$ associated with the nonhomogeneous equation $y' + Py = f$ and $y_2$ is a solution of the nonhomogeneous equation, then $y = cy_1 + y_2$ is a solution of the nonhomogeneous equation, where $c$ is any constant.

**47.** The function $f(x) = 2e^x - \frac{1}{2}(\cos x + \sin x)$ is a solution of the differential equation $y' - y = \sin x$.

# 7.5   Predator-Prey Models

Up to now, we have dealt only with population models involving a single species. But there are many instances in nature in which one species of animals feeds on another species of animals that in turn feeds on other food that is readily available. For example, wolves hunt caribou, which feed on an unlimited supply of vegetation, and sharks feed on small fish, which in turn feed on plankton. The first species is called the *predator* and the second species is called the *prey*.

Let $x(t)$ denote the number of prey and let $y(t)$ denote the number of predators at time $t$. If there are no predators and there is an unlimited supply of food, then the prey population will grow at a rate that is proportional to the current population; that is,

$$\frac{dx}{dt} = ax \qquad a > 0 \tag{1}$$

In the absence of prey, the predator population will decline at a rate that is proportional to the current population; that is,

$$\frac{dy}{dt} = -ry \qquad r > 0 \tag{2}$$

When both predators and prey are present, however, we must modify both Equations (1) and (2) to take into account the interactions of the species. It seems reasonable to assume that the number of encounters between these two species is jointly proportional to their populations, that is, the number is proportional to the product $xy$. Since these encounters are detrimental to the prey population, the rate at which the prey population changes is *decreased* by the term $bxy$, where $b$ is a positive constant. Similarly, these encounters are beneficial to the predator population, so the rate at which the predator population changes is *increased* by the term $sxy$, where $s$ is a positive constant. Thus, we are led to the following:

$$\frac{dx}{dt} = ax - bxy$$

$$\frac{dy}{dt} = -ry + sxy \tag{3}$$

where $a$, $b$, $r$ and $s$ are positive constants.

The equations in system (3) are called *predator-prey equations*. They are also called **Lotka-Volterra equations** after the mathematicians Alfred Lotka (1880–1949) and Vito Volterra (1860–1940), who independently developed mathematical models to study how two species interact. The equations are autonomous since the expression on the right-hand side of each equation does not depend explicitly on the time $t$. A **solution** of system (3) is an ordered pair of functions $(x(t), y(t))$, where $x(t)$ and $y(t)$ give the populations of prey and predators at time $t$, respectively. Although the system of equations looks simple, no exact solutions have yet been found. Nevertheless, much insight into the nature of the solutions of the system of equations can be obtained without solving them.

Observe that system (3) can be written in the form

$$\frac{dx}{dt} = x(a - by)$$

$$\frac{dy}{dt} = y(-r + sx)$$

from which we see that $x = 0$ and $y = 0$, or $(0, 0)$, and $x = r/s$ and $y = a/b$, or $\left(\frac{r}{s}, \frac{a}{b}\right)$, are solutions of system (3). These points are called **critical points** or **equilibrium points** of the system.

The solution $(0, 0)$ merely represents the fact that if there aren't any predators and prey at some point in time, then this situation will remain so forever. The solution $\left(\frac{r}{s}, \frac{a}{b}\right)$ reflects the case in which the number of predators and prey are in a state of equilibrium. The number of prey, $r/s$, is at exactly the level that will sustain the number of predators, $a/b$.

What about other, less obvious solutions? To shed some light on the nature of these solutions, let's look at the problem graphically in an example.

**EXAMPLE 1** Suppose that the population of rabbits (prey) in hundreds, $x(t)$, and the population of foxes (predators) in tens, $y(t)$, are described by the Lotka-Volterra equations with $a = 4$, $b = 1$, $r = 1$, and $s = 0.2$.

a. Write the Lotka-Volterra equations for this case.
b. Find the equilibrium points of the system.
c. Find an expression for $dy/dx$, and use it to draw a direction field for the resulting differential equation in the $xy$-plane.
d. Sketch some solution curves for the differential equation found in part (c).

**Solution**

a. The required equations are

$$\frac{dx}{dt} = 4x - xy$$

$$\frac{dy}{dt} = -y + 0.2xy$$

b. To find the equilibrium points of the system, we set $dx/dt = 0$ and $dy/dt = 0$, simultaneously, obtaining

$$4x - \quad xy = x(4 - y) \qquad = 0$$

$$-y + 0.2xy = y(-1 + 0.2x) = 0$$

from which we obtain the constant solutions (equilibrium points)

$$x = 0 \qquad \text{and} \qquad y = 0$$

or $(0, 0)$, and

$$x = \frac{1}{0.2} = 5 \qquad \text{and} \qquad y = 4$$

or $(5, 4)$. So, if at any moment in time there are no foxes and no rabbits, then there will be no predators, or prey at any time later. The other equilibrium point tells us that 500 rabbits is the exact number needed to support 40 foxes.

c. Using the Chain Rule, we have

$$\frac{dy}{dx} = \frac{\dfrac{dy}{dt}}{\dfrac{dx}{dt}} = \frac{-y + 0.2xy}{4x - xy}$$

Next, we draw a direction field for this equation using a graphing utility (see Figure 1).

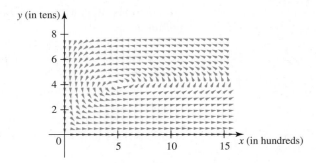

**FIGURE 1**
A direction field for Equation (1)

**d.** A few solution curves for the differential equation in part (c) are sketched in Figure 2. We have included the equilibrium points $(0, 0)$ and $(5, 4)$ in the figure. Note that the equilibrium point $(5, 4)$ lies inside the solution curves. Also, note that the solution curves are closed (a fact that we will not prove here).

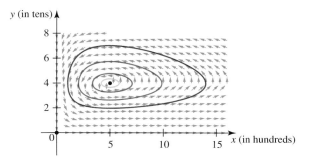

**FIGURE 2**
A few solution curves for system (3)

The solution curves shown in Figure 2 are called **phase curves,** or **phase trajectories,** of system (3) and the $xy$-plane in which the phase curves lie is called the **phase plane.** Recall that each point on the phase curve is a solution of the system (3). The next example shows that the solution points other than the equilibrium points "move" along a phase curve. A figure that consists of equilibrium points and typical phase curves is called a **phase portrait.**

**EXAMPLE 2**   Refer to the system of differential equations in Example 1:

$$\frac{dx}{dt} = 4x - xy$$

$$\frac{dy}{dt} = -y + 0.2xy$$

where $x(t)$ and $y(t)$ denote the number of rabbits (in hundreds) and foxes (in tens) at time $t$.

**a.** Suppose that at some time $t = 0$ there are 500 rabbits and 20 foxes. Draw the phase curve that corresponds to this situation.
**b.** What happens to the solution point $(x, y)$ as $t$ increases?
**c.** Use the results of parts (a) and (b) to sketch the graphs of $x(t)$ and $y(t)$.

**Solution**
**a.** We use the direction field obtained in the solution to Example 1c to help us draw the phase curve passing through the point $(5, 2)$. (See Figure 3.)

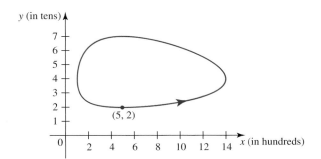

**FIGURE 3**

**b.** The phase curve is reproduced in Figure 4. At $t = 0$, $x = 5$, and $y = 2$. So

$$\left.\frac{dx}{dt}\right|_{t=0} = \left.4x - xy\right|_{(5,\,2)}$$

$$= 4(5) - (5)(2) = 10$$

Since $dx/dt > 0$, we see that the solution point $(x, y)$ moves counterclockwise around the phase curve.

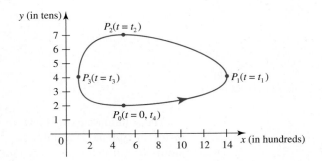

**FIGURE 4**

**c.** Refer to Figure 4. We begin at $P_0(5, 2)$ when $t = 0$ and move in a counterclockwise direction reaching $P_1(14, 4)$ at some time $t_1$, $P_2(5, 7)$ at some time $t_2$, and $P_3(1, 4)$ at some time $t_3$, and returning to $P_0(5, 2)$ at some time $t_4$. From this, we obtain the following table of values for $x$ and $y$ at different times $t$. For values of $t \geq t_4$, we simply replicate the graph obtained since the phase portrait is a closed curve.

| $t$ | 0 | $t_1$ | $t_2$ | $t_3$ | $t_4$ |
|---|---|---|---|---|---|
| $x$ | 5 | 14 | 5 | 1 | 5 |
| $y$ | 2 | 4 | 7 | 4 | 2 |

Using this table of values, we obtain the graphs of $x$ and $t$ as shown in Figure 5.

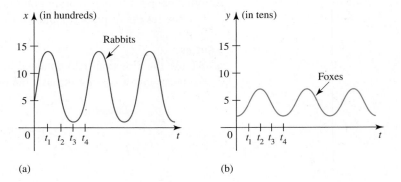

**FIGURE 5**

The graphs of the populations of (a) rabbits and (b) foxes as a function of time

Initially, there are 500 rabbits and 20 foxes. As $t$ increases, the rabbit population increases rapidly. With a plentiful supply of food available, the fox population also increases rapidly. But this rapid increase of predators starts to take a toll on the prey and the rate of increase of the population of rabbits soon slows down, the population reaching a maximum population of approximately 1400 at time $t_1$. The fox population appears to be increasing most rapidly at this time. (The graph of $y$ has an inflection point at $t = t_1$.)

From $t = t_1$ to $t = t_3$, the population of rabbits declines, the rate of decline being most rapid at $t = t_2$ (where $x$ also has an inflection point). With the decline of the rabbit population, the rate of increase of the fox population soon begins to slow down. The fox population reaches a maximum of approximately 70 at $t = t_2$, then declines rapidly until $t = t_3$, then declines less rapidly from $t = t_3$ to $t = t_4$. Finally, as the fox population declines, the rabbit population once again begins to increase (from $t = t_3$ to $t = t_4$).

The cycle then repeats over and over again. ∎

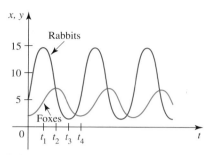

**FIGURE 6**
Graphs showing the population of rabbits, $x(t)$, in hundreds, and the population of foxes, $y(t)$, in tens

The relationship between the predator and prey populations in Examples 1 and 2 is illustrated in Figure 6, where the graphs of both $x$ and $y$ are plotted on the same set of axes. Observe that the fox population lags behind the rabbit population as both populations oscillate between their maximum and minimum values.

## 7.5  CONCEPT QUESTION

**1. a.** Let $x(t)$ and $y(t)$ denote the populations of prey and predators at time $t$, respectively. Write the Lotka-Volterra equations to model these populations.
  **b.** How are the equations modified if there are no predators? What do the resulting equation(s) say about the prey population?
  **c.** How are the equations modified if there are no prey? What do the resulting equation(s) say about the predator population?

## 7.5  EXERCISES

*In Exercises 1 and 2 you are given the Lotka-Volterra equations describing the relationship between the prey population (in hundreds) at time t, x(t), and the predator population (in tens) at time t, y(t). (a) Find the equilibrium points of the system. (b) Find an expression for dy/dx and use it to draw a direction field for the resulting differential equation in the xy-plane. (c) Sketch some solution curves for the differential equation found in part (b).*

**1.** $\dfrac{dx}{dt} = 2.4x - 1.2xy$

$\dfrac{dy}{dt} = -y + 0.8xy$

**2.** $\dfrac{dx}{dt} = 5x - 2xy$

$\dfrac{dy}{dt} = -0.6y + 0.2xy$

*In Exercises 3 and 4 you are given the phase curve associated with a system of predator-prey equations, where x(t) denotes the prey (caribou) population, in hundreds, and y(t) denotes the predator (wolves) population, in tens, at time t. (a) Describe how each population changes over time t starting from t = 0. (b) Make a rough sketch of the graphs of x and y as a function of t on the same set of axes.*

**3.**

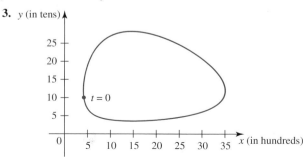

**4.**

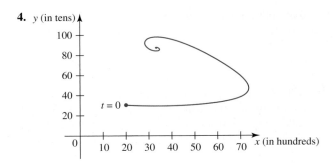

In Exercises 5 and 6 you are given the graphs of x and y as a function of time t, where x(t) denotes the prey population (in thousands) and y(t) denotes the predator population (in hundreds) at time t. Use them to sketch the associated phase curve.

**5.**

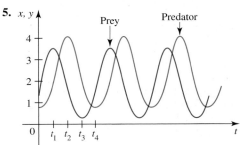

**6.**

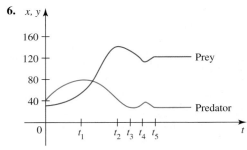

**cas** **7.** Consider the predator-prey equations

$$\frac{dx}{dt} = ax - bxy$$

$$\frac{dy}{dt} = -ry + sxy$$

where $a$, $b$, $r$, and $s$ are positive constants.

**a.** Show that

$$\frac{dy}{dx} = \frac{y(-r + sx)}{x(a - by)}$$

**b.** Show that an implicit solution of the separable differential equation in part (a) is

$$\frac{x^r y^a}{e^{sx} e^{by}} = C$$

where $C$ is a constant.

**c.** Find the equation of the phase curve passing through the point (5, 2). Then use a CAS to plot the curve of this implicit equation. Compare this curve with the phase curve shown in Figure 4 of this section.

**8.** In nature, the population of aphids (small insects that suck plant juices) is held in check by ladybugs. Assume that the population of aphids (in thousands), $x(t)$, and the population of ladybugs (in hundreds), $y(t)$, satisfy the equations

$$\frac{dx}{dt} = 2x - 1.2xy$$

$$\frac{dy}{dt} = -y + 0.8xy$$

**a.** Find the equilibrium points and interpret your results.
**b.** A direction field for the differential equation

$$\frac{dy}{dx} = \frac{-y + 0.8xy}{2x - 1.2xy}$$

is shown. Sketch a phase portrait superimposed over the direction field.

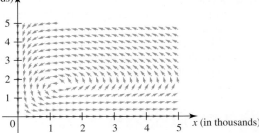

**c.** Suppose that initially there are 1000 aphids and 60 ladybugs. Draw the phase trajectory that satisfies this initial condition and use it to describe the behavior of both populations over time.
**d.** Use the result of part (c) to obtain the graphs of $x(t)$ and $y(t)$, and explain how they are related.

**9.** In the Lotka-Volterra model it was assumed that an unlimited amount of food was available to the prey. In a situation in which there is a finite amount of natural resources available to the prey, the Lotka-Volterra model can be modified to reflect this situation. Consider the following system of differential equations:

$$\frac{dx}{dt} = kx\left(1 - \frac{x}{L}\right) - axy$$

$$\frac{dy}{dt} = -by + cxy$$

where $x(t)$ and $y(t)$ represent the populations of prey and predators, respectively, and $a$, $b$, $c$, $k$, and $L$ are positive constants.

**a.** Describe what happens to the prey population in the absence of predators.

**b.** Describe what happens to the predator population in the absence of prey.

**c.** Find all the equilibrium points and explain their significance.

**10.** Refer to Exercise 9. Consider the modified Lotka-Volterra equations

$$\frac{dx}{dt} = 0.5x\left(1 - \frac{x}{150}\right) - 0.005xy$$

$$\frac{dy}{dt} = -0.2y + 0.004xy$$

where $x$ denotes the number of rabbits in tens and $y$ denotes the number of foxes at time $t$.

**a.** Using the result of Exercise 9, or otherwise, find the equilibrium points and interpret your results.

**b.** The figure shows the phase curve that starts at the point $(40, 30)$. Describe what happens to the rabbit and fox populations as $t$ increases.

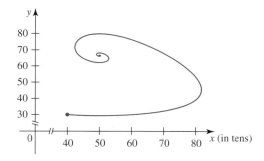

**c.** Sketch the graphs of the rabbit and fox populations as a function of time.

**11.** Consider the system of equations

$$\frac{dx}{dt} = k_1 x\left(1 - \frac{x}{L_1}\right) - axy$$

$$\frac{dy}{dt} = k_2 y\left(1 - \frac{y}{L_2}\right) - bxy$$

where $x(t)$ and $y(t)$ give the populations of two species $A$ and $B$, respectively, and $k_1$, $k_2$, $L_1$, $L_2$, $a$, and $b$ are positive constants.

**a.** Describe what happens to the population of $A$ in the absence of $B$.

**b.** Describe what happens to the population of $B$ in the absence of $A$.

**c.** Give a physical interpretation of the roles played by the terms $axy$ and $bxy$, and explain why the equations are called **competing species equations.** (Examples of competing species are trout and bass.)

**d.** Find the equilibrium points and interpret your results.

**12.** A model for the populations of trout, $x$, and bass, $y$, that compete for food and space is given by

$$\frac{dx}{dt} = 0.6x\left(1 - \frac{x}{4}\right) - 0.01xy$$

$$\frac{dy}{dt} = 0.1y\left(1 - \frac{y}{2}\right) - 0.01xy$$

where $x$ and $y$ are in thousands.

**a.** Find the equilibrium points of the system.

**b.** Plot the direction field for $dy/dx$.

**c.** Plot the phase curve that satisfies the initial condition $(6, 6)$ superimposed upon the direction field found in part (b). Does this agree with the result of part (a)?

**d.** Interpret your result.

# CHAPTER 7   REVIEW

## CONCEPT REVIEW

*In Exercises 1–8, fill in the blanks.*

**1. a.** A differential equation is one that involves the _____ or _____ of a(n) _____ function.

**b.** The order of a differential equation is the order of the _____ derivative that occurs in the equation.

**2. a.** A solution of the differential equation $dy/dx = f(x, y)$ is a function $y = y(x)$ defined on an _____ interval $I$ such that $y$ satisfies the _____ on $I$.

**b.** Graphically, the general solution of a differential equation represents a family of _____ called the _____ or _____ _____ of the differential equation.

**3.** A first-order separable differential equation is one that can be written in the form $dy/dx =$ _____; the differential equation $G(x)\, dx + H(y)\, dy = 0$ is _____.

**4. a.** The initial-value problem $dy/dt = ky$, $y(0) = y_0$ has the unique solution $y =$ _____.

**b.** The initial-value problem in part (a) is a model for exponential growth or decay; it is an exponential growth model if $k$ _____ and is an exponential decay model if $k$ _____.

**5. a.** A lineal element is a small portion of the _____ _____ at a point $(a, b)$ on a solution curve of the differential equation $y' = f(x, y)$.

**b.** A set of lineal elements drawn at various points is called a _____ field or _____ field of the differential equation.

**6. a.** The logistic differential equation has the form _____.

**b.** The solution of the logistic differential equation with initial condition $P(0) = P_0$ is _____.

**c.** If $P(0) = 0$, the population stays at _____ at all times; if $0 < P_0 < L$, then the population _____ and _____ the limiting value _____ asymptotically; if $P_0 = L$, the population at any time later remains at _____; if $P_0 > L$, then the population _____ and approaches _____ asymptotically.

**7. a.** A first-order linear differential equation is one that can be written in the form _____.

**b.** An integrating factor for the first-order linear differential equation in part (a) is _____.

**8.** To solve a first-order linear differential equation, (a) rewrite the equation in _____ form, (b) multiply both sides of the resulting equation by the _____ _____, then (c) _____ both sides of the resulting equation.

## REVIEW EXERCISES

**1.** Determine whether $y = C \cos x + \sin x$ is a solution of the differential equation $(\cos x)y' + (\sin x)y = 1$.

**2.** Determine whether $y = -2 + 3e^{3t}$ is a solution of the initial-value problem $y' - 3y = 6$, $y(0) = 1$.

*In Exercises 3–6, solve the differential equation.*

**3.** $\dfrac{dy}{dx} = 2xy^2$

**4.** $\dfrac{dy}{dx} = \dfrac{x}{y^2}$

**5.** $\dfrac{dy}{dx} = e^{y-x}$

**6.** $x\dfrac{dy}{dx} = y^2 + 1$

*In Exercises 7–10, solve the initial-value problem.*

**7.** $\dfrac{dy}{dx} = x^2 y^3$, $\quad y(0) = \dfrac{1}{2}$

**8.** $\dfrac{dy}{dx} + 3y = 6$, $\quad y(0) = 0$

**9.** $\dfrac{dy}{dx} = -\dfrac{x}{y}$, $\quad y(0) = 2$

**10.** $\dfrac{dy}{dx} = \dfrac{1 + x}{e^y}$, $\quad y(1) = 0$

**11.** Find the equation of a curve given that it passes through the point $\left(\frac{\pi}{4}, 1\right)$ and that the slope of the tangent line to the curve at any point $P(x, y)$ is given by $dy/dx = 4(y^2 + 1)$.

 **12.** Find the orthogonal trajectories of the family of curves given by $y^2 = x + C$. Use a graphing utility to draw several members of each family on the same set of axes.

**13. Bacteria Growth** A certain culture of bacteria grows at a rate that is proportional to the number present. If there are 1000 bacteria present initially and 4000 after 3 hr, find (a) an expression giving the number of bacteria in the culture after $t$ hr, (b) the number of bacteria in the culture after 6 hr, and (c) the time it takes for the number of bacteria to reach 400,000.

**14. Radioactivity** If 4 g of a radioactive substance is present at time $t = 1$ (years) and 1 g at $t = 6$, how much was present initially? What is the half-life of the substance?

**15. Radioactivity** The radioactive element radium-226 has a half-life of 1602 years. What is its decay constant?

**16. Rate of Return** A conglomerate purchased a hotel for $4.5 million and sold it five years later for $8.2 million. Find the annual rate of return (compounded continuously).

**17. Cost of Housing** The Brennans are planning to buy a house 4 years from now. Housing experts in their area have estimated that the cost of a home will increase at a rate of 3% per year compounded continuously over that 4-year period. If their predictions are correct, how much can the Brennans expect to pay for a house that currently costs $300,000?

**18. Newton's Law of Cooling** Newton's Law of Cooling (heating) states that the rate at which the temperature of an object changes is directly proportional to the difference in the temperature of the object and that of the surrounding medium. A thermometer is taken from the patio, where the temperature is 40°F, into a room where the temperature is 70°F. After 1 min, the thermometer read 52°F. How long did it take for the thermometer to reach 64°F?

**19. Newton's Law of Cooling** Refer to Exercise 18. A cup of coffee had a temperature of 200°F when it was removed from a microwave oven and placed on a counter in a room that was kept at a temperature of 70°F. The temperature of the coffee was 180°F after 5 min.

**a.** What was the temperature of the coffee after 10 min?

**b.** How long did it take for the coffee to cool to 120°F?

**20.** A motorboat is traveling in calm water when its motor is suddenly cut off. Ten seconds later, the boat's speed is 10 mph; and another 10 sec later, its speed is 4 mph. What was its speed at the instant of time when the motor was cut off if the resistance of the water is proportional to the speed of the boat?

**21. Future Value of an Annuity**  The future value $S$ of an annuity (a stream of payments made continuously) satisfies the equation

$$\frac{dS}{dt} = rS + d$$

where $r$ denotes the interest rate compounded continuously and $d$ is a positive constant giving the rate at which payments are made into the account.

   **a.** If the future value of an annuity at time $t = 0$ is $\$S_0$, find an expression for the future value of the annuity at any time $t$.

   **b.** If the future value of an annuity at $t = 0$ is \$10,000, the interest rate is 6% compounded continuously, and a constant stream of payments of \$2000 per year are made into the account, what is the future value of the annuity after 5 years?

**22.** A direction field for the differential equation $y' = y(1 - y)$ follows. Sketch the solution curves that satisfy the given initial conditions.

   **a.** $y(-1) = -0.5$
   **b.** $y(0) = 0.4$
   **c.** $y(0) = 1$
   **d.** $y(-1) = 1.4$

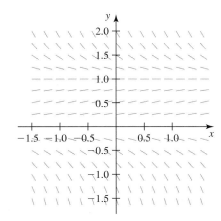

**23.** A direction field for the differential equation $y' = x^2 - y$ is shown in the figure.

   **a.** Sketch the solution curve for the initial-value problem

$$y' = x^2 - y \qquad y(0) = 1$$

   **b.** Use the graph of part (a) to estimate the value of $y$ when $x = 1$.

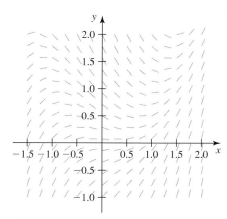

**24.** Use Euler's method with $n = 6$ to estimate $y(0.6)$, where $y$ is the solution of the initial-value problem

$$y' = y^2 - x \qquad y(0) = 1$$

**25.** Use Euler's method with $n = 5$ to estimate $y(0.5)$, where $y$ is the solution of the initial-value problem

$$y' = 2x^2y \qquad y(0) = 1$$

**26. Dissemination of Information**  Three hundred students attended the dedication ceremony of a new building on a college campus. The president of the college announced a new expansion program that included plans to make the college coeducational. The number of students who learned of the new program $t$ hr later is given by the function

$$f(t) = \frac{3000}{1 + Be^{-kt}}$$

If 600 students on campus had heard about the new program 2 hr after the ceremony, how many students had heard about the policy after 4 hr? How fast was the news spreading 4 hr after the announcement?

*In Exercises 27–30, solve the differential equation.*

**27.** $xy' + 2y = 4x^2$        **28.** $y' + 2xy = 3x$

**29.** $\dfrac{dy}{dx} + y = e^{-x} \cos 2x$    **30.** $y\dfrac{dx}{dy} + x = 2y$

*In Exercises 31–34, solve the initial-value problem.*

**31.** $3y' - y = 0, \quad y(-1) = 2$

**32.** $y' - y = 3x^2e^x, \quad y(0) = 0$

**33.** $xy' - y = x^2 \cos x, \quad y\!\left(\dfrac{\pi}{2}\right) = 0$

**34.** $xy' - y = x^2e^x, \quad y(1) = e$

**35. Trout Population** Marine biologists released 400 trout into a lake that has an estimated carrying capacity of 10,000. The trout population after the first year was 1000. Suppose that the pattern of growth of the trout population follows a logistic curve.
a. Find an equation giving the trout population after $t$ years.
b. What was the trout population after 6 years?
c. How long did it take for the population to reach 8000?

**36. Mixture Problem** A tank initially holds 200 gal of water in which 20 lb of salt has been dissolved. Brine containing 2 lb of salt per gallon enters the tank at the rate of 2 gal/min. The well-stirred solution drains from the tank at the rate of 3 gal/min. How much salt is in the tank after 20 min?

**37.** The rabbit population $x$ (in thousands) and the fox population $y$ (in hundreds) are modeled by the equations

$$\frac{dx}{dt} = 2x - 0.4xy$$

$$\frac{dy}{dt} = -1.2y + 0.3xy$$

a. What happens to the rabbit population in the absence of foxes? What happens to the fox population in the absence of rabbits?
b. Find the equilibrium points and interpret your results.
c. Find an expression for $dy/dx$ and use it to plot a phase portrait.
d. Use the phase curve corresponding to the initial populations of 4000 rabbits and 200 foxes to make a rough sketch of the graphs of $x$ and $y$ on the same set of axes.

**38.** The caribou population $x$ (in tens) and the wolf population $y$ are modeled by the equations

$$\frac{dx}{dt} = 0.5x\left(1 - \frac{x}{180}\right) - 0.005xy$$

$$\frac{dy}{dt} = -0.2y + 0.004xy$$

a. What can you say about the caribou population in the absence of wolves?
b. Find the equilibrium points and interpret your results.
c. The phase curve with initial point $(40, 30)$ is shown in the following figure. Use the phase curve to sketch the graphs of $x$ and $y$. What happens to the populations eventually?

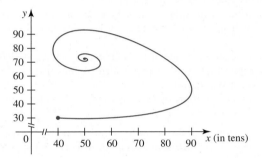

## CHALLENGE PROBLEMS

**1.** As a boat crosses the finish line in a regatta, the athlete stops rowing and allows the boat to coast to a stop. Assuming that the race takes place in a calm lake and that water resistance to the boat is directly proportional to its velocity, with constant of proportionality $k$, and that the velocity of the boat at the instant it crosses the finish line is $v_0$, show that the distance covered by the boat before it comes to a stop is $mv_0/k$, where $m$ is the combined mass of the boat and the athlete.
**Hint:** Find an expression for $s(t)$, the distance covered by the boat at time $t$ after crossing the finish line, and evaluate $\lim_{t \to \infty} s(t)$.

**2.** A *first-order homogeneous differential equation* is one of the form

$$\frac{dy}{dx} = f\left(\frac{y}{x}\right)$$

a. Show that the substitution $u = y/x$ reduces a homogeneous equation to a separable equation in the variables $u$ and $x$.
b. Solve $\dfrac{dy}{dx} = \dfrac{y - x}{y + x}$.

**3. Population Growth** Consider the logistic function

$$P(t) = \frac{L}{1 + \left(\frac{L}{P_0} - 1\right)e^{-kt}}$$

giving the population at time $t$ (see Example 2 in Section 7.3). Here, $k$ is the growth constant, $L$ is the carrying capacity of the environment, and $P_0$ is the initial population. Suppose that $P(1) = P_1$ and $P(2) = P_2$, where $P_1$ and $P_2$ are constants.

**a.** Show that

$$e^{-k} = \frac{P_0(P_2 - P_1)}{P_2(P_1 - P_0)} \quad \text{and} \quad L = \frac{P_1(P_0P_1 - 2P_0P_2 + P_1P_2)}{P_1^2 - P_0P_2}$$

**b.** The following table gives the population of the United States from the year 1900 through the year 2000.

| Year | 1900 | 1910 | 1920 | 1930 | 1940 | 1950 |
|------|------|------|------|------|------|------|
| Population (millions) | 76.21 | 92.23 | 106.02 | 123.20 | 132.16 | 151.33 |

| Year | 1960 | 1970 | 1980 | 1990 | 2000 |
|------|------|------|------|------|------|
| Population (millions) | 179.32 | 203.30 | 226.54 | 248.71 | 281.42 |

By taking $P_0 = 76.21$, $P_1 = 151.33$, and $P_2 = 281.42$, find an expression giving the population of the United States in year $t$, where $t$ is measured in 50-year intervals and $t = 0$ corresponds to 1900.

**c.** Plot the graph of $P(t)$.

**d.** Use the result of part (b) or part (c) to estimate the population of the United States in 2020.

*Source:* U.S. Census Bureau.

**4.** Find the orthogonal trajectories of the family of curves $x^2 + y^2 = 2ax$. Sketch a few members of each family.

**5. a.** Suppose that $y_1$ and $y_2$ are two different solutions of

$$\frac{dy}{dx} + P(x)y = Q(x)$$

Show that

$$y_2 = y_1(1 + Ce^{-\int [Q(x)/y_1(x)]\,dx})$$

**b.** Use the result of part (a) to solve $y' + xy = x^2 + 1$ by observing that $y = x$ is a solution.

**c.** Use the technique of Section 7.4 to verify the solution that you obtained in part (b).

**6.** The differential equation

$$\frac{dy}{dx} = P(x)y^2 + Q(x)y + R(x)$$

is called the *Riccati equation*. This equation occurs in electromagnetic theory and the study of optics. Suppose that one solution, $y_1(x)$, of the Riccati equation is known.

**a.** Show that the substitution $y = y_1 + 1/u$ reduces the Riccati equation to the first-order linear differential equation

$$\frac{du}{dx} + (2Py_1 + Q)u = -P$$

**b.** Verify that $y_1 = 1/x$ is a solution of

$$\frac{dy}{dx} = y^2 - \frac{2}{x^2}$$

Then use the result of part (a) to find the general solution.

**7. Radioactive Decay** A radioactive substance $A$ decays into another substance $B$ at a rate that is proportional to the amount present at time $t$. The new substance $B$ in turn decays into yet another substance $C$ at a rate that is proportional to the amount present at time $t$. If the amount of substance $A$ present initially is $A_0$ and there is no substance $B$ present initially, show that the amount of substance $B$ present at time $t$ is given by

$$B(t) = \frac{aA_0}{b - a}(e^{-at} - e^{-bt})$$

where $a$ and $b$ are the decay constants for substance $A$ and substance $B$, respectively, and $a \neq b$.

**8.** Show that the function defined by

$$f(x) = \frac{1}{2}\int_0^x \frac{e^{x-t}}{t}\,dt - \frac{1}{2}\int_0^x \frac{e^{t-x}}{t}\,dt$$

satisfies the differential equation

$$y'' - y = \frac{1}{x}$$

What percentage of the non-farm workforce in a country will be in the service industries one decade from now? In this chapter, we will see how a Taylor polynomial can be used to help answer this question.

Horizon International Images Limited/Alamy

# 8 Infinite Sequences and Series

**IF WE ALLOW** the number of terms of a sequence of real numbers to grow indefi-nitely, we obtain an *infinite sequence*. Infinite sequences that are *convergent* are of practical and theoretical interest. Indeed, it is the concept of a convergent sequence that allows us to define the *sum* of an *infinite series* (a series that is obtained by letting the number of terms of a series grow indefinitely). In this chapter we will see how a special type of infinite series called a *power series* affords us yet another way of representing a function. By representing a function in this manner, we are able to solve problems that we might otherwise not be able to solve.

**V** This symbol indicates that one of the following video types is available for enhanced student learning at **www.academic.cengage.com/login:**
- Chapter lecture videos
- Solutions to selected exercises

## 8.1 | Sequences

An idealized superball is dropped from a height of 1 m onto a flat surface. Suppose that each time the ball hits the surface, it rebounds to two thirds of its previous height. If we let $a_1$ denote the initial height of the ball, $a_2$ denote the maximum height attained on the first rebound, $a_3$ denote the maximum height attained on the second rebound, and so on, then we have

$$a_1 = 1, \qquad a_2 = \frac{2}{3}, \qquad a_3 = \frac{4}{9}, \qquad a_4 = \frac{8}{27}, \qquad \cdots$$

(See Figure 1.) This array of numbers, $a_1, a_2, a_3, \ldots$, is an example of an *infinite sequence,* or simply a *sequence.* If we define the function $f$ by $f(x) = \left(\frac{2}{3}\right)^{x-1}$ and allow $x$ to take on the positive integral values $x = 1, 2, 3, \ldots, n, \ldots$, then we see that the sequence $a_1, a_2, a_3, \ldots$, may be viewed as the functional values of $f$ at these numbers. Thus,

$$f(1) = 1 \qquad f(2) = \frac{2}{3} \qquad f(3) = \frac{4}{9} \qquad \cdots \qquad f(n) = \left(\frac{2}{3}\right)^{n-1} \qquad \cdots$$

$$\downarrow \qquad\qquad\quad \downarrow \qquad\qquad\quad \downarrow \qquad\qquad\qquad\qquad \downarrow$$

$$a_1 \qquad\qquad\quad a_2 \qquad\qquad\quad a_3 \qquad\qquad\qquad\qquad a_n \qquad\qquad \cdots$$

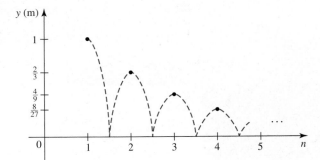

**FIGURE 1**
The ball rebounds to two thirds of its previous height upon hitting the surface.

This discussion motivates the following definition.

---

**DEFINITION    Sequence**

A **sequence** $\{a_n\}$ is a function whose domain is the set of positive integers. The functional values $a_1, a_2, a_3, \ldots, a_n, \ldots$ are the **terms** of the sequence, and the term $a_n$ is called the **$n$th term** of the sequence.

---

**Notes**

1. The sequence $\{a_n\}$ is also denoted by $\{a_n\}_{n=1}^{\infty}$.
2. Sometimes it is convenient to begin a sequence with $a_k$. In this case the sequence is $\{a_n\}_{n=k}^{\infty}$, and its terms are $a_k, a_{k+1}, a_{k+2}, \ldots, a_n, \ldots$. ∎

**EXAMPLE 1**    List the terms of the sequence.

**a.** $\left\{\dfrac{n}{n+1}\right\}$    **b.** $\left\{\dfrac{\sqrt{n}}{2^{n-1}}\right\}$    **c.** $\{(-1)^n\sqrt{n-2}\}_{n=2}^{\infty}$    **d.** $\left\{\sin\dfrac{n\pi}{3}\right\}_{n=0}^{\infty}$

**Solution**

**a.** Here, $a_n = f(n) = \dfrac{n}{n+1}$. Thus,

$$a_1 = f(1) = \frac{1}{1+1} = \frac{1}{2}, \quad a_2 = f(2) = \frac{2}{2+1} = \frac{2}{3}, \quad a_3 = f(3) = \frac{3}{3+1} = \frac{3}{4}, \quad \cdots$$

and we see that the given sequence can be written as

$$\left\{ \frac{n}{n+1} \right\} = \left\{ \frac{1}{2}, \frac{2}{3}, \frac{3}{4}, \frac{4}{5}, \dots, \frac{n}{n+1}, \dots \right\}$$

**b.** $\left\{ \dfrac{\sqrt{n}}{2^{n-1}} \right\} = \left\{ \dfrac{\sqrt{1}}{2^0}, \dfrac{\sqrt{2}}{2^1}, \dfrac{\sqrt{3}}{2^2}, \dfrac{\sqrt{4}}{2^3}, \dots, \dfrac{\sqrt{n}}{2^{n-1}}, \dots \right\}$

**c.** $\{(-1)^n \sqrt{n-2}\}_{n=2}^{\infty} = \{(-1)^2\sqrt{0}, (-1)^3\sqrt{1}, (-1)^4\sqrt{2},$

$$(-1)^5\sqrt{3}, \dots, (-1)^n\sqrt{n-2}, \dots\}$$

$$= \{0, -\sqrt{1}, \sqrt{2}, -\sqrt{3}, \dots, (-1)^n\sqrt{n-2}, \dots\}$$

Notice that $n$ starts from 2 in this example. (See Note 2 on page 726.)

**d.** $\left\{ \sin \dfrac{n\pi}{3} \right\}_{n=0}^{\infty} = \left\{ \sin 0, \sin \dfrac{\pi}{3}, \sin \dfrac{2\pi}{3}, \sin \dfrac{3\pi}{3}, \sin \dfrac{4\pi}{3}, \sin \dfrac{5\pi}{3}, \dots, \sin \dfrac{n\pi}{3}, \dots \right\}$

$$= \left\{ 0, \frac{\sqrt{3}}{2}, \frac{\sqrt{3}}{2}, 0, -\frac{\sqrt{3}}{2}, -\frac{\sqrt{3}}{2}, \dots, \sin \frac{n\pi}{3}, \dots \right\}$$

Once again, refer to Note 2. ■

We can often determine the $n$th term of a sequence by studying the first few terms of the sequence and recognizing the pattern that emerges.

**EXAMPLE 2** Find an expression for the $n$th term of each sequence.

**a.** $\left\{ 2, \dfrac{3}{\sqrt{2}}, \dfrac{4}{\sqrt{3}}, \dfrac{5}{\sqrt{4}}, \dots \right\}$    **b.** $\left\{ 1, \dfrac{1}{8}, \dfrac{1}{27}, \dfrac{1}{64}, \dots \right\}$    **c.** $\left\{ 1, -\dfrac{1}{2}, \dfrac{1}{3}, -\dfrac{1}{4}, \dots \right\}$

**Solution**

**a.** The terms of the sequence may be written in the form

$$a_1 = \frac{1+1}{\sqrt{1}}, \quad a_2 = \frac{2+1}{\sqrt{2}}, \quad a_3 = \frac{3+1}{\sqrt{3}}, \quad a_4 = \frac{4+1}{\sqrt{4}}, \quad \cdots$$

from which we see that $a_n = \dfrac{n+1}{\sqrt{n}}$.

**b.** Here,

$$a_1 = \frac{1}{1^3}, \quad a_2 = \frac{1}{2^3}, \quad a_3 = \frac{1}{3^3}, \quad a_4 = \frac{1}{4^3}, \quad \cdots$$

so $a_n = \dfrac{1}{n^3}$.

**c.** Note that $(-1)^r$ is equal to 1 if $r$ is an even integer and $-1$ if $r$ is an odd integer. Using this result, we obtain

$$a_1 = \frac{(-1)^0}{1}, \quad a_2 = \frac{(-1)^1}{2}, \quad a_3 = \frac{(-1)^2}{3}, \quad a_4 = \frac{(-1)^3}{4}, \quad \cdots$$

We conclude that the $n$th term is $a_n = (-1)^{n-1}/n$. ■

Some sequences are defined **recursively;** that is, the sequence is defined by specifying the first term or the first few terms of the sequence and a rule for calculating any other term of the sequence from the preceding term(s).

**EXAMPLE 3** List the first five terms of the recursively defined sequence $a_1 = 2$, $a_2 = 4$, and $a_{n+1} = 2a_n - a_{n-1}$ for $n \geq 2$.

**Solution**   The first two terms of the sequence are given as $a_1 = 2$ and $a_2 = 4$. To find the third term of the sequence, we put $n = 2$ in the recursion formula to obtain

$$a_3 = 2a_2 - a_1 = 2 \cdot 4 - 2 = 6$$

Next, putting $n = 3$ and $n = 4$ in succession in the recursive formula gives

$$a_4 = 2a_3 - a_2 = 2 \cdot 6 - 4 = 8 \qquad \text{and} \qquad a_5 = 2a_4 - a_3 = 2 \cdot 8 - 6 = 10 \qquad ∎$$

Since a sequence is a function, we can draw its graph. The graphs of the sequences $\{n/(n + 1)\}$ and $\{(-1)^n\}$ are shown in Figure 2. They are just the graphs of the functions $f(n) = n/(n + 1)$ and $g(n) = (-1)^n$ for $n = 1, 2, 3, \ldots$.

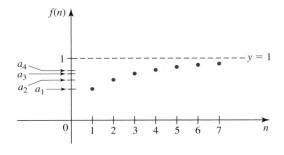

**(a)** The graph of $\left\{\dfrac{n}{n+1}\right\}$

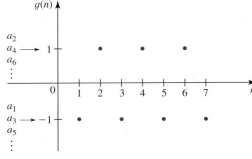

**(b)** The graph of $\{(-1)^n\}$

**FIGURE 2**

## ■ Limit of a Sequence

If you examine the graph of the sequence $\{n/(n + 1)\}$ sketched in Figure 2a, you will see that the terms of the sequence seem to get closer and closer to 1 as $n$ gets larger and larger. In this situation we say that the sequence $\{n/(n + 1)\}$ converges to the *limit* 1, written

$$\lim_{n \to \infty} \frac{n}{n + 1} = 1$$

In general, we have the following informal definition of the limit of a sequence.

**DEFINITION**   **Limit of a Sequence**

A sequence $\{a_n\}$ has the **limit** $L$, written

$$\lim_{n \to \infty} a_n = L$$

if $a_n$ can be made as close to $L$ as we please by taking $n$ sufficiently large. If $\lim_{n \to \infty} a_n$ exists, we say that the sequence **converges.** Otherwise, we say that the sequence **diverges.**

A more precise definition of the limit of a sequence follows.

---

**DEFINITION (Precise)    Limit of a Sequence**

A sequence $\{a_n\}$ **converges** and has the **limit** $L$, written

$$\lim_{n\to\infty} a_n = L$$

if for every $\varepsilon > 0$ there exists a positive integer $N$ such that $|a_n - L| < \varepsilon$ whenever $n > N$.

---

To illustrate this definition, suppose that a challenger selects an $\varepsilon > 0$. Then we must show that there exists a positive integer $N$ such that all points $(n, a_n)$ on the graph of $\{a_n\}$, where $n > N$, lie inside a band of width $2\varepsilon$ about the line $y = L$. (See Figure 3.)

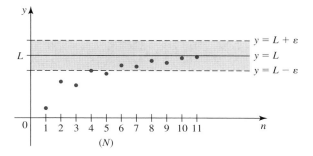

**FIGURE 3**
If $n > N$, then $L - \varepsilon < a_n < L + \varepsilon$
or, equivalently, $|a_n - L| < \varepsilon$.

To reconcile this definition with the intuitive definition of a limit, recall that $\varepsilon$ is arbitrary. Therefore, by choosing $\varepsilon$ very small, the challenger ensures that $a_n$ is "close" to $L$. Furthermore, if corresponding to *each* choice of $\varepsilon$, we can produce an $N$ such that $n > N$ implies that $|a_n - L| < \varepsilon$, then we have shown that $a_n$ can be made as close to $L$ as we please by taking $n$ sufficiently large.

Notice that the definition of the limit of a sequence is very similar to the definition of the limit of a function at infinity given in Section 3.5. This is expected, since the only difference between a function $f$ defined by $y = f(x)$ on the interval $(0, \infty)$ and the sequence $\{a_n\}$ defined by $a_n = f(n)$ is that $n$ is an integer. (See Figure 4.) This observation tells us that we can often evaluate $\lim_{n\to\infty} a_n$ by evaluating $\lim_{x\to\infty} f(x)$, where $f$ is defined on $(0, \infty)$ and $a_n = f(n)$.

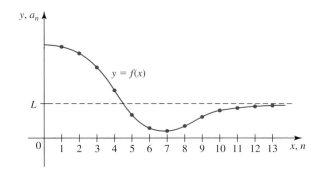

**FIGURE 4**
The graph of $\{a_n\}$ comprises
the points $(n, f(n))$ that lie
on the graph of $y = f(x)$.

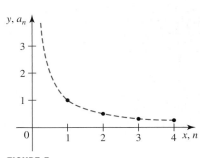

**FIGURE 5**
The graph of $\{1/n^r\}$ for $n = 1, 2, 3, 4$, and $r = 1$. The graph of $f(x) = 1/x^r$ is shown with a dashed curve.

**THEOREM 1**

If $\lim_{x \to \infty} f(x) = L$ and $\{a_n\}$ is a sequence defined by $a_n = f(n)$, where $n$ is a positive integer, then $\lim_{n \to \infty} a_n = L$.

You will be asked to prove Theorem 1 in Exercise 75.

**EXAMPLE 4** Find $\lim_{n \to \infty} \dfrac{1}{n^r}$ if $r > 0$.

**Solution** Since $a_n = 1/n^r$, we choose $f(x) = 1/x^r$, where $x > 0$. By Theorem 1 in Section 3.5 we have

$$\lim_{x \to \infty} \frac{1}{x^r} = 0$$

Using Theorem 1 of this section, we conclude that

$$\lim_{n \to \infty} \frac{1}{n^r} = 0$$

(See Figure 5.)

The converse of Theorem 1 is false. Consider, for example, the sequence $\{\sin n\pi\} = \{0\}$. This sequence evidently converges to 0, since every term of the sequence is 0. But $\lim_{x \to \infty} \sin \pi x$ does not exist. (See Figure 6.)

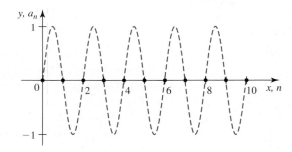

**FIGURE 6**
The graph of $\{\sin n\pi\}$ for $n = 0, 1, 2, \ldots, 10$. The graph of $f(x) = \sin \pi x$ is shown with a dashed curve.

The following limit laws for sequences are the analogs of the limit laws for functions studied in Section 1.2 and are proved in a similar manner.

**THEOREM 2**   **Limit Laws for Sequences**

Suppose that $\lim_{n \to \infty} a_n = L$ and $\lim_{n \to \infty} b_n = M$ and that $c$ is a constant. Then

1. $\lim_{n \to \infty} ca_n = cL$

2. $\lim_{n \to \infty} (a_n \pm b_n) = L \pm M$

3. $\lim_{n \to \infty} a_n b_n = LM$

4. $\lim_{n \to \infty} \dfrac{a_n}{b_n} = \dfrac{L}{M}$,   provided that $b_n \neq 0$ and $M \neq 0$

5. $\lim_{n \to \infty} a_n^p = L^p$,   if $p > 0$ and $a_n > 0$

**EXAMPLE 5** Determine whether the sequence converges or diverges.

**a.** $\left\{\dfrac{n}{n+1}\right\}$    **b.** $\{(-1)^n\}$

**Solution**

**a.** Both the numerator and the denominator of $n/(n+1)$ approach infinity as $n$ approaches infinity. So their limits do not exist, and we cannot use Law 4 of Theorem 2. But we can divide the numerator and denominator by $n$ and then apply Law 4 to evaluate the resulting limit. Thus,

$$\lim_{n\to\infty} \frac{n}{n+1} = \lim_{n\to\infty} \frac{1}{1+\dfrac{1}{n}} = 1$$

and we conclude that the sequence converges to 1. (See Figure 2a.)

**b.** The terms of the sequence are

$$-1, 1, -1, 1, \ldots$$

The sequence evidently does not approach a unique number, and we conclude that it diverges. (See Figure 2b.)

**EXAMPLE 6** Find

**a.** $\lim_{n\to\infty} \dfrac{\ln n}{n}$    **b.** $\lim_{n\to\infty} \dfrac{e^n}{n^2}$

**Solution**

**a.** Observe that both the numerator and the denominator of $(\ln n)/n$ approach infinity as $n \to \infty$. Therefore, we may not use Law 4 of Theorem 2 directly. Since $a_n = f(n) = (\ln n)/n$, we consider the function $f(x) = (\ln x)/x$. Using l'Hôpital's Rule, we find

$$\lim_{x\to\infty} \frac{\ln x}{x} = \lim_{x\to\infty} \frac{1/x}{1} = \lim_{x\to\infty} \frac{1}{x} = 0$$

Therefore, by Theorem 1 we conclude that

$$\lim_{n\to\infty} \frac{\ln n}{n} = 0$$

(See Figure 7.)

**b.** Once again both $e^n$ and $n^2$ approach infinity as $n \to \infty$. Choose $f(x) = e^x/x^2$, and use l'Hôpital's Rule twice to find that

$$\lim_{x\to\infty} \frac{e^x}{x^2} = \lim_{x\to\infty} \frac{e^x}{2x} = \lim_{x\to\infty} \frac{e^x}{2} = \infty$$

from which we see that

$$\lim_{n\to\infty} \frac{e^n}{n^2} = \infty$$

and we conclude that the sequence $\{e^n/n^2\}$ is divergent. (See Figure 8.)

The Squeeze Theorem has the following counterpart for sequences. (The proof is similar to that of the Squeeze Theorem and will be omitted.)

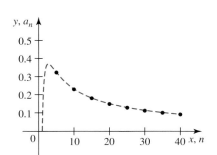

**FIGURE 7**
The graph of $\{(\ln n)/n\}$ for $n = 5, 10, 15, \ldots, 40$. The graph of $f(x) = (\ln x)/x$ is shown with a dashed curve.

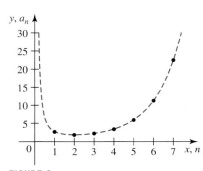

**FIGURE 8**
The graph of $\{e^n/n^2\}$. The graph of $f(x) = e^x/x^2$ is shown with a dashed curve.

**THEOREM 3    Squeeze Theorem for Sequences**

If there exists some integer $N$ such that $a_n \leq b_n \leq c_n$ for all $n \geq N$ and $\lim_{n \to \infty} a_n = \lim_{n \to \infty} c_n = L$, then $\lim_{n \to \infty} b_n = L$.

(See Figure 9.)

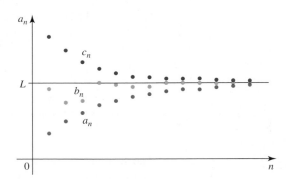

**FIGURE 9**

The sequence $\{b_n\}$ is squeezed between the sequences $\{a_n\}$ and $\{c_n\}$.

**EXAMPLE 7**    Find $\lim\limits_{n \to \infty} \dfrac{n!}{n^n}$, where $n!$ (read "$n$ factorial") is defined by

$$n! = n(n-1)(n-2) \cdots 1$$

**Solution**    Let $a_n = n!/n^n$. The first few terms of $\{a_n\}$ are

$$a_1 = \frac{1!}{1} = 1, \qquad a_2 = \frac{2!}{2^2} = \frac{2 \cdot 1}{2 \cdot 2}, \qquad a_3 = \frac{3!}{3^3} = \frac{3 \cdot 2 \cdot 1}{3 \cdot 3 \cdot 3}$$

and its $n$th term is

$$a_n = \frac{n!}{n^n} = \frac{n(n-1) \cdot \cdots \cdot 3 \cdot 2 \cdot 1}{n \cdot n \cdot \cdots \cdot n \cdot n \cdot n} = \left(\frac{n}{n}\right)\left(\frac{n-1}{n}\right) \cdot \cdots \cdot \left(\frac{3}{n}\right)\left(\frac{2}{n}\right)\left(\frac{1}{n}\right) \leq \frac{1}{n}$$

Therefore,

$$0 < a_n \leq \frac{1}{n}$$

Since $\lim_{n \to \infty} 1/n = 0$, the Squeeze Theorem implies that

$$\lim_{n \to \infty} a_n = \lim_{n \to \infty} \frac{n!}{n^n} = 0$$

The next theorem is an immediate consequence of the Squeeze Theorem.

**THEOREM 4**

If $\lim_{n \to \infty} |a_n| = 0$, then $\lim_{n \to \infty} a_n = 0$.

You are asked to prove Theorem 4 in Exercise 76.

**EXAMPLE 8** Find $\lim\limits_{n\to\infty} \dfrac{(-1)^n}{n}$.

**Solution** Since

$$\lim_{n\to\infty}\left|\frac{(-1)^n}{n}\right| = \lim_{n\to\infty}\frac{1}{n} = 0$$

we conclude by Theorem 4 that

$$\lim_{n\to\infty}\frac{(-1)^n}{n} = 0$$

The graph of the sequence $\{(-1)^n/n\}$ confirms this result. (See Figure 10.)

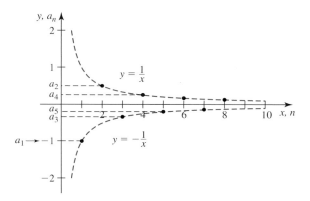

**FIGURE 10**
The terms of the sequence $\{(-1)^n/n\}$ oscillate between the graphs of $y = 1/x$ and $y = -1/x$.

If we take the composition of a function $f$ with a sequence $\{a_n\}$, we obtain another sequence $\{f(a_n)\}$. The following theorem shows how to compute the limit of the latter. The proof will be given in Appendix B.

---

**THEOREM 5**

If $\lim_{n\to\infty} a_n = L$ and the function $f$ is continuous at $L$, then

$$\lim_{n\to\infty} f(a_n) = f(\lim_{n\to\infty} a_n) = f(L)$$

---

**Note** Compare this theorem with Theorem 5 in Section 1.4.

**EXAMPLE 9** Find $\lim_{n\to\infty} e^{\sin(1/n)}$.

**Solution** Observe that $e^{\sin(1/n)} = f(a_n)$, where $f(x) = e^x$ and $a_n = \sin(1/n)$. Since

$$\lim_{n\to\infty} \sin\frac{1}{n} = 0$$

and $f$ is continuous at 0, Theorem 5 gives $\lim_{n\to\infty} e^{\sin(1/n)} = e^{\lim\limits_{n\to\infty}\sin(1/n)} = e^0 = 1.$ ∎

### ■ Bounded Monotonic Sequences

Up to now, the convergent sequences that we have dealt with had limits that are readily found. Sometimes, however, we need to show that a sequence is convergent even if its precise limit is not readily found. Our immediate goal here is to find conditions that will guarantee that a sequence converges. To do this, we need to make use of two further properties of sequences.

---

**DEFINITION**   **Monotonic Sequence**

A sequence $\{a_n\}$ is **increasing** if

$$a_1 < a_2 < a_3 < \cdots < a_n < a_{n+1} < \cdots$$

and **decreasing** if

$$a_1 > a_2 > a_3 > \cdots > a_n > a_{n+1} > \cdots$$

A sequence is **monotonic** if it is either increasing or decreasing.

---

**EXAMPLE 10**   Show that the sequence $\left\{\dfrac{n}{n+1}\right\}$ is increasing.

**Solution**   Let $a_n = n/(n+1)$. We must show that $a_n \le a_{n+1}$ for all $n \ge 1$; that is,

$$\frac{n}{n+1} \le \frac{n+1}{(n+1)+1}$$

or

$$\frac{n}{n+1} \le \frac{n+1}{n+2}$$

To show that this inequality is true, we obtain the following equivalent inequalities:

$$n(n+2) \le (n+1)(n+1) \qquad \text{Cross-multiply.}$$

$$n^2 + 2n \le n^2 + 2n + 1$$

$$0 \le 1$$

which is true for $n \ge 1$. Therefore, $a_n \le a_{n+1}$, so $\{a_n\}$ is increasing.

**Alternative Solution**   Here, $a_n = f(n) = n/(n+1)$. So consider the function $f(x) = x/(x+1)$. Since

$$f'(x) = \frac{(x+1)(1) - x(1)}{(x+1)^2} = \frac{1}{(x+1)^2} > 0 \qquad \text{if}\quad x > 0$$

we see that $f$ is increasing on $(0, \infty)$. Therefore, the given sequence is increasing.   ■

**EXAMPLE 11**   Show that the sequence $\left\{\dfrac{n}{e^n}\right\}$ is decreasing.

**Solution** We must show that $a_n \geq a_{n+1}$ for $n \geq 1$; that is,

$$\frac{n}{e^n} \geq \frac{n+1}{e^{n+1}}$$

$$ne^{n+1} \geq (n+1)e^n$$

$$ne \geq n+1 \qquad \text{Divide both sides by } e^n.$$

$$n(e-1) \geq 1$$

which is true for all $n \geq 1$, so $\{n/e^n\}$ is decreasing. ∎

Next, we explain what is meant by a *bounded* sequence.

---

**DEFINITION** **Bounded Sequence**

A sequence $\{a_n\}$ is **bounded above** if there exists a number $M$ such that

$$a_n \leq M \qquad \text{for all } n \geq 1$$

A sequence is **bounded below** if there exists a number $m$ such that

$$m \leq a_n \qquad \text{for all } n \geq 1$$

A sequence is **bounded** if it is both bounded above and bounded below.

---

For example, the sequence $\{n\}$ is bounded below by 0, but it is not bounded above. The sequence $\{n/(n+1)\}$ is bounded below by $\frac{1}{2}$ and above by 1 and is therefore bounded. (See Figure 2a.)

A bounded sequence need not be convergent. For example, the sequence $\{(-1)^n\}$ is bounded, since $-1 \leq (-1)^n \leq 1$; but it is evidently divergent. (See Figure 2b.) Also, a monotonic sequence need not be convergent. For example, the sequence $\{n\}$ is increasing and evidently divergent. However, if a sequence is *both* bounded and monotonic, then it must be convergent.

---

**THEOREM 6** **Monotone Convergence Theorem for Sequences**

Every bounded, monotonic sequence is convergent.

---

The plausibility of Theorem 6 is suggested by the sequence $\{n/(n+1)\}$ whose graph is reproduced in Figure 11. This sequence is increasing and bounded above by

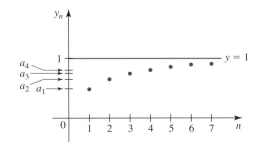

**FIGURE 11**
The increasing, bounded sequence $\{n/(n+1)\}$ is convergent.

any number $M \geq 1$. Therefore, as $n$ increases, the terms $a_n$ approach a number (which is no larger than $M$) from below. In this case the number is 1, which is also the limit of this sequence. (A proof of Theorem 6 is given at the end of this section.)

Theorem 6 can be used to find the limit of a convergent sequence indirectly, as the next example shows. It will also play an important role in infinite series (Sections 8.2–8.9).

**EXAMPLE 12** Show that $\left\{ \dfrac{2^n}{n!} \right\}$ is convergent and find its limit.

**Solution** Here, $a_n = 2^n/n!$. The first few terms of the sequence are

$$a_1 = 2, \qquad a_2 = 2, \qquad a_3 \approx 1.333333, \qquad a_4 \approx 0.666667, \qquad a_5 \approx 0.266667,$$

$$a_6 \approx 0.088889, \qquad \dots, \qquad a_{10} \approx 0.000282$$

These terms suggest that the sequence is decreasing from $n = 2$ onward. To prove this, we compute

$$\frac{a_{n+1}}{a_n} = \frac{\dfrac{2^{n+1}}{(n+1)!}}{\dfrac{2^n}{n!}} = \frac{2^{n+1} n!}{2^n (n+1)!} = \frac{2n!}{(n+1)n!} = \frac{2}{n+1} \tag{1}$$

So

$$\frac{a_{n+1}}{a_n} \leq 1 \quad \text{if} \quad n \geq 1$$

Thus, $a_{n+1} \leq a_n$ if $n \geq 1$, and this proves the assertion. Since all of the terms of the sequence are positive, $\{a_n\}$ is bounded below by 0. Therefore, the sequence is decreasing and bounded below, and Theorem 6 guarantees that it converges to a nonnegative limit $L$.

To find $L$, we first use Equation (1) to write

$$a_{n+1} = \frac{2}{n+1} a_n \tag{2}$$

Since $\lim_{n \to \infty} a_n = L$, we also have $\lim_{n \to \infty} a_{n+1} = L$. Taking the limit on both sides of Equation (2) and using Law (3) for limits of sequences, we obtain

$$L = \lim_{n \to \infty} a_{n+1} = \lim_{n \to \infty} \left( \frac{2}{n+1} a_n \right) = \lim_{n \to \infty} \frac{2}{n+1} \cdot \lim_{n \to \infty} a_n = 0 \cdot L = 0$$

We conclude that $\lim_{n \to \infty} 2^n/n! = 0$.

**Alternative Solution** Observe that

$$a_2 = \frac{2 \cdot 2}{2 \cdot 1} = 2, \quad a_3 = \frac{2 \cdot 2 \cdot 2}{3 \cdot 2 \cdot 1} = 2\left(\frac{2}{3}\right), \quad a_4 = \frac{2 \cdot 2 \cdot 2 \cdot 2}{4 \cdot 3 \cdot 2 \cdot 1} = \left(\frac{2}{4}\right)\left(\frac{2}{3}\right)2 < 2\left(\frac{2}{3}\right)^2,$$

$$a_5 = \frac{2 \cdot 2 \cdot 2 \cdot 2 \cdot 2}{5 \cdot 4 \cdot 3 \cdot 2 \cdot 1} = \left(\frac{2}{5}\right)\left(\frac{2}{4}\right)\left(\frac{2}{3}\right)2 < 2\left(\frac{2}{3}\right)^3$$

and

$$a_n = \frac{2 \cdot 2 \cdot 2 \cdot \cdots \cdot 2}{n \cdot (n-1) \cdot (n-2) \cdot \cdots \cdot 1} < 2\left(\frac{2}{3}\right)^{n-2}$$

Therefore

$$0 < a_n < 2\left(\frac{2}{3}\right)^{n-2}$$

Since $\lim_{n\to\infty}\left(\frac{2}{3}\right)^{n-2} = 0$, the Squeeze Theorem gives the desired result.    ■

The next example contains some important results that we will derive here using the Squeeze Theorem. (You will also be asked to demonstrate their validity using the properties of exponential functions in Exercise 77.)

**EXAMPLE 13**   Show that $\lim_{n\to\infty} r^n = 0$ if $|r| < 1$.

**Solution**   If $r = 0$, then each term of the sequence $\{r^n\}$ is 0, and the sequence converges to 0. Now suppose that $0 < |r| < 1$. Then $1/|r|$ is greater than 1. So there exists a positive number $p$ such that

$$\frac{1}{|r|} = 1 + p$$

Using the Binomial Theorem, we have

$$(1 + p)^n = 1 + np + \frac{n(n-1)}{2!} p^2 + \cdots + p^n > np$$

Thus,

$$0 < |r|^n = \frac{1}{(1+p)^n} < \frac{1}{np}$$

But

$$\lim_{n\to\infty} \frac{1}{np} = 0$$

so by the Squeeze Theorem

$$\lim_{n\to\infty} |r|^n = 0$$

Finally, using Theorem 4, we conclude that $\lim_{n\to\infty} r^n = 0$.    ■

If $r = 1$, then $r^n = 1$ for all $n$, and the sequence $\{r^n\}$ evidently converges to 1. If $r = -1$, then the sequence $\{r^n\} = \{(-1)^n\}$ is divergent. (See Example 5b.) If $|r| > 1$, then $|r| = 1 + p$ for some positive number $p$. Using the Binomial Theorem again, we have

$$|r|^n = (1 + p)^n > np$$

Since $p > 0$, $\lim_{n\to\infty} np = \infty$. This shows that $\{r^n\}$ diverges if $|r| > 1$.

A summary of these results follows.

**Properties of the Sequence $\{r^n\}$**

The sequence $\{r^n\}$ converges if $-1 < r \le 1$ and

$$\lim_{n\to\infty} r^n = \begin{cases} 0 & \text{if } -1 < r < 1 \\ 1 & \text{if } r = 1 \end{cases}$$

It diverges for all other values of $r$.

## ■ Proof of Theorem 6

The proof of Theorem 6 depends on the **Completeness Axiom** for the real number system, which states that *every nonempty set S of real numbers that is bounded above has a least upper bound.* Thus, if $x \leq M$ for all $x$ in $S$, then there must be a real number $b$ such that $b$ is an upper bound of $S$ ($x \leq b$ for all $x \in S$), and if $N$ is any upper bound of $S$, then $N \geq b$. For example, if $S$ is the interval $(-2, 3)$, then the number 4 (or any number greater than 3) is an upper bound of $S$ and 3 is the least upper bound of $S$. As a consequence of this axiom, it can be shown that every nonempty set of real numbers that is bounded below has a greatest lower bound. The Completeness Axiom merely states that the real number line has no gaps or holes.

**PROOF**    Suppose that $\{a_n\}$ is an increasing sequence. Since $\{a_n\}$ is bounded, the set $S = \{a_n \mid n \geq 1\}$ is bounded above, and by the Completeness Axiom it has a least upper bound $L$. (See Figure 12.) We claim that $L$ is the limit of the sequence.

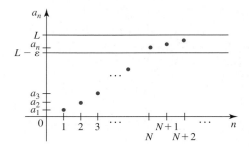

**FIGURE 12**
An increasing sequence bounded above
must converge to its least upper bound.

To show this, let $\varepsilon > 0$ be given. Then $L - \varepsilon$ is *not* an upper bound of $S$ (since $L$ is the least upper bound of $S$). Therefore, there exists an integer $N$ such that $a_N > L - \varepsilon$. But the sequence is increasing, so $a_n \geq a_N$ for every $n > N$. In other words, if $n > N$, we have $a_n > L - \varepsilon$. Since $a_n \leq L$,

$$0 \leq L - a_n < \varepsilon$$

This shows that

$$|L - a_n| < \varepsilon$$

whenever $n > N$, so $\lim_{n \to \infty} a_n = L$.

The proof is similar for the case in which $\{a_n\}$ is decreasing, except that we use the greatest lower bound instead.    ■

## 8.1    CONCEPT QUESTIONS

1. Explain each of the following terms in your own words, and give an example of each.
   a. Sequence
   b. Convergent sequence
   c. Divergent sequence
   d. Limit of a sequence

2. Explain each of the following terms in your own words, and give an example of each.
   a. Bounded sequence
   b. Monotonic sequence

## 8.1 EXERCISES

*In Exercises 1–6, write the first five terms of the sequence $\{a_n\}$ whose nth term is given.*

**1.** $a_n = \dfrac{n+1}{2n-1}$

**2.** $a_n = \dfrac{(-1)^{n+1}2^n}{n+1}$

**3.** $a_n = \sin\dfrac{n\pi}{2}$

**4.** $a_n = \dfrac{1\cdot 3\cdot 5\cdot\cdots\cdot(2n-1)}{n!}$

**5.** $a_n = \dfrac{2^n}{(2n)!}$

**6.** $a_1 = 2, \quad a_{n+1} = 3a_n + 1$

*In Exercises 7–12, find an expression for the nth term of the sequence. (Assume that the pattern continues.)*

**7.** $\left\{\dfrac{1}{2},\dfrac{2}{3},\dfrac{3}{4},\dfrac{4}{5},\dfrac{5}{6},\cdots\right\}$

**8.** $\left\{\dfrac{3}{4},\dfrac{4}{9},\dfrac{5}{16},\dfrac{6}{25},\dfrac{7}{36},\cdots\right\}$

**9.** $\left\{-1,\dfrac{1}{2},-\dfrac{1}{6},\dfrac{1}{24},-\dfrac{1}{120},\cdots\right\}$

**10.** $\{0,2,0,2,0,\ldots\}$

**11.** $\left\{\dfrac{1}{1\cdot 2},\dfrac{2}{2\cdot 3},\dfrac{3}{3\cdot 4},\dfrac{4}{4\cdot 5},\dfrac{5}{5\cdot 6},\cdots\right\}$

**12.** $\left\{\dfrac{1}{2},\dfrac{1\cdot 3}{2\cdot 4},\dfrac{1\cdot 3\cdot 5}{2\cdot 4\cdot 6},\dfrac{1\cdot 3\cdot 5\cdot 7}{2\cdot 4\cdot 6\cdot 8},\dfrac{1\cdot 3\cdot 5\cdot 7\cdot 9}{2\cdot 4\cdot 6\cdot 8\cdot 10},\cdots\right\}$

*In Exercises 13–42, determine whether the sequence $\{a_n\}$ converges or diverges. If it converges, find its limit.*

**13.** $a_n = \dfrac{2n}{n+1}$

**14.** $a_n = \sqrt{n+1}$

**15.** $a_n = 1 + 2(-1)^n$

**16.** $a_n = 1 + \dfrac{(-1)^n}{n^{3/2}}$

**17.** $a_n = \dfrac{n-1}{n} - \dfrac{2n+1}{n^2}$

**18.** $a_n = \dfrac{n^2-1}{2n^2+1}$

**19.** $a_n = \dfrac{2n^2-3n+4}{3n^2+1}$

**20.** $a_n = (-1)^n\dfrac{n+2}{3n+1}$

**21.** $a_n = \dfrac{2+(-1)^n}{n}$

**22.** $a_n = \dfrac{\sqrt{2n^2+1}}{n}$

**23.** $a_n = \dfrac{2n}{\sqrt{n+1}}$

**24.** $a_n = 1 + \left(-\dfrac{2}{e}\right)^n$

**25.** $a_n = \dfrac{n^{1/2}+n^{1/3}}{n+2n^{2/3}}$

**26.** $a_n = \cos n\pi + 2$

**27.** $a_n = \sin\dfrac{n\pi}{2}$

**28.** $a_n = \sin\left(\dfrac{n\pi}{2n+1}\right)$

**29.** $a_n = \dfrac{\sin\sqrt{n}}{\sqrt{n}}$

**30.** $a_n = \tan^{-1}n^2$

**31.** $a_n = \tanh n$

**32.** $a_n = \dfrac{\ln n^2}{\sqrt{n}}$

**33.** $a_n = \dfrac{2^n}{3^n+1}$

**34.** $a_n = \dfrac{2^n+1}{e^n}$

**35.** $a_n = \sqrt{n+1} - \sqrt{n}$

**36.** $a_n = \dfrac{n^p}{e^n}, \quad p > 0$

**37.** $a_n = \left(1+\dfrac{2}{n}\right)^{1/n}$

**38.** $a_n = \dfrac{(-2)^n}{n!}$

**39.** $a_n = \dfrac{\sin^2 n}{\sqrt{n}}$

**40.** $a_n = \dfrac{1\cdot 3\cdot 5\cdot\cdots\cdot(2n-1)}{n!}$

**41.** $a_n = \dfrac{1}{n^2} + \dfrac{2}{n^2} + \dfrac{3}{n^2} + \cdots + \dfrac{n}{n^2}$

**42.** $a_n = \dfrac{1+2+3+\cdots+n}{n+2} - \dfrac{n}{2}$

*In Exercises 43–48, (a) graph the sequence $\{a_n\}$ with a graphing utility, (b) use your graph to guess at the convergence or divergence of the sequence, and (c) use the properties of limits to verify your guess and to find the limit of the sequence if it converges.*

**43.** $a_n = \dfrac{n-1}{n+2}$

**44.** $a_n = (-1)^n\dfrac{2n+1}{n+3}$

**45.** $a_n = \dfrac{n!}{n^n}$

**46.** $a_n = 2\tan^{-1}\left(\dfrac{n+1}{n+3}\right)$

**47.** $a_n = n\sin\dfrac{1}{n}$

**48.** $a_n = \left(1-\dfrac{2}{n}\right)^n$

**49.** Evaluate

$$\lim_{n\to\infty}\dfrac{1-\left(1-\dfrac{1}{n}\right)^9}{1-\left(1-\dfrac{1}{n}\right)}$$

*Hint:* Use Theorem 1.

**50.** Evaluate

$$\lim_{n\to\infty} n\left(1-\sqrt[7]{1-\dfrac{1}{n}}\right)$$

*Hint:* Use Theorem 1.

*In Exercises 51–58, determine whether the sequence $\{a_n\}$ is monotonic. Is the sequence bounded?*

**51.** $a_n = \dfrac{3}{2n+5}$

**52.** $a_n = \dfrac{2n}{n+1}$

**53.** $a_n = 3 - \dfrac{1}{n}$

**54.** $a_n = 2 + \dfrac{(-1)^n}{n}$

**55.** $a_n = \dfrac{\sin n}{n}$

**56.** $a_n = \tan^{-1} n$

**57.** $a_n = \dfrac{n}{2^n}$

**58.** $a_n = \dfrac{\ln n}{n}$

**59. Compound Interest** If a principal of $P$ dollars is invested in an account earning interest at the rate of $r$ per year compounded monthly, then the accumulated amount $A_n$ at the end of $n$ months is

$$A_n = P\left(1 + \frac{r}{12}\right)^n$$

a. Write the first six terms of the sequence $\{A_n\}$ if $P = 10,000$ and $r = 0.105$. Interpret your results.

b. Does the sequence $\{A_n\}$ converge or diverge?

**60. Quality Control** Half a percent of the microprocessors manufactured by Alpha Corporation for use in regulating fuel consumption in automobiles are defective. It can be shown that the probability of finding at least one defective microprocessor in a random sample of $n$ microprocessors is $f(n) = 1 - (0.995)^n$. Consider the sequence $\{a_n\}$ defined by $a_n = f(n)$.

a. Write the terms $a_{10}$, $a_{100}$, and $a_{1000}$.

b. Evaluate $\lim_{n \to \infty} a_n$, and interpret your result.

**61. Annuities** An annuity is a sequence of payments made at regular intervals. Suppose that a sum of $200 is deposited at the end of each month into an account earning interest at the rate of 12% per year compounded monthly. Then the amount on deposit (called the future value of the annuity) at the end of the $n$th month is $f(n) = 20,000[(1.01)^n - 1]$. Consider the sequence $\{a_n\}$ defined by $a_n = f(n)$.

a. Find the 24th term of the sequence $\{a_n\}$, and interpret your result.

b. Evaluate $\lim_{n \to \infty} a_n$, and interpret your result.

**62. Continuously Compounded Interest** If $P$ dollars is invested in an account paying interest at the rate of $r$ per year compounded $m$ times per year, then the accumulated amount at the end of $t$ years is

$$A_m = P\left(1 + \frac{r}{m}\right)^{mt} \qquad m = 1, 2, 3, \ldots$$

a. Find the limit of the sequence $\{A_m\}$.

b. Interpret the result in part (a).
   **Note:** In this situation, interest is said to be *compounded continuously.*

c. What is the accumulated amount at the end of 3 years if $1000 is invested in an account paying interest at the rate of 10% per year compounded continuously?

 **63.** Find the limit of the sequence $\left\{\left(1 + \dfrac{2}{n}\right)^{3n}\right\}$. Confirm your results visually by plotting the graph of

$$f(x) = \left(1 + \frac{2}{x}\right)^{3x}$$

 **64.** Define the sequence $\{a_n\}$ recursively by $a_0 = 2$ and $a_{n+1} = \sqrt{a_n}$ for $n \geq 1$.

a. Show that $a_n = 2^{1/2^n}$.

b. Evaluate $\lim_{n \to \infty} a_n$.

c. Verify the result of part (b) graphically.

 **65. Newton's Method** Suppose that $A > 0$. Applying Newton's method to the solution of the equation $x^2 - A = 0$ leads to the sequence $\{x_n\}$ defined by

$$x_{n+1} = \frac{1}{2}\left(x_n + \frac{A}{x_n}\right) \qquad x_0 > 0$$

a. Show that if $L = \lim_{n \to \infty} x_n$ exists, then $L = \sqrt{A}$.
   Hint: $\lim_{n \to \infty} x_{n+1} = L$

b. Find $\sqrt{5}$ accurate to four decimal places.

 **66. Finding the Roots of an Equation** Suppose that we want to find a root of $f(x) = 0$. Newton's method provides one way of finding it. Here is another method that works under suitable conditions.

a. Write $f(x) = 0$ in the form $x = g(x)$, where $g$ is continuous. Then generate the sequence $\{x_n\}$ by the recursive formula $x_{n+1} = g(x_n)$, where $x_0$ is arbitrary.

b. Show that if the sequence $\{x_n\}$ converges to a number $r$, then $r$ is a solution of $f(x) = 0$.
   Hint: $\lim_{n \to \infty} x_{n+1} = r$

c. Use this method to find the root of $f(x) = 3x^3 - 9x + 2$ (accurate to four decimal places) that lies in the interval $(0, 1)$.
   Hint: Write $3x^3 - 9x + 2 = 0$ in the form $x = \frac{1}{9}(3x^3 + 2)$. Take $x_0 = 0$.

**67. A Floating Object** A sphere of radius 1 ft is made of wood that has a specific gravity of $\frac{2}{3}$. If the sphere is placed in water, it sinks to a depth of $h$ ft. It can be shown that $h$ satisfies the equation

$$h^3 - 3h^2 + \frac{8}{3} = 0$$

Use the method described in Exercise 66 to find $h$ accurate to three decimal places.
Hint: Show that the equation can be written in the form $h = \frac{1}{3}\sqrt{3h^3 + 8}$.

**68.** Find the limit of the sequence

$$\left\{\sqrt{2}, \sqrt{2\sqrt{2}}, \sqrt{2\sqrt{2\sqrt{2}}}, \ldots\right\}$$

Hint: Show that $a_n = 2^{(2^n - 1)/2^n} = 2^{1 - 1/2^n}$.

**69.** Consider the sequence $\{a_n\}$ defined by $a_1 = \sqrt{2}$ and $a_n = \sqrt{2 + a_{n-1}}$ for $n \geq 2$. Assuming that the sequence converges, find its limit.
Note: Using the principle of mathematical induction, it can be shown that $\{a_n\}$ is increasing and bounded by 2 and, hence, by Theorem 6 is convergent.

**70.** Show that if $\lim_{n \to \infty} a_{2n} = L$ and $\lim_{n \to \infty} a_{2n+1} = L$, then $\lim_{n \to \infty} a_n = L$.

**71.** Let the sequence $\{a_n\}$ be defined by

$$a_n = 1 + \frac{1}{2^2} + \frac{1}{3^2} + \cdots + \frac{1}{n^2}$$

  **a.** Show that $\{a_n\}$ is increasing.
  **b.** Show that $\{a_n\}$ is bounded above by establishing that $a_n < 2 - 1/n$ for $n \geq 2$.
  **Hint:** $\dfrac{1}{n^2} < \dfrac{1}{n(n-1)} = \dfrac{1}{n-1} - \dfrac{1}{n}$, for $n \geq 2$.
  **c.** Using the results of parts (a) and (b), what can you deduce about the convergence of $\{a_n\}$?

**72.** Let the sequence $\{a_n\}$ be defined by

$$a_n = \frac{1}{2+1} + \frac{1}{2^2+2} + \cdots + \frac{1}{2^n+n}$$

  **a.** Show that $\{a_n\}$ is increasing.
  **b.** Show that $\{a_n\}$ is bounded above.
  **c.** Using the results of parts (a) and (b), what can you deduce about the convergence of $\{a_n\}$?

**73.** Let the sequence $\{a_n\}$ be defined by

$$a_1 = \frac{a_0}{2+a_0}, \qquad a_2 = \frac{a_1}{2+a_1},$$

$$a_3 = \frac{a_2}{2+a_2}, \qquad \cdots, \qquad a_n = \frac{a_{n-1}}{2+a_{n-1}}, \qquad \cdots$$

  where $a_n > 0$.
  **a.** Show that $\{a_n\}$ is convergent.
  **b.** Find the limit of $\{a_n\}$.

**74.** Use the Squeeze Theorem for Sequences to prove that

$$\lim_{n \to \infty} \sqrt[n]{a} = 1 \qquad a > 0$$

  **Hint:** For $n$ sufficiently large, $1/n < a < n$.

**75.** Prove Theorem 1: If $\lim_{x \to \infty} f(x) = L$ and $\{a_n\}$ is a sequence defined by $a_n = f(n)$, where $n$ is a positive integer, then $\lim_{n \to \infty} a_n = L$.

**76.** Prove Theorem 4: If $\lim_{n \to \infty} |a_n| = 0$, then $\lim_{n \to \infty} a_n = 0$.

**77.** Prove the properties of the sequence $\{r^n\}$ given on page 737 using the results $\lim_{x \to \infty} a^x = 0$ if $0 < a < 1$ and $\lim_{x \to \infty} a^x = \infty$ if $a > 1$.

**78.** **Fibonacci Sequence** The Fibonacci sequence $\{F_n\}$ is defined by $F_1 = 1$, $F_2 = 1$, and $F_{n+1} = F_n + F_{n-1}$ for $n \geq 2$. Let $a_n = F_{n+1}/F_n$. Assuming that $\{a_n\}$ is convergent, show that

$$\lim_{n \to \infty} a_n = \frac{1}{2}(1 + \sqrt{5})$$

  **Hint:** First, show that $a_{n-1} = 1 + 1/a_{n-2}$. Then use the fact that if $\lim_{n \to \infty} a_n = L$, then $\lim_{n \to \infty} a_{n-2} = \lim_{n \to \infty} a_{n-1} = L$.
  **Note:** The number $\frac{1}{2}(1 + \sqrt{5})$, which is approximately 1.6, has the following special property: A picture with a ratio of width to height equal to this number is especially pleasing to the eye. The ancient Greeks used this "golden" ratio in designing their beautiful temples and public buildings, such as the Parthenon.

The front of the Parthenon has a ratio of width to height that is approximately 1.6 to 1.

*In Exercises 79–86, determine whether the statement is true or false. If it is true, explain why it is true. If it is false, explain why or give an example to show why it is false.*

**79.** If $\{a_n\}$ and $\{b_n\}$ are divergent, then $\{a_n + b_n\}$ is divergent.

**80.** If $\{a_n\}$ is divergent, then $\{|a_n|\}$ is divergent.

**81.** If $\{a_n\}$ converges to $L$ and $\{b_n\}$ converges to 0, then $\{a_n b_n\}$ converges to 0.

**82.** If $\{a_n\}$ converges and $\{b_n\}$ converges, then $\{a_n/b_n\}$ converges.

**83.** If $\{a_n\}$ is bounded and $\{b_n\}$ converges, then $\{a_n b_n\}$ converges.

**84.** If $\{a_n\}$ is bounded, then $\{a_n/n\}$ converges to 0.

**85.** If $\lim_{n \to \infty} a_n b_n$ exists, then both $\lim_{n \to \infty} a_n$ and $\lim_{n \to \infty} b_n$ exist.

**86.** If $\lim_{n \to \infty} |a_n|$ exists, then $\lim_{n \to \infty} a_n$ exists.

# 8.2 Series

Consider again the example involving the bouncing ball. Earlier we found a sequence describing the maximum height attained by the ball on each rebound after hitting a surface. The question that follows naturally is: How do we find the total distance traveled by the ball? To answer this question, recall that the initial height and the heights attained on each subsequent rebound are

$$1, \quad \frac{2}{3}, \quad \left(\frac{2}{3}\right)^2, \quad \left(\frac{2}{3}\right)^3, \quad \cdots$$

meters, respectively. (See Figure 1.) Observe that the distance traveled by the ball when it first hits the surface is 1 m. When it hits the surface the second time, it will have traveled a total distance of

$$1 + 2\left(\frac{2}{3}\right) \quad \text{or} \quad 1 + \frac{4}{3}$$

meters. When it hits the surface the third time, it will have traveled a distance of

$$1 + 2\left(\frac{2}{3}\right) + 2\left(\frac{2}{3}\right)^2 \quad \text{or} \quad 1 + \frac{4}{3} + \frac{8}{9}$$

meters. Continuing in this fashion, we see that the total distance traveled by the ball is

$$1 + 2\left(\frac{2}{3}\right) + 2\left(\frac{2}{3}\right)^2 + 2\left(\frac{2}{3}\right)^3 + \cdots \tag{1}$$

meters. Observe that this last expression involves the sum of infinitely many terms.

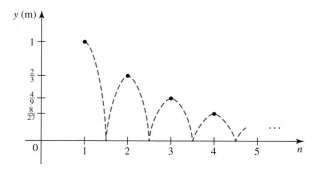

**FIGURE 1**

In general, an expression of the form

$$a_1 + a_2 + a_3 + \cdots + a_n + \cdots$$

is called an **infinite series** or, more simply, a **series.** The numbers $a_1, a_2, a_3, \ldots$ are called the **terms** of the series; $a_n$ is called the **nth term**, or **general term,** of the series; and the series itself is denoted by the symbol

$$\sum_{n=1}^{\infty} a_n$$

or simply $\Sigma\, a_n$.

*How do we define the "sum" of an infinite series, if it exists?* To answer this question, we use the same technique that we have employed several times before: using quantities that we can compute to help us define new ones. For example, in defining the slope of the tangent line to the graph of a function, we take the limit of the slope of secant lines (quantities that we can compute); and in defining the area under the graph of a function, we take the limit of the sum of the area of rectangles (again, quantities that we can compute). Here, we define the sum of an infinite series as the limit of a sequence of *finite* sums (quantities that we can compute).

We can get an inkling of how this may be done from examining the series (1) giving the total distance traveled by the ball. Define the sequence $\{S_n\}$ by

$$S_1 = 1$$

$$S_2 = 1 + 2\left(\frac{2}{3}\right)$$

$$S_3 = 1 + 2\left(\frac{2}{3}\right) + 2\left(\frac{2}{3}\right)^2$$

$$\vdots$$

$$S_n = 1 + 2\left(\frac{2}{3}\right) + 2\left(\frac{2}{3}\right)^2 + \cdots + 2\left(\frac{2}{3}\right)^{n-1}$$

giving the total vertical distance traveled by the ball when it hits the surface the first time, the second time, the third time, ... , and the $n$th time, respectively. If the series (1) has a sum $S$ (the total distance traveled by the ball), then the terms of the sequence $\{S_n\}$ form a sequence of increasingly accurate approximations to $S$. This suggests that we define

$$S = \lim_{n \to \infty} S_n$$

We will complete the solution to this problem in Example 5.

Motivated by this discussion, we define the sum of an infinite series.

---

**DEFINITION   Convergence of Infinite Series**

Given an infinite series

$$\sum_{n=1}^{\infty} a_n = a_1 + a_2 + a_3 + \cdots + a_n + \cdots$$

the **$n$th partial sum** of the series is

$$S_n = \sum_{k=1}^{n} a_k = a_1 + a_2 + a_3 + \cdots + a_n$$

If the sequence of partial sums $\{S_n\}$ **converges** to the number $S$, that is, if $\lim_{n \to \infty} S_n = S$, then the series $\Sigma\, a_n$ **converges** and has **sum** $S$, written

$$\sum_{n=1}^{\infty} a_n = a_1 + a_2 + a_3 + \cdots + a_n + \cdots = S$$

If $\{S_n\}$ diverges, then the series $\Sigma\, a_n$ **diverges.**

⚠ Be sure to note the difference between a sequence and a series. A sequence is a *succession* of terms, whereas a series is a *sum* of terms.

**EXAMPLE 1** Determine whether the series converges. If the series converges, find its sum.

**a.** $\displaystyle\sum_{n=1}^{\infty} n$　　**b.** $\displaystyle\sum_{n=1}^{\infty} \left(\frac{1}{n} - \frac{1}{n+1}\right)$

**Solution**

**a.** The $n$th partial sum of the series is

$$S_n = 1 + 2 + 3 + \cdots + n = \frac{n(n+1)}{2}$$

Since

$$\lim_{n\to\infty} S_n = \lim_{n\to\infty} \frac{n(n+1)}{2} = \infty$$

we conclude that the limit does not exist and $\sum_{n=1}^{\infty} n$ diverges.

**b.** The $n$th partial sum of the series is

$$S_n = \left(1 - \frac{1}{2}\right) + \left(\frac{1}{2} - \frac{1}{3}\right) + \left(\frac{1}{3} - \frac{1}{4}\right) + \cdots + \left(\frac{1}{n-1} - \frac{1}{n}\right) + \left(\frac{1}{n} - \frac{1}{n+1}\right)$$

Removing the parentheses, we see that all the terms of $S_n$, except for the first and last, cancel out. So

$$S_n = 1 - \frac{1}{n+1}$$

Since

$$\lim_{n\to\infty} S_n = \lim_{n\to\infty}\left(1 - \frac{1}{n+1}\right) = 1$$

we conclude that the series converges and has sum 1, that is,

$$\sum_{n=1}^{\infty} \left(\frac{1}{n} - \frac{1}{n+1}\right) = 1$$ ■

The series in Example 1b is called a **telescoping series.**

**EXAMPLE 2** Show that the series $\displaystyle\sum_{n=1}^{\infty} \frac{4}{4n^2 - 1}$ is convergent, and find its sum.

**Solution** First, we use partial fraction decomposition to rewrite the general term $a_n = 4/(4n^2 - 1)$:

$$a_n = \frac{4}{4n^2 - 1} = \frac{4}{(2n-1)(2n+1)} = \frac{2}{2n-1} - \frac{2}{2n+1}$$

Then we write the $n$th partial sum of the series as

$$S_n = \sum_{k=1}^{n} \frac{4}{4k^2 - 1} = \sum_{k=1}^{n} \left( \frac{2}{2k - 1} - \frac{2}{2k + 1} \right)$$

$$= \left( \frac{2}{1} - \frac{2}{3} \right) + \left( \frac{2}{3} - \frac{2}{5} \right) + \left( \frac{2}{5} - \frac{2}{7} \right) + \cdots + \left( \frac{2}{2n - 1} - \frac{2}{2n + 1} \right)$$

$$= 2 - \frac{2}{2n + 1} \qquad \text{This is a telescoping series.}$$

Since

$$\lim_{n \to \infty} S_n = \lim_{n \to \infty} \left( 2 - \frac{2}{2n + 1} \right) = 2$$

we conclude that the given series is convergent and has sum 2; that is,

$$\sum_{n=1}^{\infty} \frac{4}{4n^2 - 1} = 2 \qquad \blacksquare$$

## Geometric Series

Geometric series play an important role in mathematical analysis. They also arise frequently in the field of finance. The convergence or divergence of a geometric series is easily established.

---

**DEFINITION** **Geometric Series**

A series of the form

$$\sum_{n=1}^{\infty} ar^{n-1} = a + ar + ar^2 + \cdots + ar^{n-1} + \cdots \qquad a \neq 0$$

is called a **geometric series** with common ratio $r$.

---

The following theorem tells us the conditions under which a geometric series is convergent.

---

**THEOREM 1**

If $|r| < 1$, then the geometric series

$$\sum_{n=1}^{\infty} ar^{n-1} = a + ar + ar^2 + \cdots + ar^{n-1} + \cdots$$

converges, and its sum is $\displaystyle\sum_{n=1}^{\infty} ar^{n-1} = \frac{a}{1 - r}$. The series diverges if $|r| \geq 1$.

---

**PROOF** The $n$th partial sum of the series is

$$S_n = a + ar + ar^2 + \cdots + ar^{n-1}$$

Multiplying both sides of this equation by $r$ gives

$$rS_n = ar + ar^2 + ar^3 + \cdots + ar^n$$

Subtracting the second equation from the first then yields

$$(1 - r)S_n = a - ar^n = a(1 - r^n)$$

If $r \neq 1$, we can solve the last equation for $S_n$, obtaining

$$S_n = \frac{a(1 - r^n)}{1 - r}$$

From Example 13 on page 737 we know that $\lim_{n \to \infty} r^n = 0$ if $|r| < 1$, so

$$\lim_{n \to \infty} S_n = \lim_{n \to \infty} \frac{a(1 - r^n)}{1 - r} = \frac{a}{1 - r}$$

This implies that

$$\sum_{n=1}^{\infty} ar^{n-1} = \frac{a}{1 - r} \qquad |r| < 1$$

If $|r| > 1$, then the sequence $\{r^n\}$ diverges, so $\lim_{n \to \infty} S_n$ does not exist. This means that the geometric series diverges. We leave it as an exercise to show that $\{S_n\}$ diverges if $r = \pm 1$, so the series also diverges for these values of $r$. ■

**EXAMPLE 3**   Determine whether the series converges or diverges. If it converges, find its sum.

**a.** $\displaystyle\sum_{n=1}^{\infty} 3\left(-\frac{1}{2}\right)^{n-1} = 3 - \frac{3}{2} + \frac{3}{4} - \frac{3}{8} + \cdots$

**b.** $\displaystyle\sum_{n=1}^{\infty} 5\left(\frac{4}{3}\right)^{n-1} = 5 + \frac{20}{3} + \frac{80}{9} + \frac{320}{27} + \cdots$

**Solution**

**a.** This is a geometric series with $a = 3$ and common ratio $r = -\frac{1}{2}$. Since $\left|-\frac{1}{2}\right| < 1$, Theorem 1 tells us that the series converges and has sum

$$\sum_{n=1}^{\infty} 3\left(-\frac{1}{2}\right)^{n-1} = \frac{3}{1 - \left(-\frac{1}{2}\right)} = 2$$

The graphs of $\{a_n\}$ and $\{S_n\}$ for this series are shown in Figure 2a.

**b.** This is a geometric series with $a = 5$ and common ratio $r = \frac{4}{3}$. Since $\frac{4}{3} > 1$, Theorem 1 tells us that the series is divergent. The graphs of $\{a_n\}$ and $\{S_n\}$ for this series are shown in Figure 2b.

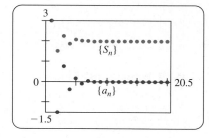

(a) The geometric series converges because $|r| < 1$.

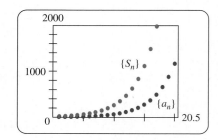

(b) The geometric series diverges because $|r| > 1$.

**FIGURE 2**

**EXAMPLE 4** Express the number $3.2\overline{14} = 3.2141414\ldots$ as a rational number.

**Solution** We rewrite the number as

$$3.2141414\ldots = 3.2 + \frac{14}{10^3} + \frac{14}{10^5} + \frac{14}{10^7} + \cdots$$

$$= \frac{32}{10} + \frac{14}{10^3}\left[1 + \frac{1}{10^2} + \frac{1}{10^4} + \cdots\right]$$

$$= \frac{32}{10} + \sum_{n=1}^{\infty}\left(\frac{14}{10^3}\right)\left(\frac{1}{10^2}\right)^{n-1}$$

The expression after the first term is a geometric series with $a = \frac{14}{1000}$ and $r = \frac{1}{100}$. Using Theorem 1, we have

$$3.2141414\ldots = \frac{32}{10} + \frac{\frac{14}{1000}}{1 - \frac{1}{100}}$$

$$= \frac{32}{10} + \frac{14}{990} = \frac{3182}{990} \qquad \blacksquare$$

**EXAMPLE 5** Complete the solution of the bouncing ball problem that was introduced at the beginning of this section. Recall that the total vertical distance traveled by the ball is given by

$$1 + 2\left(\frac{2}{3}\right) + 2\left(\frac{2}{3}\right)^2 + 2\left(\frac{2}{3}\right)^3 + \cdots$$

meters.

**Solution** If we let $d$ denote the total vertical distance traveled by the ball, then

$$d = 1 + \sum_{n=1}^{\infty}\left(\frac{4}{3}\right)\left(\frac{2}{3}\right)^{n-1}$$

The expression after the first term is a geometric series with $a = \frac{4}{3}$ and $r = \frac{2}{3}$. Using Theorem 1, we obtain

$$d = 1 + \frac{\frac{4}{3}}{1 - \frac{2}{3}} = 1 + 4 = 5$$

and conclude that the total distance traveled by the ball is 5 m. $\qquad \blacksquare$

## ■ The Harmonic Series

The series

$$\sum_{n=1}^{\infty} \frac{1}{n} = 1 + \frac{1}{2} + \frac{1}{3} + \frac{1}{4} + \cdots$$

is called the **harmonic series.** Before showing that this series is divergent, we make this observation: If a sequence $\{b_n\}$ is convergent, then any *subsequence* obtained by deleting any number of terms from the parent sequence $\{b_n\}$ must also converge to the same limit. Therefore, to show that a sequence is divergent, it suffices to produce a subsequence of the parent sequence that is divergent.

In keeping with this strategy, let us show that the subsequence

$$S_2, S_4, S_8, S_{16}, \ldots, S_{2^n}, \ldots$$

of the sequence $\{S_n\}$ of partial sums of the harmonic series is divergent. We have

$$S_2 = 1 + \frac{1}{2}$$

$$S_4 = 1 + \frac{1}{2} + \left(\frac{1}{3} + \frac{1}{4}\right) > 1 + \frac{1}{2} + \left(\frac{1}{4} + \frac{1}{4}\right) = 1 + 2\left(\frac{1}{2}\right)$$

$$S_8 = 1 + \frac{1}{2} + \left(\frac{1}{3} + \frac{1}{4}\right) + \left(\frac{1}{5} + \frac{1}{6} + \frac{1}{7} + \frac{1}{8}\right)$$

$$> 1 + \frac{1}{2} + \left(\frac{1}{4} + \frac{1}{4}\right) + \left(\frac{1}{8} + \frac{1}{8} + \frac{1}{8} + \frac{1}{8}\right) = 1 + \frac{1}{2} + \frac{1}{2} + \frac{1}{2} = 1 + 3\left(\frac{1}{2}\right)$$

$$S_{16} = 1 + \frac{1}{2} + \left(\frac{1}{3} + \frac{1}{4}\right) + \left(\frac{1}{5} + \cdots + \frac{1}{8}\right) + \left(\frac{1}{9} + \cdots + \frac{1}{16}\right)$$

$$> 1 + \frac{1}{2} + \left(\frac{1}{4} + \frac{1}{4}\right) + \underbrace{\left(\frac{1}{8} + \cdots + \frac{1}{8}\right)}_{\text{4 terms}} + \underbrace{\left(\frac{1}{16} + \cdots + \frac{1}{16}\right)}_{\text{8 terms}}$$

$$= 1 + \frac{1}{2} + \frac{1}{2} + \frac{1}{2} + \frac{1}{2} = 1 + 4\left(\frac{1}{2}\right)$$

and, in general, $S_{2^n} > 1 + n\left(\frac{1}{2}\right)$. Therefore,

$$\lim_{n \to \infty} S_{2^n} = \infty$$

so $\{S_n\}$ is divergent. This proves that the harmonic series is divergent.

## ■ The Divergence Test

The next theorem tells us that the terms of a convergent series must ultimately approach zero.

---

**THEOREM 2**

If $\sum_{n=1}^{\infty} a_n$ converges, then $\lim_{n \to \infty} a_n = 0$.

---

**PROOF**   We have $S_n = a_1 + a_2 + \cdots + a_{n-1} + a_n = S_{n-1} + a_n$, so $a_n = S_n - S_{n-1}$. Since $\sum_{n=1}^{\infty} a_n$ is convergent, the sequence $\{S_n\}$ is convergent. Let $\lim_{n \to \infty} S_n = S$. Then

$$\lim_{n \to \infty} a_n = \lim_{n \to \infty}(S_n - S_{n-1}) = \lim_{n \to \infty} S_n - \lim_{n \to \infty} S_{n-1} = S - S = 0 \qquad \blacksquare$$

The *Divergence Test* is an important consequence of Theorem 2.

---

**THEOREM 3**   **The Divergence Test**

If $\lim_{n \to \infty} a_n$ does not exist or $\lim_{n \to \infty} a_n \neq 0$, then $\sum_{n=1}^{\infty} a_n$ diverges.

The Divergence Test does *not* say that if $\lim_{n \to \infty} a_n = 0$, then $\sum_{n=1}^{\infty} a_n$ must converge. In other words, the converse of Theorem 2 is not true in general. For example, $\lim_{n \to \infty} 1/n = 0$, yet the harmonic series $\sum_{n=1}^{\infty} 1/n$ is divergent. In short, the Divergence Test rules out convergence for a series whose $n$th term does not approach zero but yields no information if $a_n$ does approach zero—that is, the series might or might not converge.

**EXAMPLE 6** Show that the following series are divergent.

**a.** $\displaystyle\sum_{n=1}^{\infty} (-1)^{n-1}$ **b.** $\displaystyle\sum_{n=1}^{\infty} \frac{2n^2 + 1}{3n^2 - 1}$

**Solution**
**a.** Here, $a_n = (-1)^{n-1}$, and

$$\lim_{n \to \infty} a_n = \lim_{n \to \infty} (-1)^{n-1}$$

does not exist. We conclude by the Divergence Test that the series diverges.
**b.** Here, $a_n = \dfrac{2n^2 + 1}{3n^2 - 1}$, and

$$\lim_{n \to \infty} a_n = \lim_{n \to \infty} \frac{2n^2 + 1}{3n^2 - 1} = \lim_{n \to \infty} \frac{2 + \dfrac{1}{n^2}}{3 - \dfrac{1}{n^2}} = \frac{2}{3} \neq 0$$

so by the Divergence Test, the series diverges. ∎

## ■ Properties of Convergent Series

The following properties of series are immediate consequences of the corresponding properties of the limits of sequences. We omit the proofs.

**THEOREM 4  Properties of Convergent Series**

If $\sum_{n=1}^{\infty} a_n = A$ and $\sum_{n=1}^{\infty} b_n = B$ are convergent and $c$ is any real number, then $\sum_{n=1}^{\infty} ca_n$ and $\sum_{n=1}^{\infty} (a_n \pm b_n)$ are also convergent, and

**a.** $\displaystyle\sum_{n=1}^{\infty} ca_n = c\sum_{n=1}^{\infty} a_n = cA$ **b.** $\displaystyle\sum_{n=1}^{\infty} (a_n \pm b_n) = \sum_{n=1}^{\infty} a_n \pm \sum_{n=1}^{\infty} b_n = A \pm B$

**EXAMPLE 7** Show that the series $\displaystyle\sum_{n=1}^{\infty} \left[ \frac{2}{n(n+1)} - \frac{4}{3^n} \right]$ is convergent, and find its sum.

**Solution**  First, consider the series $\sum_{n=1}^{\infty} 1/[n(n+1)]$. Using partial fraction decomposition, we can write this series in the form

$$\sum_{n=1}^{\infty} \frac{1}{n(n+1)} = \sum_{n=1}^{\infty} \left( \frac{1}{n} - \frac{1}{n+1} \right)$$

Using the result of Example 1, we see that

$$\sum_{n=1}^{\infty} \frac{1}{n(n+1)} = 1$$

Next, observe that $\sum_{n=1}^{\infty} \frac{4}{3^n}$ is a geometric series with $a = \frac{4}{3}$ and $r = \frac{1}{3}$, so

$$\sum_{n=1}^{\infty} \frac{4}{3^n} = \frac{\frac{4}{3}}{1 - \frac{1}{3}} = 2$$

Therefore, by Theorem 4 the given series is convergent, and

$$\sum_{n=1}^{\infty} \left[ \frac{2}{n(n+1)} - \frac{4}{3^n} \right] = 2 \sum_{n=1}^{\infty} \frac{1}{n(n+1)} - \sum_{n=1}^{\infty} \frac{4}{3^n}$$

$$= 2 \cdot 1 - 2 = 0 \qquad \blacksquare$$

## 8.2 CONCEPT QUESTIONS

1. Explain the difference between
   a. A sequence and a series
   b. A convergent sequence and a convergent series
   c. A divergent sequence and a divergent series
   d. The limit of a sequence and the sum of a series

2. Suppose that $\sum_{n=1}^{\infty} a_n = 6$.
   a. Evaluate $\lim_{n \to \infty} S_n$, where $S_n$ is the $n$th partial sum of $\sum_{n=1}^{\infty} a_n$.
   b. Find $\sum_{n=2}^{\infty} a_n$ if it is known that $a_1 = \frac{1}{2}$.

## 8.2 EXERCISES

In Exercises 1–6, find the nth partial sum $S_n$ of the telescoping series, and use it to determine whether the series converges or diverges. If it converges, find its sum.

1. $\sum_{n=2}^{\infty} \left( \frac{1}{n-1} - \frac{1}{n} \right)$

2. $\sum_{n=1}^{\infty} \left( \frac{1}{2n+3} - \frac{1}{2n+1} \right)$

3. $\sum_{n=1}^{\infty} \frac{4}{(2n+3)(2n+5)}$

4. $\sum_{n=1}^{\infty} \left( \frac{-8}{4n^2 + 4n - 3} \right)$

5. $\sum_{n=2}^{\infty} \left( \frac{1}{\ln n} - \frac{1}{\ln(n+1)} \right)$

6. $\sum_{n=1}^{\infty} \frac{2}{\sqrt{n+1} + \sqrt{n}}$

In Exercises 7–14, determine whether the geometric series converges or diverges. If it converges, find its sum.

7. $4 + \frac{8}{3} + \frac{16}{9} + \frac{32}{27} + \cdots$

8. $-\frac{1}{2} + \frac{1}{4} - \frac{1}{8} + \frac{1}{16} - \cdots$

9. $\frac{5}{3} - \frac{5}{9} + \frac{5}{27} - \frac{5}{81} + \cdots$

10. $1 + \frac{4}{3} + \frac{16}{9} + \frac{64}{27} + \cdots$

11. $\sum_{n=0}^{\infty} 2\left( -\frac{1}{\sqrt{2}} \right)^n$

12. $\sum_{n=1}^{\infty} \frac{e^n}{3^{n+1}}$

13. $\sum_{n=0}^{\infty} 2^n 3^{-n+1}$

14. $\sum_{n=1}^{\infty} (-1)^{n-1} 3^n 2^{1-n}$

In Exercises 15–22, show that the series diverges.

15. $\frac{1}{2} + \frac{2}{3} + \frac{3}{4} + \cdots$

16. $1 - \frac{3}{2} + \frac{9}{4} - \frac{27}{8} + \cdots$

17. $\sum_{n=1}^{\infty} \frac{2n}{3n+1}$

18. $\sum_{n=1}^{\infty} \frac{n^2}{2n^2 + 1}$

19. $\sum_{n=1}^{\infty} 2(1.5)^n$

20. $\sum_{n=0}^{\infty} \frac{(-1)^n 3^n}{2^{n-1}}$

21. $\sum_{n=1}^{\infty} \frac{1}{2 + 3^{-n}}$

22. $\sum_{n=1}^{\infty} \frac{n}{\sqrt{2n^2 + 1}}$

 In Exercises 23–28, (a) compute as many terms of the sequence of partial sums, $S_n$, as is necessary to convince yourself that the series converges or diverges. If it converges, estimate its sum. (b) Plot $\{S_n\}$ to give a visual confirmation of your observation in part (a). (c). If the series converges, find the exact sum. If it diverges, prove it, using the Divergence Theorem.

23. $\sum_{n=1}^{\infty} \frac{6}{n(n+1)}$

24. $\sum_{n=1}^{\infty} \frac{2n}{\sqrt{n^2 + 1}}$

25. $\sum_{n=1}^{\infty} 3\left( \frac{7}{8} \right)^{n-1}$

26. $\sum_{n=1}^{\infty} 5\left( -\frac{2}{3} \right)^{n-1}$

27. $\sum_{n=1}^{\infty} \sin n^2$

28. $\sum_{n=1}^{\infty} \left( \frac{1}{2^n} - \frac{1}{3^n} \right)$

In Exercises 29–54, determine whether the given series converges or diverges. If it converges, find its sum.

**29.** $\displaystyle\sum_{n=1}^{\infty} \frac{1}{n(n+2)}$

**30.** $\displaystyle\sum_{n=2}^{\infty} \frac{1}{n^2-1}$

**31.** $\displaystyle\sum_{n=0}^{\infty} \frac{2^n}{5^n}$

**32.** $\displaystyle\sum_{n=0}^{\infty} \frac{3^{n+1}}{5^n}$

**33.** $\displaystyle\sum_{n=0}^{\infty} \frac{(-3)^n}{2^{n+1}}$

**34.** $\displaystyle\sum_{n=1}^{\infty} 2^{-n}5^{n+1}$

**35.** $\displaystyle\sum_{n=1}^{\infty} \frac{2n-1}{3n+1}$

**36.** $\displaystyle\sum_{n=0}^{\infty} \frac{2n^2+n+1}{3n^2+2}$

**37.** $\displaystyle\sum_{n=0}^{\infty} 3(1.01)^n$

**38.** $\displaystyle\sum_{n=1}^{\infty} \frac{3^n-1}{3^{n+1}}$

**39.** $\displaystyle\sum_{n=1}^{\infty} \left[\frac{1}{2^n} - \frac{1}{n(n+1)}\right]$

**40.** $\displaystyle\sum_{n=1}^{\infty} \left[\frac{2^n}{3^{n-1}} + \frac{(-1)^{n-1}2^n}{3^{n+1}}\right]$

**41.** $\displaystyle\sum_{n=1}^{\infty} \frac{2}{1+(0.2)^n}$

**42.** $\displaystyle\sum_{n=1}^{\infty} \ln\left(\frac{n}{n+1}\right)$

**43.** $\displaystyle\sum_{n=1}^{\infty} \left[\cos\left(\frac{1}{n}\right) - \cos\left(\frac{1}{n+1}\right)\right]$

**44.** $\displaystyle\sum_{n=1}^{\infty} \frac{n!}{2^n}$

**45.** $\displaystyle\sum_{n=1}^{\infty} [2(0.1)^n + 3(-1)^n(0.2)^n]$

**46.** $\displaystyle\sum_{n=0}^{\infty} \left[\left(-\frac{3}{\pi}\right)^n + \left(\frac{e}{3}\right)^{n+1}\right]$

**47.** $\displaystyle\sum_{n=0}^{\infty} \left(\frac{2^n+3^n}{6^n}\right)$

**48.** $\displaystyle\sum_{n=1}^{\infty} \left(\frac{2^n-5^n}{3^n}\right)$

**49.** $\displaystyle\sum_{n=1}^{\infty} \tan^{-1} n$

**50.** $\displaystyle\sum_{n=1}^{\infty} \sin^2 n$

**51.** $\displaystyle\sum_{n=1}^{\infty} n \sin\frac{1}{n}$

**52.** $\displaystyle\sum_{n=1}^{\infty} \frac{\sin n}{1+e^{-n}}$

**53.** $\displaystyle\sum_{n=2}^{\infty} \frac{n}{\ln n}$

**54.** $\displaystyle\sum_{n=1}^{\infty} \left(1+\frac{2}{n}\right)^n$

In Exercises 55–58, express each number as a rational number.

**55.** $0.\overline{4} = 0.444\ldots$

**56.** $-0.\overline{23} = -0.232323\ldots$

**57.** $1.\overline{213} = 1.213213213\ldots$

**58.** $3.14\overline{234} = 3.142343434\ldots$

In Exercises 59–62, find the values of x for which the series converges, and find the sum of the series. (Hint: First show that the series is a geometric series.)

**59.** $\displaystyle\sum_{n=0}^{\infty} (-x)^n$

**60.** $\displaystyle\sum_{n=0}^{\infty} (x-2)^n$

**61.** $\displaystyle\sum_{n=1}^{\infty} 2^n(x-1)^n$

**62.** $\displaystyle\sum_{n=0}^{\infty} \frac{x^{2n}}{3^n}$

**63. Distance Traveled by a Bouncing Ball** A rubber ball is dropped from a height of 2 m onto a flat surface. Each time the ball hits the surface, it rebounds to half its previous height. Find the total distance the ball travels.

**64. Finding the Coefficient of Restitution** The *coefficient of restitution* for steel onto steel is measured by dropping a steel ball onto a steel plate. If the ball is dropped from a height $H$ and rebounds to a height $h$, then the coefficient of restitution is $\sqrt{h/H}$. Suppose that a steel ball is dropped from a height of 1 m onto a steel plate. Each time the ball hits the plate, it rebounds to $r$ times it previous height ($0 < r < 1$). If the ball travels a total distance of 2 m, find the coefficient of restitution for steel on steel.

**65. Probability of Winning a Dice Toss** Peter and Paul take turns tossing a pair of dice. The first person to throw a 7 wins. If Peter starts the game, then it can be shown that his chances of winning are

$$p = \frac{1}{6} + \left(\frac{1}{6}\right)\left(\frac{5}{6}\right)^2 + \left(\frac{1}{6}\right)\left(\frac{5}{6}\right)^4 + \cdots$$

Find $p$.

**66. Multiplier Effect of a Tax Cut** Suppose that the average wage earner saves 9% of his or her take-home pay and spends the other 91%. What is the estimated impact that a proposed $30 billion tax cut will have on the economy over the long run because of the additional spending generated by the proposed tax cut?

**Note:** This phenomenon in economics is known as the *multiplier effect*.

**67. Perpetuities** An *annuity* is a sequence of payments that are made at regular time intervals. If the payments are allowed to continue indefinitely, then it is a *perpetuity*.

a. Suppose that $P$ dollars is paid into an account at the beginning of each month and that the account earns interest at the rate of $r$ per year compounded monthly. Then the present value $V$ of the perpetuity (that is, the value of the perpetuity in today's dollars) is

$$V = P\left(1+\frac{r}{12}\right)^{-1} + P\left(1+\frac{r}{12}\right)^{-2} + \cdots + P\left(1+\frac{r}{12}\right)^{-n} + \cdots$$

Show that $V = 12P/r$.

b. Mrs. Thompson wishes to establish a fund to provide a university medical center with a monthly research grant of $150,000. If the fund will earn interest at the rate of 9% per year compounded monthly, use the result of part (a) to find the amount of the endowment she is required to make now.

**68. Residual Concentration of a Drug in the Bloodstream** Suppose that a dose of $C$ units of a certain drug is administered to a patient and that the fraction of the dose remaining in the patient's bloodstream $t$ hr after the dose is administered is given by $Ce^{-kt}$, where $k$ is a positive constant.

a. Show that the residual concentration of the drug in the bloodstream after extended treatment when a dose of $C$ units is administered at intervals of $t$ hr is given by

$$R = \frac{Ce^{-kt}}{1-e^{-kt}}$$

**b.** If the highest concentration of this particular drug that is considered safe is $S$ units, find the minimal time that must exist between doses.

Hint: $C + R \leq S$

**69. Capital Value of a Perpetuity** The capital value of a perpetuity involving payments of $P$ dollars paid at the end of each investment period into a fund that earns interest at the rate of $r$ per year compounded continuously is given by

$$A = Pe^{-r} + Pe^{-2r} + Pe^{-3r} + \cdots$$

Find an expression for $A$ that does not involve an infinite series.

**70. Sum of Areas of Nested Squares** An infinite sequence of nested squares is constructed as follows: Starting with a square with a side of length 2, each square in the sequence is constructed from the preceding square by drawing line segments connecting the midpoints of the sides of the square. Find the sum of the areas of all the squares in the sequence.

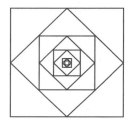

**71. Sum of Areas of Nested Triangles and Circles** An infinite sequence of nested equilateral triangles and circles is constructed as follows: Beginning with an equilateral triangle with a side of length 1, inscribe a circle followed by a triangle, followed by a circle, and so on, ad infinitum. Find the total area of the shaded regions.

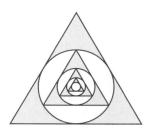

**72.** Prove or disprove: If $\Sigma\, a_n$ and $\Sigma\, b_n$ are both divergent, then $\Sigma(a_n + b_n)$ is divergent.

**73.** Suppose that $\Sigma\, a_n$ $(a_n \neq 0)$ is convergent. Prove that $\Sigma\, 1/a_n$ is divergent.

**74.** Suppose that $\Sigma\, a_n$ is convergent and $\Sigma\, b_n$ is divergent. Prove that $\Sigma(a_n + b_n)$ is divergent.

Hint: Prove by contradiction, using Theorem 4.

**75.** Suppose that $\Sigma\, a_n$ is divergent and $c \neq 0$. Prove that $\Sigma\, ca_n$ is divergent.

Hint: Prove by contradiction, using Theorem 4.

**76.** Prove that if the sequence $\{a_n\}$ converges, then the series $\Sigma(a_{n+1} - a_n)$ converges. Conversely, prove that if $\Sigma(a_{n+1} - a_n)$ converges, then $\{a_n\}$ converges.

**77.** Show that $\displaystyle\sum_{n=1}^{\infty} \frac{1}{n^2}$ converges and $\dfrac{3}{2} \leq \displaystyle\sum_{n=1}^{\infty} \frac{1}{n^2} \leq 2$.

Hint: See Exercise 71 in Section 8.1.

**78.** Prove that $\displaystyle\sum_{n=1}^{\infty} \frac{1}{2^n + 1}$ converges by showing that $\{S_n\}$ is increasing and bounded above, where $S_n$ is the $n$th partial sum of the series.

*In Exercises 79–84, determine whether the statement is true or false. If it is true, explain why it is true. If it is false, explain why or give an example to show why it is false.*

**79.** If $\lim_{n \to \infty} a_n = 0$, then $\sum_{n=1}^{\infty} a_n$ converges.

**80.** If $\lim_{n \to \infty} a_n = L$, then the telescoping series $\sum_{n=1}^{\infty}(a_{n+1} - a_n)$ converges and has sum $L - a_1$.

**81.** $\sum_{n=1}^{\infty} \sin^n x$ converges for all $x$ in $[0, 2\pi]$.

**82.** $\displaystyle\sum_{n=p}^{\infty} ar^n = \frac{ar^p}{1 - r}$ provided that $|r| < 1$.

**83.** If the sequence of partial sums of a series $\Sigma\, a_n$ is bounded above, then $\Sigma\, a_n$ must converge.

**84.** If $\Sigma(a_n + b_n)$ converges, then both $\Sigma\, a_n$ and $\Sigma\, b_n$ must converge.

---

## 8.3 The Integral Test

The convergence or divergence of a telescoping or geometric series is relatively easy to determine because we are able to find a simple formula involving a finite number of terms for the $n$th partial sum $S_n$ of these series. As we saw in Section 8.2, we can find the actual sum of a convergent series in this case by simply evaluating $\lim_{n \to \infty} S_n$. However, it is often very difficult or impossible to obtain a simple formula for the $n$th partial sum of an infinite series, and we are forced to look for alternative ways to investigate the convergence or divergence of the series.

In this and the next two sections we will develop several tests for determining the convergence or divergence of an infinite series by examining the $n$th term $a_n$ of the series. These tests will confirm the convergence of a series without yielding a value for its sum. From the practical point of view, however, this is all that is required. Once it has been ascertained that a series is convergent, we can approximate its sum to any degree of accuracy desired by adding up the terms of its $n$th partial sum $S_n$, provided that $n$ is chosen large enough. The convergence tests that are given here and in Section 8.4 apply only to series with positive terms.

## ■ The Integral Test

The Integral Test ties the convergence or divergence of an infinite series $\sum_{n=1}^{\infty} a_n$ to the convergence or divergence of the improper integral $\int_{1}^{\infty} f(x)\, dx$, where $f(n) = a_n$.

---

**THEOREM 1  The Integral Test**

Suppose that $f$ is a continuous, positive, and decreasing function on $[1, \infty)$. If $f(n) = a_n$ for $n \geq 1$, then

$$\sum_{n=1}^{\infty} a_n \qquad \text{and} \qquad \int_{1}^{\infty} f(x)\, dx$$

either both converge or both diverge.

---

**PROOF**  If you examine Figure 1a, you will see that the height of the first rectangle is $a_2 = f(2)$. Since this rectangle has width 1, the area of the rectangle is also $a_2 = f(2)$. Similarly, the area of the second rectangle is $a_3$, and so on. Comparing the sum of the areas of the first $(n - 1)$ inscribed rectangles with the area of the region under the graph of $f$ over the interval $[1, n]$, we see that

$$a_2 + a_3 + \cdots + a_n \leq \int_{1}^{n} f(x)\, dx$$

which implies that

$$S_n = a_1 + a_2 + a_3 + \cdots + a_n \leq a_1 + \int_{1}^{n} f(x)\, dx \tag{1}$$

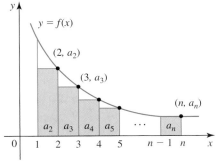

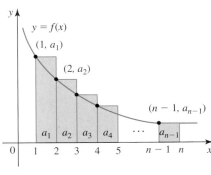

**FIGURE 1**  (a) $a_2 + a_3 + \cdots + a_n \leq \int_{1}^{n} f(x)\, dx$    (b) $\int_{1}^{n} f(x)\, dx \leq a_1 + a_2 + \cdots + a_{n-1}$

If $\int_1^\infty f(x)\, dx$ is convergent and has value $L$, then

$$S_n \le a_1 + \int_1^n f(x)\, dx \le a_1 + L$$

This shows that $\{S_n\}$ is bounded above. Also,

$$S_{n+1} = S_n + a_{n+1} \ge S_n \qquad \text{Because } a_{n+1} = f(n+1) \ge 0$$

shows that $\{S_n\}$ is increasing as well. Therefore, by Theorem 6, Section 8.1, $\{S_n\}$ is convergent. In other words, $\sum_{n=1}^\infty a_n$ is convergent.

Next, by examining Figure 1b, we can see that

$$\int_1^n f(x)\, dx \le a_1 + a_2 + \cdots + a_{n-1} = S_{n-1} \tag{2}$$

So if $\int_1^\infty f(x)\, dx$ diverges (to infinity because $f(x) \ge 0$), then $\lim_{n\to\infty} S_{n-1} = \lim_{n\to\infty} S_n = \infty$, and $\sum_{n=1}^\infty a_n$ is divergent. ∎

### Notes

1. The Integral Test simply tells us whether a series converges or diverges. If it indicates that a series converges, we may not conclude that the (finite) value of the improper integral used in conjunction with the test is the sum of the convergent series (see Exercise 54).
2. Since the convergence of an infinite series is not affected by adding or subtracting a finite number of terms to the series, we sometimes study the series $\sum_{n=N}^\infty a_n = a_N + a_{N+1} + \cdots$ rather than the series $\sum_{n=1}^\infty a_n$. In this case the series is compared with the improper integral $\int_N^\infty f(x)\, dx$, as we will see in Example 2. ∎

**EXAMPLE 1** Use the Integral Test to determine whether $\displaystyle\sum_{n=1}^\infty \frac{1}{n^2+1}$ converges or diverges.

**Solution** Here, $a_n = f(n) = 1/(n^2+1)$, so we consider the function $f(x) = 1/(x^2+1)$. Since $f$ is continuous, positive, and decreasing on $[1, \infty)$, we may use the Integral Test. Next,

$$\int_1^\infty \frac{1}{x^2+1}\, dx = \lim_{b\to\infty} \int_1^b \frac{1}{x^2+1}\, dx = \lim_{b\to\infty}\left[\tan^{-1} x\right]_1^b$$

$$= \lim_{b\to\infty}(\tan^{-1} b - \tan^{-1} 1) = \frac{\pi}{2} - \frac{\pi}{4} = \frac{\pi}{4}$$

Since $\int_1^\infty 1/(x^2+1)\, dx$ converges, we conclude that $\sum_{n=1}^\infty 1/(n^2+1)$ converges as well. ∎

**EXAMPLE 2** Use the Integral Test to determine whether $\displaystyle\sum_{n=1}^\infty \frac{\ln n}{n}$ converges or diverges.

**Solution** Here, $a_n = (\ln n)/n$, so we consider the function $f(x) = (\ln x)/x$. Observe that $f$ is continuous and positive on $[1, \infty)$. Next, we compute

$$f'(x) = \frac{x\left(\dfrac{1}{x}\right) - \ln x}{x^2} = \frac{1 - \ln x}{x^2}$$

Note that $f'(x) < 0$ if $\ln x > 1$, that is, if $x > e$. This shows that $f$ is decreasing on $[3, \infty)$. Therefore, we may use the Integral Test:

$$\int_3^\infty \frac{\ln x}{x}\, dx = \lim_{b \to \infty} \int_3^b \frac{\ln x}{x}\, dx = \lim_{b \to \infty}\left[\frac{1}{2}(\ln x)^2\right]_3^b$$

$$= \lim_{b \to \infty}\frac{1}{2}[(\ln b)^2 - (\ln 3)^2] = \infty$$

and we conclude that $\sum_{n=1}^\infty (\ln n)/n$ diverges.   ■

## ■ The $p$-Series

The following series will play an important role in our work later on.

---

**DEFINITION   $p$-Series**

A $p$-**series** is a series of the form

$$\sum_{n=1}^\infty \frac{1}{n^p} = 1 + \frac{1}{2^p} + \frac{1}{3^p} + \cdots$$

where $p$ is a constant.

---

Observe that if $p = 1$, the $p$-series is just the harmonic series $\sum_{n=1}^\infty 1/n$.

The conditions for the convergence or divergence of the $p$-series can be found by applying the Integral Test to the series.

---

**THEOREM 2   Convergence of the $p$-Series**

The $p$-series $\displaystyle\sum_{n=1}^\infty \frac{1}{n^p}$ converges if $p > 1$ and diverges if $p \le 1$.

---

**PROOF**   If $p < 0$, then $\lim_{n \to \infty}(1/n^p) = \infty$. If $p = 0$, then $\lim_{n \to \infty}(1/n^p) = 1$. In either case, $\lim_{n \to \infty}(1/n^p) \neq 0$, so the $p$-series diverges by the Divergence Test.

If $p > 0$, then the function $f(x) = 1/x^p$ is continuous, positive, and decreasing on $[1, \infty)$. In Example 2 in Section 6.6 we found that $\int_1^\infty 1/x^p\, dx$ converges if $p > 1$ and diverges if $p \le 1$. Using this result and the Integral Test, we conclude that $\sum_{n=1}^\infty 1/n^p$ converges if $p > 1$ and diverges if $0 < p \le 1$. Therefore, $\sum_{n=1}^\infty 1/n^p$ converges if $p > 1$ and diverges if $p \le 1$.   ■

**EXAMPLE 3**   Determine whether the given series converges or diverges.

**a.** $\displaystyle\sum_{n=1}^\infty \frac{1}{n^2}$     **b.** $\displaystyle\sum_{n=1}^\infty \frac{1}{\sqrt{n}}$     **c.** $\displaystyle\sum_{n=1}^\infty n^{-1.001}$

**Solution**
**a.** This is a $p$-series with $p = 2 > 1$, and hence it converges by Theorem 2.
**b.** Rewriting the series in the form $\sum_{n=1}^\infty 1/n^{1/2}$, we see that the series is a $p$-series with $p = \frac{1}{2} < 1$, and hence it diverges by Theorem 2.
**c.** We rewrite the series in the form $\sum_{n=1}^\infty 1/n^{1.001}$, which we recognize to be a $p$-series with $p = 1.001 > 1$ and conclude that the series converges.   ■

## 8.3 CONCEPT QUESTIONS

1. Consider the series $\displaystyle\sum_{n=1}^{\infty} \frac{1}{n^2} = \frac{1}{1^2} + \frac{1}{2^2} + \frac{1}{3^2} + \frac{1}{4^2} + \frac{1}{5^2} + \cdots$.
Let $f(x) = \dfrac{1}{x^2}$.
   a. Sketch a figure similar to Figure 1a for this series and function, and compute $a_1 = f(1)$, $a_2 = f(2)$, $a_3 = f(3)$, $\ldots, a_n = f(n)$.
   b. Explain why $S_n = \dfrac{1}{1^2} + \dfrac{1}{2^2} + \dfrac{1}{3^2} + \cdots + \dfrac{1}{n^2}$
   $\leq \dfrac{1}{1^2} + \displaystyle\int_1^n \frac{1}{x^2}\,dx < 1 + \int_1^{\infty} \frac{1}{x^2}\,dx$.
   c. By evaluating the improper integral in part (b), show that $S_n \leq 2$ for each $n = 1, 2, 3, \ldots$. Then use the Monotone Convergence Theorem (Section 8.1) to show that $\displaystyle\sum_{n=1}^{\infty} \frac{1}{n^2}$ converges.
   **Note:** The Swiss mathematician Leonhard Euler showed that the sum of this series is $\pi^2/6$.

2. Consider the series $\displaystyle\sum_{n=1}^{\infty} \frac{1}{n} = \frac{1}{1} + \frac{1}{2} + \frac{1}{3} + \frac{1}{4} + \frac{1}{5} + \cdots$.
Let $f(x) = \dfrac{1}{x}$.
   a. Sketch a figure similar to Figure 1b for this series and function, and compute $a_1 = f(1)$, $a_2 = f(2)$, $a_3 = f(3)$, $\ldots, a_n = f(n)$.
   b. Explain why $S_{n-1} = \dfrac{1}{1} + \dfrac{1}{2} + \dfrac{1}{3} + \cdots + \dfrac{1}{n-1} \geq \displaystyle\int_1^n \frac{1}{x}\,dx$.
   c. Show that $\displaystyle\int_1^{\infty} \frac{1}{x}\,dx$ is divergent, and conclude that $\displaystyle\sum_{n=1}^{\infty} \frac{1}{n}$ diverges.
   **Note:** This is the harmonic series that was shown to be divergent in Section 8.2.

## 8.3 EXERCISES

*In Exercises 1–8, use the Integral Test to determine whether the series is convergent or divergent.*

1. $\displaystyle\sum_{n=1}^{\infty} \frac{1}{n^4}$

2. $\displaystyle\sum_{n=1}^{\infty} \frac{3}{2n-1}$

3. $\displaystyle\sum_{n=1}^{\infty} e^{-n}$

4. $\displaystyle\sum_{n=1}^{\infty} ne^{-n}$

5. $\dfrac{1}{2} + \dfrac{1}{5} + \dfrac{1}{10} + \dfrac{1}{17} + \dfrac{1}{26} + \cdots$

6. $\dfrac{1}{3} + \dfrac{1}{7} + \dfrac{1}{11} + \dfrac{1}{15} + \dfrac{1}{19} + \cdots$

7. $\displaystyle\sum_{n=1}^{\infty} \frac{n}{(n^2+1)^{3/2}}$

8. $\displaystyle\sum_{n=2}^{\infty} \frac{1}{n\sqrt{\ln n}}$

*In Exercises 9–14, determine whether the p-series is convergent or divergent.*

9. $\displaystyle\sum_{n=1}^{\infty} \frac{1}{n^3}$

10. $\displaystyle\sum_{n=1}^{\infty} \frac{1}{n^{2/3}}$

11. $\displaystyle\sum_{n=1}^{\infty} \frac{1}{n^{1.01}}$

12. $\displaystyle\sum_{n=1}^{\infty} \frac{1}{n^e}$

13. $\displaystyle\sum_{n=1}^{\infty} n^{-\pi}$

14. $\displaystyle\sum_{n=1}^{\infty} n^{-0.98}$

*In Exercises 15–32 determine whether the given series is convergent or divergent.*

15. $\displaystyle\sum_{n=0}^{\infty} \frac{1}{\sqrt{n+1}}$

16. $\displaystyle\sum_{n=1}^{\infty} \frac{n}{\sqrt{2n^2+1}}$

17. $\displaystyle\sum_{n=1}^{\infty} \frac{1}{n\sqrt{n}}$

18. $\displaystyle\sum_{n=1}^{\infty} n^{-0.75}$

19. $\displaystyle\sum_{n=1}^{\infty} \left(\frac{1}{n\sqrt{n}} + \frac{2}{n^2}\right)$

20. $\displaystyle\sum_{n=1}^{\infty} \left[\left(\frac{2}{3}\right)^n + \frac{1}{n^{3/2}}\right]$

21. $\displaystyle\sum_{n=2}^{\infty} \frac{\ln n}{n}$

22. $\displaystyle\sum_{n=2}^{\infty} \frac{\ln n}{n^2}$

23. $\displaystyle\sum_{n=2}^{\infty} \frac{1}{n(\ln n)^2}$

24. $\displaystyle\sum_{n=1}^{\infty} \frac{e^{1/n}}{n^2}$

25. $\displaystyle\sum_{n=1}^{\infty} \frac{\sin\left(\frac{1}{n}\right)}{n^2}$

26. $\displaystyle\sum_{n=1}^{\infty} \frac{1}{\sqrt{n}+4}$

27. $\displaystyle\sum_{n=1}^{\infty} \frac{1}{4n^2-1}$

28. $\displaystyle\sum_{n=1}^{\infty} \frac{n}{2^n}$

29. $\displaystyle\sum_{n=1}^{\infty} \frac{\tan^{-1} n}{n^2+1}$

30. $\displaystyle\sum_{n=1}^{\infty} \frac{1}{e^{-n}+1}$

31. $\displaystyle\sum_{n=1}^{\infty} \frac{1}{n^2+2n+5}$

32. $\displaystyle\sum_{n=1}^{\infty} \frac{1}{2n^2+7n+3}$

*In Exercises 33 and 34, find the values of p for which the series is convergent.*

33. $\displaystyle\sum_{n=2}^{\infty} \frac{1}{n(\ln n)^p}$

34. $\displaystyle\sum_{n=1}^{\infty} \frac{\ln n}{n^p}$

35. Find the value(s) of $a$ for which the series
$\displaystyle\sum_{n=1}^{\infty} \left[\frac{a}{n+1} - \frac{1}{n+2}\right]$ converges. Justify your answer.

**36. a.** Show that if $S_n$ is the $n$th partial sum of the harmonic series, then $S_n \le 1 + \ln n$.
  **Hint:** Use Inequality (1), page 753, with $f(x) = 1/x$.
  **b.** Use part (a) to show that the sum of the first 1,000,000 terms of the harmonic series is less than 15. The harmonic series diverges very slowly!

**37. Euler's Constant**
  **a.** Show that

$$\ln(n + 1) \le 1 + \frac{1}{2} + \cdots + \frac{1}{n}$$

  and therefore,

$$0 < \ln(n + 1) - \ln n \le 1 + \frac{1}{2} + \cdots + \frac{1}{n} - \ln n$$

  Hence, deduce that the sequence $\{a_n\}$ defined by

$$a_n = 1 + \frac{1}{2} + \cdots + \frac{1}{n} - \ln n$$

  is bounded below.
  **Hint:** Use Inequality (2), page 754, with $f(x) = 1/x$.
  **b.** Show that

$$\frac{1}{n + 1} < \int_n^{n+1} \frac{1}{x}\, dx = \ln(n + 1) - \ln n$$

  and use this result to show that the sequence $\{a_n\}$ defined in part (a) is decreasing.
  **Hint:** Draw a figure similar to Figure 1.
  **c.** Use the Monotone Convergence Theorem to show that $\{a_n\}$ is convergent.
  **Note:** The number

$$\gamma = \lim_{n \to \infty} a_n = \lim_{n \to \infty} \left( 1 + \frac{1}{2} + \cdots + \frac{1}{n} - \ln n \right)$$

  whose value is $0.5772 \ldots$, is called Euler's constant.

**38. Riemann Zeta Function** The *Riemann zeta function* for real numbers is defined by

$$\xi(x) = \sum_{n=1}^{\infty} n^{-x}$$

  What is the domain of the function?

**39.** Let $a_k = f(k)$, where $f$ is a continuous, positive, and decreasing function on $[n, \infty)$, and suppose that $\sum_{n=1}^{\infty} a_n$ is convergent.
  **a.** Show, by sketching appropriate figures, that if $R_n = S - S_n$, where $S = \sum_{n=1}^{\infty} a_n$ and $S_n = \sum_{k=1}^{n} a_k$, then

$$\int_{n+1}^{\infty} f(x)\, dx \le R_n \le \int_n^{\infty} f(x)\, dx$$

  **Note:** $R_n$ is the error estimate for the Integral Test.
  **b.** Use the result of part (a) to deduce that

$$S_n + \int_{n+1}^{\infty} f(x)\, dx \le S \le S_n + \int_n^{\infty} f(x)\, dx$$

**40.** Consider the series $\displaystyle\sum_{n=1}^{\infty} \frac{1}{n^2}$, which is a convergent $p$-series $(p = 2)$.
  **a.** Use the result of Exercise 39b to show that

$$S_n + \frac{1}{n + 1} \le \sum_{n=1}^{\infty} \frac{1}{n^2} \le S_n + \frac{1}{n}$$

  where $S_n = \displaystyle\sum_{k=1}^{n} \frac{1}{k^2}$ is the $n$th partial sum of $\displaystyle\sum_{n=1}^{\infty} \frac{1}{n^2}$.
  **b.** In Exercise 77 in Section 8.2 you were asked to show that

$$\frac{3}{2} \le \sum_{n=1}^{\infty} \frac{1}{n^2} \le 2$$

  Confirm this result, using the result of part (a).
  **c.** Use the result of Exercise 39a to find the upper and lower bounds on the error incurred in approximating

$$\sum_{n=1}^{\infty} \frac{1}{n^2}$$

  using the 100th partial sum of the series.
  **cas d.** It can be shown that $\displaystyle\sum_{n=1}^{\infty} \frac{1}{n^2} = \frac{\pi^2}{6}$. Use a calculator or computer to verify this.

*In Exercises 41–44, use the result of Exercise 39 to find the maximum error if the sum of the series is approximated by $S_n$.*

**41.** $\displaystyle\sum_{n=1}^{\infty} \frac{2}{n^2};\quad S_{40}$

**42.** $\displaystyle\sum_{n=1}^{\infty} \frac{1}{n^{5/2}};\quad S_{20}$

**43.** $\displaystyle\sum_{n=1}^{\infty} \frac{1}{n^2 + 1};\quad S_{50}$

**44.** $\displaystyle\sum_{n=1}^{\infty} ne^{-n^2};\quad S_3$

*In Exercises 45–48, use the result of Exercise 39 to find the number of terms of the series that is sufficient to obtain an approximation of the sum of the series accurate to two decimal places.*

**45.** $\displaystyle\sum_{n=1}^{\infty} \frac{1}{n^2}$

**46.** $\displaystyle\sum_{n=1}^{\infty} \frac{1}{n^3}$

**47.** $\displaystyle\sum_{n=1}^{\infty} \frac{\tan^{-1} n}{1 + n^2}$

**48.** $\displaystyle\sum_{n=2}^{\infty} \frac{1}{n(\ln n)^2}$

 *In Exercises 49 and 50, use the result of Exercise 39 to find the sum of the series accurate to three decimal places using the $n$th partial sum of the series.*

**49.** $\displaystyle\sum_{n=1}^{\infty} \frac{1}{n^4}$

**50.** $\displaystyle\sum_{n=1}^{\infty} \frac{1}{n^{9/2}}$

**51. a.** Show that

$$\sum_{n=1}^{\infty} \frac{1}{n(n + 1)(n + 2)} = \sum_{n=1}^{\infty} \left[ \frac{1}{2n(n + 1)} - \frac{1}{2(n + 1)(n + 2)} \right]$$

  **b.** Use the results of part (a) to evaluate

$$\sum_{n=1}^{\infty} \frac{1}{n(n + 1)(n + 2)}$$

**52.** Evaluate $\sum_{n=1}^{\infty} \dfrac{1}{n^3}$ accurate to four decimal places by establishing parts (a) and (b) and using the results of Exercise 51.

**a.** $\displaystyle\sum_{n=1}^{\infty} \dfrac{1}{n^3} = 1 + \sum_{n=2}^{\infty} \dfrac{1}{(n-1)n(n+1)} - \sum_{n=2}^{\infty} \dfrac{1}{n^3(n^2-1)}$

**b.** $\displaystyle\sum_{n=2}^{\infty} \dfrac{1}{n^3(n^2-1)}$ can be approximated with an accuracy of
four decimal places by using six terms of the series.
**Hint:** Show that

$$\frac{1}{n^3(n^2-1)} \le \frac{2}{n^5}$$

if $n \ge 2$, and use the result of Exercise 39.

**53.** Use the Integral Test to show that $\displaystyle\sum_{n=3}^{\infty} \dfrac{1}{n(\ln n)[\ln(\ln n)]^p}$
converges if $p > 1$ and diverges if $p \le 1$.

**54.** Consider the series $\sum_{n=0}^{\infty} e^{-n}$.
**a.** Evaluate $\int_0^{\infty} e^{-x}\,dx$, and deduce from the Integral Test
that the given series is convergent.

**b.** Show that the given series is a geometric series, and find
its sum.
**c.** Conclude that although the convergence of $\int_0^{\infty} e^{-x}\,dx$
implies convergence of the infinite series, its value does
not give the sum of the infinite series.

*In Exercises 55–58, determine whether the statement is true or
false. If it is true, explain why it is true. If it is false, explain why
or give an example to show why it is false.*

**55.** Suppose that $f$ is a continuous, positive, and decreasing
function on $[1, \infty)$. If $f(n) = a_n$ for $n \ge 1$ and $\sum_{n=1}^{\infty} a_n$ is
convergent, then $\sum_{n=1}^{\infty} a_n \le a_1 + \int_1^{\infty} f(x)\,dx$.

**56.** Suppose that $f$ is a continuous, positive, and decreasing
function on $[1, \infty)$. If $f(n) = a_n$ for $n \ge 1$ and
$\int_N^{\infty} f(x)\,dx = \infty$, where $N$ is a positive integer, then $\sum_{n=1}^{\infty} a_n$
diverges.

**57.** $\displaystyle\int_1^{\infty} \dfrac{dx}{x(x+1)} < \infty$

**58.** If $\sum_{n=1}^{\infty} a_n$ is a convergent series with positive terms, then
$\sum_{n=1}^{\infty} \sqrt{a_n}$ must also converge.

---

## 8.4 The Comparison Tests

The rationale for the comparison tests is that the convergence or divergence of a given
series $\Sigma\, a_n$ can be determined by comparing its terms with the corresponding terms of
a *test series* whose convergence or divergence is known. The series that we will consider in this section have positive terms.

### The Comparison Test

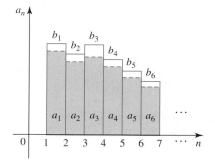

**FIGURE 1**
Each rectangle representing $a_n$ is contained in the rectangle representing $b_n$.

Suppose that the terms of a series $\Sigma\, a_n$ are smaller than the corresponding terms of a
series $\Sigma\, b_n$. This situation is illustrated in Figure 1, where the respective terms of each
series are represented by rectangles, each of width 1 and having the appropriate height.
　If $\Sigma\, b_n$ is convergent, the total area of the rectangles representing this series is finite.
Since each rectangle representing the series $\Sigma\, a_n$ is contained in a corresponding rectangle representing the terms of $\Sigma\, b_n$, the total area of the rectangles representing $\Sigma\, a_n$
must also be finite; that is, the series $\Sigma\, a_n$ must be convergent. A similar argument suggests that if all the terms of a series $\Sigma\, a_n$ are larger than the corresponding terms of a
series $\Sigma\, b_n$ that is known to be divergent, then $\Sigma\, a_n$ must itself be divergent. These
observations lead to the following theorem.

---

**THEOREM 1   The Comparison Test**

Suppose that $\Sigma\, a_n$ and $\Sigma\, b_n$ are series with positive terms.

**a.** If $\Sigma\, b_n$ is convergent and $a_n \le b_n$ for all $n$, then $\Sigma\, a_n$ is also convergent.
**b.** If $\Sigma\, b_n$ is divergent and $a_n \ge b_n$ for all $n$, then $\Sigma\, a_n$ is also divergent.

**PROOF**  Let

$$S_n = \sum_{k=1}^{n} a_k \quad \text{and} \quad T_n = \sum_{k=1}^{n} b_k$$

be the $n$th terms of the sequence of partial sums of $\Sigma\, a_n$ and $\Sigma\, b_n$, respectively. Since both series have positive terms, $\{S_n\}$ and $\{T_n\}$ are increasing.

**a.** If $\Sigma_{n=1}^{\infty} b_n$ is convergent, then there exists a number $L$ such that $\lim_{n\to\infty} T_n = L$ and $T_n \leq L$ for all $n$. Since $a_n \leq b_n$ for all $n$, we have $S_n \leq T_n$, and this implies that $S_n \leq L$ for all $n$. We have shown that $\{S_n\}$ is increasing and bounded above, so by the Monotone Convergence Theorem for Sequences of Section 8.1, $\Sigma\, a_n$ converges.

**b.** If $\Sigma_{n=1}^{\infty} b_n$ is divergent, then $\lim_{n\to\infty} T_n = \infty$, since $\{T_n\}$ is increasing. But $a_n \geq b_n$ for all $n$, and this implies that $S_n \geq T_n$, which in turn implies that $\lim_{n\to\infty} S_n = \infty$. Therefore, $\Sigma\, a_n$ diverges.  ∎

To use the Comparison Test, we need a catalog of test series whose convergence or divergence is known. For the moment we can use the geometric series and the $p$-series as test series.

**EXAMPLE 1**  Determine whether the series $\displaystyle\sum_{n=1}^{\infty} \frac{1}{n^2 + 2}$ converges or diverges.

**Solution**  Let

$$a_n = \frac{1}{n^2 + 2}$$

If $n$ is large, $n^2 + 2$ behaves like $n^2$, so $a_n$ behaves like

$$b_n = \frac{1}{n^2}$$

This observation suggests that we compare $\Sigma\, a_n$ with the test series $\Sigma\, b_n$, which is a convergent $p$-series with $p = 2$. Now,

$$0 < \frac{1}{n^2 + 2} < \frac{1}{n^2} \qquad n \geq 1$$

and the given series is indeed "smaller" than the test series $\Sigma\, 1/n^2$. Since the test series converges, we conclude by the Comparison Test that $\Sigma\, 1/(n^2 + 2)$ also converges.  ∎

**EXAMPLE 2**  Determine whether the series $\displaystyle\sum_{n=1}^{\infty} \frac{1}{3 + 2^n}$ converges or diverges.

**Solution**  Let

$$a_n = \frac{1}{3 + 2^n}$$

If $n$ is large, $3 + 2^n$ behaves like $2^n$, so $a_n$ behaves like $b_n = \left(\frac{1}{2}\right)^n$. This observation suggests that we compare $\Sigma\, a_n$ with $\Sigma\, b_n$. Now the series $\Sigma\, \frac{1}{2^n} = \Sigma\, \left(\frac{1}{2}\right)^n$ is a geometric series with $r = \frac{1}{2} < 1$, so it is convergent. Since

$$a_n = \frac{1}{3 + 2^n} < \frac{1}{2^n} = b_n \qquad n \geq 1$$

the Comparison Test tells us that the given series is convergent.  ∎

**Note**    Since the convergence or divergence of a series is not affected by the omission of a finite number of terms of the series, the condition $a_n \le b_n$ (or $a_n \ge b_n$) for all $n$ can be replaced by the condition that these inequalities hold for all $n \ge N$ for some integer $N$.    ∎

**EXAMPLE 3**    Determine whether the series $\displaystyle\sum_{n=2}^{\infty} \frac{1}{\sqrt{n} - 1}$ is convergent or divergent.

**Solution**    Let

$$a_n = \frac{1}{\sqrt{n} - 1}$$

If $n$ is large, $\sqrt{n} - 1$ behaves like $\sqrt{n}$, so $a_n$ behaves like

$$b_n = \frac{1}{\sqrt{n}}$$

Now the series

$$\sum_{n=2}^{\infty} b_n = \sum_{n=2}^{\infty} \frac{1}{\sqrt{n}} = \sum_{n=2}^{\infty} \frac{1}{n^{1/2}}$$

is a $p$-series with $p = \frac{1}{2} < 1$, so it is divergent. Since

$$a_n = \frac{1}{\sqrt{n} - 1} > \frac{1}{\sqrt{n}} = b_n \qquad \text{for } n \ge 2$$

the Comparison Test implies that the given series is divergent.    ∎

## ■ The Limit Comparison Test

Consider the series

$$\sum_{n=1}^{\infty} \frac{1}{\sqrt{n} + 1}$$

If $n$ is large, $\sqrt{n} + 1$ behaves like $\sqrt{n}$, so the $n$th term of the given series

$$a_n = \frac{1}{\sqrt{n} + 1}$$

behaves like

$$b_n = \frac{1}{\sqrt{n}}$$

Since the series $\sum_{n=1}^{\infty} b_n = \sum_{n=1}^{\infty} 1/\sqrt{n}$ is a divergent $p$-series with $p = \frac{1}{2}$, we expect the series $\sum_{n=1}^{\infty} 1/(\sqrt{n} + 1)$ to be divergent as well. But the inequality

$$a_n = \frac{1}{\sqrt{n} + 1} < \frac{1}{\sqrt{n}} = b_n \qquad n \ge 1$$

tells us that $\sum_{n=1}^{\infty} a_n$ is "smaller" than a divergent series, and this is of no help if we try to use the Comparison Test!

In situations like this, the *Limit Comparison Test* might be applicable. The rationale for this test follows: Suppose that $\Sigma\, a_n$ and $\Sigma\, b_n$ are series with positive terms and suppose that $\lim_{n\to\infty}(a_n/b_n) = L$, where $L$ is a positive constant. If $n$ is large, $a_n/b_n \approx L$ or $a_n \approx Lb_n$. It is reasonable to conjecture that the series $\Sigma\, a_n$ and $\Sigma\, b_n$ must both converge or both diverge.

---

**THEOREM 2    The Limit Comparison Test**

Suppose that $\Sigma\, a_n$ and $\Sigma\, b_n$ are series with positive terms and

$$\lim_{n\to\infty} \frac{a_n}{b_n} = L$$

where $L$ is a positive number. Then either both series converge or both diverge.

---

**PROOF**    Since $\lim_{n\to\infty}(a_n/b_n) = L > 0$, there exists an integer $N$ such that $n \geq N$ implies that

$$\left| \frac{a_n}{b_n} - L \right| < \frac{1}{2} L$$

$$\frac{1}{2} L < \frac{a_n}{b_n} < \frac{3}{2} L$$

or

$$\frac{1}{2} Lb_n < a_n < \frac{3}{2} Lb_n$$

If $\Sigma\, b_n$ converges, so does $\Sigma\, \frac{3}{2} Lb_n$. Therefore, the right side of the last inequality implies that $\Sigma\, a_n$ converges by the Comparison Test. On the other hand, if $\Sigma\, b_n$ diverges, so does $\Sigma\, \frac{1}{2} Lb_n$, and the left side of the last inequality implies by the Comparison Test that $\Sigma\, a_n$ diverges as well.    ■

**EXAMPLE 4**    Show that the series $\displaystyle\sum_{n=1}^{\infty} \frac{1}{\sqrt{n} + 1}$ is divergent.

**Solution**    As we saw earlier, $1/(\sqrt{n} + 1)$ behaves like $1/\sqrt{n}$ if $n$ is large. This suggests that we use the Limit Comparison Test with $a_n = 1/(\sqrt{n} + 1)$ and $b_n = 1/\sqrt{n}$. Thus,

$$\lim_{n\to\infty} \frac{a_n}{b_n} = \lim_{n\to\infty} \frac{\dfrac{1}{\sqrt{n} + 1}}{\dfrac{1}{\sqrt{n}}} = \lim_{n\to\infty} \frac{\sqrt{n}}{\sqrt{n} + 1} = \lim_{n\to\infty} \frac{1}{1 + \dfrac{1}{\sqrt{n}}} = 1$$

Since $\Sigma_{n=1}^{\infty} 1/\sqrt{n}$ is divergent $\left(\text{it is a } p\text{-series with } p = \frac{1}{2}\right)$, we conclude that the given series is divergent as well.    ■

**Note**    You can still use the Comparison Test to solve the problem. Simply observe that

$$\frac{1}{\sqrt{n} + 1} \geq \frac{1}{\sqrt{n} + \sqrt{n}} = \frac{1}{2\sqrt{n}} \qquad \text{for } n \geq 1$$

This suggests picking $\Sigma\, b_n$, where $b_n = 1/(2\sqrt{n})$, for the test series.    ■

**EXAMPLE 5**    Determine whether the series $\displaystyle\sum_{n=1}^{\infty} \frac{2n^2 + n}{\sqrt{4n^7 + 3}}$ converges or diverges.

**Solution**    If $n$ is large, $2n^2 + n$ behaves like $2n^2$, and $4n^7 + 3$ behaves like $4n^7$. Therefore,

$$a_n = \frac{2n^2 + n}{\sqrt{4n^7 + 3}}$$

behaves like

$$\frac{2n^2}{\sqrt{4n^7}} = \frac{2n^2}{2n^{7/2}} = \frac{1}{n^{3/2}} = b_n$$

Now

$$\lim_{n\to\infty} \frac{a_n}{b_n} = \lim_{n\to\infty} \frac{2n^2 + n}{(4n^7 + 3)^{1/2}} \cdot \frac{n^{3/2}}{1}$$

$$= \lim_{n\to\infty} \frac{2n^{7/2} + n^{5/2}}{(4n^7 + 3)^{1/2}}$$

$$= \lim_{n\to\infty} \frac{2 + \dfrac{1}{n}}{\left(4 + \dfrac{3}{n^7}\right)^{1/2}} \qquad \text{Divide numerator and}$$
$$\qquad\qquad\qquad\qquad\qquad \text{denominator by } n^{7/2}.$$

$$= 1$$

Since $\Sigma\, 1/n^{3/2}$ converges $\left(\text{it is a } p\text{-series with } p = \frac{3}{2}\right)$, the given series converges, by the Limit Comparison Test.    ■

**EXAMPLE 6**    Determine whether the series $\displaystyle\sum_{n=1}^{\infty} \frac{\sqrt{n} + \ln n}{n^2 + 1}$ converges or diverges.

**Solution**    If $n$ is large, $\sqrt{n} + \ln n$ behaves like $\sqrt{n}$. You can see this by comparing the derivatives of $f(x) = \sqrt{x}$ and $g(x) = \ln x$:

$$f'(x) = \frac{1}{2\sqrt{x}} \qquad \text{and} \qquad g'(x) = \frac{1}{x}$$

Observe that $g'(x)$ approaches zero faster than $f'(x)$ approaches zero, as $x \to \infty$. This shows that $\sqrt{x}$ grows faster than $\ln x$. Also, if $n$ is large, $n^2 + 1$ behaves like $n^2$. Therefore,

$$a_n = \frac{\sqrt{n} + \ln n}{n^2 + 1}$$

behaves like

$$\frac{\sqrt{n}}{n^2} = \frac{1}{n^{3/2}} = b_n$$

Next, we compute

$$\lim_{n \to \infty} \frac{a_n}{b_n} = \lim_{n \to \infty} \frac{n^{1/2} + \ln n}{n^2 + 1} \cdot \frac{n^{3/2}}{1}$$

$$= \lim_{n \to \infty} \frac{n^2 + n^{3/2} \ln n}{n^2 + 1}$$

$$= \lim_{n \to \infty} \frac{1 + \dfrac{\ln n}{n^{1/2}}}{1 + \dfrac{1}{n^2}} \qquad \text{Divide the numerator and denominator by } n^2.$$

In evaluating this limit, we need to compute

$$\lim_{x \to \infty} \frac{\ln x}{x^{1/2}} = \lim_{x \to \infty} \frac{\dfrac{1}{x}}{\frac{1}{2} x^{-1/2}} = \lim_{x \to \infty} \frac{2}{\sqrt{x}} = 0 \qquad \text{Use l'Hôpital's Rule.}$$

(Incidentally, this result supports the observation made earlier that $\sqrt{x}$ grows faster than $\ln x$.) Using this result, we find

$$\lim_{n \to \infty} \frac{a_n}{b_n} = \lim_{n \to \infty} \frac{1 + \dfrac{\ln n}{n^{1/2}}}{1 + \dfrac{1}{n^2}} = 1$$

Since $\Sigma \, 1/n^{3/2}$ converges $\left(\text{it is a } p\text{-series with } p = \frac{3}{2}\right)$, the given series converges, by the Limit Comparison Test. ∎

## 8.4 CONCEPT QUESTIONS

**1. a.** State the Comparison Test and the Limit Comparison Test.
  **b.** When is the Comparison Test used? When is the Limit Comparison Test used?
**2.** Let $\Sigma \, a_n$ and $\Sigma \, b_n$ be series with positive terms.
  **a.** If $\Sigma \, b_n$ is convergent and $a_n \geq b_n$ for all $n$, what can you say about the convergence or divergence of $\Sigma \, a_n$? Give examples.
  **b.** If $\Sigma \, b_n$ is divergent and $a_n \leq b_n$ for all $n$, what can you say about the convergence or divergence of $\Sigma \, a_n$? Give examples.

*In Exercises 3 and 4, let $\Sigma \, a_n$, $\Sigma \, b_n$, and $\Sigma \, c_n$ be series with positive terms.*

**3.** If $\Sigma \, a_n$ is convergent and $b_n + c_n \leq a_n$ for all $n$, what can you say about the convergence or divergence of $\Sigma \, b_n$ and $\Sigma \, c_n$?
**4.** If $\Sigma \, a_n$ is divergent and $b_n + c_n \geq a_n$ for all $n$, what can you say about the convergence or divergence of $\Sigma \, b_n$ and $\Sigma \, c_n$?

## 8.4 EXERCISES

*In Exercises 1–12, use the Comparison Test to determine whether the series is convergent or divergent.*

**1.** $\displaystyle\sum_{n=1}^{\infty} \frac{1}{2n^2 + 1}$

**2.** $\displaystyle\sum_{n=1}^{\infty} \frac{1}{n^2 + 2n}$

**3.** $\displaystyle\sum_{n=3}^{\infty} \frac{1}{n - 2}$

**4.** $\displaystyle\sum_{n=2}^{\infty} \frac{1}{n^{2/3} - 1}$

**5.** $\displaystyle\sum_{n=2}^{\infty} \frac{1}{\sqrt{n^2 - 1}}$

**6.** $\displaystyle\sum_{n=0}^{\infty} \frac{1}{\sqrt{n^3 + 1}}$

7. $\displaystyle\sum_{n=0}^{\infty} \frac{2^n}{3^n + 1}$

8. $\displaystyle\sum_{n=3}^{\infty} \frac{3^n}{2^n - 4}$

9. $\displaystyle\sum_{n=2}^{\infty} \frac{\ln n}{n}$

10. $\displaystyle\sum_{n=1}^{\infty} \frac{\cos^2 n}{n^2}$

11. $\displaystyle\sum_{n=1}^{\infty} \frac{2 + \sin n}{3^n}$

12. $\displaystyle\sum_{n=1}^{\infty} \frac{1}{n^n}$

*In Exercises 13–24, use the Limit Comparison Test to determine whether the series is convergent or divergent.*

13. $\displaystyle\sum_{n=2}^{\infty} \frac{n}{n^2 + 1}$

14. $\displaystyle\sum_{n=1}^{\infty} \frac{1}{\sqrt{n} + 2}$

15. $\displaystyle\sum_{n=2}^{\infty} \frac{n}{\sqrt{n^5 - 1}}$

16. $\displaystyle\sum_{n=1}^{\infty} \frac{2n + 1}{3n^2 - n + 1}$

17. $\displaystyle\sum_{n=1}^{\infty} \frac{3n^2 + 1}{2n^5 + n + 2}$

18. $\displaystyle\sum_{n=1}^{\infty} \frac{n^2 + 1}{n^2(n + 3)}$

19. $\displaystyle\sum_{n=2}^{\infty} \frac{1}{\sqrt{n^3 - n - 1}}$

20. $\displaystyle\sum_{n=2}^{\infty} \frac{1}{2^n - 3}$

21. $\displaystyle\sum_{n=1}^{\infty} \frac{n}{2^n - 1}$

22. $\displaystyle\sum_{n=2}^{\infty} \frac{\ln n}{n^3 - 1}$

23. $\displaystyle\sum_{n=1}^{\infty} \sin \frac{1}{n}$

24. $\displaystyle\sum_{n=1}^{\infty} \tan \frac{1}{n}$

*In Exercises 25–40, determine whether the series is convergent or divergent.*

25. $\displaystyle\sum_{n=1}^{\infty} \frac{n + 1}{(n + 2)(2n^2 + 1)}$

26. $\displaystyle\sum_{n=1}^{\infty} \frac{n}{\sqrt{n^5 + n}}$

27. $\displaystyle\sum_{n=1}^{\infty} \frac{n - 1}{n^3 + 2}$

28. $\displaystyle\sum_{n=1}^{\infty} \frac{n + 1}{2n^3 + 1}$

29. $\displaystyle\sum_{n=1}^{\infty} \frac{2^{n-1}}{n^2 + n}$

30. $\displaystyle\sum_{n=1}^{\infty} \frac{1}{n + \sqrt{n^2 - 1}}$

31. $\displaystyle\sum_{n=1}^{\infty} \frac{\sin^2 n}{n\sqrt{n + 1}}$

32. $\displaystyle\sum_{n=1}^{\infty} \frac{\tan^{-1} n}{n^3 + 1}$

33. $\displaystyle\sum_{n=2}^{\infty} \frac{1}{\ln n}$

34. $\displaystyle\sum_{n=1}^{\infty} \frac{\ln n}{n + 2}$

35. $\displaystyle\sum_{n=0}^{\infty} \frac{1}{n!}$

36. $\displaystyle\sum_{n=1}^{\infty} \frac{n^2}{n!}$

37. $\displaystyle\sum_{n=1}^{\infty} \frac{n!}{n^n}$

38. $\displaystyle\sum_{n=1}^{\infty} \frac{1}{1 + 2 + 3 + \cdots + n}$

39. $\displaystyle\sum_{n=1}^{\infty} \frac{\sqrt{n} + \ln n}{2n^2 + 3}$

40. $\displaystyle\sum_{n=1}^{\infty} \frac{2n^2 + n}{\sqrt{3n^7 + \ln n}}$

41. Let $\Sigma\, a_n$ and $\Sigma\, b_n$ be series with $0 \le a_n \le b_n$, and suppose that $\Sigma\, b_n$ is convergent with sum $T$. Then the Comparison Test implies that $\Sigma\, a_n$ also converges, say, with sum $S$. Put $R_n = S - S_n$ and $T_n = T - U_n$, where $U_n$ is the $n$th-partial sum of $\Sigma\, b_n$. Show that the remainders $R_n$ and $T_n$ satisfy $R_n \le T_n$.

*In Exercises 42–45, use the result of Exercise 41 to find an approximation of the sum of the series using its partial sum, accurate to two decimal places.*

42. $\displaystyle\sum_{n=1}^{\infty} \frac{1}{n^3 + 2n}$

43. $\displaystyle\sum_{n=1}^{\infty} \frac{\sin n + 2}{n^4}$

44. $\displaystyle\sum_{n=1}^{\infty} \frac{1}{3^n + 1}$

45. $\displaystyle\sum_{n=1}^{\infty} \frac{\tan^{-1} n}{2^n}$

46. Suppose that $\Sigma\, a_n$ is a convergent series with positive terms. Show that $\Sigma(a_n/n)$ is also convergent.

47. Suppose that $\Sigma\, a_n$ and $\Sigma\, b_n$ are convergent series with positive terms. Show that $\Sigma\, a_n b_n$ is convergent.
    **Hint:** There exists an integer $N$ such that $n \ge N$ implies that $b_n \le 1$, and therefore, $a_n b_n \le a_n$ for $n \ge N$.

48. Suppose that $\Sigma\, a_n$ is a convergent series with positive terms and $\{c_n\}$ is a sequence of positive numbers that converges to zero. Prove that $\Sigma\, a_n c_n$ is convergent.
    **Hint:** There exists an integer $N$ such that $n \ge N$ implies that $c_n < L$, where $L$ is a positive number, and therefore, $a_n c_n < L a_n$ for $n \ge N$.

49. Prove that if $a_n \ge 0$ and $\Sigma\, a_n$ converges, then $\Sigma\, a_n^2$ also converges. Is the converse true? Explain.

50. Using the result of Exercise 48 or otherwise, show that $\Sigma_{n=2}^{\infty} 1/(n^p \ln n)$ is convergent if $p > 1$.

51. **a.** Suppose that $\Sigma\, a_n$ and $\Sigma\, b_n$ are series with positive terms and $\Sigma\, b_n$ is convergent. Show that if $\lim_{n \to \infty} a_n/b_n = 0$, then $\Sigma\, a_n$ is convergent.

    **b.** Use part (a) to show that $\displaystyle\sum_{n=1}^{\infty} \frac{\ln n}{n^2}$ is convergent.

52. Give an example of a pair of series $\Sigma\, a_n$ and $\Sigma\, b_n$ with positive terms such that $\lim_{n \to \infty} a_n/b_n = 0$, $\Sigma\, b_n$ is divergent, but $\Sigma\, a_n$ is convergent. (Compare this with the result of Exercise 51.)

53. **a.** Show that if $\Sigma\, a_n$ is a convergent series with positive terms, then $\Sigma \sin a_n$ is also convergent.

    **b.** If $\Sigma\, a_n$ diverges can $\Sigma \sin a_n$ converge? Explain.

54. Prove that (a) $\displaystyle\int_1^{\infty} \frac{1}{\sqrt{x(x + 1)(x + 2)}}\, dx$ converges and

    (b) $\displaystyle\int_1^{\infty} \frac{1}{\sqrt{x(x + 1)}}\, dx$ diverges.

*In Exercises 55–58, determine whether the statement is true or false. If it is true, explain why it is true. If it is false, explain why or give an example to show why it is false.*

55. If $0 \le a_n \le b_n$ and $\Sigma\, a_n$ converges, then $\Sigma\, b_n$ diverges.

56. If $0 \le a_n \le b_n$ and $\Sigma\, b_n$ diverges, then $\Sigma\, a_n$ diverges.

57. If $a_n > 0$ and $b_n > 0$ and $\Sigma\, a_n b_n$ converges, then $\Sigma\, a_n$ and $\Sigma\, b_n$ both converge.

58. If $a_n > 0$ and $b_n > 0$ and $\Sigma \sqrt{a_n^2 + b_n^2}$ converges, then $\Sigma\, a_n$ and $\Sigma\, b_n$ both converge.

# 8.5 Alternating Series

Up to now, we have dealt mainly with series that have positive terms, and the convergence tests that we have developed are applicable only to these series. In this section and Section 8.6 we will consider series that contain both positive and negative terms. Series whose terms alternate in sign are called *alternating series*.

Examples are the *alternating harmonic series*

$$\sum_{n=1}^{\infty} \frac{(-1)^{n-1}}{n} = 1 - \frac{1}{2} + \frac{1}{3} - \frac{1}{4} + \frac{1}{5} - \frac{1}{6} + \cdots$$

and the series

$$\sum_{n=1}^{\infty} \frac{(-1)^{n} n^2}{(n+1)!} = -\frac{1}{2!} + \frac{4}{3!} - \frac{9}{4!} + \frac{16}{5!} - \frac{25}{6!} + \cdots$$

More generally, an **alternating series** is a series of the form

$$\sum_{n=1}^{\infty} (-1)^{n-1} a_n \qquad \text{or} \qquad \sum_{n=1}^{\infty} (-1)^{n} a_n$$

where $a_n$ is a positive number. We use the *Alternating Series Test* to determine convergence for these series.

---

**THEOREM 1   The Alternating Series Test**

If the alternating series

$$\sum_{n=1}^{\infty} (-1)^{n-1} a_n = a_1 - a_2 + a_3 - a_4 + a_5 - a_6 + \cdots \qquad a_n > 0$$

satisfies the conditions

**1.** $a_{n+1} \leq a_n$ for all $n$

**2.** $\displaystyle \lim_{n \to \infty} a_n = 0$

then the series converges.

---

The plausibility of Theorem 1 is suggested by Figure 1, which shows the first few terms of the sequence of partial sums $\{S_n\}$ of the alternating series

$$\sum_{n=1}^{\infty} (-1)^{n-1} a_n = a_1 - a_2 + a_3 - \cdots$$

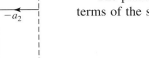

**FIGURE 1**
The terms of $\{S_n\}$ oscillate in smaller and smaller steps, and this suggests that $\lim_{n \to \infty} S_n = S$.

plotted on the number line. The point $S_2 = a_1 - a_2$ lies to the left of the number $S_1 = a_1$, since it is obtained by subtracting the positive number $a_2$ from $S_1$. But the number $S_2$ also lies to the right of the origin because $a_2 \leq a_1$. The number $S_3 = a_1 - a_2 + a_3 = S_2 + a_3$ is obtained by adding $a_3$ to $S_2$, and hence it lies to the right of $S_2$. But because $a_3 < a_2$, $S_3$ lies to the left of $S_1$. Continuing in this fashion, we see that numbers corresponding to the partial sums $\{S_n\}$ oscillate. Because $\lim_{n \to \infty} a_n = 0$, the steps get smaller and smaller. Thus, it appears that the sequence $\{S_n\}$ will approach a limit. In particular, observe that the even terms of the sequence $\{S_n\}$ are increasing, whereas the odd terms of the sequence are decreasing. This suggests that the subsequence $\{S_{2n}\}$ will approach the limit $S$ from below and the subsequence $\{S_{2n+1}\}$ will approach $S$ from above. These observations form the basis of the proof of Theorem 1.

**PROOF OF THEOREM 1** We first consider the subsequence $\{S_{2n}\}$ comprising the even terms of $\{S_n\}$. Now,

$$S_2 = a_1 - a_2 \geq 0 \qquad \text{Since } a_1 \geq a_2$$

$$S_4 = S_2 + (a_3 - a_4) \geq S_2 \qquad \text{Since } a_3 \geq a_4$$

and, in general,

$$S_{2n+2} = S_{2n} + (a_{2n+1} - a_{2n+2}) \geq S_{2n} \qquad \text{Since } a_{2n+1} \geq a_{2n+2}$$

This shows that

$$0 \leq S_2 \leq S_4 \leq \cdots \leq S_{2n} \leq \cdots$$

that is, $\{S_{2n}\}$ is increasing. Next, we write $S_{2n}$ in the form

$$S_{2n} = a_1 - (a_2 - a_3) - (a_4 - a_5) - \cdots - (a_{2n-2} - a_{2n-1}) - a_{2n}$$

and observe that every expression within the parenthesis is nonnegative (again, because $a_{n+1} \leq a_n$). Thus, we see that $S_{2n} \leq a_1$ for all $n$. This shows that the sequence $\{S_{2n}\}$ is bounded above as well. Therefore, by the Monotone Convergence Theorem for Sequences of Section 8.1, the sequence $\{S_{2n}\}$ is convergent; that is, there exists a number $S$ such that $\lim_{n \to \infty} S_{2n} = S$.

Next, we consider the subsequence $\{S_{2n+1}\}$ comprising the odd terms of $\{S_n\}$. Since $S_{2n+1} = S_{2n} + a_{2n+1}$ and $\lim_{n \to \infty} a_{2n+1} = 0$ by assumption, we have

$$\lim_{n \to \infty} S_{2n+1} = \lim_{n \to \infty} (S_{2n} + a_{2n+1})$$

$$= \lim_{n \to \infty} S_{2n} + \lim_{n \to \infty} a_{2n+1}$$

$$= S$$

Since the subsequences $\{S_{2n}\}$ and $\{S_{2n+1}\}$ of the sequence of partial sums $\{S_n\}$ both converge to $S$, we have $\lim_{n \to \infty} S_n = S$, so the series converges. ∎

**EXAMPLE 1** Show that the alternating harmonic series

$$\sum_{n=1}^{\infty} \frac{(-1)^{n-1}}{n} = 1 - \frac{1}{2} + \frac{1}{3} - \frac{1}{4} + \cdots$$

converges.

**Solution** This is an alternating series with $a_n = 1/n$, so we use the Alternating Series Test. We need to verify that (1) $a_{n+1} \leq a_n$ and (2) $\lim_{n \to \infty} a_n = 0$. But the first condition follows from the computation

$$a_{n+1} = \frac{1}{n+1} < \frac{1}{n} = a_n$$

while the second condition follows from

$$\lim_{n \to \infty} a_n = \lim_{n \to \infty} \frac{1}{n} = 0$$

Therefore, by the Alternating Series Test, the given series converges. ∎

**EXAMPLE 2**   Determine whether the series converges or diverges.

**a.** $\displaystyle\sum_{n=1}^{\infty}(-1)^{n}\frac{2n}{4n-1}$   **b.** $\displaystyle\sum_{n=1}^{\infty}(-1)^{n-1}\frac{3n}{4n^{2}-1}$

**Solution**   Since both series are alternating series we use the Alternating Series Test.

**a.** Here, $a_n = 2n/(4n-1)$. Because

$$\lim_{n\to\infty}\frac{2n}{4n-1}=\frac{1}{2}\neq 0$$

we see that condition (2) in the Alternating Series Test is not satisfied. In fact, this computation shows that

$$\lim_{n\to\infty}(-1)^{n}\frac{2n}{4n-1}$$

does not exist, and the divergence of the series follows from the Divergence Test.

**b.** Here $a_n = 3n/(4n^2 - 1)$. First we show that $a_n \geq a_{n+1}$ for all $n$. We can do this by showing that $f(x) = 3x/(4x^2 - 1)$ is decreasing for $x \geq 0$. We compute

$$f'(x) = \frac{(4x^2-1)(3)-(3x)(8x)}{(4x^2-1)^2}$$

$$= \frac{-12x^2-3}{(4x^2-1)^2}<0$$

and the desired conclusion follows. Next, we compute

$$\lim_{n\to\infty}a_n = \lim_{n\to\infty}\frac{3n}{4n^2-1}=\lim_{n\to\infty}\frac{\dfrac{3}{n}}{4-\dfrac{1}{n^2}}=0$$

Since both conditions of the Alternating Series Test are satisfied, we conclude that the series is convergent.   ■

**Notes**

1. Example 2a reminds us once again that it is a good idea to begin investigating the convergence of a series by checking for divergence using the Divergence Test.
2. Because the behavior of a finite number of terms will not affect the convergence or divergence of a series, the first condition in the Alternating Series Test can be replaced by the condition $a_{n+1} \leq a_n$ for $n \geq N$, where $N$ is some positive integer.   ■

## ■ Approximating the Sum of an Alternating Series by $S_n$

Suppose that we can show that the series $\Sigma\, a_n$ is convergent so that it has a sum $S$. If $\{S_n\}$ is the sequence of partial sums of $\Sigma\, a_n$, then $\lim_{n\to\infty} S_n = S$ or, equivalently,

$$\lim_{n\to\infty}(S - S_n) = 0$$

Thus, the sum of a convergent series can be approximated to any degree of accuracy by its $n$th partial sum $S_n$, provided that $n$ is taken large enough. To measure the accuracy of the approximation, we introduce the quantity

$$R_n = S - S_n = \sum_{k=1}^{\infty} a_k - \sum_{k=1}^{n} a_k = \sum_{k=n+1}^{\infty} a_k = a_{n+1} + a_{n+2} + a_{n+3} + \cdots$$

called the **remainder after $n$ terms** of the series $\sum_{n=1}^{\infty} a_n$. The remainder measures the error incurred when $S$ is approximated by $S_n$.

In general, it is difficult to determine the accuracy of such an approximation, but for alternating series the following theorem gives us a simple way of estimating the error.

---

**THEOREM 2    Error Estimate in Approximating an Alternating Series**

Suppose $\sum_{n=1}^{\infty}(-1)^{n-1} a_n$ is an alternating series satisfying

1. $0 \le a_{n+1} \le a_n$ for all $n$
2. $\lim\limits_{n\to\infty} a_n = 0$

If $S$ is the sum of the series, then

$$|R_n| = |S - S_n| \le a_{n+1}$$

In other words, the absolute value of the error incurred in approximating $S$ by $S_n$ is no larger than $a_{n+1}$, the first term omitted.

---

**PROOF**   We have

$$S - S_n = \sum_{k=1}^{\infty}(-1)^{k-1}a_k - \sum_{k=1}^{n}(-1)^{k-1}a_k = \sum_{k=n+1}^{\infty}(-1)^{k-1}a_k$$

$$= (-1)^n a_{n+1} + (-1)^{n+1}a_{n+2} + (-1)^{n+2}a_{n+3} + \cdots$$

$$= (-1)^n(a_{n+1} - a_{n+2} + a_{n+3} - \cdots)$$

Next,

$$a_{n+1} - a_{n+2} + a_{n+3} - a_{n+4} + \cdots$$

$$= (a_{n+1} - a_{n+2}) + (a_{n+3} - a_{n+4}) + \cdots$$

$$\ge 0 \qquad\qquad \text{Since } a_{n+1} \le a_n \text{ for all } n$$

So

$$|S - S_n| = a_{n+1} - a_{n+2} + a_{n+3} - a_{n+4} + a_{n+5} - \cdots$$

$$= a_{n+1} - (a_{n+2} - a_{n+3}) - (a_{n+4} - a_{n+5}) - \cdots$$

Since every expression within each parenthesis is nonnegative, we see that $|S - S_n| \le a_{n+1}$.    ▪

⚠ This error estimate holds only for alternating series.

**EXAMPLE 3** Show that the series $\displaystyle\sum_{n=0}^{\infty} (-1)^n \frac{1}{n!}$ is convergent, and find its sum correct to three decimal places.

**Solution** Since

$$a_{n+1} = \frac{1}{(n+1)!} = \frac{1}{n!(n+1)} < \frac{1}{n!} = a_n$$

for all $n$ and

$$\lim_{n \to \infty} a_n = \lim_{n \to \infty} \frac{1}{n!} = 0$$

we conclude that the series converges by the Alternating Series Test.

To see how many terms of the series are needed to ensure the specified accuracy of the approximation, we turn to Theorem 2. It tells us that

$$|R_n| = |S - S_n| \le a_{n+1} = \frac{1}{(n+1)!}$$

We require that $|R_n| < 0.0005$, which is satisfied if

$$\frac{1}{(n+1)!} < 0.0005 \quad \text{or} \quad (n+1)! > \frac{1}{0.0005} = 2000$$

The smallest positive integer that satisfies the last inequality is $n = 6$. Hence, the required approximation is

$$S \approx S_6 = \frac{1}{0!} - \frac{1}{1!} + \frac{1}{2!} - \frac{1}{3!} + \frac{1}{4!} - \frac{1}{5!} + \frac{1}{6!}$$

$$= 1 - 1 + \frac{1}{2} - \frac{1}{6} + \frac{1}{24} - \frac{1}{120} + \frac{1}{720}$$

$$\approx 0.368$$

## 8.5 CONCEPT QUESTIONS

1. **a.** What is an alternating series? Give an example.
   **b.** State the Alternating Series Test, and use it to determine whether the series in your example converges or diverges.
   **c.** What is the maximum error that can occur if you approximate the sum of a convergent alternating series by its $n$th partial sum?

## 8.5 EXERCISES

In Exercises 1–24, determine whether the series converges or diverges.

1. $\displaystyle\sum_{n=1}^{\infty} \frac{(-1)^{n-1}}{n+2}$

2. $\displaystyle\sum_{n=1}^{\infty} \frac{(-1)^n n}{3n-1}$

3. $\displaystyle\sum_{n=1}^{\infty} \frac{(-1)^{n+1}}{n^2}$

4. $\displaystyle\sum_{n=1}^{\infty} \frac{(-1)^{n-1} n^2}{2n^2 - 1}$

5. $\displaystyle\sum_{n=1}^{\infty} \frac{(-1)^{n-1}}{\sqrt{n}}$

6. $\displaystyle\sum_{n=1}^{\infty} \frac{(-1)^{n+1} n}{\sqrt{n^2 + 1}}$

7. $\displaystyle\sum_{n=2}^{\infty} \frac{(-1)^{n-1} \sqrt{n+1}}{n-1}$

8. $\displaystyle\sum_{n=2}^{\infty} \frac{(-1)^{n-1}}{\ln n}$

9. $\displaystyle\sum_{n=2}^{\infty} \frac{(-1)^n n}{\ln n}$

10. $\displaystyle\sum_{n=1}^{\infty} \frac{(-1)^n \ln(n+1)}{n+2}$

**11.** $\displaystyle\sum_{n=1}^{\infty} \frac{(-1)^n\, n}{2^n}$

**12.** $\displaystyle\sum_{n=1}^{\infty} \frac{(-1)^{n-1}}{ne^{-n}}$

**13.** $\displaystyle\sum_{n=0}^{\infty} \frac{(-1)^{n+1}\, e^n}{\pi^{n+1}}$

**14.** $\displaystyle\sum_{n=1}^{\infty} \frac{\cos n\pi}{n}$

**15.** $\displaystyle\sum_{n=1}^{\infty} \frac{1}{\sqrt{n}} \sin \frac{(2n-1)\pi}{2}$

**16.** $\displaystyle\sum_{n=2}^{\infty} (\ln n) \sin \frac{(2n-1)\pi}{2}$

**17.** $\displaystyle\sum_{n=1}^{\infty} \frac{\sin\left(\dfrac{n\pi}{2}\right)}{\sqrt{n^3 + 1}}$

**18.** $\displaystyle\sum_{n=1}^{\infty} (-1)^n \cos\left(\frac{\pi}{n}\right)$

**19.** $\displaystyle\sum_{n=1}^{\infty} (-1)^n\, n \sin\left(\frac{\pi}{n}\right)$

**20.** $\displaystyle\sum_{n=1}^{\infty} \frac{(-1)^n\, n!}{n^n}$

**21.** $\displaystyle\sum_{n=2}^{\infty} \frac{(-1)^n \ln n}{e^n}$

**22.** $\displaystyle\sum_{n=2}^{\infty} \frac{(-1)^{n-1}\sqrt{\ln n}}{n}$

**23.** $\displaystyle\sum_{n=1}^{\infty} \frac{(-1)^n}{\sqrt{n} + \sqrt{n+1}}$

**24.** $\displaystyle\sum_{n=1}^{\infty} \frac{(-1)^{n-1}}{\sqrt[n]{n}}$

*In Exercises 25 and 26, find the values of p for which the series is convergent.*

**25.** $\displaystyle\sum_{n=2}^{\infty} \frac{(-1)^n}{(\ln n)^p}$

**26.** $\displaystyle\sum_{n=2}^{\infty} (-1)^{n-1} \frac{(\ln n)^p}{n}$

 *In Exercises 27–30, determine the number of terms sufficient to obtain the sum of the series accurate to three decimal places.*

**27.** $\displaystyle\sum_{n=1}^{\infty} \frac{(-1)^{n-1}}{n^2 + 1}$

**28.** $\displaystyle\sum_{n=1}^{\infty} \frac{(-1)^{n-1}}{\sqrt{n}}$

**29.** $\displaystyle\sum_{n=0}^{\infty} \frac{(-2)^{n+3}}{(n+1)!}$

**30.** $\displaystyle\sum_{n=2}^{\infty} \frac{(-1)^{n-1}}{n \ln n}$

 *In Exercises 31–34, find an approximation of the sum of the series accurate to two decimal places.*

**31.** $\displaystyle\sum_{n=1}^{\infty} \frac{(-1)^n}{n^3}$

**32.** $\displaystyle\sum_{n=0}^{\infty} \frac{(-1)^n}{(2n)!}$

**33.** $\displaystyle\sum_{n=1}^{\infty} \frac{(-1)^{n-1}(n+1)}{2^n}$

**34.** $\displaystyle\sum_{n=1}^{\infty} \frac{(-1)^{n-1}}{n \cdot 2^n}$

**35.** Show that the series

$$\frac{1}{2} - \frac{1}{3} + \frac{1}{4} - \frac{1}{9} + \frac{1}{8} - \frac{1}{27} + \cdots + \frac{1}{2^n} - \frac{1}{3^n} + \cdots$$

converges, and find its sum. Why isn't the Alternating Series Test applicable?

**36.** Show that the series

$$1 - \frac{1}{4} + \frac{1}{3} - \frac{1}{16} + \frac{1}{5} - \frac{1}{36} + \cdots + \frac{1}{2n-1} - \frac{1}{(2n)^2} + \cdots$$

diverges. Why isn't the Alternating Series Test applicable?

**37. a.** Suppose that $\Sigma\, a_n$ and $\Sigma\, b_n$ are both convergent. Does it follow that $\Sigma\, a_n b_n$ must be convergent? Justify your answer.

   **b.** Suppose that $\Sigma\, a_n$ and $\Sigma\, b_n$ are both divergent. Does it follow that $\Sigma\, a_n b_n$ must be divergent? Justify your answer.

**38.** Find all values of $s$ for which $\displaystyle\sum_{n=1}^{\infty} \frac{(-1)^n}{n^s}$ converges.

**39. a.** Show that $\displaystyle\sum_{n=1}^{\infty} \frac{(-1)^n(2n+1)}{n(n+1)}$ converges.

   **b.** Find the sum of the series of part (a).

**40. a.** Show that $\displaystyle\sum_{n=0}^{\infty} \frac{(-1)^n}{n!}$ converges.

   **b.** Denote the sum of the infinite series in part (a) by $S$. Show that $S$ is irrational.
   **Hint:** Use Theorem 2.

*In Exercises 41–44, determine whether the statement is true or false. If it is true, explain why it is true. If it is false, explain why or give an example to show why it is false.*

**41.** If the alternating series $\Sigma_{n=1}^{\infty}(-1)^{n-1}\, a_n$, where $a_n > 0$, is divergent, then the series $\Sigma_{n=1}^{\infty}\, a_n$ is also divergent.

**42.** Let $\Sigma_{n=1}^{\infty}(-1)^{n+1}\, a_n$ be an alternating series, where $a_n > 0$. If $\lim_{n\to\infty} a_n = 0$, then $\Sigma_{n=1}^{\infty}(-1)^{n+1}\, a_n$ converges.

**43.** If the alternating series $\Sigma_{n=1}^{\infty}(-1)^{n+1}\, a_n$, where $a_n > 0$, converges, then both the series $\Sigma_{n=1}^{\infty} a_{2n-1}$ and $\Sigma_{n=1}^{\infty} a_{2n}$ converge.

**44.** Let $\Sigma_{n=1}^{\infty}(-1)^{n+1}\, a_n$ be an alternating series, where $a_n > 0$. If $a_{n+1} \leq a_n$ for all $n$, then $\Sigma_{n=1}^{\infty}(-1)^{n+1}\, a_n$ converges.

---

## 8.6   Absolute Convergence; the Ratio and Root Tests

### ■ Absolute Convergence

Up to now, we have considered series whose terms are all positive and series whose terms alternate between being positive and negative. Now, consider the series

$$\sum_{n=1}^{\infty} \frac{\sin 2n}{n^2} = \sin 2 + \frac{\sin 4}{2^2} + \frac{\sin 6}{3^2} + \cdots$$

With the aid of a calculator you can verify that the first term of this series is positive, the next two terms are negative, and the next term is positive. Therefore, this series is neither a series with positive terms nor an alternating series. To study the convergence of such series, we introduce the notion of *absolute convergence.*

Suppose that $\sum_{n=1}^{\infty} a_n$ is any series. Then we can form the series

$$\sum_{n=1}^{\infty} |a_n| = |a_1| + |a_2| + |a_3| + \cdots$$

by taking the absolute value of each term of the given series. Since this series contains only positive terms, we can use the tests developed in Sections 8.3 and 8.4 to determine its convergence or divergence.

---

**DEFINITION**   **Absolutely Convergent Series**

A series $\sum a_n$ is **absolutely convergent** if the series $\sum |a_n|$ is convergent.

---

Notice that if the terms of the series $\sum a_n$ are positive, then $|a_n| = a_n$. In this case absolute convergence is the same as convergence.

**EXAMPLE 1**   Show that the series

$$\sum_{n=1}^{\infty} \frac{(-1)^{n-1}}{n^2} = 1 - \frac{1}{2^2} + \frac{1}{3^2} - \frac{1}{4^2} + \cdots$$

is absolutely convergent.

**Solution**   Taking the absolute value of each term of the series, we obtain

$$\sum_{n=1}^{\infty} \left| \frac{(-1)^{n-1}}{n^2} \right| = \sum_{n=1}^{\infty} \frac{1}{n^2} = 1 + \frac{1}{2^2} + \frac{1}{3^2} + \frac{1}{4^2} + \cdots$$

which is a convergent $p$-series ($p = 2$). Hence the series is absolutely convergent.   ■

**EXAMPLE 2**   Show that the alternating harmonic series

$$\sum_{n=1}^{\infty} \frac{(-1)^{n-1}}{n} = 1 - \frac{1}{2} + \frac{1}{3} - \frac{1}{4} + \cdots$$

is not absolutely convergent.

**Solution**   Taking the absolute value of each term of the series leads to

$$\sum_{n=1}^{\infty} \left| \frac{(-1)^{n-1}}{n} \right| = \sum_{n=1}^{\infty} \frac{1}{n} = 1 + \frac{1}{2} + \frac{1}{3} + \cdots$$

which is the divergent harmonic series. This shows that the series is not absolutely convergent.   ■

In Example 2 we saw that the alternating harmonic series is not absolutely convergent; but as we proved earlier, it is convergent. Such a series is said to be *conditionally convergent.*

> **DEFINITION** Conditionally Convergent Series
>
> A series $\Sigma\, a_n$ is **conditionally convergent** if it is convergent but not absolutely convergent.

The following theorem tells us that absolute convergence is, loosely speaking, stronger than convergence.

> **THEOREM 1**
>
> If a series $\Sigma\, a_n$ is absolutely convergent, then it is convergent.

**PROOF** Using an absolute value property, we have

$$-|a_n| \le a_n \le |a_n|$$

Adding $|a_n|$ to both sides of this inequality yields

$$0 \le a_n + |a_n| \le 2|a_n|$$

If we let $b_n = a_n + |a_n|$, then the last inequality becomes $0 \le b_n \le 2|a_n|$. If $\Sigma\, a_n$ is absolutely convergent, then $\Sigma\, |a_n|$ is convergent, which in turn implies, by Theorem 4a of Section 8.2, that $\Sigma\, 2|a_n|$ is convergent. Therefore, $\Sigma\, b_n$ is convergent by the Comparison Test. Finally, since $a_n = b_n - |a_n|$, we see that $\Sigma\, a_n = \Sigma\, b_n - \Sigma\, |a_n|$ is convergent by Theorem 4b of Section 8.2. ∎

As an illustration, the series $\Sigma(-1)^{n-1}/n^2$ of Example 1 is an alternating series that can be shown to be convergent by the Alternating Series Test. Alternatively, we can show that the series is absolutely convergent (as was done in Example 1) and conclude by Theorem 1 that it must be convergent.

**EXAMPLE 3** Determine whether the series

$$\sum_{n=1}^{\infty} \frac{\sin 2n}{n^2} = \sin 2 + \frac{\sin 4}{2^2} + \frac{\sin 6}{3^2} + \cdots$$

converges or diverges.

**Solution** As was pointed out at the beginning of this section, this series contains both positive and negative terms, but it is not an alternating series because the first term is positive, the next two terms are negative, and the next term is positive.

Let's show that the series is absolutely convergent. To do this, we consider the series

$$\sum_{n=1}^{\infty} \left| \frac{\sin 2n}{n^2} \right| = \sum_{n=1}^{\infty} \frac{|\sin 2n|}{n^2}$$

Since $|\sin 2n| \le 1$ for all $n$, we see that

$$\frac{|\sin 2n|}{n^2} \le \frac{1}{n^2}$$

Now, because $\Sigma\, 1/n^2$ is a convergent $p$-series, the Comparison Test tells us that $\sum_{n=1}^{\infty} |\sin 2n|/n^2$ is convergent. This shows that the given series is absolutely convergent, and we conclude by Theorem 1 that it is convergent. ∎

## ▪ The Ratio Test

The *Ratio Test* is a test for determining whether a series is absolutely convergent. Of course, for series that contain only positive terms, the Ratio Test will just be yet another test for convergence. To gain insight into why the Ratio Test works, consider the ratios of the consecutive terms of the series $\Sigma |a_n|$:

$$\frac{|a_2|}{|a_1|}, \quad \frac{|a_3|}{|a_2|}, \quad \frac{|a_4|}{|a_3|}, \quad \cdots$$

If the terms of this sequence are ultimately less than 1, then the terms of the series $\Sigma |a_n|$ ultimately behave roughly like the terms of a geometric series $\Sigma ar^n$ with $0 < r < 1$, and we can expect the series to be convergent. On the other hand, if the terms of the series are ultimately greater than 1, then we can expect the series to be divergent.

---

**THEOREM 2    The Ratio Test**

Let $\Sigma a_n$ be a series with nonzero terms.

**a.** If $\lim\limits_{n\to\infty} \left| \dfrac{a_{n+1}}{a_n} \right| = L < 1$, then $\sum\limits_{n=1}^{\infty} a_n$ converges absolutely.

**b.** If $\lim\limits_{n\to\infty} \left| \dfrac{a_{n+1}}{a_n} \right| = L > 1$, or $\lim\limits_{n\to\infty} \left| \dfrac{a_{n+1}}{a_n} \right| = \infty$, then $\sum\limits_{n=1}^{\infty} a_n$ diverges.

**c.** If $\lim\limits_{n\to\infty} \left| \dfrac{a_{n+1}}{a_n} \right| = 1$, the test is inconclusive, and another test should be used.

---

**PROOF**

**a.** Suppose that

$$\lim_{n\to\infty} \left| \frac{a_{n+1}}{a_n} \right| = L < 1$$

Let $r$ be any number such that $0 \le L < r < 1$. Then there exists an integer $N$ such that

$$\left| \frac{a_{n+1}}{a_n} \right| < r$$

whenever $n \ge N$ or, equivalently,

$$|a_{n+1}| < |a_n| r$$

whenever $n \ge N$. Letting $n$ take on the values $N, N+1, N+2, \ldots$, successively, we obtain

$$|a_{N+1}| < |a_N| r$$

$$|a_{N+2}| < |a_{N+1}| r < |a_N| r^2$$

$$|a_{N+3}| < |a_{N+2}| r < |a_N| r^3$$

and, in general,

$$|a_{N+k}| < |a_N| r^k \qquad \text{for all } k \ge 1$$

Since the series

$$\sum_{k=1}^{\infty} |a_N| r^k = |a_N| r + |a_N| r^2 + |a_N| r^3 + \cdots \tag{1}$$

is a convergent geometric series with $0 < r < 1$ and each term of the series

$$\sum_{k=1}^{\infty} |a_{N+k}| = |a_{N+1}| + |a_{N+2}| + |a_{N+3}| + \cdots \tag{2}$$

is less than the corresponding term of the geometric series (1), the Comparison Test then implies that series (2) is convergent. Since convergence or divergence is unaffected by the omission of a finite number of terms, we see that the series $\sum_{n=1}^{\infty} |a_n|$ is also convergent.

**b.** Suppose that

$$\lim_{n \to \infty} \left| \frac{a_{n+1}}{a_n} \right| = L > 1$$

Let $r$ be any number such that $L > r > 1$. Then there exists an integer $N$ such that

$$\left| \frac{a_{n+1}}{a_n} \right| > r > 1$$

whenever $n \geq N$. This implies that $|a_{n+1}| > |a_n|$ when $n \geq N$. Thus, $\lim_{n \to \infty} a_n \neq 0$, and $\sum a_n$ is divergent by the Divergence Test.

**c.** Consider the series $\sum_{n=1}^{\infty} 1/n$ and $\sum_{n=1}^{\infty} 1/n^2$. For the first series we have

$$\lim_{n \to \infty} \left| \frac{a_{n+1}}{a_n} \right| = \lim_{n \to \infty} \frac{1}{n+1} \cdot \frac{n}{1} = \lim_{n \to \infty} \frac{1}{1 + \dfrac{1}{n}} = 1$$

and for the second series we have

$$\lim_{n \to \infty} \left| \frac{a_{n+1}}{a_n} \right| = \lim_{n \to \infty} \frac{1}{(n+1)^2} \cdot \frac{n^2}{1} = \lim_{n \to \infty} \frac{1}{\left(1 + \dfrac{1}{n}\right)^2} = 1$$

Thus,

$$\lim_{n \to \infty} \left| \frac{a_{n+1}}{a_n} \right| = 1$$

for both series. The first series is the divergent harmonic series, whereas the second series is a convergent $p$-series with $p = 2$. Thus, if $L = 1$, the series may converge or diverge, and the Ratio Test is inconclusive. ■

**EXAMPLE 4** Determine whether the series $\displaystyle\sum_{n=1}^{\infty} (-1)^{n-1} \frac{n^2 + 1}{2^n}$ is absolutely convergent, conditionally convergent, or divergent.

**Solution** We use the Ratio Test with $a_n = (-1)^{n-1}(n^2 + 1)/2^n$. We have

$$\lim_{n \to \infty} \left| \frac{a_{n+1}}{a_n} \right| = \lim_{n \to \infty} \left| \frac{(-1)^n[(n+1)^2 + 1]}{2^{n+1}} \cdot \frac{2^n}{(-1)^{n-1}(n^2 + 1)} \right|$$

$$= \lim_{n \to \infty} \frac{1}{2} \left( \frac{n^2 + 2n + 2}{n^2 + 1} \right) = \frac{1}{2} < 1$$

Therefore, by the Ratio Test, the series is absolutely convergent. ■

**EXAMPLE 5** Determine whether the series $\displaystyle\sum_{n=1}^{\infty} \frac{n!}{n^n}$ is convergent or divergent.

**Solution** Let $a_n = n!/n^n$. Then

$$\lim_{n \to \infty} \left| \frac{a_{n+1}}{a_n} \right| = \lim_{n \to \infty} \frac{a_{n+1}}{a_n} \qquad \text{Since } a_n \text{ and } a_{n+1} \text{ are positive}$$

$$= \lim_{n \to \infty} \frac{(n+1)!}{(n+1)^{n+1}} \cdot \frac{n^n}{n!}$$

$$= \lim_{n \to \infty} \frac{(n+1)n!}{(n+1)(n+1)^n} \cdot \frac{n^n}{n!}$$

$$= \lim_{n \to \infty} \left( \frac{n}{n+1} \right)^n$$

$$= \lim_{n \to \infty} \frac{1}{\left( \dfrac{n+1}{n} \right)^n} = \lim_{n \to \infty} \frac{1}{\left( 1 + \dfrac{1}{n} \right)^n} = \frac{1}{\lim_{n \to \infty}\left( 1 + \dfrac{1}{n} \right)^n} = \frac{1}{e} < 1$$

Therefore, the series converges, by the Ratio Test. ∎

**EXAMPLE 6** Determine whether the series $\displaystyle\sum_{n=1}^{\infty} (-1)^n \frac{n!}{3^n}$ is absolutely convergent, conditionally convergent, or divergent.

**Solution** Let $a_n = (-1)^n n!/3^n$. Then

$$\lim_{n \to \infty} \left| \frac{a_{n+1}}{a_n} \right| = \lim_{n \to \infty} \left| \frac{(-1)^{n+1}(n+1)!}{3^{n+1}} \cdot \frac{3^n}{(-1)^n n!} \right|$$

$$= \lim_{n \to \infty} \frac{n+1}{3} = \infty$$

and we conclude that the given series is divergent by the Ratio Test.

**Alternative Solution** Observe that for $n \geq 2$,

$$\frac{n!}{3^n} = \frac{n \cdot (n-1) \cdot \cdots \cdot 3 \cdot 2 \cdot 1}{3 \cdot 3 \cdot \cdots \cdot 3 \cdot 3 \cdot 3} \geq \frac{2 \cdot 1}{3 \cdot 3} = \frac{2}{9} \neq 0$$

Therefore, $\lim_{n \to \infty} a_n = \lim_{n \to \infty}(-1)^n n!/3^n$ does not exist, so the Divergence Test implies that the series must diverge. ∎

## The Root Test

The following test is especially useful when the $n$th term of a series involves the $n$th power. Since the proof is similar to that of the Ratio Test, it will be omitted.

---

**THEOREM 3   The Root Test**

Let $\sum_{n=1}^{\infty} a_n$ be a series.

**a.** If $\lim_{n \to \infty} \sqrt[n]{|a_n|} = L < 1$, then $\sum_{n=1}^{\infty} a_n$ converges absolutely.

**b.** If $\lim_{n \to \infty} \sqrt[n]{|a_n|} = L > 1$ or $\lim_{n \to \infty} \sqrt[n]{|a_n|} = \infty$, then $\sum_{n=1}^{\infty} a_n$ diverges.

**c.** If $\lim_{n \to \infty} \sqrt[n]{|a_n|} = 1$, the test is inconclusive, and another test should be used.

**EXAMPLE 7** Determine whether the series $\displaystyle\sum_{n=1}^{\infty} (-1)^{n-1} \frac{2^{n+3}}{(n+1)^n}$ is absolutely convergent, conditionally convergent, or divergent.

**Solution** We apply the Root Test with $a_n = (-1)^{n-1} 2^{n+3}/(n+1)^n$. We have

$$\lim_{n\to\infty} \sqrt[n]{|a_n|} = \lim_{n\to\infty} \sqrt[n]{\left|(-1)^{n-1} \frac{2^{n+3}}{(n+1)^n}\right|} = \lim_{n\to\infty} \left|\frac{2^{n+3}}{(n+1)^n}\right|^{1/n}$$

$$= \lim_{n\to\infty} \frac{2^{1+3/n}}{n+1} = 0 < 1$$

and conclude that the series is absolutely convergent.  ■

## ■ Summary of Tests for Convergence and Divergence of Series

We have developed several ways of determining whether a series is convergent or divergent. Next, we give a summary of the available tests and suggest when it might be advantageous to use each test.

---

**Summary of the Convergence and Divergence Tests for Series**

1. The **Divergence Test** often settles the question of convergence or divergence of a series $\Sigma\, a_n$ simply and quickly:

   If $\lim_{n\to\infty} a_n \neq 0$, then the series diverges.

2. If you recognize that the series is
   **a.** a **geometric series** $\sum_{n=1}^{\infty} ar^{n-1}$, then it converges with sum $a/(1-r)$ if $|r| < 1$. If $|r| \geq 1$, the series diverges.
   **b.** a **telescoping series,** then use partial fraction decomposition (if necessary) to find its $n$th partial sum $S_n$. Next determine convergence or divergence by evaluating $\lim_{n\to\infty} S_n$.
   **c.** a **$p$-series** $\sum_{n=1}^{\infty} 1/n^p$, then the series converges if $p > 1$ and diverges if $p \leq 1$.
   Sometimes a little algebraic manipulation might be required to cast the series into one of these forms. Also, a series might involve a combination (for example, a sum or difference) of these series.

3. If $f(n) = a_n$ for $n \geq 1$, where $f$ is a continuous, positive, decreasing function on $[1, \infty)$ and readily integrable, then we may use the **Integral Test:**

   $\sum_{n=1}^{\infty} a_n$ converges if $\int_1^{\infty} f(x)\, dx$ converges and diverges if $\int_1^{\infty} f(x)\, dx$ diverges.

4. If $a_n$ is positive and behaves like the $n$th term of a geometric or $p$-series for large values of $n$, then the Comparison Test or Limit Comparison Test may be used. The tests and conclusions follow:
   **a.** If $a_n \leq b_n$ for all $n$ and $\Sigma\, b_n$ converges, then $\Sigma\, a_n$ converges.
   **b.** If $a_n \geq b_n \geq 0$ for all $n$ and $\Sigma\, b_n \geq 0$ diverges, then $\Sigma\, a_n$ diverges.
   **c.** If $b_n$ is positive and $\lim_{n\to\infty}(a_n/b_n) = L > 0$, then both series converge or both diverge.
   The comparison tests can also be used on $\Sigma\, |a_n|$ to test for absolute convergence.

---

5. If the series is an **alternating series,** $\sum_{n=1}^{\infty}(-1)^n a_n$ or $\sum_{n=1}^{\infty}(-1)^{n-1} a_n$, then the Alternating Series Test should be considered:

   If $a_n \geq a_{n+1}$ for all $n$ and $\lim_{n\to\infty} a_n = 0$, then the series converges.

6. The **Ratio Test** is useful if $a_n$ involves factorials or $n$th powers. The series

   **a.** converges absolutely if $\lim\limits_{n\to\infty}\left|\dfrac{a_{n+1}}{a_n}\right| < 1$.

   **b.** diverges if $\lim\limits_{n\to\infty}\left|\dfrac{a_{n+1}}{a_n}\right| > 1$ or $\lim\limits_{n\to\infty}\left|\dfrac{a_{n+1}}{a_n}\right| = \infty$.

   The test is inconclusive if $\lim\limits_{n\to\infty}\left|\dfrac{a_{n+1}}{a_n}\right| = 1$.

7. The **Root Test** is useful if $a_n$ involves $n$th powers. The series
   **a.** converges absolutely if $\lim_{n\to\infty}\sqrt[n]{|a_n|} < 1$.
   **b.** diverges if $\lim_{n\to\infty}\sqrt[n]{|a_n|} > 1$ or $\lim_{n\to\infty}\sqrt[n]{|a_n|} = \infty$.

   The test is inconclusive if $\lim_{n\to\infty}\sqrt[n]{|a_n|} = 1$.
8. If the series $\sum a_n$ involves terms that are both positive and negative but it is not alternating, then one sometimes can prove convergence of the series by proving that $\sum |a_n|$ is convergent.

## ■ Rearrangement of Series

A series with a finite number of terms has the same sum regardless of how the terms of the series are rearranged. The situation gets a little more complicated, however, when we deal with infinite series. The following example shows that a rearrangement of a convergent series could result in a series with a different sum!

**EXAMPLE 8**    Consider the alternating harmonic series that converges to ln 2 (see Problem 57 in Exercises 8.8):

$$1 - \frac{1}{2} + \frac{1}{3} - \frac{1}{4} + \frac{1}{5} - \frac{1}{6} + \frac{1}{7} - \frac{1}{8} + \cdots = \ln 2$$

If we rearrange the series so that every positive term is followed by two negative terms, we obtain

$$1 - \frac{1}{2} - \frac{1}{4} + \frac{1}{3} - \frac{1}{6} - \frac{1}{8} + \frac{1}{5} - \frac{1}{10} - \frac{1}{12} + \cdots$$

$$= \left(1 - \frac{1}{2}\right) - \frac{1}{4} + \left(\frac{1}{3} - \frac{1}{6}\right) - \frac{1}{8} + \left(\frac{1}{5} - \frac{1}{10}\right) - \frac{1}{12} + \cdots$$

$$= \frac{1}{2} - \frac{1}{4} + \frac{1}{6} - \frac{1}{8} + \frac{1}{10} - \frac{1}{12} + \cdots$$

$$= \frac{1}{2}\left(1 - \frac{1}{2} + \frac{1}{3} - \frac{1}{4} + \frac{1}{5} - \frac{1}{6} + \cdots\right) = \frac{1}{2}\ln 2$$

Thus, rearrangement of the alternating harmonic series has a sum that is one half that of the original series!    ■

You might have noticed that the alternating harmonic series in Example 8 is *conditionally convergent*. In fact, for such series, Riemann proved the following result:

If $x$ is any real number and $\sum_{n=1}^{\infty} a_n$ is conditionally convergent, then there is a rearrangement of $\sum_{n=1}^{\infty} a_n$ that converges to $x$.

A proof of this result can be found in more advanced textbooks.

Riemann's result tells us that for conditionally convergent series, we may not rearrange their terms, lest we end up with a totally different series, that is, a series with a different sum. Actually, for conditionally convergent series, one can find rearrangements of the series that diverge to infinity, diverge to minus infinity, or oscillate between any two prescribed real numbers!

So what kind of convergent series will have rearrangements that converge to the same sum as the original series? The answer is found in the following result, which we state without proof:

If $\sum_{n=1}^{\infty} a_n$ converges absolutely and $\sum_{n=1}^{\infty} b_n$ is any rearrangement of $\sum_{n=1}^{\infty} a_n$, then $\sum_{n=1}^{\infty} b_n$ converges and $\sum_{n=1}^{\infty} a_n = \sum_{n=1}^{\infty} b_n$.

Finally, since a convergent series with positive terms is absolutely convergent, its terms can be written in any order, and the resultant series will converge and have the same sum as the original series.

**EXAMPLE 9**    Indicate the test(s) that you would use to determine whether the series converges or diverges. Explain how you arrived at your choice.

**a.** $\displaystyle\sum_{n=1}^{\infty} \frac{2n-1}{3n+1}$    **b.** $\displaystyle\sum_{n=1}^{\infty} \left[\frac{2}{3^n} - \frac{1}{n(n+1)}\right]$    **c.** $\displaystyle\sum_{n=1}^{\infty} \left(\frac{1}{n}\right)^e$

**d.** $\displaystyle\sum_{n=3}^{\infty} \frac{1}{n\sqrt{\ln n}}$    **e.** $\displaystyle\sum_{n=3}^{\infty} \frac{\ln n}{n^2}$    **f.** $\displaystyle\sum_{n=1}^{\infty} \frac{\sqrt{n^3+2}}{n^4+3n^2+1}$

**g.** $\displaystyle\sum_{n=1}^{\infty} (-1)^n \frac{\sqrt{n}}{n^2+1}$    **h.** $\displaystyle\sum_{n=1}^{\infty} \frac{n}{2^n}$    **i.** $\displaystyle\sum_{n=1}^{\infty} \frac{\sin n}{\sqrt{n^3+1}}$

**Solution**
**a.** Since
$$\lim_{n\to\infty} a_n = \lim_{n\to\infty} \frac{2n-1}{3n+1} = \frac{2}{3} \neq 0$$

we use the Divergence Test.

**b.** The series is the difference of a geometric series and a telescoping series, so we use the properties of these series to determine convergence.

**c.** Here, $a_n = \left(\dfrac{1}{n}\right)^e = \dfrac{1}{n^e}$ is a $p$-series, so we use the properties of a $p$-series to study its convergence.

**d.** The function $f(x) = \dfrac{1}{x\sqrt{\ln x}}$ is continuous, positive, and decreasing on $[3, \infty)$ and is integrable, so we choose the Integral Test.

**e.** Here,
$$a_n = \frac{\ln n}{n^2} < \frac{\sqrt{n}}{n^2} = \frac{1}{n^{3/2}} = b_n$$

and we use the Comparison Test with the test series $\sum b_n$.

**f.** $a_n = \dfrac{(n^3 + 2)^{1/2}}{n^4 + 3n^2 + 1}$ is positive and behaves like

$$b_n = \frac{(n^3)^{1/2}}{n^4} = \frac{n^{3/2}}{n^4} = \frac{1}{n^{5/2}}$$

for large values of $n$, so we use the Limit Comparison Test with test series $\sum_{n=1}^{\infty} 1/n^{5/2}$.

**g.** This is an alternating series, and we use the Alternating Series Test.

**h.** Here, $a_n = \dfrac{n}{2^n} = \left(\dfrac{n^{1/n}}{2}\right)^n$ involves the $n$th power, so the Root Test is a candidate.

In fact, here $\lim\limits_{n \to \infty} \sqrt[n]{|a_n|} = \lim\limits_{n \to \infty} \dfrac{\sqrt[n]{n}}{2} = \dfrac{1}{2} < 1$ and the series converges.

**i.** The series involves both positive and negative terms and is not an alternating series, so we use the test for absolute convergence. ∎

## 8.6  CONCEPT QUESTIONS

**1. a.** What is an absolutely convergent series? Give an example.
 **b.** What is a conditionally convergent series? Give an example.

**2. a.** State the Ratio Test and the Root Test.
 **b.** Give an example of a convergent series and an example of a divergent series for which the Ratio Test is inconclusive.
 **c.** Give an example of a convergent series and an example of a divergent series for which the Root Test is inconclusive.

## 8.6  EXERCISES

*In Exercises 1–34, determine whether the series is convergent, absolutely convergent, conditionally convergent, or divergent.*

**1.** $\displaystyle\sum_{n=1}^{\infty} \frac{(-1)^{n-1}}{\sqrt{n}}$

**2.** $\displaystyle\sum_{n=1}^{\infty} \frac{(-1)^n}{n\sqrt{n}}$

**3.** $\displaystyle\sum_{n=1}^{\infty} \frac{(-2)^{n-1}}{n^2}$

**4.** $\displaystyle\sum_{n=1}^{\infty} \frac{(-2)^n}{n!}$

**5.** $\displaystyle\sum_{n=1}^{\infty} \frac{(-1)^{n+1}}{n+1}$

**6.** $\displaystyle\sum_{n=1}^{\infty} \frac{(-1)^n n}{n^2 + 1}$

**7.** $\displaystyle\sum_{n=1}^{\infty} \frac{(-1)^n n^2}{n^2 + 3}$

**8.** $\displaystyle\sum_{n=1}^{\infty} \frac{(-1)^{n-1} n}{\sqrt{2n^2 + 1}}$

**9.** $\displaystyle\sum_{n=2}^{\infty} \frac{(-1)^n}{n \ln n}$

**10.** $\displaystyle\sum_{n=3}^{\infty} \frac{(-1)^n}{n\sqrt{\ln n}}$

**11.** $\displaystyle\sum_{n=1}^{\infty} \frac{n!}{e^n}$

**12.** $\displaystyle\sum_{n=1}^{\infty} \frac{\cos(n+1)}{n\sqrt{n}}$

**13.** $\displaystyle\sum_{n=1}^{\infty} (-1)^{n-1} \sin\left(\frac{1}{n}\right)$

**14.** $\displaystyle\sum_{n=1}^{\infty} \frac{(-1)^{n-1} \tan^{-1} n}{n^2}$

**15.** $\displaystyle\sum_{n=1}^{\infty} \frac{2^n}{n! \, n}$

**16.** $\displaystyle\sum_{n=1}^{\infty} \frac{(-5)^{n-1}}{n^2 \cdot 3^n}$

**17.** $\displaystyle\sum_{n=1}^{\infty} \frac{(-2)^n n}{(n+1)3^{n-1}}$

**18.** $\displaystyle\sum_{n=2}^{\infty} \frac{(-1)^{n+1} \ln n}{n^2 + 1}$

**19.** $\displaystyle\sum_{n=2}^{\infty} \frac{(-1)^n \ln n}{2^n}$

**20.** $\displaystyle\sum_{n=0}^{\infty} \frac{\cos n\pi}{n!}$

**21.** $\displaystyle\sum_{n=2}^{\infty} \frac{\sin\left(\dfrac{n\pi}{4}\right)}{n(\ln n)^2}$

**22.** $\displaystyle\sum_{n=1}^{\infty} \frac{(-1)^{n-1} n^5}{e^n}$

**23.** $\displaystyle\sum_{n=1}^{\infty} \frac{(-1)^{n+1} n^n}{n!}$

**24.** $\displaystyle\sum_{n=2}^{\infty} \left(\frac{\ln n}{n}\right)^n$

**25.** $\displaystyle\sum_{n=2}^{\infty} \frac{(-1)^n}{(\ln n)^n}$

**26.** $\displaystyle\sum_{n=1}^{\infty} \left(\frac{n}{2n+1}\right)^n$

**27.** $\displaystyle\sum_{n=1}^{\infty} (-1)^n \tan\left(\frac{1}{n}\right)$

**28.** $\displaystyle\sum_{n=1}^{\infty} (-1)^n n \sin\left(\frac{\pi}{n}\right)$

**29.** $\displaystyle\sum_{n=1}^{\infty} \frac{(-n)^n}{[(n+1)\tan^{-1} n]^n}$

**30.** $\displaystyle\sum_{n=2}^{\infty} \left(\sqrt[n]{n} - 1\right)^n$

Ⓥ Videos for selected exercises are available online at **www.academic.cengage.com/login**.

**31.** $\displaystyle\sum_{n=1}^{\infty} (-1)^{n-1} \frac{3 \cdot 5 \cdot 7 \cdot \cdots \cdot (2n+1)}{1 \cdot 4 \cdot 7 \cdot \cdots \cdot (3n-2)}$

**32.** $\displaystyle\sum_{n=1}^{\infty} (-1)^{n} \frac{2^n}{3 \cdot 5 \cdot 7 \cdot \cdots \cdot (2n+1)}$

**33.** $\displaystyle\sum_{n=1}^{\infty} \frac{4 \cdot 7 \cdot 10 \cdot \cdots \cdot (3n+1)}{4^n(n+1)!}$

**34.** $\displaystyle\sum_{n=1}^{\infty} \frac{(n!)^2}{(3n)!}$

**35.** Find all values of $x$ for which the series $\displaystyle\sum_{n=1}^{\infty} \frac{x^n}{n}$ (a) converges absolutely and (b) converges conditionally.

**36.** Show that the Ratio Test is inconclusive for the *p*-series.

**37.** Show that the Root Test is inconclusive for the *p*-series.

**38. a.** Show that if $\Sigma\, a_n$ converges absolutely, then $\Sigma\, a_n^2$ converges.
   **b.** Show that the converse of the result in part (a) is false by finding a series $\Sigma\, a_n$ for which $\Sigma\, a_n^2$ converges, but $\Sigma\, |a_n|$ diverges.

**39.** Show that if $\Sigma\, a_n$ diverges, then $\Sigma\, |a_n|$ diverges.

**40.** Show that if $\Sigma\, a_n$ converges absolutely, then $\Sigma\, a_n \le \Sigma\, |a_n|$.

**41.** Suppose that $\Sigma\, a_n^2$ and $\Sigma\, b_n^2$ are convergent. Show that $\Sigma\, a_n b_n$ is absolutely convergent.
   **Hint:** Show that $2|ab| \le a^2 + b^2$ by looking at $(a+b)^2$ and $(a-b)^2$.

**42.** Prove that $\displaystyle\lim_{n \to \infty} \frac{2^n n!}{n^n} = 0$.
   **Hint:** Show that $\displaystyle\sum_{n=1}^{\infty} \frac{2^n n!}{n^n}$ is convergent.

**43. a.** Show that the series $\Sigma_{n=1}^{\infty}\, np^n$, where $0 < p < 1$, is convergent.
   **b.** Show that its sum is $S = \dfrac{p}{(1-p)^2}$.
      **Hint:** Find an expression for $S_n - pS_n$.

**44. Average Number of Coin Tosses** An unbiased coin is tossed until the coin lands heads and the number of throws in the experiment is recorded. As more and more experiments are performed, the average number of tosses obtained from these experiments approaches $\Sigma_{n=1}^{\infty}\, n\left(\frac{1}{2}\right)^n$. Use the result of Exercise 43 to find this number.

**45.** Show that if $\Sigma_{n=1}^{\infty}\, |a_n|$ converges, then so does $\Sigma_{n=2}^{\infty}\, |a_n - a_{n-1}|$.

**46.** Show that if $\Sigma_{n=1}^{\infty}\, a_n$ is absolutely convergent, then $|\Sigma_{n=1}^{\infty}\, a_n| \le \Sigma_{n=1}^{\infty}\, |a_n|$.

*In Exercises 47–50, determine whether the statement is true or false. If it is true, explain why it is true. If it is false, explain why or give an example to show why it is false.*

**47.** If $\Sigma_{n=1}^{\infty}\, a_n$ and $\Sigma_{n=1}^{\infty}\, b_n$ converge absolutely, then $\Sigma_{n=1}^{\infty}(a_n + b_n)$ converges absolutely.

**48.** If $a_n > 0$ for $n \ge 1$ and $\Sigma_{n=1}^{\infty}\, a_n$ converges, then $\Sigma_{n=1}^{\infty}(-1)^n a_n$ converges.

**49.** If $\Sigma_{n=1}^{\infty}\sqrt{a_n^2 + b_n^2}$ converges, then $\Sigma_{n=1}^{\infty}\, a_n$ and $\Sigma_{n=1}^{\infty}\, b_n$ converge absolutely.

**50.** If $a_n \ne 0$ for any $n \ge 1$ and $\displaystyle\sum_{n=1}^{\infty} a_n$ converges absolutely, then $\displaystyle\sum_{n=1}^{\infty} \frac{1}{|a_n|}$ diverges.

## 8.7 Power Series

### Power Series

Until now, we have dealt with series with constant terms. In this section we will study infinite series of the form

$$\sum_{n=0}^{\infty} a_n x^n = a_0 + a_1 x + a_2 x^2 + a_3 x^3 + \cdots + a_n x^n + \cdots$$

where $x$ is a variable. More generally, we will consider series of the form

$$\sum_{n=0}^{\infty} a_n (x-c)^n = a_0 + a_1(x-c) + a_2(x-c)^2 + a_3(x-c)^3 + \cdots + a_n(x-c)^n + \cdots$$

from which $\Sigma_{n=0}^{\infty}\, a_n x^n$ may be obtained as a special case by putting $c = 0$. We may view such series as generalizations of the notion of a polynomial to an infinite series.
   Examples of power series are

$$\sum_{n=0}^{\infty} x^n = 1 + x + x^2 + x^3 + \cdots$$

$$\sum_{n=0}^{\infty} \frac{(-1)^n x^n}{n!} = 1 - x + \frac{x^2}{2!} - \frac{x^3}{3!} + \cdots$$

and

$$\sum_{n=0}^{\infty} \frac{(-1)^n \left(x - \frac{\pi}{4}\right)^{2n+1}}{(2n+1)!} = \left(x - \frac{\pi}{4}\right) - \frac{\left(x - \frac{\pi}{4}\right)^3}{3!} + \frac{\left(x - \frac{\pi}{4}\right)^5}{5!} - \cdots$$

Observe that if we truncate each of these series, we obtain a polynomial.

---

**DEFINITION   Power Series**

Let $x$ be a variable. A **power series in $x$** is a series of the form

$$\sum_{n=0}^{\infty} a_n x^n = a_0 + a_1 x + a_2 x^2 + a_3 x^3 + \cdots + a_n x^n + \cdots$$

where the $a_n$'s are constants and are called the **coefficients** of the series. More generally, a **power series in $(x - c)$**, where $c$ is a constant, is a series of the form

$$\sum_{n=0}^{\infty} a_n (x - c)^n = a_0 + a_1(x - c) + a_2(x - c)^2$$
$$+ a_3(x - c)^3 + \cdots + a_n(x - c)^n + \cdots$$

---

**Notes**

1. A power series in $(x - c)$ is also called a **power series centered at $c$** or a **power series about $c$**. Thus, a power series in $x$ is just a series centered at the origin.
2. To simplify the notation used for a power series, we have adopted the convention that $(x - c)^0 = 1$, even when $x = c$.   ■

We can view a power series as a function $f$ defined by the rule

$$f(x) = \sum_{n=0}^{\infty} a_n (x - c)^n$$

The *domain* of $f$ is the set of all $x$ for which the power series converges, and the *range of $f$* comprises the sums of the series obtained by allowing $x$ to take on all values in the domain of $f$. If a function $f$ is defined in this manner, we say that $f$ **is represented by the power series** $\sum_{n=0}^{\infty} a_n(x - c)^n$.

**EXAMPLE 1**   As an example, consider the power series

$$\sum_{n=0}^{\infty} x^n = 1 + x + x^2 + x^3 + \cdots + x^n + \cdots \tag{1}$$

Recognizing that this is a geometric series with common ratio $x$, we see that it converges for $-1 < x < 1$. Thus, the power series (1) is a rule for a function $f$ with interval $(-1, 1)$ as its domain; that is,

$$f(x) = \sum_{n=0}^{\infty} x^n = 1 + x + x^2 + x^3 + \cdots + x^n + \cdots$$

There is a simple formula for the sum of the geometric series (1), namely, $1/(1 - x)$,

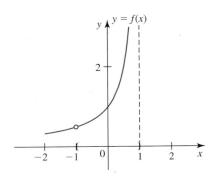

**FIGURE 1**
The function $f(x) = \sum_{n=0}^{\infty} x^n$
represents the function

$$g(x) = \frac{1}{1-x} \text{ on } (-1, 1) \text{ only.}$$

**FIGURE 2**
Observe that $S_n(x) = \sum_{k=0}^{n} x^k$
approximates $g(x)$ better and
better as $n \to \infty$ for $-1 < x < 1$.

and we see that the function *represented* by the series is the function

$$f(x) = \frac{1}{1-x} \qquad -1 < x < 1$$

Even though the domain of the function $g(x) = 1/(1-x)$ is the set of all real numbers except $x = 1$, the power series (1) represents the function $g(x) = 1/(1-x)$ only in the interval of convergence $(-1, 1)$ of the series. (See Figure 1.) Observe that the $n$th partial sum $S_n(x) = 1 + x + x^2 + \cdots + x^n$ of $\sum_{n=0}^{\infty} x^n$ approximates $g(x)$ better and better as $n$ increases for $-1 < x < 1$. But outside this interval, $S_n(x)$ diverges from $g(x)$. (See Figure 2.)

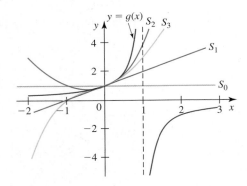

Example 1 reveals one shortcoming in representing a function by a power series. But as we will see later on, the advantages far outweigh the disadvantages.

## ■ Interval of Convergence

How do we find the domain of a function represented by a power series? Suppose that $f$ is the function represented by the power series

$$f(x) = \sum_{n=0}^{\infty} a_n(x - c)^n = a_0 + a_1(x - c) + a_2(x - c)^2 \qquad (2)$$
$$+ a_3(x - c)^3 + \cdots + a_n(x - c)^n + \cdots$$

Since $f(c) = a_0$, we see that the domain of $f$ always contains at least one number (the center of the power series) and is therefore nonempty. The following theorem, which we state without proof, tells us that the domain of a power series is always an interval with $x = c$ as its center. In the extreme cases the domain consists of the infinite interval $(-\infty, \infty)$ or just the point $x = c$, which may be regarded as a degenerate interval.

---

**THEOREM 1    Convergence of a Power Series**

Given a power series $\sum_{n=0}^{\infty} a_n(x - c)^n$, exactly one of the following is true:

**a.** The series converges only at $x = c$.
**b.** The series converges for all $x$.
**c.** There is a number $R > 0$ such that the series converges for $|x - c| < R$ and diverges for $|x - c| > R$.

---

A proof of Theorem 1 is given in Appendix B.

The number $R$ referred to in Theorem 1 is called the **radius of convergence** of the power series. The radius of convergence is $R = 0$ in case (a) and $R = \infty$ in case (b). The set of all values for which the power series converges is called the **interval of convergence** of the power series. Thus, Theorem 1 tells us that the interval of convergence of a power series centered at $c$ is (a) just the single point $c$, (b) the interval $(-\infty, \infty)$, or (c) the interval $(c - R, c + R)$. (See Figure 3.) But in the last case, Theorem 1 does not tell us whether the endpoints $x = c - R$ and $x = c + R$ are included in the interval of convergence. To determine whether they are included, we simply replace $x$ in the power series (2) by $c - R$ and $c + R$ in succession and use a convergence test on the resultant series.

**FIGURE 3**
The power series $\sum_{n=0}^{\infty} a_n(x - c)^n$ converges for $|x - c| < R$ and diverges for $|x - c| > R$.

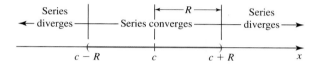

**EXAMPLE 2**  Find the radius of convergence and the interval of convergence of $\sum_{n=0}^{\infty} n!\, x^n$.

**Solution**   We can think of the given series as $\sum_{n=0}^{\infty} u_n$, where $u_n = n!\, x^n$. Applying the Ratio Test, we have

$$\lim_{n \to \infty} \left| \frac{u_{n+1}}{u_n} \right| = \lim_{n \to \infty} \left| \frac{(n + 1)!\, x^{n+1}}{n!\, x^n} \right| = \lim_{n \to \infty} (n + 1)|x| = \infty$$

whenever $x \neq 0$, and we conclude that the series diverges whenever $x \neq 0$. Therefore, the series converges only when $x = 0$, and its radius of convergence is accordingly $R = 0$.  ▪

**EXAMPLE 3**  Find the radius of convergence and the interval of convergence of

$$\sum_{n=0}^{\infty} \frac{(-1)^n x^{2n}}{(2n)!}$$

**Solution**   Let

$$u_n = \frac{(-1)^n x^{2n}}{(2n)!}$$

Then

$$\lim_{n \to \infty} \left| \frac{u_{n+1}}{u_n} \right| = \lim_{n \to \infty} \left| \frac{(-1)^{n+1} x^{2n+2}}{(2n + 2)!} \cdot \frac{(2n)!}{(-1)^n x^{2n}} \right|$$

$$= \lim_{n \to \infty} \frac{x^2}{(2n + 1)(2n + 2)} = 0 < 1$$

for each fixed value of $x$, so by the Ratio Test, the given series converges for all values of $x$. Therefore, the radius of convergence of the series is $R = \infty$, and its interval of convergence is $(-\infty, \infty)$.  ▪

**EXAMPLE 4** Find the radius of convergence and the interval of convergence of
$$\sum_{n=1}^{\infty} \frac{x^n}{n}.$$

**Solution**   Let $u_n = x^n/n$. Then
$$\lim_{n\to\infty}\left|\frac{u_{n+1}}{u_n}\right| = \lim_{n\to\infty}\left|\frac{x^{n+1}}{n+1}\cdot\frac{n}{x^n}\right| = \lim_{n\to\infty}\left(\frac{n}{n+1}\right)|x| = |x|$$

By the Ratio Test, the series converges if $|x| < 1$, that is, if $-1 < x < 1$. Therefore, the radius of convergence of the series is $R = 1$. To determine the interval of convergence of the power series, we need to examine the behavior of the series at the endpoints $x = -1$ and $x = 1$. Now, if $x = -1$, the series becomes
$$\sum_{n=1}^{\infty} \frac{(-1)^n}{n}$$

which is the convergent alternating harmonic series, and we see that $x = -1$ is in the interval of convergence of the power series. If $x = 1$, we obtain the harmonic series $\sum_{n=1}^{\infty} 1/n$, which is divergent, so $x = 1$ is not in the interval of convergence. We conclude that the interval of convergence of the given power series is $[-1, 1)$, as shown in Figure 4.   ■

**FIGURE 4**
The interval of convergence
of $\sum_{n=1}^{\infty} x^n/n$ is the interval $[-1, 1)$
with center $c = 0$ and radius $R = 1$.

**EXAMPLE 5** Find the radius of convergence and the interval of convergence of
$$\sum_{n=1}^{\infty} \frac{(x-2)^n}{n^2\cdot 3^n}.$$

**Solution**   Letting
$$u_n = \frac{(x-2)^n}{n^2\cdot 3^n}$$

we have
$$\lim_{n\to\infty}\left|\frac{u_{n+1}}{u_n}\right| = \lim_{n\to\infty}\left|\frac{(x-2)^{n+1}}{(n+1)^2 3^{n+1}}\cdot\frac{n^2\cdot 3^n}{(x-2)^n}\right|$$
$$= \lim_{n\to\infty}\left(\frac{n}{n+1}\right)^2\frac{|x-2|}{3} = \frac{|x-2|}{3}$$

By the Ratio Test, the series converges if $|x-2|/3 < 1$ or $|x-2| < 3$. The last inequality tells us that the radius of convergence of the given series is $R = 3$ and that the power series converges for $x$ in the interval $(-1, 5)$.

Next, we check the endpoints $x = -1$ and $x = 5$. If $x = -1$, the power series becomes
$$\sum_{n=1}^{\infty} \frac{(-3)^n}{n^2\cdot 3^n} = \sum_{n=1}^{\infty} \frac{(-1)^n}{n^2}$$

which is a convergent alternating series. Therefore, $x = -1$ is in the interval of convergence. Next, if $x = 5$, we obtain
$$\sum_{n=1}^{\infty} \frac{3^n}{n^2\cdot 3^n} = \sum_{n=1}^{\infty} \frac{1}{n^2}$$

**FIGURE 5**

The interval of convergence of
$$\sum_{n=1}^{\infty} \frac{(x-2)^n}{n^2 \cdot 3^n} \text{ is the interval } [-1, 5]$$
with center $c = 2$ and radius $R = 3$.

which is a convergent $p$-series. Therefore, $x = 5$ is also in the interval of convergence. We conclude, accordingly, that the interval of convergence of the given power series is $[-1, 5]$, as shown in Figure 5.  ■

**EXAMPLE 6**  Find the radius of convergence and the interval of convergence of
$$\sum_{n=0}^{\infty} \frac{(-1)^n 2^n x^n}{\sqrt{n+1}}.$$

**Solution**  Let

$$u_n = \frac{(-1)^n 2^n x^n}{\sqrt{n+1}}$$

Then

$$\lim_{n \to \infty} \left| \frac{u_{n+1}}{u_n} \right| = \lim_{n \to \infty} \left| \frac{(-1)^{n+1} 2^{n+1} x^{n+1}}{\sqrt{n+2}} \cdot \frac{\sqrt{n+1}}{(-1)^n 2^n x^n} \right|$$

$$= \lim_{n \to \infty} 2\sqrt{\frac{n+1}{n+2}} |x| = 2|x| \lim_{n \to \infty} \sqrt{\frac{1 + (1/n)}{1 + (2/n)}} = 2|x|$$

By the Ratio Test, the series converges if $2|x| < 1$ or $|x| < \frac{1}{2}$. The last inequality tells us that the radius of convergence of the power series is $R = \frac{1}{2}$, and the series converges in the interval $\left(-\frac{1}{2}, \frac{1}{2}\right)$.

Next, we check the endpoints $x = -\frac{1}{2}$ and $x = \frac{1}{2}$. If $x = -\frac{1}{2}$, the power series becomes

$$\sum_{n=0}^{\infty} \frac{(-1)^n 2^n \left(-\frac{1}{2}\right)^n}{\sqrt{n+1}} = \sum_{n=0}^{\infty} \frac{1}{\sqrt{n+1}}$$

which can be shown to be divergent by the Limit Comparison Test. (Compare it with the $p$-series $\sum_{n=1}^{\infty} 1/n^{1/2}$.) Next, if $x = \frac{1}{2}$, we have

$$\sum_{n=0}^{\infty} \frac{(-1)^n 2^n \left(\frac{1}{2}\right)^n}{\sqrt{n+1}} = \sum_{n=0}^{\infty} \frac{(-1)^n}{\sqrt{n+1}}$$

which converges, by the Alternating Series Test. Therefore, the interval of convergence of the power series is $\left(-\frac{1}{2}, \frac{1}{2}\right]$.  ■

## ■ Differentiation and Integration of Power Series

Suppose that $f$ is a function represented by a power series centered at $c$, that is,

$$f(x) = \sum_{n=0}^{\infty} a_n (x - c)^n$$

where $x$ lies in the interval of convergence of the series (domain of $f$). The following question arises naturally: Can we differentiate and integrate $f$, and if so, what are the series representations of the derivative and integral of $f$? The next theorem answers this question in the affirmative and tells us that the series representations of the derivative and integral of $f$ are found by differentiating and integrating the power series representation of $f$ term by term. (We omit its proof.)

---

**THEOREM 2    Differentiation and Integration of Power Series**

Suppose that the power series $\sum_{n=0}^{\infty} a_n(x - c)^n$ has a radius of convergence $R > 0$. Then the function $f$ defined by

$$f(x) = \sum_{n=0}^{\infty} a_n(x - c)^n = a_0 + a_1(x - c) + a_2(x - c)^2 + a_3(x - c)^3 + \cdots$$

for all $x$ in $(c - R, c + R)$ is both differentiable and integrable on $(c - R, c + R)$. Moreover, the derivative of $f$ and the indefinite integral of $f$ are

**a.** $f'(x) = a_1 + 2a_2(x - c) + 3a_3(x - c)^2 + \cdots = \sum_{n=1}^{\infty} na_n(x - c)^{n-1}$

**b.** $\int f(x)\, dx = C + a_0(x - c) + a_1 \dfrac{(x - c)^2}{2} + a_2 \dfrac{(x - c)^3}{3} + \cdots$

$$= \sum_{n=0}^{\infty} a_n \frac{(x - c)^{n+1}}{n + 1} + C$$

---

**Notes**

1. The series in parts (a) and (b) of Theorem 2 have the same radius of convergence, $R$, as the series $\sum_{n=0}^{\infty} a_n(x - c)^n$. But the interval of convergence may change. More specifically, you may lose convergence at the endpoints when you differentiate (Exercise 38) and gain convergence there when you integrate (Example 9).

2. Theorem 2 implies that a function that is represented by a power series in an interval $(c - R, c + R)$ is continuous on that interval. This follows from Theorem 1 in Section 2.1.    ■

**EXAMPLE 7**    Find a power series representation for $1/(1 - x)^2$ on $(-1, 1)$ by differentiating a power series representation of $f(x) = 1/(1 - x)$.

**Solution**    Recalling that $1/(1 - x)$ is the sum of a geometric series, we have

$$f(x) = \frac{1}{1 - x} = 1 + x + x^2 + x^3 + \cdots = \sum_{n=0}^{\infty} x^n \qquad |x| < 1$$

Differentiating both sides of this equation with respect to $x$ and using Theorem 2, we obtain

$$f'(x) = \frac{1}{(1 - x)^2} = 1 + 2x + 3x^2 + \cdots = \sum_{n=1}^{\infty} nx^{n-1} \qquad ■$$

**EXAMPLE 8**    Find a power series representation for $\ln(1 - x)$ on $(-1, 1)$.

**Solution**    We start with the equation

$$\frac{1}{1 - x} = 1 + x + x^2 + x^3 + \cdots = \sum_{n=0}^{\infty} x^n \qquad |x| < 1$$

Integrating both sides of this equation with respect to $x$ and using Theorem 2, we obtain

$$\int \frac{1}{1 - x}\, dx = \int (1 + x + x^2 + x^3 + \cdots)\, dx$$

or

$$-\ln(1-x) = x + \frac{1}{2}x^2 + \frac{1}{3}x^3 + \cdots + C$$

To determine the value of $C$, we set $x = 0$ in this equation to obtain $-\ln 1 = 0 = C$. Using this value of $C$, we see that

$$\ln(1-x) = -x - \frac{1}{2}x^2 - \frac{1}{3}x^3 - \cdots = -\sum_{n=1}^{\infty} \frac{x^n}{n} \qquad |x| < 1 \qquad \blacksquare$$

**EXAMPLE 9** Find a power series representation for $\tan^{-1} x$ by integrating a power series representation of $f(x) = 1/(1 + x^2)$.

**Solution** Observe that we can obtain a power series representation of $f$ by replacing $x$ with $-x^2$ in the equation

$$\frac{1}{1-x} = 1 + x + x^2 + \cdots \qquad |x| < 1$$

Thus,

$$\frac{1}{1+x^2} = \frac{1}{1-(-x^2)} = 1 + (-x^2) + (-x^2)^2 + (-x^2)^3 + \cdots$$

$$= 1 - x^2 + x^4 - x^6 + \cdots = \sum_{n=0}^{\infty} (-1)^n x^{2n}$$

Since the geometric series converges for $|x| < 1$, we see that this series converges for $|-x^2| < 1$, that is, $x^2 < 1$ or $|x| < 1$. Finally, integrating this equation, we have, by Theorem 2,

$$\tan^{-1} x = \int \frac{1}{1+x^2} \, dx = \int (1 - x^2 + x^4 - x^6 + \cdots) \, dx$$

$$= C + x - \frac{x^3}{3} + \frac{x^5}{5} - \frac{x^7}{7} + \cdots$$

To find $C$, we use the condition $\tan^{-1} 0 = 0$ to obtain $0 = C$. Therefore,

$$\tan^{-1} x = x - \frac{x^3}{3} + \frac{x^5}{5} - \frac{x^7}{7} + \cdots = \sum_{n=0}^{\infty} (-1)^n \frac{x^{2n+1}}{2n+1}$$

We leave it for you to show that the interval of convergence of the series is $[-1, 1]$. $\blacksquare$

## 8.7 CONCEPT QUESTIONS

1. **a.** Define a power series in $x$.
   **b.** Define a power series in $(x - c)$.
2. **a.** What is the radius of convergence of a power series?
   **b.** What is the interval of convergence of a power series?
   **c.** How do you find the radius and the interval of convergence of a power series?

3. Suppose that $\sum_{n=0}^{\infty} a_n x^n$ has radius of convergence 2. What can you say about the convergence or divergence of $\sum_{n=0}^{\infty} a_n \left(\frac{3}{2}\right)^n$?
4. Suppose that $\sum_{n=0}^{\infty} a_n (x - 2)^n$ diverges for $x = 0$. What can you say about the convergence or divergence of $\sum_{n=0}^{\infty} a_n 5^n$? What about $\sum_{n=0}^{\infty} a_n 2^n$?

## 8.7    EXERCISES

*In Exercises 1–30, find the radius of convergence and the interval of convergence of the power series.*

**1.** $\displaystyle\sum_{n=0}^{\infty} \frac{x^n}{n+1}$

**2.** $\displaystyle\sum_{n=1}^{\infty} (-1)^{n-1} n x^n$

**3.** $\displaystyle\sum_{n=1}^{\infty} \frac{x^n}{\sqrt{n}}$

**4.** $\displaystyle\sum_{n=1}^{\infty} \frac{x^n}{n^2}$

**5.** $\displaystyle\sum_{n=0}^{\infty} \frac{(2x)^n}{n!}$

**6.** $\displaystyle\sum_{n=1}^{\infty} \frac{(-1)^n x^n}{n \cdot 3^n}$

**7.** $\displaystyle\sum_{n=1}^{\infty} (nx)^n$

**8.** $\displaystyle\sum_{n=0}^{\infty} \frac{n! \, x^n}{(2n)!}$

**9.** $\displaystyle\sum_{n=2}^{\infty} \frac{x^n}{\ln n}$

**10.** $\displaystyle\sum_{n=2}^{\infty} (x \ln n)^n$

**11.** $\displaystyle\sum_{n=1}^{\infty} \frac{e^n x^n}{n}$

**12.** $\displaystyle\sum_{n=0}^{\infty} \frac{(-1)^n n! \, x^n}{2^n}$

**13.** $\displaystyle\sum_{n=1}^{\infty} \frac{(-1)^n (x-3)^n}{\sqrt{n}}$

**14.** $\displaystyle\sum_{n=1}^{\infty} \sqrt{n}(2x+3)^n$

**15.** $\displaystyle\sum_{n=1}^{\infty} \frac{(-1)^{n-1}(x-2)^n}{n \cdot 3^n}$

**16.** $\displaystyle\sum_{n=1}^{\infty} \frac{n(2x+1)^n}{2^n}$

**17.** $\displaystyle\sum_{n=0}^{\infty} \frac{(-1)^n n(x-1)^n}{n^2+1}$

**18.** $\displaystyle\sum_{n=0}^{\infty} \frac{n(x+2)^n}{(n^2+1)2^n}$

**19.** $\displaystyle\sum_{n=0}^{\infty} \frac{(-1)^n (x+2)^{2n+1}}{(2n+1)!}$

**20.** $\displaystyle\sum_{n=0}^{\infty} \frac{(-1)^n (3x+2)^{2n}}{(2n)!}$

**21.** $\displaystyle\sum_{n=1}^{\infty} \frac{2^n (x+2)^n}{n^n}$

**22.** $\displaystyle\sum_{n=1}^{\infty} \frac{(3x-1)^n}{n^3+n}$

**23.** $\displaystyle\sum_{n=2}^{\infty} \frac{(-1)^n (3x+5)^n}{n \ln n}$

**24.** $\displaystyle\sum_{n=2}^{\infty} \frac{(-1)^n (x+2)^n}{(\ln n)^n}$

**25.** $\displaystyle\sum_{n=2}^{\infty} \frac{x^n}{n(\ln n)^2}$

**26.** $\displaystyle\sum_{n=1}^{\infty} \frac{n^n (3x+5)^n}{(2n)!}$

**27.** $\displaystyle\sum_{n=1}^{\infty} \frac{2 \cdot 4 \cdot 6 \cdots 2n}{3 \cdot 5 \cdot 7 \cdots (2n+1)} x^{2n+1}$

**28.** $\displaystyle\sum_{n=1}^{\infty} \frac{(-1)^n n! \, (x-1)^n}{1 \cdot 3 \cdot 5 \cdots (2n-1)}$

**29.** $\displaystyle\sum_{n=1}^{\infty} \frac{(-1)^n 2^n n! \, x^n}{5 \cdot 8 \cdot 11 \cdots (3n+2)}$

**30.** $\displaystyle\sum_{n=1}^{\infty} \frac{(-1)^n 2 \cdot 4 \cdot 6 \cdots 2n(x-\pi)^n}{n!}$

 **31.** Consider the series $\sum_{n=0}^{\infty} x^n$ and the (sum) function $f(x) = 1/(1-x)$ represented by the series for $-1 < x < 1$.
   **a.** Find the remainder $R_n(x) = f(x) - S_n(x)$, where $S_n(x) = \sum_{k=0}^{n} x^k$ is the $n$th partial sum of $\sum_{n=0}^{\infty} x^n$ and $x$ is fixed.

   **b.** Evaluate $\lim_{n\to\infty} R_n(x)$ for each fixed $x$ in the interval $(-1, 1)$. What happens to $\lim_{n\to\infty} R_n(x)$ for $|x| > 1$?
   **c.** Plot the graphs of $R_n(x)$ for $n = 1, 2, 3, \ldots, 5$, and 20 using the viewing window $[-2, 2] \times [-10, 5]$.

 **32. A Bessel Function** The function $J_0$ defined by

$$J_0(x) = \sum_{n=0}^{\infty} \frac{(-1)^n x^{2n}}{2^{2n}(n!)^2}$$

is called the *Bessel function of order* 0.
   **a.** What is the domain of $J_0$?
   **b.** Plot the graph of $J_0$ in the viewing window $[-10, 10] \times [-0.5, 1.2]$, and plot the graphs of $S_n(x)$ for $n = 0, 1, 2, 3$, and 4 in the viewing window $[-8, 8] \times [-2, 2]$.

**33. A Bessel Function** The function $J_1$ defined by

$$J_1(x) = \sum_{n=0}^{\infty} \frac{(-1)^n x^{2n+1}}{n!(n+1)! \, 2^{2n+1}}$$

is called the *Bessel function of order* 1. What is its domain?

**34.** If $a$ is a constant, find the radius and the interval of convergence of the power series $\sum_{n=0}^{\infty} a^n (x-c)^n$.

**35.** If the radius of convergence of the power series $\sum a_n x^n$ is $R$, what is the radius of convergence of the power series $\sum a_n x^{2n}$?

**36.** Suppose that $\lim_{n\to\infty} |a_{n+1}/a_n| = L$ and $L \neq 0$. Show that the radius of convergence of the power series $\sum a_n x^n$ is $1/L$.

**37.** Suppose that $\lim_{n\to\infty} \sqrt[n]{|a_n|} = L$ and $L \neq 0$. What is the radius of convergence of the power series $\sum a_n x^n$?

**38.** Let $f(x) = \displaystyle\sum_{n=1}^{\infty} \frac{(x-2)^n}{n^2 3^n}$. Show that the domain of $f$ is $[-1, 5]$ but the domain of $f'$ is $[-1, 5)$.

**39.** Let $f(x) = \displaystyle\sum_{n=1}^{\infty} \frac{x^n}{n^2}$. Find $f'(x)$ and $f''(x)$. What are the intervals of convergence of $f$, $f'$, and $f''$?

**40.** Show that the series $\displaystyle\sum_{n=1}^{\infty} \frac{\sin(n^3 x)}{n^2}$ converges for all values of $x$, but $\displaystyle\sum_{n=1}^{\infty} \frac{d}{dx}\left[\frac{\sin(n^3 x)}{n^2}\right]$ diverges for all values of $x$. Does this contradict Theorem 2? Explain your answer.

**41.** Find the sum of the series $\sum_{n=1}^{\infty} nx^{n-1}$, $|x| < 1$.
   **Hint:** Differentiate the geometric series $\sum_{n=0}^{\infty} x^n$.

**42. a.** Find the sum of the series $\sum_{n=1}^{\infty} nx^n$, $|x| < 1$.
   **Hint:** See the hint for Exercise 41.

   **b.** Use the result of part (a) to find the sum of $\displaystyle\sum_{n=1}^{\infty} \frac{n}{2^n}$.

**43.** Suppose that the interval of convergence of the series $\sum_{n=0}^{\infty} a_n (x-c)^n$ is $(c - R, c + R]$. Prove that the series is conditionally convergent at $c + R$.

**44.** Suppose that the series $\sum_{n=0}^{\infty} a_n(x - c)^n$ is absolutely convergent at one endpoint of its interval of convergence. Prove that the series is also absolutely convergent at the other endpoint.

**45. a.** Find a power series representation for $1/(1 - t^2)$.
   **b.** Use the result of part (a) to find a power series representation of $\tanh^{-1} x$ using the relationship

$$\tanh^{-1} x = \int_0^x \frac{1}{1 - t^2} \, dt$$

What is the radius of convergence of the series?

**46.** Use the result of Example 8

$$\ln(1 - x) = -\sum_{n=1}^{\infty} \frac{x^n}{n}$$

to obtain an approximation of $\ln 1.2$ accurate to five decimal places.
**Hint:** Use Theorem 2 in Section 8.5.

**47.** Use the result of Example 9,

$$\tan^{-1} x = \sum_{n=0}^{\infty} (-1)^n \frac{x^{2n+1}}{2n + 1}$$

to obtain an approximation of $\pi$ accurate to five decimal places.
**Hint:** Use Theorem 2 in Section 8.5.

**48. Motion Along an Inclined Plane** An object of mass $m$ is thrown up an inclined plane that makes an angle of $\alpha$ with the horizontal. If air resistance proportional to the instantaneous velocity is taken into consideration, then the object reaches a maximum distance up the incline given by

$$D = \frac{mv_0}{k} - \frac{m^2 g}{k^2} (\sin \alpha) \ln\left(1 + \frac{kv_0}{mg \sin \alpha}\right)$$

where $k$ is the constant of proportionality and $g$ is the constant of acceleration due to gravity.
   **a.** Show that $D$ can be written as

$$D = \frac{1}{2} \frac{v_0^2}{g \sin \alpha} - \frac{1}{3} \frac{v_0^3}{m(g \sin \alpha)^2} k + \frac{1}{4} \frac{v_0^4}{m^2(g \sin \alpha)^3} k^2 - \cdots$$

**Hint:** Use the result of Example 8.
   **b.** Use the result of part (a) to show that in the absence of air resistance the object reaches a maximum distance of $v_0^2/(2g \sin \alpha)$ up the incline.

*In Exercises 49–52, determine whether the statement is true or false. If it is true, explain why it is true. If it is false, explain why or give an example to show why it is false.*

**49.** If the power series $\sum_{n=0}^{\infty} a_n x^n$ converges for $x = 3$, then it converges for $x = -2$.

**50.** If the power series $\sum_{n=0}^{\infty} a_n x^n$ converges for $x$ in $(-1, 1)$, then $f(x) = \sum_{n=0}^{\infty} a_n x^n$ is continuous on $(-1, 1)$.

**51.** If the interval of convergence of $\sum_{n=0}^{\infty} a_n x^n$ is $[-2, 2)$, then the interval of convergence of $\sum_{n=0}^{\infty} a_n(x - 3)^n$ is $[1, 5)$.

**52.** If the radius of convergence of $\sum_{n=0}^{\infty} a_n x^n$ is $R > 0$, then the radius of convergence of the power series in $\frac{1}{x}$, $\sum_{n=0}^{\infty} \frac{a_n}{x^n}$, is $\frac{1}{R}$.

**8.8** **Taylor and Maclaurin Series**

In Section 8.7 we saw that every power series represents a function whose domain is precisely the interval of convergence of the series. We also touched upon the converse problem: Given a function $f$ defined on an interval containing a point $c$, is there a power series centered at $c$ that represents $f$, and if so, how do we find it? There, we were able to look only at functions whose power series representations are obtained by manipulating the geometric series.

We now look at the general problem of finding power series representations for functions. The problem centers on finding the answers to two questions:

**1.** What form does the power series representation of the function $f$ take? (In other words, what does $a_n$ look like?)
**2.** What conditions will guarantee that such a power series will represent $f$?

We will consider the first question here and leave the second for Section 8.9.

■ **Taylor and Maclaurin Series**

Suppose that $f$ is a function that can be represented by a power series that is centered at $c$ and has a radius of convergence $R > 0$. If $|x - c| < R$, we have

$$f(x) = a_0 + a_1(x - c) + a_2(x - c)^2 + a_3(x - c)^3 + a_4(x - c)^4 + \cdots + a_n(x - c)^n + \cdots$$

Applying Theorem 2 of Section 8.7 repeatedly, we obtain

$$f'(x) = a_1 + 2a_2(x - c) + 3a_3(x - c)^2 + 4a_4(x - c)^3 + \cdots + na_n(x - c)^{n-1} + \cdots$$

$$f''(x) = 2a_2 + 3 \cdot 2a_3(x - c) + 4 \cdot 3a_4(x - c)^2 + \cdots + n(n - 1)a_n(x - c)^{n-2} + \cdots$$

$$f'''(x) = 3 \cdot 2a_3 + 4 \cdot 3 \cdot 2a_4(x - c) + \cdots + n(n - 1)(n - 2)a_n(x - c)^{n-3} + \cdots$$

$$\vdots$$

$$f^{(n)}(x) = n(n - 1)(n - 2)(n - 3) \cdot \cdots \cdot 2a_n + \cdots$$

$$\vdots$$

Each of these series is valid for $x$ satisfying $|x - c| < R$. Substituting $x = c$ in each of the above expressions, we obtain

$$f(c) = a_0, \qquad f'(c) = a_1, \qquad f''(c) = 2a_2,$$

$$f'''(c) = 3! \, a_3, \qquad \ldots, \qquad f^{(n)}(c) = n! \, a_n, \qquad \ldots$$

from which we find

$$a_0 = f(c), \qquad a_1 = f'(c), \qquad a_2 = \frac{f''(c)}{2!},$$

$$a_3 = \frac{f'''(c)}{3!}, \qquad \ldots, \qquad a_n = \frac{f^{(n)}(c)}{n!}, \qquad \ldots$$

We have proved that if $f$ has a power series representation, then the series must have the form given in the following theorem.

---

**THEOREM 1** **Taylor Series of $f$ at $c$**

If $f$ has a power series representation at $c$, that is, if

$$f(x) = \sum_{n=0}^{\infty} a_n(x - c)^n \qquad |x - c| < R$$

then $f^{(n)}(c)$ exists for every positive integer $n$ and

$$a_n = \frac{f^{(n)}(c)}{n!}$$

Thus,

$$f(x) = \sum_{n=0}^{\infty} \frac{f^{(n)}(c)}{n!} (x - c)^n$$

$$= f(c) + f'(c)(x - c) + \frac{f''(c)}{2!}(x - c)^2 + \frac{f'''(c)}{3!}(x - c)^3 + \cdots \qquad \textbf{(1)}$$

---

A series of this form is called the **Taylor series of the function $f$ at $c$** after the English mathematician Brook Taylor (1685–1731).

In the special case in which $c = 0$, the Taylor series becomes

$$f(x) = \sum_{n=0}^{\infty} \frac{f^{(n)}(0)}{n!} x^n = f(0) + f'(0)x + \frac{f''(0)}{2!}x^2 + \frac{f'''(0)}{3!}x^3 + \cdots \qquad \textbf{(2)}$$

This series is just the Taylor series of $f$ centered at the origin. It is called the **Maclaurin series of $f$** in honor of the Scottish mathematician Colin Maclaurin (1698–1746).

**Note**   Theorem 1 states that if a function $f$ has a power series representation at $c$, then the (unique) series must be the Taylor series at $c$. The converse is not necessarily true. Given a function $f$ with derivatives of *all* orders at $c$, we can compute the Taylor coefficients of $f$ at $c$,

$$\frac{f^{(n)}(c)}{n!} \qquad n = 0, 1, 2, \ldots$$

and, therefore, the Taylor series of $f$ at $c$ (Equation (1)). But the series that is obtained *formally* in this fashion need *not* represent $f$. Situations such as these, however, are rare. (We give an example of such a function in Exercise 75.) In view of this *we will assume, in the rest of this section, that the Taylor series of a function does represent the function, unless otherwise noted.*   ■

**EXAMPLE 1**   Let $f(x) = e^x$. Find the Maclaurin series of $f$, and determine its radius of convergence.

**Solution**   The derivatives of $f(x) = e^x$ are $f'(x) = e^x$, $f''(x) = e^x$, and, in general, $f^{(n)}(x) = e^x$, where $n \geq 1$. So

$$f(0) = 1, \qquad f'(0) = 1, \qquad f''(0) = 1, \qquad \ldots, \qquad f^{(n)}(0) = 1, \qquad \ldots$$

Therefore, if we use Equation (2), the Maclaurin series of $f$ (the Taylor series of $f$ at 0) is

$$\sum_{n=0}^{\infty} \frac{f^{(n)}(0)}{n!} x^n = \sum_{n=0}^{\infty} \frac{1}{n!} x^n = 1 + x + \frac{x^2}{2!} + \frac{x^3}{3!} + \cdots + \frac{x^n}{n!} + \cdots$$

To determine the radius of convergence of the power series, we use the ratio test with $u_n = x^n/n!$. Since

$$\lim_{n \to \infty} \left| \frac{u_{n+1}}{u_n} \right| = \lim_{n \to \infty} \left| \frac{x^{n+1}}{(n+1)!} \cdot \frac{n!}{x^n} \right| = \lim_{n \to \infty} \frac{|x|}{n+1} = 0$$

we conclude that the radius of convergence of the series is $R = \infty$.   ■

**EXAMPLE 2**   Find the Taylor series for $f(x) = \ln x$ at 1, and determine its interval of convergence.

**Solution**   We compute the values of $f$ and its derivatives at 1. Thus,

$$f(x) = \ln x \qquad\qquad\qquad f(1) = \ln 1 = 0$$

$$f'(x) = \frac{1}{x} = x^{-1} \qquad\qquad f'(1) = 1$$

$$f''(x) = -x^{-2} \qquad\qquad f''(1) = -1$$

$$f'''(x) = 2x^{-3} \qquad\qquad f'''(1) = 2$$

$$f^{(4)}(x) = -3 \cdot 2x^{-4} \qquad\qquad f^{(4)}(1) = -3 \cdot 2$$

$$\vdots \qquad\qquad\qquad\qquad \vdots$$

$$f^{(n)}(x) = (-1)^{n-1}(n-1)! \, x^{-n} \qquad f^{(n)}(1) = (-1)^{n-1}(n-1)!$$

Then using Equation (1), we obtain the Taylor series of $f(x) = \ln x$:

$$\sum_{n=0}^{\infty} \frac{f^{(n)}(1)}{n!}(x-1)^n = f(1) + f'(1)(x-1) + \frac{f''(1)}{2!}(x-1)^2 + \frac{f'''(1)}{3!}(x-1)^3 + \cdots$$

$$= (x-1) - \frac{1}{2!}(x-1)^2 + \frac{2}{3!}(x-1)^3 - \frac{3!}{4!}(x-1)^4 + \cdots$$

$$= (x-1) - \frac{(x-1)^2}{2} + \frac{(x-1)^3}{3} - \frac{(x-1)^4}{4} + \cdots$$

$$= \sum_{n=1}^{\infty} (-1)^{n-1}\frac{(x-1)^n}{n}$$

To find the interval of convergence of the series, we use the Ratio Test with $u_n = (-1)^{n-1}(x-1)^n/n$. Since

$$\lim_{n\to\infty} \left|\frac{u_{n+1}}{u_n}\right| = \lim_{n\to\infty} \left|\frac{(-1)^n(x-1)^{n+1}}{n+1} \cdot \frac{n}{(-1)^{n-1}(x-1)^n}\right|$$

$$= \lim_{n\to\infty} |x-1|\left(\frac{n}{n+1}\right) = |x-1|\lim_{n\to\infty}\frac{1}{1+\dfrac{1}{n}} = |x-1|$$

we see that the series converges for $x$ in the interval $(0, 2)$. Next, we notice that if $x = 0$, the series becomes

$$\sum_{n=1}^{\infty} \frac{(-1)^{2n-1}}{n} = -\sum_{n=1}^{\infty}\frac{1}{n}$$

Since this is the negative of the harmonic series, it is divergent. If $x = 2$, the series becomes

$$\sum_{n=1}^{\infty} \frac{(-1)^{n-1}}{n}$$

This is the alternating harmonic series and, hence, is convergent. Therefore, the Taylor series for $f(x) = \ln x$ at 1 has interval of convergence $(0, 2]$.     ■

**EXAMPLE 3**   Find the Maclaurin series of $f(x) = \sin x$, and determine its interval of convergence.

**Solution**   To find the Maclaurin series of $f(x) = \sin x$, we compute the values of $f$ and its derivatives at $x = 0$. We obtain

$$f(x) = \sin x \qquad\qquad f(0) = 0$$
$$f'(x) = \cos x \qquad\qquad f'(0) = 1$$
$$f''(x) = -\sin x \qquad\qquad f''(0) = 0$$
$$f'''(x) = -\cos x \qquad\qquad f'''(0) = -1$$
$$f^{(4)}(x) = \sin x \qquad\qquad f^{(4)}(0) = 0$$

We need not go further, since it is clear that successive derivatives of $f$ follow this same pattern. Then, using Equation (2), we obtain the Maclaurin series of $f(x) = \sin x$:

$$\sum_{n=0}^{\infty} \frac{f^{(n)}(0)}{n!} x^n = f(0) + f'(0)x + \frac{f''(0)}{2!} x^2 + \frac{f'''(0)}{3!} x^3 + \frac{f^{(4)}(0)}{4!} x^4 + \cdots$$

$$= x - \frac{x^3}{3!} + \frac{x^5}{5!} - \frac{x^7}{7!} + \cdots$$

$$= \sum_{n=0}^{\infty} \frac{(-1)^n}{(2n+1)!} x^{2n+1}$$

To find the interval of convergence of the series, we use the Ratio Test with $u_n = (-1)^n x^{2n+1}/(2n+1)!$. Since

$$\lim_{n \to \infty} \left| \frac{u_{n+1}}{u_n} \right| = \lim_{n \to \infty} \left| \frac{(-1)^{n+1} x^{2n+3}}{(2n+3)!} \cdot \frac{(2n+1)!}{(-1)^n x^{2n+1}} \right|$$

$$= \lim_{n \to \infty} \frac{|x|^2}{(2n+2)(2n+3)} = 0 < 1$$

we conclude that the interval of convergence of the series is $(-\infty, \infty)$.    ■

**EXAMPLE 4**  Find the Maclaurin series of $f(x) = \cos x$.

**Solution**  We could proceed as in Example 3, but it is easier to make use of Theorem 2 of Section 8.7 to differentiate the expression for $\sin x$ that we obtained in Example 3. Thus,

$$f(x) = \cos x = \frac{d}{dx} (\sin x) = \frac{d}{dx} \left( x - \frac{x^3}{3!} + \frac{x^5}{5!} - \frac{x^7}{7!} + \cdots \right)$$

$$= 1 - \frac{x^2}{2!} + \frac{x^4}{4!} - \frac{x^6}{6!} + \cdots$$

$$= \sum_{n=0}^{\infty} \frac{(-1)^n}{(2n)!} x^{2n}$$

Since the Maclaurin series for $\sin x$ converges for all $x$, Theorem 2 of Section 8.7 tells us that this series converges in $(-\infty, \infty)$ as well.    ■

**EXAMPLE 5**  Find the Maclaurin series for $f(x) = (1 + x)^k$, where $k$ is a real number.

**Solution**  We compute the values of $f$ and its derivatives at $x = 0$, obtaining

$$f(x) = (1 + x)^k \qquad\qquad f(0) = 1$$
$$f'(x) = k(1 + x)^{k-1} \qquad\qquad f'(0) = k$$
$$f''(x) = k(k - 1)(1 + x)^{k-2} \qquad\qquad f''(0) = k(k - 1)$$
$$f'''(x) = k(k - 1)(k - 2)(1 + x)^{k-3} \qquad\qquad f'''(0) = k(k - 1)(k - 2)$$
$$\vdots \qquad\qquad\qquad\qquad\qquad \vdots$$
$$f^{(n)}(x) = k(k - 1) \cdots (k - n + 1)(1 + x)^{k-n} \qquad f^{(n)}(0) = k(k - 1) \cdots (k - n + 1)$$

So the Maclaurin series of $f(x) = (1 + x)^k$ is

$$\sum_{n=0}^{\infty} \frac{f^{(n)}(0)}{n!} x^n = f(0) + f'(0)x + \frac{f''(0)}{2!} x^2 + \frac{f'''(0)}{3!} x^3 + \cdots$$

$$= 1 + kx + \frac{k(k-1)}{2!} x^2 + \frac{k(k-1)(k-2)}{3!} x^3 + \cdots$$

$$= \sum_{n=0}^{\infty} \frac{k(k-1)(k-2)\cdots(k-n+1)}{n!} x^n$$

Observe that if $k$ is a positive integer, then the series is infinite (by the Binomial Theorem), and so it converges for all $x$.

If $k$ is not a positive integer, then we use the Ratio Test to find the interval of convergence. Denoting the $n$th term of the series by $u_n$, we find

$$\lim_{n \to \infty} \left| \frac{u_{n+1}}{u_n} \right| = \lim_{n \to \infty} \left| \frac{k(k-1)\cdots(k-n+1)(k-n)x^{n+1}}{(n+1)!} \cdot \frac{n!}{k(k-1)\cdots(k-n+1)x^n} \right|$$

$$= \lim_{n \to \infty} \frac{|k-n|}{n+1} |x| = \lim_{n \to \infty} \frac{\left| \dfrac{k}{n} - 1 \right|}{1 + \dfrac{1}{n}} |x| = |x|$$

and we see that the series converges for $x$ in the interval $(-1, 1)$. ∎

The series in Example 5 is called the *binomial series*.

---

**The Binomial Series**

If $k$ is any real number and $|x| < 1$, then

$$(1 + x)^k = 1 + kx + \frac{k(k-1)}{2!} x^2 + \frac{k(k-1)(k-2)}{3!} x^3 + \cdots = \sum_{n=0}^{\infty} \binom{k}{n} x^n \quad \text{(3)}$$

---

**Notes**

1. The coefficients in the binomial series are referred to as **binomial coefficients** and are denoted by

$$\binom{k}{n} = \frac{k(k-1)\cdots(k-n+1)}{n!} \qquad n \geq 1, \qquad \binom{k}{0} = 1$$

2. If $k$ is a positive integer and $n > k$, then the binomial coefficient contains a factor $(k - k)$, so $\binom{k}{n} = 0$ for $n > k$. The binomial series then reduces to a polynomial of degree $k$:

$$(1 + x)^k = 1 + kx + \frac{k(k-1)}{2!} x^2 + \cdots + x^k = \sum_{n=0}^{k} \binom{k}{n} x^n$$

In other words, the expression $(1 + x)^k$ can be represented by a finite sum if $k$ is a positive integer and by an infinite series if $k$ is not a positive integer. Thus, we can view the binomial series as an extension of the Binomial Theorem to the case in which $k$ is not a positive integer.

3. Even though the binomial series always converges for $-1 < x < 1$, its convergence at the endpoints $x = -1$ or $x = 1$ depends on the value of $k$. It can be shown that the series converges at $x = 1$ if $-1 < k < 0$ and at both endpoints $x = \pm 1$ if $k \geq 0$.

4. We have derived Equation (3) under the assumption that $(1 + x)^k$ has a power series representation. In Exercise 78 we outline a procedure for deriving Equation (3) without this assumption. ∎

**EXAMPLE 6** Find a power series representation for the function $f(x) = \sqrt{1 + x}$.

**Solution** Using Equation (3) with $k = \frac{1}{2}$, we obtain

$$f(x) = (1 + x)^{1/2} = 1 + \frac{1}{2}x + \frac{\frac{1}{2}\left(\frac{1}{2} - 1\right)}{2!}x^2 + \frac{\frac{1}{2}\left(\frac{1}{2} - 1\right)\left(\frac{1}{2} - 2\right)}{3!}x^3 + \cdots$$

$$+ \frac{\frac{1}{2}\left(\frac{1}{2} - 1\right) \cdots \left(\frac{1}{2} - n + 1\right)}{n!}x^n + \cdots$$

$$= 1 + \frac{1}{2}x - \frac{1}{2 \cdot 2^2}x^2 + \frac{1 \cdot 3}{3! \cdot 2^3}x^3 + \cdots$$

$$+ (-1)^{n+1}\frac{1 \cdot 3 \cdot 5 \cdot \cdots \cdot (2n - 3)}{n! \cdot 2^n}x^n + \cdots$$

$$= 1 + \frac{1}{2}x + \sum_{n=2}^{\infty}(-1)^{n+1}\frac{1 \cdot 3 \cdot 5 \cdot \cdots \cdot (2n - 3)}{n! \cdot 2^n}x^n$$

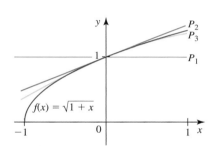

**FIGURE 1**
The graphs of $f(x) = \sqrt{1 + x}$ and the first three partial sums of the binomial series

This representation is valid for $|x| \leq 1$.

The graph of $f$ and the first three partial sums $P_1(x) = 1$, $P_2(x) = 1 + \frac{1}{2}x$, and $P_3(x) = 1 + \frac{1}{2}x - \frac{1}{8}x^2$ are shown in Figure 1. Observe that the partial sums of $f$, $P_n(x)$, approximate $f$ better and better in the interval of convergence of the series as $n$ increases. ∎

### Techniques for Finding Taylor Series

The Taylor series of a function can always be found by using Equation (1). But as Examples 7, 8, and 9 of Section 8.7 and Example 4 of this section show, it is often easier to find the series by algebraic manipulation, differentiation, or integration of some well-known series. We now elaborate further on such techniques. First, we list some common functions and their power series representations in Table 1.

**TABLE 1**

| Maclaurin Series | Interval of Convergence |
|---|---|
| 1. $\dfrac{1}{1 - x} = 1 + x + x^2 + x^3 + \cdots = \sum_{n=0}^{\infty} x^n$ | $(-1, 1)$ |
| 2. $e^x = 1 + x + \dfrac{x^2}{2!} + \dfrac{x^3}{3!} + \cdots = \sum_{n=0}^{\infty} \dfrac{x^n}{n!}$ | $(-\infty, \infty)$ |
| 3. $\sin x = x - \dfrac{x^3}{3!} + \dfrac{x^5}{5!} - \dfrac{x^7}{7!} + \cdots = \sum_{n=0}^{\infty} (-1)^n \dfrac{x^{2n+1}}{(2n + 1)!}$ | $(-\infty, \infty)$ |
| 4. $\cos x = 1 - \dfrac{x^2}{2!} + \dfrac{x^4}{4!} - \dfrac{x^6}{6!} + \cdots = \sum_{n=0}^{\infty} (-1)^n \dfrac{x^{2n}}{(2n)!}$ | $(-\infty, \infty)$ |

*(continued)*

**TABLE 1**    *(continued)*

| Maclaurin Series | Interval of Convergence |
|---|---|
| **5.** $\ln(1 + x) = x - \dfrac{x^2}{2} + \dfrac{x^3}{3} - \dfrac{x^4}{4} + \cdots = \displaystyle\sum_{n=1}^{\infty} (-1)^{n-1} \dfrac{x^n}{n}$ | $(-1, 1]$ |
| **6.** $\sin^{-1} x = x + \dfrac{x^3}{2 \cdot 3} + \dfrac{1 \cdot 3x^5}{2 \cdot 4 \cdot 5} + \cdots = \displaystyle\sum_{n=0}^{\infty} \dfrac{(2n)! \, x^{2n+1}}{(2^n \, n!)^2 (2n + 1)}$ | $[-1, 1]$ |
| **7.** $\tan^{-1} x = x - \dfrac{x^3}{3} + \dfrac{x^5}{5} - \dfrac{x^7}{7} + \cdots = \displaystyle\sum_{n=0}^{\infty} (-1)^n \dfrac{x^{2n+1}}{2n + 1}$ | $[-1, 1]$ |
| **8.** $(1 + x)^k = \displaystyle\sum_{n=0}^{\infty} \binom{k}{n} x^n = 1 + kx + \dfrac{k(k - 1)}{2!} x^2 + \dfrac{k(k - 1)(k - 2)}{3!} x^3 + \cdots$ | $(-1, 1)$ |

All of the formulas in the table except Formulas (5) and (6) have been derived in this and the previous sections. (See note on page 791.) Formula (5) follows from the result of Example 8 in Section 8.7 by replacing $-x$ by $x$. Formula (6) will be derived in Example 14.

**EXAMPLE 7**    Find the Taylor series representation of $f(x) = \dfrac{1}{1 + x}$ at $x = 2$.

**Solution**    We first rewrite $f(x)$ so that it includes the expression $(x - 2)$. Thus,

$$f(x) = \frac{1}{1 + x} = \frac{1}{3 + (x - 2)} = \frac{1}{3\left[1 + \left(\dfrac{x - 2}{3}\right)\right]} = \frac{1}{3} \cdot \frac{1}{1 + \left(\dfrac{x - 2}{3}\right)}$$

Then, using Formula (1) in Table 1 with $x$ replaced by $-(x - 2)/3$, we obtain

$$f(x) = \frac{1}{3} \left\{ \frac{1}{1 - \left[-\left(\dfrac{x - 2}{3}\right)\right]} \right\}$$

$$= \frac{1}{3} \left\{ 1 + \left[-\left(\frac{x - 2}{3}\right)\right] + \left[-\left(\frac{x - 2}{3}\right)\right]^2 + \left[-\left(\frac{x - 2}{3}\right)\right]^3 + \cdots \right\}$$

$$= \frac{1}{3} \left[ 1 - \left(\frac{x - 2}{3}\right) + \left(\frac{x - 2}{3}\right)^2 - \left(\frac{x - 2}{3}\right)^3 + \cdots \right]$$

$$= \frac{1}{3} - \frac{1}{3^2} (x - 2) + \frac{1}{3^3} (x - 2)^2 - \frac{1}{3^4} (x - 2)^3 + \cdots = \sum_{n=0}^{\infty} (-1)^n \frac{(x - 2)^n}{3^{n+1}}$$

The series converges for $|(x - 2)/3| < 1$, that is, $|x - 2| < 3$ or $-1 < x < 5$. You can verify that the series diverges at both endpoints.    ■

**EXAMPLE 8**    Find the Maclaurin series for $f(x) = x^2 \sin 2x$.

**Solution**    If we replace $x$ by $2x$ in Formula (3) in Table 1, we obtain

$$\sin 2x = (2x) - \frac{(2x)^3}{3!} + \frac{(2x)^5}{5!} - \frac{(2x)^7}{7!} + \cdots$$

$$= 2x - \frac{2^3 x^3}{3!} + \frac{2^5 x^5}{5!} - \frac{2^7 x^7}{7!} + \cdots = \sum_{n=0}^{\infty} (-1)^n \frac{2^{2n+1} x^{2n+1}}{(2n + 1)!}$$

which is valid for all $x$ in $(-\infty, \infty)$. Therefore, using Theorem 4a of Section 8.2, we obtain

$$f(x) = x^2 \sin 2x = x^2\left(2x - \frac{2^3 x^3}{3!} + \frac{2^5 x^5}{5!} - \frac{2^7 x^7}{7!} + \cdots\right)$$

$$= 2x^3 - \frac{2^3 x^5}{3!} + \frac{2^5 x^7}{5!} - \frac{2^7 x^9}{7!} + \cdots$$

$$= \sum_{n=0}^{\infty} (-1)^n \frac{2^{2n+1} x^{2n+3}}{(2n+1)!}$$

which converges for all $x$ in $(-\infty, \infty)$. ■

The next example shows how the use of trigonometric identities can help us find the Taylor series of a trigonometric function.

**EXAMPLE 9** Find the Taylor series for $f(x) = \sin x$ at $x = \pi/6$.

**Solution**   We write

$$f(x) = \sin x = \sin\left[\left(x - \frac{\pi}{6}\right) + \frac{\pi}{6}\right]$$

$$= \sin\left(x - \frac{\pi}{6}\right)\cos\frac{\pi}{6} + \cos\left(x - \frac{\pi}{6}\right)\sin\frac{\pi}{6}$$

$$= \frac{\sqrt{3}}{2}\sin\left(x - \frac{\pi}{6}\right) + \frac{1}{2}\cos\left(x - \frac{\pi}{6}\right)$$

Then using Formulas 3 and 4 with $x - (\pi/6)$ in place of $x$, we obtain

$$f(x) = \frac{\sqrt{3}}{2}\sum_{n=0}^{\infty}\frac{(-1)^n}{(2n+1)!}\left(x - \frac{\pi}{6}\right)^{2n+1} + \frac{1}{2}\sum_{n=0}^{\infty}\frac{(-1)^n}{(2n)!}\left(x - \frac{\pi}{6}\right)^{2n}$$

which converges for all $x$ in $(-\infty, \infty)$. ■

The power series representations of certain functions can also be found by adding, multiplying, or dividing the Maclaurin or Taylor series of some familiar functions as the following examples show.

**EXAMPLE 10** Find the Maclaurin series representation for $f(x) = \sinh x$.

**Solution**   We have

$$f(x) = \sinh x = \frac{1}{2}(e^x - e^{-x}) = \frac{1}{2}e^x - \frac{1}{2}e^{-x}$$

$$= \frac{1}{2}\left(1 + x + \frac{x^2}{2!} + \frac{x^3}{3!} + \cdots\right) - \frac{1}{2}\left(1 - x + \frac{x^2}{2!} - \frac{x^3}{3!} + \cdots\right)$$

$$= x + \frac{x^3}{3!} + \frac{x^5}{5!} + \cdots = \sum_{n=0}^{\infty}\frac{x^{2n+1}}{(2n+1)!}$$

Since the Maclaurin series of both $e^x$ and $e^{-x}$ converge for $x$ in $(-\infty, \infty)$, we see that this representation of $\sinh x$ is also valid for all values of $x$. ■

**EXAMPLE 11** Find the first three terms of the Maclaurin series representation for $f(x) = e^x \cos x$.

**Solution** Using Formulas (2) and (4) in Table 1, we can write

$$f(x) = e^x \cos x = \left(1 + x + \frac{x^2}{2} + \frac{x^3}{6} + \cdots\right)\left(1 - \frac{x^2}{2} + \frac{x^4}{24} - \cdots\right)$$

Multiplying and collecting like terms, we obtain

$$f(x) = (1)\left(1 - \frac{x^2}{2} + \frac{x^4}{24} - \cdots\right) + x\left(1 - \frac{x^2}{2} + \frac{x^4}{24} - \cdots\right)$$

$$+ \frac{x^2}{2}\left(1 - \frac{x^2}{2} + \frac{x^4}{24} - \cdots\right) + \frac{x^3}{6}\left(1 - \frac{x^2}{2} + \cdots\right) + \cdots$$

$$= 1 - \frac{x^2}{2} + \frac{x^4}{24} - \cdots + x - \frac{x^3}{2} + \frac{x^5}{24} - \cdots + \frac{x^2}{2} - \frac{x^4}{4}$$

$$+ \frac{x^6}{48} - \cdots + \frac{x^3}{6} - \frac{x^5}{12} + \cdots$$

$$= 1 + x - \frac{x^3}{3} + \cdots$$

**EXAMPLE 12** Find the first three terms of the Maclaurin series representation for $f(x) = \tan x$.

**Solution** Using Formulas (3) and (4) in Table 1, we have

$$f(x) = \tan x = \frac{\sin x}{\cos x} = \frac{x - \dfrac{x^3}{3!} + \dfrac{x^5}{5!} - \cdots}{1 - \dfrac{x^2}{2!} + \dfrac{x^4}{4!} - \cdots}$$

By long division we find

$$1 - \tfrac{1}{2}x^2 + \tfrac{1}{24}x^4 - \cdots \overline{)\phantom{x} \begin{array}{l} x + \tfrac{1}{3}x^3 + \tfrac{2}{15}x^5 + \cdots \\[4pt] x - \tfrac{1}{6}x^3 + \tfrac{1}{120}x^5 - \cdots \\[4pt] \underline{x - \tfrac{1}{2}x^3 + \tfrac{1}{24}x^5 - \cdots} \\[4pt] \tfrac{1}{3}x^3 - \tfrac{1}{30}x^5 + \cdots \\[4pt] \underline{\tfrac{1}{3}x^3 - \tfrac{1}{6}x^5 + \cdots} \\[4pt] \tfrac{2}{15}x^5 + \cdots \end{array}}$$

Therefore,

$$f(x) = \tan x = x + \frac{1}{3}x^3 + \frac{2}{15}x^5 + \cdots$$

In both Examples 11 and 12 we computed only the first three terms of each series. In practice, the retention of just the first few terms of a series is sufficient to obtain an acceptable approximation to the solution of a problem.

We can also use Taylor series to integrate functions whose antiderivatives cannot be found in terms of elementary functions (see page 814). Examples of such functions are $e^{-x^2}$ and $\sin x^2$. In particular, the use of Taylor series enables us to obtain approximations to definite integrals involving such functions, as illustrated in the following example.

### EXAMPLE 13

**a.** Find $\displaystyle\int e^{-x^2}\, dx$.

**b.** Find an approximation of $\displaystyle\int_0^{0.5} e^{-x^2}\, dx$ accurate to four decimal places.

**Solution**

**a.** Replacing $x$ in Formula (2) in Table 1 by $-x^2$ gives

$$e^{-x^2} = 1 - x^2 + \frac{x^4}{2!} - \frac{x^6}{3!} + \cdots = \sum_{n=0}^{\infty}(-1)^n \frac{x^{2n}}{n!}$$

Integrating both sides of this equation with respect to $x$, we obtain, by Theorem 2,

$$\int e^{-x^2}\, dx = \int \left(1 - x^2 + \frac{x^4}{2!} - \frac{x^6}{3!} + \cdots\right) dx$$

$$= C + x - \frac{1}{3}x^3 + \frac{1}{5\cdot 2!}x^5 - \frac{1}{7\cdot 3!}x^7 + \cdots$$

$$= C + \sum_{n=0}^{\infty}(-1)^n \frac{1}{(2n+1)\cdot n!}x^{2n+1}$$

Since the power series representation of $e^{-x^2}$ converges for $x$ in $(-\infty, \infty)$, this result is valid for all values of $x$.

**b.** Using the result from part (a), we obtain

$$\int_0^{0.5} e^{-x^2}\, dx = \left[x - \frac{1}{3}x^3 + \frac{1}{5\cdot 2!}x^5 - \frac{1}{7\cdot 3!}x^7 + \frac{1}{9\cdot 4!}x^9 - \frac{1}{11\cdot 5!}x^{11} + \cdots\right]_0^{1/2}$$

$$= \frac{1}{2} - \frac{1}{3}\left(\frac{1}{2}\right)^3 + \frac{1}{5\cdot 2!}\left(\frac{1}{2}\right)^5 - \frac{1}{7\cdot 3!}\left(\frac{1}{2}\right)^7 + \frac{1}{9\cdot 4!}\left(\frac{1}{2}\right)^9 - \cdots$$

$$= \frac{1}{2} - \frac{1}{24} + \frac{1}{320} - \frac{1}{5376} + \frac{1}{110592} - \cdots$$

$$\approx 0.4613$$

Since this series is alternating and its terms decrease to 0, we know, by Theorem 2 of Section 8.5, that the error incurred in the approximation does not exceed

$$\frac{1}{9\cdot 4!}\left(\frac{1}{2}\right)^9 = \frac{1}{110592} \approx 0.000009 < 0.00005$$

So the result is accurate to within four decimal places, as desired.

**EXAMPLE 14**   Find a power series representation for $\sin^{-1} x$.

**Solution**   Observe that

$$\sin^{-1} x = \int_0^x \frac{1}{\sqrt{1 - t^2}}\, dt$$

Using Equation (3) with $k = -\frac{1}{2}$ and $x = -t^2$, we have

$$\frac{1}{\sqrt{1 - t^2}} = (1 - t^2)^{-1/2} = 1 + \left(-\frac{1}{2}\right)(-t^2) + \frac{-\frac{1}{2}\left(-\frac{1}{2} - 1\right)}{2!}(-t^2)^2 + \cdots$$

$$+ \frac{-\frac{1}{2}\left(-\frac{1}{2} - 1\right)\cdots\left(-\frac{1}{2} - n + 1\right)}{n!}(-t^2)^n + \cdots$$

$$= 1 + \frac{1}{2}t^2 + \frac{1 \cdot 3}{2! \, 2^2}t^4 + \cdots + \frac{1 \cdot 3 \cdot 5 \cdots (2n - 1)}{n! \, 2^n}t^{2n} + \cdots$$

Therefore,

$$\sin^{-1} x = \int_0^x \frac{1}{\sqrt{1 - t^2}}\, dt = x + \frac{1}{2 \cdot 3}x^3 + \frac{1 \cdot 3}{2! \, 2^2 \cdot 5}x^5 + \cdots$$

$$= x + \sum_{n=1}^{\infty} \frac{1 \cdot 3 \cdot 5 \cdots (2n - 1)}{2 \cdot 4 \cdot 6 \cdots (2n)} \cdot \frac{x^{2n+1}}{2n + 1}$$

$$= \sum_{n=0}^{\infty} \frac{(2n)! \, x^{2n+1}}{(2^n \, n!)^2 (2n + 1)}$$

It can be shown that the series converges in $[-1, 1]$.   ▪

**EXAMPLE 15**   **Einstein's Special Theory of Relativity**   According to Einstein's special theory of relativity, a body of mass $m_0$ at rest has a *rest energy* $E_0 = m_0 c^2$ due to the *mass* itself. (Here, the constant $c$ denotes the speed of light.) The same body, moving at a speed $v$, has *total energy*

$$E = \frac{m_0 c^2}{\sqrt{1 - \dfrac{v^2}{c^2}}}$$

The kinetic energy, the energy of motion, is the difference between the total energy and the rest energy and is therefore given by

$$K = E - E_0 = \frac{m_0 c^2}{\sqrt{1 - \dfrac{v^2}{c^2}}} - m_0 c^2$$

Show that if $v$ is very small in comparison to $c$, the kinetic energy of the body assumes the classical form $K = \frac{1}{2}m_0 v^2$.

**Solution**   We have

$$K = \frac{m_0 c^2}{\sqrt{1 - \dfrac{v^2}{c^2}}} - m_0 c^2 = m_0 c^2\left[\left(1 - \frac{v^2}{c^2}\right)^{-1/2} - 1\right]$$

Using Equation (3) with $k = -\frac{1}{2}$ and $x = -v^2/c^2$, we obtain

$$\left(1 - \frac{v^2}{c^2}\right)^{-1/2} = 1 - \frac{1}{2}\left(-\frac{v^2}{c^2}\right) + \frac{(-\frac{1}{2})(-\frac{1}{2} - 1)}{2!}\left(-\frac{v^2}{c^2}\right)^2 + \cdots$$

$$= 1 + \frac{1}{2} \cdot \frac{v^2}{c^2} + \frac{3}{8} \cdot \frac{v^4}{c^4} + \cdots$$

Therefore, the kinetic energy is

$$K = m_0 c^2\left[\left(1 + \frac{1}{2} \cdot \frac{v^2}{c^2} + \frac{3}{8} \cdot \frac{v^4}{c^4} + \cdots\right) - 1\right]$$

$$= m_0 c^2\left(\frac{1}{2} \cdot \frac{v^2}{c^2} + \frac{3}{8} \cdot \frac{v^4}{c^4} + \cdots\right)$$

For speeds much less than the speed of light ($v$ much smaller than $c$), all the terms after the first are very small in comparison to the first and may be neglected, leading to

$$K = \frac{1}{2}m_0 v^2$$

the classical expression for kinetic energy. ■

## 8.8 CONCEPT QUESTIONS

1. **a.** What is a Taylor series? What is a Maclaurin series?
   **b.** What is the difference between a Taylor series and a Maclaurin series?
2. **a.** Suppose $f(x) = \Sigma_{n=0}^{\infty} a_n(x - c)^n$ for $x$ in $(-R, R)$, where $R > 0$. What is $f^{(n)}(c)$?
   **b.** The Taylor series of $f(x)$ at $x = 1$ is
   $$\sum_{n=0}^{\infty}(-1)^n \frac{(x - 1)^{n+1}}{n + 1}$$
   What is $f^{(5)}(1)$?
3. **a.** What is a binomial series?
   **b.** What is the $n$th term of a binomial series?
   **c.** What is the radius of convergence of a binomial series if the exponent is a nonnegative integer?

4. **a.** Consider the function $f(x) = (1 + x)^k$, where $k$ is negative. What can you say about $f(-1)$?
   **b.** Consider the function $f(x) = (1 + x)^k$, where $k$ is positive but not an integer. What can you say about the derivative or higher derivatives of $f$ at $-1$? Illustrate with an example.
   **c.** Use the results of parts (a) and (b) to explain why we can only assert, in general, that the binomial series converges for $|x| < 1$.

## 8.8 EXERCISES

**Note:** *In this exercise set, assume that all the functions have power series representations.*

*In Exercises 1–10, use Equation (1) to find the Taylor series of $f$ at the given value of $c$. Then find the radius of convergence of the series.*

1. $f(x) = e^{2x}$, $c = 0$
2. $f(x) = e^{-3x}$, $c = 0$
3. $f(x) = e^x$, $c = 2$
4. $f(x) = e^{-2x}$, $c = 3$
5. $f(x) = \sin 2x$, $c = 0$
6. $f(x) = \sin x$, $c = \dfrac{\pi}{4}$
7. $f(x) = \cos x$, $c = -\dfrac{\pi}{6}$
8. $f(x) = \dfrac{1}{x}$, $c = -1$
9. $f(x) = \ln x$, $c = 2$
10. $f(x) = \sinh x$, $c = 0$

Ⓥ Videos for selected exercises are available online at **www.academic.cengage.com/login**.

*In Exercises 11–28, use the power series representations of functions established in this section to find the Taylor series of f at the given value of c. Then find the radius of convergence of the series.*

**11.** $f(x) = \dfrac{1}{1 + x}$, $c = 1$

**12.** $f(x) = \dfrac{1}{1 + x}$, $c = -2$

**13.** $f(x) = \dfrac{1}{1 - 2x}$, $c = 1$

**14.** $f(x) = \dfrac{1}{1 + 3x}$, $c = 2$

**15.** $f(x) = \dfrac{x^2}{x^2 - 1}$, $c = 0$

**16.** $f(x) = \dfrac{1}{4 + x^2}$, $c = 0$

**17.** $f(x) = xe^{-x}$, $c = 0$

**18.** $f(x) = e^{2x}$, $c = -1$

**19.** $f(x) = x^2 \cos x$, $c = 0$

**20.** $f(x) = x \cos 3x$, $c = 0$

**21.** $f(x) = \cos^2 x$, $c = 0$

    **Hint:** $\cos^2 x = \dfrac{1}{2}(1 + \cos 2x)$

**22.** $f(x) = \sin^2 x$, $c = 0$

    **Hint:** $\sin^2 x = \dfrac{1}{2}(1 - \cos 2x)$

**23.** $f(x) = \sin x$, $c = \dfrac{\pi}{3}$

**24.** $f(x) = \cos x$, $c = \dfrac{\pi}{6}$

**25.** $f(x) = \sqrt{x} \sin^{-1} x$, $c = 0$

**26.** $f(x) = (1 + x^2)\tan^{-1} x$, $c = 0$

**27.** $f(x) = \ln(1 + x^2)$, $c = 0$

**28.** $f(x) = \ln\left(\dfrac{1 + x}{1 - x}\right)$, $c = 0$

*In Exercises 29–34, use the binomial series to find the power series representation of the function. Then find the radius of convergence of the series.*

**29.** $f(x) = \dfrac{1}{(1 + x)^2}$

**30.** $f(x) = \sqrt[3]{1 + x}$

**31.** $f(x) = \sqrt{1 - x^2}$

**32.** $f(x) = \dfrac{1}{\sqrt[3]{8 + x}}$

**33.** $f(x) = (1 - x)^{3/5}$

**34.** $f(x) = \dfrac{x}{(1 + x)^2}$

*In Exercises 35–40, find the first three terms of the Taylor series of f at the given value of c.*

**35.** $f(x) = \tan x$, $c = \dfrac{\pi}{4}$

**36.** $f(x) = \sec x$, $c = 0$

**37.** $f(x) = \sin^{-1} x$, $c = \dfrac{1}{2}$

**38.** $f(x) = \tan^{-1} x$, $c = 1$

**39.** $f(x) = e^{-x} \sin x$, $c = 0$

**40.** $f(x) = e^x \tan x$, $c = 0$

 *In Exercises 41 and 42, (a) find the power series representation for the function; (b) write the first three partial sums $P_1$, $P_2$, and $P_3$; and (c) plot the graphs of f and $P_1$, $P_2$, and $P_3$ using a viewing window that includes the interval of convergence of the power series.*

**41.** $f(x) = \sqrt[3]{1 + x}$

**42.** $f(x) = \dfrac{1}{\sqrt{9 - x}}$

**43.** Use the Maclaurin series for $e^{-x^2}$ to calculate $e^{-0.01}$ accurate to five decimal places.

**44.** Use the Maclaurin series for $\cos x$ to calculate $\cos 3°$ accurate to five decimal places.

*In Exercises 45–50, find a power series representation for the indefinite integral.*

**45.** $\displaystyle\int \dfrac{1}{1 + x^3}\, dx$

**46.** $\displaystyle\int e^{-\sqrt{x}}\, dx$

**47.** $\displaystyle\int \sin x^2\, dx$

**48.** $\displaystyle\int x \tan^{-1} x\, dx$

**49.** $\displaystyle\int \dfrac{\ln(1 + x)}{x}\, dx$

**50.** $\displaystyle\int \dfrac{\sin x}{x}\, dx$

*In Exercises 51–56, use a power series to obtain an approximation of the definite integral to four decimal places of accuracy.*

**51.** $\displaystyle\int_0^1 e^{-x^2}\, dx$

**52.** $\displaystyle\int_0^{0.5} x^2 e^{-x^2}\, dx$

**53.** $\displaystyle\int_0^{0.5} \cos x^2\, dx$

**54.** $\displaystyle\int_0^1 \sin x^2\, dx$

**55.** $\displaystyle\int_0^{0.5} x \cos x^3\, dx$

**56.** $\displaystyle\int_0^{0.5} \tan^{-1} x^3\, dx$

*In Exercises 57–62, find the sum of the given series. (Hint: Each series is the Maclaurin series of a function evaluated at an appropriate point.)*

**57.** $\displaystyle\sum_{n=1}^{\infty} (-1)^{n-1} \dfrac{1}{n}$

**58.** $\displaystyle\sum_{n=0}^{\infty} \dfrac{(-1)^n}{n!\, 2^n}$

**59.** $\displaystyle\sum_{n=0}^{\infty} (-1)^n \dfrac{\pi^{2n}}{(2n)!}$

**60.** $\displaystyle\sum_{n=0}^{\infty} (-1)^n \dfrac{\pi^{2n+1}}{(2n + 1)!\, 2^{2n+1}}$

**61.** $\displaystyle\sum_{n=1}^{\infty} (-1)^{n-1} \frac{1}{n2^n}$    **62.** $\displaystyle\sum_{n=0}^{\infty} \frac{(-1)^n}{2n+1}$

**63.** Evaluate $\displaystyle\lim_{x \to 0} \frac{\sin x - x + \frac{1}{6}x^3}{x^5}$.

**Hint:** Use the Maclaurin series representation of $\sin x$.

**64.** Evaluate $\displaystyle\lim_{x \to 0} \frac{\cos x^2 - 1 + \frac{1}{2}x^4}{x^8}$.

**Hint:** Use the Maclaurin series representation of $\cos x^2$.

**65.** Evaluate $\displaystyle\lim_{x \to 0} \frac{\tan x - x - \frac{1}{3}x^3}{x^5}$.

**Hint:** Use the result of Example 12.

**66.** Evaluate $\displaystyle\lim_{x \to 1} \frac{\ln x}{x - 1}$.

**Hint:** Use the Taylor series representation of $\ln x$ at 1.

**67. a.** Find the power series representation for $\dfrac{1}{\sqrt{1 - u^2}}$.

 **b.** Use the result of part (a) to find a power series representation of

$$\sin^{-1} x = \int_0^x \frac{1}{\sqrt{1 - t^2}}\, dt$$

 What is the radius of convergence of the series?

**68. a.** Find a power series representation of $f(x) = \sqrt[3]{1 + x^2}$.

 **b.** Use the result of part (a) to find $f^{(6)}(0)$.

**69. Force Exerted by a Charge Distribution** Suppose that a charge $Q$ is distributed uniformly along the positive $x$-axis from $x = 0$ to $x = a$ and that a negative charge $-Q$ is distributed uniformly along the negative $x$-axis from $x = -a$ to $x = 0$. If a positive charge $q$ is placed on the positive $x$-axis a distance of $x$ units ($x > a$) from the origin, then the force exerted by the charge distribution on $q$ has magnitude

$$F = \frac{qQ}{4\pi\varepsilon_0 a}\left[\frac{1}{x-a} + \frac{1}{x+a} - \frac{2}{x}\right]$$

and direction along the positive $x$-axis.

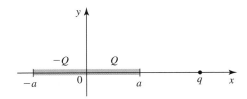

Show that if $x$ is large, then

$$F \approx \frac{qQa}{2\pi\varepsilon_0 x^3}$$

**70. Speed of a Wave** A wave of water of length $L$ travels across a body of water of depth $d$, as illustrated in the figure below. The speed of the wave is given by

$$v = \left(\frac{gL}{2\pi} \tanh \frac{2\pi d}{L}\right)^{1/2}$$

where $g$ is the constant of acceleration due to gravity.
 **a.** Show that in deep water, $v \approx \sqrt{gL/(2\pi)}$, and hence the speed of the wave is independent of the depth of the body of water.
 **b.** Show that in shallow water, $v \approx \sqrt{gd}$, so the speed of the wave is independent of the length of the wave.
 **Hint:** Find the first three nonzero terms of the Maclaurin series for $f(x) = \tanh x$.

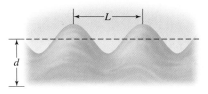

**71. Gravitational Force Between Two Masses** Suppose that a mass $M$ is distributed uniformly over a disk of radius $a$. Then it can be shown that the attractive gravitational force between the disk-shaped mass and a point mass $m$ located a distance of $x$ units above the center of the disk has magnitude

$$F = \frac{2GmM}{a^2}\left[1 - \frac{x}{\sqrt{x^2 + a^2}}\right]$$

Here, $g$ is the gravitational constant. Show that if $x$ is large, then

$$F \approx \frac{GmM}{x^2}$$

Thus, from this distance the disk "looks" like a point mass.

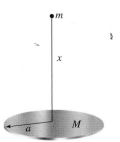

**72. Force of Attraction of a Cylinder on a Point** It can be shown that the magnitude of the force of attraction of a homogeneous right-circular cylinder upon a point $P$ on its axis is

$$F = 2\pi\sigma\left[h + \sqrt{R^2 + a^2} - \sqrt{(R+h)^2 + a^2}\right]$$

where $h$ and $a$ are the height and radius of the cylinder, $R$ is the distance between $P$ and the top of the cylinder, and $\sigma$ is the (constant) density of the solid. Show that if $R$ is large in comparison to $a$ and $h$, then

$$F \approx \frac{M}{R^2}$$

where $M = \pi a^2 h \sigma$ is the mass of the cylinder.

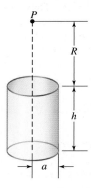

**73. Volume of Water in a Trough** A trough of length $L$ feet has a cross section in the shape of a semicircle with radius $r$ feet. When the trough is filled with water to a level that is $h$ feet as measured from the top of the trough, the volume of the water is

$$V = L\left[\frac{1}{2}\pi r^2 - r^2 \sin^{-1}\left(\frac{h}{r}\right) - h\sqrt{r^2 - h^2}\right]$$

Show that if $h$ is small in comparison to $r$ (that is, $h/r$ is small), then

$$V \approx L\left[\frac{1}{2}\pi r^2 - 2rh + \frac{1}{2}\cdot\frac{h^3}{r}\right]$$

**74.** Formula (5) in Table 1 can be used to compute the value of $\ln x$ for $-1 < x \le 1$. However, the restriction on $x$ and the slow convergence of the series limit its effectiveness from the computational point of view. A more effective formula, first obtained by the Scottish mathematician James Gregory (1638–1675), follows.

**a.** Use Formula (5) to show that

$$\ln\left(\frac{1+x}{1-x}\right) = 2\left(x + \frac{x^3}{3} + \frac{x^5}{5} + \frac{x^7}{7} + \cdots\right) \quad -1 < x < 1$$

**b.** To compute the natural logarithm of a positive number $p$, let $p = (1 + x)/(1 - x)$ and show that

$$x = \frac{p-1}{p+1} \quad -1 < x < 1$$

**c.** Use parts (a) and (b) to find ln 2 accurate to four decimal places.

**75.** Let $f$ be the function defined by

$$f(x) = \begin{cases} e^{-1/x^2} & \text{if } x \neq 0 \\ 0 & \text{if } x = 0 \end{cases}$$

Show that $f$ cannot be represented by a Maclaurin series.

**76. a.** Find the Taylor series for $f(x) = 2x^3 + 3x^2 + 1$ at $x = 1$.
  **b.** Show that the Taylor series and $f(x)$ are equal.
  **c.** What can you say about a Taylor series for a polynomial function? Justify your answer.

**77.** Show that $(1 + x)^n > 1 + nx$ for all $x > 0$ and $n > 1$.

**78.** Prove that $(1 + x)^k = \sum_{n=0}^{\infty} \binom{k}{n} x^n = \sum_{n=0}^{\infty} \frac{k!}{n!(k-n)!} x^n$, where $k$ is any real number and $|x| < 1$, by verifying the following steps.
  **a.** Let

$$f(x) = \sum_{n=0}^{\infty} \binom{k}{n} x^n = 1 + kx + \frac{k(k-1)}{2!} x^2 + \cdots$$
$$+ \frac{k(k-1)\cdots(k-n+1)}{n!} x^n + \cdots$$

Differentiate the equation with respect to $x$ to show that

$$f'(x)(1 + x) - kf(x) = 0$$

  **b.** Define the function $g$ by $g(x) = f(x)/(1 + x)^k$ and show that $g'(x) = 0$.
  **c.** Deduce that $f(x) = (1 + x)^k$.

*In Exercises 79–84, determine whether the statement is true or false. If it is true, explain why it is true. If it is false, explain why or give an example to show why it is false.*

**79.** If $P(x)$ is a polynomial function of degree $n$, then the Maclaurin series for $P$ is $P$.

**80.** Suppose that $f(x) = \sum_{n=0}^{\infty} a_n x^n$ for $x$ in $(-R, R)$, where $R > 0$ and $f$ is odd. Then $a_{2n} = 0$ for $n \ge 0$.

**81.** The function $f(x) = x^{5/3}$ has a Maclaurin series.

**82.** The Taylor series of $f(x) = (1 - x)^7$ at $x = 1$ is $(x - 1)^7$.

**83.** The Maclaurin series for $f(x) = (2 + x)^k$ is $\sum_{n=0}^{\infty} \binom{k}{n} 2^{k-n} x^n$.

**84.** If $k$ is a positive integer, then the Maclaurin series for $f(x) = (1 + x)^k$ is a polynomial of degree $k$.

In Section 8.8 we saw how Maclaurin and Taylor series of functions can be used to help us find the values of the functions they represent. We also saw how we can use these series to find the antiderivatives as well as the values of definite integrals of functions that we could not otherwise evaluate. You will recall that in each instance we were able to obtain satisfactory approximations to the actual values of these quantities by retaining just the first few terms of the series. These truncated series—the $n$th partial sums of the power series representations of the functions—are polynomials. The $n$th partial sum of the Taylor series of $f$ centered at $c$,

$$P_n(x) = \sum_{k=0}^{n} \frac{f^{(k)}(c)}{k!} (x - c)^k$$

$$= f(c) + \frac{f'(c)}{1!}(x - c) + \frac{f''(c)}{2!}(x - c)^2 + \cdots + \frac{f^{(n)}(c)}{n!}(x - c)^n$$

(1)

is called the ***n*th-degree Taylor polynomial of $f$ at $c$. If $c = 0$, we have the *n*th-degree Maclaurin polynomial of $f$.**

The accuracy with which the Taylor polynomials approximate a function $f$ in a neighborhood of $c$ is demonstrated graphically in Figure 1. Here, the function $f(x) = e^x$ with Maclaurin series

$$f(x) = 1 + x + \frac{x^2}{2!} + \frac{x^3}{3!} + \cdots + \frac{x^n}{n!} + \cdots$$

is approximated by the Maclaurin polynomials of degrees 1, 2, and 3:

$$P_1(x) = 1 + x, \quad P_2(x) = 1 + x + \frac{1}{2}x^2, \quad \text{and} \quad P_3(x) = 1 + x + \frac{1}{2}x^2 + \frac{1}{6}x^3$$

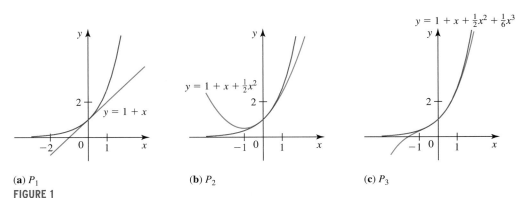

(a) $P_1$　　　　　　　　(b) $P_2$　　　　　　　　(c) $P_3$

**FIGURE 1**

As $n$ increases, $P_n(x)$ gives a better and better approximation of $f(x)$ in a neighborhood of $x = 0$.

Observe that the graph of

$$P_1(x) = 1 + x$$

is a straight line that is tangent to the graph of $f$ at $(0, 1)$. [$P_1(0) = f(0)$ and $P_1'(0) = f'(0)$]. The graph of

$$P_2(x) = 1 + x + \frac{1}{2}x^2$$

is a parabola that passes through $(0, 1)$ $[P_2(0) = f(0)]$, has a tangent line that coincides with that of $f$ at $(0, 1)$ $[P_2'(0) = f'(0)]$, and has concavity that matches that of the graph of $f$ at $(0, 1)$ $[P_2''(0) = f''(0)]$. The graph of

$$P_3(x) = 1 + x + \frac{1}{2}x^2 + \frac{1}{6}x^3$$

provides an even better approximation to the graph of $f$ than that of $P_2(x)$ near $(0, 1)$. Not only does it have the same tangent line and concavity as that of $f$ at $(0, 1)$ $[P_3'(0) = f'(0)$ and $P_3''(0) = f''(0)]$, but both $P_3$ and $f$ satisfy the condition $P_3'''(0) = f'''(0)$.

In general, you can show that if $P_n$ is the $n$th-degree Taylor polynomial of $f$ at $c$, then the derivatives of $P_n$ at $c$ agree with the derivatives of $f$ at $c$ up to and including those of order $n$ (see Exercise 48). This explains why the graph of $P_n$ more closely conforms to the graph of $f$ near $x = c$ as $n$ gets larger and larger.

Figure 2 illustrates how the Maclaurin polynomials $P_2$, $P_4$, $P_6$, and $P_8$ of $f(x) = \cos x$ approximate with increasing accuracy the function $f$. For fixed $n$ the accuracy in the approximation decreases as we move away from the center $c$.

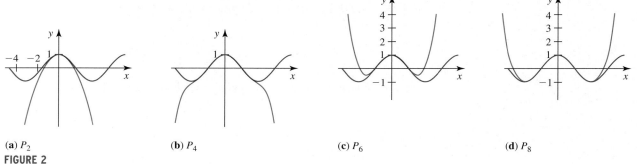

**(a)** $P_2$      **(b)** $P_4$      **(c)** $P_6$      **(d)** $P_8$

**FIGURE 2**
As $n$ increases, $P_{2n}(x)$ approximates $f(x)$ with greater and greater accuracy.

To obtain the same degree of accuracy for $x$ farther away from the center, we need to use an approximating polynomial of higher degree. Figure 3 shows the approximation of $f(x) = \cos x$ using a Maclaurin polynomial $P_{24}(x)$ of degree 24.

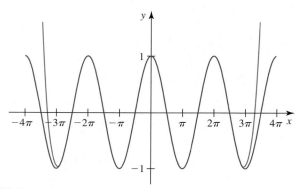

**FIGURE 3**
Approximating $f(x) = \cos x$ using a Maclaurin polynomial of degree 24

## ■ Taylor's Formula with Remainder

Two important questions arise when a function $f$ is approximated by a Taylor polynomial $P_n$:

1. How good is the approximation?
2. How large should $n$ be taken to ensure that a specified degree of accuracy is obtained?

To answer these questions, we need the following theorem, which gives the relationship between $f$ and $P_n$.

---

**THEOREM 1    Taylor's Theorem**

If $f$ has derivatives up to order $n + 1$ in an interval $I$ containing $c$, then for each $x$ in $I$, there exists a number $z$ between $x$ and $c$ such that

$$f(x) = f(c) + f'(c)(x - c) + \frac{f''(c)}{2!}(x - c)^2 + \cdots + \frac{f^{(n)}(c)}{n!}(x - c)^n + R_n(x)$$

$$= P_n(x) + R_n(x)$$

where

$$R_n(x) = \frac{f^{(n+1)}(z)}{(n + 1)!}(x - c)^{n+1} \qquad (2)$$

---

**PROOF**  Let $x$ be any point in $I$ that is different from $c$ and define

$$R_n(x) = f(x) - P_n(x)$$

where $P_n(x)$ is the $n$th-degree Taylor polynomial of $f$ at $c$. For any point $t$ in $I$, define the function $g$ by

$$g(t) = f(x) - f(t) - f'(t)(x - t) - \cdots - \frac{f^{(n)}(t)}{n!}(x - t)^n - R_n(x)\frac{(x - t)^{n+1}}{(x - c)^{n+1}}$$

If we differentiate both sides of this equation with respect to $t$, then the expression on the right side of the resulting equation will be a telescoping finite series (to see this, just write out the first several terms). Canceling like terms, we obtain the following expression for $g'(t)$:

$$g'(t) = -\frac{f^{(n+1)}(t)}{n!}(x - t)^n + (n + 1)R_n(x)\frac{(x - t)^n}{(x - c)^{n+1}}$$

We now apply Rolle's Theorem to the function $g$ defined on the interval $[c, x]$ or $[x, c]$, depending on whether $c < x$ or $c > x$. In either case we see that $g(x) = 0$. Furthermore,

$$g(c) = f(x) - f(c) - f'(c)(x - c) - \cdots - \frac{f^{(n)}(c)}{n!}(x - c)^n - R_n(x)\frac{(x - c)^{n+1}}{(x - c)^{n+1}}$$

$$= f(x) - P_n(x) - R_n(x)$$

$$= f(x) - P_n(x) - [f(x) - P_n(x)] \qquad \text{By the definition of } R_n(x)$$

$$= 0$$

Therefore, $g$ satisfies the conditions of Rolle's Theorem, so there exists a number $z$ between $c$ and $x$ such that $g'(z) = 0$. Using the expression for $g'$ obtained earlier, we have

$$g'(z) = -\frac{f^{(n+1)}(z)}{n!}(x - z)^n + (n + 1)R_n(x)\frac{(x - z)^n}{(x - c)^{n+1}} = 0$$

Solving for $R_n(x)$, we obtain

$$R_n(x) = \frac{f^{(n+1)}(z)}{(n + 1)!}(x - c)^{n+1}$$

Finally, since $g(c) = 0$, we have

$$0 = f(x) - f(c) - f'(c)(x - c) - \cdots - \frac{f^{(n)}(c)}{n!}(x - c)^n - R_n(x)$$

or

$$f(x) = f(c) + f'(c)(x - c) + \cdots + \frac{f^{(n)}(c)}{n!}(x - c)^n + R_n(x) \qquad \blacksquare$$

The expression $R_n(x)$ is called the **Taylor remainder of $f$ at $c$.** If $c = 0$, $R_n(x)$ is called the **Maclaurin remainder of $f$.** We can regard

$$R_n(x) = f(x) - P_n(x)$$

as the error that is incurred when $f(x)$ is approximated by the $n$th-degree Taylor polynomial of $f$ at $c$. Since we usually don't know the value of $z$ in Equation (2)—all we know is that it lies between $x$ and $c$—we often use Equation (2) to find a *bound* for the error in the approximation rather than attempting to find the actual error itself. Incidentally, the presence of the factor $(x - c)^{n+1}$ in Equation (2) explains why (for fixed $n$) $P_n(x)$ gives a better approximation when $x$ is closer to $c$.

**EXAMPLE 1** Let $f(x) = \ln x$.

**a.** Find the fourth-degree Taylor polynomial of $f$ at $c = 1$, and use it to approximate $\ln 1.1$.

**b.** Estimate the accuracy of the approximation that you obtained in part (a).

**Solution**

**a.** The first five derivatives of $f(x) = \ln x$ are

$$f'(x) = \frac{1}{x}, \quad f''(x) = -\frac{1}{x^2}, \quad f'''(x) = \frac{2}{x^3}, \quad f^{(4)}(x) = -\frac{3!}{x^4}, \quad \text{and} \quad f^{(5)}(x) = \frac{4!}{x^5}$$

and the values of $f(x)$ and its first four derivatives at $x = 1$ are

$$f(1) = 0, \quad f'(1) = 1, \quad f''(1) = -1, \quad f'''(1) = 2, \quad \text{and} \quad f^{(4)}(1) = -3!$$

Using Equation (1) with $n = 4$ and $c = 1$, we obtain

$$P_4(x) = f(1) + f'(1)(x - 1) + \frac{f''(1)}{2!}(x - 1)^2 + \frac{f'''(1)}{3!}(x - 1)^3 + \frac{f^{(4)}(1)}{4!}(x - 1)^4$$

$$= (x - 1) - \frac{1}{2}(x - 1)^2 + \frac{1}{3}(x - 1)^3 - \frac{1}{4}(x - 1)^4$$

Replacing $x$ by 1.1 then gives the required approximation

$$\ln 1.1 \approx 0.1 - \frac{1}{2}(0.1)^2 + \frac{1}{3}(0.1)^3 - \frac{1}{4}(0.1)^4$$

$$\approx 0.09530833$$

**b.** The error in the approximation is found by using Equation (2) with $n = 4$, $c = 1$, and $x = 1.1$. Thus,

$$R_4(1.1) = \frac{f^{(5)}(z)}{5!}(1.1 - 1)^5 = \frac{(0.1)^5}{5z^5}$$

where $1 < z < 1.1$. The largest possible value of $R_4(1.1)$ is obtained when $z = 1$. (The denominator of $R_4(1.1)$ is smallest for this value of $z$ in the interval $[1, 1.1]$.) Therefore,

$$R_4(1.1) < \frac{(0.1)^5}{5} = 0.000002$$

so the error in the approximation is less than 0.000002.

**Alternative Solution**    By replacing $x$ by $x - 1$ in Formula (5) in Section 8.8, we obtain the following power series representation of $f(x)$:

$$\ln x = (x - 1) - \frac{1}{2}(x - 1)^2 + \frac{1}{3}(x - 1)^3 - \frac{1}{4}(x - 4)^4 + \frac{1}{5}(x - 1)^5 - \cdots \qquad 0 < x \le 2$$

Therefore,

$$\ln 1.1 = 0.1 - \frac{1}{2}(0.1)^2 + \frac{1}{3}(0.1)^3 - \frac{1}{4}(0.1)^4 + \frac{1}{5}(0.1)^5 - \cdots$$

If we use just the first four terms of the series on the right to approximate $\ln 1.1$, we obtain the approximation of $\ln 1.1$ by $P_4(1.1)$. Next, since the series is an *alternating series* with terms decreasing to 0, the error in this approximation is no larger than $\frac{1}{5}(0.1)^5$, the first term that is omitted—that is, no larger than 0.000002, which is in agreement with the result that we obtained earlier.    ■

---

**EXAMPLE 2**    Let $f(x) = \sqrt{x}$.

**a.** Find the Taylor polynomial $P_2(x)$ of degree 2 at $c = 4$.
**b.** What is the maximum error incurred if $f$ is approximated by $P_2(x)$ on the interval $[3, 5]$?

**Solution**
**a.** The first two derivatives of $f(x) = \sqrt{x}$ are

$$f'(x) = \frac{1}{2}x^{-1/2} \qquad \text{and} \qquad f''(x) = -\frac{1}{4}x^{-3/2}$$

and the values of $f(x)$ and its first two derivatives at $x = 4$ are

$$f(4) = 2, \qquad f'(4) = \frac{1}{4}, \qquad \text{and} \qquad f''(4) = -\frac{1}{32}$$

Therefore, the required Taylor polynomial is

$$P_2(x) = f(4) + f'(4)(x - 4) + \frac{f''(4)}{2!}(x - 4)^2$$

$$= 2 + \frac{1}{4}(x - 4) - \frac{1}{64}(x - 4)^2$$

**b.** The Taylor remainder is

$$R_2(x) = \frac{f'''(z)}{3!}(x - 4)^3$$

where $z$ lies between 4 and $x$. But

$$f'''(x) = \frac{3}{8}x^{-5/2}$$

so

$$R_2(x) = \frac{3}{8}z^{-5/2}\frac{(x - 4)^3}{3!} = \frac{(x - 4)^3}{16z^{5/2}}$$

Now, if $x$ lies in the interval $[3, 5]$, then $3 \le x \le 5$, so $-1 \le x - 4 \le 1$, or $|x - 4| \le 1$. Furthermore, since $z > 3$, we see that

$$z^{5/2} > 3^{5/2} > 15$$

so a bound on the error incurred in approximating $f$ by $P_2$ on the interval $[3, 5]$ is

$$|R_2(x)| = \frac{|x - 4|^3}{16z^{5/2}} < \frac{1}{16 \cdot 15} < 0.0042 \qquad \blacksquare$$

**EXAMPLE 3**   Determine the degree of the Maclaurin polynomial of $f(x) = e^x$ that allows us to find the value of $\sqrt{e}$ to within an accuracy of 0.0001. Then use the polynomial to obtain the approximation.

**Solution**   We are required to estimate the value of $\sqrt{e} = e^{1/2} = f\left(\frac{1}{2}\right)$. Since $f^{(n)}(x) = e^x$ for all $n$, we see that the error in approximating $f(x)$ by $P_n(x)$ is

$$R_n(x) = \frac{f^{(n+1)}(z)}{(n + 1)!}x^{n+1} = \frac{e^z}{(n + 1)!}x^{n+1}$$

where $z$ lies between $c = 0$ and $x$. We are interested in approximating $e^{1/2}$, so we take $x = \frac{1}{2}$. Then, $0 < z < \frac{1}{2}$. Because $g(z) = e^z$ is an increasing function of $z$, we see that

$$e^z < e^{1/2} < 4^{1/2} = 2$$

Therefore,

$$R_n\left(\frac{1}{2}\right) < \frac{e^{1/2}}{(n + 1)!}\left(\frac{1}{2}\right)^{n+1} < \frac{2}{(n + 1)! \, 2^{n+1}} = \frac{1}{(n + 1)! \, 2^n}$$

Let's try $n = 4$. We obtain

$$R_4\left(\frac{1}{2}\right) < \frac{1}{5! \, 2^4} \approx 0.0005$$

Since this bound is not within the specified error bound of 0.0001, we next try $n = 5$, obtaining

$$R_5\left(\frac{1}{2}\right) < \frac{1}{6! \, 2^5} \approx 0.00004$$

This bound is less than the prescribed error bound, so we can use

$$P_5(x) = 1 + x + \frac{x^2}{2!} + \frac{x^3}{3!} + \frac{x^4}{4!} + \frac{x^5}{5!}$$

for the approximation, obtaining

$$e^{1/2} \approx P_5\left(\frac{1}{2}\right) = 1 + \frac{1}{2} + \frac{1}{2!}\left(\frac{1}{2}\right)^2 + \frac{1}{3!}\left(\frac{1}{2}\right)^3 + \frac{1}{4!}\left(\frac{1}{2}\right)^4 + \frac{1}{5!}\left(\frac{1}{2}\right)^5$$

$$\approx 1.64870 \qquad \blacksquare$$

As we saw earlier, the approximation of a function $f$ by the Taylor polynomial $P_n$ of $f$ centered at $c$ diminishes in accuracy as we move away from the center. Therefore, in approximating $f(x_0)$ by Taylor polynomials of $f$, it is best to pick the center $c$ as close to $x_0$ as possible. This is illustrated in the following example.

**EXAMPLE 4**  Suppose that we want to approximate $\cos 50°$ using the second-order Taylor polynomial $P_2(x)$ of $f(x) = \cos x$ with center at $x = \pi/4$. (Note that $\pi/4$ rad $= 45°$ is close to $50°$.)

**a.** Find $P_2(x)$.

**b.** Find the maximum error in the approximation $f(x) = P_2(x)$ when $|x - (\pi/4)| < 0.1$.

**c.** Use the results of parts (a) and (b) to find $\cos 50°$. How accurate is your estimate?

**Solution**

**a.** Since

$$f(x) = \cos x, \qquad f'(x) = -\sin x, \qquad f''(x) = -\cos x, \qquad \text{and} \qquad f'''(x) = \sin x$$

we find

$$f\left(\frac{\pi}{4}\right) = \frac{1}{\sqrt{2}}, \qquad f'\left(\frac{\pi}{4}\right) = -\frac{1}{\sqrt{2}}, \qquad \text{and} \qquad f''\left(\frac{\pi}{4}\right) = -\frac{1}{\sqrt{2}}$$

Therefore, the required Taylor polynomial is

$$P_2(x) = f\left(\frac{\pi}{4}\right) + f'\left(\frac{\pi}{4}\right)\left(x - \frac{\pi}{4}\right) + \frac{1}{2}f''\left(\frac{\pi}{4}\right)\left(x - \frac{\pi}{4}\right)^2$$

$$= \frac{1}{\sqrt{2}} - \frac{1}{\sqrt{2}}\left(x - \frac{\pi}{4}\right) - \frac{1}{2\sqrt{2}}\left(x - \frac{\pi}{4}\right)^2$$

$$= \frac{1}{\sqrt{2}}\left[1 - \left(x - \frac{\pi}{4}\right) - \frac{1}{2}\left(x - \frac{\pi}{4}\right)^2\right]$$

**b.** The error in the approximation $f(x) \approx P_2(x)$ is

$$R_2(x) = \frac{f'''(z)}{3!}\left(x - \frac{\pi}{4}\right)^3 = \frac{\sin z}{6}\left(x - \frac{\pi}{4}\right)^3$$

where $z$ lies between $\pi/4$ and $x$. Now $|\sin z| \leq 1$ for any $z$, and if $|x - (\pi/4)| < 0.1$, then

$$|R_2(x)| = \frac{|\sin z|}{6}\left|x - \frac{\pi}{4}\right|^3 < \frac{(0.1)^3}{6} \approx 0.000167 < 0.0002$$

Therefore, the maximum error in approximating $f(x)$ by $P_2(x)$ for $x$ satisfying $|x - (\pi/4)| < 0.1$ is less than 0.0002.

**c.** Since

$$5° = \frac{5\pi}{180} = \frac{\pi}{36}$$

we see that

$$50° = 45° + 5° = \frac{\pi}{4} + \frac{\pi}{36}$$

Therefore, using the result of part (a), we have

$$\cos 50° = \cos\left(\frac{\pi}{4} + \frac{\pi}{36}\right) = \frac{1}{\sqrt{2}}\left[1 - \frac{\pi}{36} - \frac{1}{2}\left(\frac{\pi}{36}\right)^2\right] \approx 0.643$$

Since $\pi/36 \approx 0.0873 < 0.1$, the results of part (b) guarantee that the error in the approximation that we just obtained is accurate to three decimal places. The true value of $\cos 50°$ is approximately 0.642787610.    ■

### ■ Representing a Function by a Series

We now turn our attention to finding the conditions under which the function $f$ has a power series representation. These conditions are spelled out in the following theorem.

---

**THEOREM 2**

Suppose that $f$ has derivatives of all order on an interval $I$ containing $c$ and that $R_n(x)$ is the Taylor remainder of $f$ at $c$. If

$$\lim_{n \to \infty} R_n(x) = 0$$

for every $x$ in $I$, then $f(x)$ is represented by the Taylor series of $f$ at $c$; that is,

$$f(x) = \sum_{n=0}^{\infty} \frac{f^{(n)}(c)}{n!}(x - c)^n$$

---

**PROOF**    As was noted earlier, $P_n(x)$ is the $n$th partial sum of the Taylor series of $f$ at $c$. By Taylor's Theorem, $P_n(x) = f(x) - R_n(x)$, and therefore,

$$\lim_{n \to \infty} P_n(x) = \lim_{n \to \infty}[f(x) - R_n(x)] = \lim_{n \to \infty} f(x) - \lim_{n \to \infty} R_n(x)$$
$$= f(x) - \lim_{n \to \infty} R_n(x) = f(x) - 0 = f(x)$$

for all $x$ in $I$. Thus, the sequence of partial sums converges to $f(x)$ for each $x$ in $I$, and the theorem is proved.    ■

Before looking at an application of Theorem 2, we state the following result, which will be used in the solution.

---

**THEOREM 3**

If $x$ is any real number, then

$$\lim_{n \to \infty} \frac{|x|^n}{n!} = 0$$

---

**PROOF**   In Example 1 in Section 8.8 we proved that the power series $\sum_{n=0}^{\infty} x^n/n!$ is absolutely convergent for every real number $x$. Since the $n$th term of a convergent series must approach zero as $n$ approaches infinity (Theorem 2 in Section 8.2), we conclude that

$$\lim_{n \to \infty} \frac{|x|^n}{n!} = 0$$  ∎

**EXAMPLE 5**   Show that the Maclaurin series $\sum_{n=0}^{\infty} x^n/n!$ of the function $f(x) = e^x$ does represent $f$.

**Solution**   We use Taylor's Theorem with $c = 0$. Since $f^{(n+1)}(x) = e^x$, we see that

$$R_n(x) = \frac{f^{(n+1)}(z)}{(n+1)!} x^{n+1} = \frac{e^z}{(n+1)!} x^{n+1}$$

where $z$ is a number between $0$ and $x$. If $x > 0$, then $e^z < e^x$, since the function $f(x) = e^x$ is increasing. Therefore,

$$0 < R_n(x) < \frac{e^x}{(n+1)!} x^{n+1}$$

By Theorem 3,

$$\lim_{n \to \infty} \frac{e^x}{(n+1)!} x^{n+1} = e^x \lim_{n \to \infty} \frac{x^{n+1}}{(n+1)!} = 0$$

so the Squeeze Theorem implies that

$$\lim_{n \to \infty} R_n(x) = 0$$

If $x < 0$, then $z < 0$, and hence $e^z < e^0 = 1$. Therefore,

$$0 < |R_n(x)| < \left| \frac{x^{n+1}}{(n+1)!} \right| = \frac{|x|^{n+1}}{(n+1)!}$$

and once again, the Squeeze Theorem implies that

$$\lim_{n \to \infty} R_n(x) = 0$$

It follows from Theorem 2 that the Maclaurin series of $f(x) = e^x$ represents the function $f$ for all $x \neq 0$. Finally, the series represents $f$ at $x = 0$, since $f(0) = e^0 = 1$, and this is also the value of the sum of the series at $0$.  ∎

**EXAMPLE 6** Show that the Maclaurin series $\sum_{n=0}^{\infty} (-1)^n \dfrac{x^{2n+1}}{(2n+1)!}$ of the function $f(x) = \sin x$ does represent $f$.

**Solution** Using Taylor's Theorem with $c = 0$, we have

$$R_n(x) = \frac{f^{(n+1)}(z)}{(n+1)!} x^{n+1}$$

where $z$ is a number between $0$ and $x$. But $f^{(n+1)}(x)$ is either $\pm\sin x$ or $\pm\cos x$ for any $n$ ($n = 0, 1, 2, \ldots$). Therefore, $|f^{(n+1)}(z)| \leq 1$, so

$$|R_n(x)| = \frac{|f^{(n+1)}(z)|}{(n+1)!} |x|^{n+1} \leq \frac{|x|^{n+1}}{(n+1)!}$$

By Theorem 3,

$$\lim_{n\to\infty} \frac{|x|^{n+1}}{(n+1)!} = 0$$

so the Squeeze Theorem implies that

$$\lim_{n\to\infty} R_n(x) = 0$$

It follows from Theorem 2 that

$$f(x) = \sum_{n=0}^{\infty} (-1)^n \frac{x^{2n+1}}{(2n+1)!}$$

as was to be shown. ■

The next example shows how a Taylor polynomial can be used to approximate an integral that involves an integrand whose antiderivative cannot be expressed as an elementary function.

**EXAMPLE 7** **Growth of the Service Industries** It has been estimated that service industries, which currently make up 30% of the nonfarm workforce in a certain country, will continue to grow at the rate of

$$R(t) = 5e^{1/(t+1)}$$

percent per decade, $t$ decades from now. Estimate the percentage of the nonfarm workforce in the service industries one decade from now.

**Solution** The percentage of the nonfarm workforce in the service industries $t$ decades from now will be given by

$$P(t) = \int 5e^{1/(t+1)} \, dt \qquad P(0) = 30$$

This integral cannot be expressed in terms of an elementary function. To obtain an approximate solution to the problem at hand, let's first make the substitution

$$u = \frac{1}{t+1}$$

So that

$$t + 1 = \frac{1}{u} \qquad \text{and} \qquad t = \frac{1}{u} - 1$$

giving

$$dt = -\frac{1}{u^2}\, du$$

The integral becomes

$$F(u) = 5 \int e^u \left(-\frac{du}{u^2}\right) = -5 \int \frac{e^u}{u^2}\, du$$

Next, let's approximate $e^u$ at $u = 0$ by a fourth-degree Taylor polynomial. Using Formula 2 in Section 8.8, we have

$$e^u \approx 1 + u + \frac{u^2}{2!} + \frac{u^3}{3!} + \frac{u^4}{4!}$$

Thus,

$$
\begin{aligned}
F(u) &\approx -5 \int \frac{1}{u^2}\left(1 + u + \frac{u^2}{2} + \frac{u^3}{6} + \frac{u^4}{24}\right) du \\
&= -5 \int \left(\frac{1}{u^2} + \frac{1}{u} + \frac{1}{2} + \frac{u}{6} + \frac{u^2}{24}\right) du \\
&= -5\left(-\frac{1}{u} + \ln u + \frac{1}{2}u + \frac{u^2}{12} + \frac{u^3}{72}\right) + C
\end{aligned}
$$

Therefore,

$$P(t) \approx -5\left[-(t+1) + \ln\left(\frac{1}{t+1}\right) + \frac{1}{2(t+1)} + \frac{1}{12(t+1)^2} + \frac{1}{72(t+1)^3}\right] + C$$

Using the condition $P(0) = 30$, we find

$$30 = P(0) \approx -5\left(-1 + \ln 1 + \frac{1}{2} + \frac{1}{12} + \frac{1}{72}\right) + C$$

or $C \approx 27.99$. So

$$P(t) \approx -5\left[-(t+1) + \ln\left(\frac{1}{t+1}\right) + \frac{1}{2(t+1)} + \frac{1}{12(t+1)^2} + \frac{1}{72(t+1)^3}\right] + 27.99$$

In particular, the percentage of the nonfarm workforce in the service industries one decade from now will be given by

$$P(1) \approx -5\left[-2 + \ln\left(\frac{1}{2}\right) + \frac{1}{4} + \frac{1}{48} + \frac{1}{576}\right] + 27.99 \approx 40.09$$

or approximately 40.1%.

## 8.9    CONCEPT QUESTIONS

1. What is the $n$th-degree Taylor polynomial of $f$ at $c$? What is the $n$th-degree Maclaurin polynomial of $f$?
2. Match each of the Taylor polynomials with the graph of the appropriate function $f$, $g$, or $h$.

   **a.** $1 + \dfrac{2}{3}(x - 1) - \dfrac{1}{9}(x - 1)^2 + \dfrac{4}{81}(x - 1)^3$

   **b.** $1 + \dfrac{1}{8}(x - 1)^2 - \dfrac{1}{8}(x - 1)^3$

   **c.** $1 - \dfrac{1}{4}(x - 1)^2 + \dfrac{1}{8}(x - 1)^3$

   **(1)**

   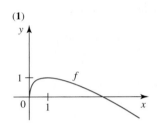

   **(2)**

   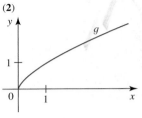

   **(3)**

   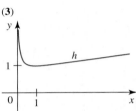

3. Write the expression for the Taylor remainder $R_n(x)$ of $f$ at $c$.
4. State the conditions under which a function $f$ will have a power series representation at $c$.

## 8.9    EXERCISES

*In Exercises 1 and 2, find the $n$th-order Taylor polynomial $P_n(x)$ at $c$ for the function $f$ and the values of $n$. Then plot the graphs of $f$ and the approximating polynomials on the same set of axes.*

1. $f(x) = e^{-x}$,    $c = 0$,    $n = 1, 2, 3$
2. $f(x) = \sin x$,    $c = 0$,    $n = 1, 2, 3, 4$

*In Exercises 3–16, find the Taylor polynomial $P_n(x)$ and the Taylor remainder $R_n(x)$ for the function $f$ and the values of $c$ and $n$.*

3. $f(x) = 2x^3 + 3x^2 + x + 1$,    $c = 1$,    $n = 4$
4. $f(x) = x^4 + 3x^3 + 2x + 3$,    $c = -1$,    $n = 4$

5. $f(x) = \sin x$,    $c = \dfrac{\pi}{2}$,    $n = 3$

6. $f(x) = \cos x$,    $c = \dfrac{\pi}{6}$,    $n = 3$

7. $f(x) = \tan x$,    $c = \dfrac{\pi}{4}$,    $n = 2$

8. $f(x) = \sqrt{x}$,    $c = 4$,    $n = 3$

9. $f(x) = \sqrt[3]{x}$,    $c = -8$,    $n = 3$

10. $f(x) = \dfrac{1}{x}$,    $c = -1$,    $n = 5$

11. $f(x) = \tan^{-1} x$,    $c = 1$,    $n = 2$

12. $f(x) = \ln x$,    $c = 4$,    $n = 3$

13. $f(x) = xe^x$,    $c = -1$,    $n = 3$

14. $f(x) = e^x \sin x$,    $c = 0$,    $n = 2$

15. $f(x) = e^x \cos 2x$,    $c = \dfrac{\pi}{6}$,    $n = 2$

16. $f(x) = \ln \sin x$,    $c = \dfrac{\pi}{6}$,    $n = 3$

*In Exercises 17–28, find the Taylor or Maclaurin polynomial $P_n(x)$ for the function $f$ with the given values of $c$ and $n$. Then give a bound on the error that is incurred if $P_n(x)$ is used to approximate $f(x)$ on the given interval.*

17. $f(x) = x^4 - 1$,    $c = 1$,    $n = 2$,    $[0.8, 1.2]$

18. $f(x) = \sin x$,    $c = \dfrac{\pi}{6}$,    $n = 5$,    $\left[0, \frac{\pi}{3}\right]$

**19.** $f(x) = \cos x$, $c = \dfrac{\pi}{4}$, $n = 4$, $\left[0, \dfrac{\pi}{2}\right]$

**20.** $f(x) = x^{1/3}$, $c = 1$, $n = 3$, $[0.8, 1.2]$

**21.** $f(x) = e^{2x}$, $c = 1$, $n = 4$, $[1, 1.1]$

**22.** $f(x) = e^{-x^2}$, $c = 0$, $n = 2$, $[0, 0.1]$

**23.** $f(x) = \sqrt{x}$, $c = 9$, $n = 3$, $[8, 10]$

**24.** $f(x) = \dfrac{1}{x}$, $c = 1$, $n = 5$, $[0.9, 1.1]$

**25.** $f(x) = \tan x$, $c = 0$, $n = 3$, $\left[0, \dfrac{\pi}{4}\right]$

**26.** $f(x) = \sec x$, $c = 0$, $n = 2$, $\left[0, \dfrac{\pi}{6}\right]$

**27.** $f(x) = \ln(x + 1)$, $c = 3$, $n = 3$, $[2, 4]$

**28.** $f(x) = \cosh x$, $c = 0$, $n = 5$, $[-1, 1]$

*In Exercises 29–38, find the Taylor polynomial of smallest degree of an appropriate function about a suitable point to approximate the given number to within the indicated accuracy.*

**29.** $e^{0.2}$, $\quad 0.0001$

**30.** $e^{-1/2}$, $\quad 0.0002$

**31.** $\sqrt{9.01}$, $\quad 0.00005$

**32.** $\sqrt[3]{-8.2}$, $\quad 0.000005$

**33.** $-\dfrac{1}{2.1}$, $\quad 0.0005$

**34.** $\ln 1.2$, $\quad 0.0001$

**35.** $\sin 0.1$, $\quad 0.00001$

**36.** $\cos 0.5$, $\quad 0.0005$

**37.** $\cos 32°$, $\quad 0.0001$

**38.** $\sin 69°$, $\quad 0.0001$

*In Exercises 39–44, prove that the given Taylor (Maclaurin) series does represent the function.*

**39.** $\displaystyle\sum_{n=0}^{\infty} (-1)^n \dfrac{x^n}{n!}$, $\quad f(x) = e^{-x}$

**40.** $\displaystyle\sum_{n=0}^{\infty} (-1)^n \dfrac{x^{2n}}{(2n)!}$, $\quad f(x) = \cos x$

**41.** $\dfrac{1}{\sqrt{2}} \displaystyle\sum_{n=0}^{\infty} (-1)^{n(n-1)/2} \dfrac{\left(x - \frac{\pi}{4}\right)^n}{n!}$, $\quad f(x) = \sin x$

**42.** $\displaystyle\sum_{n=0}^{\infty} \dfrac{x^{2n+1}}{(2n+1)!}$, $\quad f(x) = \sinh x$

**43.** $\displaystyle\sum_{n=0}^{\infty} \dfrac{x^{2n}}{(2n)!}$, $\quad f(x) = \cosh x$

**44.** $\dfrac{\sqrt{3}}{2} \displaystyle\sum_{n=0}^{\infty} (-1)^n \dfrac{\left(x - \frac{\pi}{6}\right)^{2n+1}}{(2n+1)!} + \dfrac{1}{2} \displaystyle\sum_{n=0}^{\infty} (-1)^n \dfrac{\left(x - \frac{\pi}{6}\right)^{2n}}{(2n)!}$,
$f(x) = \sin x$

**45. Growth of Service Industries** It has been estimated that service industries, which currently make up 30% of the nonfarm workforce in a certain country, will continue to grow at the rate of

$$R(t) = 6e^{1/(2t+1)}$$

percent per decade, $t$ decades from now. Estimate the percentage of the nonfarm workforce in service industries two decades from now.

**46. Concentration of Carbon Monoxide in the Air** According to a joint study conducted by a certain city's Environmental Management Department and a state government agency, the concentration of carbon monoxide (CO) in the air due to automobile exhaust $t$ years from now is given by

$$C(t) = 0.01(0.2t^2 + 4t + 64)^{2/3}$$

parts per million. Use the second Taylor polynomial of $C$ at $t = 0$ to obtain an approximation of the average level of concentration of CO in the air between $t = 0$ and $t = 2$.

**47.** Show that $y = P_1(x)$, where $P_1$ is the first-order Taylor polynomial of $f$ at $c$, is an equation of the tangent line to the graph of $f$ at the point $(c, f(c))$.

**48.** Let $P_n(x)$ be the $n$th-order Taylor polynomial of $f$ at $c$. Show that $P_n$ and $f$ have the same derivatives at $c$ up to order $n$.

**49.** Prove that

$$x - \dfrac{x^2}{2} < \ln(1 + x) < x$$

if $x > 0$.
**Hint:** Use Taylor's Theorem with $c = 0$ and $n = 1$ and $n = 2$.

**50.** Show that the error that is incurred in approximating $\sin(c + h)$ by $\sin c + h \cos c$ does not exceed $h^2/2$.

*In Exercises 51–54, determine whether the statement is true or false. If it is true, explain why it is true. If it is false, explain why or give an example to show why it is false.*

**51.** If $f$ is a polynomial function of degree $n$, then the Maclaurin polynomial of degree $n$ of $f$ is $f$ itself.

**52.** If $f(x) = e^x$ and $P_n(x)$ is the $n$th-degree Maclaurin polynomial of $f$, then $P_n(0.1) = e^{0.1}$ for some positive integer $n$.

**53.** The inequality $1 + x \le e^x$ holds for all real values of $x$.

**54.** The binomial expansion in the Binomial Theorem is the Maclaurin polynomial of $f(x) = (1 + x)^n$, where $n$ is a positive integer.

# CHAPTER 8 REVIEW

## CONCEPT REVIEW

*In Exercises 1–10, fill in the blanks.*

**1. a.** A sequence is a _____ whose domain is the set of positive _____. The term $a_n$ is called the _____ _____ of the sequence.

**b.** If $a_n$ can be made as close to the number $L$ as we please by taking $n$ sufficiently large, then $\{a_n\}$ is said to _____ to $L$.

**c.** The precise definition of a limit states that $\lim_{n\to\infty} a_n = L$ if _____ _____ $\varepsilon > 0$ there exists a _____ $N$ such that $|a_n - L| < \varepsilon$ whenever _____.

**2. a.** If $\lim_{n\to\infty} a_n = L$, $\lim_{n\to\infty} b_n = M$, and $c$ is any real number, then $\lim_{n\to\infty} ca_n = $ _____, $\lim_{n\to\infty}(a_n + b_n) = $ _____, $\lim_{n\to\infty} a_n b_n = $ _____, and $\lim_{n\to\infty} \dfrac{a_n}{b_n} = $ _____, provided that _____.

**b.** If there exists some integer $N$ such that $a_n \le b_n \le c_n$ for all $n \ge N$ and $\lim_{n\to\infty} a_n = $ _____ $= L$, then $\lim_{n\to\infty} b_n = L$.

**3. a.** A series $\sum_{n=1}^{\infty} a_n$ converges and has sum $S$ if its sequence of _____ _____ _____ converges to $S$.

**b.** The series $\sum_{n=1}^{\infty} ar^{n-1}$, $a \ne 0$, is called a _____ series. It converges if $|r| < $ _____ and diverges if $|r| \ge $ _____.

**4. a.** If $\sum_{n=1}^{\infty} a_n$ converges, then $\lim_{n\to\infty} a_n = $ _____. If $\lim_{n\to\infty} a_n$ does not exist or $\lim_{n\to\infty} a_n \ne 0$, then $\sum_{n=1}^{\infty} a_n$ _____.

**b.** If $\sum_{n=1}^{\infty} a_n = A$, $\sum_{n=1}^{\infty} b_n = B$, and $c$ is any real number, then $\sum_{n=1}^{\infty}(ca_n + b_n) = $ _____.

**5. a.** If $f$ is positive, continuous, and decreasing, and if $a_n = f(n)$, then $\sum_{n=1}^{\infty} a_n$ and $\int_1^{\infty} f(x)\, dx$ are either both _____ or _____.

**b.** The $p$-series has the form _____ and converges if _____ and diverges if _____.

**6. a.** If $\sum a_n$ and $\sum b_n$ are series with positive terms with $a_n \le b_n$ for all $n$, then $\sum b_n$ converges implies that $\sum a_n$ _____. If $\sum b_n$ diverges and _____ for all $n$, then $\sum a_n$ also diverges.

**b.** If $\sum a_n$ and $\sum b_n$ are series with positive terms and $\lim_{n\to\infty} \dfrac{a_n}{b_n} = L$, where $L$ is _____ _____ _____, then either both series _____ or both _____.

**7. a.** The series $\sum_{n=1}^{\infty}(-1)^{n-1} a_n$ is called _____ _____ series. It converges if both the conditions $a_{n+1}$ _____ $a_n$ for all $n$ and $\lim_{n\to\infty} a_n = $ _____ are satisfied.

**b.** If both the conditions in part (a) are satisfied, then the error that is incurred in approximating the sum of the alternating series by $S_n$ is no larger than _____.

**8. a.** A series $\sum a_n$ is absolutely convergent if the series _____ converges.

**b.** A series $\sum a_n$ is _____ convergent if it is convergent but not _____ convergent.

**c.** A(n) _____ convergent series is convergent.

**d.** Suppose that $\lim_{n\to\infty} |a_{n+1}/a_n| = L$. Then if $L < 1$, the series _____; if $L > 1$ or $L = \infty$, the series _____; and if $L = 1$, the Ratio Test is _____.

**e.** Suppose that $\lim_{n\to\infty} \sqrt[n]{|a_n|} = L$. Then if $L < 1$, the series _____; if $L > 1$ or $L = \infty$, the series _____; and if $L = 1$, the Root Test is _____.

**9. a.** A power series in $(x - c)$ is a series of the form _____.

**b.** For a power series in $(x - c)$, exactly one of the following is true: It converges only at _____, it converges for all _____, or it converges for $|x - c| < R$, where $R$ is the radius of convergence of the series. In the last case the series diverges for $|x - c| > R$.

**10. a.** The Taylor series of a function $f$ at $c$ has the form _____.

**b.** The $n$th-degree Taylor polynomial $P_n$ of $f$ at $c$ is the $n$th _____ _____ of the Taylor series at $c$.

**c.** Taylor's Theorem states that if $f$ has derivatives up to order $n + 1$ in an interval $I$ containing _____, then for each $x$ in _____ there exists a number $z$ between _____ _____ _____ such that $f(x) = P_n(x) + R_n(x)$ where $R_n(x) = $ _____.

**d.** If $f$ has derivatives of all order in $I$ and $\lim_{n\to\infty} R_n(x) = 0$, then $f$ is represented by the _____ _____ of $f$ at $c$.

## REVIEW EXERCISES

*In Exercises 1–8, determine whether the sequence with the given* n*th term converges or diverges. If it converges, find the limit.*

**1.** $a_n = \dfrac{n}{3n - 2}$

**2.** $a_n = \dfrac{n + 1}{2n^2}$

**3.** $a_n = 2 + 3(0.9)^n$

**4.** $a_n = 10(-1.01)^n$

**5.** $a_n = \dfrac{n}{\ln n}$

**6.** $a_n = \dfrac{\ln(n^2 + 1)}{n}$

**7.** $a_n = \dfrac{\cos n}{n}$

**8.** $a_n = \left(1 + \dfrac{3}{n}\right)^{2n}$

*In Exercises 9–12, find the sum of the series.*

**9.** $\displaystyle\sum_{n=0}^{\infty} \left(\dfrac{2}{3}\right)^n$

**10.** $\displaystyle\sum_{n=0}^{\infty} \left(\dfrac{1}{3^n} - \dfrac{1}{4^{n+1}}\right)$

**11.** $\displaystyle\sum_{n=1}^{\infty} \dfrac{1}{n(n + 3)}$

**12.** $\displaystyle\sum_{n=1}^{\infty} \left[\left(\dfrac{3}{5}\right)^n - \dfrac{1}{n(n + 1)}\right]$

*In Exercises 13–26, determine whether the series is convergent or divergent.*

**13.** $\displaystyle\sum_{n=1}^{\infty} \dfrac{n}{2n^3 + 1}$

**14.** $\displaystyle\sum_{n=1}^{\infty} \dfrac{n^3 + 1}{2n^3 - 1}$

**15.** $\displaystyle\sum_{n=1}^{\infty} \dfrac{n^3}{2^n}$

**16.** $\displaystyle\sum_{n=1}^{\infty} \dfrac{(-1)^n}{\sqrt[3]{n + 1}}$

**17.** $\displaystyle\sum_{n=1}^{\infty} \dfrac{1}{\sqrt{n^3 + n}}$

**18.** $\displaystyle\sum_{n=1}^{\infty} \dfrac{\sin n}{n^2 + 1}$

**19.** $\displaystyle\sum_{n=1}^{\infty} \dfrac{n + \cos n}{n^3 + 1}$

**20.** $\displaystyle\sum_{n=2}^{\infty} \dfrac{(-1)^n \ln n}{n}$

**21.** $\displaystyle\sum_{n=1}^{\infty} \dfrac{(-1)^n 3^n}{n \cdot 2^n}$

**22.** $\displaystyle\sum_{n=1}^{\infty} \dfrac{e^n}{(n + 1)^{2n}}$

**23.** $\displaystyle\sum_{n=2}^{\infty} \dfrac{1}{n(\ln n)^2}$

**24.** $\displaystyle\sum_{n=1}^{\infty} \dfrac{\tan^{-1} n}{\sqrt{n^2 + 1}}$

**25.** $\displaystyle\sum_{n=1}^{\infty} \dfrac{1 \cdot 3 \cdot 5 \cdot \cdots \cdot (2n - 1)}{2 \cdot 5 \cdot 8 \cdot \cdots \cdot (3n - 1)}$

**26.** $\displaystyle\sum_{n=1}^{\infty} \dfrac{1 \cdot 3 \cdot 5 \cdot \cdots \cdot (2n - 1)}{n! \, 3^n}$

*In Exercises 27–32, determine whether the series is absolutely convergent, conditionally convergent, or divergent.*

**27.** $\displaystyle\sum_{n=1}^{\infty} \dfrac{(-1)^{n-1}}{2n + 1}$

**28.** $\displaystyle\sum_{n=2}^{\infty} \dfrac{(-1)^n}{n \ln n}$

**29.** $\displaystyle\sum_{n=2}^{\infty} \dfrac{(-1)^n}{(\ln n)^{n/2}}$

**30.** $\displaystyle\sum_{n=1}^{\infty} \dfrac{(-1)^n \tan^{-1} n}{n^2 + 1}$

**31.** $\displaystyle\sum_{n=1}^{\infty} \dfrac{(-1)^n \sqrt{n}}{2n + 1}$

**32.** $\displaystyle\sum_{n=1}^{\infty} (-1)^n \dfrac{1 \cdot 3 \cdot 5 \cdot \cdots \cdot (2n + 1)}{2 \cdot 5 \cdot 8 \cdot \cdots \cdot (3n + 2)}$

**33.** Express $1.3\overline{617} = 1.3617617617\ldots$ as a rational number.

**34.** Find an approximation of the sum of the series $\displaystyle\sum_{n=1}^{\infty} \dfrac{(-1)^{n-1}}{n^3}$ accurate to three decimal places.

**35.** True or false? If $\lim_{n\to\infty} a_n \neq 0$, then $\Sigma \, a_n$ may converge conditionally but not absolutely.

**36.** True or false? If $0 \le a_n \le b_n$ and $\Sigma \, b_n$ diverges, then $\Sigma \, a_n$ diverges.

**37.** True or false? If $\Sigma \, a_n$ diverges, then $\Sigma |a_n|$ also diverges.

**38.** True or false? If $\Sigma_{n=1}^{\infty} a_n$ diverges and $a_n \ge 0$ for every $n$, then $\lim_{n\to\infty} S_n = \lim_{n\to\infty} \Sigma_{k=1}^{n} a_k = \infty$.

**39.** Find all values of $x$ for which the series $\Sigma_{n=1}^{\infty}(\cos x)^n$ converges.

**40.** Show that $\lim_{n\to\infty} nx^n = 0$ if $|x| < 1$.
**Hint:** Show that $\Sigma_{n=1}^{\infty} nx^n$ is convergent.

**41.** Show that $\Sigma_{n=1}^{\infty} a_n \; (a_n > 0)$ is convergent if and only if the sequence of partial sums of $\Sigma \, a_n$, $S_n$, are bounded for $n \ge 1$.

**42.** If $c$ is a nonzero constant and $\Sigma_{n=1}^{\infty} a_n$ diverges, prove that $\Sigma_{n=1}^{\infty} ca_n$ diverges.

*In Exercises 43–48, find the radius of convergence and the interval of convergence of the power series.*

**43.** $\displaystyle\sum_{n=0}^{\infty} \dfrac{(-1)^n x^n}{n + 1}$

**44.** $\displaystyle\sum_{n=1}^{\infty} \dfrac{n^2(x - 2)^n}{2^n}$

**45.** $\displaystyle\sum_{n=0}^{\infty} \dfrac{(-2x)^n}{n^2 + 1}$

**46.** $\displaystyle\sum_{n=1}^{\infty} \dfrac{(x + 2)^n}{n^n}$

**47.** $\displaystyle\sum_{n=2}^{\infty} \dfrac{x^n}{n(\ln n)^2}$

**48.** $\displaystyle\sum_{n=1}^{\infty} \dfrac{(-1)^n \, n!(x - 1)^n}{2 \cdot 4 \cdot 6 \cdot \cdots \cdot (2n)}$

*In Exercises 49–54, find the Taylor series of $f$ at the given value of $c$.*

**49.** $f(x) = \dfrac{x^3}{1 + x}, \quad c = 0$

**50.** $f(x) = xe^{-x^2}, \quad c = 0$

**51.** $f(x) = \cos x^2, \quad c = 0$

**52.** $f(x) = \ln x, \quad c = 3$

**53.** $f(x) = \sqrt{1 + x^2}, \quad c = 0$

**54.** $f(x) = \cos x, \quad c = \dfrac{\pi}{6}$

**55.** Find the radius of convergence of the series $\displaystyle\sum_{n=1}^{\infty} \dfrac{(2n)!}{(n!)^2} x^n$.

**56.** Find the radius of convergence of the series

$$\sum_{n=1}^{\infty} \frac{n^n}{(2n)!} (x - 1)^n.$$

**57.** Find a power series representation of $\displaystyle\int \frac{e^{-x}}{x} dx$.

**58.** Find a power series representation of $\displaystyle\int \frac{e^x - 1}{x} dx$.

**59.** Approximate $\displaystyle\int_0^{0.2} \sqrt{1 - x^2}\, dx$ to three decimal places of accuracy.

**60.** Approximate $\displaystyle\int_0^{0.1} \cos \sqrt{x}\, dx$ to three decimal places of accuracy.

*In Exercises 61 and 62, use a Taylor polynomial to approximate the number with an error of less than 0.001.*

**61.** $e^{-0.25}$               **62.** $\sin 2°$

*In Exercises 63–66, find the Taylor polynomial $P_n(x)$ and the Taylor remainder $R_n(x)$ for the given function f and the given values of c and n.*

**63.** $f(x) = \sqrt{x}$,    $c = 1$,    $n = 3$

**64.** $f(x) = \cos x$,    $c = \dfrac{\pi}{3}$,    $n = 3$

**65.** $f(x) = \csc x$,    $c = \dfrac{\pi}{2}$,    $n = 2$

**66.** $f(x) = \ln \cos x$,    $c = \dfrac{\pi}{6}$,    $n = 2$

**67.** Suppose that $\sum_{n=1}^{\infty} a_n$ and $\sum_{n=1}^{\infty} b_n$ are both convergent series with positive terms. Show that $\sum_{n=1}^{\infty} \sqrt{a_n b_n}$ also converges.
Hint: $(\sqrt{a_n} - \sqrt{b_n})^2 \geq 0$

**68.** Show that the series

$$\frac{1}{\sqrt{2} - 1} - \frac{1}{\sqrt{2} + 1} + \frac{1}{\sqrt{3} - 1} - \frac{1}{\sqrt{3} + 1} + \cdots$$

diverges. Is the Alternating Series Test applicable? Explain.

**69.** It can be shown that the magnetic field at a point $P$ a distance $y$ above the center of a circular loop of radius $R$ and carrying a steady current $I$ is directed upward and has magnitude

$$B = \frac{\mu_0 I}{2} \cdot \frac{R^2}{(R^2 + y^2)^{3/2}}$$

where $\mu_0$ is a constant called the permeability of free space (see the figure). Show that

$$B \approx \frac{\mu_0 I R^2}{2y^3}$$

if $y$ is large in comparison to $R$.

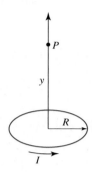

**70. Electric Field Induced by a Line Charge**  The figure below shows a straight line segment of length $2L$ that carries a uniform line charge $\lambda$. It can be shown that the electric field induced by this line charge at a distance $y$ above the origin is directed along the $y$-axis and has magnitude

$$E = \frac{1}{4\pi\varepsilon_0} \cdot \frac{2\lambda L}{y\sqrt{y^2 + L^2}}$$

where $\varepsilon_0$ is a constant called the permittivity of free space. Show that the formula becomes

$$E \approx \frac{1}{4\pi\varepsilon_0} \cdot \frac{2\lambda L}{y^2}$$

if $y$ is large in comparison to $L$. This suggests that the line charge "looks" like a point charge $q = 2\lambda L$ from this distance, so the field reduces to that induced by a point charge $q$, namely, $q/(4\pi\varepsilon_0)y^2$.

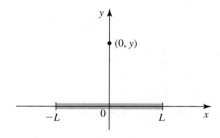

## PROBLEM-SOLVING TECHNIQUES

Although l'Hôpital's Rule is a powerful tool for evaluating limits involving an indeterminate form, it is not always the ideal choice. The following technique illustrates the usefulness of the Taylor series in solving such problems.

**EXAMPLE**  Find $\displaystyle\lim_{x\to 0}\dfrac{e^{x^2}\sin x - x\left(1 + \frac{5}{6}x^2\right)}{x^5}$.

**Solution**  Evaluating the limit, we are led to the indeterminate form $0/0$. An obvious approach is to use l'Hôpital's Rule to find the limit. But we would soon be deterred by the number of calculations (of derivatives) involved in the process. Alternatively, we can solve the problem with the aid of the power series representations of functions. Displaying terms up to those of degree five, we find

$$\lim_{x\to 0}\frac{e^{x^2}\sin x - x\left(1 + \frac{5}{6}x^2\right)}{x^5}$$

$$= \lim_{x\to 0}\frac{\left(1 + x^2 + \dfrac{x^4}{2!} + \cdots\right)\left(x - \dfrac{x^3}{6} + \dfrac{x^5}{120} - \cdots\right) - x - \dfrac{5}{6}x^3}{x^5}$$

$$= \lim_{x\to 0}\frac{x - \dfrac{x^3}{6} + \dfrac{x^5}{120} + x^3 - \dfrac{x^5}{6} + \dfrac{x^5}{2} + \cdots - x - \dfrac{5}{6}x^3}{x^5}$$

$$= \lim_{x\to 0}\frac{\dfrac{41}{120}x^5 + \cdots}{x^5} = \frac{41}{120} \approx 0.342$$

## CHALLENGE PROBLEMS

**1.** Let $x_n = \dfrac{1}{\sqrt{n^2 + 1}} + \dfrac{1}{\sqrt{n^2 + 2}} + \cdots + \dfrac{1}{\sqrt{n^2 + n}}$. Find $\lim_{n\to\infty} x_n$.

   **Hint:** Show that $y_n < x_n < z_n$, where $y_n = \dfrac{n}{\sqrt{n^2 + n}}$ and $z_n = \dfrac{n}{\sqrt{n^2 + 1}}$, and use the Squeeze Theorem.

**2.** Define $f$ by

$$f(x) = \lim_{n\to\infty}\frac{x^{2n} - 1}{x^{2n} + 1}$$

   **a.** Find a rule for $f$ that does not involve a limit.
   **b.** Sketch the graph of $f$.

**3. Cantor Set**  Start with the interval $[0, 1]$, and remove the open middle third $\left(\frac{1}{3}, \frac{2}{3}\right)$. Next, remove the open middle third from each of the two remaining closed intervals, then the open middle third from each of the four remaining closed intervals, and so on.
   **a.** Find an expression for $c_n$, the sum of the lengths of the intervals remaining after $n$ steps.
   **b.** Show that $\lim_{n\to\infty} c_n = 0$. The set $c$ remaining after all the deletions is called the **Cantor middle-third set** and can be said to have total length zero.

**4.** Find $\lim_{n\to\infty} 4n\left(\sqrt{n^2 + 1} - n\right)$.

**5.** Let $a > 0$, and define the sequence $\{x_n\}$ by

$$x_1 = \sqrt{a},\, x_2 = \sqrt{a + \sqrt{a}},\, x_3 = \sqrt{a + \sqrt{a + \sqrt{a}}},\, \ldots$$

$$x_n = \underbrace{\sqrt{a + \sqrt{a + \cdots + \sqrt{a}}}}_{n \text{ radicals}}$$

   Assuming that $\lim_{n\to\infty} x_n$ exists, what is the limit?

**6.** Find the largest term in the sequence $\{a_n\}$, where

$$a_n = \frac{2n^2}{3n^3 + 400}.$$

**7.** Show that $\displaystyle\lim_{n\to\infty}\frac{1}{n^2}\sum_{k=1}^{n}\llbracket kx \rrbracket = \frac{x}{2}$.

**8.** Let $\displaystyle\sum_{n=1}^{\infty} a_n$ be a convergent series with positive terms.

   Show that $\displaystyle\sum_{n=1}^{\infty}\frac{\sqrt{a_n}}{n}$ converges.

   **Hint:** Consider $\displaystyle\sum_{n=1}^{N}\left(\sqrt{a_n} - \frac{1}{n}\right)^2$.

**9.** Show that if $a > 1$ and $k$ is any positive integer, then

$$\lim_{n\to\infty}\frac{n^k}{a^n} = 0.$$

   **Hint:** Show that the series $\displaystyle\sum_{n=1}^{\infty}\frac{n^k}{a^n}$ converges.

**10.** Let $A_0, A_1, A_2, A_3, \ldots$ denote the areas of the regions $R_0, R_1, R_2, R_3, \ldots$ bounded by the $x$-axis and the graph of $f(x) = e^{-ax} \sin bx \ (a > 0)$ for $x \geq 0$.

**a.** Show that

$$A_n = (-1)^n \int_{n\pi/b}^{(n+1)\pi/b} e^{-ax} \sin bx \, dx \qquad n = 0, 1, 2, 3, \ldots$$

**b.** Integrate by parts to show that

$$\int e^{-ax} \sin bx \, dx = -\frac{e^{-ax}}{a^2 + b^2} (a \sin bx + b \cos bx) + C$$

**c.** Using the results of parts (a) and (b), show that

$$A_n = \frac{b}{a^2 + b^2} e^{-na\pi/b} (1 + e^{-a\pi/b}) \qquad n = 0, 1, 2, \ldots$$

**d.** Using the results of part (c), find the sum of the areas of the regions $A_0, A_1, A_2, A_3, \ldots$.

**Hint:** $\sum_{n=0}^{\infty} A_n$ is a geometric series.

**11.** Evaluate $\displaystyle\lim_{x \to 0} \frac{1 - \sqrt{1 + x^2} \cos x}{x^4}$.

**12.** Evaluate $\displaystyle\lim_{n \to \infty} \left( \frac{1 - 2 + 3 - 4 + \cdots - 2n}{\sqrt{4n^2 + 1}} \right)$.

**13.** Find the Maclaurin series of $f(x) = \dfrac{1}{1 + x + x^2 + x^3}$.

**14.** Determine whether $\sum_{n=1}^{\infty} n^{100} e^{-n} \sin n$ is convergent or divergent.

**15.** Find the values of $x$ for which the series $\displaystyle\sum_{n=1}^{\infty} \frac{(-1)^{n-1}}{n + x^2}$ converges.

**16.** Suppose that $f, f'$, and $f''$ are continuous in an interval containing $x$. Show that

$$\lim_{h \to 0} \frac{f(x + 2h) - 2f(x + h) + f(x)}{h^2} = f''(x)$$

**17.** Find the Maclaurin series of $f(x) = (1 + x)^x$ up to the $x^3$ term.

The shape assumed by the cable in a suspension bridge is a parabola. The parabola, like the ellipse and hyperbola, is a curve that is obtained as a result of the intersection of a plane and a double-napped cone, and is, accordingly, called a conic section. Conic sections appear in various fields of study such as astronomy, physics, engineering, and navigation.

James Osmond/Alamy

# 9 Conic Sections, Plane Curves, and Polar Coordinates

CONIC SECTIONS ARE curves that can be obtained by intersecting a double-napped right circular cone with a plane. Our immediate goal is to describe conics using algebraic equations. We then turn to applications of conics, which range from the design of suspension bridges to the design of satellite-signal receiving dishes and to the design of whispering galleries, in which a person standing at one spot in a gallery can hear a whisper coming from another spot in the gallery. The orbits of celestial bodies and human-made satellites can also be described by using conics.

Parametric equations afford a way of describing curves in the plane and in space. We will study these representations and use them to describe the motion of projectiles and the motion of other objects.

Polar coordinates provide an alternative way of representing points in the plane. We will see that certain curves have simpler representations with polar equations than with rectangular equations. We will also make use of polar equations to help us find the arc length of a curve, the area of a region bounded by a curve, and the area of a surface obtained by revolving a curve about a given line.

**V** This symbol indicates that one of the following video types is available for enhanced student learning at **www.academic.cengage.com/login:**
- Chapter lecture videos
- Solutions to selected exercises

## 9.1 Conic Sections

**FIGURE 1**
The reflector of a radio telescope

Figure 1 shows the reflector of a radio telescope. The shape of the surface of the reflector is obtained by revolving a plane curve called a *parabola* about its axis of symmetry. (See Figure 2a.) Figure 2b depicts the orbit of a planet $P$ around the sun, $S$. This curve is called an *ellipse*. Figure 2c depicts the trajectory of an incoming alpha particle heading toward and then repulsed by a massive atomic nucleus located at the point $F$. The trajectory is one of two branches of a *hyperbola*.

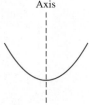

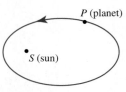

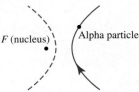

**(a)** The cross section of a radio telescope is part of a parabola.

**(b)** The orbit of a planet around the sun is an ellipse.

**(c)** The trajectory of an alpha particle in a Rutherford scattering is part of a branch of a hyperbola.

**FIGURE 2**

These curves—parabolas, ellipses, and hyperbolas—are called *conic sections* or, more simply, *conics* because they result from the intersection of a plane and a double-napped cone, as shown in Figure 3.

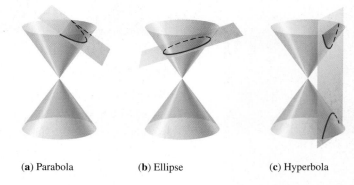

**FIGURE 3**
The conic sections

**(a)** Parabola

**(b)** Ellipse

**(c)** Hyperbola

In this section we give the geometric definition of each conic section, and we derive an equation for describing each conic section algebraically.

### ■ Parabola

We first consider a conic section called a *parabola*.

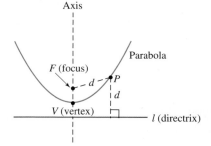

**FIGURE 4**
The distance between a point $P$ on a parabola and its focus $F$ is the same as the distance between $P$ and the directrix $l$ of the parabola.

> **DEFINITION** **Parabola**
>
> A **parabola** is the set of all points in a plane that are equidistant from a fixed point (called the **focus**) and a fixed line (called the **directrix**). (See Figure 4.)

By definition the point halfway between the focus and directrix lies on the parabola. This point $V$ is called the **vertex** of the parabola. The line passing through the focus and perpendicular to the directrix is called the **axis** of the parabola. Observe that the parabola is symmetric with respect to its axis.

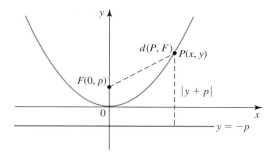

**FIGURE 5**
The parabola with focus $F(0, p)$ and directrix $y = -p$, where $p > 0$

To find an equation of a parabola, suppose that the parabola is placed so that its vertex is at the origin and its axis is along the $y$-axis, as shown in Figure 5. Further, suppose that its focus $F$ is at $(0, p)$, and its directrix is the line with equation $y = -p$. If $P(x, y)$ is any point on the parabola, then the distance between $P$ and $F$ is

$$d(P, F) = \sqrt{x^2 + (y - p)^2}$$

whereas the distance between $P$ and the directrix is $|y + p|$. By definition these distances are equal, so

$$\sqrt{x^2 + (y - p)^2} = |y + p|$$

Squaring both sides and simplifying, we obtain

$$x^2 + (y - p)^2 = |y + p|^2 = (y + p)^2$$
$$x^2 + y^2 - 2py + p^2 = y^2 + 2py + p^2$$
$$x^2 = 4py$$

---

**Standard Equation of a Parabola**

An equation of the parabola with focus $(0, p)$ and directrix $y = -p$ is

$$x^2 = 4py \tag{1}$$

---

If we write $a = 1/(4p)$, then Equation (1) becomes $y = ax^2$. Observe that the parabola opens upward if $p > 0$ and opens downward if $p < 0$. (See Figure 6.) Also, the parabola is symmetric with respect to the $y$-axis (that is, the axis of the parabola coincides with the $y$-axis), since Equation (1) remains unchanged if we replace $x$ by $-x$.

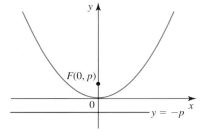

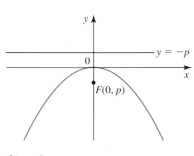

**FIGURE 6**
The parabola $x^2 = 4py$ opens upward if $p > 0$ and downward if $p < 0$.

**(a)** $p > 0$

**(b)** $p < 0$

Interchanging $x$ and $y$ in Equation (1) gives

$$y^2 = 4px \tag{2}$$

which is an equation of the parabola with focus $F(p, 0)$ and directrix $x = -p$. The parabola opens to the right if $p > 0$ and opens to the left if $p < 0$. (See Figure 7.) In both cases the axis of the parabola coincides with the $x$-axis.

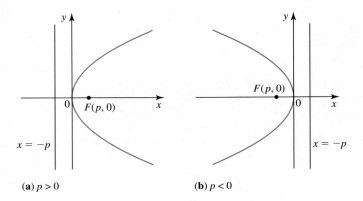

**(a)** $p > 0$        **(b)** $p < 0$

**FIGURE 7**
The parabola $y^2 = 4px$ opens to the right if $p > 0$ and to the left if $p < 0$.

**Note**   A parabola with vertex at the origin and axis of symmetry lying on the $x$-axis or $y$-axis is said to be in **standard position**. (See Figures 6 and 7.) ∎

**EXAMPLE 1**   Find the focus and directrix of the parabola $y^2 + 6x = 0$, and make a sketch of the parabola.

**Solution**   Rewriting the given equation in the form $y^2 = -6x$ and comparing it with Equation (2), we see that $4p = -6$ or $p = -\frac{3}{2}$. Therefore, the focus of the parabola is $F\left(-\frac{3}{2}, 0\right)$, and its directrix is $x = \frac{3}{2}$. The parabola is sketched in Figure 8. ∎

**EXAMPLE 2**   Find an equation of the parabola that has its vertex at the origin with axis of symmetry lying on the $y$-axis, and passes through the point $P(3, -4)$. What are the focus and directrix of the parabola?

**Solution**   An equation of the parabola has the form $y = ax^2$. To determine the value of $a$, we use the condition that the point $P(3, -4)$ lies on the parabola to obtain the equation $-4 = a(3)^2$ giving $a = -\frac{4}{9}$. Therefore, an equation of the parabola is

$$y = -\frac{4}{9}x^2$$

To find the focus of the parabola, observe that it has the form $F(0, p)$. Now

$$p = \frac{1}{4a} = \frac{1}{4\left(-\frac{4}{9}\right)} = -\frac{9}{16}$$

Therefore, the focus is $F\left(0, -\frac{9}{16}\right)$. Its directrix is $y = -p = -\left(-\frac{9}{16}\right)$, or $y = \frac{9}{16}$. The graph of the parabola is sketched in Figure 9. ∎

The parabola has many applications. For example, the cables of certain suspension bridges assume shapes that are parabolic.

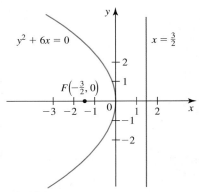

**FIGURE 8**
The parabola $y^2 + 6x = 0$

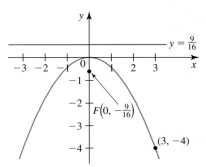

**FIGURE 9**
The parabola $y = -\frac{4}{9}x^2$

**EXAMPLE 3**  **Suspension Bridge Cables**  Figure 10 depicts a bridge, suspended by a flexible cable. If we assume that the weight of the cable is negligible in comparison to the weight of the bridge, then it can be shown that the shape of the cable is described by the equation

$$y = \frac{Wx^2}{2H}$$

where $W$ is the weight of the bridge in pounds per foot and $H$ is the tension at the lowest point of the cable in pounds (the origin). (See Exercise 89.) Suppose that the *span* of the cable is $2a$ ft and the *sag* is $h$ ft.

**a.** Find an equation describing the shape assumed by the cable in terms of $a$ and $h$.
**b.** Find the length of the cable if the span of the cable is 400 ft and the sag is 80 ft.

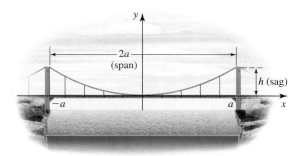

**FIGURE 10**
A bridge of length $2a$
suspended by a flexible cable

**Solution**
**a.** We can write the given equation in the form $y = kx^2$, where $k = W/(2H)$. Since the point $(a, h)$ lies on the parabola $y = kx^2$, we have

$$h = ka^2$$

or $k = h/a^2$, so the required equation is $y = hx^2/a^2$.
**b.** With $a = 200$ and $h = 80$ an equation that describes the shape of the cable is

$$y = \frac{80x^2}{200^2} = \frac{x^2}{500}$$

Next, the length of the cable is given by

$$s = 2\int_0^{200} \sqrt{1 + (y')^2}\, dx$$

But $y' = x/250$, so

$$s = 2\int_0^{200} \sqrt{1 + \left(\frac{x}{250}\right)^2}\, dx = \frac{1}{125}\int_0^{200} \sqrt{250^2 + x^2}\, dx$$

The easiest way to evaluate this integral is to use Formula 37 from the Table of Integrals found on the reference pages of this book:

$$\int \sqrt{a^2 + u^2}\, du = \frac{u}{2}\sqrt{a^2 + u^2} + \frac{a^2}{2} \ln|u + \sqrt{a^2 + u^2}| + C$$

If we let $a = 250$ and $u = x$, then

$$s = \frac{1}{125}\left[\frac{x}{2}\sqrt{250^2 + x^2} + \frac{250^2}{2}\ln|x + \sqrt{250^2 + x^2}|\right]_0^{200}$$

$$= \frac{1}{125}\left[100\sqrt{62500 + 40000} + 31250\ln|200 + \sqrt{62500 + 40000}| - 31250\ln 250\right]$$

$$= \frac{4}{5}\sqrt{102500} + 250\ln\left(\frac{200 + \sqrt{102500}}{250}\right) \approx 439$$

or 439 ft. ∎

Other applications of the parabola include the trajectory of a projectile in the absence of air resistance.

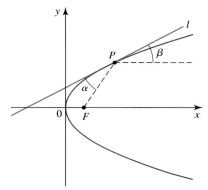

**FIGURE 11**

The reflective property states that $\alpha = \beta$.

## Reflective Property of the Parabola

Suppose that $P$ is any point on a parabola with focus $F$, and let $l$ be the tangent line to the parabola at $P$. (See Figure 11.) The reflective property states that the angle $\alpha$ that lies between $l$ and the line segment $FP$ is equal to the angle $\beta$ that lies between $l$ and the line passing through $P$ and parallel to the axis of the parabola. This property is the basis for many applications. (An outline of the proof of this property is given in Exercise 105.)

As was mentioned earlier, the reflector of a radio telescope has a shape that is obtained by revolving a parabola about its axis. Figure 12a shows a cross section of such a reflector. A radio wave coming in from a great distance may be assumed to be parallel to the axis of the parabola. This wave will strike the surface of the reflector and be reflected toward the focus $F$, where a collector is located. (The angle of incidence is equal to the angle of reflection.)

**FIGURE 12**
Applications of the reflective property of a parabola

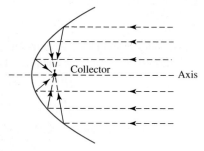

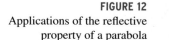

(a) A cross section of a radio telescope

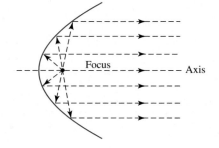

(b) A cross section of a headlight

The reflective property of the parabola is also used in the design of headlights of automobiles. Here, a light bulb is placed at the focus of the parabola. A ray of light emanating from the light bulb will strike the surface of the reflector and be reflected outward along a direction parallel to the axis of the parabola (see Figure 12b).

## Ellipses

Next, we consider a conic section called an ellipse.

> **DEFINITION**  **Ellipse**
>
> An ellipse is the set of all points in a plane the sum of whose distances from two fixed points (called the **foci**) is a constant.

Figure 13 shows an ellipse with foci $F_1$ and $F_2$. The line passing through the foci intersects the ellipse at two points, $V_1$ and $V_2$, called the **vertices** of the ellipse. The chord joining the vertices is called the **major axis,** and its midpoint is called the **center** of the ellipse. The chord passing through the center of the ellipse and perpendicular to the major axis is called the **minor axis** of the ellipse.

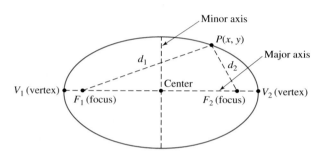

**FIGURE 13**
An ellipse with foci $F_1$ and $F_2$. A point $P(x, y)$ is on the ellipse if and only if $d_1 + d_2 =$ a constant.

**Note** We can construct an ellipse on paper in the following way: Place a piece of paper on a flat wooden board. Next, secure the ends of a piece of string to two points (the foci of the ellipse) with thumbtacks. Then trace the required ellipse with a pencil pushed against the string, as shown in Figure 14, making sure that the string is kept taut at all times. ∎

**FIGURE 14**
Drawing an ellipse on paper using thumbtacks, a string, and a pencil

To find an equation for an ellipse, suppose that the ellipse is placed so that its major axis lies along the $x$-axis and its center is at the origin, as shown in Figure 15. Then its foci $F_1$ and $F_2$ are at the points $(-c, 0)$ and $(c, 0)$, respectively. Let the sum of the distances between any point $P(x, y)$ on the ellipse and its foci be $2a > 2c > 0$. Then, by the definition of an ellipse we have

$$d(P, F_1) + d(P, F_2) = 2a$$

that is,

$$\sqrt{(x + c)^2 + y^2} + \sqrt{(x - c)^2 + y^2} = 2a$$

or

$$\sqrt{(x - c)^2 + y^2} = 2a - \sqrt{(x + c)^2 + y^2}$$

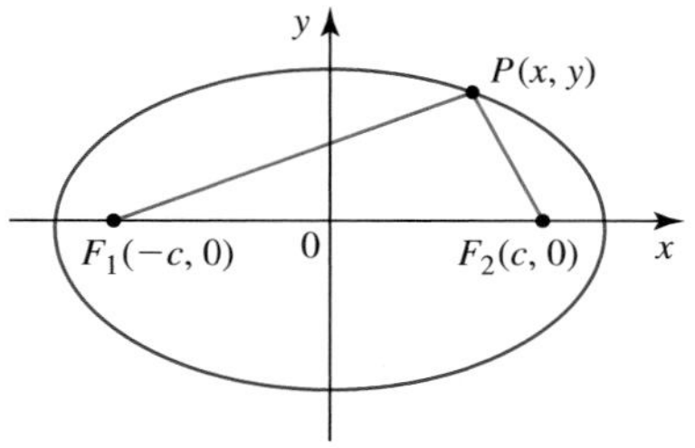

**FIGURE 15**
The ellipse with foci $F_1(-c, 0)$ and $F_2(c, 0)$

Squaring both sides of this equation, we obtain

$$x^2 - 2cx + c^2 + y^2 = 4a^2 - 4a\sqrt{(x + c)^2 + y^2} + x^2 + 2cx + c^2 + y^2$$

or, upon simplification,

$$a\sqrt{(x + c)^2 + y^2} = a^2 + cx$$

Squaring both sides again, we have

$$a^2(x^2 + 2cx + c^2 + y^2) = a^4 + 2a^2cx + c^2x^2$$

which yields

$$(a^2 - c^2)x^2 + a^2 y^2 = a^2(a^2 - c^2)$$

Recall that $a > c$, so $a^2 - c^2 > 0$. Let $b^2 = a^2 - c^2$ with $b > 0$. Then the equation of the ellipse becomes

$$b^2 x^2 + a^2 y^2 = a^2 b^2$$

or, upon dividing both sides by $a^2 b^2$, we obtain

$$\frac{x^2}{a^2} + \frac{y^2}{b^2} = 1$$

By setting $y = 0$, we obtain $x = \pm a$, which gives $(-a, 0)$ and $(a, 0)$ as the vertices of the ellipse. Similarly, by setting $x = 0$, we see that the ellipse intersects the $y$-axis at the points $(0, -b)$ and $(0, b)$. Since the equation remains unchanged if $x$ is replaced by $-x$ and $y$ is replaced by $-y$, we see that the ellipse is symmetric with respect to both axes.

Observe, too, that $b < a$, since

$$b^2 = a^2 - c^2 < a^2$$

So as the name implies, the length of the major axis, $2a$, is greater than the length of the minor axis, $2b$. Finally, observe that if the foci coincide, then $c = 0$ and $a = b$, so the ellipse is a circle with radius $r = a = b$.

Placing the ellipse so that its major axis lies along the $y$-axis and its center is at the origin leads to an equation in which the roles of $x$ and $y$ are reversed. To summarize, we have the following.

---

**Standard Equation of an Ellipse**

An equation of the ellipse with foci $(\pm c, 0)$ and vertices $(\pm a, 0)$ is

$$\frac{x^2}{a^2} + \frac{y^2}{b^2} = 1 \qquad a \geq b > 0 \tag{3}$$

and an equation of the ellipse with foci $(0, \pm c)$ and vertices $(0, \pm a)$ is

$$\frac{x^2}{b^2} + \frac{y^2}{a^2} = 1 \qquad a \geq b > 0 \tag{4}$$

where $c^2 = a^2 - b^2$. (See Figure 16.)

---

**FIGURE 16**
Two ellipses in standard position
with center at the origin

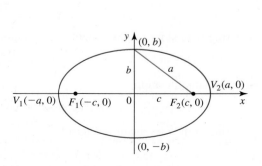

(a) The major axis is along the $x$-axis.

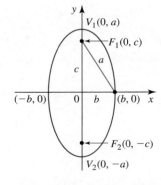

(b) The major axis is along the $y$-axis.

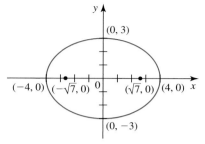

**FIGURE 17**
The ellipse $\dfrac{x^2}{16} + \dfrac{y^2}{9} = 1$

**Note** An ellipse with center at the origin and foci lying along the $x$-axis or the $y$-axis is said to be in standard position. (See Figure 16.) ■

**EXAMPLE 4** Sketch the ellipse $\dfrac{x^2}{16} + \dfrac{y^2}{9} = 1$. What are the foci and vertices?

**Solution** Here, $a^2 = 16$ and $b^2 = 9$, so $a = 4$ and $b = 3$. Setting $y = 0$ and $x = 0$ in succession gives the $x$- and $y$-intercepts as $\pm 4$ and $\pm 3$, respectively. Also, from

$$c^2 = a^2 - b^2 = 16 - 9 = 7$$

we obtain $c = \sqrt{7}$ and conclude that the foci of the ellipse are $(\pm\sqrt{7}, 0)$. Its vertices are $(\pm 4, 0)$. The ellipse is sketched in Figure 17. ■

**EXAMPLE 5** Find an equation of the ellipse with foci $(0, \pm 2)$ and vertices $(0, \pm 4)$.

**Solution** Since the foci and therefore the major axis of the ellipse lie along the $y$-axis, we use Equation (4). Here, $c = 2$ and $a = 4$, so

$$b^2 = a^2 - c^2 = 16 - 4 = 12$$

Therefore, the standard form of the equation for the ellipse is

$$\frac{x^2}{12} + \frac{y^2}{16} = 1$$

or

$$4x^2 + 3y^2 = 48$$ ■

## Reflective Property of the Ellipse

The ellipse, like the parabola, has a reflective property. To describe this property, consider an ellipse with foci $F_1$ and $F_2$ as shown in Figure 18. Let $P$ be a point on the ellipse, and let $l$ be the tangent line to the ellipse at $P$. Then the angle $\alpha$ between the line segment $F_1P$ and $l$ is equal to the angle $\beta$ between the line segment $F_2P$ and $l$. You will be asked to establish this property in Exercise 106.

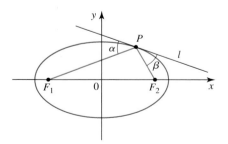

**FIGURE 18**
The reflective property
states that $\alpha = \beta$.

The reflective property of the ellipse is used to design *whispering galleries*—rooms with elliptical-shaped ceilings, in which a person standing at one focus can hear the whisper of another person standing at the other focus. A whispering gallery can be found in the rotunda of the Capitol Building in Washington, D.C. Also, Paris subway tunnels are almost elliptical, and because of the reflective property of the ellipse, whispering on one platform can be heard on the other. (See Figure 19.)

**FIGURE 19**
A cross section of a Paris subway tunnel is almost elliptical.

Yet another application of the reflective property of the ellipse can be found in the field of medicine in a procedure for removing kidney stones called *shock wave lithotripsy.* In this procedue an ellipsoidal reflector is positioned so that a transducer is at one focus and a kidney stone is at the other focus. Shock waves emanating from the transducer are reflected according to the reflective property of the ellipse onto the kidney stone, pulverizing it. This procedure obviates the necessity for surgery.

### ■ Eccentricity of an Ellipse

To measure the ovalness of an ellipse, we introduce the notion of eccentricity.

---

**DEFINITION    Eccentricity of an Ellipse**

The **eccentricity** of an ellipse is given by the ratio $e = c/a$.

---

The eccentricity of an ellipse satisfies $0 < e < 1$, since $0 < c < a$. The closer $e$ is to zero, the more circular is the ellipse.

### ■ Hyperbolas

The definition of a hyperbola is similar to that of an ellipse. The *sum* of the distances between the foci and a point on an ellipse is fixed, whereas the *difference* of these distances is fixed for a hyperbola.

---

**DEFINITION    Hyperbola**

A **hyperbola** is the set of all points in a plane the difference of whose distances from two fixed points (called the **foci**) is a constant.

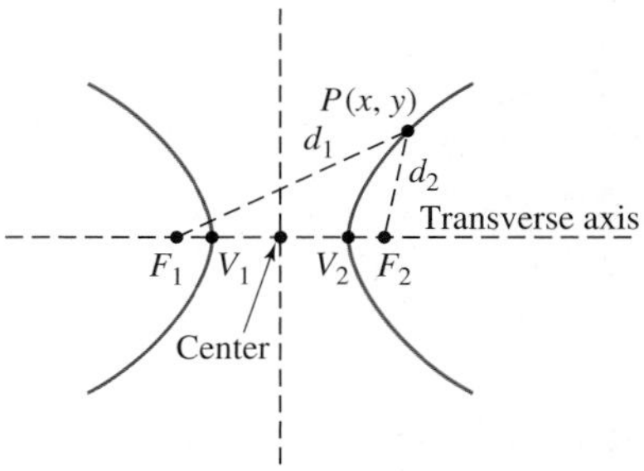

---

**FIGURE 20**
A hyperbola with foci $F_1$ and $F_2$. A point $P(x, y)$ is on the hyperbola if and only if $|d_1 - d_2|$ is a constant.

Figure 20 shows a hyperbola with foci $F_1$ and $F_2$. The line passing through the foci intersects the hyperbola at two points, $V_1$ and $V_2$, called the **vertices** of the hyperbola. The line segment joining the vertices is called the **transverse axis** of the hyperbola, and the midpoint of the transverse axis is called the **center** of the hyperbola. Observe that a hyperbola, in contrast to a parabola or an ellipse, has two separate branches.

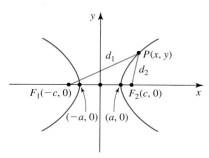

**FIGURE 21**

An equation of the hyperbola with center $(0, 0)$ and foci $(-c, 0)$ and $(c, 0)$ is $\dfrac{x^2}{a^2} - \dfrac{y^2}{b^2} = 1$

The derivation of an equation of a hyperbola is similar to that of an ellipse. Consider, for example, the hyperbola with center at the origin and foci $F_1(-c, 0)$ and $F_2(c, 0)$ on the $x$-axis. (See Figure 21.) Using the condition $d(P, F_1) - d(P, F_2) = 2a$, where $a$ is a positive constant, it can be shown that if $P(x, y)$ is any point on the hyperbola, then an equation of the hyperbola is

$$\frac{x^2}{a^2} - \frac{y^2}{b^2} = 1$$

where $b = \sqrt{c^2 - a^2}$ or $c = \sqrt{a^2 + b^2}$.

Observe that the $x$-intercepts of the hyperbola are $x = \pm a$, giving $(-a, 0)$ and $(a, 0)$ as its vertices. But there are no $y$-intercepts, since setting $x = 0$ gives $y^2 = -b^2$, which has no real solution. Also, observe that the hyperbola is symmetric with respect to both axes.

If we solve the equation

$$\frac{x^2}{a^2} - \frac{y^2}{b^2} = 1$$

for $y$, we obtain

$$y = \pm \frac{b}{a} \sqrt{x^2 - a^2}$$

Since $x^2 - a^2 \geq 0$ or, equivalently, $x \leq -a$ or $x \geq a$, we see that the hyperbola actually consists of two separate branches, as was noted earlier. Also, observe that if $x$ is large in magnitude, then $x^2 - a^2 \approx x^2$, so $y = \pm(b/a)x$. This heuristic argument suggests that both branches of the hyperbola approach the slant asymptotes $y = \pm(b/a)x$ as $x$ increases or decreases without bound. (See Figure 22.) You will be asked in Exercise 101 to demonstrate that this is true.

Finally, if the foci of a hyperbola are on the $y$-axis, then by reversing the roles of $x$ and $y$, we obtain

$$\frac{y^2}{a^2} - \frac{x^2}{b^2} = 1$$

as an equation of the hyperbola.

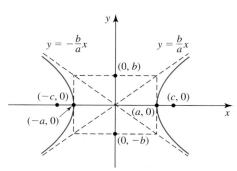

**(a)** $\dfrac{x^2}{a^2} - \dfrac{y^2}{b^2} = 1$ (The transverse axis is along the $x$-axis.)

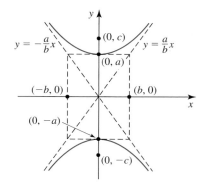

**(b)** $\dfrac{y^2}{a^2} - \dfrac{x^2}{b^2} = 1$ (The transverse axis is along the $y$-axis.)

**FIGURE 22**

Two hyperbolas in standard position with center at the origin

---

**Standard Equation of a Hyperbola**

An equation of the hyperbola with foci $(\pm c, 0)$ and vertices $(\pm a, 0)$ is

$$\frac{x^2}{a^2} - \frac{y^2}{b^2} = 1 \qquad (5)$$

where $c = \sqrt{a^2 + b^2}$. The hyperbola has asymptotes $y = \pm(b/a)x$. An equation of the hyperbola with foci $(0, \pm c)$ and vertices $(0, \pm a)$ is

$$\frac{y^2}{a^2} - \frac{x^2}{b^2} = 1 \qquad (6)$$

where $c = \sqrt{a^2 + b^2}$. The hyperbola has asymptotes $y = \pm(a/b)x$.

---

The line segment of length $2b$ joining the points $(0, -b)$ and $(0, b)$ or $(-b, 0)$ and $(b, 0)$ is called the **conjugate axis** of the hyperbola.

**EXAMPLE 6** Find the foci, vertices, and asymptotes of the hyperbola $4x^2 - 9y^2 = 36$.

**Solution** Dividing both sides of the given equation by 36 leads to the standard equation

$$\frac{x^2}{9} - \frac{y^2}{4} = 1$$

of a hyperbola. Here, $a^2 = 9$ and $b^2 = 4$, so $a = 3$ and $b = 2$. Setting $y = 0$ gives $\pm 3$ as the $x$-intercepts, so $(\pm 3, 0)$ are the vertices of the hyperbola. Also, we have $c = \sqrt{a^2 + b^2} = \sqrt{13}$, and conclude that the foci of the hyperbola are $(\pm\sqrt{13}, 0)$. Finally, the asymptotes of the hyperbola are

$$y = \pm\frac{b}{a}x = \pm\frac{2}{3}x$$

When you sketch this hyperbola, draw the asymptotes first so that you can then use them as guides for sketching the hyperbola itself. (See Figure 23.) ∎

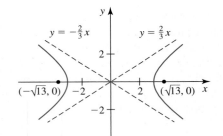

**FIGURE 23**
The graph of the hyperbola $4x^2 - 9y^2 = 36$

**EXAMPLE 7** A hyperbola has vertices $(0, \pm 3)$ and passes through the point $(2, 5)$. Find an equation of the hyperbola. What are its foci and asymptotes?

**Solution** Here, the foci lie along the $y$-axis, so the standard equation of the hyperbola has the form

$$\frac{y^2}{9} - \frac{x^2}{b^2} = 1 \qquad \text{Note that } a = 3.$$

To determine $b$, we use the condition that the hyperbola passes through the point $(2, 5)$ to write

$$\frac{25}{9} - \frac{4}{b^2} = 1$$

$$\frac{4}{b^2} = \frac{25}{9} - 1 = \frac{16}{9}$$

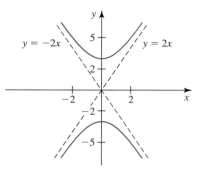

**FIGURE 24**
The graph of the hyperbola
$y^2 - 4x^2 = 9$

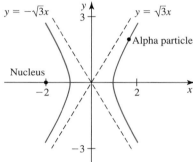

**FIGURE 25**
The trajectory of an alpha particle in a
Rutherford scattering is a branch of a
hyperbola.

or $b^2 = \frac{9}{4}$. Therefore, a required equation of the hyperbola is

$$\frac{y^2}{9} - \frac{x^2}{\frac{9}{4}} = 1$$

or, equivalently, $y^2 - 4x^2 = 9$. To find the foci of the hyperbola, we compute

$$c^2 = a^2 + b^2 = 9 + \frac{9}{4} = \frac{45}{4}$$

or $c = \pm\sqrt{45/4} = \pm 3\sqrt{5}/2$, from which we see that the foci are $\left(0, \pm\frac{3\sqrt{5}}{2}\right)$. Finally, the asymptotes are obtained by substituting $a = 3$ and $b = \frac{3}{2}$ into the equations $y = \pm(a/b)x$, giving $y = \pm 2x$. The graph of the hyperbola is shown in Figure 24. ∎

**EXAMPLE 8** **A Rutherford Scattering** A massive atomic nucleus used as a target for incoming alpha particles is located at the point $(-2, 0)$, as shown in Figure 25. Suppose that an alpha particle approaching the nucleus has a trajectory that is a branch of the hyperbola shown with asymptotes $y = \pm\sqrt{3}x$ and foci $(\pm 2, 0)$. Find an equation of the trajectory.

**Solution** The asymptotes of a hyperbola with center at the origin and foci lying on the $x$-axis have equations of the form $y = \pm(b/a)x$. Since the asymptotes of the trajectory are $y = \pm\sqrt{3}x$, we see that

$$\frac{b}{a} = \sqrt{3} \qquad \text{or} \qquad b = \sqrt{3}a$$

Next, since the foci of the hyperbola are $(\pm 2, 0)$, we know that $c = 2$. But $c^2 = a^2 + b^2$, and this gives

$$4 = a^2 + (\sqrt{3}a)^2 = a^2 + 3a^2 = 4a^2$$

or $a = 1$, so $b = \sqrt{3}$. Therefore, an equation of the trajectory is

$$\frac{x^2}{1} - \frac{y^2}{3} = 1$$

or $3x^2 - y^2 = 3$, where $x > 0$. ∎

> **DEFINITION** **Eccentricity of a Hyperbola**
> The **eccentricity** of a hyperbola is given by the ratio $e = c/a$.

Since $c > a$, the eccentricity of a hyperbola satisfies $e > 1$. The larger the eccentricity is, the flatter are the branches of the hyperbola.

## ■ Shifted Conics

By using the techniques in Section 0.4, we can obtain the equations of conics that are translated from their standard positions. In fact, by replacing $x$ by $x - h$ and $y$ by $y - k$ in their standard equations, we obtain the equation of a parabola whose vertex is translated from the origin to the point $(h, k)$ and the equation of an ellipse (or hyperbola) whose center is translated from the origin to the point $(h, k)$.

We summarize these results in Table 1. Figure 26 shows the graphs of these conics.

**TABLE 1**

| Conic | Orientation of axis | Equation of conic | | |
|-------|--------------------|-----------|----|------------|
| Parabola | Axis horizontal | $(y - k)^2 = 4p(x - h)$ | **(7)** | (See Figure 26a.) |
| Parabola | Axis vertical | $(x - h)^2 = 4p(y - k)$ | **(8)** | (See Figure 26b.) |
| Ellipse | Major axis horizontal | $\dfrac{(x - h)^2}{a^2} + \dfrac{(y - k)^2}{b^2} = 1$ | **(9)** | (See Figure 26c.) |
| Ellipse | Major axis vertical | $\dfrac{(x - h)^2}{b^2} + \dfrac{(y - k)^2}{a^2} = 1$ | **(10)** | (See Figure 26d.) |
| Hyperbola | Transverse axis horizontal | $\dfrac{(x - h)^2}{a^2} - \dfrac{(y - k)^2}{b^2} = 1$ | **(11)** | (See Figure 26e.) |
| Hyperbola | Transverse axis vertical | $\dfrac{(y - k)^2}{a^2} - \dfrac{(x - h)^2}{b^2} = 1$ | **(12)** | (See Figure 26f.) |

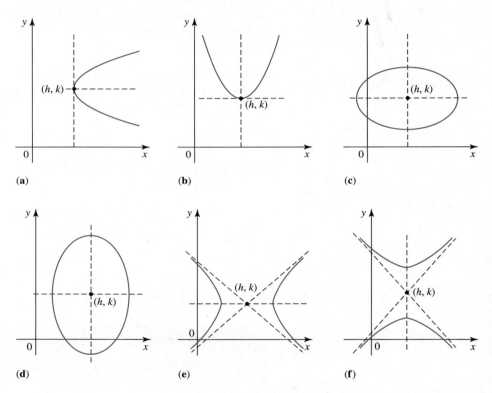

**FIGURE 26**
Shifted conics with centers at $(h, k)$

Observe that if $h = k = 0$, then each of the equations listed in Table 1 reduces to the corresponding standard equation of a conic centered at the origin, as expected.

**EXAMPLE 9**  Find the standard equation of the ellipse with foci at $(1, 2)$ and $(5, 2)$ and major axis of length 6. Sketch the ellipse.

**Solution**  Since the foci $(1, 2)$ and $(5, 2)$ have the same $y$-coordinate, we see that they lie along the line $y = 2$ parallel to the $x$-axis. The midpoint of the line segment joining $(1, 2)$ to $(5, 2)$ is $(3, 2)$, and this is the center of the ellipse. From this we can see that the distance from the center of the ellipse to each of the foci is 2, so $c = 2$. Next,

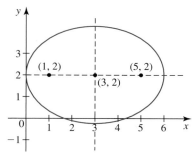

**FIGURE 27**

The ellipse $\dfrac{(x-3)^2}{9} + \dfrac{(y-2)^2}{5} = 1$

since the major axis of the ellipse is known to have length 6, we have $2a = 6$, or $a = 3$. Finally, from the relation $c^2 = a^2 - b^2$, we obtain $4 = 9 - b^2$, or $b^2 = 5$. Therefore, using Equation (9) from Table 1 with $h = 3$, $k = 2$, $a = 3$, and $b = \sqrt{5}$, we obtain the desired equation:

$$\frac{(x-3)^2}{9} + \frac{(y-2)^2}{5} = 1$$

The ellipse is sketched in Figure 27.   ■

If you expand and simplify each equation in Table 1, you will see that these equations have the general form

$$Ax^2 + By^2 + Dx + Ey + F = 0$$

where the coefficients are real numbers. Conversely, given such an equation, we can obtain an equivalent equation in the form listed in Table 1 by using the technique of completing the square. The latter can then be analyzed readily to obtain the properties of the conic that it represents.

**EXAMPLE 10**  Find the standard equation of the hyperbola

$$3x^2 - 4y^2 + 6x + 16y - 25 = 0$$

Find its foci, vertices, and asymptotes, and sketch its graph.

**Solution**   We complete the squares in $x$ and $y$:

$$3(x^2 + 2x) - 4(y^2 - 4y) = 25$$
$$3[x^2 + 2x + (1)^2] - 4[y^2 - 4y + (-2)^2] = 25 + 3 - 16$$
$$3(x + 1)^2 - 4(y - 2)^2 = 12$$

Then, dividing both sides of this equation by 12 gives the desired equation

$$\frac{(x + 1)^2}{4} - \frac{(y - 2)^2}{3} = 1$$

Comparing this equation with Equation (11) in Table 1, we see that it is an equation of a hyperbola with center $(-1, 2)$ and transverse axis parallel to the $x$-axis. We also see that $a^2 = 4$ and $b^2 = 3$, from which it follows that $c^2 = a^2 + b^2 = 4 + 3 = 7$. We can think of this hyperbola as one that is obtained by shifting a similar hyperbola, centered at the origin, one unit to the left and two units upward. Then the required foci, vertices, and asymptotes are obtained by shifting the foci, vertices, and asymptotes of this latter hyperbola accordingly. The results are as follows:

| | |
|---|---|
| **Foci** | $(-\sqrt{7} - 1, 2)$ and $(\sqrt{7} - 1, 2)$ |
| **Vertices** | $(-3, 2)$ and $(1, 2)$ |
| **Asymptotes** | $y - 2 = \pm\frac{\sqrt{3}}{2}(x + 1)$ |

The hyperbola is sketched in Figure 28.

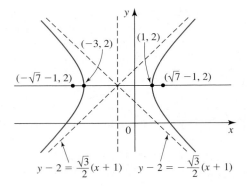

**FIGURE 28**
The hyperbola
$$3x^2 - 4y^2 + 6x + 16y - 25 = 0$$

The properties of the hyperbola are exploited in the navigational system LORAN (Long Range Navigation). This system utilizes two sets of transmitters: one set located at $F_1$ and $F_2$ and another set located at $G_1$ and $G_2$. (See Figure 29.) Suppose that synchronized signals sent out by the transmitters located at $F_1$ and $F_2$ reach a ship that is located at $P$. The difference in the times of arrival of the signals are converted by an onboard computer into the difference in the distance $d(P, F_1) - d(P, F_2)$. Using the definition of the hyperbola, we see that this places the ship on a branch of a hyperbola with foci $F_1$ and $F_2$ (Figure 29). Similarly, we see that the ship must also lie on a branch of a hyperbola with foci $G_1$ and $G_2$. Thus, the position of $P$ is given by the intersection of these two branches of the hyperbolas.

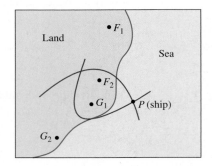

**FIGURE 29**
In the LORAN navigational system the position of a ship is the point of intersection of two branches of hyperbolas.

## 9.1 CONCEPT QUESTIONS

**1. a.** Give the definition of a parabola. What are the focus, directrix, vertex, and axis of a parabola? Illustrate with a sketch.
 **b.** Write the standard equation of (i) a parabola whose axis lies on the $y$-axis and (ii) a parabola whose axis lies on the $x$-axis. Illustrate with sketches.
**2. a.** Give the definition of an ellipse. What are the foci, vertices, center, major axis, and minor axis of an ellipse? Illustrate with a sketch.
 **b.** Write the standard equation of (i) an ellipse with foci $(\pm c, 0)$ and vertices $(\pm a, 0)$ and (ii) an ellipse with foci $(0, \pm c)$ and vertices $(0, \pm a)$. Illustrate with sketches.

**3. a.** Give the definition of a hyperbola. What are the center, the foci, and the transverse axis of the hyperbola? Illustrate with a sketch.
 **b.** Write the standard equation of (i) a hyperbola with foci $(\pm c, 0)$ and vertices $(\pm a, 0)$ and (ii) a hyperbola with foci $(0, \pm c)$ and vertices $(0, \pm a)$. Illustrate with sketches.

# 9.1  EXERCISES

*In Exercises 1–8, match the equation with one of the conics labeled (a)–(h). If the conic is a parabola, find its vertex, focus and directrix. If it is an ellipse or a hyperbola, find its vertices, foci, and eccentricity.*

**1.** $x^2 = -4y$

**2.** $y = \dfrac{x^2}{8}$

**3.** $y^2 = 8x$

**4.** $x = -\dfrac{1}{4}y^2$

**5.** $\dfrac{x^2}{9} + \dfrac{y^2}{4} = 1$

**6.** $x^2 + \dfrac{y^2}{4} = 1$

**7.** $\dfrac{x^2}{16} - \dfrac{y^2}{9} = 1$

**8.** $y^2 - \dfrac{x^2}{4} = 1$

**(a)**

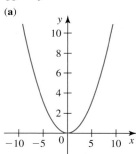

**(b)**

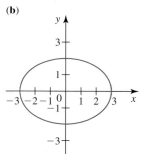

**(c)**

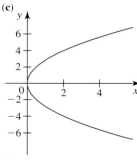

**(d)**

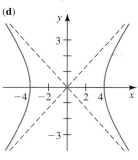

**(e)**

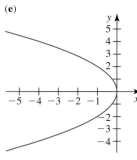

**(f)**

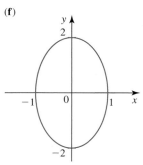

**(g)**

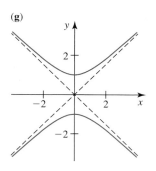

**(h)**

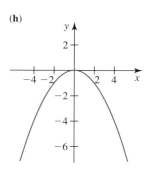

*In Exercises 9–14, find the vertex, focus, and directrix of the parabola with the given equation, and sketch the parabola.*

**9.** $y = 2x^2$

**10.** $x^2 = -12y$

**11.** $x = 2y^2$

**12.** $y^2 = -8x$

**13.** $5y^2 = 12x$

**14.** $y^2 = -40x$

*In Exercises 15–20, find the foci and vertices of the ellipse, and sketch its graph.*

**15.** $\dfrac{x^2}{4} + \dfrac{y^2}{25} = 1$

**16.** $\dfrac{x^2}{16} + \dfrac{y^2}{9} = 1$

**17.** $4x^2 + 9y^2 = 36$

**18.** $25x^2 + 16y^2 = 400$

**19.** $x^2 + 4y^2 = 4$

**20.** $2x^2 + y^2 = 4$

*In Exercises 21–26, find the vertices, foci, and asymptotes of the hyperbola, and sketch its graph using its asymptotes as an aid.*

**21.** $\dfrac{x^2}{25} - \dfrac{y^2}{144} = 1$

**22.** $\dfrac{y^2}{16} - \dfrac{x^2}{81} = 1$

**23.** $x^2 - y^2 = 1$

**24.** $4y^2 - x^2 = 4$

**25.** $y^2 - 5x^2 = 25$

**26.** $x^2 - 2y^2 = 8$

*In Exercises 27–30, find an equation of the parabola that satisfies the conditions.*

**27.** Focus $(3, 0)$, directrix $x = -3$

**28.** Focus $(0, -2)$, directrix $y = 2$

**29.** Focus $\left(-\frac{5}{2}, 0\right)$, directrix $x = \frac{5}{2}$

**30.** Focus $\left(0, \frac{3}{2}\right)$, directrix $y = -\frac{3}{2}$

*In Exercises 31–38, find an equation of the ellipse that satisfies the given conditions.*

**31.** Foci $(\pm 1, 0)$, vertices $(\pm 3, 0)$

**32.** Foci $(0, \pm 3)$, vertices $(0, \pm 5)$

**33.** Foci $(0, \pm 1)$, length of major axis 6

**34.** Vertices $(0, \pm 5)$, length of minor axis 5

**35.** Vertices $(\pm 3, 0)$, passing through $(1, \sqrt{2})$

**36.** Passes through $(1, 5)$ and $(2, 4)$ and its center is at $(0, 0)$

**37.** Passes through $\left(2, \frac{3\sqrt{3}}{2}\right)$ with vertices at $(0, \pm 5)$

**38.** $x$-intercepts $\pm 3$, $y$-intercepts $\pm \frac{1}{2}$

*In Exercises 39–44, find an equation of the hyperbola centered at the origin that satisfies the given conditions.*

**39.** foci $(\pm 5, 0)$, vertices $(\pm 3, 0)$

**40.** foci $(0, \pm 8)$, vertices $(0, \pm 4)$

**41.** foci $(0, \pm 5)$, conjugate axis of length 4

**42.** vertices $(\pm 4, 0)$ passing through $\left(5, \frac{9}{4}\right)$

**43.** vertices $(\pm 2, 0)$, asymptotes $y = \pm \frac{3}{2}x$

**44.** $y$-intercepts $\pm 1$, asymptotes $y = \pm \frac{1}{2\sqrt{2}}x$

*In Exercises 45–48, match the equation with one of the conic sections labeled (a)–(d).*

**45.** $(x + 3)^2 = -2(y - 4)$     **46.** $\dfrac{(x - 2)^2}{16} + \dfrac{(y + 3)^2}{4} = 1$

**47.** $\dfrac{(y - 3)^2}{16} - \dfrac{(x + 1)^2}{9} = 1$    **48.** $(y - 1)^2 = -4(x - 2)$

**(a)**

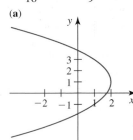

**(b)**

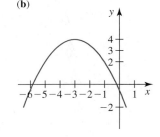

**(c)**

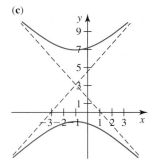

**(d)**

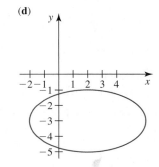

*In Exercises 49–66, find an equation of the conic satisfying the given conditions.*

**49.** Parabola, focus $(3, 1)$, directrix $x = 1$

**50.** Parabola, focus $(-2, 3)$, directrix $y = 5$

**51.** Parabola, vertex $(2, 2)$, focus $\left(\frac{3}{2}, 2\right)$

**52.** Parabola, vertex $(1, -2)$, directrix $y = 1$

**53.** Parabola, axis parallel to the $y$-axis, passes through $(-3, 2)$, $\left(0, -\frac{5}{2}\right)$, and $(1, -6)$

**54.** Parabola, axis parallel to the $x$-axis, passes through $(-6, 6)$, $(0, 0)$, and $(2, 2)$

**55.** Ellipse, foci $(\pm 1, 3)$, vertices $(\pm 3, 3)$

**56.** Ellipse, foci $(0, 2)$ and $(4, 2)$, vertices $(-1, 2)$ and $(5, 2)$

**57.** Ellipse, foci $(\pm 1, 2)$, length of major axis 8

**58.** Ellipse, foci $(1, \pm 3)$, length of minor axis 2

**59.** Ellipse, center $(2, 1)$, focus $(0, 1)$, vertex $(5, 1)$

**60.** Ellipse, foci $(2 - \sqrt{3}, -1)$ and $(2 + \sqrt{3}, -1)$, passes through $(2, 0)$

**61.** Hyperbola, foci $(-2, 2)$ and $(8, 2)$, vertices $(0, 2)$ and $(6, 2)$

**62.** Hyperbola, foci $(-4, 5)$ and $(-4, -15)$, vertices $(-4, -3)$ and $(-4, -7)$

**63.** Hyperbola, foci $(6, -3)$ and $(-4, -3)$, asymptotes $y + 3 = \pm \frac{4}{3}(x - 1)$

**64.** Hyperbola, foci $(2, 2)$ and $(2, 6)$, asymptotes $x = -2 + y$ and $x = 6 - y$

**65.** Hyperbola, vertices $(4, -2)$ and $(4, 4)$, asymptotes $y - 1 = \pm \frac{3}{2}(x - 4)$

**66.** Hyperbola, vertices $(0, -2)$ and $(4, -2)$, asymptotes $x = -y$ and $x = y + 4$

*In Exercises 67–72, find the vertex, focus, and directrix of the parabola, and sketch its graph.*

**67.** $y^2 - 2y - 4x + 9 = 0$    **68.** $y^2 - 4y - 2x - 4 = 0$

**69.** $x^2 + 6x - y + 11 = 0$    **70.** $2x^2 - 8x - y + 5 = 0$

**71.** $4y^2 - 4y - 32x - 31 = 0$

**72.** $9x^2 + 6x + 9y - 8 = 0$

*In Exercises 73–78, find the center, foci, and vertices of the ellipse, and sketch its graph.*

**73.** $(x - 1)^2 + 4(y + 2)^2 = 1$

**74.** $2x^2 + y^2 - 20x + 2y + 43 = 0$

**75.** $x^2 + 4y^2 - 2x + 16y + 13 = 0$

**76.** $2x^2 + y^2 + 12x - 6y + 25 = 0$

**77.** $4x^2 + 9y^2 - 18x - 27 = 0$

**78.** $9x^2 + 36y^2 - 36x + 48y + 43 = 0$

*In Exercises 79–84, find the center, foci, vertices, and equations of the asymptotes of the hyperbola with the given equation, and sketch its graph using its asymptotes as an aid.*

**79.** $3x^2 - 4y^2 - 8y - 16 = 0$

**80.** $4x^2 - 9y^2 - 16x - 54y + 79 = 0$

**81.** $2x^2 - 3y^2 - 4x + 12y + 8 = 0$

**82.** $4y^2 - 9x^2 + 18x + 16y + 43 = 0$

**83.** $4x^2 - 2y^2 + 8x + 8y - 12 = 0$

**84.** $4x^2 - 3y^2 - 12y - 3 = 0$

**85. Parabolic Reflectors** The following figure shows the cross section of a parabolic reflector. If the reflector is 2 ft wide at the opening and 1 ft deep, how far from the vertex should the light source be placed along the axis of symmetry of the parabola?

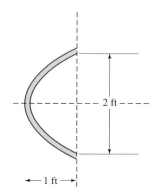

**86. Length of Suspension Bridge Cable** The figure below depicts a suspension bridge. The shape of the cable is described by the equation

$$y = \frac{hx^2}{a^2}$$

where $2a$ ft is the span of the bridge and $h$ ft is the sag. (See Example 3.) Assuming that the sag is small in comparison to the span (that is, $h/a$ is small), show that the length of the cable is

$$s \approx 2a\left(1 + \frac{2h^2}{3a^2}\right)$$

Use this approximation to estimate the length of the cable in Example 3, where $a = 200$ and $h = 80$. Compare your result with that obtained in the example.

**Hint:** Retain the first two terms of a binomial series.

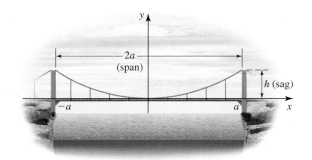

**87. Length of Suspension Bridge Cable** The cable of the suspension bridge shown in the figure below has the shape of a parabola. If the span of the bridge is 600 ft and the sag is 60 ft, what is the length of the cable?

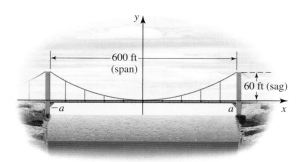

**88. Surface Area of a Satellite Dish** An 18-in. satellite dish is obtained by revolving the parabola with equation $y = \frac{4}{81}x^2$ about the $y$-axis. Find the surface area of the dish.

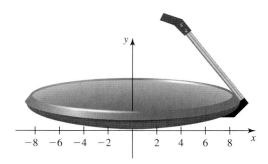

**89. Shape of a Suspension Bridge Cable** Consider a bridge of weight $W$ lb/ft suspended by a flexible cable. Assume that the weight of the cable is negligible in comparison to the weight of the bridge. The following figure shows a portion of such a structure with the lowest point of the cable located at the origin. Let $P$ be any point on the cable, and suppose that the tension of the cable at $P$ is $T$ lb and lies along the tangent at $P$ (this is the case with flexible cables).

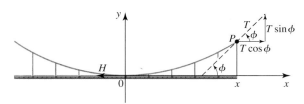

Referring to the figure, we see that

$$\frac{dy}{dx} = \tan \phi = \frac{\sin \phi}{\cos \phi} = \frac{T \sin \phi}{T \cos \phi}$$

But since the bridge is in equilibrium, the horizontal component of $T$ must be equal to $H$, the tension at the lowest point of the cable (the origin); that is, $T \cos \phi = H$. Similarly, the vertical component of $T$ must be equal to $Wx$, the load carried over that section of the cable from 0 to $P$; that is, $T \sin \phi = Wx$. Therefore,

$$\frac{dy}{dx} = \frac{Wx}{H}$$

Finally, since the lowest point of the cable is located at the origin, we have $y(0) = 0$. Solve this initial-value problem to show that the shape of the cable is a parabola.

**Note:** Observe that in a suspension bridge, the cable supports a load that is *uniformly distributed horizontally*. A cable supporting a load distributed *uniformly along its length* (for example, a cable supporting its own weight) assumes the shape of a catenary, as we saw in Section 5.8.

**90. Shape of a Suspension Bridge Cable**  Refer to Exercise 89. Suppose that the span of the cable is $2a$ ft and the sag is $h$ ft. (See Figure 10.) Show that the tension (in pounds) at the highest point of the cable has magnitude

$$T = \frac{Wa\sqrt{a^2 + 4h^2}}{2h}$$

**91. Arch of a Bridge**  A bridge spanning the Charles River has three arches that are semielliptical in shape. The base of the center arch is 24 ft across, and the maximum height of the arch is 8 ft. What is the height of the arch 6 ft from the center of the base?

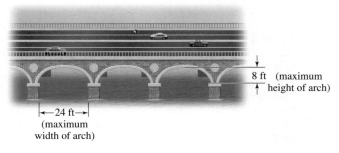

8 ft  (maximum height of arch)

|←—24 ft—→|
(maximum width of arch)

**92. a.** Find an equation of the tangent line to the parabola $y = ax^2$ at the point where $x = x_0$.
   **b.** Use the result of part (a) to show that the $x$-intercept of this tangent line is $x_0/2$.
   **c.** Use the result of part (b) to draw the tangent line.

**93.** Prove that any two distinct tangent lines to a parabola must intersect at one and only one point.

**94. a.** Show that an equation of the tangent line to the parabola $y^2 = 4px$ at the point $(x_0, y_0)$ can be written in the form
$$y_0 y = 2p(x + x_0)$$
   **b.** Use the result of part (a) to show that the $x$-intercept of this tangent line is $-x_0$.
   **c.** Use the result of part (b) to draw the tangent line for $p > 0$.

**95.** Show that an equation of the tangent line to the ellipse
$$\frac{x^2}{a^2} + \frac{y^2}{b^2} = 1$$
at the point $(x_0, y_0)$ can be written in the form
$$\frac{xx_0}{a^2} + \frac{yy_0}{b^2} = 1$$

**96.** Use the result of Exercise 95 to find an equation of the tangent line to the ellipse
$$\frac{x^2}{4} + \frac{y^2}{25} = 1$$
at the point $\left(1, \frac{5\sqrt{3}}{2}\right)$.

**97.** Show that an equation of the tangent line to the hyperbola
$$\frac{x^2}{a^2} - \frac{y^2}{b^2} = 1$$
at the point $(x_0, y_0)$ can be written in the form
$$\frac{xx_0}{a^2} - \frac{yy_0}{b^2} = 1$$

**98.** Use the result of Exercise 97 to find an equation of the tangent line to the hyperbola
$$\frac{x^2}{4} - \frac{y^2}{9} = 1$$
at the point $(4, 3\sqrt{3})$.

**99.** Show that the ellipse
$$\frac{x^2}{a^2} + \frac{2y^2}{b^2} = 1$$
and the hyperbola
$$\frac{x^2}{a^2 - b^2} - \frac{2y^2}{b^2} = 1$$
intersect at right angles.

**100.** Use the definition of a hyperbola to derive Equation (5) for a hyperbola with foci $F_1(-c, 0)$ and $F_2(c, 0)$ and vertices $V_1(a, 0)$ and $V_2(-a, 0)$.

**101.** Show that the lines $y = (b/a)x$ and $y = -(b/a)x$ are slant asymptotes of the hyperbola
$$\frac{x^2}{a^2} - \frac{y^2}{b^2} = 1$$

**102.** A transmitter $B$ is located 200 miles due east of a transmitter $A$ on a straight coastline. The two transmitters send out signals simultaneously to a ship that is located at $P$. Suppose that the ship receives the signal from $B$, 800 microseconds ($\mu$ sec) before it receives the signal from $A$.
   **a.** Assuming that radio waves travel at a speed of 980 ft/$\mu$ sec, find an equation of the hyperbola on which the ship lies (see page 838).
   **Hint:** $d(P, A) - d(P, B) = 2a$.

**b.** If the ship is sailing in a direction parallel to and 20 mi north of the coastline, locate the position of the ship at that instant of time.

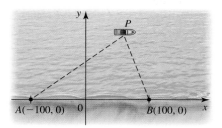

$A(-100, 0)$   0   $B(100, 0)$   $x$

**cas 103.** Use a computer algebra system (CAS) to find an approximation of the circumference of the ellipse

$$4x^2 + 25y^2 = 100$$

**cas 104.** The dwarf planet Pluto has an elliptical orbit with the sun at one focus. The length of the major axis of the ellipse is $7.33 \times 10^9$ miles, and the length of the minor axis is $7.08 \times 10^9$ miles. Use a CAS to approximate the distance traveled by the planet during one complete orbit around the sun.

**105. The Reflective Property of the Parabola** The figure shows a parabola with equation $y^2 = 4px$. The line $l$ is tangent to the parabola at the point $P(x_0, y_0)$.

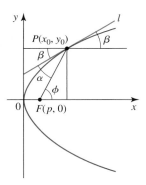

Show that $\alpha = \beta$ by establishing the following:

**a.** $\tan \beta = \dfrac{2p}{y_0}$     **b.** $\tan \phi = \dfrac{y_0}{x_0 - p}$

**c.** $\tan \alpha = \dfrac{2p}{y_0}$

Hint: $\tan \alpha = \dfrac{\tan \phi - \tan \beta}{1 + \tan \phi \tan \beta}$

**106. The Reflective Property of the Ellipse** Establish the reflective property of the ellipse by showing that $\alpha = \beta$ in Figure 18, page 831.

Hint: Use the trigonometric formula

$$\tan(\theta_1 - \theta_2) = \dfrac{\tan \theta_1 - \tan \theta_2}{1 + \tan \theta_1 \tan \theta_2}$$

**107. Reflective Property of the Hyperbola** The hyperbola also has the reflective property that the other two conics enjoy. Consider a mirror that has the shape of one branch of a hyperbola as shown in Figure (a). A ray of light aimed at a focus $F_2$ will be reflected toward the other focus, $F_1$. To establish the reflective property of the hyperbola, let $P(x_0, y_0)$ be a point on the hyperbola $x^2/a^2 - y^2/b^2 = 1$ with foci $F_1$ and $F_2$, and let $\alpha$ and $\beta$ be the angles between the lines $PF_1$ and $PF_2$, as shown in Figure (b). Show that $\alpha = \beta$.

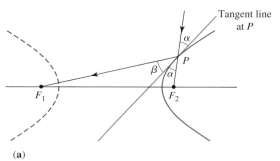

**(a)**

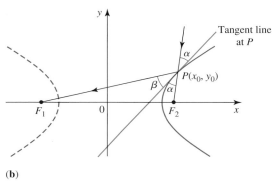

**(b)**

**108. Reflecting Telescopes** The reflective properties of the parabola and the hyperbola are exploited in designing a reflecting telescope. A hyperbolic mirror and a parabolic mirror are placed so that one focus of the hyperbola coincides with the focus $F$ of the parabola, as shown in the figure. Use the reflective properties of the two conics to explain why rays of light coming from great distances are finally focused at the eyepiece placed at $F'$, the other focus of the hyperbola.

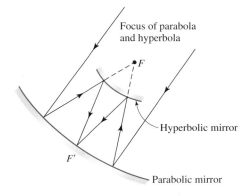

Focus of parabola and hyperbola

$F$

Hyperbolic mirror

$F'$

Parabolic mirror

*In Exercises* 109–114, *determine whether the statement is true or false. If it is true, explain why it is true. If it is false, give an example to show why it is false.*

**109.** The graph of $2x^2 - y^2 + F = 0$ is a hyperbola, provided that $F \neq 0$.

**110.** The graph of $y^4 = 16ax^2$, where $a > 0$, is a parabola.

**111.** The ellipse $b^2x^2 + a^2y^2 = a^2b^2$, where $a > b > 0$, is contained in the circle $x^2 + y^2 = a^2$ and contains the circle $x^2 + y^2 = b^2$.

**112.** The asymptotes of the hyperbola $x^2/a^2 - y^2/b^2 = 1$ are perpendicular to each other if and only if $a = b$.

**113.** If $A$ and $C$ are both positive constants, then

$$Ax^2 + Cy^2 + Dx + Ey + F = 0$$

is an ellipse.

**114.** If $A$ and $C$ have opposite signs, then

$$Ax^2 + Cy^2 + Dx + Ey + F = 0$$

is a hyperbola.

# 9.2 Plane Curves and Parametric Equations

## ■ Why We Use Parametric Equations

Figure 1a gives a bird's-eye view of a proposed training course for a yacht. In Figure 1b we have introduced an $xy$-coordinate system in the plane to describe the position of the yacht. With respect to this coordinate system the position of the yacht is given by the point $P(x, y)$, and the course itself is the graph of the rectangular equation $4x^4 - 4x^2 + y^2 = 0$, which is called a *lemniscate*. But representing the lemniscate in terms of a rectangular equation in this instance has three major drawbacks.

**FIGURE 1**     **(a)** The dots give the position of markers.          **(b)** An equation of the curve $C$ is $4x^4 - 4x^2 + y^2 = 0$.

First, the equation does not define $y$ explicitly as a function of $x$ or $x$ as a function of $y$. You can also convince yourself that this is not the graph of a function by applying the vertical and horizontal line tests to the curve in Figure 1b (see Section 0.2). Because of this, we cannot make direct use of many of the results for functions developed earlier. Second, the equation does not tell us when the yacht is at a given point $(x, y)$. Third, the equation gives no inkling as to the direction of motion of the yacht.

To overcome these drawbacks when we consider the motion of an object in the plane or plane curves that are not graphs of functions, we turn to the following representation. If $(x, y)$ is a point on a curve in the $xy$-plane, we write

$$x = f(t) \qquad y = g(t)$$

where $f$ and $g$ are functions of an auxiliary variable $t$ with (common) domain some interval $I$. These equations are called **parametric equations,** $t$ is called a **parameter,** and the interval $I$ is called a **parameter interval.**

If we think of $t$ on the closed interval $[a, b]$ as representing time, then we can interpret the parametric equations in terms of the motion of a particle as follows: At $t = a$ the particle is at the **initial point** $(f(a), g(a))$ of the curve or **trajectory** $C$. As $t$ increases from $t = a$ to $t = b$, the particle traverses the curve in a specific direction called the **orientation** of the curve, eventually ending up at the **terminal point** $(f(b), g(b))$ of the curve. (See Figure 2.)

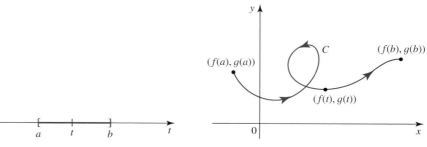

**FIGURE 2**
As $t$ increases from $a$ to $b$, the particle traces the curve from $(f(a), g(a))$ to $(f(b), g(b))$ in a specific direction.

Parameter interval is $[a, b]$.

We can also interpret the parametric equations in geometric terms as follows: We take the line segment $[a, b]$ and, by a process of stretching, bending, and twisting, make it conform geometrically to the curve $C$.

## ■ Sketching Curves Defined by Parametric Equations

Before looking at some examples, let's define the following term.

---

**DEFINITION    Plane Curve**

A plane curve is a set $C$ of ordered pairs $(x, y)$ defined by the parametric equations

$$x = f(t) \qquad \text{and} \qquad y = g(t)$$

where $f$ and $g$ are continuous functions on a parameter interval $I$.

---

**EXAMPLE 1**    Sketch the curve described by the parametric equations

$$x = t^2 - 4 \qquad \text{and} \qquad y = 2t \qquad -1 \leq t \leq 2$$

**Solution**    By plotting and connecting the points $(x, y)$ for selected values of $t$ (Table 1), we obtain the curve shown in Figure 3.

**TABLE 1**

| $t$ | $-1$ | $-\frac{1}{2}$ | $0$ | $\frac{1}{2}$ | $1$ | $2$ |
|---|---|---|---|---|---|---|
| $(x, y)$ | $(-3, -2)$ | $\left(-\frac{15}{4}, -1\right)$ | $(-4, 0)$ | $\left(-\frac{15}{4}, 1\right)$ | $(-3, 2)$ | $(0, 4)$ |

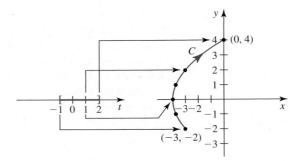

**FIGURE 3**
As $t$ increases from $-1$ to 2, the curve $C$ is traced from the initial point $(-3, -2)$ to the terminal point $(0, 4)$.

**Alternative Solution**   We eliminate the parameter $t$ by solving the second of the two given parametric equations for $t$, obtaining $t = \frac{1}{2}y$. We then substitute this value of $t$ into the first equation to obtain

$$x = \left(\frac{1}{2}y\right)^2 - 4 \qquad \text{or} \qquad x = \frac{1}{4}y^2 - 4$$

This is an equation of a parabola that has the $x$-axis as its axis of symmetry and its vertex at $(-4, 0)$. Now observe that $t = -1$ gives $(-3, -2)$ as the initial point of the curve and that $t = 2$ gives $(0, 4)$ as the terminal point of the curve. So tracing the graph from the initial point to the terminal point gives the desired curve, as obtained earlier.

We will adopt the convention here, just as we did with the domain of a function, that the parameter interval for $x = f(t)$ and $y = g(t)$ will consist of all values of $t$ for which $f(t)$ and $g(t)$ are real numbers, unless otherwise noted.

**EXAMPLE 2**   Sketch the curves represented by

**a.** $x = \sqrt{t}$ and $y = t$
**b.** $x = t$ and $y = t^2$

**Solution**
**a.** We eliminate the parameter $t$ by squaring the first equation to obtain $x^2 = t$. Substituting this value of $t$ into the second equation, we obtain $y = x^2$, which is an equation of a parabola. But note that the first parametric equation implies that $t \geq 0$, so $x \geq 0$. Therefore, the desired curve is the right portion of the parabola shown in Figure 4. Finally, note that the parameter interval is $[0, \infty)$, and as $t$ increases from 0, the desired curve starts at the initial point $(0, 0)$ and moves away from it along the parabola.

**FIGURE 4**
As $t$ increases from 0, the curve starts out at $(0, 0)$ and follows the right portion of the parabola with indicated orientation.

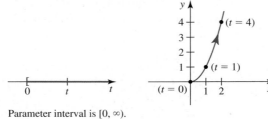

Parameter interval is $[0, \infty)$.

| $t$ | $(x, y)$ |
|---|---|
| 0 | $(0, 0)$ |
| 1 | $(1, 1)$ |
| 2 | $(\sqrt{2}, 2)$ |
| 4 | $(2, 4)$ |

**b.** Substituting the first equation into the second yields $y = x^2$. Although the rectangular equation is the same as that in part (a), the curve described by the parametric equations here is different from that of part (a), as we will now see. In this instance the parameter interval is $(-\infty, \infty)$. Furthermore, as $t$ increases from $-\infty$ to $\infty$, the curve runs along the parabola $y = x^2$ from left to right, as you can see by plotting the points corresponding to, say, $t = -1, 0$, and $1$. You can also see this by examining the parametric equation $x = t$, which tells us that $x$ increases as $t$ increases. (See Figure 5.)

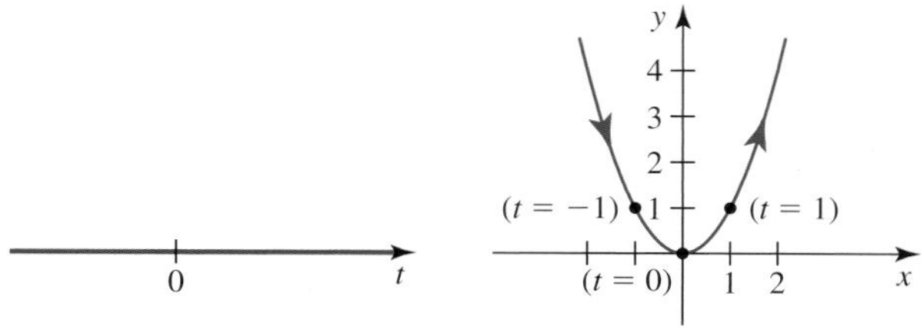

**FIGURE 5**
As $t$ increases from $-\infty$ to $\infty$, the entire parabola is traced out, from left to right.

Parameter interval is $(-\infty, \infty)$.

| $t$ | $(x, y)$ |
|-----|----------|
| $-1$ | $(-1, 1)$ |
| $0$ | $(0, 0)$ |
| $1$ | $(1, 1)$ |

For problems involving motion, it is natural to use the parameter $t$ to represent time. But other situations call for different representations or interpretations of the parameters, as the next two examples show. Here, we use an *angle* as a parameter.

**EXAMPLE 3** Describe the curves represented by the parametric equations

$$x = a \cos \theta \qquad \text{and} \qquad y = a \sin \theta \qquad a > 0$$

with parameter intervals

**a.** $[0, \pi]$
**b.** $[0, 2\pi]$
**c.** $[0, 4\pi]$

**Solution** We have $\cos \theta = x/a$ and $\sin \theta = y/a$. So

$$1 = \cos^2 \theta + \sin^2 \theta = \left(\frac{x}{a}\right)^2 + \left(\frac{y}{a}\right)^2$$

giving us

$$x^2 + y^2 = a^2$$

This tells us that each of the curves under consideration is contained in a circle of radius $a$, centered at the origin.

**a.** If $\theta = 0$, then $x = a$ and $y = 0$, giving $(a, 0)$ as the initial point on the curve. As $\theta$ increases from $0$ to $\pi$, the required curve is traced out in a counterclockwise direction, terminating at the point $(-a, 0)$. (See Figure 6a.)
**b.** Here, the curve is a complete circle that is traced out in a counterclockwise direction, starting at $(a, 0)$ and terminating at the same point (see Figure 6b).
**c.** The curve here is a circle that is traced out *twice* in a counterclockwise direction starting at $(a, 0)$ and terminating at the same point (see Figure 6c).

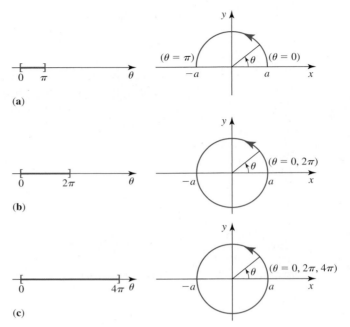

(a)

(b)

(c)

Parameter interval
**FIGURE 6**
The curve is (a) a semicircle, (b) a complete circle, and (c) a complete circle traced out twice. All curves are traced in a counterclockwise direction. ■

**EXAMPLE 4**   Describe the curve represented by

$$x = 4 \cos \theta \qquad \text{and} \qquad y = 3 \sin \theta \qquad 0 \le \theta \le 2\pi$$

**Solution**   Solving the first equation for $\cos \theta$ and the second equation for $\sin \theta$ gives

$$\cos \theta = \frac{x}{4}$$

and

$$\sin \theta = \frac{y}{3}$$

Squaring each equation and adding the resulting equations, we obtain

$$\cos^2 \theta + \sin^2 \theta = \left(\frac{x}{4}\right)^2 + \left(\frac{y}{3}\right)^2$$

Since $\cos^2 \theta + \sin^2 \theta = 1$, we end up with the rectangular equation

$$\frac{x^2}{16} + \frac{y^2}{9} = 1$$

From this we see that the curve is contained in an ellipse centered at the origin. If $\theta = 0$, then $x = 4$ and $y = 0$, giving $(4, 0)$ as the initial point of the curve. As $\theta$ increases from 0 to $2\pi$, the elliptical curve is traced out in a counterclockwise direction, termi-nating at $(4, 0)$. (See Figure 7.)

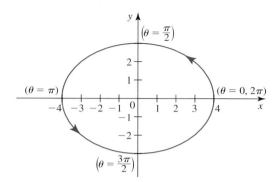

**FIGURE 7**
As $\theta$ increases from 0 to $2\pi$, the curve that is traced out in a counterclockwise direction beginning and ending at (4, 0) is an ellipse.

**EXAMPLE 5** A proposed training course for a yacht is represented by the parametric equations

$$x = \sin t \quad \text{and} \quad y = \sin 2t \quad 0 \leq t \leq 2\pi$$

where $x$ and $y$ are measured in miles.

**a.** Show that the rectangular equation of the course is $4x^4 - 4x^2 + y^2 = 0$.
**b.** Describe the course.

**Solution**
**a.** Using the trigonometric identity $\sin 2t = 2 \sin t \cos t$, we rewrite the second of the parametric equations in the form

$$y = 2 \sin t \cos t = 2x \cos t \qquad \text{Since } x = \sin t$$

Solving for $\cos t$, we have

$$\cos t = \frac{y}{2x}$$

Then, using the identity $\sin^2 t + \cos^2 t = 1$, we obtain

$$x^2 + \left(\frac{y}{2x}\right)^2 = 1$$

$$x^2 + \frac{y^2}{4x^2} = 1$$

or

$$4x^4 - 4x^2 + y^2 = 0$$

**b.** From the results of part (a) we see that the required curve is symmetric with respect to the $x$-axis, the $y$-axis, and the origin. Therefore, it suffices to concentrate first on drawing the part of the curve that lies in the first quadrant and then make use of symmetry to complete the curve. Since both $\sin t$ and $\sin 2t$ are nonnegative only for $0 \leq t \leq \frac{\pi}{2}$, we first sketch the curve corresponding to values of $t$ in $\left[0, \frac{\pi}{2}\right]$. With the help of the following table we obtain the curve shown in Figure 8. The direction of the yacht is indicated by the arrows.

| $t$ | 0 | $\frac{\pi}{6}$ | $\frac{\pi}{4}$ | $\frac{\pi}{3}$ | $\frac{\pi}{2}$ |
|-----|---|-----------------|-----------------|-----------------|-----------------|
| $(x, y)$ | $(0, 0)$ | $\left(\frac{1}{2}, \frac{\sqrt{3}}{2}\right)$ | $\left(\frac{\sqrt{2}}{2}, 1\right)$ | $\left(\frac{\sqrt{3}}{2}, \frac{\sqrt{3}}{2}\right)$ | $(1, 0)$ |

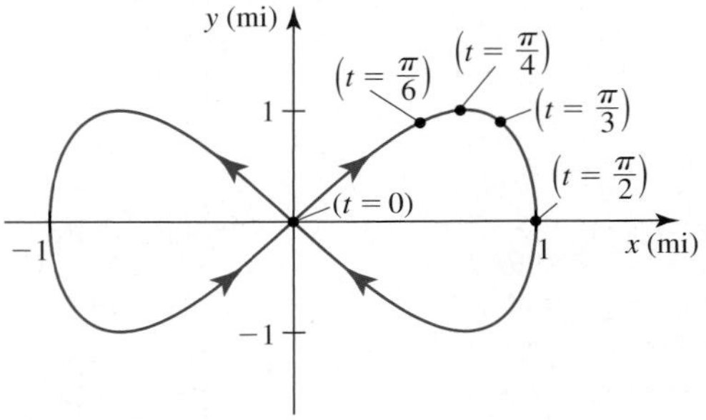

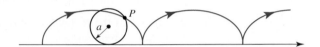

**FIGURE 8**
The training course for the yacht

**EXAMPLE 6** **Cycloids** Let $P$ be a fixed point on the rim of a wheel. If the wheel is allowed to roll along a straight line without slipping, then the point $P$ traces out a curve called a **cycloid** (see Figure 9). Suppose that the wheel has radius $a$ and rolls along the $x$-axis. Find parametric equations for the cycloid.

**FIGURE 9**
The cycloid is the curve
traced out by a fixed point $P$
on the rim of a rolling wheel.

**Solution** Suppose that the wheel rolls in a positive direction with the point $P$ initially at the origin of the coordinate system. Figure 10 shows the position of the wheel after it has rotated through $\theta$ radians. Because there is no slippage, the distance the wheel has rolled from the origin is

$$d(O, M) = \text{length of arc } PM = a\theta$$

giving its center as $C(a\theta, a)$. Also, from Figure 10 we see that the coordinates of $P(x, y)$ satisfy

$$x = d(O, M) - a \sin \theta = a\theta - a \sin \theta = a(\theta - \sin \theta)$$

and

$$y = d(C, M) - a \cos \theta = a - a \cos \theta = a(1 - \cos \theta)$$

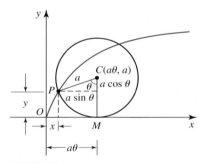

**FIGURE 10**
The position of the wheel after it has
rotated through $\theta$ radians

Although these results are derived under the tacit assumption that $0 < \theta < \frac{\pi}{2}$, it can be demonstrated that they are valid for other values of $\theta$. Therefore, the required parametric equations of the cycloid are

$$x = a(\theta - \sin \theta) \quad \text{and} \quad y = a(1 - \cos \theta) \quad -\infty < \theta < \infty$$

The cycloid provides the solution to two famous problems in mathematics:

1. *The brachistochrone problem:* Find the curve along which a moving particle (under the influence of gravity) will slide from a point $A$ to another point $B$, not directly beneath $A$, in the shortest time (see Figure 11a).
2. *The tautochrone problem:* Find the curve having the property that it takes the same time for a particle to slide to the bottom of the curve no matter where the particle is placed on the curve (see Figure 11b).

The brachistochrone problem—the problem of finding the curve of quickest descent—was advanced in 1696 by the Swiss mathematician Johann Bernoulli. Offhand, one might conjecture that such a curve should be a straight line, since it yields

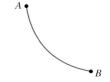

**FIGURE 11**
The cycloid provides the solution
to both the brachistochrone
and the tautochrone problem.

(**a**) The brachistochrone problem          (**b**) The tautochrone problem

the shortest distance between the two points. But the velocity of the particle moving on the straight line will build up comparatively slowly, whereas if we take a curve that is steeper near $A$, even though the path becomes longer, the particle will cover a large portion of the distance at a greater speed. The problem was solved by Johann Bernoulli, his older brother Jacob Bernoulli, Leibniz, Newton, and l'Hôpital. They found that the curve of quickest descent is an inverted arc of a cycloid (Figure 11a). As it turns out, this same curve is also the solution to the tautochrone problem.

## 9.2  CONCEPT QUESTIONS

1. What is a plane curve? Give an example of a plane curve that is not the graph of a function.

2. What is the difference between the curve $C_1$ with parametric representation $x = \cos t$ and $y = \sin t$, where $0 \le t \le 2\pi$, and the curve $C_2$ with parametric representation $x = \sin t$ and $y = \cos t$, where $0 \le t \le 2\pi$?

3. Describe the relationship between the curve $C_1$ with parametric equations $x = f(t)$ and $y = g(t)$, where $0 \le t \le 1$, and the curve $C_2$ with parametric equations $x = f(1 - t)$ and $y = g(1 - t)$, where $0 \le t \le 1$.

## 9.2  EXERCISES

In Exercises 1–28, (a) find a rectangular equation whose graph contains the curve $C$ with the given parametric equations, and (b) sketch the curve $C$ and indicate its orientation.

1. $x = 2t + 1, \quad y = t - 3$

2. $x = t - 2, \quad y = 2t - 1; \quad -1 \le t \le 5$

3. $x = \sqrt{t}, \quad y = 9 - t$

4. $x = t^2, \quad y = t - 1; \quad 0 \le t \le 3$

5. $x = t^2 + 1, \quad y = 2t^2 - 1; \quad -2 \le t \le 2$

6. $x = t^3, \quad y = 2t + 1$

7. $x = t^2, \quad y = t^3; \quad -2 \le t \le 2$

8. $x = 1 + \dfrac{1}{t}, \quad y = t + 1$

9. $x = 2 \sin \theta, \quad y = 2 \cos \theta; \quad 0 \le \theta \le 2\pi$

10. $x = \cos 2\theta, \quad y = 3 \sin \theta; \quad 0 \le \theta \le 2\pi$

11. $x = 2 \sin \theta, \quad y = 3 \cos \theta; \quad 0 \le \theta \le 2\pi$

12. $x = \cos \theta + 1, \quad y = \sin \theta - 2; \quad 0 \le \theta \le 2\pi$

13. $x = 2 \cos \theta + 2, \quad y = 3 \sin \theta - 1; \quad 0 \le \theta \le 2\pi$

14. $x = \sin \theta + 3, \quad y = 3 \cos \theta + 1; \quad 0 \le \theta \le 2\pi$

15. $x = \cos \theta, \quad y = \cos 2\theta$

16. $x = \sec \theta, \quad y = \cos \theta$

17. $x = \sec \theta, \quad y = \tan \theta; \quad -\frac{\pi}{2} < \theta < \frac{\pi}{2}$

18. $x = \cos^3 \theta, \quad y = \sin^3 \theta$

19. $x = \sin^2 \theta, \quad y = \sin^4 \theta; \quad 0 \le \theta \le \frac{\pi}{2}$

20. $x = e^t, \quad y = e^{-t}$

21. $x = -e^t, \quad y = e^{2t}$

22. $x = t^3, \quad y = 3 \ln t$

23. $x = \ln 2t, \quad y = t^2$

24. $x = e^t, \quad y = \ln t$

25. $x = \cosh t, \quad y = \sinh t$

26. $x = 3 \sinh t, \quad y = 2 \cosh t$

27. $x = (t - 1)^2, \quad y = (t - 1)^3; \quad 1 \le t \le 2$

28. $x = \dfrac{2t}{1 + t^2}, \quad y = \dfrac{1 - t^2}{1 + t^2}$

*In Exercises 29–34, the position of a particle at time t is* $(x, y)$. *Describe the motion of the particle as t varies over the time interval* $[a, b]$.

**29.** $x = t + 1$, $y = \sqrt{t}$; $[0, 4]$

**30.** $x = \sin \pi t$, $y = \cos \pi t$; $[0, 6]$

**31.** $x = 1 + \cos t$, $y = 2 + \sin t$; $[0, 2\pi]$

**32.** $x = 1 + 2 \sin 2t$, $y = 2 + 4 \sin 2t$; $[0, 2\pi]$

**33.** $x = \sin t$, $y = \sin^2 t$; $[0, 3\pi]$

**34.** $x = e^{-t}$, $y = e^{2t-1}$; $[0, \infty)$

**35. Flight Path of an Aircraft** The position $(x, y)$ of an aircraft flying in a fixed direction $t$ seconds after takeoff is given by $x = \tan(0.025\pi t)$ and $y = \sec(0.025\pi t) - 1$, where $x$ and $y$ are measured in miles. Sketch the flight path of the aircraft for $0 \le t \le \frac{40}{3}$.

**36. Trajectory of a Shell** A shell is fired from a howitzer with a muzzle speed of $v_0$ ft/sec. If the angle of elevation of the howitzer is $\alpha$, then the position of the shell after $t$ sec is described by the parametric equations

$$x = (v_0 \cos \alpha)t \quad \text{and} \quad y = (v_0 \sin \alpha)t - \frac{1}{2}gt^2$$

where $g$ is the acceleration due to gravity (32 ft/sec$^2$).
  **a.** Find the range of the shell.
  **b.** Find the maximum height attained by the shell.
  **c.** Show that the trajectory of the shell is a parabola by eliminating the parameter $t$.

**37.** Let $P_1(x_1, y_1)$ and $P_2(x_2, y_2)$ be two distinct points in the plane. Show that the parametric equations

$$x = x_1 + (x_2 - x_1)t \quad \text{and} \quad y = y_1 + (y_2 - y_1)t$$

describe (a) the line passing through $P_1$ and $P_2$ if $-\infty < t < \infty$ and (b) the line segment joining $P_1$ and $P_2$ if $0 \le t \le 1$.

**38.** Show that

$$x = a \cos t + h \quad \text{and} \quad y = b \sin t + k \quad 0 \le t \le 2\pi$$

are parametric equations of an ellipse with center at $(h, k)$ and axes of lengths $2a$ and $2b$.

**39.** Show that

$$x = a \sec t + h \quad \text{and} \quad y = b \tan t + k$$
$$t \in \left(-\frac{\pi}{2}, \frac{\pi}{2}\right) \cup \left(\frac{\pi}{2}, \frac{3\pi}{2}\right)$$

are parametric equations of a hyperbola with center at $(h, k)$ and transverse and conjugate axes of lengths $2a$ and $2b$, respectively.

**40.** Let $P$ be a point located a distance $d$ from the center of a circle of radius $r$. The curve traced out by $P$ as the circle

rolls without slipping along a straight line is called a **trochoid.** (The cycloid is the special case of a trochoid with $d = r$.) Suppose that the circle rolls along the $x$-axis in the positive direction with $\theta = 0$ when the point $P$ is at one of the lowest points on the trochoid. Show that the parametric equations of the trochoid are

$$x = r\theta - d \sin \theta \quad \text{and} \quad y = r - d \cos \theta$$

where $\theta$ is the same parameter as that for the cycloid. Sketch the trochoid for the cases in which $d < r$ and $d > r$.

**41.** The **witch of Agnesi** is the curve shown in the following figure. Show that the parametric equations of this curve are

$$x = 2a \cot \theta \quad \text{and} \quad y = 2a \sin^2 \theta$$

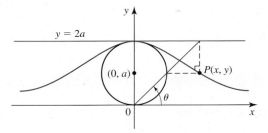

**42.** If a string is unwound from a circle of radius $a$ in such a way that it is held taut in the plane of the circle, then its end $P$ will trace a curve called the **involute of the circle.** Referring to the following figure, show that the parametric equations of the involute are

$$x = a(\cos \theta + \theta \sin \theta) \quad \text{and} \quad y = a(\sin \theta - \theta \cos \theta)$$

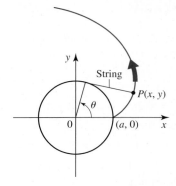

*In Exercises 43–46, use a graphing utility to plot the curve with the given parametric equations.*

**43.** $x = 2 \sin 3t$, $y = 3 \sin 1.5t$; $t \ge 0$

**44.** $x = \cos t + 5 \cos 3t$, $y = 6 \cos t - 5 \sin 3t$; $0 \le t \le 2\pi$

**45.** $x = 2 \cos t + \cos 2t$, $y = 2 \sin t - \sin 2t$; $0 \le t \le 2\pi$

**46.** $x = 3 \cos t + \cos 3t$, $y = 3 \sin t - \sin 3t$; $0 \le t \le 2\pi$

**47.** The **butterfly catastrophe curve,** which is described by the parametric equations

$$x = c(8at^3 + 24t^5) \quad \text{and} \quad y = c(-6at^2 - 15t^4)$$

occurs in the study of catastrophe theory. Plot the curve with $a = -7$ and $c = 0.03$ for $t$ in the parameter interval $[-1.629, 1.629]$.

**48.** The **swallowtail catastrophe curve,** which is described by the parametric equations

$$x = c(-2at - 4t^3) \quad \text{and} \quad y = c(at^2 + 3t^4)$$

occurs in the study of catastrophe theory. Plot the curve with $a = -2$ and $c = 0.5$ for $t$ in the parameter interval $[-1.25, 1.25]$.

**49.** The **Lissajous curves,** also known as **Bowditch curves,** have applications in physics, astronomy, and other sciences. They are described by the parametric equations

$$x = \sin(at + b\pi), \quad a \text{ a rational number, } \quad \text{and} \quad y = \sin t$$

Plot the curve with $a = 0.75$ and $b = 0$ for $t$ in the parameter interval $[0, 8\pi]$.

**50.** The **prolate cycloid** is the path traced out by a fixed point at a distance $b > a$ from the center of a rolling circle, where $a$

is the radius of the circle. The prolate cycloid is described by the parametric equations

$$x = a(t - b \sin t) \quad \text{and} \quad y = c(1 - d \cos t)$$

Plot the curve with $a = 0.1$, $b = 2$, $c = 0.25$, and $d = 2$ for $t$ in the parameter interval $[-10, 10]$.

*In Exercises 51–54, determine whether the statement is true or false. If it is true, explain why it is true. If it is false, give an example to show why it is false.*

**51.** The parametric equations $x = \cos^2 t$ and $y = \sin^2 t$, where $-\infty < t < \infty$, have the same graph as $x + y = 1$.

**52.** The graph of a function $y = f(x)$ can always be represented by a pair of parametric equations.

**53.** The curve with parametric equations $x = f(t) + a$ and $y = g(t) + b$ is obtained from the curve $C$ with parametric equations $x = f(t)$ and $y = g(t)$ by shifting the latter horizontally and vertically.

**54.** The ellipse with center at the origin and major and minor axes $a$ and $b$, respectively, can be obtained from the circle with equations $x = f(t) = \cos t$ and $y = g(t) = \sin t$ by multiplying $f(t)$ and $g(t)$ by appropriate nonzero constants.

---

## 9.3 The Calculus of Parametric Equations

### ■ Tangent Lines to Curves Defined by Parametric Equations

Suppose that $C$ is a smooth curve that is parametrized by the equations $x = f(t)$ and $y = g(t)$ with parameter interval $I$ and we wish to find the slope of the tangent line to the curve at the point $P$. (See Figure 1.) Let $t_0$ be the point in $I$ that corresponds to $P$, and let $(a, b)$ be the subinterval of $I$ containing $t_0$ corresponding to the highlighted portion of the curve $C$ in Figure 1. This subset of $C$ is the graph of a function of $x$, as you can verify using the Vertical Line Test. (The general conditions that $f$ and $g$ must satisfy for this to be true are given in Exercise 66.)

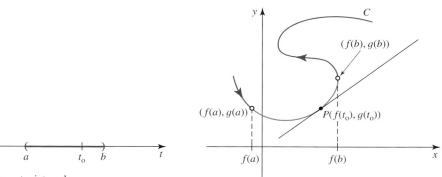

**FIGURE 1**
We want to find the slope of the tangent line to the curve at the point $P$.

Parameter interval

Let's denote this function by $F$ so that $y = F(x)$, where $f(a) < x < f(b)$. Since $x = f(t)$ and $y = g(t)$, we may rewrite this equation in the form

$$g(t) = F[f(t)]$$

Using the Chain Rule, we obtain

$$g'(t) = F'[f(t)]f'(t)$$
$$= F'(x)f'(t) \qquad \text{Replace } f(t) \text{ by } x.$$

If $f'(t) \neq 0$, we can solve for $F'(x)$, obtaining

$$F'(x) = \frac{g'(t)}{f'(t)}$$

which can also be written

$$\frac{dy}{dx} = \frac{\dfrac{dy}{dt}}{\dfrac{dx}{dt}} \qquad \text{if} \quad \frac{dx}{dt} \neq 0 \qquad\qquad \textbf{(1)}$$

The required slope of the tangent line at $P$ is then found by evaluating Equation (1) at $t_0$. Observe that Equation (1) enables us to solve the problem without eliminating $t$.

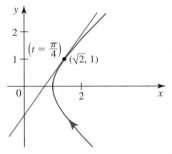

**FIGURE 2**
The tangent line to the curve at $(\sqrt{2}, 1)$

**EXAMPLE 1**   Find an equation of the tangent line to the curve

$$x = \sec t \qquad y = \tan t \qquad -\tfrac{\pi}{2} < t < \tfrac{\pi}{2}$$

at the point where $t = \pi/4$. (See Figure 2.)

**Solution**   The slope of the tangent line at any point $(x, y)$ on the curve is

$$\frac{dy}{dx} = \frac{\dfrac{dy}{dt}}{\dfrac{dx}{dt}}$$

$$= \frac{\sec^2 t}{\sec t \tan t} = \frac{\sec t}{\tan t}$$

In particular, the slope of the tangent line at the point where $t = \pi/4$ is

$$\left.\frac{dy}{dx}\right|_{t=\pi/4} = \frac{\sec \dfrac{\pi}{4}}{\tan \dfrac{\pi}{4}} = \frac{\sqrt{2}}{1} = \sqrt{2}$$

Also, when $t = \pi/4$, we have $x = \sec(\pi/4) = \sqrt{2}$ and $y = \tan(\pi/4) = 1$ giving $(\sqrt{2}, 1)$ as the point of tangency. Finally, using the point-slope form of the equation of a line, we obtain the required equation:

$$y - 1 = \sqrt{2}(x - \sqrt{2}) \qquad \text{or} \qquad y = \sqrt{2}x - 1 \qquad\qquad \blacksquare$$

## Horizontal and Vertical Tangents

A curve $C$ represented by the parametric equations $x = f(t)$ and $y = g(t)$ has a **horizontal** tangent at a point $(x, y)$ on $C$ where $dy/dt = 0$ and $dx/dt \neq 0$ and a **vertical** tangent where $dx/dt = 0$ and $dy/dt \neq 0$, so that $dy/dx$ is undefined there. Points where both $dy/dt$ and $dx/dt$ are equal to zero are candidates for horizontal or vertical tangents and may be investigated by using l'Hôpital's Rule.

**EXAMPLE 2** A curve $C$ is defined by the parametric equations $x = t^2$ and $y = t^3 - 3t$.

a. Find the points on $C$ where the tangent lines are horizontal or vertical.
b. Find the $x$- and $y$-intercepts of $C$.
c. Sketch the graph of $C$.

**Solution**
a. Setting $dy/dt = 0$ gives $3t^2 - 3 = 0$, or $t = \pm 1$. Since $dx/dt = 2t \neq 0$ at these values of $t$, we conclude that $C$ has horizontal tangents at the points on $C$ corresponding to $t = \pm 1$, that is, at $(1, -2)$ and $(1, 2)$. Next, setting $dx/dt = 0$ gives $2t = 0$, or $t = 0$. Since $dy/dt \neq 0$ for this value of $t$, we conclude that $C$ has a vertical tangent at the point corresponding to $t = 0$, or at $(0, 0)$.
b. To find the $x$-intercepts, we set $y = 0$, which gives $t^3 - 3t = t(t^2 - 3) = 0$, or $t = -\sqrt{3}, 0$, and $\sqrt{3}$. Substituting these values of $t$ into the expression for $x$ gives 0 and 3 as the $x$-intercepts. Next, setting $x = 0$ gives $t = 0$, which, when substituted into the expression for $y$, gives 0 as the $y$-intercept.
c. Using the information obtained in parts (a) and (b), we obtain the graph of $C$ shown in Figure 3.

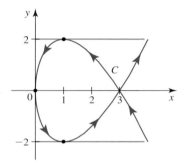

**FIGURE 3**
The graph of $x = t^2$, $y = t^3 - 3t$, and the tangent lines at $t = \pm 1$

## Finding $d^2y/dx^2$ from Parametric Equations

Suppose that the parametric equations $x = f(t)$ and $y = g(t)$ define $y$ as a twice-differentiable function of $x$ over some suitable interval. Then $d^2y/dx^2$ may be found from Equation (1) with another application of the Chain Rule.

$$\frac{d^2y}{dx^2} = \frac{d}{dx}\left(\frac{dy}{dx}\right) = \frac{\dfrac{d}{dt}\left(\dfrac{dy}{dx}\right)}{\dfrac{dx}{dt}} \qquad \text{if} \quad \frac{dx}{dt} \neq 0 \qquad (2)$$

Higher-order derivatives are found in a similar manner.

**EXAMPLE 3** Find $\dfrac{d^2y}{dx^2}$ if $x = t^2 - 4$ and $y = t^3 - 3t$.

**Solution** First, we use Equation (1) to compute

$$\frac{dy}{dx} = \frac{\dfrac{dy}{dt}}{\dfrac{dx}{dt}} = \frac{3t^2 - 3}{2t}$$

Then, using Equation (2), we obtain

$$\frac{d^2y}{dx^2} = \frac{\dfrac{d}{dt}\left(\dfrac{dy}{dx}\right)}{\dfrac{dx}{dt}} = \frac{\dfrac{d}{dt}\left(\dfrac{3t^2-3}{2t}\right)}{2t}$$

$$= \frac{\dfrac{(2t)(6t) - (3t^2 - 3)(2)}{4t^2}}{2t} \qquad \text{Use the Quotient Rule.}$$

$$= \frac{6t^2 + 6}{8t^3} = \frac{3(t^2+1)}{4t^3} \qquad\blacksquare$$

## ◼ The Length of a Smooth Curve

In Section 5.4 we showed that the length $L$ of the graph of a smooth function $f$ on an interval $[a, b]$ can be found by using the formula

$$L = \int_a^b \sqrt{1 + [f'(x)]^2}\, dx \qquad (3)$$

We now generalize this result to include curves defined by parametric equations. We begin by explaining what is meant by a *smooth* curve defined parametrically. Suppose that $C$ is represented by $x = f(t)$ and $y = g(t)$ on a parameter interval $I$. Then $C$ is **smooth** if $f'$ and $g'$ are continuous on $I$ and are not simultaneously zero, except possibly at the endpoints of $I$. A smooth curve is devoid of corners or cusps. For example, the cycloid that we discussed in Section 9.2 (see Figure 9 in that section) has sharp corners at the values $x = 2n\pi a$ and, therefore, is not smooth. However, it is smooth between these points.

Now let $P = \{t_0, t_1, \ldots, t_n\}$ be a regular partition of the parameter interval $[a, b]$. Then the point $P_k(f(t_k), g(t_k))$ lies on $C$, and the length of $C$ is approximated by the length of the polygonal curve with vertices $P_0, P_1, \ldots, P_n$. (See Figure 4.) Thus,

$$L \approx \sum_{k=1}^n d(P_{k-1}, P_k) \qquad (4)$$

where

$$d(P_{k-1}, P_k) = \sqrt{[f(t_k) - f(t_{k-1})]^2 + [g(t_k) - g(t_{k-1})]^2}$$

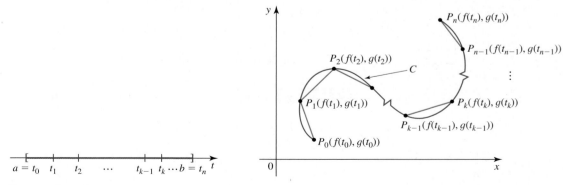

Parameter interval

**FIGURE 4**

The length of $C$ is approximated by the length of the polygonal curve (the red lines).

Now, since $f$ and $g$ both have continuous derivatives, we can use the Mean Value Theorem to write

$$f(t_k) - f(t_{k-1}) = f'(t_k^*)(t_k - t_{k-1})$$

and

$$g(t_k) - g(t_{k-1}) = g'(t_k^{**})(t_k - t_{k-1})$$

where $t_k^*$ and $t_k^{**}$ are numbers in $(t_{k-1}, t_k)$. Substituting these expressions into Equation (4) gives

$$L \approx \sum_{k=1}^{n} d(P_{k-1}, P_k) = \sum_{k=1}^{n} \sqrt{[f'(t_k^*)]^2 + [g'(t_k^{**})]^2} \, \Delta t \tag{5}$$

As in Section 5.4, we define

$$L = \lim_{n \to \infty} \sum_{k=1}^{n} d(P_{k-1}, P_k)$$

$$= \lim_{n \to \infty} \sum_{k=1}^{n} \sqrt{[f'(t_k^*)]^2 + [g'(t_k^{**})]^2} \, \Delta t \tag{6}$$

The sum in Equation (6) looks like a Riemann sum of the function $\sqrt{[f']^2 + [g']^2}$, but it is not, because $t_k^*$ is not necessarily equal to $t_k^{**}$. But it can be shown that the limit in Equation (6) is the same as that of an expression in which $t_k^* = t_k^{**}$. Therefore,

$$L = \int_a^b \sqrt{[f'(t)]^2 + [g'(t)]^2} \, dt$$

and we have the following result.

---

**THEOREM 1  Length of a Smooth Curve**

Let $C$ be a smooth curve represented by the parametric equations $x = f(t)$ and $y = g(t)$ with parameter interval $[a, b]$. If $C$ does not intersect itself, except possibly for $t = a$ and $t = b$, then the length of $C$ is

$$L = \int_a^b \sqrt{[f'(t)]^2 + [g'(t)]^2} \, dt = \int_a^b \sqrt{\left(\frac{dx}{dt}\right)^2 + \left(\frac{dy}{dt}\right)^2} \, dt \tag{7}$$

---

**Note**  Equation (7) is consistent with Equations (3) and (4) of Section 5.4. Both have the form $L = \int ds$, where $(ds)^2 = (dx)^2 + (dy)^2$.  ∎

---

**EXAMPLE 4**  Find the length of one arch of the cycloid

$$x = a(\theta - \sin \theta) \qquad y = a(1 - \cos \theta)$$

(See Example 6 in Section 9.2.)

**Solution**  One arch of the cycloid is traced out by letting $\theta$ run from 0 to $2\pi$. Now

$$\frac{dx}{d\theta} = a(1 - \cos \theta) \qquad \text{and} \qquad \frac{dy}{d\theta} = a \sin \theta$$

Therefore, using Equation (7), we find the required length to be

$$L = \int_0^{2\pi} \sqrt{\left(\frac{dx}{d\theta}\right)^2 + \left(\frac{dy}{d\theta}\right)^2} \, d\theta = \int_0^{2\pi} \sqrt{a^2(1 - \cos\theta)^2 + a^2 \sin^2\theta} \, d\theta$$

$$= \int_0^{2\pi} \sqrt{a^2 - 2a^2 \cos\theta + a^2 \cos^2\theta + a^2 \sin^2\theta} \, d\theta$$

$$= a \int_0^{2\pi} \sqrt{2(1 - \cos\theta)} \, d\theta \qquad \sin^2\theta + \cos^2\theta = 1$$

To evaluate this integral, we use the identity $\sin^2 x = \frac{1}{2}(1 - \cos 2x)$ with $\theta = 2x$. This gives $1 - \cos\theta = 2\sin^2(\theta/2)$, so

$$L = a \int_0^{2\pi} \sqrt{4 \sin^2 \frac{\theta}{2}} \, d\theta$$

$$= 2a \int_0^{2\pi} \sin \frac{\theta}{2} \, d\theta \qquad \sin\frac{\theta}{2} \geq 0 \text{ on } [0, 2\pi]$$

$$= -4a \left[ \cos \frac{\theta}{2} \right]_0^{2\pi}$$

$$= -4a(-1 - 1) = 8a \qquad \blacksquare$$

## ■ The Area of a Surface of Revolution

Recall that the formulas $S = 2\pi \int y \, ds$ and $S = 2\pi \int x \, ds$ (Formulas 11 and 12 of Section 5.4) give the area of the surface of revolution that is obtained by revolving the graph of a function about the $x$- and $y$-axes, respectively. These formulas are valid for finding the area of the surface of revolution that is obtained by revolving a curve described by parametric equations about the $x$- and the $y$-axes, provided that we replace the element of arc length $ds$ by the appropriate expression. These results, which may be derived by using the method used to derive Equation (7), are stated in the next theorem.

---

**THEOREM 2** **Area of a Surface of Revolution**

Let $C$ be a smooth curve represented by the parametric equations $x = f(t)$ and $y = g(t)$ with parameter interval $[a, b]$, and suppose that $C$ does not intersect itself, except possibly for $t = a$ and $t = b$. If $g(t) \geq 0$ for all $t$ in $[a, b]$, then the area $S$ of the surface obtained by revolving $C$ about the $x$-axis is

$$S = 2\pi \int_a^b y \sqrt{[f'(t)]^2 + [g'(t)]^2} \, dt = 2\pi \int_a^b y \sqrt{\left(\frac{dx}{dt}\right)^2 + \left(\frac{dy}{dt}\right)^2} \, dt \qquad (8)$$

If $f(t) \geq 0$ for all $t$ in $[a, b]$, then the area $S$ of the surface that is obtained by revolving $C$ about the $y$-axis is

$$S = 2\pi \int_a^b x \sqrt{[f'(t)]^2 + [g'(t)]^2} \, dt = 2\pi \int_a^b x \sqrt{\left(\frac{dx}{dt}\right)^2 + \left(\frac{dy}{dt}\right)^2} \, dt \qquad (9)$$

---

**EXAMPLE 5** Show that the surface area of a sphere of radius $r$ is $4\pi r^2$.

**Solution** We obtain this sphere by revolving the semicircle

$$x = r \cos t \qquad y = r \sin t \qquad 0 \leq t \leq \pi$$

about the *x*-axis. Using Equation (8), the surface area of the sphere is

$$S = 2\pi \int_0^\pi r \sin t \sqrt{(-r \sin t)^2 + (r \cos t)^2} \, dt$$

$$= 2\pi r \int_0^\pi \sin t \sqrt{r^2(\sin^2 t + \cos^2 t)} \, dt$$

$$= 2\pi r \int_0^\pi r \sin t \, dt \qquad \sin^2 t + \cos^2 t = 1$$

$$= 2\pi r^2 \left[ -\cos t \right]_0^\pi = 2\pi r^2 [-(-1) + 1] = 4\pi r^2 \qquad ∎$$

## 9.3    CONCEPT QUESTIONS

1. Suppose that *C* is a smooth curve with parametric equations
   $x = f(t)$ and $y = g(t)$ and parameter interval *I*. Write an
   expression for the slope of the tangent line to *C* at the point
   $(x_0, y_0)$ corresponding to $t_0$ in *I*.
2. Suppose that *C* is a smooth curve with parametric equations
   $x = f(t)$ and $y = g(t)$ and parameter interval $[a, b]$. Furthermore, suppose that *C* does not cross itself, except possibly
   for $t = a$. Write an expression giving the length of *C*.
3. Suppose that *C* is a smooth curve with parametric equations
   $x = f(t)$ and $y = g(t)$ and parameter interval $[a, b]$. Suppose,

further, that *C* does not intersect itself, except possibly for
$t = a$ and $t = b$.
   a. Write an integral giving the area of the surface obtained
      by revolving *C* about the *x*-axis assuming that $g(t) \geq 0$
      for all *t* in $[a, b]$.
   b. Write an integral giving the area of the surface obtained
      by revolving *C* about the *y*-axis assuming that $f(t) \geq 0$
      for all *t* in $[a, b]$.

## 9.3    EXERCISES

*In Exercises 1–6, find the slope of the tangent line to the curve
at the point corresponding to the value of the parameter.*

1. $x = t^2 + 1, \quad y = t^2 - t; \quad t = 1$

2. $x = t^3 - t, \quad y = t^2 - 2t + 2; \quad t = 2$

3. $x = \sqrt{t}, \quad y = \dfrac{1}{t}; \quad t = 1$

4. $x = e^{2t}, \quad y = \ln t; \quad t = 1$

5. $x = 2 \sin \theta, \quad y = 3 \cos \theta; \quad \theta = \dfrac{\pi}{4}$

6. $x = 2(\theta - \sin \theta), \quad y = 2(1 - \cos \theta); \quad \theta = \dfrac{\pi}{6}$

*In Exercises 7 and 8, find an equation of the tangent line to the
curve at the point corresponding to the value of the parameter.*

7. $x = 2t - 1, \quad y = t^3 - t^2; \quad t = 1$

8. $x = \theta \cos \theta, \quad y = \theta \sin \theta; \quad \theta = \dfrac{\pi}{2}$

*In Exercises 9 and 10, find an equation of the tangent line to the
curve at the given point. Then sketch the curve and the tangent
line(s).*

9. $x = t^2 + t, \quad y = t^2 - t^3; \quad (0, 2)$

10. $x = e^t, \quad y = e^{-t}; \quad (1, 1)$

*In Exercises 11 and 12, find the points on the curve at which the
slope of the tangent line is m.*

11. $x = 2t^2 - 1, \quad y = t^3; \quad m = 3$

12. $x = t^3, \quad y = t^2 + t; \quad m = 1$

*In Exercises 13–16, find the points on the curve at which the
tangent line is either horizontal or vertical. Sketch the curve.*

13. $x = t^2 - 4, \quad y = t^3 - 3t$

14. $x = t^3 - 3t, \quad y = t^2$

15. $x = 1 + 3 \cos t, \quad y = 2 - 2 \sin t$

16. $x = \sin t, \quad y = \sin 2t$

*In Exercises 17–24, find dy/dx and d²y/dx².*

**17.** $x = 3t^2 + 1, \quad y = 2t^3$    **18.** $x = t^3 - t, \quad y = t^3 + 2t^2$

**19.** $x = \sqrt{t}, \quad y = \dfrac{1}{t}$    **20.** $x = \sin 2t, \quad y = \cos 2t$

**21.** $x = \theta + \cos \theta, \quad y = \theta - \sin \theta$

**22.** $x = e^{-t}, \quad y = e^{2t}$

**23.** $x = \cosh t, \quad y = \sinh t$

**24.** $x = \sqrt{t^2 + 1}, \quad y = t \ln t$

**25.** Let $C$ be the curve defined by the parametric equations $x = t^2$ and $y = t^3 - 3t$ (see Example 2). Find $d^2y/dx^2$, and use this result to determine the intervals where $C$ is concave upward and where it is concave downward.

**26.** Show that the curve defined by the parametric equations $x = t^2$ and $y = t^3 - 3t$ crosses itself. Find equations of the tangent lines to the curve at that point (see Example 2).

**27.** The parametric equations of the astroid $x^{2/3} + y^{2/3} = a^{2/3}$ are $x = a \cos^3 t$ and $y = a \sin^3 t$. (Verify this!) Find an expression for the slope of the tangent line to the astroid in terms of $t$. At what points on the astroid is the slope of the tangent line equal to $-1$? Equal to 1?

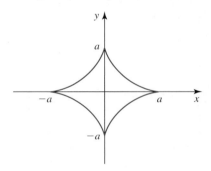

**28.** Find $dy/dx$ and $d^2y/dx^2$ if

$$x = \int_1^t \frac{\sin u}{u}\, du \quad \text{and} \quad y = \int_2^{\ln t} e^u\, du$$

**29.** The function $y = f(x)$ is defined by the parametric equations

$$x = t^5 + 5t^3 + 10t + 2 \quad \text{and} \quad y = 2t^3 - 3t^2 - 12t + 1$$
$$-2 \le t \le 2$$

Find the absolute maximum and the absolute minimum values of $f$.

**30.** Find the points on the curve with parametric equations $x = t^3 - t$ and $y = t^2$ at which the tangent line is parallel to the line with parametric equations $x = 2t$ and $y = 2t + 4$.

*In Exercises 31–36, find the length of the curve defined by the parametric equations.*

**31.** $x = 2t^2, \quad y = 3t^3; \quad 0 \le t \le 1$

**32.** $x = 2t^{3/2}, \quad y = 3t + 1; \quad 0 \le t \le 4$

**33.** $x = \sin^2 t, \quad y = \cos 2t; \quad 0 \le t \le \pi$

**34.** $x = e^t \cos t, \quad y = e^t \sin t; \quad 0 \le t \le \pi$

**35.** $x = a(\cos t + t \sin t), \quad y = a(\sin t - t \cos t); \quad 0 \le t \le \frac{\pi}{2}$

**36.** $x = (t^2 - 2)\sin t + 2t \cos t, \quad y = (2 - t^2)\cos t + 2t \sin t; \quad 0 \le t \le \pi$

**37.** Find the length of the cardioid with parametric equations

$$x = a(2 \cos t - \cos 2t) \quad \text{and} \quad y = a(2 \sin t - \sin 2t)$$

**38.** Find the length of the astroid with parametric equations

$$x = a \cos^3 t \quad \text{and} \quad y = a \sin^3 t$$

(See the figure for Exercise 27. Compare with Exercise 29 in Section 5.4.)

**39.** The position of an object at any time $t$ is $(x, y)$, where $x = \cos^2 t$ and $y = \sin^2 t$, $0 \le t \le 2\pi$. Find the distance covered by the object as $t$ runs from $t = 0$ to $t = 2\pi$.

**40.** The following figure shows the course taken by a yacht during a practice run. The parametric equations of the course are

$$x = 4\sqrt{2} \sin t \qquad y = \sin 2t \qquad 0 \le t \le 2\pi$$

where $x$ and $y$ are measured in miles. Find the length of the course.

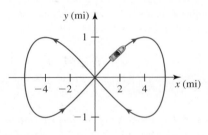

**41. Path of a Boat** Two towns, $A$ and $B$, are located directly opposite each other on the banks of a river that is 1600 ft wide and flows east with a constant speed of 4 ft/sec. A boat leaving Town $A$ travels with a constant speed of 18 ft/sec always aimed toward Town $B$. It can be shown that the path of the boat is given by the parametric equations

$$x = 800(t^{7/9} - t^{11/9}) \qquad y = 1600t \qquad 0 \le t \le 1$$

Find the distance covered by the boat in traveling from $A$ to $B$.

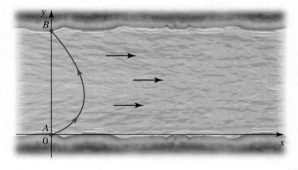

**42. Trajectory of an Electron** An electron initially located at the origin of a coordinate system is projected horizontally into a uniform electric field with magnitude $E$ and directed upward. If the initial speed of the electron is $v_0$, then its trajectory is

$$x = v_0 t \qquad y = -\frac{1}{2}\left(\frac{eE}{m}\right)t^2$$

where $e$ is the charge of the electron and $m$ is its mass. Show that the trajectory of the electron is a parabola.

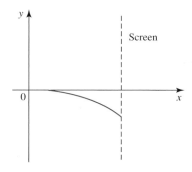

**Note:** The deflection of electrons by an electric field is used to control the direction of an electron beam in an electron gun.

**43.** Refer to Exercise 42. If a screen is placed along the vertical line $x = a$, at what point will the electron beam hit the screen?

**44.** Find the point that is located one quarter of the way along the arch of the cycloid

$$x = a(t - \sin t) \qquad y = a(1 - \cos t) \qquad 0 \le t \le 2\pi$$

as measured from the origin. What is the slope of the tangent line to the cycloid at that point? Plot the arch of the cycloid and the tangent line on the same set of axes.

**45.** The **cornu spiral** is a curve defined by the parametric equations

$$x = C(t) = \int_0^t \cos(\pi u^2/2)\,du \qquad y = S(t) = \int_0^t \sin(\pi u^2/2)\,du$$

where $C$ and $S$ are called Fresnel integrals. They are used to explain the phenomenon of light diffraction.
   **a.** Plot the spiral. Describe the behavior of the curve as $t \to \infty$ and as $t \to -\infty$.
   **b.** Find the length of the spiral from $t = 0$ to $t = a$.

**46.** Suppose that the graph of a nonnegative function $F$ on an interval $[a, b]$ is represented by the parametric equations $x = f(t)$ and $y = g(t)$ for $t$ in $[\alpha, \beta]$. Show that the area of the region under the graph of $F$ is given by

$$\int_\alpha^\beta g(t)f'(t)\,dt \qquad \text{or} \qquad \int_\beta^\alpha g(t)f'(t)\,dt$$

**47.** Use the result of Exercise 46 to find the area of the region under one arch of the cycloid $x = a(\theta - \sin \theta)$, $y = a(1 - \cos \theta)$.

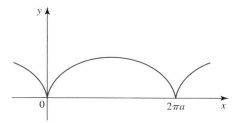

**48.** Use the result of Exercise 46 to find the area of the region enclosed by the ellipse with parametric equations $x = a \cos \theta$, $y = b \sin \theta$, where $0 \le \theta \le 2\pi$.

**49.** Use the result of Exercise 46 to find the area of the region enclosed by the astroid $x = a \cos^3 \theta$, $y = a \sin^3 \theta$. (See the figure for Exercise 27.)

**50.** Use the result of Exercise 46 to find the area of the region enclosed by the curve $x = a \sin t$, $y = b \sin 2t$.

**51.** Use the result of Exercise 46 to find the area of the region lying inside the course taken by the yacht of Exercise 40.

*In Exercises 52–57, find the area of the surface obtained by revolving the curve about the x-axis.*

**52.** $x = t$, $y = 2 - t$; $0 \le t \le 2$

**53.** $x = t^3$, $y = t^2$; $0 \le t \le 1$

**54.** $x = \sqrt{3}t^2$, $y = t - t^3$; $0 \le t \le 1$

**55.** $x = \frac{1}{3}t^3$, $y = 4 - \frac{1}{2}t^2$; $0 \le t \le 2\sqrt{2}$

**56.** $x = e^t \sin t$, $y = e^t \cos t$; $0 \le t \le \frac{\pi}{2}$

**57.** $x = t - \sin t$, $y = 1 - \cos t$; $0 \le t \le 2\pi$

*In Exercises 58–61, find the area of the surface obtained by rotating the curve about the y-axis.*

**58.** $x = t$, $y = 2t$; $0 \le t \le 4$

**59.** $x = 3t^2$, $y = 2t^3$; $0 \le t \le 1$

**60.** $x = a \cos t$, $y = b \sin t$; $-\frac{\pi}{2} \le t \le \frac{\pi}{2}$

**61.** $x = e^t - t$, $y = 4e^{t/2}$; $0 \le t \le 1$

**62.** Find the area of the surface obtained by revolving the cardioid

$$x = a(2 \cos t - \cos 2t) \qquad y = a(2 \sin t - \sin 2t)$$

about the $x$-axis.

**63.** Find the area of the surface obtained by revolving the astroid

$$x = a \cos^3 t \qquad y = a \sin^3 t$$

about the $x$-axis.

**64.** Find the areas of the surface obtained by revolving one arch of the cycloid $x = a(\theta - \sin \theta)$, $y = a(1 - \cos \theta)$ about the $x$- and $y$-axes.

**65.** Find the surface area of the torus obtained by revolving the circle $x^2 + (y - b)^2 = r^2$ $(0 < r < b)$ about the $x$-axis.

Hint: Represent the equation of the circle in parametric form: $x = r \cos t$, $y = b + r \sin t$, $0 \le t \le 2\pi$.

**66.** Show that if $f'$ is continuous and $f'(t) \ne 0$ for $a \le t \le b$, then the parametric curve defined by $x = f(t)$ and $y = g(t)$ for $a \le t \le b$ can be put in the form $y = F(x)$.

**cas** *In Exercises 67–70, (a) plot the curve defined by the parametric equations and (b) estimate the arc length of the curve accurate to four decimal places.*

**67.** $x = 2t^2$,   $y = t - t^3$;   $0 \le t \le 1$

**68.** $x = \sin(0.5t + 0.4\pi)$,   $y = \sin t$;   $0 < t \le 4\pi$

**69.** $x = 0.2(6 \cos t - \cos 6t)$,   $y = 0.2(6 \sin t - \sin 6t)$; $0 \le t \le 2\pi$

**70.** $x = 2t(1 - t^2)$,   $y = -t^2\left(1 - \dfrac{3}{2}t^2\right)$;   $-2 < t < 2$

(swallowtail castastrophe)

**cas 71.** Use a calculator or computer to approximate the area of the surface obtained by revolving the curve

$$x = 4 \sin 2t \qquad y = 2 \cos 3t \qquad 0 \le t \le \tfrac{\pi}{6}$$

about the $x$-axis.

**72. a.** Find an expression for the arc length of the curve defined by the parametric equations

$$x = f''(t)\cos t + f'(t)\sin t \qquad y = -f''(t)\sin t + f'(t)\cos t$$

where $a \le t \le b$ and $f$ has continuous third-order derivatives.

  **b.** Use the result of part (a) to find the arc length of the curve $x = 6t \cos t + 3t^2 \sin t$ and $y = -6t \sin t + 3t^2 \cos t$, where $0 \le t \le 1$.

**73.** Show that

$$x = \frac{2at}{1 + t^2} \qquad y = \frac{a(1 - t^2)}{1 + t^2}$$

where $a > 0$ and $-\infty < t < \infty$, are parametric equations of a circle. What are its center and radius?

**74.** Use the parametric representation of a circle in Exercise 73 to show that the circumference of a circle of radius $a$ is $2\pi a$.

**75.** Find parametric equations for the *Folium of Descartes*, $x^3 + y^3 = 3axy$ with parameter $t = y/x$.

**cas 76.** Use the parametric representation of the *Folium of Descartes* to estimate the length of the loop.

**cas 77.** Show that the length of the ellipse $x = a \cos t$, $y = b \sin t$, $0 \le t \le 2\pi$, where $a > b > 0$, is given by

$$L = 4a \int_0^{\pi/2} \sqrt{1 - e^2 \sin^2 t}\, dt$$

where

$$e = \frac{c}{a} = \frac{\sqrt{a^2 - b^2}}{a}$$

is the eccentricity of the ellipse.

**Note:** The integral is called an *elliptical integral of the second kind.*

**cas 78.** Use a computer or calculator and the result of Exercise 77 to estimate the circumference of the ellipse

$$\frac{x^2}{100} + \frac{y^2}{36} = 1$$

accurate to three decimal places.

*In Exercises 79–80, determine whether the statement is true or false. If it is true, explain why it is true. If it is false, give an example to show why it is false.*

**79.** If $x = f(t)$ and $y = g(t)$, $f$ and $g$ have second-order derivatives, and $f'(t) \ne 0$, then

$$\frac{d^2y}{dx^2} = \frac{f'(t)g''(t) - g'(t)f''(t)}{[f'(t)]^2}$$

**80.** The curve with parametric equations $x = f(t)$ and $y = g(t)$ is a line if and only if $f$ and $g$ are both linear functions of $t$.

# 9.4   Polar Coordinates

The curve shown in Figure 1a is a lemniscate, and the one shown in Figure 1b is called a cardioid. The rectangular equations of these curves are

$$(x^2 + y^2)^2 = 4(x^2 - y^2) \qquad \text{and} \qquad x^4 - 2x^3 + 2x^2y^2 - 2xy^2 - y^2 + y^4 = 0$$

respectively. As you can see, these equations are somewhat complicated. For example, they will not prove very helpful if we want to calculate the area enclosed by the two loops of the lemniscate shown in Figure 1a or the length of the cardioid shown in Figure 1b.

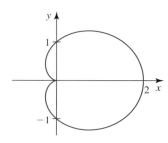

| (a) A lemniscate | (b) A cardioid |

**FIGURE 1**
A rectangular equation of the lemniscate in part (a) is $(x^2 + y^2)^2 = 4(x^2 - y^2)$, and an equation of the cardioid in part (b) is $x^4 - 2x^3 + 2x^2y^2 - 2xy^2 - y^2 + y^4 = 0$.

A question that arises naturally is: Is there a coordinate system other than the rectangular system that we can use to give a simpler representation for curves such as the lemniscate and cardioid? One such system is the *polar coordinate system*.

## The Polar Coordinate System

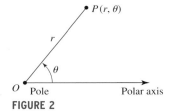

**FIGURE 2**

To construct the polar coordinate system, we fix a point $O$ called the **pole** (or **origin**) and draw a ray (half-line) emanating from $O$ called the **polar axis.** Suppose that $P$ is any point in the plane, let $r$ denote the distance from $O$ to $P$, and let $\theta$ denote the angle (in degrees or radians) between the polar axis and the line segment $OP$. (See Figure 2.) Then the point $P$ is represented by the ordered pair $(r, \theta)$, also written $P(r, \theta)$, where the numbers $r$ and $\theta$ are called the **polar coordinates** of $P$.

The **angular coordinate** $\theta$ is positive if it is measured in the counterclockwise direction from the polar axis and negative if it is measured in the clockwise direction. The **radial coordinate** $r$ may assume positive as well as negative values. If $r > 0$, then $P(r, \theta)$ is on the terminal side of $\theta$ and at a distance $r$ from the origin. If $r < 0$, then $P(r, \theta)$ lies on the ray that is opposite the terminal side of $\theta$ and at a distance of $|r| = -r$ from the pole. (See Figure 3.) Also, by convention the pole $O$ is represented by the ordered pair $(0, \theta)$ for *any* value of $\theta$. Finally, a plane that is endowed with a polar coordinate system is referred to as an $r\theta$-plane.

**FIGURE 3**

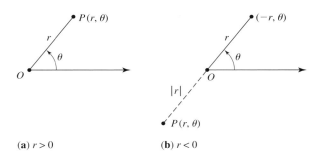

(a) $r > 0$      (b) $r < 0$

**EXAMPLE 1** Plot the following points in the $r\theta$-plane.

**a.** $\left(1, \frac{2\pi}{3}\right)$     **b.** $\left(2, -\frac{\pi}{4}\right)$     **c.** $\left(-2, \frac{\pi}{3}\right)$     **d.** $(2, -3\pi)$

**Solution**   The points are plotted in Figure 4.

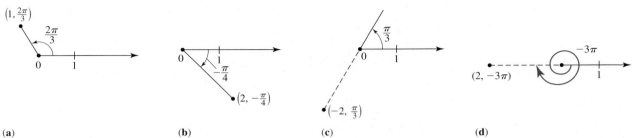

**(a)**

**(b)**

**(c)**

**(d)**

**FIGURE 4**

The points in Example 1

Unlike the representation of points in the rectangular system, the representation of points using polar coordinates is *not* unique. For example, the point $(r, \theta)$ can also be written as $(r, \theta + 2n\pi)$ or $(-r, \theta + (2n + 1)\pi)$, where $n$ is any integer. Figures 5a and 5b illustrate this for the case $n = 1$ and $n = 0$, respectively.

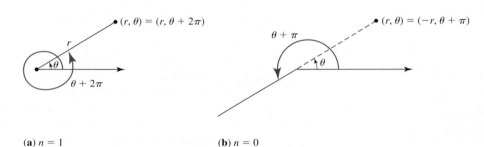

**FIGURE 5**

Representation of points using
polar coordinates is not unique.

**(a)** $n = 1$

**(b)** $n = 0$

## ■ Relationship Between Polar and Rectangular Coordinates

To establish the relationship between polar and rectangular coordinates, let's superimpose an $xy$-plane on an $r\theta$-plane in such a way that the origins coincide and the positive $x$-axis coincides with the polar axis. Let $P$ be any point in the plane other than the origin with rectangular representation $(x, y)$ and polar representation $(r, \theta)$. Figure 6a shows a situation in which $r > 0$, and Figure 6b shows a situation in which $r < 0$. If $r > 0$, we see immediately from the figure that

$$\cos \theta = \frac{x}{r} \qquad \sin \theta = \frac{y}{r}$$

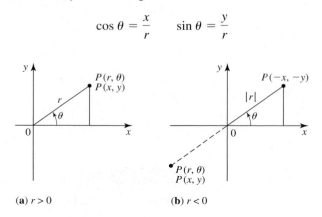

**FIGURE 6**

The relationship between polar
and rectangular coordinates

**(a)** $r > 0$

**(b)** $r < 0$

so $x = r \cos \theta$ and $y = r \sin \theta$. If $r < 0$, we see by referring to Figure 6b that

$$\cos \theta = \frac{-x}{|r|} = \frac{-x}{-r} = \frac{x}{r} \qquad \sin \theta = \frac{-y}{|r|} = \frac{-y}{-r} = \frac{y}{r}$$

so again $x = r \cos \theta$ and $y = r \sin \theta$. Finally, in either case we have

$$x^2 + y^2 = r^2 \qquad \text{and} \qquad \tan \theta = \frac{y}{x} \qquad \text{if } x \neq 0$$

---

**Relationship Between Rectangular and Polar Coordinates**

Suppose that a point $P$ (other than the origin) has representation $(r, \theta)$ in polar coordinates and $(x, y)$ in rectangular coordinates. Then

$$x = r \cos \theta \qquad \text{and} \qquad y = r \sin \theta \tag{1}$$

$$r^2 = x^2 + y^2 \qquad \text{and} \qquad \tan \theta = \frac{y}{x} \qquad \text{if } x \neq 0 \tag{2}$$

---

**EXAMPLE 2** The point $\left(4, \frac{\pi}{6}\right)$ is given in polar coordinates. Find its representation in rectangular coordinates.

**Solution** Here, $r = 4$ and $\theta = \pi/6$. Using Equation (1), we obtain

$$x = r \cos \theta = 4 \cos \frac{\pi}{6} = 4 \cdot \frac{\sqrt{3}}{2} = 2\sqrt{3}$$

$$y = r \sin \theta = 4 \sin \frac{\pi}{6} = 4 \cdot \frac{1}{2} = 2$$

Therefore, the given point has rectangular representation $(2\sqrt{3}, 2)$. ∎

**EXAMPLE 3** The point $(-1, 1)$ is given in rectangular coordinates. Find its representation in polar coordinates.

**Solution** Here, $x = -1$ and $y = 1$. Using Equation (2), we have

$$r^2 = x^2 + y^2 = (-1)^2 + 1^2 = 2$$

and

$$\tan \theta = \frac{y}{x} = -1$$

Let's choose $r$ to be positive; that is, $r = \sqrt{2}$. Next, observe that the point $(-1, 1)$ lies in the second quadrant and so we choose $\theta = 3\pi/4$ (other choices are $\theta = (3\pi/4) \pm 2n\pi$, where $n$ is an integer). Therefore, one representation of the given point is $\left(\sqrt{2}, \frac{3\pi}{4}\right)$. ∎

## ▪ Graphs of Polar Equations

The graph of a **polar equation** $r = f(\theta)$ or, more generally, $F(r, \theta) = 0$ is the set of all points $(r, \theta)$ whose coordinates satisfy the equation.

**EXAMPLE 4**   Sketch the graphs of the polar equations, and reconcile your results by finding the corresponding rectangular equations.

**a.** $r = 2$   **b.** $\theta = \dfrac{2\pi}{3}$

**Solution**

**a.** The graph of $r = 2$ consists of all points $P(r, \theta)$ where $r = 2$ and $\theta$ can assume *any* value. Since $r$ gives the distance between $P$ and the pole $O$, we see that the graph consists of all points that are located a distance of 2 units from the pole; in other words, the graph of $r = 2$ is the circle of radius 2 centered at the pole. (See Figure 7a.) To find the corresponding rectangular equation, square both sides of the given equation obtaining $r^2 = 4$. But by Equation (2), $r^2 = x^2 + y^2$, and this gives the desired equation $x^2 + y^2 = 4$. Since this is a rectangular equation of a circle with center at the origin and radius 2, the result obtained earlier has been confirmed.

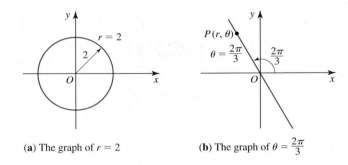

**FIGURE 7**   (a) The graph of $r = 2$   (b) The graph of $\theta = \dfrac{2\pi}{3}$

**b.** The graph of $\theta = 2\pi/3$ consists of all points $P(r, \theta)$ where $\theta = 2\pi/3$ and $r$ can assume *any* value. Since $\theta$ measures the angle the line segment $OP$ makes with the polar axis, we see that the graph consists of all points that are located on the straight line passing through the pole $O$ and making an angle of $2\pi/3$ radians with the polar axis. (See Figure 7b.) Observe that the half-line in the second quadrant consists of points for which $r > 0$, whereas the half-line in the fourth quadrant consists of points for which $r < 0$. To find the corresponding rectangular equation, we use Equation (2), $\tan \theta = y/x$, to obtain

$$\tan \frac{2\pi}{3} = \frac{y}{x} \qquad \text{or} \qquad \frac{y}{x} = -\sqrt{3}$$

or $y = -\sqrt{3}x$. This equation confirms that the graph of $\theta = 2\pi/3$ is a straight line with slope $-\sqrt{3}$.   ■

As in the case with rectangular equations, we can often obtain a sketch of the graph of a simple polar equation by plotting and connecting some points that lie on the graph.

**EXAMPLE 5**   Sketch the graph of the polar equation $r = 2 \sin \theta$. Find a corresponding rectangular equation and reconcile your results.

**Solution**   The following table shows the values of $r$ corresponding to some convenient values of $\theta$. It suffices to restrict the values of $\theta$ to those lying between 0 and $\pi$, since values of $\theta$ beyond $\pi$ will give the same points $(r, \theta)$ again.

| $\theta$ | 0 | $\frac{\pi}{6}$ | $\frac{\pi}{4}$ | $\frac{\pi}{3}$ | $\frac{\pi}{2}$ | $\frac{2\pi}{3}$ | $\frac{3\pi}{4}$ | $\frac{5\pi}{6}$ | $\pi$ |
|---|---|---|---|---|---|---|---|---|---|
| $r$ | 0 | 1 | $\sqrt{2} \approx 1.4$ | $\sqrt{3} \approx 1.7$ | 2 | $\sqrt{3} \approx 1.7$ | $\sqrt{2} \approx 1.4$ | 1 | 0 |

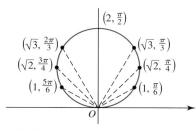

**FIGURE 8**
The graph of $r = 2\sin\theta$ is a circle. To plot the points, first draw the ray with the desired angle, then locate the point by measuring off the required distance from the pole.

The graph of $r = 2\sin\theta$ is sketched in Figure 8. To find a corresponding rectangular equation, we multiply both sides of $r = 2\sin\theta$ by $r$ to obtain $r^2 = 2r\sin\theta$ and then use the relationships $r^2 = x^2 + y^2$ (Equation (2)) and $y = r\sin\theta$ (Equation (1)), to obtain the desired equation

$$x^2 + y^2 = 2y \qquad \text{or} \qquad x^2 + y^2 - 2y = 0$$

Finally, completing the square in $y$, we have

$$x^2 + y^2 - 2y + (-1)^2 = 1$$

or

$$x^2 + (y - 1)^2 = 1$$

which is an equation of the circle with center $(0, 1)$ and radius 1, as obtained earlier.

It might have occurred to you that in the last several examples we could have obtained the graphs of the polar equations by first converting them to the corresponding rectangular equations. But as you will see, some curves are easier to graph using polar coordinates.

### ■ Symmetry

Just as the use of symmetry is helpful in graphing rectangular equations, its use is equally helpful in graphing polar equations. Three types of symmetry are illustrated in Figure 9. The test for each type of symmetry follows.

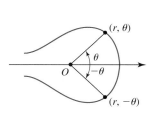

**(a)** Symmetry with respect to the polar axis

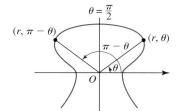

**(b)** Symmetry with respect to the line $\theta = \frac{\pi}{2}$

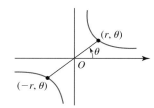

**(c)** Symmetry with respect to the pole

**FIGURE 9**
Symmetries of graphs of polar equations

---

**Tests for Symmetry**

**a.** The graph of $r = f(\theta)$ is **symmetric with respect to the polar axis** if the equation is unchanged when $\theta$ is replaced by $-\theta$.

**b.** The graph of $r = f(\theta)$ is **symmetric with respect to the vertical line** $\theta = \pi/2$ if the equation is unchanged when $\theta$ is replaced by $\pi - \theta$.

**c.** The graph of $r = f(\theta)$ is **symmetric with respect to the pole** if the equation is unchanged when $r$ is replaced by $-r$ or when $\theta$ is replaced by $\theta + \pi$.

To illustrate the use of the tests for symmetry, consider the equation $r = 2 \sin \theta$ of Example 5. Here, $f(\theta) = 2 \sin \theta$, and since

$$f(\pi - \theta) = 2 \sin(\pi - \theta) = 2(\sin \pi \cos \theta - \cos \pi \sin \theta) = 2 \sin \theta = f(\theta)$$

we conclude that the graph of $r = 2 \sin \theta$ is symmetric with respect to the vertical line $\theta = \pi/2$ (Figure 8).

**EXAMPLE 6** Sketch the graph of the polar equation $r = 1 + \cos \theta$. This is the polar form of the rectangular equation $x^4 - 2x^3 + 2x^2 y^2 - 2xy^2 - y^2 + y^4 = 0$ of the cardioid that was mentioned at the beginning of this section (Figure 1b).

**Solution** Writing $f(\theta) = 1 + \cos \theta$ and observing that

$$f(-\theta) = 1 + \cos(-\theta) = 1 + \cos \theta = f(\theta)$$

we conclude that the graph of $r = 1 + \cos \theta$ is symmetric with respect to the polar axis. In view of this, we need only to obtain that part of the graph between $\theta = 0$ and $\theta = \pi$. We can then complete the graph using symmetry.

To sketch the graph of $r = 1 + \cos \theta$ for $0 \le \theta \le \pi$, we can proceed as we did in Example 5 by first plotting some points lying on that part of the graph, or we may proceed as follows: Treat $r$ and $\theta$ as *rectangular* coordinates, and make use of our knowledge of graphing rectangular equations to obtain the graph of $r = f(\theta) = 1 + \cos \theta$ on the interval $[0, \pi]$. (See Figure 10a.) Then recalling that $\theta$ is the angular coordinate and $r$ is the radial coordinate, we see that as $\theta$ increases from 0 to $\pi$, the points on the respective rays shrink to 0. (See Figure 10b, where the corresponding points are shown.)

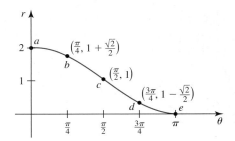

**(a)** $r = f(\theta)$, treating $r$ and $\theta$ as rectangular coordinates

**FIGURE 10**
Two steps in sketching the graph of the polar equation $r = 1 + \cos \theta$

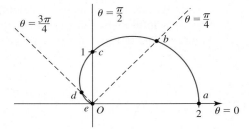

**(b)** $r = f(\theta)$, treating $r$ and $\theta$ as polar coordinates

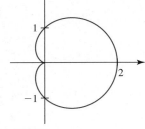

**FIGURE 11**
The graph of $r = 1 + \cos \theta$ is a cardioid.

Finally, using symmetry, we complete the graph of $r = 1 + \cos \theta$, as shown in Figure 11. It is called a **cardioid** because it is heart-shaped. ∎

**EXAMPLE 7** Sketch the graph of the polar equation $r = 2 \cos 2\theta$.

**Solution** Write $f(\theta) = 2 \cos 2\theta$, and observe that

$$f(-\theta) = 2 \cos 2(-\theta) = 2 \cos 2\theta = f(\theta)$$

and

$$f(\pi - \theta) = 2 \cos 2(\pi - \theta) = 2 \cos(2\pi - 2\theta)$$

$$= 2[\cos 2\pi \cos 2\theta + \sin 2\pi \sin 2\theta] = 2 \cos 2\theta = f(\theta)$$

Therefore, the graph of the given equation is symmetric with respect to both the polar axis and the vertical line $\theta = \pi/2$. It suffices, therefore, to obtain an accurate sketch of that part of the graph for $0 \le \theta \le \frac{\pi}{2}$ and then complete the sketch of the graph using symmetry. Proceeding as in Example 6, we first sketch the graph of $r = 2 \cos 2\theta$ for $0 \le \theta \le \frac{\pi}{2}$ treating $r$ and $\theta$ as rectangular coordinates (Figure 12a), and then transcribe the information contained in this graph onto the graph in the $r\theta$-plane for $0 \le \theta \le \frac{\pi}{2}$. (See Figure 12b.)

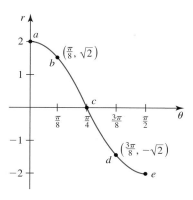

**FIGURE 12**
Two steps in sketching the graph of $r = 2 \cos 2\theta$

(**a**) $r = f(\theta)$ treating $r$ and $\theta$ as rectangular coordinates

(**b**) $r = f(\theta)$, treating $r$ and $\theta$ as polar coordinates

Finally, using the symmetry that was established earlier (Figure 13a), we complete the graph of $r = 2 \cos 2\theta$ as shown in Figure 13b. This graph is called a **four-leaved rose.**

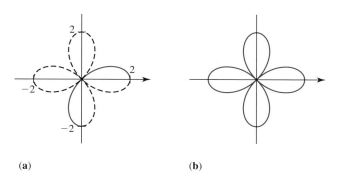

**FIGURE 13**
The graph of $r = 2 \cos 2\theta$ is a four-leaved rose.

(**a**)

(**b**)

The next example shows how the graph of a rectangular equation can be sketched more easily by first converting it to polar form.

**EXAMPLE 8** Sketch the graph of the equation $(x^2 + y^2)^2 = 4(x^2 - y^2)$ by first converting it to polar form. This is an equation of the lemniscate that was mentioned at the beginning of this section.

**Solution** To convert the given equation to polar form, we use Equations (1) and (2), obtaining

$$(r^2)^2 = 4(r^2 \cos^2 \theta - r^2 \sin^2 \theta)$$
$$= 4r^2(\cos^2 \theta - \sin^2 \theta)$$
$$r^4 = 4r^2 \cos 2\theta$$

or

$$r^2 = 4 \cos 2\theta$$

Observe that $f(\theta) = 2\sqrt{\cos 2\theta}$ is defined for $-\frac{\pi}{4} \le \theta \le \frac{\pi}{4}$ and $\frac{3\pi}{4} \le \theta \le \frac{5\pi}{4}$. Also, observe that $f(-\theta) = f(\theta)$ and $f(\pi - \theta) = f(\theta)$. (These computations are similar to those in Example 7.) So the graph of $r = 2\sqrt{\cos 2\theta}$ is symmetric with respect to the polar axis and the line $\theta = \pi/2$. The graph of $r = f(\theta)$ for $0 \le \theta \le \frac{\pi}{4}$, where $r$ and $\theta$ are treated as rectangular coordinates, is shown in Figure 14a. This leads to the part of the required graph for $0 \le \theta \le \frac{\pi}{4}$ shown in Figure 14b. Then, using symmetry, we obtain the graph of $r = 2\sqrt{\cos 2\theta}$ and, therefore, that of $(x^2 + y^2)^2 = 4(x^2 - y^2)$, as shown in Figure 15.

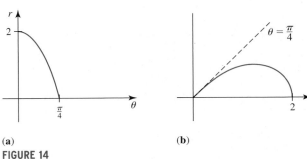

**(a)**

**FIGURE 14**

Two steps in sketching the graph of $r = 2\sqrt{\cos 2\theta}$

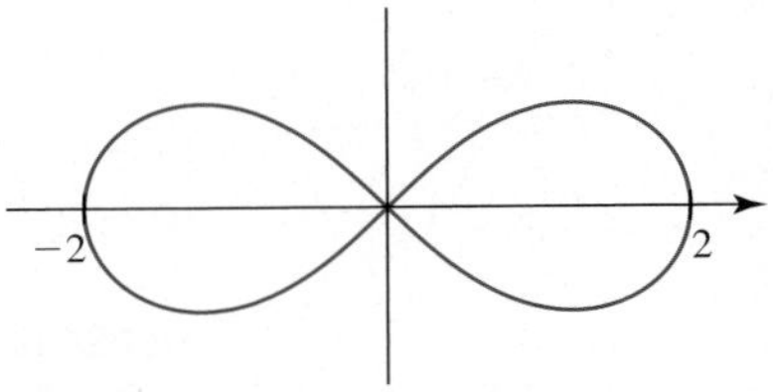

**(b)**

**FIGURE 15**

The graph of $r = 2\sqrt{\cos 2\theta}$ is a lemniscate.

## ■ Tangent Lines to Graphs of Polar Equations

To find the slope of the tangent line to the graph of $r = f(\theta)$ at the point $P(r, \theta)$, let $P(x, y)$ be the rectangular representation of $P$. Then

$$x = r \cos \theta = f(\theta) \cos \theta$$

$$y = r \sin \theta = f(\theta) \sin \theta$$

We can view these equations as parametric equations for the graph of $r = f(\theta)$ with parameter $\theta$. Then, using Equation (1) of Section 9.3, we have

$$\frac{dy}{dx} = \frac{\dfrac{dy}{d\theta}}{\dfrac{dx}{d\theta}} = \frac{\dfrac{dr}{d\theta} \sin \theta + r \cos \theta}{\dfrac{dr}{d\theta} \cos \theta - r \sin \theta} \qquad \text{if } \frac{dx}{d\theta} \ne 0 \tag{3}$$

and this gives the slope of the tangent line to the graph of $r = f(\theta)$ at any point $P(r, \theta)$.

The horizontal tangent lines to the graph of $r = f(\theta)$ are located at the points where $dy/d\theta = 0$ and $dx/d\theta \ne 0$. The vertical tangent lines are located at the points where $dx/d\theta = 0$ and $dy/d\theta \ne 0$ (so that $dy/dx$ is undefined). Also, points where both $dy/d\theta$ and $dx/d\theta$ are equal to zero are candidates for horizontal or vertical tangent lines, respectively, and may be investigated using l'Hôpital's Rule.

Equation (3) can be used to help us find the tangent lines to the graph of $r = f(\theta)$ at the pole. To see this, suppose that the graph of $f$ passes through the pole when $\theta = \theta_0$. Then $f(\theta_0) = 0$. If $f'(\theta_0) \ne 0$, then Equation (3) reduces to

$$\frac{dy}{dx} = \frac{f'(\theta_0) \sin \theta_0 + f(\theta_0) \cos \theta_0}{f'(\theta_0) \cos \theta_0 - f(\theta_0) \sin \theta_0} = \frac{\sin \theta_0}{\cos \theta_0} = \tan \theta_0$$

This shows that $\theta = \theta_0$ is a tangent line to the graph of $r = f(\theta)$ at the pole $(0, \theta_0)$. The following summarizes this discussion.

$\theta = \theta_0$ is a tangent line to the graph of $r = f(\theta)$ at the pole if $f(\theta_0) = 0$ and $f'(\theta_0) \neq 0$.

**EXAMPLE 9** Consider the cardioid $r = 1 + \cos \theta$ of Example 6.

**a.** Find the slope of the tangent line to the cardioid at the point where $\theta = \pi/6$.
**b.** Find the points on the cardioid where the tangent lines are horizontal and where the tangent lines are vertical.

**Solution**
**a.** The slope of the tangent line to the cardioid $r = 1 + \cos \theta$ at any point $P(r, \theta)$ is given by

$$\frac{dy}{dx} = \frac{\dfrac{dr}{d\theta} \sin \theta + r \cos \theta}{\dfrac{dr}{d\theta} \cos \theta - r \sin \theta} = \frac{(-\sin \theta)(\sin \theta) + (1 + \cos \theta)\cos \theta}{(-\sin \theta)(\cos \theta) - (1 + \cos \theta)\sin \theta}$$

$$= \frac{(\cos^2 \theta - \sin^2 \theta) + \cos \theta}{-2 \sin \theta \cos \theta - \sin \theta} = -\frac{\cos 2\theta + \cos \theta}{\sin 2\theta + \sin \theta}$$

At the point on the cardioid where $\theta = \pi/6$, the slope of the tangent line is

$$\frac{dy}{dx}\bigg|_{\theta = \pi/6} = -\frac{\cos\left(\dfrac{\pi}{3}\right) + \cos\left(\dfrac{\pi}{6}\right)}{\sin\left(\dfrac{\pi}{3}\right) + \sin\left(\dfrac{\pi}{6}\right)} = -\frac{\dfrac{1}{2} + \dfrac{\sqrt{3}}{2}}{\dfrac{\sqrt{3}}{2} + \dfrac{1}{2}} = -1$$

**b.** Observe that $dy/d\theta = 0$ if

$$\cos 2\theta + \cos \theta = 0$$

$$2 \cos^2 \theta + \cos \theta - 1 = 0$$

$$(2 \cos \theta - 1)(\cos \theta + 1) = 0$$

that is, if $\cos \theta = \frac{1}{2}$ or $\cos \theta = -1$. This gives

$$\theta = \frac{\pi}{3}, \quad \pi, \quad \text{or} \quad \frac{5\pi}{3}$$

Next, $dx/d\theta = 0$ if

$$\sin 2\theta + \sin \theta = 0$$

$$2 \sin \theta \cos \theta + \sin \theta = 0$$

$$\sin \theta (2 \cos \theta + 1) = 0$$

that is, if $\sin \theta = 0$ or $\cos \theta = -\frac{1}{2}$. This gives

$$\theta = 0, \quad \pi, \quad \frac{2\pi}{3}, \quad \text{or} \quad \frac{4\pi}{3}$$

In view of the remarks following Equation (3), we see that $\theta = \pi/3$ and $\theta = 5\pi/3$ give rise to horizontal tangents. To investigate the candidate $\theta = \pi$, where both $dy/d\theta$ and $dx/d\theta$ are equal to zero, we use l'Hôpital's Rule. Thus,

$$\lim_{\theta \to \pi^-} \frac{dy}{dx} = -\lim_{\theta \to \pi^-} \frac{\cos 2\theta + \cos \theta}{\sin 2\theta + \sin \theta}$$

$$= -\lim_{\theta \to \pi^-} \frac{-2 \sin 2\theta - \sin \theta}{2 \cos 2\theta + \cos \theta} = 0$$

Similarly, we see that

$$\lim_{\theta \to \pi^+} \frac{dy}{dx} = 0$$

Therefore, $\theta = \pi$ also gives rise to a horizontal tangent. Thus, the horizontal tangent lines occur at

$$\left(\tfrac{3}{2}, \tfrac{\pi}{3}\right), \quad (0, \pi), \quad \text{and} \quad \left(\tfrac{3}{2}, \tfrac{5\pi}{3}\right)$$

The vertical tangent lines occur at $\theta = 0$, $2\pi/3$, and $4\pi/3$. The points are $(2, 0)$, $\left(\tfrac{1}{2}, \tfrac{2\pi}{3}\right)$, and $\left(\tfrac{1}{2}, \tfrac{4\pi}{3}\right)$. These tangent lines are shown in Figure 16. ∎

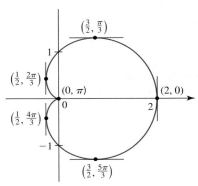

**FIGURE 16**
The horizontal and vertical tangents to the graph of $r = 1 + \cos \theta$

**EXAMPLE 10** Find the tangent lines of $r = \cos 2\theta$ at the origin.

**Solution** Setting $f(\theta) = \cos 2\theta = 0$, we find that

$$2\theta = \frac{\pi}{2}, \quad \frac{3\pi}{2}, \quad \frac{5\pi}{2}, \quad \text{or} \quad \frac{7\pi}{2}$$

or

$$\theta = \frac{\pi}{4}, \quad \frac{3\pi}{4}, \quad \frac{5\pi}{4}, \quad \text{or} \quad \frac{7\pi}{4}$$

Next, we compute $f'(\theta) = -2 \sin 2\theta$. Since $f'(\theta) \neq 0$ for each of these values of $\theta$, we see that $\theta = \pi/4$ and $\theta = 3\pi/4$ (that is, $y = x$ and $y = -x$) are tangent lines to the graph of $r = \cos 2\theta$ at the pole (see Figure 17). ∎

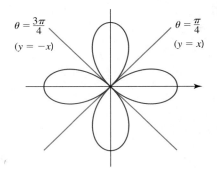

**FIGURE 17**
The tangent lines to the graph of $r = \cos 2\theta$ at the origin

## 9.4 CONCEPT QUESTIONS

1. Let $P(r, \theta)$ be a point in the plane with polar coordinates $r$ and $\theta$. Find all possible representations of $P(r, \theta)$.
2. Suppose that $P$ has representation $(r, \theta)$ in polar coordinates and $(x, y)$ in rectangular coordinates. Express (a) $x$ and $y$ in terms of $r$ and $\theta$ and (b) $r$ and $\theta$ in terms of $x$ and $y$.
3. Explain how you would determine whether the graph of $r = f(\theta)$ is symmetric with respect to (a) the polar axis, (b) the vertical line $\theta = \pi/2$, and (c) the pole.

4. Suppose that $r = f(\theta)$, where $f$ is differentiable.
   a. Write an expression for $dy/dx$.
   b. How do you find the points on the graph of $r = f(\theta)$ where the tangent lines are horizontal and where the tangent lines are vertical?
   c. How do you find the tangent lines to the graph of $r = f(\theta)$ (if they exist) at the pole?

## 9.4  EXERCISES

*In Exercises 1–8, plot the point with the polar coordinates. Then find the rectangular coordinates of the point.*

**1.** $\left(4, \frac{\pi}{4}\right)$

**2.** $\left(2, \frac{\pi}{6}\right)$

**3.** $\left(4, \frac{3\pi}{2}\right)$

**4.** $(6, 3\pi)$

**5.** $\left(-\sqrt{2}, \frac{\pi}{4}\right)$

**6.** $\left(-1, \frac{\pi}{3}\right)$

**7.** $\left(-4, -\frac{3\pi}{4}\right)$

**8.** $\left(5, -\frac{5\pi}{6}\right)$

*In Exercises 9–16, plot the point with the rectangular coordinates. Then find the polar coordinates of the point taking $r > 0$ and $0 \le \theta < 2\pi$.*

**9.** $(2, 2)$

**10.** $(1, -1)$

**11.** $(0, 5)$

**12.** $(3, -4)$

**13.** $(-\sqrt{3}, -\sqrt{3})$

**14.** $(2\sqrt{3}, -2)$

**15.** $(5, -12)$

**16.** $(3, -1)$

*In Exercises 17–24, sketch the region comprising points whose polar coordinates satisfy the given conditions.*

**17.** $r \ge 1$

**18.** $r > 1$

**19.** $0 \le r \le 2$

**20.** $1 \le r < 2$

**21.** $0 \le \theta \le \frac{\pi}{4}$

**22.** $0 \le r \le 3, \quad 0 \le \theta \le \frac{\pi}{3}$

**23.** $1 \le r \le 3, \quad -\frac{\pi}{6} \le \theta \le \frac{\pi}{6}$

**24.** $2 < r < 4, \quad -\frac{\pi}{2} < \theta < \frac{\pi}{2}$

*In Exercises 25–32, convert the polar equation to a rectangular equation.*

**25.** $r \cos \theta = 2$

**26.** $r \sin \theta = -3$

**27.** $2r \cos \theta + 3r \sin \theta = 6$

**28.** $r \sin \theta = 2r \cos \theta$

**29.** $r^2 = 4r \cos \theta$

**30.** $r^2 = \sin 2\theta$

**31.** $r = \dfrac{1}{1 - \sin \theta}$

**32.** $r = \dfrac{3}{4 - 5 \cos \theta}$

*In Exercises 33–38, convert the rectangular equation to a polar equation.*

**33.** $x = 4$

**34.** $x + 2y = 3$

**35.** $x^2 + y^2 = 9$

**36.** $x^2 - y^2 = 1$

**37.** $xy = 4$

**38.** $y^2 - x^2 = 4\sqrt{x^2 + y^2}$

*In Exercises 39–64, sketch the curve with the polar equation.*

**39.** $r = 3$

**40.** $r = -2$

**41.** $\theta = \dfrac{\pi}{3}$

**42.** $\theta = -\dfrac{\pi}{6}$

**43.** $r = 3 \cos \theta$

**44.** $r = -4 \sin \theta$

**45.** $r = 3 \cos \theta - 2 \sin \theta$

**46.** $r = 2 \sin \theta + 4 \cos \theta$

**47.** $r = 1 + \cos \theta$

**48.** $r = 1 + \sin \theta$

**49.** $r = 4(1 - \sin \theta)$

**50.** $r = 3 - 3 \cos \theta$

**51.** $r = 2 \csc \theta$

**52.** $r = -3 \sec \theta$

**53.** $r = \theta, \quad \theta \ge 0$   (spiral)

**54.** $r = \dfrac{1}{\theta}$   (spiral)

**55.** $r = e^{\theta}, \quad \theta \ge 0$   (logarithmic spiral)

**56.** $r^2 = \dfrac{1}{\theta}$   (lituus)

**57.** $r^2 = 4 \sin 2\theta$   (lemniscate)

**58.** $r = 1 - 2 \cos \theta$   (limaçon)

**59.** $r = 3 + 2 \sin \theta$   (limaçon)

**60.** $r = \sin 2\theta$   (four-leaved rose)

**61.** $r = \sin 3\theta$   (three-leaved rose)

**62.** $r = 2 \cos 4\theta$   (eight-leaved rose)

**63.** $r = 4 \sin 4\theta$   (eight-leaved rose)

**64.** $r = 2 \sin 5\theta$   (five-leaved rose)

*In Exercises 65–72, find the slope of the tangent line to the curve with the polar equation at the point corresponding to the given value of $\theta$.*

**65.** $r = 4 \cos \theta, \quad \theta = \dfrac{\pi}{3}$

**66.** $r = 3 \sin \theta, \quad \theta = \dfrac{\pi}{4}$

**67.** $r = \sin \theta + \cos \theta, \quad \theta = \dfrac{\pi}{4}$

**68.** $r = 1 + 3 \cos \theta, \quad \theta = \dfrac{\pi}{2}$

**69.** $r = \theta, \quad \theta = \pi$

**70.** $r = \sin 3\theta, \quad \theta = \dfrac{\pi}{3}$

**71.** $r^2 = 4 \cos 2\theta, \quad \theta = \dfrac{\pi}{6}$

**72.** $r = 2 \sec \theta, \quad \theta = \dfrac{\pi}{4}$

*In Exercises 73–78, find the points on the curve with the given polar equation where the tangent line is horizontal or vertical.*

**73.** $r = 4 \cos \theta$

**74.** $r = \sin \theta + \cos \theta$

**75.** $r = \sin 2\theta$

**76.** $r^2 = 4 \cos 2\theta$

**77.** $r = 1 + 2 \cos \theta$

**78.** $r = 1 + \sin \theta$

**79.** Show that the rectangular equation

$$x^4 - 2x^3 + 2x^2y^2 - 2xy^2 - y^2 + y^4 = 0$$

is an equation of the cardioid with polar equation $r = 1 + \cos \theta$.

**80.** Show that the polar equation $r = a \sin \theta + b \cos \theta$, where $a$ and $b$ are nonzero, represents a circle. What are the center and radius of the circle?

**81. a.** Show that the distance between the points with polar coordinates $(r_1, \theta_1)$ and $(r_2, \theta_2)$ is given by

$$d = \sqrt{r_1^2 + r_2^2 - 2r_1r_2 \cos(\theta_1 - \theta_2)}$$

**b.** Find the distance between the points with polar coordinates $\left(4, \frac{2\pi}{3}\right)$ and $\left(2, \frac{\pi}{3}\right)$.

**82.** Show that the curves with polar equations $r = a \sin \theta$ and $r = a \cos \theta$ intersect at right angles.

 **83. a.** Plot the graphs of the cardioids $r = a(1 + \cos \theta)$ and $r = a(1 - \cos \theta)$.
**b.** Show that the cardioids intersect at right angles except at the pole.

**84.** Let $\psi$ be the angle between the radial line $OP$ and the tangent line to the curve with polar equation $r = f(\theta)$ at $P$ (see the figure). Show that

$$\tan \psi = r \frac{d\theta}{dr}$$

**Hint:** Observe that $\psi = \phi - \theta$. Then use the trigonometric identity

$$\tan(a - b) = \frac{\tan a - \tan b}{1 + \tan a \tan b}$$

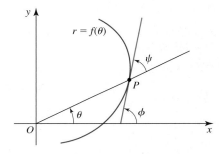

 *In Exercises 85–92, use a graphing utility to plot the curve with the polar equation.*

**85.** $r = \cos \theta (4 \sin^2 \theta - 1), \quad 0 \le \theta < 2\pi$

**86.** $r = 3 \sin \theta \cos^2 \theta, \quad 0 \le \theta < 2\pi$

**87.** $r = 0.3\left[1 + 2 \sin\left(\frac{\theta}{2}\right)\right], \quad 0 \le \theta < 4\pi$
(nephroid of Freeth)

**88.** $r = \dfrac{1 - 10 \cos \theta}{1 + 10 \cos \theta}, \quad 0 \le \theta < 2\pi$

**89.** $r^2 = 0.8(1 - 0.8 \sin^2 \theta), \quad 0 \le \theta < 2\pi$ (hippopede curve)

**90.** $r^2 = \dfrac{\frac{1}{4} \sin^2 \theta - 3.6 \cos^2 \theta}{\sin^2 \theta - \cos^2 \theta}, \quad 0 \le \theta < 2\pi$ (devil's curve)

**91.** $r = \dfrac{0.1}{\cos 3\theta}, \quad 0 \le \theta < \pi$ (epi-spiral)

**92.** $r = \dfrac{\sin \theta}{\theta}, \quad -6\pi \le \theta < 6\pi$ (cochleoid)

*In Exercises 93–95, determine whether the statement is true or false. If it is true, explain why it is true. If it is false, give an example to show why it is false.*

**93.** If $P(r_1, \theta_1)$ and $P(r_2, \theta_2)$ represent the same point in polar coordinates, then $r_1 = r_2$.

**94.** If $P(r_1, \theta_1)$ and $P(r_2, \theta_2)$ represent the same point in polar coordinates, then $\theta_1 = \theta_2$.

**95.** The graph of $r = f(\theta)$ has a horizontal tangent line at a point on the graph if $dy/d\theta = 0$, where $y = f(\theta)\sin \theta$.

---

## 9.5 Areas and Arc Lengths in Polar Coordinates

In this section we see how the use of polar equations to represent curves such as lemniscates and cardioids will simplify the task of finding the areas of the regions enclosed by these curves as well as the lengths of these curves.

### ■ Areas in Polar Coordinates

To develop a formula for finding the area of a region bounded by a curve defined by a polar equation, we need the formula for the area of a sector of a circle

$$A = \frac{1}{2}r^2\theta \tag{1}$$

where $r$ is the radius of the circle and $\theta$ is the central angle measured in radians. (See Figure 1.) This formula follows by observing that the area of a sector is $\theta/(2\pi)$ times that of the area of a circle; that is,

$$A = \frac{\theta}{2\pi} \cdot \pi r^2 = \frac{1}{2}r^2\theta$$

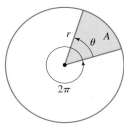

**FIGURE 1**
The area of a sector of a circle is $A = \frac{1}{2}r^2\theta$.

Now let $R$ be a region bounded by the graph of the polar equation $r = f(\theta)$ and the rays $\theta = \alpha$ and $\theta = \beta$, where $f$ is a nonnegative continuous function and $0 \le \beta - \alpha < 2\pi$, as shown in Figure 2a. Let $P$ be a regular partition of the interval $[\alpha, \beta]$:

$$\alpha = \theta_0 < \theta_1 < \theta_2 < \cdots < \theta_n = \beta$$

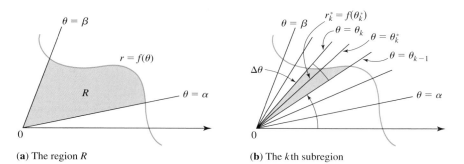

**FIGURE 2**   (**a**) The region $R$                 (**b**) The $k$th subregion

The rays $\theta = \theta_k$ divide $R$ into $n$ subregions $R_1, R_2, \ldots, R_n$ of area $\Delta A_1, \Delta A_2, \ldots, \Delta A_n$, respectively. If we choose $\theta_k^*$ in the interval $[\theta_{k-1}, \theta_k]$, then the area of $\Delta A_k$ of the $k$th subregion bounded by the rays $\theta = \theta_{k-1}$ and $\theta = \theta_k$ is approximated by the sector of a circle with central angle

$$\Delta\theta = \frac{\beta - \alpha}{n}$$

and radius $f(\theta_k^*)$ (highlighted in Figure 2b). Using Equation (1), we have

$$\Delta A_k \approx \frac{1}{2}[f(\theta_k^*)]^2 \Delta\theta$$

Therefore, an approximation of the area $A$ of $R$ is

$$A = \sum_{k=1}^{n} \Delta A_k \approx \sum_{k=1}^{n} \frac{1}{2}[f(\theta_k^*)]^2 \Delta\theta \tag{2}$$

But the sum in Equation (2) is a Riemann sum of the continuous function $\frac{1}{2}f^2$ over the interval $[\alpha, \beta]$. Therefore, it is true, although we will not prove it here, that

$$A = \lim_{n \to \infty} \sum_{k=1}^{n} \frac{1}{2}[f(\theta_k^*)]^2 \Delta\theta = \int_{\alpha}^{\beta} \frac{1}{2}[f(\theta)]^2 \, d\theta$$

---

**THEOREM 1   Area Bounded by a Polar Curve**

Let $f$ be a continuous, nonnegative function on $[\alpha, \beta]$ where $0 \le \beta - \alpha < 2\pi$. Then the area $A$ of the region bounded by the graphs of $r = f(\theta)$, $\theta = \alpha$, and $\theta = \beta$ is given by

$$A = \int_{\alpha}^{\beta} \frac{1}{2}[f(\theta)]^2 \, d\theta = \int_{\alpha}^{\beta} \frac{1}{2} r^2 \, d\theta$$

---

**Note**   When you determine the limits of integration, keep in mind that the region $R$ is swept out in a counterclockwise direction by the ray emanating from the origin, starting at the angle $\alpha$ and terminating at the angle $\beta$.

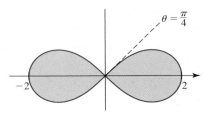

**FIGURE 3**
The region enclosed by the lemniscate
$r^2 = 4 \cos 2\theta$

**EXAMPLE 1** Find the area of the region enclosed by the lemniscate $r^2 = 4 \cos 2\theta$. This lemniscate has rectangular equation $x^4 + 2x^2y^2 - 4x^2 + 4y^2 + y^4 = 0$, as you can verify.

**Solution** The lemniscate is shown in Figure 3. Making use of symmetry, we see that the required area $A$ is four times that of the area swept out by the ray emanating from the origin as $\theta$ increases from 0 to $\pi/4$. In other words,

$$A = 4 \int_0^{\pi/4} \frac{1}{2} r^2 \, d\theta = 8 \int_0^{\pi/4} \cos 2\theta \, d\theta$$

$$= \left[ 4 \sin 2\theta \right]_0^{\pi/4} = 4 \qquad \blacksquare$$

**EXAMPLE 2** Find the area of the region enclosed by the cardioid $r = 1 + \cos \theta$.

**Solution** The graph of the cardioid $r = 1 + \cos \theta$, sketched previously in Example 6 in Section 9.4, is reproduced in Figure 4. Observe that the ray emanating from the origin sweeps out the required region exactly once as $\theta$ increases from 0 to $2\pi$. Therefore, the required area $A$ is

$$A = \int_0^{2\pi} \frac{1}{2} r^2 \, d\theta = \int_0^{2\pi} \frac{1}{2} (1 + \cos \theta)^2 \, d\theta$$

$$= \frac{1}{2} \int_0^{2\pi} (1 + 2 \cos \theta + \cos^2 \theta) \, d\theta$$

$$= \frac{1}{2} \int_0^{2\pi} \left( 1 + 2 \cos \theta + \frac{1 + \cos 2\theta}{2} \right) d\theta$$

$$= \frac{1}{2} \int_0^{2\pi} \left( \frac{3}{2} + 2 \cos \theta + \frac{1}{2} \cos 2\theta \right) d\theta$$

$$= \frac{1}{2} \left[ \frac{3}{2} \theta + 2 \sin \theta + \frac{1}{4} \sin 2\theta \right]_0^{2\pi} = \frac{3}{2} \pi \qquad \blacksquare$$

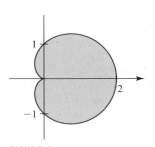

**FIGURE 4**
The region enclosed by the
cardioid $r = 1 + \cos \theta$

**EXAMPLE 3** Find the area inside the smaller loop of the limaçon $r = 1 + 2 \cos \theta$.

**Solution** We first sketch the limaçon $r = 1 + 2 \cos \theta$ (Figure 5). Observe that the region of interest is swept out by the ray emanating from the origin as $\theta$ runs from $2\pi/3$ to $4\pi/3$. We can also take advantage of symmetry by observing that the required area is double the area of the smaller loop lying below the polar axis. Since this region is swept out by the ray emanating from the origin as $\theta$ runs from $2\pi/3$ to $\pi$, we see that the required area is

$$A = 2 \int_{2\pi/3}^{\pi} \frac{1}{2} r^2 \, d\theta = \int_{2\pi/3}^{\pi} r^2 \, d\theta$$

$$= \int_{2\pi/3}^{\pi} (1 + 2 \cos \theta)^2 \, d\theta$$

$$= \int_{2\pi/3}^{\pi} (1 + 4 \cos \theta + 4 \cos^2 \theta) \, d\theta$$

$$= \int_{2\pi/3}^{\pi} \left[ 1 + 4 \cos \theta + 4 \left( \frac{1 + \cos 2\theta}{2} \right) \right] d\theta$$

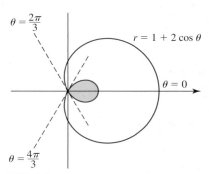

**FIGURE 5**
The limaçon $r = 1 + 2 \cos \theta$

$$= \int_{2\pi/3}^{\pi} (3 + 4 \cos \theta + 2 \cos 2\theta) \, d\theta$$

$$= \left[ 3\theta + 4 \sin \theta + \sin 2\theta \right]_{2\pi/3}^{\pi}$$

$$= 3\pi - \left( 2\pi + 4 \cdot \frac{\sqrt{3}}{2} - \frac{\sqrt{3}}{2} \right) = \pi - \frac{3\sqrt{3}}{2} \qquad \blacksquare$$

## ■ Area Bounded by Two Graphs

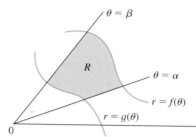

**FIGURE 6**
$R$ is the region bounded by the graphs of $r = f(\theta)$ and $r = g(\theta)$ for $\alpha \le \theta \le \beta$.

Consider the region $R$ bounded by the graphs of the polar equations $r = f(\theta)$ and $r = g(\theta)$, and the rays $\theta = \alpha$ and $\theta = \beta$, where $f(\theta) \ge g(\theta) \ge 0$ and $0 \le \beta - \alpha < 2\pi$. (See Figure 6.) From the figure we can see that the area $A$ of $R$ is found by subtracting the area of the region inside $r = g(\theta)$ from the area of the region inside $r = f(\theta)$. Using Theorem 1, we obtain the following theorem.

> **THEOREM 2    Area Bounded by Two Polar Curves**
>
> Let $f$ and $g$ be continuous on $[\alpha, \beta]$, where $0 \le g(\theta) \le f(\theta)$ and $0 \le \beta - \alpha < 2\pi$. Then the area $A$ of the region bounded by the graphs of $r = g(\theta)$, $r = f(\theta)$, $\theta = \alpha$, and $\theta = \beta$ is given by
>
> $$A = \frac{1}{2} \int_{\alpha}^{\beta} \{ [f(\theta)]^2 - [g(\theta)]^2 \} \, d\theta$$

**EXAMPLE 4** Find the area of the region that lies outside the circle $r = 3$ and inside the cardioid $r = 2 + 2 \cos \theta$.

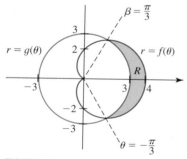

**FIGURE 7**
$R$ is the region outside the circle $r = 3$ and inside the cardioid $r = 2 + 2 \cos \theta$.

**Solution**   We first sketch the circle $r = 3$ and the cardioid $r = 2 + 2 \cos \theta$. The required region is shown shaded in Figure 7.

To find the points of intersection of the two curves, we solve the two equations simultaneously. We have $2 + 2 \cos \theta = 3$ or $\cos \theta = \frac{1}{2}$, which gives $\theta = \pm\pi/3$. Since the region of interest is swept out by the ray emanating from the origin as $\theta$ varies from $-\pi/3$ to $\pi/3$, we see that the required area is, by Theorem 2,

$$A = \frac{1}{2} \int_{\alpha}^{\beta} \{ [f(\theta)]^2 - [g(\theta)]^2 \} \, d\theta$$

where $f(\theta) = 2 + 2 \cos \theta = 2(1 + \cos \theta)$, $g(\theta) = 3$, $\alpha = -\pi/3$, and $\beta = \pi/3$. If we take advantage of symmetry, we can write

$$A = 2 \left( \frac{1}{2} \right) \int_{0}^{\pi/3} \{ [2(1 + \cos \theta)]^2 - 3^2 \} \, d\theta$$

$$= \int_{0}^{\pi/3} (4 + 8 \cos \theta + 4 \cos^2 \theta - 9) \, d\theta$$

$$= \int_{0}^{\pi/3} \left( -5 + 8 \cos \theta + 4 \cdot \frac{1 + \cos 2\theta}{2} \right) d\theta$$

$$= \int_{0}^{\pi/3} (-3 + 8 \cos \theta + 2 \cos 2\theta) \, d\theta$$

$$= \left[ -3\theta + 8 \sin \theta + \sin 2\theta \right]_{0}^{\pi/3}$$

$$= \left( -\pi + 8 \left( \frac{\sqrt{3}}{2} \right) + \frac{\sqrt{3}}{2} \right) = \frac{9\sqrt{3}}{2} - \pi \qquad \blacksquare$$

## ■ Arc Length in Polar Coordinates

To find the length of a curve $C$ defined by a polar equation $r = f(\theta)$ for $\alpha \leq \theta \leq \beta$, we use Equation (1) in Section 9.4 to write the parametric equations

$$x = r \cos \theta = f(\theta) \cos \theta \qquad \text{and} \qquad y = r \sin \theta = f(\theta) \sin \theta \qquad \alpha \leq \theta \leq \beta$$

for the curve, regarding $\theta$ as the parameter. Then

$$\frac{dx}{d\theta} = f'(\theta) \cos \theta - f(\theta) \sin \theta \qquad \text{and} \qquad \frac{dy}{d\theta} = f'(\theta) \sin \theta + f(\theta) \cos \theta$$

Therefore,

$$\left(\frac{dx}{d\theta}\right)^2 + \left(\frac{dy}{d\theta}\right)^2 = [f'(\theta)]^2 \cos^2 \theta - 2f'(\theta)f(\theta) \cos \theta \sin \theta + [f(\theta)]^2 \sin^2 \theta$$
$$+ [f'(\theta)]^2 \sin^2 \theta + 2f'(\theta)f(\theta) \cos \theta \sin \theta + [f(\theta)]^2 \cos^2 \theta$$
$$= [f'(\theta)]^2 + [f(\theta)]^2 \qquad \text{sin}^2 \theta + \cos^2 \theta = 1$$

Consequently, if $f'$ is continuous, then Theorem 1 in Section 9.3 gives the arc length of $C$ as

$$L = \int_\alpha^\beta \sqrt{\left(\frac{dx}{d\theta}\right)^2 + \left(\frac{dy}{d\theta}\right)^2}\, d\theta = \int_\alpha^\beta \sqrt{[f'(\theta)]^2 + [f(\theta)]^2}\, d\theta$$

---

**THEOREM 3    Arc Length**

Let $f$ be a function with a continuous derivative on an interval $[\alpha, \beta]$. If the graph $C$ of $r = f(\theta)$ is traced exactly once as $\theta$ increases from $\alpha$ to $\beta$, then the length $L$ of $C$ is given by

$$L = \int_\alpha^\beta \sqrt{[f'(\theta)]^2 + [f(\theta)]^2}\, d\theta = \int_\alpha^\beta \sqrt{\left(\frac{dr}{d\theta}\right)^2 + r^2}\, d\theta$$

---

**EXAMPLE 5**    Find the length of the cardioid $r = 1 + \cos \theta$.

**Solution**    The cardioid is shown in Figure 8. Observe that the cardioid is traced exactly once as $\theta$ runs from $\theta$ to $2\pi$. However, we can also take advantage of symmetry to see that the required length is twice that of the length of the cardioid lying above the polar axis. Thus,

$$L = 2 \int_0^\pi \sqrt{\left(\frac{dr}{d\theta}\right)^2 + r^2}\, d\theta$$

But $r = 1 + \cos \theta$, so

$$\frac{dr}{d\theta} = -\sin \theta$$

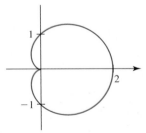

**FIGURE 8**
The cardioid $r = 1 + \cos \theta$

Therefore,

$$L = 2 \int_0^\pi \sqrt{(-\sin \theta)^2 + (1 + \cos \theta)^2}\, d\theta$$

$$= 2 \int_0^\pi \sqrt{\sin^2 \theta + 1 + 2 \cos \theta + \cos^2 \theta}\, d\theta$$

$$= 2 \int_0^\pi \sqrt{2 + 2 \cos \theta} \, d\theta \qquad \sin^2 \theta + \cos^2 \theta = 1$$

$$= 2\sqrt{2} \int_0^\pi \sqrt{1 + \cos \theta} \, d\theta = 2\sqrt{2} \int_0^\pi \sqrt{2 \cos^2 \frac{\theta}{2}} \, d\theta$$

$$= 4 \int_0^\pi \left| \cos \frac{\theta}{2} \right| d\theta = 4 \int_0^\pi \cos \frac{\theta}{2} \, d\theta \qquad \cos \frac{\theta}{2} \geq 0 \text{ on } [0, \pi]$$

$$= \left[ 4(2) \sin \frac{\theta}{2} \right]_0^\pi = 8 \qquad \blacksquare$$

## Area of a Surface of Revolution

The formulas for finding the area of a surface obtained by revolving a curve defined by a polar equation about the polar axis or about the line $\theta = \pi/2$ can be derived by using Equations (8) and (9) of Section 9.3 and the equations $x = r \cos \theta$ and $y = r \sin \theta$.

---

**THEOREM 4**   **Area of a Surface of a Revolution**

Let $f$ be a function with a continuous derivative on an interval $[\alpha, \beta]$. If the graph $C$ of $r = f(\theta)$ is traced exactly once as $\theta$ increases from $\alpha$ to $\beta$, then the area of the surface obtained by revolving $C$ about the indicated line is given by

**a.** $S = 2\pi \displaystyle\int_\alpha^\beta r \sin \theta \sqrt{\left(\frac{dr}{d\theta}\right)^2 + r^2} \, d\theta$    (about the polar axis)

**b.** $S = 2\pi \displaystyle\int_\alpha^\beta r \cos \theta \sqrt{\left(\frac{dr}{d\theta}\right)^2 + r^2} \, d\theta$    (about the line $\theta = \pi/2$)

---

**Note**   In using Theorem 4, we must choose $[\alpha, \beta]$ so that the surface is only traced once when $C$ is revolved about the line.    $\blacksquare$

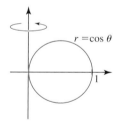

(a)

(b)

**FIGURE 9**

The solid obtained by revolving the circle $r = \cos \theta$ (a) about the line $\theta = \pi/2$ is a torus (b).

**EXAMPLE 6**   Find the area $S$ of the surface obtained by revolving the circle $r = \cos \theta$ about the line $\theta = \pi/2$. (See Figure 9.)

**Solution**   Observe that the circle is traced exactly once as $\theta$ increases from 0 to $\pi$. Therefore, using Theorem 4 with $r = \cos \theta$, $\alpha = 0$, and $\beta = \pi$, we obtain

$$S = 2\pi \int_\alpha^\beta f(\theta) \cos \theta \sqrt{\left(\frac{dr}{d\theta}\right)^2 + r^2} \, d\theta$$

$$= 2\pi \int_0^\pi \cos \theta (\cos \theta) \sqrt{(-\sin \theta)^2 + (\cos \theta)^2} \, d\theta$$

$$= 2\pi \int_0^\pi \cos^2 \theta \, d\theta = \pi \int_0^\pi (1 + \cos 2\theta) \, d\theta$$

$$= \pi \left[ \theta + \frac{\sin 2\theta}{2} \right]_0^\pi = \pi^2 \qquad \blacksquare$$

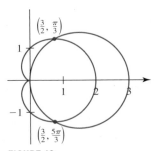

**FIGURE 10**
The graphs of the cardioid
$r = 1 + \cos\theta$ and the circle
$r = 3\cos\theta$

## ■ Points of Intersection of Graphs in Polar Coordinates

In Example 4 we were able to find the points of intersection of two curves with representations in polar coordinates by solving a system of two equations simultaneously. This is not always the case. Consider for example, the graphs of the cardioid $r = 1 + \cos\theta$ and the circle $r = 3\cos\theta$ shown in Figure 10. Solving the two equations simultaneously, we obtain

$$3\cos\theta = 1 + \cos\theta$$

$$\cos\theta = \frac{1}{2}$$

(3)

or $\theta = \pi/3$ and $5\pi/3$. Therefore, the points of intersection are $\left(\frac{3}{2}, \frac{\pi}{3}\right)$ and $\left(\frac{3}{2}, \frac{5\pi}{3}\right)$. But one glance at Figure 10 shows the pole as a third point of intersection that is not revealed in our calculation. To see how this can happen, think of the cardioid as being traced by the point $(r, \theta)$ satisfying

$$r = f(\theta) = 1 + \cos\theta \qquad 0 \le \theta \le 2\pi$$

with $\theta$ as a parameter. If we think of $\theta$ as representing time, then as $\theta$ runs from $\theta = 0$ through $\theta = 2\pi$, the point $(r, \theta)$ starts at $(2, 0)$ and traverses the cardioid in a counterclockwise direction, eventually returning to the point $(2, 0)$. (See Figure 11a.) Similarly, the circle is traced *twice* in the counterclockwise direction, by the point $(r, \theta)$, where

$$r = g(\theta) = 3\cos\theta \qquad 0 \le \theta \le 2\pi$$

and the parameter $\theta$, once again representing time, runs from $\theta = 0$ through $\theta = 2\pi$ (see Figure 11b).

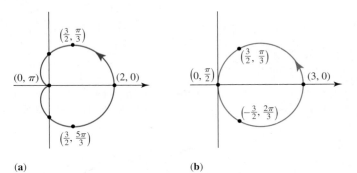

**FIGURE 11**     (a)                                (b)

Observe that the point tracing the cardioid arrives at the point $\left(\frac{3}{2}, \frac{\pi}{3}\right)$ on the cardioid at precisely the same time that the point tracing the circle arrives at the point $\left(\frac{3}{2}, \frac{\pi}{3}\right)$ on the circle. A similar observation holds at the point $\left(\frac{3}{2}, \frac{5\pi}{3}\right)$ on each of the two curves. These are the points of intersection found earlier.

Next, observe that the point tracing the cardioid arrives at the origin when $\theta = \pi$. But the point tracing the circle first arrives at the origin when $\theta = \pi/2$ and then again when $\theta = 3\pi/2$. In other words, these two points arrive at the origin at *different* times, so there is no (common) value of $\theta$ corresponding to the origin that satisfies both Equations (3) simultaneously. Thus, although the origin is a point of intersection of the two curves, this fact will not show up in the solution of the system of equations. For this reason it is recommended that we sketch the graphs of polar equations when finding their points of intersection.

**EXAMPLE 7** Find the points of inte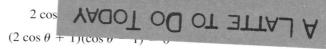

**Solution** We solve the system of eq

We set $\cos \theta = \cos 2\theta$ and use the id

$$2 \cos$$

$$(2 \cos \theta + 1)(\cos \theta - 1) =$$

So

$$\cos \theta = -\frac{1}{2} \quad \text{or} \quad \cos \theta = 1$$

that is,

$$\theta = \frac{2\pi}{3}, \quad \frac{4\pi}{3}, \quad \text{or} \quad 0$$

These values of $\theta$ give $\left(-\frac{1}{2}, \frac{2\pi}{3}\right)$, $\left(-\frac{1}{2}, \frac{4\pi}{3}\right)$, and $(1, 0)$ as the points of intersection. Since both graphs also pass through the pole, we conclude that the pole is also a point of intersection. (See Figure 12.)

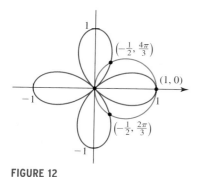

**FIGURE 12**

## 9.5 CONCEPT QUESTIONS

**1. a.** Let $f$ be nonnegative and continuous on $[\alpha, \beta]$, where $0 \le \beta - \alpha < 2\pi$. Write an integral giving the area of the region bounded by the graphs of $r = f(\theta)$, $\theta = \alpha$, and $\theta = \beta$. Make a sketch of the region.
   **b.** If $f$ and $g$ are continuous on $[\alpha, \beta]$ and $0 \le g(\theta) \le f(\theta)$, where $0 \le \alpha \le \beta \le 2\pi$, write an integral giving the area of the region bounded by the graphs of $r = g(\theta)$, $r = f(\theta)$, $\theta = \alpha$, and $\theta = \beta$. Make a sketch of the region.

**2.** Suppose that $f$ has a continuous derivative on an interval $[\alpha, \beta]$. If the graph $C$ of $r = f(\theta)$ is traced exactly once as $\theta$ increases from $\alpha$ to $\beta$, write an integral giving the length of $C$.

**3.** Suppose that $f$ is a function with a continuous derivative on $[\alpha, \beta]$ and the graph $C$ of $r = f(\theta)$ is traced exactly once as $\theta$ increases from $\alpha$ to $\beta$. Write an integral giving the area of the surface obtained by revolving $C$ about (a) the polar axis, $y \ge 0$, and (b) the line $\theta = \pi/2$, $x \ge 0$.

## 9.5 EXERCISES

**1. a.** Find a rectangular equation of the circle $r = 4 \cos \theta$, and use it to find its area.
   **b.** Find the area of the circle of part (a) by integration.

**2. a.** By finding a rectangular equation, show that the polar equation $r = 2 \cos \theta - 2 \sin \theta$ represents a circle. Then find the area of the circle.
   **b.** Find the area of the circle of part (a) by integration.

*In Exercises 3–8, find the area of the region bounded by the curve and the rays.*

**3.** $r = \theta, \quad \theta = 0, \quad \theta = \pi$

**4.** $r = \dfrac{1}{\theta}, \quad \theta = \dfrac{\pi}{6}, \quad \theta = \dfrac{\pi}{3}$

**5.** $r = e^{\theta}, \quad \theta = -\dfrac{\pi}{2}, \quad \theta = 0$

**6.** $r = e^{-2\theta}, \quad \theta = 0, \quad \theta = \dfrac{\pi}{4}$

**7.** $r = \sqrt{\cos \theta}, \quad \theta = 0, \quad \theta = \dfrac{\pi}{2}$

**8.** $r = \cos 2\theta, \quad \theta = 0, \quad \theta = \dfrac{\pi}{16}$

*In Exercises 9–12, find the area of the shaded region.*

**9.**

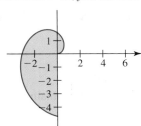

$r = \theta$

**10.**

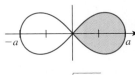

$r = a\sqrt{\cos 2\theta}$

**11.**

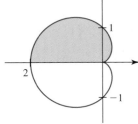

$r = 1 - \cos \theta$

**12.**

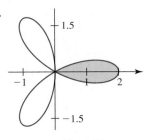

$r = 2 \cos 3\theta$

*In Exercises 13–18, sketch the curve, and find the area of the region enclosed by it.*

**13.** $r = 3 \sin \theta$

**14.** $r = 2(1 - \cos \theta)$

**15.** $r^2 = \sin \theta$

**16.** $r^2 = 3 \sin 2\theta$

**17.** $r = 2 \sin 2\theta$

**18.** $r = 2 \sin 3\theta$

*In Exercises 19–22, find the area of the region enclosed by one loop of the curve.*

**19.** $r = \cos 2\theta$

**20.** $r = 2 \cos 3\theta$

**21.** $r = \sin 4\theta$

**22.** $r = 2 \cos 4\theta$

*In Exercises 23–24, find the area of the region described.*

**23.** The inner loop of the limaçon $r = 1 + 2 \cos \theta$

**24.** Between the loops of the limaçon $r = 1 + 2 \sin \theta$

*In Exercises 25–28, find the area of the shaded region.*

**25.**

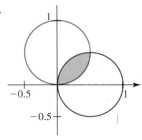

$r = \sin \theta, r = \cos \theta$

**26.**

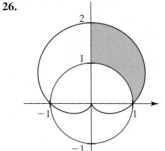

$r = 1, r = 1 + \sin \theta$

**27.**

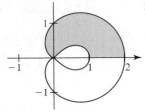

$r = 1 + \cos \theta, r = \sqrt{\cos 2\theta}$

**28.**

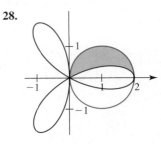

$r = 2 \cos 3\theta, r = 2 \cos \theta$

*In Exercises 29–34, find all points of intersection of the given curves.*

**29.** $r = 1$ and $r = 1 + \cos \theta$

**30.** $r = 3$ and $r = 2 + 2 \cos \theta$

**31.** $r = 2$ and $r = 4 \cos 2\theta$

**32.** $r = 1$ and $r^2 = 2 \cos 2\theta$

**33.** $r = \sin \theta$ and $r = \sin 2\theta$

**34.** $r = \cos \theta$ and $r = \cos 2\theta$

*In Exercises 35–40, find the area of the region that lies outside the first curve and inside the second curve.*

**35.** $r = 1 + \cos \theta$, $r = 3 \cos \theta$

**36.** $r = 1 - \sin \theta$, $r = 1$

**37.** $r = 4 \cos \theta$, $r = 2$

**38.** $r = 3 \sin \theta$, $r = 2 - \sin \theta$

**39.** $r = 1 - \cos \theta$, $r = \dfrac{3}{2}$

**40.** $r = 2 \cos 3\theta$, $r = 1$

*In Exercises 41–46, find the area of the region that is enclosed by both of the curves.*

**41.** $r = 1$, $r = 2 \sin \theta$

**42.** $r = \cos \theta$, $r = \sqrt{3} \sin \theta$

**43.** $r = \sin \theta$, $r = 1 - \sin \theta$

**44.** $r = \cos \theta$, $r = 1 - \cos \theta$

**45.** $r^2 = 4 \cos 2\theta$, $r = \sqrt{2}$

**46.** $r = \sqrt{3} \sin \theta$, $r = 1 + \cos \theta$

*In Exercises 47–54, find the length of the given curve.*

**47.** $r = 5 \sin \theta$

**48.** $r = 2\theta$; $0 \le \theta \le 2\pi$

**49.** $r = e^{-\theta}$; $0 \le \theta \le 4\pi$

**50.** $r = 1 + \sin \theta$; $0 \le \theta \le 2\pi$

**51.** $r = \sin^3 \dfrac{\theta}{3}$; $0 \le \theta \le \pi$

**52.** $r = \cos^2 \dfrac{\theta}{2}$

**53.** $r = a \sin^4 \dfrac{\theta}{4}$

**54.** $r = \sec \theta;$   $0 \le \theta \le \dfrac{\pi}{3}$

*In Exercises 55–60, find the area of the surface obtained by revolving the given curve about the given line.*

**55.** $r = 4 \cos \theta$ about the polar axis

**56.** $r = 2 \cos \theta$ about the line $\theta = \dfrac{\pi}{2}$

**57.** $r = 2 + 2 \cos \theta$ about the polar axis

**58.** $r^2 = \cos 2\theta$ about the polar axis

**59.** $r^2 = \cos 2\theta$ about the line $\theta = \dfrac{\pi}{2}$

**60.** $r = e^{a\theta}$,   $0 \le \theta \le \dfrac{\pi}{2}$ about the line $\theta = \dfrac{\pi}{2}$

*In Exercises 61 and 62, find the area of the region enclosed by the given curve. (Hint: Convert the rectangular equation to a polar equation.)*

**61.** $(x^2 + y^2)^3 = 16x^2 y^2$

**62.** $x^4 + y^4 = 4(x^2 + y^2)$

**63.** Let $P$ be a point other than the origin lying on the curve $r = f(\theta)$. If $\psi$ is the angle between the tangent line to the curve at $P$ and the radial line $OP$, then $\tan \psi = \dfrac{r}{dr/d\theta}$. (See Section 9.4, Exercise 84.)

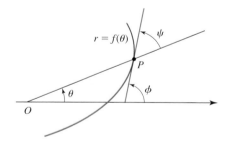

**a.** Show that the angle between the tangent line to the logarithmic spiral $r = e^{m\theta}$ and the radial line at the point of tangency is a constant.

**b.** Suppose the curve with polar equation $r = f(\theta)$ has the property that at any point on the curve, the angle $\psi$ between the tangent line to the curve at that point and the radial line from the origin to that point is a constant. Show that $f(\theta) = Ce^{m\theta}$, where $C$ and $m$ are constants.

**64.** Find the length of the logarithmic spiral $r = ae^{m\theta}$ between the point $(r_0, \theta_0)$ and the point $(r, \theta)$, and use this result to deduce that the length of a logarithmic spiral is proportional to the difference between the radial coordinates of the points.

**65.** Show that the length of the parabola $y = (1/2p)x^2$ on the interval $[0, a]$ is the same as the length of the spiral $r = p\theta$ for $0 \le r \le a$.

**cas 66.** Plot the curve $r = \sin(3 \cos \theta)$, and find an approximation of the area enclosed by the curve accurate to four decimal places.

**cas** *In Exercises 67–69, (a) plot the curve, and (b) find an approximation of its length accurate to two decimal places.*

**67.** $r = \sqrt{1 + \theta^2}$, where $0 \le \theta \le 2\pi$   (involute of a circle)

**68.** $r = 0.2\sqrt{\theta} + 1$, where $0 \le \theta \le 6\pi$   (parabolic spiral)

**69.** $r = 3 \sin \theta \cos^2 \theta$, where $0 \le \theta \le \pi$   (bifolia)

**70. a.** Let $f$ be a function with a continuous derivative in an interval $[\alpha, \beta]$. If the graph $C$ of $r = f(\theta)$ is traced exactly once as $\theta$ increases from $\alpha$ to $\beta$, show that the rectangular coordinates of the centroid of $C$ are

$$\bar{x} = \frac{\displaystyle\int_\alpha^\beta r \cos \theta \sqrt{(r')^2 + r^2}\, d\theta}{\displaystyle\int_\alpha^\beta \sqrt{(r')^2 + r^2}\, d\theta}$$

and

$$\bar{y} = \frac{\displaystyle\int_\alpha^\beta r \sin \theta \sqrt{(r')^2 + r^2}\, d\theta}{\displaystyle\int_\alpha^\beta \sqrt{(r')^2 + r^2}\, d\theta}$$

**Hint:** See the directions for Exercises 47 and 48 in Section 5.7.

**b.** Use the result of part (a) to find the centroid of the upper semicircle $r = a$, where $a > 0$ and $0 \le \theta \le \pi$.

**cas 71. a.** Plot the curve with polar equation $r = 2 \cos^3 \theta$ where $-\dfrac{\pi}{2} \le \theta \le \dfrac{\pi}{2}$.

**b.** Find the Cartesian coordinates of the centroid of the region bounded by the curve of part (a).

**cas 72. a.** Plot the graphs of $r = 1 + \cos \theta$ and $r = 3 \cos \theta$ for $0 \le \theta \le 2\pi$, treating $r$ and $\theta$ as rectangular coordinates.

**b.** Refer to page 880. Reconcile your results with the discussion of finding the points of intersection of graphs in polar coordinates.

**cas** *In Exercises 73 and 74, (a) find the polar representation of the curve given in rectangular coordinates, (b) plot the curve, and (c) find the area of the region enclosed by a loop (or loops) of the curve.*

**73.** $x^3 - 3xy + y^3 = 0$   (folium of Descartes)

**74.** $(x^2 + y^2)^{1/2} - \cos\left[4 \tan^{-1}\left(\dfrac{y}{x}\right)\right] = 0$   (rhodenea)

*In Exercises 75 and 76, determine whether the statement is true or false. If it is true, explain why it is true. If it is false, give an example to show why it is false.*

**75.** If there exists a $\theta_0$ such that $f(\theta_0) = g(\theta_0)$, then the graphs of $r = f(\theta)$ and $r = g(\theta)$ have at least one point of intersection.

**76.** If $P$ is a point of intersection of the graphs of $r = f(\theta)$ and $r = g(\theta)$, then there must exist a $\theta_0$ such that $f(\theta_0) = g(\theta_0)$.

## 9.6 Conic Sections in Polar Coordinates

In Section 9.1 we obtained representations of the conic sections—the parabola, the ellipse, and the hyperbola—in terms of rectangular equations. In this section we will show that all three types of conic sections can be represented by a single polar equation. As you saw in the preceding sections, some problems can be solved more easily using polar coordinates rather than rectangular coordinates.

We begin by proving the following theorem, which gives an equivalent definition of each conic section in terms of its focus and directrix. As a corollary, we will obtain the desired representation of the conic sections in polar form.

---

**THEOREM 1**

Let $F$ be a fixed point, let $l$ be a fixed line in the plane, and let $e$ be a fixed positive number. Then the set of all points $P$ in the plane satisfying

$$\frac{d(P, F)}{d(P, l)} = e$$

is a conic section. The point $F$ is the **focus** of the conic section, and the line $l$ is its **directrix**. The number $e$, which is the ratio of the distance between $P$ and $F$ and the distance between $P$ and $l$, is called the **eccentricity** of the conic. The conic is an ellipse if $e < 1$, a parabola if $e = 1$, or a hyperbola if $e > 1$.

---

The three types of conics are illustrated in Figure 1.

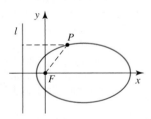

(a) $\dfrac{d(P, F)}{d(P, l)} = e < 1$ (ellipse)

(b) $\dfrac{d(P, F)}{d(P, l)} = e = 1$ (parabola)

(c) $\dfrac{d(P, F)}{d(P, l)} = e > 1$ (hyperbola)

**FIGURE 1**

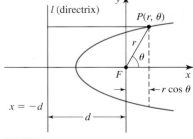

**FIGURE 2**

**PROOF** Observe that if $e = 1$, then $d(P, F) = d(P, l)$. That is, the distance between a point on the curve and the focus is equal to the distance between the point and the directrix. But this is just the definition of a parabola, so the curve is a conic section.

In what follows, we will assume that $e \neq 1$. Refer to Figure 2, where we have placed the focus $F$ at the origin and the directrix $l$ parallel to and $d$ units to the left of the $y$-axis. Therefore, the directrix has equation $x = -d$, where $d > 0$. If $P(r, \theta)$ is any point lying on the curve, then you can see from Figure 2 that

$$d(P, F) = r \qquad \text{and} \qquad d(P, l) = d + r \cos \theta$$

Therefore, the condition $d(P, F)/d(P, l) = e$ or, equivalently, $d(P, F) = e \cdot d(P, l)$, implies that

$$r = e(d + r \cos \theta) \tag{1}$$

Converting this equation to rectangular coordinates gives

$$\sqrt{x^2 + y^2} = e(d + x)$$

which, upon squaring, yields

$$x^2 + y^2 = e^2(d + x)^2 = e^2(d^2 + 2dx + x^2)$$

$$(1 - e^2)x^2 - 2de^2x + y^2 = e^2d^2$$

or

$$x^2 - \left(\frac{2e^2d}{1 - e^2}\right)x + \frac{y^2}{1 - e^2} = \frac{e^2d^2}{1 - e^2}$$

Completing the square in $x$, we obtain

$$\left(x - \frac{e^2d}{1 - e^2}\right)^2 + \frac{y^2}{1 - e^2} = \frac{e^2d^2}{1 - e^2} + \frac{e^4d^2}{(1 - e^2)^2} = \frac{e^2d^2}{(1 - e^2)^2} \tag{2}$$

Now, if $e < 1$, then $1 - e^2 > 0$. Dividing both sides by $e^2d^2/(1 - e^2)^2$, we can write Equation (2) in the form

$$\frac{(x - h)^2}{a^2} + \frac{y^2}{b^2} = 1$$

where

$$h = \frac{e^2d}{1 - e^2}, \qquad a^2 = \frac{e^2d^2}{(1 - e^2)^2}, \qquad \text{and} \qquad b^2 = \frac{e^2d^2}{1 - e^2} \tag{3}$$

This is an equation of an ellipse centered at the point $(h, 0)$ on the $x$-axis.

Next, we compute

$$c^2 = a^2 - b^2 = \frac{e^4d^2}{(1 - e^2)^2} \tag{4}$$

from which we obtain

$$c = \frac{e^2d}{1 - e^2} = h$$

Recalling that the foci of an ellipse are located at a distance $c$ from its center, we have shown that $F$ is indeed the focus of the ellipse. It also follows from Equations (3) and (4) that the eccentricity of the ellipse is given by

$$e = \frac{c}{a} \tag{5}$$

where $c^2 = a^2 - b^2$.

If $e > 1$, then $1 - e^2 < 0$. Proceeding in a similar manner as before, we can write Equation (2) in the form

$$\frac{(x - h)^2}{a^2} - \frac{y^2}{b^2} = 1$$

which is an equation of a hyperbola. We also see that the eccentricity of the hyperbola is

$$e = \frac{c}{a} \tag{6}$$

where $c^2 = a^2 + b^2$.

If we solve Equation (1) for $r$, we obtain the polar equation

$$r = \frac{ed}{1 - e \cos \theta}$$

of the conic shown in Figure 2. If the directrix is chosen so that it lies to the right of the focus, say, $x = d$, where $d > 0$, then the polar equation of the conic is

$$r = \frac{ed}{1 + e \cos \theta}$$

Similarly, we can show that if the directrix $y = \pm d$ is chosen to be parallel to the polar axis, then the polar equation of the conic is

$$r = \frac{ed}{1 \pm e \sin \theta}$$

(See Exercises 28–30.) ◼

The conics are illustrated in Figure 3.

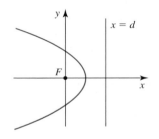

**(a)** $r = \dfrac{ed}{1 + e \cos \theta}$

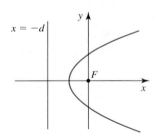

**(b)** $r = \dfrac{ed}{1 - e \cos \theta}$

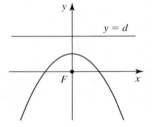

**(c)** $r = \dfrac{ed}{1 + e \sin \theta}$

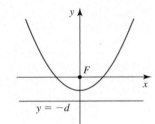
**(d)** $r = \dfrac{ed}{1 - e \sin \theta}$

**FIGURE 3**
Polar equations of conics

---

**THEOREM 2**

A polar equation of the form

$$r = \frac{ed}{1 \pm e \cos \theta} \qquad \text{or} \qquad r = \frac{ed}{1 \pm e \sin \theta}$$

represents a conic section with eccentricity $e$. The conic is a parabola if $e = 1$, an ellipse if $e < 1$, and a hyperbola if $e > 1$.

---

**EXAMPLE 1**    Find a polar equation of a parabola that has its focus at the pole and the line $y = 2$ as its directrix.

**Solution**   Since this conic section is a parabola, we see that $e = 1$. Next, observe that its directrix, $y = 2$, is parallel to and lies above the polar axis. So letting $d = 2$ and referring to Figure 3c, we see that a required equation of the parabola is

$$r = \frac{2}{1 + \sin \theta}$$

**EXAMPLE 2**   A conic has polar equation

$$r = \frac{15}{3 + 2 \cos \theta}$$

Find the eccentricity and the directrix of the conic section, and sketch the conic section.

**Solution**   We begin by rewriting the given equation in standard form by dividing both its numerator and denominator by 3, obtaining

$$r = \frac{5}{1 + \frac{2}{3} \cos \theta}$$

Then using Theorem 2, we see that $e = \frac{2}{3}$. Since $ed = 5$, we have

$$d = \frac{5}{e} = \frac{5}{\frac{2}{3}} = \frac{15}{2}$$

Since $e < 1$, we conclude that the conic section is an ellipse with focus at the pole and major axis lying along the polar axis. Its directrix has rectangular equation $x = \frac{15}{2}$. Setting $\theta = 0$ and $\theta = \pi$ successively gives $r = 3$ and $r = 15$, giving the vertices of the ellipse in polar coordinates as $(3, 0)$ and $(15, \pi)$. The center of the ellipse is the midpoint $(6, \pi)$ in polar coordinates of the line segment joining the vertices. Since the length of the major axes of the ellipse is 18, we have $2a = 18$, or $a = 9$. Finally, since $e = c/a$, we find that

$$c = ae = 9\left(\frac{2}{3}\right) = 6$$

So

$$b^2 = a^2 - c^2 = 81 - 36 = 45$$

or

$$b = 3\sqrt{5}$$

The graph of the conic is sketched in Figure 4.

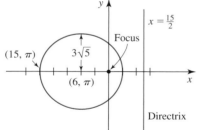

**FIGURE 4**
The graph of $r = \dfrac{15}{3 + 2 \cos \theta}$

**EXAMPLE 3**   Sketch the graph of the polar equation

$$r = \frac{20}{2 + 3 \sin \theta}$$

**Solution**   By dividing the numerator and the denominator of the given equation by 2, we obtain the equation

$$r = \frac{10}{1 + \frac{3}{2} \sin \theta}$$

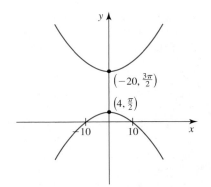

**FIGURE 5**

The graph of $r = \dfrac{20}{2 + 3 \sin \theta}$

in standard form. We see that $e = \frac{3}{2}$, so the equation represents a hyperbola with one focus at the pole. Comparing this equation with the equation associated with Figure 3c, we see that the transverse axis of the hyperbola lies along the line $\theta = \pi/2$. To find the vertices of the hyperbola, we set $\theta = \pi/2$ and $\theta = 3\pi/2$ successively, giving $\left(4, \frac{\pi}{2}\right)$ and $\left(-20, \frac{3\pi}{2}\right)$ as the required vertices in polar coordinates. The center of the hyperbola in polar coordinates is the midpoint $\left(12, \frac{\pi}{2}\right)$ of the line segment joining the vertices. The $x$-intercepts (we superimpose the Cartesian system over the polar system) are found by setting $\theta = 0$ and $\theta = \pi$, giving the $x$-intercepts as 10 and $-10$. The required graph may be sketched in two steps; first, we sketch the lower branch of the hyperbola, making use of the $x$-intercepts that we just found. Then, using symmetry, we sketch the upper branch of the hyperbola. (See Figure 5.) ■

## ■ Eccentricity of a Conic

As we saw in Theorem 1, the nature of a conic section is determined by its eccentricity $e$. To see in greater detail the role that is played by the eccentricity of a conic, let's first consider the case in which $e < 1$, so that the conic under consideration is an ellipse. Now by Equation (5) we have

$$e = \frac{c}{a} = \frac{\sqrt{a^2 - b^2}}{a}$$

If $e$ is close to 0, then $\sqrt{a^2 - b^2}$ is close to 0, or $a$ is close to $b$. This means that the ellipse is almost circular (see Figure 6a). On the other hand, if $e$ is close to 1, then $\sqrt{a^2 - b^2} \approx a$, $a^2 - b^2 \approx a^2$, or $b$ is small. This means that the ellipse is very flat (see Figure 6b).

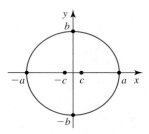

(**a**) $e$ is close to 0.

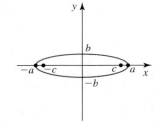

(**b**) $e$ is close to 1.

**FIGURE 6**
The ellipse is almost circular if $e$ is close to 0 and is very flat if $e$ is close to 1.

If $e = 1$, then the conic is a parabola. We leave it to you to perform a similar analysis in the case in which $e > 1$ (so that the conic is a hyperbola).

In Figure 7 we show two hyperbolas: In part (a) the eccentricity $e$ is close to but greater than 1. In part (b) the eccentricity $e$ is much larger than 1.

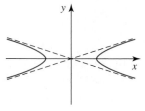

(**a**)

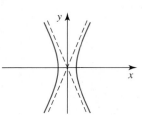

(**b**)

**FIGURE 7**
The eccentricity of the hyperbola in part (a) is close to 1, whereas the eccentricity of the hyperbola in part (b) is much larger than 1.

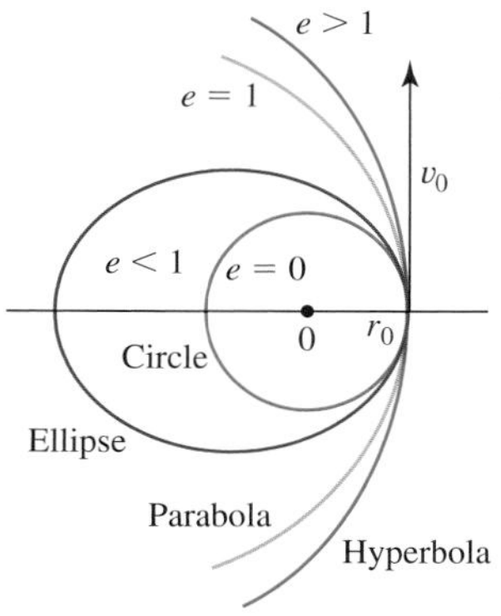

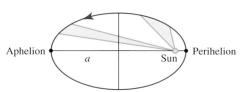

*e > 1*

*e = 1*

$v_0$

*e < 1*  *e = 0*

Circle  0  $r_0$

Ellipse

Parabola

Hyperbola

**FIGURE 8**
The speed $v_0$ determines
the orbit of the body.

# Motion of Celestial Bodies

In the last few sections we have seen numerous applications of conics. Yet another important application of the conics arises in the motion of celestial bodies.

Figure 8 shows a body a distance $r_0$ from the origin 0 moving with a speed $v_0$ and in a direction perpendicular to the line passing through 0 and $v_0$. It can be shown, although we will not do so here, that the orbit of the body about the origin depends on the magnitude of $v_0$. For the planets in the solar system (with the sun at the origin) and for certain comets such as Halley's comet, the speed $v_0$ is such that they remain captive and will never leave the system; their orbits are ellipses. However, if the speed $v_0$ of a body is sufficiently large, then its orbit about the sun is a parabola ($e = 1$) or a branch of a hyperbola ($e > 1$). In both these cases the body makes but a single pass about the sun!

The orbits of the planets about the sun, moreover, are described by Kepler's Laws.

> **Kepler's Laws**
> 1. Planets move in orbits that are ellipses with the sun at one focus.
> 2. The line from the sun to a planet sweeps out equal areas in equal times. (See Figure 9.)
> 3. The square of a planet's period is proportional to the cube of the length of the semimajor axis of its orbit.

Aphelion  $a$  Sun  Perihelion

**FIGURE 9**
Equal areas are swept out in equal times, $T^2 \propto a^3$, where $T$ is the period.

The positions of a planet that are closest to and farthest from the sun are called the *perihelion* and *aphelion,* respectively.

**EXAMPLE 4** **The Orbit of Halley's Comet** Halley's comet has an elliptical orbit with an eccentricity of 0.967. Its perihelion distance (shortest distance from the sun) is $8.9 \times 10^7$ km.

**a.** Find a polar equation for the orbit.
**b.** Find the distance of the comet from the sun when it is at the aphelion.

**Solution**
**a.** Suppose that the axis is horizontal as shown in Figure 10. Then the polar equation can be chosen to have the form

$$r = \frac{ed}{1 + e \cos \theta}$$

The distance of the comet from the sun when it is at the perihelion is given by

$$a - c = a - ea = a(1 - e)$$

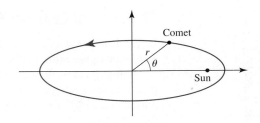

**FIGURE 10**
In actuality the trajectory is much flatter.

But we are given that at the perihelion the distance from the sun is $8.9 \times 10^7$ km and $e = 0.967$. So

$$a(1 - 0.967) = 8.9 \times 10^7$$

or

$$a = \frac{8.9 \times 10^7}{1 - 0.967} \approx 2.697 \times 10^9$$

Next, from Equation (3) we see that

$$ed = a(1 - e^2)$$
$$= (2.697 \times 10^9)(1 - 0.967^2) = 1.75 \times 10^8$$

So the required equation is

$$r = \frac{1.75 \times 10^8}{1 + 0.967 \cos \theta}$$

**b.** The aphelion distance (farthest distance from the sun) is

$$a + c = a + ea = a(1 + e) \approx (2.697 \times 10^9)(1 + 0.967) \approx 5.305 \times 10^9$$

kilometers. ∎

## 9.6 CONCEPT QUESTIONS

1. Consider the polar equations

$$r = \frac{ed}{1 \pm e \cos \theta} \quad \text{and} \quad r = \frac{ed}{1 \pm e \sin \theta}$$

Explain the role of the numbers $d$ and $e$. Illustrate each with a sketch.

2. Give a classification of the conics in terms of their eccentricities.

3. Identify the conic:

**a.** $r = \dfrac{3}{1 + 2 \sin \theta}$  **b.** $r = \dfrac{6}{3 + \cos \theta}$

**c.** $r = \dfrac{2}{3(1 + \cos \theta)}$  **d.** $r = \dfrac{5}{3 - 2 \sin \theta}$

## 9.6 EXERCISES

*In Exercises 1–8, write a polar equation of the conic that has a focus at the origin and the given properties. Identify the conic.*

1. Eccentricity 1, directrix $x = -2$

2. Eccentricity $\frac{1}{3}$, directrix $x = 3$

3. Eccentricity $\frac{1}{2}$, directrix $y = -2$

4. Eccentricity 1, directrix $y = -3$

5. Eccentricity $\frac{3}{2}$, directrix $x = 1$

6. Eccentricity $\frac{5}{4}$, directrix $y = -2$

**7.** Eccentricity 0.4, directrix $y = 0.4$

**8.** Eccentricity $\frac{1}{2}$, directrix $r = -2 \sec \theta$

*In Exercises 9–20, (a) find the eccentricity and an equation of the directrix of the conic, (b) identify the conic, and (c) sketch the curve.*

**9.** $r = \dfrac{8}{6 + 2 \sin \theta}$

**10.** $r = \dfrac{8}{6 - 2 \sin \theta}$

**11.** $r = \dfrac{10}{4 + 6 \cos \theta}$

**12.** $r = \dfrac{10}{4 - 6 \cos \theta}$

**13.** $r = \dfrac{5}{2 + 2 \cos \theta}$

**14.** $r = \dfrac{5}{2 - 2 \sin \theta}$

**15.** $r = \dfrac{1}{3 - 2 \cos \theta}$

**16.** $r = \dfrac{12}{3 + \cos \theta}$

**17.** $r = \dfrac{1}{1 - \sin \theta}$

**18.** $r = \dfrac{1}{1 + \cos \theta}$

**19.** $r = -\dfrac{6}{\sin \theta - 2}$

**20.** $r = -\dfrac{2}{\cos \theta - 3}$

*In Exercises 21–26, use Equation (5) or Equation (6) to find the eccentricity of the conic with the given rectangular equation.*

**21.** $\dfrac{x^2}{9} + \dfrac{y^2}{16} = 1$

**22.** $\dfrac{x^2}{5} - \dfrac{y^2}{3} = 1$

**23.** $x^2 - y^2 = 1$

**24.** $9x^2 + 25y^2 = 225$

**25.** $x^2 - 9y^2 + 2x - 54y = 105$

**26.** $2x^2 + y^2 + 4x - 6y + 7 = 0$

**27.** Show that the parabolas with polar equations

$$r = \frac{c}{1 + \sin \theta} \quad \text{and} \quad r = \frac{d}{1 - \sin \theta}$$

intersect at right angles.

**28.** Show that a conic with focus at the origin, eccentricity $e$, and directrix $x = d$ has polar equation

$$r = \frac{ed}{1 + e \cos \theta}$$

**29.** Show that a conic with focus at the origin, eccentricity $e$, and directrix $y = d$ has polar equation

$$r = \frac{ed}{1 + e \sin \theta}$$

**30.** Show that a conic with focus at the origin, eccentricity $e$, and directrix $y = -d$ has polar equation

$$r = \frac{ed}{1 - e \sin \theta}$$

**31. a.** Show that the polar equation of an ellipse with one focus at the pole and major axis lying along the polar axis is given by

$$r = \frac{a(1 - e^2)}{1 - e \cos \theta}$$

where $e$ is the eccentricity of the ellipse and $2a$ is the length of its major axis.

**b.** The planets revolve about the sun in elliptical orbits with the sun at one focus. The points on the orbit where a planet is nearest to and farthest from the sun are called the *perihelion* and the *aphelion* of the orbit, respectively. Use the result of part (a) to show that the perihelion distance (minimum distance from the planet to the sun) is $a(1 - e)$.

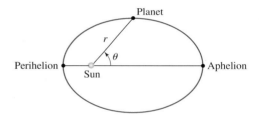

*In Exercises 32 and 33, use the results of Exercise 31 to find a polar equation describing the approximate orbit of the given planet and to find the perihelion and aphelion distances.*

**32.** Earth: $e = 0.017$, $a = 92.957 \times 10^6$ mi

**33.** Saturn: $e = 0.056$, $a = 1.427 \times 10^9$ km

**34.** The dwarf planet Pluto revolves about the sun in an elliptical orbit. The eccentricity of the orbit is 0.249, and its perihelion distance is $4.43 \times 10^9$ km. Use the results of Exercise 31 to find a polar equation for the orbit of Pluto and find its aphelion distance.

**35.** The planet Mercury revolves about the sun in an elliptical orbit. Its perihelion distance is approximately $4.6 \times 10^7$ km, and its aphelion distance is approximately $7.0 \times 10^7$ km. Use the results of Exercise 31 to estimate the eccentricity of Mercury's orbit.

# CHAPTER 9 REVIEW

## CONCEPT REVIEW

*In Exercises 1–16, fill in the blanks.*

**1. a.** A parabola is the set of all points in the plane that are _____ from a fixed _____ and a fixed _____. The fixed _____ is called the _____, and the fixed _____ is called the _____.

   **b.** The point halfway between the focus and the directrix of a parabola is called its _____. The axis of the parabola is the line passing through the _____ and perpendicular to the _____.

**2. a.** An equation of a parabola with focus $(0, p)$ and directrix $y = -p$ is _____.

   **b.** An equation of a parabola with focus _____ and directrix _____ is $y^2 = 4px$.

**3. a.** An ellipse is the set of all points in a plane the _____ of whose distances from two fixed points (called the _____) is a _____.

   **b.** The vertices of an ellipse are the points of intersection of the line passing through the _____ and the ellipse. The chord joining the vertices is called the _____, and its midpoint is called the _____ of the ellipse. The chord passing through the center of the ellipse and perpendicular to the major axis is called the _____ _____ of the ellipse.

**4. a.** An equation of the ellipse with foci $(\pm c, 0)$ and vertices $(\pm a, 0)$ is _____, where $c^2 = $ _____.

   **b.** An equation of the ellipse with foci _____ and vertices _____ is $x^2/b^2 + y^2/a^2 = 1$.

**5. a.** A hyperbola is the set of all points in a plane the _____ of whose distances from two fixed points (called the _____) is a _____.

   **b.** The line passing through the foci intersects the hyperbola at two points called the _____ of the hyperbola. The line segment joining the vertices is called the _____ axis of the hyperbola, and the midpoint of the _____ axis is called the _____ of the hyperbola. A hyperbola has _____ _____ branches.

**6. a.** An equation of a hyperbola with foci $(\pm c, 0)$ and vertices $(\pm a, 0)$ is _____, where $c^2 = $ _____. The hyperbola has asymptotes _____.

   **b.** An equation of the hyperbola with foci _____ and vertices _____ is $y^2/a^2 - x^2/b^2 = 1$, where $c^2 = $ _____. The hyperbola has asymptotes _____.

**7.** A plane curve is a set $C$ of ordered pairs $(x, y)$ defined by the parametric equations _____, where $f$ and $g$ are continuous functions on an interval $I$; $I$ is called the _____ interval.

**8. a.** If $x = f(t)$ and $y = g(t)$, where $f$ and $g$ are differentiable and $f'(t) \neq 0$, then $dy/dx = $ _____.

   **b.** If $x = f(t)$ and $y = g(t)$ define $y$ as a twice-differentiable function of $x$ over some suitable interval, then $d^2y/dx^2 = $ _____.

**9. a.** A curve $C$ represented by $x = f(t)$ and $y = g(t)$ on a parameter interval $I$ is smooth if _____ and _____ are continuous on $I$ and are not _____ _____, except possibly at the _____ of $I$.

   **b.** If $C$ is a smooth curve represented by $x = f(t)$ and $y = g(t)$ with parameter interval $[a, b]$, then the length of $C$ is $L = $ _____.

**10.** If $C$ is a smooth curve as described in Question 9b, $C$ does not intersect itself, except possibly at _____, and $g(t) \geq 0$, then the area of the surface obtained by revolving $C$ about the $x$-axis is $S = $ _____. If $f(t) \geq 0$ for all $t$ in $[a, b]$, then the area of the surface obtained by revolving $C$ about the $y$-axis is $S = $ _____.

**11. a.** The rectangular coordinates $(x, y)$ of a point $P$ are related to the polar coordinates of $P$ by the equations $x = $ _____ and $y = $ _____.

   **b.** The polar coordinates $(r, \theta)$ of a point $P$ are related to the rectangular coordinates of $P$ by the equations $r^2 = $ _____ and $\tan \theta = $ _____.

**12.** The horizontal tangent lines to the graph of $r = f(\theta)$ are located at the points where $dy/d\theta$ _____ and $dx/d\theta$ _____. The vertical tangent lines are located at the points where $dx/d\theta$ _____ and $dy/d\theta$ _____. Horizontal and vertical tangent lines may also be located at points where $dy/d\theta$ and $dx/d\theta$ are both equal to _____.

**13. a.** If $f$ is nonnegative and continuous on $[\alpha, \beta]$, where $0 \leq \alpha < \beta \leq 2\pi$, then the area of the region bounded by the graphs of $r = f(\theta)$, $\theta = \alpha$, and $\theta = \beta$ is given by _____.

   **b.** Let $f$ and $g$ be continuous on $[\alpha, \beta]$, where $0 \leq g(\theta) \leq f(\theta)$ and $0 \leq \alpha < \beta \leq 2\pi$. Then the area of the region bounded by the graphs of $r = g(\theta)$, $r = f(\theta)$, $\theta = \alpha$, and $\theta = \beta$ is given by _____.

**14.** Suppose that $f$ has a continuous derivative on $[\alpha, \beta]$. If the graph $C$ of $r = f(\theta)$ is traced exactly _____ as $\theta$ increases from $\alpha$ to $\beta$, then the length of $C$ is given by _____.

**15.** Let $F$ be a fixed point, let $l$ be a fixed line in the plane, and let $e$ be a fixed positive number. Then a conic section defined by the equation _____ is an ellipse if $e$ satisfies

_____, a parabola if $e$ satisfies _____, and a hyperbola if $e$ satisfies _____.

**16.** A conic section can be represented by a polar equation of the form _____ or _____. It is an ellipse, a parabola, or a hyperbola depending on whether $e$ satisfies _____, _____, or _____, respectively.

## REVIEW EXERCISES

*In Exercises 1–6, find the vertices and the foci of the conic and sketch its graph.*

**1.** $\dfrac{x^2}{4} + \dfrac{y^2}{9} = 1$

**2.** $\dfrac{(x-1)^2}{2} + \dfrac{(y+1)^2}{4} = 1$

**3.** $x^2 - 9y^2 = 9$

**4.** $y^2 - 2y - 8x - 15 = 0$

**5.** $y^2 - 9x^2 + 8y + 7 = 0$

**6.** $4x^2 + 25y^2 - 16x + 50y - 59 = 0$

*In Exercises 7–12, find a rectangular equation of the conic satisfying the given conditions.*

**7.** parabola, focus $(-2, 0)$, directrix $x = 2$

**8.** parabola, vertex $(-2, 2)$, directrix $y = 4$

**9.** ellipse, vertices $(\pm 7, 0)$, foci $(\pm 2, 0)$

**10.** ellipse, foci $(\pm 2, 3)$, major axis has length 8

**11.** hyperbola, foci $\left(0, \pm\frac{3}{2}\sqrt{5}\right)$, vertices $(0, \pm 3)$

**12.** hyperbola, vertices $(-2, 0)$ and $(2, 0)$, asymptotes $y = \pm\dfrac{3}{2}x$

**13.** Show that if $m$ is any real number, then there is exactly one line of slope $m$ that is tangent to the parabola $x^2 = 4py$ and its equation is $y = mx - pm^2$.

**14.** Show that if $m$ is any real number, then there are exactly two lines of slope $m$ that are tangent to the ellipse $x^2/a^2 + y^2/b^2 = 1$ and their equations are $y = mx \pm \sqrt{a^2m^2 + b^2}$.

*In Exercises 15–18, (a) find a rectangular equation whose graph contains the curve $C$ with the given parametric equations, and (b) sketch the curve $C$ and indicate its orientation.*

**15.** $x = 1 + 2t$, $y = 3 - 2t$

**16.** $x = e^t$, $y = e^{-2t}$

**17.** $x = 1 + 2\sin t$, $y = 3 + 2\cos t$

**18.** $x = \cos^3 t$, $y = 4\sin^3 t$

*In Exercises 19–22, find the slope of the tangent line to the curve at the point corresponding to the value of the parameter.*

**19.** $x = t^3 + 1$, $y = 2t^2 - 1$; $t = 1$

**20.** $x = \sqrt{t+1}$, $y = \sqrt{16 - t}$; $t = 0$

**21.** $x = te^{-t}$, $y = \dfrac{1}{t^2 + 1}$; $t = 0$

**22.** $x = 1 - \sin^2 t$, $y = \cos^3 t$; $t = \dfrac{\pi}{4}$

*In Exercises 23 and 24, find $dy/dx$ and $d^2y/dx^2$.*

**23.** $x = t^3 + 1$, $y = t^4 + 2t^2$

**24.** $x = e^t \sin t$, $y = e^t \cos t$

*In Exercises 25 and 26, find the points on the curve with the given parametric equations at which the tangent lines are vertical or horizontal.*

**25.** $x = t^3 - 4t$, $y = t^2 + 2$

**26.** $x = 1 - 2\cos t$, $y = 1 - 2\sin t$

*In Exercises 27 and 28, find the length of the curve defined by the given parametric equations.*

**27.** $x = \dfrac{1}{6}t^6$, $y = 2 - \dfrac{1}{4}t^4$; $0 \le t \le \sqrt[4]{8}$

**28.** $x = \sqrt{3}t^2$, $y = t - t^3$; $-1 \le t \le 1$

**29.** The position of a body at time $t$ is $(x, y)$, where $x = e^{-t}\cos t$ and $y = e^{-t}\sin t$. Find the distance covered by the body as $t$ runs from 0 to $\pi/2$.

**30.** The course taken by an oceangoing racing boat during a practice run is described by the parametric equations

$$x = \sqrt{3}(t - 1)^2 \quad \text{and} \quad y = (t - 1) - (t - 1)^3$$
$$0 \le t \le 2$$

where $x$ and $y$ are measured in miles. Sketch the path of the boat, and find the length of the course.

*In Exercises 31 and 32, find the area of the surface obtained by revolving the given curve about the x-axis.*

**31.** $x = t^2$, $y = \dfrac{t}{3}(3 - t^2)$; $0 \le t \le \sqrt{3}$

**32.** $x = \ln(\sec t + \tan t) - \sin t$, $y = \cos t$; $0 \le t \le \frac{\pi}{3}$

*In Exercises 33–38, sketch the curve with the given polar equation.*

**33.** $r = 2\sin\theta$

**34.** $r = 3 - 4\cos\theta$

**35.** $r = 2 \cos 5\theta$    **36.** $r = e^{-\theta}$

**37.** $r^2 = \cos 2\theta$    **38.** $r = 2 \sin \theta \cos^2 \theta$

*In Exercises 39 and 40, find the slope of the tangent line to the curve with the given polar equation at the point corresponding to the given value of $\theta$.*

**39.** $r = e^{2\theta}, \quad \theta = \dfrac{\pi}{2}$    **40.** $r = 2 - \sin \theta, \quad \theta = \dfrac{\pi}{2}$

*In Exercises 41 and 42, find the points of intersection of the given curves.*

**41.** $r = \sin \theta, \quad r = 1 - \sin \theta$

**42.** $r = \cos \theta, \quad r = \cos 2\theta$

**43.** Find the area of the region enclosed by the curve with polar equation $r = 2 + \cos \theta$.

**44.** Find the area of the region enclosed by the curve with polar equation $r = 1 + \sin \theta$.

**45.** Find the area of the region that is enclosed between the petals of the curves with polar equations $r = 2 \sin 2\theta$ and $r = 2 \cos 2\theta$.

**46.** Find the area of the region that is enclosed between the curves with polar equations $r = 3 + 2 \sin \theta$ and $r = 2$.

*In Exercises 47 and 48, find the length of the given curve.*

**47.** $r = \theta^2, \quad 0 \le \theta \le 2\pi$

**48.** $r = 2(\sin \theta + \cos \theta), \quad 0 \le \theta \le 2\pi$

*In Exercises 49 and 50, find the area of the surface obtained by revolving the curve about the given line.*

**49.** $r = 2 \sin \theta$ about the polar axis

**50.** $r = \sqrt{\cos 2\theta}, \quad 0 \le \theta \le \frac{\pi}{4},$ about the line $\theta = \dfrac{\pi}{2}$

*In Exercises 51 and 52, sketch the curve with the given equation.*

**51.** $r = \dfrac{1}{1 + \sin \theta}$    **52.** $r = \dfrac{16}{3 - 5 \cos \theta}$

*In Exercises 53–54, plot the curve with the parametric equations.*

 **53.** $x = 0.15(2 \cos t + 3 \cos 2t), \quad y = 0.15(2 \sin t - 3 \sin 2t);$ $0 \le t \le 2\pi$ (hypotrochoid)

 **54.** $x = 0.24(-7t^3 + 3t^5), \quad y = 0.09(14t^2 - 5t^4);$ $-1.629 \le t \le 1.629$ (butterfly catastrophe)

*In Exercises 55–57, plot the curve with the polar equation.*

 **55.** $r = 0.1e^{0.1\theta}$ where $0 \le \theta \le \dfrac{15\pi}{2}$

 **56.** $r = \dfrac{0.1}{\cos 4\theta}$ where $0 \le \theta \le 2\pi$ (epi-spiral)

**57.** $r = \dfrac{1}{2 \sinh \theta}$ where $-2\pi \le \theta \le 2\pi$ (spiral of Poinsot)

**58.** An ant crawls along the curve $x = \frac{1}{2}t^2, y = \frac{1}{3}(2t + 1)^{3/2}$ starting at the point $\left(0, \frac{1}{3}\right)$ and ending at the point $\left(\frac{1}{2}, \sqrt{3}\right)$, where $x$ and $y$ are measured in feet. Find the distance traveled by the ant.

**59.** An egg has the shape of a solid obtained by revolving the upper half of the ellipse $x^2 + 2y^2 = 2$ about the $x$-axis. What is the surface area of the egg?

**60.** A piston is attached to a crankshaft by means of a connecting rod of length $L$, as shown in the figure. If the disk is of radius $r$, find the parametric equations giving the position of the point $P$ using the angle $\theta$ as a parameter.

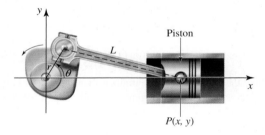

## CHALLENGE PROBLEMS

**1. a.** In the following figure, the axes of an $xy$-coordinate system have been rotated about the origin through an angle of $\theta$ to produce a new $x'y'$-coordinate system.

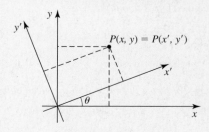

Show that

$$x = x' \cos \theta - y' \sin \theta \qquad y = x' \sin \theta + y' \cos \theta$$

**b.** Show that the equation $Ax^2 + Bxy + Cy^2 + F = 0$, where $B \ne 0$, will have the form $(A'x')^2 + (C'y')^2 + F = 0$ in the $x'y'$-coordinate system obtained by rotating the $xy$-system through an angle $\theta$ given by

$$\cot 2\theta = \frac{A - C}{B}$$

**c.** Sketch the ellipse $2x^2 + \sqrt{3}xy + y^2 - 20 = 0$.

**2. a.** Show that the area of the ellipse

$$Ax^2 + Bxy + Cy^2 + F = 0$$

where $B^2 - 4AC < 0$, is given by

$$S = -\frac{2\pi F}{\sqrt{4AC - B^2}}$$

**b.** Find the area of the ellipse $2x^2 + \sqrt{3}xy + y^2 = 20$.

**3.** Find the length of the curve with parametric equations

$$x = \int_1^t \frac{\cos u}{u}\, du \qquad y = \int_1^t \frac{\sin u}{u}\, du$$

between the origin and the nearest point from the vertical tangent line.

**4.** Find the area of the surface obtained by revolving one branch of the lemniscate $r = a\sqrt{\cos 2\theta}$ about the line $\theta = \pi/4$.

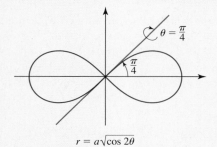

$r = a\sqrt{\cos 2\theta}$

**5.** The curve with equation $x^3 + y^3 = 3axy$, where $a$ is a nonzero constant, is called the **folium of Descartes**.

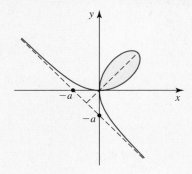

**a.** Show that the polar equation of the curve is

$$r = \frac{3a \sec\theta \tan\theta}{1 + \tan^3\theta}$$

**b.** Find the area of the region enclosed by the loop of the curve.

**6.** Find the rectangular coordinates of the centroid of the region that is completely enclosed by the curve $r = 3\cos^3\theta$.

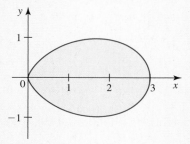

**7.** An ant is placed at each corner of a square with sides of length $a$. Starting at the same instant of time, all four ants begin to move counterclockwise at the same speed and in such a way that each ant moves toward the next at all times. The resulting path of each ant is a spiral curve that converges to the center of the square.

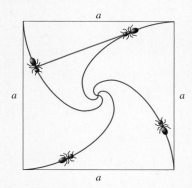

**a.** Taking the pole to be the center of the square, find the polar equation describing the path taken by one of the ants. Hint: The line passing through the position of two adjacent ants is tangent to the path of one of them.

**b.** Find the distance traveled by an ant as it moves from a corner of the square to its center.

The photograph shows an Olympian throwing a javelin. Once the javelin has been launched, its direction of motion and speed at any point in its flight through the air can be described by a *vector* at that point. In this chapter we will study vectors—objects that possess both magnitude and direction—and look at some of their applications.

# 10 Vectors and the Geometry of Space

**IN THIS CHAPTER** we will study *vectors,* quantities that have both direction and magnitude. Vectors can be used to describe the position, velocity, and acceleration of a body moving in a plane or in space. Since a force is determined by the direction along which it acts and by its magnitude, we can also represent a force by a vector.

There are two ways in which vectors can be *multiplied* together. These operations on vectors enable us to find the work that is done by a force in moving an object from one point in space to another point, to find the angle between two lines, to compute the volume of a parallelepiped, and to find the torque exerted by a person pushing on a leveraged-impact lug wrench when changing a tire—just to mention a few applications.

Vectors also facilitate the algebraic representation of lines and planes in space. Using such representations, we can easily find the distance between a point in space and a plane and the distance between two skew lines (lines that are neither parallel nor intersect). Finally, we introduce two alternative coordinate systems in space: the *cylindrical coordinate system* and the *spherical coordinate system*. Each of these systems enables us to obtain relatively simple algebraic representation of certain surfaces.

**V** This symbol indicates that one of the following video types is available for enhanced student learning at **www.academic.cengage.com/login:**
- Chapter lecture videos
- Solutions to selected exercises

**V** 🖉 This symbol indicates that step-by-step video lessons for hand-drawing certain complex figures are available.

## 10.1 Vectors in the Plane

### ■ Vectors

Some physical quantities, such as force and velocity, possess both magnitude (size) and direction. These quantities are called **vectors** and can be represented by arrows or directed line segments. The arrow points in the direction of the vector, and the length of the arrow gives the magnitude of the vector.

Figure 1a gives an aerial view of a tugboat trying to free a cruise liner that has run aground in shallow waters. The magnitude and direction of the force exerted by the tugboat are represented by the vector shown in the figure.

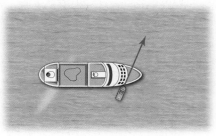

**FIGURE 1**

(**a**) The vector represents the force exerted by a tugboat on a ship.

(**b**) The vectors represent the velocity of blood cells flowing through an artery.

In Figure 1b the vectors (arrows) give the magnitude and direction of blood cells flowing through an artery. Observe that the lengths of the vectors vary; this reflects the fact that the blood cells near the central axis have a greater velocity than those near the walls of the artery.

Vectors are customarily denoted by lowercase boldface type such as $\mathbf{v}$ and $\mathbf{w}$. However, if a vector $\mathbf{v}$ is defined by a directed line segment from the **initial point** $A$ of the vector to the **terminal point** $B$ of the vector, then it is written $\mathbf{v} = \overrightarrow{AB}$. (See Figure 2.)*

Two vectors, $\mathbf{v}$ and $\mathbf{w}$, that have the same magnitude and direction are said to be **equal,** written $\mathbf{v} = \mathbf{w}$. Thus, the vectors $\mathbf{v} = \overrightarrow{AB}$ and $\mathbf{w} = \overrightarrow{CD}$ shown in Figure 3 are equal.

### ■ Scalar Multiples

In contrast to a vector, a **scalar** is a quantity that has magnitude but no direction. Real numbers and complex numbers are examples of scalars. In this text, however, the term *scalar* will always refer to a real number. A vector can be multiplied by a scalar. If $c \neq 0$ is a scalar and $\mathbf{v}$ is a vector, then the **scalar multiple** of $c$ and $\mathbf{v}$ is a vector $c\mathbf{v}$. The magnitude of $c\mathbf{v}$ is $|c|$ times the magnitude of $\mathbf{v}$, and the direction of $c\mathbf{v}$ is the same as that of $\mathbf{v}$ if $c > 0$ and opposite that of $\mathbf{v}$ if $c < 0$. (See Figure 4.) Observe that two nonzero vectors are **parallel** if they are scalar multiples of one another.

For convenience we define the **zero vector,** denoted by $\mathbf{0}$, to be the vector with length zero and having *no* direction. If $c = 0$, then $c\mathbf{v} = \mathbf{0}$ for any vector $\mathbf{v}$.

**FIGURE 2**
$\mathbf{v}$ is the directed line segment from $A$ to $B$.

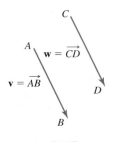

$\mathbf{v} = \mathbf{w}$

**FIGURE 3**
$\mathbf{v}$ and $\mathbf{w}$ have the same length and direction.

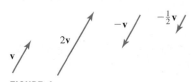

**FIGURE 4**
Scalar multiples of $\mathbf{v}$

*If the vector $\mathbf{v}$ is handwritten, it is more convenient to write it in the form $\vec{v}$.

## ■ Vector Addition: The Parallelogram Law

Two vectors may be added together. To see how, consider the two nonzero vectors **v** and **w** shown in Figure 5a. Translate the vector **w** (move **w** without changing its magnitude or direction) so that the initial point of **w** coincides with the terminal point of **v**. (See Figure 5b.) Then the **sum** of **v** and **w**, written **v** + **w**, is the vector represented by the arrow with tail at the initial point of **v** and head at the terminal point of **w** (Figure 5c). If you examine Figure 5d, you will see that the line segment representing the vector **v** + **w** coincides with the diagonal of the parallelogram determined by **v** and **w**. For this reason we say that vector addition obeys the Parallelogram Law. Try to translate the vector **v** instead of **w**, and convince yourself that the result is the same.

(**a**) The vectors **v** and **w**

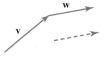

(**b**) **w** translated

(**c**) **v** + **w**

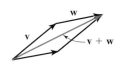

(**d**) **v** + **w** lies on the diagonal of the parallelogram determined by **v** and **w**.

**FIGURE 5**
Geometric construction of **v** + **w**

The difference of two vectors **v** and **w**, written **v** − **w**, is defined by

$$\mathbf{v} - \mathbf{w} = \mathbf{v} + (-\mathbf{w}) \qquad \text{Difference of two vectors}$$

To describe this operation geometrically, consider once again the two vectors **v** and **w** of Figure 5a, which are reproduced in Figure 6a. If we translate **w**, reverse it to obtain −**w**, and then use the parallelogram law to add **v** to −**w**, we obtain **v** − **w**, as shown in Figure 6b.

**FIGURE 6**
Geometric construction of **v** − **w**

(**a**) The vectors **v** and **w**

(**b**) The vector **v** − **w**

## ■ Vectors in the Coordinate Plane

Just as the introduction of a rectangular coordinate system in the plane enabled us to describe geometric objects in algebraic terms, we will see that the introduction of a rectangular coordinate system in a "vector space" will enable us to represent vectors algebraically.

**EXAMPLE 1** Let **a** be a vector with initial point $A(0, 0)$ and terminal point $B(3, 2)$, and let **b** be a vector with initial point $C(1, 3)$ and terminal point $D(4, 5)$. Show that **a** = **b**.

**Solution** The vectors $\mathbf{a} = \overrightarrow{AB}$ and $\mathbf{b} = \overrightarrow{CD}$ are shown in Figure 7. To show that **a** = **b**, we need to show that both vectors have the same length and direction. Using the distance formula, we find

$$\text{length of } \overrightarrow{AB} = \sqrt{(3 - 0)^2 + (2 - 0)^2} = \sqrt{9 + 4} = \sqrt{13}$$

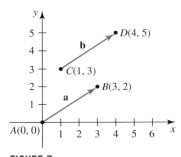

**FIGURE 7**
**a** = **b** because both vectors have the same length and direction.

and

$$\text{length of } \overrightarrow{CD} = \sqrt{(4-1)^2 + (5-3)^2} = \sqrt{9+4} = \sqrt{13}$$

so **a** and **b** have the same length. Next, we find

$$\text{slope of } \overrightarrow{AB} = \frac{2-0}{3-0} = \frac{2}{3}$$

and

$$\text{slope of } \overrightarrow{CD} = \frac{5-3}{4-1} = \frac{2}{3}$$

so **a** and **b** have the same direction. This proves that **a** = **b**.   ∎

In Example 1 we saw that the vector **b** may be represented by the vector **a** that has its initial point at the origin. In general, it is true that any vector in the plane can be represented by such a vector. To see this, suppose that $\mathbf{b} = \overrightarrow{P_1 P_2}$ is any vector with initial point $P_1(x_1, y_1)$ and terminal point $P_2(x_2, y_2)$. (See Figure 8.) Let $a_1 = x_2 - x_1$ and $a_2 = y_2 - y_1$. Then the vector $\mathbf{a} = \overrightarrow{OP}$ with $O(0, 0)$ and $P(a_1, a_2)$ is the required vector, since the length of **b** is

$$\sqrt{a_1^2 + a_2^2} = \sqrt{(x_2 - x_1)^2 + (y_2 - y_1)^2}$$

which is also the length of $\mathbf{b} = \overrightarrow{P_1 P_2}$. Similarly, the slope of **a** is

$$\frac{a_2}{a_1} = \frac{y_2 - y_1}{x_2 - x_1} \qquad x_1 \neq x_2$$

which is also the slope of **b**. (We leave the proof of the case in which $x_1 = x_2$ to you.)

The vector **a** with initial point at the origin and terminal point $P(a_1, a_2)$ is called the **position vector** of the point $P(a_1, a_2)$ and is denoted by $\langle a_1, a_2 \rangle$. Thus, we have shown that any vector in a coordinate plane is equal to a position vector. Since the zero vector has length zero, its terminal point must coincide with its initial point; therefore, it is equal to the position vector of the point $(0, 0)$.

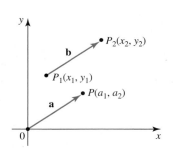

**FIGURE 8**
Any vector **b** in the plane can be represented by a vector **a** emanating from the origin.

---

**DEFINITIONS    A Vector in the Coordinate Plane**

A **vector** in the plane is an ordered pair $\mathbf{a} = \langle a_1, a_2 \rangle$ of real numbers, $a_1$ and $a_2$, called the **scalar components** of **a**. The **zero vector** is $\mathbf{0} = \langle 0, 0 \rangle$.

---

We have also established the following result.

---

**DEFINITION**

Given the points $P_1(x_1, y_1)$ and $P_2(x_2, y_2)$, the vector $\overrightarrow{P_1 P_2}$ is represented by the **position vector**

$$\mathbf{a} = \overrightarrow{P_1 P_2} = \langle x_2 - x_1, y_2 - y_1 \rangle \qquad (1)$$

---

Thus, the components of a vector are found by subtracting the respective coordinates of its initial point from the coordinates of its terminal point.

**EXAMPLE 2** Find the vector **a** with initial point $A(-1, -2)$ and terminal point $B(3, 2)$.

**Solution** Using Equation (1), we find the vector **a** to be

$$\mathbf{a} = \langle 3 - (-1), 2 - (-2) \rangle = \langle 4, 4 \rangle$$ ∎

## ■ Length of a Vector

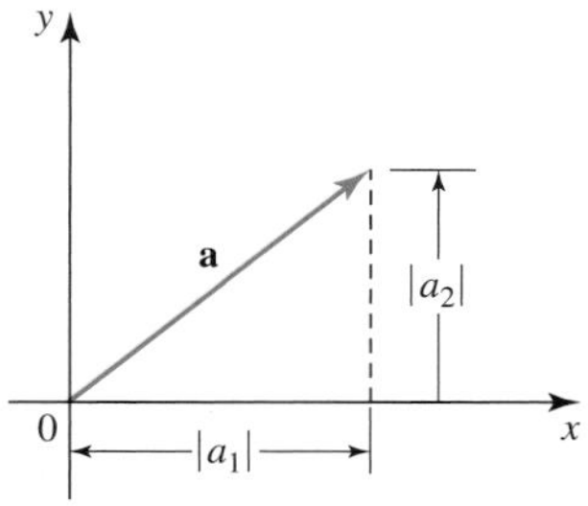

The **length** or the **magnitude** of a vector $\mathbf{a} = \langle a_1, a_2 \rangle$, denoted by the symbol $|\mathbf{a}|$, is found by using the Pythagorean Theorem. (See Figure 9.)

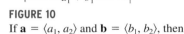

> **DEFINITION**
>
> The **length** or **magnitude** of $\mathbf{a} = \langle a_1, a_2 \rangle$ is
> $$|\mathbf{a}| = \sqrt{a_1^2 + a_2^2} \qquad (2)$$

**FIGURE 9**
The length of **a** is $|\mathbf{a}| = \sqrt{a_1^2 + a_2^2}$.

## ■ Vector Addition in the Coordinate Plane

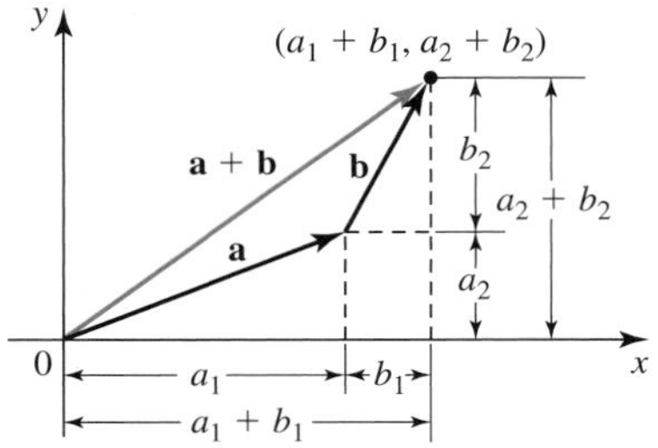

Vector addition is carried out componentwise. To add the two vectors $\mathbf{a} = \langle a_1, a_2 \rangle$ and $\mathbf{b} = \langle b_1, b_2 \rangle$, we add their components. (See Figure 10.)

> **Parallelogram Law for Vector Addition**
>
> If $\mathbf{a} = \langle a_1, a_2 \rangle$ and $\mathbf{b} = \langle b_1, b_2 \rangle$, then
> $$\mathbf{a} + \mathbf{b} = \langle a_1 + b_1, a_2 + b_2 \rangle \qquad (3)$$

**FIGURE 10**
If $\mathbf{a} = \langle a_1, a_2 \rangle$ and $\mathbf{b} = \langle b_1, b_2 \rangle$, then
$\mathbf{a} + \mathbf{b} = \langle a_1 + b_1, a_2 + b_2 \rangle$.

**EXAMPLE 3** If $\mathbf{a} = \langle 3, -2 \rangle$, and $\mathbf{b} = \langle -1, 3 \rangle$, then

$$\mathbf{a} + \mathbf{b} = \langle 3, -2 \rangle + \langle -1, 3 \rangle = \langle 3 + (-1), -2 + 3 \rangle = \langle 2, 1 \rangle$$ ∎

> **Scalar Multiplication**
>
> If $\mathbf{a} = \langle a_1, a_2 \rangle$ and $c$ is a scalar, then
> $$c\mathbf{a} = \langle ca_1, ca_2 \rangle \qquad (4)$$
>
> (See Figure 11.)

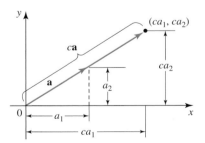

**FIGURE 11**
If $\mathbf{a} = \langle a_1, a_2 \rangle$, then $c\mathbf{a} = \langle ca_1, ca_2 \rangle$.

Recall that the **difference** of **a** and **b** is defined by $\mathbf{a} - \mathbf{b} = \mathbf{a} + (-\mathbf{b})$, so if $\mathbf{a} = \langle a_1, a_2 \rangle$ and $\mathbf{b} = \langle b_1, b_2 \rangle$, then

$$\mathbf{a} - \mathbf{b} = \mathbf{a} + (-\mathbf{b}) = \mathbf{a} + (-1)\mathbf{b} = \langle a_1, a_2 \rangle + \langle -b_1, -b_2 \rangle$$

$$= \langle a_1 - b_1, a_2 - b_2 \rangle \quad \text{Vector subtraction}$$

**EXAMPLE 4** Let $\mathbf{a} = \langle 1, -2 \rangle$ and $\mathbf{b} = \langle -2, 5 \rangle$. Find

**a.** $\mathbf{a} + \mathbf{b}$ **b.** $\mathbf{a} - \mathbf{b}$ **c.** $5\mathbf{a}$ **d.** $3\mathbf{a} + 2\mathbf{b}$ **e.** $|3\mathbf{a} + 2\mathbf{b}|$

**Solution**

a. $\mathbf{a} + \mathbf{b} = \langle 1, -2 \rangle + \langle -2, 5 \rangle = \langle 1 - 2, -2 + 5 \rangle = \langle -1, 3 \rangle$

b. $\mathbf{a} - \mathbf{b} = \langle 1, -2 \rangle - \langle -2, 5 \rangle = \langle 1 - (-2), -2 - 5 \rangle = \langle 3, -7 \rangle$

c. $5\mathbf{a} = 5\langle 1, -2 \rangle = \langle 5(1), 5(-2) \rangle = \langle 5, -10 \rangle$

d. $3\mathbf{a} + 2\mathbf{b} = 3\langle 1, -2 \rangle + 2\langle -2, 5 \rangle = \langle 3, -6 \rangle + \langle -4, 10 \rangle = \langle -1, 4 \rangle$

e. $|3\mathbf{a} + 2\mathbf{b}| = |\langle -1, 4 \rangle|$      Use part (d).

$$= \sqrt{(-1)^2 + 4^2}$$
$$= \sqrt{1 + 16} = \sqrt{17}$$  ∎

## Properties of Vectors

The operations of vector addition and scalar multiplication obey the following rules.

---

**THEOREM 1    Rules for Vector Addition and Scalar Multiplication**

Suppose that $\mathbf{a}$, $\mathbf{b}$, and $\mathbf{c}$ are vectors and that $c$ and $d$ are scalars. Then

1. $\mathbf{a} + \mathbf{b} = \mathbf{b} + \mathbf{a}$
2. $(\mathbf{a} + \mathbf{b}) + \mathbf{c} = \mathbf{a} + (\mathbf{b} + \mathbf{c})$
3. $\mathbf{a} + \mathbf{0} = \mathbf{0} + \mathbf{a} = \mathbf{a}$
4. $\mathbf{a} + (-\mathbf{a}) = \mathbf{0}$
5. $c(\mathbf{a} + \mathbf{b}) = c\mathbf{a} + c\mathbf{b}$
6. $c(d\mathbf{a}) = (cd)\mathbf{a}$
7. $(c + d)\mathbf{a} = c\mathbf{a} + d\mathbf{a}$
8. $1\mathbf{a} = \mathbf{a}$

---

We will prove the first of these rules and leave the proofs of the others as exercises.

**PROOF OF 1**    Let $\mathbf{a} = \langle a_1, a_2 \rangle$ and $\mathbf{b} = \langle b_1, b_2 \rangle$. Then

$$\mathbf{a} + \mathbf{b} = \langle a_1, a_2 \rangle + \langle b_1, b_2 \rangle = \langle a_1 + b_1, a_2 + b_2 \rangle$$
$$= \langle b_1 + a_1, b_2 + a_2 \rangle = \mathbf{b} + \mathbf{a}$$  ∎

## Unit Vectors

A **unit vector** is a vector of length 1. Unit vectors are primarily used as indicators of direction. For example, if $\mathbf{a}$ is a nonzero vector, then the vector

$$\mathbf{u} = \frac{\mathbf{a}}{|\mathbf{a}|}$$

is a unit vector having the same direction as $\mathbf{a}$. (See Figure 12.) Furthermore, by writing $\mathbf{a}$ in the form

$$\mathbf{a} = |\mathbf{a}|\left(\frac{\mathbf{a}}{|\mathbf{a}|}\right) = |\mathbf{a}|\,\mathbf{u} \tag{5}$$

Magnitude of $\mathbf{a}$ ⎯⎯⎯⎯ ⎿ ⎾ ⎯⎯⎯⎯ Direction of $\mathbf{a}$

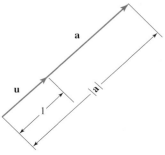

**FIGURE 12**
The unit vector $\mathbf{u}$ indicates the direction of $\mathbf{a}$.

the two properties of magnitude and direction that define a vector are clearly displayed.

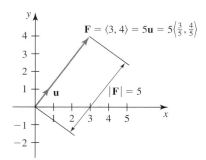

**FIGURE 13**
The vector $\mathbf{F} = \langle 3, 4 \rangle$ can be written in the alternate form $\mathbf{F} = 5\mathbf{u}$, where $\mathbf{u}$ is the unit vector in the direction of $\mathbf{F}$.

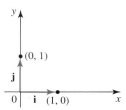

**FIGURE 14**
The unit vectors $\mathbf{i}$ and $\mathbf{j}$ point in the positive $x$- and $y$-direction, respectively.

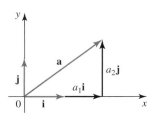

**FIGURE 15**
Any vector in the plane can be expressed in terms of the standard basis vectors $\mathbf{i}$ and $\mathbf{j}$.

**EXAMPLE 5** Let $\mathbf{F} = \langle 3, 4 \rangle$ be a vector describing a force acting on a particle. Express $\mathbf{F}$ in terms of its magnitude (in dynes) and a unit vector having the same direction as $\mathbf{F}$.

**Solution** The magnitude of $\mathbf{F}$ is

$$|\mathbf{F}| = \sqrt{3^2 + 4^2} = \sqrt{9 + 16} = 5$$

or 5 dynes. Its direction is

$$\mathbf{u} = \frac{\mathbf{F}}{|\mathbf{F}|} = \frac{1}{5}\langle 3, 4 \rangle = \left\langle \frac{3}{5}, \frac{4}{5} \right\rangle$$

So

$$\mathbf{F} = \langle 3, 4 \rangle = 5\left\langle \frac{3}{5}, \frac{4}{5} \right\rangle$$

(See Figure 13.)

## Standard Basis Vectors

There are two unit vectors in the coordinate plane that are singled out for a special role. They are the vectors $\mathbf{i}$ and $\mathbf{j}$ defined by

$$\mathbf{i} = \langle 1, 0 \rangle \qquad \text{and} \qquad \mathbf{j} = \langle 0, 1 \rangle$$

The vector $\mathbf{i}$ points in the positive $x$-direction, whereas the vector $\mathbf{j}$ points in the positive $y$-direction. (See Figure 14.)

Let $\mathbf{a} = \langle a_1, a_2 \rangle$ be a vector in the coordinate plane. Then

$$\mathbf{a} = \langle a_1, a_2 \rangle = \langle a_1, 0 \rangle + \langle 0, a_2 \rangle \qquad \text{By the definition of vector addition}$$

$$= a_1 \langle 1, 0 \rangle + a_2 \langle 0, 1 \rangle \qquad \text{By the definition of scalar multiplication}$$

$$= a_1 \mathbf{i} + a_2 \mathbf{j}$$

This shows that any vector in the plane can be expressed in terms of the vectors $\mathbf{i}$ and $\mathbf{j}$. (See Figure 15.) For this reason the vectors $\mathbf{i}$ and $\mathbf{j}$ are referred to as **standard basis vectors**. The vectors $a_1 \mathbf{i}$ and $a_2 \mathbf{j}$ are the **horizontal** and **vertical vector components** of $\mathbf{a}$. We also say that $\mathbf{a}$ is *resolved* into a (vector) sum of $a_1 \mathbf{i}$ and $a_2 \mathbf{j}$.

**EXAMPLE 6** Let $\mathbf{F} = \langle 3, 4 \rangle$ be the force vector of Example 5. Express $\mathbf{F}$ in terms of the standard basis vectors $\mathbf{i}$ and $\mathbf{j}$, and identify the horizontal and vertical vector components of $\mathbf{F}$.

**Solution** Since

$$\mathbf{F} = \langle 3, 4 \rangle$$
$$= 3\mathbf{i} + 4\mathbf{j}$$

the horizontal vector component of $\mathbf{F}$ is $3\mathbf{i}$, and the vertical vector component of $\mathbf{F}$ is $4\mathbf{j}$.

**Note** By using standard basis vectors, we are able to express any vector in the coordinate plane in two ways:

$$\mathbf{a} = \langle a_1, a_2 \rangle \qquad \text{and} \qquad \mathbf{a} = a_1 \mathbf{i} + a_2 \mathbf{j}$$

We will use these representations interchangeably.

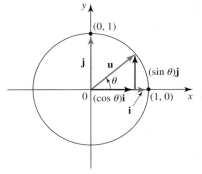

**FIGURE 16**
Every unit vector **u** can be expressed
in the form **u** = (cos θ)**i** + (sin θ)**j**.

## ■ Angular Form of the Unit Vector

Let θ be the angle that the unit vector **u** makes with the positive x-axis. (See Figure 16.)
Then resolving **u** into a sum of horizontal and vertical vector components gives

$$\mathbf{u} = (\cos\theta)\mathbf{i} + (\sin\theta)\mathbf{j} \tag{6}$$

**EXAMPLE 7**   Find an expression for the vector **a** of length 5 that makes an angle of
π/6 radians with the positive axis.

**Solution**   Using Equation (6), we see that the unit vector making an angle of π/6 with
the positive axis is

$$\mathbf{u} = \left(\cos\frac{\pi}{6}\right)\mathbf{i} + \left(\sin\frac{\pi}{6}\right)\mathbf{j} = \frac{\sqrt{3}}{2}\mathbf{i} + \frac{1}{2}\mathbf{j} = \frac{1}{2}(\sqrt{3}\mathbf{i} + \mathbf{j})$$

Therefore, the required expression is

$$\mathbf{a} = 5\mathbf{u} = \frac{5}{2}(\sqrt{3}\mathbf{i} + \mathbf{j})$$    ■

**EXAMPLE 8**   **Finding the True Course and Ground Speed of an Airplane**   An airplane,
on level flight, is headed in a direction that makes an angle of 45° with the north (measured in a clockwise direction) and has an airspeed of 500 mph. It is subjected to a tail-
wind blowing at 80 mph in a direction that makes an angle of 75° with the north. (See
Figure 17.) The **true course** and **ground speed** of the airplane are given by the direction and magnitude of the resultant **v** + **w**, where **v** is the velocity of the plane and **w**
is the velocity of the wind. (See Figure 18.) Find the true course and ground speed of
the airplane.

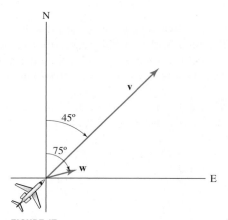

**FIGURE 17**
**v** is the velocity of the plane and **w** is the
velocity of the wind.

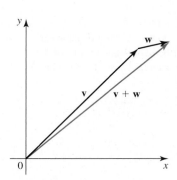

**FIGURE 18**
**v** + **w** gives the true course and
ground speed of the airplane.

**Solution**   With respect to the coordinate system shown in Figure 17, we can represent
**v** and **w** as

$$\mathbf{v} = (500\cos 45°)\mathbf{i} + (500\sin 45°)\mathbf{j}$$    Velocity of the plane

and

$$\mathbf{w} = (80 \cos 15°)\mathbf{i} + (80 \sin 15°)\mathbf{j} \qquad \text{Velocity of the wind}$$

Therefore,

$$\mathbf{v} + \mathbf{w} = [(500 \cos 45°)\mathbf{i} + (500 \sin 45°)\mathbf{j}] + [(80 \cos 15°)\mathbf{i} + (80 \sin 15°)\mathbf{j}]$$

$$= (500 \cos 45° + 80 \cos 15°)\mathbf{i} + (500 \sin 45° + 80 \sin 15°)\mathbf{j}$$

$$\approx 430.8\mathbf{i} + 374.3\mathbf{j}$$

The magnitude of $\mathbf{v} + \mathbf{w}$ is

$$|\mathbf{v} + \mathbf{w}| \approx \sqrt{(430.8)^2 + (374.3)^2} \approx 570.7$$

and this gives the ground speed of the airplane. To find its true course, we compute the unit vector $\mathbf{u}$ having the same direction as $\mathbf{v} + \mathbf{w}$. Thus,

$$\mathbf{u} = \frac{\mathbf{v} + \mathbf{w}}{|\mathbf{v} + \mathbf{w}|} \approx \frac{1}{570.7}(430.8\mathbf{i} + 374.3\mathbf{j}) \approx 0.7549\mathbf{i} + 0.6559\mathbf{j}$$

If we write $\mathbf{u}$ in the form $\mathbf{u} = (\cos\theta)\mathbf{i} + (\sin\theta)\mathbf{j}$ then we see that

$$\cos\theta \approx 0.7549 \qquad \text{or} \qquad \theta \approx \cos^{-1} 0.7549 \approx 41.0°$$

Since $\theta$ is measured from the positive $x$-axis, we conclude that the true course of the airplane is approximately $(90 - 41)°$, or $49°$ with the north. ∎

## 10.1 CONCEPT QUESTIONS

1. **a.** What is a vector? Give examples.
   **b.** What is the scalar multiple of a vector $\mathbf{v}$ and a scalar $c$? Give a geometric interpretation.
   **c.** How are two nonzero vectors added? Illustrate geometrically.
   **d.** What is the difference of the vectors $\mathbf{v}$ and $\mathbf{w}$? Illustrate geometrically.
2. **a.** What is a vector in the $xy$-plane? Give an example.
   **b.** What is the position vector $\overrightarrow{P_1P_2}$ of the vector with initial point $P_1(a_1, a_2)$ and terminal point $P_2(b_1, b_2)$?
   **c.** What is the length of the vector of part (b)?

3. State the rule for vector addition and scalar multiplication.
4. **a.** What is a unit vector?
   **b.** If $\mathbf{a} \neq \mathbf{0}$, write $\mathbf{a}$ in terms of its magnitude and direction.
   **c.** What are the standard basis vectors in the $xy$-plane?
   **d.** If the vector $\mathbf{a}$ has magnitude 3 and makes an angle of $2\pi/3$ radians with the positive $x$-axis, what are its horizontal and vertical vector components? Make a sketch.

## 10.1 EXERCISES

1. Which quantity is a vector and which is a scalar? Explain.
   **a.** The amount of water in a swimming pool
   **b.** The speed and direction of a jet stream (a current of rapidly moving air found in the upper levels of the atmosphere) at a certain point
   **c.** The population of Los Angeles
   **d.** The initial speed and direction of a bullet as it leaves a gun

2. Classify each of the following as a scalar or a vector.
   **a.** temperature  **b.** momentum
   **c.** specific heat  **d.** weight
   **e.** work  **f.** density

3. State whether the expression makes sense. Explain your answer.
   **a.** $\mathbf{a} + c$  **b.** $|\mathbf{a}|\,\mathbf{b}$
   **c.** $2\mathbf{a} + \mathbf{b} + 0$  **d.** $\dfrac{\mathbf{a}}{\mathbf{b}}$
   **e.** $\dfrac{\mathbf{a}}{|\mathbf{b}|}$, $\mathbf{b} \neq \mathbf{0}$  **f.** $\dfrac{|\mathbf{a}|\,\mathbf{b} - |\mathbf{b}|\,\mathbf{a}}{|\mathbf{a}|^2}$, $\mathbf{a} \neq \mathbf{0}$

*In Exercises 4–7, show that the vector **a** and the position vector **b** are equal.*

**4.**

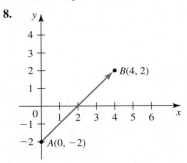

**5.**

**6.**

**7.**

*In Exercises 8–11, find the vector $\overrightarrow{AB}$. Then sketch the position vector that is equal to $\overrightarrow{AB}$.*

**8.**

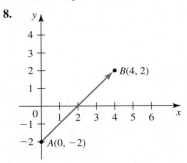

**9.**

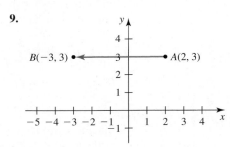

**10.**

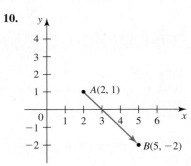

**11.**

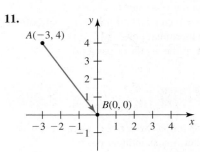

*In Exercises 12–15, use the geometric interpretation of scalar multiplication and the parallelogram law of vector addition to sketch the indicated vector.*

**12.** $\mathbf{a} + 2\mathbf{b}$

**13.** $2\mathbf{a} - 3\mathbf{b}$

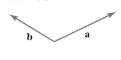

**14.** $\dfrac{1}{2}\mathbf{a} + \mathbf{b}$

**15.** $(\mathbf{a} + 2\mathbf{b}) + \mathbf{c}$

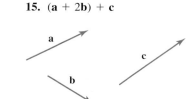

*In Exercises 16–19, express the vector* **v** *in terms of the vectors shown.*

**16.**

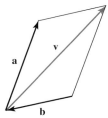

**17.**

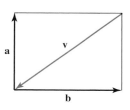

**18.**

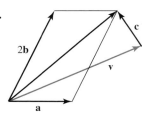

**19.**

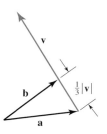

*In Exercises 20–23, find the vector $\overrightarrow{AB}$. Sketch $\overrightarrow{AB}$ and the position vector that is equal to $\overrightarrow{AB}$.*

**20.** $A(1, 3), B(3, 4)$

**21.** $A(3, 4), B(1, 3)$

**22.** $A\left(-\dfrac{1}{2}, -\dfrac{3}{2}\right), B\left(2, \dfrac{1}{2}\right)$

**23.** $A(0.1, 0.5), B(-0.2, 0.4)$

**24.** Suppose that $\overrightarrow{AB} = \langle 2, 3 \rangle$ and $A(-1, 1)$. Find $B$.

**25.** Suppose that $\overrightarrow{AB} = \langle -1, 4 \rangle$ and $B(0, 2)$. Find $A$.

*In Exercises 26–31, find* $2\mathbf{a}, \mathbf{a} + \mathbf{b}, \mathbf{a} - \mathbf{b}, and \ |2\mathbf{a} + \mathbf{b}|.$

**26.** $\mathbf{a} = \langle 1, 3 \rangle$ and $\mathbf{b} = \langle -2, 1 \rangle$

**27.** $\mathbf{a} = \langle -1, 2 \rangle$ and $\mathbf{b} = \langle 3, 1 \rangle$

**28.** $\mathbf{a} = 2\mathbf{i} - \mathbf{j}$ and $\mathbf{b} = 3\mathbf{i} + \mathbf{j}$

**29.** $\mathbf{a} = 3\mathbf{i} - 2\mathbf{j}$ and $\mathbf{b} = 2\mathbf{i}$

**30.** $\mathbf{a} = \langle 1, 2.4 \rangle$ and $\mathbf{b} = \langle -1, 0.4 \rangle$

**31.** $\mathbf{a} = \dfrac{1}{2}\mathbf{i} + \dfrac{3}{2}\mathbf{j}$ and $\mathbf{b} = \dfrac{3}{4}\mathbf{i} - \dfrac{1}{4}\mathbf{j}$

**32.** If $\mathbf{a} = \langle a_1, a_2 \rangle$ and $\mathbf{b} = \langle b_1, b_2 \rangle$ and if $c$ is a scalar, what are $\mathbf{a} - \mathbf{b}$ and $c(\mathbf{a} + 2\mathbf{b})$?

*In Exercises 33 and 34, find* $2\mathbf{a} - 3\mathbf{b}$ *and* $\frac{1}{2}\mathbf{a} + \frac{1}{3}\mathbf{b}.$

**33.** $\mathbf{a} = 2\mathbf{i}$ and $\mathbf{b} = -6\mathbf{j}$

**34.** $\mathbf{a} = 2\mathbf{i} - 4\mathbf{j}$ and $\mathbf{b} = 2\mathbf{i} - \mathbf{j}$

**35.** Let $\mathbf{u} = \langle -1, 3 \rangle$, $\mathbf{v} = \langle 2, 4 \rangle$, and $\mathbf{w} = \langle 6, 4 \rangle$. Find $a$ and $b$ if $a\mathbf{u} + b\mathbf{v} = \mathbf{w}$.

**36.** Let $\mathbf{u} = \langle 2, 1 \rangle$, $\mathbf{v} = \langle 3, -1 \rangle$ and $\mathbf{w} = \langle 2, 4 \rangle$. Find $a$ and $b$ if $a\mathbf{u} - b\mathbf{v} = 2\mathbf{w}$.

*In Exercises 37–42, determine whether* **b** *is parallel to* $\mathbf{a} = 3\mathbf{i} - 2\mathbf{j}.$

**37.** $\mathbf{b} = 9\mathbf{i} - 6\mathbf{j}$

**38.** $\mathbf{b} = -\dfrac{3}{2}\mathbf{i} + \mathbf{j}$

**39.** $\mathbf{b} = \mathbf{i} - \mathbf{j}$

**40.** $\mathbf{b} = 6\mathbf{i} - 5\mathbf{j}$

**41.** $\mathbf{b} = \langle -1, 2 \rangle + 4\langle 1, -1 \rangle$

**42.** $\mathbf{b} = 6\langle -1, 2 \rangle + 4\langle 3, -4 \rangle$

**43.** Determine the value of $c$ such that $\mathbf{a} = c\mathbf{i} - 2\mathbf{j}$ and $\mathbf{b} = -4\mathbf{i} + 3\mathbf{j}$ are parallel.

**44.** Prove that $\mathbf{a} = \langle a_1, a_2 \rangle$ and $\mathbf{b} = \langle b_1, b_2 \rangle$ are parallel if and only if $a_1 b_2 - a_2 b_1 = 0$.

*In Exercises 45–48, find a unit vector that has (a) the same direction as* **a** *and (b) a direction opposite to that of* **a**.

**45.** $\mathbf{a} = \langle 2, 1 \rangle$

**46.** $\mathbf{a} = -3\mathbf{i} + 4\mathbf{j}$

**47.** $\mathbf{a} = \langle -\sqrt{3}, 1 \rangle$

**48.** $\mathbf{a} = \langle 0, 3 \rangle$

*In Exercises 49–52, find a vector* **a** *with the given length and having the same direction as* **b**.

**49.** $|\mathbf{a}| = 5$;  $\mathbf{b} = \langle 1, 1 \rangle$

**50.** $|\mathbf{a}| = 2$;  $\mathbf{b} = -3\mathbf{i} + 4\mathbf{j}$

**51.** $|\mathbf{a}| = \sqrt{3}$;  $\mathbf{b} = 3\mathbf{i}$

**52.** $|\mathbf{a}| = 4$;  $\mathbf{b} = \langle \sqrt{3}, -1 \rangle$

**53.** Let $\mathbf{a} = \langle -3, 4 \rangle$ and $\mathbf{b} = \langle 1, 2 \rangle$. Find a vector with length 3 and having the same direction as $2\mathbf{a} - 3\mathbf{b}$.

**54.** Let $\mathbf{a} = \langle 1, -2 \rangle$ and $\mathbf{b} = \langle -1, 3 \rangle$. Find a vector with length 4 and having a direction opposite to that of $\mathbf{a} - 3\mathbf{b}$.

*In Exercises 55–58,* **F** *represents a force acting on a particle. Express* **F** *in terms of its magnitude and direction. What are the horizontal and vertical vector components of* **F**?

**55.** $\mathbf{F} = \langle 3, 1 \rangle$   **56.** $\mathbf{F} = \langle -3, 4 \rangle$

**57.** $\mathbf{F} = \sqrt{3}\mathbf{i} + 6\mathbf{j}$   **58.** $\mathbf{F} = \langle 0, -5 \rangle$

*In Exercises 59–62, find a vector* **a** *that has the given length and makes an angle θ with the positive x-axis.*

**59.** $|\mathbf{a}| = 2; \quad \theta = 0$   **60.** $|\mathbf{a}| = 5; \quad \theta = \dfrac{\pi}{3}$

**61.** $|\mathbf{a}| = 3; \quad \theta = \dfrac{5\pi}{3}$   **62.** $|\mathbf{a}| = 1; \quad \theta = \dfrac{\pi}{2}$

**63. Production Planning** The Acrosonic Company manufactures two different loudspeaker systems in two locations. Suppose that it produced $a_1$ Model $A$ systems and $b_1$ Model $B$ systems in Location I last year. Then we can record this data by writing the production vector $\mathbf{v}_1 = \langle a_1, b_1 \rangle$. Suppose further that the company also produced $a_2$ Model $A$ systems and $b_2$ Model $B$ systems in Location II in the same year. Then we can record this using the vector $\mathbf{v}_2 = \langle a_2, b_2 \rangle$.

**a.** Find $\mathbf{v}_1 + \mathbf{v}_2$, and interpret your result.

**b.** For the next year the company wishes to boost the production of both speaker systems by 10%. Write a vector reflecting the desired level of output.

**64. Pulling a Sled** A child's sled is pulled with a constant force **F** of magnitude 10 lb that makes an angle of 30° with the horizontal. Find the horizontal and vertical vector components $\mathbf{F}_1$ and $\mathbf{F}_2$ of the force.

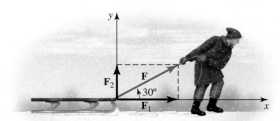

**65. Velocity of a Shell** A shell is fired from a howitzer at an angle of elevation of 45° and with an initial speed of 800 ft/sec. Find the horizontal and vertical vector components of its velocity.

**66. Towing a Cruise Ship** The following figure gives an aerial view of two tugboats attempting to free a cruise ship that ran aground during a storm. Tugboat I exerts a force of magnitude 3000 lb, and Tugboat II exerts a force of magnitude 2400 lb.

**a.** Find expressions for the forces $\mathbf{F}_1$ and $\mathbf{F}_2$.

**b.** Find the resultant force **F** acting on the cruise ship.

**c.** Find the angle θ and the magnitude of **F** if **F** acts along the positive x-axis.

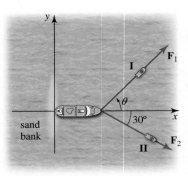

**67.** An object located at the origin is acted upon by three forces, $\mathbf{F}_1$, $\mathbf{F}_2$, and $\mathbf{F}_3$, as shown in the following figure. Find $\mathbf{F}_3$ if the object is in equilibrium.

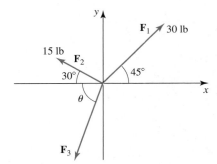

**68.** A model airplane on display in a hobby store is suspended by two wires attached to the ceiling as shown in the following figure. The model airplane weighs 2 lb. Find the tensions $\mathbf{T}_1$ and $\mathbf{T}_2$ of the wires.

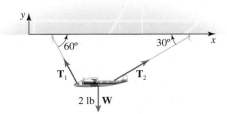

**69.** A river $\frac{1}{2}$ mi wide flows parallel to the shore at the rate of 5 mph. If a motorboat can move at 10 mph in still water, at what angle with respect to the shore should the boat be pointed to move in a direction perpendicular to the shore? How long will the boat take to cross the river?

70. **Finding the True Course and Ground Speed of an Airplane** An airplane on level flight has an airspeed of 300 mph and is headed in a direction that makes an angle of 30° with the north (measured in a clockwise direction). The plane is subjected to a headwind blowing at 60 mph in a direction that makes an angle of 270° with the north. Find the true course and ground speed of the plane.

71. **Finding the True Course and Ground Speed of an Airplane** An airplane pilot wishes to maintain a true course in the direction that makes an angle of 60° with the north (measured in a clockwise direction) and a ground speed of 240 mph when the wind is blowing directly west at 40 mph. Find the required airspeed and compass heading of the airplane.

72. Use vectors to prove that the line segment joining the midpoints of two sides of a triangle is parallel to the third side and half its length.

73. Use vectors to prove that the diagonals of a parallelogram bisect each other.

74. The opposite sides of a quadrilateral are parallel and of equal length. Use vectors to prove that the other two sides must be parallel and of equal length.

*In Exercises 75–82, prove the stated property if $\mathbf{a} = \langle a_1, a_2 \rangle$, $\mathbf{b} = \langle b_1, b_2 \rangle$, $\mathbf{c} = \langle c_1, c_2 \rangle$, and c and d are scalars.*

75. $(\mathbf{a} + \mathbf{b}) + \mathbf{c} = \mathbf{a} + (\mathbf{b} + \mathbf{c})$

76. $\mathbf{a} + \mathbf{0} = \mathbf{a}$

77. $\mathbf{a} + (-\mathbf{a}) = \mathbf{0}$

78. $c(\mathbf{a} + \mathbf{b}) = c\mathbf{a} + c\mathbf{b}$

79. $c(d\mathbf{a}) = (cd)\mathbf{a}$

80. $(c + d)\mathbf{a} = c\mathbf{a} + d\mathbf{a}$

81. $1\mathbf{a} = \mathbf{a}$

82. $c\mathbf{a} = \mathbf{0}$ if and only if $c = 0$ or $\mathbf{a} = \mathbf{0}$

*In Exercises 83–88, determine whether the statement is true or false. If it is true, explain why. If it is false, explain why or give an example that shows it is false.*

83. $\mathbf{v} - \mathbf{v} = 0$.

84. If $|\mathbf{v}| = |\mathbf{w}|$, then $\mathbf{v} = \mathbf{w}$.

85. If $\mathbf{u}$ is a unit vector having the same direction as $\mathbf{v}$, then $\mathbf{v} = |\mathbf{v}|\mathbf{u}$.

86. If $\mathbf{v} = a\mathbf{i} + b\mathbf{j}$, $\mathbf{w} = b\mathbf{i} - a\mathbf{j}$, and $\mathbf{v} = \mathbf{w}$, then $a = b = 0$.

87. If $\mathbf{v} = a\mathbf{i} + b\mathbf{j}$, where not both $a$ and $b$ are equal to zero, then

$$\mathbf{u} = \pm\left(\frac{a}{\sqrt{a^2 + b^2}}\mathbf{i} + \frac{b}{\sqrt{a^2 + b^2}}\mathbf{j}\right)$$

are two unit vectors having the same direction as $\mathbf{v}$.

88. If $P_1(x_1, y_1)$ and $P_2(x_2, y_2)$ are distinct points and the vector $\overrightarrow{P_1P_2}$ is parallel to $\mathbf{v} = \langle 1, 2 \rangle$, then there exists a nonzero constant $c$ such that $x_2 = x_1 + c$ and $y_2 = y_1 + 2c$.

## 10.2 Coordinate Systems and Vectors in 3-Space

### ▪ Coordinate Systems in Space

The plane curve $C$ shown in Figure 1a gives the path taken by an aircraft as it taxis to the runway. The position of the plane may be specified by the coordinates of the point $P(x, y)$ lying on the curve $C$. Figure 1b shows the flight path of the plane shortly *after* takeoff. Because the aircraft is now in the air, we also need to specify its altitude when giving its position. This can be done by introducing an axis that is perpendicular to the $x$- and $y$-axes at the origin. The position of the plane may then be specified by giving the *three* **coordinates** $x$, $y$, and $z$ of the point $P$ represented by the **ordered triple** $(x, y, z)$. Here, the number $z$ gives the altitude of the plane.

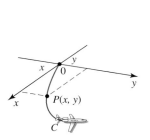

(a) The path of a plane taxiing on the ground

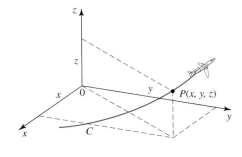

(b) The plane's path after takeoff

**FIGURE 1**

The three positive axes that we have just drawn in Figure 1b are part of a three-dimensional coordinate system. Figure 2 shows a **three-dimensional rectangular coordinate system** along with the points $A(2, 4, 5)$, $B(3, -4, -2)$, and $C(-2, -3, 3)$.

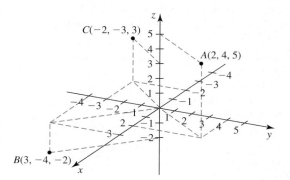

**FIGURE 2**
The points $A(2, 4, 5)$, $B(3, -4, -2)$, and $C(-2, -3, 3)$

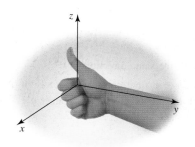

**FIGURE 3**
The right-handed system

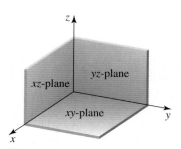

**FIGURE 4**
The three coordinate planes

The coordinate systems shown in Figure 1 and Figure 2 are **right-handed:** If you start by pointing the fingers of your right hand in the direction of the positive $x$-axis and then curl them toward the positive $y$-axis, your thumb will point in the positive direction of the $z$-axis (see Figure 3).

The three coordinate axes determine the three coordinate planes: The $xy$-plane is determined by the $x$- and $y$-axes, the $yz$-plane is determined by the $y$- and $z$-axes, and the $xz$-plane is determined by the $x$- and $z$-axes. (See Figure 4.) These coordinate planes divide 3-space into eight **octants.** The first octant is the one determined by the positive axes.

Just as an equation in $x$ and $y$ represents a curve in the plane, an equation in $x$, $y$, and $z$ represents a *surface* in 3-space. The simplest surfaces in 3-space, other than the coordinate planes, are the planes that are parallel to the coordinate planes.

**EXAMPLE 1**   Sketch the surface represented by the equation

**a.** $x = 3$      **b.** $z = 4$

**Solution**

**a.** The equation $x = 3$ tells us that the surface consists of the set of points in 3-space whose $x$-coordinate is held fast at 3 while $y$ and $z$ are allowed to range over all real numbers, written $\{(x, y, z) \mid x = 3\}$. This surface is the plane that is parallel to the $yz$-plane and located three units in front of it. (See Figure 5.)

**b.** Similarly, we see that the equation $z = 4$ represents the set $\{(x, y, z) \mid z = 4\}$ and is the plane that is parallel to the $xy$-plane and located four units above it.

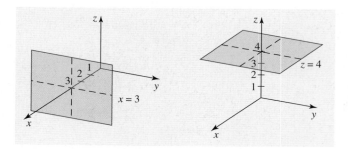

**FIGURE 5**
The planes $x = 3$ and $z = 4$

Keep in mind the dimensions you are working in. In the $xy$-plane (2-space) the equation $x = 3$ represents the vertical line parallel to the $y$-axis; in the $xyz$-plane (3-space) the equation $x = 3$ represents a plane parallel to the $yz$-plane, as we have just seen.

In general, if $k$ is a constant, then $x = k$ represents a plane that is parallel to the $yz$-plane; $y = k$ represents a plane that is parallel to the $xz$-plane; and $z = k$ represents a plane that is parallel to the $xy$-plane.

## ■ The Distance Formula

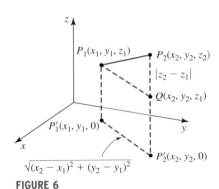

**FIGURE 6**

To find a formula for the distance between two points $P_1(x_1, y_1, z_1)$ and $P_2(x_2, y_2, z_2)$ in 3-space, refer to Figure 6. First, apply the distance formula in 2-space to see that the distance between $P'_1(x_1, y_1, 0)$ and $P'_2(x_2, y_2, 0)$, the respective projections of $P_1(x_1, y_1, z_1)$ and $P_2(x_2, y_2, z_2)$ onto the $xy$-plane, is

$$d(P'_1, P'_2) = \sqrt{(x_2 - x_1)^2 + (y_2 - y_1)^2}$$

But this is also the distance $d(P_1, Q)$ between the points $P_1(x_1, y_1, z_1)$ and $Q(x_2, y_2, z_1)$. Then applying the Pythagorean Theorem to the right triangle $P_1QP_2$, we have

$$[d(P_1, P_2)]^2 = [d(P_1, Q)]^2 + [d(P_2, Q)]^2 = (x_2 - x_1)^2 + (y_2 - y_1)^2 + (z_2 - z_1)^2$$

which is equivalent to the following.

---

**The Distance Formula**

$$d(P_1, P_2) = \sqrt{(x_2 - x_1)^2 + (y_2 - y_1)^2 + (z_2 - z_1)^2} \qquad \textbf{(1)}$$

---

**EXAMPLE 2** Find the distance between $(3, -2, 1)$ and $(1, 0, 3)$.

**Solution** Using the distance formula (1) with $P_1(3, -2, 1)$ and $P_2(1, 0, 3)$, we find that the required distance is

$$d = \sqrt{(1 - 3)^2 + [0 - (-2)]^2 + (3 - 1)^2}$$
$$= \sqrt{4 + 4 + 4} = \sqrt{12} = 2\sqrt{3} \qquad \blacksquare$$

## ■ The Midpoint Formula

The formula for finding the coordinates of the midpoint of the line segment joining two points $P_1(x_1, y_1, z_1)$ and $P_2(x_2, y_2, z_2)$ in 3-space is just an extension of that for finding the coordinates of the midpoint of the line segment joining two points in the plane. (See Exercise 77.)

---

**The Midpoint Formula**

$$\left( \frac{x_1 + x_2}{2}, \frac{y_1 + y_2}{2}, \frac{z_1 + z_2}{2} \right) \qquad \textbf{(2)}$$

---

**EXAMPLE 3**   Find the midpoint of the line segment joining $(3, -2, 1)$ and $(1, 0, 3)$.

**Solution**   Using the midpoint formula (2) with $P_1(3, -2, 1)$ and $P_2(1, 0, 3)$, we find that the midpoint is

$$\left(\frac{3 + 1}{2}, \frac{-2 + 0}{2}, \frac{1 + 3}{2}\right) \quad \text{or} \quad (2, -1, 2) \qquad \blacksquare$$

As the next example illustrates, we can also use the distance formula to help us find an equation of a sphere.

**EXAMPLE 4**

**a.** Find an equation of the sphere with center $C(h, k, l)$ and radius $r$.
**b.** Find an equation of the sphere that has a diameter with endpoints $(3, -2, 1)$ and $(1, 0, 3)$.

**Solution**
**a.** The sphere is the set of all points $P(x, y, z)$ whose distance from $C(h, k, l)$ is $r$, or, equivalently, the square of the distance from $P$ to $C$ is $r^2$. Using the distance formula, we see that an equation of the sphere is

$$(x - h)^2 + (y - k)^2 + (z - l)^2 = r^2$$

**b.** From Example 2 we see that the distance between $(3, -2, 1)$ and $(1, 0, 3)$ is $2\sqrt{3}$, so the radius of the sphere is $\frac{1}{2}(2\sqrt{3})$, or $\sqrt{3}$. Next, from Example 3 we see that the midpoint of the line segment joining $(3, -2, 1)$ and $(1, 0, 3)$ is $(2, -1, 2)$. This point is the center of the sphere. Finally, using the result of part (a), we obtain the equation of the sphere:

$$(x - 2)^2 + (y + 1)^2 + (z - 2)^2 = 3 \qquad \blacksquare$$

The equation that we obtained in Example 4a is called the *standard equation* of a sphere.

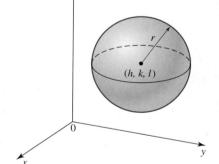

**FIGURE 7**
The sphere with center $C(h, k, l)$ and radius $r$

---

**The Standard Equation of a Sphere with Center $(h, k, l)$ and Radius $r$**

$$(x - h)^2 + (y - k)^2 + (z - l)^2 = r^2 \qquad (3)$$

---

The graph of this equation appears in Figure 7.

**EXAMPLE 5**   Show that $x^2 + y^2 + z^2 - 4x + 2y + 6z + 5 = 0$ is an equation of a sphere, and find its center and radius.

**Solution**   By completing the squares in $x$, $y$, and $z$, we can write the given equation in the form

$$[x^2 - 4x + (-2)^2] + (y^2 + 2y + 1) + (z^2 + 6z + 9) = -5 + 4 + 1 + 9$$

or

$$(x - 2)^2 + (y + 1)^2 + (z + 3)^2 = 3^2$$

Comparing this equation with Equation (3), we conclude that it is an equation of the sphere of radius 3 with center at $(2, -1, -3)$. $\qquad \blacksquare$

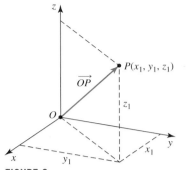

**FIGURE 8**
The position vector $\overrightarrow{OP}$ has initial point $O$ and terminal point $P$.

## Vectors in 3-Space

A vector in 3-space is an ordered triple of real numbers

$$\mathbf{a} = \langle a_1, a_2, a_3 \rangle$$

where $a_1$, $a_2$, and $a_3$ are the **components** of the vector. In particular, the **position vector** of a point $P(x_1, y_1, z_1)$ is the vector $\overrightarrow{OP} = \langle x_1, y_1, z_1 \rangle$ with initial point at the origin and terminal point $P(x_1, y_1, z_1)$. (See Figure 8.)

The basic definitions and operations of vectors in 3-space are natural generalizations of those of vectors in the plane.

---

**DEFINITION    Vectors in 3-Space**

If $\mathbf{a} = \langle a_1, a_2, a_3 \rangle$ and $\mathbf{b} = \langle b_1, b_2, b_3 \rangle$ are vectors in 3-space and $c$ is a scalar, then

1. $\mathbf{a} = \mathbf{b}$    if and only if    $a_1 = b_1, a_2 = b_2$ and $a_3 = b_3$    Equality

2. $\mathbf{a} + \mathbf{b} = \langle a_1 + b_1, a_2 + b_2, a_3 + b_3 \rangle$    Vector addition

3. $c\mathbf{a} = \langle ca_1, ca_2, ca_3 \rangle$    Scalar multiplication

4. $|\mathbf{a}| = \sqrt{a_1^2 + a_2^2 + a_3^2}$    Length

---

Also, the rules of vector addition and scalar multiplication stated in Theorem 1 of Section 10.1 are valid for vectors in 3-space. The proofs are similar.

The following representation of a vector $\mathbf{a}$ in 3-space is a natural extension of the representation of vectors in the plane.

---

The vector with initial point $P_1(x_1, y_1, z_1)$ and terminal point $P_2(x_2, y_2, z_2)$ is

$$\overrightarrow{P_1P_2} = \langle x_2 - x_1, y_2 - y_1, z_2 - z_1 \rangle \tag{4}$$

---

Thus, we can find the components of a vector by subtracting the respective coordinates of its initial point from the coordinates of its terminal point, as illustrated in Figure 9. The vectors $\overrightarrow{OP_1}$ and $\overrightarrow{OP_2}$ are the position vectors of the point $P_1(x_1, y_1, z_1)$ and $P_2(x_2, y_2, z_2)$. As a natural extension of the definition of vector subtraction in 2-space into 3-space, we have

$$\overrightarrow{P_1P_2} = \overrightarrow{OP_2} - \overrightarrow{OP_1} = \langle x_2, y_2, z_2 \rangle - \langle x_1, y_1, z_1 \rangle = \langle x_2 - x_1, y_2 - y_1, z_2 - z_1 \rangle$$

**FIGURE 9**
$\overrightarrow{P_1P_2} = \langle x_2 - x_1, y_2 - y_1, z_2 - z_1 \rangle$
is represented by the position vector $\overrightarrow{OP}$ of the point $P(x_2 - x_1, y_2 - y_1, z_2 - z_1)$.

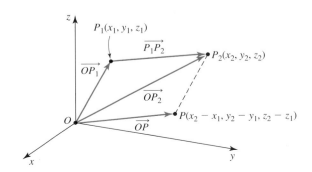

By considering the parallelogram $OPP_2P_1$ in Figure 9, you can convince yourself that $\overrightarrow{P_1P_2}$ is represented by the position vector $\overrightarrow{OP}$ of the point $(x_2 - x_1, y_2 - y_1, z_2 - z_1)$.

**EXAMPLE 6** Let $P(2, -1, 2)$ and $Q(1, 4, 5)$ be two points in 3-space.

**a.** Find the vector $\overrightarrow{PQ}$.
**b.** Find $|\overrightarrow{PQ}|$.
**c.** Find a unit vector having the same direction as $\overrightarrow{PQ}$.

**Solution**
**a.** Using Equation (4) with $P_1 = P$ and $P_2 = Q$, we have

$$\overrightarrow{PQ} = \langle 1 - 2, 4 - (-1), 5 - 2 \rangle = \langle -1, 5, 3 \rangle$$

**b.** Using the result of part (a), we have

$$|\overrightarrow{PQ}| = \sqrt{(-1)^2 + 5^2 + 3^2} = \sqrt{35}$$

**c.** Using the results of parts (a) and (b), we obtain the unit vector

$$\mathbf{u} = \frac{\overrightarrow{PQ}}{|\overrightarrow{PQ}|} = \frac{1}{\sqrt{35}} \langle -1, 5, 3 \rangle$$

The vector $\overrightarrow{PQ}$, the position vector $\mathbf{a}$ which equals $\overrightarrow{PQ}$, and the unit (position) vector $\mathbf{u}$ are shown in Figure 10.

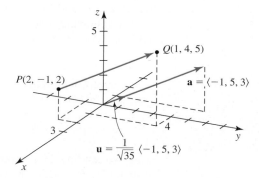

**FIGURE 10**
The vector $\overrightarrow{PQ}$, its equivalent position vector $\mathbf{a}$, and the (position) unit vector $\mathbf{u}$

## Standard Basis Vectors in Space

In Section 10.1 we saw that any vector in the plane can be expressed in terms of the standard basis vectors $\mathbf{i} = \langle 1, 0 \rangle$ and $\mathbf{j} = \langle 0, 1 \rangle$. In three-dimensional space, the 3-space vectors

$$\mathbf{i} = \langle 1, 0, 0 \rangle, \qquad \mathbf{j} = \langle 0, 1, 0 \rangle, \qquad \text{and} \qquad \mathbf{k} = \langle 0, 0, 1 \rangle$$

form a basis for the space, in the sense that any vector in the space can be expressed in terms of these vectors. In fact, if $\mathbf{a} = \langle a_1, a_2, a_3 \rangle$ is a vector in 3-space, we can write

$$\mathbf{a} = \langle a_1, a_2, a_3 \rangle = \langle a_1, 0, 0 \rangle + \langle 0, a_2, 0 \rangle + \langle 0, 0, a_3 \rangle$$

$$= a_1 \langle 1, 0, 0 \rangle + a_2 \langle 0, 1, 0 \rangle + a_3 \langle 0, 0, 1 \rangle$$

$$= a_1 \mathbf{i} + a_2 \mathbf{j} + a_3 \mathbf{k}$$

The standard basis vectors **i**, **j**, and **k** are shown in Figure 11a. Figure 11b shows the vector **a** and its three vector components $a_1\mathbf{i}$, $a_2\mathbf{j}$, and $a_3\mathbf{k}$ in the $x$-, $y$-, and $z$-directions, respectively.

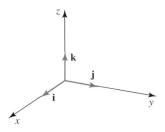

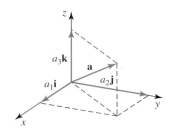

**FIGURE 11**

(a) The standard basis vectors **i**, **j**, and **k**

(b) The vectors $a_1\mathbf{i}$, $a_2\mathbf{j}$, and $a_3\mathbf{k}$ are the vector components of **a** in the $x$-, $y$-, and $z$-directions.

**EXAMPLE 7**  Write $\mathbf{a} = \langle -1, 2, -3 \rangle$ and $\mathbf{b} = \langle 2, 0, 4 \rangle$ in terms of the standard basis vectors **i**, **j**, and **k**. Then compute $2\mathbf{a} - 3\mathbf{b}$.

**Solution**  We have

$$\mathbf{a} = \langle -1, 2, -3 \rangle = -\mathbf{i} + 2\mathbf{j} - 3\mathbf{k}$$

and

$$\mathbf{b} = \langle 2, 0, 4 \rangle = 2\mathbf{i} + 0\mathbf{j} + 4\mathbf{k} = 2\mathbf{i} + 4\mathbf{k}$$

Next, we find

$$2\mathbf{a} - 3\mathbf{b} = 2(-\mathbf{i} + 2\mathbf{j} - 3\mathbf{k}) - 3(2\mathbf{i} + 4\mathbf{k})$$
$$= -2\mathbf{i} + 4\mathbf{j} - 6\mathbf{k} - 6\mathbf{i} - 12\mathbf{k}$$
$$= -8\mathbf{i} + 4\mathbf{j} - 18\mathbf{k} \qquad \text{Add components.}$$

which can also be written $\langle -8, 4, -18 \rangle$.

## 10.2  CONCEPT QUESTIONS

**1. a.** What is a right-handed rectangular coordinate system in a three-dimensional space? Plot the point $(-2, 3, 4)$ in this system.
  **b.** What is the distance between the points $P_1(a_1, b_1, c_1)$ and $P_2(a_2, b_2, c_2)$?
  **c.** What is the point midway between $P_1$ and $P_2$ of part (b)?
**2.** What is the standard equation of a sphere with center $C(h, k, l)$ and radius $r$? What happens if $r < 0$? If $r = 0$?

**3. a.** If $\mathbf{a} = \langle a_1, a_2, a_3 \rangle$ and $\mathbf{b} = \langle b_1, b_2, b_3 \rangle$, what is the sum of **a** and **b**, the scalar multiple of **a** by the scalar $c$, and the length of **a**?
  **b.** If $P_1(a_1, a_2, a_3)$ and $P_2(b_1, b_2, b_3)$ are points in 3-space, what is $\overrightarrow{P_1 P_2}$? $\overrightarrow{P_2 P_1}$? How are the two vectors related?
**4. a.** What are the standard basis vectors in a three-dimensional space? Make a sketch.
  **b.** If $P_1(a_1, a_2, a_3)$ and $P_2(b_1, b_2, b_3)$ are points in 3-space, write $\overrightarrow{P_1 P_2}$ in terms of the standard basis vectors.

## 10.2  EXERCISES

*In Exercises 1–6, plot the given points in a three-dimensional coordinate system.*

**1.** $(3, 2, 4)$      **2.** $(2, 3, 2)$

**3.** $(3, -1, 4)$      **4.** $(0, 2, 4)$

**5.** $(-3, -2, 4)$      **6.** $(-2, 0, 4)$

*In Exercises 7 and 8, find the coordinates of the indicated points.*

**7.**

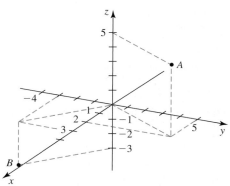

**8.**

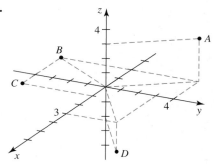

*In Exercises 9–12, sketch the plane in three-dimensional space represented by the equation.*

**9.** $y = 5$      **10.** $x = -4$

**11.** $z = 4$      **12.** $z = -4$

*In Exercises 13–16, describe the region in three-dimensional space represented by the inequality.*

**13.** $x \geq 3$      **14.** $y \leq -4$

**15.** $z > 3$      **16.** $y < 4$

**17. Air Traffic Control** Suppose that the control tower of a municipal airport is located at the origin of a three-dimensional coordinate system with orientation as shown in the figure. At an instant of time, Plane $A$ is 1000 ft west and 2000 ft south of the tower and flying at an altitude of 3000 ft, and Plane $B$ is 4000 ft east and 1000 ft north of the tower and flying at an altitude of 1000 ft.

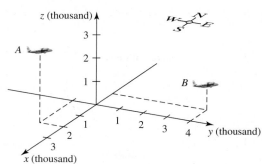

**a.** Write the coordinates of Plane $A$ and Plane $B$.
**b.** How far apart are the planes at that instant of time?

**18.** After holding a short conversation with each other, Jack and Jill proceeded to return to their downtown offices. Jack walked 1 block east, then 2 blocks north, then took the elevator to the thirtieth floor. At the moment Jack emerged from the elevator, Jill had walked $1\frac{1}{2}$ blocks south and $1\frac{1}{4}$ blocks west. If the length of a city block is 1000 ft and a story in Jack's office building is 10 ft high, how far apart are Jack and Jill at that moment?

*In Exercises 19–22, find the length of each side of the triangle $ABC$ and determine whether the triangle is an isosceles triangle, a right triangle, both, or neither.*

**19.** $A(0, 1, 2)$, $B(4, 3, 3)$, $C(3, 4, 5)$

**20.** $A(3, 4, 1)$, $B(4, 4, 6)$, $C(3, 1, 2)$

**21.** $A(-1, 0, 1)$, $B(1, 1, -1)$, $C(1, 1, 1)$

**22.** $A(-1, 5, 2)$, $B(1, -1, 2)$, $C(-3, 1, -2)$

*In Exercises 23 and 24, determine whether the given points are collinear.*

**23.** $A(2, 3, 2)$, $B(-4, 0, 5)$, and $C(4, 4, 1)$

**24.** $A(-1, 3, -2)$, $B(2, 1, -1)$, and $C(8, -3, 1)$

*In Exercises 25 and 26, find the midpoint of the line segment joining the given points.*

**25.** $(2, 4, -6)$ and $(-4, 2, 4)$      **26.** $\left(\frac{1}{2}, -4, 2\right)$ and $\left(\frac{3}{2}, 2, -4\right)$

*In Exercises 27–30, find the standard equation of the sphere with center $C$ and radius $r$.*

**27.** $C(2, 1, 3)$;  $r = 3$      **28.** $C(3, 2, 0)$;  $r = \sqrt{3}$

**29.** $C(3, -1, 2)$;  $r = 4$      **30.** $C(1, \sqrt{2}, -2)$;  $r = 5$

**31.** Find an equation of the sphere that has the points $(2, -3, 4)$ and $(3, 2, 1)$ at opposite ends of its diameter.

**32.** Find an equation of the sphere centered at the point $(2, -3, 4)$ and tangent to the $xy$-plane.

**33.** Find an equation of the sphere that contains the point $(1, 3, 5)$ and is centered at the point $(-1, 2, 4)$.

**34.** Find an equation of the sphere centered at the point $(2, 3, 6)$ and tangent to the sphere with equation $x^2 + y^2 + z^2 = 9$.

*In Exercises 35–40, find the center and the radius of the sphere that has the given equation.*

**35.** $x^2 + y^2 + z^2 - 2x - 4y - 6z + 10 = 0$

**36.** $x^2 + y^2 + z^2 + 4x - 5y + 2z + 5 = 0$

**37.** $x^2 + y^2 + z^2 - 4x + 6y = 0$

**38.** $x^2 + y^2 + z^2 = y$

**39.** $2x^2 + 2y^2 + 2z^2 - 6x - 4y + 2z = 1$

**40.** $3x^2 + 3y^2 + 3z^2 = 6z + 1$

*In Exercises 41–44, describe the region in 3-space satisfying the inequality or inequalities.*

**41.** $x^2 + y^2 + z^2 < 4$

**42.** $x^2 + y^2 + z^2 - 2x - 4y + 2z + 5 \geq 0$

**43.** $1 \leq x^2 + y^2 + z^2 \leq 9$

**44.** $x^2 + y^2 + z^2 \leq 4, \quad z \geq 0$

*In Exercises 45–48, find the vector $\overrightarrow{AB}$. Then sketch the position vector that is equal to $\overrightarrow{AB}$.*

**45.**

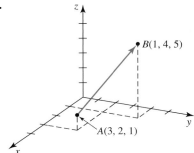

**46.**

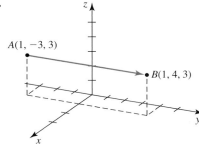

**47.**

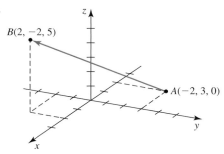

**48.**

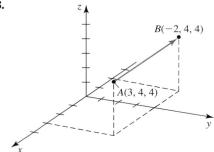

*In Exercises 49 and 50, find the vector $\overrightarrow{AB}$. Then sketch $\overrightarrow{AB}$ and the position vector that is equal to $\overrightarrow{AB}$.*

**49.** $A(2, 1, 0)$, $B(1, 4, 5)$

**50.** $A(-2, -1, 2)$, $B(1, 3, 4)$

**51.** Suppose that $\overrightarrow{AB} = \langle -1, 3, 4 \rangle$ and $B(2, -3, 1)$. Find $A$.

**52.** Suppose that $\overrightarrow{AB} = \langle 3, 0, 4 \rangle$ and $A(-1, -2, -4)$. Find the midpoint of the line segment joining $A$ and $B$.

*In Exercises 53–56, find $\mathbf{a} + \mathbf{b}$, $2\mathbf{a} - 3\mathbf{b}$, $|3\mathbf{a}|$, $|-2\mathbf{b}|$, and $|\mathbf{a} - \mathbf{b}|$.*

**53.** $\mathbf{a} = \langle -1, 2, 0 \rangle$ and $\mathbf{b} = \langle 2, 3, -1 \rangle$

**54.** $\mathbf{a} = 2\mathbf{i} - \mathbf{j} + \mathbf{k}$ and $\mathbf{b} = 3\mathbf{i} + 2\mathbf{k}$

**55.** $\mathbf{a} = \langle 0, 2.1, 3.4 \rangle$ and $\mathbf{b} = \langle 1, 4.1, -5.6 \rangle$

**56.** $\mathbf{a} = -2\mathbf{i} + 4\mathbf{k}$ and $\mathbf{b} = 2\mathbf{j} - \mathbf{k}$

**57.** Let $\mathbf{u} = \langle -1, 3, -2 \rangle$, $\mathbf{v} = \langle 2, 1, 4 \rangle$, and $\mathbf{w} = \langle 3, -2, 1 \rangle$. Find $a$, $b$, and $c$ if $a\mathbf{u} + b\mathbf{v} + c\mathbf{w} = \langle 2, 0, 2 \rangle$.

**58.** Refer to Exercise 57. Show that $\langle 2, 0, 2 \rangle$ cannot be written in the form $a\mathbf{u} + b\mathbf{v}$ for any choice of $a$ and $b$.

*In Exercises 59–62, determine whether $\mathbf{b}$ is parallel to $\mathbf{a} = \mathbf{i} - 2\mathbf{j} + 5\mathbf{k}$.*

**59.** $\mathbf{b} = \langle 3, -6, 15 \rangle$

**60.** $\mathbf{b} = \dfrac{1}{3}\mathbf{i} - \dfrac{2}{3}\mathbf{j} + \dfrac{5}{3}\mathbf{k}$

**61.** $\mathbf{b} = 2\mathbf{i} - 3\mathbf{j} + 10\mathbf{k}$

**62.** $\mathbf{b} = \langle -2, 4, -10 \rangle$

*In Exercises 63–66, find a unit vector that has* (a) *the same direction as* **a** *and* (b) *a direction opposite to that of* **a**.

**63.** $\mathbf{a} = \langle 1, 2, 2 \rangle$      **64.** $\mathbf{a} = -3\mathbf{i} + 4\mathbf{j} + 5\mathbf{k}$

**65.** $\mathbf{a} = -\mathbf{i} + 3\mathbf{j} - \mathbf{k}$      **66.** $\mathbf{a} = -2\langle 0, -3, 4 \rangle$

*In Exercises 67–70, find a vector* **a** *that has the given length and the same direction as* **b**.

**67.** $|\mathbf{a}| = 10$;   $\mathbf{b} = \langle 1, 1, 1 \rangle$    **68.** $|\mathbf{a}| = 2$;   $\mathbf{b} = \mathbf{i} - 2\mathbf{j} + 3\mathbf{k}$

**69.** $|\mathbf{a}| = 3$;   $\mathbf{b} = 2\mathbf{i} + 4\mathbf{j}$    **70.** $|\mathbf{a}| = 4$;   $\mathbf{b} = \langle -1, 0, 1 \rangle$

**71.** Let $\mathbf{a} = \langle 3, -1, 2 \rangle$ and $\mathbf{b} = \langle 1, 0, -1 \rangle$. Find a vector that has length 2 and the same direction as $\mathbf{a} - 2\mathbf{b}$.

**72.** Let $\mathbf{a} = \langle 1, 0, -2 \rangle$ and $\mathbf{b} = \langle 3, 4, 1 \rangle$. Find a vector that has length $|2\mathbf{a} + \mathbf{b}|$ and a direction opposite to that of $\mathbf{a} - \mathbf{b}$.

*In Exercises 73 and 74,* **F** *represents a force acting on a particle. Express* **F** *in terms of its magnitude and direction.*

**73.** $\mathbf{F} = \langle -3, 4, 5 \rangle$      **74.** $\mathbf{F} = 2\mathbf{i} + 3\mathbf{j} - 4\mathbf{k}$

**75.** Refer to the figure below. Show that

$$|\mathbf{a} - \mathbf{b}|^2 = |\mathbf{a}|^2 + |\mathbf{b}|^2 - 2|\mathbf{a}||\mathbf{b}| \cos \theta$$

where $\theta$ is the angle between **a** and **b**.

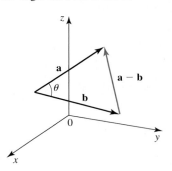

**76.** Refer to the figure below. Show that

$$|\mathbf{a} + \mathbf{b}|^2 = |\mathbf{a}|^2 + |\mathbf{b}|^2 + 2|\mathbf{a}||\mathbf{b}| \cos \theta$$

where $\theta$ is the angle between **a** and **b**.

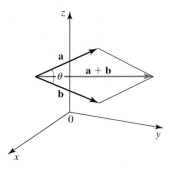

**77.** Prove that the midpoint of the line segment joining the points $P_1(x_1, y_1, z_1)$ and $P_2(x_2, y_2, z_2)$ is

$$\left( \frac{x_1 + x_2}{2}, \frac{y_1 + y_2}{2}, \frac{z_1 + z_2}{2} \right)$$

**78.** A person standing on a bridge watches a canoe go by. The canoe moves at a constant speed of 5 ft/sec in a direction parallel to the $y$-axis. Find a formula for the distance between the spectator and the canoe. How fast is this distance changing when the canoe is 60 ft ($y = 60$) from the bridge?

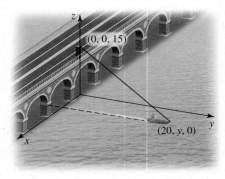

**79.** **Newton's Law of Gravitation** Newton's law of gravitation states that every particle of matter in the universe attracts every other particle with a force whose magnitude is proportional to the product of the masses of the particles and inversely proportional to the square of the distance between them. Show that if a particle of mass $m_1$ is located at a point $A$ and a particle of mass $m_2$ is located at a point $B$, then the force of attraction exerted by the particle located at $A$ on the particle located at $B$ is

$$\mathbf{F} = \frac{Gm_1 m_2}{|\overrightarrow{BA}|^3} \overrightarrow{BA}$$

where $G$ is a positive constant.

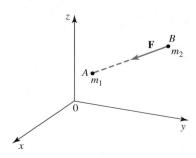

**80.** Refer to Exercise 79. Suppose that two particles of mass $m_1$ and $m_2$ are located at the origin and the point $(d, 0)$, respectively, in a two-dimensional coordinate system. Further, suppose that a third particle of mass $m$ is located at the point $P(x, y)$.

a. Write a vector $\mathbf{F}$ giving the force exerted by the particles of mass $m_1$ and $m_2$, respectively, on $m$.

b. Where should the mass $m$ be located so that the system is in equilibrium?

81. **Coulomb's Law** Coulomb's law states that the force of attraction or repulsion between two point charges (that is, charged bodies whose sizes are small in comparison to the distance between them) is directly proportional to the product of the charges and inversely proportional to the square of the distance between them. Show that if the charges $q_1$ and $q_2$ are located at the points $A$ and $B$ (see the figure below), respectively, then the force $\mathbf{F}_1$ exerted by the charge located at $A$ on the charge located at $B$ is given by

$$\mathbf{F}_1 = \frac{kq_1q_2}{|\overrightarrow{AB}|^3}\overrightarrow{AB}$$

and the force $\mathbf{F}_2$ exerted by the charge located at $B$ on the charge located at $A$ is given by

$$\mathbf{F}_2 = \frac{kq_1q_2}{|\overrightarrow{BA}|^3}\overrightarrow{BA}$$

where $k$ is the constant of proportionality.

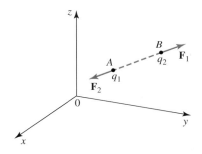

82. Refer to Exercise 81. Suppose that point charges $q_1, q_2, \ldots, q_n$ are placed at the points $P_1, P_2, \ldots, P_n$, respectively.

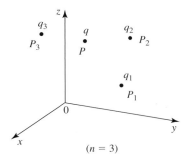

$(n = 3)$

According to the principle of superposition, the total force $\mathbf{F}$ exerted by these charges on the charge $q$ located at the point $P$ is given by

$$\mathbf{F} = kq\left(\frac{q_1}{|\overrightarrow{P_1P}|^3}\overrightarrow{P_1P} + \frac{q_2}{|\overrightarrow{P_2P}|^3}\overrightarrow{P_2P} + \cdots + \frac{q_n}{|\overrightarrow{P_nP}|^3}\overrightarrow{P_nP}\right)$$

$$= kq\sum_{i=1}^{n}\frac{q_i}{|\overrightarrow{P_iP}|^3}\overrightarrow{P_iP}$$

where $k$ is a constant of proportionality.

a. Four equal charges are placed at the points $P_1(d, 0, 0)$, $P_2(0, d, 0)$, $P_3(-d, 0, 0)$, and $P_4(0, -d, 0)$, where $d > 0$ (see the following figure). Find the total force $\mathbf{F}$ exerted by these charges on the charge $q_0$ placed at the point $(0, 0, z)$ on the $z$-axis.

b. Does the result of part (a) agree with what you might expect if $z$ is very large in comparison to $d$?

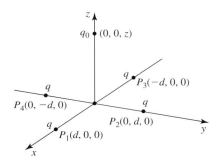

*In Exercises 83–86, determine whether the statement is true or false. If it is true, explain why. If it is false, explain why or give an example that shows it is false.*

83. The equation $x^2 + 2y^2 + z^2 + Ex + Fy + G = 0$, where $E$, $F$, and $G$ are constants, cannot be that of a sphere.

84. The set of points $(x, y, z)$ satisfying $(x - 1)^2 + (y - 2)^2 + (z - 3)^2 \leq 1$ and $(x - 1)^2 + (y - 2)^2 + (z - 3)^2 > 2$ is the empty set.

85. If $P_1$, $P_2$, $Q_1$ and $Q_2$ are points in 3-space and $\overrightarrow{P_1P_2} = \overrightarrow{Q_1Q_2}$, then $P_1 = Q_1$ and $P_2 = Q_2$.

86. If $|c\mathbf{a}| = 1$ and $\mathbf{a} \neq \mathbf{0}$, then $c = 1/|\mathbf{a}|$.

## 10.3    The Dot Product

### ■ Finding the Dot Product

So far, we have looked at two operations involving vectors: vector addition and scalar multiplication. In vector addition, two vectors are combined to yield another vector, and in scalar multiplication a scalar and a vector are combined to yield another vector. In this section we will look at another way of combining two vectors. This operation, called the *dot product,* combines two vectors to yield a *scalar.* As we shall see shortly, the dot product plays a role in the computation of many quantities: the length of a vector, the angle between two vectors, and the work done by a force in moving an object from one point to another, just to mention a few.

---

**DEFINITION    Dot Product**

Let $\mathbf{a} = \langle a_1, a_2, a_3 \rangle$ and $\mathbf{b} = \langle b_1, b_2, b_3 \rangle$ be any two vectors in space. Then the **dot product** of $\mathbf{a}$ and $\mathbf{b}$ is the number $\mathbf{a} \cdot \mathbf{b}$ defined by

$$\mathbf{a} \cdot \mathbf{b} = a_1 b_1 + a_2 b_2 + a_3 b_3$$

---

Thus, we can find the dot product of the two vectors $\mathbf{a}$ and $\mathbf{b}$ by adding the products of their corresponding components.

**Notes**

1. The dot product of two vectors is also called the **inner product,** or **scalar product,** of the two vectors.
2. The definition just given pertains to the dot product of two three-dimensional vectors. For vectors in two-dimensional space, the definition is

$$\mathbf{a} \cdot \mathbf{b} = \langle a_1, a_2 \rangle \cdot \langle b_1, b_2 \rangle = a_1 b_1 + a_2 b_2 \qquad ■$$

**EXAMPLE 1**    Find the dot product of each pair of vectors:

**a.** $\mathbf{a} = \langle 1, 3 \rangle$    and    $\mathbf{b} = \langle -1, 2 \rangle$        **b.** $\mathbf{a} = \langle 1, -2, 4 \rangle$    and    $\mathbf{b} = \langle -1, -2, 3 \rangle$

**Solution**
**a.** $\mathbf{a} \cdot \mathbf{b} = \langle 1, 3 \rangle \cdot \langle -1, 2 \rangle$
$\qquad = (1)(-1) + (3)(2) = 5$
**b.** $\mathbf{a} \cdot \mathbf{b} = \langle 1, -2, 4 \rangle \cdot \langle -1, -2, 3 \rangle$
$\qquad = (1)(-1) + (-2)(-2) + (4)(3) = 15 \qquad ■$

The dot product obeys the following rules.

---

**Properties of the Dot Product**

Let $\mathbf{a}$, $\mathbf{b}$, and $\mathbf{c}$ be vectors in 2- or 3-space and let $c$ be a scalar. Then

1. $\mathbf{a} \cdot \mathbf{b} = \mathbf{b} \cdot \mathbf{a}$
2. $\mathbf{a} \cdot (\mathbf{b} + \mathbf{c}) = \mathbf{a} \cdot \mathbf{b} + \mathbf{a} \cdot \mathbf{c}$
3. $(c\mathbf{a}) \cdot \mathbf{b} = c(\mathbf{a} \cdot \mathbf{b}) = \mathbf{a} \cdot (c\mathbf{b})$
4. $\mathbf{a} \cdot \mathbf{a} = |\mathbf{a}|^2$
5. $\mathbf{0} \cdot \mathbf{a} = 0$

---

We will prove Properties 1 and 4 here and leave the proofs of the other three as an exercise.

**PROOF OF 1** Let $\mathbf{a} = \langle a_1, a_2, a_3 \rangle$ and $\mathbf{b} = \langle b_1, b_2, b_3 \rangle$. Then

$$\mathbf{a} \cdot \mathbf{b} = \langle a_1, a_2, a_3 \rangle \cdot \langle b_1, b_2, b_3 \rangle$$
$$= a_1 b_1 + a_2 b_2 + a_3 b_3 = b_1 a_1 + b_2 a_2 + b_3 a_3$$
$$= \langle b_1, b_2, b_3 \rangle \cdot \langle a_1, a_2, a_3 \rangle = \mathbf{b} \cdot \mathbf{a}$$

**PROOF OF 4** Let $\mathbf{a} = \langle a_1, a_2, a_3 \rangle$. Then

$$\mathbf{a} \cdot \mathbf{a} = \langle a_1, a_2, a_3 \rangle \cdot \langle a_1, a_2, a_3 \rangle$$
$$= a_1^2 + a_2^2 + a_3^2 = |\mathbf{a}|^2$$

Property 4 of dot products gives us the following formula for computing the length of a vector $\mathbf{a}$ in terms of the dot product of $\mathbf{a}$ with itself. Thus,

$$|\mathbf{a}| = \sqrt{\mathbf{a} \cdot \mathbf{a}} \tag{1}$$

**EXAMPLE 2** Let $\mathbf{a} = \langle 1, 2, 4 \rangle$, $\mathbf{b} = \langle -1, -2, 3 \rangle$, and $\mathbf{c} = \langle 3, 1, 2 \rangle$. Compute

**a.** $(\mathbf{a} + \mathbf{b}) \cdot \mathbf{c}$      **b.** $(3\mathbf{a}) \cdot \mathbf{c}$      **c.** $(\mathbf{a} \cdot \mathbf{b})\mathbf{c}$      **d.** $|\mathbf{a} - 2\mathbf{b}|$

**Solution**
**a.** $\mathbf{a} + \mathbf{b} = \langle 1, 2, 4 \rangle + \langle -1, -2, 3 \rangle = \langle 0, 0, 7 \rangle$. Therefore,

$$(\mathbf{a} + \mathbf{b}) \cdot \mathbf{c} = \langle 0, 0, 7 \rangle \cdot \langle 3, 1, 2 \rangle = 0(3) + 0(1) + 7(2) = 14$$

**b.** $3\mathbf{a} = 3\langle 1, 2, 4 \rangle = \langle 3, 6, 12 \rangle$. Therefore,

$$(3\mathbf{a}) \cdot \mathbf{c} = \langle 3, 6, 12 \rangle \cdot \langle 3, 1, 2 \rangle = 3(3) + 6(1) + 12(2) = 39$$

**c.** $\mathbf{a} \cdot \mathbf{b} = \langle 1, 2, 4 \rangle \cdot \langle -1, -2, 3 \rangle = 1(-1) + 2(-2) + 4(3) = 7$. Therefore,

$$(\mathbf{a} \cdot \mathbf{b})\mathbf{c} = 7\langle 3, 1, 2 \rangle = \langle 21, 7, 14 \rangle$$

**d.** $\mathbf{a} - 2\mathbf{b} = \langle 1, 2, 4 \rangle - 2\langle -1, -2, 3 \rangle = \langle 1, 2, 4 \rangle - \langle -2, -4, 6 \rangle = \langle 3, 6, -2 \rangle$. Therefore,

$$|\mathbf{a} - 2\mathbf{b}| = \sqrt{(\mathbf{a} - 2\mathbf{b}) \cdot (\mathbf{a} - 2\mathbf{b})} = \sqrt{\langle 3, 6, -2 \rangle \cdot \langle 3, 6, -2 \rangle}$$
$$= \sqrt{9 + 36 + 4} = \sqrt{49} = 7$$

## The Angle Between Two Vectors

The **angle between two nonzero vectors** is the angle $\theta$ between their corresponding position vectors, where $0 \le \theta \le \pi$. (See Figure 1.)

**Notes**
**1.** If two vectors are parallel, then $\theta = 0$ or $\theta = \pi$.
**2.** The angle between the zero vector and another vector is not defined.

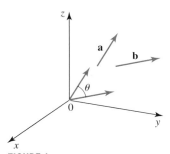

**FIGURE 1**
The angle between $\mathbf{a}$ and $\mathbf{b}$ is the angle between their corresponding position vectors.

---

**THEOREM 1**    **The Angle Between Two Vectors**

Let $\theta$ be the angle between two nonzero vectors $\mathbf{a}$ and $\mathbf{b}$. Then

$$\cos \theta = \frac{\mathbf{a} \cdot \mathbf{b}}{|\mathbf{a}||\mathbf{b}|} \tag{2}$$

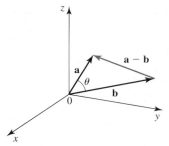

**FIGURE 2**
The angle between **a** and **b** is $\theta$.

**PROOF**   Consider the triangle determined by the vectors **a**, **b**, and **a** − **b** as shown in Figure 2. Using the law of cosines, we have

$$|\mathbf{a} - \mathbf{b}|^2 = |\mathbf{a}|^2 + |\mathbf{b}|^2 - 2|\mathbf{a}||\mathbf{b}|\cos\theta \qquad c^2 = a^2 + b^2 - 2ab\cos C$$

But

$$|\mathbf{a} - \mathbf{b}|^2 = (\mathbf{a} - \mathbf{b}) \cdot (\mathbf{a} - \mathbf{b}) \qquad \text{Equation (1)}$$

$$= \mathbf{a} \cdot \mathbf{a} - \mathbf{a} \cdot \mathbf{b} - \mathbf{b} \cdot \mathbf{a} + \mathbf{b} \cdot \mathbf{b}$$

$$= |\mathbf{a}|^2 - 2\mathbf{a} \cdot \mathbf{b} + |\mathbf{b}|^2 \qquad \mathbf{a} \cdot \mathbf{b} = \mathbf{b} \cdot \mathbf{a}$$

so we have

$$|\mathbf{a}|^2 - 2\mathbf{a} \cdot \mathbf{b} + |\mathbf{b}|^2 = |\mathbf{a}|^2 + |\mathbf{b}|^2 - 2|\mathbf{a}||\mathbf{b}|\cos\theta$$

$$-2\mathbf{a} \cdot \mathbf{b} = -2|\mathbf{a}||\mathbf{b}|\cos\theta$$

or

$$\cos\theta = \frac{\mathbf{a} \cdot \mathbf{b}}{|\mathbf{a}||\mathbf{b}|}$$   ■

**Note**   Because of Equation (2), the dot product of two vectors **a** and **b** can also be defined by the equation $\mathbf{a} \cdot \mathbf{b} = |\mathbf{a}||\mathbf{b}|\cos\theta$, where $\theta$ is the angle between **a** and **b**.   ■

**EXAMPLE 3**   Find the angle between the vectors $\mathbf{a} = \langle 2, 1, 3 \rangle$ and $\mathbf{b} = \langle 3, -2, 2 \rangle$.

**Solution**   We have

$$|\mathbf{a}| = \sqrt{2^2 + 1^2 + 3^2} = \sqrt{14} \qquad |\mathbf{b}| = \sqrt{3^2 + (-2)^2 + 2^2} = \sqrt{17}$$

and

$$\mathbf{a} \cdot \mathbf{b} = 2(3) + 1(-2) + 3(2) = 10$$

so upon using Equation (2), we have

$$\cos\theta = \frac{\mathbf{a} \cdot \mathbf{b}}{|\mathbf{a}||\mathbf{b}|} = \frac{10}{\sqrt{14}\sqrt{17}}$$

and

$$\theta = \cos^{-1}\left(\frac{10}{\sqrt{14}\,\sqrt{17}}\right) \approx 49.6°$$   ■

## ■ Orthogonal Vectors

Two nonzero vectors **a** and **b** are said to be **perpendicular,** or **orthogonal,** if the angle between them is a right angle. Now, suppose that **a** and **b** are orthogonal so that the angle between them is $\pi/2$. Then Equation (2) gives

$$\mathbf{a} \cdot \mathbf{b} = |\mathbf{a}||\mathbf{b}|\cos\frac{\pi}{2} = 0$$

Conversely, if $\mathbf{a} \cdot \mathbf{b} = 0$, then $\cos\theta = 0$ (because $|\mathbf{a}|$ and $|\mathbf{b}|$ are both nonzero), so $\theta = \pi/2$. We have proved the following.

**THEOREM 2**

Two nonzero vectors **a** and **b** are orthogonal if and only if $\mathbf{a} \cdot \mathbf{b} = 0$.

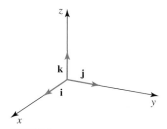

**FIGURE 3**
The standard basis vectors **i**, **j**, and **k** are mutually orthogonal.

The standard basis vectors $\mathbf{i} = \langle 1, 0, 0 \rangle$, $\mathbf{j} = \langle 0, 1, 0 \rangle$, and $\mathbf{k} = \langle 0, 0, 1 \rangle$ are mutually orthogonal; that is, any two of them are orthogonal. This is evident if you examine Figure 3.

For example,

$$\mathbf{i} \cdot \mathbf{j} = \langle 1, 0, 0 \rangle \cdot \langle 0, 1, 0 \rangle = 0 \quad \text{and} \quad \mathbf{j} \cdot \mathbf{k} = \langle 0, 1, 0 \rangle \cdot \langle 0, 0, 1 \rangle = 0$$

**EXAMPLE 4** Determine whether the vectors $\mathbf{a} = 2\mathbf{i} + 3\mathbf{j} + 3\mathbf{k}$ and $\mathbf{b} = 3\mathbf{i} - 4\mathbf{j} + 2\mathbf{k}$ are orthogonal.

**Solution** We compute

$$\mathbf{a} \cdot \mathbf{b} = (2\mathbf{i} + 3\mathbf{j} + 3\mathbf{k})(3\mathbf{i} - 4\mathbf{j} + 2\mathbf{k})$$
$$= 2(3) + 3(-4) + 3(2) = 0$$

and conclude that $\mathbf{a}$ and $\mathbf{b}$ are indeed orthogonal. ■

## Direction Cosines

We can describe the direction of a nonzero vector $\mathbf{a}$ by giving the angles $\alpha$, $\beta$, and $\gamma$ that $\mathbf{a}$ makes with the positive $x$-, $y$-, and $z$-axes, respectively. (See Figure 4.) These angles are called the **direction angles** of $\mathbf{a}$. The cosines of these angles, $\cos \alpha$, $\cos \beta$, and $\cos \gamma$, are called the **direction cosines** of the vector $\mathbf{a}$.

Let $\mathbf{a} = a_1\mathbf{i} + a_2\mathbf{j} + a_3\mathbf{k}$ be a nonzero vector in 3-space. Then

$$\mathbf{a} \cdot \mathbf{i} = (a_1\mathbf{i} + a_2\mathbf{j} + a_3\mathbf{k}) \cdot \mathbf{i} = a_1 \quad \mathbf{i} \cdot \mathbf{i} = 1, \quad \mathbf{j} \cdot \mathbf{i} = \mathbf{k} \cdot \mathbf{i} = 0$$

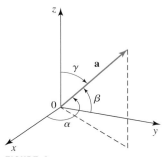

**FIGURE 4**
The direction angles of a vector

So

$$\cos \alpha = \frac{\mathbf{a} \cdot \mathbf{i}}{|\mathbf{a}||\mathbf{i}|} = \frac{a_1}{|\mathbf{a}|}$$

Similarly,

$$\cos \beta = \frac{a_2}{|\mathbf{a}|} \quad \text{and} \quad \cos \gamma = \frac{a_3}{|\mathbf{a}|}$$

By squaring and adding the three direction cosines, we obtain

$$\cos^2 \alpha + \cos^2 \beta + \cos^2 \gamma = \frac{a_1^2}{|\mathbf{a}|^2} + \frac{a_2^2}{|\mathbf{a}|^2} + \frac{a_3^2}{|\mathbf{a}|^2} = \frac{|\mathbf{a}|^2}{|\mathbf{a}|^2} = 1$$

**THEOREM 3**

The three direction cosines of a nonzero vector $\mathbf{a} = a_1\mathbf{i} + a_2\mathbf{j} + a_3\mathbf{k}$ in 3-space are

$$\cos \alpha = \frac{a_1}{|\mathbf{a}|} \quad \cos \beta = \frac{a_2}{|\mathbf{a}|} \quad \cos \gamma = \frac{a_3}{|\mathbf{a}|} \tag{3}$$

The direction cosines satisfy

$$\cos^2 \alpha + \cos^2 \beta + \cos^2 \gamma = 1 \tag{4}$$

**Notes**

1. If $\mathbf{a} = a_1\mathbf{i} + a_2\mathbf{j} + a_3\mathbf{k}$ is nonzero, then the unit vector having the same direction as $\mathbf{a}$ is

$$\mathbf{u} = \frac{\mathbf{a}}{|\mathbf{a}|} = \frac{a_1}{|\mathbf{a}|}\mathbf{i} + \frac{a_2}{|\mathbf{a}|}\mathbf{j} + \frac{a_3}{|\mathbf{a}|}\mathbf{k}$$

$$= (\cos\alpha)\mathbf{i} + (\cos\beta)\mathbf{j} + (\cos\gamma)\mathbf{k} \qquad (5)$$

This shows that the direction cosines of $\mathbf{a}$ are the components of the unit vector in the direction of $\mathbf{a}$. This augments the statement made earlier that the direction cosines of a vector define the direction of that vector.

2. From Equation (5) we see that

$$\mathbf{a} = |\mathbf{a}|\big[(\cos\alpha)\mathbf{i} + (\cos\beta)\mathbf{j} + (\cos\gamma)\mathbf{k}\big]$$

$$\uparrow \qquad\qquad\qquad \uparrow$$

Magnitude          Direction ∎

---

**EXAMPLE 5**   Find the direction angles of the vector $\mathbf{a} = 2\mathbf{i} + 3\mathbf{j} + \mathbf{k}$.

**Solution**   We have

$$|\mathbf{a}| = \sqrt{2^2 + 3^2 + 1^2} = \sqrt{14}$$

so by Equation (3),

$$\cos\alpha = \frac{2}{\sqrt{14}} \qquad \cos\beta = \frac{3}{\sqrt{14}} \qquad \cos\gamma = \frac{1}{\sqrt{14}}$$

Therefore,

$$\alpha = \cos^{-1}\left(\frac{2}{\sqrt{14}}\right) \approx 58° \qquad \beta = \cos^{-1}\left(\frac{3}{\sqrt{14}}\right) \approx 37° \qquad \gamma = \cos^{-1}\left(\frac{1}{\sqrt{14}}\right) \approx 74°$$

∎

## ■ Vector Projections and Components

Figure 5a depicts a child pulling a sled with a constant force represented by the vector $\mathbf{F}$. The force $\mathbf{F}$ can be expressed as the sum of two forces: a horizontal component $\mathbf{F}_1$ and a vertical component $\mathbf{F}_2$, as shown in Figure 5b.

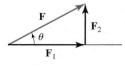

**FIGURE 5**          (a) $\mathbf{F}$ makes an angle $\theta$ with the line of motion.          (b) $\mathbf{F} = \mathbf{F}_1 + \mathbf{F}_2$

Observe that $\mathbf{F}_1$ acts in the direction of motion, whereas $\mathbf{F}_2$ acts in a direction perpendicular to the direction of motion. We will see why it is useful to look at $\mathbf{F}$ in this way when we study the work done by $\mathbf{F}$ in moving the sled.

More generally, we are interested in the component of one vector $\mathbf{b}$ in the direction of another nonzero vector $\mathbf{a}$. The vector that is obtained by projecting $\mathbf{b}$ onto the

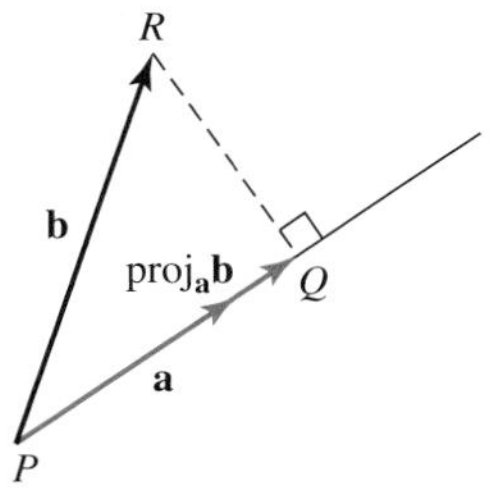

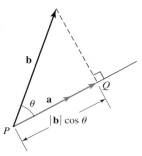

**FIGURE 6**
$\text{proj}_{\mathbf{a}}\mathbf{b}:\left(\overrightarrow{PQ}\right)$ is the vector projection of **b** onto **a**.

**FIGURE 7**
$\text{comp}_{\mathbf{a}}\mathbf{b} = |\mathbf{b}|\cos\theta$

line containing the vector **a** is called the **vector projection of b onto a** (also called the **vector component of b along a**) and denoted by

$$\text{proj}_{\mathbf{a}}\mathbf{b}$$

(See Figure 6.)

The **scalar projection of b onto a** (also called the **scalar component of b along a**) is the length of $\text{proj}_{\mathbf{a}}\mathbf{b}$ if the projection has the same direction as **a** and the negative of the length of $\text{proj}_{\mathbf{a}}\mathbf{b}$ if the projection has the opposite direction. We denote this scalar projection by

$$\text{comp}_{\mathbf{a}}\mathbf{b}$$

As you can see from Figure 7, it is just the number $|\mathbf{b}|\cos\theta$. Observe that if $\frac{\pi}{2} < \theta \leq \pi$, then $|\mathbf{b}|\cos\theta$ is negative. We encourage you to make a sketch of this situation.

Since

$$\cos\theta = \frac{\mathbf{b}\cdot\mathbf{a}}{|\mathbf{b}||\mathbf{a}|}$$

we can write

$$|\mathbf{b}|\cos\theta = \frac{|\mathbf{b}|(\mathbf{b}\cdot\mathbf{a})}{|\mathbf{b}||\mathbf{a}|} = \frac{\mathbf{b}\cdot\mathbf{a}}{|\mathbf{a}|}$$

Therefore, the scalar component of **b** along **a** is

$$\text{comp}_{\mathbf{a}}\mathbf{b} = \frac{\mathbf{b}\cdot\mathbf{a}}{|\mathbf{a}|} \tag{6}$$

**Note**  Writing

$$\frac{\mathbf{b}\cdot\mathbf{a}}{|\mathbf{a}|} = \mathbf{b}\cdot\left(\frac{\mathbf{a}}{|\mathbf{a}|}\right)$$

we see that the scalar component of **b** along **a** can also be calculated by taking the dot product of **b** with the unit vector in the direction of **a**.  ■

The vector projection of **b** onto **a** is the scalar component of **b** along **a** times the direction of **a**. (See Figures 6 and 7.) Thus, we have

$$\text{proj}_{\mathbf{a}}\mathbf{b} = \left(\frac{\mathbf{b}\cdot\mathbf{a}}{|\mathbf{a}|}\right)\frac{\mathbf{a}}{|\mathbf{a}|} = \left(\frac{\mathbf{b}\cdot\mathbf{a}}{|\mathbf{a}|^2}\right)\mathbf{a} = \left(\frac{\mathbf{b}\cdot\mathbf{a}}{\mathbf{a}\cdot\mathbf{a}}\right)\mathbf{a} \tag{7}$$

(See Figure 8.)

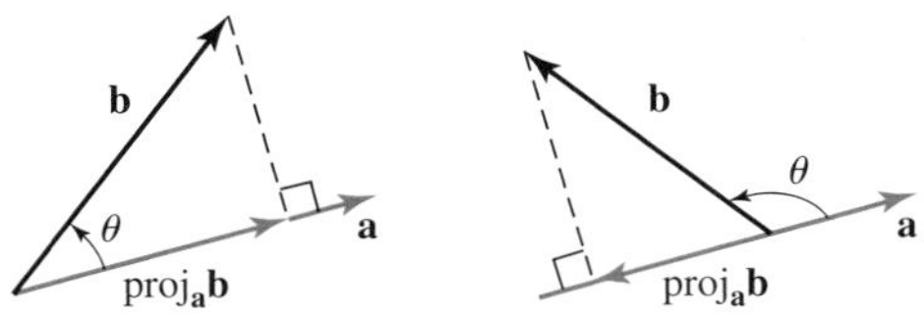

**FIGURE 8**
$\text{proj}_{\mathbf{a}}\mathbf{b}$ points in the same direction as **a** if $\theta$ is acute and points in the opposite direction as **a** if $\theta$ is obtuse.

(**a**) $\theta$ is acute.

(**b**) $\theta$ is obtuse.

**EXAMPLE 6** Let $\mathbf{b} = 2\mathbf{i} + 3\mathbf{j} - 4\mathbf{k}$, and let $\mathbf{a} = 3\mathbf{i} - 2\mathbf{j} + \mathbf{k}$. Find the scalar component of $\mathbf{b}$ along $\mathbf{a}$ and the vector projection of $\mathbf{b}$ onto $\mathbf{a}$.

**Solution**   The scalar component of $\mathbf{b}$ along $\mathbf{a}$ is

$$\text{comp}_{\mathbf{a}}\mathbf{b} = \frac{\mathbf{b} \cdot \mathbf{a}}{|\mathbf{a}|} = \frac{2(3) + (3)(-2) + (-4)(1)}{\sqrt{3^2 + (-2)^2 + 1^2}} = -\frac{4}{\sqrt{14}}$$

Next, we compute the unit vector in the direction of $\mathbf{a}$. Thus,

$$\frac{\mathbf{a}}{|\mathbf{a}|} = \frac{3\mathbf{i} - 2\mathbf{j} + \mathbf{k}}{\sqrt{14}} = \frac{1}{\sqrt{14}}(3\mathbf{i} - 2\mathbf{j} + \mathbf{k})$$

Therefore,

$$\text{proj}_{\mathbf{a}}\mathbf{b} = \left(\frac{\mathbf{b} \cdot \mathbf{a}}{|\mathbf{a}|}\right)\frac{\mathbf{a}}{|\mathbf{a}|}$$

$$= -\frac{4}{\sqrt{14}} \cdot \frac{1}{\sqrt{14}}(3\mathbf{i} - 2\mathbf{j} + \mathbf{k})$$

$$= -\frac{6}{7}\mathbf{i} + \frac{4}{7}\mathbf{j} - \frac{2}{7}\mathbf{k} \qquad\blacksquare$$

Using vector projections, we can express any vector $\mathbf{b}$ as the sum of a vector parallel to a vector $\mathbf{a}$ and a vector perpendicular to $\mathbf{a}$. In fact, from Figure 9 we see that

$$\mathbf{b} = \text{proj}_{\mathbf{a}}\mathbf{b} + (\mathbf{b} - \text{proj}_{\mathbf{a}}\mathbf{b})$$

$$= \underbrace{\left(\frac{\mathbf{b} \cdot \mathbf{a}}{\mathbf{a} \cdot \mathbf{a}}\right)\mathbf{a}}_{\text{Parallel to } \mathbf{a}} + \underbrace{\left[\mathbf{b} - \left(\frac{\mathbf{b} \cdot \mathbf{a}}{\mathbf{a} \cdot \mathbf{a}}\right)\mathbf{a}\right]}_{\text{Orthogonal to } \mathbf{a}} \qquad (8)$$

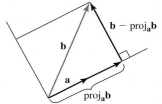

**FIGURE 9**
The vector $\mathbf{b}$ can be written as the sum of a vector parallel to $\mathbf{a}$ and a vector orthogonal to $\mathbf{a}$.

**EXAMPLE 7** Write $\mathbf{b} = 3\mathbf{i} - \mathbf{j} + 2\mathbf{k}$ as the sum of a vector parallel to $\mathbf{a} = 2\mathbf{i} - \mathbf{j} + \mathbf{k}$ and a vector perpendicular to $\mathbf{a}$.

**Solution**   Using Equation (8) with

$$\mathbf{b} \cdot \mathbf{a} = (3)(2) + (-1)(-1) + (2)(1) = 9$$

and

$$\mathbf{a} \cdot \mathbf{a} = 2^2 + (-1)^2 + 1^2 = 6$$

gives

$$\mathbf{b} = \left(\frac{\mathbf{b} \cdot \mathbf{a}}{\mathbf{a} \cdot \mathbf{a}}\right)\mathbf{a} + \left[\mathbf{b} - \left(\frac{\mathbf{b} \cdot \mathbf{a}}{\mathbf{a} \cdot \mathbf{a}}\right)\mathbf{a}\right]$$

$$= \frac{9}{6}(2\mathbf{i} - \mathbf{j} + \mathbf{k}) + \left[3\mathbf{i} - \mathbf{j} + 2\mathbf{k} - \frac{9}{6}(2\mathbf{i} - \mathbf{j} + \mathbf{k})\right]$$

$$= \underbrace{\left(3\mathbf{i} - \frac{3}{2}\mathbf{j} + \frac{3}{2}\mathbf{k}\right)}_{\text{Parallel to } \mathbf{a}} + \underbrace{\left(\frac{1}{2}\mathbf{j} + \frac{1}{2}\mathbf{k}\right)}_{\text{Perpendicular to } \mathbf{a}} \qquad\blacksquare$$

### ■ Work

One application of vector projections lies in the computation of the work done by a force. Recall that the work $W$ done by a constant force $\mathbf{F}$ acting along the line of motion in moving an object a distance $d$ is given by $W = |\mathbf{F}|\, d$. But if the constant force $\mathbf{F}$ acts in a direction that is different from the direction of motion, as in the case we mentioned earlier of a child pulling a sled, then work is done only by that component of the force in the direction of motion.

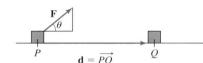

**FIGURE 10**
The work done by $\mathbf{F}$ in moving an object from $P$ to $Q$ is $\mathbf{F} \cdot \mathbf{d}$.

To derive an expression for the work done in this situation, suppose that $\mathbf{F}$ moves an object from $P$ to $Q$. (See Figure 10.) Then, letting $\mathbf{d}$ denote the **displacement vector** $\overrightarrow{PQ}$, we see that the work done by $\mathbf{F}$ in moving an object from $P$ to $Q$ is given by

$$W = \left( |\mathbf{F}| \cos\theta \right) |\mathbf{d}| \qquad \text{component of } \mathbf{F} \text{ along } \mathbf{d} \cdot \text{distance moved}$$

$$= |\mathbf{F}||\mathbf{d}| \cos\theta$$

$$= \mathbf{F} \cdot \mathbf{d} \tag{9}$$

Thus, the work done by a constant force $\mathbf{F}$ in moving an object through a displacement $\mathbf{d}$ is the dot product of $\mathbf{F}$ and $\mathbf{d}$.

**EXAMPLE 8**   A force $\mathbf{F} = 2\mathbf{i} + 3\mathbf{j} + 4\mathbf{k}$ moves a particle along the line segment from the point $P(1, 2, 1)$ to the point $Q(3, 6, 5)$. (See Figure 11.) Find the work done by the force if $|\mathbf{F}|$ is measured in newtons and $|\mathbf{d}|$ is measured in meters.

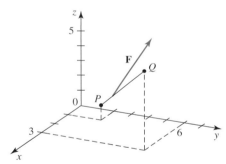

**FIGURE 11**
The force $\mathbf{F}$ moves the particle from the point $P$ to the point $Q$.

**Solution**   The displacement vector is $\mathbf{d} = \overrightarrow{PQ} = \langle 2, 4, 4 \rangle = 2\mathbf{i} + 4\mathbf{j} + 4\mathbf{k}$. Therefore, using Equation (9), we see that the work done by $\mathbf{F}$ is

$$\mathbf{F} \cdot \mathbf{d} = (2\mathbf{i} + 3\mathbf{j} + 4\mathbf{k}) \cdot (2\mathbf{i} + 4\mathbf{j} + 4\mathbf{k})$$

$$= 4 + 12 + 16 = 32$$

or 32 joules.    ■

---

### 10.3    CONCEPT QUESTIONS

**1. a.** What is the dot product of the vectors $\mathbf{a} = \langle a_1, a_2, a_3 \rangle$ and $\mathbf{b} = \langle b_1, b_2, b_3 \rangle$?
   **b.** State the properties of the dot product.
   **c.** Express $|\mathbf{a}|$ in terms of the dot product.
   **d.** What is the angle between two nonzero vectors $\mathbf{a}$ and $\mathbf{b}$ in terms of the dot product?
**2. a.** What does it mean for two nonzero vectors to be orthogonal?
   **b.** State the condition(s) for two nonzero vectors to be orthogonal.

**3. a.** What are the direction cosines of a vector $\mathbf{a}$ in 3-space?
   **b.** Write the nonzero vector $\mathbf{a}$ in terms of its magnitude and its direction cosines.
**4. a.** What is the scalar component of $\mathbf{b}$ along the nonzero vector $\mathbf{a}$? Illustrate with a diagram.
   **b.** What is the vector projection of $\mathbf{b}$ onto $\mathbf{a}$? Illustrate.
**5.** What is the work done by a constant force $\mathbf{F}$ in moving an object along a straight line from the point $P$ to the point $Q$?

## 10.3 EXERCISES

1. State whether the expression makes sense. Explain.

   a. $(\mathbf{a} \cdot \mathbf{b})\mathbf{c}$         b. $\mathbf{a} \cdot (\mathbf{b} \cdot \mathbf{c})$

   c. $\mathbf{a} + \mathbf{b} \cdot \mathbf{c}$         d. $\mathbf{a} \cdot \mathbf{b} - |\mathbf{a}||\mathbf{b}|$

   e. $|\mathbf{a}|\mathbf{b} + (\mathbf{a} \cdot \mathbf{b})\mathbf{a}$         f. $\mathbf{a} \cdot \left(\dfrac{\mathbf{b}}{|\mathbf{b}|}\right)$

2. a. If $\mathbf{a} \cdot \mathbf{a} = 0$ and $\mathbf{a} \cdot \mathbf{b} = 0$, what can you say about $\mathbf{b}$?

   b. If $\mathbf{a} \cdot \mathbf{a} \neq 0$ and $\mathbf{a} \cdot \mathbf{b} = 0$, what can you say about $\mathbf{b}$?

*In Exercises 3–8, find* $\mathbf{a} \cdot \mathbf{b}$.

3. $\mathbf{a} = \langle 1, 3 \rangle$,   $\mathbf{b} = \langle 2, -1 \rangle$

4. $\mathbf{a} = \langle 1, 3, -2 \rangle$,   $\mathbf{b} = \langle -1, 1, -1 \rangle$

5. $\mathbf{a} = 2\mathbf{i} + 3\mathbf{j}$,   $\mathbf{b} = \mathbf{i} - 2\mathbf{j}$

6. $\mathbf{a} = 2\mathbf{i} - 3\mathbf{j} + \mathbf{k}$,   $\mathbf{b} = -\mathbf{i} + 2\mathbf{j} - 2\mathbf{k}$

7. $\mathbf{a} = \langle 0, 1, -3 \rangle$,   $\mathbf{b} = \langle 10, \pi, -\pi \rangle$

8. $\mathbf{a} = 2\mathbf{i} + 3\mathbf{k}$,   $\mathbf{b} = -\mathbf{i} + 0.2\mathbf{j} - \sqrt{3}\mathbf{k}$

*In Exercises 9–16,* $\mathbf{a} = \langle 1, -3, 2 \rangle$, $\mathbf{b} = \langle -2, 4, 1 \rangle$, *and* $\mathbf{c} = \langle 2, -4, 1 \rangle$. *Find the indicated quantity.*

9. $\mathbf{a} \cdot (\mathbf{b} + \mathbf{c})$

10. $\mathbf{b} \cdot (\mathbf{a} - \mathbf{c})$

11. $(2\mathbf{a} + 3\mathbf{b}) \cdot (3\mathbf{c})$

12. $(\mathbf{a} - \mathbf{b}) \cdot (\mathbf{a} + \mathbf{b})$

13. $(\mathbf{a} \cdot \mathbf{b})\mathbf{c}$

14. $(\mathbf{a} \cdot \mathbf{b})\mathbf{c} - (\mathbf{b} \cdot \mathbf{c})\mathbf{a}$

15. $|\mathbf{a} - \mathbf{b}|^2 + |\mathbf{a} + \mathbf{b}|^2$

16. $\left(\dfrac{\mathbf{a} \cdot \mathbf{b}}{\mathbf{b} \cdot \mathbf{b}}\right)\mathbf{b}$

*In Exercises 17–22, find the angle between the vectors.*

17. $\mathbf{a} = \langle 2, 1 \rangle$,   $\mathbf{b} = \langle 3, 4 \rangle$

18. $\mathbf{a} = \mathbf{i} + 3\mathbf{j}$,   $\mathbf{b} = -\mathbf{i} + 2\mathbf{j}$

19. $\mathbf{a} = \langle 1, 1, 1 \rangle$,   $\mathbf{b} = \langle 2, 3, -6 \rangle$

20. $\mathbf{a} = \mathbf{i} + 2\mathbf{j} + \mathbf{k}$,   $\mathbf{b} = 8\mathbf{i} - 4\mathbf{j} - 3\mathbf{k}$

21. $\mathbf{a} = -2\mathbf{j} + 3\mathbf{k}$,   $\mathbf{b} = \mathbf{i} + \mathbf{j} + 2\mathbf{k}$

22. $\mathbf{a} = \langle -2, 1, 1 \rangle$,   $\mathbf{b} = \langle -3, 2, 1 \rangle$

23. Find the value of $c$ such that the angle between $\mathbf{a} = \langle 1, c \rangle$ and $\mathbf{b} = \langle 1, 2 \rangle$ is $45°$.

24. Find the value(s) of $c$ such that the angle between $\mathbf{a} = \mathbf{i} + c\mathbf{j} + 2\mathbf{k}$ and $\mathbf{b} = -\mathbf{i} + 2\mathbf{j} - \mathbf{k}$ is $60°$.

*In Exercises 25–30, determine whether the vectors are orthogonal, parallel, or neither.*

25. $\mathbf{a} = \langle 1, 2 \rangle$,   $\mathbf{b} = \langle 3, 0 \rangle$

26. $\mathbf{a} = \langle 3, -4 \rangle$,   $\mathbf{b} = \langle 4, 3 \rangle$

27. $\mathbf{a} = \mathbf{i} - 2\mathbf{j} + \mathbf{k}$,   $\mathbf{b} = 3\mathbf{i} + 2\mathbf{j} - 2\mathbf{k}$

28. $\mathbf{a} = 2\mathbf{i} + 4\mathbf{j} - \mathbf{k}$,   $\mathbf{b} = 6\mathbf{i} + 12\mathbf{j} - 3\mathbf{k}$

29. $\mathbf{a} = \langle 2, 3, -1 \rangle$,   $\mathbf{b} = \langle 2, -1, 1 \rangle$

30. $\mathbf{a} = \langle 2, 3, -3 \rangle$,   $\mathbf{b} = \langle \frac{4}{3}, 2, -2 \rangle$

31. Find a value of $c$ such that $\langle c, 2, -1 \rangle$ and $\langle 2, 3, c \rangle$ are orthogonal.

32. Find a unit vector that is orthogonal to both $\mathbf{a} = \mathbf{i} + \mathbf{j} + \mathbf{k}$ and $\mathbf{b} = -2\mathbf{i} + \mathbf{k}$.

*In Exercises 33–36, find the direction cosines and direction angles of the vector.*

33. $\mathbf{a} = \langle 1, 2, 3 \rangle$        34. $\mathbf{a} = 2\mathbf{i} + 2\mathbf{j} - \mathbf{k}$

35. $\mathbf{a} = -\mathbf{i} + 3\mathbf{j} + 5\mathbf{k}$        36. $\mathbf{a} = \langle 3, -4, 5 \rangle$

37. A vector has direction angles $\alpha = \pi/3$ and $\gamma = \pi/4$. Find the direction angle $\beta$.

38. Find a unit vector whose direction angles are all equal.

*In Exercises 39–44, find* (a) $\text{proj}_{\mathbf{a}}\mathbf{b}$ *and* (b) $\text{proj}_{\mathbf{b}}\mathbf{a}$.

39. $\mathbf{a} = \langle 2, 3 \rangle$,   $\mathbf{b} = \langle 1, 4 \rangle$

40. $\mathbf{a} = -\mathbf{i} + 2\mathbf{j}$,   $\mathbf{b} = -3\mathbf{i} + 4\mathbf{j}$

41. $\mathbf{a} = 2\mathbf{i} + \mathbf{j} + 4\mathbf{k}$,   $\mathbf{b} = 3\mathbf{i} + \mathbf{k}$

42. $\mathbf{a} = \langle 1, 2, 0 \rangle$,   $\mathbf{b} = \langle -3, 0, -4 \rangle$

43. $\mathbf{a} = \langle -3, 4, -2 \rangle$,   $\mathbf{b} = \langle 0, 1, 0 \rangle$

44. $\mathbf{a} = \langle -1, 3, -2 \rangle$,   $\mathbf{b} = \langle 0, 3, 1 \rangle$

*In Exercises 45–48, write* $\mathbf{b}$ *as the sum of a vector parallel to* $\mathbf{a}$ *and a vector perpendicular to* $\mathbf{a}$.

45. $\mathbf{a} = \langle 1, 3 \rangle$,   $\mathbf{b} = \langle 2, 4 \rangle$

46. $\mathbf{a} = -\mathbf{i} + 2\mathbf{j}$,   $\mathbf{b} = 2\mathbf{i} + 3\mathbf{j}$

47. $\mathbf{a} = \mathbf{i} + 2\mathbf{j} + 3\mathbf{k}$,   $\mathbf{b} = 2\mathbf{i} - \mathbf{j} + \mathbf{k}$

48. $\mathbf{a} = \mathbf{i} + 2\mathbf{k}$,   $\mathbf{b} = 2\mathbf{i} - \mathbf{j}$

*In Exercises 49 and 50, find the work done by the force* $\mathbf{F}$ *in moving a particle from the point P to the point Q.*

49. $\mathbf{F} = 2\mathbf{i} + 3\mathbf{j} - \mathbf{k}$;   $P(-1, -2, 2)$;   $Q(2, 1, 5)$

50. $\mathbf{F} = \langle 1, 4, 5 \rangle$;   $P(2, 3, 1)$;   $Q(-1, 2, -4)$

51. Find the angle between a diagonal of a cube and one of its edges.

52. Find the angle between a diagonal of a cube and a diagonal of one of its sides.

**53.** Refer to the following figure. Find the angles $\theta$ and $\psi$.

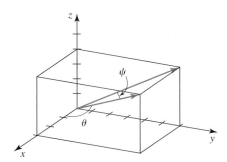

**54. Bond Angle of a Molecule of Methane** The following figure gives the configuration of a molecule of methane, $CH_4$. The four hydrogen atoms are located at the vertices of a regular tetrahedron and the carbon atom is located at the centroid. The *bond angle* for the molecule is the angle between the line segments joining the carbon atom to two of the hydrogen atoms. Show that this angle is approximately $109.5°$
**Hint:** The centroid of the tetrahedron is $\left(\frac{k}{2}, \frac{k}{2}, \frac{k}{2}\right)$.

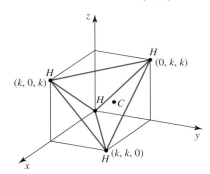

**55.** A passenger in an airport terminal pulls his luggage with a constant force of magnitude 24 lb. If the handle makes an angle of $30°$ with the horizontal surface, find the work done by the passenger in pulling the luggage a distance of 50 ft.

**56.** A child pulls a toy wagon up a straight incline that makes an angle of $15°$ with the horizontal. If the handle makes an

angle of $30°$ with the incline and she exerts a constant force of 15 lb on the handle, find the work done in pulling the wagon a distance of 30 ft along the incline.

**57.** The following figure gives an aerial view of two tugboats attempting to free a cruise ship that ran aground during a storm. Tugboat I exerts a force of magnitude 3000 lb, and Tugboat II exerts a force of magnitude 2400 lb. If the resultant force acts along the positive $x$-axis and the cruise ship is towed a distance of 100 ft in that direction, find the work done by each tugboat. (See Exercise 66 in Section 10.1.)

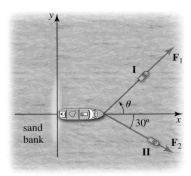

**58.** Prove Property 2 of the dot product:
$\mathbf{a} \cdot (\mathbf{b} + \mathbf{c}) = \mathbf{a} \cdot \mathbf{b} + \mathbf{a} \cdot \mathbf{c}$.

**59.** Prove Property 3 of the dot product:
$(c\mathbf{a}) \cdot \mathbf{b} = c(\mathbf{a} \cdot \mathbf{b}) = \mathbf{a} \cdot (c\mathbf{b})$.

**60.** Prove that if $\mathbf{a}$ is orthogonal to both $\mathbf{b}$ and $\mathbf{c}$, then $\mathbf{a}$ is orthogonal to $p\mathbf{b} + q\mathbf{c}$ for any scalars $p$ and $q$. Give a geometric interpretation of this result.

**61.** Let $\mathbf{a}$ and $\mathbf{b}$ be nonzero vectors. Then the vector $\mathbf{b} - \text{proj}_{\mathbf{a}}\mathbf{b}$ is called the *vector component of* $\mathbf{b}$ *orthogonal to* $\mathbf{a}$.
   **a.** Make a sketch of the vectors $\mathbf{a}$, $\mathbf{b}$, $\text{proj}_{\mathbf{a}}\mathbf{b}$ and $\mathbf{b} - \text{proj}_{\mathbf{a}}\mathbf{b}$.
   **b.** Show that $\mathbf{b} - \text{proj}_{\mathbf{a}}\mathbf{b}$ is orthogonal to $\mathbf{a}$.

**62.** Refer to Exercise 61. Let $\mathbf{a} = \langle 2, 1 \rangle$ and $\mathbf{b} = \langle 4, 5 \rangle$.
   **a.** Find $\mathbf{b} - \text{proj}_{\mathbf{a}}\mathbf{b}$.
   **b.** Sketch the vectors $\mathbf{a}$, $\mathbf{b}$, $\text{proj}_{\mathbf{a}}\mathbf{b}$ and $\mathbf{b} - \text{proj}_{\mathbf{a}}\mathbf{b}$.
   **c.** Show that $\mathbf{b} - \text{proj}_{\mathbf{a}}\mathbf{b}$ is orthogonal to $\mathbf{a}$.

**63. a.** Show that the vector $\mathbf{n} = \langle a, b \rangle$ is orthogonal to the line $ax + by + c = 0$.

**b.** Use the result of part (a) to show that the distance from a point $P_1(x_1, y_1)$ to the line $ax + by + c = 0$ is

$$\frac{|ax_1 + by_1 + c|}{\sqrt{a^2 + b^2}}$$

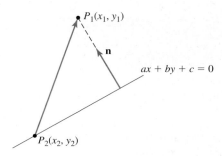

**Hint:** Let $P_2(x_2, y_2)$ be a point on the line, and consider the scalar projection of $\overrightarrow{P_1P_2}$ onto $\mathbf{n}$.

**c.** Use the formula found in part (b) to find the distance from the point $(1, -3)$ to the line $2x + 3y - 6 = 0$.

**64.** Prove that $(\mathbf{a} \cdot \mathbf{b})\mathbf{c} - (\mathbf{b} \cdot \mathbf{c})\mathbf{a}$ is orthogonal to $\mathbf{b}$, where $\mathbf{a}$, $\mathbf{b}$, and $\mathbf{c}$ are any three vectors.

*In Exercises 65–68, determine whether the statement is true or false. If it is true, explain why. If it is false, explain why or give an example that shows it is false.*

**65.** $(\mathbf{a} \cdot \mathbf{b})^2 = |\mathbf{a}|^2 |\mathbf{b}|^2$

**66.** If $\mathbf{a} \neq \mathbf{0}$, and $\mathbf{a} \cdot \mathbf{b} = \mathbf{a} \cdot \mathbf{c}$, then $\mathbf{b} = \mathbf{c}$.

**67.** If $\mathbf{u}$, $\mathbf{v}$, and $\mathbf{w}$ are nonzero vectors, and both $\mathbf{u}$ and $\mathbf{v}$ are orthogonal to $\mathbf{w}$, then $2\mathbf{u} + 3\mathbf{v}$ is orthogonal to $\mathbf{w}$.

**68.** If $\mathbf{a}$ and $\mathbf{b}$ are nonzero vectors, then $\text{proj}_\mathbf{a}\mathbf{b} = \mathbf{0}$ if and only if $\mathbf{a}$ is orthogonal to $\mathbf{b}$.

# 10.4    The Cross Product

## ■ The Cross Product of Two Vectors

In the preceding section we saw how an operation called the dot product combines two vectors to yield a *scalar.* In this section we will look at yet another operation on vectors. This operation, called the *cross product,* combines two vectors to yield a vector.

> **DEFINITION    The Cross Product of Two Vectors in Space**
>
> Let $\mathbf{a} = a_1\mathbf{i} + a_2\mathbf{j} + a_3\mathbf{k}$ and $\mathbf{b} = b_1\mathbf{i} + b_2\mathbf{j} + b_3\mathbf{k}$ be any two vectors in space. Then the **cross product** of $\mathbf{a}$ and $\mathbf{b}$ is the vector
>
> $$\mathbf{a} \times \mathbf{b} = (a_2b_3 - a_3b_2)\mathbf{i} + (a_3b_1 - a_1b_3)\mathbf{j} + (a_1b_2 - a_2b_1)\mathbf{k} \qquad (1)$$

The cross product is used in computing quantities as diverse as the volume of a parallelepiped and the rate of rotation of an incompressible fluid.

Before giving a geometric interpretation of the cross product of two vectors, let's find a simpler way to remember the cross product. Recall that a **determinant of order 2** is defined by

$$\begin{vmatrix} a & b \\ c & d \end{vmatrix} = ad - bc$$

For example,

$$\begin{vmatrix} 2 & 1 \\ 3 & -4 \end{vmatrix} = 2(-4) - 1(3) = -11$$

A **determinant of order 3** is defined in terms of second-order determinants as follows:

$$\begin{vmatrix} a_1 & a_2 & a_3 \\ b_1 & b_2 & b_3 \\ c_1 & c_2 & c_3 \end{vmatrix} = a_1 \begin{vmatrix} b_2 & b_3 \\ c_2 & c_3 \end{vmatrix} - a_2 \begin{vmatrix} b_1 & b_3 \\ c_1 & c_3 \end{vmatrix} + a_3 \begin{vmatrix} b_1 & b_2 \\ c_1 & c_2 \end{vmatrix}$$

In this definition the determinant is said to be *expanded about the first row.* Observe that each term on the right involves the product of a term from the first row and a second-order determinant that is obtained by deleting the row and column containing that term. Also note how the signs of the terms alternate.

As an example,

$$\begin{vmatrix} 1 & 2 & 4 \\ 3 & -1 & 2 \\ 4 & 0 & 3 \end{vmatrix} = 1 \begin{vmatrix} -1 & 2 \\ 0 & 3 \end{vmatrix} - 2 \begin{vmatrix} 3 & 2 \\ 4 & 3 \end{vmatrix} + 4 \begin{vmatrix} 3 & -1 \\ 4 & 0 \end{vmatrix}$$

$$= 1(-3 - 0) - 2(9 - 8) + 4(0 + 4) = 11$$

As a mnemonic device for remembering the expression for the cross product of $\mathbf{a}$ and $\mathbf{b}$, where $\mathbf{a} = a_1\mathbf{i} + a_2\mathbf{j} + a_3\mathbf{k}$ and $\mathbf{b} = b_1\mathbf{i} + b_2\mathbf{j} + b_3\mathbf{k}$, let's expand the following expression as if it were a determinant. (Technically it is not a determinant, because $\mathbf{i}$, $\mathbf{j}$, and $\mathbf{k}$ are not real numbers.) Thus,

$$\begin{vmatrix} \mathbf{i} & \mathbf{j} & \mathbf{k} \\ a_1 & a_2 & a_3 \\ b_1 & b_2 & b_3 \end{vmatrix} = \mathbf{i} \begin{vmatrix} a_2 & a_3 \\ b_2 & b_3 \end{vmatrix} - \mathbf{j} \begin{vmatrix} a_1 & a_3 \\ b_1 & b_3 \end{vmatrix} + \mathbf{k} \begin{vmatrix} a_1 & a_2 \\ b_1 & b_2 \end{vmatrix}$$

$$= (a_2b_3 - a_3b_2)\mathbf{i} + (a_3b_1 - a_1b_3)\mathbf{j} + (a_1b_2 - a_2b_1)\mathbf{k}$$

Comparing the last expression with Equation (1), we are led to the following result.

Let $\mathbf{a} = a_1\mathbf{i} + a_2\mathbf{j} + a_3\mathbf{k}$ and $\mathbf{b} = b_1\mathbf{i} + b_2\mathbf{j} + b_3\mathbf{k}$. Then

$$\mathbf{a} \times \mathbf{b} = \begin{vmatrix} \mathbf{i} & \mathbf{j} & \mathbf{k} \\ a_1 & a_2 & a_3 \\ b_1 & b_2 & b_3 \end{vmatrix} \qquad\qquad (2)$$

 Note the order in which the scalar components of the vectors are written.

**EXAMPLE 1**  Let $\mathbf{a} = 2\mathbf{i} + \mathbf{j} - \mathbf{k}$ and $\mathbf{b} = -3\mathbf{i} - 2\mathbf{j} + \mathbf{k}$. Find $\mathbf{a} \times \mathbf{b}$ and $\mathbf{b} \times \mathbf{a}$.

**Solution**

$$\mathbf{a} \times \mathbf{b} = \begin{vmatrix} \mathbf{i} & \mathbf{j} & \mathbf{k} \\ 2 & 1 & -1 \\ -3 & -2 & 1 \end{vmatrix} = \mathbf{i} \begin{vmatrix} 1 & -1 \\ -2 & 1 \end{vmatrix} - \mathbf{j} \begin{vmatrix} 2 & -1 \\ -3 & 1 \end{vmatrix} + \mathbf{k} \begin{vmatrix} 2 & 1 \\ -3 & -2 \end{vmatrix}$$

$$= -\mathbf{i} + \mathbf{j} - \mathbf{k}$$

and

$$\mathbf{b} \times \mathbf{a} = \begin{vmatrix} \mathbf{i} & \mathbf{j} & \mathbf{k} \\ -3 & -2 & 1 \\ 2 & 1 & -1 \end{vmatrix} = \mathbf{i} \begin{vmatrix} -2 & 1 \\ 1 & -1 \end{vmatrix} - \mathbf{j} \begin{vmatrix} -3 & 1 \\ 2 & -1 \end{vmatrix} + \mathbf{k} \begin{vmatrix} -3 & -2 \\ 2 & 1 \end{vmatrix}$$

$$= \mathbf{i} - \mathbf{j} + \mathbf{k}$$ ∎

Note that $\mathbf{b} \times \mathbf{a} = -\mathbf{a} \times \mathbf{b}$ in Example 1. This is true in general if we recall the property of determinants that states that if two rows of a determinant are interchanged, then the sign of the determinant is changed.

## Geometric Properties of the Cross Product

The cross product $\mathbf{a} \times \mathbf{b}$ being a vector, has both magnitude and direction. The following theorem tells us the direction of the vector $\mathbf{a} \times \mathbf{b}$.

---

### THEOREM 1

Let $\mathbf{a}$ and $\mathbf{b}$ be nonzero vectors in 3-space. Then $\mathbf{a} \times \mathbf{b}$ is orthogonal to both $\mathbf{a}$ and $\mathbf{b}$.

---

**PROOF**  Let $\mathbf{a} = a_1\mathbf{i} + a_2\mathbf{j} + a_3\mathbf{k}$ and $\mathbf{b} = b_1\mathbf{i} + b_2\mathbf{j} + b_3\mathbf{k}$. Then by Equation (1) or by expanding Equation (2), we have

$$\mathbf{a} \times \mathbf{b} = (a_2b_3 - a_3b_2)\mathbf{i} + (a_3b_1 - a_1b_3)\mathbf{j} + (a_1b_2 - a_2b_1)\mathbf{k}$$

Therefore,

$$(\mathbf{a} \times \mathbf{b}) \cdot \mathbf{a} = [(a_2b_3 - a_3b_2)\mathbf{i} + (a_3b_1 - a_1b_3)\mathbf{j} + (a_1b_2 - a_2b_1)\mathbf{k}]$$
$$\cdot (a_1\mathbf{i} + a_2\mathbf{j} + a_3\mathbf{k})$$
$$= (a_2b_3 - a_3b_2)a_1 + (a_3b_1 - a_1b_3)a_2 + (a_1b_2 - a_2b_1)a_3$$
$$= a_1a_2b_3 - a_1b_2a_3 + b_1a_2a_3 - a_1a_2b_3 + a_1b_2a_3 - b_1a_2a_3 = 0$$

which shows that $\mathbf{a} \times \mathbf{b}$ is orthogonal to $\mathbf{a}$. Similarly, by showing that $(\mathbf{a} \times \mathbf{b}) \cdot \mathbf{b} = 0$, we prove that $\mathbf{a} \times \mathbf{b}$ is orthogonal to $\mathbf{b}$. Therefore, $\mathbf{a} \times \mathbf{b}$ is orthogonal to both $\mathbf{a}$ and $\mathbf{b}$. ∎

Let $\mathbf{a}$ and $\mathbf{b}$ be vectors in 3-space, and suppose that $\mathbf{a}$ and $\mathbf{b}$ have the same initial point. Then Theorem 1 tells us that $\mathbf{a} \times \mathbf{b}$ has a direction that is perpendicular to the plane determined by $\mathbf{a}$ and $\mathbf{b}$. (See Figure 1.) The direction of $\mathbf{a} \times \mathbf{b}$ is determined by the right-hand rule: Point the fingers of your open right hand in the direction of $\mathbf{a}$, then curl them towards the vector $\mathbf{b}$. Your thumb will then point in the direction of $\mathbf{a} \times \mathbf{b}$.

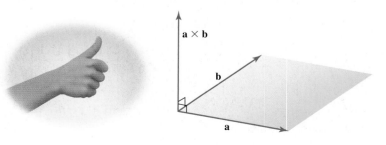

**FIGURE 1**
The vector $\mathbf{a} \times \mathbf{b}$ is orthogonal to both $\mathbf{a}$ and $\mathbf{b}$ with direction determined by the right-hand rule.

The next theorem gives the magnitude of $\mathbf{a} \times \mathbf{b}$.

---

**THEOREM 2**

Let $\mathbf{a}$ and $\mathbf{b}$ be vectors in 3-space. Then

$$|\mathbf{a} \times \mathbf{b}| = |\mathbf{a}||\mathbf{b}| \sin \theta$$

where $\theta$ is the angle between $\mathbf{a}$ and $\mathbf{b}$ and $0 \le \theta \le \pi$.

---

**PROOF** Let $\mathbf{a} = a_1\mathbf{i} + a_2\mathbf{j} + a_3\mathbf{k}$ and $\mathbf{b} = b_1\mathbf{i} + b_2\mathbf{j} + b_3\mathbf{k}$. Then, from Equation (1) we have

$$\mathbf{a} \times \mathbf{b} = (a_2b_3 - a_3b_2)\mathbf{i} + (a_3b_1 - a_1b_3)\mathbf{j} + (a_1b_2 - a_2b_1)\mathbf{k}$$

Next, using Property (4) of dot products we have

$$
\begin{aligned}
|\mathbf{a} \times \mathbf{b}|^2 &= (\mathbf{a} \times \mathbf{b}) \cdot (\mathbf{a} \times \mathbf{b}) \\
&= a_2^2 b_3^2 - 2a_2 a_3 b_2 b_3 + a_3^2 b_2^2 + a_3^2 b_1^2 - 2a_1 a_3 b_1 b_3 + a_1^2 b_3^2 \\
&\quad + a_1^2 b_2^2 - 2a_1 a_2 b_1 b_2 + a_2^2 b_1^2 \\
&= (a_1^2 + a_2^2 + a_3^2)(b_1^2 + b_2^2 + b_3^2) - (a_1 b_1 + a_2 b_2 + a_3 b_3)^2 \\
&= |\mathbf{a}|^2 |\mathbf{b}|^2 - (\mathbf{a} \cdot \mathbf{b})^2 \\
&= |\mathbf{a}|^2 |\mathbf{b}|^2 - |\mathbf{a}|^2 |\mathbf{b}|^2 \cos^2 \theta \qquad \text{Use Equation (2) of Section 10.3.} \\
&= |\mathbf{a}|^2 |\mathbf{b}|^2 (1 - \cos^2 \theta) = |\mathbf{a}|^2 |\mathbf{b}|^2 \sin^2 \theta
\end{aligned}
$$

Finally, taking the square root on both sides and observing that $\sin \theta \ge 0$ for $0 \le \theta \le \pi$, we obtain

$$|\mathbf{a} \times \mathbf{b}| = |\mathbf{a}||\mathbf{b}| \sin \theta \qquad \blacksquare$$

We can combine the results of Theorems 1 and 2 to express the vector $\mathbf{a} \times \mathbf{b}$ in the following form.

---

**An Alternative Definition of $\mathbf{a} \times \mathbf{b}$**

$$\mathbf{a} \times \mathbf{b} = \underbrace{(|\mathbf{a}||\mathbf{b}| \sin \theta)}_{\text{Length of } \mathbf{a} \times \mathbf{b}} \overset{\displaystyle \uparrow}{\mathbf{n}}_{\text{Direction of } \mathbf{a} \times \mathbf{b}}$$

where $\theta$ is the angle between $\mathbf{a}$ and $\mathbf{b}$, and $\mathbf{n}$ is a unit vector orthogonal to both $\mathbf{a}$ and $\mathbf{b}$. (See Figure 2.)

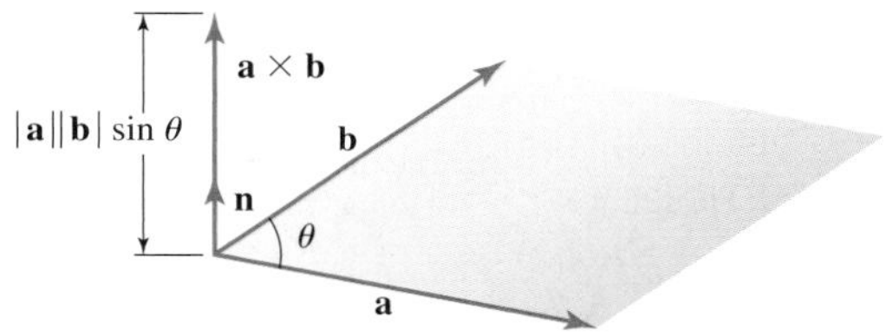

---

**FIGURE 2**
The vector $\mathbf{a} \times \mathbf{b}$ has length $|\mathbf{a}||\mathbf{b}| \sin \theta$ and direction given by $\mathbf{n}$, the unit vector perpendicular to the plane determined by $\mathbf{a}$ and $\mathbf{b}$.

**Note** Since $\mathbf{b} \times \mathbf{a} = -\mathbf{a} \times \mathbf{b}$ we see that the vector $\mathbf{b} \times \mathbf{a}$ has the same length as $\mathbf{a} \times \mathbf{b}$ but points in the direction opposite to that of $\mathbf{a} \times \mathbf{b}$. ■

**EXAMPLE 2** Let $\mathbf{a} = 2\mathbf{i} + 3\mathbf{j}$ and $\mathbf{b} = 2\mathbf{j} + \mathbf{k}$.

**a.** Find a unit vector $\mathbf{n}$ that is orthogonal to both $\mathbf{a}$ and $\mathbf{b}$.
**b.** Express $\mathbf{a} \times \mathbf{b}$ in terms of $|\mathbf{a} \times \mathbf{b}|$ and $\mathbf{n}$.

**Solution**
**a.** A vector that is orthogonal to both $\mathbf{a}$ and $\mathbf{b}$ is

$$\mathbf{a} \times \mathbf{b} = \begin{vmatrix} \mathbf{i} & \mathbf{j} & \mathbf{k} \\ 2 & 3 & 0 \\ 0 & 2 & 1 \end{vmatrix} = 3\mathbf{i} - 2\mathbf{j} + 4\mathbf{k}$$

The length of $\mathbf{a} \times \mathbf{b}$ is

$$|\mathbf{a} \times \mathbf{b}| = \sqrt{3^2 + (-2)^2 + 4^2} = \sqrt{29}$$

Therefore, a unit vector that is orthogonal to both $\mathbf{a}$ and $\mathbf{b}$ is

$$\mathbf{n} = \frac{\mathbf{a} \times \mathbf{b}}{|\mathbf{a} \times \mathbf{b}|} = \frac{3}{\sqrt{29}}\mathbf{i} - \frac{2}{\sqrt{29}}\mathbf{j} + \frac{4}{\sqrt{29}}\mathbf{k}$$

**b.** We can write $\mathbf{a} \times \mathbf{b}$ as

$$\mathbf{a} \times \mathbf{b} = |\mathbf{a} \times \mathbf{b}|\mathbf{n}$$

$$= \sqrt{29}\left( \frac{3}{\sqrt{29}}\mathbf{i} - \frac{2}{\sqrt{29}}\mathbf{j} + \frac{4}{\sqrt{29}}\mathbf{k} \right)$$

The vectors $\mathbf{a}$, $\mathbf{b}$, $\mathbf{a} \times \mathbf{b}$ and $\mathbf{n}$ are shown in Figure 3. ■

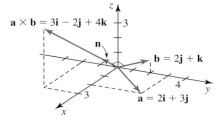

**FIGURE 3**
The vector $\mathbf{a} \times \mathbf{b}$ and the unit vector $\mathbf{n}$ are orthogonal to both $\mathbf{a}$ and $\mathbf{b}$.

**EXAMPLE 3** **An Application in Mechanics** Figure 4a depicts a force $\mathbf{F}$ applied at a point (the terminal point of the position vector $\mathbf{r}$) on a wrench. This force produces a torque that acts along the axis of the bolt and has the effect of driving the bolt forward. To derive an expression for the torque, we recall that the magnitude of the torque $\boldsymbol{\tau}$ is given by

$|\boldsymbol{\tau}| = $ the length of the moment arm $\times$ the magnitude of the vertical component of $\mathbf{F}$

$$= |\mathbf{r}||\mathbf{F}| \sin \theta$$

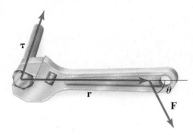
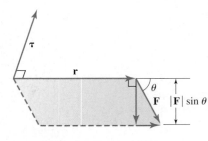

**FIGURE 4**

(a) The force $\mathbf{F}$ applied on the wrench produces a torque that acts along the axis of the bolt.

(b) The torque $\boldsymbol{\tau} = \mathbf{r} \times \mathbf{F}$

(See Figure 4b.) Since the direction of the torque is along the axis of the bolt (which is orthogonal to the plane determined by $\mathbf{r}$ and $\mathbf{F}$), we conclude that $\boldsymbol{\tau} = \mathbf{r} \times \mathbf{F}$. For example, if a force of magnitude 3 lb is applied to the wrench at a point $1\frac{1}{2}$ ft from the bolt at an angle of $60°$, then the magnitude of the torque exerted on the bolt will be

$$\left(\frac{3}{2}\right)(3) \sin 60° \qquad \text{or} \qquad \frac{9\sqrt{3}}{4} \text{ ft-lb} \qquad \blacksquare$$

The following test for determining whether two vectors are parallel is an immediate consequence of Theorem 2.

---

**Test for Parallel Vectors**

Two nonzero vectors **a** and **b** are parallel if and only if $\mathbf{a} \times \mathbf{b} = \mathbf{0}$.

---

**PROOF** Two nonzero vectors **a** and **b** are parallel if and only if $\theta = 0$ or $\pi$. In either case, $|\mathbf{a} \times \mathbf{b}| = |\mathbf{a}||\mathbf{b}| \sin \theta = 0$, so $\mathbf{a} \times \mathbf{b} = \mathbf{0}$. $\qquad \blacksquare$

## ■ Finding the Area of a Triangle

Consider the parallelogram determined by the vectors **a** and **b** shown in Figure 5a. The altitude of the parallelogram is $|\mathbf{b}| \sin \theta$ and the length of its base is $|\mathbf{a}|$, so its area is

$$A = |\mathbf{a}||\mathbf{b}| \sin \theta = |\mathbf{a} \times \mathbf{b}| \qquad \text{By Theorem 2}$$

Thus, the *length of the cross product* $\mathbf{a} \times \mathbf{b}$ *and the area of the parallelogram determined by* **a** *and* **b** have the same numerical value. (See Figure 5b.)

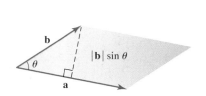

(a) The parallelogram determined by **a** and **b**

(b) The length of $\mathbf{a} \times \mathbf{b}$ is numerically equal to the area of the parallelogram determined by **a** and **b**.

**FIGURE 5**

Therefore, the area of the triangle determined by **a** and **b** is $\frac{1}{2}|\mathbf{a} \times \mathbf{b}|$.

**EXAMPLE 4** Find the area of the triangle with vertices $P(3, -3, 0)$, $Q(1, 2, 2)$, and $R(1, -2, 5)$.

**Solution** The area of $\triangle PQR$ is half the area of the parallelogram determined by the vectors $\overrightarrow{PQ}$ and $\overrightarrow{PR}$. Now $\overrightarrow{PQ} = \langle -2, 5, 2 \rangle$ and $\overrightarrow{PR} = \langle -2, 1, 5 \rangle$, so

$$\overrightarrow{PQ} \times \overrightarrow{PR} = \begin{vmatrix} \mathbf{i} & \mathbf{j} & \mathbf{k} \\ -2 & 5 & 2 \\ -2 & 1 & 5 \end{vmatrix}$$

$$= \mathbf{i} \begin{vmatrix} 5 & 2 \\ 1 & 5 \end{vmatrix} - \mathbf{j} \begin{vmatrix} -2 & 2 \\ -2 & 5 \end{vmatrix} + \mathbf{k} \begin{vmatrix} -2 & 5 \\ -2 & 1 \end{vmatrix}$$

$$= 23\mathbf{i} + 6\mathbf{j} + 8\mathbf{k}$$

Therefore, the area of the parallelogram is

$$|\overrightarrow{PQ} \times \overrightarrow{PR}| = \sqrt{23^2 + 6^2 + 8^2} = \sqrt{629} \approx 25.1$$

so the area of the required triangle is $\frac{1}{2}\sqrt{629}$ or approximately 12.5. (See Figure 6.)

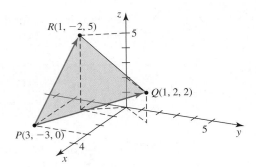

**FIGURE 6**
The triangle with vertices
at $P(3, -3, 0)$, $Q(1, 2, 2)$,
and $R(1, -2, 5)$

## ▤ Properties of the Cross Product

The cross product obeys the following rules.

> **THEOREM 3** **Properties of the Cross Product**
>
> If **a**, **b**, and **c** are vectors and $c$ is a scalar, then
>
> 1. $\mathbf{a} \times \mathbf{b} = -\mathbf{b} \times \mathbf{a}$
> 2. $\mathbf{a} \times (\mathbf{b} + \mathbf{c}) = \mathbf{a} \times \mathbf{b} + \mathbf{a} \times \mathbf{c}$
> 3. $(\mathbf{a} + \mathbf{b}) \times \mathbf{c} = \mathbf{a} \times \mathbf{c} + \mathbf{b} \times \mathbf{c}$
> 4. $c(\mathbf{a} \times \mathbf{b}) = (c\mathbf{a}) \times \mathbf{b} = \mathbf{a} \times (c\mathbf{b})$
> 5. $\mathbf{a} \times \mathbf{0} = \mathbf{0} \times \mathbf{a} = \mathbf{0}$
> 6. $\mathbf{a} \times \mathbf{a} = \mathbf{0}$
> 7. $\mathbf{a} \cdot (\mathbf{b} \times \mathbf{c}) = (\mathbf{a} \times \mathbf{b}) \cdot \mathbf{c}$
> 8. $\mathbf{a} \times (\mathbf{b} \times \mathbf{c}) = (\mathbf{a} \cdot \mathbf{c})\mathbf{b} - (\mathbf{a} \cdot \mathbf{b})\mathbf{c}$

**PROOF** Each of these properties may be proved by applying the definition of the cross product. For example, to prove Property 1, let $\mathbf{a} = a_1\mathbf{i} + a_2\mathbf{j} + a_3\mathbf{k}$ and $b = b_1\mathbf{i} + b_2\mathbf{j} + b_3\mathbf{k}$. Then

$$\mathbf{a} \times \mathbf{b} = \begin{vmatrix} \mathbf{i} & \mathbf{j} & \mathbf{k} \\ a_1 & a_2 & a_3 \\ b_1 & b_2 & b_3 \end{vmatrix} = (a_2b_3 - a_3b_2)\mathbf{i} - (a_1b_3 - a_3b_1)\mathbf{j} + (a_1b_2 - a_2b_1)\mathbf{k}$$

and

$$\mathbf{b} \times \mathbf{a} = \begin{vmatrix} \mathbf{i} & \mathbf{j} & \mathbf{k} \\ b_1 & b_2 & b_3 \\ a_1 & a_2 & a_3 \end{vmatrix} = (b_2a_3 - b_3a_2)\mathbf{i} - (b_1a_3 - b_3a_1)\mathbf{j} + (b_1a_2 - b_2a_1)\mathbf{k}$$

So $\mathbf{a} \times \mathbf{b} = -\mathbf{b} \times \mathbf{a}$. See Example 1 for an illustration of this property. ▤

The proofs of the other properties are left as exercises.

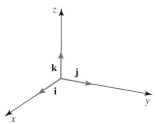

**FIGURE 7**
Using the right-hand rule, we can see the validity of the relationships in Example 5.

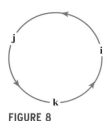

**FIGURE 8**
The cross product of two consecutive vectors in the counterclockwise direction is the next vector and has a positive direction; in the clockwise direction it is the next vector and has a negative direction. For example, $\mathbf{i} \times \mathbf{j} = \mathbf{k}$ and $\mathbf{j} \times \mathbf{i} = -\mathbf{k}$.

**EXAMPLE 5**    By direct computation or using Property 6 of Theorem 3, we can show that

$$\mathbf{i} \times \mathbf{i} = \mathbf{0}, \qquad \mathbf{j} \times \mathbf{j} = \mathbf{0}, \qquad \text{and} \qquad \mathbf{k} \times \mathbf{k} = \mathbf{0} \tag{3}$$

Next, by direct computation, we can verify that

$$\mathbf{i} \times \mathbf{j} = \mathbf{k}, \qquad \mathbf{j} \times \mathbf{k} = \mathbf{i}, \qquad \text{and} \qquad \mathbf{k} \times \mathbf{i} = \mathbf{j} \tag{4}$$

so by Property 1 of Theorem 3 we also have

$$\mathbf{j} \times \mathbf{i} = -\mathbf{k}, \qquad \mathbf{k} \times \mathbf{j} = -\mathbf{i}, \qquad \text{and} \qquad \mathbf{i} \times \mathbf{k} = -\mathbf{j} \tag{5}$$

These results are also evident if you interpret each of the cross products (3)–(5) geometrically while looking at Figure 7.    ∎

You can use a simple mnemonic device to help remember the cross products (3)–(5). Consider the circle shown in Figure 8. The cross product of two consecutive vectors in the counterclockwise direction is the next vector, and its direction is positive. Likewise, the cross product of two consecutive vectors in the clockwise direction is the next vector but with a negative direction.

## The Scalar Triple Product

Suppose that **a**, **b**, and **c** are three vectors in three-dimensional space. The dot product of **a** and **b** × **c**, **a** · (**b** × **c**), is called the **scalar triple product.** If we write $\mathbf{a} = a_1\mathbf{i} + a_2\mathbf{j} + a_3\mathbf{k}$, $\mathbf{b} = b_1\mathbf{i} + b_2\mathbf{j} + b_3\mathbf{k}$, and $\mathbf{c} = c_1\mathbf{i} + c_2\mathbf{j} + c_3\mathbf{k}$, then by direct computation,

$$\mathbf{a} \cdot (\mathbf{b} \times \mathbf{c}) = \begin{vmatrix} a_1 & a_2 & a_3 \\ b_1 & b_2 & b_3 \\ c_1 & c_2 & c_3 \end{vmatrix} \tag{6}$$

(We leave it as an exercise for you to verify this computation.)

The geometric significance of the scalar triple product can be seen by examining the parallelepiped determined by the vectors **a**, **b**, and **c**. (See Figure 9.) The base of the parallelepiped is a parallelogram with adjacent sides determined by **b** and **c** with area $|\mathbf{b} \times \mathbf{c}|$. If $\theta$ is the angle between **a** and **b** × **c**, then the height of the parallelepiped is given by $h = |\mathbf{a}||\cos\theta|$. Therefore, the volume of the parallelepiped is

$$V = |\mathbf{b} \times \mathbf{c}||\mathbf{a}||\cos\theta| \qquad \text{area of base · height}$$

$$= |\mathbf{a} \cdot (\mathbf{b} \times \mathbf{c})| \qquad \text{By Equation (2) of Section 10.3}$$

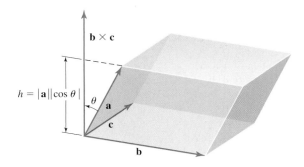

**FIGURE 9**
The volume $V$ of the parallelepiped is equal to $|\mathbf{a} \cdot (\mathbf{b} \times \mathbf{c})|$.

We have established the following result.

---

**THEOREM 4** **Geometric Interpretation of the Scalar Triple Product**

The volume $V$ of the parallelepiped determined by the vectors $\mathbf{a}$, $\mathbf{b}$, and $\mathbf{c}$ is given by

$$V = |\mathbf{a} \cdot (\mathbf{b} \times \mathbf{c})|$$

---

**EXAMPLE 6** Find the volume of the parallelepiped determined by the vectors $\mathbf{a} = \mathbf{i} + 2\mathbf{j} + 3\mathbf{k}$, $\mathbf{b} = \mathbf{i} - \mathbf{j} + \mathbf{k}$, and $\mathbf{c} = 3\mathbf{i} + \mathbf{j} - 2\mathbf{k}$.

**Solution** By Theorem 4 the volume of the parallelepiped is $V = |\mathbf{a} \cdot (\mathbf{b} \times \mathbf{c})|$. But by Equation (6),

$$\mathbf{a} \cdot (\mathbf{b} \times \mathbf{c}) = \begin{vmatrix} 1 & 2 & 3 \\ 1 & -1 & 1 \\ 3 & 1 & -2 \end{vmatrix} = 1 \begin{vmatrix} -1 & 1 \\ 1 & -2 \end{vmatrix} - 2 \begin{vmatrix} 1 & 1 \\ 3 & -2 \end{vmatrix} + 3 \begin{vmatrix} 1 & -1 \\ 3 & 1 \end{vmatrix}$$

$$= 1(1) - 2(-5) + 3(4) = 23$$

Therefore, the required volume is $|23|$, or 23. ■

Because the volume of a parallelepiped is zero if and only if the vectors $\mathbf{a}$, $\mathbf{b}$, and $\mathbf{c}$ forming the adjacent sides of the parallelepiped are coplanar, we have the following result.

---

**Test for Coplanar Vectors**

The vectors $\mathbf{a} = a_1\mathbf{i} + a_2\mathbf{j} + a_3\mathbf{k}$, $\mathbf{b} = b_1\mathbf{i} + b_2\mathbf{j} + b_3\mathbf{k}$, and $\mathbf{c} = c_1\mathbf{i} + c_2\mathbf{j} + c_3\mathbf{k}$ are coplanar if and only if

$$\mathbf{a} \cdot (\mathbf{b} \times \mathbf{c}) = \begin{vmatrix} a_1 & a_2 & a_3 \\ b_1 & b_2 & b_3 \\ c_1 & c_2 & c_3 \end{vmatrix} = 0$$

---

## 10.4 CONCEPT QUESTIONS

**1. a.** Give the definition of $\mathbf{a} \times \mathbf{b}$.
   **b.** What is the length or magnitude of $\mathbf{a} \times \mathbf{b}$ in terms of the angle between $\mathbf{a}$ and $\mathbf{b}$? Give a geometric interpretation of $|\mathbf{a} \times \mathbf{b}|$.
   **c.** What is the direction of $\mathbf{a} \times \mathbf{b}$?
   **d.** Write $\mathbf{a} \times \mathbf{b}$ in terms of $\mathbf{a}$, $\mathbf{b}$, and $\theta$ (the angle between $\mathbf{a}$ and $\mathbf{b}$).

**2. a.** How would you use the cross product of two nonzero vectors $\mathbf{a}$ and $\mathbf{b}$ to determine whether $\mathbf{a}$ and $\mathbf{b}$ are parallel?
   **b.** What can you say about the three nonzero vectors $\mathbf{a}$, $\mathbf{b}$, and $\mathbf{c}$ if $(\mathbf{c} \times \mathbf{b}) \cdot \mathbf{a} = 0$?

## 10.4 EXERCISES

1. State whether the expression makes sense. Explain.
   **a.** $(\mathbf{a} \times \mathbf{b}) \cdot \mathbf{c}$       **b.** $\mathbf{a} \times (\mathbf{b} \cdot \mathbf{c})$
   **c.** $(\mathbf{a} + \mathbf{b}) \cdot (\mathbf{c} \times \mathbf{d})$       **d.** $(\mathbf{a} \times \mathbf{b}) \cdot (\mathbf{c} \times \mathbf{d})$
   **e.** $\mathbf{a} \times [(\mathbf{b} \cdot \mathbf{c})\mathbf{d}]$       **f.** $(\mathbf{a} \times \mathbf{b}) \times (\mathbf{c} \times \mathbf{d})$

2. If $\mathbf{a} \cdot \mathbf{b} = 0$ and $\mathbf{a} \times \mathbf{b} = \mathbf{0}$ what can you say about $\mathbf{a}$ and $\mathbf{b}$?

*In Exercises 3–10, find $\mathbf{a} \times \mathbf{b}$.*

3. $\mathbf{a} = \mathbf{i} + \mathbf{j}, \quad \mathbf{b} = 2\mathbf{j} + 3\mathbf{k}$

4. $\mathbf{a} = \langle 0, 1, 2 \rangle, \quad \mathbf{b} = \langle 2, 1, 3 \rangle$

5. $\mathbf{a} = \langle 1, -2, 1 \rangle, \quad \mathbf{b} = \langle 3, 1, -2 \rangle$

6. $\mathbf{a} = 2\mathbf{i} - 3\mathbf{j} + 4\mathbf{k}, \quad \mathbf{b} = -\mathbf{i} - 2\mathbf{j} + 3\mathbf{k}$

7. $\mathbf{a} = 2\mathbf{i} + 3\mathbf{k}, \quad \mathbf{b} = -3\mathbf{i} + 2\mathbf{j} - \mathbf{k}$

8. $\mathbf{a} = \langle 0, 1, 0 \rangle, \quad \mathbf{b} = \langle -3, 0, 2 \rangle$

9. $\mathbf{a} = 2\mathbf{i} + \mathbf{j} - 3\mathbf{k}, \quad \mathbf{b} = \frac{2}{3}\mathbf{i} + \frac{1}{3}\mathbf{j} - \mathbf{k}$

10. $\mathbf{a} = \langle 1, 1, 2 \rangle, \quad \mathbf{b} = \langle \frac{1}{2}, 2, -\frac{1}{2} \rangle$

*In Exercises 11 and 12, find $\mathbf{a} \times \mathbf{b}$ and $\mathbf{b} \times \mathbf{a}$.*

11. $\mathbf{a} = \mathbf{i} + 2\mathbf{j} + 3\mathbf{k}, \quad \mathbf{b} = 2\mathbf{i} - \mathbf{j} - \mathbf{k}$

12. $\mathbf{a} = \langle 1, 1, 2 \rangle, \quad \mathbf{b} = \langle -1, 3, -1 \rangle$

*In Exercises 13 and 14, find two vectors that are orthogonal to both $\mathbf{a}$ and $\mathbf{b}$.*

13. $\mathbf{a} = 2\mathbf{i} - 3\mathbf{j} + 4\mathbf{k}, \quad \mathbf{b} = -\mathbf{i} + \mathbf{j} - 2\mathbf{k}$

14. $\mathbf{a} = \langle 1, -2, 1 \rangle, \quad \mathbf{b} = \langle 2, 3, -4 \rangle$

*In Exercises 15 and 16, find two unit vectors that are orthogonal to both $\mathbf{a}$ and $\mathbf{b}$.*

15. $\mathbf{a} = -3\mathbf{i} + \mathbf{j} - 2\mathbf{k}, \quad \mathbf{b} = \mathbf{i} + \mathbf{j} + \mathbf{k}$

16. $\mathbf{a} = \langle -1, 1, -1 \rangle, \quad \mathbf{b} = \langle 0, 3, 4 \rangle$

*In Exercises 17–20, find the area of the triangle with the given vertices.*

17. $P(1, 0, 0), Q(0, 1, 0), R(0, 0, 1)$

18. $P(1, 1, 1), Q(1, 2, 1), R(2, 2, 3)$

19. $P(1, -1, 2), Q(2, 3, 1), R(-2, 3, 4)$

20. $P(0, 0, 0), Q(1, 3, 2), R(-1, -2, 3)$

*In Exercises 21–26, let $\mathbf{a} = \mathbf{i} - \mathbf{j} + \mathbf{k}$, $\mathbf{b} = 2\mathbf{i} + 3\mathbf{j} - \mathbf{k}$, and $\mathbf{c} = -\mathbf{i} + \mathbf{j} + 2\mathbf{k}$. Find the indicated quantity.*

21. $(2\mathbf{a}) \times \mathbf{b}$

22. $(\mathbf{a} + \mathbf{b}) \times \mathbf{c}$

23. $(\mathbf{a} \times \mathbf{b}) \times \mathbf{c}$

24. $\mathbf{a} \cdot (\mathbf{b} \times \mathbf{c})$

25. $(\mathbf{a} \times \mathbf{b}) \cdot \mathbf{c}$

26. $(3\mathbf{a}) \times (\mathbf{a} - 2\mathbf{b} + 3\mathbf{c})$

*In Exercises 27 and 28, find the volume of the parallelepiped determined by the vectors, $\mathbf{a}$, $\mathbf{b}$, and $\mathbf{c}$.*

27. $\mathbf{a} = \mathbf{i} + \mathbf{j}, \mathbf{b} = \mathbf{j} - 2\mathbf{k}, \mathbf{c} = \mathbf{i} + 2\mathbf{j} + 3\mathbf{k}$

28. $\mathbf{a} = \langle 1, 3, 2 \rangle, \mathbf{b} = \langle 2, -1, 3 \rangle, \mathbf{c} = \langle 1, -1, -2 \rangle$

*In Exercises 29 and 30, find the volume of the parallelepiped with adjacent edges PQ, PR, and PS.*

29. $P(0, 0, 0), Q(3, -2, 1), R(1, 2, 2), S(1, 1, 4)$

30. $P(1, 1, 1), Q(2, 1, 3), R(-1, 0, 3), S(4, -1, 2)$

31. Find the height of a parallelepiped determined by $\mathbf{a} = \mathbf{i} + 2\mathbf{j} + \mathbf{k}, \mathbf{b} = 2\mathbf{i} + \mathbf{j} - \mathbf{k}$, and $\mathbf{c} = \mathbf{i} + \mathbf{j} + 3\mathbf{k}$ if its base is determined by $\mathbf{a}$ and $\mathbf{b}$.

32. Find $c$ such that $\mathbf{a} = 2\mathbf{i} + 3\mathbf{j} + \mathbf{k}, \mathbf{b} = \mathbf{i} + 2\mathbf{j} + 3\mathbf{k}$, and $\mathbf{c} = \mathbf{i} - 3\mathbf{j} + c\mathbf{k}$ are coplanar.

33. Determine whether the vectors $\mathbf{a} = \mathbf{i} + 2\mathbf{j} + 4\mathbf{k}$, $\mathbf{b} = -2\mathbf{i} + 3\mathbf{j} - \mathbf{k}$, and $\mathbf{c} = \mathbf{j} + \mathbf{k}$ are coplanar.

34. Find the value of $c$ such that the vectors $\mathbf{u} = \mathbf{i} - 2\mathbf{j} + 3\mathbf{k}$, $\mathbf{v} = \mathbf{i} + 2\mathbf{j} - \mathbf{k}$, and $\mathbf{w} = 2\mathbf{i} + 3\mathbf{j} + c\mathbf{k}$ are coplanar.

35. Determine whether the points $P(1, 0, 1), Q(2, 3, 1)$, $R(-1, 2, -3)$, and $S(\frac{2}{3}, -1, 1)$ are coplanar.

36. Find the area of the parallelogram shown in the figure in terms of $a$, $b$, $c$, and $d$.

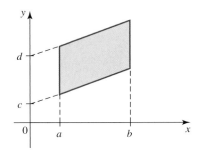

**37.** A force of magnitude 50 lb is applied at the end of an 18-in.-long leveraged-impact lug wrench in the direction as shown in the figure. Find the magnitude of the torque about $P$.

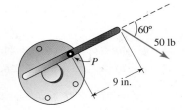

**38.** A 15-lb force is applied to a stapler at the point shown. Find the magnitude of the torque about $P$.

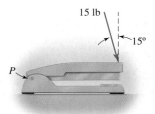
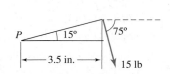

**39. Force on a Proton Moving Through a Magnetic Field** The force $\mathbf{F}$ acting on a charge $q$ moving with velocity $\mathbf{v}$ in a magnetic field $\mathbf{B}$ is given by $\mathbf{F} = q\mathbf{v} \times \mathbf{B}$, where $q$ is measured in coulombs, $|\mathbf{B}|$ in tesla, and $|\mathbf{v}|$ in meters per second. Suppose that a proton beam moves through a region of space where there is a uniform magnetic field $\mathbf{B} = 2\mathbf{k}$. The protons have velocity

$$\mathbf{v} = \left(\frac{3}{2} \times 10^5\right)\mathbf{i} + \left(\frac{3\sqrt{3}}{2} \times 10^5\right)\mathbf{k}$$

Find the force on a proton if its charge is $q = 1.6 \times 10^{-19}$ C.

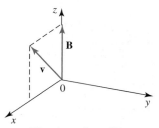

Directions of $\mathbf{v}$ and $\mathbf{B}$

**40.** Find $(\mathbf{a} \times \mathbf{b}) \times \mathbf{c}$ and $\mathbf{a} \times (\mathbf{b} \times \mathbf{c})$ given that $\mathbf{a} = 2\mathbf{i} + \mathbf{j}$, $\mathbf{b} = 3\mathbf{i} + \mathbf{j} - \mathbf{k}$, and $\mathbf{c} = \mathbf{i} + \mathbf{k}$. Does the associative law hold for vector products?

**41.** Let $\mathbf{a} = a_1\mathbf{i} + a_2\mathbf{j} + a_3\mathbf{k}$, $\mathbf{b} = b_1\mathbf{i} + b_2\mathbf{j} + b_3\mathbf{k}$ and $\mathbf{c} = c_1\mathbf{i} + c_2\mathbf{j} + c_3\mathbf{k}$. Prove that

$$\mathbf{a} \cdot (\mathbf{b} \times \mathbf{c}) = (\mathbf{a} \times \mathbf{b}) \cdot \mathbf{c} = \begin{vmatrix} a_1 & a_2 & a_3 \\ b_1 & b_2 & b_3 \\ c_1 & c_2 & c_3 \end{vmatrix}$$

**42.** Prove that $(\mathbf{a} + \mathbf{b}) \times (\mathbf{a} - \mathbf{b}) = 2(\mathbf{b} \times \mathbf{a})$.

**43.** Prove that $\mathbf{a} \times (\mathbf{b} \times \mathbf{c}) + \mathbf{b} \times (\mathbf{c} \times \mathbf{a}) + \mathbf{c} \times (\mathbf{a} \times \mathbf{b}) = \mathbf{0}$.

**44.** Prove that

$$(\mathbf{a} \times \mathbf{b}) \cdot (\mathbf{c} \times \mathbf{d}) = \begin{vmatrix} \mathbf{a} \cdot \mathbf{c} & \mathbf{b} \cdot \mathbf{c} \\ \mathbf{a} \cdot \mathbf{d} & \mathbf{b} \cdot \mathbf{d} \end{vmatrix}$$

**45.** Prove Lagrange's identity:

$$|\mathbf{a} \times \mathbf{b}|^2 = |\mathbf{a}|^2|\mathbf{b}|^2 - (\mathbf{a} \cdot \mathbf{b})^2$$

**46.** Refer to the following figure.

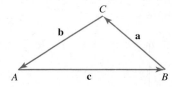

    **a.** Show that $\mathbf{a} + \mathbf{b} + \mathbf{c} = \mathbf{0}$.

    **b.** Show that $\mathbf{a} \times \mathbf{b} = \mathbf{b} \times \mathbf{c} = \mathbf{c} \times \mathbf{a}$, and hence deduce the law of sines for plane triangles:

$$\frac{\sin A}{a} = \frac{\sin B}{b} = \frac{\sin C}{c} \quad a = |\mathbf{a}|, \quad b = |\mathbf{b}|, \quad c = |\mathbf{c}|$$

*In Exercises 47–52, prove the given property of the cross-product.*

**47.** $\mathbf{a} \times (\mathbf{b} + \mathbf{c}) = \mathbf{a} \times \mathbf{b} + \mathbf{a} \times \mathbf{c}$

**48.** $c(\mathbf{a} \times \mathbf{b}) = (c\mathbf{a}) \times \mathbf{b}$

**49.** $\mathbf{a} \times \mathbf{0} = \mathbf{0} \times \mathbf{a} = \mathbf{0}$

**50.** $\mathbf{a} \times \mathbf{a} = \mathbf{0}$

**51.** $(\mathbf{a} + \mathbf{b}) \times \mathbf{c} = \mathbf{a} \times \mathbf{c} + \mathbf{b} \times \mathbf{c}$

**52.** $\mathbf{a} \times (\mathbf{b} \times \mathbf{c}) = (\mathbf{a} \cdot \mathbf{c})\mathbf{b} - (\mathbf{a} \cdot \mathbf{b})\mathbf{c}$

**53. Angular Velocity** Consider a rigid body rotating about a fixed axis with a constant angular speed $\omega$. The *angular velocity* is represented by a vector $\boldsymbol{\omega}$ of magnitude $\omega$ lying along the axis of rotation as shown in the figure. If we place the origin 0 on the axis of rotation and let $\mathbf{R}$ denote the position vector of a particle in the body, then the velocity $\mathbf{v}$ of the particle is given by

$$\mathbf{v} = \boldsymbol{\omega} \times \mathbf{R}$$

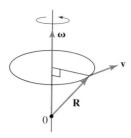

Suppose that the axis of rotation is parallel to the vector $2\mathbf{i} + 2\mathbf{j} + \mathbf{k}$. What is the speed of a particle at the instant it passes through the point $(3, 5, 2)$?

54. Find a unit vector in the plane that contains the vectors $\mathbf{a} = \mathbf{i} + 3\mathbf{j} + 2\mathbf{k}$ and $\mathbf{b} = -\mathbf{i} + 2\mathbf{j} + 4\mathbf{k}$, and is perpendicular to the vector $\mathbf{c} = 3\mathbf{i} - \mathbf{j} + 2\mathbf{k}$.

55. Find $a$ and $b$ if $\mathbf{a} = \langle -1, a, 3 \rangle$ and $\mathbf{b} = \langle 2, 3, b \rangle$ are parallel.

56. Find $s$ and $t$ such that $(\mathbf{a} \times \mathbf{b}) \times \mathbf{a} = s\mathbf{a} + t\mathbf{b}$, where $\mathbf{a} = \langle 2, 1, 3 \rangle$ and $\mathbf{b} = \langle 1, 3, 4 \rangle$.

*In Exercises 57–62, determine whether the statement is true or false. If it is true, explain why. If it is false, explain why or give an example that shows it is false.*

57. $\mathbf{a} \times \mathbf{b} + \mathbf{b} \times \mathbf{a} = \mathbf{0}$

58. $[(\mathbf{a} \times \mathbf{b}) \times \mathbf{c}] \times [(\mathbf{a} \times \mathbf{b}) \times \mathbf{c}] = \mathbf{0}$

59. $\mathbf{a} \cdot (\mathbf{a} \times \mathbf{b}) = 0$

60. $\mathbf{a} \times \mathbf{b} = \mathbf{a} \times \mathbf{c}$ if and only if $\mathbf{b} = \mathbf{c}$

61. $(\mathbf{a} - \mathbf{b}) \times (\mathbf{a} + \mathbf{b}) = 2\mathbf{a} \times \mathbf{b}$

62. If $\mathbf{a} \neq \mathbf{0}$, $\mathbf{a} \cdot \mathbf{b} = \mathbf{a} \cdot \mathbf{c}$, and $\mathbf{a} \times \mathbf{b} = \mathbf{a} \times \mathbf{c}$ then $\mathbf{b} = \mathbf{c}$.

## 10.5 Lines and Planes in Space

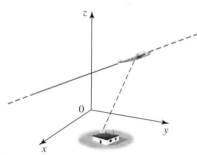

**FIGURE 1**
The path of the airplane is a straight line.

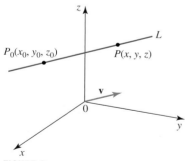

**FIGURE 2**
The line $L$ passes through $P_0$ and is parallel to the vector $\mathbf{v}$.

### ■ Equations of Lines in Space

Figure 1 depicts an airplane flying in a straight line above a ground radar station. How fast is the distance between the airplane and the radar station changing at any time? How close to the radar station does the airplane get? To answer questions such as these, we need to be able to describe the path of the airplane. More specifically, we want to be able to represent a line in space algebraically.

In this section we will see how lines as well as planes in space can be described in algebraic terms. We begin by considering a line in space. Such a line is uniquely determined by specifying its direction and a point through which it passes. The direction may be specified by a vector that has the same direction as the line. So suppose that the line $L$ passes through the point $P_0(x_0, y_0, z_0)$ and has the same direction as the vector $\mathbf{v} = \langle a, b, c \rangle$. (See Figure 2.)

Let $P(x, y, z)$ be *any* point on $L$. Then the vector $\overrightarrow{P_0P}$ is parallel to $\mathbf{v}$. But two vectors are parallel if and only if one is a scalar multiple of the other. Therefore, there exists some number $t$, called a *parameter,* such that

$$\overrightarrow{P_0P} = t\mathbf{v}$$

or, since $\overrightarrow{P_0P} = \langle x - x_0, y - y_0, z - z_0 \rangle$, we have

$$\langle x - x_0, y - y_0, z - z_0 \rangle = t\langle a, b, c \rangle = \langle ta, tb, tc \rangle$$

Equating the corresponding components of the two vectors then yields

$$x - x_0 = ta, \qquad y - y_0 = tb, \qquad \text{and} \qquad z - z_0 = tc$$

Solving these equations for $x$, $y$, and $z$, respectively, gives the following standard *parametric equations* of the line $L$.

---

**DEFINITION** Parametric Equations of a Line

The parametric equations of the line passing through the point $P_0(x_0, y_0, z_0)$ and parallel to the vector $\mathbf{v} = \langle a, b, c \rangle$ are

$$x = x_0 + at, \qquad y = y_0 + bt, \qquad \text{and} \qquad z = z_0 + ct \qquad \textbf{(1)}$$

Each value of the parameter $t$ corresponds to a point $P(x, y, z)$ on $L$. As $t$ takes on all values in the parameter interval $(-\infty, \infty)$, the line $L$ is traced out. (See Figure 3.)

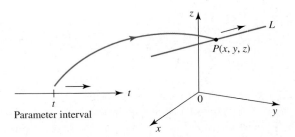

**FIGURE 3**
As $t$ runs through all values in the parameter interval $(-\infty, \infty)$, $L$ is traced out.

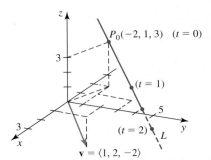

**FIGURE 4**
The line $L$ and some points on $L$ corresponding to selected values of $t$. Note the orientation of the line.

**EXAMPLE 1** Find parametric equations for the line passing through the point $P_0(-2, 1, 3)$ and parallel to the vector $\mathbf{v} = \langle 1, 2, -2 \rangle$.

**Solution** We use Equation (1) with $x_0 = -2, y_0 = 1, z_0 = 3, a = 1, b = 2$, and $c = -2$, obtaining

$$x = -2 + t, \qquad y = 1 + 2t, \qquad \text{and} \qquad z = 3 - 2t$$

The line $L$ in question is sketched in Figure 4. ∎

Suppose that the vector $\mathbf{v} = \langle a, b, c \rangle$ defines the direction of a line $L$. Then the numbers $a$, $b$, and $c$ are called the **direction numbers** of $L$. Observe that if a line $L$ is described by a set of parametric equations (1), then the direction numbers of $L$ are precisely the coefficients of $t$ in each of the parametric equations.

There is another way of describing a line in space. We start with the parametric equations of the line $L$,

$$x = x_0 + at, \qquad y = y_0 + bt, \qquad \text{and} \qquad z = z_0 + ct$$

If the direction numbers $a$, $b$, and $c$ are all nonzero, then we can solve each of these equations for $t$. Thus,

$$t = \frac{x - x_0}{a}, \qquad t = \frac{y - y_0}{b}, \qquad \text{and} \qquad t = \frac{z - z_0}{c}$$

which gives the following *symmetric equations* of $L$.

---

**DEFINITION** **Symmetric Equations of a Line**

The **symmetric equations** of the line $L$ passing through the point $P_0(x_0, y_0, z_0)$ and parallel to the vector $\mathbf{v} = \langle a, b, c \rangle$ are

$$\frac{x - x_0}{a} = \frac{y - y_0}{b} = \frac{z - z_0}{c} \tag{2}$$

---

**Note** Suppose $a = 0$ and both $b$ and $c$ are not equal to zero, then the parametric equations of the line take the form

$$x = x_0, \qquad y = y_0 + bt, \qquad \text{and} \qquad z = z_0 + ct$$

and the line lies in the plane $x = x_0$ (parallel to the $yz$-plane). Solving the second and third equations for $t$ leads to

$$x = x_0, \qquad \frac{y - y_0}{b} = \frac{z - z_0}{c}$$

which are the symmetric equations of the line. We leave it to you to consider and interpret the other cases. ◼

### EXAMPLE 2

**a.** Find parametric equations and symmetric equations for the line $L$ passing through the points $P(-3, 3, -2)$ and $Q(2, -1, 4)$.
**b.** At what point does $L$ intersect the $xy$-plane?

**Solution**

**a.** The direction of $L$ is the same as that of the vector $\overrightarrow{PQ} = \langle 5, -4, 6 \rangle$. Since $L$ passes through $P(-3, 3, -2)$, we can use Equation (1) with $a = 5$, $b = -4$, $c = 6$, $x_0 = -3$, $y_0 = 3$, and $z_0 = -2$, to obtain the parametric equations

$$x = -3 + 5t, \qquad y = 3 - 4t, \qquad \text{and} \qquad z = -2 + 6t$$

Next, using Equation (2), we obtain the following symmetric equations for $L$:

$$\frac{x + 3}{5} = \frac{y - 3}{-4} = \frac{z + 2}{6}$$

**b.** At the point where the line intersects the $xy$-plane, we have $z = 0$. So setting $z = 0$ in the third parametric equation, we obtain $t = \frac{1}{3}$. Substituting this value of $t$ into the other parametric equations gives the required point as $\left(-\frac{4}{3}, \frac{5}{3}, 0\right)$. (See Figure 5.)

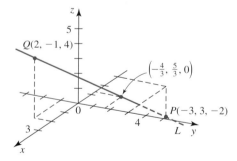

**FIGURE 5**
The line $L$ intersects the $xy$-plane at the point $\left(-\frac{4}{3}, \frac{5}{3}, 0\right)$.

◼

Suppose that $L_1$ and $L_2$ are lines having the same directions as the vectors $\mathbf{v}_1$ and $\mathbf{v}_2$, respectively. Then $L_1$ is **parallel** to $L_2$ if $\mathbf{v}_1$ is parallel to $\mathbf{v}_2$.

### EXAMPLE 3 Let $L_1$ be the line with parametric equations

$$x = 1 + 2t, \qquad y = 2 - 3t, \qquad \text{and} \qquad z = 2 + t$$

and let $L_2$ be the line with parametric equations

$$x = 3 - 4t, \qquad y = 1 + 4t, \qquad \text{and} \qquad z = -3 + 4t$$

**a.** Show that the lines $L_1$ and $L_2$ are not parallel to each other.
**b.** Do the lines $L_1$ and $L_2$ intersect? If so, find their point of intersection.

**Solution**

**a.** By inspection the direction numbers of $L_1$ are $2, -3$, and $1$. Therefore, $L_1$ has the same direction as the vector $\mathbf{v}_1 = \langle 2, -3, 1 \rangle$. Similarly, we see that $L_2$ has direction given by the vector $\mathbf{v}_2 = \langle -4, 4, 4 \rangle = -4\langle 1, -1, -1 \rangle$. Since $\mathbf{v}_1$ is not a scalar multiple of $\mathbf{v}_2$, the vectors are not parallel, so $L_1$ and $L_2$ are not parallel as well.

**b.** Suppose that $L_1$ and $L_2$ intersect at the point $P_0(x_0, y_0, z_0)$. Then there must exist parameter values $t_1$ and $t_2$ such that

$$x_0 = 1 + 2t_1, \qquad y_0 = 2 - 3t_1, \qquad \text{and} \qquad z_0 = 2 + t_1$$

$t_1$ corresponds to $P_0$ on $L_1$.

and

$$x_0 = 3 - 4t_2, \qquad y_0 = 1 + 4t_2, \qquad \text{and} \qquad z_0 = -3 + 4t_2$$

$t_2$ corresponds to $P_0$ on $L_2$.

This leads to the system of three linear equations

$$1 + 2t_1 = 3 - 4t_2$$
$$2 - 3t_1 = 1 + 4t_2$$
$$2 + t_1 = -3 + 4t_2$$

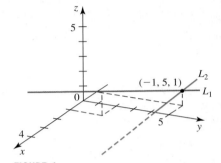

**FIGURE 6**
The lines $L_1$ and $L_2$ intersect at the point $(-1, 5, 1)$.

that must be satisfied by $t_1$ and $t_2$. Adding the first two equations gives $3 - t_1 = 4$, or $t_1 = -1$. Substituting this value of $t_1$ into either the first or the second equation then gives $t_2 = 1$. Finally, substituting these values of $t_1$ and $t_2$ into the third equation gives $2 - 1 = -3 + 4(1) = 1$, which shows that the third equation is also satisfied by these values. We conclude that $L_1$ and $L_2$ do indeed intersect at a point.

To find the point of intersection, substitute $t_1 = -1$ into the parametric equations defining $L_1$, or, equivalently substitute $t_2 = 1$ into the parametric equations defining $L_2$. In both cases we find that $x_0 = -1$, $y_0 = 5$, and $z_0 = 1$, so the point of intersection is $(-1, 5, 1)$. (See Figure 6.) ∎

**FIGURE 7**
The lines $L_1$ and $L_2$ are skew lines.

Two lines in space are said to be **skew** if they do not intersect and are not parallel. (See Figure 7.)

---

**EXAMPLE 4** **Flight Path of Two Airplanes** As two planes fly by each other, their flight paths are given by the straight lines

$$L_1: \quad x = 1 - t \qquad y = -2 - 3t \qquad z = 4 + t$$

and

$$L_2: \quad x = 2 - 2t \qquad y = -4 + 3t \qquad z = 1 + 4t$$

Show that the lines are skew and, therefore, that there is no danger of the planes colliding.

**Solution** The directions of $L_1$ and $L_2$ are given by the directions of the vectors $\mathbf{v}_1 = \langle -1, -3, 1 \rangle$ and $\mathbf{v}_2 = \langle -2, 3, 4 \rangle$, respectively. Since one vector is not a scalar multiple of the other, the lines $L_1$ and $L_2$ are not parallel. Next, suppose that the two lines do intersect at some point $P_0(x_0, y_0, z_0)$. Then

$$x_0 = 1 - t_1 \qquad y_0 = -2 - 3t_1 \qquad z_0 = 4 + t_1$$

and

$$x_0 = 2 - 2t_2 \qquad y_0 = -4 + 3t_2 \qquad z_0 = 1 + 4t_2$$

for some $t_1$ and $t_2$. Equating the values of $x_0$, $y_0$, and $z_0$ then gives

$$1 - \ t_1 = \ \ 2 - 2t_2$$
$$-2 - 3t_1 = -4 + 3t_2$$
$$4 + \ t_1 = \ \ 1 + 4t_2$$

Solving the first two equations for $t_1$ and $t_2$ yields $t_1 = \frac{1}{9}$ and $t_2 = \frac{5}{9}$. Substituting these values of $t_1$ and $t_2$ into the third equation gives $\frac{37}{9} = \frac{29}{9}$, a contradiction. This shows that there are no values of $t_1$ and $t_2$ that satisfy the three equations simultaneously. Thus, $L_1$ and $L_2$ do not intersect. We have shown that $L_1$ and $L_2$ are skew lines, so there is no possibility of the planes colliding.   ◼

## ■ Equations of Planes in Space

A plane in space is uniquely determined by specifying a point $P_0(x_0, y_0, z_0)$ lying in the plane and a vector $\mathbf{n} = \langle a, b, c \rangle$ that is **normal** (perpendicular) to it. (See Figure 8.) To find an equation of the plane, let $P(x, y, z)$ be *any* point in the plane. Then the vector $\overrightarrow{P_0P}$ must be orthogonal to $\mathbf{n}$. But two vectors are orthogonal if and only if their dot product is equal to zero. Therefore, we must have

$$\mathbf{n} \cdot \overrightarrow{P_0P} = 0 \qquad (3)$$

Since $\overrightarrow{P_0P} = \langle x - x_0, y - y_0, z - z_0 \rangle$, we can also write Equation (3) as

$$\langle a, b, c \rangle \cdot \langle x - x_0, y - y_0, z - z_0 \rangle = 0$$

or

$$a(x - x_0) + b(y - y_0) + c(z - z_0) = 0$$

**FIGURE 8**
The vector $\overrightarrow{P_0P}$ lying in the plane must be orthogonal to the normal $\mathbf{n}$ so that $\mathbf{n} \cdot \overrightarrow{P_0P} = 0$.

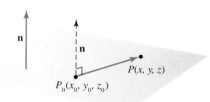

---

**DEFINITION   The Standard Form of the Equation of a Plane**

The **standard form of the equation of a plane** containing the point $P_0(x_0, y_0, z_0)$ and having the normal vector $\mathbf{n} = \langle a, b, c \rangle$ is

$$a(x - x_0) + b(y - y_0) + c(z - z_0) = 0 \qquad (4)$$

---

**EXAMPLE 5**   Find an equation of the plane containing the point $P_0(3, -3, 2)$ and having a normal vector $\mathbf{n} = \langle 4, 2, 3 \rangle$. Find the $x$-, $y$-, and $z$-intercepts, and make a sketch of the plane.

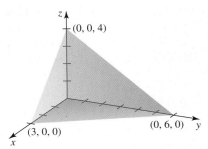

**FIGURE 9**
The portion of the plane
$4x + 2y + 3z = 12$ in the
first octant

**Solution** We use Equation (4) with $a = 4$, $b = 2$, $c = 3$, $x_0 = 3$, $y_0 = -3$, and $z_0 = 2$, obtaining

$$4(x - 3) + 2(y + 3) + 3(z - 2) = 0$$

or

$$4x + 2y + 3z = 12$$

To find the $x$-intercept, we note that any point on the $x$-axis must have both its $y$- and $z$-coordinates equal to zero. Setting $y = z = 0$ in the equation of the plane, we find that $x = 3$. Therefore, 3 is the $x$-intercept. Similarly, we find that the $y$- and $z$-intercepts are 6 and 4, respectively. By connecting the points $(3, 0, 0)$, $(0, 6, 0)$, and $(0, 0, 4)$ with straight line segments, we obtain a sketch of that portion of the plane lying in the first octant. (See Figure 9.) ∎

**EXAMPLE 6** Find an equation of the plane containing the points $P(3, -1, 1)$, $Q(1, 4, 2)$, and $R(0, 1, 4)$.

**Solution** To use Equation (4), we need to find a vector normal to the plane in question. Observe that both of the vectors $\overrightarrow{PQ} = \langle -2, 5, 1 \rangle$ and $\overrightarrow{PR} = \langle -3, 2, 3 \rangle$ lie in the plane, so the vector $\overrightarrow{PQ} \times \overrightarrow{PR}$ is normal to the plane. Denoting this vector by **n**, we have

$$\mathbf{n} = \overrightarrow{PQ} \times \overrightarrow{PR} = \begin{vmatrix} \mathbf{i} & \mathbf{j} & \mathbf{k} \\ -2 & 5 & 1 \\ -3 & 2 & 3 \end{vmatrix} = 13\mathbf{i} + 3\mathbf{j} + 11\mathbf{k}$$

Finally, using the point $P(3, -1, 1)$ in the plane (any of the other two points will also do) and the normal vector **n** just found, with $a = 13$, $b = 3$, $c = 11$, $x_0 = 3$, $y_0 = -1$, and $z_0 = 1$, Equation (4) gives

$$13(x - 3) + 3(y + 1) + 11(z - 1) = 0$$

or, upon simplification,

$$13x + 3y + 11z = 47$$

The plane is sketched in Figure 10.

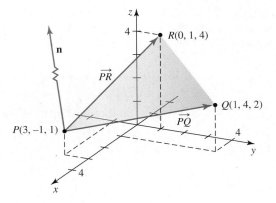

**FIGURE 10**
The normal to the plane
is $\mathbf{n} = \overrightarrow{PQ} \times \overrightarrow{PR}$.

By expanding Equation (4) and regrouping the terms, as we did in Examples 5 and 6, we obtain the **general form** of the equation of a plane in space,

$$ax + by + cz = d \tag{5}$$

where $d = ax_0 + by_0 + cz_0$. Conversely, given $ax + by + cz = d$ with $a$, $b$, and $c$ not all equal to zero, we can choose numbers $x_0$, $y_0$, and $z_0$ such that $ax_0 + by_0 + cz_0 = d$. For example, if $c \neq 0$, we can pick $x_0$ and $y_0$ arbitrarily and solve the equation $ax_0 + by_0 + cz_0 = d$ for $z_0$, obtaining $z_0 = (d - ax_0 - by_0)/c$. Therefore, with these choices of $x_0$, $y_0$, and $z_0$, Equation (5) takes the form

$$ax + by + cz = ax_0 + by_0 + cz_0$$

or

$$a(x - x_0) + b(y - y_0) + c(z - z_0) = 0$$

which we recognize to be an equation of the plane containing the point $(x_0, y_0, z_0)$ and having a normal vector $\mathbf{n} = \langle a, b, c \rangle$. (See Equation (4).) An equation of the form $ax + by + cz = d$, with $a$, $b$, and $c$ not all zero, is called a **linear equation in the three variables** $x$, $y$, and $z$.

---

**THEOREM 1**

Every plane in space can be represented by a linear equation $ax + by + cz = d$, where $a$, $b$, and $c$ are not all equal to zero. Conversely, every linear equation $ax + by + cz = d$ represents a plane in space having a normal vector $\langle a, b, c \rangle$.

---

**Note** Notice that the coefficients of $x$, $y$, and $z$ are precisely the components of the normal vector $\mathbf{n} = \langle a, b, c \rangle$. Thus, we can write a normal vector to a plane by simply inspecting its equation. ∎

### ▪ Parallel and Orthogonal Planes

Two planes with normal vectors $\mathbf{m}$ and $\mathbf{n}$ are **parallel** to each other if $\mathbf{m}$ and $\mathbf{n}$ are parallel; the planes are orthogonal if $\mathbf{m}$ and $\mathbf{n}$ are orthogonal. (See Figure 11.)

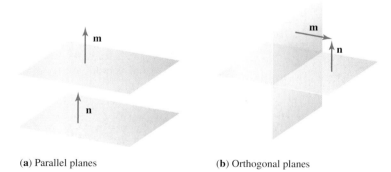

**FIGURE 11**
Two planes are parallel if $\mathbf{m}$ and $\mathbf{n}$ are parallel and orthogonal if $\mathbf{m}$ and $\mathbf{n}$ are orthogonal.

(a) Parallel planes          (b) Orthogonal planes

**EXAMPLE 7** Find an equation of the plane containing $P(2, -1, 3)$ and parallel to the plane defined by $2x - 3y + 4z = 6$.

**Solution** By Theorem 1 the normal vector of the given plane is $\mathbf{n} = \langle 2, -3, 4 \rangle$. Since the required plane is parallel to the given plane, it also has $\mathbf{n}$ as a normal vector. Therefore, using Equation (4), we obtain

$$2(x - 2) - 3(y + 1) + 4(z - 3) = 0$$

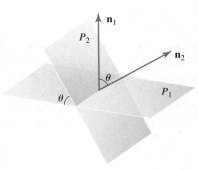

**FIGURE 12**
The angle between two planes is the angle between their normal vectors.

or

$$2x - 3y + 4z = 19$$

as an equation of the plane.   ∎

## The Angle Between Two Planes

Two distinct planes in space are either parallel to each other or intersect in a straight line. If they do intersect, then the **angle between the two planes** is defined to be the acute angle between their normal vectors (see Figure 12).

**EXAMPLE 8**   Find the angle between the two planes defined by $3x - y + 2z = 1$ and $2x + 3y - z = 4$.

**Solution**   The normal vectors of these planes are

$$\mathbf{n}_1 = \langle 3, -1, 2 \rangle \quad \text{and} \quad \mathbf{n}_2 = \langle 2, 3, -1 \rangle$$

Therefore, the angle $\theta$ between the planes is given by

$$\cos \theta = \frac{\mathbf{n}_1 \cdot \mathbf{n}_2}{|\mathbf{n}_1||\mathbf{n}_2|} \qquad \text{Use Equation (2) of Section 10.3.}$$

$$= \frac{\langle 3, -1, 2 \rangle \cdot \langle 2, 3, -1 \rangle}{\sqrt{9 + 1 + 4} \sqrt{4 + 9 + 1}} = \frac{3(2) + (-1)(3) + 2(-1)}{\sqrt{14} \sqrt{14}} = \frac{1}{14}$$

or

$$\theta = \cos^{-1} \left( \frac{1}{14} \right) \approx 86°$$   ∎

**EXAMPLE 9**   Find parametric equations for the line of intersection of the planes defined by $3x - y + 2z = 1$ and $2x + 3y - z = 4$.

**Solution**   We need the direction of the line of intersection $L$ as well as a point on $L$. To find the direction of $L$, we observe that a vector $\mathbf{v}$ is parallel to $L$ if and only if it is orthogonal to the normal vectors of both planes. (See Figure 13 for the general case.) In other words, $\mathbf{v} = \mathbf{n}_1 \times \mathbf{n}_2$, where $\mathbf{n}_1$ and $\mathbf{n}_2$ are the normal vectors of the two planes. Here, the normal vectors are $\mathbf{n}_1 = \langle 3, -1, 2 \rangle$ and $\mathbf{n}_2 = \langle 2, 3, -1 \rangle$, so the vector $\mathbf{v}$ is given by

$$\mathbf{v} = \mathbf{n}_1 \times \mathbf{n}_2 = \begin{vmatrix} \mathbf{i} & \mathbf{j} & \mathbf{k} \\ 3 & -1 & 2 \\ 2 & 3 & -1 \end{vmatrix} = -5\mathbf{i} + 7\mathbf{j} + 11\mathbf{k}$$

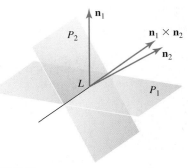

**FIGURE 13**
The vector $\mathbf{n}_1 \times \mathbf{n}_2$ has the same direction as $L$, the line of intersection of the two planes.

To find a point on $L$, let's set $z = 0$ in both of the equations defining the planes. (This will give us the point where $L$ intersects the $xy$-plane.) We obtain

$$3x - y = 1 \quad \text{and} \quad 2x + 3y = 4$$

Solving these equations simultaneously gives $x = \frac{7}{11}$ and $y = \frac{10}{11}$. Finally, by using Equation (1), the required parametric equations are

$$x = \frac{7}{11} - 5t, \qquad y = \frac{10}{11} + 7t, \qquad \text{and} \qquad z = 11t$$   ∎

## ■ The Distance Between a Point and a Plane

To find a formula for the distance between a point and a plane, suppose that $P_1(x_1, y_1, z_1)$ is a point *not* lying in the plane $ax + by + cz = d$. Let $P_0(x_0, y_0, z_0)$ be any point lying in the plane. Then, as you can see in Figure 14, the distance $D$ between $P_1$ and the plane is given by the length of the vector projection of $\overrightarrow{P_0P_1}$ onto the normal vector $\mathbf{n} = \langle a, b, c \rangle$ of the plane. Equivalently, $D$ is the absolute value of the scalar component of $\overrightarrow{P_0P_1}$ along $\mathbf{n}$. Using Equation (6) of Section 10.3 (and taking the absolute value), we obtain

$$D = \frac{|\overrightarrow{P_0P_1} \cdot \mathbf{n}|}{|\mathbf{n}|}$$

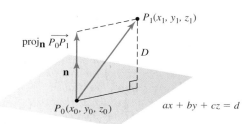

**FIGURE 14**
The distance from $P_1$ to the plane is the length of $\text{proj}_{\mathbf{n}} \overrightarrow{P_0P_1}$.

But $\overrightarrow{P_0P_1} = \langle x_1 - x_0, y_1 - y_0, z_1 - z_0 \rangle$, so we have

$$D = \frac{|\langle x_1 - x_0, y_1 - y_0, z_1 - z_0 \rangle \cdot \langle a, b, c \rangle|}{\sqrt{a^2 + b^2 + c^2}}$$

$$= \frac{|a(x_1 - x_0) + b(y_1 - y_0) + c(z_1 - z_0)|}{\sqrt{a^2 + b^2 + c^2}}$$

$$= \frac{|ax_1 + by_1 + cz_1 - (ax_0 + by_0 + cz_0)|}{\sqrt{a^2 + b^2 + c^2}}$$

Since $P_0(x_0, y_0, z_0)$ lies in the plane, its coordinates must satisfy the equation of the plane, that is, $ax_0 + by_0 + cz_0 = d$; so we can write $D$ in the following form:

$$D = \frac{|ax_1 + by_1 + cz_1 - d|}{\sqrt{a^2 + b^2 + c^2}} \tag{6}$$

**EXAMPLE 10**   Find the distance between the point $(-2, 1, 3)$ and the plane $2x - 3y + z = 1$.

**Solution**   Using Equation (6) with $x_1 = -2$, $y_1 = 1$, $z_1 = 3$, $a = 2$, $b = -3$, $c = 1$ and $d = 1$, we obtain

$$D = \frac{|2(-2) - 3(1) + 1(3) - 1|}{\sqrt{2^2 + (-3)^2 + 1^2}} = \frac{5}{\sqrt{14}} = \frac{5\sqrt{14}}{14} \qquad ■$$

## 10.5 CONCEPT QUESTIONS

**1. a.** Write the parametric equations of the line passing through the point $P_0(x_0, y_0, z_0)$ and having the same direction as the vector $\mathbf{v} = \langle a, b, c \rangle$.

**b.** What are the symmetric equations of the line of part (a)?

**c.** Write the parametric and symmetric equations of the line that passes through the point $P(x_0, y_0, z_0)$ and has the same direction as the vector $\mathbf{v} = \langle a, b, 0 \rangle$.

**2.** If you are given two lines $L_1$ and $L_2$ in space, how do you determine whether they are (a) parallel to each other, (b) perpendicular to each other, or (c) skew?

**3. a.** Write the standard form of an equation of the plane containing the point $P_0(x_0, y_0, z_0)$ and having the normal vector $\mathbf{n} = \langle a, b, c \rangle$.

**b.** What is the general form of the equation of a plane in space?

**4. a.** What is the angle between two planes in space? How do you find it?

**b.** Write the formula giving the distance between a point and a plane in space.

## 10.5 EXERCISES

*In Exercises 1 and 2 describe in your own words the strategy you might adopt to solve the problem. For example, to find the parametric equations of the line passing through two distinct points, you might use this strategy: Let P and Q denote the two points, and write the vector $\overrightarrow{PQ}$ that gives the direction numbers of the line. Then using this information and either P or Q, write the desired equations using Equation (1).*

**1.** Find parametric equations of a line, given that the line

   **a.** Passes through a given point and is parallel to a given line.

   **b.** Passes through a given point and is perpendicular to two distinct lines passing through that point.

   **c.** Passes through a given point lying in a given plane and is perpendicular to the plane.

   **d.** Is the intersection of two given nonparallel planes.

**2.** Find an equation of a plane, given that the plane

   **a.** Contains three distinct points.

   **b.** Contains a given line and a point not lying on the line.

   **c.** Contains a given point and is parallel to a given plane.

   **d.** Contains two intersecting nonparallel lines.

   **e.** Contains two parallel and distinct lines.

*In Exercises 3–6, find parametric and symmetric equations for the line passing through the point P that is parallel to the vector $\mathbf{v}$.*

**3.** $P(1, 3, 2)$;  $\mathbf{v} = \langle 2, 4, 5 \rangle$

**4.** $P(1, -4, 2)$;  $\mathbf{v} = 2\mathbf{i} - 3\mathbf{j} + \mathbf{k}$

**5.** $P(3, 0, -2)$;  $\mathbf{v} = 2\mathbf{i} - \mathbf{j} + 3\mathbf{k}$

**6.** $P(0, 1, 3)$;  $\mathbf{v} = \langle 2, -3, 4 \rangle$

*In Exercises 7–10, find parametric and symmetric equations for the line passing through the given points.*

**7.** $(2, 1, 4)$ and $(1, 3, 7)$

**8.** $(3, -2, 1)$ and $(3, 4, 4)$

**9.** $\left(-1, -2, -\frac{1}{2}\right)$ and $\left(1, \frac{3}{2}, -3\right)$

**10.** $\left(\frac{1}{2}, -\frac{1}{3}, \frac{1}{4}\right)$ and $\left(\frac{1}{2}, -\frac{1}{3}, \frac{3}{4}\right)$

**11.** Find parametric and symmetric equations of the line passing through the point $(1, 2, -1)$ and parallel to the line with parametric equations $x = -1 + t$, $y = 2 + 2t$, and $z = -2 - 3t$. At what points does the line intersect the coordinate planes?

**12.** Find parametric equations of the line passing through the point $(-1, 3, -2)$ and parallel to the line with symmetric equation

$$\frac{x - 2}{3} = \frac{y + 1}{-3} = z + 2$$

At what point does the line intersect the $yz$-plane?

**13.** Determine whether the point $(-3, 6, 1)$ lies on the line $L$ passing through the point $(-1, 4, 3)$ and parallel to the vector $\mathbf{v} = -\mathbf{i} + \mathbf{j} - \mathbf{k}$.

**14.** Find parametric equations of the line that is parallel to the line with equation

$$\frac{x - 1}{4} = \frac{y + 4}{5} = \frac{z + 1}{2}$$

and contains the point of intersection of the lines

$$L_1: \quad x = 4 + t \qquad y = 5 + t \qquad z = -1 + 2t$$

$$L_2: \quad x = 6 + 2t \qquad y = 11 + 4t \qquad z = -3 + t$$

*In Exercises 15–18, determine whether the lines $L_1$ and $L_2$ are parallel, are skew, or intersect each other. If they intersect, find the point of intersection.*

**15.** $L_1: x = -1 + 3t, y = -2 + 3t, z = 3 + t$
    $L_2: x = 1 + 4t, y = -2 + 6t, z = 4 + t$

**16.** $L_1: x = 1 - 2t, y = -1 - 3t, z = -2 + t$
    $L_2: x = -3 + t, y = -2 + 2t, z = 3 - t$

**17.** $L_1$: $\dfrac{x-2}{4} = \dfrac{z-1}{-1}$, $y = 3$

$\quad L_2$: $\dfrac{x-2}{2} = \dfrac{y-3}{2} = z - 1$

**18.** $L_1$: $\dfrac{x-4}{-1} = \dfrac{y+1}{6} = z - 4$

$\quad L_2$: $\dfrac{x-1}{2} = \dfrac{y-1}{-4} = \dfrac{z-5}{-1}$

*In Exercises 19–22, determine whether the lines $L_1$ and $L_2$ intersect. If they do intersect, find the angle between them.*

**19.** $L_1$: $x = 1 - t, y = 3 - 2t, z = t$
$\quad L_2$: $x = 2 + 3t, y = 3 + 2t, z = 1 + t$

**20.** $L_1$: $x = 2 + 3t, y = -2 - 2t, z = 3 + t$
$\quad L_2$: $x = -3 - t, y = 1 + t, z = -4 + 5t$

**21.** $L_1$: $\dfrac{x-1}{-3} = \dfrac{y+2}{2} = \dfrac{z+1}{4}$

$\quad L_2$: $\dfrac{x+2}{2} = \dfrac{y-4}{4} = z - 3$

**22.** $L_1$: $\dfrac{x+4}{3} = \dfrac{y+1}{-2} = \dfrac{z-3}{3}$

$\quad L_2$: $\dfrac{x+32}{6} = \dfrac{y-8}{-2}, z = 4$

*In Exercises 23–26, find an equation of the plane that has the normal vector $\mathbf{n}$ and passes through the given point.*

**23.** $(2, 1, 5)$; $\mathbf{n} = \langle 1, 2, 4 \rangle$

**24.** $(-1, 3, -2)$; $\mathbf{n} = \mathbf{i} - 2\mathbf{j} + \mathbf{k}$

**25.** $(1, 3, 0)$; $\mathbf{n} = 2\mathbf{i} - 4\mathbf{k}$

**26.** $(3, 0, 3)$; $\mathbf{n} = \langle 0, 0, 1 \rangle$

*In Exercises 27–30, find an equation of the plane that passes through the given point and is parallel to the given plane.*

**27.** $(3, 6, -2)$; $2x + 3y - z = 4$

**28.** $(2, -1, 0)$; $x - 2y - 3z = 1$

**29.** $(-1, -2, -3)$; $x - 3z = 1$

**30.** $(0, 2, -1)$; $\dfrac{1}{2}x - \dfrac{1}{3}y + \dfrac{1}{4}z = 2$

*In Exercises 31 and 32, find an equation of the plane that passes through the three given points.*

**31.** $(1, 0, -2), (1, 3, 2), (2, 3, 0)$

**32.** $(2, 3, -1), (1, -2, 3), (-1, 2, 4)$

*In Exercises 33–36, find an equation of the plane that passes through the given point and contains the given line.*

**33.** $(1, 3, 2)$; $x = 1 + t, y = -1 - 2t, z = 3 + 2t$

**34.** $(-1, 2, 3)$; $x = -1 + 2t, y = -2 + 3t, z = 3 - t$

**35.** $(3, -4, 5)$; $\dfrac{x-2}{2} = \dfrac{y+1}{-3} = \dfrac{z+3}{5}$

**36.** $(1, 3, 0)$; $\dfrac{x+4}{-3} = \dfrac{y-3}{5}, z = 2$

*In Exercises 37 and 38, find an equation of the plane passing through the given points and perpendicular to the given plane.*

**37.** $(2, 1, 1)$ and $(-1, 3, 2)$; $2x + 3y - 4z = 3$

**38.** $(-1, 3, 0)$ and $(2, -1, 4)$; $3x - 4y + 5z = 1$

*In Exercises 39–42, determine whether the planes are parallel, orthogonal, or neither. If they are neither parallel nor orthogonal, find the angle between them.*

**39.** $x + 2y + z = 1$, $2x - 3y + 4z = 3$

**40.** $2x - y + 4z = 7$, $6x - 3y + 12z = -1$

**41.** $3x - y + 2z = 2$, $2x + 3y + z = 4$

**42.** $4x - 4y + 2z = 7$, $3x + 2y - 2z = 5$

*In Exercises 43 and 44, find the angle between the plane and the line.*

**43.** $x + y + 2z = 6$; $x = 1 + t$, $y = 2 + t$, $z = -1 + t$

**44.** $2x - 3y + 4z = 12$; $\dfrac{x-1}{2} = \dfrac{y+1}{3} = \dfrac{z}{2}$

*In Exercises 45 and 46, find parametric equations for the line of intersection of the planes.*

**45.** $2x - 3y + 4z = 3$, $x + 4y - 2z = 7$

**46.** $3x + y - 2z = 4$, $2x - y - 3z = 6$

**47.** Find parametric equations of the line that passes through the point $(2, 3, -1)$ and is perpendicular to the plane $2x + 4y - 3z = 4$.

**48.** Find an equation of the plane that passes through the point $(3, -2, 4)$ and is perpendicular to the line

$$\frac{x+1}{-2} = \frac{y-2}{3} = \frac{z+4}{4}$$

**49.** Find an equation of the plane that contains the lines given by

$$x = -1 + 2t \qquad y = 2 - 3t \qquad z = 1 + t$$
$$x = 2 - t \qquad y = 1 - 2t \qquad z = 5 - 3t$$

**50.** Find an equation of the plane that passes through the point $(3, 4, 1)$ and contains the line of intersection of the planes $x - y + 2z = 1$ and $2x + 3y - z = 2$.

**51.** Find an equation of the plane that is orthogonal to the plane $3x + 2y - 4z = 7$ and contains the line of intersection of the planes $2x - 3y + z = 3$ and $x + 2y - 3z = 5$.

**52.** Find an equation of the plane that is parallel to the line of intersection of the planes $x - y + 2z = 3$ and $2x + 3y - z = 4$ and contains the points $(2, 3, 5)$ and $(3, 4, 1)$.

*In Exercises 53 and 54, find the point of intersection, if any, of the plane and the line.*

**53.** $2x + 3y - z = 9$; $\quad x = 2 + 3t, \quad y = -1 + t,$
$\quad z = 3 - 2t$

**54.** $x - y + 2z = 13$; $\quad \dfrac{x + 1}{3} = \dfrac{y - 2}{4} = z + 1$

*In Exercises 55 and 56, find the distance between the point and the plane.*

**55.** $(3, 1, 2)$; $\quad 2x - 3y + 4z = 7$

**56.** $(-1, 3, -2)$; $\quad 3x + y + z = 2$

*In Exercises 57 and 58, show that the two planes are parallel and find the distance between them.*

**57.** $x + 2y - 4z = 1$, $\quad x + 2y - 4z = 7$

**58.** $2x - 3y + z = 2$, $\quad 4x - 6y + 2z = 8$

**59.** Let $P$ be a point that is not on the line $L$. Show that the distance $D$ between the point $P$ and the line $L$ is

$$D = \frac{|\overrightarrow{QP} \times \mathbf{u}|}{|\mathbf{u}|}$$

where $\mathbf{u}$ is a vector having the same direction as $L$, and $Q$ is any point on $L$.

*In Exercises 60 and 61, use the result of Exercise 59 to find the distance between the point and the line.*

**60.** $(3, 4, 6)$; $\quad x = 2 + t, y = 1 - 2t, z = 3 - t$

**61.** $(1, -2, 3)$; $\quad \dfrac{x + 2}{3} = \dfrac{y - 1}{1} = \dfrac{z + 3}{2}$

**62.** Find the distance between the point $(1, 4, 2)$ and the line passing through the points $(-1, 3, -1)$ and $(1, 2, 3)$.

**63.** Show that the distance $D$ between the parallel planes $ax + by + cz = d_1$ and $ax + by + cz = d_2$ is

$$D = \frac{|d_1 - d_2|}{\sqrt{a^2 + b^2 + c^2}}$$

*In Exercises 64 and 65, find the distance between the skew lines. (Hint: Use the result of Exercise 63.)*

**64.** $x = 1 + 5t, y = -1 + 2t, z = 2 - 3t$ $\quad$ and
$\quad x = 1 + t, y = -1 - 2t, z = 1 + t$

**65.** $\dfrac{x - 1}{-2} = \dfrac{y - 4}{-6} = \dfrac{z - 3}{-2}$ $\quad$ and $\quad x - 2 = \dfrac{y + 2}{-5} = \dfrac{z - 1}{-3}$

**66.** Find the distance between the line given by

$$x = 1 + 3t, \qquad y = 2 - 6t, \qquad \text{and} \qquad z = -1 + 2t$$

and the plane that passes through the point $(2, 3, 1)$ and is perpendicular to the line containing the points $(2, 1, 4)$ and $(4, 4, 10)$.

*In Exercises 67–72, determine whether the statement is true or false. If it is true, explain why. If it is false, explain why or give an example that shows it is false.*

**67.** If the lines $L_1$ and $L_2$ are both perpendicular to the line $L_3$, then $L_1$ must be parallel to $L_2$.

**68.** If the lines $L_1$ and $L_2$ do not intersect, then they must be parallel to each other.

**69.** If the planes $P_1$ and $P_2$ are both perpendicular to the plane $P_3$, then $P_1$ must be perpendicular to $P_2$.

**70.** If the planes $P_1$ and $P_2$ are both parallel to a line $L$, then $P_1$ and $P_2$ must be parallel to each other.

**71.** There always exists a unique plane passing through a given point and a given line.

**72.** Given any two lines that are not coincident, there is a plane containing the two lines.

---

# 10.6 Surfaces in Space

In Section 10.5 we saw that the graph of a linear equation in three variables is a plane in space. In general, the graph of an equation in three variables, $F(x, y, z) = 0$, is a surface in 3-space. In this section we will study surfaces called *cylinders* and *quadric surfaces*.

The paraboloidal surface shown in Figure 1a is an example of a quadric surface. A uniformly rotating liquid acquires this shape as a result of the interaction between the force of gravity and centrifugal force. As was explained in Section 9.1, this surface is ideal for radio and optical telescope mirrors. (See Figure 1b.) Mathematically, a paraboloid is obtained by revolving a parabola about its axis of symmetry. (See Figure 1c.)

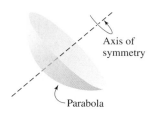

Axis of
symmetry

Parabola

(**a**) Surface of rotating liquid    (**b**) Surface of a radio telescope    (**c**) Surface obtained by revolving a
parabola about its axis

**FIGURE 1**

## ■ Traces

Just as we can use the $x$- and $y$-intercepts of a plane curve to help us sketch the graph of a plane curve, so can we use the traces of a surface in the coordinate planes to help us sketch the surface itself. The **trace** of a surface $S$ in a plane is the intersection of the surface and the plane. In particular, the traces of $S$ in the $xy$-plane, the $yz$-plane, and the $xz$-plane are called the **$xy$-trace,** the **$yz$-trace** and the **$xz$-trace,** respectively.

To find the $xy$-traces, we set $z = 0$ and sketch the graph of the resulting equation in the $xy$-plane. The other traces are obtained in a similar manner. Of course, if the surface does not intersect the plane, there is no trace in that plane.

**EXAMPLE 1**    Consider the plane with equation $4x + 2y + 3z = 12$. (See Example 5 in Section 10.5.) Find the traces of the plane in the coordinate planes, and sketch the plane.

**Solution**    To find the $xy$-trace, we first set $z = 0$ in the given equation to obtain the equation $4x + 2y = 12$. Then we sketch the graph of this equation in the $xy$-plane. (See Figure 2a.) To find the $yz$-trace, we set $x = 0$ to obtain the equation $2y + 3z = 12$, whose graph in the $yz$-plane gives the required trace. (See Figure 2b.) The $xz$-trace is obtained in a similar manner. (See Figure 2c.) The graph of the plane in the first octant is sketched in Figure 2d.

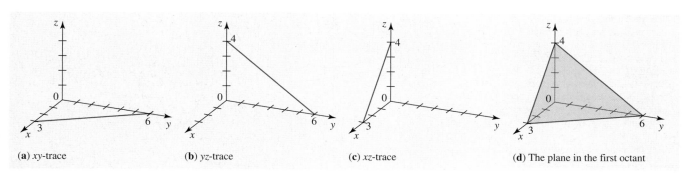

(**a**) $xy$-trace            (**b**) $yz$-trace            (**c**) $xz$-trace            (**d**) The plane in the first octant

**FIGURE 2**

The traces of the plane $4x + 2y + 3z = 12$ in the coordinate planes are shown in parts (a)–(c).    ■

Sometimes it is useful to obtain the traces of a surface in planes that are parallel to the coordinate planes, as illustrated in the next example.

**EXAMPLE 2**  Let $S$ be the surface defined by $z = x^2 + y^2$.

**a.** Find the traces of $S$ in the coordinate planes.
**b.** Find the traces of $S$ in the plane $z = k$, where $k$ is a constant.
**c.** Sketch the surface $S$.

**Solution**

**a.** Setting $z = 0$ gives $x^2 + y^2 = 0$, from which we see that the $xy$-trace is the origin $(0, 0)$. (See Figure 3a.) Next, setting $x = 0$ gives $z = y^2$, from which we see that the $yz$-trace is a parabola. (See Figure 3b.) Finally, setting $y = 0$ gives $z = x^2$, so the $xz$-trace is also a parabola. (See Figure 3c.)

**b.** Setting $z = k$, we obtain $x^2 + y^2 = k$, from which we see that the trace of $S$ in the plane $z = k$ is a circle of radius $\sqrt{k}$ centered at the point of intersection of the plane and the $z$-axis, provided that $k > 0$. (See Figure 3d.) Observe that if $k = 0$, the trace is the point $(0, 0)$ (degenerate circle) obtained in part (a).

**c.** The graph of $z = x^2 + y^2$ sketched in Figure 3e is called a *circular paraboloid* because its traces in planes parallel to the coordinate planes are either circles or parabolas.

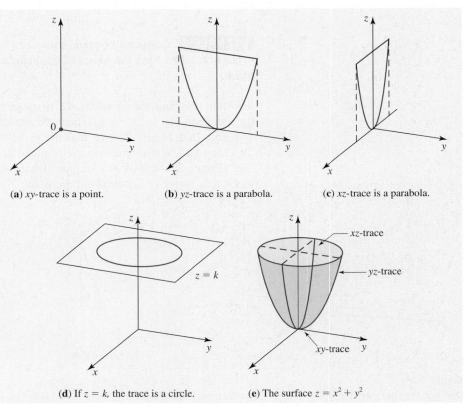

**(a)** $xy$-trace is a point.

**(b)** $yz$-trace is a parabola.

**(c)** $xz$-trace is a parabola.

**(d)** If $z = k$, the trace is a circle.

**(e)** The surface $z = x^2 + y^2$

**FIGURE 3**
The traces of the surface $S$

## ■ Cylinders

We now turn our attention to a class of surfaces called *cylinders*.

---

**DEFINITION    Cylinder**

Let $C$ be a curve in a plane, and let $l$ be a line that is not parallel to that plane. Then the set of all points generated by letting a line traverse $C$ while parallel to $l$ at all times is called a cylinder. The curve $C$ is called the **directrix** of the cylinder, and each line through $C$ parallel to $l$ is called a **ruling** of the cylinder. (See Figure 4.)

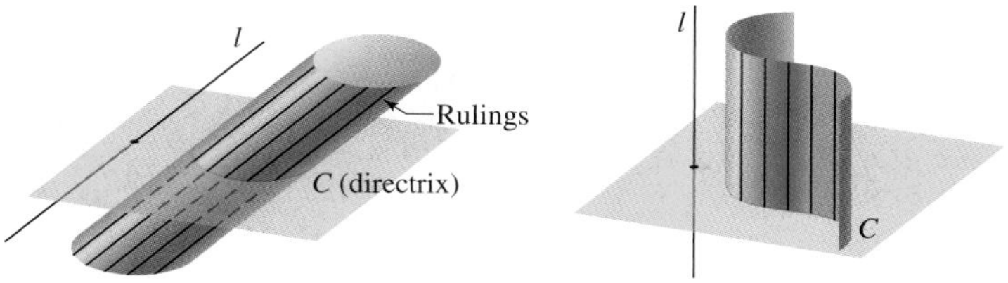

---

**FIGURE 4**
Two cylinders. The curve $C$ is the directrix. The rulings are parallel to $l$.

Cylinders in which the directrix lies in a coordinate plane and the rulings are perpendicular to that plane have relatively simple algebraic representations. Consider, for example, the surface $S$ with equation $f(x, y) = 0$. The $xy$-trace of $S$ is the graph $C$ of the equation $f(x, y) = 0$ in the $xy$-plane. (See Figure 5a.) Next, observe that if $(x, y, 0)$ is any point on $C$, then the point $(x, y, z)$ must satisfy the equation $f(x, y) = 0$ for *any* value of $z$ (since $z$ is not present in the equation). But all such points lie on the line perpendicular to the $xy$-plane and pass through the point $(x, y, 0)$. This shows that the surface $S$ is a cylinder with directrix $f(x, y) = 0$ and rulings that are parallel to the direction of the axis of the missing variable $z$. (See Figure 5b.)

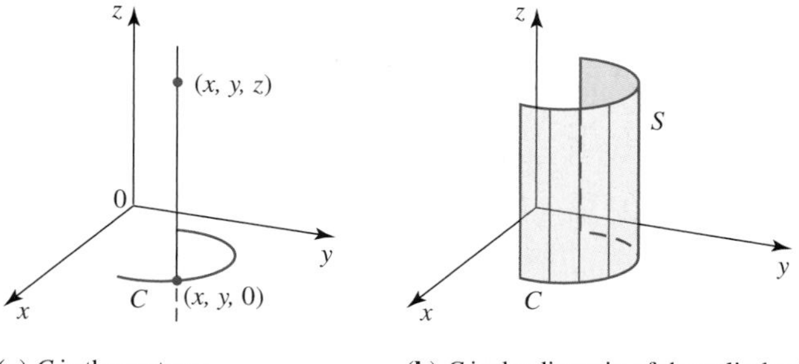

**FIGURE 5**
The surface $f(x, y) = 0$ is a cylinder with directrix $C$ defined by $f(x, y) = 0$ in the $xy$-plane and with rulings parallel to the direction of the axis of the missing variable $z$.

**(a)** $C$ is the $xy$-trace.

**(b)** $C$ is the directrix of the cylinder $S$. The rulings are parallel to the $z$-axis.

**EXAMPLE 3**   Sketch the graph of $y = x^2 - 4$.

**Solution**   The given equation has the form $f(x, y) = 0$, where $f(x, y) = x^2 - y - 4$. Therefore, its graph is a cylinder with directrix given by the graph of $y = x^2 - 4$ in the $xy$-plane and rulings parallel to the $z$-axis (corresponding to the variable missing in the equation). The graph of $y = x^2 - 4$ in the $xy$-plane is the parabola shown in Figure 6a. The required cylinder is shown in Figure 6b. It is called a **parabolic cylinder.**

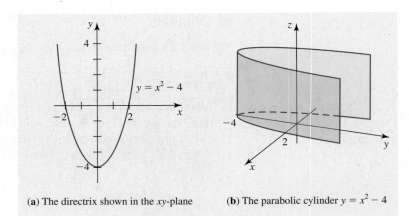

(a) The directrix shown in the $xy$-plane    (b) The parabolic cylinder $y = x^2 - 4$

**FIGURE 6**
The graph of $y = x^2 - 4$ is sketched in two steps.

**EXAMPLE 4** Sketch the graph of $\dfrac{y^2}{4} + \dfrac{z^2}{9} = 1$.

**Solution**  The given equation has the form $f(y, z) = 0$, where

$$f(y, z) = \frac{y^2}{4} + \frac{z^2}{9} - 1$$

Its graph is a cylinder with directrix given by

$$\frac{y^2}{4} + \frac{z^2}{9} = 1$$

and rulings parallel to the $x$-axis. The graph of

$$\frac{y^2}{4} + \frac{z^2}{9} = 1$$

in the $yz$-plane is the ellipse shown in Figure 7a. The required cylinder is shown in Figure 7b. It is called an **elliptic cylinder.**

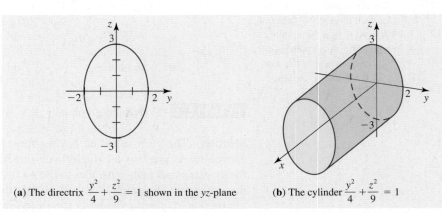

(a) The directrix $\dfrac{y^2}{4} + \dfrac{z^2}{9} = 1$ shown in the $yz$-plane    (b) The cylinder $\dfrac{y^2}{4} + \dfrac{z^2}{9} = 1$

**FIGURE 7**

**EXAMPLE 5** Sketch the graph of $z = \cos x$.

**Solution** The given equation has the form $f(x, z) = 0$, where $f(x, z) = z - \cos x$. Therefore, its graph is a cylinder with directrix given by the graph of $z = \cos x$ in the $xz$-plane and rulings parallel to the $y$-axis. The graph of the directrix in the $xz$-plane is shown in Figure 8a, and the graph of the cylinder is sketched in Figure 8b.

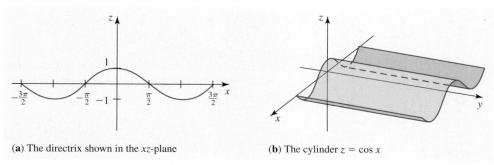

**(a)** The directrix shown in the $xz$-plane     **(b)** The cylinder $z = \cos x$

**FIGURE 8**

Note that, while an equation in two variables represents a curve in 2-space, the same equation represents a cylinder when we are working in 3-space. For example, the equation $x^2 + y^2 = 1$ represents a circle in the plane, but the same equation represents a right circular cylinder in 3-space.

## Quadric Surfaces

The equation of a sphere given in Section 10.2 and the equations in Examples 1, 2, and 3 in this section are special cases of the second-degree equation in $x$, $y$, and $z$

$$Ax^2 + By^2 + Cz^2 + Dxy + Exz + Fyz + Gx + Hy + Iz + J = 0$$

where $A, B, C, \dots, J$ are constants. The graph of this equation is a **quadric surface.** By making a suitable translation and/or rotation of the coordinate system, a quadric surface can always be put in standard position with respect to a new coordinate system. (See Figure 9.) With respect to the new system the equation will assume one of the two standard forms

$$\overline{A}X^2 + \overline{B}Y^2 + \overline{C}Z^2 + \overline{J} = 0 \qquad \text{or} \qquad \overline{A}X^2 + \overline{B}Y^2 + \overline{I}Z = 0$$

For this reason we will restrict our study of quadric surfaces to those represented by the equations

$$AX^2 + BY^2 + CZ^2 + J = 0 \qquad \text{or} \qquad AX^2 + BY^2 + IZ = 0$$

When we sketch the following quadric surfaces, we will find it useful to look at their traces in the coordinate planes as well as planes that are parallel to the coordinate planes.

In the remainder of this section, unless otherwise noted, $a$, $b$, and $c$ denote positive real numbers.

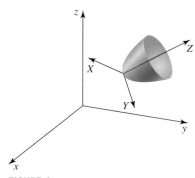

**FIGURE 9**
By translating and rotating the $xyz$-system, we have the $XYZ$-system in which the paraboloid is in standard position with respect to the latter.

**Ellipsoids** The graph of the equation

$$\frac{x^2}{a^2} + \frac{y^2}{b^2} + \frac{z^2}{c^2} = 1$$

is an ellipsoid because its traces in the planes parallel to the coordinate planes are ellipses. In fact, its trace in the plane $z = k$, where $-c < k < c$, is the ellipse

$$\frac{x^2}{a^2} + \frac{y^2}{b^2} = 1 - \frac{k^2}{c^2}$$

and, in particular, its trace in the $xy$-plane is the ellipse

$$\frac{x^2}{a^2} + \frac{y^2}{b^2} = 1$$

shown in Figure 10a.

Similarly, you may verify that its traces in the planes $x = k$ $(-a < k < a)$ and $y = k$ $(-b < k < b)$ are ellipses and, in particular, that its $yz$- and $xz$-traces are the ellipses

$$\frac{y^2}{b^2} + \frac{z^2}{c^2} = 1 \qquad \text{and} \qquad \frac{x^2}{a^2} + \frac{z^2}{c^2} = 1$$

respectively. (See Figures 10b–c.) The ellipsoid is sketched in Figure 10d.

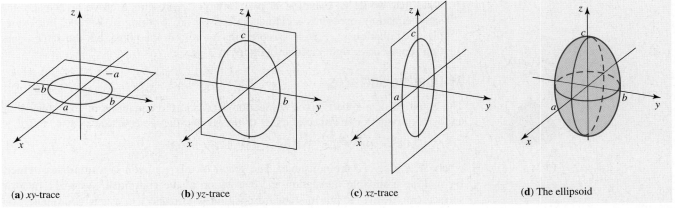

(a) $xy$-trace  (b) $yz$-trace  (c) $xz$-trace  (d) The ellipsoid

**FIGURE 10**
The traces in the coordinate planes and the ellipsoid $\dfrac{x^2}{a^2} + \dfrac{y^2}{b^2} + \dfrac{z^2}{c^2} = 1$

Note that if $a = b = c$, then the ellipsoid is in fact a sphere of radius $a$ with center at the origin.

### Hyperboloids of One Sheet
The graph of the equation

$$\frac{x^2}{a^2} + \frac{y^2}{b^2} - \frac{z^2}{c^2} = 1$$

is a **hyperboloid of one sheet**. The $xy$-trace of this surface is the ellipse

$$\frac{x^2}{a^2} + \frac{y^2}{b^2} = 1$$

(Figure 11a) whereas both the *yz*- and *xz*-traces are hyperbolas (Figures 11b–c), as you may verify. The trace of the surface in the plane $z = k$ is an ellipse

$$\frac{x^2}{a^2} + \frac{y^2}{b^2} = 1 + \frac{k^2}{c^2}$$

As $|k|$ increases, the ellipses grow larger and larger. The hyperboloid is sketched in Figure 11d.

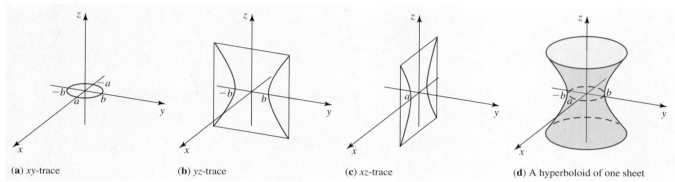

| **(a)** *xy*-trace | **(b)** *yz*-trace | **(c)** *xz*-trace | **(d)** A hyperboloid of one sheet |

**FIGURE 11**
The traces in the coordinate plane and the hyperboloid of one sheet $\dfrac{x^2}{a^2} + \dfrac{y^2}{b^2} - \dfrac{z^2}{c^2} = 1$

The *z*-axis is called the **axis of the hyperboloid.** Note that the orientation of the axis of the hyperboloid is associated with the term that has a minus sign in front of it. Thus, if the minus sign had been in front of the term involving *x*, then the surface would have been a hyperboloid of one sheet with the *x*-axis as its axis.

## Hyperboloids of Two Sheets   The graph of the equation

$$-\frac{x^2}{a^2} - \frac{y^2}{b^2} + \frac{z^2}{c^2} = 1$$

is a **hyperboloid of two sheets.** The *xz*- and *yz*-traces are the hyperbolas

$$-\frac{x^2}{a^2} + \frac{z^2}{c^2} = 1 \qquad \text{and} \qquad -\frac{y^2}{b^2} + \frac{z^2}{c^2} = 1$$

sketched in Figures 12a–b. The trace of the surface in the plane $z = k$ is an ellipse

$$\frac{x^2}{a^2} + \frac{y^2}{b^2} = \frac{k^2}{c^2} - 1$$

provided that $|k| > c$. There are no values of *x* and *y* that satisfy the equation if $|k| < c$, so the surface is made up of two parts, as shown in Figure 12c: one part lying on or above the plane $z = c$ and the other part lying on or below the plane $z = -c$.

The axis of the hyperboloid is the *z*-axis. Observe that the sign associated with the variable *z* is positive. Had the positive sign been in front of one of the other variables, then the surface would have been a hyperboloid of two sheets with its axis along the axis associated with that variable.

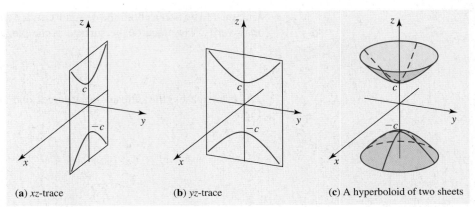

**(a)** $xz$-trace      **(b)** $yz$-trace     **(c)** A hyperboloid of two sheets

**FIGURE 12**
The traces in the $xz$- and $yz$-planes and the hyperboloid of two sheets $-\dfrac{x^2}{a^2} - \dfrac{y^2}{b^2} + \dfrac{z^2}{c^2} = 1$

## Cones   The graph of the equation

$$\frac{x^2}{a^2} + \frac{y^2}{b^2} - \frac{z^2}{c^2} = 0$$

is a double-napped **cone.** The $xz$- and $yz$-traces are the lines $z = \pm(c/a)x$ and $z = \pm(c/b)y$, respectively. (See Figures 13a–b.) The trace in the plane $z = k$ is an ellipse,

$$\frac{x^2}{a^2} + \frac{y^2}{b^2} = \frac{k^2}{c^2}$$

As $|k|$ increases, so do the lengths of the axes of the resulting ellipses. The traces in planes parallel to the other two coordinate planes are hyperbolas. The cone is sketched in Figure 13c. The **axis of the cone** is the $z$-axis.

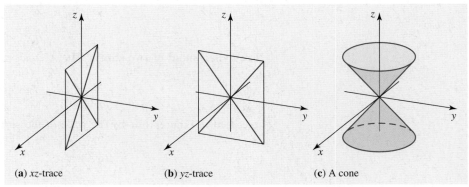

**(a)** $xz$-trace     **(b)** $yz$-trace     **(c)** A cone

**FIGURE 13**
The traces in the $xz$- and $yz$-planes and the cone $\dfrac{x^2}{a^2} + \dfrac{y^2}{b^2} - \dfrac{z^2}{c^2} = 0$

## Paraboloids   The graph of the equation

$$\frac{x^2}{a^2} + \frac{y^2}{b^2} = cz$$

where $c$ is a real number, is called an **elliptic paraboloid** because its traces in planes parallel to the $xy$-coordinate plane are ellipses and its traces in planes parallel to the other two coordinate planes are parabolas. If $a = b$, the surface is called a **circular paraboloid.** We will let you verify these statements. The graph of an elliptic paraboloid with $c > 0$ is sketched in Figure 14a. The **axis** of the paraboloid is the $z$-axis, and its **vertex** is the origin.

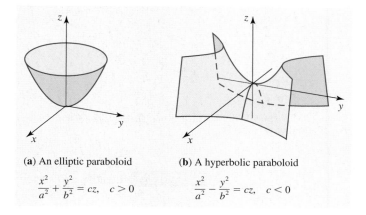

(**a**) An elliptic paraboloid

$$\frac{x^2}{a^2} + \frac{y^2}{b^2} = cz, \quad c > 0$$

(**b**) A hyperbolic paraboloid

$$\frac{x^2}{a^2} - \frac{y^2}{b^2} = cz, \quad c < 0$$

**FIGURE 14**

## Hyperbolic Paraboloids    The graph of the equation

$$\frac{x^2}{a^2} - \frac{y^2}{b^2} = cz$$

where $c$ is a real number, is called a **hyperbolic paraboloid** because the $xz$- and $yz$-traces are parabolas and the traces in planes parallel to the $xy$-plane are hyperbolas. The graph of a hyperbolic paraboloid with $c < 0$ is shown in Figure 14b.

**EXAMPLE 6**   Identify and sketch the surface $12x^2 - 3y^2 + 4z^2 + 12 = 0$.

**Solution**    Rewriting the equation in the standard form

$$-\frac{x^2}{1} + \frac{y^2}{4} - \frac{z^2}{3} = 1$$

we see that it represents a hyperboloid of two sheets with the $y$-axis as its axis.

To sketch the surface, observe that the surface intersects the $y$-axis at the points $(0, -2, 0)$ and $(0, 2, 0)$, as you can verify by setting $x = z = 0$ in the given equation. Next, let's find the trace in the plane $y = k$. We obtain

$$\frac{x^2}{1} + \frac{z^2}{3} = \frac{k^2}{4} - 1$$

In particular, the trace in the plane $y = 6$ is the ellipse

$$\frac{x^2}{1} + \frac{z^2}{3} = 8 \qquad \text{or} \qquad \frac{x^2}{8} + \frac{z^2}{24} = 1$$

A sketch of this trace is shown in Figure 15a. The completed sketch of the hyperboloid of two sheets is shown in Figure 15b.

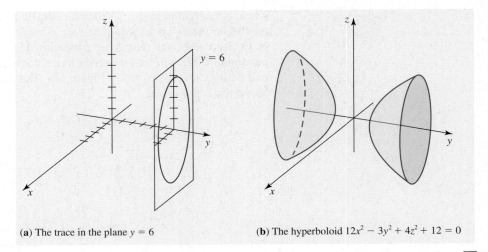

**V** ✎ **FIGURE 15**
Steps in sketching a
hyperboloid of two sheets

(a) The trace in the plane $y = 6$

(b) The hyperboloid $12x^2 - 3y^2 + 4z^2 + 12 = 0$

**EXAMPLE 7** Identify and sketch the surface $4x - 3y^2 - 12z^2 = 0$.

**Solution** Rewriting the equation in the standard form,

$$4x = 3y^2 + 12z^2$$

we see that it represents a paraboloid with the $x$-axis as its axis. To sketch the surface, let's find the trace in the plane $x = k$. We obtain

$$4k = 3y^2 + 12z^2$$

Letting $k = 3$, we see that the trace in the plane $x = 3$ is the ellipse with equation $12 = 3y^2 + 12z^2$ or, in standard form,

$$\frac{y^2}{4} + \frac{z^2}{1} = 1$$

A sketch of this trace is shown in Figure 16a. The completed sketch of the paraboloid is shown in Figure 16b.

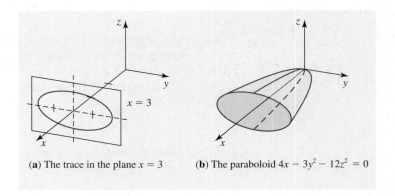

(a) The trace in the plane $x = 3$

(b) The paraboloid $4x - 3y^2 - 12z^2 = 0$

We now give a summary of the quadric surfaces and their general shapes, and we also suggest an aid for sketching these surfaces. Note that in many instances, finding the intercepts and using a judiciously chosen trace will be sufficient to help you obtain a good sketch of the surface.

| Equation | Surface (computer generated) | Aid for Sketching the Figure |
|---|---|---|
| **Ellipsoid**<br><br>$$\frac{x^2}{a^2} + \frac{y^2}{b^2} + \frac{z^2}{c^2} = 1$$<br><br>**Note:** All signs are positive. | | Find $x$-, $y$-, and $z$-intercepts, and then sketch.<br><br> |
| **Hyperboloid of One Sheet**<br><br>$$\frac{x^2}{a^2} + \frac{y^2}{b^2} - \frac{z^2}{c^2} = 1$$<br><br>**Notes:**<br>**1.** One sign is negative.<br>**2.** The axis lies along the coordinate axis associated with the variable with the negative coefficient. | | Sketch the trace on the plane $z = k$ (in this case) for an appropriate value of $k$ and for $z = 0$. Then use symmetry.<br><br> |
| **Hyperboloid of Two Sheets**<br><br>$$-\frac{x^2}{a^2} - \frac{y^2}{b^2} + \frac{z^2}{c^2} = 1$$<br><br>**Notes:**<br>**1.** Two signs are negative.<br>**2.** The axis lies along the coordinate axis associated with the variable with the positive coefficient. | | Sketch the trace on the plane $z = k$ (in this case) for an appropriate value of $k$. Find the $z$-intercept (in this case) and use symmetry.<br><br> |

*(continued)*

| Equation | Surface (computer generated) | Aid for Sketching the Figure |
|---|---|---|
| **Cone** $$\frac{x^2}{a^2} + \frac{y^2}{b^2} - \frac{z^2}{c^2} = 0$$ **Notes:** 1. One sign is negative. 2. The constant term is zero. 3. The axis lies along the coordinate axis associated with the variable with the negative coefficient. | | Sketch the trace on the plane $z = k$ (in this case) for an appropriate value of $k$. Then use symmetry. |
| **Paraboloids** $$\frac{x^2}{a^2} + \frac{y^2}{b^2} = cz$$ **Notes:** 1. There are two positive signs. 2. The axis lies along the coordinate axis associated with the variable of degree 1. 3. It opens upward if $c > 0$ and opens downward if $c < 0$. | | Sketch the trace on the plane $z = k$ (in this case) for an appropriate value of $k$. |

| Equation | Surface (computer generated) | Aid for Sketching the Figure |
|---|---|---|
| **Hyperbolic Paraboloid** $$\frac{x^2}{a^2} - \frac{y^2}{b^2} = cz$$ **Note:** There is one positive and one negative sign. | 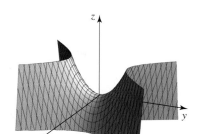 | **a.** For the case $c < 0$, sketch the parabolas $$z = \frac{x^2}{ca^2} \quad\text{and}\quad z = -\frac{y^2}{cb^2}$$ 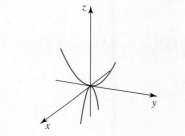 **b.** Sketch the hyperbola $$\frac{x^2}{a^2} - \frac{y^2}{b^2} = ck$$ for an appropriate value of $k$. 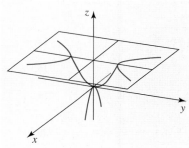 **c.** Complete your sketch. 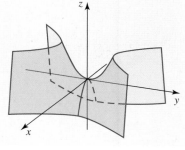 |

## 10.6 CONCEPT QUESTIONS

1. What is the trace of a surface in a plane? Illustrate by showing the trace of the surface $z = x^2 + y^2$ in the plane $z = 4$.
2. What is a cylinder? Illustrate by sketching the cylinder $x = y^2 - 4$.
3. **a.** What is a quadric surface?
   **b.** Write a standard equation for (1) an ellipsoid, (2) a hyperboloid of one sheet, (3) a hyperboloid of two sheets, (4) a cone, (5) an elliptic paraboloid, and (6) a hyperbolic paraboloid.

## 10.6 EXERCISES

*In Exercises 1–12, sketch the graph of the cylinder with the given equation.*

1. $x^2 + y^2 = 4$
2. $y^2 + z^2 = 9$
3. $x^2 + z^2 = 16$
4. $y = 4x^2$
5. $z = 4 - x^2$
6. $y = z^2 - 9$
7. $9x^2 + 4y^2 = 36$
8. $x^2 + 4z^2 = 16$
9. $yz = 1$
10. $y^2 - x^2 = 1$
11. $z = \cos y$
12. $y = \sec x, \quad -\frac{\pi}{2} < x < \frac{\pi}{2}$

*In Exercises 13–20, match each equation with one of the graphs labeled* (a)–(h).

13. $x^2 + \dfrac{y^2}{16} + \dfrac{z^2}{4} = 1$
14. $x^2 - \dfrac{y^2}{9} + z^2 = 9$
15. $-2x^2 - 2y^2 + z^2 = 1$
16. $x^2 - y^2 + z^2 = 0$
17. $x^2 + \dfrac{z^2}{4} = y$
18. $x^2 - \dfrac{y^2}{4} = -z$
19. $\dfrac{x^2}{4} + \dfrac{y^2}{9} + \dfrac{z^2}{25} = 1$
20. $x^2 - z^2 = -y$

(c)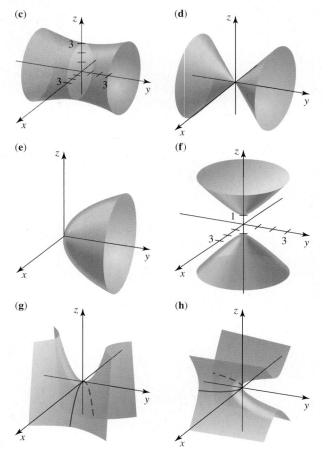

(d)

(e)

(f)

(g)

(h)

(a)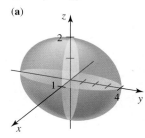

(b)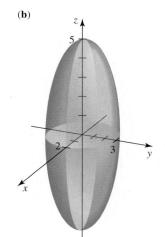

*In Exercises 21–44, write the given equation in standard form and sketch the surface represented by the equation.*

21. $4x^2 + y^2 + z^2 = 4$
22. $4x^2 + 4y^2 + z^2 = 16$
23. $9x^2 + 4y^2 + z^2 = 36$
24. $36x^2 + 100y^2 + 225z^2 = 900$
25. $4x^2 + 4y^2 - z^2 = 4$

**26.** $9x^2 + 9z^2 - 4y^2 = 36$   **27.** $x^2 + 4y^2 - z^2 = 4$

**28.** $9x^2 + 9y^2 - 4z^2 = 36$   **29.** $z^2 - x^2 - y^2 = 1$

**30.** $y^2 - x^2 - 9z^2 = 9$   **31.** $4x^2 - y^2 + 2z^2 + 4 = 0$

**32.** $4x^2 - 3y^2 + 12z^2 + 12 = 0$

**33.** $x^2 + y^2 - z^2 = 0$   **34.** $y^2 + z^2 = x^2$

**35.** $9x^2 + 4y^2 - z^2 = 0$   **36.** $x^2 - 4y^2 - 16z^2 = 0$

**37.** $x^2 + y^2 = z$   **38.** $y^2 + z^2 = x$

**39.** $x^2 + 9y^2 = z$   **40.** $x^2 + z^2 + y = 1$

**41.** $z = x^2 + y^2 + 4$   **42.** $z = x^2 + 4y^2 - 4$

**43.** $y^2 - x^2 = z$   **44.** $x^2 - y^2 = z$

*In Exercises 45–50, sketch the region bounded by the surfaces with the given equations.*

**45.** $x + 3y + 2z = 6$,   $x = 0$,   $y = 0$,   and   $z = 0$

**46.** $z = \sqrt{x^2 + y^2}$   and   $z = 2$

**47.** $x^2 + y^2 = 4$,   $x + z = 2$,   $x = 0$,   $y = 0$,   and   $z = 0$

**48.** $y^2 + z^2 = 1$,   $x^2 + z^2 = 1$,   $x = 0$,   $y = 0$,   and   $z = 0$

**49.** $z = \sqrt{x^2 + y^2}$   and   $z = 9 - x^2 - y^2$

**50.** $z = x^2 + y^2$   and   $z = 2 - x^2 - y^2$

*In Exercises 51 and 52, find an equation of the surface satisfying the conditions. Identify the surface.*

**51.** The set of all points equidistant from the point $(-3, 0, 0)$ and the plane $x = 3$.

**52.** The set of all points whose distance from the $y$-axis is twice its distance from the $xz$-plane.

**53.** Show that the curve of intersection of the surfaces $2x^2 + y^2 - 3z^2 + 2y = 6$ and $4x^2 + 2y^2 - 6z^2 - 4x = 4$ lies in a plane.

**54.** Show that the straight lines
  $L_1$: $x = a + t, y = b + t, z = b^2 - a^2 + 2(b - a)t$ and
  $L_2$: $x = a + t, y = b - t, z = b^2 - a^2 - 2(b + a)t$,
  passing through each point $(a, b, b^2 - a^2)$ on the hyperbolic paraboloid $z = y^2 - x^2$, both lie entirely on the surface.
  **Note:** This shows that the hyperbolic paraboloid is a **ruled surface,** that is, a surface that can be swept out by a line moving in space. The only other quadric surfaces that are ruled surfaces are cylinders, cones, and hyperboloids of one sheet.

**cas** *In Exercises 55–60, use a computer algebra system (CAS) to plot the surface with the given equation.*

**55.** $2x^2 + 3y^2 + 6z^2 = 36$   **56.** $-x^2 + 4y^2 + z^2 = 2$

**57.** $-2x^2 - 9y^2 + z^2 = 1$   **58.** $-x^2 - 3y^2 + z^2 = 0$

**59.** $-x^2 + y^2 - z = 0$

**60.** $x^2 - 2x + 4y^2 - 16y - z = 0$

*In Exercises 61–64, determine whether the statement is true or false. If it is true, explain why. If it is false, explain why or give an example that shows it is false.*

**61.** The graph of $y = x + 3$ in 3-space is a line lying in the $xy$-plane.

**62.** The graph of $9x^2 + 4z^2 = 1$ and $y = 0$ is the graph of an ellipse lying in the $xz$-plane.

**63.** The surface $\dfrac{x^2}{a^2} + \dfrac{y^2}{b^2} + \dfrac{z^2}{c^2} = 4$ is an ellipsoid obtained by stretching the ellipsoid $\dfrac{x^2}{a^2} + \dfrac{y^2}{b^2} + \dfrac{z^2}{c^2} = 1$ by a factor of 2 in each of the $x$-, $y$-, and $z$-directions.

**64.** The surface $\dfrac{x^2}{a^2} + \dfrac{y^2}{b^2} = c(z - z_0)$ is obtained by translating the paraboloid $\dfrac{x^2}{a^2} + \dfrac{y^2}{b^2} = cz$ vertically.

# 10.7 Cylindrical and Spherical Coordinates

Just as certain curves in the plane are described more easily by using polar coordinates than by using rectangular coordinates, there are some surfaces in space that can be described more conveniently by using coordinates other than rectangular coordinates. In this section we will look at two such coordinate systems.

## The Cylindrical Coordinate System

The **cylindrical coordinate system** is just an extension of the polar coordinate system in the plane to a three-dimensional system in space obtained by adding the (perpendicular) $z$-axis to the system (see Figure 1).

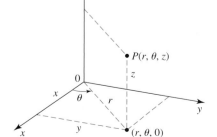

**FIGURE 1**
The cylindrical coordinate system

A point $P$ in this system is represented by the ordered triple $(r, \theta, z)$, where $r$ and $\theta$ are the polar coordinates of the projection of $P$ onto the $xy$-plane and $z$ is the directed distance from $(r, \theta, 0)$ to $P$.

The relationship between rectangular coordinates and cylindrical coordinates can be seen by examining Figure 1. If $P$ has representation $(x, y, z)$ in terms of rectangular coordinates, then we have the following equations for converting cylindrical coordinates to rectangular coordinates and vice versa.

---

**Converting Cylindrical to Rectangular Coordinates**

$$x = r \cos \theta \qquad y = r \sin \theta \qquad z = z \qquad\qquad \textbf{(1)}$$

---

**Converting Rectangular to Cylindrical Coordinates**

$$r^2 = x^2 + y^2 \qquad \tan \theta = \frac{y}{x} \qquad z = z \qquad\qquad \textbf{(2)}$$

---

**EXAMPLE 1**   The point $\left(3, \frac{\pi}{4}, 3\right)$ is expressed in cylindrical coordinates. Find its rectangular coordinates.

**Solution**   We are given that $r = 3$, $\theta = \pi/4$, and $z = 3$. Using the equations in (1), we have

$$x = r \cos \theta = 3 \cos \frac{\pi}{4} = \frac{3\sqrt{2}}{2}$$

$$y = r \sin \theta = 3 \sin \frac{\pi}{4} = \frac{3\sqrt{2}}{2}$$

and

$$z = 3$$

Therefore, the rectangular coordinates of the given point are $\left(\frac{3\sqrt{2}}{2}, \frac{3\sqrt{2}}{2}, 3\right)$. (See Figure 2.)

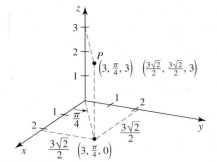

**FIGURE 2**
The point $P$ can be written as $\left(\frac{3\sqrt{2}}{2}, \frac{3\sqrt{2}}{2}, 3\right)$ in rectangular coordinates.

**EXAMPLE 2**   The point $(-\sqrt{2}, \sqrt{2}, 2)$ is expressed in rectangular coordinates. Find its cylindrical coordinates.

**Solution**   We are given that $x = -\sqrt{2}$, $y = \sqrt{2}$, and $z = 2$. Using the equations in (2), we have

$$r^2 = x^2 + y^2 = (-\sqrt{2})^2 + (\sqrt{2})^2 = 2 + 2 = 4$$

and

$$\tan \theta = \frac{y}{x} = \frac{\sqrt{2}}{-\sqrt{2}} = -1$$

So $r = \pm 2$, and

$$\theta = \tan^{-1}(-1) + n\pi = \frac{3}{4}\pi + n\pi$$

where $n$ is an integer; and $z = 2$.

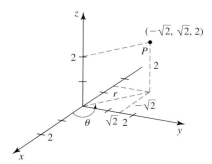

**FIGURE 3**
The point $P$ can be written as $\left(2, \frac{3\pi}{4}, 2\right)$ or $\left(-2, \frac{7\pi}{4}, 2\right)$, among others, in cylindrical coordinates.

We have two choices for $r$ and infinitely many choices for $\theta$. For example, we have the representations

$$\left(2, \frac{3\pi}{4}, 2\right) \qquad \text{If we pick } r > 0$$

and

$$\left(-2, \frac{7\pi}{4}, 2\right) \qquad \text{If we pick } r < 0$$

The point is shown in Figure 3. Note that neither the combination $r = 2$ and $\theta = 7\pi/4$ nor $r = -2$ and $\theta = 3\pi/4$ in Example 2 will do. (Why?) ∎

Cylindrical coordinates are especially useful in describing surfaces that are symmetric about the $z$-axis. For example, the circular cylinder with rectangular equation $x^2 + y^2 = c^2$ has the simple representation $r = c$ in the cylindrical coordinate system. (See Figure 4.)

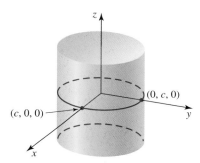

**FIGURE 4**
The circular cylinder has the simple representation $r = c$ in cylindrical coordinates.

**EXAMPLE 3**  Find an equation in cylindrical coordinates of the surface with the given rectangular equation.

**a.** $x^2 + y^2 = 9z$ **b.** $x^2 + y^2 = z^2$ **c.** $9x^2 + 9y^2 + 4z^2 = 36$

**Solution**  In each case we use the relationship $r^2 = x^2 + y^2$.
**a.** We obtain $r^2 = 9z$ as the required equation.
**b.** Here, $r^2 = z^2$.
**c.** Here we have $9(x^2 + y^2) + 4z^2 = 36$,
$$9r^2 + 4z^2 = 36$$

The surfaces are shown in Figure 5.

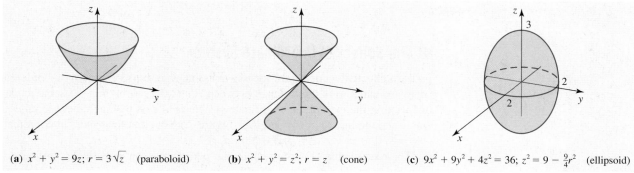

**(a)** $x^2 + y^2 = 9z$; $r = 3\sqrt{z}$  (paraboloid) **(b)** $x^2 + y^2 = z^2$; $r = z$  (cone) **(c)** $9x^2 + 9y^2 + 4z^2 = 36$; $z^2 = 9 - \frac{9}{4}r^2$  (ellipsoid)

**FIGURE 5** ∎

**EXAMPLE 4**  Find an equation in rectangular coordinates for the surface with the given cylindrical equation.

**a.** $\theta = \dfrac{\pi}{4}$ **b.** $r^2 \cos 2\theta - z^2 = 4$

**Solution**

**a.** Using the equations in (2), we have

$$\frac{y}{x} = \tan\theta = \tan\frac{\pi}{4} = 1$$

or

$$y = x$$

**b.** First, we use the trigonometric identity $\cos 2\theta = \cos^2\theta - \sin^2\theta$ to rewrite the given equation in the form

$$r^2(\cos^2\theta - \sin^2\theta) - z^2 = 4$$

$$r^2\cos^2\theta - r^2\sin^2\theta - z^2 = 4$$

Using the equations in (1), we then obtain

$$x^2 - y^2 - z^2 = 4$$

The surfaces are shown in Figure 6. Note that Figure 6a shows only the part of the plane in the first octant.

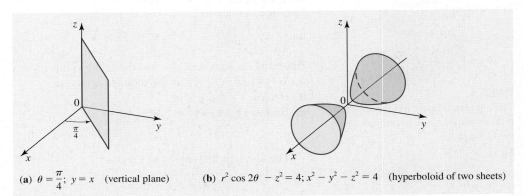

**(a)** $\theta = \dfrac{\pi}{4}$; $y = x$ (vertical plane)       **(b)** $r^2\cos 2\theta - z^2 = 4$; $x^2 - y^2 - z^2 = 4$ (hyperboloid of two sheets)

**FIGURE 6**

## ■ The Spherical Coordinate System

In the **spherical coordinate system** a point $P$ is represented by an ordered triple $(\rho, \theta, \phi)$, where $\rho$ is the distance between $P$ and the origin, $\theta$ is the same angle as the one used in the cylindrical coordinate system, and $\phi$ is the angle between the positive $z$-axis and the line segment $OP$. (See Figure 7.) Note that the spherical coordinates satisfy $\rho \geq 0$, $0 \leq \theta < 2\pi$, and $0 \leq \phi \leq \pi$.

The relationship between rectangular coordinates and spherical coordinates can be seen by examining Figure 7. If $P$ has representation $(x, y, z)$ in terms of rectangular coordinates, then

$$x = r\cos\theta \quad \text{and} \quad y = r\sin\theta$$

Since

$$r = \rho\sin\phi \quad \text{and} \quad z = \rho\cos\phi$$

we have the following equations for converting spherical coordinates to rectangular coordinates and vice versa.

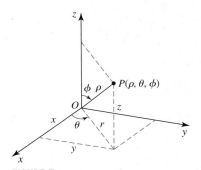

**FIGURE 7**
The spherical coordinate system

---

**Converting Spherical to Rectangular Coordinates**

$$x = \rho \sin \phi \cos \theta \qquad y = \rho \sin \phi \sin \theta \qquad z = \rho \cos \phi \tag{3}$$

---

**Converting Rectangular to Spherical Coordinates**

$$\rho^2 = x^2 + y^2 + z^2 \qquad \tan \theta = \frac{y}{x} \qquad \cos \phi = \frac{z}{\rho} \tag{4}$$

---

**EXAMPLE 5**   The point $\left(3, \frac{\pi}{3}, \frac{\pi}{4}\right)$ is expressed in spherical coordinates. Find its rectangular coordinates.

**Solution**   Using the equations in (3) with $\rho = 3$, $\theta = \pi/3$, and $\phi = \pi/4$, we have

$$x = \rho \sin \phi \cos \theta = 3 \sin \frac{\pi}{4} \cos \frac{\pi}{3} = 3\left(\frac{\sqrt{2}}{2}\right)\left(\frac{1}{2}\right) = \frac{3\sqrt{2}}{4}$$

$$y = \rho \sin \phi \sin \theta = 3 \sin \frac{\pi}{4} \sin \frac{\pi}{3} = 3\left(\frac{\sqrt{2}}{2}\right)\left(\frac{\sqrt{3}}{2}\right) = \frac{3\sqrt{6}}{4}$$

and

$$z = \rho \cos \phi = 3 \cos \frac{\pi}{4} = 3\left(\frac{\sqrt{2}}{2}\right) = \frac{3\sqrt{2}}{2}$$

Thus, in terms of rectangular coordinates the given point is $\left(\frac{3\sqrt{2}}{4}, \frac{3\sqrt{6}}{4}, \frac{3\sqrt{2}}{2}\right)$.   ∎

**EXAMPLE 6**   The point $(\sqrt{3}, 3, 2)$ is given in rectangular coordinates. Find its spherical coordinates.

**Solution**   We use the equations in (4). First, we have

$$\rho^2 = x^2 + y^2 + z^2 = (\sqrt{3})^2 + 3^2 + 2^2 = 3 + 9 + 4 = 16$$

so $\rho = 4$. (Remember that $\rho \geq 0$.) Next, from

$$\tan \theta = \frac{y}{x} = \frac{3}{\sqrt{3}} = \sqrt{3}$$

we see that $\theta = \pi/3$. Finally, from

$$\cos \phi = \frac{z}{\rho} = \frac{2}{4} = \frac{1}{2}$$

we see that $\phi = \pi/3$. Therefore, in terms of spherical coordinates, the given point is $\left(4, \frac{\pi}{3}, \frac{\pi}{3}\right)$.   ∎

Spherical coordinates are particularly useful in describing surfaces that are symmetric about the origin. For example, the sphere with rectangular equation $x^2 + y^2 + z^2 = c^2$ has the simple representation $\rho = c$ in the spherical coordinate system. (See Figure 8a.) Also, shown in Figures 8b–8c are surfaces described by the equations $\theta = c$ and $\phi = c$, where $0 < c < \frac{\pi}{2}$.

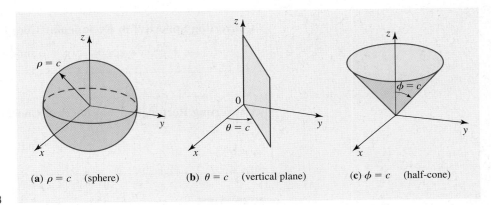

**(a)** $\rho = c$    (sphere)          **(b)** $\theta = c$    (vertical plane)          **(c)** $\phi = c$    (half-cone)

**FIGURE 8**

**EXAMPLE 7**   Find an equation in spherical coordinates for the paraboloid with rectangular equation $4z = x^2 + y^2$.

**Solution**   Using the equations in (3), we obtain

$$4\rho \cos \phi = \rho^2 \sin^2 \phi \cos^2 \theta + \rho^2 \sin^2 \phi \sin^2 \theta$$
$$= \rho^2 \sin^2 \phi (\cos^2 \theta + \sin^2 \theta)$$
$$= \rho^2 \sin^2 \phi$$

or

$$\rho \sin^2 \phi = 4 \cos \phi$$

**EXAMPLE 8**   Find an equation in rectangular coordinates for the surface with spherical equation $\rho = 4 \cos \phi$.

**Solution**   Multiplying both sides of the given equation by $\rho$ gives

$$\rho^2 = 4\rho \cos \phi$$

Then, using the equations in (4), we have

$$x^2 + y^2 + z^2 = 4z$$

or, upon completing the square in $z$, we obtain

$$x^2 + y^2 + (z - 2)^2 = 4$$

which is an equation of the sphere with center $(0, 0, 2)$ and radius 2.

## 10.7   CONCEPT QUESTIONS

1. Sketch the cylindrical coordinate system, and use it as an aid to help you give the equations (a) for converting cylindrical coordinates to rectangular coordinates and (b) for converting rectangular coordinates to cylindrical coordinates.

2. Sketch the spherical coordinate system, and use it as an aid to help you give the equations (a) for converting spherical coordinates to rectangular coordinates and (b) for converting rectangular coordinates to spherical coordinates.

## 10.7  EXERCISES

*In Exercises 1–6, the point is expressed in cylindrical coordinates. Write it in terms of rectangular coordinates.*

**1.** $\left(3, \frac{\pi}{2}, 2\right)$      **2.** $(4, 0, -3)$

**3.** $\left(\sqrt{2}, \frac{\pi}{4}, \sqrt{3}\right)$      **4.** $\left(2, \frac{\pi}{3}, 5\right)$

**5.** $\left(3, -\frac{\pi}{6}, 2\right)$      **6.** $(1, \pi, \pi)$

*In Exercises 7–12, the point is expressed in rectangular coordinates. Write it in terms of cylindrical coordinates.*

**7.** $(2, 0, 3)$      **8.** $(3, 3, 3)$

**9.** $(1, \sqrt{3}, 5)$      **10.** $(\sqrt{2}, -\sqrt{2}, 4)$

**11.** $(\sqrt{3}, 1, -2)$      **12.** $(\sqrt{3}, -1, 4)$

*In Exercises 13–18, the point is expressed in spherical coordinates. Write it in terms of rectangular coordinates.*

**13.** $(5, 0, 0)$      **14.** $\left(2, \frac{\pi}{2}, \frac{\pi}{6}\right)$

**15.** $\left(2, 0, \frac{\pi}{4}\right)$      **16.** $\left(3, \frac{\pi}{4}, \frac{3\pi}{4}\right)$

**17.** $\left(5, \frac{\pi}{6}, \frac{\pi}{4}\right)$      **18.** $\left(1, \pi, \frac{\pi}{2}\right)$

*In Exercises 19–24, the point is expressed in rectangular coordinates. Write it in terms of spherical coordinates.*

**19.** $(-2, 0, 0)$      **20.** $(1, 1, 1)$

**21.** $(\sqrt{3}, 0, 1)$      **22.** $(-2, 2\sqrt{3}, 4)$

**23.** $(0, 2\sqrt{3}, 2)$      **24.** $(\sqrt{3}, 1, 2\sqrt{3})$

*In Exercises 25–30, the point is expressed in cylindrical coordinates. Write it in terms of spherical coordinates.*

**25.** $\left(2, \frac{\pi}{4}, 0\right)$      **26.** $\left(2, \frac{\pi}{2}, -2\right)$

**27.** $\left(4, \frac{\pi}{3}, -4\right)$      **28.** $(12, \pi, 5)$

**29.** $\left(4, \frac{\pi}{6}, 6\right)$      **30.** $\left(12, \frac{\pi}{2}, 5\right)$

*In Exercises 31–36, the point is expressed in spherical coordinates. Write it in terms of cylindrical coordinates.*

**31.** $(3, 0, 0)$      **32.** $\left(5, \frac{\pi}{6}, \frac{\pi}{2}\right)$

**33.** $\left(2, \frac{3\pi}{2}, \frac{\pi}{2}\right)$      **34.** $\left(4, -\frac{\pi}{6}, \frac{\pi}{6}\right)$

**35.** $\left(1, \frac{\pi}{4}, \frac{\pi}{3}\right)$      **36.** $\left(5, \frac{\pi}{4}, \frac{3\pi}{4}\right)$

**37.** Find the distance between $\left(2, \frac{\pi}{3}, 0\right)$ and $(1, \pi, 2)$, where the points are given in cylindrical coordinates.

**38.** Find the distance between $\left(4, \frac{\pi}{2}, \frac{2\pi}{3}\right)$ and $\left(3, \pi, \frac{\pi}{2}\right)$, where the points are given in spherical coordinates.

*In Exercises 39–58, identify the surface whose equation is given.*

**39.** $r = 2$      **40.** $z = 2$

**41.** $\rho = 2$      **42.** $\theta = \dfrac{\pi}{6}$ (spherical coordinates)

**43.** $\phi = \dfrac{\pi}{4}$      **44.** $z = 4r^2$

**45.** $z = 4 - r^2$      **46.** $r = 6 \sin \theta$

**47.** $\rho \cos \phi = 3$      **48.** $\rho \sin \phi = 3$

**49.** $r \sec \theta = 4$      **50.** $r = -\csc \theta$

**51.** $z = r^2 \sin^2 \theta$      **52.** $\rho = 4 \cos \phi$

**53.** $r^2 + z^2 = 16$      **54.** $3r^2 - 4z^2 = 12$

**55.** $\rho = 2 \csc \phi \sec \theta$      **56.** $\rho^2(\sin^2 \phi - 2 \cos^2 \phi) = 1$

**57.** $r^2 - 3r + 2 = 0$      **58.** $\rho^2 - 4\rho + 3 = 0$

*In Exercises 59–66, write the given equation (a) in cylindrical coordinates and (b) in spherical coordinates.*

**59.** $x^2 + y^2 + z^2 = 4$      **60.** $x^2 - y^2 + z^2 = 4$

**61.** $x^2 + y^2 = 2z$      **62.** $x^2 + y^2 = 9$

**63.** $2x + 3y - 4z = 12$      **64.** $x^2 + y^2 = 4y$

**65.** $x^2 + z^2 = 4$      **66.** $x^2 - y^2 - z^2 = 1$

*In Exercises 67–70, sketch the region described by the inequalities.*

**67.** $r \le z \le 2$      **68.** $r^2 \le z \le 4 - r^2$

**69.** $0 \le \theta \le 2\pi, \quad 0 \le \phi \le \frac{\pi}{6}, \quad 0 \le \rho \le a \sec \phi$

**70.** $0 \le \phi \le \frac{\pi}{4}, \quad \rho \le 2$

**71. Spherical Coordinate System for the Earth** A spherical coordinate system for the earth can be set up as follows. Let the origin of the system be at the center of the earth, and choose the positive $x$-axis to pass through the point of intersection of the equator and the prime meridian and the positive $z$-axis to pass through the North Pole. Recall that the parallels of latitude are measured from 0° to 90° degrees north and south of the equator and the meridians of longitude are measured from 0° to 180° east and west of the prime meridian.

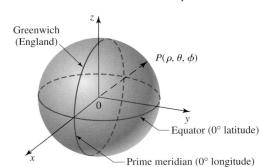

**a.** Express the locations of Los Angeles (latitude 34.06° North, longitude 118.25° West) and Paris (latitude 48.52° North, longitude 2.20° East) in terms of spherical coordinates. Take the radius of the earth to be 3960 miles.

V  Videos for selected exercises are available online at **www.academic.cengage.com/login**.

**b.** Express the points found in part (a) in terms of rectangular coordinates.

**c.** Find the great-circle distance between Los Angeles and Paris. (A *great circle* is the circle obtained by intersecting a sphere with a plane passing through the center of the sphere.)

**72.** A **geodesic** on a surface is the curve that minimizes the distance between any two points on the surface. Suppose a cylinder of radius $r$ is oriented so that its axis coincides with the $z$-axis of a cylindrical coordinate system. If $P_1(r, \theta_1, z_1)$ and $P_2(r, \theta_2, z_2)$ are two points on the cylinder, show that the length of the geodesic joining $P_1$ to $P_2$ is

$$\sqrt{r^2(\theta_2 - \theta_1)^2 + (z_2 - z_1)^2}$$

*In Exercises 73–76, determine whether the statement is true or false. If it is true, explain why. If it is false, explain why or give an example that shows it is false.*

**73.** The representation of a point in the cylindrical coordinate system is not unique.

**74.** The equation $\theta = \pi/4$ in cylindrical coordinates represents the plane with rectangular equation $y = x$.

**75.** The equation $\theta = c$, where $c$ is a constant, in cylindrical coordinates and the equation $\theta = c$ in spherical coordinates represent different surfaces.

**76.** The surface defined by the spherical equation $\phi = \pi/4$ is the same as the surface defined by the rectangular equation $x^2 + y^2 = z^2$.

# CHAPTER 10 REVIEW

## CONCEPT REVIEW

*In Exercises 1–14, fill in the blanks.*

**1. a.** A vector is a quantity that possesses both _____ and _____.

**b.** A vector can be represented by an _____; the _____ points in the _____ of the vector, and the _____ of the arrow gives its magnitude.

**c.** The vector $\mathbf{v} = \overrightarrow{AB}$ has _____ point _____ and _____ point _____.

**d.** Two vectors $\mathbf{v}$ and $\mathbf{w}$ are equal if they have the same _____ and _____.

**2. a.** The scalar multiple of a scalar $c$ and a vector $\mathbf{v}$ is the vector _____ whose magnitude is _____ times that of $\mathbf{v}$ and the direction is the same as that of $\mathbf{v}$ if _____ and opposite that of $\mathbf{v}$ if _____.

**b.** The vector $\mathbf{v} + \mathbf{w}$ is represented by the arrow with tail at the _____ point of $\mathbf{v}$ and head at the _____ point of $\mathbf{w}$.

**c.** The vector $\mathbf{v} - \mathbf{w}$ is defined to be the vector _____.

**3. a.** A vector in the plane is an ordered pair $\mathbf{a} = $ _____ of real numbers _____ and _____, called the _____ components of $\mathbf{a}$. The zero vector is $\mathbf{0} = $ _____.

**b.** If $\mathbf{a} = \langle a_1, a_2 \rangle$ and $\mathbf{b} = \langle b_1, b_2 \rangle$, then $\mathbf{a} + \mathbf{b} = $ _____. If $\mathbf{a} = \langle a_1, a_2 \rangle$ and $c$ is a scalar, then $c\mathbf{a} = $ _____.

**4. a.** If $\mathbf{a} \neq \mathbf{0}$, then a unit vector having the same direction as $\mathbf{a}$ is $\mathbf{u} = $ _____.

**b.** The vectors $\mathbf{i} = $ _____ and $\mathbf{j} = $ _____ are called the _____ _____ vectors. If $\mathbf{a}$ is any vector in the plane, then $\mathbf{a}$ can be expressed in terms of $\mathbf{i}$ and $\mathbf{j}$ as $\mathbf{a} = $ _____.

**c.** If $\theta$ is the angle that the unit vector $\mathbf{u}$ makes with the positive $x$-axis, then $\mathbf{u}$ can be written in terms of $\theta$ as $\mathbf{u} = $ _____.

**5. a.** The standard equation of a sphere with center _____ and radius _____ is $(x - h)^2 + (y - k)^2 + (z - l)^2 = r^2$.

**b.** The midpoint of the line segment joining the points $P_1(x_1, y_1, z_1)$ and $P_2(x_2, y_2, z_2)$ is _____.

**6. a.** The vector with initial point $P_1(x_1, y_1, z_1)$ and terminal point $P_2(x_2, y_2, z_2)$ is $\overrightarrow{P_1P_2} = $ _____.

**b.** A vector $\mathbf{a} = \langle a_1, a_2, a_3 \rangle$ in 3-space can be written in terms of the basis vectors $\mathbf{i} = $ _____, $\mathbf{j} = $ _____, and $\mathbf{k} = $ _____ as $\mathbf{a} = $ _____.

**7. a.** The dot product of $\mathbf{a} = \langle a_1, a_2, a_3 \rangle$ and $\mathbf{b} = \langle b_1, b_2, b_3 \rangle$ is $\mathbf{a} \cdot \mathbf{b} = $ _____ and is a _____.

**b.** The magnitude of a vector $\mathbf{a}$ can be written in terms of the dot product as $|\mathbf{a}| = $ _____.

**c.** The angle between two nonzero vectors $\mathbf{a}$ and $\mathbf{b}$ is given by $\cos \theta = $ _____.

**8. a.** Two nonzero vectors $\mathbf{a}$ and $\mathbf{b}$ are orthogonal if and only if _____.

**b.** The angles $\alpha$, $\beta$, and $\gamma$ that a nonzero vector $\mathbf{a}$ makes with the positive $x$-, $y$-, and $z$-axes, respectively, are called the _____ angles of $\mathbf{a}$.

**c.** The direction cosines of a nonzero vector $\mathbf{a} = a\mathbf{i} + b\mathbf{j} + c\mathbf{k}$ satisfy _____.

**9. a.** The vector obtained by projecting $\mathbf{b}$ onto the line containing the vector $\mathbf{a}$ is called the _____ _____ of $\mathbf{b}$ onto $\mathbf{a}$; it is also called the _____ _____ of $\mathbf{b}$ along $\mathbf{a}$.

**b.** The length of $\text{proj}_{\mathbf{a}}\mathbf{b}$ is called the _____ _____ of $\mathbf{b}$ along $\mathbf{a}$.

**c.** $\text{proj}_{\mathbf{a}}\mathbf{b}$ = _____.

**d.** The work done by a constant force $\mathbf{F}$ in moving an object along a straight line from $P$ to $Q$ is
$W$ = _____.

**10. a.** If $\mathbf{a}$ and $\mathbf{b}$ are nonzero vectors in 3-space, then $\mathbf{a} \times \mathbf{b}$ is _____ to both $\mathbf{a}$ and $\mathbf{b}$.

**b.** If $0 \le \theta \le \pi$ is the angle between $\mathbf{a}$ and $\mathbf{b}$, then $|\mathbf{a} \times \mathbf{b}|$ = _____.

**c.** Two nonzero vectors $\mathbf{a}$ and $\mathbf{b}$ are parallel if and only if $\mathbf{a} \times \mathbf{b}$ = _____.

**11. a.** The scalar triple product of $\mathbf{a}$, $\mathbf{b}$, and $\mathbf{c}$ is _____.

**b.** The volume $V$ of the parallelepiped determined by $\mathbf{a}$, $\mathbf{b}$, and $\mathbf{c}$ is $V$ = _____.

**12. a.** The parametric equations of the line passing through $P_0(x_0, y_0, z_0)$ and parallel to $\mathbf{v} = \langle a, b, c \rangle$ are _____.

**b.** The symmetric equations of the line passing through $P_0(x_0, y_0, z_0)$ and parallel to $\mathbf{v} = \langle a, b, c \rangle$ are _____.

**13. a.** The standard form of the equation of the plane containing the points $P_0(x_0, y_0, z_0)$ and having the normal vector $\mathbf{n} = \langle a, b, c \rangle$ is _____.

**b.** The linear equation $ax + by + cz = d$ represents a _____ in space having the normal vector _____. The acute angle between two intersecting planes is the angle between their _____ _____.

**14. a.** If a point has rectangular coordinates $(x, y, z)$ and cylindrical coordinates $(r, \theta, z)$, then $x$ = _____, $y$ = _____, and $z$ = _____; and $r^2$ = _____, $\tan \theta$ = _____, and $z$ = _____.

**b.** If a point has rectangular coordinates $(x, y, z)$ and spherical coordinates $(\rho, \theta, \phi)$, then $x$ = _____, $y$ = _____, and $z$ = _____; and $\rho^2$ = _____, $\tan \theta$ = _____, $\cos \phi$ = _____.

## REVIEW EXERCISES

*In Exercises 1–17, let* $\mathbf{a} = 2\mathbf{i} - \mathbf{j} + 3\mathbf{k}$, $\mathbf{b} = \mathbf{i} + 2\mathbf{j} - \mathbf{k}$, *and* $\mathbf{c} = 3\mathbf{i} - 2\mathbf{j} + \mathbf{k}$. *Find the given quantities.*

**1.** $2\mathbf{a} - 3\mathbf{b}$

**2.** $\mathbf{a} \cdot (\mathbf{b} + \mathbf{c})$

**3.** $|3\mathbf{a} + 2\mathbf{b}|$

**4.** $|\mathbf{a}| + |\mathbf{c}|$

**5.** $\mathbf{a} \times \mathbf{c}$

**6.** $|\mathbf{b} \times (\mathbf{c} \times \mathbf{a})|$

**7.** $\mathbf{a} \cdot (\mathbf{b} \times \mathbf{c})$

**8.** $|\mathbf{a} \times \mathbf{a}|$

**9.** $\mathbf{a} \times (\mathbf{b} + \mathbf{c})$

**10.** The angle between $\mathbf{a}$ and $\mathbf{b}$

**11.** Two unit vectors having the same direction as $\mathbf{c}$

**12.** A vector having twice the magnitude of $\mathbf{a}$ and direction opposite to that of $\mathbf{a}$

**13.** The direction cosines of $\mathbf{b}$

**14.** The scalar projection of $\mathbf{b}$ onto $\mathbf{a}$

**15.** The vector projection of $\mathbf{b}$ onto $\mathbf{a}$

**16.** The scalar projection of $\mathbf{b} \times \mathbf{c}$ onto $\mathbf{a}$

**17.** The volume of the parallelepiped determined by $\mathbf{a}$, $\mathbf{b}$, and $\mathbf{c}$

**18.** Which of the following are legitimate operations?
**a.** $\mathbf{a} \cdot (\mathbf{b} - \mathbf{c})$
**b.** $\mathbf{a} \times (\mathbf{b} \cdot \mathbf{c})$
**c.** $(|\mathbf{a}|\mathbf{b} - |\mathbf{b}|\mathbf{a} \times \mathbf{c})$

**19.** Show that $\mathbf{a} = 2\mathbf{i} - 3\mathbf{j} + 4\mathbf{k}$ and $\mathbf{b} = 3\mathbf{i} + 6\mathbf{j} + 3\mathbf{k}$ are orthogonal.

**20.** Find the value of $x$ such that $3\mathbf{i} + x\mathbf{j} - 2\mathbf{k}$ and $2x\mathbf{i} - 3\mathbf{j} + 6\mathbf{k}$ are orthogonal.

**21.** Find $c$ such that $\mathbf{a} = 2\mathbf{i} + 3\mathbf{j} + \mathbf{k}$, $\mathbf{b} = \mathbf{i} + 2\mathbf{j} + 3\mathbf{k}$, and $\mathbf{c} = \mathbf{i} - 3\mathbf{j} + c\mathbf{k}$ are coplanar.

**22.** Find two unit vectors that are orthogonal to $\langle 3, 1, -2 \rangle$ and $\langle 2, 3, -2 \rangle$.

**23.** Find the acute angle between two diagonals of a cube.

**24.** Find the volume of the parallelepiped with adjacent sides $AB$, $AC$, and $AD$, where $A(2, -1, 1)$, $B(2, 0, 1)$, $C(4, -1, 1)$, and $D(5, -2, 0)$.

**25. a.** Find a vector perpendicular to the plane passing through the points $P(-1, 2, -2)$, $Q(2, 3, 1)$, and $R(3, 2, 1)$.
**b.** What is the area of the triangle with vertices $P$, $Q$, and $R$?

**26.** A force $\mathbf{F} = \mathbf{i} + 2\mathbf{j} + 4\mathbf{k}$ moves an object along the line segment from $(-3, -1, 1)$ to $(2, 1, 1)$. Find the work done by the force if the distance is measured in feet and the force is measured in pounds.

**27.** A constant force has a magnitude of 20 newtons and acts in the direction of the vector $\mathbf{a} = 2\mathbf{i} - \mathbf{j} + 3\mathbf{k}$. If this force moves a particle along the line segment from $(1, 2, 1)$ to $(2, 1, 4)$ and the distance is measured in meters, find the work done by the force.

**28.** Two men wish to push a crate in the $x$-direction as shown in the figure. If one man pushes with a force $\mathbf{F}_1$ of magnitude 80 N in the direction indicated in the figure and the second man pushes in the indicated direction, find the force $\mathbf{F}_2$ with which he must push.

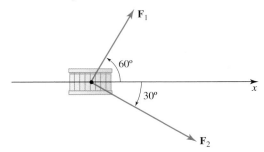

**29.** Two forces $\mathbf{F}_1$ and $\mathbf{F}_2$ with magnitude 10 N and 8 N, respectively, are applied to a bar as shown in the figure. Find the resultant torque about the point 0.

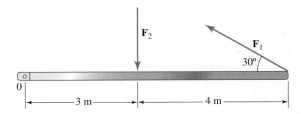

*In Exercises 30–33, find (a) parametric equations and (b) symmetric equations for the line satisfying the given conditions.*

**30.** Passes through $(2, 3, 1)$ and has the direction of $\mathbf{v} = \mathbf{i} - 2\mathbf{j} + 3\mathbf{k}$

**31.** Passes through $(-1, 2, -4)$ and $(2, -1, 3)$

**32.** Passes through $(2, -1, 3)$ and is parallel to the line with parametric equations $x = 1 - 2t$, $y = 2 + 3t$, $z = -1 - t$

**33.** Passes through $(1, 2, 4)$ and is perpendicular to $\mathbf{u} = \langle 1, -2, 1 \rangle$ and $\mathbf{v} = \langle 3, 2, 5 \rangle$

*In Exercises 34–37, find an equation of the plane satisfying the given conditions.*

**34.** Passes through $(-1, 2, 3)$ and has a normal vector $2\mathbf{i} - 3\mathbf{j} + 5\mathbf{k}$

**35.** Passes through $(-2, 4, 3)$ and is parallel to the plane with equation $2x + 4y - 3z = 12$

**36.** Passes through $(-2, 1, 1)$, $(2, -2, 4)$, and $(3, 1, 5)$

**37.** Passes through $(3, 2, 2)$ and is parallel to the $xz$-plane

**38.** Find the point of intersection (if any) of the line with parametric equations $x = 1 + 2t$, $y = -1 + t$, and $z = 2 + 3t$ and the plane $2x + 3y - 4z = 6$

**39.** Find the distance between the point $(2, 1, 4)$ and the plane $2x - 3y + 4z = 12$

**40.** Determine whether the lines with parametric equations $x = 3 - 3t$, $y = -1 + 2t$, $z = 3 - 2t$, and $x = 1 + 2t$, $y = 2 - 3t$, $z = 2 + t$ are parallel, are skew, or intersect. If they intersect, find the point of intersection.

**41.** Show that the lines with symmetric equations

$$\frac{x - 1}{-2} = \frac{y - 3}{-1} = \frac{z}{2}$$

and

$$\frac{x - 2}{-3} = \frac{y - 1}{1} = \frac{z + 1}{3}$$

intersect, and find the angle between the two lines.

**42.** Determine whether the planes $2x + 3y - 4z = 12$ and $2x - 3y - 5z = 8$ are parallel, perpendicular, or neither. If they are neither parallel nor perpendicular, find the angle between them.

*In Exercises 43 and 44, find the distance between the parallel planes.*

**43.** $x + 2y - 3z = 2$;   $2x + 4y - 6z = 6$

**44.** $2x - 3y - z = 2$;   $6x - 9y - 3z = 10$

**45.** Find the distance between the point $(3, 4, 5)$ and the plane $2x + 4y - 3z = 12$.

**46.** Find the curve of intersection of the plane $x + z = 5$ and the cylinder $x^2 + y^2 = 9$.

*In Exercises 47–50, describe and sketch the region in 3-space defined by the inequality or inequalities.*

**47.** $x^2 + y^2 \leq 4$

**48.** $1 \leq x^2 + z^2 \leq 4$

**49.** $y \leq x$,   $0 \leq x \leq 1$,   $0 \leq y \leq 1$,   $0 \leq z \leq 1$

**50.** $9x^2 + 4y^2 = 36$,   $0 \leq z \leq 2$

*In Exercises 51–58, identify and sketch the surface represented by the given equation.*

**51.** $2x - y = 6$

**52.** $x = 9 - y^2$

**53.** $x = y^2 + z^2$

**54.** $9x^2 + 4y^2 - z^2 = 36$

**55.** $4x^2 + 9z^2 = y^2$

**56.** $225x^2 + 100y^2 + 36z^2 = 900$

**57.** $x^2 - z^2 = y$

**58.** $z = \sin y$

**59.** The point $(1, 1, \sqrt{2})$ is expressed in rectangular coordinates. Write it in terms of cylindrical and spherical coordinates.

**60.** The point $\left(2, \frac{\pi}{6}, 4\right)$ is expressed in cylindrical coordinates. Write it in terms of rectangular and spherical coordinates.

**61.** The point $\left(2, \frac{\pi}{4}, \frac{\pi}{3}\right)$ is expressed in spherical coordinates. Write it in terms of rectangular and cylindrical coordinates.

*In Exercises 62–66, identify the surface whose equation is given.*

**62.** $z = -2$

**63.** $\theta = \frac{\pi}{3}$   (spherical coordinates)

**64.** $\phi = \frac{\pi}{3}$

**65.** $r = 2 \sin \theta$

**66.** $\rho = 2 \sec \phi$

*In Exercises 67–70, write the given equation (a) in cylindrical coordinates and (b) in spherical coordinates.*

**67.** $x^2 + y^2 = 2$

**68.** $x^2 + y^2 + z^2 = 9$

**69.** $x^2 + y^2 + 2z^2 = 1$

**70.** $x^2 + y^2 + z^2 = 2y$

*In Exercises 71 and 72, sketch the region described by the given inequalities.*

**71.** $0 \leq r \leq z$,   $0 \leq \theta \leq \frac{\pi}{2}$

**72.** $0 \leq \phi \leq \frac{\pi}{3}$,   $\rho \leq 2$

## CHALLENGE PROBLEMS

1. Prove the Cauchy-Schwarz inequality $|\mathbf{a} \cdot \mathbf{b}| \le |\mathbf{a}||\mathbf{b}|$, without using trigonometry, by demonstrating the following:

   **a.** $|t\mathbf{a} + \mathbf{b}|^2 = |\mathbf{a}|^2t^2 + 2(\mathbf{a} \cdot \mathbf{b})t + |\mathbf{b}|^2 \ge 0$ for all values of $t$

   **b.** $4(\mathbf{a} \cdot \mathbf{b})^2 - 4|\mathbf{a}|^2|\mathbf{b}|^2 \le 0$

   **Hint:** Recall the relationship between the discriminant $b^2 - 4ac$ of the quadratic equation $at^2 + bt + c = 0$ and the nature of the roots of the equation.

2. Use vectors to prove that if $A$ and $B$ are endpoints of a diameter of a circle and $C$ is any other point on the circle, then the triangle $\triangle ABC$ is a right triangle.

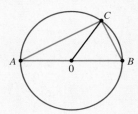

3. Let $\mathbf{a} = a_1\mathbf{i} + a_2\mathbf{j} + a_3\mathbf{k}$ and $\mathbf{b} = b_1\mathbf{i} + b_2\mathbf{j} + b_3\mathbf{k}$ be nonparallel vectors in three-dimensional space, and let $\mathbf{c} = x\mathbf{i} + y\mathbf{j} + z\mathbf{k}$ be a vector that is perpendicular to both $\mathbf{a}$ and $\mathbf{b}$.

   **a.** Show that $x$, $y$, and $z$ satisfy the system of two equations

   $$a_1x + a_2y + a_3z = 0$$
   $$b_1x + b_2y + b_3z = 0$$

   **b.** Solving the system in part (a) for $x$ and $y$ in terms of $z$, show that

   $$x = \frac{a_2b_3 - a_3b_2}{a_1b_2 - a_2b_1} z \quad \text{and} \quad y = \frac{a_3b_1 - a_1b_3}{a_1b_2 - a_2b_1} z$$

   **c.** Show that $\mathbf{c}$ can be written in the form

   $$\mathbf{c} = x\mathbf{i} + y\mathbf{j} + z\mathbf{k}$$
   $$= (a_2b_3 - a_3b_2)\mathbf{i} + (a_3b_1 - a_1b_3)\mathbf{j} + (a_1b_2 - a_2b_1)\mathbf{k}$$
   $$= \mathbf{a} \times \mathbf{b}$$

   (This is the definition of the cross-product of $\mathbf{a}$ and $\mathbf{b}$ given in Section 10.4.)

4. The area of a triangle with sides of lengths $a$, $b$, and $c$ is given by

   $$A = \sqrt{s(s - a)(s - b)(s - c)}$$

where $s = \frac{1}{2}(a + b + c)$ is the semiperimeter of the triangle. Derive this formula, known as Heron's formula.

**Hint:** Let $a$ and $b$ denote two sides of the triangle. Then $A = \frac{1}{2}|\mathbf{a} \times \mathbf{b}|$. Use the result of Exercise 44 in Section 10.4.

5. **a.** Let $\mathbf{a}$, $\mathbf{b}$, and $\mathbf{c}$ be noncoplanar vectors and let $\mathbf{v}$ be an arbitrary vector. Show that there exist constants $\alpha$, $\beta$, and $\gamma$ such that $\mathbf{v} = \alpha\mathbf{a} + \beta\mathbf{b} + \gamma\mathbf{c}$.

   **Hint:** To find $\alpha$, take the dot product of $\mathbf{v}$ with $\mathbf{b} \times \mathbf{c}$.

   **b.** Let $\mathbf{a} = \langle 1, 3, 1 \rangle$, $\mathbf{b} = \langle 2, -1, 1 \rangle$, and $\mathbf{c} = \langle 3, 1, 2 \rangle$. Express $\mathbf{v} = \langle 3, 2, 4 \rangle$ in terms of $\mathbf{a}$, $\mathbf{b}$, and $\mathbf{c}$ as suggested in part (a).

6. **a.** Consider the portion of the plane $ax + by + cz = d$ lying in the first octant, where $a$, $b$, and $c$, are positive real constants. Show that its area is given by

   $$\frac{d^2\sqrt{a^2 + b^2 + c^2}}{2abc}$$

   **b.** Use the result of part (a) to find the area of the plane $x + 2y + 3z = 6$ that lies in the first octant.

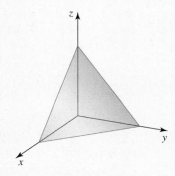

7. Find the points of intersection of the line

   $$x - 2 = \frac{y}{2} = \frac{z - 24}{16}$$

   and the elliptic paraboloid $z = 4x^2 + y^2$.

8. **a.** Let $\mathbf{a}$, $\mathbf{b}$, and $\mathbf{c}$ be vectors in 3-space. Show that there exist scalars $s$ and $t$ such that

   $$\mathbf{a} \times (\mathbf{b} \times \mathbf{c}) = s\mathbf{b} + t\mathbf{c}$$

   **b.** Let $\mathbf{a} = \langle 1, -2, 4 \rangle$, $\mathbf{b} = \langle 2, 3, 2 \rangle$, and $\mathbf{c} = \langle -2, 4, 5 \rangle$. Use the result of part (a) to write $\mathbf{a} \times (\mathbf{b} \times \mathbf{c})$ in the form $s\mathbf{b} + t\mathbf{c}$, where $s$ and $t$ are scalars.

Anthony Reynolds/Corbis

The "human cannonball" is a popular attraction at circuses. The trajectory of the person shot out of the cannon can be described by a *vector function*—a function whose domain is a set of real numbers and whose range is a set of vectors. We will look at an exercise involving a human cannonball in Section 11.4.

# 11 Vector-Valued Functions

**IN THIS CHAPTER** we will study functions whose values are vectors in the plane or in space. These vector-valued functions can be used to describe plane curves and space curves, and they also allow us to study the motion of objects along such curves.

We will also develop formulas for computing the arc length of plane and space curves and for finding the *curvature* of a curve. (The curvature measures the rate at which a curve bends.)

We end the chapter by demonstrating how vector calculus can be used to prove Kepler's laws of planetary motion.

**V** This symbol indicates that one of the following video types is available for enhanced student learning at **www.academic.cengage.com/login**:
- Chapter lecture videos
- Solutions to selected exercises

# 11.1 Vector-Valued Functions and Space Curves

In Section 9.2 we saw that the position of an object such as a boat or a car moving in the $xy$-plane can be described by a pair of parametric equations

$$x = f(t) \qquad y = g(t)$$

where $f$ and $g$ are continuous functions on a parameter interval $I$.

Using vector notation, we can denote the position of the object in an equivalent and somewhat abbreviated form via its *position vector* **r** as follows: For each $t$ in $I$, the position vector **r** of the object is the vector with initial point at the origin and terminal point $(f(t), g(t))$. In other words,

$$\mathbf{r}(t) = \langle f(t), g(t) \rangle = f(t)\mathbf{i} + g(t)\mathbf{j} \qquad t \in I$$

As $t$ takes on increasing values, the terminal point of $\mathbf{r}(t)$ traces the path of the object which is a plane curve $C$. This is illustrated in Figure 1 for the parameter interval $I = [a, b]$.

**FIGURE 1**
As $t$ increases from $a$ to $b$, the terminal point of **r** traces the curve $C$.

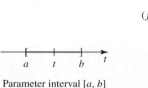
Parameter interval $[a, b]$

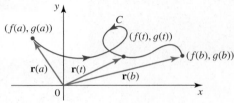

Similarly, in 3-space we can describe the position of an object such as a plane or a satellite using the parametric equations

$$x = f(t) \qquad y = g(t) \qquad z = h(t)$$

where $f$, $g$, and $h$ are continuous functions on a parameter interval $I$. Equivalently, we can describe its position using the *position vector* **r** defined by

$$\mathbf{r}(t) = \langle f(t), g(t), h(t) \rangle = f(t)\mathbf{i} + g(t)\mathbf{j} + h(t)\mathbf{k} \qquad t \in I$$

As $t$ takes on increasing values, the terminal point of $\mathbf{r}(t)$ traces the path of the object, which is a **space curve** $C$. (See Figure 2.)

**FIGURE 2**
As $t$ increases from $a$ to $b$, the terminal point of **r** traces the curve $C$.

Parameter interval $[a, b]$

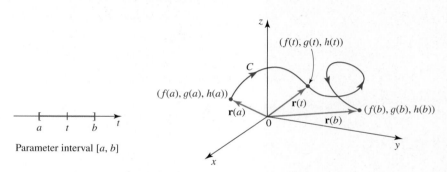

The function **r** is called a **vector-valued function,** or **vector function,** of a real variable $t$ because its value $\mathbf{r}(t)$ is a vector and its **domain** (parameter interval) is a subset of the real numbers.

> **DEFINITION**   **Vector Function**
>
> A **vector-valued function**, or **vector function**, is a function **r** defined by
>
> $$\mathbf{r}(t) = f(t)\mathbf{i} + g(t)\mathbf{j} + h(t)\mathbf{k}$$
>
> where the component functions $f$, $g$, and $h$ of **r** are real-valued functions of the parameter $t$ lying in a **parameter interval $I$.**

Unless otherwise specified, the parameter interval will be taken to be the intersection of the domains of the real-valued functions $f$, $g$, and $h$.

**EXAMPLE 1**   Find the domain (parameter interval) of the vector function

$$\mathbf{r}(t) = \left\langle \frac{1}{t}, \sqrt{t - 1}, \ln t \right\rangle$$

**Solution**   The component functions of **r** are $f(t) = 1/t$, $g(t) = \sqrt{t - 1}$, and $h(t) = \ln t$. Observe that $f$ is defined for all values of $t$ except $t = 0$, $g$ is defined for all $t \geq 1$, and $h$ is defined for all $t > 0$. Therefore, $f$, $g$, and $h$ are all defined if $t \geq 1$, and we conclude that the domain of **r** is $[1, \infty)$. ■

## Curves Defined by Vector Functions

As was mentioned earlier, a plane or space curve is the curve traced out by the terminal point of $\mathbf{r}(t)$ of a vector function **r** as $t$ takes on all values in a parameter interval.

**EXAMPLE 2**   Sketch the curve defined by the vector function

$$\mathbf{r}(t) = \langle 3 \cos t, -2 \sin t \rangle \qquad 0 \leq t \leq 2\pi$$

**Solution**   The parametric equations for the curve are

$$x = 3 \cos t \qquad \text{and} \qquad y = -2 \sin t$$

Solving the first equation for $\cos t$ and the second equation for $\sin t$ and using the identity $\cos^2 t + \sin^2 t = 1$, we obtain the rectangular equation

$$\frac{x^2}{9} + \frac{y^2}{4} = 1$$

The curve described by this equation is the ellipse shown in Figure 3. As $t$ increases from 0 to $2\pi$, the terminal point of **r** traces the ellipse in a clockwise direction.

**FIGURE 3**
As $t$ increases from 0 to $2\pi$, the terminal point of the vector $\mathbf{r}(t)$ traces the ellipse $\dfrac{x^2}{9} + \dfrac{y^2}{4} = 1$ in a clockwise direction, starting and ending at $(3, 0)$.

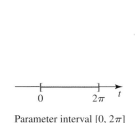

Parameter interval $[0, 2\pi]$

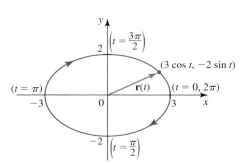

■

**EXAMPLE 3** Sketch the curve defined by the vector function

$$\mathbf{r}(t) = (2 - 4t)\mathbf{i} + (-1 + 3t)\mathbf{j} + (3 + 2t)\mathbf{k} \qquad 0 \le t \le 1$$

**Solution** The parametric equations for the curve are

$$x = 2 - 4t \qquad y = -1 + 3t \qquad z = 3 + 2t$$

which are parametric equations of the line passing through the point $(2, -1, 3)$ with direction numbers $-4$, $3$, and $2$. Because the parameter interval is the closed interval $[0, 1]$, we see that the curve is a straight line segment: Its initial point $(2, -1, 3)$ is the terminal point of the vector $\mathbf{r}(0) = 2\mathbf{i} - \mathbf{j} + 3\mathbf{k}$, and its terminal point $(-2, 2, 5)$ is the terminal point of the vector $\mathbf{r}(1) = -2\mathbf{i} + 2\mathbf{j} + 5\mathbf{k}$. (See Figure 4.)

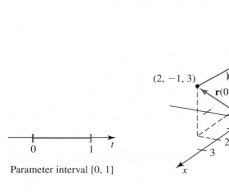

**FIGURE 4**
As $t$ increases from 0 to 1, the tip of $\mathbf{r}(t)$ traces the straight line segment from $(2, -1, 3)$ to $(-2, 2, 5)$.

Parameter interval $[0, 1]$

**EXAMPLE 4** Sketch the curve defined by the vector function

$$\mathbf{r}(t) = 3\mathbf{i} + t\mathbf{j} + (4 - t^2)\mathbf{k} \qquad -2 \le t \le 2$$

**Solution** The parametric equations for the curve are

$$x = 3 \qquad y = t \qquad z = 4 - t^2$$

Eliminating $t$ from the second and third equations, we obtain

$$z = 4 - y^2$$

Since the $x$-coordinate of any point on the curve must always be 3, as implied by the equation $x = 3$, we conclude that the desired curve is contained in the parabola $z = 4 - y^2$, which lies in the plane $x = 3$. In fact, as $t$ runs from $-2$ to 2, the terminal point of $\mathbf{r}$ traces the part of the parabola starting at the point $(3, -2, 0)$ [since $\mathbf{r}(-2) = 3\mathbf{i} - 2\mathbf{j}$] and ending at the point $(3, 2, 0)$ [since $\mathbf{r}(2) = 3\mathbf{i} + 2\mathbf{j}$], as shown in Figure 5.

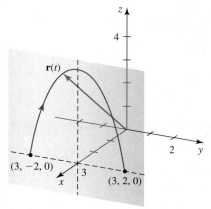

**FIGURE 5**
As $t$ increases from $-2$ to 2, the terminal point of $\mathbf{r}(t)$ traces the part of the parabola lying in the plane $x = 3$ from the point $(3, -2, 0)$ to the point $(3, 2, 0)$.

**EXAMPLE 5** Sketch the curve defined by the vector function

$$\mathbf{r}(t) = 2\cos t\,\mathbf{i} + 2\sin t\,\mathbf{j} + t\mathbf{k} \qquad 0 \le t \le 2\pi$$

**Solution** The parametric equations for the curve are

$$x = 2\cos t \qquad y = 2\sin t \qquad z = t$$

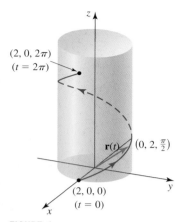

**FIGURE 6**

As $t$ increases from 0 to $2\pi$, the terminal point of $\mathbf{r}(t)$ traces the helix beginning at $(2, 0, 0)$ and terminating at $(2, 0, 2\pi)$.

From the first two equations we obtain

$$\left(\frac{x}{2}\right)^2 + \left(\frac{y}{2}\right)^2 = \cos^2 t + \sin^2 t = 1 \qquad \text{or} \qquad x^2 + y^2 = 4$$

This says that the curve lies on the right circular cylinder of radius 2, whose axis is the $z$-axis. At $t = 0$, $\mathbf{r}(0) = 2\mathbf{i}$, and this gives $(2, 0, 0)$ as the starting point of the curve. Since $z = t$, the $z$-coordinate of the point on the curve increases (linearly) as $t$ increases, and the curve spirals upward around the cylinder in a counterclockwise direction, terminating at the point $(2, 0, 2\pi)$ $[\mathbf{r}(2\pi) = 2\mathbf{i} + 2\pi\mathbf{k}]$. The curve, called a **helix,** is shown in Figure 6. ■

**EXAMPLE 6** Find a vector function that describes the curve of intersection of the cylinder $x^2 + y^2 = 4$ and the plane $x + y + 2z = 4$. (See Figure 7.)

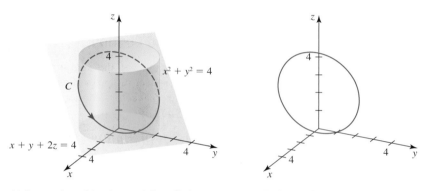

**FIGURE 7**

(**a**) Intersection of the plane and the cylinder    (**b**) Curve of intersection

**Solution** If $P(x, y, z)$ is any point on the curve of intersection $C$, then the $x$- and $y$-coordinates lie on the right circular cylinder of radius 2 and axis lying along the $z$-axis. Therefore,

$$x = 2 \cos t \qquad \text{and} \qquad y = 2 \sin t$$

To find the $z$-coordinate of the point, we substitute these values of $x$ and $y$ into the equation of the plane, obtaining

$$2 \cos t + 2 \sin t + 2z = 4 \qquad \text{or} \qquad z = 2 - \cos t - \sin t$$

So a required vector function is

$$\mathbf{r}(t) = 2 \cos t\mathbf{i} + 2 \sin t\mathbf{j} + (2 - \cos t - \sin t)\mathbf{k} \qquad 0 \le t \le 2\pi \qquad ■$$

You might have noticed that the space curves in Examples 4, 5, and 6 are relatively easy to sketch by hand. This is partly because they are relatively simple and partly because they lie in a plane. For more complicated curves we turn to computers.

**EXAMPLE 7** Use a computer to plot the curve represented by

$$\mathbf{r}(t) = (0.2 \sin 20t + 0.8) \cos t\mathbf{i} + (0.2 \sin 20t + 0.8) \sin t\mathbf{j} + 0.2 \cos 20t\mathbf{k}$$

$$0 \le t \le 2\pi$$

**Solution**  The curve is shown in Figure 8.

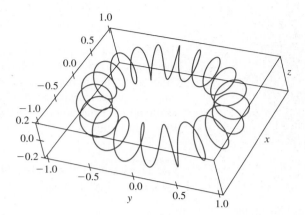

**FIGURE 8**
The curve in Example 7 is
called a **toroidal spiral**
because it lies on a torus.

## Limits and Continuity

Because the range of the vector function **r** is a subset of vectors in two- or three-dimensional space, the properties of vectors given in Chapter 10 can be used to study the properties of vector functions. For example, we add two vector functions componentwise. Thus, if

$$\mathbf{r}_1(t) = f_1(t)\mathbf{i} + g_1(t)\mathbf{j} + h_1(t)\mathbf{k} \qquad \text{and} \qquad \mathbf{r}_2(t) = f_2(t)\mathbf{i} + g_2(t)\mathbf{j} + h_2(t)\mathbf{k}$$

then

$$(\mathbf{r}_1 + \mathbf{r}_2)(t) = \mathbf{r}_1(t) + \mathbf{r}_2(t) = [f_1(t) + f_2(t)]\mathbf{i} + [g_1(t) + g_2(t)]\mathbf{j} + [h_1(t) + h_2(t)]\mathbf{k}$$

Similarly, if $c$ is a scalar, then the scalar multiple of **r** by $c$ is

$$(c\mathbf{r})(t) = c\mathbf{r}(t) = cf(t)\mathbf{i} + cg(t)\mathbf{j} + ch(t)\mathbf{k}$$

Next, because the components $f$, $g$, and $h$ of the vector function **r** are real-valued functions, we can investigate the notions of limits and continuity involving **r** using the properties of such functions. As you might expect, the limit of $\mathbf{r}(t)$ is defined in terms of the limits of its component functions.

---

**DEFINITION**  **The Limit of a Vector Function**

Let **r** be a function defined by $\mathbf{r}(t) = f(t)\mathbf{i} + g(t)\mathbf{j} + h(t)\mathbf{k}$. Then

$$\lim_{t \to a} \mathbf{r}(t) = \left[\lim_{t \to a} f(t)\right]\mathbf{i} + \left[\lim_{t \to a} g(t)\right]\mathbf{j} + \left[\lim_{t \to a} h(t)\right]\mathbf{k}$$

provided that the limits of the component functions exist.

---

To obtain a geometric interpretation of $\lim_{t \to a} \mathbf{r}(t)$, suppose that the limit exists. Let $\lim_{t \to a} f(t) = L_1$, $\lim_{t \to a} g(t) = L_2$, and $\lim_{t \to a} h(t) = L_3$, and let $\mathbf{L} = L_1\mathbf{i} + L_2\mathbf{j} + L_3\mathbf{k}$. Then, by definition, $\lim_{t \to a} \mathbf{r}(t) = \mathbf{L}$. This says that as $t$ approaches $a$, the vector $\mathbf{r}(t)$ approaches the constant vector **L**. (See Figure 9.)

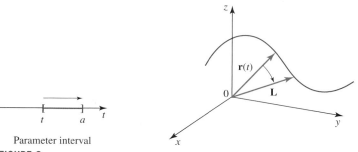

Parameter interval

**FIGURE 9**

$\lim_{t \to a} \mathbf{r}(t) = \mathbf{L}$ means that as $t$ approaches $a$, $\mathbf{r}(t)$ approaches $\mathbf{L}$.

**EXAMPLE 8**   Find $\lim_{t \to 0} \mathbf{r}(t)$, where $\mathbf{r}(t) = \sqrt{t + 2}\,\mathbf{i} + t \cos 2t\,\mathbf{j} + e^{-t}\mathbf{k}$.

**Solution**

$$\lim_{t \to 0} \mathbf{r}(t) = \left[\lim_{t \to 0} \sqrt{t + 2}\right]\mathbf{i} + \left[\lim_{t \to 0} t \cos 2t\right]\mathbf{j} + \left[\lim_{t \to 0} e^{-t}\right]\mathbf{k}$$

$$= \sqrt{2}\,\mathbf{i} + \mathbf{k}$$

The notion of continuity is extended to vector functions via the following definition.

---

**DEFINITION   Continuity of a Vector Function**

A vector function $\mathbf{r}$ is continuous at $a$ if

$$\lim_{t \to a} \mathbf{r}(t) = \mathbf{r}(a)$$

A vector function $\mathbf{r}$ **is continuous on an interval** $I$ if it is continuous at every number in $I$.

---

It follows from this definition that a vector function is continuous at $a$ if and only if each of its component functions is continuous at $a$.

**EXAMPLE 9**   Find the interval(s) on which the vector function $\mathbf{r}$ defined by

$$\mathbf{r}(t) = \sqrt{t}\,\mathbf{i} + \left(\frac{1}{t^2 - 1}\right)\mathbf{j} + \ln t\,\mathbf{k}$$

is continuous.

**Solution**   The component functions of $\mathbf{r}$ are $f(t) = \sqrt{t}$, $g(t) = 1/(t^2 - 1)$, and $h(t) = \ln t$. Observe that $f$ is continuous for $t \geq 0$, $g$ is continuous for all values of $t$ except $t = \pm 1$, and $h$ is continuous for $t > 0$. Therefore, $\mathbf{r}$ is continuous on the intervals $(0, 1)$ and $(1, \infty)$.

## 11.1 CONCEPT QUESTIONS

**1. a.** What is a vector-valued function?
 **b.** Give an example of a vector function. What is the parameter interval of the function that you picked?
**2.** Let $\mathbf{r}(t)$ be a vector function defined by $\mathbf{r}(t) = \langle f(t), g(t), h(t) \rangle$.
 **a.** Define $\lim_{t \to a} \mathbf{r}(t)$.
 **b.** Give an example of a vector function $\mathbf{r}(t)$ such that $\lim_{t \to 1} \mathbf{r}(t)$ does not exist.

**3. a.** What does it mean for a vector function $\mathbf{r}(t)$ to be continuous at $a$? Continuous on an interval $I$?
 **b.** Give an example of a function $\mathbf{r}(t)$ that is defined on the interval $(-1, 1)$ but fails to be continuous at 0.

## 11.1 EXERCISES

*In Exercises 1–6, find the domain of the vector function.*

**1.** $\mathbf{r}(t) = t\mathbf{i} + \dfrac{1}{t}\mathbf{j}$

**2.** $\mathbf{r}(t) = \cos t\mathbf{i} + 2 \sin t\mathbf{j} + \sqrt{t + 1}\,\mathbf{k}$

**3.** $\mathbf{r}(t) = \left\langle \sqrt{t}, \dfrac{1}{t - 1}, \ln t \right\rangle$ 　 **4.** $\mathbf{r}(t) = \left\langle \dfrac{1}{\sqrt{t - 1}}, e^{-t} \right\rangle$

**5.** $\mathbf{r}(t) = \ln t\mathbf{i} + \cosh t\mathbf{j} + \tanh t\mathbf{k}$

**6.** $\mathbf{r}(t) = \sqrt[3]{t}\,\mathbf{i} + e^{1/t}\mathbf{j} + \dfrac{1}{t + 2}\mathbf{k}$

*In Exercises 7–12, match the vector functions with the curves labeled (a)–(f). Explain your choices.*

**7.** $\mathbf{r}(t) = t^2\mathbf{i} + t^2\mathbf{j} + t^2\mathbf{k}$

**8.** $\mathbf{r}(t) = 2 \cos 2t\mathbf{i} + t\mathbf{j} + 2 \sin 2t\mathbf{k}$

**9.** $\mathbf{r}(t) = t\mathbf{i} + t\mathbf{j} + \left(\dfrac{1}{t^2 + 1}\right)\mathbf{k}$

**10.** $\mathbf{r}(t) = t \sin t\mathbf{i} + t \cos t\mathbf{j} + t\mathbf{k}, \quad 0 \le t \le 10\pi$

**11.** $\mathbf{r}(t) = 2 \cos t\mathbf{i} + 3 \sin t\mathbf{j} + e^{0.1t}\mathbf{k}, \quad t \ge 0$

**12.** $\mathbf{r}(t) = \cos t\mathbf{i} + \sin t\mathbf{j} + \sin 3t\mathbf{k}$

**(a)** 　　　　　　　　　　**(b)**

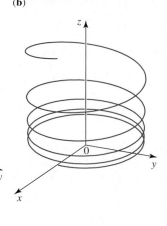

**(c)** 　　　　　　　　　　**(d)**

**(e)** 　　　　　　　　　　**(f)**

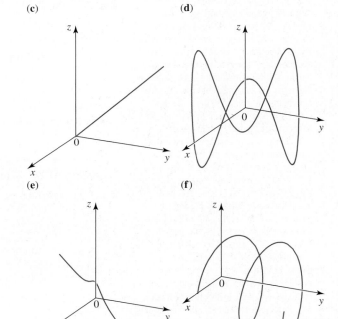

*In Exercises 13–26, sketch the curve with the given vector function, and indicate the orientation of the curve.*

**13.** $\mathbf{r}(t) = 2t\mathbf{i} + (3t + 1)\mathbf{j}, \quad -1 \le t \le 2$

**14.** $\mathbf{r}(t) = \sqrt{t}\,\mathbf{i} + (4 - t)\mathbf{j}, \quad t \ge 0$

**15.** $\mathbf{r}(t) = \langle t^2, t^3 \rangle, \quad -1 \le t \le 2$

**16.** $\mathbf{r}(t) = 2 \sin t\mathbf{i} + 3 \cos t\mathbf{j}, \quad 0 \le t \le 2\pi$

**17.** $\mathbf{r}(t) = e^t\mathbf{i} + e^{2t}\mathbf{j}, \quad -\infty < t < \infty$

**18.** $\mathbf{r}(t) = \langle 1 + 2 \cos t, 3 + 2 \sin t \rangle, \quad 0 \le t \le 2\pi$

**19.** $\mathbf{r}(t) = (1 + t)\mathbf{i} + (2 - t)\mathbf{j} + (3 - 2t)\mathbf{k}, \quad -\infty < t < \infty$

**20.** $\mathbf{r}(t) = (2 + t)\mathbf{i} + (3 - 2t)\mathbf{j} + (2 + 4t)\mathbf{k}, \quad 0 \le t \le 1$

**21.** $\mathbf{r}(t) = \langle t, t^2, t^3 \rangle, \quad t \ge 0$

**22.** $\mathbf{r}(t) = 2\cos t\mathbf{i} + 4\sin t\mathbf{j} + 3\mathbf{k}, \quad 0 \le t \le 2\pi$

**23.** $\mathbf{r}(t) = 2\cos t\mathbf{i} + 4\sin t\mathbf{j} + t\mathbf{k}, \quad 0 \le t \le 2\pi$

**24.** $\mathbf{r}(t) = t\mathbf{i} + 2t\mathbf{j} + \sin 2t\mathbf{k}, \quad -\infty < t < \infty$

**25.** $\mathbf{r}(t) = \langle t\cos t, t\sin t, t\rangle, \quad -\infty < t < \infty$
Hint: Show that it lies on a cone.

**26.** $\mathbf{r}(t) = e^t\cos t\mathbf{i} + e^t\sin t\mathbf{j} + e^t\mathbf{k}, \quad -\infty < t < \infty$

**cas** *In Exercises 27–30 use a computer to graph the curve described by the function.*

**27.** $\mathbf{r}(t) = 2\sin \pi t\mathbf{i} + 3\cos \pi t\mathbf{j} + 0.1t\mathbf{k}, \quad 0 \le t \le 10$

**28.** $\mathbf{r}(t) = (t^2 - t + 1)\mathbf{i} + (t^2 + 1)\mathbf{j} + t^3\mathbf{k}, \quad -3 \le t \le 3$

**29.** $\mathbf{r}(t) = \sin 3t\cos t\mathbf{i} + \sin 3t\sin t\mathbf{j} + \dfrac{t}{2\pi}\mathbf{k}, \quad -2\pi \le t \le 2\pi$
(rotating sine wave)

**30.** $\mathbf{r}(t) = \dfrac{1}{2}\sin t\mathbf{i} + \dfrac{1}{2}\cos t\mathbf{j} + \dfrac{t^2}{100\pi^2}\mathbf{k}, \quad 0 \le t \le 10\pi$
(Fresnel integral spiral)

**31. a.** Show that the curve

$$\mathbf{r}(t) = \sqrt{1 - 0.09\cos^2 10t}\,\cos t\mathbf{i}$$
$$+ \sqrt{1 - 0.09\cos^2 10t}\,\sin t\mathbf{j} + 0.3\cos 10t\mathbf{k}$$

lies on a sphere.

 **b.** Graph the curve described by $\mathbf{r}(t)$ for $0 \le t \le 2\pi$.

**32. a.** Show that the curve

$$\mathbf{r}(t) = (1 + \cos 12t)\cos t\mathbf{i} + (1 + \cos 12t)\sin t\mathbf{j}$$
$$+ (1 + \cos 12t)\mathbf{k}$$

lies on a cone.

 **b.** Graph the curve described by $\mathbf{r}(t)$ for $0 \le t \le 2\pi$.

*In Exercises 33–36, find a vector function describing the curve of intersection of the two surfaces.*

**33.** The cylinder $x^2 + y^2 = 1$ and the plane $x + y + 2z = 1$

**34.** The cylinder $x^2 + y^2 = 4$ and the surface $z = xy$

**35.** The cone $z = \sqrt{x^2 + y^2}$ and the plane $x + y + z = 1$

**36.** The paraboloid $z = x^2 + y^2$ and the sphere $x^2 + y^2 + z^2 = 1$

*In Exercises 37–42, find the given limit.*

**37.** $\lim\limits_{t \to 0} [(t^2 + 1)\mathbf{i} + \cos t\mathbf{j} - 3\mathbf{k}]$

**38.** $\lim\limits_{t \to 0} \left\langle e^{-t}, \dfrac{\sin t}{t}, \cos t\right\rangle$

**39.** $\lim\limits_{t \to 2} \left[\sqrt{t}\mathbf{i} + \left(\dfrac{t^2 - 4}{t - 2}\right)\mathbf{j} + \left(\dfrac{t}{t^2 + 1}\right)\mathbf{k}\right]$

**40.** $\lim\limits_{t \to 0^+} \left[\cos t\mathbf{i} + \dfrac{\tan t}{t}\mathbf{j} + t\ln t\mathbf{k}\right]$

**41.** $\lim\limits_{t \to \infty} \left\langle e^{-t}, \dfrac{1}{t}, \dfrac{2t^2}{t^2 + 1}\right\rangle$

**42.** $\lim\limits_{t \to -\infty} \left[\left(\dfrac{t - 1}{2t + 1}\right)\mathbf{i} + e^{2t}\mathbf{j} + \tan^{-1} t\mathbf{k}\right]$

*In Exercises 43–48 find the interval(s) on which the vector function is continuous.*

**43.** $\mathbf{r}(t) = \sqrt{t + 1}\mathbf{i} + \dfrac{1}{t}\mathbf{j}$

**44.** $\mathbf{r}(t) = \sin t\mathbf{i} + \cos t\mathbf{j} - \tan^{-1} t\mathbf{k}$

**45.** $\mathbf{r}(t) = \left\langle \dfrac{\cos t - 1}{t}, \dfrac{\sqrt{t}}{1 + 2t}, te^{-1/t}\right\rangle$

**46.** $\mathbf{r}(t) = \left(\dfrac{2t}{t^2 - 4}\right)\mathbf{i} + \sin^{-1} t\mathbf{j} + \sqrt[3]{t}\mathbf{k}$

**47.** $\mathbf{r}(t) = e^{-t}\mathbf{i} + \cos\sqrt{4 - t}\mathbf{j} + \dfrac{1}{t^2 - 1}\mathbf{k}$

**48.** $\mathbf{r}(t) = \dfrac{1}{\sqrt{t}}\mathbf{i} + \tan t\mathbf{j} + e^{-t}\cos t\mathbf{k}$

**49. Trajectory of a Plane** An airplane is circling an airport in a holding pattern. Suppose that the airport is located at the origin of a three-dimensional coordinate system and that the trajectory of the plane traveling at a constant speed is described by

$$\mathbf{r}(t) = 44{,}000\cos 60t\mathbf{i} + 44{,}000\sin 60t\mathbf{j} + 10{,}000\mathbf{k}$$

where the distance is measured in feet and the time is measured in hours. What is the distance covered by the plane over a 2-min period?

**50. Temperature at a Point** Suppose that the temperature at a point $(x, y, z)$ in 3-space is $T(x, y, z) = x^2 + 2y^2 + 3z^2$ and that the position of a particle at time $t$ is described by $\mathbf{r}(t) = \langle t, t^2, e^t\rangle$. What is the temperature at the point occupied by the particle when $t = 1$?

*In Exercises 51–54, suppose that $\mathbf{u}$ and $\mathbf{v}$ are vector functions such that $\lim\limits_{t \to a} \mathbf{u}(t)$ and $\lim\limits_{t \to a} \mathbf{v}(t)$ exist and $c$ is a constant. Prove the given property.*

**51.** $\lim\limits_{t \to a} [\mathbf{u}(t) + \mathbf{v}(t)] = \lim\limits_{t \to a} \mathbf{u}(t) + \lim\limits_{t \to a} \mathbf{v}(t)$

**52.** $\lim\limits_{t \to a} c\mathbf{u}(t) = c\lim\limits_{t \to a} \mathbf{u}(t)$

**53.** $\lim\limits_{t \to a} [\mathbf{u}(t) \cdot \mathbf{v}(t)] = \lim\limits_{t \to a} \mathbf{u}(t) \cdot \lim\limits_{t \to a} \mathbf{v}(t)$

**54.** $\lim\limits_{t \to a} [\mathbf{u}(t) \times \mathbf{v}(t)] = \lim\limits_{t \to a} \mathbf{u}(t) \times \lim\limits_{t \to a} \mathbf{v}(t)$

**55. a.** Prove that if $\mathbf{r}$ is a vector function that is continuous at $a$, then $|\mathbf{r}|$ is also continuous at $a$.
**b.** Show that the converse is false by exhibiting a vector function $\mathbf{r}$ such that $|\mathbf{r}|$ is continuous at $a$ but $\mathbf{r}$ is not continuous at $a$.

**56.** Evaluate

$$\lim_{h \to 0} \left\langle \frac{(t + h)^2 - t^2}{h}, \frac{\cos(t + h) - \cos t}{h}, \frac{e^{t+h} - e^t}{h} \right\rangle$$

**57.** Evaluate

$$\lim_{t \to 0} \left[ \frac{\sin t}{t} \mathbf{i} + \frac{1 - \cos t}{t^2} \mathbf{j} + \frac{\ln(1 + t^2)}{\cos t - e^{-t}} \mathbf{k} \right]$$

**58. a.** Find a vector function describing the curve of intersection of the plane $x + y + 2z = 2$ and the paraboloid $z = x^2 + y^2$.

**b.** Find the point(s) on the curve of part (a) that are closest to and farthest from the origin.

*In Exercises 59–62, determine whether the statement is true or false. If it is true, explain why. If it is false, explain why or give an example to show that it is false.*

**59.** The curve defined by $\mathbf{r}_1(t) = t^2 \mathbf{i} + t^2 \mathbf{j} + t^2 \mathbf{k}$ is the same as the curve defined by $\mathbf{r}_2(\theta) = \theta \mathbf{i} + \theta \mathbf{j} + \theta \mathbf{k}$.

**60.** If $f$, $g$, and $h$ are linear functions of $t$ for $t$ in $(-\infty, \infty)$, then $\mathbf{r}(t) = f(t)\mathbf{i} + g(t)\mathbf{j} + h(t)\mathbf{k}$ defines a line in 3-space.

**61.** The curve defined by $\mathbf{r}(t) = f(t)\mathbf{i} + g(t)\mathbf{j} + c\mathbf{k}$, where $c$ is a constant, is a curve lying in the plane $z = c$.

**62.** If $\mathbf{r}$ is continuous on an interval $I$ and if $a$ is any number in $I$, then $\lim_{t \to a} \mathbf{r}(t) = \mathbf{r}(a)$.

---

## 11.2    Differentiation and Integration of Vector-Valued Functions

### ■ The Derivative of a Vector Function

The derivative of a vector function is defined in much the same way as the derivative of a real-valued function of a real variable.

> **DEFINITION    Derivative of a Vector Function**
> The derivative of a **vector function r** is the vector function $\mathbf{r}'$ defined by
>
> $$\mathbf{r}'(t) = \frac{d\mathbf{r}}{dt} = \lim_{h \to 0} \frac{\mathbf{r}(t + h) - \mathbf{r}(t)}{h}$$
>
> provided that the limit exists.

To obtain a geometric interpretation of this derivative, let $\mathbf{r}$ be a vector function, and let $C$ be the curve traced by the tip of $\mathbf{r}$. Let $t$ be a fixed but otherwise arbitrary number in the parameter interval $I$. If $h > 0$, then the vector $\mathbf{r}(t + h) - \mathbf{r}(t)$ lies on the secant line passing through the points $P$ and $Q$, the terminal points of the vectors $\mathbf{r}(t)$ and $\mathbf{r}(t + h)$, respectively. (See Figure 1.)

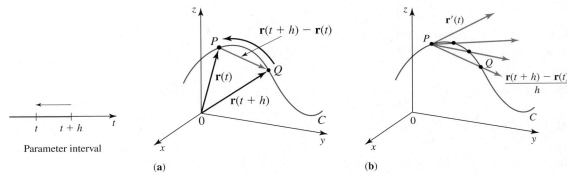

**FIGURE 1**
As $h$ approaches 0, $Q$ approaches $P$ along $C$, and the vector $\dfrac{\mathbf{r}(t + h) - \mathbf{r}(t)}{h}$ approaches the tangent vector $\mathbf{r}'(t)$.

The vector $[\mathbf{r}(t + h) - \mathbf{r}(t)]/h$, which is a scalar multiple of $\mathbf{r}(t + h) - \mathbf{r}(t)$, also lies on the secant line. (See Figure 1b.) As $h$ approaches 0, the number $t + h$ approaches $t$ along the parameter interval, and the point $Q$, in turn, approaches the point $P$ along the curve $C$. As a consequence, the vector $[\mathbf{r}(t + h) - \mathbf{r}(t)]/h$ approaches the fixed vector $\mathbf{r}'(t)$, which lies on the tangent line to the curve at $P$. In other words, the derivative $\mathbf{r}'$ of the vector $\mathbf{r}$ may be interpreted as the **tangent vector** to the curve defined by $\mathbf{r}$ at the point $P$, provided that $\mathbf{r}'(t) \neq \mathbf{0}$. If we divide $\mathbf{r}'(t)$ by its length, we obtain the **unit tangent vector**

$$\mathbf{T}(t) = \frac{\mathbf{r}'(t)}{|\mathbf{r}'(t)|}$$

which has unit length and the direction of $\mathbf{r}'$.

The following theorem tells us that the derivative $\mathbf{r}'$ of a vector function can be found by differentiating the components of $\mathbf{r}$.

---

**THEOREM 1   Differentiation of Vector Functions**

Let $\mathbf{r}(t) = f(t)\mathbf{i} + g(t)\mathbf{j} + h(t)\mathbf{k}$, where $f$, $g$, and $h$ are differentiable functions of $t$. Then

$$\mathbf{r}'(t) = f'(t)\mathbf{i} + g'(t)\mathbf{j} + h'(t)\mathbf{k}$$

---

**PROOF**   We compute

$$\mathbf{r}'(t) = \lim_{\Delta t \to 0} \frac{\mathbf{r}(t + \Delta t) - \mathbf{r}(t)}{\Delta t} \qquad \text{We use } \Delta t \text{ instead of } h \text{ so as not to confuse the increment}$$
$$\text{of } t \text{ with the component function } h.$$

$$= \lim_{\Delta t \to 0} \left[ \frac{f(t + \Delta t)\mathbf{i} + g(t + \Delta t)\mathbf{j} + h(t + \Delta t)\mathbf{k} - [f(t)\mathbf{i} + g(t)\mathbf{j} + h(t)\mathbf{k}]}{\Delta t} \right]$$

$$= \lim_{\Delta t \to 0} \left[ \frac{f(t + \Delta t) - f(t)}{\Delta t}\mathbf{i} + \frac{g(t + \Delta t) - g(t)}{\Delta t}\mathbf{j} + \frac{h(t + \Delta t) - h(t)}{\Delta t}\mathbf{k} \right]$$

$$= \left[ \lim_{\Delta t \to 0} \frac{f(t + \Delta t) - f(t)}{\Delta t} \right]\mathbf{i} + \left[ \lim_{\Delta t \to 0} \frac{g(t + \Delta t) - g(t)}{\Delta t} \right]\mathbf{j} + \left[ \lim_{\Delta t \to 0} \frac{h(t + \Delta t) - h(t)}{\Delta t} \right]\mathbf{k}$$

$$= f'(t)\mathbf{i} + g'(t)\mathbf{j} + h'(t)\mathbf{k} \qquad \blacksquare$$

---

**EXAMPLE 1**

**a.** Find the derivative of $\mathbf{r}(t) = (t^2 + 1)\mathbf{i} + e^{-t}\mathbf{j} - \sin 2t\,\mathbf{k}$.

**b.** Find the point of tangency and the unit tangent vector at the point on the curve corresponding to $t = 0$.

**Solution**

**a.** Using Theorem 1, we obtain

$$\mathbf{r}'(t) = 2t\mathbf{i} - e^{-t}\mathbf{j} - 2\cos 2t\,\mathbf{k}$$

**b.** Since $\mathbf{r}(0) = \mathbf{i} + \mathbf{j}$, we see that the point of tangency is $(1, 1, 0)$. Next, since $\mathbf{r}'(0) = -\mathbf{j} - 2\mathbf{k}$, we find the unit tangent vector at $(1, 1, 0)$ to be

$$\mathbf{T}(0) = \frac{\mathbf{r}'(0)}{|\mathbf{r}'(0)|} = \frac{-\mathbf{j} - 2\mathbf{k}}{\sqrt{1 + 4}} = -\frac{1}{\sqrt{5}}\mathbf{j} - \frac{2}{\sqrt{5}}\mathbf{k} \qquad \blacksquare$$

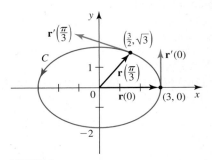

**FIGURE 2**
The vectors $\mathbf{r}'(0)$ and $\mathbf{r}'(\pi/3)$ are tangent to the curve at the points $(3, 0)$ and $\left(\frac{3}{2}, \sqrt{3}\right)$, respectively.

**EXAMPLE 2** Find the tangent vectors to the plane curve $C$ defined by the vector function $\mathbf{r}(t) = 3 \cos t\mathbf{i} + 2 \sin t\mathbf{j}$ at the points where $t = 0$ and $t = \pi/3$. Make a sketch of $C$, and display the position vectors $\mathbf{r}(0)$ and $\mathbf{r}(\pi/3)$ and the tangent vectors $\mathbf{r}'(0)$ and $\mathbf{r}'(\pi/3)$.

**Solution** The tangent vector to the curve $C$ at any point is given by

$$\mathbf{r}'(t) = -3 \sin t\mathbf{i} + 2 \cos t\mathbf{j}$$

In particular, the tangent vectors at the points where $t = 0$ and $t = \pi/3$ are

$$\mathbf{r}'(0) = 2\mathbf{j} \quad \text{and} \quad \mathbf{r}'\left(\frac{\pi}{3}\right) = -\frac{3\sqrt{3}}{2}\mathbf{i} + \mathbf{j}$$

These vectors are shown emanating from their points of tangency at $(3, 0)$ and $\left(\frac{3}{2}, \sqrt{3}\right)$ in Figure 2. ∎

**EXAMPLE 3** Find parametric equations for the tangent line to the helix with parametric equations

$$x = 3 \cos t \qquad y = 2 \sin t \qquad z = t$$

at the point where $t = \pi/6$.

**Solution** The vector function that describes the helix is

$$\mathbf{r}(t) = 3 \cos t\mathbf{i} + 2 \sin t\mathbf{j} + t\mathbf{k}$$

The tangent vector at any point on the helix is

$$\mathbf{r}'(t) = -3 \sin t\mathbf{i} + 2 \cos t\mathbf{j} + \mathbf{k}$$

In particular, the tangent vector at the point $\left(\frac{3\sqrt{3}}{2}, 1, \frac{\pi}{6}\right)$, where $t = \pi/6$, is

$$\mathbf{r}'\left(\frac{\pi}{6}\right) = -\frac{3}{2}\mathbf{i} + \sqrt{3}\mathbf{j} + \mathbf{k}$$

Finally, we observe that the required tangent line passes through the point $\left(\frac{3\sqrt{3}}{2}, 1, \frac{\pi}{6}\right)$ and has the same direction as the tangent vector $\mathbf{r}'(\pi/6)$. Using Equation (1) of Section 10.5, we see that the parametric equations of this line are

$$x = \frac{3\sqrt{3}}{2} - \frac{3}{2}t, \qquad y = 1 + \sqrt{3}t, \qquad \text{and} \qquad z = \frac{\pi}{6} + t \qquad ∎$$

## Higher-Order Derivatives

Higher-order derivatives of vector functions are obtained by successive differentiation of the lower-order derivatives of the function. For example, the **second derivative** of $\mathbf{r}(t)$ is

$$\mathbf{r}''(t) = \frac{d}{dt}\mathbf{r}'(t) = f''(t)\mathbf{i} + g''(t)\mathbf{j} + h''(t)\mathbf{k}$$

**EXAMPLE 4** Find $\mathbf{r}''(t)$ if $\mathbf{r}(t) = 2e^{3t}\mathbf{i} + \ln t\mathbf{j} + \sin t\mathbf{k}$.

**Solution** We have

$$\mathbf{r}'(t) = 6e^{3t}\mathbf{i} + \frac{1}{t}\mathbf{j} + \cos t\mathbf{k}$$

and

$$\mathbf{r}''(t) = 18e^{3t}\mathbf{i} - \frac{1}{t^2}\mathbf{j} - \sin t\mathbf{k}$$ ■

## ▨ Rules of Differentiation

The following theorem gives the rules of differentiation for vector functions. As you might expect, some of the rules are similar to the differentiation rules of Chapter 2.

---

**THEOREM 2    Rules of Differentiation**

Suppose that $\mathbf{u}$ and $\mathbf{v}$ are differentiable vector functions, $f$ is a differentiable real-valued function, and $c$ is a scalar. Then

1. $\dfrac{d}{dt}[\mathbf{u}(t) \pm \mathbf{v}(t)] = \mathbf{u}'(t) \pm \mathbf{v}'(t)$

2. $\dfrac{d}{dt}[c\mathbf{u}(t)] = c\mathbf{u}'(t)$

3. $\dfrac{d}{dt}[f(t)\mathbf{u}(t)] = f'(t)\mathbf{u}(t) + f(t)\mathbf{u}'(t)$

4. $\dfrac{d}{dt}[\mathbf{u}(t) \cdot \mathbf{v}(t)] = \mathbf{u}'(t) \cdot \mathbf{v}(t) + \mathbf{u}(t) \cdot \mathbf{v}'(t)$

5. $\dfrac{d}{dt}[\mathbf{u}(t) \times \mathbf{v}(t)] = \mathbf{u}'(t) \times \mathbf{v}(t) + \mathbf{u}(t) \times \mathbf{v}'(t)$

6. $\dfrac{d}{dt}[\mathbf{u}(f(t))] = \mathbf{u}'(f(t))f'(t)$      Chain Rule

---

We will prove Rule 4 and leave the proofs of the other rules as exercises.

**PROOF**    Let

$$\mathbf{u}(t) = f_1(t)\mathbf{i} + g_1(t)\mathbf{j} + h_1(t)\mathbf{k} \qquad \text{and} \qquad \mathbf{v}(t) = f_2(t)\mathbf{i} + g_2(t)\mathbf{j} + h_2(t)\mathbf{k}$$

Then

$$\mathbf{u}(t) \cdot \mathbf{v}(t) = f_1(t)f_2(t) + g_1(t)g_2(t) + h_1(t)h_2(t)$$

Therefore

$$\begin{aligned}
\frac{d}{dt}[\mathbf{u}(t) \cdot \mathbf{v}(t)] &= [f'_1(t)f_2(t) + g'_1(t)g_2(t) + h'_1(t)h_2(t)] \\
&\quad + [f_1(t)f'_2(t) + g_1(t)g'_2(t) + h_1(t)h'_2(t)] \\
&= \mathbf{u}'(t) \cdot \mathbf{v}(t) + \mathbf{u}(t) \cdot \mathbf{v}'(t)
\end{aligned}$$ ■

**EXAMPLE 5**   Suppose that $\mathbf{v}$ is a differentiable vector function of constant length $c$. Show that $\mathbf{v} \cdot \mathbf{v}' = 0$. In other words, the vector $\mathbf{v}$ and its tangent vector $\mathbf{v}'$ must be orthogonal.

**Solution**   The condition on $\mathbf{v}$ implies that

$$\mathbf{v} \cdot \mathbf{v} = |\mathbf{v}|^2 = c^2$$

Differentiating both sides of this equation with respect to $t$ and using Rule 4 of differentiation, we obtain

$$\frac{d}{dt}(\mathbf{v} \cdot \mathbf{v}) = \mathbf{v} \cdot \mathbf{v}' + \mathbf{v}' \cdot \mathbf{v} = \frac{d}{dt}(c^2) = 0$$

But $\mathbf{v}' \cdot \mathbf{v} = \mathbf{v} \cdot \mathbf{v}'$, so we have

$$2\mathbf{v} \cdot \mathbf{v}' = 0 \qquad \text{or} \qquad \mathbf{v} \cdot \mathbf{v}' = 0 \qquad \blacksquare$$

The result of Example 5 has the following geometric interpretation: If a curve lies on a sphere with center at the origin, then the tangent vector $\mathbf{r}'(t)$ is always perpendicular to the position vector $\mathbf{r}(t)$.

**EXAMPLE 6**   Let $\mathbf{r}(s) = 2 \cos 2s\mathbf{i} + 3 \sin 2s\mathbf{j} + 4s\mathbf{k}$, where $s = f(t) = t^2$. Find $\dfrac{d\mathbf{r}}{dt}$.

**Solution**   Using the Chain Rule, we obtain

$$\frac{d}{dt}[\mathbf{r}(s)] = \frac{d}{ds}(2 \cos 2s\mathbf{i} + 3 \sin 2s\mathbf{j} + 4s\mathbf{k})\left(\frac{ds}{dt}\right)$$

$$= (-4 \sin 2s\mathbf{i} + 6 \cos 2s\mathbf{j} + 4\mathbf{k})(2t)$$

$$= -8t \sin 2t^2\mathbf{i} + 12t \cos 2t^2\mathbf{j} + 8t\mathbf{k} \qquad \text{Replace } s \text{ by } t^2. \quad \blacksquare$$

## ■ Integration of Vector Functions

As with the differentiation of vector functions, integration of vector functions is done component-wise, so we have the following definitions.

---

**DEFINITIONS**   **Integration of Vector Functions**
Let $\mathbf{r}(t) = f(t)\mathbf{i} + g(t)\mathbf{j} + h(t)\mathbf{k}$, where $f$, $g$, and $h$ are integrable. Then

**1.** The **indefinite integral of r with respect to $t$** is

$$\int \mathbf{r}(t)\, dt = \left[\int f(t)\, dt\right]\mathbf{i} + \left[\int g(t)\, dt\right]\mathbf{j} + \left[\int h(t)\, dt\right]\mathbf{k}$$

**2.** The **definite integral of r** over the interval $[a, b]$ is

$$\int_a^b \mathbf{r}(t)\, dt = \left[\int_a^b f(t)\, dt\right]\mathbf{i} + \left[\int_a^b g(t)\, dt\right]\mathbf{j} + \left[\int_a^b h(t)\, dt\right]\mathbf{k}$$

---

**EXAMPLE 7**   Find $\int \mathbf{r}(t)\, dt$ if $\mathbf{r}(t) = (t + 1)\mathbf{i} + \cos 2t\mathbf{j} + e^{3t}\mathbf{k}$.

**Solution**

$$\int \mathbf{r}(t)\, dt = \int [(t + 1)\mathbf{i} + \cos 2t\mathbf{j} + e^{3t}\mathbf{k}]\, dt$$

$$= \left[\int (t + 1)\, dt\right]\mathbf{i} + \left[\int \cos 2t\, dt\right]\mathbf{j} + \left[\int e^{3t}\, dt\right]\mathbf{k}$$

$$= \left(\frac{1}{2}t^2 + t + C_1\right)\mathbf{i} + \left(\frac{1}{2}\sin 2t + C_2\right)\mathbf{j} + \left(\frac{1}{3}e^{3t} + C_3\right)\mathbf{k}$$

where $C_1$, $C_2$, and $C_3$ are constants of integration. We can rewrite the last expression as

$$\left(\frac{1}{2}t^2 + t\right)\mathbf{i} + \frac{1}{2}\sin 2t\mathbf{j} + \frac{1}{3}e^{3t}\mathbf{k} + C_1\mathbf{i} + C_2\mathbf{j} + C_3\mathbf{k}$$

or, upon letting $\mathbf{C} = C_1\mathbf{i} + C_2\mathbf{j} + C_3\mathbf{k}$,

$$\int \mathbf{r}(t)\, dt = \left(\frac{1}{2}t^2 + t\right)\mathbf{i} + \frac{1}{2}\sin 2t\mathbf{j} + \frac{1}{3}e^{3t}\mathbf{k} + \mathbf{C}$$

where $\mathbf{C}$ is a constant (vector) of integration. ■

**Note** In general, the indefinite integral of $\mathbf{r}$ can be written as

$$\int \mathbf{r}(t)\, dt = \mathbf{R}(t) + \mathbf{C}$$

where $\mathbf{C}$ is an arbitrary constant vector and $\mathbf{R}'(t) = \mathbf{r}(t)$. ■

**EXAMPLE 8** Find the antiderivative of $\mathbf{r}'(t) = \cos t\mathbf{i} + e^{-t}\mathbf{j} + \sqrt{t}\mathbf{k}$ satisfying the initial condition $\mathbf{r}(0) = \mathbf{i} + 2\mathbf{j} + 3\mathbf{k}$.

**Solution** We have

$$\mathbf{r}(t) = \int \mathbf{r}'(t)\, dt = \int (\cos t\mathbf{i} + e^{-t}\mathbf{j} + t^{1/2}\mathbf{k})\, dt$$

$$= \sin t\mathbf{i} - e^{-t}\mathbf{j} + \frac{2}{3}t^{3/2}\mathbf{k} + \mathbf{C}$$

where $\mathbf{C}$ is a constant (vector) of integration. To determine $\mathbf{C}$, we use the condition $\mathbf{r}(0) = \mathbf{i} + 2\mathbf{j} + 3\mathbf{k}$ to obtain

$$\mathbf{r}(0) = 0\mathbf{i} - \mathbf{j} + 0\mathbf{k} + \mathbf{C} = \mathbf{i} + 2\mathbf{j} + 3\mathbf{k}$$

from which we find $\mathbf{C} = \mathbf{i} + 3\mathbf{j} + 3\mathbf{k}$. Therefore,

$$\mathbf{r}(t) = \sin t\mathbf{i} - e^{-t}\mathbf{j} + \frac{2}{3}t^{3/2}\mathbf{k} + \mathbf{i} + 3\mathbf{j} + 3\mathbf{k}$$

$$= (1 + \sin t)\mathbf{i} + (3 - e^{-t})\mathbf{j} + \left(3 + \frac{2}{3}t^{3/2}\right)\mathbf{k}$$ ■

**EXAMPLE 9** Evaluate $\int_0^1 \mathbf{r}(t)\, dt$ if $\mathbf{r}(t) = t^2\mathbf{i} + \dfrac{1}{t+1}\mathbf{j} + e^{-t}\mathbf{k}$.

**Solution**

$$\int_0^1 \mathbf{r}(t)\, dt = \int_0^1 \left(t^2\mathbf{i} + \frac{1}{t+1}\mathbf{j} + e^{-t}\mathbf{k}\right) dt$$

$$= \left[\int_0^1 t^2 dt\right]\mathbf{i} + \left[\int_0^1 \frac{1}{t+1}\, dt\right]\mathbf{j} + \left[\int_0^1 e^{-t}\, dt\right]\mathbf{k}$$

$$= \left[\frac{1}{3}t^3\right]_0^1 \mathbf{i} + \left[\ln(t+1)\right]_0^1 \mathbf{j} + \left[-e^{-t}\right]_0^1 \mathbf{k}$$

$$= \frac{1}{3}\mathbf{i} + \ln 2\mathbf{j} + \left(1 - \frac{1}{e}\right)\mathbf{k}$$ ■

## 11.2    CONCEPT QUESTIONS

1.  a.  What is the derivative of a vector function?
    b.  If $\mathbf{r}(t) = f(t)\mathbf{i} + g(t)\mathbf{j} + h(t)\mathbf{k}$, what is $\mathbf{r}'(t)$?
    c.  Give an example of a function $\mathbf{r}(t)$ such that $\mathbf{r}'(0)$ does not exist.

2.  If $\mathbf{w}(t) = \mathbf{u}(f(t)) \times \mathbf{v}(f(t))$, what is $\mathbf{w}'(t)$? Assume that $\mathbf{u}$, $\mathbf{v}$, and $f$ are all differentiable.

3.  Let $\mathbf{r}(t) = f(t)\mathbf{i} + g(t)\mathbf{j} + h(t)\mathbf{k}$.
    a.  What is the indefinite integral of $\mathbf{r}$ with respect to $t$?
    b.  What is the definite integral of $\mathbf{r}$ over the interval $[a, b]$?

## 11.2    EXERCISES

*In Exercises 1–8, find $\mathbf{r}'(t)$ and $\mathbf{r}''(t)$.*

1.  $\mathbf{r}(t) = t\mathbf{i} + t^2\mathbf{j} + t^3\mathbf{k}$

2.  $\mathbf{r}(t) = \sqrt{t}\mathbf{i} + \dfrac{1}{t}\mathbf{j} + \ln t\mathbf{k}$

3.  $\mathbf{r}(t) = \langle t^2 - 1, \sqrt{t^2 + 1} \rangle$

4.  $\mathbf{r}(t) = \langle t \cos t, t \sin t, \tan t \rangle$

5.  $\mathbf{r}(t) = \langle t \cos t - \sin t, t \sin t + \cos t \rangle$

6.  $\mathbf{r}(t) = e^{-t}\mathbf{i} + te^t\mathbf{j} + e^{-2t}\mathbf{k}$

7.  $\mathbf{r}(t) = e^{-t} \sin t\mathbf{i} + e^{-t} \cos t\mathbf{j} + \tan^{-1} t\mathbf{k}$

8.  $\mathbf{r}(t) = \langle \sin^{-1} t, \sec t, \ln|t| \rangle$

*In Exercises 9–16, (a) find $\mathbf{r}(a)$ and $\mathbf{r}'(a)$ at the given value of a. (b) Sketch the curve defined by $\mathbf{r}$ and the vectors $\mathbf{r}(a)$ and $\mathbf{r}'(a)$ on the same set of axes.*

9.  $\mathbf{r}(t) = \sqrt{t}\mathbf{i} + (t - 4)\mathbf{j};\quad a = 2$

10. $\mathbf{r}(t) = \sin t\mathbf{i} + \cos t\mathbf{j};\quad a = \dfrac{\pi}{4}$

11. $\mathbf{r}(t) = \langle 4 \cos t, 2 \sin t \rangle;\quad a = \dfrac{\pi}{3}$

12. $\mathbf{r}(t) = t^2\mathbf{i} + t^3\mathbf{j};\quad a = 1$

13. $\mathbf{r}(t) = (2 + 3t)\mathbf{i} + (1 - 2t)\mathbf{j};\quad a = 1$

14. $\mathbf{r}(t) = \langle e^t, e^{-2t} \rangle;\quad a = 0$

15. $\mathbf{r}(t) = \sec t\mathbf{i} + 2 \tan t\mathbf{j};\quad a = \dfrac{\pi}{4}$

16. $\mathbf{r}(t) = b \cos^3 t\mathbf{i} + b \sin^3 t\mathbf{j};\quad a = \dfrac{\pi}{4}$

*In Exercises 17–20, find the unit tangent vector $\mathbf{T}(t)$ at the point corresponding to the given value of the parameter t.*

17. $\mathbf{r}(t) = t\mathbf{i} + 2t\mathbf{j} + 3t\mathbf{k};\quad t = 1$

18. $\mathbf{r}(t) = \langle e^t, te^{-t}, (t + 1)e^{2t} \rangle;\quad t = 0$

19. $\mathbf{r}(t) = 2 \sin 2t\mathbf{i} + 3 \cos 2t\mathbf{j} + 3\mathbf{k};\quad t = \dfrac{\pi}{6}$

20. $\mathbf{r}(t) = t \sin t\mathbf{i} + t \cos t\mathbf{j} + t\mathbf{k};\quad t = \dfrac{\pi}{2}$

*In Exercises 21–26, find parametric equations for the tangent line to the curve with the given parametric equations at the point with the indicated value of t.*

21. $x = t,\quad y = t^2,\quad z = t^3;\quad t = 1$

22. $x = 1 + t,\quad y = t^2 - 4,\quad z = \sqrt{t};\quad t = 4$

23. $x = \sqrt{t + 2},\quad y = \dfrac{1}{t + 1},\quad z = \dfrac{2}{t^2 + 4};\quad t = 2$

24. $x = 2 \cos t,\quad y = t^2,\quad z = 2 \sin t;\quad t = \dfrac{\pi}{4}$

25. $x = t \cos t,\quad y = t \sin t,\quad z = te^t;\quad t = \dfrac{\pi}{6}$

26. $x = e^{-t} \cos t,\quad y = e^{-t} \sin t,\quad z = \sin^{-1}t;\quad t = 0$

*In Exercises 27–34, find or evaluate the integral.*

27. $\displaystyle\int (t\mathbf{i} + 2t^2\mathbf{j} + 3\mathbf{k})\, dt$

28. $\displaystyle\int_0^1 (t\mathbf{i} + t^2\mathbf{j} + t^3\mathbf{k})\, dt$

29. $\displaystyle\int \left( \sqrt{t}\mathbf{i} + \dfrac{1}{t}\mathbf{j} - t^{3/2}\mathbf{k} \right) dt$

30. $\displaystyle\int_1^2 \left[ \sqrt{t - 1}\mathbf{i} + \dfrac{1}{\sqrt{t}}\mathbf{j} + (2t - 1)^5\mathbf{k} \right] dt$

31. $\displaystyle\int (\sin 2t\mathbf{i} + \cos 2t\mathbf{j} + e^{-t}\mathbf{k})\, dt$

32. $\displaystyle\int (te^t\mathbf{i} + 2\mathbf{j} - \sec^2 t\mathbf{k})\, dt$

33. $\displaystyle\int (t \cos t\mathbf{i} + t \sin t^2\mathbf{j} - te^{t^2}\mathbf{k})\, dt$

34. $\displaystyle\int \left[ \dfrac{1}{1 + t^2}\mathbf{i} + \dfrac{t}{1 + 2t^2}\mathbf{j} - \dfrac{1}{\sqrt{1 - t^2}}\mathbf{k} \right] dt$

*In Exercises 35–40, find* $\mathbf{r}(t)$ *satisfying the given conditions.*

**35.** $\mathbf{r}'(t) = 2\mathbf{i} + 4t\mathbf{j} - 6t^2\mathbf{k}$;   $\mathbf{r}(0) = \mathbf{i} + \mathbf{k}$

**36.** $\mathbf{r}'(t) = 2\sin 2t\mathbf{i} + 3\cos 2t\mathbf{j} + t\mathbf{k}$;   $\mathbf{r}(0) = \mathbf{i} + 2\mathbf{j} + \dfrac{1}{2}\mathbf{k}$

**37.** $\mathbf{r}'(t) = 2e^{2t}\mathbf{i} + 3e^{-t}\mathbf{j} + e^t\mathbf{k}$;   $\mathbf{r}(0) = \mathbf{i} - \mathbf{j} + \mathbf{k}$

**38.** $\mathbf{r}'(t) = \sqrt{t+1}\,\mathbf{i} + \dfrac{t}{t^2+1}\mathbf{j} + \dfrac{1}{t}\mathbf{k}$;   $\mathbf{r}(3) = \mathbf{i} + \mathbf{j} + 2\mathbf{k}$

**39.** $\mathbf{r}''(t) = \sqrt{t}\,\mathbf{i} + \sec^2 t\mathbf{j} + e^t\mathbf{k}$;   $\mathbf{r}'(0) = \mathbf{i} + \mathbf{k}$, $\mathbf{r}(0) = 2\mathbf{i} + \mathbf{j} - \mathbf{k}$

**40.** $\mathbf{r}''(t) = 3\cos 2t\mathbf{i} + 4\sin 2t\mathbf{j} + \mathbf{k}$;   $\mathbf{r}'(0) = \mathbf{i} + 2\mathbf{j}$, $\mathbf{r}(0) = 2\mathbf{i} + \mathbf{j} - \mathbf{k}$

*In Exercises 41–46, let* $\mathbf{u}(t) = t^2\mathbf{i} - 2t\mathbf{j} + 2\mathbf{k}$, $\mathbf{v}(t) = \cos t\mathbf{i} + \sin t\mathbf{j} + t^2\mathbf{k}$, *and* $f(t) = e^{2t}$.

**41.** Show that $\dfrac{d}{dt}[\mathbf{u}(t) + \mathbf{v}(t)] = \mathbf{u}'(t) + \mathbf{v}'(t)$.

**42.** Show that $\dfrac{d}{dt}[3\mathbf{u}(t)] = 3\mathbf{u}'(t)$.

**43.** Show that $\dfrac{d}{dt}[f(t)\mathbf{u}(t)] = f'(t)\mathbf{u}(t) + f(t)\mathbf{u}'(t)$.

**44.** Show that $\dfrac{d}{dt}[\mathbf{u}(t) \cdot \mathbf{v}(t)] = \mathbf{u}'(t) \cdot \mathbf{v}(t) + \mathbf{u}(t) \cdot \mathbf{v}'(t)$.

**45.** Show that $\dfrac{d}{dt}[\mathbf{u}(t) \times \mathbf{v}(t)] = \mathbf{u}'(t) \times \mathbf{v}(t) + \mathbf{u}(t) \times \mathbf{v}'(t)$.

**46.** Show that $\dfrac{d}{dt}[\mathbf{u}(f(t))] = \mathbf{u}'[f(t)]f'(t)$.

*In Exercises 47–52, suppose* $\mathbf{u}$ *and* $\mathbf{v}$ *are differentiable vector functions, f is a differentiable real-valued function, and c is a scalar. Prove each rule.*

**47.** $\dfrac{d}{dt}[\mathbf{u}(t) + \mathbf{v}(t)] = \mathbf{u}'(t) + \mathbf{v}'(t)$

**48.** $\dfrac{d}{dt}[\mathbf{u}(t) - \mathbf{v}(t)] = \mathbf{u}'(t) - \mathbf{v}'(t)$

**49.** $\dfrac{d}{dt}[c\mathbf{u}(t)] = c\mathbf{u}'(t)$

**50.** $\dfrac{d}{dt}[f(t)\mathbf{u}(t)] = f'(t)\mathbf{u}(t) + f(t)\mathbf{u}'(t)$

**51.** $\dfrac{d}{dt}[\mathbf{u}(t) \times \mathbf{v}(t)] = \mathbf{u}'(t) \times \mathbf{v}(t) + \mathbf{u}(t) \times \mathbf{v}'(t)$

**52.** $\dfrac{d}{dt}[\mathbf{u}(f(t))] = \mathbf{u}'(f(t))f'(t)$

**53.** Prove that $\dfrac{d}{dt}[\mathbf{r}(t) \times \mathbf{r}'(t)] = \mathbf{r}(t) \times \mathbf{r}''(t)$.

**54.** Prove that

$$\dfrac{d}{dt}[\mathbf{r}(t) \cdot (\mathbf{u}(t) \times \mathbf{v}(t))]$$
$$= \mathbf{r}'(t) \cdot [\mathbf{u}(t) \times \mathbf{v}(t)] + \mathbf{r}(t) \cdot [\mathbf{u}'(t) \times \mathbf{v}(t)]$$
$$+ \mathbf{r}(t) \cdot [\mathbf{u}(t) \times \mathbf{v}'(t)]$$

*In Exercises 55–58, find the indicated derivative.*

**55.** $\dfrac{d}{dt}\left[\mathbf{r}(-t) + \mathbf{r}\left(\dfrac{1}{t}\right)\right]$

**56.** $\dfrac{d}{dt}[\mathbf{r}(2t) \cdot \mathbf{r}(t^2)]$

**57.** $\dfrac{d}{dt}[\mathbf{r}(t) \cdot (\mathbf{r}'(t) \times \mathbf{r}''(t))]$

**58.** $\dfrac{d}{dt}\{\mathbf{u}(t) \times [\mathbf{v}(t) \times \mathbf{w}(t)]\}$

*In Exercises 59 and 60, suppose that* $\mathbf{u}$ *and* $\mathbf{v}$ *are integrable on* $[a, b]$ *and that c is a scalar. Prove each property.*

**59.** $\displaystyle\int_a^b [\mathbf{u}(t) + \mathbf{v}(t)]\,dt = \int_a^b \mathbf{u}(t)\,dt + \int_a^b \mathbf{v}(t)\,dt$

**60.** $\displaystyle\int_a^b c\mathbf{u}(t)\,dt = c\int_a^b \mathbf{u}(t)\,dt$

**61. a.** Suppose that $\mathbf{r}$ is integrable on $[a, b]$ and that $\mathbf{c}$ is a constant vector. Prove that

$$\int_a^b \mathbf{c} \cdot \mathbf{r}(t)\,dt = \mathbf{c} \cdot \int_a^b \mathbf{r}(t)\,dt$$

**b.** Verify this property directly for the vector function

$$\mathbf{r}(t) = \sin t\mathbf{i} + \cos t\mathbf{j} + t\mathbf{k},$$
$$\mathbf{c} = 2\mathbf{i} + 3\mathbf{j} - \mathbf{k}, \quad \text{and} \quad a = 0, \quad b = \pi$$

*In Exercises 62–65, determine whether the statement is true or false. If it is true, explain why. If it is false, explain why or give an example to show that it is false.*

**62.** If $\mathbf{c}$ is a constant vector, then $\dfrac{d}{dt}(\mathbf{c}) = 0$.

**63.** $\dfrac{d}{dt}(|\mathbf{u}|^2) = 2\mathbf{u} \cdot \mathbf{u}'$

**64.** If $\mathbf{r}'(t) = \mathbf{0}$, then $\mathbf{r}(t) = \mathbf{c}$, where $\mathbf{c}$ is an arbitrary constant vector.

**65.** If $\mathbf{r}$ is differentiable and $\mathbf{r}(t) \cdot \mathbf{r}'(t) = 0$ for all $t$, then $\mathbf{r}$ must have constant length.

## 11.3    Arc Length and Curvature

### ■ Arc Length

In Section 9.3 we saw that the length of the plane curve given by the parametric equations $x = f(t)$ and $y = g(t)$, where $a \leq t \leq b$, is

$$L = \int_a^b \sqrt{\left(\frac{dx}{dt}\right)^2 + \left(\frac{dy}{dt}\right)^2}\, dt = \int_a^b \sqrt{[f'(t)]^2 + [g'(t)]^2}\, dt$$

Now, suppose that $C$ is described by the vector function $\mathbf{r}(t) = f(t)\mathbf{i} + g(t)\mathbf{j}$ instead. Then

$$\mathbf{r}'(t) = f'(t)\mathbf{i} + g'(t)\mathbf{j}$$

and

$$|\mathbf{r}'(t)| = \sqrt{\mathbf{r}'(t) \cdot \mathbf{r}'(t)} = \sqrt{[f'(t)]^2 + [g'(t)]^2}$$

from which we see that $L$ can also be written in the form

$$L = \int_a^b |\mathbf{r}'(t)|\, dt$$

A similar formula for calculating the length of a space curve is contained in the following theorem.

---

**THEOREM 1    Arc Length of a Space Curve**

Let $C$ be a curve given by the vector function

$$\mathbf{r}(t) = f(t)\mathbf{i} + g(t)\mathbf{j} + h(t)\mathbf{k} \qquad a \leq t \leq b$$

where $f'$, $g'$, and $h'$ are continuous. If $C$ is traversed exactly once as $t$ increases from $a$ to $b$, then its length is given by

$$L = \int_a^b \sqrt{[f'(t)]^2 + [g'(t)]^2 + [h'(t)]^2}\, dt = \int_a^b |\mathbf{r}'(t)|\, dt$$

---

**EXAMPLE 1**    Find the length of the arc of the helix $C$ given by the vector function $\mathbf{r}(t) = 2 \cos t\mathbf{i} + 2 \sin t\mathbf{j} + t\mathbf{k}$, where $0 \leq t \leq 2\pi$, as shown in Figure 1.

**Solution**    We first compute

$$\mathbf{r}'(t) = -2 \sin t\mathbf{i} + 2 \cos t\mathbf{j} + \mathbf{k}$$

Then, using Theorem 1, we see that the length of the arc in question is

$$L = \int_0^{2\pi} |\mathbf{r}'(t)|\, dt = \int_0^{2\pi} \sqrt{4 \sin^2 t + 4 \cos^2 t + 1}\, dt$$

$$= \int_0^{2\pi} \sqrt{5}\, dt = 2\sqrt{5}\pi \qquad ■$$

**FIGURE 1**
The length of the arc of the helix for $0 \leq t \leq 2\pi$ is $2\sqrt{5}\pi$.

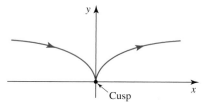

**FIGURE 2**
The curve defined by $\mathbf{r}(t) = t^3\mathbf{i} + t^2\mathbf{j}$
is smooth everywhere except at $(0, 0)$.

## Smooth Curves

A curve that is defined by a vector function $\mathbf{r}$ on a parameter interval $I$ is said to be **smooth** if $\mathbf{r}'(t)$ is continuous and $\mathbf{r}'(t) \neq \mathbf{0}$ for all $t$ in $I$ with the possible exception of the endpoints. For example, the plane curve defined by $\mathbf{r}(t) = t^3\mathbf{i} + t^2\mathbf{j}$ is smooth everywhere except at the point $(0, 0)$ corresponding to $t = 0$. To see this, we compute $\mathbf{r}'(t) = 3t^2\mathbf{i} + 2t\mathbf{j}$ and note that $\mathbf{r}'(0) = \mathbf{0}$. The curve is shown in Figure 2. The point $(0, 0)$ where the curve has a sharp corner is called a **cusp.**

## Arc Length Parameter

The curve $C$ described by the vector function $\mathbf{r}(t)$ with parameter $t$ in some parameter interval $I$ is said to be **parametrized** by $t$. A curve $C$ can have more than one parametrization. For example, the helix represented by the vector function

$$\mathbf{r}_1(t) = 2\cos t\,\mathbf{i} + 3\sin t\,\mathbf{j} + t\mathbf{k} \qquad 2\pi \leq t \leq 4\pi$$

with parameter $t$ is also represented by the function

$$\mathbf{r}_2(u) = 2\cos e^u\mathbf{i} + 3\sin e^u\mathbf{j} + e^u\mathbf{k} \qquad \ln 2\pi \leq u \leq \ln 4\pi$$

with parameter $u$, where $t$ and $u$ are related by $t = e^u$.

A useful parametrization of a curve $C$ is obtained by using the arc length of $C$ as its parameter. To see how this is done, we need the following definition.

---

**DEFINITION**   **Arc Length Function**

Suppose that $C$ is a smooth curve described by $\mathbf{r}(t) = f(t)\mathbf{i} + g(t)\mathbf{j} + h(t)\mathbf{k}$, where $a \leq t \leq b$. Then the **arc length function** $s$ is defined by

$$s(t) = \int_a^t |\mathbf{r}'(u)|\, du \tag{1}$$

---

Thus, $s(t)$ is the length of that part of $C$ (shown in red) between $\mathbf{r}(a)$ and $\mathbf{r}(t)$. (See Figure 3.) Because $s(a) = 0$, we see that the length $L$ of $C$ from $t = a$ to $t = b$ is

$$s(b) = \int_a^b |\mathbf{r}'(t)|\, dt$$

**FIGURE 3**
The arc length function $s(t)$ gives the length of that part of $C$ corresponding to the parameter interval $[a, t]$.

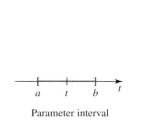

Parameter interval

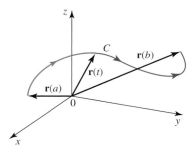

If we differentiate both sides of Equation (1) with respect to $t$ and use the Fundamental Theorem of Calculus, Part 2, we obtain

$$\frac{ds}{dt} = |\mathbf{r}'(t)| \tag{2}$$

or, in differential form,

$$ds = |\mathbf{r}'(t)|\, dt \tag{3}$$

The following example shows how to parametrize a curve in terms of its arc length.

**EXAMPLE 2**   Find the arc length function $s(t)$ for the circle $C$ in the plane described by

$$\mathbf{r}(t) = 2\cos t\mathbf{i} + 2\sin t\mathbf{j} \qquad 0 \le t \le 2\pi$$

Then use your result to find a parametrization of $C$ in terms of $s$.

**Solution**   We first compute $\mathbf{r}'(t) = -2\sin t\mathbf{i} + 2\cos t\mathbf{j}$, and then compute

$$|\mathbf{r}'(t)| = \sqrt{4\sin^2 t + 4\cos^2 t} = 2$$

Using Equation (1), we obtain

$$s(t) = \int_0^t |\mathbf{r}'(u)|\, du = \int_0^t 2\, du = 2t \qquad 0 \le t \le 2\pi$$

Writing $s$ for $s(t)$, we have $s = 2t$, where $0 \le t \le 2\pi$, which when solved for $t$, yields $t = t(s) = s/2$. Substituting this value of $t$ into the equation for $\mathbf{r}(t)$ gives

$$\mathbf{r}(t(s)) = 2\cos\left(\frac{s}{2}\right)\mathbf{i} + 2\sin\left(\frac{s}{2}\right)\mathbf{j}$$

Finally, since $s(0) = 0$ and $s(2\pi) = 4\pi$, we see that the parameter interval for this parametrization by the arc length $s$ is $[0, 4\pi]$. (See Figure 4.)

**FIGURE 4**
The curve $C$ is described by
$\mathbf{r}(t) = 2\cos t\mathbf{i} + 2\sin t\mathbf{j}$,
where $0 \le t \le 2\pi$, and
$\mathbf{r}(t(s)) = 2\cos(s/2)\mathbf{i} + 2\sin(s/2)\mathbf{j}$,
where $0 \le s \le 4\pi$.

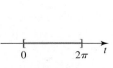

Parameter interval for $\mathbf{r}(t)$

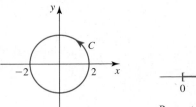

Parameter interval for $\mathbf{r}(s)$

One reason for using the arc length of a curve $C$ as the parameter stems from the fact that its tangent vector $\mathbf{r}'(s)$ has unit length; that is, $\mathbf{r}'(s)$ is a unit tangent vector. Consider the circle of Example 2. Here,

$$\mathbf{r}'(s) = -\sin\left(\frac{s}{2}\right)\mathbf{i} + \cos\left(\frac{s}{2}\right)\mathbf{j}$$

so

$$|\mathbf{r}'(s)| = \sqrt{\sin^2\left(\frac{s}{2}\right) + \cos^2\left(\frac{s}{2}\right)} = 1$$

# Curvature

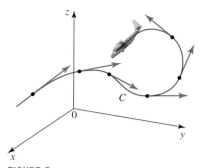

**FIGURE 5**
The unit tangent vector $\mathbf{T}(s)$ turns faster along the stretch of the path where the turn is sharper.

Figure 5 depicts the flight path $C$ of an aerobatic plane as it executes a maneuver. Suppose that the smooth curve $C$ is defined by the vector function $\mathbf{r}(s)$, where $s$ is the arc length parameter. Then the unit tangent vector function $\mathbf{T}(s) = \mathbf{r}'(s)$ gives the direction of the plane at the point on $C$ corresponding to the parameter value $s$.

In Figure 5 we have drawn the unit tangent vector $\mathbf{T}(s)$ to $C$ corresponding to several values of $s$. Observe that $\mathbf{T}(s)$ turns rather slowly along a stretch of the flight path that is relatively straight but turns more quickly along a stretch of the curve where the plane executes a sharp turn.

To measure how quickly a curve bends, we introduce the notion of the *curvature* of a curve. Specifically, we define the curvature at a point on a curve $C$ to be the magnitude of the rate of change of the unit tangent vector with respect to arc length at that point.

---

**DEFINITION   Curvature**

Let $C$ be a smooth curve defined by $\mathbf{r}(s)$, where $s$ is the arc length of the parameter. Then the **curvature** of $C$ at $s$ is

$$\kappa(s) = \left| \frac{d\mathbf{T}}{ds} \right| = |\mathbf{T}'(s)|$$

where $\mathbf{T}$ is the unit tangent vector.

---

**Note**   The Greek letter $\kappa$ is read "kappa." ∎

Although the use of the arc length parameter $s$ provides us with a natural way for defining the curvature of a curve, it is generally easier to find the curvature in terms of the parameter $t$. To see how this is done, let's apply the Chain Rule (Rule 6 in Section 11.2) to write

$$\frac{d\mathbf{T}}{dt} = \frac{d\mathbf{T}}{ds} \frac{ds}{dt}$$

Then

$$\kappa(s) = \left| \frac{d\mathbf{T}}{ds} \right| = \frac{\left| \dfrac{d\mathbf{T}}{dt} \right|}{\left| \dfrac{ds}{dt} \right|}$$

Since $ds/dt = |\mathbf{r}'(t)|$ by Equation (2), we are led to the following formula:

$$\kappa(t) = \frac{|\mathbf{T}'(t)|}{|\mathbf{r}'(t)|} \tag{4}$$

**EXAMPLE 3**   Find the curvature of a circle of radius $a$.

**Solution**   Without loss of generality we may take the circle $C$ with center at the origin. This circle is represented by the vector function

$$\mathbf{r}(t) = a \cos t\, \mathbf{i} + a \sin t\, \mathbf{j} \qquad 0 \le t \le 2\pi$$

Now

$$\mathbf{r}'(t) = -a \sin t\mathbf{i} + a \cos t\mathbf{j}$$

so

$$|\mathbf{r}'(t)| = \sqrt{a^2 \sin^2 t + a^2 \cos^2 t} = a$$

Therefore,

$$\mathbf{T}(t) = \frac{\mathbf{r}'(t)}{|\mathbf{r}'(t)|} = -\sin t\mathbf{i} + \cos t\mathbf{j}$$

Next, we compute

$$\mathbf{T}'(t) = -\cos t\mathbf{i} - \sin t\mathbf{j}$$

and

$$|\mathbf{T}'(t)| = \sqrt{\cos^2 t + \sin^2 t} = 1$$

Finally, using Equation (4), we obtain

$$\kappa(t) = \frac{|\mathbf{T}'(t)|}{|\mathbf{r}'(t)|} = \frac{1}{a}$$

Therefore, the curvature at every point on the circle of radius $a$ is $1/a$. This result agrees with our intuition: A big circle has a small curvature and vice versa. ■

The following formula expresses the curvature in terms of the vector function $\mathbf{r}$ and its derivatives.

---

**THEOREM 2  Formula for Finding Curvature**

Let $C$ be a smooth curve given by the vector function $\mathbf{r}$. Then the curvature of $C$ at any point on $C$ corresponding to $t$ is given by

$$\kappa(t) = \frac{|\mathbf{r}'(t) \times \mathbf{r}''(t)|}{|\mathbf{r}'(t)|^3}$$

---

**PROOF**  We begin by recalling that

$$\mathbf{T}(t) = \frac{\mathbf{r}'(t)}{|\mathbf{r}'(t)|}$$

Since $|\mathbf{r}'(t)| = ds/dt$, we have

$$\mathbf{r}'(t) = \frac{ds}{dt}\mathbf{T}(t)$$

Differentiating both sides of this equation with respect to $t$ and using Rule 3 in Section 11.2, we obtain

$$\mathbf{r}''(t) = \frac{d^2s}{dt^2}\mathbf{T}(t) + \frac{ds}{dt}\mathbf{T}'(t)$$

Next, we use the fact that $\mathbf{T} \times \mathbf{T} = \mathbf{0}$ (Property 6 of Theorem 3 in Section 10.4) to obtain

$$\mathbf{r}'(t) \times \mathbf{r}''(t) = \left(\frac{ds}{dt}\right)^2 (\mathbf{T}(t) \times \mathbf{T}'(t))$$

Also, $|\mathbf{T}(t)| = 1$ for all $t$ implies that $\mathbf{T}(t)$ and $\mathbf{T}'(t)$ are orthogonal. (See Example 5 in Section 11.2.) Therefore, using Theorem 2 in Section 10.4, we have

$$|\mathbf{r}'(t) \times \mathbf{r}''(t)| = \left(\frac{ds}{dt}\right)^2 |\mathbf{T}(t) \times \mathbf{T}'(t)| = \left(\frac{ds}{dt}\right)^2 |\mathbf{T}(t)| |\mathbf{T}'(t)| = \left(\frac{ds}{dt}\right)^2 |\mathbf{T}'(t)|$$

Upon solving for $|\mathbf{T}'(t)|$, we obtain

$$|\mathbf{T}'(t)| = \frac{|\mathbf{r}'(t) \times \mathbf{r}''(t)|}{\left(\dfrac{ds}{dt}\right)^2} = \frac{|\mathbf{r}'(t) \times \mathbf{r}''(t)|}{|\mathbf{r}'(t)|^2}$$

from which we deduce that

$$\kappa(t) = \frac{|\mathbf{T}'(t)|}{|\mathbf{r}'(t)|} = \frac{|\mathbf{r}'(t) \times \mathbf{r}''(t)|}{|\mathbf{r}'(t)|^3} \qquad\blacksquare$$

**EXAMPLE 4**  Find the curvature of the "twisted cubic" described by the vector function

$$\mathbf{r}(t) = t\mathbf{i} + \frac{1}{2} t^2 \mathbf{j} + \frac{1}{3} t^3 \mathbf{k}$$

**Solution**  Since

$$\mathbf{r}'(t) = \mathbf{i} + t\mathbf{j} + t^2 \mathbf{k}$$

and

$$\mathbf{r}''(t) = \mathbf{j} + 2t\mathbf{k}$$

we have

$$\mathbf{r}'(t) \times \mathbf{r}''(t) = \begin{vmatrix} \mathbf{i} & \mathbf{j} & \mathbf{k} \\ 1 & t & t^2 \\ 0 & 1 & 2t \end{vmatrix} = t^2 \mathbf{i} - 2t\mathbf{j} + \mathbf{k}$$

so

$$|\mathbf{r}'(t) \times \mathbf{r}''(t)| = \sqrt{t^4 + 4t^2 + 1}$$

Also,

$$|\mathbf{r}'(t)| = \sqrt{1 + t^2 + t^4} = \sqrt{t^4 + t^2 + 1}$$

Therefore,

$$\kappa(t) = \frac{|\mathbf{r}'(t) \times \mathbf{r}''(t)|}{|\mathbf{r}'(t)|^3} = \frac{\sqrt{t^4 + 4t^2 + 1}}{(t^4 + t^2 + 1)^{3/2}} \qquad\blacksquare$$

If a plane curve $C$ happens to be contained in the graph of a function defined by $y = f(x)$, then we can use the following formula to compute its curvature.

---

**THEOREM 3    Formula for the Curvature of the Graph of a Function**

If $C$ is the graph of a twice differentiable function $f$, then the curvature at the point $(x, y)$ where $y = f(x)$ is given by

$$\kappa(x) = \frac{|f''(x)|}{[1 + [f'(x)]^2]^{3/2}} = \frac{|y''|}{[1 + (y')^2]^{3/2}} \tag{5}$$

---

**PROOF**    Using $x$ as the parameter, we can represent $C$ by the vector function $\mathbf{r}(x) = x\mathbf{i} + f(x)\mathbf{j} + 0\mathbf{k}$. Differentiating $\mathbf{r}(x)$ with respect to $x$ successively, we obtain

$$\mathbf{r}'(x) = \mathbf{i} + f'(x)\mathbf{j} + 0\mathbf{k} \qquad \text{and} \qquad \mathbf{r}''(x) = 0\mathbf{i} + f''(x)\mathbf{j} + 0\mathbf{k}$$

from which we obtain

$$\mathbf{r}'(x) \times \mathbf{r}''(x) = \begin{vmatrix} \mathbf{i} & \mathbf{j} & \mathbf{k} \\ 1 & f'(x) & 0 \\ 0 & f''(x) & 0 \end{vmatrix} = f''(x)\mathbf{k}$$

and

$$|\mathbf{r}'(x) \times \mathbf{r}''(x)| = |f''(x)|$$

Also,

$$|\mathbf{r}'(x)| = \sqrt{1 + [f'(x)]^2}$$

Therefore,

$$\kappa(x) = \frac{|\mathbf{r}'(x) \times \mathbf{r}''(x)|}{|\mathbf{r}'(x)|^3} = \frac{|f''(x)|}{[1 + [f'(x)]^2]^{3/2}} \qquad \blacksquare$$

---

**EXAMPLE 5**

**a.** Find the curvature of the parabola $y = \frac{1}{4}x^2$ at the points where $x = 0$ and $x = 1$.
**b.** Find the point(s) where the curvature is largest.

**Solution**
**a.** We first compute $y' = \frac{1}{2}x$ and $y'' = \frac{1}{2}$. Then using Theorem 3, we find the curvature at any point $(x, y)$ on the parabola $y = \frac{1}{2}x^2$ to be

$$\kappa(x) = \frac{|y''|}{[1 + (y')^2]^{3/2}} = \frac{\frac{1}{2}}{\left(1 + \frac{1}{4}x^2\right)^{3/2}} = \frac{4}{(4 + x^2)^{3/2}}$$

In particular, the curvature at the point $(0, 0)$, where $x = 0$, is

$$\kappa(0) = \frac{4}{(4 + x^2)^{3/2}}\bigg|_{x=0} = \frac{1}{2}$$

and the curvature at the point $\left(1, \frac{1}{4}\right)$, where $x = 1$, is

$$\kappa(1) = \frac{4}{(4 + x^2)^{3/2}}\bigg|_{x=1} = \frac{4}{5^{3/2}} \approx 0.358$$

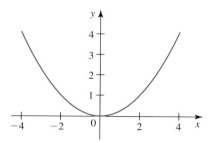

**FIGURE 6**
The graph of $y = \frac{1}{4}x^2$.

**b.** To find the value of $x$ at which $\kappa$ is largest, we compute

$$\kappa'(x) = \frac{d}{dx}\left[4(4 + x^2)^{-3/2}\right] = -6(4 + x^2)^{-5/2}(2x) = -\frac{12x}{(4 + x^2)^{5/2}}$$

Setting $\kappa'(x) = 0$ yields the sole critical point $x = 0$. We leave it to you to show that $x = 0$ does give the absolute maximum value of $\kappa(x)$.

The graph of $y = \frac{1}{4}x^2$ is shown in Figure 6. ■

## Radius of Curvature

Suppose that $C$ is a plane curve with curvature $\kappa$ at the point $P$. Then the reciprocal of the curvature, $\rho = 1/\kappa$, is called the **radius of curvature** of $C$ at $P$. The radius of curvature at any point $P$ on a curve $C$ is the radius of the circle that best "fits" the curve at that point. This circle, which lies on the concave side of the curve and shares a common tangent line with the curve at $P$, is called the **circle of curvature** or **osculating circle**. (See Figure 7.)

The center of the circle is called the **center of curvature**. As an example, the curvature of the parabola $y = \frac{1}{4}x^2$ of Example 5 at the point $(0, 0)$ was found to be $\frac{1}{2}$. Therefore, the radius of curvature of the parabola at $(0, 0)$ is $\rho = 1/(1/2) = 2$. The circle of curvature is shown in Figure 8. Its equation is $x^2 + (y - 2)^2 = 4$.

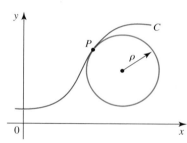

**FIGURE 7**
The radius of curvature at $P$ is the radius of the circle that best fits the curve $C$ at $P$.

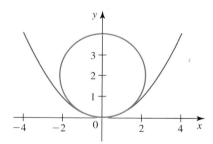

**FIGURE 8**
The circle of curvature is tangent to the parabola.

## 11.3 CONCEPT QUESTIONS

1. Give the formula for finding the arc length of the curve $C$ defined by $\mathbf{r}(t) = \langle f(t), g(t), h(t) \rangle$ for $a \leq t \leq b$. What condition, if any, must be imposed on $C$?

2. **a.** What is a smooth curve?
   **b.** Give an example of a curve in 3-space that is not smooth.

3. **a.** What is the arc length function associated with $\mathbf{r}(t) = \langle f(t), g(t), h(t) \rangle$, where $a \leq t \leq b$?
   **b.** If a curve is parametrized in terms of its arc length, what is the unit tangent vector $\mathbf{T}(s)$? What is $\mathbf{T}(t)$, where $t$ is not the arc length parameter?

4. **a.** What is the curvature of a smooth curve $C$ at $s$, where $s$ is the arc length parameter?
   **b.** If $t$ is not the arc length parameter, what is the curvature of $C$ at $t$?
   **c.** What is the radius of curvature of a curve $C$ at a point $P$ on $C$?

## 11.3 EXERCISES

*In Exercises 1–8, find the length of the curve.*

**1.** $\mathbf{r}(t) = t\mathbf{i} + 2t\mathbf{j} + 3t\mathbf{k}, \quad 0 \le t \le 4$

**2.** $\mathbf{r}(t) = \langle 5t, 3t^2, 4t^2 \rangle, \quad 0 \le t \le 2$

**3.** $\mathbf{r}(t) = 4 \sin t\mathbf{i} + 3t\mathbf{j} + 4 \cos t\mathbf{k}, \quad 0 \le t \le 2\pi$

**4.** $\mathbf{r}(t) = a \cos t\mathbf{i} + a \sin t\mathbf{j} + bt\mathbf{k}, \quad 0 \le t \le 2\pi$

**5.** $\mathbf{r}(t) = \langle e^t \cos t, e^t \sin t, e^t \rangle, \quad 0 \le t \le 2\pi$

**6.** $\mathbf{r}(t) = t^2\mathbf{i} + t \cos t\mathbf{j} + t \sin t\mathbf{k}, \quad 0 \le t \le 1$

**7.** $\mathbf{r}(t) = 2t\mathbf{i} + t^2\mathbf{j} + \ln t\mathbf{k}, \quad 1 \le t \le e$

**8.** $\mathbf{r}(t) = (\cos t + t \sin t)\mathbf{i} + (\sin t - t \cos t)\mathbf{j} + t^2\mathbf{k},$
$\quad 0 \le t \le \frac{\pi}{2}$

 *In Exercises 9 and 10, use a calculator or computer to graph the curve represented by $\mathbf{r}(t)$, and find the length of the curve for t defined on the indicated interval.*

**9.** $\mathbf{r}(t) = t \sin t\mathbf{i} + t \cos t\mathbf{j} + t\mathbf{k}; \quad [0, 2\pi]$

**10.** $\mathbf{r}(t) = 2 \sin t\mathbf{i} + 2 \cos t\mathbf{j} + \frac{1}{2}t^2\mathbf{k}; \quad [0, 2\pi]$

*In Exercises 11–14, find the arc length function s(t) for the curve defined by $\mathbf{r}(t)$. Then use this result to find a parametrization of C in terms of s.*

**11.** $\mathbf{r}(t) = (1 + t)\mathbf{i} + (1 + 2t)\mathbf{j} + 3t\mathbf{k}, \quad t \ge 0$

**12.** $\mathbf{r}(t) = 4 \sin t\mathbf{i} + 4 \cos t\mathbf{j} + 3t\mathbf{k}, \quad t \ge 0$

**13.** $\mathbf{r}(t) = e^t \cos t\mathbf{i} + e^t \sin t\mathbf{j} + e^t\mathbf{k}, \quad t \ge 0$

**14.** $\mathbf{r}(t) = a \cos^3 t\mathbf{i} + a \sin^3 t\mathbf{j} + \mathbf{k}, \quad 0 \le t \le \frac{\pi}{2}$

*In Exercises 15–20, use Theorem 2 to find the curvature of the curve.*

**15.** $\mathbf{r}(t) = 2t\mathbf{i} + 2t\mathbf{j} + \mathbf{k}$

**16.** $\mathbf{r}(t) = t\mathbf{i} + \mathbf{j} + t^2\mathbf{k}$

**17.** $\mathbf{r}(t) = t\mathbf{i} + \frac{1}{2}t^2\mathbf{j} + t^2\mathbf{k}$

**18.** $\mathbf{r}(t) = (1 - t)\mathbf{i} + (1 + t)\mathbf{j} + 3t^2\mathbf{k}$

**19.** $\mathbf{r}(t) = 2 \sin t\mathbf{i} + 2 \cos t\mathbf{j} + 2t\mathbf{k}$

**20.** $\mathbf{r}(t) = \langle e^t \cos t, e^t \sin t, e^t \rangle$

*In Exercises 21–26, use Theorem 3 to find the curvature of the curve.*

**21.** $y = x^3 + 1$

**22.** $y = x^4$

**23.** $y = \sin 2x$

**24.** $y = \ln x$

**25.** $y = e^{-x^2}$

**26.** $y = \sec x$

**27.** Find the point(s) on the graph of $y = e^{-x^2}$ at which the curvature is zero.

**28.** Find an equation of the circle of curvature for the graph of $f(x) = x + (1/x)$ at the point $(1, 2)$. Sketch the graph of $f$ and the circle of curvature.

*In Exercises 29–32, find the point(s) on the curve at which the curvature is largest.*

**29.** $y = e^x$

**30.** $y = \ln x$

**31.** $xy = 1$

**32.** $4x^2 + 9y^2 = 36$

*In Exercises 33–36, match the curve with the graph of its curvature $y = \kappa(x)$ in (a)–(d).*

**33.**

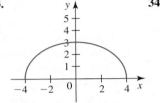

**34.**

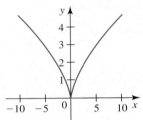

**35.**

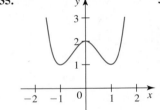

**36.**

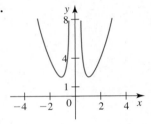

**(a)**

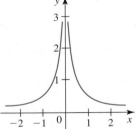

**(b)**

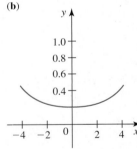

**(c)**

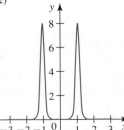

**(d)**

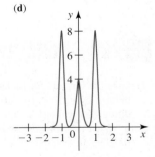

 *In Exercises 37 and 38, find the curvature function $\kappa(x)$ of the curve. Then use a calculator or computer to graph both the curve and its curvature function $\kappa(x)$ on the same set of axes.*

**37.** $y = e^{-x^2}$

**38.** $y = \ln(1 + x^2)$

**39.** Suppose that $C$ is a smooth curve with parametric equations $x = f(t)$, $y = g(t)$. Using Theorem 2, show that the curvature at the point $(x, y)$ corresponding to any value of $t$ is

$$\kappa(t) = \frac{|f'(t)g''(t) - g'(t)f''(t)|}{\{[f'(t)]^2 + [g'(t)]^2\}^{3/2}}$$

*In Exercises 40 and 41, use the formula in Exercise 39 to find the curvature of the curve.*

**40.** $x = \cos t$, $\quad y = t \sin t$

**41.** $x = t - \sin t$, $\quad y = 1 - \cos t$

**42. a.** The curvature of the curve $C$ at $P$ shown in the figure is 2. Sketch the osculating circle at $P$. (Use the tangent line shown at $P$ as an aid.)
**b.** What is the curvature of $C$ at $Q$?

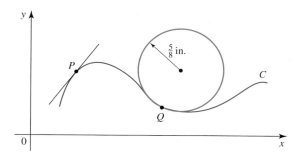

**43. a.** Find the curvature at the point $(x, y)$ on the ellipse

$$\frac{x^2}{9} + \frac{y^2}{4} = 1$$

**b.** Find the curvature and the equation of the osculating circle at the points $(3, 0)$ and $(0, 2)$.
**c.** Sketch the graph of the ellipse and the osculating circles of part (b).

**44.** Find the curvature $\kappa(t)$ for the curve with parametric equations

$$x = t^2 \quad \text{and} \quad y = t^3$$

What happens to $\kappa(t)$ as $t$ approaches 0?
**Note:** The curve is not smooth at $t = 0$.

**45.** The spiral of cornu is defined by the parametric equations

$$x = \int_0^t \cos\left(\frac{\pi u^2}{2}\right) du \quad y = \int_0^t \sin\left(\frac{\pi u^2}{2}\right) du$$

and was encountered in Exercise 45 of Section 9.3. Its graph follows.

**a.** Find $\dfrac{dy}{dx}$ and $\dfrac{d^2y}{dx^2}$.
**b.** Find the curvature of the spiral.
**Note:** The curvature $\kappa(t)$ increases from 0 at a constant rate with respect to $t$ as $t$ increases from $t = 0$. This property of the spiral

of cornu makes the curve useful in highway design: It provides a gradual transition from a straight road (zero curvature) to a curved road (with positive curvature), such as an exit ramp.

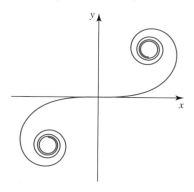

**46.** Suppose that the curve $C$ is described by a polar equation $r = f(\theta)$. Show that the curvature at the point $(r, \theta)$ is given by

$$\kappa(\theta) = \frac{|2(r')^2 - rr'' + r^2|}{[(r')^2 + r^2]^{3/2}}$$

**Hint:** Represent $C$ by $\mathbf{r}(\theta) = r \cos \theta \mathbf{i} + r \sin \theta \mathbf{j}$.

*In Exercises 47 and 48, use the formula in Exercise 46 to find the curvature of the curve.*

**47.** $r = 1 + \sin \theta$    **48.** $r = e^\theta$

**49.** Show that the curvature at every point on the helix

$$x = a \cos t \quad y = a \sin t \quad z = bt$$

where $a > 0$, is given by $\kappa(t) = a/(a^2 + b^2)$.

**50.** Find the curvature at the point $(x, y, z)$ on an elliptic helix with parametric equations

$$x = a \cos t \quad y = b \sin t \quad z = ct$$

where $a$, $b$, and $c$ are positive and $a \neq b$.

**51.** Find the arc length of $\mathbf{r}(t) = t \cos t \mathbf{i} + t \sin t \mathbf{j} + t \mathbf{k}$, where $0 \leq t \leq 2\pi$.

**52.** Find the curvature of the graph of $x^3 + y^3 = 9xy$ (folium of Descartes) at the point $(2, 4)$ accurate to four decimal places.

*In Exercises 53–58, determine whether the statement is true or false. If it is true, explain why. If it is false, explain why or give an example to show that it is false.*

**53.** If $C$ is a smooth curve in the $xy$-plane defined by $\mathbf{r}(t) = x(t)\mathbf{i} + y(t)\mathbf{j}$ on a parameter interval $I$, then $dy/dx$ is defined at every point on the curve.

**54.** The curve defined by $\mathbf{r}(t) = \langle t, |t| \rangle$ is smooth.

**55.** If the graph of a twice differentiable function $f$ has an inflection point at $a$, then the curvature at the point $(a, f(a))$ is zero.

**56.** If $C$ is the curve defined by the parametric equations

$$x = 1 - 2t \qquad y = 2 + 3t \qquad z = 4t$$

then $d\mathbf{T}/ds = 0$, where $\mathbf{T}$ is the unit tangent vector to $C$.

**57.** The radius of curvature of the plane curve $y = \sqrt{a^2 - x^2}$ is constant at each point on the curve.

**58.** If $\mathbf{r}'(t)$ is continuous for all $t$ in an interval $I$, then $\mathbf{r}$ defines a smooth curve.

## 11.4 Velocity and Acceleration

### ■ Velocity, Acceleration, and Speed

The curve $C$ in Figure 1 is the flight path of a fighter plane. We can represent $C$ by the vector function

$$\mathbf{r}(t) = f(t)\mathbf{i} + g(t)\mathbf{j} + h(t)\mathbf{k} \qquad t \in I$$

where we think of the parameter interval $I$ as a time interval and use $\mathbf{r}(t)$ to indicate the position of the plane at time $t$.

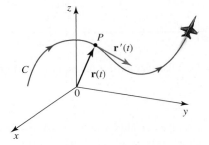

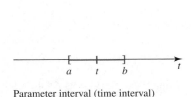

**FIGURE 1**
The position vector $\mathbf{r}(t)$ gives the position of a fighter plane at time $t$, and its derivative $\mathbf{r}'(t)$ gives the plane's velocity at time $t$.

Parameter interval (time interval)

From Sections 11.2 and 11.3 we know that the vector $\mathbf{r}'(t)$ has the following properties:

**1.** $\mathbf{r}'(t)$ is tangent to $C$ at the point $P$ corresponding to time $t$.

**2.** $|\mathbf{r}'(t)| = \dfrac{ds}{dt}$.

Since $ds/dt$ is the rate of change of the distance (measured along the arc) with respect to time, it measures the *speed* of the plane. Thus, the vector $\mathbf{r}'(t)$ gives both the speed and the direction of the plane. In other words, it makes sense to define the *velocity vector* of the plane at time $t$ to be $\mathbf{r}'(t)$, the rate of change of its position vector with respect to time. Similarly, we define the *acceleration vector* of the plane at time $t$ to be $\mathbf{r}''(t)$, the rate of change of its velocity vector with respect to time.

To gain insight into the nature of the acceleration vector, let's refer to Figure 2. Here, $t$ is fixed, and $h$ is a small number. The vector $\mathbf{r}'(t)$ is tangent to the flight path at the tip of the position vector $\mathbf{r}(t)$, and $\mathbf{r}'(t + h)$ is tangent to the flight path at the tip of $\mathbf{r}(t + h)$. The vector

$$\frac{\mathbf{r}'(t + h) - \mathbf{r}'(t)}{h}$$

points in the general direction in which the plane is turning. Therefore, the acceleration vector

$$\mathbf{r}''(t) = \frac{d}{dt}\mathbf{r}'(t) = \lim_{h \to 0} \frac{\mathbf{r}'(t + h) - \mathbf{r}'(t)}{h}$$

points toward the concave side of the flight path as long as the direction of $\mathbf{r}'(t)$ is changing, in agreement with our intuition.

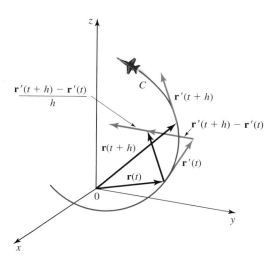

**FIGURE 2**
To find $\mathbf{r}'(t + h) - \mathbf{r}'(t)$, translate $\mathbf{r}'(t + h)$ so that its tail is at the tip of $\mathbf{r}(t)$.

Let's summarize these definitions.

---

**DEFINITIONS** Velocity, Acceleration, and Speed

Let $\mathbf{r}(t) = f(t)\mathbf{i} + g(t)\mathbf{j} + h(t)\mathbf{k}$ be the position vector of an object. If $f$, $g$, and $h$ are twice differentiable functions of $t$, then the **velocity vector** $\mathbf{v}(t)$, **acceleration vector** $\mathbf{a}(t)$, and **speed** $|\mathbf{v}(t)|$ of the object at time $t$ are defined by

$$\mathbf{v}(t) = \mathbf{r}'(t) = f'(t)\mathbf{i} + g'(t)\mathbf{j} + h'(t)\mathbf{k}$$

$$\mathbf{a}(t) = \mathbf{r}''(t) = f''(t)\mathbf{i} + g''(t)\mathbf{j} + h''(t)\mathbf{k}$$

$$|\mathbf{v}(t)| = |\mathbf{r}'(t)| = \sqrt{[f'(t)]^2 + [g'(t)]^2 + [h'(t)]^2}$$

---

**EXAMPLE 1** The position of an object moving in a plane is given by

$$\mathbf{r}(t) = t^2\mathbf{i} + t\mathbf{j} \qquad t \geq 0$$

Find its velocity, acceleration, and speed when $t = 2$. Sketch the path of the object and the vectors $\mathbf{v}(2)$ and $\mathbf{a}(2)$.

**Solution** The velocity and acceleration vectors of the object are

$$\mathbf{v}(t) = \mathbf{r}'(t) = 2t\mathbf{i} + \mathbf{j}$$

and

$$\mathbf{a}(t) = \mathbf{r}''(t) = 2\mathbf{i}$$

Therefore, its velocity, acceleration, and speed when $t = 2$ are

$$\mathbf{v}(2) = 4\mathbf{i} + \mathbf{j}$$

$$\mathbf{a}(2) = 2\mathbf{i}$$

and

$$|\mathbf{v}(2)| = \sqrt{16 + 1} = \sqrt{17}$$

respectively.

To sketch the path of the object, observe that the parametric equations of the curve described by $\mathbf{r}(t)$ are $x = t^2$ and $y = t$. By eliminating $t$ from these equations, we obtain the rectangular equation $x = y^2$, where $y \geq 0$, which tells us that the path of the object is contained in the graph of the parabola $x = y^2$. This path together with the vectors $\mathbf{v}(2)$ and $\mathbf{a}(2)$ is shown in Figure 3. ∎

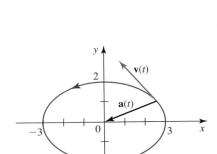

**FIGURE 3**
The path of the object $C$ and the vectors $\mathbf{v}(2)$ and $\mathbf{a}(2)$

**EXAMPLE 2**   Find the velocity vector, speed, and acceleration vector of an object that moves along the plane curve $C$ described by the position vector

$$\mathbf{r}(t) = 3 \cos t\,\mathbf{i} + 2 \sin t\,\mathbf{j}$$

**Solution**   The velocity vector is

$$\mathbf{v}(t) = -3 \sin t\,\mathbf{i} + 2 \cos t\,\mathbf{j}$$

The speed of the object at time $t$ is

$$|\mathbf{v}(t)| = \sqrt{9 \sin^2 t + 4 \cos^2 t}$$

Finally, the acceleration vector is

$$\mathbf{a}(t) = -3 \cos t\,\mathbf{i} - 2 \sin t\,\mathbf{j} = -\mathbf{r}(t)$$

which shows the acceleration is directed toward the origin (see Figure 4). ∎

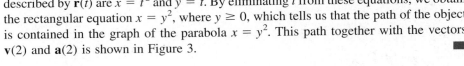

**FIGURE 4**
The acceleration vector $\mathbf{a}$ points toward the origin.

**EXAMPLE 3**   Find the velocity vector, acceleration vector, and speed of a particle with position vector

$$\mathbf{r}(t) = \sqrt{t}\,\mathbf{i} + t^2\mathbf{j} + e^{2t}\mathbf{k} \qquad t \geq 0$$

**Solution**   The required quantities are

$$\mathbf{v}(t) = \mathbf{r}'(t) = \frac{1}{2} t^{-1/2}\mathbf{i} + 2t\mathbf{j} + 2e^{2t}\mathbf{k} = \frac{1}{2\sqrt{t}}\mathbf{i} + 2t\mathbf{j} + 2e^{2t}\mathbf{k}$$

$$\mathbf{a}(t) = \mathbf{r}''(t) = -\frac{1}{4} t^{-3/2}\mathbf{i} + 2\mathbf{j} + 4e^{2t}\mathbf{k} = -\frac{1}{4\sqrt{t^3}}\mathbf{i} + 2\mathbf{j} + 4e^{2t}\mathbf{k}$$

and

$$|\mathbf{v}(t)| = \sqrt{\frac{1}{4t} + 4t^2 + 4e^{4t}} = \frac{\sqrt{1 + 16t^3 + 16te^{4t}}}{2\sqrt{t}}$$ ∎

Suppose that we are given the velocity or acceleration vector of a moving object. Then it is possible to find the position vector of the object by integration, as is shown in the next example.

**EXAMPLE 4** A moving object has an initial position and an initial velocity given by the vectors $\mathbf{r}(0) = \mathbf{i} + 2\mathbf{j} + \mathbf{k}$ and $\mathbf{v}(0) = \mathbf{i} + 2\mathbf{k}$. Its acceleration at time $t$ is $\mathbf{a}(t) = 6t\mathbf{i} + \mathbf{j} + 2\mathbf{k}$. Find its velocity and position at time $t$.

**Solution**  Since $\mathbf{v}'(t) = \mathbf{a}(t)$, we can obtain $\mathbf{v}(t)$ by integrating both sides of this equation with respect to $t$. Thus,

$$\mathbf{v}(t) = \int \mathbf{a}(t)\, dt = \int (6t\mathbf{i} + \mathbf{j} + 2\mathbf{k})\, dt = 3t^2\mathbf{i} + t\mathbf{j} + 2t\mathbf{k} + \mathbf{C}$$

Letting $t = 0$ in this expression and using the initial condition $\mathbf{v}(0) = \mathbf{i} + 2\mathbf{k}$, we obtain

$$\mathbf{v}(0) = \mathbf{C} = \mathbf{i} + 2\mathbf{k}$$

Therefore, the velocity of the object at any time $t$ is

$$\mathbf{v}(t) = (3t^2\mathbf{i} + t\mathbf{j} + 2t\mathbf{k}) + \mathbf{i} + 2\mathbf{k}$$
$$= (3t^2 + 1)\mathbf{i} + t\mathbf{j} + 2(t + 1)\mathbf{k}$$

Next, integrating the equation $\mathbf{r}'(t) = \mathbf{v}(t)$ with respect to $t$ gives

$$\mathbf{r}(t) = \int \mathbf{v}(t)\, dt = \int [(3t^2 + 1)\mathbf{i} + t\mathbf{j} + 2(t + 1)\mathbf{k}]\, dt$$

$$= (t^3 + t)\mathbf{i} + \frac{1}{2}t^2\mathbf{j} + (t^2 + 2t)\mathbf{k} + \mathbf{D}$$

Letting $t = 0$ in $\mathbf{r}(t)$ and using the initial condition $\mathbf{r}(0) = \mathbf{i} + 2\mathbf{j} + \mathbf{k}$, we have

$$\mathbf{r}(0) = \mathbf{D} = \mathbf{i} + 2\mathbf{j} + \mathbf{k}$$

Therefore, the position of the object at any time $t$ is

$$\mathbf{r}(t) = (t^3 + t)\mathbf{i} + \frac{1}{2}t^2\mathbf{j} + (t^2 + 2t)\mathbf{k} + (\mathbf{i} + 2\mathbf{j} + \mathbf{k})$$

$$= (t^3 + t + 1)\mathbf{i} + \left(\frac{1}{2}t^2 + 2\right)\mathbf{j} + (t^2 + 2t + 1)\mathbf{k}$$

$$= (t^3 + t + 1)\mathbf{i} + \left(\frac{1}{2}t^2 + 2\right)\mathbf{j} + (t + 1)^2\mathbf{k} \qquad ∎$$

## ■ Motion of a Projectile

A projectile of mass $m$ is fired from a height $h$ with an initial velocity $\mathbf{v}_0$ and an angle of elevation $\alpha$. If we describe the position of the projectile at any time $t$ by the position vector $\mathbf{r}(t)$, then its initial position may be described by the vector

$$\mathbf{r}(0) = h\mathbf{j}$$

and its initial velocity by the vector

$$\mathbf{v}(0) = \mathbf{v}_0 = (v_0 \cos \alpha)\mathbf{i} + (v_0 \sin \alpha)\mathbf{j} \qquad v_0 = |\mathbf{v}_0| \qquad \textbf{(1)}$$

(See Figure 5.)

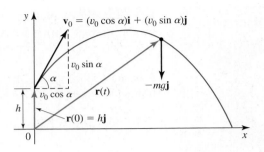

**FIGURE 5**
The initial position of the projectile is $\mathbf{r}(0) = h\mathbf{j}$, and its initial velocity is $\mathbf{v}_0 = (v_0 \cos \alpha)\mathbf{i} + (v_0 \sin \alpha)\mathbf{j}$.

If we assume that air resistance is negligible and that the only external force acting on the projectile is due to gravity, then the force acting on the projectile during its flight is

$$\mathbf{F} = -mg\mathbf{j}$$

where $g$ is the acceleration due to gravity (32 ft/sec$^2$ or 9.8 m/sec$^2$). By Newton's Second Law of Motion this force is equal to $m\mathbf{a}$, where $\mathbf{a}$ is the acceleration of the projectile. Therefore,

$$m\mathbf{a} = -mg\mathbf{j}$$

giving the acceleration of the projectile as

$$\mathbf{a}(t) = -g\mathbf{j}$$

To find the velocity of the projectile at any time $t$, we integrate the last equation with respect to $t$ to obtain

$$\mathbf{v}(t) = \int -g\mathbf{j}\, dt = -gt\mathbf{j} + \mathbf{C}$$

Setting $t = 0$ and using the initial condition $\mathbf{v}(0) = \mathbf{v}_0$, we obtain

$$\mathbf{v}(0) = \mathbf{C} = \mathbf{v}_0$$

Therefore, the velocity of the projectile at any time $t$ is

$$\mathbf{v}(t) = -gt\mathbf{j} + \mathbf{v}_0$$

Integrating this equation then gives

$$\mathbf{r}(t) = \int (-gt\mathbf{j} + \mathbf{v}_0)\, dt = -\frac{1}{2} gt^2\mathbf{j} + \mathbf{v}_0 t + \mathbf{D}$$

Setting $t = 0$ and using the initial condition $\mathbf{r}(0) = h\mathbf{j}$, we obtain

$$\mathbf{r}(0) = \mathbf{D} = h\mathbf{j}$$

Therefore, the position of the projectile at any time $t$ is

$$\mathbf{r}(t) = -\frac{1}{2} gt^2\mathbf{j} + \mathbf{v}_0 t + h\mathbf{j}$$

or, upon using Equation (1),

$$\mathbf{r}(t) = -\frac{1}{2} gt^2\mathbf{j} + [(v_0 \cos \alpha)\mathbf{i} + (v_0 \sin \alpha)\mathbf{j}]t + h\mathbf{j}$$

$$= (v_0 \cos \alpha)t\mathbf{i} + \left[h + (v_0 \sin \alpha)t - \frac{1}{2} gt^2\right]\mathbf{j}$$

---

**DEFINITION  Position Function for a Projectile**
The trajectory of a projectile fired from a height $h$ with an initial speed $v_0$ and an angle of elevation $\alpha$ is given by the position vector function

$$\mathbf{r}(t) = (v_0 \cos \alpha)t\mathbf{i} + \left[ h + (v_0 \sin \alpha)t - \frac{1}{2}gt^2 \right]\mathbf{j} \qquad (2)$$

where $g$ is the constant of acceleration due to gravity.

---

**EXAMPLE 5  Motion of a Projectile**  A shell is fired from a gun located on a hill 100 m above a level terrain. The muzzle speed of the gun is 500 m/sec, and its angle of elevation is 30°.

**a.** Find the range of the shell.
**b.** What is the maximum height attained by the shell?
**c.** What is the speed of the shell at impact?

**Solution**  Using Equation (2) with $h = 100$, $v_0 = 500$, $\alpha = 30°$, and $g = 9.8$, we see that the position of the shell at any time $t$ is given by

$$\mathbf{r}(t) = (500 \cos 30°)t\mathbf{i} + [100 + (500 \sin 30°)t - 4.9t^2]\mathbf{j}$$

$$= 250\sqrt{3}\,t\mathbf{i} + (100 + 250t - 4.9t^2)\mathbf{j}$$

The corresponding parametric equations are

$$x = 250\sqrt{3}\,t \qquad \text{and} \qquad y = 100 + 250t - 4.9t^2$$

**a.** We first find the time when the shell strikes the ground by solving the equation

$$4.9t^2 - 250t - 100 = 0$$

obtained by setting $y = 0$. Using the quadratic formula, we have

$$t = \frac{250 \pm \sqrt{62{,}500 + 1960}}{9.8} \approx 51.4 \qquad \text{We reject the negative root.}$$

or 51.4 sec. Substituting this value of $t$ into the expression for $x$ we find that the range of the shell is approximately

$$250\sqrt{3}\,(51.4) \approx 22{,}257$$

or 22,257 m.

**b.** The height of the shell at any time $t$ is given by

$$y = 100 + 250t - 4.9t^2$$

To find the maximum value of $y$, we solve

$$y' = 250 - 9.8t = 0$$

to obtain $t \approx 25.5$. Since $y'' = -9.8 < 0$, the Second Derivative Test implies that at approximately 25.5 sec into flight, the shell attains its maximum height

$$y\Big|_{t \approx 25.5} \approx 100 + 250(25.5) - 4.9(25.5)^2 \approx 3289$$

or 3289 m.

**c.** By differentiating the position function

$$\mathbf{r}(t) = 250\sqrt{3}\,t\mathbf{i} + (100 + 250t - 4.9t^2)\mathbf{j}$$

we obtain the velocity of the shell at any time $t$. Thus,

$$\mathbf{v}(t) = \mathbf{r}'(t) = 250\sqrt{3}\mathbf{i} + (250 - 9.8t)\mathbf{j}$$

From part (a) we know that the time of impact is $t \approx 51.4$. So at the time of impact the velocity of the shell is

$$\mathbf{v}(51.4) \approx 250\sqrt{3}\mathbf{i} + [250 - 9.8(51.4)]\mathbf{j}$$

$$= 250\sqrt{3}\mathbf{i} - 253.7\mathbf{j}$$

Therefore, its speed at impact is

$$|\mathbf{v}(51.4)| = \sqrt{(250\sqrt{3})^2 + (253.7)^2} \approx 502$$

or 502 m/sec. The trajectory of the shell is shown in Figure 6.

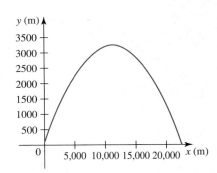

**FIGURE 6**
The trajectory of the shell

## 11.4    CONCEPT QUESTIONS

**1. a.** What are the velocity, acceleration, and speed of an object with position vector $\mathbf{r}(t)$?
  **b.** Give the expressions for the quantities in part (a) if the position vector has the form $\mathbf{r}(t) = f(t)\mathbf{i} + g(t)\mathbf{j} + h(t)\mathbf{k}$.

**2.** A projectile of mass $m$ is fired from a height $h$ with an initial velocity $\mathbf{v}_0$ and an angle of elevation $\alpha$. Write a vector representing
  **a.** Its initial position.
  **b.** Its initial velocity in terms of $v_0 = |\mathbf{v}_0|$ and $\alpha$.
  **c.** Its velocity at time $t$.
  **d.** Its position at time $t$.

## 11.4    EXERCISES

*In Exercises 1–6, find the velocity, acceleration, and speed of an object with the position function for the given value of t. Sketch the path of the object and its velocity and acceleration vectors.*

**1.** $\mathbf{r}(t) = t\mathbf{i} + (4 - t^2)\mathbf{j}$;  $t = 1$

**2.** $\mathbf{r}(t) = \langle t^2 - 4, 2t \rangle$;  $t = 1$

**3.** $\mathbf{r}(t) = \cos t\mathbf{i} + 3 \sin t\mathbf{j}$;  $t = \dfrac{\pi}{4}$

**4.** $\mathbf{r}(t) = e^t\mathbf{i} + e^{-t}\mathbf{j}$;  $t = 0$

**5.** $\mathbf{r}(t) = \cos t\mathbf{i} + \sin t\mathbf{j} + t\mathbf{k}$;  $t = \dfrac{\pi}{2}$

**6.** $\mathbf{r}(t) = \langle t, t^2, t^3 \rangle$;  $t = 1$

*In Exercises 7–12, find the velocity, acceleration, and speed of an object with the given position vector.*

**7.** $\mathbf{r}(t) = t\mathbf{i} + t^2\mathbf{j} + (t^2 - 4)\mathbf{k}$

**8.** $\mathbf{r}(t) = \langle \sqrt{t}, 1 + \sqrt{t}, t \rangle$

**9.** $\mathbf{r}(t) = t\mathbf{i} + t^2\mathbf{j} + \dfrac{1}{t}\mathbf{k}$

**10.** $\mathbf{r}(t) = e^t\mathbf{i} + e^{-t}\mathbf{j} + t^2\mathbf{k}$

**11.** $\mathbf{r}(t) = e^t\langle \cos t, \sin t, 1 \rangle$

**12.** $\mathbf{r}(t) = t \cos t\mathbf{i} + t \sin t\mathbf{j} + t^2\mathbf{k}$

*In Exercises 13–18, find the velocity and position vectors of an object with the given acceleration, initial velocity, and position.*

**13.** $\mathbf{a}(t) = -32\mathbf{k}$,  $\mathbf{v}(0) = \mathbf{i} + 2\mathbf{j}$,  $\mathbf{r}(0) = 128\mathbf{k}$

**14.** $\mathbf{a}(t) = 2\mathbf{i} + t\mathbf{k}$,  $\mathbf{v}(0) = \mathbf{k}$,  $\mathbf{r}(0) = \mathbf{0}$

**15.** $\mathbf{a}(t) = \mathbf{i} - t\mathbf{j} + (1 + t)\mathbf{k}$,  $\mathbf{v}(0) = \mathbf{i} + \mathbf{k}$,  $\mathbf{r}(0) = \mathbf{j} + \mathbf{k}$

**16.** $\mathbf{a}(t) = \langle e^t, 0, e^{-t} \rangle$,  $\mathbf{v}(0) = \langle 1, 2, 0 \rangle$,  $\mathbf{r}(0) = \langle 3, 1, 2 \rangle$

**17.** $\mathbf{a}(t) = -\cos t\mathbf{i} - \sin t\mathbf{j} + \mathbf{k}$,  $\mathbf{v}(0) = 2\mathbf{k}$, $\mathbf{r}(0) = \mathbf{i}$

**18.** $\mathbf{a}(t) = \langle \cosh t, \sinh t, 0 \rangle$,  $\mathbf{v}(0) = \langle 0, 1, 1 \rangle$,  $\mathbf{r}(0) = \langle 1, 0, 0 \rangle$

**19.** An object moves with a constant speed. Show that the velocity and acceleration vectors associated with this motion are orthogonal.
Hint: Study Example 5 in Section 11.2.

**20.** Suppose that the acceleration of a moving object is always **0**. Show that its motion is rectilinear (that is, along a straight line).

**21.** A particle moves in three-dimensional space in such a way that its velocity is always orthogonal to its position vector. Show that its trajectory lies on a sphere centered at the origin.

**22.** A particle moves in three-dimensional space in such a way that its velocity is always parallel to its position vector. Show that its trajectory lies on a straight line passing through the origin.

**23. Motion of a Projectile** A projectile is fired from ground level with an initial speed of 1500 ft/sec and an angle of elevation of 30°.
   **a.** Find the range of the projectile.
   **b.** What is the maximum height attained by the projectile?
   **c.** What is the speed of the projectile at impact?

**24. Motion of a Projectile** Rework Exercise 23 if the projectile is fired with an angle of elevation of 60°.

**25. Motion of a Projectile** Rework Exercise 23 if the projectile is fired with an angle of elevation of 30° from a height of 200 ft above a level terrain.

**26. Motion of a Projectile** A shell is fired from a gun situated on a hill 500 ft above level ground. If the angle of elevation of the gun is 0° and the muzzle speed of the shell is 2000 ft/sec, when and where will the shell strike the ground?

**27. Motion of a Projectile** A mortar shell is fired with a muzzle speed of 500 ft/sec. Find the angle of elevation of the mortar if the shell strikes a target located 1200 ft away.

**28. Path of a Baseball** A baseball player throws a ball at an angle of 45° with the horizontal. If the ball lands 250 ft away, what is the initial speed of the ball? (Ignore the height of the player.)

**29.** An object moves in a circular path described by the position vector

$$\mathbf{r}(t) = a \cos \omega t \mathbf{i} + a \sin \omega t \mathbf{j}$$

where $\omega = d\theta/dt$ is the constant angular velocity of the object.
   **a.** Find the velocity vector, and show that it is orthogonal to $\mathbf{r}(t)$.
   **b.** Find the acceleration vector, and show that it always points toward the center of the circle.
   **c.** Find the speed and the magnitude of the acceleration vector of the object.

**30.** An object located at the origin is to be projected at an initial speed of $v_0$ m/sec and an angle of elevation of $\alpha$ so that it will strike a target located at the point $(r, 0)$. Neglecting air resistance, find the required angle $\alpha$.

**31. Human Cannonball** The following figure shows the trajectory of a "human cannonball" who will be shot out of a cannon located at ground level onto a net. If the angle of elevation of the cannon is 60° and the initial speed of the man is

$v_0$ ft/sec, determine the range of values of $v_0$ that will allow the man to land on the net. Neglect air resistance.

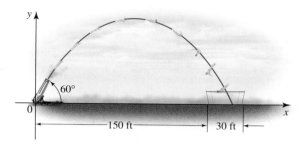

**32. Cycloid Motion** A particle of charge $Q$ is released at rest from the origin in a region of uniform electric and magnetic fields described by $\mathbf{E} = E\mathbf{k}$ and $\mathbf{B} = B\mathbf{i}$.

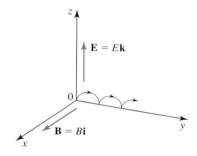

   **a.** Use the *Lorentz Force Law,* $\mathbf{F} = Q[\mathbf{E} + (\mathbf{v} \times \mathbf{B})]$ where $\mathbf{v}$ is the velocity of the particle, to show that

$$\mathbf{F} = QB\frac{dz}{dt}\mathbf{j} + Q\left(E - B\frac{dy}{dt}\right)\mathbf{k}$$

   **b.** Use the result of part (a) and Newton's Second Law of Motion to show that the equations of motion of the particle take the form

$$\frac{d^2x}{dt^2} = 0 \qquad \frac{d^2y}{dt^2} = \omega\frac{dz}{dt} \qquad \frac{d^2z}{dt^2} = \omega\left(\frac{E}{B} - \frac{dy}{dt}\right)$$

   where $\omega = \dfrac{QB}{m}$ and $m$ is the mass of the particle.

   **c.** Show that the general solution of the system in part (b) is $x(t) = C_1 t + C_2$, $y(t) = C_3 \cos \omega t + C_4 \sin \omega t + (E/B)t + C_5$, and $z(t) = C_4 \cos \omega t - C_3 \sin \omega t + C_6$.

   **d.** Use the initial conditions

$$x(0) = \frac{dx}{dt}(0) = 0, \qquad y(0) = \frac{dy}{dt}(0) = 0,$$

$$\text{and} \qquad z(0) = \frac{dz}{dt}(0) = 0$$

   to determine $C_1, C_2, \ldots, C_6$ and hence show that the trajectory of the particle is the cycloid $x(t) = 0$, $y(t) = (E/(\omega B))(\omega t - \sin \omega t)$, and $z(t) = (E/(\omega B))(1 - \cos \omega t)$. (See Section 9.2.)

**33. Newton's Law of Inertia** As a model train moves along a straight track at a constant speed of $v_0$ ft/sec, a ball bearing is ejected vertically from the train at an initial speed of $v_1$ ft/sec. Show that at some later time the ball bearing will return to the location on the train from which it was released.

**Note:** This experiment demonstrates Newton's Law of Inertia.

**34.** Let $\mathbf{r}(t)$ be the position vector of a moving particle and let $r(t) = |\mathbf{r}(t)|$.

 **a.** Show that $\mathbf{r} \cdot \mathbf{r}' = rr'$.
 **b.** What can you say about the orbit of the particle if $\mathbf{v} = \mathbf{r}'$ is perpendicular to $\mathbf{r}$?
 **c.** What can you say about the relationship between the velocity vector and the position vector of the particle if the orbit is circular?

**35.** A particle has position given by $\mathbf{r}(t) = \langle t, t^2, t^3 \rangle$ at time $t$ for $0 \le t \le 1$. At time $t = 1$ the particle departs the curve and flies off along the line tangent to the curve at $\mathbf{r}(1)$. If the particle maintains a constant speed given by $|\mathbf{v}(1)|$, what is the trajectory of the particle for $t \ge 1$? What is its position at $t = 2$?

**36. Motion of a Projectile** Suppose that a projectile is fired from the origin of a two-dimensional coordinate system at an angle of elevation of $\alpha$ and an initial speed of $v_0$.

 **a.** Show that the position function $\mathbf{r}(t)$ of the projectile is equivalent to the parametric equations $x(t) = (v_0 \cos \alpha)t$ and $y(t) = (v_0 \sin \alpha)t - (1/2)gt^2$.
 **b.** Eliminate $t$ in the equations in part (a) to find an equation in $x$ and $y$ describing the trajectory of the projectile. What is the shape of the trajectory?

*In Exercises 37–38, determine whether the statement is true or false. If it is true, explain why. If it is false, explain why or give an example to show that it is false.*

**37.** If $\mathbf{r}(t)$ gives the position of a particle at time $t$, then $\mathbf{r}'(t) = |\mathbf{r}'(t)|\mathbf{T}(t)$, where $\mathbf{T}(t)$ is the unit tangent vector to the curve described by $\mathbf{r}$ at $t$ and $\mathbf{r}'(t) \ne \mathbf{0}$.

**38.** If a particle moves in such a way that its speed is always constant, then its acceleration is zero.

## 11.5 Tangential and Normal Components of Acceleration

### ■ The Unit Normal

Suppose that $C$ is a smooth space curve described by the vector function $\mathbf{r}(t)$. Then, as we saw earlier,

$$\mathbf{T}(t) = \frac{\mathbf{r}'(t)}{|\mathbf{r}'(t)|} \qquad \mathbf{r}'(t) \ne \mathbf{0}$$

is the *unit tangent vector* to the curve $C$ at the point corresponding to $t$. Since $|\mathbf{T}(t)| = 1$ for every $t$, the result of Example 5 in Section 11.2 tells us that the vector $\mathbf{T}'(t)$ is orthogonal to $\mathbf{T}(t)$. Therefore, if $\mathbf{r}'$ is also smooth, we can *normalize* $\mathbf{T}'(t)$ to obtain a *unit* vector that is orthogonal to $\mathbf{T}(t)$. This vector

$$\mathbf{N}(t) = \frac{\mathbf{T}'(t)}{|\mathbf{T}'(t)|}$$

is called the **principal unit normal vector** (or simply the **unit normal**) to the curve $C$ at the point corresponding to $t$. (See Figure 1.)

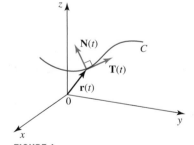

**FIGURE 1**
At each point on the curve $C$, the unit normal vector $\mathbf{N}(t)$ is orthogonal to $\mathbf{T}(t)$ and points in the direction the curve is turning.

**EXAMPLE 1** Let $C$ be the helix defined by

$$\mathbf{r}(t) = 2 \cos t\,\mathbf{i} + 2 \sin t\,\mathbf{j} + t\mathbf{k} \qquad t \ge 0$$

Find $\mathbf{T}(t)$ and $\mathbf{N}(t)$. Sketch $C$ and the vectors $\mathbf{T}(\pi/2)$ and $\mathbf{N}(\pi/2)$.

**Solution** Since

$$\mathbf{r}'(t) = -2 \sin t\,\mathbf{i} + 2 \cos t\,\mathbf{j} + \mathbf{k}$$

and

$$|\mathbf{r}'(t)| = \sqrt{4 \sin^2 t + 4 \cos^2 t + 1} = \sqrt{5}$$

we have

$$\mathbf{T}(t) = \frac{\mathbf{r}'(t)}{|\mathbf{r}'(t)|} = \frac{1}{\sqrt{5}}(-2 \sin t\mathbf{i} + 2 \cos t\mathbf{j} + \mathbf{k})$$

Next, differentiating $\mathbf{T}$, we obtain

$$\mathbf{T}'(t) = \frac{1}{\sqrt{5}}(-2 \cos t\mathbf{i} - 2 \sin t\mathbf{j}) = -\frac{2}{\sqrt{5}}(\cos t\mathbf{i} + \sin t\mathbf{j})$$

Since

$$|\mathbf{T}'(t)| = \frac{2}{\sqrt{5}}\sqrt{\cos^2 t + \sin^2 t} = \frac{2}{\sqrt{5}}$$

it follows that

$$\mathbf{N}(t) = \frac{\mathbf{T}'(t)}{|\mathbf{T}'(t)|} = -(\cos t\mathbf{i} + \sin t\mathbf{j})$$

In particular, at $t = \pi/2$ we have

$$\mathbf{T}\!\left(\frac{\pi}{2}\right) = \frac{1}{\sqrt{5}}(-2 \sin t\mathbf{i} + 2 \cos t\mathbf{j} + \mathbf{k})\Big|_{t=\pi/2} = -\frac{2}{\sqrt{5}}\mathbf{i} + \frac{1}{\sqrt{5}}\mathbf{k}$$

and

$$\mathbf{N}\!\left(\frac{\pi}{2}\right) = -(\cos t\mathbf{i} + \sin t\mathbf{j})\Big|_{t=\pi/2} = -\mathbf{j}$$

The curve $C$ and the unit vectors $\mathbf{T}(\pi/2)$ and $\mathbf{N}(\pi/2)$ are shown in Figure 2. Note that, in general, the principal normal vector $\mathbf{N}(t)$ is parallel to the $xy$-plane and points toward the $z$-axis.  ■

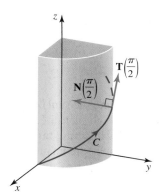

**FIGURE 2**
The unit vectors $\mathbf{T}(\pi/2)$ and $\mathbf{N}(\pi/2)$ at the point $\left(0, 2, \frac{\pi}{2}\right)$ on the helix

## Tangential and Normal Components of Acceleration

Let's return to the study of the motion of an object moving along the curve $C$ described by the vector function $\mathbf{r}$ defined on the parameter interval $I$. Recall that the speed $v$ of the object at any time $t$ is $v = |\mathbf{v}(t)| = |\mathbf{r}'(t)|$. But

$$\mathbf{T} = \frac{\mathbf{r}'(t)}{|\mathbf{r}'(t)|}$$

so we can write

$$\mathbf{v}(t) = \mathbf{r}'(t) = |\mathbf{r}'(t)|\mathbf{T} = v\mathbf{T} \tag{1}$$

which expresses the velocity of the object in terms of its speed and direction. (See Figure 3.)

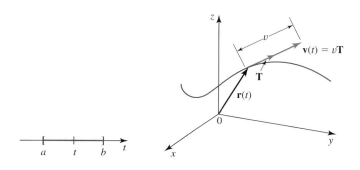

**FIGURE 3**
The velocity of the object at time $t$ is $\mathbf{v}(t) = v\mathbf{T}$.

The acceleration of the object at time $t$ is

$$\mathbf{a} = \mathbf{v}' = \frac{d}{dt}(v\mathbf{T}) = v'\mathbf{T} + v\mathbf{T}'$$

To obtain an expression for $\mathbf{T}'$, recall that

$$\mathbf{N} = \frac{\mathbf{T}'}{|\mathbf{T}'|}$$

so $\mathbf{T}' = |\mathbf{T}'|\mathbf{N}$. Now we need an expression for $|\mathbf{T}'|$. But from Equation (4) in Section 11.3 we have

$$\kappa = \frac{|\mathbf{T}'|}{|\mathbf{r}'|}$$

where $\kappa$ is the curvature of $C$. This gives

$$|\mathbf{T}'| = \kappa|\mathbf{r}'| = \kappa v$$

so $\mathbf{T}' = |\mathbf{T}'|\mathbf{N} = \kappa v\mathbf{N}$.

Therefore,

$$\mathbf{a} = v'\mathbf{T} + v(\kappa v\mathbf{N})$$
$$= v'\mathbf{T} + \kappa v^2\mathbf{N} \tag{2}$$

This result shows that the acceleration vector $\mathbf{a}$ can be resolved into the sum of two vectors—one along the tangential direction and the other along the normal direction. The magnitude of the acceleration along the tangential direction is called the **tangential scalar component of acceleration** and is denoted by $a_{\mathrm{T}}$, whereas the magnitude of the acceleration along the normal direction is called the **normal scalar component of acceleration** and is denoted by $a_{\mathrm{N}}$. Thus,

$$\mathbf{a} = a_{\mathrm{T}}\mathbf{T} + a_{\mathrm{N}}\mathbf{N} \tag{3}$$

where

$$a_{\mathrm{T}} = v' \quad \text{and} \quad a_{\mathrm{N}} = \kappa v^2 \tag{4}$$

(See Figure 4.)

The following theorem gives formulas for calculating $a_{\mathrm{T}}$ and $a_{\mathrm{N}}$ directly from $\mathbf{r}$ and its derivatives.

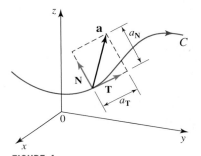

**FIGURE 4**
The acceleration $\mathbf{a}$ has a component $a_{\mathrm{T}}\mathbf{T}$ in the tangential direction and a component $a_{\mathrm{N}}\mathbf{N}$ in the normal direction.

---

**THEOREM 1**    **Tangential and Normal Components of Acceleration**

Let $\mathbf{r}(t)$ be the position vector of an object moving along a smooth curve $C$. Then

$$\mathbf{a} = a_{\mathrm{T}}\mathbf{T} + a_{\mathrm{N}}\mathbf{N}$$

where

$$a_{\mathrm{T}} = \frac{\mathbf{r}'(t) \cdot \mathbf{r}''(t)}{|\mathbf{r}'(t)|} \quad \text{and} \quad a_{\mathrm{N}} = \frac{|\mathbf{r}'(t) \times \mathbf{r}''(t)|}{|\mathbf{r}'(t)|}$$

---

**PROOF**    If we take the dot product of $\mathbf{v}$ and $\mathbf{a}$ as given by Equations (1) and (2), we obtain

$$\mathbf{v} \cdot \mathbf{a} = (v\mathbf{T}) \cdot (v'\mathbf{T} + \kappa v^2\mathbf{N})$$
$$= vv'\mathbf{T} \cdot \mathbf{T} + \kappa v^3\mathbf{T} \cdot \mathbf{N}$$

But $\mathbf{T} \cdot \mathbf{T} = |\mathbf{T}|^2 = 1$, since $\mathbf{T}$ is a unit vector, and $\mathbf{T} \cdot \mathbf{N} = 0$, since $\mathbf{T}$ and $\mathbf{N}$ are orthogonal. Therefore,

$$\mathbf{v} \cdot \mathbf{a} = vv'$$

or, in view of Equation (4),

$$a_{\mathbf{T}} = v' = \frac{\mathbf{v} \cdot \mathbf{a}}{v} = \frac{\mathbf{r}'(t) \cdot \mathbf{r}''(t)}{|\mathbf{r}'(t)|}$$

Next, using Equation (4) and the formula for curvature (Theorem 2 in Section 11.3), we have

$$a_{\mathbf{N}} = \kappa v^2 = \frac{|\mathbf{r}'(t) \times \mathbf{r}''(t)|}{|\mathbf{r}'(t)|^3} |\mathbf{r}'(t)|^2 = \frac{|\mathbf{r}'(t) \times \mathbf{r}''(t)|}{|\mathbf{r}'(t)|} \qquad \blacksquare$$

**EXAMPLE 2** A particle moves along a curve described by the vector function $\mathbf{r}(t) = t\mathbf{i} + t^2\mathbf{j} + t^3\mathbf{k}$. Find the tangential scalar and normal scalar components of acceleration of the particle at any time $t$.

**Solution** We begin by computing

$$\mathbf{r}'(t) = \mathbf{i} + 2t\mathbf{j} + 3t^2\mathbf{k}$$

$$\mathbf{r}''(t) = 2\mathbf{j} + 6t\mathbf{k}$$

Then, using Theorem 1, we obtain

$$a_{\mathbf{T}} = \frac{\mathbf{r}'(t) \cdot \mathbf{r}''(t)}{|\mathbf{r}'(t)|} = \frac{4t + 18t^3}{\sqrt{1 + 4t^2 + 9t^4}}$$

Next, we compute

$$\mathbf{r}'(t) \times \mathbf{r}''(t) = \begin{vmatrix} \mathbf{i} & \mathbf{j} & \mathbf{k} \\ 1 & 2t & 3t^2 \\ 0 & 2 & 6t \end{vmatrix} = 6t^2\mathbf{i} - 6t\mathbf{j} + 2\mathbf{k}$$

Then, using Theorem 1, we have

$$a_{\mathbf{N}} = \frac{|\mathbf{r}'(t) \times \mathbf{r}''(t)|}{|\mathbf{r}'(t)|} = \frac{\sqrt{36t^4 + 36t^2 + 4}}{\sqrt{1 + 4t^2 + 9t^4}} = 2\sqrt{\frac{9t^4 + 9t^2 + 1}{9t^4 + 4t^2 + 1}} \qquad \blacksquare$$

**EXAMPLE 3** **Motion of a Projectile** Refer to Example 5 in Section 11.4. The position function of a shell is given by

$$\mathbf{r}(t) = 250\sqrt{3}\,t\mathbf{i} + (100 + 250t - 4.9t^2)\mathbf{j}$$

**a.** Find the tangential and normal scalar components of acceleration of the shell at any time $t$.
**b.** Find $a_{\mathbf{T}}(t)$ and $a_{\mathbf{N}}(t)$ for $t = 0$, 12.75, 25.5, and 38.25.
**c.** Is the shell accelerating or decelerating in the tangential direction at the values of $t$ specified in part (b)?

**Solution**
**a.** $\mathbf{r}'(t) = 250\sqrt{3}\mathbf{i} + (250 - 9.8t)\mathbf{j}$

$\mathbf{r}''(t) = -9.8\mathbf{j}$

The tangential scalar component acceleration of the shell is

$$a_T(t) = \frac{\mathbf{r}'(t) \cdot \mathbf{r}''(t)}{|\mathbf{r}'(t)|} = \frac{-9.8(250 - 9.8t)}{\sqrt{96.04t^2 - 4900t + 250{,}000}}$$

Next

$$\mathbf{r}'(t) \times \mathbf{r}''(t) = \begin{vmatrix} \mathbf{i} & \mathbf{j} & \mathbf{k} \\ 250\sqrt{3} & 250 - 9.8t & 0 \\ 0 & -9.8 & 0 \end{vmatrix} = -2450\sqrt{3}\,\mathbf{k}$$

So the normal scalar component of acceleration of the shell is

$$a_N(t) = \frac{|\mathbf{r}'(t) \times \mathbf{r}''(t)|}{|\mathbf{r}'(t)|} = \frac{2450\sqrt{3}}{\sqrt{96.04t^2 - 4900t + 250{,}000}}$$

**b.** The values of $a_T(t)$ and $a_N(t)$ for the specified values of $t$ are shown in Table 1.

**TABLE 1**

| $t$ | 0 | 12.75 | 25.5 | 38.25 |
|-----|------|-------|------|-------|
| $a_T(t)$ | −4.9 | −2.7 | 0 | 2.7 |
| $a_N(t)$ | 8.5 | 9.4 | 9.8 | 9.4 |

**c.** Since $a_T(0) = -4.9 < 0$, the shell is decelerating at $t = 0$. Since $a_T(12.75) \approx -2.7 < 0$, the shell is decelerating at $t = 12.75$ but not by as much as it was at $t = 0$. Since $a_T(25.5) \approx 0$, the shell is neither accelerating nor decelerating at $t = 25.5$ (when the shell is at its maximum height). Finally, since $a_T(38.25) \approx 2.7 > 0$, the shell is accelerating at $t = 38.25$ as it continues to plunge toward the earth.　■

## ■ Kepler's Laws of Planetary Motion

We close this chapter by demonstrating how calculus can be used to derive Kepler's Laws of Planetary Motion. After laboring for more than 20 years analyzing the empirical data obtained by the Danish astronomer Tycho Brahe, the German astronomer Johannes Kepler (1571–1630) formulated the following three laws describing the motion of the planets around the sun.

**Kepler's Laws**

1. The orbit of each planet is an ellipse with the sun at one focus. (See Figure 5a.)
2. The line joining the sun to a planet sweeps out equal areas in equal intervals of time. (See Figure 5b.)
3. The square of the period of revolution $T$ of a planet is proportional to the cube of the length of the major axis $a$ of its orbit; that is, $T^2 = ka^3$, where $k$ is a constant.

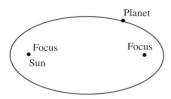

**(a)** The orbit of a planet is an ellipse.

**FIGURE 5**

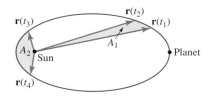

**(b)** If the time intervals $[t_1, t_2]$ and $[t_3, t_4]$ are of equal length, then the area of $A_1$ is equal to the area of $A_2$.

Sir Isaac Newton proved these laws approximately 50 years after they were formulated. He showed that they were consequences of his own Law of Universal Gravitation and the Second Law of Motion. We will prove Kepler's First Law and leave the derivation of the other two laws as exercises. (See Exercises 37 and 38.)

### ▇ Derivation of Kepler's First Law

We begin by showing that the orbit of a planet lies in a plane. Let's place the sun at the origin of a coordinate system. By Newton's Law of Gravitation the force **F** of gravitational attraction exerted by the sun on the planet is given by

$$\mathbf{F} = -\frac{GMm}{r^2}\mathbf{u}$$

where $M$ and $m$ are the masses of the sun and the planet, respectively; **r** is the position vector of the planet; $G$ is the gravitational constant; $r = |\mathbf{r}|$; and $\mathbf{u} = \mathbf{r}/|\mathbf{r}|$ is the unit vector having the same direction as **r**. This force, which is always directed toward a fixed point $O$, is an example of a **central force.** But by Newton's Second Law of Motion the acceleration, **a**, of the planet is related to the force, **F**, to which it is subjected by

$$\mathbf{F} = m\mathbf{a}$$

Equating these two expressions for **F** gives

$$m\mathbf{a} = -\frac{GMm}{r^2}\mathbf{u}$$

or, upon dividing through by $m$,

$$\mathbf{a} = -\frac{GM}{r^2}\mathbf{u}$$

Next, we will show that for any central force, **r** and **a** satisfy $\mathbf{r} \times \mathbf{a} = \mathbf{0}$. To see this, we compute

$$\mathbf{r} \times \mathbf{a} = \mathbf{r} \times \left(-\frac{GM}{r^2}\mathbf{u}\right) = -\frac{GM}{r^2}(\mathbf{r} \times \mathbf{u})$$

$$= -\frac{GM}{r^2}\left(\mathbf{r} \times \frac{\mathbf{r}}{|\mathbf{r}|}\right) = -\frac{GM}{r^3}(\mathbf{r} \times \mathbf{r}) = \mathbf{0}$$

Using this result, we have

$$\frac{d}{dt}(\mathbf{r} \times \mathbf{v}) = \mathbf{r}' \times \mathbf{v} + \mathbf{r} \times \mathbf{v}' = \mathbf{v} \times \mathbf{v} + \mathbf{r} \times \mathbf{a} = \mathbf{0} + \mathbf{0} = \mathbf{0}$$

Integrating both sides of this equation with respect to $t$ yields

$$\mathbf{r} \times \mathbf{v} = \mathbf{c}$$

where $\mathbf{c}$ is a constant vector. By the definition of the cross-product, $\mathbf{c}$ is orthogonal to both $\mathbf{r}$ and $\mathbf{v}$, and we conclude that both $\mathbf{r}(t)$ and $\mathbf{v}(t)$ lie in a fixed plane containing the point $O$. This shows that the orbit of the planet is a plane curve, as was claimed earlier. (See Figure 6.)

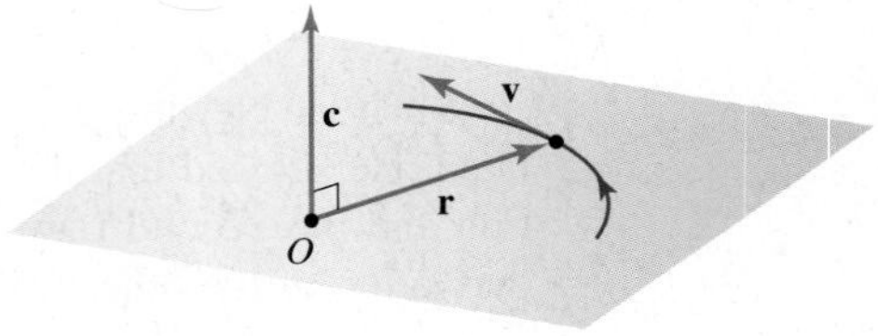

**FIGURE 6**
The orbit of the planet lies in the plane passing through the origin and orthogonal to $\mathbf{c}$.

To show that this curve is an ellipse with the sun at one focus, we observe that

$$\mathbf{c} = \mathbf{r} \times \mathbf{v} = \mathbf{r} \times \mathbf{r}' = (r\mathbf{u}) \times (r\mathbf{u})'$$

$$= (r\mathbf{u}) \times (r\mathbf{u}' + r'\mathbf{u}) = r^2(\mathbf{u} \times \mathbf{u}') + rr'(\mathbf{u} \times \mathbf{u})$$

$$= r^2(\mathbf{u} \times \mathbf{u}') \qquad \mathbf{u} \times \mathbf{u} = \mathbf{0}.$$

Therefore,

$$\mathbf{a} \times \mathbf{c} = \left(-\frac{GM}{r^2}\mathbf{u}\right) \times [r^2(\mathbf{u} \times \mathbf{u}')] = -GM[\mathbf{u} \times (\mathbf{u} \times \mathbf{u}')]$$

$$= -GM[(\mathbf{u} \cdot \mathbf{u}')\mathbf{u} - (\mathbf{u} \cdot \mathbf{u})\mathbf{u}'] \qquad \text{See Theorem 3 in Section 10.4.}$$

Since $\mathbf{u} \cdot \mathbf{u} = |\mathbf{u}|^2 = 1$, we see that $\mathbf{u} \cdot \mathbf{u}' = 0$. (See Example 5 in Section 11.2.) So the last equation reduces to

$$\mathbf{a} \times \mathbf{c} = GM\mathbf{u}'$$

But $\mathbf{a} \times \mathbf{c}$ can also be written as

$$\mathbf{a} \times \mathbf{c} = \mathbf{v}' \times \mathbf{c} = \frac{d}{dt}(\mathbf{v} \times \mathbf{c}) \qquad \text{Remember that } \mathbf{c} \text{ is a constant vector.}$$

Therefore,

$$\frac{d}{dt}(\mathbf{v} \times \mathbf{c}) = GM\mathbf{u}' = \frac{d}{dt}(GM\mathbf{u})$$

Integrating both sides of this equation with respect to $t$ gives

$$\mathbf{v} \times \mathbf{c} = GM\mathbf{u} + \mathbf{b} \tag{5}$$

where $\mathbf{b}$ is a constant vector that depends on the initial conditions. If we take the dot product of both sides of the last equation with $\mathbf{r}$, we have

$$\mathbf{r} \cdot (\mathbf{v} \times \mathbf{c}) = \mathbf{r} \cdot (GM\mathbf{u} + \mathbf{b}) = GM\mathbf{r} \cdot \mathbf{u} + \mathbf{r} \cdot \mathbf{b}$$

$$= GM(r\mathbf{u} \cdot \mathbf{u}) + \mathbf{r} \cdot \mathbf{b} = GMr + \mathbf{r} \cdot \mathbf{b}$$

But

$$\mathbf{r} \cdot (\mathbf{v} \times \mathbf{c}) = \mathbf{c} \cdot (\mathbf{r} \times \mathbf{v}) \qquad \text{See Theorem 3 in Section 10.4.}$$

$$= \mathbf{c} \cdot \mathbf{c} = |\mathbf{c}|^2$$

So

$$GMr + \mathbf{r} \cdot \mathbf{b} = |\mathbf{c}|^2 = c^2 \tag{6}$$

where $c = |\mathbf{c}|$. If $\mathbf{b} = \mathbf{0}$, then Equation (6) reduces to

$$GMr = c^2$$

or

$$r = \frac{c^2}{GM}$$

and the orbit of the planet is a circle. Note that none of the orbits of the planets in our solar system has such an orbit. So we may assume that $\mathbf{b} \neq \mathbf{0}$ for planets in our solar system. In this case, letting $\theta$ be the angle between $\mathbf{r}$ and $\mathbf{b}$ and writing $|\mathbf{c}| = c$, we can write Equation (6) in the form

$$GMr + |\mathbf{r}\|\mathbf{b}| \cos \theta = c^2$$

$$GMr + rb \cos \theta = c^2$$

or

$$r = \frac{c^2}{GM + b \cos \theta} \tag{7}$$

Dividing both the numerator and denominator of Equation (7) by $GM$, we obtain

$$r = \frac{\dfrac{c^2}{GM}}{1 + \left(\dfrac{b}{GM}\right)\cos \theta} = \frac{\dfrac{ec^2}{b}}{1 + e \cos \theta}$$

where $e = b/GM$. Finally, if we write $d = c^2/b$, we obtain the equation

$$r = \frac{ed}{1 + e \cos \theta} \tag{8}$$

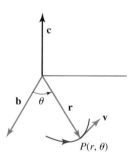

**FIGURE 7**
In the plane determined by $\mathbf{r}$ and $\mathbf{c}$, $r$ and $\theta$ are polar coordinates.

Since $\mathbf{v} \times \mathbf{c}$ and $\mathbf{r}$ are both orthogonal to $\mathbf{c}$, we see from Equation (5) that $\mathbf{b}$ is orthogonal to $\mathbf{c}$. Therefore, in the plane determined by $\mathbf{b}$ and $\mathbf{r}$ (see Figure 7), we can regard $\mathbf{r}$ and $\theta$ as polar coordinates of a point $P$ on the orbit of the planet. Comparing Equation (8) with that of Theorem 2 in Section 9.6, we see that it is the polar equation of a conic section with focus at the origin and eccentricity $e$. Since the orbit of the planet is a closed curve, we know that the conic must be an ellipse. This completes the proof of Kepler's First Law.

## 11.5   CONCEPT QUESTIONS

**1. a.** What are the unit tangent and the unit normal vectors at a point on a curve? Illustrate with a sketch.
   **b.** Suppose that a curve is described by the vector function $\mathbf{r}(t)$. Give formulas for computing the quantities in part (a).

**2. a.** What are the tangential and normal components of acceleration of an acceleration vector? Illustrate with a sketch.
   **b.** Write expressions for the quantities in part (a) in terms of $\mathbf{r}(t)$ and its derivatives.

## 11.5 EXERCISES

*In Exercises 1–4, find the unit tangent and unit normal vector* $\mathbf{T}(t)$ *and* $\mathbf{N}(t)$ *for the curve C defined by* $\mathbf{r}(t)$. *Sketch the graph of C, and show* $\mathbf{T}(t)$ *and* $\mathbf{N}(t)$ *for the given value of t.*

1. $\mathbf{r}(t) = t\mathbf{i} + 2t^2\mathbf{j}; \quad t = 1$

2. $\mathbf{r}(t) = t^2\mathbf{i} - 2t\mathbf{j}; \quad t = 1$

3. $\mathbf{r}(t) = t^2\mathbf{i} + t^3\mathbf{j}; \quad t = 1$

4. $\mathbf{r}(t) = (2 + \cos t)\mathbf{i} + (3 - \sin t)\mathbf{j}; \quad t = \dfrac{\pi}{4}$

*In Exercises 5–10, find the unit tangent and unit normal vectors* $\mathbf{T}(t)$ *and* $\mathbf{N}(t)$ *for the curve C defined by* $\mathbf{r}(t)$.

5. $\mathbf{r}(t) = \mathbf{i} + t\mathbf{j} + t^2\mathbf{k}$

6. $\mathbf{r}(t) = t\mathbf{i} + t^2\mathbf{j} + \dfrac{2}{3}t^3\mathbf{k}$

7. $\mathbf{r}(t) = \langle \sin 2t, \cos 2t, 3t \rangle$

8. $\mathbf{r}(t) = 2 \cos t\mathbf{i} + \mathbf{j} + 2 \sin t\mathbf{k}$

9. $\mathbf{r}(t) = e^t\langle \cos t, \sin t, 1 \rangle$     10. $\mathbf{r}(t) = 2t\mathbf{i} + t^2\mathbf{j} + \ln t\mathbf{k}$

*In Exercises 11–18, find the scalar tangential and normal components of acceleration of a particle with the given position vector.*

11. $\mathbf{r}(t) = t\mathbf{i} + (t^2 + 4)\mathbf{j}$     12. $\mathbf{r}(t) = (2t^2 - 1)\mathbf{i} + 2t\mathbf{j}$

13. $\mathbf{r}(t) = t\mathbf{i} + t^2\mathbf{j} + t^3\mathbf{k}$

14. $\mathbf{r}(t) = t^2\mathbf{i} + t^3\mathbf{j} + t^2\mathbf{k}; \quad t > 0$

15. $\mathbf{r}(t) = 2 \sin t\mathbf{i} + 2 \cos t\mathbf{j} + t\mathbf{k}$

16. $\mathbf{r}(t) = \cos^2 t\mathbf{i} + \sin^2 t\mathbf{j} + t\mathbf{k}$

17. $\mathbf{r}(t) = e^t\langle \cos t, \sin t, 1 \rangle$     18. $\mathbf{r}(t) = \langle t \cos t, t \sin t, 4 \rangle$

19. The accompanying figure shows the path of an object moving in the plane and its acceleration vector **a**, its unit tangent vector **T**, and its unit normal vector **N** at the points A and B.
    a. Sketch the vectors $a_\mathbf{T}\mathbf{T}$ and $a_\mathbf{N}\mathbf{N}$ at A and B.
    b. Is the particle accelerating or decelerating at A? At B?

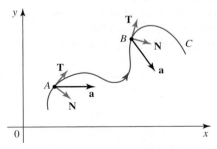

20. At a certain instant of time, the position, velocity, and acceleration of a particle moving in the plane are $\mathbf{r} = 4\mathbf{i} + 2\mathbf{j}$, $\mathbf{v} = 3\mathbf{i} + 4\mathbf{j}$, and $\mathbf{a} = 5\mathbf{i} - 5\mathbf{j}$, respectively.
    a. Sketch **r**, **v**, and **a**.
    b. Is the particle accelerating or decelerating at that instant of time? Explain.
    c. Verify your assertion by computing $a_\mathbf{T}$.

21. At a certain instant of time, the velocity and acceleration of a particle are $\mathbf{v} = 2\mathbf{i} + 3\mathbf{j} - 6\mathbf{k}$ and $\mathbf{a} = -6\mathbf{i} - 4\mathbf{j} + 3\mathbf{k}$, respectively.
    a. Find $a_\mathbf{T}$ and $a_\mathbf{N}$.
    b. Is the particle accelerating or decelerating?

22. At a certain instant of time, the velocity and acceleration of a particle at that time are $\mathbf{v} = \langle 2, 3, 6 \rangle$ and $\mathbf{a} = \langle -6, -4, 3 \rangle$, respectively.
    a. Find $a_\mathbf{T}$ and $a_\mathbf{N}$.
    b. Is the particle accelerating or decelerating?

23. The position of a particle at time t is $\mathbf{r}(t) = \langle \cos t^2, \sin t^2 \rangle$.
    a. Show that the path of the particle is a circular orbit with center at the origin.
    b. Show that $\mathbf{r} \cdot \mathbf{a} \le 0$, where **a** is the acceleration vector of the particle.
    Hint: Show that $\mathbf{a} \cdot \mathbf{r} + \mathbf{v} \cdot \mathbf{v} = 0$.

24. **Trajectory of a Shell** A shell is fired from a howitzer with a muzzle speed of $v_0$ m/sec at angle of elevation of $\alpha$. What are the scalar tangential and normal components of acceleration of the shell?

25. A particle moves along a curve C with a constant speed. Show that the acceleration of the particle is always normal to C.

26. An object moves along a curve C with a constant speed. Show that the magnitude of the acceleration of the object is directly proportional to the curvature of C.

27. Suppose that a particle moves along a plane curve that is the graph of a function f whose second derivative exists. Show that its normal component of acceleration is zero when the particle is at an inflection point of the graph of f.

*Let C be a smooth curve defined by* $\mathbf{r}(t)$, *and let* $\mathbf{T}(t)$ *and* $\mathbf{N}(t)$ *be the unit tangent vector and unit normal vector to C corresponding to t. The plane determined by* **T** *and* **N** *is called the* **osculating plane***. In Exercises 28 and 29, find an equation of the osculating plane of the curve described by* $\mathbf{r}(t)$ *at the point corresponding to the given value of t.*

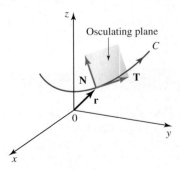

28. $\mathbf{r}(t) = t\mathbf{i} + 2t^2\mathbf{j} + t^3\mathbf{k}; \quad t = 1$

29. $\mathbf{r}(t) = \langle e^t, e^{-t}, \sqrt{2}t \rangle; \quad t = 0$

Let $C$ be a smooth curve defined by $\mathbf{r}(t)$, and let $\mathbf{T}(t)$ and $\mathbf{N}(t)$ be the unit tangent vector and unit normal vector to $C$ corresponding to $t$. The vector $\mathbf{B}$ defined by $\mathbf{B} = \mathbf{T} \times \mathbf{N}$ is orthogonal to $\mathbf{T}$ and $\mathbf{N}$ and is called the **unit binormal vector.** The vectors $\mathbf{T}$, $\mathbf{N}$, and $\mathbf{B}$ form a right-handed set of orthogonal unit vectors, called the **TNB frame,** that move along $C$ as $t$ varies. (Also, see Challenge Problems, page 1027.) In Exercises 30 and 31, find $\mathbf{B}$ for the curve described by $\mathbf{r}(t)$.

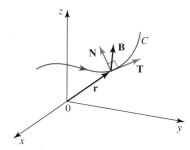

**30.** $\mathbf{r}(t) = t\mathbf{i} + 2t^2\mathbf{j} + t^3\mathbf{k}$

**31.** $\mathbf{r}(t) = 2\cosh t\,\mathbf{i} + 2\sinh t\,\mathbf{j} + 2t\mathbf{k}$

**32.** Refer to Exercises 30 and 31. Show that $\mathbf{B}$ can be expressed in terms of $r$ and its derivatives by the formula

$$\mathbf{B} = \frac{\mathbf{r}' \times \mathbf{r}''}{|\mathbf{r}' \times \mathbf{r}''|}$$

**33.** Rework Exercise 30 using the formula for $\mathbf{B}$ in Exercise 32.

**34.** Let $\mathbf{T}$, $\mathbf{N}$, and $\mathbf{B}$ be the unit tangent, unit normal, and unit binormal, respectively, associated with a smooth curve $C$ described by $\mathbf{r}(t)$.
   **a.** Show that $d\mathbf{B}/ds$ is orthogonal to $\mathbf{T}$ and to $\mathbf{B}$.
   **b.** Use the result of part (a) to show that $d\mathbf{B}/ds = -\tau\mathbf{N}$ for some scalar $\tau(t)$. (The number $\tau(t)$ is called the **torsion** of the curve. It measures the rate at which the curve twists out of its osculating plane (see page 1022). We define $\tau$ to be equal to 0 for a straight line.)
   **c.** Use the result of part (b) to show that the torsion of a plane curve is zero.

The torsion of a curve defined by $\mathbf{r}(t)$ is given by

$$\tau = \frac{(\mathbf{r}' \times \mathbf{r}'') \cdot \mathbf{r}'''}{|\mathbf{r}' \times \mathbf{r}''|^2}$$

In Exercises 35 and 36, find the torsion of the curve defined by $\mathbf{r}(t)$. (Also, see Challenge Problems, page 1027.)

**35.** $\mathbf{r}(t) = \cos t\,\mathbf{i} + \sin t\,\mathbf{j} + t\mathbf{k}$

**36.** $\mathbf{r}(t) = (t - \sin t)\mathbf{i} + (1 - \cos t)\mathbf{j} + t\mathbf{k}$

**37. Kepler's Second Law** Prove Kepler's Second Law using the following steps. (All notation is the same as that used in the text).
   **a.** Show that if $A(t)$ is the area swept out by the radius vector $\mathbf{r}(t)$ in the time interval $[t_0, t]$, then

$$\frac{dA}{dt} = \frac{1}{2}r^2\frac{d\theta}{dt}$$

(See the figure below.)
   **b.** Show that $\mathbf{c} = r^2\dfrac{d\theta}{dt}\mathbf{k}$, so $r^2\dfrac{d\theta}{dt} = c$.
   **c.** Conclude that $\dfrac{dA}{dt} = \dfrac{1}{2}c$, so the rate at which the area is swept out is constant. This is precisely Kepler's Second Law.

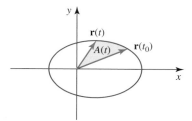

**38. Kepler's Third Law** Prove Kepler's Third Law by using the following steps. In addition, suppose that the lengths of the major and minor axes of the elliptical orbit are $2a$ and $2b$, respectively. (All notation is the same as that used in the text.)
   **a.** Use the result of part (c) Exercise 37 and the fact that the area of the ellipse is $\pi ab$ square units to show that $T = 2\pi ab/c$.
   **b.** Show that $c^2/(GM) = ed = b^2/a$.
   **c.** Using the result of parts (a) and (b), show that $T^2 = ka^3$, where $k = 4\pi^2/(GM)$.

**39. Period of the Earth's Orbit** The period of the earth's orbit about the sun is approximately 365.26 days. Also, the mass of the sun is approximately $1.99 \times 10^{30}$ kg, and the gravitational constant is $G = 6.67 \times 10^{-11}$ Nm²/kg². Find the length of the major axis of the earth's orbit.

**40. Artificial Satellites** A communications relay satellite is to be placed in *geosynchronous* orbit; that is, its circular orbit about the earth is to have a period of revolution of 24 hr so that the satellite appears to be stationary in the sky. Use the fact that the moon has a period of 27.32 days in a circular orbit of radius 238,850 mi from the center of the earth to determine the radius of the satellite's orbit.

**41. Motion of a Projectile** A projectile is fired from a height $h$ with an initial speed $v_0$ and an angle of elevation $\alpha$.
   **a.** What are the scalar tangential and normal components of acceleration of the projectile?
   **b.** What are the scalar tangential and normal components of acceleration of the projectile when the projectile is at its maximum height?

**42. Trajectory of a Shell** A shell is fired from a gun located on a hill 50 m above a level terrain. The muzzle speed of the gun is 500 m/sec, and its angle of elevation is 45°.
   **a.** Find the scalar tangential and normal components of acceleration of the shell.
   Hint: Use the result of Exercise 41.
   **b.** When is the shell accelerating, and when is it decelerating?

**43.** Derive the following formula for calculating the radius of curvature $\rho$ of a curve $C$ represented by $\mathbf{r}(t) = x(t)\mathbf{i} + y(t)\mathbf{j} + z(t)\mathbf{k}$:

$$\rho = \frac{(x'(t))^2 + (y'(t))^2 + (z'(t))^2}{\sqrt{(x''(t))^2 + (y''(t))^2 + (z''(t))^2 - (v'(t))^2}}$$

**44.** Use the result of Exercise 43 to find the radius of curvature of the space curve with position vector $\mathbf{r}(t) = t\mathbf{i} + \sin t\mathbf{j} + \cos t\mathbf{k}$.

*In Exercises 45–48, determine whether the statement is true or false. If it is true, explain why it is true. If it is false, explain why or give an example to show why it is false.*

**45.** If $|\mathbf{r}'(t)| = c$, where $c$ is a nonzero constant, then the unit normal to the curve $C$ defined by $\mathbf{r}(t)$ is given by $\mathbf{N} = \mathbf{r}''/|\mathbf{r}''|$.

**46.** If motion takes place along the $x$-axis, then $a_\mathbf{T} = d^2x/dt^2$.

**47.** If $\mathbf{r}(t)$ is the position vector of a particle with respect to time $t$ and $\mathbf{r}(s)$ is the position vector of the particle with respect to arc length, then $\mathbf{r}''(t)$ is a scalar multiple of $\mathbf{r}''(s)$.

**48.** If $\mathbf{r}(t)$ is the position vector of a particle moving along a smooth curve $C$, then $\mathbf{a} = v'\mathbf{T} + (v^2/\rho)\mathbf{N}$, where $\rho$ is the radius of curvature.

# CHAPTER 11 REVIEW

## CONCEPT REVIEW

*In Exercises 1–11, fill in the blanks.*

**1. a.** A vector function is a function of the form $\mathbf{r}(t) = $ _____, where $f$, $g$, and $h$ are _____ functions of a variable _____, called a _____.
   **b.** The domain of $\mathbf{r}(t)$, called the _____ _____ is a subset of the _____ _____.

**2. a.** A space curve is traced out by the _____ point of $\mathbf{r}(t)$ as $t$ takes on all values in the _____ interval $[a, b]$.
   **b.** The terminal point of $\mathbf{r}(a)$ corresponds to the _____ point of the curve, and the terminal point of $\mathbf{r}(b)$ gives the _____ point of the curve.

**3. a.** If $\mathbf{r}(t) = \langle f(t), g(t), h(t) \rangle$, then $\lim_{t \to a} \mathbf{r}(t)$ exists if and only if _____, _____, and _____ exist.
   **b.** A vector function $\mathbf{r}$ is continuous at $a$ if $\lim_{t \to a} \mathbf{r}(t) = $ _____. The function is _____ on $I$ if it is continuous at each point in $I$.

**4. a.** The derivative of $\mathbf{r}$ is $\mathbf{r}'(t) = $ _____, provided that the limit exists.
   **b.** If $\mathbf{r}(t) = \langle f(t), g(t), h(t) \rangle$, then $\mathbf{r}'(t) = $ _____.

**5. a.** If $\mathbf{u}$ and $\mathbf{v}$ are differentiable, then $\dfrac{d}{dt}[\mathbf{u}(t) \cdot \mathbf{v}(t)] = $ _____, and $\dfrac{d}{dt}[\mathbf{u}(t) \times \mathbf{v}(t)] = $ _____.
   **b.** If $\mathbf{u}$ is differentiable and $f$ is differentiable, then $\dfrac{d}{dt}[\mathbf{u}(f(t))] = $ _____.

**6.** If $f$, $g$, and $h$ are integrable and $\mathbf{r}(t) = \langle f(t), g(t), h(t) \rangle$, then
   **a.** $\int \mathbf{r}(t)\, dt = $ _____.
   **b.** $\int_a^b \mathbf{r}(t)\, dt = $ _____.

**7.** The length of the curve $C$, $\mathbf{r}(t) = \langle f(t), g(t), h(t) \rangle$, where $a \le t \le b$, is given by $L = $ _____.

**8. a.** A curve $C$ described by $\mathbf{r}(t)$ with parameter $t$ is said to be _____ by $t$.
   **b.** The arc length function $s$ associated with a smooth curve $C$ described by $\mathbf{r}(t)$ is $s(t) = $ _____.
   **c.** The curve $C$ has arc length parametrization if it is parametrized by the _____ _____ function $s(t)$.

**9. a.** If a smooth curve $C$ is described by $\mathbf{r}(s)$, where $s$ is the arc length parameter, then the curvature of $C$ is $\kappa(s) = $ _____.
   **b.** If $C$ is parametrized by $t$, then $\kappa(t) = $ _____.
   **c.** The curvature of $C$ is also given by $\kappa(t) = $ _____.
   **d.** If $C$ is a plane curve, then $\kappa(x) = $ _____.
   **e.** For a plane curve $C$, the reciprocal of the curvature at the point $P$ is called the _____ _____ _____ of $C$ at $P$, and the circle with this _____ that shares a common _____ _____ with the curve at $P$ is called the _____ _____ _____.

**10.** If the position vector of a particle is $\mathbf{r}(t)$, then its velocity is _____, its speed is _____, and its acceleration is _____. The acceleration vector points toward the _____ side of the trajectory of the particle.

**11. a.** If $C$ is a smooth curve described by the vector function $\mathbf{r}(t)$, then the unit tangent vector is $\mathbf{T}(t) = $ _____, and the principal unit normal vector is $\mathbf{N}(t) = $ _____.
   **b.** The acceleration of a particle can be resolved into the sum of two vectors—one along the direction of _____ and the other along the direction of _____. In fact, $\mathbf{a} = a_\mathbf{T}\mathbf{T} + a_\mathbf{N}\mathbf{N}$, where $a_\mathbf{T} = $ _____ and $a_\mathbf{N} = $ _____; the former is called the scalar _____ component of acceleration, and the latter is called the scalar _____ component of acceleration.
   **c.** In terms of $\mathbf{r}$ and its derivatives, $a_\mathbf{T} = $ _____, and $a_\mathbf{N} = $ _____.

# REVIEW EXERCISES

*In Exercises 1–4, sketch the curve with the given vector equation, and indicate the orientation of the curve.*

1. $\mathbf{r}(t) = (2 + 3t)\mathbf{i} + (2t - 1)\mathbf{j}$

2. $\mathbf{r}(t) = t^3\mathbf{i} + t^2\mathbf{j}; \quad 0 \le t \le 2$

3. $\mathbf{r}(t) = (\cos t - 1)\mathbf{i} + (\sin t + 2)\mathbf{j} + 2\mathbf{k}$

4. $\mathbf{r}(t) = 2 \cos t\mathbf{i} + 3 \sin t\mathbf{j} + t^2\mathbf{k}; \quad 0 \le t \le 2\pi$

5. Find the domain of $\mathbf{r}(t) = \dfrac{1}{\sqrt{5 - t}}\mathbf{i} + \dfrac{\sin t}{t}\mathbf{j} + \ln(1 + t)\mathbf{k}$.

6. Find $\lim\limits_{t\to 0^+} \mathbf{r}(t)$, where $\mathbf{r}(t) = \dfrac{\sqrt{t}}{1 + t^2}\mathbf{i} + \dfrac{t^2}{\sin t}\mathbf{j} + \dfrac{e^t - 1}{t}\mathbf{k}$.

7. Find the interval in which

$$\mathbf{r}(t) = \sqrt{t + 1}\mathbf{i} + \frac{e^t}{\sqrt{2 - t}}\mathbf{j} + \frac{t^2}{(t - 1)^2}\mathbf{k}$$

   is continuous.

8. Find $\mathbf{r}'(t)$ if $\mathbf{r}(t) = \left[\displaystyle\int_0^t \cos^2 u\, du\right]\mathbf{i} + \left[\displaystyle\int_0^{t^2} \sin u\, du\right]\mathbf{j}$.

*In Exercises 9–12, find $\mathbf{r}'(t)$ and $\mathbf{r}''(t)$.*

9. $\mathbf{r}(t) = \sqrt{t}\mathbf{i} + t^2\mathbf{j} + \dfrac{1}{t + 1}\mathbf{k}$

10. $\mathbf{r}(t) = e^{-t}\mathbf{i} + t \cos t\mathbf{j} + t \sin t\mathbf{k}$

11. $\mathbf{r}(t) = (t^2 + 1)\mathbf{i} + 2t\mathbf{j} + \ln t\mathbf{k}$

12. $\mathbf{r}(t) = \langle t \sin t, t \cos t, e^{2t} \rangle$

*In Exercises 13 and 14, find parametric equations for the tangent line to the curve with the given parametric equations at the point with the given value of t.*

13. $x = t^2 + 1, \quad y = 2t - 3, \quad z = t^3 + 1; \quad t = 0$

14. $x = t \cos t - \sin t, \quad y = t \sin t + \cos t, \quad z = t^2; \quad t = \dfrac{\pi}{2}$

*In Exercises 15 and 16, evaluate the integral.*

15. $\displaystyle\int \left( \sqrt{t}\mathbf{i} + e^{-2t}\mathbf{j} + \dfrac{1}{t + 1}\mathbf{k} \right) dt$

16. $\displaystyle\int_0^1 (2t\mathbf{i} + t^2\mathbf{j} + t^{3/2}\mathbf{k})\, dt$

*In Exercises 17 and 18, find $\mathbf{r}(t)$ for the vector function $\mathbf{r}'(t)$ or $\mathbf{r}''(t)$ and the given initial condition(s).*

17. $\mathbf{r}'(t) = 2\sqrt{t}\mathbf{i} + 3 \cos 2\pi t\mathbf{j} - e^{-t}\mathbf{k}; \quad \mathbf{r}(0) = \mathbf{i} + 2\mathbf{j}$

18. $\mathbf{r}''(t) = 2\mathbf{i} + t\mathbf{j} + e^{-t}\mathbf{k}; \quad \mathbf{r}'(0) = \mathbf{i} + \mathbf{k},$
    $\mathbf{r}(0) = 2\mathbf{i} + \mathbf{j} + 3\mathbf{k}$

*In Exercises 19 and 20, find the unit tangent and the unit normal vectors for the curve C defined by $\mathbf{r}(t)$ for the given value of t.*

19. $\mathbf{r}(t) = t\mathbf{i} + t^2\mathbf{j} + t^3\mathbf{k}; \quad t = 1$

20. $\mathbf{r}(t) = 2 \cos t\mathbf{i} + 2 \sin t\mathbf{j} + e^t\mathbf{k}; \quad t = 0$

*In Exercises 21 and 22, find the length of the curve.*

21. $\mathbf{r}(t) = 2 \sin 2t\mathbf{i} + 2 \cos 2t\mathbf{j} + 3t\mathbf{k}; \quad 0 \le t \le 2$

22. $\mathbf{r}(t) = \sqrt{2}t\mathbf{i} + \dfrac{1}{2}t^2\mathbf{j} + \ln t\mathbf{k}; \quad 1 \le t \le 2$

*In Exercises 23 and 24, find the curvature of the curve.*

23. $\mathbf{r}(t) = t\mathbf{i} + t^2\mathbf{j} + t^3\mathbf{k}$

24. $\mathbf{r}(t) = t \sin t\mathbf{i} + t \cos t\mathbf{j} + t\mathbf{k}$

*In Exercises 25 and 26, find the curvature of the plane curve, and determine the point on the curve at which the curvature is largest.*

25. $y = x - \dfrac{1}{4}x^2$      26. $y = e^{-x}$

*In Exercises 27 and 28, find the velocity, acceleration, and speed of the object with the given position vector.*

27. $\mathbf{r}(t) = 2t\mathbf{i} + e^{-2t}\mathbf{j} + \cos t\mathbf{k}$

28. $\mathbf{r}(t) = te^{-t}\mathbf{i} + \cos 2t\mathbf{j} + \sin 2t\mathbf{k}$

*In Exercises 29 and 30, find the velocity and position vectors of an object with the given acceleration and the given initial velocity and position.*

29. $\mathbf{a}(t) = t\mathbf{i} + \dfrac{1}{3}t^2\mathbf{j} + 3\mathbf{k}; \quad \mathbf{v}(0) = 2\mathbf{i} + 3\mathbf{j} + \mathbf{k}, \quad \mathbf{r}(0) = \mathbf{0}$

30. $\mathbf{a}(t) = e^t\mathbf{i} + e^{-t}\mathbf{j} + t\mathbf{k}; \quad \mathbf{v}(0) = 2\mathbf{i}, \quad \mathbf{r}(0) = \mathbf{i} + \mathbf{k}$

*In Exercises 31–34, find the scalar tangential and normal components of acceleration of a particle with the given position vector.*

31. $\mathbf{r}(t) = \mathbf{i} + t\mathbf{j} + t^2\mathbf{k}$

32. $\mathbf{r}(t) = 2 \cos t\mathbf{i} + 3 \sin t\mathbf{j} + t\mathbf{k}$

33. $\mathbf{r}(t) = \cos t\mathbf{i} + \sin 2t\mathbf{j}$

34. $\mathbf{r}(t) = \sqrt{2}t\mathbf{i} + e^t\mathbf{j} + e^{-t}\mathbf{k}$

35. **A Shot Put**  In a track and field meet, a shot putter heaves a shot at an angle of 45° with the horizontal. As the shot leaves her hand, it is at a height of 7 ft and moving at a speed of 40 ft/sec. Set up a coordinate system so that the shot putter is at the origin.
    a. What is the position of the shot at time $t$?
    b. How far is her put?

## CHALLENGE PROBLEMS

1. **a.** Show that the curve $C$ defined by the vector function
$$\mathbf{r}(t) = (a_1 t^2 + b_1 t + c)\mathbf{i} + (a_2 t^2 + b_2 t + c)\mathbf{j} + (a_3 t^2 + b_3 t + c)\mathbf{k}$$
lies in a plane.
   **b.** Show that the plane of part (a) can be written in the form
$$\begin{vmatrix} x - c & y - c & z - c \\ a_1 & a_2 & a_3 \\ b_1 & b_2 & b_3 \end{vmatrix} = 0$$

2. **Tracking Planes in a Holding Pattern at an Airport** Suppose that an airport is located at the origin of a three-dimensional coordinate system and two airplanes are circling the airport in a holding pattern at an altitude of 2 mi. The planes fly at a constant speed of 300 mph along circular paths of radius 10 mi and are separated by 90°, as shown in the figure.

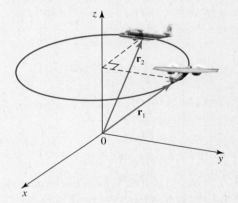

   **a.** Show that the position vectors of the planes are
$$\mathbf{r}_1(t) = 10 \cos 30t\,\mathbf{i} + 10 \sin 30t\,\mathbf{j} + 2\mathbf{k}$$
and
$$\mathbf{r}_2(t) = -10 \sin 30t\,\mathbf{i} + 10 \cos 30t\,\mathbf{j} + 2\mathbf{k}$$
respectively.
   **b.** Let $\mathbf{r} = \mathbf{r}_2 - \mathbf{r}_1$. Interpret your results.
   **c.** Find $\mathbf{r}'$, and interpret your result.
   **d.** Find $\mathbf{r}''$, and interpret your result.

3. **Hitting a Moving Target** A target is located at a height $h$ over level ground, and a gun, located at ground level and at a distance $d$ from the point directly below the target, is aimed directly at the target. Suppose that the gun is fired at the instant the target is released.
   **a.** Show that the bullet will hit the target if its initial speed $v_0$ satisfies
$$v_0 \geq \sqrt{\frac{g(d^2 + h^2)}{2h}}$$
   **b.** Assuming that the condition in part (a) is satisfied, find the distance the target has fallen before it was hit.

4. **Motion of a Projectile** A projectile of mass $m$ is fired from the origin of a coordinate system at an angle of elevation $\alpha$. Assume that air resistance acting on the projectile is proportional to its velocity. Then by Newton's Second Law of Motion the motion of the projectile is described by the equation
$$m\mathbf{r}'' = -mg\mathbf{j} - k\mathbf{r}' \tag{1}$$
where $\mathbf{r}(t)$ is the position vector of the projectile and $k > 0$ is the constant of proportionality.
   **a.** By integrating Equation (1), obtain the equation
$$\mathbf{r}' + \frac{k}{m}\mathbf{r} = -gt\mathbf{j} + \mathbf{v}_0 \tag{2}$$
where $\mathbf{v}_0 = \mathbf{v}(0) = \mathbf{r}'(0)$.
   **b.** Multiply both sides of Equation (2) by $e^{(k/m)t}$, and show that the left-hand side of the resulting equation can be written as $\dfrac{d}{dt}[e^{(k/m)t}\,\mathbf{r}(t)]$. Make use of this observation to find an expression for $\mathbf{r}(t)$.

5. **Motion of a Projectile** Refer to Exercise 4.
   **a.** If the initial speed of the projectile is $v_0$, show that the position function $\mathbf{r}(t)$ is equivalent to the parametric equations
$$x(t) = \frac{mv_0 \cos \alpha}{k}(1 - e^{-(k/m)t})$$
$$y(t) = \left(\frac{m^2 g}{k^2} + \frac{mv_0 \sin \alpha}{k}\right)(1 - e^{-(k/m)t}) - \frac{mg}{k}t$$
   **b.** Solve the first equation in part (a) for $t$ to obtain
$$t = \frac{m}{k}\ln\frac{mv_0 \cos \alpha}{mv_0 \cos \alpha - kx}$$
Then substitute this value into the second equation in part (a) to obtain
$$y = \left(\frac{mg}{kv_0 \cos \alpha} + \tan \alpha\right)x + \frac{m^2 g}{k^2}\ln\left(1 - \frac{kx}{mv_0 \cos \alpha}\right) \tag{3}$$

    **c.** Suppose that a projectile of weight 1600 lb is fired from the origin with an initial speed of 1200 mph and at an angle of elevation of 30°. Draw the trajectories of the projectile for values of $k$ equal to 0.01, 0.1, 0.5, and 1, using the viewing rectangle $[0, 100{,}000] \times [0, 15{,}000]$. Comment on the shape of the trajectories.
   **d.** Expand the expression for $y$ in Equation (3) as a power series to show that
$$y = (\tan \alpha)x - \frac{1}{2}\frac{g}{(v_0 \cos \alpha)^2}x^2 - \frac{1}{3}\frac{kg}{m(v_0 \cos \alpha)^3}x^3 - \cdots$$
and hence deduce that if $k$ is very small, then the trajectory of the projectile is almost parabolic.

**6.** A particle moves in a circular orbit in the plane given by $\mathbf{r}(t) = R\cos t\,\mathbf{i} + R\sin t\,\mathbf{j}$, where $R$ is a constant. At a certain instant of time, the particle is to be released so that it will strike a target located at the point $(a, b)$, where $a^2 + b^2 > R^2$. Find the time at which the particle is to be released.

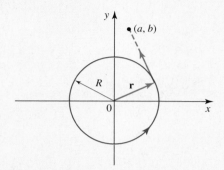

**7. Coriolis Acceleration** Consider the motion of a particle in the $xy$-plane in which the position of the particle is given in polar coordinates $r$ and $\theta$.

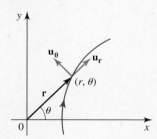

**a.** If $\mathbf{u_r}$ and $\mathbf{u_\theta}$ are unit vectors that point in the direction of the position vector and at right angles to it (in the direction of increasing $\theta$), respectively, show that

$$\mathbf{u_r} = \cos\theta\,\mathbf{i} + \sin\theta\,\mathbf{j}$$

$$\mathbf{u_\theta} = -\sin\theta\,\mathbf{i} + \cos\theta\,\mathbf{j}$$

**b.** If $\mathbf{r} = r\mathbf{u_r}$ is the position vector of a particle located at $(r, \theta)$, show that its velocity vector is given by

$$\mathbf{v}(t) = \mathbf{r}'(t) = \frac{dr}{dt}\mathbf{u_r} + r\frac{d\theta}{dt}\mathbf{u_\theta}$$

and its acceleration vector is given by

$$\mathbf{a}(t) = \frac{d\mathbf{v}}{dt} = \left[\frac{d^2r}{dt^2} - r\left(\frac{d\theta}{dt}\right)^2\right]\mathbf{u_r} + \left[r\frac{d^2\theta}{dt^2} + 2\frac{dr}{dt}\frac{d\theta}{dt}\right]\mathbf{u_\theta}$$

**Note:** The fourth term in the expression for $\mathbf{a}(t)$

$$2\frac{dr}{dt}\frac{d\theta}{dt}\mathbf{u_\theta}$$

is called the Coriolis acceleration. It is due partly to the change in the direction of the radial component of velocity and partly to the change in the transverse component of velocity.

**8. Kepler's Second Law of Planetary Motion** Use the result of Exercise 7(b) to prove Kepler's Second Law of Planetary Motion: The radius vector in a central force field (that is, one in which the force is always directed radially toward or away from the origin) sweeps over area at a constant rate.
**Hint:** Use Newton's Second Law of Motion, $\mathbf{F} = m\mathbf{a}$, to show that $mr^2(d\theta/dt) = C$, where $C$ is constant, and use the fact that the area swept out by $\mathbf{r}$ is $\dfrac{dA}{dt} = \dfrac{1}{2}r^2\dfrac{d\theta}{dt}$.

**9. Coriolis Acceleration** A turntable rotates at a constant angular velocity of 30 rev/min. An ant walks from the center of the turntable outward toward the edge at a speed of 2 cm/sec (relative to the turntable).
**a.** What are the speed and the magnitude of the acceleration of the ant 3 sec later?
**b.** What is the magnitude of the Coriolis acceleration at that time?
**Hint:** Use the results of Challenge Problem 7.

 **10. Path of a Boat** The path of a boat is given by

$$\mathbf{r}(t) = \frac{1}{5}(t^2 - 4t + 8)\mathbf{i} + \frac{1}{5}(3t^2 - 6t + 4)\mathbf{j} \qquad 0 \le t \le 3$$

The shoreline lies along the positive $x$-axis. All distances are measured in miles and time is measured in minutes.
**a.** Plot the path of the boat.
**b.** At what time is the boat closest to the shoreline? What is the distance of the boat from the shoreline at that time?
**c.** What is the velocity, speed, and acceleration of the boat at the time it is closest to the shoreline?

**11.** The binormal vector $\mathbf{B}$ is defined by $\mathbf{B} = \mathbf{T} \times \mathbf{N}$ (see Exercises 30 and 31 on page 1023).
**a.** Show that $d\mathbf{T}/ds = \kappa\mathbf{N}$.
**b.** Show that $\mathbf{N} = \mathbf{B} \times \mathbf{T}$ and $\mathbf{T} = \mathbf{N} \times \mathbf{B}$.
**Hint:** Use Theorem 3, page 936.
**c.** Use the result of part (b) and the result of Exercise 34, page 1023, to show that $d\mathbf{N}/ds = -\kappa\mathbf{T} + \tau\mathbf{B}$.
(The three formulas $d\mathbf{T}/ds = \kappa\mathbf{N}$, $d\mathbf{N}/ds = -\kappa\mathbf{T} + \tau\mathbf{B}$, and $d\mathbf{B}/ds = -\tau\mathbf{N}$ are called the **Frenet-Serret formulas.**)

**12.** Refer to Exercises 35 and 36, page 1023. Use the Frenet-Serret formulas to derive the formula

$$\tau = \frac{(\mathbf{r}' \times \mathbf{r}'') \cdot \mathbf{r}'''}{|\mathbf{r}' \times \mathbf{r}''|^2}$$

Shaun Botterill/Getty Images

The rules for the new International America's cup class include a formula that governs the basic yacht dimensions. This formula balances the rated length, the sail area, and the displacement of the yacht. It is an example of an expression involving three variables. We will use this formula in this chapter.

# 12 Functions of Several Variables

**UP TO NOW** we have dealt primarily with functions involving one independent variable. In this chapter we consider functions involving two or more independent variables. The related notions of limits, continuity, differentiability, and optimization of a function of one variable have their counterparts in the case of a function of several variables, and we will develop these concepts in this chapter. As we will see, many real-life applications of mathematics involve more than one independent variable.

**V** This symbol indicates that one of the following video types is available for enhanced student learning at **www.academic.cengage.com/login:**
  • Chapter lecture videos          • Solutions to selected exercises

**V** ✎ This symbol indicates that step-by-step video lessons for hand-drawing certain complex figures are available.

## 12.1    Functions of Two or More Variables

### ■ Functions of Two Variables

Up to now, we have dealt only with functions of one variable. In practice, however, we often encounter situations in which one quantity depends on two or more quantities. For example, consider the following:

- The volume $V$ of a right circular cylinder depends on its radius $r$ and its height $h$ ($V = \pi r^2 h$).
- The volume $V$ of a rectangular box depends on its length $l$, width $w$, and height $h$ ($V = lwh$).
- The revenue $R$ from the sale of commodities $A$, $B$, $C$, and $D$ at the unit prices of 10, 14, 20, and 30 dollars, respectively, depends on the number of units $x$, $y$, $z$, and $w$ of commodities $A$, $B$, $C$, and $D$ sold ($R = 10x + 14y + 20z + 30w$).

Just as we used a function of one variable to describe the dependency of one variable on another, we can use the notion of a function of several variables to describe the dependency of one variable on several variables. We begin with the definition of a function of two variables.

---

**DEFINITION    Function of Two Variables**

Let $D = \{(x, y) \mid x, y \in R\}$ be a subset of the $xy$-plane. A **function** $f$ **of two variables** is a rule that assigns to each ordered pair of real numbers $(x, y)$ in $D$ a unique real number $z$. The set $D$ is called the **domain** of $f$, and the set of corresponding values of $z$ is called the **range** of $f$.

---

The number $z$ is usually written $z = f(x, y)$. The variables $x$ and $y$ are **independent variables,** and $z$ is the **dependent variable.**

As in the case of a function of a single variable, a function of two or more variables can be described verbally, numerically, graphically, or algebraically.

---

**EXAMPLE 1**    **Home Mortgage Payments**    In a typical housing loan, the borrower makes periodic payments toward reducing indebtedness to the lender, who charges interest at a fixed rate on the unpaid portion of the debt. In practice, the borrower is required to repay the lender in periodic installments, usually of the same size over a fixed term, so that the loan (principal plus interest charges) is amortized at the end of the term. Table 1 gives the monthly loan repayment on a loan of $1000, $f(t, r)$, where $t$ is the term of the loan in years and $r$ is the interest rate per annum (%/year) compounded monthly. Referring to the table, we see that the monthly installment for a 30-year loan of $1000 when the current interest rate is 7%/year is given by $f(30, 7) = 6.6530$ (dollars). Therefore, if the amount borrowed is $350,000, the monthly repayment is 350(6.6530), or $2328.55.

**TABLE 1**

| | $r$ / $t$ | 6 | $6\frac{1}{4}$ | $6\frac{1}{2}$ | $6\frac{3}{4}$ | 7 | $7\frac{1}{4}$ | $7\frac{1}{2}$ | $7\frac{3}{4}$ | 8 |
|---|---|---|---|---|---|---|---|---|---|---|
| | **Interest rate %/year** | | | | | | | | | |
| | 5 | 19.3328 | 19.4493 | 19.5661 | 19.6835 | 19.8012 | 19.9194 | 20.0379 | 20.1570 | 20.2764 |
| | 10 | 11.1021 | 11.2280 | 11.3548 | 11.4824 | 11.6108 | 11.7401 | 11.8702 | 12.0011 | 12.1328 |
| | 15 | 8.4386 | 8.5742 | 8.7111 | 8.8491 | 8.9883 | 9.1286 | 9.2701 | 9.4128 | 9.5565 |
| Term of the loan (years) | 20 | 7.1643 | 7.3093 | 7.4557 | 7.6036 | 7.7530 | 7.9038 | 8.0559 | 8.2095 | 8.3644 |
| | 25 | 6.4430 | 6.5967 | 6.7521 | 6.9091 | 7.0678 | 7.2281 | 7.3899 | 7.5533 | 7.7182 |
| | 30 | 5.9955 | 6.1572 | 6.3207 | 6.4860 | 6.6530 | 6.8218 | 6.9921 | 7.1641 | 7.3376 |
| | 35 | 5.7019 | 5.8708 | 6.0415 | 6.2142 | 6.3886 | 6.5647 | 6.7424 | 6.9218 | 7.1026 |
| | 40 | 5.5021 | 5.6774 | 5.8546 | 6.0336 | 6.2143 | 6.3967 | 6.5807 | 6.7662 | 6.9531 |

Although the monthly installments based on a $1000 loan are displayed in the form of a table for selected values of $t$ and $r$ in Example 1, an algebraic expression for computing $f(t, r)$ also exists:

$$f(t, r) = \frac{10r}{12\left[1 - \left(1 + \dfrac{0.01r}{12}\right)^{-12t}\right]}$$

But as in the case of a single variable, we are primarily interested in functions that can be described by an equation relating the **dependent variable** $z$ to the **independent variables** $x$ and $y$. Also, as in the case of a single variable, unless otherwise specified, the domain of a function of two variables is the set of all points $(x, y)$ for which $z = f(x, y)$ is a real number.

**EXAMPLE 2**  Let $f(x, y) = x^2 - xy + 2y$. Find the domain of $f$, and evaluate $f(1, 2)$, $f(2, 1)$, $f(t, 2t)$, $f(x^2, y)$, and $f(x + y, x - y)$.

**Solution**  Since $x^2 - xy + 2y$ is a real number whenever $(x, y)$ is an ordered pair of real numbers, we see that the domain of $f$ is the entire $xy$-plane. Next, we have

$$f(1, 2) = 1^2 - (1)(2) + 2(2) = 3$$
$$f(2, 1) = 2^2 - (2)(1) + 2(1) = 4$$
$$f(t, 2t) = t^2 - (t)(2t) + 2(2t) = -t^2 + 4t$$
$$f(x^2, y) = (x^2)^2 - (x^2)(y) + 2y = x^4 - x^2y + 2y$$

and

$$f(x + y, x - y) = (x + y)^2 - (x + y)(x - y) + 2(x - y)$$
$$= x^2 + 2xy + y^2 - x^2 + y^2 + 2x - 2y$$
$$= 2(y^2 + xy + x - y)$$

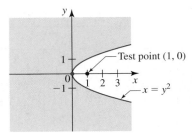

(a) The domain of $f(x, y) = \sqrt{y^2 - x}$

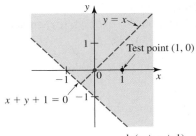

(b) The domain of $g(x, y) = \dfrac{\ln(x + y + 1)}{y - x}$

**FIGURE 1**

**EXAMPLE 3** Find and sketch the domain of the function:

**a.** $f(x, y) = \sqrt{y^2 - x}$ **b.** $g(x, y) = \dfrac{\ln(x + y + 1)}{y - x}$

**Solution**

**a.** $f(x, y)$ is a real number provided that $y^2 - x \geq 0$. Therefore, the domain of $f$ is

$$D = \{(x, y) \mid y^2 - x \geq 0\}$$

To sketch the region $D$, we first draw the curve $y^2 - x = 0$, or $y^2 = x$, which is a parabola (Figure 1a). Observe that this curve divides the $xy$-plane into two regions: the points satisfying $y^2 - x > 0$ and the points satisfying $y^2 - x < 0$. To determine the region of interest, we pick a point in one of the regions, say, the point $(1, 0)$. Substituting the coordinates $x = 1$ and $y = 0$ into the inequality $y^2 - x > 0$, we obtain $0 - 1 > 0$, which is false. This shows that the test point is *not* contained in the required region. Therefore, the region that does not contain the test point together with the curve $x = y^2$ is the required domain (Figure 1a).

**b.** Because the logarithmic function is defined only for positive numbers, we must have $x + y + 1 > 0$. Furthermore, the denominator of the expression cannot be zero, so $y - x \neq 0$, or $y \neq x$. Therefore, the domain of $g$ is

$$D = \{(x, y) \mid x + y + 1 > 0 \quad \text{and} \quad y \neq x\}$$

To sketch the domain of $D$, we first draw the graph of the equation

$$x + y + 1 = 0$$

which is a straight line. The dashed line is used to indicate that points on the line are not included in $D$. This line divides the $xy$-plane into two half-planes. If we pick the test point $(1, 0)$ and substitute the coordinates $x = 1$ and $y = 0$ into the inequality $x + y + 1 > 0$, we obtain $2 > 0$, which is true. This computation tells us that the upper half-plane containing the test point satisfies the inequality $x + y + 1 > 0$. Next, because $y \neq x$, all the points lying on the line $y = x$ in this half-plane must be excluded from $D$. Again, we indicate this with a dashed line (Figure 1b). ■

## Graphs of Functions of Two Variables

Just as the graph of a function of one variable enables us to visualize the function, so too does the graph of a function of two variables.

> **DEFINITION** **Graph of a Function of Two Variables**
>
> Let $f$ be a function of two variables with domain $D$. The graph of $f$ is the set
>
> $$S = \{(x, y, z) \mid z = f(x, y), (x, y) \in D\}$$

**FIGURE 2**
The graph of $f$ is the surface $S$ consisting of all points $(x, y, z)$, where $z = f(x, y)$ and $(x, y) \in D$.

Since each ordered triple $(x, y, z)$ may be represented as a point in three-dimensional space, $R^3$, the set $S$ is a surface in space (see Figure 2).

**EXAMPLE 4**   Sketch the graph of $f(x, y) = \sqrt{9 - x^2 - y^2}$. What is the range of $f$?

**Solution**   The domain of $f$ is $D = \{(x, y) \mid x^2 + y^2 \le 9\}$, the disk with radius 3, centered at the origin. Writing $z = f(x, y)$, we have

$$z = \sqrt{9 - x^2 - y^2}$$
$$z^2 = 9 - x^2 - y^2$$

or

$$x^2 + y^2 + z^2 = 9$$

The last equation represents a sphere of radius 3 centered at the origin. Since $z \ge 0$, we see that the graph of $f$ is just an upper hemisphere (Figure 3). Furthermore, $z$ must be less than or equal to 3, so the range of $f$ is $[0, 3]$.

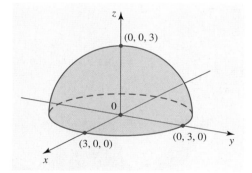

Ⓥ ✏ **FIGURE 3**
The graph of $f(x, y) = \sqrt{9 - x^2 - y^2}$ is the upper hemisphere of radius 3, centered at the origin.

∎

## ■ Computer Graphics

The graph of a function of two variables can be sketched with the aid of a graphing utility. In most cases the techniques that are used involve plotting the traces of a surface in the vertical planes $x = k$ and $y = k$ for equally spaced values of $k$. The program uses a "hidden line" routine that determines what parts of certain traces should be eliminated to give the illusion of the surface in three dimensions. In the next example we sketch the graph of a function of two variables and then show a computer-generated version of it.

**EXAMPLE 5**   Let $f(x, y) = x^2 + 4y^2$.

**a.** Sketch the graph of $f$.        **b.** Use a CAS to plot the graph of $f$.

**Solution**
**a.** We recognize that the graph of the function is the surface $z = x^2 + 4y^2$, which is the elliptic paraboloid

$$\frac{x^2}{1} + \frac{y^2}{\left(\dfrac{1}{2}\right)^2} = z$$

Using the drawing skills developed in Section 10.6, we obtain the sketch shown in Figure 4a.

**b.** The computer-generated graph of $f$ is shown in Figure 4(b).

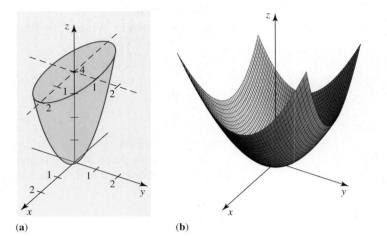

The graph of $f(x, y) = x^2 + 4y^2$

**(a)**                          **(b)**

Figure 5 shows the computer-generated graphs of several functions.

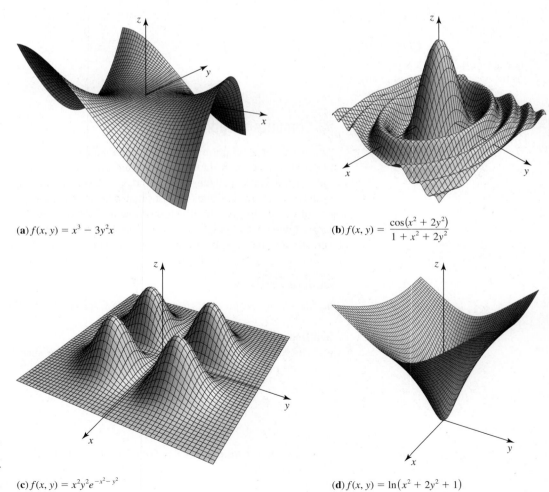

**(a)** $f(x, y) = x^3 - 3y^2x$

**(b)** $f(x, y) = \dfrac{\cos(x^2 + 2y^2)}{1 + x^2 + 2y^2}$

**FIGURE 5**
Some computer-generated
graphs of functions of
two variables     **(c)** $f(x, y) = x^2 y^2 e^{-x^2 - y^2}$

**(d)** $f(x, y) = \ln(x^2 + 2y^2 + 1)$

## ◼ Level Curves

We can visualize the graph of a function of two variables by using *level curves*. To define the level curve of a function $f$ of two variables, let $z = f(x, y)$ and consider the trace of $f$ in the plane $z = k$ ($k$, a constant), as shown in Figure 6a. If we project this trace onto the $xy$-plane, we obtain a curve $C$ with equation $f(x, y) = k$, called a *level curve* of $f$ (Figure 6b).

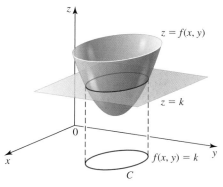

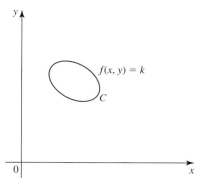

**(a)** The level curve $C$ with equation $f(x, y) = k$ is the projection of the trace of $f$ in the plane $z = k$ onto the $xy$-plane.

**(b)** The level curve $C$

**FIGURE 6**

---

> **DEFINITION  Level Curves**
>
> The **level curves** of a function $f$ of two variables are the curves in the $xy$-plane with equations $f(x, y) = k$, where $k$ is a constant in the range of $f$.

---

Notice that the level curve with equation $f(x, y) = k$ is the set of all points in the domain of $f$ corresponding to the points on the surface $z = f(x, y)$ having the same height or depth $k$. By drawing the level curves corresponding to several admissible values of $k$, we obtain a *contour map*. The map enables us to visualize the surface represented by the graph of $z = f(x, y)$: We simply lift or depress the level curve to see the "cross sections" of the surface. Figure 7a shows a hill, and Figure 7b shows a contour map associated with that hill.

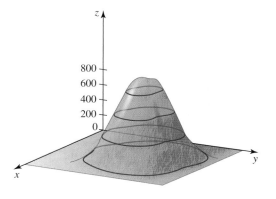

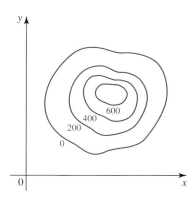

**FIGURE 7**    **(a)** A hill

**(b)** A contour map of the hill

**EXAMPLE 6** Sketch a contour map for the surface described by $f(x, y) = x^2 + y^2$, using the level curves corresponding to $k = 0, 1, 4, 9$, and 16.

**Solution**    The level curve of $f$ corresponding to each value of $k$ is a circle $x^2 + y^2 = k$ of radius $\sqrt{k}$, centered at the origin. For example, if $k = 4$, the level curve is the circle with equation $x^2 + y^2 = 4$, centered at the origin and having radius 2. The required contour map of $f$ comprises the origin and the four concentric circles shown in Figure 8a. The graph of $f$ is the paraboloid shown in Figure 8b.

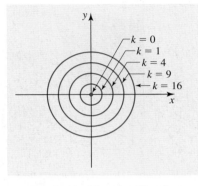

**⊙ ✎ FIGURE 8**            (a) Contour map for $f(x, y) = x^2 + y^2$            (b) The graph of $z = x^2 + y^2$

**EXAMPLE 7** Sketch a contour map for the hyperbolic paraboloid defined by $f(x, y) = y^2 - x^2$.

**Solution**    The level curve corresponding to each value of $k$ is the graph of the equation $y^2 - x^2 = k$. For $k > 0$ the level curves have equations

$$\frac{y^2}{k} - \frac{x^2}{k} = 1$$

or

$$\frac{y^2}{(\sqrt{k})^2} - \frac{x^2}{(\sqrt{k})^2} = 1$$

These curves are a family of hyperbolas with asymptotes $y = \pm x$ and vertices $(0, \pm\sqrt{k})$. For example, if $k = 4$, then the level curve is the hyperbola

$$\frac{y^2}{4} - \frac{x^2}{4} = 1$$

with vertices $(0, \pm 2)$.

If $k < 0$, the level curves have equations $y^2 - x^2 = k$ or $x^2 - y^2 = -k$, which can be put in the standard form

$$\frac{x^2}{(\sqrt{-k})^2} - \frac{y^2}{(\sqrt{-k})^2} = 1$$

and represent a family of hyperbolas with asymptotes $y = \pm x$. The contour map comprising the level curves corresponding to $k = 0, \pm 2, \pm 4, \pm 6$, and $\pm 8$ is sketched in Figure 9a. The graph of $z = y^2 - x^2$ is shown in Figure 9b.

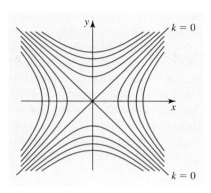
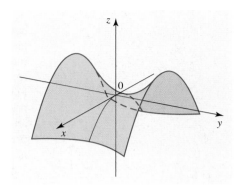

**FIGURE 9**    (**a**) Contour map for $f(x, y) = y^2 - x^2$        (**b**) The graph of $z = y^2 - x^2$

Figure 10 shows some computer-generated graphs of functions of two variables and their corresponding level curves.

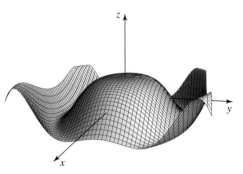

(**a**) Graph of $f(x, y) = \cos\left(\dfrac{x^2 + 2y^2}{4}\right)$

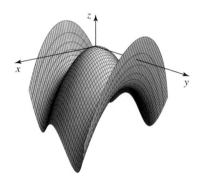

(**b**) Graph of $f(x, y) = y^4 - 8y^2 + 4x^2$

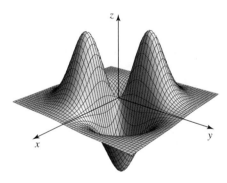

(**c**) Graph of $f(x, y) = -xye^{-x^2 - y^2}$

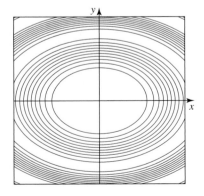

(**d**) Level curves of $f(x, y) = \cos\left(\dfrac{x^2 + 2y^2}{4}\right)$

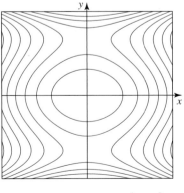

(**e**) Level curves of $f(x, y) = y^4 - 8y^2 + 4x^2$

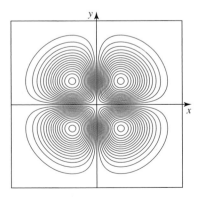

(**f**) Level curves of $f(x, y) = -xye^{-x^2 - y^2}$

**FIGURE 10**
The graphs of some functions and their level curves

Aside from their use in constructing topographic maps of mountain ranges, level curves are found in many areas of practical interest. For example, if $T(x, y)$ denotes the temperature at a location within the continental United States with longitude $x$ and latitude $y$ at a certain time of day, then the temperature at the point $(x, y)$ is the height (or depth) of the surface with equation $z = T(x, y)$. In this context the level curve $T(x, y) = k$ is a curve superimposed on the map of the United States connecting all points that have the same temperature at a given time (Figure 11). These level curves are called **isotherms.** Similarly, if $P(x, y)$ measures the barometric pressure at the location $(x, y)$, then the level curves of the function $P$ are called **isobars.** All points on an isobar $P(x, y) = k$ have the same barometric pressure at a given time.

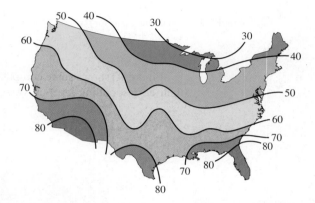

**FIGURE 11**
Isotherms: level curves connecting points that have the same temperature

## Functions of Three Variables and Level Surfaces

A function $f$ of three variables is a rule that assigns to each ordered triple $(x, y, z)$ in a domain $D = \{(x, y, z) \mid x, y, z \in R\}$ a unique real number $w$ denoted by $f(x, y, z)$. For example, the volume $V$ of a rectangular box of length $x$, width $y$, and height $z$ can be described by the function $f$ defined by $f(x, y, z) = xyz$.

**EXAMPLE 8** Find the domain of the function $f$ defined by

$$f(x, y, z) = \sqrt{x + y - z} + xe^{yz}$$

**Solution** $f(x, y, z)$ is a real number provided that $x + y - z \geq 0$ or, equivalently, $z \leq x + y$. Therefore, the domain of $f$ is

$$D = \{(x, y, z) \mid z \leq x + y\}$$

This is the half-space consisting of all points lying on or below the plane $z = x + y$. ∎

Since the graph of a function of three variables is composed of the points $(x, y, z, w)$, where $w = f(x, y, z)$, lying in four-dimensional space, we cannot draw the graphs of such functions. But by examining the **level surfaces,** which are the surfaces with equations

$$f(x, y, z) = k \qquad k, \text{ a constant}$$

we are often able to gain some insight into the nature of $f$.

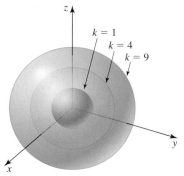

**FIGURE 12**
The level surfaces of
$f(x, y, z) = x^2 + y^2 + z^2$
corresponding to $k = 1, 4, 9$

**EXAMPLE 9** Find the level surfaces of the function $f$ defined by

$$f(x, y, z) = x^2 + y^2 + z^2$$

**Solution** The required level surfaces of $f$ are the graphs of the equations $x^2 + y^2 + z^2 = k$, where $k \geq 0$. These surfaces are concentric spheres of radius $\sqrt{k}$ centered at the origin (see Figure 12). Observe that $f$ has the same value for all points $(x, y, z)$ lying on any such sphere. ∎

## 12.1 CONCEPT QUESTIONS

1. What is a function of two variables? Give an example of one by stating its rule, domain, and range.
2. What is the graph of a function of two variables? Illustrate with a sketch.
3. What is a level curve of a function of two variables? Illustrate with a sketch.

4. What is a level surface of a function of three variables? If $w = T(x, y, z)$ gives the temperature of a point $(x, y, z)$ in three-dimensional space, what does the level surface $w = k$ describe?

## 12.1 EXERCISES

1. Let $f(x, y) = x^2 + 3xy - 2x + 3$. Find
   **a.** $f(1, 2)$     **b.** $f(2, 1)$
   **c.** $f(2h, 3k)$     **d.** $f(x + h, y)$
   **e.** $f(x, y + k)$

2. Let $g(x, y) = \dfrac{2xy}{2x^2 + 3y^2}$. Find

   **a.** $g(-1, 2)$     **b.** $g(2, -1)$
   **c.** $g(u, -v)$     **d.** $g(2, a)$
   **e.** $g(u + v, v)$

3. Let $f(x, y, z) = \sqrt{x^2 + 2y^2 + 3z^2}$. Find
   **a.** $f(1, 2, 3)$     **b.** $f(0, 2, -1)$
   **c.** $f(t, -t, t)$     **d.** $f(u, u - 1, u + 1)$
   **e.** $f(-x, x, -2x)$

4. Let $g(r, s, t) = re^{s/t}$. Find
   **a.** $g(2, 0, 3)$     **b.** $g(1, \ln 3, 1)$
   **c.** $g(-1, -1, -1)$     **d.** $g(t, t, t)$
   **e.** $g(r + h, s + k, t + l)$

*In Exercises 5–14, find the domain and the range of the function.*

5. $f(x, y) = x + 3y - 1$     6. $g(x, y) = x^2 + 2y^2 + 3$

7. $f(u, v) = \dfrac{uv}{u - v}$     8. $h(x, y) = \sqrt{x - 2y}$

9. $g(x, y) = \sqrt{4 - x^2 - y^2}$     10. $h(x, y) = \ln(xy - 1)$

11. $f(x, y, z) = \sqrt{9 - x^2 - y^2 - z^2}$

12. $g(x, y, z) = \dfrac{x}{y + z}$

13. $h(u, v, w) = \tan u + v \cos w$

14. $f(x, y, z) = \dfrac{1}{\sqrt{4 - x^2 - y^2 - z^2}}$

*In Exercises 15–22, find and sketch the domain of the function.*

15. $f(x, y) = \sqrt{y} - \sqrt{x}$     16. $g(x, y) = \dfrac{xy}{2x - y}$

17. $f(u, v) = \dfrac{uv}{u^2 - v^2}$     18. $h(x, y) = \sqrt{xy - 1}$

19. $f(x, y) = x \ln y + y \ln x$

20. $h(x, y) = \dfrac{\ln(y - x)}{\sqrt{x - y + 1}}$

21. $f(x, y, z) = \sqrt{9 - x^2 - y^2 - z^2}$

22. $g(x, y, z) = \dfrac{\sqrt{4 - x^2 - y^2}}{z - 3}$

*In Exercises 23–30, sketch the graph of the function.*

**23.** $f(x, y) = 4$

**24.** $f(x, y) = 6 - 2x + 3y$

**25.** $f(x, y) = x^2 + y^2$

**26.** $g(x, y) = y^2$

**27.** $h(x, y) = 9 - x^2 - y^2$

**28.** $f(x, y) = \sqrt{x^2 + y^2}$

**29.** $f(x, y) = \frac{1}{2}\sqrt{36 - 9x^2 - 36y^2}$

**30.** $f(x, y) = \cos x$

**31.** The figure shows the contour map of a hill. The numbers in the figure are measured in feet. Use the figure to answer the questions below.

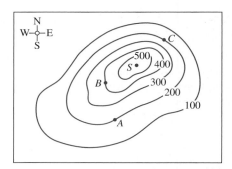

**a.** What is the altitude of the point on the hill corresponding to the point $A$? The point $B$?

**b.** If you start out from the point on the hill corresponding to point $A$ and move north, will you be ascending or descending? What if you move east from the point on the hill corresponding to point $B$?

**c.** Is the hill steeper at the point corresponding to $A$ or at the point corresponding to $C$? Explain.

**32.** A contour map of a function $f$ is shown in the figure. Use it to estimate the value of $f$ at $P$ and $Q$.

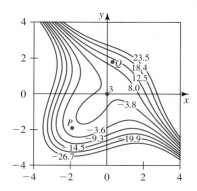

*In Exercises 33–38, match the function with one of the graphs labeled* a *through* f.

**(a)**                    **(b)**

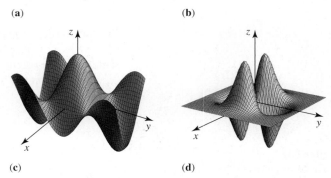

**(c)**                    **(d)**

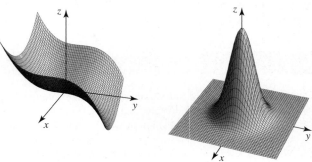

**(e)**                    **(f)**

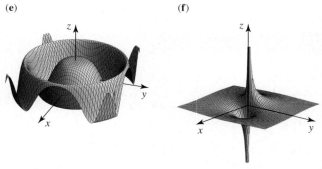

**33.** $f(x, y) = 2x^2 - y^3$

**34.** $f(x, y) = \cos(x^2 + y^2)$

**35.** $f(x, y) = \cos\dfrac{x}{2}\cos y$

**36.** $f(x, y) = (x^2 - y^2)e^{-x^2 - y^2}$

**37.** $f(x) = e^{-x^2 - y^2}$

**38.** $f(x, y) = -\dfrac{x}{2(x^2 + y^2)}$

**cas** *In Exercises 39–42, use a computer or calculator to plot the graph of the function.*

**39.** $f(x, y) = 3x^2 - 3y^2 + 2$

**40.** $f(x, y) = (4x^2 + 9y^2)e^{-x^2 - y^2}$

**41.** $f(x, y) = \cos x + \cos y$

**42.** $f(x, y) = \dfrac{1 - 2\sin(x^2 + y^2)}{x^2 + y^2}$

*In Exercises 43–52, sketch the level curves $f(x, y) = k$ of the function for the indicated values of k.*

**43.** $f(x, y) = 2x + 3y;$ $k = -2, -1, 0, 1, 2$

**44.** $f(x, y) = x^2 + 4y^2;$ $k = 0, 1, 2, 3, 4$

**45.** $f(x, y) = xy;$ $k = -2, -1, 0, 1, 2$

**46.** $f(x, y) = \sqrt{16 - x^2 - y^2};$ $k = 0, 1, 2, 3, 4$

**47.** $f(x, y) = \dfrac{x + y}{x - y};$ $k = -2, 0, 1, 2$

**48.** $f(x, y) = y^2 - x^2;$ $k = -2, -1, 0, 1, 2$

**49.** $f(x, y) = \ln(x + y);$ $k = -2, -1, 0, 1, 2$

**50.** $f(x, y) = \dfrac{x}{y};$ $k = -2, -1, 0, 1, 2$

**51.** $f(x, y) = y - x^2;$ $k = -2, -1, 0, 1, 2$

**52.** $f(x, y) = x - \sin y;$ $k = -2, -1, 0, 1, 2$

*In Exercises 53–56, describe the level surfaces of the function.*

**53.** $f(x, y, z) = 2x + 4y - 3z + 1$

**54.** $f(x, y, z) = 2x^2 + 3y^2 + 6z^2$

**55.** $f(x, y, z) = x^2 + y^2 - z^2$

**56.** $f(x, y, z) = -x^2 - y^2 + z + 2$

*In Exercises 57–62, match the graph of the surface with one of the contour maps labeled a through f.*

**(a)**

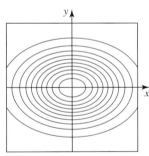

**(b)**

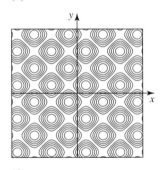

**(c)**

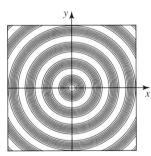

**(d)**

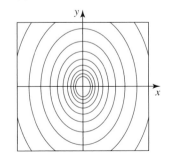

**(e)**

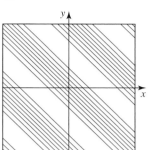

**(f)**

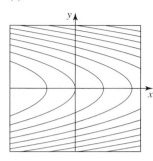

**57.** $f(x, y) = e^{1 - 2x^2 - 4y^2}$

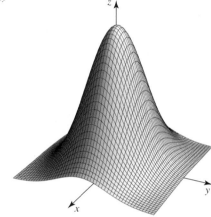

**58.** $f(x, y) = x + y^2$

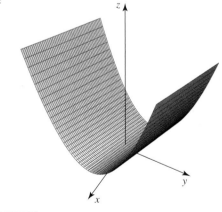

**59.** $f(x, y) = \cos\sqrt{x^2 + y^2}$

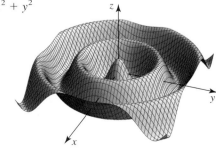

**60.** $f(x, y) = \sin x + \sin y$

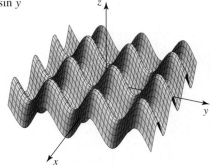

**61.** $f(x, y) = \sin(x + y)$

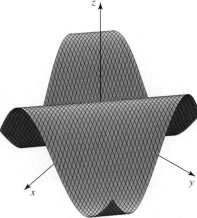

**62.** $f(x, y) = \ln(2x^2 + y^2)$

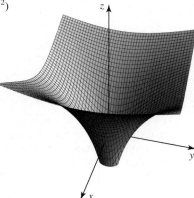

**cas** *In Exercises 63–66, (a) use a computer or calculator to plot the graph of the function f, and (b) plot some level curves of f and compare them with the graph obtained in part (a).*

**63.** $f(x, y) = |x| + |y|$

**64.** $f(x, y) = \dfrac{xy}{\sqrt{x^2 + y^2}}$

**65.** $f(x, y) = \dfrac{xy(x^2 - y^2)}{x^2 + y^2}$

**66.** $f(x, y) = ye^{1 - x^2 - y^2}$

**67.** Find an equation of the level curve of $f(x, y) = \sqrt{x^2 + y^2}$ that contains the point $(3, 4)$.

**68.** Find an equation of the level surface of $f(x, y, z) = 2x^2 + 3y^2 - z$ that contains the point $(-1, 2, -3)$.

**69.** Can two level curves of a function $f$ of two variables $x$ and $y$ intersect? Explain.

**70.** A *level set* of $f$ is the set $S = \{(x, y) \mid f(x, y) = k$, where $k$ is in the range of $f\}$. Let

$$f(x, y) = \begin{cases} 0 & \text{if } x^2 + y^2 < 1 \\ x^2 + y^2 - 1 & \text{if } x^2 + y^2 \geq 1 \end{cases}$$

Sketch the level set of $f$ for $k = 0$ and $3$.

**71.** Refer to Exercise 70. Let

$$f(x, y) = \begin{cases} 1 - \sqrt{x^2 + y^2} & \text{if } x^2 + y^2 < 1 \\ x^2 + y^2 - 1 & \text{if } x^2 + y^2 \geq 1 \end{cases}$$

(a) Sketch the graph of $f$ and (b) describe the level set of $f$ for $k = 0, \frac{1}{2}, 1$, and $3$.

**72. Body Mass** The body mass index (BMI) is used to identify, evaluate, and treat overweight and obese adults. The BMI value for an adult of weight $w$ (in kilograms) and height $h$ (in meters) is defined to be

$$M = f(w, h) = \frac{w}{h^2}$$

According to federal guidelines, an adult is overweight if he or she has a BMI value between 25 and 29.9 and is "obese" if the value is greater than or equal to 30.

    **a.** What is the BMI of an adult who weighs in at 80 kg and stands 1.8 m tall?

    **b.** What is the maximum weight of an adult of height 1.8 m who is not classified as overweight or obese?

**73. Poiseuille's Law** Poiseuille's Law states that the resistance $R$, measured in dynes, of blood flowing in a blood vessel of length $l$ and radius $r$ (both in centimeters) is given by

$$R = f(l, r) = \frac{kl}{r^4}$$

where $k$ is the viscosity of blood (in dyne-sec/cm$^2$). What is the resistance, in terms of $k$, of blood flowing through an arteriole with radius 0.1 cm and length 4 cm?

**74. Surface Area of a Human Body** An empirical formula by E.F. Dubois relates the surface area $S$ of a human body (in square meters) to its weight $W$ (in kilograms) and its height $h$ (in centimeters). The formula, given by

$$S = 0.007184W^{0.425}H^{0.725}$$

is used by physiologists in metabolism studies.

    **a.** Find the domain of the function $S$.

    **b.** What is the surface area of a human body that weighs 70 kg and has a height of 178 cm?

**75. Cobb-Douglas Production Function** Economists have found that the output of a finished product, $f(x, y)$, is sometimes described by the function

$$f(x, y) = ax^b y^{1-b}$$

where $x$ stands for the amount of money expended for labor, $y$ stands for the amount expended on capital, and $a$ and $b$ are positive constants with $0 < b < 1$.
   **a.** If $p$ is a positive number, show that $f(px, py) = pf(x, y)$.
   **b.** Use the result of part (a) to show that if the amount of money expended for labor and capital are both increased by $r$ percent, then the output is also increased by $r$ percent.

**76. Continuous Compound Interest** If a principal of $P$ dollars is deposited in an account earning interest at the rate of $r$/year compounded continuously, then the accumulated amount at the end of $t$ years is given by

$$A = f(P, r, t) = Pe^{rt}$$

dollars. Find the accumulated amount at the end of 3 years if $10,000 is deposited in an account earning interest at the rate of 10%/year.

**77. Home Mortgages** Suppose a home buyer secures a bank loan of $A$ dollars to purchase a house. If the interest rate charged is $r$/year and the loan is to be amortized in $t$ years, then the principal repayment at the end of $i$ months is given by

$$B = f(A, r, t, i) = A\left[\frac{\left(1 + \frac{r}{12}\right)^i - 1}{\left(1 + \frac{r}{12}\right)^{12t} - 1}\right] \qquad 0 \le i \le 12t$$

Suppose the Blakelys borrow $280,000 from a bank to help finance the purchase of a house and the bank charges interest at a rate of 6%/year, compounded monthly. If the Blakelys agree to repay the loan in equal installments over 30 years, how much will they owe the bank after the sixtieth payment (5 years)? The 240th payment (20 years)?

**78. Wilson Lot-Size Formula** The Wilson lot-size formula in economics states that the optimal quantity $Q$ of goods for a store to order is given by

$$Q = f(C, N, h) = \sqrt{\frac{2CN}{h}}$$

where $C$ is the cost of placing an order, $N$ is the number of items the store sells per week, and $h$ is the weekly holding cost for each item. Find the most economical quantity of ten-speed bicycles to order if it costs the store $20 to place an order and $5 to hold a bicycle for a week and the store expects to sell 40 bicycles a week.

**79. Force Generated by a Centrifuge** A centrifuge is a machine designed for the specific purpose of subjecting materials to a sustained centrifugal force. The magnitude of a centrifugal force $F$ in dynes is given by

$$F = f(M, S, R) = \frac{\pi^2 S^2 MR}{900}$$

where $S$ is in revolutions per minute (rpm), $M$ is the mass in grams, and $R$ is the radius in centimeters. Find the centrifugal force generated by an object revolving at the rate of 600 rpm in a circle of radius 10 cm. Express your answer as a multiple of the force of gravity. (Recall that 1 gram of force is equal to 980 dynes.)

**80. Temperature of a Thin Metal Plate** A thin metal plate located in the $xy$-plane has a temperature of

$$T(x, y) = \frac{120}{1 + 2x^2 + y^2}$$

degrees Celsius at the point $(x, y)$. Describe the isotherms of $T$, and sketch those corresponding to $T = 120, 60, 40$, and 20.

**81. International America's Cup Class** Drafted by an international committee in 1989, the rules for the new International America's cup class includes a formula that governs the basic yacht dimensions. The formula $f(L, S, D) \le 42$, where

$$f(L, S, D) = \frac{L + 1.25S^{1/2} - 9.80D^{1/3}}{0.388}$$

balances the rated length $L$ (in meters), the rated sail area $S$ (in square meters) and the displacement $D$ (in cubic meters). All changes in the basic dimensions are tradeoffs. For example, if you want to pick up speed by increasing the sail area, you must pay for it by decreasing the length or increasing the displacement, both of which slow the boat down. Show that yacht $A$ of rated length 20.95 m, rated sail area 277.3 m$^2$, and displacement 17.56 m$^3$, and the longer and heavier yacht $B$ with $L = 21.87$, $S = 311.78$, and $D = 22.48$ both satisfy the formula.

**82. Ideal Gas Law** According to the *ideal gas law*, the volume $V$ of an ideal gas is related to its pressure $P$ and temperature $T$ by the formula

$$V = \frac{kT}{P}$$

where $k$ is a positive constant. Describe the level curves of $V$, and give a physical interpretation of your result.

**83. Newton's Law of Gravitation** According to Newton's Law of Gravitation a body of mass $m_1$ located at the origin of an $xyz$-coordinate system attracts another body of mass $m_2$ located at the point $(x, y, z)$ with a force of magnitude given by

$$F = \frac{Gm_1m_2}{x^2 + y^2 + z^2}$$

where $G$ is the universal constant of gravitation. Describe the level surfaces of $F$, and give a physical interpretation of your result.

**84. Equipotential Curves** Consider the crescent-shaped region $R$ (in the following figure) that lies inside the disk

$$D_1 = \{(x, y) \mid (x - 2)^2 + y^2 \le 4\}$$

and outside the disk

$$D_2 = \{(x, y) \mid (x - 1)^2 + y^2 \le 1\}$$

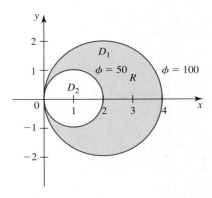

If the electrostatic potential along the inner circle is kept at 50 volts and the electrostatic potential along the outer circle is kept at 100 volts, then the electrostatic potential at any point $(x, y)$ in the region $R$ is given by

$$\phi(x, y) = 150 - \frac{200x}{x^2 + y^2}$$

Show that the equipotential curves of $\phi$ are arcs of circles that have their centers on the positive $x$-axis and pass through the origin. Sketch the equipotential curve corresponding to a potential of 75 volts.

**85. The Doppler Effect** Suppose that a sound with frequency $f$ is emitted by an object moving along a straight line with speed $u$ and that a listener is traveling along the same line in the opposite direction with speed $v$. Then the frequency $F$ heard by the listener is given by

$$F = \left(\frac{c - v}{c + u}\right)f$$

where $c$ is the speed of sound in still air, about 1100 ft/sec. (This phenomenon is called the **Doppler effect.**) Suppose a railroad train is traveling at 100 ft/sec (approximately 68 mph) in still air and the frequency of a note emitted by the locomotive whistle is 500 Hz. What is the frequency of the note heard by a passenger in a train moving at 50 ft/sec in the opposite direction to the first train?

**86.** A function $f(x, y)$ is *homogeneous of degree n* if it satisfies the equation $f(tx, ty) = t^n f(x, y)$ for all $t$. Show that

$$f(x, y) = \frac{xy - y^2}{2x + y}$$

is homogeneous of degree 1.

*In Exercises 87–90, determine whether the statement is true or false. If it is true, explain why it is true. If it is false, give an example to show why it is false.*

**87.** $f$ is a function of $x$ and $y$ if and only if for any two points $P_1(x_1, y_1)$ and $P_2(x_2, y_2)$ in the domain of $f$, $f(x_1, y_1) = f(x_2, y_2)$ implies that $P_1(x_1, y_1) = P_2(x_2, y_2)$.

**88.** The equation $x^2 + y^2 + z^2 = 4$ defines at least two functions of $x$ and $y$.

**89.** The level curves of a function $f$ of two variables, $f(x, y) = k$, exist for all values of $k$.

**90.** The level surfaces of the function $f(x, y, z) = ax + by + cz + d$ consist of a family of parallel planes that are orthogonal to the vector $\mathbf{n} = a\mathbf{i} + b\mathbf{j} + c\mathbf{k}$.

## 12.2   Limits and Continuity

### ■ An Intuitive Definition of a Limit

Figure 1 shows the graph of a function $f$ of two variables. This figure suggests that $f(x, y)$ is close to the number $L$ if the point $(x, y)$ is close to the point $(a, b)$.

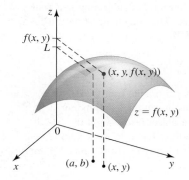

**FIGURE 1**
The functional value $f(x, y)$ is close to $L$ if $(x, y)$ is close to $(a, b)$.

> **DEFINITION   Limit of a Function of Two Variables at a Point**
>
> Let $f$ be a function that is defined for all points $(x, y)$ close to the point $(a, b)$ with the possible exception of $(a, b)$ itself. Then the **limit of $f(x, y)$ as $(x, y)$ approaches $(a, b)$** is $L$, written
>
> $$\lim_{(x, y) \to (a, b)} f(x, y) = L$$
>
> if $f(x, y)$ can be made as close to $L$ as we please by restricting $(x, y)$ to be sufficiently close to $(a, b)$.

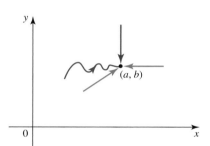

**FIGURE 2**
There are infinitely many paths the point $(x, y)$ could take in approaching the point $(a, b)$.

At first glance, there appears to be little difference between this definition and the definition of the limit of a function of one variable, with the exception that the points $(x, y)$ and $(a, b)$ lie in the plane. But there are subtle differences. In the case of a function of one variable, the point $x$ can approach the point $x = a$ from only two directions: from the left and from the right. As a consequence, the function $f$ has a limit $L$ as $x$ approaches $a$ if and only if $f(x)$ approaches $L$ from the left ($\lim_{x \to a^-} f(x) = L$) and from the right ($\lim_{x \to a^+} f(x) = L$), a fact that we observed in Section 1.1.

The situation is a little more complicated in the case of a function of two variables because there are infinitely many ways in which we can approach a point $(a, b)$ in the plane (Figure 2). Thus, if $f$ has a limit $L$ as $(x, y)$ approaches $(a, b)$, then $f(x, y)$ must approach $L$ along *every* possible path leading to $(a, b)$.

To see why this is true, suppose that

$$f(x, y) \to L_1 \quad \text{as} \quad (x, y) \to (a, b)$$

along a path $C_1$ and that

$$f(x, y) \to L_2 \quad \text{as} \quad (x, y) \to (a, b)$$

along another path $C_2$, where $L_1 \neq L_2$. Then no matter how close $(x, y)$ is to $(a, b)$, $f(x, y)$ will assume values that are close to $L_1$ and also values that are close to $L_2$ depending on whether $(x, y)$ is on $C_1$ or on $C_2$. Therefore, $f(x, y)$ cannot be made as close as we please to a unique number $L$ by restricting $(x, y)$ to be sufficiently close to $(a, b)$; that is, $\lim_{(x, y) \to (a, b)} f(x, y)$ cannot exist.

An immediate consequence of this observation is the following criterion for demonstrating that a limit does *not* exist.

> **Technique for Showing That $\lim_{(x, y) \to (a, b)} f(x, y)$ Does Not Exist**
>
> If $f(x, y)$ approaches two different numbers as $(x, y)$ approaches $(a, b)$ along two different paths, then $\lim_{(x, y) \to (a, b)} f(x, y) = L$ does not exist.

**EXAMPLE 1**   Show that $\displaystyle \lim_{(x, y) \to (0, 0)} \frac{x^2 - y^2}{x^2 + y^2}$ does not exist.

**Solution**   The function $f(x, y) = (x^2 - y^2)/(x^2 + y^2)$ is defined everywhere except at $(0, 0)$. Let's approach $(0, 0)$ along the $x$-axis (see Figure 3). On the path $C_1$, $y = 0$, so

$$\lim_{\substack{(x, y) \to (0, 0) \\ \text{along } C_1}} f(x, y) = \lim_{x \to 0} f(x, 0) = \lim_{x \to 0} \frac{x^2}{x^2} = \lim_{x \to 0} 1 = 1$$

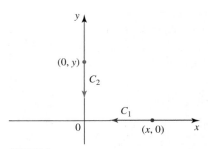

**FIGURE 3**

A point on $C_1$ has the form $(x, 0)$, and a point on $C_2$ has the form $(0, y)$.

Next, let's approach $(0, 0)$ along the $y$-axis. On the path $C_2$, $x = 0$ (Figure 3), so

$$\lim_{\substack{(x, y) \to (0, 0) \\ \text{along } C_2}} f(x, y) = \lim_{y \to 0} f(0, y) = \lim_{y \to 0} \frac{-y^2}{y^2} = \lim_{y \to 0}(-1) = -1$$

Since $f(x, y)$ approaches two different numbers as $(x, y)$ approaches $(0, 0)$ along two different paths, we conclude that the given limit does not exist. ■

**EXAMPLE 2** Show that $\displaystyle\lim_{(x, y) \to (0, 0)} \frac{xy}{x^2 + y^2}$ does not exist.

**Solution**  The function $f(x, y) = xy/(x^2 + y^2)$ is defined everywhere except at $(0, 0)$. Let's approach $(0, 0)$ along the $x$-axis (Figure 4). On the path $C_1$, $y = 0$, so

$$\lim_{\substack{(x, y) \to (0, 0) \\ \text{along } C_1}} f(x, y) = \lim_{x \to 0} f(x, 0) = \lim_{x \to 0} \frac{0}{x^2} = \lim_{x \to 0} 0 = 0$$

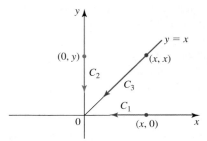

**FIGURE 4**

$f(x, y) \to 0$ as $(x, y) \to (0, 0)$ along $C_1$ and $C_2$, but $f(x, y) \to \frac{1}{2}$ as $(x, y) \to (0, 0)$ along $C_3$, so $\lim_{(x, y) \to (0, 0)} f(x, y)$ does not exist.

Similarly, you can show that $f(x, y)$ also approaches 0 as $(x, y)$ approaches $(0, 0)$ along the $y$-axis, path $C_2$ (Figure 4).

Now consider yet another approach to $(0, 0)$, this time along the line $y = x$ (Figure 4). On the path $C_3$, $y = x$, so

$$\lim_{\substack{(x, y) \to (0, 0) \\ \text{along } C_3}} f(x, y) = \lim_{x \to 0} f(x, x) = \lim_{x \to 0} \frac{x^2}{x^2 + x^2} = \lim_{x \to 0} \frac{1}{2} = \frac{1}{2}$$

Since $f(x, y)$ approaches two different numbers as $(x, y)$ approaches $(0, 0)$ along two different paths, we conclude that the given limit does not exist.

The graph of $f$ shown in Figure 5 confirms this result visually. Notice the ridge that occurs above the line $y = x$ because $f(x, y) = \frac{1}{2}$ for all points $(x, y)$ on that line except at the origin.

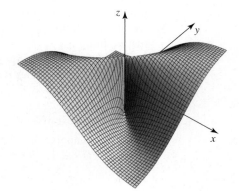

**FIGURE 5**

The graph of $f(x, y) = \dfrac{xy}{x^2 + y^2}$

■

Although the method of Examples 1 and 2 is effective in demonstrating when a limit does not exist, it cannot be used to prove the existence of the limit of a function at a point. Using this method, we would have to show that $f(x, y)$ approaches a unique number $L$ as $(x, y)$ approaches the point along *every* path, which is clearly an impossible task. Fortunately, the Limit Laws for a function of a single variable can be extended to functions of two or more variables. For example, the Sum Law, the Product Law, the Quotient Law, and so forth, all hold. So does the Squeeze Theorem.

**EXAMPLE 3** Evaluate

**a.** $\lim_{(x, y)\to(1, 2)}(x^3y^2 - x^2y + x^2 - 2x + 3y)$

**b.** $\lim_{(x, y)\to(2, 4)} \sqrt[3]{\dfrac{8xy}{2x + y}}$

**Solution**

**a.** We have

$$\lim_{(x, y)\to(1, 2)} (x^3y^2 - x^2y + x^2 - 2x + 3y) = (1)^3(2)^2 - (1)^2(2) + (1)^2 - 2(1) + 3(2)$$
$$= 7$$

**b.** We have

$$\lim_{(x, y)\to(2, 4)} \sqrt[3]{\frac{8xy}{2x + y}} = \sqrt[3]{\lim_{(x, y)\to(2, 4)} \frac{8xy}{2x + y}}$$
$$= \sqrt[3]{\frac{8(2)(4)}{2(2) + 4}} = \sqrt[3]{8} = 2 \qquad \blacksquare$$

The next example utilizes the Squeeze Theorem to show the existence of a limit.

**EXAMPLE 4** Find $\lim_{(x, y)\to(0, 0)} \dfrac{2x^2y}{x^2 + y^2}$ if it exists.

**Solution** Observe that the numerator of the rational function has degree 3, whereas the denominator has degree 2. This suggests that when $x$ and $y$ are both close to zero, the numerator is much smaller than the denominator, and we suspect that the limit might exist and that it is equal to zero.

To prove our assertion, we observe that $y^2 \geq 0$, so $x^2/(x^2 + y^2) \leq 1$. Therefore,

$$0 \leq \left|\frac{2x^2y}{x^2 + y^2}\right| = \frac{2x^2|y|}{x^2 + y^2} \leq 2|y|$$

Let $f(x, y) = 0$, $g(x, y) = \left|\dfrac{2x^2y}{x^2 + y^2}\right|$, and $h(x, y) = 2|y|$. Then

$$\lim_{(x, y)\to(0, 0)} f(x, y) = \lim_{(x, y)\to(0, 0)} 0 = 0 \quad \text{and} \quad \lim_{(x, y)\to(0, 0)} h(x, y) = \lim_{(x, y)\to(0, 0)} 2|y| = 0$$

By the Squeeze Theorem,

$$\lim_{(x, y)\to(0, 0)} g(x, y) = \lim_{(x, y)\to(0, 0)} \left|\frac{2x^2y}{x^2 + y^2}\right| = 0$$

and this, in turn, implies that

$$\lim_{(x, y)\to(0, 0)} \frac{2x^2y}{x^2 + y^2} = 0 \qquad \blacksquare$$

## ■ Continuity of a Function of Two Variables

The definition of continuity for a function of two variables is similar to that for a function of one variable.

> **DEFINITION    Continuity at a Point**
>
> Let $f$ be a function that is defined for all points $(x, y)$ close to the point $(a, b)$. Then $f$ is **continuous at the point** $(a, b)$ if
>
> $$\lim_{(x,\,y)\to(a,\,b)} f(x, y) = f(a, b)$$

Thus, $f$ is continuous at $(a, b)$ if $f(x, y)$ approaches $f(a, b)$ as $(x, y)$ approaches $(a, b)$ along any path. Loosely speaking, a function $f$ is continuous at a point $(a, b)$ if the graph of $f$ does not have a hole, gap, or jump at $(a, b)$. If $f$ is not continuous at $(a, b)$, then $f$ is said to be **discontinuous** there. For example, the functions $f$, $g$, and $h$ whose graphs are shown in Figure 6 are discontinuous at the indicated points.

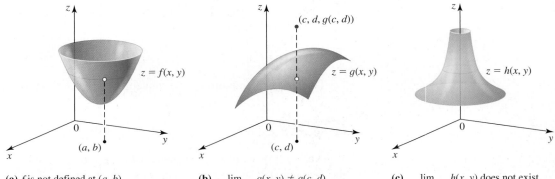

**FIGURE 6**    **(a)** $f$ is not defined at $(a, b)$.    **(b)** $\displaystyle\lim_{(x,\,y)\to(c,\,d)} g(x, y) \neq g(c, d)$    **(c)** $\displaystyle\lim_{(x,\,y)\to(0,\,0)} h(x, y)$ does not exist.

## ■ Continuity on a Set

Let's digress a little to introduce some terminology. We define the **$\delta$-neighborhood** about $(a, b)$ to be the set

$$N_\delta = \{(x, y)\mid \sqrt{(x - a)^2 + (y - b)^2} < \delta\}$$

Thus, $N_\delta$ is just the set of all points lying inside the circle of radius $\delta$ centered at $(a, b)$ (see Figure 7).

Let $R$ be a plane region. A point $(a, b)$ is said to be an **interior point** of $R$ if there exists a $\delta$-neighborhood about $(a, b)$ that lies entirely in $R$ (Figure 8). A point $(a, b)$ is called a **boundary point** of $R$ if every $\delta$-neighborhood of $R$ contains points in $R$ and also points not in $R$.

A region $R$ is said to be an **open region** if every point of $R$ is an interior point of $R$. A region is **closed** if it contains all of its boundary points. Finally, a region that contains some but not all of its boundary points is neither open nor closed. For example, the regions

$$A = \left\{(x, y)\,\Big|\, \frac{x^2}{9} + \frac{y^2}{4} < 1\right\}, \qquad B = \left\{(x, y)\,\Big|\, \frac{x^2}{9} + \frac{y^2}{4} \leq 1\right\}$$

and

$$C = \left\{(x, y)\,\Big|\, \frac{x^2}{9} + \frac{y^2}{4} \leq 1; y \geq 0\right\} \bigcup \left\{(x, y)\,\Big|\, \frac{x^2}{9} + \frac{y^2}{4} < 1; y < 0\right\}$$

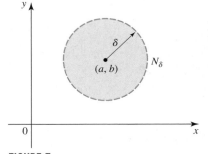

**FIGURE 7**
The $\delta$-neighborhood about $(a, b)$

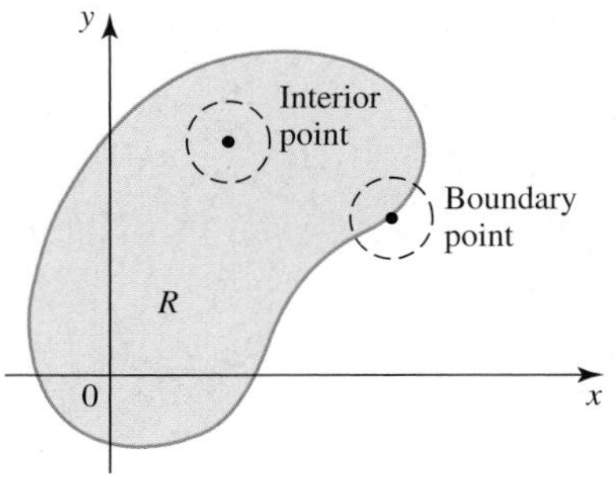

**FIGURE 8**
An interior point and a boundary point of $R$

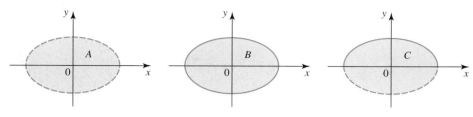

**FIGURE 9**
Every point in $A$ is an interior point; $B$ contains all of its boundary points;
$C$ contains some but not all of its boundary points.

shown in Figure 9a–c are open, closed, and neither open nor closed, respectively.

As we mentioned in Section 1.4, continuity is a "local" concept. The following definition explains what we mean by continuity on a region.

---

**DEFINITION   Continuity on a Region**

Let $R$ be a region in the plane. Then $f$ is **continuous on $R$** if $f$ is continuous at every point $(x, y)$ in $R$. If $(a, b)$ is a boundary point, the condition for continuity is modified to read

$$\lim_{(x, y) \to (a, b)} f(x, y) = f(a, b)$$

where $(x, y) \in R$, that is, $(x, y)$ is restricted to approach $(a, b)$ along paths lying inside $R$.

---

**EXAMPLE 5**  Show that the function $f$ defined by $f(x, y) = \sqrt{9 - x^2 - y^2}$ is continuous on the closed region $R = \{(x, y) \mid x^2 + y^2 \leq 9\}$, which is the set of all points lying on and inside the circle of radius 3 centered at $(0, 0)$ in the $xy$-plane.

**Solution**   Observe that the set $R$ is precisely the domain of $f$. Now, if $(a, b)$ is any interior point of $R$, then

$$\lim_{(x, y) \to (a, b)} f(x, y) = \lim_{(x, y) \to (a, b)} \sqrt{9 - x^2 - y^2}$$

$$= \sqrt{\lim_{(x, y) \to (a, b)} (9 - x^2 - y^2)}$$

$$= \sqrt{9 - a^2 - b^2}$$

$$= f(a, b)$$

This shows that $f$ is continuous at $(a, b)$.

Next, if $(c, d)$ is a boundary point of $R$ and $(x, y)$ is restricted to lie inside $R$, we obtain

$$\lim_{(x, y) \to (c, d)} f(x, y) = f(c, d)$$

as before, thus showing that $f$ is continuous at $(c, d)$ as well.

The graph of $f$ is the upper hemisphere of radius 3 centered at the origin together with the circle in the $xy$-plane having equation $x^2 + y^2 = 9$. (See Figure 10.)

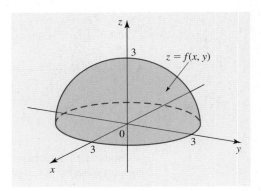

**FIGURE 10**
The graph of $f(x, y) = \sqrt{9 - x^2 - y^2}$
has no holes, gaps, or jumps.

The following theorem summarizes the properties of continuous functions of two variables. The proofs of these properties follow from the limit laws and will be omitted.

---

**THEOREM 1  Properties of Continuous Functions of Two Variables**

If $f$ and $g$ are continuous at $(a, b)$, then the following functions are also continuous at $(a, b)$.

**a.** $f \pm g$     **b.** $fg$     **c.** $cf$   $c$, a constant     **d.** $f/g$   $g(a, b) \neq 0$

---

A consequence of Theorem 1 is that *polynomial* and *rational* functions are continuous.

A **polynomial function** of two variables is a function whose rule can be expressed as a finite sum of terms of the form $cx^m y^n$, where $c$ is a constant and $m$ and $n$ are nonnegative integers. For example, the function $f$ defined by

$$f(x, y) = 2x^2 y^5 - 3xy^3 + 8xy^2 - 3y + 4$$

is a polynomial function in the two variables $x$ and $y$. A **rational function** is the quotient of two polynomial functions. For example, the function $g$ defined by

$$g(x, y) = \frac{x^3 + xy + y^2}{x^2 - y^2}$$

is a rational function.

---

**THEOREM 2  Continuity of Polynomial and Rational Functions**

A polynomial function is continuous everywhere (that is, in the whole plane). A rational function is continuous at all points in its domain (that is, at all points where its denominator is defined and not equal to zero).

---

**EXAMPLE 6**  Determine where the function is continuous:

**a.** $f(x, y) = \dfrac{xy(x^2 - y^2)}{x^2 + y^2}$     **b.** $g(x, y) = \dfrac{1}{y - x^2}$

**Solution**

**a.** The function $f$ is a rational function and is therefore continuous everywhere except at $(0, 0)$, where its denominator is equal to zero (Figure 11).

**b.** The function $g$ is a rational function and is continuous everywhere except along the curve $y = x^2$, where its denominator is equal to zero (Figure 12).

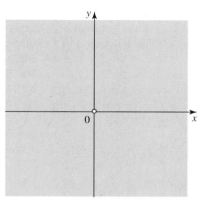

**FIGURE 11**
The graph of $f$ has a hole at the origin.

(**a**) The domain of $f$

(**b**) The graph of $z = \dfrac{xy(x^2 - y^2)}{x^2 + y^2}$

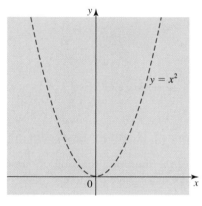

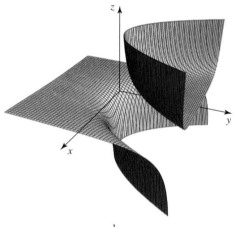

**FIGURE 12**
As $(x, y)$ approaches the curve $y = x^2$ from the region $y > x^2$, $z = f(x, y)$ approaches infinity; as $(x, y)$ approaches the curve $y = x^2$ from the region $y < x^2$, $z$ approaches minus infinity.

(**a**) The domain of $g$

(**b**) The graph of $z = \dfrac{1}{y - x^2}$

The next theorem tells us that the composite function of two continuous functions is also a continuous function.

---

**THEOREM 3  Continuity of a Composite Function**

If $f$ is continuous at $(a, b)$ and $g$ is continuous at $f(a, b)$, then the composite function $h = g \circ f$ defined by $h(x, y) = g(f(x, y))$ is continuous at $(a, b)$.

---

**EXAMPLE 7** Determine where the function is continuous:

**a.** $F(x, y) = \sin xy$    **b.** $G(x, y) = \dfrac{\frac{1}{2}\cos(2x^2 + y^2)}{1 + 2x^2 + y^2}$

**Solution**

**a.** We can view the function $F$ as the composition $g \circ f$ of the functions $f$ and $g$ defined by $f(x, y) = xy$ and $g(t) = \sin t$. Thus,

$$F(x, y) = g(f(x, y)) = \sin (f(x, y)) = \sin xy$$

Since $f$ is continuous on the whole plane and $g$ is continuous on $(-\infty, \infty)$, we conclude that $F$ is continuous everywhere. The graph of $F$ is shown in Figure 13a.

**b.** The function $G$ is the quotient of $p(x, y) = \frac{1}{2} \cos(2x^2 + y^2)$ and $q(x, y) = 1 + 2x^2 + y^2$. The function $p$ in turn involves the composition of $g(t) = \frac{1}{2} \cos t$ and $f(x, y) = 2x^2 + y^2$. Since both $f$ and $g$ are continuous everywhere, we see that $p$ is continuous everywhere. The function $q$ is continuous everywhere as well and is never zero. Therefore, by Theorem 3, $G$ is continuous everywhere. The graph of $G$ is shown in Figure 13b.

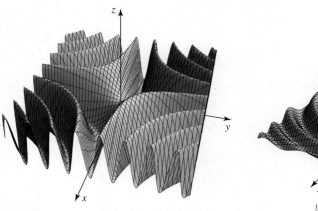

**FIGURE 13**    **(a)** $F(x, y) = \sin xy$ is continuous everywhere.    **(b)** $G(x, y) = \dfrac{\frac{1}{2}\cos(2x^2 + y^2)}{1 + 2x^2 + y^2}$ is continuous everywhere.

## ■ Functions of Three or More Variables

The notions of the limit of a function of three or more variables and that of the continuity of a function of three or more variables parallel those of a function of two variables. For example, if $f$ is a function of three variables, then we write

$$\lim_{(x, y, z) \to (a, b, c)} f(x, y, z) = L$$

to mean that there exists a number $L$ such that $f(x, y, z)$ can be made as close to $L$ as we please by restricting $(x, y, z)$ to be sufficiently close to $(a, b, c)$.

**EXAMPLE 8**   Evaluate   $\displaystyle\lim_{(x, y, z) \to (\frac{\pi}{2}, 0, 1)} \frac{e^{2y}(\sin x + \cos y)}{1 + y^2 + z^2}$.

**Solution**

$$\lim_{(x, y, z) \to (\frac{\pi}{2}, 0, 1)} \frac{e^{2y}(\sin x + \cos y)}{1 + y^2 + z^2} = \frac{e^0[\sin (\pi/2) + \cos 0]}{1 + 0 + 1} = \frac{2}{2} = 1$$

A function $f$ of three variables is continuous at $(a, b, c)$ if

$$\lim_{(x, y, z) \to (a, b, c)} f(x, y, z) = f(a, b, c)$$

**EXAMPLE 9** Determine where $f(x, y, z) = \dfrac{\ln z}{\sqrt{1 - x^2 - y^2 - z^2}}$ is continuous.

**Solution** We require that $z > 0$ and $1 - x^2 - y^2 - z^2 > 0$; that is, $z > 0$ and $x^2 + y^2 + z^2 < 1$. So $f$ is continuous on the set $\{(x, y, z) \mid x^2 + y^2 + z^2 < 1$ and $z > 0\}$, which is the set of points above the $xy$-plane and inside the upper hemisphere with center at the origin and radius 1. ∎

## The $\varepsilon$-$\delta$ Definition of a Limit (Optional)

The notion of the limit of a function of two variables given earlier can be made more precise as follows.

> **DEFINITION** Limit of $f(x, y)$
>
> Let $f$ be a function of two variables that is defined for all points $(x, y)$ on a disk with center at $(a, b)$ with the possible exception of $(a, b)$ itself. Then
> $$\lim_{(x, y) \to (a, b)} f(x, y) = L$$
> if for every $\varepsilon > 0$, there exists a $\delta > 0$ such that
> $$|f(x, y) - L| < \varepsilon \qquad \text{whenever} \qquad 0 < \sqrt{(x - a)^2 + (y - b)^2} < \delta$$

Geometrically speaking, $f$ has the limit $L$ at $(a, b)$ if given *any* $\varepsilon > 0$, we can find a circle of radius $\delta$ centered at $(a, b)$ such that $L - \varepsilon < f(x, y) < L + \varepsilon$ for all interior points $(x, y) \neq (a, b)$ of the circle (Figure 14).

**EXAMPLE 10** Prove that $\lim_{(x, y) \to (a, b)} x = a$.

**Solution** Let $\varepsilon > 0$ be given. We need to show that there exists a $\delta > 0$ such that
$$|f(x, y) - a| < \varepsilon$$
whenever $(x, y) \neq (a, b)$ is in the $\delta$-neighborhood about $(a, b)$. To find such a $\delta$, consider
$$|f(x, y) - a| = |x - a| = \sqrt{(x - a)^2} \le \sqrt{(x - a)^2 + (y - b)^2}$$
Thus, if we pick $\delta = \varepsilon$, we see that $\delta > 0$ and that $\sqrt{(x - a)^2 + (y - b)^2} < \delta$ implies that $|f(x, y) - a| < \varepsilon$ as was to be shown. Since $\varepsilon$ is arbitrary, the proof is complete. ∎

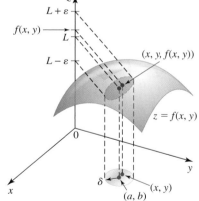

**FIGURE 14**
$f(x, y)$ lies in the interval
$(L - \varepsilon, L + \varepsilon)$ whenever
$(x, y) \neq (a, b)$ is in the
$\delta$-neighborhood of $(a, b)$.

**EXAMPLE 11** Prove that $\lim_{(x, y) \to (0, 0)} \dfrac{2x^2 y}{x^2 + y^2} = 0$. (See Example 4.)

**Solution** Let $\varepsilon > 0$ be given. Consider
$$|f(x, y) - 0| = \left| \frac{2x^2 y}{x^2 + y^2} \right| = 2|y| \left( \frac{x^2}{x^2 + y^2} \right) \qquad (x, y) \neq (0, 0)$$
$$\le 2|y| = 2\sqrt{y^2} \le 2\sqrt{x^2 + y^2}$$

If we pick $\delta = \varepsilon/2$, then $\delta > 0$, and $\sqrt{x^2 + y^2} < \delta$ implies that $|f(x, y) - 0| < \varepsilon$. Since $\varepsilon$ is arbitrary, the proof is complete. ∎

## 12.2 CONCEPT QUESTIONS

1. **a.** Explain what it means for a function $f$ of two variables to have a limit at $(a, b)$.
   **b.** Describe a technique that you could use to show that the limit of $f(x, y)$ as $(x, y)$ approaches $(a, b)$ does not exist.
2. Explain what it means for a function of two variables to be continuous (a) at a point $(a, b)$ and (b) on a region in the plane.
3. Determine whether each function $f$ is continuous or discontinuous. Explain your answer.
   **a.** $f(P, T)$ measures the volume of a balloon ascending into the sky as a function of the atmospheric pressure $P$ and the air temperature $T$.

**b.** $f(H, W)$ measures the surface area of a human body as a function of its height $H$ and weight $W$.
**c.** $f(d, t)$ measures the fare as a function of distance $d$ and time $t$ for taking a cab from O'Hare Airport to downtown Chicago.
**d.** $f(T, P)$ measures the volume of a certain mass of gas as a function of the temperature $T$ and the pressure $P$.

4. Suppose $f$ has the property that it is not defined at the point $(1, 2)$ but $\lim_{(x, y)\to(1, 2)} f(x, y) = 3$. Can you define $f(1, 2)$ so that $f$ is continuous at $(1, 2)$? If so, what should the value of $f(1, 2)$ be?

## 12.2 EXERCISES

*In Exercises 1–12, show that the limit does not exist.*

1. $\lim\limits_{(x, y)\to(0, 0)} \dfrac{x^2 - y^2}{2x^2 + y^2}$

2. $\lim\limits_{(x, y)\to(0, 0)} \dfrac{2x^2 - 3xy + 4y^2}{2x^2 + 3y^2}$

3. $\lim\limits_{(x, y)\to(0, 0)} \dfrac{3xy}{3x^2 + y^2}$

4. $\lim\limits_{(x, y)\to(0, 0)} \dfrac{xy^2}{x^2 + y^4}$

5. $\lim\limits_{(x, y)\to(0, 0)} \dfrac{2xy}{\sqrt{x^4 + y^4}}$

6. $\lim\limits_{(x, y)\to(0, 0)} \dfrac{\sin xy}{x^2 + y^2}$

7. $\lim\limits_{(x, y)\to(1, 0)} \dfrac{2xy - 2y}{x^2 + y^2 - 2x + 1}$

8. $\lim\limits_{(x, y)\to(0, 0)} \dfrac{xy^3 \cos x}{2x^2 + y^6}$

9. $\lim\limits_{(x, y, z)\to(0, 0, 0)} \dfrac{xy + yz + xz}{x^2 + y^2 + z^2}$

10. $\lim\limits_{(x, y, z)\to(0, 0, 0)} \dfrac{2xyz}{x^3 + y^3 + z^3}$

11. $\lim\limits_{(x, y, z)\to(0, 0, 0)} \dfrac{xz^2 + 2y^2}{x^2 + 2y^2 + z^4}$
    **Hint:** Approach $(0, 0, 0)$ along the curve with parametric equations $x = t^2$, $y = t^2$, $z = t$.

12. $\lim\limits_{(x, y, z)\to(0, 0, 0)} \dfrac{xy}{x^2 + y^2 + z^2}$

*In Exercises 13–26, find the given limit.*

13. $\lim\limits_{(x, y)\to(1, 2)} (x^2 + 2y^2)$

14. $\lim\limits_{(x, y)\to(1, -1)} (2x^2 + xy + 3y + 1)$

15. $\lim\limits_{(x, y)\to(1, 2)} \dfrac{2x^2 - 3y^3 + 4}{3 - xy}$

16. $\lim\limits_{(x, y)\to(-1, 3)} \dfrac{x + 2y^2}{(x - 1)(y + 1)}$

17. $\lim\limits_{(x, y)\to(1, -2)} \dfrac{3xy}{2x^2 - y^2}$

18. $\lim\limits_{(x, y)\to(1, \frac{1}{2})} x^2 \sin \pi(2x + y)$

19. $\lim\limits_{(x, y)\to(0^+, 0^+)} \dfrac{e^{\sqrt{x+y}}}{x + y - 1}$

20. $\lim\limits_{(x, y)\to(0, 1)} \dfrac{\sin^{-1}\left(\dfrac{x}{y}\right)}{1 + \dfrac{x}{y}}$

21. $\lim\limits_{(x, y)\to(1, 1)} \dfrac{\tan^{-1}\left(\dfrac{x}{y}\right)}{\cos^{-1}(x - 2y)}$

22. $\lim\limits_{(x, y)\to(0, 1)} e^{-x} \sin^{-1}(y - x)$

23. $\lim\limits_{(x, y)\to(2, 1)} \ln(x^2 - 3y)$

24. $\lim\limits_{(x, y)\to(3, 4)} e^{\sqrt{x^2 + y^2}}$

25. $\lim\limits_{(x, y, z)\to(1, 2, 3)} \dfrac{xy + yz + xz}{xyz - 3}$

26. $\lim\limits_{(x, y, z)\to(0, 3, 1)} [e^{\sin \pi x} + \ln(\cos \pi(y - z))]$

*In Exercises 27–30, use polar coordinates to find the limit. Hint: If $x = r \cos \theta$ and $y = r \sin \theta$, then $(x, y) \to (0, 0)$ if and only if $r \to 0^+$.*

27. $\lim\limits_{(x, y)\to(0, 0)} \dfrac{x^3 + y^3}{x^2 + y^2}$

28. $\lim\limits_{(x, y)\to(0, 0)} \dfrac{\sin(2x^2 + 2y^2)}{x^2 + y^2}$

29. $\lim\limits_{(x, y)\to(0, 0)} (x^2 + y^2) \ln(x^2 + y^2)$

30. $\lim\limits_{(x, y)\to(0, 0)} \dfrac{\tan(2x^2 + 2y^2)}{\tanh(3x^2 + 3y^2)}$

*In Exercises 31–40, determine where the function is continuous.*

31. $f(x, y) = \dfrac{2xy}{2x + 3y - 1}$

32. $f(x, y) = \dfrac{x^3 + xy + y^3}{x^2 + y^2}$

33. $g(x, y) = \sqrt{x + y} - \sqrt{x - y}$

34. $h(x, y) = \sin(2x + 3y)$    **35.** $F(x, y) = \sqrt{x}e^{x/y}$

**36.** $G(x, y) = \ln(2x - y)$

**37.** $f(x, y, z) = \dfrac{xyz}{x^2 + y^2 + z^2 - 4}$

**38.** $g(x, y, z) = \sqrt{x} + \cos\sqrt{y + z}$

**39.** $h(x, y, z) = x \ln(yz - 1)$

**40.** $F(x, y, z) = x \tan \dfrac{y}{z}$

**41.** Let

$$f(x, y) = \begin{cases} \dfrac{\sin xy}{xy} & \text{if } xy \neq 0 \\ 1 & \text{if } xy = 0 \end{cases}$$

    **a.** Determine all the points where $f$ is continuous.

**cas** **b.** Plot the graph of $f$. Does the graph give a visual confirmation of your conclusion in part (a)?

**42.** Let

$$f(x, y) = \begin{cases} \dfrac{x}{\sin x} + y & \text{if } x \neq 0 \\ 1 + y & \text{if } x = 0 \end{cases}$$

    **a.** Determine all the points where $f$ is continuous.

**cas** **b.** Plot the graph of $f$. Does the graph give a visual confirmation of your conclusion in part (a)?

*In Exercises 43–48, find $h(x, y) = g(f(x, y))$, and determine where $h$ is continuous.*

**43.** $f(x, y) = x^2 - xy + y^2$, $g(t) = t \cos t + \sin t$

**44.** $f(x, y) = x^3 + xy - xy^2 + y^3$, $g(t) = te^{-t}$

**45.** $f(x, y) = 2x - y$, $g(t) = \dfrac{t + 2}{t - 1}$

**46.** $f(x, y) = x - 2y + 3$, $g(t) = \sqrt{t} + \dfrac{1}{t}$

**47.** $f(x, y) = x \tan y$, $g(t) = \cos t$

**48.** $f(x, y) = y \ln x$, $g(t) = e^{t^2}$

**49.** Use the precise definition of a limit to prove that $\lim_{(x, y) \to (a, b)} c = c$ where $c$ is a constant.

**50.** Use the precise definition of a limit to prove that $\lim_{(x, y) \to (a, b)} y = b$.

**51.** Use the precise definition of a limit to prove that

$$\lim_{(x, y) \to (0, 0)} \frac{3xy^3}{x^2 + y^2} = 0$$

**52.** Use the precise definition of a limit to prove that if $\lim_{(x, y) \to (a, b)} f(x, y) = L$ and $c$ is a constant, then $\lim_{(x, y) \to (a, b)} cf(x, y) = cL$.

*In Exercises 53–58, determine whether the statement is true or false. If it is true, explain why it is true. If it is false, give an example to show why it is false.*

**53.** If $\lim_{(x, y) \to (a, b)} f(x, y) = L$, then $\lim_{(x, y) \to (a, b) \text{ along } C} f(x, y) = L$, where $C$ is any path leading to $(a, b)$.

**54.** If $\lim_{(x, y) \to (a, b)} f(x, y) = L$ and $f$ is defined at $(a, b)$, then $f(a, b) = L$.

**55.** If $f(x, y) = g(x)h(y)$, where $g$ and $h$ are continuous at $a$ and $b$, respectively, then $f$ is continuous at $(a, b)$.

**56.** If $f(1, 3) = 4$, then $\lim_{(x, y) \to (1, 3)} f(x, y) = 4$.

**57.** If $f$ is continuous at $(3, -1)$ and $f(3, -1) = 2$, then $\lim_{(x, y) \to (3, -1)} f(x, y) = 2$.

**58.** If $f$ is continuous at $(a, b)$ and $g$ is continuous at $f(a, b)$, then $\lim_{(x, y) \to (a, b)} g(f(x, y)) = g(f(a, b))$.

---

# 12.3 Partial Derivatives

## ▮ Partial Derivatives of Functions of Two Variables

For a function of one variable $x$, there is no ambiguity when we speak of the rate of change of $f(x)$ with respect to $x$. The situation becomes more complicated, however, when we study the rate of change of a function of two or more variables. For example, for the function of two variables defined by the equation $z = f(x, y)$, *both* the independent variables $x$ and $y$ may be allowed to vary in some arbitrary fashion, thus making it unclear what we mean by the phrase "the rate of change of $z$ with respect to $x$ and $y$."

One way of getting around this difficulty is to hold one variable constant and consider the rate of change of $f$ with respect to the other variable. This approach might be familiar to anyone who has used the expression "everything else being equal" while debating the merits of a complicated issue.

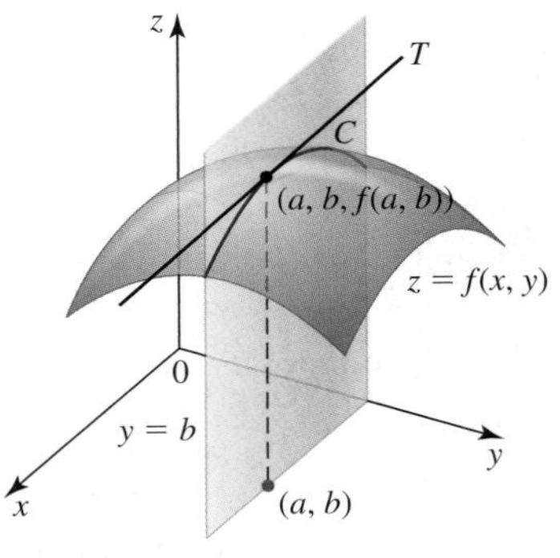

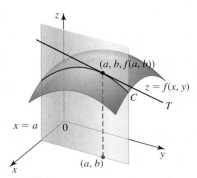

**FIGURE 1**

$$\lim_{h \to 0} \frac{f(a + h, b) - f(a, b)}{h}$$ measures

the slope of $T$ and the rate of change of $f(x, y)$ in the $x$-direction when $x = a$ and $y = b$.

**FIGURE 2**

$$\lim_{h \to 0} \frac{f(a, b + h) - f(a, b)}{h}$$ measures

the slope of $T$ and the rate of change of $f(x, y)$ in the $y$-direction when $x = a$ and $y = b$.

Specifically, suppose that $(a, b)$ is a point in the domain of $f$. Fix $y = b$. Then the function that is defined by $z = f(x, b)$ is a function of the single variable $x$. Its graph is the curve $C$ formed by the intersection of the vertical plane $y = b$ and the surface $z = f(x, y)$ (Figure 1).

Therefore, the quantity

$$\lim_{h \to 0} \frac{f(a + h, b) - f(a, b)}{h} \tag{1}$$

if it exists, measures both the slope of the tangent line $T$ to the curve $C$ at the point $(a, b, f(a, b))$ as well as the rate of change of $f(x, y)$ with respect to $x$ (in the $x$-direction) with $y$ held constant when $x = a$ and $y = b$.

Similarly, the quantity

$$\lim_{h \to 0} \frac{f(a, b + h) - f(a, b)}{h} \tag{2}$$

if it exists, measures the slope of the tangent line $T$ to the curve $C$ (formed by the intersection of the vertical plane $x = a$ and the surface $z = f(x, y)$ at $(a, b, f(a, b))$, and the rate of change of $f(x, y)$ with respect to $y$ (in the $y$-direction) with $x$ held constant when $x = a$ and $y = b$ (Figure 2).

In expressions (1) and (2) the point $(a, b)$ is fixed but otherwise arbitrary. Therefore, we may replace $(a, b)$ by $(x, y)$, leading to the following definitions.

**DEFINITIONS** **Partial Derivatives of a Function of Two Variables**

Let $z = f(x, y)$. Then the **partial derivative of $f$ with respect to $x$** is

$$\frac{\partial f}{\partial x} = \lim_{h \to 0} \frac{f(x + h, y) - f(x, y)}{h}$$

and the **partial derivative of $f$ with respect to $y$** is

$$\frac{\partial f}{\partial y} = \lim_{h \to 0} \frac{f(x, y + h) - f(x, y)}{h}$$

provided that each limit exists.

## Computing Partial Derivatives

The partial derivatives of $f$ can be calculated by using the following rules.

**Computing Partial Derivatives**

To compute $\partial f / \partial x$, treat $y$ as a constant and differentiate in the usual manner with respect to $x$ (an operation that we denote by $\partial / \partial x$).

To compute $\partial f / \partial y$, treat $x$ as a constant and differentiate in the usual manner with respect to $y$ (an operation that we denote by $\partial / \partial y$).

**EXAMPLE 1**   Find $\dfrac{\partial f}{\partial x}$ and $\dfrac{\partial f}{\partial y}$ if $f(x, y) = 2x^2y^3 - 3xy^2 + 2x^2 + 3y^2 + 1$.

**Solution**   To compute $\partial f/\partial x$, we think of the variable $y$ as a constant and differentiate with respect to $x$. Let's write

$$f(x, y) = 2x^2y^3 - 3xy^2 + 2x^2 + 3y^2 + 1$$

where the variable $y$ to be treated as a constant is shown in color. Then

$$\frac{\partial f}{\partial x} = 4xy^3 - 3y^2 + 4x$$

To compute $\partial f/\partial y$, we think of the variable $x$ as a constant and differentiate with respect to $y$. In this case,

$$f(x, y) = 2x^2y^3 - 3xy^2 + 2x^2 + 3y^2 + 1$$

and

$$\frac{\partial f}{\partial y} = 6x^2y^2 - 6xy + 6y$$ ∎

Before looking at more examples, let's introduce some alternative notations for the partial derivatives of a function. If $z = f(x, y)$, then

$$\frac{\partial}{\partial x} f(x, y) = \frac{\partial f}{\partial x} = f_x = z_x \quad \text{and} \quad \frac{\partial}{\partial y} f(x, y) = \frac{\partial f}{\partial y} = f_y = z_y$$

**EXAMPLE 2**   Find $f_x$ and $f_y$ if $f(x, y) = x \cos xy^2$.

**Solution**   To compute $f_x$, we think of the variable $y$ as a constant and differentiate with respect to $x$. Thus,

$$f(x, y) = x \cos xy^2$$

and

$$f_x = \frac{\partial}{\partial x} (x \cos xy^2) = x \frac{\partial}{\partial x} (\cos xy^2) + (\cos xy^2) \frac{\partial}{\partial x} (x) \qquad \text{Use the Product Rule.}$$

$$= x(-\sin xy^2) \frac{\partial}{\partial x} (xy^2) + \cos xy^2 \qquad \text{Use the Chain Rule on the first term.}$$

$$= -xy^2 \sin xy^2 + \cos xy^2$$

Next, to compute $f_y$, we treat $x$ as a constant and differentiate with respect to $y$. Thus,

$$f(x, y) = x \cos xy^2$$

and

$$f_y = \frac{\partial}{\partial y} (x \cos xy^2) = x \frac{\partial}{\partial y} (\cos xy^2) + (\cos xy^2) \frac{\partial}{\partial y} (x)$$

$$= x(-\sin xy^2) \frac{\partial}{\partial y} (xy^2) + 0 = -2x^2y \sin xy^2$$ ∎

**EXAMPLE 3** Let $f(x, y) = 4 - 2x^2 - y^2$. Find the slope of the tangent line at the point $(1, 1, 1)$ on the curve formed by the intersection of the surface $z = f(x, y)$ and

**a.** the plane $y = 1$          **b.** the plane $x = 1$

**Solution**

**a.** The slope of the tangent line at any point on the curve formed by the intersection of the plane $y = 1$ and the surface $z = 4 - 2x^2 - y^2$ is given by

$$\frac{\partial f}{\partial x} = \frac{\partial}{\partial x}(4 - 2x^2 - y^2) = -4x$$

In particular, the slope of the required tangent line is

$$\left.\frac{\partial f}{\partial x}\right|_{(1,1)} = -4(1) = -4$$

**b.** The slope of the tangent line at any point on the curve formed by the intersection of the plane $x = 1$ and the surface $z = 4 - 2x^2 - y^2$ is given by

$$\frac{\partial f}{\partial y} = \frac{\partial}{\partial y}(4 - 2x^2 - y^2) = -2y$$

In particular, the slope of the required tangent line is

$$\left.\frac{\partial f}{\partial y}\right|_{(1,1)} = -2(1) = -2$$

(See Figure 3.)

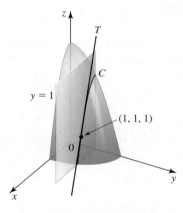

**FIGURE 3**          **(a)** The slope of the tangent line is −4.          **(b)** The slope of the tangent line is −2.

**EXAMPLE 4** **Electrostatic Potential** Figure 4 shows a crescent-shaped region $R$ that lies inside the disk $D_1 = \{(x, y) \mid (x - 2)^2 + y^2 \leq 4\}$ and outside the disk $D_2 = \{(x, y) \mid (x - 1)^2 + y^2 \leq 1\}$. Suppose that the electrostatic potential along the inner circle is kept at 50 volts and the electrostatic potential along the outer circle is kept at 100 volts. Then the electrostatic potential at any point $(x, y)$ in $R$ is given by

$$U(x, y) = 150 - \frac{200x}{x^2 + y^2}$$

volts.

**a.** Compute $U_x(x, y)$ and $U_y(x, y)$.

**b.** Compute $U_x(3, 1)$ and $U_y(3, 1)$ and interpret your results.

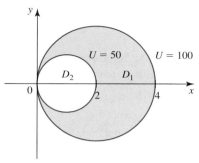

**FIGURE 4**

The electrostatic potential inside the crescent-shaped region is $U(x, y)$.

**Solution**

**a.** $U_x(x, y) = \dfrac{\partial}{\partial x}\left[150 - \dfrac{200x}{x^2 + y^2}\right] = -\dfrac{\partial}{\partial x}\left(\dfrac{200x}{x^2 + y^2}\right)$

$$= -\frac{(x^2 + y^2)\dfrac{\partial}{\partial x}(200x) - 200x\dfrac{\partial}{\partial x}(x^2 + y^2)}{(x^2 + y^2)^2}$$

$$= -\frac{200(x^2 + y^2) - 200x(2x)}{(x^2 + y^2)^2} = \frac{200(x^2 - y^2)}{(x^2 + y^2)^2}$$

$$U_y(x, y) = \frac{\partial}{\partial y}\left[150 - \frac{200x}{x^2 + y^2}\right] = -\frac{\partial}{\partial y}\left(\frac{200x}{x^2 + y^2}\right)$$

$$= -200x\frac{\partial}{\partial y}(x^2 + y^2)^{-1}$$

$$= -200x(-1)(x^2 + y^2)^{-2}\frac{\partial}{\partial y}(x^2 + y^2)$$

$$= 200x(x^2 + y^2)^{-2}(2y) = \frac{400xy}{(x^2 + y^2)^2}$$

**b.** $U_x(3, 1) = \dfrac{200(9 - 1)}{(9 + 1)^2} = 16$  and  $U_y(3, 1) = \dfrac{400(3)(1)}{(9 + 1)^2} = 12$

This tells us that the rate of change of the electrostatic potential at the point $(3, 1)$ in the $x$-direction is 16 volts per unit change in $x$ with $y$ held fixed at 1, and the rate of change of the electrostatic potential at the point $(3, 1)$ in the $y$-direction is 12 volts per unit change in $y$ with $x$ held fixed at 3.  ■

**EXAMPLE 5**  **A Production Function**  The production function of a certain country is given by

$$f(x, y) = 20x^{2/3}y^{1/3}$$

billion dollars, when $x$ billion dollars of labor and $y$ billion dollars of capital are spent.

**a.** Compute $f_x(x, y)$ and $f_y(x, y)$.

**b.** Compute $f_x(125, 27)$ and $f_y(125, 27)$, and interpret your results.

**c.** Should the government encourage capital investment rather than investment in labor to increase the country's productivity?

**Solution**

**a.** $f_x(x, y) = \dfrac{\partial}{\partial x}(20x^{2/3}y^{1/3}) = (20)\left(\dfrac{2}{3}x^{-1/3}\right)(y^{1/3}) = \dfrac{40}{3}\left(\dfrac{y}{x}\right)^{1/3}$

$f_y(x, y) = \dfrac{\partial}{\partial y}(20x^{2/3}y^{1/3}) = (20x^{2/3})\left(\dfrac{1}{3}y^{-2/3}\right) = \dfrac{20}{3}\left(\dfrac{x}{y}\right)^{2/3}$

**b.** $f_x(125, 27) = \dfrac{40}{3}\left(\dfrac{27}{125}\right)^{1/3} = \dfrac{40}{3}\left(\dfrac{3}{5}\right) = 8$

This says that the production is increasing at the rate of $8 billion per billion dollar increase in labor expenditure when the labor expenditure stands at $125 billion (capital expenditure held constant at $27 billion).

Next,

$$f_y(125, 27) = \dfrac{20}{3}\left(\dfrac{125}{27}\right)^{2/3} = \dfrac{20}{3}\left(\dfrac{25}{9}\right) = 18\dfrac{14}{27}$$

This tells us that production is increasing at the rate of approximately $18.5 billion per billion dollar increase in capital outlay when the capital expenditure stands at $27 billion (with labor expenditure held constant at $125 billion).

**c.** Yes. Since a unit increase in capital expenditure results in a greater increase in production than a unit increase in labor, the government should encourage spending on capital rather than on labor.    ■

Sometimes we have available only the contour map of a function $f$. In such instances we can use the contour map to help us estimate the partial derivatives of $f$ at a specified point, as the following example shows.

**EXAMPLE 6**    Figure 5 shows the contour map of a function $f$. Use it to estimate $f_x(3, 1)$ and $f_y(3, 1)$.

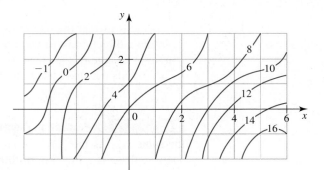

**FIGURE 5**
A contour map of $f$

**Solution**    To estimate $f_x(3, 1)$, we start at the point $(3, 1)$, where the value of $f$ at $(3, 1)$ can be read off from the contour map: $f(3, 1) = 8$. Then we proceed along the positive $x$-axis until we arrive at the point on the next level curve whose location is approximately $(3.8, 1)$. Using the definition of the partial derivative, we find

$$f_x(3, 1) \approx \dfrac{f(3.8, 1) - f(3, 1)}{3.8 - 3} = \dfrac{10 - 8}{0.8} = 2.5$$

Similarly, starting at the point $(3, 1)$ and moving along the positive $y$-axis, we find

$$f_y(3, 1) \approx \dfrac{f(3, 3) - f(3, 1)}{3 - 1} = \dfrac{6 - 8}{2} = -1$$    ■

## ■ Implicit Differentiation

**EXAMPLE 7** Suppose $z$ is a differentiable function of $x$ and $y$ that is defined implicitly by $x^2 + y^3 - z + 2yz^2 = 5$. Find $\partial z/\partial x$ and $\partial z/\partial y$.

**Solution** Differentiating the given equation implicitly with respect to $x$, we find

$$\frac{\partial}{\partial x}(x^2 + y^3 - z + 2yz^2) = \frac{\partial}{\partial x}(5)$$

$$2x - \frac{\partial z}{\partial x} + 2y\left(2z\frac{\partial z}{\partial x}\right) = 0 \qquad \text{Remember that } y \text{ is treated as a constant.}$$

$$\frac{\partial z}{\partial x}(4yz - 1) + 2x = 0$$

and

$$\frac{\partial z}{\partial x} = \frac{2x}{1 - 4yz}$$

Next, differentiating the given equation with respect to $y$, we obtain

$$\frac{\partial}{\partial y}(x^2 + y^3 - z + 2yz^2) = \frac{\partial}{\partial y}(5)$$

$$3y^2 - \frac{\partial z}{\partial y} + 2y\left(2z\frac{\partial z}{\partial y}\right) + 2z^2 = 0$$

$$3y^2 - \frac{\partial z}{\partial y}(1 - 4yz) + 2z^2 = 0$$

and

$$\frac{\partial z}{\partial y} = \frac{3y^2 + 2z^2}{1 - 4yz} \qquad ■$$

## ■ Partial Derivatives of Functions of More Than Two Variables

The partial derivatives of a function of more than two variables are defined in much the same way as the partial derivatives of a function of two variables. For example, suppose that $f$ is a function of three variables defined by $w = f(x, y, z)$. Then the partial derivative of $f$ with respect to $x$ is defined as

$$\frac{\partial w}{\partial x} = \frac{\partial f}{\partial x} = \lim_{h \to 0} \frac{f(x + h, y, z) - f(x, y, z)}{h}$$

where $y$ and $z$ are held fixed, provided that the limit exists. The other two partial derivatives, $\partial f/\partial y$ and $\partial f/\partial z$, are defined in a similar manner.

> **Finding the Partial Derivative of a Function of More Than Two Variables**
>
> To find the partial derivative of a function of more than two variables with respect to a certain variable, say $x$, we treat all the other variables as if they are constants and differentiate with respect to $x$ in the usual manner.

**EXAMPLE 8** Find

**a.** $f_x$ if $f(x, y, z) = x^2 y + y^2 z + xz$         **b.** $h_w$ if $h(x, y, z, w) = \dfrac{xw^2}{y + \sin zw}$

**Solution**

**a.** To find $f_x$, we treat $y$ and $z$ as constants and differentiate $f$ with respect to $x$ to obtain

$$f_x = \frac{\partial}{\partial x}(x^2 y + y^2 z + xz) = 2xy + z$$

**b.** To find $h_w$, we treat $x$, $y$, and $z$ as constants and differentiate $h$ with respect to $w$, obtaining

$$h_w = \frac{\partial}{\partial w}\left(\frac{xw^2}{y + \sin zw}\right)$$

$$= \frac{(y + \sin zw)\dfrac{\partial}{\partial w}(xw^2) - xw^2 \dfrac{\partial}{\partial w}(y + \sin zw)}{(y + \sin zw)^2} \qquad \text{Use the Quotient Rule.}$$

$$= \frac{(y + \sin zw)(2xw) - xw^2\left[0 + \cos zw \cdot \dfrac{\partial}{\partial w}(zw)\right]}{(y + \sin zw)^2} \qquad \text{Use the Chain Rule.}$$

$$= \frac{2xw(y + \sin zw) - xw^2 z \cos zw}{(y + \sin zw)^2} = \frac{xw(2y + 2\sin zw - wz \cos zw)}{(y + \sin zw)^2} \qquad \blacksquare$$

## ■ Higher-Order Derivatives

Consider the function $z = f(x, y)$ of two variables. Each of the partial derivatives $\partial f/\partial x$ and $\partial f/\partial y$ are functions of $x$ and $y$. Therefore, we may take the partial derivatives of these functions to obtain the four **second-order partial derivatives**

$$\frac{\partial^2 f}{\partial x^2} = \frac{\partial}{\partial x}\left(\frac{\partial f}{\partial x}\right), \quad \frac{\partial^2 f}{\partial y\, \partial x} = \frac{\partial}{\partial y}\left(\frac{\partial f}{\partial x}\right), \quad \frac{\partial^2 f}{\partial x\, \partial y} = \frac{\partial}{\partial x}\left(\frac{\partial f}{\partial y}\right), \quad \text{and} \quad \frac{\partial^2 f}{\partial y^2} = \frac{\partial}{\partial y}\left(\frac{\partial f}{\partial y}\right)$$

(See Figure 6.)

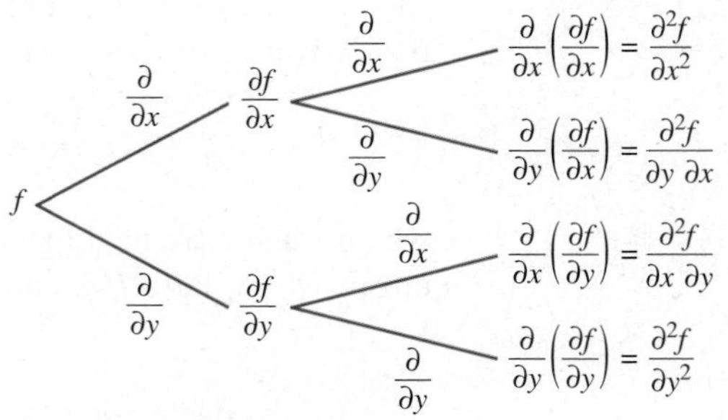

**FIGURE 6**
The differential operators are shown on the limbs of the tree diagram.

Before we turn to an example, let's introduce some additional notation for the second-order partial derivatives of $f$:

$$\frac{\partial^2 f}{\partial x^2} = f_{xx} \qquad \frac{\partial^2 f}{\partial y\, \partial x} = f_{xy} \qquad \frac{\partial^2 f}{\partial x\, \partial y} = f_{yx} \qquad \frac{\partial^2 f}{\partial y^2} = f_{yy}$$

Note the order in which the derivatives are taken: Using the notation $\partial^2 f/(\partial y\,\partial x)$, we differentiate first with respect to $x$—the independent variable that appears first when read from *right to left*. In the notation $f_{xy}$ we also differentiate first with respect to $x$—the independent variable that appears first when read from *left to right*. The derivatives $f_{xy}$ and $f_{yx}$ are called **mixed partial derivatives.**

**Note**   If $f$ is defined by the equation $z = f(x, y)$, then the four partial derivatives of $f$ are also written

$$z_{xx} \quad z_{xy} \quad z_{yx} \quad \text{and} \quad z_{yy}$$

**EXAMPLE 9**   Find the second-order partial derivatives of $f(x, y) = 2xy^2 - 3x^2 + xy^3$.

**Solution**   We first compute the first-order partial derivatives

$$f_x = \frac{\partial}{\partial x}(2xy^2 - 3x^2 + xy^3) = 2y^2 - 6x + y^3$$

and

$$f_y = \frac{\partial}{\partial y}(2xy^2 - 3x^2 + xy^3) = 4xy + 3xy^2$$

Then differentiating each of these functions, we obtain

$$f_{xx} = \frac{\partial}{\partial x} f_x = \frac{\partial}{\partial x}(2y^2 - 6x + y^3) = -6$$

$$f_{xy} = \frac{\partial}{\partial y} f_x = \frac{\partial}{\partial y}(2y^2 - 6x + y^3) = 4y + 3y^2$$

$$f_{yx} = \frac{\partial}{\partial x} f_y = \frac{\partial}{\partial x}(4xy + 3xy^2) = 4y + 3y^2$$

$$f_{yy} = \frac{\partial}{\partial y} f_y = \frac{\partial}{\partial y}(4xy + 3xy^2) = 4x + 6xy$$

Notice that the mixed derivatives $f_{xy}$ and $f_{yx}$ in Example 9 are equal. The following theorem, which we state without proof, gives the conditions under which this is true.

---

**THEOREM 1    Clairaut's Theorem**

If $f(x, y)$ and its partial derivatives $f_x, f_y, f_{xy}$, and $f_{yx}$ are continuous on an open region $R$, then

$$f_{xy}(x, y) = f_{yx}(x, y)$$

for all $(x, y)$ in $R$.

---

A function $u$ of two variables $x$ and $y$ is called a **harmonic function** if $u_{xx} + u_{yy} = 0$ for all $(x, y)$ in the domain of $u$. Harmonic functions are used in the study of heat conduction, fluid flow, and potential theory. The *partial differential equation* $u_{xx} + u_{yy} = 0$ is called **Laplace's equation,** named for Pierre Laplace (1749–1827).

---

**Historical Biography**

SPL/Photo Researchers, Inc.

**ALEXIS CLAUDE CLAIRAUT**
(1713-1765)

Alexis Claude Clairaut was one of twenty children born to his mother but the only to survive to adulthood. His father, a mathematics teacher in Paris, educated his son at home with very high standards: Alexis was taught to read using Euclid's *Elements*. As a result of both nature and nurture, Clairaut turned out to be a very precocious mathematician. He studied calculus by the age of 10 and wrote an original mathematical paper at 13. At 18 he published his first paper; he also became the youngest member ever admitted to the prestigious Academie des Sciences. Clairaut excelled in many areas of mathematics, including geometry, calculus, and celestial mechanics. He was the first to prove the prediction by Isaac Newton and the astronomer Christiaan Huygens that the earth is an oblate ellipsoid. Clairaut also accurately predicted the return of Halley's comet in 1759, a prediction that made him famous. Clairaut developed the notation for partial derivatives that we still use today, and he was the first to prove that mixed second-order partial derivatives of a function at a point are equal if the derivatives are continuous at that point.

**EXAMPLE 10** Show that the function $u(x, y) = e^x \cos y$ is harmonic in the $xy$-plane.

**Solution** We find

$$u_x = \frac{\partial}{\partial x}(e^x \cos y) = e^x \cos y, \qquad u_y = \frac{\partial}{\partial y}(e^x \cos y) = -e^x \sin y$$

$$u_{xx} = \frac{\partial}{\partial x}(e^x \cos y) = e^x \cos y, \qquad u_{yy} = \frac{\partial}{\partial y}(-e^x \sin y) = -e^x \cos y$$

Therefore,

$$u_{xx} + u_{yy} = e^x \cos y - e^x \cos y = 0$$

This holds for all $(x, y)$ in the plane, so $u$ is harmonic there. ■

Partial derivatives of order three and higher are defined in a similar manner. For example,

$$f_{xxx} = \frac{\partial}{\partial x}f_{xx} \qquad f_{xxy} = \frac{\partial}{\partial y}f_{xx} \qquad \text{and} \qquad f_{xyx} = \frac{\partial}{\partial x}f_{xy}$$

Also, Theorem 1 is valid for mixed derivatives of higher order. For example, if the third partial derivatives of $f$ are continuous, then the order in which the differentiation is taken does not matter.

**EXAMPLE 11** Let $f(x, y, z) = xe^{yz}$. Compute $f_{xzy}$ and $f_{yxz}$.

**Solution** We have

$$f_x = \frac{\partial}{\partial x}(xe^{yz}) = e^{yz}$$

$$f_{xz} = \frac{\partial}{\partial z}f_x = \frac{\partial}{\partial z}(e^{yz}) = ye^{yz}$$

so

$$f_{xzy} = \frac{\partial}{\partial y}f_{xz} = \frac{\partial}{\partial y}(ye^{yz}) = e^{yz} + yze^{yz} = (1 + yz)e^{yz}$$

Next, we have

$$f_y = \frac{\partial}{\partial y}(xe^{yz}) = xze^{yz}$$

$$f_{yx} = \frac{\partial}{\partial x}f_y = \frac{\partial}{\partial x}(xze^{yz}) = ze^{yz}$$

so

$$f_{yxz} = \frac{\partial}{\partial z}f_{yx} = \frac{\partial}{\partial z}(ze^{yz}) = e^{yz} + yze^{yz} = (1 + yz)e^{yz}$$

Observe that both $f_{xzy}$ and $f_{yxz}$ are continuous everywhere and are equal. ■

## 12.3 CONCEPT QUESTIONS

1. **a.** Define the partial derivatives of a function of two variables, $x$ and $y$, with respect to $x$ and with respect to $y$.
   **b.** Give a geometric and a physical interpretation of $f_x(x, y)$.
2. Let $f$ be a function of $x$ and $y$. Describe a procedure for finding $f_x$ and $f_y$.

3. Suppose $F(x, y, z) = 0$ defines $x$ implicitly as a function of $y$ and $z$; that is, $x = f(y, z)$. Describe a procedure for finding $\partial x / \partial z$. Illustrate with an example of your choice.
4. If $f$ is a function of $x$ and $y$, give a condition that will guarantee that $f_{xy}(x, y) = f_{yx}(x, y)$ for all $(x, y)$ in some open region.

## 12.3 EXERCISES

1. Let $f(x, y) = x^2 + 2y^2$.
   **a.** Find $f_x(2, 1)$ and $f_y(2, 1)$.
   **b.** Interpret the numbers in part (a) as slopes.
   **c.** Interpret the numbers in part (a) as rates of change.

2. Let $f(x, y) = 9 - x^2 + xy - 2y^2$.
   **a.** Find $f_x(1, 2)$ and $f_y(1, 2)$.
   **b.** Interpret the numbers in part (a) as slopes.
   **c.** Interpret the numbers in part (a) as rates of change.

3. Determine the sign of $\partial f / \partial x$ and $\partial f / \partial y$ at the points $P$, $Q$, and $R$ on the graph of the function $f$ shown in the figure.

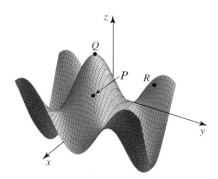

**(b)**

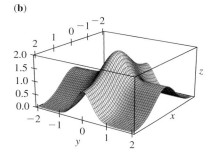

**(c)**

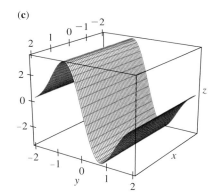

4. The graphs of a function $f$ and its partial derivatives $f_x$ and $f_y$ are labeled (a), (b), and (c). Identify the graphs of $f$, $f_x$, and $f_y$, and give a reason for your answer.

**(a)**

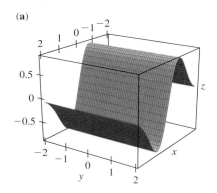

5. The figure below shows the contour map of the function $T$ (measured in degrees Fahrenheit) giving the temperature at each point $(x, y)$ on an 8 in. $\times$ 5 in. rectangular metal plate. Use it to estimate the rate of change of the temperature at the point $(3, 2)$ in the positive $x$-direction and in the positive $y$-direction.

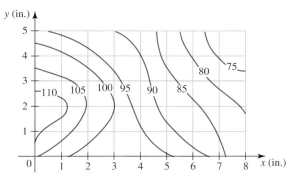

**V** Videos for selected exercises are available online at **www.academic.cengage.com/login**.

*In Exercises 6–29, find the first partial derivatives of the function.*

**6.** $f(x, y) = 3x - 4y + 2$     **7.** $f(x, y) = 2x^2 - 3xy + y^2$

**8.** $z = 2x^3 - 3x^2y^3 + xy^2 - 2x$

**9.** $z = x\sqrt{y}$     **10.** $f(x, y) = (2x^2 - y^3)^4$

**11.** $g(r, s) = \sqrt{r + s^2}$     **12.** $h(u, v) = \ln(u^2 + v^2)$

**13.** $f(x, y) = xe^{y/x}$     **14.** $f(x, y) = e^x \cos y + e^y \sin x$

**15.** $z = \tan^{-1}(x^2 + y^2)$     **16.** $f(x, y) = \sqrt{3 - 2x^2 - y^2}$

**17.** $g(u, v) = \dfrac{uv}{u^2 + v^3}$

**18.** $f(x, y) = \sinh xy$     **19.** $g(x, y) = x^2 \cosh \dfrac{x}{y}$

**20.** $z = \ln(e^x + y^2)$     **21.** $f(x, y) = y^x$

**22.** $f(x, y) = \displaystyle\int_x^y \cos t \, dt$     **23.** $f(x, y) = \displaystyle\int_x^y te^{-t} \, dt$

**24.** $f(x, y, z) = 2x^3 + 3xy + 2yz - z^2$

**25.** $g(x, y, z) = \sqrt{xyz}$     **26.** $f(u, v, w) = ue^v - ve^u + we^u$

**27.** $u = xe^{y/z} - z^2$     **28.** $u = x \sin \dfrac{y}{x + z}$

**29.** $f(r, s, t) = rs \ln st$

*In Exercises 30–33, use implicit differentiation to find $\partial z/\partial x$ and $\partial z/\partial y$.*

**30.** $x^2y + xz + yz^2 = 8$     **31.** $xe^y + ye^{-x} + e^z = 10$

**32.** $2 \cos(x + 2y) + \sin yz - 1 = 0$

**33.** $\ln(x^2 + z^2) + yz^3 + 2x^2 = 10$

*In Exercises 34–39, find the second partial derivatives of the function.*

**34.** $f(x, y) = x^4 - 2x^2y^3 + y^4 - 3x$

**35.** $g(x, y) = x^3y^2 + xy^3 - 2x + 3y + 1$

**36.** $z = xe^{2y} + ye^{2x}$

**37.** $w = \cos(2u - v) + \sin(2u + v)$

**38.** $z = \sqrt{x^2 + y^2}$     **39.** $h(x, y) = \tan^{-1} \dfrac{y}{x}$

*In Exercises 40–45, find the indicated partial derivative.*

**40.** $f(x, y) = x^3 + y^3 - 3x^2y^2 + 2x + 3y + 4;$   $f_{xxx}$

**41.** $f(x, y) = x^4 - 2x^2y^2 + xy^3 + 2y^4;$   $f_{xyx}$

**42.** $f(x, y, z) = \ln(x^2 + y^2 + z^2);$   $f_{yxz}$

**43.** $z = x \cos y + y \sin x;$   $\dfrac{\partial^3 z}{\partial x \, \partial y \, \partial x}$

**44.** $p = e^{uvw};$   $\dfrac{\partial^3 p}{\partial u \, \partial w \, \partial v}$

**45.** $h(x, y, z) = e^x \cos(y + 2z);$   $h_{zzy}$

*In Exercises 46–49, show that the mixed partial derivatives $f_{xy}$ and $f_{yx}$ are equal.*

**46.** $f(x, y) = x^2 + 2x^2y + y^3$     **47.** $f(x, y) = x \sin^2 y + y^2 \cos x$

**48.** $f(x, y) = e^{-2x} \cos 3y$     **49.** $f(x, y) = \tan^{-1}(x^2 + y^3)$

*In Exercises 50–53, show that the mixed partial derivatives $f_{xyz}$, $f_{yxz}$, and $f_{zyx}$ are equal.*

**50.** $f(x, y, z) = x^2y^3 - y^2z^3$

**51.** $f(x, y, z) = \sqrt{9 - x^2 - 2y^2 - z^2}$

**52.** $f(x, y, z) = \ln(x + 2y + 3z)$

**53.** $f(x, y, z) = e^{-x} \cos yz$

**54.** The figure shows the contour map of a function $f$. Use it to determine the sign of (a) $f_x$, (b) $f_y$, (c) $f_{xx}$, (d) $f_{xy}$, and (e) $f_{yy}$ at the point $P$.

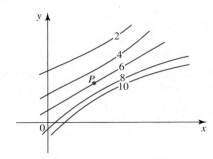

*In Exercises 55 and 56, show that the function satisfies the one-dimensional heat equation $u_t = c^2 u_{xx}$.*

**55.** $u = e^{-t} \sin \dfrac{x}{c}$     **56.** $u = e^{-c^2k^2t} \cos kx$

*In Exercises 57 and 58, show that the function satisfies the one-dimensional wave equation $u_{tt} = c^2 u_{xx}$.*

**57.** $u = \cos(x - ct) + 2 \sin(x + ct)$

**58.** $u = \sin(kct) \sin(kx)$

*In Exercises 59–64, show that the function satisfies the two-dimensional Laplace's equation $u_{xx} + u_{yy} = 0$.*

**59.** $u = 3x^2y - y^3$     **60.** $u = \dfrac{x}{x^2 + y^2}$

**61.** $u = \ln\sqrt{x^2 + y^2}$     **62.** $u = e^{-x} \cos y + e^{-y} \cos x$

**63.** $u = \tan^{-1} \dfrac{y}{x}$

**64.** $u = \cosh y \sin x + \sinh y \cos x$

*In Exercises 65 and 66, show that the function satisfies the three-dimensional Laplace's equation $u_{xx} + u_{yy} + u_{zz} = 0$.*

**65.** $u = x^2 + 3xy + 2y^2 - 3z^2 + 4xyz$

**66.** $u = (x^2 + y^2 + z^2)^{-1/2}$

**67.** Show that the function $z = \sqrt{x^2 + y^2}\,\tan^{-1}\dfrac{y}{x}$ satisfies the equation $x\dfrac{\partial z}{\partial x} + y\dfrac{\partial z}{\partial y} = z$.

**68.** Show that the function $u = 20x^2 \cos\dfrac{y}{x}$ satisfies the equation $x\dfrac{\partial u}{\partial x} + y\dfrac{\partial u}{\partial y} = 2u$.

**69.** According to the ideal gas law, the volume $V$ (in liters) of an ideal gas is related to its pressure $P$ (in pascals) and temperature $T$ (in kelvins) by the formula

$$V = \frac{kT}{P}$$

where $k$ is a constant. Compute $\partial V/\partial T$ and $\partial V/\partial P$ if $k = 8.314$, $T = 300$, and $P = 125$, and interpret your results.

**70.** Refer to Exercise 69. Show that

$$\frac{\partial V}{\partial T} \cdot \frac{\partial T}{\partial P} \cdot \frac{\partial P}{\partial V} = -1$$

**71.** The total resistance $R$ (in ohms) of three resistors with resistances of $R_1$, $R_2$, and $R_3$ ohms connected in parallel is given by the formula

$$\frac{1}{R} = \frac{1}{R_1} + \frac{1}{R_2} + \frac{1}{R_3}$$

Find $\partial R/\partial R_1$ and interpret your result.

**72.** The height of a hill (in feet) is given by

$$h(x, y) = 20(16 - 4x^2 - 3y^2 + 2xy + 28x - 18y)$$

where $x$ is the distance (in miles) east and $y$ the distance (in miles) north of Bolton. If you are at a point on the hill 1 mile north and 1 mile east of Bolton, what is the rate of change of the height of the hill (a) in a northerly direction and (b) in an easterly direction?

**73. Profit Versus Inventory and Floor Space** The monthly profit (in dollars) of the Barker Department Store depends on its level of inventory $x$ (in thousands of dollars) and the floor space $y$ (in thousands of square feet) available for display of its merchandise, as given by the equation

$$P(x, y) = -0.02x^2 - 15y^2 + xy + 39x + 25y - 15,000$$

Find $\partial P/\partial x$ and $\partial P/\partial y$ when $x = 5000$ and $y = 200$, and interpret your result.

**74. Steady-State Temperature** Consider the upper half-disk $H = \{(x, y)\,|\,x^2 + y^2 \le 1, y \ge 0\}$ (see the figure). If the temperature at points on the upper boundary is kept at $100°C$ and the temperature at points on the lower boundary is kept at $50°C$, then the steady-state temperature at any point $(x, y)$ inside the half-disk is given by

$$T(x, y) = 100 - \frac{100}{\pi}\tan^{-1}\frac{1 - x^2 - y^2}{2y}$$

**a.** Compute $T_x\!\left(\tfrac{1}{2}, \tfrac{1}{2}\right)$, and interpret your result.

**b.** Find the rate of change of the temperature at the point $P\!\left(\tfrac{1}{2}, \tfrac{1}{2}\right)$ in the $y$-direction.

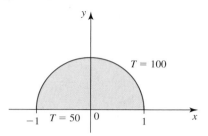

**75. Electric Potential** A charge $Q$ (in coulombs) located at the origin of a three-dimensional coordinate system produces an electric potential $V$ (in volts) given by

$$V(x, y, z) = \frac{kQ}{\sqrt{x^2 + y^2 + z^2}}$$

where $k$ is a positive constant and $x$, $y$, and $z$ are measured in meters. Find the rate of change of the potential at the point $P(1, 2, 3)$ in the $x$-direction.

**76. Surface Area of a Human** The formula

$$S = 0.007184W^{0.425}H^{0.725}$$

gives the surface area $S$ of a human body (in square meters) in terms of its weight $W$ (in kilograms) and its height $H$ (in centimeters). Compute $\partial S/\partial W$ and $\partial S/\partial H$ when $W = 70$ and $H = 180$, and interpret your results.

**77. Arson Study** A study of arson for profit conducted for a certain city found that the number of suspicious fires is approximated by the formula

$$N(x, y) = \frac{120\sqrt{1000 + 0.03x^2 y}}{(5 + 0.2y)^2}$$

$$0 \le x \le 150, \quad 5 \le y \le 35$$

where $x$ denotes the number of persons per census tract and $y$ denotes the level of reinvestment in conventional mortgages by the city's ten largest banks measured in cents per dollars deposited. Compute $\partial N/\partial x$ and $\partial N/\partial y$ when $x = 100$ and $y = 20$, and interpret your results.

**78. Production Functions** The productivity of a Central American country is given by the function

$$f(x, y) = 20x^{3/4}y^{1/4}$$

when $x$ units of labor and $y$ units of capital are used.

**a.** What are the marginal productivity of labor and the marginal productivity of capital when the amounts expended on labor and capital are 256 units and 16 units, respectively?

**b.** Should the government encourage capital investment rather than increased expenditure on labor at this time to increase the country's productivity?

**79. Wind Chill Factor** A formula that meteorologists use to calculate the wind chill temperature (the temperature that you would feel in still air that is the same as the actual temperature when the presence of wind is taken into consideration) is

$$T = f(t, s) = 35.74 + 0.6125t - 35.75s^{0.16} + 0.4275ts^{0.16}$$

$$s \geq 1$$

where $t$ is the air temperature in degrees Fahrenheit and $s$ is the wind speed in mph.

a. What is the wind chill temperature when the actual air temperature is 32°F and the wind speed is 20 mph?

b. What is the rate of change of the wind chill temperature with respect to the wind speed if the temperature is 32°F and the wind speed is 20 mph?

**80. Wind Chill Factor** The wind chill temperature is the temperature that you would feel in still air that is the same as the actual temperature when the presence of wind is taken into consideration. The following table gives the wind chill temperature $T = f(t, s)$ in degrees Fahrenheit in terms of the actual air temperature $t$ in degrees Fahrenheit and the wind speed $s$ in mph.

| | | | Wind speed (mph) | | | | |
|---|---|---|---|---|---|---|---|
| $t$ \ $s$ | 10 | 15 | 20 | 25 | 30 | 35 | 40 |
| **30** | 21.2 | 19.0 | 17.4 | 16.0 | 14.9 | 13.9 | 13.0 |
| **32** | 23.7 | 21.6 | 20.0 | 18.7 | 17.6 | 16.6 | 15.8 |
| **34** | 26.2 | 24.2 | 22.6 | 21.4 | 20.3 | 19.4 | 18.6 |
| **36** | 28.7 | 26.7 | 25.2 | 24.0 | 23.0 | 22.2 | 21.4 |
| **38** | 31.2 | 29.3 | 27.9 | 26.7 | 25.7 | 24.9 | 24.2 |
| **40** | 33.6 | 31.8 | 20.5 | 29.4 | 28.5 | 27.7 | 26.9 |

(Actual air temperature (°F))

a. Estimate the rate of change of the wind chill temperature $T$ with respect to the actual air temperature when the wind speed is constant at 25 mph and the actual air temperature is 34°F.

Hint: Show that it is given by

$$\frac{\partial T}{\partial t}(34, 25) \approx \frac{f(36, 25) - f(34, 25)}{2}$$

b. Estimate the rate of change of the wind chill temperature $T$ with respect to the wind speed when the actual air temperature is constant at 34°F and the wind speed is 25 mph.

*Source: National Weather Service.*

**81.** Let $f$ be a function of two variables.

a. Put $g(x) = f(x, b)$, and use the definition of the derivative of a function of one variable to show that $f_x(a, b) = g'(a)$.

b. Put $h(y) = f(a, y)$, and show that $f_y(a, b) = h'(b)$.

**82. a.** Use the result of Exercise 81 to find $f_x\left(1, \frac{\pi}{2}\right)$ if $f(x, y) = x^2 \cos xy$.

b. Verify the result of part (a) by evaluating $f_x(x, y)$ at $\left(1, \frac{\pi}{2}\right)$.

**cas** *In Exercises 83 and 84, use the result of Exercise 81 and a calculator or computer to find the partial derivative.*

**83.** $f_x(2, 1)$ if $f(x, y) = \ln\left(e^{xy} + \cos\sqrt{x^2 + y^2}\right)$

**84.** $f_y(2, 1)$ if $f(x, y) = \dfrac{\sin \pi xy}{\left(1 + \sqrt{x^2 + y^3}\right)^{3/2}}$

**85. Cobb-Douglas Production Function** Show that the Cobb-Douglas production function $P = kx^\alpha y^{1-\alpha}$, where $0 < \alpha < 1$, satisfies the equation

$$x \frac{\partial P}{\partial x} + y \frac{\partial P}{\partial y} = P$$

**Note:** This equation is called **Euler's equation.**

**86.** Let $S$ be the surface with equation $z = f(x, y)$, where $f$ has continuous first-order partial derivatives and $P(x_0, y_0, z_0)$ is a point on $S$ (see the figure). Let $C_1$ and $C_2$ be the curves obtained by the intersection of the surface $S$ with the planes $x = x_0$ and $y = y_0$, respectively. Let $T_1$ and $T_2$ be the tangent lines to the curves $C_1$ and $C_2$ at $P$. Then the *tangent plane* to the surface $S$ at the point $P$ is the plane that contains both tangent lines $T_1$ and $T_2$.

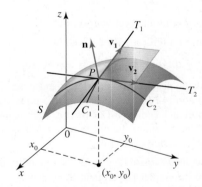

a. Show that the vectors $\mathbf{v}_1 = \mathbf{i} + f_x(x_0, y_0)\mathbf{k}$ and $\mathbf{v}_2 = \mathbf{j} + f_y(x_0, y_0)\mathbf{k}$ are parallel to $T_1$ and $T_2$, respectively.

b. Using the result of part (a), find a vector $\mathbf{n}$ that is normal to both $\mathbf{v}_1$ and $\mathbf{v}_2$.

c. Use the result of part (b) to show that an equation of the tangent plane to $S$ at $P$ is

$$z - z_0 = f_x(x_0, y_0)(x - x_0) + f_y(x_0, y_0)(y - y_0)$$

**87.** Use the result of Exercise 86 to find an equation of the tangent plane to the paraboloid $z = x^2 + \frac{1}{4}y^2$ at the point $(1, 2, 2)$.

**88. Engine Efficiency** The efficiency of an internal combustion engine is given by

$$E = \left(1 - \frac{v}{V}\right)^{0.4}$$

where $V$ and $v$ are the respective maximum and minimum volumes of air in each cylinder.

**a.** Show that $\partial E / \partial V > 0$, and interpret your result.

**b.** Show that $\partial E / \partial v < 0$, and interpret your result.

**89.** A semi-infinite strip has faces that are insulated. If the edges $x = 0$ and $x = \pi$ of the strip are kept at temperature zero and the base of the strip is kept at temperature 1, then the steady-state temperature (that is, the temperature after a long time) is given by

$$T(x, y) = \frac{2}{\pi} \tan^{-1} \frac{\sin x}{\sinh y}$$

Find $\dfrac{\partial T}{\partial x}\left(\frac{\pi}{2}, 1\right)$ and $\dfrac{\partial T}{\partial y}\left(\frac{\pi}{2}, 1\right)$, and interpret your results.

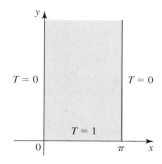

**90.** Let

$$f(x, y) = \begin{cases} \dfrac{xy(x^2 - y^2)}{x^2 + y^2} & \text{if } (x, y) \neq (0, 0) \\ 0 & \text{if } (x, y) = (0, 0) \end{cases}$$

**a.** Find $f_x(x, y)$ and $f_y(x, y)$ for $(x, y) \neq (0, 0)$.

**b.** Use the definition of partial derivatives to find $f_x(0, 0)$ and $f_y(0, 0)$.

**c.** Show that $f_{xy}(0, 0) = -1$ and $f_{yx}(0, 0) = 1$.

**d.** Does the result of part (c) contradict Theorem 1? Explain.

**91.** Does there exist a function $f$ of two variables $x$ and $y$ with continuous second-order partial derivatives such that $f_x(x, y) = e^{2x}(2 \cos xy - y \sin xy)$ and $f_y(x, y) = -ye^{2x} \sin xy$? Explain.

**92.** Show that if a function $f$ of two variables $x$ and $y$ has continuous third-order partial derivatives, then $f_{xyx} = f_{yxx} = f_{xxy}$.

*In Exercises 93–96, determine whether the statement is true or false. If it is true, explain why it is true. If it is false, give an example to show why it is false.*

**93.** If $z = f(x, y)$ has a partial derivative with respect to $x$ at the point $(a, b)$, then

$$\frac{\partial f}{\partial x}(a, b) = \lim_{x \to a} \frac{f(x, b) - f(a, b)}{x - a}$$

**94.** If $\partial f / \partial y \, (a, b) = 0$, then the tangent line to the curve formed by the intersection of the plane $x = a$ and the surface $z = f(x, y)$ at the point $(a, b, f(a, b))$ is horizontal; that is, it is parallel to the $xy$-plane.

**95.** If $f_{xx}(x, y)$ is defined for all $x$ and $y$ and $f_{xx}(a, b) < 0$ for all $x$ in the interval $(a, b)$, then the curve $C$ formed by the intersection of the plane $y = b$ and the surface $z = f(x, y)$ is concave downward on $(a, b)$.

**96.** If $f(x, y) = \ln xy$, then $f_{xy}(x, y) = f_{yx}(x, y)$ for all $(x, y)$ in $D = \{(x, y) \mid xy > 0\}$.

---

## 12.4 Differentials

### ■ Increments

Recall that if $f$ is a function of one variable defined by $y = f(x)$, then the *increment* in $y$ is defined to be

$$\Delta y = f(x + \Delta x) - f(x)$$

where $\Delta x$ is an increment in $x$ (Figure 1a). The increment of a function of two or more variables is defined in an analogous manner. For example, if $z$ is a function of two variables defined by $z = f(x, y)$, then the **increment** in $z$ produced by increments of $\Delta x$ and $\Delta y$ in the independent variables $x$ and $y$, respectively, is defined to be

$$\Delta z = f(x + \Delta x, y + \Delta y) - f(x, y) \tag{1}$$

(See Figure 1b.)

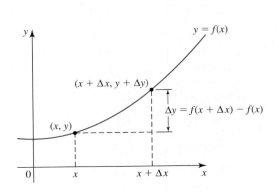

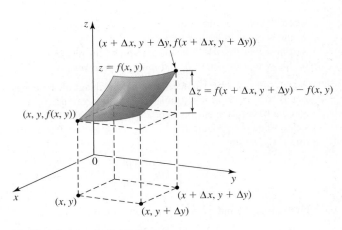

**(a)** The increment $\Delta y$ is the change in $y$ as $x$ changes from $x$ to $x + \Delta x$.

**(b)** The increment $\Delta z$ is the change in $z$ as $x$ changes from $x$ to $x + \Delta x$ and $y$ changes from $y$ to $y + \Delta y$.

**FIGURE 1**

**EXAMPLE 1**  Let $z = f(x, y) = 2x^2 - xy$. Find $\Delta z$. Then use your result to find the change in $z$ if $(x, y)$ changes from $(1, 1)$ to $(0.98, 1.03)$.

**Solution**   Using Equation (1), we obtain

$$\Delta z = f(x + \Delta x, y + \Delta y) - f(x, y)$$

$$= [2(x + \Delta x)^2 - (x + \Delta x)(y + \Delta y)] - (2x^2 - xy)$$

$$= 2x^2 + 4x\,\Delta x + 2(\Delta x)^2 - xy - x\,\Delta y - y\,\Delta x - \Delta x\,\Delta y - 2x^2 + xy$$

$$= (4x - y)\,\Delta x - x\,\Delta y + 2(\Delta x)^2 - \Delta x\,\Delta y$$

Next, to find the increment in $z$ if $(x, y)$ changes from $(1, 1)$ to $(0.98, 1.03)$, we note that $\Delta x = 0.98 - 1 = -0.02$ and $\Delta y = 1.03 - 1 = 0.03$. Therefore, using the result obtained earlier with $x = 1$, $y = 1$, $\Delta x = -0.02$, and $\Delta y = 0.03$, we obtain

$$\Delta z = [4(1) - 1](-0.02) - (1)(0.03) + 2(-0.02)^2 - (-0.02)(0.03)$$

$$= -0.0886$$

You can verify the correctness of this result by computing the quantity $f(0.98, 1.03) - f(1, 1)$.    ∎

## ■ The Total Differential

Recall from Section 2.10 that if $f$ is a function of one variable defined by $y = f(x)$, then the differential of $f$ at $x$ is defined by

$$dy = f'(x)\,dx$$

where $dx = \Delta x$ is the differential in $x$. Furthermore,

$$\Delta y \approx dy \tag{2}$$

if $\Delta x$ is small (see Figure 2).

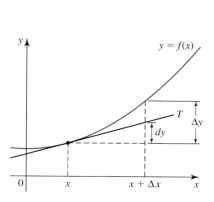

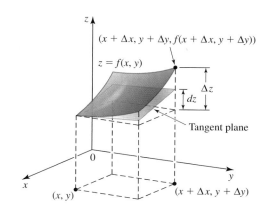

(a) Relationship between $dy$ and $\Delta y$

(b) Relationship between $dz$ and $\Delta z$. The tangent plane is the analog of the tangent line $T$ in the one-variable case.

**FIGURE 2**

For an analog of this result for a function of two variables, we begin with the following definition.

---

**DEFINITION Differentials**

Let $z = f(x, y)$, and let $\Delta x$ and $\Delta y$ be increments of $x$ and $y$, respectively. The **differentials** $dx$ and $dy$ of the independent variables $x$ and $y$ are

$$dx = \Delta x \qquad \text{and} \qquad dy = \Delta y$$

The **differential** $dz$, or **total differential,** of the dependent variable $z$ is

$$dz = \frac{\partial f}{\partial x}\, dx + \frac{\partial f}{\partial y}\, dy = f_x(x, y)\, dx + f_y(x, y)\, dy$$

---

Later in this section, we will show that

$$\Delta z = dz + \varepsilon_1\, \Delta x + \varepsilon_2\, \Delta y$$

where $\varepsilon_1$ and $\varepsilon_2$ are functions of $\Delta x$ and $\Delta y$ that approach 0 as $\Delta x$ and $\Delta y$ approach 0. This implies that

$$\Delta z \approx dz \qquad\qquad\qquad\qquad \textbf{(3)}$$

if both $\Delta x$ and $\Delta y$ are small.

Figure 2b shows the geometric relationship between $\Delta z$ and $dz$. Observe that as $x$ changes from $x$ to $x + \Delta x$ and $y$ changes from $y$ to $y + \Delta y$, $\Delta z$ measures the change in the height of the graph of $f$, whereas $dz$ measures the change in the height of the tangent plane.*

---

*For now, we will rely on our intuitive definition of the tangent plane. We will define the tangent plane in Section 12.7.

**EXAMPLE 2**    Let $z = f(x, y) = 2x^2 - xy$.

**a.** Find the differential $dz$.
**b.** Compute the value of $dz$ if $(x, y)$ changes from $(1, 1)$ to $(0.98, 1.03)$, and compare your result with the value of $\Delta z$ obtained in Example 1.

**Solution**

**a.** $dz = \dfrac{\partial f}{\partial x} dx + \dfrac{\partial f}{\partial y} dy = (4x - y)\, dx - x\, dy$

**b.** Here $x = 1$, $y = 1$, $dx = \Delta x = -0.02$, and $dy = \Delta y = 0.03$. Therefore,

$$dz = [4(1) - 1](-0.02) - 1(0.03) = -0.09$$

The value of $\Delta z$ obtained in Example 1 was $-0.0886$, so $dz$ is a good approximation of $\Delta z$ in this case. Observe that it is easier to compute $dz$ than to compute $\Delta z$.    ■

**EXAMPLE 3**    A storage tank has the shape of a right circular cylinder. Suppose that the radius and height of the tank are measured at 1.5 ft and 5 ft, respectively, with a possible error of 0.05 ft and 0.1 ft, respectively. Use differentials to estimate the maximum error in calculating the capacity of the tank.

**Solution**    The capacity (volume) of the tank is $V = \pi r^2 h$. The error in calculating the capacity of the tank is given by

$$\Delta V \approx dV = \frac{\partial V}{\partial r} dr + \frac{\partial V}{\partial h} dh = 2\pi rh\, dr + \pi r^2\, dh$$

Since the errors in the measurement of $r$ and $h$ are at most 0.05 ft and 0.1 ft, respectively, we have $dr = 0.05$ and $dh = 0.1$. Therefore, taking $r = 1.5$, $h = 5$, $dr = 0.05$, and $dh = 0.1$, we obtain

$$dV = 2\pi rh\, dr + \pi r^2\, dh$$
$$\approx 2\pi(1.5)(5)(0.05) + \pi(1.5)^2(0.1) = 0.975\pi$$

Thus, the maximum error in calculating the volume of the storage tank is approximately $0.975\pi$, or 3.1, ft$^3$.    ■

**EXAMPLE 4**    **The Error in Computing the Range of a Projectile**    If a projectile is fired with an angle of elevation $\theta$ and initial speed of $v$ ft/sec, then its range (in feet) is

$$R = \frac{v^2 \sin 2\theta}{g}$$

where $g$ is the constant of acceleration due to gravity. (See Figure 3.) Suppose that a projectile is launched with an initial speed of 2000 ft/sec at an angle of elevation of $\pi/12$ radians and that the maximum percentage errors in the measurement of $v$ and $\theta$ are 0.5% and 1%, respectively.

**a.** Estimate the maximum error in the computation of the range of the projectile.
**b.** Find the maximum percentage error in computing the range of this projectile.

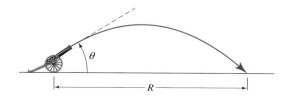

**FIGURE 3**
We want to find the range $R$ of a
projectile fired with an angle of
elevation $\theta$ and initial speed of $v$ ft/sec.

**Solution**

**a.** The error in the computation of $R$ is

$$\Delta R \approx dR = \frac{\partial R}{\partial v}\,dv + \frac{\partial R}{\partial \theta}\,d\theta = \frac{2v \sin 2\theta}{g}\,dv + \frac{2v^2 \cos 2\theta}{g}\,d\theta$$

The maximum error in the computation of $v$ is $(0.005)(2000)$ or 10 ft/sec; that is, $|dv| \le 10$. Also, the maximum error in the computation of $\theta$ is $(0.01)(\pi/12)$ radians. In other words, $|d\theta| \le 0.01(\pi/12)$. Therefore, the maximum error in computing the range of the projectile is approximately

$$|\Delta R| \approx |dR| \le \frac{2v \sin 2\theta}{g}|dv| + \frac{2v^2 \cos 2\theta}{g}|d\theta|$$

$$= \frac{2(2000) \sin\left(\dfrac{\pi}{6}\right)}{32}(10) + \frac{2(2000)^2 \cos\left(\dfrac{\pi}{6}\right)}{32}\left(\frac{0.01\pi}{12}\right)$$

$$\approx 1192$$

or approximately 1192 ft.

**b.** Using $v = 2000$ and $\theta = \pi/12$, we find the range of the projectile to be

$$R = \frac{v^2 \sin 2\theta}{g} = \frac{(2000)^2 \sin\left(\dfrac{\pi}{6}\right)}{32} = 62{,}500$$

Therefore, the maximum percentage error in computing the range of the projectile is

$$100\left|\frac{\Delta R}{R}\right| \approx 100\left(\frac{1192}{62{,}500}\right)$$

or approximately 1.91%.  ■

## ■ Error in Approximating $\Delta z$ by $dz$

The following theorem tells us that $dz$ gives a good approximation of $\Delta z$ if $\Delta x$ and $\Delta y$ are small, provided that both $f_x$ and $f_y$ are continuous.

---

**THEOREM 1**

Let $f$ be a function defined on an open region $R$. Suppose that the points $(x, y)$ and $(x + \Delta x, y + \Delta y)$ are in $R$ and that $f_x$ and $f_y$ are continuous at $(x, y)$. Then

$$\Delta z = f_x(x, y)\,\Delta x + f_y(x, y)\,\Delta y + \varepsilon_1\,\Delta x + \varepsilon_2\,\Delta y$$

where $\varepsilon_1$ and $\varepsilon_2$ are functions of $\Delta x$ and $\Delta y$ such that

$$\lim_{(\Delta x, \Delta y)\to(0, 0)} \varepsilon_1 = 0 \qquad \text{and} \qquad \lim_{(\Delta x, \Delta y)\to(0, 0)} \varepsilon_2 = 0$$

**PROOF** Fix $x$ and $y$. By adding and subtracting $f(x + \Delta x, y)$ to $\Delta z$, we have

$$\Delta z = f(x + \Delta x, y + \Delta y) - f(x, y)$$
$$= [f(x + \Delta x, y) - f(x, y)] + [f(x + \Delta x, y + \Delta y) - f(x + \Delta x, y)]$$
$$= \Delta z_1 + \Delta z_2$$

where $\Delta z_1$ is the change in $z$ as $(x, y)$ changes from $(x, y)$ to $(x + \Delta x, y)$ and $\Delta z_2$ is the change in $z$ as $(x, y)$ changes from $(x + \Delta x, y)$ to $(x + \Delta x, y + \Delta y)$. (See Figure 4a.)

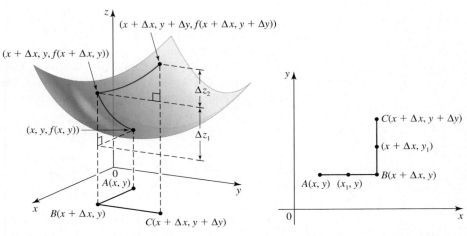

**FIGURE 4**

**(a)** $\Delta z_1 = f(x + \Delta x, y) - f(x, y)$ and
$\Delta z_2 = f(x + \Delta x, y + \Delta y) - f(x + \Delta x, y)$

**(b)** The points $A$, $B$, and $C$ shown in the $xy$-plane.

On the interval between $A$ and $B$, $y$ is constant, so the function $g$ defined by $g(t) = f(t, y)$ for $x \le t \le x + \Delta x$ is a function of one variable. (See Figure 4b.) Therefore, by the Mean Value Theorem, there exists a point $(x_1, y)$ with $x < x_1 < x + \Delta x$ such that

$$g(x + \Delta x) - g(x) = g'(x_1) \Delta x$$

Since $g'(x_1) = f_x(x_1, y)$, we have

$$\Delta z_1 = f(x + \Delta x, y) - f(x, y) = g(x + \Delta x) - g(x)$$
$$= g'(x_1) \Delta x = f_x(x_1, y) \Delta x \qquad x < x_1 < x + \Delta x$$

Next, on the interval between $B$ and $C$, both $x$ and $\Delta x$ are constant, so the function $h$ defined by $h(t) = f(x + \Delta x, t)$ for $y \le t \le y + \Delta y$ is a function of one variable. (See Figure 4b.) Therefore, by the Mean Value Theorem there exists a point $(x + \Delta x, y_1)$ with $y < y_1 < y + \Delta y$ such that

$$h(y + \Delta y) - h(y) = h'(y_1) \Delta y$$

Since $h'(y_1) = f_y(x + \Delta x, y_1)$, we have

$$\Delta z_2 = f(x + \Delta x, y + \Delta y) - f(x + \Delta x, y)$$
$$= h(y + \Delta y) - h(y) = h'(y_1) \Delta y = f_y(x + \Delta x, y_1) \Delta y$$

Therefore,

$$\Delta z = \Delta z_1 + \Delta z_2$$
$$= f_x(x_1, y) \Delta x + f_y(x + \Delta x, y_1) \Delta y$$

Adding and subtracting $f_x(x, y) \Delta x + f_y(x, y) \Delta y$ to the right-hand side of the previous equation and rearranging terms, we obtain

$$\Delta z = f_x(x, y) \Delta x + f_y(x, y) \Delta y + [f_x(x_1, y) - f_x(x, y)] \Delta x + [f_y(x + \Delta x, y_1) - f_y(x, y)] \Delta y$$

$$= f_x(x, y) \Delta x + f_y(x, y) \Delta y + \varepsilon_1 \Delta x + \varepsilon_2 \Delta y$$

where

$$\varepsilon_1 = f_x(x_1, y) - f_x(x, y)$$

and

$$\varepsilon_2 = f_y(x + \Delta x, y_1) - f_y(x, y)$$

Observe that as $(\Delta x, \Delta y) \to (0, 0)$, $x_1 \to x$ and $y_1 \to y$. Therefore, the continuity of $f_x$ and $f_y$ implies that

$$\lim_{(\Delta x, \Delta y) \to (0, 0)} \varepsilon_1 = 0 \quad \text{and} \quad \lim_{(\Delta x, \Delta y) \to (0, 0)} \varepsilon_2 = 0$$

and this proves the result. ■

**Note**  Observe that the conclusion of Theorem 1 can be written as

$$\Delta z - dz = \varepsilon_1 \Delta x + \varepsilon_2 \Delta y$$

Therefore, if $\Delta x$ and $\Delta y$ are both small, then

$$\Delta z - dz = (\text{small number})(\text{small number}) + (\text{small number})(\text{small number})$$

and this quantity is a *very* small number, which accounts for the closeness of the approximation. Compare this with the case of a function of one variable discussed in Section 2.10. ■

### ■ Differentiability of a Function of Two Variables

The conclusion of Theorem 1 can be written as

$$\Delta z = dz + \varepsilon_1 \Delta x + \varepsilon_2 \Delta y \tag{4}$$

where $\varepsilon_1 \to 0$ and $\varepsilon_2 \to 0$ as $(\Delta x, \Delta y) \to (0, 0)$. We *define* a function of two variables to be *differentiable* if $z = f(x, y)$ satisfies Equation (4).

---

**DEFINITION    Differentiability of a Function of Two Variables**

Let $z = f(x, y)$. The function $f$ is differentiable at $(a, b)$ if $\Delta z$ can be expressed in the form

$$\Delta z = f_x(a, b) \Delta x + f_y(a, b) \Delta y + \varepsilon_1 \Delta x + \varepsilon_2 \Delta y$$

where $\varepsilon_1 \to 0$ and $\varepsilon_2 \to 0$ as $(\Delta x, \Delta y) \to (0, 0)$. The function $f$ is differentiable in a region $R$ if it is differentiable at each point of $R$.

---

**EXAMPLE 5**  Show that the function $f$ defined by $f(x, y) = 2x^2 - xy$ is differentiable in the plane.

**Solution**  Write $z = f(x, y) = 2x^2 - xy$, and let $(x, y)$ be any point in the plane. Then using the result of Example 1, we have

$$\Delta z = (4x - y) \Delta x - x \Delta y + 2(\Delta x)^2 - \Delta x \Delta y$$

Since $f_x = 4x - y$ and $f_y = -x$, we can write

$$\Delta z = f_x \Delta x + f_y \Delta y + \varepsilon_1 \Delta x + \varepsilon_2 \Delta y$$

where $\varepsilon_1 = 2 \Delta x$ and $\varepsilon_2 = -\Delta x$. Since $\varepsilon_1 \to 0$ and $\varepsilon_2 \to 0$ as $(\Delta x, \Delta y) \to (0, 0)$, it follows that $f$ is differentiable at $(x, y)$. But $(x, y)$ is any point in the plane, so $f$ is differentiable in the plane. ∎

The next theorem, which is an immediate consequence of Theorem 1, guarantees when a function of two variables is differentiable.

---

**THEOREM 2    Criterion for Differentiability**

Let $f$ be a function of the variables $x$ and $y$. If $f_x$ and $f_y$ exist and are continuous on an open region $R$, then $f$ is differentiable in $R$.

---

For the function $f(x, y) = 2x^2 - xy$ of Example 5, we have $f_x(x, y) = 4x - y$ and $f_y(x, y) = -x$, both of which are continuous everywhere. Therefore, by Theorem 2 we conclude that $f$ is differentiable in the plane, as demonstrated earlier.

 Remember that the mere existence of the partial derivatives $f_x$ and $f_y$ of a function $f$ at a point $(x, y)$ is not enough to guarantee the differentiability of $f$ at $(x, y)$. (See Exercise 43.)

## ■ Differentiability and Continuity

Just as a differentiable function of one variable is continuous, the following theorem shows that a differentiable function of two variables is also continuous.

---

**THEOREM 3    Differentiable Functions Are Continuous**

Let $f$ be a function of two variables. If $f$ is differentiable at $(a, b)$, then $f$ is continuous at $(a, b)$.

---

**PROOF**    Using the result of Theorem 1, we have

$$\Delta z = f(a + \Delta x, b + \Delta y) - f(a, b)$$
$$= f_x(a, b) \Delta x + f_y(a, b) \Delta y + \varepsilon_1 \Delta x + \varepsilon_2 \Delta y$$

Writing $x = a + \Delta x$ and $y = b + \Delta y$, we have

$$f(x, y) - f(a, b) = [f_x(a, b) + \varepsilon_1](x - a) + [f_y(a, b) + \varepsilon_2](y - b)$$

Noting that $\varepsilon_1 \to 0$ and $\varepsilon_2 \to 0$ as $(\Delta x, \Delta y) \to (0, 0)$, we see that

$$f(x, y) - f(a, b) \to 0 \quad \text{as} \quad (\Delta x, \Delta y) \to (0, 0)$$

Equivalently,

$$\lim_{(x, y) \to (a, b)} f(x, y) = f(a, b)$$

Therefore, $f$ is continuous at $(a, b)$. ∎

## ■ Functions of Three or More Variables

The notions of differentiability and the differential of functions of more than two variables are similar to those of functions of two variables. For example, suppose that $f$ is a function of three variables that is defined by $w = f(x, y, z)$. Then the increment $\Delta w$ of $w$ corresponding to increments of $\Delta x$, $\Delta y$, and $\Delta z$ of $x$, $y$, and $z$, respectively, is

$$\Delta w = f(x + \Delta x, y + \Delta y, z + \Delta z) - f(x, y, z)$$

The function $f$ is **differentiable** at $(x, y, z)$ if $\Delta w$ can be written in the form

$$\Delta w = f_x(x, y, z)\, \Delta x + f_y(x, y, z)\, \Delta y + f_z(x, y, z)\, \Delta z + \varepsilon_1\, \Delta x + \varepsilon_2\, \Delta y + \varepsilon_3\, \Delta z$$

where $\varepsilon_1, \varepsilon_2$, and $\varepsilon_3$ are functions of $\Delta x, \Delta y$, and $\Delta z$ that approach zero as $(\Delta x, \Delta y, \Delta z) \to (0, 0, 0)$.

The **differential** $dw$ of the dependent variable $w$ is defined to be

$$dw = \frac{\partial w}{\partial x}\, dx + \frac{\partial w}{\partial y}\, dy + \frac{\partial w}{\partial z}\, dz$$

where $dx = \Delta x$, $dy = \Delta y$, and $dz = \Delta z$ are the differentials of the independent variables, $x$, $y$, and $z$. If $f$ has continuous partial derivatives and $dx$, $dy$, and $dz$ are all small, then $\Delta w \approx dw$.

**EXAMPLE 6** **Maximum Error in Calculating Centrifugal Force** A centrifuge is a machine designed for the specific purpose of subjecting materials to a sustained centrifugal force. The magnitude of a centrifugal force $F$ in dynes is given by

$$F = f(M, S, R) = \frac{\pi^2 S^2 MR}{900}$$

where $S$ is in revolutions per minute (rpm), $M$ is the mass in grams, and $R$ is the radius in centimeters. If the maximum percentage errors in the measurement of $M$, $S$, and $R$ are 0.1%, 0.4%, and 0.2%, respectively, use differentials to estimate the maximum percentage error in calculating $F$.

**Solution** The error in calculating $F$ is $\Delta F$, and

$$\Delta F \approx dF = \frac{\partial F}{\partial M}\, dM + \frac{\partial F}{\partial S}\, dS + \frac{\partial F}{\partial R}\, dR$$

$$= \frac{\pi^2 S^2 R}{900}\, dM + \frac{2\pi^2 SMR}{900}\, dS + \frac{\pi^2 S^2 M}{900}\, dR$$

Therefore,

$$\frac{\Delta F}{F} \approx \frac{dF}{F} = \frac{dM}{M} + 2\frac{dS}{S} + \frac{dR}{R}$$

and

$$\left| \frac{\Delta F}{F} \right| \approx \left| \frac{dF}{F} \right| \leq \left| \frac{dM}{M} \right| + 2\left| \frac{dS}{S} \right| + \left| \frac{dR}{R} \right|$$

Since

$$\left| \frac{dM}{M} \right| \leq 0.001, \qquad \left| \frac{dS}{S} \right| \leq 0.004, \qquad \text{and} \qquad \left| \frac{dR}{R} \right| \leq 0.002$$

we have

$$\left|\frac{dF}{F}\right| \le 0.001 + 2(0.004) + 0.002 = 0.011$$

Thus, the maximum percentage error in calculating the centrifugal force is approximately 1.1%. ∎

## 12.4 CONCEPT QUESTIONS

1. If $z = f(x, y)$, what is the differential of $x$? The differential of $y$? What is the total differential of $z$?
2. Let $z = f(x, y)$. What is the relationship between the actual change $\Delta z$, when $x$ changes from $x$ to $x + \Delta x$ and $y$ changes from $y$ to $y + \Delta y$, and the total differential $dz$ of $f$ at $(x, y)$?
3. **a.** What does it mean for a function $f$ of two variables $x$ and $y$ to be differentiable at $(a, b)$? To be differentiable in a region $R$?

   **b.** Give a condition that guarantees that a function $f$ of two variables $x$ and $y$ is differentiable in an open region $R$.

   **c.** If a function $f$ of two variables $x$ and $y$ is differentiable at $(a, b)$, what can you say about the continuity of $f$ at $(a, b)$?

## 12.4 EXERCISES

1. Let $z = 2x^2 + 3y^2$, and suppose that $(x, y)$ changes from $(2, -1)$ to $(2.01, -0.98)$.
   **a.** Compute $\Delta z$.     **b.** Compute $dz$.
   **c.** Compare the values of $\Delta z$ and $dz$.

2. Let $z = x^2 - 2xy + 3y^2$, and suppose that $(x, y)$ changes from $(2, 1)$ to $(1.97, 1.02)$.
   **a.** Compute $\Delta z$.     **b.** Compute $dz$.
   **c.** Compare the values of $\Delta z$ and $dz$.

*In Exercises 3–20, find the differential of the function.*

3. $z = 3x^2y^3$

4. $z = x^4 - 2x^2y^2 + 3xy^2 + y^3$

5. $z = \dfrac{x + y}{x - y}$

6. $w = \dfrac{xy}{1 + x^2}$

7. $z = (2x^2y + 3y^3)^3$

8. $z = \sqrt{2x^2 + 3y^2}$

9. $w = ye^{x^2 - y^2}$

10. $z = \ln(2x + 3y)$

11. $w = x^2 \ln(x^2 + y^2)$

12. $z = x^2 \sin 2y$

13. $z = e^{2x} \cos 3y$

14. $w = \tan^{-1}\left(\dfrac{y}{x}\right)$

15. $w = x^2 + xy + z^2$

16. $w = \sqrt{x^2 + xy + z^2}$

17. $w = x^2 e^{-yz}$

18. $w = e^{-x^2} \sin(2y + 3z)$

19. $w = x^2 e^y + y \ln z$

20. $w = x \cosh yz$

*In Exercises 21–24, use differentials to approximate the change in f due to the indicated change in the independent variables.*

21. $f(x, y) = x^4 - 3x^2y^2 + y^3 - 2y + 4$;   $(x, y)$ changes from $(2, 2)$ to $(1.98, 2.01)$.

22. $f(x, y) = \sqrt{2x + 3y} - \dfrac{x}{y}$;   $(x, y)$ changes from $(3, 1)$ to $(2.96, 1.02)$.

23. $f(x, y, z) = \ln(2x - y) + e^{2xz}$;   $(x, y, z)$ changes from $(2, 3, 0)$ to $(2.01, 2.97, 0.04)$.

24. $f(x, y, z) = x^2y \cos \pi z$;   $(x, y, z)$ changes from $(1, 3, 2)$ to $(0.98, 2.97, 2.01)$.

25. The dimensions of a closed rectangular box are measured as 30 in., 40 in., and 60 in., with a maximum error of 0.2 in. in each measurement. Use differentials to estimate the maximum error in calculating the volume of the box.

26. Use differentials to estimate the maximum error in calculating the surface area of the box of Exercise 25.

27. A piece of land is triangular in shape. Two of its sides are measured as 80 and 100 ft, and the included angle is measured as $\pi/3$ rad. If the sides are measured with a maximum error of 0.3 ft and the angle is measured with a maximum error of $\pi/180$ rad, what is the approximate maximum error in the calculated area of the land?

28. **Production Functions** The productivity of a certain country is given by the function

$$f(x, y) = 30x^{4/5}y^{1/5}$$

when $x$ units of labor and $y$ units of capital are utilized. What is the approximate change in the number of units produced if the amount expended on labor is decreased from 243 to 240 units and the amount expended on capital is increased from 32 units to 35 units?

29. The pressure $P$ (in pascals), the volume $V$ (in liters), and the temperature $T$ (in kelvins) of an ideal gas are related by the

equation $PV = 8.314T$. Use differentials to find the approximate change in the pressure of the gas if its volume increases from 20 L to 20.2 L and its temperature decreases from 300 K to 295 K.

30. Consider the ideal gas law equation $PV = 8.314T$ of Exercise 29. If $T$ and $P$ are measured with maximum errors of 0.6% and 0.4%, respectively, determine the maximum percentage error in calculating the value of $V$.

31. **Surface Area of Humans** The surface area $S$ of humans is related to their weight $W$ and height $H$ by the formula $S = 0.1091W^{0.425}H^{0.725}$. If $W$ and $H$ are measured with maximum errors of 3% and 2%, respectively, find the approximate maximum percentage error in the measurement of $S$.

32. **Specific Gravity** The specific gravity of an object with density greater than that of water can be determined by using the formula

$$S = \frac{A}{A - W}$$

where $A$ and $W$ are the weights of the object in air and in water, respectively. If the measurements of an object are $A = 2.2$ lb and $W = 1.8$ lb with maximum errors of 0.02 lb and 0.04 lb, respectively, find the approximate maximum error in calculating $S$.

33. **Flow of Blood** The flow of blood through an arteriole measured in cm³/sec is given by

$$F = \frac{\pi PR^4}{8kL}$$

where $L$ is the length of the arteriole in centimeters, $R$ is the radius in centimeters, $P$ is the difference in pressure between the two ends of the arteriole in dyne-sec/cm², and $k$ is the viscosity of blood in dyne-sec/cm². Find the approximate maximum percentage error in measuring the flow of blood if an error of at most 1% is made in measuring the length of the arteriole and an error of at most 2% is made in measuring its radius. Assume that $P$ and $k$ are constant.

34. The figure below shows two long, parallel wires that are at a distance of $d$ m apart, carrying currents of $I_1$ and $I_2$ amps. It can be shown that the force of attraction per unit length between the two wires as a result of magnetic fields generated by the currents is given by

$$f = \frac{\mu_0}{2\pi}\left(\frac{I_1 I_2}{D}\right)$$

teslas per meter, where $\mu_0$ ($4\pi \times 10^{-7}$ N/amp²) is a constant called the *permeability of free space*. Use differentials to find the approximate percentage change in $f$ if $I_1$ increases by 2%, $I_2$ decreases by 2%, and $D$ decreases by 5%.

35. **Error in Measuring the Period of a Pendulum** The period $T$ of a simple pendulum executing small oscillations is given by $T = 2\pi\sqrt{L/g}$, where $L$ is the length of the pendulum and $g$ is the constant of acceleration due to gravity. If $T$ is computed by using $L = 4$ ft and $g = 32$ ft/sec², find the approximate percentage error in $T$ if the true values for $L$ and $g$ are 4.05 ft and 32.2 ft/sec².

36. **Error in Calculating the Power of a Battery** Suppose that the source of current in an electric circuit is a battery. Then the power output $P$ (in watts) obtained if the circuit has a resistance of $R$ ohms is given by

$$P = \frac{E^2 R}{(R + r)^2}$$

where $E$ is the electromotive force (EMF) in volts and $r$ is the internal resistance of the battery. Estimate the maximum percentage error in calculating the power if an EMF of 12 volts is applied in a circuit with a resistance of 100 ohms, the internal resistance of the battery is 5 ohms, and the possible maximum percentage errors in measuring $E$, $R$, and $r$ are 2%, 3%, and 1%, respectively.

37. **Error in Measuring the Resistance of a Circuit** The total resistance $R$ (in ohms) of three resistors with resistances of $R_1$, $R_2$, and $R_3$ ohms connected in parallel is given by

$$\frac{1}{R} = \frac{1}{R_1} + \frac{1}{R_2} + \frac{1}{R_3}$$

If $R_1$, $R_2$, and $R_3$ are measured as 20, 30, and 50 ohms, respectively, with a maximum error of 0.5 in each measurement, estimate the maximum error in the calculated value of $R$.

38. A container with a constant cross section of $A$ ft² is filled with water to a height of $h$ ft. The water is then allowed to flow out through an orifice of cross section $a$ in.² located at the base of the container. It can be shown that the time (in seconds) that it takes to empty the tank is given by

$$T = f(A, a, h) = \frac{A}{a}\sqrt{\frac{2h}{g}}$$

where $g$ is the constant of acceleration. Suppose that the measurements of $A$, $a$, and $h$ are 5 ft², 2 in.², and 16 ft with errors of 0.05 ft², $-0.04$ in.², and 0.2 ft, respectively. Find the error in computing $T$. (Take $g$ to be 32 ft/sec².)

**39. Suspension Bridge Cables** The supports of a cable of a suspension bridge are at the same level and at a distance of $L$ ft apart. The supports are $a$ feet higher than the lowest point of the cable (see the figure). If the weight of the cable is negligible and the bridge has a uniform weight of $W$ lb/ft, then the tension (in lb) in the cable at its lowest point is given by

$$H = \frac{WL^2}{8a}$$

If $W$, $L$, and $a$ are measured with possible maximum errors of 1%, 2%, and 2%, respectively, determine the maximum percentage error in calculating $H$.

**40. Flight of a Projectile** A projectile is fired with a muzzle velocity of $v$ ft/sec at an angle $\alpha$ radians above the horizontal. If the launch site is located at a height of $h$ ft above the target (see the figure below), then the time of the flight of the projectile in seconds is given by

$$T = \frac{v \sin \alpha + \sqrt{(v \sin \alpha)^2 + 2gh}}{g}$$

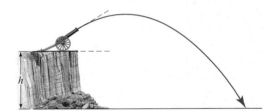

Suppose that the projectile is fired with an initial speed of 800 ft/sec at an angle of elevation of $\pi/4$ radians from a site

that is located 400 ft above the target. If the initial speed of the projectile, the angle of elevation of the cannon, and the height of the site above the target are measured with maximum possible percentage errors of 0.05%, 0.02%, and 0.5%, respectively, find the maximum error in computing the time of flight of the projectile. (Take $g$ to be 32 ft/sec².)

*In Exercises 41 and 42, show that the function is differentiable in the plane. (See Example 5.)*

**41.** $f(x, y) = x^2 - y^2$

**42.** $f(x, y) = 2xy - y^2$

**43.** Let $f$ be defined by

$$f(x, y) = \begin{cases} \dfrac{xy}{x^2 + y^2} & \text{if } (x, y) \neq 0 \\ 0 & \text{if } (x, y) = (0, 0) \end{cases}$$

Show that $f_x(0, 0)$ and $f_y(0, 0)$ both exist but that $f$ is not differentiable at $(0, 0)$.
**Hint:** Use the result of Theorem 3.

*In Exercises 44–47, determine whether the statement is true or false. If it is true, explain why it is true. If it is false, give an example to show why it is false.*

**44.** If $z = f(x, y)$ and $dz = 0$ for all $x$ and $y$ and for all differentials $dx$ and $dy$, then $f_x(x, y) = 0$ and $f_y(x, y) = 0$ for all $x$ and $y$.

**45.** If $f(x, y)$ is differentiable at $(a, b)$, then $f(a, b) = \lim_{(x, y) \to (a, b)} f(x, y)$.

**46.** If $F(x, y) = f(x) + g(y)$, where $f$ and $g$ are differentiable in the interval $(a, b)$, then $F$ is differentiable on $R = \{(x, y) \,|\, a < x < b, a < y < b\}$.

**47.** The function

$$f(x, y) = \begin{cases} x^2 + y^2 & \text{if } (x, y) \neq (0, 0) \\ 1 & \text{if } (x, y) = (0, 0) \end{cases}$$

is differentiable everywhere.

# 12.5 The Chain Rule

## The Chain Rule for Functions Involving One Independent Variable

In this section we extend the Chain Rule to functions of two or more variables. First, let's recall the Chain Rule for functions of one variable: If $y$ is a differentiable function of $x$ and $x$ is a differentiable function of $t$ (so that $y$ is a function of $t$), then

$$\frac{dy}{dt} = \frac{dy}{dx}\frac{dx}{dt}$$

This rule is easily recalled by using the diagram shown in Figure 1.

We begin by looking at the Chain Rule for the case in which a variable $w$ depends on two *intermediate* variables $x$ and $y$, which in turn depend on a third variable $t$ (so $w$ is a function of one independent variable $t$).

**FIGURE 1**
To find $dy/dt$, compute $dy/dx$ ($y$ depends on $x$), compute $dx/dt$ ($x$ depends on $t$), and then multiply the two quantities together.

**THEOREM 1**   **The Chain Rule for Functions Involving One Independent Variable**

Let $w = f(x, y)$, where $f$ is a differentiable function of $x$ and $y$. If $x = g(t)$ and $y = h(t)$, where $g$ and $h$ are differentiable functions of $t$, then $w$ is a differentiable function of $t$, and

$$\frac{dw}{dt} = \frac{\partial w}{\partial x}\frac{dx}{dt} + \frac{\partial w}{\partial y}\frac{dy}{dt}$$

**Note**   Observe that the derivative of $w$ with respect to $t$ is written with an ordinary d ($d$) rather than a curly d ($\partial$), since $w$ is a function of the single variable $t$.   ◼

**PROOF**   Let $t$ change from $t$ to $t + \Delta t$. This produces a change

$$\Delta x = g(t + \Delta t) - g(t)$$

in $x$ from $x$ to $x + \Delta x$ and a change

$$\Delta y = h(t + \Delta t) - h(t)$$

in $y$ from $y$ to $y + \Delta y$. Since $g$ and $h$ are differentiable, they are continuous at $t$, so both $\Delta x$ and $\Delta y$ approach zero as $\Delta t$ approaches zero.

Next, observe that the changes of $\Delta x$ in $x$ and $\Delta y$ in $y$ in turn produce a change $\Delta w$ in $w$ from $w$ to $w + \Delta w$. Since $f$ is differentiable, we have

$$\Delta w = \frac{\partial w}{\partial x}\Delta x + \frac{\partial w}{\partial y}\Delta y + \varepsilon_1 \Delta x + \varepsilon_2 \Delta y$$

where $\varepsilon_1 \to 0$ and $\varepsilon_2 \to 0$ as $(\Delta x, \Delta y) \to (0, 0)$. Dividing both sides of this equation by $\Delta t$, we have

$$\frac{\Delta w}{\Delta t} = \frac{\partial w}{\partial x}\frac{\Delta x}{\Delta t} + \frac{\partial w}{\partial y}\frac{\Delta y}{\Delta t} + \varepsilon_1\frac{\Delta x}{\Delta t} + \varepsilon_2\frac{\Delta y}{\Delta t}$$

Letting $\Delta t \to 0$, we have

$$\frac{dw}{dt} = \lim_{\Delta t \to 0}\frac{\Delta w}{\Delta t}$$

$$= \frac{\partial w}{\partial x}\lim_{\Delta t \to 0}\frac{\Delta x}{\Delta t} + \frac{\partial w}{\partial y}\lim_{\Delta t \to 0}\frac{\Delta y}{\Delta t} + \lim_{\Delta t \to 0}\varepsilon_1\lim_{\Delta t \to 0}\frac{\Delta x}{\Delta t} + \lim_{\Delta t \to 0}\varepsilon_2\lim_{\Delta t \to 0}\frac{\Delta y}{\Delta t}$$

$$= \frac{\partial w}{\partial x}\frac{dx}{dt} + \frac{\partial w}{\partial y}\frac{dy}{dt} + 0 \cdot \frac{dx}{dt} + 0 \cdot \frac{dy}{dt}$$

$$= \frac{\partial w}{\partial x}\frac{dx}{dt} + \frac{\partial w}{\partial y}\frac{dy}{dt}$$   ◼

The tree diagram in Figure 2 will help you recall this version of the Chain Rule. There are two "limbs" on this tree leading from $w$ to $t$. To find $dw/dt$, multiply the partial derivatives along each limb, and then add the products of these partial derivatives.

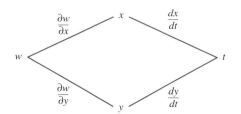

**FIGURE 2**
$w$ depends on $t$ via $x$ and $y$.

**EXAMPLE 1**  Let $w = x^2 y - xy^3$, where $x = \cos t$ and $y = e^t$. Find $dw/dt$ and its value when $t = 0$.

**Solution**  Observe that $w$ is a function of $x$ and $y$ and that both these variables are functions of $t$. Thus, we have the situation depicted in the schematic in Figure 2. Using the Chain Rule, we have

$$\frac{dw}{dt} = \frac{\partial w}{\partial x}\frac{dx}{dt} + \frac{\partial w}{\partial y}\frac{dy}{dt}$$

$$= (2xy - y^3)(-\sin t) + (x^2 - 3xy^2)e^t$$

$$= y(y^2 - 2x)\sin t + x(x - 3y^2)e^t$$

To find the value of $dw/dt$ when $t = 0$, we first observe that if $t = 0$, then $x = \cos 0 = 1$ and $y = e^0 = 1$. So

$$\left.\frac{dw}{dt}\right|_{t=0} = 0 + 1(1 - 3)e^0 = -2$$    ∎

The Chain Rule in Theorem 1 can be extended to the case involving a function of any finite number of intermediate variables. For example, if $w = f(x_1, x_2, \ldots, x_n)$, where $f$ is a differentiable function of $x_1, x_2, \ldots, x_n$ and $x_1 = f_1(t), x_2 = f_2(t), \ldots, x_n = f_n(t)$, where $f_1, f_2, \ldots, f_n$ are differentiable functions of $t$, then

$$\frac{dw}{dt} = \frac{\partial w}{\partial x_1}\frac{dx_1}{dt} + \frac{\partial w}{\partial x_2}\frac{dx_2}{dt} + \cdots + \frac{\partial w}{\partial x_n}\frac{dx_n}{dt}$$

This is easier to recall if you look at Figure 3, which shows the dependency of the variables involved: Multiply the derivatives along each limb leading from $w$ to $t$, and add the products of these derivatives.

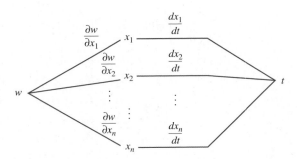

**FIGURE 3**
$w$ depends on $t$ via $x_1, x_2, \ldots, x_n$.

**EXAMPLE 2**  **Tracking a Missile Cruiser**    Figure 4 depicts an AWACS (Airborne Warning and Control System) aircraft tracking a missile cruiser. The flight path of the plane is described by the parametric equations

$$x = 20\cos 12t, \qquad y = 20\sin 12t, \qquad z = 3$$

and the course of the missile cruiser is given by

$$x = 30 + 20t, \qquad y = 40 + 10t^2, \qquad z = 0$$

where $0 \leq t \leq 1$, and $x$, $y$, and $z$ are measured in miles and $t$ in hours. How fast is the distance between the AWACS plane and the missile cruiser changing when $t = 0$?

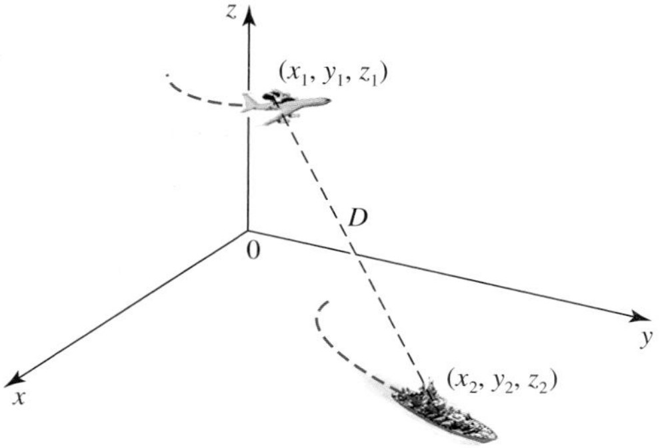

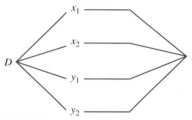

**FIGURE 4**
An AWACS aircraft tracking a missile cruiser.

**FIGURE 5**
$D$ depends on $t$ via the variables $x_1$, $x_2$, $y_1$, and $y_2$.

**Solution**   At time $t$ the position of the AWACS plane is given by the point $(x_1, y_1, z_1)$ and the position of the missile cruiser is given by the point $(x_2, y_2, z_2)$, so the distance $D$ between the plane and the cruiser is

$$D = \sqrt{(x_2 - x_1)^2 + (y_2 - y_1)^2 + (z_2 - z_1)^2}$$
$$= \sqrt{(x_2 - x_1)^2 + (y_2 - y_1)^2 + 9}$$

We want to compute $dD/dt$ when $t = 0$. To find $dD/dt$, we note that $D$ is a function of the four variables $x_1$, $x_2$, $y_1$, and $y_2$—all of which are functions of the single variable $t$ (Figure 5). By the Chain Rule we have

$$\frac{dD}{dt} = \frac{\partial D}{\partial x_1}\frac{dx_1}{dt} + \frac{\partial D}{\partial x_2}\frac{dx_2}{dt} + \frac{\partial D}{\partial y_1}\frac{dy_1}{dt} + \frac{\partial D}{\partial y_2}\frac{dy_2}{dt}$$

But

$$\frac{\partial D}{\partial x_1} = \frac{-(x_2 - x_1)}{\sqrt{(x_2 - x_1)^2 + (y_2 - y_1)^2 + 9}}, \qquad \frac{\partial D}{\partial x_2} = \frac{x_2 - x_1}{\sqrt{(x_2 - x_1)^2 + (y_2 - y_1)^2 + 9}}$$

$$\frac{\partial D}{\partial y_1} = \frac{-(y_2 - y_1)}{\sqrt{(x_2 - x_1)^2 + (y_2 - y_1)^2 + 9}}, \qquad \frac{\partial D}{\partial y_2} = \frac{y_2 - y_1}{\sqrt{(x_2 - x_1)^2 + (y_2 - y_1)^2 + 9}}$$

$$\frac{dx_1}{dt} = -240 \sin 12t, \qquad \frac{dx_2}{dt} = 20, \qquad \frac{dy_1}{dt} = 240 \cos 12t, \qquad \text{and} \qquad \frac{dy_2}{dt} = 20t$$

If $t = 0$, $x_1 = 20$, $y_1 = 0$, $x_2 = 30$, and $y_2 = 40$, then

$$\sqrt{(x_2 - x_1)^2 + (y_2 - y_1)^2 + 9} = \sqrt{(30 - 20)^2 + (40 - 0)^2 + 9} = \sqrt{1709}$$

Thus,

$$\frac{\partial D}{\partial x_1} = -\frac{10}{\sqrt{1709}} \approx -0.24, \qquad \frac{\partial D}{\partial x_2} = \frac{10}{\sqrt{1709}} \approx 0.24,$$

$$\frac{\partial D}{\partial y_1} = -\frac{40}{\sqrt{1709}} \approx -0.97, \qquad \frac{\partial D}{\partial y_2} = \frac{40}{\sqrt{1709}} \approx 0.97$$

and

$$\frac{dx_1}{dt} = 0, \qquad \frac{dx_2}{dt} = 20, \qquad \frac{dy_1}{dt} = 240, \qquad \frac{dy_2}{dt} = 0$$

Therefore, when $t = 0$,

$$\frac{dD}{dt} = (-0.24)(0) + (0.24)(20) + (-0.97)(240) + (0.97)(0)$$

$$= -228$$

that is, the distance between the AWACS aircraft and the missile cruiser is decreasing at the rate of 228 mph at that instant of time. ■

### ■ The Chain Rule for Functions Involving Two Independent Variables

We now look at the Chain Rule for the case in which a variable $w$ depends on two intermediate variables $x$ and $y$, each of which in turn depends on two variables $u$ and $v$ (so that $w$ is a function of two independent variables $u$ and $v$). More specifically, we have the following theorem.

---

**THEOREM 2    The Chain Rule for Functions Involving Two Independent Variables**

Let $w = f(x, y)$, where $f$ is a differentiable function of $x$ and $y$. Suppose that $x = g(u, v)$ and $y = h(u, v)$ and the partial derivatives $\partial g/\partial u$, $\partial g/\partial v$, $\partial h/\partial u$, and $\partial h/\partial v$ exist. Then

$$\frac{\partial w}{\partial u} = \frac{\partial w}{\partial x}\frac{\partial x}{\partial u} + \frac{\partial w}{\partial y}\frac{\partial y}{\partial u}$$

and

$$\frac{\partial w}{\partial v} = \frac{\partial w}{\partial x}\frac{\partial x}{\partial v} + \frac{\partial w}{\partial y}\frac{\partial y}{\partial v}$$

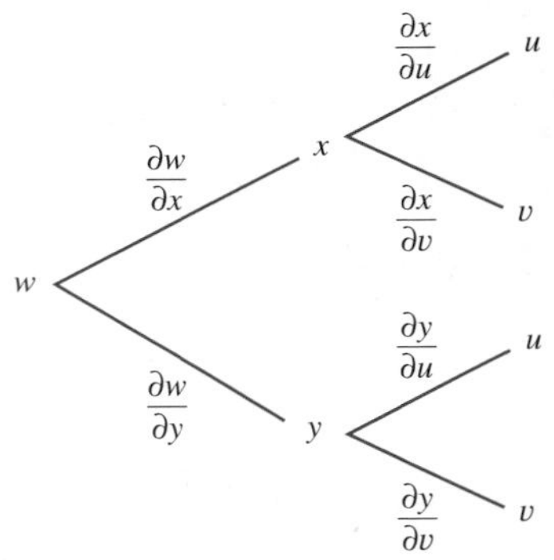

---

**FIGURE 6**
$w$ depends on $u$ and $v$ via $x$ and $y$.

**PROOF**    For $\partial w/\partial u$ we think of $v$ as a constant, so $g$ and $h$ are differentiable functions of $u$. Then the result follows from Theorem 1. The expression $\partial w/\partial v$ is derived in a similar manner. ■

The tree diagram shown in Figure 6 will help you to recall the Chain Rule given in Theorem 2.

To obtain $\partial w/\partial u$, observe that $w$ is connected to $u$ by two "limbs," one from $w$ to $u$ via $x$ and the other from $w$ to $u$ via $y$. Multiply the partial derivatives along each of these limbs, and add the product of these partial derivatives together to get $\partial w/\partial u$. The expression for $\partial w/\partial v$ is found in a similar manner.

**EXAMPLE 3**    Let $w = 2x^2y$, where $x = u^2 + v^2$ and $y = u^2 - v^2$. Find $\partial w/\partial u$ and $\partial w/\partial v$.

**Solution**    Observe that $w$ is a function of $x$ and $y$ and that both of these variables are functions of $u$ and $v$. Thus, we have the situation depicted in Figure 6. Using the Chain Rule (Theorem 2), we have

$$\frac{\partial w}{\partial u} = \frac{\partial w}{\partial x}\frac{\partial x}{\partial u} + \frac{\partial w}{\partial y}\frac{\partial y}{\partial u}$$

$$= 4xy(2u) + 2x^2(2u) = 4xu(2y + x)$$

and

$$\frac{\partial w}{\partial v} = \frac{\partial w}{\partial x}\frac{\partial x}{\partial v} + \frac{\partial w}{\partial y}\frac{\partial y}{\partial v}$$

$$= 4xy(2v) + 2x^2(-2v) = 4xv(2y - x) \qquad \blacksquare$$

## ■ The General Chain Rule

The Chain Rule in Theorem 2 can be extended to the case involving any finite number of intermediate variables and any finite number of independent variables. For example, if $w = f(x_1, x_2, \ldots, x_n)$, where $f$ is a differentiable function of $n$ intermediate variables, $x_1, x_2, \ldots, x_n$, and $x_1 = f_1(t_1, t_2, \ldots, t_m)$, $x_2 = f_2(t_1, t_2, \ldots, t_m)$, ..., $x_n = f_n(t_1, t_2, \ldots, t_m)$, where $f_1, f_2, \ldots, f_n$ are differentiable functions of $m$ variables, $t_1, t_2, \ldots, t_m$, then

$$\frac{\partial w}{\partial t_1} = \frac{\partial w}{\partial x_1}\frac{\partial x_1}{\partial t_1} + \frac{\partial w}{\partial x_2}\frac{\partial x_2}{\partial t_1} + \cdots + \frac{\partial w}{\partial x_n}\frac{\partial x_n}{\partial t_1}$$

$$\frac{\partial w}{\partial t_2} = \frac{\partial w}{\partial x_1}\frac{\partial x_1}{\partial t_2} + \frac{\partial w}{\partial x_2}\frac{\partial x_2}{\partial t_2} + \cdots + \frac{\partial w}{\partial x_n}\frac{\partial x_n}{\partial t_2}$$

$$\vdots$$

$$\frac{\partial w}{\partial t_m} = \frac{\partial w}{\partial x_1}\frac{\partial x_1}{\partial t_m} + \frac{\partial w}{\partial x_2}\frac{\partial x_2}{\partial t_m} + \cdots + \frac{\partial w}{\partial x_n}\frac{\partial x_n}{\partial t_m}$$

(See Figure 7.)

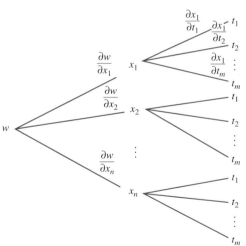

**FIGURE 7**
$w$ depends on $t_1, t_2, \ldots, t_m$ via $x_1, x_2, \ldots, x_n$.

**EXAMPLE 4** Let $w = x^2 y + y^2 z^3$, where $x = r\cos s$, $y = r\sin s$, and $z = re^s$. Find the value of $\partial w/\partial s$ when $r = 1$ and $s = 0$.

**Solution** Observe that $w$ is a function of $x$, $y$, and $z$, which in turn are functions of $r$ and $s$ (Figure 8).

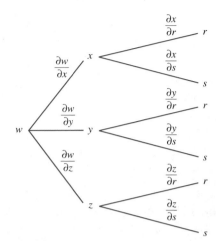

**FIGURE 8**
$w$ depends on $r$ and $s$ via $x$, $y$, and $z$.

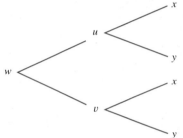

**FIGURE 9**
$w$ depends on $x$ and $y$ via the intermediate variables $u$ and $v$.

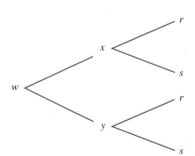

**FIGURE 10**
$w$ depends on $r$ and $s$ via the intermediate variables $x$ and $y$.

Multiplying the partial derivatives on the limbs that connect $w$ to $s$ on the tree diagram and adding the products of these derivatives, we obtain

$$\frac{\partial w}{\partial s} = \frac{\partial w}{\partial x}\frac{\partial x}{\partial s} + \frac{\partial w}{\partial y}\frac{\partial y}{\partial s} + \frac{\partial w}{\partial z}\frac{\partial z}{\partial s}$$

$$= 2xy(-r\sin s) + (x^2 + 2yz^3)(r\cos s) + 3y^2 z^2(re^s)$$

When $r = 1$ and $s = 0$, we have $x = 1$, $y = 0$, and $z = 1$, so

$$\frac{\partial w}{\partial s} = 2(1)(0)(0) + (1)(1) + 3(0)(1)(1) = 1$$ ∎

**EXAMPLE 5** If $w = f(x^2 - y^2, y^2 - x^2)$ and $f$ is differentiable, show that $w$ satisfies the equation

$$y\frac{\partial w}{\partial x} + x\frac{\partial w}{\partial y} = 0$$

**Solution** Introduce the intermediate variables $u = x^2 - y^2$ and $v = y^2 - x^2$. Then $w = g(x, y) = f(u, v)$ (Figure 9).
Using the Chain Rule, we have

$$\frac{\partial w}{\partial x} = \frac{\partial w}{\partial u}\frac{\partial u}{\partial x} + \frac{\partial w}{\partial v}\frac{\partial v}{\partial x} = \frac{\partial w}{\partial u}(2x) + \frac{\partial w}{\partial v}(-2x)$$

and

$$\frac{\partial w}{\partial y} = \frac{\partial w}{\partial u}\frac{\partial u}{\partial y} + \frac{\partial w}{\partial v}\frac{\partial v}{\partial y} = \frac{\partial w}{\partial u}(-2y) + \frac{\partial w}{\partial v}(2y)$$

Therefore,

$$y\frac{\partial w}{\partial x} + x\frac{\partial w}{\partial y} = \left(2xy\frac{\partial w}{\partial u} - 2xy\frac{\partial w}{\partial v}\right) + \left(-2xy\frac{\partial w}{\partial u} + 2xy\frac{\partial w}{\partial v}\right) = 0$$ ∎

**EXAMPLE 6** Let $w = f(x, y)$, where $f$ has continuous second-order partial derivatives, and let $x = r^2 + s^2$ and $y = 2rs$. Find $\partial^2 w/\partial r^2$.

**Solution** We begin by calculating $\partial w/\partial r$. Using the Chain Rule, we have

$$\frac{\partial w}{\partial r} = \frac{\partial w}{\partial x}\frac{\partial x}{\partial r} + \frac{\partial w}{\partial y}\frac{\partial y}{\partial r} = \frac{\partial w}{\partial x}(2r) + \frac{\partial w}{\partial y}(2s)$$

(See Figure 10.) Next, we apply the Product Rule to $\partial w/\partial r$ to obtain

$$\frac{\partial^2 w}{\partial r^2} = \frac{\partial}{\partial r}\left(2r\frac{\partial w}{\partial x} + 2s\frac{\partial w}{\partial y}\right)$$

$$= 2\frac{\partial w}{\partial x} + 2r\frac{\partial}{\partial r}\left(\frac{\partial w}{\partial x}\right) + 2s\frac{\partial}{\partial r}\left(\frac{\partial w}{\partial y}\right) \quad \textbf{(1)}$$

To compute the partial derivatives appearing in the last two terms of Equation (1), we observe that since $w$ is a function of $r$ and $s$ via the intermediate variables $x$ and $y$, the same is true of $\partial w/\partial x$ and $\partial w/\partial y$ (Figure 11).

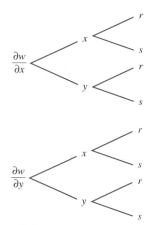

**FIGURE 11**
Both $\partial w/\partial x$ and $\partial w/\partial y$ depend on $r$ and $s$ via the intermediate variables $x$ and $y$.

Using the Chain Rule once again, we have

$$\frac{\partial}{\partial r}\left(\frac{\partial w}{\partial x}\right) = \frac{\partial}{\partial x}\left(\frac{\partial w}{\partial x}\right)\frac{\partial x}{\partial r} + \frac{\partial}{\partial y}\left(\frac{\partial w}{\partial x}\right)\frac{\partial y}{\partial r}$$

$$= \frac{\partial^2 w}{\partial x^2}(2r) + \frac{\partial^2 w}{\partial y\,\partial x}(2s)$$

and

$$\frac{\partial}{\partial r}\left(\frac{\partial w}{\partial y}\right) = \frac{\partial}{\partial x}\left(\frac{\partial w}{\partial y}\right)\frac{\partial x}{\partial r} + \frac{\partial}{\partial y}\left(\frac{\partial w}{\partial y}\right)\frac{\partial y}{\partial r}$$

$$= \frac{\partial^2 w}{\partial x\,\partial y}(2r) + \frac{\partial^2 w}{\partial y^2}(2s)$$

Substituting these expressions into Equation (1) and observing that $f_{xy} = f_{yx}$ because they are continuous, we have

$$\frac{\partial^2 w}{\partial r^2} = 2\frac{\partial w}{\partial x} + 2r\left(2r\frac{\partial^2 w}{\partial x^2} + 2s\frac{\partial^2 w}{\partial y\,\partial x}\right) + 2s\left(2r\frac{\partial^2 w}{\partial x\,\partial y} + 2s\frac{\partial^2 w}{\partial y^2}\right)$$

$$= 2\frac{\partial w}{\partial x} + 4r^2\frac{\partial^2 w}{\partial x^2} + 8rs\frac{\partial^2 w}{\partial x\,\partial y} + 4s^2\frac{\partial^2 w}{\partial y^2} \qquad \blacksquare$$

## Implicit Differentiation

The Chain Rule for a function of several variables can be used to find the derivative of a function implicitly. We will consider two situations.

First, suppose that the equation $F(x, y) = 0$, where $F$ is a differentiable function, defines a differentiable function $f$ of $x$ via the equation $y = f(x)$. If we differentiate both sides of $w = F(x, y) = 0$ with respect to $x$, we obtain

$$\frac{\partial w}{\partial x} = \frac{\partial F}{\partial x} + \frac{\partial F}{\partial y}\frac{dy}{dx} = 0$$

(see Figure 12) which implies that

$$\frac{dy}{dx} = -\frac{\dfrac{\partial F}{\partial x}}{\dfrac{\partial F}{\partial y}} = -\frac{F_x}{F_y} \quad \text{if } F_y \neq 0$$

Let's summarize this result.

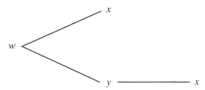

**FIGURE 12**
Tree diagram showing dependency of $w$ on $x$ directly and via $y$

---

**THEOREM 3 Implicit Differentiation: One Independent Variable**

Suppose that the equation $F(x, y) = 0$, where $F$ is differentiable, defines $y$ implicitly as a differentiable function of $x$. Then

$$\frac{dy}{dx} = -\frac{F_x(x, y)}{F_y(x, y)} \qquad \text{if } F_y(x, y) \neq 0 \tag{2}$$

**EXAMPLE 7** Find $\dfrac{dy}{dx}$ if $x^3 + xy + y^2 = 4$.

**Solution**    The given equation can be rewritten as

$$F(x, y) = x^3 + xy + y^2 - 4 = 0$$

Then Equation (2) immediately gives

$$\frac{dy}{dx} = -\frac{F_x}{F_y} = -\frac{3x^2 + y}{x + 2y}$$

∎

As a second application of the Chain Rule to implicit differentiation, suppose that the equation $F(x, y, z) = 0$, where $F$ is a differentiable function, defines a differentiable function $f$ of $x$ and $y$ via the equation $z = f(x, y)$. Differentiating both sides of $w = F(x, y, z) = 0$ with respect to $x$, we obtain

$$\frac{\partial w}{\partial x} = \frac{\partial F}{\partial x} + \frac{\partial F}{\partial z}\frac{\partial z}{\partial x} = F_x + F_z\frac{\partial z}{\partial x} = 0$$

(see Figure 13) which gives

$$\frac{\partial z}{\partial x} = -\frac{F_x}{F_z}$$

provided that $F_z \neq 0$.

Similarly, we see that

$$\frac{\partial z}{\partial y} = -\frac{F_y}{F_z} \quad \text{if } F_z \neq 0$$

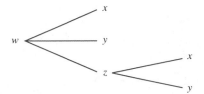

**FIGURE 13**
$w$ depends on $x$ and $y$ directly and via $z$.

---

**THEOREM 4    Implicit Differentiation: Two Independent Variables**

Suppose the equation $F(x, y, z) = 0$, where $F$ is differentiable, defines $z$ implicitly as a differentiable function of $x$ and $y$. Then

$$\frac{\partial z}{\partial x} = -\frac{F_x(x, y, z)}{F_z(x, y, z)} \quad \text{and} \quad \frac{\partial z}{\partial y} = -\frac{F_y(x, y, z)}{F_z(x, y, z)} \quad \text{if } F_z(x, y, z) \neq 0 \ \textbf{(3)}$$

---

**EXAMPLE 8** Find $\dfrac{\partial z}{\partial x}$ and $\dfrac{\partial z}{\partial y}$ if $2x^2z - 3xy^2 + yz - 8 = 0$.

**Solution**    Here, $F(x, y, z) = 2x^2z - 3xy^2 + yz - 8 = 0$, and Equation (3) gives

$$\frac{\partial z}{\partial x} = -\frac{F_x(x, y, z)}{F_z(x, y, z)} = -\frac{4xz - 3y^2}{2x^2 + y} = \frac{3y^2 - 4xz}{2x^2 + y}$$

and

$$\frac{\partial z}{\partial y} = -\frac{F_y(x, y, z)}{F_z(x, y, z)} = -\frac{-6xy + z}{2x^2 + y} = \frac{6xy - z}{2x^2 + y}$$

∎

## 12.5 CONCEPT QUESTIONS

1. Suppose that $w = f(x, y)$, $x = g(t)$, and $y = h(t)$, where $f$, $g$, and $h$ are differentiable functions. Write an expression for $dw/dt$. Illustrate with a tree diagram.
2. Suppose that $w = f(x, y)$, $x = g(u, v)$, and $y = h(u, v)$, where $f$, $g$, and $h$ are differentiable functions. Write an expression for $\partial w/\partial v$. Illustrate with a tree diagram.
3. Suppose that $w = f(x_1, x_2, \ldots, x_n)$, $x_1 = f_1(t_1, t_2, \ldots, t_m)$, $x_2 = f_2(t_1, t_2, \ldots, t_m) \ldots$, $x_n = f_n(t_1, t_2, \ldots, t_m)$, where $f, f_1$,

$f_2, \ldots, f_n$ are differentiable functions. Write an expression for $\partial w/\partial t_i$, where $1 \le i \le m$. Illustrate with a tree diagram.

4. **a.** Suppose that $F(x, y) = 0$ defines $y$ implicitly as a function of $x$ and $F$ is differentiable. Write an expression for $dy/dx$. Illustrate with a tree diagram.
   **b.** Suppose that $F(x, y, z) = 0$ defines $z$ implicitly as a function of $x$ and $y$ and $F$ is differentiable. Write an expression for $\partial z/\partial x$. Illustrate with a tree diagram.

## 12.5 EXERCISES

In Exercises 1–8, use the Chain Rule to find $dw/dt$.

1. $w = x^2 - y^2$, $x = t^2 + 1$, $y = t^3 + t$
2. $w = \sqrt{x^2 + 2y^2}$, $x = \sqrt{t}$, $y = \sqrt{2t + 1}$
3. $w = r \cos s + s \sin r$, $r = e^{-2t}$, $s = t^3 - 2t$
4. $w = \ln(x + y^2)$, $x = \tan t$, $y = \sec t$
5. $w = 2x^3y^2z$, $x = t$, $y = \cos t$, $z = t \sin t$
6. $w = pe^{qr}$, $p = \sqrt{t}$, $q = \sin 2t$, $r = \dfrac{t}{t^2 + 1}$
7. $w = \tan^{-1} xz + \dfrac{z}{y}$, $x = t$, $y = t^2$, $z = \sinh t$
8. $w = x\sqrt{y^2 + z^2}$, $x = \dfrac{1}{t}$, $y = e^{-t} \cos t$, $z = e^{-t} \sin t$

In Exercises 9–14, use the Chain Rule to find $\partial w/\partial u$ and $\partial w/\partial v$.

9. $w = x^3 + y^3$, $x = u^2 + v^2$, $y = 2uv$
10. $w = \sin xy$, $x = (u + v)^3$, $y = \sqrt{v}$
11. $w = e^x \cos y$, $x = \ln(u^2 + v^2)$, $y = \sqrt{uv}$
12. $w = x \ln y + 2^y$, $x = \ln u$, $y = ue^v$
13. $w = x \tan^{-1} yz$, $x = \sqrt{u}$, $y = e^{-2v}$, $z = v \cos u$
14. $w = x \cosh y + y \sinh z$, $x = u^2 - v^2$, $y = \ln(u + 1)$, $z = \dfrac{u}{v}$

In Exercises 15–18, write the Chain Rule for finding the indicated derivative with the aid of a tree diagram.

15. $w = f(r, s, u, v)$, $r = g(t)$, $s = h(t)$, $u = p(t)$, $v = q(t)$; $\dfrac{dw}{dt}$
16. $w = f(x, y)$, $x = g(u, v, t)$, $y = h(u, v, t)$; $\dfrac{\partial w}{\partial v}$
17. $w = f(x, y, z)$, $x = g(r, s, t)$, $y = h(r, s, t)$, $z = p(r, s, t)$; $\dfrac{\partial w}{\partial t}$
18. $w = f(x, y)$, $x = g(u, v, r, s)$, $y = h(u, v, r, s)$; $\dfrac{\partial w}{\partial r}$

In Exercises 19–26, use the Chain Rule to find the indicated derivative.

19. $w = x^2 + xy + y^2 + z^3$, $x = 2t$, $y = e^t$, $z = \cos 2t$; $\dfrac{dw}{dt}$
20. $z = x\sqrt{y} + \sqrt{x}$, $x = 2s + t$, $y = s^2 - 7t$; $\dfrac{\partial z}{\partial t}$ if $s = 4$ and $t = 1$
21. $u = \dfrac{x}{x^2 + y^2}$, $x = \sec 2t$, $y = \tan t$; $\dfrac{du}{dt}\bigg|_{t=0}$
22. $w = \dfrac{u}{\sqrt{u^2 + v^2}}$, $u = x + 2y + 3z$, $v = x \cos \pi(y + z)$; $\dfrac{\partial w}{\partial x}$ and $\dfrac{\partial w}{\partial z}$ if $x = 0$, $y = 1$, and $z = 1$
23. $u = x \csc yz$, $x = rs$, $y = s^2t$, $z = \dfrac{s}{t^2}$; $\dfrac{\partial u}{\partial s}$ and $\dfrac{\partial u}{\partial t}$
24. $w = \cos(2x + 3y)$, $x = r^2st$, $y = s^2tu$; $\dfrac{\partial w}{\partial r}$ and $\dfrac{\partial w}{\partial u}$
25. $w = \dfrac{x^2y}{z^2}$, $x = re^{st}$, $y = se^{rt}$, $z = e^{rst}$; $\dfrac{\partial w}{\partial r}$ and $\dfrac{\partial w}{\partial t}$ if $r = 1$, $s = 2$, and $t = 0$
26. $w = \dfrac{x + y}{x + z}$, $x = r \cos s$, $y = r \sin s$, $z = s \tan t$; $\dfrac{\partial w}{\partial r}$ and $\dfrac{\partial w}{\partial t}$
27. Given the system
$$\begin{cases} x = u^2 + v^2 \\ y = u^2 - v^2 \end{cases}$$
find $\partial u/\partial x$, $\partial u/\partial y$, $\partial v/\partial x$ and $\partial v/\partial y$.
28. Given the system
$$\begin{cases} x = \dfrac{1}{2}(u^2 - v^2) \\ y = uv \end{cases}$$
find $\partial u/\partial x$, $\partial u/\partial y$, $\partial v/\partial x$ and $\partial v/\partial y$.

*In Exercises 29–32, use Equation (2) to find dy/dx.*

**29.** $x^3 - 2xy + y^3 = 4$    **30.** $x^4 + 2x^2y^2 - 3xy - x = 5$

**31.** $2x^2 + 3\sqrt{xy} - 2y = 4$    **32.** $x \sec y + y \cos x = 1$

*In Exercises 33–36, use Equation (3) to find ∂z/∂x and ∂z/∂y.*

**33.** $x^2 + xy - x^2z + yz^2 = 0$

**34.** $x^2 + y^2 + z^2 - xy - yz - xz = 1$

**35.** $xe^y + ye^{xz} + x^2e^{x/y} = 10$

**36.** $\ln(x^2 + y^2) + x \ln z - \cos(xyz) = 0$

**37.** Find $dy/dx$ if $x^3 + y^3 - 3axy = 0$, $a > 0$.

**38.** The radius of a right circular cylinder is increasing at the rate of 0.1 cm/sec while its height is decreasing at the rate of 0.2 cm/sec. Find the rate at which the volume of the cylinder is changing when its radius is 60 cm and its height is 130 cm.

**39.** The radius of a right circular cone is decreasing at the rate of 0.2 in./min while its height is increasing at the rate of 0.1 in./min. Find the rate at which the area of its lateral surface is changing when its radius is 10 in. and its height is 18 in.

**40.** The pressure $P$ (in pascals), the volume $V$ (in liters), and the temperature $T$ (in kelvins) of 1 mole of an ideal gas are related by the equation $PV = 8.314T$. Find the rate at which the pressure of the gas is changing when its volume is 20 L and is increasing at the rate of 0.2 L/sec and its temperature is 300 K and is increasing at the rate of 0.3 K/sec.

**41.** Car $A$ is approaching an intersection from the north, and car $B$ is approaching the same intersection from the east. At a certain instant of time car $A$ is 0.4 mile from the intersection and approaching it at 45 mph, while car $B$ is 0.3 mile from the intersection and approaching it at a speed of 30 mph. How fast is the distance between the two cars changing?

**42.** The position of boat $A$ at time $t$ is given by the parametric equations

$$x_1 = -5t, \qquad y_1 = 5t$$

and the position of boat $B$ at time $t$ is given by

$$x_2 = 5t, \qquad y_2 = 2t + t^2$$

where $0 < t < 15$, and $x_1, y_1, x_2, y_2$ are measured in feet and $t$ is measured in seconds. How fast is the distance between the two boats changing when $t = 10$?

**43.** The total resistance $R$ (in ohms) of $n$ resistors with resistances $R_1, R_2, \ldots, R_n$ ohms connected in parallel is given by the formula

$$\frac{1}{R} = \frac{1}{R_1} + \frac{1}{R_2} + \cdots + \frac{1}{R_n}$$

Show that

$$\frac{\partial R}{\partial R_k} = \left(\frac{R}{R_k}\right)^2$$

**44. The Doppler Effect** Suppose a sound with frequency $f$ is emitted by an object moving along a straight line with speed $u$ and a listener is traveling along the same line in the opposite direction with speed $v$. Then the frequency $F$ heard by the listener is given by

$$F = \left(\frac{c - v}{c + u}\right)f$$

where $c$ is the speed of sound in still air—about 1100 ft/sec. (This phenomenon is called the **Doppler effect.**) Suppose that a railroad train is traveling at 100 ft/sec in still air and accelerating at the rate of 3 ft/sec$^2$ and that a note emitted by the locomotive whistle is 500 Hz. If a passenger is on a train that is moving at 50 ft/sec in the direction opposite to that of the first train and accelerating at the rate of 5 ft/sec$^2$, how fast is the frequency of the note he hears changing?

**45. Rate of Change in Temperature** The temperature at a point $(x, y, z)$ is given by

$$T(x, y, z) = \frac{60}{1 + x^2 + y^2 + z^2}$$

where $T$ is measured in degrees Fahrenheit and $x$, $y$, and $z$ are measured in feet. Suppose the position of a flying insect is

$$\mathbf{r}(t) = 2t\mathbf{i} + t^2\mathbf{j} + t^3\mathbf{k} \qquad 0 \le t \le 5$$

where $t$ is measured in seconds and the distance is measured in feet. Find the rate of change in temperature that the insect experiences at $t = 2$.

**46.** If $z = f(x, y)$, where $x = r \cos \theta$ and $y = r \sin \theta$, show that

$$\left(\frac{\partial z}{\partial x}\right)^2 + \left(\frac{\partial z}{\partial y}\right)^2 = \left(\frac{\partial z}{\partial r}\right)^2 + \frac{1}{r^2}\left(\frac{\partial z}{\partial \theta}\right)^2$$

**47.** If $u = f(x, y)$, where $x = e^r \cos \theta$ and $y = e^r \sin \theta$, show that

$$\left(\frac{\partial u}{\partial x}\right)^2 + \left(\frac{\partial u}{\partial y}\right)^2 = e^{-2r}\left[\left(\frac{\partial u}{\partial r}\right)^2 + \left(\frac{\partial u}{\partial \theta}\right)^2\right]$$

**48.** If $u = f(x, y)$, where $x = e^r \cos \theta$ and $y = e^r \sin \theta$, show that

$$\frac{\partial^2 u}{\partial x^2} + \frac{\partial^2 u}{\partial y^2} = e^{-2r}\left[\frac{\partial^2 u}{\partial r^2} + \frac{\partial^2 u}{\partial \theta^2}\right]$$

**49.** If $z = f(x, y)$, where $x = u - v$ and $y = v - u$, show that

$$\frac{\partial z}{\partial u} + \frac{\partial z}{\partial v} = 0$$

**50.** If $z = f(x, y)$, where $x = u + v$ and $y = u - v$, show that

$$\left(\frac{\partial z}{\partial x}\right)^2 - \left(\frac{\partial z}{\partial y}\right)^2 = \frac{\partial z}{\partial u}\frac{\partial z}{\partial v}$$

**51.** If $z = f(x + at) + g(x - at)$, show that $z$ satisfies the wave equation

$$\frac{\partial^2 z}{\partial t^2} = a^2 \frac{\partial^2 z}{\partial x^2}$$

*Hint:* Let $u = x + at$ and $v = x - at$.

**52.** If $z = f(x^2 + y^2)$, show that

$$y\left(\frac{\partial z}{\partial x}\right) - x\left(\frac{\partial z}{\partial y}\right) = 0$$

*Hint:* Let $u = x^2 + y^2$.

**53.** If $z = f(u, v)$, where $u = g(x, y)$ and $v = h(x, y)$, show that

$$\frac{\partial^2 z}{\partial x^2} = \frac{\partial^2 z}{\partial u^2}\left(\frac{\partial u}{\partial x}\right)^2 + \left(\frac{\partial^2 z}{\partial v \, \partial u} + \frac{\partial^2 z}{\partial u \, \partial v}\right)\frac{\partial u}{\partial x}\frac{\partial v}{\partial x}$$
$$+ \frac{\partial^2 z}{\partial v^2}\left(\frac{\partial v}{\partial x}\right)^2 + \frac{\partial z}{\partial u}\frac{\partial^2 u}{\partial x^2} + \frac{\partial z}{\partial v}\frac{\partial^2 v}{\partial x^2}$$

Assume that all second-order partial derivatives are continuous.

**54.** If $z = f(u, v)$, where $u = g(x, y)$ and $v = h(x, y)$, show that

$$\frac{\partial^2 z}{\partial y \, \partial x} = \frac{\partial^2 z}{\partial u^2}\frac{\partial u}{\partial x}\frac{\partial u}{\partial y} + \frac{\partial^2 z}{\partial v \, \partial u}\frac{\partial u}{\partial x}\frac{\partial v}{\partial y} + \frac{\partial^2 z}{\partial u \, \partial v}\frac{\partial u}{\partial y}\frac{\partial v}{\partial x}$$
$$+ \frac{\partial^2 z}{\partial v^2}\frac{\partial v}{\partial x}\frac{\partial v}{\partial y} + \frac{\partial z}{\partial u}\frac{\partial^2 u}{\partial y \, \partial x} + \frac{\partial z}{\partial v}\frac{\partial^2 v}{\partial y \, \partial x}$$

Assume that all second-order partial derivatives are continuous.

**55.** A function $f$ is *homogeneous of degree n* if $f(tx, ty) = t^n f(x, y)$ for every $t$, where $n$ is an integer. Show that if $f$ is homogeneous of degree $n$, then

$$x\frac{\partial f}{\partial x} + y\frac{\partial f}{\partial y} = nf$$

*Hint:* Differentiate both sides of the given equation with respect to $t$.

*In Exercises 56–59, find the degree of homogeneity of f and show that f satisfies the equation*

$$x\frac{\partial f}{\partial x} + y\frac{\partial f}{\partial y} = nf \qquad \text{See Exercise 55.}$$

**56.** $f(x, y) = 2x^3 + 4x^2y + y^3$

**57.** $f(x, y) = \dfrac{xy^2}{\sqrt{x^2 + y^2}}$     **58.** $f(x, y) = \tan^{-1}\left(\dfrac{y}{x}\right)$

**59.** $f(x, y) = e^{x/y}$

**60.** Suppose that the functions $u = f(x, y)$ and $v = g(x, y)$ satisfy the *Cauchy-Riemann equations*

$$\frac{\partial u}{\partial x} = \frac{\partial v}{\partial y} \quad \text{and} \quad \frac{\partial u}{\partial y} = -\frac{\partial v}{\partial x}$$

If $x = r\cos\theta$ and $y = r\sin\theta$, show that $u$ and $v$ satisfy

$$\frac{\partial u}{\partial r} = \frac{1}{r}\frac{\partial v}{\partial \theta} \quad \text{and} \quad \frac{\partial v}{\partial r} = -\frac{1}{r}\frac{\partial u}{\partial \theta}$$

the polar coordinate form of the Cauchy-Riemann equations.

**61.** Show that the functions $u = \ln\sqrt{x^2 + y^2}$ and $v = \tan^{-1}\left(\dfrac{y}{x}\right)$ satisfy the Cauchy-Riemann equations (see Exercise 60).

**62. a.** Let $P(a, b)$ be a point on the curve defined by the equation $f(x, y) = 0$. Show that if the curve has a tangent line at $P(a, b)$, then an equation of the tangent line can be written in the form

$$f_x(a, b)(x - a) + f_y(a, b)(y - b) = 0$$

**b.** Find an equation of the tangent line to the ellipse

$$\frac{x^2}{4} + \frac{y^2}{9} = 1$$

at the point $\left(1, \frac{3\sqrt{3}}{2}\right)$.

**63. a.** Use implicit differentiation to find an expression for $d^2y/dx^2$ given the implicit equation $f(x, y) = 0$. (Assume that $f$ has continuous second partial derivatives.)

**b.** Use the result of part (a) to find $d^2y/dx^2$ if $x^3 + y^3 - 3xy = 0$. What is its domain?

**64. Course Taken by a Yacht** The following figure depicts a bird's-eye view of the course taken by a yacht during an outing. The pier is located at the origin and the course is described by the equation

$$x^3 + y^3 - 9xy = 0 \qquad x \geq 0, y \geq 0$$

where $x$ and $y$ are measured in miles. When the yacht was at the point $(2, 4)$, it was sailing in an easterly direction at the rate of 16 mph. How fast was it moving in the northerly direction at that instant of time?

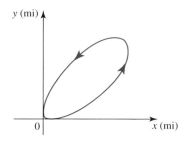

*In Exercises 65–66, determine whether the statement is true or false. If it is true, explain why it is true. If it is false, give an example to show why it is false.*

**65.** If $F(x, y) = 0$, where $F$ is differentiable, then

$$\frac{dx}{dy} = -\frac{F_y(x, y)}{F_x(x, y)}$$

provided that $F_x(x, y) \neq 0$.

**66.** If $z = \cos xy$ for $x > 0$ and $y > 0$ and $xy \neq n\pi$, where $n$ is an integer, then

$$\frac{\partial z}{\partial x} = \frac{1}{\partial x/\partial z}$$

provided that $\partial x/\partial z \neq 0$.

## 12.6 Directional Derivatives and Gradient Vectors

To study the heat conduction properties of a certain material, heat is applied to one corner of a thin rectangular sheet of that material. Suppose that the heated corner of the sheet is located at the origin of the $xy$-coordinate plane, as shown in Figure 1 and that the temperature at any point $(x, y)$ on the sheet is given by $T = f(x, y)$.

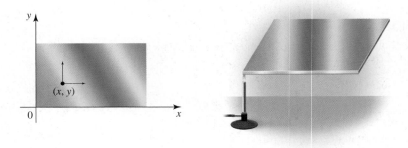

**FIGURE 1**
The temperature at the point
$(x, y)$ is $T = f(x, y)$.

From our previous work we can find the rate at which the temperature is changing at the point $(x, y)$ in the $x$-direction by computing $\partial f / \partial x$. Similarly, $\partial f / \partial y$ gives the rate of change of $T$ in the $y$-direction. But how fast does the temperature change if we move in a direction other than those just mentioned?

In this section we will attempt to answer questions of this nature. More generally, we will be interested in the problem of finding the rate of change of a function $f$ in a specified direction.

### ■ The Directional Derivative

Let's look at the problem from an intuitive point of view. Suppose that $f$ is a function defined by the equation $z = f(x, y)$, and let $P(a, b)$ be a point in the domain $D$ of $f$. Furthermore, let **u** be a unit (position) vector having a specified direction. Then the vertical plane containing the line $L$ passing through $P(a, b)$ and having the same direction as **u** will intersect the surface $z = f(x, y)$ along a curve $C$ (Figure 2). Intuitively, we see that the rate of change of $z$ at the point $P(a, b)$ with respect to the distance measured along $L$ is given by the slope of the tangent line $T$ to the curve $C$ at the point $P'(a, b, f(a, b))$.

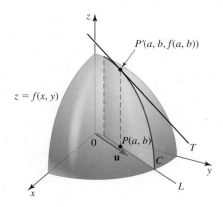

**FIGURE 2**
The rate of change of $z$ at $P(a, b)$
with respect to the distance measured
along $L$ is given by the slope of $T$.

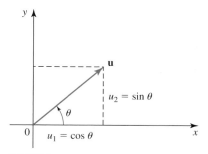

**FIGURE 3**
Any direction in the plane can be specified in terms of a unit vector **u**.

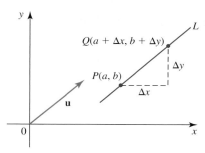

**FIGURE 4**
The point $Q(a + \Delta x, b + \Delta y)$ lies on $L$ and is distinct from $P(a, b)$.

**FIGURE 5**
The secant line $S$ passes through the points $P'$ and $Q'$ on the curve $C$.

Let's find the slope of $T$. First, observe that **u** may be specified by writing $\mathbf{u} = u_1\mathbf{i} + u_2\mathbf{j}$ for appropriate components $u_1$ and $u_2$. Equivalently, we may specify **u** by giving the angle $\theta$ that it makes with the positive $x$-axis, in which case $u_1 = \cos\theta$ and $u_2 = \sin\theta$ (Figure 3).

Next, let $Q(a + \Delta x, b + \Delta y)$ be any point distinct from $P(a, b)$ lying on the line $L$ passing through $P$ and having the same direction as **u** (Figure 4).

Since the vector $\overrightarrow{PQ}$ is parallel to **u**, it must be a scalar multiple of **u**. In other words, there exists a nonzero number $h$ such that

$$\overrightarrow{PQ} = h\mathbf{u} = hu_1\mathbf{i} + hu_2\mathbf{j}$$

But $\overrightarrow{PQ}$ is also given by $\Delta x\mathbf{i} + \Delta y\mathbf{j}$, and therefore,

$$\Delta x = hu_1, \qquad \Delta y = hu_2, \qquad \text{and} \qquad h = \sqrt{(\Delta x)^2 + (\Delta y)^2}$$

So the point $Q$ can be expressed as $Q(a + hu_1, b + hu_2)$. Therefore, the slope of the secant line $S$ passing through the points $P'$ and $Q'$ (see Figure 5) is given by

$$\frac{\Delta z}{h} = \frac{f(a + hu_1, b + hu_2) - f(a, b)}{h} \tag{1}$$

Observe that Equation (1) also gives the average rate of change of $z = f(x, y)$ from $P(a, b)$ to $Q(a + \Delta x, b + \Delta y) = Q(a + hu_1, b + hu_2)$ in the direction of **u**.

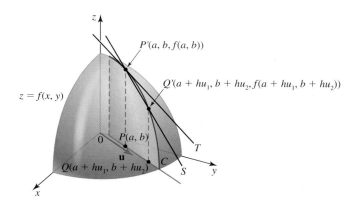

If we let $h$ approach zero in Equation (1), we see that the slope of the secant line $S$ approaches the slope of the tangent line at $P'$. Also, the average rate of change of $z$ approaches the (instantaneous) rate of change of $z$ at $(a, b)$ in the direction of **u**. This limit, whenever it exists, is called the *directional derivative of $f$* at $(a, b)$ in the direction of **u**. Since the point $P(a, b)$ is arbitrary, we can replace it by $P(x, y)$ and define the directional derivative of $f$ at any point as follows.

**DEFINITION   Directional Derivative**

Let $f$ be a function of $x$ and $y$ and let $\mathbf{u} = u_1\mathbf{i} + u_2\mathbf{j}$ be a unit vector. Then the directional derivative of $f$ at $(x, y)$ in the direction of **u** is

$$D_{\mathbf{u}}f(x, y) = \lim_{h \to 0} \frac{f(x + hu_1, y + hu_2) - f(x, y)}{h} \tag{2}$$

if this limit exists.

**Note**    If $\mathbf{u} = \mathbf{i}$ ($u_1 = 1$ and $u_2 = 0$), then Equation (2) gives

$$D_{\mathbf{i}}f(x, y) = \lim_{h \to 0} \frac{f(x + h, y) - f(x, y)}{h} = f_x(x, y)$$

That is, the directional derivative of $f$ in the $x$-direction is the partial derivative of $f$ in the $x$-direction, as expected. Similarly, you can show that $D_{\mathbf{j}}f(x, y) = f_y(x, y)$.    ■

The following theorem helps us to compute the directional derivatives of functions without appealing directly to the definition of the directional derivative. More specifically, it gives the directional derivative of $f$ in terms of its partial derivatives $f_x$ and $f_y$.

---

**THEOREM 1**    If $f$ is a differentiable function of $x$ and $y$, then $f$ has a directional derivative in the direction of any unit vector $\mathbf{u} = u_1\mathbf{i} + u_2\mathbf{j}$ and

$$D_{\mathbf{u}}f(x, y) = f_x(x, y)u_1 + f_y(x, y)u_2 \qquad (3)$$

---

**PROOF**    Fix the point $(a, b)$. Then the function $g$ defined by

$$g(h) = f(a + hu_1, b + hu_2)$$

is a function of the single variable $h$. By the definition of the derivative,

$$g'(0) = \lim_{h \to 0} \frac{g(h) - g(0)}{h} = \lim_{h \to 0} \frac{f(a + hu_1, b + hu_2) - f(a, b)}{h}$$

$$= D_{\mathbf{u}}f(a, b)$$

Next, observe that $g$ may be written as $g(h) = f(x, y)$ where $x = a + hu_1$ and $y = b + hu_2$. Therefore, by the Chain Rule we have

$$g'(h) = \frac{\partial f}{\partial x}\frac{dx}{dh} + \frac{\partial f}{\partial y}\frac{dy}{dh} = f_x(x, y)u_1 + f_y(x, y)u_2$$

In particular, when $h = 0$, we have $x = a$, $y = b$, so

$$g'(0) = f_x(a, b)u_1 + f_y(a, b)u_2$$

Comparing this expression for $g'(0)$ with the one obtained earlier, we conclude that

$$D_{\mathbf{u}}f(a, b) = f_x(a, b)u_1 + f_y(a, b)u_2$$

Finally, since $(a, b)$ is arbitrary, we may replace it by $(x, y)$ and the result follows.

■

**EXAMPLE 1**    Find the directional derivative of $f(x, y) = 4 - 2x^2 - y^2$ at the point $(1, 1)$ in the direction of the unit vector $\mathbf{u}$ that makes an angle of $\pi/3$ radians with the positive $x$-axis.

**Solution**    Here

$$\mathbf{u} = \cos\left(\frac{\pi}{3}\right)\mathbf{i} + \sin\left(\frac{\pi}{3}\right)\mathbf{j} = \frac{1}{2}\mathbf{i} + \frac{\sqrt{3}}{2}\mathbf{j}$$

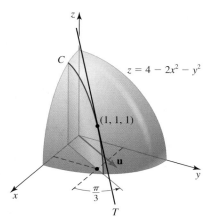

**FIGURE 6**
The slope of the tangent line to the curve $C$ at $(1, 1, 1)$ is $\approx -3.732$.

so $u_1 = \frac{1}{2}$ and $u_2 = \frac{\sqrt{3}}{2}$. Using Equation (3), we find that

$$D_{\mathbf{u}}f(x, y) = f_x(x, y)u_1 + f_y(x, y)u_2$$

$$= (-4x)\left(\frac{1}{2}\right) + (-2y)\left(\frac{\sqrt{3}}{2}\right) = -(2x + \sqrt{3}y)$$

In particular,

$$D_{\mathbf{u}}f(1, 1) = -(2 + \sqrt{3}) \approx -3.732$$

(See Figure 6.) ∎

**EXAMPLE 2** Find the directional derivative of $f(x, y) = e^x \cos 2y$ at the point $\left(0, \frac{\pi}{4}\right)$ in the direction of $\mathbf{v} = 2\mathbf{i} + 3\mathbf{j}$.

**Solution** The unit vector $\mathbf{u}$ that has the same direction as $\mathbf{v}$ is

$$\mathbf{u} = \frac{\mathbf{v}}{|\mathbf{v}|} = \frac{2}{\sqrt{13}}\mathbf{i} + \frac{3}{\sqrt{13}}\mathbf{j}$$

Using Equation (3) with $u_1 = 2/\sqrt{13}$ and $u_2 = 3/\sqrt{13}$, we have

$$D_{\mathbf{u}}f(x, y) = f_x(x, y)u_1 + f_y(x, y)u_2$$

$$= (e^x \cos 2y)\left(\frac{2}{\sqrt{13}}\right) + (-2e^x \sin 2y)\left(\frac{3}{\sqrt{13}}\right)$$

In particular,

$$D_{\mathbf{u}}f\left(0, \frac{\pi}{4}\right) = \left(e^0 \cos \frac{\pi}{2}\right)\left(\frac{2}{\sqrt{13}}\right) - 2(e^0)\left(\sin \frac{\pi}{2}\right)\left(\frac{3}{\sqrt{13}}\right) = -\frac{6}{\sqrt{13}} = -\frac{6\sqrt{13}}{13}$$

∎

## ▪ The Gradient of a Function of Two Variables

The directional derivative $D_{\mathbf{u}}f(x, y)$ can be written as the dot product of the unit vector

$$\mathbf{u} = u_1\mathbf{i} + u_2\mathbf{j}$$

and the vector

$$f_x(x, y)\mathbf{i} + f_y(x, y)\mathbf{j}$$

Thus,

$$D_{\mathbf{u}}f(x, y) = (u_1\mathbf{i} + u_2\mathbf{j}) \cdot [f_x(x, y)\mathbf{i} + f_y(x, y)\mathbf{j}] = f_x(x, y)u_1 + f_y(x, y)u_2$$

The vector $f_x(x, y)\mathbf{i} + f_y(x, y)\mathbf{j}$ plays an important role in many other computations and is given a special name.

---

**DEFINITION** Gradient of a Function of Two Variables

Let $f$ be a function of two variables $x$ and $y$. The **gradient** of $f$ is the vector function

$$\nabla f(x, y) = f_x(x, y)\mathbf{i} + f_y(x, y)\mathbf{j}$$

**Notes**

1. $\nabla f$ is read "del f."
2. $\nabla f(x, y)$ is sometimes written **grad** $f(x, y)$.

**EXAMPLE 3**  Find the gradient of $f(x, y) = x \sin y + y \ln x$ at the point $(e, \pi)$.

**Solution**  Since

$$f_x(x, y) = \sin y + \frac{y}{x} \text{ and } f_y(x, y) = x \cos y + \ln x$$

we have

$$\nabla f(x, y) = f_x(x, y)\mathbf{i} + f_y(x, y)\mathbf{j}$$

$$= \left( \sin y + \frac{y}{x} \right)\mathbf{i} + (x \cos y + \ln x)\mathbf{j}$$

So the gradient of $f$ at $(e, \pi)$ is

$$\nabla f(e, \pi) = \left( \sin \pi + \frac{\pi}{e} \right)\mathbf{i} + (e \cos \pi + \ln e)\mathbf{j}$$

$$= \frac{\pi}{e}\mathbf{i} + (1 - e)\mathbf{j}$$

Theorem 1 can be rewritten in terms of the gradient of $f$ as follows.

---

**THEOREM 2**

If $f$ is a differentiable function of $x$ and $y$, then $f$ has a directional derivative in the direction of any unit vector $\mathbf{u}$, and

$$D_{\mathbf{u}}f(x, y) = \nabla f(x, y) \cdot \mathbf{u} \qquad (4)$$

---

To give a geometric interpretation of Equation (4), suppose that $(a, b)$ is a fixed point in the $xy$-plane. Then

$$D_{\mathbf{u}}f(a, b) = \nabla f(a, b) \cdot \mathbf{u} = \frac{\nabla f(a, b) \cdot \mathbf{u}}{|\mathbf{u}|} \qquad \text{since } |\mathbf{u}| = 1$$

so by Equation (6) of Section 10.3 we see that $D_{\mathbf{u}}f(a, b)$ can be viewed as the scalar component of $\nabla f(a, b)$ along $\mathbf{u}$ (Figure 7).

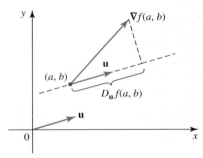

**FIGURE 7**
The directional derivative of $f$ at $(a, b)$ in the direction of $\mathbf{u}$ is the scalar component of the gradient of $f$ at $(a, b)$ along $\mathbf{u}$.

**EXAMPLE 4**  Let $f(x, y) = x^2 - 2xy$.

**a.** Find the gradient of $f$ at the point $(1, -2)$.
**b.** Use the result of (a) to find the directional derivative of $f$ at $(1, -2)$ in the direction from $P(-1, 2)$ to $Q(2, 3)$.

**Solution**
**a.** The gradient of $f$ at any point $(x, y)$ is

$$\nabla f(x, y) = (2x - 2y)\mathbf{i} - 2x\mathbf{j}$$

**b.** The gradient of $f$ at the point $(1, -2)$ is

$$\nabla f(1, -2) = (2 + 4)\mathbf{i} - 2\mathbf{j} = 6\mathbf{i} - 2\mathbf{j}$$

The desired direction is given by the direction of the vector $\overrightarrow{PQ} = 3\mathbf{i} + \mathbf{j}$. A unit vector that has the same direction as $\overrightarrow{PQ}$ is

$$\mathbf{u} = \frac{3}{\sqrt{10}}\mathbf{i} + \frac{1}{\sqrt{10}}\mathbf{j}$$

Using Equation (4), we obtain

$$D_{\mathbf{u}}f(1, -2) = \nabla f(1, -2) \cdot \mathbf{u}$$

$$= (6\mathbf{i} - 2\mathbf{j}) \cdot \left(\frac{3}{\sqrt{10}}\mathbf{i} + \frac{1}{\sqrt{10}}\mathbf{j}\right)$$

$$= \frac{18}{\sqrt{10}} - \frac{2}{\sqrt{10}} = \frac{16}{\sqrt{10}} \approx 5.1$$

or a change in $f$ of 5.1 per unit change in the direction of the vector $\mathbf{u}$. The gradient vector $\nabla f(1, -2)$, the unit vector $\mathbf{u}$, and the geometrical interpretation of $D_{\mathbf{u}}f(1, -2)$ are shown in Figure 8.

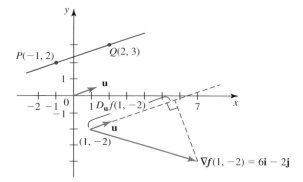

**FIGURE 8**
$D_{\mathbf{u}}f(1, -2)$ viewed as the scalar component of $\nabla f(1, -2)$ along $\mathbf{u}$.

## ■ Properties of the Gradient

The following theorem gives some important properties of the gradient of a function.

---

**THEOREM 3    Properties of the Gradient**

Suppose $f$ is differentiable at the point $(x, y)$.

1. If $\nabla f(x, y) = \mathbf{0}$, then $D_{\mathbf{u}}f(x, y) = 0$ for every $\mathbf{u}$.
2. The maximum value of $D_{\mathbf{u}}f(x, y)$ is $|\nabla f(x, y)|$, and this occurs when $\mathbf{u}$ has the same direction as $\nabla f(x, y)$.
3. The minimum value of $D_{\mathbf{u}}f(x, y)$ is $-|\nabla f(x, y)|$, and this occurs when $\mathbf{u}$ has the direction of $-\nabla f(x, y)$.

---

**PROOF**  Suppose $\nabla f(x, y) = \mathbf{0}$. Then for any $\mathbf{u} = u_1\mathbf{i} + u_2\mathbf{j}$, we have

$$D_{\mathbf{u}}f(x, y) = \nabla f(x, y) \cdot \mathbf{u} = (0\mathbf{i} + 0\mathbf{j}) \cdot (u_1\mathbf{i} + u_2\mathbf{j}) = 0$$

Next, if $\nabla f(x, y) \neq \mathbf{0}$, then

$$D_{\mathbf{u}}f(x, y) = \nabla f(x, y) \cdot \mathbf{u} = |\nabla f(x, y)|\,|\mathbf{u}|\cos\theta = |\nabla f(x, y)|\cos\theta$$

where $\theta$ is the angle between $\nabla f(x, y)$ and $\mathbf{u}$. Since the maximum value of $\cos\theta$ is 1 and this occurs when $\theta = 0$, we see that the maximum value of $D_{\mathbf{u}}f(x, y)$ is $|\nabla f(x, y)|$

and this occurs when both $\nabla f(x, y)$ and $\mathbf{u}$ have the same direction. Similarly, Property (3) is proved by observing that $\cos \theta$ has a minimum value of $-1$ when $\theta = \pi$. ■

**Notes**
1. Property (2) of Theorem 3 tells us that $f$ *increases* most rapidly in the direction of $\nabla f(x, y)$. This direction is called the direction of steepest ascent.
2. Property (3) of Theorem 3 says that $f$ *decreases* most rapidly in the direction of $-\nabla f(x, y)$. This direction is called the direction of steepest descent. ■

**EXAMPLE 5**   **Quickest Descent**   Suppose a hill is described mathematically by using the model $z = f(x, y) = 300 - 0.01x^2 - 0.005y^2$, where $x$, $y$, and $z$ are measured in feet. If you are at the point $(50, 100, 225)$ on the hill, in what direction should you aim your toboggan if you want to achieve the quickest descent? What is the maximum rate of decrease of the height of the hill at this point?

**Solution**   The gradient of the "height" function is

$$\nabla f(x, y) = f_x(x, y)\mathbf{i} + f_y(x, y)\mathbf{j} = -0.02x\mathbf{i} - 0.01y\mathbf{j}$$

Therefore, the direction of greatest *increase* in $z$ when you are at the point $(50, 100, 225)$ is given by the direction of

$$\nabla f(50, 100) = -\mathbf{i} - \mathbf{j}$$

So by pointing the toboggan in the direction of the vector

$$-\nabla f(50, 100) = -(-\mathbf{i} - \mathbf{j}) = \mathbf{i} + \mathbf{j}$$

you will achieve the quickest descent.
    The maximum rate of *decrease* of the height of the hill at the point $(50, 100, 225)$ is

$$|\nabla f(50, 100)| = |-\mathbf{i} - \mathbf{j}| = \sqrt{2}$$

or approximately $1.41$ ft/ft. The graph of $f$ and the direction of greatest descent are shown in Figure 9. ■

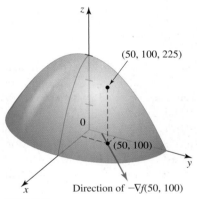
(50, 100, 225)

0

(50, 100)

$x$   Direction of $-\nabla f(50, 100)$

**FIGURE 9**
The direction of greatest descent is in the direction of $-\nabla f(50, 100)$.

**EXAMPLE 6**   **Path of a Heat-Seeking Object**   A heat-seeking object is located at the point $(2, 3)$ on a metal plate whose temperature at a point $(x, y)$ is $T(x, y) = 30 - 8x^2 - 2y^2$. Find the path of the object if it moves continuously in the direction of maximum increase in temperature at each point.

**Solution**   Let the path of the object be described by the position function

$$\mathbf{r}(t) = x(t)\mathbf{i} + y(t)\mathbf{j}$$

where

$$\mathbf{r}(0) = 2\mathbf{i} + 3\mathbf{j}$$

Since the object moves in the direction of maximum increase in temperature, its velocity vector at time $t$ has the same direction as the gradient of $T$ at time $t$. Therefore, there exists a scalar function of $t$, $k$, such that $\mathbf{v}(t) = k\nabla T(x, y)$. But

$$\mathbf{v}(t) = \mathbf{r}'(t) = \frac{dx}{dt}\mathbf{i} + \frac{dy}{dt}\mathbf{j}$$

and $\nabla T = -16x\mathbf{i} - 4y\mathbf{j}$. So we have

$$\frac{dx}{dt}\mathbf{i} + \frac{dy}{dt}\mathbf{j} = -16kx\mathbf{i} - 4ky\mathbf{j}$$

or, equivalently, the system

$$\frac{dx}{dt} = -16kx \qquad \frac{dy}{dt} = -4ky$$

Therefore

$$\frac{dy}{dx} = \frac{\dfrac{dy}{dt}}{\dfrac{dx}{dt}} = \frac{-4ky}{-16kx} \qquad \text{or} \qquad \frac{dy}{dx} = \frac{y}{4x}$$

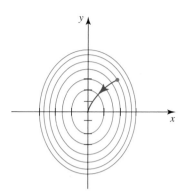

**FIGURE 10**
The path of the heat-seeking object

This is a first-order separable differential equation. The solution of this equation is $x = Cy^4$, where $C$ is a constant. (See Section 7.1.) Using the initial condition $y(2) = 3$, we have $2 = C(3^4)$, or $C = 2/(81)$. So

$$x = \frac{2y^4}{81}$$

The path of the heat-seeking object is shown in Figure 10. ∎

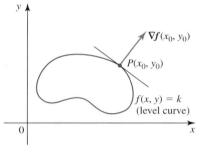

**FIGURE 11**
$\nabla f(x_0, y_0)$ is perpendicular to the level curve $f(x, y) = k$ at $P(x_0, y_0)$.

In Figure 10 observe that at each point where the path intersects a level curve that is part of the contour map of $T$, the gradient vector $\nabla T$ is perpendicular to the level curve at that point. To see why this makes sense, refer to Figure 11, where we show the level curve $f(x, y) = k$ of a function $f$ for some $k$ and a point $P(x_0, y_0)$ lying on the curve. If we move away from $P(x_0, y_0)$ along the level curve then the values of $f$ remain constant (at $k$). It seems reasonable to conjecture that by moving away in a direction that is perpendicular to the tangent line to the level curve at $P(x_0, y_0)$, $f$ will increase at the fastest rate. But this direction is given by the direction of $\nabla f(x_0, y_0)$. This will be demonstrated in Section 12.7.

## ■ Functions of Three Variables

The definitions of the directional derivative and the gradient of a function of three or more variables are similar to those for a function of two variables. Also, the algebraic results that are obtained for the case of a function of two variables carry over to the higher-dimensional case and are summarized in the following theorem.

**THEOREM 4** **Directional Derivative and Gradient of a Function of Three Variables**

Let $f$ be a differentiable function of $x$, $y$, and $z$, and let $\mathbf{u} = u_1\mathbf{i} + u_2\mathbf{j} + u_3\mathbf{k}$ be a unit vector. The directional derivative of $f$ in the direction of $\mathbf{u}$ is given by

$$D_{\mathbf{u}}f(x, y, z) = f_x(x, y, z)u_1 + f_y(x, y, z)u_2 + f_z(x, y, z)u_3$$

The **gradient of $f$** is

$$\nabla f(x, y, z) = f_x(x, y, z)\mathbf{i} + f_y(x, y, z)\mathbf{j} + f_z(x, y, z)\mathbf{k}$$

We also write

$$D_{\mathbf{u}}f(x, y, z) = \nabla f(x, y, z) \cdot \mathbf{u}$$

The properties of the gradient given in Theorem 3 for a function of two variables are also valid for a function of three or more variables. For example, the direction of greatest increase of $f$ coincides with that of the gradient of $f$ and has magnitude $|\nabla f(x, y, z)|$.

**EXAMPLE 7**    **Electric Potential**    Suppose a point charge $Q$ (in coulombs) is located at the origin of a three-dimensional coordinate system. This charge produces an electric potential $V$ (in volts) given by

$$V(x, y, z) = \frac{kQ}{\sqrt{x^2 + y^2 + z^2}}$$

where $k$ is a positive constant and $x$, $y$, and $z$ are measured in meters.

**a.** Find the rate of change of the potential at the point $P(1, 2, 3)$ in the direction of the vector $\mathbf{v} = 2\mathbf{i} + \mathbf{j} - 2\mathbf{k}$.

**b.** In which direction does the potential increase most rapidly at $P$, and what is the rate of increase?

**Solution**

**a.** We begin by computing the gradient of $V$. Since

$$V_x = \frac{\partial}{\partial x} [kQ(x^2 + y^2 + z^2)^{-1/2}] = kQ\left(-\tfrac{1}{2}\right)(x^2 + y^2 + z^2)^{-3/2}(2x)$$

$$= -\frac{kQx}{(x^2 + y^2 + z^2)^{3/2}}$$

and by symmetry

$$V_y = -\frac{kQy}{(x^2 + y^2 + z^2)^{3/2}} \quad \text{and} \quad V_z = -\frac{kQz}{(x^2 + y^2 + z^2)^{3/2}}$$

we obtain

$$\nabla V(x, y, z) = V_x\mathbf{i} + V_y\mathbf{j} + V_z\mathbf{k}$$

$$= -\frac{kQ}{(x^2 + y^2 + z^2)^{3/2}} (x\mathbf{i} + y\mathbf{j} + z\mathbf{k})$$

In particular,

$$\nabla V(1, 2, 3) = -\frac{kQ}{14^{3/2}} (\mathbf{i} + 2\mathbf{j} + 3\mathbf{k})$$

A unit vector $\mathbf{u}$ that has the same direction as $\mathbf{v} = 2\mathbf{i} + \mathbf{j} - 2\mathbf{k}$ is

$$\mathbf{u} = \tfrac{1}{3}(2\mathbf{i} + \mathbf{j} - 2\mathbf{k})$$

By Theorem 4 the rate of change of $V$ at $P(1, 2, 3)$ in the direction of $\mathbf{v}$ is

$$D_{\mathbf{u}}V(1, 2, 3) = \nabla V(1, 2, 3) \cdot \mathbf{u} = -\frac{kQ}{14^{3/2}} (\mathbf{i} + 2\mathbf{j} + 3\mathbf{k}) \cdot \frac{(2\mathbf{i} + \mathbf{j} - 2\mathbf{k})}{3}$$

$$= -\frac{kQ}{(3)(14)\sqrt{14}} (2 + 2 - 6) = \frac{kQ}{21\sqrt{14}} = \frac{\sqrt{14}kQ}{294}$$

In other words, the potential is increasing at the rate of $\sqrt{14}kQ/294$ volts/m.

**b.** The maximum rate of change of $V$ occurs in the direction of the gradient of $V$, that is, in the direction of the vector $-(\mathbf{i} + 2\mathbf{j} + 3\mathbf{k})$. Observe that this vector points toward the origin from $P(1, 2, 3)$. The maximum rate of change of $V$ at $P(1, 2, 3)$ is given by

$$|\nabla V(1, 2, 3)| = \left| -\frac{kQ}{14^{3/2}} (\mathbf{i} + 2\mathbf{j} + 3\mathbf{k}) \right|$$

$$= \frac{kQ}{14^{3/2}} \sqrt{1 + 4 + 9} = \frac{kQ}{14}$$

or $kQ/14$ volts/m.

## 12.6 CONCEPT QUESTIONS

**1. a.** Let $f$ be a function of $x$ and $y$, and let $\mathbf{u} = u_1\mathbf{i} + u_2\mathbf{j}$ be a unit vector. Define the directional derivative of $f$ in the direction of $\mathbf{u}$. Why is it necessary to use a unit vector to indicate the direction?

**b.** If $f$ is a differentiable function of $x$, $y$, and $z$ and $\mathbf{u} = u_1\mathbf{i} + u_2\mathbf{j} + u_3\mathbf{k}$ is a unit vector, express $D_{\mathbf{u}}f(x, y, z)$ in terms of the partial derivatives of $f$ and the components of $\mathbf{u}$.

**2. a.** What is the gradient of a function $f(x, y)$ of two variables $x$ and $y$?

**b.** What is the gradient of a function $f(x, y, z)$ of three variables $x$, $y$, and $z$?

**c.** If $f$ is a differentiable function of $x$ and $y$ and $\mathbf{u}$ is a unit vector, write $D_{\mathbf{u}}f(x, y)$ in terms of $f$ and $\mathbf{u}$.

**d.** If $f$ is a differentiable function of $x$, $y$, and $z$ and $\mathbf{u}$ is a unit vector, write $D_{\mathbf{u}}f(x, y, z)$ in terms of $f$ and $\mathbf{u}$.

**3. a.** If $f$ is a differentiable at $(x, y)$, what can you say about $D_{\mathbf{u}}f(x, y)$ if $\nabla f(x, y) = \mathbf{0}$?

**b.** What is the maximum (minimum) value of $D_{\mathbf{u}}f(x, y)$, and when does it occur?

## 12.6 EXERCISES

*In Exercises 1–4, find the directional derivative of the function $f$ at the point $P$ in the direction of the unit vector that makes the angle $\theta$ with the positive x-axis.*

**1.** $f(x, y) = x^3 - 2x^2 + y^3$; $\quad P(1, 2)$, $\quad \theta = \dfrac{\pi}{6}$

**2.** $f(x, y) = \sqrt{y^2 - x^2}$; $\quad P(4, 5)$, $\quad \theta = \dfrac{3\pi}{4}$

**3.** $f(x, y) = (x + 1)e^y$; $\quad P(3, 0)$, $\quad \theta = \dfrac{\pi}{2}$

**4.** $f(x, y) = \sin xy$; $\quad P(1, 0)$, $\quad \theta = -\dfrac{\pi}{4}$

*In Exercises 5–10, find the gradient of $f$ at the point $P$.*

**5.** $f(x, y) = 2x + 3xy - 3y + 4$; $\quad P(2, 1)$

**6.** $f(x, y) = \dfrac{1}{x^2 + y^2}$; $\quad P(1, 2)$

**7.** $f(x, y) = x \sin y + y \cos x$; $\quad P\left(\frac{\pi}{4}, \frac{\pi}{2}\right)$

**8.** $f(x, y, z) = \dfrac{x + y}{x + z}$; $\quad P(1, 2, 3)$

**9.** $f(x, y, z) = xe^{yz}$; $\quad P(1, 0, 2)$

**10.** $f(x, y, z) = \ln(x^2 + y^2 + z^2)$; $\quad P(1, 1, 1)$

*In Exercises 11–28, find the directional derivative of the function $f$ at the point $P$ in the direction of the vector $\mathbf{v}$.*

**11.** $f(x, y) = x^3 - x^2y^2 + xy + y^2$; $\quad P(1, -1)$, $\quad \mathbf{v} = \mathbf{i} - 2\mathbf{j}$

**12.** $f(x, y) = x^3 - y^3$; $\quad P(2, 1)$, $\quad \mathbf{v} = \dfrac{1}{\sqrt{2}}(\mathbf{i} + \mathbf{j})$

**13.** $f(x, y) = \dfrac{y}{x}$; $\quad P(3, 1)$, $\quad \mathbf{v} = -\mathbf{i}$

**14.** $f(x, y) = \sqrt{x^2 + y^2 + 1}$; $\quad P(2, 2)$, $\quad \mathbf{v} = 3\mathbf{i} + 4\mathbf{j}$

**15.** $f(x, y) = \dfrac{x + y}{x - y}$; $\quad P(2, 1)$, $\quad \mathbf{v} = -\mathbf{i} + 3\mathbf{j}$

**16.** $f(x, y) = xe^{xy}$; $\quad P(2, 0)$, $\quad \mathbf{v} = 2\mathbf{i} - \mathbf{j}$

**17.** $f(x, y) = x \sin^2 y$; $\quad P\left(-1, \frac{\pi}{4}\right)$, $\quad \mathbf{v} = -2\mathbf{i} + 3\mathbf{j}$

**18.** $f(x, y) = \tan^{-1} \dfrac{y}{x}$; $\quad P(1, 1)$, $\quad \mathbf{v} = \mathbf{i} - \mathbf{j}$

**19.** $f(x, y, z) = x^2y^3z^4$; $\quad P(3, -2, 1)$, $\quad \mathbf{v} = \mathbf{i} + \mathbf{j} + \mathbf{k}$

**20.** $f(x, y, z) = x^2 + 2xy^2 + 2yz^3$; $\quad P(2, 1, -1)$, $\quad \mathbf{v} = \mathbf{i} + 2\mathbf{j} + 2\mathbf{k}$

**21.** $f(x, y, z) = \sqrt{xyz}$; $\quad P(4, 2, 2)$, $\quad \mathbf{v} = 2\mathbf{i} - 4\mathbf{j} + 4\mathbf{k}$

**22.** $f(x, y, z) = \sqrt{xy^2 + 6y^2z^2}$; $\quad P(2, 3, -1)$, $\quad \mathbf{v} = 2\mathbf{i} - \mathbf{k}$

**23.** $f(x, y, z) = x^2 e^{yz}$;   $P(2, 3, 0)$,   $\mathbf{v} = \mathbf{i} - 2\mathbf{j} + 3\mathbf{k}$

**24.** $f(x, y, z) = \ln(x^2 + y^2 + z^2)$;   $P(1, 2, -1)$,
$\mathbf{v} = -3\mathbf{i} + 2\mathbf{j} + \mathbf{k}$

**25.** $f(x, y, z) = x^2 y \cos 2z$;   $P\left(-1, 2, \frac{\pi}{4}\right)$,   $\mathbf{v} = \mathbf{i} - \mathbf{j} + \mathbf{k}$

**26.** $f(x, y, z) = e^x(2 \cos y + 3 \sin z)$;   $P\left(1, \frac{\pi}{6}, \frac{\pi}{6}\right)$,
$\mathbf{v} = 2\mathbf{i} - \mathbf{j} + 3\mathbf{k}$

**27.** $f(x, y, z) = x \tan^{-1}\left(\dfrac{y}{z}\right)$;   $P(3, -2, 2)$,   $\mathbf{v} = \mathbf{i} + 2\mathbf{j} - \mathbf{k}$

**28.** $f(x, y, z) = x^2 \sin^{-1} yz$;   $P(2, 1, 0)$,   $\mathbf{v} = \dfrac{1}{\sqrt{3}}(\mathbf{i} - \mathbf{j} + \mathbf{k})$

*In Exercises 29–32, find the directional derivative of the function
f at the point P in the direction from P to the point Q.*

**29.** $f(x, y) = x^3 + y^3$;   $P(1, 2)$,   $Q(2, 5)$

**30.** $f(x, y) = xe^{-y}$;   $P(2, 0)$,   $Q(-1, 2)$

**31.** $f(x, y, z) = x \sin(2y + 3z)$;   $P\left(1, \frac{\pi}{4}, -\frac{\pi}{12}\right)$,   $Q\left(3, \frac{\pi}{2}, -\frac{\pi}{4}\right)$

**32.** $f(x, y, z) = \dfrac{x + y}{y + z}$;   $P(2, 1, 1)$,   $Q(3, 2, -2)$

*In Exercises 33–36, find a vector giving the direction in which
the function f increases most rapidly at the point P. What is the
maximum rate of increase?*

**33.** $f(x, y) = \sqrt{2x + 3y^2}$;   $P(3, 2)$

**34.** $f(x, y) = e^{-2x} \cos y$;   $P\left(0, \frac{\pi}{4}\right)$

**35.** $f(x, y, z) = x^3 + 2xz + 2yz^2 + z^3$;   $P(-1, 3, 2)$

**36.** $f(x, y, z) = \ln(x^2 + 2y^2 + 3z^2)$;   $P(1, 2, -1)$

*In Exercises 37–40, find a vector giving the direction in which
the function f decreases most rapidly at the point P. What is the
maximum rate of decrease?*

**37.** $f(x, y) = \tan^{-1}(2x + y)$;   $P(0, 0)$

**38.** $f(x, y) = xe^{-y^2}$;   $P(1, 0)$

**39.** $f(x, y, z) = \dfrac{x}{y} + \dfrac{y}{z}$;   $P(1, -1, 2)$

**40.** $f(x, y, z) = \sqrt{xy} \cos z$;   $P\left(4, 1, \frac{\pi}{4}\right)$

**41.** The height of a hill (in feet) is given by

$$h(x, y) = 20(16 - 4x^2 - 3y^2 + 2xy + 28x - 18y)$$

where $x$ is the distance (in miles) east and $y$ the distance (in
miles) north of Bolton. In what direction is the slope of the
hill steepest at the point 1 mile north and 1 mile east of
Bolton? What is the steepest slope at that point?

**42.** **Path of Steepest Ascent** The following figure shows the contour map of a hill with its summit denoted by $S$. Draw the
curve from $P$ to $S$ that is associated with the path you will
take to reach the summit by ascending the direction of the
greatest increase in altitude.
**Hint:** Study Figure 10.

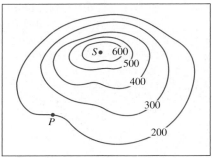

**Note:** This path is called the *path of steepest ascent*.

**43.** **Path of Steepest Descent** The figure shows a topographical
map of a 620-ft hill with contours at 100-ft intervals.

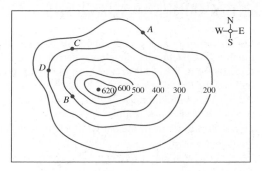

**a.** If you start from $A$ and proceed in a southwesterly direction, will you be ascending, descending, or neither
ascending nor descending? What if you start from $B$?

**b.** If you start from $C$ and proceed in a westerly direction,
will you be ascending, descending, or neither ascending
nor descending?

**c.** If you start from $D$, in what direction should you proceed
to have the steepest ascent?

**d.** If you want to climb to the summit of the hill using the
gentlest ascent, would you start from the east or the west?

**44.** **Steady-State Temperature** Consider the upper half-disk
$H = \{(x, y) \mid x^2 + y^2 \le 1, y \ge 0\}$ (see the figure). If the
temperature at points on the upper boundary is kept at
100°C and the temperature at points on the lower boundary
is kept at 50°C, then the steady-state temperature at any
point $(x, y)$ inside the half-disk is given by

$$T(x, y) = 100 - \frac{100}{\pi} \tan^{-1} \frac{1 - x^2 - y^2}{2y}$$

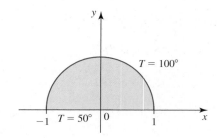

Find the rate of change of the temperature at the point $P\left(\frac{1}{2}, \frac{1}{2}\right)$ in the direction of the vector $\mathbf{v} = 2\mathbf{i} + 3\mathbf{j}$.

**45. Steady-State Temperature** Consider the upper half-disk $H = \{(x, y) \mid x^2 + y^2 \leq 1, y \geq 0\}$ (see the figure). If the temperature at points on the upper boundary is kept at $100°C$ and the temperature at points on the lower boundary is kept at $0°C$, then the steady-state temperature at any point $(x, y)$ inside the half-disk is given by

$$T(x, y) = \frac{200}{\pi} \tan^{-1} \frac{2y}{1 - x^2 - y^2}$$

  **a.** Find the gradient of $T$ at the point $\left(\frac{\sqrt{7}}{4}, \frac{1}{4}\right)$, and interpret your result.
  **b.** Sketch the isothermal curve of $T$ passing through the point $\left(\frac{\sqrt{7}}{4}, \frac{1}{4}\right)$ and the gradient vector $\nabla T$ at that point on the same coordinate system.

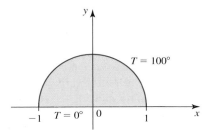

**46.** The temperature at a point $P(x, y, z)$ of a solid ball of radius 4 with center at the origin is given by $T(x, y, z) = xy + yz + xz$. Find the direction in which $T$ is increasing most rapidly at $P(1, 1, 2)$.

**47.** Let $T(x, y, z)$ represent the temperature at a point $P(x, y, z)$ of a region $R$ in space. If the isotherms of $T$ are concentric spheres, show that the temperature gradient $\nabla T$ points either toward or away from the center of the spheres.
  **Hint:** Recall that the isotherms of $T$ are the sets on which $T$ is constant.

**48.** Suppose the temperature at the point $(x, y)$ on a thin sheet of metal is given by

$$T(x, y) = \frac{100(1 + 3x + 2y)}{1 + 2x^2 + 3y^2}$$

degrees Fahrenheit. In what direction will the temperature be increasing most rapidly at the point $(1, 2)$? In what direction will it be decreasing most rapidly?

**49.** The temperature (in degrees Fahrenheit) at a point $(x, y)$ on a metal plate is

$$T(x, y) = 90 - 6x^2 - 2y^2$$

An insect located at the point $(1, 1)$ crawls in the direction in which the temperature drops most rapidly.
  **a.** Find the path of the insect.
  **b.** Sketch a few level curves of $T$ and the path found in part (a).

**50. Cobb-Douglas Production Function** The output of a finished product is given by the production function

$$f(x, y) = 100x^{0.6}y^{0.4}$$

where $x$ stands for the number of units of labor and $y$ stands for the number of units of capital. Currently, the amount being spent on labor is 500 units, and the amount being spent on capital is 250 units. If the manufacturer wishes to expand production by injecting an additional 10 units into labor, how much more should be put into capital to maximize the increase in output?

**51.** The figure shows the contour map of a function $f$ of two variables $x$ and $y$. Use it to estimate the directional derivative of $f$ at $P_0$ in the indicated direction.

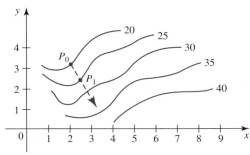

**Hint:** If $\mathbf{u}$ is a unit vector having indicated direction, then

$$D_\mathbf{u} f(P_0) \approx \frac{f(P_1) - f(P_0)}{d(P_0, P_1)}$$

when $f(P_i)$ is the value of $f$ at $P_i$ ($i = 0, 1$) and $d(P_0, P_1)$ is the distance between $P_0$ and $P_1$.

**52.** A rectangular metal plate of dimensions 8 in. $\times$ 4 in. is placed on a rectangular coordinate system with one corner at the origin and the longer side along the positive $x$-axis. The figure shows the contour map of the function $f$ describing the temperature of the plate in degrees Fahrenheit. Use the contour map to estimate the rate of change of the temperature at the point $(3, 1)$ in the direction from the point $(3, 1)$ toward the point $(5, 4)$.

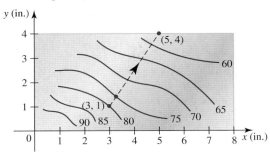

**Hint:** See Exercise 51.

**53.** Suppose that $f$ is differentiable, and suppose that the directional derivative of $f$ at the origin attains a maximum value of 5 in the direction of the vector from the origin to the point $(-3, 4)$. Find $\nabla f(0, 0)$.

**54.** Find unit vector(s) $\mathbf{u} = \langle u_1, u_2 \rangle$ such that the directional derivative of $f(x, y) = x^2 + e^{-xy}$ at the point $(1, 0)$ in the direction of $\mathbf{u}$ has value 1.

**cas** In Exercises 55 and 56, (a) plot several level curves of each pair of functions f and g using the same viewing window, and (b) show analytically that each level curve of f intersects all level curves of g at right angles.

**55.** $f(x, y) = x^2 - y^2$, $g(x, y) = xy$

**56.** $f(x, y) = e^x \cos y$, $g(x, y) = e^x \sin y$

**57.** Let $f(x, y) = x^2 + y^2$, and let $g(x, y) = x^2 - y^2$. Find the direction in which f increases most rapidly and the direction in which g increases most rapidly at $(0, 0)$. Is Theorem 3 applicable here?

In Exercises 58–62, determine whether the statement is true or false. If it is true, explain why it is true. If it is false, give an example to show why it is false.

**58.** If f is differentiable at each point, then the directional derivative of f exists in all directions.

**59.** If f is differentiable at each point, then the value of the directional derivative at any point in a given direction depends only on the direction and the partial derivatives $f_x$ and $f_y$ at that point.

**60.** If $f_x(a, b) = 0$ and $f_y(a, b) = 0$, then $\nabla f(a, b) = \mathbf{0}$.

**61.** The maximum value of $D_{\mathbf{u}} f(x, y)$ is $\sqrt{f_x^2(x, y) + f_y^2(x, y)}$.

**62.** If $\nabla f$ is known, then we can determine f completely.

# 12.7 Tangent Planes and Normal Lines

One compelling reason for studying the tangent line to a curve is that the curve may be approximated by its tangent line near a point of tangency (Figure 1). Answers to questions about the curve near a point of tangency may be obtained indirectly by analyzing the tangent line, a relatively simple task, rather than by studying the curve itself. As you might recall, both the approximation of the change in a function using its differential and Newton's method for finding the zeros of a function are based on this observation.

Our motivation for studying tangent planes to a surface in space is the same as that for studying tangent lines to a curve: Near a point of tangency, a surface may be approximated by its tangent plane (Figure 1). We will show later that approximating the change in $z = f(x, y)$ using the differential is tantamount to approximating this change by the change in $z$ on the tangent plane.

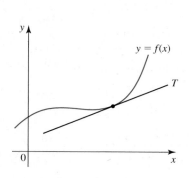

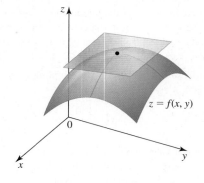

**FIGURE 1**
Near a point of tangency, the tangent line approximates the curve, and the tangent plane approximates the surface.

**(a)** The tangent line to a curve

**(b)** The tangent plane to a surface

## ■ Geometric Interpretation of the Gradient

We begin by looking at the geometric interpretation of the gradient of a function. This vector will play a central role in our effort to find the tangent plane to a surface.

Suppose that the temperature $T$ at any point $(x, y)$ in the plane is given by the function $f$; that is, $T = f(x, y)$. Then the level curve $f(x, y) = c$, where $c$ is a constant, gives the set of points in the plane that have temperature $c$ (Figure 2). Recall that such a curve is called an isothermal curve.

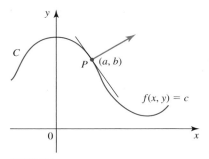

**FIGURE 2**
The level curve $C$ defined by
$f(x, y) = c$ is an isothermal curve.

If we are at the point $P(a, b)$, in what direction should we move if we want to experience the greatest increase in temperature? Since the temperature remains constant if we move along $C$, it seems reasonable to conjecture that proceeding in the direction perpendicular to the tangent line to $C$ at $P$ will result in the greatest increase in temperature. But as we saw in Section 12.6, the function $f$ (and hence the temperature) increases most rapidly in the direction given by its gradient $\nabla f(a, b)$. These observations suggest that the gradient $\nabla f(a, b)$ is perpendicular to the tangent line to the level curve $f(x, y) = c$ at $P$. That this is indeed the case can be demonstrated as follows:

Suppose that the curve $C$ is represented by the vector function

$$\mathbf{r}(t) = g(t)\mathbf{i} + h(t)\mathbf{j}$$

where $g$ and $h$ are differentiable functions, $a = g(t_0)$ and $b = h(t_0)$, and $t_0$ lies in the parameter interval (Figure 3). Since the point $(x, y) = (g(t), h(t))$ lies on $C$, we have

$$f(g(t), h(t)) = c$$

for all $t$ in the parameter interval.

**FIGURE 3**
The curve $C$ may be represented by
$\mathbf{r}(t) = x\mathbf{i} + y\mathbf{j} = g(t)\mathbf{i} + h(t)\mathbf{j}$.

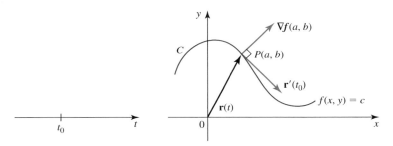

Differentiating both sides of this equation with respect to $t$ and using the Chain Rule for a function of two variables, we obtain

$$\frac{\partial f}{\partial x}\frac{dx}{dt} + \frac{\partial f}{\partial y}\frac{dy}{dt} = 0$$

Recalling that

$$\nabla f(x, y) = \frac{\partial f}{\partial x}\mathbf{i} + \frac{\partial f}{\partial y}\mathbf{j} \qquad \text{and} \qquad \mathbf{r}'(t) = \frac{dx}{dt}\mathbf{i} + \frac{dy}{dt}\mathbf{j}$$

we can write this last equation in the form

$$\nabla f(x, y) \cdot \mathbf{r}'(t) = 0$$

In particular, when $t = t_0$, we have

$$\nabla f(a, b) \cdot \mathbf{r}'(t_0) = 0$$

Thus, if $\mathbf{r}'(t_0) \neq \mathbf{0}$, the vector $\nabla f(a, b)$ is orthogonal to the tangent vector $\mathbf{r}'(t_0)$ at $P(a, b)$. Loosely speaking, we have demonstrated the following:

$\nabla f$ is orthogonal to the level curve $f(x, y) = c$ at $P$.     See Figure 3.

**EXAMPLE 1**  Let $f(x, y) = x^2 - y^2$. Find the level curve of $f$ passing through the point $(5, 3)$. Also, find the gradient of $f$ at that point, and make a sketch of both the level curve and the gradient vector.

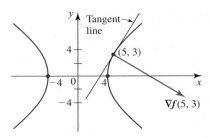

**FIGURE 4**
The gradient $\nabla f(5, 3)$ is orthogonal to the level curve $x^2 - y^2 = 16$ at $(5, 3)$.

**Solution** Since $f(5, 3) = 25 - 9 = 16$, the required level curve is the hyperbola $x^2 - y^2 = 16$. The gradient of $f$ at any point $(x, y)$ is

$$\nabla f(x, y) = 2x\mathbf{i} - 2y\mathbf{j}$$

and, in particular, the gradient of $f$ at $(5, 3)$ is

$$\nabla f(5, 3) = 10\mathbf{i} - 6\mathbf{j}$$

The level curve and $\nabla f(5, 3)$ are shown in Figure 4.  ∎

**EXAMPLE 2** Refer to Example 1. Find equations of the normal line and the tangent line to the curve $x^2 - y^2 = 16$ at the point $(5, 3)$.

**Solution** We think of the curve $x^2 - y^2 = 16$ as the level curve $f(x, y) = k$ of the function $f(x, y) = x^2 - y^2$ for $k = 16$. From Example 1 we see that $\nabla f(5, 3) = 10\mathbf{i} - 6\mathbf{j}$. Since this gradient is normal to the curve $x^2 - y^2 = 16$ at $(5, 3)$ (see Figure 4), we see that the slope of the required normal line is

$$m_1 = -\frac{6}{10} = -\frac{3}{5}$$

Therefore, an equation of the normal line is

$$y - 3 = -\frac{3}{5}(x - 5) \qquad \text{or} \qquad y = -\frac{3}{5}x + 6$$

The slope of the required tangent line is

$$m_2 = -\frac{1}{m_1} = \frac{5}{3}$$

so an equation of the tangent line is

$$y - 3 = \frac{5}{3}(x - 5) \qquad \text{or} \qquad y = \frac{5}{3}x - \frac{16}{3}$$  ∎

Next, suppose that $F(x, y, z) = k$ is the level surface $S$ of a differentiable function $F$ defined by $T = F(x, y, z)$. You may think of the function $F$ as giving the temperature at any point $(x, y, z)$ in space and interpret the following argument in terms of this application.

Suppose that $P(a, b, c)$ is a point on $S$ and let $C$ be a smooth curve on $S$ passing through $P$. Then $C$ can be described by the vector function

$$\mathbf{r}(t) = f(t)\mathbf{i} + g(t)\mathbf{j} + h(t)\mathbf{k}$$

where $f(t_0) = a$, $g(t_0) = b$, $h(t_0) = c$, and $t_0$ is a point in the parameter interval (Figure 5).

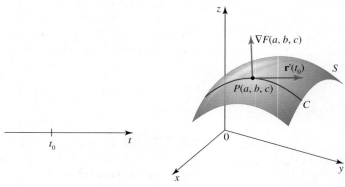

**FIGURE 5**
The curve $C$ is described by
$\mathbf{r}(t) = f(t)\mathbf{i} + g(t)\mathbf{j} + h(t)\mathbf{k}$ with
$P(a, b, c)$ corresponding to $t_0$.

Since the point $(x, y, z) = (f(t), g(t), h(t))$ lies on $S$, we have

$$F(f(t), g(t), h(t)) = k$$

for all $t$ in the parameter interval. If $\mathbf{r}$ is differentiable, then we can use the Chain Rule to differentiate both sides of this equation to obtain

$$\frac{\partial F}{\partial x}\frac{dx}{dt} + \frac{\partial F}{\partial y}\frac{dy}{dt} + \frac{\partial F}{\partial z}\frac{dz}{dt} = 0$$

This is the same as

$$[F_x(x, y, z)\mathbf{i} + F_y(x, y, z)\mathbf{j} + F_z(x, y, z)\mathbf{k}] \cdot \left[\frac{dx}{dt}\mathbf{i} + \frac{dy}{dt}\mathbf{j} + \frac{dz}{dt}\mathbf{k}\right] = 0$$

or, in an even more abbreviated form,

$$\nabla F(x, y, z) \cdot \mathbf{r}'(t) = 0$$

In particular, at $t = t_0$ we have

$$\nabla F(a, b, c) \cdot \mathbf{r}'(t_0) = 0$$

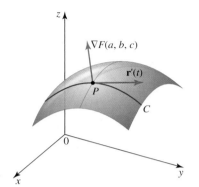

**FIGURE 6**
The gradient $\nabla F(a, b, c)$ is orthogonal to the tangent vector of *every* curve on $S$ passing through $P(a, b, c)$.

This shows that if $\mathbf{r}'(t_0) \neq \mathbf{0}$, then the gradient vector $\nabla F(a, b, c)$ is orthogonal to the tangent vector $\mathbf{r}'(t_0)$ to $C$ at $P$ (Figure 6). Since this argument holds for any differentiable curve passing through $P(a, b, c)$ on $S$, we have shown that $\nabla F(a, b, c)$ is orthogonal to the tangent vector of *every* curve on $S$ passing through $P$. Thus, loosely speaking, we have demonstrated the following result.

> $\nabla F$ is orthogonal to the level surface $F(x, y, z) = 0$ at $P$.

**Note**    Interpreting the function $F$ as giving the temperature at any point $(x, y, z)$ in space as was suggested earlier, we see that the level surface $F(x, y, z) = k$ gives all points $(x, y, z)$ in space whose temperature is $k$. The result that was just derived simply states that if you are at any point on this surface, then moving away from it in a direction of $\nabla F$ (perpendicular to the surface at that point) will result in the greatest increase in temperature.    ■

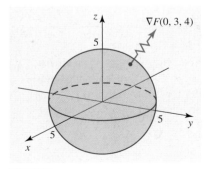

**FIGURE 7**
The gradient $\nabla F(0, 3, 4)$ is orthogonal to the level surface $x^2 + y^2 + z^2 = 25$ at $(0, 3, 4)$.

**EXAMPLE 3**    Let $F(x, y, z) = x^2 + y^2 + z^2$. Find the level surface that contains the point $(0, 3, 4)$. Also, find the gradient of $F$ at that point, and make a sketch of both the level surface and the gradient vector.

**Solution**    Since $F(0, 3, 4) = 0 + 9 + 16 = 25$, the required level surface is the sphere $x^2 + y^2 + z^2 = 25$ with center at the origin and radius 5. The gradient of $F$ at any point $(x, y, z)$ is

$$\nabla F(x, y, z) = 2x\mathbf{i} + 2y\mathbf{j} + 2z\mathbf{k}$$

so the gradient of $F$ at $(0, 3, 4)$ is

$$\nabla F(0, 3, 4) = 6\mathbf{j} + 8\mathbf{k}$$

The level surface and $\nabla F(0, 3, 4)$ are sketched in Figure 7.    ■

## ■ Tangent Planes and Normal Lines

We are now in a position to define a tangent plane to a surface in space. But before doing so, let's digress a little to talk about the representation of surfaces in space. Up to now, we have assumed that a surface in space is described by a function $f$ with explicit representation $z = f(x, y)$.

Another way of describing a surface in space is via a function that is represented implicitly by the equation

$$F(x, y, z) = 0 \qquad\qquad (1)$$

Here, $F$ is a function of the three variables $x$, $y$, and $z$ described by the equation $w = F(x, y, z)$. Thus, we can think of Equation (1) as representing a *level surface* of $F$.

For a surface $S$ that is given explicitly by $z = f(x, y)$, we define

$$F(x, y, z) = z - f(x, y)$$

This shows that we can also view $S$ as the level surface of $F$ given by Equation (1). For example, the surface described by $z = x^2 + 2y^2 + 1$ can be viewed as the level surface of $F$ defined by $F(x, y, z) = 0$, where $F(x, y, z) = z - x^2 - 2y^2 - 1$.

To define a tangent plane, let $S$ be a surface described by $F(x, y, z) = 0$, and let $P(a, b, c)$ be a point on $S$. Then, as we saw earlier, the gradient $\nabla F(a, b, c)$ at $P$ is orthogonal to the tangent vector of *every* curve on $S$ passing through $P$ (Figure 8). This suggests that we define the *tangent plane* to $S$ at $P$ to be the plane passing through $P$ and containing all these tangent vectors. Equivalently, the tangent plane should have $\nabla F(a, b, c)$ as a normal vector.

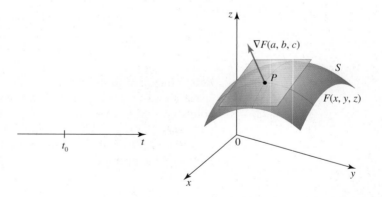

**FIGURE 8**
The tangent plane to $S$ at $P$ contains the tangent vectors to all curves on $S$ passing through $P$.

---

**DEFINITIONS   Tangent Plane and Normal Line**

Let $P(a, b, c)$ be a point on the surface $S$ described by $F(x, y, z) = 0$, where $F$ is differentiable at $P$, and suppose that $\nabla F(a, b, c) \neq \mathbf{0}$. Then the **tangent plane** to $S$ at $P$ is the plane that passes through $P$ and has normal vector $\nabla F(a, b, c)$. The **normal line** to $S$ at $P$ is the line that passes through $P$ and has the same direction as $\nabla F(a, b, c)$.

---

Using Equation (4) from Section 10.5, we see that an equation of the tangent plane is

$$F_x(a, b, c)(x - a) + F_y(a, b, c)(y - b) + F_z(a, b, c)(z - c) = 0 \qquad\qquad (2)$$

and using Equation (2) of Section 10.5, we see that the equations of the normal line (in symmetric form) are

$$\frac{x - a}{F_x(a, b, c)} = \frac{y - b}{F_y(a, b, c)} = \frac{z - c}{F_z(a, b, c)} \tag{3}$$

**EXAMPLE 4** Find equations of the tangent plane and normal line to the ellipsoid with equation $4x^2 + y^2 + 4z^2 = 16$ at the point $(1, 2, \sqrt{2})$.

**Solution** The given equation can be written in the form $F(x, y, z) = 0$, where $F(x, y, z) = 4x^2 + y^2 + 4z^2 - 16$. The partial derivatives of $F$ are

$$F_x(x, y, z) = 8x, \qquad F_y(x, y, z) = 2y, \qquad \text{and} \qquad F_z(x, y, z) = 8z$$

In particular, at the point $(1, 2, \sqrt{2})$

$$F_x(1, 2, \sqrt{2}) = 8, \qquad F_y(1, 2, \sqrt{2}) = 4, \qquad \text{and} \qquad F_z(1, 2, \sqrt{2}) = 8\sqrt{2}$$

Then, using Equation (2), we find that an equation of the tangent plane to the ellipsoid at $(1, 2, \sqrt{2})$ is

$$8(x - 1) + 4(y - 2) + 8\sqrt{2}(z - \sqrt{2}) = 0$$

or $2x + y + 2\sqrt{2}z = 8$. Next, using Equation (3), we obtain the following parametric equations of the normal line:

$$\frac{x - 1}{8} = \frac{y - 2}{4} = \frac{z - \sqrt{2}}{8\sqrt{2}} \qquad \text{or} \qquad \frac{x - 1}{2} = y - 2 = \frac{z - \sqrt{2}}{2\sqrt{2}}$$

The tangent plane and normal line are shown in Figure 9. ▪

**EXAMPLE 5** Find equations of the tangent plane and normal line to the graph of the function $f$ defined by $f(x, y) = 4x^2 + y^2 + 2$ at the point where $x = 1$ and $y = 1$.

**Solution** Here, the surface is defined by

$$z = f(x, y) = 4x^2 + y^2 + 2$$

and we recognize it to be a paraboloid. This equation can be rewritten in the form

$$F(x, y, z) = z - f(x, y) = 0$$

where $F(x, y, z) = z - 4x^2 - y^2 - 2$. The partial derivatives of $F$ are

$$F_x(x, y, z) = -8x, \qquad F_y(x, y, z) = -2y, \qquad \text{and} \qquad F_z(x, y, z) = 1$$

If $x = 1$ and $y = 1$, then $z = f(1, 1) = 4 + 1 + 2 = 7$. At the point $(1, 1, 7)$ we have

$$F_x(1, 1, 7) = -8, \qquad F_y(1, 1, 7) = -2, \qquad \text{and} \qquad F_z(1, 1, 7) = 1$$

Then, using Equation (2), we find an equation of the tangent plane to the paraboloid at $(1, 1, 7)$ to be

$$-8(x - 1) - 2(y - 1) + 1(z - 7) = 0$$

or $8x + 2y - z = 3$. Next, using Equation (3), we find that the parametric equations of the normal line at $(1, 1, 7)$ are

$$\frac{x - 1}{-8} = \frac{y - 1}{-2} = \frac{z - 7}{1}$$

The tangent plane and normal line are shown in Figure 10. ▪

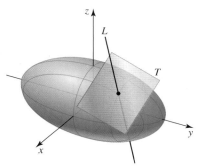

**FIGURE 9**
The tangent plane and normal line to the ellipsoid $4x^2 + y^2 + 4z^2 = 16$ at $(1, 2, \sqrt{2})$.

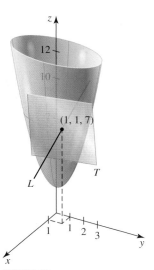

**FIGURE 10**
The tangent plane and normal line to the paraboloid $z = 4x^2 + y^2 + 2$ at $(1, 1, 7)$.

## Using the Tangent Plane of $f$ to Approximate the Surface $z = f(x, y)$

We conclude this section by showing that in approximating the change $\Delta z$ in $z = f(x, y)$ as $(x, y)$ changes from $(a, b)$ to $(a + \Delta x, b + \Delta y)$ by the differential $dz = f_x(a, b)\, \Delta x + f_y(a, b)\, \Delta y$, we are in effect using the tangent plane of $f$ near $P(a, b)$ to approximate the surface $z = f(x, y)$ near $P(a, b)$.

We begin by finding an expression for the tangent plane to the surface $z = f(x, y)$ at $(a, b)$. Writing $z = f(x, y)$ in the form $F(x, y, z) = z - f(x, y) = 0$ we see that

$$F_x(a, b, c) = -f_x(a, b), \qquad F_y(a, b, c) = -f_y(a, b), \qquad \text{and} \qquad F_z(a, b, c) = 1$$

Using Equation (2), we find that the required equation is

$$-f_x(a, b)(x - a) - f_y(a, b)(y - b) + (z - c) = 0$$

or

$$z - f(a, b) = f_x(a, b)\, \Delta x + f_y(a, b)\, \Delta y \qquad {\scriptstyle c = f(a, b)} \qquad \textbf{(4)}$$

But the expression on the right is just the differential of $f$ at $(a, b)$. So Equation (4) implies that $dz = z - f(a, b)$; that is, $dz$ represents the change in height of the tangent plane (Figure 11).

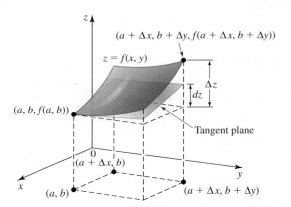

**FIGURE 11**
The relationship between $\Delta z$ and $dz$

By Theorem 1 of Section 12.4 we have

$$\Delta z = f_x(a, b)\, \Delta x + f_y(a, b)\, \Delta y + \varepsilon_1\, \Delta x + \varepsilon_2\, \Delta y$$

or

$$\Delta z - dz = \varepsilon_1\, \Delta x + \varepsilon_2\, \Delta y$$

where $\varepsilon_1$ and $\varepsilon_2$ are functions of $\Delta x$ and $\Delta y$ that approach 0 as $\Delta x$ and $\Delta y$ approach 0. Therefore, as was pointed out in Section 12.4, $\Delta z \approx dz$ if $\Delta x$ and $\Delta y$ are small. Recalling the meaning of $\Delta z$ (see Figure 11), we see that we are using the tangent plane at $(a, b)$ to approximate the surface $z = f(x, y)$ when $(x, y)$ is close to $(a, b)$.

## 12.7    CONCEPT QUESTIONS

**1. a.** Consider the level curve $f(x, y) = c$, where $f$ is differentiable and $c$ is a constant. What can you say about $\nabla f$ at a point $P$ on the level curve? Illustrate with a figure.
   **b.** Repeat part (a) for a level surface $F(x, y, z) = c$.

**2. a.** Define the tangent plane to a surface $S$ described by $F(x, y, z) = 0$ at the point $P(a, b, c)$ on $S$. Illustrate with a figure and give an equation of the tangent plane.
   **b.** Repeat part (a) for the normal line to $S$ at $P$.

## 12.7 EXERCISES

*In Exercises 1–4, sketch (a) the level curve of the function f that passes through the point P and (b) the gradient of f at P.*

**1.** $f(x, y) = y^2 - x^2;$ $P(1, 2)$

**2.** $f(x, y) = 4x^2 + y^2;$ $P(\frac{\sqrt{3}}{2}, 1)$

**3.** $f(x, y) = x^2 + y;$ $P(1, 3)$

**4.** $f(x, y) = 2x + 3y;$ $P(-3, 4)$

*In Exercises 5–8, find equations of the normal and tangent lines to the curve at the given point.*

**5.** $\dfrac{x^2}{9} + \dfrac{y^2}{16} = 1;$ $(\frac{3\sqrt{3}}{2}, 2)$

**6.** $x^4 - x^2 + y^2 = 0;$ $(\frac{1}{2}, \frac{\sqrt{3}}{4})$

**7.** $x^4 + 2x^2y^2 + y^4 - 9x^2 + 9y^2 = 0;$ $(\sqrt{5}, -1)$

**8.** $2x + y - e^{x-y} = 2;$ $(1, 1)$

*In Exercises 9–14, sketch (a) the level surface of the function F that passes through the point P and (b) the gradient of F at P.*

**9.** $F(x, y, z) = x^2 + y^2 + z^2;$ $P(1, 2, 2)$

**10.** $F(x, y, z) = z - x^2 - y^2;$ $P(1, 1, 2)$

**11.** $F(x, y, z) = x^2 + y^2;$ $P(0, 2, 4)$

**12.** $F(x, y, z) = 2x + 3y + z;$ $P(2, 3, 1)$

**13.** $F(x, y, z) = -x^2 + y^2 - z^2;$ $P(1, 3, 2)$

**14.** $F(x, y, z) = xy;$ $P(2, \frac{1}{2}, 0)$

*In Exercises 15–32, find equations for the tangent plane and the normal line to the surface with the equation at the given point.*

**15.** $x^2 + 4y^2 + 9z^2 = 17;$ $P(2, 1, 1)$

**16.** $2x^2 - y^2 + 3z^2 = 2;$ $P(2, -3, 1)$

**17.** $x^2 - 2y^2 - 4z^2 = 4;$ $P(4, -2, -1)$

**18.** $x^2 + y^2 + z^2 - 2xy + 4xz - x + y = 12;$ $P(1, 0, 2)$

**19.** $xy + yz + xz = 11;$ $P(1, 2, 3)$

**20.** $xyz = -4;$ $P(2, -1, 2)$

**21.** $z = 9x^2 + 4y^2;$ $P(-1, 2, 25)$

**22.** $z = y^2 - 2x^2;$ $P(2, 4, 8)$

**23.** $xz^2 + yx^2 + y^2 - 2x + 3y + 6 = 0;$ $P(-2, 1, 3)$

**24.** $x^3 - xy^2 + z^3 - 2x + 6 = 0;$ $P(1, 2, -1)$

**25.** $z = xe^y;$ $P(2, 0, 2)$

**26.** $z = e^x \sin \pi y;$ $P(0, 1, 0)$

**27.** $z = \ln(xy + 1);$ $P(3, 0, 0)$

**28.** $z = \ln \dfrac{x}{y};$ $P(2, 2, 0)$

**29.** $z = \tan^{-1}\left(\dfrac{y}{x}\right);$ $P\left(1, 1, \dfrac{\pi}{4}\right)$

**30.** $z - x \cos y = 0;$ $P\left(2, \dfrac{\pi}{3}, 1\right)$

**31.** $\sin xy + 3z = 3;$ $P(0, 3, 1)$

**32.** $e^x(\cos y + 1) - 2z = -2;$ $P(0, 0, 2)$

**33.** Show that an equation of the tangent plane to the ellipsoid

$$\frac{x^2}{a^2} + \frac{y^2}{b^2} + \frac{z^2}{c^2} = 1$$

at the point $(x_0, y_0, z_0)$ can be written as

$$\frac{xx_0}{a^2} + \frac{yy_0}{b^2} + \frac{zz_0}{c^2} = 1$$

**34.** Show that an equation of the tangent plane to the hyperboloid

$$\frac{x^2}{a^2} + \frac{y^2}{b^2} - \frac{z^2}{c^2} = 1$$

at the point $(x_0, y_0, z_0)$ can be written as

$$\frac{xx_0}{a^2} + \frac{yy_0}{b^2} - \frac{zz_0}{c^2} = 1$$

**35.** Find an equation of the tangent plane to the hyperboloid of two sheets

$$\frac{x^2}{a^2} - \frac{y^2}{b^2} - \frac{z^2}{c^2} = 1$$

at the point $(x_0, y_0, z_0)$, and express it in a form similar to that of Exercise 34.

**36.** Show that an equation of the tangent plane to the elliptic paraboloid

$$\frac{x^2}{a^2} + \frac{y^2}{b^2} = cz$$

at the point $(x_0, y_0, z_0)$ can be written as

$$\frac{2xx_0}{a^2} + \frac{2yy_0}{b^2} = c(z + z_0)$$

**37.** Find the points on the sphere $x^2 + y^2 + z^2 = 14$ at which the tangent plane is parallel to the plane $x + 2y + 3z = 12$.

**38.** Find the points on the hyperboloid of two sheets $-x^2 - 2y^2 + z^2 = 4$ at which the tangent plane is parallel to the plane $2x + 2y + 4z = 1$.

**39.** Find the points on the hyperboloid of one sheet $2x^2 - y^2 + z^2 = 1$ at which the normal line is parallel to the line passing through the points $(-1, 1, 2)$ and $(3, 3, 3)$.

**40.** Find the points on the surface $x^2 + 4y^2 + 3z^2 - 2xy = 16$ at which the tangent plane is horizontal.

**41.** Two surfaces are *tangent* to each other at a point $P$ if and only if they have a common tangent plane at that point. Show that the elliptic paraboloid $2x^2 + y^2 - z - 5 = 0$ and the sphere $x^2 + y^2 + z^2 - 6x - 8y - z + 17 = 0$ are tangent to each other at the point $(1, 2, 1)$.

**42.** Two surfaces are *orthogonal* to each other at a point of intersection $P$ if and only if their normal lines at $P$ are orthogonal. Show that the sphere $x^2 + y^2 + z^2 - 17 = 0$ and the elliptic paraboloid $2x^2 - y + 2z^2 + 2 = 0$ are orthogonal to each other at the point $(1, 4, 0)$.

**43.** Show that any line that is tangent to the ellipse $(x^2/a^2) + (y^2/b^2) = 1$ has the equation

$$(b \cos \theta)x + (a \sin \theta)y = ab$$

where $\theta$ lies in the interval $[0, 2\pi)$.

**44.** Suppose that two surfaces $F(x, y, z) = 0$ and $G(x, y, z) = 0$ intersect along a curve $C$ and that $P(x_0, y_0, z_0)$ is a point on $C$. Show that the vector $\nabla F(x_0, y_0, z_0) \times \nabla G(x_0, y_0, z_0)$ is parallel to the tangent line to $C$ at $P$. Illustrate with a sketch.

**45.** Refer to Exercise 44. Let $C$ be the intersection of the sphere $x^2 + y^2 + z^2 = 2$ and the paraboloid $z = x^2 + y^2$. Find the parametric equations of the tangent line to $C$ at the point $\left(-\frac{\sqrt{2}}{2}, \frac{\sqrt{2}}{2}, 1\right)$.

*In Exercises 46–49, determine whether the statement is true or false. If it is true, explain why it is true. If it is false, give an example to show why it is false.*

**46.** The tangent line at a point $P$ on the level curve $f(x, y) = c$ is orthogonal to $\nabla f$ at $P$.

**47.** The line with equations

$$\frac{x - 2}{4} = \frac{y + 1}{6} = -\frac{z}{2}$$

is perpendicular to the plane with equation $2x + 3y - z = 4$.

**48.** If an equation of the tangent plane at the point $P_0(x_0, y_0, z_0)$ on the surface described by $F(x, y, z) = 0$ is $ax + by + cz = d$, then $\nabla F(x_0, y_0, z_0) = k\langle a, b, c \rangle$ for some scalar $k$.

**49.** The vector equation of the normal line passing through the point $P_0(x_0, y_0, z_0)$ on the surface with equation $F(x, y, z) = 0$ is $\mathbf{r}(t) = \langle x_0, y_0, z_0 \rangle + t \nabla F(x_0, y_0, z_0)$.

## 12.8  Extrema of Functions of Two Variables

### ■ Relative and Absolute Extrema

In Chapter 3 we saw that the solution of a problem often reduces to finding the extreme values of a function of one variable. A similar situation arises when we solve problems involving a function of two or more variables.

For example, suppose that the Scandi Company manufactures computer desks in both assembled and unassembled versions. Then its weekly profit $P$ is a function of the number of assembled units, $x$, and the number of unassembled units, $y$, sold per week; that is, $P = f(x, y)$. A question of paramount importance to the manufacturer is: How many assembled desks and how many unassembled desks should the company manufacture per week to maximize its weekly profit? Mathematically, the problem is solved by finding the values of $x$ and $y$ that will make $f(x, y)$ a maximum.

In this section and the next section we will focus our attention on finding the extrema of a function of two variables. As in the case of a function of one variable, we distinguish between the relative (or local) extrema and the absolute extrema of a function of two variables.

---

**DEFINITION    Relative Extrema of a Function of Two Variables**

Let $f$ be a function defined on a region $R$ containing the point $(a, b)$. Then $f$ has a **relative maximum** at $(a, b)$ if $f(x, y) \leq f(a, b)$ for all points $(x, y)$ in an open disk containing $(a, b)$. The number $f(a, b)$ is called a **relative maximum value.**

Similarly, $f$ has a **relative minimum** at $(a, b)$ with **relative minimum value** $f(a, b)$ if $f(x, y) \geq f(a, b)$ for all points $(x, y)$ in an open disk containing $(a, b)$.

Loosely speaking, $f$ has a relative maximum at $(a, b)$ if the point $(a, b, f(a, b))$ is the highest point on the graph of $f$ when compared to all nearby points. A similar interpretation holds for a relative minimum.

If the inequalities in this last definition hold for all points $(x, y)$ in the domain of $f$, then $f$ has an **absolute maximum (absolute minimum)** at $(a, b)$ with **absolute maximum value (absolute minimum value)** $f(a, b)$. Figure 1 shows the graph of a function defined on a domain $D$ with relative maxima at $(a, b)$ and $(e, g)$ and a relative minimum at $(c, d)$. The absolute maximum of $f$ occurs at $(e, g)$, and the absolute minimum of $f$ occurs at $(h, i)$.

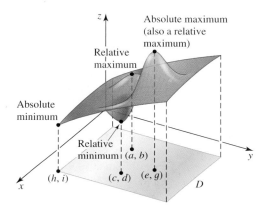

**FIGURE 1**
The relative and absolute extrema of the function $f$ over the domain $D$

## ■ Critical Points—Candidates for Relative Extrema

Figure 2a shows the graph of a function $f$ with a relative maximum at a point $(a, b)$ lying inside the domain of $f$. As you can see, the tangent plane to the surface $z = f(x, y)$ at the point $(a, b, f(a, b))$ is horizontal. This means that all the directional derivatives of $f$ at $(a, b)$, if they exist, must be zero. In particular, $f_x(a, b) = 0$ and $f_y(a, b) = 0$. Next, Figure 2b shows the graph of a function with a relative maximum at a point $(a, b)$. Note that both $f_x(a, b)$ and $f_y(a, b)$ do not exist because the surface $z = f(x, y)$ has a point $(a, b, f(a, b))$ that looks like a jagged mountain peak.

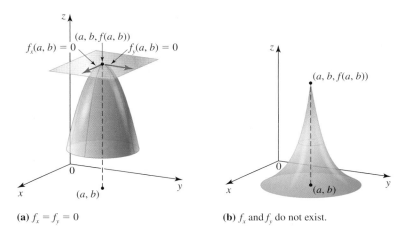

**(a)** $f_x = f_y = 0$     **(b)** $f_x$ and $f_y$ do not exist.

**FIGURE 2**
At a relative extremum of $f$, either $f_x = f_y = 0$ or one or both partial derivatives do not exist.

You are encouraged to draw similar graphs of functions having relative minima at points lying inside the domain of the functions. All of these points are critical points of a function of two variables.

> **DEFINITION** **Critical Points of a Function**
>
> Let $f$ be defined on an open region $R$ containing the point $(a, b)$. We call $(a, b)$ a **critical point** of $f$ if
>
> **a.** $f_x$ and/or $f_y$ do not exist at $(a, b)$ or
> **b.** both $f_x(a, b) = 0$ and $f_y(a, b) = 0$.

The next theorem tells us that the relative extremum of a function $f$ defined on an open region can occur only at a critical point of $f$.

> **THEOREM 1** **The Critical Points of $f$ Are Candidates for Relative Extrema**
>
> If $f$ has a relative extremum (relative maximum or relative minimum) at a point $(a, b)$ in the domain of $f$, then $(a, b)$ must be a critical point of $f$.

**PROOF** If either $f_x$ or $f_y$ does not exist at $(a, b)$, then $(a, b)$ is a critical point of $f$. So suppose that both $f_x(a, b)$ and $f_y(a, b)$ exist. Let $g(x) = f(x, b)$. If $f$ has a relative extremum at $(a, b)$, then $g$ has a relative extremum at $a$, so by Theorem 1 of Section 3.1, $g'(a) = 0$. But

$$g'(a) = \lim_{h \to 0} \frac{f(a + h, b) - f(a, b)}{h} = f_x(a, b)$$

Therefore, $f_x(a, b) = 0$. Similarly, by considering the function $f(y) = f(a, y)$, we obtain $f_y(a, b) = 0$. Thus, $(a, b)$ is a critical point of $f$. ∎

**EXAMPLE 1** Let $f(x, y) = x^2 + y^2 - 4x - 6y + 17$. Find the critical point of $f$, and show that $f$ has a relative minimum at that point.

**Solution** To find the critical point of $f$, we compute

$$f_x(x, y) = 2x - 4 = 2(x - 2) \qquad \text{and} \qquad f_y(x, y) = 2y - 6 = 2(y - 3)$$

Observe that both $f_x$ and $f_y$ are continuous for all values of $x$ and $y$. Setting $f_x$ and $f_y$ equal to zero, we find that $x = 2$ and $y = 3$, so $(2, 3)$ is the only critical point of $f$. Next, to show that $f$ has a relative minimum at this point, we complete the squares in $x$ and $y$ and write $f(x, y)$ in the form

$$f(x, y) = (x - 2)^2 + (y - 3)^2 + 4$$

Notice that $(x - 2)^2 \geq 0$ and $(y - 3)^2 \geq 0$, so $f(x, y) \geq 4$ for all $(x, y)$ in the domain of $f$. Therefore, $f(2, 3) = 4$ is a relative minimum value of $f$. In fact, we have shown that 4 is the absolute minimum value of $f$. The graph of $f$ shown in Figure 3 confirms this result. ∎

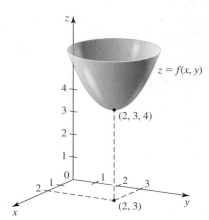

**FIGURE 3**
The function $f$ has a relative minimum at $(2, 3)$.

**EXAMPLE 2** Let $f(x, y) = 3 - \sqrt{x^2 + y^2}$. Show that $(0, 0)$ is the only critical point of $f$ and that $f(0, 0) = 3$ is a relative maximum value of $f$.

**Solution** The partial derivatives of $f$ are

$$f_x(x, y) = -\frac{x}{\sqrt{x^2 + y^2}} \qquad \text{and} \qquad f_y(x, y) = -\frac{y}{\sqrt{x^2 + y^2}}$$

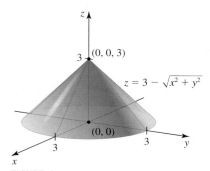

**FIGURE 4**
The function $f$ has a relative maximum at $(0, 0)$.

Since both $f_x(x, y)$ and $f_y(x, y)$ are undefined at $(0, 0)$, we conclude that $(0, 0)$ is a critical point of $f$. Also, $f_x(x, y)$ and $f_y(x, y)$ are not both equal to zero at any point. This tells us that $(0, 0)$ is the only critical point of $f$. Finally, since $\sqrt{x^2 + y^2} \geq 0$ for all values of $x$ and $y$, we see that $f(x, y) \leq 3$ for all points $(x, y)$. We conclude that $f(0, 0) = 3$ is a relative (indeed, the absolute) maximum of $f$. The graph of $f$ shown in Figure 4 confirms this result. ∎

As in the case of a function of one variable, a critical point of a function of two variables is only a candidate for a relative extremum of the function. A critical point need not give rise to a relative extremum, as the following example shows.

**EXAMPLE 3** Show that the point $(0, 0)$ is a critical point of $f(x, y) = y^2 - x^2$ but that it does not give rise to a relative extremum of $f$.

**Solution** The partial derivatives of $f$,

$$f_x(x, y) = -2x \qquad \text{and} \qquad f_y(x, y) = 2y$$

are continuous everywhere. Since $f_x$ and $f_y$ are both equal to zero at $(0, 0)$, we conclude that $(0, 0)$ is a critical point of $f$ and that it is the only candidate for a relative extremum of $f$. But notice that for points on the $x$-axis we have $y = 0$, so $f(x, y) = -x^2 < 0$ if $x \neq 0$; and for points on the $y$-axis we have $x = 0$, so $f(x, y) = y^2 > 0$ if $y \neq 0$. Therefore, every open disk containing $(0, 0)$ has points where $f$ takes on positive values as well as points where $f$ takes on negative values. This shows that $f(0, 0) = 0$ cannot be a relative extremum of $f$. The graph of $f$ is shown in Figure 5. The point $(0, 0, 0)$ is called a *saddle point*.

**FIGURE 5**
The point $(0, 0)$ is a critical point of $f(x, y) = y^2 - x^2$, but it does not give rise to a relative extremum of $f$.

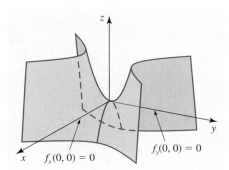

∎

As we have observed, the critical point $(0, 0)$ in Example 3 does not yield a relative maximum or minimum. In general, a critical point of a differentiable function of two variables that does not give rise to a relative extremum is called a **saddle point.** A saddle point is the analog of an inflection point for the case of a function of one variable.

### ■ The Second Derivative Test for Relative Extrema

In Examples 1 and 3 we were able to determine, either by inspection or with the help of simple algebraic manipulations, whether $f$ did or did not possess a relative extremum at a critical point. For more complicated functions, the following test may be used. This test is the analog of the Second Derivative Test for a function of one variable. Its proof will be omitted.

---

**THEOREM 2** **The Second Derivative Test for a Function of Two Variables**

Suppose that $f$ has continuous second-order partial derivatives on an open region containing a critical point $(a, b)$ of $f$. Let

$$D(x, y) = f_{xx}(x, y)f_{yy}(x, y) - f_{xy}^2(x, y)$$

**a.** If $D(a, b) > 0$ and $f_{xx}(a, b) < 0$, then $f(a, b)$ is a **relative maximum value.**

**b.** If $D(a, b) > 0$ and $f_{xx}(a, b) > 0$, then $f(a, b)$ is a **relative minimum value.**

**c.** If $D(a, b) < 0$, then $(a, b, f(a, b))$ is a **saddle point.**

**d.** If $D(a, b) = 0$, then the test is inconclusive.

---

**EXAMPLE 4** Find the relative extrema of $f(x, y) = x^3 + y^2 - 2xy + 7x - 8y + 2$.

**Solution** First, we find the critical points of $f$. Since

$$f_x(x, y) = 3x^2 - 2y + 7 \quad \text{and} \quad f_y(x, y) = 2y - 2x - 8$$

are both continuous for all values of $x$ and $y$, the only critical points of $f$, if any, are found by solving the system of equations $f_x(x, y) = 0$ and $f_y(x, y) = 0$, that is, by solving

$$3x^2 - 2y + 7 = 0 \quad \text{and} \quad 2y - 2x - 8 = 0$$

From the second equation we obtain $y = x + 4$, which upon substitution into the first equation yields

$$3x^2 - 2x - 1 = 0 \quad \text{or} \quad (3x + 1)(x - 1) = 0$$

Therefore, $x = -\frac{1}{3}$ or $x = 1$. Substituting each of these values of $x$ into the expression for $y$ gives $y = \frac{11}{3}$ and $y = 5$, respectively. Therefore, the critical points of $f$ are $\left(-\frac{1}{3}, \frac{11}{3}\right)$ and $(1, 5)$.

Next, we use the Second Derivative Test to determine the nature of each of these critical points. We begin by computing $f_{xx}(x, y) = 6x$, $f_{yy}(x, y) = 2$, $f_{xy}(x, y) = -2$, and

$$D(x, y) = f_{xx}(x, y)f_{yy}(x, y) - f_{xy}^2(x, y)$$
$$= (6x)(2) - (-2)^2 = 4(3x - 1)$$

To test the point $\left(-\frac{1}{3}, \frac{11}{3}\right)$, we compute

$$D\left(-\tfrac{1}{3}, \tfrac{11}{3}\right) = 4(-1 - 1) = -8 < 0$$

from which we deduce that $\left(-\frac{1}{3}, \frac{11}{3}\right)$ gives rise to the saddle point $\left(-\frac{1}{3}, \frac{11}{3}, -\frac{373}{27}\right)$ of $f$. Next, to test the critical point $(1, 5)$, we compute

$$D(1, 5) = 4(3 - 1) = 8 > 0$$

which indicates that $(1, 5)$ gives a relative extremum of $f$. Since

$$f_{xx}(1, 5) = 6(1) = 6 > 0$$

we see that $(1, 5)$ yields a relative minimum of $f$. Its value is

$$f(1, 5) = (1)^3 + (5)^2 - 2(1)(5) + 7(1) - 8(5) + 2$$
$$= -15$$

The graph and contour map of $f$ are shown in Figures 6a and 6b.

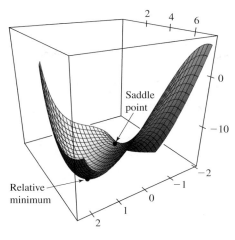

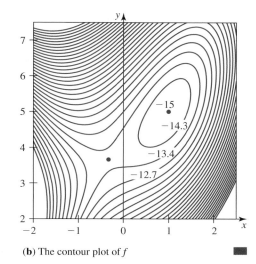

**FIGURE 6**    (a) The graph of $f(x, y) = x^3 + y^2 - 2xy + 7x - 8y + 2$    (b) The contour plot of $f$

**EXAMPLE 5** **Priority Mail Regulations**    Postal regulations specify that the combined length and girth of a package sent by priority mail may not exceed 108 in. Find the dimensions of a rectangular package with the greatest possible volume satisfying these regulations.

**FIGURE 7**
The combined length and girth of the package is $x + 2y + 2z$ inches.

**Solution**    Let the length, width, and height of the package be $x$, $y$, and $z$ inches respectively, as shown in Figure 7. Then the volume of the package is $V = xyz$. Observe that the combined length and girth of the package is $(x + 2y + 2z)$ inches. Clearly, we should let this quantity be as large as possible, that is, we should let

$$x + 2y + 2z = 108$$

With the help of this equation we can express $V$ as a function of two variables. For example, solving the equation for $x$ in terms of $y$ and $z$, we obtain

$$x = 108 - 2y - 2z$$

which, upon substitution into the expression for $V$, gives

$$V = f(y, z) = (108 - 2y - 2z)yz = 108yz - 2y^2z - 2yz^2$$

To find the critical points of $f$, we set

$$f_y = 108z - 4yz - 2z^2 = 2z(54 - 2y - z) = 0$$

and

$$f_z = 108y - 2y^2 - 4yz = 2y(54 - y - 2z) = 0$$

Since $y$ and $z$ are both nonzero (otherwise, $V$ would be zero), we are led to the system

$$54 - 2y - \phantom{2}z = 0$$
$$54 - \phantom{2}y - 2z = 0$$

Multiplying the second equation by 2 gives $108 - 2y - 4z = 0$. Then subtracting this equation from the first equation gives $-54 + 3z = 0$, or $z = 18$. Substituting this value of $z$ into either equation in the system then yields $y = 18$. Therefore, the only critical point of $f$ is $(18, 18)$.

We could use the Second Derivative Test to show that the point $(18, 18)$ gives a relative maximum of $V$, or, as in this situation, we can simply argue from physical considerations that $V$ must attain an absolute maximum at $(18, 18)$. Finally, from the equation

$$x = 108 - 2y - 2z$$

found earlier, we see that when $y = z = 18$,

$$x = 108 - 2(18) - 2(18) = 36$$

Therefore, the dimensions of the required package are 18 in. $\times$ 18 in. $\times$ 36 in. ■

## ■ Finding the Absolute Extremum Values of a Continuous Function on a Closed Set

Recall that if $f$ is a continuous function of one variable on a closed interval $[a, b]$, then the Extreme Value Theorem guarantees that $f$ has an absolute maximum value and an absolute minimum value. The analog of this theorem for a function of two variables follows.

---

**THEOREM 3    The Extreme Value Theorem for Functions of Two Variables**

If $f$ is continuous on a closed, bounded set $D$ in the plane, then $f$ attains an absolute maximum value $f(a, b)$ at some point $(a, b)$ in $D$ and an absolute minimum value $f(c, d)$ at some point $(c, d)$ in $D$.

---

The following procedure for finding the extreme values of a function of two variables is the analog of the one for finding the extreme values of a function of one variable discussed in Section 3.1.

---

**Finding the Absolute Extremum Values of $f$ on a Closed, Bounded Set $D$**

1. Find the values of $f$ at the critical points of $f$ in $D$.
2. Find the extreme values of $f$ on the boundary of $D$.
3. The absolute maximum value of $f$ and the absolute minimum value of $f$ are precisely the largest and the smallest numbers found in Steps 1 and 2.

---

The justification for this procedure is similar to that for a function of one variable on a closed interval $[a, b]$: If an absolute extremum of $f$ occurs in the interior of $D$, then it must also be a relative extremum of $f$, and hence it must occur at a critical point of $f$. Otherwise, the absolute extremum of $f$ must occur at a boundary point of $D$.

**EXAMPLE 6**  Find the absolute maximum and the absolute minimum values of the function $f(x, y) = 2x^2 + y^2 - 4x - 2y + 3$ on the rectangle

$$D = \{(x, y) \mid 0 \leq x \leq 3, 0 \leq y \leq 2\}$$

**Solution**  Since $f$ is a polynomial, it is continuous on the closed, bounded set $D$. Therefore, Theorem 3 guarantees the existence of an absolute maximum and an absolute minimum value of $f$ on $D$.

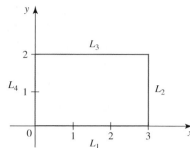

**FIGURE 8**
The boundary of $D$ consists of the four line segments $L_1$, $L_2$, $L_3$, and $L_4$.

**Step 1** To find the critical point of $f$ in $D$, we set $f_x = 4x - 4 = 0$ and $f_y = 2y - 2 = 0$. Solving this system of equations gives $(1, 1)$ as the only critical point of $f$. The value of $f$ at this point is $f(1, 1) = 0$.

**Step 2** Next, we look for extreme values of $f$ on the boundary of $D$. We can think of this boundary as being made up of four line segments $L_1$, $L_2$, $L_3$, and $L_4$, as shown in Figure 8.

**On $L_1$:** Here $y = 0$, so we have

$$f(x, 0) = 2x^2 - 4x + 3 \qquad 0 \le x \le 3$$

To find the extreme values of the continuous function $f(x, 0)$ of *one* variable on the closed bounded interval $[0, 3]$, we use the method of Section 3.1. Setting

$$f'(x, 0) = 4x - 4 = 0$$

gives $x = 1$ as the only critical number of $f(x, 0)$ in $(0, 3)$. Evaluating $f(x, 0)$ at $x = 1$, as well as at the endpoints of the interval $[0, 3]$, gives $f(0, 0) = 3$, $f(1, 0) = 1$, and $f(3, 0) = 9$. Thus, $f$ has the absolute minimum value of 1 and the absolute maximum value of 9 on $L_1$.

**On $L_2$:** Here $x = 3$, so we have

$$f(3, y) = y^2 - 2y + 9 \qquad 0 \le y \le 2$$

Setting $f'(3, y) = 2y - 2 = 0$ yields $y = 1$ as the only critical number of $f(3, y)$ in $(0, 2)$. Evaluating $f(3, y)$ at the endpoints of $[0, 2]$ and at the critical number $y = 1$ gives $f(3, 0) = 9$, $f(3, 1) = 8$, and $f(3, 2) = 9$. We see that $f$ has the absolute minimum value of 8 and the absolute maximum value of 9 on $L_2$.

**On $L_3$:** Here $y = 2$, so we have

$$f(x, 2) = 2x^2 - 4x + 3 \qquad 0 \le x \le 3$$

Setting $f'(x, 2) = 4x - 4 = 0$ gives $x = 1$ as the only critical number of $f(x, 2)$ in $(0, 3)$. Since $f(0, 2) = 3$, $f(1, 2) = 1$, and $f(3, 2) = 9$, we see that $f(x, 2)$ has the absolute minimum value of 1 and the absolute maximum value of 9 on $L_3$.

**On $L_4$:** Here $x = 0$, so we have

$$f(0, y) = y^2 - 2y + 3 \qquad 0 \le y \le 2$$

Setting $f'(0, y) = 2y - 2 = 0$ gives $y = 1$ as the only critical number of $f(0, y)$ in $(0, 2)$. Since $f(0, 0) = 3$, $f(0, 1) = 2$, and $f(0, 2) = 3$, we see that $f(0, y)$ has the absolute minimum value of 2 and the absolute maximum value of 3 on $L_4$.

**Step 3** Table 1 summarizes the results of our computations. Comparing the value of $f$ obtained at the various points, we conclude that the absolute minimum value of $f$ on $D$ is 0 attained at the critical point $(1, 1)$ of $f$ and that the absolute maximum value of $f$ on $D$ is 9 attained at the boundary points $(3, 0)$ and $(3, 2)$.

**TABLE 1**

| | Critical point | Boundary point on $L_1$ | | Boundary point on $L_2$ | | | Boundary point on $L_3$ | | Boundary point on $L_4$ | | |
|---|---|---|---|---|---|---|---|---|---|---|---|
| $(x, y)$ | $(1, 1)$ | $(1, 0)$ | $(3, 0)$ | $(3, 1)$ | $(3, 0)$ | $(3, 2)$ | $(1, 2)$ | $(3, 2)$ | $(0, 1)$ | $(0, 0)$ | $(0, 2)$ |
| **Extreme value:** $f(x, y)$ | 0 | 1 | 9 | 8 | 9 | 9 | 1 | 9 | 2 | 3 | 3 |

## 12.8 CONCEPT QUESTIONS

**1. a.** What does it mean when one says that $f$ has a relative maximum (relative minimum) at a point $(a, b)$ in the domain of $f$? What is $f(a, b)$ called in each case?

**b.** What does it mean when one says that $f$ has an absolute maximum (absolute minimum) at $(a, b)$? What is $f(a, b)$ called in each case?

**2. a.** What is a critical point of a function $f(x, y)$?

**b.** What role does a critical point of a function play in the determination of the relative extrema of the function?

**c.** State the Second Derivative Test for a function of two variables.

**3. a.** What can you say about the existence of a maximum value and a minimum value of a continuous function of two variables defined on a closed, bounded set on the plane?

**b.** Describe a strategy for finding the absolute extreme values of a continuous function on a closed, bounded set in the plane.

## 12.8 EXERCISES

*In Exercises 1–22, find and classify the relative extrema and saddle points of the function.*

**1.** $f(x, y) = x^2 + y^2 - 2x + 4y$

**2.** $f(x, y) = 2x^2 + y^2 - 6x + 2y + 1$

**3.** $f(x, y) = -x^2 - 3y^2 + 4x - 6y + 8$

**4.** $f(x, y) = -2x^2 - 3y^2 + 6x - 4y - 6$

**5.** $f(x, y) = x^2 + 3xy + 3y^2$

**6.** $f(x, y) = x^2 + 3xy + 2y^2 + 1$

**7.** $f(x, y) = 2x^2 + y^2 - 2xy - 8x - 2y + 2$

**8.** $f(x, y) = x^2 + 3y^2 - 6xy - 2x + 4y$

**9.** $f(x, y) = x^2 + 2y^2 + x^2y + 3$

**10.** $f(x, y) = x^2 - y^2 + 2xy^2 + 1$

**11.** $f(x, y) = x^2 + 5y^2 + x^2y + 2y^3$

**12.** $f(x, y) = x^3 - 3xy + y^3 + 3$

**13.** $f(x, y) = x^2 - 6x - x\sqrt{y} + y$

**14.** $f(x, y) = xy(3 - x - y)$

**15.** $f(x, y) = \dfrac{x^2y^2 - 2y - 4x}{xy}$

**16.** $f(x, y) = -\dfrac{4y}{x^2 + y^2 + 1}$

**17.** $f(x, y) = e^{-x^2 - y^2}$

**18.** $f(x, y) = e^x \sin y, \quad x \geq 0, \quad 0 \leq y \leq 2\pi$

**19.** $f(x, y) = x \sin y, \quad x \geq 0, \quad 0 \leq y \leq 2\pi$

**20.** $f(x, y) = xe^x \sin y, \quad x \geq 0, \quad 0 \leq y \leq 2\pi$

**21.** $f(x, y) = e^{-x} \cos y, \quad x \geq 0, \quad 0 \leq y \leq 2\pi$

**22.** $f(x, y) = \sin x + \sin y, \quad 0 \leq x \leq 2\pi, \quad 0 \leq y \leq 2\pi$

*In Exercises 23 and 24, (a) use the graph and the contour map of $f$ to estimate the relative extrema and saddle point(s) of $f$, and (b) verify your guess analytically.*

**23.** $f(x, y) = x^3 - 3xy^2 + y^4$

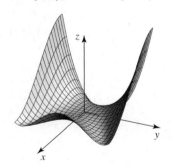
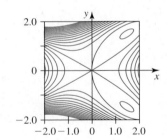

**24.** $f(x, y) = 3xy^2 - x^3$

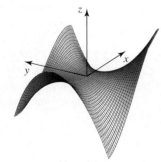
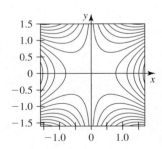

**cas** In Exercises 25–28, plot the graph and the contour map of f, and use them to estimate the relative maximum and minimum values and the saddle point(s) of f. Then find these values and saddle point(s) analytically.

**25.** $f(x, y) = \dfrac{1}{2}x^4 - 2x^3 + 4xy + y^2$

**26.** $f(x, y) = (x^2 + y^2)e^{-y}$

**27.** $f(x, y) = xy - \dfrac{2}{x} - \dfrac{4}{y} + 8$

**28.** $f(x, y) = 6xy^2 - 2x^3 - 3y^4$  ("Monkey Saddle")

**cas** In Exercises 29–32, use a graphing calculator or computer to find the critical points of f correct to three decimal places. Then use these results to find the relative extrema of f. Plot the graph of f.

**29.** $f(x, y) = 2x^4 - 8x^2 + y^2 + 4x - 2y - 5$

**30.** $f(x, y) = 2 - 2x^2 + 5xy + 2y - y^4$

**31.** $f(x, y) = -x^4 - y^4 + 2x^2y + x^2 + y - 2$

**32.** $f(x, y) = x^4 - 2x^2 + x + y^2 + e^{-y}$

In Exercises 33–40, find the absolute extrema of the function on the set D.

**33.** $f(x, y) = 2x + 3y - 6$;
$D = \{(x, y) \mid 0 \le x \le 2, -2 \le y \le 3\}$

**34.** $f(x, y) = x^2 + xy + y^2$;
$D = \{(x, y) \mid -2 \le x \le 2, -1 \le y \le 1\}$

**35.** $f(x, y) = 3x + 4y - 12$;  $D$ is the closed triangular region with vertices $(0, 0)$, $(3, 0)$, and $(3, 4)$.

**36.** $f(x, y) = 3x^2 + 2xy + y^2$;  $D$ is the closed triangular region with vertices $(-2, -1)$, $(1, -1)$, and $(1, 2)$.

**37.** $f(x, y) = xy - x^2$;  $D$ is the region bounded by the parabola $y = x^2$ and the line $y = 4$.

**38.** $f(x, y) = 4x^2 + y^2$;  $D$ is the region bounded by the parabola $y = 4 - x^2$ and the x-axis.

**39.** $f(x, y) = x^2 + 4y^2 + 3x - 1$;  $D = \{(x, y) \mid x^2 + y^2 \le 4\}$

**40.** $f(x, y) = 4x^2 + y^2 + 2x - y$;  $D = \{(x, y) \mid 4x^2 + y^2 \le 1\}$

**41.** Find the shortest distance from the origin to the plane $x + 2y + z = 4$.
Hint: The square of the distance from the origin to any point $(x, y, z)$ on the plane is $d^2 = x^2 + y^2 + z^2 = x^2 + y^2 + (4 - x - 2y)^2$.
Minimize $d^2 = f(x, y) = x^2 + y^2 + (4 - x - 2y)^2$.

**42.** Find the point on the plane $x + 2y - z = 5$ that is closest to the point $(2, 3, -1)$.
Hint: Study the hint of Exercise 41.

**43.** Find the points on the surface $z^2 = xy - x + 4y + 21$ that are closest to the origin. What is the shortest distance from the origin to the surface?

**44.** Find the points on the surface $xy^2z = 4$ that are closest to the origin. What is the shortest distance from the origin to the surface?

**45.** Find three positive real numbers whose sum is 500 and whose product is as large as possible.

**46.** Find the dimensions of an open rectangular box of maximum volume that can be constructed from 48 ft$^2$ of cardboard.

**47.** Find the dimensions of a closed rectangular box of maximum volume that can be constructed from 48 ft$^2$ of cardboard.

**48.** An open rectangular box having a volume of 108 in$^3$. is to be constructed from cardboard. Find the dimensions of such a box if the amount of cardboard used in its construction is to be minimized.

**49.** Find the dimensions of the rectangular box of maximum volume with faces parallel to the coordinate planes that can be inscribed in the ellipsoid

$$\frac{x^2}{4} + \frac{y^2}{9} + \frac{z^2}{16} = 1$$

**50.** Solve the problem posed in Exercise 49 for the general case of an ellipsoid with equation

$$\frac{x^2}{a^2} + \frac{y^2}{b^2} + \frac{z^2}{c^2} = 1$$

where $a$, $b$, and $c$ are positive real numbers.

**51.** Find the dimensions of the rectangular box of maximum volume lying in the first octant with three of its faces lying in the coordinate planes and one vertex lying in the plane $2x + 3y + z = 6$. What is the volume of such a box?

**52.** Solve the problem posed in Exercise 51 for the general case of a plane with equation

$$\frac{x}{a} + \frac{y}{b} + \frac{z}{c} = 1$$

where $a$, $b$, and $c$ are positive real numbers.

**53.** An open rectangular box is to have a volume of 12 ft$^3$. If the material for its base costs three times as much (per square foot) as the material for its sides, what are the dimensions of the box that can be constructed at a minimum cost?

**54.** A closed rectangular box is to have a volume of 16 ft$^3$. If the material for its base costs twice as much (per square foot) as the material for its top and sides, find the dimensions of the box that can be constructed at a minimum cost.

**55. Locating a Radio Station** The following figure shows the locations of three neighboring communities. The operators of a newly proposed radio station have decided that the site $P(x, y)$ for the station should be chosen so that the sum of the squares of the distances from the site to each community is minimized. Find the location of the proposed radio station.

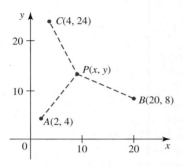

**56. Parcel Post Regulations** Postal regulations specify that a parcel sent by parcel post may have a combined length and girth of no more than 130 inches. Find the dimensions of a cylindrical package of greatest volume that can be sent through the mail. What is the volume of such a package?
**Hint:** The length plus the girth is $2\pi r + l$.

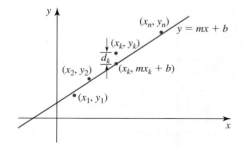

**57.** Suppose a relationship exists between two quantities $x$ and $y$ and that we have obtained the following data relating $y$ to $x$:

| $x$ | $x_1$ | $x_2$ | $\cdots$ | $x_n$ |
|---|---|---|---|---|
| $y$ | $y_1$ | $y_2$ | $\cdots$ | $y_n$ |

The figure below shows the points $(x_1, y_1)$, $(x_2, y_2)$, ..., $(x_n, y_n)$ plotted in the $xy$-plane. (This figure is called a **scatter diagram.**) If the data points are scattered about a straight line, as in this illustration, then it is reasonable to describe the relationship between $x$ and $y$ in terms of a linear equation $y = mx + b$.

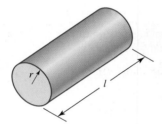

One criterion for determining the straight line that "best" fits the data calls for minimizing the sum of the squares of the deviations $\sum_{k=1}^{n} d_k^2$, where $d_k = y_k - (mx_k + b) = y_k - mx_k - b$. This sum is a function of $m$ and $b$; that is,

$$g(m, b) = \sum_{k=1}^{n} (y_k - mx_k - b)^2$$

and it is minimized with respect to the variables $m$ and $b$ by solving the system comprising the equations $g_m(a, b) = 0$ and $g_b(a, b) = 0$ for $m$ and $b$. Show that this leads to the system

$$\left(\sum_{k=1}^{n} x_k\right)m + nb = \sum_{k=1}^{n} y_k$$

$$\left(\sum_{k=1}^{n} x_k^2\right)m + \left(\sum_{k=1}^{n} x_k\right)b = \sum_{k=1}^{n} x_k y_k$$

This method of determining the equation $y = mx + b$ is called the **method of least squares,** and the line with equation $y = mx + b$ is called a **least squares** or **regression line.**

**58.** Use the method of least squares (Exercise 57) to find the straight line $y = mx + b$ that best fits the data points $(1, 3)$, $(2, 5)$, $(3, 5)$, $(4, 7)$, and $(5, 8)$. Plot the scatter diagram, and sketch the graph of the regression line on the same set of axes.

**59. Information Security Software Sales** Refer to Exercise 57. As online attacks persist, spending on information security software continues to rise. The following table gives the forecast for the worldwide sales (in billions of dollars) of information security software through 2007 ($x = 0$ corresponds to 2002).

| Year, $x$ | 0 | 1 | 2 | 3 | 4 | 5 |
|---|---|---|---|---|---|---|
| Spending, $y$ | 6.8 | 8.3 | 9.8 | 11.3 | 12.8 | 14.9 |

**a.** Find an equation of the least-squares line for these data.
**b.** Use the result of part (a) to forecast the spending on information security software in 2010, assuming that this trend continues.
*Source:* International Data Corp.

**60. Male Life Expectancy At 65** Refer to Exercise 57. The projections of male life expectancy at age 65 in the United States are summarized in the following table ($x = 0$ corresponds to 2000).

| Year, $x$ | 0 | 10 | 20 | 30 | 40 | 50 |
|---|---|---|---|---|---|---|
| Years beyond 65, $y$ | 15.9 | 16.8 | 17.6 | 18.5 | 19.3 | 20.3 |

**a.** Find an equation of the least-squares line for these data.

**b.** Use the result of (a) to estimate the life expectancy at 65 of a male in 2040. How does this result compare with the given data for that year?

**c.** Use the result of (a) to estimate the life expectancy at 65 of a male in 2030.

*Source:* U.S. Census Bureau.

**61. Operations Management Consulting Spending** Refer to Exercise 57. The following table gives the projected operations management consulting spending (in billions of dollars) from 2005 through 2010. Here, $x = 5$ corresponds to 2005.

| Year, $x$ | 5 | 6 | 7 | 8 | 9 | 10 |
|---|---|---|---|---|---|---|
| Spending, $y$ | 40 | 43.2 | 47.4 | 50.5 | 53.7 | 56.8 |

**a.** Find an equation of the least-squares line for these data.

**b.** Use the results of part (a) to estimate the average rate of change of operations management consulting spending from 2005 through 2010.

**c.** Use the results of part (a) to estimate the amount of spending on operations management consulting in 2011, assuming that the trend continues.

*Source:* Kennedy Information.

**62.** Let $f(x, y) = x^2 - y^2 + 2xy + 2$.

**a.** Show that $f$ has no maximum or minimum values.

**cas b.** Find the maximum and minimum values of $f$ in the region $D = \{(x, y) \mid x^2 + 4y^2 \leq 4\}$.

**Hint:** On the boundary of $D$, let $x = 2 \cos t$, $y = \sin t$ for $0 \leq t \leq 2\pi$.

**63.** Let $f(x, y) = -3x^2 + 6x - 4y^2 - 4y - 3$.

**a.** Show that $f$ has no minimum value.

**cas b.** Find the maximum and minimum values of $f$ in the region $D = \{(x, y) \mid x^2 + y^2 \leq 1\}$.

**Hint:** On the boundary of $D$, let $x = \cos t$, $y = \sin t$ for $0 \leq t \leq 2\pi$.

**64.** Let $f(x, y) = Ax^2 + 2Bxy + Cy^2$, where $B^2 - 4AC \neq 0$. Find conditions in terms of $A$, $B$, and $C$ such that $f$ has a relative minimum at $(0, 0)$; a relative maximum at $(0, 0)$; and a saddle point at $(0, 0)$.

*In Exercises 65–68, determine whether the statement is true or false. If it is true, explain why it is true. If it is false, give an example to show why it is false.*

**65.** If $f(x, y)$ has a relative maximum at $(a, b)$, then $f_x(a, b) = 0$ and $f_y(a, b) = 0$.

**66.** Let $h(x, y) = f(x) + g(y)$, where $f$ and $g$ have second-order derivatives. If the graph of $f$ is concave upward on $(-\infty, \infty)$ and the graph of $g$ is concave downward on $(-\infty, \infty)$, then $h$ cannot have a relative maximum or a relative minimum at any point.

**67.** If $\nabla f(a, b) = \mathbf{0}$, then $f$ has a relative extremum at $(a, b)$.

**68.** If $f(x, y)$ has continuous second-order partial derivatives and $f_{xx}(x, y) + f_{yy}(x, y) = 0$ and $f_{xy}(x, y) \neq 0$ for all $(x, y)$, then $f$ cannot have a relative extremum.

---

# 12.9 Lagrange Multipliers

## ■ Constrained Maxima and Minima

Many practical optimization problems involve maximizing or minimizing an objective function subject to one or more constraints, or side conditions. In Example 5 of Section 12.8 we discussed the problem of maximizing the (volume) function

$$V = f(x, y, z) = xyz$$

subject to the constraint

$$g(x, y, z) = x + 2y + 2z = 108$$

In this case the constraint expresses the condition that the combined length plus girth of a package is 108 in. (the maximum allowed by postal regulations).

As another example, consider a problem encountered in the construction of an AC transformer. Here, we are required to find the cross-shaped iron core of largest surface area that can be inserted into a coil of radius $a$ (Figure 1). In terms of $x$ and $y$ we see that the surface area of the iron core is

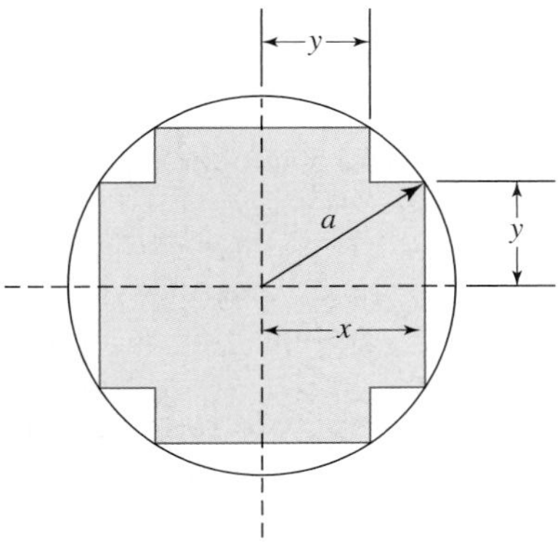

$$S = 4xy + 4y(x - y) = 8xy - 4y^2$$

**FIGURE 1**
We want to find the core of largest surface area that can be inserted into a coil of radius $a$.

Next, observe that $x$ and $y$ must satisfy the equation $x^2 + y^2 = a^2$. Therefore, the problem is equivalent to one of maximizing the objective function

$$f(x, y) = 8xy - 4y^2$$

subject to the constraint

$$g(x, y) = x^2 + y^2 = a^2$$

We will complete the solution of this problem in Example 2.

Figure 2a shows the graph of a function $f$ defined by the equation $z = f(x, y)$. Observe that $f$ has an absolute minimum at $(0, 0)$ and an absolute minimum value of $0$. However, if the independent variables $x$ and $y$ are subjected to a constraint of the form $g(x, y) = k$, then the points $(x, y, z)$ that satisfy both $z = f(x, y)$ and $g(x, y) = k$ lie on the curve $C$, the intersection of the surface $z = f(x, y)$ and the cylinder $g(x, y) = k$ (Figure 2b). From the figure you can see that the absolute minimum of $f$ subject to the constraint $g(x, y) = k$ occurs at the point $(a, b)$. Furthermore, $f$ has the **constrained** absolute minimum value $f(a, b)$ rather than the unconstrained absolute minimum value of $0$ at $(0, 0)$.

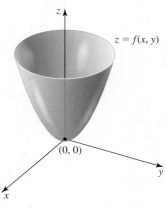

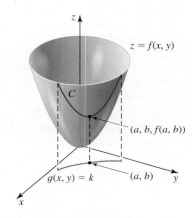

**FIGURE 2**

The function $f$ has an unconstrained minimum value of $0$, but it has a constrained minimum value of $f(a, b)$ when subjected to the constraint $g(x, y) = k$.

**(a)** $f$ is not subject to any constraints.

**(b)** $f$ is subject to a constraint.

The problem that we discussed at the beginning of this section (maximizing the volume of a box subject to a given constraint) was first solved in Section 12.8. Recall the method of solution that we used:

First, we solved the constraint equation

$$g(x, y, z) = x + 2y + 2z = 108$$

for $x$ in terms of $y$ and $z$. We then substituted this expression for $x$ into the equation

$$V = f(x, y, z) = xyz$$

thereby obtaining an expression for $V$ involving the variables $y$ and $z$ and satisfying the constraint equation. Next, we found the maximum of $V$ by treating $V$ as an unconstrained function of $y$ and $z$.

The major drawback of this method is that it relies on our ability to solve the constraint equation $g(x, y) = k$ for one variable explicitly in terms of the other (or $g(x, y, z) = k$ for one variable explicitly in terms of the other two variables in the case of a constraint involving three variables). This might not always be possible or convenient. Moreover, even when we are able to solve the constraint equation $g(x, y) = k$ for $y$ explicitly in terms of $x$, the resulting function of one variable that is obtained by substituting this expression for $y$ into the objective function $f(x, y)$ might turn out to be unnecessarily complicated.

### ■ The Method of Lagrange Multipliers

We will now consider a method, called the **method of Lagrange multipliers** (named after the French mathematician Joseph Lagrange, 1736–1813), which obviates the need to solve the constraint equation for one variable in terms of the other variables. To see how this method works, let's reexamine the problem of finding the absolute minimum of the objective function $f$ subject to the constraint $g(x, y) = k$ that we considered earlier. Figure 3a shows the level curves of $f$ drawn in the $xyz$-coordinate system. These level curves are reproduced in the $xy$-plane in Figure 3b.

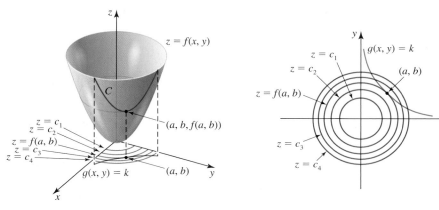

**FIGURE 3**      **(a)** The level curves of $f$ in the $xyz$-plane                    **(b)** The level curves of $f$ in the $xy$-plane

Observe that the level curves of $f$ with equations $f(x, y) = c$, where $c < f(a, b)$, have no points in common with the graph of the constraint equation $g(x, y) = k$ (for example, the level curves $f(x, y) = c_1$ and $f(x, y) = c_2$ shown in Figure 3). Thus, points lying on these curves are not candidates for the constrained minimum of $f$.

On the other hand, the level curves of $f$ with equation $f(x, y) = c$, where $c \geq f(a, b)$, do intersect the graph of the constraint equation $g(x, y) = k$ (such as the level curves of $f(x, y) = c_3$ and $f(x, y) = c_4$). These points of intersection are candidates for the constrained minimum of $f$.

Finally, observe that the larger $c$ is for $c \geq f(a, b)$, the larger the value $f(x, y)$ is for $(x, y)$ lying on the level curve $g(x, y) = k$. This observation suggests that we can find the constrained minimum of $f$ by choosing the smallest value of $c$ so that the level curve $f(x, y) = c$ still intersects the curve $g(x, y) = k$. At such a point $(a, b)$ the level curve of $f$ just touches the graph of the constraint equation $g(x, y) = k$. That is, the two curves have a common tangent at $(a, b)$ (see Figure 3b). Equivalently, their normal lines at this point coincide. Putting it yet another way, the gradient vectors $\nabla f(a, b)$ and $\nabla g(a, b)$ have the same direction, so $\nabla f(a, b) = \lambda \nabla g(a, b)$ for some scalar $\lambda$ (lambda).

A similar result holds for the problem of maximizing or minimizing a function $f$ of three variables defined by $w = f(x, y, z)$ and subject to the constraint $g(x, y, z) = k$. In this situation, $f$ has a constrained maximum or constrained minimum at a point $(a, b, c)$ where the level surface $f(x, y, z) = f(a, b, c)$ is tangent to the level surface $g(x, y, z) = k$. But this means that the normals of these surfaces, and therefore their gradient vectors, at the point $(a, b, c)$ must be parallel to each other. Thus, there is a scalar $\lambda$ such that $\nabla f(a, b, c) = \lambda \nabla g(a, b, c)$.

These geometric arguments suggest the following theorem.

---

**THEOREM 1    Lagrange's Theorem**

Let $f$ and $g$ have continuous first partial derivatives in some region $D$ in the plane. If $f$ has an extremum at a point $(a, b)$ on the smooth constraint curve $g(x, y) = c$ lying in $D$ and $\nabla g(a, b) \neq 0$, then there is a real number $\lambda$ such that

$$\nabla f(a, b) = \lambda \nabla g(a, b)$$

---

The number $\lambda$ in Theorem 1 is called a **Lagrange multiplier.**

**PROOF**    Suppose that the smooth curve $C$ described by $g(x, y) = c$ is represented by the vector function

$$\mathbf{r}(t) = x(t)\mathbf{i} + y(t)\mathbf{j}, \qquad \mathbf{r}'(t) \neq \mathbf{0}$$

where $x'$ and $y'$ are continuous on an open interval $I$ (Figure 4). Then the values assumed by $f$ on $C$ are given by

$$h(t) = f(x(t), y(t))$$

for $t$ in $I$. Suppose that $f$ has an extreme value at $(a, b)$. If $t_0$ is the point in $I$ corresponding to the point $(a, b)$, then $h$ has an extreme value at $t_0$. Therefore, $h'(t_0) = 0$. Using the Chain Rule, we have

$$
\begin{aligned}
h'(t_0) &= f_x(x(t_0), y(t_0))x'(t_0) + f_y(x(t_0), y(t_0))y'(t_0) \\
&= f_x(a, b)x'(t_0) + f_y(a, b)y'(t_0) \\
&= \nabla f(a, b) \cdot \mathbf{r}'(t_0) = 0
\end{aligned}
$$

This shows that $\nabla f(a, b)$ is orthogonal to $\mathbf{r}'(t_0)$. But as we demonstrated in Section 12.7, $\nabla g(a, b)$ is orthogonal to $\mathbf{r}'(t_0)$. Therefore, the gradient vectors $\nabla f(a, b)$ and $\nabla g(a, b)$ are parallel, so there is a number $\lambda$ such that $\nabla f(a, b) = \lambda \nabla g(a, b)$.

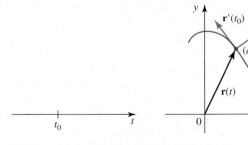

(a) The parameter interval $I$

(b) The smooth curve $C$ is represented by the vector function $\mathbf{r}(t)$.

**FIGURE 4**

The proof of Lagrange's Theorem for functions of three variables is similar to that for functions of two variables. In the case involving three variables, level surfaces rather than level curves are involved. Lagrange's Theorem leads to the following procedure for finding the constrained extremum values of functions. We state it for the case of functions of three variables.

### The Method of Lagrange Multipliers

Suppose $f$ and $g$ have continuous first partial derivatives. To find the maximum and minimum values of $f$ subject to the constraint $g(x, y, z) = k$ (assuming that these extreme values exist and that $\nabla g \neq \mathbf{0}$ on $g(x, y, z) = k$):

1. Solve the equations

$$\nabla f(x, y, z) = \lambda \nabla g(x, y, z) \qquad \text{and} \qquad g(x, y, z) = k$$

   for $x$, $y$, $z$, and $\lambda$.

2. Evaluate $f$ at each solution point found in Step 1. The largest value yields the constrained maximum of $f$, and the smallest value yields the constrained minimum of $f$.

**Note**   Since

$$\nabla f(x, y, z) = f_x(x, y, z)\mathbf{i} + f_y(x, y, z)\mathbf{j} + f_z(x, y, z)\mathbf{k}$$

and

$$\nabla g(x, y, z) = g_x(x, y, z)\mathbf{i} + g_y(x, y, z)\mathbf{j} + g_z(x, y, z)\mathbf{k}$$

we see, by equating like components, that the vector equation

$$\nabla f(x, y, z) = \lambda \nabla g(x, y, z)$$

is equivalent to the three scalar equations

$$f_x(x, y, z) = \lambda g_x(x, y, z), \quad f_y(x, y, z) = \lambda g_y(x, y, z), \quad \text{and} \quad f_z(x, y, z) = \lambda g_z(x, y, z)$$

These scalar equations together with the constraint equation $g(x, y, z) = k$ give a system of four equations to be solved for the four unknowns $x$, $y$, $z$, and $\lambda$.  ∎

**EXAMPLE 1**   Find the maximum and minimum values of the function $f(x, y) = x^2 - 2y$ subject to $x^2 + y^2 = 9$.

**Solution**   The constraint equation is $g(x, y) = x^2 + y^2 = 9$. Since

$$\nabla f(x, y) = 2x\mathbf{i} - 2\mathbf{j} \qquad \text{and} \qquad \nabla g(x, y) = 2x\mathbf{i} + 2y\mathbf{j}$$

the equation $\nabla f(x, y) = \lambda \nabla g(x, y)$ becomes

$$2x\mathbf{i} - 2\mathbf{j} = \lambda(2x\mathbf{i} + 2y\mathbf{j}) = 2\lambda x\mathbf{i} + 2\lambda y\mathbf{j}$$

Equating like components and rewriting the constraint equation lead to the following system of three equations in the three variables $x$, $y$, and $\lambda$:

$$2x = 2\lambda x \tag{1a}$$

$$-2 = 2\lambda y \tag{1b}$$

$$x^2 + y^2 = 9 \tag{1c}$$

From Equation (1a) we have

$$2x(1 - \lambda) = 0$$

so $x = 0$, or $\lambda = 1$. If $x = 0$, then Equation (1c) gives $y = \pm 3$. If $\lambda = 1$, then Equation (1b) gives $y = -1$, which upon substitution into Equation (1c) yields $x = \pm 2\sqrt{2}$. Therefore, $f$ has possible extreme values at the points $(0, -3)$, $(0, 3)$, $(-2\sqrt{2}, -1)$, and $(2\sqrt{2}, -1)$. Evaluating $f$ at each of these points gives

$$f(0, -3) = 6, \quad f(0, 3) = -6, \quad f(-2\sqrt{2}, -1) = 10, \quad \text{and} \quad f(2\sqrt{2}, -1) = 10$$

We conclude that the maximum value of $f$ on the circle $x^2 + y^2 = 9$ is 10, attained at the points $(-2\sqrt{2}, -1)$ and $(2\sqrt{2}, -1)$, and that the minimum value of $f$ on the circle is $-6$, attained at the point $(0, 3)$.

Figure 5 shows the graph of the constraint equation $x^2 + y^2 = 9$ and some level curves of the objective function $f$. Observe that the extreme values of $f$ are attained at the points where the level curves of $f$ are tangent to the graph of the constraint equation.

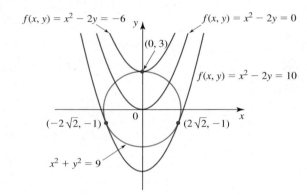

**FIGURE 5**

The extreme values of $f$ occur at the points where the level curves of $f$ are tangent to the graph of the constraint equation (the circle).

**EXAMPLE 2**    Complete the solution to the problem posed at the beginning of this section: Find the cross-shaped iron core of largest surface area that can be inserted into a coil of radius $a$ (Figure 6).

**Solution**    Recall that the problem reduces to one of finding the largest value of the objective function $f(x, y) = 8xy - 4y^2$ subject to the constraint $g(x, y) = x^2 + y^2 = a^2$. Since

$$\nabla f(x, y) = 8y\mathbf{i} + (8x - 8y)\mathbf{j} \quad \text{and} \quad \nabla g(x, y) = 2x\mathbf{i} + 2y\mathbf{j}$$

the equation $\nabla f(x, y) = \lambda \nabla g(x, y)$ becomes

$$8y\mathbf{i} + (8x - 8y)\mathbf{j} = \lambda(2x\mathbf{i} + 2y\mathbf{j}) = 2\lambda x\mathbf{i} + 2\lambda y\mathbf{j}$$

Equating like components and rewriting the constraint equation, we get the following system of three equations in the three variables $x$, $y$, and $\lambda$:

$$8y = 2\lambda x \tag{2a}$$

$$8x - 8y = 2\lambda y \tag{2b}$$

$$x^2 + y^2 = a^2 \tag{2c}$$

Solving Equation (2a) for $y$, we obtain $y = \frac{1}{4}\lambda x$. Substituting this expression for $y$ into Equation (2b) gives

$$8x - 2\lambda x = \frac{1}{2}\lambda^2 x$$

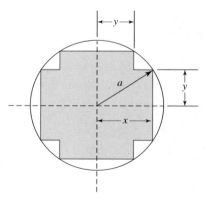

**FIGURE 6**

A cross-shaped iron core of largest surface area is to be inserted into the coil.

or

$$x(\lambda^2 + 4\lambda - 16) = 0$$

Observe that $x \neq 0$; otherwise, Equation (2a) implies that $y = 0$, so Equation (2c) becomes $0 = a^2$, which is impossible. So we have $\lambda^2 + 4\lambda - 16 = 0$. Using the quadratic formula, we obtain

$$\lambda = \frac{-4 \pm \sqrt{16 + 64}}{2} = -2 \pm 2\sqrt{5}$$

Observe that $\lambda$ must be positive; otherwise, Equation (2a) implies that $x$ or $y$ must be negative. So we choose $\lambda = -2 + 2\sqrt{5} \approx 2.4721$. Next, substituting $y = \frac{1}{4}\lambda x$ into Equation (2c) gives

$$x^2 + \frac{1}{16}\lambda^2 x^2 = a^2$$

$$x^2 \left(1 + \frac{\lambda^2}{16}\right) = a^2$$

$$x^2 \left(\frac{\lambda^2 + 16}{16}\right) = a^2$$

or

$$x = \frac{4a}{\sqrt{\lambda^2 + 16}} \approx \frac{4}{\sqrt{(2.4721)^2 + 16}} a \qquad \text{Recall that } \lambda \approx 2.4721.$$

$$\approx 0.8507a$$

Finally,

$$y = \frac{1}{4}\lambda x \approx \frac{1}{4}(2.4721)(0.8507a) \approx 0.5258a$$

Therefore, the core will have the largest surface area if $x \approx 0.8507a$ and $y \approx 0.5258a$, where $a$ is the radius of the coil.

Figure 7 shows the graph of the constraint equation $x^2 + y^2 = a^2$ (the circle of radius $a$ centered at the origin) and several level curves of the objective function $f$. Once again, observe that the maximum value of $f$, $f(0.8507a, 0.5258a) \approx 2.4725a^2$, occurs at the point $(0.8507a, 0.5258a)$, where the level curve of $f$ is tangent to the graph of the constraint equation. ∎

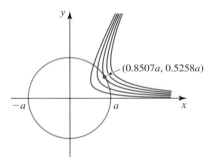

**FIGURE 7**
The maximum value of $f$ occurs at the point where the level curve of $f$ is tangent to the level curve of the constraint equation.

(0.8507a, 0.5258a)

**EXAMPLE 3**  Find the dimensions of a rectangular package having the greatest possible volume and satisfying the postal regulation that specifies that the combined length and girth of the package may not exceed 108 inches. (See Example 5 in Section 12.8.)

**Solution**  Recall that to solve this problem, we need to find the largest value of the volume function $f(x, y, z) = xyz$ subject to the constraint $g(x, y, z) = x + 2y + 2z = 108$. To solve this problem using the method of Lagrange multipliers, observe that

$$\nabla f(x, y, z) = yz\mathbf{i} + xz\mathbf{j} + xy\mathbf{k} \qquad \text{and} \qquad \nabla g(x, y, z) = \mathbf{i} + 2\mathbf{j} + 2\mathbf{k}$$

so the equation $\nabla f(x, y, z) = \lambda \nabla g(x, y, z)$ becomes

$$yz\mathbf{i} + xz\mathbf{j} + xy\mathbf{k} = \lambda(\mathbf{i} + 2\mathbf{j} + 2\mathbf{k})$$

Equating components and rewriting the constraint equation give the following system of four equations in the four variables $x$, $y$, $z$, and $\lambda$:

$$yz = \lambda \tag{3a}$$

$$xz = 2\lambda \tag{3b}$$

$$xy = 2\lambda \tag{3c}$$

$$x + 2y + 2z = 108 \tag{3d}$$

Substituting Equation (3a) into Equation (3b) yields

$$xz = 2yz \qquad \text{or} \qquad z(x - 2y) = 0$$

Since $z \neq 0$, we have $x = 2y$. Next, substituting Equation (3a) into Equation (3c) gives

$$xy = 2yz \qquad \text{or} \qquad y(x - 2z) = 0$$

Since $y \neq 0$, we have $x = 2z$. Equating the two expressions for $x$ just obtained gives

$$2y = 2z \qquad \text{or} \qquad y = z$$

Finally, substituting the expressions for $x$ and $y$ into Equation (3d) gives

$$2z + 2z + 2z = 108 \qquad \text{or} \qquad z = 18$$

So $y = 18$ and $x = 2(18) = 36$. Therefore, the dimensions of the package are 18 in. $\times$ 18 in. $\times$ 36 in., as was obtained before. ∎

## Interpreting Our Results

Geometrically, this problem is one of finding the point on the plane $x + 2y + 2z = 108$ at which $f(x, y, z) = xyz$ has the largest value. The point $(36, 18, 18)$ is precisely the point at which the level surface $xyz = f(36, 18, 18) = 11{,}664$ is tangent to the plane $x + 2y + 2z = 108$.

**EXAMPLE 4** Find the dimensions of the open rectangular box of maximum volume that can be constructed from a rectangular piece of cardboard having an area of 48 ft$^2$. What is the volume of the box?

**Solution** Let the length, width, and height of the box (in feet) be $x$, $y$, and $z$, as shown in Figure 8. Then the volume of the box is $V = xyz$. The area of the bottom of the box plus the area of the four sides is

$$xy + 2xz + 2yz$$

square feet, and this is equal to the area of the cardboard; that is,

$$xy + 2xz + 2yz = 48$$

Thus, the problem is one of maximizing the objective function

$$f(x, y, z) = xyz$$

subject to the constraint

$$g(x, y, z) = xy + 2xz + 2yz = 48$$

Since

$$\nabla f(x, y, z) = yz\mathbf{i} + xz\mathbf{j} + xy\mathbf{k}$$

**FIGURE 8**

An open rectangular box of maximum volume is to be constructed from a piece of cardboard. What are the dimensions of the box?

and

$$\nabla g(x, y, z) = (y + 2z)\mathbf{i} + (x + 2z)\mathbf{j} + (2x + 2y)\mathbf{k}$$

the equation $\nabla f(x, y, z) = \lambda \nabla g(x, y, z)$ becomes

$$yz\mathbf{i} + xz\mathbf{j} + xy\mathbf{k} = \lambda[(y + 2z)\mathbf{i} + (x + 2z)\mathbf{j} + (2x + 2y)\mathbf{k}]$$

Equating like components and rewriting the constraint equation give the following system of four equations in the unknowns $x$, $y$, $z$, and $\lambda$:

$$yz = \lambda(y + 2z) \tag{4a}$$

$$xz = \lambda(x + 2z) \tag{4b}$$

$$xy = \lambda(2x + 2y) \tag{4c}$$

$$xy + 2xz + 2yz = 48 \tag{4d}$$

Multiplying Equations (4a), (4b), and (4c) by $x$, $y$, and $z$, respectively, gives

$$xyz = \lambda(xy + 2xz) \tag{5a}$$

$$xyz = \lambda(xy + 2yz) \tag{5b}$$

$$xyz = \lambda(2xz + 2yz) \tag{5c}$$

From Equations (5a) and (5b), we obtain

$$\lambda(xy + 2xz) = \lambda(xy + 2yz) \tag{6}$$

Observe that $\lambda \neq 0$; otherwise, Equations (4a), (4b), and (4c) would imply that $yz = xz = xy = 0$, thus contradicting Equation (4d). Dividing both sides of (6) by $\lambda$ and simplifying give

$$2xz = 2yz \qquad \text{or} \qquad 2z(x - y) = 0$$

Now $z \neq 0$; otherwise, Equation (4a) would imply that $\lambda = 0$, which, as was observed earlier, is impossible. Therefore, $x = y$.

Next, from Equations (5b) and (5c), we have

$$\lambda(xy + 2yz) = \lambda(2xz + 2yz) \tag{7}$$

Dividing both sides of Equation (7) by $\lambda$ and simplifying, we get

$$xy = 2xz \qquad \text{or} \qquad x(y - 2z) = 0$$

Since $x \neq 0$, we have $y = 2z$. Finally, substituting $x = y = 2z$ into Equation (4d) gives

$$4z^2 + 4z^2 + 4z^2 = 48$$

or $z = 2$ (we reject the negative root, since $z$ must be positive). Therefore, $x = y = 4$, and the dimensions of the box are 4 ft $\times$ 4 ft $\times$ 2 ft. Its volume is 32 ft$^3$. ■

## Interpreting Our Results

Geometrically, this problem is one of finding the point on the surface $xy + 2xz + 2yz = 48$ at which $f(x, y, z) = xyz$ has the largest value. The point $(4, 4, 2)$ is precisely the point at which the level surface $xyz = f(4, 4, 2) = 32$ is tangent to the surface $xy + 2xz + 2yz = 48$.

The next example shows how the method of Lagrange multipliers can be used to help find the absolute extreme values of a function on a closed, bounded set.

**EXAMPLE 5** Find the absolute extreme values of $f(x, y) = 2x^2 + y^2 - 2y + 1$ subject to the constraint $x^2 + y^2 \leq 4$.

**Solution** The inequality $x^2 + y^2 \leq 4$ defines the disk $D$, which is a closed, bounded set with boundary given by the circle $x^2 + y^2 = 4$. So, following the procedure given in Section 12.8, we first find the critical number(s) of $f$ inside $D$. Setting

$$f_x(x, y) = 4x = 0$$

$$f_y(x, y) = 2y - 2 = 2(y - 1) = 0$$

simultaneously gives $(0, 1)$ as the only critical point of $f$ in $D$.

Next, we find the critical numbers of $f$ on the boundary of $D$ using the method of Lagrange multipliers. Writing $g(x, y) = x^2 + y^2 = 4$, we have

$$\nabla f(x, y) = 4x\mathbf{i} + 2(y - 1)\mathbf{j} \quad \text{and} \quad \nabla g(x, y) = 2x\mathbf{i} + 2y\mathbf{j}$$

The equation $\nabla f(x, y) = \lambda g(x, y)$ and the constraint equation give the system

$$4x = 2\lambda x \tag{8a}$$

$$2(y - 1) = 2\lambda y \tag{8b}$$

$$x^2 + y^2 = 4 \tag{8c}$$

Equation (8a) gives

$$2x(\lambda - 2) = 0$$

that is, $x = 0$ or $\lambda = 2$. If $x = 0$, then Equation (8c) gives $y = \pm 2$. Next, if $\lambda = 2$, then Equation (8b) gives

$$2(y - 1) = 4y \quad \text{or} \quad y = -1$$

in which case $x = \pm\sqrt{3}$. So $f$ has the critical points $(0, -2)$, $(0, 2)$, $(-\sqrt{3}, -1)$ and $(\sqrt{3}, -1)$ on the boundary of $D$.

Finally, we construct the following table.

| $(x, y)$ | $f(x, y) = 2x^2 + y^2 - 2y + 1$ |
|---|---|
| $(0, 1)$ | 0 |
| $(-\sqrt{3}, -1)$ | 10 |
| $(\sqrt{3}, -1)$ | 10 |
| $(0, -2)$ | 9 |
| $(0, 2)$ | 1 |

From the table we see that $f$ has an absolute minimum value of 0 attained at $(0, 1)$ and an absolute maximum value of 10 attained at $(-\sqrt{3}, -1)$ and $(\sqrt{3}, -1)$. ∎

## Optimizing a Function Subject to Two Constraints

Some applications involve maximizing or minimizing an objective function $f$ subject to two or more constraints. Consider, for example, the problem of finding the extreme values of $f(x, y, z)$ subject to the two constraints

$$g(x, y, z) = k \quad \text{and} \quad h(x, y, z) = l$$

It can be shown that if $f$ has an extremum at $(a, b, c)$ subject to these constraints, then there are real numbers (Lagrange multipliers) $\lambda$ and $\mu$ such that

$$\nabla f(a, b, c) = \lambda \nabla g(a, b, c) + \mu \nabla h(a, b, c) \tag{9}$$

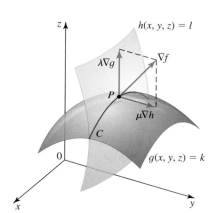

**FIGURE 9**

If $f$ has an extreme value at $P(a, b, c)$, then $\nabla f(a, b, c) = \lambda \nabla g(a, b, c) + \mu \nabla h(a, b, c)$.

Geometrically, we are looking for the extreme values of $f(x, y, z)$ on the curve of intersection of the level surfaces $g(x, y, z) = k$ and $h(x, y, z) = l$. Condition (9) is a statement that at an extremum point $(a, b, c)$, the gradient of $f$ must lie in the plane determined by the gradient of $g$ and the gradient of $h$. (See Figure 9.) The vector equation (9) is equivalent to three scalar equations. When combined with the two constraint equations, this leads to a system of five equations that can be solved for the five unknowns $x$, $y$, $z$, $\lambda$, and $\mu$.

**EXAMPLE 6** Find the maximum and minimum values of the function $f(x, y, z) = 3x + 2y + 4z$ subject to the constraints $x - y + 2z = 1$ and $x^2 + y^2 = 4$.

**Solution** Write the constraint equations in the form

$$g(x, y, z) = x - y + 2z = 1 \quad \text{and} \quad h(x, y, z) = x^2 + y^2 = 4$$

Then the equation $\nabla f(x, y, z) = \lambda \nabla g(x, y, z) + \mu \nabla h(x, y, z)$ becomes

$$3\mathbf{i} + 2\mathbf{j} + 4\mathbf{k} = \lambda(\mathbf{i} - \mathbf{j} + 2\mathbf{k}) + \mu(2x\mathbf{i} + 2y\mathbf{j})$$

$$= (\lambda + 2\mu x)\mathbf{i} + (-\lambda + 2\mu y)\mathbf{j} + 2\lambda\mathbf{k}$$

Equating like components and rewriting the constraint equations lead to the following system of five equations in the five variables, $x$, $y$, $z$, $\lambda$, and $\mu$:

$$3 = \lambda + 2\mu x \tag{10a}$$

$$2 = -\lambda + 2\mu y \tag{10b}$$

$$4 = 2\lambda \tag{10c}$$

$$x - y + 2z = 1 \tag{10d}$$

$$x^2 + y^2 = 4 \tag{10e}$$

From Equation (10c) we have $\lambda = 2$. Next, substituting this value of $\lambda$ into Equations (10a) and (10b) gives

$$3 = 2 + 2\mu x \quad \text{or} \quad 1 = 2\mu x \tag{11a}$$

and

$$2 = -2 + 2\mu y \quad \text{or} \quad 4 = 2\mu y \tag{11b}$$

Solving Equations (11a) and (11b) for $x$ and $y$ gives $x = 1/(2\mu)$ and $y = 2/\mu$. Substituting these values of $x$ and $y$ into Equation (10e) yields

$$\left(\frac{1}{2\mu}\right)^2 + \left(\frac{2}{\mu}\right)^2 = 4$$

$$1 + 16 = 16\mu^2 \quad \text{or} \quad \mu^2 = \frac{17}{16}$$

Therefore, $\mu = \pm\sqrt{17}/4$, so $x = \pm 2/\sqrt{17}$ and $y = \pm 8/\sqrt{17}$. Using Equation (10d), we have

$$z = \frac{1}{2}(1 - x + y) = \frac{1}{2}\left(1 \mp \frac{2}{\sqrt{17}} \pm \frac{8}{\sqrt{17}}\right)$$

$$= \frac{1}{2}\left(1 \pm \frac{6}{\sqrt{17}}\right)$$

The value of $f$ at the point $\left(\frac{2}{\sqrt{17}}, \frac{8}{\sqrt{17}}, \frac{1}{2} + \frac{3}{\sqrt{17}}\right)$ is

$$3\left(\frac{2}{\sqrt{17}}\right) + 2\left(\frac{8}{\sqrt{17}}\right) + 4\left(\frac{1}{2} + \frac{3}{\sqrt{17}}\right) = 2 + \frac{34}{\sqrt{17}} = 2(1 + \sqrt{17})$$

and the value of $f$ at the point $\left(-\frac{2}{\sqrt{17}}, -\frac{8}{\sqrt{17}}, \frac{1}{2} - \frac{3}{\sqrt{17}}\right)$ is

$$3\left(-\frac{2}{\sqrt{17}}\right) + 2\left(-\frac{8}{\sqrt{17}}\right) + 4\left(\frac{1}{2} - \frac{3}{\sqrt{17}}\right) = 2 - \frac{34}{\sqrt{17}} = 2(1 - \sqrt{17})$$

Therefore, the maximum value of $f$ is $2(1 + \sqrt{17})$, and the minimum value of $f$ is $2(1 - \sqrt{17})$. ∎

## 12.9  CONCEPT QUESTIONS

1. What is a constrained maximum (minimum) value problem? Illustrate with examples.
2. Describe the method of Lagrange multipliers for finding the extrema of $f(x, y)$ subject to the constraint $g(x, y) = c$. State the method for the case in which $f$ and $g$ are functions of three variables.
3. The figure at the right shows the contour map of a function $f$ and the curve of the equation $g(x, y) = 4$. Use the figure to obtain estimates of the maximum and minimum values of $f$ subject to the constraint $g(x, y) = 4$.

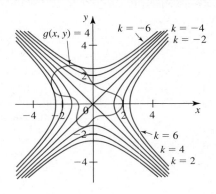

## 12.9  EXERCISES

*In Exercises 1–4, use the method of Lagrange multipliers to find the extrema of the function f subject to the given constraint. Sketch the graph of the constraint equation and several level curves of f. Include the level curves that touch the graph of the constraint equation at the points where the extrema occur.*

1. $f(x, y) = 3x + 4y$;  $x^2 + y^2 = 1$
2. $f(x, y) = x^2 + y^2$;  $2x + 4y = 5$
3. $f(x, y) = x^2 + y^2$;  $xy = 1$
4. $f(x, y) = xy$;  $x^2 + y^2 = 4$

*In Exercises 5–16, use the method of Lagrange multipliers to find the extrema of the function f subject to the given constraint.*

5. $f(x, y) = xy$;  $2x + 3y = 6$
6. $f(x, y) = x^2 - y^2$;  $x^2 + y^2 = 1$
7. $f(x, y) = xy$;  $x^2 + 4y^2 = 1$
8. $f(x, y) = 8x + 9y$;  $4x^2 + 9y^2 = 36$
9. $f(x, y) = x^2 + xy + y^2$;  $x^2 + y^2 = 8$
10. $f(x, y) = x^2 + y^2$;  $x^4 + y^4 = 1$
11. $f(x, y, z) = x + 2y + z$;  $x^2 + 4y^2 - z = 0$

12. $f(x, y, z) = x + y + z$;  $x^2 + y^2 + z^2 = 1$
13. $f(x, y, z) = x + 2y - 2z$;  $x^2 + 2y^2 + 4z^2 = 1$
14. $f(x, y, z) = x^2 + y^2 + z^2$;  $y - x = 1$
15. $f(x, y, z) = xyz$;  $x^2 + 2y^2 + \frac{1}{2}z^2 = 6$
16. $f(x, y, z) = xy + xz$;  $x^2 + y^2 + z^2 = 8$

*In Exercises 17–20, use the method of Lagrange multipliers to find the extrema of the function subject to the given constraints.*

17. $f(x, y, z) = 2x + y$;  $x + y + z = 1$,  $y^2 + z^2 = 9$
18. $f(x, y, z) = x + y + z$;  $x^2 + y^2 = 1$,  $x + z = 2$
19. $f(x, y, z) = yz + xz$;  $xz = 1$,  $y^2 + z^2 = 1$
20. $f(x, y, z) = x^2 + y^2 + z^2$;  $2x + y + z = 2$, $x - 2y + 3z = -4$

*In Exercises 21–22, use the method of Lagrange multipliers to find the extrema of the function subject to the inequality constraint.*

21. $f(x, y) = 3x^2 + 2y^2 - 2x - 1$;  $x^2 + y^2 \le 9$
22. $f(x, y) = x^2 y$;  $4x^2 + y^2 \le 4$

**23.** Find the point on the plane $x + 2y + z = 4$ that is closest to the origin.

**24.** Find the maximum and minimum distances from the origin to the curve $5x^2 + 6xy + 5y^2 - 10 = 0$.

**25.** Find the point on the plane $x + 2y - z = 5$ that is closest to the point $(2, 3, -1)$.

**26.** Find the points on the surface $z^2 = xy - x + 4y + 21$ that are closest to the origin. What is the shortest distance from the origin to the surface?

**27.** Find the points on the surface $xy^2z = 4$ that are closest to the origin. What is the shortest distance from the origin to the surface?

**28.** Find three positive real numbers whose sum is 500 and whose product is as large as possible.

**29.** Find the dimensions of a closed rectangular box of maximum volume that can be constructed from 48 ft$^2$ of cardboard.

**30.** Find the dimensions of an open rectangular box of maximum volume that can be constructed from 12 ft$^2$ of cardboard.

**31.** An open rectangular box having a volume of 108 in$^3$. is to be constructed from cardboard. Find the dimensions of such a box if the amount of cardboard used in its construction is to be minimized.

**32.** Find the dimensions of the rectangular box of maximum volume with faces parallel to the coordinate planes that can be inscribed in the ellipsoid
$$\frac{x^2}{4} + \frac{y^2}{9} + \frac{z^2}{16} = 1$$

**33.** Solve the problem posed in Exercise 32 for the general case of an ellipsoid with equation
$$\frac{x^2}{a^2} + \frac{y^2}{b^2} + \frac{z^2}{c^2} = 1$$
where $a$, $b$, and $c$ are positive real numbers.

**34.** Find the dimensions of the rectangular box of maximum volume lying in the first octant with three of its faces lying in the coordinate planes and one vertex lying in the plane $2x + 3y + z = 6$. What is the volume of the box?

**35.** Solve the problem posed in Exercise 34 for the general case of a plane with equation
$$\frac{x}{a} + \frac{y}{b} + \frac{z}{c} = 1$$
where $a$, $b$, and $c$ are positive real numbers.

**36.** An open rectangular box is to have a volume of 12 ft$^3$. If the material for its base costs three times as much (per square foot) as the material for its sides, what are the dimensions of the box that can be constructed at the minimum cost?

**37.** A rectangular box is to have a volume of 16 ft$^3$. If the material for its base costs twice as much (per square foot) as the material for its top and sides, find the dimensions of the box that can be constructed at the minimum cost.

**38. Maximizing Profit** The total daily profit (in dollars) realized by Weston Publishing in publishing and selling its dictionaries is given by the profit function
$$P(x, y) = -0.005x^2 - 0.003y^2 - 0.002xy + 14x + 12y - 200$$
where $x$ stands for the number of deluxe editions and $y$ denotes the number of standard editions sold daily. Weston's management decides that publication of these dictionaries should be restricted to a total of exactly 400 copies per day. How many deluxe copies and how many standard copies should be published each day to maximize Weston's daily profit?

**39. Cobb-Douglas Production Function** Suppose $x$ units of labor and $y$ units of capital are required to produce
$$f(x, y) = 100x^{3/4}y^{1/4}$$
units of a certain product. If each unit of labor costs \$200 and each unit of capital costs \$300 and a total of \$60,000 is available for production, determine how many units of labor and how many units of capital should be used to maximize production.

**40. a.** Find the distance between the point $P(x_1, y_1)$ and the line $ax + by + c = 0$ using the method of Lagrange multipliers.
   **b.** Use the result of part (a) to find the distance between the point $(2, -1)$ and the line $2x + 3y - 6 = 0$.

**41.** Let $f(x, y) = x - y$ and $g(x, y) = x + x^5 - y$.
   **a.** Use the method of Lagrange multipliers to find the point(s) where $f$ may have a relative maximum or relative minimum subject to the constraint $g(x, y) = 1$.
   **b.** Plot the graph of $g$ and the level curves of $f(x, y) = k$ for $k = -2, -1, 0, 1, 2$, using the viewing window $[-4, 4] \times [-4, 4]$. Then use this to explain why the point(s) found in part (a) does not give rise to a relative maximum or a relative minimum of $f$
   **c.** Verify the observation made in part (b) analytically.

**42.** Let $f(x, y) = x^2 - y^2$, and let $g(x, y) = x + y$.
   **a.** Show that $f$ has no maximum or minimum values when subjected to the constraint $g(x, y) = 1$.
   **b.** What happens when you try to use the method of Lagrange multipliers to find the extrema of $f$ subject to $g(x, y) = 1$? Does this contradict Theorem 1?

**43.** Find the point on the line of intersection of the planes $x + 2y - 3z = 9$ and $2x - 3y + z = 4$ that is closest to the origin.

**44.** Find the shortest distance from the origin to the curve with equation $y = (x - 1)^{3/2}$. Explain why the method of Lagrange multipliers fails to give the solution.

**45. a.** Find the maximum distance from the origin to the Folium of Descartes, $x^3 + y^3 - 3axy = 0$, where $a > 0$, $x \geq 0$ and $y \geq 0$, using symmetry.
   **b.** Verify the result of part (a), using the method of Lagrange multipliers.

*In Exercises 46 and 47, use the fact that a vector in n-space has the form $v = \langle v_1, v_2, \ldots, v_n \rangle$ and the gradient of a function of n variables, $f(x_1, x_2, \ldots, x_n)$, is defined by $\nabla f = \langle f_{x_1}, f_{x_2}, \ldots, f_{x_n} \rangle$. Also, assume that Theorem 1 holds for the n-dimensional case.*

**46. a.** Find the maximum value of

$$f(x_1, x_2, \ldots, x_n, y_1, y_2, \ldots, y_n)$$

$$= \sum_{i=1}^{n} x_i y_i = x_1 y_1 + x_2 y_2 + \cdots + x_n y_n$$

subject to the constraints

$$\sum_{i=1}^{n} x_i^2 = x_1^2 + x_2^2 + \cdots + x_n^2 = 1$$

and

$$\sum_{i=1}^{n} y_i^2 = y_1^2 + y_2^2 + \cdots + y_n^2 = 1$$

**b.** Use the result of part (a) to show that if $a_1, a_2, \ldots, a_n$, $b_1, b_2, \ldots, b_n$ are any numbers, then

$$\sum_{i=1}^{n} a_i b_i \leq \sqrt{\sum_{i=1}^{n} a_i^2} \sqrt{\sum_{i=1}^{n} b_i^2}$$

**Hint:** Put $x_i = \dfrac{a_i}{\sqrt{\sum_{i=1}^{n} a_i^2}}$ and $y_i = \dfrac{b_i}{\sqrt{\sum_{i=1}^{n} b_i^2}}$.

**Note:** This inequality is called the Cauchy-Schwarz Inequality. (Compare this with Exercise 14 in the Challenge Problems for Chapter 4.)

**47. a.** Let $p$ and $q$ be positive numbers satisfying $(1/p) + (1/q) = 1$. Find the minimum value of

$$f(x, y) = \frac{x^p}{p} + \frac{y^q}{q} \qquad x > 0, \quad y > 0$$

subject to the constraint $xy = c$, where $c$ is a constant.

**b.** Use the result of part (a) to show that if $x$ and $y$ are positive numbers, then

$$\frac{x^p}{p} + \frac{y^q}{q} \geq xy$$

where $p > 0$ and $q > 0$ and $(1/p) + (1/q) = 1$.

**48. a.** Let $x_1, x_2, \ldots, x_n$ be positive numbers. Find the maximum value of

$$f(x_1, x_2, \ldots, x_n) = \sqrt[n]{x_1 x_2 \cdots x_n}$$

subject to the constraint $x_1 + x_2 + \cdots + x_n = c$, where $c$ is a constant.

**b.** Use the result of part (a) to show that if $x_1, x_2, \ldots, x_n$ are positive numbers, then

$$\sqrt[n]{x_1 x_2 \cdots x_n} \leq \frac{x_1 + x_2 + \cdots + x_n}{n}$$

This shows that the geometric mean of $n$ positive numbers cannot exceed the arithmetic mean of the numbers.

**49. Snell's Law of Refraction** According to Fermat's Principle in optics, the path $POQ$ taken by a ray of light (see the figure below) in traveling across the plane separating two optical media is such that the time taken is minimal. Using this principle, derive Snell's Law of Refraction:

$$\frac{v_1}{\sin \theta_1} = \frac{v_2}{\sin \theta_2}$$

where $\theta_1$ is the angle of incidence, $\theta_2$ is the angle of refraction, and $v_1$ and $v_2$ are the speeds of light in the two media.

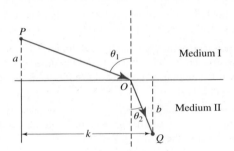

**Hint:** Show that the time taken by the ray of light in traveling from $P$ to $Q$ is

$$t = \frac{a}{v_1 \cos \theta_1} + \frac{b}{v_2 \cos \theta_2}$$

Then minimize $t = f(\cos \theta_1, \cos \theta_2)$ subject to $a \tan \theta_1 + b \tan \theta_2 = k$, where $k$ is a constant.

*In Exercises 50–52, determine whether the statement is true or false. If it is true, explain why it is true. If it is false, give an example to show why it is false.*

**50.** Suppose $f$ and $g$ have continuous first partial derivatives in some region $D$ in the plane. If $f$ has an extremum at a point $(a, b)$ subject to the constraint $g(x, y) = c$, then there exists a constant $\lambda$ such that $(a, b)$ is a critical point of $F = f + \lambda g$; that is, $F_x(a, b) = 0$, $F_y(a, b) = 0$, and $F_\lambda(a, b) = 0$.

**51.** If $(a, b)$ gives rise to a (constrained) extremum of $f$ subject to the constraint $g(x, y) = 0$, then $(a, b)$ also gives rise to an unconstrained extremum of $f$.

**52.** If $(a, b)$ gives rise to a (constrained) extremum of $f$ subject to the constraint $g(x, y) = 0$, then $f_x(a, b) = 0$ and $f_y(a, b) = 0$ simultaneously.

# CHAPTER 12  REVIEW

## CONCEPT REVIEW

*In Exercises 1–17, fill in the blanks.*

**1. a.** A function $f$ of two variables, $x$ and $y$, is a _____ that assigns to each ordered pair _____ in the domain of $f$, exactly one real number $f(x, y)$.

  **b.** The number $z = f(x, y)$ is called a _____ variable, and $x$ and $y$ are _____ variables. The totality of the numbers $z$ is called the _____ of the function $f$.

  **c.** The graph of $f$ is the set $S =$ _____.

**2. a.** The curves in the $xy$-plane with equation $f(x, y) = k$, where $k$ is a constant in the range of $f$, are called the _____ _____ of $f$.

  **b.** A level surface of a function $f$ of three variables is the graph of the equation _____, where $k$ is a constant in the range of _____.

**3.** $\lim_{(x, y) \to (a, b)} f(x, y) = L$ means there exists a number _____ such that $f(x, y)$ can be made as close to _____ as we please by restricting $(x, y)$ to be sufficiently close to _____.

**4.** If $f(x, y)$ approaches $L_1$ as $(x, y)$ approaches $(a, b)$ along one path, and $f(x, y)$ approaches $L_2$ as $(x, y)$ approaches $(a, b)$ along another path with $L_1 \neq L_2$, then $\lim_{(x, y) \to (a, b)} f(x, y)$ _____ _____ exist.

**5. a.** $f(x, y)$ is continuous at $(a, b)$ if $\lim_{(x, y) \to (a, b)} f(x, y) =$ _____.

  **b.** $f(x, y)$ is continuous on a region $R$ if $f$ is continuous at every point $(x, y)$ in _____.

**6. a.** A polynomial function is continuous _____; a rational function is continuous at all points in its _____.

  **b.** If $f$ is continuous at $(a, b)$ and $g$ is continuous at $f(a, b)$, then the composite function $h = g \circ f$ is continuous at _____.

**7. a.** The partial derivative of $f(x, y)$ with respect to $x$ is _____ if the limit exists. The partial derivative $(\partial f/\partial x)(a, b)$ gives the slope of the tangent line to the curve obtained by the intersection of the plane _____ and the graph of $z = f(x, y)$ at _____; it also measures the rate of change of $f(x, y)$ in the _____-direction with $y$ held _____ at _____.

  **b.** To compute $\partial f/\partial x$ where $f$ is a function of $x$ and $y$, treat _____ as a constant and differentiate with respect to _____ in the usual manner.

**8.** If $f(x, y)$ and its partial derivatives $f_x, f_y, f_{xy}$, and $f_{yx}$ are continuous on an open region $R$, then $f_{xy}(x, y) =$ _____ for all $(x, y)$ in $R$.

**9. a.** The total differential $dz$ of $z = f(x, y)$ is $dz =$ _____.

  **b.** If $\Delta z = f(x + \Delta x, y + \Delta y) - f(x, y)$, then $\Delta z \approx$ _____.

  **c.** $\Delta z = f_x(x, y)\, \Delta x + f_y(x, y)\, \Delta y + \varepsilon_1\, \Delta x + \varepsilon_2\, \Delta y$, where $\varepsilon_1$ and $\varepsilon_2$ are functions of _____ and _____ such that $\lim_{(\Delta x, \Delta y) \to (0, 0)} \varepsilon_1 =$ _____ and $\lim_{(\Delta x, \Delta y) \to (0, 0)} \varepsilon_2 =$ _____.

  **d.** The function $z = f(x, y)$ is differentiable at $(a, b)$ if $\Delta z$ can be expressed in the form $\Delta z =$ _____, where _____ and _____ as $(\Delta x, \Delta y) \to$ _____.

**10. a.** If $f$ is a function of $x$ and $y$, and $f_x$ and $f_y$ are continuous on an open region $R$, then $f$ is _____ in $R$.

  **b.** If $f$ is differentiable at $(a, b)$, then $f$ is _____ at $(a, b)$.

**11. a.** If $w = f(x, y)$, $x = g(t)$, and $y = h(t)$, then under suitable conditions the Chain Rule gives $dw/dt =$ _____.

  **b.** If $w = f(x, y)$, $x = g(u, v)$, and $y = h(u, v)$, then $\partial w/\partial u =$ _____.

  **c.** If $F(x, y) = 0$, where $F$ is differentiable, then $dy/dx =$ _____, provided that _____.

  **d.** If $F(x, y, z) = 0$, where $F$ is differentiable, and $F$ defines $z$ implicitly as a function of $x$ and $y$, then $\partial z/\partial x =$ _____ and $\partial z/\partial y =$ _____, provided that _____.

**12. a.** If $f$ is a function of $x$ and $y$ and $\mathbf{u} = u_1 \mathbf{i} + u_2 \mathbf{j}$ is a unit vector, then the directional derivative of $f$ in the direction of $\mathbf{u}$ is $D_{\mathbf{u}} f(x, y) =$ _____ if the limit exists.

  **b.** The directional derivative $D_{\mathbf{u}} f(a, b)$ measures the rate of change of $f$ at _____ in the direction of _____.

  **c.** If $f$ is differentiable, then $D_{\mathbf{u}} f(x, y) =$ _____.

  **d.** The gradient of $f(x, y)$ is $\nabla f(x, y) =$ _____.

  **e.** In terms of the gradient, $D_{\mathbf{u}} f(x, y) =$ _____.

**13. a.** The maximum value of $D_{\mathbf{u}} f(x, y)$ is _____, and this occurs when $\mathbf{u}$ has the same direction as _____.

  **b.** The minimum value of $D_{\mathbf{u}} f(x, y)$ is _____, and this occurs when $\mathbf{u}$ has the direction of _____.

**14. a.** $\nabla f$ is _____ to the level curve $f(x, y) = c$ at $P$.

  **b.** $\nabla F$ is _____ to the level surface $F(x, y, z) = 0$ at $P$.

  **c.** The tangent plane to the surface $F(x, y, z) = 0$ at the point $P(a, b, c)$ is _____; the normal line passing through $P(a, b, c)$ has symmetric equations _____.

**15. a.** If $f(x, y) \leq f(a, b)$ for all points in an open disk containing $(a, b)$, then $f$ has a _____ _____ at $(a, b)$.

  **b.** If $f(x, y) \geq f(a, b)$ for all points in the domain of $f$, then $f$ has an _____ _____ at $(a, b)$.

**c.** If $f$ is defined on an open region $R$ containing the point $(a, b)$, then $(a, b)$ is a critical point of $f$ if (1) $f_x$ and/or $f_y$ _____ _____ _____ at $(a, b)$ or (2) both $f_x(a, b)$ and $f_y(a, b)$ equal _____.

**d.** If $f$ has a relative extremum at $(a, b)$, then $(a, b)$ must be a _____ _____ of $f$.

**e.** To determine whether a critical point of $f$ does give rise to a relative extremum, we use the _____ _____ _____.

**16. a.** If $f$ is continuous on a closed, bounded set $D$ in the plane, then $f$ has an absolute maximum value _____ at some point _____ in $D$, and $f$ has an absolute minimum value _____ at some point _____ in $D$.

**b.** To find the absolute extreme values of $f$ on a closed, bounded set $D$, (1) find the values of $f$ at the _____ _____ _____ _____ in $D$, (2) find the extreme values of $f$ on the _____ of $D$. Then the largest and smallest values found in (1) and (2) give the _____ value of $f$ and the _____ value of $f$ on $D$.

**17. a.** If $f(x, y)$ has an extremum at a point $(a, b)$ lying on the curve with equation $g(x, y) = c$, then the extremum is called a _____ extremum.

**b.** If $f$ has an extremum at $(a, b)$ subject to the constraint $g(x, y) = c$, then $\nabla f(a, b) =$ _____, where $\lambda$ is a real number called a Lagrange _____.

**c.** To find the maximum and minimum values of $f$ subject to the constraint $g(x, y) = c$, we solve the system of equations $\nabla f(x, y) =$ _____ and $g(x, y) = c$ for $x$, $y$, and $\lambda$. We then evaluate _____ at each of the _____ _____ found in the last step. The largest value yields the constrained _____ of $f$, and the smallest value yields the constrained _____ of $f$.

## REVIEW EXERCISES

*In Exercises 1–4, find and sketch the domain of the function.*

**1.** $f(x, y) = \dfrac{\sqrt{9 - x^2 - y^2}}{x^2 + y^2}$

**2.** $f(x, y) = \dfrac{\ln(x - 2y - 4)}{y + x}$

**3.** $f(x, y) = \sin^{-1} x + \tan^{-1} y$

**4.** $f(x, y) = \ln(xy - 1)$

*In Exercises 5 and 6, sketch the graph of the function.*

**5.** $f(x, y) = 4 - x^2 - y^2$

**6.** $f(x, y) = \sqrt{1 - x^2 - y^2}$

*In Exercises 7–10, sketch several level curves for the function.*

**7.** $f(x, y) = x^2 + 2y$

**8.** $f(x, y) = y^2 - x^2$

**9.** $f(x, y) = e^{x^2 + y^2}$

**10.** $f(x, y) = \ln xy$

*In Exercises 11–14, find the limit or show that it does not exist.*

**11.** $\displaystyle\lim_{(x, y) \to (0, 0)} \dfrac{\sqrt{xy + 4}}{2y + 3}$

**12.** $\displaystyle\lim_{(x, y) \to (0, 0)} \dfrac{x^2 y^2}{x^4 + 3y^4}$

**13.** $\displaystyle\lim_{(x, y) \to (1, 0^+)} \dfrac{x^2 y + x^3}{\sqrt{x} + \sqrt{y}}$

**14.** $\displaystyle\lim_{(x, y, z) \to (0, 0, 0)} \dfrac{x^2 - 2y^2 + 3z^2}{x^2 + y^2 + z^2}$

*In Exercises 15 and 16, determine where the function is continuous.*

**15.** $f(x, y) = \dfrac{\ln(x - y)}{(x^2 + y^2)^{3/2}}$

**16.** $f(x, y, z) = e^{x/y} \cos z + \sqrt{x - y}$

*In Exercises 17–22, find the first partial derivatives of the function.*

**17.** $f(x, y) = 2x^2 y - \sqrt{x}$

**18.** $f(x, y) = \dfrac{xy^2}{x^2 + y^2}$

**19.** $f(r, s) = re^{-(r^2 + s^2)}$

**20.** $f(u, v) = e^{2u} \cos(u^2 + v^2)$

**21.** $f(x, y, z) = \dfrac{x^2 - y^2}{z^2 - x^2}$

**22.** $f(r, s, t) = r \cos st + s \sin\left(\dfrac{s}{t}\right)$

*In Exercises 23–26, find the second partial derivatives of the function.*

**23.** $f(x, y) = x^4 - 2x^2 y^3 + y^2 - 2$

**24.** $f(x, y) = e^{-xy} \cos(2x + 3y)$

**25.** $f(x, y, z) = x^2 yz^3$

**26.** $f(u, v, w) = ue^{-v} \sin w$

**27.** If $u = \sqrt{x^2 + y^2 + z^2}$, show that

$$\frac{\partial^2 u}{\partial x^2} + \frac{\partial^2 u}{\partial y^2} + \frac{\partial^2 u}{\partial z^2} = \frac{2}{u}$$

**28.** Show that the function $u = e^{-t} \cos(x/c)$ satisfies the one-dimensional heat equation $u_t = c^2 u_{xx}$.

*In Exercises 29 and 30, show that the function satisfies Laplace's equation $u_{xx} + u_{yy} + u_{zz} = 0$.*

**29.** $u = 2z^2 - x^2 - y^2$

**30.** $u = z \tan^{-1}\dfrac{y}{x}$

**31.** Find $dz$ if $z = x^2 \tan^{-1} y^3$.

**32.** Use differentials to approximate the change in
$f(x, y) = x^2 - 3xy + y^2$ if $(x, y)$ changes from $(2, -1)$ to
$(1.9, -0.8)$.

**33.** Use differentials to approximate $(2.01)^2\sqrt{(1.98)^2 + (3.02)^3}$.

**34. Estimating Changes in Profit** The total daily profit function (in
dollars) of Weston Publishing Company realized in publishing and selling its English language dictionaries is given by

$$P(x, y) = -0.0005x^2 - 0.003y^2 - 0.002xy + 14x + 12y - 200$$

where $x$ denotes the number of deluxe copies and $y$ denotes
the number of standard copies published and sold daily.
Currently, the number of deluxe and standard copies of the
dictionaries published and sold daily are 1000 and 1700,
respectively. Determine the approximate daily change in the
total daily profit if the number of deluxe copies is increased
to 1050 and the number of standard copies is decreased to
1650 per day.

**35.** Does a function $f$ such that $\nabla f = -y\mathbf{i} + x\mathbf{j}$ exist? Explain.

**36.** According to Ohm's Law, $R = V/I$, where $R$ is the resistance in ohms, $V$ is the electromotive force in volts, and $I$
is the current in amperes. If the errors in the measurements
made in a certain experiment in $V$ and $I$ are 2% and 1%,
respectively, use differentials to estimate the maximum
percentage error in the calculated value of $R$.

**37.** Let $z = x^2 y - \sqrt{y}$, where $x = e^{2t}$ and $y = \cos t$. Use the
Chain Rule to find $dz/dt$.

**38.** Let $w = e^x \cos y + y \sin e^x$, where $x = u^2 - v^2$ and
$y = \sqrt{uv}$. Use the Chain Rule to find $\partial w/\partial u$ and $\partial w/\partial v$.

**39.** Use partial differentiation to find $dy/dx$ if
$x^3 - 3x^2y + 2xy^2 + 2y^3 = 9$.

**40.** Find $\partial z/\partial x$ and $\partial z/\partial y$ if $x^3 z^2 + yz^3 = \cos xz$.

*In Exercises 41–44, find the gradient of the function $f$ at the
indicated point.*

**41.** $f(x, y) = \sqrt{x^2 + y^2}$;   $P(1, 2)$

**42.** $f(x, y) = e^{-x} \tan y$;   $P\left(0, \frac{\pi}{4}\right)$

**43.** $f(x, y, z) = xy^2 - yz^2 + zx^2$;   $P(2, 1, -3)$

**44.** $f(x, y, z) = x \ln y + y \ln z$;   $P(2, 1, 1)$

*In Exercises 45–48, find the directional derivative of the function
$f$ at the point $P$ in the indicated direction.*

**45.** $f(x, y) = x^3 y^2 - xy^3$;   $P(2, -1)$, in the direction of
$\mathbf{v} = 3\mathbf{i} - 4\mathbf{j}$.

**46.** $f(x, y) = e^{-x^2} \cos y$;   $P\left(0, \frac{\pi}{2}\right)$, in the direction from $P(1, 3)$
to $Q(3, 1)$.

**47.** $f(x, y, z) = x\sqrt{y^2 + z^2}$;   $P(2, 3, 4)$ in the direction of
$\mathbf{v} = \mathbf{i} - 2\mathbf{j} + 2\mathbf{k}$.

**48.** $f(x, y, z) = x^2 \ln y + xy^2 e^z$;   $P(2, 1, 0)$ in the direction of
$\mathbf{v} = \langle 3, -1, 2 \rangle$.

**49.** Find the direction in which $f(x, y) = \sqrt{x} + xy^2$ increases
most rapidly at the point $(4, 1)$. What is the maximum rate
of increase?

**50.** Find the direction in which $f(x, y, z) = xe^{yz}$ decreases most
rapidly at the point $(4, 3, 0)$. What is the greatest rate of
decrease?

*In Exercises 51–54, find equations for the tangent plane and the
normal line to the surface with the equation at the given point.*

**51.** $2x^2 + 4y^2 + 9z^2 = 27$;   $P(1, 2, 1)$

**52.** $x^2 + 2y^2 - 3z^2 = 19$;   $P(2, 3, 1)$

**53.** $z = x^2 + 3xy^2$;   $P(3, 1, 18)$

**54.** $z = xe^{-y}$;   $P(1, 0, 1)$

**55.** Let $f(x, y) = x^2 + y^2$, and let $g(x, y) = y^2/x^2$.
**cas a.** Plot several level curves of $f$ and $g$ using the same viewing window.
   **b.** Show analytically that each level curve of $f$ intersects all
   level curves of $g$ at right angles.

**56.** Show that if $\nabla f(x_0, y_0) \neq 0$, then an equation of the tangent
line to the level curve $f(x, y) = f(x_0, y_0)$ at the point $(x_0, y_0)$
is

$$f_x(x_0, y_0)(x - x_0) + f_y(x_0, y_0)(y - y_0) = 0$$

*In Exercises 57–60, find the relative extrema and saddle points
of the function.*

**57.** $f(x, y) = x^2 + xy + y^2 - 5x + 8y + 5$

**58.** $f(x, y) = 8x^3 - 6xy + y^3$

**59.** $f(x, y) = x^3 - 3xy + y^2$

**60.** $f(x, y) = \dfrac{2}{x} + \dfrac{4}{y} - xy$

*In Exercises 61 and 62, find the absolute extrema of the function
on the set D.*

**61.** $f(x, y) = x^2 + xy^2 - y^3$;
   $D = \{(x, y) \mid -1 \le x \le 1, \ 0 \le y \le 2\}$

**62.** $f(x, y) = (x^2 + 3y^2)e^{-x}$;   $D = \{(x, y) \mid x^2 + y^2 \le 9\}$

*In Exercises 63–66, use the method of Lagrange multipliers to
find the extrema of the function $f$ subject to the constraints.*

**63.** $f(x, y) = xy^2$;   $x^2 + y^2 = 4$

**64.** $f(x, y) = \dfrac{1}{x} + \dfrac{1}{y}$;   $\dfrac{1}{x^2} + \dfrac{1}{y^2} = 9$

**65.** $f(x, y, z) = xy + yz + xz$;   $x + 2y + 3z = 1$

**66.** $f(x, y, z) = 3x^2 + 2y^2 + z^2$;   $x + y + z = 1$,
   $2x - y + z = 2$

**67.** Let $f(x, y) = Ax^2 + Bxy + Cy^2 + Dx + Ey + F$. Show that if $f$ has a relative maximum or a relative minimum at a point $(x_0, y_0)$, then $x_0$ and $y_0$ must satisfy the system of equations

$$2Ax + By + D = 0$$

$$Bx + 2Cy + E = 0$$

simultaneously.

**68.** Let $f(x, y) = x^2 + 2Bxy + y^2$, where $B > 0$. For what value of $B$ does $f$ have a relative minimum at $(0, 0)$? A saddle point at $(0, 0)$? Are there any values of $B$ such that $f$ has a relative maximum at $(0, 0)$?

**69.** Find the point on the paraboloid

$$z = \frac{x^2}{4} + \frac{y^2}{25}$$

that is closest to the point $(3, 0, 0)$.

**70. Isothermal Curves** Consider the upper half-disk $H = \{(x, y) \mid x^2 + y^2 \le 1, y \ge 0\}$ (see the figure). If the temperature at points on the upper boundary is kept at $100°C$ and the temperature at points on the lower boundary is kept at $50°C$, then the steady-state temperature $T(x, y)$ at any point inside the disk is given by

$$T(x, y) = 100 - \frac{100}{\pi} \tan^{-1} \frac{1 - x^2 - y^2}{2y}$$

Show that the isothermal curves $T(x, y) = k$ are arcs of circles that pass through the points $x = \pm 1$. Sketch the isothermal curve corresponding to a temperature of $75°C$.

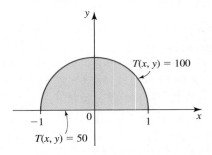

*In Exercises 71 and 72, determine whether the statement is true or false. If it is true, explain why it is true. If it is false, give an example to show why it is false.*

**71.** The directional derivative of $f(x, y)$ at the point $(a, b)$ in the positive $x$-direction is $f_x(a, b)$.

**72.** If we know the gradient of $f(x, y, z)$ at the point $P(a, b, c)$, then we can compute the directional derivative of $f$ in any direction at $P$.

## CHALLENGE PROBLEMS

**1.** Find and sketch the domain of

$$f(x, y) = \sqrt{36 - 4x^2 - 9y^2} + \ln(x^2 - 2x + y^2)$$

$$+ \frac{1}{\sqrt{4x^2 + 16x + 4y^2 + 15}}$$

**2.** Describe the domain of

$$H(x, y, z) = \sqrt{x - a} + \sqrt{b - x} + \sqrt{y - c}$$

$$+ \sqrt{d - y} + \sqrt{z - e} + \sqrt{f - z}$$

where $a \le b$, $c \le d$, and $e \le f$.

**3.** Suppose $f$ has continuous second partial derivatives in $x$ and $y$. Then the *second-order directional derivative* of $f$ in the direction of the unit vector $\mathbf{u} = u_1\mathbf{i} + u_2\mathbf{j}$ is defined to be

$$D_{\mathbf{u}}^2 f(x, y) = D_{\mathbf{u}}(D_{\mathbf{u}} f)$$

  **a.** Find an expression in terms of the partial derivatives of $f$ for $D_{\mathbf{u}}^2 f$.
  **b.** Find $D_{\mathbf{u}}^2 f(1, 0)$ if $f(x, y) = xy^2 + e^{xy}$ and $\mathbf{u}$ has the same direction as $\mathbf{v} = 2\mathbf{i} - 3\mathbf{j}$.

**4.** Consider the quadratic polynomial function

$$f(x, y) = Ax^2 + 2Bxy + Cy^2 + 2Dx + 2Ey + F$$

Find conditions on the coefficients of $f$ such that $f$ has (a) a relative maximum and (b) a relative minimum. What are the coordinates of the point in terms of the coefficients of $f$?

**5.** Let $a$, $b$, and $c$ denote the sides of a triangle of area $A$, and let $\alpha$, $\beta$, and $\gamma$ denote the angles opposite them. If $A = f(a, b, c)$, show that

$$\frac{\partial f}{\partial a} = \frac{1}{2} R \cos \alpha$$

where $R$ is the radius of the circumscribing circle.

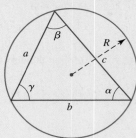

6. Linda has 24 feet of fencing with which to enclose a triangular flower garden. What should the lengths of the sides of the garden be if the area is to be as large as possible?
   **Hint:** Heron's formula states that the area of a triangle with sides $a$, $b$, and $c$ is given by $A = \sqrt{s(s-a)(s-b)(s-c)}$, where $s = \frac{1}{2}(a+b+c)$ is the semiperimeter.

7. Find the directional derivative at $\left(1, 2, \frac{\pi}{4}\right)$ of the function $f(x, y, z) = x^2 + y \cos z$ in the direction of increasing $t$ along the curve in three-dimensional space described by the position vector $\mathbf{r}(t) = \langle t, t^2, t^3 \rangle$ at $\mathbf{r}(1)$.

8. Let
$$f(x, y) = \begin{cases} \dfrac{xy(x^2 - y^2)}{x^2 + y^2} & (x, y) \neq (0, 0) \\ 0 & (x, y) = (0, 0) \end{cases}$$

   Use the definition of partial derivatives to show that $f_{xy}(0, 0) = -1$ and $f_{yx}(0, 0) = 1$.

9. Show that Laplace's equation $\dfrac{\partial^2 u}{\partial x^2} + \dfrac{\partial^2 u}{\partial y^2} + \dfrac{\partial^2 u}{\partial z^2} = 0$ in cylindrical coordinates takes the form
$$\frac{\partial^2 u}{\partial r^2} + \frac{1}{r} \cdot \frac{\partial u}{\partial r} + \frac{1}{r^2} \cdot \frac{\partial^2 u}{\partial \theta^2} + \frac{\partial^2 u}{\partial z^2} = 0$$

10. Consider the problem of determining the maximum and the minimum distances from the point $(x_0, y_0, z_0)$ to the ellipsoid
$$\frac{x^2}{a^2} + \frac{y^2}{b^2} + \frac{z^2}{c^2} = 1$$

   **a.** Show that the solutions are
$$x = \frac{a^2 x_0}{a^2 - \lambda}, \qquad y = \frac{b^2 y_0}{b^2 - \lambda}, \qquad z = \frac{c^2 z_0}{c^2 - \lambda}$$

   where $\lambda$ satisfies
$$\frac{a^2 x_0^2}{(a^2 - \lambda)^2} + \frac{b^2 y_0^2}{(b^2 - \lambda)^2} + \frac{c^2 z_0^2}{(c^2 - \lambda)^2} = 1$$

   **b.** Use the result of part (a) to solve the problem with $a = 2$, $b = 3$, $c = 1$, and $(x_0, y_0, z_0) = (3, 2, 4)$.

A lawn sprinkler sprays water in a circular pattern. If we know the amount of the water per hour that the sprinkler delivers to any point within the circular region, can we find the total amount of water accumulated per hour in that part of the lawn? We will consider such problems in this chapter.

# 13 Multiple Integrals

**IN THIS CHAPTER** we extend the notion of the integral of a function of one variable to the integral of a function of two or three variables. Applications of double and triple integrals include finding the area of a surface, finding the center of mass of a planar object, and finding the centroid of a solid. We end the chapter by looking at how certain multiple integrals can be more easily evaluated by a change of variables.

**V** This symbol indicates that one of the following video types is available for enhanced student learning at **www.academic.cengage.com/login:**
- Chapter lecture videos
- Solutions to selected exercises

## 13.1  Double Integrals

### An Introductory Example

Suppose a piece of straight, thin wire of length $(b - a)$ is placed on the $x$-axis of a coordinate system, as shown in Figure 1. Further suppose that the wire has linear mass density given by $f(x)$ at $x$ for $a \le x \le b$, where $f$ is a nonnegative continuous function on $[a, b]$. Let $P = \{x_0, x_1, x_2, \ldots, x_n\}$, where $a = x_0$ and $b = x_n$, be a regular partition of $[a, b]$. Then the continuity of $f$ tells us that, $f(x) \approx f(c_k)$ for each $x$ in the $k$th subinterval $[x_{k-1}, x_k]$, where $c_k$ is an evaluation point in $[x_{k-1}, x_k]$, provided $n$ is large enough. Therefore, the mass of the piece of wire lying on $[x_{k-1}, x_k]$ is

$$\Delta m_k \approx f(c_k)\, \Delta x \qquad \Delta x = \frac{b - a}{n}$$

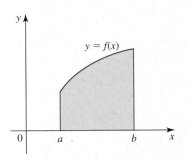

**FIGURE 1**
The mass of a straight wire of length $(b - a)$ is given by $\int_a^b f(x)\, dx$, where $f(x)$ is the density of the wire at any point $x$ for $a \le x \le b$.

This leads to the definition of the mass of the wire as

$$m = \lim_{n \to \infty} \sum_{k=1}^{n} \Delta m_k = \lim_{n \to \infty} \sum_{k=1}^{n} f(c_k)\, \Delta x = \int_a^b f(x)\, dx$$

Thus, the mass of the curve has the same numerical value as that of the area under the graph of the (nonnegative) density function $f$ shown in Figure 1.

Now let's consider a thin rectangular plate occupying the region

$$R = \{(x, y) \mid a \le x \le b, c \le y \le d\}$$

(See Figure 2.) If the plate is homogeneous (having a constant mass density of $k$ g/cm$^2$), then its mass is given by

$$m = k(b - a)(d - c) \qquad \text{mass density} \cdot \text{area}$$

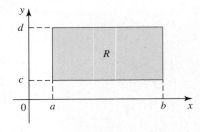

**FIGURE 2**       (a) A thin rectangular plate       (b) The plate placed in the $xy$-plane

Observe that $m$ has the same numerical value as that of the volume of the rectangular box bounded above by the graph of the constant function $f(x, y) = k$ and below by $R$. (See Figure 3a.) Next, instead of being constant, suppose that the mass density of the

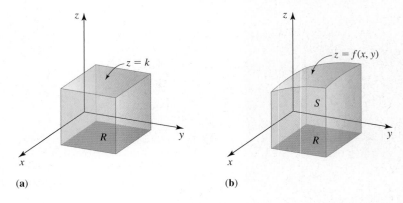

**FIGURE 3**
The mass of the plate $R$ is numerically equal to that of the volume of the solid region lying directly above $R$ and below the surface $z = f(x, y)$.

(a)                                      (b)

plate is given by the mass density function $f$. Then it seems reasonable to conjecture that the mass of the plate is given by the "volume" of the solid region $S$ lying directly above $R$ and below the graph of $z = f(x, y)$. (See Figure 3b.) We will show in Section 13.4 that this is indeed the case.

## ■ Volume of a Solid Between a Surface and a Rectangle

We will now show that the volume of a solid $S$ can be defined as a limit of Riemann sums. Suppose that $f$ is a nonnegative continuous function* of two variables that is defined on a rectangle

$$R = [a, b] \times [c, d] = \{(x, y) \mid a \leq x \leq b, c \leq y \leq d\}$$

and suppose that $f(x, y) \geq 0$ on $R$. Let

$$a = x_0 < x_1 < \cdots < x_{i-1} < x_i < \cdots < x_m = b$$

be a regular partition of the interval $[a, b]$ into $m$ subintervals of length $\Delta x = (b - a)/m$, and let

$$c = y_0 < y_1 < \cdots < y_{j-1} < y_j < \cdots < y_n = d$$

be a regular partition of the interval $[c, d]$ into $n$ subintervals of length $\Delta y = (d - c)/n$. The grid comprising segments of the vertical lines $x = x_i$ for $0 \leq i \leq m$ and the horizontal lines $y = y_j$ for $0 \leq j \leq n$ partition $R$ into $N = mn$ subrectangles $R_{11}, R_{12}, \ldots, R_{ij}, \ldots, R_{mn}$, where $R_{ij} = [x_{i-1}, x_i] \times [y_{j-1}, y_j] = \{(x, y) \mid x_{i-1} \leq x \leq x_i, y_{j-1} \leq y \leq y_j\}$ as shown in Figure 4. The area of each subrectangle is $\Delta A = \Delta x \, \Delta y$. This partition is called a **regular partition** of $R$.

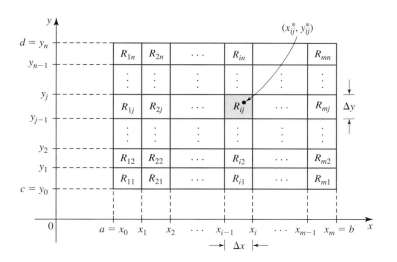

**FIGURE 4**
A partition $P = \{R_{ij}\}$ of $R$

The partition $P = \{R_{11}, R_{12}, \ldots, R_{ij}, \ldots, R_{mn}\}$ divides the solid $S$ between the graph of $z = f(x, y)$ and $R$ into $N = mn$ solids; the solid $S_{ij}$ is bounded below by $R_{ij}$ and bounded above by the part of the surface $z = f(x, y)$ that lies directly above $R_{ij}$. (See Figure 5.)

---

*As in the case of the integral of a function of one variable, these assumptions will simplify the discussion.

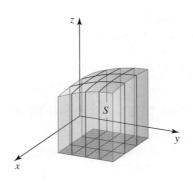
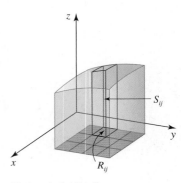

(a) The solid $S$ is the union of $N = mn$
solids (shown here with $m = 3$, $n = 4$)

(b) A typical solid $S_{ij}$

**FIGURE 5**

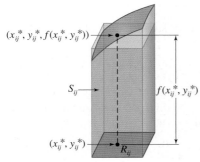

**FIGURE 6**
The volume of $S_{ij}$ is approximated
by the volume of the parallelepiped
with base $R_{ij}$ and height $f(x_{ij}^*, y_{ij}^*)$.

Let $(x_{ij}^*, y_{ij}^*)$ be an evaluation point in $R_{ij}$. Then the parallelepiped with base $R_{ij}$, height $f(x_{ij}^*, y_{ij}^*)$, and volume

$$f(x_{ij}^*, y_{ij}^*) \, \Delta A$$

gives an approximation of the volume of $S_{ij}$. (See Figure 6.)

Therefore, the volume $V$ of $S$ is approximated by the volume of the sum of $N = mn$ parallelepipeds; that is,

$$V \approx \sum_{i=1}^{m} \sum_{j=1}^{n} f(x_{ij}^*, y_{ij}^*) \, \Delta A \tag{1}$$

If we take $m$ and $n$ to be larger and larger, then, intuitively, we can expect the approximation (1) to improve. This suggests the following definition.

> **DEFINITION    Volume Under the Graph of $z = f(x, y)$**
>
> Let $f$ be defined on the rectangle $R$ and suppose $f(x, y) \geq 0$ on $R$. Then the volume $V$ of the solid $S$ that lies directly above $R$ and below the surface $z = f(x, y)$ is
>
> $$V = \lim_{m, n \to \infty} \sum_{i=1}^{m} \sum_{j=1}^{n} f(x_{ij}^*, y_{ij}^*) \, \Delta A \tag{2}$$
>
> if the limit exists.

Because of the assumption that $f$ be continuous, it can be shown that the limit in Equation (2) always exists regardless of how the evaluation points $(x_{ij}^*, y_{ij}^*)$ in $R_{ij}$, for $1 \leq i \leq m$ and $1 \leq j \leq n$, are chosen.

**EXAMPLE 1** Approximate the volume $V$ of the solid lying under the graph of the elliptic paraboloid $z = 8 - 2x^2 - y^2$ and above the rectangle $R = \{(x, y) \mid 0 \leq x \leq 1, 0 \leq y \leq 2\}$. Use the partition $P$ of $R$ that is obtained by dividing $R$ into four subrectangles with the lines $x = \frac{1}{2}$ and $y = 1$, and choose the evaluation point $(x_{ij}^*, y_{ij}^*)$ to be the upper right-hand corner of $R_{ij}$. (See Figure 7.)

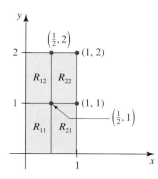
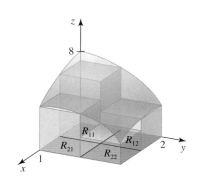

**FIGURE 7**

(a) The region $R$ is divided into four subrectangles

(b) The solid lying under the graph of $z = 8 - 2x^2 - y^2$ and above $R$

**Solution**   Here,

$$\Delta x = \frac{1 - 0}{2} = \frac{1}{2} \quad \text{and} \quad \Delta y = \frac{2 - 0}{2} = 1$$

so $\Delta A = \left(\frac{1}{2}\right)(1) = \frac{1}{2}$. Also, $x_0 = 0$, $x_1 = \frac{1}{2}$, and $x_2 = 1$, and $y_0 = 0$, $y_1 = 1$, and $y_2 = 2$. Taking $(x_{11}^*, y_{11}^*) = (x_1, y_1) = \left(\frac{1}{2}, 1\right)$, $(x_{12}^*, y_{12}^*) = (x_1, y_2) = \left(\frac{1}{2}, 2\right)$, $(x_{21}^*, y_{21}^*) = (x_2, y_1) = (1, 1)$, and $(x_{22}^*, y_{22}^*) = (x_2, y_2) = (1, 2)$, we have

$$V \approx \sum_{i=1}^{2} \sum_{j=1}^{2} f(x_{ij}^*, y_{ij}^*) \, \Delta A = f(x_{11}^*, y_{11}^*) \, \Delta A + f(x_{12}^*, y_{12}^*) \, \Delta A + f(x_{21}^*, y_{21}^*) \, \Delta A + f(x_{22}^*, y_{22}^*) \, \Delta A$$

$$= f\left(\frac{1}{2}, 1\right) \Delta A + f\left(\frac{1}{2}, 2\right) \Delta A + f(1, 1) \, \Delta A + f(1, 2) \, \Delta A$$

$$= \left(\frac{13}{2}\right)\left(\frac{1}{2}\right) + \left(\frac{7}{2}\right)\left(\frac{1}{2}\right) + (5)\left(\frac{1}{2}\right) + (2)\left(\frac{1}{2}\right) = \frac{17}{2} \qquad \blacksquare$$

The approximations to the volume in Example 1 get better and better as $m$ and $n$ increase, as shown in Figure 8.

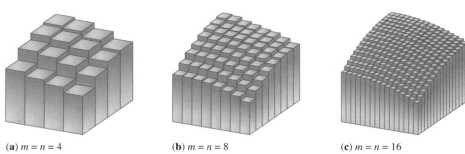

(a) $m = n = 4$        (b) $m = n = 8$        (c) $m = n = 16$

**FIGURE 8**

The approximation of $V$ using the sum of the volumes of 16 parallelepipeds in (a), 64 parallelepipeds in (b), and 256 parallelepipeds in (c).

**Note**   Suppose that the mass density of a rectangular plate $R = \{(x, y) \mid 0 \le x \le 1, 0 \le y \le 2\}$ is $f(x, y) = 8 - 2x^2 - y^2$ g/cm$^2$. Then the result of Example 1 tells us that the mass of the plate is approximately $\frac{17}{2}$ g.    $\blacksquare$

## The Double Integral Over a Rectangular Region

Thus far, we have assumed that $f(x, y) \geq 0$ on the rectangle $R$. This condition was imposed so that we could give a simple geometric interpretation for the limit in Equation (2). In the general situation we have the following.

---

**DEFINITION    Riemann Sum**

Let $f$ be a continuous function of two variables defined on a rectangle $R$, and let $P = \{R_{ij}\}$ be a regular partition of $R$. A **Riemann sum of $f$** over $R$ with respect to the partition $P$ is a sum of the form

$$\sum_{i=1}^{m} \sum_{j=1}^{n} f(x_{ij}^*, y_{ij}^*) \, \Delta A \tag{3}$$

where $(x_{ij}^*, y_{ij}^*)$ is an evaluation point in $R_{ij}$.

---

**DEFINITION    Double Integral of $f$ Over a Rectangle $R$**

Let $f$ be a continuous function of two variables defined on a rectangle $R$. The **double integral of $f$** over $R$ is

$$\iint_R f(x, y) \, dA = \lim_{m, n \to \infty} \sum_{i=1}^{m} \sum_{j=1}^{n} f(x_{ij}^*, y_{ij}^*) \, \Delta A \tag{4}$$

if this limit exists for all choices of the evaluation point $(x_{ij}^*, y_{ij}^*)$ in $R_{ij}$.

---

**Notes**

1. If the double integral of $f$ over $R$ exists, then $f$ is said to be **integrable over $R$.** It can be shown, although we will not do so here, that if $f$ is continuous on $R$, then $f$ is integrable over $R$.
2. If $f$ is integrable, then the Riemann sum (3) is an approximation of the double integral (4).
3. If $f(x, y) \geq 0$ on $R$, then $\iint_R f(x, y) \, dA$ gives the volume of the solid lying directly above $R$ and below the surface $z = f(x, y)$.
4. We use the (double) integral sign to denote the limit, whenever it exists, because it is related to the definite integral, as you will see in Section 13.2.    ■

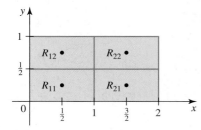

**FIGURE 9**
The partition
$P = \{R_{11}, R_{12}, R_{21}, R_{22}\}$ of $R$

**EXAMPLE 2**  Find an approximation for $\iint_R (x - 4y) \, dA$, where $R = \{(x, y) \,|\, 0 \leq x \leq 2, 0 \leq y \leq 1\}$, using the Riemann sum of $f(x, y) = x - 4y$ over $R$ with $m = n = 2$ and taking the evaluation point $(x_{ij}^*, y_{ij}^*)$ to be the center of $R_{ij}$.

**Solution**   Here,

$$\Delta x = \frac{2 - 0}{2} = 1 \qquad \Delta y = \frac{1 - 0}{2} = \frac{1}{2}$$

and $x_0 = 0$, $x_1 = 1$, $x_2 = 2$, $y_0 = 0$, $y_1 = \frac{1}{2}$, $y_2 = 1$. The partition $P$ is shown in Figure 9. Using Equation (3) with $f(x, y) = x - 4y$, $\Delta A = \Delta x \, \Delta y = (1)\left(\frac{1}{2}\right) = \frac{1}{2}$, and

$(x_{11}^*, y_{11}^*) = \left(\frac{1}{2}, \frac{1}{4}\right)$, $(x_{12}^*, y_{12}^*) = \left(\frac{1}{2}, \frac{3}{4}\right)$, $(x_{21}^*, y_{21}^*) = \left(\frac{3}{2}, \frac{1}{4}\right)$, and $(x_{22}^*, y_{22}^*) = \left(\frac{3}{2}, \frac{3}{4}\right)$, we obtain

$$\iint_R f(x, y)\, dA \approx \sum_{i=1}^{2} \sum_{j=1}^{2} f(x_{ij}^*, y_{ij}^*)\, \Delta A$$

$$= f(x_{11}^*, y_{11}^*)\, \Delta A + f(x_{12}^*, y_{12}^*)\, \Delta A + f(x_{21}^*, y_{21}^*)\, \Delta A + f(x_{22}^*, y_{22}^*)\, \Delta A$$

$$= f\left(\frac{1}{2}, \frac{1}{4}\right)\frac{1}{2} + f\left(\frac{1}{2}, \frac{3}{4}\right)\frac{1}{2} + f\left(\frac{3}{2}, \frac{1}{4}\right)\frac{1}{2} + f\left(\frac{3}{2}, \frac{3}{4}\right)\frac{1}{2}$$

$$= \left(-\frac{1}{2}\right)\left(\frac{1}{2}\right) + \left(-\frac{5}{2}\right)\left(\frac{1}{2}\right) + \left(\frac{1}{2}\right)\left(\frac{1}{2}\right) + \left(-\frac{3}{2}\right)\left(\frac{1}{2}\right) = -2 \quad \blacksquare$$

## Double Integrals Over General Regions

Next we will extend the definition of the double integral to more general functions and regions. Suppose that $f$ is a bounded function defined on a bounded plane region $D$. If you like, you can think of $f$ as a mass density function for a thin plate occupying a nonrectangular region $D$ (in which case $f(x, y) \geq 0$ on $D$) and of what follows as a way of finding the mass of the plate. Since $D$ is bounded, it can be enclosed in a rectangle $R$. Let $Q$ be a regular partition of $R$ into subrectangles $R_{11}, R_{12}, \ldots, R_{ij}, \ldots, R_{mn}$. (See Figure 10.)

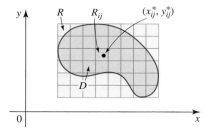

**FIGURE 10**
A partition $Q$ of $R$

Let's define the function

$$f_D(x, y) = \begin{cases} f(x, y) & \text{if } (x, y) \text{ is in } D \\ 0 & \text{if } (x, y) \text{ is in } R \text{ but not in } D \end{cases}$$

Note that $f_D$ takes on the same value as $f$ if $(x, y)$ is in $D$, but it takes on the value zero if $(x, y)$ lies outside $D$. (See Figure 11.)

Now let $(x_{ij}^*, y_{ij}^*)$ be an evaluation point in the subrectangle $R_{ij}$ of $Q$ for $1 \leq i \leq m$ and $1 \leq j \leq n$. Then the sum

$$\sum_{i=1}^{m} \sum_{j=1}^{n} f_D(x_{ij}^*, y_{ij}^*)\, \Delta A$$

is a **Riemann sum of $f$ over $D$** with respect to the partition $Q$. Taking the limit of these sums as $m, n \to \infty$ gives the **double integral of $f$ over $D$**. Thus,

$$\iint_D f(x, y)\, dA = \lim_{m, n \to \infty} \sum_{i=1}^{m} \sum_{j=1}^{n} f_D(x_{ij}^*, y_{ij}^*)\, \Delta A \tag{5}$$

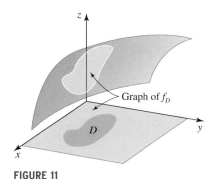

**FIGURE 11**
$f_D(x, y) = f(x, y)$ if $(x, y)$ lies in $D$, but $f_D(x, y) = 0$ if $(x, y)$ lies outside $D$.

if the limit exists. Again, it can be shown that if $f$ is continuous, then the limit (5) always exists regardless of how the evaluation points $(x_{ij}^*, y_{ij}^*)$ in $R_{ij}$ are chosen.

**Notes**

1. If $f(x, y) \geq 0$ on $D$, then $\iint_D f(x, y)\, dA$ gives the volume of the solid lying directly above $D$ and below the surface $z = f(x, y)$.

2. If $\rho(x, y) \geq 0$ on $D$, where $\rho$ is a mass density function, then $\iint_D \rho(x, y)\, dA$ gives the mass of the thin plate occupying the plane region $D$ in the $xy$-plane. This will be demonstrated in Section 13.4. $\blacksquare$

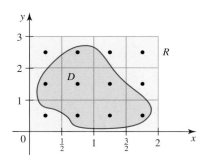

**FIGURE 12**
$f_D(x, y) = f(x, y)$ if $(x, y)$ lies in $D$, but $f_D(x, y) = 0$ if $(x, y)$ lies outside $D$.

**EXAMPLE 3** Find an approximation for $\iint_D (x + 2y)\, dA$, where $D$ is the region shown in Figure 12, using the Riemann sum of $f(x, y) = x + 2y$ over $D$ with respect to the partition $Q$ obtained by dividing the rectangle $\{(x, y) \mid 0 \leq x \leq 2, 0 \leq y \leq 3\}$

into 12 subrectangles by taking $m = 4$ and $n = 3$ and taking the evaluation points to be the center of $R_{ij}$.

**Solution** Here,

$$\Delta A = (\Delta x)(\Delta y) = \left(\frac{2 - 0}{4}\right)\left(\frac{3 - 0}{3}\right) = \frac{1}{2}$$

Next, define

$$f_D(x, y) = \begin{cases} f(x, y) & \text{if } (x, y) \text{ is in } D \\ 0 & \text{if } (x, y) \text{ is not in } D \end{cases}$$

Then

$$\iint\limits_D (x + 2y) \, dA \approx \sum_{i=1}^{4} \sum_{j=1}^{3} f_D(x_{ij}^*, y_{ij}^*) \, \Delta A \qquad f(x, y) = x + 2y$$

$$= \left[ f_D\left(\frac{1}{4}, \frac{1}{2}\right) + f_D\left(\frac{1}{4}, \frac{3}{2}\right) + f_D\left(\frac{1}{4}, \frac{5}{2}\right) + f_D\left(\frac{3}{4}, \frac{1}{2}\right) + f_D\left(\frac{3}{4}, \frac{3}{2}\right) + f_D\left(\frac{3}{4}, \frac{5}{2}\right) \right.$$

$$\left. + f_D\left(\frac{5}{4}, \frac{1}{2}\right) + f_D\left(\frac{5}{4}, \frac{3}{2}\right) + f_D\left(\frac{5}{4}, \frac{5}{2}\right) + f_D\left(\frac{7}{4}, \frac{1}{2}\right) + f_D\left(\frac{7}{4}, \frac{3}{2}\right) + f_D\left(\frac{7}{4}, \frac{5}{2}\right) \right] \Delta A$$

$$= \frac{1}{2}\left[ f\left(\frac{1}{4}, \frac{3}{2}\right) + f\left(\frac{3}{4}, \frac{1}{2}\right) + f\left(\frac{3}{4}, \frac{3}{2}\right) + f\left(\frac{3}{4}, \frac{5}{2}\right) + f\left(\frac{5}{4}, \frac{1}{2}\right) + f\left(\frac{5}{4}, \frac{3}{2}\right) + f\left(\frac{7}{4}, \frac{1}{2}\right) \right]$$

$$= \frac{1}{2}\left\{ \left[\frac{1}{4} + 2\left(\frac{3}{2}\right)\right] + \left[\frac{3}{4} + 2\left(\frac{1}{2}\right)\right] + \left[\frac{3}{4} + 2\left(\frac{3}{2}\right)\right] + \left[\frac{3}{4} + 2\left(\frac{5}{2}\right)\right] + \left[\frac{5}{4} + 2\left(\frac{1}{2}\right)\right] \right.$$

$$\left. + \left[\frac{5}{4} + 2\left(\frac{3}{2}\right)\right] + \left[\frac{7}{4} + 2\left(\frac{1}{2}\right)\right] \right\} = 11.875 \qquad \blacksquare$$

## ■ Properties of Double Integrals

Double integrals have many of the properties that single integrals enjoy. We list some of them in the following theorem, the proof of which will be omitted.

---

**THEOREM 1    Properties of the Definite Integral**

Let $f$ and $g$ be defined on a suitably restricted region $D$, so that both $\iint_D f(x, y) \, dA$ and $\iint_D g(x, y) \, dA$ exist, and let $c$ be a constant. Then

**1.** $\displaystyle\iint\limits_D cf(x, y) \, dA = c\iint\limits_D f(x, y) \, dA$

**2.** $\displaystyle\iint\limits_D [f(x, y) \pm g(x, y)] \, dA = \iint\limits_D f(x, y) \, dA \pm \iint\limits_D g(x, y) \, dA$

**3.** If $f(x, y) \geq 0$ on $D$, then $\displaystyle\iint\limits_D f(x, y) \, dA \geq 0$

**4.** If $f(x, y) \geq g(x, y)$ on $D$, then $\displaystyle\iint\limits_D f(x, y) \, dA \geq \iint\limits_D g(x, y) \, dA$

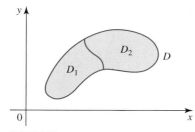

**FIGURE 13**
$D = D_1 \cup D_2$ where $D_1 \cap D_2 = \emptyset$.

**5.** If $D = D_1 \cup D_2$, where $D_1$ and $D_2$ are two nonoverlapping subregions with the possible exception of their common boundaries, then

$$\iint_D f(x, y)\, dA = \iint_{D_1} f(x, y)\, dA + \iint_{D_2} f(x, y)\, dA$$

(See Figure 13.)

## 13.1 CONCEPT QUESTIONS

**1.** Let $f(x, y) = x + 2y$.
   **a.** Complete the table of values for $f(x, y)$ in the following table.

| y \ x | 1 | $\frac{3}{2}$ | 2 | $\frac{5}{2}$ | 3 | $\frac{7}{2}$ | 4 |
|---|---|---|---|---|---|---|---|
| **0** | | | | | | | |
| $\frac{1}{4}$ | | | | | | | |
| $\frac{1}{2}$ | | | | | | | |
| $\frac{3}{4}$ | | | | | | | |
| **1** | | | | | | | |

   **b.** Use the table of values from part (a) to estimate the volume of the solid lying under the graph of $z = x + 2y$ and above the rectangular region $R = [0, 1] \times [1, 4]$ using a regular partition with $m = 2$ and $n = 3$ and choosing the evaluation point $(x_{ij}^*, y_{ij}^*)$ to be the lower left-hand corner of $R_{ij}$.
   **c.** Repeat part (b), this time choosing the evaluation point $(x_{ij}^*, y_{ij}^*)$ to be the center of $R_{ij}$.

**2.** **a.** Let $f$ be a continuous function defined on the rectangular region $[a, b] \times [c, d]$. Define $\iint_R f(x, y)\, dA$.
   **b.** Suppose that $f(x, y) = k$, where $k$ is a constant. Find $\iint_R k\, dA$ for $R = [a, b] \times [c, d]$ using your definition from part (a).

## 13.1 EXERCISES

*In Exercises 1–4, find an approximation for the volume V of the solid lying under the graph of the elliptic paraboloid $z = 8 - 2x^2 - y^2$ and above the rectangular region $R = \{(x, y) \mid 0 \le x \le 1, 0 \le y \le 2\}$. Use a regular partition P of R with $m = n = 2$, and choose the evaluation point $(x_{ij}^*, y_{ij}^*)$ as indicated in each exercise.*

**1.** The lower left-hand corner of $R_{ij}$

**2.** The upper left-hand corner of $R_{ij}$

**3.** The lower right-hand corner of $R_{ij}$

**4.** The center of $R_{ij}$

*In Exercises 5–8, find the Riemann sum $\sum_{i=1}^{m} \sum_{j=1}^{n} f(x_{ij}^*, y_{ij}^*) \Delta A$ of f over the region R with respect to the regular partition P with the indicated values of m and n.*

**5.** $f(x, y) = 2x + 3y$;   $R = [0, 1] \times [0, 3]$;   $m = 2, n = 3$; $(x_{ij}^*, y_{ij}^*)$ is the lower left-hand corner of $R_{ij}$

**6.** $f(x, y) = x^2 - 2y$;   $R = [1, 5] \times [1, 3]$;   $m = 4, n = 2$; $(x_{ij}^*, y_{ij}^*)$ is the upper right-hand corner of $R_{ij}$

**7.** $f(x, y) = x^2 + 2y^2$;   $R = [-1, 3] \times [0, 4]$;   $m = 4, n = 4$; $(x_{ij}^*, y_{ij}^*)$ is the center of $R_{ij}$

**8.** $f(x, y) = 2xy$;   $R = [-1, 1] \times [-2, 2]$;   $m = 4, n = 4$; $(x_{ij}^*, y_{ij}^*)$ is the center of $R_{ij}$

**9.** The figure on the following page shows a region $D$ enclosed by a rectangular region $R$ and a partition $Q$ of $R$ into subrectangles with $m = 5$ and $n = 3$. Suppose that $f$ is continuous on $D$ and the values of $f$ at the evaluation points of $Q$ that lie in $D$ are as shown in the figure (next to the evaluation points). Define

$$f_D(x, y) = \begin{cases} f(x, y) & \text{if } (x, y) \text{ is in } D \\ 0 & \text{if } (x, y) \text{ is in } R \text{ but not in } D \end{cases}$$

Compute $\sum_{i=1}^{5} \sum_{j=1}^{3} f_D(x_{ij}^*, y_{ij}^*) \, \Delta A$.

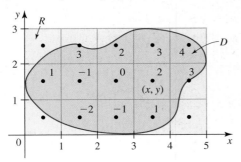

**10. Volume of Water in a Pond** The following figure depicts a pond that is 40 ft long and 20 ft wide. The depth of the pond is measured at the center of each subrectangle in the imaginary partition of the rectangle that is superimposed over the aerial view of the pond. These measurements (in feet) are shown in the figure. Estimate the volume of water in the pond.
**Hint:** See Exercise 9.

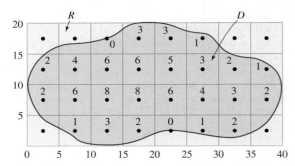

**11.** The figure shows the contour map of a function $f$ on the set $R = \{(x, y) \mid 0 \le x \le 2, 0 \le y \le 2\}$. Estimate $\iint_R f(x, y) \, dA$ using a Riemann sum with $m = n = 2$ and choosing the evaluation point $(x_{ij}^*, y_{ij}^*)$ to be the center of $R_{ij}$.

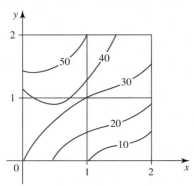

**12. Room Temperature** The figure represents a part of a room with a fireplace located at the origin. The curves shown are the level curves of the temperature function $T$, and are called *isothermals* because the temperature is the same at all points on an isothermal. Estimate the average temperature in this

part of the room using a regular partition with $m = n = 3$ and choosing the evaluation point $(x_{ij}^*, y_{ij}^*)$ to be the center of $R_{ij}$.

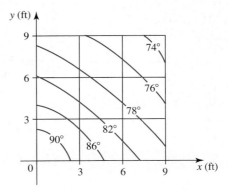

*In Exercises 13–16, find the double integral by interpreting it as the volume of a solid.*

**13.** $\displaystyle\iint_R 2 \, dA$, where $R = [-1, 3] \times [2, 5]$

**14.** $\displaystyle\iint_R 2x \, dA$, where $R = \{(x, y) \mid 0 \le x \le 2, 0 \le y \le 1\}$

**15.** $\displaystyle\iint_R (6 - 2y) \, dA$, where $R = \{(x, y) \mid 0 \le x \le 4, 0 \le y \le 2\}$

**16.** $\displaystyle\iint_R \sqrt{9 - x^2 - y^2} \, dA$, where
$R = \{(x, y) \mid x^2 + y^2 \le 9, x \ge 0, y \ge 0\}$

*In Exercises 17 and 18, the double integral gives the volume of a solid. Describe the solid.*

**17.** $\displaystyle\iint_R (4 - x^2) \, dA$, where $R = \{(x, y) \mid 0 \le y \le x, 0 \le x \le 2\}$

**18.** $\displaystyle\iint_R \left(3 - \frac{1}{2}x - \frac{3}{4}y\right) dA$, where
$R = \{(x, y) \mid 2x + 3y \le 12, x \ge 0, y \ge 0\}$

*In Exercises 19 and 20, the expression is the limit of a Riemann sum of a function $f$ over a rectangle $R$. Write this expression as a double integral over $R$.*

**19.** $\displaystyle\lim_{m, n \to \infty} \sum_{i=1}^{m} \sum_{j=1}^{n} (3 - 2x_{ij}^* + y_{ij}^*) \, \Delta A$,    $R = [-1, 2] \times [1, 3]$

**20.** $\displaystyle\lim_{m, n \to \infty} \sum_{i=1}^{m} \sum_{j=1}^{n} \sqrt{(x_{ij}^*)^2 + 2(y_{ij}^*)^2} \, \Delta A$,    $R = [0, 1] \times [0, 2]$

**cas** *In Exercises 21 and 22, use a computer algebra system (CAS) to obtain an approximate value of the double integral using a regu-*

*lar partition with the given value of m and n and choosing the evaluation point $(x_{ij}^*, y_{ij}^*)$ to be the center of $R_{ij}$.*

**21.** $\displaystyle\iint_R \sqrt{1 + x^2 + y^2}\, dA$, where $R = [0, 1] \times [0, 1]$;
$m = 10, n = 10$

**22.** $\displaystyle\iint_R \frac{1}{1 + e^{xy}}\, dA$, where $R = [0, 1] \times [1, 3]$;
$m = 10, n = 20$

**23.** Use Property 4 (in Theorem 1) of the double integral to show that if $f$ and $|f|$ are integrable over $D$, then

$$\left|\iint_D f(x, y)\, dA\right| \le \iint_D |f(x, y)|\, dA$$

**24.** Use a geometric argument and Theorem 1 to show that if $f(x, y) = k$, where $k$ is a constant, then $\iint_R k\, dA = k \cdot \text{area of } R$.

**25.** Let $R = \{(x, y)\,|\,0 \le x \le 1, 0 \le y \le 1\}$. Show that $0 \le \iint_R e^{-x} \cos y\, dA \le 1$.

**26.** Let $R = \left[0, \frac{1}{2}\right] \times \left[0, \frac{1}{2}\right]$. Show that $0 \le \iint_R \sin(2x + 3y)\, dA \le \frac{1}{4}$.

*In Exercises 27–30, determine whether the statement is true or false. If it is true, explain why. If it is false, explain why or give an example that shows it is false.*

**27.** If $f$ and $g$ are continuous on $D$, then

$$\iint_D [2f(x, y) - 3g(x, y)]\, dA = 2\iint_D f(x, y)\, dA - 3\iint_D g(x, y)\, dA$$

**28.** If $f$ and $g$ are continuous on $D$, then

$$\iint_D [f(x, y)g(x, y)]\, dA = \left[\iint_D f(x, y)\, dA\right]\left[\iint_D g(x, y)\, dA\right]$$

**29.** $\displaystyle\iint_R \frac{\sqrt{x^2 + xy + y^2 + 1}}{\cos(x^2 + y^2)}\, dA \ge \pi$, where $R = \{(x, y)\,|\,x^2 + y^2 \le 1\}$

**30.** If $f$ is nonnegative and continuous on
$D = \{(x, y)\,|\,0 \le x \le 1, 0 \le y \le 1\}$ and
$E = \left\{(x, y)\,|\,0 \le x \le 1, \frac{1}{2} \le y \le 1\right\}$, then

$$\iint_D f(x, y)\, dA \ge \iint_E f(x, y)\, dA$$

## 13.2   Iterated Integrals

### ■ Iterated Integrals Over Rectangular Regions

Just as it is difficult to find the value of an integral of a function of one variable directly from its definition, the task is even harder in the case of double integrals. Fortunately, as you will see, the value of a double integral can be found by evaluating two single integrals.

We begin by looking at the simple case in which $f$ is a continuous function defined on the rectangular region $R = \{(x, y)\,|\,a \le x \le b, c \le y \le b\}$ shown in Figure 1b.

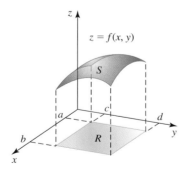

**FIGURE 1**

(a) The graph of $f$

(b) The domain $R$ of $f$

If we fix $x$, then $f(x, y)$ is a function of the single variable $y$ for $c \le y \le d$. As such, we can integrate the function with respect to $y$ over the interval $[c, d]$. This operation is called *partial integration with respect to y* and is the reverse of the operation of partial differentiation studied in Chapter 12. The result is the number

$$\int_c^d f(x, y) \, dy$$

that depends on the value of $x$ in $[a, b]$. In other words, the rule

$$A(x) = \int_c^d f(x, y) \, dy \qquad a \le x \le b \tag{1}$$

defines a function $A$ of $x$ on $[a, b]$. If we integrate the function $A$ with respect to $x$ over $[a, b]$, we obtain

$$\int_a^b A(x) \, dx = \int_a^b \left[ \int_c^d f(x, y) \, dy \right] dx \tag{2}$$

The integral on the right-hand side of Equation (2) is usually written in the form

$$\int_a^b \int_c^d f(x, y) \, dy \, dx \tag{3}$$

without the brackets and is called an *iterated* or *repeated integral*.

Similarly, by holding $y$ fixed and integrating the resulting function with respect to $x$ over $[a, b]$, we obtain a function of $y$ on the interval $[c, d]$. If this function is then integrated with respect to $y$ over $[c, d]$, we obtain the iterated integral

$$\int_c^d \int_a^b f(x, y) \, dx \, dy = \int_c^d \left[ \int_a^b f(x, y) \, dx \right] dy \tag{4}$$

Observe that when we evaluate an iterated integral, we work *from the inside out*.

**EXAMPLE 1**   Evaluate the iterated integrals:

**a.** $\displaystyle \int_1^2 \int_0^1 3x^2 y \, dx \, dy$    **b.** $\displaystyle \int_0^1 \int_1^2 3x^2 y \, dy \, dx$

**Solution**
**a.** By definition,

$$\int_1^2 \int_0^1 3x^2 y \, dx \, dy = \int_1^2 \left[ \int_0^1 3x^2 y \, dx \right] dy$$

Now the integral inside the brackets is found by integrating with respect to $x$ while treating $y$ as a constant. This gives

$$\int_0^1 3x^2 y \, dx = \left[ x^3 y \right]_{x=0}^{x=1} = y$$

Therefore,

$$\int_1^2 \int_0^1 3x^2 y \, dx \, dy = \int_1^2 y \, dy$$

$$= \left[ \frac{1}{2} y^2 \right]_1^2 = \frac{3}{2}$$

**b.** Here, we first integrate with respect to $y$ and then with respect to $x$, obtaining

$$\int_0^1 \int_1^2 3x^2 y \, dy \, dx = \int_0^1 \left[ \int_1^2 3x^2 y \, dy \right] dx$$

$$= \int_0^1 \left[ \frac{3}{2} x^2 y^2 \right]_{y=1}^{y=2} dx$$

$$= \int_0^1 \frac{9}{2} x^2 \, dx = \left[ \frac{3}{2} x^3 \right]_0^1 = \frac{3}{2}$$

### Fubini's Theorem for Rectangular Regions

Observe that the two iterated integrals in Example 1 are equal. Thus, the example seems to suggest that the order of integration of the iterated integrals does not matter. To see why this might be true for continuous functions, consider the special case in which $f$ is nonnegative. Let's calculate the volume $V$ of the solid $S$ lying under the graph of $z = f(x, y)$ and above the rectangular region $R = \{(x, y) \mid a \le x \le b, c \le y \le d\}$.

Using the method of cross sections of Section 5.2, we see that

$$V = \int_a^b A(x) \, dx$$

where $A(x)$ is the area of the cross section of $S$ in the plane perpendicular to the $x$-axis at $x$. (See Figure 2a.) But from the figure, you can see that $A(x)$ is the area under the graph $C$ of the function defined by $g(y) = f(x, y)$ for $c \le y \le d$, where $x$ is fixed. So

$$A(x) = \int_c^d g(y) \, dy = \int_c^d f(x, y) \, dy \qquad x \text{ fixed}$$

Therefore,

$$V = \int_a^b A(x) \, dx = \int_a^b \left[ \int_c^d f(x, y) \, dy \right] dx$$

Similarly, using cross sections perpendicular to the $y$-axis (Figure 2b), you can show that

$$V = \int_c^d \left[ \int_a^b f(x, y) \, dx \right] dy$$

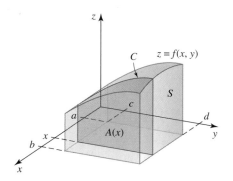

(**a**) $A(x)$ is the area of a cross section of $S$ in the plane perpendicular to the $x$-axis.

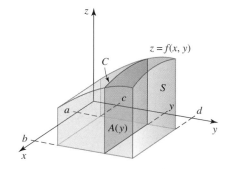

(**b**) $A(y)$ is the area of a cross section of $S$ in the plane perpendicular to the $y$-axis.

**FIGURE 2**

Now, by definition,

$$V = \iint_R f(x, y)\, dA$$

Therefore, we have shown that

$$\iint_R f(x, y)\, dA = \int_a^b \int_c^d f(x, y)\, dy\, dx = \int_c^d \int_a^b f(x, y)\, dx\, dy$$

This discussion suggests the following theorem, which is named after the Italian mathematician Guido Fubini (1879–1943). Its proof lies outside the scope of this book and will be omitted.

---

**THEOREM 1    Fubini's Theorem for Rectangular Regions**

Let $f$ be continuous over the rectangle $R = \{(x, y) \mid a \le x \le b, c \le y \le d\}$. Then

$$\iint_R f(x, y)\, dA = \int_a^b \int_c^d f(x, y)\, dy\, dx = \int_c^d \int_a^b f(x, y)\, dx\, dy$$

---

Fubini's Theorem provides us with a practical method for finding double integrals by expressing them in terms of iterated integrals that we can evaluate by integrating with respect to one variable at a time. It also states that the order in which the integration is carried out does not matter, an important option, as you will see later on. Finally, observe that Fubini's Theorem holds for *any* continuous function; $f(x, y)$ may assume negative as well as positive values on $R$.

**EXAMPLE 2**  Evaluate $\iint_R (1 - 2xy^2)\, dA$, where

$$R = \{(x, y) \mid 0 \le x \le 2, -1 \le y \le 1\}$$

**Solution**  Using Fubini's Theorem, we obtain

$$\iint_R (1 - 2xy^2)\, dA = \int_{-1}^1 \int_0^2 (1 - 2xy^2)\, dx\, dy$$

$$= \int_{-1}^1 \left[ x - x^2 y^2 \right]_{x=0}^{x=2}\, dy$$

$$= \int_{-1}^1 (2 - 4y^2)\, dy = \left[ 2y - \frac{4}{3}y^3 \right]_{-1}^1$$

$$= \left( 2 - \frac{4}{3} \right) - \left( -2 + \frac{4}{3} \right) = \frac{4}{3}$$

We leave it for you to verify that

$$\iint_R (1 - 2xy^2)\, dA = \int_0^2 \int_{-1}^1 (1 - 2xy^2)\, dy\, dx = \frac{4}{3}$$

as well.

**EXAMPLE 3**  Find the volume of the solid lying under the elliptic paraboloid $z = 8 - 2x^2 - y^2$ and above the rectangular region $R = \{(x, y) \mid 0 \le x \le 1, 0 \le y \le 2\}$. (See Figure 3.) Compare with Example 1 in Section 13.1.

**Solution**  Using Fubini's Theorem, we see that the required volume is

$$V = \iint\limits_R (8 - 2x^2 - y^2) \, dA = \int_0^2 \int_0^1 (8 - 2x^2 - y^2) \, dx \, dy$$

$$= \int_0^2 \left[ 8x - \frac{2}{3}x^3 - xy^2 \right]_{x=0}^{x=1} dy$$

$$= \int_0^2 \left( \frac{22}{3} - y^2 \right) dy = \left[ \frac{22}{3}y - \frac{1}{3}y^3 \right]_0^2 = 12$$

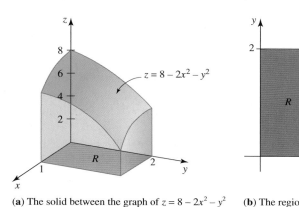

**FIGURE 3**

(a) The solid between the graph of $z = 8 - 2x^2 - y^2$ and the rectangular region $R$

(b) The region $R$

## Iterated Integrals Over Nonrectangular Regions

Fubini's Theorem is valid for regions that are more general than rectangular regions. More specifically, it is valid for the two types of regions that we will now describe. A plane region $R$ is said to be **y-simple** if it lies between two functions of $x$; that is,

$$R = \{(x, y) \mid a \le x \le b, g_1(x) \le y \le g_2(x)\}$$

where $g_1$ and $g_2$ are continuous on $[a, b]$. (See Figure 4.)

An **x-simple** region $R$ is one that lies between two functions of $y$; that is,

$$R = \{(x, y) \mid c \le y \le d, h_1(y) \le x \le h_2(y)\}$$

where $h_1$ and $h_2$ are continuous on $[c, d]$. (See Figure 5.)

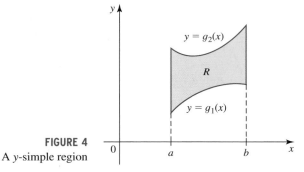

**FIGURE 4**
A y-simple region

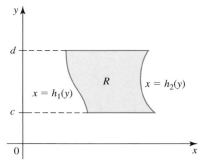

**FIGURE 5**
An x-simple region

The following theorem tells us that a double integral over a $y$-simple or an $x$-simple region can be found by evaluating an iterated integral.

---

**THEOREM 2** Fubini's Theorem for General Regions

Let $f$ be continuous on a region $R$.

**1.** If $R$ is a $y$-simple region, then

$$\iint_R f(x, y)\, dA = \int_a^b \int_{g_1(x)}^{g_2(x)} f(x, y)\, dy\, dx$$

**2.** If $R$ is an $x$-simple region, then

$$\iint_R f(x, y)\, dA = \int_c^d \int_{h_1(y)}^{h_2(y)} f(x, y)\, dx\, dy$$

---

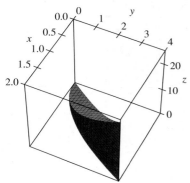

**FIGURE 6**
The graph of the solid $S$

**EXAMPLE 4** Find the volume of the solid $S$ lying under the graph of the surface $z = x^3 + 4y$ and above the region $R$ in the $xy$-plane bounded by the line $y = 2x$ and the parabola $y = x^2$. (See Figure 6.)

**Solution** First, we make a sketch of the region $R$. (See Figure 7a). We see that $R$ can be viewed as a $y$-simple region; that is,

$$R = \{(x, y) \mid 0 \le x \le 2, x^2 \le y \le 2x\}$$

where $g_1(x) = x^2$ and $g_2(x) = 2x$. Observe that if we integrate over a $y$-simple region, we integrate with respect to $y$ first. The appropriate limits of integration can be found by drawing a vertical arrow as shown in Figure 7a. The arrow begins at the lower boundary of the region described by $y = g_1(x) = x^2$, giving the lower limit of integration as $g_1(x) = x^2$, and terminates at the upper boundary of the region described by $y = g_2(x) = 2x$, giving the upper limit of integration as $g_2(x) = 2x$. To find the limits for integrating with respect to $x$, observe that a vertical line sweeping from left to right meets the extreme left point of $R$ when $x = 0$ (the lower limit of integration) and meets the extreme right point of $R$ when $x = 2$ (the upper limit of integration). Using Fubini's Theorem for general regions, we obtain

$$V = \iint_R f(x, y)\, dA = \int_0^2 \int_{x^2}^{2x} (x^3 + 4y)\, dy\, dx$$

$$= \int_0^2 \left[ x^3 y + 2y^2 \right]_{y=x^2}^{y=2x} dx = \int_0^2 \left[ (2x^4 + 8x^2) - (x^5 + 2x^4) \right] dx$$

$$= \int_0^2 (8x^2 - x^5)\, dx = \left[ \frac{8}{3} x^3 - \frac{1}{6} x^6 \right]_0^2 = \frac{32}{3}$$

**Alternative Solution** We can view the region $R$ as an $x$-simple region

$$R = \left\{ (x, y) \mid 0 \le y \le 4, \frac{y}{2} \le x \le \sqrt{y} \right\}$$

where $h_1(y) = y/2$ and $h_2(y) = \sqrt{y}$ are obtained by solving $y = 2x$ and $y = x^2$ for $x$ in terms of $y$, respectively. (See Figure 7b.) If we integrate over an $x$-simple region, we

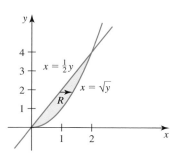

**FIGURE 7**    (**a**) The region $R$ viewed as a $y$-simple region    (**b**) The region $R$ viewed as an $x$-simple region

integrate with respect to $x$ first. A horizontal arrow starting from the left boundary of $R$ described by $h_1(y) = y/2$ and terminating at the right boundary of $R$ described by $h_2(y) = \sqrt{y}$ gives the lower and upper limits of integration with respect to $x$. The limits of integration with respect to $y$ are found by letting a horizontal line sweep through the region. This line meets the lowest point of $R$ when $y = 0$ (the lower limit of integration) and the highest point of $R$ when $y = 4$ (the upper limit of integration). Once again using Fubini's Theorem, we obtain

$$V = \iint_R f(x, y) \, dA = \int_0^4 \int_{y/2}^{\sqrt{y}} (x^3 + 4y) \, dx \, dy$$

$$= \int_0^4 \left[ \frac{1}{4} x^4 + 4xy \right]_{x=y/2}^{x=\sqrt{y}} dy = \int_0^4 \left[ \left( \frac{1}{4} y^2 + 4y^{3/2} \right) - \left( \frac{1}{64} y^4 + 2y^2 \right) \right] dy$$

$$= \int_0^4 \left( -\frac{7}{4} y^2 + 4y^{3/2} - \frac{1}{64} y^4 \right) dy = \left[ -\frac{7}{12} y^3 + \frac{8}{5} y^{5/2} - \frac{1}{320} y^5 \right]_0^4 = \frac{32}{3}$$

as before.    ■

**EXAMPLE 5**    Evaluate $\iint_R (2x - y) \, dA$, where $R$ is the region bounded by the parabola $x = y^2$ and the straight line $x - y = 2$.

**Solution**    The region $R$ is shown in Figure 8. It is both $y$-simple and $x$-simple. But observe that it is more convenient to view it as an $x$-simple region because the lower boundary of $R$ consists of two curves when viewed as a $y$-simple region. In fact, viewing $R$ as a $y$-simple region (Figure 8a) and using Fubini's Theorem, we find

$$\iint_R (2x - y) \, dA = \int_0^1 \int_{-\sqrt{x}}^{\sqrt{x}} (2x - y) \, dy \, dx + \int_1^4 \int_{x-2}^{\sqrt{x}} (2x - y) \, dy \, dx$$

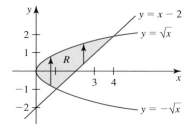

**FIGURE 8**    (**a**) $R$ viewed as a $y$-simple region    (**b**) $R$ viewed as an $x$-simple region

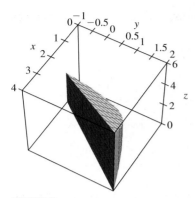

**FIGURE 9**
The solid $S$

On the other hand, viewing $R$ as an $x$-simple region (Figure 8b), we have

$$\iint_R (2x - y)\, dA = \int_{-1}^{2} \int_{y^2}^{y+2} (2x - y)\, dx\, dy$$

$$= \int_{-1}^{2} \left[x^2 - xy\right]_{x=y^2}^{x=y+2} dy = \int_{-1}^{2} \left\{ \left[(y+2)^2 - y(y+2)\right] - \left[y^4 - y^3\right] \right\} dy$$

$$= \int_{-1}^{2} (4 + 2y + y^3 - y^4)\, dy = \left[4y + y^2 + \frac{1}{4}y^4 - \frac{1}{5}y^5\right]_{-1}^{2} = \frac{243}{20}$$

which is easier to evaluate.

The double integral $\iint_R (2x - y)\, dA$ gives the volume of the solid $S$ shown in Figure 9. ∎

Example 5 shows that it is sometimes easier to integrate in one order rather than the other because of the shape of $R$. In certain instances the nature of the function dictates the order of integration, as the next example shows.

**EXAMPLE 6** Evaluate $\int_0^1 \int_y^1 \frac{\sin x}{x}\, dx\, dy$.

**Solution** Because

$$\int \frac{\sin x}{x}\, dx$$

cannot be expressed in terms of elementary functions, the given integral cannot be evaluated as it stands. So let's attempt to evaluate it by reversing the order of integration. We begin by using Fubini's Theorem to express the iterated integral as a double integral. The order of integration of the given integral suggests that

$$\int_0^1 \int_y^1 \frac{\sin x}{x}\, dx\, dy = \iint_R \frac{\sin x}{x}\, dA$$

where $R = \{(x, y) \mid 0 \le y \le 1,\ y \le x \le 1\}$ is viewed as an $x$-simple region (see Figure 10a).

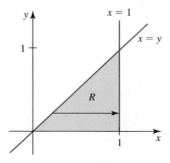

(**a**) $R$ viewed as an $x$-simple region

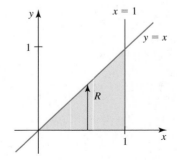

(**b**) $R$ viewed as a $y$-simple region

**FIGURE 10**

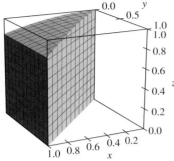

**FIGURE 11**
The solid $S$ represented by the double integral $\displaystyle\int_0^1\int_y^1 \frac{\sin x}{x}\, dx\, dy$

Viewing $R$ as a $y$-simple region (Figure 10b), we find, again by Fubini's Theorem, that

$$\int_0^1\int_y^1 \frac{\sin x}{x}\, dx\, dy = \iint_R \frac{\sin x}{x}\, dA$$

$$= \int_0^1\int_0^x \frac{\sin x}{x}\, dy\, dx = \int_0^1 \left[\frac{y\sin x}{x}\right]_{y=0}^{y=x} dx$$

$$= \int_0^1 \sin x\, dx = \left[-\cos x\right]_0^1 = -\cos 1 + 1 \approx 0.46$$

The double integral $\displaystyle\int_0^1\int_y^1 \frac{\sin x}{x}\, dx\, dy$ gives the volume of the solid $S$ shown in Figure 11. ∎

## 13.2 CONCEPT QUESTIONS

1. Suppose that $f$ is continuous on the rectangular region $R = [a, b] \times [c, d]$.
   a. Explain the difference between the iterated integrals

   $$\int_a^b \left[\int_c^d f(x, y)\, dy\right] dx \quad \text{and} \quad \int_c^d \left[\int_a^b f(x, y)\, dx\right] dy$$

   b. Give a geometric interpretation of each of the iterated integrals in part (a), where $f$ is nonnegative.
   c. What does Fubini's Theorem say about the two iterated integrals in part (a)?

2. a. What is a $y$-simple region, and what is an $x$-simple region?
   b. Express $\iint_R f(x, y)\, dA$ as an iterated integral if $R$ is a $y$-simple region. As an $x$-simple region.
   c. Explain why it is sometimes advantageous to reverse the order of integration of an iterated integral.

## 13.2 EXERCISES

*In Exercises 1–12, evaluate the iterated integral.*

**1.** $\displaystyle\int_0^1\int_0^2 (x + 2y)\, dy\, dx$

**2.** $\displaystyle\int_{-1}^1\int_0^3 (3x^2 + y)\, dx\, dy$

**3.** $\displaystyle\int_0^2\int_1^4 y\sqrt{x}\, dy\, dx$

**4.** $\displaystyle\int_0^1\int_0^1 \frac{x}{1 + xy}\, dy\, dx$

**5.** $\displaystyle\int_0^\pi\int_0^\pi \cos(x + y)\, dy\, dx$

**6.** $\displaystyle\int_0^{\pi/2}\int_0^{\ln 2} e^{-x}\sin y\, dx\, dy$

**7.** $\displaystyle\int_0^4\int_0^{\sqrt{x}} 2xy\, dy\, dx$

**8.** $\displaystyle\int_0^{1/2}\int_0^{\sqrt{1-x}} 2xy\, dy\, dx$

**9.** $\displaystyle\int_0^1\int_0^{\sqrt{1-y^2}} x\, dx\, dy$

**10.** $\displaystyle\int_0^1\int_0^{\sqrt{1-x^2}} (x + y)\, dy\, dx$

**11.** $\displaystyle\int_{-1}^1\int_x^{2x} e^{x+y}\, dy\, dx$

**12.** $\displaystyle\int_0^\pi\int_{e^{-2x}}^{e^{\cos x}} \frac{\ln y}{y}\, dy\, dx$

*In Exercises 13–32, evaluate the double integral.*

**13.** $\displaystyle\iint_R (x + y^2)\, dA$, where

$R = \{(x, y)\,|\, 0 \le x \le 1, -1 \le y \le 2\}$

**14.** $\displaystyle\iint_R (3x^2 + 2xy^3)\, dA$, where

$R = \{(x, y)\,|\, -1 \le x \le 2, 0 \le y \le 2\}$

**15.** $\displaystyle\iint_R (x\cos y + y\sin x)\, dA$, where

$R = \left\{(x, y)\,|\, 0 \le x \le \frac{\pi}{2}, 0 \le y \le \frac{\pi}{4}\right\}$

**16.** $\displaystyle\iint_R ye^{xy}\, dA$, where $R = \{(x, y)\,|\, 0 \le x \le 1, 0 \le y \le 1\}$

**17.** $\iint\limits_R (x + 2y)\, dA$, where $R = \{(x, y)\,|\,0 \le x \le 1, 0 \le y \le x\}$

**18.** $\iint\limits_R \sqrt{1 - x^2}\, dA$, where $R = \{(x, y)\,|\,0 \le x \le 1, 0 \le y \le x\}$

**19.** $\iint\limits_R (x^3 + 2y)\, dA$, where

$R = \{(x, y)\,|\,0 \le x \le 2, x^2 \le y \le 2x\}$

**20.** $\iint\limits_R xy\, dA$, where

$R = \{(x, y)\,|\,-1 \le x \le 2, -x^2 \le y \le 1 + x^2\}$

**21.** $\iint\limits_R (1 + 2x + 2y)\, dA$, where

$R = \{(x, y)\,|\,0 \le y \le 1, y \le x \le 2y\}$

**22.** $\iint\limits_R (x^2 + y^2)\, dA$, where

$R = \{(x, y)\,|\,0 \le y \le 1, -y - 1 \le x \le y - 1\}$

**23.** $\iint\limits_R x \cos y\, dA$, where

$R = \left\{(x, y)\,|\,0 \le y \le \frac{\pi}{2}, 0 \le x \le \sin y\right\}$

**24.** $\iint\limits_R \dfrac{1}{xy}\, dA$, where $R = \{(x, y)\,|\,1 \le y \le e, y \le x \le y^2\}$

**25.** $\iint\limits_R x^2 y\, dA$, where $R$ is the region bounded by the graphs of

$y = x$, $y = 2x$, $x = 1$, and $x = 2$

**26.** $\iint\limits_R xy\, dA$, where $R$ is the region bounded by the graphs of

$y = x^3$, $y = 1$, and $x = 0$

**27.** $\iint\limits_R (\sin x - y)\, dA$, where $R$ is the region bounded by the graphs of $y = \cos x$, $y = 0$, $x = 0$, and $x = \pi/2$

**28.** $\iint\limits_R (x^2 + y)\, dA$, where $R$ is the region bounded by the graphs of $y = x^2 + 2$, $x = 0$, $x = 1$ and $y = 0$

**29.** $\iint\limits_R 4x^3\, dA$, where $R$ is the region bounded by the graphs of

$y = (x - 1)^2$ and $y = -x + 3$

**30.** $\iint\limits_R 2xy^2\, dA$, where $R$ is the region bounded by the graphs of

$x = y^2$ and $x = 3 - 2y^2$

**31.** $\iint\limits_R ye^x\, dA$, where $R$ is the triangular region with vertices

$(0, 0)$, $(4, 4)$, and $(6, 0)$

**32.** $\iint\limits_R y\, dA$, where $R$ is the half-disk defined by the inequalities

$x^2 + y^2 \le 1$ and $y \ge 0$

*In Exercises 33–38, find the volume of the solid shown in the figure.*

**33.**　　　　　　　　　　**34.**

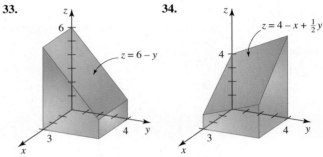

**35.**　　　　　　　　　　**36.**

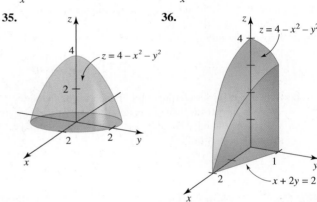

**37.**　　　　　　　　　　**38.**

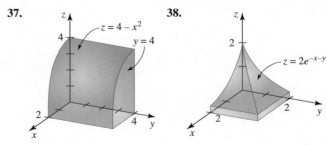

*In Exercises 39–46, find the volume of the solid.*

**39.** The solid under the plane $z = 4 - 2x - y$ and above the region $R = \{(x, y)\,|\,0 \le x \le 1, 0 \le y \le 2\}$ lying in the $xy$-plane

**40.** The solid under the plane $z = x + 2y$ and above the triangular region in the $xy$-plane bounded by the lines $y = 2x$, $y = 0$, and $x = 2$

**41.** The solid under the surface $z = xy$ and above the triangular region in the $xy$-plane bounded by the lines $y = 2x$, $y = -x + 6$, and $y = 0$

**42.** The solid under the surface $z = x^2 + y$ and above the region in the $xy$-plane bounded by the parabolas $y = x^2$ and $y = 2 - x^2$

**43.** The solid under the paraboloid $z = x^2 + y^2$ and above the region in the $xy$-plane bounded by the line $y = x$ and the parabola $y = x^2$

**44.** The solid under the paraboloid $z = x^2 + 3y^2$ and above the region in the $xy$-plane bounded by the graphs of $y = \sqrt{x}$, $y = 0$, and $x = 4$

**45.** The solid bounded by the cylinder $y^2 + z^2 = 9$ and the planes $x = 0$, $y = 0$, $z = 0$, and $2x + y = 2$

**46.** The solid bounded by the cylinder $x^2 + y^2 = 4$ and the planes $z = 4 - y$ and $z = 0$

*In Exercises 47–54, sketch the region of integration for the iterated integral, and reverse the order of integration.*

**47.** $\displaystyle\int_0^1 \int_0^{1-x} f(x, y)\, dy\, dx$  **48.** $\displaystyle\int_0^1 \int_{2x}^2 f(x, y)\, dy\, dx$

**49.** $\displaystyle\int_0^1 \int_{y^2}^{\sqrt[3]{y}} f(x, y)\, dx\, dy$  **50.** $\displaystyle\int_{-2}^2 \int_{-\sqrt{4-y^2}}^{4-y^2} f(x, y)\, dx\, dy$

**51.** $\displaystyle\int_{-1}^{5/2} \int_{y^2-4}^{(3/2)y-3/2} f(x, y)\, dx\, dy$

**52.** $\displaystyle\int_{-1}^1 \int_{x^2}^{3-2x^2} f(x, y)\, dy\, dx$

**53.** $\displaystyle\int_1^e \int_0^{\ln x} f(x, y)\, dy\, dx$

**54.** $\displaystyle\int_0^{\pi/4} \int_0^{\tan x} f(x, y)\, dy\, dx$

*In Exercises 55–60, evaluate the integral by reversing the order of integration.*

**55.** $\displaystyle\int_0^1 \int_{2y}^2 e^{-x^2}\, dx\, dy$  **56.** $\displaystyle\int_0^2 \int_{y/2}^1 e^{y/x}\, dx\, dy$

**57.** $\displaystyle\int_0^4 \int_{\sqrt{x}}^2 \sin y^3\, dy\, dx$  **58.** $\displaystyle\int_0^2 \int_{x^2}^4 x \cos y^2\, dy\, dx$

**59.** $\displaystyle\int_0^4 \int_{\sqrt{y}}^2 \frac{1}{\sqrt{x^3 + 1}}\, dx\, dy$

**60.** $\displaystyle\int_0^1 \int_{\tan^{-1} y}^{\pi/4} \sec^2 x\sqrt{1 + \sec^2 x}\, dx\, dy$

**61.** Suppose that $f(x, y) = g(x)h(y)$ and let $R = \{(x, y)\,|\,a \le x \le b, c \le y \le d\}$. Show that

$$\iint_R f(x, y)\, dA = \left[\int_a^b g(x)\, dx\right]\left[\int_c^d h(y)\, dy\right]$$

**62.** Suppose that $f(x, y)$ has continuous second-order partial derivatives. Find

$$\iint_R f_{xy}(x, y)\, dA$$

where $R = \{(x, y)\,|\,a \le x \le b, c \le y \le d\}$.

**63.** The following figure depicts a semicircular metal plate whose density at the point $(x, y)$ is $(1 + y)$ slugs/ft². What is the mass of the plate?

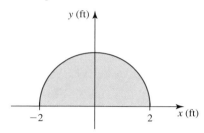

**64. Population Density** The population density (number of people per square mile) of a coastal town is described by the function

$$f(x, y) = \frac{10{,}000e^y}{1 + 0.5|x|} \qquad -10 \le x \le 10, \qquad -4 \le y \le 0$$

where $x$ and $y$ are measured in miles. Find the population inside the rectangular area described by

$$R = \{(x, y)\,|-5 \le x \le 5, -2 \le y \le 0\}$$

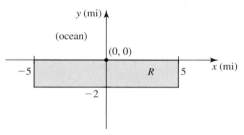

**65. Population Density** Refer to Exercise 64. Find the average population density inside the rectangular area $R$.

**66. Population Density** The population density (number of people per square mile) of a certain city is given by the function

$$f(x, y) = \frac{50{,}000|xy|}{(x^2 + 20)(y^2 + 36)}$$

where the origin $(0, 0)$ gives the location of the government center. Find the population inside the rectangular area described by $R = \{(x, y)\,|-15 \le x \le 15, -20 \le y \le 20\}$.

**cas 67. a.** Plot the region $R$ bounded by the graphs of $y = \cos x$ and $y = x^2 + x$ and the $y$-axis.
  **b.** Find the $x$-coordinate of the point of intersection of the graphs of $y = \cos x$ and $y = x^2 + x$ for $x > 0$ accurate to three decimal places.
  **c.** Estimate $\iint_R x\, dA$.

**cas** **68. a.** Plot the region $R$ bounded by the graphs of $y = e^{-2x}$ and $y = x\sqrt{1 - x^2}$. Then find the $x$-coordinates of the points of intersection of the two graphs accurate to three decimal places.
  **b.** Estimate $\iint_R x^{1/3} y^{2/3} \, dA$.

**cas** *In Exercises 69–72, use a calculator or computer to compute the iterated integral accurate to four decimal places.*

**69.** $\displaystyle\int_0^1 \int_0^2 x^2 y^3 \cos(x + y) \, dy \, dx$

**70.** $\displaystyle\int_0^1 \int_1^2 \frac{xy}{\sqrt{x^2 + y^2}} \, dy \, dx$

**71.** $\displaystyle\int_0^1 \int_0^{1-x} \sqrt{1 + x^2 + y^3} \, dy \, dx$

**72.** $\displaystyle\int_0^2 \int_{-\sqrt{4-x^2}}^{4-x^2} \frac{e^{xy}}{1 + x^2 + y^2} \, dy \, dx$

*In Exercises 73–78, determine whether the statement is true or false. If it is true, explain why. If it is false, explain why or give an example that shows it is false.*

**73.** If $f$ is continuous on $R = [a, b] \times [c, d]$, then
$$\iint_R f(x, y) \, dA = \int_a^b \left[ \int_c^d f(x, y) \, dy \right] dx = \int_c^d \left[ \int_a^b f(x, y) \, dx \right] dy$$

**74.** If $f$ is continuous on $R = [a, b] \times [c, d]$, then
$$\int_c^d \left[ \int_a^b f(x, y) \, dx \right] dy = \int_d^c \left[ \int_b^a f(x, y) \, dx \right] dy$$

**75.** If $f$ is a nonnegative continuous function on the interval $[a, b]$, then the area under the graph of $f$ on $[a, b]$ is $\int_a^b \left[ \int_0^{f(x)} dy \right] dx$.

**76.** If $f$ is continuous on $R = [0, 1] \times [0, 1]$, then
$$\int_0^1 \left[ \int_0^y f(x, y) \, dx \right] dy = \int_0^1 \left[ \int_0^x f(x, y) \, dy \right] dx$$

**77.** $\displaystyle\int_0^2 \int_{-1}^1 x \cos(y^2) \, dx \, dy \neq 0$

**78.** $\displaystyle\int_0^1 \int_0^1 (\sqrt{x} + y)\cos(\sqrt{xy}) \, dx \, dy \leq 1.2$

## 13.3    Double Integrals in Polar Coordinates

### Polar Rectangles

Some double integrals are easier to evaluate if they are expressed in terms of polar coordinates. This is especially true when the region of integration is a **polar rectangle,**
$$R = \{(r, \theta) \mid a \leq r \leq b, \alpha \leq \theta \leq \beta\}$$

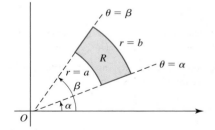

**FIGURE 1**
A polar rectangle is bounded by circular arcs and rays.

(See Figure 1.) Observe that $R$ is a part of an annular ring with inner radius $r = a$ and outer radius $r = b$. Therefore, its area is the difference between the area of the circular sector of radius $b$ and central angle $\Delta\theta = \beta - \alpha$, and the area of the circular sector of radius $a$ and the same central angle $\Delta\theta$. Since the area of a circular sector of radius $r$ and central angle $\theta$ is $\frac{1}{2} r^2 \theta$, we see that the area of $R$ is

$$A = \frac{1}{2} b^2 \Delta\theta - \frac{1}{2} a^2 \Delta\theta = \frac{1}{2} (b^2 - a^2) \Delta\theta \tag{1}$$

$$= \frac{1}{2} (b + a)(b - a) \Delta\theta = \bar{r} \, \Delta r \, \Delta\theta$$

where $\Delta r = b - a$ and $\bar{r} = \frac{1}{2}(b + a)$ is the *average radius* of the polar rectangle.

### Double Integrals Over Polar Rectangles

To define a double integral over a polar rectangle $R$, suppose $f$ is a continuous function on $R$. We start by taking a regular partition

$$a = r_0 < r_1 < r_2 < \cdots < r_{i-1} < r_i < \cdots < r_m = b$$

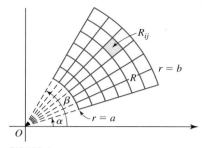

**FIGURE 2**
A polar partition of the polar region $R$
with $m = 6$ and $n = 6$

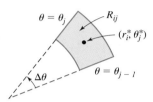

**FIGURE 3**
A polar subrectangle $R_{ij}$ and
its center $(r_i^*, \theta_j^*)$

of $[a, b]$ into $m$ subintervals of equal length $\Delta r = (b - a)/m$, and a regular partition

$$\alpha = \theta_0 < \theta_1 < \theta_2 < \cdots < \theta_{j-1} < \theta_j < \cdots < \theta_n = \beta$$

of $[\alpha, \beta]$ into $n$ subintervals of equal length $\Delta \theta = (\beta - \alpha)/n$. Then the circles $r = r_i$ and the rays $\theta = \theta_j$ determine a **polar partition** $P$ of $R$ into $N = mn$ polar rectangles $R_{11}, R_{12}, \ldots, R_{ij}, \ldots, R_{mn}$, where $R_{ij} = \{(r, \theta) \mid r_{i-1} \leq r \leq r_i, \theta_{j-1} \leq \theta \leq \theta_j\}$, as shown in Figure 2. Figure 3 shows a typical polar subrectangle $R_{ij}$ enlarged for the sake of clarity. The center of $R_{ij}$ is the point $(r_i^*, \theta_j^*)$, where $r_i^*$ is the average radius of $R_{ij}$, and $\theta_j^*$ is the average angle of $R_{ij}$. In other words, $r_i^* = \frac{1}{2}(r_{i-1} + r_i)$ and $\theta_j^* = \frac{1}{2}(\theta_{j-1} + \theta_j)$. Observe that the center of $R_{ij}$, when expressed in terms of rectangular coordinates, takes the form $(r_i^* \cos \theta_j^*, r_i^* \sin \theta_j^*)$. Also, from Equation (1) we see that the area of $R_{ij}$ is $\Delta A_i = r_i^* \Delta r \Delta \theta$. Therefore, the Riemann sum of $f$ over the polar partition $P$ is

$$\sum_{i=1}^{m} \sum_{j=1}^{n} f(r_i^* \cos \theta_j^*, r_i^* \sin \theta_j^*) \Delta A_i = \sum_{i=1}^{m} \sum_{j=1}^{n} f(r_i^* \cos \theta_j^*, r_i^* \sin \theta_j^*) r_i^* \Delta r \Delta \theta$$

$$= \sum_{i=1}^{m} \sum_{j=1}^{n} g(r_i^*, \theta_j^*) \Delta r \Delta \theta$$

where $g(r, \theta) = rf(r \cos \theta, r \sin \theta)$. But the last sum is just a Riemann sum associated with the double integral

$$\int_{\alpha}^{\beta} \int_{a}^{b} g(r, \theta) \, dr \, d\theta$$

Therefore, we have

$$\iint_R f(x, y) \, dA = \lim_{m, n \to \infty} \sum_{i=1}^{m} \sum_{j=1}^{n} f(r_i^* \cos \theta_j^*, r_i^* \sin \theta_j^*) \Delta A$$

$$= \lim_{m, n \to \infty} \sum_{i=1}^{m} \sum_{j=1}^{n} g(r_i^*, \theta_j^*) \Delta r \Delta \theta$$

$$= \int_{\alpha}^{\beta} \int_{a}^{b} g(r, \theta) \, dr \, d\theta = \int_{\alpha}^{\beta} \int_{a}^{b} f(r \cos \theta, r \sin \theta) \, r \, dr \, d\theta$$

> **Transforming a Double Integral Over a Polar Rectangle to Polar Coordinates**
>
> Let $f$ be continuous on a polar rectangle $R = \{(r, \theta) \mid 0 \leq a \leq r \leq b, \alpha \leq \theta \leq \beta\}$, where $0 \leq \beta - \alpha \leq 2\pi$. Then
>
> $$\iint_R f(x, y) \, dA = \int_{\alpha}^{\beta} \int_{a}^{b} f(r \cos \theta, r \sin \theta) \, r \, dr \, d\theta \qquad (2)$$

Thus, we formally transform a double integral over a polar rectangle from rectangular to polar coordinates by substituting

$$x = r \cos \theta, \qquad y = r \sin \theta, \qquad dA = r \, dr \, d\theta$$

and inserting the appropriate limits.

⚠ Do not forget the factor $r$ on the right-hand side of Equation (2). You can remember the expression for $dA$ by making a sketch of the "infinitesimal polar rectangle" shown in Figure 4. The polar rectangle is similar to an ordinary rectangle with sides of length $r \, d\theta$ and $dr$, and therefore, it has "area" $dA = (r \, d\theta) \, dr = r \, dr \, d\theta$.

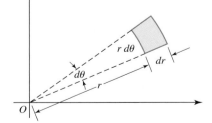

**FIGURE 4**
The infinitesimal polar rectangle has
"area" $dA = r \, dr \, d\theta$.

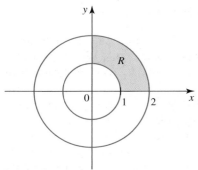

**FIGURE 5**
The region
$R = \{(r, \theta) \mid 1 \le r \le 2, 0 \le \theta \le \frac{\pi}{2}\}$

**EXAMPLE 1** Evaluate $\iint_R (2x + 3y) \, dA$, where $R$ is the region in the first quadrant bounded by the circles $x^2 + y^2 = 1$ and $x^2 + y^2 = 4$.

**Solution** The region $R$ is a polar rectangle that can also be described in terms of polar coordinates by

$$R = \left\{(r, \theta) \mid 1 \le r \le 2, 0 \le \theta \le \frac{\pi}{2}\right\}$$

(See Figure 5.) Using Equation (2), we obtain

$$\iint_R (2x + 3y) \, dA = \int_0^{\pi/2} \int_1^2 (2r \cos \theta + 3r \sin \theta) \, r \, dr \, d\theta$$

$$= \int_0^{\pi/2} \int_1^2 (2r^2 \cos \theta + 3r^2 \sin \theta) \, dr \, d\theta$$

$$= \int_0^{\pi/2} \left[\frac{2}{3} r^3 \cos \theta + r^3 \sin \theta\right]_{r=1}^{r=2} d\theta$$

$$= \int_0^{\pi/2} \left(\frac{14}{3} \cos \theta + 7 \sin \theta\right) d\theta$$

$$= \left[\frac{14}{3} \sin \theta - 7 \cos \theta\right]_0^{\pi/2} = \frac{35}{3}$$

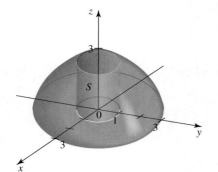

**FIGURE 6**
The solid $S$ lies above the disk
$x^2 + y^2 \le 1$ and under the
hemisphere $z = \sqrt{9 - x^2 - y^2}$.

**EXAMPLE 2** Find the volume of the solid $S$ that lies below the hemisphere $z = \sqrt{9 - x^2 - y^2}$, above the $xy$-plane, and inside the cylinder $x^2 + y^2 = 1$.

**Solution** The solid $S$ is shown in Figure 6. It lies between the hemisphere $z = \sqrt{9 - x^2 - y^2}$ and the circular disk centered at the origin with radius 1. A polar representation of $R$ is

$$R = \{(r, \theta) \mid 0 \le r \le 1, 0 \le \theta \le 2\pi\}$$

Also, in polar coordinates we can write $z = \sqrt{9 - x^2 - y^2} = \sqrt{9 - r^2}$. Therefore, the required volume is given by

$$V = \iint_R f(x, y) \, dA = \int_0^{2\pi} \int_0^1 \sqrt{9 - r^2} \, r \, dr \, d\theta$$

$$= \int_0^{2\pi} \left[-\frac{1}{3}(9 - r^2)^{3/2}\right]_{r=0}^{r=1} d\theta$$

$$= \frac{1}{3}(27 - 16\sqrt{2}) \int_0^{2\pi} d\theta = \frac{2\pi}{3}(27 - 16\sqrt{2})$$

or approximately 9.16.

**Note** You can appreciate the role played by polar coordinates in Example 2 by observing that in rectangular coordinates,

$$V = \int_{-1}^{1} \int_{-\sqrt{1-y^2}}^{\sqrt{1-y^2}} \sqrt{9 - x^2 - y^2} \, dx \, dy$$

which is not easy to evaluate.

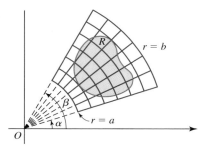

**FIGURE 7**
An inner polar partition of the region R

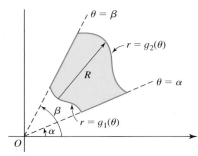

**FIGURE 8**
The polar region
$R = \{(r, \theta) \mid \alpha \leq \theta \leq \beta,$
$g_1(\theta) \leq r \leq g_2(\theta)\}$. Observe
that r runs from the curve
$r = g_1(\theta)$ to the curve $r = g_2(\theta)$
as indicated by the arrow.

## Double Integrals Over General Regions

The results obtained thus far can be extended to more general regions. If the bounded region R is such a region, then we can transform the double integral $\iint_R f(x, y)\, dA$ into one involving polar coordinates by expressing it as a limit of Riemann sums associated with the function

$$f_R(x, y) = \begin{cases} f(x, y) & \text{if } (x, y) \text{ is in } R \\ 0 & \text{if } (x, y) \text{ is outside } R \end{cases}$$

(See Figure 7.)

We will not pursue the details. Instead, we will state the result for the type of region that occurs most frequently in practice: A region is **r-simple** if it is bounded by the graphs of two functions of $\theta$. The r-simple region described by

$$R = \{(r, \theta) \mid \alpha \leq \theta \leq \beta, g_1(\theta) \leq r \leq g_2(\theta)\}$$

where $g_1$ and $g_2$ are continuous on $[\alpha, \beta]$, is shown in Figure 8.

> **Transforming a Double Integral Over a Polar Region to Polar Coordinates**
>
> Let f be continuous on a polar region of the form
>
> $$R = \{(r, \theta) \mid \alpha \leq \theta \leq \beta, g_1(\theta) \leq r \leq g_2(\theta)\}$$
>
> where $0 \leq \beta - \alpha \leq 2\pi$. Then
>
> $$\iint_R f(x, y)\, dA = \int_\alpha^\beta \int_{g_1(\theta)}^{g_2(\theta)} f(r\cos\theta, r\sin\theta)\, r\, dr\, d\theta \qquad (3)$$

**Note**   $\theta$-simple regions (regions that are bounded by the graphs of functions of r) will be considered in Exercise 44. ∎

**EXAMPLE 3**   Use a double integral to find the area enclosed by one loop of the three-leaved rose $r = \sin 3\theta$.

**Solution**   The graph of $r = \sin 3\theta$ is shown in Figure 9. Observe that a loop of the rose is described by the region

$$R = \left\{ (r, \theta) \mid 0 \leq \theta \leq \frac{\pi}{3}, 0 \leq r \leq \sin 3\theta \right\}$$

and may be viewed as being r-simple, where $g_1(\theta) = 0$ and $g_2(\theta) = \sin 3\theta$. Taking $f(x, y) = 1$ in Equation (3), we see that the required area is given by

$$A = \iint_R dA = \int_0^{\pi/3} \int_0^{\sin 3\theta} r\, dr\, d\theta$$

$$= \int_0^{\pi/3} \left[ \frac{1}{2} r^2 \right]_{r=0}^{r=\sin 3\theta} d\theta$$

$$= \frac{1}{2} \int_0^{\pi/3} \sin^2 3\theta\, d\theta = \frac{1}{4} \int_0^{\pi/3} (1 - \cos 6\theta)\, d\theta \qquad \sin^2\theta = \frac{1 - \cos 2\theta}{2}$$

$$= \frac{1}{4} \left[ \theta - \frac{1}{6} \sin 6\theta \right]_{\theta=0}^{\theta=\pi/3} = \frac{\pi}{12}$$

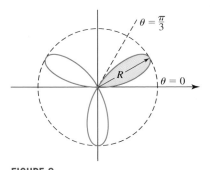

**FIGURE 9**
The region R viewed as an r-simple region

or approximately 0.26. ∎

**EXAMPLE 4** Evaluate $\iint_R y \, dA$, where $R$ is the region in the first quadrant that is outside the circle $r = 2$ and inside the cardioid $r = 2(1 + \cos \theta)$.

**Solution**    The required region

$$R = \left\{ (r, \theta) \,\middle|\, 0 \le \theta \le \tfrac{\pi}{2}, \, 2 \le r \le 2(1 + \cos \theta) \right\}$$

is sketched in Figure 10 and may be viewed as being $r$-simple. Recalling that $y = r \sin \theta$ and using Equation (3), we obtain

$$\iint_R y \, dA = \int_0^{\pi/2} \int_2^{2(1+\cos\theta)} r(\sin\theta) \, r \, dr \, d\theta = \int_0^{\pi/2} \int_2^{2(1+\cos\theta)} r^2(\sin\theta) \, dr \, d\theta$$

$$= \int_0^{\pi/2} \left[ \frac{1}{3} r^3 \sin\theta \right]_{r=2}^{r=2(1+\cos\theta)} d\theta$$

$$= \frac{8}{3} \int_0^{\pi/2} \left[ (1 + \cos\theta)^3 \sin\theta - \sin\theta \right] d\theta$$

$$= \frac{8}{3} \left[ -\frac{1}{4}(1 + \cos\theta)^4 + \cos\theta \right]_0^{\pi/2} = \frac{22}{3} \qquad \blacksquare$$

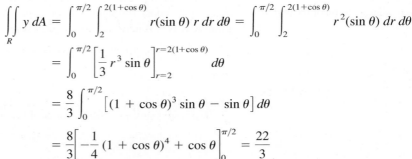

**FIGURE 10**
The polar region
$R = \left\{ (r, \theta) \,\middle|\, 0 \le \theta \le \tfrac{\pi}{2}, \right.$
$\left. 2 \le r \le 2(1 + \cos\theta) \right\}$

**EXAMPLE 5** Find the volume of the solid that lies below the paraboloid $z = 4 - x^2 - y^2$, above the $xy$-plane, and inside the cylinder $(x - 1)^2 + y^2 = 1$.

**Solution**    The solid $S$ under consideration is shown in Figure 11a. It lies above the disk $R$ bounded by the circle with center $(1, 0)$ and radius 1 shown in Figure 11b. This unit circle has polar equation $r = 2 \cos \theta$, as you can verify by replacing $x$ and $y$ in the rectangular equation of the circle by $x = r \cos \theta$ and $y = r \sin \theta$. Therefore,

$$R = \left\{ (r, \theta) \,\middle|\, -\tfrac{\pi}{2} \le \theta \le \tfrac{\pi}{2}, \, 0 \le r \le 2\cos\theta \right\}$$

and may be viewed as being $r$-simple, where $g_1(\theta) = 0$ and $g_2(\theta) = 2\cos\theta$. Using the relationship $x^2 + y^2 = r^2$ and taking advantage of symmetry, we see that the required volume is

$$V = \iint_R (4 - x^2 - y^2) \, dA = \int_{-\pi/2}^{\pi/2} \int_0^{2\cos\theta} (4 - r^2) r \, dr \, d\theta = 2 \int_0^{\pi/2} \int_0^{2\cos\theta} (4r - r^3) \, dr \, d\theta$$

$$= 2 \int_0^{\pi/2} \left[ 2r^2 - \frac{1}{4} r^4 \right]_{r=0}^{r=2\cos\theta} d\theta = 8 \int_0^{\pi/2} (2\cos^2\theta - \cos^4\theta) \, d\theta$$

$$= 8 \int_0^{\pi/2} \left[ 1 + \cos 2\theta - \left( \frac{1 + \cos 2\theta}{2} \right)^2 \right] d\theta \qquad \cos^2\theta = \frac{1 + \cos 2\theta}{2}$$

$$= 8 \int_0^{\pi/2} \left[ \frac{3}{4} + \frac{1}{2} \cos 2\theta - \frac{1 + \cos 4\theta}{8} \right] d\theta$$

$$= 8 \left[ \frac{5}{8} \theta + \frac{1}{4} \sin 2\theta - \frac{1}{32} \sin 4\theta \right]_0^{\pi/2} = \frac{5\pi}{2}$$

or approximately 7.85.

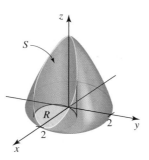

**FIGURE 11**    (**a**) The solid $S$        (**b**) The region $R$ is $r$-simple.

## 13.3 CONCEPT QUESTIONS

**1. a.** What is a polar rectangle? Illustrate with a sketch.
   **b.** Suppose $f$ is continuous on a polar rectangle
   $R = \{(r, \theta) \mid a \leq r \leq b, \alpha \leq \theta \leq \beta\}$, where
   $0 \leq \beta - \alpha \leq 2\pi$. Write $\iint_R f(x, y)\, dA$ in terms of polar
   coordinates.

**2. a.** What is an $r$-simple region? Illustrate with a sketch.
   **b.** Suppose that $f$ is continuous on a region of the form
   $R = \{(r, \theta) \mid \alpha \leq \theta \leq \beta, g_1(\theta) \leq r \leq g_2(\theta)\}$ where
   $0 \leq \beta - \alpha \leq 2\pi$. Write $\iint_R f(x, y)\, dA$ in terms of polar
   coordinates.

## 13.3 EXERCISES

*In Exercises 1–4, determine whether to use polar coordinates or rectangular coordinates to evaluate the integral $\iint_R f(x, y)\, dA$, where $f$ is a continuous function. Then write an expression for the (iterated) integral.*

**1.**

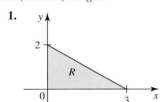

**2.**

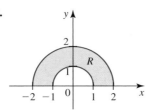

**3.**

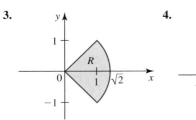

**4.**

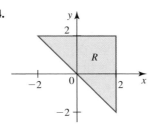

*In Exercises 5–8, sketch the region of integration associated with the integral.*

**5.** $\displaystyle\int_0^\pi \int_1^4 f(r \cos \theta, r \sin \theta)\, r\, dr\, d\theta$

**6.** $\displaystyle\int_0^\pi \int_0^{4 \sin \theta} f(r \cos \theta, r \sin \theta)\, r\, dr\, d\theta$

**7.** $\displaystyle\int_{\pi/4}^{\pi/2} \int_0^{2\sqrt{2}} f(r \cos \theta, r \sin \theta)\, r\, dr\, d\theta$

**8.** $\displaystyle\int_0^{2\pi} \int_0^{1+\cos \theta} f(r \cos \theta, r \sin \theta)\, r\, dr\, d\theta$

*In Exercises 9–16, evaluate the integral by changing to polar coordinates.*

**9.** $\displaystyle\iint_R 3y\, dA$, where $R$ is the disk of radius 2 centered at the origin

**10.** $\displaystyle\iint_R (x + 2y)\, dA$, where $R$ is the region in the first quadrant bounded by the circle $x^2 + y^2 = 9$

**11.** $\displaystyle\iint_R xy\, dA$, where $R$ is the region in the first quadrant bounded by the circle $x^2 + y^2 = 4$ and the lines $x = 0$ and $x = y$

**12.** $\displaystyle\iint_R \sqrt{x^2 + y^2}\, dA$, where $R$ is the region in the first quadrant bounded by the circle $x^2 + y^2 = 4$ and the lines $y = 0$ and $y = \sqrt{3}x$

**13.** $\displaystyle\iint_R \frac{y^2}{x^2 + y^2}\, dA$, where $R$ is the annular region bounded by the circles $x^2 + y^2 = 1$ and $x^2 + y^2 = 2$

**14.** $\displaystyle\iint_R \sin(x^2 + y^2)\, dA$, where $R$ is the region in the first

quadrant bounded by the circles $x^2 + y^2 = 1$ and $x^2 + y^2 = 9$

**15.** $\displaystyle\iint_R y\, dA$, where $R$ is the smaller of the two regions bounded

by the circle $x^2 + y^2 = 2x$ and the line $y = x$

**16.** $\displaystyle\iint_R (x + y)\, dA$, where $R$ is the region in the first quadrant

bounded by the circles $x^2 + y^2 = 4$ and $x^2 + y^2 = 2y$

*In Exercises 17–26, use polar coordinates to find the volume of the solid region T.*

**17.** $T$ lies below the paraboloid $z = x^2 + y^2$, above the $xy$-plane, and inside the cylinder $x^2 + y^2 = 4$.

**18.** $T$ lies below the paraboloid $z = 9 - x^2 - y^2$, above the $xy$-plane, and inside the cylinder $x^2 + y^2 = 1$.

**19.** $T$ lies below the cone $z = \sqrt{x^2 + y^2}$, above the $xy$-plane, and inside the cylinder $x^2 + y^2 = 4$.

**20.** $T$ lies below the cone $z = \sqrt{x^2 + y^2}$, above the $xy$-plane, inside the cylinder $x^2 + y^2 = 4$, and outside the cylinder $x^2 + y^2 = 1$.

**21.** $T$ lies under the plane $3x + 4y + z = 12$, above the $xy$-plane, and inside the cylinder $x^2 + y^2 = 2x$.

**22.** $T$ lies under the paraboloid $z = x^2 + y^2$, above the $xy$-plane, and inside the cylinder $x^2 + y^2 = 2y$.

**23.** $T$ is bounded by the paraboloid $z = 9 - 2x^2 - 2y^2$ and the plane $z = 1$.

**24.** $T$ is bounded by the paraboloids $z = 5x^2 + 5y^2$ and $z = 12 - x^2 - y^2$.

**25.** $T$ is below the sphere $x^2 + y^2 + z^2 = 2$ and above the cone $z = \sqrt{x^2 + y^2}$.

**26.** $T$ is inside the sphere $x^2 + y^2 + z^2 = 4$ and inside the cylinder $x^2 + y^2 = 2y$.

*In Exercises 27–32, use a double integral to find the area of the region R.*

**27.** $R$ is bounded by the circle $r = 3 \cos \theta$.

**28.** $R$ is bounded by one loop of the four-leaved rose $r = \cos 2\theta$.

**29.** $R$ is bounded by the cardioid $r = 3 - 3 \sin \theta$.

**30.** $R$ is bounded by the lemniscate $r^2 = 4 \cos 2\theta$.

**31.** $R$ is outside the circle $r = a$ and inside the circle $r = 2a \sin \theta$.

**32.** $R$ is inside the circle $r = 3 \sin \theta$ and outside the cardioid $r = 1 + \sin \theta$.

*In Exercises 33–40, evaluate the integral by changing to polar coordinates.*

**33.** $\displaystyle\int_{-2}^{2} \int_0^{\sqrt{4-x^2}} \sqrt{x^2 + y^2}\, dy\, dx$

**34.** $\displaystyle\int_0^3 \int_0^{\sqrt{9-x^2}} (x^2 + y^2)^{3/2}\, dy\, dx$

**35.** $\displaystyle\int_{-1}^{1} \int_0^{\sqrt{1-y^2}} \frac{1}{1 + x^2 + y^2}\, dx\, dy$

**36.** $\displaystyle\int_1^3 \int_0^x \frac{1}{\sqrt{x^2 + y^2}}\, dy\, dx$

**37.** $\displaystyle\int_{-2}^{2} \int_0^{\sqrt{4-x^2}} e^{x^2+y^2}\, dy\, dx$

**38.** $\displaystyle\int_0^1 \int_0^{\sqrt{1-y^2}} \cos(x^2 + y^2)\, dx\, dy$

**39.** $\displaystyle\int_0^2 \int_{-\sqrt{2x-x^2}}^{\sqrt{2x-x^2}} x\, dy\, dx$

**40.** $\displaystyle\int_0^1 \int_0^{\sqrt{1-x^2}} \tan^{-1}\!\left(\frac{y}{x}\right) dy\, dx$

*In Exercises 41 and 42, write the sum of the double integrals as a simple double integral using polar coordinates. Then evaluate the resulting integral.*

**41.** $\displaystyle\int_0^{\sqrt{2}} \int_0^x xy\, dy\, dx + \int_{\sqrt{2}}^2 \int_0^{\sqrt{4-x^2}} xy\, dy\, dx$

**42.** $\displaystyle\int_0^1 \int_{\sqrt{1-x^2}}^{\sqrt{4-x^2}} \sqrt{x^2 + y^2}\, dy\, dx + \int_1^2 \int_0^{\sqrt{4-x^2}} \sqrt{x^2 + y^2}\, dy\, dx$

**43. a.** Suppose that $f$ is continuous on the region $R$ bounded by the lines $y = x$, $y = -x$, and $y = 1$. Show that

$$\iint_R f(x, y)\, dA = \int_{\pi/4}^{3\pi/4} \int_0^{\csc \theta} f(r \cos \theta, r \sin \theta)\, r\, dr\, d\theta$$

**b.** Use the result of part (a) to evaluate

$$\int_0^1 \int_{-y}^{y} \sqrt{x^2 + y^2}\, dx\, dy$$

**44.** A region is **θ-simple** if it is bounded by the graphs of two functions of $r$. A $\theta$-simple region is described by

$$R = \{(r, \theta) \mid a \le r \le b, g_1(r) \le \theta \le g_2(r)\}$$

where $g_1$ and $g_2$ are continuous on $[a, b]$. It can be shown that if $f$ is continuous on $R$, then

$$\iint_R f(x, y)\, dA = \int_a^b \int_{g_1(r)}^{g_2(r)} f(r \cos \theta, r \sin \theta)\, r\, d\theta\, dr$$

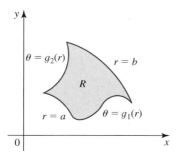

Use this formula to find the area of the smaller region bounded by the spiral $r\theta = 1$, the circles $r = 1$ and $r = 2$, and the polar axis.

45. The integral $I = \int_{-\infty}^{\infty} e^{-x^2/2}\, dx$ occurs in the study of probability and statistics. Show that $I = \sqrt{2\pi}$ by verifying the following steps.

a. Sketch the regions $R_1 = \{(x, y) \mid x^2 + y^2 \le a^2,\ x \ge 0, y \ge 0\}$, $R_2 = \{(x, y) \mid 0 \le x \le a, 0 \le y \le a\}$, and $R_3 = \{(x, y) \mid x^2 + y^2 \le 2a^2, x \ge 0, y \ge 0\}$ on the same plane. Observe that $R_1$ lies inside $R_2$ and that $R_2$ lies inside $R_3$.

b. Show that

$$\iint_{R_1} f(r, \theta)\, dA = \frac{\pi}{4}\left(1 - e^{-a^2}\right)$$

where $f(r, \theta) = e^{-r^2}$, and that

$$\iint_{R_3} f(r, \theta)\, dA = \frac{\pi}{4}\left(1 - e^{-2a^2}\right)$$

c. By considering $\iint_{R_2} f(x, y)\, dA$, where $f(x, y) = e^{-x^2 - y^2}$, and using the results of part (b), show that

$$\frac{\pi}{4}\left(1 - e^{-a^2}\right) < \left(\int_0^a e^{-x^2}\, dx\right)^2 < \frac{\pi}{4}\left(1 - e^{-2a^2}\right)$$

d. Show that $\int_0^{\infty} e^{-x^2}\, dx = \lim_{a\to\infty} \int_0^a e^{-x^2}\, dx = \sqrt{\pi}/2$, and hence deduce the result $I = \sqrt{2\pi}$.

*Use the results of Exercise 45 to evaluate the integrals in Exercises 46 and 47.*

46. $\displaystyle\int_0^{\infty} x^2 e^{-x^2}\, dx$

47. $\displaystyle\int_0^{\infty} \frac{e^{-x}}{\sqrt{x}}\, dx$

48. **Water Delivered by a Water Sprinkler** A lawn sprinkler sprays water in a circular pattern. It delivers water to a depth of $f(r) = 0.1re^{-0.1r}$ ft/hr at a distance of $r$ ft from the sprinkler.

a. Find the total amount of water that is accumulated in an hour in a circular region of radius 50 ft centered at the sprinkler.

b. What is the average amount of water that is delivered to the region in part (a) in an hour?

**Hint:** The average value of $f$ over a region

$$D = \frac{1}{A(D)} \iint_D f(x, y)\, dA$$

where $A(D)$, is the area of $D$.

*In Exercises 49 and 50, determine whether the statement is true or false. If it is true, explain why. If it is false, explain why or give an example that shows it is false.*

49. If $R = \{(r, \theta) \mid \alpha \le \theta \le \beta, 0 \le r \le g(\theta)\}$, where $0 \le \beta - \alpha \le 2\pi$ and $f(r\cos\theta, r\sin\theta) = 1$ for all $(r, \theta)$ in $R$, then $\int_\alpha^\beta \int_0^{g(\theta)} f(r\cos\theta, r\sin\theta)\, r\, dr\, d\theta$ gives the area of $R$.

50. If $R$ is the triangular region, whose vertices in rectangular coordinates are $(0, 0)$, $(1, 0)$, and $(1, 1)$, then $\iint_R f(x, y)\, dA = \int_0^{\pi/4} \int_0^{\sec\theta} f(r\cos\theta, r\sin\theta)\, r\, dr\, d\theta$, where $r$ and $\theta$ are polar coordinates.

## 13.4 Applications of Double Integrals

### Mass of a Lamina

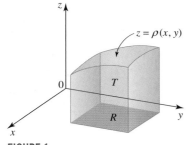

**FIGURE 1**
The mass of the plate $R$ is numerically equal to the volume of the solid $T$.

We mentioned in Section 13.1 that the mass of a thin rectangular plate $R$ lying in the $xy$-plane and having mass density $\rho(x, y)$ at a point $(x, y)$ in $R$ is given by the volume of the solid region $T$ lying directly above $R$ and bounded above by $z = \rho(x, y)$. (See Figure 1.) We will now show that this is the case. In fact, we will demonstrate that the mass of a lamina occupying a region $R$ in the $xy$-plane and having mass density $\rho(x, y)$ at a point $(x, y)$, where $\rho$ is a nonnegative continuous function, is given by $\iint_R \rho(x, y)\, dA$. The double integral also gives the volume of the solid region lying directly above $R$ and bounded above by the surface $z = \rho(x, y)$. (See Figure 2.)

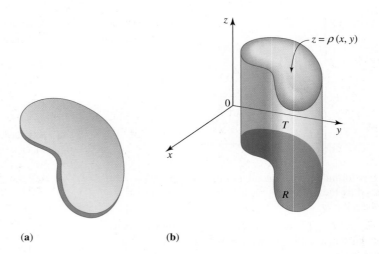

**FIGURE 2**
The mass of the lamina in part (a)
is numerically equal to the
volume of the solid $T$ in part (b).

(a)                (b)

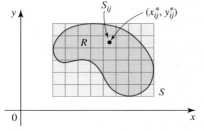

**FIGURE 3**
$P = \{S_{11}, S_{12}, \ldots, S_{ij}, \ldots, S_{mn}\}$ is a
partition of $S$.

Let $S$ be a rectangle containing $R$, and let $P = \{S_{11}, S_{12}, \ldots, S_{ij}, \ldots, S_{mn}\}$ be a regular partition of $S$. (See Figure 3.) Define

$$\rho_R(x, y) = \begin{cases} \rho(x, y) & \text{if } (x, y) \text{ is in } R \\ 0 & \text{if } (x, y) \text{ is inside } S \text{ but outside } R \end{cases}$$

Let $(x_{ij}^*, y_{ij}^*)$ be a point in $S_{ij}$ that also lies in $R$. If both $m$ and $n$ are large (so that the dimensions of $S_{ij}$ are small), then the continuity of $\rho$ implies that $\rho(x, y)$ is approximately equal to $\rho(x_{ij}^*, y_{ij}^*)$ for all points $(x, y)$ in $S_{ij}$. Therefore, the mass of that piece of $R$ lying in $S_{ij}$ with area $\Delta A$ is approximately

$$\rho(x_{ij}^*, y_{ij}^*) \, \Delta A \qquad \text{constant density} \cdot \text{area}$$

Summing the masses of all such pieces gives an approximation of the mass of $R$:

$$\sum_{i=1}^{m} \sum_{j=1}^{n} \rho(x_{ij}^*, y_{ij}^*) \, \Delta A$$

We can expect the approximation to improve as both $m$ and $n$ get larger and larger. Therefore, it is reasonable to define the mass of the lamina as the limiting value of the sums of this form. But each of these sums is just the Riemann sum of $\rho_R$ over $S$. This leads to the following definition.

---

**DEFINITION    Mass of a Lamina**

Suppose that a lamina occupies a region $R$ in the plane and the mass density of the lamina at a point $(x, y)$ in $R$ is $\rho(x, y)$, where $\rho$ is a continuous density function. Then the mass of the lamina is given by

$$m = \iint_R \rho(x, y) \, dA \qquad\qquad \textbf{(1)}$$

---

**Note**    We obtain other physical interpretations of the double integral $\iint_R f(x, y) \, dA$ by letting $f$ represent various types of densities. For example, if an electric charge is spread over a plane surface $R$ and the charge density (charge per unit area) at a point $(x, y)$ in $R$ is $\sigma(x, y)$, then the total charge on the surface is given by

$$Q = \iint\limits_{R} \sigma(x, y)\, dA \tag{2}$$

For another example, suppose that the population density (number of people per unit area) at a point $(x, y)$ in a plane region $R$ is $\delta(x, y)$; then the total population in the region is given by

$$N = \iint\limits_{R} \delta(x, y)\, dA \tag{3}$$

$\blacksquare$

**EXAMPLE 1** Find the mass of a lamina occupying a triangular region $R$ with vertices $(0, 0)$, $(2, 0)$, and $(0, 2)$ if its mass density at a point $(x, y)$ in $R$ is $\rho(x, y) = x + 2y$.

**Solution** The region $R$ is shown in Figure 4. Viewing $R$ as a $y$-simple region and using Equation (1), we see that the required mass is given by

$$m = \iint\limits_{R} \rho(x, y)\, dA = \int_0^2 \int_0^{2-x} (x + 2y)\, dy\, dx$$

$$= \int_0^2 \left[ xy + y^2 \right]_{y=0}^{y=2-x} dx = \int_0^2 \left[ x(2 - x) + (2 - x)^2 \right] dx$$

$$= \int_0^2 (4 - 2x)\, dx = \left[ 4x - x^2 \right]_0^2 = 4$$

$\blacksquare$

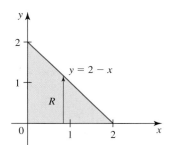

**FIGURE 4**
The region $R$ is both $x$-simple and $y$-simple. Here, we view it as $y$-simple.

**EXAMPLE 2** **Electric Charge Over a Region** An electric charge is spread over a region $R$ lying in the first quadrant and inside the circle $x^2 + y^2 = 4$. Find the total charge on $R$ if the charge density (measured in coulombs per square meter) at a point $(x, y)$ in $R$ is directly proportional to the square of the distance between the point and the origin.

**Solution** The region $R$ is shown in Figure 5. The charge density function is given by $\sigma(x, y) = k(x^2 + y^2)$, where $k$ is the constant of proportionality. Viewing $R$ as a $y$-simple region and using Equation (2), we see that the total charge on $R$ is given by

$$Q = \iint\limits_{R} \sigma(x, y)\, dA = \int_0^2 \int_0^{\sqrt{4-x^2}} k(x^2 + y^2)\, dy\, dx$$

or, changing to polar coordinates,

$$Q = \int_0^{\pi/2} \int_0^2 (kr^2)\, r\, dr\, d\theta = k \int_0^{\pi/2} \int_0^2 r^3\, dr\, d\theta = k \int_0^{\pi/2} \left[ \frac{1}{4} r^4 \right]_{r=0}^{r=2} d\theta$$

$$= 4k \int_0^{\pi/2} d\theta = 2\pi k$$

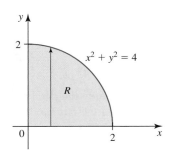

**FIGURE 5**
The region $R$ is both $x$-simple and $y$-simple. Here, we view it as $y$-simple.

or $2\pi k$ coulombs.

$\blacksquare$

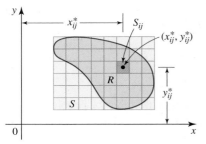

**FIGURE 6**
The region $R$ contained in a rectangle $S$

## Moments and Center of Mass of a Lamina

We considered the moments and the center of mass of a homogeneous lamina in Section 5.7. Using double integrals, we can now find the moments and center of mass of a lamina with *variable* density. Suppose that a lamina with continuous mass density function $\rho$ occupies a region $R$ in the $xy$-plane. (See Figure 6.)

Let $S$ be a rectangle containing $R$, and let $P = \{S_{11}, S_{12}, \ldots, S_{ij}, \ldots, S_{mn}\}$ be a regular partition of $S$. Choose $(x_{ij}^*, y_{ij}^*)$ to be any evaluation point in $S_{ij}$. If $m$ and $n$ are large, then the mass of the part of the lamina occupying the subrectangle $S_{ij}$ is approximately $\rho(x_{ij}^*, y_{ij}^*)\, \Delta A$. Consequently, the moment of this part of the lamina with respect to the $x$-axis is approximately

$$[\rho(x_{ij}^*, y_{ij}^*)\, \Delta A]y_{ij}^* \qquad \text{mass} \cdot \text{moment arm}$$

Adding up these $mn$ moments and taking the limit of the resulting sum as $m$ and $n$ approach infinity, we obtain the moment of the lamina with respect to the $x$-axis. A similar argument gives the moment of the lamina about the $y$-axis. These formulas and the formula for the center of mass of a lamina follow.

> **DEFINITION** **Moments and Center of Mass of a Lamina**
>
> Suppose that a lamina occupies a region $R$ in the $xy$-plane and the mass density of the lamina at a point $(x, y)$ in $R$ is $\rho(x, y)$, where $\rho$ is a continuous density function. Then the **moments of mass** of the lamina with respect to the $x$- and $y$-axes are
>
> $$M_x = \iint\limits_R y\rho(x, y)\, dA \qquad \text{and} \qquad M_y = \iint\limits_R x\rho(x, y)\, dA \qquad \text{(4a)}$$
>
> Furthermore, the **center of mass** of the lamina is located at the point $(\bar{x}, \bar{y})$, where
>
> $$\bar{x} = \frac{M_y}{m} = \frac{1}{m}\iint\limits_R x\rho(x, y)\, dA \qquad \bar{y} = \frac{M_x}{m} = \frac{1}{m}\iint\limits_R y\rho(x, y)\, dA \qquad \text{(4b)}$$
>
> where the mass of the lamina is given by
>
> $$m = \iint\limits_R \rho(x, y)\, dA$$

**Note** If the density function $\rho$ is constant on $R$, then the point $(\bar{x}, \bar{y})$ is also called the *centroid* of the region $R$. (See Section 5.7.) ∎

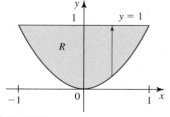

**FIGURE 7**
The region $R$ occupied by the lamina viewed as being $y$-simple

**EXAMPLE 3** A lamina occupies a region $R$ in the $xy$-plane bounded by the parabola $y = x^2$ and the line $y = 1$. (See Figure 7.) Find the center of mass of the lamina if its mass density at a point $(x, y)$ is directly proportional to the distance between the point and the $x$-axis.

**Solution** The mass density of the lamina is given by $\rho(x, y) = ky$, where $k$ is the constant of proportionality. Since $R$ is symmetric with respect to the $y$-axis and the density of the lamina is directly proportional to the distance from the $x$-axis, we see that the center of mass is located on the $y$-axis. Thus, $\bar{x} = 0$. To find $\bar{y}$, we view $R$ as being $y$-simple and first compute

$$m = \iint_R \rho(x, y)\, dA = \int_{-1}^{1} \int_{x^2}^{1} ky\, dy\, dx = k \int_{-1}^{1} \left[ \frac{1}{2} y^2 \right]_{y=x^2}^{y=1} dx$$

$$= \frac{k}{2} \int_{-1}^{1} (1 - x^4)\, dx = \frac{k}{2} \left[ x - \frac{1}{5} x^5 \right]_{-1}^{1} = \frac{4k}{5}$$

Then using Equation (4b), we obtain

$$\bar{y} = \frac{1}{m} \iint_R y\rho(x, y)\, dA = \frac{5}{4k} \int_{-1}^{1} \int_{x^2}^{1} y(ky)\, dy\, dx$$

$$= \frac{5}{4} \int_{-1}^{1} \int_{x^2}^{1} y^2\, dy\, dx = \frac{5}{4} \int_{-1}^{1} \left[ \frac{1}{3} y^3 \right]_{y=x^2}^{y=1} dx$$

$$= \frac{5}{12} \int_{-1}^{1} (1 - x^6)\, dx = \frac{5}{12} \left[ x - \frac{1}{7} x^7 \right]_{-1}^{1} = \frac{5}{7}$$

Therefore, the center of mass of the lamina is located at $\left( 0, \frac{5}{7} \right)$. ∎

## Moments of Inertia

The moments of mass of a lamina, $M_x$ and $M_y$, are called the **first moments** of the lamina with respect to the $x$- and $y$-axes. We can also consider the **second moment** or **moment of inertia** of a lamina about an axis. We begin by recalling that the moment of inertia of a particle of mass $m$ with respect to an axis is defined to be

$$I = mr^2 \qquad \text{mass} \cdot \text{the square of the distance of the moment arm}$$

To understand the physical significance of the moment of inertia of a particle, suppose that a particle of mass $m$ rotates with constant angular velocity $\omega$ about a stationary axis. (See Figure 8.) The velocity of the particle is $v = r\omega$, where $r$ is the distance of the particle from the axis. The kinetic energy of the particle is

$$\frac{1}{2} mv^2 = \frac{1}{2} mr^2 \omega^2 = \frac{1}{2} I\omega^2 \qquad I = mr^2$$

This tells us that the moment of inertia $I$ of the particle with respect to the axis plays the same role in rotational motion that the mass $m$ of a particle plays in rectilinear motion. Since the mass $m$ is a measure of the inertia or resistance to rectilinear motion (the larger $m$ is, the greater the energy needed), we see that the moment of inertia $I$ is a measure of the resistance of the particle to rotational motion.

To define the moment of inertia of a lamina occupying a region $R$ in the $xy$-plane and having mass density described by a continuous function $\rho$, we proceed as before by enclosing $R$ with a rectangle and partitioning the latter using a regular partition. The moment of inertia of the piece of the lamina occupying the subrectangle $R_{ij}$ about the $x$-axis is approximately $[\rho(x_{ij}^*, y_{ij}^*) \Delta A](y_{ij}^*)^2$, where $(x_{ij}^*, y_{ij}^*)$ is a point in $R_{ij}$. Taking the limit of the sum of the second moments as $m$ and $n$ approach infinity, we obtain the **moment of inertia** of the lamina **with respect to the $x$-axis.** In a similar manner we obtain the **moment of inertia** of a lamina **with respect to the $y$-axis.**

The formulas for these quantities and the formulas for the moment of inertia of a lamina with **respect to the origin** (the sum of the moments with respect to $x$ and with respect to $y$) follow.

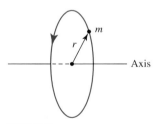

**FIGURE 8**
A particle of mass $m$ rotating about a stationary axis

**DEFINITION    Moments of Inertia of a Lamina**

The **moment of inertia** of a lamina with respect to the **x-axis**, the **y-axis**, and the **origin** are, respectively, as follows:

$$I_x = \lim_{m, n \to \infty} \sum_{i=1}^{m} \sum_{j=1}^{n} (y_{ij}^*)^2 \rho(x_{ij}^*, y_{ij}^*) \, \Delta A = \iint_R y^2 \rho(x, y) \, dA \qquad \text{(5a)}$$

$$I_y = \lim_{m, n \to \infty} \sum_{i=1}^{m} \sum_{j=1}^{n} (x_{ij}^*)^2 \rho(x_{ij}^*, y_{ij}^*) \, \Delta A = \iint_R x^2 \rho(x, y) \, dA \qquad \text{(5b)}$$

$$I_0 = \lim_{m, n \to \infty} \sum_{i=1}^{m} \sum_{j=1}^{n} [(x_{ij}^*)^2 + (y_{ij}^*)^2] \rho(x_{ij}^*, y_{ij}^*) \, \Delta A$$

$$= \iint_R (x^2 + y^2) \rho(x, y) \, dA = I_x + I_y \qquad \text{(5c)}$$

**EXAMPLE 4**    Find the moments of inertia with respect to the x-axis, the y-axis, and the origin of a thin homogeneous disk of mass $m$ and radius $a$, centered at the origin.

**Solution**    Since the disk is homogeneous, its density is constant and given by $\rho(x, y) = m/(\pi a)^2$. Using Equation (5a), we see that the moment of inertia of the disk about the x-axis is given by

$$I_x = \iint_R y^2 \rho(x, y) \, dA = \frac{m}{\pi a^2} \int_0^{2\pi} \int_0^a (r \sin \theta)^2 \, r \, dr \, d\theta$$

$$= \frac{m}{\pi a^2} \int_0^{2\pi} \int_0^a r^3 \sin^2 \theta \, dr \, d\theta = \frac{m}{\pi a^2} \int_0^{2\pi} \left[ \frac{1}{4} r^4 \sin^2 \theta \right]_{r=0}^{r=a} d\theta$$

$$= \frac{ma^2}{4\pi} \int_0^{2\pi} \sin^2 \theta \, d\theta = \frac{ma^2}{8\pi} \int_0^{2\pi} (1 - \cos 2\theta) \, d\theta$$

$$= \frac{ma^2}{8\pi} \left[ \theta - \frac{1}{2} \sin 2\theta \right]_0^{2\pi} = \frac{1}{4} ma^2$$

By symmetry we see that $I_y = I_x = \frac{1}{4} ma^2$. Finally, using Equation (5c), we see that the moment of inertia of the disk about the origin is given by

$$I_0 = I_x + I_y = \frac{1}{4} ma^2 + \frac{1}{4} ma^2 = \frac{1}{2} ma^2 \qquad \blacksquare$$

## ▮ Radius of Gyration of a Lamina

If we imagine that the mass of a lamina is concentrated at a point at a distance $R$ from the axis, then the moment of inertia of this "point mass" would be the same as the moment of inertia of the lamina. (See Figure 9.) The distance $R$ is called the **radius of gyration of the lamina with respect to the axis.** Thus, if the mass of the lamina is $m$ and its moment of inertia with respect to the axis is $I$, then

$$mR^2 = I$$

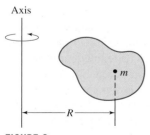

**FIGURE 9**
$R$ is the radius of gyration of the lamina with respect to the axis.

from which we see that

$$R = \sqrt{\frac{I}{m}} \tag{6}$$

**EXAMPLE 5** Find the radius of gyration of the disk of Example 4 with respect to the $y$-axis.

**Solution** Using the result of Example 4, we have $I_y = \frac{1}{4}ma^2$. Therefore, using Equation (6), we see that the radius of gyration of the disk about the $y$-axis is

$$\bar{\bar{x}} = \sqrt{\frac{I_y}{m}} = \sqrt{\frac{\frac{1}{4}ma^2}{m}} = \frac{1}{2}a$$

**Note** In Example 5 we have used the customary notation $\bar{\bar{x}}$ for the radius of gyration of a lamina with respect to the $y$-axis. The radius of gyration of a lamina with respect to the $x$-axis is denoted by $\bar{\bar{y}}$.

## 13.4 CONCEPT QUESTIONS

1. A lamina occupies a region $R$ in the plane. If the mass density of the lamina is $\rho(x, y)$, write an integral giving (a) the mass of the lamina, (b) the moments of mass of the lamina with respect to the $x$- and $y$-axes, and (c) the center of mass of the lamina.

2. A lamina occupies a region $R$ in the plane.
   **a.** Write an integral giving the moment of inertia of the lamina with respect to the $x$-axis, the $y$-axis, and the origin.

   **b.** Write an integral giving the moment of inertia of the lamina with respect to a line $L$.
   **Hint:** Let $d(x, y)$ denote the distance between a point $(x, y)$ in $R$ and the line $L$.

3. What is the radius of gyration of a lamina with respect to an axis? Illustrate with a sketch.

## 13.4 EXERCISES

*In Exercises 1–12, find the mass and the center of mass of the lamina occupying the region $R$ and having the given mass density.*

1. $R$ is the rectangular region with vertices $(0, 0)$, $(3, 0)$, $(3, 2)$, and $(0, 2)$;  $\rho(x, y) = y$

2. $R$ is the rectangular region with vertices $(0, 0)$, $(3, 0)$, $(3, 1)$, and $(0, 1)$;  $\rho(x, y) = x^2 + y^2$

3. $R$ is the triangular region with vertices $(0, 0)$, $(2, 1)$, and $(4, 0)$;  $\rho(x, y) = x$

4. $R$ is the triangular region with vertices $(1, 0)$, $(1, 1)$, and $(0, 1)$;  $\rho(x, y) = x + y$

5. $R$ is the region bounded by the graphs of the equations $y = \sqrt{x}$, $y = 0$, and $x = 4$;  $\rho(x, y) = xy$

6. $R$ is the region bounded by the parabola $y = 4 - x^2$ and the $x$-axis;  $\rho(x, y) = y$

7. $R$ is the region bounded by the graphs of $y = e^x$, $y = 0$, $x = 0$, and $x = 1$;  $\rho(x, y) = 2xy$

8. $R$ is the region bounded by the graphs of $y = \ln x$, $y = 0$, and $x = e$;  $\rho(x, y) = y/x$

9. $R$ is the region bounded by the graphs of $y = \sin x$, $y = 0$, $x = 0$, and $x = \pi$;  $\rho(x, y) = y$

10. $R$ is the region in the first quadrant bounded by the circle $x^2 + y^2 = 1$;  $\rho(x, y) = x + y$

11. $R$ is the region bounded by the circle $r = 2 \cos \theta$; $\rho(r, \theta) = r$

12. $R$ is the region bounded by the cardioid $r = 1 + \cos \theta$; $\rho(r, \theta) = 3$

13. An electric charge is spread over a rectangular region $R = \{(x, y) \mid 0 \le x \le 3, 0 \le y \le 1\}$. Find the total charge on $R$ if the charge density at a point $(x, y)$ in $R$ (measured in coulombs per square meter) is $\sigma(x, y) = 2x^2 + y^3$.

14. **Electric Charge on a Disk** An electric charge is spread over the half-disk $H$ described by $x^2 + y^2 = 4$, $y \ge 0$. Find the total charge on $H$ if the charge density at any point $(x, y)$ in $H$ (measured in coulombs per square meter) is $\sigma(x, y) = \sqrt{x^2 + y^2}$.

15. **Temperature of a Hot Plate** An 8-in. hot plate is described by the set $S = \{(x, y) \mid x^2 + y^2 \le 16\}$. The temperature at the

point $(x, y)$ is $T(x, y) = 400 \cos(0.1\sqrt{x^2 + y^2})$, measured in degrees Fahrenheit. What is the average temperature of the hot plate?

16. **Population Density of a City**  The population density (number of people per square mile) of a certain city is

$$\sigma(x, y) = 3000e^{-(x^2+y^2)}$$

where $x$ and $y$ are measured in miles. Find the population within a 1-mi radius of the town hall, located at the origin.

*In Exercises 17–20, find the moments of inertia $I_x$, $I_y$, and $I_0$ and the radii of gyration $\bar{\bar{x}}$ and $\bar{\bar{y}}$ for the lamina occupying the region $R$ and having uniform density $\rho$.*

17. $R$ is the rectangular region with vertices $(0, 0)$, $(a, 0)$, $(a, b)$, and $(0, b)$.

18. $R$ is the triangular region with vertices $(0, 0)$, $(a, 0)$, and $(0, b)$.

19. $D$ is the half-disk $H = \{(x, y) \mid x^2 + y^2 \le R^2, y \ge 0\}$.

20. $R$ is the region bounded by the ellipse $\dfrac{x^2}{a^2} + \dfrac{y^2}{b^2} = 1$.

*In Exercises 21–24, find the moments of inertia $I_x$, $I_y$, and $I_0$ and the radii of gyration $\bar{\bar{x}}$ and $\bar{\bar{y}}$ for the lamina.*

21. The lamina of Exercise 1

22. The lamina of Exercise 3

23. The lamina of Exercise 5

24. The lamina of Exercise 10

25. A thin metal plate has the shape of the region $R$ inside the circle $x^2 + y^2 = 4$, below the line $y = x$, to the right of the line $x = 1$, and above the $x$-axis. Its density is $\rho(x, y) = y/x$ for $(x, y)$ in $R$. Find the mass of the plate.

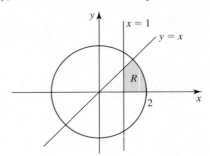

26. Find the rectangular coordinates of the centroid of the region lying between the circles $r = 2 \cos \theta$ and $r = 4 \cos \theta$.

*In Exercises 27–29, determine whether the statement is true or false. If it is true, explain why. If it is false, explain why or give an example that shows it is false.*

27. A piece of metal is laminated from two thin sheets of metal with mass density $\rho_1(x, y)$ and $\rho_2(x, y)$. If it occupies a region $R$ in the plane, then the mass of the laminate is $\iint_R \rho_1(x, y) \, dA + \iint_R \rho_2(x, y) \, dA$.

28. If the region occupied by a lamina is symmetric with respect to both the $x$- and $y$-axes, then the center of mass of the lamina must be located at the origin.

29. If a lamina occupies a region $R$ in the plane, then its center of mass must be located in $R$.

## 13.5    Surface Area

In Section 5.4 we saw that the area of a surface of revolution can be found by evaluating a simple integral. We now turn our attention to the problem of finding the area of more general surfaces. More specifically, we will consider surfaces that are graphs of functions of two variables. As you will see, the area of these surfaces can be found by using double integrals.

### ■ Area of a Surface $z = f(x, y)$

For simplicity we will consider the case in which $f$ is defined in an open set containing a rectangular region $R = [a, b] \times [c, d] = \{(x, y) \mid a \le x \le b, c \le y \le d\}$ and $f(x, y) \ge 0$ on $R$. Furthermore, we assume that $f$ has continuous first-order partial derivatives in that region. We wish first to define what we mean by the *area* of the surface $S$ with equation $z = f(x, y)$ (Figure 1) and then to find a formula that will enable us to calculate this area.

Let $P$ be a regular partition of $R$ into $N = mn$ subrectangles $R_{11}, R_{12}, \ldots, R_{mn}$. Corresponding to the subrectangle $R_{ij}$, there is the part $S_{ij}$ of $S$ (called a **patch**) that lies

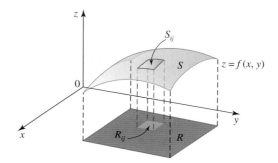

**FIGURE 1**
The surface $S$ is the graph of
$z = f(x, y)$ for $(x, y)$ in $R$.

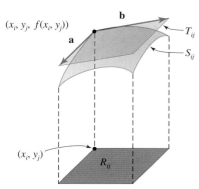

**FIGURE 2**
The tangent plane determined by $\mathbf{a}$ and
$\mathbf{b}$ approximates $S$ well if $R_{ij}$ is small.

directly above $R_{ij}$ with area denoted by $\Delta S_{ij}$. Since the subrectangles $R_{ij}$ are nonoverlapping except for their common boundaries, so are the patches $S_{ij}$ of $S$, so the area of $S$ is given by

$$A = \sum_{i=1}^{m} \sum_{j=1}^{n} \Delta S_{ij} \qquad (1)$$

Next, let's find an approximation of $\Delta S_{ij}$. Let $(x_i, y_j)$ be the corner of $R_{ij}$ closest to the origin, and let $(x_i, y_j, f(x_i, y_j))$ be the point directly above it. If you refer to Figure 2, you can see that $\Delta S_{ij}$ is approximated by the area of $\Delta T_{ij}$ of the parallelogram $T_{ij}$ that is part of the tangent plane to $S$ at the point $(x_i, y_j, f(x_i, y_j))$ and lying directly above $R_{ij}$. To find a formula for $\Delta T_{ij}$, let $\mathbf{a}$ and $\mathbf{b}$ be vectors that have initial point $(x_i, y_j, f(x_i, y_j))$ and lie along the sides of the approximating parallelogram. Now from Section 12.3 we see that the slopes of the tangent lines passing through $(x_i, y_j, f(x_i, y_j))$ and having the directions of $\mathbf{a}$ and $\mathbf{b}$ are given by $f_x(x_i, y_j)$ and $f_y(x_i, y_j)$, respectively. Therefore,

$$\mathbf{a} = \Delta x \mathbf{i} + f_x(x_i, y_j) \Delta x \mathbf{k} \qquad \text{and} \qquad \mathbf{b} = \Delta y \mathbf{j} + f_y(x_i, y_j) \Delta y \mathbf{k}$$

From Section 10.4 we have $\Delta T_{ij} = |\mathbf{a} \times \mathbf{b}|$. But

$$\mathbf{a} \times \mathbf{b} = \begin{vmatrix} \mathbf{i} & \mathbf{j} & \mathbf{k} \\ \Delta x & 0 & f_x(x_i, y_j) \Delta x \\ 0 & \Delta y & f_y(x_i, y_j) \Delta y \end{vmatrix}$$

$$= -f_x(x_i, y_j) \Delta x \Delta y \mathbf{i} - f_y(x_i, y_j) \Delta x \Delta y \mathbf{j} + \Delta x \Delta y \mathbf{k}$$

$$= [-f_x(x_i, y_j) \mathbf{i} - f_y(x_i, y_j) \mathbf{j} + \mathbf{k}] \Delta A$$

where $\Delta A = \Delta x \Delta y$ is the area of $R_{ij}$. Therefore,

$$\Delta T_{ij} = |\mathbf{a} \times \mathbf{b}| = \sqrt{[f_x(x_i, y_j)]^2 + [f_y(x_i, y_j)]^2 + 1} \, \Delta A \qquad (2)$$

If we approximate $\Delta S_{ij}$ by $\Delta T_{ij}$, then Equation (1) becomes

$$A \approx \sum_{i=1}^{m} \sum_{j=1}^{n} \Delta T_{ij}$$

Intuitively, we see that the approximation should get better and better as both $m$ and $n$ get larger and larger. This suggests that we define

$$A = \lim_{m, n \to \infty} \sum_{i=1}^{m} \sum_{j=1}^{n} \sqrt{[f_x(x_i, y_j)]^2 + [f_y(x_i, y_j)]^2 + 1} \, \Delta A$$

Using the definition of the double integral, we obtain the following result, which is stated for the general case in which $R$ is not necessarily rectangular and $f(x, y)$ is not necessarily positive.

**Formula for Finding the Area of a Surface $z = f(x, y)$**

Let $f$ be defined on a region $R$ in the $xy$-plane and suppose that $f_x$ and $f_y$ are continuous. The area $A$ of the surface $z = f(x, y)$ is

$$A = \iint_R \sqrt{[f_x(x, y)]^2 + [f_y(x, y)]^2 + 1} \; dA \qquad (3)$$

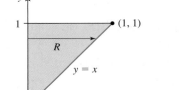

**FIGURE 3**
The region
$R = \{(x, y) \mid 0 \le x \le y, 0 \le y \le 1\}$
viewed as an $x$-simple region

**EXAMPLE 1**  Find the area of the part of the surface with equation $z = 2x + y^2$ that lies directly above the triangular region $R$ in the $xy$-plane with vertices $(0, 0)$, $(1, 1)$, and $(0, 1)$.

**Solution**  The region $R$ is shown in Figure 3. It is both a $y$-simple and an $x$-simple region. Viewed as an $x$-simple region

$$R = \{(x, y) \mid 0 \le x \le y, 0 \le y \le 1\}$$

Using Equation (3) with $f(x, y) = 2x + y^2$, we see that the required area is

$$A = \iint_R \sqrt{[f_x(x, y)]^2 + [f_y(x, y)]^2 + 1} \; dA$$

$$= \iint_R \sqrt{2^2 + (2y)^2 + 1} \; dA = \int_0^1 \int_0^y \sqrt{4y^2 + 5} \; dx \, dy$$

$$= \int_0^1 \left[ x\sqrt{4y^2 + 5} \right]_{x=0}^{x=y} dy = \int_0^1 y\sqrt{4y^2 + 5} \; dy$$

$$= \left[ \frac{1}{8} \cdot \frac{2}{3} (4y^2 + 5)^{3/2} \right]_0^1 = \frac{1}{12} (27 - 5\sqrt{5})$$

or approximately 1.32.   ∎

**EXAMPLE 2**  Find the surface area of the part of the paraboloid $z = 9 - x^2 - y^2$ that lies above the plane $z = 5$.

**Solution**  The paraboloid is sketched in Figure 4a. The paraboloid intersects the plane $z = 5$ along the circle $x^2 + y^2 = 4$. Therefore, the surface of interest lies directly above the disk $R = \{(x, y) \mid x^2 + y^2 \le 4\}$ shown in Figure 4b. Using Equation (3) with $f(x, y) = 9 - x^2 - y^2$, we find the required area to be

$$A = \iint_R \sqrt{[f_x(x, y)]^2 + [f_y(x, y)]^2 + 1} \; dA$$

$$= \iint_R \sqrt{(-2x)^2 + (-2y)^2 + 1} \; dA$$

$$= \iint_R \sqrt{4x^2 + 4y^2 + 1} \; dA$$

**Historical Biography**

GASPARD MONGE
(1746-1818)

In 1789, at the beginning of the French Revolution, Gaspard Monge was one of the best-known mathematicians in France. In addition to doing theoretical work in descriptive geometry, Monge applied his skills to construction projects, general architecture, and military applications. Before the revolution he was appointed examiner of naval cadets. This position took him away from his professorship in Mézières, but he used his salary to pay other people to fulfill his teaching duties. With this arrangement in place, in 1796 Monge embarked on a prolonged absence from France. He traveled first to Italy, where he became friendly with Napoleon Bonaparte. Two years later, he joined Bonaparte's expeditionary force to Egypt. While in Egypt, Monge carried out many technical and scientific tasks, including the establishment of the Institut d'Egypt in Cairo. Monge returned to Paris in 1799, where he resumed his teaching duties and returned to his research. He received numerous awards for his work and accepted an appointment as a senator for life during Napoleon's military dictatorship. However, after the defeat of Napoleon in 1815, life became difficult for Monge. He was expelled from the Institut de France, and his life was continually threatened. Monge died in Paris in 1818. He is known today primarily for his application of the calculus to the study of curvature of surfaces, and he is considered the father of differential geometry.

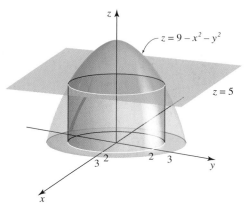

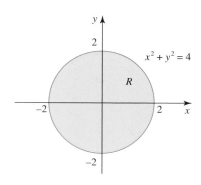

(a) The part of the paraboloid that lies above the plane $z = 5$    (b) The disk $R = \{(x,y) \mid x^2 + y^2 \le 4\}$
**FIGURE 4**

Changing to polar coordinates, we have

$$A = \int_0^{2\pi} \int_0^2 \sqrt{4r^2 + 1}\, r\, dr\, d\theta$$

$$= \int_0^{2\pi} \left[ \frac{1}{8} \cdot \frac{2}{3} (4r^2 + 1)^{3/2} \right]_{r=0}^{r=2} d\theta$$

$$= \int_0^{2\pi} \left[ \frac{1}{12} (17^{3/2} - 1) \right] d\theta = 2\pi \left( \frac{1}{12} \right)(17\sqrt{17} - 1) = \frac{1}{6}\pi(17\sqrt{17} - 1)$$

or approximately 36.2.

## Area of Surfaces with Equations $y = g(x, z)$ and $x = h(y, z)$

Formulas for finding the area of surfaces that are graphs of $y = g(x, z)$ and $x = h(y, z)$ are developed in a similar manner.

**Formulas for Finding the Area of Surfaces in the Form $y = g(x, z)$ and $x = h(y, z)$.**

Let $g$ be defined on a region $R$ in the $xz$-plane, and suppose that $g_x$ and $g_z$ are continuous. The area $A$ of the surface $y = g(x, z)$ is

$$A = \iint_R \sqrt{[g_x(x, z)]^2 + [g_z(x, z)]^2 + 1}\, dA \tag{4}$$

Let $h$ be defined on a region $R$ in the $yz$-plane, and suppose that $h_y$ and $h_z$ are continuous. The area $A$ of the surface $x = h(y, z)$ is

$$A = \iint_R \sqrt{[h_y(y, z)]^2 + [h_z(y, z)]^2 + 1}\, dA \tag{5}$$

These situations are depicted in Figure 5.

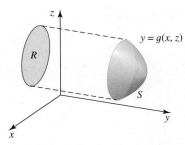

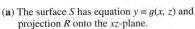

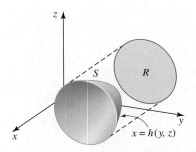

**FIGURE 5**

(a) The surface $S$ has equation $y = g(x, z)$ and projection $R$ onto the $xz$-plane.

(b) The surface $S$ has equation $x = h(y, z)$ and projection $R$ onto the $yz$-plane.

**EXAMPLE 3**  Find the area of that part of the plane $y + z = 2$ inside the cylinder $x^2 + z^2 = 1$.

**Solution**  The surface $S$ of interest is sketched in Figure 6a. The projection of $S$ onto the $xz$-plane is the disk $R = \{(x, z) \mid x^2 + z^2 \le 1\}$ shown in Figure 6b. Using Equation (4) with $g(x, z) = 2 - z$, we see that the area of $S$ is

$$A = \iint_R \sqrt{[g_x(x, z)]^2 + [g_z(x, z)]^2 + 1} \, dA$$

$$= \iint_R \sqrt{0^2 + (-1)^2 + 1} \, dA = \sqrt{2} \iint_R 1 \, dA = \sqrt{2}\pi$$

upon observing that the area of $R$ is $\pi$.

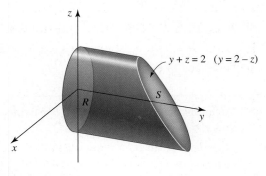

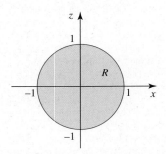

**FIGURE 6**    (a) The surface $S$

(b) The projection $R$ of $S$ onto the $xz$-plane

## 13.5  CONCEPT QUESTIONS

1. Write an integral giving the area of the surface $z = f(x, y)$ defined over a region $R$ in the $xy$-plane.

2. Write an integral giving the area of the surface $x = f(y, z)$ defined over a region $R$ in the $yz$-plane.

## 13.5   EXERCISES

*In Exercises 1–14, find the area of the surface S.*

**1.** $S$ is the part of the plane $2x + 3y + z = 12$ that lies above the rectangular region $R = \{(x, y) \mid 0 \le x \le 2, 0 \le y \le 1\}$.

**2.** $S$ is the part of the plane $3x + 2y + z = 6$ that lies above the triangular region with vertices $(0, 0)$, $(1, 3)$, and $(0, 3)$.

**3.** $S$ is the part of the surface $z = \frac{1}{2}x^2 + y$ that lies above the triangular region with vertices $(0, 0)$, $(1, 0)$, and $(1, 1)$.

**4.** $S$ is the part of the surface $z = 2 - x^2 + y$ that lies above the triangular region with vertices $(0, -1)$, $(1, 0)$, and $(0, 1)$.

**5.** $S$ is the part of the paraboloid $z = 9 - x^2 - y^2$ that lies above the $xy$-plane.

**6.** $S$ is the part of the paraboloid $y = 9 - x^2 - z^2$ that lies between the planes $y = 0$ and $y = 5$.

**7.** $S$ is the part of the sphere $x^2 + y^2 + z^2 = 9$ that lies above the plane $z = 2$.

**8.** $S$ is the part of the hyperbolic paraboloid $z = y^2 - x^2$ that lies above the annular region $A = \{(x, y) \mid 1 \le x^2 + y^2 \le 4\}$.

**9.** $S$ is the part of the surface $x = yz$ that lies inside the cylinder $y^2 + z^2 = 16$.

**10.** $S$ is the part of the sphere $x^2 + y^2 + z^2 = 9$ that lies to the right of the $xz$-plane and inside the cylinder $x^2 + z^2 = 4$.

**11.** $S$ is the part of the sphere $x^2 + y^2 + z^2 = 8$ that lies inside the cone $z^2 = x^2 + y^2$.

**12.** $S$ is the part of the hyperbolic paraboloid $y = x^2 - z^2$ that lies in the first octant and inside the cylinder $x^2 + z^2 = 4$.

**13.** $S$ is the part of the sphere $x^2 + y^2 + z^2 = a^2$ that lies inside the cylinder $x^2 - ax + y^2 = 0$.

**14.** $S$ comprises the parts of the cylinder $x^2 + z^2 = 1$ that lie within the cylinder $y^2 + z^2 = 1$.

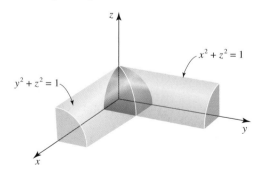

**Hint:** The figure shows the intersection of the two cylinders in the first octant. Use symmetry.

**15.** Let $S$ be the part of the plane $ax + by + cz = d$ lying in the first octant whose projection onto the $xy$-plane is a region $R$. Prove that the area of $S$ is $(1/c) \sqrt{a^2 + b^2 + c^2} \, A(R)$, where $A(R)$ is the area of $R$.

**16. a.** Let $S$ be the part of the sphere $x^2 + y^2 + z^2 = a^2$ that lies above the region $R = \{(x, y) \mid x^2 + y^2 \le b^2, 0 \le b \le a\}$ in the $xy$-plane. Show that the area of $S$ is $2\pi a(a - \sqrt{a^2 - b^2})$.

**b.** Use the result of part (a) to deduce that the area of a sphere of radius $a$ is $4\pi a^2$.

**cas** *In Exercises 17–20, use a calculator or a computer to approximate the area of the surface S, accurate to four decimal places.*

**17.** $S$ is the part of the paraboloid $z = x^2 + y^2$ that lies above the square region $R = \{(x, y) \mid 0 \le x \le 2, 0 \le y \le 2\}$.

**18.** $S$ is the part of the paraboloid $z = 9 - x^2 - y^2$ that lies above the square region $R = \{(x, y) \mid -2 \le x \le 2, -2 \le y \le 2\}$.

**19.** $S$ is the part of the surface $z = e^{-x^2 - y^2}$ that lies inside the cylinder $x^2 + y^2 \le 4$.

**20.** $S$ is the part of the surface $z = \sin(x^2 + y^2)$ that lies above the disk $x^2 + y^2 \le 1$.

*In Exercises 21–24, write a double integral that gives the surface area of the part of the graph of f that lies above the region R. Do not evaluate the integral.*

**21.** $f(x, y) = 3x^2 y^2$;   $R = \{(x, y) \mid -1 \le x \le 1, -1 \le y \le 1\}$

**22.** $f(x, y) = x^2 - 3xy + y^2$;   $R$ is the triangular region with vertices $(0, 0)$, $(1, 1)$, and $(0, 1)$

**23.** $f(x, y) = \dfrac{1}{2x + 3y}$;   $R = \{(x, y) \mid 0 \le x \le 2, 0 \le y \le x\}$

**24.** $f(x, y) = e^{-xy}$;   $R = \{(x, y) \mid 0 \le x \le 1, 0 \le y \le 2\}$

*In Exercises 25 and 26, determine whether the statement is true or false. If it is true, explain why. If it is false, explain why or give an example that shows it is false.*

**25.** If $f(x, y) = \sqrt{4 - x^2 - y^2}$, then $\iint_R \sqrt{f_x^2 + f_y^2 + 1} \, dA = 8\pi$, where $R = \{(x, y) \mid 0 \le x^2 + y^2 \le 4\}$.

**26.** If $z = f(x, y)$ is defined over a region $R$ in the $xy$-plane, then $\iint_R \sqrt{f_x^2 + f_y^2 + 1} \, dA \ge A(R)$, where $A(R)$ denotes the area of $R$. (Assume that $f_x$ and $f_y$ exist.)

## 13.6 Triple Integrals

### Triple Integrals Over a Rectangular Box

Just as the mass of a piece of straight, thin wire of linear mass density $\delta(x)$, where $a \le x \le b$, is given by the single integral $\int_a^b \delta(x)\,dx$, and the mass of a thin plate $D$ of mass density $\sigma(x, y)$ is given by the double integral $\iint_D \sigma(x, y)\,dA$, we will now see that the mass of a solid object $T$ with mass density $\rho(x, y, z)$ is given by a *triple integral*.

Let's consider the simplest case in which the solid takes the form of a rectangular box:

$$B = [a, b] \times [c, d] \times [p, q] = \{(x, y, z)\,|\,a \le x \le b, c \le y \le d, p \le z \le q\}$$

Suppose that the mass density of the solid is $\rho(x, y, z)$ g/m$^3$, where $\rho$ is a positive continuous function defined on $B$. Let

$$a = x_0 < x_1 < \cdots < x_{i-1} < x_i < \cdots < x_l = b$$
$$c = y_0 < y_1 < \cdots < y_{j-1} < y_j < \cdots < y_m = d$$
$$p = z_0 < z_1 < \cdots < z_{k-1} < z_k < \cdots < z_n = q$$

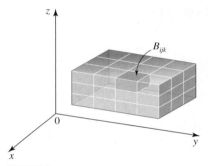

**FIGURE 1**
A partition $P = \{B_{ijk}\}$ of $B$

be regular partitions of the intervals $[a, b]$, $[c, d]$, and $[p, q]$ of length $\Delta x = (b - a)/l$, $\Delta y = (d - c)/m$, and $\Delta z = (q - p)/n$, respectively. The planes $x = x_i$, for $1 \le i \le l$, $y = y_j$, for $1 \le j \le m$, and $z = z_k$, for $1 \le k \le n$, parallel to the $yz$-, $xz$-, and $xy$-coordinate planes divide the box $B$ into $N = lmn$ boxes $B_{111}, B_{112}, \ldots, B_{ijk}, \ldots, B_{lmn}$, as shown in Figure 1. The volume of $B_{ijk}$ is $\Delta V = \Delta x\,\Delta y\,\Delta z$.

Let $(x_{ijk}^*, y_{ijk}^*, z_{ijk}^*)$ be an arbitrary point in $B_{ijk}$. If $l$, $m$, and $n$ are large (so that the dimensions of $B_{ijk}$ are small), then the continuity of $\rho$ implies that $\rho(x, y, z)$ does not vary appreciably from $\rho(x_{ijk}^*, y_{ijk}^*, z_{ijk}^*)$, whenever $(x, y, z)$ is in $B_{ijk}$. Therefore, we can approximate the mass of $B_{ijk}$ by

$$\rho(x_{ijk}^*, y_{ijk}^*, z_{ijk}^*)\,\Delta V \qquad \text{constant mass density} \cdot \text{volume}$$

where $\Delta V = \Delta x\,\Delta y\,\Delta z$. Adding up the masses of the $N$ boxes, we see that the mass of the box $B$ is approximately

$$\sum_{i=1}^{l} \sum_{j=1}^{m} \sum_{k=1}^{n} \rho(x_{ijk}^*, y_{ijk}^*, z_{ijk}^*)\,\Delta V \tag{1}$$

We expect the approximation to improve as $l$, $m$, and $n$ get larger and larger. Therefore, it is reasonable to define the mass of the box $B$ as

$$\lim_{l, m, n \to \infty} \sum_{i=1}^{l} \sum_{j=1}^{m} \sum_{k=1}^{n} \rho(x_{ijk}^*, y_{ijk}^*, z_{ijk}^*)\,\Delta V \tag{2}$$

The expression in (1) is an example of a *Riemann sum* of a function of three variables over a box and the corresponding limit in (2) is the *triple integral* of $f$ over $B$. More generally, we have the following definitions. Notice that no assumption regarding the sign of $f(x, y, z)$ is made in these definitions.

---

**DEFINITION**  Triple Integral of $f$ Over a Rectangular Box $B$

Let $f$ be a continuous function of three variables defined on a rectangular box $B$, and let $P = \{B_{ijk}\}$ be a partition of $B$.

1. A **Riemann sum of $f$ over $B$** with respect to the partition $P$ is a sum of the form

$$\sum_{i=1}^{l} \sum_{j=1}^{m} \sum_{k=1}^{n} f(x_{ijk}^{*}, y_{ijk}^{*}, z_{ijk}^{*}) \Delta V$$

where $(x_{ijk}^{*}, y_{ijk}^{*}, z_{ijk}^{*})$ is a point in $B_{ijk}$.

2. The **triple integral of $f$ over $B$** is

$$\iiint_{B} f(x, y, z) \, dV = \lim_{l, m, n \to \infty} \sum_{i=1}^{l} \sum_{j=1}^{m} \sum_{k=1}^{n} f(x_{ijk}^{*}, y_{ijk}^{*}, z_{ijk}^{*}) \Delta V$$

if the limit exists for all choices of $(x_{ijk}^{*}, y_{ijk}^{*}, z_{ijk}^{*})$ in $B_{ijk}$.

---

As in the case of double integrals, a triple integral may be found by evaluating an appropriate iterated integral.

---

**THEOREM 1**

Let $f$ be continuous on the rectangular box

$$B = \{(x, y, z) \mid a \leq x \leq b, c \leq y \leq d, p \leq z \leq q\}$$

Then

$$\iiint_{B} f(x, y, z) \, dV = \int_{p}^{q} \int_{c}^{d} \int_{a}^{b} f(x, y, z) \, dx \, dy \, dz \qquad (3)$$

---

The iterated integral in Equation (3) is evaluated by first integrating with respect to $x$ while holding $y$ and $z$ constant, then integrating with respect to $y$ while holding $z$ constant, and finally integrating with respect to $z$. The triple integral in Equation (3) can also be expressed as any one of five other iterated integrals, each with a different order of integration. For example, we can write

$$\iiint_{B} f(x, y, z) \, dV = \int_{a}^{b} \int_{p}^{q} \int_{c}^{d} f(x, y, z) \, dy \, dz \, dx$$

where the iterated integral is evaluated by successively integrating with respect to $y$, $z$, and then $x$. (Remember, we work "from the inside out.")

**EXAMPLE 1**  Evaluate $\iiint_{B} (x^2 y + yz^2) \, dV$, where

$$B = \{(x, y, z) \mid -1 \leq x \leq 1, 0 \leq y \leq 3, 1 \leq z \leq 2\}$$

**Solution**   We can express the given integral as one of six integrals. For example, if we choose to integrate with respect to $x$, $y$, and $z$, in that order, then we obtain

$$\iiint_B (x^2y + yz^2)\, dV = \int_1^2 \int_0^3 \int_{-1}^1 (x^2y + yz^2)\, dx\, dy\, dz$$

$$= \int_1^2 \int_0^3 \left[\frac{1}{3}x^3y + xyz^2\right]_{x=-1}^{x=1}\, dy\, dz$$

$$= \int_1^2 \int_0^3 \left[\frac{2}{3}y + 2yz^2\right]\, dy\, dz$$

$$= \int_1^2 \left[\frac{1}{3}y^2 + y^2z^2\right]_{y=0}^{y=3}\, dz$$

$$= \int_1^2 (3 + 9z^2)\, dz = \left[3z + 3z^3\right]_1^2 = 24 \quad \blacksquare$$

## ■ Triple Integrals Over General Bounded Regions in Space

We can extend the definition of the triple integral to more general regions using the same technique that we used for double integrals. Suppose that $T$ is a bounded solid region in space. Then it can be enclosed in a rectangular box $B = [a, b] \times [c, d] \times [p, q]$. Let $P$ be a regular partition of $B$ into $N = lmn$ boxes with sides of length $\Delta x = (b - a)/l$, $\Delta y = (d - c)/m$, $\Delta z = (q - p)/n$, and volume $\Delta V = \Delta x\, \Delta y\, \Delta z$. Thus, $P = \{B_{111}, B_{112}, \ldots, B_{ijk}, \ldots, B_{lmn}\}$. (See Figure 2.)

Define

$$F(x, y, z) = \begin{cases} f(x, y, z) & \text{if } (x, y, z) \text{ is in } T \\ 0 & \text{if } (x, y, z) \text{ is in } B \text{ but not in } T \end{cases}$$

Then a **Riemann sum of $f$ over $T$** with respect to the partition $P$ is given by

$$\sum_{i=1}^l \sum_{j=1}^m \sum_{k=1}^n F(x_{ijk}^*, y_{ijk}^*, z_{ijk}^*)\, \Delta V$$

where $(x_{ijk}^*, y_{ijk}^*, z_{ijk}^*)$ is an arbitrary point in $B_{ijk}$ and $\Delta V$ is the volume of $B_{ijk}$. If we take the limit of these sums as $l$, $m$, $n$ approach infinity, we obtain the **triple integral of $f$ over $T$.** Thus,

$$\iiint_T f(x, y, z)\, dV = \lim_{l, m, n \to \infty} \sum_{i=1}^l \sum_{j=1}^m \sum_{k=1}^n F(x_{ijk}^*, y_{ijk}^*, z_{ijk}^*)\, \Delta V$$

provided that the limit exists for all choices of $(x_{ijk}^*, y_{ijk}^*, z_{ijk}^*)$ in $T$.

**Notes**

1. If $f$ is continuous and the surface bounding $T$ is "sufficiently smooth," it can be shown that $f$ is integrable over $T$.

2. The properties of double integrals that are listed in Theorem 1, Section 13.1, with the necessary modifications are also enjoyed by triple integrals.   ■

## ■ Evaluating Triple Integrals Over General Regions

We will now restrict our attention to certain types of regions. A region $T$ is **$z$-simple** if it lies between the graphs of two continuous functions of $x$ and $y$, that is, if

$$T = \{(x, y, z)\,|\,(x, y) \in R, k_1(x, y) \le z \le k_2(x, y)\}$$

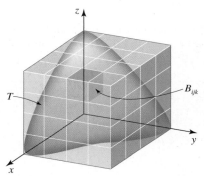

**FIGURE 2**
The box $B_{ijk}$ is a typical element of the partition of $B$.

where $R$ is the projection of $T$ onto the $xy$-plane. (See Figure 3.) If $f$ is continuous on $T$, then

$$\iiint_T f(x, y, z)\, dV = \iint_R \left[ \int_{k_1(x, y)}^{k_2(x, y)} f(x, y, z)\, dz \right] dA \tag{4}$$

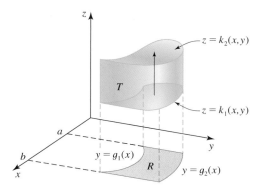

**FIGURE 3**
A $z$-simple region $T$ is bounded by the surfaces $z = k_1(x, y)$ and $z = k_2(x, y)$.

The iterated integral on the right-hand side of Equation (4) is evaluated by first integrating with respect to $z$ while holding $x$ and $y$ constant. The resulting double integral is then evaluated by using the method of Section 13.2. For example, if $R$ is $y$-simple, as shown in Figure 3, then

$$R = \{(x, y)\,|\,a \le x \le b, g_1(x) \le y \le g_2(x)\}$$

in which case Equation (4) becomes

$$\iiint_T f(x, y, z)\, dV = \int_a^b \int_{g_1(x)}^{g_2(x)} \int_{k_1(x, y)}^{k_2(x, y)} f(x, y, z)\, dz\, dy\, dx$$

To determine the "limits of integration" with respect to $z$, notice that $z$ runs from the lower surface $z = k_1(x, y)$ to the upper surface $z = k_2(x, y)$ as indicated by the arrow in Figure 3.

**EXAMPLE 2** Evaluate $\iiint_T z\, dV$ where $T$ is the solid in the first octant bounded by the graphs of $z = 1 - x^2$ and $y = x$.

**Solution** The solid $T$ is shown in Figure 4a. The solid is $z$-simple because it is bounded below by the graph of $z = k_1(x, y) = 0$ and above by $z = k_2(x, y) = 1 - x^2$.

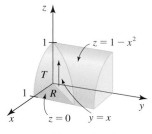

(**a**) The solid $T$ is $z$-simple.

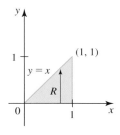

(**b**) The projection of the solid $T$ onto $R$ in the $xy$-plane is $y$-simple.

**FIGURE 4**

The projection of $T$ onto the $xy$-plane is the set $R$ that is sketched in Figure 4b. Regarding $R$ as a $y$-simple region, we obtain

$$\iiint_T z \, dV = \iint_R \left[ \int_{z=k_1(x, y)}^{z=k_2(x, y)} z \, dz \right] dA = \int_0^1 \int_0^x \int_0^{1-x^2} z \, dz \, dy \, dx$$

$$= \int_0^1 \int_0^x \left[ \frac{1}{2} z^2 \right]_{z=0}^{z=1-x^2} dy \, dx$$

$$= \frac{1}{2} \int_0^1 \int_0^x (1 - x^2)^2 \, dy \, dx$$

$$= \frac{1}{2} \int_0^1 \left[ (1 - x^2)^2 y \right]_{y=0}^{y=x} dx$$

$$= \frac{1}{2} \int_0^1 x(1 - x^2)^2 \, dx = \left[ \left( \frac{1}{2} \right) \left( -\frac{1}{2} \right) \left( \frac{1}{3} \right) (1 - x^2)^3 \right]_0^1 = \frac{1}{12} \quad \blacksquare$$

There are two other simple regions besides the $z$-simple region just considered. An **$x$-simple region $T$** is one that lies between the graphs of two continuous functions of $y$ and $z$. In other words, $T$ may be described as

$$T = \{(x, y, z) \mid (y, z) \in R, k_1(y, z) \leq x \leq k_2(y, z)\}$$

where $R$ is the projection of $T$ onto the $yz$-plane. (See Figure 5.) Here, we have

$$\iiint_T f(x, y, z) \, dV = \iint_R \left[ \int_{k_1(y, z)}^{k_2(y, z)} f(x, y, z) \, dx \right] dA \tag{5}$$

The (double) integral over the plane region $R$ is evaluated by integrating with respect to $y$ or $z$ first depending on whether $R$ is $y$-simple or $z$-simple.

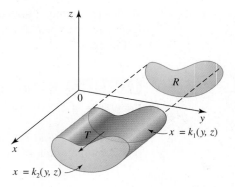

**FIGURE 5**
An $x$-simple region $T$ is bounded by the surfaces $x = k_1(y, z)$ and $x = k_2(y, z)$.

A **$y$-simple region $T$** lies between the graphs of two continuous functions of $x$ and $z$. In other words, $T$ may be described as

$$T = \{(x, y, z) \mid (x, z) \in R, k_1(x, z) \leq y \leq k_2(x, z)\}$$

where $R$ is the projection of $T$ onto the $xz$-plane. (See Figure 6.) In this case we have

$$\iiint_T f(x, y, z) \, dV = \iint_R \left[ \int_{k_1(x, z)}^{k_2(x, z)} f(x, y, z) \, dy \right] dA \tag{6}$$

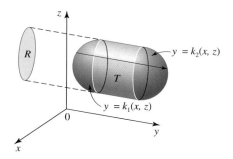

**FIGURE 6**
A $y$-simple region $T$ is bounded by the surfaces $y = k_1(x, z)$ and $y = k_2(x, z)$.

Again depending on whether $R$ is an $x$-simple or $z$-simple plane region, the double integration is carried out first with respect to $x$ or $z$.

**EXAMPLE 3**   Evaluate $\iiint_T \sqrt{x^2 + z^2}\, dV$, where $T$ is the region bounded by the cylinder $x^2 + z^2 = 1$ and the planes $y + z = 2$ and $y = 0$.

**Solution**   The solid $T$ is shown in Figure 7a. Although $T$ can be viewed as an $x$-simple or $z$-simple region, it is easier to view it as a $y$-simple region. (Try It!) In this case we see that $T$ is bounded to the left by the graph of the function $y = k_1(x, z) = 0$ and to the right by the graph of the function $y = k_2(x, z) = 2 - z$. The projection of $T$ onto the $xz$-plane is the set $R$, which is sketched in Figure 7b. We have

$$\iiint_T \sqrt{x^2 + z^2}\, dV = \iint_R \left[ \int_{k_1(x, z)}^{k_2(x, z)} \sqrt{x^2 + z^2}\, dy \right] dA$$

$$= \iint_R \left[ \int_0^{2-z} \sqrt{x^2 + z^2}\, dy \right] dA$$

$$= \iint_R \left[ \sqrt{x^2 + z^2}\, y \right]_{y=0}^{y=2-z} dA$$

$$= \iint_R \sqrt{x^2 + z^2}\, (2 - z)\, dA$$

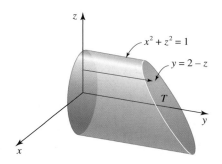

(**a**) The solid $T$ is viewed as being $y$-simple.

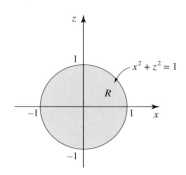

(**b**) The projection of the solid $T$ onto $R$ in the $xz$-plane

**FIGURE 7**

Since $R$ is a circular region, it is more convenient to use polar coordinates when integrating over $R$. So letting $x = r \cos \theta$ and $z = r \sin \theta$, we have

$$\iint\limits_{R} \sqrt{x^2 + z^2}\,(2 - z)\,dA = \int_0^{2\pi} \int_0^1 r(2 - r \sin \theta)\, r\, dr\, d\theta$$

$$= \int_0^{2\pi} \int_0^1 (2r^2 - r^3 \sin \theta)\, dr\, d\theta$$

$$= \int_0^{2\pi} \left[ \frac{2}{3} r^3 - \frac{1}{4} r^4 \sin \theta \right]_{r=0}^{r=1} d\theta$$

$$= \int_0^{2\pi} \left( \frac{2}{3} - \frac{1}{4} \sin \theta \right) d\theta$$

$$= \left[ \frac{2}{3} \theta + \frac{1}{4} \cos \theta \right]_0^{2\pi} = \frac{4\pi}{3}$$

Therefore,

$$\iiint\limits_{T} \sqrt{x^2 + z^2}\,dV = \frac{4\pi}{3}$$

## Volume, Mass, Center of Mass, and Moments of Inertia

Before looking at other examples, let's list some applications of triple integrals. Let $f(x, y, z) = 1$ for all points in a solid $T$. Then the triple integral of $f$ over $T$ gives the **volume** $V$ of $T$; that is,

$$V = \iiint\limits_{T} dV \tag{7}$$

We also have the following.

**DEFINITIONS** Mass, Center of Mass, and Moments of Inertia for Solids in Space

Suppose that $\rho(x, y, z)$ gives the mass density at the point $(x, y, z)$ of a solid $T$. Then the **mass** $m$ of $T$ is

$$m = \iiint\limits_{T} \rho(x, y, z)\, dV \tag{8}$$

The **moments** of $T$ about the three coordinate planes are

$$M_{yz} = \iiint\limits_{T} x\rho(x, y, z)\, dV \tag{9a}$$

$$M_{xz} = \iiint\limits_{T} y\rho(x, y, z)\, dV \tag{9b}$$

$$M_{xy} = \iiint\limits_{T} z\rho(x, y, z)\, dV \tag{9c}$$

The **center of mass** of $T$ is located at the point $(\bar{x}, \bar{y}, \bar{z})$, where

$$\bar{x} = \frac{M_{yz}}{m}, \qquad \bar{y} = \frac{M_{xz}}{m}, \qquad \bar{z} = \frac{M_{xy}}{m} \tag{10}$$

and the **moments of inertia** of $T$ about the three coordinate axes are

$$I_x = \iiint_T (y^2 + z^2)\rho(x, y, z)\, dV \tag{11a}$$

$$I_y = \iiint_T (x^2 + z^2)\rho(x, y, z)\, dV \tag{11b}$$

$$I_z = \iiint_T (x^2 + y^2)\rho(x, y, z)\, dV \tag{11c}$$

If the mass density is constant, then the center of mass of a solid is called the **centroid** of $T$.

**EXAMPLE 4**   Let $T$ be the solid tetrahedron bounded by the plane $x + y + z = 1$ and the three coordinate planes $x = 0$, $y = 0$, and $z = 0$. Find the mass of $T$ if the mass density of $T$ is directly proportional to the distance between a base of $T$ and a point on $T$.

**Solution**   The solid $T$ is shown in Figure 8a. It is $x$-, $y$-, and $z$-simple. For example, it can be viewed as being $x$-simple if you observe that it is bounded by the surface $x = k_1(y, z) = 0$ and the surface $x = k_2(y, z) = 1 - y - z$. (Solve the equation $x + y + z = 1$ for $x$.)

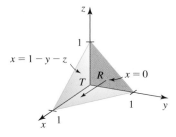

(a) The solid $T$ viewed as an $x$-simple region

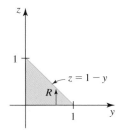

(b) The projection of the solid $T$ onto the $yz$-plane viewed as a $z$-simple region

**FIGURE 8**

The projection of $T$ onto the $yz$-plane is the set $R$ shown in Figure 8b. Observe that the upper boundary of $R$ lies along the line that is the intersection of $x + y + z = 1$ and the plane $x = 0$ and hence has equation $y + z = 1$ or $z = 1 - y$. If we take the base of $T$ as the face of the tetrahedron lying on the $xy$-plane (actually, by symmetry, any face will do), then the mass density function for $T$ is $\rho(x, y, z) = kz$, where $k$ is the constant of proportionality. Using Equation (8), we see that the required mass is

$$
\begin{aligned}
m &= \iiint_T \rho(x, y, z) \, dV = \iiint_T kz \, dV \\
&= k \int_0^1 \int_0^{1-y} \int_0^{1-y-z} z \, dx \, dz \, dy \qquad \text{View } T \text{ as } x\text{-simple.} \\
&= k \int_0^1 \int_0^{1-y} \left[ zx \right]_{x=0}^{x=1-y-z} dz \, dy = k \int_0^1 \int_0^{1-y} \left[ (1-y)z - z^2 \right] dz \, dy \qquad \text{View } R \text{ as } z\text{-simple.} \\
&= k \int_0^1 \left[ \frac{1}{2}(1-y)z^2 - \frac{1}{3}z^3 \right]_{z=0}^{z=1-y} dy \\
&= k \int_0^1 \frac{1}{6}(1-y)^3 \, dy = k \left[ \left( \frac{1}{6} \right)\left( -\frac{1}{4} \right)(1-y)^4 \right]_0^1 = \frac{k}{24}
\end{aligned}
$$

$\blacksquare$

**EXAMPLE 5** Let $T$ be the solid that is bounded by the parabolic cylinder $y = x^2$ and the planes $z = 0$ and $y + z = 1$. Find the center of mass of $T$, given that it has uniform density $\rho(x, y, z) = 1$.

**Solution** The solid $T$ is shown in Figure 9a. It is $x$-, $y$-, and $z$-simple. Let's choose to view it as being $z$-simple. (You are also encouraged to solve the problem by viewing $T$ as $x$-simple and $y$-simple.) In this case we see that $T$ lies between the $xy$-plane $z = k_1(x, y) = 0$ and the plane $z = k_2(x, y) = 1 - y$. The projection of $T$ onto the $xy$-plane is the region $R$ shown in Figure 9b. As a first step toward finding the center of mass of $T$, let's find the mass of $T$. Using Equation (8), we have

$$
\begin{aligned}
m &= \iiint_T \rho(x, y, z) \, dV = \iiint_T dV \\
&= \int_{-1}^1 \int_{x^2}^1 \int_0^{1-y} dz \, dy \, dx = \int_{-1}^1 \int_{x^2}^1 \left[ z \right]_{z=0}^{z=1-y} dy \, dx \\
&= \int_{-1}^1 \int_{x^2}^1 (1-y) \, dy \, dx = \int_{-1}^1 \left[ y - \frac{1}{2}y^2 \right]_{y=x^2}^{y=1} dx \\
&= \int_{-1}^1 \left( \frac{1}{2} - x^2 + \frac{1}{2}x^4 \right) dx = \left[ \frac{1}{2}x - \frac{1}{3}x^3 + \frac{1}{10}x^5 \right]_{-1}^1 = \frac{8}{15}
\end{aligned}
$$

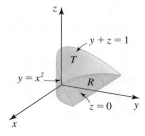

(a) The solid $T$ is viewed as a $z$-simple region.

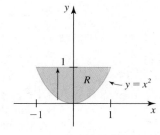

(b) The projection $R$ of $T$ onto the $xy$-plane viewed as being $y$-simple

**FIGURE 9**

By symmetry we see that $\bar{x} = 0$. Next, using Equations (9b) and (10), we have

$$\bar{y} = \frac{1}{m} \iiint\limits_T y\rho(x, y, z) \, dV = \frac{15}{8} \iiint\limits_T y \, dV$$

$$= \frac{15}{8} \int_{-1}^{1} \int_{x^2}^{1} \int_{0}^{1-y} y \, dz \, dy \, dx = \frac{15}{8} \int_{-1}^{1} \int_{x^2}^{1} \left[yz\right]_{z=0}^{z=1-y} \, dy \, dx$$

$$= \frac{15}{8} \int_{-1}^{1} \int_{x^2}^{1} (y - y^2) \, dy \, dx = \frac{15}{8} \int_{-1}^{1} \left[\frac{1}{2} y^2 - \frac{1}{3} y^3\right]_{y=x^2}^{y=1} \, dx$$

$$= \frac{15}{8} \int_{-1}^{1} \left(\frac{1}{6} - \frac{1}{2} x^4 + \frac{1}{3} x^6\right) \, dx = 2\left(\frac{15}{8}\right) \int_{0}^{1} \left(\frac{1}{6} - \frac{1}{2} x^4 + \frac{1}{3} x^6\right) \, dx$$

The integrand is an even function.

$$= \frac{15}{4} \left[\frac{1}{6} x - \frac{1}{10} x^5 + \frac{1}{21} x^7\right]_{0}^{1} = \frac{3}{7}$$

Similarly, you can verify that

$$\bar{z} = \frac{1}{m} \iiint\limits_T z\rho(x, y, z) \, dV = \frac{15}{8} \iiint\limits_T z \, dV \qquad \text{Use Equation (9c).}$$

$$= \frac{15}{8} \int_{-1}^{1} \int_{x^2}^{1} \int_{0}^{1-y} z \, dz \, dy \, dx = \frac{2}{7}$$

Therefore, the center of mass of $T$ is located at the point $\left(0, \frac{3}{7}, \frac{2}{7}\right)$. ∎

**EXAMPLE 6**  Find the moments of inertia about the three coordinate axes for the solid rectangular parallelepiped of constant density $k$ shown in Figure 10.

**Solution**  Using Equation (11a) with $\rho(x, y, z) = k$, we obtain

$$I_x = \iiint\limits_T (y^2 + z^2)k \, dV$$

$$= \int_{-c/2}^{c/2} \int_{-b/2}^{b/2} \int_{-a/2}^{a/2} k(y^2 + z^2) \, dx \, dy \, dz$$

Observe that the integrand is an even function of $x$, $y$, and $z$. Taking advantage of symmetry, we can write

$$I_x = 8k \int_{0}^{c/2} \int_{0}^{b/2} \int_{0}^{a/2} (y^2 + z^2) \, dx \, dy \, dz = 8k \int_{0}^{c/2} \int_{0}^{b/2} \left[(y^2 + z^2)x\right]_{x=0}^{x=a/2} \, dy \, dz$$

$$= 4ka \int_{0}^{c/2} \int_{0}^{b/2} (y^2 + z^2) \, dy \, dz = 4ka \int_{0}^{c/2} \left[\frac{1}{3} y^3 + z^2 y\right]_{y=0}^{y=b/2} \, dz$$

$$= 4ka \int_{0}^{c/2} \left(\frac{b^3}{24} + \frac{bz^2}{2}\right) \, dz = 4ka \left(\frac{b^3}{24} z + \frac{b}{6} z^3\right)\Big|_{z=0}^{z=c/2}$$

$$= 4ka \left(\frac{b^3 c}{48} + \frac{bc^3}{48}\right) = \frac{kabc}{12} (b^2 + c^2)$$

$$= \frac{1}{12} m(b^2 + c^2) \qquad m = kabc = \text{mass of the solid}$$

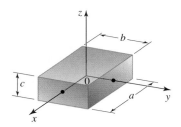

**FIGURE 10**
The center of the solid is placed at the origin.

Similarly, we find

$$I_y = \frac{1}{12}\,m(a^2 + c^2) \qquad \text{and} \qquad I_z = \frac{1}{12}\,m(a^2 + b^2) \qquad \blacksquare$$

## 13.6 CONCEPT QUESTIONS

**1. a.** Define the Riemann sum of $f$ over a rectangular box $B$.
   **b.** Define the triple integral of $f$ over $B$.

**2.** Suppose that $f$ is continuous on the rectangular box
   $B = [a, b] \times [c, d] \times [p, q]$.
   **a.** Explain how you would evaluate $\iiint_B f(x, y, z)\, dV$.
   **b.** Write all iterated integrals that are associated with the triple integral of part (a).

**3. a.** What is a $z$-simple region in space? An $x$-simple region? A $y$-simple region?
   **b.** Write the integral $\iiint_T f(x, y, z)\, dV$, where $T$ is a $z$-simple region. An $x$-simple region. A $y$-simple region.

## 13.6 EXERCISES

*In Exercises 1–4, evaluate the integral $\iiint_B f(x, y, z)\, dV$ using the indicated order of integration.*

**1.** $f(x, y, z) = x + y + z$; $B = \{(x, y, z)\,|\,0 \le x \le 2, 0 \le y \le 1, 0 \le z \le 3\}$. Integrate (a) with respect to $x$, $y$, and $z$, in that order, and (b) with respect to $z$, $y$, and $x$, in that order.

**2.** $f(x, y, z) = xyz$; $B = \{(x, y, z)\,|\,-1 \le x \le 1, 0 \le y \le 2, -2 \le z \le 6\}$. Integrate (a) with respect to $y$, $x$, and $z$, in that order, and (b) with respect to $x$, $z$, and $y$, in that order.

**3.** $f(x, y, z) = xy^2 + yz^2$; $B = \{(x, y, z)\,|\,0 \le x \le 2, -1 \le y \le 1, 0 \le z \le 3\}$. Integrate (a) with respect to $z$, $y$, and $x$, in that order, and (b) with respect to $x$, $y$, and $z$, in that order.

**4.** $f(x, y, z) = xy^2 \cos z$; $B = \{(x, y, z)\,|\,0 \le x \le 2, 0 \le y \le 3, 0 \le z \le \frac{\pi}{2}\}$. Integrate (a) with respect to $y$, $z$, and $x$, in that order, and (b) with respect to $y$, $x$, and $z$, in that order.

*In Exercises 5–10, evaluate the iterated integral.*

**5.** $\displaystyle\int_0^1 \int_0^x \int_0^{x+y} x\, dz\, dy\, dx$

**6.** $\displaystyle\int_0^1 \int_0^z \int_0^y 2xz\, dx\, dy\, dz$

**7.** $\displaystyle\int_0^{\pi/2} \int_1^2 \int_0^{\sqrt{1-z}} y \cos x\, dy\, dz\, dx$

**8.** $\displaystyle\int_{-1}^1 \int_0^2 \int_0^{\sqrt{4-z^2}} y^2 z\, dx\, dz\, dy$

**9.** $\displaystyle\int_0^4 \int_0^1 \int_0^x 2\sqrt{y}\,e^{-x^2}\, dz\, dx\, dy$

**10.** $\displaystyle\int_1^e \int_1^x \int_0^{1/(xy)} 2 \ln y\, dz\, dy\, dx$

*In Exercises 11–14, the figure shows the region of integration for $\iiint_T f(x, y, z)\, dV$. Express the triple integral as an iterated integral in six different ways using different orders of integration.*

**11.**      **12.**

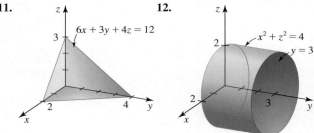

**13.**      **14.**

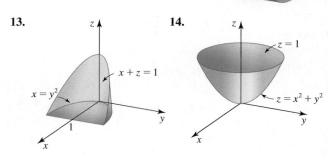

*In Exercises 15–22, evaluate the integral $\iiint_T f(x, y, z)\, dV$.*

**15.** $f(x, y, z) = x$; $T$ is the tetrahedron bounded by the planes $x = 0$, $y = 0$, $z = 0$, and $x + y + z = 1$

**16.** $f(x, y, z) = y$; $T$ is the region bounded by the planes $x = 0$, $y = 0$, $z = 0$, and $2x + 3y + z = 6$

**17.** $f(x, y, z) = 2z$; $T$ is the region bounded by the cylinder $y = x^3$ and the planes $y = x$, $z = 2x$, and $z = 0$

**18.** $f(x, y, z) = x + 2y$; $T$ is the region bounded by the cylinder $y = \sqrt{x}$ and the planes $y = x$, $z = 2x$, and $z = 0$

**19.** $f(x, y, z) = y$; $T$ is the region bounded by the paraboloid $y = x^2 + z^2$ and the plane $y = 4$

**20.** $f(x, y, z) = z$; $T$ is the region bounded by the parabolic cylinder $y = x^2$ and the planes $y + z = 1$ and $z = 0$

**21.** $f(x, y, z) = z$; $T$ is the region bounded by the cylinder $x^2 + z^2 = 4$ and the planes $x = 2y$, $y = 0$, and $z = 0$

**22.** $f(x, y, z) = \sqrt{x^2 + z^2}$; $T$ is the region bounded by the paraboloids $y = x^2 + z^2$ and $y = 8 - x^2 - z^2$

*In Exercises 23–28, sketch the solid bounded by the graphs of the equations, and then use a triple integral to find the volume of the solid.*

**23.** $3x + 2y + z = 6$, $x = 0$, $y = 0$, $z = 0$

**24.** $y = 2z$, $y = x^2$, $y = 4$, $z = 0$

**25.** $x = 4 - y^2$, $x + z = 4$, $x = 0$, $z = 0$

**26.** $z = 1 - x^2$, $y = x$, $y = 2 - x$, $z = 0$

**27.** $z = x^2 + y^2$, $z = 8 - x^2 - y^2$

**28.** $x^2 + z^2 = 4$, $y^2 + z^2 = 4$

**29.** Find the volume of the tetrahedron with vertices $(0, 0, 0)$, $(1, 0, 0)$, $(0, 3, 0)$, and $(0, 0, 2)$.

**30.** Find the volume of the tetrahedron with vertices $(0, 0, 0)$, $(1, 0, 0)$, $(1, 0, 1)$, and $(1, 1, 0)$.

*In Exercises 31–34, sketch the solid whose volume is given by the iterated integral.*

**31.** $\int_0^1 \int_0^{1-y} \int_0^{1-x-y} dz\, dx\, dy$    **32.** $\int_0^1 \int_0^{1-y} \int_0^{2-2z} dx\, dz\, dy$

**33.** $\int_{-2}^2 \int_0^{4-y^2} \int_0^{y+2} dz\, dx\, dy$    **34.** $\int_0^1 \int_{-\sqrt{1-y}}^{\sqrt{1-y}} \int_0^y dz\, dx\, dy$

*In Exercises 35–38, express the triple integral $\iiint_T f(x, y, z)\, dV$ as an iterated integral in six different ways using different orders of integration.*

**35.** $T$ is the solid bounded by the planes $x + 2y + 3z = 6$, $x = 0$, $y = 0$, and $z = 0$.

**36.** $T$ is the tetrahedron bounded by the planes $z = 0$, $x = 0$, $y = 0$, $y = 2 - 2z$, and $z = 1 - x$.

**37.** $T$ is the solid bounded by the circular cylinder $x^2 + y^2 = 1$ and the planes $z = 0$ and $z = 2$.

**38.** $T$ is the solid bounded by the parabolic cylinder $y = x^2$ and the planes $z = 0$ and $z = 4 - y$.

**39.** Let $f(x, y, z) = x + y + z$ and let $B = \{(x, y, z)\,|\,0 \le x \le 4, 0 \le y \le 4, 0 \le z \le 4\}$.

    **a.** Use a Riemann sum with $m = n = p = 2$, and choose the evaluation point $(x_{ijk}^*, y_{ijk}^*, z_{ijk}^*)$ to be the midpoint

of the subrectangles $R_{ijk}(1 \le i, j, k \le 2)$ to estimate $\iiint_B f(x, y, z)\, dV$.

    **b.** Find the exact value of $\iiint_B f(x, y, z)\, dV$.

**40.** Let $f(x, y, z) = \sqrt{x^2 + y^2 + z^2}$ and let $B = \{(x, y, z)\,|\,0 \le x \le 4, 0 \le y \le 2, 0 \le z \le 1\}$.

    **a.** Use a Riemann sum with $m = n = p = 2$, and choose the evaluation point $(x_{ijk}^*, y_{ijk}^*, z_{ijk}^*)$ to be the midpoint of the subrectangles $R_{ijk}(1 \le i, j, k \le 2)$ to estimate $\iiint_B f(x, y, z)\, dV$.

**cas**  **b.** Use a computer algebra system to estimate $\iiint_B f(x, y, z)\, dV$ accurate to four decimal places.

**cas** *In Exercises 41 and 42, use a computer algebra system to estimate the triple integral accurate to four decimal places.*

**41.** $\displaystyle\int_{-1}^1 \int_0^2 \int_1^2 \frac{\cos xy}{\sqrt{1 + xyz^2}}\, dx\, dy\, dz$

**42.** $\displaystyle\int_0^1 \int_0^{1-x} \int_0^{1-x^2} xe^{yz}\, dz\, dy\, dx$

*In Exercises 43–46, find the center of mass of the solid $T$ having the given mass density.*

**43.** $T$ is the tetrahedron bounded by the planes $x = 0$, $y = 0$, $z = 0$, and $x + y + z = 1$. The mass density at a point $P$ of $T$ is directly proportional to the distance between $P$ and the $yz$-plane.

**44.** $T$ is the wedge bounded by the planes $x = 0$, $y = 0$, $z = 0$, $z = -\frac{2}{3}y + 2$ and $x = 1$. The mass density at a point $P$ of $T$ is directly proportional to the distance between $P$ and the $xy$-plane.

**45.** $T$ is the solid bounded by the cylinder $y^2 + z^2 = 4$ and the planes $x = 0$ and $x = 3$. The mass density at a point $P$ of $T$ is directly proportional to the distance between $P$ and the $yz$-plane.

**46.** $T$ is the solid bounded by the parabolic cylinder $z = 1 - x^2$ and the planes $y + z = 1$, $y = 0$, and $z = 0$. $T$ has uniform mass density $\rho(x, y, z) = k$, where $k$ is a constant.

*In Exercises 47–50, set up, but do not evaluate, the iterated integral giving the mass of the solid $T$ having mass density given by the function $\rho$.*

**47.** $T$ is the solid bounded by the cylinder $x^2 + z^2 = 1$ in the first octant and the plane $z + y = 1$; $\rho(x, y, z) = xy + z^2$

**48.** $T$ is the solid bounded by the ellipsoid $36x^2 + 9y^2 + 4z^2 = 36$ and the planes $y = 0$ and $z = 0$; $\rho(x, y, z) = \sqrt{yz}$

**49.** $T$ is the solid bounded by the parabolic cylinder $z = 1 - y^2$ and the planes $2x + y = 2$, $y = 0$, and $z = 0$; $\rho(x, y, z) = \sqrt{x^2 + y^2 + z^2}$

**50.** $T$ is the upper hemisphere bounded by the sphere $x^2 + y^2 + z^2 = 1$ and the plane $z = 0$; $\rho(x, y, z) = \sqrt{1 + x^2 + y^2}$

**51.** Let $T$ be a cube bounded by the planes $x = 0$, $x = 1$, $y = 0$, $y = 1$, $z = 0$, and $z = 1$. Find the moments of inertia of $T$ with respect to the coordinate axes if $T$ has constant mass density $k$.

**52.** Let $T$ be a rectangular box bounded by the planes $x = 0$, $x = a$, $y = 0$, $y = b$, $z = 0$, and $z = c$. Find the moments of inertia of $T$ with respect to the coordinate axes if $T$ has constant mass density $k$.

**53.** Let $T$ be the solid bounded by the planes $x + y + z = 1$, $x = 0$, $y = 0$, and $z = 0$. Find the moments of inertia of $T$ with respect to the $x$-, $y$-, and $z$-axes if $T$ has mass density given by $\rho(x, y, z) = x$.

**54.** Let $T$ be the solid bounded by the cylinder $y = x^2$ and the planes $y = x$, $z = 0$, and $z = x$. Find the moments of inertia of $T$ with respect to the coordinate axes if $T$ has mass density given by $\rho(x, y, z) = z$.

*The average value of a function f of three variables over a solid region T is defined to be*

$$f_{av} = \frac{1}{V(T)} \iiint_T f(x, y, z)\, dV$$

*where V(T) is the volume of T. Use this definition in Exercises 55–58.*

**55.** Find the average value of $f(x, y, z) = x + y + z$ over the rectangular box $T$ bounded by the planes $x = 0$, $x = 1$, $y = 0$, $y = 2$, $z = 0$, and $z = 3$.

**56.** Find the average value of $f(x, y, z) = x^2 + y^2 + z^2$ over the tetrahedron bounded by the planes $x + y + z = 1$, $x = 0$, $y = 0$, and $z = 0$.

**57.** Find the average value of $f(x, y, z) = xyz$ over the solid region lying inside the spherical ball of radius 2 with center at the origin and in the first octant.

**58. Average Temperature in a Room** A rectangular room can be described by the set $B = \{(x, y, z) \mid 0 \le x \le 20, 0 \le y \le 40, 0 \le z \le 9\}$. If the temperature (in degrees Fahrenheit) at a point $(x, y, z)$ in the room is given by $f(x, y, z) = 60 + 0.2x + 0.1y + 0.2z$, what is the average temperature in the room?

**59.** Find the region $T$ that will make the value of $\iiint_T (1 - 2x^2 - 3y^2 - z^2)^{1/3}\, dV$ as large as possible.

**60.** Find the values of $a$ and $b$ that will maximize $\iiint_T (4 - x^2 - y^2 - z^2)\, dV$, where $T = \{(x, y, z) \mid 1 \le a \le x^2 + y^2 + z^2 \le b \le 2\}$.

*In Exercises 61–64, determine whether the statement is true or false. If it is true, explain why. If it is false, explain why or give an example that shows it is false.*

**61.** If $B = [-1, 1] \times [-2, 2] \times [-3, 3]$, then $\iiint_B \sqrt{x^2 + y^2 + z^2}\, dV > 0$.

**62.** If $T$ is a solid sphere of radius $a$ centered at the origin, then $\iiint_T x\, dV = 0$.

**63.** $12 \le \displaystyle\int_1^2 \int_1^3 \int_1^4 \sqrt{1 + x^2 + y^2 + z^2}\, dz\, dy\, dx \le 6\sqrt{30}$

**64.** $\displaystyle\iiint_T k\, dV = \frac{28\pi k}{3}$, where $T = \{(x, y, z) \mid 1 \le (x - 1)^2 + (y - 2)^2 + (z + 1)^2 \le 4\}$ and $k$ is a constant

---

## 13.7 Triple Integrals in Cylindrical and Spherical Coordinates

Just as some double integrals are easier to evaluate by using polar coordinates, we will see that some triple integrals are easier to evaluate by using cylindrical or spherical coordinates.

### Cylindrical Coordinates

Let $T$ be a **z-simple region** described by

$$T = \{(x, y, z) \mid (x, y) \in R, h_1(x, y) \le z \le h_2(x, y)\}$$

where $R$ is the projection of $T$ onto the $xy$-plane. (See Figure 1.) As we saw in Section 13.6, if $f$ is continuous on $T$, then

$$\iiint_T f(x, y, z)\, dV = \iint_R \left[ \int_{h_1(x, y)}^{h_2(x, y)} f(x, y, z)\, dz \right] dA \qquad (1)$$

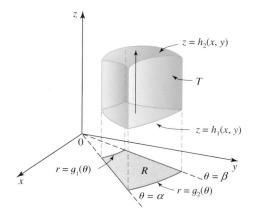

**FIGURE 1**
$T$ viewed as a $z$-simple region

Now suppose that the region $R$ can be described in polar coordinates by

$$R = \{(r, \theta) \,|\, \alpha \le \theta \le \beta, g_1(\theta) \le r \le g_2(\theta)\}$$

Then, since $x = r \cos \theta$, $y = r \sin \theta$, and $z = z$ in cylindrical coordinates, we use Equation (2) in Section 13.3 to obtain the following formula.

---

**Triple Integral in Cylindrical Coordinates**

$$\iiint_T f(x, y, z) \, dV = \int_\alpha^\beta \int_{g_1(\theta)}^{g_2(\theta)} \int_{h_1(r \cos \theta, \, r \sin \theta)}^{h_2(r \cos \theta, \, r \sin \theta)} f(r \cos \theta, r \sin \theta, z) \, r \, dz \, dr \, d\theta \quad (2)$$

---

**Note**   As an aid to remembering Equation (2), observe that the element of volume in cylindrical coordinates is $dV = r \, dz \, dr \, d\theta$, as is suggested by Figure 2.

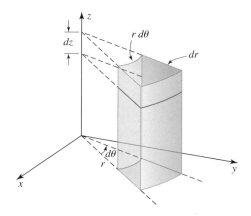

**FIGURE 2**
The element of volume in cylindrical coordinates is $dV = r \, dz \, dr \, d\theta$.

**EXAMPLE 1**   A solid $T$ is bounded by the cone $z = \sqrt{x^2 + y^2}$ and the plane $z = 2$. (See Figure 3.) The mass density at any point of the solid is proportional to the distance between the axis of the cone and the point. Find the mass of $T$.

**Solution**   The solid $T$ is described by

$$T = \left\{(x, y, z) \,|\, (x, y) \in R, \sqrt{x^2 + y^2} \le z \le 2\right\}$$

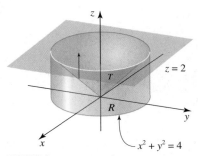

**FIGURE 3**

The arrow runs from the lower surface $z = h_1(x, y) = \sqrt{x^2 + y^2}$ to the upper surface $z = h_2(x, y) = 2$ of $T$.

where $R = \{(x, y) \mid 0 \le x^2 + y^2 \le 4\}$. In cylindrical coordinates,

$$T = \{(r, \theta, z) \mid 0 \le \theta \le 2\pi, 0 \le r \le 2, r \le z \le 2\}$$

and

$$R = \{(r, \theta) \mid 0 \le \theta \le 2\pi, 0 \le r \le 2\}$$

Since the density of the solid at $(x, y, z)$ is proportional to the distance from the $z$-axis to the point in question, we see that the density function is

$$\rho(x, y, z) = k\sqrt{x^2 + y^2} = kr$$

where $k$ is the constant of proportionality. Therefore, if we use Equation (8) in Section 13.6, the mass of $T$ is

$$m = \iiint_T \rho(x, y, z) \, dV = \iiint_T k\sqrt{x^2 + y^2} \, dV$$

$$= \int_0^{2\pi} \int_0^2 \int_r^2 (kr) \, r \, dz \, dr \, d\theta$$

$$= k \int_0^{2\pi} \int_0^2 \left[ r^2 z \right]_{z=r}^{z=2} dr \, d\theta = k \int_0^{2\pi} \int_0^2 (2r^2 - r^3) \, dr \, d\theta$$

$$= k \int_0^{2\pi} \left[ \frac{2}{3} r^3 - \frac{1}{4} r^4 \right]_{r=0}^{r=2} d\theta = \frac{4}{3} k \int_0^{2\pi} d\theta = \frac{8}{3} \pi k \quad \blacksquare$$

**EXAMPLE 2**   Find the centroid of a homogeneous solid hemisphere of radius $a$.

**Solution**   The solid $T$ is shown in Figure 4. In rectangular coordinates we can write

$$T = \left\{ (x, y, z) \mid (x, y) \in R, 0 \le z \le \sqrt{a^2 - x^2 - y^2} \right\}$$

where

$$R = \{(x, y) \mid 0 \le x^2 + y^2 \le a^2\}$$

In cylindrical coordinates we have

$$T = \left\{ (r, \theta, z) \mid 0 \le \theta \le 2\pi, 0 \le r \le a, 0 \le z \le \sqrt{a^2 - r^2} \right\}$$

and

$$R = \{(r, \theta) \mid 0 \le \theta \le 2\pi, 0 \le r \le a\}$$

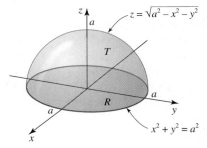

**FIGURE 4**

A homogeneous solid hemisphere of radius $a$

By symmetry the centroid lies on the $z$-axis. Therefore, it suffices to find $\bar{z} = M_{xy}/V$, where $V$, the volume of $T$, is $\frac{1}{2} \cdot \frac{4}{3}\pi a^3$, or $\frac{2}{3}\pi a^3$. Using Equation (9c) in Section 13.6, with $\rho(x, y, z) = 1$, we obtain

$$M_{xy} = \iiint_T z \, dV = \int_0^{2\pi} \int_0^a \int_0^{\sqrt{a^2 - r^2}} z \, r \, dz \, dr \, d\theta$$

$$= \int_0^{2\pi} \int_0^a \left[ \frac{1}{2} z^2 \right]_{z=0}^{z=\sqrt{a^2 - r^2}} r \, dr \, d\theta = \frac{1}{2} \int_0^{2\pi} \int_0^a (a^2 - r^2) \, r \, dr \, d\theta$$

$$= \frac{1}{2} \int_0^{2\pi} \left[ \frac{1}{2} a^2 r^2 - \frac{1}{4} r^4 \right]_{r=0}^{r=a} d\theta$$

$$= \frac{1}{2} \left( \frac{1}{4} a^4 \right) \int_0^{2\pi} d\theta = \frac{1}{8} a^4 (2\pi) = \frac{1}{4} \pi a^4$$

Therefore,

$$\bar{z} = \frac{M_{xy}}{V} = \frac{\pi a^4}{4} \cdot \frac{3}{2\pi a^3} = \frac{3}{8} a$$

so the centroid is located at the point $\left(0, 0, \frac{3a}{8}\right)$. ∎

### ■ Spherical Coordinates

When the region of integration is bounded by portions of spheres and cones, a triple integral is generally easier to evaluate if it is expressed in terms of spherical coordinates. Recall from Section 10.7 that the relationship between spherical coordinates $\rho, \phi, \theta$ and rectangular coordinates $x, y, z$ is given by

$$x = \rho \sin \phi \cos \theta, \qquad y = \rho \sin \phi \sin \theta, \qquad z = \rho \cos \phi \qquad (3)$$

(See Figure 5.)

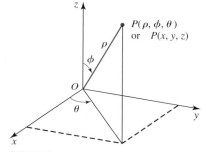

**FIGURE 5**
The point $P$ has representation $(\rho, \phi, \theta)$ in spherical coordinates and $(x, y, z)$ in rectangular coordinates.

To see the role played by spherical coordinates in integration, let's consider the simplest case in which the region of integration is a **spherical wedge** (the analog of a rectangular box)

$$T = \{(\rho, \phi, \theta) \mid a \le \rho \le b, c \le \phi \le d, \alpha \le \theta \le \beta\}$$

where $a \ge 0, 0 \le d - c \le \pi$, and $0 \le \beta - \alpha \le 2\pi$. To integrate over such a region, let

$$a = \rho_0 < \rho_1 < \cdots < \rho_{i-1} < \rho_i < \cdots < \rho_l = b$$

$$c = \phi_0 < \phi_1 < \cdots < \phi_{j-1} < \phi_j < \cdots < \phi_m = d$$

$$\alpha = \theta_0 < \theta_1 < \cdots < \theta_{k-1} < \theta_k < \cdots < \theta_n = \beta$$

be regular partitions of the intervals $[a, b]$, $[c, d]$, and $[\alpha, \beta]$, respectively, where $\Delta\rho = (b - a)/l$, $\Delta\phi = (d - c)/m$ and $\Delta\theta = (\beta - \alpha)/n$. The concentric spheres $\rho_i$, where $1 \le i \le l$, half-cones $\phi = \phi_j$, where $1 \le j \le m$, and the half-planes $\theta = \theta_k$, where $1 \le k \le n$, divide the spherical wedge $T$ into $N = lmn$ spherical wedges $T_{111}, T_{112}, \ldots, T_{lmn}$. A typical wedge $T_{ijk}$ comprising the spherical partition $P = \{T_{ijk}\}$ is shown in Figure 6.

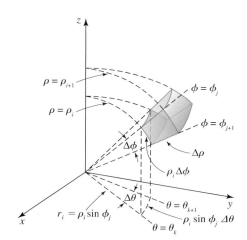

**FIGURE 6**
A typical spherical wedge in the partition $P$ of the solid $T$

If you refer to Figure 6, you will see that $T_{ijk}$ is approximately a rectangular box with dimensions $\Delta\rho$, $\rho_i \Delta\phi$ (the arc of a circle with radius $\rho_i$ that subtends an angle of

$\Delta\phi$) and $\rho_i \sin \phi_j \Delta\theta$ (the arc of a circle with radius $\rho_i \sin \phi_j$ and subtending an angle of $\Delta\theta$). Thus, its volume $\Delta V$ is

$$\Delta V = \rho_i^2 \sin \phi_j \, \Delta\rho \, \Delta\phi \, \Delta\theta$$

Therefore, an approximation to a Riemann sum of $f$ over $T$ is

$$\sum_{i=1}^{l} \sum_{j=1}^{m} \sum_{k=1}^{n} f(\rho_i^* \sin \phi_j^* \cos \theta_k^*, \rho_i^* \sin \phi_j^* \sin \theta_k^*, \rho_i^* \cos \phi_j^*)\rho_i^{*2} \sin \phi_j^* \, \Delta\rho \, \Delta\phi \, \Delta\theta$$

But this is a Riemann sum of the function

$$F(\rho, \phi, \theta) = f(\rho \sin \phi \cos \theta, \rho \sin \phi \sin \theta, \rho \cos \phi)\rho^2 \sin \phi$$

and its limit is the triple integral

$$\int_{\alpha}^{\beta} \int_{c}^{d} \int_{a}^{b} F(\rho, \phi, \theta) \, \rho^2 \sin \phi \, d\rho \, d\phi \, d\theta$$

Therefore, we have the following formula for transforming a triple integral in rectangular coordinates into one involving spherical coordinates.

---

**Triple Integral in Spherical Coordinates**

$$\iiint_T f(x, y, z) \, dV = \int_{\alpha}^{\beta} \int_{c}^{d} \int_{a}^{b} f(\rho \sin \phi \cos \theta, \rho \sin \phi \sin \theta, \rho \cos \phi)\rho^2 \sin \phi \, d\rho \, d\phi \, d\theta \qquad \textbf{(4)}$$

where $T$ is the spherical wedge

$$T = \{(\rho, \phi, \theta) \mid a \leq \rho \leq b, c \leq \phi \leq d, \alpha \leq \theta \leq \beta\}$$

---

Equation (4) states that to transform a triple integral in rectangular coordinates to one in spherical coordinates, make the substitutions

$$x = \rho \sin \phi \cos \theta, \quad y = \rho \sin \phi \sin \theta, \quad z = \rho \cos \phi, \quad \text{and} \quad x^2 + y^2 + z^2 = \rho^2$$

then make the appropriate change in the limits of integration, and replace $dV$ by $\rho^2 \sin \phi \, d\rho \, d\phi \, d\theta$. This element of volume can be recalled with the help of Figure 7.

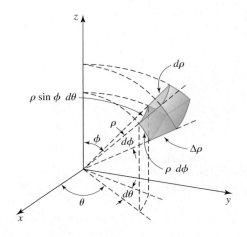

**FIGURE 7**
The element of volume in spherical coordinates is $dV = \rho^2 \sin \phi \, d\rho \, d\phi \, d\theta$.

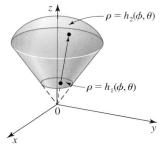

**FIGURE 8**
A $\rho$-simple region is bounded by the surfaces $\rho = h_1(\phi, \theta)$ and $\rho = h_2(\phi, \theta)$

Equation (4) can be extended to include more general regions. For example, if $T$ is **$\rho$-simple,** that is, if the region $T$ can be described by

$$T = \{(\rho, \phi, \theta) \,|\, h_1(\phi, \theta) \le \rho \le h_2(\phi, \theta), c \le \phi \le d, \alpha \le \theta \le \beta\}$$

then

$$\iiint_T f(x, y, z)\, dV$$

$$= \int_\alpha^\beta \int_c^d \int_{h_1(\phi, \theta)}^{h_2(\phi, \theta)} f(\rho \sin \phi \cos \theta, \rho \sin \phi \sin \theta, \rho \cos \phi)\rho^2 \sin \phi \, d\rho \, d\phi \, d\theta \qquad (5)$$

Observe that $\rho$-simple regions are precisely those regions that lie between two surfaces $\rho = h_1(\phi, \theta)$ and $\rho = h_2(\phi, \theta)$, as shown in Figure 8. To find the limits of integration with respect to $\rho$, we draw a radial line emanating from the origin. The line first intersects the surface, $\rho = h_1(\phi, \theta)$, giving the lower limit of integration, and then intersects the surface $\rho = h_2(\phi, \theta)$, giving the upper limit of integration.

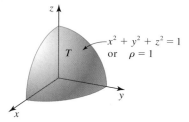

**FIGURE 9**
$T$ is the part of the ball $x^2 + y^2 + z^2 \le 1$ lying in the first octant.

**EXAMPLE 3** Evaluate $\iiint_T x \, dV$, where $T$ is the part of the region in the first octant lying inside the sphere $x^2 + y^2 + z^2 = 1$.

**Solution** The solid $T$ is shown in Figure 9. Since the boundary of $T$ is part of a sphere, let's use spherical coordinates. In terms of spherical coordinates we can write

$$T = \left\{(\rho, \phi, \theta) \,|\, 0 \le \rho \le 1, 0 \le \phi \le \tfrac{\pi}{2}, 0 \le \theta \le \tfrac{\pi}{2}\right\}$$

Furthermore, $x = \rho \sin \phi \cos \theta$. Therefore, using Equation (4), we obtain

$$\iiint_T x \, dV = \int_0^{\pi/2} \int_0^{\pi/2} \int_0^1 (\rho \sin \phi \cos \theta)\rho^2 \sin \phi \, d\rho \, d\phi \, d\theta$$

$$= \int_0^{\pi/2} \int_0^{\pi/2} \int_0^1 \rho^3 \sin^2 \phi \cos \theta \, d\rho \, d\phi \, d\theta$$

$$= \int_0^{\pi/2} \int_0^{\pi/2} \left[\tfrac{1}{4}\rho^4 \sin^2 \phi \cos \theta\right]_{\rho=0}^{\rho=1} d\phi \, d\theta = \tfrac{1}{4}\int_0^{\pi/2} \int_0^{\pi/2} \sin^2 \phi \cos \theta \, d\phi \, d\theta$$

$$= \tfrac{1}{8}\int_0^{\pi/2} \int_0^{\pi/2} (1 - \cos 2\phi)\cos \theta \, d\phi \, d\theta = \tfrac{1}{8}\int_0^{\pi/2} \cos \theta \left[\phi - \tfrac{1}{2}\sin 2\phi\right]_{\phi=0}^{\phi=\pi/2} d\theta$$

$$= \tfrac{\pi}{16}\int_0^{\pi/2} \cos \theta \, d\theta = \tfrac{\pi}{16}\sin \theta \Big|_0^{\pi/2} = \tfrac{\pi}{16} \qquad \blacksquare$$

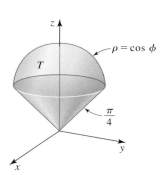

**FIGURE 10**
The solid $T$ is bounded below by part of a cone and above by part of a sphere.

**EXAMPLE 4** Find the center of mass of the solid $T$ of uniform density bounded by the cone $z = \sqrt{x^2 + y^2}$ and the sphere $x^2 + y^2 + z^2 = z$. (See Figure 10.)

**Solution** We first express the given equations in terms of spherical coordinates. The equation of the cone is

$$\rho \cos \phi = \sqrt{\rho^2 \sin^2 \phi \cos^2 \theta + \rho^2 \sin^2 \phi \sin^2\theta} = \rho \sin \phi$$

which simplifies to $\cos\phi = \sin\phi$, $\tan\phi = 1$, or $\phi = \pi/4$. Next, we see that the equation of the sphere is

$$\rho^2 = \rho\cos\phi \qquad \text{or} \qquad \rho = \cos\phi$$

Therefore, the solid under consideration can be described by

$$T = \left\{(\rho,\phi,\theta)\,|\,0 \le \rho \le \cos\phi, 0 \le \phi \le \tfrac{\pi}{4}, 0 \le \theta \le 2\pi\right\}$$

Let the uniform density of $T$ be $k$. Then the mass of $T$ is

$$m = k\iiint_T dV = k\int_0^{2\pi}\int_0^{\pi/4}\int_0^{\cos\phi} \rho^2 \sin\phi\, d\rho\, d\phi\, d\theta \qquad {\scriptstyle h_1(\phi,\theta) = 0,\, h_2(\phi,\theta) = \cos\phi}$$

$$= k\int_0^{2\pi}\int_0^{\pi/4}\left[\frac{1}{3}\rho^3\sin\phi\right]_{\rho=0}^{\rho=\cos\phi} d\phi\, d\theta$$

$$= \frac{k}{3}\int_0^{2\pi}\int_0^{\pi/4} \cos^3\phi\sin\phi\, d\phi\, d\theta = \frac{k}{3}\int_0^{2\pi}\left[-\frac{1}{4}\cos^4\phi\right]_{\phi=0}^{\phi=\pi/4} d\theta$$

$$= \frac{k}{16}\int_0^{2\pi} d\theta = \frac{\pi k}{8}$$

By symmetry the center of mass lies on the $z$-axis, so it suffices to find $\bar{z} = M_{xy}/m$. Using Equation (9c) in Section 13.6, with $\rho(x, y, z) = 1$, we obtain

$$M_{xy} = \iiint_T kz\, dV = k\int_0^{2\pi}\int_0^{\pi/4}\int_0^{\cos\phi} (\rho\cos\phi)\rho^2\sin\phi\, d\rho\, d\phi\, d\theta$$

$$= k\int_0^{2\pi}\int_0^{\pi/4}\left[\frac{1}{4}\rho^4\cos\phi\sin\phi\right]_{\rho=0}^{\rho=\cos\phi} d\phi\, d\theta$$

$$= \frac{k}{4}\int_0^{2\pi}\int_0^{\pi/4} \cos^5\phi\sin\phi\, d\phi\, d\theta = \frac{k}{4}\int_0^{2\pi}\left[-\frac{1}{6}\cos^6\phi\right]_{\phi=0}^{\phi=\pi/4} d\theta$$

$$= -\frac{k}{24}\left(\frac{(\sqrt{2})^6}{2^6} - 1\right)\int_0^{2\pi} d\theta = \frac{7k}{192}\int_0^{2\pi} d\theta = \frac{7k\pi}{96}$$

Therefore,

$$\bar{z} = \frac{M_{xy}}{m} = \frac{7k\pi}{96}\cdot\frac{8}{\pi k} = \frac{7}{12}$$

so the center of mass is located at $\left(0, 0, \frac{7}{12}\right)$. ∎

## 13.7 CONCEPT QUESTIONS

1. Write the triple integral $\iiint_T f(x, y, z)\, dV$ in cylindrical coordinates if

$$T = \{(r, \theta, z)\,|\,\alpha \le \theta \le \beta, g_1(\theta) \le r \le g_2(\theta),$$

$$h_1(r\cos\theta, r\sin\theta) \le z \le h_2(r\cos\theta, r\sin\theta)\}$$

2. Write the triple integral $\iiint_T f(x, y, z)\, dV$ in spherical coordinates if

$$T = \{(\rho,\phi,\theta)\,|\,h_1(\phi,\theta) \le \rho \le h_2(\phi,\theta), c \le \phi \le d, \alpha \le \theta \le \beta\}$$

3. Write the element of volume $dV$ in (a) cylindrical coordinates and (b) spherical coordinates.

## 13.7    EXERCISES

*In Exercises 1–4, sketch the solid whose volume is given by the integral, and evaluate the integral.*

**1.** $\int_0^{\pi/2} \int_0^3 \int_0^{r^2} r \, dz \, dr \, d\theta$

**2.** $\int_0^{2\pi} \int_1^2 \int_0^{2-r} r \, dz \, dr \, d\theta$

**3.** $\int_0^{2\pi} \int_0^{\pi/2} \int_0^2 \rho^2 \sin \phi \, d\rho \, d\phi \, d\theta$

**4.** $\int_0^{2\pi} \int_0^{\pi/4} \int_0^{2 \sec \phi} \rho^2 \sin \phi \, d\rho \, d\phi \, d\theta$

*In Exercises 5–18, solve the problem using cylindrical coordinates.*

**5.** Evaluate $\iiint_T \sqrt{x^2 + y^2} \, dV$, where $T$ is the solid bounded by the cylinder $x^2 + y^2 = 1$ and the planes $z = 1$ and $z = 3$.

**6.** Evaluate $\iiint_T e^{x^2 + y^2} \, dV$, where $T$ is the solid bounded by the cylinder $x^2 + y^2 = 4$ and the planes $z = 0$ and $z = 4$.

**7.** Evaluate $\iiint_T y \, dV$, where $T$ is the part of the solid in the first octant lying inside the paraboloid $z = 4 - x^2 - y^2$.

**8.** Evaluate $\iiint_T x \, dV$, where $T$ is the part of the solid in the first octant bounded by the paraboloid $z = x^2 + y^2$ and the plane $z = 4$.

**9.** Evaluate $\iiint_T (x^2 + y^2) \, dV$, where $T$ is the solid bounded by the cone $z = 4 - \sqrt{x^2 + y^2}$ and the $xy$-plane.

**10.** Evaluate $\iiint_T y^2 \, dV$, where $T$ is the solid that lies within the cylinder $x^2 + y^2 = 1$ and between the $xy$-plane and the paraboloid $z = 2x^2 + 2y^2$.

**11.** Find the volume of the solid bounded above by the sphere $x^2 + y^2 + z^2 = 9$ and below by the paraboloid $8z = x^2 + y^2$.

**12.** Find the volume of the solid bounded by the paraboloids $z = x^2 + y^2$ and $z = 12 - 2x^2 - 2y^2$.

**13.** A solid is bounded by the cylinder $x^2 + y^2 = 4$ and the planes $z = 0$ and $z = 3$. Find the center of mass of the solid if the mass density at any point is directly proportional to its distance from the $xy$-plane.

**14.** A solid is bounded by the cone $z = \sqrt{x^2 + y^2}$ and the plane $z = 4$. Find its center of mass if the mass density at $P(x, y, z)$ is directly proportional to the distance between $P$ and the $z$-axis.

**15.** Find the center of mass of a homogeneous solid bounded by the paraboloid $z = 4 - x^2 - y^2$ and $z = 0$.

**16.** Find the center of mass of a homogeneous solid bounded by the paraboloids $z = x^2 + y^2$ and $z = 36 - 3x^2 - 3y^2$.

**17.** Find the moment of inertia about the $z$-axis of a homogeneous solid bounded by the cone $z = \sqrt{x^2 + y^2}$ and the paraboloid $z = x^2 + y^2$.

**18.** Find the moment of inertia about the $z$-axis of a solid bounded by the cylinder $x^2 + y^2 = 4$ and the planes $z = 0$ and $z = 3$ if the mass density at any point on the solid is directly proportional to its distance from the $xy$-plane.

*In Exercises 19–24, solve the problem by using spherical coordinates.*

**19.** Evaluate $\iiint_B \sqrt{x^2 + y^2 + z^2} \, dV$, where $B$ is the unit ball $x^2 + y^2 + z^2 \leq 1$.

**20.** Evaluate $\iiint_B e^{(x^2 + y^2 + z^2)^{3/2}} dV$, where $B$ is the part of the unit ball $x^2 + y^2 + z^2 \leq 1$ lying in the first octant.

**21.** Evaluate $\iiint_T y \, dV$, where $T$ is the solid bounded by the hemisphere $z = \sqrt{1 - x^2 - y^2}$ and the $xy$-plane.

**22.** Evaluate $\iiint_T x^2 \, dV$, where $T$ is the part of the unit ball $x^2 + y^2 + z^2 \leq 1$ lying in the first octant.

**23.** Evaluate $\iiint_T xz \, dV$, where $T$ is the solid bounded above by the sphere $x^2 + y^2 + z^2 = 4$ and below by the cone $z = \sqrt{x^2 + y^2}$.

**24.** Evaluate $\iiint_T z \, dV$, where $T$ is the solid bounded above by the sphere $x^2 + y^2 + z^2 = 4$ and below by the cone $z = \sqrt{x^2 + y^2}$.

**25.** Find the volume of the solid that is bounded above by the plane $z = 1$ and below by the cone $z = \sqrt{x^2 + y^2}$.

**26.** Find the volume of the solid bounded by the cone $z = \sqrt{x^2 + y^2}$, the cylinder $x^2 + y^2 = 4$, and the plane $z = 0$.

**27.** Find the volume of the solid lying outside the cone $z = \sqrt{x^2 + y^2}$ and inside the upper hemisphere $x^2 + y^2 + z^2 \leq 1$.

**28.** Find the volume of the solid lying above the cone $\phi = \pi/6$ and below the sphere $\rho = 4 \cos \phi$.

**29.** Find the centroid of a homogeneous solid hemisphere of radius $a$.

**30.** Find the centroid of the solid of Exercise 28.

**31.** Find the mass of a solid hemisphere of radius $a$ if the mass density at any point on the solid is directly proportional to its distance from the base of the solid.

**32.** Find the center of mass of the solid of Exercise 31.

**33.** Find the mass of the solid bounded by the cone $z = \sqrt{x^2 + y^2}$ and the plane $z = 2$ if the mass density at any point on the solid is directly proportional to the square of its distance from the origin.

34. Find the center of mass of the solid of Exercise 33.

35. Find the moment of inertia about the $z$-axis of the solid of Exercise 28, assuming that it has constant mass density.

36. Find the moment of inertia with respect to the axis of symmetry for a solid hemisphere of radius $a$ if the density at a point is directly proportional to its distance from the center of the base.

37. Find the moment of inertia with respect to a diameter of the base of a homogeneous solid hemisphere of radius $a$.

38. Show that the average distance from the center of a circle of radius $a$ to other points of the circle is $2a/3$ and that the average distance from the center of a sphere of radius $a$ to other points of the sphere is $3a/4$.

39. Let $T$ be a uniform solid of mass $m$ bounded by the spheres $\rho = a$ and $\rho = b$, where $0 < a < b$. Show that the moment of inertia of $T$ about a diameter of $T$ is

$$I = \frac{2m}{5}\left(\frac{b^5 - a^5}{b^3 - a^3}\right)$$

40. a. Use the result of Exercise 39 to find the moment of inertia of a uniform solid ball of mass $m$ and radius $b$ about a diameter of the ball.

    b. Use the result of Exercise 39 to find the moment of inertia of a hollow spherical shell of mass $m$ and radius $b$ about a diameter of the shell.

    Hint: Find $\lim_{a \to b^-} I$.

*In Exercises 41 and 42, evaluate the integral by using cylindrical coordinates.*

41. $\displaystyle\int_{-1}^{1}\int_{0}^{\sqrt{1-x^2}}\int_{0}^{\sqrt{4-x^2-y^2}} z\, dz\, dy\, dx$

42. $\displaystyle\int_{-1}^{1}\int_{-\sqrt{1-x^2}}^{\sqrt{1-x^2}}\int_{\sqrt{x^2+y^2}}^{2-x^2-y^2} (x^2 + y^2)^{3/2}\, dz\, dy\, dx$

*In Exercises 43 and 44, evaluate the integral by using spherical coordinates.*

43. $\displaystyle\int_{0}^{1}\int_{0}^{\sqrt{1-x^2}}\int_{\sqrt{x^2+y^2}}^{\sqrt{2-x^2-y^2}} (x^2 + y^2 + z^2)^{3/2}\, dz\, dy\, dx$

44. $\displaystyle\int_{-3}^{3}\int_{-\sqrt{9-x^2}}^{\sqrt{9-x^2}}\int_{4}^{\sqrt{25-x^2-y^2}} (x^2 + y^2 + z^2)^{-1/2}\, dz\, dy\, dx$

45. The temperature (in degrees Fahrenheit) at a point $(x, y, z)$ of a solid ball of radius 3 in. centered at the origin is given by $T(x, y, z) = 20(x^2 + y^2 + z^2)$. What is the average temperature of the ball?

*In Exercises 46–50, determine whether the statement is true or false. If it is true, explain why. If it is false, explain why or give an example that shows it is false.*

46. The volume of the solid bounded above by the paraboloid $z = 4 - x^2 - y^2$ and below by the $xy$-plane in cylindrical coordinates is $\int_0^{2\pi}\int_0^2\int_0^{4-r^2} dz\, dr\, d\theta$.

47. $\int_0^{\pi/2}\int_0^{2\pi}\int_0^2 \rho^2 \sin\phi\, d\rho\, d\theta\, d\phi = \frac{16\pi}{3}$

48. If $T = \{(\rho, \phi, \theta) \,|\, a < \rho < b, 0 \le \phi \le \frac{\pi}{2}, 0 \le \theta \le \frac{\pi}{2}\}$, then $\iiint_T dV = \frac{\pi}{6}(b^3 - a^3)$.

49. If $T$ is a solid with constant density $k$, then its moment of inertia about the $z$-axis is given by $I_z = k\iiint_T \rho^2 \sin^2\phi\, dV$.

50. If $T = \{(\rho, \phi, \theta) \,|\, 0 < \rho < a, 0 \le \phi \le \frac{\pi}{2}, 0 \le \theta \le 2\pi\}$, then $\iiint_T \rho \cos\theta\, dV = 0$.

---

## 13.8 Change of Variables in Multiple Integrals

We often use a change of variable (a substitution) when we integrate a function of one variable to transform the given integral into one that is easier to evaluate. For example, using the substitution $x = \sin\theta$, we find

$$\int_0^1 \sqrt{1 - x^2}\, dx = \int_0^{\pi/2} \cos^2\theta\, d\theta = \frac{1}{2}\int_0^{\pi/2}(1 + \cos 2\theta)\, d\theta$$

$$= \frac{\pi}{4}$$

Observe that the interval of integration is $[0, 1]$ if we integrate with respect to $x$, and it changes to $\left[0, \frac{\pi}{2}\right]$ if we integrate with respect to $\theta$. More generally, the substitution $x = g(u)$ [so $dx = g'(u)\, du$] enables us to write

$$\int_a^b f(x)\, dx = \int_c^d f(g(u))g'(u)\, du \tag{1}$$

where $a = g(c)$ and $b = g(d)$.

As you have also seen on many occasions, a change of variables can be used to help us to evaluate integrals involving a function of two or more variables. For example, in evaluating a double integral $\iint_R f(x, y)\, dA$, where $R$ is a circular region, it is often helpful to use the substitution

$$x = r\cos\theta \qquad y = r\sin\theta$$

to transform the original integral into one involving polar coordinates. In this instance we have

$$\iint_R f(x, y)\, dA = \iint_D f(r\cos\theta, r\sin\theta)\, r\, dr\, d\theta$$

where $D$ is in the region in the $r\theta$-plane that corresponds to the region $R$ in the $xy$-plane.

These examples raise the following questions:

**1.** If an integral $\iint f(x, y)\, dA$ cannot be readily found when we are integrating with respect to the variables $x$ and $y$, can we find a substitution $x = g(u, v)$, $y = h(u, v)$ that transforms this integral into one involving the variables $u$ and $v$ that is more convenient to evaluate?
**2.** What form does the latter integral take?

## ■ Transformations

The substitutions that are used to change an integral involving the variables $x$ and $y$ into one involving the variables $u$ and $v$ are determined by a **transformation** or function $T$ from the $uv$-plane to the $xy$-plane. This function associates with each point $(u, v)$ in a region $S$ in the $uv$-plane exactly one point $(x, y)$ in the $xy$-plane. (See Figure 1.) The point $(x, y)$, called the **image** of the point $(u, v)$ under the transformation $T$, is written $(x, y) = T(u, v)$ and is defined by the equations

$$x = g(u, v) \qquad y = h(u, v) \tag{2}$$

where $g$ and $h$ are functions of two variables. The totality of all points in the $xy$-plane that are images of all points in $S$ is called the **image of $S$** and denoted by $T(S)$. Figure 1 gives a geometric visualization of a transformation $T$ that maps a region $S$ in the $uv$-plane onto a region $R$ in the $xy$-plane.

**FIGURE 1**
$T$ maps the region $S$ in the $uv$-plane onto the region $R$ in the $xy$-plane.

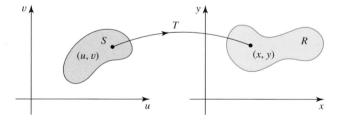

A transformation $T$ is **one-to-one** if no two distinct points in the $uv$-plane have the same image. In this case it may be possible to solve Equation (2) for $u$ and $v$ in terms of $x$ and $y$ to obtain the equations

$$u = G(x, y) \qquad v = H(x, y)$$

which defines the **inverse transformation** $T^{-1}$ from the $xy$-plane to the $uv$-plane.

**EXAMPLE 1**   Let $T$ be a transformation defined by the equations

$$x = u + v \qquad y = v$$

Find the image of the rectangular region $S = \{(u, v) \mid 0 \le u \le 2, 0 \le v \le 1\}$ under the transformation $T$.

**Solution**   Let's see how the sides of the rectangle $S$ are transformed by $T$. Referring to Figure 2a, observe $0 \le u \le 2$ and $v = 0$ on $S_1$. Using the given equations describing $T$, we see that $x = u$ and $y = 0$. This shows that $S_1$ is mapped onto the line segment $0 \le x \le 2$ and $y = 0$ (labeled $T(S_1)$ in Figure 2b). On $S_2$, $u = 2$ and $0 \le v \le 1$, so $x = 2 + y$, for $0 \le y \le 1$. This gives the image of $S_2$ under $T$ as the line segment $T(S_2)$. On $S_3$, $0 \le u \le 2$ and $v = 1$, so $x = u + 1$ and $y = 1$, which means that $S_3$ is mapped onto the line segment $T(S_3)$ described by $1 \le x \le 3$, $y = 1$. Finally, on $S_4$, $u = 0$ and $0 \le v \le 1$, and this gives the image of $S_4$ as the line segment $x = y$, for $0 \le y \le 1$. Observe that as the perimeter of $S$ is traced in a counterclockwise direction, so too is the boundary of the image $R = T(S)$ of $S$. The image of $S$ under $T$ is the region inside and on the parallelogram $R$.

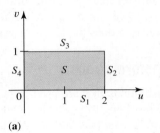

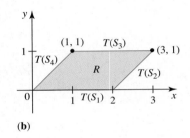

**FIGURE 2**
The region $S$ in part (a) is transformed onto the region $R$ in part (b) by $T$.

**(a)**                                      **(b)**

## ■ Change of Variables in Double Integrals

To see how a double integral is changed under the transformation $T$ defined by Equation (2), let's consider the effect that $T$ has on the area of a small rectangular region $S$ in the $uv$-plane with vertices $(u_0, v_0)$, $(u_0 + \Delta u, v_0)$, $(u_0 + \Delta u, v_0 + \Delta v)$, and $(u_0, v_0 + \Delta v)$ as shown in Figure 3a. The image of $S$ is the region $R = T(S)$ in the $xy$-plane shown in Figure 3b. The lower left-hand corner point of $S$, $(u_0, v_0)$, is mapped onto the point $(x_0, y_0) = T(u_0, v_0) = (g(u_0, v_0), h(u_0, v_0))$ by $T$. On the side $L_1$ of $S$, $u_0 \le u \le u_0 + \Delta u$ and $v = v_0$. Therefore, the image $T(L_1)$ of $L_1$ under $T$ is the curve with equations

$$x = g(u, v_0) \qquad y = h(u, v_0)$$

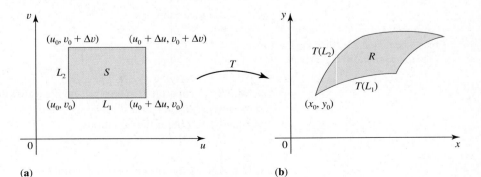

**FIGURE 3**
The transformation $T$ maps $S$ onto $R$.       **(a)**                              **(b)**

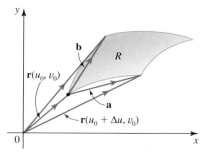

**FIGURE 4**
The vector
$\mathbf{a} = \mathbf{r}(u_0 + \Delta u, v_0) - \mathbf{r}(u_0, v_0)$

or, in vector form,

$$\mathbf{r}(u, v_0) = g(u, v_0)\mathbf{i} + h(u, v_0)\mathbf{j}$$

with parameter interval $[u_0, u_0 + \Delta u]$. As you can see from Figure 4, the vector

$$\mathbf{a} = \mathbf{r}(u_0 + \Delta u, v_0) - \mathbf{r}(u_0, v_0)$$

provides us with an approximation of $\mathbf{T}(L_1)$. Similarly, we see that the vector

$$\mathbf{b} = \mathbf{r}(u_0, v_0 + \Delta v) - \mathbf{r}(u_0, v_0)$$

provides us with an approximation of $\mathbf{T}(L_2)$.

But we can write

$$\mathbf{a} = \left[ \frac{\mathbf{r}(u_0 + \Delta u, v_0) - \mathbf{r}(u_0, v_0)}{\Delta u} \right] \Delta u$$

If $\Delta u$ is small, as we have assumed, then the term inside the brackets is approximately equal to $\mathbf{r}_u(u_0, v_0)$. So

$$\mathbf{a} \approx \Delta u \, \mathbf{r}_u(u_0, v_0)$$

Similarly, we see that

$$\mathbf{b} \approx \Delta v \, \mathbf{r}_v(u_0, v_0)$$

This suggests that we can approximate $R$ by the parallelogram having $\Delta u \, \mathbf{r}_u(u_0, v_0)$ and $\Delta v \, \mathbf{r}_v(u_0, v_0)$ as adjacent sides. (See Figure 5.) The area of this parallelogram is $|\mathbf{a} \times \mathbf{b}|$, or

$$|(\Delta u \, \mathbf{r}_u) \times (\Delta v \, \mathbf{r}_v)| = |\mathbf{r}_u \times \mathbf{r}_v| \, \Delta u \, \Delta v$$

where the partial derivatives are evaluated at $(u_0, v_0)$. But

$$\mathbf{r}_u = g_u\mathbf{i} + h_u\mathbf{j} = \frac{\partial x}{\partial u}\mathbf{i} + \frac{\partial y}{\partial u}\mathbf{j}$$

where the partial derivatives are evaluated at $(u_0, v_0)$. Similarly,

$$\mathbf{r}_v = g_v\mathbf{i} + h_v\mathbf{j} = \frac{\partial x}{\partial v}\mathbf{i} + \frac{\partial y}{\partial v}\mathbf{j}$$

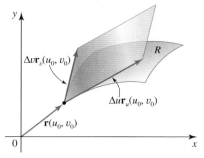

**FIGURE 5**
The image region $R$ is approximated by the parallelogram with sides $\Delta u \, \mathbf{r}_u(u_0, v_0)$ and $\Delta v \, \mathbf{r}_v(u_0, v_0)$.

So

$$\mathbf{r}_u \times \mathbf{r}_v = \begin{vmatrix} \mathbf{i} & \mathbf{j} & \mathbf{k} \\ \dfrac{\partial x}{\partial u} & \dfrac{\partial y}{\partial u} & 0 \\ \dfrac{\partial x}{\partial v} & \dfrac{\partial y}{\partial v} & 0 \end{vmatrix} = \begin{vmatrix} \dfrac{\partial x}{\partial u} & \dfrac{\partial y}{\partial u} \\ \dfrac{\partial x}{\partial v} & \dfrac{\partial y}{\partial v} \end{vmatrix} \mathbf{k} = \begin{vmatrix} \dfrac{\partial x}{\partial u} & \dfrac{\partial x}{\partial v} \\ \dfrac{\partial y}{\partial u} & \dfrac{\partial y}{\partial v} \end{vmatrix} \mathbf{k}$$

Before proceeding, let's define the following determinant, which is named after the German mathematician Carl Jacobi (1804–1851).

---

**DEFINITION   The Jacobian**

The Jacobian of the transformation $T$ defined by $x = g(u, v)$ and $y = h(u, v)$ is

$$\frac{\partial(x, y)}{\partial(u, v)} = \begin{vmatrix} \dfrac{\partial x}{\partial u} & \dfrac{\partial x}{\partial v} \\ \dfrac{\partial y}{\partial u} & \dfrac{\partial y}{\partial v} \end{vmatrix} = \frac{\partial x}{\partial u}\frac{\partial y}{\partial v} - \frac{\partial y}{\partial u}\frac{\partial x}{\partial v}$$

In terms of the Jacobian we can write the approximation of the area $\Delta A$ of $R$ as

$$\Delta A \approx |\mathbf{r}_u \times \mathbf{r}_v| \Delta u \, \Delta v = \left| \frac{\partial(x, y)}{\partial(u, v)} \right| \Delta u \, \Delta v \tag{3}$$

where the Jacobian is evaluated at $(u_0, v_0)$.

Now let $R$ be the image (in the $xy$-plane) under $T$ of the region $S$ in the $uv$-plane; that is, let $R = T(S)$ as shown in Figure 6. Enclose $S$ by a rectangle, and partition the latter into $mn$ rectangles $S_{ij}$, where $1 \le i \le m$, $1 \le j \le n$. The images $S_{ij}$ are transformed onto images $R_{ij}$ in the $xy$-plane, as shown in Figure 6.

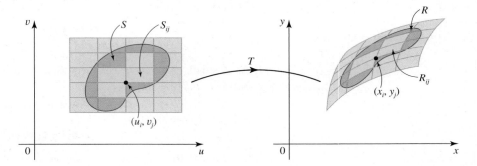

**FIGURE 6**
The images $S_{ij}$ in the $uv$-plane
are transformed onto the
images $R_{ij}$ in the $xy$-plane.

Suppose that $f$ is continuous on $R$, and define $F$ by

$$F_R(x, y) = \begin{cases} f(x, y) & \text{if } (x, y) \in R \\ 0 & \text{if } (x, y) \notin R \end{cases}$$

Using the approximation in Equation (3) on each subrectangle $R_{ij}$, we can write the double integral of $f$ over $R$ as

$$\iint\limits_R f(x, y) \, dA = \lim_{m, n \to \infty} \sum_{i=1}^m \sum_{j=1}^n F_R(x_i, y_j) \, \Delta A$$

$$= \lim_{m, n \to \infty} \sum_{i=1}^m \sum_{j=1}^n F_R(g(u_i, v_j), h(u_i, v_j)) \left| \frac{\partial(x, y)}{\partial(u, v)} \right| \Delta u \, \Delta v$$

where the Jacobian is evaluated at $(u_i, v_j)$. But the sum on the right is the Riemann sum associated with the integral

$$\iint\limits_S f(g(u, v), h(u, v)) \left| \frac{\partial(x, y)}{\partial(u, v)} \right| du \, dv$$

This discussion suggests the following result. Its proof can be found in books on advanced calculus.

---

**THEOREM 1** **Change of Variables in Double Integrals**

Let $T$ be a one-to-one transformation defined by $x = g(u, v)$, $y = h(u, v)$ that maps a region $S$ in the $uv$-plane onto a region $R$ in the $xy$-plane. Suppose that the boundaries of both $R$ and $S$ consist of finitely many piecewise smooth, simple, closed curves. Furthermore, suppose that the first-order partial derivatives of $g$ and $h$ are continuous functions. If $f$ is continuous on $R$ and the Jacobian of $T$ is nonzero, then

$$\iint\limits_R f(x, y) \, dA = \iint\limits_S f(g(u, v), h(u, v)) \left| \frac{\partial(x, y)}{\partial(u, v)} \right| du \, dv \tag{4}$$

**Note** Theorem 1 tells us that we can formally transform an integral $\iint_R f(x, y)\, dA$ involving the variables $x$ and $y$ into an integral involving the variables $u$ and $v$ by replacing $x$ by $g(u, v)$ and $y$ by $h(u, v)$ and the area element $dA$ in $x$ and $y$ by the area element

$$dA = \left| \frac{\partial(x, y)}{\partial(u, v)} \right| du\, dv$$

in $u$ and $v$. If you compare Equation (4) with Equation (1), you will see that the absolute value of the Jacobian of $T$ plays the same role as the derivative $g'(u)$ of the "transformation" $g$ defined by $x = g(u)$ in the one-dimensional case. ∎

**EXAMPLE 2** Use the transformation $T$ defined by the equations $x = u + v$, $y = v$ to evaluate $\iint_R (x + y)\, dA$, where $R$ is the parallelogram shown in Figure 2b. (See Example 1.)

**Solution** Recall that the transformation $T$ maps the much simpler rectangular region $S = \{(u, v) \mid 0 \le u \le 2, 0 \le v \le 1\}$ onto $R$ and that this is precisely the reason for choosing this transformation. The Jacobian of $T$ is

$$\frac{\partial(x, y)}{\partial(u, v)} = \begin{vmatrix} \dfrac{\partial x}{\partial u} & \dfrac{\partial x}{\partial v} \\[2mm] \dfrac{\partial y}{\partial u} & \dfrac{\partial y}{\partial v} \end{vmatrix} = \begin{vmatrix} 1 & 1 \\ 0 & 1 \end{vmatrix} = 1$$

Using Theorem 1, we obtain

$$\iint_R (x + y)\, dA = \iint_S [(u + v) + v](1)\, du\, dv$$

$$= \int_0^1 \int_0^2 (u + 2v)\, du\, dv = \int_0^1 \left[ \frac{1}{2} u^2 + 2uv \right]_{u=0}^{u=2} dv$$

$$= \int_0^1 (2 + 4v)\, dv = \left[ 2v + 2v^2 \right]_0^1 = 4 \qquad ∎$$

In Example 2 the transformation $T$ was chosen so that the region $S$ in the $uv$-plane corresponding to the region $R$ could be described more simply. This made it easier to evaluate the transformed integral. In other instances the transformation is chosen so that the corresponding integrand in $u$ and $v$ is easier to integrate than the original integrand in the variables $x$ and $y$, as the following example shows.

**EXAMPLE 3** Evaluate

$$\iint_R \cos\left( \frac{x - y}{x + y} \right) dA$$

where $R$ is the trapezoidal region with vertices $(1, 0)$, $(2, 0)$, $(0, 2)$, and $(0, 1)$.

**Solution** As it stands, this integral is difficult to evaluate. But observe that the form of the integrand suggests that we make the substitution

$$u = x - y \qquad v = x + y$$

These equations define a transformation $T^{-1}$ from the $xy$-plane to the $uv$-plane. If we solve these equations for $x$ and $y$ in terms of $u$ and $v$, we obtain the transformation $T$ from the $uv$-plane to the $xy$-plane defined by

$$x = \frac{1}{2}(u + v) \qquad y = \frac{1}{2}(v - u)$$

The given region $R$ is shown in Figure 7.

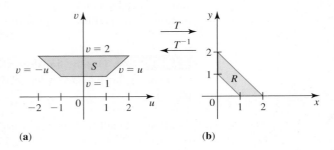

**FIGURE 7**
$T$ maps $S$ onto $R$, and
$T^{-1}$ maps $R$ onto $S$.

To find the region $S$ in the $uv$-plane that is mapped onto $R$ under the transformation $T$, observe that the sides of $R$ lie on the lines

$$y = 0, \qquad y + x = 2, \qquad x = 0, \qquad \text{and} \qquad y + x = 1$$

Using the equations defining $T^{-1}$, we see that the sides of $S$ corresponding to these sides of $R$ are

$$v = u, \qquad v = 2, \qquad v = -u, \qquad \text{and} \qquad v = 1$$

The region $S$ is shown in Figure 7a.

The Jacobian of $T$ is

$$\frac{\partial(x, y)}{\partial(u, v)} = \begin{vmatrix} \dfrac{\partial x}{\partial u} & \dfrac{\partial x}{\partial v} \\ \dfrac{\partial y}{\partial u} & \dfrac{\partial y}{\partial v} \end{vmatrix} = \begin{vmatrix} \dfrac{1}{2} & \dfrac{1}{2} \\ -\dfrac{1}{2} & \dfrac{1}{2} \end{vmatrix} = \frac{1}{2}$$

If we use Theorem 1 while viewing $S$ as a $u$-simple region, we find

$$\iint_R \cos\left(\frac{x - y}{x + y}\right) dA = \iint_S \cos\left(\frac{u}{v}\right) \left|\frac{\partial(x, y)}{\partial(u, v)}\right| du\, dv$$

$$= \int_1^2 \int_{-v}^{v} \cos\left(\frac{u}{v}\right) \cdot \left(\frac{1}{2}\right) du\, dv = \frac{1}{2} \int_1^2 \left[v \sin\left(\frac{u}{v}\right)\right]_{u=-v}^{u=v} dv$$

$$= \sin 1 \int_1^2 v\, dv = \frac{3}{2} \sin 1 \qquad \blacksquare$$

The next example shows how the formula for integration in polar coordinates can be derived with the help of Theorem 1.

**EXAMPLE 4**   Suppose that $f$ is continuous on a polar rectangle

$$R = \{(r, \theta) \,|\, a \le r \le b, \alpha \le \theta \le \beta\}$$

in the *xy*-plane. Show that

$$\iint_R f(x, y)\, dA = \iint_S f(r \cos \theta, r \sin \theta)\, r\, dr\, d\theta$$

where *S* is the region in the *rθ*-plane mapped onto *R* under the transformation *T* defined by

$$x = g(r, \theta) = r \cos \theta \qquad y = h(r, \theta) = r \sin \theta$$

**Solution**   Observe that *T* maps the *r*-simple region

$$S = \{(r, \theta) \,|\, a \le r \le b, \alpha \le \theta \le \beta\}$$

onto the polar rectangle *R* as shown in Figure 8. The Jacobian of *T* is

$$\frac{\partial(x, y)}{\partial(r, \theta)} = \begin{vmatrix} \dfrac{\partial x}{\partial r} & \dfrac{\partial x}{\partial \theta} \\[2mm] \dfrac{\partial y}{\partial r} & \dfrac{\partial y}{\partial \theta} \end{vmatrix} = \begin{vmatrix} \cos \theta & -r \sin \theta \\ \sin \theta & r \cos \theta \end{vmatrix}$$

$$= r \cos^2 \theta + r \sin^2 \theta = r > 0$$

Using Theorem 1, we obtain

$$\iint_R f(x, y)\, dA = \iint_S f(g(r, \theta), h(r, \theta)) \left| \frac{\partial(x, y)}{\partial(r, \theta)} \right| dr\, d\theta$$

$$= \int_\alpha^\beta \int_{g_1(\theta)}^{g_2(\theta)} f(r \cos \theta, r \sin \theta)\, r\, dr\, d\theta$$

as was to be shown.

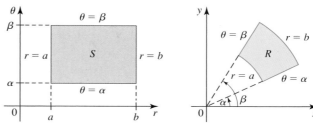

**FIGURE 8**
*T* maps the region *S* onto the polar rectangle *R*.

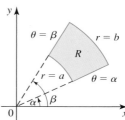

## Change of Variables in Triple Integrals

The results for a change of variables for double integrals can be extended to the case involving triple integrals. Let *T* be a transformation from the *uvw*-space to the *xyz*-space defined by the equations

$$x = g(u, v, w), \qquad y = h(u, v, w), \qquad z = k(u, v, w)$$

and suppose that *T* maps a region *S* in *uvw*-space onto a region *R* in *xyz*-space. The Jacobian of *T* is

$$\frac{\partial(x, y, z)}{\partial(u, v, w)} = \begin{vmatrix} \dfrac{\partial x}{\partial u} & \dfrac{\partial x}{\partial v} & \dfrac{\partial x}{\partial w} \\[2mm] \dfrac{\partial y}{\partial u} & \dfrac{\partial y}{\partial v} & \dfrac{\partial y}{\partial w} \\[2mm] \dfrac{\partial z}{\partial u} & \dfrac{\partial z}{\partial v} & \dfrac{\partial z}{\partial w} \end{vmatrix}$$

The following is the analog of Equation (4) for triple integrals.

---

**Change of Variables in Triple Integrals**

$$\iiint\limits_{R} f(x, y, z)\, dV = \iiint\limits_{S} f(g(u, v, w), h(u, v, w), k(u, v, w)) \left| \frac{\partial(x, y, z)}{\partial(u, v, w)} \right| du\, dv\, dw \qquad (5)$$

---

**EXAMPLE 5**  Use Equation (5) to derive the formula for changing a triple integral in rectangular coordinates to one in spherical coordinates.

**Solution**  The required transformation is defined by the equations

$$x = \rho \sin \phi \cos \theta, \qquad y = \rho \sin \phi \sin \theta, \qquad z = \rho \cos \phi$$

where $\rho$, $\phi$, and $\theta$ are spherical coordinates. The Jacobian of $T$ is

$$\frac{\partial(x, y, z)}{\partial(\rho, \phi, \theta)} = \begin{vmatrix} \sin \phi \cos \theta & \rho \cos \phi \cos \theta & -\rho \sin \phi \sin \theta \\ \sin \phi \sin \theta & \rho \cos \phi \sin \theta & \rho \sin \phi \cos \theta \\ \cos \phi & -\rho \sin \phi & 0 \end{vmatrix}$$

Expanding the determinant by the third row, we find

$$\frac{\partial(x, y, z)}{\partial(\rho, \phi, \theta)} = \cos \phi \begin{vmatrix} \rho \cos \phi \cos \theta & -\rho \sin \phi \sin \theta \\ \rho \cos \phi \sin \theta & \rho \sin \phi \cos \theta \end{vmatrix} + \rho \sin \phi \begin{vmatrix} \sin \phi \cos \theta & -\rho \sin \phi \sin \theta \\ \sin \phi \sin \theta & \rho \sin \phi \cos \theta \end{vmatrix}$$

$$= \cos \phi (\rho^2 \cos \phi \sin \phi \cos^2 \theta + \rho^2 \cos \phi \sin \phi \sin^2 \theta) + \rho \sin \phi (\rho \sin^2 \phi \cos^2 \theta + \rho \sin^2 \phi \sin^2 \theta)$$

$$= \rho^2 \cos^2 \phi \sin \phi + \rho^2 \sin^3 \phi = \rho^2 \sin \phi$$

Since $0 \leq \phi \leq \pi$, we see that $\sin \phi \geq 0$, so

$$\left| \frac{\partial(x, y, z)}{\partial(\rho, \phi, \theta)} \right| = |\rho^2 \sin \phi| = \rho^2 \sin \phi$$

Using Equation (5), we obtain

$$\iiint\limits_{R} f(x, y, z)\, dV = \iiint\limits_{S} f(\rho \sin \phi \cos \theta, \rho \sin \phi \sin \theta, \rho \cos \phi)\rho^2 \sin \phi\, d\rho\, d\phi\, d\theta$$

which is Equation (4) in Section 13.7, the formula for integrating a triple integral in spherical coordinates. ∎

---

## 13.8  CONCEPT QUESTIONS

**1. a.** Let $T$ be a transformation defined by $x = g(u, v)$ and $y = h(u, v)$. What is the Jacobian of $T$?

  **b.** Write the Jacobian of the transformation $T$ given by $x = g(u, v, w)$, $y = h(u, v, w)$, and $z = k(u, v, w)$.

**2. a.** Let $T$ be the one-to-one transformation defined by $x = g(u, v)$ and $y = h(u, v)$ that maps a region $S$ in the $uv$-plane onto a region $R$ in the $xy$-plane. Write the formula for transforming the integral $\iint_{R} f(x, y)\, dA$ into an integral involving $u$ and $v$ over the region $S$.

  **b.** Repeat part (a) for the case of a triple integral.

## 13.8  EXERCISES

*In Exercises 1–6, sketch the image $R = T(S)$ of the set $S$ under the transformation $T$ defined by the equations $x = g(u, v)$, $y = h(u, v)$.*

**1.** $S = \{(u, v) \mid 0 \le u \le 2, 0 \le v \le 1\}; \quad x = u - v, y = v$

**2.** $S = \{(u, v) \mid 0 \le u \le 1, 0 \le v \le 2\}; \quad x = u + v,$
$y = u - v$

**3.** $S$ is the triangular region with vertices $(0, 0)$, $(1, 1)$, $(0, 1)$;
$x = u + 2v, y = 2v$.

**4.** $S$ is the trapezoidal region with vertices $(-2, 0)$, $(-1, 0)$, $(0, 1)$, $(0, 2)$; $\quad x = u + v, y = u - v$

**5.** $S = \{(u, v) \mid u^2 + v^2 \le 1, u \ge 0, v \ge 0\}; \quad x = u^2 - v^2;$
$y = 2uv$

**6.** $S = \{(u, v) \mid 1 \le u \le 2, 0 \le v \le \frac{\pi}{2}\};$
$x = u \cos v, y = u \sin v$

*In Exercises 7–12, find the Jacobian of the transformation $T$ defined by the equations.*

**7.** $x = 2u + v, \quad y = u^2 - v$

**8.** $x = u^2 - v^2, \quad y = 2uv$

**9.** $x = e^u \cos 2v, \quad y = e^u \sin 2v$

**10.** $x = u \ln v, \quad y = v \ln u$

**11.** $x = u + v + w, \quad y = u - v + w, \quad z = u - 2v + 3w$

**12.** $x = 2u + w, \quad y = u^2 - v^2, \quad z = u + v^2 - 2w^2$

*In Exercises 13–20, evaluate the integral using the transformation $T$.*

**13.** $\iint_R (x + y)\, dA$, where $R$ is the parallelogram bounded by the lines with equations $y = -2x$, $y = \frac{1}{2}x - \frac{15}{2}$, $y = -2x + 10$, and $y = \frac{1}{2}x$; $\quad T$ is defined by $x = u + 2v$ and $y = v - 2u$

**14.** $\iint_R (2x + 3y)\, dA$, where $R$ is the parallelogram bounded by the lines with equations $y = 2x$, $y = \frac{1}{2}x + 3$, $y = 2x + 3$, and $y = \frac{1}{2}x$; $\quad T$ is defined by $x = u - 2v$ and $y = 2u - v$

**15.** $\iint_R 2xy\, dA$, where $R$ is the region in the first quadrant bounded by the ellipse $4x^2 + 9y^2 = 36$; $\quad T$ is defined by $x = 3u$ and $y = 2v$

**16.** $\iint_R \cos(x^2 - xy + y^2)\, dA$, where $R$ is the region bounded by the ellipse $x^2 - xy + y^2 = 2$; $\quad T$ is defined by $x = \sqrt{2}u - \sqrt{2/3}v$ and $y = \sqrt{2}u + \sqrt{2/3}v$

**17.** $\iint_R \sqrt{1 - \dfrac{x^2}{4} - \dfrac{y^2}{9}}\, dA$, where $R$ is the region bounded by

the ellipse $\dfrac{x^2}{4} + \dfrac{y^2}{9} = 1$; $\quad T$ is defined by $x = 2u$ and $y = 3v$

**18.** $\iint_R xy^2\, dA$, where $R$ is the region in the first quadrant bounded by the hyperbolas $xy = 1$ and $xy = 2$ and the lines $y = x$ and $y = 2x$; $\quad T$ is defined by $x = \dfrac{u}{v}$ and $y = v$

**19.** $\iint_R \dfrac{1}{\sqrt{x^2 + y^2}}\, dA$, where $R = \{(x, y) \mid x^2 + y^2 \le 1, y \ge 0\}$; $T$ is defined by $x = u^2 - v^2$ and $y = 2uv$, where $u, v \ge 0$.

**20.** $\iint_R y \sin x\, dA$, where $R$ is the region bounded by the graphs of $x = y^2$, $x = 0$, and $y = 1$; $\quad T$ is defined by $x = u^2$ and $y = v$

*In Exercises 21–26, evaluate the integral by making a suitable change of variables.*

**21.** $\iint_R (2x + y)\, dA$, where $R$ is the parallelogram bounded by the lines $x + y = -1$, $x + y = 3$, $2x - y = 0$, and $2x - y = 4$

**22.** $\iint_R (x + y) \sin(2x - y)\, dA$, where $R$ is the parallelogram bounded by the lines $y = -x$, $y = -x + 1$, $y = 2x$, and $y = 2x - 2$

**23.** $\iint_R e^{(x-y)/(x+y)}\, dA$, where $R$ is the triangular region bounded by the lines $x = 0$, $y = 0$, and $x + y = 1$

**24.** $\iint_R e^{(x+y)/(x-y)}\, dA$, where $R$ is the trapezoidal region with vertices $(-2, 0)$, $(-1, 0)$, $(0, 1)$, and $(0, 2)$

**25.** $\iint_R xy\, dA$, where $R$ is the region in the first quadrant bounded by the ellipse $\dfrac{x^2}{a^2} + \dfrac{y^2}{b^2} = 1$

**26.** $\iint_R \ln(4x^2 + 25y^2 + 1)\, dA$, where $R$ is the region bounded by the ellipse $4x^2 + 25y^2 = 1$

**27.** Find the volume $V$ of the solid $E$ enclosed by the ellipsoid

$$\frac{x^2}{a^2} + \frac{y^2}{b^2} + \frac{z^2}{c^2} = 1$$

**Hint:** $V = \iiint_E dV$. Use the transformation $x = au$, $y = bv$, and $z = cw$.

**28.** Let $E$ be the solid enclosed by the ellipsoid

$$\frac{x^2}{a^2} + \frac{y^2}{b^2} + \frac{z^2}{c^2} = 1$$

Find the mass of $E$ if it has constant mass density $\delta$.
**Hint:** Use the transformation of Exercise 27.

**29.** Find the moment of inertia, $I_x$, of the lamina that has constant mass density $\rho$ and occupies the disk $x^2 + y^2 - ax \le 0$ about the $x$-axis.

**30.** Show that the moment of inertia of the solid of Exercise 28 about the $z$-axis is $I_z = \frac{1}{5}m(a^2 + b^2)$, where $m = \frac{4}{3}\pi\delta abc$ is the mass of the solid.

**31.** Use Formula (5) to find the formula for changing a triple integral in rectangular coordinates to one in cylindrical coordinates.

*In Exercises 32 and 33, determine whether the statement is true or false. If it is true, explain why. If it is false, explain why or give an example that shows it is false.*

**32.** If $T$ is defined by $x = g(u, v)$ and $y = h(u, v)$ and maps a region $S$ in the $uv$-plane onto a region $R$ in the $xy$-plane, then the area of $R$ is the same as the area of $S$.

**33.** If $T$ is defined by $x = g(u, v)$, $y = h(u, v)$ and maps a region $S$ onto a region $R$, then

$$\iint_R (x^2 + y^2) \, dx \, dy = \iint_S (u^2 + v^2) \left| \frac{\partial(x, y)}{\partial(u, v)} \right| du \, dv$$

# CHAPTER 13 REVIEW

## CONCEPT QUESTIONS

*In Exercises 1–12, fill in the blanks.*

**1. a.** If $f$ is a continuous function defined on a rectangle $R = [a, b] \times [c, d]$, then the Riemann sum of $f$ over $R$ with respect to a partition $P = \{R_{ij}\}$ is _____, where $(x_{ij}^*, y_{ij}^*)$ is a point in $R_{ij}$.
  **b.** The double integral $\iint_R f(x, y) \, dA = $ _____ if the limit exists for all choices of _____ in $R_{ij}$.
  **c.** If $f(x, y) \ge 0$ on $R$, then $\iint_R f(x, y) \, dA$ gives the _____ of the solid lying directly above $R$ and below the surface _____.
  **d.** If $D$ is a bounded region that is not rectangular, then $\iint_D f(x, y) \, dA = $ _____, where $f_D(x, y) = $ _____ if $(x, y)$ is in $D$ and $f_D(x, y) = $ _____ if $(x, y)$ is not in $D$.

**2.** The following properties hold for double integrals:
  **a.** $\iint_D cf(x, y) \, dA = $ _____
  **b.** $\iint_D [f(x, y) \pm g(x, y)] \, dA = $ _____
  **c.** If $f(x, y) \ge 0$ on $D$, then $\iint_D f(x, y) \, dA$_____
  **d.** If $f(x, y) \ge g(x, y)$ on $D$, then $\iint_D f(x, y) \, dA$_____.
  **e.** If $D = D_1 \cup D_2$ and $D_1 \cap D_2 = \varnothing$, then $\iint_D f(x, y) \, dA = $ _____.

**3. a.** If $R = [a, b] \times [c, d]$, then the two iterated integrals of $f$ over $R$ are _____ and _____.
  **b.** Fubini's Theorem for a rectangular region $R = [a, b] \times [c, d]$ states that $\iint_R f(x, y) \, dA$ is equal to the _____ integrals in part (a).

**4. a.** A $y$-simple region has the form $R = $ _____, where $g_1$ and $g_2$ are continuous functions on $[a, b]$.

  **b.** An $x$-simple region has the form $R = $ _____, where $h_1$ and $h_2$ are continuous functions on $[c, d]$.
  **c.** Fubini's Theorem for the $y$-simple region $R$ of part (a), states that $\iint_R f(x, y) \, dA = $ _____. If $R$ is the $x$-simple region $R$ of part (b), then $\iint_R f(x, y) \, dA = $ _____.

**5. a.** A polar rectangle is a set of the form $R = $ _____.
  **b.** If $f$ is continuous on a polar rectangle $R$, then $\iint_R f(x, y) \, dA = $ _____.
  **c.** An $r$-simple region is a set of the form $R = $ _____.
  **d.** If $f$ is continuous on an $r$-simple region $R$, then $\iint_R f(x, y) \, dA = $ _____.

**6.** If a lamina occupies a region $R$ in the plane and the mass density of the lamina is $\rho(x, y)$, then
  **a.** The mass of the lamina is given by $m = $ _____.
  **b.** The moments of the lamina with respect to the $x$- and $y$-axes are $M_x = $ _____ and $M_y = $ _____. The coordinates of the center of mass of the lamina are $\bar{x} = $ _____ and $\bar{y} = $ _____.
  **c.** The moments of inertia of the lamina with respect to the $x$-axis, the $y$-axis, and the origin are $I_x = $ _____, $I_y = $ _____, and $I_0 = $ _____, respectively.
  **d.** If the moment of inertia of a lamina with respect to an axis is $I$, then its radius of gyration with respect to the axis is $R = $ _____.

**7. a.** If $f_x$ and $f_y$ are continuous on a region $R$ in the $xy$-plane, then the area of the surface $z = f(x, y)$ over $R$ is $A = $ _____.
  **b.** If $g$ is defined in a region $R$ in the $xz$-plane, then the area of the surface $y = g(x, z)$ is $A = $ _____.

**c.** If $h$ is defined in a region $R$ in the $yz$-plane, then the area of the surface $x = h(y, z)$ is $A =$ _____.

**8. a.** If $f$ is a continuous function defined on a rectangular box $B = [a, b] \times [c, d] \times [p, q]$, then the Riemann sum of $f$ over $B$ with respect to a partition $P = \{B_{ijk}\}$ is _____, where $(x_{ijk}^*, y_{ijk}^*, z_{ijk}^*)$ is a point in $B_{ijk}$.

**b.** The triple integral $\iiint_B f(x, y, z)\, dV =$ _____ if the limit exists for all choices of $(x_{ijk}^*, y_{ijk}^*, z_{ijk}^*)$ in $B_{ijk}$.

**c.** If $f$ is continuous on a bounded solid region $T$ in space, $B$ is a rectangular box that contains $T$, $Q = \{B_{111}, B_{112}, \ldots, B_{ijk}, \ldots, B_{lmn}\}$ is a partition of $B$, $F$ is a function defined by

$$F(x, y, z) = \begin{cases} f(x, y, z) & \text{if } (x, y, z) \text{ is in } T \\ 0 & \text{if } (x, y, z) \text{ is in } B \text{ but not in } T \end{cases}$$

then a Riemann sum of $f$ over $T$ is _____.

**d.** The triple integral of $f$ over $T$ is $\iiint_T f(x, y, z)\, dV =$ _____ provided that the limit exists for all choices of $(x_{ijk}^*, y_{ijk}^*, z_{ijk}^*)$ in $T$.

**9. a.** If $f$ is continuous on $B = [a, b] \times [c, d] \times [p, q]$, then $\iiint_B f(x, y, z)\, dV$ is equal to any of six iterated integrals depending on the _____ of integration. If we integrate with respect to $x$, $y$, and $z$, in that order, then $\iiint_B f(x, y, z)\, dV =$ _____.

**b.** If $f$ is continuous on a $z$-simple region $T = \{(x, y, z) \,|\, (x, y) \in R, k_1(x, y) \le z \le k_2(x, y)\}$, where $R$ is the projection of $T$ onto the $xy$-plane, then $\iiint_T f(x, y, z)\, dV =$ _____.

**10.** If $\rho(x, y, z)$ gives the density at the point $(x, y, z)$ of a solid $T$, then

**a.** The mass of $T$ is $m =$ _____.

**b.** The moment of $T$ about the $yz$-plane is $M_{yz} =$ _____, the moment of $T$ about the $xz$-plane is $M_{xz} =$ _____,

and the moment of $T$ about the $xy$-plane is $M_{xy} =$ _____.

**c.** The _____ _____ _____ of $T$ is located at the point $(\bar{x}, \bar{y}, \bar{z})$, where $\bar{x} =$ _____, $\bar{y} =$ _____, and $\bar{z} =$ _____.

**d.** The moments of inertia of $T$ about the $x$-, $y$-, and $z$-axes are $I_x =$ _____, $I_y =$ _____, and $I_z =$ _____.

**11. a.** If $T$ is a $z$-simple region described by $T = \{(x, y, z) \,|\, (x, y) \in R, h_1(x, y) \le z \le h_2(x, y)\}$, where $R = \{(r, \theta) \,|\, \alpha \le \theta \le \beta, g_1(\theta) \le r \le g_2(\theta)\}$, then in terms of cylindrical coordinates, $\iiint_T f(x, y, z)\, dV =$ _____.

**b.** If $T = \{(\rho, \phi, \theta) \,|\, a \le \rho \le b, c \le \phi \le d, \alpha \le \theta \le \beta\}$ is a spherical wedge, then in terms of spherical coordinates, $\iiint_T f(x, y, z)\, dV =$ _____.

**c.** If $T$ is $\rho$-simple, $T = \{(\rho, \phi, \theta) \,|\, h_1(\phi, \theta) \le \rho \le h_2(\phi, \theta), c \le \phi \le d, \alpha \le \theta \le \beta\}$, then $\iiint_T f(x, y, z)\, dV =$ _____.

**12. a.** If $T$ is a transformation defined by $x = g(u, v)$ and $y = h(u, v)$, then the Jacobian of $T$ is $\dfrac{\partial(x, y)}{\partial(u, v)} =$ _____.

**b.** If $T$ maps $S$ in the $uv$-plane onto a region $R$ in the $xy$-plane, then the formula for transforming the integral $\iint_R f(x, y)\, dx\, dy$ into one involving $u$ and $v$ is $\iint_R f(x, y)\, dx\, dy =$ _____.

**c.** If $T$ maps $S$ in $uvw$-space onto $R$ in $xyz$-space and is defined by $x = g(u, v, w)$, $y = h(u, v, w)$, and $z = k(u, v, w)$, then the Jacobian $T$ is $\dfrac{\partial(x, y, z)}{\partial(u, v, w)} =$ _____, and the change of variable formula for triple integrals is $\iiint_R f(x, y, z)\, dx\, dy\, dz =$ _____.

## REVIEW EXERCISES

*In Exercises 1–8 evaluate the iterated integral.*

**1.** $\displaystyle\int_0^2 \int_{-1}^2 (2x + 3xy^2)\, dx\, dy$

**2.** $\displaystyle\int_0^\pi \int_0^1 x \sin xy\, dy\, dx$

**3.** $\displaystyle\int_0^1 \int_x^{\sqrt{x}} (2x + 3y)\, dy\, dx$

**4.** $\displaystyle\int_0^1 \int_0^{\sqrt{1-y^2}} 2y\, dx\, dy$

**5.** $\displaystyle\int_0^2 \int_y^2 \frac{1}{4 + y^2}\, dx\, dy$

**6.** $\displaystyle\int_1^e \int_0^{1/x} \sqrt{\ln x}\, dy\, dx$

**7.** $\displaystyle\int_0^2 \int_0^{\sqrt{z}} \int_0^x (x + 2z)\, dy\, dx\, dz$

**8.** $\displaystyle\int_1^2 \int_x^3 \int_0^y \frac{y}{y + z}\, dz\, dy\, dx$

*In Exercises 9–12, sketch the region of integration for the iterated integral.*

**9.** $\displaystyle\int_1^2 \int_{\ln x}^{\sqrt[3]{x}} f(x, y)\, dy\, dx$

**10.** $\displaystyle\int_0^1 \int_0^{\sin^{-1} y} f(x, y)\, dx\, dy$

**11.** $\displaystyle\int_0^\pi \int_0^{1 + \cos \theta} f(r, \theta)\, r\, dr\, d\theta$

**12.** $\displaystyle\int_{-\sqrt{2}}^{\sqrt{2}} \int_{y^2}^2 \int_0^{2-x} f(x, y, z)\, dz\, dx\, dy$

*In Exercises 13 and 14, reverse the order of integration, and evaluate the resulting integral.*

**13.** $\displaystyle\int_0^1 \int_y^1 \sin x^2\, dx\, dy$

**14.** $\displaystyle\int_0^1 \int_y^{\sqrt{y}} \frac{\cos x}{x}\, dx\, dy$

*In Exercises 15–26, evaluate the multiple integral.*

**15.** $\iint\limits_R (x^2 + 3y^2)\, dA$, where

$R = \{(x, y)\,|-1 \le x \le 1, 0 \le y \le 2\}$

**16.** $\iint\limits_R (x + y)\, dA$, where

$R = \{(x, y)\,|0 \le x \le 1, 0 \le y \le \sqrt{1 - x^2}\}$

**17.** $\iint\limits_R y\, dA$, where $R$ is the region bounded by the parabola

$x = y^2$ and the line $x - 2y = 3$

**18.** $\iint\limits_R (x + 2y)\, dA$, where $R$ is the region bounded by the

graphs of $x = 4 - y^2$, $x = 0$, and $y = 0$

**19.** $\iint\limits_R x\, dA$, where $R$ is the region in the first quadrant bounded

by the ellipse $4x^2 + 9y^2 = 36$

**20.** $\iint\limits_R \ln x\, dy\, dx$, where $R$ is the region bounded by the graphs

of $y = 1/x$, $y = x$, and $x = e$

**21.** $\iiint\limits_T xy\, dV$, where

$T = \{(x, y, z)\,|0 \le x \le 1, 0 \le y \le x^2, 0 \le z \le x + y\}$

**22.** $\iiint\limits_T z\, dV$, where $R$ is the tetrahedron bounded by the planes

$x + 2y + z = 6$, $x = 0$, $y = 0$, and $z = 0$

**23.** $\iiint\limits_T xyz\, dV$, where $T$ is the region bounded by the hemi-

sphere $z = \sqrt{1 - x^2 - y^2}$ and the plane $z = 0$

**24.** $\iiint\limits_T z\, dV$, where $T$ is the region bounded by the cylinder

$x^2 + z^2 = 1$ and the planes $y = x$, $y = 2x$, and $z = 0$

**25.** $\iiint\limits_T x^2 z\, dV$, where $T$ is the region bounded above by the

paraboloid $y = 1 - x^2 - z^2$, above the plane $z = 0$, and to
the left by the plane $y = 0$

**26.** $\iiint\limits_T \dfrac{1}{\sqrt{x^2 + y^2 + z^2}}\, dV$, where $T$ is the region bounded

above by the hemisphere $z = \sqrt{1 - x^2 - y^2}$ and below by
the plane $z = 0$

*In Exercises 27–32, find the volume of the solid.*

**27.** The solid under the surface $z = xy^2$ and above the rectangu-
lar region $R = \{(x, y)\,|0 \le x \le 1, 1 \le y \le 2\}$

**28.** The solid under the paraboloid $z = 4 - x^2 - y^2$ and above
the triangular region in the $xy$-plane with vertices $(0, 0)$,
$(1, 1)$, and $(0, 1)$

**29.** The solid bounded by the paraboloid $z = x^2 + y^2$, the cylinder $x^2 + y^2 = 1$, and the plane $z = 0$

**30.** The solid under the paraboloid $z = 9 - x^2 - y^2$ and above
the circular region $x^2 + y^2 \le 4$ in the $xy$-plane

**31.** The solid under the surface $z = e^{-(x^2 + y^2)}$, within the cylinder
$x^2 + y^2 = 1$ and above the plane $z = 0$

**32.** The solid bounded above by the paraboloid
$z = 4 - x^2 - y^2$ and below by the cone $z = \sqrt{x^2 + y^2}$

*In Exercises 33–36, find the mass and the center of mass of
the lamina occupying the region D and having the given mass
density.*

**33.** $D$ is the region in the first quadrant bounded by the graphs
of $y = x$ and $y = x^3$;   $\rho(x, y) = y$

**34.** $D$ is the region bounded by the parabola $y = x^2$ and the line
$y = 4$;   $\rho(x, y) = x^2 y$

**35.** $D$ is the region in the first quadrant bounded by the circle
$x^2 + y^2 = 1$;   $\rho(x, y) = \sqrt{x^2 + y^2}$

**36.** $D$ is the region bounded by the semicircle $y = \sqrt{4 - x^2}$
and the $x$-axis;   $\rho(x, y) = x^2 y$

*In Exercises 37 and 38, find the moments of inertia $I_x$, $I_y$, and $I_0$
of the lamina occupying the region D and having the given mass
density.*

**37.** $D$ is the region bounded by the triangle with vertices $(0, 0)$,
$(0, 1)$, and $(1, 1)$;   $\rho(x, y) = x^2 + y^2$

**38.** $D$ is the region bounded by the graphs of $y = x$ and $y = x^2$;
$\rho(x, y) = x$

*In Exercises 39 and 40, find the area of the surface S.*

**39.** $S$ is the part of the plane $2x + 3y + z = 6$ in the first octant.

**40.** $S$ is the part of the paraboloid $z = x^2 + y^2$ below the plane
$z = 4$.

*In Exercises 41 and 42, evaluate the integral by changing to
cylindrical or spherical coordinates.*

**41.** $\displaystyle\int_0^2 \int_0^{\sqrt{4-x^2}} \int_0^1 (x^2 + y^2)^{3/2}\, dz\, dy\, dx$

**42.** $\displaystyle\int_0^3 \int_0^{\sqrt{9-x^2}} \int_0^{\sqrt{9-x^2-y^2}} z\sqrt{x^2 + y^2 + z^2}\, dz\, dy\, dx$

**43.** Express the triple integral $\iiint_T f(x, y, z)\, dV$ as an iterated
integral in six different ways using different orders of integration, where $T$ is the tetrahedron bounded by the planes
$2x + 3y + z = 6$, $x = 0$, $y = 0$, and $z = 0$.

**44.** Set up, but do not evaluate, the iterated integral giving the
mass of the solid bounded by the cone $z = \sqrt{x^2 + y^2}$ and

the sphere $x^2 + y^2 + z^2 = 8$ if the density of the solid at any point $P$ is $\rho(x, y, z) = \sqrt{1 + xz}$.

**45.** Find the Jacobian of the transformation $T$ defined by the equations $x = u + w^2$, $y = 2u^2 + v$, and $z = u^2 - v^2 + 2w$.

**46.** Use the transformation $x = u/v$ and $y = v$ to evaluate $\iint_R y \cos xy \, dy \, dx$, where $R$ is the region bounded by the hyperbolas $xy = 1$ and $xy = 4$ and the lines $y = 1$ and $y = 4$.

**47.** Evaluate $\iint_R e^{(x-y)/(x+y)} \, dA$, where $R$ is the triangular region bounded by the lines $y = x$, $x + y = 2$, and $y = 0$.

*In Exercises 48–53, state whether the statement is true or false. Give a reason for your answer.*

**48.** $\displaystyle\int_a^b \int_a^b f(x)f(y) \, dx \, dy = \left[\int_a^b f(x) \, dx\right]^2$

**49.** $\displaystyle\int_0^1 \int_{-2}^3 (x + \cos xy) \, dx \, dy = \int_{-2}^3 \int_0^1 (x + \cos xy) \, dy \, dx$

**50.** $\displaystyle\int_0^1 \int_0^y f(x, y) \, dx \, dy = \int_0^1 \int_0^x f(x, y) \, dy \, dx$

**51.** If $\iint_D f(x, y) \, dA \geq 0$, then $f(x, y) \geq 0$ for all $(x, y)$ in $D$.

**52.** $\displaystyle\int_{-1}^1 \int_0^3 x^3 \sin y^2 \, dy \, dx = 0$

**53.** $\displaystyle\int_0^1 \int_1^3 [\sqrt{x} + \cos^2(xy)] \, dx \, dy \leq 6$

# CHALLENGE PROBLEMS

**1. a.** Use the definition of the double integral as a limit of a Riemann sum to compute $\iint_R (3x^2 + 2y) \, dA$, where $R = \{(x, y) \mid 0 \leq x \leq 2, 0 \leq y \leq 1\}$.
   **Hint:** Take $\Delta x = 2/m$, and $\Delta y = 1/n$, so that $x_i = 2i/m$, where $1 \leq i \leq m$, and $y_j = j/n$, where $1 \leq j \leq n$.
   **b.** Verify the result of part (a) by evaluating an appropriate iterated integral.

**2.** The following figure shows a triangular lamina. Its mass density at $(x, y)$ is $f(x, y) = \cos(y^2)$. Find its mass.

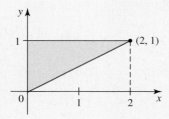

**3.** Show that the area of the parallelogram shown in the figure is $(b - a)(d - c)$, where $a \leq b$ and $c \leq d$.

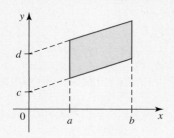

**4.** Using the result of Problem 3, show that the area of the parallelogram determined by the vectors $\mathbf{a} = \langle a_1, a_2 \rangle$ and $\mathbf{b} = \langle b_1, b_2 \rangle$ is $|a_1b_2 - a_2b_1|$.

**5. Monte Carlo Integration** This is a method that is used to find the area of complicated bounded regions in the $xy$-plane. To describe the method, suppose that $D$ is such a region completely enclosed by a rectangle $R = \{(x, y) \mid a \leq x \leq b, c \leq y \leq d\}$, as shown in the figure. Using a random number generator, we then pick points in $R$. If $A(D)$ denotes the area of $D$, then

$$\frac{A(D)}{A(R)} \approx \frac{N(D)}{n}$$

where $N(D)$ denotes the number of points landing in $D$, $A(R) = (b - a)(d - c)$, and $n$ is the number of points picked. Then

$$A(D) \approx \frac{(b - a)(d - c)}{n} N(D)$$

Use Monte Carlo integration with $n = 5000$ to estimate the area of the disk of radius 5.

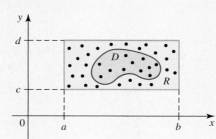

6. The expression $\sum_{i=1}^{m} \sum_{j=1}^{n} (x_i^2 + y_j^3) \Delta x \Delta y$, where $x_i = i/m$, $i = 1, 2, \ldots, m$, and $y_j = 1 + (j/n), j = 1, 2, \ldots, n$, is the Riemann sum of a function $f(x, y)$ over a region associated with a regular pattern.
   a. Write a double integral corresponding to this Riemann sum.
   b. Write an iterated integral corresponding to this Riemann sum.

7. a. Suppose that $f(x, y)$ is continuous in the triangular region $R = \{(x, y) \mid x \leq b, y \geq a, y \leq x\}$. Show that
$$\int_a^b \left[ \int_a^x f(x, y) \, dy \right] dx = \int_a^b \left[ \int_y^b f(x, y) \, dx \right] dy$$
   b. Use the result of part (a) to evaluate
$$\int_0^1 \left[ \int_y^1 \sin x^2 \, dx \right] dy$$

8. Let $f$ be a continuous function of one variable. Show that
$$\int_a^x \int_a^y \int_a^z f(t) \, dt \, dz \, dy = \frac{1}{2} \int_a^x (x - t)^2 f(t) \, dt$$
   **Hint:** Use the result of Exercise 7.

9. a. Let $R$ be a region in the $xy$-plane that is symmetric with respect to the $y$-axis, and let $f$ be a function that satisfies the condition $f(-x, y) = -f(x, y)$. Show that $\iint_R f(x, y) \, dA = 0$.
   b. Use the result of part (a) to show that if a lamina with uniform density $\rho$ that occupies a plane region that is symmetric with respect to a straight line $L$, then the centroid of the lamina lies on $L$.

10. In Exercise 6 in the Challenge Problems for Chapter 10, you were asked to show that the area of the portion of the plane $ax + by + cz = d$, where $a, b,$ and $c$ are positive constants, in the first octant is given by
$$\frac{d^2 \sqrt{a^2 + b^2 + c^2}}{2abc}$$
Derive this formula again, this time using integration. Show that the result can also be written as
$$\frac{A(R)}{c} \sqrt{a^2 + b^2 + c^2}$$

where $A(R)$ is the area of the region $R$ in the $xy$-plane.

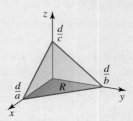

11. A thin rectangular metal plate has dimensions $a$ ft by $b$ ft and a constant density of $k$ slugs/ft$^2$. The plate is placed in the $xy$-plane as shown in the figure and is allowed to rotate about the $z$-axis at a constant angular velocity of $\omega$ radians/sec.

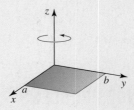

   a. Show that the kinetic energy of the plate is given by
$$E = \frac{k\omega^2}{2} \int_0^b \int_0^a (x^2 + y^2) \, dx \, dy = \frac{1}{3} (a^2 + b^2) m\omega^2$$
   where $m = kab$.
   **Hint:** The kinetic energy of a particle of mass $m$ slugs and velocity $v$ ft/sec is $\frac{1}{2}mv^2$ ft-lb.
   b. Show that $E = \frac{1}{2} I \omega^2$, where $I = \frac{2}{3}(a^2 + b^2)m$.

12. The Schwartz inequality for functions of one variable holds for multiple integrals. (See Exercise 14 in the Challenge Problems for Chapter 4.) Thus,
$$\left| \iint_D f(x, y)g(x, y) \, dA \right| \leq \sqrt{\iint_D [f(x, y)]^2 \, dA \iint_D [g(x, y)]^2 \, dA}$$

   a. Use Schwartz's inequality to prove that
$$\left| \iint_D \sqrt{4x^2 - y^2} \, dA \right| \leq \frac{2\sqrt{3}}{3}$$
   where $D$ is the triangle with vertices $A(0, 0)$, $B(1, 2)$, and $C(1, 0)$.
   b. Find the exact value of the integral in part (a). How accurate is the estimate?

A vector field in a region in three-dimensional space is a vector-valued function that assigns a vector to each point in the region. Vector fields are used in aerodynamics to model the speed and direction of air flow around an airplane. The photograph shows the air flow from the wing of an agricultural plane. The air flow was made visible by a technique that uses colored smoke rising from the ground. The *wingtip vortex*, a tube of circulating air that is left behind by the wing as it generates lift, exerts a powerful influence on the flow field behind the plane. This is the reason that the Federal Aviation Administration (FAA) requires aircraft to maintain set distances behind each other when they land.

NASA Langley Research Center

# 14 Vector Analysis

A *VECTOR FIELD* is a function that assigns a vector to each point in a region. The study of vector fields is motivated by many physical fields such as force fields and velocity fields. Gravitational and electric fields are examples of force fields, and the flow of water through a channel and the flow of air around an airfoil are examples of velocity fields.

The calculus of vector fields enables us to calculate many quantities of interest associated with force fields and velocity fields. For example, using the notion of the line integral, which is a generalization of the definite integral, we are able to calculate the work done by a force field in moving a body from one point to another along a curve. Using *surface integrals*, which are generalizations of double integrals, we can calculate the flux (flow of fluids and gases) across a surface.

The calculations involving line integrals and surface integrals are facilitated by the theorems of Green and Stokes and the Divergence Theorem, all of which may be regarded as analogs of the Fundamental Theorem of Calculus in higher dimensions.

**V** This symbol indicates that one of the following video types is available for enhanced student learning at **www.academic.cengage.com/login:**
• Chapter lecture videos       • Solutions to selected exercises

## 14.1    Vector Fields

Figure 1 shows the airflow around an airfoil in a wind tunnel. The smooth curves, traced by the individual air particles and made visible by kerosene smoke, are called **stream-lines.**

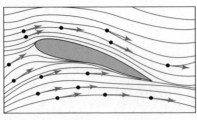

**FIGURE 1**
A vector field associated with the airflow around an airfoil

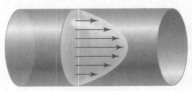

**FIGURE 2**
A vector field associated with the flow of blood in an artery

To facilitate the analysis of this flow, we can associate a tangent vector with each point on a streamline. The direction of the vector indicates the direction of flow of the air particle, and the length of the vector gives the speed of the particle. If we assign a tangent vector to each point on every streamline, we obtain what is called a *vector field* associated with this flow.

Another example of a vector field arises in the study of the flow of blood through an artery. Here, the vectors give the direction of flow and the speed of the blood cells (see Figure 2).

---

**DEFINITION    Vector Field in Two-Dimensional Space**

Let $R$ be a region in the plane. A **vector field in $R$** is a vector-valued function **F** that associates with each point $(x, y)$ in $R$ a two-dimensional vector

$$\mathbf{F}(x, y) = P(x, y)\mathbf{i} + Q(x, y)\mathbf{j}$$

where $P$ and $Q$ are functions of two variables defined on $R$.

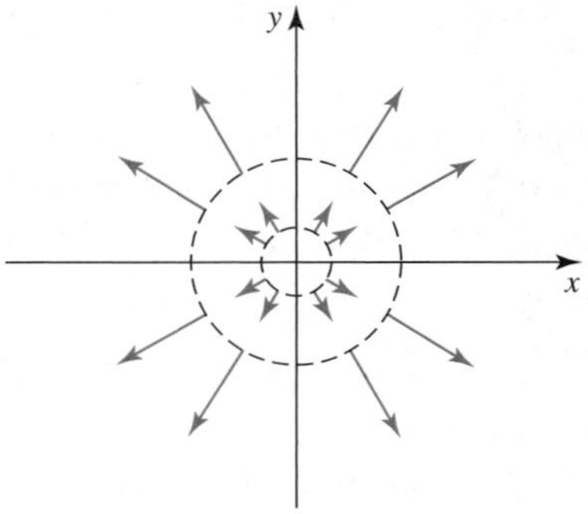

---

**EXAMPLE 1**    A vector field **F** in $R^2$ (two-dimensional space) is defined by $\mathbf{F}(x, y) = x\mathbf{i} + y\mathbf{j}$. Describe **F**, and sketch a few vectors representing the vector field.

**Solution**    The vector-valued function **F** associates with each point $(x, y)$ in $R^2$ its position vector $\mathbf{r} = x\mathbf{i} + y\mathbf{j}$. This vector points directly away from the origin and has length

$$|\mathbf{F}(x, y)| = |\mathbf{r}| = \sqrt{x^2 + y^2} = r$$

which is equal to the distance of $(x, y)$ from the origin. As an aid to sketching some vectors representing **F**, observe that each point on a circle of radius $r$ centered at the origin is associated with a vector of length $r$. Figure 3 shows a few vectors representing this vector field. ∎

**FIGURE 3**
Some vectors representing the vector field $\mathbf{F}(x, y) = x\mathbf{i} + y\mathbf{j}$

**EXAMPLE 2**    A vector field **F** in $R^2$ is defined by $\mathbf{F}(x, y) = -y\mathbf{i} + x\mathbf{j}$. Describe **F**, and sketch a few vectors representing the vector field.

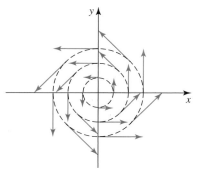

**FIGURE 4**
Some vectors representing the
vector field $\mathbf{F}(x, y) = -y\mathbf{i} + x\mathbf{j}$

**Solution**  Let $\mathbf{r} = x\mathbf{i} + y\mathbf{j}$ be the position vector of the point $(x, y)$. Then

$$\mathbf{F} \cdot \mathbf{r} = (-y\mathbf{i} + x\mathbf{j}) \cdot (x\mathbf{i} + y\mathbf{j})$$
$$= -yx + xy = 0$$

and this shows that $\mathbf{F}$ is orthogonal to the vector $\mathbf{r}$. This means that $\mathbf{F}(x, y)$ is tangent to the circle of radius $r = |\mathbf{r}|$ with center at the origin. Furthermore,

$$|\mathbf{F}(x, y)| = \sqrt{(-y)^2 + x^2} = \sqrt{x^2 + y^2} = r$$

gives the length of the position vector. Therefore, $\mathbf{F}$ associates with each point $(x, y)$ a vector of length equal to the distance between the origin and $(x, y)$ and direction that is perpendicular to the position vector of $(x, y)$. A few vectors representing this vector field are sketched in Figure 4. As in Example 1, this task is facilitated by first sketching a few concentric circles centered at the origin.  ∎

The "spin" vector field of Example 2 is used to describe phenomena as diverse as whirlpools and the motion of a ferris wheel. It is called a **velocity field.**

The definition of vector fields in three-dimensional space is similar to that in two-dimensional vector fields.

> **DEFINITION**  **Vector Field in Three-Dimensional Space**
> Let $T$ be a region in space. A **vector field in $T$** is a vector-valued function $\mathbf{F}$ that associates with each point $(x, y, z)$ in $T$ a three-dimensional vector
> $$\mathbf{F}(x, y, z) = P(x, y, z)\mathbf{i} + Q(x, y, z)\mathbf{j} + R(x, y, z)\mathbf{k}$$
> where $P$, $Q$, and $R$ are functions of three variables defined on $T$.

Important applications of vector fields in three-dimensional space occur in the form of *gravitational* and *electric fields,* as described in the following examples.

**EXAMPLE 3**  **Gravitational Field**  Suppose that an object $O$ of mass $M$ is located at the origin of a three-dimensional coordinate system. We can think of this object as inducing a **force field $\mathbf{F}$** in space. The effect of this **gravitational field** is to attract any object placed in the vicinity of $O$ toward it with a force that is governed by Newton's Law of Gravitation. To find an expression for $\mathbf{F}$, suppose that an object of mass $m$ is located at a point $(x, y, z)$ with position vector $\mathbf{r} = x\mathbf{i} + y\mathbf{j} + z\mathbf{k}$. Then, according to Newton's Law of Gravitation, the force of attraction of the object $O$ of mass $M$ on the object of mass $m$ has magnitude

$$\frac{GmM}{|\mathbf{r}|^2}$$

and direction given by the unit vector $-\mathbf{r}/|\mathbf{r}|$, where $G$ is the gravitational constant. Therefore, we can write

$$\mathbf{F}(x, y, z) = -\frac{GM}{|\mathbf{r}|^3}\mathbf{r}$$

$$= -\frac{GMx}{(x^2 + y^2 + z^2)^{3/2}}\mathbf{i} - \frac{GMy}{(x^2 + y^2 + z^2)^{3/2}}\mathbf{j} - \frac{GMz}{(x^2 + y^2 + z^2)^{3/2}}\mathbf{k}$$

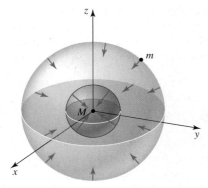

**FIGURE 5**
A gravitational force field

The force exerted by the gravitational field $\mathbf{F}$ on a particle of mass $m$ with position vector $\mathbf{r}$ is $m\mathbf{F}$. The vector field $\mathbf{F}$ is sketched in Figure 5.

Observe that all the arrows point toward the origin and that the lengths of the arrows decrease as one moves farther away from the origin. Physically, $\mathbf{F}(x, y, z)$ is the force per unit mass that would be exerted on a test mass placed at the point $P(x, y, z)$. ■

**EXAMPLE 4** **Electric Field** Suppose that a charge of $Q$ coulombs is located at the origin of a three-dimensional coordinate system. Then, according to **Coulomb's Law,** the electric force exerted by this charge on a charge of $q$ coulombs located at a point $(x, y, z)$ with position vector $\mathbf{r} = x\mathbf{i} + y\mathbf{j} + z\mathbf{k}$ has magnitude

$$\frac{k|q\|Q|}{}$$

(where $k$, the electrical constant, depends on the units used) and direction given by the unit vector $\mathbf{r}/|\mathbf{r}|$ for like charges $Q$ and $q$ (repulsion). Therefore, we can write the **electric field $\mathbf{E}$** that is induced by $Q$ as

$$\mathbf{E}(x, y, z) = \frac{kQ}{|\mathbf{r}|^3}\,\mathbf{r}$$

$$= \frac{kQx}{(x^2 + y^2 + z^2)^{3/2}}\,\mathbf{i} + \frac{kQy}{(x^2 + y^2 + z^2)^{3/2}}\,\mathbf{j} + \frac{kQz}{(x^2 + y^2 + z^2)^{3/2}}\,\mathbf{k}$$

The force exerted by the electric field $\mathbf{E}$ on a charge of $q$ coulombs, located at $(x, y, z)$, is $q\mathbf{E}$. Physically, $\mathbf{E}(x, y, z)$ is the force per unit charge that would be exerted on a test charge placed at the point $P(x, y, z)$. ■

## ■ Conservative Vector Fields

Recall from our work in Section 12.6 that if $f$ is a scalar function of three variables, then the *gradient* of $f$, written $\nabla f$ or grad $f$, is defined by

$$\nabla f(x, y, z) = f_x(x, y, z)\mathbf{i} + f_y(x, y, z)\mathbf{j} + f_z(x, y, z)\mathbf{k}$$

If $f$ is a function of two variables, then

$$\nabla f(x, y) = f_x(x, y)\mathbf{i} + f_y(x, y)\mathbf{j}$$

Since $\nabla f$ assigns to each point $(x, y, z)$ the vector $\nabla f(x, y, z)$, we see that $\nabla f$ is a vector field that associates with each point in its domain a vector giving the direction of greatest increase of $f$. (See Section 12.6.) The vector field $\nabla f$ is called the **gradient vector field** of $f$.

**EXAMPLE 5** Find the gradient vector field of $f(x, y, z) = x^2 + xy + y^2 z^3$.

**Solution** The required gradient vector field is given by

$$\nabla f(x, y, z) = \frac{\partial f}{\partial x}\mathbf{i} + \frac{\partial f}{\partial y}\mathbf{j} + \frac{\partial f}{\partial z}\mathbf{k}$$

$$= \frac{\partial}{\partial x}(x^2 + xy + y^2 z^3)\mathbf{i} + \frac{\partial}{\partial y}(x^2 + xy + y^2 z^3)\mathbf{j} + \frac{\partial}{\partial z}(x^2 + xy + y^2 z^3)\mathbf{k}$$

$$= (2x + y)\mathbf{i} + (x + 2yz^3)\mathbf{j} + 3y^2 z^2\mathbf{k} \qquad ■$$

Before we proceed further, it should be pointed out that vector fields in both two- and three-dimensional space can be plotted with the help of most computer algebra systems. The computer often scales the lengths of the vectors but still gives a good visual representation of the vector field. The vector fields of Examples 1 and 2 and two examples of vector fields in 3-space are shown in Figures 6a–6d.

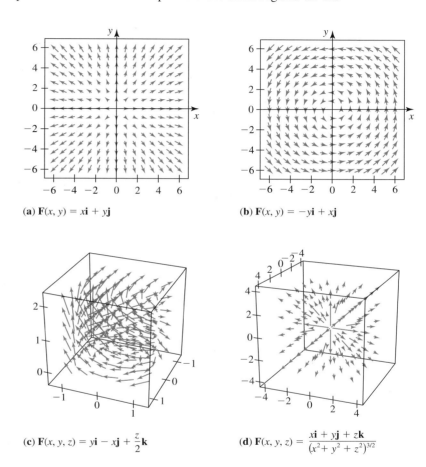

(a) $\mathbf{F}(x, y) = x\mathbf{i} + y\mathbf{j}$

(b) $\mathbf{F}(x, y) = -y\mathbf{i} + x\mathbf{j}$

(c) $\mathbf{F}(x, y, z) = y\mathbf{i} - x\mathbf{j} + \dfrac{z}{2}\mathbf{k}$

(d) $\mathbf{F}(x, y, z) = \dfrac{x\mathbf{i} + y\mathbf{j} + z\mathbf{k}}{(x^2 + y^2 + z^2)^{3/2}}$

**FIGURE 6**
Some computer-generated graphs of vector fields

Not all vector fields are gradients of scalar functions, but those that are play an important role in the physical sciences.

---

**DEFINITION** Conservative Vector Field

A vector field $\mathbf{F}$ in a region $R$ is **conservative** if there exists a scalar function $f$ defined in $R$ such that

$$\mathbf{F} = \nabla f$$

The function $f$ is called a **potential function** for $\mathbf{F}$.

---

The reason for using the words *conservative* and *potential* in this definition will be apparent when we discuss the law of conservation of energy in Section 14.4.

Vector fields of the form

$$F(x, y, z) = \frac{k}{|\mathbf{r}|^3} \mathbf{r}$$

are called **inverse square fields.** The gravitational and electric fields in Examples 3 and 4 are inverse square fields. The next example shows that these fields are conservative.

**EXAMPLE 6** Find the gradient vector field of the function

$$f(x, y, z) = -\frac{k}{\sqrt{x^2 + y^2 + z^2}}$$

and hence deduce that the inverse square field **F** is conservative.

**Solution** The gradient vector field of $f$ is given by

$$\nabla f(x, y, z) = f_x(x, y, z)\mathbf{i} + f_y(x, y, z)\mathbf{j} + f_z(x, y, z)\mathbf{k}$$

$$= \frac{kx}{(x^2 + y^2 + z^2)^{3/2}}\mathbf{i} + \frac{ky}{(x^2 + y^2 + z^2)^{3/2}}\mathbf{j} + \frac{kz}{(x^2 + y^2 + z^2)^{3/2}}\mathbf{k}$$

$$= \frac{k}{|\mathbf{r}|^3}\mathbf{r}$$

where $\mathbf{r} = x\mathbf{i} + y\mathbf{j} + z\mathbf{k}$. This shows that the inverse square field

$$F(x, y, z) = \frac{k}{|\mathbf{r}|^3}\mathbf{r}$$

is the gradient of the potential function $f$ and is therefore conservative.  ■

**Note** In Example 6 we were able to show that an inverse square field **F** is conservative because we were *given* a potential function $f$ such that $\mathbf{F} = \nabla f$. In Section 14.4 we will learn how to find the potential function $f$ for a conservative vector field. We will also learn how to determine whether a vector field is conservative without knowing its potential function.  ■

## 14.1  CONCEPT QUESTIONS

**1. a.** What is a vector field in the plane? In space? Give examples of each.
   **b.** Give three examples of vector fields with a physical interpretation.

**2. a.** What is a conservative vector field? Give an example.
   **b.** What is a potential function? Give an example.

## 14.1   EXERCISES

*In Exercises 1–6, match the vector field with one of the plots labeled* (a)–(f).

**(a)**

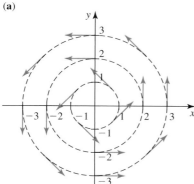

**(b)**

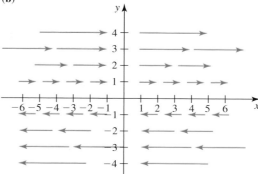

**(c)**

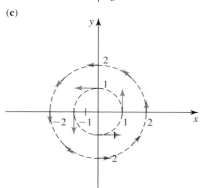

**(d)**

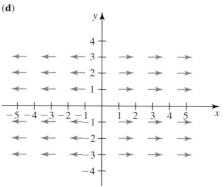

**(e)**

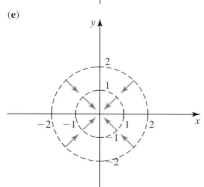

**(f)**

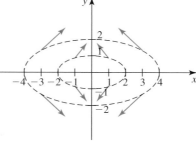

**1.** $\mathbf{F}(x, y) = y\mathbf{i}$

**2.** $\mathbf{F}(x, y) = \dfrac{x}{|x|}\mathbf{i}$

**3.** $\mathbf{F}(x, y) = -\dfrac{y}{x^2 + y^2}\mathbf{i} + \dfrac{x}{x^2 + y^2}\mathbf{j}$

**4.** $\mathbf{F}(x, y) = -\dfrac{y}{\sqrt{x^2 + y^2}}\mathbf{i} + \dfrac{x}{\sqrt{x^2 + y^2}}\mathbf{j}$

**5.** $\mathbf{F}(x, y) = -\dfrac{x}{\sqrt{x^2 + y^2}}\mathbf{i} - \dfrac{y}{\sqrt{x^2 + y^2}}\mathbf{j}$

**6.** $\mathbf{F}(x, y) = -\dfrac{1}{2}x\mathbf{i} + y\mathbf{j}$

*In Exercises 7–18, sketch several vectors associated with the vector field* **F**.

**7.** $\mathbf{F}(x, y) = 2\mathbf{i}$

**8.** $\mathbf{F}(x, y) = \mathbf{i} + \mathbf{j}$

**9.** $\mathbf{F}(x, y) = -x\mathbf{i} - y\mathbf{j}$

**10.** $\mathbf{F}(x, y) = y\mathbf{i} - x\mathbf{j}$

**11.** $\mathbf{F}(x, y) = x\mathbf{i} - 2y\mathbf{j}$     **12.** $\mathbf{F}(x, y) = x\mathbf{i} + 3y\mathbf{j}$

**13.** $\mathbf{F}(x, y) = \dfrac{x}{\sqrt{x^2 + y^2}}\mathbf{i} + \dfrac{y}{\sqrt{x^2 + y^2}}\mathbf{j}$

**14.** $\mathbf{F}(x, y) = \dfrac{y}{\sqrt{x^2 + y^2}}\mathbf{i} - \dfrac{x}{\sqrt{x^2 + y^2}}\mathbf{j}$

**15.** $\mathbf{F}(x, y, z) = c\mathbf{j}, \quad c$ a constant

**16.** $\mathbf{F}(x, y, z) = z\mathbf{k}$     **17.** $\mathbf{F}(x, y, z) = \mathbf{i} + \mathbf{j} + \mathbf{k}$

**18.** $\mathbf{F}(x, y, z) = x\mathbf{i} + y\mathbf{j} + z\mathbf{k}$

*In Exercises 19–22, match the vector field with one of the plots labeled* (a)–(d).

**19.** $\mathbf{F}(x, y, z) = \mathbf{i} + \mathbf{j} + 2\mathbf{k}$     **20.** $\mathbf{F}(x, y, z) = x\mathbf{i} + y\mathbf{j} + 2\mathbf{k}$

**21.** $\mathbf{F}(x, y, z) = -x\mathbf{i} - y\mathbf{j} - z\mathbf{k}$

**22.** $\mathbf{F}(x, y, z) = \dfrac{x}{\sqrt{x^2 + y^2 + z^2}}\mathbf{i} + \dfrac{y}{\sqrt{x^2 + y^2 + z^2}}\mathbf{j}$
$+ \dfrac{z}{\sqrt{x^2 + y^2 + z^2}}\mathbf{k}$

**(a)**                                    **(b)**

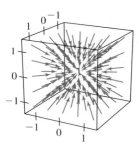

**(c)**                                    **(d)**

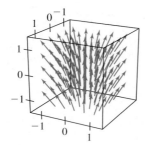

**cas** *In Exercises 23–26, use a computer algebra system to plot the vector field.*

**23.** $\mathbf{F}(x, y) = \dfrac{1}{10}(x + y)\mathbf{i} + \dfrac{1}{10}(x - y)\mathbf{j}$

**24.** $\mathbf{F}(x, y) = 2xy\mathbf{i} + 2x^2y\mathbf{j}$

**25.** $\mathbf{F}(x, y, z) = \dfrac{1}{5}(-y\mathbf{i} + x\mathbf{j} + z\mathbf{k})$

**26.** $\mathbf{F}(x, y, z) = -\dfrac{x\mathbf{i} + y\mathbf{j} + z\mathbf{k}}{\sqrt{x^2 + y^2 + z^2}}$

*In Exercises 27–32, find the gradient vector field of the scalar function f. (That is, find the conservative vector field* **F** *for the potential function f of* **F**.)

**27.** $f(x, y) = x^2y - y^3$     **28.** $f(x, y) = e^{-2x}\sin 3y$

**29.** $f(x, y, z) = xyz$     **30.** $f(x, y, z) = xy^2 - yz^3$

**31.** $f(x, y, z) = y\ln(x + z)$     **32.** $f(x, y, z) = \tan^{-1}(xyz)$

**33. Velocity of a Particle** A particle is moving in a velocity field

$$\mathbf{V}(x, y, z) = 2x\mathbf{i} + (x + 3y)\mathbf{j} + z^2\mathbf{k}$$

At time $t = 2$ the particle is located at the point $(1, 3, 2)$.
**a.** What is the velocity of the particle at $t = 2$?
**b.** What is the approximate location of the particle at $t = 2.01$?

**34. Velocity of Flow** The following figure shows a lateral section of a tube through which a liquid is flowing. The velocity of flow may vary from point to point, but it is independent of time.
**a.** Assuming that the flow is from right to left, sketch vectors emanating from the indicated points representing the speed and direction of fluid flow. Give a reason for your answer. (The answer is not unique.)

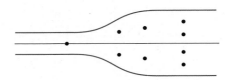

**b.** Explain why it is a bad idea to seek shelter in a tunnel when a tornado is approaching.

**35.** Show that the vector field $\mathbf{F}(x, y) = y\mathbf{i}$ is not a gradient vector field of a scalar function $f$.
**Hint:** If $\mathbf{F}$ is a gradient vector field of $f$, then $\partial f/\partial x = y$ and $\partial f/\partial y = 0$. Show that $f$ cannot exist.

**36.** Is $\mathbf{F}(x, y) = -y\mathbf{i} + x\mathbf{j}$ a gradient vector field of a scalar function $f$? Explain your answer.

*In Exercises 37–40, determine whether the statement is true or false. If it is true, explain why. If it is false, explain why or give an example that shows it is false.*

**37.** If $\mathbf{F}$ is a vector field in the plane, then $\mathbf{G} = c\mathbf{F}$ defined by $\mathbf{G}(x, y) = c\mathbf{F}(x, y)$, where $c$ is a constant, is also a vector field.

**38.** If $\mathbf{F}$ is a velocity field in space, then $|\mathbf{F}(x, y, z)|$ gives the speed of a particle at the point $(x, y, z)$, and $\mathbf{F}(x, y, z)/|\mathbf{F}(x, y, z)|$, where $|\mathbf{F}(x, y, z)| \neq 0$, is a unit vector giving its direction.

**39.** A constant vector field $\mathbf{F}(x, y, z) = a\mathbf{i} + b\mathbf{j} + c\mathbf{k}$ is a gradient vector field.

**40.** All the vectors of the vector field $\mathbf{F}(x, y) = x^2\mathbf{i} + y^2\mathbf{j}$ point outward in a radial direction from the origin.

## 14.2 Divergence and Curl

In this section we will look at two ways of measuring the rate of change of a vector field **F**: the **divergence** of **F** at a point $P$ and the **curl** of **F** at $P$. The divergence and curl of a vector field play a very important role in describing fluid flow, heat conduction, and electromagnetism.

### ■ Divergence

Suppose that **F** is a vector field in 2- or 3-space and $P$ is a point in its domain. For the purpose of this discussion, let's also suppose that the vector field **F** describes the flow of a fluid in 2- or 3-space. Then the divergence of **F** at $P$, written div $\mathbf{F}(P)$, measures the rate per unit area (or volume) at which the fluid departs or accumulates at $P$. Let's consider several examples.

### EXAMPLE 1

**a.** Figure 1a shows the vector field $\mathbf{F}(x, y) = x\mathbf{i} + y\mathbf{j}$ described in Example 1 of Section 14.1. Let $P$ be a point in the plane, and let $N$ be a neighborhood of $P$ with center $P$. Referring to Figure 1b, observe that an arrow entering $N$ along a streamline is matched by one that emerges from $N$ and has a greater length (because it is located farther away from the origin). This shows that more fluid leaves than enters a neighborhood of $P$. We will show in Example 2a that the vector field **F** is "divergent" at $P$; that is, the divergence of **F** at $P$ is positive.

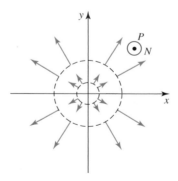 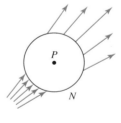

(**a**) The vector field $\mathbf{F}(x, y) = x\mathbf{i} + y\mathbf{i}$

(**b**) Flow through a neighborhood of $P$ (enlarged and not to scale)

**FIGURE 1**

**b.** Figure 2a shows the vector field $\mathbf{F}(x, y) = y\mathbf{i}$ for $x \geq 0$ and $y \geq 0$. Observe that the streamlines are parallel to the $x$-axis and that the lengths of the arrows on each horizontal line are constant. We can think of **F** as describing the flow of a river near one side of a riverbank. The velocity of flow is near zero close to the bank (the $x$-axis) and increases as we move away from it. You can see from Figure 2b that the amount of fluid flowing into the neighborhood $N$ of $P$ is matched by the same amount that exits $N$. Consequently, we expect the "divergence" at $P$ to be zero. We will show that this is the case in Example 2b.

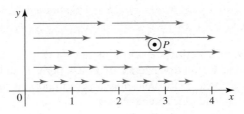

(a) The vector field $\mathbf{F}(x, y) = y\mathbf{i}$

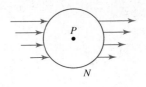

(b) Flow through a neighborhood of $P$ (enlarged and not to scale)

**FIGURE 2**

**c.** Figure 3a shows the vector field

$$\mathbf{F}(x, y) = \frac{1}{x + 1}\mathbf{i}$$

for $x \geq 0$ and $y \geq 0$. Observe that the streamlines are parallel to the $x$-axis and that the lengths of the arrows on each horizontal line get smaller as $x$ increases. From Figure 3b you can see that the "flow" into a neighborhood $N$ of $P$ is greater than the flow that emerges from $N$. In this case, more fluid enters the neighborhood than leaves it, and the "divergence" is negative. We will show that our intuition is correct in Example 2.

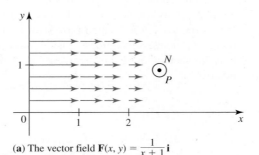

(a) The vector field $\mathbf{F}(x, y) = \frac{1}{x+1}\mathbf{i}$

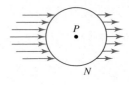

(b) Flow through a neighborhood of $P$ (enlarged and not to scale)

**FIGURE 3**

Up to now, we have looked at the notion of "divergence" intuitively. The divergence of a vector field can be defined as follows.

---

**DEFINITION** **Divergence of a Vector Field**

Let $\mathbf{F}(x, y, z) = P\mathbf{i} + Q\mathbf{j} + R\mathbf{k}$ be a vector field in space, where $P$, $Q$, and $R$ have first-order partial derivatives in some region $T$. The divergence of $\mathbf{F}$ is the scalar function defined by

$$\operatorname{div} \mathbf{F} = \frac{\partial P}{\partial x} + \frac{\partial Q}{\partial y} + \frac{\partial R}{\partial z} \tag{1}$$

---

(We will justify this definition of divergence in Section 14.8.) In two-dimensional space,

$$\mathbf{F}(x, y) = P\mathbf{i} + Q\mathbf{j} \qquad \text{and} \qquad \operatorname{div} \mathbf{F} = \frac{\partial P}{\partial x} + \frac{\partial Q}{\partial y}$$

As an aid to remembering Equation (1), let's introduce the vector differential operator $\nabla$ (read "del") defined by

$$\nabla = \frac{\partial}{\partial x}\mathbf{i} + \frac{\partial}{\partial y}\mathbf{j} + \frac{\partial}{\partial z}\mathbf{k}$$

If we let $\nabla$ operate on a scalar function $f(x, y, z)$, we obtain

$$\nabla f(x, y, z) = \left(\frac{\partial}{\partial x}\mathbf{i} + \frac{\partial}{\partial y}\mathbf{j} + \frac{\partial}{\partial z}\mathbf{k}\right)f(x, y, z)$$

$$= \frac{\partial}{\partial x}f(x, y, z)\mathbf{i} + \frac{\partial}{\partial y}f(x, y, z)\mathbf{j} + \frac{\partial}{\partial z}f(x, y, z)\mathbf{k}$$

$$= \frac{\partial f}{\partial x}(x, y, z)\mathbf{i} + \frac{\partial f}{\partial y}(x, y, z)\mathbf{j} + \frac{\partial f}{\partial z}(x, y, z)\mathbf{k}$$

which is the gradient of $f$. If we take the "dot product" of $\nabla$ with the vector field $\mathbf{F}(x, y, z) = P\mathbf{i} + Q\mathbf{j} + R\mathbf{k}$, we obtain

$$\nabla \cdot \mathbf{F} = \left(\frac{\partial}{\partial x}\mathbf{i} + \frac{\partial}{\partial y}\mathbf{j} + \frac{\partial}{\partial z}\mathbf{k}\right) \cdot (P\mathbf{i} + Q\mathbf{j} + R\mathbf{k})$$

$$= \frac{\partial}{\partial x}P + \frac{\partial}{\partial y}Q + \frac{\partial}{\partial z}R = \frac{\partial P}{\partial x} + \frac{\partial Q}{\partial y} + \frac{\partial R}{\partial z}$$

which is the divergence of the vector field $\mathbf{F}$. Thus, we can write the divergence of $\mathbf{F}$ symbolically as

$$\text{div } \mathbf{F} = \nabla \cdot \mathbf{F} \qquad (2)$$

Let's apply the definition of divergence to the vector fields that we discussed in Example 1.

**EXAMPLE 2** Find the divergence of (a) $\mathbf{F}(x, y) = x\mathbf{i} + y\mathbf{j}$, (b) $\mathbf{F}(x, y) = y\mathbf{i}$, and (c) $\mathbf{F}(x, y) = \dfrac{1}{x + 1}\mathbf{i}$. Reconcile your results with the intuitive observations that were made in Example 1.

**Solution**

a. $\text{div } \mathbf{F} = \dfrac{\partial}{\partial x}(x) + \dfrac{\partial}{\partial y}(y) = 1 + 1 = 2$. Here, div $\mathbf{F} > 0$, as expected.

b. Here, $\mathbf{F} = y\mathbf{i} + 0\mathbf{j}$, so $\text{div } \mathbf{F} = \dfrac{\partial}{\partial x}(y) + \dfrac{\partial}{\partial y}(0) = 0$. In this case, div $\mathbf{F} = 0$, as was observed in Example 1b.

c. With $\mathbf{F} = (x + 1)^{-1}\mathbf{i} + 0\mathbf{j}$ we find

$$\text{div } \mathbf{F} = \frac{\partial}{\partial x}(x + 1)^{-1} + \frac{\partial}{\partial y}(0) = -(x + 1)^{-2} = -\frac{1}{(x + 1)^2}$$

and div $\mathbf{F} < 0$, as we concluded intuitively in Example 1c. ■

We turn now to an example involving a vector field whose streamlines are not so easily visualized.

**EXAMPLE 3** Find the divergence of $\mathbf{F}(x, y, z) = xyz\mathbf{i} + x^2y^2z\mathbf{j} + xy^2\mathbf{k}$ at the point $(1, -1, 2)$.

**Solution**

$$\text{div } \mathbf{F} = \frac{\partial}{\partial x}(xyz) + \frac{\partial}{\partial y}(x^2y^2z) + \frac{\partial}{\partial z}(xy^2)$$

$$= yz + 2x^2yz$$

In particular, at the point $(1, -1, 2)$ we find

$$\text{div } \mathbf{F}(1, -1, 2) = (-1)(2) + 2(1)^2(-1)(2) = -6 \qquad\blacksquare$$

The divergence of the vector field $\mathbf{F}(x, y) = y\mathbf{i}$ of Examples 1b and 2b is zero. In general, if div $\mathbf{F} = 0$, then $\mathbf{F}$ is called **incompressible.** In electromagnetic theory a vector field $\mathbf{F}$ that satisfies $\nabla \cdot \mathbf{F} = 0$ is called **solenoidal.** For example, the electric field $\mathbf{E}$ in Example 4 is solenoidal. We will study the *divergence* of a vector field in greater detail in Section 14.8.

**EXAMPLE 4** Show that the divergence of the electric field $\mathbf{E}(x, y, z) = \dfrac{kQ}{|\mathbf{r}|^3}\mathbf{r}$, where $\mathbf{r} = x\mathbf{i} + y\mathbf{j} + z\mathbf{k}$, is zero.

**Solution** We first write

$$\mathbf{E}(x, y, z) = \frac{kQx}{(x^2 + y^2 + z^2)^{3/2}}\mathbf{i} + \frac{kQy}{(x^2 + y^2 + z^2)^{3/2}}\mathbf{j} + \frac{kQz}{(x^2 + y^2 + z^2)^{3/2}}\mathbf{k}$$

Then

$$\text{div } \mathbf{E} = kQ\left\{\frac{\partial}{\partial x}\left[\frac{x}{(x^2 + y^2 + z^2)^{3/2}}\right] + \frac{\partial}{\partial y}\left[\frac{y}{(x^2 + y^2 + z^2)^{3/2}}\right] + \frac{\partial}{\partial z}\left[\frac{z}{(x^2 + y^2 + z^2)^{3/2}}\right]\right\}$$

But

$$\frac{\partial}{\partial x}\left[\frac{x}{(x^2 + y^2 + z^2)^{3/2}}\right] = \frac{\partial}{\partial x}[x(x^2 + y^2 + z^2)^{-3/2}]$$

$$= (x^2 + y^2 + z^2)^{-3/2} + x \cdot \left(-\frac{3}{2}\right)(x^2 + y^2 + z^2)^{-5/2}(2x)$$

$$= (x^2 + y^2 + z^2)^{-5/2}[(x^2 + y^2 + z^2) - 3x^2]$$

$$= \frac{-2x^2 + y^2 + z^2}{(x^2 + y^2 + z^2)^{5/2}}$$

Similarly, we find

$$\frac{\partial}{\partial y}\left[\frac{y}{(x^2 + y^2 + z^2)^{3/2}}\right] = \frac{x^2 - 2y^2 + z^2}{(x^2 + y^2 + z^2)^{5/2}}$$

and

$$\frac{\partial}{\partial z}\left[\frac{z}{(x^2 + y^2 + z^2)^{3/2}}\right] = \frac{x^2 + y^2 - 2z^2}{(x^2 + y^2 + z^2)^{5/2}}$$

Therefore,

$$\text{div } \mathbf{E} = kQ\left[\frac{-2x^2 + y^2 + z^2}{(x^2 + y^2 + z^2)^{5/2}} + \frac{x^2 - 2y^2 + z^2}{(x^2 + y^2 + z^2)^{5/2}} + \frac{x^2 + y^2 - 2z^2}{(x^2 + y^2 + z^2)^{5/2}}\right] = 0 \qquad\blacksquare$$

Axis

**FIGURE 4**
A paddle wheel

## Curl

We now turn our attention to the other measure of the rate of change of a vector field **F**. Let **F** be a vector field in 3-space, and let $P$ be a point in its domain. Once again, let's think of the vector field as one that describes the flow of fluid. Suppose that a small paddle wheel, like the one shown in Figure 4, is immersed in the fluid at $P$. Then the curl of **F**, written curl **F**, is a measure of the tendency of the fluid to rotate the device about its vertical axis at $P$. Later, we will show that the paddle wheel will rotate most rapidly if its axis coincides with the direction of curl **F** at $P$ and that its maximum rate of rotation at $P$ is given by the length of curl **F** at $P$.

**EXAMPLE 5**

**a.** Consider the vector field $\mathbf{F}(x, y, z) = y\mathbf{i}$ for $x \geq 0$ similar to that of Example 1b. This field is shown in Figure 5a. Notice that the positive $z$-axis points vertically out of the page. Suppose that a paddle wheel is planted at a point $P$. Referring to Figure 5b, you can see that the arrows in the upper half of the circle with center at $P$ are longer than those in the lower half. This shows that the net clockwise flow of the fluid is greater than the net counterclockwise flow. This will cause the paddle to rotate in a clockwise direction, as we will show in Example 6.

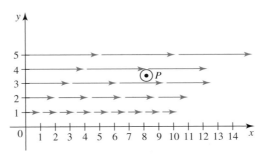

**(a)** The vector field $\mathbf{F}(x, y, z) = y\mathbf{i}$

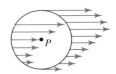

**(b)** Flow through a neighborhood of $P$ at which a paddle wheel is located (enlarged and not to scale)

**FIGURE 5**

**b.** Consider the vector field $\mathbf{F}(x, y, z) = -y\mathbf{i} + x\mathbf{j}$ shown in Figure 6a. Observe that it is similar to the spin vector field of Example 2 in Section 14.1. Again, the positive $z$-axis points vertically out of the page. If a paddle wheel is placed at the origin, it is easy to see that it will rotate in a counterclockwise direction. Next, suppose that the paddle wheel is planted at a point $P$ other than the origin. If you refer to Figure 6b, you can see that the circle with center at $P$ is divided into two arcs by the points of tangency of the two half-lines starting from the origin. Notice that the arc farther from the origin is longer than the one closer to the origin and that the flow on the larger arc is counterclockwise, whereas the flow on the shorter arc is clockwise. Furthermore, the arrows emanating from the longer arc are longer than those emanating from the shorter arc. This shows that the amount of fluid flowing in the counterclockwise direction is greater than that flowing in the clockwise direction. Therefore, the paddle wheel will rotate in a counterclockwise direction, as we will show in Example 6.

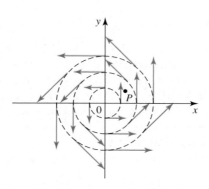

(**a**) The vector field $\mathbf{F}(x, y, z) = -y\mathbf{i} + x\mathbf{j}$

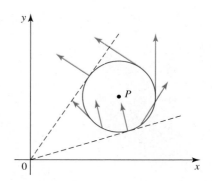

(**b**) Flow through a neighborhood of $P$ at which a paddle wheel is located (enlarged and not to scale)

**FIGURE 6**

**c.** Consider the vector field $\mathbf{F}(x, y, z) = x\mathbf{i} + y\mathbf{j}$ shown in Figure 7a. Note that it is similar to the vector field in Example 1 in Section 14.1. Suppose that a paddle wheel is placed at a point $P$. Then referring to Figure 7(b) and using an argument involving symmetry, you can convince yourself that the paddle wheel will not rotate. Again, we will show in Example 6 that this is true.

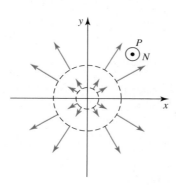

(**a**) The vector field $\mathbf{F}(x, y, z) = x\mathbf{i} + y\mathbf{j}$

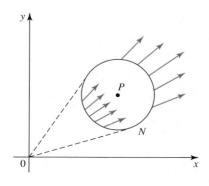

(**b**) Flow through a neighborhood of $P$ at which a paddle wheel is located (enlarged and not to scale)

**FIGURE 7**

The following definition provides us with an exact way to measure the curl of a vector field.

---

**DEFINITION** **Curl of a Vector Field**

Let $\mathbf{F}(x, y, z) = P\mathbf{i} + Q\mathbf{j} + R\mathbf{k}$ be a vector field in space, where $P$, $Q$, and $R$ have first-order partial derivatives in some region $T$. The curl of $\mathbf{F}$ is the vector field defined by

$$\text{curl } \mathbf{F} = \nabla \times \mathbf{F} = \left( \frac{\partial R}{\partial y} - \frac{\partial Q}{\partial z} \right)\mathbf{i} + \left( \frac{\partial P}{\partial z} - \frac{\partial R}{\partial x} \right)\mathbf{j} + \left( \frac{\partial Q}{\partial x} - \frac{\partial P}{\partial y} \right)\mathbf{k}$$

---

(We will justify this definition in Section 14.9.)

As in the case of the cross product of two vectors, we can remember the expression for the curl of a vector field by writing it (formally) in determinant form:

$$\text{curl } \mathbf{F} = \nabla \times \mathbf{F} = \begin{vmatrix} \mathbf{i} & \mathbf{j} & \mathbf{k} \\ \dfrac{\partial}{\partial x} & \dfrac{\partial}{\partial y} & \dfrac{\partial}{\partial z} \\ P & Q & R \end{vmatrix}$$

$$= \left( \frac{\partial R}{\partial y} - \frac{\partial Q}{\partial z} \right) \mathbf{i} + \left( \frac{\partial P}{\partial z} - \frac{\partial R}{\partial x} \right) \mathbf{j} + \left( \frac{\partial Q}{\partial x} - \frac{\partial P}{\partial y} \right) \mathbf{k}$$

Let's apply this definition to the vector fields that we discussed in Example 5.

**EXAMPLE 6** Find the curl of (a) $\mathbf{F}(x, y, z) = y\mathbf{i}$ for $x \geq 0$, (b) $\mathbf{F}(x, y, z) = -y\mathbf{i} + x\mathbf{j}$, and (c) $\mathbf{F}(x, y, z) = x\mathbf{i} + y\mathbf{j}$. Reconcile your results with the intuitive observations that were made in Example 5.

**Solution**

a.

$$\text{curl } \mathbf{F} = \nabla \times \mathbf{F} = \begin{vmatrix} \mathbf{i} & \mathbf{j} & \mathbf{k} \\ \dfrac{\partial}{\partial x} & \dfrac{\partial}{\partial y} & \dfrac{\partial}{\partial z} \\ y & 0 & 0 \end{vmatrix}$$

$$= \left[ \frac{\partial}{\partial y}(0) - \frac{\partial}{\partial z}(0) \right] \mathbf{i} - \left[ \frac{\partial}{\partial x}(0) - \frac{\partial}{\partial z}(y) \right] \mathbf{j} + \left[ \frac{\partial}{\partial x}(0) - \frac{\partial}{\partial y}(y) \right] \mathbf{k} = -\mathbf{k}$$

This shows that curl $\mathbf{F}$ is a (unit) vector that points vertically into the page. Applying the right-hand rule, we see that this result tells us that at any point in the vector field, the paddle wheel will rotate in a clockwise direction, as was observed earlier.

b.

$$\text{curl } \mathbf{F} = \nabla \times \mathbf{F} = \begin{vmatrix} \mathbf{i} & \mathbf{j} & \mathbf{k} \\ \dfrac{\partial}{\partial x} & \dfrac{\partial}{\partial y} & \dfrac{\partial}{\partial z} \\ -y & x & 0 \end{vmatrix}$$

$$= \left[ \frac{\partial}{\partial y}(0) - \frac{\partial}{\partial z}(x) \right] \mathbf{i} - \left[ \frac{\partial}{\partial x}(0) - \frac{\partial}{\partial z}(-y) \right] \mathbf{j} + \left[ \frac{\partial}{\partial x}(x) - \frac{\partial}{\partial y}(-y) \right] \mathbf{k}$$

$$= 2\mathbf{k}$$

The result tells us that curl $\mathbf{F}$ points vertically out of the page, so the paddle wheel will rotate in a counterclockwise direction when placed at any point in the vector field $\mathbf{F}$.

c.

$$\text{curl } \mathbf{F} = \nabla \times \mathbf{F} = \begin{vmatrix} \mathbf{i} & \mathbf{j} & \mathbf{k} \\ \dfrac{\partial}{\partial x} & \dfrac{\partial}{\partial y} & \dfrac{\partial}{\partial z} \\ x & y & 0 \end{vmatrix}$$

$$= \left[ \frac{\partial}{\partial y}(0) - \frac{\partial}{\partial z}(y) \right] \mathbf{i} - \left[ \frac{\partial}{\partial x}(0) - \frac{\partial}{\partial z}(x) \right] \mathbf{j} + \left[ \frac{\partial}{\partial x}(y) - \frac{\partial}{\partial y}(x) \right] \mathbf{k} = \mathbf{0}$$

This shows that a paddle wheel placed at any point in $\mathbf{F}$ will not rotate, as observed earlier.

The vector field **F** in Example 6c has the property that curl **F** = **0** at any point $P$. In general, if curl **F** = **0** at a point $P$, then **F** is said to be **irrotational** at $P$. This means that there are no vortices or whirlpools there.

### EXAMPLE 7

**a.** Find curl **F** if $\mathbf{F}(x, y, z) = xy\mathbf{i} + xz\mathbf{j} + xyz^2\mathbf{k}$.
**b.** What is curl $\mathbf{F}(-1, 2, 1)$?

**Solution**
**a.** By definition,

$$
\text{curl } \mathbf{F} = \nabla \times \mathbf{F} = \begin{vmatrix} \mathbf{i} & \mathbf{j} & \mathbf{k} \\ \dfrac{\partial}{\partial x} & \dfrac{\partial}{\partial y} & \dfrac{\partial}{\partial z} \\ xy & xz & xyz^2 \end{vmatrix}
$$

$$
= \left[ \frac{\partial}{\partial y}(xyz^2) - \frac{\partial}{\partial z}(xz) \right]\mathbf{i} - \left[ \frac{\partial}{\partial x}(xyz^2) - \frac{\partial}{\partial z}(xy) \right]\mathbf{j} + \left[ \frac{\partial}{\partial x}(xz) - \frac{\partial}{\partial y}(xy) \right]\mathbf{k}
$$

$$
= (xz^2 - x)\mathbf{i} - yz^2\mathbf{j} + (z - x)\mathbf{k}
$$

$$
= x(z^2 - 1)\mathbf{i} - yz^2\mathbf{j} + (z - x)\mathbf{k}
$$

**b.** curl $\mathbf{F}(-1, 2, 1) = (-1)(1^2 - 1)\mathbf{i} - (2)(1^2)\mathbf{j} + [1 - (-1)]\mathbf{k} = -2\mathbf{j} + 2\mathbf{k}$ ∎

The *div* and *curl* of vector fields enjoy some algebraic properties as illustrated in the following examples. Other properties can be found in the exercises at the end of this section.

### EXAMPLE 8 Let $f$ be a scalar function, and let **F** be a vector field. If $f$ and the components of **F** have first-order partial derivatives, show that

$$
\text{div}(f\mathbf{F}) = f \text{ div } \mathbf{F} + \mathbf{F} \cdot \nabla f
$$

**Solution** Let's write $\mathbf{F} = P\mathbf{i} + Q\mathbf{j} + R\mathbf{k}$, where $P$, $Q$, and $R$ are functions of $x$, $y$, and $z$. Then

$$
f\mathbf{F} = f(P\mathbf{i} + Q\mathbf{j} + R\mathbf{k}) = fP\mathbf{i} + fQ\mathbf{j} + fR\mathbf{k}
$$

so the left-hand side of the given equation reads

$$
\text{div}(f\mathbf{F}) = \nabla \cdot (f\mathbf{F}) = \left( \frac{\partial}{\partial x}\mathbf{i} + \frac{\partial}{\partial y}\mathbf{j} + \frac{\partial}{\partial z}\mathbf{k} \right) \cdot (fP\mathbf{i} + fQ\mathbf{j} + fR\mathbf{k})
$$

$$
= \frac{\partial}{\partial x}(fP) + \frac{\partial}{\partial y}(fQ) + \frac{\partial}{\partial z}(fR)
$$

$$
= f\frac{\partial P}{\partial x} + \frac{\partial f}{\partial x}P + f\frac{\partial Q}{\partial y} + \frac{\partial f}{\partial y}Q + f\frac{\partial R}{\partial z} + \frac{\partial f}{\partial z}R
$$

$$
= f\left( \frac{\partial P}{\partial x} + \frac{\partial Q}{\partial y} + \frac{\partial R}{\partial z} \right) + \left( \frac{\partial f}{\partial x}P + \frac{\partial f}{\partial y}Q + \frac{\partial f}{\partial z}R \right)
$$

$$
= f(\nabla \cdot \mathbf{F}) + (\nabla f) \cdot \mathbf{F}
$$

$$
= f \text{ div } \mathbf{F} + \mathbf{F} \cdot \nabla f
$$

which is equal to the right-hand side. ∎

**EXAMPLE 9**   Let $\mathbf{F} = P\mathbf{i} + Q\mathbf{j} + R\mathbf{k}$ be a vector field in space, and suppose that $P$, $Q$, and $R$ have continuous second-order partial derivatives. Show that

$$\text{div curl } \mathbf{F} = 0$$

**Solution**   Direct computation shows that

$$\text{div curl } \mathbf{F} = \nabla \cdot (\nabla \times \mathbf{F})$$

$$= \left( \frac{\partial}{\partial x}\mathbf{i} + \frac{\partial}{\partial y}\mathbf{j} + \frac{\partial}{\partial z}\mathbf{k} \right) \cdot \left[ \left( \frac{\partial R}{\partial y} - \frac{\partial Q}{\partial z} \right)\mathbf{i} + \left( \frac{\partial P}{\partial z} - \frac{\partial R}{\partial x} \right)\mathbf{j} + \left( \frac{\partial Q}{\partial x} - \frac{\partial P}{\partial y} \right)\mathbf{k} \right]$$

$$= \frac{\partial}{\partial x}\left( \frac{\partial R}{\partial y} - \frac{\partial Q}{\partial z} \right) + \frac{\partial}{\partial y}\left( \frac{\partial P}{\partial z} - \frac{\partial R}{\partial x} \right) + \frac{\partial}{\partial z}\left( \frac{\partial Q}{\partial x} - \frac{\partial P}{\partial y} \right)$$

$$= \frac{\partial^2 R}{\partial x\, \partial y} - \frac{\partial^2 Q}{\partial x\, \partial z} + \frac{\partial^2 P}{\partial y\, \partial z} - \frac{\partial^2 R}{\partial y\, \partial x} + \frac{\partial^2 Q}{\partial z\, \partial x} - \frac{\partial^2 P}{\partial z\, \partial y}$$

$$= 0$$

Here we have used the fact that the mixed derivatives are equal because, by assumption, they are continuous.   ■

## 14.2   CONCEPT QUESTIONS

**1. a.** Define the divergence of a vector field $\mathbf{F}$ and give a formula for finding it.

   **b.** Define the curl of a vector field $\mathbf{F}$, and give a formula for finding it.

   **c.** Suppose that $\mathbf{F}$ is the velocity vector field associated with the airflow around an airfoil. Give an interpretation of $\nabla \cdot \mathbf{F}$ and $\nabla \times \mathbf{F}$.

**2. a.** What is meant by a vector field $\mathbf{F}$ that is incompressible? Give a physical example of an (almost) incompressible field.

   **b.** Repeat part (a) for an irrotational vector field.

## 14.2   EXERCISES

*In Exercises 1–4, you are given the vector field $\mathbf{F}$ and a plot of the vector field in the xy-plane. (The z-component of $\mathbf{F}$ is 0.) (a) By studying the plot of $\mathbf{F}$, determine whether div $\mathbf{F}$ is positive, negative, or zero. Justify your answer. (b) Find div $\mathbf{F}$, and reconcile your result with your answer in part (a). (c) By studying the plot of $\mathbf{F}$, determine whether a paddle wheel planted at a point in the field will rotate clockwise, rotate counterclockwise, or not rotate at all. Justify your answer. (d) Find curl $\mathbf{F}$, and reconcile your result with your answer in part (c).*

**1.** $\mathbf{F}(x, y, z) = \dfrac{x}{|x|}\mathbf{i}, \quad x \neq 0$

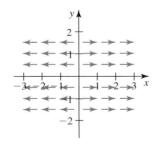

**2.** $\mathbf{F}(x, y, z) = -x\mathbf{j}$

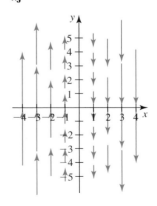

**3.** $F(x, y, z) = \dfrac{x}{\sqrt{x^2 + y^2}}\mathbf{i} + \dfrac{y}{\sqrt{x^2 + y^2}}\mathbf{j}$

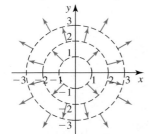

**4.** $F(x, y, z) = -\dfrac{y}{\sqrt{x^2 + y^2}}\mathbf{i} + \dfrac{x}{\sqrt{x^2 + y^2}}\mathbf{j}$

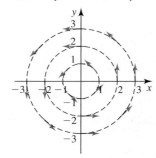

*In Exercises 5–12, find* (a) *the divergence and* (b) *the curl of the vector field* **F**.

**5.** $F(x, y, z) = yz\mathbf{i} + xz\mathbf{j} + xy\mathbf{k}$

**6.** $F(x, y, z) = x^2 y\mathbf{i} - xy^2\mathbf{j} + xyz\mathbf{k}$

**7.** $F(x, y, z) = x^2 y^3\mathbf{i} + xz^2\mathbf{k}$

**8.** $F(x, y, z) = yz^2\mathbf{i} + x^2 z\mathbf{j}$

**9.** $F(x, y, z) = \sin x\mathbf{i} + x\cos y\mathbf{j} + \sin z\mathbf{k}$

**10.** $F(x, y, z) = x\cos y\mathbf{i} + y\tan x\mathbf{j} + \sec z\mathbf{k}$

**11.** $F(x, y, z) = e^{-x}\cos y\mathbf{i} + e^{-x}\sin y\mathbf{j} + \ln z\mathbf{k}$

**12.** $F(x, y, z) = e^{xyz}\mathbf{i} + \cos(x + y)\mathbf{j} - \ln(x + z)\mathbf{k}$

*In Exercises 13–15, let* **F** *be a vector field, and let f be a scalar field. Determine whether each expression is meaningful. If so, state whether the expression represents a scalar field or a vector field.*

**13. a.** $\nabla \times f$        **b.** $\nabla \cdot f$
     **c.** $\nabla \times (\nabla f)$        **d.** grad **F**

**14. a.** $\mathrm{div}(\nabla f)$        **b.** $\mathrm{grad}(\nabla f)$
     **c.** $\nabla \times (\mathrm{grad}\, f)$        **d.** curl(curl **F**)

**15. a.** $\nabla \times (\nabla \times \mathbf{F})$        **b.** $\nabla \cdot (\nabla f)$
     **c.** $\nabla \cdot (\nabla \cdot \mathbf{F})$        **d.** $\nabla \times [\nabla \times (\nabla f)]$

**16.** Find div **F** if $\mathbf{F} = \mathrm{grad}\, f$, where $f(x, y, z) = 2xy^2 z^3$.

**17.** Show that the vector field $\mathbf{F}(x, y, z) = f(y, z)\mathbf{i} + g(x, z)\mathbf{j} + h(x, y)\mathbf{k}$, where $f$, $g$, and $h$ are differentiable, is incompressible.

**18.** Show that the vector field $\mathbf{F}(x, y, z) = f(x)\mathbf{i} + g(y)\mathbf{j} + h(z)\mathbf{k}$, where $f$, $g$ and $h$ are differentiable, is irrotational.

*In Exercises 19–26, prove the property for vector fields* **F** *and* **G** *and scalar fields f and g. Assume that the appropriate partial derivatives exist and are continuous.*

**19.** $\mathrm{div}(\mathbf{F} + \mathbf{G}) = \mathrm{div}\,\mathbf{F} + \mathrm{div}\,\mathbf{G}$

**20.** $\mathrm{curl}(\mathbf{F} + \mathbf{G}) = \mathrm{curl}\,\mathbf{F} + \mathrm{curl}\,\mathbf{G}$

**21.** $\mathrm{curl}(f\mathbf{F}) = f\,\mathrm{curl}\,\mathbf{F} + (\nabla f) \times \mathbf{F}$

**22.** $\mathrm{curl}(\nabla f) = \mathbf{0}$

**23.** $\mathrm{div}(\mathbf{F} \times \mathbf{G}) = \mathbf{G} \cdot \mathrm{curl}\,\mathbf{F} - \mathbf{F} \cdot \mathrm{curl}\,\mathbf{G}$

**24.** $\mathrm{div}(\nabla f \times \nabla g) = 0$

**25.** $\nabla \times (\nabla \times \mathbf{F}) = \nabla(\nabla \cdot \mathbf{F}) - \nabla^2 \mathbf{F}$, where
$$\nabla^2 \mathbf{F} = \left(\frac{\partial^2}{\partial x^2} + \frac{\partial^2}{\partial y^2} + \frac{\partial^2}{\partial z^2}\right)\mathbf{F}$$

**26.** $\nabla \times [\nabla f + (\nabla \times \mathbf{F})] = \nabla \times (\nabla \times \mathbf{F})$
  **Hint:** Use the results of Exercises 20 and 22.

**27.** Show that there is no vector field **F** in space such that
  $\mathrm{curl}\,\mathbf{F} = xy\mathbf{i} - yz\mathbf{j} + xy\mathbf{k}$.
  **Hint:** See Example 9.

**28.** Find the value of the constant $c$ such that the vector field
$$\mathbf{G}(x, y, z) = (2x + 3y + z^2)\mathbf{i} + (cy - z)\mathbf{j} + (x - y + 2z)\mathbf{k}$$
  is the curl of some vector field **F**.

**29.** Show that $\mathbf{F} = (\cos x)y\mathbf{i} + (\sin y)x\mathbf{j}$ is not a gradient vector field.
  **Hint:** See Exercise 22.

**30.** Let $f$ be a differentiable function, $\mathbf{r} = x\mathbf{i} + y\mathbf{j} + z\mathbf{k}$, and $r = |\mathbf{r}|$.
  **a.** Find $\mathrm{curl}[f(r)\mathbf{r}]$ by interpreting it geometrically.
  **b.** Verify your answer to part (a) analytically.

*In Exercises 31–34, let* $\mathbf{r} = x\mathbf{i} + y\mathbf{j} + z\mathbf{k}$ *and* $r = |\mathbf{r}|$.

**31.** Show that $\nabla r = \mathbf{r}/r$.

**32.** Show that $\nabla(1/r) = -\mathbf{r}/r^3$.

**33.** Show that $\nabla(\ln r) = \mathbf{r}/r^2$.

**34.** Show that $\nabla r^n = nr^{n-2}\mathbf{r}$.

*In Exercises 35–38, the differential operator* $\nabla^2$ *(called the* ***Laplacian****) is defined by* $\nabla^2 = \nabla \cdot \nabla = \dfrac{\partial^2}{\partial x^2} + \dfrac{\partial^2}{\partial y^2} + \dfrac{\partial^2}{\partial z^2}$.

*It acts on f to produce the function* $\nabla^2 f = \dfrac{\partial^2 f}{\partial x^2} + \dfrac{\partial^2 f}{\partial y^2} + \dfrac{\partial^2 f}{\partial z^2}$.

*Assume that f and g have second-order partial derivatives.*

**35.** Show that $\nabla \cdot (\nabla f) = \nabla^2 f$.

**36.** Show that $\nabla^2(fg) = f\nabla^2 g + g\nabla^2 f + 2\nabla f \cdot \nabla g$.

**37.** Show that $\nabla^2 r^3 = 12r$, where $r = |\mathbf{r}|$ and $\mathbf{r} = x\mathbf{i} + y\mathbf{j} + z\mathbf{k}$.

**38.** Show that $\nabla^2\left(\dfrac{1}{r}\right) = 0$, where $r = |\mathbf{r}|$ and $\mathbf{r} = x\mathbf{i} + y\mathbf{j} + z\mathbf{k}$.

**39. Angular Velocity of a Particle**  A particle located at the point $P$ is rotating about the $z$-axis on a circle of radius $R$ that lies in the plane $z = h$, as shown in the figure. Suppose that the angular speed of the particle is a constant $\omega$. Then this rotational motion can be described by the vector $\mathbf{w} = \omega\mathbf{k}$, which gives the **angular velocity** of $P$.

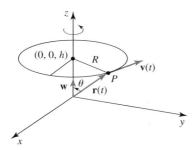

**a.** Show that the velocity $\mathbf{v}(t)$ of $P$ is given by $\mathbf{v} = \mathbf{w} \times \mathbf{r}$.
   **Hint:** The position of $P$ is $\mathbf{r}(t) = R \cos \omega t\mathbf{i} + R \sin \omega t\mathbf{j} + h\mathbf{k}$.
**b.** Show that $\mathbf{v} = -\omega y\mathbf{i} + \omega x\mathbf{j}$.
**c.** Show that curl $\mathbf{v} = 2\mathbf{w}$. This shows that the angular velocity of $P$ is one half the curl of its tangential velocity.

**40. Maxwell's Equations**  Maxwell's equations relating the electric field $\mathbf{E}$ and the magnetic field $\mathbf{H}$, where $c$ is the speed of light, are given by

$$\nabla \cdot \mathbf{E} = 0, \qquad \nabla \cdot \mathbf{H} = 0,$$

$$\nabla \times \mathbf{E} = -\frac{1}{c}\frac{\partial \mathbf{H}}{\partial t}, \qquad \nabla \times \mathbf{H} = \frac{1}{c}\frac{\partial \mathbf{E}}{\partial t}$$

Show that

**a.** $\nabla^2\mathbf{E} = \dfrac{1}{c^2}\dfrac{\partial^2\mathbf{E}}{\partial t^2}$

**b.** $\nabla^2\mathbf{H} = \dfrac{1}{c^2}\dfrac{\partial^2\mathbf{H}}{\partial t^2}$

   **Hint:** Use Exercise 25.

*In Exercises 41–48, determine whether the statement is true or false. If it is true, explain why. If it is false, explain why or give an example that shows it is false.*

**41.** If $\mathbf{F}$ is a nonconstant vector field, then div $\mathbf{F} \neq 0$.

**42.** If $\mathbf{F}(x, y) \neq \mathbf{0}$ and div $\mathbf{F} = 0$ for all $x$ and $y$, then the streamlines of $\mathbf{F}$ must be closed curves.

**43.** If the streamlines of a vector field $\mathbf{F}$ are straight lines, then div $\mathbf{F} = 0$.

**44.** If the streamlines of a vector field $\mathbf{F}$ are concentric circles, then curl $\mathbf{F} = \mathbf{0}$.

**45.** If the streamlines of a vector field $\mathbf{F}$ are straight lines, then curl $\mathbf{F} = \mathbf{0}$.

**46.** The curl of a "spin" field is never equal to $\mathbf{0}$.

**47.** There is no nonzero vector field $\mathbf{F}$ such that div $\mathbf{F} = 0$ and curl $\mathbf{F} = \mathbf{0}$, simultaneously.

**48.** If curl $\mathbf{F} = \mathbf{0}$, then $\mathbf{F}$ must be a constant vector field.

---

## 14.3  Line Integrals

### ■ Line Integrals

Once again recall that the mass of a thin, straight wire of length $(b - a)$ and linear mass density $f(x)$ is given by

$$m = \int_a^b f(x)\, dx$$

which has the same numerical value as the area under the graph of $f$ on $[a, b]$. (See Figure 1.)

Instead of being straight, suppose that the wire takes the shape of a plane curve $C$ described by the parametric equations $x = x(t)$ and $y = y(t)$, where $a \leq t \leq b$, or, equivalently, by the vector equation $\mathbf{r}(t) = x(t)\mathbf{i} + y(t)\mathbf{j}$ with parameter interval $[a, b]$. (See Figure 2a.) Furthermore, suppose that the linear mass density of the wire is given by a continuous function $f(x, y)$. Then one might conjecture that the mass of the curved wire should be numerically equal to the area of the region under the graph of $z = f(x, y)$ with $(x, y)$ lying on $C$. (See Figure 2b.)

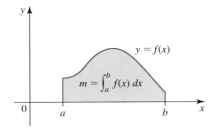

**FIGURE 1**
The mass of a wire of length $(b - a)$ and linear mass density $f(x)$ is $\int_a^b f(x)\, dx$.

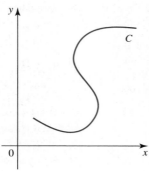

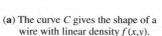

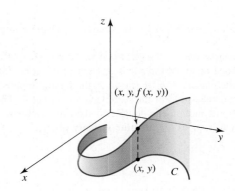

**FIGURE 2**

(a) The curve $C$ gives the shape of a wire with linear density $f(x,y)$.

(b) The region under the graph of $f$ along $C$

But how do we define this area, and how do we compute it? As we will now see, this area can be defined in terms of an integral called a *line integral,* even though the term "curve integral" would seem more appropriate.

Let $C$ be a smooth plane curve defined by the parametric equations

$$x = x(t), \qquad y = y(t), \qquad a \le t \le b$$

or, equivalently, by the vector equation $\mathbf{r}(t) = x(t)\mathbf{i} + y(t)\mathbf{j}$, and let $P$ be a regular partition of the parameter interval $[a, b]$ with partition points

$$a = t_0 < t_1 < t_2 < \cdots < t_n = b$$

If $x_k = x(t_k)$ and $y_k = y(t_k)$, then the points $P_k(x_k, y_k)$ divide $C$ into $n$ subarcs $\widehat{P_0P_1}$, $\widehat{P_1P_2}, \ldots, \widehat{P_{n-1}P_n}$ of lengths $\Delta s_1, \Delta s_2, \ldots, \Delta s_n$, respectively. (See Figure 3.) Next, we pick any evaluation point $t_k^*$ in the subinterval $[t_{k-1}, t_k]$. This point is mapped onto the point $P_k^*(x_k^*, y_k^*)$ lying in the subarc $\widehat{P_{k-1}P_k}$. If $f$ is any function of two variables with domain that contains the curve $C$, then we can evaluate $f$ at the point $(x_k^*, y_k^*)$, obtaining $f(x_k^*, y_k^*)$. If $f$ is positive, we can think of the product $f(x_k^*, y_k^*)\,\Delta s_k$ as representing the area of a curved panel with a curved base of length $\Delta s_k$ and constant height $f(x_k^*, y_k^*)$. (See Figure 4.) This panel is an approximation of the area under the curve $z = f(x, y)$ on the subarc $\widehat{P_{k-1}P_k}$. Therefore, the sum

$$\sum_{k=1}^{n} f(x_k^*, y_k^*)\,\Delta s_k \tag{1}$$

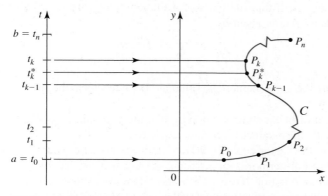

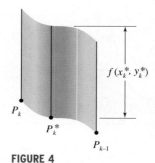

**FIGURE 3**

(a) A parameter interval

(b) The point $P_k(x_k, y_k)$ corresponds to the point $t_k$.

**FIGURE 4**
The product $f(x_k^*, y_k^*)\,\Delta s_k$ gives the area of a curved panel with a curved base of length $\Delta s_k$ and with constant height.

gives an approximation of the area under the curve $z = f(x, y)$ and along the curve $C$. If we let $n \to \infty$, then it seems reasonable to expect that this sum will approach the area under the curve $z = f(x, y)$ along the curve $C$. This observation suggests the following definition.

> **DEFINITION** Line Integral
>
> If $f$ is defined in a region containing a smooth curve $C$ with parametric representation $\mathbf{r}(t)$, where $a \le t \le b$, then the **line integral of $f$ along $C$** is
>
> $$\int_C f(x, y) \, ds = \lim_{n \to \infty} \sum_{k=1}^{n} f(x_k^*, y_k^*) \Delta s_k \tag{2}$$
>
> provided that the limit exists.

**Note** Observe that over a small piece of a curved wire represented by the segment $\overparen{P_{k-1}P_k}$, the linear density of the wire does not vary by much. Therefore, we may assume that the linear mass density of the wire in the segment $\overparen{P_{k-1}P_k}$ is approximately $f(x_k^*, y_k^*)$, so the mass of this segment is approximately $f(x_k^*, y_k^*) \Delta s_k$. (This is also the area of a typical panel.) Adding the masses of all the segments of the wire leads to the sum (1). Taking the limit as $n \to \infty$ in (1) then gives the mass of the wire. ■

In general, it can be shown that if $f$ is continuous, then the limit in Equation (2) always exists, and the line integral can be evaluated as an ordinary definite integral with respect to a single variable by using the following formula.

$$\int_C f(x, y) \, ds = \int_a^b f(x(t), y(t)) \sqrt{[x'(t)]^2 + [y'(t)]^2} \, dt \tag{3}$$

**Notes**
1. Equation (3) is easier to remember by observing that the element of arc length is given by $ds = |\mathbf{r}'(t)| \, dt = \sqrt{[x'(t)]^2 + [y'(t)]^2} \, dt$.
2. If $C$ is given by the interval $[a, b]$, then $C$ is just the line segment joining $(a, 0)$ to $(b, 0)$. So $C$ can be described by the parametric equations $x = t$ and $y = 0$, where $a \le t \le b$. In this case, Equation (3) becomes $\int_C f(x, y) \, ds = \int_a^b f(t, 0) \, dt = \int_a^b g(x) \, dx$, where $g(x) = f(x, 0)$. So the line integral reduces to an integral of a function defined on an interval $[a, b]$, as expected. ■

**EXAMPLE 1** Evaluate $\int_C (1 + xy) \, ds$, where $C$ is the quarter-circle described by $\mathbf{r}(t) = \cos t \mathbf{i} + \sin t \mathbf{j}$, $0 \le t \le \frac{\pi}{2}$, as shown in Figure 5.

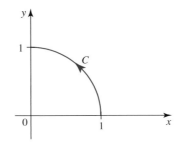

**FIGURE 5**
The curve $C$ is described by $\mathbf{r}(t) = \cos t \mathbf{i} + \sin t \mathbf{j}$, where $0 \le t \le \frac{\pi}{2}$.

**Solution**    Here, $x(t) = \cos t$ and $y(t) = \sin t$, so $x'(t) = -\sin t$ and $y'(t) = \cos t$. Therefore, using Equation (3), we obtain

$$\int_C (1 + xy)\, ds = \int_0^{\pi/2} (1 + \cos t \sin t)\sqrt{[x'(t)]^2 + [y'(t)]^2}\, dt$$

$$= \int_0^{\pi/2} (1 + \cos t \sin t)\sqrt{(-\sin t)^2 + (\cos t)^2}\, dt$$

$$= \int_0^{\pi/2} (1 + \cos t \sin t)\, dt = \left[t + \frac{1}{2}\sin^2 t\right]_0^{\pi/2}$$

$$= \frac{\pi}{2} + \frac{1}{2} = \frac{1}{2}(\pi + 1)$$

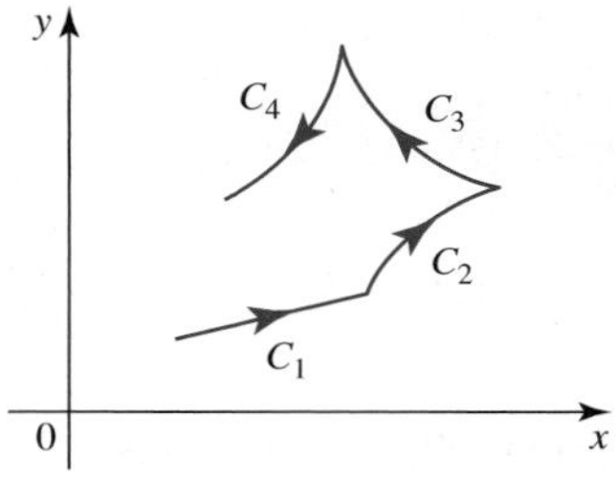

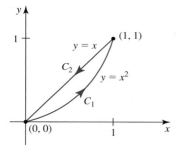

**FIGURE 6**
A piecewise-smooth curve composed of four smooth curves ($n = 4$)

A curve is **piecewise-smooth** if it is made up of a finite number of smooth curves $C_1, C_2, \ldots, C_n$ connected at consecutive endpoints as shown in Figure 6. If $f$ is continuous in a region containing $C$, then it can be shown that

$$\int_C f(x, y)\, ds = \int_{C_1} f(x, y)\, ds + \int_{C_2} f(x, y)\, ds + \cdots + \int_{C_n} f(x, y)\, ds$$

**FIGURE 7**
$C$ is composed of two smooth curves $C_1$ and $C_2$.

**EXAMPLE 2**    Evaluate $\int_C 2x\, ds$, where $C$ consists of the arc $C_1$ of the parabola $y = x^2$ from $(0, 0)$ to $(1, 1)$ followed by the line segment $C_2$ from $(1, 1)$ to $(0, 0)$.

**Solution**    The curve $C$ is shown in Figure 7. $C_1$ can be parametrized by taking $x = t$, where $t$ is a parameter. Thus,

$$C_1: \quad x(t) = t, \qquad y(t) = t^2, \qquad 0 \le t \le 1$$

Therefore,

$$\int_{C_1} 2x\, ds = \int_0^1 2x\sqrt{[x'(t)]^2 + [y'(t)]^2}\, dt$$

$$= 2\int_0^1 t\sqrt{1 + 4t^2}\, dt$$

$$= \left[2\left(\frac{1}{8}\right)\left(\frac{2}{3}\right)(1 + 4t^2)^{3/2}\right]_0^1 = \frac{5\sqrt{5} - 1}{6}$$

$C_2$ can be parametrized by taking $x = 1 - t$. Thus,

$$C_2: \quad x(t) = 1 - t, \qquad y(t) = 1 - t, \qquad 0 \le t \le 1$$

Therefore,

$$\int_{C_2} 2x\, ds = \int_0^1 2x\sqrt{[x'(t)]^2 + [y'(t)]^2}\, dt$$

$$= 2\int_0^1 (1 - t)\sqrt{1 + 1}\, dt$$

$$= \left[2\sqrt{2}\left(t - \frac{1}{2}t^2\right)\right]_0^1 = \sqrt{2}$$

Putting these results together, we have

$$\int_C 2x \, ds = \int_{C_1} 2x \, ds + \int_{C_2} 2x \, ds = \frac{5\sqrt{5} - 1}{6} + \sqrt{2}$$

As we saw earlier, the mass of a thin wire represented by $C$ that has linear mass density $\rho(x, y)$ is given by

$$m = \int_C \rho(x, y) \, ds$$

The **center of mass** of the wire is located at the point $(\bar{x}, \bar{y})$, where

$$\bar{x} = \frac{1}{m} \int_C x\rho(x, y) \, ds \qquad \bar{y} = \frac{1}{m} \int_C y\rho(x, y) \, ds \tag{4}$$

**EXAMPLE 3**  **The Mass and Center of Mass of a Wire**  A thin wire has the shape of a semicircle of radius $a$. The linear mass density of the wire is proportional to the distance from the diameter that joins the two endpoints of the wire. Find the mass of the wire and the location of its center of mass.

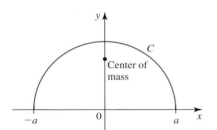

**FIGURE 8**
The curve $C$ has parametric equations
$x = a \cos t$ and $y = a \sin t$, where
$0 \le t \le \pi$.

**Solution**  If the wire is placed on a coordinate system as shown in Figure 8, then it coincides with the curve $C$ described by the parametric equations $x = a \cos t$ and $y = a \sin t$, where $0 \le t \le \pi$. Its linear mass density is given by $\rho(x, y) = ky$, where $k$ is a positive constant. Since $x'(t) = -a \sin t$ and $y'(t) = a \cos t$, we see that the mass of the wire is

$$m = \int_C \rho(x, y) \, ds = \int_C ky \, ds = \int_0^\pi ky\sqrt{[x'(t)]^2 + [y'(t)]^2} \, dt$$

$$= \int_0^\pi ka \sin t \sqrt{(-a \sin t)^2 + (a \cos t)^2} \, dt$$

$$= ka^2 \int_0^\pi \sin t \, dt = \left[-ka^2 \cos t\right]_0^\pi = 2ka^2$$

Next, we note that by symmetry, $\bar{x} = 0$. Using Equation (4), we obtain

$$\bar{y} = \frac{1}{m} \int_C y\rho(x, y) \, ds = \frac{1}{2ka^2} \int_0^\pi ky^2 \, ds = \frac{1}{2a^2} \int_0^\pi a(a \sin t)^2 \, dt$$

$$= \frac{a}{2} \int_0^\pi \sin^2 t \, dt = \frac{a}{4} \int_0^\pi (1 - \cos 2t) \, dt$$

$$= \frac{a}{4}\left[t - \frac{1}{2} \sin 2t\right]_0^\pi = \frac{1}{4} \pi a$$

Therefore, the center of mass of the curve is located at $\left(0, \frac{\pi a}{4}\right)$. (See Figure 8.)

## Line Integrals with Respect to Coordinate Variables

The line integrals that we have dealt with up to now are taken *with respect to arc length*. Two other line integrals are obtained by replacing $\Delta s_k$ in Equation (2) by $\Delta x_k = x(t_k) - x(t_{k-1})$ and $\Delta y_k = y(t_k) - y(t_{k-1})$. In the first instance we have the **line**

**integral of $f$ along $C$ with respect to $x$,**

$$\int_C f(x, y)\, dx = \lim_{n \to \infty} \sum_{k=1}^{n} f(x_k^*, y_k^*)\, \Delta x_k$$

and in the second instance we have the **line integral of $f$ along $C$ with respect to $y$,**

$$\int_C f(x, y)\, dy = \lim_{n \to \infty} \sum_{k=1}^{n} f(x_k^*, y_k^*)\, \Delta y_k$$

Line integrals with respect to both coordinate variables can also be evaluated as ordinary definite integrals with respect to a single variable. In fact, since $x = x(t)$ and $y = y(t)$, we see that $dx = x'(t)\, dt$ and $dy = y'(t)\, dt$. This leads to the following formulas:

$$\int_C f(x, y)\, dx = \int_a^b f(x(t), y(t)) x'(t)\, dt \tag{5a}$$

$$\int_C f(x, y)\, dy = \int_a^b f(x(t), y(t)) y'(t)\, dt \tag{5b}$$

Thus, if $P$ and $Q$ are continuous functions of $x$ and $y$, then

$$\int_C P(x, y)\, dx + Q(x, y)\, dy = \int_C P(x, y)\, dx + \int_C Q(x, y)\, dy$$

can be evaluated as an ordinary integral of a single variable using the formula

$$\int_C P(x, y)\, dx + Q(x, y)\, dy = \int_a^b [P(x(t), y(t)) x'(t) + Q(x(t), y(t)) y'(t)]\, dt \tag{6}$$

**EXAMPLE 4**   Evaluate $\int_C y\, dx + x^2\, dy$, where (a) $C$ is the line segment $C_1$ from $(1, -1)$ to $(4, 2)$, (b) $C$ is the arc $C_2$ of the parabola $x = y^2$ from $(1, -1)$ to $(4, 2)$, and (c) $C$ is the arc $C_3$ of the parabola $x = y^2$ from $(4, 2)$ to $(1, -1)$. (See Figure 9.)

**Solution**
**a.** $C_1$ can be described by the parametric equations

$$x = 1 + 3t, \qquad y = -1 + 3t, \qquad 0 \le t \le 1$$

(See Section 9.5.) We have $dx = 3\, dt$ and $dy = 3\, dt$, so Equation (6) gives

$$\int_{C_1} y\, dx + x^2\, dy = \int_0^1 (-1 + 3t)(3\, dt) + (1 + 3t)^2 (3\, dt)$$

$$= 27 \int_0^1 (t^2 + t)\, dt = 27 \left[ \frac{1}{3} t^3 + \frac{1}{2} t^2 \right]_0^1 = \frac{45}{2}$$

**b.** A parametric representation of $C_2$ is obtained by letting $y = t$. Thus,

$$C_2: \quad x = t^2, \qquad y = t, \qquad -1 \le t \le 2$$

Then $dx = 2t\, dt$ and $dy = dt$, so Equation (6) gives

$$\int_{C_2} y\, dx + x^2\, dy = \int_{-1}^2 t(2t\, dt) + (t^2)^2\, dt$$

$$= \int_{-1}^2 (2t^2 + t^4)\, dt = \left[ \frac{2}{3} t^3 + \frac{1}{5} t^5 \right]_{-1}^2 = \frac{63}{5}$$

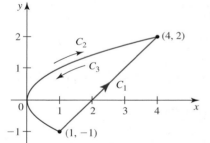

**FIGURE 9**
The curves $C_1$, $C_2$, and $C_3$

**c.** $C_3$ can be parametrized by taking $y = -t$. Thus,

$$C_3: \quad x = t^2, \qquad y = -t, \qquad -2 \leq t \leq 1$$

Then $dx = 2t\, dt$ and $dy = -dt$, so Equation (6) gives

$$\int_{C_3} y\, dx + x^2\, dy = \int_{-2}^{1} (-t)(2t\, dt) + (t^2)^2\, (-dt)$$

$$= -1 \int_{-2}^{1} (2t^2 + t^4)\, dt = -\left[ \frac{2}{3} t^3 + \frac{1}{5} t^5 \right]_{-2}^{1} = -\frac{63}{5} \quad \blacksquare$$

Example 4 sheds some light on the nature of line integrals. First of all, the results of parts (a) and (b) suggest that the value of a line integral depends not only on the endpoints, but also on the curve joining these points. Second, the results of parts (b) and (c) seem to suggest that reversing the direction in which a curve is traced changes the sign of the value of the line integral.

This latter observation turns out to be true in the general case. For example, suppose that the **orientation** of the curve $C$ (the direction in which it is traced as $t$ increases) is reversed. Let $-C$ denote precisely the curve $C$ with its orientation reversed (so that the curve is traced from $B$ to $A$ instead of from $A$ to $B$ as shown in Figure 10). Then

$$\int_{-C} P\, dx + Q\, dy = -\int_{C} P\, dx + Q\, dy$$

In contrast, note that the value of a line integral taken with respect to arc length does *not* change sign when $C$ is reversed. These results follow because the terms $x'(t)$ and $y'(t)$ change sign but $ds$ does not when the orientation of $C$ is reversed.

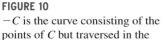

**FIGURE 10**
$-C$ is the curve consisting of the points of $C$ but traversed in the opposite direction.

### ■ Line Integrals in Space

The line integrals in two-dimensional space that we have just considered can be extended to line integrals in three-dimensional space. Suppose that $C$ is a smooth space curve described by the parametric equations

$$x = x(t), \qquad y = y(t), \qquad z = z(t), \qquad a \leq t \leq b$$

or, equivalently, by the vector equation $\mathbf{r}(t) = x(t)\mathbf{i} + y(t)\mathbf{j} + z(t)\mathbf{k}$, and let $f$ be a function of three variables that is defined and continuous on some region containing $C$. We define the **line integral of $f$ along $C$** (with respect to arc length) by

$$\int_{C} f(x, y, z)\, ds = \lim_{n \to \infty} \sum_{k=1}^{n} f(x_k^*, y_k^*, z_k^*)\, \Delta s_k$$

This integral can be evaluated as an ordinary integral by using the following formula, which is the analog of Equation (3) for the three-dimensional case.

$$\int_{C} f(x, y, z)\, ds = \int_{a}^{b} f(x(t), y(t), z(t)) \sqrt{\left(\frac{dx}{dt}\right)^2 + \left(\frac{dy}{dt}\right)^2 + \left(\frac{dz}{dt}\right)^2}\, dt \qquad \text{(7)}$$

If we make use of vector notation, Equation (7) can be written in the equivalent form

$$\int_{C} f(x, y, z)\, ds = \int_{a}^{b} f(\mathbf{r}(t)) |\mathbf{r}'(t)|\, dt$$

**EXAMPLE 5**    Evaluate $\int_C kz\, ds$, where $k$ is a constant and $C$ is the circular helix with parametric equations $x = \cos t$, $y = \sin t$, and $z = t$, where $0 \le t \le 2\pi$.

**Solution**    With $x'(t) = -\sin t$, $y'(t) = \cos t$, and $z'(t) = 1$, Equation (7) gives

$$\int_C kz\, ds = \int_0^{2\pi} kt \sqrt{[x'(t)]^2 + [y'(t)]^2 + [z'(t)]^2}\, dt$$

$$= \int_0^{2\pi} kt \sqrt{\sin^2 t + \cos^2 t + 1}\, dt$$

$$= \sqrt{2}k \int_0^{2\pi} t\, dt = \sqrt{2}k \left[ \frac{1}{2} t^2 \right]_0^{2\pi} = 2\sqrt{2}k\pi^2 \qquad \blacksquare$$

**Note**    In Example 5, suppose that $C$ represents a thin wire whose linear mass density is directly proportional to its height. Then our calculations tell us that its mass is $2\sqrt{2}k\pi^2$ units. $\qquad \blacksquare$

Line integrals along a curve $C$ in space with respect to $x$, $y$, and $z$ are defined in much the same way as line integrals along a curve in two-dimensional space. For example, the **line integral of $f$ along $C$ with respect to $x$** is given by

$$\int_C f(x, y, z)\, dx = \lim_{n \to \infty} \sum_{k=1}^{n} f(x_k^*, y_k^*, z_k^*)\, \Delta x_k$$

so

$$\int_C f(x, y, z)\, dx = \int_a^b f(x(t), y(t), z(t))x'(t)\, dt \qquad \textbf{(8)}$$

If the line integrals with respect to $x$, $y$, and $z$ occur together, we have

$$\int_C P(x, y, z)\, dx + Q(x, y, z)\, dy + R(x, y, z)\, dz$$

$$= \int_a^b \left[ P(x(t), y(t), z(t)) \frac{dx}{dt} + Q(x(t), y(t), z(t)) \frac{dy}{dt} + R(x(t), y(t), z(t)) \frac{dz}{dt} \right] dt \qquad \textbf{(9)}$$

**EXAMPLE 6**    Evaluate $\int_C y\, dx + z\, dy + x\, dz$, where $C$ consists of part of the twisted cubic $C_1$ with parametric equations $x = t$, $y = t^2$, and $z = t^3$, where $0 \le t \le 1$, followed by the line segment $C_2$ from $(1, 1, 1)$ to $(0, 1, 0)$.

**Solution**    The curve $C$ is shown in Figure 11. Integrating along $C_1$, we have $dx = dt$, $dy = 2t\, dt$, and $dz = 3t^2\, dt$. Therefore,

$$\int_{C_1} y\, dx + z\, dy + x\, dz = \int_0^1 t^2\, dt + t^3(2t\, dt) + t(3t^2)\, dt$$

$$= \int_0^1 (t^2 + 3t^3 + 2t^4)\, dt$$

$$= \left[ \frac{1}{3} t^3 + \frac{3}{4} t^4 + \frac{2}{5} t^5 \right]_0^1 = \frac{89}{60}$$

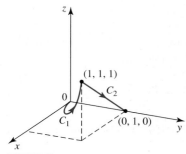

**FIGURE 11**
The curve $C$ is composed of $C_1$ and $C_2$ traversed in the directions shown.

Next, we write the parametric equations of the line segment from $(1, 1, 1)$ to $(0, 1, 0)$.

On $C_2$:    $x = 1 - t$,      $y = 1$,      $z = 1 - t$,      $0 \le t \le 1$

Then $dx = -dt$, $dy = 0$, and $dz = -dt$. Therefore,

$$\int_{C_2} y\,dx + z\,dy + x\,dz = \int_0^1 1(-dt) + (1-t)(0) + (1-t)(-dt)$$

$$= \int_0^1 (t-2)\,dt = \left[\frac{1}{2}t^2 - 2t\right]_0^1 = -\frac{3}{2}$$

Finally, putting these results together, we have

$$\int_C y\,dx + z\,dy + x\,dz = \frac{89}{60} - \frac{3}{2} = -\frac{1}{60}$$ ∎

## ■ Line Integrals of Vector Fields

Up to now, we have considered line integrals involving a scalar function $f$. We now turn our attention to the study of line integrals of vector fields. Suppose that we want to find the work done by a continuous force field $\mathbf{F}$ in moving a particle from a point $A$ to a point $B$ along a smooth curve $C$ in space. Let $C$ be represented parametrically by

$$x = x(t), \qquad y = y(t), \qquad z = z(t), \qquad a \le t \le b$$

or, equivalently, by the vector equation $\mathbf{r}(t) = x(t)\mathbf{i} + y(t)\mathbf{j} + z(t)\mathbf{k}$ with parameter interval $[a, b]$. Take a regular partition $P$ of the parameter interval $[a, b]$ with partition points

$$a = t_0 < t_1 < t_2 < \cdots < t_n = b$$

If $x_k = x(t_k)$, $y_k = y(t_k)$, and $z_k = z(t_k)$, then the points $P_k(x_k, y_k, z_k)$ divide $C$ into $n$ subarcs $\widehat{P_0P_1}$, $\widehat{P_1P_2}, \ldots, \widehat{P_{n-1}P_n}$ of lengths $\Delta s_1$, $\Delta s_2, \ldots, \Delta s_n$, respectively. (See Figure 12.) Furthermore, because $\mathbf{r}$ is smooth, the unit tangent vector $\mathbf{T}(t)$ at any point on the subarc $\widehat{P_{k-1}P_k}$ will not exhibit an appreciable change in direction and may be approximated by $\mathbf{T}(t_k^*)$. Also, because $\mathbf{F}$ is continuous, the force $\mathbf{F}(x(t), y(t), z(t))$ for $t_{k-1} \le t \le t_k$ is approximated by $\mathbf{F}(x_k^*, y_k^*, z_k^*)$. Therefore, we can approximate the work done by $\mathbf{F}$ in moving the particle along the curve from $P_{k-1}$ to $P_k$ by the work done by the component of the constant force $\mathbf{F}(x_k^*, y_k^*, z_k^*)$ in the direction of the line segment (approximated by $\mathbf{T}(t_k^*)$) from $P_{k-1}$ to $P_k$, that is, by

$$\Delta W_k = \mathbf{F}(x_k^*, y_k^*, z_k^*) \cdot \mathbf{T}(t_k^*)\,\Delta s_k \qquad \text{Constant force in the direction of } \mathbf{T}(x_k^*) \text{ times displacement}$$

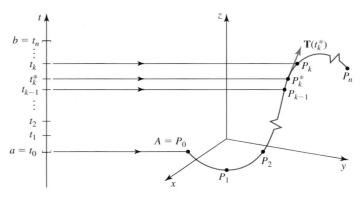

**(a)** A parameter interval

**(b)** The partition $P$ of $[a, b]$ breaks the curve $C$ into $n$ subarcs.

**FIGURE 12**

Here, we have used the fact that the length of the line segment from $P_{k-1}$ to $P_k$ is approximately $\Delta s_k$. So the total work done by $\mathbf{F}$ in moving the particle from $A$ to $B$ is

$$W \approx \sum_{k=1}^{n} \mathbf{F}(x_k^*, y_k^*, z_k^*) \cdot \mathbf{T}(t_k^*)\, \Delta s_k$$

This approximation suggests that we define the **work** $W$ done by the force field $\mathbf{F}$ as

$$W = \lim_{n \to \infty} \sum_{k=1}^{n} \mathbf{F}(x_k^*, y_k^*, z_k^*) \cdot \mathbf{T}(t_k^*)\, \Delta s_k = \int_C \mathbf{F} \cdot \mathbf{T}\, ds \tag{10}$$

Since $\mathbf{T}(t) = \mathbf{r}'(t)/|\mathbf{r}'(t)|$, Equation (10) can also be written in the form

$$W = \int_a^b \left[ \mathbf{F}(\mathbf{r}(t)) \cdot \frac{\mathbf{r}'(t)}{|\mathbf{r}'(t)|} \right] |\mathbf{r}'(t)|\, dt$$

$$= \int_a^b \mathbf{F}(\mathbf{r}(t)) \cdot \mathbf{r}'(t)\, dt$$

The last integral is usually written in the form $\int_C \mathbf{F} \cdot d\mathbf{r}$. In words, it says that the work done by a force is given by the line integral of the tangential component of the force with respect to arc length. Although this integral was defined in the context of work done by a force, integrals of this type occur frequently in many other areas of physics and engineering.

---

**DEFINITION** Line Integral of Vector Fields

Let $\mathbf{F}$ be a continuous vector field defined in a region that contains a smooth curve $C$ described by a vector function $\mathbf{r}(t)$, $a \le t \le b$. Then the **line integral of $\mathbf{F}$ along $C$** is

$$\int_C \mathbf{F} \cdot d\mathbf{r} = \int_C \mathbf{F} \cdot \mathbf{T}\, ds = \int_a^b \mathbf{F}(\mathbf{r}(t)) \cdot \mathbf{r}'(t)\, dt \tag{11}$$

---

**Note** We remind you that $d\mathbf{r}$ is an abbreviation for $\mathbf{r}'(t)\, dt$ and that $\mathbf{F}(\mathbf{r}(t))$ is an abbreviation for $\mathbf{F}(x(t), y(t), z(t))$. ∎

**EXAMPLE 7** Find the work done by the force field $\mathbf{F}(x, y, z) = -y\mathbf{i} + x\mathbf{j} + z\mathbf{k}$ in moving a particle along the helix $C$ described by the parametric equations $x = \cos t$, $y = \sin t$, and $z = t$ from $(1, 0, 0)$ to $\left(0, 1, \frac{\pi}{2}\right)$. (See Figure 13.)

**Solution** Since $x(t) = \cos t$, $y(t) = \sin t$, and $z(t) = t$, we see that

$$\mathbf{F}(\mathbf{r}(t)) = \mathbf{F}(x(t), y(t), z(t)) = -y\mathbf{i} + x\mathbf{j} + z\mathbf{k}$$

$$= -\sin t\,\mathbf{i} + \cos t\,\mathbf{j} + t\mathbf{k}$$

Furthermore, observe that the vector equation of $C$ is

$$\mathbf{r}(t) = x(t)\mathbf{i} + y(t)\mathbf{j} + z(t)\mathbf{k} = \cos t\,\mathbf{i} + \sin t\,\mathbf{j} + t\mathbf{k} \qquad 0 \le t \le \frac{\pi}{2}$$

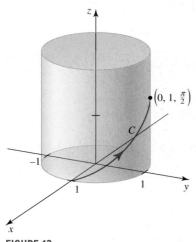

$\left(0, 1, \frac{\pi}{2}\right)$

$C$

**FIGURE 13**
The curve $C$ is described by
$\mathbf{r}(t) = \cos t\,\mathbf{i} + \sin t\,\mathbf{j} + t\mathbf{k}$, $0 \le t \le \frac{\pi}{2}$.

from which we have

$$\mathbf{r}'(t) = -\sin t\mathbf{i} + \cos t\mathbf{j} + \mathbf{k}$$

Therefore, the work done by the force is

$$W = \int_C \mathbf{F} \cdot d\mathbf{r} = \int_0^{\pi/2} \mathbf{F}(\mathbf{r}(t)) \cdot \mathbf{r}'(t) \, dt$$

$$= \int_0^{\pi/2} (-\sin t\mathbf{i} + \cos t\mathbf{j} + t\mathbf{k}) \cdot (-\sin t\mathbf{i} + \cos t\mathbf{j} + \mathbf{k}) \, dt$$

$$= \int_0^{\pi/2} (\sin^2 t + \cos^2 t + t) \, dt = \int_0^{\pi/2} (1 + t) \, dt = \left[ t + \frac{1}{2} t^2 \right]_0^{\pi/2} = \frac{\pi}{2}\left( 1 + \frac{\pi}{4} \right)$$

∎

We close this section by pointing out the relationship between line integrals of vector fields and line integrals of scalar fields with respect to the coordinate variables. Suppose that a vector field $\mathbf{F}$ in space is defined by $\mathbf{F} = P(x, y, z)\mathbf{i} + Q(x, y, z)\mathbf{j} + R(x, y, z)\mathbf{k}$. Then by Equation (11) we have

$$\int_C \mathbf{F} \cdot d\mathbf{r} = \int_a^b \mathbf{F}(\mathbf{r}(t)) \cdot \mathbf{r}'(t) \, dt = \int_a^b (P\mathbf{i} + Q\mathbf{j} + R\mathbf{k}) \cdot (x'(t)\mathbf{i} + y'(t)\mathbf{j} + z'(t)\mathbf{k}) \, dt$$

$$= \int_a^b [P(x(t), y(t), z(t))x'(t) + Q(x(t), y(t), z(t))y'(t) + R(x(t), y(t), z(t))z'(t)] \, dt$$

But the integral on the right is just the line integral of Equation (9). Therefore, we have shown that

$$\int_C \mathbf{F} \cdot d\mathbf{r} = \int_C P \, dx + Q \, dy + R \, dz \qquad \text{where} \quad \mathbf{F} = P\mathbf{i} + Q\mathbf{j} + R\mathbf{k} \qquad \textbf{(12)}$$

You are urged to rework Example 7 with the aid of Equation (12).

As a consequence of Equation (12), we have the result

$$\int_{-C} \mathbf{F} \cdot d\mathbf{r} = -\int_C \mathbf{F} \cdot d\mathbf{r}$$

(see page 1243). This result also follows from the equation

$$\int_{-C} \mathbf{F} \cdot d\mathbf{r} = -\int_C \mathbf{F} \cdot \mathbf{T} \, ds$$

and we observe that even though line integrals with respect to arc length do not change sign when the direction traversed is reversed, the unit vector $\mathbf{T}$ does change sign when $C$ is replaced by $-C$.

**EXAMPLE 8**  Let $\mathbf{F}(x, y) = -\frac{1}{8}(x - y)\mathbf{i} - \frac{1}{8}(x + y)\mathbf{j}$ be the force field shown in Figure 14. Find the work done on a particle that moves along the quarter-circle of radius 1 centered at the origin (a) in a counterclockwise direction from $(1, 0)$ to $(0, 1)$ and (b) in a clockwise direction from $(0, 1)$ to $(1, 0)$.

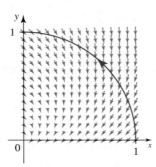

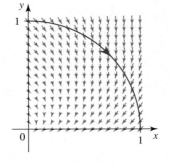

**FIGURE 14**

The force field
$\mathbf{F}(x, y) = -\frac{1}{8}(x - y)\mathbf{i} - \frac{1}{8}(x + y)\mathbf{j}$

**(a)** The direction of the path goes against the direction of **F**.

**(b)** The direction of the path has the same direction as the direction of **F**.

**Solution**

**a.** The path of the particle may be represented by $\mathbf{r}(t) = \cos t\mathbf{i} + \sin t\mathbf{j}$ for $0 \leq t \leq \frac{\pi}{2}$. Since $x = \cos t$ and $y = \sin t$, we find

$$\mathbf{F}(\mathbf{r}(t)) = -\frac{1}{8}(\cos t - \sin t)\mathbf{i} - \frac{1}{8}(\cos t + \sin t)\mathbf{j}$$

and

$$\mathbf{r}'(t) = -\sin t\mathbf{i} + \cos t\mathbf{j}$$

Therefore, the work done by the force on the particle is

$$\int_C \mathbf{F} \cdot d\mathbf{r} = \int_0^{\pi/2} \mathbf{F}(\mathbf{r}(t)) \cdot \mathbf{r}'(t)\, dt = \frac{1}{8}\int_0^{\pi/2} (\cos t \sin t - \sin^2 t - \cos^2 t - \sin t \cos t)\, dt$$

$$= -\frac{1}{8}\int_0^{\pi/2} dt = -\frac{\pi}{16}$$

**b.** Here, we can represent the path by $\mathbf{r}(t) = \sin t\mathbf{i} + \cos t\mathbf{j}$ for $0 \leq t \leq \frac{\pi}{2}$. Then $x = \sin t$ and $y = \cos t$. So

$$\int_C \mathbf{F} \cdot d\mathbf{r} = \int_0^{\pi/2} \mathbf{F}(\mathbf{r}(t)) \cdot \mathbf{r}'(t)\, dt = \frac{1}{8}\int_0^{\pi/2} (-\sin t \cos t + \cos^2 t + \sin^2 t + \sin t \cos t)\, dt$$

$$= \frac{1}{8}\int_0^{\pi/2} dt = \frac{\pi}{16} \qquad \blacksquare$$

In Example 8, observe that the work done by **F** on the particle in part (a) is negative because the force field opposes the motion of the particle.

## 14.3 CONCEPT QUESTIONS

**1. a.** Define the line integral of a function $f(x, y, z)$ along a smooth curve $C$ with parametric representation $\mathbf{r}(t)$, where $a \leq t \leq b$.
  **b.** Write a formula for evaluating the line integral of part (a).
**2. a.** Define the line integral of a function $f(x, y, z)$ along a smooth curve with respect to $x$, with respect to $y$, and with respect to $z$.

  **b.** Write formulas for evaluating the line integrals for the integrals of part (a).
  **c.** Write a formula for evaluating $\int_C P\, dx + Q\, dy + R\, dz$.
**3. a.** Define the line integral of a vector field **F** along a smooth curve $C$.
  **b.** If **F** is a force field, what does the line integral in part (a) represent?

## 14.3 EXERCISES

*In Exercises 1–22, evaluate the line integral over the given curve C.*

**1.** $\int_C (x + y)\, ds$;   $C: \mathbf{r}(t) = 3t\mathbf{i} + 4t\mathbf{j}$,   $0 \le t \le 1$

**2.** $\int_C (x^2 + 2y)\, ds$;   $C: \mathbf{r}(t) = t\mathbf{i} + (t + 1)\mathbf{j}$,   $0 \le t \le 2$

**3.** $\int_C y\, ds$;   $C: \mathbf{r}(t) = 2t\mathbf{i} + t^3\mathbf{j}$, $0 \le t \le 1$

**4.** $\int_C (x + y^3)\, ds$;   $C: \mathbf{r}(t) = t^3\mathbf{i} + t\mathbf{j}$,   $0 \le t \le 1$

**5.** $\int_C (xy^2 + yx^2)\, ds$, where $C$ is the upper semicircle $y = \sqrt{4 - x^2}$

**6.** $\int_C (x^2 + y^2)\, ds$, where $C$ is the right half of the circle $x^2 + y^2 = 9$

**7.** $\int_C 2xy\, ds$, where $C$ is the line segment joining $(-2, -1)$ to $(1, 3)$

**8.** $\int_C (x^2 + 2y)\, ds$, where $C$ is the line segment joining $(-1, 1)$ to $(0, 3)$

**9.** $\int_C (x + 3y^2)\, dx$;   $C: \mathbf{r}(t) = (-1 + 2t)\mathbf{i} + (1 + 3t)\mathbf{j}$, $0 \le t \le 1$

**10.** $\int_C (x + 3y^2)\, dy$;   $C: \mathbf{r}(t) = (-1 + 2t)\mathbf{i} + (1 + 3t)\mathbf{j}$, $0 \le t \le 1$

**11.** $\int_C xy\, dx + (x + y)\, dy$, where $C$ consists of the line segment from $(1, 2)$ to $(3, 4)$ and the line segment from $(3, 4)$ to $(4, 0)$

**12.** $\int_C (y - x)\, dx + y^2\, dy$, where $C$ consists of the line segment from $(0, 0)$ to $(1, 0)$, and the line segment from $(1, 0)$ to $(2, 4)$

**13.** $\int_C y\, dx + x\, dy$, where $C$ consists of the arc of the parabola $y = 4 - x^2$ from $(-2, 0)$ to $(0, 4)$ and the line segment from $(0, 4)$ to $(2, 0)$

**14.** $\int_C (2x + y)\, dx + 2y\, dy$, where $C$ consists of the elliptical path $9x^2 + 16y^2 = 144$ from $(4, 0)$ to $(0, 3)$ and the circular path $x^2 + y^2 = 9$ from $(0, 3)$ to $(-3, 0)$

**15.** $\int_C xyz\, ds$;   $C: \mathbf{r}(t) = (1 + t)\mathbf{i} + 2t\mathbf{j} + (1 - t)\mathbf{k}$, $0 \le t \le 1$

**16.** $\int_C xyz^2\, ds$, where $C$ is the line segment joining $(1, 1, 0)$ to $(2, 3, 1)$

**17.** $\int_C xy^2\, ds$;   $C: \mathbf{r}(t) = \cos 2t\mathbf{i} + \sin 2t\mathbf{j} + 3t\mathbf{k}$,   $0 \le t \le \frac{\pi}{2}$

**18.** $\int_C (8x + 27z)\, ds$;   $C: \mathbf{r}(t) = t\mathbf{i} + 2t^2\mathbf{j} + 3t^3\mathbf{k}$,   $0 \le t \le 1$

**19.** $\int_C (x + y)\, dx + xy\, dy + y\, dz$;   $C: \mathbf{r}(t) = e^t\mathbf{i} + e^{-t}\mathbf{j} + 2e^{2t}\mathbf{k}$, $0 \le t \le 1$

**20.** $\int_C x\, dx - y^2\, dy + yz\, dz$;   $C: \mathbf{r}(t) = t\mathbf{i} + \cos t\mathbf{j} + \sin t\mathbf{k}$, $0 \le t \le \frac{\pi}{4}$

**21.** $\int_C xy\, dx - yz\, dy + x^2\, dz$, where $C$ consists of the line segment from $(0, 0, 0)$ to $(1, 1, 0)$ and the line segment from $(1, 1, 0)$ to $(2, 3, 5)$

**22.** $\int_C (x + y + z)\, dx + (x - y)\, dy + xz\, dz$, where $C$ consists of the line segment from $(0, 0, 0)$ to $(1, 1, 1)$ and the line segment from $(1, 1, 1)$ to $(-1, -2, 3)$

**23.** A thin wire has the shape of a semicircle of radius $a$. Find the mass and the location of the center of mass of the wire if it has a constant linear mass density $k$.

**24.** A thin wire in the shape of a quarter-circle $\mathbf{r}(t) = a \cos t\mathbf{i} + a \sin t\mathbf{j}$, $0 \le t \le \frac{\pi}{2}$, has linear mass density $\pi(x, y) = k(x + y)$, where $k$ is a positive constant. Find the mass and the location of the center of mass of the wire.

**25.** A thin wire has the shape of a semicircle $x^2 + y^2 = a^2$, $y \ge 0$. Find the center of mass of the wire if the linear mass density of the wire at any point is proportional to its distance from the line $y = a$.

**26.** A thin wire of constant linear mass density $k$ takes the shape of an arch of the cycloid $x = a(t - \sin t)$, $y = a(1 - \cos t)$, $0 \le t \le 2\pi$. Determine the mass of the wire, and find the location of its center of mass.

**27.** A thin wire of constant linear mass density $k$ has the shape of the astroid $x = \cos^3 t$, $y = \sin^3 t$, $0 \le t \le \frac{\pi}{2}$. Determine the location of its center of mass.

**28.** A thin wire has the shape of the helix $x = a \cos t$, $y = a \sin t$, $z = bt$, $0 \le t \le 3\pi$. Find the mass and the center of mass of the wire if it has constant linear mass density $k$.

Hint: $\bar{x} = \dfrac{1}{m} \displaystyle\int_C x\rho(x, y, z)\, ds$, $\bar{y} = \dfrac{1}{m} \displaystyle\int_C y\rho(x, y, z)\, ds$,

$\bar{z} = \dfrac{1}{m} \displaystyle\int_C z\rho(x, y, z)\, ds$, where $m = \displaystyle\int_C \rho(x, y, z)\, ds$

**29.** The vector field $\mathbf{F}(x, y) = (x - y)\mathbf{i} + (x + y)\mathbf{j}$ is shown in the figure. A particle is moved from the point $(-2, 0)$ to the point $(2, 0)$ along the upper semicircle of radius 2 with center at the origin.

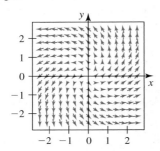

**a.** By inspection, determine whether the work done by $\mathbf{F}$ on the particle is positive, zero, or negative.
**b.** Find the work done by $\mathbf{F}$ on the particle.

**30.** The vector field $\mathbf{F}(x, y) = -\dfrac{y}{\sqrt{x^2 + y^2}}\mathbf{i} + \dfrac{x}{\sqrt{x^2 + y^2}}\mathbf{j}$ is shown in the figure. A particle is moved once around the circle of radius 2 with center at the origin in the counterclockwise direction.

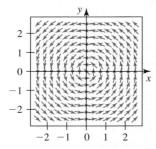

**a.** By inspection, determine whether the work done by $\mathbf{F}$ on the particle is positive, zero, or negative.
**b.** Find the work done by $\mathbf{F}$ on the particle.

*In Exercises 31–36, find the work done by the force field $\mathbf{F}$ on a particle that moves along the curve C.*

**31.** $\mathbf{F}(x, y) = (x^2 + y^2)\mathbf{i} + xy\mathbf{j}$;  $C: \mathbf{r}(t) = t^2\mathbf{i} + t^3\mathbf{j}$, $0 \le t \le 1$

**32.** $\mathbf{F}(x, y) = \ln x\mathbf{i} + y^2\mathbf{j}$;  $C: \mathbf{r}(t) = t\mathbf{i} + t^2\mathbf{j}$, $1 \le t \le 2$

**33.** $\mathbf{F}(x, y) = xe^y\mathbf{i} + y\mathbf{j}$, where $C$ is the part of the parabola $y = x^2$ from $(-1, 1)$ to $(2, 4)$

**34.** $\mathbf{F}(x, y) = x\mathbf{i} + (y + 1)\mathbf{j}$, where $C$ is an arch of the cycloid $x = t - \sin t$, $y = 1 - \cos t$, $0 \le t \le 2\pi$

**35.** $\mathbf{F}(x, y, z) = x^2\mathbf{i} + y^2\mathbf{j} + z^2\mathbf{k}$;  $C: \mathbf{r}(t) = t\mathbf{i} + t^2\mathbf{j} + t^3\mathbf{k}$, $0 \le t \le 1$

**36.** $\mathbf{F}(x, y, z) = (x + 2y)\mathbf{i} + 2z\mathbf{j} + (x - y)\mathbf{k}$, where $C$ is the line segment from $(-1, 3, 2)$ to $(1, -2, 4)$.

**37. Walking up a Spiral Staircase** A spiral staircase is described by the parametric equations

$$x = 5 \cos t, \qquad y = 5 \sin t, \qquad z = \frac{16}{\pi}t, \qquad 0 \le t \le \frac{\pi}{2}$$

where the distance is measured in feet. If a 90-lb girl walks up the staircase, what is the work done by her against gravity in walking to the top of the staircase?
**Note:** You can also obtain the answer using elementary physics.

**38.** A particle is moved along a path from $(0, 0)$ to $(1, 2)$ by the force $\mathbf{F} = 2xy^2\mathbf{i} + 3yx^2\mathbf{j}$. Which of the following polygonal paths results in the least work?
**a.** The path from $(0, 0)$ to $(1, 0)$ to $(1, 2)$
**b.** The path from $(0, 0)$ to $(0, 2)$ to $(1, 2)$
**c.** The path from $(0, 0)$ to $(1, 2)$

**39. Newton's Second Law of Motion** Suppose that the position of a particle of varying mass $m(t)$ in 3-space at time $t$ is $\mathbf{r}(t)$. According to Newton's Second Law of Motion, the force acting on the particle at $\mathbf{r}(t)$ is

$$\mathbf{F}(\mathbf{r}(t)) = \frac{d}{dt}[m(t)\mathbf{v}(t)]$$

**a.** Show that $\mathbf{F}(\mathbf{r}(t)) \cdot \mathbf{r}'(t) = m'(t)v^2(t) + m(t)v(t)v'(t)$, where $v = |\mathbf{r}'|$ is the speed of the particle.
**b.** Show that if $m$ is constant, then the work done by the force in moving the particle along its path from $t = a$ to $t = b$ is

$$W = \frac{m}{2}[v^2(b) - v^2(a)]$$

**Note:** The function $W(t) = \frac{1}{2}mv^2(t)$ is the kinetic energy of the particle.

**40. Work Done by an Electric Field** Suppose that a charge of $Q$ coulombs is located at the origin of a three-dimensional coordinate system. This charge induces an electric field

$$\mathbf{E}(x, y, z) = \frac{cQ}{|\mathbf{r}|^3}\mathbf{r}$$

where $\mathbf{r} = x\mathbf{i} + y\mathbf{j} + z\mathbf{k}$ and $c$ is a constant (see Example 4 in Section 14.1). Find the work done by the electric field on a particle of charge $q$ coulombs as it is moved along the path $C: \mathbf{r}(t) = t\mathbf{i} + 2t\mathbf{j} + (1 + 4t)\mathbf{k}$, where $0 \le t \le 1$.

**41. Work Done by an Electric Field**  The electric field **E** at any point $(x, y, z)$ induced by a point charge $Q$ located at the origin is given by

$$\mathbf{E} = \frac{Q\mathbf{r}}{4\pi\varepsilon_0|\mathbf{r}|^3}$$

where $\mathbf{r} = \langle x, y, z \rangle$ and $\varepsilon_0$ is a positive constant called the permittivity of free space.

**a.** Find the work done by the field when a particle of charge $q$ coulombs is moved from $A(2, 1, 0)$ to $D(0, 5, 5)$ along the indicated paths.

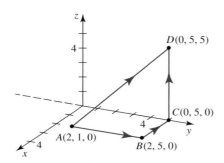

**(i)**  The straight line segment from $A$ to $D$.
**(ii)** The polygonal path from $A(2, 1, 0)$ to $B(2, 5, 0)$ to $C(0, 5, 0)$ and then to $D(0, 5, 5)$.

**b.** Is there any difference in the work done in part (i) and part (ii)?

**42. Magnitude of a Magnetic Field**  The following figure shows a long straight wire that is carrying a steady current $I$. This current induces a magnetic field **B** whose direction is *circumferential*; that is, it circles around the wire. Ampere's Law states that

$$\int_C \mathbf{B} \cdot d\mathbf{r} = \mu_0 I$$

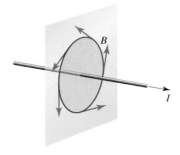

In words, the line integral of the tangential component of the magnetic field around a closed loop $C$ is proportional to the current $I$ passing through any surface bounded by the loop. The constant $\mu_0$ is called the permeability of free space. By taking the loop to be a circle of radius $r$ centered on the wire, show that the magnitude $B = |\mathbf{B}|$ of the magnetic field at a distance $r$ from the center of the wire is

$$B = \frac{\mu_0 I}{2\pi r}$$

**cas** *In Exercises 43 and 44, plot the graph of the vector field* **F** *and the curve C on the same set of axes. Guess at whether the line integral of* **F** *over C is positive, negative, or zero. Verify your answer by evaluating the line integral.*

**43.** $\mathbf{F}(x, y) = \dfrac{1}{2}(x - y)\mathbf{i} + \dfrac{1}{2}(x + y)\mathbf{j}$;   $C$ is the curve
$\mathbf{r}(t) = 2\sin t\,\mathbf{i} + 2\cos t\,\mathbf{j}$,   $0 \le t \le \pi$

**44.** $\mathbf{F}(x, y) = \dfrac{1}{4}x\mathbf{i} + \dfrac{1}{2}y\mathbf{j}$;   $C$ is the curve $\mathbf{r}(t) = t\mathbf{i} + (1 - t^2)\mathbf{j}$,
$-1 \le t \le 1$

*In Exercises 45–48, determine whether the statement is true or false. If it is true, explain why. If it is false, explain why or give an example that shows it is false.*

**45.** If $\mathbf{F}(x, y) = x\mathbf{i} + y\mathbf{j}$, then $\int_C \mathbf{F} \cdot d\mathbf{r} = 0$, where $C$ is any circular path centered at the origin.

**46.** If $f(x, y)$ is continuous and $C$ is a smooth curve, then $\int_C f(x, y)\, ds = -\int_{-C} f(x, y)\, ds$.

**47.** If $C$ is a smooth curve defined by $\mathbf{r}(t) = x(t)\mathbf{i} + y(t)\mathbf{j}$ with $a \le t \le b$, then $\int_C xy\, dy = \frac{1}{2}xy^2 \Big|_{t=a}^{t=b}$.

**48.** If $f(x, y)$ is continuous and $C$ is a smooth curve defined by $\mathbf{r}(t) = x(t)\mathbf{i} + y(t)\mathbf{j}$ with $a \le t \le b$, then $\left[\int_C f(x, y)\, ds\right]^2 = \left[\int_C f(x, y)\, dx\right]^2 + \left[\int_C f(x, y)\, dy\right]^2$.

## 14.4    Independence of Path and Conservative Vector Fields

The gravitational field possesses an important property that we will demonstrate in the following example.

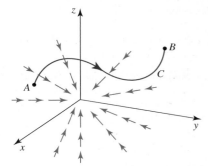

**FIGURE 1**
The particle moves from $A$ to $B$ along the path $C$ in the gravitational field.

**EXAMPLE 1**    **Work Done on a Particle by a Gravitational Field**    Consider the gravitational field $\mathbf{F}$ induced by an object of mass $M$ located at the origin (see Example 3, Section 14.1):

$$\mathbf{F}(x, y, z) = -\frac{GM}{|\mathbf{r}|^3}\mathbf{r}$$

$$= -\frac{GMx}{(x^2 + y^2 + z^2)^{3/2}}\mathbf{i} - \frac{GMy}{(x^2 + y^2 + z^2)^{3/2}}\mathbf{j} - \frac{GMz}{(x^2 + y^2 + z^2)^{3/2}}\mathbf{k}$$

Suppose that a particle with mass $m$ moves in the gravitational field $\mathbf{F}$ from the point $A(x(a), y(a), z(a))$ to the point $B(x(b), y(b), z(b))$ along a smooth curve $C$ defined by

$$\mathbf{r}(t) = x(t)\mathbf{i} + y(t)\mathbf{j} + z(t)\mathbf{k}$$

with parameter interval $[a, b]$. (See Figure 1.) What is the work $W$ done by $\mathbf{F}$ on the particle?

**Solution**    To find $W$, we note that the particle moving in the gravitational field $\mathbf{F}$ is subjected to a force of $m\mathbf{F}$, so the work done by the force on the particle is

$$W = \int_C m\mathbf{F} \cdot d\mathbf{r} = \int_a^b m\mathbf{F}(\mathbf{r}(t)) \cdot \mathbf{r}'(t)\, dt$$

$$= -GMm \int_a^b \left[ \frac{x}{(x^2 + y^2 + z^2)^{3/2}}\mathbf{i} + \frac{y}{(x^2 + y^2 + z^2)^{3/2}}\mathbf{j} + \frac{z}{(x^2 + y^2 + z^2)^{3/2}}\mathbf{k} \right]$$

$$\cdot \left[ \frac{dx}{dt}\mathbf{i} + \frac{dy}{dt}\mathbf{j} + \frac{dz}{dt}\mathbf{k} \right] dt$$

$$= -GMm \int_a^b \left[ \frac{x}{(x^2 + y^2 + z^2)^{3/2}}\frac{dx}{dt} + \frac{y}{(x^2 + y^2 + z^2)^{3/2}}\frac{dy}{dt} + \frac{z}{(x^2 + y^2 + z^2)^{3/2}}\frac{dz}{dt} \right] dt$$

But the expression inside the brackets can be written as

$$\frac{d}{dt}f(x, y, z) = \frac{\partial f}{\partial x}\frac{dx}{dt} + \frac{\partial f}{\partial y}\frac{dy}{dt} + \frac{\partial f}{\partial z}\frac{dz}{dt}$$

where

$$f(x, y, z) = \frac{1}{\sqrt{x^2 + y^2 + z^2}}$$

as you can verify. (Also, see Example 6 in Section 14.1.) Using this result, we can write

$$W = -GMm \int_a^b \frac{d}{dt}\left( \frac{1}{\sqrt{x^2 + y^2 + z^2}} \right) dt = -\frac{GMm}{\sqrt{x^2 + y^2 + z^2}}\Bigg|_{t=a}^{t=b}$$

$$= -GMm\, f(x, y, z)\Bigg|_{t=a}^{t=b} = -GMm[f(x(b), y(b), z(b)) - f(x(a), y(a), z(a))]\quad\blacksquare$$

**Note**   Don't worry about finding the potential function

$$f(x, y, z) = \frac{1}{\sqrt{x^2 + y^2 + z^2}}$$

for the gravitational field **F**. We will develop a systematic method for finding potential functions $f$ of gradient fields $\nabla f$ later in this section.    ■

Example 1 shows that the work done on a particle by a gravitational field **F** depends *only* on the initial point $A$ and the endpoint $B$ of a curve $C$ and *not* on the curve itself. We say that the value of the line integral along the path $C$ is *independent of the path.* (A path is a piecewise-smooth curve.)

More generally, we say that the line integral $\int_C \mathbf{F} \cdot d\mathbf{r}$ is **independent of path** if

$$\int_{C_1} \mathbf{F} \cdot d\mathbf{r} = \int_{C_2} \mathbf{F} \cdot d\mathbf{r}$$

for any two paths $C_1$ and $C_2$ that have the same initial and terminal points.

Observe that the gravitational field **F** happens to be a *conservative* vector field with potential function $f$; that is, $\mathbf{F} = \nabla f$. Also, Example 1 seems to suggest that if $\mathbf{F} = \nabla f$ is a gradient vector field with potential function $f$, then

$$\int_C \mathbf{F} \cdot d\mathbf{r} = \int_C \nabla f \cdot d\mathbf{r} = f(x(b), y(b), z(b)) - f(x(a), y(a), z(a)) \qquad \textbf{(1)}$$

This expression reminds us of Part 2 of the Fundamental Theorem of Calculus which states that

$$\int_a^b F'(x) \, dx = F(b) - F(a)$$

where $F$ is continuous on $[a, b]$. The Fundamental Theorem of Calculus, Part 2, tells us that if the derivative of $F$ in the interior of the interval $[a, b]$ is known, then the integral of $F'$ over $[a, b]$ is given by the difference of the values of $F$ (an antiderivative of $F'$) at the endpoints of $[a, b]$. If we think of $\nabla f$ as some kind of derivative of $f$, then Equation (1) says that if we know the "derivative" of $f$, then the line integral of $\nabla f$ is given by the difference of the values of the potential function $f$ ("antiderivative" of $\nabla f$) at the endpoints of the curve $C$.

We now show that Equation (1) is indeed true for all conservative vector fields. We state and prove the result for a function $f$ of two variables and a curve $C$ in the plane.

---

**THEOREM 1   Fundamental Theorem for Line Integrals**

Let $\mathbf{F}(x, y) = \nabla f(x, y)$ be a conservative vector field in an open region $R$, where $f$ is a differentiable potential function for **F**. If $C$ is any piecewise-smooth curve lying in $R$ given by

$$\mathbf{r}(t) = x(t)\mathbf{i} + y(t)\mathbf{j} \qquad a \le t \le b$$

then

$$\int_C \mathbf{F} \cdot d\mathbf{r} = \int_C \nabla f \cdot d\mathbf{r} = f(x(b), y(b)) - f(x(a), y(a))$$

**PROOF**    We will give the proof for a smooth curve $C$. Since $\mathbf{F}(x, y) = \nabla f = f_x(x, y)\mathbf{i} + f_y(x, y)\mathbf{j}$, we see that

$$\int_C \mathbf{F} \cdot d\mathbf{r} = \int_C \nabla f \cdot d\mathbf{r} = \int_a^b \nabla f \cdot \frac{d\mathbf{r}}{dt}\,dt$$

$$= \int_a^b \left[ \frac{\partial f}{\partial x}\frac{dx}{dt} + \frac{\partial f}{\partial y}\frac{dy}{dt} \right] dt$$

$$= \int_a^b \frac{d}{dt}[f(x(t), y(t))]\,dt \qquad \text{Use the Chain Rule.}$$

$$= f(x(t), y(t)) \Big|_{t=a}^{t=b} = f(x(b), y(b)) - f(x(a), y(a))\qquad \blacksquare$$

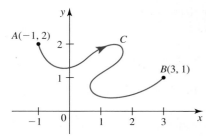

**FIGURE 2**
$C$ is a piecewise smooth curve joining $A$ to $B$.

**EXAMPLE 2**    Let $\mathbf{F}(x, y) = 2xy\mathbf{i} + x^2\mathbf{j}$ be a force field.

**a.** Prove that $\mathbf{F}$ is conservative by showing that it is the gradient of the potential function $f(x, y) = x^2 y$.

**b.** Use the Fundamental Theorem for Line Integrals to evaluate $\int_C \mathbf{F} \cdot d\mathbf{r}$, where $C$ is any piecewise-smooth curve joining the point $A(-1, 2)$ to the point $B(3, 1)$ (See Figure 2.)

**Solution**

**a.** Since $\nabla f(x, y) = \dfrac{\partial}{\partial x}(x^2 y)\mathbf{i} + \dfrac{\partial}{\partial y}(x^2 y)\mathbf{j} = 2xy\mathbf{i} + x^2\mathbf{j} = \mathbf{F}(x, y)$, we conclude that $\mathbf{F}$ is indeed conservative.

**b.** Thanks to the Fundamental Theorem for Line Integrals, we do not need to know the rule defining the curve $C$; the integral depends only on the coordinates of the endpoints $A$ and $B$ of the curve. We have

$$\int_C \mathbf{F} \cdot d\mathbf{r} = f(3, 1) - f(-1, 2) = x^2 y \Big|_{(-1, 2)}^{(3, 1)}$$

$$= (3)^2(1) - (-1)^2(2) = 7 \qquad \blacksquare$$

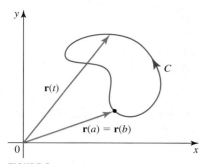

**FIGURE 3**
On the closed curve $C$, the tip of $\mathbf{r}(t)$ starts at $\mathbf{r}(a)$, traverses $C$, and ends up back at $\mathbf{r}(b) = \mathbf{r}(a)$.

## Line Integrals Along Closed Paths

A path is **closed** if its terminal point coincides with its initial point. If a curve $C$ has parametric representation $\mathbf{r}(t)$ with parameter interval $[a, b]$, then $C$ is closed if $\mathbf{r}(a) = \mathbf{r}(b)$. (See Figure 3.)

The following theorem gives an alternative method for determining whether a line integral is independent of path.

**THEOREM 2**

Suppose that $\mathbf{F}$ is a continuous vector field in a region $R$. Then $\int_C \mathbf{F} \cdot d\mathbf{r}$ is independent of path if and only if $\int_C \mathbf{F} \cdot d\mathbf{r} = 0$ for every closed path $C$ in $R$.

**PROOF** Suppose that $\int_C \mathbf{F} \cdot d\mathbf{r}$ is independent of path in $R$, and let $C$ be any closed path in $R$. We can pick any two points $A$ and $B$ on $C$ and regard $C$ as being made up of the path from $A$ to $B$ and the path from $B$ to $A$. (See Figure 4a.) Then

$$\int_C \mathbf{F} \cdot d\mathbf{r} = \int_{C_1} \mathbf{F} \cdot d\mathbf{r} + \int_{C_2} \mathbf{F} \cdot d\mathbf{r} = \int_{C_1} \mathbf{F} \cdot d\mathbf{r} - \int_{-C_2} \mathbf{F} \cdot d\mathbf{r}$$

where $-C_2$ is the path $C_2$ traversed in the opposite direction. But both $C_1$ and $-C_2$ have the same initial point $A$ and the same terminal point $B$. Since the line integral is assumed to be independent of path, we have

$$\int_{C_1} \mathbf{F} \cdot d\mathbf{r} = \int_{-C_2} \mathbf{F} \cdot d\mathbf{r}$$

and this implies that $\int_C \mathbf{F} \cdot d\mathbf{r} = 0$.

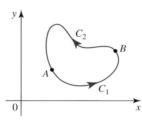 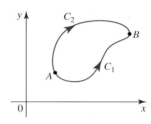

**FIGURE 4**
$C$ is a closed path in an open region $R$.

(a) $C$ is made up of $C_1$ and $C_2$.  (b) $C$ is made up of $C_1$ and $-C_2$.

Conversely, suppose that $\int_C \mathbf{F} \cdot d\mathbf{r} = 0$ for every closed path $C$ in $R$. Let $A$ and $B$ be any two points in $R$ and let $C_1$ and $C_2$ be any two paths in $R$ connecting $A$ to $B$, respectively. (See Figure 4b.) Let $C$ be the closed path composed of $C_1$ followed by $-C_2$. Then

$$0 = \int_C \mathbf{F} \cdot d\mathbf{r} = \int_{C_1} \mathbf{F} \cdot d\mathbf{r} + \int_{-C_2} \mathbf{F} \cdot d\mathbf{r} = \int_{C_1} \mathbf{F} \cdot d\mathbf{r} - \int_{C_2} \mathbf{F} \cdot d\mathbf{r}$$

so $\int_{C_1} \mathbf{F} \cdot d\mathbf{r} = \int_{C_2} \mathbf{F} \cdot d\mathbf{r}$, which shows that the line integral is independent of path. ∎

As a consequence of Theorem 1, we see that if a body moves along a closed path that ends where it began, then the work done by a conservative force field on the body is zero.

### ■ Independence of Path and Conservative Vector Fields

The Fundamental Theorem for Line Integrals tells us that the line integral of a *conservative* vector field is independent of path. A question that arises naturally is: Is a vector field whose integral is independent of path necessarily a conservative vector field? To answer this question, we need to consider regions that are both *open* and *connected*. A region is **open** if it doesn't contain any of its boundary points. It is **connected** if any two points in the region can be joined by a path that lies in the region. (See Figure 5.) The following theorem provides an answer to part of the first question that we raised.

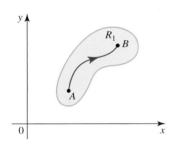

(a) The plane region $R_1$ is connected.

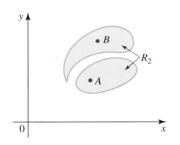

(b) The region $R_2$ is not connected, since it is impossible to find a path from $A$ to $B$ lying strictly within $R_2$.

**FIGURE 5**

**THEOREM 3    Independence of Path and Conservative Vector Fields**

Let $\mathbf{F}$ be a continuous vector field in an open, connected region $R$. The line integral $\int_C \mathbf{F} \cdot d\mathbf{r}$ is independent of path if and only if $\mathbf{F}$ is conservative, that is, if and only if $\mathbf{F} = \nabla f$ for some scalar function $f$.

**PROOF**   If $\mathbf{F}$ is conservative, then the Fundamental Theorem for Line Integrals implies that the line integral is independent of path. We will prove the converse for the case in which $R$ is a plane region; the proof for the three-dimensional case is similar. Suppose that the integral is independent of path in $R$. Let $(x_0, y_0)$ be a fixed point in $R$, and let $(x, y)$ be any point in $R$. If $C$ is any path from $(x_0, y_0)$ to $(x, y)$, we define the function $f$ by

$$f(x, y) = \int_C \mathbf{F} \cdot d\mathbf{r} = \int_{(x_0, y_0)}^{(x, y)} \mathbf{F} \cdot d\mathbf{r}$$

Since $R$ is open, there exists a disk contained in $R$ with center $(x, y)$. Pick any point $(x_1, y)$ in the disk with $x_1 < x$. Now, by assumption, the line integral is independent of path, so we can choose $C$ to be the path consisting of any path $C_1$ from $(x_0, y_0)$ to $(x_1, y)$ followed by the horizontal line segment $C_2$ from $(x_1, y)$ to $(x, y)$, as shown in Figure 6. Then

$$f(x, y) = \int_{C_1} \mathbf{F} \cdot d\mathbf{r} + \int_{C_2} \mathbf{F} \cdot d\mathbf{r} = \int_{(x_0, y_0)}^{(x_1, y)} \mathbf{F} \cdot d\mathbf{r} + \int_{C_2} \mathbf{F} \cdot d\mathbf{r}$$

Since the first of the two integrals on the right does not depend on $x$, we have

$$\frac{\partial}{\partial x} f(x, y) = 0 + \frac{\partial}{\partial x} \int_{C_2} \mathbf{F} \cdot d\mathbf{r}$$

If we write $\mathbf{F}(x, y) = P(x, y)\mathbf{i} + Q(x, y)\mathbf{j}$, then

$$\int_{C_2} \mathbf{F} \cdot d\mathbf{r} = \int_{C_2} P(x, y)\, dx + Q(x, y)\, dy$$

Now $C_2$ can be represented parametrically by $x(t) = t$, $y(t) = y$, where $x_1 \le t \le x$ and $y$ is a constant. This gives $dx = x'(t)\, dt = dt$ and $dy = 0$ since $y$ is constant on $C_2$. Therefore,

$$\frac{\partial}{\partial x} f(x, y) = \frac{\partial}{\partial x} \int_{C_2} P(x, y)\, dx + Q(x, y)\, dy$$

$$= \frac{\partial}{\partial x} \int_{x_1}^{x} P(t, y)\, dt = P(x, y)$$

upon using the Fundamental Theorem of Calculus, Part 1. Similarly, by choosing $C$ to be the path with a vertical line segment as shown in Figure 7, we can show that

$$\frac{\partial}{\partial y} f(x, y) = Q(x, y)$$

Therefore,

$$\mathbf{F} = P\mathbf{i} + Q\mathbf{j} = \frac{\partial f}{\partial x}\mathbf{i} + \frac{\partial f}{\partial y}\mathbf{j} = \nabla f$$

that is, $\mathbf{F}$ is conservative.   ▪

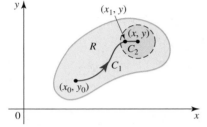

**FIGURE 6**
The path $C$ consists of an arbitrary path $C_1$ from $(x_0, y_0)$ to $(x_1, y)$ followed by the horizontal line segment from $(x_1, y)$ to $(x, y)$.

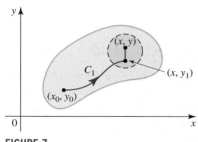

**FIGURE 7**
The path $C$ consists of an arbitrary path from $(x_0, y_0)$ to $(x, y_1)$ followed by the vertical line segment from $(x, y_1)$ to $(x, y)$.

## ■ Determining Whether a Vector Field Is Conservative

Although Theorem 3 provides us with a good characterization of conservative vector fields, it does not help us to determine whether a vector field is conservative, since it is not practical to evaluate the line integral of **F** over all possible paths. Before stating a criterion for determining whether a vector field is conservative, we look at a condition that must be satisfied by a conservative vector field.

---

**THEOREM 4**

If $\mathbf{F}(x, y) = P(x, y)\mathbf{i} + Q(x, y)\mathbf{j}$ is a conservative vector field in an open region $R$ and both $P$ and $Q$ have continuous first-order partial derivatives in $R$, then

$$\frac{\partial Q}{\partial x} = \frac{\partial P}{\partial y}$$

at each point $(x, y)$ in $R$.

---

**PROOF**   Because $\mathbf{F} = P\mathbf{i} + Q\mathbf{j}$ is conservative in $R$, there exists a function $f$ such that $\mathbf{F} = \nabla f$, that is,

$$P\mathbf{i} + Q\mathbf{j} = f_x\mathbf{i} + f_y\mathbf{j}$$

This equation is equivalent to the two equations

$$P = f_x \qquad \text{and} \qquad Q = f_y$$

Since $P_y$ and $Q_x$ are continuous by assumption, it follows from Clairaut's Theorem in Section 12.3 that

$$\frac{\partial P}{\partial y} = f_{xy} = f_{yx} = \frac{\partial Q}{\partial x}$$

■

The converse of Theorem 4 holds only for a certain type of region. To describe this region, we need the notion of a *simple curve*. A plane curve described by $\mathbf{r} = \mathbf{r}(t)$ is a **simple curve** if it does not intersect itself anywhere except possibly at its endpoints; that is, $\mathbf{r}(t_1) \neq \mathbf{r}(t_2)$ if $a < t_1 < t_2 < b$. (See Figure 8.)

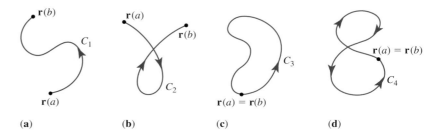

**FIGURE 8**
$C_1$ is simple, $C_2$ is not simple, $C_3$ is simple and closed, and $C_4$ is closed but not simple.

**(a)**　　**(b)**　　**(c)**　　**(d)**

A connected region $R$ in the plane is a **simply-connected region** if every simple closed curve $C$ in $R$ encloses only points that are in $R$. As is illustrated in Figure 9, a simply-connected region not only is connected, but also does not have any hole(s).

**FIGURE 9**
$R_1$ is simply-connected; $R_2$ is not
simply-connected because the simple
closed curve shown encloses points
outside $R_2$; $R_3$ is not simply-connected
because it is not connected.

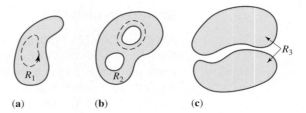

(a)        (b)        (c)

The following theorem, which is a partial converse of Theorem 4, gives us a test to determine whether a vector field on a simply-connected region in the plane is conservative.

---

**THEOREM 5    Test for a Conservative Vector Field in the Plane**

Let $\mathbf{F} = P\mathbf{i} + Q\mathbf{j}$ be a vector field in an open simply-connected region $R$ in the plane. If $P$ and $Q$ have continuous first-order partial derivatives on $R$ and

$$\frac{\partial Q}{\partial x} = \frac{\partial P}{\partial y} \qquad\qquad (2)$$

for all $(x, y)$ in $R$, then $\mathbf{F}$ is conservative in $R$.

---

The proof of this theorem can be found in advanced calculus books.

**EXAMPLE 3**  Determine whether the vector field $\mathbf{F}(x, y) = (x^2 - 2xy + 1)\mathbf{i} + (y^2 - x^2)\mathbf{j}$ is conservative.

**Solution**    Here, $P(x, y) = x^2 - 2xy + 1$ and $Q(x, y) = y^2 - x^2$. Since

$$\frac{\partial P}{\partial y} = -2x = \frac{\partial Q}{\partial x}$$

for all $(x, y)$ in the plane, which is open and simply-connected, we conclude by Theorem 5 that $\mathbf{F}$ is conservative. ◼

**EXAMPLE 4**  Determine whether the vector field $\mathbf{F}(x, y) = 2xy^2\mathbf{i} + x^2y\mathbf{j}$ is conservative.

**Solution**    Here, $P(x, y) = 2xy^2$ and $Q(x, y) = x^2y$. So

$$\frac{\partial P}{\partial y} = 4xy \qquad \text{and} \qquad \frac{\partial Q}{\partial x} = 2xy$$

Since $\partial P/\partial y \neq \partial Q/\partial x$ except along the $x$- or $y$-axis, we see that Equation (2) of Theorem 5 is not satisfied for all points $(x, y)$ in any open simply-connected region in the plane. Therefore, $\mathbf{F}$ is not conservative. ◼

## ◼ Finding a Potential Function

Once we have ascertained that a vector field $\mathbf{F}$ is conservative, how do we go about finding a potential function $f$ for $\mathbf{F}$? One such technique is utilized in the following example.

**EXAMPLE 5** Let $\mathbf{F}(x, y) = 2xy\mathbf{i} + (1 + x^2 - y^2)\mathbf{j}$.

**a.** Show that $\mathbf{F}$ is conservative, and find a potential function $f$ such that $\mathbf{F} = \nabla f$.

**b.** If $\mathbf{F}$ is a force field, find the work done by $\mathbf{F}$ in moving a particle along any path from $(1, 0)$ to $(2, 3)$.

**Solution**

**a.** Here, $P(x, y) = 2xy$ and $Q(x, y) = 1 + x^2 - y^2$. Since

$$\frac{\partial P}{\partial y} = 2x = \frac{\partial Q}{\partial x}$$

for all points in the plane, we see that $\mathbf{F}$ is conservative. Therefore, there exists a function $f$ such that $\mathbf{F} = \nabla f$. In this case the equation reads

$$2xy\mathbf{i} + (1 + x^2 - y^2)\mathbf{j} = \frac{\partial f}{\partial x}\mathbf{i} + \frac{\partial f}{\partial y}\mathbf{j}$$

This vector equation is equivalent to the system of scalar equations

$$\frac{\partial f}{\partial x} = 2xy \tag{3}$$

$$\frac{\partial f}{\partial y} = 1 + x^2 - y^2 \tag{4}$$

Integrating Equation (3) with respect to $x$, (so that $y$ is treated as a constant), we have

$$f(x, y) = x^2 y + g(y) \tag{5}$$

where $g(y)$ is the constant of integration. (Remember that $y$ is treated as a constant, so the most general expression of a constant here involves a function of $y$.) To determine $g(y)$, we differentiate Equation (5) with respect to $y$, obtaining

$$\frac{\partial f}{\partial y} = x^2 + g'(y) \tag{6}$$

Comparing Equation (6) with Equation (4) leads to

$$x^2 + g'(y) = 1 + x^2 - y^2$$

or

$$g'(y) = 1 - y^2 \tag{7}$$

Integrating Equation (7) with respect to $y$ gives

$$g(y) = y - \frac{1}{3}y^3 + C$$

where $C$ is a constant. Finally, substituting $g(y)$ into Equation (5) gives

$$f(x, y) = x^2 y + y - \frac{1}{3}y^3 + C$$

the desired potential function.

**b.** Since $\mathbf{F}$ is conservative, we know that the work done by $\mathbf{F}$ in moving a particle from $(1, 0)$ to $(2, 3)$ is independent of the path connecting these two points.

Using Equation (1), we see that the work done by $\mathbf{F}$ is

$$W = \int_C \mathbf{F} \cdot d\mathbf{r} = \int_C \nabla f \cdot d\mathbf{r} = f(2, 3) - f(1, 0)$$

$$= \left[ (2^2)(3) + 3 - \frac{1}{3}(3^3) \right] - \left[ (1^2)(0) + 0 - \frac{1}{3}(0) \right] = 6 \qquad \blacksquare$$

**Note**  In Example 5a you may also integrate Equation (4) first with respect to $y$ and proceed in a similar manner. $\qquad \blacksquare$

The following theorem provides us with a test to determine whether a vector field in space is conservative. Theorem 6 is an extension of Theorem 5, and its proof will be omitted.

---

**THEOREM 6    Test for a Conservative Vector Field in Space**

Let $\mathbf{F} = P\mathbf{i} + Q\mathbf{j} + R\mathbf{k}$ be a vector field in an open, simply connected region $D$ in space. If $P$, $Q$, and $R$ have continuous first-order partial derivatives in space, then $\mathbf{F}$ is conservative if curl $\mathbf{F} = \mathbf{0}$ for all points in $D$. Equivalently, $\mathbf{F}$ is conservative if

$$\frac{\partial R}{\partial y} = \frac{\partial Q}{\partial z}, \qquad \frac{\partial R}{\partial x} = \frac{\partial P}{\partial z}, \qquad \text{and} \qquad \frac{\partial Q}{\partial x} = \frac{\partial P}{\partial y}$$

---

The following example illustrates how to find a potential function for a conservative vector field in space.

**EXAMPLE 6**  Let $\mathbf{F}(x, y, z) = 2xyz^2\mathbf{i} + x^2z^2\mathbf{j} + 2x^2yz\mathbf{k}$.

**a.** Show that $\mathbf{F}$ is conservative, and find a function $f$ such that $\mathbf{F} = \nabla f$.
**b.** If $\mathbf{F}$ is a force field, find the work done by $\mathbf{F}$ in moving a particle along any path from $(0, 1, 0)$ to $(1, 2, -1)$.

**Solution**
**a.** We compute

$$\text{curl } \mathbf{F} = \begin{vmatrix} \mathbf{i} & \mathbf{j} & \mathbf{k} \\ \dfrac{\partial}{\partial x} & \dfrac{\partial}{\partial y} & \dfrac{\partial}{\partial z} \\ 2xyz^2 & x^2z^2 & 2x^2yz \end{vmatrix}$$

$$= (2x^2z - 2x^2z)\mathbf{i} - (4xyz - 4xyz)\mathbf{j} + (2xz^2 - 2xz^2)\mathbf{k}$$

$$= \mathbf{0}$$

Since curl $\mathbf{F} = \mathbf{0}$ for all points in $R^3$, we see that $\mathbf{F}$ is a conservative vector field by Theorem 6. Therefore, there exists a function $f$ such that $\mathbf{F} = \nabla f$. In this case the equation reads

$$2xyz^2\mathbf{i} + x^2z^2\mathbf{j} + 2x^2yz\mathbf{k} = \frac{\partial f}{\partial x}\mathbf{i} + \frac{\partial f}{\partial y}\mathbf{j} + \frac{\partial f}{\partial z}\mathbf{k}$$

This vector equation is equivalent to the system of three scalar equations

$$\frac{\partial f}{\partial x} = 2xyz^2 \qquad (8)$$

$$\frac{\partial f}{\partial y} = x^2z^2 \qquad (9)$$

$$\frac{\partial f}{\partial z} = 2x^2yz \qquad (10)$$

Integrating Equation (8) with respect to $x$ (so that $y$ and $z$, are treated as constants), we have

$$f(x, y, z) = x^2yz^2 + g(y, z) \qquad (11)$$

where $g(y, z)$ is the constant of integration. To determine $g(y, z)$, we differentiate Equation (11) with respect to $y$, obtaining

$$\frac{\partial f}{\partial y} = x^2z^2 + \frac{\partial g}{\partial y} \qquad (12)$$

Comparing Equation (12) with Equation (9) leads to

$$x^2z^2 + \frac{\partial g}{\partial y} = x^2z^2$$

or

$$\frac{\partial g}{\partial y} = 0 \qquad (13)$$

Integrating Equation (13) with respect to $y$ (so that $z$, is treated as a constant), we obtain $g(y, z) = h(z)$, so

$$f(x, y, z) = x^2yz^2 + h(z) \qquad (14)$$

Differentiating Equation (14) with respect to $z$, and comparing the result with Equation (10), we have

$$\frac{\partial f}{\partial z} = 2x^2yz + h'(z) = 2x^2yz$$

Therefore, $h'(z) = 0$ and $h(z) = C$, where $C$ is a constant. Finally, substituting the value of $h(z)$ into Equation (14) gives

$$f(x, y, z) = x^2yz^2 + C$$

as the desired potential function.

**b.** Since $\mathbf{F}$ is conservative, we know that the work done by $\mathbf{F}$ in moving a particle from $(0, 1, 0)$ to $(1, 2, -1)$ is independent of the path connecting these two points. Therefore, the work done by $\mathbf{F}$ is

$$W = \int_C \mathbf{F} \cdot d\mathbf{r} = \int_C \nabla f \cdot d\mathbf{r} = f(1, 2, -1) - f(0, 1, 0)$$

$$= (1)^2(2)(-1)^2 - 0 = 2$$

■

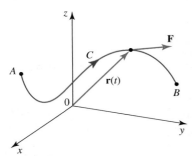

**FIGURE 10**
The path $C$ of a body from $A = \mathbf{r}(a)$
to $B = \mathbf{r}(b)$

## Conservation of Energy

The Fundamental Theorem for Line Integrals can be used to derive one of the most important laws of physics: the Law of Conservation of Energy. Suppose that a body of mass $m$ is moved from $A$ to $B$ along a piecewise-smooth curve $C$ such that its position at any time $t$ is given by $\mathbf{r}(t)$, $a \le t \le b$, and suppose that the body is subjected to the action of a continuous conservative force field $\mathbf{F}$. (See Figure 10.) To find the work done by the force on the body, we use Newton's Second Law of Motion to write $\mathbf{F} = m\mathbf{a} = m\mathbf{v}'(t) = m\mathbf{r}''(t)$, where $\mathbf{v}(t) = \mathbf{r}'(t)$ and $\mathbf{a}(t) = \mathbf{r}''(t)$ are the velocity and acceleration of the body at any time $t$, respectively. The work done by the force $\mathbf{F}$ on the body as it is moved from $A$ to $B$ along $C$ is

$$W = \int_C \mathbf{F} \cdot d\mathbf{r} = \int_a^b \mathbf{F}(\mathbf{r}(t)) \cdot \mathbf{r}'(t)\, dt$$

$$= \int_a^b m\mathbf{r}''(t) \cdot \mathbf{r}'(t)\, dt$$

$$= \frac{m}{2} \int_a^b \frac{d}{dt}[\mathbf{r}'(t) \cdot \mathbf{r}'(t)]\, dt \qquad \text{Use Theorem 2 in Section 11.2.}$$

$$= \frac{m}{2} \int_a^b \frac{d}{dt} |\mathbf{r}'(t)|^2\, dt$$

$$= \frac{m}{2} \Big[|\mathbf{r}'(t)|^2\Big]_a^b \qquad \text{Use the Fundamental Theorem for Line Integrals.}$$

$$= \frac{m}{2} \left(|\mathbf{r}'(b)|^2 - |\mathbf{r}'(a)|^2\right)$$

$$= \frac{1}{2} m|\mathbf{v}(b)|^2 - \frac{1}{2} m|\mathbf{v}(a)|^2 \qquad \text{Since } \mathbf{v}(t) = \mathbf{r}'(t)$$

Since the kinetic energy $K$ of a particle of mass $m$ and speed $v$ is $\frac{1}{2}mv^2$, we can write

$$W = K(B) - K(A) \tag{15}$$

which says that the work done by the force field on the body as it moves from $A$ to $B$ along $C$ is equal to the change in kinetic energy of the body at $A$ and $B$.

Since $\mathbf{F}$ is conservative, there is a scalar function $f$ such that $\mathbf{F} = \nabla f$. The **potential energy** $P$ of a body at the point $(x, y, z)$ in a conservative force field is defined to be $P(x, y, z) = -f(x, y, z)$, so we have $\mathbf{F} = -\nabla P$. Consequently, the work done by $\mathbf{F}$ on the body as it is moved from $A$ to $B$ along $C$ is given by

$$W = \int_C \mathbf{F} \cdot d\mathbf{r} = -\int_C \nabla P \cdot d\mathbf{r}$$

$$= \Big[-P(\mathbf{r}(t))\Big]_a^b = -[P(\mathbf{r}(b)) - P(\mathbf{r}(a))]$$

$$= P(A) - P(B)$$

Comparing this equation with Equation (15), we see that

$$P(A) + K(A) = P(B) + K(B)$$

which states that as the body moves from one point to another in a conservative force field, then the sum of its potential energy and kinetic energy remains constant. This is the **Law of Conservation of Energy** and is the reason why certain vector fields are called *conservative*.

## 14.4 CONCEPT QUESTIONS

1. State the Fundamental Theorem for Line Integrals.
2. **a.** Explain what it means for the line integral $\int_C \mathbf{F} \cdot d\mathbf{r}$ to be independent of path?
   **b.** If $\int_C \mathbf{F} \cdot d\mathbf{r}$ is independent of path for all paths $C$ in an open, connected region $R$, what can you say about $\mathbf{F}$?

3. **a.** How do you determine whether a vector field $\mathbf{F} = P(x, y)\mathbf{i} + Q(x, y)\mathbf{j}$ is conservative?
   **b.** How do you determine whether a vector field $\mathbf{F} = P(x, y, z)\mathbf{i} + Q(x, y, z)\mathbf{j} + R(x, y, z)\mathbf{k}$ is conservative?

## 14.4 EXERCISES

*In Exercises 1–10, determine whether* $\mathbf{F}$ *is conservative. If so, find a function f such that* $\mathbf{F} = \nabla f$.

1. $\mathbf{F}(x, y) = (4x + 3y)\mathbf{i} + (3x - 2y)\mathbf{j}$
2. $\mathbf{F}(x, y) = (2x^2 + 4y)\mathbf{i} + (2x - 3y^2)\mathbf{j}$
3. $\mathbf{F}(x, y) = (2x + y^2)\mathbf{i} + (x^2 + y)\mathbf{j}$
4. $\mathbf{F}(x, y) = (x^2 + y^2)\mathbf{i} + 2xy\mathbf{j}$
5. $\mathbf{F}(x, y) = y^2 \cos x\mathbf{i} + (2y \sin x + 3)\mathbf{j}$
6. $\mathbf{F}(x, y) = (x \cos y + \sin y)\mathbf{i} + (\cos y - x \sin y)\mathbf{j}$
7. $\mathbf{F}(x, y) = (e^{-x} - 2y \cos 2x)\mathbf{i} + (\sin 2x + ye^{-x})\mathbf{j}$
8. $\mathbf{F}(x, y) = (\tan y + 2xy)\mathbf{i} + (x \sec^2 y + x^2)\mathbf{j}$
9. $\mathbf{F}(x, y) = \left(x^2 + \dfrac{y}{x}\right)\mathbf{i} + (y^2 + \ln x)\mathbf{j}$
10. $\mathbf{F}(x, y) = (e^x \cos y + y \sec^2 x)\mathbf{i} + (\tan x - e^x \cos y)\mathbf{j}$

*In Exercises 11–18, (a) show that* $\mathbf{F}$ *is conservative and find a function f such that* $\mathbf{F} = \nabla f$, *and (b) use the result of part (a) to evaluate* $\int_C \mathbf{F} \cdot d\mathbf{r}$, *where C is any path from* $A(x_0, y_0)$ *to* $B(x_1, y_1)$.

11. $\mathbf{F}(x, y) = (2y + 1)\mathbf{i} + (2x + 3)\mathbf{j}$; $A(0, 0)$ and $B(-1, 1)$
12. $\mathbf{F}(x, y) = (x - 2y)\mathbf{i} + (y - 2x)\mathbf{j}$; $A(0, 0)$ and $B(1, 1)$
13. $\mathbf{F}(x, y) = (2xy^2 + 2y)\mathbf{i} + (2x^2y + 2x)\mathbf{j}$; $A(-1, 1)$ and $B(1, 2)$
14. $\mathbf{F}(x, y) = 2xy^3\mathbf{i} + (3x^2y^2 + 1)\mathbf{j}$; $A(1, 1)$ and $B(2, 0)$
15. $\mathbf{F}(x, y) = xe^{2y}\mathbf{i} + x^2e^{2y}\mathbf{j}$; $A(0, 0)$ and $B(-1, 1)$
16. $\mathbf{F}(x, y) = 2x \sin y\mathbf{i} + x^2 \cos y\mathbf{j}$; $A(0, 0)$ and $B\left(1, \frac{\pi}{2}\right)$
17. $\mathbf{F}(x, y) = e^x \sin y\mathbf{i} + (e^x \cos y + y)\mathbf{j}$; $A(0, 0)$ and $B(0, \pi)$
18. $\mathbf{F}(x, y) = (x + \tan^{-1} y)\mathbf{i} + \dfrac{x + y}{1 + y^2}\mathbf{j}$; $A(0, 0)$ and $B(1, 1)$

*In Exercises 19 and 20, evaluate* $\int_C \mathbf{F} \cdot d\mathbf{r}$ *for the vector field* $\mathbf{F}$ *and the path C. (Hint: Show that* $\mathbf{F}$ *is conservative, and pick a simpler path.)*

19. $\mathbf{F}(x, y) = (2xy^2 + \cos y)\mathbf{i} + (2x^2y - x \sin y)\mathbf{j}$
    $C: \mathbf{r}(t) = (1 - \cos t)\mathbf{i} + \sin t\mathbf{j}$, $0 \le t \le \pi$
20. $\mathbf{F}(x, y) = (e^y - y^2 \sin x)\mathbf{i} + (xe^y + 2y \cos x)\mathbf{j}$
    $C: \mathbf{r}(t) = 4 \cos t\mathbf{i} + 3 \sin t\mathbf{j}$, $0 \le t \le \pi$

*In Exercises 21 and 22, find the work done by the force field* $\mathbf{F}$ *on a particle moving along a path from P to Q.*

21. $\mathbf{F}(x, y) = 2\sqrt{y}\mathbf{i} + \dfrac{x}{\sqrt{y}}\mathbf{j}$; $A(1, 1)$, $B(2, 9)$
22. $\mathbf{F}(x, y) = -e^{-x} \cos y\mathbf{i} - e^{-x} \sin y\mathbf{j}$; $A(0, 0)$, $B(1, \pi)$
23. Show that the line integral $\int_C yz\, dx + xz\, dy + xyz\, dz$ is not independent of path.
24. Show that the following line integral is not independent of path: $\int_C e^{-y} \sin z\, dx - xe^{-y} \sin z\, dy - xe^{-y} \cos z\, dz$

*In Exercises 25–32, determine whether* $\mathbf{F}$ *is conservative. If so, find a function f such that* $\mathbf{F} = \nabla f$.

25. $\mathbf{F}(x, y, z) = yz\mathbf{i} + xz\mathbf{j} + xy\mathbf{k}$
26. $\mathbf{F}(x, y, z) = 2xy^2z\mathbf{i} + 2x^2yz\mathbf{j} + x^2y^2\mathbf{k}$
27. $\mathbf{F}(x, y, z) = 2xy\mathbf{i} + (x^2 + z^2)\mathbf{j} + xy\mathbf{k}$
28. $\mathbf{F}(x, y, z) = \sin y\mathbf{i} + (x \cos y + \cos z)\mathbf{j} + \sin z\mathbf{k}$
29. $\mathbf{F}(x, y, z) = e^x \cos z\mathbf{i} + z \sinh y\mathbf{j} + (\cosh y - e^x \sin z)\mathbf{k}$
30. $\mathbf{F}(x, y, z) = ze^{xz}\mathbf{i} + \ln z\mathbf{j} + \left(xe^{xz} + \dfrac{y}{z}\right)\mathbf{k}$
31. $\mathbf{F}(x, y, z) = z \cos(x + y)\mathbf{i} + z \sin(x + y)\mathbf{j} + \cos(x + y)\mathbf{k}$
32. $\mathbf{F}(x, y, z) = \dfrac{1}{yz}\mathbf{i} - \dfrac{x}{y^2z}\mathbf{j} - \dfrac{x}{yz^2}\mathbf{k}$

*In Exercises 33–36, (a) show that* $\mathbf{F}$ *is conservative, and find a function f such that* $\mathbf{F} = \nabla f$, *and (b) use the result of part (a) to evaluate* $\int_C \mathbf{F} \cdot d\mathbf{r}$, *where C is any curve from* $A(x_0, y_0, z_0)$ *to* $B(x_1, y_1, z_1)$.

33. $\mathbf{F}(x, y, z) = yz^2\mathbf{i} + xz^2\mathbf{j} + 2xyz\mathbf{k}$; $A(0, 0, 1)$ and $B(1, 3, 2)$
34. $\mathbf{F}(x, y, z) = 2xy^2z^3\mathbf{i} + 2x^2yz^3\mathbf{j} + 3x^2y^2z^2\mathbf{k}$; $A(0, 0, 0)$ and $B(1, 1, 1)$
35. $\mathbf{F}(x, y, z) = \cos y\mathbf{i} + (z^2 - x \sin y)\mathbf{j} + 2yz\mathbf{k}$; $A(1, 0, 0)$ and $B(2, 2\pi, 1)$
36. $\mathbf{F}(x, y, z) = e^y\mathbf{i} + (xe^y + \ln z)\mathbf{j} + \left(\dfrac{y}{z}\right)\mathbf{k}$; $A(0, 1, 1)$ and $B(1, 0, 2)$
37. Evaluate $\int_C (2xy^2 - 3)\, dx + (2x^2y + 1)\, dy$, where $C$ is the curve $x^4 - 6xy^3 - 4y^2 = 0$ from $(0, 0)$ to $(2, 1)$.

V Videos for selected exercises are available online at **www.academic.cengage.com/login**.

**38.** Evaluate $\int_C (3x^2y + e^y)\,dx + (x^3 + xe^y - 2y)\,dy$, where $C$ is the curve of Exercise 37.

**39.** Let

$$\mathbf{E}(x, y, z) = \frac{kQ}{|\mathbf{r}|^3}\,\mathbf{r}$$

where $k$ is a constant, and let $\mathbf{r} = x\mathbf{i} + y\mathbf{j} + z\mathbf{k}$ be the electric field induced by a charge $Q$ located at the origin. (See Example 4 in Section 14.1.) Find the work done by $\mathbf{E}$ in moving a charge of $q$ coulombs from the point $A(1, 3, 2)$ along any path to the point $B(2, 4, 1)$.

**40.** Find the work that is done by the force field $\mathbf{F}(x, y, z) = y^2z\mathbf{i} + 2xyz\mathbf{j} + xy^2\mathbf{k}$ on a particle moving along a path from $P(1, 1, 1)$ to $Q(2, 1, 3)$.

**41.** Let

$$\mathbf{F}(x, y) = \frac{y}{x^2 + y^2}\,\mathbf{i} - \frac{x}{x^2 + y^2}\,\mathbf{j}$$

**a.** Show that $\dfrac{\partial Q}{\partial x} = \dfrac{\partial P}{\partial y}$.

**b.** Show that $\int_C \mathbf{F} \cdot d\mathbf{r}$ is not independent of path by computing $\int_{C_1} \mathbf{F} \cdot d\mathbf{r}$ and $\int_{C_2} \mathbf{F} \cdot d\mathbf{r}$, where $C_1$ and $C_2$ are the upper and lower semicircles of radius 1, centered at the origin, from $(1, 0)$ to $(-1, 0)$.

**c.** Do your results contradict Theorem 5? Explain.

**42.** Let

$$\mathbf{F}(x, y, z) = \frac{y}{(y^2 + z^2)^2}\,\mathbf{j} + \frac{z}{(y^2 + z^2)^2}\,\mathbf{k}$$

**a.** Show that curl $\mathbf{F} = \mathbf{0}$.

**b.** Is $\mathbf{F}$ conservative? Explain.

*In Exercises 43–48, determine whether the statement is true or false. If it is true, explain why. If it is false, explain why or give an example that shows it is false.*

**43.** The region $R = \{(x, y) \mid 0 < x^2 + y^2 < 1\}$ is simply-connected.

**44.** If $\mathbf{F}$ is a nonconservative vector field, then $\int_C \mathbf{F} \cdot d\mathbf{r} \neq 0$ whenever $C$ is a closed path.

**45.** If $\mathbf{F}$ has continuous first-order partial derivatives in space and $C$ is any smooth curve, then $\int_C \nabla f \cdot d\mathbf{r}$ depends only on the endpoints of $C$.

**46.** If $\mathbf{F} = P\mathbf{i} + Q\mathbf{j}$ is in an open connected region $R$ and $\dfrac{\partial Q}{\partial x} = \dfrac{\partial P}{\partial y}$ for all $(x, y)$ in $R$, then $\int_C \mathbf{F} \cdot d\mathbf{r} = 0$ for any smooth curve $C$ in $R$.

**47.** If $\mathbf{F}(x, y)$ is continuous and $C$ is a smooth curve, then $\int_C \mathbf{F} \cdot d\mathbf{r} = -\int_{-C} \mathbf{F} \cdot d\mathbf{r}$.

**48.** If $\mathbf{F}$ has first-order partial derivatives in a simply-connected region $R$, then $\int_C \mathbf{F} \cdot d\mathbf{r} = 0$ for every closed path in $R$.

---

## 14.5 Green's Theorem

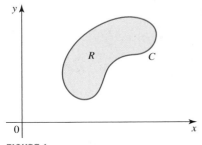

**FIGURE 1**
A plane region $R$ bounded by a simple closed plane curve $C$

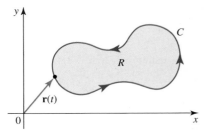

**FIGURE 2**
The curve $C$ traversed in the positive or counterclockwise direction

### ■ Green's Theorem for Simple Regions

Green's Theorem, named after the English mathematical physicist George Green (1793–1841), relates a line integral around a simple closed plane curve $C$ to a double integral over the plane region $R$ bounded by $C$. (See Figure 1.)

Before stating Green's Theorem, however, we need to explain what is meant by the *orientation* of a simple closed curve. Suppose that $C$ is defined by the vector function $\mathbf{r}(t)$, where $a \leq t \leq b$. Then $C$ is traversed in the **positive** or **counterclockwise** direction if the region $R$ is always on the left as the terminal point of $\mathbf{r}(t)$ traces the boundary curve $C$. (See Figure 2.)

**THEOREM 1    Green's Theorem**

Let $C$ be a piecewise-smooth, simple closed curve that bounds a region $R$ in the plane. If $P$ and $Q$ have continuous partial derivatives on an open set that contains $R$, then

$$\oint_C P\,dx + Q\,dy = \iint_R \left[\frac{\partial Q}{\partial x} - \frac{\partial P}{\partial y}\right] dA \qquad (1)$$

where the line integral over $C$ is taken in the positive (counterclockwise) direction.

**Note**  The notation

$$\oint_C P\,dx + Q\,dy \qquad \text{or} \qquad \oint_C P\,dx + Q\,dy$$

is sometimes used to indicate that the line integral over a simple closed curved $C$ is taken in the positive, or counterclockwise, direction. ∎

Since it is not easy to prove Green's Theorem for general regions, we will prove it only for the special case in which the region $R$ is both a $y$-simple and an $x$-simple region. (See Section 13.2.) Such regions are called **simple** or **elementary regions.**

**PROOF OF GREEN'S THEOREM FOR SIMPLE REGIONS**  Let $R$ be a simple region with boundary $C$ as shown in Figure 3. Since

$$\oint_C P\,dx + Q\,dy = \oint_C P\,dx + \oint_C Q\,dy$$

we can consider each integral on the right separately. Since $R$ is a $y$-simple region, it can be described as

$$R = \{(x, y) \mid a \le x \le b, f_1(x) \le y \le f_2(x)\}$$

where $f_1$ and $f_2$ are continuous on $[a, b]$. Observe that the boundary $C$ of $R$ consists of the curves $C_1$ and $C_2$ that are the graphs of the functions $f_1$ and $f_2$ as shown in the figure. Therefore,

$$\oint_C P\,dx = \int_{C_1} P\,dx + \int_{C_2} P\,dx$$

where $C_1$ and $C_2$ are oriented as shown in Figure 3.

Observe that the point $(x, f_1(x))$ traces $C_1$ as $x$ increases from $a$ to $b$, whereas the point $(x, f_2(x))$ traces $C_2$ as $x$ *decreases* from $b$ to $a$. Therefore,

$$
\begin{aligned}
\oint_C P\,dx &= \oint_{C_1} P\,dx + \oint_{C_2} P\,dx \\
&= \int_a^b P(x, f_1(x))\,dx + \int_b^a P(x, f_2(x))\,dx \\
&= \int_a^b P(x, f_1(x))\,dx - \int_a^b P(x, f_2(x))\,dx \\
&= \int_a^b [P(x, f_1(x)) - P(x, f_2(x))]\,dx
\end{aligned}
\tag{2}
$$

Next, we find

$$
\begin{aligned}
\iint_R \frac{\partial P}{\partial y}\,dA &= \int_a^b \int_{f_1(x)}^{f_2(x)} \frac{\partial P}{\partial y}(x, y)\,dy\,dx \\
&= \int_a^b [P(x, f_2(x)) - P(x, f_1(x))]\,dx
\end{aligned}
\tag{3}
$$

where the last equality is obtained with the aid of the Fundamental Theorem of Calculus. Comparing Equation (3) with Equation (2), we see that

$$\oint_C P\,dx = -\iint_R \frac{\partial P}{\partial y}\,dA \tag{4}$$

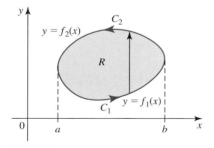

**FIGURE 3**
The simple region $R$ viewed as a $y$-simple region

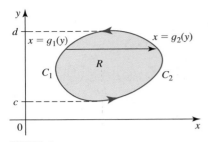

**FIGURE 4**
The simple region $R$ viewed as an $x$-simple region

By viewing $R$ as an $x$-simple region (Figure 4),

$$R = \{(x, y) \,|\, c \le y \le d, g_1(y) \le x \le g_2(y)\}$$

you can show in a similar manner that

$$\oint_C Q \, dy = \iint_R \frac{\partial Q}{\partial x} \, dA \qquad (5)$$

(See Exercise 48.) Adding Equation (4) and Equation (5), we obtain Equation (1), the conclusion of Green's Theorem for the case of a simple region. ∎

**EXAMPLE 1** Evaluate $\oint_C x^2 \, dx + (xy + y^2) \, dy$, where $C$ is the boundary of the region $R$ bounded by the graphs of $y = x$ and $y = x^2$ and is oriented in a positive direction.

**Solution** The region $R$ is shown in Figure 5. Observe that $R$ is simple. Using Green's Theorem with $P(x, y) = x^2$ and $Q(x, y) = xy + y^2$, we have

$$\oint_C x^2 \, dx + (xy + y^2) \, dy = \iint_R \left[ \frac{\partial Q}{\partial x} - \frac{\partial P}{\partial y} \right] dA = \int_0^1 \int_{x^2}^{x} (y - 0) \, dy \, dx$$

$$= \int_0^1 \left[ \frac{1}{2} y^2 \right]_{y=x^2}^{y=x} dx = \frac{1}{2} \int_0^1 (x^2 - x^4) \, dx$$

$$= \frac{1}{2} \left( \frac{1}{3} x^3 - \frac{1}{5} x^5 \right) \Big|_0^1 = \frac{1}{15}$$ ∎

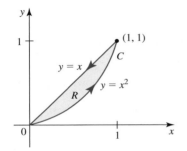

**FIGURE 5**
The curve $C$ is the boundary of the region $R$.

**EXAMPLE 2** Evaluate $\oint_C (y^2 + \tan x) \, dx + (x^3 + 2xy + \sqrt{y}) \, dy$, where $C$ is the circle $x^2 + y^2 = 4$ and is oriented in a positive direction.

**Solution** The simple region $R$ bounded by $C$ is the disk $R = \{(x, y) \,|\, x^2 + y^2 \le 4\}$ shown in Figure 6. Using Green's Theorem with $P(x, y) = y^2 + \tan x$ and $Q(x, y) = x^3 + 2xy + \sqrt{y}$, we find

$$\frac{\partial Q}{\partial x} = \frac{\partial}{\partial x} (x^3 + 2xy + \sqrt{y}) = 3x^2 + 2y \quad \text{and} \quad \frac{\partial P}{\partial y} = \frac{\partial}{\partial y} (y^2 + \tan x) = 2y$$

and so

$$\oint_C (y^2 + \tan x) \, dx + (x^3 + 2xy + \sqrt{y}) \, dy = \iint_R \left[ \frac{\partial Q}{\partial x} - \frac{\partial P}{\partial y} \right] dA = \iint_R 3x^2 \, dA$$

$$= 3 \int_0^{2\pi} \int_0^2 (r \cos \theta)^2 r \, dr \, d\theta \qquad \text{Use polar coordinates.}$$

$$= 3 \int_0^{2\pi} \int_0^2 r^3 \cos^2 \theta \, dr \, d\theta$$

$$= 3 \int_0^{2\pi} \left[ \frac{1}{4} r^4 \cos^2 \theta \right]_{r=0}^{r=2} d\theta$$

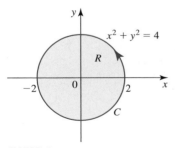

**FIGURE 6**
The region $R$ is the disk bounded by the circle $x^2 + y^2 \le 4$.

$$= 12 \int_0^{2\pi} \cos^2 \theta \, d\theta$$

$$= 6 \int_0^{2\pi} (1 + \cos 2\theta) \, d\theta$$

$$= 6 \left[ \theta + \frac{1}{2} \sin 2\theta \right]_0^{2\pi} = 12\pi \qquad \blacksquare$$

The results obtained in Examples 1 and 2 can be verified by evaluating the given line integrals directly without the benefit of Green's Theorem, but this entails much more work than evaluating the corresponding double integrals. In certain situations, however, the opposite is true; that is, it is easier to evaluate a line integral than it is to evaluate the corresponding double integral. This fact is exploited in the following formulas based on Green's Theorem for finding the area of a plane region.

---

**THEOREM 2   Finding Area Using Line Integrals**

Let $R$ be a plane region bounded by a piecewise-smooth simple closed curve $C$. Then the area of $R$ is given by

$$A = \oint_C x \, dy = -\oint_C y \, dx = \frac{1}{2} \oint_C x \, dy - y \, dx \qquad (6)$$

---

**PROOF**   Taking $P(x, y) = 0$ and $Q(x, y) = x$, Green's Theorem gives

$$\oint_C x \, dy = \iint_R \left[ \frac{\partial Q}{\partial x} - \frac{\partial P}{\partial y} \right] dA = \iint_R 1 \, dA = A$$

Similarly, by taking $P(x, y) = -y$ and $Q(x, y) = 0$, we have

$$\oint_C -y \, dx = \iint_R \left[ \frac{\partial Q}{\partial x} - \frac{\partial P}{\partial y} \right] dA = \iint_R 1 \, dA = A$$

Finally, with $P(x, y) = -\frac{1}{2}y$ and $Q(x, y) = \frac{1}{2}x$, we have

$$\oint_C -\frac{1}{2} y \, dx + \frac{1}{2} x \, dy = \oint_C P \, dx + Q \, dy = \iint_R \left[ \frac{\partial Q}{\partial x} - \frac{\partial P}{\partial y} \right] dA = \iint_R \left( \frac{1}{2} + \frac{1}{2} \right) dA = A \qquad \blacksquare$$

**EXAMPLE 3**   Find the area enclosed by the ellipse $\dfrac{x^2}{a^2} + \dfrac{y^2}{b^2} = 1$.

**Solution**   The ellipse $C$ can be represented by the parametric equations $x = a \cos t$ and $y = b \sin t$, where $0 \le t \le 2\pi$. Also observe that the ellipse is traced in the counterclockwise direction as $t$ increases from $0$ to $2\pi$. Using Equation (6), we have

$$A = \frac{1}{2} \oint_C x \, dy - y \, dx = \frac{1}{2} \int_0^{2\pi} (a \cos t)(b \cos t) \, dt - (b \sin t)(-a \sin t) \, dt$$

$$= \frac{ab}{2} \int_0^{2\pi} dt = \pi ab \qquad \blacksquare$$

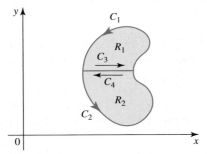

**FIGURE 7**
The region $R$ is the union of two simple regions $R_1$ and $R_2$.

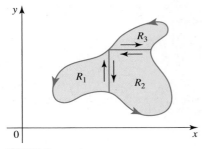

**FIGURE 8**
The region $R$ is a union of three simple regions $R_1$, $R_2$, and $R_3$.

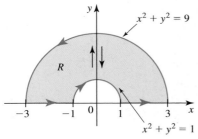

**FIGURE 9**
The region $R$ is divided into two simple regions by the crosscut that lies on the $y$-axis.

## ■ Green's Theorem for More General Regions

So far, we have proved Green's Theorem for the case in which $R$ is a simple region, but the theorem can be extended to the case in which the region $R$ is a finite union of simple regions. For example, the region $R$ shown in Figure 7 is not simple, but it can be written as $R = R_1 \cup R_2$, where $R_1$ and $R_2$ are both simple. The boundary of $R_1$ is $C_1 \cup C_3$, and the boundary of $R_2$ is $C_2 \cup C_4$, where $C_3$ and $C_4$ are paths along the crosscut traversed in the indicated directions.

Applying Green's Theorem to each of the regions $R_1$ and $R_2$ gives

$$\oint_{C_1 \cup C_3} P\, dx + Q\, dy = \iint_{R_1} \left[ \frac{\partial Q}{\partial x} - \frac{\partial P}{\partial y} \right] dA$$

and

$$\oint_{C_2 \cup C_4} P\, dx + Q\, dy = \iint_{R_2} \left[ \frac{\partial Q}{\partial x} - \frac{\partial P}{\partial y} \right] dA$$

Adding these two equations and observing that the line integrals along $C_3$ and $C_4$ cancel each other, we obtain

$$\oint_{C_1 \cup C_3} P\, dx + Q\, dy + \oint_{C_2 \cup C_4} P\, dx + Q\, dy = \oint_{C_1 \cup C_2} P\, dx + Q\, dy = \iint_{R} \left[ \frac{\partial Q}{\partial x} - \frac{\partial P}{\partial y} \right] dA$$

which is Green's Theorem for the region $R = R_1 \cup R_2$ with boundary $C = C_1 \cup C_2$.

A similar argument enables us to establish Green's Theorem for the general case in which $R$ is the union of any finite number of nonoverlapping, except perhaps for the common boundaries, simple regions (see Figure 8).

**EXAMPLE 4** Evaluate $\oint_C (e^x + y^2)\, dx + (x^2 + 3xy)\, dy$, where $C$ is the positively oriented closed curve lying on the boundary of the semiannular region $R$ bounded by the upper semicircles $x^2 + y^2 = 1$ and $x^2 + y^2 = 9$ and the $x$-axis as shown in Figure 9.

**Solution** The region $R$ is not simple, but it can be divided into two simple regions by means of the crosscut that is the intersection of $R$ and the $y$-axis. Also notice that in polar coordinates,

$$R = \{(r, \theta)\,|\, 1 \leq r \leq 3, 0 \leq \theta \leq \pi\}$$

Using Green's Theorem with $P(x, y) = e^x + y^2$ and $Q(x, y) = x^2 + 3xy$, we have

$$\frac{\partial Q}{\partial x} = \frac{\partial}{\partial x}(x^2 + 3xy) = 2x + 3y \qquad \text{and} \qquad \frac{\partial P}{\partial y} = \frac{\partial}{\partial y}(e^x + y^2) = 2y$$

and so

$$\oint_C (e^x + y^2)\, dx + (x^2 + 3xy)\, dy = \iint_R \left[ \frac{\partial Q}{\partial x} - \frac{\partial P}{\partial y} \right] dA = \iint_R (2x + y)\, dA$$

$$= \int_0^\pi \int_1^3 (2r \cos \theta + r \sin \theta) r\, dr\, d\theta \qquad \text{Use polar coordinates.}$$

$$= \int_0^\pi (2 \cos \theta + \sin \theta) \left[ \frac{1}{3} r^3 \right]_1^3 d\theta$$

$$= \frac{26}{3} \left[ 2 \sin \theta - \cos \theta \right]_0^\pi = \frac{52}{3} \qquad ■$$

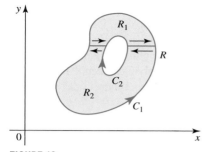

**FIGURE 10**
The annular region $R$ can be divided into two simple regions using two crosscuts.

Green's Theorem can be extended to even more general regions. Recall that a region $R$ is simply-connected if for every simple closed curve $C$ that lies in $R$, the region bounded by $C$ is also in $R$. Thus, as was noted earlier, a simply-connected region "has no holes." For example, a rectangle is simply-connected, but an annulus (a ring bounded by two concentric circles) is not. Also, **multiply-connected regions** may have one or more holes in them and also may have boundaries that consist of two or more simple closed curves. For example, the annular region $R$ shown in Figure 10 has a boundary $C$ consisting of two simple closed curves $C_1$ and $C_2$. Observe that $C$ is traversed in the positive direction provided that $C_1$ is traversed in the counterclockwise direction and $C_2$ is traversed in the clockwise direction (so that the region $R$ always lies to the left as the curve is traced).

The region $R$ can be divided into two simple regions, $R_1$ and $R_2$, by means of two crosscuts, as shown in Figure 10. Applying Green's Theorem to each of these subregions of $R$, we obtain

$$\iint_R \left[ \frac{\partial Q}{\partial x} - \frac{\partial P}{\partial y} \right] dA = \iint_{R_1} \left[ \frac{\partial Q}{\partial x} - \frac{\partial P}{\partial y} \right] dA + \iint_{R_2} \left[ \frac{\partial Q}{\partial x} - \frac{\partial P}{\partial y} \right] dA$$

$$= \int_{\partial R_1} P \, dx + Q \, dy + \int_{\partial R_2} P \, dx + Q \, dy$$

where $\partial R_1$ and $\partial R_2$ denote the boundaries of $R_1$ and $R_2$, respectively. Since the line integrals along the crosscuts are traversed in opposite directions, they cancel out, and we have

$$\iint_R \left[ \frac{\partial Q}{\partial x} - \frac{\partial P}{\partial y} \right] dA = \oint_{C_1} P \, dx + Q \, dy + \oint_{C_2} P \, dx + Q \, dy = \oint_C P \, dx + Q \, dy$$

which is Green's Theorem for the region $R$. Observe that the second line integral above is traversed in the clockwise direction.

**EXAMPLE 5** Let $C$ be a smooth, simple, closed curve that does not pass through the origin. Show that

$$\oint_C -\frac{y}{x^2 + y^2} \, dx + \frac{x}{x^2 + y^2} \, dy$$

is equal to zero if $C$ does not enclose the origin but is equal to $2\pi$ if $C$ encloses the origin.

**Solution** Suppose that $C$ does not enclose the origin. (See Figure 11.) Using Green's Theorem with $P(x, y) = -y/(x^2 + y^2)$ and $Q(x, y) = x/(x^2 + y^2)$ so that

$$\frac{\partial Q}{\partial x} = \frac{(x^2 + y^2)(1) - x(2x)}{(x^2 + y^2)^2} = \frac{y^2 - x^2}{(x^2 + y^2)^2}$$

and

$$\frac{\partial P}{\partial y} = \frac{(x^2 + y^2)(-1) - (-y)(2y)}{(x^2 + y^2)^2} = \frac{y^2 - x^2}{(x^2 + y^2)^2} = \frac{\partial Q}{\partial x}$$

we obtain

$$\oint_C -\frac{y}{x^2 + y^2} \, dx + \frac{x}{x^2 + y^2} \, dy = \iint_R \left[ \frac{\partial Q}{\partial x} - \frac{\partial P}{\partial y} \right] dA = \iint_R 0 \, dA = 0$$

Here, $R$ denotes the region enclosed by $C$.

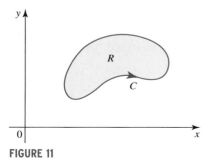

**FIGURE 11**
$C$ does not enclose the origin.

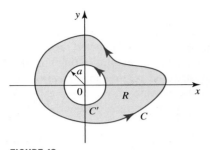

**FIGURE 12**
*C* encloses the origin.

Next, suppose that *C encloses* the origin. Since *P* and *Q* are *not* continuous in the region enclosed by *C*, Green's Theorem is not directly applicable. Let *C'* be a counterclockwise-oriented circle with center at the origin and radius *a* chosen small enough so that *C'* lies inside *C*. (See Figure 12.) Then both *P* and *Q* have continuous partial derivatives in the annular region bounded by *C* and *C'*. Applying Green's Theorem to the multiply-connected region *R* with its positively oriented boundary $C \cup (-C')$, we obtain

$$\oint_C P\,dx + Q\,dy + \oint_{-C'} P\,dx + Q\,dy = \iint_R \left[ \frac{\partial Q}{\partial x} - \frac{\partial P}{\partial y} \right] dA = \iint_R 0\,dA = 0$$

or, upon reversing the direction of traversal of the second line integral,

$$\oint_C P\,dx + Q\,dy - \oint_{C'} P\,dx + Q\,dy = 0$$

Therefore,

$$\oint_C P\,dx + Q\,dy = \oint_{C'} P\,dx + Q\,dy$$

Up to this point, we have shown that the required line integral is equal to the line integral taken over the circle *C'* in the counterclockwise direction. To evaluate this integral, we represent the circle by the parametric equations $x = a \cos t$ and $y = a \sin t$, where $0 \le t \le 2\pi$. We obtain

$$\oint_{C'} -\frac{y}{x^2 + y^2}\,dx + \frac{x}{x^2 + y^2}\,dy = \int_0^{2\pi} -\frac{(a \sin t)(-a \sin t)}{(a \cos t)^2 + (a \sin t)^2}\,dt + \frac{(a \cos t)(a \cos t)}{(a \cos t)^2 + (a \sin t)^2}\,dt$$

$$= \int_0^{2\pi} 1\,dt = 2\pi$$

Therefore,

$$\oint_C -\frac{y}{x^2 + y^2}\,dx + \frac{x}{x^2 + y^2}\,dy = 2\pi \qquad \blacksquare$$

## ■ Vector Form of Green's Theorem

The vector form of Green's Theorem has two useful versions: one involving the curl of a vector field and another involving the divergence of a vector field.

Suppose that the curve *C*, the plane region *R*, and the functions *P* and *Q* satisfy the hypothesis of Green's Theorem. Let $\mathbf{F} = P\mathbf{i} + Q\mathbf{j}$ be a vector field. Then

$$\oint_C \mathbf{F} \cdot \mathbf{T}\,ds = \oint_C P\,dx + Q\,dy$$

Recalling that *P* and *Q* are functions of *x* and *y*, we have

$$\text{curl } \mathbf{F} = \nabla \times \mathbf{F} = \begin{vmatrix} \mathbf{i} & \mathbf{j} & \mathbf{k} \\ \dfrac{\partial}{\partial x} & \dfrac{\partial}{\partial y} & \dfrac{\partial}{\partial z} \\ P & Q & 0 \end{vmatrix} = \left( \frac{\partial Q}{\partial x} - \frac{\partial P}{\partial y} \right)\mathbf{k}$$

Remember that *P* and *Q* are functions of *x* and *y*.

so

$$(\text{curl }\mathbf{F}) \cdot \mathbf{k} = \left(\frac{\partial Q}{\partial x} - \frac{\partial P}{\partial y}\right)\mathbf{k} \cdot \mathbf{k} = \frac{\partial Q}{\partial x} - \frac{\partial P}{\partial y}$$

Therefore, Green's Theorem can be written in the vector form

$$\oint_C \mathbf{F} \cdot \mathbf{T}\, ds = \iint_R \text{curl }\mathbf{F} \cdot \mathbf{k}\, dA \tag{7}$$

Equation (7) states that the line integral of the tangential component of $\mathbf{F}$ around a closed curve $C$ is equal to the double integral of the normal component to $R$ of curl $\mathbf{F}$ over the region $R$ enclosed by $C$.

Next, let the curve $C$ be represented by the vector equation $\mathbf{r}(t) = x(t)\mathbf{i} + y(t)\mathbf{j}$, $a \le t \le b$. Then the outer unit normal vector to $C$ is

$$\mathbf{n}(t) = \frac{y'(t)}{|\mathbf{r}'(t)|}\mathbf{i} - \frac{x'(t)}{|\mathbf{r}'(t)|}\mathbf{j}$$

which you can verify by showing that $\mathbf{n}(t) \cdot \mathbf{T}(t) = 0$, where

$$\mathbf{T}(t) = \frac{x'(t)}{|\mathbf{r}'(t)|}\mathbf{i} + \frac{y'(t)}{|\mathbf{r}'(t)|}\mathbf{j}$$

is the unit tangent vector to $C$. (See Figure 13.) We have

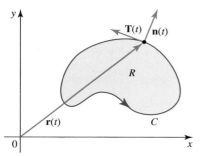

**FIGURE 13**
$\mathbf{n}(t)$ is the outer normal vector to $C$.

$$\oint_C \mathbf{F} \cdot \mathbf{n}\, ds = \int_a^b (\mathbf{F} \cdot \mathbf{n})(t)|\mathbf{r}'(t)|\, dt$$

$$= \int_a^b \left[\frac{P(x(t), y(t))y'(t)}{|\mathbf{r}'(t)|} - \frac{Q(x(t), y(t))x'(t)}{|\mathbf{r}'(t)|}\right]|\mathbf{r}'(t)|\, dt$$

$$= \int_a^b P(x(t), y(t))y'(t)\, dt - \int_a^b Q(x(t), y(t))x'(t)\, dt$$

$$= \oint_C P\, dy - Q\, dx$$

But by Green's Theorem,

$$\oint_C P\, dy - Q\, dx = \iint_R \left[\frac{\partial}{\partial x}(P) - \frac{\partial}{\partial y}(-Q)\right] dA$$

$$= \iint_R \left(\frac{\partial P}{\partial x} + \frac{\partial Q}{\partial y}\right) dA$$

Observing that the integrand of the last integral is just the divergence of $\mathbf{F}$, we obtain the second vector form of Green's Theorem:

$$\oint_C \mathbf{F} \cdot \mathbf{n}\, ds = \iint_R \text{div }\mathbf{F}\, dA \tag{8}$$

Equation (8) states that the line integral of the normal component of $\mathbf{F}$ around a closed curve $C$ is equal to the double integral of the divergence of $\mathbf{F}$ over $R$.

## 14.5    CONCEPT QUESTIONS

1. State Green's Theorem.

2. Write three line integrals that give the area of a region bounded by a piecewise smooth curve $C$.

## 14.5    EXERCISES

In Exercises 1–4, evaluate the line integral (a) directly and (b) by using Green's Theorem, where $C$ is positively oriented.

1. $\oint_C 2xy\, dx + 3xy^2\, dy$, where $C$ is the square with vertices $(0, 0)$, $(1, 0)$, $(1, 1)$, and $(0, 1)$

2. $\oint_C x^2\, dx + xy\, dy$, where $C$ is the triangle with vertices $(0, 0)$, $(1, 0)$, and $(0, 1)$

3. $\oint_C y^2\, dx + (x^2 + 2xy)\, dy$, where $C$ is the boundary of the region bounded by the graphs of $y = x$ and $y = x^3$ lying in the first quadrant

4. $\oint_C 2x\, dx - 3y\, dy$, where $C$ is the circle $x^2 + y^2 = a^2$

In Exercises 5–16, use Green's Theorem to evaluate the line integral along the positively oriented closed curve $C$.

5. $\oint_C x^3\, dx + xy\, dy$, where $C$ is the triangle with vertices $(0, 0)$, $(1, 1)$, and $(0, 1)$

6. $\oint_C (x^2 + y^2)\, dx - 2xy\, dy$, where $C$ is the square with vertices $(\pm 1, \pm 1)$

7. $\oint_C (x^2 y + x^3)\, dx + 2xy\, dy$, where $C$ is the boundary of the region bounded by the graphs of $y = x$ and $y = x^2$

8. $\oint_C (-y^3 + \cos x)\, dx + e^{y^2}\, dy$, where $C$ is the boundary of the region bounded by the parabolas $y = x^2$ and $x = y^2$

9. $\oint_C (y^2 + \cos x)\, dx + (x - \tan^{-1} y)\, dy$, where $C$ is the boundary of the region bounded by the graphs of $y = 4 - x^2$ and $y = 0$

10. $\oint_C x^2 y\, dx + y^3\, dy$, where $C$ consists of the line segment from $(-1, 0)$ to $(1, 0)$ and the upper half of the circle $x^2 + y^2 = 1$

11. $\oint_C (x^2 - y)\, dx + \sqrt{1 + y^2}\, dy$, where $C$ is the astroid $x^{2/3} + y^{2/3} = a^{2/3}$

12. $\oint_C 6xy\, dx + (3x^2 + \ln(1 + y))\, dy$, where $C$ is the cardioid $r = 1 + \cos\theta$

13. $\oint_C (x + e^x \sin y)\, dx + (x + e^x \cos y)\, dy$, where $C$ is the ellipse $\dfrac{x^2}{9} + \dfrac{y^2}{4} = 1$

14. $\oint_C \dfrac{y}{1 + x^2}\, dx + (x + \tan^{-1} x)\, dy$, where $C$ is the right-hand loop of the lemniscate $r^2 = \cos 2\theta$

15. $\oint_C (-y\, dx + x\, dy)$, where $C$ is the boundary of the annular region formed by circles $x^2 + y^2 = 1$ and $x^2 + y^2 = 4$

16. $\oint_C 3x^2 y\, dx + (x^3 + x)\, dy$, where $C$ is the boundary of the region lying between the ellipse $\dfrac{x^2}{4} + \dfrac{y^2}{9} = 1$ and the circle $x^2 + y^2 = 1$

17. Use Green's Theorem to find the work done by the force $\mathbf{F}(x, y) = (x^2 - y^2)\mathbf{i} + 2xy\mathbf{j}$ in moving a particle in the positive direction once around the triangle with vertices $(0, 0)$, $(1, 0)$, and $(0, 1)$.

18. Use Green's Theorem to find the work done by the force $\mathbf{F}(x, y) = 3y\mathbf{i} - 2x\mathbf{j}$ in moving a particle once around the ellipse $\dfrac{x^2}{4} + \dfrac{y^2}{9} = 1$ in the clockwise direction.

In Exercises 19–22, use one of the formulas on page 1267 to find the area of the indicated region.

19. The region enclosed by the astroid $x^{2/3} + y^{2/3} = a^{2/3}$

20. The region bounded by an arc of the cycloid $x = a(t - \sin t)$, $y = a(1 - \cos t)$, and the $x$-axis

21. The region enclosed by the curve $x = a \sin t$ and $y = b \sin 2t$

22. The region enclosed by the curve $x = \cos t$ and $y = 4 \sin^3 t$, where $0 \le t \le 2\pi$

**23. a.** Plot the curve $C$ defined by $x = t(1 - t^2)$ and $y = t^2(1 - t^3)$, where $0 \le t \le 1$.
 **b.** Find the area of the region enclosed by the curve $C$.

**24. a.** Plot the deltoid defined by $x = \frac{1}{4}(2 \cos t + \cos 2t)$ and $y = \frac{1}{4}(2 \sin t - \sin 2t)$, where $0 \le t \le 2\pi$.
 **b.** Find the area of the region enclosed by the deltoid.

**25. Swallowtail Catastrophe**
 **a.** Plot the swallowtail catastrophe defined by $x = 2t(1 - t^2)$ and $y = \frac{1}{2}t^2(3t^2 - 2)$, where $-1 \le t \le 1$.
 **b.** Find the area of the region enclosed by the swallowtail catastrophe.

**26.** Refer to the following figure. Suppose that $\displaystyle\int_{C_2} \mathbf{F} \cdot d\mathbf{r} = 3\pi$, where $\mathbf{F}(x, y) = P(x, y)\mathbf{i} + Q(x, y)\mathbf{j}$, and that $\left(\dfrac{\partial Q}{\partial x} - \dfrac{\partial P}{\partial y}\right) = 6$ for all $(x, y)$ in the region $R$ bounded by the circles $C_1$ and $C_2$, and oriented in a counterclockwise direction. Use Green's Theorem to find $\displaystyle\int_{C_1} \mathbf{F} \cdot d\mathbf{r}$.

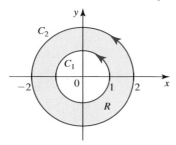

**27.** Refer to the figure below. Suppose that $\displaystyle\oint_{C_2} \mathbf{F} \cdot d\mathbf{r} = 2\pi$ and $\displaystyle\oint_{C_3} \mathbf{F} \cdot d\mathbf{r} = 3\pi$, where $\mathbf{F}(x, y) = P(x, y)\mathbf{i} + Q(x, y)\mathbf{j}$, and that $\left(\dfrac{\partial Q}{\partial x} - \dfrac{\partial P}{\partial y}\right) = 6$ for all $(x, y)$ in the region $R$ lying inside the curve $C_1$ and outside the curves $C_2$ and $C_3$. Use Green's Theorem to find $\displaystyle\oint_{C_1} \mathbf{F} \cdot d\mathbf{r}$.

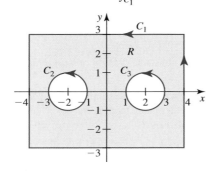

**28.** Evaluate $\int_{C_1} (x^2 + 2y)\, dx + \left(4x + e^{y^2}\right) dy$, where $C_1$ is the semi-elliptical path from $A$ to $B$ shown in the figure.
**Hint:** Use Green's Theorem, noting that $C_1 \cup C_2$, where $C_2$ is the straight path from $(-3, 0)$ to $(3, 0)$, is a closed path.

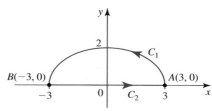

**29.** Evaluate $\int_{C_1}(x^2 + 2y)\, dx + (3x - \sinh y)\, dy$, where $C_1$ is the path $ABCDEF$ shown in the figure.
**Hint:** See the hint in Exercise 28.

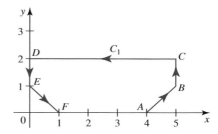

**30. a.** Let $C$ be the line segment joining the points $(x_1, y_1)$ and $(x_2, y_2)$. Show that $\int_C -y\, dx + x\, dy = x_1 y_2 - x_2 y_1$.
 **b.** Use the result of part (a) to show that the area of a polygon with vertices $(x_1, y_1), (x_2, y_2), \ldots, (x_n, y_n)$ (appearing in the counterclockwise order) is

$$A = \frac{1}{2}[(x_1 y_2 - x_2 y_1) + (x_2 y_3 - x_3 y_2) + \cdots$$
$$+ (x_{n-1} y_n - x_n y_{n-1}) + (x_n y_1 - x_1 y_n)]$$

*In Exercises 31 and 32, use the result of Exercise 30 to find the area of the shaded region.*

**31.** 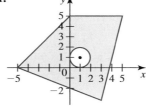    **32.**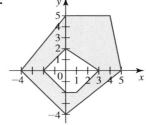

*In Exercises 33 and 34, use the result of Exercise 30 to find the area of the polygon.*

**33.** Pentagon with vertices $(0, 0)$, $(2, 0)$, $(3, 1)$, $(1, 3)$, and $(-1, 1)$.

**34.** Hexagon with vertices $(0, 0)$, $(3, 0)$, $(4, 1)$, $(2, 4)$, $(0, 3)$, and $(-2, 1)$.

**35.** Let $R$ be a plane region of area $A$ bounded by a piecewise-smooth simple closed curve $C$. Use Green's Theorem to show that the centroid of $R$ is $(\bar{x}, \bar{y})$, where

$$\bar{x} = \frac{1}{2A} \oint_C x^2 \, dy \qquad \bar{y} = -\frac{1}{2A} \oint_C y^2 \, dx$$

*In Exercises 36 and 37, use the result of Exercise 35 to find the centroid of the region.*

**36.** The triangle with vertices $(0, 0)$, $(1, 0)$, and $(1, 1)$.

**37.** The region bounded by the graphs of $y = 0$ and $y = 9 - x^2$.

**38.** A plane lamina with constant density $\rho$ has the shape of a region bounded by a piecewise-smooth simple closed curve $C$. Show that its moments of inertia about the axes are

$$I_x = -\frac{\rho}{3} \oint_C y^3 \, dx \qquad I_y = \frac{\rho}{3} \oint_C x^3 \, dy$$

**39.** Use the result of Exercise 38 to find the moment of inertia of a circular lamina of radius $a$ and constant density $\rho$ about a diameter.

**40.** Show that if $f$ and $g$ have continuous derivatives, then

$$\oint_C f(x) \, dx + g(y) \, dy = 0$$

for every piecewise-smooth simple closed curve $C$.

**41.** Let $C$ be a piecewise-smooth simple closed curve that encloses a region $R$ of area $A$. Show that

$$\oint_C (ay + b) \, dx + (cx + d) \, dy = (c - a)A$$

**42.** Let $C$ be a piecewise-smooth simple closed curve that does not pass through the origin. Evaluate

$$\oint_C \frac{x}{x^2 + y^2} \, dx + \frac{y}{x^2 + y^2} \, dy$$

(a) where $C$ does not enclose the origin and (b) where $C$ encloses the origin.

**43.** Let $P(x, y) = -\dfrac{y}{x^2 + y^2}$ and $Q(x, y) = \dfrac{x}{x^2 + y^2}$.

**a.** Show that $\oint_C (P \, dx + Q \, dy) \neq 0$, where $C$ is the circle of radius 1 centered at the origin.

**b.** Verify that $\dfrac{\partial P}{\partial y} = \dfrac{\partial Q}{\partial x}$.

**c.** Do parts (a) and (b) contradict each other? Explain.

**44.** Let $R$ be the region bounded by the circles of radius 1 and 3 centered at the origin, and let $C$ be the circle of radius 2 centered at the origin described by $\mathbf{r}(t) = 2\cos t\mathbf{i} + 2\sin t\mathbf{j}$, where $0 \le t \le 2\pi$. Let

$$P(x, y) = -\frac{y}{x^2 + y^2} \qquad \text{and} \qquad Q(x, y) = \frac{x}{x^2 + y^2}$$

**a.** Show that $\dfrac{\partial P}{\partial y} = \dfrac{\partial Q}{\partial x}$ in $R$ but $\oint_C P \, dx + Q \, dy \neq 0$.

**b.** Does this contradict Green's Theorem? Explain.

**45. a.** Use Green's Theorem to show that

$$\oint_C (\cos x + x^3 y) \, dx + (x^4 + e^y) \, dy = 0$$

where $C$ is the boundary of the square with vertices $(-1, -1)$, $(1, -1)$, $(1, 1)$, and $(-1, 1)$.

**b.** Note that

$$\frac{\partial}{\partial x}(x^4 + e^y) \neq \frac{\partial}{\partial y}(\cos x + x^3 y)$$

Does this contradict Theorem 4 of Section 14.4? Explain.

**c.** Evaluate the line integral of part (a), taking $C$ to be the boundary of the square with vertices $(0, 0)$, $(1, 0)$, $(1, 1)$, and $(0, 1)$.

**46.** Can Green's Theorem be applied to evaluate

$$\oint_C \frac{x}{\sqrt{(x - 2)^2 + y^2}} \, dx + \frac{y}{\sqrt{(x - 2)^2 + y^2}} \, dy$$

where $C$ is the circle of radius 1 centered at the origin? Explain.

**47.** Show that if $P(y)$ and $Q(x)$ have continuous derivatives, then

$$\oint_C P(y) \, dx + Q(x) \, dy = 2\big[Q(t) + P(t)\big]_{t=-1}^{t=1}$$

where $C$ is the rectangular path that is traced in a counter-clockwise direction with vertices $(-1, -1)$, $(1, -1)$, $(1, 1)$, and $(-1, 1)$.

**48.** Refer to the proof of Green's Theorem. Show that by viewing $R$ as an $x$-simple region, we have

$$\oint_C Q \, dy = \iint_R \frac{\partial Q}{\partial x} \, dA$$

*In Exercises 49–51, determine whether the statement is true or false. If it is true, explain why. If it is false, explain why or give an example that shows it is false.*

**49.** If $a$ and $b$ are constants, then $\oint_C a \, dx + b \, dy \neq 0$, where $C$ is a simple closed curve.

**50.** If $C$ is a piecewise-smooth simple closed curve that bounds a region $R$ in the plane, then $\oint_C xy^2 \, dx + (x^2 y + x) \, dy$ is equal to the area of $R$.

**51.** The work done by the force field $\mathbf{F}(x, y) = -\frac{1}{2}y\mathbf{i} + \frac{1}{2}x\mathbf{j}$ on a particle that moves once around a piecewise-smooth simple closed curve in a counterclockwise direction is numerically equal to the area of the region bounded by the curve.

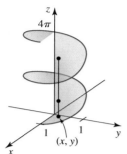

**FIGURE 1**
The helicoid shown here
is *not* the graph of a
function $z = f(x, y)$.

## 14.6 Parametric Surfaces

### ■ Why We Use Parametric Surfaces

In Chapter 12 we studied surfaces that are graphs of functions of two variables. However, not every surface is the graph of a function $z = f(x, y)$. Consider, for example, the *helicoid* shown in Figure 1. Observe that the point $(x, y)$ in the $xy$-plane is associated with *more than one* point on the helicoid, so this surface cannot be the graph of a function $z = f(x, y)$.

Just as we found it useful to describe a curve in the plane (and in space) as the image of a line under a vector-valued function **r** rather than as the graph of a function, we will now see that a similar situation exists for surfaces. Instead of a single parameter, however, we will use two parameters and view a surface in space as the image of a plane region. More specifically, we have the following.

---

**DEFINITION  Parametric Surface**

Let

$$\mathbf{r}(u, v) = x(u, v)\mathbf{i} + y(u, v)\mathbf{j} + z(u, v)\mathbf{k}$$

be a vector-valued function defined for all points $(u, v)$ in a region $D$ in the $uv$-plane. The set of all points $(x, y, z)$ in $R^3$ satisfying the **parametric equations**

$$x = x(u, v), \qquad y = y(u, v), \qquad z = z(u, v)$$

as $(u, v)$ ranges over $D$ is called a **parametric surface $S$ represented by r**. The region $D$ is called the **parameter domain.**

---

Thus, as $(u, v)$ ranges over $D$, the tip of the vector $\mathbf{r}(u, v)$ traces out the surface $S$ (see Figure 2). Put another way, we can think of **r** as mapping each point $(u, v)$ in $D$ onto a point $(x(u, v), y(u, v), z(u, v))$ on $S$ in such a way that the plane region $D$ is bent, twisted, stretched, and/or shrunk to yield the surface $S$.

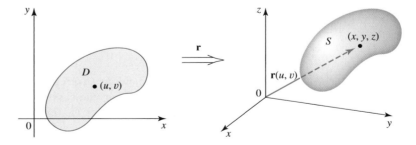

**FIGURE 2**
The function **r** maps $D$
onto the surface $S$.

**EXAMPLE 1**  Identify and sketch the surface represented by

$$\mathbf{r}(u, v) = 2 \cos u\mathbf{i} + 2 \sin u\mathbf{j} + v\mathbf{k}$$

with parameter domain $D = \{(u, v) | 0 \leq u \leq 2\pi, 0 \leq v \leq 3\}$.

**Solution**  The parametric equations for the surface are

$$x = 2 \cos u, \qquad y = 2 \sin u, \qquad z = v$$

Eliminating the parameters $u$ and $v$ in the first two equations, we obtain

$$x^2 + y^2 = 4\cos^2 u + 4\sin^2 u = 4$$

Observe that the variable $z$ is missing in this equation, so it represents a cylinder with the $z$-axis as its axis. (See Section 10.6.) Furthermore, the trace in the $xy$-plane is a circle of radius 2, and we conclude that the cylinder is a circular cylinder. Finally, because $0 \le v \le 3$, the third equation $z = v$ tells us that $0 \le z \le 3$. Thus, the required surface is the truncated cylinder shown in Figure 3.

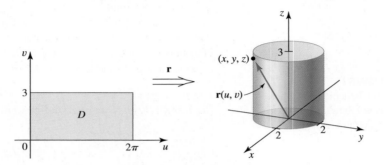

**FIGURE 3**
The function **r** "bends" the rectangular region $D$ into a cylinder.

There is another way of visualizing the way **r** maps the domain $D$ onto a surface $S$. If we fix $u$ by setting $u = u_0$, where $u_0$ is a constant, and allow $v$ to vary so that the points $(u_0, v)$ lie in $D$, then we obtain a vertical line segment $L_1$ lying in $D$. When restricted to $L_1$, the function **r** becomes a function involving one parameter $v$ whose domain is the parameter interval $L_1$. Therefore, $\mathbf{r}(u_0, v)$ maps $L_1$ onto a curve $C_1$ lying on $S$ (see Figure 4).

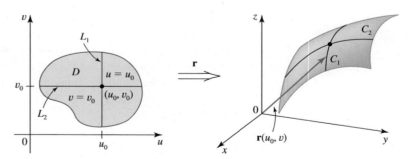

**FIGURE 4**
**r** maps $L_1$ onto $C_1$ and $L_2$ onto $C_2$.

Similarly, by holding $v$ fixed, say, $v = v_0$, where $v_0$ is a constant, the tip of the resulting vector $\mathbf{r}(u, v_0)$ traces the curve $C_2$ as $u$ is allowed to assume values in the parameter interval $L_2$. The curves $C_1$ and $C_2$ are called **grid curves.**

By way of illustration, if we set $u = u_0$ in Example 1, then both $x = 2\cos u_0$ and $y = 2\sin u_0$ are constant. So the vertical line $u = u_0$ is mapped onto the vertical line segment $(2\cos u_0, 2\sin u_0, v)$, $0 \le v \le 3$. Similarly, you can verify that a horizontal line segment $v = v_0$ in $D$ is mapped onto a circle on the cylinder at a height of $v_0$ units from the $xy$-plane.

**EXAMPLE 2**    Use a computer algebra system (CAS) to generate the surface represented by

$$\mathbf{r}(u, v) = \sin u \cos v \mathbf{i} + \sin u \sin v \mathbf{j} + \cos u \mathbf{k}$$

with parameter domain $D = \{(u, v) \mid 0 \le u \le \pi, 0 \le v \le 2\pi\}$. Identify the curves on the surface that correspond to the curves with $u$ held constant and those with $v$ held constant.

**Solution**   The required surface is the unit sphere centered at the origin. (See Figure 5a.) You can verify that this is the case by eliminating $u$ and $v$ in the parametric equations

$$x = \sin u \cos v, \qquad y = \sin u \sin v, \qquad z = \cos u$$

to obtain the rectangular equation $x^2 + y^2 + z^2 = 1$ for the sphere. Fixing $u = u_0$, where $u_0$ is a constant, leads to the equations

$$x = \sin u_0 \cos v, \qquad y = \sin u_0 \sin v, \qquad z = \cos u_0$$

We have

$$x^2 + y^2 = \sin^2 u_0 \cos^2 v + \sin^2 u_0 \sin^2 v$$
$$= \sin^2 u_0 (\cos^2 v + \sin^2 v) = \sin^2 u_0$$

The system of equations

$$\left.\begin{array}{l} x^2 + y^2 = \sin^2 u_0 \\[4pt] z = \cos u_0 \end{array}\right\}$$

for a fixed $u_0$ lying in $[0, \pi]$ or, equivalently, the vector-valued function

$$\mathbf{r}(u_0, v) = \sin u_0 \cos v\,\mathbf{i} + \sin u_0 \sin v\,\mathbf{j} + \cos u_0\mathbf{k}$$

represents a circle of radius $\sin u_0$ on the sphere that is parallel to the *xy*-plane. Thus, if we think of the sphere as a globe then the horizontal line segments in the domain of **r** are mapped onto the latitudinal lines, or *parallels*. (See Figure 5b.) Similarly, we can show that the vertical line segments in the domain of **r** with $v = v_0$, where $v_0$ is a constant, are mapped by

$$\mathbf{r}(u, v_0) = \sin u \cos v_0\,\mathbf{i} + \sin u \sin v_0\,\mathbf{j} + \cos u\mathbf{k}$$

onto the *meridians of longitude*—great circles on the surface of the globe passing through the poles.

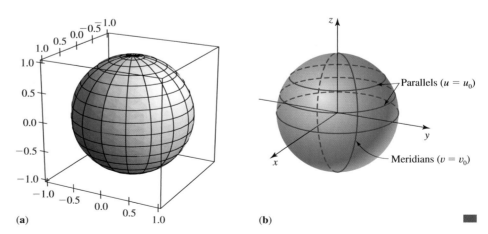

**FIGURE 5**   (a)          (b)

## ■ Finding Parametric Representations of Surfaces

We now turn our attention to finding vector-valued function representations of surfaces. We begin by showing that if a surface is the graph of a function $f(x, y)$, then it has a simple parametric representation.

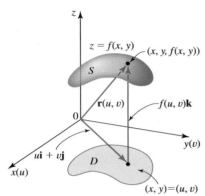

**FIGURE 6**
The vector $\mathbf{r}(u, v) = u\mathbf{i} + v\mathbf{j} + f(u, v)\mathbf{k}$ by the rule for vector addition.

**EXAMPLE 3**

a. Find a parametric representation for the graph of a function $f(x, y)$.
b. Use the result of part (a) to find a parametric representation for the elliptic paraboloid $z = 4x^2 + y^2$.

**Solution**

a. Suppose that $S$ is the graph of $z = f(x, y)$ defined on a domain $D$ in the $xy$-plane. (See Figure 6.) We simply pick $x$ and $y$ to be the parameters; in other words, we write the desired parametric equations as

$$x = x(u, v) = u, \qquad y = y(u, v) = v, \qquad z = z(u, v) = f(u, v)$$

and take the domain of $f$ to be the parameter domain. Equivalently, we obtain the vector-valued representation by writing

$$\mathbf{r}(u, v) = u\mathbf{i} + v\mathbf{j} + f(u, v)\mathbf{k}$$

b. The surface is the graph of the function $f(x, y) = 4x^2 + y^2$. So we can let $x$ and $y$ be the parameters. Thus, the required parametric equations are

$$x = u, \qquad y = v, \qquad z = 4u^2 + v^2$$

and the corresponding vector-valued function is

$$\mathbf{r}(u, v) = u\mathbf{i} + v\mathbf{j} + (4u^2 + v^2)\mathbf{k}$$

The parameter domain is $D = \{(u, v) \mid -\infty < u < \infty, -\infty < v < \infty\}$. ▪

**EXAMPLE 4**  Find a parametric representation for the cone $x = \sqrt{y^2 + z^2}$.

**Solution**  The surface is the graph of the function $f(y, z) = \sqrt{y^2 + z^2}$. So we can let $y$ and $z$ be the parameters. Thus, the required parametric equations are

$$x = \sqrt{u^2 + v^2}, \qquad y = u, \qquad z = v$$

and the corresponding vector-valued function is

$$\mathbf{r}(u, v) = \sqrt{u^2 + v^2}\,\mathbf{i} + u\mathbf{j} + v\mathbf{k}$$

The parameter domain is $D = \{(u, v) \mid -\infty < u < \infty, -\infty < v < \infty\}$. ▪

**EXAMPLE 5**

a. Find a parametric representation of the plane that passes through the point $P_0$ with position vector $\mathbf{r_0}$ and contains two nonparallel vectors $\mathbf{a}$ and $\mathbf{b}$.
b. Using the result of part (a), find a parametric representation of the plane passing through the point $P_0(3, -1, 1)$ and containing the vectors $\mathbf{a} = -2\mathbf{i} + 5\mathbf{j} + \mathbf{k}$ and $\mathbf{b} = -3\mathbf{i} + 2\mathbf{j} + 3\mathbf{k}$. (This is the plane in Example 6 in Section 10.5.)

**Solution**

a. Let $P$ be a point lying on the plane, and let $\mathbf{r} = \overrightarrow{OP}$. Since $\overrightarrow{P_0P}$ lies in the plane determined by $\mathbf{a}$ and $\mathbf{b}$, there exist real numbers $u$ and $v$ such that $\overrightarrow{P_0P} = u\mathbf{a} + v\mathbf{b}$. (See Figure 7.) Furthermore, we see that $\mathbf{r} = \mathbf{r_0} + \overrightarrow{P_0P} = \mathbf{r_0} + u\mathbf{a} + v\mathbf{b}$. Finally, since any point on the plane is located at the tip of $\mathbf{r}$ for an appropriate choice of $u$ and $v$, we see that the required representation is

$$\mathbf{r}(u, v) = \mathbf{r_0} + u\mathbf{a} + v\mathbf{b}$$

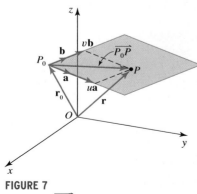

**FIGURE 7**
$\mathbf{r} = \mathbf{r_0} + \overrightarrow{P_0P} = \mathbf{r_0} + u\mathbf{a} + v\mathbf{b}$

The parameter domain is $D = \{(u, v) \mid -\infty < u < \infty, -\infty < v < \infty\}$.

**b.** The required representation is

$$\mathbf{r}(u, v) = (3\mathbf{i} - \mathbf{j} + \mathbf{k}) + u(-2\mathbf{i} + 5\mathbf{j} + \mathbf{k}) + v(-3\mathbf{i} + 2\mathbf{j} + 3\mathbf{k})$$

$$= (-2u - 3v + 3)\mathbf{i} + (5u + 2v - 1)\mathbf{j} + (u + 3v + 1)\mathbf{k}$$

with domain $D = \{(u, v) \mid -\infty < u < \infty, -\infty < v < \infty\}$. ∎

**Note** The representation in Example 5b is by no means unique. For example, an equation of the plane in question is $13x + 3y + 11z = 47$. (See Example 6 in Section 10.5.) Solving this equation for $z$ in terms of $x$ and $y$, we obtain $z = f(x, y) = \frac{1}{11}(47 - 13x - 3y)$. Thus, the plane is the graph of the function $f$, and this observation leads us to the representation

$$\mathbf{r}(u, v) = u\mathbf{i} + v\mathbf{j} + \left(\frac{47 - 13u - 3v}{11}\right)\mathbf{k}$$ ∎

The next two examples involve surfaces that are not graphs of functions.

**EXAMPLE 6** Find a parametric representation for the cone $x^2 + y^2 = z^2$.

**Solution** The cone has a simple representation $r^2 = z^2$ in cylindrical coordinates. This suggests that we choose $r$ and $\theta$ as parameters. Writing $u$ for $r$ and $v$ for $\theta$, we have

$$x = u \cos v, \qquad y = u \sin v, \qquad z = u$$

as the required parametric equations. In vector form we have

$$\mathbf{r}(u, v) = u \cos v\,\mathbf{i} + u \sin v\,\mathbf{j} + u\mathbf{k}$$ ∎

**EXAMPLE 7** Find a parametric representation for the helicoid shown in Figure 1.

**Solution** Recall that the parametric equations for a helix are

$$x = a \cos \theta, \qquad y = a \sin \theta, \qquad z = \theta$$

where $\theta$ and $z$ are in cylindrical coordinates. This suggests that we let $u$ denote $r$ and $v$ denote $\theta$. Then the parametric equations for the helicoid are

$$x = u \cos v, \qquad y = u \sin v, \qquad z = v$$

with parameter domain $D = \{(u, v) \mid -1 \le u \le 1, 0 \le v \le 4\pi\}$. In vector form we have

$$\mathbf{r}(u, v) = u \cos v\,\mathbf{i} + u \sin v\,\mathbf{j} + v\mathbf{k}$$ ∎

We now turn our attention to finding the parametric representation for surfaces of revolution. Suppose that a surface $S$ is obtained by revolving the graph of the function $y = f(x)$ for $a \le x \le b$ about the $x$-axis, where $f(x) \ge 0$. (See Figure 8.) Letting $u$ denote $x$ and $v$ denote the angle shown in the figure, we see that if $(x, y, z)$ is any point on $S$, then

$$x = u, \qquad y = f(u) \cos v, \qquad z = f(u) \sin v \tag{1}$$

or, equivalently,

$$r(u, v) = u\mathbf{i} + f(u) \cos v\,\mathbf{j} + f(u) \sin v\,\mathbf{k}$$

The parameter domain is $D = \{(u, v) \mid a \le u \le b, 0 \le v \le 2\pi\}$.

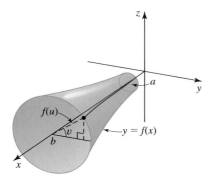

**FIGURE 8**
$S$ is obtained by revolving the graph of $f$ between $x = a$ and $x = b$ about the $x$-axis.

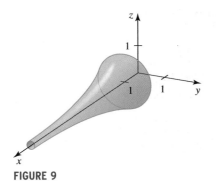

**FIGURE 9**
Gabriel's Horn

**EXAMPLE 8** **Gabriel's Horn** Find a parametric representation for the surface obtained by revolving the graph of $f(x) = 1/x$, where $1 \le x < \infty$, about the x-axis.

**Solution** Using Equation (1), we obtain the parametric equations

$$x = u, \qquad y = \frac{1}{u}\cos v, \qquad z = \frac{1}{u}\sin v$$

with parametric domain $D = \{(u, v) \mid 1 \le u < \infty, 0 \le v \le 2\pi\}$. The resulting surface is a portion of Gabriel's Horn as shown in Figure 9. ■

## ■ Tangent Planes to Parametric Surfaces

Suppose that $S$ is a parametric surface represented by the vector function

$$\mathbf{r}(u, v) = x(u, v)\mathbf{i} + y(u, v)\mathbf{j} + z(u, v)\mathbf{k}$$

and $P_0$ is a point on the surface $S$ represented by the vector $\mathbf{r}(u_0, v_0)$, where $(u_0, v_0)$ is a point in the parameter domain $D$ of $\mathbf{r}$. If we fix $u$ by putting $u = u_0$ and allow $v$ to vary, then the tip of $\mathbf{r}(u_0, v)$ traces the curve $C_1$ lying on $S$. (See Figure 10.) The tangent vector to $C_1$ at $P_0$ is given by

$$\mathbf{r}_v(u_0, v_0) = \frac{\partial x}{\partial v}(u_0, v_0)\mathbf{i} + \frac{\partial y}{\partial v}(u_0, v_0)\mathbf{j} + \frac{\partial z}{\partial v}(u_0, v_0)\mathbf{k}$$

Similarly, by holding $v$ fast, $v = v_0$, and allowing $u$ to vary, the tip of $\mathbf{r}(u, v_0)$ traces the curve $C_2$ lying on $S$, with tangent vector at $P_0$ given by

$$\mathbf{r}_u(u_0, v_0) = \frac{\partial x}{\partial u}(u_0, v_0)\mathbf{i} + \frac{\partial y}{\partial u}(u_0, v_0)\mathbf{j} + \frac{\partial z}{\partial u}(u_0, v_0)\mathbf{k}$$

If $\mathbf{r}_u(u, v) \times \mathbf{r}_v(u, v) \ne \mathbf{0}$ for each $(u, v)$ in the parameter domain of $\mathbf{r}$, then the surface $S$ is said to be **smooth.** For a smooth surface the **tangent plane** to $S$ at $P_0$ is the plane that contains the tangent vectors $\mathbf{r}_u(u_0, v_0)$ and $\mathbf{r}_v(u_0, v_0)$ and thus has a normal vector given by

$$\mathbf{n} = \mathbf{r}_u(u_0, v_0) \times \mathbf{r}_v(u_0, v_0)$$

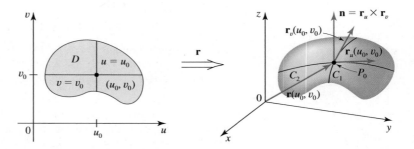

**FIGURE 10**

**EXAMPLE 9** Find an equation of the tangent plane to the helicoid

$$\mathbf{r}(u, v) = u\cos v\mathbf{i} + u\sin v\mathbf{j} + v\mathbf{k}$$

of Example 7 at the point where $u = \frac{1}{2}$ and $v = \frac{\pi}{4}$.

**Solution**  We start by finding the partial derivatives of $\mathbf{r}$. Thus,

$$\mathbf{r}_u(u, v) = \cos v\mathbf{i} + \sin v\mathbf{j}$$

$$\mathbf{r}_v(u, v) = -u \sin v\mathbf{i} + u \cos v\mathbf{j} + \mathbf{k}$$

So

$$\mathbf{r}_u\left(\frac{1}{2}, \frac{\pi}{4}\right) = \frac{\sqrt{2}}{2}\mathbf{i} + \frac{\sqrt{2}}{2}\mathbf{j}$$

$$\mathbf{r}_v\left(\frac{1}{2}, \frac{\pi}{4}\right) = -\frac{1}{2} \cdot \frac{\sqrt{2}}{2}\mathbf{i} + \frac{1}{2} \cdot \frac{\sqrt{2}}{2}\mathbf{j} + \mathbf{k} = -\frac{\sqrt{2}}{4}\mathbf{i} + \frac{\sqrt{2}}{4}\mathbf{j} + \mathbf{k}$$

A normal vector to the tangent plane is

$$\mathbf{n} = \mathbf{r}_u\left(\frac{1}{2}, \frac{\pi}{4}\right) \times \mathbf{r}_v\left(\frac{1}{2}, \frac{\pi}{4}\right) = \begin{vmatrix} \mathbf{i} & \mathbf{j} & \mathbf{k} \\ \dfrac{\sqrt{2}}{2} & \dfrac{\sqrt{2}}{2} & 0 \\ -\dfrac{\sqrt{2}}{4} & \dfrac{\sqrt{2}}{4} & 1 \end{vmatrix} = \frac{\sqrt{2}}{2}\mathbf{i} - \frac{\sqrt{2}}{2}\mathbf{j} + \frac{1}{2}\mathbf{k}$$

Since any normal vector will do, let's take $\mathbf{n} = \sqrt{2}\mathbf{i} - \sqrt{2}\mathbf{j} + \mathbf{k}$.

Next note that the point $\left(\frac{1}{2}, \frac{\pi}{4}\right)$ in the parameter domain is mapped onto the point with coordinates

$$x_0 = \frac{1}{2}\cos\frac{\pi}{4} = \frac{1}{2} \cdot \frac{\sqrt{2}}{2} = \frac{\sqrt{2}}{4}, \qquad y_0 = \frac{1}{2}\sin\frac{\pi}{4} = \frac{\sqrt{2}}{4}, \qquad z_0 = \frac{\pi}{4}$$

Therefore, an equation of the required tangent plane at $\left(\frac{\sqrt{2}}{4}, \frac{\sqrt{2}}{4}, \frac{\pi}{4}\right)$ is

$$\sqrt{2}\left(x - \frac{\sqrt{2}}{4}\right) - \sqrt{2}\left(y - \frac{\sqrt{2}}{4}\right) + 1\left(z - \frac{\pi}{4}\right) = 0$$

$$\sqrt{2}x - \sqrt{2}y + z - \frac{\pi}{4} = 0$$

or

$$4\sqrt{2}x - 4\sqrt{2}y + 4z - \pi = 0 \qquad \blacksquare$$

## ■ Area of a Parametric Surface

In Section 13.5, we learned how to find the area of a surface that is the graph of a function $z = f(x, y)$. We now take on the task of finding the areas of parametric surfaces, which are more general than the surfaces (graphs) defined by functions.

For simplicity, let's assume that the parametric surface $S$ defined by

$$\mathbf{r}(u, v) = x(u, v)\mathbf{i} + y(u, v)\mathbf{j} + z(u, v)\mathbf{k}$$

has parameter domain $R$ that is a rectangle. (See Figure 11.) Let $P$ be a regular partition of $R$ into $n = mn$ subrectangles $R_{11}, R_{12}, \ldots, R_{mn}$. The subrectangle $R_{ij}$ is mapped by $\mathbf{r}$ onto the patch $S_{ij}$ with area denoted by $\Delta S_{ij}$. Since the subrectangles $R_{ij}$ are nonoverlapping, except for their common boundaries, so are the patches $S_{ij}$, and so the area of $S$ is given by

$$S = \sum_{i=1}^{m}\sum_{j=1}^{n}\Delta S_{ij}$$

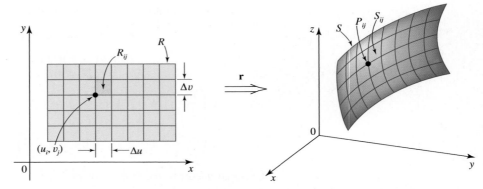

**FIGURE 11**
The subrectangle $R_{ij}$ is
mapped onto the patch $S_{ij}$.

Next, let's find an approximation of $\Delta S_{ij}$. Let $(u_i, v_j)$ be the corner of $R_{ij}$ closest to the origin with image the point $P_{ij}$ represented by $\mathbf{r}(u_i, v_j)$ as shown in Figure 12. For the sake of clarity, both $R_{ij}$ and $S_{ij}$ are shown enlarged. The sides of $S_{ij}$ with corner represented by $\mathbf{r}(u_i, v_j)$ are approximated by $\mathbf{a}$ and $\mathbf{b}$, where $\mathbf{a} = \mathbf{r}(u_i + \Delta u, v_j) - \mathbf{r}(u_i, v_j)$ and $\mathbf{b} = \mathbf{r}(u_i, v_j + \Delta v) - \mathbf{r}(u_i, v_j)$. So $\Delta S_{ij}$ may be approximated by the area of the parallelogram with $\mathbf{a}$ and $\mathbf{b}$ as adjacent sides, that is,

$$\Delta S_{ij} \approx |\mathbf{a} \times \mathbf{b}|$$

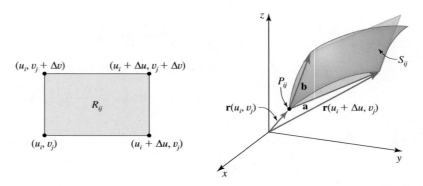

**FIGURE 12**

But we can write

$$\mathbf{a} = \left[ \frac{\mathbf{r}(u_i + \Delta u, v_j) - \mathbf{r}(u_i, v_j)}{\Delta u} \right] \Delta u$$

If $\Delta u$ is small, as we assume, then the term inside the brackets is approximately equal to $\mathbf{r}_u(u_i, v_j)$. So

$$\mathbf{a} \approx \Delta u\, \mathbf{r}_u(u_i, v_j)$$

Similarly, we see that

$$\mathbf{b} \approx \Delta v\, \mathbf{r}_v(u_i, v_j)$$

Therefore,

$$\Delta S_{ij} \approx |[\Delta u\, \mathbf{r}_u(u_i, v_j)] \times [\Delta v\, \mathbf{r}_v(u_i, v_j)]|$$
$$= |\mathbf{r}_u(u_i, v_j) \times \mathbf{r}_v(u_i, v_j)|\, \Delta u\, \Delta v$$

and the area of $S$ may be approximated by

$$\sum_{i=1}^{m} \sum_{j=1}^{n} |\mathbf{r}_u(u_i, v_j) \times \mathbf{r}_v(u_i, v_j)|\, \Delta u\, \Delta v$$

Intuitively, the approximation gets better and better as $m$ and $n$ get larger and larger. But the double sum is the Riemann sum of $|\mathbf{r}_u \times \mathbf{r}_v|$, and we are led to define the area of $S$ as

$$\lim_{m,\,n \to \infty} \sum_{i=1}^{m} \sum_{j=1}^{n} |\mathbf{r}_u(u_i, v_j) \times \mathbf{r}_v(u_i, v_j)|\, \Delta u\, \Delta v$$

Alternatively, we have the following definition.

---

**DEFINITION    Surface Area** (Parametric Form)

Let $S$ be a smooth surface represented by the equation

$$\mathbf{r}(u, v) = x(u, v)\mathbf{i} + y(u, v)\mathbf{j} + z(u, v)\mathbf{k}$$

with parameter domain $D$. If $S$ is covered just once as $(u, v)$ varies throughout $D$, then the **surface area** of $S$ is

$$A(S) = \iint_D |\mathbf{r}_u \times \mathbf{r}_v|\, dA \qquad (2)$$

---

**EXAMPLE 10**    Find the surface area of a sphere of radius $a$.

**Solution**    The sphere centered at the origin with radius $a$ is represented by the equation

$$\mathbf{r}(u, v) = a \sin u \cos v\, \mathbf{i} + a \sin u \sin v\, \mathbf{j} + a \cos u\, \mathbf{k}$$

with parameter domain $D = \{(u, v)\,|\, 0 \le u \le \pi, 0 \le v \le 2\pi\}$. We find

$$\mathbf{r}_u(u, v) = a \cos u \cos v\, \mathbf{i} + a \cos u \sin v\, \mathbf{j} - a \sin u\, \mathbf{k}$$

$$\mathbf{r}_v(u, v) = -a \sin u \sin v\, \mathbf{i} + a \sin u \cos v\, \mathbf{j}$$

and so

$$\mathbf{r}_u \times \mathbf{r}_v = \begin{vmatrix} \mathbf{i} & \mathbf{j} & \mathbf{k} \\ a \cos u \cos v & a \cos u \sin v & -a \sin u \\ -a \sin u \sin v & a \sin u \cos v & 0 \end{vmatrix}$$

$$= a^2 \sin^2 u \cos v\, \mathbf{i} + a^2 \sin^2 u \sin v\, \mathbf{j} + a^2 \sin u \cos u\, \mathbf{k}$$

Therefore

$$|\mathbf{r}_u \times \mathbf{r}_v| = \sqrt{a^4 \sin^4 u \cos^2 v + a^4 \sin^4 u \sin^2 v + a^4 \sin^2 u \cos^2 u}$$

$$= \sqrt{a^4 \sin^4 u + a^4 \sin^2 u \cos^2 u} = \sqrt{a^4 \sin^2 u}$$

$$= a^2 \sin u$$

since $\sin u \ge 0$ for $0 \le u \le \pi$. Using Equation (2), the area of the sphere is

$$A = \iint_D |\mathbf{r}_u \times \mathbf{r}_v|\, dA = \int_0^{2\pi} \int_0^{\pi} a^2 \sin u\, du\, dv$$

$$= \int_0^{2\pi} \left[ -a^2 \cos u \right]_{u=0}^{u=\pi} dv = \int_0^{2\pi} 2a^2\, dv = 2a^2(2\pi) = 4\pi a^2 \qquad \blacksquare$$

**EXAMPLE 11**  Find the area of one complete turn of the helicoid of width one represented by the equation $\mathbf{r}(u, v) = u \cos v \mathbf{i} + u \sin v \mathbf{j} + v \mathbf{k}$ with parameter domain $D = \{(u, v) \mid 0 \le u \le 1, 0 \le v \le 2\pi\}$. (Refer to Figure 1.)

**Solution**  We first find

$$\mathbf{r}_u = \cos v \mathbf{i} + \sin v \mathbf{j}$$

$$\mathbf{r}_v = -u \sin v \mathbf{i} + u \cos v \mathbf{j} + \mathbf{k}$$

and so

$$\mathbf{r}_u \times \mathbf{r}_v = \begin{vmatrix} \mathbf{i} & \mathbf{j} & \mathbf{k} \\ \cos v & \sin v & 0 \\ -u \sin v & u \cos v & 1 \end{vmatrix}$$

$$= \sin v \mathbf{i} - \cos v \mathbf{j} + (u \cos^2 v + u \sin^2 v)\mathbf{k}$$

$$= \sin v \mathbf{i} - \cos v \mathbf{j} + u\mathbf{k}$$

Therefore,

$$|\mathbf{r}_u \times \mathbf{r}_v| = \sqrt{\sin^2 v + \cos^2 v + u^2} = \sqrt{1 + u^2}$$

So, the required area is

$$A = \iint_D |\mathbf{r}_u \times \mathbf{r}_v|\, dA = \int_0^{2\pi} \int_0^1 \sqrt{1 + u^2}\, du\, dv$$

$$= \int_0^{2\pi} \left[ \frac{u}{2} \sqrt{1 + u^2} + \frac{1}{2} \ln(u + \sqrt{1 + u^2}) \right]_{u=0}^{u=1} dv$$

$$= \int_0^{2\pi} \left[ \frac{1}{2} \sqrt{2} + \frac{1}{2} \ln(1 + \sqrt{2}) \right] dv \qquad \text{Use Formula 37 from the Table of Integrals.}$$

$$= \pi[\sqrt{2} + \ln(1 + \sqrt{2})] \approx 7.212 \qquad \blacksquare$$

## 14.6  CONCEPT QUESTIONS

**1. a.** Define a parametric surface.
   **b.** What are the grid curves of a parametric surface?
**2. a.** What is a smooth surface?
   **b.** Explain how you would find an equation of the tangent plane to a smooth surface with representation $\mathbf{r}(u, v)$ at the point represented by $\mathbf{r}(u_0, v_0)$.

**3.** Write a double integral giving the area of a surface $S$ defined by a vector function $\mathbf{r}(u, v)$, where $(u, v)$ lies in the parameter domain $D$ of $\mathbf{r}$.

## 14.6 EXERCISES

*In Exercises 1–4, match the equation with one of the graphs labeled (a)–(d). Give a reason for your choice.*

**1.** $\mathbf{r}(u, v) = 2 \cos u \mathbf{i} + 2 \sin u \mathbf{j} + v \mathbf{k}$

**2.** $\mathbf{r}(u, v) = u \cos v \mathbf{i} + u \sin v \mathbf{j} + u \mathbf{k}$

**3.** $\mathbf{r}(u, v) = u \cos v \mathbf{i} + u \sin v \mathbf{j} + u^2 \mathbf{k}$

**4.** $\mathbf{r}(u, v) = u \cos v \mathbf{i} + u \sin v \mathbf{j} + v \mathbf{k}$

(a)  (b)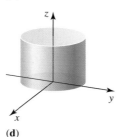

(c)  (d)

*In Exercises 5–8, find an equation in rectangular coordinates, and then identify and sketch the surface.*

**5.** $\mathbf{r}(u, v) = (u - v)\mathbf{i} + 3v\mathbf{j} + (u + v)\mathbf{k}$

**6.** $\mathbf{r}(u, v) = (u^2 + v^2)\mathbf{i} + u\mathbf{j} + v\mathbf{k}$

**7.** $\mathbf{r}(u, v) = 3 \sin u \mathbf{i} + 2 \cos u \mathbf{j} + v \mathbf{k}, \quad 0 \le v \le 2$

**8.** $\mathbf{r}(u, v) = 2 \cos v \cos u \mathbf{i} + 2 \cos v \sin u \mathbf{j} + 3 \sin v \mathbf{k}$

**cas** *In Exercises 9–14, use a computer algebra system (CAS) to graph the surface represented by the vector function.*

**9.** $\mathbf{r}(u, v) = (u + v)\mathbf{i} + (u - v)\mathbf{j} + (u^2 + v^2)\mathbf{k}; \quad -1 \le u \le 1, \\ -1 \le v \le 1$

**10.** $\mathbf{r}(u, v) = u\mathbf{i} + (v - 1)\mathbf{j} + (v^3 - v)\mathbf{k}; \quad 0 \le u \le 1, \\ -2 \le v \le 1$

**11.** $\mathbf{r}(u, v) = \cos u \sin v \mathbf{i} + \sin u \sin v \mathbf{j} + (1 + \cos v)\mathbf{k}; \\ 0 \le u \le 2\pi, \quad 0 \le v \le 2\pi$

**12.** $\mathbf{r}(u, v) = v \cos u \sin v \mathbf{i} + v \sin u \sin v \mathbf{j} + u \cos v \mathbf{k}; \\ 0 \le u \le 2\pi, \quad 0 \le v \le \pi$

**13.** $\mathbf{r}(u, v) = v \cos u \mathbf{i} + v \sin u \mathbf{j} + v^2 \mathbf{k}; \quad 0 \le u \le 2\pi, \\ 0 \le v \le 1$

**14.** $\mathbf{r}(u, v) = \left[ 2 \cos u + v \cos\left(\dfrac{u}{2}\right) \right]\mathbf{i} \\ + \left[ 2 \sin u + v \cos\left(\dfrac{u}{2}\right) \right]\mathbf{j} + v \sin\left(\dfrac{u}{2}\right)\mathbf{k}; \\ 0 \le u \le 2\pi, \quad -\tfrac{1}{2} \le v \le \tfrac{1}{2}$

**Note:** This is a representation for the **Möbius strip.**

*In Exercises 15–22, find a vector representation for the surface.*

**15.** The plane that passes through the point $(2, 1, -3)$ and contains the vectors $2\mathbf{i} + \mathbf{j} - \mathbf{k}$ and $\mathbf{i} - 2\mathbf{j} - \mathbf{k}$

**16.** The plane $2x + 3y + z = 6$

**17.** The lower half of the sphere $x^2 + y^2 + z^2 = 1$

**18.** The upper half of the ellipsoid $9x^2 + 4y^2 + 36z^2 = 36$

**19.** The part of the cylinder $x^2 + y^2 = 4$ between $z = -1$ and $z = 3$

**20.** The part of the cylinder $9y^2 + 4z^2 = 36$ between $x = 0$ and $x = 3$

**21.** The part of the paraboloid $z = 9 - x^2 - y^2$ inside the cylinder $x^2 + y^2 = 4$

**22.** The part of the plane $z = x + 2$ that lies inside the cylinder $x^2 + y^2 = 1$

*In Exercises 23–26, find a vector equation for the surface obtained by revolving the graph of the function about the indicated axis. Graph the surface.*

**23.** $y = \sqrt{x}, \quad 0 \le x \le 4; \quad x\text{-axis}$

**24.** $y = e^{-x}, \quad 0 \le x \le 1; \quad x\text{-axis}$

**25.** $x = 9 - y^2, \quad 0 \le y \le 3; \quad y\text{-axis}$

**26.** $y = \cos z, \quad -\pi \le z \le \pi; \quad z\text{-axis}$

*In Exercises 27–32, find an equation of the tangent plane to the parametric surface represented by* $\mathbf{r}$ *at the specified point.*

**27.** $\mathbf{r}(u, v) = (u + v)\mathbf{i} + (u - v)\mathbf{j} + v^2 \mathbf{k}; \quad (2, 0, 1)$

**28.** $\mathbf{r}(u, v) = u\mathbf{i} + (u^2 + v^2)\mathbf{j} + v\mathbf{k}; \quad (2, 5, 1)$

**29.** $\mathbf{r}(u, v) = u \cos v \mathbf{i} + 2u \sin v \mathbf{j} + u^2 \mathbf{k}; \quad u = 1, \quad v = \pi$

**30.** $\mathbf{r}(u, v) = \cos u \sin v \mathbf{i} + \sin u \sin v \mathbf{j} + \cos v \mathbf{k}; \quad u = \dfrac{\pi}{2}, \\ v = \dfrac{\pi}{4}$

**31.** $\mathbf{r}(u, v) = ue^v \mathbf{i} + ve^u \mathbf{j} + uv \mathbf{k}; \quad u = 0, \quad v = \ln 2$

**32.** $\mathbf{r}(u, v) = uv\mathbf{i} + u \ln v \mathbf{j} + v \mathbf{k}; \quad u = 1, \quad v = 1$

*In Exercises 33–40, find the area of the surface.*

**33.** The part of the plane $\mathbf{r}(u, v) = (u + 2v - 1)\mathbf{i} + (2u + 3v + 1)\mathbf{j} + (u + v + 2)\mathbf{k}; \quad 0 \le u \le 1, \quad 0 \le v \le 2$

**34.** The part of the plane $2x + 3y - z = 1$ that lies above the rectangular region $[1, 2] \times [1, 3]$

**35.** The part of the plane $z = 8 - 2x - 3y$ that lies inside the cylinder $x^2 + y^2 = 4$

**36.** The part of the paraboloid $\mathbf{r}(u, v) = u \cos v\mathbf{i} + u \sin v\mathbf{j} + u^2\mathbf{k}$; $\quad 0 \le u \le 3, \quad 0 \le v \le 2\pi$

**37.** The part of the cone $\mathbf{r}(u, v) = u \cos v\mathbf{i} + u \sin v\mathbf{j} + u\mathbf{k}$; $1 \le u \le 2, \quad 0 \le v \le \frac{\pi}{2}$

**38.** The part of the sphere $\mathbf{r}(u, v) = a \sin u \cos v\mathbf{i} + a \sin u \sin v\mathbf{j} + a \cos u\mathbf{k}$ that lies in the first octant

**39.** The surface $\mathbf{r}(u, v) = \sin u \cos v\mathbf{i} + \sin u \sin v\mathbf{j} + u\mathbf{k}$; $0 \le u \le \pi, \quad 0 \le v \le 2\pi$

**40.** The part of the surface $\mathbf{r}(u, v) = u^2\mathbf{i} + uv\mathbf{j} + \frac{1}{2}v^2\mathbf{k}$; $0 \le u \le 1, \quad 0 \le v \le 2$

**cas 41. a.** Show that the vector equation $\mathbf{r}(u, v) = a \sin u \cos v\mathbf{i} + b \sin u \sin v\mathbf{j} + c \cos u\mathbf{k}$, where $0 \le u \le \pi$ and $0 \le v \le 2\pi$, represents the ellipsoid

$$\frac{x^2}{a^2} + \frac{y^2}{b^2} + \frac{z^2}{c^2} = 1$$

**b.** Use a CAS to graph the ellipsoid with $a = 3$, $b = 4$, and $c = 5$.

**c.** Use a CAS to find the approximate surface area of the ellipsoid of part (b).

**cas 42. a.** Show that the vector equation $\mathbf{r}(u, v) = a \sin^3 u \cos^3 v\mathbf{i} + a \sin^3 u \sin^3 v\mathbf{j} + a \cos^3 u\mathbf{k}$, where $0 \le u \le \pi$ and $0 \le v \le 2\pi$, represents the **astroidal sphere** $x^{2/3} + y^{2/3} + z^{2/3} = a^{2/3}$.

**b.** Use a CAS to graph the astroidal sphere with $a = 1$.

**c.** Find the area of the astroidal sphere with $a = 1$.

**43.** Find the area of the part of the cone $z = \sqrt{x^2 + y^2}$ that is cut off by the cylinder $x^2 + (y - 1)^2 = 1$.

**44.** In Section 12.7 we showed that the tangent plane to the graph $S$ of a function $f(x, y)$ at the point $(a, b, f(a, b))$ is given by the equation

$$z - f(a, b) = f_x(a, b)(x - a) + f_y(a, b)(y - b)$$

(See Equation (4) in Section 12.7.) Show that parametrizing $S$ by $\mathbf{r}(x, y) = x\mathbf{i} + y\mathbf{j} + f(x, y)\mathbf{k}$ yields the same tangent plane.

**45.** In Section 5.4 we defined the area of the surface of revolution obtained by revolving the graph of a nonnegative smooth function $f$ on $[a, b]$ about the $x$-axis as

$$S = 2\pi \int_a^b f(x)\sqrt{1 + [f'(x)]^2}\, dx$$

Use Equation (1) to derive this formula.

**46.** If the circle with center at $(a, 0, 0)$ and radius $b$, where $0 < b < a$, in the $xz$-plane is revolved about the $z$-axis, we obtain a torus represented parametrically by

$$x = (a + b \cos v)\cos u$$

$$y = (a + b \cos v)\sin u$$

$$z = b \sin v$$

with parametric domain $D = \{(u, v) \mid 0 \le u \le 2\pi,\ 0 \le v \le 2\pi\}$. (See the figure below.) Find the surface area of the torus.

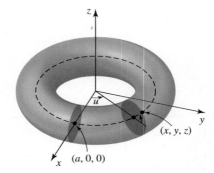

*In Exercises 47 and 48, determine whether the statement is true or false. If it is true, explain why. If it is false, explain why or give an example that shows it is false.*

**47.** The surface described by $\mathbf{r}(u, v) = u \cos v\mathbf{i} + u \sin v\mathbf{j} + u\mathbf{k}$, where $-2 \le u \le 2$ and $0 \le v \le 2\pi$, is smooth.

**48.** If $\mathbf{r}(u, v) = 2 \sin u \cos v\mathbf{i} + 2 \sin u \sin v\mathbf{j} + 2 \cos u\mathbf{k}$, where $0 \le u \le \frac{\pi}{2}$ and $0 \le v \le \frac{\pi}{2}$, then

$$\int_0^{\pi/2} \int_0^{\pi/2} |\mathbf{r}_u \times \mathbf{r}_v|\, du\, dv = 2\pi$$

# 14.7 Surface Integrals

## ■ Surface Integrals of Scalar Fields

As we saw in Section 13.1, the mass of a thin plate lying in a plane region can be found by evaluating the double integral $\iint_R \sigma(x, y)\, dA$, where $\sigma(x, y)$ is the mass density of the plate at any point $(x, y)$ in $R$. Now, instead of a flat plate, let's suppose that we have a plate that takes the form of a curved surface. How do we determine the mass of this plate?

For simplicity, let's suppose that the thin plate has the shape of the surface $S$ that is the graph of a continuous function $g$ of two variables defined by $z = g(x, y)$. To further simplify our discussion, suppose that the domain of $g$ is a rectangular region $R = \{(x, y) \mid a \le x \le b, c \le y \le d\}$. A typical surface is shown in Figure 1.

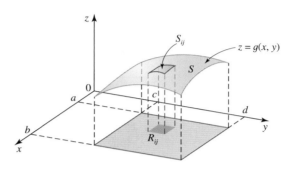

**FIGURE 1**
A thin plate that takes the shape of a surface $S$ defined by $z = g(x, y)$

Let the mass density of the plate at any point on $S$ be $\sigma(x, y, z)$, where $\sigma$ is a nonnegative continuous function defined on an open region containing $S$, and let $P = \{R_{ij}\}$ be a partition of $R$ into $N = mn$ subrectangles. Corresponding to each subrectangle $R_{ij}$, there is a part of $S$, $S_{ij}$, that lies directly above $R_{ij}$ with area $\Delta S_{ij}$. Then

$$\Delta S_{ij} = \sqrt{[g_x(x_i, y_j)]^2 + [g_y(x_i, y_j)]^2 + 1}\ \Delta A \qquad (1)$$

where $(x_i, y_j)$ is the corner of $R_{ij}$ closest to the origin and $\Delta A$ is the area of $R_{ij}$. If $m$ and $n$ are large so that the dimensions of $R_{ij}$ are small, then the continuity of $g$ and $\sigma$ implies that $\sigma(x, y, z)$ does not differ appreciably from $\sigma(x_i, y_j, g(x_i, y_j))$. Therefore, the mass of the part of the plate that lies on $S$ and directly above $R_{ij}$ is

$$\Delta m_{ij} \approx \sigma(x_i, y_j, g(x_i, y_j))\, \Delta S_{ij} \qquad \text{Constant mass density} \cdot \text{surface area}$$

Using Equation (1), we see that the mass of the plate is approximately

$$\sum_{i=1}^{m} \sum_{j=1}^{n} \sigma(x_i, y_j, g(x_i, y_j)) \sqrt{[g_x(x_i, y_j)]^2 + [g_y(x_i, y_j)]^2 + 1}\ \Delta A$$

The approximation should improve as $m$ and $n$ approach infinity. This suggests that we define the mass of the plate to be

$$\lim_{n,\, m \to \infty} \sum_{i=1}^{m} \sum_{j=1}^{n} \sigma(x_i, y_j, g(x_i, y_j)) \sqrt{[g_x(x_i, y_j)]^2 + [g_y(x_i, y_j)]^2 + 1}\ \Delta A$$

Using the definition of the double integral, we see that the required mass, $m$, is

$$m = \iint_S \sigma(x, y, z)\, dS = \iint_R \sigma(x, y, g(x, y)) \sqrt{[g_x(x, y)]^2 + [g_y(x, y)]^2 + 1}\ dA \qquad (2)$$

if we assume that both $g_x$ and $g_y$ are continuous on $R$.

The integral that appears in Equation (2) is a *surface integral*. More generally, we can define the surface integral of a function $f$ over nonrectangular regions as follows.

> **DEFINITION    Surface Integral of a Scalar Function**
>
> Let $f$ be a function of three variables defined in a region in space containing a surface $S$. The **surface integral of $f$ over $S$** is
>
> $$\iint_S f(x, y, z)\, dS = \lim_{n,\, m \to \infty} \sum_{i=1}^{m} \sum_{j=1}^{n} f(x_i, y_j, g(x_i, y_j))\, \Delta S_{ij}$$

We also have the following formulas for evaluating a surface integral depending on the way $S$ is defined.

> **THEOREM 1    Evaluation of Surface Integrals**
>              (for Surfaces That Are Graphs)
>
> **1.** If $S$ is defined by $z = g(x, y)$ and the projection of $S$ onto the $xy$-plane is $R$ (Figure 2a), then
>
> $$\iint_S f(x, y, z)\, dS = \iint_R f(x, y, g(x, y))\sqrt{[g_x(x, y)]^2 + [g_y(x, y)]^2 + 1}\; dA \quad \textbf{(3)}$$
>
> **2.** If $S$ is defined by $y = g(x, z)$ and the projection of $S$ onto the $xz$-plane is $R$ (Figure 2b), then
>
> $$\iint_S f(x, y, z)\, dS = \iint_R f(x, g(x, z), z)\sqrt{[g_x(x, z)]^2 + [g_z(x, z)]^2 + 1}\; dA \quad \textbf{(4)}$$
>
> **3.** If $S$ is defined by $x = g(y, z)$ and the projection of $S$ onto the $yz$-plane is $R$ (Figure 2c), then
>
> $$\iint_S f(x, y, z)\, dS = \iint_R f(g(y, z), y, z)\sqrt{[g_y(y, z)]^2 + [g_z(y, z)]^2 + 1}\; dA \quad \textbf{(5)}$$

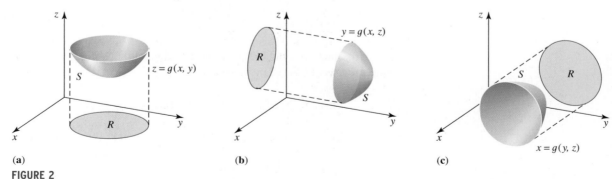

**FIGURE 2**
The surfaces $S$ and their projections onto the coordinate planes

**Note**    If we take $f(x, y, z) = 1$, then each of the formulas gives the area of $S$.    ∎

**EXAMPLE 1**    Evaluate $\iint_S x\, dS$, where $S$ is the part of the plane $2x + 3y + z = 6$ that lies in the first octant.

**Solution**    The plane $S$ is shown in Figure 3a, and its projection onto the $xy$-plane is shown in Figure 3b. Using Equation (3) with $f(x, y, z) = x$ and $z = g(x, y) = 6 - 2x - 3y$, we have

$$\iint_S f(x, y, z)\, dS = \iint_S x\, dS = \iint_R x\sqrt{[g_x(x, y)]^2 + [g_y(x, y)]^2 + 1}\, dA$$

$$= \iint_R x\sqrt{(-2)^2 + (-3)^2 + 1}\, dA = \sqrt{14} \iint_R x\, dA$$

$$= \sqrt{14} \int_0^3 \int_0^{2-(2/3)x} x\, dy\, dx \qquad \text{View } R \text{ as } y\text{-simple.}$$

$$= \sqrt{14} \int_0^3 \left[xy\right]_{y=0}^{y=2-(2/3)x} dx$$

$$= \sqrt{14} \int_0^3 \left(2x - \frac{2}{3}x^2\right) dx = \sqrt{14}\left[x^2 - \frac{2}{9}x^3\right]_0^3 = 3\sqrt{14}$$

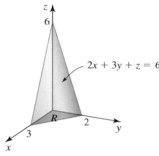

(a) The surface $S$

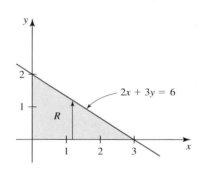

(b) The projection $R$ of $S$ onto the $xy$-plane viewed as $y$-simple

**FIGURE 3**

**EXAMPLE 2**    Find the mass of the surface $S$ composed of the part of the paraboloid $y = x^2 + z^2$ between the planes $y = 1$ and $y = 4$ if the density at a point $P$ on $S$ is inversely proportional to the distance between $P$ and the axis of symmetry of $S$.

**Solution**    The surface $S$ is shown in Figure 4a, and its projection onto the $xz$-plane is shown in Figure 4b. Using Equation (4) with $f(x, y, z) = \sigma(x, y, z) = k(x^2 + z^2)^{-1/2}$, where $k$ is the constant of proportionality and $y = g(x, z) = x^2 + z^2$, we have

$$m = \iint_S \sigma(x, y, z)\, dS = \iint_S k(x^2 + z^2)^{-1/2}\, dS$$

$$= k\iint_R (x^2 + z^2)^{-1/2}\sqrt{[g_x(x, z)]^2 + [g_z(x, z)]^2 + 1}\, dA$$

$$= k\iint_R (x^2 + z^2)^{-1/2}\sqrt{(2x)^2 + (2z)^2 + 1}\, dA$$

$$= k\iint_R (x^2 + z^2)^{-1/2}\sqrt{4x^2 + 4z^2 + 1}\, dA$$

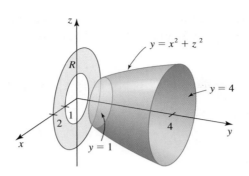

**FIGURE 4**      (**a**) The surface $S$          (**b**) The projection $R$ of $S$ onto the $xz$-plane

Changing to polar coordinates, $x = r \cos \theta$ and $z = r \sin \theta$, we obtain

$$m = k \int_0^{2\pi} \int_1^2 \left(\frac{1}{r}\right)\sqrt{4r^2 + 1}\, r\, dr\, d\theta = 2k \int_0^{2\pi} \int_1^2 \sqrt{r^2 + \left(\tfrac{1}{2}\right)^2}\, dr\, d\theta$$

$$= 2k \int_0^{2\pi} \left[\frac{r}{2}\sqrt{r^2 + \tfrac{1}{4}} + \frac{1}{8}\ln\left|r + \sqrt{r^2 + \tfrac{1}{4}}\right|\right]_{r=1}^{r=2} d\theta \qquad \text{Use Formula 37 from the Table of Integrals.}$$

$$= k \left[\sqrt{17} - \frac{1}{2}\sqrt{5} + \frac{1}{4}\ln\left(\frac{4 + \sqrt{17}}{2 + \sqrt{5}}\right)\right] \int_0^{2\pi} d\theta$$

$$= k\pi \left[2\sqrt{17} - \sqrt{5} + \frac{1}{2}\ln\left(\frac{4 + \sqrt{17}}{2 + \sqrt{5}}\right)\right]$$
■

## ■ Parametric Surfaces

If a surface $S$ is represented by a vector equation

$$\mathbf{r}(u, v) = x(u, v)\mathbf{i} + y(u, v)\mathbf{j} + z(u, v)\mathbf{k}$$

with parameter domain $D$, then an element of surface area is given by

$$|\mathbf{r}_u(u, v) \times \mathbf{r}_v(u, v)|\, dA$$

as we saw in Section 14.6. This leads to the following formula for evaluating a surface integral in which the surface is defined parametrically.

---

**THEOREM 2    Evaluation of Surface Integrals** (for Parametric Surfaces)

If $f$ is a continuous function in a region that contains a smooth surface $S$ with parametric representation

$$\mathbf{r}(u, v) = x(u, v)\mathbf{i} + y(u, v)\mathbf{j} + z(u, v)\mathbf{k} \qquad (u, v) \in D$$

then the **surface integral of $f$ over $S$** is

$$\iint_S f(x, y, z)\, dS = \iint_D f(\mathbf{r}(u, v))|\mathbf{r}_u \times \mathbf{r}_v|\, dA \tag{6}$$

where $f(\mathbf{r}(u, v)) = f(x(u, v), y(u, v), z(u, v))$.

**Note** You can show that if $S$ is the graph of a function $z = g(x, y)$, then Equation (3) follows from Equation (6) by putting $\mathbf{r}(u, v) = u\mathbf{i} + v\mathbf{j} + g(u, v)\mathbf{k}$. (See Exercise 46.) ∎

**EXAMPLE 3** Evaluate $\displaystyle\iint_S \frac{x - y}{\sqrt{2z + 1}} \, dS$, where $S$ is the surface represented by $\mathbf{r}(u, v) = (u + v)\mathbf{i} + (u - v)\mathbf{j} + (u^2 + v^2)\mathbf{k}$, where $0 \leq u \leq 1$ and $0 \leq v \leq 2$.

**Solution** We first find

$$\mathbf{r}_u(u, v) = \mathbf{i} + \mathbf{j} + 2u\mathbf{k}$$

$$\mathbf{r}_v(u, v) = \mathbf{i} - \mathbf{j} + 2v\mathbf{k}$$

$$\mathbf{r}_u \times \mathbf{r}_v = \begin{vmatrix} \mathbf{i} & \mathbf{j} & \mathbf{k} \\ 1 & 1 & 2u \\ 1 & -1 & 2v \end{vmatrix}$$

$$= 2[(u + v)\mathbf{i} + (u - v)\mathbf{j} - \mathbf{k}]$$

so

$$|\mathbf{r}_u \times \mathbf{r}_v| = 2\sqrt{(u + v)^2 + (u - v)^2 + 1} = 2\sqrt{2u^2 + 2v^2 + 1}$$

Therefore,

$$\iint_S \frac{x - y}{\sqrt{2z + 1}} \, dS = \int_0^2 \int_0^1 \frac{(u + v) - (u - v)}{\sqrt{2(u^2 + v^2) + 1}} \cdot 2\sqrt{2u^2 + 2v^2 + 1} \, du \, dv$$

$$= 4 \int_0^2 \int_0^1 v \, du \, dv$$

$$= 4 \int_0^2 \left[ uv \right]_{u=0}^{u=1} dv$$

$$= 4 \int_0^2 v \, dv = 2v^2 \Big|_0^2 = 8$$ ∎

## Oriented Surfaces

One of the most important applications of surface integrals involves the computation of the flux of a vector field across an *oriented* surface. Before explaining the notion of *flux,* however, we need to elaborate on the meaning of *orientation.*

A surface $S$ is **orientable** or **two-sided** if it has a unit normal vector $\mathbf{n}$ that varies *continuously* over $S$, that is, if the components of $\mathbf{n}$ are continuous at each point $(x, y, z)$ on $S$. Closed surfaces (surfaces that are boundaries of solids) such as spheres are examples of orientable surfaces. There are two possible choices of $\mathbf{n}$ for orientable surfaces: the **unit inner normal** that points *inward* from $S$ and the **unit outer normal** that points *outward* from $S$ (see Figure 5). By convention, however, the **positive orientation** for a closed surface $S$ is the one for which the unit normal vector points outward from $S$.

An example of a nonorientable surface is the Möbius strip, which can be constructed by taking a long, rectangular strip of paper, giving it a half-twist, and then taping the short edges together to produce the surface shown in Figure 6. If you take a unit normal $\mathbf{n}$ starting at $P$ (see Figure 6), then you can move it along the surface in such a way that upon returning to the starting point (and without crossing any edges), it will point in a direction precisely opposite to its initial direction. This shows that $\mathbf{n}$ does not vary continuously on a Möbius strip, and accordingly, the strip is not orientable.

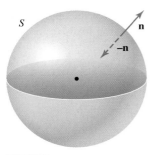

**FIGURE 5**
Unit inner and outer normals to an (orientable) closed surface $S$

**FIGURE 6**
The Möbius strip can be constructed by using a rectangular strip of paper.

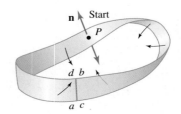

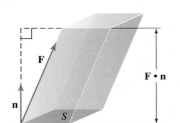

**FIGURE 7**
If $S$ is flat and $F$ is constant, then the flux is equal to the volume of the prism.

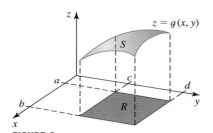

**FIGURE 8**
A smooth surface $S$ defined by $z = g(x, y)$ for $(x, y)$ in $R$

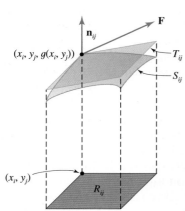

**FIGURE 9**

## Surface Integrals of Vector Fields

Suppose that $\mathbf{F}$ is a continuous vector field defined in a region $R$ in space. We can think of $\mathbf{F}(x, y, z)$ as giving the velocity of a fluid at a point $(x, y, z)$ in $R$, and $S$ as a smooth, oriented surface lying in $R$. If $S$ is flat and $\mathbf{F}$ is a constant field, then the flux, or rate of flow (volume of fluid crossing $S$ per unit time), is equal to

$$\mathbf{F} \cdot \mathbf{n}\, A(S)$$

(the *normal component* of $\mathbf{F}$ with respect to $S$ times the area of $S$). Geometrically, the flux is given by the volume of fluid in the prism in Figure 7.

More generally, suppose that $S$ is the graph of a function of two variables defined by $z = g(x, y)$, where, for simplicity, we assume that the domain of $g$ is a rectangular region $R = \{(x, y) \mid a \le x \le b, c \le y \le d\}$. (See Figure 8.) Let $P = \{R_{ij}\}$ be a partition of $R$ into $N = mn$ subrectangles. Corresponding to each subrectangle $R_{ij}$ there is the part of $S$ that lies directly above $R_{ij}$ with area $\Delta S_{ij}$. As in Section 13.5, let $(x_i, y_j)$ be the corner of $R_{ij}$ closest to the origin, and let $(x_i, y_j, g(x_i, y_j))$ be the point directly above it, as shown in Figure 9. Let $\mathbf{n}_{ij}$ denote the unit normal vector to $S$ at $(x_i, y_j, g(x_i, y_j))$. If $m$ and $n$ are large so that the dimensions of $R_{ij}$ are small, then the continuity of $\mathbf{F}$ implies that $\mathbf{F}(x, y, z)$ does not differ appreciably from $\mathbf{F}(x_i, y_j, g(x_i, y_j))$ on $R_{ij}$.

Furthermore, the continuity of $g$ implies that $S_{ij}$ may be approximated by $T_{ij}$, the parallelogram that is part of the tangent plane to $S$ at the point $(x_i, y_j, g(x_i, y_j))$ lying directly above $R_{ij}$. But the flux of $\mathbf{F}$ across (the flat) $T_{ij}$ is approximately

$$\mathbf{F}(x_i, y_j, g(x_i, y_j)) \cdot \mathbf{n}_{ij}(\Delta T_{ij})$$

where $\Delta T_{ij}$ is the area of $T_{ij}$. Since $\Delta T_{ij} \approx \Delta S_{ij}$, we see that the flux of $\mathbf{F}$ across $S$ may be approximated by the sum

$$\sum_{i=1}^{m} \sum_{j=1}^{n} \mathbf{F}(x_i, y_j, g(x_i, y_j)) \cdot \mathbf{n}_{ij}\, \Delta S_{ij}$$

$$= \sum_{i=1}^{m} \sum_{j=1}^{n} \mathbf{F}(x_i, y_j, g(x_i, y_j)) \cdot \mathbf{n}_{ij} \sqrt{[g_x(x_i, y_j)]^2 + [g_y(x_i, y_j)]^2 + 1}\, \Delta A$$

This last equality follows upon using Equation (1). We can expect that the approximation will get better as the partition $P$ becomes finer. This observation leads to the following definition.

> **DEFINITION**    **Surface Integral of a Vector Field**
>
> Let $\mathbf{F}$ be a continuous vector field defined in a region containing an oriented surface $S$ with unit normal vector $\mathbf{n}$. The **surface integral of $\mathbf{F}$ across $S$ in the direction of $\mathbf{n}$** is
>
> $$\iint_S \mathbf{F} \cdot d\mathbf{S} = \iint_S \mathbf{F} \cdot \mathbf{n}\, dS = \lim_{m, n \to \infty} \sum_{i=1}^{m} \sum_{j=1}^{n} \mathbf{F}(x_i, y_j, f(x_i, y_j)) \cdot \mathbf{n}_{ij}\, \Delta S_{ij}$$

Thus, the surface integral (also called **flux integral**) of a vector field **F** across an oriented surface $S$ is the integral of the *normal* component of **F** over $S$. If the fluid has density $\rho(x, y, z)$ at $(x, y, z)$, then the flux integral

$$\iint_S \rho \mathbf{F} \cdot \mathbf{n} \, dS$$

gives the *mass* of the fluid flowing across $S$ per unit time.

To obtain a formula for finding the flux integral in terms of $g(x, y)$, recall from Section 12.6 that the normal to the surface $z = g(x, y)$ is given by $\nabla G$, where $G(x, y, z) = z - g(x, y)$. Therefore, the unit normal to $S$ is

$$\mathbf{n} = \frac{\nabla G(x, y, z)}{|\nabla G(x, y, z)|} = \frac{-g_x(x, y)\mathbf{i} - g_y(x, y)\mathbf{j} + \mathbf{k}}{\sqrt{[g_x(x, y)]^2 + [g_y(x, y)]^2 + 1}}$$

Furthermore, in Section 13.5 we showed that the "element of area" $dS$ is given by

$$dS = \sqrt{[g_x(x, y)]^2 + [g_y(x, y)]^2 + 1} \, dA$$

so

$$\iint_S \mathbf{F} \cdot \mathbf{n} \, dS = \iint_D \frac{\mathbf{F} \cdot [-g_x(x, y)\mathbf{i} - g_y(x, y)\mathbf{j} + \mathbf{k}]}{\sqrt{[g_x(x, y)]^2 + [g_y(x, y)]^2 + 1}} \cdot \sqrt{[g_x(x, y)]^2 + [g_y(x, y)]^2 + 1} \, dA$$

$$= \iint_D \mathbf{F} \cdot [-g_x(x, y)\mathbf{i} - g_y(x, y)\mathbf{j} + \mathbf{k}] \, dA$$

where $D$ is the projection of $S$ onto the $xy$-plane.

If $\mathbf{F}(x, y, z) = P(x, y, z)\mathbf{i} + Q(x, y, z)\mathbf{j} + R(x, y, z)\mathbf{k}$, then we can write

$$\iint_S \mathbf{F} \cdot \mathbf{n} \, dS = \iint_D (-Pg_x - Qg_y + R) \, dA$$

---

**THEOREM 3  Evaluation of Surface Integrals** (for Graphs)

If $\mathbf{F} = P\mathbf{i} + Q\mathbf{j} + R\mathbf{k}$ is a continuous vector field in a region that contains a smooth oriented surface $S$ given by $z = g(x, y)$ and $D$ is its projection onto the $xy$-plane, then

$$\iint_S \mathbf{F} \cdot d\mathbf{S} = \iint_D (-Pg_x - Qg_y + R) \, dA \tag{7}$$

---

Before looking at the next example, we note the following property of surface integrals: If $S = S_1 \cup S_2 \cup \cdots \cup S_n$, where each of the surfaces is smooth and intersect only along their boundaries, then

$$\iint_S \mathbf{F} \cdot d\mathbf{S} = \iint_{S_1} \mathbf{F} \cdot d\mathbf{S} + \cdots + \iint_{S_n} \mathbf{F} \cdot d\mathbf{S}$$

**EXAMPLE 4**  Evaluate $\iint_S \mathbf{F} \cdot d\mathbf{S}$, where $\mathbf{F}(x, y, z) = x\mathbf{i} + y\mathbf{j} + z\mathbf{k}$ and $S$ is the surface that is composed of the part of the paraboloid $z = 1 - x^2 - y^2$ lying above the $xy$-plane and the disk $D = \{(x, y) \mid 0 \le x^2 + y^2 \le 1\}$.

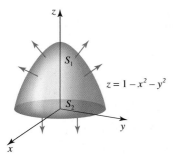

**FIGURE 10**
The part of the paraboloid
$z = 1 - x^2 - y^2$ that lies above
the $xy$-plane and is oriented so that
the unit normal vector $\mathbf{n}$ points
upward. The unit normal for the
disk $D = S_2$ points downward.

**Solution**   The (closed) surface $S$ together with a few vectors from the vector field $\mathbf{F}$ is shown in Figure 10. Writing the equation of the surface $S_1$ in the form $g(x, y) = 1 - x^2 - y^2$, we find that $g_x = -2x$ and $g_y = -2y$. Observe that the projection of $S$ onto the $xy$-plane is $D = \{(x, y) \mid 0 \le x^2 + y^2 \le 1\}$. Also, $P(x, y, z) = x$, $Q(x, y, z) = y$, and $R(x, y, z) = z$. So using Equation (7), we obtain

$$\iint_{S_1} \mathbf{F} \cdot d\mathbf{S} = \iint_D (-Pg_x - Qg_y + R) \, dA$$

$$= \iint_D [-x(-2x) - y(-2y) + z] \, dA$$

$$= \iint_D (2x^2 + 2y^2 + z) \, dA$$

$$= \iint_D [2x^2 + 2y^2 + (1 - x^2 - y^2)] \, dA \qquad z = 1 - x^2 - y^2$$

$$= \iint_D (1 + x^2 + y^2) \, dA$$

$$= \int_0^{2\pi} \int_0^1 (1 + r^2) r \, dr \, d\theta \qquad \text{Use polar coordinates.}$$

$$= \int_0^{2\pi} \left[ \frac{1}{2} r^2 + \frac{1}{4} r^4 \right]_{r=0}^{r=1} d\theta = \int_0^{2\pi} \frac{3}{4} \, d\theta = \frac{3}{2}\pi$$

Next, observe that the normal for the surface $S_2$ is $\mathbf{n} = -\mathbf{k}$. (Remember that the normal for a closed surface, by convention, points outward.) So we have

$$\iint_{S_2} \mathbf{F} \cdot d\mathbf{S} = \iint_{S_2} \mathbf{F} \cdot (-\mathbf{k}) \, dS = \iint_D (-z) \, dA = \iint_D 0 \, dA = 0$$

because $z = 0$ on $S_2$. Therefore,

$$\iint_S \mathbf{F} \cdot d\mathbf{S} = \iint_{S_1} \mathbf{F} \cdot d\mathbf{S} + \iint_{S_2} \mathbf{F} \cdot d\mathbf{S} = \frac{3}{2}\pi + 0 = \frac{3\pi}{2} \qquad \blacksquare$$

**Notes**
1. If the vector field $\mathbf{F}$ of Example 4 describes the velocity of a fluid flowing through the paraboloidal surface $S$, then the integral $\iint_S \mathbf{F} \cdot \mathbf{n} \, dS$ that we have just evaluated tells us that the fluid is flowing out through $S$ at the rate of $3\pi/2$ cubic units per unit time.
2. In Example 4, if we had wanted the paraboloid to be oriented so that the normal pointed downward, then we would simply have picked the normal to be $-\mathbf{n}$. In this case the fluid flows into $S$ at the rate of $3\pi/2$ cubic units per unit time.   $\blacksquare$

## ■ Parametric Surfaces

If an oriented surface $S$ is a smooth surface represented by a vector equation

$$\mathbf{r}(u, v) = x(u, v)\mathbf{i} + y(u, v)\mathbf{j} + z(u, v)\mathbf{k}$$

with parameter domain $D$, then the normal to $S$ is

$$\mathbf{n} = \frac{\mathbf{r}_u \times \mathbf{r}_v}{|\mathbf{r}_u \times \mathbf{r}_v|}$$

Therefore,

$$\iint_S \mathbf{F} \cdot d\mathbf{S} = \iint_S \mathbf{F} \cdot \mathbf{n}\, dS = \iint_S \mathbf{F} \cdot \frac{\mathbf{r}_u \times \mathbf{r}_v}{|\mathbf{r}_u \times \mathbf{r}_v|}\, dS$$

$$= \iint_D \left[ \mathbf{F}(\mathbf{r}(u, v)) \cdot \frac{\mathbf{r}_u \times \mathbf{r}_v}{|\mathbf{r}_u \times \mathbf{r}_v|} \right] |\mathbf{r}_u \times \mathbf{r}_v|\, dA$$

$$= \iint_D \mathbf{F}(\mathbf{r}(u, v)) \cdot (\mathbf{r}_u \times \mathbf{r}_v)\, dA$$

---

**THEOREM 4   Evaluation of Surface Integrals of a Vector Field**
**(for Parametric Surfaces)**

If $\mathbf{F}$ is a continuous vector field in a region that contains a smooth, oriented surface $S$ with parametric representation

$$\mathbf{r}(u, v) = x(u, v)\mathbf{i} + y(u, v)\mathbf{j} + z(u, v)\mathbf{k} \qquad (u, v) \in D$$

then the surface integral of $f$ over $S$ is

$$\iint_S \mathbf{F} \cdot d\mathbf{S} = \iint_D \mathbf{F}(\mathbf{r}(u, v)) \cdot (\mathbf{r}_u \times \mathbf{r}_v)\, dA \tag{8}$$

where

$$\mathbf{F}(\mathbf{r}(u, v)) = \mathbf{F}(x(u, v), y(u, v), z(u, v))$$

---

**EXAMPLE 5**   Find the flux of the vector field $\mathbf{F}(x, y, z) = y\mathbf{i} + x\mathbf{j} + 2z\mathbf{k}$ across the unit sphere $x^2 + y^2 + z^2 = 1$.

**Solution**   The unit sphere has parametric representation

$$\mathbf{r}(u, v) = \sin u \cos v\,\mathbf{i} + \sin u \sin v\,\mathbf{j} + \cos u\,\mathbf{k}$$

with parameter domain $D = \{(u, v)\,|\,0 \le u \le \pi, 0 \le v \le 2\pi\}$. Proceeding as in Example 10 in Section 14.6 and taking $a = 1$, we find

$$\mathbf{r}_u \times \mathbf{r}_v = \sin^2 u \cos v\,\mathbf{i} + \sin^2 u \sin v\,\mathbf{j} + \sin u \cos u\,\mathbf{k}$$

Therefore,

$$\mathbf{F}(\mathbf{r}(u, v)) \cdot (\mathbf{r}_u \times \mathbf{r}_v) = (\sin u \sin v\,\mathbf{i} + \sin u \cos v\,\mathbf{j} + 2 \cos u\,\mathbf{k})$$
$$\cdot (\sin^2 u \cos v\,\mathbf{i} + \sin^2 u \sin v\,\mathbf{j} + \sin u \cos u\,\mathbf{k})$$
$$= \sin^3 u \sin v \cos v + \sin^3 u \cos v \sin v + 2 \cos^2 u \sin u$$
$$= 2(\sin^3 u \sin v \cos v + \cos^2 u \sin u)$$

Using Equation (8), the flux across the sphere is

$$\iint_S \mathbf{F} \cdot d\mathbf{S} = \iint_D \mathbf{F}(\mathbf{r}(u, v)) \cdot (\mathbf{r}_u \times \mathbf{r}_v) \, dA$$

$$= 2 \int_0^{2\pi} \int_0^{\pi} (\sin^3 u \sin v \cos v + \cos^2 u \sin u) \, du \, dv$$

$$= 2 \int_0^{\pi} \sin^3 u \, du \int_0^{2\pi} \sin v \cos v \, dv + 2 \int_0^{\pi} \cos^2 u \sin u \, du \int_0^{2\pi} dv$$

The first term on the right is equal to zero because

$$\int_0^{2\pi} \sin v \cos v \, dv = \frac{1}{2} \sin^2 v \Big|_0^{2\pi} = 0$$

so

$$\iint_S \mathbf{F} \cdot d\mathbf{S} = 2 \int_0^{\pi} \cos^2 u \sin u \, du \int_0^{2\pi} dv$$

$$= 2 \left( -\frac{1}{3} \cos^3 u \right) \Big|_0^{\pi} \cdot 2\pi$$

$$= \frac{8\pi}{3} \qquad \blacksquare$$

We have used an example involving fluid flow to illustrate the concept of the surface integral of a vector field. But these integrals have wider applications in the physical sciences. For example, if $\mathbf{E}$ is the electric field induced by an electric charge of $q$ coulombs located at the origin of a three-dimensional coordinate system, then by Coulomb's Law (Example 4 in Section 14.1),

$$\mathbf{E} = \frac{q}{4\pi\varepsilon_0} \cdot \frac{\mathbf{r}}{|\mathbf{r}|^3}$$

where $\varepsilon_0$ is a constant called the *permittivity of free space*. If $S$ is a sphere of radius $r$ centered at the origin, then the surface integral

$$\iint_S \mathbf{E} \cdot \mathbf{n} \, dS$$

is the flux of $\mathbf{E}$ passing through $S$.

Yet another application of surface integrals can be found in the study of heat flow. Suppose that the temperature at a point $(x, y, z)$ in a homogeneous body is $T(x, y, z)$. Empirical results suggest that heat will flow from points at higher temperatures to those at lower temperatures. Since the temperature gradient $\nabla T$ points in the direction of maximum increase of the temperature, we see that the flow of heat can be described by the vector field

$$\mathbf{q} = -k\nabla T$$

where $k$ is a constant of proportionality known as the thermal conductivity of the body. The rate at which heat flows across a surface $S$ in the body is given by the surface integral

$$\iint_S \mathbf{q} \cdot \mathbf{n} \, dS = -k \iint_S \nabla T \cdot \mathbf{n} \, dS$$

**EXAMPLE 6** **Rate of Flow of Heat Across a Sphere** The temperature at a point $P(x, y, z)$ in a medium with thermal conductivity $k$ is inversely proportional to the distance between $P$ and the origin. Find the rate of flow of heat across a sphere $S$ of radius $a$, centered at the origin.

**Solution** We have

$$T(x, y, z) = \frac{c}{\sqrt{x^2 + y^2 + z^2}}$$

where $c$ is the constant of proportionality. Then the flow of heat is

$$\mathbf{q} = -k\nabla T = -k\left[ -\frac{cx}{(x^2 + y^2 + z^2)^{3/2}}\mathbf{i} - \frac{cy}{(x^2 + y^2 + z^2)^{3/2}}\mathbf{j} - \frac{cz}{(x^2 + y^2 + z^2)^{3/2}}\mathbf{k} \right]$$

$$= \frac{ck}{(x^2 + y^2 + z^2)^{3/2}}(x\mathbf{i} + y\mathbf{j} + z\mathbf{k})$$

The outward unit normal to the sphere $x^2 + y^2 + z^2 = a^2$ at the point $(x, y, z)$ is

$$\mathbf{n} = \frac{1}{a}(x\mathbf{i} + y\mathbf{j} + z\mathbf{k})$$

So the rate at which heat flows across $S$ is

$$\iint_S \mathbf{q} \cdot \mathbf{n}\, dS = \iint_S \frac{ck}{(x^2 + y^2 + z^2)^{3/2}}(x\mathbf{i} + y\mathbf{j} + z\mathbf{k}) \cdot \left[ \frac{1}{a}(x\mathbf{i} + y\mathbf{j} + z\mathbf{k}) \right] dS$$

$$= \frac{ck}{a} \iint_S \frac{1}{\sqrt{x^2 + y^2 + z^2}}\, dS$$

$$= \frac{ck}{a^2} \iint_S dS \qquad \text{Since } x^2 + y^2 + z^2 = a^2 \text{ on } S$$

$$= \frac{ck}{a^2} A(S) = \frac{ck}{a^2}(4\pi a^2) = 4\pi ck \qquad \blacksquare$$

## 14.7 CONCEPT QUESTIONS

1. **a.** Define the surface integral of a scalar function $f$ over a surface that is the graph of a function $z = f(x, y)$.
   **b.** How do you evaluate the integral of part (a)?
   **c.** How do you evaluate the surface integral if the surface is represented by a vector function $\mathbf{r}(u, v)$?

2. What is an orientable surface? Give an example of a surface that is not orientable.

3. **a.** Define the surface (flux) integral of a vector field $\mathbf{F}$ over an oriented surface $S$ with a unit normal $\mathbf{n}$.
   **b.** How do you evaluate the surface integral if the surface is the graph of a function $z = g(x, y)$?
   **c.** How do you evaluate the surface integral if the surface is represented by the vector function $\mathbf{r}(u, v)$?

## 14.7 EXERCISES

*In Exercises 1–14, evaluate $\iint_S f(x, y, z)\, dS$.*

1. $f(x, y, z) = x + y$;  $S$ is the part of the plane $3x + 2y + z = 6$ in the first octant

2. $f(x, y, z) = xy$;  $S$ is the part of the plane $2x + 3y + z = 6$ in the first octant

**3.** $f(x, y, z) = y$;   $S$ is the part of the surface
$z = 2x + y^2$ above the rectangular region
$R = \{(x, y) \mid 0 \le x \le 2, 0 \le y \le 1\}$

**4.** $f(x, y, z) = xz^2$;   $S$ is the part of the surface
$y = x^2 + 2z$ to the right of the square region
$R = \{(x, z) \mid 0 \le x \le 1, 0 \le z \le 1\}$

**5.** $f(x, y, z) = x + 2y + z$;   $S$ is the part of the plane
$y + z = 4$ inside the cylinder $x^2 + y^2 = 1$

**6.** $f(x, y, z) = xz$;   $S$ is the part of the plane $y + z = 4$
inside the cylinder $x^2 + y^2 = 4$

**7.** $f(x, y, z) = x^2z$;   $S$ is the part of the cone $z = \sqrt{x^2 + y^2}$
inside the cylinder $x^2 + y^2 = 1$

**8.** $f(x, y, z) = x + 2y + 3z$;   $S$ is the part of the cone
$x = \sqrt{y^2 + z^2}$ between the planes $x = 1$ and $x = 4$

**9.** $f(x, y, z) = xyz$;   $S$ is the part of the cylinder $x^2 + y^2 = 4$
in the first octant between $z = 0$ and $z = 4$

**10.** $f(x, y, z) = z^2$;   $S$ is the hemisphere $z = \sqrt{9 - x^2 - y^2}$

**11.** $f(x, y, z) = x + \dfrac{y}{\sqrt{4z + 5}}$;   $S$ is the surface with
vector representation $\mathbf{r}(u, v) = u\mathbf{i} + v\mathbf{j} + (v^2 - 1)\mathbf{k}$,
$0 \le u \le 1, -1 \le v \le 1$

**12.** $f(x, y, z) = x + z$;   $S$ is the surface with vector representation
$\mathbf{r}(u, v) = u \sin v\mathbf{i} + u \cos v\mathbf{j} + u^2\mathbf{k}, 0 \le u \le 1, 0 \le v \le \frac{\pi}{2}$

**13.** $f(x, y, z) = z\sqrt{1 + x^2 + y^2}$;   $S$ is the helicoid with
representation $\mathbf{r}(u, v) = u \cos v\mathbf{i} + u \sin v\mathbf{j} + v\mathbf{k}$,
$0 \le u \le 1, 0 \le v \le 2\pi$

**14.** $f(x, y, z) = z$;   $S$ is the part of the torus with vector
representation $\mathbf{r}(u, v) = (a + b \cos v)\cos u\mathbf{i} +$
$(a + b \cos v)\sin u\mathbf{j} + b \sin v\mathbf{k}, 0 \le u \le 2\pi, 0 \le v \le \frac{\pi}{2}$

*In Exercises 15–18, find the mass of the surface S having the
given mass density.*

**15.** $S$ is the part of the plane $x + 2y + 3z = 6$ in the first
octant; the density at a point $P$ on $S$ is directly proportional
to the square of the distance between $P$ and the $yz$-plane.

**16.** $S$ is the part of the paraboloid $z = x^2 + y^2$ between the
planes $z = 1$ and $z = 4$; the density at a point $P$ on $S$ is
directly proportional to the distance between $P$ and the axis
of symmetry of $S$.

**17.** $S$ is the hemisphere $x^2 + y^2 + z^2 = 4, z \ge 0$; the density at
a point $P$ on $S$ is directly proportional to the distance
between $P$ and the $xy$-plane.

**18.** $S$ is the part of the sphere $x^2 + y^2 + z^2 = 1$ that lies above
the cone $z = \sqrt{x^2 + y^2}$; the density at a point $P$ on $S$ is
directly proportional to the distance between $P$ and the
$xy$-plane.

*In Exercises 19–28, evaluate $\iint_S \mathbf{F} \cdot d\mathbf{S}$, that is, find the flux of
$\mathbf{F}$ across S. If S is closed, use the positive (outward) orientation.*

**19.** $\mathbf{F}(x, y, z) = 2x\mathbf{i} + 2y\mathbf{j} + z\mathbf{k}$;   $S$ is the part of the paraboloid
$z = 4 - x^2 - y^2$ above the $xy$-plane; $\mathbf{n}$ points upward

**20.** $\mathbf{F}(x, y, z) = 3x\mathbf{i} + 3y\mathbf{j} + 2\mathbf{k}$;   $S$ is the part of the paraboloid $z = x^2 + y^2$ between the planes $z = 0$ and $z = 4$;
$\mathbf{n}$ points downward

**21.** $\mathbf{F}(x, y, z) = x\mathbf{i} + y\mathbf{j} + z\mathbf{k}$;   $S$ is the part of the plane
$3x + 2y + z = 6$ in the first octant; $\mathbf{n}$ points upward

**22.** $\mathbf{F}(x, y, z) = 6z\mathbf{i} + 2x\mathbf{j} + y\mathbf{k}$;   $S$ is the part of the plane
$2x + 3y + 6z = 12$ in the first octant; $\mathbf{n}$ points upward

**23.** $\mathbf{F}(x, y, z) = -y\mathbf{i} + x\mathbf{j} + 2z\mathbf{k}$;   $S$ is the hemisphere
$z = \sqrt{4 - x^2 - y^2}$; $\mathbf{n}$ points upward

**24.** $\mathbf{F}(x, y, z) = x\mathbf{i} + y\mathbf{j} + z\mathbf{k}$;   $S$ is the hemisphere
$z = \sqrt{9 - x^2 - y^2}$; $\mathbf{n}$ points upward
**Hint:** First evaluate $\iint_{\bar{S}} \mathbf{F} \cdot \mathbf{n} \, dS$, where $\bar{S}$ is the part of the hemisphere $z = \sqrt{9 - x^2 - y^2}$ inside the cylinder $x^2 + y^2 = a^2$, where
$0 < a < 3$. Then take the limit as $a \to 3^-$.

**25.** $\mathbf{F}(x, y, z) = 2\mathbf{i} + 3\mathbf{j} + \mathbf{k}$;   $S$ is the part of the cone
$z = \sqrt{x^2 + y^2}$ inside the cylinder $x^2 + y^2 = 1$; $\mathbf{n}$ points
upward

**26.** $\mathbf{F}(x, y, z) = x\mathbf{i} + 2y\mathbf{j} + z\mathbf{k}$;   $S$ is the part of the cone
$y = \sqrt{x^2 + z^2}$ inside the cylinder $x^2 + z^2 = 1$; $\mathbf{n}$ points
to the right

**27.** $\mathbf{F}(x, y, z) = y^3\mathbf{i} + x^2\mathbf{j} + z\mathbf{k}$;   $S$ is the boundary of the
cylindrical solid bounded by $x^2 + y^2 = 9, z = 0$, and $z = 3$

**28.** $\mathbf{F}(x, y, z) = x^2\mathbf{i} + xy\mathbf{j} + xz\mathbf{k}$;   $S$ is the surface of the tetrahedron with vertices $(0, 0, 0), (1, 0, 0), (0, 2, 0)$, and
$(0, 0, 3)$

*In Exercises 29 and 30, a thin sheet has the shape of the surface
S. If its density (mass per unit area) at the point $(x, y, z)$ is
$\rho(x, y, z)$, then its **center of mass** is $(\bar{x}, \bar{y}, \bar{z})$, where*

$$\bar{x} = \frac{1}{m}\iint_S x\rho(x, y, z) \, dS, \qquad \bar{y} = \frac{1}{m}\iint_S y\rho(x, y, z) \, dS,$$

$$\bar{z} = \frac{1}{m}\iint_S z\rho(x, y, z) \, dS$$

*and m is the mass of the sheet. Find the center of mass of the
sheet.*

**29.** $S$ is the upper hemisphere $x^2 + y^2 + z^2 = a^2, z \ge 0$,
$\rho(x, y, z) = k$, where $k$ is a constant.

**30.** $S$ is the part of the paraboloid $z = 4 - \frac{1}{2}x^2 - \frac{1}{2}y^2, z \ge 0$,
$\rho(x, y, z) = k$, where $k$ is a constant.

In Exercises 31 and 32, a thin sheet has the shape of a surface S. If its density (mass per unit area) at the point $(x, y, z)$ is $\rho(x, y, z)$, then its moment of inertia about the z-axis is $I_z = \iint_S (x^2 + y^2)\rho(x, y, z)\, dS$.

**31.** Show that the moment of inertia of a spherical shell of uniform density about its diameter is $\frac{2}{3}ma^2$, where $m$ is its mass and $a$ is its radius.

**32.** Find the moment of inertia of the conical shell $z^2 = x^2 + y^2$, where $0 \le z \le 2$, if it has constant density $k$.

In Exercises 33 and 34 the electric charge density at a point $(x, y, z)$ on S is $\sigma(x, y, z)$, and the total charge on S is given by $Q = \iint_S \sigma(x, y, z)\, dS$.

**33. Electric Charge** Find the total charge on the part of the plane $2x + 3y + z = 6$ in the first octant if the charge density at a point $P$ on the surface is directly proportional to the square of the distance between $P$ and the xy-plane.

**34. Electric Charge** Find the total charge on the part of the hemisphere $z = \sqrt{25 - x^2 - y^2}$ that lies directly above the plane region $R = \{(x, y)\,|\,x^2 + y^2 \le 9\}$ if the charge density at a point $P$ on the surface is directly proportional to the square of the distance between $P$ and the xy-plane.

**35. Flow of a Fluid** The flow of a fluid is described by the vector field $\mathbf{F}(x, y, z) = 2x\mathbf{i} + 2y\mathbf{j} + 3z\mathbf{k}$. Find the rate of flow of the fluid upward through the surface $S$ that is the part of the plane $2x + 3y + z = 6$ in the first octant.

**36. Flow of a Liquid** The flow of a liquid is described by the vector field $\mathbf{F}(x, y, z) = x\mathbf{i} + y\mathbf{j} + 3z\mathbf{k}$. If the mass density of the fluid is 1000 (in appropriate units), find the rate of flow (mass per unit time) upward of the liquid through the surface $S$ that is part of the paraboloid $z = 9 - x^2 - y^2$ above the xy-plane.
Hint: The flux is $\iint_S \rho \mathbf{F} \cdot \mathbf{n}\, dS$, where $\rho$ is the mass density function.

**37. Rate of Flow of Heat** The temperature at a point $(x, y, z)$ in a homogeneous body with thermal conductivity $k = 5$ is $T(x, y, z) = x^2 + y^2$. Find the rate of flow of heat across the cylindrical surface $x^2 + y^2 = 1$ between the planes $z = 0$ and $z = 1$.

**38. Rate of Flow of Heat** The temperature at a point $P(x, y, z)$ in a medium with thermal conductivity $k$ is proportional to the square of the distance between $P$ and the origin. Find the rate of flow of heat across a sphere $S$ of radius $a$, centered at the origin.

**39. a.** Suppose that $\mathbf{F} = P\mathbf{i} + Q\mathbf{j} + R\mathbf{k}$ is a continuous vector field in a region that contains a smooth oriented surface $S$ given by $y = g(x, z)$ and $D$ is its projection onto the

xz-plane. Write a double integral similar to Equation (7) that gives $\iint_S \mathbf{F} \cdot d\mathbf{S}$.

**b.** Use the result of part (a) to evaluate $\iint_S \mathbf{F} \cdot d\mathbf{S}$, where $\mathbf{F}(x, y, z) = y\mathbf{i} + z\mathbf{j} - 3yz^2\mathbf{k}$ and $S$ is the part of the cylinder $x^2 + y^2 = 4$ that lies in the first octant between $z = 0$ and $z = 3$ and oriented away from the origin.

**40. Flux of an Electric Field** Find the flux of the electric field

$$\mathbf{E} = \frac{q}{4\pi\varepsilon_0}\frac{\mathbf{r}}{|\mathbf{r}|^3}$$

across the sphere $x^2 + y^2 + z^2 = a^2$. Is the flux independent of the radius of the sphere? Give a physical interpretation.

**41.** Suppose that the density at each point of a thin spherical shell of radius $R$ is proportional to the (linear) distance from the point to a fixed point on the sphere. Find the total mass of the shell.

**42. Mass of a Ramp** Suppose that the density at each point of a spiral ramp represented by the vector equation $\mathbf{r}(u, v) = u \cos v\mathbf{i} + u \sin v\mathbf{j} + v\mathbf{k}$, where $0 \le u \le 3$ and $0 \le v \le 6\pi$, is proportional to the distance of the point from the central axis of the ramp. What is the total mass of the ramp?

**43.** Suppose that $f$ is a nonnegative real-valued function defined on the interval $[a, b]$ and $f$ has a continuous derivative in $(a, b)$. Show that the area of the surface of revolution $S$ obtained by revolving the graph of $f$ about the x-axis is given by

$$2\pi \int_a^b f(x)\sqrt{1 + [f'(x)]^2}\, dx$$

Hint: First find a parametric representation of $S$ (see Section 14.6).

**44.** Find the flux of $\mathbf{F}(x, y, z) = 2xz\mathbf{i} + yz\mathbf{j} - z^2\mathbf{k}$ out of a unit cube $T = \{(x, y, z)\,|\,0 \le x \le 1, 0 \le y \le 1, 0 \le z \le 1\}$.
Hint: The flux out of the cube is the sum of the fluxes across the sides of the cube.

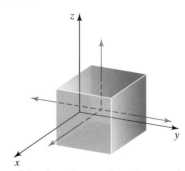

Four of the six unit normal vectors are shown.

**45. a.** Let $f$ be a function of three variables defined on a region in space containing a surface $S$. Suppose that $S$ is the graph of the function $z = g(x, y)$ that is represented implicitly by the equation $F(x, y, z) = 0$, where $F$ is differentiable. Show that

$$\iint_S f(x, y, z)\, dS = \iint_D \frac{f\sqrt{F_x^2 + F_y^2 + F_z^2}}{|F_z|}\, dA$$

where $D$ is the projection of $S$ onto the $xy$-plane.

**b.** Re-solve Example 1 using the result of part (a).

**46.** Show that if the surface $S$ is the graph of a function $z = g(x, y)$, then Equation (3) follows from Equation (6) by putting $\mathbf{r}(u, v) = u\mathbf{i} + v\mathbf{j} + g(u, v)\mathbf{k}$.

**47.** Let $\mathbf{F}$ and $\mathbf{G}$ be continuous vector fields defined on a smooth, oriented surface $S$. If $a$ and $b$ are constants, show that

$$\iint_S (a\mathbf{F} + b\mathbf{G}) \cdot d\mathbf{S} = a\iint_S \mathbf{F} \cdot d\mathbf{S} + b\iint_S \mathbf{F} \cdot d\mathbf{S}$$

*In Exercises 48 and 49, determine whether the statement is true or false. If it is true, explain why. If it is false, explain why or give an example that shows it is false.*

**48.** If $f(x, y, z) \geq 0$, then $\iint_S f\, dS \geq A(S)$, where $A(S)$ is the area of $S$.

**49.** If $\mathbf{F}$ is a constant vector field and $S$ is a sphere, then $\iint_S \mathbf{F} \cdot d\mathbf{S} = 0$.

# 14.8  The Divergence Theorem

Recall that Green's Theorem can be written in the form

$$\oint_C \mathbf{F} \cdot \mathbf{n}\, ds = \iint_R \operatorname{div} \mathbf{F}\, dA \qquad \text{Equation (8) in Section 14.5}$$

where $\mathbf{F}$ is a vector field in the plane, $C$ is an oriented, piecewise-smooth, simple closed curve that bounds a region $R$, and $\mathbf{n}$ is the outer normal vector to $C$. The theorem states that the line integral of the normal component of a vector field in two-dimensional space around a simple closed curve is equal to the double integral of the divergence of the vector field over the plane region bounded by the curve.

## ■ The Divergence Theorem

The *Divergence Theorem* generalizes this result to the case involving vector fields in three-dimensional space. This theorem, also called *Gauss's Theorem* in honor of the German mathematician Karl Friedrich Gauss (1777–1855), relates the surface integral of the normal component of a vector field $\mathbf{F}$ in three-dimensional space over a closed surface $S$ to the triple integral of the divergence of $\mathbf{F}$ over the solid region $T$ bounded by $S$.

Although the Divergence Theorem is true for very general surfaces, we will restrict our attention to the case in which the solid regions $T$ are simultaneously $x$-, $y$-, and $z$-simple. These regions are called **simple solid regions.** Examples are regions bounded by spheres, ellipsoids, cubes, and tetrahedrons.

---

**THEOREM 1   The Divergence Theorem**

Let $T$ be a simple solid region bounded by a closed piecewise-smooth surface $S$, and let $\mathbf{n}$ be the unit *outer* normal to $S$. If $\mathbf{F} = P\mathbf{i} + Q\mathbf{j} + R\mathbf{k}$ is a vector field, where $P$, $Q$, and $R$ have continuous partial derivatives on an open region containing $T$, then

$$\iint_S \mathbf{F} \cdot d\mathbf{S} = \iint_S \mathbf{F} \cdot \mathbf{n}\, dS = \iiint_T \operatorname{div} \mathbf{F}\, dV \qquad (1)$$

In words, the surface integral of the normal component of $\mathbf{F}$ over a closed surface $S$ is equal to the volume integral of the divergence of $\mathbf{F}$ over the solid $T$ bounded by $S$.

**PROOF**  Recall that if $\mathbf{F} = P\mathbf{i} + Q\mathbf{j} + R\mathbf{k}$, then

$$\nabla \cdot \mathbf{F} = \frac{\partial P}{\partial x} + \frac{\partial Q}{\partial y} + \frac{\partial R}{\partial z}$$

Therefore, the right-hand side of Equation (1) takes the form

$$\iiint_T \operatorname{div} \mathbf{F}\, dV = \iiint_T \frac{\partial P}{\partial x}\, dV + \iiint_T \frac{\partial Q}{\partial y}\, dV + \iiint_T \frac{\partial R}{\partial z}\, dV$$

Next, if $\mathbf{n}$ is the unit outer normal vector to $S$, then the left-hand side of Equation (1) assumes the form

$$\iint_S \mathbf{F} \cdot \mathbf{n}\, dS = \iint_S (P\mathbf{i} + Q\mathbf{j} + R\mathbf{k}) \cdot \mathbf{n}\, dS$$

$$= \iint_S P\mathbf{i} \cdot \mathbf{n}\, dS + \iint_S Q\mathbf{j} \cdot \mathbf{n}\, dS + \iint_S R\mathbf{k} \cdot \mathbf{n}\, dS$$

By equating the last two expressions, we see that the Divergence Theorem will be proved if we can show that

$$\iint_S P\mathbf{i} \cdot \mathbf{n}\, dS = \iiint_T \frac{\partial P}{\partial x}\, dV \tag{2}$$

$$\iint_S Q\mathbf{j} \cdot \mathbf{n}\, dS = \iiint_T \frac{\partial Q}{\partial y}\, dV \tag{3}$$

and

$$\iint_S R\mathbf{k} \cdot \mathbf{n}\, dS = \iiint_T \frac{\partial R}{\partial z}\, dV \tag{4}$$

**PROOF OF EQUATION (4)**  Because $T$ is $z$-simple, it can be described by the set

$$T = \{(x, y, z) \mid (x, y) \in D,\ k_1(x, y) \leq z \leq k_2(x, y)\}$$

where $D$ is the projection of $T$ onto the $xy$-plane, and $k_1$ and $k_2$ are continuous functions of $x$ and $y$. (See Figure 1.) Using Equation (4) in Section 13.6, we obtain

$$\iiint_T \frac{\partial R}{\partial z}\, dV = \iint_D \left[ \int_{k_1(x, y)}^{k_2(x, y)} \frac{\partial R}{\partial z}\, dz \right] dA$$

$$= \iint_D [R(x, y, k_2(x, y)) - R(x, y, k_1(x, y))]\, dA \tag{5}$$

that gives an alternative expression for the right-hand side of Equation (4).

Next, observe that the boundary of $T$ may consist of up to six surfaces (Figure 1). On each of the vertical sides, $\mathbf{k} \cdot \mathbf{n} = 0$, so

$$\iint_{S_3} R\mathbf{k} \cdot \mathbf{n}\, dS = \iint_{S_4} R\mathbf{k} \cdot \mathbf{n}\, dS = \iint_{S_5} R\mathbf{k} \cdot \mathbf{n}\, dS = \iint_{S_6} R\mathbf{k} \cdot \mathbf{n}\, dS = 0$$

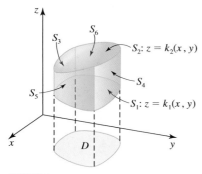

**FIGURE 1**
$T$ viewed as a $z$-simple region

Therefore,

$$\iint_S R\mathbf{k} \cdot \mathbf{n} \, dS = \iint_{S_1} R\mathbf{k} \cdot \mathbf{n} \, dS + \iint_{S_2} R\mathbf{k} \cdot \mathbf{n} \, dS \tag{6}$$

To evaluate the integrals on the right-hand side of Equation (6), observe that the outer normal $\mathbf{n}$ points downward on $S_1$. Writing $g_1(x, y, z) = z - k_1(x, y)$, we find

$$\mathbf{n} = -\frac{\nabla g_1(x, y, z)}{|\nabla g_1(x, y, z)|} = \frac{\dfrac{\partial k_1}{\partial x}\mathbf{i} + \dfrac{\partial k_1}{\partial y}\mathbf{j} - \mathbf{k}}{\sqrt{\left(\dfrac{\partial k_1}{\partial x}\right)^2 + \left(\dfrac{\partial k_1}{\partial y}\right)^2 + 1}}$$

Therefore, using Equation (6), we obtain

$$\iint_{S_1} R\mathbf{k} \cdot \mathbf{n} \, dS = \iint_D R(x, y, k_1(x, y)) \cdot \frac{-1}{\sqrt{\left(\dfrac{\partial k_1}{\partial x}\right)^2 + \left(\dfrac{\partial k_1}{\partial y}\right)^2 + 1}} \sqrt{\left(\dfrac{\partial k_1}{\partial x}\right)^2 + \left(\dfrac{\partial k_1}{\partial y}\right)^2 + 1} \, dA$$

$$= -\iint_D R(x, y, k_1(x, y)) \, dA$$

On $S_2$ the outer normal $\mathbf{n}$ points upward. Writing $g_2(x, y, z) = z - k_2(x, y)$, we see that

$$\mathbf{n} = \frac{\nabla g_2(x, y, z)}{|\nabla g_2(x, y, z)|} = \frac{-\dfrac{\partial k_2}{\partial x}\mathbf{i} - \dfrac{\partial k_2}{\partial y}\mathbf{j} + \mathbf{k}}{\sqrt{\left(\dfrac{\partial k_2}{\partial x}\right)^2 + \left(\dfrac{\partial k_2}{\partial y}\right)^2 + 1}}$$

so

$$\iint_{S_2} R\mathbf{k} \cdot \mathbf{n} \, dS = \iint_D R(x, y, k_2(x, y)) \cdot \frac{1}{\sqrt{\left(\dfrac{\partial k_2}{\partial x}\right)^2 + \left(\dfrac{\partial k_2}{\partial y}\right)^2 + 1}} \sqrt{\left(\dfrac{\partial k_2}{\partial x}\right)^2 + \left(\dfrac{\partial k_2}{\partial y}\right)^2 + 1} \, dA$$

$$= \iint_D R(x, y, k_2(x, y)) \, dA$$

Therefore, Equation (6) becomes

$$\iint_S R\mathbf{k} \cdot \mathbf{n} \, dS = \iint_D [R(x, y, k_2(x, y)) - R(x, y, k_1(x, y))] \, dA$$

Comparing this with Equation (5), we have

$$\iint_S R\mathbf{k} \cdot \mathbf{n} \, dS = \iiint_T \frac{\partial R}{\partial z} \, dV$$

so Equation (4) is established. Equations (2) and (3) are proved in a similar manner by viewing $R$ as $x$-simple and $y$-simple, respectively. ∎

**EXAMPLE 1** Compute $\iint_S \mathbf{F} \cdot \mathbf{n} \, dS$ given that

$$\mathbf{F}(x, y, z) = (x + \sin z)\mathbf{i} + (2y + \cos x)\mathbf{j} + (3z + \tan y)\mathbf{k}$$

and $S$ is the unit sphere $x^2 + y^2 + z^2 = 1$.

**Solution**   To evaluate the integral directly would be a difficult task. Applying the Divergence Theorem, we have

$$\iint_S \mathbf{F} \cdot \mathbf{n} \, dS = \iiint_T \text{div } \mathbf{F} \, dV$$

But

$$\nabla \cdot \mathbf{F} = \frac{\partial}{\partial x}(x + \sin z) + \frac{\partial}{\partial y}(2y + \cos x) + \frac{\partial}{\partial z}(3z + \tan y) = 1 + 2 + 3 = 6$$

and the solid $T$ is the unit ball $B$ bounded by the unit sphere $x^2 + y^2 + z^2 = 1$. Therefore,

$$\iint_S \mathbf{F} \cdot \mathbf{n} \, dS = \iiint_B \nabla \cdot \mathbf{F} \, dV = \iiint_B 6 \, dV = 6V(B) = 6\left[\frac{4}{3}\pi(1)^3\right] = 8\pi \quad \blacksquare$$

**EXAMPLE 2**   Let $T$ be the solid bounded by the cylinder $x^2 + y^2 = 4$ and the planes $z = 0$ and $z = 3$, and let $S$ be the surface of $T$. Calculate the outward flux of the vector field $\mathbf{F}(x, y, z) = xy^2\mathbf{i} + yz^2\mathbf{j} + zx^2\mathbf{k}$ over $S$.

**Solution**   The surface $S$ is shown in Figure 2. The flux of $\mathbf{F}$ over $S$ is given by $\iint_S \mathbf{F} \cdot \mathbf{n} \, dS$, which by the Divergence Theorem can also be found by evaluating $\iiint_T \nabla \cdot \mathbf{F} \, dV$. Now

$$\text{div } \mathbf{F} = \frac{\partial}{\partial x}(xy^2) + \frac{\partial}{\partial y}(yz^2) + \frac{\partial}{\partial z}(zx^2) = y^2 + z^2 + x^2$$

Therefore,

$$\iint_S \mathbf{F} \cdot \mathbf{n} \, dS = \iiint_T \text{div } \mathbf{F} \, dV = \iiint_T (x^2 + y^2 + z^2) \, dV$$

Using cylindrical coordinates to evaluate the triple integral, we have

$$\iint_S \mathbf{F} \cdot \mathbf{n} \, dS = \int_0^{2\pi} \int_0^2 \int_0^3 (r^2 + z^2)r \, dz \, dr \, d\theta$$

$$= \int_0^{2\pi} \int_0^2 \left[r^3 z + \frac{1}{3}rz^3\right]_{z=0}^{z=3} dr \, d\theta$$

$$= \int_0^{2\pi} \int_0^2 (3r^3 + 9r) \, dr \, d\theta = \int_0^{2\pi} \left[\frac{3}{4}r^4 + \frac{9}{2}r^2\right]_{r=0}^{r=2} d\theta$$

$$= \int_0^{2\pi} 30 \, d\theta = 60\pi \quad \blacksquare$$

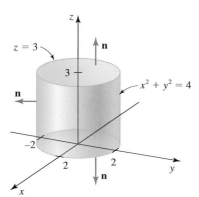

**FIGURE 2**
The surface $S$ and some of the outer normals to $S$

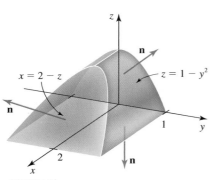

**FIGURE 3**
The surface $S$ and some of the outer normals to $S$

**EXAMPLE 3**   Let $T$ be the region bounded by the parabolic cylinder $z = 1 - y^2$ and the planes $z = 0$, $x = 0$, and $x + z = 2$, and let $S$ be the surface of $T$. If $\mathbf{F}(x, y, z) = xy^2\mathbf{i} + \left(\frac{1}{3}y^3 - \cos xz\right)\mathbf{j} + xe^y\mathbf{k}$, find $\iint_S \mathbf{F} \cdot \mathbf{n} \, dS$.

**Solution**   The surface $S$ is shown in Figure 3. We first compute

$$\nabla \cdot \mathbf{F} = \frac{\partial}{\partial x}(xy^2) + \frac{\partial}{\partial y}\left(\frac{1}{3}y^3 - \cos xz\right) + \frac{\partial}{\partial z}(xe^y) = y^2 + y^2 = 2y^2$$

Viewing $T$ as an $x$-simple region, we use the Divergence Theorem to obtain

$$\iint_S \mathbf{F} \cdot \mathbf{n} \, dS = \iiint_T \operatorname{div} \mathbf{F} \, dV = \iiint_T 2y^2 \, dV$$

$$= \int_{-1}^{1} \int_0^{1-y^2} \int_0^{2-z} 2y^2 \, dx \, dz \, dy = \int_{-1}^{1} \int_0^{1-y^2} \left[2y^2 x\right]_{x=0}^{x=2-z} dz \, dy$$

$$= \int_{-1}^{1} \int_0^{1-y^2} 2y^2(2 - z) \, dz \, dy = \int_{-1}^{1} \left[-y^2(2 - z)^2\right]_{z=0}^{z=1-y^2} dy$$

$$= \int_{-1}^{1} \left[-y^2(1 + y^2)^2 + 4y^2\right] dy = \int_{-1}^{1} (3y^2 - 2y^4 - y^6) \, dy$$

$$= 2 \int_0^{1} (3y^2 - 2y^4 - y^6) \, dy$$

$$= 2\left[y^3 - \frac{2}{5} y^5 - \frac{1}{7} y^7\right]_0^1 = \frac{32}{35} \qquad \blacksquare$$

The Divergence Theorem was stated for simple solid regions. But it can be extended to include regions that are finite unions of simple solid regions. For example, let $T$ be the region that lies between the closed surfaces $S_1$ and $S_2$ with $S_1$ lying inside $S_2$. Then the boundary of $T$ is $S = S_1 \cup S_2$, as shown in Figure 4. If $\mathbf{n}_1$ and $\mathbf{n}_2$ denote the outward normals of $S_1$ and $S_2$, respectively, then the normal to $S$ is given by $\mathbf{n} = -\mathbf{n}_1$ on $S_1$ and by $\mathbf{n} = \mathbf{n}_2$ on $S_2$. Applying the Divergence Theorem to $T$, we obtain

$$\iiint_T \operatorname{div} \mathbf{F} \, dV = \iint_S \mathbf{F} \cdot \mathbf{n} \, dS = \iint_{S_1} \mathbf{F} \cdot \mathbf{n} \, dS + \iint_{S_2} \mathbf{F} \cdot \mathbf{n} \, dS$$

$$= \iint_{S_1} \mathbf{F} \cdot (-\mathbf{n}_1) \, dS + \iint_{S_2} \mathbf{F} \cdot \mathbf{n}_2 \, dS$$

$$= -\iint_{S_1} \mathbf{F} \cdot \mathbf{n}_1 \, dS + \iint_{S_2} \mathbf{F} \cdot \mathbf{n}_2 \, dS$$

$$= -\iint_{S_1} \mathbf{F} \cdot d\mathbf{S} + \iint_{S_2} \mathbf{F} \cdot d\mathbf{S} \qquad (7)$$

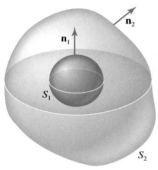

**FIGURE 4**
The region $T$ lies between $S_1$ and $S_2$.

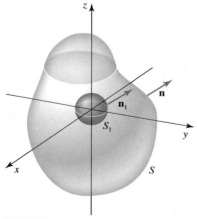

**FIGURE 5**
The region $T$ lies between the sphere $S_1$ and the surface $S$.

**EXAMPLE 4**    **Flux of an Electric Field**    Consider the electric field

$$\mathbf{E} = \frac{q}{4\pi\varepsilon_0} \frac{\mathbf{r}}{|\mathbf{r}|^3}$$

induced by a point charge $q$ placed at the origin of a three-dimensional coordinate system, where $\mathbf{r} = x\mathbf{i} + y\mathbf{j} + z\mathbf{k}$. (See Example 4 in Section 14.1.) Find the flux of $\mathbf{E}$ across a smooth surface $S$ that encloses the origin.

**Solution**    The Divergence Theorem is not immediately applicable because $\mathbf{E}$ is not continuous at the origin. To avoid this difficulty, let's construct a sphere with center at the origin and a radius $a$ that is small enough to ensure that the sphere lies completely inside $S$. (See Figure 5.) If we denote this sphere by $S_1$, then $\mathbf{E}$ satisfies the conditions of the Divergence Theorem for the solid $T$ that lies between $S_1$ and $S$. Using Equation (7), we obtain

$$\iiint_T \text{div } \mathbf{E} \, dV = -\iint_{S_1} \mathbf{E} \cdot d\mathbf{S} + \iint_S \mathbf{E} \cdot d\mathbf{S}$$

But we showed that div $\mathbf{E} = 0$ in Example 4 in Section 14.2. So we have

$$\iint_S \mathbf{E} \cdot d\mathbf{S} = \iint_{S_1} \mathbf{E} \cdot d\mathbf{S} = \iint_{S_1} \mathbf{E} \cdot \mathbf{n} \, dS$$

To evaluate the integral on the right, note that the unit normal to the sphere $S_1$ is $\mathbf{n} = \mathbf{r}/|\mathbf{r}|$. Therefore,

$$\mathbf{E} \cdot \mathbf{n} = \frac{q}{4\pi\varepsilon_0} \frac{\mathbf{r}}{|\mathbf{r}|^3} \cdot \left(\frac{\mathbf{r}}{|\mathbf{r}|}\right) = \frac{q}{4\pi\varepsilon_0} \frac{\mathbf{r} \cdot \mathbf{r}}{|\mathbf{r}|^4}$$

$$= \frac{q}{4\pi\varepsilon_0 |\mathbf{r}|^2} \qquad \mathbf{r} \cdot \mathbf{r} = |\mathbf{r}|^2$$

$$= \frac{q}{4\pi\varepsilon_0 a^2}$$

because $|\mathbf{r}| = a$ on the sphere $S_1$. Therefore, we have

$$\iint_S \mathbf{E} \cdot d\mathbf{S} = \iint_{S_1} \mathbf{E} \cdot d\mathbf{S} = \frac{q}{4\pi\varepsilon_0 a^2} \iint_{S_1} dS = \frac{q}{4\pi\varepsilon_0 a^2} A(S_1)$$

$$= \frac{q}{4\pi\varepsilon_0 a^2}(4\pi a^2) = \frac{q}{\varepsilon_0} \qquad \blacksquare$$

The result in Example 4 shows that the flux across *any* closed surface that contains the charge $q$ is $q/\varepsilon_0$. This is intuitively clear, since any closed surface enclosing the charge $q$ would trap the same number of field lines.

Furthermore, by using the principle of superposition (the field induced by several electric charges is the vector sum of the fields due to the individual charges), it can be shown that for any closed surface $S$,

$$\iint_S \mathbf{E} \cdot \mathbf{n} \, dS = \frac{Q}{\varepsilon_0}$$

where $Q$ is the total charge enclosed by $S$. This is one of the most important laws in electrostatics and is known as **Gauss's Law**.

## ◼ Interpretation of Divergence

For a physical interpretation of the divergence of a vector field $\mathbf{F}$, we can think of $\mathbf{F}$ as representing the velocity field associated with the flow of a fluid. Let $P_0(x_0, y_0, z_0)$ be a point in the fluid, and let $B_r$ be a ball with radius $r$, centered at $P_0$, and having the sphere $S_r$ for its boundary, as shown in Figure 6. If $r$ is small, then the continuity of

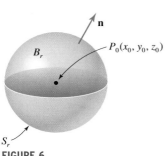

**FIGURE 6**

$B_r$ is a ball of radius $r$ centered at $P_0(x_0, y_0, z_0)$.

div $\mathbf{F}$ guarantees that $(\text{div } \mathbf{F})(P) \approx (\text{div } \mathbf{F})(P_0)$ for all points $P$ in $B_r$. Therefore, using the Divergence Theorem, we have

$$\iint_{S_r} \mathbf{F} \cdot \mathbf{n} \, dS = \iiint_{B_r} \text{div} \cdot \mathbf{F} \, dV \approx \iiint_{B_r} (\text{div } \mathbf{F})(P_0) \, dV$$

$$= (\text{div } \mathbf{F})(P_0) \iiint_{B_r} dV = (\text{div } \mathbf{F})(P_0) V(B_r)$$

where $V(B_r) = \frac{4}{3}\pi r^3$. This approximation improves as $r \to 0$, and we have

$$(\text{div } \mathbf{F})(P_0) = \lim_{r \to 0} \frac{1}{V(B_r)} \iint_{B_r} \mathbf{F} \cdot \mathbf{n} \, dS \tag{8}$$

Equation (8) tells us that we can interpret div $\mathbf{F}(P_0)$ as the rate of flow outward of the fluid per unit volume at $P_0$—hence the term *divergence*. In general, if div $\mathbf{F}(P) > 0$, the net flow is outward near $P$, and $P$ is called a **source**. If div $\mathbf{F}(P) < 0$, the net flow is inward near $P$, and $P$ is called a **sink**. Finally, if the fluid is incompressible and there are no sources or sinks present, then no fluid exits or enters $B_r$ and, accordingly, div $\mathbf{F}(P) = 0$ at every point $P$.

## 14.8 CONCEPT QUESTIONS

1. State the Divergence Theorem.
2. Suppose that the vector field $\mathbf{F}$ is associated with the flow of a fluid and $P$ is a point in the domain of $\mathbf{F}$. Give a physical interpretation for div $\mathbf{F}(P)$. What happens to the flow at $P$ if div $\mathbf{F}(P) > 0$? If div $\mathbf{F}(P) < 0$? If div $\mathbf{F}(P) = 0$?

## 14.8 EXERCISES

*In Exercises 1–4, verify the Divergence Theorem for the given vector field $\mathbf{F}$ and region $T$.*

1. $\mathbf{F}(x, y, z) = x\mathbf{i} + y\mathbf{j} + z\mathbf{k}$; $T$ is the cube bounded by the planes $x = 0$, $x = 1$, $y = 0$, $y = 1$, $z = 0$, and $z = 1$

2. $\mathbf{F}(x, y, z) = 2xy\mathbf{i} - y^2\mathbf{j} + 3yz\mathbf{k}$; $T$ is the cube bounded by the planes $x = 0$, $x = 2$, $y = 0$, $y = 2$, $z = 0$, and $z = 2$

3. $\mathbf{F}(x, y, z) = y\mathbf{i} + z\mathbf{j} - 3yz^2\mathbf{k}$; $T$ is the region bounded by the cylinder $x^2 + y^2 = 4$ in the first octant between $z = 0$ and $z = 3$

4. $\mathbf{F}(x, y, z) = x\mathbf{i} + y\mathbf{j} + 2z^2\mathbf{k}$; $T$ is the region bounded by the paraboloid $z = x^2 + y^2$ and the plane $z = 1$

*In Exercises 5–18, use the Divergence Theorem to find the flux of $\mathbf{F}$ across $S$; that is, calculate $\iint_S \mathbf{F} \cdot \mathbf{n} \, dS$.*

5. $\mathbf{F}(x, y, z) = xy^2\mathbf{i} + 2yz\mathbf{j} - 3x^2y^3\mathbf{k}$; $S$ is the surface of the cube bounded by the planes $x = \pm 1$, $y = \pm 1$, and $z = \pm 1$

6. $\mathbf{F}(x, y, z) = 2xz\mathbf{i} - y^2\mathbf{j} + yz\mathbf{k}$; $S$ is the surface of the rectangular box bounded by the planes $x = 0$, $x = 2$, $y = 0$, $y = 3$, $z = -1$, and $z = 1$

7. $\mathbf{F}(x, y, z) = (x^3 + \cos y)\mathbf{i} + (y^3 + \sin xz)\mathbf{j} + (z^3 + 2e^{-x})\mathbf{k}$; $S$ is the surface of the region bounded by the cylinder $y^2 + z^2 = 1$ and the planes $x = 0$ and $x = 3$

8. $\mathbf{F}(x, y, z) = \sin y\mathbf{i} + (x^2y + e^z)\mathbf{j} + (2x^2z + e^{-x})\mathbf{k}$; $S$ is the surface of the region bounded by the cylinder $z = 4 - x^2$ and the planes $z = 0$, $y = 0$ and $y + z = 5$

9. $\mathbf{F}(x, y, z) = 2xy\mathbf{i} + y^2\mathbf{j} + (x^2 + yz)\mathbf{k}$; $S$ is the surface of the tetrahedron bounded by the planes $x + y + z = 1$, $x = 0$, $y = 0$, and $z = 0$

10. $\mathbf{F}(x, y, z) = x^2\mathbf{i} + xz^2\mathbf{j} + (2xz + \sin xy)\mathbf{k}$; $S$ is the surface of the tetrahedron bounded by the planes $x + 2y + 3z = 6$, $x = 0$, $y = 0$, and $z = 0$

11. $\mathbf{F}(x, y, z) = x\mathbf{i} + 2y\mathbf{j} + 3z\mathbf{k}$; $S$ is the sphere $x^2 + y^2 + z^2 = 9$

12. $\mathbf{F}(x, y, z) = (x + y^2)\mathbf{i} + (\sqrt{x^2 + z^2} + y)\mathbf{j} + \cos xy \, \mathbf{k}$; $S$ is the surface of the region bounded by the cylinder $x^2 + z^2 = 1$ and the planes $y = 0$ and $y = 1$

13. $\mathbf{F}(x, y, z) = xz\mathbf{i} - yz\mathbf{j} + xy\mathbf{k}$; $S$ is the ellipsoid $9x^2 + 4y^2 + 36z^2 = 36$

**14.** $\mathbf{F}(x, y, z) = (x + ye^z)\mathbf{i} - (y + \tan xz)\mathbf{j} + (x + \cosh y)\mathbf{k}$;
$S$ is the surface of the region bounded by the cone
$y = \sqrt{x^2 + z^2}$ and the plane $y = 4$

**15.** $\mathbf{F}(x, y, z) = (x^3 + 1)\mathbf{i} + (yz^2 + \cos xz)\mathbf{j} + (2y^2z + e^{\tan x})\mathbf{k}$;
$S$ is the sphere $x^2 + y^2 + z^2 = 1$

**16.** $\mathbf{F}(x, y, z) = x^3\mathbf{i} - 3yz^2\mathbf{j} + 2z^3\mathbf{k}$; $S$ is the surface of the
region bounded by the paraboloid $y = 9 - x^2 - z^2$ and the
$xz$-plane

**17.** $\mathbf{F}(x, y, z) = xz\mathbf{i} + x^2y\mathbf{j} + (y^2z + 1)\mathbf{k}$; $S$ is the surface of
the region that lies between the cylinders $x^2 + y^2 = 1$ and
$x^2 + y^2 = 4$ and between the planes $z = 1$ and $z = 3$

**18.** $\mathbf{F}(x, y, z) = yz^2\mathbf{i} + (y^3 + xz)\mathbf{j} - (y^2z + 10x)\mathbf{k}$; $S$ is the
surface of the region between the spheres $x^2 + y^2 + z^2 = 1$
and $x^2 + y^2 + z^2 = 4$

*In Exercises 19–25, assume that $S$ and $T$ satisfy the conditions of
the Divergence Theorem.*

**19.** Show that the volume of $T$ is given by $V(T) = \frac{1}{3}\iint_S \mathbf{r} \cdot \mathbf{n}\, dS$,
where $\mathbf{r} = x\mathbf{i} + y\mathbf{j} + z\mathbf{k}$.

**20.** Show that $\iint_S \mathbf{a} \cdot \mathbf{n}\, dS = 0$, where $\mathbf{a}$ is a constant vector.

**21.** Show that if $\mathbf{F}$ has continuous second-order derivatives, then

$$\iint_S \operatorname{curl} \mathbf{F} \cdot \mathbf{n}\, dS = 0$$

**22.** Show that if $f$ has continuous second-order partial derivatives, then

$$\iiint_T \nabla^2 f\, dV = \iint_S D_{\mathbf{n}} f\, dS$$

where $\nabla^2 f = \dfrac{\partial^2 f}{\partial x^2} + \dfrac{\partial^2 f}{\partial y^2} + \dfrac{\partial^2 f}{\partial z^2}$ and $D_{\mathbf{n}} f$ is the directional
derivative of $f$ in the direction of an outer normal $\mathbf{n}$ of $S$.

**23.** Show that $\iiint_T \nabla f\, dV = \iint_S f\mathbf{n}\, dS$.
**Hint:** Apply the Divergence Theorem to $\mathbf{F} = f\mathbf{c}$, where $\mathbf{c}$ is a constant vector.

**24.** Show that if $f$ and $g$ have continuous second-order partial
derivatives, then

$$\iiint_T (f\nabla^2 g + \nabla f \cdot \nabla g)\, dV = \iint_S (f\nabla g) \cdot \mathbf{n}\, dS$$

**25.** Show that if $f$ and $g$ have continuous second-order partial
derivatives, then

$$\iiint_T (f\nabla^2 g - g\nabla^2 f)\, dV = \iint_S (f\nabla g - g\nabla f) \cdot \mathbf{n}\, dS$$

**26.** Find the flux of the vector field

$$\mathbf{F}(x, y, z) = \frac{x\mathbf{i} + y\mathbf{j} + z\mathbf{k}}{(x^2 + y^2 + z^2)^{3/2}}$$

across the ellipsoid $(x^2/9) + (y^2/16) + (z^2/4) = 1$.

*In Exercises 27–29, determine whether the statement is true or
false. If it is true, explain why. If it is false, explain why or give
an example that shows it is false.*

**27.** If $\operatorname{div} \mathbf{F} = 0$, then $\iint_S \mathbf{F} \cdot d\mathbf{S} = 0$ for every closed surface $S$.

**28.** If $\mathbf{F}$ is a constant vector field and $S$ is a cube, then
$\iint_S \mathbf{F} \cdot d\mathbf{S} = 0$.

**29.** If $|\mathbf{F}(x, y, z)| \le 1$ for all points $(x, y, z)$ in a solid region $T$
bounded by a closed surface $S$, then $\iiint_T \operatorname{div} \mathbf{F}\, dV \le A(S)$,
where $A(S)$ is the area of $S$.

# 14.9 Stokes' Theorem

## ■ Stokes' Theorem

In this section we consider another generalization of Green's Theorem to higher dimensions. We start with the following version of Green's Theorem:

$$\oint_C \mathbf{F} \cdot \mathbf{T}\, ds = \iint_R \operatorname{curl} \mathbf{F} \cdot \mathbf{k}\, dA \qquad \text{Equation (7) in Section 14.5}$$

where the plane curve $C$ is an oriented piecewise-smooth simply closed curve that
bounds a region $R$. The theorem states that the line integral of the tangential component of a vector field in two-dimensional space around a closed curve is equal to the
double integral of the normal component to $R$ of the curl of the vector field over the
plane region bounded by the curve.

Stokes' Theorem generalizes this version of Green's Theorem to three-dimensional
space. Named after the English mathematical physicist George G. Stokes (1819–1903),

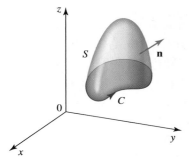

**FIGURE 1**
The curve $C$ is a boundary of the surface $S$.

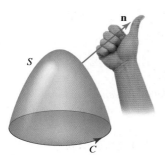

**FIGURE 2**
The orientation of $S$ induces an orientation on $C$.

Stokes' Theorem relates the line integral of the tangential component of a vector field in three-dimensional space around a simple closed curve in space to the surface integral of the normal component of the curl of the vector field over any surface that has the closed curve as its boundary. (See Figure 1.)

The orientation of the surface induces an orientation on $C$ that is determined by using the right-hand rule: Imagine grasping the normal vector $\mathbf{n}$ to $S$ with your right hand in such a way that your thumb points in the direction of $\mathbf{n}$. Then your fingers will point toward the positive direction of $C$. (See Figure 2.)

---

**THEOREM 1 Stokes' Theorem**

Let $S$ be an oriented piecewise-smooth surface that has a unit normal vector $\mathbf{n}$ and is bounded by a simple, closed, positively oriented curve $C$. If $\mathbf{F} = P\mathbf{i} + Q\mathbf{j} + R\mathbf{k}$ is a vector field, where $P$, $Q$, and $R$ have continuous partial derivatives in an open region containing $S$, then

$$\oint_C \mathbf{F} \cdot d\mathbf{r} = \oint_C \mathbf{F} \cdot \mathbf{T} \, ds = \iint_S \operatorname{curl} \mathbf{F} \cdot d\mathbf{S} = \iint_S \operatorname{curl} \mathbf{F} \cdot \mathbf{n} \, dS \qquad (1)$$

---

In words, the line integral of the tangential component of $\mathbf{F}$ around a simple closed curve $C$ is equal to the surface integral of the normal component of the curl of $\mathbf{F}$ over any surface $S$ with $C$ as its boundary.

Stokes' Theorem provides us with the following physical interpretation: If $\mathbf{F}$ is a force field, then the work done by $\mathbf{F}$ along $C$ is equal to the flux of curl $\mathbf{F}$ across $S$. The proof of Stokes' Theorem can be found in more advanced textbooks.

**EXAMPLE 1** Verify Stokes' Theorem for the case in which $\mathbf{F}(x, y, z) = 3z\mathbf{i} + 2x\mathbf{j} + y^2\mathbf{k}$, $S$ is the part of the paraboloid $z = 4 - x^2 - y^2$ with $z \geq 0$, and $C$ is the trace of $S$ on the $xy$-plane.

**Solution** The surface $S$ and the curve $C$ are sketched in Figure 3. We begin by calculating

$$\operatorname{curl} \mathbf{F} = \begin{vmatrix} \mathbf{i} & \mathbf{j} & \mathbf{k} \\ \dfrac{\partial}{\partial x} & \dfrac{\partial}{\partial y} & \dfrac{\partial}{\partial z} \\ 3z & 2x & y^2 \end{vmatrix} = 2y\mathbf{i} + 3\mathbf{j} + 2\mathbf{k}$$

Next, writing $g(x, y) = 4 - x^2 - y^2$, we find that $g_x = -2x$ and $g_y = -2y$. Also, observe that the projection of $S$ onto the $xy$-plane is $R = \{(x, y) \mid x^2 + y^2 \leq 4\}$. So using Equation (7) of Section 14.7 with $P(x, y) = 2y$, $Q(x, y) = 3$, and $R(x, y) = 2$, we obtain

$$\iint_S \operatorname{curl} \mathbf{F} \cdot d\mathbf{S} = \iint_R (-Pg_x - Qg_y + R) \, dA$$

$$= \iint_R (4xy + 6y + 2) \, dA$$

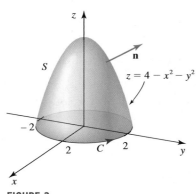

**FIGURE 3**
The outer normal $\mathbf{n}$ to $S$ induces the positive direction for $C$ as shown.

Changing to polar coordinates, we find

$$\iint_S \text{curl } \mathbf{F} \cdot d\mathbf{S} = \int_0^{2\pi} \int_0^2 (4r^2 \cos \theta \sin \theta + 6r \sin \theta + 2)r \, dr \, d\theta$$

$$= \int_0^{2\pi} \int_0^2 (4r^3 \cos \theta \sin \theta + 6r^2 \sin \theta + 2r) \, dr \, d\theta$$

$$= \int_0^{2\pi} \left[ r^4 \cos \theta \sin \theta + 2r^3 \sin \theta + r^2 \right]_{r=0}^{r=2} d\theta$$

$$= \int_0^{2\pi} [16 \cos \theta \sin \theta + 16 \sin \theta + 4) \, d\theta$$

$$= \left[ 8 \sin^2 \theta - 16 \cos \theta + 4\theta \right]_0^{2\pi} = 8\pi$$

To evaluate the line integral on the left-hand side of Equation (1), we observe that $C$ can be parametrized by

$$\mathbf{r}(t) = 2 \cos t\mathbf{i} + 2 \sin t\mathbf{j} + 0\mathbf{k} \qquad 0 \le t \le 2\pi$$

Therefore, with $\mathbf{F}(\mathbf{r}(t)) = 0\mathbf{i} + 4 \cos t\mathbf{j} + 4 \sin^2 t\mathbf{k}$ we have

$$\oint_C \mathbf{F} \cdot d\mathbf{r} = \int_0^{2\pi} \mathbf{F}(\mathbf{r}(t)) \cdot \mathbf{r}'(t) \, dt$$

$$= \int_0^{2\pi} (4 \cos t\mathbf{j} + 4 \sin^2 t\mathbf{k}) \cdot (-2 \sin t\mathbf{i} + 2 \cos t\mathbf{j}) \, dt$$

$$= \int_0^{2\pi} 8 \cos^2 t \, dt = 4 \int_0^{2\pi} (1 + \cos 2t) \, dt$$

$$= 4 \left[ t + \frac{1}{2} \sin 2t \right]_0^{2\pi} = 8\pi$$

which is the same as the surface integral. This verifies the solution. ■

**EXAMPLE 2**  Evaluate $\oint_C \mathbf{F} \cdot d\mathbf{r}$, where $\mathbf{F}(x, y, z) = \cos z\mathbf{i} + x^2\mathbf{j} + 2y\mathbf{k}$ and $C$ is the curve of intersection of the plane $x + z = 2$ and the cylinder $x^2 + y^2 = 1$.

**Solution**  The curve $C$ is an ellipse, as shown in Figure 4. We can evaluate $\oint_C \mathbf{F} \cdot d\mathbf{r}$ directly (see Exercise 29). But it is easier to use Stokes' Theorem. Thus,

$$\oint_C \mathbf{F} \cdot d\mathbf{r} = \oint_C \mathbf{F} \cdot \mathbf{T} \, ds = \iint_S \text{curl } \mathbf{F} \cdot d\mathbf{S}$$

where $S$ is the elliptic plane region lying in the plane $x + z = 2$ and enclosed by $C$. Of all the surfaces that have $C$ as their boundary, this choice of $S$ is clearly the most convenient for our purpose.

We first find

$$\text{curl } \mathbf{F} = \begin{vmatrix} \mathbf{i} & \mathbf{j} & \mathbf{k} \\ \dfrac{\partial}{\partial x} & \dfrac{\partial}{\partial y} & \dfrac{\partial}{\partial z} \\ \cos z & x^2 & 2y \end{vmatrix} = 2\mathbf{i} - \sin z\mathbf{j} + 2x\mathbf{k}$$

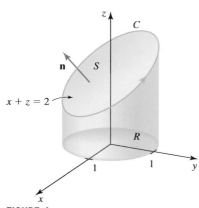

**FIGURE 4**
The surface $S$ is enclosed by $C$ and the disk $R$ is the projection of $S$ onto the $xy$-plane.

Then, writing $g(x, y) = -x + 2$, we find $g_x = -1$ and $g_y = 0$. Next, observe that the projection of $S$ onto the $xy$-plane is $R = \{(x, y) \mid x^2 + y^2 \leq 1\}$. So using Equation (7) of Section 14.7 with $P(x, y) = 2$, $Q(x, y) = -\sin z$, and $R(x, y) = 2x$, we obtain

$$\iint_S \text{curl } \mathbf{F} \cdot d\mathbf{S} = \iint_R (-Pg_x - Qg_y + R) \, dA$$

$$= \iint_R (2 + 2x) \, dx$$

$$= 2 \iint_R (1 + x) \, dA$$

Changing to polar coordinates, we obtain

$$\iint_S \text{curl } \mathbf{F} \cdot d\mathbf{S} = 2 \int_0^{2\pi} \int_0^1 (1 + r \cos \theta) r \, dr \, d\theta = 2 \int_0^{2\pi} \int_0^1 (r + r^2 \cos \theta) \, dr \, d\theta$$

$$= 2 \int_0^{2\pi} \left[ \frac{1}{2} r^2 + \frac{1}{3} r^3 \cos \theta \right]_{r=0}^{r=1} d\theta = 2 \int_0^{2\pi} \left( \frac{1}{2} + \frac{1}{3} \cos \theta \right) d\theta$$

$$= 2 \left[ \frac{1}{2} \theta + \frac{1}{3} \sin \theta \right]_0^{2\pi} = 2\pi$$

so by Stokes' Theorem we have

$$\oint_C \mathbf{F} \cdot d\mathbf{r} = \iint_S \text{curl } \mathbf{F} \cdot d\mathbf{S} = 2\pi \qquad \blacksquare$$

**EXAMPLE 3**   Evaluate $\iint_S \text{curl } \mathbf{F} \cdot d\mathbf{S}$, where $\mathbf{F}(x, y, z) = yz\mathbf{i} - xz\mathbf{j} + z^3\mathbf{k}$ and $S$ is the part of the sphere $x^2 + y^2 + z^2 = 8$ lying inside the cone $z = \sqrt{x^2 + y^2}$. (See Figure 5.)

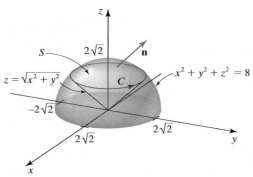

**FIGURE 5**
The surface $S$ is enclosed by $C$; the outer normal $\mathbf{n}$ induces the positive direction of $C$ shown.

**Solution**   The boundary of $S$ is the curve whose equation is obtained by solving the equations $x^2 + y^2 + z^2 = 8$ and $z = \sqrt{x^2 + y^2}$ simultaneously. Squaring the second equation and substituting this result into the first equation give $2z^2 = 8$, or $z = 2$ (since $z > 0$). Therefore, the boundary of $S$ is the circle $C$ with equations $x^2 + y^2 = 4$ and

$z = 2$. Since $C$ is easily parametrized, we can use Stokes' Theorem to help us find the value of the given surface integral by evaluating the line integral $\oint_C \mathbf{F} \cdot d\mathbf{r}$. A vector equation of $C$ is

$$\mathbf{r}(t) = 2 \cos t\mathbf{i} + 2 \sin t\mathbf{j} + 2\mathbf{k} \qquad 0 \le t \le 2\pi$$

so

$$\mathbf{r}'(t) = -2 \sin t\mathbf{i} + 2 \cos t\mathbf{j}$$

Furthermore, we have

$$\mathbf{F}(\mathbf{r}(t)) = (2 \sin t)(2)\mathbf{i} - (2 \cos t)(2)\mathbf{j} + (2^3)\mathbf{k} = 4 \sin t\mathbf{i} - 4 \cos t\mathbf{j} + 8\mathbf{k}$$

Therefore,

$$\iint_S \operatorname{curl} \mathbf{F} \cdot d\mathbf{S} = \oint_C \mathbf{F} \cdot d\mathbf{r}$$

$$= \oint_C \mathbf{F}(\mathbf{r}(t)) \cdot \mathbf{r}'(t) \, dt$$

$$= \int_0^{2\pi} (4 \sin t\mathbf{i} - 4 \cos t\mathbf{j} + 8\mathbf{k}) \cdot (-2 \sin t\mathbf{i} + 2 \cos t\mathbf{j}) \, dt$$

$$= \int_0^{2\pi} (-8 \sin^2 t - 8 \cos^2 t) \, dt$$

$$= -8 \int_0^{2\pi} dt = -16\pi \qquad \blacksquare$$

## ■ Interpretation of Curl

Just as the divergence theorem can be used to give a physical interpretation of the divergence of a vector field, Stokes' Theorem can be used to give a physical interpretation of the curl vector. Once again, we think of $\mathbf{F}$ as representing the velocity field in fluid flow. Let $P_0(x_0, y_0, z_0)$ be a point in the fluid, and let $S$ be a circular disk with radius $r$ centered at $P_0$ and boundary $C_r$, as shown in Figure 6. Let $\mathbf{n}$ be a unit vector normal to $S_r$ at $P_0$. Applying Stokes' Theorem to the vector field $\mathbf{F}$ on the surface $S_r$, we obtain

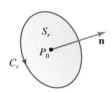

**FIGURE 6**
$C_r$ is a circular disk of radius $r$ at $P_0(x_0, y_0, z_0)$.

$$\oint_C \mathbf{F} \cdot d\mathbf{r} = \oint_C \mathbf{F} \cdot \mathbf{T} \, ds = \iint_{S_r} \operatorname{curl} \mathbf{F} \cdot d\mathbf{S} \tag{2}$$

Since $\mathbf{F} \cdot \mathbf{T}$ is the component of $\mathbf{F}$ tangent to $C_r$, we see that $\oint_{C_r} \mathbf{F} \cdot d\mathbf{r}$ is a measure of the tendency of the fluid to move around $C_r$, and accordingly, this line integral is called the **circulation of F around $C_r$.** By taking $r$ small, we see that the circulation of $\mathbf{F}$ around $C_r$ is a measure of the tendency of the field to rotate around the axis determined by $\mathbf{n}$.

Next, for small $r$ the continuity of curl $\mathbf{F}$ implies that $(\operatorname{curl} \mathbf{F})(P) \approx (\operatorname{curl} \mathbf{F})(P_0)$ for all points $P$ in $S_r$. Therefore, if $r$ is small, we can write

$$\iint_{S_r} \operatorname{curl} \mathbf{F} \cdot d\mathbf{S} = \iint_{S_r} (\operatorname{curl} \mathbf{F})(P_0) \cdot \mathbf{n} \, dS$$

$$\approx (\operatorname{curl} \mathbf{F})(P_0) \cdot \mathbf{n} \iint_{S_r} dS = (\operatorname{curl} \mathbf{F})(P_0) \cdot \mathbf{n}(\pi r^2)$$

So for small $r$, Equation (2) gives

$$\oint_C \mathbf{F} \cdot d\mathbf{r} \approx (\text{curl } \mathbf{F})(P_0) \cdot \mathbf{n}(\pi r^2)$$

The approximation improves as $r \to 0$, and we have

$$(\text{curl } \mathbf{F})(P_0) \cdot \mathbf{n} = \lim_{r \to 0} \frac{1}{\pi r^2} \oint_C \mathbf{F} \cdot d\mathbf{r} \tag{3}$$

Equation (3) gives the relationship between the curl and the circulation. It tells us that we can think of $|\text{curl } \mathbf{F}(P_0)|$ as a measure of the magnitude of the tendency of the fluid to rotate about the axis determined by $\mathbf{n}$. It also tells us that we can think of $\text{curl } \mathbf{F}(P_0)$ as determining the axis about which the circulation of $\mathbf{F}$ is greatest near $P_0$.

We now summarize the types of line and surface integrals and the major theorems associated with these integrals.

## ■ Summary of Line and Surface Integrals

---

**Line Integrals**

---

**a. Element of arc length:**   $ds = |\mathbf{r}'(t)| \, dt$

$$= \sqrt{[x'(t)]^2 + [y'(t)]^2 + [z'(t)]^2} \, dt$$

**b. Line integral of a scalar function:**   $\displaystyle \int_C f(x, y, z) \, ds = \int_a^b f(x(t), y(t), z(t)) |\mathbf{r}'(t)| \, dt$

**c. Line integral of a vector field:**   $\displaystyle \int_C \mathbf{F} \cdot d\mathbf{r} = \int_C \mathbf{F} \cdot \mathbf{T} \, ds = \int_a^b \mathbf{F}(\mathbf{r}(t)) \cdot \mathbf{r}'(t) \, dt$

---

**Surface Integrals**

---

**a. Element of surface area:**
   **(i)** If $S$ is the graph of $z = g(x, y)$, then $dS = \sqrt{g_x^2 + g_y^2 + 1} \, dA$.
   **(ii)** If $S$ is represented parametrically by $\mathbf{r}(u, v)$, then $dS = |\mathbf{r}_u \times \mathbf{r}_v| \, dA$.

**b. Surface integral of a scalar function:**
   **(i)** If $S$ is the graph of $z = g(x, y)$, then

$$\iint_S f(x, y, z) \, dS = \iint_R f(x, y, g(x, y)) \sqrt{g_x^2 + g_y^2 + 1} \, dA$$

   **(ii)** If $S$ is represented parametrically by $\mathbf{r}(u, v)$, then

$$\iint_S f(\mathbf{r}(u, v)) \, dS = \iint_R f(x(u, v), y(u, v), z(u, v)) |\mathbf{r}_u \times \mathbf{r}_v| \, dA$$

**c. Surface integral of a vector field:**
   **(i)** If $S$ is the graph of $z = g(x, y)$, then

$$\iint_S \mathbf{F} \cdot d\mathbf{S} = \iint_S \mathbf{F} \cdot \mathbf{n} \, dS = \iint_R \mathbf{F} \cdot (-g_x \mathbf{i} - g_y \mathbf{j} + \mathbf{k}) \, dA$$

   **(ii)** If $S$ is represented parametrically by $\mathbf{r}(u, v)$, then

$$\iint_S \mathbf{F} \cdot d\mathbf{S} = \iint_S \mathbf{F} \cdot \mathbf{n} \, dS = \iint_D \mathbf{F} \cdot (\mathbf{r}_u \times \mathbf{r}_v) \, dA$$

### ■ Summary of Major Theorems Involving Line Integrals and Surface Integrals

**1. Fundamental Theorem for Line Integrals**

$$\int_C \nabla f \cdot d\mathbf{r} = f(\mathbf{r}(b)) - f(\mathbf{r}(a))$$

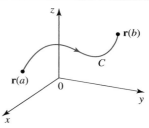

**2. Green's Theorem**

$$\oint_C P\, dx + Q\, dy = \iint_R \left( \frac{\partial Q}{\partial x} - \frac{\partial P}{\partial y} \right) dA$$

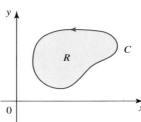

**3. Divergence Theorem**

$$\iint_S \mathbf{F} \cdot d\mathbf{S} = \iiint_T \operatorname{div} \mathbf{F}\, dV$$

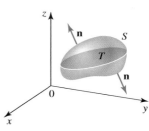

**4. Stokes' Theorem**

$$\oint_C \mathbf{F} \cdot d\mathbf{r} = \iint_S \operatorname{curl} \mathbf{F} \cdot d\mathbf{S}$$

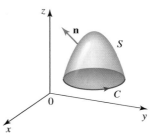

## 14.9  CONCEPT QUESTIONS

1. State Stokes' Theorem.
2. Suppose that $\mathbf{F} = P\mathbf{i} + Q\mathbf{j} + R\mathbf{k}$ is a vector field in three-dimensional space such that $P$, $Q$, and $R$ have continuous partial derivatives. Let $S_1$ be the hemisphere $z = \sqrt{1 - x^2 - y^2}$, and let $S_2$ be the paraboloid $z = 1 - x^2 - y^2$. Explain why

$$\iint_{S_1} \operatorname{curl} \mathbf{F} \cdot d\mathbf{S} = \iint_{S_2} \operatorname{curl} \mathbf{F} \cdot d\mathbf{S}$$

## 14.9    EXERCISES

*In Exercises 1–4, verify Stokes' Theorem for the given vector field **F** and the surface S, oriented with the normal pointing upward.*

**1.** $\mathbf{F}(x, y, z) = 2z\mathbf{i} + 3x\mathbf{j} - 2y\mathbf{k}$;    S is the part of the paraboloid $z = 9 - x^2 - y^2$ with $z \geq 0$

**2.** $\mathbf{F}(x, y, z) = 2y\mathbf{i} - 2x\mathbf{j} + z\mathbf{k}$;    S is the part of the plane $z = 1$ lying within the cone $z = \sqrt{x^2 + y^2}$

**3.** $\mathbf{F}(x, y, z) = y\mathbf{i} + z\mathbf{j} + x\mathbf{k}$;    S is the part of the plane $2x + 2y + z = 6$ lying in the first octant

**4.** $\mathbf{F}(x, y, z) = (x + 2y)\mathbf{i} + yz^2\mathbf{j} + y^2z\mathbf{k}$;    S is the hemisphere $z = \sqrt{1 - x^2 - y^2}$

*In Exercises 5–10, use Stokes' Theorem to evaluate $\iint_S \operatorname{curl} \mathbf{F} \cdot d\mathbf{S}$.*

**5.** $\mathbf{F}(x, y, z) = 2y\mathbf{i} + xz^2\mathbf{j} + x^2ye^z\mathbf{k}$;    S is the hemisphere $z = \sqrt{4 - x^2 - y^2}$ oriented with normal pointing upward

**6.** $\mathbf{F}(x, y, z) = 5yz\mathbf{i} + 2xz\mathbf{j} - 3xz^2\mathbf{k}$;    S is the part of the paraboloid $z = x^2 + y^2$ lying below the plane $z = 4$ and oriented with normal pointing downward

**7.** $\mathbf{F}(x, y, z) = xyz\mathbf{i} + 2x\mathbf{j} + \tan^{-1} y^2\mathbf{k}$;    S is the part of the hemisphere $z = \sqrt{4 - x^2 - y^2}$ lying inside the cylinder $x^2 + y^2 = 1$ and oriented with normal pointing upward

**8.** $\mathbf{F}(x, y, z) = xy\mathbf{i} + yz\mathbf{j} + xz\mathbf{k}$;    S is the part of the cylinder $z = \sqrt{1 - x^2}$ lying in the first octant between $y = 0$ and $y = 1$ and oriented with normal pointing in the positive x-direction

**9.** $\mathbf{F}(x, y, z) = z \sin x\mathbf{i} + 2x\mathbf{j} + e^x \cos z\mathbf{k}$;    S is the part of the ellipsoid $9x^2 + 9y^2 + 4z^2 = 36$ lying above the xy-plane and oriented with normal pointing upward

**10.** $\mathbf{F}(x, y, z) = yz^2\mathbf{i} - xz\mathbf{j} + z^3\mathbf{k}$;    S is the part of the cone $z = \sqrt{x^2 + y^2}$ between the planes $z = 1$ and $z = 3$ and oriented with normal pointing upward

*In Exercises 11–16, use Stokes' Theorem to evaluate $\oint_C \mathbf{F} \cdot d\mathbf{r}$.*

**11.** $\mathbf{F}(x, y, z) = (y - z)\mathbf{i} + (z - x)\mathbf{j} + (x - y)\mathbf{k}$;    C is the boundary of the part of the plane $2x + 3y + z = 6$ in the first octant, oriented in a counterclockwise direction when viewed from above

**12.** $\mathbf{F}(x, y, z) = y\mathbf{i} + z\mathbf{j} + x\mathbf{k}$;    C is the boundary of the triangle with vertices $(0, 0, 0)$, $(1, 0, 0)$, and $(0, 1, 1)$ oriented in a counterclockwise direction when viewed from above

**13.** $\mathbf{F}(x, y, z) = 3xz\mathbf{i} + e^{xz}\mathbf{j} + 2xy\mathbf{k}$;    C is the circle obtained by intersecting the cylinder $x^2 + z^2 = 1$ with the plane $y = 3$ oriented in a counterclockwise direction when viewed from the right

**14.** $\mathbf{F}(x, y, z) = 2z\mathbf{i} + xy\mathbf{j} + 4y\mathbf{k}$;    C is the ellipse obtained by intersecting the plane $y + z = 4$ with the cylinder $x^2 + y^2 = 4$, oriented in a counterclockwise direction when viewed from above

**15.** $\mathbf{F}(x, y, z) = xe^y\mathbf{i} + ye^x\mathbf{j} + (xyz)\mathbf{k}$;    C is the path consisting of straight line segments joining the points $(0, 0, 0)$, $(0, 1, 0)$, $(2, 1, 0)$, $(2, 0, 0)$, $(2, 0, 1)$, $(0, 0, 1)$, and $(0, 0, 0)$ in that order

**16.** $\mathbf{F}(x, y, z) = \left(\dfrac{y}{1 + x^2}\right)\mathbf{i} + \tan^{-1} x\mathbf{j} + xy\mathbf{k}$;    C is the curve obtained by intersecting the cylinder $x^2 + y^2 = 1$ with the hyperbolic paraboloid $z = y^2 - x^2$, oriented in a counterclockwise direction when viewed from above

**17.** Find the work done by the force field $\mathbf{F}(x, y, z) = (e^x + z)\mathbf{i} + (x^2 + \cosh y)\mathbf{j} + (y^2 + z^3)\mathbf{k}$ on a particle when it is moved along the triangular path that is obtained by intersecting the plane $2x + 2y + z = 2$ with the coordinate planes and oriented in a counterclockwise direction when viewed from above.

**18.** Find the work done by the force field $\mathbf{F}(x, y, z) = xy^2\mathbf{i} + (x/z)\mathbf{j} + (2x + y)\mathbf{k}$ on a particle when it is moved along the rectangular path with vertices $A(0, 0, 3)$, $B(2, 0, 3)$, $C(2, 4, 3)$, $D(0, 4, 3)$, and $A(0, 0, 3)$ in that order.

**19. Ampere's Law**    A steady current in a long wire produces a magnetic field that is tangent to any circle that lies in the plane perpendicular to the wire and whose center lies on the wire. (See the figure below.)

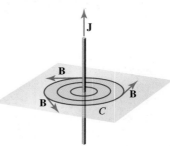

Let **J** denote the vector that points in the direction of the current and has magnitude measured in amperes per square meter. This vector is called the electric current density. One of Maxwell's equations states that curl $\mathbf{B} = \mu_0\mathbf{J}$, where **B** denotes the magnetic field intensity and $\mu_0$ is a constant called the permeability of free space. Using Stokes' Theorem, show that $\oint_C \mathbf{B} \cdot d\mathbf{r} = \mu_0 I$, where C is any closed curve enclosing the curve and I is the net current that passes through any surface bounded by C. This is Ampere's Law. (See Exercise 42 in Section 14.3.)

**20.** Let S be a sphere, and suppose that **F** satisfies the conditions of Stokes' Theorem. Show that $\iint_S \operatorname{curl} \mathbf{F} \cdot d\mathbf{S} = 0$.

21. Let $f$ and $g$ have continuous partial derivatives, and let $C$ and $S$ satisfy the conditions of Stokes' Theorem. Show that:

a. $\oint_C (f \nabla g) \cdot d\mathbf{r} = \iint_S (\nabla f \times \nabla g) \cdot d\mathbf{S}$

b. $\oint_C (f \nabla f) \cdot d\mathbf{r} = 0$    c. $\oint_C (f \nabla g + g \nabla f) \cdot d\mathbf{r} = 0$

22. Evaluate $\oint_C (2xy + z^2)\, dx + (x^2 - 1)\, dy + 2xz\, dz$, where $C$ is the curve

$$\mathbf{r}(t) = (1 + \cos t)\mathbf{i} + (1 + \sin t)\mathbf{j} + (1 - \sin t - \cos t)\mathbf{k}$$
$$0 \le t \le 2\pi$$

23. Let $\mathbf{F}(x, y, z) = f(r)\mathbf{r}$, where $\mathbf{r} = x\mathbf{i} + y\mathbf{j} + z\mathbf{k}$, $f$ is a differentiable function, and $r = |\mathbf{r}|$. Evaluate $\oint_C \mathbf{F} \cdot d\mathbf{r}$, where $C$ is the boundary of the triangle with vertices $(4, 2, 0)$, $(1, 5, 2)$, and $(1, -1, 5)$, traced in a counterclockwise direction when viewed from above the plane.

24. Let $\mathbf{F}(x, y, z) = xy\mathbf{i} + (4x - yz)\mathbf{j} + (xy - \sqrt{z})\mathbf{k}$, and let $C$ be a circle of radius $r$ lying in the plane $x + y + z = 5$. (See the following figure.) If $\oint_C \mathbf{F} \cdot d\mathbf{r} = \sqrt{3}\pi$, where $C$ is oriented in the counterclockwise direction when viewed from above the plane, what is the value of $r$?

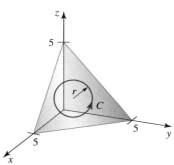

25. Use Stokes' Theorem to evaluate $\oint_C e^x \cos z\, dx + 2xy^2\, dy + \cot^{-1} y\, dz$, where $C$ is the circle $x^2 + y^2 = 4$ and $z = 0$. **Hint:** Find a surface $S$ with $C$ as its boundary and such that $C$ is oriented counterclockwise when viewed from above.

26. Find $\iint_S \text{curl } \mathbf{F} \cdot d\mathbf{S}$ if $\mathbf{F}(x, y, z) = (x + y - z + 2)\mathbf{i} + (y \cos z + 4)\mathbf{j} + xz\mathbf{k}$, where $S$ is the surface with the normal pointing outward and having the boundary $x^2 + y^2 = 1$, $z = 0$, shown in the figure.

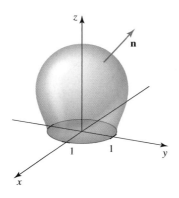

27. Let $S$ be the oriented piecewise-smooth surface that is the boundary of a simple solid region $T$. If $\mathbf{F} = P\mathbf{i} + Q\mathbf{j} + R\mathbf{k}$ is a vector field, where $P$, $Q$, and $R$ have continuous second-order partial derivatives in an open region containing $T$, show that

$$\iint_S \text{curl } \mathbf{F} \cdot d\mathbf{S} = 0$$

28. Suppose that $f$ has continuous second-order partial derivatives in a simply connected set $D$. Use Stokes' Theorem to show that

$$\oint_C \nabla f \cdot d\mathbf{r} = 0$$

for any simple piecewise-smooth closed curve $C$ lying in $D$.

29. Refer to Example 2. Evaluate $\oint_C \mathbf{F} \cdot d\mathbf{r}$ directly (that is, without using Stokes' Theorem), where $F(x, y, z) = \cos z\mathbf{i} + x^2\mathbf{j} + 2y\mathbf{k}$ and $C$ is the curve of intersection of the plane $x + z = 2$ and the cylinder $x^2 + y^2 = 1$.

30. Refer to Exercise 16. Evaluate $\oint_C \mathbf{F} \cdot d\mathbf{r}$ directly (that is, without using Stokes' Theorem), where
$$F(x, y, z) = \left(\frac{y}{1 + x^2}\right)\mathbf{i} + \tan^{-1} x\mathbf{j} + xy\mathbf{k}$$ and $C$ is the curve obtained by intersecting the cylinder $x^2 + y^2 = 1$ with the hyperbolic paraboloid $z = y^2 - x^2$, oriented in a counterclockwise direction when viewed from above.

*In Exercises 31 and 32, determine whether the statement is true or false. If it is true, explain why. If it is false, explain why or give an example that shows it is false.*

31. If $\mathbf{F} = P\mathbf{i} + Q\mathbf{j} + R\mathbf{k}$, where $P$, $Q$, and $R$ have continuous partial derivatives in three-dimensional space and $S_1$ and $S_2$ are the upper and lower hemisphere $z = \sqrt{4 - x^2 - y^2}$ and $z = -\sqrt{4 - x^2 - y^2}$, respectively, then $\iint_{S_1} \text{curl } \mathbf{F} \cdot d\mathbf{S} = \iint_{S_2} \text{curl } \mathbf{F} \cdot d\mathbf{S}$.

32. If $\mathbf{F}$ has continuous partial derivatives on an open region containing an oriented surface $S$ bounded by a piecewise-smooth simple closed curve $C$ and curl $\mathbf{F}$ is tangent to $S$, then $\oint_C \mathbf{F} \cdot d\mathbf{r} = 0$.

# CHAPTER 14   REVIEW

## CONCEPT REVIEW

*In Exercises 1–18, fill in the blanks.*

**1.** A vector field on a region $R$ is a function $\mathbf{F}$ that associates with each point $(x, y)$ in two-dimensional space a two-dimensional _____; if $R$ is a subset of three-dimensional space, then $\mathbf{F}$ associates each point $(x, y, z)$ with a three-dimensional _____.

**2.** A vector field $\mathbf{F}$ is conservative if there exists a scalar function $f$ such that $\mathbf{F} = $ _____. The function $f$ is called a _____ function for $\mathbf{F}$.

**3. a.** The divergence of $\mathbf{F}$ (describing the flow of fluid) at a point $P$ measures the rate of flow per unit area (or volume) at which the fluid _____ or _____ at $P$.

 **b.** The divergence of $\mathbf{F} = P\mathbf{i} + Q\mathbf{j} + R\mathbf{k}$ in three-dimensional space is defined by div $\mathbf{F} = $ _____; if div $\mathbf{F}(P) < 0$, more fluid _____ a neighborhood of $P$ than _____ from it; if div $\mathbf{F}(P) = 0$, the amount of fluid entering a neighborhood of $P$ _____ the amount departing from it; if div $\mathbf{F}(P) > 0$, more fluid _____ a neighborhood of $P$ than _____ it.

**4. a.** If $\mathbf{F}$ describes fluid flow, then the curl of $\mathbf{F}$ measures the tendency of the fluid to _____ a paddle wheel.

 **b.** The curl of $\mathbf{F} = P\mathbf{i} + Q\mathbf{j} + R\mathbf{k}$ is defined by curl $\mathbf{F} = $ _____.

**5. a.** If $C$ is a smooth curve, then the line integral of $f$ along $C$ is $\int_C f(x, y)\, ds = $ _____.

 **b.** The formula for evaluating the line integral of part (a) is $\int_C f(x, y)\, ds = $ _____.

 **c.** The line integral of $f$ along $C$ with respect to $x$ is defined to be $\int_C f(x, y)\, dx = $ _____ with formula $\int_C f(x, y)\, dx = $ _____, where $\mathbf{r}(t) = x(t)\mathbf{i} + y(t)\mathbf{j}$, $a \le t \le b$.

 **d.** The line integral of $f$ along $C$ with respect to $y$ is defined to be $\int_C f(x, y)\, dy = $ _____ with formula $\int_C f(x, y)\, dy = $ _____.

**6.** The formula for evaluating the line integral of $f$ along a curve $C$ (with respect to arc length) parametrized by $\mathbf{r}(t) = x(t)\mathbf{i} + y(t)\mathbf{j} + z(t)\mathbf{k}$, $a \le t \le b$, is $\int_C f(x, y, z)\, ds = $ _____.

**7. a.** If $\mathbf{F}$ is a continuous vector field and $C$ is a smooth curve described by $\mathbf{r}(t)$, $a \le t \le b$, then the formula for evaluating the line integral of $\mathbf{F}$ along $C$ is $\int_C \mathbf{F} \cdot d\mathbf{r} = $ _____.

 **b.** If $\mathbf{F}$ is a force field, then $\int_C \mathbf{F} \cdot d\mathbf{r}$ gives the _____ done by $\mathbf{F}$ on a particle as it moves along $C$ from $t = a$ to $t = b$.

**8. a.** If $\int_{C_1} \mathbf{F} \cdot d\mathbf{r} = \int_{C_2} \mathbf{F} \cdot d\mathbf{r}$ for any two paths having the same initial and terminal points, then the line integral $\int_C \mathbf{F} \cdot d\mathbf{r}$ is _____ _____ _____.

 **b.** The Fundamental Theorem for Line Integrals states that if $\mathbf{F} = \nabla f$, where $f$ is a _____ function for $\mathbf{F}$ and $C$ is any piecewise-smooth curve described by $\mathbf{r}(t)$, $a \le t \le b$, then $\int_C \mathbf{F} \cdot d\mathbf{r} = \int_C \nabla f \cdot d\mathbf{r} = $ _____.

**9. a.** The line integral $\int_C \mathbf{F} \cdot d\mathbf{r}$ is independent of path if and only if $\int_C \mathbf{F} \cdot d\mathbf{r} = 0$ for every _____ path $C$.

 **b.** If $\mathbf{F}$ is continuous on an open, _____ region $R$, then the line integral $\int_C \mathbf{F} \cdot d\mathbf{r}$ is independent of path if and only if $\mathbf{F}$ is _____.

**10. a.** If $\mathbf{F}(x, y) = P(x, y)\mathbf{i} + Q(x, y)\mathbf{j}$ is a conservative vector field in an open region $R$ and both $P$ and $Q$ have continuous first-order partial derivatives in $R$, then _____ at each point in $R$.

 **b.** If $\mathbf{F} = P\mathbf{i} + Q\mathbf{j}$ is defined on an open, _____ region $R$ in the plane and $P$ and $Q$ have continuous first-order derivatives on $R$ and _____, for all $(x, y)$ in $R$, then $\mathbf{F}$ is conservative in $R$.

**11.** If $\mathbf{F} = P\mathbf{i} + Q\mathbf{j} + R\mathbf{k}$, where $P$, $Q$, and $R$ have continuous first-order partial derivatives in space, then $\mathbf{F}$ is conservative if and only if _____ $= \mathbf{0}$, or, in terms of the partial derivatives of $P$, $Q$, and $R$, $\dfrac{\partial R}{\partial y} = $ _____, $\dfrac{\partial R}{\partial x} = $ _____, and $\dfrac{\partial Q}{\partial x} = $ _____.

**12.** Green's Theorem states that if $C$ is a _____, simple _____ curve that bounds a region $R$ in the plane and $P$ and $Q$ have continuous partial derivatives on an open set containing $R$, then $\int_C P\, dx + Q\, dy = $ _____.

**13.** If $R$ is a plane region bounded by a piecewise-smooth, simple closed curve $C$, then the area of $R$ is given by $A = $ _____ $ = $ _____ $ = $ _____.

**14.** If $\mathbf{F} = P\mathbf{i} + Q\mathbf{j}$, then Green's Theorem has the vector forms $\oint_C \mathbf{F} \cdot \mathbf{T}\, ds = \oint_C P\, dx + Q\, dy = $ _____ and $\oint_C \mathbf{F} \cdot \mathbf{n}\, ds = $ _____.

**15. a.** A parametric surface $S$ can be represented by the vector equation $\mathbf{r}(u, v) = $ _____, where $(u, v)$ lies in its _____ domain; this vector equation is equivalent to the three _____ _____ $x = x(u, v)$, $y = y(u, v)$, and $z = z(u, v)$.

 **b.** If $S$ is the graph of the function $z = f(x, y)$, then a vector representation of $S$ is $\mathbf{r}(u, v) = $ _____.

**c.** If $S$ is a surface obtained by revolving the graph of a nonnegative function $f(x)$, where $a \leq x \leq b$, about the $x$-axis, then a vector representation of $S$ is $\mathbf{r}(u, v) = $ _____ with parameter domain $D = $ _____.

**16.** If a parametric surface $S$ is represented by $\mathbf{r}(u, v) = x(u, v)\mathbf{i} + y(u, v)\mathbf{j} + z(u, v)\mathbf{k}$ with parameter domain $D$, then its surface area is $A(S) = $ _____.

**17. a.** If $S$ is defined by $z = f(x, y)$ and the projection of $S$ onto the $xy$-plane is $R$, then $\iint_S F(x, y, z) \, dS = $ _____.

**b.** If $S$ is defined by $\mathbf{r}(u, v) = x(u, v)\mathbf{i} + y(u, v)\mathbf{j} + z(u, v)\mathbf{k}$ with parameter domain $D$, then $\iint_S F(x, y, z) \, dS = $ _____.

**18. a.** The positive orientation of a surface $S$ is the one for which the unit normal vector points _____ from $S$.

**b.** The flux of a vector field $\mathbf{F}$ across an oriented surface $S$ in the direction of the unit normal $\mathbf{n}$ is _____.

## REVIEW EXERCISES

*In Exercises 1–4, find* (a) *the divergence and* (b) *the curl of the vector field* **F**.

**1.** $\mathbf{F}(x, y, z) = xy^2\mathbf{i} + yz^2\mathbf{j} + zx^2\mathbf{k}$

**2.** $\mathbf{F}(x, y, z) = xy \cos y\mathbf{i} + y \sin x\mathbf{j} + xz\mathbf{k}$

**3.** $\mathbf{F}(x, y, z) = e^x \sin y\mathbf{i} + e^x \cos y\mathbf{j} + e^z\mathbf{k}$

**4.** $\mathbf{F}(x, y, z) = \ln(x^2 + y^2)\mathbf{i} + x \cos y\mathbf{j} + z^2\mathbf{k}$

*In Exercises 5–14, evaluate the line integral.*

**5.** $\displaystyle\int_C y \, ds$, where $C$ is the arch of the parabola $y = \sqrt{x}$ from $(1, 1)$ to $(4, 2)$

**6.** $\displaystyle\int_C (1 - x^2) \, ds$, where $C$ is the upper semicircle centered at the origin and joining $(0, -1)$ to $(0, 1)$

**7.** $\displaystyle\int_C xy^2 \, ds$, where $C$ is the curve given by $\mathbf{r}(t) = \sin t\mathbf{i} + \cos t\mathbf{j} + t\mathbf{k}, \ 0 \leq t \leq \frac{\pi}{2}$

**8.** $\displaystyle\int_C xyz \, ds$, where $C$ is the line segment from $(1, 1, 0)$ to $(2, 3, 4)$

**9.** $\displaystyle\int_C x^2y \, dx + x^3y \, dy$, where $C$ is the graph of $y = \sqrt[3]{x}$ from $(1, 1)$ to $(8, 2)$

**10.** $\displaystyle\int_C xy \, dx - xy^2 dy$, where $C$ is the quarter-circle from $(0, -1)$ to $(1, 0)$, centered at the origin

**11.** $\displaystyle\int_C yz \, dx - y \cos x \, dy + y \, dz$, where $C$ is the curve $x = t$, $y = \cos t, z = \sin t, 0 \leq t \leq \frac{\pi}{2}$

**12.** $\displaystyle\int_C xe^{-y} \, dx + \cos y \, dy + z^2 \, dz$,

$C: \mathbf{r}(t) = t\mathbf{i} + t^2\mathbf{j} + t^3\mathbf{k}, \ 0 \leq t \leq 1$

**13.** $\displaystyle\int_C xy \, dx + e^{-y} \, dy + ze^x \, dz$, where $C$ is the line segment joining $(0, 0, 0)$ to $(1, 1, 2)$

**14.** $\displaystyle\int_C z \, dx + x \, dy + x^2 \, dz$, where $C$ consists of the line segment joining $(0, 0, 0)$ to $(1, 0, 0)$ and the line segment from $(1, 0, 0)$ to $(2, 1, 3)$.

*In Exercises 15 and 16, find the work done by the force field* **F** *in moving a particle along the curve C.*

**15.** $\mathbf{F}(x, y, z) = xy\mathbf{i} + (y + z)\mathbf{j} + z^2\mathbf{k}$, where $C$ is the line segment from $(1, 1, 1)$ to $(2, 3, 5)$

**16.** $\mathbf{F}(x, y, z) = yz\mathbf{i} + z\mathbf{j} + x\mathbf{k}$, where $C$ is part of the helix given by $x = 2t, y = 2 \sin t, z = 2 \cos t, 0 \leq t \leq \frac{\pi}{2}$

*In Exercises 17 and 18, show that* **F** *is a conservative vector field and find a function f such that* $\mathbf{F} = \nabla f$.

**17.** $\mathbf{F}(x, y) = (4xy + 3y^2)\mathbf{i} + (2x^2 + 6xy)\mathbf{j}$

**18.** $\mathbf{F}(x, y, z) = (y^2 + 2xz)\mathbf{i} + (2xy - z^2)\mathbf{j} + (x^2 - 2yz)\mathbf{k}$

*In Exercises 19 and 20, show that* **F** *is conservative and use this result to evaluate* $\int_C \mathbf{F} \cdot \mathbf{T} \, ds$ *for the given curve C.*

**19.** $\mathbf{F}(x, y) = (2xy + y^3)\mathbf{i} + (x^2 + 3xy^2)\mathbf{j}$; $C$ is the elliptical path $9x^2 + 25y^2 = 225$ from $(-5, 0)$ to $(0, 3)$ traversed in a counterclockwise direction

**20.** $\mathbf{F}(x, y, z) = (2xy + yz^2)\mathbf{i} + (x^2 + xz^2)\mathbf{j} + 2xyz\mathbf{k}$; $C$ is the twisted cubic $x = t, y = t^2, z = t^3$ from $(0, 0, 0)$ to $(1, 1, 1)$

*In Exercises 21–24, use Green's Theorem to evaluate the line integral along the positively oriented closed curve C.*

**21.** $\displaystyle\oint_C (y^2 + \sec x) \, dx + (x^2 + y^5) \, dy$, where $C$ is the boundary of the region enclosed by the graphs of $y = 4 - x^2$ and $y = -x + 2$

**22.** $\oint_C xy\,dx + (x^2 + \sqrt{y})\,dy$, where $C$ is the boundary of the region enclosed by the graphs of $y = \sqrt{x}$, $y = 0$, and $x = 4$

**23.** $\oint_C (x^2y + e^x)\,dx + (e^{-y} - xy^2)\,dy$, where $C$ is the circle $x^2 + y^2 = 1$

**24.** $\oint_C (2y + \cosh x)\,dx + (x - \sinh y)\,dy$, where $C$ is the ellipse $9x^2 + 4y^2 = 36$

*In Exercises 25–28, evaluate the surface integral.*

**25.** $\iint_S (y + xz)\,dS$, where $S$ is the part of the plane

$2x + 2y + 3z = 6$ in the first octant

**26.** $\iint_S z\,dS$, where $S$ is the part of the paraboloid

$z = 4 - x^2 - y^2$ inside the cylinder $x^2 + y^2 = 1$

**27.** $\iint_S \mathbf{F} \cdot \mathbf{n}\,dS$, where $\mathbf{F}(x, y, z) = x\mathbf{i} + y\mathbf{j} + z\mathbf{k}$ and $S$ is the

part of the paraboloid $y = 1 - x^2 - z^2$ lying to the right of the $xz$-plane; $\mathbf{n}$ points to the right

**28.** $\iint_S \mathbf{F} \cdot \mathbf{n}\,dS$, where $\mathbf{F}(x, y, z) = y\mathbf{i} - x\mathbf{j} + z\mathbf{k}$ and $S$ is the

part of the paraboloid $z = 5 - x^2 - y^2$ lying above the plane $z = 1$; $\mathbf{n}$ points upward

*In Exercises 29 and 30, find the mass of the surface $S$ having the given mass density.*

**29.** $S$ is the part of the plane $x + y + z = 1$ in the first octant; the density at a point $P$ on $S$ is directly proportional to the square of the distance between $P$ and the $yz$-plane.

**30.** $S$ is the part of the paraboloid $z = x^2 + y^2$ lying inside the cylinder $x^2 + y^2 = 2$; the density at a point $P$ on $S$ is directly proportional to the distance between $P$ and the $xy$-plane.

*In Exercises 31 and 32, use the Divergence Theorem to find the flux of $\mathbf{F}$ across $S$; that is, calculate $\iint_S \mathbf{F} \cdot \mathbf{n}\,dS$.*

**31.** $\mathbf{F}(x, y, z) = x\mathbf{i} - y\mathbf{j} + z\mathbf{k}$; $S$ is the surface of the cylinder $x^2 + y^2 = 4$ bounded by the planes $z = 0$ and $z = 3$; $\mathbf{n}$ points outward

**32.** $\mathbf{F}(x, y, z) = y^2\mathbf{i} + yz^2\mathbf{j} + z^3\mathbf{k}$; $S$ is the unit sphere centered at the origin; $\mathbf{n}$ points outward

*In Exercises 33 and 34, use Stokes' Theorem to evaluate $\iint_S \text{curl } \mathbf{F} \cdot \mathbf{n}\,dS$.*

**33.** $\mathbf{F}(x, y, z) = (x + y^2 - 2)\mathbf{i} - 2xy\mathbf{j} - (x^2 + yz^2)\mathbf{k}$; $S$ is the part of the paraboloid $z = 4 - x^2 - y^2$ above the $xy$-plane with an outward normal

**34.** $\mathbf{F}(x, y, z) = y^2z\mathbf{i} + 2xz\mathbf{j} + \cos z\mathbf{k}$; $S$ is the part of the sphere $x^2 + y^2 + z^2 = 16$ below the plane $z = 2$ with an outward normal

*In Exercises 35 and 36, use Stokes' Theorem to evaluate $\oint_C \mathbf{F} \cdot \mathbf{T}\,ds$.*

**35.** $\mathbf{F}(x, y, z) = (2x + y)\mathbf{i} - (3x + z)\mathbf{j} + (y - z)\mathbf{k}$; $C$ is the curve obtained by intersecting the plane $2x + y + z = 6$ with the coordinate planes, oriented clockwise when viewed from the top

**36.** $\mathbf{F}(x, y, z) = y\mathbf{i} + xz\mathbf{j} + xyz\mathbf{k}$; $C$ is the curve obtained by intersecting the surface $z = x^2y$ with the planes $x = 0$, $x = 1$, $y = 0$, and $y = 1$, oriented counterclockwise when viewed from above

**37.** Find the work done by the force field $\mathbf{F}(x, y) = 2xy^3\mathbf{i} + 3x^2y^2\mathbf{j}$ when a particle is moved from $(0, 0)$ to $(2, 4)$ along the path shown in the figure.

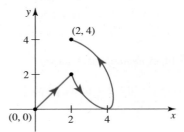

**38.** Find the work done by the force field $\mathbf{F}(x, y, z) = 2xy\mathbf{i} + (x^2 + 2yz^2)\mathbf{j} + 2y^2z\mathbf{k}$ when a particle is moved from $(2, 0, 0)$ to $(0, 3, 0)$ along the path shown in the figure.

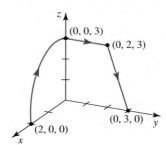

**39.** Let

$$\mathbf{F}(x, y) = \frac{x - y}{x^2 + y^2}\mathbf{i} + \frac{x + y}{x^2 + y^2}\mathbf{j}$$

Evaluate $\oint_C \mathbf{F} \cdot d\mathbf{r}$, where $C$ is the path shown in the figure.

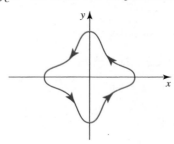

**40.** Let $\mathbf{F}(x, y, z) = \left(4y - \frac{1}{2}z^2\right)\mathbf{i} + 2xz\mathbf{j} - x^2\mathbf{k}$, and let $C$ be a simple closed curve lying in the plane $2x + 2y + z = 6$ (see the following figure) and oriented in the counterclockwise direction when viewed from above the plane. If $\oint_C \mathbf{F} \cdot d\mathbf{r} = -24$, what is the area of the region enclosed by $C$?

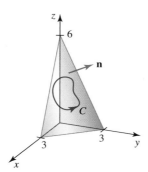

*In Exercises 41–48, state whether the statement is true or false. (Assume that all differentiability conditions are met.) Give a reason for your answer.*

**41.** If $\mathbf{F}$ is a vector field, then curl (curl $\mathbf{F}$) is a vector field.

**42.** If $\mathbf{F}$ and $\mathbf{G}$ are vector fields, then div ($\mathbf{F} \times \mathbf{G}$) is a scalar field.

**43.** If $f$ is a scalar field, then $\nabla \cdot [\nabla \times (\nabla f)]$ is undefined.

**44.** If $\mathbf{F}$ is a vector field and $f$ is a scalar field, then div$[\nabla f \times ($curl $\mathbf{F})]$ is undefined.

**45.** If $f$ has continuous partial derivatives at all points and $\nabla f = \mathbf{0}$, then $f$ is a constant function.

**46.** If $\nabla \times \mathbf{F} = \mathbf{0}$, then $\mathbf{F}$ is a constant vector field.

**47.** If div $\mathbf{F} = 0$, then $\iint_S \mathbf{F} \cdot \mathbf{n} \, dS = 0$ for every closed surface $S$.

**48.** If $\oint_C \mathbf{F} \cdot d\mathbf{r} = 0$ for every closed path $C$, then curl $\mathbf{F} = \mathbf{0}$.

## CHALLENGE PROBLEMS

**1.** Find and sketch the domain of the vector-valued function

$$\mathbf{F}(x, y) = \frac{1}{\sqrt{4 - x^2 - 4y^2}}\mathbf{i} + \frac{1}{\sqrt{4x^2 + 4y^2 - 1}}\mathbf{j}$$

**2.** Find the domain of

$$\mathbf{F}(x, y, z) = \sqrt{|x| - y - 1}\,\mathbf{i}$$
$$+ \ln(1 - |x| - y)\mathbf{j} + \frac{\ln \ln(z - 1)}{\sqrt{z - 3}}\mathbf{k}$$

**3.** The curve with equation $x^3 + y^3 = 3axy$, where $a$ is a nonzero constant, is called the **folium of Descartes.**

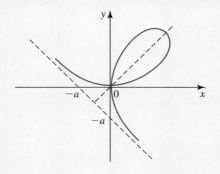

**a.** Show that a parametric representation of this curve is

$$x = \frac{3at}{1 + t^3} \qquad y = \frac{3at^2}{1 + t^3}$$

**Hint:** Use the parameter $t = y/x$.

**b.** Find the area of the region enclosed by the loop of the curve.

**4.** Let

$$P(x, y) = \frac{x - y}{x^2 + y^2} \qquad \text{and} \qquad Q(x, y) = \frac{x + y}{x^2 + y^2}$$

Evaluate $\oint_C P \, dx + Q \, dy$, where $C$ is the curve shown in the figure.

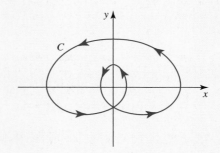

**5.** Let $\mathbf{F}(x, y, z) = y \cos x\mathbf{i} + (x + \sin x)\mathbf{j} + \cos z\mathbf{k}$, and let $C$ be the curve represented by $\mathbf{r}(t) = (1 + \cos t)\mathbf{i} + (1 + \sin t)\mathbf{j} + (1 - \sin t - \cos t)\mathbf{k}$ for $0 \le t \le 2\pi$. Evaluate $\oint_C \mathbf{F} \cdot d\mathbf{r}$.

**6.** A differential equation of the form

$$P(x, y)\, dx + Q(x, y)\, dy = 0$$

is called *exact* if $\dfrac{\partial Q}{\partial x} = \dfrac{\partial P}{\partial y}$ for all $(x, y)$. Show that the equation

$$(2y^2 + 6xy - 2)\, dx + (4xy + 3x^2 + 3)\, dy = 0$$

is exact and solve it.

**7.** Let $T$ be a one-to-one transformation defined by $x = g(u, v)$, $y = h(u, v)$ that maps a region $S$ in the $uv$-plane onto a region $R$ in the $xy$-plane. Use Green's Theorem to prove the change of variables formula

$$\iint\limits_{R} dA = \iint\limits_{S} \left| \frac{\partial(x, y)}{\partial(u, v)} \right| du\, dv$$

for the case in which $f(x, y) = 1$. (Compare with Formula (4) in Section 13.8.)

# The Real Number Line, Inequalities, and Absolute Value

## ■ The Real Number Line

The real number system is made up of the set of real numbers together with the usual operations of addition, subtraction, multiplication, and division.

We can represent real numbers geometrically by points on a **real number,** or **coordinate, line.** This line can be constructed as follows. Arbitrarily select a point on a straight line to represent the number 0. This point is called the **origin.** If the line is horizontal, then a point at a convenient distance to the right of the origin is chosen to represent the number 1. This determines the scale for the number line. Each positive real number lies at an appropriate distance to the right of the origin, and each negative real number lies at an appropriate distance to the left of the origin (see Figure 1).

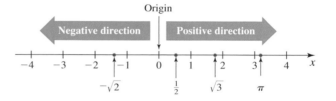

**FIGURE 1**
The real number line

A *one-to-one correspondence* is set up between the set of all real numbers and the set of points on the number line; that is, exactly one point on the line is associated with each real number. Conversely, exactly one real number is associated with each point on the line. The real number that is associated with a point on the real number line is called the **coordinate** of that point.

## ■ Intervals

Throughout this book we often restrict our attention to subsets of the set of real numbers. For example, if $x$ denotes the number of cars rolling off a plant assembly line each day, then $x$ must be nonnegative—that is, $x \geq 0$. Further, suppose that management decides that the daily production must not exceed 200 cars. Then, $x$ must satisfy the inequality $0 \leq x \leq 200$.

More generally, we will be interested in the following subsets of real numbers: open intervals, closed intervals, and half-open intervals. The set of all real numbers that lie *strictly* between two fixed numbers $a$ and $b$ is called an **open interval** $(a, b)$. It consists of all real numbers $x$ that satisfy the inequalities $a < x < b$, and it is called "open" because neither of its endpoints is included in the interval. A **closed interval** contains *both* of its endpoints. Thus, the set of all real numbers $x$ that satisfy the inequalities

$a \le x \le b$ is the closed interval $[a, b]$. Notice that square brackets are used to indicate that the endpoints are included in this interval. **Half-open intervals** contain only *one* of their endpoints. Thus, the interval $[a, b)$ is the set of all real numbers $x$ that satisfy $a \le x < b$, whereas the interval $(a, b]$ is described by the inequalities $a < x \le b$. Examples of these **finite intervals** are illustrated in Table 1.

**TABLE 1** Finite Intervals

| Interval | Graph | Example |
|---|---|---|
| Open: $(a, b)$ | | $(-2, 1)$ |
| Closed: $[a, b]$ | | $[-1, 2]$ |
| Half-open: $(a, b]$ | | $\left(\frac{1}{2}, 3\right]$ |
| Half-open: $[a, b)$ | | $\left[-\frac{1}{2}, 3\right)$ |

In addition to finite intervals, we will encounter **infinite intervals.** Examples of infinite intervals are the half lines $(a, \infty)$, $[a, \infty)$, $(-\infty, a)$, and $(-\infty, a]$ defined by the set of all real numbers that satisfy $x > a$, $x \ge a$, $x < a$, and $x \le a$, respectively. The symbol $\infty$, called *infinity,* is not a real number. It is used here only for notational purposes. The notation $(-\infty, \infty)$ is used for the set of all real numbers $x$, since by definition the inequalities $-\infty < x < \infty$ hold for any real number $x$. Infinite intervals are illustrated in Table 2.

**TABLE 2** Infinite Intervals

| Interval | Graph | Example |
|---|---|---|
| $(a, \infty)$ | | $(2, \infty)$ |
| $[a, \infty)$ | | $[-1, \infty)$ |
| $(-\infty, a)$ | | $(-\infty, 1)$ |
| $(-\infty, a]$ | | $\left(-\infty, -\frac{1}{2}\right]$ |

## ◼ Inequalities

The following properties may be used to solve one or more inequalities involving a variable.

**Properties of Inequalities**

If $a$, $b$, and $c$, are any real numbers, then

|  |  | Example |
|---|---|---|
| **Property 1** | If $a < b$ and $b < c$, then $a < c$. | $2 < 3$ and $3 < 8$, so $2 < 8$. |
| **Property 2** | If $a < b$, then $a + c < b + c$. | $-5 < -3$, so $-5 + 2 < -3 + 2$; that is, $-3 < -1$. |
| **Property 3** | If $a < b$ and $c > 0$, then $ac < bc$. | $-5 < -3$, and since $2 > 0$, we have $(-5)(2) < (-3)(2)$; that is, $-10 < -6$. |
| **Property 4** | If $a < b$ and $c < 0$, then $ac > bc$. | $-2 < 4$, and since $-3 < 0$, we have $(-2)(-3) > (4)(-3)$; that is, $6 > -12$. |

Similar properties hold if each inequality sign, $<$, between $a$ and $b$ and between $b$ and $c$ is replaced by $\geq$, $>$, or $\leq$. Note that Property 4 says that an inequality sign is reversed if the inequality is multiplied by a negative number.

A real number is a *solution of an inequality* involving a variable if a true statement is obtained when the variable is replaced by that number. The set of all real numbers satisfying the inequality is called the *solution set*. We often use interval notation to describe the solution set.

**EXAMPLE 1**   Find the set of real numbers that satisfy $-1 \leq 2x - 5 < 7$.

**Solution**   Add 5 to each member of the given double inequality, obtaining

$$4 \leq 2x < 12$$

Next, multiply each member of the resulting double inequality by $\frac{1}{2}$, yielding

$$2 \leq x < 6$$

Thus, the solution is the set of all values of $x$ lying in the interval $[2, 6)$.   ■

**EXAMPLE 2**   Solve the inequality $x^2 + 2x - 8 < 0$.

**Solution**   Observe that $x^2 + 2x - 8 = (x + 4)(x - 2)$, so the given inequality is equivalent to the inequality $(x + 4)(x - 2) < 0$. Since the product of two real numbers is negative if and only if the two numbers have opposite signs, we solve the inequality $(x + 4)(x - 2) < 0$ by studying the signs of the two factors $x + 4$ and $x - 2$. Now, $x + 4 > 0$ if $x > -4$, and $x + 4 < 0$ if $x < -4$. Similarly, $x - 2 > 0$ if $x > 2$, and $x - 2 < 0$ if $x < 2$. These results are summarized graphically in Figure 2.

**FIGURE 2**
Sign diagram for $(x + 4)(x - 2)$

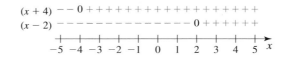

From Figure 2 we see that the two factors $x + 4$ and $x - 2$ have opposite signs if and only if $x$ lies strictly between $-4$ and $2$. Therefore, the required solution is the interval $(-4, 2)$. ∎

**EXAMPLE 3**  Solve the inequality $\dfrac{x + 1}{x - 1} \geq 0$.

**Solution**  The quotient $(x + 1)/(x - 1)$ is strictly positive if and only if both the numerator and the denominator have the same sign. The signs of $x + 1$ and $x - 1$ are shown in Figure 3.

**FIGURE 3**

Sign diagram for $\dfrac{x + 1}{x - 1}$

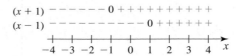

From Figure 3 we see that $x + 1$ and $x - 1$ have the same sign if and only if $x < -1$ or $x > 1$. The quotient $(x + 1)/(x - 1)$ is equal to zero if and only if $x = -1$. Therefore, the required solution is the set of all $x$ in the intervals $(-\infty, -1]$ and $(1, \infty)$. ∎

## ■ Absolute Value

> **DEFINITION**  **Absolute Value**
>
> The **absolute value** of a number $a$ is denoted by $|a|$ and is defined by
>
> $$|a| = \begin{cases} a & \text{if } a \geq 0 \\ -a & \text{if } a < 0 \end{cases}$$

**FIGURE 4**

The absolute value of a number

Since $-a$ is a positive number when $a$ is negative, it follows that the absolute value of a number is always nonnegative. For example, $|5| = 5$ and $|-5| = -(-5) = 5$. Geometrically, $|a|$ is the distance between the origin and the point on the number line that represents the number $a$. (See Figure 4.)

**Absolute Value Properties**

If $a$ and $b$ are any real numbers, then

|  |  |  | **Example** |
|---|---|---|---|
| **Property 5** | $|-a| = |a|$ | | $|-3| = -(-3) = 3 = |3|$ |
| **Property 6** | $|ab| = |a||b|$ | | $|(2)(-3)| = |-6| = 6 = (2)(3)$ |
| | | | $= |2||-3|$ |
| **Property 7** | $\left|\dfrac{a}{b}\right| = \dfrac{|a|}{|b|}$ | $(b \neq 0)$ | $\left|\dfrac{(-3)}{(-4)}\right| = \left|\dfrac{3}{4}\right| = \dfrac{3}{4} = \dfrac{|-3|}{|-4|}$ |
| **Property 8** | $|a + b| \leq |a| + |b|$ | | $|8 + (-5)| = |3| = 3$ |
| | | | $\leq |8| + |-5| = 13$ |

Property 8 is called the **triangle inequality.** To prove the triangle inequality, note that

$$-|a| < a < |a| \qquad \text{and} \qquad -|b| < b < |b|$$

Adding the respective numbers in these inequalities, we have

$$-(|a| + |b|) \leq a + b \leq |a| + |b|$$

which is equivalent to

$$|a + b| \leq |a| + |b|$$

as was to be shown.

**EXAMPLE 4**   Evaluate each of the following expressions:

**a.** $|\pi - 5| + 3$   **b.** $|\sqrt{3} - 2| + |2 - \sqrt{3}|$

**Solution**
**a.** Since $\pi - 5 < 0$, we see that $|\pi - 5| = -(\pi - 5)$. Therefore,

$$|\pi - 5| + 3 = -(\pi - 5) + 3 = 8 - \pi$$

**b.** Since $\sqrt{3} - 2 < 0$, we see that $|\sqrt{3} - 2| = -(\sqrt{3} - 2)$. Next, observe that $2 - \sqrt{3} > 0$, so $|2 - \sqrt{3}| = 2 - \sqrt{3}$. Therefore,

$$|\sqrt{3} - 2| + |2 - \sqrt{3}| = -(\sqrt{3} - 2) + (2 - \sqrt{3})$$
$$= 4 - 2\sqrt{3} = 2(2 - \sqrt{3}) \qquad \blacksquare$$

**EXAMPLE 5**   Solve the inequalities $|x| \leq 5$ and $|x| \geq 5$.

**Solution**   First, we consider the inequality $|x| \leq 5$. If $x \geq 0$, then $|x| = x$, so $|x| \leq 5$ implies $x \leq 5$ in this case. On the other hand, if $x < 0$, then $|x| = -x$, so $|x| \leq 5$ implies $-x \leq 5$ or $x \geq -5$. Thus, $|x| \leq 5$ means $-5 \leq x \leq 5$. (See Figure 5a.) To obtain an alternative solution, observe that $|x|$ is the distance from the point $x$ to zero, so the inequality $|x| \leq 5$ implies immediately that $-5 \leq x \leq 5$.

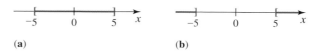

**FIGURE 5**          (a)                    (b)

Next, the inequality $|x| \geq 5$ states that the distance from $x$ to zero is greater than or equal to 5. This observation yields the result $x \geq 5$ or $x \leq -5$. (See Figure 5b.)   $\blacksquare$

**EXAMPLE 6**   Solve the inequality $|2x - 3| \leq 1$.

**Solution**   The inequality $|2x - 3| \leq 1$ is equivalent to the inequalities $-1 \leq 2x - 3 \leq 1$. (See Example 5.) Thus, $2 \leq 2x \leq 4$ and $1 \leq x \leq 2$. The solution is therefore given by the set of all $x$ in the interval $[1, 2]$. (See Figure 6.)   $\blacksquare$

**EXAMPLE 7**   Solve $|2x + 3| \geq 5$.

**Solution**   The inequality $|2x + 3| \geq 5$ is equivalent to $2x + 3 \geq 5$ or $2x + 3 \leq -5$. (See Example 5 with $x$ replaced by $2x + 3$.) The first inequality gives $x \geq 1$, and the second gives $x \leq -4$. So the solution is $\{x \mid x \leq -4 \quad \text{or} \quad x \geq 1\} = (-\infty, -4] \cup [1, \infty)$.   $\blacksquare$

**FIGURE 6**
$|2x - 3| \leq 1$

**EXAMPLE 8** If $|x - 2| < 0.1$ and $|y - 3| < 0.2$, find an upper bound for $|x + y - 5|$.

**Solution** We have

$$|x + y - 5| = |(x - 2) + (y - 3)|$$

$$\leq |x - 2| + |y - 3| \qquad \text{Use the Triangle Inequality.}$$

$$< 0.1 + 0.2 = 0.3$$

So $|x + y - 5| < 0.3$. ∎

## EXERCISES

*In Exercises 1–6, show the interval on a number line.*

**1.** $(3, 6)$

**2.** $(-2, 5]$

**3.** $[-1, 4)$

**4.** $\left[-\dfrac{6}{5}, -\dfrac{1}{2}\right]$

**5.** $(0, \infty)$

**6.** $(-\infty, 5]$

*In Exercises 7–10, determine whether the statement is true or false.*

**7.** $-3 < -20$

**8.** $-5 \leq -5$

**9.** $\dfrac{2}{3} > \dfrac{5}{6}$

**10.** $-\dfrac{5}{6} < -\dfrac{11}{12}$

*In Exercises 11–28, find the values of x that satisfy the inequality (inequalities).*

**11.** $2x + 4 < 8$

**12.** $-6 > 4 + 5x$

**13.** $-4x \geq 20$

**14.** $-12 \leq -3x$

**15.** $-6 < x - 2 < 4$

**16.** $0 \leq x + 1 \leq 4$

**17.** $x + 1 > 4$  or  $x + 2 < -1$

**18.** $x + 1 > 2$  or  $x - 1 < -2$

**19.** $x + 3 > 1$  and  $x - 2 < 1$

**20.** $x - 4 \leq 1$  and  $x + 3 > 2$

**21.** $(x + 3)(x - 5) \leq 0$

**22.** $(2x - 4)(x + 2) \geq 0$

**23.** $(2x - 3)(x - 1) \geq 0$

**24.** $(3x - 4)(2x + 2) \leq 0$

**25.** $\dfrac{x + 3}{x - 2} \geq 0$

**26.** $\dfrac{2x - 3}{x + 1} \geq 4$

**27.** $\dfrac{x - 2}{x - 1} \leq 2$

**28.** $\dfrac{2x - 1}{x + 2} \leq 4$

*In Exercises 29–38, evaluate the expression.*

**29.** $|-6 + 2|$

**30.** $4 + |-4|$

**31.** $\dfrac{|-12 + 4|}{|16 - 12|}$

**32.** $\left|\dfrac{0.2 - 1.4}{1.6 - 2.4}\right|$

**33.** $\sqrt{3}|-2| + 3|-\sqrt{3}|$

**34.** $|-1| + \sqrt{2}|-2|$

**35.** $|\pi - 1| + 2$

**36.** $|\pi - 6| - 3$

**37.** $|\sqrt{2} - 1| + |3 - \sqrt{2}|$

**38.** $|2\sqrt{3} - 3| - |\sqrt{3} - 4|$

*In Exercises 39–44, suppose that a and b are real numbers other than zero and that a > b. State whether the inequality is true or false.*

**39.** $b - a > 0$

**40.** $\dfrac{a}{b} > 1$

**41.** $a^2 > b^2$

**42.** $\dfrac{1}{a} > \dfrac{1}{b}$

**43.** $a^3 > b^3$

**44.** $-a < -b$

*In Exercises 45–50, determine whether the statement is true for all real numbers a and b.*

**45.** $|-a| = a$

**46.** $|b^2| = b^2$

**47.** $|a - 4| = |4 - a|$

**48.** $|a + 1| = |a| + 1$

**49.** $|a + b| = |a| + |b|$

**50.** $|a - b| = |a| - |b|$

*In Exercises 51–54, solve the equation for x.*

**51.** $|3x| = 4$

**52.** $|2x + 4| = 1$

**53.** $|x + 2| = |2x + 3|$

**54.** $\left|\dfrac{3x + 1}{x + 2}\right| = 3$

*In Exercises 55–64, solve the inequality.*

**55.** $|x| < 4$

**56.** $|x| > 3$

**57.** $|x - 2| < 1$

**58.** $|x - 4| < 0.1$

**59.** $|x + 3| \geq 2$

**60.** $|3x - 2| > 1$

**61.** $|2x + 3| \leq 0.2$

**62.** $|3x - 2| < 4$

**63.** $1 \leq |x| \leq 3$

**64.** $0 < |x - 2| < \frac{1}{3}$

**65.** If $|x - 1| < 0.2$ and $|y - 4| < 0.2$, find an upper bound for $|x + y - 5|$.

**66.** Prove that $|x - y| \geq |x| - |y|$.
Hint: Write $x = (x - y) + y$ and use the Triangle Inequality.

# Proofs of Theorems

In this appendix we give the proofs of some of the theorems that appear in the body of the text.

## ■ Chapter 1

### Theorem 1   Limit Laws

(Section 1.2)

#### PRODUCT LAW

If $\lim_{x \to a} f(x) = L$ and $\lim_{x \to a} g(x) = M$, then $\lim_{x \to a} [f(x)g(x)] = LM$.

**PROOF**   Let $\varepsilon > 0$ be given. We want to show that there exists a $\delta > 0$ such that

$$0 < |x - a| < \delta \Rightarrow |f(x)g(x) - LM| < \varepsilon.$$

We begin by considering

$$
\begin{aligned}
|f(x)g(x) - LM| &= |f(x)g(x) - Lg(x) + Lg(x) - LM| \\
&= |[f(x) - L]g(x) + L[g(x) - M]| \\
&\leq |[f(x) - L]g(x)| + |L[g(x) - M]| \quad \text{Use the Triangle Inequality.} \\
&= |f(x) - L||g(x)| + |L||g(x) - M| \quad\quad (1)
\end{aligned}
$$

Thus, our goal will be achieved if we can show that the expression on the right of Equation (1) is less than $\varepsilon$ whenever $0 < |x - a| < \delta$. Since $\lim_{x \to a} g(x) = M$, there exists a $\delta_1 > 0$ such that

$$0 < |x - a| < \delta_1 \Rightarrow |g(x) - M| < 1$$

and therefore,

$$|g(x)| = |g(x) - M + M| = |[g(x) - M] + M| \leq |g(x) - M| + |M| < 1 + |M| \quad (2)$$

Also, because $\lim_{x \to a} g(x) = M$, there exists a $\delta_2 > 0$ such that

$$0 < |x - a| < \delta_2 \Rightarrow |g(x) - M| < \frac{\varepsilon}{2(1 + |L|)} \quad (3)$$

Next, since $\lim_{x \to a} f(x) = L$, there exists a $\delta_3 > 0$ such that

$$0 < |x - a| < \delta_3 \Rightarrow |f(x) - L| < \frac{\varepsilon}{2(1 + |M|)} \quad (4)$$

If we let $\delta$ denote the smallest of the three numbers $\delta_1$, $\delta_2$, and $\delta_3$, then $\delta > 0$, and if $0 < |x - a| < \delta$, then we have $0 < |x - a| < \delta_1$, $0 < |x - a| < \delta_2$, and

$0 < |x - a| < \delta_3$, so all three Inequalities (2)–(4) hold simultaneously. Therefore, if $0 < |x - a| < \delta$, then by Equation (1),

$$|f(x)g(x) - LM| \le \frac{\varepsilon}{2(1 + |M|)} \cdot (1 + |M|) + |L| \cdot \frac{\varepsilon}{2(1 + |L|)}$$

$$< \frac{\varepsilon}{2} + \frac{\varepsilon}{2} = \varepsilon$$

This completes our proof.                                                               ■

## CONSTANT MULTIPLE LAW

$$\lim_{x \to a} [cf(x)] = c \lim_{x \to a} f(x) \quad \text{for every } c$$

**PROOF**    Put $g(x) = c$ in the Product Law, obtaining

$$\lim_{x \to a} [cf(x)] = \lim_{x \to a} [g(x)f(x)] = [\lim_{x \to a} g(x)][\lim_{x \to a} f(x)]$$

$$= c \lim_{x \to a} f(x) \qquad \text{Use Law 1 in Section 1.2.}$$

as was to be shown.                                                                     ■

## QUOTIENT LAW

$$\lim_{x \to a} \frac{f(x)}{g(x)} = \frac{L}{M} \quad \text{provided that } M \ne 0$$

**PROOF**    We can write

$$\frac{f(x)}{g(x)} = f(x) \cdot \frac{1}{g(x)}$$

and thus use the Product Law established earlier to help us with the proof. Let's first show that

$$\lim_{x \to a} \frac{1}{g(x)} = \frac{1}{M}$$

Let $\varepsilon > 0$ be given. We need to show that there exists a $\delta > 0$ such that

$$0 < |x - a| < \delta \Rightarrow \left| \frac{1}{g(x)} - \frac{1}{M} \right| < \varepsilon$$

We consider

$$\left| \frac{1}{g(x)} - \frac{1}{M} \right| = \left| \frac{M - g(x)}{Mg(x)} \right| = \frac{|M - g(x)|}{|M||g(x)|} \tag{5}$$

We want to show that the denominator of the last expression is bounded away from 0 when $x$ is close to $a$. To do this, observe that if $\lim_{x \to a} g(x) = M$, then there exists a $\delta_1 > 0$ such that

$$0 < |x - a| < \delta_1 \Rightarrow |g(x) - M| < \frac{|M|}{2} \qquad \text{Remember that } M \ne 0, \text{ so } |M| > 0.$$

Then

$$|M| = |M - g(x) + g(x)| \leq |M - g(x)| + |g(x)|$$

$$< \frac{|M|}{2} + |g(x)|$$

and this implies that if $0 < |x - a| < \delta_1$, then

$$|g(x)| > \frac{|M|}{2} \tag{6}$$

Thus, if $x$ satisfies $0 < |x - a| < \delta_1$ then Inequality (6) implies that

$$\frac{1}{|M| \, |g(x)|} < \frac{1}{|M|} \cdot \frac{2}{|M|} = \frac{2}{|M|^2}$$

Also, since $\lim_{x \to a} g(x) = M$, there exists a $\delta_2 > 0$ such that

$$0 < |x - a| < \delta_2 \Rightarrow |g(x) - M| < \frac{|M|^2 \varepsilon}{2} \tag{7}$$

If we let $\delta$ denote the smaller of the two numbers $\delta_1$ and $\delta_2$, then $\delta > 0$ and if $0 < |x - a| < \delta$, then $0 < |x - a| < \delta_1$ and $0 < |x - a| < \delta_2$, so both Inequalities (6) and (7) hold simultaneously. Therefore, if $0 < |x - a| < \delta$, then Equation (5) implies that

$$\left| \frac{1}{g(x)} - \frac{1}{M} \right| = \frac{|M - g(x)|}{|M| \, |g(x)|} < \frac{2}{|M|^2} \cdot \frac{|M|^2 \varepsilon}{2} = \varepsilon$$

This establishes that

$$\lim_{x \to a} \frac{1}{g(x)} = \frac{1}{M}$$

Thus, using the Product Law proved earlier, we have

$$\lim_{x \to a} \frac{f(x)}{g(x)} = \lim_{x \to a} f(x) \cdot \frac{1}{g(x)} = L \left( \frac{1}{M} \right) = \frac{L}{M}$$

and the desired result follows.

## ROOT LAW

If $\lim_{x \to a} f(x) = L$, then $\lim_{x \to a} \sqrt[n]{f(x)} = \sqrt[n]{L}$, provided that $L > 0$ if $n$ is even.

**PROOF** First, we prove that

$$\lim_{x \to a} \sqrt{x} = \sqrt{a} \qquad a > 0$$

Let $\varepsilon > 0$ be given. We have

$$|\sqrt{x} - \sqrt{a}| = \left| \frac{\sqrt{x} - \sqrt{a}}{1} \cdot \frac{\sqrt{x} + \sqrt{a}}{\sqrt{x} + \sqrt{a}} \right| = \frac{|x - a|}{\sqrt{x} + \sqrt{a}}$$

Let's agree to pick $\delta \le \frac{3}{4}a$. Then $\delta > 0$. If $|x - a| < \delta$, then $-\frac{3}{4}a < x - a < \frac{3}{4}a$ or $\frac{1}{4}a < x < \frac{7}{4}a$, and so $\sqrt{x} + \sqrt{a} > \frac{\sqrt{a}}{2} + \sqrt{a} = \frac{3\sqrt{a}}{2}$. Let's pick $\delta$ to be the smaller of $\frac{3}{4}a$ and $\frac{3\sqrt{a}}{2}\varepsilon$. Then $0 < |x - a| < \delta$ implies

$$|\sqrt{x} - \sqrt{a}| = \frac{|x - a|}{\sqrt{x} + \sqrt{a}} < \frac{\dfrac{3\sqrt{a}}{2}\varepsilon}{\dfrac{3\sqrt{a}}{2}} = \varepsilon$$

Since $\varepsilon$ is arbitrary, the result follows.

To complete the proof of the Root Law, we apply Theorem 4 (Section 1.4) to the special case where $f(x) = \sqrt[n]{x}$, with $n$ a positive integer. Then, assuming that all roots exist, we have

$$f(g(x)) = \sqrt[n]{g(x)}$$

and

$$f\left[\lim_{x \to a} g(x)\right] = \sqrt[n]{\lim_{x \to a} g(x)}$$

So, Theorem 4 gives

$$\lim_{x \to a} \sqrt[n]{g(x)} = \sqrt[n]{\lim_{x \to a} g(x)}$$

and this completes the proof of the Root Law.    ▬

## Theorem 3    The Squeeze Theorem    (Section 1.2)

Suppose that $f(x) \le g(x) \le h(x)$ for all $x$ in an open interval containing $a$, except possibly at $a$ and that

$$\lim_{x \to a} f(x) = L = \lim_{x \to a} h(x)$$

Then $\lim_{x \to a} g(x) = L$.

**PROOF**   Let $\varepsilon > 0$ be given. Since $\lim_{x \to a} f(x) = L$, there exists a $\delta_1 > 0$ such that

$$0 < |x - a| < \delta_1 \Rightarrow |f(x) - L| < \varepsilon$$

or, equivalently,

$$L - \varepsilon < f(x) < L + \varepsilon \tag{8}$$

Next, since $\lim_{x \to a} h(x) = L$, there exists a $\delta_2 > 0$ such that

$$0 < |x - a| < \delta_2 \Rightarrow |h(x) - L| < \varepsilon$$

or, equivalently,

$$L - \varepsilon < h(x) < L + \varepsilon \tag{9}$$

If we let $\delta$ denote the smaller of the two numbers $\delta_1$ and $\delta_2$, then $\delta > 0$, and if $0 < |x - a| < \delta$, then $0 < |x - a| < \delta_1$ and $0 < |x - a| < \delta_2$, so both Inequalities (8) and (9) hold simultaneously. Therefore, if $0 < |x - a| < \delta$, then

$$L - \varepsilon < f(x) \le g(x) \le h(x) < L + \varepsilon$$

which implies that

$$L - \varepsilon < g(x) < L + \varepsilon$$

or, equivalently, $|g(x) - L| < \varepsilon$. Therefore, $\lim_{x \to a} g(x) = L$.    ▬

## Theorem 4                                                          (Section 1.2)

Suppose that $f(x) \leq g(x)$ for all $x$ in an open interval containing $a$, except possibly at $a$ and that

$$\lim_{x \to a} f(x) = L \quad \text{and} \quad \lim_{x \to a} g(x) = M$$

Then $L \leq M$.

**PROOF**   Suppose, to the contrary, that $L > M$. Now, since $\lim_{x \to a} f(x) = L$ and $\lim_{x \to a} g(x) = M$, by the Sum Law (Theorem 1a),

$$\lim_{x \to a} [g(x) - f(x)] = \lim_{x \to a} g(x) - \lim_{x \to a} f(x) = M - L$$

Since $L - M > 0$, we can take $\varepsilon = L - M$ and find a $\delta > 0$ such that

$$|[g(x) - f(x)] - (M - L)| < L - M$$

or

$$g(x) - f(x) - M + L < L - M$$

or

$$g(x) < f(x)$$

But this contradicts the condition in the hypothesis, which states that $g(x) \geq f(x)$. Therefore, the assumption $L > M$ must be false, and we conclude that $L \leq M$, as was to be shown.   ■

## Theorem 4    Limit of a Composite Function                        (Section 1.4)

If the function $f$ is continuous at $L$ and $\lim_{x \to a} g(x) = L$, then

$$\lim_{x \to a} f(g(x)) = f(L)$$

**PROOF**   Let $\varepsilon > 0$ be given. We want to show that there exists a $\delta > 0$ such that

$$0 < |x - a| < \delta \Rightarrow |f(g(x)) - f(L)| < \varepsilon$$

Now, since $f$ is continuous at $L$, we have

$$\lim_{y \to L} f(y) = f(L) \tag{10}$$

so there exists a $\delta_1 > 0$ such that

$$0 < |y - L| < \delta_1 \Rightarrow |f(y) - f(L)| < \varepsilon$$

Next, since $\lim_{x \to a} g(x) = L$, there exists a $\delta > 0$ such that

$$0 < |x - a| < \delta \Rightarrow |g(x) - L| < \delta_1 \tag{11}$$

Combining (10) and (11), we see that

$$0 < |x - a| < \delta \Rightarrow |g(x) - L| < \delta_1 \Rightarrow |f(g(x)) - f(L)| < \varepsilon$$

as was to be shown.   ■

# ■ Chapter 2

## Theorem 1   The Chain Rule
(Section 2.6)

If $f$ is differentiable at $x$ and $g$ is differentiable at $f(x)$, then the composition $h = g \circ f$ defined by $h(x) = g[f(x)]$ is differentiable, and

$$h'(x) = g'[f(x)]f'(x)$$

Also, if we write $u = f(x)$ and $y = g(u) = g[f(x)]$, then

$$\frac{dy}{dx} = \frac{dy}{du}\frac{du}{dx}$$

**PROOF**   According to the definition of the derivative, we have, for fixed $a$,

$$f'(a) = \lim_{\Delta x \to 0} \frac{f(a + \Delta x) - f(a)}{\Delta x} = \lim_{\Delta x \to 0} \frac{\Delta y}{\Delta x} \tag{12}$$

where $\Delta y = f(a + \Delta x) - f(a)$. Let us define a function of $\Delta x$ by

$$\varepsilon(\Delta x) = \begin{cases} \dfrac{\Delta y}{\Delta x} - f'(a) & \text{if } \Delta x \neq 0 \\ 0 & \text{if } \Delta x = 0 \end{cases}$$

Then

$$\lim_{\Delta x \to 0} \varepsilon(\Delta x) = \lim_{\Delta x \to 0}\left(\frac{\Delta y}{\Delta x} - f'(a)\right) = 0$$

by Equation (12). So $\varepsilon$ is a continuous function of $\Delta x$. Therefore, for a differentiable function $f$, we can write

$$\Delta y = f'(a)\,\Delta x + \varepsilon(\Delta x)\,\Delta x \tag{13}$$

where $\lim_{\Delta x \to 0} \varepsilon(\Delta x) = 0$.

Now suppose that $u = f(x)$, where $f$ is differentiable at $a$ and $y = g(u)$ is differentiable at $b = f(a)$. If $\Delta x$ is an increment in $x$ and $\Delta u$ and $\Delta y$ are the corresponding increments in $u$ and $y$, then using Equation (13), we have

$$\Delta u = f'(a)\,\Delta x + \varepsilon_1(\Delta x)\,\Delta x = [f'(a) + \varepsilon_1(\Delta x)]\,\Delta x \tag{14}$$

where $\varepsilon_1(\Delta x) \to 0$ as $\Delta x \to 0$. Similarly,

$$\Delta y = g'(b)\,\Delta u + \varepsilon_2(\Delta u)\,\Delta u = [g'(b) + \varepsilon_2(\Delta u)]\,\Delta u \tag{15}$$

where $\varepsilon_2(\Delta u) \to 0$ as $\Delta u \to 0$. Substituting the expression for $\Delta u$ in Equation (14) into Equation (15), we obtain

$$\Delta y = [g'(b) + \varepsilon_2(\Delta u)][f'(a) + \varepsilon_1(\Delta x)]\,\Delta x$$

or

$$\frac{\Delta y}{\Delta x} = [g'(b) + \varepsilon_2(\Delta u)][f'(a) + \varepsilon_1(\Delta x)] \tag{16}$$

Now, if $\Delta x \to 0$, then Equation (14) implies that $\Delta u \to 0$. So both $\varepsilon_1(\Delta x)$ and $\varepsilon_2(\Delta u)$ approach 0 as $\Delta x \to 0$. Therefore, from Equation (16) we have

$$\frac{dy}{dx} = \lim_{\Delta x \to 0} \frac{\Delta y}{\Delta x} = \lim_{\Delta x \to 0} [g'(b) + \varepsilon_2(\Delta u)][f'(a) + \varepsilon_1(\Delta x)]$$

$$= g'(b)f'(a) = g'(f(a))f'(a)$$

Since $a$ is arbitrary, the proof is complete.   ■

## Theorem 1   Continuity and Differentiability of Inverse Functions

**(Sections 1.4 and 2.7)**

Let $f$ be one-to-one so that it has an inverse $f^{-1}$.

1.  If $f$ is continuous on its domain, then $f^{-1}$ is continuous on its domain.
2.  If $f$ is differentiable at $c$ and $f'(c) \neq 0$, then $f^{-1}$ is differentiable at $f(c)$.

**PROOF OF (1)**   We first show that $f$ is monotonic. Suppose, on the contrary, that $f$ is neither increasing nor decreasing. Then there exists numbers $x_1$, $x_2$, $x_3$ in $(a, b)$ with $x_1 < x_2 < x_3$ such that $f(x_2)$ does not lie between $f(x_1)$ and $f(x_3)$.

There are two cases: (i) $f(x_3)$ lies between $f(x_1)$ and $f(x_2)$ or (ii) $f(x_1)$ lies between $f(x_2)$ and $f(x_3)$. In case (i) the Intermediate Value Theorem implies that there exists a $c$ satisfying $x_1 < c < x_2$ such that $f(c) = f(x_3)$. In case (ii) the Intermediate Value Theorem implies that there exists a $c$ satisfying $x_2 < c < x_3$ such that $f(c) = f(x_1)$. In either case we have shown that $f$ is not one-to-one. This is a contradiction. Thus, $f$ is indeed monotonic.

Without loss of generality, let us assume that $f$ is increasing on $(a, b)$. Let $y_0$ be any number in the domain of $f^{-1}$. Then $f^{-1}(y_0) = x_0$, where $x_0$ is a number in $(a, b)$ such that $f(x_0) = y_0$. We want to show that $f^{-1}$ is continuous at $y_0$. So let $\varepsilon > 0$ be given and be sufficiently small so that the interval $(x_0 - \varepsilon, x_0 + \varepsilon)$ is contained in the interval $(a, b)$. Since $f$ is increasing, we see that the interval $I = (x_0 - \varepsilon, x_0 + \varepsilon)$ is mapped onto the interval $J = (f(x_0 - \varepsilon), f(x_0 + \varepsilon))$.

The function $f^{-1}$ maps $J$ onto $I$. Now let $\delta$ denote the smaller of the numbers $\delta_1 = y_0 - f(x_0 - \varepsilon)$ and $\delta_2 = f(x_0 + \varepsilon) - y_0$. Then the interval $(y_0 - \delta, y_0 + \delta)$ is contained in $J$ and so is mapped onto the interval $I$ by $f^{-1}$. (See Figure 1.) Thus, we have found a $\delta > 0$ such that $0 < |y - y_0| < \delta$ implies that $|f^{-1}(y) - f^{-1}(y_0)| < \varepsilon$. This shows that $f^{-1}$ is continuous at $y_0$. Since $y_0$ is an arbitrary number in the domain of $f^{-1}$, we have shown that $f^{-1}$ is continuous in its domain.

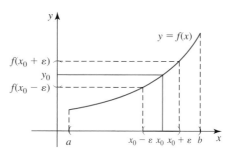

**FIGURE 1**

## Chapter 3

## Note on the Definition of Concavity of a Function

**(Section 3.4)**

If the graph of $f$ is concave upward on an open interval $I$, then it lies above its tangent lines, and if the graph is concave downward on $I$, then it lies below all of its tangent lines.

**PROOF**   Suppose that the graph of $f$ is concave upward on an interval $I = (a, b)$. Then $f'$ is increasing on $I$. If $c$ is a number in $I$, then an equation of the tangent line to the graph of $f$ at $(c, f(c))$ is

$$L(x) = f(c) + f'(c)(x - c)$$

If $x$ is any number in the interval $(a, c)$, then the directed distance from the point $(x, f(x))$ on the graph of $f$ to the point $(x, L(x))$ on the tangent line is given by

$$D(x) = f(x) - L(x) = f(x) - [f(c) + f'(c)(x - c)] \tag{17}$$
$$= f(x) - f(c) - f'(c)(x - c)$$

Now, by the Mean Value Theorem there exists a number $z$ in $(x, c)$ such that

$$f'(z) = \frac{f(x) - f(c)}{x - c} \tag{18}$$

Substituting Equation (18) into Equation (17) gives

$$D(x) = f'(z)(x - c) - f'(c)(x - c) = [f'(z) - f'(c)](x - c)$$

Now, $[f'(z) - f'(c)] < 0$ because $f'$ is increasing. Furthermore, $(x - c) < 0$ because $x$ lies in $(a, c)$. So $D(x) > 0$ for all $x$ in $(a, c)$, and this tells us that the graph of $f$ lies above the tangent line at $x$ for all $x$ in $(a, c)$. In a similar manner you can show that the graph of $f$ also lies above the tangent line at $x$ for all $x$ on $(c, b)$. This completes the proof for the case in which $f$ is concave upward. The proof for the case in which $f$ is concave downward is similar. ■

## Theorem 1   l'Hôpital's Rule   (Section 3.8)

Suppose that $f$ and $g$ are differentiable on an open interval $I$ that contains $a$, with the possible exception of $a$ itself, and $g'(a) \neq 0$ for all $x$ in $I$. If $\lim\limits_{x \to a} \dfrac{f(x)}{g(x)}$ has an indeterminate form of type $0/0$ or $\infty/\infty$, then

$$\lim_{x \to a} \frac{f(x)}{g(x)} = \lim_{x \to a} \frac{f'(x)}{g'(x)}$$

provided that the limit on the right exists or is infinite.

**PROOF for the Case $\lim\limits_{x \to a} \dfrac{f(x)}{g(x)} = \dfrac{0}{0}$**   First, we need to show that if $f$ and $g$ are continuous on $[a, b]$ and differentiable on $(a, b)$ and $g'(x) \neq 0$ for all $x$ in $(a, b)$, then there exists a number $c$ in $(a, b)$ such that

$$\frac{f'(c)}{g'(c)} = \frac{f(b) - f(a)}{g(b) - g(a)} \tag{19}$$

This is called Cauchy's Mean Value Theorem.

To prove this, first observe that $g(a) \neq g(b)$; otherwise, an application of Rolle's Theorem implies that there exists a $c$ in $(a, b)$ such that $g'(c) = 0$, contradicting the assumption that $g'(x) \neq 0$ for all $x$ in $(a, b)$. Put

$$h(x) = f(x) - f(a) - \frac{f(b) - f(a)}{g(b) - g(a)} [g(x) - g(a)] \tag{20}$$

Then Equation (19) follows by applying Rolle's Theorem to Equation (20). To prove l'Hôpital's Rule, suppose that $\lim_{x \to a} f(x) = 0$ and $\lim_{x \to a} g(x) = 0$. Define

$$F(x) = \begin{cases} f(x) & \text{if } x \neq a \\ 0 & \text{if } x = a \end{cases} \quad \text{and} \quad G(x) = \begin{cases} g(x) & \text{if } x \neq a \\ 0 & \text{if } x = a \end{cases}$$

Then $F$ is continuous on $I$, since $f$ is continuous on $\{x \in I \mid x \neq a\}$ and $\lim_{x \to a} F(x) = \lim_{x \to a} f(x) = 0 = F(a)$. Similarly, we see that $G$ is also continuous on $I$.

For any $x$ in $I$ and $x > a$, $F$ and $G$ are continuous on $[a, x]$ and differentiable on $(a, x)$, and $G'(x) \neq 0$, since $G'(x) = g'(x)$. So, by Cauchy's Mean Value Theorem there exists a $z$ such that $a < z < x$, and

$$\frac{F'(z)}{G'(z)} = \frac{F(x) - F(a)}{G(x) - G(a)} = \frac{F(x)}{G(x)}$$

because, by definition, $F(a) = 0$ and $G(a) = 0$. Now, if we let $x$ approach $a$ from the right, then $z \to a^+$ because $a < z < x$. Therefore,

$$\lim_{x \to a^+} \frac{f(x)}{g(x)} = \lim_{x \to a^+} \frac{F(x)}{G(x)} = \lim_{z \to a^+} \frac{F'(z)}{G'(z)} = \lim_{z \to a^+} \frac{f'(z)}{g'(z)} = L$$

where

$$L = \lim_{x \to a^+} \frac{f'(x)}{g'(x)}$$

which we assume to exist. In a similar manner we can show that

$$\lim_{x \to a^-} \frac{f(x)}{g(x)} = L$$

So

$$\lim_{x \to a} \frac{f(x)}{g(x)} = L = \lim_{x \to a} \frac{f'(x)}{g'(x)}$$

as was to be shown.   ■

## ■ Chapter 8

### Theorem 5   (Section 8.1)

If $\lim_{n \to \infty} a_n = L$ and the function $f$ is continuous at $L$, then

$$\lim_{n \to \infty} f(a_n) = f(\lim_{n \to \infty} a_n) = f(L)$$

**PROOF**   Let $\varepsilon > 0$ be given. We want to show that there exists a positive integer $N$ such that $n > N$ implies that $|f(a_n) - f(L)| < \varepsilon$. Since $f$ is continuous at $L$, there exists a $\delta > 0$ such that $0 < |x - L| < \delta$ implies that $|f(x) - f(L)| < \varepsilon$. Next, since $\lim_{n \to \infty} a_n = L$, there exists a positive integer $N$ such that $n > N \Rightarrow |a_n - L| < \delta$. Now suppose that $n > N$. Then $0 < |a_n - L| < \delta$, which implies that $|f(a_n) - f(L)| < \varepsilon$ and thus completes the proof.   ■

### Theorem 1   Convergence of Power Series   (Section 8.7)

Given a power series $\sum_{n=0}^{\infty} a_n(x - c)^n$, exactly one of the following is true:

**a.** The series converges only at $x = c$.
**b.** The series converges for all $x$.
**c.** There is a number $R > 0$ such that the series converges for $|x - c| < R$ and diverges for $|x - c| > R$.

**PROOF**    It suffices to prove the theorem for the special case where $c = 0$. The general case then follows if we replace $x$ by $x - c$. So, let us prove the following result.

Given a power series $\sum_{n=0}^{\infty} a_n x^n$, exactly one of the following is true:

a. The series converges only at $x = 0$.
b. The series converges for all $x$.
c. There is a number $R > 0$ such that the series converges for $|x| < R$ and diverges for $|x| > R$.

We begin by establishing the following results:

1. If a power series $\sum_{n=0}^{\infty} a_n x^n$ converges at $x = b$, where $b \neq 0$, then it converges for all $x$ satisfying $|x| < |b|$.

2. If a power series $\sum_{n=0}^{\infty} a_n x^n$ diverges at $x = c$, where $c \neq 0$, then it diverges for all $x$ satisfying $|x| > |c|$.

**PROOF OF (1) AND (2)**    Suppose $\sum_{n=0}^{\infty} a_n x^n$ converges at $b$. Then $\lim_{n \to \infty} a_n b^n = 0$. So there exists a positive integer $N$ such that $n \geq N$ implies that $|a_n b^n| < 1$. Therefore, if $n \geq N$, then we have

$$|a_n x^n| = \left| a_n x^n \frac{b^n}{b^n} \right| = |a_n b^n| \left| \frac{x}{b} \right|^n < \left| \frac{x}{b} \right|^n$$

If $|x| < |b|$, then $|x/b| < 1$ and so $\sum_{n=0}^{\infty} |x/b|^n$ is a convergent geometric series. Therefore, using the Comparison Test, we see that $\sum_{n=0}^{\infty} |a_n x^n|$ is convergent, and this shows that $\sum_{n=0}^{\infty} a_n x^n$ is absolutely convergent and, therefore, convergent. Next, suppose $\sum_{n=0}^{\infty} a_n x^n$ diverges at $x = c$. If $x$ is any number satisfying $|x| > |c|$, then part (1) of the theorem shows that $\sum_{n=0}^{\infty} a_n x^n$ cannot converge because, otherwise, $\sum_{n=0}^{\infty} a_n c^n$ would converge, a contradiction. Therefore, $\sum_{n=0}^{\infty} a_n x^n$ diverges if $|x| > |c|$.

We are now in the position to prove the theorem. Suppose that neither case (a) nor case (b) is true. Then there exists nonzero numbers $b$ and $d$ such that $\sum_{n=0}^{\infty} a_n x^n$ converges at $x = b$ and diverges at $x = d$. Let $S = \{x \mid \sum_{n=0}^{\infty} a_n x^n \text{ converges}\}$. Then $S$ is nonempty since $b \in S$. Furthermore, our earlier result shows that $\sum_{n=0}^{\infty} a_n x^n$ diverges if $|x| > |d|$ and so $|x| \leq |d|$ for all $x$ in $S$. Thus $|d|$ is an upper bound for $S$. By the Completeness Axiom (see Section 8.1), $S$ has a least upper bound $R$. Now, if $|x| > R$, then $x \notin S$ and so $\sum_{n=0}^{\infty} a_n x^n$ diverges. If $|x| < R$, then $|x|$ is not an upper bound for $S$ and so there exists a $b$ in $S$ satisfying $|b| > |x|$. Since $b \in S$, $\sum_{n=0}^{\infty} a_n b^n$ converges, and this implies that $\sum_{n=0}^{\infty} a_n x^n$ converges.    ■

## ■ Chapter 12

### Theorem 1    Clairaut's Theorem                               (Section 12.3)

If $f(x, y)$ and its partial derivatives $f_x, f_y, f_{xy}$, and $f_{yx}$ are continuous on an open region $R$, then

$$f_{xy}(x, y) = f_{yx}(x, y)$$

for all $(x, y)$ in $R$.

**PROOF**    Let $(a, b)$ be any point in $R$. Fix it, and let $h$ and $k$ be nonzero numbers such that the point $(a + h, b + k)$ also lie in $R$. Define the function $F$ by

$$F(h, k) = [f(a + h, b + k) - f(a + h, b)] - [f(a, b + k) - f(a, b)] \qquad (1)$$

If we let $g(x) = f(x, b + k) - f(x, b)$, then Equation (1) can be written

$$F(h, k) = g(a + h) - g(a)$$

Applying the Mean Value Theorem to the function $g$ on the interval $[a, a + h]$ (or $[a + h, a]$), we see that there exists a number $c$ lying between $a$ and $a + h$ such that

$$g(a + h) - g(a) = g'(c)h = h[f_x(c, b + k) - f_x(c, b)]$$

If we apply the Mean Value Theorem again to the function $f_x$ on the interval $[b, b + k]$ (or $[b + k, b]$), we see that there exists a number $d$ lying between $b$ and $b + k$ such that

$$f_x(c, b + k) - f_x(c, b) = f_{xy}(c, d)k$$

Therefore,

$$F(h, k) = hkf_{xy}(c, d)$$

If we let $(h, k) \to (0, 0)$, then $(c, d)$ approaches $(a, b)$, and the continuity of $f_{xy}$ implies that

$$\lim_{(h, k) \to (0, 0)} \frac{F(h, k)}{hk} = f_{xy}(a, b)$$

Similarly, if we write

$$F(h, k) = [f(a + h, b + k) - f(a, b + k)] - [f(a + h, b) - f(a, b)]$$

and use the Mean Value Theorem twice, the continuity of $f_{yx}$ at $(a, b)$ implies that

$$\lim_{(h, k) \to (0, 0)} \frac{F(h, k)}{hk} = f_{yx}(a, b)$$

Therefore, $f_{xy}(a, b) = f_{yx}(a, b)$. Finally, since $(a, b)$ is any point in $R$, we see that this equality holds for any $(x, y)$ in $R$, and the theorem is proved.  ■

# The Definition of the Logarithm as an Integral

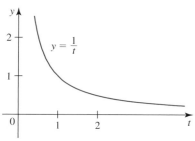

**FIGURE 1**
The function $f(t) = 1/t$ is continuous on $(0, \infty)$.

In Section 0.8, we gave an intuitive introduction of the exponential and logarithmic functions based on a numerical and graphical approach. In this appendix, we use the Fundamental Theorem of Calculus to help us define the logarithmic function. Then we use this definition to define the exponential function. This alternative approach, independent of the intuitive approach used earlier, puts the definition of these functions on a firm mathematical footing.

## ■ The Natural Logarithm Function

Recall that the Fundamental Theorem of Calculus, Part 1, states that if $f$ is a continuous function on an open interval $I$ and if $a$ is any number in $I$, then we can define a differentiable function $F$ by

$$F(x) = \int_a^x f(t)\, dt \qquad x \in I$$

Now consider the function $f$ defined by $f(t) = 1/t$ on the interval $(0, \infty)$. (See Figure 1.) Since $f$ is continuous on $(0, \infty)$, the Fundamental Theorem of Calculus, Part 1, guarantees that we can define a differentiable function on $(0, \infty)$ as follows.

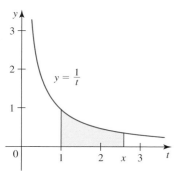

(a) If $x > 1$, $\ln x = \int_1^x \frac{1}{t}\, dt$

---

**DEFINITION   The Natural Logarithmic Function**
The **natural logarithmic function,** denoted by **ln,** is the function defined by

$$\ln x = \int_1^x \frac{1}{t}\, dt \tag{1}$$

for all $x > 0$.

---

The expression $\ln x$, read "ell-en of $x$," is called the **natural logarithm of $x$** because it has all the properties of logarithmic functions, as we shall see.

If $x > 1$, we can interpret $\ln x$ as the area of the region under the graph of $y = 1/t$ on the interval $[1, x]$. (See Figure 2.) For $x = 1$ we have

$$\ln 1 = \int_1^1 \frac{1}{t}\, dt = 0$$

If $0 < x < 1$, then

$$\ln x = \int_1^x \frac{1}{t}\, dt = -\int_x^1 \frac{1}{t}\, dt < 0$$

so $\ln x$ can be interpreted as the *negative* of the area of the region under the graph of $y = 1/t$ on the interval $[x, 1]$ (Figure 2b).

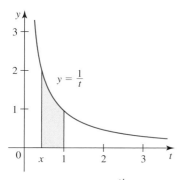

(b) If $0 < x < 1$, $\ln x = -\int_x^1 \frac{1}{t}\, dt$
**FIGURE 2**
$\ln x$ interpreted in terms of area

## ■ The Derivative of $\ln x$

Recall that the Fundamental Theorem of Calculus, Part 1, states that if $f$ is continuous on an open interval $I$ and the function $F$ is defined by

$$F(x) = \int_a^x f(t)\, dt \qquad a \in I$$

then $F'(x) = f(x)$. Applying this theorem to the function $f(t) = 1/t$ gives

$$\frac{d}{dx} \ln x = \frac{d}{dx} \int_1^x \frac{1}{t}\, dt = \frac{1}{x} \qquad x > 0 \tag{2}$$

Next, using the Chain Rule, we see that if $u$ is a differentiable function of $x$, then

$$\frac{d}{dx} \ln u = \frac{1}{u} \frac{du}{dx} \qquad u > 0 \tag{3}$$

## ■ Laws of Logarithms

The laws for differentiating the logarithmic function can be used to prove the following familiar laws of logarithms.

---

**THEOREM 1**   **Laws of Logarithms**

Let $x$ and $y$ be positive numbers and let $r$ be a rational number. Then

**a.** $\ln 1 = 0$ 　　　　　　　　　　　**b.** $\ln xy = \ln x + \ln y$

**c.** $\ln \dfrac{x}{y} = \ln x - \ln y$ 　　　　**d.** $\ln x^r = r \ln x$

---

**PROOF**

**a.** Law (a) was proved on page A 19.

**b.** Define the function $F(x) = \ln ax$, where $a$ is a positive constant. Then, using Equation (3), we have

$$F'(x) = \frac{d}{dx} (\ln ax) = \frac{1}{ax} \frac{d}{dx} (ax) = \frac{a}{ax} = \frac{1}{x}$$

But by Equation (2) we have

$$\frac{d}{dx} \ln x = \frac{1}{x}$$

Therefore, $F(x)$ and $\ln x$ have the same derivative and, by Theorem 1 of Section 4.1, must differ by a constant; that is,

$$F(x) = \ln ax = \ln x + C$$

Letting $x = 1$ in this equation and recalling that $\ln 1 = 0$, we have

$$\ln a = \ln 1 + C = C$$

Therefore,

$$\ln ax = \ln x + \ln a$$

Since $a$ can be any positive number, we have shown that

$$\ln xy = \ln x + \ln y$$

c. Using the result of part (b) with $x = 1/y$, we have

$$\ln \frac{1}{y} + \ln y = \ln\left(\frac{1}{y} \cdot y\right) = \ln 1 = 0$$

so

$$\ln \frac{1}{y} = -\ln y$$

Using the result of part (b) once again, we obtain

$$\ln \frac{x}{y} = \ln\left(x \cdot \frac{1}{y}\right) = \ln x + \ln \frac{1}{y} = \ln x - \ln y$$

as desired.

d. Define the functions $F$ and $G$ by $F(x) = \ln x^r$ and $G(x) = r \ln x$, respectively. Then using Equation (3), we have

$$F'(x) = \frac{1}{x^r} \cdot rx^{r-1} = \frac{r}{x}$$

Next, using Equation (2), we find

$$G'(x) = \frac{r}{x}$$

Therefore, $F$ and $G$ must differ by a constant; that is,

$$\ln x^r = r \ln x + C$$

Letting $x = 1$ in this equation gives

$$\ln 1 = r \ln 1 + C$$

or $C = 0$, so

$$\ln x^r = r \ln x$$

as was to be shown.  ▐

## ▐ The Graph of the Natural Logarithmic Function

To help us draw the graph of the natural logarithmic function, we first note that $f(x) = \ln x$ has the following properties:

1. The domain of $f$ is $(0, \infty)$, by definition.
2. $f$ is continuous on $(0, \infty)$, since it is differentiable there.
3. $f$ is increasing on $(0, \infty)$, since $f'(x) = \frac{1}{x} > 0$ on $(0, \infty)$.
4. The graph of $f$ is concave downward on $(0, \infty)$ since $f''(x) = -\frac{1}{x^2} < 0$ on $(0, \infty)$.

Next, using the Trapezoidal Rule or Simpson's Rule, we have

$$f(2) = \ln 2 = \int_1^2 \frac{1}{t} \, dt \approx 0.693$$

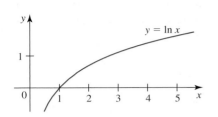

**FIGURE 3**
The graph of the natural logarithmic function $y = \ln x$

Then, using Theorem 1d, we obtain the following table of values.

| $x$ | 4 | 8 | $\frac{1}{2}$ | $\frac{1}{4}$ | $\frac{1}{8}$ |
|---|---|---|---|---|---|
| $f(x)$ | 1.386 | 2.079 | $-0.693$ | $-1.386$ | $-2.079$ |

Using the properties of $f(x) = \ln x$, the sample values just obtained, and the results

$$\lim_{x \to 0^+} \ln x = -\infty \qquad \text{and} \qquad \lim_{x \to \infty} \ln x = \infty \tag{4}$$

which we will establish at the end of this section, we sketch the graph of $f(x) = \ln x$, as shown in Figure 3.

## The Natural Exponential Function

Since the natural logarithmic function defined by $y = \ln x$ is continuous and increasing on the interval $(0, \infty)$, it is one-to-one there and, hence, has an inverse function. This inverse function is called the *natural exponential function* and is defined as follows.

> **DEFINITION    The Natural Exponential Function**
>
> The **natural exponential function,** denoted by **exp,** is the function satisfying the conditions:
>
> **1.** $\ln(\exp x) = x$ for all $x$ in $(-\infty, \infty)$.
> **2.** $\exp(\ln x) = x$ for all $x$ in $(0, \infty)$.
>
> Equivalently,
>
> $$\exp(x) = y \qquad \text{if and only if} \qquad \ln y = x$$

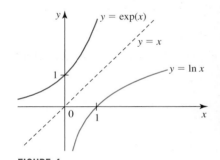

**FIGURE 4**
The graph of $y = \exp(x)$ is obtained by reflecting the graph of $y = \ln x$ with respect to the line $y = x$.

That the domain of exp is $(-\infty, \infty)$ and its range is $(0, \infty)$ follows because the range of ln is $(-\infty, \infty)$ and its domain is $(0, \infty)$. The graph of $y = \exp(x)$ can be obtained by reflecting the graph of $y = \ln x$ about the line $y = x$. (See Figure 4.)

## The Number $e$

We begin by recalling that the natural logarithmic function ln is continuous and one-to-one and that its range is $(-\infty, \infty)$. Therefore, by the Intermediate Value Theorem there must be a unique real number $x_0$ such that $\ln x_0 = 1$. Let's denote $x_0$ by $e$. In view of the definition of ln, the number $e$ can be defined as follows.

> **DEFINITION    The Number $e$**
>
> The number $e$ is the number such that
>
> $$\ln e = \int_1^e \frac{1}{t} \, dt = 1$$

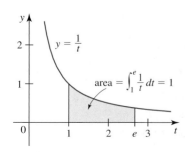

**FIGURE 5**
The area of the region under the graph of $f(t) = 1/t$ on $[1, e]$ is 1.

Figure 5 gives a geometric interpretation of the number $e$. It can be shown that the number $e$ is irrational and has the following approximation:

$$e \approx 2.718281828$$

You can verify this using a graphing calculator. Plot the graphs of the functions $y_1 = \ln x$ and $y_2 = 1$. Then use the function for finding the intersection of two curves to estimate the $x$-coordinate of the point of intersection.

## ■ Defining the Natural Exponential Function

Using Law (c) of logarithms, we see that if $r$ is a *rational* number, then

$$\ln e^r = r \ln e = r(1) = r$$

Equivalently, $e^r = y$ if and only if $\ln y = r$. The equation $\ln e^r = r$ can be used to motivate the definition of $e^x$ for every *real* number $x$.

---

**DEFINITION**   $e^x$

If $x$ is any real number, then

$$e^x = y \qquad \text{if and only if} \qquad \ln y = x$$

---

Now, by definition of the natural exponential function we have

$$\exp(x) = y \qquad \text{if and only if} \qquad \ln y = x$$

Comparing this definition with the definition of $e^x$ gives the following rule for defining the natural exponential function.

---

**DEFINITION**   **The Natural Exponential Function**

The natural exponential function, exp, is defined by the rule

$$\exp(x) = e^x$$

---

In view of this, we have the following theorem, which gives us another way of expressing the fact that exp and ln are inverse functions.

---

**THEOREM 1**

**a.** $\ln e^x = x$,   for $x \in (-\infty, \infty)$     **b.** $e^{\ln x} = x$,   for $x \in (0, \infty)$

---

The graph of the natural exponential function $y = e^x$ was sketched earlier (Figure 4). We summarize the important properties of this function.

---

**Properties of the Natural Exponential Function**

**1.** The domain of $f(x) = e^x$ is $(-\infty, \infty)$, and its range is $(0, \infty)$.
**2.** The function $f(x) = e^x$ is continuous and increasing on $(-\infty, \infty)$.
**3.** The graph of $f(x) = e^x$ is concave upward on $(-\infty, \infty)$.
**4.** $\lim\limits_{x \to -\infty} e^x = 0$ and $\lim\limits_{x \to \infty} e^x = \infty$.

## ■ The Laws of Exponents

The following laws of exponents are useful when working with exponential functions.

---

**THEOREM 2   Laws of Exponents**

Let $x$ and $y$ be real numbers and $r$ be a rational number. Then

**a.** $e^x e^y = e^{x+y}$ **b.** $\dfrac{e^x}{e^y} = e^{x-y}$ **c.** $(e^x)^r = e^{rx}$

---

**PROOF**   We will prove Law (a). The proofs of the other two laws are similar and will be omitted. We have

$$\ln(e^x e^y) = \ln e^x + \ln e^y = x + y = \ln e^{x+y}$$

Since the natural logarithmic function is one-to-one, we see that

$$e^x e^y = e^{x+y}$$ ■

## ■ The Derivatives of Exponential Functions

Since the inverse function of a differentiable function is itself differentiable, we see that the natural exponential function is differentiable. In fact, as the following theorem shows, the natural exponential function is its own derivative!

---

**THEOREM 3   The Derivatives of Exponential Functions**

Let $u$ be a differentiable function of $x$. Then

**a.** $\dfrac{d}{dx} e^x = e^x$ **b.** $\dfrac{d}{dx} e^u = e^u \dfrac{du}{dx}$

---

**PROOF**

**a.** Let $y = e^x$, so that $\ln y = x$. Differentiating both sides of the last equation implicitly with respect to $x$ gives

$$\frac{1}{y} \frac{dy}{dx} = 1 \qquad \text{or} \qquad \frac{dy}{dx} = y = e^x$$

**b.** This follows from part (a) by using the Chain Rule. ■

## ■ Logarithmic Functions with Base $a$

If $a$ is a positive real number with $a \neq 1$, then the function $f$ defined by $f(x) = a^x$ is one-to-one on $(-\infty, \infty)$, and its range is $(0, \infty)$. Therefore, it has an inverse on $(0, \infty)$. This function is called the logarithmic function with base $a$ and is denoted by $\log_a$.

---

**DEFINITION   Logarithmic Function with Base $a$**

**The logarithmic function with base $a$,** denoted by $\log_a$, is the function satisfying the relationship

$$y = \log_a x \qquad \text{if and only if} \qquad x = a^y$$

Observe that if $a = e$, then this definition reduces to the relationship between the natural logarithmic function ln and the natural exponential function exp.

## ■ The Definition of the Number $e$ as a Limit

If we use the definition of the derivative as a limit to compute $f'(1)$, where $f(x) = \ln x$, we obtain

$$f'(1) = \lim_{h \to 0} \frac{f(1 + h) - f(1)}{h}$$

$$= \lim_{h \to 0} \frac{\ln(1 + h) - \ln 1}{h} = \lim_{h \to 0} \frac{\ln(1 + h)}{h} \qquad \ln 1 = 0$$

$$= \lim_{h \to 0} \ln(1 + h)^{1/h}$$

$$= \ln\left[\lim_{h \to 0}(1 + h)^{1/h}\right] \qquad \text{Use the continuity of ln.}$$

But

$$f'(1) = \left[\frac{d}{dx} \ln x\right]_{x=1} = \left[\frac{1}{x}\right]_{x=1} = 1$$

so

$$\ln\left[\lim_{h \to 0}(1 + h)^{1/h}\right] = 1$$

or

$$\lim_{h \to 0}(1 + h)^{1/h} = e$$

We now prove Equation (4). First, using Law (d) of logarithms with $x = 2$, we have $\ln 2^r = r \ln 2$, where $r$ is any rational number. If we pick $r = n$, where $n$ is a positive integer, then

$$\lim_{n \to \infty} \ln 2^n = \lim_{n \to \infty} n(\ln 2) = \infty$$

since $\ln 2 > 0$. But, as we established earlier, $\ln x$ is an increasing function, and this allows us to conclude that

$$\lim_{x \to \infty} \ln x = \infty$$

Next, let's put $u = 1/x$. Then $u \to \infty$ as $x \to 0^+$. Therefore,

$$\lim_{x \to 0^+} \ln x = \lim_{u \to \infty} \ln \frac{1}{u} = \lim_{u \to \infty}(-\ln u) = -\infty$$

and (4) is proved.

## CHAPTER 0

### 0.1 Self-Check Diagnostic Test • page 2

**1.** $y = -\frac{7}{3}x + \frac{2}{3}$   **2.** $y = \frac{3}{2}x - \frac{13}{2}$   **3.** No, they do not.

**4.** $y = -\frac{3}{2}x + 6$   **5.** $y = -\frac{3}{4}x - \frac{7}{4}$

### Exercises 0.1 • page 13

**1.** 6   **3.** $-1.1$

**5. a.** $m_1 > 0, m_2 > 0, m_3 = 0, m_4 < 0$   **b.** $L_4, L_3, L_2, L_1$

**7.** $-22$   **9.** 1   **11.** $\sqrt{3}/3$   **13.** $\sqrt{3}$   **15.** $135°$

**17.** $60°$   **19.** $150°$

**21.** 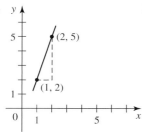   **23.**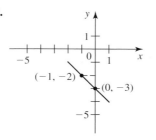

**25.** Parallel   **27.** Perpendicular   **29.** $-6$   **31.** 5

**35.** No   **37.** $y = \frac{3}{4}x + 2, m = \frac{3}{4}, b = 2$

**39.** $y = -\frac{A}{B}x + \frac{C}{B}, m = -\frac{A}{B}, b = \frac{C}{B}$

**41.** $y = \frac{\sqrt{6}}{3}x - \frac{4\sqrt{3}}{3}, m = \frac{\sqrt{6}}{3}, b = -\frac{4\sqrt{3}}{3}$

**43.** $60°$   **45.** $135°$   **47.** $y = 2x - 11$

**49.** $y = 4x - 4$   **51.** $x = 2$   **53.** $y = 3x - 5$

**55.** $y = -\frac{2}{3}x - 3$   **57.** $y = -\frac{1}{3}x - \frac{14}{3}$

**59.** $y = \frac{\sqrt{3}}{3}x + \left(3 - \frac{2\sqrt{3}}{3}\right)$   **61.** Perpendicular

**63.** Perpendicular   **65.** $(2, 1)$

**67. a.** $(x - 2)^2 + (y + 3)^2 = 25$   **b.** $x^2 + y^2 = 13$

　　**c.** $(x - 2)^2 + (y + 3)^2 = 34$

　　**d.** $(x + a)^2 + (y - a)^2 = 4a^2$

**71.** $\frac{x}{-4} + \frac{y}{-1} = 1$   **73.** $2\sqrt{5}$   **75.** $\arctan\left(\frac{1}{6}\right) \approx 9.5°$

**77. a.** $y = \frac{12}{7}x$   **b.** $y = -\frac{12}{7}x + 6$

　　**c.** $y = 1.2x - 4.2$   **d.** $y = -\frac{6}{7}x + \frac{57}{7}$

**79. a.**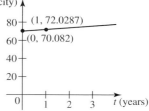

**b.** $1.9467, 70.082$   **d.** 2005

**81. a.** 88.8 tons

**b.**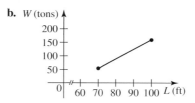

**85.** True   **87.** True

### 0.2 Self-Check Diagnostic Test • page 16

**1.** $2; 1; 4$   **2.** $2x + h + 2$   **3.** $\left[-\frac{1}{2}, 1\right) \cup (1, \infty)$

**4.** $(-\infty, \infty); [-1, \infty)$

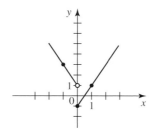

**5.** Odd

### Exercises 0.2 • page 23

*Abbreviations: D, domain; R, range.*

**1.** $f(0) = 4, f(-4) = -8, f(a) = 3a + 4, f(-a) = -3a + 4,$
　　$f(a + 1) = 3a + 7, f(2a) = 6a + 4, f(\sqrt{a}) = 3\sqrt{a} + 4,$
　　$f(x + 1) = 3x + 7$

**3.** $g(-2) = -8, g(\sqrt{3}) = -3 + 2\sqrt{3}, g(a^2) = -a^4 + 2a^2,$
　　$g(a + h) = -a^2 - 2ah - h^2 + 2a + 2h, 1/(g(3)) = -\frac{1}{3}$

**5.** $f(-1) = -1, f(0) = 0, f(x^2) = 2x^6 - x^2,$
　　$f(\sqrt{x}) = 2x^{3/2} - \sqrt{x}, f\left(\frac{1}{x}\right) = \frac{2}{x^3} - \frac{1}{x}$

**7.** $f(-2) = 5, f(0) = 1, f(1) = 1$     **9.** $x + 1$

**11.** $1 - 2x - h$     **13. a.** 4  **b.** $-2$

**15.** $(-\infty, 0) \cup (0, \infty)$     **17.** $\left(-\infty, -\frac{1}{2}\right) \cup \left(-\frac{1}{2}, 1\right) \cup (1, \infty)$

**19.** $[-3, 3]$     **21.** $[2, 4]$

**23.** $[-2, -1) \cup (-1, 0) \cup (0, 1) \cup (1, 2]$

**25.** $(-\infty, -1) \cup (1, \infty)$

**27. a.** $-2$  **b. (i)** 2  **(ii)** 1  **c.** $[0, 6]$  **d.** $[-2, 6]$     **29.** Yes

**31.** $D: (-\infty, \infty), R: (-\infty, \infty)$     **33.** $D: [1, \infty), R: [0, \infty)$

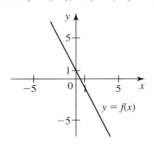 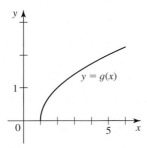

**35.** $D: (-\infty, -1] \cup [1, \infty), R: [0, \infty)$

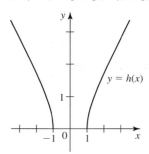

**37.** $D: (-\infty, \infty), R: [0, \infty)$

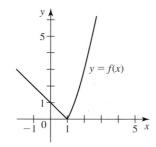

**39.** No     **41.** No     **43.** Yes     **45.** Even     **47.** Odd

**49.** Even     **51.** Neither     **53.** Even

**55. a.** 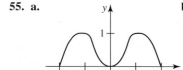     **b.**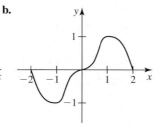

**57.** They stopped from 9:00 to 9:15 A.M. and from 12:00 to 1:00 P.M. They traveled at constant rates between stops.

**59.**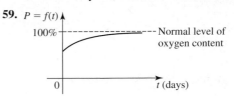

**61.** 4.9623%, 70.0923%

**63. a.** \$82.95   **b.** \$85.65, \$90.25, \$96.75

**c.**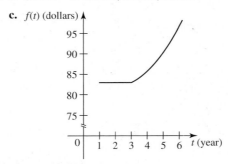

**65. a.** 130 tons, 100 tons, 40 tons

**b.**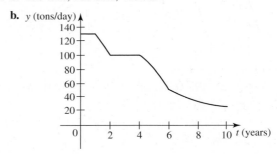

**69.** True     **71.** False

## 0.3 Self-Check Diagnostic Test • page 27

**1.** $\frac{4}{3}$     **2.** Odd     **4.** $a^2 \cos^2 \theta$     **5.** $\theta = \pi/3, \pi$, and $5\pi/3$

## Exercises 0.3 • page 37

*Abbreviations: D, domain; R, range; A, amplitude; P, period.*

**1.** $5\pi/6$     **3.** $11\pi/6$     **5.** $-2\pi/3$     **7.** $-5\pi/12$

**9.** $60°$     **11.** $150°$     **13.** $-90°$     **15.** $-585°$

**17.** $\sqrt{3}/2, \frac{1}{2}, \sqrt{3}$     **19.** $-\frac{1}{2}, -\sqrt{3}, -2$

**21.** $0, 0$, undefined     **23.** $2, -2\sqrt{3}/3, -\sqrt{3}$

**25.** $\cos \theta = -\frac{4}{5}, \tan \theta = -\frac{3}{4}, \cot \theta = -\frac{4}{3}, \sec \theta = -\frac{5}{4}, \csc \theta = \frac{5}{3}$

**27.** $0, \sqrt{2}/2, -\sqrt{3}/2, 0, \cos a$

**29.** $\left\{\frac{\pi}{2} + 2n\pi \,|\, n \text{ is an integer}\right\}$

**31. a.** Odd  **b.** Odd  **c.** Odd

**43.** $D = (-\infty, \infty)$, $R = [-2, 2]$

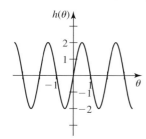

**45.** $A = 1$, $P = 2\pi$         **47.** $A = 1$, $P = 2\pi$

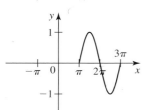    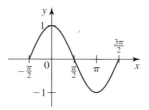

**49.** $A = 1$, $P = 2\pi$         **51.** $A = 2$, $P = \pi$

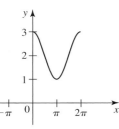    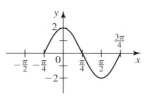

**53.** $A = 1$, $P = \pi$         **55.** $A = 2$, $P = 2\pi/3$

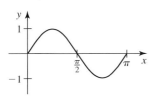    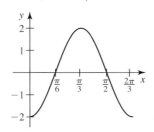

**57.** $A = 3$, $P = \pi$

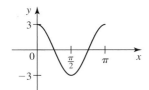

**59.** $x = \pi/4, 5\pi/4$     **61.** $t = \pi$

**63.** $x = \pi/4, \pi/2, 5\pi/4, 3\pi/2$     **65.** $x = 0, \pi/3, 5\pi/3$

**67.** About 2.4 min

**69.** True

**71.** True

**73.** False

## 0.4 Self-Check Diagnostic Test • page 39

**1.** $2x + \dfrac{1}{x+1}$, $(-\infty, -1) \cup (-1, \infty)$; $2x - \dfrac{1}{x+1}$,

$(-\infty, -1) \cup (-1, \infty)$; $\dfrac{2x}{x+1}$, $(-\infty, -1) \cup (-1, \infty)$;

$2x^2 + 2x$, $(-\infty, -1) \cup (-1, \infty)$

**2.** $\dfrac{\sqrt{x+1}}{\sqrt{x+1}+1}$; $[-1, \infty)$     **3.** $f(x) = 3x^2 + 1$; $g(x) = \dfrac{10}{\sqrt{x}}$

**4.** $g(x) = 3\sqrt{x-2} - 5$     **5.** $f(x) = 2x^2 + x - 1$

## Exercises 0.4 • page 48

*Abbreviations: D, domain; R, range.*

**1. a.** $x^2 + 3x - 1$, $D = (-\infty, \infty)$

  **b.** $-x^2 + 3x + 1$, $D = (-\infty, \infty)$

  **c.** $3x^3 - 3x$, $D = (-\infty, \infty)$

  **d.** $\dfrac{3x}{x^2 - 1}$, $D = (-\infty, -1) \cup (-1, 1) \cup (1, \infty)$

**3. a.** $\sqrt{x+1} + \sqrt{x-1}$, $D = [1, \infty)$

  **b.** $\sqrt{x+1} - \sqrt{x-1}$, $D = [1, \infty)$

  **c.** $\sqrt{x^2 - 1}$, $D = [1, \infty)$   **d.** $\dfrac{\sqrt{x+1}}{\sqrt{x-1}}$, $D = (1, \infty)$

**5.** $4x^2 + 12x + 9$, $D = (-\infty, \infty)$; $2x^2 + 3$, $D = (-\infty, \infty)$

**7.** $\dfrac{x-1}{x+1}$, $D = (-\infty, -1) \cup (-1, 1) \cup (1, \infty)$; $\dfrac{1+x}{1-x}$,

$D = (-\infty, 0) \cup (0, 1) \cup (1, \infty)$

**9.** 10

**11. a.** $g \circ f = \begin{cases} x^2 + 2x + 1 & \text{if } x < 0 \\ x^2 - 2x + 1 & \text{if } x \ge 0 \end{cases}$

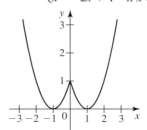

  **b.** $f \circ g = \begin{cases} x^2 + 1 & \text{if } x < 0 \\ x^2 - 1 & \text{if } x \ge 0 \end{cases}$

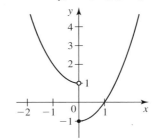

**13. a.** $8 + \sqrt{2}$  **b.** $34$  **c.** $36$  **d.** $18 + \sqrt{3}$

**15.** $\sqrt{2x^2 - 1}$     **17.** $f(x) = 3x^2 + 4$, $g(x) = x^{3/2}$

**19.** $f(x) = \sqrt{x^2 - 4}$, $g(x) = 1/x$

**21.** $f(t) = t^2$, $g(t) = \sin t$

**23. a.** $f(x) = \sqrt{x}$, $g(x) = 1 - x$, $h(x) = \sqrt{x}$

  **b.** $f(x) = x^3$, $g(x) = \sin x$, $h(x) = 2x + 3$

**25. a.** $4$  **b.** $5$  **c.** $1$  **d.** $3$  **e.** $2$  **f.** $6$

**27.** Curve 1: $y = f(x) + 1$; curve 2: $y = f(x) - 1$

**29.** Curve 1: $y = f\left(\frac{x}{2}\right)$; curve 2: $y = f(2x)$

**31.** $g(x) = x^3 + x + 2$     **33.** $g(x) = x + 3 + \dfrac{1}{\sqrt{x + 3}}$

**35.** $g(x) = \dfrac{3\sqrt{x}}{x^2 + 1}$     **37.** $g(x) = \dfrac{x}{2} \sin \dfrac{x}{2}$

**39.** $g(x) = \sqrt{8x - 4x^2} + 1$

**41. a.**

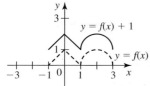

**b.**

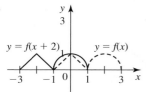

**c.**     **d.**

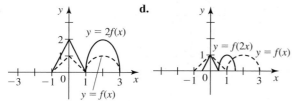

**e.**     **f.**

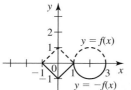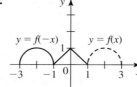

**g.**

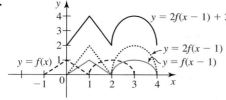

**h.**

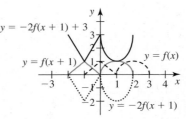

**43.**

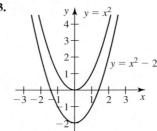

**45.**

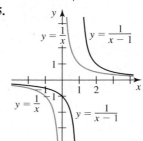

**47.**

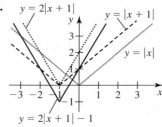

**49.**

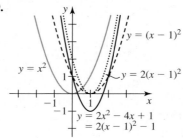

**51.**

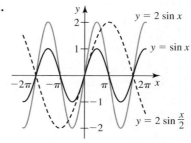

**53.**

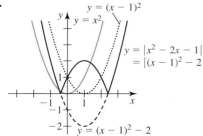

**55. a.** Graph $f(x)$ for $x \geq 0$ and $f(-x)$ for $x < 0$.

**b.**

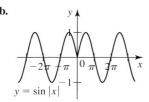

**57. a.** $2x + 5$   **b.** $\frac{4}{3}x - 4$   **59.** $3x - 1$

**61. a.** $\begin{cases} \sqrt{-x} - \sin x & \text{if } -2\pi \leq x < 0 \\ \sqrt{x} + \sin x & \text{if } 0 \leq x \leq 2\pi \end{cases}$

   **b.** $\begin{cases} -\sqrt{-x} + \sin x & \text{if } -2\pi \leq x < 0 \\ \sqrt{x} + \sin x & \text{if } 0 \leq x \leq 2\pi \end{cases}$

**63. a.** $D(t) = 0.33t^2 + 1.1t + 16.9$, $D(4) = 26.58$

   **b.** $P(t) = \dfrac{1.21t^2 + 6t + 14.5}{1.54t^2 + 7.1t + 31.4}$, $P(4) \approx 0.69$

**65. a.** $(g \circ f)(0) = 26$   **b.** $(g \circ f)(6) = 42$

**67. a.** $3.5t^2 + 2.4t + 71.2$   **b.** 71,200; 109,900

**69.** True   **71.** True   **73.** True

## Exercises 0.5 • page 56

**1. a.**

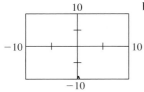

**b.**

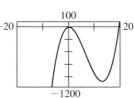

**3. a.**

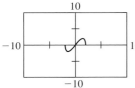

**b.**

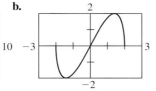

**5.**

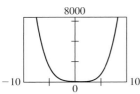

**7.**

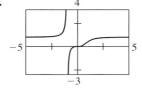

**9.**

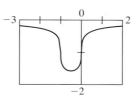

**11.**

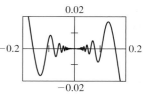

**13.**

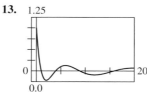

**15.**

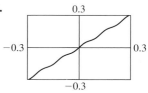

**17.** $-1.47569$   **19.** $-1.11769, 0.35855$

**21.** $-0.45662, 1.25873$

**23.** $(-2.33712, 2.41174), (6.05141, -2.50154)$

**25.** $(-1.02193, -6.34606), (1.2414, -1.59306),$
   $(5.78053, 7.93912)$

**27.** $(-2.51746, -1.16879), (0.91325, 1.58299)$

**29. a.**

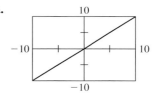

   **b.**

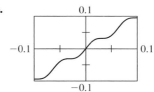

**31. a.**

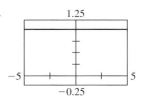

   **b.** No; $f$ is not defined at $x = 0$.

**33. a.**

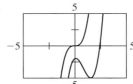

   **b.**

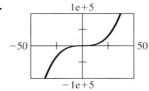

## 0.6 Self-Check Diagnostic Test • page 57

**1. a.** $f(x) = 3x - 1$   **b.** $f(x) = 2x^4 - 3x^2 + 7$

   **c.** $f(x) = \dfrac{x^2}{x^2 - 9}$   **d.** $f(x) = x^{5/7}$

   **e.** $f(x) = \sqrt{x - 1}$   **f.** $f(x) = 2 \sin x$

**2. a.** $V(0) = C$; the initial book value of the asset

   **b.** $\dfrac{C - S}{n}$ dollars/year

**3.** $V(x) = 4x^3 - 48x^2 + 144x$

## Exercises 0.6 • page 69

**1. a.** Polynomial, 3   **b.** Power   **c.** Rational

   **d.** Rational   **e.** Algebraic   **f.** Trigonometric

**3. a.** 59.7 million   **b.** 152.54 million

**5.** 648,000; 902,000; 1,345,200; 1,762,800

**7.** 4.6%, 8.51%, 15.91%

**11. a.** $\dfrac{100x}{7960 + x}$   **b.** 74.8%

**13. a.** 
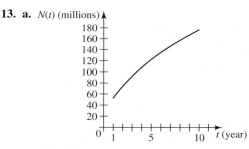

   **b.** 157 million

**15. a.** $A(t) = 0.010716t^2 + 0.8212t + 313.4$

   **b.** 
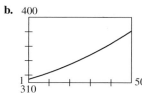

   **c.** 338 ppmv   **d.** 387 ppmv

**17. a.** $f(t) = -0.425t^3 + 3.65714t^2 + 4.01786t + 43.6643$

   **b.** 
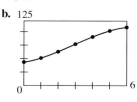

   **c.** $43.66 million; $77.16 million; $107.63 million

**19. a.** $f(t) = 0.00125t^4 - 0.005093t^3 - 0.024306t^2 + 0.128624t + 1.70992$

   **b.** 
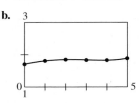

   **c.** 1.71, 1.81, 1.85, 1.84, 1.83, 1.89

**21.** $f(x) = 40x - x^2$, (0, 40)

**23.** $V = 4x^3 - 46x^2 + 120x$; (0, 4)

**25.** $f(x) = 28x - \left(\frac{\pi}{2} + 2\right)x^2$; $\left(0, \frac{56}{4 + \pi}\right)$

**27.** $f(x) = -2x + 52 - \dfrac{50}{x}$, (1, 25)

**29. a.** $R(x) = -4x^2 + 520x + 12,000$

   **b.** $26,400   **c.** $28,800

## 0.7 Self-Check Diagnostic Test • page 73

**1.** No   **2.** 1

**3.** $\dfrac{x + 2}{3}$

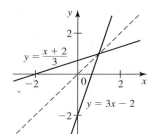

**4. a.** $-\dfrac{\pi}{4}$   **b.** $\dfrac{5\pi}{6}$   **5.** $\dfrac{x}{\sqrt{1 - x^2}}$

## Exercises 0.7 • page 82

**7.** Yes   **9.** No   **11.** Yes   **13.** Yes

**15.** 2   **17.** 0   **19.** $\pi/6$

**21.**

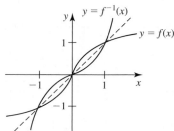

**23.** $\sqrt[3]{x-1}$

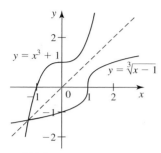

**25.** $\sqrt{9-x^2}$, $x \geq 0$

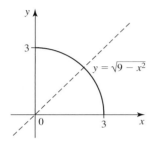

**27.** $\frac{1}{2}(1 + \sin^{-1} x)$

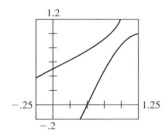

**29.** $x^3 + 1$

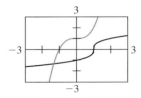

**31.** $\dfrac{1 - \sqrt{1 - 4x^2}}{2x}$, $-\frac{2}{5} \leq x \leq \frac{2}{5}$

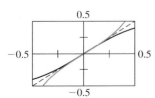

**33.** $\begin{cases} \dfrac{x + 1}{2} & \text{if } x < 1 \\ x^2 & \text{if } 1 \leq x < 2 \\ \sqrt{2x + 12} & \text{if } x \geq 2 \end{cases}$   Domain: $(-\infty, \infty)$

**35.** 0   **37.** $\pi/6$   **39.** $\pi/3$   **41.** $\pi/3$   **43.** $\pi/3$
**45.** $-\pi/6$   **47.** $\sqrt{2}/2$   **49.** $\sqrt{1 - x^2}$   **51.** $x$

**53.** $\dfrac{1}{\sqrt{x^2 - 1}}$

**55. a.** $f^{-1}(F) = \frac{5}{9}(F - 32)$   **b.** $[-459.67, \infty)$

**57. a.** $f^{-1}(p) = \dfrac{10}{9}\left[\left(\dfrac{p}{10.72}\right)^{10/3} - 10\right]$   **b.** 7.58

**61.** True   **63.** True   **65.** False

## 0.8 Self-Check Diagnostic Test • page 84

**1.** $3x$   **2.** 3.6620   **3.** $\left(0, \dfrac{1}{e}\right) \cup \left(\dfrac{1}{e}, \infty\right)$

**4.** $e^{x/2} - 1$, $(-\infty, \infty)$   **5.** $\ln \dfrac{x\sqrt{x + 1}}{\cos x}$

## Exercises 0.8 • page 93

**1. a.** 1.7917   **b.** 0.4055   **3. a.** 3.4011   **b.** 2.0149
**5.** $\ln 2 + \frac{1}{2} \ln 3 - \ln 5$   **7.** $\frac{1}{3} \ln x + \frac{2}{3} \ln y - \frac{1}{2} \ln z$
**9.** $\frac{1}{3} \ln(x + 1) - \frac{1}{3} \ln(x - 1)$   **11.** $\ln 2$

**13.** $\ln \dfrac{8}{\sqrt{x + 1}}$   **15. a.** 3   **b.** $x^2$   **17. a.** 9   **b.** $\sqrt{x}$

**19.** $(-\infty, \infty)$   **21.** $[0, \infty)$
**23.** $\left(-\frac{1}{2}, \infty\right)$   **25.** $\left(2k\pi - \frac{\pi}{2}, 2k\pi + \frac{\pi}{2}\right)$, $k = 0, \pm 1, \pm 2, \ldots$
**27. a.** $x = 2$   **b.** $x = -\frac{3}{2}$
**29. a.** $x = \ln \frac{5}{2} - 2$   **b.** $x = e^2 - 1$
**31. a.** $x = 5 \ln \frac{3}{8}$   **b.** $x = \ln 2$ or $x = \ln 3$
**33.** Odd   **35.** Odd
**37.**

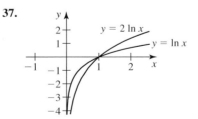

**39.**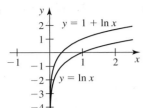

**41.**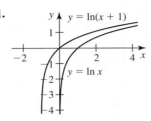

**43. a.**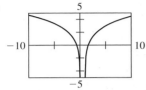

**b.** $x > 1$

The graph of $f$ is the right branch and the graph of $g$ consists of both branches.

**45.**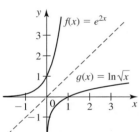

**47.**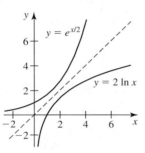

**49.** $\ln(x - 1)$

**51.** $\ln\left(\dfrac{x + 1}{x - 1}\right)$

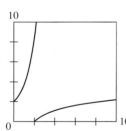

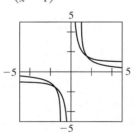

**53. a.**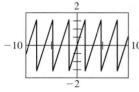

**b.** Yes

**55.** No    **57.** 0.83% per decade

**59.** 0.094 percent; 0.075 percent

**61. a.** 180.7 per decade    **b.** 52

**63. a.**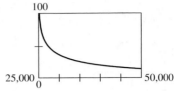

**b.** \$35,038.78 per year

**65.** $0.4732p_0$; mountaineers experience difficulty in breathing at very high altitudes because the air is more rarefied owing to a decrease in atmospheric pressure.

**67.** 153,024; 235,181

**69. a.** 50   **b.** 70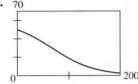

**71. a.** \$8052.55   **b.** \$8144.47   **c.** \$8193.08

**d.** \$8226.54   **e.** \$8243.04

**73.** \$2.58 million   **75.** True   **77.** True   **79.** False

## Chapter 0 Review Exercises • page 96

**1.** $-\frac{7}{3}$   **3.** $-2$   **5.** $y = -4$   **7.** $y = -\frac{4}{3}x + \frac{1}{3}$

**9.** $y = -\frac{3}{5}x + \frac{12}{5}$   **11.** $y = -1$

**13. a–b.**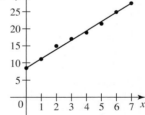

**c.** $y = 2.7x + 8.5$   **d.** 30.1 million

**15.** $f(0) = 0, f\left(\frac{\pi}{6}\right) = \sqrt{3}/3, f\left(\frac{\pi}{4}\right) = 1, f\left(\frac{\pi}{3}\right) = \sqrt{3}, f(\pi) = 0$

**17.** $(-\infty, -2) \cup (-2, 2) \cup (2, \infty)$

**19.** $[1, 2) \cup (2, \infty)$   **21.** $(-\infty, \infty)$   **23.** $(0, \infty)$

**25.** Domain $(-\infty, \infty)$, range $[1, 2]$

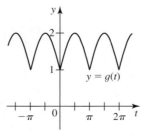

**27.** Even    **29.** Even    **31. a.** $330°$    **b.** $-450°$    **c.** $-315°$

**33. a.** $\theta = \pi/3, 5\pi/3$    **b.** $\theta = 5\pi/6, 11\pi/6$

**35. a.** $\pi/4, \pi/2, 5\pi/4, 3\pi/2$    **b.** $0, 2\pi/3, \pi, 4\pi/3$

**37.** $|x|, (-\infty, \infty)$

**39.** $f(x) = x^2, g(x) = \cos x, h(x) = 1 + \sqrt{x+2}$

**41.** $e^{2/5}$    **43.** 9    **45.** $(\ln 4)^2$    **47.** $\ln \frac{3}{4}$    **49.** $1 + \sqrt{2}$

**51.** $1, 2$    **53.** $\tan 1$    **55.** $\frac{1}{2}\ln(y-2)$

**57.** $3 \ln x + \frac{1}{2} \ln y - \ln z$    **59.** $\ln \dfrac{x^5}{y^2(x+y)^2}$

**61.**

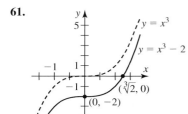

**63.**

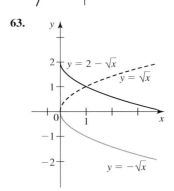

**65.**

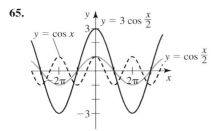

**67.**

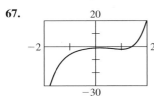

**69.** $-2.18271, 0.59237, 1.89858$

**71.** $-1.08659, -1, 1.26456$    **73.** $116\frac{2}{3}$ mg

**75.** $f(4) = 0, f(5) = 20, f(6) \approx 34, f(7) \approx 48, f(8) \approx 64,$
$f(9) \approx 80, f(10) \approx 98, f(11) \approx 116, f(12) \approx 136$

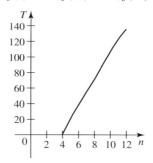

**77. a.** $f(r) = \pi r^2$    **b.** $g(t) = 2t$

   **c.** $h(t) = f[g(t)] = \pi(2t)^2 = 4\pi t^2$    **d.** $3600\pi$ ft$^2$

**79. a.**

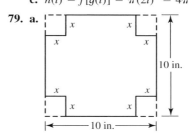

   **b.** $V(x) = 4x^3 - 40x^2 + 100x$    **c.** 64 in.$^3$

**81. a.**

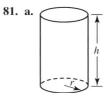

   **b.** $C(r) = 16\pi r^2 + \dfrac{256\pi}{r}$    **c.** \$603.19    **83.** \$337,653

# CHAPTER 1

## Exercises 1.1 • page 109

**1. a.** 2    **b.** $-1$    **c.** Does not exist    **3. a.** 2    **b.** 2    **c.** 2

**5. a.** $\infty$    **b.** 1    **c.** Does not exist

**7. a.** True    **b.** True    **c.** False    **d.** True    **e.** True    **f.** False

**9.**

| $x$ | $f(x)$ |
|---|---|
| 0.9 | $-0.90909$ |
| 0.99 | $-0.99010$ |
| 0.999 | $-0.99900$ |
| 1.001 | $-1.00100$ |
| 1.01 | $-1.01010$ |
| 1.1 | $-1.11111$ |

$-1$

**11.**

| $x$ | $f(x)$ |
|---|---|
| 1.9 | 0.25158 |
| 1.99 | 0.25016 |
| 1.999 | 0.25002 |
| 2.001 | 0.24998 |
| 2.01 | 0.24984 |
| 2.1 | 0.24846 |

$\frac{1}{4}$

**13.**

| $x$ | $f(x)$ |
|-----|--------|
| 1.9 | $-0.06370$ |
| 1.99 | $-0.06262$ |
| 1.999 | $-0.06251$ |
| 2.001 | $-0.06249$ |
| 2.01 | $-0.06238$ |
| 2.1 | $-0.06135$ |

$-0.0625$

**15.**

| $x$ | $f(x)$ |
|-----|--------|
| $-0.1$ | 0.95163 |
| $-0.01$ | 0.99502 |
| $-0.001$ | 0.99950 |
| 0.001 | 1.00050 |
| 0.01 | 1.00502 |
| 0.1 | 1.05171 |

1

**17.**

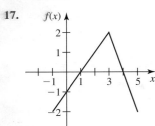

**a.** 2 **b.** 2 **c.** 2

**19.**

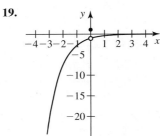

**a.** $-1$ **b.** $-1$ **c.** $-1$

**21.**

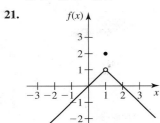

**a.** 1 **b.** 1 **c.** 1

**23.** 2 **25.** $-1$ **27.** 3

**29.**

| $x$ | $f(x)$ |
|-----|--------|
| $-0.01$ | 0 |
| $-0.001$ | 0 |
| $-0.0001$ | 0 |
| 0 | 0 |
| 0.0001 | $-0.3056$ |
| 0.001 | 0.8269 |
| 0.01 | $-0.5064$ |

Does not exist; 0;
does not exist

**35.** 2.7183

**37.**

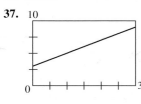

7

**31. a.**

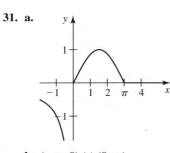

**b.** $(-\infty, 0) \cup (0, \infty)$
**c.** $(-\infty, 0) \cup (0, \infty)$
**d.** $(-\infty, \infty)$

**39.**

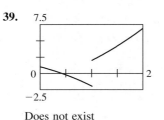

Does not exist

**41.**

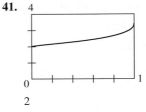

**43.** False **45.** False

## Exercises 1.2 • page 123

**1.** 10 **3.** 0 **5.** 1 **7.** $-\frac{1}{3}$ **9.** $4 - 2\sqrt{2}$ **11.** 4
**13.** $\frac{1}{2}$ **15.** $-\frac{2}{3}$ **17.** 1 **19.** $2\sqrt{2}/\pi$ **21.** 1
**23.** 16 **25.** 1 **27.** $-7$ **29.** 11 **31.** 2 **33.** 1
**35.** 0 **37.** Incorrect **39.** $f(x) = 1, g(x) = x, a = 0$
**41.** 4 **43.** $\infty$ **45.** 2 **47.** $\frac{\sqrt{6}}{2}$ **49.** $-\frac{3}{2}$ **51.** 3
**53.** $\frac{1}{2}$ **55.** Does not exist **57.** $\sqrt{3}/6$ **59.** $\frac{1}{2}$ **61.** 6
**63.** 1 **65.** $\frac{1}{3}$ **67.** $\sqrt{2}$ **69.** 0 **71.** $-\frac{1}{2}$
**73.** $-\sqrt{2}/2$ **75.** 1 **77.** $-1$
**79. a.** 12 **b.** 12

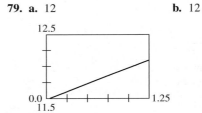

**c.** 12

| $x$ | $f(x)$ |
|-----|--------|
| 0.99 | 11.995 |
| 0.999 | 11.999 |
| 1 | undefined |
| 1.001 | 12.001 |
| 1.01 | 12.005 |

**81. a.** $[0, c)$ **b.** 0 **83.** 0
**85. a.** 1, 2 **b.** No, $\lim_{x \to -1^-} f(x) \neq \lim_{x \to -1^+} f(x)$
**87.** 1, 1, yes **89.** Yes, 1 **93.** No **95.** No
**99.** False **101.** False

## Exercises 1.3 • page 133

**1.** 0.003 **3.** 0.005 **5.** 0.02 **7.** 0.0007
**9.** 0.01 **31.** False **33.** True

## Exercises 1.4 • page 145

**1.** Nowhere **3.** At $\pm 1$ **5.** At 0 **7.** None
**9.** 2 **11.** $\pm 2$ **13.** 0, 2 **15.** $-2, 0$
**17.** $0, \pm 1, \pm 2, \ldots$ **19.** None **21.** $-1$ **23.** 0
**25.** $\pm \pi/4, \pm 3\pi/4, \pm 5\pi/4, \ldots$ **27.** 3
**29.** $a = \frac{8}{3}, b = \frac{4}{3}$ **31.** 2 **33.** Yes **35.** No
**37.** $(-\infty, \infty)$ **39.** $(-\infty, \infty)$ **41.** $[-3, 3]$
**43.** $(-3, 0)$ and $(0, 3)$ **45.** $[0, \infty)$ **47.** $(2, \infty)$

**49.** 5   **51.** $f(0) = -\frac{2}{5}$   **53.** $f(0) = \frac{1}{2}$   **55.** $f(0) = 1$

**57.** No   **59.** 3   **61.** 3

**69. b.** 0.57926   **71.** No   **73.** 1.34

**75.**

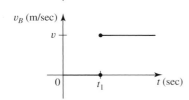

**77. c.** $\frac{1}{2}$ sec and $\frac{7}{2}$ sec

**83. a.** No   **b.** No

**89.**

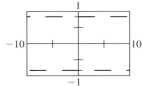

   $f$ is continuous at all numbers except $n\pi$, $n$ an integer

**97.** False   **99.** True

## Exercises 1.5 • page 155

**1.** $-0.15$ mph per thousand cars, $-0.3$ mph per thousand cars

**3.** Rising at 3.08%/hr; falling at 21.15%/hr

**5. a.** At $t_1$, the velocity of Car $A$ is greater, but the acceleration of Car $B$ is greater.
   **b.** At $t_2$, both cars have the same velocity, but Car $B$ has greater acceleration.

**7. a.** 0   **b.** 0   **c.** $y = 5$

**9. a.** $2h + 8$   **b.** 8   **c.** $y = 8x - 9$

**11. a.** $h^2 + 6h + 12$   **b.** 12   **c.** $y = 12x - 16$

**13. a.** $-\dfrac{1}{1 + h}$   **b.** $-1$   **c.** $y = -x + 2$

**15.** 4   **17.** 13   **19.** $-1$

**21. a.** 1.25 ft/sec, 1.125 ft/sec, 1.025 ft/sec, 1.0025 ft/sec, 1.00025 ft/sec
   **b.** 1 ft/sec

**23. a.** 48 ft/sec, 56 ft/sec, 62.4 ft/sec
   **b.** 64 ft/sec
   **c.** $-32$ ft/sec, falling
   **d.** $t = 8$ sec

**25. a.** 820 ft   **b.** 20.5 ft/sec   **c.** 40.5 ft/sec

**27. a.** $3\pi$ units$^2$/unit   **b.** $4\pi$ units$^2$/unit

**29. a.** $-\$2.005$ per thousand tents, $-\$2.001$ per thousand tents
   **b.** Decrease of $2 per thousand tents

**31.** 30.8 ft/sec

**33. a.**

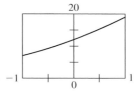

   **b.** 12

**35.** $f(x) = x^5$, $a = 1$   **37.** $f(x) = x^2 + \sqrt{x}$, $a = 4$

**39.** $f(x) = x^4$, $a = 1$   **41.** True   **43.** False

## Chapter 1 Concept Review • page 158

**1. a.** $L, f, L, a$   **b.** right   **c.** exist, $L$   **d.** $\varepsilon > 0, \delta > 0$

**3.** $\lim_{x \to a} g(x) = L$

**5. a.** $(-\infty, \infty)$   **b.** its domain   **c.** continuous

**7. a.** $m_{\tan} = \lim\limits_{h \to 0} \dfrac{f(a + h) - f(a)}{h}$   **b.** $y - f(a) = m_{\tan}(x - a)$

## Chapter 1 Review Exercises • page 159

**1. a.** 0   **b.** 0   **c.** 0

**3.**
   **a.** 2   **b.** 2   **c.** 2

**5.**
   **a.** 0   **b.** 0   **c.** 0

**7.** 34   **9.** $2\sqrt{3}$   **11.** 9   **13.** $\frac{1}{2}$   **15.** $\frac{7}{8}$   **17.** $-\frac{1}{16}$

**19.** 0   **21.** 6   **23.** $\infty$   **25.** 0   **27.** 1

**31.** At $b$ and $c$   **33. a.** Yes   **b.** Yes   **c.** No   **35.** None

**37.** None   **39.** $n\pi$, $n$ an integer   **41.** $x \le 0$ and 1

**43.** $\frac{1}{2}$   **49.** No   **53. a.** $[0, c)$   **b.** 0

**55. a.** 7 ft/sec; 6 ft/sec; 5.2 ft/sec; 5.02 ft/sec   **b.** 5 ft/sec

## Chapter 1 Challenge Problems • page 162

**1.** $\frac{1}{3}$   **3.** $\pi/4$   **7.** 2   **11.** At 0, 1, and $\frac{3}{2}$

# CHAPTER 2

## 2.1 Exercises • page 172

**1.** $0, (-\infty, \infty)$     **3.** $3, (-\infty, \infty)$     **5.** $6x - 1, (-\infty, \infty)$

**7.** $6x^2 + 1, (-\infty, \infty)$     **9.** $\dfrac{1}{2\sqrt{x+1}}, (-1, \infty)$

**11.** $-\dfrac{1}{(x+2)^2}, (-\infty, -2) \cup (-2, \infty)$

**13.** $-\dfrac{6}{(2x+1)^2}, \left(-\infty, -\tfrac{1}{2}\right) \cup \left(-\tfrac{1}{2}, \infty\right)$

**15.** $y = 4x - 3$     **17.** $y = 6x - 4$     **19.** $y = \dfrac{\sqrt{3}}{6}x + \dfrac{\sqrt{3}}{3}$

**21. a.** $y = -x + 2$

**b.**

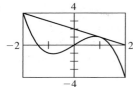

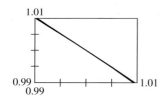

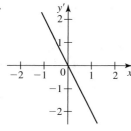

**23.** $-3$     **25.** $\tfrac{1}{2}$     **27.** (c)     **29.** (b)

**31.**

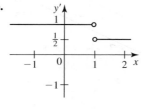

**33.**

**35.**

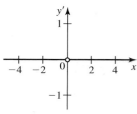

**37. a.** $f'(h)$ is measured in degrees Fahrenheit per foot and gives the instantaneous rate of change of the temperature at a given height $h$.

**b.** Negative

**c.** $-0.05°F$

**39. a.** $C'(x)$, measured in dollars per unit, gives the instantaneous rate of change of the total manufacturing cost $C$ when $x$ units of a certain product are produced.

**b.** Positive

**c.** $\$20$

**41. a.** $2x - 2$   **b.** $(1, 0)$

**c.**

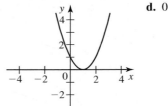

**d.** $0$

**43.** $0$     **45.** $\pm 2$     **47.** $\pm 2$

**53.** $y'$ ($\$$ million/year)

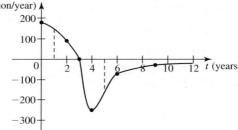

$\$150$ million per year, $-\$160$ million per year

**55. a.** The average rate of change of $f(x) = x^3$ over the interval $[x, x + h]$

**b.** $f'(x) = 3x^2$

**c.**

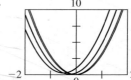

**57. a.**

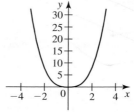

**b.** $x \in (-\infty, \infty)$

**c.** $f'(x) = \begin{cases} -3x^2 & \text{if } x < 0 \\ 3x^2 & \text{if } x \geq 0 \end{cases}$

**61. a.** 0   **b.**

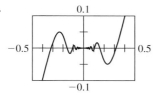

**65.** True    **67.** False    **69.** False

## 2.2 Exercises • page 183

**1.** 0    **3.** $6x$    **5.** $2.1x^{1.1}$    **7.** $\dfrac{3}{2\sqrt{x}} + 2e^x$    **9.** $-\dfrac{84}{x^{13}}$

**11.** $2x - 2$    **13.** $2\pi r + 2\pi$    **15.** $0.06x - 0.4$

**17.** $10x^4 - 12x^3 + 3x^2 + 8x$    **19.** $2x - 4 - \dfrac{3}{x^2}$

**21.** $16x^3 - \frac{15}{2}x^{3/2}$    **23.** $-\dfrac{3}{x^2} - \dfrac{8}{x^3}$

**25.** $-\dfrac{16}{t^5} + \dfrac{9}{t^4} - \dfrac{2}{t^2}$    **27.** $0.002x - 0.4 - \dfrac{200}{x^2}$

**29.** $2 - \dfrac{5}{2\sqrt{x}} + e^{x+1}$    **31.** $\dfrac{1}{3x^{2/3}} - \dfrac{1}{2x^{3/2}}$

**33. a.** 20   **b.** $-4$   **c.** 20

**35. a.** $y = 5x - 4$   **b.**

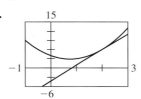

**37. a.** $y = -x + 1$   **b.**

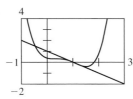

**39.** $(-2, 10)$ and $(1, -17)$    **41.** $\left(-\frac{1}{2}, -3\right)$ and $\left(\frac{1}{2}, 3\right)$

**43. a.** $y = 9x - 15,\ y = -\frac{1}{9}x + \frac{29}{9}$

**b.**

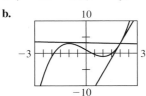

**45.** $\pm 3$

**47. a.** $\left(1, -\frac{13}{12}\right)$ and $(0, 0)$

    **b.** $(0, 0),\ \left(2, -\frac{8}{3}\right),$ and $\left(-1, -\frac{5}{12}\right)$

    **c.** $(0, 0),\ \left(4, \frac{80}{3}\right),$ and $\left(-3, \frac{81}{4}\right)$

**49.** $\left(-1, \frac{20}{3}\right)$ and $\left(1, \frac{10}{3}\right)$    **51.** $y = 2x - 4$ and $y = 6x - 16$

**53.** 3    **55.** 11    **57.** $2x(7x^5 - 1)$    **59.** 1 P.M.

**61. a.** 1.94% per decade, 2.48% per decade

    **b.** 3.87%, 6.08%

**63. a.** $3.6t^2 - 10.62t + 80$

    **b.** $-\$3.42$ per person per year, $\$32.58$ per person per year

**65. a.** 12

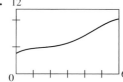

    **b.** 0.3835, 1.0489, 1.7311

    **c.** The number of Alzheimer's patients will be increasing at the rate of approximately 0.3835 million patients per decade at the beginning of 2010, 1.0489 million patients per decade in 2020, and 1.7311 million patients per decade in 2030.

**67. a.** $-0.0123t^4 + 0.2639t^3 - 2.2601t^2 + 5.6608t + 143.6$

    **b.** 150

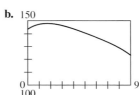

    **c.** 5.6608, $-0.6064$, $-6.7602$

**69.** $\dfrac{3\pi}{R}\sqrt{\dfrac{r}{g}}$    **71.** $A = \frac{1}{4},\ B = \frac{1}{2},\ C = \frac{1}{4}$

**75.** False    **77.** True

## 2.3 Exercises • page 193

**1.** $xe^x(x + 2)$    **3.** $\dfrac{(2t^2 + 7t + 2)e^t}{2\sqrt{t}}$    **5.** $2e^x(e^x + 1)$

**7.** $-\dfrac{1}{(x - 1)^2}$    **9.** $-\dfrac{7}{(3x - 2)^2}$    **11.** $\dfrac{1 - x^2 e^x}{(1 + xe^x)^2}$

**13.** $3x^2 + 2x - 1$    **15.** $2x + xe^x + e^x - 4e^{2x}$

**17.** $\dfrac{-3x^2 + 1}{\sqrt{x}(x^2 + 1)^2}$    **19.** $\dfrac{2x(x + 2)}{(x^2 + x + 1)^2}$

**21.** $\dfrac{(x^2 - x - 5)e^x}{(x - 2)^2}$    **23.** $-\dfrac{x^2 + 2x + 2}{x^2(x + 2)^2}$

**25.** $\dfrac{3\sqrt{3}x + 2\sqrt{x} + \sqrt{3}}{2\sqrt{x}(3x - 1)^2}$    **27.** $\dfrac{ad - bc}{(cx + d)^2}$

**29.** $\dfrac{1 + e^x + x^2 e^x + e^{2x}}{(1 - xe^x)^2}$    **31.** 10

**33.** 29    **35.** $\left(1, \dfrac{1}{e}\right)$    **37.** $(1, e+1)$

**39. a.** $y = \frac{1}{4}ex + \frac{1}{4}e$

**b.**

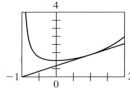

**41. a.** $y = -\frac{13}{3}x - \frac{14}{3}$

**b.**

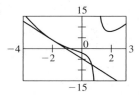

**43. a.** $y = 7x - 5$, $y = -\frac{1}{7}x + \frac{15}{7}$

**b.**

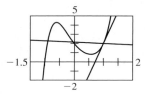

**45.** 8    **47.** $-9$    **49.** $-2$    **51.** $56x^6 - 12x^2 + 4$

**53.** $\dfrac{e^x(x^2 - 2x + 2)}{x^3}$    **55.** $6x - 4$

**57.** $\dfrac{\sqrt{x}(4x^2 - 20x + 15)}{4e^x}$    **59. a.** 44    **b.** 10

**61.** $f'(x) = 8x^3 - 8x$, $f''(x) = 24x^2 - 8$, $f'''(x) = 48x$,
$f^{(4)}(x) = 48$, $f^{(n)}(x) = 0$ for $n \geq 5$

**63.** 0.0375 part per million per year, 0.006 part per million per year

**65. a.** 120

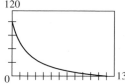

**b.** $f'(0) \approx -57.5266$, $f'(2) = -14.6165$

**67.** $h$    **69.** $f''(0) = 0$, yes    **71.** False    **73.** True

**75.** True

## 2.4 Exercises • page 204

**1.** 7.44 ft, 7.44 ft/sec, 7.44 ft/sec

**3.** $-2$ ft, $-8$ ft/sec, 8 ft/sec

**5.** $\frac{4}{5}$ ft, $-\frac{6}{25}$ ft/sec, $\frac{6}{25}$ ft/sec

**7.** 0 ft, 0 ft/sec, 0 ft/sec

**9. a.** Never

**b.** Always positive

**c.**

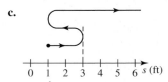

**11. a.** $s(1) = 9$ ft

**b.** Positive when $0 < t < 1$, negative when $t > 1$

**c.**

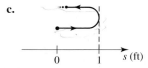

**13. a.** $s(0) = 1$ ft, $s(1) = 3$ ft, $s(2) = 1$ ft

**b.** Positive when $0 < t < 1$ and when $t > 2$, negative when $1 < t < 2$

**c.**

**15. a.** $s(1) = 1$ ft

**b.** Positive when $0 < t < 1$, negative when $t > 1$

**c.**

**17. a.** $6(2t - 3)$

**b.** $a\left(\frac{3}{2}\right) = 0$, $a(t) < 0$ if $0 < t < \frac{3}{2}$, $a(t) > 0$ if $t > \frac{3}{2}$

**19. a.** $\dfrac{4t(t^2 - 3)}{(t^2 + 1)^3}$

**b.** $a(0) = a(\sqrt{3}) = 0$, $a(t) < 0$ if $0 < t < \sqrt{3}$, $a(t) > 0$ if $t > \sqrt{3}$

**21. a.** 100 (kg · m)/sec    **b.** 250 J

**23. a.** Ascending at $t = t_0$, stationary at $t = t_1$, descending at $t = t_2$

**b.** Positive at $t = t_0$, 0 at $t = t_1$, positive at $t = t_2$

**25. a.** 1.65 sec    **b.** $-14.14$ m/sec

**27. a.** $\frac{1}{16}t^3 - \frac{3}{2}t^2 + 8t$    **b.** $v(0) = 0$, $v(8) = 0$, $v(16) = 0$

**c.** 64 ft

**29. a.** $60°$    **b.** $(200\sqrt{3}, 300)$    **c.** 300 ft

**d.** $400\sqrt{3}$ ft    **e.** $120°$

**31. a.** 3,000    **b.** $10.6°$

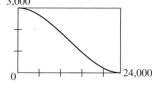

**33.** 22 ft/sec$^2$, 0.88 ft/sec$^2$    **35.** $120, $120.06

**37. a.** $10,000 - 200x$   **b.** 200, 0, $-200$

**39. a.** 2.38 cm   **b.** 0.00227 cm/cm

## 2.5 Exercises • page 213

**1.** $-4 \sin x - 2$    **3.** $(\sec t)(3 \sec t - 4 \tan t)$

**5.** $e^u(\cot u - \csc^2 u)$    **7.** $\cos 2x$    **9.** $-\sin \theta$

**11.** $e^{-x}(\cos x - \sin x)$    **13.** $\dfrac{1 + \sec x - x \sec x \tan x}{(1 + \sec x)^2}$

**15.** $\dfrac{e^{2x} \sec x(\tan x + 1) - \csc^2 x - \cot x}{e^x}$

**17.** $\dfrac{\cos x - \sin x - 1}{(1 - \cos x)^2}$    **19.** $-\dfrac{2}{(\sin \theta - \cos \theta)^2}$

**21.** $e^x \sin x(2 \cos x + \sin x)$    **23.** $2e^x \cos x$

**25.** $-5 \cos x + x \sin x$    **27.** $-\dfrac{4x^2 \cos x + 4x \sin x + \cos x}{4x\sqrt{x}}$

**29. a.** $y = \dfrac{\sqrt{3}}{2}x + \dfrac{6 - \sqrt{3}\pi}{12}$    **b.**

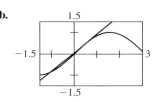

**31. a.** $y = 2\sqrt{3}x + \dfrac{6 - 2\sqrt{3}\pi}{3}$

**b.**

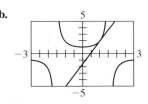

**33.** $\dfrac{\sqrt{2}\pi(8 + \pi)}{16}$    **35.** $-1$    **37.** $2k\pi, k = 0, \pm 1, \pm 2, \ldots$

**39.** $\dfrac{(2k + 1)\pi}{2}, k = 0, \pm 1, \pm 2, \ldots$

**41.** $f'(x) = \cos x, f''(x) = -\sin x, f'''(x) = -\cos x,$
$f^{(4)}(x) = \sin x, \ldots$

**43.** 2 ft, $-3$ ft/sec, 3 ft/sec, $-2$ ft/sec$^2$

**45.** $-\csc x \cot x$    **51.** True

## 2.6 Exercises • page 226

**1.** $6(2x + 4)^2$    **3.** $-\dfrac{2x}{3(x^2 + 1)^{4/3}}$

**5.** $\dfrac{e^x - \sin x}{2\sqrt{e^x + \cos x}}$    **7.** $10(2x + 1)^4$

**9.** $(2x - 1)e^{x^2 - x}$    **11.** $6\left(t + \dfrac{2}{t}\right)^5\left(1 - \dfrac{2}{t^2}\right)$

**13.** $u^2(2u^2 - 1)^3(22u^2 - 3)$    **15.** $2xe^{-2x}(1 - x)$

**17.** $\dfrac{1 - 2u^2}{\sqrt{1 - u^2}}$    **19.** $-\dfrac{20t^2 + 40t + 9}{t^4(1 + 2t)^4}$

**21.** $\dfrac{5s\left(1 + \sqrt{1 + s^2}\right)^4}{\sqrt{1 + s^2}}$    **23.** $\dfrac{e^x(2e^x + 3)}{(1 + e^{-x})^2}$    **25.** $3 \cos 3x$

**27.** $\pi \sec^2(\pi t - 1)$    **29.** $3 \cos x \sin^2 x$

**31.** $2 \cos 2x + \dfrac{\sec^2 \sqrt{x}}{2\sqrt{x}}$    **33.** $3 \sin x \cos x(\sin x - \cos x)$

**35.** $\dfrac{4 \sin 3x \cos 3x}{\sqrt[3]{1 + \sin^2 3x}}$    **37.** $-\dfrac{3(2x - \pi \sec \pi x \tan \pi x)}{(x^2 - \sec \pi x)^4}$

**39.** $-\dfrac{\sin x}{\sqrt{1 + 2 \cos x}}$    **41.** $-\dfrac{6 \sin 3x}{(1 - \cos 3x)^2}$

**43.** $-e^{\cos x} \sin x$    **45.** $\dfrac{\cos 2x + \sin 2x}{\sqrt{\sin 2x - \cos 2x}}$

**47.** $\dfrac{4}{(1 - x)^2} \sin\left(\dfrac{1 + x}{1 - x}\right) \cos\left(\dfrac{1 + x}{1 - x}\right)$

**49.** $-\dfrac{2(1 + x^2) \sin 2x + x \cos 2x}{(1 + x^2)^{3/2}}$

**51.** $\sec^2 x \, (3 \sec^2 3x + 2 \tan x \tan 3x)$    **53.** $\cos x \cos(\sin x)$

**55.** $-3\pi \cos \pi x \cos^2(\sin \pi x) \sin(\sin \pi x)$

**57.** $[1 + (3 \ln 5)x]5^{3x}$    **59.** $6(2^x + 3^{-x})^5(2^x \ln 2 - 3^{-x} \ln 3)$

**61.** $\left(\dfrac{e}{x} + 1\right)x^e e^x$    **63.** $-\ln 2(\csc^2 x)2^{\cot x}$

**65.** $48x(6x^2 - 1)(2x^2 - 1)^2$

**67.** $2(\cos 2t + 2t^2 \sin t^2 - \cos t^2)$

**69. a.** $y = \dfrac{3\sqrt{2}}{2}x - \dfrac{\sqrt{2}}{2}$    **b.**

**71. a.** $y = \dfrac{1}{e}$   **b.**

**73.** 300    **75.** No    **77.** $y = (\ln 2)x + 2$

**79.** $-4, \frac{1}{2}$, does not exist

**81.** $a \cdot \cos x \cdot f'(\sin x) - b \cdot \sin x \cdot g'(\cos x)$

**83.** $2x[f'(x^2 + 1) + g'(x^2 - 1)]$    **85.** $-4$

**87. a.** $f'(x) = \begin{cases} -\dfrac{2}{(2 - x^2)^{3/2}} & \text{if } x < 0 \\[2mm] \dfrac{2}{(2 - x^2)^{3/2}} & \text{if } x > 0 \end{cases}$

$y = -2x - 1, \; y = 2x - 1$

**b.**

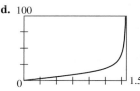

**89.** 0.6% per year, 0.4% per year, 25.9%

**91. a.** 0.83% per decade   **b.** $-0.18\%$ per decade per decade

**93. a.** 80 mg   **b.** $-\dfrac{4 \ln 2}{7} \cdot 2^{-t/140}$ mg/day

**95.** $\frac{3}{4}$ ft, $-1$ ft/sec, 1 ft/sec, $-3$ ft/sec$^2$

**97.** \$0 per share per day, \$28 per share

**99.** $\dfrac{6u_0\sigma^6}{r^7}\left[2\left(\dfrac{\sigma}{r}\right)^6 - 1\right]$    **101.** $\sqrt{3}/3$ ft/sec, $-10\sqrt{3}/9$ ft/sec$^2$

**103. a.** 1.94 sec   **b.** 7.57 m/sec

**105.** $\dfrac{k_1(EC - q_0)}{C(k_1 + k_2 t)^2}\left(\dfrac{k_1}{k_1 + k_2 t}\right)^{(1 - Ck_2)/(Ck_2)}$

**107.** $-\dfrac{\sigma}{2\varepsilon_0}\left(\dfrac{r}{\sqrt{r^2 + R^2}} - 1\right)$

**109. b.** 6.1 ft/sec, 12.3 ft/sec   **c.** $\infty$

     **d.** 100

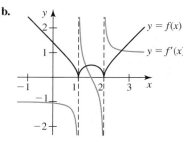

**111.** $\left(2 - \dfrac{1}{x^2}\right)\sin\dfrac{1}{x} - \dfrac{2}{x}\cos\dfrac{1}{x}$ $(x \neq 0)$, no

**113.** $\dfrac{x + 1}{|x + 1|}, \; x \neq -1$    **115.** $\dfrac{\sin 2x}{2|\sin x|}, \; x \neq k\pi, \; k$ an integer

**117. a.** $\dfrac{(2x - 3)(x - 1)(x - 2)}{2|x^2 - 3x + 2|^{3/2}}, \; x \neq 1, 2$

     **b.**

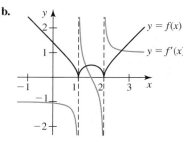

**119.** False    **121.** True

## 2.7 Exercises • page 240

**1.** $-\dfrac{2x}{y}$   **3.** $-\dfrac{y(y + 2x)}{x(2y + x)}$   **5.** $\dfrac{3x^2 - 1}{6y^2 + 1}$   **7.** $-\dfrac{y^2}{x^2}$

**9.** $\dfrac{x^4 + 2x^2y^2 + x^2y - y^3 + y^4}{x(x^2 - y^2)}$ or $\dfrac{3x^2 + 2x + y^2 - y}{x - 2xy - 2y}$

**11.** $\dfrac{x + 1}{2 - y}$   **13.** $-\sqrt{y/x}$   **15.** $\dfrac{\cos(x + y)}{2y - \cos(x + y)}$

**17.** $\dfrac{y - 6x^2 \tan(x^3 + y^3)\sec^2(x^3 + y^3)}{6y^2 \tan(x^3 + y^3)\sec^2(x^3 + y^3) - x}$

**19.** $\dfrac{-y\sqrt{1 + \cos^2 y}}{\cos y \sin y + x\sqrt{1 + \cos^2 y}}$    **21.** $\dfrac{3x^2 - e^{2y}}{2xe^{2y} + 2}$

**23.** $y = \dfrac{\sqrt{3}}{6}x - \dfrac{2\sqrt{3}}{3}$    **25.** $y = x - 2$    **27.** 1

**29.** $\sqrt{3}/3$    **31.** $y = -\dfrac{1}{e}x + 1$    **33.** $2y/x^2$

**35.** $-\dfrac{\sin x \sin^2 y + \cos y \cos^2 x}{\sin^3 y}$    **37. a.** $\dfrac{1}{2\sqrt{x}}$   **b.** $\dfrac{1}{2\sqrt{x}}$

**39. b.** $\frac{1}{2}$    **41. b.** 1    **43. b.** $\frac{1}{36}$    **45. b.** $-\frac{25}{4}$    **47.** $\frac{1}{3}$

**49.** $\dfrac{3}{\sqrt{1 - 9x^2}}$    **51.** $\dfrac{2x}{1 + x^4}$    **53.** $\tan^{-1} 3t + \dfrac{3t}{1 + 9t^2}$

**55.** $\dfrac{1}{|u|\sqrt{4u^2 - 1}}$    **57.** $-\dfrac{1}{\sqrt{1 - x^2}}$    **59.** $\dfrac{1 - x}{1 + x^2} + \cot^{-1} x$

**61.** $2x \tan^{-1} x + 1$    **63.** $\dfrac{1}{t^2 + 1}$    **65.** $\dfrac{2\cos 2x}{1 + \sin^2 2x}$

**67.** $\dfrac{2e^{2x}}{\sqrt{1 - e^{4x}}}$    **69.** $\dfrac{2x}{(1 - x^4)^{3/2}}$

**71.** $-\dfrac{1}{|\theta|(\sec^{-1}\theta)^2\sqrt{\theta^2 - 1}}$    **73.** $y = \dfrac{\sqrt{3}}{2}x + 2\sqrt{3}$

**75.** $y = 2x - \frac{1}{2}$    **77.** $y = -\frac{9}{13}x + \frac{40}{13}$

**79.** $y = \dfrac{\pi + 2\sqrt{3}}{6}x - \dfrac{\sqrt{3}}{6}$

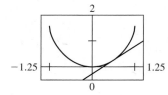

**81. a.** $y = -\frac{1}{4}x + \frac{3}{2}, \; y = 4x - \frac{45}{4}$

     **b.**

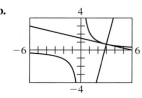

**83. a.** $y = \frac{15}{2}x + 18$, $y = -\frac{2}{15}x + \frac{41}{15}$

**b.**

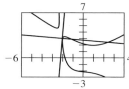

**85. a.** $\dfrac{y - x^2}{y^2 - x}$  **b.** $y = -x + 3$  **c.** $(0, 0)$, $\left(\sqrt[3]{2}, \sqrt[3]{4}\right)$

**87.** $\frac{3}{25}$ ft/sec

**93. b.**

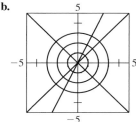

**95. b.**

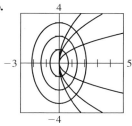

**97. b.**

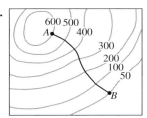

**99.** $17.9°$/sec  **103.** True

## 2.8 Exercises • page 250

**1.** $\dfrac{2}{2x + 3}$  **3.** $\dfrac{1}{2x}$  **5.** $\dfrac{1}{u(u + 1)}$

**7.** $(\ln x)^2 + 2\ln x$  **9.** $\dfrac{x(1 - \ln x) + 1}{x(x + 1)^2}$  **11.** $\dfrac{1}{x \ln x}$

**13.** $\dfrac{\ln x + 1}{x \ln x}$  **15.** $\dfrac{\cos(\ln x)}{x}$  **17.** $2x \ln \cos x - x^2 \tan x$

**19.** $\tan u$  **21.** $\dfrac{2 \cos t + \sin t + 1}{(\sin t + 1)(\cos t + 2)}$

**23.** $\dfrac{2x + 1}{(x^2 + x + 1)\ln 2}$  **25.** $\dfrac{t}{(t^2 + 1)\ln 10}$  **27.** $y = x - 1$

**29.** $2(2x + 1)(3x^2 - 4)^2(24x^2 + 9x - 8)$

**31.** $-\dfrac{x^2 - 2x - 1}{3(x - 1)^{2/3}(x^2 + 1)^{4/3}}$  **33.** $[x(\ln x + 1)^2 + 1]x^{x-1}$

**35.** $3^x \ln 3$  **37.** $\left[\dfrac{1}{x(x + 2)} - \dfrac{\ln(x + 2)}{x^2}\right](x + 2)^{1/x}$

**39.** $\dfrac{\cos x \ln(\cos x) - x \sin x}{2 \cos x}(\sqrt{\cos x})^x$

**41.** $y(\ln x + 1)$  **43.** $\dfrac{x + y}{x - y}$

**45. a.** $\dfrac{ab}{M + m - at} - g$  **b.** $-b \ln \dfrac{M}{M + m} - \dfrac{gm}{a}; \dfrac{ab}{M} - g$

**47.** 0.0580%/kg; 0.0133%/kg

**51. a.** $0.3307 \, p_0$  **b.** $-0.0000455 \, p_0$ atm/m

**53.** False  **55.** False

## 2.9 Exercises • page 257

**1.** $\frac{3}{2}$  **3.** $-6$  **5.** $1$

**7. a.** $\dfrac{dV}{dt} = 3x^2 \dfrac{dx}{dt}$  **b.** 150 in.$^3$/sec  **9.** $-3$ units/sec

**11. a.** $-2 < x < 2$  **b.** $x = \pm 2$  **c.** $x < -2$ or $x > 2$

**13.** 188.5 ft$^2$/sec  **17.** 19.2 ft/sec  **19.** 29 packs per week

**21.** $-6.96$ ft/sec  **23.** 10 ft/sec  **25.** 20.1 in.$^3$/sec

**27.** 17 ft/sec  **29.** $-23$ km/h  **31.** $\frac{1}{7}$ L/sec

**33.** $9.1 \times 10^{-32}$ kg/sec  **35.** 196.8 ft/sec

**37. a.** $x^2 - (6 \cos \theta)x - 40 = 0$  **b.** $-1205.5$ m/sec

**39.** $-80.15$ ft/sec  **41.** 0.04 in.$^3$/sec

## 2.10 Exercises • page 270

**1. a.** 0.02, 0.0804  **b.** 0.08  **c.** 0.0004

**3. a.** 0.1, 0.033150  **b.** 0.033333  **c.** $-0.000183$

**5.** $dx$  **7.** $-dx$  **9.** $\frac{57}{4} dx$  **11.** $-\dfrac{\sqrt{2}}{2} dx$  **13.** $-dx$

**15.** $2 \, dx$  **17.** $2 \, dx$  **19.** $7x - 4$  **21.** $x - 1$

**23.** 1.975, 2.025

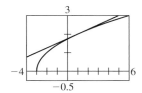

**25.** 1.006  **27.** 1.9885  **29.** 6%  **31.** $2000

**33.** $-40\%$  **35.** 3%  **37.** 15%  **39.** 1.75 ft

**41.** 18.1%  **43.** 0.68 hr  **45.** $\phi = 45°$  **47.** 2.78 kg

**49.** True  **51.** False

## Chapter 2 Concept Review • page 273

**1. a.** $f'(x) = \lim\limits_{h \to 0} \dfrac{f(x + h) - f(x)}{h}$  **b.** limit

**c.** tangent line; $(a, f(a))$  **d.** $f(x)$; $x$; $(a, f(a))$

**e.** $y = f'(a)(x - a) + f(a)$

**3. a.** 0   **b.** $nx^{n-1}$

**5.** $dy/dx$   **7.** $C'$; $R'$; $P'$; $\overline{C}'$

**9. a.** $n[f(x)]^{n-1} \cdot f'(x)$

   **b.** $\cos f(x) \cdot f'(x)$; $-\sin f(x) \cdot f'(x)$; $\sec^2 f(x) \cdot f'(x)$;
   $\sec f(x) \tan f(x) \cdot f'(x)$; $-\csc f(x) \cot f(x) \cdot f'(x)$;
   $-\csc^2 f(x) \cdot f'(x)$

**11.** $\dfrac{1}{u \ln a}\dfrac{du}{dx}$   **13.** $y$; $dy/dt$; $a$

**15. a.** $x_2 - x_1$   **b.** $f(x + \Delta x) - f(x)$

## Chapter 2 Review Exercises • page 274

**1.** $2x - 2$

**3.**

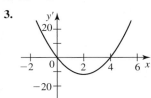

**5. a.** The rate of change of the amount of money on deposit
   with respect to the interest rate, measured in dollars per
   unit change in interest.

   **b.** Positive   **c.** \$607.75

**7.** $f(x) = 3x^{3/2}$, $a = 4$   **9.** $2x^5 - 8x^3 + 2x$

**11.** $4t + \dfrac{4}{t^2} - \dfrac{1}{t\sqrt{t}}$   **13.** $\dfrac{3}{(2t+1)^2}$   **15.** $\dfrac{1 - 3u^2}{2\sqrt{u}(u^2+1)^2}$

**17.** $-\sin\theta - 2\cos\theta$   **19.** $(1 - x^2)\sin x + 3x\cos x$

**21.** $\dfrac{\cos t - t\sin t + \sin t - t\sin t \tan t - t\sec t}{(1 + \tan t)^2}$

**23.** $-\frac{3}{2}(3t^2 + 2)(t^3 + 2t + 1)^{-5/2}$

**25.** $\frac{1}{2}\sqrt{s^3 + s + 1}(11s^3 + 5s + 2)$

**27.** $\dfrac{t + 2}{(\sqrt{t} + 1)^3}$   **29.** $-2\sin(2x + 1)$

**31.** $2x + \dfrac{2x\cos 2x - \sin 2x}{x^2}$   **33.** $-\dfrac{2}{x^2}\sec^2\dfrac{2}{x}$

**35.** $-3\cot^2 x \csc^2 x$   **37.** $-\dfrac{\theta\sin\theta + 2\cos\theta}{\theta^3}$

**39.** $\dfrac{1}{2(x+1)}$   **41.** $\dfrac{\ln x + 2}{2\sqrt{x}}$   **43.** $5e^{-x}(\cos 2x - \sin 2x)$

**45.** $-\dfrac{y(y + x\ln y)}{x(x + y\ln x)}$   **47.** $\dfrac{2(1-x)}{x}$   **49.** $\dfrac{2e^x + 3}{2(1 + e^{-x})^{3/2}}$

**51.** $-\csc x \cot x \cdot e^{\csc x}$   **53.** $e^{e^x + x}$   **55.** $\dfrac{ye^{-x} - e^{y^2}}{e^{-x} + 2xye^{y^2}}$

**57.** $3^{x\cot x}(\cot x - x\csc^2 x)\ln 3$   **59.** $\sec^{-1} x + \dfrac{x}{|x|\sqrt{x^2 - 1}}$

**61.** $\dfrac{x}{(x^2 + 2)\sqrt{x^2 + 1}}$   **63.** $\dfrac{-1}{2\sqrt{x - x^2}\,[1 + (\cos^{-1}\sqrt{x})^2]}$

**65.** $-\frac{47}{1156}$   **67.** $6x + 2 - \dfrac{2}{x^3}$   **69.** $\dfrac{3x - 2}{(2x - 1)^{3/2}}$

**71.** $-2\cos 2x$   **73.** $\dfrac{1}{2}\csc^2\dfrac{\theta}{2}\cot\dfrac{\theta}{2}$

**75.** $-2(\csc^2 t)(1 - t\cot t)$   **77.** $-\frac{1}{27}$   **79.** 10

**81.** $\dfrac{1}{3}\left[\dfrac{f(x)}{g(x)}\right]^{-2/3}\dfrac{g(x)f'(x) - f(x)g'(x)}{[g(x)]^2}$   **83.** $\dfrac{3x}{2y}$   **85.** $-\dfrac{y^3}{x^3}$

**87.** $-\dfrac{2x^2 + 2xy + y^2}{x^2 + 2xy + 2y^2}$   **89.** $\dfrac{\sin y - \sin(x + y)}{\sin(x + y) - x\cos y}$

**91.** $-\dfrac{y}{x}$   **93.** $10x^4 - 3x^2$   **95.** $\frac{3}{16}\,dx$

**97.** $\frac{7}{3}\,dx$   **99.** $\dfrac{6 - \sqrt{3}\pi}{12}\,dx$   **101.** $y = 2e$

**103.** $y = -x + 2$, $y = x$   **105.** $-\dfrac{2x}{y^5}$

**107.** $-\dfrac{\sqrt{3}}{2}x + \dfrac{1}{12}(\sqrt{3}\pi + 9)$   **109. a.** $(1, 2)$   **b.** $y = 2x$

**111.** 0   **113.** $\frac{20}{3}t^{-1/3} - \frac{5}{3}t^{2/3}$, $-\frac{20}{9}t^{-4/3} - \frac{10}{9}t^{-1/3}$

**115.** 30 cm/sec, $-200$ cm/sec/cm

**117.** 0.0107 m²/kg   **119.** 0.05164

**121. a.** $\dfrac{\pi r^2}{3}$   **b.** $\dfrac{2\pi rh}{3}$   **125.** $\frac{15}{4}$   **127.** 7.7 ft/sec

## Chapter 2 Challenge Problems • page 279

**1.** 64

**3. b.** $(-1)^n n!\left[\dfrac{1}{(x + 2)^{n+1}} + \dfrac{1}{(x - 1)^{n+1}}\right]$

**5.** $\dfrac{1 \cdot 3 \cdot 5 \cdot \cdots \cdot 17}{2^{10}}(1 - x)^{-21/2}(39 - x)$

**9.** $\dfrac{xf'(\sqrt{1 + x^2})}{\sqrt{1 + x^2}}$   **13.** $-10$   **17.** No

# CHAPTER 3

## Exercises 3.1 • page 291

*Abbreviations: abs. max., absolute maximum; abs. min., absolute minimum.*

**1.** Abs. max. $f(1) = 3$

**3.** Abs. max. $f(2n) = 1$, abs. min. $f(2n + 1) = 0$, $n$, an integer

**5.** Abs. max. $f(1) = 37$, abs. min. $f(5) = -5$

**7.**

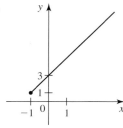

Abs. min. $f(-1) = 1$

**9.**

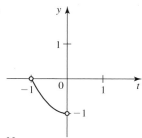

None

**11.**

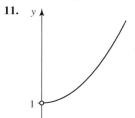

None

**13.**

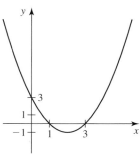

Abs. min. $f(2) = -1$

**15.**

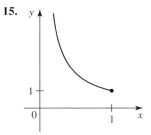

Abs. min. $f(1) = 1$

**17.**

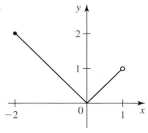

Abs. max. $f(-2) = 2$,
abs. min. $f(0) = 0$

**19.**

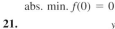

Abs. max. $f\left(\frac{\pi}{2}\right) = 2$

**21.**

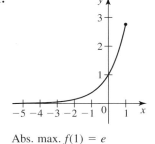

Abs. max. $f(1) = e$

**23.**

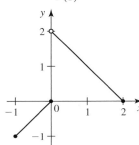

Abs. min. $f(-1) = -1$

**25.** None    **27.** $-1$    **29.** $\pm\sqrt{2}$    **31.** 0, 3    **33.** 0

**35.** $\pm 1$    **37.** $n\pi/4$, $n = 0, \pm 1, \pm 2, \ldots$    **39.** 0

**41.** Abs. max. $f(2) = 0$, abs. min. $f\left(\frac{1}{2}\right) = -\frac{9}{4}$

**43.** Abs. max. $h(2) = 21$, abs. min. $h(-3) = h(0) = 1$

**45.** Abs. max. $g(-2) = 17$, abs. min. $g(-1) = 0$

**47.** Abs. max. $f(1) = \frac{1}{2}$, abs. min. $f(-1) = -\frac{1}{2}$

**49.** Abs. max. $g(2) = 2$, abs. min. $g(4) = \frac{4}{3}$

**51.** Abs. max. $f(9) = 3$, abs. min. $f(1) = -1$

**53.** Abs. max. $f(0) = f(2) = 0$, abs. min. $f(\pm 1) = -3$

**55.** Abs. max. $f\left(\frac{\pi}{4}\right) = 5$, abs. min. $f(0) = f\left(\frac{\pi}{2}\right) = 2$

**57.** Abs. max. $f(1) = 1/e$, abs. min. $f(-1) = -e$

**59.** Abs. max. $f(2) = -0.61$, abs. min. $f(1) = -1$

**61.** 1667 dozen

**63.** Highest at 6 A.M. and 10 A.M., lowest at 7 A.M.

**67.** 10,000    **71.** 7.4 lb    **73.** $52.79/ft$^2$

**75. a.**

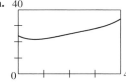

**b.** 21.5%

**77.** Halfway up the side of the cylinder

**79.** $\approx$93 ft    **81.** $0.85a$, $0.53a$

**83. a.** $\tan^{-1}\mu$    **b.** $\dfrac{\mu W}{\sqrt{\mu^2 + 1}}$ lb    **c.**

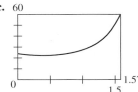

**85.** No    **89.** No abs. max., abs. min. 0

**91.**

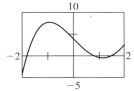

Abs. max. $f(-0.9) \approx 7.8$, abs. min. $f(-2) = -4.16$

**93.**

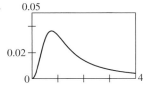

Abs. max. $f(0.8) \approx 0.037$, abs. min. $f(0) = 0$

**95. a.**

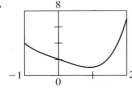

**b.** Abs. max. 7, abs. min. $2 - \frac{9}{8}\sqrt[3]{3/4} \approx 0.978$

**97. a.**

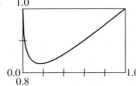

**b.** Abs. max. 1, abs. min. $\sqrt{2}(2 - \sqrt{2}) \approx 0.828$

**99.** False    **101.** False

## Exercises 3.2 • page 302

**1.** 2    **3.** $-\dfrac{1 + \sqrt{7}}{3}$    **5.** $\pm\sqrt{2}/2$    **7.** $\pi/2$    **9.** 1

**11.** $\sqrt{3}$    **13.** $\dfrac{2 + 2\sqrt{3}}{3}$    **15.** $2\ln\left(\dfrac{2e^2}{e^2 - 1}\right)$

**17.** There is at least one instant during the 30-min flight when the plane is neither climbing nor descending.

**19.** $c = 4$; the aircraft attains the highest altitude at $t = 4$.

**21.** $f$ is not differentiable on $(-1, 1)$.

**23.** No; $f$ is not differentiable at $x = 1$.

**31. b.**

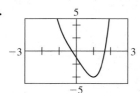

**39.** (3, 3)    **41. b.** $\frac{1}{2}$

**43. a.** 1.325    **b.**

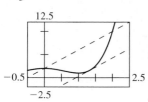

**45. a.** 0.691    **b.** 1.25

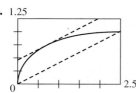

**47.** False    **49.** True    **51.** True

## Exercises 3.3 • page 311

*Abbreviations: rel. max., relative maximum; rel. min., relative minimum.*

**1. a.** Increasing on $(-\infty, -2)$, constant on $(-2, 2)$, decreasing on $(2, \infty)$

   **b.** Rel. max. $f(x) = 2$, $x \in [-2, 2]$

**3. a.** Increasing on $(-1, \infty)$, decreasing on $(-\infty, -1)$

   **b.** Rel. min. $f(-1) = 0$

**5. a.** Increasing on $(-\infty, -1)$ and $(-1, \infty)$    **b.** None

**7. a.** Increasing on $(-2.5, 2.5)$, decreasing on $(-\infty, -2.5)$ and $(2.5, \infty)$

   **b.** Rel. max. at 2.5, rel. min. at $-2.5$

**9. a.** Increasing on $(1, \infty)$, decreasing on $(-\infty, 1)$

   **b.** Rel. min. $f(1) = -1$

**11. a.** Increasing on $(-\infty, -\sqrt{2})$ and $(\sqrt{2}, \infty)$, decreasing on $(-\sqrt{2}, \sqrt{2})$

   **b.** Rel. max. $f(-\sqrt{2}) = 1 + 4\sqrt{2}$, rel. min. $f(\sqrt{2}) = 1 - 4\sqrt{2}$

**13. a.** Increasing on $(-\infty, -2)$ and $(1, \infty)$, decreasing on $(-2, 1)$

   **b.** Rel. max. $f(-2) = 25$, rel. min. $f(1) = -2$

**15. a.** Increasing on $(3, \infty)$, decreasing on $(-\infty, 3)$

   **b.** Rel. min. $f(3) = -21$

**17. a.** Increasing on $(-\infty, \infty)$    **b.** None

**19. a.** Increasing on $(-\infty, 0)$ and $\left(\frac{4}{5}, \infty\right)$, decreasing on $\left(0, \frac{4}{5}\right)$

   **b.** Rel. max. $f(0) = 0$, rel. min. $f\left(\frac{4}{5}\right) = -1.10592$

**21. a.** Increasing on $(-\infty, -1)$ and $(1, \infty)$, decreasing on $(-1, 0)$ and $(0, 1)$

   **b.** Rel. max. $f(-1) = -2$, rel. min. $f(1) = 2$

**23. a.** Increasing on $(-\infty, 0)$ and $(2, \infty)$, decreasing on $(0, 1)$ and $(1, 2)$

   **b.** Rel. max. $f(0) = 0$, rel. min. $f(2) = 4$

**25. a.** Decreasing on $(-\infty, -2)$, $(-2, 2)$, and $(2, \infty)$

   **b.** None

**27. a.** Increasing on $(-\infty, 0)$ and $\left(\frac{6}{5}, \infty\right)$, decreasing on $\left(0, \frac{6}{5}\right)$

   **b.** Rel. max. $f(0) = 0$, rel. min. $f\left(\frac{6}{5}\right) \approx -2.03$

**29. a.** Increasing on $\left(0, \frac{3}{4}\right)$, decreasing on $\left(\frac{3}{4}, 1\right)$

   **b.** Rel. max. $f\left(\frac{3}{4}\right) = 3\sqrt{3}/16$

**31. a.** Increasing on $\left(\frac{\pi}{3}, \frac{5\pi}{3}\right)$, decreasing on $\left(0, \frac{\pi}{3}\right)$ and $\left(\frac{5\pi}{3}, 2\pi\right)$

   **b.** Rel. max. $f\left(\frac{5\pi}{3}\right) \approx 6.97$, rel. min. $f\left(\frac{\pi}{3}\right) \approx -0.68$

**33. a.** Increasing on $\left(\frac{\pi}{2}, \pi\right)$ and $\left(\frac{3\pi}{2}, 2\pi\right)$, decreasing on $\left(0, \frac{\pi}{2}\right)$ and $\left(\pi, \frac{3\pi}{2}\right)$

   **b.** Rel. max. $f(\pi) = 1$, rel. min. $f\left(\frac{\pi}{2}\right) = f\left(\frac{3\pi}{2}\right) = 0$

**35. a.** Increasing on $(0, 2)$, decreasing on $(-\infty, 0)$ and $(2, \infty)$

   **b.** Rel. max. $f(2) = 4e^{-2}$, rel. min. $f(0) = 0$

**37. a.** Increasing on $(e, \infty)$, decreasing on $(0, 1)$ and $(1, e)$

   **b.** Rel. min. $f(e) = 2e$

**39.** Rel. max. $f\left(-\dfrac{\sqrt{3}}{2}\right) = \sqrt{3} - \dfrac{\pi}{3}$,

   rel. min. $f\left(\dfrac{\sqrt{3}}{2}\right) = \dfrac{\pi}{3} - \sqrt{3}$

**41. a.** Increasing on $(20.2, 20.6)$ and $(21.7, 21.8)$, constant on $(19.6, 20.2)$ and $(20.6, 21.1)$, decreasing on $(21.1, 21.7)$ and $(21.8, 22.7)$

**43.** Increasing on $(1, 4)$, decreasing on $(0, 1)$, rel. min. $f(1) = 32$

**45.** Decreasing on $(0, 500)$, increasing on $(500, \infty)$, $\overline{C}(500) = 35$

**47.** Rising on $(0, 1)$, $(7, 13)$, and $(19, 24)$; falling on $(1, 7)$ and $(13, 19)$; rel. max. $(1, 12.4)$ and $(13, 12.4)$, rel. min. $(7, 2.8)$ and $(19, 2.8)$

**49. a.** 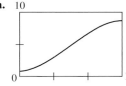   **51. a.**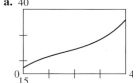

**53. b.** 4505 cases per year, 272 cases per year

**55.** Increasing on $(0, e)$, decreasing on $(e, \infty)$

**57. a.** 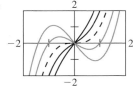   **b.** $a \le 0$

**63.** $a = -4$, $b = 24$

**65.** No. $f$ is not continuous on an interval containing $x = 0$.

**69. a.**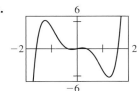

   **b.**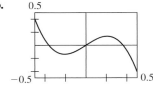

   Increasing on $(-\infty, -1.2)$, $(-0.2, 0.2)$, and $(1.2, \infty)$; decreasing on $(-1.2, -0.2)$ and $(0.2, 1.2)$

**71. a.**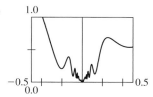

**73.** True      **75.** False      **77.** False

## Exercises 3.4 • page 324

*Abbreviations: CU, concave upward; CD, concave downward; IP, inflection point; rel. max., relative maximum; rel. min., relative minimum.*

**1.** CU on $(0, \infty)$, CD on $(-\infty, 0)$, IP $(0, 0)$

**3.** CU on $(-\infty, -4)$ and $(4, \infty)$, CD on $(-4, 4)$

**5.** CD on $(-\infty, -2)$, $(-2, 2)$, and $(2, \infty)$

**7.** CU on $(-1, 0)$ and $(1, \infty)$, CD on $(-\infty, -1)$ and $(0, 1)$, IP at $x = -1, 0, 1$

**9.** (b)      **11.** CU on $(0, \infty)$, CD on $(-\infty, 0)$, IP $(0, 0)$

**13.** CU on $(-\infty, 0)$ and $(1, \infty)$, CD on $(0, 1)$, IP $(0, 0)$ and $(1, -1)$

**15.** CU on $(-\infty, 0)$, CD on $(0, \infty)$, IP $(0, 1)$

**17.** CU on $(-\infty, -1)$ and $(0, \infty)$, CD on $(-1, 0)$, IP $\left(-1, -\frac{4}{15}\right)$ and $(0, 0)$

**19.** CD on $(-1, 0)$ and $(0, 1)$

**21.** CU on $(-\infty, 0)$ and $(0, \infty)$

**23.** CU on $(-1, 0)$ and $(1, \infty)$, CD on $(-\infty, -1)$ and $(0, 1)$, IP $(0, 0)$

**25.** CU on $\left(\frac{\pi}{2}, \pi\right)$, CD on $\left(0, \frac{\pi}{2}\right)$, IP $\left(\frac{\pi}{2}, 0\right)$

**27.** CU on $\left(\frac{3\pi}{4}, \frac{7\pi}{4}\right)$, CD on $\left(0, \frac{3\pi}{4}\right)$ and $\left(\frac{7\pi}{4}, 2\pi\right)$, IP $\left(\frac{3\pi}{4}, 0\right)$ and $\left(\frac{7\pi}{4}, 0\right)$

**29.** CU on $\left(-\pi, -\frac{3\pi}{4}\right)$, $\left(-\frac{\pi}{2}, -\frac{\pi}{4}\right)$, $\left(0, \frac{\pi}{4}\right)$, and $\left(\frac{\pi}{2}, \frac{3\pi}{4}\right)$, CD on $\left(-\frac{3\pi}{4}, -\frac{\pi}{2}\right)$, $\left(-\frac{\pi}{4}, 0\right)$, $\left(\frac{\pi}{4}, \frac{\pi}{2}\right)$, and $\left(\frac{3\pi}{4}, \pi\right)$, IP $\left(-\frac{\pi}{2}, 0\right)$, $(0, 0)$, and $\left(\frac{\pi}{2}, 0\right)$

**31.** CD on $(-\infty, 0)$ and $(0, \infty)$

**33.** 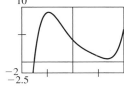   **a.** CU on $(-0.4, \infty)$, CD on $(-\infty, -0.4)$   **b.** IP $(-0.4, 6.4)$

**35.** 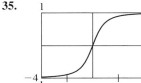   **a.** CU on $(-\infty, 0)$, CD on $(0, \infty)$   **b.** IP $(0, 0)$

**37.** Rel. max. $h(-1) = -\frac{22}{3}$, rel. min. $h(5) = -\frac{130}{3}$

**39.** Rel. min. $f(3) = -27$

**41.** Rel. max. $f\left(-\frac{\sqrt{2}}{2}\right) = -2\sqrt{2}$, rel. min. $f\left(\frac{\sqrt{2}}{2}\right) = 2\sqrt{2}$

**43.** Rel. min. $g(2) = 2 - 2\ln 2$

**45.** Rel. max. $f\left(\frac{\pi}{4}\right) = \sqrt{2}$    **47.** Rel. max. $f\left(\frac{\pi}{3}\right) = 3\sqrt{3}/2$

**49.** IP $(0, 0)$

**51.** 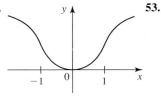   **53.**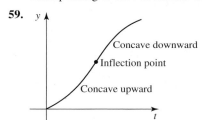

**55.** **a.** $D_1'(t) > 0,\ D_2'(t) > 0,\ D_1''(t) > 0,\ D_2''(t) < 0$

 **b.** With the promotion the deposits will increase at an increasing rate; without it they will increase at a decreasing rate.

**57.** The restoration process is working at its peak at the time $t_0$ corresponding to the $t$-coordinate of $Q$.

**59.**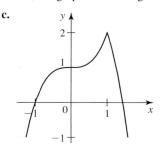

 $f(t)$ increases at an increasing rate before the IP and a decreasing rate after the IP.

**61.** **b.** Sales continued to accelerate.

**67.** **a.** 506,000 in 1999, 125,480 in 2005

 **b.** The number of measles deaths was dropping.

 **c.** April 2002; decreasing at about 41,000 deaths annually

**69.** **a.** 71.6 kg/yr, 63.1 kg/yr   **b.** 5.7 years

**73.** **a.** CU on $(0, 1)$, CD on $(-\infty, 0)$ and $(1, \infty)$

 **b.** No; the graph has no tangent line at $(1, 2)$.

 **c.**

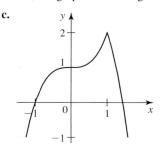

**75.** $c \geq \frac{3}{2}$   **77.** No   **83.** False   **85.** True

## Exercises 3.5 • page 343

*Abbreviations: HA, horizontal asymptote(s); VA, vertical asymptote(s).*

**1.** **a.** $-\infty$  **b.** $\infty$  **c.** $\infty$  **d.** $-\infty$    **3.** **a.** $-\infty$  **b.** 0  **c.** 0

**5.** $\infty$   **7.** $-\infty$   **9.** $\infty$   **11.** $\infty$   **13.** $\infty$   **15.** $\infty$

**17.** $\infty$   **19.** $\infty$   **21.** $\frac{3}{2}$   **23.** 0   **25.** $\frac{1}{9}$   **27.** $-\frac{2}{3}$

**29.** $-\frac{1}{6}$   **31.** $2\sqrt{3}/3$   **33.** $\frac{2}{3}$   **35.** 0   **37.** $-\infty, 1, 0, 1$

**39.** **b.** 0   **c.** 0.5

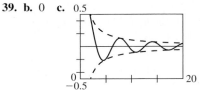

**41.** **a.** $\frac{1}{2}$  **b.** $\frac{1}{2}$  **c.** $\frac{1}{2}$

**43.** **a.** 0.7  **b.** 0.707  **c.** $\sqrt{2}/2$

**45.** HA $y = \pm 1$    **47.** HA $y = -1$, VA $x = \pm 1$

**49.** HA $y = 0$, VA $x = -2$    **51.** HA $y = 1$, VA $x = -1$

**53.** HA $y = 0$, VA $x = -2, 3$    **55.** HA $y = 1$, VA $t = \pm 2$

**57.** 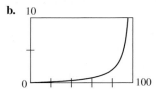   **59.**

**61.** **a.** $\lim_{x \to 100^-} C(x) = \infty$; the cost increases dramatically as the amount of pollutant removed approaches 100%.

 **b.** 10

**63.** **a.** 25,000   **b.** 30

**65.** **67.**

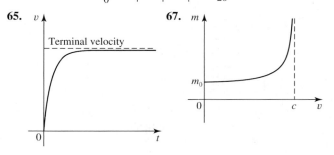

**69.**

| $n$ | 1 | 10 | $10^2$ | $10^3$ | $10^4$ | $10^5$ | $10^6$ |
|---|---|---|---|---|---|---|---|
| $\left(1 + \dfrac{1}{n}\right)^n$ | 2 | 2.59374 | 2.70481 | 2.71692 | 2.71815 | 2.71827 | 2.71828 |

**71.** $33,201.17     **73.** $48,635.23

**75. a.** 0.6  **b.** 0.577  **c.** $\sqrt{3}/3$

**77. b.** $\approx$25,067 mph  **c.**

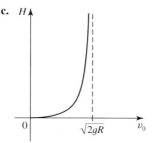

**81.** 1    **89.** False    **91.** False    **93.** True

## Exercises 3.6 • page 358

**1.**

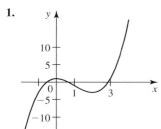

**3.**

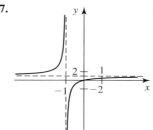

**5.**

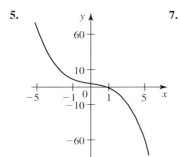

**7.**

**9.**

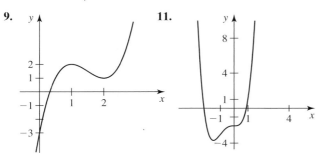

**11.**

**13.**

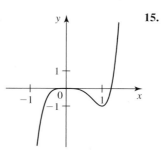

**15.**

**17.**

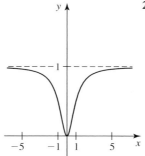

**19.**

**21.**

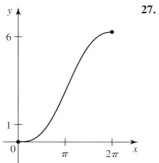

**23.**

**25.**

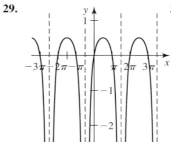

**27.**

**29.**

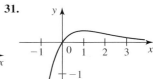

**31.**

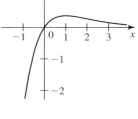

**33.**

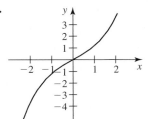

**35.**

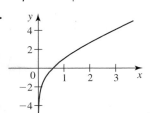

**37.**

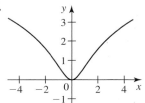

**39.** $y = u$    **41.** $y = \frac{1}{2}x - \frac{1}{2}$

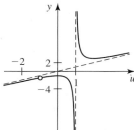
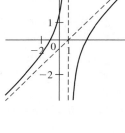

**45.**

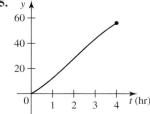

The average worker's efficiency increases until 10 A.M. and then begins to decrease.

**47.**

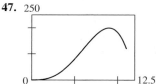

The rate of increase of the ozone level is maximum at about 1 P.M.

**49.**

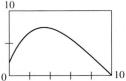

The amount of salt increases to a peak level of approximately 7.43 lb after approximately 3.4 min, and then it declines.

**51.** $f(v)$

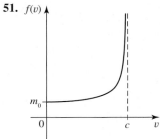

A particle's mass increases as its speed increases.

**53.**

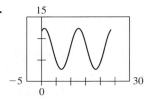

At 12 A.M. the water level is $\approx 11.8$ ft. At 7 A.M. and 7 P.M. it has a rel. min. of $\approx 2.8$ ft; at 1 A.M. and 1 P.M. it has a rel. max. of $\approx 12.4$ ft.

**55.**

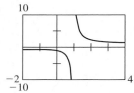

**57.**

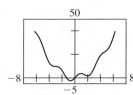

**59. a.**

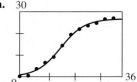

**b.** 1.5 in./hr, 0.5 in./hr

**c.** 2:24 A.M. on February 7, 1.6 in./hr

**61.**

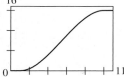

## Exercises 3.7 • page 369

**1.** 50, 50　　**3.** 18, 3　　**5.** 25 m × 25 m

**7.** 100 ft × $66\frac{2}{3}$ ft　　**9.** 12 in. × 6 in. × 2 in.

**11.** 7.6 in. × 7.6 in. × 3.8 in.

**13.** Radius $36/\pi$ in., length 36 in., volume $46{,}656/\pi$ in.$^3$

**15.** 1.39 ft × 2.08 ft × 0.83 ft　　**17.** $2\sqrt{2} \times \sqrt{2}$　　**19.** $\frac{3}{2} \times 1$

**21.** $(-2, 1)$　　**23.** $(-1.6180, -0.6180)$, $(0.6180, 1.6180)$

**25.** $4\sqrt{2} \times 2\sqrt{2}$　　**27.** $(-2.834, -4.032)$　　**29.** 20 trees/acre

**31.** $\frac{2\sqrt{3}}{3}$ ft × $\frac{2\sqrt{3}}{3}$ ft × $\frac{10\sqrt{3}}{9}$ ft; $\frac{40\sqrt{3}}{27}$ ft$^3$

**35.** Width 11.5 in., height 19.9 in.

**37.** $\dfrac{2\pi(3 - \sqrt{6})}{3}$ rad, $128\pi\sqrt{3}$ in.$^3$

**39.** $\approx 6.7$ ft, $\approx 6.7$ ft　　**41.** $\sqrt[3]{32/\pi}$, $2\sqrt[3]{32/\pi}$

**43.** $\approx 8.6$ mi from $O$

**45. a.** $D(x) = \sqrt{x^2 + \dfrac{(x+1)^2}{x} - \dfrac{4(x+1)}{\sqrt{x}} + 4}$

　　**b.** $\approx 0.45$ mi

**47.** $r$, $\dfrac{E^2}{4r}$ watts　　**49.** $C$　　**53.** Radius $R/2$, height $H/2$

**55.** $y = -2x + 4$　　**57.** $5\sqrt{2}/2$ ft　　**59.** Yes

**61.** Radius 1.5 ft, height 3 ft

**63. a.** $\frac{3}{2}u$ ft/sec

　　**b.**

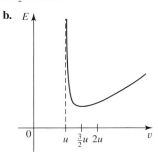

**65.** 655 ft　　**67.** $\overline{PR} = \dfrac{3\sqrt{7}}{7}$ mi, $\dfrac{30 + \sqrt{7}}{12}$ hr

## Exercises 3.8 • page 387

**1.** $\frac{1}{2}$　　**3.** 12　　**5.** 1　　**7.** 1　　**9.** 2　　**11.** $\frac{1}{2}$　　**13.** $\frac{1}{3}$

**15.** $\infty$　　**17.** 0　　**19.** 0　　**21.** $\frac{9}{4}$　　**23.** $\infty$　　**25.** $-\frac{1}{2}$

**27.** 2　　**29.** 0　　**31.** $-2$　　**33.** $-\infty$　　**35.** $\frac{1}{2}$　　**37.** $-1$

**39.** 0　　**41.** 1　　**43.** $e^2$　　**45.** 1　　**47.** $\infty$　　**49.** 1

**51.** 0　　**53.** $-\infty$　　**55.** $\frac{1}{3}$　　**57.** $\frac{5}{2}$　　**61.** $\dfrac{Vt}{L}$　　**63.** $a$

**69.** 3　　**71.** $\frac{1}{2}$　　**73.** False

## Exercises 3.9 • page 394

**1.** 0.7709　　**3.** 1.1219, 2.5745　　**5.** $(0.8767, 0.7686)$

**7.** $(0.4502, 0.4502)$　　**9.** 1.37880　　**11.** 0.75488

**13.** 0.87122　　**15.** $-1.347296$　　**17.** 0.739085

**19.** 1.76322　　**21.** All terms are $x_0$.　　**23.** 1.8794

**25.** 2.4495　　**27.** 2.1147　　**29.** $t \approx 41.08569$

**31.** $(0.786, 0.666)$

**33.** $f'(1)$ is not defined; using $x_0 = 0$ leads to $x_1 = 1$.

**35. b.** 0.7351　　**c.** $\approx 2.74$ mi

　　**d.**

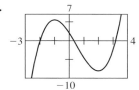

**39.** 6.7%/year

**41. c.**

**Chapter 3 Concept Review • page 397**

**1. a.** $f(x) \le f(c)$; absolute maximum value

　　**b.** $f(c) \le f(x)$; open interval

**3.** Continuous; absolute maximum value; absolute minimum value

**5. a.** $f(x_1) < f(x_2)$

　　**b.** increasing; decreasing

　　**c.** $< 0$

**7. a.** values; arbitrarily large; $a$

　　**b.** values; to $L$; sufficiently large

　　**c.** arbitrarily large; decreases

**9. a.** $M > 0$; $\delta > 0$; $f(x) > M$

　　**b.** $M > 0$; $N$; $x < N$; $f(x) > M$

**Chapter 3 Review Exercises • page 397**

*Abbreviations: abs. max., absolute maximum; abs. min., absolute minimum; CU, concave upward; CD, concave downward; IP, inflection point; HA, horizontal asymptote(s); VA, vertical asymptote(s).*

**1.** Abs. max. $f(2) = 1$, abs. min. $f(-1) = -8$

**3.** Abs. max. $h(2) = -16$, abs. min. $h(4) = -32$

**5.** Abs. max. $f(3) = \frac{107}{9}$, abs. min. $f(1) = 3$

**7.** Abs. min. $f(3) = -3$

**9.** Abs. max. $f\left(\frac{7\pi}{4}\right) = \sqrt{2}$, abs. min. $f\left(\frac{3\pi}{4}\right) = -\sqrt{2}$

**11.** Abs. min. $f\left(\frac{\pi}{3}\right) = \dfrac{-3\sqrt{3} + \pi}{6}$

**13.** Absolute minimum $f(1) = 0$, absolute maximum $f(e) = 1/e$

**15.** $-1$  **17.** $\sqrt{3}$  **19.** $\cos^{-1}\left(\frac{2}{\pi}\right)$

**21.** $f$ is not continuous on $[-2, 0]$.

**23. a.** Increasing on $(-\infty, \infty)$  **b.** None
  **c.** CU on $(1, \infty)$, CD on $(-\infty, 1)$  **d.** $\left(1, -\frac{17}{3}\right)$

**25. a.** Increasing on $(-1, 0)$ and on $(1, \infty)$, decreasing on $(-\infty, -1)$ and on $(0, 1)$
  **b.** Rel. max. $(0, 0)$, rel. min. $(-1, -1)$ and $(1, -1)$
  **c.** CU on $\left(-\infty, -\frac{\sqrt{3}}{3}\right)$ and on $\left(\frac{\sqrt{3}}{3}, \infty\right)$, CD on $\left(-\frac{\sqrt{3}}{3}, \frac{\sqrt{3}}{3}\right)$
  **d.** $\left(-\frac{\sqrt{3}}{3}, -\frac{5}{9}\right), \left(\frac{\sqrt{3}}{3}, -\frac{5}{9}\right)$

**27. a.** Increasing on $(-\infty, 0)$ and on $(2, \infty)$, decreasing on $(0, 1)$ and on $(1, 2)$
  **b.** Rel. max. $(0, 0)$, rel. min. $(2, 4)$
  **c.** CU on $(1, \infty)$, CD on $(-\infty, 1)$
  **d.** None

**29. a.** Decreasing on $(-\infty, \infty)$  **b.** None
  **c.** CU on $(1, \infty)$, CD on $(-\infty, 1)$  **d.** $(1, 0)$

**31. a.** Increasing on $(-\infty, -1)$ and on $(-1, \infty)$  **b.** None
  **c.** CU on $(-\infty, -1)$, CD on $(-1, \infty)$  **d.** None

**35.** Decreasing on $\left(0, \frac{1}{2}\right)$, increasing on $\left(\frac{1}{2}, \infty\right)$

**37.** $-\infty$  **39.** $-\infty$  **41.** $\infty$  **43.** 3  **45.** 0

**47.** HA $y = 0$, VA $x = -\frac{3}{2}$

**49.** 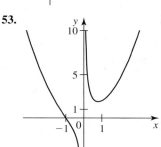  **51.**

**53.** 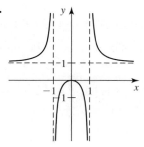  **55.**

**57.** 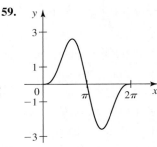  **59.**

**61.** 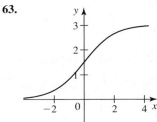  **63.**

**65.** 0  **67.** $\frac{2}{3}$  **69.** 0  **71.** 0

**73.** $\frac{1}{2}$  **75.** $6885.64

**79. a.** Increasing on $(12.7, 30)$, decreasing on $(0, 12.7)$
  **b.** $P(12.7) \approx 7.9$
  **c.** The smallest percentage of women over 65 in the workforce was about 7.9% in late 1982.

**81.** $10\sqrt{2} \times 4\sqrt{2}$  **83.** $(2, 1)$

**85. a.** 30,000

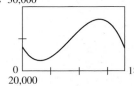

**87.** Maximum when $t = 2, 6, 10, 14, \ldots$; minimum when $t = 0, 4, 8, 12, \ldots$

**89. a.** 43 mg/day  **b.** $t = 1$  **c.** 125 mg

**91.** 1 ft $\times$ 2 ft $\times$ 2 ft  **93.** 1 ft

**95.** $-0.7709$  **97.** $0.8767$

**99.** $a = -4$, $b = 11$  **101.** No

## Chapter 3 Challenge Problems • page 401

**1.** $f\left(\frac{1}{2}\right) = \frac{1}{4}$

**3.** Highest point: $(-1, 2)$, lowest point: $(1, -2)$

**17.** $\dfrac{a_{n-1} - b_{n-1}}{n}$

# CHAPTER 4

## 4.1 Exercises • page 412

**1.** $\frac{1}{2}x^2 + 2x + C$　　**3.** $\frac{1}{3}x^3 - x^2 + 3x + C$

**5.** $\frac{1}{5}x^{10} + 3e^x + 4x + C$　　**7.** $\frac{2}{3}x^{3/2} + 6\sqrt{x} + C$

**9.** $e^t + \frac{t^{e+1}}{e+1} + C$　　**11.** $6\sqrt{u} + C$

**13.** $3x + \frac{2}{x} - \frac{1}{3x^3} + C$　　**15.** $\frac{2}{5}x^{5/2} - \frac{4}{3}x^{3/2} + 6x^{1/2} + C$

**17.** $\frac{2^x}{\ln 2} + \ln|x| + C$　　**19.** $-3\cos x - 4\sin x + C$

**21.** $-\cot x + \frac{2}{3}x^{3/2} + C$　　**23.** $-\csc x + C$

**25.** $\tan x + 2\cot x + C$　　**27.** $\tan x - \cot x + C$

**29.** $\tan x + \sec x + C$

**31. a.** $\frac{1}{2}x^2 - 3x + C$　**b.**

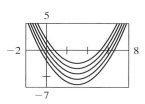

**33. a.** $x^2 - \cos x + C$　**b.**

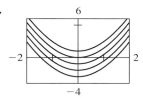

**35.** $x^2 + x + 1$　　**37.** $2\sqrt{x} - 2$　　**39.** $e^x - \cos x$

**41.** $3x^2 - 4x + 5$　　**43.** $\frac{1}{2}x^4 + x^3 + x^2 + 3x + 3$

**45.** $-4t^{1/2} + 4t - 7$　　**47.** $\frac{1}{3}x^3 - x^2 + 3x - \frac{1}{3}$

**49.** Graph 1 is $f$.　　**51.** $2t^3 - 2t^2 + t - 2$

**53.** $t^3 - 2t^2 + 4t - 2$　　**55.** $-\sin t + 2\cos t + t + 1$

**57.** $0.1t^2 + 3t$

**59. a.** $-16t^2 + 400$　**b.** 5 sec　**c.** $-160$ ft/sec

**65.** 2.15 m, 0.82 m　　**67.** $-\frac{88}{9}$ ft/sec², 396 ft

**69.** 0.924 ft/sec²　　**71.** Branch $A$　　**73.** $10{,}000x - 100x^2$

**75. a.** $0.001547x^3 - 0.1506x^2 + 4.9x - 53.09$

　　**b.** 0.96%, 3.42%

**77.** $1.0974t^3 - 0.0915t^4 + 34$

**79.** $-\frac{1}{25}\left(2\sqrt{5}t - \frac{1}{100}t^2\right) + 20$

**81.** True　　**83.** False　　**85.** False　　**87.** True

## 4.2 Exercises • page 425

**1.** $\frac{1}{12}(2x + 3)^6 + C$　　**3.** $\sqrt{x^2 + 1} + C$　　**5.** $\frac{1}{4}\tan^4 x + C$

**7.** $\frac{1}{5}(x^2 + 1)^5 + C$　　**9.** $\frac{5}{16}(2x - 4)^{8/5} + C$

**11.** $\frac{1}{8}(2x^2 + 3)^6 + C$　　**13.** $-\frac{1}{3}\sqrt{(1 - 2x)^3} + C$

**15.** $\frac{1}{10}(s^4 - 1)^{5/2} + C$　　**17.** $-\frac{1}{2}e^{-x^2} + C$

**19.** $\frac{1}{2}(e^{2x} + 4x - e^{-2x}) + C$　　**21.** $-\dfrac{1}{2^x \ln 2} + C$

**23.** $e^{-1/x} + C$　　**25.** $4\sin\frac{x}{2} + C$　　**27.** $\dfrac{1}{2\pi}\sin \pi x^2 + C$

**29.** $-\dfrac{1}{2\pi}\cos^2 \pi x + C$ or $\dfrac{1}{2\pi}\sin^2 \pi x + C$　　**31.** $\frac{1}{3}\sec 3x + C$

**33.** $\cos u^{-1} + C$　　**35.** $\frac{1}{6}\tan^2 3x + C$

**37.** $\sqrt{2 + \sin 2t} + C$　　**39.** $\dfrac{1}{1 - \sec x} + C$

**41.** $\frac{1}{2}x - \dfrac{1}{4\pi}\sin 2\pi x + C$　　**43.** $-\ln(1 + e^{-x}) + C$

**45.** $\dfrac{2(\ln x)^{3/2}}{3\sqrt{\ln 10}} + C$　　**47.** $\frac{1}{2}[\ln(e^x + 1)]^2 + C$

**49.** $\frac{1}{2}\ln|2x + 3| + C$　　**51.** $\ln|\ln x| + C$

**53.** $\ln|1 + \sin x| + C$　　**55.** $\ln|\sec\theta + \tan\theta| + \sin\theta + C$

**57.** $\ln|2 + x\ln x| + C$　　**59.** $\sin^{-1}\left(\frac{1}{4}x\right) + C$

**61.** $\dfrac{1}{18}\sec^{-1}\left(\dfrac{x^2}{9}\right) + C$　　**63.** $x - 2\tan^{-1}x + C$

**65.** $-\sin^{-1}\left(\dfrac{\cos x}{2}\right) + C$　　**67.** $e^{\tan^{-1}x} + C$

**69.** $\tan^{-1}\left(\dfrac{\sqrt{x}}{2}\right) + C$　　**71.** $\dfrac{1}{2}\tan^{-1}\left(\dfrac{x - 2}{2}\right) + C$

**73.** $\frac{2}{15}(3x + 8)\sqrt{(x - 4)^3} + C$

**75.** $\frac{1}{9}(x^2 + 1)^{9/2} - \frac{1}{7}(x^2 + 1)^{7/2} + C$

**77.** $\frac{2}{3}(x + 1)^{3/2} - \frac{2}{3}x^{3/2} + C$　　**79.** $\frac{1}{3}(1 + x^2)^{3/2} + \frac{2}{3}$

**81.** $-\frac{1}{3}(16 - t^2)^{3/2} + \frac{64}{3}$　　**83.** 80.04 years

**85.** $\dfrac{1.2}{\pi}\left(1 - \cos\dfrac{\pi t}{2}\right)$

**87.** $v(t) = -2\sin 2t + \frac{3}{2}\cos 2t$, $s(t) = \cos 2t + \frac{3}{4}\sin 2t - 1$

**89.** True

## 4.3 Exercises • page 442

**1. a.**

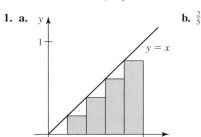

**b.** $\frac{2}{5}$

**3. a.**

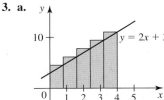

**b.** $\frac{156}{5}$

**5. a.**

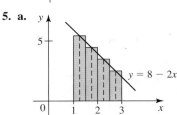

**b.** 8

**7. a.**

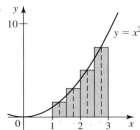

**b.** $\frac{69}{8} = 8.625$

**9. a.**

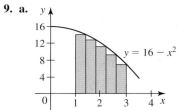

**b.** 21.68

**11. a.**

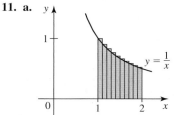

**b.** 0.72

**13.** 10    **15.** 25    **17.** 55    **19.** 6.15

**21.** $\displaystyle\sum_{k=1}^{30} 2k$    **23.** $\displaystyle\sum_{k=1}^{11}(2k+1)$    **25.** $\displaystyle\sum_{k=1}^{5}\left(\frac{2k}{5}+1\right)$

**27.** $\displaystyle\frac{1}{n}\sum_{k=1}^{n}\left[2\left(\frac{k}{n}\right)^3 - 1\right]$    **29.** $\displaystyle\frac{1}{n}\sum_{k=1}^{n}\sin\left(1+\frac{k}{n}\right)$

**31.** 120    **33.** 275    **35.** 13,695

**37.** $\dfrac{4n^3 + 12n^2 + 11n}{3}$    **39.** 1    **41.** $\frac{15}{2}$    **43.** $\frac{13}{3}$

**45.** 6    **47.** $\frac{1}{3}$    **49.** 9    **51.** $\frac{14}{3}$

**53. a.** $\displaystyle\lim_{n\to\infty}\frac{32}{n^5}\sum_{k=1}^{n}k^4$

**b.** $\dfrac{16(n+1)(2n+1)(3n^2+3n-1)}{15n^4}$    **c.** $\frac{32}{5}$

**55. a.** $\displaystyle\lim_{n\to\infty}\sum_{k=1}^{n}\left[\left(2+\frac{3k}{n}\right)^4 + 2\left(2+\frac{3k}{n}\right)^2 + \left(2+\frac{3k}{n}\right)\right]\left(\frac{3}{n}\right)$

**b.** $\dfrac{3(2357n^4 + 3270n^3 + 1200n^2 - 27)}{10n^4}$    **c.** $\frac{7071}{10}$

**59.** 9400 ft$^2$    **63.** False    **65.** False

## 4.4 Exercises • page 459

**1. a.** 14    **b.** 4    **c.** 6

**3. a.**

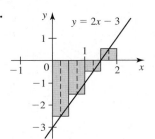

**b.** −2

**5. a.**

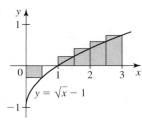

**b.** 0.83

**7.** 2    **9.** −4    **11.** $-\frac{5}{3}$    **13.** $\displaystyle\int_{-3}^{-1}(4x-3)\,dx$

**15.** $\displaystyle\int_{1}^{2}\frac{2x}{x^2+1}\,dx$    **17.** $\displaystyle\lim_{n\to\infty}\sum_{k=1}^{n}\left(1+c_k^3\right)^{1/3}\Delta x$

**19. a.** −6    **b.** $-\frac{19}{2}$    **c.** 13    **d.** $-\frac{5}{2}$

**21.**

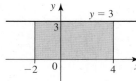

18

**23.**

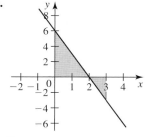

$\frac{9}{2}$

**25.**

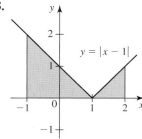

$y = |x - 1|$

$\frac{5}{2}$

**27.**

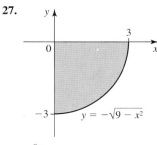

$y = -\sqrt{9 - x^2}$

$-\frac{9\pi}{4}$

**29. a.** 2  **b.** 1  **c.** 6  **d.** −13

**31. a.** 3  **b.** −7  **c.** 19    **33.** 0

**41.** $\sqrt{3} \le \int_1^2 \sqrt{1 + 2x^3}\, dx \le \sqrt{17}$

**43.** $3 \le \int_{-1}^2 (x^2 - 2x + 2)\, dx \le 15$

**45.** $0.367 \le \int_0^1 e^{-x^2}\, dx \le 1$

**47. a.**

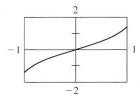

**53. a.**

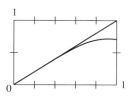

**57. a.** $1 \le \int_0^1 \sqrt{1 + x^2}\, dx \le \sqrt{2}$

  **b.** $1 \le \int_0^1 \sqrt{1 + x^2}\, dx \le \dfrac{1 + \sqrt{2}}{2}$

**59.** 4    **61.** No    **63.** True

**65.** False    **67.** True    **69.** False

## 4.5 Exercises • page 475

**1. a.** $x^2$  **b.** $\frac{1}{3}x^3 - \frac{8}{3}$  **c.** $x^2$    **3.** $\sqrt{3x + 5}$

**5.** $\dfrac{1}{x^2 + 1}$    **7.** $-\sin 2x$    **9.** $\dfrac{\sin \sqrt{x}}{2x}$

**11.** $-\dfrac{\sin x \cos^2 x}{\cos x + 1}$    **13.** $x(9x - 4)\ln x$

**15.** 20    **17.** $-\frac{22}{3}$    **19.** 21

**21.** $2 + 3 \ln 3$    **23.** 2    **25.** $\frac{32}{3}$

**27.** $208\sqrt{2}/105$    **29.** $\dfrac{3 - \sqrt{3}}{3}$

**31.** $\frac{4}{3}$    **33.** $2\sqrt{3}/3$    **35.** $\frac{19}{6}$    **37.** $\frac{121}{5}$

**39.** $\frac{6561}{2}$    **41.** $\frac{3}{4}(4\sqrt[3]{4} - 1)$    **43.** $\frac{1}{3}$

**45.** $\frac{1}{2}[\ln 2 - \ln(e^{-2} + 1)]$    **47.** $2 - \sqrt{2}$

**49.** 0    **51.** 1    **53.** $\frac{1}{2}e(e^{e-1} - 1)$

**55.** $\dfrac{2}{\ln 3} + \dfrac{1}{4}$    **57.** $\dfrac{\pi}{12}$    **59.** $\dfrac{\pi}{12}$

**61.** $\dfrac{\pi}{16}$    **63. b.** 0.14342    **65.** 6

**67.** $\frac{32}{3}$    **69.** $2\left(\sqrt{e} - \dfrac{1}{e}\right)$

**71. a.**

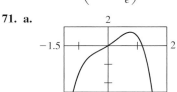

  **b.** 0, 1.165  **c.** 1.026

**73.** $\frac{1}{5}$    **75.** $\frac{56}{3}$    **77.** $\frac{1}{2}$    **79.** $\dfrac{\sqrt{10} - 1}{3}$

**81. a.** $\dfrac{\sqrt{21}-3}{3}$   **b.**

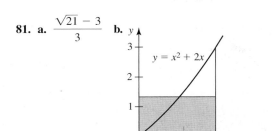

**83. a.** $\frac{769}{225}$   **b.**

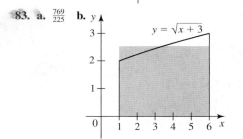

**85. a.** 4.5 ft   **b.** 15.75 ft

**87.** 46%; 24%   **93.** $\frac{80}{3}$ cm/sec

**95.** 150.937 pollutant standard index

**97.** 8373 wolves, 50,804 caribou

**99.** 343.45 ppmv/year

**101.** 39.16 million barrels

**103.** 43.3 sec   **105.** 15.54

**109.** Maxima at $(2n-1)\pi$, minima at $2n\pi$, $n = 1, 2, \ldots$

**111.** 3   **113.** 3   **115.** 0   **117. b.** $\pi$

**125.** True   **127.** True

## 4.6 Exercises • page 489

**1. a.** 2.75   **b.** $\frac{8}{3}$   Exact value: $\frac{8}{3}$

**3. a.** 3.7708   **b.** $\frac{15}{4}$   Exact value: $\frac{15}{4}$

**5. a.** 4.3766   **b.** 4.3328   Exact value: $\frac{13}{3}$

**7. a.** 0.3129   **b.** 0.3161   Exact value: 0.316060

**9.** 0.5523   **11.** 0.8806   **13.** 1.1643

**15.** 3.2411   **17.** 2.2955   **19.** 1.9101

**21. a.** $\frac{3}{4}$   **b.** 0   **23. a.** $\frac{9}{1024}$   **b.** $\frac{81}{262,144}$

**25. a.** $\pi^3/1728$   **b.** $\pi^5/1,866,240$

**27.** 0.0026; $1.63 \times 10^{-5}$   **29.** 13

**31.** 27   **33.** 28   **35.** 8

**37.** 4   **39.** 4   **41.** 52.82 ft/sec

**43.** 474.77 million barrels

**45.** 1922.4 ft³/sec   **47.** False

## Chapter 4 Concept Review • page 491

**1. a.** $F' = f$   **b.** $F(x) + C$   **3. a.** unknown   **b.** function

**5. a.** $\displaystyle\int_a^b f(x)\,dx$   **b.** minus

**7. a.** $f(x)$   **b.** $F(b) - F(a)$; antiderivative   **c.** $\displaystyle\int_a^b f'(x)\,dx$

**9.** $f(c) = \dfrac{1}{b-a}\displaystyle\int_a^b f(x)\,dx$

## Chapter 4 Review Exercises • page 492

**1.** $\frac{1}{2}x^4 - \frac{4}{3}x^3 + \frac{3}{2}x^2 + 4x + C$   **3.** $\frac{3}{8}x^{8/3} - \frac{10}{7}x^{7/5} + C$

**5.** $\frac{3}{5}x^{5/3} + 3x + \frac{2}{3}x^{-3} + C$   **7.** $\frac{1}{8}(1+2t)^4 + C$

**9.** $\frac{1}{27}(3t-4)^9 + C$   **11.** $\frac{1}{3}x^3 + 2x - \dfrac{1}{x} + C$

**13.** $-\dfrac{1}{4(3x^2+2x)^2} + C$   **15.** $-\frac{1}{5}\cos^5 t + C$

**17.** $-2\sqrt{1-\sin\theta} + C$   **19.** $-\frac{1}{2}\csc x^2 + C$

**21.** $\frac{1}{5}\ln|5x-3| + C$   **23.** $\frac{2}{3}x - \frac{1}{9}\ln|3x+2| + C$

**25.** $\dfrac{2^{t^2}}{2\ln 2} + C$   **27.** $-\cos\ln x + C$

**29.** $-2\ln|\cos\sqrt{x}| + C$   **31.** $\frac{1}{4}(\tan^{-1}2x)^2 + C$

**33.** 16   **35.** $\frac{1}{8}$   **37.** 2   **39.** $\frac{155}{6}$

**41.** $\frac{1}{2}(\sqrt{2}-1)$   **43.** $\dfrac{\sqrt{3}-\sqrt{2}}{2}$

**45.** $2(1-\ln 2)$   **47.** $\frac{1}{2}(e^2 - 1)$

**49.** 2   **51.** $(\ln 2)^2$   **53.** $\dfrac{2xe^{x^2}}{x^4 + 1}$

**55.** $3x^2\sin x^3 - 2x\sin x^2$   **57.** 1.1502

**59. a.** $\frac{1}{128}$   **b.** $\frac{7}{24,576}$   **61.** $\frac{2}{3}x^{3/2} - \cos x + 3$

**63. a.** $-16t^2 + 128$   **b.** $2\sqrt{2}$ sec   **c.** $-64\sqrt{2}$ ft/sec

**65.** 352 ft   **67.** 0

**69.** $-0.000002x^3 - 0.02x^2 + 200x - 80,000$

**71. a.** $11.9\sqrt{1+0.91t}$   **b.** 25.6 million   **73.** $6/(5\pi)$ L/sec

**75. a.** 64.45°F   **b.** 64.63°F   **77. b.** 0.500005

## Chapter 4 Challenge Problems • page 495

**1.** $|b| - |a|$   **3.** $20\sqrt{2}$   **5.** $-\frac{4}{5}$

**7.** $\dfrac{1}{\sin a}\tan^{-1}\left(\dfrac{x + \cos a}{\sin a}\right) + C$   **9.** $f(a)$   **11. b.** 0

**13. b.** $\dfrac{\sin x}{2\sqrt{x}} + \dfrac{\sin(1/x^2)}{x^2}$   **15. b.** Yes   **23.** $f(x) = x$

# CHAPTER 5

## Exercises 5.1 • page 506

**1.** $\frac{40}{3}$    **3.** $\frac{40}{3}$    **5.** $\frac{13}{12}$

**7.** $\int_{2010}^{2050} [f(t) - g(t)]\, dt$ billion barrels

**9.**

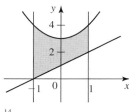

$\frac{14}{3}$

**11.**

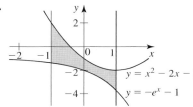

$y = x^2 - 2x - 1$

$y = -e^x - 1$

$\dfrac{3e^2 + 2e - 3}{3e}$

**13.**

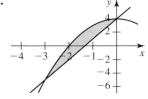

$\frac{9}{2}$

**15.**

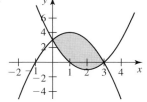

9

**17.**

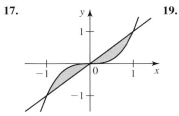

$\frac{1}{2}$

**19.**

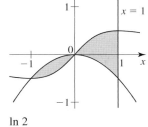

$x = 1$

ln 2

**21.**

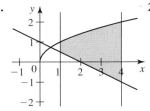

$\frac{65}{12}$

**23.**

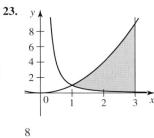

8

**25.**

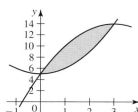

9

**27.**

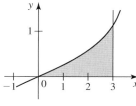

$4 - \sqrt{7}$

**29.**

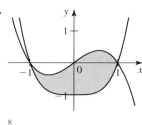

$\frac{21}{2}$

**31.**

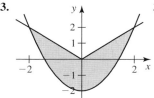

$\frac{8}{5}$

**33.**

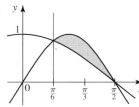

$\frac{20}{3}$

**35.**

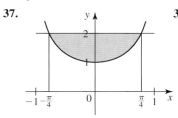

$\frac{1}{4}$

**37.**

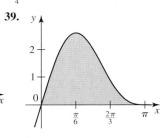

$\pi - 2$

**39.**

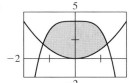

4

**41.** $4\sqrt{2}/3$    **43.** 11    **45. a.** 8    **b.** 8

**47.** $A = \int_0^b [g(x) - f(x)]\, dx$

**49. a.**

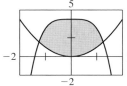

**b.** $-1.25$ and $1.25$, $7.48$

**51. a.**

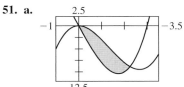

**b.** 0 and 2.25, 7.59

**53. a.** 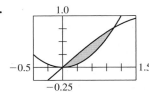   **b.** 0 and 0.88, 0.14

**55.** 30 ft/sec   **57.** 3,661,581 ft$^2$

**59. a.** 300   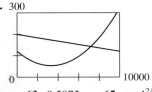   **b.** 342,000

**61.** $\frac{1}{12}$   **63.** 0.5875   **65. a.** $4^{2/3}$   **b.** 0.69

**67. a.** $(m - 1)/(m + 1)$   **b.** 0, 1

**69. a.** 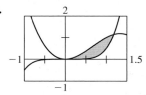   **b.** 0.78540

**71.** False   **73.** True

## Exercises 5.2 • page 522

**1.** $2\pi/3$   **3.** $153\pi/5$   **5.** $\pi(1 - (\pi/4))$   **7.** $\pi/10$

**9.** $64\pi/15$   **11.** $47\pi/84$   **13.** $32\pi/5$   **15.** $16\pi/15$

**17.** $(\pi/2)(e^2 - 1)$   **19.** $\pi/2$   **21.** $16\pi/3$   **23.** $64\pi/3$

**25.** $128\pi/15$   **27.** $\pi^2/4$   **29.** $3\pi/10$

**31.** $(\pi/2)(e^2 - 1)$   **33.** $19\pi/48$

**35. a.**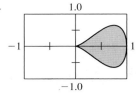

**b.** (0, 0) and (1.18, 1.14), 1.08

**37.** $64\pi/15$   **39.** $1088\pi/15$   **41.** $104\pi/15$

**43.** 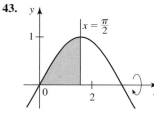   **45.**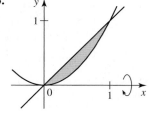

**47.** $(\pi/15)a^3$   **49.** $2\pi/5$   **51.** $71\pi/30$

**53.** $(\pi h/3)(R^2 + rR + r^2)$   **55.** $(\pi h^2/3)(3r - h)$

**57.** 32   **59.** $81\sqrt{3}/40$

**61. a.**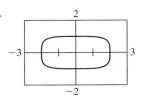

**b.** 12.6   **c.** 10.9832

**63. a.** 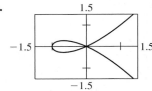   **b.** $\pi/24$

**65.** $\frac{32}{2}$   **67.** $\frac{2}{3}a^3$   **69. b.** $\pi r^2 h$   **71.** 33.52 ft$^3$

## Exercises 5.3 • page 533

**1.** $8\pi/3$   **3.** $\pi/6$   **5.** $48\pi/5$

**7.** $8\pi$   **9.** $8\pi/3$

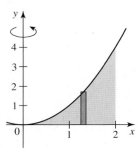

**11.** $2\pi$   **13.** $\pi \ln 5$

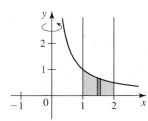   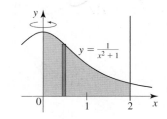

**15.** $18\pi$   **17.** $8\pi$

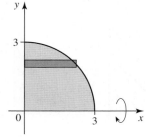   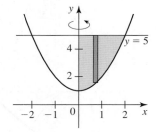

**19.** $\pi/3$   **21.** $19\pi/48$

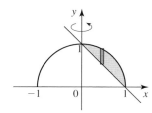

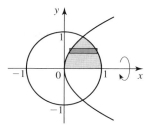

**23. a.**

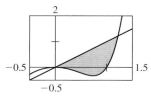

**b.** $(0, 0)$ and $(1.22, 1.22)$   **c.** 3.67

**25.** $16\pi/3$   **27.** $9\pi/4$

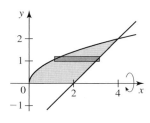

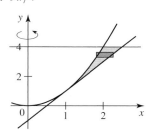

**29.** $\pi/15$   **31.** $32\pi/3$

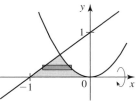

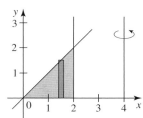

**33.** $128\pi/3$   **35.** $8\pi/15$

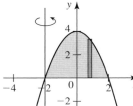

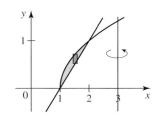

**37.**

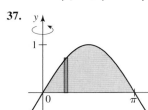

**39.**

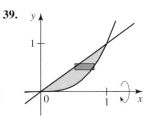

**43.** $\frac{4}{3}\pi a^2 b$   **45.** $2\pi^2 a^2 b$

**47.** $10\pi$ ft$^3$   **49.** 296,231,243 ft$^3$

## Exercises 5.4 • page 546

**1.** $\frac{1}{27}(80\sqrt{10} - 13\sqrt{13})$   **3.** $\frac{8}{27}(10\sqrt{10} - 1)$

**5.** $\sqrt{73}$   **7.** $3\sqrt{5}$   **9.** $\frac{2}{27}(37\sqrt{37} - 1)$   **11.** 45

**13.** $\frac{2}{27}(37\sqrt{37} - 1)$   **15.** $\frac{3}{4}$   **17.** 0.8814

**19.** $\displaystyle\int_{-1}^{2}\sqrt{1 + 4x^2}\,dx$   **21.** $\displaystyle\int_{-1}^{2}\sqrt{1 + \frac{4x^2}{(x^2 + 1)^4}}\,dx$

**23.** $\displaystyle\int_{0}^{\pi/4}\sqrt{1 + \sec^4 x}\,dx$

**25. a.**

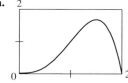

**b.** $\displaystyle\int_{0}^{2}\sqrt{1 + 4x^4(3 - 2x)^2}\,dx$   **c.** 4.2008

**27. a.**

**b.** $\displaystyle\int_{0}^{4}\sqrt{1 + \left(1 - \frac{1}{\sqrt{x}}\right)^2}\,dx$   **c.** 4.8086

**29.** $6a$   **31.** 0.6325   **33.** $5\sqrt{5}\pi$

**35.** $(\pi/27)(10\sqrt{10} - 1)$   **37.** $(\pi/6)(17\sqrt{17} - 1)$

**39.** $\dfrac{45\sqrt{13}\pi}{4}$   **41.** $\pi/4$   **43.** $(\pi/16)(16\ln 2 + 15)$

**45.** $2\pi\displaystyle\int_{0}^{\pi/2}\sin x\sqrt{1 + \cos^2 x}\,dx$   **47.** 21.4018

**51.** $4\pi$   **53.** $\frac{8}{3}$ mi   **55.** 24,223.5 ft   **57.** 30.73 in.

## Exercises 5.5 • page 554

**1.** 200 ft-lb   **3.** $-500$ ft-lb

**5.** 6 ft-lb   **7.** 18 J   **9.** $-\dfrac{2}{\pi}$ J

**11.** $\frac{17}{6}$ J   **13.** $\frac{7}{24}$ ft-lb   **15.** 360 ft-lb

**17.** 15,600 ft-lb   **19.** 56,458 ft-lb

**21.** 1740 ft-lb   **23.** 121,919 J   **25.** 128,252 ft-lb

**27.** 5799 ft-lb   **29.** 1405 ft-lb   **31.** $1.91 \times 10^{13}$ J

**33.** $3.46 \times 10^8$ ft-lb

**35.** $\dfrac{qQ}{4\pi\varepsilon_0}\left(\dfrac{1}{\sqrt{a^2+R^2}}-\dfrac{1}{\sqrt{b^2+R^2}}\right)$

**37.** $nRT\ln\dfrac{V_1-nb}{V_0-nb}+an^2\dfrac{V_0-V_1}{V_0V_1}$

### Exercises 5.6 • page 563

**1. a.** 187.2 lb   **b.** 93.6 lb   **c.** 31.2 lb

**3.** 1040 lb    **5.** 6760 lb    **7.** 1064.96 lb    **9.** 3057.6 lb

**11.** 374.4 lb    **13.** 561.6 lb    **15.** 12.0 lb    **17.** 7841 lb

**19. a.** 8387 lb   **b.** 36,741 lb

**21.** 477.9 lb    **23.** 610,509 lb

### Exercises 5.7 • page 574

**1.** $\frac{7}{6}$ m    **3.** $\frac{1}{3}$ m    **5.** $\left(-\frac{5}{12},\frac{3}{2}\right)$    **7.** $\left(\frac{5}{6},\frac{7}{9}\right)$    **9.** $\left(1,\frac{2}{3}\right)$

**11.** $\left(0,\frac{8}{5}\right)$    **13.** $\left(0,\frac{1}{5}\right)$    **15.** $\left(1,\frac{2}{5}\right)$    **17.** $\left(5,\frac{10}{7}\right)$

**19.** $\left(\frac{9}{20},\frac{9}{20}\right)$    **21.** $\left(\frac{16}{35},\frac{16}{35}\right)$    **23.** $\left(1,\frac{13}{5}\right)$    **25.** $\left(0,\frac{4}{3}\right)$

**27.** $\left(0,-\dfrac{20}{3(8+\pi)}\right)$    **29.** $\left(\frac{37}{18},\frac{23}{18}\right)$    **31.** $\left(\dfrac{\pi-4}{\pi+4},0\right)$

**33.** $\left(\dfrac{\pi-2}{\pi+2},0\right)$    **35.** $\left(\dfrac{a}{3},\dfrac{b}{3}\right)$    **39.** $(1.44,0.36)$

**41.** $72\pi^2$    **43.** $8\pi$    **45.** $\left(0,\dfrac{4R}{3\pi}\right)$

**47.** $\left(0,\dfrac{2a}{\pi}\right)$    **49.** $\left(\dfrac{1}{2},\dfrac{\pi}{8}\right)$

### Exercises 5.8 • page 584

**1. a.** 3.6269   **b.** 27.3082   **c.** 0.0993

**3. a.** 1   **b.** 0.6481   **c.** 1.3333

**5. a.** 0.4812   **b.** $-0.4812$   **c.** 0.8047

**17.** $\operatorname{csch}x=\frac{3}{4}$, $\cosh x=\frac{5}{3}$, $\operatorname{sech}x=\frac{3}{5}$, $\tanh x=\frac{4}{5}$, $\coth x=\frac{5}{4}$

**19.** $3\cosh 3x$    **21.** $-3\operatorname{sech}^2(1-3x)$    **23.** $e^{2t}$

**25.** $\tanh x$    **27.** $\dfrac{2u\sinh u^2}{\cosh^2(\cosh u^2)}$

**29.** $12t\cosh(3t^2+1)\sinh(3t^2+1)$

**31.** $\sinh v^2+2v^2\cosh v^2$    **33.** $2e^{2x}\operatorname{sech}^2(e^{2x}+1)$

**35.** $-\frac{2}{3}(\cosh x-\sinh x)^{2/3}$    **37.** $-\operatorname{csch}x$

**39.** $\dfrac{1}{1+\cosh x}$

**41.** $\dfrac{(1+\tanh 2t)\cdot\dfrac{1}{\sqrt{t^2-1}}-2\cosh^{-1}t\,\operatorname{sech}^2 2t}{(1+\tanh 2t)^2}$

**43.** $\dfrac{3}{\sqrt{1+9x^2}}$    **45.** $\dfrac{1}{\sqrt{\cosh^{-1}2x}\,\sqrt{4x^2-1}}$

**47.** $-\dfrac{1}{(2x+1)\sqrt{-2x}}$    **49.** $\cosh^{-1}x^2+\dfrac{2x^2}{\sqrt{x^4-1}}$

**51.** $-\dfrac{1}{2x\sqrt{1-x}}$    **53.** $\dfrac{9(x-1)}{\sqrt{9x^2-1}}$

**55.** $\frac{1}{2}\sinh(2x+3)+C$    **57.** $\frac{2}{3}(\sinh x)^{3/2}+C$

**59.** $\frac{1}{3}\ln|\sinh 3x|+C$    **61.** $\ln(1+\cosh x)+C$

**63.** $\pi a^3\left(\dfrac{b}{a}+\dfrac{1}{2}\sinh\dfrac{2b}{a}\right)$    **65.** $\dfrac{T_0^2}{bW}\sinh\dfrac{Wb}{T_0}$

**67. a.** $0,\ -2$ ft/sec

**b.**

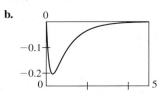

**69.** $\pi[(b-a)+\frac{1}{2}(\sinh 2b-\sinh 2a)]$

**71.** $\left(0,\dfrac{e^{2a}+4a-e^{-2a}}{8(e^a-e^{-a})}\right)$

**77.** True

**79.** True

### Chapter 5 Concept Review • page 586

**1. a.** $\displaystyle\int_a^b[f(x)-g(x)]\,dx$

**b.** $\displaystyle\int_a^b|f(x)-g(x)|\,dx$

**3.** $\displaystyle\pi\int_a^b\{[f(x)]^2-[g(x)]^2\}\,dx$

**5. a.** $\displaystyle 2\pi\int_a^b xf(x)\,dx$

**b.** $\displaystyle 2\pi\int_c^d yf(y)\,dy$

**7.** $\displaystyle\int_a^x\sqrt{1+[f'(t)]^2}\,dt$; $[a,b]$; $\sqrt{1+(y')^2}\,dx$; $(dx)^2+(dy)^2$

**9.** $\displaystyle\int_a^b F(x)\,dx$

**11. a.** $\displaystyle\int_a^b\rho f(x)\,dx$

**b.** $\displaystyle\frac{1}{2}\int_a^b\rho[f(x)]^2\,dx$; $\displaystyle\int_a^b\rho xf(x)\,dx$

**c.** $\dfrac{M_y}{m}$; $\dfrac{M_x}{m}$

## Chapter 5 Review Exercises • page 587

**1.**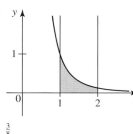

$\frac{3}{8}$

**3.**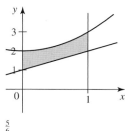

$\frac{5}{6}$

**5.**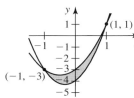

$\frac{4}{3}$

**7.**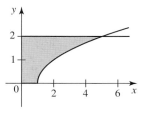

$\frac{14}{3}$

**9.**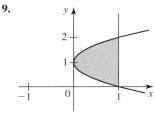

$\frac{4}{3}$

**11.**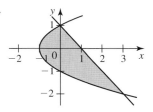

$\frac{9}{2}$

**13.**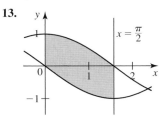

2

**15.**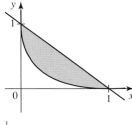

$\frac{1}{3}$

**17.** $\frac{1}{2}(e^4 + 2\ln 2 - e^2)$    **19.** $8\pi$    **21.** $8\pi/21$

**23.** $7\pi/30$    **25.** $\pi$    **27.** $2\coth 2x$    **29.** $\dfrac{\operatorname{sech}^2 x}{\sqrt{2 - \operatorname{sech}^2 x}}$

**31.** $\dfrac{-1}{2\sqrt{x - x^2}\left[1 + (\cos^{-1}\sqrt{x})^2\right]}$    **33.** $-\frac{1}{2}x + \frac{1}{4}\sinh 2x + C$

**35.** $\frac{1}{2}\cosh 2t + C$    **37.** $\frac{1}{6}$    **39.** $722$ m$^3$

**41.** $36$    **43.** $\frac{1}{4}(3 + 2\ln 2)$    **45.** $14\pi/3$

**47.** $2\pi\displaystyle\int_1^2 \dfrac{(x^3 + 1)\sqrt{4x^6 + x^4 - 4x^3 + 1}}{x^3}\,dx$

**51.** $\frac{2}{3}$ ft-lb    **53.** $367{,}566$ ft-lb

**55.** deep end: 30,576 lb; shallow end: 5616 lb; other sides: 41,080 lb

**57.** $\left(1, -\frac{2}{5}\right)$    **59.** $\left(1, -\frac{2}{5}\right)$

## Chapter 5 Challenge Problems • page 590

**1.** $a = 3$, $b = 5$    **3.** $y = \dfrac{\sqrt{3}}{3}x^2$    **5.** $f(x) = \frac{2}{3}(1 + x^2)^{3/2}$

**7.** $52$    **9.** $\dfrac{4\pi ab^2(2\sqrt{3} - 3)}{9}$

# CHAPTER 6

### Exercises 6.1 • page 601

**1.** $\frac{1}{4}(2x - 1)e^{2x} + C$    **3.** $\sin x - x\cos x + C$

**5.** $\frac{1}{4}x^2(2\ln 2x - 1) + C$    **7.** $-(x^2 + 2x + 2)e^{-x} + C$

**9.** $(x^2 - 2)\sin x + 2x\cos x + C$

**11.** $x\tan^{-1}x - \frac{1}{2}\ln(1 + x^2) + C$

**13.** $\frac{2}{9}t\sqrt{t}(3\ln t - 2) + C$

**15.** $x\tan x + \ln|\cos x| + C$

**17.** $\frac{1}{13}e^{2x}(3\sin 3x + 2\cos 3x) + C$

**19.** $\frac{1}{4}[\sin(2u + 1) - 2u\cos(2u + 1)] + C$

**21.** $x\tan x + \ln|\cos x| - \frac{1}{2}x^2 + C$

**23.** $2[(x - 2)\sin\sqrt{x} + 2\sqrt{x}\cos\sqrt{x}] + C$

**25.** $\frac{1}{4}\sec^3\theta\tan\theta + \frac{3}{8}(\sec\theta\tan\theta + \ln|\sec\theta + \tan\theta|) + C$

**27.** $x^3\cosh x - 3x^2\sinh x + 6x\cosh x - 6\sinh x + C$

**29.** $-e^{-x}\ln(e^x + 1) - \ln(1 + e^{-x}) + C$

**31.** $4\sqrt{1 - x} - 2\ln\left|\dfrac{\sqrt{1 - x} + 1}{\sqrt{1 - x} - 1}\right| - 2\sqrt{1 - x}\ln x + C$

**33.** $\frac{1}{9}(2e^3 + 1)$    **35.** $\frac{1}{6}(\pi + 6 - 3\sqrt{3})$

**37.** $\dfrac{3\sqrt{e} - 4}{2e}$    **39.** $\dfrac{4\pi - 3\sqrt{3}}{6}$    **41.** $\dfrac{2\ln 2 + \pi - 4}{2}$

**43.** $\frac{1}{2}\ln\frac{3}{2} + \frac{\pi}{36}(9 - 4\sqrt{3})$    **45.** $e - 2$    **47.** $\frac{1}{8}(\pi - 4\ln 2)$

**49. a.**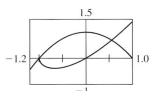

   $-1$, $0.555$    **b.** $1.251$

**51.** $\pi$    **53.** $\frac{\pi}{5}(2e^{\pi/2} - 3)$    **55.** $2\pi(\pi + 2)$

**57.** $P(t) = 820 - 40(t + 20)e^{-0.05t}$; 92.2 million metric tons

**59.** $3te^{-4t}$    **61.** $\frac{1}{3}e^{-20t}(\sin 60t + 3\cos 60t) + q_0 - 1$

**63.** $v_0 t - \dfrac{1}{2}gt^2 + \dfrac{ms}{r}\left\{\left(1 - \dfrac{r}{m}t\right)\left[\ln\left(1 - \dfrac{r}{m}t\right) - 1\right] + 1\right\}$

**65.** $(c_2 - c_1)\left(\dfrac{r_1}{r_2 - r_1} + \dfrac{1}{\ln r_1 - \ln r_2}\right) + c_2$

**67.** $16$    **69.** True

## Exercises 6.2 • page 611

**1.** $\frac{1}{4}\sin^4 x + C$    **3.** $\frac{1}{4}\left(\frac{1}{3}\sin^6 2x - \frac{1}{4}\sin^8 2x\right) + C$

**5.** $\frac{1}{3}\cos^3 x - \cos x + C$    **7.** $\pi/2$    **9.** $\frac{3}{8}$

**11.** $\frac{1}{16}(2x - \sin 2x) + C$    **13.** $\pi/16$

**15.** $\frac{1}{64}\left[12x^2 + 8\sin(2x^2) + \sin(4x^2)\right] + C$

**17.** $\frac{1}{8}(2x^2 - \cos 2x - 2x\sin 2x) + C$

**19.** $\dfrac{4 - \pi}{4}$    **21.** $\dfrac{1}{2}\sec^4\dfrac{x}{2} - 2\tan^2\dfrac{x}{2} - \ln\cos^2\dfrac{x}{2} + C$

**23.** $\dfrac{1}{4\pi}\tan^4(\pi x) + C$    **25.** $\frac{1}{9}\tan^3 3x + \frac{1}{15}\tan^5 3x + C$

**27.** $\frac{2}{3}\tan^{3/2}\theta + \frac{2}{7}\tan^{7/2}\theta + C$    **29.** $-\frac{1}{2}\cot 2x - x + C$

**31.** $-\frac{1}{2}(\cot x \csc x + \ln|\csc x + \cot x|) + C$

**33.** $-\left(\cot t + \frac{1}{3}\cot^3 t\right) + C$

**35.** $-\frac{1}{5}\cot^5 t + \frac{1}{3}\cot^3 t - \cot t - t + C$

**37.** $\frac{4}{3}$    **39.** $-\frac{1}{3}$    **41.** $\frac{1}{4}\sin 2\theta + \frac{1}{12}\sin 6\theta + C$

**43.** $\dfrac{\ln|\sin 2\theta|}{2} - \dfrac{\sin^2 2\theta}{4} + C$    **45.** $\frac{1}{2}\tan^4\sqrt{t} + C$

**47.** $\frac{1}{2}\sin 2x + C$    **49.** $\frac{1}{2}$    **51.** $\frac{1}{2}$

**53. a.**

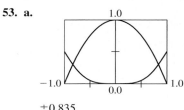

1.0
−1.0   0.0   1.0
±0.835

**b.** 1.165

**55.** $\dfrac{\pi(3\pi - 8)}{12}$    **57.** $\left(\frac{1}{2}(\pi - 2), \frac{\pi}{8}\right)$    **59.** $8/\pi, 0$

**61.** $8\pi/15$    **63.** $-\dfrac{2I_0\sin\alpha}{\pi}$    **65.** $0.12RI_0^2 T$

**67.**

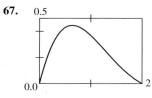

0.5
0.0   2

**a.** 0.55   **b.** 2.63

## Exercises 6.3 • page 619

**1.** $-\sqrt{9 - x^2} + C$    **3.** $-\dfrac{(4 - x^2)^{3/2}}{3} + C$

**5.** $-\dfrac{1}{2}\ln\left|\dfrac{\sqrt{4 + x^2} + 2}{x}\right| + C$    **7.** $-\dfrac{\sqrt{x^2 + 4}}{4x} + C$

**9.** $-\frac{1}{15}(3x^2 + 2)(1 - x^2)^{3/2} + C$

**11.** $\frac{1}{3}(x^2 - 18)\sqrt{x^2 + 9} + C$    **13.** $-\dfrac{x}{9\sqrt{x^2 - 9}} + C$

**15.** $\sqrt{16x^2 - 9} - 3\sec^{-1}\left(\frac{4}{3}x\right) + C$

**17.** $-\dfrac{(1 - x^2)^{3/2}}{3x^3} + C$    **19.** $-\dfrac{1}{2}\ln\left|\dfrac{\sqrt{9x^2 + 4} + 2}{x}\right| + C$

**21.** $\frac{1}{3}(4\pi + 3\sqrt{3})$    **23.** $\frac{1}{2}(\sqrt{3} - \sqrt{2})$

**25.** $2\sin^{-1}\left(\frac{1}{2}e^x\right) + \frac{1}{2}e^x\sqrt{4 - e^{2x}} + C$    **27.** $\ln\dfrac{2\sqrt{3} + 3}{3}$

**29.** $\sin^{-1}(t - 1) + C$

**31.** $\dfrac{1}{16}\tan^{-1}\left(\dfrac{x + 2}{2}\right) + \dfrac{x + 2}{8(x^2 + 4x + 8)} + C$

**33.** $-\dfrac{1}{2}\ln\dfrac{\sqrt{2} + 1}{\sqrt{3} + 2}$    **35.** $\dfrac{\pi}{4}b$    **37.** $\dfrac{4\pi}{3}(2\pi - 3\sqrt{3})$

**39.** $2 + \ln\left(\dfrac{2\sqrt{3}}{3} - \dfrac{\sqrt{3}}{3}\right) - \sqrt{2} - \ln(\sqrt{2} - 1)$

**41.** $\dfrac{8(8\pi + 9\sqrt{3})}{3}$ or $\approx 108.59$ lb

**43. a.**

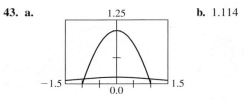

1.25
−1.5   0.0   1.5
±0.953

**b.** 1.114

**45.** $\dfrac{2\pi a^2 b}{\sqrt{a^2 - b^2}}\left[\sin^{-1}\left(\dfrac{\sqrt{a^2 - b^2}}{a}\right) + \dfrac{b\sqrt{a^2 - b^2}}{a^2}\right]$

**47.** $\left\{\frac{544}{10}[\tan^{-1}(0.25) - \tan^{-1}(-2.25)] + 28\right\}$ or $\approx 104$ PSI

**49.** $\dfrac{2\sqrt{5}k}{5\,a^2}$

**55. b.** $\sqrt{1 + x^2}\tan^{-1} x - \ln(\sqrt{1 + x^2} + x) + C$

## Exercises 6.4 • page 630

**1. a.** $\dfrac{A}{x} + \dfrac{B}{x - 5}$    **b.** $\dfrac{A}{x + 1} + \dfrac{B}{3x - 2}$

**3. a.** $\dfrac{A}{x} + \dfrac{B}{x^2} + \dfrac{C}{x + 1}$    **b.** $\dfrac{A}{x + 4} + \dfrac{B}{x - 1}$

**5. a.** $\dfrac{A}{x + 2} + \dfrac{B}{x - 2} + \dfrac{Cx + D}{x^2 + 4}$    **b.** $\dfrac{A}{2x - 1} + \dfrac{Bx + C}{x^2 + 4}$

**7.** $\dfrac{1}{4}\ln\left|\dfrac{x - 4}{x}\right| + C$    **9.** $\ln\dfrac{|t|^3}{(t + 1)^2} + C$    **11.** $\frac{1}{4}\ln\frac{5}{3}$

**13.** $\frac{1}{3}\ln|(x - 2)(x + 1)^2| + C$    **15.** $\ln\left|\dfrac{(x + 3)^3(x - 2)}{(x + 2)^2}\right| + C$

**17.** $2x + \ln|x(x - 1)^2| + C$    **19.** $\ln\left|\dfrac{x^3}{x - 1}\right| - \dfrac{2}{x - 1} + C$

**21.** $\ln|x(x+1)^3| - \dfrac{2}{x} + C$

**23.** $\ln\left|\dfrac{(v-1)^2}{v}\right| - \dfrac{v}{(v-1)^2} + C$

**25.** $2\ln\left|\dfrac{x}{(x+1)^2}\right| + \dfrac{2}{x+1} + x + C$

**27.** $\ln|x| - \dfrac{1}{2}\ln|x^2-1| - \dfrac{1}{2(x^2-1)} + C$

**29.** $\frac{1}{6}(31\ln|x-1| + 8\ln|x+2| - 3\ln|x+3|) + C$

**31.** $x + \tan^{-1}x + \ln\left|\dfrac{x+1}{x^2+1}\right| + C$

**33.** $\dfrac{5}{2}\ln(x^2+1) - 3\tan^{-1}x - \dfrac{1}{x^2+1} + C$

**35.** $\dfrac{1}{2}\ln\dfrac{(x+1)^2}{x^2-4x+6} + C$

**37.** $\dfrac{1}{6}\left[\ln\dfrac{x^2-x+1}{(x+1)^2} + 2\sqrt{3}\tan^{-1}\dfrac{\sqrt{3}(2x-1)}{3}\right] + C$

**39.** $2\ln 2 + \dfrac{\pi}{4} - \dfrac{1}{2}$

**41.** $2\sqrt{3}\tan^{-1}\left[\dfrac{\sqrt{3}}{3}(2x+1)\right] + \dfrac{1}{x^2+x+1} + C$

**43.** $-\dfrac{1}{2(2x^3 - x^2 + 8x + 4)} + C$

**45.** $\ln\left|\dfrac{\cos x}{\cos x + 1}\right| + \sec x + C$

**47.** $\dfrac{1}{3}\ln\left|\dfrac{e^t-1}{e^t+2}\right| + C$

**49.** $e^t + \dfrac{1}{6}\ln\left|\dfrac{(e^t+1)^3(e^t-1)}{(e^t+2)^{16}}\right| + C$

**51.** $3x^{1/3} + \dfrac{1}{2}\ln\dfrac{x^{2/3} - x^{1/3} + 1}{(x^{1/3}+1)^2}$

$\qquad - \sqrt{3}\tan^{-1}\left[\dfrac{\sqrt{3}}{3}(2x^{1/3}-1)\right] + C$

**53.** $\tan\dfrac{x}{2} + C$

**55.** $\dfrac{2\sqrt{15}}{15}\tan^{-1}\dfrac{\sqrt{15}(8\sin x + \cos x + 1)}{15(1+\cos x)} + C$

**57.** $\ln 2$     **59.** $\frac{1}{2}(\ln|\sin x + \cos x| + x) + C$     **61.** $\ln\frac{4}{3}$

**63. a.**     3                      **b.** 0.892

2.144

**65.** $\dfrac{2\pi}{3}(1 + 3\ln 3 - 6\ln 2)$     **67.** $2\ln 3 - 1$

**69. a.** $\dfrac{5}{3}\ln|x+2| - \dfrac{10}{7}\ln|x+3|$

$\qquad - \dfrac{1}{42}\left[5\ln(x^2+x+1) + 2\sqrt{3}\tan^{-1}\dfrac{2x+1}{\sqrt{3}}\right] + C$

**b.** $\dfrac{5}{3(x+2)} - \dfrac{10}{7(x+3)} - \dfrac{5x+4}{21(x^2+x+1)}$

**c.** $\dfrac{1}{42}\left\{-2\sqrt{3}\tan^{-1}\left(\dfrac{2x+1}{\sqrt{3}}\right) + 70\ln(x+2)\right.$

$\qquad\qquad \left. - 5[12\ln(x+3) + \ln(x^2+x+1)]\right\}$

**71.** 9231     **73.** False     **75.** False

## Exercises 6.5 • page 639

**1.** $\frac{1}{15}(3x-1)(1+2x)^{3/2} + C$

**3.** $\dfrac{1}{8}\left(1 + 2x - \dfrac{1}{1+2x} - 2\ln|1+2x|\right) + C$

**5.** $-\dfrac{\sqrt{3+2x}}{x} + \dfrac{\sqrt{3}}{3}\ln\left|\dfrac{\sqrt{3+2x}-\sqrt{3}}{\sqrt{3+2x}+\sqrt{3}}\right| + C$

**7.** $-\dfrac{\sqrt{3}}{3}\ln\left|\dfrac{\sqrt{3+2x^2}+\sqrt{3}}{\sqrt{2}x}\right| + C$

**9.** $\sqrt{2-x^2} - \sqrt{2}\ln\left|\dfrac{\sqrt{2}+\sqrt{2-x^2}}{x}\right| + C$

**11.** $\sqrt{x^2-3} - \sqrt{3}\cos^{-1}\dfrac{\sqrt{3}}{|x|} + C$

**13.** $\dfrac{e^x}{\sqrt{1-e^{2x}}} + C$

**15.** $\frac{1}{16}\left[(8x^2-1)\cos^{-1}2x - 2x\sqrt{1-4x^2}\right] + C$

**17.** $\frac{1}{2}\sin(x^2+1) - \frac{1}{2}x^2\cos(x^2+1) + C$     **19.** $\pi/4$

**21.** $-\frac{1}{13}e^{-2x}(2\sin 3x + 3\cos 3x) + C$

**23.** $-\frac{1}{8}e^{-2x}(4x^3 + 6x^2 + 6x + 3) + C$

**25.** $-\tan^{-1}(\cos x) + C$

**27.** $\dfrac{x^4}{16}(4\ln 5x - 1) + C$

**29.** $\dfrac{x-3}{2}\sqrt{6x-x^2} + \dfrac{9}{2}\cos^{-1}\left(\dfrac{3-x}{3}\right) + C$

**31.** $\dfrac{8\sqrt{3}}{9}\cos^{-1}\left(\dfrac{4-3x}{4}\right) - \dfrac{x+4}{6}\sqrt{8x-3x^2} + C$     **33.** $\frac{4}{3}$

**35.** $-2(\cos x - 1)e^{\cos x} + C$     **37.** $\frac{1}{9}(2e^3+1)$     **39.** $\frac{3}{16}\pi^2$

**41.** $\left(\dfrac{\pi^2-4}{4\pi}, \dfrac{3}{8}\right)$     **43.** 28,284     **45.** 44; 49

**47.** $\dfrac{\pi}{32}[18\sqrt{5} - \ln(2+\sqrt{5})]$

**53.** *Maple:* $\dfrac{2(x+2)^{5/2}}{5} - \dfrac{4(x+2)^{3/2}}{3}$;

*Mathematica:* $\frac{2}{15}(2+x)^{3/2}(-4+3x)$;

TI-89: $\dfrac{2(x+2)^{3/2}(3x-4)}{15}$

**55.** *Maple:* $2\sqrt{x+2} - \sqrt{2}\,\text{arctanh}\left(\dfrac{\sqrt{x+2}\sqrt{2}}{2}\right)$;

*Mathematica:* $2\sqrt{2+x} - \sqrt{2}\,\text{arctanh}\,\dfrac{\sqrt{2+x}}{\sqrt{2}}$;

TI-89:

$\dfrac{\sqrt{2}\ln(|\sqrt{x+2}-\sqrt{2}|) - \sqrt{2}\ln(|\sqrt{x+2}+\sqrt{2}|) + 4\sqrt{x+2}}{2}$

**57.** *Maple and TI-89:* $\frac{1}{4}\cos^3 x \sin x + \frac{3}{8}\cos x \sin x + \frac{3}{8}x$;

*Mathematica:* $\frac{3}{8}x + \frac{1}{4}\sin 2x + \frac{1}{32}\sin 4x$

**59.** *Maple:* $x^5 e^x - 5x^4 e^x + 20x^3 e^x - 60x^2 e^x + 120xe^x - 120e^x$;

*Mathematica:* $e^x(-120 + 120x - 60x^2 + 20x^3 - 5x^4 + x^5)$;

TI-89: $(x^5 - 5x^4 + 20x^3 - 60x^2 + 120x - 120)\,e^x$

**61.** *Maple:* $\frac{2}{3}(e^x + 1)^{3/2} - 2\sqrt{e^x + 1}$;

*Mathematica:* $\frac{2}{3}(-2 + e^x)\sqrt{1 + e^x}$;

TI-89: $\dfrac{2(e^x - 2)\sqrt{e^x + 1}}{3}$

**63.** $-\frac{3}{5}(x+3)(2-x)^{2/3} + C$    **65.** $-\sin(1/x) + C$

**67.** $1 + \dfrac{\pi}{6} - \dfrac{\sqrt{3}}{2}$    **69.** $\dfrac{\sqrt{3}}{36}\pi$

**71.** $\ln\left(\ln x + \sqrt{1 + (\ln x)^2}\right) + C$    **73.** $\dfrac{2(8 - 3\sqrt{3})}{3}$

**75.** $\dfrac{3^{x^3}}{\ln 3} + C$    **77.** $\frac{1}{4}\left[(2x^2 - 1)\sin^{-1} x + x\sqrt{1 - x^2}\right] + C$

**79.** $2\ln\dfrac{\sqrt{5} - 1}{2} + \dfrac{\sqrt{5}}{2}$    **81.** $\frac{1}{2}(1 - \ln 2)$

**83.** $x\tan x + \ln|\cos x| - \frac{1}{2}x^2 + C$    **85.** $\sqrt{2}(2 - \sqrt{3})$

**87.** $2\ln(\sqrt{x+1} + 1) + C$    **89.** $\ln\left|x - 3 + \sqrt{x^2 - 6x}\right| + C$

**91.** $-\frac{1}{6}\cot^3 2x + \frac{1}{2}\cot 2x + x + C$

**93.** $\sqrt{x^2 + 9} - 3\ln\left|\dfrac{3 + \sqrt{x^2 + 9}}{x}\right| + C$

**95.** $\dfrac{1}{3}\ln|x - 1| - \dfrac{1}{6}\ln(x^2 + x + 1)$

$\quad - \dfrac{\sqrt{3}}{3}\tan^{-1}\left[\dfrac{\sqrt{3}}{3}(2x + 1)\right] + C$

**97.** $\dfrac{1}{4}\ln\left|\dfrac{x^2 + x + 1}{x^2 - x + 1}\right|$

$\quad + \dfrac{\sqrt{3}}{6}\left[\tan^{-1}\dfrac{\sqrt{3}(2x + 1)}{3} + \tan^{-1}\dfrac{\sqrt{3}(2x - 1)}{3}\right] + C$

**99.** $\frac{1}{2}e^{e^{x^2}} + C$

## Exercises 6.6 • page 653

**1.** $2\sqrt{3}$   **3.** $\frac{1}{2}$   **5.** $\pi$   **7.** $\frac{1}{2}$   **9.** 100   **11.** $2\sqrt{3}/3$

**13.** $1/(2e^2)$   **15.** Diverges   **17.** Diverges   **19.** $\pi/2$

**21.** $\pi/2$   **23.** 0   **25.** 3   **27.** $-\frac{9}{2}$   **29.** $3\sqrt[3]{3}$

**31.** Diverges   **33.** $-\frac{1}{4}$   **35.** $\sqrt{2}$   **37.** 4

**39.** Diverges   **41.** $-4$   **43.** Convergent

**45.** Convergent   **47.** Divergent   **49.** 1   **51.** $\frac{32}{105}\pi$

**53.** $\pi/2$   **55.** $\pi/2$   **57.** $3\pi/4$   **59.** $\dfrac{qQ}{4\pi a\varepsilon_0}$

**61.** $\dfrac{qQ}{4\pi\varepsilon_0 c}\ln\left|\dfrac{\sqrt{a^2 + c^2} + c}{a}\right|$    **63. b.** \$100,000

**65.** 2.3%    **67.** 1; $2(\sqrt{2} - 1)$

**69.** Converges if $p > 1$, diverges if $p \le 1$

**73. a.** $\displaystyle\int_1^\infty t^{-7/2}\,dt$   **c.** Converges

**75.** $\dfrac{1}{s}$, $s > 0$    **77.** $\dfrac{1}{s^2}$, $s > 0$

**81.** False    **83.** True    **85.** True    **87.** True

## Chapter 6 Concept Review • page 656

**1.** Product; $uv - \displaystyle\int v\,du$; $u$; easily integrated;

$f(x)g(x)\Big|_a^b - \displaystyle\int_a^b g(x)f'(x)\,dx$

**3. a.** $\sec x$; odd   **b.** $\tan x$; even

**5. a.** $a\sin\theta$   **b.** $a\tan\theta$   **c.** $a\sec\theta$

**7.** $\displaystyle\lim_{a\to-\infty}\int_a^b f(x)\,dx$; $\displaystyle\lim_{b\to\infty}\int_a^b f(x)\,dx$; $\displaystyle\int_{-\infty}^c f(x)\,dx + \int_c^\infty f(x)\,dx$;

$\displaystyle\lim_{c\to b^-}\int_a^c f(x)\,dx$; $\displaystyle\int_a^c f(x)\,dx + \int_c^b f(x)\,dx$

## Chapter 6 Review Exercises • page 657

**1.** $2(x - \ln|x + 1|) + C$    **3.** $-\frac{1}{3}(x^2 + 18)\sqrt{9 - x^2} + C$

**5.** $\frac{1}{9}x^3(3\ln x - 1) + C$    **7.** $-\cot\theta - \csc\theta + C$

**9.** $\dfrac{1}{27}\left(\ln\left|\dfrac{x}{x + 3}\right| + \dfrac{6}{x + 3} - \dfrac{3}{x}\right) + C$

**11.** $\dfrac{x}{2}\sqrt{x^2 - 4} - 2\ln\left|x + \sqrt{x^2 - 4}\right| + C$

**13.** $\frac{1}{4}\left[(2\theta^2 - 1)\sin^{-1}\theta + \theta\sqrt{1 - \theta^2}\right] + C$

**15.** $-\frac{1}{2}(e^\pi + 1)$

**17.** $\dfrac{1}{4}\left[\ln\dfrac{x^8}{(x + 1)^2(x^2 + 1)^3} - 2\tan^{-1}x\right] + C$

**19.** $\frac{1}{2}\left(\frac{1}{7}\tan^7 2x + \frac{1}{9}\tan^9 2x\right) + C$

**21.** $\dfrac{x \sin x + \cos x - 1}{\sin x} + C$　**23.** $\frac{1}{4}(\ln x)^4 + C$

**25.** $2 \tan^{-1}\sqrt{4x - 1} + C$　**27.** $\tan x\big(\ln|\tan x| - 1\big) + C$

**29.** $\frac{1}{8}(2 \cos 2x - \cos 4x) + C$

**31.** $\frac{1}{2}\sin^{-1}(2x + 3) + C$

**33.** $-\frac{1}{6}\sin t \cos^5 t + \frac{1}{24}\sin t \cos^3 t + \frac{1}{16}\sin t \cos t + \frac{1}{16}t + C$

**35.** $x + 2\sqrt{x} + 2\ln|\sqrt{x} - 1| + C$

**37.** $\frac{1}{2}x^2 \cos^{-1} 2x - \frac{1}{8}x\sqrt{1 - 4x^2} + 16 \sin^{-1} 2x + C$

**39.** $\dfrac{x}{2} - \dfrac{1}{4}e^{-2x} + C$

**41.** $\frac{1}{2}\ln\big(2x + 1 + \sqrt{4x^2 + 4x + 10}\big) + C$　**43.** 1

**45.** Divergent　**47.** $-\frac{9}{2}$　**49.** $\frac{3}{2}$

**51.** $\dfrac{x}{8}(3 + 2x^2)\sqrt{3 + x^2} - \dfrac{9}{8}\ln|x + \sqrt{3 + x^2}| + C$

**53.** $\ln|\ln(x + 1)| + C$　**55.** $\frac{1}{3}\tan x \sec^2 x + \frac{2}{3}\tan x + C$

**57.** $e^x f(x) + C$　**59.** 6　**61.** $6\pi$

**63. a.**

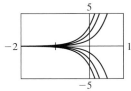

**65.** $\dfrac{2\pi}{9}(2e^3 + 1)$

**67.** $\frac{1}{2}\ln(\sqrt{3} + 2) + \sqrt{3}$　**69.** 1187.99 ft

## Chapter 6 Challenge Problems • page 660

**1.** $1/e$　**3. b.** $\frac{1}{2}$　**5.** $\pi/6$　**19.** 72.6 ft

# CHAPTER 7

## Exercises 7.1 • page 675

**5. a.** 3

**b.**

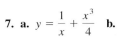

**c.** $y = 2e^{3x}$, yes

**7. a.** $y = \dfrac{1}{x} + \dfrac{x^3}{4}$　**b.**

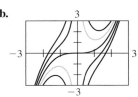

**9.** $y = Cx^2$　**11.** $y = Ce^{x^3/3}$　**13.** $y = Ce^{-2/x} - \frac{3}{2}$

**15.** $y = \sin^{-1}(\tan x + C)$　**17.** $y = -\dfrac{2}{(\ln x)^2 + C}$

**19.** $y = \frac{2}{3} + \frac{1}{3}e^{3x^2/2}$　**21.** $y^{3/2} = \frac{1}{2}(x^3 + 1)$

**23.** $y = \ln(x^3 + e)$　**25.** $I = 2(1 - e^{-2t})$

**27.** $f(x) = \sqrt{x^3 + 8}$

**29.** $y^2 - x^2 = C$　**31.** $y^2 + 2x = C$

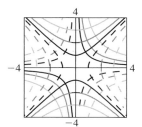
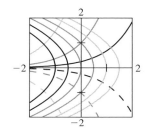

**33.** $\frac{1}{3}$　**35.** 200　**37.** $\frac{1}{2}$ in.　**39.** $y = \dfrac{50}{4t + 1}; \dfrac{50}{9}$ g

**41.** 57.8 years　**43.** 14,176 years old　**45.** After 3.6 min

**47.** 1.05 mph

**49. a.** $P(t) = (P_0 + I/k)e^{kt} - I/k$　**b.** 304.9 million

**51. a.** $\dfrac{dv}{dt} = g - kv$　**b.** $v(t) = \dfrac{g}{k}(1 - e^{-kt})$　**c.** $\dfrac{g}{k}$

**53. a.** $P(t) = \dfrac{1}{(1 - 0.01kt)^{100}}$

**b.** 1e+199

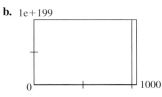

**c.** The population grows without bound after a finite period of time.

**55. a.** $C_0 e^{-kT}$　**b.** $C_0(e^{-kT} + e^{-2kT})$

**c.** $C_0(e^{-kT} + e^{-2kT} + \cdots + e^{-NkT})$　**d.** $\dfrac{C_0 e^{-kT}}{1 - e^{-kT}}$ g/mL

**57. a.** 2974, 5319, 6955　**b.** 3.6 weeks

**c.** 10,000

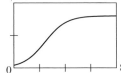

**59.** False　**61.** True

## Exercises 7.2 • page 687

**1.** (a)

**3.** (b)

**5.** 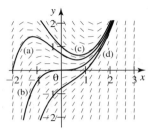   **7.**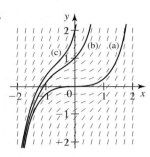

**9.** 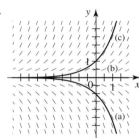   **11.**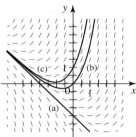

**13.** 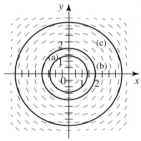   **15.**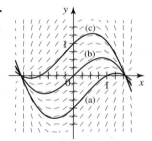

**17. a.** 2.88   **b.** 3.04

**19. a.** 3.19   **b.** 3.26

**21. a.** 0.83   **b.** 0.82

**23. a.** 1.78   **b.** 1.79

**25. a.** 1.34   **b.** 1.37

**27. a, c.**

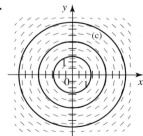

**b.** $x^2 + y^2 = 16$

**29. a, b.**

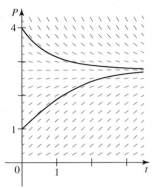

**c.** $\lim_{t \to \infty} P(t) \approx 2.72$

**31. a.** 73.6 ft/sec   **b.** $v(t) = 160 - 30e^{-t/5}$; $v(2) \approx 72.9$ ft/sec

**33.** False   **35.** True

## Exercises 7.3 • page 698

**1. a.** 0.02   **b.** 1000   **3. a.** 0.5   **b.** 500

**5. a.** 100   **b.** $P(t) = 0$, $P(t) = 100$

**c–e.**

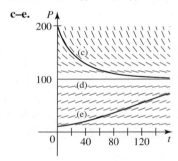

**7. a.** 0.02   **b.** 4000   **c.** 10

**11. a.** 86.12%   **b.** 1970

**13. a.** $P(t) = \dfrac{100}{1 + 9\left(\frac{11}{51}\right)^{t/30}}$   **b.** 70 days

**17. a, c.**

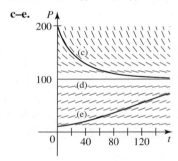

If the initial fish population is 100, then it will be gone after 5 weeks. If the initial population is 300, then it increases to 600 over time. If it is 700, then it decreases to 600 over time.

**b.** $P(t) = 200$, $P(t) = 600$

**19. a, c.**

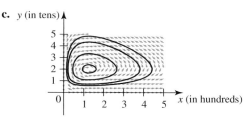

**b.** $P(t) = 1000$

**21.** 173    **23.** True    **25.** True

## Exercises 7.4 • page 710

**1.** No    **3.** No    **5.** $y = \frac{1}{4}e^{2x} + Ce^{-2x}$

**7.** $y = \frac{1}{4}x^3 + \frac{C}{x}$    **9.** $y = \frac{1}{3}x^2 \sin 3x + Cx^2$

**11.** $y = \frac{2}{3}x(\ln x)^3 + Cx$    **13.** $y = \frac{t^2}{2(t+1)} + \frac{C}{t+1}$

**15.** $y = \frac{x}{2}e^{-2x} + \frac{Ce^{-2x}}{x}$    **17.** $y = 1 - 2e^{-x}$

**19.** $y = \frac{1}{2}\left(1 + e^{-x^2}\right)$    **21.** $y = \frac{(\ln x)^2 + 2}{2(x+1)}$

**25.** $y = \frac{\sqrt{3}x^2}{\sqrt{9 - 8x^3}}$    **27.** $f(x) = x \ln x + x$

**29. a.** $y(t) = 90(1 - e^{-t/15})$    **b.** 33 min

**31. a.** $y(t) = 80 - t - \frac{(80 - t)^4}{512,000}$    **b.** 34.7 lb

   **c.** 35 lb    **d.** 37.8 lb

**33.** $Q(t) = 0.24 - 0.19e^{-5t}$, $I(t) = 0.95e^{-5t}$

**35. a.** $8(1 - e^{-4t})$    **b.** 7.85 ft/sec    **c.** 0.17 sec

**37.** 29.5 ft; 8 ft/sec

**39.** $I = \frac{E_0 R}{R^2 + \omega^2 L^2}\left[\cos \omega t + \frac{L\omega}{R}\sin \omega t - e^{-(R/L)t}\right]$

**43.** True    **45.** True    **47.** True

## Exercises 7.5 • page 717

**1. a.** $(0, 0)$ and $\left(\frac{5}{4}, 2\right)$    **b.** $\frac{-y + 0.8xy}{2.4x - 1.2xy}$

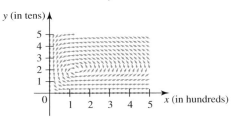

**c.**

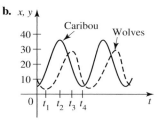

**3. a.** At the start there are approximately 480 caribou and 100 wolves. The caribou population fluctuates between a minimum of ≈480 and a maximum of ≈3500; the wolf population fluctuates between ≈30 and ≈290. The caribou population leads the wolf population.

**b.**

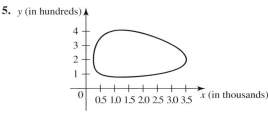

**5.**

**7. c.** $xy^4 - 3.9830e^{0.2x}e^y = 0$

**9. a.** The system of differential equations reduces to the single differential equation $\frac{dx}{dt} = kx\left(1 - \frac{x}{L}\right)$. If $x(0) = 0$, then the prey population stays at zero at all times. If $0 < x(0) < L$, then the prey population approaches the carrying capacity $L$ of the environment. If $x(0) = L$, the prey population at any time $t > 0$ remains at $L$. If $x(0) > L$, then the prey population decreases and approaches $L$.

**b.** The system of differential equations reduces to $\frac{dy}{dt} = -by$ with solution $y = y_0 e^{-bt}$, where $y_0 = y(0)$ is the initial predator population. Since $y \to 0$ as $t \to \infty$, the predator population eventually dies out.

**c.** $(0, 0)$, $(L, 0)$, and $\left(\frac{b}{c}, \frac{k(cL - b)}{acL}\right)$; If there are no predators or prey at any time $t$, the situation will persist forever. If there are exactly $L$ prey and no predators at some time, then the prey population will remain at $L$ forever. If at least one of the populations is not equal to zero, then we conclude that a prey population of $b/c$ is exactly the number that will support a stable predator population of $\frac{k(cL - b)}{acL}$.

**11. a.** The system reduces to $\dfrac{dx}{dt} = k_1 x\left(1 - \dfrac{x}{L_1}\right)$. If $x(0) = 0$,
then the population of species $A$ stays at zero at all
times. If $0 < x(0) < L_1$, then the population of
species $A$ approaches $L_1$. If $x(0) = L_1$, the population
remains at $L_1$. If $x(0) > L_1$, then the population
decreases and approaches $L_1$.

**b.** The system reduces to $\dfrac{dy}{dt} = k_2 y\left(1 - \dfrac{y}{L_2}\right)$. The population of species $B$ behaves in a manner similar to that of
the population of species $A$ described in part (a).

**c.** The terms $axy$ and $bxy$ arise from the interaction of the
species. The rates of change of the populations decline as
a result of their competition for resources.

**d.** $(0, 0)$, $(0, L_2)$, $(L_1, 0)$, and
$$\left(\frac{k_2 L_1(aL_2 - k_1)}{abL_1 L_2 - k_1 k_2}, \; \frac{k_1 L_2(bL_1 - k_2)}{abL_1 L_2 - k_1 k_2}\right)$$

## Chapter 7 Concept Review • page 719

**1. a.** derivative; differential; unknown   **b.** highest

**3.** $g(x)h(y)$; separable

**5. a.** tangent line   **b.** slope; direction

**7. a.** $\dfrac{dy}{dx} + P(x)y = Q(x)$   **b.** $u(x) = e^{\int P(x)\,dx}$

## Chapter 7 Review Exercises • page 720

**1.** Yes   **3.** $y = -\dfrac{1}{x^2 + C}$   **5.** $e^y = \dfrac{e^x}{1 + Ce^x}$

**7.** $y^2 = \dfrac{3}{12 - 2x^3}$   **9.** $x^2 + y^2 = 4$   **11.** $y = \dfrac{1 + \tan 4x}{1 - \tan 4x}$

**13. a.** $x(t) = 1000 \cdot 4^{t/3}$   **b.** 16,000   **c.** 13 hr

**15.** $-0.000433$   **17.** \$338,249

**19. a.** 163°F   **b.** 28.6 min

**21. a.** $S(t) = \dfrac{1}{r}[(rS_0 + d)e^{rt} - d]$   **b.** \$25,160.55

**23. a.**

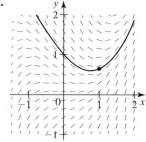

**b.** 0.6

**25.** 1.061   **27.** $y = x^2 + (C/x^2)$

**29.** $y = \frac{1}{2}e^{-x}\sin 2x + Ce^{-x}$   **31.** $y = 2e^{(x+1)/3}$

**33.** $y = x\sin x - x$

**35. a.** $P(t) = \dfrac{10,000}{1 + 24\left(\frac{3}{8}\right)^t}$   **b.** 9374   **c.** 4.7 years

**37. a.** The fox population dies out; the rabbit population grows
indefinitely.

**b.** $(0, 0)$, $(4, 5)$. There are no rabbits and no foxes or 4000
rabbits will support a stable population of 500 foxes.

**c.** $\dfrac{-1.2y + 0.3xy}{2x - 0.4xy}$

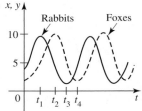

**d.**

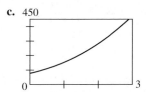

## Chapter 7 Challenge Problems • page 722

**3. b.** $P(t) \approx \dfrac{1168.60}{1 + 14.334e^{-0.7572t}}$

**c.** 450

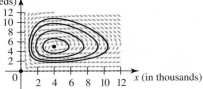

**d.** 351.06 million

**5. b.** $y_1 = x$, $y_2 = x + Ce^{-x^2/2}$

# CHAPTER 8

*Abbreviations:* C, convergent; AC, absolutely convergent;
CC, conditionally convergent; D, divergent; R, radius of
convergence; I, interval of convergence

## Exercises 8.1 • page 739

**1.** $2, 1, \frac{4}{5}, \frac{5}{7}, \frac{2}{3}$   **3.** $1, 0, -1, 0, 1$   **5.** $1, \frac{1}{6}, \frac{1}{90}, \frac{1}{2520}, \frac{1}{113,400}$

**7.** $a_n = \dfrac{n}{n + 1}$   **9.** $a_n = \dfrac{(-1)^n}{n!}$   **11.** $a_n = \dfrac{1}{n + 1}$

**13.** 2    **15.** D    **17.** 1    **19.** $\frac{2}{3}$    **21.** 0

**23.** D    **25.** 0    **27.** D    **29.** 0    **31.** 1

**33.** 0    **35.** 0    **37.** 1    **39.** 0    **41.** $\frac{1}{2}$

**43. a.** 1.0

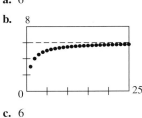

**b.** 1    **c.** 1

**45. a.** 1.25

**b.** 0    **c.** 0

**47. a.** 1.0

**b.** 1    **c.** 1

**49.** 9    **51.** Monotonic, bounded    **53.** Monotonic, bounded

**55.** Not monotonic, bounded    **57.** Monotonic, bounded

**59. a.** 10088, 10176, 10265, 10355, 10445, 10537

**b.** Diverges

**61. a.** 5394.69    **b.** $\infty$

**63.** $e^6$    **65. b.** 2.2361    **67.** 1.226    **69.** 2

**71. c.** $\{a_n\}$ converges    **73. b.** 0    **79.** False

**81.** True    **83.** False    **85.** False

## Exercises 8.2 • page 750

**1.** 1    **3.** $\frac{2}{5}$    **5.** $1/\ln 2$    **7.** 12    **9.** $\frac{5}{4}$

**11.** $2(2 - \sqrt{2})$    **13.** 9

**23. a.** 6

**b.** 8

**c.** 6

**25. a.** 24

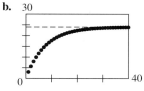

**b.** 30

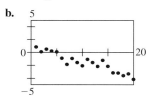

**c.** 24

**27. a.** diverges

**b.** 5

**c.** D

**29.** $\frac{3}{4}$    **31.** $\frac{5}{3}$    **33.** D    **35.** D    **37.** D    **39.** 0

**41.** D    **43.** $\cos 1 - 1$    **45.** $-\frac{5}{18}$    **47.** $\frac{7}{2}$    **49.** D

**51.** D    **53.** D    **55.** $\frac{4}{9}$    **57.** $\frac{404}{333}$

**59.** $|x| < 1, \dfrac{1}{1 + x}$    **61.** $\dfrac{1}{2} < x < \dfrac{3}{2}, \dfrac{2(x - 1)}{3 - 2x}$

**63.** 6 m    **65.** $\frac{6}{11}$    **67. b.** \$20 million

**69.** $A = \dfrac{P}{e^r - 1}$    **71.** $\dfrac{3\sqrt{3} - \pi}{9}$

**79.** False    **81.** False    **83.** False

## Exercises 8.3 • page 756

**1.** C    **3.** C    **5.** C    **7.** C    **9.** C    **11.** C

**13.** C    **15.** D    **17.** C    **19.** C    **21.** D    **23.** C

**25.** C    **27.** C    **29.** C    **31.** C    **33.** $p > 1$

**35.** $a = 1$    **41.** 0.05    **43.** $(\pi/2) - \tan^{-1} 50$

**45.** $n \geq 200$    **47.** $n \geq 314$    **49.** 1.082

**51. b.** $\frac{1}{4}$    **55.** True    **57.** True

## Exercises 8.4 • page 763

**1.** C    **3.** D    **5.** D    **7.** C    **9.** D    **11.** C

**13.** D    **15.** C    **17.** C    **19.** C    **21.** C    **23.** D

**25.** C    **27.** C    **29.** D    **31.** C    **33.** D    **35.** C

**37.** C    **39.** C    **43.** 3.06    **45.** 0.99

**49.** The converse is false.    **53. b.** Yes

**55.** False    **57.** False

## Exercises 8.5 • page 769

**1.** C    **3.** C    **5.** C    **7.** C    **9.** D    **11.** C
**13.** C    **15.** C    **17.** C    **19.** D    **21.** C    **23.** C
**25.** $p > 0$    **27.** 44    **29.** 10    **31.** $-0.90$    **33.** 0.56
**35.** $\frac{1}{2}$; $a_{n+1} \le a_n$ is not satisfied.    **37. a.** No  **b.** No
**39. b.** $-1$    **41.** True    **43.** False

## Exercises 8.6 • page 779

**1.** CC    **3.** D    **5.** CC    **7.** D    **9.** CC    **11.** D
**13.** CC    **15.** C    **17.** AC    **19.** AC    **21.** AC
**23.** D    **25.** AC    **27.** CC    **29.** AC    **31.** AC
**33.** C    **35. a.** $-1 < x < 1$  **b.** $x = -1$
**47.** True    **49.** True

## Exercises 8.7 • page 788

**1.** $R = 1, I = [-1, 1)$    **3.** $R = 1, I = [-1, 1)$
**5.** $R = \infty, I = (-\infty, \infty)$    **7.** $R = 0, I = \{0\}$
**9.** $R = 1, I = [-1, 1)$    **11.** $R = \frac{1}{e}, I = \left[-\frac{1}{e}, \frac{1}{e}\right)$
**13.** $R = 1, I = (2, 4]$    **15.** $R = 3, I = (-1, 5]$
**17.** $R = 1, I = (0, 2]$    **19.** $R = \infty, I = (-\infty, \infty)$
**21.** $R = \infty, I = (-\infty, \infty)$
**23.** $R = \frac{1}{3}, I = \left(-2, -\frac{4}{3}\right]$
**25.** $R = 1, I = [-1, 1]$
**27.** $R = 1, I = (-1, 1)$
**29.** $R = \frac{3}{2}, I = \left(-\frac{3}{2}, \frac{3}{2}\right]$
**31. a.** $\dfrac{x^{n+1}}{1 - x}$    **b.** 0; the limit does not exist.
**c.**

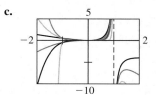

**33.** $(-\infty, \infty)$    **35.** $\sqrt{R}$    **37.** $1/L$
**39.** $\displaystyle\sum_{n=1}^{\infty} \frac{x^{n-1}}{n}$; $\displaystyle\sum_{n=2}^{\infty} \frac{(n-1)x^{n-2}}{n}$; $[-1, 1], [-1, 1), (-1, 1)$
**41.** $\dfrac{1}{(1 - x)^2}$    **45. a.** $\displaystyle\sum_{n=0}^{\infty} t^{2n}$  **b.** $\displaystyle\sum_{n=0}^{\infty} \frac{x^{2n+1}}{2n + 1}$, 1
**47.** 3.14159    **49.** True    **51.** True

## Exercises 8.8 • page 801

**1.** $\displaystyle\sum_{n=0}^{\infty} \frac{2^n x^n}{n!}$, $\infty$    **3.** $\displaystyle\sum_{n=0}^{\infty} \frac{e^2}{n!} (x - 2)^n$, $\infty$

**5.** $2x - \dfrac{2^3}{3!}x^3 + \dfrac{2^5}{5!}x^5 - \cdots + \dfrac{(-1)^k 2^{2k+1}}{(2k + 1)!} x^{2k+1} + \cdots$, $\infty$

**7.** $\dfrac{\sqrt{3}}{2} \displaystyle\sum_{n=0}^{\infty} \frac{(-1)^n}{(2n)!} \left(x + \frac{\pi}{6}\right)^{2n}$
   $+ \dfrac{1}{2} \displaystyle\sum_{n=0}^{\infty} \frac{(-1)^n}{(2n + 1)!} \left(x + \frac{\pi}{6}\right)^{2n+1}$, $\infty$

**9.** $\ln 2 + \displaystyle\sum_{n=1}^{\infty} \frac{(-1)^{n-1}}{n2^n} (x - 2)^n$, 2

**11.** $\displaystyle\sum_{n=0}^{\infty} \frac{(-1)^n}{2^{n+1}} (x - 1)^n$, 2    **13.** $\displaystyle\sum_{n=0}^{\infty} (-1)^{n+1} 2^n (x - 1)^n$, $\frac{1}{2}$

**15.** $-\displaystyle\sum_{n=0}^{\infty} x^{2n+2}$, 1    **17.** $\displaystyle\sum_{n=0}^{\infty} \frac{(-1)^n x^{n+1}}{n!}$, $\infty$

**19.** $\displaystyle\sum_{n=0}^{\infty} \frac{(-1)^n x^{2n+2}}{(2n)!}$, $\infty$    **21.** $1 + \displaystyle\sum_{n=1}^{\infty} \frac{(-1)^n 2^{2n-1} x^{2n}}{(2n)!}$, $\infty$

**23.** $\dfrac{1}{2} \displaystyle\sum_{n=0}^{\infty} \frac{(-1)^n \left(x - \frac{\pi}{3}\right)^{2n+1}}{(2n + 1)!} + \dfrac{\sqrt{3}}{2} \displaystyle\sum_{n=0}^{\infty} \frac{(-1)^n \left(x - \frac{\pi}{3}\right)^{2n}}{(2n)!}$, $\infty$

**25.** $\displaystyle\sum_{n=0}^{\infty} \frac{(2n)! x^{3/2} x^{2n}}{(2^n n!)^2 (2n + 1)}$, 1    **27.** $\displaystyle\sum_{n=1}^{\infty} (-1)^{n-1} \frac{x^{2n}}{n}$, 1

**29.** $\displaystyle\sum_{n=0}^{\infty} (-1)^n (n + 1)x^n$, 1

**31.** $1 - \dfrac{1}{2}x^2 - \displaystyle\sum_{n=2}^{\infty} \frac{1 \cdot 3 \cdot 5 \cdot \cdots \cdot (2n - 3)}{n! \, 2^n} x^{2n}$, 1

**33.** $1 - \dfrac{3}{5}x - 3 \displaystyle\sum_{n=2}^{\infty} \frac{2 \cdot 7 \cdot 12 \cdot \cdots \cdot (5n - 8)}{n! \, 5^n} x^n$, 1

**35.** $f(x) = 1 + 2\left(x - \dfrac{\pi}{4}\right) + 2\left(x - \dfrac{\pi}{4}\right)^2 + \cdots$

**37.** $f(x) = \dfrac{\pi}{6} + \dfrac{2\sqrt{3}}{3}\left(x - \dfrac{1}{2}\right) + \dfrac{2\sqrt{3}}{9}\left(x - \dfrac{1}{2}\right)^2 + \cdots$

**39.** $f(x) = x - x^2 + \dfrac{1}{3}x^3 - \cdots$

**41. a.** $f(x) = 1 + \dfrac{1}{3}x$
   $+ \displaystyle\sum_{n=2}^{\infty} (-1)^{n-1} \frac{2 \cdot 5 \cdot 8 \cdot \cdots \cdot (3n - 4)}{n! \, 3^n} x^n$, $R = 1$
**b.** $P_1(x) = 1, P_2(x) = 1 + \frac{1}{3}x, P_3(x) = 1 + \frac{1}{3}x - \frac{1}{9}x^2$
**c.**

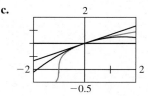

**43.** 0.99005

**45.** $\displaystyle\sum_{n=0}^{\infty} (-1)^n \frac{x^{3n+1}}{3n+1} + C$    **47.** $\displaystyle\sum_{n=0}^{\infty} \frac{(-1)^n x^{4n+3}}{(2n+1)!\,(4n+3)} + C$

**49.** $\displaystyle\sum_{n=1}^{\infty} \frac{(-1)^{n-1} x^n}{n^2} + C$    **51.** 0.7468    **53.** 0.4969

**55.** 0.1248    **57.** $\ln 2$    **59.** $\cos \pi = -1$

**61.** $\ln 3 - \ln 2$    **63.** $\frac{1}{120}$    **65.** $\frac{2}{15}$

**67. a.** $\dfrac{1}{\sqrt{1-u^2}} = 1 + \dfrac{1}{2} u^2 + \dfrac{1\cdot 3}{2!\,2^2} u^4 + \cdots$

$\qquad + \dfrac{1\cdot 3\cdot 5 \cdot \cdots \cdot (2n-1)}{n!\,2^n} u^{2n} + \cdots$

**b.** $\displaystyle\int_0^x \dfrac{dt}{\sqrt{1-t^2}} = \sum_{n=0}^{\infty} \dfrac{(2n)!}{(n!\,2^n)^2 (2n+1)} x^{2n+1}, \; R = 1$

**79.** True    **81.** False    **83.** True

## Exercises 8.9 • page 816

**1.** $P_1(x) = 1 - x$, $P_2(x) = 1 - x + \frac{1}{2} x^2$,
$P_3(x) = 1 - x + \frac{1}{2} x^2 - \frac{1}{6} x^3$

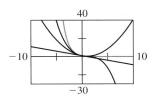

**3.** $P_4(x) = 7 + 13(x-1) + 9(x-1)^2 + 2(x-1)^3$, $R_4(x) = 0$

**5.** $P_3(x) = 1 - \dfrac{1}{2}\left(x - \dfrac{\pi}{2}\right)^2$, $R_3(x) = \dfrac{\sin z}{24}\left(x - \dfrac{\pi}{2}\right)^4$,

$z$ lies between $x$ and $\dfrac{\pi}{2}$

**7.** $P_2(x) = 1 + 2\left(x - \dfrac{\pi}{4}\right) + 2\left(x - \dfrac{\pi}{4}\right)^2$,

$R_2(x) = \dfrac{\sec^2 z(2\tan^2 z + \sec^2 z)}{3}\left(x - \dfrac{\pi}{4}\right)^3$,

$z$ lies between $x$ and $\dfrac{\pi}{4}$

**9.** $P_3(x) = -2 + \dfrac{1}{12}(x+8) + \dfrac{1}{288}(x+8)^2$

$\qquad + \dfrac{5}{20{,}736}(x+8)^3$,

$R_3(x) = -\dfrac{10}{243 z^{11/3}}(x+8)^4$, $z$ lies between $x$ and $-8$

**11.** $P_2(x) = \dfrac{\pi}{4} + \dfrac{1}{2}(x-1) - \dfrac{1}{4}(x-1)^2$,

$R_2(x) = \dfrac{3z^2 - 1}{3(1+z^2)^3}(x-1)^3$, $z$ lies between $x$ and $1$

**13.** $P_3(x) = -\dfrac{1}{e} + \dfrac{1}{2e}(x+1)^2 + \dfrac{1}{3e}(x+1)^3$,

$R_3(x) = \dfrac{(z+4)e^z}{24}(x+1)^4$, $z$ lies between $x$ and $-1$

**15.** $P_2(x) = \dfrac{1}{2} e^{\pi/6} + \left(\dfrac{1}{2} - \sqrt{3}\right) e^{\pi/6}\left(x - \dfrac{\pi}{6}\right)$

$\qquad - \left(\dfrac{3}{4} + \sqrt{3}\right) e^{\pi/6}\left(x - \dfrac{\pi}{6}\right)^2$,

$R_2(x) = \dfrac{e^z(2\sin 2z - 11\cos 2z)}{6}\left(x - \dfrac{\pi}{6}\right)^3$,

$z$ lies between $x$ and $\dfrac{\pi}{6}$

**17.** $4(x-1) + 6(x-1)^2$, $4(1.2)(1.2 - 1)^3$

**19.** $\dfrac{\sqrt{2}}{2} - \dfrac{\sqrt{2}}{2}\left(x - \dfrac{\pi}{4}\right) - \dfrac{\sqrt{2}}{4}\left(x - \dfrac{\pi}{4}\right)^2$

$\qquad + \dfrac{\sqrt{2}}{12}\left(x - \dfrac{\pi}{4}\right)^3 + \dfrac{\sqrt{2}}{48}\left(x - \dfrac{\pi}{4}\right)^4$, $\dfrac{\sin \frac{\pi}{2}}{120}\left(\dfrac{\pi}{4}\right)^5$

**21.** $e^2 + 2e^2(x-1) + 2e^2(x-1)^2 + \dfrac{4}{3} e^2 (x-1)^3$

$\qquad + \dfrac{2}{3} e^2 (x-1)^4$, $\dfrac{32 e^{2(1.1)}}{5!}(0.1)^5$

**23.** $3 + \dfrac{1}{6}(x-9) - \dfrac{1}{216}(x-9)^2 + \dfrac{1}{3888}(x-9)^3$,

$\qquad \dfrac{5}{128} \cdot \dfrac{1}{8^{7/2}}(1)^4$

**25.** $x + \dfrac{1}{3} x^3$, $\dfrac{10\left(\frac{\pi}{4}\right)^4}{3}$

**27.** $\ln 4 + \dfrac{1}{4}(x-3) - \dfrac{1}{32}(x-3)^2 + \dfrac{1}{192}(x-3)^3$, $\dfrac{6(1)^4}{3^4 4!}$

**29.** $f(x) = e^x$, $c = 0$, $P_3(0.2) \approx 1.2213$

**31.** $f(x) = \sqrt{x}$, $c = 9$, $P_1(9.01) \approx 3.00167$

**33.** $f(x) = \dfrac{1}{x}$, $c = -2$, $P_2(-2.1) \approx -0.476$

**35.** $f(x) = \sin x$, $c = 0$, $P_3(0.1) \approx 0.09983$

**37.** $f(x) = \cos x$, $c = \dfrac{\pi}{6}$, $P_2\left(\dfrac{8\pi}{45}\right) \approx 0.8480$    **45.** 48%

**51.** True    **53.** True

## Chapter 8 Concept Review • page 818

**1. a.** function; integers; $n$th term

   **b.** converge

   **c.** for every; positive integer; $n > N$

**3. a.** partial sums $\{S_n\}$

   **b.** geometric; 1; 1

**5. a.** convergent; divergent **b.** $\sum_{n=1}^{\infty} \dfrac{1}{n^p}$; $p > 1$; $p \le 1$

**7. a.** an alternating; $\le$; $0$ **b.** $a_{n+1}$

**9. a.** $\sum_{n=0}^{\infty} a_n(x - c)^n$ **b.** $x = c$; $x$

## Chapter 8 Review Exercises • page 819

**1.** $\frac{1}{3}$ **3.** 2 **5.** D **7.** 0 **9.** 3 **11.** $\frac{11}{18}$ **13.** C

**15.** C **17.** C **19.** C **21.** D **23.** C **25.** C

**27.** CC **29.** AC **31.** CC **33.** $\frac{6802}{4995}$ **35.** False

**37.** True **39.** $x \ne k\pi$, $k = 0, \pm 1, \pm 2, \ldots$

**43.** $R = 1$, $I = (-1, 1]$ **45.** $R = \frac{1}{2}$, $I = \left[-\frac{1}{2}, \frac{1}{2}\right]$

**47.** $R = 1$, $I = [-1, 1]$ **49.** $\sum_{n=0}^{\infty} (-1)^n x^{n+3}$

**51.** $\sum_{n=0}^{\infty} (-1)^n \dfrac{x^{4n}}{(2n)!}$

**53.** $1 + \dfrac{1}{2}x^2 + \sum_{n=2}^{\infty} (-1)^{n-1} \dfrac{1 \cdot 3 \cdot 5 \cdot \,\cdots\, \cdot (2n - 3)}{n! \, 2^n} x^{2n}$

**55.** $\frac{1}{4}$ **57.** $\ln|x| + \sum_{n=1}^{\infty} (-1)^n \dfrac{x^n}{n! \, n} + C$

**59.** 0.199 **61.** 0.779

**63.** $P_3(x) = 1 + \frac{1}{2}(x - 1) - \frac{1}{8}(x - 1)^2 + \frac{1}{16}(x - 1)^3$,

$R_3(x) = \dfrac{-5(x - 1)^4}{128 z^{7/2}}$, $z$ lies between $x$ and 1

**65.** $P_2(x) = 1 + \dfrac{1}{2}\left(x - \dfrac{\pi}{2}\right)^2$,

$R_2(x) = -\dfrac{\csc z \cot z(\cot^2 z + 5 \csc^2 z)}{6}\left(x - \dfrac{\pi}{2}\right)^3$,

$z$ lies between $x$ and $\dfrac{\pi}{2}$

## Chapter 8 Challenge Problems • page 821

**1.** 1 **3. a.** $c_n = \left(\frac{2}{3}\right)^n$ **5.** $\frac{1}{2}(1 + \sqrt{1 + 4a})$ **11.** $\frac{1}{3}$

**13.** $\sum_{n=0}^{\infty} x^{4n} - \sum_{n=0}^{\infty} x^{4n+1}$ **15.** $(-\infty, \infty)$

**17.** $P_3 = 1 + x^2 - \frac{1}{2}x^3$, $|x| < 1$

# CHAPTER 9

## Exercises 9.1 • page 839

**1.** Parabola (h), vertex $(0, 0)$, focus $(0, -1)$, directrix $y = 1$

**3.** Parabola (c), vertex $(0, 0)$, focus $(2, 0)$, directrix $x = -2$

**5.** Ellipse (b), vertices $(\pm 3, 0)$, foci $(\pm \sqrt{5}, 0)$, eccentricity $\frac{\sqrt{5}}{3}$

**7.** Hyperbola (d), vertices $(\pm 4, 0)$, foci $(\pm 5, 0)$, eccentricity $\frac{5}{4}$

**9.** Vertex $(0, 0)$,
focus $\left(0, \frac{1}{8}\right)$,
directrix $y = -\frac{1}{8}$

**11.** Vertex $(0, 0)$,
focus $\left(\frac{1}{8}, 0\right)$,
directrix $x = -\frac{1}{8}$

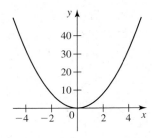

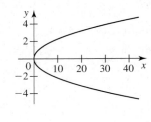

**13.** Vertex $(0, 0)$,
focus $\left(\frac{3}{5}, 0\right)$,
directrix $x = -\frac{3}{5}$

**15.** Foci $(0, \pm\sqrt{21})$,
vertices $(0, \pm 5)$

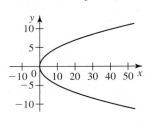

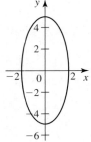

**17.** Foci $(\pm\sqrt{5}, 0)$,
vertices $(\pm 3, 0)$

**19.** Foci $(\pm\sqrt{3}, 0)$,
vertices $(\pm 2, 0)$

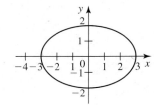

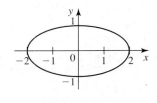

**21.** Foci $(\pm 13, 0)$,
vertices $(\pm 5, 0)$,
asymptotes $y = \pm\frac{12}{5}x$

**23.** Foci $(\pm\sqrt{2}, 0)$,
vertices $(\pm 1, 0)$,
asymptotes $y = \pm x$

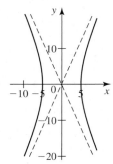

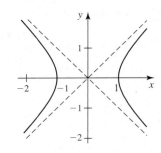

**25.** Foci $(0, \pm\sqrt{30})$, vertices $(0, \pm5)$, asymptotes $y = \pm\sqrt{5}x$

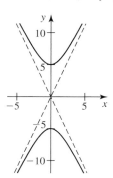

**27.** $y^2 = 12x$     **29.** $y^2 = -10x$     **31.** $\dfrac{x^2}{9} + \dfrac{y^2}{8} = 1$

**33.** $\dfrac{x^2}{8} + \dfrac{y^2}{9} = 1$     **35.** $\dfrac{x^2}{9} + \dfrac{4y^2}{9} = 1$     **37.** $\dfrac{73x^2}{400} + \dfrac{y^2}{25} = 1$

**39.** $\dfrac{x^2}{9} - \dfrac{y^2}{16} = 1$     **41.** $\dfrac{y^2}{21} - \dfrac{x^2}{4} = 1$     **43.** $\dfrac{x^2}{4} - \dfrac{y^2}{9} = 1$

**45.** (b)     **47.** (c)     **49.** $(y - 1)^2 = 4(x - 2)$

**51.** $(y - 2)^2 = -2(x - 2)$     **53.** $(x + 3)^2 = -2(y - 2)$

**55.** $\dfrac{x^2}{9} + \dfrac{(y - 3)^2}{8} = 1$     **57.** $\dfrac{x^2}{16} + \dfrac{(y - 2)^2}{15} = 1$

**59.** $\dfrac{(x - 2)^2}{9} + \dfrac{(y - 1)^2}{5} = 1$     **61.** $\dfrac{(x - 3)^2}{9} - \dfrac{(y - 2)^2}{16} = 1$

**63.** $\dfrac{(x - 1)^2}{9} - \dfrac{(y + 3)^2}{16} = 1$     **65.** $\dfrac{(y - 1)^2}{9} - \dfrac{(x - 4)^2}{4} = 1$

**67.** Vertex $(2, 1)$,
focus $(3, 1)$,
directrix $x = 1$

**69.** Vertex $(-3, 2)$,
focus $\left(-3, \frac{9}{4}\right)$,
directrix $y = \frac{7}{4}$

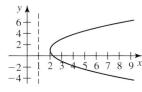
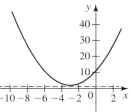

**71.** Vertex $\left(-1, \frac{1}{2}\right)$,
focus $\left(1, \frac{1}{2}\right)$,
directrix $x = -3$

**73.** Center $(1, -2)$,
foci $\left(1 \pm \frac{\sqrt{3}}{2}, -2\right)$,
vertices $(0, -2)$ and $(2, -2)$

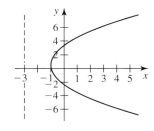
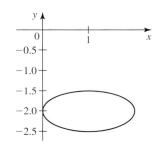

**75.** Center $(1, -2)$, foci $(1 \pm \sqrt{3}, -2)$, vertices $(-1, -2)$ and $(3, -2)$

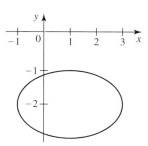

**77.** Center $\left(\frac{9}{4}, 0\right)$, foci $\left(\frac{9}{4} \pm \frac{\sqrt{105}}{4}, 0\right)$, vertices $\left(\frac{9}{4} \pm \frac{3\sqrt{21}}{4}, 0\right)$

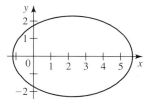

**79.** Center $(0, -1)$, foci $(\pm\sqrt{7}, -1)$, vertices $(\pm2, -1)$,
asymptotes $y = \pm\frac{\sqrt{3}}{2}x - 1$

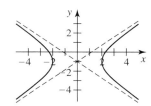

**81.** Center $(1, 2)$, foci $(1, 2 \pm \sqrt{15})$, vertices $(1, 2 \pm \sqrt{6})$,
asymptotes $y - 2 = \pm\frac{\sqrt{6}}{3}(x - 1)$

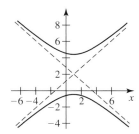

**83.** Center $(-1, 2)$, foci $(-1 \pm \sqrt{6}, 2)$, vertices $(-1 \pm \sqrt{2}, 2)$,
asymptotes $y - 2 = \pm\sqrt{2}(x + 1)$

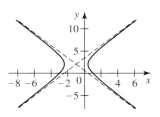

**85.** $\frac{1}{4}$ ft     **87.** 616 ft     **91.** 6.93 ft     **103.** 23.013

**109.** True     **111.** True     **113.** False

## Exercises 9.2 • page 851

**1. a.** $x - 2y - 7 = 0$
  **b.**
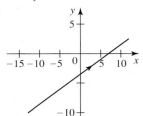

**3. a.** $y = 9 - x^2,\ x \geq 0$
  **b.**
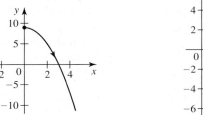

**5. a.** $y = 2x - 3,\ 1 \leq x \leq 5$
  **b.**
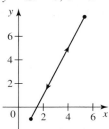

**7. a.** $x = y^{2/3},\ -8 \leq y \leq 8$
  **b.**
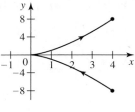

**9. a.** $x^2 + y^2 = 4$
  **b.**
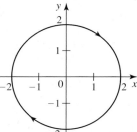

**11. a.** $\frac{1}{4}x^2 + \frac{1}{9}y^2 = 1$
  **b.**
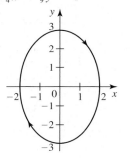

**13. a.** $\frac{1}{4}(x - 2)^2 + \frac{1}{9}(y + 1)^2 = 1$
  **b.**

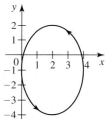

**15. a.** $y = 2x^2 - 1,\ -1 \leq x \leq 1$
  **b.**
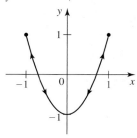

**17. a.** $x^2 - y^2 = 1,\ x \geq 1$
  **b.**

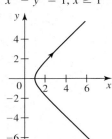

**19. a.** $y = x^2,\ 0 \leq x \leq 1$
  **b.**

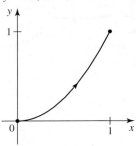

**21. a.** $y = x^2,\ x < 0$
  **b.**
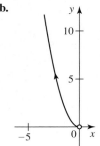

**23. a.** $y = \frac{1}{4}e^{2x}$
  **b.**

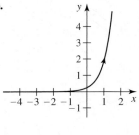

**25. a.** $x^2 - y^2 = 1,\ x \geq 1$
  **b.**

**27. a.** $y = x^{3/2},\ 0 \leq x \leq 1$
  **b.**
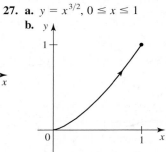

**29.** As $t$ increases, the particle moves along the parabola $y = \sqrt{x - 1}$ from $(1, 0)$ to $(5, 2)$.

**31.** The particle starts out at $(2, 2)$ and travels once counterclockwise along the circle of radius 1 centered at $(1, 2)$.

**33.** The particle starts at $(0, 0)$ and moves to the right along the parabola $y = x^2$ to $(1, 1)$, then back to $(-1, 1)$, then again to $(1, 1)$, and finally back to $(0, 0)$.

**35.**

**43.**
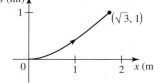
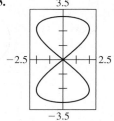

**45.**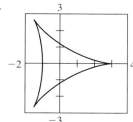

**47.**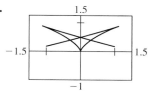

**49.**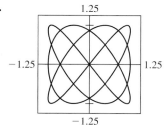

**51.** False   **53.** True

### Exercises 9.3 • page 859

**1.** $\frac{1}{2}$   **3.** $-2$   **5.** $-\frac{3}{2}$   **7.** $y = \frac{1}{2}x - \frac{1}{2}$

**9.** $y = 5x + 2$

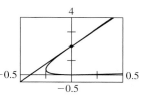

**11.** $(31, 64)$

**13.** Horizontal at $(-3, \pm 2)$, vertical at $(-4, 0)$

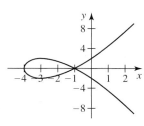

**15.** Horizontal at $(1, 0)$ and $(1, 4)$, vertical at $(4, 2)$ and $(-2, 2)$

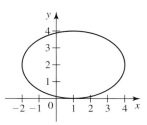

**17.** $\frac{dy}{dx} = t, \frac{d^2y}{dx^2} = \frac{1}{6t}$   **19.** $\frac{dy}{dx} = -\frac{2}{t^{3/2}}, \frac{d^2y}{dx^2} = \frac{6}{t^2}$

**21.** $\frac{dy}{dx} = \frac{1 - \cos\theta}{1 - \sin\theta}, \frac{d^2y}{dx^2} = \frac{\sin\theta + \cos\theta - 1}{(1 - \sin\theta)^3}$

**23.** $\frac{dy}{dx} = \coth t, \frac{d^2y}{dx^2} = -\frac{1}{\sinh^3 t}$

**25.** Concave downward on $(-\infty, 0)$, concave upward on $(0, \infty)$

**27.** $dy/dx = -\tan t$, $m = -1$ at $\left(\frac{\sqrt{2}}{4}a, \frac{\sqrt{2}}{4}a\right)$ and $\left(-\frac{\sqrt{2}}{4}a, -\frac{\sqrt{2}}{4}a\right)$, $m = 1$ at $\left(-\frac{\sqrt{2}}{4}a, \frac{\sqrt{2}}{4}a\right)$ and $\left(\frac{\sqrt{2}}{4}a, -\frac{\sqrt{2}}{4}a\right)$

**29.** Absolute maximum $f(-14) = 8$, absolute minimum $f(94) = -19$

**31.** $\frac{1}{243}(97^{3/2} - 64)$   **33.** $2\sqrt{5}$   **35.** $\frac{1}{8}a\pi^2$   **37.** $16a$

**39.** $4\sqrt{2}$   **41.** Approximately 1639 ft

**43.** $\left(a, -\frac{1}{2}\left(\frac{eE}{mv_0^2}\right)a^2\right)$

**45. a.**

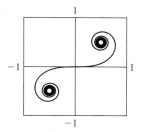

As $t \to \infty$, the curve spirals about and converges to the point $\left(\frac{1}{2}, \frac{1}{2}\right)$. As $t \to -\infty$, the curve spirals about and converges to the point $\left(-\frac{1}{2}, -\frac{1}{2}\right)$.

**b.** $a$

**47.** $3\pi a^2$   **49.** $3\pi a^2/8$   **51.** $32\sqrt{2}/3$

**53.** $\frac{2(247\sqrt{13} + 64)\pi}{1215}$   **55.** $148\pi/5$   **57.** $64\pi/3$

**59.** $\frac{24(\sqrt{2} + 1)\pi}{5}$   **61.** $\pi(e^2 + 2e - 6)$   **63.** $12\pi a^2/5$

**65.** $4\pi^2 rb$

**67.** 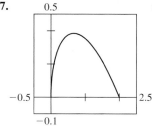   **69.**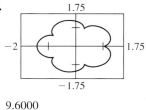

9.6000

2.2469

**71.** 33.66   **73.** Center $(0, 0)$, radius $a$

**75.** $x = \frac{3at}{t^3 + 1}, y = \frac{3at^2}{t^3 + 1}$   **79.** False

## Exercises 9.4 • page 873

**1.**

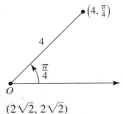

$(4, \frac{\pi}{4})$

$(2\sqrt{2}, 2\sqrt{2})$

**3.**

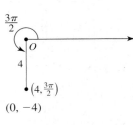

$\frac{3\pi}{2}$

$(4, \frac{3\pi}{2})$

$(0, -4)$

**5.**

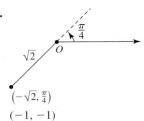

$(-\sqrt{2}, \frac{\pi}{4})$

$(-1, -1)$

**7.**

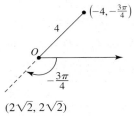

$(-4, -\frac{3\pi}{4})$

$(2\sqrt{2}, 2\sqrt{2})$

**9.**

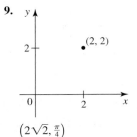

$(2\sqrt{2}, \frac{\pi}{4})$

**11.**

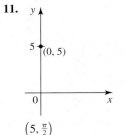

$(5, \frac{\pi}{2})$

**13.**

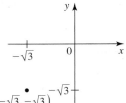

$(-\sqrt{3}, -\sqrt{3})$

$(\sqrt{6}, \frac{5\pi}{4})$

**15.**

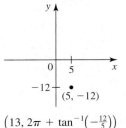

$(5, -12)$

$(13, 2\pi + \tan^{-1}(-\frac{12}{5}))$

**17.**

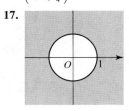

**19.**

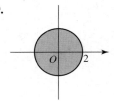

**21.**

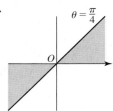

$\theta = \frac{\pi}{4}$

**23.**

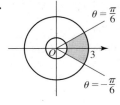

$\theta = \frac{\pi}{6}$

$\theta = -\frac{\pi}{6}$

**25.** $x = 2$    **27.** $2x + 3y = 6$    **29.** $x^2 + y^2 - 4x = 0$

**31.** $x^2 - 2y - 1 = 0$    **33.** $r = 4 \sec \theta$    **35.** $r = 3$

**37.** $r^2 = 8 \csc 2\theta$

**39.**

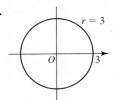

$r = 3$

**41.**

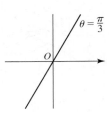

$\theta = \frac{\pi}{3}$

**43.**

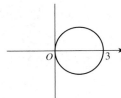

**45.**

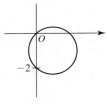

**47.**

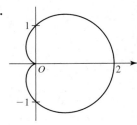

**49.**

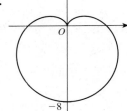

**51.**

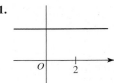

**53.**

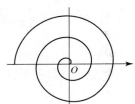

**55.**

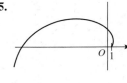

**57.**

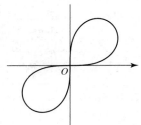

**59.**

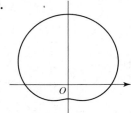

**61.**

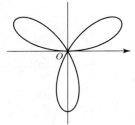

**63.**

**65.** $\sqrt{3}/3$    **67.** $-1$    **69.** $\pi$    **71.** 0

**73.** Horizontal at $\left(2\sqrt{2}, \frac{\pi}{4}\right)$ and $\left(-2\sqrt{2}, \frac{3\pi}{4}\right)$, vertical at the pole and $(4, 0)$

**75.** Horizontal at $(0, 0)$, $(\sin(2\tan^{-1}\sqrt{2}), \tan^{-1}\sqrt{2})$, $(\sin(2\tan^{-1}(-\sqrt{2})), \pi + \tan^{-1}(-\sqrt{2}))$, $(0, \pi)$, $(\sin(2\tan^{-1}\sqrt{2}), \pi + \tan^{-1}\sqrt{2})$, and $(\sin(2\tan^{-1}(-\sqrt{2})), 2\pi + \tan^{-1}(-\sqrt{2}))$;
vertical at $\left(\sin\left(2\sin^{-1}\frac{\sqrt{3}}{3}\right), \sin^{-1}\frac{\sqrt{3}}{3}\right)$, $\left(0, \frac{\pi}{2}\right)$, $\left(\sin\left(2\sin^{-1}\left(-\frac{\sqrt{3}}{3}\right)\right), \pi + \sin^{-1}\left(-\frac{\sqrt{3}}{3}\right)\right)$, $\left(\sin\left(2\sin^{-1}\frac{\sqrt{3}}{3}\right), \pi + \sin^{-1}\frac{\sqrt{3}}{3}\right)$, $\left(0, \frac{3\pi}{2}\right)$, and $\left(\sin\left(2\sin^{-1}\left(-\frac{\sqrt{3}}{3}\right)\right), 2\pi + \sin^{-1}\left(-\frac{\sqrt{3}}{3}\right)\right)$

**77.** Horizontal at $\left(1 + \dfrac{\sqrt{33}-1}{4}, \cos^{-1}\dfrac{\sqrt{33}-1}{8}\right)$, $\left(1 - \dfrac{\sqrt{33}+1}{4}, \cos^{-1}\dfrac{-\sqrt{33}-1}{8}\right)$, $\left(1 - \dfrac{\sqrt{33}+1}{4}, 2\pi - \cos^{-1}\left(\dfrac{-1-\sqrt{33}}{8}\right)\right)$, and $\left(1 + \dfrac{\sqrt{33}-1}{4}, 2\pi - \cos^{-1}\dfrac{\sqrt{33}-1}{8}\right)$, vertical at $(3, 0)$, $\left(\frac{1}{2}, \cos^{-1}\left(-\frac{1}{4}\right)\right)$, $(-1, \pi)$, and $\left(\frac{1}{2}, 2\pi - \cos^{-1}\left(-\frac{1}{4}\right)\right)$

**81. b.** $2\sqrt{3}$

**83. a.**

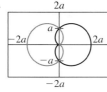

**85.**

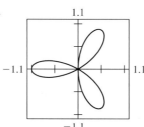

**87.**

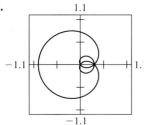

**89.**

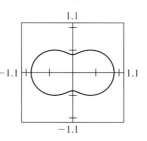

**91.**

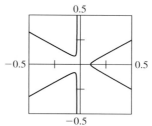

**93.** False    **95.** False

## Exercises 9.5 • page 881

**1. a.** $4\pi$   **b.** $4\pi$    **3.** $\frac{1}{6}\pi^3$

**5.** $\dfrac{e^\pi - 1}{4e^\pi}$    **7.** $\frac{1}{2}$    **9.** $9\pi^3/16$    **11.** $3\pi/4$

**13.**

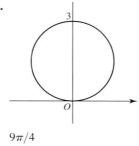

$9\pi/4$

**15.**

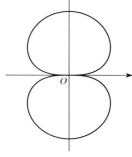

2

**17.**

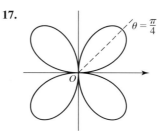

$2\pi$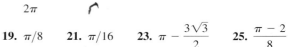

**19.** $\pi/8$    **21.** $\pi/16$    **23.** $\pi - \dfrac{3\sqrt{3}}{2}$    **25.** $\dfrac{\pi - 2}{8}$

**27.** $\dfrac{3\pi - 1}{4}$    **29.** $\left(1, \frac{\pi}{2}\right)$ and $\left(1, \frac{3\pi}{2}\right)$

**31.** $\left(2, \frac{\pi}{6}\right)$, $\left(2, \frac{\pi}{3}\right)$, $\left(2, \frac{2\pi}{3}\right)$, $\left(2, \frac{5\pi}{6}\right)$, $\left(2, \frac{7\pi}{6}\right)$, $\left(2, \frac{4\pi}{3}\right)$, $\left(2, \frac{5\pi}{3}\right)$, and $\left(2, \frac{11\pi}{6}\right)$

**33.** $\left(\frac{\sqrt{3}}{2}, \frac{\pi}{3}\right)$, $\left(-\frac{\sqrt{3}}{2}, \frac{5\pi}{3}\right)$, and the pole    **35.** $\pi$

**37.** $\dfrac{4\pi + 6\sqrt{3}}{3}$    **39.** $\dfrac{4\pi + 9\sqrt{3}}{8}$    **41.** $\dfrac{4\pi - 3\sqrt{3}}{6}$

**43.** $\dfrac{7\pi - 12\sqrt{3}}{12}$    **45.** $\dfrac{2\pi + 12 - 6\sqrt{3}}{3}$    **47.** $5\pi$

**51.** $\dfrac{4\pi - 3\sqrt{3}}{8}$    **53.** $16a/3$

**57.** $128\pi/5$    **59.** $2\sqrt{2}\pi$    **61.** $2\pi$

**b.** 22.01

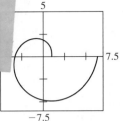

**69. a.**     **b.** 5.37

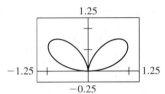

**71. a.**    **b.** $\left(\dfrac{21}{20}, 0\right)$

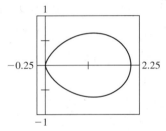

**73. a.** $r = \dfrac{3\cos\theta\sin\theta}{\cos^3\theta + \sin^3\theta},\ -\dfrac{\pi}{4} < \theta < \dfrac{3\pi}{4}$

**b.**     **c.** $\dfrac{3}{2}$

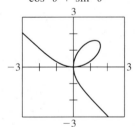

**75.** True

## Exercises 9.6 • page 890

**1.** $r = \dfrac{2}{1 - \cos\theta}$, parabola

**3.** $r = \dfrac{2}{2 - \sin\theta}$, ellipse

**5.** $r = \dfrac{3}{2 + 3\cos\theta}$, hyperbola

**7.** $r = \dfrac{4}{25 + 10\sin\theta}$, ellipse

**9. a.** $e = \frac{1}{3}, y = 4$
**b.** Ellipse
**c.**

**11. a.** $e = \frac{3}{2}, x = \frac{5}{3}$
**b.** Hyperbola
**c.**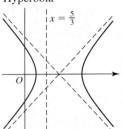

**13. a.** $e = 1, x = \frac{5}{2}$
**b.** Parabola
**c.**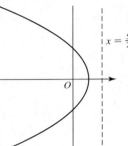

**15. a.** $e = \frac{2}{3}, x = -\frac{1}{2}$
**b.** Ellipse
**c.**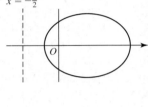

**17. a.** $e = 1, y = -1$
**b.** Parabola
**c.**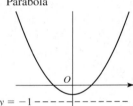

**19. a.** $e = \frac{1}{2}, y = -6$
**b.** Ellipse
**c.**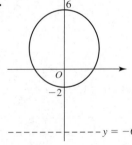

**21.** $\sqrt{7}/4$    **23.** $\sqrt{2}$    **25.** $\sqrt{10}/3$

**33.** $r = \dfrac{1.423 \times 10^9}{1 - 0.056\cos\theta}$, perihelion $1.347 \times 10^9$ km, aphelion $1.507 \times 10^9$ km

**35.** 0.207

## Chapter 9 Concept Review • page 892

**1. a.** equidistant; point; line; point; focus; line; directrix
**b.** vertex; focus (or vertex); directrix

**3. a.** sum; foci; constant
**b.** foci; major axis; center; minor axis

**5. a.** difference; foci; constant
**b.** vertices; transverse; transverse; center; two separate

**7.** $x = f(t), y = g(t)$; parameter

**9. a.** $f'(t)$; $g'(t)$; simultaneously zero; endpoints

**b.** $\int_a^b \sqrt{[f'(t)]^2 + [g'(t)]^2}\, dt = \int_a^b \sqrt{\left(\frac{dx}{dt}\right)^2 + \left(\frac{dy}{dt}\right)^2}\, dt$

**11. a.** $r\cos\theta$; $r\sin\theta$

**b.** $x^2 + y^2$; $\dfrac{y}{x}$  $(x \neq 0)$

**13. a.** $A = \dfrac{1}{2}\int_\alpha^\beta r^2\, d\theta = \dfrac{1}{2}\int_\alpha^\beta [f(\theta)]^2\, d\theta$

**b.** $A = \dfrac{1}{2}\int_\alpha^\beta \{[f(\theta)]^2 - [g(\theta)]^2\}\, d\theta$

**15.** $\dfrac{d(P,F)}{d(P,l)} = e$; $0 < e < 1$; $e = 1$; $e > 1$

## Chapter 9 Review Exercises • page 893

**1.** Vertices $(0, \pm 3)$, foci $(0, \pm\sqrt{5})$

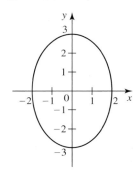

**3.** Vertices $(\pm 3, 0)$, foci $(\pm\sqrt{10}, 0)$

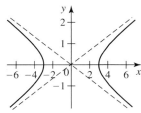

**5.** Vertices $(0, -1)$ and $(0, -7)$, foci $(0, -4 \pm \sqrt{10})$

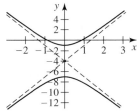

**7.** $y^2 = -8x$   **9.** $\dfrac{x^2}{49} + \dfrac{y^2}{45} = 1$   **11.** $y^2 - 4x^2 = 9$

**15. a.** $y = 4 - x$

**b.**
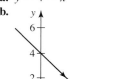

**17. a.** $(x - 1)^2 + (y - 3)^2 = 4$

**b.**
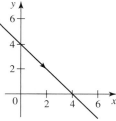

**19.** $\frac{4}{3}$   **21.** $0$   **23.** $\dfrac{dy}{dx} = \dfrac{4(t^2 + 1)}{3t}$, $\dfrac{d^2y}{dx^2} = \dfrac{4(t^2 - 1)}{9t^4}$

**25.** Vertical at $\left(\pm\frac{16\sqrt{3}}{9}, \frac{10}{3}\right)$, horizontal at $(0, 2)$

**27.** $\frac{13}{3}$   **29.** $\sqrt{2}(1 - e^{-\pi/2})$   **31.** $3\pi$

**33.**
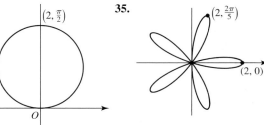

**35.**

**37.**
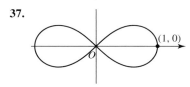

**39.** $-2$   **41.** $\left(\frac{1}{2}, \frac{\pi}{6}\right)$, $\left(\frac{1}{2}, \frac{5\pi}{6}\right)$, and the pole   **43.** $9\pi/2$

**45.** $2(\pi - 2)$   **47.** $\frac{8}{3}[(\pi^2 + 1)^{3/2} - 1]$   **49.** $4\pi^2$

**51.**
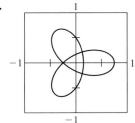

**53.**

**55.**

**57.**
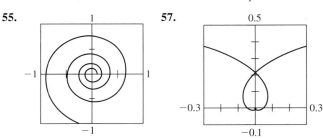

**59.** $\pi(\pi + 2)$

### Chapter 9 Challenge Problems • page 894

**1. c.**

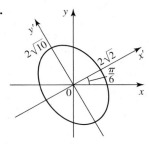

**3.** $\ln \dfrac{\pi}{2}$   **5. b.** $3a^2/2$

**7. a.** $r = \dfrac{\sqrt{2}}{2}\, ae^{(\pi/4)-\theta}$ (for an ant starting at the northeast corner)

**b.** $a$

# CHAPTER 10

## Exercises 10.1 • page 905

**1. a.** Scalar **b.** Vector **c.** Scalar **d.** Vector

**3. a.** No **b.** Yes **c.** No **d.** No **e.** Yes **f.** Yes

**9.** $\langle -5, 0 \rangle$   **11.** $\langle 3, -4 \rangle$

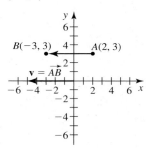 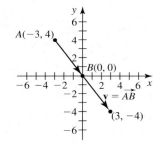

**13.** 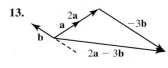 **15.**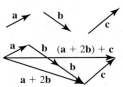

**17.** $\mathbf{v} = -(\mathbf{a} + \mathbf{b})$   **19.** $\mathbf{v} = 3(\mathbf{b} - \mathbf{a})$

**21.** $\langle -2, -1 \rangle$   **23.** $\langle -0.3, -0.1 \rangle$

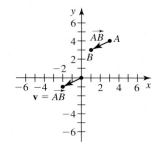 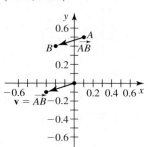

**25.** $(1, -2)$

**27.** $2\mathbf{a} = \langle -2, 4 \rangle$, $\mathbf{a} + \mathbf{b} = \langle 2, 3 \rangle$, $\mathbf{a} - \mathbf{b} = \langle -4, 1 \rangle$, $2\mathbf{a} + \mathbf{b} = \langle 1, 5 \rangle$, $|2\mathbf{a} + \mathbf{b}| = \sqrt{26}$

**29.** $2\mathbf{a} = 6\mathbf{i} - 4\mathbf{j}$, $\mathbf{a} + \mathbf{b} = 5\mathbf{i} - 2\mathbf{j}$, $\mathbf{a} - \mathbf{b} = \mathbf{i} - 2\mathbf{j}$, $2\mathbf{a} + \mathbf{b} = 8\mathbf{i} - 4\mathbf{j}$, $|2\mathbf{a} + \mathbf{b}| = 4\sqrt{5}$

**31.** $2\mathbf{a} = \mathbf{i} + 3\mathbf{j}$, $\mathbf{a} + \mathbf{b} = \frac{5}{4}\mathbf{i} + \frac{5}{4}\mathbf{j}$, $\mathbf{a} - \mathbf{b} = -\frac{1}{4}\mathbf{i} + \frac{7}{4}\mathbf{j}$, $2\mathbf{a} + \mathbf{b} = \frac{7}{4}\mathbf{i} + \frac{11}{4}\mathbf{j}$, $|2\mathbf{a} + \mathbf{b}| = \frac{\sqrt{170}}{4}$

**33.** $2\mathbf{a} - 3\mathbf{b} = 4\mathbf{i} + 18\mathbf{j}$, $\frac{1}{2}\mathbf{a} + \frac{1}{3}\mathbf{b} = \mathbf{i} - 2\mathbf{j}$

**35.** $a = -1.6$, $b = 2.2$   **37.** Parallel   **39.** Not parallel

**41.** Parallel   **43.** $\frac{8}{3}$

**45. a.** $\left\langle \frac{2\sqrt{5}}{5}, \frac{\sqrt{5}}{5} \right\rangle$ **b.** $\left\langle -\frac{2\sqrt{5}}{5}, -\frac{\sqrt{5}}{5} \right\rangle$

**47. a.** $\left\langle -\frac{\sqrt{3}}{2}, \frac{1}{2} \right\rangle$ **b.** $\left\langle \frac{\sqrt{3}}{2}, -\frac{1}{2} \right\rangle$

**49.** $\left\langle \frac{5\sqrt{2}}{2}, \frac{5\sqrt{2}}{2} \right\rangle$   **51.** $\sqrt{3}\mathbf{i}$   **53.** $\left\langle -\frac{27\sqrt{85}}{85}, \frac{6\sqrt{85}}{85} \right\rangle$

**55.** $\mathbf{F} = \sqrt{10}\left\langle \frac{3\sqrt{10}}{10}, \frac{\sqrt{10}}{10} \right\rangle$, 3, 1   **57.** $\sqrt{39}\left( \frac{\sqrt{13}}{13}\mathbf{i} + \frac{2\sqrt{39}}{13}\mathbf{j} \right)$, $\sqrt{3}$, 6

**59.** $2\mathbf{i}$   **61.** $\dfrac{3}{2}\mathbf{i} - \dfrac{3\sqrt{3}}{2}\mathbf{j}$

**63. a.** $\langle a_1 + a_2, b_1 + b_2 \rangle$; the company produced $a_1 + a_2$ Model $A$ systems and $b_1 + b_2$ Model $B$ systems

**b.** $\langle 1.1(a_1 + a_2), 1.1(b_1 + b_2) \rangle$

**65.** $400\sqrt{2}\mathbf{i}$ ft/sec, $400\sqrt{2}\mathbf{j}$ ft/sec   **67.** 30 lb

**69.** $60°$, $\dfrac{\sqrt{3}}{30}$ hr   **71.** 275.4 mph, E 25.8° N

**83.** False   **85.** True   **87.** False

## Exercises 10.2 • page 916

**1.** 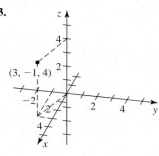 **3.**

**5.** $(-3, -2, 4)$

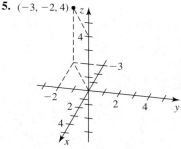

**7.** $A(2, 5, 5)$, $B(3, -3, -3)$

**9.** 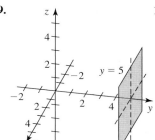 **11.**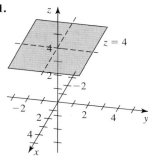

**13.** The subspace of three-dimensional space that lies on or in front of the plane $x = 3$

**15.** The subspace of three-dimensional space that lies strictly above the plane $z = 3$

**17. a.** $A(2, -1, 3)$, $B(-1, 4, 1)$ (in units of 1000 ft)

**b.** 6164 ft

**19.** $\sqrt{21}$, $\sqrt{6}$, $\sqrt{27}$, right **21.** 3, 2, $\sqrt{5}$, right

**23.** Collinear **25.** $(-1, 3, -1)$

**27.** $(x - 2)^2 + (y - 1)^2 + (z - 3)^2 = 9$

**29.** $(x - 3)^2 + (y + 1)^2 + (z - 2)^2 = 16$

**31.** $\left(x - \frac{5}{2}\right)^2 + \left(y + \frac{1}{2}\right)^2 + \left(z - \frac{5}{2}\right)^2 = \frac{35}{4}$

**33.** $(x + 1)^2 + (y - 2)^2 + (z - 4)^2 = 6$

**35.** $(1, 2, 3)$, 2 **37.** $(2, -3, 0)$, $\sqrt{13}$ **39.** $\left(\frac{3}{2}, 1, -\frac{1}{2}\right)$, 2

**41.** All points inside the sphere with radius 2 and center $(0, 0, 0)$

**43.** All points lying on or between two concentric spheres with radii 1 and 3 and center $(0, 0, 0)$

**45.** $\langle -2, 2, 4 \rangle$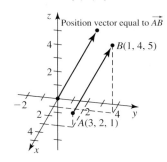

**47.** $\langle 4, -5, 5 \rangle$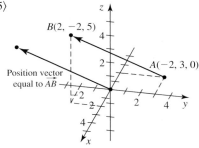

**49.** $\langle -1, 3, 5 \rangle$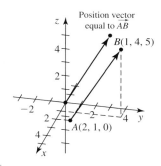

**51.** $(3, -6, -3)$

**53.** $\mathbf{a} + \mathbf{b} = \langle 1, 5, -1 \rangle$, $2\mathbf{a} - 3\mathbf{b} = \langle -8, -5, 3 \rangle$, $|3\mathbf{a}| = 3\sqrt{5}$, $|-2\mathbf{b}| = 2\sqrt{14}$, $|\mathbf{a} - \mathbf{b}| = \sqrt{11}$

**55.** $\mathbf{a} + \mathbf{b} = \langle 1, 6.2, -2.2 \rangle$, $2\mathbf{a} - 3\mathbf{b} = \langle -3, -8.1, 23.6 \rangle$, $|3\mathbf{a}| \approx 11.99$, $|-2\mathbf{b}| \approx 14.02$, $|\mathbf{a} - \mathbf{b}| \approx 9.27$

**57.** $a = \frac{4}{35}$, $b = \frac{16}{35}$, $c = \frac{2}{5}$

**59.** Parallel **61.** Not parallel

**63. a.** $\langle \frac{1}{3}, \frac{2}{3}, \frac{2}{3} \rangle$ **b.** $\langle -\frac{1}{3}, -\frac{2}{3}, -\frac{2}{3} \rangle$

**65. a.** $\frac{\sqrt{11}}{11}(-\mathbf{i} + 3\mathbf{j} - \mathbf{k})$ **b.** $\frac{\sqrt{11}}{11}(\mathbf{i} - 3\mathbf{j} + \mathbf{k})$

**67.** $\langle \frac{10\sqrt{3}}{3}, \frac{10\sqrt{3}}{3}, \frac{10\sqrt{3}}{3} \rangle$ **69.** $\frac{3\sqrt{5}}{5}\mathbf{i} + \frac{6\sqrt{5}}{5}\mathbf{j}$

**71.** $\langle \frac{\sqrt{2}}{3}, -\frac{\sqrt{2}}{3}, \frac{4\sqrt{2}}{3} \rangle$ **73.** $5\sqrt{2}\langle -\frac{3\sqrt{2}}{10}, \frac{2\sqrt{2}}{5}, \frac{\sqrt{2}}{2} \rangle$

**83.** True **85.** False

## Exercises 10.3 • page 928

**1. a.** Yes **b.** No **c.** No **d.** Yes **e.** Yes **f.** Yes

**3.** $-1$ **5.** $-4$ **7.** $4\pi$ **9.** 4 **11.** $-75$

**13.** $\langle -24, 48, -12 \rangle$ **15.** 70 **17.** 26.6° **19.** 94.7°

**21.** 63.1° **23.** $\frac{1}{3}$ **25.** Neither **27.** Neither

**29.** Orthogonal **31.** $-6$

**33.** $\cos \alpha = \sqrt{14}/14$, $\alpha \approx 74.5°$, $\cos \beta = \sqrt{14}/7$, $\beta \approx 57.7°$, $\cos \gamma = 3\sqrt{14}/14$, $\gamma \approx 36.7°$

**35.** $\cos \alpha = -\sqrt{35}/35$, $\alpha \approx 99.7°$, $\cos \beta = 3\sqrt{35}/35$, $\beta \approx 59.5°$, $\cos \gamma = \sqrt{35}/7$, $\gamma \approx 32.3°$

**37.** $\pi/3$ or $2\pi/3$

**39. a.** $\langle \frac{28}{13}, \frac{42}{13} \rangle$ **b.** $\langle \frac{14}{17}, \frac{56}{17} \rangle$

**41. a.** $\frac{20}{21}\mathbf{i} + \frac{10}{21}\mathbf{j} + \frac{40}{21}\mathbf{k}$ **b.** $3\mathbf{i} + \mathbf{k}$

**43. a.** $\langle -\frac{12}{29}, \frac{16}{29}, -\frac{8}{29} \rangle$ **b.** $\langle 0, 4, 0 \rangle$

**45.** $\mathbf{b} = \langle \frac{7}{5}, \frac{21}{5} \rangle + \langle \frac{3}{5}, -\frac{1}{5} \rangle$

**47.** $\mathbf{b} = \left( \frac{3}{14}\mathbf{i} + \frac{3}{7}\mathbf{j} + \frac{9}{14}\mathbf{k} \right) + \left( \frac{25}{14}\mathbf{i} - \frac{10}{7}\mathbf{j} + \frac{5}{14}\mathbf{k} \right)$

**49.** 12 **51.** 54.7° **53.** $\theta \approx 67.4°$, $\psi \approx 22.6°$

**55.** 1039.2 ft-lb

**57.** 274,955 ft-lb by Tugboat I, 207,846 ft-lb by Tugboat II

**61. a.**

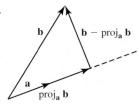

**63. c.** $\sqrt{13}$    **65.** False    **67.** True

## Exercises 10.4 • page 939

**1. a.** Yes  **b.** No  **c.** Yes  **d.** Yes  **e.** Yes  **f.** Yes

**3.** $3\mathbf{i} - 3\mathbf{j} + 2\mathbf{k}$    **5.** $3\mathbf{i} + 5\mathbf{j} + 7\mathbf{k}$    **7.** $-6\mathbf{i} - 7\mathbf{j} + 4\mathbf{k}$

**9.** 0    **11.** $\mathbf{i} + 7\mathbf{j} - 5\mathbf{k}, -\mathbf{i} - 7\mathbf{j} + 5\mathbf{k}$

**13.** $2\mathbf{i} - \mathbf{k}, -2\mathbf{i} + \mathbf{k}$    **15.** $\pm\dfrac{\sqrt{26}}{26}(3\mathbf{i} + \mathbf{j} - 4\mathbf{k})$

**17.** $\sqrt{3}/2$    **19.** $\sqrt{401}/2$    **21.** $-4\mathbf{i} + 6\mathbf{j} + 10\mathbf{k}$

**23.** $\mathbf{i} - \mathbf{j} + \mathbf{k}$    **25.** 15    **27.** 5    **29.** 21    **31.** $\sqrt{3}$

**33.** Yes    **35.** Yes    **37.** 32.5 ft-lb

**39.** $-4.8 \times 10^{-14}\mathbf{j}$ newtons    **53.** $3\sqrt{2}$

**55.** $a = -\frac{3}{2}, b = -6$    **57.** True    **59.** True    **61.** True

## Exercises 10.5 • page 950

**1. a.** The direction of the required line is the same as the direction of the given line and so can be obtained from the parametric equations of the latter. Use this information with the given point to write down the required equation.
  **b.** Obtain the vectors $\mathbf{v}_1$ and $\mathbf{v}_2$ from the two given lines. Find the vector $\mathbf{n} = \mathbf{v}_1 \times \mathbf{v}_2$. Then the required line has the direction of $\mathbf{n}$ and contains the given point.
  **c.** A vector parallel to the required line is $\mathbf{n}$, a vector normal to the plane. It can be obtained from the equation of the plane. Use $\mathbf{n}$ and the given point to write down the required parametric equation.
  **d.** Obtain $\mathbf{n}_1$ and $\mathbf{n}_2$, the normals to the two planes, from the given equations of the planes. The direction of the required line is given by $\mathbf{v} = \mathbf{n}_1 \times \mathbf{n}_2$. Find a point on the required line by setting one variable, say $z$, equal to 0 and solving the resulting simultaneous equations in two variables.

**3.** $x = 1 + 2t, \ y = 3 + 4t, z = 2 + 5t$;
$\dfrac{x-1}{2} = \dfrac{y-3}{4} = \dfrac{z-2}{5}$

**5.** $x = 3 + 2t, y = -t, z = -2 + 3t; \dfrac{x-3}{2} = \dfrac{y}{-1} = \dfrac{z+2}{3}$

**7.** $x = 2 - t, y = 1 + 2t, z = 4 + 3t$;
$\dfrac{x-2}{-1} = \dfrac{y-1}{2} = \dfrac{z-4}{3}$

**9.** $x = -1 + 4t, y = -2 + 7t, z = -\dfrac{1}{2} - 5t$;
$\dfrac{x+1}{4} = \dfrac{y+2}{7} = \dfrac{z+\frac{1}{2}}{-5}$

**11.** $x = 1 + t, y = 2 + 2t, z = -1 - 3t$;
$\dfrac{x-1}{1} = \dfrac{y-2}{2} = \dfrac{z+1}{-3}$; xy-plane: $\left(\frac{2}{3}, \frac{4}{3}, 0\right)$,
xz-plane: $(0, 0, 2)$, yz-plane: $(0, 0, 2)$

**13.** Yes    **15.** Intersect at $(5, 4, 5)$

**17.** Intersect at $(2, 3, 1)$    **19.** Intersect, 49.1°

**21.** Skew    **23.** $x + 2y + 4z = 24$

**25.** $x - 2z = 1$    **27.** $2x + 3y - z = 26$

**29.** $x - 3z = 8$    **31.** $6x - 4y + 3z = 0$

**33.** $6x - y - 4z = -5$    **35.** $9x + 11y + 3z = -2$

**37.** $11x + 10y + 13z = 45$    **39.** Orthogonal

**41.** Neither, 69.1°    **43.** 70.5°

**45.** $x = \frac{17}{4} - 10t, y = 8t, z = -\frac{11}{8} + 11t$

**47.** $x = 2 + 2t, y = 3 + 4t, z = -1 - 3t$

**49.** $11x + 5y - 7z = -8$    **51.** $6x - 7y + z = 11$

**53.** $(5, 0, 1)$    **55.** $4\sqrt{29}/29$    **57.** $2\sqrt{21}/7$

**61.** $6\sqrt{42}/7$    **65.** $\sqrt{6}/2$    **67.** False

**69.** False    **71.** False

## Exercises 10.6 • page 966

**1.** 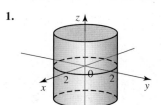    **3.**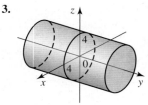

**5.** 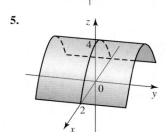    **7.**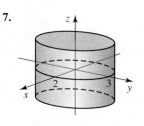

**9.** 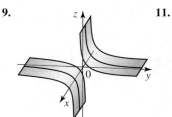    **11.**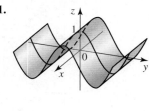

**13.** (a)   **15.** (f)   **17.** (e)   **19.** (b)

**21.** $\dfrac{x^2}{1^2} + \dfrac{y^2}{2^2} + \dfrac{z^2}{2^2} = 1$   **23.** $\dfrac{x^2}{2^2} + \dfrac{y^2}{3^2} + \dfrac{z^2}{6^2} = 1$

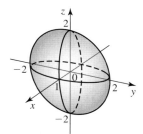   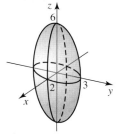

**25.** $x^2 + y^2 - \dfrac{z^2}{2^2} = 1$   **27.** $\dfrac{x^2}{2^2} + \dfrac{y^2}{1^2} - \dfrac{z^2}{2^2} = 1$

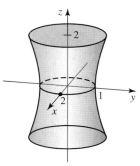   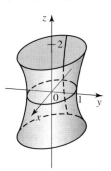

**29.** $-x^2 - y^2 + z^2 = 1$   **31.** $-\dfrac{x^2}{1^2} + \dfrac{y^2}{2^2} - \dfrac{z^2}{(\sqrt{2})^2} = 1$

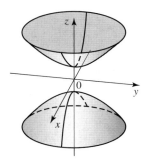   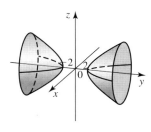

**33.** $x^2 + y^2 - z^2 = 0$   **35.** $\dfrac{x^2}{1^2} + \dfrac{y^2}{\left(\frac{3}{2}\right)^2} - \dfrac{z^2}{3^2} = 0$

or $\dfrac{x^2}{(2)^2} + \dfrac{y^2}{(3)^2} - \dfrac{z^2}{(6)^2} = 0$

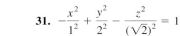

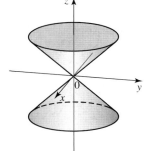   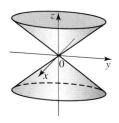

**37.** $x^2 + y^2 = z$   **39.** $\dfrac{x^2}{3^2} + \dfrac{y^2}{1^2} = \dfrac{z}{3^2}$

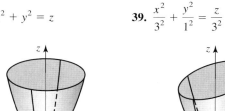   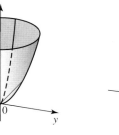   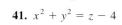   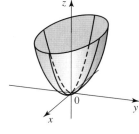

**41.** $x^2 + y^2 = z - 4$   **43.** $y^2 - x^2 = z$

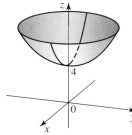   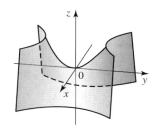

**45.**   **47.**

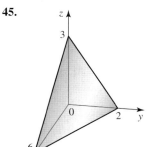   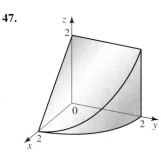

**49.**

**51.** $y^2 + z^2 = -12x$, a paraboloid

**53.** $x + y = 2$, a plane

**55.**

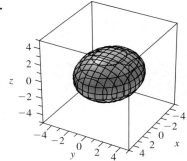

**57.**

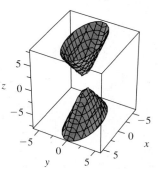

**59.**

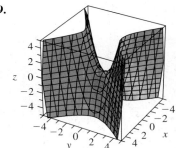

**61.** False    **63.** True

### Exercises 10.7 • page 973

**1.** $(0, 3, 2)$    **3.** $(1, 1, \sqrt{3})$    **5.** $\left(\frac{3\sqrt{3}}{2}, -\frac{3}{2}, 2\right)$

**7.** $(2, 0, 3)$    **9.** $\left(2, \frac{\pi}{3}, 5\right)$    **11.** $\left(2, \frac{\pi}{6}, -2\right)$

**13.** $(0, 0, 5)$    **15.** $(\sqrt{2}, 0, \sqrt{2})$    **17.** $\left(\frac{5\sqrt{6}}{4}, \frac{5\sqrt{2}}{4}, \frac{5\sqrt{2}}{2}\right)$

**19.** $\left(2, \pi, \frac{\pi}{2}\right)$    **21.** $\left(2, 0, \frac{\pi}{3}\right)$    **23.** $\left(4, \frac{\pi}{2}, \frac{\pi}{3}\right)$

**25.** $\left(2, \frac{\pi}{4}, \frac{\pi}{2}\right)$    **27.** $\left(4\sqrt{2}, \frac{\pi}{3}, \frac{3\pi}{4}\right)$    **29.** $\left(2\sqrt{13}, \frac{\pi}{6}, \cos^{-1}\left(\frac{3\sqrt{13}}{13}\right)\right)$

**31.** $(0, 0, 3)$    **33.** $\left(2, \frac{3\pi}{2}, 0\right)$    **35.** $\left(\frac{\sqrt{3}}{2}, \frac{\pi}{4}, \frac{1}{2}\right)$    **37.** $\sqrt{11}$

**39.** Circular cylinder with radius 2 and central axis the $z$-axis

**41.** Sphere with center the origin and radius 2

**43.** Upper half of a right circular cone with vertex the origin and axis the positive $z$-axis

**45.** Paraboloid with vertex $(0, 0, 4)$ and axis the $z$-axis, opening downward

**47.** Plane parallel to the $xy$-plane and three units above it

**49.** Circular cylinder with radius 2 and central axis the line parallel to the $z$-axis passing through $(2, 0, 0)$

**51.** Parabolic cylinder

**53.** Sphere with radius 4 centered at the origin

**55.** Plane parallel to the $yz$-plane passing through $(2, 0, 0)$

**57.** Two circular cylinders with radii 1 and 2 and axis the $z$-axis

**59. a.** $r^2 + z^2 = 4$    **b.** $\rho = 2$

**61. a.** $r^2 - 2z = 0$    **b.** $\rho \sin^2 \phi - 2 \cos \phi = 0$

**63. a.** $r(2 \cos \theta + 3 \sin \theta) - 4z = 12$
   **b.** $\rho(2 \sin \phi \cos \theta + 3 \sin \phi \sin \theta - 4 \cos \phi) = 12$

**65. a.** $r^2 \cos^2 \theta + z^2 = 4$
   **b.** $\rho^2(\sin^2 \phi \cos^2 \theta + \cos^2 \phi) = 4$

**67.**

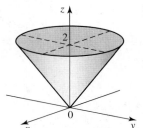

**69.**

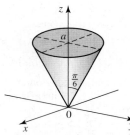

**71. a.** $(3960, 241.75°, 55.94°)$, $(3960, 2.20°, 41.48°)$
   **b.** $(-1552.8, -2889.9, 2217.8)$, $(2621.0, 100.7, 2966.8)$
   **c.** $5663$ mi

**73.** True    **75.** True

### Chapter 10 Concept Review • page 974

**1. a.** direction; magnitude    **b.** arrow; arrow; direction; length
   **c.** initial; $A$; terminal; $B$    **d.** direction; magnitude

**3. a.** $\langle a_1, a_2 \rangle$; $a_1$; $a_2$; scalar; $\langle 0, 0 \rangle$
   **b.** $\langle a_1 + b_1, a_2 + b_2 \rangle$; $\langle ca_1, ca_2 \rangle$

**5. a.** $(h, k, l)$; $r$
   **b.** $\left(\dfrac{x_1 + x_2}{2}, \dfrac{y_1 + y_2}{2}, \dfrac{z_1 + z_2}{2}\right)$

**7. a.** $a_1 b_1 + a_2 b_2 + a_3 b_3$; scalar
   **b.** $\sqrt{\mathbf{a} \cdot \mathbf{a}}$
   **c.** $\dfrac{\mathbf{a} \cdot \mathbf{b}}{|\mathbf{a}||\mathbf{b}|}$

**9. a.** vector projection; vector component
   **b.** scalar component
   **c.** $\left(\dfrac{\mathbf{b} \cdot \mathbf{a}}{|\mathbf{a}|^2}\right)\mathbf{a}$
   **d.** $\mathbf{F} \cdot \overrightarrow{PQ}$

**11. a.** $\mathbf{a} \cdot (\mathbf{b} \times \mathbf{c})$    **b.** $|\mathbf{a} \cdot (\mathbf{b} \times \mathbf{c})|$

**13. a.** $a(x - x_0) + b(y - y_0) + c(z - z_0) = 0$
   **b.** Plane; $\mathbf{n} = \langle a, b, c \rangle$; normal vectors

## Chapter 10 Review Exercises • page 975

**1.** $\mathbf{i} - 8\mathbf{j} + 9\mathbf{k}$ **3.** $\sqrt{114}$ **5.** $5\mathbf{i} + 7\mathbf{j} - \mathbf{k}$

**7.** $-20$ **9.** $12\mathbf{j} + 4\mathbf{k}$

**11.** $\dfrac{3\sqrt{14}}{14}\mathbf{i} - \dfrac{\sqrt{14}}{7}\mathbf{j} + \dfrac{\sqrt{14}}{14}\mathbf{k}, \ -\dfrac{3\sqrt{14}}{14}\mathbf{i} + \dfrac{\sqrt{14}}{7}\mathbf{j} - \dfrac{\sqrt{14}}{14}\mathbf{k}$

**13.** $\cos \alpha = \sqrt{6}/6, \cos \beta = \sqrt{6}/3, \cos \gamma = -\sqrt{6}/6$

**15.** $-\dfrac{3}{7}\mathbf{i} + \dfrac{3}{14}\mathbf{j} - \dfrac{9}{14}\mathbf{k}$ **17.** 20 **21.** $-22$ **23.** $70.5°$

**25. a.** $3\mathbf{i} + 3\mathbf{j} - 4\mathbf{k}$ **b.** $\frac{1}{2}\sqrt{34}$ **27.** 64 J **29.** $11\mathbf{k}$

**31. a.** $x = -1 + 3t, y = 2 - 3t, z = -4 + 7t$
**b.** $\dfrac{x+1}{3} = \dfrac{y-2}{-3} = \dfrac{z+4}{7}$

**33. a.** $x = 1 + 6t, y = 2 + t, z = 4 - 4t$
**b.** $\dfrac{x-1}{6} = \dfrac{y-2}{1} = \dfrac{z-4}{-4}$

**35.** $2x + 4y - 3z = 3$ **37.** $y = 2$ **39.** $5\sqrt{29}/29$

**41.** $32.7°$ **43.** $\sqrt{14}/14$ **45.** $5\sqrt{29}/29$

**47.** All points lying on or inside the (infinite) circular cylinder of radius 2 with axis the $z$-axis

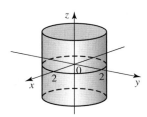

**49.** All points lying on or inside the prism shown

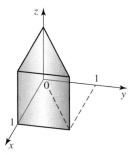

**51.** A plane perpendicular to the $xy$-plane

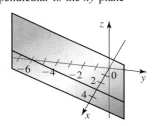

**53.** A paraboloid

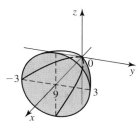

**55.** An elliptic cone with axis the $y$-axis

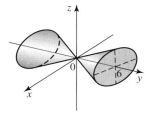

**57.** A hyperbolic paraboloid.

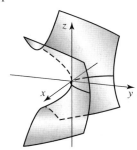

**59.** $\left(\sqrt{2}, \frac{\pi}{4}, \sqrt{2}\right), \left(2, \frac{\pi}{4}, \frac{\pi}{4}\right)$ **61.** $\left(\frac{\sqrt{6}}{2}, \frac{\sqrt{6}}{2}, 1\right), \left(\sqrt{3}, \frac{\pi}{4}, 1\right)$

**63.** The half-plane containing the $z$-axis making an angle of $\pi/3$ with the positive $x$-axis

**65.** A right circular cylinder with radius 1 and axis parallel to the $z$-axis passing through $(0, 1, 0)$

**67. a.** $r^2 = 2$ **b.** $\rho = \sqrt{2}\csc\phi$

**69. a.** $r^2 + 2z^2 = 1$ **b.** $\rho^2(1 + \cos^2\phi) = 1$

**71.** $0 \le r \le z, 0 \le \theta \le \frac{\pi}{2}$

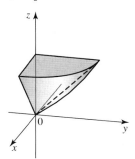

## Chapter 10 Challenge Problems • page 977

**5. b.** $\mathbf{v} = -9\langle 1, 3, 1 \rangle - 15\langle 2, -1, 1 \rangle + 14\langle 3, 1, 2 \rangle$

**7.** $(1, -2, 8)$ and $(3, 2, 40)$

# CHAPTER 11

## Exercises 11.1 • page 986

**1.** $(-\infty, 0) \cup (0, \infty)$    **3.** $(0, 1) \cup (1, \infty)$    **5.** $(0, \infty)$

**7.** (c)    **9.** (e)    **11.** (b)

**13.**
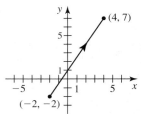

**15.**
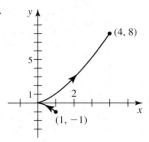

**17.**
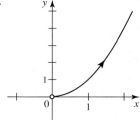

**19.**
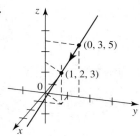

**21.**
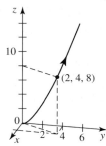

**23.**
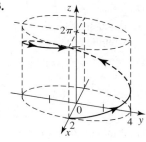

**25.**
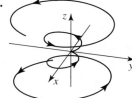

**27.**
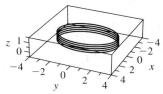

**29.**
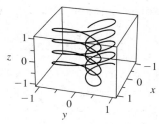

**31. b.**
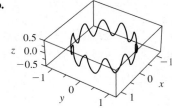

**33.** $\mathbf{r}(t) = \left\langle \cos t, \sin t, \dfrac{1 - \cos t - \sin t}{2} \right\rangle, 0 \le t \le 2\pi$

**35.** $\mathbf{r}(t) = \left\langle t, \dfrac{2t - 1}{2t - 2}, \dfrac{-2t^2 + 2t - 1}{2t - 2} \right\rangle$

**37.** $\mathbf{i} + \mathbf{j} - 3\mathbf{k}$    **39.** $\sqrt{2}\mathbf{i} + 4\mathbf{j} + \frac{2}{5}\mathbf{k}$    **41.** $\langle 0, 0, 2 \rangle$

**43.** $[-1, 0)$ and $(0, \infty)$    **45.** $(0, \infty)$

**47.** $(-\infty, -1), (-1, 1),$ and $(1, 4]$    **49.** 88,000 ft

**57.** $\mathbf{i} + \frac{1}{2}\mathbf{j}$    **59.** False    **61.** True

## Exercises 11.2 • page 994

**1.** $\mathbf{r}'(t) = \mathbf{i} + 2t\mathbf{j} + 3t^2\mathbf{k}, \mathbf{r}''(t) = 2\mathbf{j} + 6t\mathbf{k}$

**3.** $\mathbf{r}'(t) = \left\langle 2t, \dfrac{t}{\sqrt{t^2 + 1}} \right\rangle, \mathbf{r}''(t) = \left\langle 2, \dfrac{1}{(t^2 + 1)^{3/2}} \right\rangle$

**5.** $\mathbf{r}'(t) = \langle -t \sin t, t \cos t \rangle,$
$\mathbf{r}''(t) = \langle -\sin t - t \cos t, \cos t - t \sin t \rangle$

**7.** $\mathbf{r}'(t) = (\cos t - \sin t)e^{-t}\mathbf{i} - (\cos t + \sin t)e^{-t}\mathbf{j} + \dfrac{1}{t^2 + 1}\mathbf{k},$

$\mathbf{r}''(t) = -2e^{-t} \cos t\mathbf{i} + 2e^{-t} \sin t\mathbf{j} - \dfrac{2t}{(t^2 + 1)^2}\mathbf{k}$

**9. a.** $\mathbf{r}(2) = \sqrt{2}\mathbf{i} - 2\mathbf{j}, \mathbf{r}'(2) = \dfrac{\sqrt{2}}{4}\mathbf{i} + \mathbf{j}$

**b.**
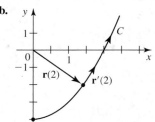

**11. a.** $r(\pi/3) = \langle 2, \sqrt{3} \rangle$, $r'(\pi/3) = \langle -2\sqrt{3}, 1 \rangle$
   **b.**

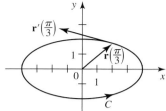

**13. a.** $r(1) = 5i - j$, $r'(1) = 3i - 2j$
   **b.**

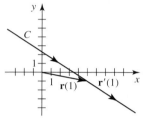

**15. a.** $r(\pi/4) = \sqrt{2}i + 2j$, $r'(\pi/4) = \sqrt{2}i + 4j$
   **b.**

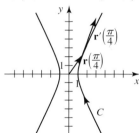

**17.** $\dfrac{\sqrt{14}}{14}i + \dfrac{\sqrt{14}}{7}j + \dfrac{3\sqrt{14}}{14}k$    **19.** $\dfrac{2\sqrt{31}}{31}i - \dfrac{3\sqrt{93}}{31}j$

**21.** $x = 1 + t$, $y = 1 + 2t$, $z = 1 + 3t$

**23.** $x = 2 + \frac{1}{4}t$, $y = \frac{1}{3} - \frac{1}{9}t$, $z = \frac{1}{4} - \frac{1}{8}t$

**25.** $x = \dfrac{\sqrt{3}\pi}{12} + \left( \dfrac{\sqrt{3}}{2} - \dfrac{\pi}{12} \right)t$, $y = \dfrac{\pi}{12} + \left( \dfrac{1}{2} + \dfrac{\sqrt{3}\pi}{12} \right)t$,

$z = \dfrac{\pi}{6}e^{\pi/6} + \left( 1 + \dfrac{\pi}{6} \right)e^{\pi/6}t$

**27.** $\frac{1}{2}t^2 i + \frac{2}{3}t^3 j + 3t k + C$

**29.** $\frac{2}{3}t^{3/2}i + \ln|t|j - \frac{2}{5}t^{5/2}k + C$

**31.** $-\frac{1}{2}\cos 2t i + \frac{1}{2}\sin 2t j - e^{-t}k + C$

**33.** $(\cos t + t \sin t)i - \frac{1}{2}\cos t^2 j - \frac{1}{2}e^{t^2} + C$

**35.** $(2t + 1)i + 2t^2 j - (2t^3 - 1)k$

**37.** $r(t) = e^{2t}i - (3e^{-t} - 2)j + e^t k$

**39.** $r(t) = \left( \frac{4}{15}t^{5/2} + t + 2 \right)i - (\ln|\cos t| - 1)j + (e^t - 2)k$

**55.** $-r'(-t) - \dfrac{1}{t^2}r'\left( \dfrac{1}{t} \right)$

**57.** $r'(t) \cdot [r'(t) \times r''(t)] + r(t) \cdot [r'(t) \times r'''(t)]$

**63.** True    **65.** True

**Exercises 11.3 • page 1004**

**1.** $4\sqrt{14}$    **3.** $10\pi$    **5.** $\sqrt{3}(e^{2\pi} - 1)$    **7.** $e^2$

**9.**

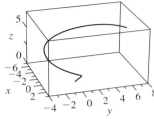

$\pi\sqrt{4\pi^2 + 2} + \ln\left( \sqrt{2\pi^2 + 1} + \sqrt{2}\pi \right)$

**11.** $s(t) = \sqrt{14}t$, $t \geq 0$;

$r(t(s)) = \left( 1 + \dfrac{\sqrt{14}}{14}s \right)i + \left( 1 + \dfrac{2\sqrt{14}}{14}s \right)j + \left( \dfrac{3\sqrt{14}}{14}s \right)k$,

$s \geq 0$

**13.** $s(t) = \sqrt{3}(e^t - 1)$;

$r(t(s)) = \left( 1 + \dfrac{\sqrt{3}}{3}s \right)\cos\left( \ln\left( 1 + \dfrac{\sqrt{3}}{3}s \right) \right)i$

$\qquad + \left( 1 + \dfrac{\sqrt{3}}{3}s \right)\sin\left( \ln\left( 1 + \dfrac{\sqrt{3}}{3}s \right) \right)j$

$\qquad + \left( 1 + \dfrac{\sqrt{3}}{3}s \right)k$,    $s \geq 0$

**15.** 0    **17.** $\dfrac{\sqrt{5}}{(1 + 5t^2)^{3/2}}$    **19.** $\frac{1}{4}$    **21.** $\dfrac{6|x|}{(1 + 9x^4)^{3/2}}$

**23.** $\dfrac{4|\sin 2x|}{(1 + 4\cos^2 2x)^{3/2}}$    **25.** $\dfrac{2|2x^2 - 1|e^{2x^2}}{(e^{2x^2} + 4x^2)^{3/2}}$

**27.** $\left( -\dfrac{\sqrt{2}}{2}, e^{-1/2} \right)$, $\left( \dfrac{\sqrt{2}}{2}, e^{-1/2} \right)$    **29.** $\left( \ln\dfrac{\sqrt{2}}{2}, \dfrac{\sqrt{2}}{2} \right)$

**31.** $(-1, -1)$, $(1, 1)$    **33.** (b)    **35.** (d)

**37.** $\kappa(x) = \dfrac{2|2x^2 - 1|e^{2x^2}}{(e^{2x^2} + 4x^2)^{3/2}}$

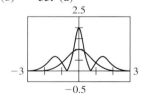

**41.** 0 for $t \neq 2n\pi$, where $n$ is an integer

**43. a.** $\dfrac{162}{(81 - 5x^2)^{3/2}}$

   **b.** $\kappa(3) = \frac{3}{4}$, osculating circle at $(3, 0)$ has equation
   $\left( x - \frac{5}{3} \right)^2 + y^2 = \frac{16}{9}$; $\kappa(0) = \frac{2}{9}$, osculating circle at $(0, 2)$
   has equation $x^2 + \left( y + \frac{5}{2} \right)^2 = \frac{81}{4}$

   **c.**

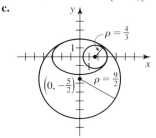

**45. a.** $\dfrac{dy}{dx} = \tan\dfrac{1}{2}\pi t^2$, $\dfrac{d^2y}{dx^2} = \dfrac{\pi t}{\cos^3\frac{1}{2}\pi t^2}$   **b.** $\pi t$

**47.** $\dfrac{3\sqrt{2}}{4\sqrt{1 + \sin\theta}}$

**51.** $\sqrt{2}\left[\pi\sqrt{1 + 4\pi^2} + \frac{1}{2}\ln(2\pi + \sqrt{1 + 4\pi^2})\right]$

**53.** True   **55.** True   **57.** True

## Exercises 11.4 • page 1012

**1.** $\mathbf{v}(1) = \mathbf{i} - 2\mathbf{j}$, $\mathbf{a}(1) = -2\mathbf{j}$, $|\mathbf{v}(1)| = \sqrt{5}$

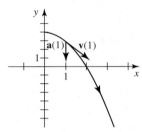

**3.** $\mathbf{v}\!\left(\dfrac{\pi}{4}\right) = -\dfrac{\sqrt{2}}{2}\mathbf{i} + \dfrac{3\sqrt{2}}{2}\mathbf{j}$, $\mathbf{a}\!\left(\dfrac{\pi}{4}\right) = -\dfrac{\sqrt{2}}{2}\mathbf{i} - \dfrac{3\sqrt{2}}{2}\mathbf{j}$,
$\left|\mathbf{v}\!\left(\dfrac{\pi}{4}\right)\right| = \sqrt{5}$

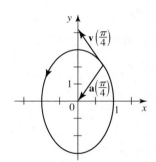

**5.** $\mathbf{v}(\pi/2) = -\mathbf{i} + \mathbf{k}$, $\mathbf{a}(\pi/2) = -\mathbf{j}$, $|\mathbf{v}(\pi/2)| = \sqrt{2}$

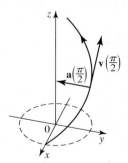

**7.** $\mathbf{v}(t) = \mathbf{i} + 2t\mathbf{j} + 2t\mathbf{k}$, $\mathbf{a}(t) = 2\mathbf{j} + 2\mathbf{k}$, $|\mathbf{v}(t)| = \sqrt{8t^2 + 1}$

**9.** $\mathbf{v}(t) = \mathbf{i} + 2t\mathbf{j} - \dfrac{1}{t^2}\mathbf{k}$, $\mathbf{a}(t) = 2\mathbf{j} + \dfrac{2}{t^3}\mathbf{k}$,
$|\mathbf{v}(t)| = \dfrac{\sqrt{4t^6 + t^4 + 1}}{t^2}$

**11.** $\mathbf{v}(t) = e^t\langle\cos t - \sin t, \cos t + \sin t, 1\rangle$,
$\mathbf{a}(t) = e^t\langle-2\sin t, 2\cos t, 1\rangle$, $|\mathbf{v}(t)| = \sqrt{3}\,e^t$

**13.** $\mathbf{v}(t) = \mathbf{i} + 2\mathbf{j} - 32t\mathbf{k}$, $\mathbf{r}(t) = t\mathbf{i} + 2t\mathbf{j} + (128 - 16t^2)\mathbf{k}$

**15.** $\mathbf{v}(t) = (t + 1)\mathbf{i} - \frac{1}{2}t^2\mathbf{j} + \left(\frac{1}{2}t^2 + t + 1\right)\mathbf{k}$,
$\mathbf{r}(t) = \left(\frac{1}{2}t^2 + t\right)\mathbf{i} + \left(1 - \frac{1}{6}t^3\right)\mathbf{j} + \left(\frac{1}{6}t^3 + \frac{1}{2}t^2 + t + 1\right)\mathbf{k}$

**17.** $\mathbf{v}(t) = -\sin t\,\mathbf{i} + (\cos t - 1)\mathbf{j} + (t + 2)\mathbf{k}$,
$\mathbf{r}(t) = \cos t\,\mathbf{i} + (\sin t - t)\mathbf{j} + \left(\frac{1}{2}t^2 + 2t\right)\mathbf{k}$

**23. a.** 60,892 ft   **b.** 8789 ft   **c.** 1500 ft/sec

**25. a.** 61,185 ft   **b.** 8989 ft   **c.** 1504 ft/sec

**27.** 4.4°

**29. a.** $-a\omega\sin\omega t\,\mathbf{i} + a\omega\cos\omega t\,\mathbf{j}$
**b.** $-\omega^2\mathbf{r}(t)$
**c.** $|\mathbf{v}(t)| = a\omega$, $|\mathbf{a}(t)| = a\omega^2$

**31.** 74.4 ft/sec $\leq v_0 \leq$ 81.6 ft/sec

**35.** $\mathbf{r}(t) = \langle 1 + t, 1 + 2t, 1 + 3t\rangle$, $\mathbf{r}(2) = \langle 3, 5, 7\rangle$

**37.** True

## Exercises 11.5 • page 1022

**1.** $\mathbf{T}(t) = \dfrac{1}{\sqrt{1 + 16t^2}}\mathbf{i} + \dfrac{4t}{\sqrt{1 + 16t^2}}\mathbf{j}$,
$\mathbf{N}(t) = -\dfrac{4t}{\sqrt{1 + 16t^2}}\mathbf{i} + \dfrac{1}{\sqrt{1 + 16t^2}}\mathbf{j}$

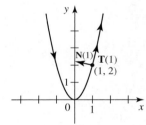

**3.** $\mathbf{T}(t) = \dfrac{2}{\sqrt{4 + 9t^2}}\mathbf{i} + \dfrac{3t}{\sqrt{4 + 9t^2}}\mathbf{j}$,
$\mathbf{N}(t) = -\dfrac{3t}{\sqrt{4 + 9t^2}}\mathbf{i} + \dfrac{2}{\sqrt{4 + 9t^2}}\mathbf{j}$

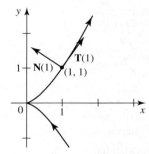

**5.** $\mathbf{T}(t) = \dfrac{1}{\sqrt{1 + 4t^2}}\mathbf{j} + \dfrac{2t}{\sqrt{1 + 4t^2}}\mathbf{k}$,

$\mathbf{N}(t) = -\dfrac{2t}{\sqrt{1 + 4t^2}}\mathbf{j} + \dfrac{1}{\sqrt{1 + 4t^2}}\mathbf{k}$

**7.** $\mathbf{T}(t) = \left\langle \dfrac{2\sqrt{13}}{13}\cos 2t, -\dfrac{2\sqrt{13}}{13}\sin 2t, \dfrac{3\sqrt{13}}{13} \right\rangle$,

$\mathbf{N}(t) = \langle -\sin 2t, -\cos 2t, 0 \rangle$

**9.** $\mathbf{T}(t) = \left\langle \dfrac{\sqrt{3}}{3}(\cos t - \sin t), \dfrac{\sqrt{3}}{3}(\cos t + \sin t), \dfrac{\sqrt{3}}{3} \right\rangle$,

$\mathbf{N}(t) = \left\langle -\dfrac{\sqrt{2}}{2}(\sin t + \cos t), \dfrac{\sqrt{2}}{2}(\cos t - \sin t), 0 \right\rangle$

**11.** $a_{\mathbf{T}} = \dfrac{4t}{\sqrt{1 + 4t^2}}$, $a_{\mathbf{N}} = \dfrac{2}{\sqrt{1 + 4t^2}}$

**13.** $a_{\mathbf{T}} = \dfrac{18t^3 + 4t}{\sqrt{9t^4 + 4t^2 + 1}}$, $a_{\mathbf{N}} = \dfrac{2\sqrt{9t^4 + 9t^2 + 1}}{\sqrt{9t^4 + 4t^2 + 1}}$

**15.** $a_{\mathbf{T}} = 0$, $a_{\mathbf{N}} = 2$

**17.** $a_{\mathbf{T}} = \sqrt{3}e^t$, $a_{\mathbf{N}} = \sqrt{2}e^t$

**19. a.**

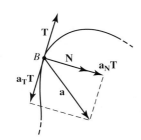

**b.** Accelerating at $A$, decelerating at $B$

**21. a.** $a_{\mathbf{T}} = -6$, $a_{\mathbf{N}} = 5$  **b.** Decelerating

**29.** $x - y - \sqrt{2}z = 0$  **31.** $\dfrac{\sqrt{2}}{2}(-\tanh t\mathbf{i} + \mathbf{j} - \operatorname{sech} t\mathbf{k})$

**33.** $\dfrac{1}{\sqrt{36t^4 + 9t^2 + 4}}(6t^2\mathbf{i} - 3t\mathbf{j} + 2\mathbf{k})$

**35.** $\frac{1}{2}$  **39.** $2.99 \times 10^{11}$ m

**41. a.** $a_{\mathbf{T}} = \dfrac{g(gt - v_0 \sin \alpha)}{\sqrt{(v_0 \cos \alpha)^2 + (v_0 \sin \alpha - gt)^2}}$,

$a_{\mathbf{N}} = \dfrac{gv_0 \cos \alpha}{\sqrt{(v_0 \cos \alpha)^2 + (v_0 \sin \alpha - gt)^2}}$

**b.** $a_{\mathbf{T}} = 0$, $a_{\mathbf{N}} = g$

**45.** True  **47.** False

## Chapter 11 Concept Review • page 1024

**1. a.** $\langle f(t), g(t), h(t) \rangle$; real-valued; $t$; parameter
**b.** parameter interval; real numbers
**3. a.** $\lim\limits_{t\to a} f(t)$; $\lim\limits_{t\to a} g(t)$; $\lim\limits_{t\to a} h(t)$  **b.** $\mathbf{r}(a)$; continuous
**5. a.** $\mathbf{u}'(t) \cdot \mathbf{v}(t) + \mathbf{u}(t) \cdot \mathbf{v}'(t)$; $\mathbf{u}'(t) \times \mathbf{v}(t) + \mathbf{u}(t) \times \mathbf{v}'(t)$
**b.** $\mathbf{u}'(f(t))f'(t)$

**7.** $\displaystyle\int_a^b \sqrt{[f'(t)]^2 + [g'(t)]^2 + [h'(t)]^2}\, dt$

**9. a.** $\left|\dfrac{d\mathbf{T}}{ds}\right|$  **b.** $\dfrac{|\mathbf{T}'(t)|}{|\mathbf{r}'(t)|}$  **c.** $\dfrac{|\mathbf{r}'(t) \times \mathbf{r}''(t)|}{|\mathbf{r}'(t)|^3}$  **d.** $\dfrac{|y''|}{[1 + (y')^2]^{3/2}}$

**e.** radius of curvature; radius; tangent line; circle of curvature

**11. a.** $\dfrac{\mathbf{r}'(t)}{|\mathbf{r}'(t)|}$; $\dfrac{\mathbf{T}'(t)}{|\mathbf{T}'(t)|}$  **b.** $\mathbf{T}$; $\mathbf{N}$; $v'$; $\kappa v^2$; tangential; normal

**c.** $\dfrac{\mathbf{r}'(t) \cdot \mathbf{r}''(t)}{|\mathbf{r}'(t)|}$; $\dfrac{|\mathbf{r}'(t) \times \mathbf{r}''(t)|}{|\mathbf{r}'(t)|}$

## Chapter 11 Review Exercises • page 1025

**1.**

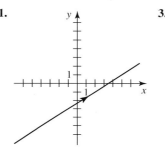

**3.**

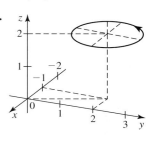

**5.** $(-1, 0)$ and $(0, 5)$  **7.** $[-1, 1)$ and $(1, 2)$

**9.** $\mathbf{r}'(t) = \dfrac{1}{2\sqrt{t}}\mathbf{i} + 2t\mathbf{j} - \dfrac{1}{(t + 1)^2}\mathbf{k}$,

$\mathbf{r}''(t) = -\dfrac{1}{4t^{3/2}}\mathbf{i} + 2\mathbf{j} + \dfrac{2}{(t + 1)^3}\mathbf{k}$

**11.** $\mathbf{r}'(t) = 2t\mathbf{i} + 2\mathbf{j} + \dfrac{1}{t}\mathbf{k}$, $\mathbf{r}''(t) = 2\mathbf{i} - \dfrac{1}{t^2}\mathbf{k}$

**13.** $x = 1$, $y = -3 + 2t$, $z = 1$

**15.** $\frac{2}{3}t^{3/2}\mathbf{i} - \frac{1}{2}e^{-2t}\mathbf{j} + \ln|t + 1|\mathbf{k} + \mathbf{C}$

**17.** $\left(\dfrac{4}{3}t^{3/2} + 1\right)\mathbf{i} + \left(\dfrac{3}{2\pi}\sin 2\pi t + 2\right)\mathbf{j} + (e^{-t} - 1)\mathbf{k}$

**19.** $\mathbf{T}(1) = \dfrac{\sqrt{14}}{14}(\mathbf{i} + 2\mathbf{j} + 3\mathbf{k})$,

$\mathbf{N}(1) = \dfrac{\sqrt{266}}{266}(-11\mathbf{i} - 8\mathbf{j} + 9\mathbf{k})$

**21.** 10  **23.** $\dfrac{2\sqrt{1 + 9t^2 + 9t^4}}{(1 + 4t^2 + 9t^4)^{3/2}}$

**25.** $\dfrac{4}{(x^2 - 4x + 8)^{3/2}}$, $(2, 1)$

**27.** $\mathbf{v}(t) = 2\mathbf{i} - 2e^{-2t}\mathbf{j} - \sin t\mathbf{k}$,

$\mathbf{a}(t) = \mathbf{r}''(t) = 4e^{-2t}\mathbf{j} - \cos t\mathbf{k}$,

$|\mathbf{v}(t)| = \sqrt{4 + 4e^{-4t} + \sin^2 t}$

**29.** $\mathbf{v}(t) = \left(\frac{1}{2}t^2 + 2\right)\mathbf{i} + \left(\frac{1}{3}t^3 + 3\right)\mathbf{j} + (3t + 1)\mathbf{k}$,

$\mathbf{r}(t) = \left(\frac{1}{6}t^3 + 2t\right)\mathbf{i} + \left(\frac{1}{36}t^4 + 3t\right)\mathbf{j} + \left(\frac{3}{2}t^2 + t\right)\mathbf{k}$

**31.** $a_T = \dfrac{4t}{\sqrt{1 + 4t^2}}$, $a_N = \dfrac{2}{\sqrt{1 + 4t^2}}$

**33.** $a_T = \dfrac{\sin t \cos t - 8 \sin 2t \cos 2t}{\sqrt{\sin^2 t + 4 \cos^2 2t}}$,

$a_N = \dfrac{2|2 \sin t \sin 2t + \cos t \cos 2t|}{\sqrt{\sin^2 t + 4 \cos^2 2t}}$

**35. a.** $20\sqrt{2}\,t\mathbf{i} + (7 + 20\sqrt{2}\,t - 16t^2)\mathbf{j}$   **b.** 56.2 ft

### Chapter 11 Challenge Problems • page 1026

**3. b.** $\dfrac{g(d^2 + h^2)}{2v_0^2}$

**5. c.** 15000

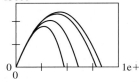

**9. a.** $|\mathbf{v}(3)| \approx 18.96$ cm/sec, $|\mathbf{a}(3)| \approx 60.54$ cm/sec$^2$
   **b.** $4\pi$ cm/sec$^2$

# CHAPTER 12

### Exercises 12.1 • page 1039

*Abbreviations: D, domain; R, range.*

**1. a.** 8   **b.** 9   **c.** $4h^2 + 18hk - 4h + 3$

   **d.** $x^2 + 2xh + h^2 + 3xy + 3hy - 2x - 2h + 3$

   **e.** $x^2 + 3xy + 3xk - 2x + 3$

**3. a.** 6   **b.** $\sqrt{11}$   **c.** $\sqrt{6}|t|$   **d.** $\sqrt{6u^2 + 2u + 5}$

   **e.** $\sqrt{15}|x|$

**5.** $D = \{(x, y) \mid -\infty < x < \infty, -\infty < y < \infty\}$,
$R = \{z \mid -\infty < z < \infty\}$

**7.** $D = \{(u, v) \mid u \neq v\}$, $R = \{z \mid -\infty < z < \infty\}$

**9.** $D = \{(x, y) \mid x^2 + y^2 \leq 4\}$, $R = \{z \mid 0 \leq z \leq 2\}$

**11.** $D = \{(x, y, z) \mid x^2 + y^2 + z^2 \leq 9\}$, $R = \{z \mid 0 \leq z \leq 3\}$

**13.** $D = \{(u, v, w) \mid u \neq \frac{\pi}{2} + n\pi, n \text{ an integer}\}$,
$R = \{z \mid -\infty < z < \infty\}$

**15.** $\{(x, y) \mid x \geq 0 \text{ and } y \geq 0\}$    **17.** $\{(u, v) \mid u \neq v, u \neq -v\}$

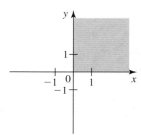
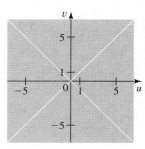

**19.** $\{(x, y) \mid x > 0, y > 0\}$    **21.** $\{(x, y, z) \mid x^2 + y^2 + z^2 \leq 9\}$

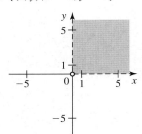
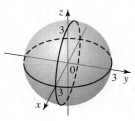

**23.**
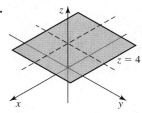
   **25.**
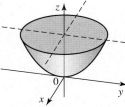

**27.**
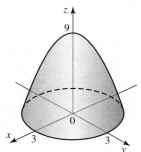
   **29.**
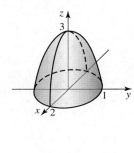

**31. a.** 200 ft, 400 ft   **b.** ascending, ascending   **c.** $C$

**33.** (c)    **35.** (a)    **37.** (d)

**39.**
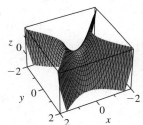
   **41.**
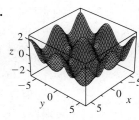

**43.**
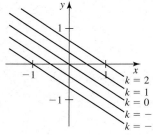

**45.**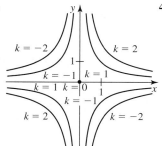

**47.**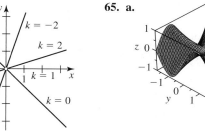

**49.**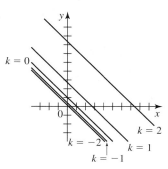

**51.**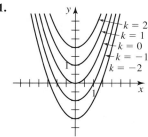

**53.** A family of parallel planes with normal vector $\langle 2, 4, -3 \rangle$

**55.** A cone with vertex the origin and axis the $z$-axis (if $k = 0$), a family of hyperboloids of one sheet with axis the $z$-axis (if $k > 0$), and a family of hyperboloids of two sheets with axis the $z$-axis (if $k < 0$)

**57.** (a)      **59.** (c)      **61.** (e)

**63. a.** 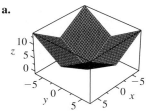      **b.**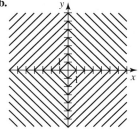

**65. a.** 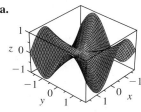      **b.**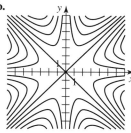

**67.** $\sqrt{x^2 + y^2} = 5$      **69.** No

**71. a.**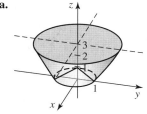

**b.** For $k = 0$, $x^2 + y^2 = 1$; for $k = \frac{1}{2}$, $x^2 + y^2 = \frac{1}{4}$ or $x^2 + y^2 = \frac{3}{2}$; for $k = 1$, $(0, 0)$ or $x^2 + y^2 = 2$; for $k = 3$, $x^2 + y^2 = 4$

**73.** 40,000$k$ dynes      **77.** \$260,552.20, \$151,210.04

**79.** 40.28 $g$

**83.** A family of concentric spheres centered at the origin

**85.** $\approx$435 Hz      **87.** False      **89.** False

## Exercises 12.2 • page 1054

**13.** 9      **15.** $-18$      **17.** 3      **19.** $-1$      **21.** $\frac{1}{4}$      **23.** 0

**25.** $\frac{11}{3}$      **27.** 0      **29.** 0      **31.** $\{(x, y) \mid 2x + 3y \neq 1\}$

**33.** $\{(x, y) \mid x \geq 0, |y| \leq x\}$      **35.** $\{(x, y) \mid x \geq 0, y \neq 0\}$

**37.** $\{(x, y) \mid x^2 + y^2 + z^2 \neq 4\}$      **39.** $\{(x, y, z) \mid yz > 1\}$

**41. a.** The entire plane      **b.**

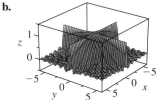

**43.** $(x^2 - xy + y^2) \cos(x^2 - xy + y^2) + \sin(x^2 - xy + y^2)$
The entire plane

**45.** $\dfrac{2x - y + 2}{2x - y - 1}$; $\{(x, y) \mid 2x - y \neq 1\}$

**47.** $\cos(x \tan y)$; $\{(x, y) \mid y \neq \frac{\pi}{2} + n\pi, n$ an integer$\}$

**53.** True      **55.** True      **57.** True

## Exercises 12.3 • page 1065

**1. a.** 4, 4

**b.** $f_x(2, 1) = 4$ says that the slope of the tangent line to the curve of intersection of the surface $z = x^2 + 2y^2$ and the plane $y = 1$ at the point $(2, 1, 6)$ is 4. $f_y(2, 1) = 4$ says that the slope of the tangent line to the curve of intersection of the surface $z = x^2 + 2y^2$ and the plane $x = 2$ at the point $(2, 1, 6)$ is 4.

**c.** $f_x(2, 1) = 4$ says that the rate of change of $f(x, y)$ with respect to $x$ with $y$ fixed at 1 is 4 units per unit change in $x$. $f_y(2, 1) = 4$ says that the rate of change of $f(x, y)$ with respect to $y$ with $x$ fixed at 2 is 4 units per unit change in $y$.

**3.** At $P$, $\dfrac{\partial f}{\partial x} < 0$ and $\dfrac{\partial f}{\partial y} < 0$. At $Q$, $\dfrac{\partial f}{\partial x} = \dfrac{\partial f}{\partial y} = 0$. At $R$, $\dfrac{\partial f}{\partial x} < 0$ and $\dfrac{\partial f}{\partial y} > 0$.

**5.** $-7.1°\text{F/in.}$, $-3.8°\text{F/in.}$

**7.** $f_x(x, y) = 4x - 3y$, $f_y(x, y) = -3x + 2y$

**9.** $\dfrac{\partial z}{\partial x} = \sqrt{y}$, $\dfrac{\partial z}{\partial y} = \dfrac{x}{2\sqrt{y}}$

**11.** $g_r(r, s) = \dfrac{1}{2\sqrt{r}}$, $g_s(r, s) = 2s$

**13.** $f_x(x, y) = e^{y/x}\left(1 - \dfrac{y}{x}\right)$, $f_y(x, y) = e^{y/x}$

**15.** $\dfrac{\partial z}{\partial x} = \dfrac{2x}{1 + (x^2 + y^2)^2}$, $\dfrac{\partial z}{\partial y} = \dfrac{2y}{1 + (x^2 + y^2)^2}$

**17.** $g_u(u, v) = \dfrac{v(v^3 - u^2)}{(u^2 + v^3)^2}$, $g_v(u, v) = \dfrac{u(u^2 - 2v^3)}{(u^2 + v^3)^2}$

**19.** $g_x(x, y) = 2x\cosh\dfrac{x}{y} + \dfrac{x^2}{y}\sinh\dfrac{x}{y}$, $g_y(x, y) = -\dfrac{x^3}{y^2}\sinh\dfrac{x}{y}$

**21.** $f_x(x, y) = y^x \ln y$, $f_y(x, y) = xy^{x-1}$

**23.** $f_x(x, y) = -xe^{-x}$, $f_y(x, y) = ye^{-y}$

**25.** $g_x(x, y, z) = \dfrac{\sqrt{xyz}}{2x}$, $g_y(x, y, z) = \dfrac{\sqrt{xyz}}{2y}$, $g_z(x, y, z) = \dfrac{\sqrt{xyz}}{2z}$

**27.** $\dfrac{\partial u}{\partial x} = e^{y/z}$, $\dfrac{\partial u}{\partial y} = \dfrac{x}{z}e^{y/z}$, $\dfrac{\partial u}{\partial z} = -\dfrac{xy}{z^2}e^{y/z} - 2z$

**29.** $f_r(r, s, t) = s\ln st$, $f_s(r, s, t) = r\ln st + r$, $f_t(r, s, t) = rs/t$

**31.** $\dfrac{\partial z}{\partial x} = \dfrac{ye^{-x} - e^y}{e^z}$, $\dfrac{\partial z}{\partial y} = -\dfrac{xe^y + e^{-x}}{e^z}$

**33.** $\dfrac{\partial z}{\partial x} = -\dfrac{2x(2x^2 + 2z^2 + 1)}{z(3yz^3 + 3x^2yz + 2)}$, $\dfrac{\partial z}{\partial y} = -\dfrac{z^2(x^2 + z^2)}{3yz^3 + 3x^2yz + 2}$

**35.** $g_{xx}(x, y) = 6xy^2$, $g_{yy} = 2x^3 + 6xy$, $g_{xy} = g_{yx} = 6x^2y + 3y^2$

**37.** $\dfrac{\partial^2 w}{\partial u^2} = -4\cos(2u - v) - 4\sin(2u + v)$, $\dfrac{\partial^2 w}{\partial v^2} = -\cos(2u - v) - \sin(2u + v)$, $\dfrac{\partial^2 w}{\partial u\,\partial v} = \dfrac{\partial^2 w}{\partial v\,\partial u} = 2\cos(2u - v) - 2\sin(2u + v)$

**39.** $h_{xx}(x, y) = \dfrac{2xy}{(x^2 + y^2)^2}$, $h_{yy}(x, y) = -\dfrac{2xy}{(x^2 + y^2)^2}$, $h_{yx}(x, y) = h_{xy}(x, y) = \dfrac{y^2 - x^2}{(x^2 + y^2)^2}$

**41.** $-8y$    **43.** $-\sin x$    **45.** $4e^x\sin(y + 2z)$

**69.** $\partial V/\partial T = 0.066512$, $\partial V/\partial P = -0.1596288$

**71.** $\left(\dfrac{R}{R_1}\right)^2$    **73.** \$39 per \$1000; $-\$975$ per 1000 ft$^2$

**75.** $-\sqrt{14kQ}/196$ volts per meter

**77.** $\partial N/\partial x \approx 1.06$, $\partial N/\partial y \approx -2.85$

**79. a.** $\approx 19.70°\text{F}$

**b.** $\approx -0.285°$ for each 1 mph increase in wind speed

**83.** $\approx 0.9872$    **87.** $2x + y - z = 2$

**89.** $0, -\dfrac{4e}{\pi(e^2 + 1)}$    **91.** No    **93.** True    **95.** True

## Exercises 12.4 • page 1078

**1. a.** $-0.0386$  **b.** $-0.04$

**3.** $dz = 6xy^3\,dx + 9x^2y^2\,dy$

**5.** $dz = -\dfrac{2y}{(x - y)^2}\,dx + \dfrac{2x}{(x - y)^2}\,dy$

**7.** $dz = 12xy^3(2x^2 + 3y^2)^2\,dx + 3y^2(2x^2 + 9y^2)(2x^2 + 3y^2)^2\,dy$

**9.** $dw = 2xye^{x^2 - y^2}\,dx + (1 - 2y^2)e^{x^2 - y^2}\,dy$

**11.** $dw = \left[2x\ln(x^2 + y^2) + \dfrac{2x^3}{x^2 + y^2}\right]dx + \dfrac{2x^2y}{x^2 + y^2}\,dy$

**13.** $dz = 2e^{2x}\cos 3y\,dx - 3e^{2x}\sin 3y\,dy$

**15.** $dw = (2x + y)\,dx + x\,dy + 2z\,dz$

**17.** $dw = 2xe^{-yz}\,dx - x^2ze^{-yz}\,dy - x^2ye^{-yz}\,dz$

**19.** $dw = 2xe^y\,dx + (x^2e^y + \ln z)\,dy + \dfrac{y}{z}\,dz$

**21.** $-0.06$    **23.** $0.21$    **25.** 1080 in.$^3$    **27.** 58.3 ft$^2$

**29.** $-3.3256$ Pa    **31.** 2.73%    **33.** 9%    **35.** 0.3%

**37.** $0.19\ \Omega$    **39.** 7%    **45.** True    **47.** False

## Exercises 12.5 • page 1089

**1.** $4xt - 6yt^2 - 2y$

**3.** $-2(\cos s + s \cos r)e^{-2t} + (\sin r - r \sin s)(3t^2 - 2)$

**5.** $2x^2y[3yz - 2xz \sin t + xy(\sin t + t \cos t)]$

**7.** $\dfrac{-2tz + y^2z - 2tx^2z^3 + \cosh t(y + xy^2 + x^2yz^2)}{y^2(1 + x^2z^2)}$

**9.** $\dfrac{\partial w}{\partial u} = 6(x^2u + y^2v),\ \dfrac{\partial w}{\partial v} = 6(x^2v + y^2u)$

**11.** $\dfrac{\partial w}{\partial u} = e^x\left(\dfrac{2u \cos y}{u^2 + v^2} - \dfrac{\sqrt{uv}\,\sin y}{2u}\right),$

$\dfrac{\partial w}{\partial v} = e^x\left(\dfrac{2v \cos y}{u^2 + v^2} - \dfrac{\sqrt{uv}\,\sin y}{2v}\right)$

**13.** $\dfrac{\partial w}{\partial u} = \dfrac{(\tan^{-1} yz)\sqrt{u}}{2u} - \dfrac{xyv \sin u}{1 + y^2z^2},$

$\dfrac{\partial w}{\partial v} = \dfrac{x(y \cos u - 2ze^{-2v})}{1 + y^2z^2}$

**15.** $\dfrac{dw}{dt} = \dfrac{\partial w}{\partial r}\dfrac{dr}{dt} + \dfrac{\partial w}{\partial s}\dfrac{ds}{dt} + \dfrac{\partial w}{\partial u}\dfrac{du}{dt} + \dfrac{\partial w}{\partial v}\dfrac{dv}{dt}$

**17.** $\dfrac{\partial w}{\partial t} = \dfrac{\partial w}{\partial x}\dfrac{\partial x}{\partial t} + \dfrac{\partial w}{\partial y}\dfrac{\partial y}{\partial t} + \dfrac{\partial w}{\partial z}\dfrac{\partial z}{\partial t}$

**19.** $4x + 2y + xe^t + 2ye^t - 6z^2 \sin 2t$   **21.** 0

**23.** $\dfrac{\partial u}{\partial s} = \dfrac{\csc yz(rt^2 - 2xzst^3 \cot yz - xy \cot yz)}{t^2},$

$\dfrac{\partial u}{\partial t} = \dfrac{sx \csc yz \cot yz\,(2y - st^3z)}{t^3}$

**25.** $\dfrac{\partial w}{\partial r} = 4,\ \dfrac{\partial w}{\partial t} = 2$

**27.** $\dfrac{\partial u}{\partial x} = \dfrac{1}{4u},\ \dfrac{\partial u}{\partial y} = \dfrac{1}{4u},\ \dfrac{\partial v}{\partial x} = \dfrac{1}{4v},\ \dfrac{\partial v}{\partial y} = -\dfrac{1}{4v}$

**29.** $\dfrac{3x^2 - 2y}{2x - 3y^2}$   **31.** $\dfrac{8x\sqrt{xy} + 3y}{4\sqrt{xy} - 3x}$

**33.** $\dfrac{\partial z}{\partial x} = \dfrac{2x + y - 2xz}{x^2 - 2yz},\ \dfrac{\partial z}{\partial y} = \dfrac{x + z^2}{x^2 - 2yz}$

**35.** $\dfrac{\partial z}{\partial x} = -\dfrac{ye^y + y^2ze^{xz} + x(2y + x)e^{x/y}}{xy^2e^{xz}},$

$\dfrac{\partial z}{\partial y} = \dfrac{x^3e^{x/y} - xy^2e^y - y^2e^{xz}}{xy^3e^{xz}}$

**37.** $-\dfrac{x^2 - ay}{y^2 - ax}$   **39.** $\approx -13.2$ in.$^2$/min

**41.** $-54$ mph   **45.** $\approx -1.53°$F/sec   **57.** 2   **59.** 0

**63. a.** $-\dfrac{f_x^2 f_{yy} - 2f_x f_y f_{xy} + f_y^2 f_{xx}}{f_y^3}$   **b.** $\dfrac{2xy}{(x - y^2)^3},\ \{(x, y)\,|\,x \neq y^2\}$

**65.** True

## Exercises 12.6 • page 1101

**1.** $6 - \sqrt{3}/2$   **3.** 4   **5.** $5\mathbf{i} + 3\mathbf{j}$

**7.** $\dfrac{4 - \sqrt{2}\pi}{4}\mathbf{i} + \dfrac{\sqrt{2}}{2}\mathbf{j}$

**9.** $\mathbf{i} + 2\mathbf{j}$   **11.** $-2\sqrt{5}/5$   **13.** $\frac{1}{9}$   **15.** $7\sqrt{10}/5$

**17.** $-4\sqrt{13}/13$   **19.** $-76\sqrt{3}$   **21.** $\frac{1}{6}$

**23.** $20\sqrt{14}/7$   **25.** $-4\sqrt{3}/3$   **27.** $\dfrac{\sqrt{6}}{24}(3 - \pi)$

**29.** $39\sqrt{10}/10$   **31.** $\dfrac{12\sqrt{2}}{\sqrt{13\pi^2 + 576}}$   **33.** $\mathbf{i} + 6\mathbf{j},\ \sqrt{74}/6$

**35.** $7\mathbf{i} + 8\mathbf{j} + 34\mathbf{k},\ 3\sqrt{141}$   **37.** $-2\mathbf{i} - \mathbf{j},\ \sqrt{5}$

**39.** $4\mathbf{i} + 2\mathbf{j} - \mathbf{k},\ \sqrt{21}/4$   **41.** $\mathbf{i} - \mathbf{j},\ 440\sqrt{2}$

**43. a.** Ascending, descending   **b.** Neither   **c.** East   **d.** East

**45. a.** $\dfrac{100\sqrt{7}}{\pi}\mathbf{i} + \dfrac{500}{\pi}\mathbf{j}$   **b.**

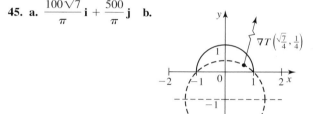

**49. a.** $x = y^3$   **b.**

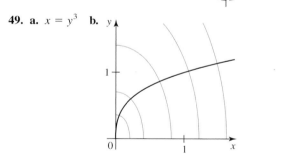

**51.** 6.25   **53.** $-3\mathbf{i} + 4\mathbf{j}$

**55. a.**

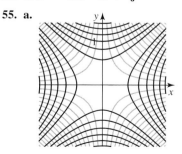

**57.** $f$ increases most rapidly in a direction along any line emanating from the origin and moving away from it; $g$ increases most rapidly in the $x$-direction (positive or negative). No.

**59.** True   **61.** True

## Exercises 12.7 • page 1111

**1.**

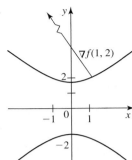

**3.**

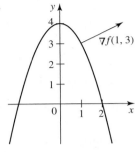

**5.** $y = \dfrac{\sqrt{3}}{4}x + \dfrac{7}{8}, y = -\dfrac{4\sqrt{3}}{3}x + 8$

**7.** $y = -\dfrac{7\sqrt{5}}{5}x + 6, y = \dfrac{\sqrt{5}}{7}x - \dfrac{12}{7}$

**9.**

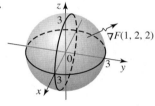

**11.**

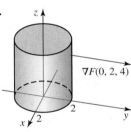

**13.**

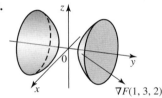

**15.** $2x + 4y + 9z = 17, \dfrac{x-2}{2} = \dfrac{y-1}{4} = \dfrac{z-1}{9}$

**17.** $x + y + z = 1, x - 4 = y + 2 = z + 1$

**19.** $5x + 4y + 3z = 22, \dfrac{x-1}{5} = \dfrac{y-2}{4} = \dfrac{z-3}{3}$

**21.** $18x - 16y + z = -25, \dfrac{x+1}{-18} = \dfrac{y-2}{16} = \dfrac{z-25}{-1}$

**23.** $x + 3y - 4z = -11, x + 2 = \dfrac{y-1}{3} = \dfrac{z-3}{-4}$

**25.** $x + 2y - z = 0, x - 2 = \dfrac{y}{2} = \dfrac{z-2}{-1}$

**27.** $3y - z = 0, x = 3, \dfrac{y}{3} = \dfrac{z}{-1}$

**29.** $x - y + 2z = \dfrac{\pi}{2}, x - 1 = \dfrac{y-1}{-1} = \dfrac{z-\frac{\pi}{4}}{2}$

**31.** $x + z = 1, \dfrac{x}{1} = \dfrac{z-1}{1}, y = 3$

**35.** $\dfrac{xx_0}{a^2} - \dfrac{yy_0}{b^2} - \dfrac{zz_0}{c^2} = 1$   **37.** $(-1, -2, -3)$ and $(1, 2, 3)$

**39.** $\left(-\dfrac{2\sqrt{5}}{5}, \dfrac{2\sqrt{5}}{5}, -\dfrac{\sqrt{5}}{5}\right)$ and $\left(\dfrac{2\sqrt{5}}{5}, -\dfrac{2\sqrt{5}}{5}, \dfrac{\sqrt{5}}{5}\right)$

**45.** $x = -\dfrac{\sqrt{2}}{2} + t, y = \dfrac{\sqrt{2}}{2} + t, z = 1$

**47.** True   **49.** True

## Exercises 12.8 • page 1120

**1.** Relative minimum $f(1, -2) = -5$

**3.** Relative maximum $f(2, -1) = 15$

**5.** Relative minimum $f(0, 0) = 0$

**7.** Relative minimum $f(5, 6) = -24$

**9.** Relative minimum $f(0, 0) = 3$, saddle points $(-2, -1, 5)$ and $(2, -1, 5)$

**11.** Relative minimum $f(0, 0) = 0$, relative maximum $f\left(0, -\dfrac{5}{3}\right) = \dfrac{125}{27}$, saddle points $(-2, -1, 3)$ and $(2, -1, 3)$

**13.** Relative minimum $f(4, 4) = -12$

**15.** Relative minimum $f(-1, -2) = 6$

**17.** Relative maximum $f(0, 0) = 1$

**19.** Saddle points $(0, 0, 0)$, $(0, \pi, 0)$, and $(0, 2\pi, 0)$

**21.** None

**23.** Relative minima at $\left(\dfrac{3}{2}, -\dfrac{3}{2}\right)$ and $\left(\dfrac{3}{2}, \dfrac{3}{2}\right)$, saddle point $(0, 0, 0)$

**25.** Relative minima $f(-1, 2) = -\dfrac{3}{2}$ and $f(4, -8) = -64$, saddle point $(0, 0, 0)$

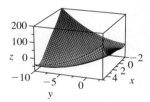
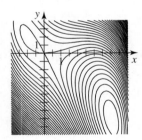

**27.** Relative minimum $f(-1, -2) = 14$

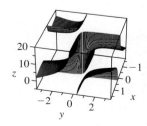
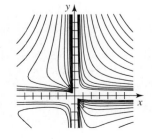

**29.** Relative minima $f(-1.526, 1) \approx -19.888$ and
$f(1.267, 1) \approx -8.620$, saddle point $(0.259, 1, -5.492)$

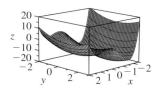

**31.** Relative maxima $f(\pm 1.225, 1) \approx 0.250$,
saddle point $(0, 0.630, -1.528)$

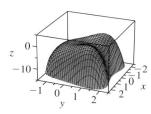

**33.** Absolute minimum $-12$, absolute maximum 7

**35.** Absolute minimum $-12$, absolute maximum 13

**37.** Absolute minimum $-12$, absolute maximum 4

**39.** Absolute minimum $-\frac{13}{4}$, absolute maximum $\frac{63}{4}$

**41.** $2\sqrt{6}/3$    **43.** $(2, -3, \pm 1)$, $\sqrt{14}$    **45.** $\frac{500}{3}, \frac{500}{3}, \frac{500}{3}$

**47.** $2\sqrt{2}$ ft $\times$ $2\sqrt{2}$ ft $\times$ $2\sqrt{2}$ ft

**49.** $4\sqrt{3}/3 \times 2\sqrt{3} \times 8\sqrt{3}/3$

**51.** $1 \times \frac{2}{3} \times 2$, $\frac{4}{3}$    **53.** 2 ft $\times$ 2 ft $\times$ 3 ft    **55.** $\left(\frac{26}{3}, 12\right)$

**59. a.** $y = 1.59t + 6.69$    **b.** \$19.41 billion

**61. a.** $y = 3.39x + 23.19$    **b.** \$3.39 billion/yr    **c.** \$60.5 billion

**63. b.** Maximum 1, minimum $-13.5833$

**65.** False    **67.** False

## Exercises 12.9 • page 1134

**1.** Minimum $-5$, maximum 5

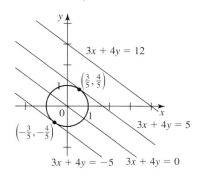

**3.** Minimum 2

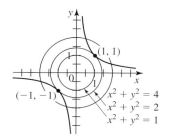

**5.** Maximum $\frac{3}{2}$    **7.** Minimum $-\frac{1}{4}$, maximum $\frac{1}{4}$

**9.** Minimum 4, maximum 12    **11.** Minimum $-\frac{1}{2}$

**13.** Minimum $-2$, maximum 2

**15.** Minimum $-2\sqrt{2}$, maximum $2\sqrt{2}$

**17.** Minimum $2 - 3\sqrt{5}$, maximum $2 + 3\sqrt{5}$

**19.** Minimum $\frac{1}{2}$, maximum $\frac{3}{2}$    **21.** Minimum $-\frac{4}{3}$, maximum 32

**23.** $\left(\frac{2}{3}, \frac{4}{3}, \frac{2}{3}\right)$    **25.** $\left(\frac{4}{3}, \frac{5}{3}, -\frac{1}{3}\right)$

**27.** $(-2^{1/4}, \pm 2^{3/4}, -2^{1/4})$ and $(2^{1/4}, \pm 2^{3/4}, 2^{1/4})$, $2\sqrt[4]{2}$

**29.** $2\sqrt{2}$ ft $\times$ $2\sqrt{2}$ ft $\times$ $2\sqrt{2}$ ft    **31.** 6 in. $\times$ 6 in. $\times$ 3 in.

**33.** $\dfrac{2\sqrt{3}}{3} a \times \dfrac{2\sqrt{3}}{3} b \times \dfrac{2\sqrt{3}}{3} c$    **35.** $\dfrac{a}{3} \times \dfrac{b}{3} \times \dfrac{c}{3}, \dfrac{abc}{27}$

**37.** $\frac{2}{3}\sqrt[3]{36}$ ft $\times$ $\frac{2}{3}\sqrt[3]{36}$ ft $\times$ $\sqrt[3]{36}$ ft

**39.** 225 units of labor and 50 units of capital

**41. a.** $(0, -1)$

**b.**

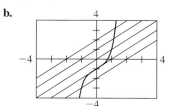

**43.** $\left(\frac{3}{2}, -\frac{3}{2}, -\frac{7}{2}\right)$    **45. a.** $\dfrac{3\sqrt{2}}{2} a$    **47. a.** $c$    **51.** False

## Chapter 12 Concept Review • page 1137

**1. a.** rule; $(x, y)$    **b.** real; real; range

   **c.** $\{(x, y, z) \mid z = f(x, y), (x, y) \in D\}$

**3.** $L; L; (a, b)$

**5. a.** $f(a, b)$    **b.** $R$

**7. a.** $\lim\limits_{h \to 0} \dfrac{f(x + h, y) - f(x, y)}{h}$; $y = b$; $(a, b, f(a, b))$;

$x$; constant; $y = b$

**b.** $y$; $x$

**9. a.** $f_x\, dx + f_y\, dy$  **b.** $dz$  **c.** $\Delta x$; $\Delta y$; 0; 0

**d.** $f_x(a, b)\, \Delta x + f_y(a, b)\, \Delta y + \varepsilon_1\, \Delta x + \varepsilon_2\, \Delta y$; $\varepsilon_1 \to 0$;
$\varepsilon_2 \to 0$; $(0, 0)$

**11. a.** $\dfrac{\partial w}{\partial x}\dfrac{dx}{dt} + \dfrac{\partial w}{\partial y}\dfrac{dy}{dt}$

**b.** $\dfrac{\partial w}{\partial x}\dfrac{\partial x}{\partial u} + \dfrac{\partial w}{\partial y}\dfrac{\partial y}{\partial u}$

**c.** $-\dfrac{F_x(x, y)}{F_y(x, y)}$; $F_y(x, y) \neq 0$

**d.** $-\dfrac{F_x(x, y, z)}{F_z(x, y, z)}$; $-\dfrac{F_y(x, y, z)}{F_z(x, y, z)}$; $F_z(x, y, z) \neq 0$

**13. a.** $|\nabla f(x, y)|$; $\nabla f(x, y)$  **b.** $-|\nabla f(x, y)|$; $-\nabla f(x, y)$

**15. a.** Relative maximum  **b.** Absolute minimum

**c.** Does not exist; 0  **d.** Critical point

**e.** Second Derivative Test

**17. a.** Constrained  **b.** $\lambda \nabla g(x, y)$; multiplier

**c.** $\lambda \nabla g(x, y)$; $f(x, y)$; critical points; maximum; minimum

## Chapter 12 Review Exercises • page 1138

*Abbreviations: D, domain; R, range.*

**1.** $D = \{(x, y) \mid 0 < x^2 + y^2 \le 9\}$

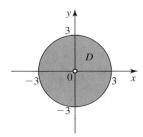

**3.** $D = \{(x, y) \mid -1 \le x \le 1\}$

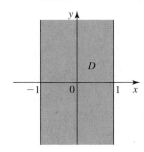

**5.**

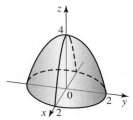

**7.**

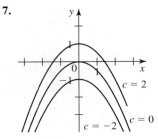

**9.**

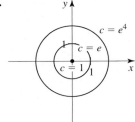

**11.** $\frac{2}{3}$   **13.** 1    **15.** $\{(x, y) \mid y < x\}$

**17.** $f_x(x, y) = 4xy - \dfrac{1}{2\sqrt{x}}$, $f_y(x, y) = 2x^2$

**19.** $f_r(r, s) = (1 - 2r^2)e^{-(r^2 + s^2)}$, $f_s(r, s) = -2rse^{-(r^2 + s^2)}$

**21.** $f_x(x, y, z) = \dfrac{2x(z^2 - y^2)}{(z^2 - x^2)^2}$, $f_y(x, y, z) = \dfrac{2y}{x^2 - z^2}$,

$f_z(x, y, z) = \dfrac{2z(y^2 - x^2)}{(z^2 - x^2)^2}$

**23.** $f_{xx}(x, y) = 4(3x^2 - y^3)$, $f_{yy}(x, y) = -12x^2y + 2$,

$f_{xy}(x, y) = f_{yx}(x, y) = -12xy^2$

**25.** $f_{xx}(x, y, z) = 2yz^3$, $f_{yy}(x, y, z) = 0$, $f_{zz}(x, y, z) = 6x^2yz$,

$f_{xy}(x, y, z) = f_{yx}(x, y, z) = 2xz^3$,

$f_{xz}(x, y, z) = f_{zx}(x, y, z) = 6xyz^2$,

$f_{yz}(x, y, z) = f_{zy}(x, y, z) = 3x^2z^2$

**31.** $2x \tan^{-1} y^3\, dx + \dfrac{3x^2 y^2}{1 + y^6}\, dy$

**33.** 22.853    **35.** No

**37.** $4xye^{2t} + \left(\dfrac{1}{2\sqrt{y}} - x^2\right)\sin t$    **39.** $\dfrac{3x^2 - 6xy + 2y^2}{3x^2 - 4xy - 6y^2}$

**41.** $\dfrac{\sqrt{5}}{5}(\mathbf{i} + 2\mathbf{j})$    **43.** $-11\mathbf{i} - 5\mathbf{j} + 10\mathbf{k}$    **45.** $\frac{127}{5}$

**47.** $\frac{29}{15}$    **49.** $\frac{5}{4}\mathbf{i} + 8\mathbf{j}$, $\sqrt{1049}/4$

**51.** $2x + 8y + 9z = 27$, $\dfrac{x - 1}{2} = \dfrac{y - 2}{8} = \dfrac{z - 1}{9}$

**53.** $9x + 18y - z = 27$, $\dfrac{x - 3}{9} = \dfrac{y - 1}{18} = \dfrac{z - 18}{-1}$

**55. a.**

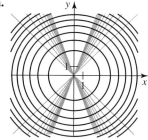

**57.** Relative minimum $f(6, -7) = -38$

**59.** Relative minimum $f\left(\frac{3}{2}, \frac{9}{4}\right) = -\frac{27}{16}$, saddle point $(0, 0, 0)$

**61.** Minimum $-11$, maximum $\frac{31}{27}$

**63.** Minimum $-\dfrac{16\sqrt{3}}{9}$, maximum $\dfrac{16\sqrt{3}}{9}$

**65.** Maximum $\frac{1}{8}$    **69.** $(2, 0, 1)$    **71.** True

### Chapter 12 Challenge Problems • page 1140

**1.** $\{(x, y, z) \mid 4x^2 + 9y^2 \le 36,\ x^2 - 2x + y^2 > 0,$ and $4x^2 + 16y + 4y^2 > -15\}$

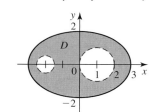

**3. a.** $D_{\mathbf{u}}^2 f(x, y) = f_{xx}u_1^2 + 2f_{xy}u_1 u_2 + f_{yy}u_2^2$  **b.** $\frac{15}{13}$

**7.** $\dfrac{\sqrt{14}}{7}(1 - \sqrt{2})$

# CHAPTER 13

### Exercises 13.1 • page 1151

**1.** $\frac{29}{2}$  **3.** $\frac{25}{2}$  **5.** $\frac{21}{2}$  **7.** 204  **9.** 15  **11.** 129

**13.** 24  **15.** 32

**17.** The wedge bounded above by the cylinder $z = 4 - x^2$ and below by the triangular base $R = \{(x, y) \mid 0 \le y \le x, 0 \le x \le 2\}$

**19.** $\displaystyle\iint_R (3 - 2x + y)\, dA$

**21.** 1.28079    **27.** True    **29.** True

### Exercises 13.2 • page 1161

**1.** 5  **3.** $10\sqrt{2}$  **5.** $-4$  **7.** $\frac{64}{3}$  **9.** $\frac{1}{3}$

**11.** $\dfrac{2e^6 - 3e^5 + 3e - 2}{6e^3}$  **13.** $\frac{9}{2}$  **15.** $\frac{1}{32}\pi^2(2\sqrt{2} + 1)$

**17.** $\frac{2}{3}$  **19.** $\frac{32}{5}$  **21.** $\frac{13}{6}$  **23.** $\frac{1}{6}$  **25.** $\frac{93}{10}$

**27.** $\dfrac{4 - \pi}{8}$  **29.** $\frac{72}{5}$  **31.** $4e^6 - 15e^4 - 1$  **33.** 48

**35.** $8\pi$  **37.** $\frac{64}{3}$  **39.** 4  **41.** 40  **43.** $\frac{3}{35}$

**45.** $\sqrt{5} + \frac{9}{2}\sin^{-1}\frac{2}{3} + \frac{1}{6}(5\sqrt{5} - 27)$

**47.**

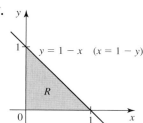

$\displaystyle\int_0^1 \int_0^{1-y} f(x, y)\, dx\, dy$

**49.**

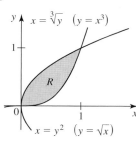

$\displaystyle\int_0^1 \int_{x^3}^{\sqrt{x}} f(x, y)\, dy\, dx$

**51.**

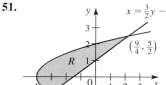

$\displaystyle\int_{-4}^{-3} \int_{-\sqrt{x+4}}^{\sqrt{x+4}} f(x, y)\, dy\, dx + \int_{-3}^{9/4} \int_{(2/3)x+1}^{\sqrt{x+4}} f(x, y)\, dy\, dx$

**53.**

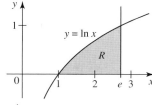

$\displaystyle\int_0^1 \int_{e^y}^{e} f(x, y)\, dx\, dy$

**55.** $\dfrac{e^4 - 1}{4e^4}$  **57.** $\dfrac{1 - \cos 8}{3}$  **59.** $\frac{4}{3}$

**63.** $\frac{2}{3}(3\pi + 8)$ slugs  **65.** 2166 people per square mile

**67. a.**

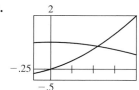

**b.** 0.550  **c.** 0.062

**69.** $-0.8784$  **71.** 0.5610  **73.** True

**75.** True  **77.** False

## Exercises 13.3 • page 1169

**1.** Rectangular, $\displaystyle\int_0^3\int_0^{-(2/3)x+2} f(x,y)\,dy\,dx$

**3.** Polar, $\displaystyle\int_{-\pi/4}^{\pi/4}\int_0^{\sqrt{2}} f(r\cos\theta,\,r\sin\theta)\,r\,dr\,d\theta$

**5.** 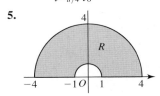   **7.**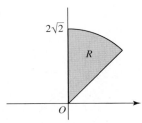

**9.** 0   **11.** 1   **13.** $\pi/2$   **15.** $\frac{1}{6}$   **17.** $8\pi$

**19.** $16\pi/3$   **21.** $9\pi$   **23.** $16\pi$   **25.** $\frac{4}{3}(\sqrt{2}-1)\pi$

**27.** $9\pi/4$   **29.** $27\pi/2$   **31.** $\dfrac{2\pi+3\sqrt{3}}{6}a^2$

**33.** $8\pi/3$   **35.** $(\pi/2)\ln 2$   **37.** $(\pi/2)(e^4-1)$

**39.** $\pi$   **41.** $\displaystyle\int_0^{\pi/4}\int_0^2 (r\cos\theta)(r\sin\theta)\,r\,dr\,d\theta,\ 1$

**43. b.** $\frac{1}{3}[\sqrt{2}+\ln(\sqrt{2}+1)]$

**45. a.**

**47.** $\sqrt{\pi}$   **49.** True

## Exercises 13.4 • page 1177

**1.** $m=6,\ (\bar{x},\bar{y})=\left(\frac{3}{2},\frac{4}{3}\right)$   **3.** $m=4,\ (\bar{x},\bar{y})=\left(\frac{7}{3},\frac{1}{3}\right)$

**5.** $m=\frac{32}{3},\ (\bar{x},\bar{y})=\left(3,\frac{8}{7}\right)$

**7.** $m=\frac{1}{4}(e^2+1),\ (\bar{x},\bar{y})=\left(\dfrac{e^2-1}{e^2+1},\dfrac{8(2e^3+1)}{27(e^2+1)}\right)$

**9.** $m=\frac{\pi}{4},\ (\bar{x},\bar{y})=\left(\frac{\pi}{2},\frac{16}{9\pi}\right)$   **11.** $m=\frac{32}{9},\ (\bar{x},\bar{y})=\left(\frac{6}{5},0\right)$

**13.** $\frac{75}{4}$ coulombs   **15.** 384.14°F

**17.** $I_x=\frac{1}{3}\rho ab^3,\ I_y=\frac{1}{3}\rho a^3 b,\ I_0=\frac{1}{3}\rho ab(a^2+b^2),\ \bar{\bar{x}}=\frac{\sqrt{3}}{3}a,$
$\bar{\bar{y}}=\frac{\sqrt{3}}{3}b$

**19.** $I_x=\frac{1}{8}\pi\rho R^4,\ I_y=\frac{1}{8}\pi\rho R^4,\ I_0=\frac{1}{4}\pi\rho R^4,\ \bar{\bar{x}}=\frac{1}{2}R,\ \bar{\bar{y}}=\frac{1}{2}R$

**21.** $I_x=12,\ I_y=18,\ I_0=30,\ \bar{\bar{x}}=\sqrt{3},\ \bar{\bar{y}}=\sqrt{2}$

**23.** $I_x=16,\ I_y=\frac{512}{5},\ I_0=\frac{592}{5},\ \bar{\bar{x}}=\frac{4\sqrt{15}}{5},\ \bar{\bar{y}}=\frac{\sqrt{6}}{2}$

**25.** $\ln 2-\frac{1}{4}$   **27.** True   **29.** False

## Exercises 13.5 • page 1183

**1.** $2\sqrt{14}$   **3.** $\frac{1}{3}(3\sqrt{3}-2\sqrt{2})$

**5.** $(\pi/6)(37\sqrt{37}-1)$   **7.** $6\pi$

**9.** $(2\pi/3)(17\sqrt{17}-1)$   **11.** $16\pi(2-\sqrt{2})$

**13.** $2a^2(\pi-2)$   **17.** 13.0046   **19.** 14.3290

**21.** $\displaystyle\int_{-1}^1\int_{-1}^1 \sqrt{36x^2y^4+36x^4y^2+1}\,dy\,dx$

**23.** $\displaystyle\int_0^2\int_0^x \dfrac{\sqrt{13+(2x+3y)^4}}{(2x+3y)^2}\,dy\,dx$

**25.** True

## Exercises 13.6 • page 1194

**1.** 18   **3.** 4   **5.** $\frac{3}{8}$   **7.** $-\frac{1}{4}$

**9.** $\dfrac{16(e-1)}{3e}$

**11.** $\displaystyle\int_0^2\int_0^{4-2x}\int_0^{(12-6x-3y)/4} f(x,y,z)\,dz\,dy\,dx,$

$\displaystyle\int_0^4\int_0^{(4-y)/2}\int_0^{(12-6x-3y)/4} f(x,y,z)\,dz\,dx\,dy,$

$\displaystyle\int_0^4\int_0^{(12-3y)/4}\int_0^{(12-3y-4z)/6} f(x,y,z)\,dx\,dz\,dy,$

$\displaystyle\int_0^3\int_0^{(12-4z)/3}\int_0^{(12-3y-4z)/6} f(x,y,z)\,dx\,dy\,dz,$

$\displaystyle\int_0^2\int_0^{(6-3x)/2}\int_0^{(12-6x-4z)/3} f(x,y,z)\,dy\,dz\,dx,$

$\displaystyle\int_0^3\int_0^{(6-2z)/3}\int_0^{(12-6x-4z)/3} f(x,y,z)\,dy\,dx\,dz$

**13.** $\displaystyle\int_0^1\int_{-\sqrt{x}}^{\sqrt{x}}\int_0^{1-x} f(x,y,z)\,dz\,dy\,dx,$

$\displaystyle\int_{-1}^1\int_{y^2}^1\int_0^{1-x} f(x,y,z)\,dz\,dx\,dy,$

$\displaystyle\int_0^1\int_0^{1-x}\int_{-\sqrt{x}}^{\sqrt{x}} f(x,y,z)\,dy\,dz\,dx,$

$\displaystyle\int_0^1\int_0^{1-z}\int_{-\sqrt{x}}^{\sqrt{x}} f(x,y,z)\,dy\,dx\,dz,$

$\displaystyle\int_{-1}^1\int_0^{1-y^2}\int_{y^2}^{1-z} f(x,y,z)\,dx\,dz\,dy,$

$\displaystyle\int_0^1\int_{-\sqrt{1-z}}^{\sqrt{1-z}}\int_{y^2}^{1-z} f(x,y,z)\,dx\,dy\,dz$

**15.** $\frac{1}{24}$   **17.** $\frac{1}{3}$   **19.** $64\pi/3$   **21.** 1

**23.**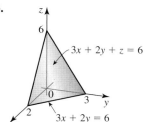

$3x + 2y + z = 6$

$3x + 2y = 6$

6

**25.**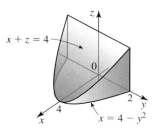

$x + z = 4$

$x = 4 - y^2$

$\frac{128}{5}$

**27.**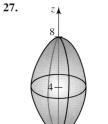

$16\pi$

**29.** 1

**31.**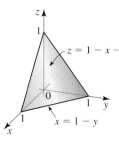

$z = 1 - x - y$

$x = 1 - y$

**33.**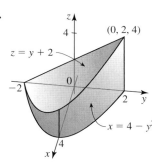

$z = y + 2$

$(0, 2, 4)$

$x = 4 - y^2$

**35.** $\int_0^6 \int_0^{(6-x)/2} \int_0^{(6-x-2y)/3} f(x, y, z) \, dz \, dy \, dx,$

$\int_0^6 \int_0^{(6-x)/3} \int_0^{(6-x-3z)/2} f(x, y, z) \, dy \, dz \, dx,$

$\int_0^3 \int_0^{6-2y} \int_0^{(6-x-2y)/3} f(x, y, z) \, dz \, dx \, dy,$

$\int_0^3 \int_0^{(6-2y)/3} \int_0^{6-2y-3z} f(x, y, z) \, dx \, dz \, dy,$

$\int_0^2 \int_0^{6-3z} \int_0^{(6-x-3z)/2} f(x, y, z) \, dy \, dx \, dz,$

$\int_0^2 \int_0^{(6-3z)/2} \int_0^{6-2y-3z} f(x, y, z) \, dx \, dy \, dz$

**37.** $\int_{-1}^1 \int_{-\sqrt{1-x^2}}^{\sqrt{1-x^2}} \int_0^2 f(x, y, z) \, dz \, dy \, dx,$

$\int_{-1}^1 \int_{-\sqrt{1-y^2}}^{\sqrt{1-y^2}} \int_0^2 f(x, y, z) \, dz \, dx \, dy,$

$\int_{-1}^1 \int_0^2 \int_{-\sqrt{1-x^2}}^{\sqrt{1-x^2}} f(x, y, z) \, dy \, dz \, dx,$

$\int_0^2 \int_{-1}^1 \int_{-\sqrt{1-x^2}}^{\sqrt{1-x^2}} f(x, y, z) \, dy \, dx \, dz,$

$\int_{-1}^1 \int_0^2 \int_{-\sqrt{1-y^2}}^{\sqrt{1-y^2}} f(x, y, z) \, dx \, dz \, dy,$

$\int_0^2 \int_{-1}^1 \int_{-\sqrt{1-y^2}}^{\sqrt{1-y^2}} f(x, y, z) \, dx \, dy \, dz$

**39. a.** 384  **b.** 384   **41.** 0.4439

**43.** $\left(\frac{2}{5}, \frac{1}{5}, \frac{1}{5}\right)$   **45.** $(2, 0, 0)$

**47.** $\int_0^1 \int_0^{1-y} \int_0^{\sqrt{1-z^2}} (xy + z^2) \, dx \, dz \, dy$

**49.** $\int_0^1 \int_0^{(2-y)/2} \int_0^{1-y^2} \sqrt{x^2 + y^2 + z^2} \, dz \, dx \, dy$

**51.** $I_x = \frac{2}{3}k, I_y = \frac{2}{3}k, I_z = \frac{2}{3}k$   **53.** $I_x = \frac{1}{180}, I_y = \frac{1}{90}, I_z = \frac{1}{90}$

**55.** 3   **57.** $1/\pi$   **59.** $T = \{(x, y, z) \mid 2x^2 + 3y^2 + z^2 \le 1\}$

**61.** True   **63.** True

## Exercises 13.7 • page 1203

**1.**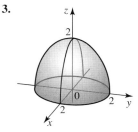

$81\pi/8$

**3.**

$16\pi/3$

**5.** $4\pi/3$   **7.** $\frac{64}{15}$   **9.** $512\pi/5$   **11.** $40\pi/3$

**13.** $(0, 0, 2)$   **15.** $\left(0, 0, \frac{4}{3}\right)$   **17.** $\frac{1}{15}\pi\rho$   **19.** $\pi$

**21.** 0   **23.** 0   **25.** $\pi/3$   **27.** $\sqrt{2}\pi/3$

**29.** $\left(0, 0, \frac{3}{8}a\right)$   **31.** $\frac{1}{4}ka^4\pi$   **33.** $\frac{48}{5}k\pi$   **35.** $\frac{67}{15}k\pi$

**37.** $\frac{4}{15}ka^5\pi$   **41.** $7\pi/8$   **43.** $\frac{1}{3}(2 - \sqrt{2})\pi$

**45.** 108°F   **47.** True   **49.** True

## Exercises 13.8 • page 1213

**1.**

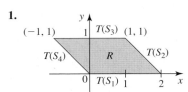

**3.**

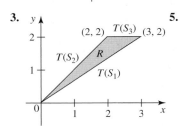

**5.**

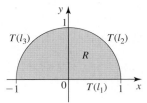

**7.** $-2(u + 1)$    **9.** $2e^{2u}$    **11.** $-4$    **13.** $45$

**15.** $0$    **17.** $4\pi$    **19.** $\pi$    **21.** $\frac{32}{3}$    **23.** $\dfrac{e^2 - 1}{4e}$

**25.** $\frac{1}{8}a^2 b^2$    **27.** $\frac{4}{3}\pi abc$    **29.** $\frac{1}{64}\pi \rho a^4$

**31.** $\displaystyle\iiint_R f(x, y, z)\, dV = \iiint_S f(r\cos\theta, r\sin\theta, z)\, r\, dz\, dr\, d\theta$

**33.** False

## Chapter 13 Concept Review • page 1214

**1. a.** $\displaystyle\sum_{i=1}^{m}\sum_{j=1}^{n} f(x_{ij}^*, y_{ij}^*)\, \Delta A$

**b.** $\displaystyle\lim_{m,\,n\to\infty}\sum_{i=1}^{m}\sum_{j=1}^{n} f(x_{ij}^*, y_{ij}^*)\, \Delta A;\ (x_{ij}^*, y_{ij}^*)$

**c.** volume; $z = f(x, y)$

**d.** $\displaystyle\lim_{m,\,n\to\infty}\sum_{i=1}^{m}\sum_{j=1}^{n} f_D(x_{ij}^*, y_{ij}^*)\, \Delta A;\ f(x, y);\ 0$

**3. a.** $\displaystyle\int_a^b\int_c^d f(x, y)\, dy\, dx;\ \int_c^d\int_a^b f(x, y)\, dx\, dy$

**b.** iterated

**5. a.** $\{(r, \theta)\,|\,a \le r \le b;\ \alpha \le \theta \le \beta\}$

**b.** $\displaystyle\int_\alpha^\beta\int_a^b f(r\cos\theta, r\sin\theta)\, r\, dr\, d\theta$

**c.** $\{(r, \theta)\,|\,\alpha \le \theta \le \beta,\ g_1(\theta) \le r \le g_2(\theta)\}$

**d.** $\displaystyle\int_\alpha^\beta\int_{g_1(\theta)}^{g_2(\theta)} f(r\cos\theta, r\sin\theta)\, r\, dr\, d\theta$

**7. a.** $\displaystyle\iint_R \sqrt{(f_x)^2 + (f_y)^2 + 1}\, dA$

**b.** $\displaystyle\iint_R \sqrt{(g_x)^2 + (g_z)^2 + 1}\, dA$

**c.** $\displaystyle\iint_R \sqrt{(h_y)^2 + (h_z)^2 + 1}\, dA$

**9. a.** order; $\displaystyle\int_p^q\int_c^d\int_a^b f(x, y, z)\, dx\, dy\, dz$

**b.** $\displaystyle\iint_R \left[\int_{k_1(x,\,y)}^{k_2(x,\,y)} f(x, y, z)\, dz\right] dA$

**11. a.** $\displaystyle\int_\alpha^\beta\int_{g_1(\theta)}^{g_2(\theta)}\int_{h_1(r\cos\theta,\,r\sin\theta)}^{h_2(r\cos\theta,\,r\sin\theta)} f(r\cos\theta, r\sin\theta, z)\, r\, dz\, dr\, d\theta$

**b.** $\displaystyle\int_\alpha^\beta\int_c^d\int_a^b f(\rho\sin\phi\cos\theta, \rho\sin\phi\sin\theta,$ $\rho\cos\phi)\rho^2 \sin\phi\, d\rho\, d\phi\, d\theta$

**c.** $\displaystyle\int_\alpha^\beta\int_c^d\int_{h_1(\phi,\,\theta)}^{h_2(\phi,\,\theta)} f(\rho\sin\phi\cos\theta,$ $\rho\sin\phi\sin\theta, \rho\cos\phi)\rho^2 \sin\phi\, d\rho\, d\phi\, d\theta$

## Chapter 13 Review Exercises • page 1215

**1.** $18$    **3.** $\frac{23}{60}$    **5.** $\dfrac{\pi}{4} - \dfrac{\ln 2}{2}$    **7.** $\frac{8}{15}(5 + \sqrt{2})$

**9.**

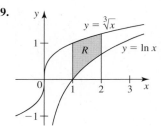

**11.**

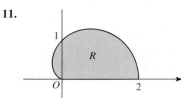

**13.** $\frac{1}{2}(1 - \cos 1)$    **15.** $\frac{52}{3}$    **17.** $\frac{32}{3}$    **19.** $6$    **21.** $\frac{19}{168}$

**23.** $0$    **25.** $\frac{4}{105}$    **27.** $\frac{7}{6}$    **29.** $\pi/2$    **31.** $\dfrac{\pi(e - 1)}{e}$

**33.** $m = \frac{2}{21}$, $(\bar{x}, \bar{y}) = \left(\frac{21}{32}, \frac{21}{40}\right)$    **35.** $m = \frac{\pi}{6}$, $(\bar{x}, \bar{y}) = \left(\frac{3}{2\pi}, \frac{3}{2\pi}\right)$

**37.** $I_x = \frac{2}{9}$, $I_y = \frac{4}{45}$, $I_0 = \frac{14}{45}$    **39.** $3\sqrt{14}$    **41.** $16\pi/5$

**43.** $\displaystyle\int_0^3 \int_0^{(6-2x)/3} \int_0^{6-2x-3y} f(x, y, z)\, dz\, dy\, dx,$

$\displaystyle\int_0^2 \int_0^{(6-3y)/2} \int_0^{6-2x-3y} f(x, y, z)\, dz\, dx\, dy,$

$\displaystyle\int_0^3 \int_0^{6-2x} \int_0^{(6-2x-z)/3} f(x, y, z)\, dy\, dz\, dx,$

$\displaystyle\int_0^6 \int_0^{(6-z)/2} \int_0^{(6-2x-z)/3} f(x, y, z)\, dy\, dx\, dz,$

$\displaystyle\int_0^2 \int_0^{6-3y} \int_0^{(6-3y-z)/2} f(x, y, z)\, dx\, dz\, dy,$

$\displaystyle\int_0^6 \int_0^{(6-z)/3} \int_0^{(6-3y-z)/2} f(x, y, z)\, dx\, dy\, dz$

**45.** $2 - 16uvw - 4uw$  **47.** $e - 1$

**49.** True  **51.** False  **53.** True

## Chapter 13 Challenge Problems • page 1217

**1. a.** 10  **7. b.** $\frac{1}{2}(1 - \cos 1)$

# CHAPTER 14

## Exercises 14.1 • page 1225

**1.** (b)  **3.** (c)  **5.** (e)

**7.**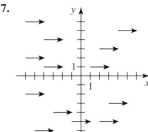

**9.**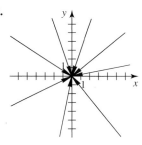

**11.**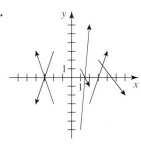

**13.**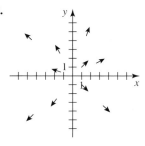

**15.**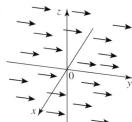

**17.**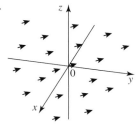

**19.** (c)  **21.** (a)

**23.** 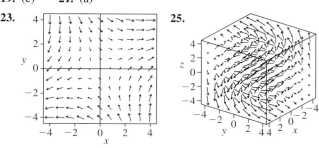  **25.**

**27.** $2xy\mathbf{i} + (x^2 - 3y^2)\mathbf{j}$  **29.** $yz\mathbf{i} + xz\mathbf{j} + xy\mathbf{k}$

**31.** $\dfrac{y}{x + z}\mathbf{i} + \ln(x + z)\mathbf{j} + \dfrac{y}{x + z}\mathbf{k}$

**33. a.** $2\mathbf{i} + 10\mathbf{j} + 4\mathbf{k}$  **b.** $(1.02, 3.1, 2.04)$

**37.** True  **39.** True

## Exercises 14.2 • page 1235

**1. a.** 0  **b.** 0  **c.** Will not rotate  **d.** 0

**3. a.** Positive  **b.** $\dfrac{1}{\sqrt{x^2 + y^2}}$  **c.** Will not rotate  **d.** 0

**5. a.** 0  **b.** 0

**7. a.** $2x(y^3 + z)$  **b.** $-z^2\mathbf{j} - 3x^2y^2\mathbf{k}$

**9. a.** $\cos x - x \sin y + \cos z$  **b.** $\cos y\mathbf{k}$

**11. a.** $1/z$  **b.** 0

**13. a.** No  **b.** No  **c.** Yes, a vector field  **d.** No

**15. a.** Yes, a vector field  **b.** Yes, a scalar field  **c.** No  **d.** Yes, a vector field

**41.** False  **43.** False  **45.** False  **47.** False

## Exercises 14.3 • page 1249

**1.** $\frac{35}{2}$  **3.** $\dfrac{13\sqrt{13} - 8}{54}$  **5.** $\frac{32}{3}$  **7.** 5  **9.** 42

**11.** $\frac{22}{3}$  **13.** 0  **15.** $\sqrt{6}/2$  **17.** 0

**19.** $\dfrac{1}{2}e^2 + 4e + \dfrac{1}{e} - \dfrac{9}{2}$  **21.** $\frac{7}{2}$  **23.** $\pi k a, \left(0, \dfrac{2a}{\pi}\right)$

**25.** $\left(0, \dfrac{a(4 - \pi)}{2(\pi - 2)}\right)$  **27.** $\left(\frac{2}{5}, \frac{2}{5}\right)$

**29. a.** Negative  **b.** $-4\pi$

**31.** $\frac{23}{24}$  **33.** $\frac{1}{2}(e^4 - e + 15)$  **35.** 1  **37.** $-720$ ft-lb

**41. a. (i)** $\dfrac{(2\sqrt{5} - \sqrt{2})qQ}{40\pi\varepsilon_0}$

   **(ii)** $\dfrac{(2\sqrt{5} - \sqrt{2})qQ}{40\pi\varepsilon_0}$

**b.** No

**43.**

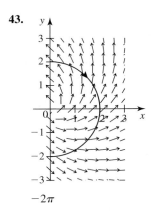

**45.** True    **47.** False

### Exercises 14.4 • page 1263

**1.** $2x^2 + 3xy - y^2 + C$    **3.** No    **5.** $y^2 \sin x + 3y + C$

**7.** No    **9.** $\frac{1}{3}x^3 + y \ln x + \frac{1}{3}y^3 + C$

**11. a.** $2xy + x + 3y + C$    **b.** 0

**13. a.** $x^2y^2 + 2xy + C$    **b.** 9

**15. a.** $\frac{1}{2}x^2e^{2y} + C$    **b.** $\frac{1}{2}e^2$

**17. a.** $e^x \sin y + \frac{1}{2}y^2 + C$    **b.** $\frac{1}{2}\pi^2$    **19.** 2    **21.** 10

**25.** $xyz + C$    **27.** No    **29.** $e^x \cos z + z \cosh y + C$

**31.** No    **33. a.** $xyz^2 + C$    **b.** 12

**35. a.** $x \cos y + yz^2 + C$    **b.** $2\pi + 1$    **37.** $-1$

**39.** $qkQ\left(\dfrac{\sqrt{14}}{14} - \dfrac{\sqrt{21}}{21}\right)$    **41. c.** No    **43.** False

**45.** False    **47.** True

### Exercises 14.5 • page 1272

**1. a.** 0  **b.** 0    **3. a.** $\frac{4}{15}$  **b.** $\frac{4}{15}$    **5.** $\frac{1}{3}$    **7.** $\frac{1}{12}$

**9.** $-\frac{352}{15}$    **11.** $\frac{3}{8}\pi a^2$    **13.** $6\pi$    **15.** $6\pi$    **17.** $\frac{2}{3}$

**19.** $\frac{3}{8}\pi a^2$    **21.** $\frac{8}{3}ab$

**23. a.**

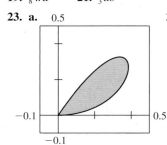

**b.** $\frac{7}{120}$

**25. a.**

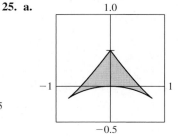

**b.** $\frac{32}{105}$

**27.** $288 - 7\pi$    **29.** $-12$    **31.** $47.5 - \pi$    **33.** 7

**37.** $\left(0, \frac{18}{5}\right)$    **39.** $\frac{1}{4}\rho a^4\pi$    **43. c.** No    **45. b.** No  **c.** $\frac{3}{4}$

**49.** False    **51.** True

### Exercises 14.6 • page 1285

**1.** (b)    **3.** (a)

**5.** $3x + 2y - 3z = 0$, a plane

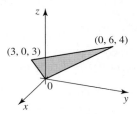

**7.** $\frac{1}{9}x^2 + \frac{1}{4}y^2 = 1, 0 \le z \le 2$, a cylinder with an elliptical cross section and axis the $z$-axis, bounded below by the plane $z = 0$ and above by the plane $z = 2$

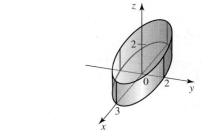

**9.** 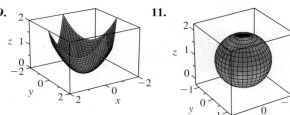 **11.**

**13.**

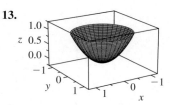

**15.** $\mathbf{r}(u, v) = (2 + 2u + v)\mathbf{i} + (1 + u - 2v)\mathbf{j} - (3 + u + v)\mathbf{k}$

**17.** $\mathbf{r}(u, v) = \cos v \cos u\mathbf{i} + \cos v \sin u\mathbf{j} - \sin v\mathbf{k}$ with domain $D = \{(u, v) \mid 0 \le u \le 2\pi, 0 \le v \le \pi\}$

**19.** $\mathbf{r}(u, v) = 2 \cos u\mathbf{i} + 2 \sin u\mathbf{j} + v\mathbf{k}$ with domain $D = \{(u, v) \mid 0 \le u \le 2\pi, -1 \le v \le 3\}$

**21.** $\mathbf{r}(u, v) = v \cos u\mathbf{i} + v \sin u\mathbf{j} + (9 - v^2)\mathbf{k}$ with domain $D = \{(u, v) \mid 0 \le u \le 2\pi, 0 \le v \le 2\}$

**23.** $\mathbf{r}(u, v) = u\mathbf{i} + \sqrt{u}\cos v\mathbf{j} + \sqrt{u}\sin v\mathbf{k}$ with domain $D = \{(u, v) \mid 0 \le u \le 4, 0 \le v \le 2\pi\}$

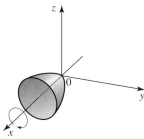

**25.** $\mathbf{r}(u, v) = (9 - u^2)\cos v\mathbf{i} + u\mathbf{j} + (9 - u^2)\sin v\mathbf{k}$ with domain $D = \{(u, v) \mid 0 \le u \le 3, 0 \le v \le 2\pi\}$

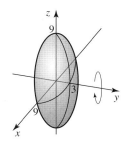

**27.** $x - y - z = 1$    **29.** $2x + z = -1$
**31.** $(\ln 2)x - 2z = 0$    **33.** $2\sqrt{3}$    **35.** $4\sqrt{14}\pi$,
**37.** $\dfrac{3\sqrt{2}\pi}{4}$    **39.** $[2\sqrt{2} + \ln(3 + 2\sqrt{2})]\pi$
**41. b.**     **c.** 199.455

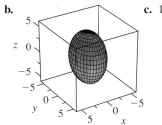

**43.** $\sqrt{2}\pi$    **47.** False

## Exercises 14.7 • page 1297

**1.** $5\sqrt{14}$    **3.** $\frac{1}{6}(27 - 5\sqrt{5})$    **5.** $4\sqrt{2}\pi$    **7.** $\sqrt{2}\pi/5$
**9.** 32    **11.** $\frac{1}{4}[2\sqrt{5} + \ln(\sqrt{5} + 2)]$    **13.** $\frac{8}{3}\pi^2$
**15.** $18\sqrt{14}k$    **17.** $8k\pi$    **19.** $40\pi$    **21.** 18
**23.** $32\pi/3$    **25.** $\pi$    **27.** $27\pi$    **29.** $\left(0, 0, \frac{a}{2}\right)$
**33.** $18\sqrt{14}k$    **35.** 42    **37.** $-20\pi$
**39. a.** $\iint\limits_{D}(-Pg_x + Q - Rg_z)\, dA$
   **b.** 15
**41.** $\frac{16}{3}\pi kR^3$    **45. b.** $3\sqrt{14}$    **49.** True

## Exercises 14.8 • page 1306

**5.** $\frac{8}{3}$    **7.** $63\pi/2$    **9.** $\frac{5}{24}$    **11.** $216\pi$    **13.** 0
**15.** $8\pi/5$    **17.** $27\pi$    **27.** True    **29.** True

## Exercises 14.9 • page 1314

**5.** $-8\pi$    **7.** $2\pi$    **9.** $8\pi$    **11.** $-36$    **13.** $-6\pi$
**15.** $-\frac{1}{2}e^2 + 2e - \frac{3}{2}$    **17.** 2    **23.** 0    **25.** $8\pi$
**29.** $2\pi$    **31.** False

## Chapter 14 Concept Review • page 1316

**1.** vector; vector
**3. a.** departs, accumulates
   **b.** $\dfrac{\partial P}{\partial x} + \dfrac{\partial Q}{\partial y} + \dfrac{\partial R}{\partial z}$; enters; departs; equals; departs; enters
**5. a.** $\lim\limits_{n\to\infty} \sum\limits_{k=1}^{n} f(x_k^*, y_k^*)\, \Delta s_k$
   **b.** $\displaystyle\int_a^b f(x(t), y(t))\sqrt{[x'(t)]^2 + [y'(t)]^2}\, dt$
   **c.** $\lim\limits_{n\to\infty} \sum\limits_{k=1}^{n} f(x_k^*, y_k^*)\, \Delta x_k$; $\displaystyle\int_a^b f(x(t), y(t))x'(t)\, dt$
   **d.** $\lim\limits_{n\to\infty} \sum\limits_{k=1}^{n} f(x_k^*, y_k^*)\, \Delta y_k$; $\displaystyle\int_a^b f(x(t), y(t))y'(t)\, dt$
**7. a.** $\displaystyle\int_a^b \mathbf{F} \cdot \mathbf{T}\, ds = \int_a^b \mathbf{F}(\mathbf{r}(t)) \cdot \mathbf{r}'(t)\, dt$
   **b.** work
**9. a.** closed    **b.** connected; conservative
**11.** curl $\mathbf{F}$; $\partial Q/\partial z$; $\partial P/\partial z$; $\partial P/\partial y$
**13.** $\displaystyle\oint_C x\, dy$; $-\displaystyle\oint_C y\, dx$; $\dfrac{1}{2}\displaystyle\oint_C x\, dy - y\, dx$
**15. a.** $x(u, v)\mathbf{i} + y(u, v)\mathbf{j} + z(u, v)\mathbf{k}$; parameter; parametric equations
   **b.** $u\mathbf{i} + v\mathbf{j} + f(u, v)\mathbf{k}$
   **c.** $u\mathbf{i} + f(u)\cos v\mathbf{j} + f(u)\sin v\mathbf{k}$; $\{(u, v) \mid a \le u \le b, 0 \le v \le 2\pi\}$
**17. a.** $\iint\limits_{R} F(x, y, f(x, y))\sqrt{[f_x(x, y)]^2 + [f_y(x, y)]^2 + 1}\, dA$
   **b.** $\iint\limits_{D} F(\mathbf{r}(u, v))|\mathbf{r}_u \times \mathbf{r}_v|\, dA$

## Chapter 14 Review Exercises • page 1317

**1. a.** $y^2 + z^2 + x^2$    **b.** $-2yz\mathbf{i} - 2xz\mathbf{j} - 2xy\mathbf{k}$
**3. a.** $e^z$    **b.** 0    **5.** $\frac{1}{12}(17\sqrt{17} - 5\sqrt{5})$    **7.** $\sqrt{2}/3$
**9.** $\dfrac{54{,}229}{110}$    **11.** $\dfrac{3\pi + 10}{12}$    **13.** $\frac{16}{3} - e^{-1}$    **15.** $\frac{109}{2}$

**17.** $2x^2y + 3xy^2 + C$    **19.** 0    **21.** $-\frac{171}{10}$    **23.** $-\pi/2$

**25.** $\dfrac{9\sqrt{17}}{4}$    **27.** $3\pi/2$    **29.** $\dfrac{\sqrt{3}}{12}k$    **31.** $12\pi$

**33.** 0    **35.** 0    **37.** 256    **39.** 0    **41.** True

**43.** False    **45.** True    **47.** True

### Chapter 14 Challenge Problems • page 1319

**1.** $\left\{(x,y) \left| \dfrac{x^2}{4} + \dfrac{y^2}{1} < 1 \text{ and } x^2 + y^2 > \dfrac{1}{4}\right.\right\}$

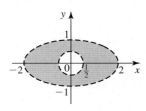

**3. b.** $\frac{3}{2}a^2$    **5.** $\pi$

## APPENDIX A

### Exercises • page A 6

**1.**

**3.**

**5.**

**7.** False    **9.** False    **11.** $(-\infty, 2)$    **13.** $(-\infty, -5]$

**15.** $(-4, 6)$    **17.** $(-\infty, -3) \cup (3, \infty)$    **19.** $(-2, 3)$

**21.** $[-3, 5]$    **23.** $(-\infty, 1] \cup [\frac{3}{2}, \infty)$

**25.** $(-\infty, -3] \cup (2, \infty)$    **27.** $(-\infty, 0] \cup (1, \infty)$    **29.** 4

**31.** 2    **33.** $5\sqrt{3}$    **35.** $\pi + 1$    **37.** 2    **39.** False

**41.** False    **43.** True    **45.** False    **47.** True

**49.** False    **51.** $x = \pm\frac{4}{3}$    **53.** $x = -1$ or $x = -\frac{5}{3}$

**55.** $(-4, 4)$    **57.** $(1, 3)$    **59.** $(-\infty, -5] \cup [-1, \infty)$

**61.** $[-1.6, -1.4]$    **63.** $[-3, -1] \cup [1, 3]$    **65.** 0.4

# ALGEBRA

## Arithmetic Operations

$$\frac{a+b}{c} = \frac{a}{c} + \frac{b}{c}$$

$$\frac{a}{b} + \frac{c}{d} = \frac{ad+bc}{bd}$$

$$\frac{\left(\dfrac{a}{b}\right)}{\left(\dfrac{c}{d}\right)} = \left(\frac{a}{b}\right)\left(\frac{d}{c}\right) = \frac{ad}{bc}$$

## Exponents and Radicals

$$x^m x^n = x^{m+n} \qquad \frac{x^m}{x^n} = x^{m-n} \qquad (x^m)^n = x^{mn}$$

$$x^{-n} = \frac{1}{x^n} \qquad (xy)^n = x^n y^n \qquad \left(\frac{x}{y}\right)^n = \frac{x^n}{y^n}$$

$$x^{n/m} = \sqrt[m]{x^n} \qquad \sqrt[n]{xy} = \sqrt[n]{x}\,\sqrt[n]{y} \qquad \sqrt[n]{\frac{x}{y}} = \frac{\sqrt[n]{x}}{\sqrt[n]{y}}$$

## Factoring

$$x^2 - y^2 = (x-y)(x+y)$$

$$x^3 - y^3 = (x-y)(x^2 + xy + y^2)$$

$$x^3 + y^3 = (x+y)(x^2 - xy + y^2)$$

## Binomial Theorem

$$(x+y)^2 = x^2 + 2xy + y^2$$

$$(x-y)^2 = x^2 - 2xy + y^2$$

$$(x+y)^3 = x^3 + 3x^2 y + 3xy^2 + y^3$$

$$(x-y)^3 = x^3 - 3x^2 y + 3xy^2 - y^3$$

$$(x+y)^n = x^n + nx^{n-1}y + \frac{n(n-1)}{2}x^{n-2}y^2 + \cdots$$

$$+ \binom{n}{k}x^{n-k}y^k + \cdots + nxy^{n-1} + y^n$$

where $\dbinom{n}{k} = \dfrac{n(n-1)\cdots(n-k+1)}{1\cdot 2\cdot 3\cdot\cdots\cdot k}$

## Quadratic Formula

If $ax^2 + bx + c = 0$, then $x = \dfrac{-b \pm \sqrt{b^2 - 4ac}}{2a}$.

## Inequalities and Absolute Value

If $a < b$ and $b < c$, then $a < c$.

If $a < b$, then $a + c < b + c$.

If $a < b$ and $c > 0$, then $ca < cb$.

If $a < b$ and $c < 0$, then $ca > cb$.

If $a > 0$, then

$|x| = a$ if and only if $x = a$ or $x = -a$

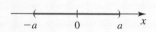

$|x| < a$ if and only if $-a < x < a$

$|x| > a$ if and only if $x > a$ or $x < -a$

# GEOMETRY

## Geometric Formulas

**Formulas for area $A$, circumference $C$, and volume $V$:**

| Triangle | Circle | Sector of Circle |
|---|---|---|

$A = \frac{1}{2}bh = \frac{1}{2}ab\sin\theta \qquad A = \pi r^2 \qquad \frac{1}{2}r^2\theta$ ($\theta$ in radians)

$\qquad\qquad\qquad\qquad\qquad C = 2\pi r \qquad s = r\theta$

**Parallelogram**     **Trapezoid**

$A = bh \qquad\qquad A = \frac{1}{2}(a+b)h$

**Sphere**    **Cylinder**    **Cone**

$V = \frac{4}{3}\pi r^3 \qquad V = \pi r^2 h \qquad V = \frac{1}{3}\pi r^2 h$

$A = 4\pi r^2 \qquad\qquad\qquad\quad A = \pi r\sqrt{r^2 + h^2}$

(lateral surface area)

## Distance and Midpoint Formulas

Distance between $P_1 = (x_1, y_1)$ and $P_2 = (x_2, y_2)$:

$$d = \sqrt{(x_2 - x_1)^2 + (y_2 - y_1)^2}$$

Midpoint of $\overline{P_1 P_2}$:

$$\left( \frac{x_1 + x_2}{2}, \frac{y_1 + y_2}{2} \right)$$

## Lines

Slope of the line through $P_1 = (x_1, y_1)$ and $P_2 = (x_2, y_2)$:

$$m = \frac{y_2 - y_1}{x_2 - x_1}$$

Slope-intercept equation of the line with slope $m$ and $y$-intercept $b$:

$$y = mx + b$$

Point-slope equation of the line through $P_1 = (x_1, y_1)$ with slope $m$:

$$y - y_1 = m(x - x_1)$$

## Equation of a Circle

Circle with center $(h, k)$ and radius $r$:

$$(x - h)^2 + (y - k)^2 = r^2$$

## TRIGONOMETRY

## Angle Measurement

$$\pi \text{ radians} = 180° \qquad 1° = \frac{\pi}{180} \text{ rad} \qquad 1 \text{ rad} = \frac{180°}{\pi}$$

## Right Triangle Definitions

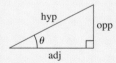

$$\sin \theta = \frac{\text{opp}}{\text{hyp}} \qquad \cos \theta = \frac{\text{adj}}{\text{hyp}} \qquad \tan \theta = \frac{\text{opp}}{\text{adj}}$$

$$\csc \theta = \frac{\text{hyp}}{\text{opp}} \qquad \sec \theta = \frac{\text{hyp}}{\text{adj}} \qquad \cot \theta = \frac{\text{adj}}{\text{opp}}$$

## Trigonometric Functions

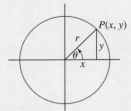

$$\sin \theta = \frac{y}{r} \qquad \cos \theta = \frac{x}{r} \qquad \tan \theta = \frac{y}{x}$$

$$\csc \theta = \frac{r}{y} \qquad \sec \theta = \frac{r}{x} \qquad \cot \theta = \frac{x}{y}$$

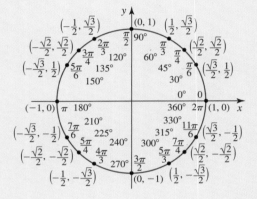

## Graphs of Trigonometric Functions

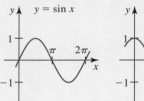
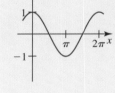
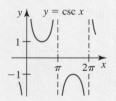
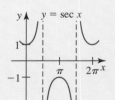
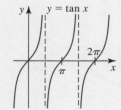
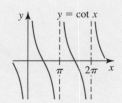

## Fundamental Identities

$$\csc \theta = \frac{1}{\sin \theta}$$

$$\sec \theta = \frac{1}{\cos \theta}$$

$$\tan \theta = \frac{\sin \theta}{\cos \theta}$$

$$\cot \theta = \frac{\cos \theta}{\sin \theta}$$

$$\sin^2 \theta + \cos^2 \theta = 1$$

$$1 + \tan^2 \theta = \sec^2 \theta$$

$$1 + \cot^2 \theta = \csc^2 \theta$$

$$\sin(-\theta) = -\sin \theta$$

$$\cos(-\theta) = \cos \theta$$

$$\tan(-\theta) = -\tan \theta$$

$$\sin\left(\frac{\pi}{2} - \theta\right) = \cos \theta$$

$$\cos\left(\frac{\pi}{2} - \theta\right) = \sin \theta$$

$$\tan\left(\frac{\pi}{2} - \theta\right) = \cot \theta$$

## The Law of Sines

$$\frac{\sin A}{a} = \frac{\sin B}{b} = \frac{\sin C}{c}$$

## The Law of Cosines

$$a^2 = b^2 + c^2 - 2bc \cos A$$

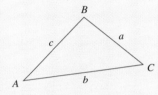

## Addition and Subtraction Formulas

$$\sin(x + y) = \sin x \cos y + \cos x \sin y$$

$$\sin(x - y) = \sin x \cos y - \cos x \sin y$$

$$\cos(x + y) = \cos x \cos y - \sin x \sin y$$

$$\cos(x - y) = \cos x \cos y + \sin x \sin y$$

$$\tan(x + y) = \frac{\tan x + \tan y}{1 - \tan x \tan y}$$

$$\tan(x - y) = \frac{\tan x - \tan y}{1 + \tan x \tan y}$$

## Double-Angle Formulas

$$\sin 2x = 2 \sin x \cos x$$

$$\cos 2x = \cos^2 x - \sin^2 x = 2 \cos^2 x - 1 = 1 - 2 \sin^2 x$$

$$\tan 2x = \frac{2 \tan x}{1 - \tan^2 x}$$

## Half-Angle Formulas

$$\sin^2 x = \frac{1 - \cos 2x}{2} \qquad \cos^2 x = \frac{1 + \cos 2x}{2}$$